Greenland
(Kalatdlit-Nunat)

Alaska

Yukon   N.W.T.   Nunavut

B.C.   Alta.   Sask.   Man.   Ont.   Que.   [Labrador]

Nfld

P.E.I.   St.Pierre
and Miquelon

N.B.   N.S.

Wash.   Mont.   N.Dak.   Minn.   Vt.   Maine
Oreg.   Idaho   Wyo.   S.Dak.   Wis.   Mich.   N.Y.   N.H.
Nev.   Utah   Colo.   Nebr.   Iowa   Ill.   Ind.   Ohio   Pa.   Mass.
Calif.   Kans.   Mo.   Ky.   W.Va.   Va.   R.I.
Ariz.   N.Mex.   Okla.   Ark.   Tenn.   N.C.   Conn.
N.J.
Del.
Md.
Tex.   La.   Miss.   Ala.   Ga.   S.C.
Fla.

Mexico

Flora of North America

Contributors to Volume 17

Adjoa Richardson Ahedor
Dirk C. Albach
Kerry A. Barringer
Richard E. Brainerd
Judith M. Canne-Hilliker†
Robert S. Capers
L. Turner Collins
Alison E. L. Colwell
Bradley M. Daugherty
J. Mark Egger
Wayne J. Elisens
Reidar Elven
Naomi S. Fraga
Craig C. Freeman
Galina Gussarova
Neil A. Harriman
John F. Hays
James Henrickson

Noel H. Holmgren
Larry D. Hufford
Brian R. Keener
Kim R. Kersh
Richard V. Lansdown
Donald H. Les
Deborah Q. Lewis
David C. Michener
Kamal I. Mohamed
Jeffery J. Morawetz
David F. Murray
Lytton J. Musselman
Allan D. Nelson
Guy L. Nesom
Eliane Meyer Norman
Richard G. Olmstead
Nick Otting
Michael S. Park

Robert E. Preston
Richard K. Rabeler
Christopher P. Randle
Bruce W. Robart
Llorenç Sáez Goñalons
Alexey Shipunov
Heidi Solstad
Bruce A. Sorrie
Gordon C. Tucker
Melissa C. Tulig
Simon Uribe-Convers
Jay B. Walker
Justin K. Williams
Barbara L. Wilson
George Yatskievych
Elizabeth H. Zacharias
Peter F. Zika
Scott Zona

Editors for Volume 17

Ariel S. Buback
Assisting Technical Editor

Tammy M. Charron
Managing Editor

Craig C. Freeman
Co-Lead Editor, Co-Taxon Editor
for Linderniaceae, Mazaceae,
Orobanchaceae, Pedaliaceae,
Phrymaceae, Plantaginaceae, and
Scrophulariaceae

Kanchi Gandhi
Nomenclatural Editor

Cassandra L. Howard
Senior Technical Editor

Robert W. Kiger
Bibliographic Editor

Deborah Q. Lewis
Taxon Editor for Paulowniaceae
and Tetrachondraceae

Richard K. Rabeler
Co-Lead Editor, Co-Taxon Editor
for Linderniaceae, Mazaceae,
Orobanchaceae, Pedaliaceae,
Phrymaceae, Plantaginaceae, and
Scrophulariaceae

John L. Strother
Reviewing Editor

James L. Zarucchi
Editorial Director

Volume 17 Composition

Tammy M. Charron
Production Coordinator and
Managing Editor

Kristin Pierce
Compositor and Editorial Assistant

Oxford University Press is a department of the University of Oxford.
It furthers the University's objective of excellence in research,
scholarship, and education by publishing worldwide.

Oxford New York

Auckland Cape Town Dar es Salaam Hong Kong Karachi Kuala Lumpur
Madrid Melbourne Mexico City Nairobi New Delhi Shanghai Taipei Toronto

With offices in

Argentina Austria Brazil Chile Czech Republic France Greece Guatemala Hungary Italy
Japan Poland Portugal Singapore South Korea Switzerland Thailand Turkey Ukraine Vietnam

Oxford is a registered trademark of Oxford University Press in the UK and certain other countries.

Published by Oxford University Press, Inc.
198 Madison Avenue, New York, New York 10016
www.oup.com

Library of Congress Cataloging-in-Publication Data
(Revised for Volume 17)
Flora of North America North of Mexico
edited by Flora of North America Editorial Committee.
Includes bibliographical references and indexes.
Contents: v. 1. Introduction—v. 2. Pteridophytes and gymnosperms—
v. 3. Magnoliophyta: Magnoliidae and Hamamelidae—
v. 22. Magnoliophyta: Alismatidae, Arecidae, Commelinidae (in part), and Zingiberidae—
v. 26. Magnoliophyta: Liliidae: Liliales and Orchidales—
v. 23. Magnoliophyta: Commelinidae (in part): Cyperaceae—
v. 25. Magnoliophyta: Commelinidae (in part): Poaceae, part 2—
v. 4. Magnoliophyta: Caryophyllidae (in part): part 1—
v. 5. Magnoliophyta: Caryophyllidae (in part): part 2—
v. 19, 20, 21. Magnoliophyta: Asteridae (in part): Asteraceae, parts 1–3—
v. 24. Magnoliophyta: Commelinidae (in part): Poaceae, part 1—
v. 27. Bryophyta, part 1—
v. 8. Magnoliophyta: Paeoniaceae to Ericaceae—
v. 7. Magnoliophyta: Salicaceae to Brassicaceae—
v. 28. Bryophyta, part 2—
v. 9. Magnoliophyta: Picramniaceae to Rosaceae—
v. 6. Magnoliophyta: Cucurbitaceae to Droseraceae—
v. 12. Magnoliophyta: Vitaceae to Garryaceae—
v. 17 Magnoliophyta: Tetrachondraceae to Orobanchaceae

ISBN: 978-0-19-086851-2 (v. 17)
1. Botany—North America.
2. Botany—United States.
3. Botany—Canada.
I. Flora of North America Editorial Committee.
QK110.F55 2002 581.97 92-30459

1 2 3 4 5 6 7 8 9
Printed in the United States of America on acid-free paper

Contents

Noel and Patricia Holmgren, Big Flat, head of Long Canyon, Grand County, Utah, May 20, 2001.
Photo by and courtesy of Tori Butt.

This volume of the Flora of North America North of Mexico is dedicated
to Noel Herman Holmgren (b. 1937) and Patricia Kern Holmgren (b. 1940),
preeminent students of the western North America flora, and esteemed scholars and colleagues.

FOUNDING MEMBER INSTITUTIONS

Flora of North America Association

Agriculture and Agri-Food Canada
Ottawa, Ontario

Arnold Arboretum
Jamaica Plain, Massachusetts

Canadian Museum of Nature
Ottawa, Ontario

Carnegie Museum of
Natural History
Pittsburgh, Pennsylvania

Field Museum of Natural History
Chicago, Illinois

Fish and Wildlife Service
United States Department of
the Interior
Washington, D.C.

Harvard University Herbaria
Cambridge, Massachusetts

Hunt Institute for Botanical
Documentation
Carnegie Mellon University
Pittsburgh, Pennsylvania

Jacksonville State University
Jacksonville, Alabama

Jardin Botanique de Montréal
Montréal, Québec

Kansas State University
Manhattan, Kansas

Missouri Botanical Garden
St. Louis, Missouri

New Mexico State University
Las Cruces, New Mexico

The New York Botanical Garden
Bronx, New York

New York State Museum
Albany, New York

Northern Kentucky University
Highland Heights, Kentucky

Université de Montréal
Montréal, Québec

University of Alaska
Fairbanks, Alaska

University of Alberta
Edmonton, Alberta

The University of British Columbia
Vancouver, British Columbia

University of California
Berkeley, California

University of California
Davis, California

University of Idaho
Moscow, Idaho

University of Illinois
Urbana-Champaign, Illinois

University of Iowa
Iowa City, Iowa

The University of Kansas
Lawrence, Kansas

University of Michigan
Ann Arbor, Michigan

University of Oklahoma
Norman, Oklahoma

University of Ottawa
Ottawa, Ontario

University of Southwestern
Louisiana
Lafayette, Louisiana

The University of Texas
Austin, Texas

University of Western Ontario
London, Ontario

University of Wyoming
Laramie, Wyoming

Utah State University
Logan, Utah

For their support of the preparation of this volume,
we gratefully acknowledge and thank:

Franklinia Foundation

The Philecology Foundation

The Andrew W. Mellon Foundation

The David and Lucile Packard Foundation

an anonymous foundation

Chanticleer Foundation

The Stanley Smith Horticultural Trust

Hall Family Charitable Fund

WEM Foundation

William T. Kemper Foundation

For sponsorship of illustrations included in this volume,
we express sincere appreciation to:

American Penstemon Society
  Volume 17 Frontispiece,
    Penstemon grahamii, Plantaginaceae

Arizona Native Plant Society, Tucson, Arizona
  Castilleja lanata, Orobanchaceae

Deborah Bell, Washington, D.C.
  Paulownia tomentosa, Paulowniaceae

James Duemmel, Bellingham, Washington
  Castilleja hispida var. *hispida*,
    Orobanchaceae

Tracy Fuentes, Seattle, Washington
  Castilleja chlorotica, Orobanchaceae
  Castilleja levisecta, Orobanchaceae
  Castilleja victoriae, Orobanchaceae

Kirk Garanflo, Hoffman Estates, Illinois
  Pedicularis canadensis, Orobanchaceae
  Scrophularia marilandica, Scrophulariaceae

Marilee Henry and Jef Thorson, Bothwell, Washington
  Castilleja cryptantha, Orobanchaceae

Noel and Patricia Holmgren, New York, New York
  Penstemon franklinii, Plantaginaceae
  Penstemon rhizomatosus, Plantaginaceae

Ida Bomholt Dyrholm Jacobsen, Nuussuaq, Greenland
  Pedicularis groenlandica, Orobanchaceae

Rhoda Love, Eugene, Oregon
  Castilleja pruinosa, Orobanchaceae

Wendy S. McClure, Poulsbo, Washington
  Castilleja, three species, Orobanchaceae—in
    memory of Margaret Ashworth

Nancy R. Morin, Point Arena, California
  Penstemon oklahomensis,
    Plantaginaceae—in honor of
      Craig C. Freeman

Richard K. Rabeler, Ann Arbor, Michigan
  Chaenorhinum minus subsp. *minus*,
    Plantaginaceae

Judith H. Reynolds, Albuquerque, New Mexico
  Penstemon cardinalis var. *cardinalis*,
    Plantaginaceae
  Penstemon dasyphyllus, Plantaginaceae
  Penstemon pinifolius, Plantaginaceae

Jennifer Richards, Miami, Florida
  Bacopa monnieri, Plantaginaceae

Washington Native Plant Society
  Central Puget Sound Chapter
    Castilleja, one species,
      Orobanchaceae
  Central Washington Chapter
    Castilleja elmeri, Orobanchaceae
  Columbia Basin Chapter
    Castilleja thompsonii,
      Orobanchaceae
  Koma Kulshan Chapter
    Castilleja parviflora var. *oreopola*,
      Orobanchaceae
    Castilleja rupicola, Orobanchaceae
  Northeast Washington Chapter
    Castilleja, two species,
      Orobanchaceae
  Okanogan Chapter
    Castilleja cervina, Orobanchaceae
  Olympic Peninsula Chapter
    Castilleja miniata var. *miniata*
      Orobanchaceae
  Salal Chapter
    Castilleja, two species,
      Orobanchaceae
  South Sound Chapter
    Castilleja, one species,
      Orobanchaceae
  Wenatchee Valley Chapter
    Castilleja, one species,
      Orobanchaceae

Project Staff — past and present
involved with the preparation of Volume 17

Barbara Alongi, *Illustrator*
Bobbi Angell, *Illustrator*
Mike Blomberg, *Imaging*
Ariel S. Buback, *Assisting Technical Editor*
Trisha K. Distler, *GIS Analyst*
Pat Harris†, *Editorial Assistant*
Linny Heagy, *Illustrator*
Suzanne E. Hirth, *Editorial Assistant*
Cassandra L. Howard, *Senior Technical Editor*
Ruth T. King, *Editorial Assistant*
Marjorie C. Leggitt, *Illustrator*
John Myers, *Illustrator and Illustration Compositor*
Kristin Pierce, *Editorial Assistant and Compositor*
Andrew C. Pryor, *Assisting Technical Editor*
Heidi H. Schmidt, *Managing Editor* (2007–2017)
Hong Song, *Programmer*
Yevonn Wilson-Ramsey, *Illustrator*

Contributors to Volume 17

Adjoa Richardson Ahedor
Rose State College
Midwest City, Oklahoma

Dirk C. Albach
Carl von Ossietzky-Universität
Oldenburg, Germany

Kerry A. Barringer
Jersey City, New Jersey

Richard E. Brainerd
Corvallis, Oregon

Judith M. Canne-Hilliker†
University of Guelph
Guelph, Ontario

Robert S. Capers
University of Connecticut
Storrs, Connecticut

L. Turner Collins
Springfield, Missouri

Alison E. L. Colwell
University of California
Berkeley, California

Bradley M. Daugherty
Eastern Illinois University
Charleston, Illinois

J. Mark Egger
Seattle, Washington

Wayne J. Elisens
University of Oklahoma
Norman, Oklahoma

Reidar Elven
Natural History Museum
University of Oslo
Oslo, Norway

Naomi S. Fraga
Rancho Santa Ana Botanic Garden
Claremont, California

Craig C. Freeman
The University of Kansas
Lawrence, Kansas

Galina Gussarova
Tromsø Museum
University of Tromsø
Tromsø, Norway

Neil A. Harriman
University of Wisconsin Oshkosh
Oshkosh, Wisconsin

John F. Hays
Gainesville, Florida

James Henrickson
The University of Texas at Austin
Austin, Texas

Noel H. Holmgren
The New York Botanical Garden
Bronx, New York

Larry D. Hufford
Washington State University
Pullman, Washington

Brian R. Keener
The University of West Alabama
Livingston, Alabama

Kim R. Kersh
University of California
Berkeley, California

Richard V. Lansdown
Stoud, England

Donald H. Les
University of Connecticut
Storrs, Connecticut

Deborah Q. Lewis
Iowa State University
Ames, Iowa

David C. Michener
Matthaei Botanical Gardens
  and Nichols Arboretum
Ann Arbor, Michigan

Kamal I. Mohamed
State University of New York
Oswego, New York

Jeffery J. Morawetz
Rancho Santa Ana Botanic Garden
Claremont, California

David F. Murray
University of Alaska
Museum of the North
University of Alaska Fairbanks
Fairbanks, Alaska

Lytton J. Musselman
Old Dominion University
Norfolk, Virginia

Allan D. Nelson
Tarleton State University
Stephenville, Texas

Guy L. Nesom
Fort Worth, Texas

Eliane Meyer Norman
Stetson University
DeLand, Florida

Richard G. Olmstead
University of Washington
Seattle, Washington

Nick Otting
Corvallis, Oregon

Michael S. Park
University of California
Berkeley, California

Robert E. Preston
ICF International
Sacramento, California

Richard K. Rabeler
University of Michigan
Ann Arbor, Michigan

Christopher P. Randle
Sam Houston State University
Huntsville, Texas

Bruce W. Robart
University of Pittsburgh at
Johnstown
Johnstown, Pennsylvania

Llorenç Sáez Goñalons
Universitat Autònoma de
Barcelona
Barcelona, Spain

Alexey Shipunov
Minot State University
Minot, South Dakota

Heidi Solstad
Natural History Museum
University of Oslo
Oslo, Norway

Bruce A. Sorrie
North Carolina Botanical Garden
Chapel Hill, North Carolina

Gordon C. Tucker
Eastern Illinois University
Charleston, Illinois

Melissa C. Tulig
The New York Botanical Garden
Bronx, New York

Simon Uribe-Convers
University of Missouri-St. Louis
St. Louis, Missouri

Jay B. Walker
Union High School
Sapulpa, Oklahoma

Justin K. Williams
Sam Houston State University
Huntsville, Texas

Barbara L. Wilson
Oregon State University
Corvallis, Oregon

George Yatskievych
The University of Texas at Austin
Austin, Texas

Elizabeth H. Zacharias
Princeton, New Jersey

Peter F. Zika
University of Washington
Seattle, Washington

Scott Zona
Florida International University
Miami, Florida

Taxonomic Reviewers

Wayne J. Elisens
University of Oklahoma
Norman, Oklahoma

L. Dwayne Estes
Austin Peay State University
Clarkesville, Tennessee

Stephen C. Meyers
Oregon State University
Corvallis, Oregon

Jeffery J. Morawetz
Rancho Santa Ana Botanic Garden
Claremont, California

Sergei L. Mosyakin
N. G. Kholodny Institute of
  Botany
Kiev, Ukraine

Guy L. Nesom
Fort Worth, Texas

Richard G. Olmstead
University of Washington
Seattle, Washington

C. Thomas Philbrick
Western Connecticut State
  University
Danbury, Connecticut

Leila M. Shultz
Utah State University
Logan, Utah

John L. Strother
University of California
Berkeley, California

Regional Reviewers

ALASKA / YUKON

Bruce Bennett
Yukon Department of
  Environment
Whitehorse, Yukon

Robert Lipkin
Alaska Natural Heritage Program
University of Alaska
Anchorage, Alaska

David F. Murray
University of Alaska
  Museum of the North
University of Alaska Fairbanks
Fairbanks, Alaska

Carolyn Parker
University of Alaska
  Museum of the North
University of Alaska Fairbanks
Fairbanks, Alaska

Mary Stensvold
Sitka, Alaska

PACIFIC NORTHWEST

Edward R. Alverson
The Nature Conservancy
Eugene, Oregon

Curtis R. Björk
University of British Columbia
Vancouver, British Columbia

David E. Giblin
University of Washington
Seattle, Washington

Richard R. Halse
Oregon State University
Corvallis, Oregon

Linda Jennings
University of British Columbia
Vancouver, British Columbia

Aaron Liston
Oregon State University
Corvallis, Oregon

Frank Lomer
Vancouver, British Columbia

Frédéric Coursol
Mirabel, Québec

William J. Crins
*Ontario Ministry of Natural
Resources
Peterborough, Ontario*

Marian Munro
*Nova Scotia Museum of Natural
History
Halifax, Nova Scotia*

Michael J. Oldham
*Natural Heritage Information
Centre
Peterborough, Ontario*

NORTHEASTERN UNITED STATES

Ray Angelo
*New England Botanical Club
Cambridge, Massachusetts*

Tom S. Cooperrider
*Kent State University
Kent, Ohio*

Arthur Haines
Canton, Maine

Michael A. Homoya
*Indiana Department of Natural
Resources
Indianapolis, Indiana*

Robert F. C. Naczi
*The New York Botanical Garden
Bronx, New York*

Anton A. Reznicek
*University of Michigan
Ann Arbor, Michigan*

Edward G. Voss†
*University of Michigan
Ann Arbor, Michigan*

Kay Yatskievych
Austin, Texas

SOUTHEASTERN UNITED STATES

Mac H. Alford
*University of Southern Mississippi
Hattiesburg, Mississippi*

J. Richard Carter Jr.
*Valdosta State University
Valdosta, Georgia*

L. Dwayne Estes
*Austin Peay State University
Clarksville, Tennessee*

W. John Hayden
*University of Richmond
Richmond, Virginia*

Wesley Knapp
*North Carolina Natural Heritage
Program
Raleigh, North Carolina*

John B. Nelson
*University of South Carolina
Columbia, South Carolina*

Chris Reid
*Louisiana State University
Baton Rouge, Louisiana*

Bruce A. Sorrie
*University of North Carolina
Chapel Hill, North Carolina*

Dan Spaulding
*Anniston Museum of Natural
Science
Anniston, Alabama*

R. Dale Thomas
Seymour, Tennessee

Lowell E. Urbatsch
*Louisiana State University
Baton Rouge, Louisiana*

Theo Witsel
*Arkansas Natural Heritage
Commission
Little Rock, Arkansas*

B. Eugene Wofford
*University of Tennessee
Knoxville, Tennessee*

FLORIDA

Loran C. Anderson
*Florida State University
Tallahassee, Florida*

Bruce F. Hansen
*University of South Florida
Tampa, Florida*

Richard P. Wunderlin
*University of South Florida
Tampa, Florida*

Preface for Volume 17

Since the publication of *Flora of North America* Volume 12 (the twentieth volume in the *Flora* series) in late 2016, the membership of the Flora of North America Association [FNAA] Board of Directors has changed. Kerry A. Barringer, Wayne J. Elisens, and Heidi H. Schmidt, the former Managing Editor, have departed the board. The new board members include Dale E. Johnson and the current Managing Editor, Tammy M. Charron. As a result of a reorganization finalized in 2003, the FNAA Board of Directors succeeded the former Editorial Committee; for the sake of continuity of citation, authorship of *Flora* volumes is to be cited as "Flora of North America Editorial Committee, eds."

Most of the editorial process for this volume was done at The University of Kansas in Lawrence, University of Michigan in Ann Arbor, and Missouri Botanical Garden in St. Louis. Final processing and composition took place at the Missouri Botanical Garden; this included pre-press processing, typesetting and layout, plus coordination for all aspects of planning, executing, and scanning the illustrations. Other aspects of production, such as art panel composition plus labeling and occurrence map generation, were carried out elsewhere in North America.

Illustrations published in this volume were executed by five very talented artists: Barbara Alongi, Bobbi Angell, Linny Heagy, John Myers, and Yevonn Wilson-Ramsey. Barbara Alongi prepared illustrations for Orobanchaceae in part (*Aureolaria, Euphrasia, Orthocarpus, Rhinanthus, Striga* & *Triphysaria*), Pedaliaceae, *Erythranthe* (Phrymaceae), Plantaginaceae in greater part (excluding those prepared by Myers & Wilson-Ramsey cited below), Scrophulariaceae in part (*Bontia, Limosella* & *Myoporum*), and Tetrachondraceae. Bobbi Angell created the frontispiece depicting *Penstemon grahamii* (Plantaginaceae). Linny Heagy illustrated taxa of *Agalinis* (Orobanchaceae) and Scrophulariaceae in greater part (*Buddleja, Capraria, Emorya, Leucophyllum* & *Verbascum*). John Myers illustrated Mazaceae, Orobanchaceae in part (*Boschniakia, Castilleja, Chloropyron, Conopholis, Cordylanthus, Dicranostegia, Epifagus, Kopsiopsis, Orobanche* & *Pedicularis*), and Plantaginaceae in part (*Keckiella, Leucospora, Penstemon, Schistophragma* & *Tonella*). Yevonn Wilson-Ramsey prepared illustrations for Linderniaceae, Orobanchaceae in part (*Bartsia, Bellardia, Brachystigma, Buchnera, Dasistoma, Macranthera, Melampyrum, Odontites, Schwalbea* & *Seymeria*), Paulowniaceae, Phrymaceae (except *Erythranthe*), Plantaginaceae in part (*Bacopa, Callitriche, Chaenorhinum, Hippuris, Lagotis, Mecardonia, Veronica* & *Veronicastrum*), and *Scrophularia* (Scrophulariaceae). In addition to preparing various illustrations, John Myers composed and labeled all of the line drawings that appear in this volume.

Starting with Volume 8, published in 2009, the circumscription and ordering of some families within the *Flora* have been modified so that they mostly reflect that of the Angiosperm Phylogeny Group [APG] rather than the previously followed Cronquist organizational structure. The groups of families found in this and future volumes in the series are mostly ordered following E. M. Haston et al. (2007); since APG views of relationships and circumscriptions have evolved, and will change further through time, some discrepancies in organization will occur. Volume 30 of the *Flora of North America* will contain a comprehensive index to the published volumes.

Support from many institutions and by numerous individuals has enabled the *Flora* to be produced. Members of the Flora of North America Association remain deeply thankful to the many people who continue to help create, encourage, and sustain the *Flora*.

Introduction

Scope of the Work

Flora of North America North of Mexico is a synoptic account of the plants of North America north of Mexico: the continental United States of America (including the Florida Keys and Aleutian Islands), Canada, Greenland (Kalâtdlit-Nunât), and St. Pierre and Miquelon. The *Flora* is intended to serve both as a means of identifying plants within the region and as a systematic conspectus of the North American flora.

The *Flora* will be published in 30 volumes. Volume 1 contains background information that is useful for understanding patterns in the flora. Volume 2 contains treatments of ferns and gymnosperms. Families in volumes 3–26, the angiosperms, were first arranged according to the classification system of A. Cronquist (1981) with some modifications, and starting with Volume 8, the circumscriptions and ordering of families generally follow those of the Angiosperm Phylogeny Group [APG] (see E. Haston et al. 2007). Bryophytes are being covered in volumes 27–29. Volume 30 will contain the cumulative bibliography and index.

The first two volumes were published in 1993, Volume 3 in 1997, and Volumes 22, 23, and 26, the first three of five volumes covering the monocotyledons, appeared in 2000, 2002, and 2002, respectively. Volume 4, the first part of the Caryophyllales, was published in late 2003. Volume 25, the second part of the Poaceae, was published in mid 2003, and Volume 24, the first part, was published in January 2007. Volume 5, completing the Caryophyllales plus Polygonales and Plumbaginales, was published in early 2005. Volumes 19–21, treating Asteraceae, were published in early 2006. Volume 27, the first of two volumes treating mosses in North America, was published in late 2007. Volume 8, Paeoniaceae to Ericaceae, was published in September 2009, and Volume 7, Salicaceae to Brassicaceae, appeared in 2010. In 2014, Volume 28 was published, completing the treatment of mosses for the flora area, and at the end of 2014, Volume 9, Picramniaceae to Rosaceae was published. Volume 6, which covered Cucurbitaceae to Droseraceae, was published in 2015. Volume 12, Vitaceae to Garryaceae, was published in late 2016. The correct bibliographic citation for the *Flora* is: Flora of North America Editorial Committee, eds. 1993+. Flora of North America North of Mexico. 21+ vols. New York and Oxford.

Volume 17 treats 952 species in 95 genera contained in 9 families. For additional statistics please refer to Table 1 on p. xx.

Contents · General

The *Flora* includes accepted names, selected synonyms, literature citations, identification keys, descriptions, phenological information, summaries of habitats and geographic ranges, and other biological observations. Each volume contains a bibliography and an index to the taxa included in that volume. The treatments, written and reviewed by experts from throughout the systematic botanical community, are based on original observations of herbarium specimens and, whenever possible, on living plants. These observations are supplemented by critical reviews of the literature.

Table 1. *Statistics for Volume 17 of Flora of North America.*

Family	Total Genera	Endemic Genera	Introduced Genera	Total Species	Endemic Species	Introduced Species	Conservation Taxa
Tetrachondraceae	1	0	0	1	0	0	0
Plantaginaceae	45	7	11	460	322	59	68
Scrophulariaceae	9	0	3	45	10	23	4
Linderniaceae	3	0	1	10	4	4	0
Pedaliaceae	2	0	2	2	0	2	0
Mazaceae	1	0	1	2	0	2	0
Phrymaceae	6	0	1	139	113	1	35
Paulowniaceae	1	0	1	1	0	1	0
Orobanchaceae	27	5	3	292	200	14	56
Totals	**95**	**12**	**23**	**952**	**649**	**106**	**163**

Italic = introduced

Basic Concepts

Our goal is to make the *Flora* as clear, concise, and informative as practicable so that it can be an important resource for both botanists and nonbotanists. To this end, we are attempting to be consistent in style and content from the first volume to the last. Readers may assume that a term has the same meaning each time it appears and that, within groups, descriptions may be compared directly with one another. Any departures from consistent usage will be explicitly noted in the treatments (see References).

Treatments are intended to reflect current knowledge of taxa throughout their ranges worldwide, and classifications are therefore based on all available evidence. Where notable differences of opinion about the classification of a group occur, appropriate references are mentioned in the discussion of the group.

Documentation and arguments supporting significantly revised classifications are published separately in botanical journals before publication of the pertinent volume of the *Flora*. Similarly, all new names and new combinations are published elsewhere prior to their use in the *Flora*. No nomenclatural innovations will be published intentionally in the *Flora*.

Taxa treated in full include extant and recently extinct or extirpated native species, named hybrids that are well established (or frequent), introduced plants that are naturalized, and cultivated plants that are found frequently outside cultivation. Taxa mentioned only in discussions include waifs known only from isolated old records and some non-native, economically important or extensively cultivated plants, particularly when they are relatives of native species. Excluded names and taxa are listed at the ends of appropriate sections, for example, species at the end of genus, genera at the end of family.

Treatments are intended to be succinct and diagnostic but adequately descriptive. Characters and character states used in the keys are repeated in the descriptions. Descriptions of related taxa at the same rank are directly comparable.

With few exceptions, taxa are presented in taxonomic sequence. If an author is unable to produce a classification, the taxa are arranged alphabetically and the reasons are given in the discussion.

Treatments of hybrids follow that of one of the putative parents. Hybrid complexes are treated at the ends of their genera, after the descriptions of species.

We have attempted to keep terminology as simple as accuracy permits. Common English equivalents usually have been used in place of Latin or Latinized terms or other specialized terminology, whenever the correct meaning could be conveyed in approximately the same space, for example, "pitted" rather than "foveolate," but "striate" rather than "with fine longitudinal lines." See *Categorical Glossary for the Flora of North America Project* (R. W. Kiger and D. M. Porter 2001; also available online at http://huntbot.andrew.cmu.edu) for standard definitions of generally used terms. Very specialized terms are defined, and sometimes illustrated, in the relevant family or generic treatments.

References

Authoritative general reference works used for style are *The Chicago Manual of Style,* ed. 14 (University of Chicago Press 1993); *Webster's New Geographical Dictionary* (Merriam-Webster 1988); and *The Random House Dictionary of the English Language,* ed. 2, unabridged (S. B. Flexner and L. C. Hauck 1987). *B-P-H/S. Botanico-Periodicum-Huntianum/Supplementum* (G. D. R. Bridson and E. R. Smith 1991), *BPH-2: Periodicals with Botanical Content* (Bridson 2004), and *BPH Online* [http://fmhibd.library.cmu.edu/HIBD-DB/bpho/findrecords.php] (Bridson and D. W. Brown) have been used for abbreviations of serial titles, and *Taxonomic Literature*, ed. 2 (F. A. Stafleu and R. S. Cowan 1976–1988) and its supplements by Stafleu et al. (1992–2009) have been used for abbreviations of book titles.

Graphic Elements

All genera and more than 25 percent of the species in this volume are illustrated. The illustrations may show diagnostic traits or complex structures. Most illustrations have been drawn from herbarium specimens selected by the authors. Data on specimens that were used and parts that were illustrated have been recorded. This information, together with the archivally preserved original drawings, is deposited in the Missouri Botanical Garden Library and is available for scholarly study.

Specific Information in Treatments

Keys

Dichotomous keys are included for all ranks below family if two or more taxa are treated. More than one key may be given to facilitate identification of sterile material or for flowering versus fruiting material.

Nomenclatural Information

Basionyms of accepted names, with author and bibliographic citations, are listed first in synonymy, followed by any other synonyms in common recent use, listed in alphabetical order, without bibliographic citations.

The last names of authors of taxonomic names have been spelled out. The conventions of *Authors of Plant Names* (R. K. Brummitt and C. E. Powell 1992) have been used as a guide for including first initials to discriminate individuals who share surnames.

If only one infraspecific taxon within a species occurs in the flora area, nomenclatural information (literature citation, basionym with literature citation, relevant other synonyms) is given for the species, as is information on the number of infraspecific taxa in the species and their distribution worldwide, if known. A description and detailed distributional information are given only for the infraspecific taxon.

Descriptions

Character states common to all taxa are noted in the description of the taxon at the next higher rank. For example, if sexual condition is dioecious for all species treated within a genus, that character state is given in the generic description. Characters used in keys are repeated in the descriptions. Characteristics are given as they occur in plants from the flora area. Characteristics that occur only in plants from outside the flora area may be given within square brackets, or instead may be noted in the discussion following the description. In families with one genus and one or more species, the family description is given as usual, the genus description is condensed, and the species are described as usual. Any special terms that may be used when describing members of a genus are presented and explained in the genus description or discussion.

In reading descriptions, the reader may assume, unless otherwise noted, that: the plants are green, photosynthetic, and reproductively mature; woody plants are perennial; stems are erect; roots are fibrous; leaves are simple and petiolate; flowers are bisexual, radially symmetric, and pediceled; perianth parts are hypogynous, distinct, and free; and ovaries are superior. Because measurements and elevations are almost always approximate, modifiers such as "about," "circa," or "±" are usually omitted.

Unless otherwise noted, dimensions are length × width. If only one dimension is given, it is length or height. All measurements are given in metric units. Measurements usually are based on dried specimens but these should not differ significantly from the measurements actually found in fresh or living material.

Chromosome numbers generally are given only if published and vouchered counts are available from North American material or from an adjacent region. No new counts are published intentionally in the *Flora*. Chromosome counts from nonsporophyte tissue have been converted to the $2n$ form. The base number ($x =$) is given for each genus. This represents the lowest known haploid count for the genus unless evidence is available that the base number differs.

Flowering time and often fruiting time are given by season, sometimes qualified by early, mid, or late, or by months. Elevations over 200 m generally are rounded to the nearest 100 m; those 100 m and under are rounded to the nearest 10 m. Mean sea level is shown as 0 m, with the understanding that this is approximate. Elevation often is omitted from herbarium specimen labels, particularly for collections made where the topography is not remarkable, and therefore precise elevation is sometimes not known for a given taxon.

The term "introduced" is defined broadly to refer to plants that were released deliberately or accidentally into the flora and that now are naturalized, that is, exist as wild plants in areas in which they were not recorded as native in the past. The distribution of introduced taxa are often poorly documented and changing, so the distribution statements for those taxa may not be fully accurate.

If a taxon is globally rare or if its continued existence is threatened in some way, the words "of conservation concern" appear before the statements of elevation and geographic range.

Criteria for taxa of conservation concern are based on NatureServe's (formerly The Nature Conservancy)—see http://www.natureserve.org—designations of global rank (G-rank) G1 and G2:

G1 Critically imperiled globally because of extreme rarity (5 or fewer occurrences or fewer than 1000 individuals or acres) or because of some factor(s) making it especially vulnerable to extinction.

G2 Imperiled globally because of rarity (5–20 occurrences or fewer than 3000 individuals or acres) or because of some factor(s) making it very vulnerable to extinction throughout its range.

The occurrence of species and infraspecific taxa within political subunits of the *Flora* area is depicted by dots placed on the outline map to indicate occurrence in a state or province. The Nunavut boundary on the maps has been provided by the GeoAccess Division, Canada Centre for Remote Sensing, Earth Science. Authors are expected to have seen at least one specimen documenting each geographic unit record (except in rare cases when undoubted literature reports may be used) and have been urged to examine as many specimens as possible from throughout the range of each taxon. Additional information about taxon distribution may be presented in the discussion.

Distributions are stated in the following order: Greenland; St. Pierre and Miquelon; Canada (provinces and territories in alphabetic order); United States (states in alphabetic order); Mexico (11 northern states may be listed specifically, in alphabetic order); West Indies; Bermuda; Central America (Belize, Costa Rica, El Salvador, Guatemala, Honduras, Nicaragua, Panama); South America; Europe, or Eurasia; Asia (including Indonesia); Africa; Atlantic Islands; Indian Ocean Islands; Pacific Islands; Australia; Antarctica.

Discussion

The discussion section may include information on taxonomic problems, distributional and ecological details, interesting biological phenomena, and economic uses.

Selected References

Major references used in preparation of a treatment or containing critical information about a taxon are cited following the discussion. These, and other works that are referred to in discussion or elsewhere, are included in Literature Cited at the end of the volume.

CAUTION

The Flora of North America Editorial Committee **does not encourage, recommend, promote, or endorse** any of the folk remedies, culinary practices, or various utilizations of any plant described within this volume. Information about medicinal practices and/or ingestion of plants, or of any part or preparation thereof, has been included only for historical background and as a matter of interest. Under no circumstances should the information contained in these volumes be used in connection with medical treatment. Readers are strongly cautioned to remember that many plants in the flora are toxic or can cause unpleasant or adverse reactions if used or encountered carelessly.

Key to boxed codes following accepted names:

C of conservation concern
E endemic to the flora area
F illustrated
I introduced to the flora area
W weedy, based mostly on R. H. Callihan et al. (1995) and/or D. T. Patterson et al. (1989)

Flora of North America

Phylogeny and Classification of Lamiales with Emphasis on Scrophulariaceae in the Broad Sense

Richard G. Olmstead

Introduction

Sorting relationships in Lamiales is one of the thorniest problems in angiosperm systematics (W. S. Judd and R. G. Olmstead 2004; D. E. Soltis et al. 2011). Relationships among several clades now recognized as families (for example, Angiosperm Phylogeny Group 2016) remain uncertain (Olmstead et al. 2001; B. Oxelman et al. 2005; B. Schäferhoff et al. 2010). Results of phylogenetic studies (for example, L. A. McDade et al. 2008; Olmstead et al. 2009; H. E. Marx et al. 2010; J. R. McNeal et al. 2013; M. Perret et al. 2013; Li Bo et al. 2016) have converged on a classification at the family level that is likely to remain stable. Figure 1 depicts a conservative estimate of relationships at the family level derived from multigene DNA studies (Schäferhoff et al.; N. F. Refulio-Rodriguez and Olmstead 2014).

Traditional circumscriptions of Lamiales often included only Lamiaceae, Verbenaceae, and a few very small segregate or isolated families (G. L. Stebbins 1974; A. L. Takhtajan 1980, 1997) and sometimes included Boraginaceae (for example, A. Cronquist 1981). Some late twentieth-century classifications recognized additional orders that now are included in Lamiales: Callitrichales, Hippuridales, Plantaginales, and Scrophulariales (for example, Cronquist; Takhtajan 1997). R. F. Thorne (1992b) recognized suborder Lamiineae, which was similar to Lamiales of Cronquist, minus Boraginaceae, in a broadly circumscribed Scrophulariales. The Scrophulariales of Thorne, renamed Lamiales in subsequent treatments (Thorne 2000b; Thorne and J. L. Reveal 2007), was similar in circumscription to Lamiales recognized today (Angiosperm Phylogeny Group

2016), with minor differences; Thorne did not include Plocospermataceae Hutchinson and Tetrachondraceae in his Scrophulariales. Recognition of an order Lamiales corresponding to the group with the present circumscription derives from early molecular systematic studies of *rbc*L sequences, in which the clade was first identified and named Lamiales (R. G. Olmstead et al. 1992, 1993) and expanded in many subsequent studies (for example, Olmstead et al. 2001; B. Bremer et al. 2002; B. Oxelman et al. 2005; B. Schäferhoff et al. 2010; N. F. Refulio-Rodriguez and Olmstead 2014).

Adjustments in family circumscriptions in Lamiales include large-scale changes; for example, about one half of Verbenaceae have been transferred to Lamiaceae (P. D. Cantino et al. 1992; S. J. Wagstaff and R. G. Olmstead 1997; R. M. Harley et al. 2004). Small-scale changes have also been made, such as consolidating Avicenniaceae Miquel into Acanthaceae (A. E. Schwarzbach and L. A. McDade 2002), or identifying isolated lineages and recognizing them as distinct families (for example, Schlegeliaceae Reveal from Bignoniaceae, R. E. Spangler and Olmstead 1999; Olmstead et al. 2009; Thomandersiaceae Sreemadhavan from Acanthaceae, A. H. Wortley et al. 2007).

Perhaps the most sweeping changes at the family level have been made to the traditional concept of Scrophulariaceae (for example, A. Cronquist 1981; R. F. Thorne 1992b; A. L. Takhtajan 1997), wherein nine families (Figure 1) have been formed from the traditionally circumscribed family (R. G. Olmstead et al. 2001; B. Oxelman et al. 2005; R. Rahmanzadeh et al. 2005; D. C. Tank et al. 2006;

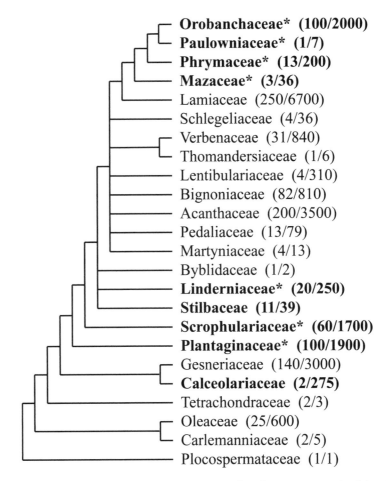

FIGURE 1. Phylogenetic relationships among clades of Lamiales representing families as recognized by APG III. Families consisting all or in part of Scrophulariaceae in the broad sense are in bold; of those, the ones present in North America are indicated by an asterisk (*). Estimated numbers of genera/species are indicated in parentheses.

J. L. Reveal 2011). Additionally, molecular evidence (Olmstead et al.; Oxelman et al.; B. Schäferhoff et al. 2010; I. C. Gormley et al. 2015) now supports separation of New World Martyniaceae from Old World Pedaliaceae, contrary to some twentieth-century treatments (for example, G. L. Stebbins 1974; Cronquist). The splitting of Scrophulariaceae resulted in enlarging some traditionally small families, including the expansion of Orobanchaceae to accommodate the hemiparasitic members of Scrophulariaceae (J. R. McNeal et al. 2013), expanding Plantaginaceae from two or three genera and about 300 species to more than 100 genera and about 1900 species (D. C. Albach et al. 2005), and transforming the monogeneric Phrymaceae by the addition of a dozen genera and about 200 species (P. M. Beardsley and Olmstead 2002; W. R. Barker et al. 2012).

Splitting Scrophulariaceae

In most twentieth-century classifications (for example, A. L. Takhtajan 1980, 1997; A. Cronquist 1981; R. F. Thorne 1992b), families in Lamiales as circumscribed here were defined by the presence of some distinctive trait or traits (for example, explosive capsules with retinacula in Acanthaceae, winged seeds lacking endosperm in Bignoniaceae, ovules reduced to two per carpel and separated by a false septum in Lamiaceae and Verbenaceae, achlorophyllous parasites in Orobanchaceae, wind pollination in Plantaginaceae, and anthers adherent in pairs in Gesneriaceae Richard & Jussieu). Scrophulariaceae, the largest of the families in those classifications, was defined by a combination of traits (typically bilaterally symmetric flowers with didynamous stamens and two multi-ovulate carpels forming a capsular fruit) that are individually found in many other families. Despite what may be seen in hindsight as evidence of either polyphyly or paraphyly for a large group without a distinctive defining trait, there is little evidence in writings accompanying the angiosperm classifications from the nineteenth and twentieth centuries that suggests Scrophulariaceae were not considered to be a natural group, apart from inclusion or exclusion of Buddlejaceae, Globulariaceae de Candolle, Lentibulariaceae, Plantaginaceae, Selaginaceae Choisy, and Salpiglossaceae Hutchinson (the last now in Solanaceae).

The earliest molecular evidence to hint that Scrophulariaceae, as traditionally defined, were not monophyletic, came from a study aimed at placing the aquatic genera *Callitriche* and *Hippuris*. That study (R. G. Olmstead and P. A. Reeves 1995) found that the two aquatic genera were sister groups and nested within a group of genera assigned to Scrophulariaceae in the broad sense, including *Antirrhinum*, *Digitalis*, *Plantago*, and *Veronica*. A second group of genera, including *Buddleja*, *Scrophularia*, *Selago* Linnaeus, and *Verbascum*, was distant from the clade containing the aquatic genera. At about the same time, a study of parasitic plants in Orobanchaceae and Scrophulariaceae found that many hemiparasitic plants assigned to Scrophulariaceae in the broad sense were more closely related to the holoparasitic Orobanchaceae than they were to other Scrophulariaceae (C. W. dePamphilis et al. 1997; A. D. Wolfe and dePamphilis 1998). An effort to combine data from these two studies with additional sampling of genera of Scrophulariaceae in the broad sense, and representatives of most families of Lamiales, identified six distinct branches comprising, wholly or in large part, Scrophulariaceae in the broad sense: Calceolariaceae Olmstead, *Mimulus*, Orobanchaceae, Plantaginaceae, Scrophulariaceae in the narrow sense, and Stilbaceae Kunth (Olmstead et al. 2001).

Subsequent studies added additional unsampled genera and further clarified relationships of segregate groups of Scrophulariaceae among Lamiales. *Mimulus* and related genera were assigned to Phrymaceae, along with the Asian genera *Lancea* Hooker f. & Thompson and *Mazus*, as subfam. Mazoideae (named, but not validly published in P. M. Beardsley and R. G. Olmstead 2002), and a group of four Australian genera (Beardsley and W. R. Barker 2005). Beardsley and co-workers (Beardsley and Olmstead; Beardsley and Barker) showed that *Mimulus* was a broadly paraphyletic ancestral group from which the other genera were derived and set the stage for splitting *Mimulus* into monophyletic genera (Barker et al. 2012). Linderniaceae was recognized as another previously unsampled lineage comprising about 15 genera (B. Oxelman et al. 2005; R. Rahmanzadeh et al. 2005). Mazoideae has not been found to occur consistently with Phrymaceae in phylogenies and should not be

included with that family (Oxelman et al.; D. C. Albach et al. 2009; Xia Z. et al. 2009). A clade comprising *Rehmannia* Liboschitz ex Fischer & C. A. Meyer and *Triaenophora* (J. D. Hooker) Solereder occupies an isolated position close to Orobanchaceae and Phrymaceae (Oxelman et al.; B. Schäferhoff et al. 2010). J. L. Reveal (2011) described new families to accommodate each of the last two groups, Mazaceae and Rehmanniaceae Reveal, respectively; the latter was included in Orobanchaceae by Angiosperm Phylogeny Group (2016). Much of this work was contemporaneous with the writing of the treatment of Scrophulariaceae for *The Families and Genera of Vascular Plants* (E. Fischer 2004). In that volume, genera of Scrophulariaceae in the broad sense are included in one family; the genera are organized along lines of the families that were then becoming recognized.

Synopsis of Current Classification

Seven of the nine families derived all, or in large part, from the former Scrophulariales are represented in the flora of North America. Mazaceae and Paulowniaceae are represented by introduced species only; the rest have substantial native representation in North America.

Plantaginaceae contain the largest number of representatives in the flora area and are the largest clade worldwide to emerge from Scrophulariaceae in the broad sense, with about 100 genera and 1900 species. Some formerly recognized families (for example, A. Cronquist 1981) belong here, including Callitrichaceae Link, Globulariaceae, Hippuridaceae Vest, and Plantaginaceae in the narrow sense. Plantaginaceae have been divided into 12 tribes (D. C. Albach et al. 2005). Of these, nine are represented in North America; Angelonieae Pennell, Digitalideae Dumortier, and Russelieae Pennell are represented by introduced species. All genera of Cheloneae D. Don are represented in the flora area, except for *Uroskinnera* Lindley (Central America) and *Pennellianthus* Crosswhite (East Asia) (A. D. Wolfe et al. 2002). Plantaginaceae are primarily Northern Hemisphere in distribution; Angelonieae is South American and is sister to Gratioleae D. Don, which is widely distributed in Northern and Southern hemispheres (Albach et al.). This inclusive clade (Angelonieae + Gratioleae) is sister to the rest of the family, in which the predominantly North American *Chelone* is sister to the other mostly Northern Hemisphere tribes.

Scrophulariaceae in the narrow sense are primarily a Southern Hemisphere group with the greatest generic diversity in southern Africa (P. Kornhall et al. 2001) and another center of diversity in Australia (B. Oxelman et al. 2005). Though much smaller than the traditional family, this is still a sizable group with about 60 genera and about 1700 species. The former families Myoporaceae R. Brown (distributed primarily in Australia) and Buddlejaceae belong here (R. G. Olmstead and P. A. Reeves 1995; Oxelman et al. 1999, 2005). Despite the Southern Hemisphere origins of Scrophulariaceae, Scrophularieae Dumortier (primarily *Scrophularia* and *Verbascum*) has diversified extensively in the Northern Hemisphere with over 500 species. *Buddleja* also has diversified in the Northern Hemisphere with significant diversity in the New World as far north as the southwest United States, and in the eastern Himalayas. A synopsis of Scrophulariaceae (D. C. Tank et al. 2006) recognized eight tribes; five of these occur in the flora area. Four tribes, Aptosimeae Bentham & Hooker f., Hemimerideae Bentham, Limoselleae Dumortier, and Teedieae Bentham, are entirely or predominantly southern African in distribution. Only the Australian Myoporeae Reichenbach and their sister group, Leucophylleae Miers, of North America and Central America are absent from Africa. Despite the traditional image of bilaterally symmetric corollas, most Scrophulariaceae have radially symmetric corollas; bilateral symmetry is derived in *Scrophularia*, Leucophylleae, and *Eremophila* R. Brown and relatives (Myoporeae).

Linderniaceae are predominantly an African group and one of the last of the Scrophulariaceae segregate families to be recognized (R. Rahmanzadeh et al. 2005); the family comprises 20 genera and about 250 species. Two genera of Linderniaceae are native to North America: *Micranthemum* with three native species and *Lindernia* with six species (D. Q. Lewis 2000). *Torenia* is represented in the flora area by one introduced species.

Mazaceae include *Mazus*, an East Asian and Australian genus of about 33 species, with two adventive species in the flora area, along with *Lancea* (two species in China) and *Dodartia* Linnaeus (one species in Russia). Similarities with *Mimulus* led to classification of *Mazus* and *Lancea* with *Mimulus* in Scrophulariaceae tribes Gratioleae (R. Wettstein 1891–1893) or Mimuleae Dumortier (F. W. Pennell 1920c) and reflect the close relationship inferred by P. M. Beardsley and R. G. Olmstead (2002), who assigned them to Mazoideae (an unpublished name) in Phrymaceae. Subsequent research (B. Oxelman et al. 2005) showed that relationship to be close but not to be monophyletic.

Phrymaceae received the largest proportional increase of any family in Lamiales following recircumscription, going from a single species, *Phryma leptostachya*, recognized in many late twentieth-century classifications (for example, R. F. Thorne 1992b; A. L. Takhtajan

1997), to 13 genera and nearly 200 species (P. M. Beardsley and R. G. Olmstead 2002; W. R. Barker et al. 2012). Most of the increase came with the recognition that the clade that included *Mimulus* in the broad sense also encompassed several smaller genera, including *Phryma* nested within a paraphyletic *Mimulus* (Beardsley and Olmstead; Beardsley and Barker 2005), and that Phrymaceae was the only available name at the rank of family. Distribution of Phrymaceae is predominantly within the flora area with greatest diversity in western North America; Phrymaceae also have a second center of diversity in Australia. The type of *Mimulus* is the eastern North American *M. ringens*, which is not a part of the western North America radiation. Redrawing of generic boundaries in Phrymaceae (Barker et al.) led to the reclassification of all western North American species into *Diplacus* and *Erythranthe*, and unispecific *Mimetanthe*, which shares some synapomorphies with *Diplacus*. The phylogenetic work by Beardsley and co-workers (Beardsley and Olmstead; Beardsley et al. 2004) implicitly suggested that *Diplacus* and *Erythranthe* represent separate radiations in western North America from ancestors outside of the region. Acceptance of *Diplacus* and *Erythranthe* is consistent with the pattern of diversification.

Paulowniaceae comprises a single genus, *Paulownia*, an East Asian genus of seven species of large trees assigned to Scrophulariaceae by G. Bentham (1876); it was placed in Bignoniaceae in some classifications (A. L. Takhtajan 1980; A. Cronquist 1981; R. F. Thorne 1983). J. E. Armstrong (1985) examined morphological and anatomical characters and argued that the preponderance of evidence allied *Paulownia* with Scrophulariaceae, which swayed most late twentieth-century classifications (Thorne 1992b; Takhtajan 1997). Molecular phylogenetic evidence showed that it does not belong with Bignoniaceae, Scrophulariaceae, or any other family; it forms an isolated lineage near Lamiaceae, Orobanchaceae, and Phrymaceae. T. Nakai (1949) had previously established the family Paulowniaceae. One species, *P. tomentosa*, is naturalized in North America.

The remaining families segregated from Scrophulariaceae in the broad sense do not occur naturally in the flora area. Calceolariaceae Olmstead, with two genera and about 275 species, occur primarily in Andean South America and in New Zealand with two species of *Jovellana* Ruiz & Pavón (U. Molau 1988; S. Andersson 2006; A. Cosacov et al. 2009). Stilbaceae are native to South Africa; the family has a core of about five genera with ericoid foliage and small radially symmetric flowers. Molecular phylogenetic studies have expanded Stilbaceae by including *Retzia* Thunberg (B. Bremer et al. 1994) and former members of Scrophulariaceae, *Halleria* Linnaeus

(R. G. Olmstead et al. 2001), and Bowkerieae Barringer (B. Oxelman et al. 2005); the family now includes 11 genera and about 40 species. *Rehmannia* and *Triaenophora* represent a pair of isolated genera comprising about 11 species formerly assigned to Scrophulariaceae in the broad sense, but found to belong close to Orobanchaceae (Oxelman et al.; D. C. Albach et al. 2009; Xia Z. et al. 2009), and described as Rehmanniaceae by J. L. Reveal (2011).

Orobanchaceae are the second largest family of Scrophulariaceae in the broad sense, with about 100 genera and about 1850 to 2100 species (the range is due to widely varying species estimates for *Euphrasia* and *Pedicularis*), and includes the holoparasitic species that have comprised this family in traditional classifications (A. Cronquist 1981; R. F. Thorne 1992b; A. L. Takhtajan 1997) and hemiparasitic species that have been included traditionally in Scrophulariaceae in the broad sense. The non-parasitic *Lindenbergia* Lehmann is included in Orobanchaceae and is sister to the parasitic members; excluding *Lindenbergia* would result in a less well-supported and possibly non-monophyletic group (R. G. Olmstead et al. 2001; A. D. Wolfe et al. 2005; J. R. Bennett and S. Mathews 2006; J. R. McNeal et al. 2013). *Rehmannia* and *Triaenophora*, which share the corolla aestivation character diagnostic of the family, were included in Orobanchaceae by Angiosperm Phylogeny Group (2016). Parasitism is a nearly universal characteristic of Orobanchaceae; it does not provide the defining trait. The so-called rhinanthoid corolla aestivation, wherein the lateral lobes forming the abaxial corolla lip are external in bud, is a shared trait that unites non-parasitic and parasitic members.

J. R. McNeal et al. (2013) identified six major clades of Orobanchaceae (excluding Rehmanniaceae), with *Lindenbergia* composing one of them. The other five major clades are parasitic, and all have representatives in the flora area, even though an origin in Asia is inferred (A. D. Wolfe et al. 2005; McNeal et al.). One clade, centered on *Castilleja* and *Pedicularis*, is predominantly New World in distribution, even though *Pedicularis* is mostly Asian. Conventional circumscriptions of Orobanchaceae restricted the family to the holoparasites and recognized a relationship to the hemiparasitic Scrophulariaceae (for example, A. Cronquist 1981). Phylogenetic studies (C. W. dePamphilis et al. 1997; McNeal et al.) have shown that there are at least three origins of holoparasitism, and each is in a different one of the parasitic clades. There are other lineages in which photosynthetic activity is limited to a brief portion of the life history of the plant; they are often referred to as holoparasites while still meeting the definition of a hemiparasite (McNeal et al.).

TETRACHONDRACEAE Skottsberg ex Wettstein

• Tetrachondra Family

Richard K. Rabeler

Craig C. Freeman

Herbs, perennial, sometimes annual, not fleshy, autotrophic. **Stems** prostrate to ascending. **Leaves** cauline, opposite, simple; stipules reduced to ridges; petiole absent; blade not fleshy, not leathery, margins entire or finely serrate. **Inflorescences** terminal or axillary, dichasial or monochasial cymes, sometimes flowers solitary in axils. **Flowers** bisexual, perianth and androecium partly hypogynous [hypogynous]; sepals 4 or 5(or 6), proximally connate, calyx radially symmetric; petals 4(or 5), proximally connate, corolla radially symmetric, regular, short-funnelform [subrotate]; stamens 4(or 5), adnate to corolla, equal, staminode 0; pistil 1, 2-carpellate, ovary partly inferior [superior], 2[–4]-locular, placentation axile [basal]; ovules anatropous or amphitropous, unitegmic, tenuinucellate; style 1; stigma 1. **Fruits** capsules [nutlets], dehiscence septicidal. **Seeds** [4]70–120+, yellow, angled to angled-globular; embryo straight, endosperm abundant. $x = 10, 11$.

Genera 2, species 3 (1 in the flora): c, se United States, Mexico, West Indies, Central America, South America, Pacific Islands (New Zealand); introduced elsewhere in Pacific Islands, Australia.

Works by B. Oxelman et al. (1999) and S. J. Wagstaff (2004) are followed in circumscribing Tetrachondraceae as comprising two formerly disparate genera, *Polypremum* and *Tetrachondra* Petrie ex Oliver. Oxelman et al. found evidence from *rbc*L and *ndh*F analyses supporting their membership in a clade within the Lamiales. This was confirmed by B. Schäferhoff et al. (2010) from analyses using *trn*K/*mat*K, *trn*L-F, and *rps*16 sequence data.

Polypremum and *Tetrachondra* have generally been placed in different families. *Tetrachondra*, a genus of two species with sessile leaves and flowers with a superior ovary and gynobasic style, known only from Argentina, Chile, and New Zealand, was first placed in Crassulaceae (where its two-carpellate ovary was anomalous) by W. W. Hamilton in 1885 (S. J. Wagstaff 2004). C. Skottsberg (1912) thought it might be best treated as a unigeneric family near the Lamiaceae; A. Cronquist (1981) placed it within the Lamiaceae. *Polypremum* has often been placed in the Buddlejaceae, Loganiaceae, or Rubiaceae (see G. K. Rogers 1986).

A. L. Takhtajan (2009) followed L. Watson and M. J. Dallwitz (http://delta-intkey.com) in recognizing Watson's unigeneric, unpublished Polypremaceae, which Reveal published in 2011.

SELECTED REFERENCES Skottsberg, C. 1912. *Tetrachondra patagonica* n. sp. und die systematische Stellung der Gattung. Bot. Jahrb. Syst. 48(suppl.): 17–26. Wagstaff, S. J. 2004. Tetrachondraceae. In: K. Kubitzki et al., eds. 1990+. The Families and Genera of Vascular Plants. 14+ vols. Berlin, etc. Vol. 7, pp. 441–444.

1. POLYPREMUM Linnaeus, Sp. Pl. 1: 111. 1753; Gen. Pl. ed. 5, 50. 1754 • [Greek *polys*, many, and *premnon*, stump or stem, alluding to diffuse much-branched habit]

Richard K. Rabeler

Perennials or annuals, taprooted. **Stems** glabrous or sparsely scabrous along ridges. **Cymes:** bracts absent. **Pedicels** absent; bracteoles present. **Flowers:** calyx urceolate, lobes lanceolate; corolla white; stamens: filaments glabrous; stigma capitate. **Seeds:** wings absent. $x = 10, 11$.

Species 1: c, se United States, Mexico, West Indies, Central America, South America; introduced in Pacific Islands, Australia.

Polypremum is sometimes confused with *Loeflingia* and *Scleranthus* of the Caryophyllaceae; the opposite, linear-acuminate leaves and axillary flowers are suggestive of those taxa. The relatively small, but conspicuous, white petals (absent or at most rudimentary in *Loeflingia* and *Scleranthus*) and the two-lobed capsule (three-valved capsule in *Loeflingia*, utricle in *Scleranthus*) distinguish *Polypremum*.

1. Polypremum procumbens Linnaeus, Sp. Pl. 1: 111. 1753 • Juniper leaf, rustweed [F][W]

Stems (5–)8–20(–33) cm, much-branched distally. **Leaves** sometimes forming dense, overwintering rosettes; blade 1-veined, linear to narrowly lanceolate, (5–)10–25(–42) × (0.5–)1(–2) mm, margins ciliate proximally, apex acuminate. **Bracteoles** leaflike. **Calyx lobes** keeled, (2–)3 mm, margins scarious, especially medially, apex attenuate. **Corollas** 1–2 mm, usually ½–¾ length of calyx lobes, lobe apex rounded. **Stamens** included. **Capsules** 2-lobed, 2 × 1.5–2 mm, slightly flattened. **Seeds** 0.3 mm, shiny. $2n = 20$ (Guyana), 22.

Flowering May–Nov. Pine woods and barrens, roadsides, sand dunes, sandy fields, waste places; 0–1000 m; Ala., Ark., Del., Fla., Ga., Ill., Ky., La., Md., Miss., Mo., N.C., Okla., S.C., Tenn., Tex., Va.; Mexico; West Indies; Central America; South America; introduced in Pacific Islands, Australia.

Collections of *Polypremum procumbens* were gathered from ship ballast piles in New Jersey (in 1865, 1875) and Pennsylvania (in 1865, 1868), and there was an 1873 report from Long Island, New York.

Since a report from Hawaii (F. R. Fosberg 1962), populations of *Polypremum procumbens* have been discovered in widely scattered locations, often near roads or runways, in the Pacific Basin. Specimens have been examined from the Territory of Guam, Kwajalein Atoll of the Republic of Marshall Islands, the Republic of Palau, Wallis Island, and one site in eastern Australia.

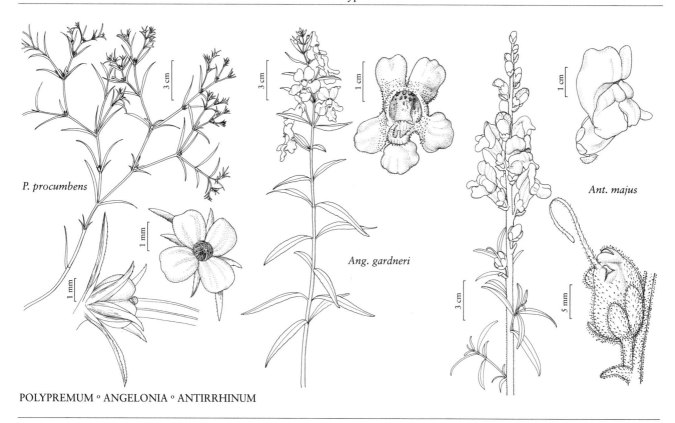

P. procumbens

Ang. gardneri

Ant. majus

POLYPREMUM ∘ ANGELONIA ∘ ANTIRRHINUM

PLANTAGINACEAE Jussieu

• Plantain Family

Craig C. Freeman

Richard K. Rabeler

Wayne J. Elisens

Herbs, subshrubs, or shrubs, annual or perennial, sometimes biennial, rarely fleshy (*Plantago maritima*), autotrophic. **Stems** prostrate, decumbent, ascending, or erect, sometimes arching, creeping, sprawling, spreading, scrambling, pendent, climbing, or reclining, sometimes absent. **Leaves** deciduous or persistent, basal, basal and cauline, or cauline, opposite, subopposite, alternate, helical, or whorled, sometimes opposite proximally, alternate distally, simple; stipules absent; petiole present or absent; blade fleshy or not, leathery or not, margins entire to subentire, toothed, or lobed. **Inflorescences** axillary or terminal, spikes, racemes, cymes, corymbs, thyrses, whorls, panicles, spiciform, or flowers 1–4. **Flowers** bisexual, rarely unisexual (*Callitriche, Hippuris, Littorella*), perianth and androecium hypogynous (epigynous in *Hippuris*); sepals (2–)4 or 5, basally or proximally connate, sometimes distinct, rarely a minute rim (*Hippuris*) or 0 (*Callitriche*), calyx radially or bilaterally symmetric; petals 0 (*Callitriche, Hippuris*) or (3 or) 4 or 5, connate, corolla radially or bilaterally symmetric, ± bilabiate, bilabiate and personate, or regular, rarely unilabiate (*Synthyris*), ± rotate to salverform, ± cylindric, tubular, funnelform, urceolate, ellipsoid, globular, ovoid, or ligulate; stamens (1 or)2–4 or 5, adnate to corolla or free, didynamous or equal, staminodes 0 or 1(–3); pistil 1, 2-carpellate (1 in *Hippuris*), ovary superior (inferior in *Hippuris*), 2-locular, sometimes 1-locular or 4-locular (*Callitriche*), placentation axile, sometimes parietal (*Dopatrium*), parietal centrally and axile distally and proximally (*Synthyris*), apical (*Hippuris*), basal (*Callitriche, Littorella*), or free-central (*Plantago*); ovules anatropous, hemitropous, hemianatropous, or campylotropous, unitegmic, tenuinucellate; styles [0]1 (2 in *Callitriche*); stigma 1 (2 in *Callitriche*). **Fruits** capsules, dehiscence loculicidal, septicidal, or poricidal, or nutlets (*Littorella*), pyxides (*Plantago*), schizocarps (*Callitriche*), drupes (*Hippuris*), or drupelike (*Lagotis*). **Seeds** 1–300, white, tan, brown, yellow, gray, black, maroon, or red, ovoid to ellipsoid, cylindric, globular, oblong, conic, disciform, patelliform, or angled; embryo straight, rarely curved, endosperm present.

Genera ca. 100, species ca. 1900 (45 genera, 460 species in the flora): nearly worldwide, apparently absent in much of tropical South America, and arid northern Africa and northern Australia.

The concept of Plantaginaceae has expanded considerably since the molecular study by R. G. Olmstead and P. A. Reeves (1995) reported that many of the genera at that time placed in the Scrophulariaceae did not cluster with *Scrophularia* and instead formed a clade with genera that usually had been considered to be distantly related and placed in different families, including *Callitriche*, *Hippuris*, and *Plantago*. As treated here, Plantaginaceae include Callitrichaceae Link and Hippuridaceae Vest. Subsequent studies (for example, D. C. Albach et al. 2005; B. Oxelman et al. 2005) have confirmed this basic alignment; as yet, there is no comprehensive phylogenetic classification of the family. Albach et al. suggested that the genera of Plantaginaceae might be placed in 12 tribes, noting molecular support for at least six of them. Species in the flora area are distributed among nine of the 12 tribes: Angelonieae Pennell (genus 1), Antirrhineae Dumortier (genera 2–18), Callitricheae Dumortier (genera 19 and 20), Cheloneae D. Don (genera 21–27), Digitalideae Dumortier (genus 28), Gratioleae D. Don (genera 29–38), Plantagineae Dumortier (genera 39 and 40), Russelieae Pennell (genus 41), and Veroniceae Duby (genera 42–45). Angelonieae, Digitalideae, and Russelieae are represented in the flora area only by introduced species. Within the family, genera are arranged alphabetically within tribes.

In the flora area, 20 of the 45 genera contain one species each; *Penstemon* contains nearly half of all North American Plantaginaceae species. *Chelone*, *Chionophila*, *Holmgrenanthe*, *Howelliella*, *Nothochelone*, *Sophronanthe*, and *Tonella* are endemic. Genera that include popular ornamental plants are *Angelonia*, *Antirrhinum*, *Bacopa*, *Chelone*, *Digitalis*, *Linaria*, *Penstemon*, *Russelia*, *Veronica*, and *Veronicastrum*. Ten genera are represented only by introduced species: *Angelonia*, *Antirrhinum*, *Chaenorhinum*, *Cymbalaria*, *Digitalis*, *Kickxia*, *Limnophila*, *Linaria*, *Misopates*, and *Russelia*. *Linaria*, *Plantago*, *Scoparia*, and *Veronica* contain globally important pest species, some of which are listed as noxious weeds in parts of the flora area.

SELECTED REFERENCES Albach, D. C. et al. 2004. Evolution of Veroniceae: A phylogenetic perspective. Ann. Missouri Bot. Gard. 91: 275–302. Albach, D. C. et al. 2004b. A new classification of the tribe Veroniceae—Problems and a possible solution. Taxon 53: 429–452. Albach, D. C., H. M. Meudt, and B. Oxelman. 2005. Piecing together the "new" Plantaginaceae. Amer. J. Bot. 92: 297–315. Erbar, C. and P. Leins. 2004. Callitrichaceae. In: K. Kubitzki et al., eds. 1990+. The Families and Genera of Vascular Plants. 14+ vols. Berlin etc. Vol. 7, pp. 50–56. Ghebrehiwet, M., B. Bremer, and M. Thulin. 2000. Phylogeny of the tribe Antirrhineae (Scrophulariaceae) based on morphological and *ndh*F sequence data. Pl. Syst. Evol. 220: 223–239. Leins, P. and C. Erbar. 2004. Hippuridaceae. In: K. Kubitzki et al., eds. 1990+. The Families and Genera of Vascular Plants. 14+ vols. Berlin etc. Vol. 7, pp. 163–166. Pennell, F. W. 1946. Reconsideration of the *Bacopa-Herpestis* problem of the Scrophulariaceae. Proc. Acad. Nat. Sci. Philadelphia 98: 83–98. Sutton, D. A. 1988. A Revision of the Tribe Antirrhineae. London. Wolfe, A. D. et al. 1997. Using restriction-site variation of PCR-amplified cpDNA genes for phylogenetic analysis of tribe Cheloneae (Scrophulariaceae). Amer. J. Bot. 84: 555–564. Wolfe, A. D. et al. 2002. A phylogenetic and biogeographic analysis of the Cheloneae (Scrophulariaceae) based on ITS and *mat*K sequence data. Syst. Bot. 27: 138–148.

1. Corolla lobes 0; stamens 1; sepals 0 or minute rims at summits of ovaries.
   2. Leaves opposite; fruits schizocarps; ovaries superior; styles 2 . . . . 19. *Callitriche*, p. 49
   2. Leaves whorled; fruits drupes; ovaries inferior; styles 1 . . . . .20. *Hippuris*, p. 55
1. Corolla lobes 3–5, rarely 0; stamens 2–4(or 5); sepals 2–5.
   3. Corolla tubes spurred, ± saccate, or gibbous, or with rounded sacs at bases of median lobes.
      4. Corolla tubes with rounded sacs at bases of median lobes . . . . .1. *Angelonia*, p. 15
      4. Corolla tube bases spurred, ± saccate, or gibbous, sometimes obscurely so.
         5. Corolla tube bases ± saccate or gibbous adaxially, sometimes obscurely so.
            6. Corollas: middle lobes of abaxial lips folded lengthwise, enclosing stamens and styles; stamens included . . . . .23. *Collinsia*, p. 62
            6. Corollas: middle lobes of abaxial lips not folded lengthwise, not enclosing stamens and styles; stamens exserted . . . . .27. *Tonella* (in part), p. 256
         5. Corolla tube bases spurred or gibbous abaxially.
            7. Corolla tube bases spurred.
               8. Inflorescences terminal, racemes.

9. Corolla abaxial lips as long as or slightly longer than adaxials; filaments usually hairy proximally; capsules 9–12 mm . . . . 10. *Linaria*, p. 27

9. Corolla abaxial lips much longer than adaxials; filaments glabrous; capsules 2–4.8 mm . . . . 16. *Nuttallanthus*, p. 40

8. Inflorescences axillary or terminal, flowers solitary.

10. Stems ascending or erect; leaf blades linear, lanceolate, or oblanceolate; calyx lobes glandular-pubescent.

11. Pedicels 5–20 mm. . . . . 3. *Chaenorhinum*, p. 17

11. Pedicels 1–4 mm. . . . . 18. *Sairocarpus* (in part), p. 43

10. Stems pendent, decumbent, prostrate, or erect; leaf blades oblong-ovate, orbiculate, cordate, or reniform; calyx lobes glabrous or villous.

12. Perennials; stems glabrous; calyx lobes glabrous . . . . 4. *Cymbalaria*, p. 19

12. Annuals; stems villous to glandular-hairy; calyx lobes villous . . . . 9. *Kickxia*, p. 25

7. Corolla tube bases gibbous.

13. Inflorescences terminal, racemes.

14. Corollas strongly bilabiate, not personate. . . . . 8. *Howelliella*, p. 24

14. Corollas bilabiate and personate.

15. Corollas 25–45 mm; capsules 10–15 mm . . . . 2. *Antirrhinum*, p. 16

15. Corollas 9–18 mm; capsules 4–11 mm.

16. Calyx lobes linear, usually longer than corolla tubes in flower; seed wings present . . . . 13. *Misopates*, p. 36

16. Calyx lobes ovate to lanceolate, shorter than or as long as corolla tubes in flower; seed wings absent. . . . 18. *Sairocarpus* (in part), p. 43

13. Inflorescences axillary, flowers solitary.

17. Corollas bilabiate; stamens 2; staminodes 2 or 3 . . . . 14. *Mohavea*, p. 37

17. Corollas bilabiate and personate; stamens 4; staminodes 0 or 1.

18. Locules unequal . . . . 18. *Sairocarpus* (in part), p. 43

18. Locules equal.

19. Stems erect, ascending, or sprawling, glabrous or hairy; bracts absent; pedicels twining; pollen sacs 2 . . . . 15. *Neogaerrhinum*, p. 39

19. Stems erect, glandular-hairy; bracts present; pedicels not twining, recurved in fruit; pollen sacs 1 . . . . 17. *Pseudorontium*, p. 42

[3. Shifted to left margin.—Ed.]

3. Corolla tubes not spurred, gibbous, or saccate.

20. Shrubs or subshrubs.

21. Stamens 2 . . . . 44. *Veronica* (in part), p. 305

21. Stamens 4.

22. Distal leaf blades needlelike or scalelike; capsules densely packed with white, membranous hairs; inflorescences axillary, cymes. 41. *Russelia*, p. 294

22. Distal leaf blades not needlelike or scalelike; capsules not densely packed with white, membranous hairs; inflorescences terminal, panicles, thyrses, corymbs, racemes, or spikelike racemes.

23. Staminodes 0; bracteoles absent; leaves whorled . . . . 6. *Gambelia*, p. 21

23. Staminodes 1; bracteoles usually present; leaves opposite, subopposite, or whorled, distals rarely alternate.

24. Stamen filament bases eglandular-hairy; nectaries hypogynous discs . . . . 24. *Keckiella*, p. 75

24. Stamen filament bases glabrous or glandular-puberulent proximally, rarely pubescent distally; nectaries epistaminal . . . . 26. *Penstemon* (in part), p. 82

20. Herbs.

25. Fruits pyxides or nutlets; leaves basal only, rarely cauline; corolla lobes 4; stamens free.

26. Flowers unisexual; fruits nutlets; leaf blade margins entire. . . . . 39. *Littorella*, p. 280

26. Flowers bisexual; fruits pyxides; leaf blade margins entire or toothed . . . . . . 40. *Plantago*, p. 281

[25. Shifted to left margin.—Ed.]

25. Fruits capsules, rarely drupelike; leaves basal and cauline or cauline only, rarely basal only; corolla lobes 0 or 3–5; stamens adnate to corolla, rarely inserted on receptacle.
   27. Stamens 2 or 3.
      28. Ovaries 1-locular; leaves basal or basal and cauline.
         29. Leaf blade margins entire; petioles absent; sepals 5; inflorescences of solitary flowers; annuals . . . . .30. *Dopatrium*, p. 263
         29. Leaf blade margins toothed or deeply incised to pinnatifid; petioles present; sepals 2–4(or 5); inflorescences racemes; perennials . . . . 43. *Synthyris*, p. 296
      28. Ovaries 2-locular; leaves cauline, sometimes basal and cauline.
         30. Sepals 2; corolla lobes 3; fruits drupelike . . . . .42. *Lagotis*, p. 294
         30. Sepals 4 or 5; corolla lobes 4 or 5; fruits capsules.
            31. Bracts absent; stems prostrate; corollas radially symmetric, rarely bilaterally symmetric, regular. . . . . .29. *Bacopa* (in part), p. 260
            31. Bracts present; stems creeping to decumbent, ascending, or erect; corollas bilaterally symmetric, weakly bilabiate, bilabiate, or bilabiate and personate, sometimes regular.
               32. Corolla lobes 5.
                  33. Leaf blades not leathery; pollen sacs perpendicular to filaments, connectives dilated . . . . .31. *Gratiola*, p. 264
                  33. Leaf blades leathery; pollen sacs parallel to filaments, connectives not dilated . . . . .37. *Sophronanthe*, p. 277
               32. Corolla lobes 4.
                  34. Leaves opposite, distals sometimes alternate; stamen filaments glabrous . . . . .44. *Veronica* (in part), p. 305
                  34. Leaves whorled, rarely opposite; stamen filaments hairy proximally . . . . 45. *Veronicastrum*, p. 322
   27. Stamens 4(or 5).
      35. Leaves alternate.
         36. Staminodes 0; inflorescences terminal, racemes; bracts present. . . . . .28. *Digitalis*, p. 258
         36. Staminodes 1; inflorescences axillary, flowers solitary; bracts absent.
            37. Leaf blade margins entire; corollas blue to violet, pink, or red.
               38. Annuals; seed wings present . . . . 5. *Epixiphium*, p. 20
               38. Perennials; seed wings absent . . . . .12. *Maurandella*, p. 35
            37. Leaf blade margins dentate or spinulose; corollas pale yellow to yellow or ochroleucous.
               39. Leaf blade margins spinulose; stems erect; ovaries 1-locular . . . . 7. *Holmgrenanthe*, p. 22
               39. Leaf blade margins dentate; stems pendent; ovaries 2-locular . . . . 11. *Mabrya*, p. 33
      35. Leaves opposite or whorled, distals sometimes alternate.
         40. Plants paludal or aquatic; leaves dimorphic, margins of submerged leaves pinnatifid, margins of aerial leaves entire, serrate, or pinnatifid . . . . 33. *Limnophila*, p. 271
         40. Plants terrestrial or, if paludal or aquatic, leaves monomorphic and margins of leaves rarely pinnatifid.
            41. Corollas radially symmetric.
               42. Corolla lobes 5, rarely 4, throats not densely pilose internally; bracteoles present . . . . .29. *Bacopa* (in part), p. 260
               42. Corolla lobes 4, throats densely pilose internally; bracteoles absent . . . . .36. *Scoparia*, p. 276
            41. Corollas bilaterally symmetric, rarely nearly radially symmetric.
               43. Seeds 2 or 4; stamens equal; annuals; stigmas linear . . . . 27. *Tonella* (in part), p. 256
               43. Seeds (2–)5–150; stamens didynamous, rarely equal; perennials, sometimes annuals; stigmas capitate, 2-lobed, or cuneate.
                  44. Staminodes 1.
                     45. Bracteoles absent; leaf blade margins entire; seed wings present . . . . .22. *Chionophila*, p. 61
                     45. Bracteoles present; leaf blade margins toothed or entire; seed wings absent or present.

[46. Shifted to left margin.—Ed.]

46. Bracteoles ± as large as calyx lobes, nearly surrounding calyces of flowers they subtend . . . . 21. *Chelone* (in part), p. 57
46. Bracteoles smaller than calyx lobes, not surrounding calyces of flowers they subtend.
    47. Inflorescences spikes, racemes, or flowers solitary; calyces bilaterally symmetric; capsule dehiscence loculicidal . . . . 38. *Stemodia* (in part), p. 279
    47. Inflorescences cymes or thyrses; calyces radially symmetric or nearly so; capsule dehiscence septicidal.
        48. Stamen filaments hirsute proximally, puberulent distally; seed wings present; nectaries hypogynous discs . . . . 25. *Nothochelone*, p. 80
        48. Stamen filaments glabrous or glandular-puberulent proximally, rarely pubescent distally; seed wings absent, sometimes narrow; nectaries epistaminal . . . . 26. *Penstemon* (in part), p. 82

[44. Shifted to left margin.—Ed.]

44. Staminodes 0.
    49. Leaf blade margins pinnatifid to bipinnatifid; bracteoles absent.
        50. Corollas pale blue to pale lavender; seeds white, parallel-ridged . . . . 32. *Leucospora*, p. 270
        50. Corollas pink or purple; seeds yellow or brown, spirally ridged . . . . 35. *Schistophragma*, p. 275
    49. Leaf blade margins entire, crenate, serrate, dentate, or denticulate; bracteoles present.
        51. Leaf blade margins entire; sepals distinct, outer wider than inner . . . . 29. *Bacopa* (in part), p. 260
        51. Leaf blade margins toothed; sepals basally or proximally connate, outer calyx lobes ± as wide as inner.
            52. Inflorescences spikelike cymes; bracteoles ± as large as calyx lobes, nearly surrounding calyces of flowers they subtend; calyces radially symmetric . . . . 21. *Chelone* (in part), p. 57
            52. Inflorescences spikes, racemes, or flowers solitary; bracteoles smaller than calyx lobes, not surrounding calyces of flowers they subtend; calyces bilaterally symmetric.
                53. Stems glabrous; petioles present. . . . . 34. *Mecardonia*, p. 273
                53. Stems hairy; petioles absent . . . . 38. *Stemodia* (in part), p. 279

1. ANGELONIA Bonpland in A. von Humboldt and A. J. A. Bonpland, Pl. Aequinoct. 2: 92, plate 108. 1812 • [Latin rendering of Venezuelan common name *angelon*] ☐I

Kerry A. Barringer

Neil A. Harriman

Herbs, annual or perennial. **Stems** erect, hairy [glabrous]. **Leaves** cauline, opposite [alternate distally]; petiole absent; blade not fleshy, not leathery, margins serrate to subentire [entire]. **Inflorescences** terminal, racemes [thyrses or flowers solitary]; bracts present. **Pedicels** present; bracteoles absent [present]. **Flowers** bisexual; sepals 5, basally connate, calyx slightly bilaterally symmetric, campanulate to rotate, lobes lanceolate; corolla purple, mauve, or white, bilaterally symmetric, bilabiate, short-tubular [globular], not spurred, tube with a pair of rounded sacs at base of median lobe abaxially, lobes 5, abaxial 3, adaxial 2, adaxial lip with concave palate at base containing a cylindric, 2-fid tooth [ridge]; stamens 4, basally adnate to corolla, didynamous, filaments glandular-hairy [glabrous]; staminode 0; ovary 2-locular, placentation axile; stigma punctiform. **Fruits** capsules, dehiscence loculicidal. **Seeds** 10–50, light brown to brown, prismatic, wings absent. $x = 10$.

Species 25 (1 in the flora): introduced, Florida; Mexico, West Indies, Central America, South America; introduced also in Asia, Africa, Pacific Islands.

1. Angelonia gardneri Hooker, Bot. Mag. 66: plate 3754. 1839 F I

Stems slightly 4-angled, 30–50 cm, stipitate-glandular, especially on angles. **Leaves:** blade narrowly lanceolate, 40–90 × 5–15 mm, smaller distally, base narrowed, apex acute, abaxial surface glandular-pubescent, adaxial glabrous, sometimes sparsely glandular-pubescent when young. **Racemes** glandular-pubescent; bracts 5–20 × 4–10 mm, larger in fruit, flowers 1 per axil. **Pedicels** strongly recurved in fruit, 9–12 mm, glandular-pubescent. **Flowers:** sepals 3–4 × 2–3 mm, margins pale; corolla tube white to purple with darker spots within, 3–5 mm, throat 9–12 mm diam., palate concave, 3–5 mm; filaments 3–4 mm. **Capsules** globular, 3–8 mm, glabrous; style and stigma persistent. **Seeds** 1–1.5 mm, reticulate-alveolate. $2n = 20$ (India).

Flowering May–Oct. Roadsides, disturbed ground; 0–100 m; introduced; Fla.; South America (Brazil); introduced also elsewhere in South America, Asia, Africa, Pacific Islands.

Mature fruits are seldom found in the flora area. Purple, pink, white, and variegated corolla forms of *Angelonia gardneri* and the Mexican *A. angustifolia* Bentham, distinguished by its glabrous stems and larger flowers, are grown as annuals in temperate North America, Europe, and Asia.

Angelonia gardneri has not been collected outside of cultivation since the 1940s; it has become much more common in cultivation and is very likely to escape.

2. ANTIRRHINUM Linnaeus, Sp. Pl. 2: 612. 1753; Gen. Pl. ed. 5, 268. 1754

• Snapdragon [Greek *anti*, like or resembling, and *rhinos*, nose, alluding to shape of corolla] I

Kerry A. Barringer

Neil A. Harriman

Herbs [shrubs], perennial or annual. **Stems** erect, filiform, twining branches absent, glabrous or hairy. **Leaves** cauline, opposite proximally, alternate distally; petiole absent [present]; blade not fleshy, not leathery, margins entire. **Inflorescences** terminal [axillary], racemes [flowers solitary]; bracts present. **Pedicels** present; bracteoles absent [present]. **Flowers** bisexual; sepals 5, basally connate, calyx slightly bilaterally symmetric, cupulate, lobes ovate to lanceolate; corolla mauve, purple, red, yellow, or white, bilaterally symmetric, bilabiate and personate, tubular [funnelform], 25–45 mm, tube base gibbous abaxially, not spurred, lobes 5, abaxial 3, adaxial 2; stamens 4, basally adnate to corolla, didynamous, filaments glabrous, pollen sacs 2 per filament; staminode 0 or 1, minute; ovary 2-locular, placentation axile; stigma capitate. **Fruits** capsules, 10–15 mm, locules unequal, dehiscence poricidal. **Seeds** 20–100, light brown to brown, ovoid, reticulate, wings absent. $x = 8$.

Species 19 (1 in the flora): introduced; Europe (Mediterranean region); introduced also in temperate regions nearly worldwide.

Antirrhinum is sometimes treated as a larger genus including both the species of Mediterranean Europe and the species of western North America (D. M. Thompson 1988). Here, the North American species are treated in five genera, *Howelliella*, *Mohavea*, *Neogaerrhinum*, *Pseudorontium*, and *Sairocarpus*, believed to be more closely related to one another than to the European species (D. A. Sutton 1988). There is not yet enough evidence to resolve these relationships; DNA sequence data seem to indicate that the European and North American species are sister groups and that there are at least two groups of closely related species among the American taxa (R. K. Oyama and D. A. Baum 2004; M. Fernández-Mazuecos et al. 2013).

Antirrhinum is distinguished from the segregate genera by the radially symmetric seeds, terminal (versus oblique) styles, and flowers more than 2 cm.

SELECTED REFERENCE Thompson, D. M. 1988. Systematics of *Antirrhinum* (Scrophulariaceae) in the New World. Syst. Bot. Monogr. 22: 1–142.

1. Antirrhinum majus Linnaeus, Sp. Pl. 2: 617. 1753

• Muflier commun

Stems terete, 3–8(–15) dm, glabrous or sparsely stipitate-glandular proximally, stipitate-glandular distally. **Leaves:** blade narrowly elliptic to lanceolate, 50–70 × 5–20 mm, glabrous or sparsely stipitate-glandular proximally. **Inflorescences** stipitate-glandular, sometimes glabrous; bracts similar to distal leaves. **Pedicels** 1–7 mm, stipitate-glandular. **Flowers:** sepals 5–10 mm, stipitate-glandular; corolla palate yellow; filaments 20–40 mm. **Capsules** 7–10 mm wide. *2n* = 16 (Europe).

Flowering May–Oct. Disturbed ground; 0–2000 m; introduced; B.C., Ont., Que.; Calif., Conn., D.C., Ill., Iowa, La., Mass., Mich., Mo., N.Y., Ohio, Oreg., Pa., Utah, Vt., Va., Wash., Wis.; sw Europe; introduced also in Mexico, West Indies, Central America, South America, Asia, Africa, Australia.

Antirrhinum majus is a popular garden plant grown as an annual; it occasionally escapes but is short-lived. Some cultivars have been developed with different growth forms, corolla colors, or open-throated flowers.

3. CHAENORHINUM (de Candolle) Reichenbach, Consp. Regn. Veg., 123. 1828 (as Chaenarrhinum) • Dwarf-snapdragon [Greek *chaino*, to gape, and *rhis*, snout, alluding to open throat of corolla as compared to *Antirrhinum* and *Linaria*] I

Richard K. Rabeler

Craig C. Freeman

Linaria Miller sect. *Chaenorhinum* de Candolle in J. Lamarck and A. P. de Candolle, Fl. Franç. ed. 3, 6: 410. 1815

Herbs, annual [perennial]. **Stems** erect or ascending, glandular-pubescent, sometimes glabrate. **Leaves** cauline, opposite proximally, alternate distally; petiole present proximally, sometimes absent distally; blade oblanceolate to lanceolate, not fleshy, not leathery, margins entire. **Inflorescences** axillary [terminal], flowers solitary [racemes]; bracts present. **Pedicels** 5–20 mm; bracteoles absent [present]. **Flowers** bisexual; sepals 5, proximally connate, calyx bilaterally symmetric, urceolate-spreading, lobes narrowly oblanceolate to narrowly lanceolate, glandular-pubescent; corolla white to grayish or lilac [blue, pink, pale yellow], usually with 2 purple spots or stripes internally on palate, bilaterally symmetric, bilabiate and personate, cylindric, tube base not gibbous, spurred abaxially, lobes 5, abaxial 3, adaxial 2; stamens 4, adnate to corolla, didynamous, filaments glabrous; staminode 0[1]; ovary 2-locular, placentation axile; stigma capitate. **Fruits** capsules, dehiscence poricidal. **Seeds** 40–60, blackish brown to brown [black, grayish], ovoid to ellipsoid, wings absent. *x* = 7.

Species 23 (1 in the flora): introduced; Europe, Asia (Georgia, Turkmenistan), n Africa; introduced also in e Asia.

Contrary to usual practice, de Candolle did not double the r in his spelling of *Chaenorhinum*; the orthography was changed in later works, including de Candolle's own (including the 1829 reissue of the 1815 volume), as some authors thought it should be corrected (for example, D. A. Sutton 1988).

Capsule shape, wall thickness, and dehiscence; seed surface ridging; and corolla lobe margins are features used in placing *Chaenorhinum* species into either three sections (D. A. Sutton 1988) or three genera (F. Speta 1980). In the latter scheme, *C. minus*, along with nine other species, is treated as *Microrrhinum* (Endlicher) Fourreau. Recent molecular studies (N. Yousefi et al. 2016) demonstrated that the *Chaenorhinum* clade consists of two main subclades mostly corresponding to sect. *Chaenorhinum* and sect. *Microrrhinum* (including two of the recent segregate genera of Speta, *Albraunia* Speta and *Holzneria* Speta; see Speta 1982); however, *C. minus* probably belongs to the third, early-branching subclade, which is sister to the rest of the genus. Thus, it is better to accept *Chaenorhinum* in a wider sense, including *Microrrhinum*.

Chaenorhinum origanifolium (Linnaeus) Kosteletzky, a perennial native to southwestern Mediterranean Europe, is sometimes grown in rock gardens; it might be expected to escape locally.

SELECTED REFERENCES Arnold, R. M. 1982. Floral biology of *Chaenorrhinum minus* (Scrophulariaceae) a self-compatible annual. Amer. Midl. Naturalist 108: 317–324. Widrlechner, M. P. 1983. Historical and phenological observations on the spread of *Chaenorrhinum minus* across North America. Canad. J. Bot. 61: 179–187.

1. Chaenorhinum minus (Linnaeus) Lange in H. M. Willkomm and J. M. C. Lange, Prodr. Fl. Hispan. 2: 577. 1870 (as Chaenorrhinum) • Lesser toadflax, chénorhinum mineur [F] [I] [W]

Antirrhinum minus Linnaeus, Sp. Pl. 2: 617. 1753

Subspecies 4 (1 in the flora): introduced; Europe, sw Asia; introduced also in e Asia.

1a. Chaenorhinum minus (Linnaeus) Lange subsp. **minus** [F] [I] [W]

Stems much-branched from or near base, sometimes un-branched, often zig-zag, (4–) 8–28(–40) cm. **Leaves:** blade 5–15(–30) × 1–4(–5) mm, base tapered, apex obtuse to acute, surfaces glandular-pubescent. **Pedicels** ascending. **Flowers:** calyx lobes accrescent, abaxial 1.5–2.5 × 0.3–0.4 mm, adaxial 2.8–3.5 × 0.4–0.8 mm, margins entire, apex obtuse; corolla 8–11 mm (including 1–2.3 mm spur), sparsely glandular-pubescent externally, glandular-pubescent internally on abaxial surface, especially along ridges of palate and into throat, throat 2–3 mm diam., abaxial lobes spreading, adaxial projecting; stamens included, anthers opposite, navic-ular, marginally coherent, pollen sacs of longer pair of stamens 0.3–0.4 mm, of shorter pair of stamens 0.1–0.2 mm, glabrous; ovary glandular-pubescent; style 1.5–2 mm, glandular-pubescent proximally. **Capsules** obovoid to ellipsoid, 4–5.4 × 2–3.6 mm, glandular-pubescent distally. **Seeds** 0.6–0.9 mm, prominently ribbed longitudinally. $2n = 14, 28$ (both Europe).

Flowering May–Oct. Gravelly railroad rights-of-way, road shoulders, urban areas, streambeds; 0–1600 m; introduced; Alta., B.C., Man., N.B., Nfld. and Labr. (Nfld.), N.W.T., N.S., Ont., P.E.I., Que., Sask.; Ala., Ark., Colo., Conn., Del., Ga., Idaho, Ill., Ind., Iowa, Kans., Ky., La., Maine, Md., Mass., Mich., Minn., Mo., Mont., Nebr., N.H., N.J., N.Y., N.C., N.Dak., Ohio, Okla., Oreg., Pa., R.I., Tenn., Tex., Vt., Va., Wash., W.Va., Wis.; Europe; sw Asia; introduced also in e Asia (Russian Far East).

Subspecies *minus* is the only member of *Chaenorhinum minus* to have become widely established as a weed; the other three subspecies occur in Europe, one each in Corsica, Crete, and Turkey.

First collected in Camden, New Jersey (1874), and St. John, New Brunswick (1881), *Chaenorhinum minus* likely came to North America as a contaminant in ship ballast, with subsequent dispersal along railroad lines (M. P. Widrlechner 1983). R. M. Arnold (1981, 1982) noted that self-compatibility, a short generation time, drought tolerance, and seed dispersal enhanced by passing trains allowed *C. minus* to be a successful colonizer of railroad rights-of-way, where it was once common but is now scarce because of herbicide use.

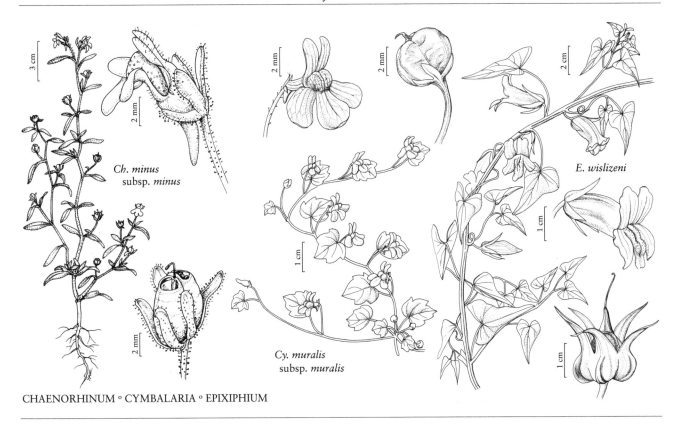

Ch. minus subsp. *minus*

Cy. muralis subsp. *muralis*

E. wislizeni

CHAENORHINUM ∘ CYMBALARIA ∘ EPIXIPHIUM

4. CYMBALARIA Hill, Brit. Herb., 113, plate 17 [upper left center]. 1756 • [Latin *cymbalum*, rounded, concave, and -*aria*, resemblance, alluding to leaf shape] I

Wayne J. Elisens

Herbs, perennial. **Stems** pendent or prostrate, glabrous [villous]. **Leaves** cauline, alternate [opposite]; petiole present; blade reniform to orbiculate, fleshy, not leathery, margins lobed [entire]. **Inflorescences** axillary, flowers solitary; bracts absent. **Pedicels** present, often elongating in fruit; bracteoles absent. **Flowers** bisexual; sepals 5, basally connate, calyx radially symmetric, campanulate, lobes linear to lanceolate, glabrous; corolla lilac to violet [blue, white], bilaterally symmetric, bilabiate and personate, tubular, tube base not gibbous, spurred abaxially, lobes 5, abaxial 3, adaxial 2; stamens 4, basally adnate to corolla, didynamous, filaments glabrous; staminode 0; ovary 2-locular, placentation axile; stigma scarcely clavate, entire. **Fruits** capsules, dehiscence loculicidal [indehiscent]. **Seeds** [6–]15–40, dark brown to black, ovoid to oblong, wings absent. $x = 7$.

Species 9 (1 in the flora): introduced; Europe (Mediterranean region), sw Asia; introduced in Mexico, Central America, South America, Asia, Afric, Atlantic Islands, Pacific Islands, Australia.

Cymbalaria is characterized by personate, spurred corollas and lobed, reniform to orbiculate leaf blades with palmate venation. Generic delimitation has been consistent since R. Wettstein (1891–1893) treated *Cymbalaria* as distinct from *Linaria*. *Cymbalaria muralis* has been included in molecular phylogenetic analyses of Antirrhineae and Plantaginaceae. Using multiple plastid and nuclear molecular markers, *Cymbalaria* is strongly supported as a monophyletic clade sister to *Asarina procumbens* Miller (M. Ghebrehiwet et al. 2000; D. C. Albach et al. 2005;

M. Fernández-Mazuecos et al. 2013; P. Carnicero et al. 2017). Several investigations support sister group status for *Asarina* Miller and *Cymbalaria* and the New World Maurandyinae clade represented by *Epixiphium*, *Lophospermum* D. Don, *Maurandya* Ortega, *Maurandella*, and *Rhodochiton* Zuccarini ex Otto & A. Dietrich.

1. **Cymbalaria muralis** P. Gaertner, B. Meyer & Scherbius, Oekon. Fl. Wetterau 2: 397. 1800 • Kenilworth ivy F I W

Antirrhinum cymbalaria Linnaeus, Sp. Pl. 2: 612. 1753; *Linaria cymbalaria* (Linnaeus) Miller

Subspecies 3 (1 in the flora): introduced; Europe; introduced also in Mexico, Central America, South America, Asia, Africa, Atlantic Islands, Pacific Islands, Australia.

With its five- to nine-lobed leaves and glabrous stems and leaves, *Cymbalaria muralis* is distinctive within *Cymbalaria*. Three subspecies are recognized; only subsp. *muralis* occurs in North America (D. A. Sutton 1988); it is distributed widely throughout the world. Subspecies *visianii* (Kümmerle ex Jávorka) D. A. Webb and subsp. *pubescens* (J. Presl & C. Presl) D. A. Webb occur in Italy and the Balkan Peninsula. The pedicels of *C. muralis* elongate and recurve during fruit maturation, which results in significant amounts of short-range seed dispersal (T. Junghans and E. Fischer 2007).

1a. **Cymbalaria muralis** P. Gaertner, B. Meyer & Scherbius subsp. **muralis** F I W

Perennials 10–60 cm. **Leaves:** petiole 10–22 mm; blade 5–40 × 6–60 mm, surfaces glabrous or with scattered hairs. **Pedicels** ascending, 10–35 mm in flower, recurved, to 90 mm in fruit. **Flowers:** sepals equal, lobes 1.5–3 × 0.5–1 mm, apex acute; corolla tube 3–5 mm, glabrous, palate inflated, abaxial plicae yellow, lobes: abaxial spreading, adaxial erect, subequal, 2–3 mm, apex round; stamens included, filaments incurved, abaxial 2–4 mm, adaxial 3–5 mm, pollen sacs oblong; ovary glabrous, locules subequal; style included, 3–4 mm, base persistent in fruit; stigma straight. **Capsules** globular, 3–5 mm. **Seeds** 0.5–1 mm, cristate-tuberculate. *2n* = 14.

Flowering May–Oct. Rock walls, rocky disturbed sites; 0–400 m; introduced; B.C., Man., N.B., Ont., Que.; Ark., Calif., Colo., Conn., Del., Ill., Ind., Ky., Md., Mass., Mich., Mo., Nebr., N.J., N.C., Ohio, Oreg., Pa., S.C., S.Dak., Tenn., Vt., Va., Wash., W.Va., Wis.; s Europe; introduced also in Mexico, Central America, South America, Asia, Africa, Atlantic Islands, Pacific Islands, Australia.

Subspecies *muralis* occurs to 2000 m elevation outside the flora area.

5. **EPIXIPHIUM** (Engelmann ex A. Gray) Munz, Proc. Calif. Acad. Sci., ser. 4, 15: 380. 1926 • [Greek *epi-*, upon, and *xiphos*, sword, alluding to sword-shaped persistent style]

Wayne J. Elisens

Maurandya Ortega subg. *Epixiphium* Engelmann ex A. Gray, Proc. Amer. Acad. Arts 7: 377. 1868 (as Maurandia)

Herbs, annual, taprooted. **Stems** climbing or scrambling, glabrous. **Leaves** cauline, alternate; petiole twining; blade fleshy, not leathery, margins entire. **Inflorescences** axillary, flowers solitary; bracts absent. **Pedicels** present; bracteoles absent. **Flowers** bisexual; sepals 5, distinct, lanceolate, calyx radially symmetric, campanulate; corolla blue to violet, bilaterally symmetric, bilabiate, tubular, tube base not spurred or gibbous, lobes 5, abaxial 3, adaxial 2; stamens 4, basally adnate to corolla, didynamous, filaments basally hairy; staminode 1, filamentous; ovary 2-locular, placentation axile; stigma 2-lobed. **Fruits** capsules, dehiscence loculicidal. **Seeds** 100–200, dark brown, ovoid-ellipsoid, wings present. *x* = 12.

Species 1: sw, sc United States, n Mexico.

Epixiphium is defined by distinctive characteristics in Antirrhineae: annual life cycle, taproots, keeled sepals, indurate capsules with regular transverse dehiscence, persistent style bases, and winged seeds. Based on this morphological distinctness, *Epixiphium* has been recognized either as a section or subgenus within *Maurandya* or as a genus. The latter view is adopted here based on the number of unique morphological characteristics. In phylogenetic studies based on morphological data, *Epixiphium* is either part of a trichotomy with *Lophospermum* D. Don and *Rhodochiton* Zuccarini ex Otto & A. Dietrich and sister to a *Maurandya* and *Maurandella* clade (M. Ghebrehiwet et al. 2000) or is basal within a *Maurandya* clade (W. J. Elisens 1985). Molecular ITS data placed *Epixiphium* in the *Cymbalaria* clade (M. Fernández-Mazuecos et al. 2013), sister to six other genera in a subclade that was sister to *Asarina* and *Cymbalaria*. *Epixiphium wislizeni* has not been included in phylogenetic studies using molecular data.

1. **Epixiphium wislizeni** (Engelmann ex A. Gray) Munz, Proc. Calif. Acad. Sci., ser. 4, 15: 380. 1926 • Sand snapdragon [F]

Maurandya wislizeni Engelmann ex A. Gray in W. H. Emory, Rep. U.S. Mex. Bound. 2(1): 111. 1859 (as Maurandia); *Asarina wislizeni* (Engelmann ex A. Gray) Pennell

Annuals 30–120 cm, vines. **Leaves:** petiole 14–54 mm; blade hastate to broadly sagittate, 21–75 × 8–48 mm, surfaces glabrous. **Pedicels** ascending, 3–9 mm, thickened in fruit. **Flowers:** sepals 13–18 × 2–4 mm, basally keeled, membranous in flower, 20–32 × 8–12 mm, indurate in fruit; corolla tube 17–22 mm, glabrous or sparsely hairy, throat open, palate not inflated, abaxial plicae white spotted with blue or violet, lobes: abaxial reflexed, adaxial erect, equal, 7–12 mm; stamens included, filaments incurved, hairy at base, abaxial 11–13 mm, adaxial 13–15 mm, pollen sacs oblong; ovary glabrous, locules subequal; style included, 13–15 mm, indurate in fruit, stigma recurved. **Capsules** ovoid, compressed apically, 11–15 mm, indurate. **Seeds** 3–4 mm, surface tuberculate, circumalate. $2n = 24$.

Flowering Apr–Nov. Active and stabilized siliceous and gypseous dunes and sandy soils; 1100–2100 m; Ariz., Calif., N.Mex., Tex.; Mexico (Chihuahua).

Epixiphium wislizeni is easily identified in fruit and by its occurrence on unconsolidated, sandy soils. When vegetative or in flower, it resembles *Maurandella antirrhiniflora*. In Texas, *E. wislizeni* is known from the trans-Pecos region.

6. **GAMBELIA** Nuttall, Proc. Acad. Nat. Sci. Philadelphia 4: 7. 1848 • Greenbright [For William Gambel, 1823–1849, American naturalist] [C]

Wayne J. Elisens

Shrubs. Stems pendent or erect, glabrous or glandular-hairy. **Leaves** persistent, cauline, whorled [opposite]; petiole present; blade not fleshy, not leathery, margins entire, distal leaf blade not needlelike or scalelike. **Inflorescences** terminal [axillary], racemes; bracts present. **Pedicels** present; bracteoles absent. **Flowers** bisexual; sepals 5, basally connate, calyx radially symmetric, campanulate, lobes lanceolate [ovate]; corolla red, bilaterally symmetric, bilabiate and personate [open-throated], tubular, tube base not spurred or gibbous, lobes 5, abaxial 3, adaxial 2; stamens 4, basally adnate to corolla, didynamous, filaments basally hairy, distally glabrous; staminode 0; ovary 2-locular, placentation axile; stigma 2-lobed. **Fruits** capsules, dehiscence loculicidal. **Seeds** 40–80[–120], dark brown to black, ovoid [oblong], wings absent. $x = 15$.

Species 4 (1 in the flora): California, w Mexico.

Gambelia is defined by its shrubby habit, whorled (rarely opposite) leaves, and red corollas. Generic delimitation has varied considerably with species placed in *Gambelia* (D. A. Sutton 1988; E. Fischer 2004) and *Saccularia* Kellogg (W. Rothmaler 1943) or combined with South American species in *Galvezia* Dombey ex Jussieu (P. A. Munz 1926). Most authors recognize *Gambelia* as exclusively North American and *Galvezia* as comprising species in South America and the Galapagos Islands (Sutton; Fischer). *Gambelia speciosa* was included in molecular phylogenetic analyses of Antirrhineae by M. Ghebrehiwet et al. (2000) and M. Fernández-Mazuecos et al. (2013). The latter investigation demonstrated strong support for recognition of a North American *Gambelia* nested in an *Antirrhinum* clade distinct from South American species of *Galvezia* nested within a *Galvezia* clade. The extent of isozyme, morphological, and molecular divergence (Sutton; W. J. Elisens and A. D. Nelson 1993; Ghebrehiwet et al.) among species in this clade is consistent with recognition of *Gambelia* for North American species and *Galvezia* for South American species.

SELECTED REFERENCE Elisens, W. J. and A. D. Nelson. 1993. Morphological and isozyme divergence in *Gambelia* (Scrophulariaceae): Species delimitation and biogeographic relationships. Syst. Bot. 18: 454–468.

1. **Gambelia speciosa** Nuttall, Proc. Acad. Nat. Sci. Philadelphia 4: 7. 1848 • Shrub snapdragon [C][F]

Galvezia speciosa (Nuttall) A. Gray

Plants 60–150 cm. **Leaves:** petiole 3–8 mm; blade elliptic to ovate, 15–45 × 4–26 mm, glabrous or glandular-hairy. **Pedicels** ascending, 10–20 mm. **Flowers:** sepals equal, lanceolate, 6–10 × 1–3 mm, apex acuminate, glandular-hairy; corolla tube 15–18 mm, glandular-hairy, palate inflated, abaxial plicae absent; lobes: abaxial spreading, adaxial erect, subequal, 2–5 mm, apex round; stamens included to slightly exserted, filaments incurved, abaxial 10–12 mm, adaxial 11–13 mm, pollen sacs oblong; ovary glandular-hairy, locules subequal; style included, 10–12 mm, base persistent in fruit, stigma slightly recurved. **Capsules** ovoid to globular, 6–9 mm. **Seeds** 1–2 mm, irregularly cristate. **2*n* = 30.**

Flowering Mar–Jun. Rock walls in coastal canyons, cliffs; of conservation concern; 0–500 m; Calif.; Mexico (Baja California).

Gambelia speciosa is unique among the four species in *Gambelia* by its personate corolla and its geographic distribution on San Clemente and Santa Catalina islands (California) and Guadalupe Island (Mexico). Populations within and among islands are variable in indument.

7. HOLMGRENANTHE   Elisens, Syst. Bot. Monogr. 5: 54, fig. 24. 1985 • Rocklady [For Arthur Herman Holmgren, 1912–1992, Noel Herman Holmgren, b. 1937, and Patricia Kern Holmgren, b. 1940, American botanists, and Greek *anthos*, flower]   [C][E]

Wayne J. Elisens

Herbs, perennial; caudex woody. **Stems** erect, hairy. **Leaves** cauline, alternate; petiole present; blade not fleshy, leathery, margins spinulose. **Inflorescences** axillary, flowers solitary; bracts absent. **Pedicels** present; bracteoles absent. **Flowers** bisexual; sepals 5, distinct, lanceolate, calyx radially symmetric, urceolate; corolla pale yellow, bilaterally symmetric, weakly bilabiate, tubular, tube base not spurred or gibbous, lobes 5, abaxial 3, adaxial 2; stamens 4, basally adnate to corolla, didynamous, filaments glandular-hairy; staminode 1, filamentous; ovary 1-locular, placentation axile; stigma conical. **Fruits** capsules, dehiscence loculicidal. **Seeds** 30–70, tan, polygonal-pyramidal, wings absent.

Species 1: California.

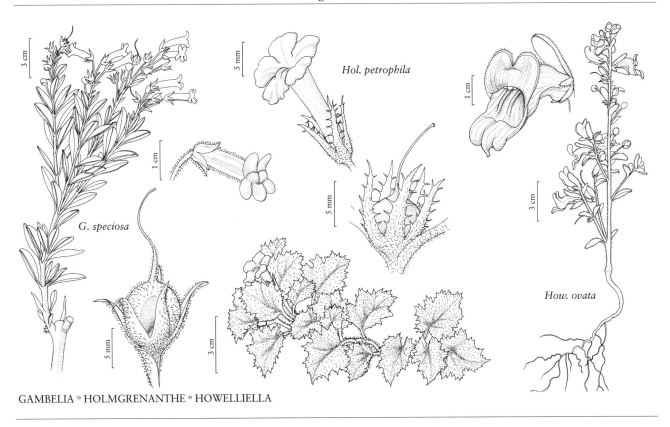

G. speciosa

Hol. petrophila

How. ovata

GAMBELIA ° HOLMGRENANTHE ° HOWELLIELLA

Holmgrenanthe is defined by characteristics unique or uncommon in Antirrhineae: cespitose habit, leaves and sepals with spinulose margins, a one-locular ovary with a T-shaped septum, and foveolate, pyramidal seeds. Based on its morphological distinctness, *Holmgrenanthe* has been recognized as monotypic since its delimitation (D. A. Sutton 1988; E. Fischer 2004). In phylogenetic studies based on morphological data, *H. petrophila* is basal to other genera (M. Ghebrehiwet et al. 2000) or is part of an unresolved basal trichotomy among genera (W. J. Elisens 1985) within the *Maurandya* clade (Maurandyinae). Bayesian and maximum likelihood analyses of nuclear ITS data in Antirrhineae (M. Fernández-Mazuecos et al. 2013) placed *Holmgrenanthe* in a *Cymbalaria* clade comprising two subclades, one including *Holmgrenanthe* and six other genera in subtribe Maurandyinae and the other *Asarina* and *Cymbalaria*.

1. **Holmgrenanthe petrophila** (Coville & C. V. Morton) Elisens, Syst. Bot. Monogr. 5: 54. 1985 [C][E][F]

Maurandya petrophila Coville & C. V. Morton, J. Wash. Acad. Sci. 25: 292, fig. [p. 292]. 1935 (as Maurandia); *Asarina petrophila* (Coville & C. V. Morton) Pennell; *Maurandella petrophila* (Coville & C. V. Morton) Rothmaler

Plants 5–12 cm, cespitose. **Stems** brittle. **Leaves:** petiole 12–27 mm; blade orbiculate to reniform, 12–35 × 14–27 mm, apex spinulose. **Pedicels** ascending, 1–4 mm, glandular-puberulent to glandular-villous. **Flowers:** sepals falcate, 9–13 × 2–3 mm, margins spinulose, glandular-villous; corolla tube 20–24 mm, glandular-villous, throat open, palate not inflated, abaxial plicae dark yellow with ligulate hairs; lobes: abaxial projecting, adaxial erect, equal, rounded, 7–12 mm; stamens included, filaments incurved, yellow at base, abaxial 12–14 mm, adaxial 7–9 mm, pollen sacs oblong; ovary with a T-shaped septum; style included, terete, 9–10 mm, stigma recurved. **Capsules** globular, 8–10 mm. **Seeds** 2–3 mm, surface foveolate.

Flowering Apr–Jun. Calcareous canyon walls; of conservation concern; 700–1800 m; Calif.

Holmgrenanthe petrophila is known from the limestone walls of Fall and Titus canyons in the Grapevine Mountains in Death Valley National Monument, Inyo County.

8. HOWELLIELLA Rothmaler, Leafl. W. Bot. 7: 115. 1954 • [For John Thomas Howell, 1903–1994, California botanist] E

Kerry A. Barringer

Herbs, annual. **Stems** erect or ascending, filiform, twining branches absent, glandular-pubescent. **Leaves** cauline, opposite proximally, alternate distally; petiole present proximally, nearly absent distally; blade not fleshy, not leathery, margins entire. **Inflorescences** terminal, racemes; bracts present. **Pedicels** present, not twining; bracteoles absent. **Flowers** bisexual; sepals 5, basally connate, calyx bilaterally symmetric, tubular to cupulate, lobes ovate or lanceolate, adaxial lobe larger; corolla creamy white to pale pink, bilaterally symmetric, strongly bilabiate, tubular, tube base gibbous abaxially, not spurred, lobes 5, abaxial 3, adaxial 2; stamens 4, basally adnate to corolla, didynamous, filaments glandular-hairy distally, pollen sacs 2 per filament; staminode 0; ovary 2-locular, placentation axile; stigma punctiform. **Fruits** capsules, locules equal to subequal, dehiscence poricidal. **Seeds** 15–50, brown to black, ovoid to subconical, wings absent. *x* = 16.

Species 1: California.

Howelliella is rare and closely related to *Sairocarpus*, but the palate does not block the mouth of the corolla tube, the floor of the corolla tube has two longitudinal folds, and the abaxial lip is relatively small. The corolla tube is slightly curved in *Howelliella*; it is straight in related species. Morphological and molecular studies agree that *Howelliella* is most closely related to *S. subcordatus* and *S. vexillocalyculatus*, which also have greatly enlarged adaxial calyx lobes (D. M. Thompson 1988; R. K. Oyama and D. A. Baum 2004; M. Fernández-Mazuecos et al. 2013).

1. **Howelliella ovata** (Eastwood) Rothmaler, Leafl. W. Bot. 7: 115. 1954 • Oval-leafed snapdragon E F

Antirrhinum ovatum Eastwood, Bull. Torrey Bot. Club 32: 213. 1905

Stems 10–50 cm, branches spreading or ascending. **Leaves:** blade ovate, 10–50 × 5–20 mm, apex glandular, glandular-hairy. **Pedicels** 2–4 mm. **Flowers** chasmogamous; calyx lobes unequal, lateral and abaxial lanceolate, 4–5 mm, apex acute, adaxial ovate, 11–12 mm, apex obtuse, gland-tipped; corolla slightly curved, 17–22 mm, glandular-hairy externally, mouth 6–8 mm diam., abaxial lip soon withering, palate yellow, slightly convex, not blocking mouth. **Capsules** ovoid, 8–10 mm, puberulent, with 3 apical pores. **Seeds** sometimes asymmetric, 1 mm, reticulate-ridged. *2n* = 32.

Flowering May–Jul. Slopes on clay soils, burned or disturbed sites; (200–)500–1000 m; Calif.

Howelliella ovata is found only on clay soil in limited areas of Kern, Monterey, San Benito, San Luis Obispo, and Ventura counties. D. M. Thompson (1988) suggested that it needs disturbance to prosper, especially fire followed by heavy rain, and pointed out that the spreading, lateral branches sometimes help support the main stem. They are stout and strong and do not resemble the filiform, twining branchlets found in some species of *Sairocarpus*.

9. **KICKXIA** Dumortier, Fl. Belg., 35. 1827 • Cancerwort, fluellin [For Jean Jacques Kickx, 1842–1887, Belgian botanist] 丄

Wayne J. Elisens

Herbs, annual [perennial]. **Stems** decumbent or prostrate to erect [climbing], villous to glandular-hairy [glabrous]. **Leaves** cauline, alternate, sometimes opposite proximally; petiole present; blade oblong-ovate to orbiculate or cordate, not fleshy, not leathery, margins entire or dentate [lobed]. **Inflorescences** axillary, flowers solitary; bracts absent. **Pedicels** present, spreading; bracteoles absent. **Flowers** bisexual; sepals 5, basally connate, equal, calyx radially symmetric, campanulate, lobes lanceolate [linear], villous; corolla yellow [white], often tinged blue to violet [red], bilaterally symmetric, bilabiate and personate, tubular, tube base not gibbous, spurred abaxially, lobes 5, abaxial 3, spreading, adaxial 2, erect, subequal, apex round; stamens 4, basally adnate to corolla, didynamous, included, filaments incurved, sparsely hairy or glabrous, anthers coherent, ciliate, pollen sacs oblong; staminode 0; ovary 2-locular, villous, locules subequal, placentation axile; stigma capitate. **Fruits** capsules, dehiscence poricidal [indehiscent]. **Seeds** [6–]20–40[–60], dark brown to black, oblong-ovoid [reniform], wings absent. $x = 9$.

Species 9 or 46 (2 in the flora): introduced; Eurasia, n Africa, Atlantic Islands; introduced also in Mexico, West Indies, Central America, South America, Pacific Islands, Australia.

Kickxia is characterized by personate, long-spurred corollas, coherent anthers, and poricidal capsule dehiscence. Early treatments proposed that *Kickxia* was congeneric with, or closely related to, *Linaria* (G. Bentham 1876; P. A. Munz 1926); later taxonomic treatments and phylogenetic analyses have supported generic rank for *Kickxia* (W. Rothmaler 1943; D. A. Sutton 1988; M. Ghebrehiwet 2001). Morphology- and molecular-based phylogenies support *Anarrhinum* Desfontaines ($x = 10$) as the most likely sister group to *Kickxia* ($x = 9$) and their basal position in Antirrhineae as the *Anarrhinum* clade (Ghebrehiwet et al. 2000; P. Vargas et al. 2004).

Although usually considered in a broad sense (D. A. Sutton 1988), N. Yousefi et al. (2016) proposed, on the basis of ITS and *trnl* sequence data, that *Kickxia* sect. *Valvatae* (Wettstein) Janchen be recognized at the genus level. *Kickxia* then becomes a genus of nine species, with 37 being transferred to *Nanorrhinum* Betsche.

SELECTED REFERENCE Ghebrehiwet, M. 2001. Taxonomy, phylogeny, and biogeography of *Kickxia* and *Nanorrhinum* (Scrophulariaceae). Nordic J. Bot. 20: 655–690.

1. Leaf blade bases: proximal truncate or rounded to cuneate, distal hastate to sagittate; sepal lobes not accrescent . . . . 1. *Kickxia elatine*
1. Leaf blade bases rounded to cordate; sepal lobes accrescent . . . . 2. *Kickxia spuria*

1. **Kickxia elatine** (Linnaeus) Dumortier, Fl. Belg., 35. 1827 • Sharpleaf cancerwort F I W

Antirrhinum elatine Linnaeus, Sp. Pl. 2: 612. 1753

Plants 4–120 cm. **Leaves:** petiole 2–6 mm; blade: proximal 15–35 × 10–30 mm, base truncate or rounded to cuneate, distal 4–25 × 2–14 mm, base hastate to sagittate, villous with scattered glandular hairs. **Pedicels** 10–22 mm. **Flowers:** sepal lobes lanceolate, 3–6 × 1–2 mm, not accrescent, apex acuminate, villous; corolla tube 1–2 mm, palate yellow, inflated, hairy, spurs 5–6 mm, abaxial lobes yellow to violet, adaxial blue to violet, 2–3 mm; abaxial filaments 1.5–2 mm, adaxial 1–1.5 mm; style included, 1.3–1.5 mm; stigma straight. **Capsules** ovoid-globular, 3–5 mm. **Seeds** 1–1.5 mm, cristate-tuberculate. $2n = 36$.

Flowering May–Oct(–Dec). Gravelly or sandy disturbed sites, roadsides, stream banks, gravel bars, glades; 0–900 m; introduced; B.C., Ont.; Ala., Ark., Calif., Conn., Del., Ga., Ill., Ind., Kans., Ky., La., Md., Mass., Mich., Mo., N.J., N.Y., N.C., Ohio, Okla., Oreg., Pa., R.I., S.C., Tenn., Tex., Va., Wash., W.Va., Wis.; Eurasia; n Africa; introduced also in Mexico, Central America, South America, Atlantic Islands, Australia.

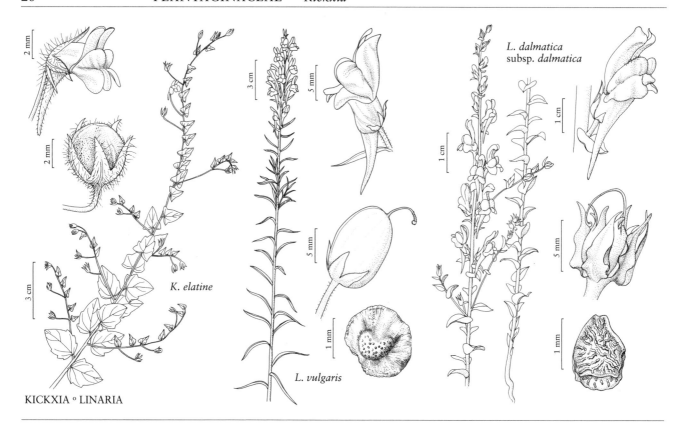

K. elatine

L. vulgaris

L. dalmatica subsp. dalmatica

KICKXIA ∘ LINARIA

Kickxia elatine differs from *K. spuria* in leaf and calyx characters. Two subspecies of *K. elatine* have been described [subsp. *elatine* and subsp. *crinita* (Mabille) W. Greuter]; they are poorly defined (D. A. Sutton 1988; M. Ghebrehiwet 2001) and differentiated on the basis of indument density.

2. Kickxia spuria (Linnaeus) Dumortier, Fl. Belg., 35. 1827 • Roundleaf cancerwort I W

Antirrhinum spurium Linnaeus, Sp. Pl. 2: 613. 1753

Plants 10–80 cm. **Leaves:** petiole 1–4 mm; blade 4–40 × 3–30 mm, base rounded to cordate, villous with scattered glandular hairs. **Pedicels** 6–12 mm. **Flowers:** sepal lobes broadly lanceolate to ovate, 5–6 × 3–4 mm, accrescent, apex acute, villous; corolla tube 3–5 mm, palate yellow, inflated, hairy, spurs 6–8 mm, abaxial lobes yellow, adaxial violet, 3–4 mm; abaxial filaments 2–2.5 mm, adaxial 1.5–2 mm; style included, 1.2–1.5 mm; stigma straight. **Capsules** ovoid-globular, 3–5 mm. **Seeds** 1–1.5 mm, cristate-tuberculate. $2n = 18$.

Flowering May–Nov. Gravelly or sandy disturbed sites, gravel bars, stream banks, roadsides; 40–700 m; introduced; B.C., Ont.; Ala., Calif., Fla., Ky., Mo., Mont., N.J., N.Y., N.C., Ohio, Oreg., Pa., R.I., Tenn., Wis.; Europe; Asia; n Africa; Atlantic Islands; introduced also in South America, Australia.

The accrescent sepal lobes of *Kickxia spuria* is a unique characteristic in *Kickxia*. D. A. Sutton (1988) recognized two subspecies that he characterized as poorly defined [subsp. *spuria* and subsp. *integrifolia* (Brotero) R. Fernandes]; these were not differentiated by M. Ghebrehiwet (2001).

10. LINARIA Miller, Gard. Dict. Abr. ed. 4, vol. 2. 1754 • Toadflax [Latin *linum*, flax, and *-aria*, resemblance, alluding to leaf similarity] ⊡

Llorenç Sáez Goñalons

Herbs, perennial or annual; caudex woody or herbaceous, shoot buds arising from roots. **Stems** prostrate to erect, glabrous or glandular-pubescent in inflorescence. **Leaves** cauline, usually whorled proximally, alternate distally; petiole absent; blade sometimes ± fleshy, not leathery, margins entire. **Inflorescences** terminal, racemes, compact to elongate; bracts present. **Pedicels** present; bracteoles absent. **Flowers** bisexual; sepals 5, proximally connate, calyx ± bilaterally symmetric to nearly radially symmetric, 5-lobed, lobes linear to linear-lanceolate, lanceolate, ovate-lanceolate, or ovate; corolla violet, blue, purple, red, yellow, or white, bilaterally symmetric, bilabiate, subcylindric, tube base not gibbous, spurred abaxially, lobes 5, abaxial 2, adaxial 3, abaxial lip as long as or slightly longer than adaxial, adaxial erect-patent to patent, with palate ± occluding mouth of tube; stamens 4, basally adnate to corolla, didynamous, filaments usually hairy proximally; staminode 1, scalelike; ovary 2-locular, placentation axile; styles simple or 2-fid; stigma capitate, clavate, linear, 2-lobed, or emarginate. **Fruits** capsules, 9–12 mm, dehiscence poricidal, each loculus wall later splitting into valves. **Seeds** 3–122, gray, brown, or black, globular, reniform, trigonous, tetrahedral, or disciform, wings present or absent. $x = 6$.

Species 150 (10 in the flora): introduced; Europe (especially Mediterranean region), w Asia, n Africa; introduced also in temperate regions of s South America (Argentina), Pacific Islands (New Zealand), Australia.

In this treatment, the species are arranged following the infrageneric classification proposed by D. A. Sutton (1988), with 150 species in seven sections; our ten belong to four of these sections.

North American species of *Linaria canadensis* (Linnaeus) Dumont de Courset, *L. floridana* Chapman, and *L. texana* Scheele included in sect. *Leptoplectron* Pennell, an invalidly published sectional name, have been included in *Nuttallanthus*, which differs from *Linaria* in having the abaxial lip of the corolla greatly exceeding the adaxial lip, a weakly developed palate, and prismatic seeds usually with prominent longitudinal ridges. Phylogenetic studies of *Linaria* and *Nuttallanthus* based on nuclear ribosomal DNA ITS sequences (M. Fernández-Mazuecos et al. 2013) concluded that on the basis of morphological traits and phylogenetic results, it would probably be appropriate to circumscribe *Nuttallanthus* species as a section (sect. *Lectoplectron*) of *Linaria*.

Linaria species often are cultivated and some (such as *L. bipartita*, *L. dalmatica*, *L. triornithophora*, and *L. vulgaris*) become naturalized; how long they may persist is unknown.

Although *Linaria supina* (Linnaeus) Chazelles de Prizy subsp. *supina* (lesser butter-and-eggs) has been collected in several states (New Jersey, New York, and Pennsylvania), all of the collections were made before 1900, and *L. supina* has yet to become established in the flora area. It would key closest to *L. vulgaris*, having shorter fertile stems [5–24 cm versus (25–)35–90 (–120) cm], smaller corollas (15–25 mm versus 27–33 mm), and globular (versus oblong-globular or ovoid) capsules. The report from Massachusetts by C. H. Knowlton and W. Deane (1923) was based on a cultivated plant (10 July 1863, *Boott s.n.*, GH), and the report from California is an error (http://ucjeps.berkeley.edu/cgi-bin/get_IJM.pl?tid=11372).

1. Styles 2-fid; stigmas with 2 discrete areas; seed wings absent; annuals [sect. *Versicolores*].
    2. Racemes: rachis glandular-hairy, hairs 0.1–0.7 mm; capsule loculi subequal to unequal . . . . .10. *Linaria maroccana*
    2. Racemes: rachis glabrous or sparsely glandular-hairy, hairs 0.1–0.4 mm; capsule loculi equal, sometimes subequal.
        3. Corollas purple, mauve, or violet, usually with yellow palate . . . . .8. *Linaria bipartita*
        3. Corollas yellow. . . . . .9. *Linaria spartea*
1. Styles simple; stigmas entire; seed wings present or absent; perennials (some plants may bloom first year).
    4. Seeds disciform, subglobular, elliptic-globular, or deltate-globular, with prominent encircling wing.
        5. Corollas 27–32(–33) mm, white, pale yellow, or bright yellow, sometimes with orange palate [sect. *Vulgaris*] . . . . 1. *Linaria vulgaris*
        5. Corollas 35–57 mm, violet, blue-violet, reddish purple, or pink, rarely white, with yellow, white, or lilac palate [sect. *Pelisserianae*] . . . . 7. *Linaria triornithophora*
    4. Seeds trigonous, subtrigonous, or ± tetrahedral, wings absent [sect. *Speciosae*].
        6. Spurs 2–5 mm; corollas white or pale lilac . . . . 6. *Linaria repens*
        6. Spurs 5–30 mm; corollas yellow or violet to purple.
            7. Corollas violet to purple, with yellow or lilac palate . . . . 5. *Linaria purpurea*
            7. Corollas pale to bright yellow.
                8. Corollas 16–20(–23) mm; spurs 7–9(–10) mm. .2. *Linaria genistifolia*
                8. Corollas 27–52 mm; spurs 11–30 mm.
                    9. Corollas: adaxial lip sinuses 2–3.5(–4) mm; calyx lobes (1.9–)3–9.5 × (0.9–)1–3.5 mm in flower . . . . .3. *Linaria dalmatica*
                    9. Corollas: adaxial lip sinuses 4–6 mm; calyx lobes 7–14 × 3.5–7 mm in flower . . . . .4. *Linaria grandiflora*

1. Linaria vulgaris Miller, Gard. Dict. ed. 8, Linaria no. 1. 1768 • Yellow or common toadflax, butter-and-eggs, linaire vulgaire F I W

Antirrhinum linaria Linnaeus

Perennials, reproducing vegetatively by adventitious buds from roots. **Fertile stems** erect or suberect, (25–)35–90 (–120) cm; sterile stems to 18 cm, usually absent. **Leaves** of fertile stems: blade linear to linear-lanceolate, 4–68 × 1–7 (–15) mm, apex acute. **Racemes** 11–31(–50)-flowered, usually dense, rachis glabrous or glandular-hairy, hairs 0.1–0.2 mm; bracts linear to lanceolate, 4–20 (–25) × 1–2 mm. **Pedicels** erect, 3–7.7 mm in flower, 3.5–8(–13) mm in fruit. **Calyx lobes** lanceolate to ovate, 2.5–4.5 × 0.8–1.5 mm in flower, 2.5–5(–6) × 1–2(–2.5) mm in fruit, apex acute. **Corollas** white, pale yellow, or bright yellow, sometimes with orange palate, 27–32(–33) mm; tube 5.5–9(–10) mm wide, spurs straight or curved, 11–15 mm, slightly shorter or subequal to rest of corolla, abaxial lip sinus 3–3.5 mm, adaxial lip sinus (2.2–)2.4–4(–4.5) mm. **Styles** simple; stigma entire. **Capsules** oblong-globular or ovoid, 5–10(–11) × 3–6 mm, glabrous; loculi equal. **Seeds** gray, brown, or black, disciform, subglobular, with prominent encircling wing, 1.6–2.5(–3) × 1.5–2.3 mm, ± plane or biconvex; disc gray, brown, or black, ± reniform, tuberculate; wing ± entire, 0.4–0.7 mm wide, membranous, smooth. 2*n* = 12.

Flowering May–Oct. Disturbed places, roadsides, railroad rights-of-way, rocky slopes, cultivated or abandoned fields, usually in moist or dry soils; 0–2300 (–3100) m; introduced; Alta., B.C., Man., N.B., Nfld. and Labr., N.W.T., N.S., Ont., P.E.I., Que., Sask., Yukon; Ala., Alaska, Ariz., Ark., Calif., Colo., Conn., Del., D.C., Ga., Idaho, Ill., Ind., Iowa, Kans., Ky., La., Maine, Md., Mass., Mich., Minn., Mo., Mont., Nebr., Nev., N.H., N.J., N.Mex., N.Y., N.C., N.Dak., Ohio, Okla., Oreg., Pa., R.I., S.Dak., Tenn., Tex., Utah, Vt., Va., Wash., W.Va., Wis., Wyo.; Europe; c Asia; introduced also in temperate regions of s South America (Argentina), Pacific Islands (New Zealand), Australia.

Linaria vulgaris was introduced into the northeastern United States (New England) before 1672 as an ornamental and medicinal plant (M. L. Fernald 1905b). In part from vigorous formation of adventitious shoots from the roots (T. S. Bakshi and R. T. Coupland 1960), *L. vulgaris* is now considered a noxious weed in nine states and four Canadian provinces (E. A. Lehnhoff 2008).

Hybrids between *Linaria repens* and *L. vulgaris* are known as *L.* ×*sepium* G. J. Allman; this hybrid presents an intermediate morphology between *L. repens* and *L. vulgaris* with regard to corolla and spur size (16–21 mm and 6–9.5 mm, respectively). Specimens examined suggest that this hybrid is relatively common in Newfoundland.

Hybrids between *Linaria dalmatica* subsp. *dalmatica* and *L. vulgaris* (*L.* ×*hybrida* Schur) are found in California (Lassen Creek, Modoc County), Colorado (Jefferson County), Idaho (Coeur d'Alene Lake, Kootenai County), and Montana (Beaverhead-Deerlodge National Forest) (S. M. Ward et al. 2009). In these localities, both potential parental species are sympatric or nearly sympatric. Ward et al. confirmed that hybridization occurs between populations of *L. dalmatica* and *L. vulgaris*, and that the hybrid progeny are viable and fertile.

SELECTED REFERENCE Saner, M. A. et al. 1995. The biology of Canadian weeds. 105. *Linaria vulgaris* Mill. Canad. J. Pl. Sci. 75: 525–537.

2. **Linaria genistifolia** (Linnaeus) Miller, Gard. Dict. ed. 8, Linaria no. 14. 1768 • Broomleaf toadflax [I]

Antirrhinum genistifolium Linnaeus, Sp. Pl. 2: 616. 1753

Subspecies 6 (1 in the flora): introduced; Europe.

2a. **Linaria genistifolia** (Linnaeus) Miller subsp. **genistifolia** [I]

Perennials, reproducing vegetatively by rhizomes. **Fertile stems** usually erect, 18–60 cm; sterile stems usually absent. **Leaves** of fertile stems: blade lanceolate or linear-lanceolate, flat, 5–45(–85) × 2–10(–35) mm, apex acute. **Racemes** 5–26-flowered, lax; bracts lanceolate, 3–10 × 1–2 mm. **Pedicels** erect-patent to erect, 2–4 mm in flower, 3–8 mm in fruit. **Calyx lobes** ovate-lanceolate to lanceolate, 2.5–5 × 0.7–1.5 mm in flower, 3–6 × 0.8–1.7 mm in fruit, apex acute. **Corollas** pale to bright yellow, 16–20(–23) mm; tube 4–5 mm wide, spurs straight, 7–9(–10) mm, slightly shorter or subequal to rest of corolla, abaxial lip sinus 2–3 mm, adaxial lip sinus 2–3.5 mm. **Styles** simple; stigma entire. **Capsules** subglobular, 4–5 × 4–4.8 mm, glabrous; loculi equal. **Seeds** black or blackish brown, subtrigonous or ± tetrahedral, 0.8–1.2(–1.4) × 0.6–1 mm, with longitudinal marginal ridges and anastomosed ridges or tubercles on faces; wing absent. $2n$ = 12 (Europe).

Flowering Jul–Sep. Roadsides, disturbed places; 0–2100 m; introduced; Mass., Minn., N.Y., Ohio, Oreg., Utah, Wash.; se Europe.

Collections of subsp. *genistifolia* on ballast are known from New York (pre-1900) and Oregon (1912).

3. **Linaria dalmatica** (Linnaeus) Miller, Gard. Dict. ed. 8, Linaria no. 13. 1768 • Dalmatian toadflax, linaire á feuilles larges [F][I][W]

Antirrhinum dalmaticum Linnaeus, Sp. Pl. 2: 616. 1753; *Linaria genistifolia* (Linnaeus) Miller subsp. *dalmatica* (Linnaeus) Maire & Petitmengin

Perennials, reproducing vegetatively by adventitious buds or stolons. **Fertile stems** erect, 49–150 cm; sterile stems to 15 cm. **Leaves** of fertile stems: blade ovate to lanceolate, flat, 10–65 × 3–31 mm, apex acute. **Racemes** 1–35-flowered, lax or dense; bracts ovate to lanceolate, 3–20 × 2–12 mm. **Pedicels** erect-patent to erect, 3–22 mm in flower, 4–32 mm in fruit. **Calyx lobes** ovate to linear-lanceolate, (1.9–)3–9.5 × (0.9–)1–3.5 mm in flower, 4–10(–13) × 1.5–5 mm in fruit, apex acute. **Corollas** pale to bright yellow, (27–)28–49 mm; tube 6–11 mm wide, spurs straight or curved, 11–30 mm, slightly shorter, subequal to, or longer than rest of corolla, abaxial lip sinus (1.8–)2–2.5 mm, adaxial lip sinus 2–3.5(–4) mm. **Styles** simple; stigma entire. **Capsules** subglobular, 4–7 × 4–7 mm, glabrous; loculi equal, rarely subequal. **Seeds** brown, brown-gray, or black, subtrigonous or ± tetrahedral, 1–2 × 0.8–1.6 mm, with longitudinal obtuse marginal ridges and anastomosed obtuse or truncate ridges on faces; wing absent.

Subspecies 2 (2 in the flora): introduced; Europe, sw Asia; introduced also in s South America (Argentina), Australia.

Linaria dalmatica was originally introduced into North America as an ornamental plant and is currently classified as a noxious weed in seven states in the United States and two Canadian provinces (https://plants.usda.gov/core/profile?symbol=LIDA).

Subspecies of *Linaria dalmatica* can be difficult to recognize because characteristics partly overlap and intermediate specimens may be found.

Linaria genistifolia and *L. dalmatica* are closely related, and the latter is sometimes treated as a subspecies of *L. genistifolia* by several authors. Although further research is required in this species complex (including *L. grandiflora*, among others), the author recognizes *L. genistifolia* and *L. dalmatica* as independent species. Nevertheless, published references to *L. genistifolia* could in fact correspond to *L. dalmatica*. In order to obtain a conclusive identification for either species, it is necessary to study flowering specimens.

SELECTED REFERENCES Alex, J. F. 1962. The taxonomy, history, and distribution of *Linaria dalmatica*. Canad. J. Bot. 40: 295–307. Vujnovic, K. and R. W. Wein. 1997. The biology of Canadian weeds. 106. *Linaria dalmatica* (L.) Mill. Canad. J. Pl. Sci. 77: 483–491.

1. Pedicels 4–10(–13) mm in fruit; corollas (27–)28–38(–42) mm; spurs 11–24 mm, shorter, subequal to, or longer than rest of corolla . . . . 3a. *Linaria dalmatica* subsp. *dalmatica*
1. Pedicels (7–)12–32 mm in fruit; corollas 30–49 mm; spurs 18–30 mm, subequal to or longer than rest of corolla . . . . 3b. *Linaria dalmatica* subsp. *macedonica*

3a. Linaria dalmatica (Linnaeus) Miller subsp. dalmatica F I

Bracts usually ± equal to or longer than fruiting pedicels. **Pedicels** 3–8 mm in flower, 4–10(–13) mm in fruit. **Calyx lobes** ovate to lanceolate, 3–9 × 1–3.5 mm in flower, 4–10(–13) × 1.5–5 mm in fruit. **Corollas** (27–)28–38(–42) mm; spurs straight or curved, 11–24 mm, shorter, subequal to, or longer than rest of corolla. **2n** = 12.

Flowering Apr–Sep. Roadsides, disturbed places, railroad rights-of-way, lawns, waste grounds, stream banks, rangelands, croplands, pastures, yellow-pine forests, pinyon/juniper woodlands, sagebrush scrub, usually well-drained soils; 0–3100 m; introduced; Alta., B.C., Man., N.B., Nfld. and Labr. (Nfld.), N.S., Ont., Que., Sask., Yukon; Ariz., Calif., Colo., Conn., Idaho, Ill., Ind., Kans., Maine, Mass., Mich., Minn., Mont., Nebr., Nev., N.H., N.Mex., N.Dak., Ohio, Okla., Oreg., Pa., R.I., S.C., S.Dak., Utah, Vt., Wash., Wyo.; Europe; sw Asia; introduced also in s South America (Argentina), Australia.

Hybrids between subsp. *dalmatica* and *Linaria vulgaris* are known as *L.* ×*hybrida* Schur; see under 1. *L. vulgaris*.

3b. Linaria dalmatica (Linnaeus) Miller subsp. macedonica (Grisebach) D. A. Sutton, Revis. Antirrhineae, 321. 1988 I

Linaria macedonica Grisebach, Spic. Fl. Rumel. 2: 19. 1844; *L. dalmatica* var. *macedonica* (Grisebach) Vandas

Bracts usually shorter than fruiting pedicels. **Pedicels** 7–22 mm in flower, (7–)12–32 mm in fruit. **Calyx lobes** ovate to linear-lanceolate, (1.9–)3–9.5 × (0.9–)1–3 mm in flower, 4–9(–11) × 1.5–4 mm in fruit. **Corollas** 30–49 mm; spurs straight, 18–30 mm, subequal to or longer than rest of corolla. **2n** = 12 (Europe).

Flowering May–Sep. Disturbed places, open banks; 50–2000 m; introduced; Ont.; Colo., Idaho, Mich., N.H., Utah, Wash., Wyo.; s Europe.

J. F. Alex (1962) noted that, after his study of over 200 specimens from North America, he was convinced that var. *macedonica*, native to the mountains of the western Balkan Peninsula, was not found in the flora area. Nevertheless, specimens have been seen during this study that, according to the identification key provided by D. A. Sutton (1988), are referable to subsp. *macedonica*. Whether those specimens with longer corollas and longer fruiting pedicels are extreme phenotypic variations within the typical *Linaria dalmatica* in North America, or they correspond to an independent introduction of subsp. *macedonica*, requires additional study; meanwhile, it is preferable to consider the specimens of *L. dalmatica* with such different extreme characters as subsp. *macedonica*.

4. Linaria grandiflora Desfontaines, Ann. Mus. Natl. Hist. Nat. 11: 51, plate 2. 1808 I

Perennials, reproducing vegetatively by adventitious buds or stolons. **Fertile stems** erect, (30–)60–74(–100) cm; sterile stems to 13.5 cm. **Leaves** of fertile stems: blade ovate to lanceolate, flat, 12–48(–65) × 3–18(–30) mm, apex acute. **Racemes** 1–36-flowered, dense in flower, lax in fruit; bracts ovate to lanceolate, 7–14(–20) × 2–7 mm. **Pedicels** erect-patent to erect, (0.5–)2–6(–15) mm in flower, 4–15(–21) mm in fruit. **Calyx lobes** ovate, sometimes ovate-lanceolate, 7–14 × 3.5–7 mm in flower, 7–14 × 3.5–9 mm in fruit, apex acute. **Corollas** pale to bright yellow, (40–)41–52 mm; tube (6–)7.5–12 mm wide, spurs straight or curved, 19–22(–23) mm, slightly shorter or subequal to rest of corolla, abaxial lip sinus 2.5–3.5(–4) mm, adaxial lip sinus 4–6 mm. **Styles** simple; stigma entire. **Capsules** globular, 6–8(–11) × 6–10.7 mm, glabrous; loculi equal. **Seeds** black, subtrigonous or tetrahedral, 0.9–1.5 mm, with longitudinal acute marginal ridges and anastomosed acute ridges on faces; wing absent.

Flowering Jul–Sep. Roadsides, disturbed places; 600–2100 m; introduced; Colo., Idaho, Wash.; se Europe; sw Asia.

Linaria grandiflora is closely related to *L. dalmatica* and sometimes has been included within *L. dalmatica*. One specimen collected in Washington (Kittitas County, west of Cle Elum, 18 June 1962, *C. L. Hitchcock & C. V. Muhlick 22329*, NY) exhibits discordant features (for example, pedicels 15–21 mm). According to D. A. Sutton (1988), *L. grandiflora* bears pedicels 0.5–4 mm, although he indicated that plants called *L. pancicii* Janka (which were included in the synonymy of *L. grandiflora*) have long pedicels. For the rest of the characteristics (corolla size, adaxial lip sinus length, calyx lobe size and shape), that specimen is fairly assimilable in *L. grandiflora*.

5. Linaria purpurea (Linnaeus) Miller, Gard. Dict. ed. 8,
Linaria no. 5. 1768 • Purple toadflax

Antirrhinum purpureum Linnaeus, Sp. Pl. 2: 613. 1753

Perennials, from taproot, not reproducing vegetatively by stolons. **Fertile stems** erect or suberect, to 71(–140) cm; sterile stems to 18 cm. **Leaves** of fertile stems: blade oblanceolate to linear, usually flat, 5–45 (–60) × 0.8–4(–8) mm, apex acute or subobtuse. **Racemes** 1–117-flowered, dense; bracts linear, 2–5(–5.5) × 0.3–1 mm. **Pedicels** erect, 1–3 mm in flower, 2–4(–5) mm in fruit. **Calyx lobes** linear to linear-lanceolate, 1.5–3 × 0.5–1 mm in flower, 2–3.5 × 0.7–1.2 mm in fruit, apex acute or subacute. **Corollas** violet to purple, with yellow or lilac palate, 9–13 (–17) mm; tube 1.5–2.5 mm wide, spurs curved, 5–7 (–9) mm, subequal to rest of corolla, abaxial lip sinus (0.6–)0.8–1.5(–2) mm, adaxial lip sinus 1 mm. **Styles** simple; stigma entire. **Capsules** subglobular, 2.7–4 × 2.5–3.7 mm, glabrous; loculi equal. **Seeds** black or blackish brown, subtrigonous or ± tetrahedral, 0.8–1.2 × 0.6–1 mm, with longitudinal marginal ridges and anastomosed ridges or tubercles on faces; wing absent. **2*n*** = 12 (Europe).

Flowering Jun–Oct. Disturbed places, railroad rights-of-way, beach foreshores; 0–1900 m; introduced; B.C.; Calif., Oreg., Wash.; s Europe (Italy); introduced also in s South America (Argentina), n Europe, Australia.

6. Linaria repens (Linnaeus) Miller, Gard. Dict. ed. 8,
Linaria no. 6. 1768 • Striped toadflax

Antirrhinum repens Linnaeus, Sp. Pl. 2: 614. 1753

Perennials, reproducing vegetatively by rhizomes. **Fertile stems** erect, 12–68 cm; sterile stems to 15 cm. **Leaves** of fertile stems: blade narrowly elliptic, oblanceolate, or linear, flat, sometimes revolute distally, 4–50 × (0.5–)0.7–6 mm, apex acute. **Racemes** 1–43-flowered, dense in flower, lax in fruit; bracts linear, 2–11(–12) × 0.3–1.5 mm. **Pedicels** erect, 2–5 mm in flower, 2.5–12(–14) mm in fruit. **Calyx lobes** linear-lanceolate, 2.5–3.5 × 0.5–1 mm in flower, 3–4 × 0.7–1.1 mm in fruit, apex acute. **Corollas** white or pale lilac, 9–14 mm; tube 2–3.5(–4) mm wide, spurs straight, 2–5 mm, much shorter than rest of corolla, abaxial lip sinus 2–2.5(–3) mm, adaxial lip sinus 1–2.5 mm. **Styles** simple; stigma entire. **Capsules** subglobular,

3–4 × 2.8–4 mm, glabrous; loculi equal. **Seeds** gray or black, trigonous or ± tetrahedral, 1.2–1.8(–2) × 0.6–1 mm, with longitudinal marginal ridges and anastomosed ridges, sometimes with tubercles, on faces; wing absent. **2*n*** = 12 (Europe).

Flowering Jul–Sep. Roadsides, railroad rights-of-way, waste places; 0–200 m; introduced; N.B., Nfld. and Labr. (Nfld.), N.S.; Maine, Mass., N.J., N.Y.; w Europe.

Pre-1900 collections of *Linaria repens* on ballast are known from New Jersey and New York. *Linaria repens* is known from an old collection in Nova Scotia and was once collected from St. Pierre and Miquelon; the population did not persist.

Linaria repens is polymorphic, easily recognized by the spur much shorter than the rest of the corolla. Hybrids between it and *L. vulgaris* (known as *L. ×sepium*) are relatively common in Newfoundland (see comments under 1. *L. vulgaris*).

7. Linaria triornithophora (Linnaeus) Willdenow, Enum.
Pl., 639. 1809 • Yellow-throated purple toadflax

Antirrhinum triornithophorum Linnaeus, Sp. Pl. 2: 613. 1753

Perennials, not reproducing vegetatively by stolons. **Fertile stems** erect or suberect, 50–130 cm; sterile stems to 20 cm. **Leaves** of fertile stems: blade lanceolate to ovate-lanceolate, rarely elliptic, flat, 25–75 × 5–30 mm, apex acute or obtuse. **Racemes** 4–30-flowered, lax; bracts linear, 6–38 × 1.5–9 mm. **Pedicels** erect, 7–35 mm in flower, 15–38 mm in fruit. **Calyx lobes** ovate-lanceolate, 5–9 × 1.2–2 mm in flower, 6–10 × 1.5–2 mm in fruit, apex acute. **Corollas** violet, blue-violet, reddish purple, or pink, rarely white, with yellow, white, or lilac palate, 35–57 mm; tube 6–11 mm wide, spurs straight, 16–27 mm, subequal to rest of corolla, abaxial lip sinus 2.5–6 mm, adaxial lip sinus 4–8 mm. **Styles** simple; stigma entire. **Capsules** depressed-globular, 4–7.5 × 6–9 mm, glabrous; loculi equal. **Seeds** blackish brown to dark brown, disciform, elliptic-globular, or deltate-globular, with prominent encircling wing, 2–2.3 × 2–2.3 mm, plano-convex; disc blackish brown to dark brown, deltate-globular or densely tuberculate, sometimes with short ridges; wing entire, 0.1–0.2 mm wide, ± thick towards disc, not tuberculate. **2*n*** = 12.

Flowering Mar–Sep. Disturbed places; 20–60 m; introduced; Oreg.; s Europe (nw Iberian Peninsula).

Linaria triornithophora is known from Lincoln County.

8. **Linaria bipartita** (Ventenat) Willdenow, Enum. Pl., 640. 1809 • Clovenlip toadflax I

Antirrhinum bipartitum Ventenat, Descr. Pl. Nouv., plate 82. 1802

Annuals, not reproducing vegetatively by stolons. **Fertile stems** erect, 8–65 cm; sterile stems to 11 cm. **Leaves** of fertile stems: blade linear to linear-oblong, flat, 5–57 × 0.5–3.5 mm, apex acute or obtuse. **Racemes** 1–35-flowered, lax or ± dense in flower, lax in fruit, rachis glabrous or sparsely glandular-hairy, hairs 0.1–0.3 mm; bracts linear to linear-lanceolate, rarely lanceolate, 4–8 × 0.5–2.5 mm. **Pedicels** erect, 1–7 mm in flower, 2.5–9 mm in fruit. **Calyx lobes** lanceolate, 3–5 × 0.7–1.5 mm in flower, 3–6 × 1–2 mm in fruit, apex acute or subacute. **Corollas** purple, mauve, or violet, usually with yellow palate, 17–24 mm; tube 2–3.5 mm wide, spurs straight or curved, 8–14 mm, usually longer than rest of corolla, abaxial lip sinus 0.7–1.5 mm, adaxial lip sinus 1.5–3.5 mm. **Styles** 2-fid; stigma with 2 discrete areas. **Capsules** oblong-ovoid, 4.5–7 × 3.8–6.3 mm, glabrous or sparsely glandular-pubescent; loculi equal or subequal. **Seeds** dark gray to black, reniform or subtrigonous, 0.5–0.7 × 0.3–0.5 mm, with ± conspicuous transverse ridges; wing absent. $2n = 12$ (Africa).

Flowering Apr–Sep. Waste ground, roadsides; 0–1200 m; introduced; Que.; Ariz., Calif., N.J., Va.; n Africa (Morocco).

Glabrous plants (without hairs on the inflorescence axis, calyx bracts, and pedicels) are included in *Linaria bipartita*. This is the main feature to distinguish it from *L. maroccana* in absence of mature fruits. Single collections have been found that include both glabrous plants and ones with sparsely hairy calyces and pedicels on the same sheet (for example, New Jersey, Somerset County, Watchung, 11 July 1931, *Moldenke 1900*, NY), suggesting that *L. bipartita* may be polymorphic with respect to this feature.

Some specimens of *Linaria bipartita* have been reported as *L. incarnata* (Ventenat) Sprengel, which is easily separable from *L. bipartita* by its densely hairy inflorescences [multicellular glandular hairs (0.1–)0.3–1 mm, with purple to violet transverse walls]. At present, there is no specimen evidence of *L. incarnata* occurring in North America.

9. **Linaria spartea** (Linnaeus) Chazelles de Prizy, Suppl. Dict. Jard. 2: 38. 1790 • Ballast toadflax I

Antirrhinum sparteum Linnaeus, Sp. Pl. 2: 1197. 1753

Annuals, not reproducing vegetatively by stolons. **Fertile stems** ascending to erect, 5–31(–65) cm; sterile stems to 8(–25) cm. **Leaves** of fertile stems: blade linear, flat, 3–35 (–48) × 0.5–2.5 mm, apex acute or obtuse. **Racemes** 1–9(–25)-flowered, lax, rachis glabrous or sparsely glandular-hairy, hairs 0.1–0.4 mm; bracts linear, 3–5 × 0.5–1 mm. **Pedicels** erect, 4–7 mm in flower, 5–10 mm in fruit. **Calyx lobes** linear to linear-lanceolate, 3.5–5.5 × 0.7–1.5 mm in flower, 4–6.5 × 0.8–1.8(–2.2) mm in fruit, apex acute or subacute. **Corollas** yellow, 19–26(–30) mm; tube 2.5–3.5 mm wide, spurs straight or curved, 9–14 (–15) mm, subequal to or longer than rest of corolla, abaxial lip sinus 1–2.5 mm, adaxial lip sinus 2–4 mm. **Styles** 2-fid; stigma with 2 discrete areas. **Capsules** oblong-ovoid, (2–)3–4.5(–5) × (1.8–)2.2–3.5(–4) mm, glabrous or sparsely glandular-pubescent; loculi equal or subequal. **Seeds** gray or dark gray, reniform, 0.4–0.7 × 0.3–0.5 mm, with conspicuous transverse ridges; wing absent. $2n = 12$ (Europe).

Flowering Aug–Sep. Dry fields; 0–200 m; introduced; Conn.; w Europe (s France, Portugal, Spain).

Linaria spartea is known from Fairfield County.

10. **Linaria maroccana** Hooker f., Bot. Mag. 98: plate 5983. 1872 • Moroccan toadflax I

Annuals, not reproducing vegetatively by stolons. **Fertile stems** erect, 7–70 cm; sterile stems to 13.5 cm. **Leaves** of fertile stems: blade linear, flat, 5–35 × 1–5 mm, apex acute or subobtuse. **Racemes** 1–41-flowered, lax or dense, rachis glandular-hairy, hairs 0.1–0.7 mm; bracts linear to lanceolate, 4–9 × 0.5–2 mm. **Pedicels** erect, 4–9 mm in flower, 5–13 mm in fruit. **Calyx lobes** lanceolate or linear-lanceolate, 3–4.5 × 1–2 mm in flower, 3.5–6 × 1–2.5 mm in fruit, apex acute or subacute. **Corollas** purple or red, with yellow or red palate, 22–29 mm; tube 2–3.5 mm wide, spurs straight or curved, 12–17 mm, longer than rest of corolla, abaxial lip sinus 1–2 mm, adaxial lip sinus 3–6 mm. **Styles** 2-fid; stigma with 2 discrete areas. **Capsules** oblong, 4–6 × 3–5.2 mm, glabrous or sparsely glandular-pubescent; loculi subequal to unequal. **Seeds** dark gray to black, reniform or trigonous-pyriform, 0.5–0.7 × 0.3–0.5 mm, with ± conspicuous transverse ridges; wing absent. $2n = 12$ (Africa).

Flowering Mar–Aug. Roadsides, grasslands, dry fields; 0–1200 m; introduced; Que., Sask.; Ariz., Calif., Conn., Fla., Maine, Mass., N.H., N.Y., Va.; n Africa (Morocco); introduced also in n Europe, Australia.

Linaria maroccana has also been collected in British Columbia, Oregon, and Washington; the populations have not persisted.

The identity of the plants that have been called *Linaria maroccana* in North America is controversial. Typical specimens of *L. maroccana* have fruits strongly asymmetric, with the adaxial loculus continuing beyond the stylar base and curving over the abaxial loculus. The American plants are unusual in having fruits with unequal loculi, with the adaxial loculus usually not curving over the abaxial loculus. The specimens studied probably correspond to cultivated forms (sometimes sold under the name *L. maroccana* hort.) of hybrid origin between *L. maroccana* and species of sect. *Versicolores* [*L. bipartita*, *L. gharbensis* Battandier & Pitard, and *L. incarnata* (Ventenat) Sprengel]. A possible name for these plants would be *L. versicolor* (Jacquin) Chazelles (*Antirrhinum versicolor* Jacquin); this issue is, at present, somewhat confusing. The identity of this taxon is not well established, and this name is not currently used in any modern treatment of *Linaria*. See D. A. Sutton (1988) for more information about the name *L. versicolor*.

Linaria pinifolia (Poiret) Thellung, a perennial species endemic to northern Africa, has been reported (for example, http://data.canadensys.net/vascan/search?lang=en; B. G. Baldwin et al. 2012) from North America. All the examined specimens labeled as *L. pinifolia* appear to be assignable to *L. maroccana*.

11. MABRYA Elisens, Syst. Bot. Monogr. 5: 57, fig. 26. 1985 • [For Tom J. Mabry, 1932–2015, American botanist and phytochemist] [C]

Wayne J. Elisens

Herbs, perennial; caudex woody. **Stems** pendent [erect], glandular-hairy. **Leaves** cauline, alternate; petiole present; blade fleshy, not leathery, margins broadly dentate [crenate]. **Inflorescences** axillary, flowers solitary; bracts absent. **Pedicels** present; bracteoles absent. **Flowers** bisexual; sepals 5, basally connate, calyx radially symmetric, campanulate, lobes ovate; corolla ochroleucous to yellow [pink, red], bilaterally symmetric, bilabiate, tubular, tube base not spurred or gibbous, lobes 5, abaxial 3, adaxial 2; stamens 4, basally adnate to corolla, didynamous, filaments glandular-hairy; staminode 1, filamentous; ovary 2-locular, placentation axile; stigma 2-lobed. **Fruits** capsules, dehiscence loculicidal. **Seeds** 20–300, dark brown [tan], oblong [ellipsoid, globular], wings absent [present]. $x = 12$.

Species 5 (1 in the flora): Arizona, n Mexico.

Mabrya is defined by several vegetative characteristics including pendent or erect, brittle stems, straight petioles, and glandular-hairy stems, leaves, and flowers. Species of *Mabrya* were placed previously in *Maurandya* Ortega by most authors; they have been recognized in *Mabrya* by W. J. Elisens (1985), D. A. Sutton (1988), and E. Fischer (2004). The phylogenetic relationships of *Mabrya* based on morphological characters are either unresolved (M. Ghebrehiwet et al. 2000) or poorly resolved (Elisens) because of character polymorphism and lack of synapomorphies for the genus. Molecular ITS data placed *Holmgrenanthe* in the *Cymbalaria* clade (M. Fernández-Mazuecos et al. 2013), in a subclade of seven genera that was sister to *Asarina* and *Cymbalaria*; *Mabrya* was sister to a node shared by *Lophospermum* and *Maurandya*.

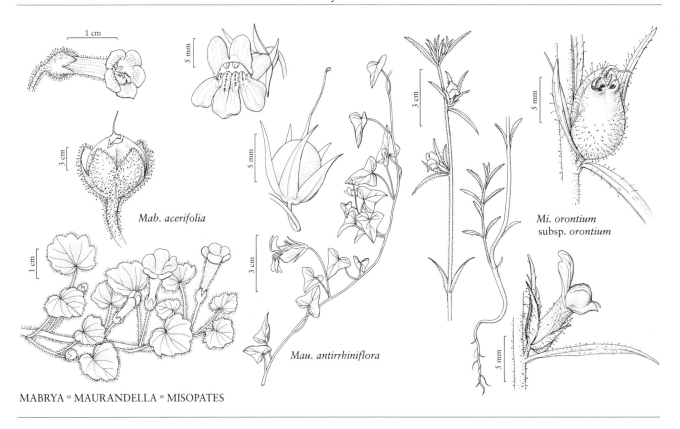

Mab. acerifolia

Mau. antirrhiniflora

Mi. orontium subsp. orontium

MABRYA ° MAURANDELLA ° MISOPATES

1. Mabrya acerifolia (Pennell) Elisens, Syst. Bot. Monogr. 5: 58. 1985 • Brittlestem [C] [E] [F]

Maurandya acerifolia Pennell, J. Wash. Acad. Sci. 19: 69. 1924; *Asarina acerifolia* (Pennell) Pennell

Plants 10–40 cm, cespitose. **Leaves:** petiole 12–45 mm; blade reniform, 12–24 × 15–35 mm, surfaces glandular-hairy. **Pedicels** ascending in flower, 8–25 mm, thickened and recurved in fruit. **Flowers:** sepals equal, 6–8 × 3–5 mm, apex acute, glandular-hairy; corolla tube 14–21 mm, glandular-hairy, throat open, palate not inflated, abaxial plicae dark yellow, lobes equal, 2–4 mm, apex round; stamens included, filaments incurved, hairs at base and apex clavate, abaxial 9–11 mm, adaxial 11–13 mm, pollen sacs oblong; ovary glabrous, locules subequal; style included, 10–12 mm, base persistent in fruit; stigma recurved. **Capsules** globular, 7–9 mm. **Seeds** 1–2 mm, tuberculate to cristate. $2n = 24$.

Flowering Feb–May. Rock walls, canyons, cliffs, roadcuts; of conservation concern; 200–1000 m; Ariz.

Mabrya acerifolia is easily identified vegetatively and in flower or fruit by its cespitose habit on rock walls, dense covering of glandular hairs, reniform leaves with dentate margins, ochroleucous to yellow corollas with yellow plicae, and globular capsules on recurved pedicels. The species is endemic to south-central Arizona in Maricopa and Pinal counties within a 30 km radius of the type locality.

12. MAURANDELLA (A. Gray) Rothmaler, Feddes Repert. Spec. Nov. Regni Veg. 52: 26. 1943 • Snapdragon vine [Genus *Maurandya* and Latin *-ella*, diminutive, alluding to presence of personate corolla in *Maurandella*]

Wayne J. Elisens

Antirrhinum Linnaeus [unranked] *Maurandella* A. Gray, Proc. Amer. Acad. Arts 7: 375. 1868

Herbs, perennial; caudex woody. **Stems** climbing, glabrous. **Leaves** cauline, alternate; petiole twining; blade fleshy, not leathery, margins entire. **Inflorescences** axillary, flowers solitary; bracts absent. **Pedicels** present; bracteoles absent. **Flowers** bisexual; sepals 5, distinct, lanceolate, calyx bilaterally symmetric, campanulate; corolla blue to violet, pink, or red, bilaterally symmetric, bilabiate and personate, tubular, tube base not spurred or gibbous, lobes 5, abaxial 3, adaxial 2; stamens 4, basally adnate to corolla, didynamous, filaments glandular-hairy; staminode 1, filamentous; ovary 2-locular, placentation axile; stigma 2-lobed. **Fruits** capsules, dehiscence loculicidal. **Seeds** 80–150, dark brown to black, ovoid to oblong-polygonal, wings absent. $x = 12$.

Species 2 (1 in the flora): sw, sc United States, Mexico.

Maurandella most commonly has been recognized as a genus segregated from *Maurandya* Ortega or as a section within *Maurandya*, based on distinctive characteristics: personate corolla, globular capsule, V-shaped septum, and marked asymmetry of the locules. The V-shaped septum results in unequal locule volumes, which is unique in Antirrhineae. Based on this unique combination of characteristics, *Maurandella* is recognized here as a genus distinct from *Epixiphium* and *Maurandya*.

Initial molecular phylogenetic studies indicated a sister group relationship between species of *Maurandella* and *Maurandya* (M. Ghebrehiwet et al. 2000); they lacked the taxon sampling needed to resolve relationships among closely related species in *Epixiphium*, *Maurandella*, and *Maurandya* (C. E. Freeman and R. Scogin 1999; P. Vargas et al. 2004). The more complete ITS sampling by M. Fernández-Mazuecos et al. (2013) placed both genera in the *Cymbalaria* clade, but not as sister taxa; *Maurandya* was sister to *Lophospermum*, whereas *Maurandella* was nested in a more basal position within a seven-taxon subclade comprising genera in subtribe Maurandyinae.

Maurandella hederifolia Rothmaler is known from northern Mexico.

1. **Maurandella antirrhiniflora** (Humboldt & Bonpland ex Willdenow) Rothmaler, Feddes Repert. Spec. Nov. Regni Veg. 52: 27. 1943 • Roving sailor F

Maurandya antirrhiniflora Humboldt & Bonpland ex Willdenow, Hort. Berol. 2: plate 83. 1806 (as Maurandia); *Antirrhinum antirrhiniflora* (Humboldt & Bonpland ex Willdenow) Hitchcock; *Asarina antirrhiniflora* (Humboldt & Bonpland ex Willdenow) Pennell

Perennials vines, 50–300 cm. **Leaves** glabrous; petiole 5–45 mm; blade hastate to sagittate, 5–30 × 4–35 mm. **Pedicels** ascending, winged, 10–45 mm, glabrous. **Flowers:** sepals subequal, adaxial lateral pair falcate, 9–14 × 2–3 mm, apex acute, surfaces glabrous; corolla tube 13–17 mm, glabrous externally, throat closed, palate inflated and streaked with white, yellow, or violet, puberulent to villous, lobes rounded, 4–8 mm, abaxial lobes projecting, adaxial erect; stamens included, filaments incurved, yellow-villous basally, abaxial 9–12 mm, adaxial 11–14 mm, pollen sacs oblong; ovary septum V-shaped, locules unequal; style terete, included, 10–13 mm, stigma recurved. **Capsules** globular, 7–10 mm, glabrous, larger locule dehiscing first. **Seeds** 1–1.5 mm, surfaces cristate-tuberculate. $2n = 24$.

Flowering Feb–Dec. Canyons, rocky slopes, arroyos, coastal dunes, desert flats on siliceous, calcareous, gypseous substrates; 0–1800(–2600) m; Ariz., Calif., Nev., N.Mex., Tex.; Mexico.

Maurandella antirrhiniflora has been treated either as one of two species in *Maurandella* (along with *M. hederifolia*) (W. Rothmaler 1943; D. A. Sutton 1988), or as one species with *M. hederifolia* treated at subspecific rank (W. J. Elisens 1985). *Maurandella antirrhiniflora* is recognized here without infraspecific variants to be consistent with the pattern of interspecific variation in Antirrhineae. Variable floral coloration is present in central Arizona and California, with pink and red corolla morphs occurring in discrete populations from the widespread blue to violet morphs.

13. MISOPATES Rafinesque, Autik. Bot., 158. 1840 • [Greek plant name used by Dioscorides; probably *misos*, to hate, and *pateo*, to trample, alluding to erect stems (in contrast to low lying habit of *Orontium aquaticum*)] Ⓘ

Craig C. Freeman

Herbs, annual. **Stems** ascending or erect, glandular-pubescent. **Leaves** cauline, opposite proximally, alternate distally; petiole absent; blade not fleshy, not leathery, margins entire. **Inflorescences** terminal, racemes; bracts present. **Pedicels** present; bracteoles absent. **Flowers** bisexual; sepals 5, basally connate, calyx bilaterally symmetric, short-cupulate, lobes linear, usually longer than corolla tube in flower; corolla pink or purple with darker veins, rarely white, bilaterally symmetric, bilabiate and personate, cylindric or urceolate, 10–15 mm, tube base gibbous abaxially, not spurred, lobes 5, abaxial 3, adaxial 2; stamens 4, basally adnate to corolla, didynamous, filaments glabrous; staminode 1, lingulate; ovary 2-locular, placentation axile; stigma capitate. **Fruits** capsules, (5–)7–11 mm, dehiscence poricidal (adaxial locule, smaller adaxial locule sometimes indehiscent). **Seeds** 200–250, dark brown, angled to obovoid, wings present. $x = 8$.

Species 7 (1 in the flora): introduced; s Europe, Asia, n Africa; introduced also in South America, s Africa, Pacific Islands, Australia.

Using molecular data, P. Vargas et al. (2004) and M. Fernández-Mazuecos et al. (2013) found *Misopates* to be monophyletic and sister to the Old World genera *Acanthorrhinum* Rothmaler and *Pseudomisopates* Güemes. *Misopates* is distinguished from related genera in the flora area by its annual habit, calyx lobes usually longer than the corolla tubes, and capsules with unequal locules.

1. **Misopates orontium** (Linnaeus) Rafinesque, Autik. Bot., 158. 1840 • Lesser snapdragon, weasel's-snout, muflier rubicond Ⓕ Ⓘ Ⓦ

Antirrhinum orontium Linnaeus, Sp. Pl. 2: 617. 1753

Subspecies 2 (1 in the flora): introduced; Europe, sw, sc Asia, n Africa; introduced also in South America, s Africa, Pacific Islands, Australia.

Subspecies *gibbosum* (Wallich) D. A. Sutton, which occurs in the eastern Mediterranean, southwestern and south-central Asia, and northeastern Africa, has more slender stems, smaller flowers, and more lax inflorescences than subsp. *orontium* (D. A. Sutton 1988). Subspecies *gibbosum* does not appear to have been introduced in the flora area.

1a. **Misopates orontium** (Linnaeus) Rafinesque subsp. **orontium** Ⓕ Ⓘ Ⓦ

Stems simple or branched, 10–70 cm. **Leaves:** blade linear to narrowly lanceolate or narrowly oblanceolate, 12–55 × 2–10 mm, glandular. **Inflorescences** 2–25 cm, densely glandular-hairy; bracts linear, 15–25 × 0.9–1.2 mm. **Pedicels** ascending, 0.5–2.5 mm in flower, to 4 mm in fruit. **Flowers:** calyx lobes accrescent, abaxial 10–15 × 1–2 mm in fruit, adaxial 15–20 × 1–2.5 mm in fruit; corolla sparsely glandular-pubescent externally, tube 6–8 mm, internally glandular-pubescent abaxially, hairs yellow, palate densely white-lanate, throat 3.5–4.5 mm diam., abaxial lip: middle lobe narrower than laterals; stamens included; staminode

0.2 mm; style persistent, densely glandular-pubescent. **Capsules** obliquely ovoid or oblong-ovoid, 4–7 mm wide, glandular-pubescent. **Seeds** 0.9–1.2 mm, papillate, abaxial face keeled, adaxial with sinuate marginal ridges, wings narrow. $2n = 16$ (Portugal).

Flowering May–Oct. Disturbed sites, including roadsides, rail yards, construction sites, abandoned areas, gravel bars; 0–1000 m; introduced; B.C., Ont., Que.; Alaska, Calif., Conn., Fla., Idaho, Ill., Ky., Maine, Mich., N.J., N.Y., Ohio, Oreg., Pa., Utah, Va., Wash.; Europe; Asia; n Africa; introduced also in South America (Argentina, Bolivia, Ecuador), s Africa (Republic of South Africa), Pacific Islands (Hawaii, New Zealand), Australia.

Subspecies *orontium* is widely, sporadically naturalized, mostly in temperate regions of the world.

14. MOHAVEA A. Gray in War Department [U.S.], Pacif. Railr. Rep. 4(5): 122. 1857 • [Alluding to Mohave River]

Kerry A. Barringer

Herbs, annual. **Stems** erect or ascending, filiform, twining branches absent, densely glandular-villous. **Leaves** cauline, opposite proximally, alternate distally; petiole present; blade fleshy, not leathery, margins entire. **Inflorescences** axillary, flowers solitary; bracts absent. **Pedicels** present; bracteoles absent. **Flowers** bisexual; sepals 5, basally connate, calyx bilaterally symmetric, cupulate, lobes lanceolate, adaxial largest; corolla yellow or white, bilaterally symmetric, bilabiate, tubular, tube base usually gibbous abaxially, not spurred, lobes 5, abaxial 3, adaxial 2, palate partially blocking mouth; stamens 2, basally adnate to corolla, filaments glabrous or sparsely hairy, pollen sacs 1 per filament; staminodes 2 or 3, linear, adaxial minute or absent; ovary 2-locular, placentation axile; stigma punctiform. **Fruits** capsules, locules equal, dehiscence poricidal. **Seeds** 20–50, dark brown to black, oblong, wings present, abaxial, cupulate. $x = 15$.

Species 2 (2 in the flora): sw United States, nw Mexico.

Mohavea is similar to *Pseudorontium*, which also has asymmetric, winged seeds. Seeds of *Mohavea* are notched at the proximal end. Plants of the genus can be distinguished from others in Antirrhineae by their relatively large, acute or obtuse corolla lobes, more or less radially symmetric fruits, and two fertile anthers, each with a single pollen sac. R. K. Oyama and D. A. Baum (2004) and M. Fernández-Mazuecos et al. (2013) found evidence that *Mohavea* nests among species treated in *Sairocarpus*.

1. Corollas yellow, 15–20 mm; pedicels 3–5 mm in flower . . . . 1. *Mohavea breviflora*
1. Corollas pale yellow to white, 25–35 mm; pedicels 5–6 mm in flower . . . . 2. *Mohavea confertiflora*

1. Mohavea breviflora Coville, Contr. U.S. Natl. Herb. 4: 168, plate 17. 1893 • Desert snapdragon E

Antirrhinum mohavea D. J. Keil

Stems 5–20(–40) cm, branched. **Leaves:** blade narrowly lanceolate to elliptic, 15–40 × 3–15 mm. **Pedicels** 3–5 mm in flower, 5–10 mm in fruit. **Flowers:** calyx abaxial and lateral lobes 5–9 mm, adaxial lobe 7–10 mm; corolla yellow, sometimes lightly spotted with maroon, 15–20 mm, palate convex, yellow, marked with maroon. **Seeds** 2–2.5 mm. $2n = 30$.

Flowering Feb–Apr. Desert, gravelly slopes, washes, roadsides; -40–1900 m; Ariz., Calif., Nev., Utah.

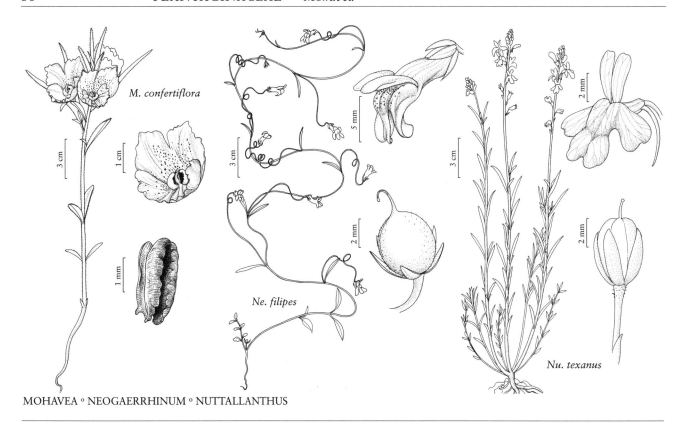

MOHAVEA ∘ NEOGAERRHINUM ∘ NUTTALLANTHUS

2. Mohavea confertiflora (Bentham ex A. de Candolle) A. Heller, Muhlenbergia 8: 48. 1912

• Ghost flower F

Antirrhinum confertiflorum Bentham ex A. de Candolle in A. P. de Candolle and A. L. P. P. de Candolle, Prodr. 10: 592. 1846; *Mohavea viscida* A. Gray

Stems 5–60 cm, simple or branched. **Leaves:** blade linear to narrowly lanceolate, proximal sometimes elliptic, 20–65 × 3–15 mm. **Pedicels** 5–6 mm in flower, 6–16 mm in fruit. **Flowers:** calyx abaxial lobe 13–15 mm, adaxial and lateral lobes 10–13 mm; corolla pale yellow to white, often marked with maroon, 25–35 mm, palate convex, yellow and spotted with maroon or maroon. **Seeds** 1.5–2 mm. $2n = 30$.

Flowering Mar–Apr. Desert, rocky slopes, washes; -40–1200 m; Ariz., Calif., Nev.; Mexico (Baja California, Sonora).

The pale, maroon-spotted flowers of *Mohavea confertiflora* resemble those of *Mentzelia involucrata*, which sometimes grows nearby. *Mentzelia* flowers produce nectar, but *Mohavea* flowers do not; R. J. Little (1980) suggested that *M. confertiflora* flowers mimic *Mentzelia* to attract pollinators. Also, the maroon spots on the palate of *M. confertiflora* flowers are said to resemble the color and form of the abdomen of female *Xeralictus* bees as they are exposed when the females bury their heads in the *Mentzelia* flowers to collect nectar. Male *Xeralictus* bees often attempt to mate with the females at these times, and they have been observed trying to mate with the palates of *M. confertiflora* flowers as well (D. Yanega, http://entmuseum.ucr.edu/bug_spotlight/posted%20 Images-pages/33.htm).

L. C. Hileman et al. (2003) found that the symmetry of the adaxial corolla lip and the abortion of the lateral and adaxial stamens, characteristics that distinguish the genus, were controlled by regulation of the expression of two genes. They suggested that these changes are adaptations to the pollination of this species.

15. NEOGAERRHINUM Rothmaler, Feddes Repert. Spec. Nov. Regni Veg. 52: 31. 1943

• [Greek *neo-*, new, *gaea*, earth or world, and *rhinum*, nose, alluding to being native to the New World]

Kerry A. Barringer

Herbs, annual. **Stems** erect, ascending, or sprawling, filiform, twining branches absent, glabrous or hairy. **Leaves** cauline, opposite proximally, alternate distally; petiole present proximally, absent distally; blade not fleshy, not leathery, margins entire. **Inflorescences** axillary, flowers solitary; bracts absent. **Pedicels** present, elongate, twining; bracteoles absent. **Flowers** bisexual, cleistogamous and chasmogamous; sepals 5, basally connate, calyx slightly bilaterally symmetric, cupulate, lobes lanceolate; corolla purple or yellow, bilaterally symmetric, bilabiate and personate, tubular, tube base gibbous abaxially, not spurred, lobes 5, abaxial 3, adaxial 2; stamens 4, basally adnate to corolla, didynamous, filaments glabrous or sparsely glandular-hairy distally, pollen sacs 2 per filament; staminode 1, minute; ovary 2-locular, placentation axile; stigma punctiform or subcapitate. **Fruits** capsules, locules equal, dehiscence poricidal. **Seeds** 20–100, black, ovoid to subglobular, wings absent. *x* = 15.

Species 2 (2 in the flora): sw United States, nw Mexico.

Neogaerrhinum is closely related to the *Sairocarpus* species that have filiform, prehensile branches; in *Neogaerrhinum*, pedicels are elongate and twining, and fruits are symmetric and usually drooping.

In both species of *Neogaerrhinum*, the cleistogamous flowers have white corollas and are smaller than the chasmogamous flowers. Only the chasmogamous flowers are used in the key and descriptions below.

1. Corollas yellow; calyx lobes 3–4.5 mm; capsules puberulent . . . . 1. *Neogaerrhinum filipes*
1. Corollas lavender to purple; calyx lobes 4–8 mm; capsules glabrous . . . . 2. *Neogaerrhinum strictum*

1. Neogaerrhinum filipes (A. Gray) Rothmaler, Feddes Repert. Spec. Nov. Regni Veg. 52: 31. 1943 [F]

Antirrhinum filipes A. Gray in J. C. Ives, Rep. Colorado R. 4: 19. 1861; *A. cooperi* A. Gray; *Asarina filipes* (A. Gray) Pennell

Stems sprawling, 10–70(–100) cm, glabrous or hairy proximally. **Leaves:** blade ovate to lanceolate, 6–50 × 2–15 mm, glabrous. **Pedicels** 30–100 mm. **Flowers:** calyx lobes equal or subequal, 3–4.5 mm; corolla yellow, 10–15 mm, mouth 3–3.5 mm diam., abaxial lobes reflexed, often purple near base, adaxial lobes projecting, golden yellow, palate yellow with maroon spots. **Capsules** often pendent, globular, 3–5 mm, puberulent, with 2, irregular pores. **Seeds** spotted with white, ovoid to subglobular, 1 mm, abaxial surface longitudinally ribbed. 2*n* = 30.

Flowering Mar–May. Desert washes and shrublands; 60–2000 m; Ariz., Calif., Nev., Utah; Mexico (Sonora).

2. Neogaerrhinum strictum (Hooker & Arnott) Rothmaler, Feddes Repert. Sp. Nov. Regni Veg. 52: 31. 1943 • Kellogg's snapdragon

Maurandya stricta Hooker & Arnott, Bot. Beechy Voy., 375. 1839 (as Maurandia); *Antirrhinum hookerianum* Pennell ex Millspaugh; *Asarina stricta* (Hooker & Arnott) Pennell

Stems erect or ascending, 10–80 cm, hairy proximally. **Leaves:** blade ovate proximally, linear-lanceolate distally, 10–50 × 1–15 mm, hairy or glabrous. **Pedicels** 20–90 mm. **Flowers:** calyx lobes equal, 4–8 mm; corolla lavender to purple, usually with darker veins, 10–14 mm, mouth 4–5 mm diam., abaxial lobes reflexed, purple, adaxial lobes reflexed, lavender to purple, palate white with purple veins. **Capsules** usually erect, globular to broadly ovoid, 5–7 mm, glabrous, with 2 pores elongating into slits. **Seeds** not spotted, ovoid, 0.8 mm, abaxial surface with blocklike tubercles. 2*n* = 30.

Flowering Apr–Jun. Disturbed sites, chaparral, coastal sage; 10–1000 m; Calif.; Mexico (Baja California).

Neogaerrhinum strictum grows from the San Francisco Bay area to northern Mexico. D. M. Thompson (1988) found that it can grow abundantly following a fire.

The name *Antirrhinum kelloggii* Greene has been used for this species, but it is invalid.

16. NUTTALLANTHUS D. A. Sutton, Revis. Antirrhineae, 455, fig. 122. 1988 • Toadflax [For Thomas Nuttall, 1786–1859, British naturalist and plant collector, and Greek *anthos*, flower]

Craig C. Freeman

Herbs, annual or biennial. **Stems** heteromorphic, sterile prostrate or decumbent to ascending, glabrous, fertile erect, glabrous or glandular-pubescent distally. **Leaves** cauline, whorled on sterile stems, alternate, sometimes whorled proximally, on fertile stems; petiole absent or essentially so; blade not fleshy, not leathery, margins entire. **Inflorescences** terminal, spikelike racemes; bracts present. **Pedicels** present; bracteoles absent. **Flowers** bisexual; sepals 5, proximally connate, calyx ± symmetric, short-cupulate, lobes linear-lanceolate or lanceolate to narrowly ovate or oblong; corolla white to blue or pale violet, bilaterally symmetric, bilabiate and personate, tubular, tube base not gibbous, spurred abaxially, lobes 5, abaxial 3, adaxial 2, abaxial lip much longer than adaxial and strongly arched at palate; stamens 4, proximally adnate to corolla, didynamous, filaments glabrous; staminode 0 or 1, minute; ovary 2-locular, placentation axile; stigma capitate. **Fruits** capsules, 2–4.8 mm, dehiscence poricidal by 4 or 5 irregular valves in each locule. **Seeds** 100–200, black or gray, angled, wings absent. $x = 6$.

Species 4 (3 in the flora): North America, Mexico, South America; introduced in West Indies, e Europe, Pacific Islands.

Species of *Nuttallanthus* native to the New World have been included historically in *Linaria*; F. W. Pennell (1919, 1935) included them in the invalidly published sect. *Lectoplectron* Pennell. Sutton erected *Nuttallanthus* to accommodate these species based on their corollas with larger abaxial lips relative to adaxial lips, weakly developed palates, and less well-developed spurs, and their angled, four- to seven-ridged seeds. M. Fernández-Mazeucos et al. (2013) presented molecular evidence for the monophyly of *Nuttallanthus* as a member of one of three basal clades within *Linaria*. Support for two of the three basal clades, including one containing *Nuttallanthus*, was sufficiently low that the authors concluded that further studies were needed to determine whether *Nuttallanthus* should be included in *Linaria* as a section. Consequently, *Nuttallanthus* is maintained here as a distinct genus. *Nuttallanthus subandinus* D. A. Sutton occurs in South America.

Nuttallanthus species usually grow in well-drained, sandy soil in native plant communities that experience disturbances and also in disturbed habitats, including former cropland. The species sometimes grow in mixed populations; the North American species are cross incompatible (P. T. Crawford and W. J. Elisens 2006).

In the descriptions, corolla length is the distance from the distal point of the medial lobe of the abaxial lip to the distal point of the spur. In addition to chasmogamous flowers, cleistogamous flowers are produced early and late in the reproductive cycle. Fruits produced by cleistogamous flowers are smaller and contain fewer seeds than do fruits produced by chasmogamous flowers (P. T. Crawford and W. J. Elisens 2006).

SELECTED REFERENCES Crawford, P. T. 2003. Biosystematics of North American Species of *Nuttallanthus* (Lamiales). Ph.D. dissertation. University of Oklahoma. Crawford, P. T. and W. J. Elisens. 2006. Genetic variation and reproductive system among North American species of *Nuttallanthus* (Plantaginaceae). Amer. J. Bot. 93: 582–591.

1. Corolla spurs 0.1–0.5 mm; pedicels 6–14 mm in fruit, ascending, sometimes erect, hairs to 0.3 mm . . . . 2. *Nuttallanthus floridanus*
1. Corolla spurs 2–11 mm; pedicels 1.8–8(–9) mm in fruit, erect, hairs to 0.1 mm.
   2. Edges of seeds sharp, faces obscurely tuberculate; corollas 8–14 mm, spurs 2–7 mm . . . . 1. *Nuttallanthus canadensis*
   2. Edges of seeds rounded, rarely angled or irregularly dentate, faces prominently pointed-tuberculate, rarely with rounded ridges and scattered, rounded tubercles; corollas (11–)14–22 mm, spurs 4.5–11 mm . . . . 3. *Nuttallanthus texanus*

1. **Nuttallanthus canadensis** (Linnaeus) D. A. Sutton, Revis. Antirrhineae, 457. 1988 • Canada toadflax, linaire du Canada E W

Antirrhinum canadense Linnaeus, Sp. Pl. 2: 618. 1753; *Linaria canadensis* (Linnaeus) Dumont de Courset

Fertile stems 1–4(–7), simple, rarely distally branched, 11–70 cm. **Leaves:** blades of sterile-stem leaves narrowly elliptic to obovate, 2–12 × 0.5–3 mm, blades of fertile-stem leaves linear, 5–43 × 0.5–2.2 mm. **Racemes** 1–18 cm; bracts narrowly oblanceolate or lanceolate to narrowly elliptic, 1.1–3 mm. **Pedicels** erect, 1.8–5.5 mm in fruit, sparsely glandular-pubescent, sometimes glabrous, hairs to 0.1 mm. **Flowers:** calyx lobes linear-lanceolate to narrowly ovate, 2.1–3.5 × 0.4–1 mm, proximally sparsely glandular-pubescent, sometimes glabrous; corolla white to blue, 8–14 mm, spurs straight or curved, 2–7 mm, abaxial lip 2–4.5 mm, adaxial 1.2–2(–3) mm. **Capsules** oblong-ovoid, 2.6–3.9 × 2.6–3.3 mm. **Seeds** black or gray, 0.3–0.5 mm, edges sharp, faces obscurely tuberculate. $2n = 12$.

Flowering Feb–Jul(–Sep). Sandy prairies, woodlands, roadsides, fallow fields, disturbed sites; 0–300 m; B.C., N.B., N.S., Ont., P.E.I., Que.; Ala., Ark., Calif., Conn., Del., D.C., Fla., Ga., Ill., Ind., Iowa, Kans., La., Maine, Md., Mass., Mich., Minn., Miss., Mo., Mont., Nebr., N.H., N.J., N.Y., N.C., N.Dak., Ohio, Okla., Pa., R.I., S.C., S.Dak., Tenn., Tex., Vt., Va., Wash., W.Va., Wis.; introduced in e Europe (Russia).

Nuttallanthus canadensis and *N. texanus* are sympatric through much of their ranges. In Texas, where they sometimes occur in mixed populations, R. Kral (1955) observed that *N. canadensis* bloomed and set fruit earlier than did *N. texanus*.

2. **Nuttallanthus floridanus** (Chapman) D. A. Sutton, Revis. Antirrhineae, 461. 1988 • Apalachicola or Florida toadflax E W

Linaria floridana Chapman, Fl. South. U.S., 290. 1860

Fertile stems 1–5, simple, sometimes distally branched, 9–45 cm. **Leaves:** blades of sterile-stem leaves elliptic to obovate, 1–6 × 0.5–2.8 mm, blades of fertile-stem leaves linear, 5–32 × 0.4–0.9(–1.4) mm. **Racemes** 1–11 cm; bracts narrowly lanceolate to narrowly elliptic, 1–3.5(–5) mm. **Pedicels** ascending, sometimes erect, 6–14 mm in fruit, sparsely to densely glandular-pubescent, hairs to 0.3 mm. **Flowers:** calyx lobes lanceolate, 1.3–2.8 × 0.3–0.7 mm, sparsely to densely glandular-pubescent; corolla blue, 5–9 mm, spurs straight, 0.1–0.5 mm, abaxial lip 3–4.5 mm, adaxial 1–3.4 mm. **Capsules** oblong-globular, 2–3.7 × 1.8–3 mm. **Seeds** black, 0.3–0.5 mm, edges rounded, faces obscurely tuberculate.

Flowering Mar–Apr. Sandy woodlands, scrublands, sandhills, coastal dunes; 0–100 m; Ala., Fla., Ga., Miss.

3. **Nuttallanthus texanus** (Scheele) D. A. Sutton, Revis. Antirrhineae, 460. 1988 • Texas toadflax F W

Linaria texana Scheele, Linnaea 21: 761. 1849; *L. canadensis* (Linnaeus) Dumont de Courset var. *texana* (Scheele) Pennell

Fertile stems 1–13(–30), distally branched, sometimes simple, 17–70 cm. **Leaves:** blades of sterile-stem leaves elliptic or oblong-elliptic to obovate, 2–18 × 0.5–3 mm, blades of fertile-stem leaves linear to narrowly elliptic, 7–34 × (0.5–)1–3.1 mm. **Racemes** (2–)4–20 cm; bracts narrowly oblanceolate or lanceolate to narrowly elliptic, (0.7–)2–4 mm. **Pedicels** erect, 2–8(–9) mm in fruit, glabrous or sparsely glandular-pubescent, hairs to 0.1 mm. **Flowers:** calyx lobes lanceolate to oblong, (2–)2.5–4.2 × 0.8–1.6 mm, proximally sparsely glandular-pubescent, sometimes glabrous; corolla blue to pale violet, (11–)14–22 mm, spurs curved, sometimes straight, 4.5–11 mm, abaxial lip 5–11 mm, adaxial 3–6 mm. **Capsules** oblong-ovoid, 2.6–4.8 × 2.5–4 mm. **Seeds** gray, 0.3–0.5 mm, edges rounded, rarely angled or irregularly dentate, faces prominently pointed-tuberculate, rarely with rounded ridges and scattered, rounded tubercles. $2n = 12, 24$.

Flowering Feb–Jul. Sandy prairies, woodlands, roadsides, fallow fields, rocky bluffs, disturbed sites; 0–1800 m; Alta., B.C., Sask.; Ala., Ariz., Ark., Calif., Colo., Fla., Ga., Ill., Kans., Ky., La., Minn., Miss., Mo., Mont., Nebr., N.Mex., N.C., N.Dak., Okla., Oreg., S.C., S.Dak., Tenn., Tex., Utah, Va., Wash., Wyo.; Mexico; South America; introduced in West Indies (Dominican Republic), Pacific Islands (Hawaii).

Sutton observed that some specimens from the northwestern United States and southwestern Canada have seeds with dentate ridges, similar to those of the South American *Nuttallanthus subandinus*. Sutton did not explicitly associate the North American plants with that species, stating that further study was needed. Among specimens examined for this treatment were five from Kansas, Louisiana, and Oklahoma with dentate-ridged seeds apparently similar to those illustrated in Sutton's monograph. In other features, these specimens most resemble *N. texanus*.

P. T. Crawford (2003) also reported two seed morphotypes in *Nuttallanthus texanus*. Among plants from California, Oklahoma, Texas, and the southeastern United States were individuals bearing seeds with ridges rounded and faces densely covered with acute tubercles. Among plants from Arkansas, Louisiana, Oklahoma, and Texas were individuals bearing seeds with more sharply angled ridges and faces with rounded ridges and scattered, rounded tubercles. From his description and photographs, this second type does not match the dentate-ridged seed type discussed above. Pending further study, plants with all three seed types are included here in *N. texanus*.

17. PSEUDORONTIUM Rothmaler, Feddes Repert. Spec. Nov. Regni Veg. 52: 33. 1943

• [Greek *pseudo*, false, and genus *Orontium*, alluding to resemblance of seeds]

Kerry A. Barringer

Neil A. Harriman

Antirrhinum Linnaeus sect. *Pseudorontium* A. Gray, Proc. Amer. Acad. Arts 12: 81. 1876

Herbs, annual. **Stems** erect, branching, filiform, twining branches absent, glandular-hairy, viscid. **Leaves** cauline, opposite proximally, alternate distally; petiole present; blade not fleshy, not leathery, margins entire. **Inflorescences** axillary, flowers solitary; bracts present, alternate. **Pedicels** present, not twining, recurved in fruit; bracteoles absent. **Flowers** bisexual; sepals 5, basally connate, calyx slightly bilaterally symmetric, cupulate, lobes linear to narrowly oblanceolate; corolla purple with darker veins, bilaterally symmetric, bilabiate and personate, tubular, tube base gibbous abaxially, not spurred, lobes 5, abaxial 2, adaxial 3; stamens 4, basally adnate to corolla, didynamous, filaments glabrous or sparsely hairy, pollen sacs 1 per filament; staminode 0 or 1, capitate; ovary 2-locular, placentation axile; stigma subcapitate. **Fruits** capsules, locules equal, dehiscence poricidal. **Seeds** 30–70, black or dark brown, ellipsoid, wings present, cupulate. $x = 13$.

Species 1: sw United States, nw Mexico.

Pseudorontium is similar to *Mohavea* in having winged seeds, but R. K. Oyama and D. A. Baum (2004) and P. Vargas et al. (2004) found *Pseudorontium* positioned well outside the taxa usually included in *Antirrhinum* in both the narrow (D. A. Sutton 1988) and broad (D. M. Thompson 1988) senses. ITS results in M. Fernández-Mazuecos et al. (2013) placed *Pseudorontium* and *Mohavea* in different clades; *Pseudorontium* was sister to *Galvezia*. The pendent, globular capsules, eccentrically winged seeds, and lack of twining branches distinguish *P. cyathiferum* from species of *Sairocarpus*.

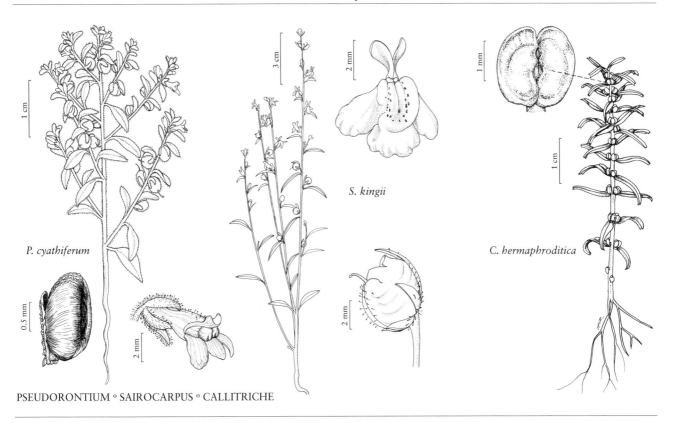

S. kingii

P. cyathiferum

C. hermaphroditica

PSEUDORONTIUM ○ SAIROCARPUS ○ CALLITRICHE

1. Pseudorontium cyathiferum (Bentham) Rothmaler, Feddes Repert. Spec. Nov. Regni Veg. 52: 33. 1943

• Dog's mouth F

Antirrhinum cyathiferum Bentham, Bot. Voy. Sulphur, 40, plate 19. 1844; *A. chytrospermum* A. Gray

Stems 15–35(–70) cm. **Leaves:** petiole 4–15 mm; blade ovate to broadly ovate, 30–40 × 20–25 mm, smaller distally. **Pedicels** 1–3 mm, longer in fruit, glandular-hairy. **Flowers:** calyx lobes equal, 4–5 mm in flower, 6.5–9 mm in fruit, glandular-hairy; corolla 8–10 mm, abaxial lip cream, often with darker veins, palate white with yellow spots, furrowed, adaxial lip cream with deep purple veins; staminode 0–1 mm. **Capsules** pendent, globular, 5–8 mm, sparsely glandular-hairy. **Seeds** 1.5–2.5 mm, ridged and tuberculate dorsally, winged adaxially. $2n = 26$.

Flowering Jan–Apr. Rocky slopes, washes; 0–800 m; Ariz., Calif.; Mexico (Baja California, Sonora).

Plants of *Pseudorontium cyathiferum* will flower and set fruit at any time of the year if moisture is available. Sand grains often stick to the viscid hairs, covering the plants.

18. SAIROCARPUS D. A. Sutton, Revis. Antirrhineae, 461, figs. 123–125, 126.1, 126.2, 126.4. 1988 • Snapdragon [Greek *sairos*, curling back lips to show teeth, and *karpos*, fruit, perhaps alluding to recurved teeth of capsules]

Kerry A. Barringer

Herbs, annual, biennial, or perennial. **Stems** erect [ascending or sprawling], filiform, twining branches often present, glabrous, hairy, or glandular-hairy. **Leaves** cauline, sometimes basal, opposite or alternate proximally, alternate distally; petiole present or absent; blade linear to oblanceolate, not fleshy, not leathery, margins entire. **Inflorescences** terminal or axillary,

racemes or flowers solitary; bracts present or absent. **Pedicels** 1–25(–30) mm; bracteoles absent.
Flowers bisexual, cleistogamous or chasmogamous; sepals 5, basally connate, calyx bilaterally
symmetric, tubular or cupulate, lobes ovate to lanceolate, shorter or as long as corolla tube in
flower, adaxial largest, glabrous or hairy to glandular-hairy; corolla white to purple, pink, red,
or tan, bilaterally symmetric, bilabiate and personate, tubular, 5.5–18 mm, tube base usually
gibbous, not spurred (not gibbous, spurred in *S. cornutus*), lobes 5, abaxial 3, adaxial 2; stamens
4, basally adnate to corolla, didynamous, filaments glabrous or glandular-hairy, pollen sacs
2 per filament; staminode 0; ovary 2-locular, placentation axile; stigma punctiform. **Fruits**
capsules, 2.5–11 mm, locules unequal, dehiscence poricidal. **Seeds** 5–40, brown to black, ovoid
to oblong, wings absent. x = 8.

Species 12 (9 in the flora): w United States, n Mexico.

Sairocarpus is sometimes treated as a subgenus of *Antirrhinum* (D. M. Thompson 1988;
M. Wetherwax and Thompson 2012). For a discussion of the generic segregates, see
2. *Antirrhinum*.

Sairocarpus is a New World genus distinguished from the Old World *Antirrhinum* by
its smaller flowers, disjunct distribution, and base chromosome number x = 8 following
R. K. Oyama and D. A. Baum (2004) and P. Vargas et al. (2004). Additional study of generic
limits may be warranted since more complete ITS sampling, including all *Sairocarpus* species in
the flora area (M. Fernández-Mazuecos et al. 2013), revealed that *Sairocarpus* is polyphyletic
if *Gambelia*, *Howelliella*, *Mohavea*, and *Neogaerrhinum* are recognized. Other New World-
only segregates of *Antirrhinum* are *Gambelia*, *Howelliella*, *Mohavea*, *Neogaerrhinum*, and
Pseudorontium; they can be distinguished from *Sairocarpus* by their fruits with equal locules.
Plants of *Mohavea* and *Pseudorontium* also have distinctive, winged seeds.

Some species of *Sairocarpus* have filiform, twining branches, usually on the distal parts of the
stems. These branches wrap around nearby objects giving additional support to these weak-
stemmed plants.

Cleistogamous and chasmogamous flowers are produced in some species of *Sairocarpus*.
The cleistogamous flowers usually form early in the season and are smaller and paler than the
chasmogamous flowers. In species with twining branches, cleistogamous flowers are usually
borne close to the main stem. Fruits from cleistogamous flowers are usually smaller and have
fewer seeds than those from chasmogamous flowers. Only chasmogamous flowers are described
below.

1. Calyx lobes unequal.
  2. Stems viscid; branches not twining; plants annuals or perennials; corollas pale pink to
    red. . . . . 4. *Sairocarpus multiflorus*
  2. Stems not viscid; branches twining; plants annuals; corollas white to tan or light
    purple.
    3. Corollas 5.5–7 mm; leaf blades elliptic to linear . . . . .3. *Sairocarpus kingii*
    3. Corollas 8–17 mm; leaf blades ovate to narrowly elliptic.
      4. Corollas white to tan, not veined; stems hairy . . . . .6. *Sairocarpus subcordatus*
      4. Corollas light purple to white, often dark-veined; stems glabrous or
        glandular-hairy. . . . . 7. *Sairocarpus vexillocalyculatus*
1. Calyx lobes equal to subequal.
  5. Stems glandular-hairy.
    6. Leaf blades linear to oblanceolate; stems self-supporting; branches not twining;
      capsules: abaxial locules indehiscent or with 1 pore . . . . 1. *Sairocarpus cornutus*
    6. Leaf blades ovate; stems not self-supporting; branches twining; capsules: abaxial
      locules with 1 pore . . . . .5. *Sairocarpus nuttallianus*
  5. Stems glabrous, or basally hairy and otherwise glabrous.

[7. Shifted to left margin.—Ed.]

7. Branches not twining; capsules: abaxial locules with 2 pores; plants perennials; corollas
pink to pale pink . . . . 8. *Sairocarpus virga*
7. Branches twining; capsules: abaxial locules indehiscent or with 1 pore; plants annuals;
corollas white to light purple.
    8. Inflorescences racemes; pedicels 1–5 mm; corolla palates purple-spotted. . . . . 2. *Sairocarpus coulterianus*
    8. Inflorescences flowers solitary; pedicels 4–25(–30) mm; corolla palates purple-veined
    . . . . 9. *Sairocarpus watsonii*

1. **Sairocarpus cornutus** (Bentham) D. A. Sutton, Revis. Antirrhineae, 468. 1988 • Spurred snapdragon E

Antirrhinum cornutum Bentham, Pl. Hartw., 328. 1849

Annuals. Stems 10–60 cm, self-supporting, glandular-hairy; branches not twining. **Leaves** opposite proximally, alternate distally; blade linear to oblanceolate, 7–43 × 3–6 (–10) mm, surfaces glandular-hairy. **Inflorescences** axillary, flowers solitary. **Pedicels** 1–4 mm. **Flowers** chasmogamous; calyx lobes equal or subequal, glandular-hairy, hairs long, eglandular, adaxial lobe 3–7 × 1 mm; corolla white to pale lavender, often violet-veined, 9–14 mm, base spurred, mouth 2–3 mm diam., palate sometimes purple-veined, rounded or angular, 3.5–6.5 mm diam., glabrous or hairy. **Capsules** ovoid, 5–7 mm, glandular-hairy, abaxial locule indehiscent or with 1 pore. **Seeds** black to dark brown, 1 mm, tuberculate and longitudinally ridged. $2n = 32$.

Subspecies 2 (2 in the flora): California.

1. Corolla veins violet, palates rounded, hairy, without yellow patches . . . .
. . . . 1a. *Sairocarpus cornutus* subsp. *cornutus*
1. Corolla veins not contrasting, palates angular, glabrous or hairy, often with 2 yellow patches
. . . . 1b. *Sairocarpus cornutus* subsp. *leptaleus*

1a. **Sairocarpus cornutus** (Bentham) D. A. Sutton subsp. **cornutus** E

Leaves: blade linear to oblanceolate. **Pedicels** 1–2(–4) mm. **Flowers:** calyx lobes equal, 4–5.5 mm; corolla white, veins violet, 9–11 mm, palate rounded, hairy. **Seeds:** ridges not reticulate.

Flowering Jun–Aug. Dry stream margins, disturbed areas, often on serpentine; 50–1200 m; Calif.

Subspecies *cornutus* is known from the Inner North Coast Ranges, the western Cascade Ranges, and the northern Sacramento Valley.

1b. **Sairocarpus cornutus** (Bentham) D. A. Sutton subsp. **leptaleus** (A. Gray) Barringer, Phytoneuron 2013-34: 1. 2013 E

Antirrhinum leptaleum A. Gray, Proc. Amer. Acad. Arts 7: 373. 1868; *A. cornutum* Bentham var. *leptaleum* (A. Gray) Munz

Leaves: blade oblanceolate. **Pedicels** 1–2 mm. **Flowers:** calyx lobes subequal, 3–7 mm; corolla white to pale lavender, veins not contrasting, 7–14 mm, palate angular, glabrous or hairy, often with 2 yellow patches. **Seeds:** ridges reticulate.

Flowering Jun–Aug. Rocky washes, disturbed areas, foothills; 200–2100 m; Calif.

Subspecies *leptaleus* is known from the Sierra Nevada, especially in the foothill region.

2. **Sairocarpus coulterianus** (Bentham ex A. de Candolle) D. A. Sutton, Revis. Antirrhineae, 468. 1988 • Coulter's snapdragon

Antirrhinum coulterianum Bentham ex A. de Candolle in A. P. de Candolle and A. L. P. P. de Candolle, Prodr. 10: 592. 1846; *A. coulterianum* subsp. *orcuttianum* (A. Gray) Pennell

Annuals. Stems 12–150 cm, seldom self-supporting, basally hairy, otherwise glabrous; branches twining. **Leaves** opposite proximally, basal rosette sometimes present, alternate distally; blade lanceolate to linear, 20–50 (–110) × 1–5(–25) mm, surfaces glabrous or slightly villous. **Inflorescences** terminal, racemes. **Pedicels** 1–5 mm. **Flowers** cleistogamous and chasmogamous; calyx lobes equal to subequal, glandular-hairy, adaxial lobe 3–6 × 0.5–1 mm; corolla white to light purple, 9–12 mm, base gibbous, mouth 1–1.5(–2) mm diam., palate purple-spotted, rounded, 4–8 mm diam., glandular-hairy. **Capsules** narrowly ovoid, 5–10 mm, sparsely glandular-hairy, abaxial locule indehiscent. **Seeds** black, 1 mm, longitudinally ridged, reticulate. $2n = 30$.

Flowering Apr–Jul. Coastal and desert scrub, burned slopes; 0–2700 m; Calif.; Mexico (Baja California).

The relatively large, rounded palates make the flowers of *Sairocarpus coulterianus* distinctive. Plants growing in chaparral and coastal sage habitats usually have basal rosettes. Sometimes, the racemes develop an elongate, prehensile tip. *Sairocarpus coulterianus* is known from the southern Outer South Coastal Ranges, southwestern mainland California, and the northwestern edge of the Sonoran Desert.

3. **Sairocarpus kingii** (S. Watson) D. A. Sutton, Revis. Antirrhineae, 470. 1988 • Least snapdragon E F

Antirrhinum kingii S. Watson, Botany (Fortieth Parallel), 215, plate 21, figs. 1–4. 1871

Annuals. Stems 3–45 cm, not self-supporting, glabrous; branches twining. **Leaves** opposite proximally, alternate distally; blade elliptic to linear, 5–35 × 1–7 mm, surfaces glabrous. **Inflorescences** axillary, flowers solitary. **Pedicels** 1–12 mm. **Flowers** chasmogamous and cleistogamous; calyx lobes unequal, glandular-hairy, adaxial lobe 4–5.5 × 1.5–2 mm; corolla white, purple-veined, 5.5–7 mm, base gibbous, mouth 3–3.5 mm diam., palate purple-veined, angular, 2.5–3 mm diam., minutely papillate. **Capsules** ovoid to globular, 3–4.5 mm, glandular-hairy, abaxial locule with 1 pore. **Seeds** black, 1 mm, reticulate, ridged, or tuberculate. $2n = 30$.

Flowering Apr–Jul. Washes, scree, rocky slopes; 500–2300 m; Ariz., Calif., Idaho, Nev., Oreg., Utah.

4. **Sairocarpus multiflorus** D. A. Sutton, Revis. Antirrhineae, 467. 1988 • Sierra snapdragon E

Antirrhinum multiflorum Pennell in L. Abrams and R. S. Ferris, Ill. Fl. Pacific States 3: 789. 1951 (not *A.* ×*multiflorum* J. Vick 1868), based on *A. glandulosum* Lindley, Edwards's Bot. Reg. 22: plate 1893. 1836, not Lejeune 1813; *A. thompsonii* D. J. Keil

Annuals or perennials. Stems 10–150 cm, self-supporting, glandular-hairy, viscid; branches not twining. **Leaves** subopposite proximally, alternate distally; blade linear to lanceolate, 10–65 × 2–10 mm, surfaces glandular-hairy. **Inflorescences** terminal, racemes. **Pedicels** 2–4(–10) mm. **Flowers** chasmogamous; calyx lobes unequal, densely glandular-hairy, adaxial lobe 5–13.5 × 2.5–5.5 mm; corolla pale pink to red with tan-brown withered area on abaxial lip, 13–18 mm, base gibbous, mouth 3.5–5 mm diam., palate white, not veined, rounded, 5–8.5 mm diam., puberulent. **Capsules** ovoid, 7–10 mm, glandular-hairy, abaxial locule with 1 pore. **Seeds** black, 1 mm, ridged, reticulate. $2n = 32$.

Flowering Apr–Aug. Rocky, disturbed, or burned areas; 10–2200 m; Calif.

5. **Sairocarpus nuttallianus** (Bentham ex A. de Candolle) D. A. Sutton, Revis. Antirrhineae, 464. 1988 • Violet snapdragon

Antirrhinum nuttallianum Bentham ex A. de Candolle in A. P. de Candolle and A. L. P. P. de Candolle, Prodr. 10: 592. 1846; *A. nuttallianum* subsp. *subsessile* (A. Gray) D. M. Thompson; *A. nuttallianum* var. *subsessile* (A. Gray) Jepson; *A. pusillum* Brandegee; *A. subsessile* A. Gray; *Sairocarpus pusillus* (Brandegee) D. A. Sutton

Annuals, rarely biennials. Stems 6–200 cm, not self-supporting, glandular-hairy; branches twining. **Leaves** opposite proximally, alternate distally; blade ovate, 2–60 × 1–50 mm, surfaces glandular-hairy. **Inflorescences** axillary, flowers solitary. **Pedicels** 2–20(–25)mm. **Flowers** cleistogamous and chasmogamous; calyx lobes equal, glandular hairy, adaxial lobe 3–6 × 1–3.5 mm; corolla pale purple to purple, sometimes dark-veined, 7–12 mm, base slightly gibbous, mouth 2.5–3.5 mm diam., palate white, purple-veined, rounded, 2.5–6 mm diam., puberulent. **Capsules** ovoid, 3–11 mm, glandular-hairy, abaxial locule with 1 pore. **Seeds** brown, 0.5–1 mm, ridged longitudinally. $2n = 32$.

Flowering Mar–Aug. Stabilized coastal dunes, rocky or disturbed areas; 0–1300 m; Ariz., Calif.; Mexico (Baja California, Baja California Sur, Sonora).

Plants of *Sairocarpus nuttallianus* are unique in having gold-colored hairs in the mouth of the corolla. D. M. Thompson (1988) recognized two intergrading subspecies based on degree of hairiness and slight differences in seed sculpturing, but the differences are minor and inconsistent.

6. **Sairocarpus subcordatus** (A. Gray) D. A. Sutton, Revis. Antirrhineae, 477. 1988 • Dimorphic snapdragon E

Antirrhinum subcordatum A. Gray, Proc. Amer. Acad. Arts 20: 306. 1885

Annuals. Stems 8–90 cm, not self-supporting, hairy; branches twining. **Leaves** opposite proximally, alternate distally; blade ovate, 10–60 × 5–45 mm, surfaces glabrous, glandular-hairy, or eglandular-hairy. **Inflorescences** axillary, racemes or flowers solitary. **Pedicels** 1–3 mm. **Flowers** chasmogamous; calyx lobes unequal, hairy or glandular-hairy, adaxial lobe 8–14 × 5–10 mm; corolla white to tan, 13–17 mm, base gibbous, mouth 2.5–3.5 mm diam., palate not veined, slightly 2-lobed, 5–9 mm diam., tomentose. **Capsules** narrowly ovoid, 5–6 mm, glandular-hairy, abaxial locule with 1 pore. **Seeds** dark brown to black, 1 mm, tuberculate, reticulate-ridged. $2n = 30$.

Flowering May–Jun. Open slopes on serpentine, under shrubs; 150–800 m; Calif.

Sairocarpus subcordatus is similar to *S. vexillocalyculatus*; it has relatively broader leaves and white to tan corollas with abaxial lobes reflexed. The species is known from the northern and central Inner North Coast Ranges.

7. **Sairocarpus vexillocalyculatus** (Kellogg) D. A. Sutton, Revis. Antirrhineae, 472. 1988 • Wiry snapdragon E

Antirrhinum vexillocalyculatum Kellogg, Proc. Calif. Acad. Sci. 1: 27. 1855 (as vexillo-calyculatum)

Annuals. Stems 7–170 cm, not self-supporting, glabrous, eglandular-hairy, or glandular-hairy; branches twining. **Leaves** opposite proximally, alternate distally; blade ovate to narrowly elliptic, 10–60 × 1–20 mm, surfaces glabrous or hairy. **Inflorescences** axillary, racemes or flowers solitary. **Pedicels** 1–4 mm. **Flowers** cleistogamous and chasmogamous; calyx lobes unequal, glandular-hairy, adaxial lobe 3.5–14 × 1–9 mm; corolla light purple to white, often dark-veined, 8–17 mm, base gibbous, mouth 2–4 mm diam., palate sometimes purple-veined, convex, 3.5–8 mm diam., puberulent. **Capsules** ovoid, 4–8 mm, glandular-hairy, abaxial locule with 1 pore. **Seeds** dark brown to black, 0.7–1.5 mm, tuberculate, reticulate-ridged. $2n = 30$.

Subspecies 3 (3 in the flora): w United States.

Subspecies *breweri* and *vexillocalyculatus* intergrade where their distributions overlap in the coastal ranges of California. Subspecies *intermedius* is disjunct in the Sierra Nevada.

1. Stems proximally glabrous or eglandular-hairy; corollas 11–17 mm, adaxial lips 7–17 mm . . . . .7a. *Sairocarpus vexillocalyculatus* subsp. *vexillocalyculatus*
1. Stems proximally glandular-hairy and, sometimes, eglandular-hairy; corollas 8–15 mm, adaxial lips 2–5.5 mm.
  2. Inflorescence branches with 1 leaf at each proximal node . . . .7b. *Sairocarpus vexillocalyculatus* subsp. *breweri*
  2. Inflorescence branches with 2 leaves at each proximal node . . . .7c. *Sairocarpus vexillocalyculatus* subsp. *intermedius*

7a. **Sairocarpus vexillocalyculatus** (Kellogg) D. A. Sutton subsp. **vexillocalyculatus** E

Antirrhinum elmeri Rothmaler; *Sairocarpus elmeri* (Rothmaler) D. A. Sutton

Stems 7–170 cm, proximally glabrous or eglandular-hairy. **Inflorescences:** branches with 2 leaves at each proximal node. **Flowers:** calyx adaxial lobe 7–14 mm; corolla light purple, without darker veins, 11–17 mm, adaxial lip 7–17 mm.

Flowering Jun–Aug. Gravelly slopes, disturbed sites, often on serpentine; 50–1200 m; Calif.

Subspecies *vexillocalyculatus* grows in the same general area as *Sairocarpus subcordatus*. *Sairocarpus subcordatus* can be distinguished from subsp. *vexillocalyculatus* by its broader leaves, white to tan corollas, and reflexed abaxial corolla lobes.

7b. **Sairocarpus vexillocalyculatus** (Kellogg) D. A. Sutton subsp. **breweri** (A. Gray) Barringer, Phytoneuron 2013-34: 2. 2013 • Brewer's snapdragon E

Antirrhinum breweri A. Gray, Proc. Amer. Acad. Arts 7: 374. 1868; *A. vagans* A. Gray var. *breweri* (A. Gray) Jepson; *A. vexillocalyculatum* Kellogg subsp. *breweri* (A. Gray) D. M. Thompson; *Sairocarpus breweri* (A. Gray) D. A. Sutton

Stems 10–70(–100) cm, proximally glandular-hairy and, sometimes, eglandular-hairy. **Inflorescences:** branches with 1 leaf at each proximal node. **Flowers:** calyx adaxial lobe

3.5–5.5 mm; corolla white to light purple, often with darker veins, 8–15 mm, adaxial lip 2–5.5 mm.

Flowering Jun–Sep. Gravelly lower slopes, disturbed areas, often on serpentine; 50–2000 m; Calif., Oreg.

Subspecies *breweri* and *Sairocarpus cornutus* both have white to purple corollas and grow in the same general area; *S. cornutus* has a deeper spur at the base of the corolla and is self-supporting, without twining branches.

7c. Sairocarpus vexillocalyculatus (Kellogg) D. A. Sutton subsp. **intermedius** (D. M. Thompson) Barringer, Phytoneuron 2013-34: 2. 2013 E

Antirrhinum vexillocalyculatum Kellogg subsp. *intermedium* D. M. Thompson, Syst. Bot. Monogr. 22: 84, figs. 14H, 32. 1988 (as vexillo-calyculatum)

Stems 10–100 cm, proximally glandular-hairy and eglandular-hairy. **Inflorescences:** branches with 2 leaves at each proximal node. **Flowers:** calyx adaxial lobe 4.5–6.5(–10) mm; corolla light purple, dark-veined on palate, 10–14 mm, adaxial lip 2.5–5.5 mm.

Flowering Jun–Sep. Gravelly slopes, disturbed sites; 100–1400 m; Calif.

Subspecies *intermedius* is known from the northern and central Sierra Nevada.

8. Sairocarpus virga (A. Gray) D. A. Sutton, Revis. Antirrhineae, 466. 1988 • Tall or twig-like snapdragon E

Antirrhinum virga A. Gray, Proc. Amer. Acad. Arts 7: 373. 1868

Perennials. Stems 40–220 cm, self-supporting, glabrous; branches not twining. **Leaves** alternate; blade linear, 50–120 × 3–10 mm, surfaces glabrous. **Inflorescences** terminal, racemes. **Pedicels** 2–6 mm. **Flowers** chasmogamous; calyx lobes equal, glabrous, adaxial lobe 6–8 × 1.5–2.5 mm; corolla pink to pale pink, 13–18 mm, base gibbous, mouth 3–5 mm diam., palate not veined, rounded, 5–7.5 mm diam., puberulent. **Capsules** globular-ovoid, 7–9 mm, glabrous, abaxial locule with 2 pores. **Seeds** black, 1–1.5 mm, ridges reticulate. $2n = 32$.

Flowering Jun–Jul. Openings in chaparral, rocky areas, often on serpentine; 200–2000 m; Calif.

Stems of *Sairocarpus virga* regrow quickly after fires (D. M. Thompson 1988).

Sairocarpus virga is known from the southern High and Inner North Coast ranges.

9. Sairocarpus watsonii (Vasey & Rose) D. A. Sutton, Revis. Antirrhineae, 471. 1988 • Watson's snapdragon

Antirrhinum watsonii Vasey & Rose, Proc. U.S. Natl. Mus. 11: 533. 1889 (as watsoni); *A. kingii* S. Watson var. *watsonii* (Vasey & Rose) Munz

Annuals. Stems 8–85 cm, not self-supporting, basally hairy, otherwise glabrous; branches twining. **Leaves** opposite proximally, alternate distally; blade linear, 5–55 × 1–4 mm, surfaces glabrous. **Inflorescences** axillary, flowers solitary. **Pedicels** 4–25(–30) mm. **Flowers** cleistogamous and chasmogamous; calyx lobes equal, glandular-hairy, adaxial lobe 2.5–4 × 1–1.5 mm; corolla light purple, purple-veined, 5.5–7 mm, base barely gibbous, mouth 2–2.5 mm diam., palate purple-veined, angular, 2.5–4 mm diam., minutely papillate. **Capsules** globular-ovoid, 2.5–6 mm, glandular-hairy, abaxial locule with 1 pore. **Seeds** black, 0.6–1 mm, tuberculate. $2n = 30$.

Flowering Mar–Apr. Gravelly slopes; (0–) 200–1000 m; Ariz.; Mexico (Baja California, Baja California Sur, Sonora).

The only North American collections of *Sairocarpus watsonii* are from Organ Pipe National Monument in Pima County. *Sairocarpus watsonii* is similar to *S. kingii*, which grows farther north and has unequal sepals.

19. CALLITRICHE Linnaeus, Sp. Pl. 2: 969. 1753; Gen. Pl. ed. 5, 5. 1754

• Water-starwort [Greek *kallos*, beautiful, and *trichos*, hair, presumably alluding to fine leaves of some growth forms]

Richard V. Lansdown

Herbs, perennial, submersed, amphibious, or terrestrial. **Stems** creeping or supported by water, glabrous or bearing scales. **Leaves** cauline, opposite, decussate; petiole present or absent; blade not fleshy, not leathery, margins entire [toothed], notched [pointed]. **Inflorescences** axillary, pistillate and staminate flowers sometimes in same axil; bracts present or absent. **Pedicels** present or absent; bracteoles present or absent. **Flowers** unisexual; sepals 0; petals 0. **Staminate flowers** solitary; stamen 1, anthers [3 or]4-locular, filaments glabrous; staminode 0. **Pistillate flowers** usually abaxial to staminate, ovary superior, 4-locular, placentation basal; styles 2, erect, ascending, spreading, recurved, deflexed, or reflexed, stigma simple, terete. **Fruits** schizocarps, each separating into mericarps. **Mericarps** 4, white, maroon, brown, grayish brown, or black, reniform, wings present, partial, or absent. $x = 3, 5$.

Species ca. 75 (13 in the flora): nearly worldwide; mainly montane where native in tropical regions.

Callitriche has been included in Haloragaceae, a family now included in Saxifragales (Angiosperm Phylogeny Group 2016), and in the monogeneric Callitrichaceae, which is nested within Plantaginaceae (D. C. Albach et al. 2005).

C. F. Hegelmaier (1864) divided the genus into two sections: *Callitriche* (as *Eucallitriche* Hegelmaier) and *Pseudocallitriche* Hegelmaier. N. C. Fassett (1951b) emended the diagnosis and established sect. *Microcallitriche* Fassett. H. D. Schotsman (1967) divided the genus into five groups; one conforms to *Pseudocallitriche*, three to *Callitriche*, and one to *Microcallitriche*. In a comprehensive study of taxa from North America, South America, and the Mediterranean, including cytological analysis, C. T. Philbrick and D. H. Les (2000) found support for only sections *Callitriche* and *Pseudocallitriche*. All these infrageneric separations are based on a small proportion of the described taxa and almost exclusively cover taxa from Europe and the Americas.

Review of the full range of described taxa of *Callitriche* shows that there may be other means of subdividing the genus. For example, three Australasian terrestrial species (*C. capricorni* R. Mason, *C. muelleri* Sonder, and *C. sonderi* Hegelmaier) have trilocular anthers; in these species and *C. cycloptera* Schotsman, the filament arises from the pedicel (H. D. Schotsman 1985). *Callitriche fassettii*, *C. hermaphroditica*, *C. stenoptera*, and *C. transvolgensis* Tzvelev form a closely related group relatively distinct from their nearest relatives. Current concepts of both *C. heterophylla* and *C. palustris* include taxa that have been recognized at subspecific rank. It is likely that a comprehensive review of morphological data combined with molecular study would enable recognition of more useful groupings; to date, intermediates have been found between all subgeneric groupings.

Descriptions here refer to aquatic or amphibious forms except for *Callitriche pedunculosa*, *C. peploides*, and *C. terrestris*, which occur only as terrestrial forms. Reliable characters by which to distinguish *Callitriche* taxa are few. Useful characters include wings on mericarps and numbers and distributions of flowers. The presence, absence, or extent of wings on mericarps is important. Here, wing applies to a clearly differentiated structure arising from the exocarp of a mericarp. All species may have solitary flowers in pairs of axils; some species may have various combinations of pistillate and staminate flowers.

Callitriche is of considerable secondary importance in freshwater wetland systems providing habitat for many kinds of animals.

SELECTED REFERENCES Fassett, N. C. 1951b. *Callitriche* in the New World. Rhodora 53: 137–155, 161–182, 185–194, 209–222. Hegelmaier, C. F. 1864. Monographie der Gattung *Callitriche*. Stuttgart. Miller, N. G. 2001. The Callitrichaceae in the southeastern United States. Harvard Pap. Bot. 5: 277–301. Philbrick, C. T. 1989. Systematic Studies of North American *Callitriche* (Callitrichaceae). Ph.D. dissertation. University of Connecticut. Philbrick, C. T. and D. H. Les. 2000. Phylogenetic studies in *Callitriche*: Implications for interpretation of ecological, karyological and pollination system evolution. Aquatic Bot. 68: 123–141.

1. Leaves not connate at bases (leaf bases not joined by a ridge of tissue extending across nodes); leaf and stem scales absent; bracts absent; leaves 1-veined.
    2. Pedicels 0.5–1 mm in fruit; schizocarps consistently shorter than wide; mericarps black . . . . 2. *Callitriche fassettii*
    2. Pedicels 0–0.2 mm in fruit; schizocarps as long as or shorter than wide, sometimes longer than wide; mericarps brown, dark brown, or gray-brown.
        3. Mericarps dark brown, wings 0.2–0.8 mm wide, as wide as or wider than mericarp body. . . . . 3. *Callitriche hermaphroditica*
        3. Mericarps brown or gray-brown, wings 0.05–0.1 mm wide, width less than mericarp body. . . . . 11. *Callitriche stenoptera*
1. Leaves connate at bases; leaf and stem scales present; bracts present or absent; leaves 1+-veined.
    4. Styles reflexed, appressed to ovary; pollen colorless . . . . 1. *Callitriche brutia*
    4. Styles erect, ascending, spreading, reflexed, recurved, or deflexed; pollen yellow.
        5. Schizocarps shorter than wide; bracts absent.
            6. Mericarps swollen at bases. . . . . .9. *Callitriche peploides*
            6. Mericarps not swollen.
                7. Schizocarps (0.5–)0.6–0.7(–0.8) mm; wings 0.01–0.03 mm wide; pedicels 0.4–0.6 mm in fruit . . . . 12. *Callitriche terrestris*
                7. Schizocarps 0.7–1 mm; wings 0.06–0.2 mm wide; pedicels 0–5 mm in fruit.
                    8. Mericarp wings straight. . . . . 6. *Callitriche marginata*
                    8. Mericarp wings curled, giving appearance of thickened margins . . . . 8. *Callitriche pedunculosa*
        5. Schizocarps as long as or longer than wide; bracts present, sometimes caducous.
            9. Mericarps pale grayish brown . . . . 10. *Callitriche stagnalis*
            9. Mericarps black.
                10. Mericarps not winged, winged only at apex, or winged throughout but widest at apex.
                    11. Schizocarps ± round, as long as wide. . . . . 4. *Callitriche heterophylla*
                    11. Schizocarps obovoid, usually widest distal to middle, longer than wide, sometimes as long as wide . . . . 7. *Callitriche palustris*
                10. Mericarps evenly winged throughout.
                    12. Schizocarps as long as or longer than wide; pedicels 0–25 mm in fruit . . . . .5. *Callitriche longipedunculata*
                    12. Schizocarps ± as long as wide; pedicels 0–0.5 mm in fruit . . . . . 13. *Callitriche trochlearis*

1. Callitriche brutia Petagna, Inst. Bot. 2: 10. 1787

Varieties 2 (1 in the flora): North America, Europe, Africa; introduced to Australia.

Variety *brutia* is known from Europe and Africa plus is introduced to Australia; it differs from var. *hamulata* in chromosome number (2*n* = 28) and pedicellate fruit.

1a. Callitriche brutia Petagna var. hamulata (Kützing ex W. D. J. Koch) Lansdown, Watsonia 26: 113. 2006
• Narrowleaf water-starwort

Callitriche hamulata Kützing ex W. D. J. Koch, Syn. Fl. Germ. Helv., 246. 1837; *C. intermedia* Hoffmann; *C. intermedia* subsp. *hamulata* (Kützing ex W. D. J. Koch) A. R. Clapham

Leaves connate at base, ± linear, tapering strongly from near base or spatulate, 5.4–18 × 0.3–3 mm, 1+-veined. **Stem and leaf scales** present.

Inflorescences: bracts caducous. **Pedicels** 0 mm, rarely less than 2 mm in fruit. **Flowers** usually solitary; styles reflexed, appressed to ovary; pollen colorless. **Schizocarps** 1.1–1.5(–1.6) × 1.1–1.4 mm, as long as wide; mericarps blackish, not swollen, winged throughout, wings straight, 0.03–0.3 mm wide. $2n = 38$.

Flowering Jul–Sep. Fast-flowing streams, backwaters, lakes, pools; 0–1600 m; Greenland; B.C.; Oreg., Wash.; Europe.

Variety *hamulata* is considered native in Greenland and is introduced in other parts of the flora area; the earliest specimen of var. *hamulata* known from western North America is from Oregon in 1956 (*Steward 7238*, MO).

2. **Callitriche fassettii** Schotsman, Acta Bot. Neerl. 15: 477, figs. 1, 2. 1966 • Fassett's water-starwort C E

Callitriche autumnalis Linnaeus var. *bicarpellaris* Fenley ex Jepson; *C. hermaphroditica* Linnaeus var. *bicarpellaris* (Fenley ex Jepson) H. Mason

Leaves not connate at base, ± linear, tapering from base, 12 × 0.3–0.5 mm, 1-veined. **Stem and leaf scales** absent. **Inflorescences:** bracts absent. **Pedicels** 0.5–1 mm in fruit. **Flowers** solitary or 1+ per axil; styles recurved, ± reflexed; pollen colorless. **Schizocarps** 1.1–1.8 × 1.5–2.5 mm, shorter than wide; mericarps black, not swollen, winged throughout, wings straight, 0.1–0.3 mm wide.

Flowering Apr–May(–Jul). Vernal and permanent pools, running water; of conservation concern; 0–1600 m; Calif., Nebr., Oreg.

Callitriche fassettii is widespread in California and adjacent Oregon. The record from Nebraska is based on a single specimen, which could be due to introduction, or could involve an identification or recording error and requires confirmation.

3. **Callitriche hermaphroditica** Linnaeus, Cent. Pl. I, 31. 1755 • Autumnal or northern water-starwort, callitriche hermaphrodiate F

Callitriche bifida (Linnaeus) Morong; *C. palustris* Linnaeus var. *bifida* Linnaeus

Leaves not connate at base, ± linear, tapering from base, (7–)7.6–11.5 × 0.9–1.4 mm, 1-veined. **Stem and leaf scales** absent. **Inflorescences:** bracts absent. **Pedicels** 0 mm in fruit. **Flowers** solitary; styles erect, later recurved; pollen colorless. **Schizocarps** 1.1–1.9 × 1.2–2.4 mm, shorter

than or as long as wide, sometimes longer than wide; mericarps dark brown, not swollen, winged throughout, wings straight, 0.2–0.8 mm wide, as wide as or wider than mericarp body. $2n = 6$.

Flowering Jul–Sep(–Nov). Lakes, ponds, backwaters, ditches; 0–3000 m; Greenland; Alta., B.C., Nfld. and Labr. (Nfld.), N.W.T., Ont., Que., Sask.; Alaska, Colo., Idaho, Mont., Nebr., N.J., N.Y., Oreg., Utah, Vt., Wis., Wyo.; Eurasia.

Infraspecific divisions have been based on the size of fruit (for example, R. V. Lansdown 2006); assignment of North American populations to these taxa would be premature.

Callitriche autumnalis Linnaeus is a superfluous renaming of *C. hermaphroditica*, with which it is homotypic (R. V. Lansdown and C. E. Jarvis 2004).

Callitriche fassettii, *C. hermaphroditica*, and *C. stenoptera* are morphologically very close; however, *C. hermaphroditica* always has sessile fruit, whereas the fruit of both the other species may be pedicellate. The mericarp wing of *C. stenoptera* is consistently very narrow and that of *C. fassettii* is slightly wider than that of *C. hermaphroditica*, which is generally wide, particularly at the apex.

4. **Callitriche heterophylla** Pursh, Fl. Amer. Sept. 1: 3. 1813 F

Callitriche anceps Fernald; *C. bolanderi* Hegelmaier; *C. deflexa* A. Braun ex Hegelmaier var. *austinii* (Engelmann) Hegelmaier; *C. heterophylla* subsp. *bolanderi* (Hegelmaier) Calder & Roy L. Taylor; *C. heterophylla* var. *bolanderi* (Hegelmaier) Fassett

Leaves connate at base, ± linear, tapering from near base or spatulate, 3.7–24.7 × 0.3–3.5 mm, 1+-veined. **Stem and leaf scales** present. **Inflorescences:** bracts caducous. **Pedicels** 0 mm in fruit. **Flowers** usually solitary; styles erect or spreading; pollen yellow. **Schizocarps** ± round, (0.4–)0.5–0.9(–1.2) × 0.5–1(–1.2) mm, as long as wide; mericarps black, not swollen, not winged or winged only at apex, wings straight, 0.05–0.1 mm wide.

Flowering Feb–Nov. Fast-flowing streams, backwaters, ditches, swamps, *Sphagnum* bogs, lakes, ponds, springs, seepages, seasonally damp soils in shade; 0–3000 m; Greenland; Alta., B.C., Man., N.B., Nfld. and Labr. (Nfld.), N.S., Ont., Que., Yukon; Ala., Alaska, Ariz., Ark., Calif., Conn., Del., Fla., Ga., Idaho, Ill., Ind., Iowa, Kans., Ky., La., Maine, Md., Mass., Miss., Mo., Mont., Nev., N.H., N.J., N.Mex., N.Y., N.C., Ohio, Okla., Oreg., Pa., R.I., S.C., Tenn., Tex., Vt., Va., Wash., Wis.; Mexico; West Indies (Antilles); Central America; South America; introduced in Pacific Islands (New Zealand).

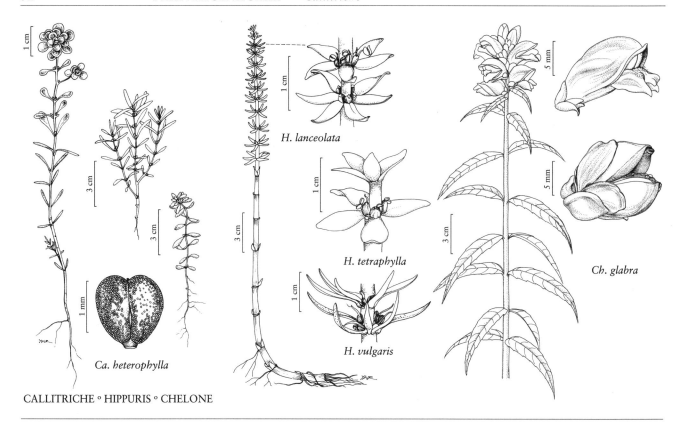

H. lanceolata

H. tetraphylla

H. vulgaris

Ch. glabra

Ca. heterophylla

CALLITRICHE ∘ HIPPURIS ∘ CHELONE

Two varieties have been recognized in *Callitriche heterophylla*, with var. *bolanderi* distinguished on the basis of its schizocarps being larger (for example, N. C. Fassett 1951b); however, the size of the schizocarps overlaps widely, and measurements of schizocarps from throughout North America show that there is no clear geographical or ecological separation between plants with smaller fruit and those with larger fruit. Therefore, there is no justification for maintaining the distinction.

Populations of small plants with linear leaves from parts of Canada and Alaska have been separated as *Callitriche anceps*. However, these appear to represent the lower extremes of variation in *C. heterophylla*, and morphological characters do not justify maintaining the distinction.

5. **Callitriche longipedunculata** Morong, Bull. Torrey Bot. Club 18: 236. 1891 • Longstock water-starwort
Ⓒ Ⓔ

Callitriche marginata Torrey var. *longipedunculata* (Morong) Jepson

Leaves connate at base, spatulate, 2.8–5 × 1.8–3.3 mm, 1+-veined. **Stem and leaf scales** present. **Inflorescences:** bracts caducous. **Pedicels** 0–25 mm in fruit. **Flowers** usually 1 staminate and 1 pistillate in 1 axil opposed by 1(+) pistillate; styles erect and spreading

to recurved; pollen yellow. **Schizocarps** 0.9–1.4 × 0.8–1.2 mm, as long as or longer than wide; mericarps black, not swollen, evenly winged throughout, wings straight, 0.03–0.1 mm wide.

Flowering Mar–May. Vernal pools; of conservation concern; 0–700 m; Calif.

Callitriche longipedunculata is known from most of California, including the Inner North Coast Ranges, Cascade Range Foothills, Great Central Valley, San Francisco Bay Area, and western Pensisular Ranges (R. V. Lansdown and R. E. Preston, http://ucjeps. berkeley.edu/cgi-bin/get_IJM.pl?tid=16691).

6. **Callitriche marginata** Torrey in War Department [U.S.], Pacif. Railr. Rep. 4(5): 135. 1857 • Winged water-starwort

Callitriche sepulta S. Watson

Leaves connate at base, narrowly spatulate, 3.5–7.1(–7.2) × 0.7–1.4 mm, 1+-veined. **Stem and leaf scales** present. **Inflorescences:** bracts absent. **Pedicels** 1.2–3.3 mm in fruit. **Flowers** usually 1 pistillate and 1 staminate in each axil, sometimes 1 or 2 axils with 2 pistillate, rarely solitary; styles sharply deflexed; pollen yellow. **Schizocarps** 0.7–0.9 × 1.1–1.4 mm, shorter than wide; mericarps maroon or white, not swollen, winged throughout, wings straight, 0.1–0.2 mm wide. **2*n*** = 20.

Flowering Apr–Aug. Lakes, reservoirs, floodplain pools, streams, vernal pools; 0–1000 m; B.C.; Calif., Oreg.; Mexico (Baja California).

Callitriche marginata is known throughout the California floristic province northward to southeastern British Columbia. There are no records from Washington State and British Columbia; occurrences may be remnants of a previously more widespread population (M. Miller 2001).

SELECTED REFERENCE Miller, M. 2001. Stewardship Account for Winged Water Starwort *Callitriche marginata*. Prepared for the B.C. Conservation Data Centre and the Garry Oak Ecosystems Recovery Team. Victoria.

7. **Callitriche palustris** Linnaeus, Sp. Pl. 2: 969. 1753 • Vernal water-starwort, callitriche des marais

Callitriche palustris var. *stenocarpa* (Hegelmaier) Jepson; *C. palustris* var. *verna* (Linnaeus) Fenley ex Jepson; *C. stenocarpa* Hegelmaier; *C. verna* Linnaeus

Leaves connate at base, ± linear, tapering strongly from near base or spatulate, 2.5–9 × 0.3–5 mm, 1+-veined. **Stem and leaf scales** present. **Inflorescences:** bracts caducous, sometimes persistent. **Pedicels** 0–0.2 mm in fruit. **Flowers** usually 1 staminate and 1 pistillate in 1 axil opposed by 1 pistillate; styles erect or spreading; pollen yellow. **Schizocarps** obovoid, usually widest distal to middle, 0.8–1.4 × 0.6–1.4 mm, longer than wide, sometimes as long as wide; mericarps black, not swollen, not winged, winged only at apex, or winged throughout but widest at apex, wings straight, 0.03–0.1 mm wide. $2n = 20$.

Flowering May–Oct. Fast-flowing rivers and streams, backwaters, ditches, swamps, *Sphagnum* bogs, lakes, ponds, springs, seepages; 0–3600 m; Greenland; St. Pierre and Miquelon; Alta., B.C., Man., N.B., Nfld. and Labr. (Nfld.), N.W.T., Nunavut, Ont., P.E.I., Que., Sask., Yukon; Alaska, Ariz., Calif., Colo., Conn., Del., Idaho, Ill., Iowa, Maine, Mass., Mich., Minn., Mont., Nebr., Nev., N.H., N.J., N.Mex., N.Y., N.Dak., Oreg., Pa., R.I., S.Dak., Tex., Utah, Vt., Wash., Wis., Wyo.; Mexico (Chihuahua, San Luis Potosí); Eurasia; introduced in Australia.

Callitriche palustris has been included on the Composite List of Weeds maintained by the Weed Science Society of America (http://wssa.net/wssa/weed/composite-list-of-weeds/); its inclusion seems unwarranted.

8. **Callitriche pedunculosa** Nuttall, Trans. Amer. Philos. Soc., n. s. 5: 140. 1835 • Nuttall's water-starwort [E]

Leaves connate at base, obovate-oblanceolate to spatulate, (2.2–)2.7–5.7 × 0.7–2 mm, 2+-veined. **Stem and leaf scales** present. **Inflorescences:** bracts absent. **Pedicels** 0–5 mm in fruit. **Flowers** usually 1 staminate and 1 pistillate in 1 axil opposed by 1 pistillate, rarely both pairs at an axil with 1 staminate and 1 pistillate; styles loosely ascending or ± reflexed; pollen yellow. **Schizocarps** 0.8–1 × 0.9–1.3 mm, shorter than wide; mericarps black, not swollen, winged throughout, wings curled, giving appearance of thickened margins, 0.06–0.15 mm wide. $2n = 20$.

Flowering Feb–Jun(Sep–Oct). Seasonally wet sandy soils, in or shaded by woodlands; 0–100 m; Ala., La., Miss., Tex.

Callitriche nuttallii Torrey is an illegitimate, superfluous name that applies here. Torrey proposed it as a replacement for *C. pedunculosa* Nuttall based on the mistaken belief that the name by Nuttall was a later homonym of one by Arnott; the purported name by Arnott does not exist.

9. **Callitriche peploides** Nuttall, Trans. Amer. Philos. Soc., n. s. 5: 141. 1835 [E]

Callitriche peploides var. *semialata* Fassett

Leaves connate at base, spatulate-obovate, 1.4–2.3 × 0.4–1.5 mm, 2+-veined. **Stem and leaf scales** present. **Inflorescences:** bracts absent. **Pedicels** 0 mm in fruit. **Flowers** not solitary, usually 1 staminate and 1 pistillate in 1 axil opposed by 1 pistillate; styles erect to reflexed; pollen yellow. **Schizocarps** 0.4–0.7 × 0.7–0.9 mm, shorter than wide; mericarps black, swollen at base, not winged or winged only at apex, wings straight, 0–0.03 mm wide. $2n = 10$.

Flowering (Jan–)Feb–Apr(–May), Oct–Dec. Seasonally damp hollows, sandy soils in woodlands, gardens, waste grounds, rocky areas; 0–100 m; Ala., Ark., Calif., Fla., Ga., La., Miss., Tex.; introduced in Asia (China, Taiwan), Indian Ocean Islands (Madagascar, Réunion), Pacific Islands (New Zealand).

Three varieties of *Callitriche peploides* have been recognized (N. C. Fassett 1951b). A collection reported as a weed of a container in a nursery in San Bernardino County, California (T. C. Fuller 1967) is referable to var. *semialata*. The characters used to distinguish these varieties do not appear reliable, and it seems best to disregard them (R. V. Lansdown and M. J. M. Christenhusz 2011).

10. Callitriche stagnalis Scopoli, Fl. Carniol. ed. 2, 2: 251. 1772 • Pond water-starwort, callitriche des eaux stagnantes [I] [W]

Leaves connate at base, obovate-spatulate to ± round, 4–9 × 1–4 mm, 1+-veined. **Stem and leaf scales** present. **Inflorescences:** bracts persistent. **Pedicels** 0–0.2 mm in fruit. **Flowers** solitary; styles erect; pollen yellow. **Schizocarps** 1.2–1.8 × 1–1.7 mm, ± as long as wide; mericarps pale grayish brown, not swollen, winged throughout, wings straight, 0.1–0.5 mm wide. $2n = 10$.

Flowering May–Oct. Standing water, stream margins, backwaters, ponds, pools, ditches; 0–1500 m; introduced; St. Pierre and Miquelon; B.C., Nfld. and Labr. (Nfld.), Que.; Ala., Calif., Conn., Maine, Md., Mass., Mo., Mont., N.J., N.Y., Oreg., Pa., Va., Wash., Wis.; Europe; introduced also in Australia.

SELECTED REFERENCE Philbrick, C. T., R. A. Aakjar Jr., and R. L. Stuckey. 1998. Invasion and spread of *Callitriche stagnalis* (Callitrichaceae) in North America. Rhodora 100: 25–38.

11. Callitriche stenoptera Lansdown, Novon 19: 367, fig. 1. 2009 [E]

Leaves not connate at base, ± linear tapering from base, 5.9–11.3 × 0.3–1.1 mm, 1-veined. **Stem and leaf scales** absent. **Inflorescences:** bracts absent. **Pedicels** 0–0.2 mm in fruit. **Flowers** usually solitary or 1 pistillate and 1 staminate in an axil; styles reflexed or strongly curved; pollen colorless. **Schizocarps** 0.8–1.2 × 0.9–1.3 mm, as long as or shorter than wide; mericarps brown or gray-brown, not swollen, winged throughout, wings straight, 0.05–0.1 mm wide, width less than mericarp body.

Flowering Jun–Sep. Lakes, pools, ponds, ditches, streams, backwaters; 0–3000 m; Alta., B.C., Nfld. and Labr. (Nfld.), Ont., Sask.; Ariz., Colo., Idaho, Minn., Mont., N.Mex., Oreg., Utah, Wyo.

Callitriche stenoptera most closely resembles *C. fassettii* and *C. hermaphroditica* but is distinguished from them by the fruit on which the wing is less than 0.1 mm wide, even at the fruit apex.

12. Callitriche terrestris Rafinesque, Med. Repos., hexade 2, 5: 358. 1808 (as terrestre) • Terrestrial water-starwort

Leaves connate at base, obovate-oblanceolate to spatulate, 1.5–3.9 × 0.8–1.4(–1.5) mm, 2+-veined. **Stem and leaf scales** present. **Inflorescences:** bracts absent. **Pedicels** 0.4–0.6 mm in fruit. **Flowers** usually solitary or 1 staminate and 1 pistillate opposed by 1 pistillate; styles deflexed; pollen yellow. **Schizocarps** (0.5–)0.6–0.7(–0.8) × (0.5–)0.8–1 mm, shorter than wide; mericarps black, not swollen, winged only at apex, wings straight, 0.01–0.03 mm wide. $2n = 10$.

Flowering Apr–Jul(–Aug), Nov–Dec. Seasonally wet hollows, damp soils, cultivated fields, rocky grounds, paths, tracks, sandy bars; 0–600 m; N.B.; Ark., Conn., D.C., Ill., Ind., Kans., Ky., La., Mass., Miss., Mo., N.Y., N.C., N.Dak., Ohio, Pa., Tenn., Tex., Va.; Mexico; Central America; South America; introduced in Asia (China, Japan).

13. Callitriche trochlearis Fassett, Rhodora 53: 194. 1951 • Effluent water-starwort [C] [E]

Leaves connate at base, ± parallel-sided, 5.2–9.1 × 0.1–0.9 mm, 1+-veined. **Stem and leaf scales** present. **Inflorescences:** bracts persistent. **Pedicels** 0–0.5 mm in fruit. **Flowers** usually 1 staminate and 1 pistillate in each axil; styles spreading; pollen yellow. **Schizocarps** 1–1.2 × 1.1–1.3 mm, ± as long as wide; mericarps black, not swollen, evenly winged throughout, wings straight, 0.05–0.1 mm wide. $2n = 40$.

Flowering Apr–Jul. Vernal and perennial pools, ditches; of conservation concern; 0–2000 m; Calif., Oreg.

Callitriche trochlearis occurs in northwestern and central-western California plus western Oregon.

20. HIPPURIS Linnaeus, Sp. Pl. 1: 4. 1753; Gen. Pl. ed. 5, 4. 1754 • Mare's tail [Greek *hippos*, horse, and *oura*, tail, alluding to appearance of stem and leaves]

Reidar Elven

David F. Murray

Heidi Solstad

Herbs, perennial; rhizomatous, emergent aquatics in fresh or brackish water. **Stems** erect, glabrous. **Leaves** cauline, whorled; petiole absent; blade not fleshy, not leathery (fleshy or leathery in *H. tetraphylla*), margins entire. **Inflorescences** axillary, flowers solitary; bracts absent. **Pedicels** present (proximal) or absent (distal); bracteoles absent. **Flowers** bisexual or unisexual; calyx a minute rim adhering to summit of inferior ovary; petals 0; stamen 1, adnate to ovary, filaments glabrous; staminode 0; ovary 1-locular, placentation apical; stigma linear along surfaces of style. **Fruits** drupes. **Seeds** 1, brownish, globular, wings absent. $x = 8$.

Species 4 (4 in the flora): North America, South America, Eurasia; introduced in Australia.

Leaf characteristics of *Hippuris* used here are derived from whorls on the emergent portions of the stems; morphology of submerged leaves differs sharply from that of emergent shoots.

M. E. McCully and H. M. Dale (1961) proposed that the taxa treated below all could be expressions of phenotypic plasticity of *Hippuris vulgaris* developed in different regimes of salts and photoperiod; this was not accepted by E. Hultén (1973), nor is it accepted here. Number of leaves in a whorl varies among plants and even on the same stem. Nevertheless, there are clear limits and discontinuities in leaf number and shape among taxa, which are well-correlated with less variable characters as well as with ecology and geography.

Hippuris has been placed in Halagoraceae or in Hippuridaceae as a monogeneric family. Molecular phylogenetic studies now place it in Plantaginaceae (D. C. Albach et al. 2005).

SELECTED REFERENCE   McCully, M. E. and H. M. Dale. 1961. Heterophylly in *Hippuris*, a problem in identification. Canad. J. Bot. 39: 1099–1116.

1. Flowers unisexual; leaves 2–10 mm, midveins often conspicuous, lateral veins absent;
   stems 15–100 mm; rhizomes 1 mm diam. . . . . 1. *Hippuris montana*
1. Flowers bisexual; leaves 3–35 mm, midveins inconspicuous, lateral veins present, sometimes
   obscure; stems 80–500 mm; rhizomes (2–)3–7 mm diam.
   2. Leaves on mid portions of emergent shoots in whorls of (7 or)8 or 9(–12), tips often
      curled in dried plants; filaments longer than anthers . . . . 2. *Hippuris vulgaris*
   2. Leaves on mid portions of emergent shoots in whorls of 3–6(or 7), tips not curled in
      dried plants; filaments equal to or shorter than anthers.
      3. Leaves on mid portions of emergent shoots in whorls of (5 or)6(or 7), linear to
         narrowly oblong or lanceolate, 0.5–1.5 mm wide, apices subacute. . . . . 3. *Hippuris lanceolata*
      3. Leaves on mid portions of emergent shoots in whorls of 3–5(or 6), oblanceolate
         or oblong to broadly obovate, 2–8 mm wide, apices obtuse, rounded, or blunt
         . . . . 4. *Hippuris tetraphylla*

1. **Hippuris montana** Ledebour ex Reichenbach, Iconogr. Bot. Pl. Crit. 1: 71, plate 86, fig. 181. 1823 [E]

Stems 15–100 mm. **Rhizomes** 1 mm diam. **Leaves** on mid portions of emergent shoots in whorls of 5–7, linear, 2–10 × 0.2–0.5 mm, midvein often conspicuous, lateral veins absent, apex acute, tip translucent, callous, not curled in dried plants. **Flowers** unisexual, staminate in leaf whorls proximal to pistillate; filaments longer than anthers. **Drupes** 1.2 × 1 mm. $2n = 16$.

Flowering summer. Shallow streams, stream banks, bogs, seeps, upper montane and alpine zones; 100–1400 m; Alta., B.C., N.W.T., Yukon; Alaska, Wash.

Hippuris montana differs from other species of *Hippuris* by its diminutive size and the tendency for the plants to be woven into the moss carpet; it is probably often overlooked by collectors.

The single occurrence reported by N. N. Tzvelev (1980) in the Russian Far East (lower Amur River) of an otherwise North American endemic has not been confirmed.

2. **Hippuris vulgaris** Linnaeus, Sp. Pl. 1: 4. 1753
   • Hippuride vulgaire [F][W]

Stems 100–400 mm. **Rhizomes** (2–)3–5 mm diam. **Leaves** on mid portions of emergent shoots in whorls of (7 or)8 or 9(–12), linear to narrowly oblong or lanceolate, 3–35 × 0.5–2.5 mm, midvein inconspicuous, lateral veins present, sometimes obscure, apex subacute to acute or attenuate, tip often curled in dried plants. **Flowers** bisexual; filaments longer than anthers. **Drupes** 1.5–2 × 0.8–1 mm. $2n = 32$.

Flowering summer. Shallow freshwater pools, pond margins; 0–2900 m; St. Pierre and Miquelon; Alta., B.C., Man., N.B., Nfld. and Labr., N.W.T., N.S., Nunavut, Ont., P.E.I., Que., Sask., Yukon; Alaska, Ariz., Calif., Colo., Ill., Ind., Maine, Mass., Mich., Minn., Mont., Nebr., Nev., N.H., N.Mex., N.Y., N.Dak., Oreg., S.Dak., Utah, Vt., Wash., Wis., Wyo.; s South America; Eurasia; introduced in Australia.

Hippuris vulgaris is the most common and widespread species of *Hippuris*; it is largely absent from the Canadian Arctic Archipelago and Greenland. All specimens seen by the authors from that region are *H. lanceolata*.

The distribution of *Hippuris vulgaris* is bipolar, occurring also in southern South America (Patagonia: Argentina and Chile) and Australia; it exists in some areas as a naturalized introduction, possibly from being used in aquaria and ornamental pools. In Australia, *H. vulgaris* is monitored for its potential to become noxious by spreading rapidly in shallow waterways.

3. **Hippuris lanceolata** Retzius, Observ. Bot. 3: 7, plate 1. 1783 • Hippuride à feuilles lancéolées [F]

Stems 100–500 mm. **Rhizomes** 4–7 mm diam. **Leaves** on mid portions of emergent shoots in whorls of (5 or)6(or 7), linear to narrowly oblong or lanceolate, 5–20 × 0.5–1.5 mm, midvein inconspicuous, lateral veins present, sometimes obscure, apex subacute, tip not curled in dried plants. **Flowers** bisexual; filaments equal to or shorter than anthers. **Drupes** 1.8–2 × 0.6–1.2 mm. $2n = 32$ (Russian Far East).

Flowering summer. Shallow fresh and brackish pools, pond margins; 0–300 m; Greenland; Man., Nfld. and Labr. (Labr.), N.W.T., Nunavut, Ont., Que., Yukon; Alaska; Eurasia.

N. N. Tzvelev (1980) speculated that *Hippuris lanceolata* arose from hybridization between *H. tetraphylla* and *H. vulgaris* or their precursors. Although *H. lanceolata* is intermediate in some features, and is often misplaced with either *H. tetraphylla* or *H. vulgaris*, it is fertile and there is no indication of pollen abortion or failure of seed set. No transitional plants have been seen; hybrid origin appears unlikely. The range of *H. lanceolata* extends well north of that of either putative parent, especially that of *H. vulgaris*. *Hippuris lanceolata* is the sole species of *Hippuris* in some areas of the Arctic.

4. **Hippuris tetraphylla** Linnaeus f., Suppl. Pl., 81. 1782
   • Hippuride à quatre feuilles [F]

Stems 80–450 mm. **Rhizomes** 3–5 mm diam. **Leaves** on mid portions of emergent shoots in whorls of 3–5(or 6), oblanceolate or oblong to broadly obovate, 6–15 × 2–8 mm, midvein inconspicuous, lateral veins present, sometimes obscure, apex obtuse, rounded, or blunt, tip not curled in dried plants. **Flowers** bisexual; filaments equal to or shorter than anthers. **Drupes** 1.8–2 × 0.6–1.2 mm. $2n = 32$.

Flowering summer. Saline or brackish lagoons and pools of maritime coastlines; 0 m; B.C., Man., Nfld. and Labr. (Labr.), N.W.T., Nunavut, Ont., Que., Yukon; Alaska; Eurasia.

The leaves of *Hippuris tetraphylla* tend to be fleshy or leathery, which is typical of obligate halophytes.

21. CHELONE Linnaeus, Sp. Pl. 2: 611. 1753; Gen. Pl. ed. 5, 267. 1754 • Turtlehead
[Greek *chelon*, tortoise, alluding to fancied resemblance between flower back and tortoise
back] E

Allan D. Nelson

Herbs, perennial; rhizomes producing 1–12 aerial stems. **Stems** erect, glabrous or glabrate.
Leaves cauline, opposite; petiole present or absent; blade not fleshy, not leathery, margins
serrate to dentate. **Inflorescences** axillary and terminal, spikelike cymes; bracts ovate to broadly
lanceolate, margins ciliate or not. **Pedicels** absent or present; bracteoles ± as large as calyx
lobes, nearly surrounding calyx of flowers they subtend. **Flowers** bisexual; sepals 5, proximally
connate, calyx radially symmetric, campanulate, lobes ovate, outer lobes ± as wide as inner;
corolla white, light pink-red, or dark purple throughout or white and pink to purple distally,
bilaterally symmetric, bilabiate, tubular, tube base not spurred or gibbous, throat not densely
pilose internally, lobes 5, abaxial 3, ovate to rounded with middle abaxial one elevated into
villous to lanate, bearded palate nearly closing throat, adaxial 2; stamens 4(or 5), proximally
adnate to corolla, didynamous, filaments pubescent to villous; staminode (0 or)1, spatulate;
ovary 2-locular, placentation axile; stigma capitate. **Fruits** capsules, dehiscence septicidal. **Seeds**
ca. 50, tan to light brown, darker towards center, globular to ovoid, wings present. *x* = 14.

Species 4 (4 in the flora): c, e North America.

Chelone is a member of Cheloneae, which in North America includes *Chionophila*, *Collinsia*,
Keckiella, *Nothochelone*, *Penstemon*, and *Tonella*. These genera share a cymose inflorescence,
the presence of a staminode, simple hairs, and stems that contain pith (A. D. Wolfe et al. 2002).
In phylogenetic analyses using molecular (Wolfe et al.) and morphological (A. D. Nelson 1995)
data, species of *Chelone* occur in a clade with *N. nemorosa*. *Chelone* has a reduced cyme with
relatively large bracteoles, adaxially ridged corollas, triangular pollen amb (grain shape from
polar view), rugulate-reticulate pollen exine sculpting, and circumalate (surrounded by a wing)
seeds; *Nothochelone* has a more-branched cyme with relatively small bracteoles, abaxially
ridged corollas, circular pollen amb, reticulate pollen exine sculpting, and asymmetric seed
wings (Nelson). Both genera have hypogynous disc nectaries (Wolfe et al.).

Species of *Chelone* occur in wetland habitats. Their seeds float readily and are likely dispersed
by water. Flowers are pollinated by bumblebees (*Bombus* spp.; F. W. Pennell 1935). Some leaf
and seed herbivores use species of *Chelone* as a host (N. E. Stamp 1984, 1987). Common
examples of leaf herbivores include the Baltimore checkerspot (*Euphydryas phaeton*) on
C. glabra as well as sawfly larvae (*Macrophya nigra* and *Tenthredo grandis*), which are known
from *C. glabra* and *C. obliqua*. One common larval seed predator known from *C. glabra* and
C. obliqua is an agromyzid dipteran (*Phytomyza cheloniae*).

Species and cultivars of *Chelone* are available from nurseries and florists. Plants can escape
from cultivation (T. S. Cooperrider 1969) and may have become naturalized outside their native
range, especially in New England. *Chelone* was used medicinally by Native Americans and
pioneers (C. S. Rafinesque 1828[–1830]; D. E. Moerman 1998).

Purported examples of inter- and intraspecific hybridization in *Chelone* were reported by
A. D. Nelson and W. J. Elisens (1999). Rare individuals of *C. cuthbertii* or *C. lyonii* with narrow
leaf bases might be the result of hybridization with *C. glabra* or *C. obliqua* (Nelson 1995).
Rare individuals of *C. lyonii* with pink-tipped staminodes might result from hybridization with
C. cuthbertii (Nelson). Individuals of *C. glabra* with completely pink, red, or purple corollas
might be the result of hybridization with sympatric species of *Chelone* that have similar corollas
(Nelson); this seems to occur rarely based on morphologic and isozyme variation (Nelson and
Elisens).

Chelone glabra and *C. obliqua* may be difficult to distinguish because all qualitative characters exhibit some overlap. Staminode colors usually are preserved in herbarium specimens; corolla and beard colors are often difficult to determine from dried plants.

SELECTED REFERENCES Nelson, A. D. 1995. Polyploid Evolution in *Chelone* (Scrophulariaceae). Ph.D. dissertation. University of Oklahoma. Nelson, A. D. and W. J. Elisens. 1999. Polyploid evolution and biogeography in *Chelone* (Scrophulariaceae): Morphological and isozyme evidence. Amer. J. Bot. 86: 1487–1501.

1. Petioles 0–3 mm; staminode apices purple; corollas pink-red to purple . . . . 1. *Chelone cuthbertii*
1. Petioles (0–)2–40 mm; staminode apices white to light pink or green, rarely purple; corollas completely dark pink, pink-red, red, or purple with abaxial surfaces sometimes paler to white, or completely white to white in tube and distally green-white, pink, red, or purple.
  2. Petioles (2–)10–40 mm; leaf blade bases rounded to truncate; bracts 2–7 mm; staminode apices white to light pink . . . . 2. *Chelone lyonii*
  2. Petioles (0–)2–20 mm; leaf blade bases cuneate; bracts 4–23 mm; staminode apices white or green, rarely purple.
    3. Corollas completely white to white in tube and distally green-white, pink, red, or purple; palates white-bearded, rarely green-yellow-bearded; staminode apices green. . . . . .3. *Chelone glabra*
    3. Corollas dark pink to red to purple, sometimes paler to white abaxially; palates yellow-bearded, rarely white-bearded; staminode apices white, rarely green or purple. . . . . .4. *Chelone obliqua*

1. Chelone cuthbertii Small, Fl. S.E. U.S., 1058, 1337. 1903 • Cuthbert's turtlehead E

Stems 23–100 cm. **Leaves:** petiole 0–3 mm; blade broadly lanceolate to ovate, 49–115 × 10–36 mm, base rounded, margins once-serrate, teeth 2–7 per cm, abaxial surface glabrous or pilose, adaxial usually glabrous. **Cymes** 25–70 mm; bracts 3–8 × 3–8 mm, apex obtuse to acute. **Flowers:** calyx lobes 5–9 × 3–6 mm, margins not or sparsely ciliate; corolla pink-red to purple, tube 11–19 mm, abaxial lobes 8–17 × 3–11 mm, adaxial slightly keeled; palate yellow-bearded; adaxial filaments 11–23 mm; staminode 6–15 mm, apex purple; style 16–24 mm. *2n* = 28.

Flowering Jul–Oct. Bogs, streamsides, wet streamheads, swamps; 0–1200 m; Ga., N.C., S.C., Va.

Chelone cuthbertii occurs in the coastal plain and Appalachian regions. The species can be distinguished from others of the genus by its short petioles and purple-tipped staminodes.

Chelone cuthbertii is rare throughout its range, making it especially vulnerable to extinction. Drainage of wetlands and bog succession threaten the habitat where the species is found (NatureServe, www.natureserve.org/explorer).

2. Chelone lyonii Pursh, Fl. Amer. Sept. 2: 737. 1813 (as lyoni) • Lyon's or pink turtlehead E

Stems 35–100 cm. **Leaves:** petiole (2–)10–40 mm; blade broadly lanceolate to ovate, 37–137 × (20–)30–55(–80) mm, base rounded to truncate, margins once-serrate, teeth 3–8 per cm, abaxial surface glabrous or pilose to slightly villous, adaxial glabrate or mostly glabrous. **Cymes** 27–71 mm; bracts 2–7 × 2–8 mm, apex obtuse to acute, sometimes acuminate. **Flowers:** calyx lobes 5–11 × 3–7 mm, margins sparsely to densely ciliate; corolla pink-red to purple, tube 15–21 mm, abaxial lobes 10–12(–14) × 5–12 mm, adaxial strongly keeled; palate yellow-bearded; adaxial filaments 16–23 mm; staminode (8–)10–15 mm, apex white to light pink; style 20–30 mm. *2n* = 28.

Flowering Jun–Oct. Stream banks, cove and spruce-fir forests, balds; 60–2000 m; Ala., Ga., Miss., N.C., S.C., Tenn.

Garden escapes of *Chelone lyonii* are reported from Connecticut, Maine, Massachusetts, and New York. *Chelone lyonii* can be distinguished from other members of the genus by its longer petioles, rounded to truncate leaf bases, and white to light pink staminodes. It is sometimes confused with *C. obliqua*, especially where *C. lyonii* is found with leaves narrowed to the base. *Chelone lyonii* can be distinguished from *C. obliqua* by wider leaf blades, longer petioles, more strongly keeled corollas, shorter abaxial corolla lips, and longer staminodes. *Chelone lyonii* has been proposed as a diploid progenitor for the allopolyploid *C. obliqua* (A. D. Nelson 1995; Nelson and W. J. Elisens 1999).

3. Chelone glabra Linnaeus, Sp. Pl. 2: 611. 1753

• White turtlehead, galane glabre E F

Chelone chlorantha Pennell & Wherry; *C. glabra* var. *dilatata* Fernald & Wiegand; *C. glabra* var. *elatior* Rafinesque; *C. glabra* subsp. *linifolia* (N. Coleman) Pennell; *C. glabra* var. *linifolia* N. Coleman

Stems 20–230 cm. **Leaves:** petiole (0–)2–10(–20) mm; blade broadly elliptic to narrowly elliptic, 17–230 × 6–54 mm, base cuneate, margins once-serrate, teeth 1–6 per cm, abaxial surface glabrous or pilose, rarely tomentose, adaxial glabrate or mostly glabrous. **Cymes** 30–115 mm; bracts 4–23 × 3–10 mm, apex acute to acuminate, rarely obtuse. **Flowers:** calyx lobes 5–11 × 3–8 mm, margins not or sparsely, rarely densely, ciliate; corolla completely white to white in tube and distally green-white, pink, red, or purple, tube 13–20 mm, abaxial lobes 6–16 × 5–15 mm, adaxial slightly keeled; palate white-bearded, rarely green-yellow-bearded; adaxial filaments (10–)15–24 mm; staminode 4–12(–16) mm, apex green; style 15–30 mm. $2n = 28$.

Flowering Jul–Nov. Bogs, fens, marshes, swamps, seeps, stream banks, wet meadows and woods, margins of ponds and lakes; 0–2000 m; St. Pierre and Miquelon; Man., N.B., Nfld. and Labr. (Nfld.), N.S., Ont., P.E.I., Que.; Ala., Ark., Conn., Del., D.C., Ga., Ill., Ind., Iowa, Ky., Maine, Md., Mass., Mich., Minn., Miss., Mo., N.H., N.J., N.Y., N.C., Ohio, Pa., R.I., S.C., Tenn., Vt., Va., W.Va., Wis.

A tetraploid population reported for *Chelone glabra* (T. S. Cooperrider 1970) likely is based on a misidentified specimen of *C. obliqua* var. *erwiniae*.

Chelone glabra has been proposed as a diploid progenitor for the allopolyploid *C. obliqua* (A. D. Nelson 1995; Nelson and W. J. Elisens 1999).

Infraspecific variants of *Chelone glabra* as proposed by F. W. Pennell (1935) are not recognized here because morphological variation within and among populations appears to be independent of geography. Qualitative and quantitative characters used to distinguish varieties have been shown to be highly variable; recognition of varieties is unwarranted (A. D. Nelson 1995; Nelson and W. J. Elisens 1999).

Chelone glabra in central Maine is visited exclusively by two species of bumblebees (*Bombus fervidus* and *B. vagans*) (B. Heinrich 1975). American Indians and pioneers used *C. glabra* as a tonic, laxative, and treatment for jaundice and internal parasites, and as an ointment to relieve itching and inflammation (C. S. Rafinesque 1828[–1830]); it is also used as an ornamental in bog gardens.

4. Chelone obliqua Linnaeus, Syst. Nat. ed. 12, 2: 408. 1767 E

Stems 25–180 cm. **Leaves:** petiole 3–20 mm; blade broadly elliptic to narrowly elliptic, 45–197 × 8–35(–50) mm, base cuneate, margins once- or twice-serrate, teeth 1–7 per cm, abaxial surface glabrous or pilose, rarely villous, adaxial usually glabrous. **Cymes** 38–86 mm; bracts 4–10(–17) × 3–8 mm, apex obtuse to acute or acuminate. **Flowers:** calyx lobes 7–10 × 3–7 mm, margins not or sparsely to densely ciliate; corolla dark pink to red to purple, sometimes paler to white abaxially, tube 14–22 mm, abaxial lobes (10–)12–19 × 5–15 mm, adaxial slightly keeled; palate yellow-bearded, rarely white-bearded; adaxial filaments (13–)16–27 mm; staminode 4–12(–14) mm, apex white, rarely green or purple; style 16–34 mm.

Varieties 3 (3 in the flora): c, e United States.

Chelone obliqua can be identified by its completely dark pink, red, or purple corollas with abaxial surfaces sometimes paler to white, yellow beards, and white staminodes. The species may be difficult to distinguish from other species of *Chelone* when rarely its beards are white or staminodes green- or purple-tipped. This may be due to rare intra- and interspecific hybridization (A. D. Nelson and W. J. Elisens 1999).

Chelone obliqua is allopolyploid (A. D. Nelson 1995; Nelson and W. J. Elisens 1999) and has a recombinant phenotype representing all three extant diploid species. Variation within *C. obliqua* reflects multiple independent origins (Nelson; Nelson and Elisens); some of the rare color variants could be due to this rather than hybridization (Nelson).

Chelone obliqua comprises two known chromosome races, $2n = 4x = 56$ and $2n = 6x = 84$. Tetraploids are found in the Blue Ridge Province; hexaploids are found in the Interior Low and Ozark plateaus as well as Central Lowland provinces and the Coastal Plain Province. Within each of these three physiographic provinces, distinct genotypes occur (A. D. Nelson 1995; Nelson and W. J. Elisens 1999; NatureServe, www.natureserve.org/explorer). Seven populations of *C. obliqua* were examined for this treatment, and it appears from this limited sample that the cytotypes might have regional ranges and minor morphological variation that support the varieties of *C. obliqua* as proposed initially by Pennell and Wherry and later treated as subspecies by Pennell. With the exception of calyx lobe margin indument, new characters are used to distinguish varieties because those used by Pennell had wide quantitative ranges or were highly variable qualitative characters.

1. Staminodes 4–8(–12) mm; calyx lobe margins densely ciliate; Interior Low and Ozark plateaus, Central Lowland provinces . . . . .4b. *Chelone obliqua* var. *speciosa*
1. Staminodes (6–)8–12(–14) mm; calyx lobe margins not or sparsely ciliate; Coastal Plain and Blue Ridge provinces.
    2. Abaxial corolla lobes (12–)15–19 mm; mid-cauline leaf blades 53–80(–117) mm; Coastal Plain Province . . . . .4a. *Chelone obliqua* var. *obliqua*
    2. Abaxial corolla lobes 12–15(–16) mm; mid-cauline leaf blades (60–)80–197 mm; Blue Ridge Province. . . . 4c. *Chelone obliqua* var. *erwiniae*

4a. Chelone obliqua Linnaeus var. **obliqua** • Red turtlehead E

Leaves: mid-cauline blade 53–80 (–117) mm. **Flowers:** calyx lobe margins not or sparsely ciliate; corolla: abaxial lobes (12–) 15–19 mm; staminode (6–)8–12 (–14) mm. *2n* = 84.

Flowering Jul–Oct. Along streams, marshes, swamps, seeps, springs, wet meadows and woods, pond and lake margins; 0–100 m; Ala., Ga., Ky., Md., Miss., N.C., S.C., Tenn., Va.

Variety *obliqua* is considered to be vulnerable to extinction throughout the Coastal Plain Province; specific threats to populations have not been assessed (NatureServe, www.natureserve.org/explorer).

4b. Chelone obliqua Linnaeus var. **speciosa** Pennell & Wherry, Bartonia 10: 19, plate 2, fig. 1. 1929 • Rose or purple turtlehead E

Chelone obliqua subsp. *speciosa* (Pennell & Wherry) Pennell

Leaves: mid-cauline blade (69–) 88–170 mm. **Flowers:** calyx lobe margins densely ciliate; corolla: abaxial lobes 10–16 mm; staminode 4–8(–12) mm. *2n* = 84.

Flowering Jul–Oct. Along streams, marshes, swamps, seeps, springs, wet meadows and woods, pond and lake margins; 90–200 m; Ark., Ill., Ind., Iowa, Ky., Mich., Minn., Mo.

Variety *speciosa* occurs in the Interior Low and Ozark plateaus as well as Central Lowland provinces, where it is vulnerable to extinction. NatureServe (www.natureserve.org/explorer) estimates 21 to 80 populations of var. *speciosa* throughout its range and cites activities affecting wetland hydrology as threats to these populations.

4c. Chelone obliqua Linnaeus var. **erwiniae** Pennell & Wherry, Bartonia 10: 19. 1929 • Erwin's or red turtlehead E

Chelone obliqua subsp. *erwiniae* (Pennell & Wherry) Pennell

Leaves: mid-cauline blade (60–) 80–197 mm. **Flowers:** calyx lobe margins not or sparsely ciliate; corolla: abaxial lobes 12–15(–16) mm; staminode (6–)8–12(–14) mm. *2n* = 56.

Flowering Jul–Oct. Along streams, marshes, swamps, seeps, springs, wet meadows and woods, pond and lake margins; 1000–1700 m; N.C., S.C.

Variety *erwiniae* is considered vulnerable to extinction in the Blue Ridge Province; threats to populations have not been assessed (NatureServe, www.natureserve.org/explorer). N. E. Stamp (1987) identified two common larval seed predators, *Endothenia herbesana* and *Phytomyza cheloniae*, which were observed to attack 25% of the capsules from a population and destroy 21% of its seeds. Insect defoliation prior to development of inflorescences reduces floral and seed output significantly in *Chelone obliqua* var. *erwiniae* and *C. glabra* (Stamp 1984). Severe herbivory after flower buds appeared caused decreased seed output, and plants that were defoliated twice in one year often produced no flowers (Stamp 1984). Because of severe levels of herbivory and seed predation observed in var. *erwiniae*, it may reproduce via seed infrequently, relying on clonal reproduction by rhizomes (Stamp 1987).

22. CHIONOPHILA Bentham in A. P. de Candolle and A. L. P. P. de Candolle, Prodr. 10: 331. 1846 • Snowlover [Greek *chion*, snow, and *philios*, loving, alluding to high-elevation habitats] E

Craig C. Freeman

Pentstemonopsis Rydberg

Herbs, perennial; caudex woody or herbaceous. **Stems** erect, glabrous or puberulent. **Leaves** basal and cauline, cauline smaller, opposite, sometimes alternate distally; petiole present or absent; blade not fleshy, ± leathery or not, margins entire. **Inflorescences** terminal, spikelike, secund racemes; bracts present. **Pedicels** present or absent; bracteoles absent. **Flowers** bisexual; sepals 5, proximally connate, calyx radially symmetric, cylindric to funnelform, lobes triangular to narrowly ovate; corolla greenish white, creamy white, or pale lavender, bilaterally symmetric, bilabiate, tubular-funnelform, tube base not spurred or gibbous, throat not densely pilose internally, lobes 5, abaxial 3, adaxial 2; stamens 4, proximally adnate to corolla, didynamous, filaments glabrous; staminode 1, ± filiform; ovary 2-locular, placentation axile; stigma capitate. **Fruits** capsules, dehiscence septicidal. **Seeds** 10–20, tan or brown, ellipsoid to fusiform or ellipsoid-disciform, wings present. *x* = 8.

Species 2 (2 in the flora): w United States.

Chionophila shares morphological similarities with *Penstemon*, most notably an epistaminal nectary of glandular hairs. Both have a base chromosome number of eight. R. M. Straw (1966) hypothesized that they are sister taxa. Relationships among genera in Cheloneae remain equivocal (A. D. Wolfe et al. 1997, 2002, 2006; S. L. Datwyler and Wolfe 2004). *Chionophila* is distinguished from *Penstemon* by a more prominent calyx tube, spikelike racemes, prominently winged seeds, and absence of bracteoles.

1. Verticillasters continuous; calyx tubes 8–9 mm . . . . 1. *Chionophila jamesii*
1. Verticillasters interrupted; calyx tubes 1–2 mm . . . . 2. *Chionophila tweedyi*

1. Chionophila jamesii Bentham in A. P. de Candolle and A. L. P. P. de Candolle, Prodr. 10: 331. 1846 • Rocky Mountain snowlover E F

Stems 1 or 2(or 3), (3–)5–12 (–15) cm, puberulent or retrorsely hairy, sometimes glabrate. **Leaves:** basal and proximal cauline, blade oblanceolate to narrowly oblanceolate or spatulate, 12–78 × 2–18 mm, surfaces glabrous or glabrate; cauline 1–3 pairs, blade narrowly lanceolate to linear, 8–28 × 1–3 mm. **Racemes** 1–5 cm, verticillasters 2–7, continuous, sparsely puberulent and, usually, sparsely glandular-puberulent; bracts ovate to lanceolate, proximal ones 8–19 × 4–7 mm. **Pedicels** 0–4 mm, glabrous or sparsely glandular-pubescent. **Flowers:** calyx tube 8–9 mm, sparsely glandular-puberulent, lobes triangular, 1.5–2.5 × 2–2.5 mm; corolla greenish white or creamy white, 10–15 mm, glabrous externally, palate and proximal parts of abaxial limb densely white-lanate, hairs to 1.5 mm, tube 3–4 mm, pollen sacs 0.5–0.6 mm, explanate; staminode 5–7 mm; style 10–12 mm. **Capsules** 8–9.5 × 4.5–6 mm. **Seeds** tan to brown, ellipsoid to fusiform with tail on each end, 3–4.2 mm. *2n* = 16.

Flowering Jun–Aug. Gravelly slopes, alpine meadows, subalpine bogs; 3300–4100 m; Colo., N.Mex., Wyo.

Chionophila jamesii occurs in the central Rocky Mountains from the Medicine Bow Mountains of south-central Wyoming to the Culebra Range in Taos County in north-central New Mexico.

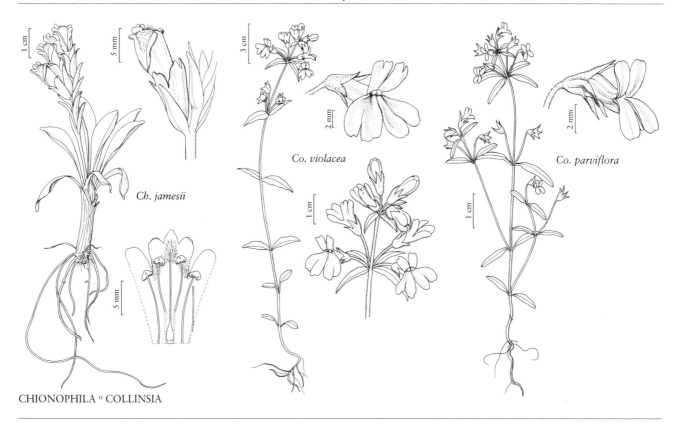

Ch. jamesii

Co. violacea

Co. parviflora

CHIONOPHILA ○ COLLINSIA

2. Chionophila tweedyi (Canby & Rose) L. F. Henderson, Bull. Torrey Bot. Club 27: 352. 1900 • Tweedy's snowlover E

Penstemon tweedyi Canby & Rose, Bot. Gaz. 15: 66. 1890; *Pentstemonopsis tweedyi* (Canby & Rose) Rydberg

Stems 1–4, (5–)10–18(–25) cm, glabrous or sparsely glandular-pubescent. **Leaves:** basal and proximal cauline, blade oblanceolate, (8–)25–90 × (1–)3–13 mm, surfaces glabrous; cauline 1–4 pairs, blade narrowly lanceolate to linear, 2–22 × 1–3 mm.

Racemes 1–7 cm, verticillasters 4–10, interrupted, glandular-pubescent; bracts lanceolate, proximal ones 3–9 × 1–3 mm. **Pedicels** 0–7 mm, glandular-pubescent. **Flowers:** calyx tube 1–2 mm, glandular-pubescent, lobes triangular to narrowly ovate, 2.5–3 × 1.4–2 mm; corolla pale lavender, 9–14 mm, glabrous externally, palate and proximal parts of abaxial limb glandular-pubescent, hairs to 0.1 mm, tube 1–2 mm, pollen sacs 0.3–0.4 mm, explanate; staminode 4–5 mm; style 6–7 mm. **Capsules** 6.5–9 × 4–5 mm. **Seeds** brown, elliptic-disciform, 2.4–3 mm.

Flowering Jun–Aug. Talus slopes, meadows, usually near timber lines; 2100–3200 m; Idaho, Mont.

Chionophila tweedyi is known from central Idaho and southwestern Montana.

23. COLLINSIA Nuttall, J. Acad. Nat. Sci. Philadelphia 1: 190, plate 9. 1817 • Chinese-houses, blue-eyed Mary [For Zaccheus Collins, 1764–1831, Philadelphia botanist]

Michael S. Park

Herbs, annuals. **Stems** erect to ascending or decumbent, glabrous or hairy, often glandular. **Leaves** cauline, opposite; petiole present on proximals, absent on distals; blade not fleshy (fleshy in *C. callosa*), not leathery, margins entire, sometimes toothed. **Inflorescences** terminal, interrupted whorls or flowers solitary; bracts present. **Pedicels** present; bracteoles present, distalmost sometimes absent. **Flowers** bisexual; sepals 5, connate, calyx bilaterally symmetric, urceolate or cup-shaped to campanulate, lobes deltate to ovate, inner face glabrous or white-hairy (*C. antonina*); corolla white, pink, purple, or blue with markings on adaxial lobes, bilaterally symmetric, bilabiate, tubular to funnelform, tube base not spurred abaxially, ± saccate or gibbous (forming a prominent or obscure pouch) adaxially, lobes 5, abaxial 3, 2 as wings, middle lobe of abaxial lip (keel) folded lengthwise, enclosing stamens and styles, adaxial 2, distal portion spreading to reflexed (banner); stamens 4, adnate near base of throat, didynamous, included, filaments glabrous or hairy, spur 0 or 1; staminode 1, glandlike; ovary 2-loculed, placentation axile; stigma minutely 2-lobed. **Fruits** capsules, dehiscence septicidal and loculicidal. **Seeds** 2–16, reddish brown, brown, or cinnamon red, oblong, ellipsoid, ovoid, or globular, rarely prismatic or hourglass-shaped, margins thickened and inrolled proximally, or thin and not inrolled, hollow near attachment to placenta, wings absent, rarely present. *x* = 7.

Species 21 (21 in the flora): North America, nw Mexico.

The flowers in *Collinsia* resemble those of Fabaceae subfam. Faboideae. The two lobes of the adaxial corolla lip are partially connate and distally reflexed; these lobes, the banner, often contain taxon-specific markings. The three lobes of the abaxial lip are more complexly connate: the two lateral lobes are referred to as wings; the central lobe, the keel, is folded, curved along its distal portion, and concealed by the wings. The keel encloses the stamens, style, and stigma.

In species with purple or blue corollas, some plants produce white or pink corollas. The frequency of such plants is usually very low in most species. The frequency of such plants in *Collinsia heterophylla* is much higher in the South Coast Ranges of California, and this trait appears to have reached fixation in some populations.

In taxa that are conspicuously glandular, a reaction occurs when the glands are ruptured, either when handled roughly in the field or when pressed for preservation as a specimen. The ruptured glands will cause a stain that is initially iron yellow-orange and later becomes purplish.

SELECTED REFERENCES Baldwin, B. G., S. Kalisz, and W. S. Armbruster. 2011. Phylogenetic perspectives on diversification, biogeography, and floral evolution of *Collinsia* and *Tonella*. Amer. J. Bot. 98: 731–753. Ganders, F. R. and G. R. Krause. 1986. Systematics of *Collinsia parviflora* and *C. grandiflora* (Scrophulariaceae). Madroño 33: 63–70. Newsom, V. M. 1929. A revision of the genus *Collinsia* (Scrophulariaceae). Bot. Gaz. 87: 260–301. Tunbridge, N. D., C. Sears, and E. Elle. 2011. Variation in floral morphology and ploidy among populations of *Collinsia parviflora* and *Collinsia grandiflora*. Botany (Ottawa) 89: 19–33.

1. Corollas ± uniformly dark purple, rarely pale pinkish purple, banner lobe bases with 2-crested, crescent-shaped folds extending 1–1.5 mm from throat openings . . . . .9. *Collinsia greenei*
1. Corollas uniformly pale, or with pale regions, especially on throats and bases of banners, banner lobe bases with folds not 2-crested, not crescent-shaped, extending less than 1 mm from throat openings, or folds absent.
   2. Flowers crowded; pedicels shorter than calyces, not or scarcely visible.
      3. Stamens: basal spurs of adaxial filaments 1. . . . . 20. *Collinsia heterophylla*
      3. Stamens: basal spurs of adaxial filaments 0 or 1.
         4. Banner lengths 0.7–0.9+ times wings.
            5. Corollas ± uniformly whitish to pinkish, banner lobes toothed; leaf blade margins crenate . . . . .18. *Collinsia bartsiifolia* (in part)
            5. Corollas tinged bluish, distally violet to magenta, banner purple-dotted near base, banner lobes notched; leaf blade margins entire, rarely serrate
            . . . . 21. *Collinsia concolor*
         4. Banner lengths less than 0.8 times wings.
            6. Banner lengths 0.5–0.8 times wings. . . . . .18. *Collinsia bartsiifolia* (in part)
            6. Banner lengths 0.1–0.5 times wings.

7. Leaf blade abaxial surfaces glabrous; inflorescences: flower whorls 1 per branch; banner reflexed portion 1 mm, shorter than basal portion, brownish, not red-banded; wings sparsely and finely glandular, not hairy; sand dunes . . . . .17. *Collinsia corymbosa*
7. Leaf blade abaxial surfaces densely hairy; inflorescences: flower whorls 2–8 on main axes, fewer on lateral branches; banner reflexed portion 2–3.5 mm, equal to basal portion, red-banded; wings usually hairy adaxially; rocky habitats . . . . .19. *Collinsia tinctoria*

[2. Shifted to left margin.—Ed.]

2. Flowers not crowded proximally; pedicels longer than calyces, visible, distalmost sometimes not or scarcely visible (*C. multicolor*).
   8. Calyx lobes: inner face white-hairy; siliceous shale screes . . . . 6. *Collinsia antonina*
   8. Calyx lobes: inner face glabrous; rocky, gravelly, sandy, or loamy soils, rarely on siliceous shale screes.
      9. Calyx lobe apices acute to acuminate.
         10. Stamens: adaxial filaments glabrous; calyx lobes equal to capsules.
            11. Corollas 8–15 mm, wings 2–6 mm wide, widely obovate, throats strongly angled to tubes. . . . . .3. *Collinsia grandiflora*
            11. Corollas 4–8 mm, wings 1(–3) mm wide, oblong, throats barely angled to tubes . . . . .4. *Collinsia parviflora*
         10. Stamens: adaxial filaments hairy; calyx lobes surpassing capsules.
            12. Keels sparsely long-hairy near tips; capsules with red blotches; seeds round, margins thin, not inrolled. . . . . 12. *Collinsia sparsiflora*
            12. Keels glabrous or sparsely glandular; capsules without red blotches; seeds oblong to oval, margins thickened, inrolled.
               13. Banner bases white, with maroon dots and lines, rarely unmarked . . . . 5. *Collinsia multicolor*
               13. Banner bases yellow, with maroon or orange spots.
                  14. Pedicels glandular abaxially, glabrate adaxially; corolla wings and keels bluish, wings notched less than 0.1 times whole lengths, banner lengths (0.7–)0.8–1 times wings, bases with maroon spots; seeds 2–4, 2–3 mm . . . . .1. *Collinsia verna*
                  14. Pedicels glandular abaxially and adaxially; corolla wings and keels violet, wings notched 0.2 times whole lengths, banner lengths 0.6–0.7(–0.8) times wings, bases with orange spots; seeds 6–12, 1–1.5 mm . . . . .2. *Collinsia violacea*
      9. Calyx lobe apices subacute, obtuse, or rounded.
         15. Inflorescences ± eglandular. . . . . 7. *Collinsia parryi*
         15. Inflorescences conspicuously glandular.
            16. Pedicels pendent and/or sigmoid in fruit, sometimes reflexed; distalmost bracts 0–2 mm.
               17. Wings surpassing keel by 1–2 mm. . . . . 14. *Collinsia wrightii*
               17. Wings equal to keel.
                  18. Leaf blades linear, lengths usually 6+ times widths . . . . 15. *Collinsia torreyi*
                  18. Leaf blades elliptic to ovate, lengths usually 2–5 times widths . . . . 16. *Collinsia latifolia*
            16. Pedicels spreading to ascending; distalmost bracts 2+ mm.
               19. Leaf blades oblong to ovate, lanceolate, or oblanceolate, lengths usually less than 6 times widths.
                  20. Annuals fleshy; distal leaf blade bases clasping; seeds 6–8; calyces campanulate to urceolate, lobes equal to capsules. . . . . 8. *Collinsia callosa*
                  20. Annuals not fleshy; distal leaf blade bases tapered; seeds 2; calyces campanulate, lobes surpassing capsules. . . . . 13. *Collinsia childii*
               19. Leaf blades linear to narrowly oblanceolate, lengths usually 6+ times widths.

21. Corollas 8–12(–15) mm, pouches prominent, throats strongly angled to tubes, banner reflexed portion lengths 1.5–2 times throats; seeds round to oblong, margins thin, scarcely inrolled . . . . .10. *Collinsia linearis*

21. Corollas 4–8 mm, pouches ± hidden by calyces, throats barely angled to tubes, banner reflexed portion lengths 1 times throats; seeds oblong to ovate, margins thickened, inrolled . . . . 11. *Collinsia rattanii*

1. Collinsia verna Nuttall, J. Acad. Nat. Sci. Philadelphia 1: 190, plate 9. 1817 [E]

Collinsia bicolor Rafinesque

Annuals 10–30 cm. **Stems** erect to ascending. **Leaf blades** ovate to elliptic or lanceolate, base cuneate to subcordate, margins shallowly and coarsely serrate. **Inflorescences** ± glandular, scaly-hairy; nodes 1–6(–8)-flowered; flowers not crowded proximally, often crowded distally; distalmost bracts linear, 5–6(+) mm. **Pedicels** ascending to reflexed, usually longer than calyx, visible, glandular abaxially, glabrate adaxially. **Flowers:** calyx lobes deltate, surpassing capsule, apex acute to acuminate; corolla blue, banner white to pale lilac, base yellow with small maroon spots, wings and keel bluish, 8–15 mm, keel sparsely glandular; banner length (0.7–)0.8–1 times wings, lobe base without folds; wings widely obovate, notched less than 0.1 times whole length; throat angled to tube, longer than diam., pouch arched, slightly expanded; stamens: abaxial filaments glabrous or sparsely hairy at base, adaxials hairy, basal spur 0. **Capsules** without red blotches. **Seeds** 2–4, oblong to oval, 2–3 mm, margins thickened, inrolled. *2n* = 14.

Flowering Apr–Jun. Moist woodlands; 80–1000 m; Ont.; Ark., Ill., Ind., Iowa, Kans., Ky., Mich., Mo., N.Y., Ohio, Okla., Pa., Tenn., Tex., Va., W.Va., Wis.

2. Collinsia violacea Nuttall, Trans. Amer. Philos. Soc., n. s. 5: 179. 1835 • Violet blue-eyed Mary [E][F]

Annuals 10–35(–60) cm. **Stems** erect to ascending. **Leaf blades** oblong to lanceolate, margins entire or weakly serrate. **Inflorescences** glabrous or glandular to scaly-hairy; nodes 1–6(–8)-flowered; flowers not crowded proximally, sometimes crowded distally; distalmost bracts linear, 5–6(+) mm. **Pedicels** ascending to reflexed, usually longer than calyx, visible, glandular abaxially and adaxially. **Flowers:** calyx lobes deltate, surpassing capsule, apex acuminate; corolla violet, banner pale violet to white, base yellow with dark orange spot, wings and keel violet, 10–15 mm, keel glabrous or sparsely glandular; banner length 0.6–0.7(–0.8) times wings, lobe base without folds; wings narrowly obcordate, notched 0.2 times whole length; throat slightly angled to tube, longer than diam., pouch arched, slightly expanded; stamens: abaxial filaments glabrous, adaxials hairy, basal spur 0. **Capsules** without red blotches. **Seeds** 6–12, oblong to oval, 1–1.5 mm, margins thickened, inrolled. *2n* = 14.

Flowering Apr–Jun. Sandy or rocky soils, dry open areas, woodlands; 10–300 m; Ark., Ill., Kans., Mo., Okla., Tex.

3. Collinsia grandiflora Lindley, Bot. Reg. 13: plate 1107. 1827 • Large-flowered blue-eyed Mary [E]

Collinsia parviflora Lindley var. *diehlii* (M. E. Jones) Pennell; *C. parviflora* var. *grandiflora* (Lindley) Ganders & G. R. Krause

Annuals (4–)6–35 cm. **Stems** erect to ascending. **Leaf blades** narrowly oblong to lanceolate, margins subentire. **Inflorescences** glabrous or finely glandular to scaly-hairy; nodes 1–6(–8)-flowered; flowers not crowded proximally, sometimes crowded distally; distalmost bracts linear, 5–6(+) mm. **Pedicels** ascending to reflexed, proximals usually longer than calyx, visible. **Flowers:** calyx lobes ± deltate, equal to capsule, apex acuminate; corolla bluish, banner pale at center, 8–15 mm, keel glabrous; banner length 0.8–1 times wings, lobe base without folds; banner lobes and wings widely obovate, usually 2–6 mm wide; throat strongly angled to tube, longer than diam., pouch prominent, angular; stamens: filaments glabrous, basal spur 0. **Seeds** (3 or)4, oblong, 2–2.5 mm, margins thickened, inrolled. *2n* = 14, 28, 42.

Flowering (Mar–)May–Jul. Gravelly or grassy margins of coniferous or open oak woodlands, moss-covered rock outcrops, other open areas; 0–1800 m; B.C.; Calif., Oreg., Wash.

Collinsia grandiflora occurs mostly in the coastal ranges. The distinction between *C. grandiflora* and *C. parviflora* is usually clear in California where corolla lobe shape and size are mostly well correlated. The distinction is much less clear in British Columbia, Oregon, and Washington.

An alternative to the treatment here would be to follow F. R. Ganders and G. R. Krause (1986), who suggested that *Collinsia grandiflora* and *C. parviflora* be treated as one species with two intergrading varieties.

4. Collinsia parviflora Lindley, Bot. Reg. 13: plate 1082. 1827 • Small-flowered blue-eyed Mary E F

Collinsia grandiflora Lindley var. *pusilla* A. Gray

Annuals 3–40 cm. **Stems** erect to ascending. **Leaf blades** ± linear-lanceolate, obovate, or narrowly elliptic, margins subentire. **Inflorescences** glabrous or sparsely and finely glandular; proximal nodes 1-flowered, distals 3–5(–7)-flowered; flowers not crowded proximally, sometimes crowded distally; distalmost bracts linear, 5–6 mm. **Pedicels** ascending to reflexed, longer than calyx, visible. **Flowers:** calyx lobes ± deltate, equal to capsule, apex sharply acute to acuminate; corolla blue, banner whitish or blue-tipped, 4–8 mm, glabrous; banner length 0.8–1 times wings, lobe base without folds; banner lobes and wings blue, sometimes purplish, oblong, 1(–3) mm wide; throat barely angled to tube, tube and throat white, narrowed to lips, pouch angular, ± hidden by calyx; stamens: filaments glabrous, basal spur 0. **Seeds** (3 or)4, oblong, 2–2.5 mm, margins thickened, inrolled. $2n$ = 14, 28, 42.

Flowering Mar–Jul. Forests, grasslands, meadows, eroded banks, bedrock depressions, scree slopes, shrublands, shaded shorelines; 0–3500 m; Alta., B.C., Man., N.S., Ont., Sask., Yukon; Alaska, Ariz., Calif., Colo., Idaho, Mass., Mich., Mont., Nebr., Nev., N.Mex., N.Dak., Oreg., Pa., S.Dak., Utah, Vt., Wash., Wyo.

Collinsia parviflora is the closest relative of *C. grandiflora* and is primarily a plant of moist montane habitats with well-drained, rocky or sandy soil. However, it occurs in a wide range of habitats across its entire range. The species is also the most widespread taxon within *Collinsia*. Some plants from the western coastal ranges may be difficult to separate from *C. grandiflora*.

Collinsia parviflora is frequently confused with *C. wrightii*. The corollas of *C. wrightii* are distinctly purplish; those of *C. parviflora* are bright blue. The acute to acuminate sepals of *C. parviflora* contrast with the blunt, rounded tips of sepals of *C. wrightii*.

5. Collinsia multicolor Lindley & Paxton, Paxton's Fl. Gard. 2: 89, plate 55. 1851 • San Francisco collinsia C E

Annuals 30–60 cm. **Stems** ascending. **Leaf blades:** middle and distal lanceolate-deltate, margins coarsely serrate. **Inflorescences** ± glandular; proximal nodes 1- or 2-flowered, distals 2–4-flowered; flowers not crowded proximally, sometimes crowded distally; distalmost bracts linear, 3–5 mm. **Pedicels** ascending to spreading, proximalmost sometimes longer than calyx, distalmost sometimes shorter than calyx, visible or distalmost not or scarcely visible. **Flowers:** calyx lobes deltate, surpassing capsule, apex acute; corolla mostly white to pale lilac, banner base white with maroon dots and lines, rarely unmarked, wings and keel lavender to bluish purple, 12–18 mm, usually glabrous; banner length 0.7–0.8 times wings, lobe base without folds; banner lobes and wings obovate, notched; keel sometimes sparsely glandular-hairy; tube longer than diam., adaxial pouch rounded, slightly gibbous, not prominent; stamens: abaxial filaments glabrous, adaxials hairy, basal spur 0(or 1). **Capsules** without red blotches. **Seeds** 8(–12), oblong, 2–2.5 mm, margins thickened, inrolled.

Flowering Mar–May. Moist, ± shady scrub, woodlands; of conservation concern; 0–300 m; Calif.

Collinsia multicolor is known from the Santa Cruz Mountains. The flowers of *C. multicolor* are similar to those of *C. heterophylla*, including markings at the base of the banner; *C. multicolor* lacks the curved basal spurs at the bases of the adaxial filaments, and its banner lobes and wings are notched. In *C. multicolor*, the adaxial side of the corolla tube is rounded and slightly gibbous, unlike the tube of *C. heterophylla*, which is saccate basally.

6. Collinsia antonina Hardham, Leafl. W. Bot. 10: 133. 1964 • San Antonio collinsia C E

Collinsia antonina subsp. *purpurea* Hardham

Annuals 4–15 cm. **Stems** erect. **Leaf blades** oblong, margins crenate. **Inflorescences** ± finely scaly, usually sparsely, finely glandular; nodes 1–3-flowered; flowers not crowded; distalmost bracts linear, 2–3 mm. **Pedicels** ascending to spreading, proximalmost longer than calyx, distalmost equal to calyx, visible. **Flowers:** calyx lobes lanceolate, slightly surpassing capsule, apex blunt to rounded, inner face white-hairy; corolla purple, lobes purple, rarely white, throat white with red-purple spots at base of banner, 4.5–8 mm, glabrous; banner length 1 times wings, lobe base without folds; stamens: abaxial filaments glabrous, adaxials sparsely hairy, basal spur 0. **Seeds** 6–8, oblong, 1.5–2 mm, margins thickened, inrolled. $2n = 14$.

Flowering Mar–Apr. Margins of oak scrub on screes; of conservation concern; 200–400 m; Calif.

Collinsia antonina is geographically narrowly endemic, known only from Monterey County. It occurs on scree derived from whitish siliceous shale of the Monterey Formation at the edge of woodlands near the shade of *Quercus john-tuckeri*. It is morphologically similar to *C. parryi*, which lacks the coarse white hairs on the inner face of the sepals. DNA studies (B. G. Baldwin et al. 2011) show a more distant relationship between *C. antonina* and *C. parryi* than suspected from morphology alone.

7. **Collinsia parryi** A. Gray in A. Gray et al., Syn. Fl. N. Amer. 2(1): 257. 1878 • Parry's blue-eyed Mary [E]

Annuals 10–40 cm. **Stems** erect to ascending. **Leaf blades** ± lanceolate, margins entire or crenate. **Inflorescences** ± eglandular; nodes 1–3(–5)-flowered; flowers not crowded; distalmost bracts linear, 2–3 mm. **Pedicels** ascending to spreading, usually longer than calyx, visible. **Flowers:** calyx lobes ovate, equal to capsule, apex obtuse to subacute or obscurely rounded; corolla blue-violet to lavender, rarely white, 4–10 mm, glabrous; banner length 1 times wings, lobe base without folds; stamens: abaxial filaments glabrous, adaxials sparsely spreading-hairy, basal spur 0. **Seeds** 8–12, oblong, 1–1.5 mm, margins thickened, inrolled. $2n = 14$.

Flowering Apr–May(–Jun). Open chaparral, sagebrush scrub, mixed woodlands; 500–1600 m; Calif.

Collinsia parryi is most closely related to *C. concolor*, which has larger flowers arranged in tiers of whorls; their ranges are largely allopatric. *Collinsia parryi* occurs most commonly on the drier, leeward sides of the Peninsular and Transverse ranges. B. G. Baldwin et al. (2011) sampled chloroplast DNA, ribosomal DNA, and introns of nuclear-coding DNA and showed that many individuals of *C. parryi* had zero sequence-divergence from *C. concolor*. This result suggests a recent diversification of these taxa from an ancestor that was most like *C. concolor*.

8. **Collinsia callosa** Parish, Erythea 7: 96. 1899 • Desert mountain blue-eyed Mary [E]

Annuals 4–25 cm, fleshy. **Stems** erect to ascending. **Leaf blades** oblong to ovate, length usually less than 6 times width, base of distals clasping, margins usually entire. **Inflorescences** glandular; nodes 1–3-flowered; flowers not crowded; distalmost bracts ovate, 2–3 mm. **Pedicels** ascending to spreading, longer than calyx, visible. **Flowers:** calyx campanulate to urceolate, lobes narrowly deltate to lanceolate, equal to capsule, apex subacute to rounded; corolla lavender-blue, rarely pink, lobe base white, keel tip purple, 7–9 mm; banner length 0.8–1 times wings, lobe base without folds; stamens: filaments glabrous, adaxials rarely sparsely hairy, basal spur 0(or 1). **Seeds** 6–8, oblong to hourglass-shaped, 1.8–2.5 mm, margins thickened, inrolled.

Flowering Apr–Jun. Disturbed, rocky slopes, open chaparral, sagebrush scrub, pinyon/juniper or pine woodlands; 1000–2300 m; Calif.

Collinsia callosa occurs primarily on the eastern sides of the southernmost Coast Ranges, Transverse Ranges, and southern Sierra Nevada. A report of this species from Nye County, Nevada (J. T. Kartesz 1987) has not been verified by the author.

9. **Collinsia greenei** A. Gray, Proc. Amer. Acad. Arts 10: 75. 1874 • Greene's blue-eyed Mary [E] [F]

Annuals 10–30 cm. **Stems** erect to ascending. **Leaf blades** narrowly lanceolate to ovate or oblanceolate, margins entire or serrate. **Inflorescences** glandular; nodes 1–5-flowered; flowers crowded or not; distalmost bracts linear, 2–3 mm. **Pedicels** ascending to spreading, proximalmost sometimes longer than calyx, distalmost equal to or shorter than calyx, visible or not. **Flowers:** calyx lobes lanceolate to ovate, surpassing capsule, apex subacute to rounded; corolla ± uniformly dark purple, rarely pale pinkish purple, 10–15 mm, sparsely glandular; banner length 0.5 times wings, base with 2-crested, crescent-shaped folds extending 1–1.5 mm from throat opening; stamens: filaments glabrous, adaxials sometimes hairy, basal spur 0. **Seeds** 2–4, oval, 2–3 mm, margins thin, not inrolled. $2n = 14$.

Flowering Apr–Jul(–Aug). Open chaparral or coniferous forests, serpentine slopes; 300–2500 m; Calif.

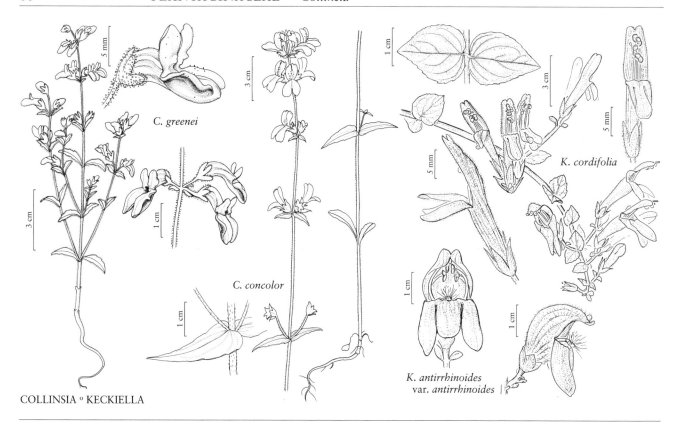

COLLINSIA ° KECKIELLA

Collinsia greenei occurs on ophiolites, most frequently on soil derived from serpentinite and similarly altered ultramafic rock. Within its range, only *C. rattanii* occurs on these substrates; *C. latifolia*, *C. parviflora*, and *C. wrightii* are not on highly mafic soil.

The dark, nearly uniformly purple corollas of *Collinsia greenei* are distinctive, and the crescent-shaped flap of tissue on the adaxial lobes is unique. Other taxa have folds that bulge outward but are neither doubly crested nor crescent-shaped.

10. Collinsia linearis A. Gray, Proc. Amer. Acad. Arts 15: 50. 1879 • Narrowleaf blue-eyed Mary [E]

Collinsia rattanii A. Gray var. *linearis* (A. Gray) Newsom

Annuals 10–40 cm. **Stems** erect to ascending. **Leaf blades** linear to narrowly oblanceolate, length usually 6+ times width, margins entire, subentire, or ± dentate. **Inflorescences** finely scaly and spreading-glandular; nodes 1–3(–5)-flowered; flowers not crowded; distal-most bracts linear, 2–5 mm. **Pedicels** ascending, longer than calyx, visible. **Flowers:** calyx lobes lanceolate to ovate, equal to capsule, apex subacute to slightly rounded; corolla white to blue-purple, 8–12(–15) mm, length 2+ times calyx, glandular; banner length 0.8–1 times wings, base with folds bulging 0.5 mm away from throat opening, reflexed portion length 1.5–2 times throat; throat strongly angled to tube, as wide as long, pouch ± square, prominent; stamens: abaxial filaments glabrous, adaxials hairy at base, basal spur 0(or 1). **Seeds** 2–4(–6), round to oblong, 1.5–2.2 mm, margins thin, scarcely inrolled. $2n = 14$.

Flowering (Apr–)May–Jul. Open coniferous forests; 200–2000 m; Calif., Oreg.

Collinsia linearis and *C. rattanii* occupy similar habitats, often growing in rocky soil derived from mafic, volcanic rock. They form a lineage but may not be reciprocally monophyletic (B. G. Baldwin et al. 2011). Other than differences in corolla size and shape, and, possibly, seed morphology, they are extremely similar. Corolla shape (floor of throat angled 45° to 60° from abaxial side of calyx and reflexed portion of adaxial lip length 1.5 to 2 times throat) and smaller corolla size are characteristic of *C. linearis*. Some plants from this lineage, mostly from Oregon, are difficult to assign to either *C. linearis* or *C. rattanii*.

Specimens annotated as *Collinsia linearis* from British Columbia have not been verified.

11. **Collinsia rattanii** A. Gray, Proc. Amer. Acad. Arts 15: 50. 1879 (as rattani) • Sticky blue-eyed Mary [E]

Collinsia rattanii subsp. *glandulosa* (Howell) Pennell

Annuals 8–40 cm. **Stems** erect to ascending. **Leaf blades** ± linear, length usually 6+ times width, margins entire, rarely crenate. **Inflorescences** scaly and spreading-glandular; nodes 1–3(–5)-flowered; flowers not crowded; distalmost bracts linear, 2–5 mm. **Pedicels** ascending, longer than calyx, visible. **Flowers:** calyx lobes lanceolate to ovate, equal to capsule, apex subacute to slightly rounded; corolla purplish lavender, rarely white, 4–8 mm, glabrous or keel sometimes sparsely glandular; banner length 0.8–1 times wings, lobe base with folds bulging less than 0.5 mm away from throat opening, reflexed portion length 1 times throat; throat barely angled to tube, pouch ± hidden by calyx; stamens: abaxial filaments glabrous, adaxials hairy at base, basal spur 0(or 1). **Seeds** (3 or)4–6, oblong to ovate, 1.5–2 mm, margins thickened, inrolled. $2n = 14$.

Flowering May–Jul(–Aug). Open coniferous forests; 60–1500 m; Calif., Oreg., Wash.

Collinsia rattanii is best distinguished from *C. linearis* by corolla shape (floor of throat nearly parallel to abaxial side of calyx and reflexed portion of adaxial lip about equal to throat) and smaller size.

12. **Collinsia sparsiflora** Fischer & C. A. Meyer, Index Seminum (St. Petersburg) 2: 33. 1836 • Few-flowered collinsia [E]

Annuals 5–30 cm. **Stems** erect to ascending. **Leaf blades** usually linear to oblong, margins entire. **Inflorescences** glabrous or finely hairy, eglandular; nodes 1- or 2(or 3)-flowered; flowers not crowded; distalmost bracts linear to narrowly lanceolate, 2–4(+) mm. **Pedicels** ascending to spreading, usually longer than calyx, visible. **Flowers:** calyx lobes narrowly triangular to lanceolate, surpassing capsule, apex sharply acute to acuminate; corolla lavender to purple, rarely white, 5–20 mm, keel sparsely long-hairy near tip; banner length 0.8–0.9 times wings, lobe base without folds; banner lobes and wings oblong to obovate, entire or notched; stamens: abaxial filaments glabrous or sparsely hairy at base, adaxials sparsely spreading-hairy on proximal ½, basal spur 0. **Capsules** with red blotches. **Seeds** 4–12, round, flattened, 2–3 mm, margins thin, not inrolled.

Varieties 2 (2 in the flora): w United States.

Collinsia sparsiflora occurs in open habitats and woodland edges; it is easily recognized by the red blotches on fruits and the frequent occurrence of a single flower per node. In immature plants, the presence of a red band spanning the circumference of the base of the calyx lobes along with the relatively frequent occurrence of a single flower per node distinguishes *C. sparsiflora*.

V. M. Newsom (1929) recognized five varieties in *Collinsia sparsiflora*.

Two varieties are recognized here to accommodate the general pattern of locally distinct small- and large-flowered populations of *Collinsia sparsiflora*. Where these taxa are sympatric, the author has not found intermediates that suggest local interbreeding.

1. Corollas 9–20 mm, angles between corolla tube-throat and calyx 45–70°; seeds 2.5–3 mm. . . . . .12a. *Collinsia sparsiflora* var. *sparsiflora*
1. Corollas 5–9 mm, angles between corolla tube-throat and calyx 30–45°; seeds 2–2.3(–2.5) mm . . . . 12b. *Collinsia sparsiflora* var. *collina*

12a. **Collinsia sparsiflora** Fischer & C. A. Meyer var. **sparsiflora** [E]

Collinsia bruceae M. E. Jones; *C. sparsiflora* var. *arvensis* (Greene) Jepson; *C. sparsiflora* var. *bruceae* (M. E. Jones) Newsom

Flowers: corolla 9–20 mm, angle between corolla tube-throat and calyx 45–70°. **Seeds** 2.5–3 mm. $2n = 14$.

Flowering Mar–May. Grassy, sometimes disturbed, places, drying meadows, chaparral, oak woodlands, dry mixed woodlands; 0–2000 m; Calif., Oreg., Wash.

In the southern portion of the interior North Coast Ranges of California, plants of var. *sparsiflora* with corollas 14–20 mm and strongly declined occur on mafic substrates of volcanic origin. V. M. Newsom (1929) treated these plants as var. *arvensis*.

V. M. Newsom (1929) treated plants of var. *sparsiflora* with corollas 9–13 mm in the Cascade Ranges of southern Oregon and northern California as var. *bruceae*. The shape of the corolla of var. *bruceae* is intermediate to that of the smaller-flowered var. *collina* with which it is sympatric. These plants are not as tall as those of the Coastal Ranges of California.

12b. Collinsia sparsiflora Fischer & C. A. Meyer var. **collina** (Jepson) Newsom, Bot. Gaz. 87: 286. 1929 [E]

Collinsia parviflora Lindley var. *collina* Jepson, Man. Fl. Pl. Calif., 904. 1925; *C. solitaria* Kellogg

Flowers: corolla 5–9 mm, angle between corolla tube-throat and calyx 30–45°. **Seeds** 2–2.3(–2.5) mm. $2n = 14$.

Flowering Mar–May. Disturbed grassy fields, road banks, open chaparral, open oak and dry mixed woodlands; 100–1200 m; Calif.

Plants of var. *collina* are diminutive in many respects compared to those of var. *sparsiflora*.

13. Collinsia childii Parry ex A. Gray in A. Gray et al., Syn. Fl. N. Amer. 2(1): 257. 1878 • Child's blue-eyed Mary [E]

Annuals 8–35 cm, not fleshy. **Stems** erect to ascending. **Leaf blades** oblong to lanceolate or oblanceolate, length usually less than 6 times width, base of distals tapered, margins entire or serrulate. **Inflorescences** densely glandular; nodes 2–5-flowered; flowers not crowded; distalmost bracts linear, 2–3(+) mm. **Pedicels** spreading to ascending, longer than calyx, visible. **Flowers:** calyx campanulate, lobes lanceolate, surpassing capsule, apex subacute to rounded; corolla whitish or pale lavender, banner base with central field of purple spots, 6–9 (–11) mm, glabrate; banner length 0.9–1 times wings, lobe base without folds; stamens: filaments glabrous, basal spur 0. **Seeds** 2, ovate to oval, 2–3 mm, margins thickened, inrolled.

Flowering (Apr–)May–Jul. Shaded slopes, mixed oak-conifer woodlands; 1000–2200 m; Calif.

Collinsia childii occurs principally on shaded, rocky soil. It is most common in the Sierra Nevada and Transverse ranges in the interior of southern California. There are also isolated populations in the South Coast Ranges.

14. Collinsia wrightii S. Watson, Proc. Amer. Acad. Arts 24: 84. 1889 • Wright's blue-eyed Mary [E]

Collinsia brachysiphon Eastwood; *C. monticola* Davidson; *C. torreyi* A. Gray var. *brachysiphon* (Eastwood) Jepson; *C. torreyi* var. *brevicarinata* Newsom; *C. torreyi* var. *wrightii* (S. Watson) I. M. Johnston

Annuals 5–25 cm. **Stems** erect. **Leaf blades** linear to elliptic, margins entire or serrate. **Inflorescences** densely glandular; nodes (1–)3–6-flowered; flowers not crowded; distalmost bracts linear, 0–2 mm. **Pedicels** ascending to spreading, sometimes reflexed, pendent and/or sigmoid in fruit, usually longer than calyx, visible. **Flowers:** calyx lobes narrowly deltate to lanceolate, equal to capsule, apex subacute to rounded; corolla blue-violet to purple, banner white, cream, or pale lilac, 4–6(–9) mm, sparsely glandular; banner length 0.7–0.8 times wings, lobe base without folds, wings surpassing keel by 1–2 mm; stamens: abaxial filaments glabrous, adaxials glabrous or hairy at base, basal spur 0. **Seeds** 2, ovate to oblong, sometimes compressed, 2–2.5 mm, margins thickened, inrolled.

Flowering May–Aug. Coniferous forests, usually in sandy, granitic soils; 800–4000 m; Calif.

A report of *Collinsia wrightii* (as *C. torreyi* var. *wrightii*) from western Nevada (J. T. Kartesz 1987) has not been verified by the author.

Collinsia wrightii is often confused with *C. parviflora* and less frequently with *C. torreyi*. The corolla of *C. wrightii* is purplish as in *C. torreyi* and not the bright blue of *C. parviflora*. *Collinsia wrightii* and *C. parviflora* differ most evidently in plant stature and corolla size; *C. wrightii* is generally more diminutive.

The large-flowered form of *Collinsia wrightii*, corolla (6–)8–9 mm, has been treated as *C. torreyi* var. *brevicarinata* and occurs only near Kings Canyon and Sequoia national parks. The flowers are similar in size to those of *C. torreyi* as treated here but differ in corolla shape (wings surpassing tip of keel by 1.5–2 mm) and leaf shape and length. The seeds of the large-flowered form are compressed; those of the small-flowered form are not.

A phylogenetic study (B. G. Baldwin et al. 2011) showed that *Collinsia wrightii* and *C. torreyi* belong to divergent lineages and supports the recognition of these taxa as distinct species based on differences in the lengths of the wings relative to the tips of the keels.

15. Collinsia torreyi A. Gray, Proc. Amer. Acad. Arts 7: 378. 1868 • Torrey's blue-eyed Mary [E]

Annuals 5–25 cm. Stems erect. Leaf blades linear, length usually 6+ times width, margins entire. Inflorescences densely glandular; nodes (1–)3–6-flowered; flowers not crowded; distalmost bracts linear, 0–2 mm. Pedicels ascending to spreading, sometimes reflexed, pendent and/or sigmoid in fruit, usually longer than calyx, visible. Flowers: calyx lobes lanceolate to ovate, equal to capsule, apex subacute to rounded; corolla blue-violet to purple, banner white, cream, or pale lilac, 6–9 mm, sparsely glandular; banner length 0.9–1 times wings, lobe base with folds bulging 0.5 mm away from throat opening at base of each lobe, wings equal to keel; stamens: abaxial filaments glabrous, adaxials glabrous or hairy at base, basal spur 0. Seeds 2, ovate to oblong, sometimes compressed, 2–3 mm, margins thickened, inrolled. $2n = 42$.

Flowering May–Aug. Mixed oak-conifer forests; 1000–3000 m; Calif., Nev.

Collinsia torreyi occurs only in the Sierra Nevada. In flower, it is superficially similar to *C. linearis*, with linear leaves and similarly colored and shaped flowers. In fruit, differences in the pedicels are distinct: S-shaped in *C. torreyi* and straight in *C. linearis*. Their ranges do not overlap.

16. Collinsia latifolia (Newsom) B. G. Baldwin, Kalisz & Armbruster, Amer. J. Bot. 98: 747. 2011 • Broad-leafed collinsia [E]

Collinsia torreyi A. Gray var. *latifolia* Newsom, Bot. Gaz. 87: 299. 1929

Annuals 5–25 cm. Stems erect to ascending. Leaf blades elliptic to ovate, length usually 2–5 times width, margins entire or serrate. Inflorescences densely glandular; nodes (1–)3–6-flowered; flowers not crowded; distalmost bracts linear, 0–2 mm. Pedicels ascending to spreading, sometimes reflexed, pendent and/or sigmoid in fruit, usually longer than calyx, visible. Flowers: calyx lobes lanceolate to ovate, equal to capsule, apex subacute to rounded; corolla blue-violet to purple, banner white, cream, or pale lilac, 6–9 mm, sparsely glandular; banner length 0.7–0.8(–0.9) times wings, lobe base with folds bulging 0.5 mm away from throat opening at base of each lobe, wings equal to keel; stamens: abaxial filaments glabrous, adaxials glabrous or hairy at base, basal spur 0. Seeds 2, ovate to oblong, often curled toward attachment side, 2–2.5 mm, margins thickened, inrolled.

Flowering Jun–Aug. Mixed oak-conifer forests, openings near montane chaparral; 1000–2500 m; Calif., Oreg.

Collinsia latifolia occurs on volcanic and metamorphic substrates in northern California and southern Oregon in the Klamath and Cascade ranges, and in the North Coast Ranges of California. Its flowers are nearly identical to those of *C. torreyi*, which differs in leaf shape: linear in *C. torreyi* and elliptic to ovate in *C. latifolia*.

17. Collinsia corymbosa Herder, Gartenflora 1868: 33, plate 568. 1868; Index Seminum (St. Petersburg) 1866: 32. 1868 • Round-headed Chinese-houses [C] [E]

Annuals 5–25 cm. Stems decumbent. Leaf blades lanceolate to ovate, margins crenate, abaxial surface glabrous, adaxial subglabrous or finely gray-hairy. Inflorescences sparsely and finely glandular; whorl 1 per branch; nodes (1–)3–10-flowered; flowers crowded; distalmost bracts ovate, 5–9 mm. Pedicels ascending to spreading, shorter than calyx, not or scarcely visible. Flowers: calyx lobes oblong to ovate, surpassing capsule, apex rounded; corolla usually whitish, 14–22 mm, wings sparsely and finely glandular, not hairy; banner length 0.1–0.3(–0.4) times wings, lobe base without folds, reflexed portion 1 mm, shorter than basal portion, brownish, not red-banded; stamens: filaments hairy, basal spur 0. Seeds 8–16, oblong to oval, 2–2.5 mm, margins thickened, inrolled. $2n = 14$.

Flowering Apr–Jun. Coastal sand dunes; of conservation concern; 0–20 m; Calif.

Collinsia corymbosa is known from Mendocino County. Specimens from other sites identified as *C. corymbosa* are *C. bartsiifolia* var. *hirsuta*. A phylogenetic study using DNA showed evidence of a close relationship between *C. corymbosa* and *C. bartsiifolia* (B. G. Baldwin et al. 2011).

18. Collinsia bartsiifolia Bentham in A. P. de Candolle and A. L. P. P. de Candolle, Prodr. 10: 318. 1846 (as bartsiaefolia) • White collinsia E

Annuals 5–35 cm. **Stems** erect to ascending or decumbent. **Leaf blades** lanceolate-oblong to oblong, 1–4 cm, margins crenate. **Inflorescences** sparsely and obscurely glandular or shaggy-hairy or densely glandular; whorls (1 or)2–7 on main axis, sometimes fewer on branches; nodes (2 or)3–6-flowered; flowers crowded; distalmost bracts linear, 3–5(+) mm. **Pedicels** ascending to spreading, shorter than calyx, not or scarcely visible. **Flowers:** calyx lobes oblong to ovate, equal to capsule, apex rounded; corolla whitish to pinkish, banner lobes and wings rarely purple-tipped and/or with purple spots at base, 8–20 mm, glabrous or glandular; banner length 0.5–0.9+ times wings, lobe base without folds, tube longer than diam., hairy inside; banner lobes ± oblong, toothed, backs often touching or parallel, wings ± oblong to obovate, entire or notched; stamens: abaxial filaments glabrous or hairy, adaxials hairy, basal spur 0(or 1). **Seeds** 8–16, oblong to ovate, sometimes prismatic or hourglass-shaped, 0.8–1.5 mm, margins thickened, inrolled.

Varieties 4 (4 in the flora): California.

1. Corollas 13–20 mm.
  2. Banner lengths 0.7–0.9+ times wings; plants (10–)20–35 cm. . . . .
       . . . . 18a. *Collinsia bartsiifolia* var. *bartsiifolia*
  2. Banner lengths 0.5–0.7 times wings; plants 5–20 cm. . . . . . 18d. *Collinsia bartsiifolia* var. *hirsuta*
1. Corollas 8–14 mm.
  3. Corollas glabrous, usually white to pale lavender, wings often pinkish, obovate, notably notched; inflorescences not or scarcely glandular. . . 18b. *Collinsia bartsiifolia* var. *davidsonii*
  3. Corollas glandular, pink to pinkish lavender, wings usually purple-tipped with purple spots at base, oblong to oblanceolate, entire or shallowly notched; inflorescences usually notably densely glandular . . . .
       . . . . .18c. *Collinsia bartsiifolia* var. *stricta*

18a. Collinsia bartsiifolia Bentham var. **bartsiifolia** E

Plants (10–)20–35 cm. **Inflorescences** eglandular. **Flowers:** corolla usually white to pale lavender, 15–20 mm, glabrous; banner length 0.7–0.9+ times wings, wings obovate, notched. **2***n* = 14.

Flowering Mar–Jun. Open, sandy places; 0–600 m; Calif.

Variety *bartsiifolia* seems to be in decline due to habitat loss in the Central Valley. Coastal populations are on sand dunes and are identical to those of the interior, which are found in sandy places but not on dunes. There are few modern records in either region. Some populations are, in part, threatened by off-road vehicle activity.

18b. Collinsia bartsiifolia Bentham var. **davidsonii** (Parish) Newsom, Bot. Gaz. 87: 272. 1929 (as bartsiaefolia) E

Collinsia davidsonii Parish, Zoë 4: 147. 1893

Plants 5–20(–25) cm. **Inflorescences** not or scarcely glandular. **Flowers:** corolla usually white to pale lavender, wings and keel often pinkish, 9–14 mm, glabrous; banner length 0.7–0.9+ times wings, wings obovate, notched. **2***n* = 14.

Flowering Apr–Jun. Open, sandy places, rarely on serpentine soils; 500–1300 m; Calif.

Variety *davidsonii* is known from drier regions east of the South Coast and Transverse ranges. Some collections suggest a relationship between precipitation and corolla size; others are of intermediate size and difficult to separate from var. *bartsiifolia*. The two varieties, in general, are not sympatric.

Some populations of var. *davidsonii* near New Idria and San Benito Mountain in San Benito County occur on serpentine soil. In those populations, banners are slightly reduced relative to those of plants in nearby populations found in sandy habitats.

18c. Collinsia bartsiifolia Bentham var. **stricta** (Greene) Newsom, Bot. Gaz. 87: 273. 1929 (as bartsiaefolia) E

Collinsia stricta Greene, Pittonia 2: 23. 1889

Plants 12–35 cm. **Inflorescences** usually notably densely glandular. **Flowers:** corolla pink to pinkish lavender, banner lobes and wings usually purple-tipped with purple spots at base, 8–12 mm, glandular; banner length 0.6–0.8 times wings, wings oblong to oblanceolate, shallowly notched or entire. **2***n* = 14.

Flowering May–Jul. Chaparral, open oak and dry pine-oak woodlands; 700–1300 m; Calif.

There have been no collections or reports of var. *stricta* since 1975 (*Heckard, Chuang & Baciagalupi 4054*, JEPS); it may be extinct. The few specimens of this variety are from the west slope of the Sierra Nevada, either in the foothills or the lower fringe of *Pinus ponderosa* woodland. Compared to other varieties, var. *stricta* is more glandular and the corolla is more darkly hued, especially distally. Purplish coloration is relatively rare in other varieties. The throat is slightly elongated relative to the other varieties.

18d. Collinsia bartsiifolia Bentham var. **hirsuta** (Kellogg) Pennell in L. Abrams and R. S. Ferris, Ill. Fl. Pacific States 3: 776. 1951 (as bartsiaefolia) [E]

Collinsia hirsuta Kellogg, Proc. Calif. Acad. Sci. 2: 110, fig. 34. 1863

Plants 5–20 cm. **Inflorescences** eglandular. **Flowers:** corolla usually white to pale lavender, 13–18 mm, eglandular; banner length 0.5–0.6(–0.7) times wings. *2n* = 14.

Flowering Apr–Jun. Sand dunes, sandy sites, rarely rocky slopes; 0–100 m; Calif.

Variety *hirsuta* is presumed extinct on the San Francisco Peninsula. It is known from a single collection at Point Reyes (Marin County) in 1906.

The type collection of *Collinsia hirsuta* is unknown and presumed to have been lost in 1906. The depiction of corollas is inconsistent in the illustration accompanying the original description. In the detail of a single corolla, slight reduction of the banner is depicted, but in the habit, there is near equivalence in length between the banner and wings. The original publication states that the description was based on plants collected within the vicinity of the meeting (at the California Academy of Sciences in San Francisco). Most known collections are from western San Francisco; these conform to the variety as described here.

Some treatments have classified plants of var. *hirsuta* (as treated here) as *Collinsia corymbosa* (V. M. Newsom 1929, cited as transitional to *C. bartsiifolia* var. *bartsiifolia* by E. C. Neese 1993b). The reduction of the corolla banner in var. *hirsuta* is not extreme like that of *C. corymbosa*. Plants of the former have vegetative characteristics and the inflorescence structure of *C. bartsiifolia* var. *bartsiifolia*.

19. Collinsia tinctoria Hartweg ex Bentham, Pl. Hartw., 328. 1849 • Tincture plant [E]

Annuals 20–60 cm. **Stems** ascending. **Leaf blades** usually lanceolate-deltate, margins entire or serrate, abaxial surface densely hairy, adaxial glabrous or glabrate, usually strongly whitish-mottled. **Inflorescences** glandular; whorls 2–8 on main axis, fewer on lateral branches; nodes 3–10-flowered; flowers crowded; distalmost bracts linear, 5–6 mm. **Pedicels** ascending to spreading, shorter than calyx, not or scarcely visible. **Flowers:**

calyx lobes linear to lanceolate, surpassing capsule, apex subacute to rounded; corolla white to yellowish or pale lavender, rarely purple, 12–20 mm, wings usually long-hairy on adaxial surface, keel glandular, hairy; adaxial pouch projecting 2–4 mm from tube base; banner length 0.4–0.5 times wings, lobe base without folds, reflexed portion 2–3.5 mm, equal to basal portion, red-banded; wings usually red-dotted; stamens: abaxial filaments glabrous, rarely hairy, adaxials hairy, basal spur 0 or 1. **Seeds** 4–8, oval, flattened, 2–2.5 mm, margins thin, not inrolled. *2n* = 14.

Flowering May–Aug. Rocky habitats, openings in dry mixed pine-oak woodlands or coniferous forests; 100–2500 m; Calif.

Collinsia tinctoria is often on scree slopes and in full sun. Most populations are on the western Sierra Nevada and the Cascade Ranges. There are a few populations in the southern North Coast Ranges (Sonoma County) and at Mount Diablo (Contra Costa County). The reduction of the banner and the rearward projection of the pouch beyond the calyx are two floral traits that distinguish it. The mottled leaves of *C. tinctoria* are unique in *Collinsia*.

20. Collinsia heterophylla Graham, Bot. Mag. 65: plate 3695. 1838 • Purple Chinese-houses

Annuals 10–50 cm. **Stems** erect to ascending. **Leaf blades** lanceolate-deltate, margins serrate. **Inflorescences** glabrous or hairy, ± glandular; whorls 2–7 on main axis; nodes 2–7-flowered; flowers crowded; distalmost bracts linear to lanceolate, 5–6 mm. **Pedicels** ascending to spreading, shorter than calyx, not or scarcely visible. **Flowers:** calyx lobes linear to ovate, equal to capsule, apex subacute to acute; corolla red-purple, rarely white, banner white to lavender or tipped dark violet, maroon spots near center and forming horizontal lines near base, wings whitish to rose purple, keel usually with darker red tip, 10–20 mm, usually glabrous; tube hairy inside, as wide as long, saccate basally, adaxial pouch prominent and ± square; banner length 0.6–0.9 times wings, lobe base without folds; stamens: abaxial filaments glabrous, adaxials hairy, basal spur 1. **Seeds** 6–12, ovate, 1.5–2 mm, margins slightly thickened, inrolled. *2n* = 14.

Varieties 2 (2 in the flora): California, nw Mexico.

Collinsia heterophylla is the most widespread and frequently encountered species in California. There is structure in phylogenetic analysis of northern and southern populations (that is, paraphyly of southern populations) suggesting northward expansion and evidence for introgression with *C. tinctoria* (B. G. Baldwin et al. 2011) in the southern Sierra Nevada

(Kern County and possibly Fresno County). This species is often confused with other collinsias that have tiered whorls of flowers, most frequently with *C. bartsiifolia* and *C. concolor*, which both lack the bold horizontal line present at the base of the adaxial corolla lip of *C. heterophylla*. Plants with reduced corolla pigmentation, nearly white, are uncommon outside of the South Coast Ranges and are often misidentified as *C. bartsiifolia. Collinsia bicolor* Bentham, which pertains here, is a later homonym of *C. bicolor* Rafinesque, a synonym of 1. *C. verna*.

1. Corollas (13–)15–20 mm; leaf blade abaxial surfaces glabrous, midveins rarely hairy; banner lengths 0.8–0.9 times wings, reflexed portions 4+ times basal portions . . . .
. . . . 20a. *Collinsia heterophylla* var. *heterophylla*
1. Corollas 10–15(–18) mm; leaf blade abaxial surfaces sparsely hairy; banner lengths 0.6–0.8 times wings, reflexed portions 2–3 times basal portions . . . .
. . . . 20b. *Collinsia heterophylla* var. *austromontana*

20a. Collinsia heterophylla Graham var. **heterophylla**

Leaf blades: abaxial surface glabrous, midvein rarely hairy. **Flowers:** corolla (13–)15–20 mm, banner length 0.8–0.9 times wings, reflexed portion 4+ times basal portion.

Flowering Mar–Jun. Chaparral, open mixed woodlands, oak woodlands; 0–1300 m; Calif.; Mexico (Baja California).

Variety *heterophylla* is frequently grown in gardens and appears to be capable of escaping. Reports of *Collinsia heterophylla* from the eastern United States may be from such escapes as waifs.

20b. Collinsia heterophylla Graham var. **austromontana** (Newsom) Munz, Man. S. Calif. Bot., 464, 600. 1935 • Downy Chinese-houses E

Collinsia bicolor Rafinesque var. *austromontana* Newsom, Bot. Gaz. 87: 277. 1929; *C. austromontana* (Newsom) Pennell

Leaf blades: abaxial surface sparsely hairy. **Flowers:** corolla 10–15(–18) mm, banner length 0.6–0.8 times wings, reflexed portion 2–3 times basal portion.

Flowering May–Jul(–Aug). Chaparral, mixed woodlands; 300–1700 m; Calif.

Variety *austromontana* is narrowly distributed in the Transverse and northern Peninsular ranges of southern California; its range does not overlap with that of var. *heterophylla*. The defining characteristics, reduction of the banner and hairiness of the abaxial leaf surface, vary. Some plants (with banner size near the limit for var. *austromontana* and limited leaf hairiness) might be misidentified as var. *heterophylla*.

21. Collinsia concolor Greene, Erythea 3: 49. 1895
• Chinese-houses F

Annuals 15–40 cm. **Stems** erect to ascending. **Leaf blades** narrowly oblong to widely lanceolate, margins entire, rarely serrate. **Inflorescences** finely hairy to shaggy, usually finely glandular; whorls 2–5 on main axis, fewer on lateral branches; nodes (2 or)3–7-flowered; flowers crowded; distalmost bracts linear, 5–6 mm. **Pedicels** ascending to spreading, shorter than calyx, not or scarcely visible. **Flowers:** calyx lobes oblong to lanceolate, surpassing capsule, apex subacute to acute; corolla tinged bluish, distally violet to magenta, banner purple-dotted near base, 11–15 mm, banner usually sparsely hairy, keel usually sparsely hairy at tip; banner length 0.7–0.9 times wings, lobe base without folds, tube longer than diam., hairy inside, lobes obovate, notched; stamens: abaxial filaments glabrous, rarely hairy, adaxials hairy, basal spur 0(or 1). **Seeds** 8–16, round to oval, flattened, 1.5–2 mm, margins slightly thickened, inrolled. $2n = 14$.

Flowering Apr–Jun. Openings and margins of chaparral, oak or pinyon-juniper woodlands; 300–1700 m; Calif.; Mexico (Baja California).

Collinsia concolor grows in dry habitats of the Peninsular Ranges. Flowers in tiers of dense whorls with bluish-tinged (or blue-blotched) corollas, a triangular region of purple spots in the adaxial lip, and lack of curved appendages at the bases of the filaments are unique to it. It is frequently confused with *C. heterophylla* because of morphological similarity and range overlap.

24. KECKIELLA Straw, Brittonia 19: 203. 1967 • [For David Daniels Keck, 1903–1995, California botanist, and *ella*, honor]

David C. Michener

Noel H. Holmgren

Lepidostemon Lemaire, Ill. Hort. 9: plate 315. 1862, not Hooker f. & Thomson 1861 [Brassicaceae]

Subshrubs or shrubs. Stems erect to spreading, sometimes climbing, glabrous or hairy. **Leaves** drought-deciduous, cauline, opposite, subopposite, or whorled; petiole present or absent; blade fleshy or not, leathery or not, margins entire or toothed, sometimes relatively small, distal leaf blade not needlelike or scalelike. **Inflorescences** terminal, panicles, thyrses, corymbs, or spikelike racemes; bracts present. **Pedicels** present; bracteoles present, sometimes much reduced. **Flowers** bisexual; sepals 5, basally connate, calyx weakly bilaterally symmetric, short-tubular, lobes lanceolate to ovate, sometimes oblanceolate; corolla red, reddish orange, pink, white, cream, yellow, brownish yellow, or purplish brown, sometimes with red-purple nectar guides, bilaterally symmetric, strongly bilabiate, ± tubular, tube base not spurred or gibbous, lobes 5, abaxial 3, adaxial 2, adaxial lip hooded; stamens 4, basally adnate to corolla, didynamous, filaments basally eglandular-hairy, pollen sacs explanate; staminode 1, filamentous; nectary a hypogynous disc; ovary 2-locular, placentation axile; stigma capitate. **Fruits** capsules, dehiscence septicidal, sometimes also loculicidal distally, not densely packed with white membranous hairs. **Seeds** 20–100, brown, ovoid, wings essentially absent. $x = 8$.

Species 7 (7 in the flora): w United States, nw Mexico.

Keckiella is nearly endemic to the California Floristic Province (P. H. Raven and D. I. Axelrod 1978). In the flora area, taxa range beyond California into Arizona and Nevada; two occur in Mexico (Baja California Peninsula).

All taxa of *Keckiella* were first described in *Penstemon*, where they were long recognized as related to each other and anomalous within *Penstemon*. *Keckiella* taxa were treated as *Keckia* Straw before the prior use of that name for a fossil alga was noticed. The relationship of *Keckiella* and *Penstemon* with the rest of Cheloneae was clarified by A. D. Wolfe et al. (2002). This confirmed the classification implied by D. D. Keck (1936) and proposed by R. M. Straw (1966). The two genera are now recognized as distinct evolutionary lineages based on morphology, rust fungi relationships, phytochemistry, and DNA evidence (D. B. O. Savile 1968b, 1979; D. C. Michener 1982; O. Mistretta and R. Scogin 1989; Wolfe et al.; C. E. Freeman et al. 2003).

Among *Keckiella* species, gross morphological and anatomical characteristics of leaf and wood correlate with habitat; floral characters, other than the distinctive nectary structure that helps define *Keckiella*, reflect the pollination syndrome (D. C. Michener 1981, 1982). Leaf morphology is typical of drought-deciduous plants (Michener 1981). Varieties are recognized in part on foliage and calyx indument. Leaf primordia of all species are (weakly) pubescent with glandular and non-glandular hairs, making indument a challenging character for interpretation.

Pollination biology and habitat tolerances appear to have been important in species diversification. Bee pollination is fundamental in *Keckiella*, with floral size among the species scaled to a range of native bees; hummingbird pollination is polyphyletic in parallel with *Penstemon* (D. C. Michener 1982; C. E. Freeman et al. 2003; Paul Wilson et al. 2007). Natural hybrids have been found repeatedly for two species pairs (*K. antirrhinoides* var. *antirrhinoides* and *K. cordifolia*; *K. breviflora* var. *glabrisepala* and *K. lemmonii*) where the ranges overlap. The pollination syndromes are not effective barriers to extraneous cross pollination: even the hummingbird-pollinated species are visited by small pollen-foraging bees that may promote some pollination or hybridization separate from the hummingbirds (Michener). Leaf flavonoids do not support the hummingbird-pollinated species being each other's closest relatives (O. Mistretta and R. Scogin 1989).

Species of *Keckiella* are widespread but only locally common in their habitat. They usually occur in unstable sites such as slopes, scree, and crevices in rocky faces from near sea level to 3000 m. These local micro-sites may be less fire prone than the immediately surrounding communities. Fire ecology may be important in the evolution of the species; root gnarls found only in *K. antirrhinoides* var. *antirrhinoides* and *K. cordifolia* may help these species resprout following the fires typical of their low-elevation habitats (D. C. Michener 1981).

The eco-evolutionary segregation of *Keckiella* from its relatives may have been centered on the Klamath region during the Tertiary (A. D. Wolfe et al. 2002). C. E. Freeman et al. (2003) recognized three clades within *Keckiella*: the basal-most *K. rothrockii* lineage, a northern clade (*K. breviflora*, *K. corymbosa*, *K. lemmonii*, and *K. ternata*), ranging from southern to northern California and through almost the entire 3000 m elevational range of the genus, and a southern clade (*K. antirrhinoides* and *K. cordifolia*) nearly restricted to lower elevations in southern California and Baja California, Mexico. They argued that speciation in these clades likely reflects repeated regional and elevational migration during the Pleistocene, presumably including the ecological scenarios that led to the modern species in fire-dominated vegetation types.

SELECTED REFERENCES Freeman, C. E. et al. 2003. Inferred phylogeny in *Keckiella* (Scrophulariaceae) based on noncoding chloroplast and nuclear ribosomal DNA sequences. Syst. Bot. 28: 782–790. Michener, D. C. 1982. Studies on the Evolution of *Keckiella* (Scrophulariaceae). Ph.D. dissertation. Claremont Graduate School. Mistretta, O. and R. Scogin. 1989. Foliar flavonoid aglycones of the genus *Keckiella* (Scrophulariaceae). Biochem. Syst. & Ecol. 17: 455–457.

1. Corollas pink, red, or reddish orange, tubes plus indistinct throats 16–25 mm.
    2. Leaves whorled (3s), sometimes opposite; calyces 3.8–7.2 mm; corolla adaxial lips 5.5–9.5 mm. . . . . 5. *Keckiella ternata*
    2. Leaves opposite; calyces 6.4–13 mm; corolla adaxial lips 9–21 mm.
        3. Leaf bases rounded, truncate, or cordate; pollen sacs 1.1–1.5 mm . . . . 3. *Keckiella cordifolia*
        3. Leaf bases wedge-shaped; pollen sacs 0.9–1 mm. . . . . 7. *Keckiella corymbosa*
1. Corollas white, cream, yellow, brownish yellow, or purplish brown, tubes plus distinct throats 4–11 mm.
    4. Corollas 10–15 mm, tubes plus throats longer than adaxial lips, adaxial lips 2.6–6 mm.
        5. Inflorescences spikelike racemes; staminodes glabrous, usually included; stems densely short-hairy when young, not glaucous . . . . 1. *Keckiella rothrockii*
        5. Inflorescences panicles; staminodes densely yellow-hairy, exserted; stems glabrous when young, glaucous. . . . . .6. *Keckiella lemmonii*
    4. Corollas (12–)15–23 mm, tubes plus throats shorter than adaxial lips, adaxial lips 8–15 mm.
        6. Staminodes densely yellow-hairy; corollas yellow; pollen sacs 1.1–1.8 mm. . . . . . .
            . . . . 2. *Keckiella antirrhinoides*
        6. Staminodes glabrous; corollas white or cream, lobes sometimes rose tinged; pollen sacs 0.6–0.8 mm. . . . . 4. *Keckiella breviflora*

1. Keckiella rothrockii (A. Gray) Straw, Brittonia 19: 203. 1967 • Rothrock's keckiella E

Penstemon rothrockii A. Gray in A. Gray et al., Syn. Fl. N. Amer. 2(1): 260. 1878 (as Pentstemon)

Stems erect, 3–6 dm, densely short-hairy when young, not glaucous. **Leaves** subopposite or whorled (3s), subsessile; blade oblanceolate or lanceolate to widely obovate, 5–16 mm, margins entire or serrulate, sometimes with 3–7 undulations instead of teeth. **Inflorescences** spikelike racemes, short-hairy. **Flowers:** calyx 4–7 mm, lobes lanceolate; corolla brownish yellow, pale yellow, or cream, purple- or reddish brown-lined, 10–15 mm, tube plus distinct throat 7–11 mm, longer than adaxial lip, adaxial lip 4–5 mm; pollen sacs 0.8–1.1 mm; staminode glabrous, usually included.

Varieties 2 (2 in the flora): w United States.

1. Corollas 10–12 mm, glabrescent; leaves canescent . . . . . . 1a. *Keckiella rothrockii* var. *rothrockii*
1. Corollas 13–15 mm, sparsely hairy; leaves glabrescent . . . . . 1b. *Keckiella rothrockii* var. *jacintensis*

1a. Keckiella rothrockii (A. Gray) Straw var. **rothrockii** E

Leaves canescent. **Corollas** 10–12 mm, glabrescent.

Flowering Jun–Aug. Sagebrush steppes, juniper-pinyon woodlands; 1900–3200 m; Calif., Nev.

Variety *rothrockii* is known from the southern high Sierra Nevada into adjacent southwestern Great Basin ranges of central California and the Transverse Ranges of southern California.

1b. Keckiella rothrockii (A. Gray) Straw var. **jacintensis** (Abrams) N. H. Holmgren, Brittonia 44: 481. 199 E

Penstemon jacintensis Abrams, Bull. Torrey Bot. Club 33: 445. 1906 (as Pentstemon); *Keckiella rothrockii* subsp. *jacintensis* (Abrams) Straw; *P. rothrockii* A. Gray subsp. *jacintensis* (Abrams) D. D. Keck

Leaves glabrescent. **Corollas** 13–15 mm, sparsely hairy.

Flowering Jun–Aug. Sagebrush steppes, juniper-pinyon woodlands; 1900–3200 m; Calif.

Variety *jacintensis* is geographically isolated in the San Jacinto Mountains in Riverside County.

2. Keckiella antirrhinoides (Bentham ex A. de Candolle) Straw, Brittonia 19: 203. 1967 • Snapdragon penstemon F

Penstemon antirrhinoides Bentham ex A. de Candolle in A. P. de Candolle and A. L. P. P. de Candolle, Prodr. 10: 594. 1846 (as Pentstemon)

Stems spreading to erect, 6–25 dm, canescent, rarely glabrous when young. **Leaves** opposite, on short shoots as axillary clusters on older stems; blade oblanceolate or lanceolate to narrowly obovate or ovate, 5–20 mm, margins entire or 2-toothed, sometimes relatively small. **Inflorescences** panicles, finely short-hairy and sparsely glandular. **Flowers:** calyx 3–9 mm, lobes lanceolate to ovate; corolla yellow drying blackish, 15–23 mm, tube plus distinct throat 6–10 mm, shorter than adaxial lip, adaxial lip 8–15 mm; pollen sacs 1.1–1.8 mm; staminode densely yellow-hairy, exserted.

Varieties 2 (2 in the flora): sw United States, nw Mexico.

1. Calyces 3–6 mm, lobes widely ovate, apices obtuse to acute; plants finely hairy; pollen sacs 1.4–1.8 mm . . . . 2a. *Keckiella antirrhinoides* var. *antirrhinoides*
1. Calyces 5.5–9 mm, lobes lanceolate, apices acute to acuminate; plants canescent; pollen sacs 1.1–1.5 mm . . . . 2b. *Keckiella antirrhinoides* var. *microphylla*

2a. Keckiella antirrhinoides (Bentham ex A. de Candolle) Straw var. **antirrhinoides** F

Plants finely hairy. **Flowers:** calyx 3–6 mm, lobes widely ovate, apex obtuse to acute; pollen sacs 1.4–1.8 mm.

Flowering Apr–May. Chaparral, oak forests; 100–1600 m; Calif.; Mexico (Baja California).

Variety *antirrhinoides* is found primarily in mountains of southern California away from the immediate coast in Riverside County.

2b. Keckiella antirrhinoides (Bentham ex A. de Candolle) Straw var. **microphylla** (A. Gray) N. H. Holmgren, Brittonia 44: 481. 1992

Penstemon microphyllus A. Gray in War Department [U.S.], Pacif. Railr. Rep. 4(5): 119. 1857 (as Pentstemon); *Keckiella antirrhinoides* subsp. *microphylla* (A. Gray) Straw; *P. antirrhinoides* Bentham subsp. *microphyllus* (A. Gray) D. D. Keck

Plants canescent. **Flowers:** calyx 5.5–9 mm, lobes lanceolate, apex acute to acuminate; pollen sacs 1.1–1.5 mm.

Flowering Apr–Jun. Juniper-pinyon woodlands, Joshua-tree scrub; 400–1800 m; Ariz., Calif., Nev.; Mexico (Baja California).

In the flora area, var. *microphylla* is found mainly in the interior Peninsular Ranges and the Desert Province of southern California.

3. Keckiella cordifolia (Bentham) Straw, Brittonia 19: 203. 1967 • Heartleaf keckiella [F]

Penstemon cordifolius Bentham, Scroph. Ind., 7. 1835 (as Pentstemon cordifolium)

Stems climbing, 10–30 dm, glabrous or short-hairy when young. **Leaves** opposite; blade ovate, 20–65 mm, base rounded, truncate, or cordate, margins usually 3–11-toothed. **Inflorescences** panicles, glandular and hairy. **Flowers:** calyx 7–13 mm, lobes lanceolate; corolla red to reddish orange, 31–43 mm, tube plus indistinct throat 18–25 mm, adaxial lip 11–21 mm; pollen sacs 1.1–1.5 mm; staminode densely yellow-hairy, included. $2n = 16$.

Flowering May–Jul. Chaparral, forests; 0–1500 m; Calif.; Mexico (Baja California).

In the flora area, *Keckiella cordifolia* occurs primarily in the South Coast Ranges, the San Bernardino Mountains, the Peninsular Ranges, and the Channel Islands. It grows in wooded, riparian habitats and is the most mesic of keckiellas.

Individuals with yellow corollas are presumed to be mutants in which the red anthocyanins are not expressed. Individual plants may develop as weakly sprawling to climbing shrubs. The relatively long stems are not self-supporting.

4. Keckiella breviflora (Lindley) Straw, Brittonia 19: 203. 1967 • Bush beardtongue [E]

Penstemon breviflorus Lindley, Edwards's Bot. Reg. 23: plate 1946. 1837 (as Pentstemon)

Stems erect, spreading with age, 5–20 dm, glabrous when young, glaucous. **Leaves** opposite, subsessile; blade oblanceolate or lanceolate, 10–40 mm, margins 4–12-toothed. **Inflorescences** thyrses, glabrous or sticky-hairy. **Flowers:** calyx 4–8 mm, lobes lanceolate to ovate; corolla white or cream, lobes sometimes rose tinged, lined purplish or pinkish, 12–18 mm, tube plus distinct throat 4–8 mm, shorter than adaxial lip, adaxial lip 8–12 mm; pollen sacs 0.6–0.8 mm; staminode glabrous, slightly exserted.

Varieties 2 (2 in the flora): w United States.

1. Calyces glandular . . . . .4a. *Keckiella breviflora* var. *breviflora*
1. Calyces glabrous . . . . .4b. *Keckiella breviflora* var. *glabrisepala*

4a. Keckiella breviflora (Lindley) Straw var. **breviflora** [E]

Calyces glandular. $2n = 16$.

Flowering May–Jul. Rocky slopes, forests, chaparral; 30–2700 m; Calif., Nev.

4b. Keckiella breviflora (Lindley) Straw var. **glabrisepala** (D. D. Keck) N. H. Holmgren in A. Cronquist et al., Intermount. Fl. 4: 369. 1984 [E]

Penstemon breviflorus Lindley subsp. *glabrisepalus* D. D. Keck, Madroño 3: 207. 1936; *Keckiella breviflora* subsp. *glabrisepala* (D. D. Keck) Straw

Calyces glabrous.

Flowering May–Jul. Rocky slopes, forests, chaparral; 100–2700 m; Calif., Nev.

Putative hybrids between var. *glabrisepala* and *Keckiella lemmonii* were first reported by D. D. Keck (1936) based on staminode indument. The same characteristics are found in this variety in Alpine County, California, outside the current range of *K. lemmonii*, suggesting potential widespread introgression (D. C. Michener 1982).

An 1887 collection by E. Palmer (*71*, Los Angeles Bay, Gulf of California, Lower California, NY) seems misplaced as to location.

5. **Keckiella ternata** (Torrey ex A. Gray) Straw, Brittonia 19: 204. 1967 • Scarlet keckiella

Penstemon ternatus Torrey ex A. Gray in W. H. Emory, Rep. U.S. Mex. Bound. 2(1): 115. 1859 (as Pentstemon)

Stems spreading to erect, 5–25 dm, glabrous when young, glaucous. **Leaves** whorled (3s), sometimes opposite; blade linear to narrowly oblanceolate or lanceolate, 15–60 mm, margins 4–10-toothed. **Inflorescences** panicles, glabrous or glandular-hairy. **Flowers:** calyx 3.8–7.2 mm, lobes lanceolate to ovate; corolla red, 21–31 mm, tube plus indistinct throat 16–24 mm, adaxial lip 5.5–9.5 mm; pollen sacs 0.8–1.1 mm; staminode densely yellow-hairy, included.

Varieties 2 (2 in the flora): California, nw Mexico.

1. Calyces glabrous, 4.5–7.2 mm ....
.................5a. *Keckiella ternata* var. *ternata*
1. Calyces glandular, 3.8–5.5 mm ....
...........5b. *Keckiella ternata* var. *septentrionalis*

5a. **Keckiella ternata** (Torrey ex A. Gray) Straw var. **ternata**

Calyces 4.5–7.2 mm, glabrous. **2n** = 16.

Flowering Jun–Sep. Juniper-pinyon woodlands, chaparral, forests; 300–2700 m; Calif.; Mexico (Baja California).

In the flora area, var. *ternata* extends from the San Gabriel Mountains through the Peninsular Ranges and San Jacinto Mountains in southern California.

5b. **Keckiella ternata** (Torrey ex A. Gray) Straw var. **septentrionalis** (Munz & I. M. Johnston) N. H. Holmgren, Brittonia 44: 481. 1992 E

Penstemon ternatus Torrey ex A. Gray var. *septentrionalis* Munz & I. M. Johnston, Bull. S. Calif. Acad. Sci. 23: 28. 1924; *Keckiella ternata* subsp. *septentrionalis* (Munz & I. M. Johnston) Straw; *P. ternatus* subsp. *septentrionalis* (Munz & I. M. Johnston) D. D. Keck

Calyces 3.8–5.5 mm, glandular.

Flowering Jun–Sep. Mixed-hardwood forests, chaparral; 900–1900 m; Calif.

Variety *septentrionalis* is essentially restricted to the Tehachapi and Transverse ranges of central and southern California.

Variety *septentrionalis* is regionally sympatric with the relatively smaller-flowered and pubescent *Keckiella breviflora* var. *breviflora*. D. C. Michener (1982) proposed that *K. breviflora* var. *breviflora* may represent complex introgression between the two species.

6. **Keckiella lemmonii** (A. Gray) Straw, Brittonia 19: 203. 1967 • Lemmon's keckiella E

Penstemon lemmonii A. Gray in W. H. Brewer et al., Bot. California 1: 557. 1876 (as Pentstemon lemmoni)

Stems erect, virgate with age, 5–15 dm, glabrous when young, glaucous. **Leaves** opposite; blade lanceolate to ovate, 10–65 mm, margins usually 2–12-toothed. **Inflorescences** panicles, glandular-hairy. **Flowers:** calyx 3.5–6.7 mm, lobes lanceolate; corolla purplish brown, abaxial lip pale yellow, brown-purple lined, 11–15 mm, tube plus distinct throat 5.5–9.5 mm, longer than adaxial lip, adaxial lip 2.6–6 mm; pollen sacs 0.6–0.9 mm; staminode densely yellow-hairy, exserted. **2n** = 16.

Flowering Jun–Aug. Rocky slopes, coniferous and mixed forests, chaparral; 40–1900 m; Calif., Nev.

Keckiella lemmonii is essentially restricted to the mountains of northern California from the North Coast Ranges through the Klamath Mountains and the northern high Sierra Nevada.

7. **Keckiella corymbosa** (Bentham ex A. de Candolle) Straw, Brittonia 19: 203. 1967 • Redwood keckiella E

Penstemon corymbosus Bentham ex A. de Candolle in A. P. de Candolle and A. L. P. P. de Candolle, Prodr. 10: 593. 1846 (as Pentstemon)

Stems spreading to erect, 3–6 dm, glabrous or hairy when young. **Leaves** opposite; blade oblanceolate or lanceolate to narrowly obovate or ovate, 10–35 mm, base wedge-shaped, margins entire or 3–5-toothed. **Inflorescences** corymbs, glandular-hairy and densely coarse-hairy. **Flowers:** calyx 6.4–11 mm, lobes lanceolate, sometimes ovate to oblanceolate; corolla pink to red, 22–40 mm, tube plus indistinct throat 17–22 mm, adaxial lip 9–15 mm; pollen sacs 0.9–1 mm; staminode densely yellow-hairy, included. **2n** = 16.

Flowering Jun–Aug(–Oct). Rocky slopes, coniferous or hardwood forests, chaparral; 40–2000 m; Calif.

Keckiella corymbosa occurs in the coastal ranges from the Central Coast to the North Coast regions of California. D. D. Keck (1936) argued that indument variation has no coherent pattern within *K. corymbosa*. D. C. Michener (1982) proposed that widespread introgression involving multiple species might be responsible.

25. NOTHOCHELONE (A. Gray) Straw, Brittonia 18: 85. 1966 • Woodland beardtongue [Greek *notho-*, spurious, and generic name *Chelone*] E

Allan D. Nelson

Chelone Linnaeus subg. *Nothochelone* A. Gray in A. Gray et al., Syn. Fl. N. Amer. 2(1): 259. 1878

Herbs, perennial; caudex unbranched, woody. **Stems** arching or reclining, glabrous or puberulent. **Leaves** cauline, opposite; petiole absent or present; blade not fleshy, not leathery, margins subentire to serrate or dentate. **Inflorescences** axillary or terminal, cymes; bracts present. **Pedicels** present; bracteoles smaller than calyx lobes, not surrounding calyx of flower they subtend. **Flowers** bisexual; sepals 5, connate proximally, distinct nearly to base, calyx radially symmetric, campanulate, lobes ovate to lanceolate; corolla pink, pinkish to bluish purple or maroon-red, often paler abaxially, bilaterally symmetric, bilabiate, tubular, tube base not spurred or gibbous, throat not densely pilose internally, lobes 5, abaxial 3, adaxial 2; stamens 4, adnate to corolla, didynamous, filaments hirsute proximally, puberulent distally; staminode 1, straplike; nectary a hypogynous disc; ovary 2-locular, placentation axile; stigma capitate. **Fruits** capsules, dehiscence septicidal. **Seeds** ca. 50, tan to brown, flattened, wings present. $x = 15$.

Species 1: w North America.

Nothochelone is a member of Cheloneae, which in North America includes *Chelone*, *Chionophila*, *Collinsia*, *Keckiella*, *Penstemon*, and *Tonella*. These genera share cymose inflorescences, presence of a staminode, simple hairs, and stems that contain pith (A. D. Wolfe et al. 2002). In phylogenetic analyses using molecular (Wolfe et al.) and morphological (A. D. Nelson 1995) data, *N. nemorosa* occurs in a clade with species of *Chelone*. *Nothochelone* has a more branched cyme with relatively small bracteoles, abaxially ridged corollas, circular pollen amb (grain shape from polar view), reticulate pollen exine sculpting, and asymmetric seed wings; *Chelone* has smaller cymes with relatively large bracteoles, adaxially ridged corollas, triangular pollen amb, rugulate-reticulate pollen exine sculpting, and circumalate seeds (Nelson). Both genera have hypogynous disc nectaries (Wolfe et al.). *Chionophila* shares a clade with sister genera *Chelone* and *Nothochelone*. *Chionophila* has spikelike racemes without bracteoles, no ridges on the corolla, circular pollen amb, reticulate pollen sculpting, seeds with wings, and epistaminal nectaries.

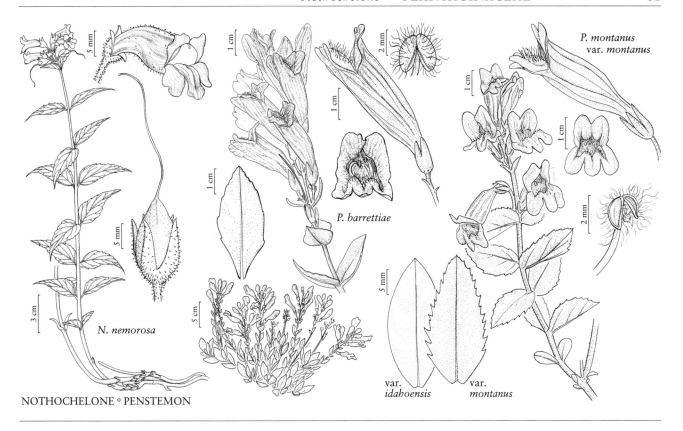

N. nemorosa

P. barrettiae

P. montanus
var. montanus

var. idahoensis   var. montanus

NOTHOCHELONE ○ PENSTEMON

1. Nothochelone nemorosa (Douglas) Straw, Brittonia 18: 85. 1966 [E] [F]

Chelone nemorosa Douglas, Bot. Reg. 14: plate 1211. 1828; *Penstemon nemorosus* (Douglas) Trautvetter

Stems clustered, 22–120 cm. **Leaves:** petiole 0–2(–5) mm; blade ovate to lanceolate or elliptic, (18–)40–140 × (4–)15–40 mm, base cuneate to rounded, margins with (0 or)1–5(–7) teeth per cm, apex acute to acuminate, surfaces glabrous or puberulent-pubescent mostly along veins. **Cymes** 3–38 cm. **Pedicels** 3–27 mm; bracteoles lanceolate, 3–10 × 1–2 mm. **Flowers:** calyx lobes (1–)3–13 × 1–3(–4) mm; corolla tube (14–)17–23 × 8–11 mm, abaxial lip (3–)4–10 × 1–5 mm, adaxial keeled, shorter than abaxial, 2.4–5 mm; adaxial filaments 12–20 mm; staminode 7–15 mm, coarsely bearded for entire length, hairs tan to orange; ovary glabrous, septum complete, locules equal, subtended by nectariferous disc; style terete, 15–28 mm, glabrous. **Capsules** ovoid, 9–15 ×5–9 mm. **Seeds** 2–4 × 0.5–1.1 mm, including wings. $2n$ = 30.

Flowering Jun–Aug. Cliffs, rocky places, trails in moist, humid, conifer and mixed evergreen forests; 0–2700(–3100) m; B.C.; Calif., Oreg., Wash.

Nothochelone nemorosa has sometimes been included in *Chelone* or *Penstemon*; studies of anatomy (R. M. Straw 1966; A. D. Nelson 1995) and molecular variation (A. D. Wolfe et al. 1997, 2002) support its separation in a monospecific genus.

26. PENSTEMON Schmidel, Icon. Pl. ed. Keller, [2]. 1763 • Beardtongue [Greek *pente*, five, and *stemon*, stamen, alluding to the conspicuous nature of the staminode]

Craig C. Freeman

Shrubs, subshrubs, or herbs, perennial; caudex woody, sometimes herbaceous (sect. *Erianthera*, sect. *Coerulei*, sect. *Glabri*, sect. *Penstemon*) or rhizomelike. **Stems** prostrate, erect, ascending, or decumbent, glabrous or variously hairy, sometimes also glandular. **Leaves** deciduous in shrubby species, rarely persistent, basal and cauline, sometimes only cauline, rarely only basal, opposite, rarely whorled, alternate, or subalternate distally; petiole present or absent; blade leathery or not, margins entire or variously toothed (1-pinnatifid in *P. dissectus*), distal leaf blade not needlelike or scalelike. **Inflorescences** terminal, thyrses, sometimes racemiform or paniculiform due to expansion or reduction of cymose lateral branches and/or number of verticillasters; bracts present. **Pedicels** present, rarely absent; bracteoles smaller than calyx lobes, not surrounding calyx of flowers they subtend. **Flowers** bisexual; sepals 5, proximally connate, calyx essentially radially symmetric, short-campanulate, rarely short-tubular, lobes ± ovate, lanceolate, round, elliptic, oblong, obovate, oblanceolate, truncate, linear, or orbiculate; corolla ± white, pink, red, blue, purple, lilac, violet, or crimson, rarely yellow or orange, bilaterally symmetric, rarely nearly radially symmetric, ± bilabiate or bilabiate and personate, rarely nearly regular, salverform, tubular, funnelform, ventricose, ampliate, ventricose-ampliate, tubular-funnelform, or tubular-salverform, tube base not spurred or gibbous, lobes 5, abaxial 3, adaxial 2; stamens 4, proximally adnate to corolla, didynamous, filaments glabrous or glandular-puberulent proximally, rarely pubescent proximally and/or distally; staminode 1, threadlike to straplike; nectaries epistaminal; ovary 2-locular, placentation axile; stigma capitate. **Fruits** capsules, dehiscence septicidal, not densely packed with white, membranous hairs. **Seeds** (2–)5–40(–100+), tan, brown, or black, angled, rarely reniform, patelliform, disciform, rounded, or angled-elongate, wings absent, sometimes narrow. $x = 8$.

Species ca. 280 (239 in the flora): North America, Mexico, Central America (Guatemala).

Penstemon is nearly endemic to North America, with three species that range south into Guatemala; it is the third largest genus in number of species in the flora area after *Carex* (Cyperaceae) and *Astragalus* (Fabaceae). Some species, especially in the western United States, have exceedingly narrow ranges.

Orthography of the genus name varied from the 1700s into the early 1900s and is summarized by F. S. Crosswhite (1965d).

Monographic studies of most parts of the genus were carried out by F. W. Pennell (1920b, 1921, 1935, 1940), D. D. Keck (1932, 1936b, 1937, 1937b, 1938, 1940b, 1945), Keck and A. Cronquist (1957), and F. S. Crosswhite (1965, 1965b, 1965c, 1966, 1967, 1967b, 1967c). Many western species were studied for floristic treatments in the Intermountain region (N. H. Holmgren 1984, 2017) and California (Holmgren 1993; M. Wetherwax and Holmgren 2017, http://ucjeps.berkeley.edu/eflora/eflora_display.php?tid=11396).

Penstemon shares morphological similarities with *Chionophila*, most notably an epistaminal nectary of glandular hairs. Both genera have a base chromosome number of $x = 8$. R. M. Straw (1966) hypothesized the genera to be sister taxa. Relationships among genera in Cheloneae remain equivocal (A. D. Wolfe et al. 1997, 2002, 2006; S. L. Datwyler and Wolfe 2004). *Penstemon* is distinguished from *Chionophila* by shorter calyx tubes (the calyx lobes usually distinct nearly to their bases), thyrsoid inflorescences, bracteoles in the inflorescences, and unwinged or, rarely, narrowly winged seeds.

The only broad molecular survey of *Penstemon* (A. D. Wolfe et al. 2006) found limited support for morphology-based infrageneric groups, and it lacked sufficient resolution to test the monophyly of large parts of the genus. C. A. Wessinger et al. (2016) obtained better species-level resolution in a study that focused on several sections of the genus. Both studies demonstrated that some groups are monophyletic or could be made so with minor adjustments and suggested that the infrageneric classification that draws heavily on floral characters may need extensive revision. A major overhaul of the infrageneric classification of *Penstemon* must await studies to resolve topological incongruencies between gene trees and to provide greater resolution within and among clades, issues likely due to hybridization and the recent, rapid radiation of the genus. Except for shifts of some species among subgenera or sections, a morphology-based classification with two subgenera and 16 sections is followed here.

Published chromosome counts exist for about half of the species in the flora area. Most *Penstemon* species counted are diploid ($2n = 16$); published tetraploid, hexaploid, octoploid, and dodecaploid counts exist for 24 species in sect. *Penstemon*, sect. *Saccanthera*, and sect. *Spectabiles* in the flora area. Most polyploids are assumed to be allopolyploids owing to the high incidence of hybridization; that hypothesis has not been tested.

Although the taxonomy of *Penstemon* traditionally places heavy emphasis on flower morphology, not all authors have used descriptive terms consistently. An overview of important diagnostic terms as applied in this treatment follows.

Basic corolla parts are the tube, throat, and limb (with abaxial and adaxial lobes). Some authors use the term tube to refer to the entire corolla proximal to the limb. In most species, the basal part of the corolla, proximal to the point of adnation of the staminode, is cylindric and narrower than the rest of the corolla; that part of the corolla is herein called the tube. In most species, the corolla expands near the point of adnation of the staminode to the corolla, forming the throat; the throat extends from tube to limb. Relative to the tube, the throat can be unexpanded with essentially parallel surfaces (corolla salverform), slightly expanded with nearly parallel sides (corolla tubular), expanded with straight and slightly diverging surfaces (corolla funnelform), inflated abaxially (corolla ventricose), inflated adaxially (corolla ampliate), or inflated abaxially and adaxially (corolla ventricose-ampliate). Most species have distinctly bilaterally symmetric corollas; some (for example, *Penstemon centranthifolius*, *P. cyathophorus*, *P. eatonii*, *P. goodrichii*) have corollas nearly radially symmetric. Corollas with the abaxial surface of the throat strongly uparched and closing or nearly closing the orifice are described as personate.

Nectar guides are present in corollas of most species, especially those presumably pollinated by hymenopterans. Members of subg. *Dasanthera* basically lack nectar guides; the two prominent ridges in the abaxial surface of the throat usually are lighter in color compared to the rest of the corolla. This condition also occurs in some species in sect. *Saccanthera*.

The taxonomy of *Penstemon* relies heavily on stamen morphology, which varies among species in size, orientation, vesture, dehiscence, and shape. Differences among species within sections often are subtle. Characters should be determined from mature anthers that have dehisced and shed their pollen. Pollen sac length is the distance from the distal tip of the sac to its point of confluence with the opposite pollen sac (the connective). Orientation refers to the relative position of paired pollen sacs at dehiscence: parallel (sacs parallel to each other and basically parallel to the filament); divergent (distal tips spreading from each other and the filament); and opposite (distal tips essentially opposite each other and in line with the point of attachment of the filament).

Sutures of dehisced pollen sacs vary from smooth to papillose (with papillae mostly less than 0.1 mm) or denticulate (teeth 0.1+ mm). Vestiture of the sides of the anthers varies from absent to densely lanate, the hairs reaching 4 mm and obscuring the external surface in most members of subg. *Dasanthera*.

Dehiscence patterns usually are consistent within a species with few exceptions. Complete dehiscence occurs when both pollen sacs open entirely from their apices proximally to, and across, the connective. Complete dehiscence can result in anthers that remain more or less boat-shaped (navicular) or become essentially flat (explanate). Two basic patterns of incomplete dehiscence exist: the proximal portion of each pollen sac can remain indehiscent (pollen sacs dehiscent distally), sometimes with only the connective not splitting, or the distal portion of each sac can remain indehiscent (pollen sacs dehiscent proximally). Anthers of the former type typically have been described simply as dehiscing incompletely proximally; the pollen sacs can remain straight, sometimes they twist sigmoidally. Anthers that dehisce incompletely distally usually are described as saccate because of their pouchlike appearance.

The diversity of floral forms in *Penstemon* has stimulated great interest in the evolution of floral morphology (R. M. Straw 1955, 1956, 1956b; J. Walker-Larsen and L. D. Harder 2001; M. C. Castellanos et al. 2004, 2006), pollination ecology and pollination syndromes (Straw 1963; F. S. Crosswhite and C. D. Crosswhite 1966; R. Schmid 1976; H. R. Lawson et al. 1989; R. R. Clinebell and P. Bernhardt 1998; R. S. Lange and P. E. Scott 1999; V. J. Tepedino et al. 1999, 2006, 2007; J. D. Thomson et al. 2000; Paul Wilson et al. 2004), and reproductive biology (R. L. Nielson 1998; J. Chari and Wilson 2001; G. Dieringer and L. Cabrera R. 2002; J. L. Hawk and Tepedino 2007). Studies show that flower color is a good predictor of pollen vector (red or pink, salverform or tubular flowers pollinated by hummingbirds; purple, blue, or white, tubular, funnelform, or ventricose flowers pollinated by hymenopterans). Hummingbird pollination has evolved independently nearly a dozen times in *Penstemon*.

Penstemons are popular as ornamentals. Readers interested in growing penstemons should consult R. Nold (1999) or D. Lindgren and E. Wilde (2003). Nearly two-dozen species of *Penstemon* throughout North America were used by Native American tribes for drugs, food, fiber, dyes, and ceremonies (D. E. Moerman 1998).

SELECTED REFERENCES   Clark, D. V. 1971. Speciation in *Penstemon*. Ph.D. dissertation. University of Montana.   Keck, D. D. 1936b. Studies in *Penstemon* III. Madroño 3: 248–250.   Straw, R. M. 1956. Adaptive morphology of the *Penstemon* flower. Phytomorphology 6: 112–119.   Straw, R. M. 1956b. Floral isolation in *Penstemon*. Amer. Naturalist 90: 47–53. Straw, R. M. 1963. Bee-fly pollination of *Penstemon ambiguus*. Ecology 44: 818–819. Straw, R. M. 1966. A redefinition of *Penstemon* (Scrophulariaceae). Brittonia 18: 80–95. Wessinger, C. A. et al. 2016. Multiplexed shotgun genotyping resolves species relationships within the North American genus *Penstemon*. Amer. J. Bot. 103: 912–922. Wilson, Paul et al. 2004. A multivariate search for pollination syndromes among penstemons. Oikos 104: 345–361. Wolfe, A. D. et al. 2006. Phylogeny, taxonomic affinities, and biogeography of *Penstemon* (Plantaginaceae) based on ITS and cpDNA sequence data. Amer. J. Bot. 93: 1699–1713.

Key to Subgenera of *Penstemon*

1. Staminodes included, length ⅓–¾ times corolla throats; pollen sacs explanate, sides usually densely lanate (subexplanate and sides glabrous in *P. personatus*). . . . 26a. *Penstemon* subg. *Dasanthera*, p. 85
1. Staminodes included to exserted, if included, length ¾+ times corolla throats; pollen sacs navicular, navicular-sigmoid, saccate, or explanate, sides glabrous or hairy, if lanate, hairs not concealing surfaces, or if nearly concealing surfaces, pollen sacs not explanate . . . . .26b. *Penstemon* subg. *Penstemon*, p. 92

26a. Penstemon Schmidel subg. Dasanthera (Rafinesque) Pennell, Contr. U.S. Natl. Herb. 20: 327. 1920 E

Dasanthera Rafinesque, J. Phys. Chim. Hist. Nat. Arts 89: 99. 1819

Flowers: pollen sacs usually explanate, dehiscing completely, connective splitting, sides usually densely lanate, (subexplanate, sides glabrous in *P. personatus*), sutures smooth; staminode included, length usually ⅓–¾ times corolla throat. **Seeds** angled, wings absent or narrow.

Species 10 (10 in the flora): w North America.

Subgenus *Dasanthera* appears to be the basal lineage in the clade corresponding to *Penstemon* (A. D. Wolfe et al. 2006). As circumscribed here, subg. *Dasanthera* comprises a single section, sect. *Erianthera*. Morphologic and molecular data indicate that the persistent-leaved lineage of the Sierra-Cascade is derived from the mostly deciduous-leaved lineage of the Rocky Mountains (A. D. Every 1977; S. L. Datwyler and Wolfe 2004).

SELECTED REFERENCES Datwyler, S. L. and A. D. Wolfe. 2004. Phylogenetic relationships and morphological evolution in *Penstemon* subg. *Dasanthera* (Veronicaceae). Syst. Bot. 29: 165–176. Every, A. D. 1977. Biosystematics of *Penstemon* Subgenus *Dasanthera*—A Naturally Hybridizing Species Complex. Ph.D. dissertation. University of Washington.

26a.1. Penstemon Schmidel (subg. Dasanthera) sect. Erianthera G. Don, Gen. Hist. 4: 639. 1837 (as Pentstemon) E

Herbs, shrubs, or subshrubs. Stems glabrous or retrorsely hairy, sometimes glandular-pubescent distally, glaucous or not. **Leaves** cauline, opposite, rarely subopposite (*P. lyallii*), leathery, sometimes not (*P. lyallii*, *P. personatus*), glabrous or hairy, glaucous or not; cauline petiolate, short-petiolate, or sessile, blade round, orbiculate, ovate, obovate, lanceolate, oblanceolate, spatulate, elliptic, or oblong, rarely linear, margins entire or toothed. **Thyrses** continuous or interrupted, ± secund, rarely cylindric or conic, axis hairy, rarely glabrous, cymes 2 per node; peduncles and pedicels spreading or ascending to erect. **Flowers:** calyx lobes: margins entire or erose, ± scarious or herbaceous, glandular-pubescent, rarely glabrous or inconspicuously glandular; corolla lilac, lavender, blue, violet, purple, pink, red, scarlet, or white, bilaterally symmetric, strongly bilabiate, personate or not, funnelform, rarely tubular-funnelform, glabrous externally, rarely glandular-pubescent, hairy internally abaxially, rarely glabrous, throat gradually inflated, not constricted at orifice, 2-ridged abaxially; stamens included or longer pair exserted, filaments glabrous, pollen sacs opposite; staminode 0.1–0.4 mm diam., tip straight to recurved, glabrous or distal 10–50% hairy, hairs to 1.5 mm; style glabrous. **Capsules** glabrous. **Seeds** brown, 0.8–4 mm.

Species 10 (10 in the flora): w North America.

Morphologic and molecular data support inclusion of *Penstemon personatus* in sect. *Erianthera* (A. D. Wolfe et al. 2006). D. D. Keck (1936b) placed *P. personatus* in sect. *Cryptostemon* D. D. Keck; that name is invalid, because it lacked a diagnosis.

1. Pollen sac sides glabrous . . . . 9. *Penstemon personatus*
1. Pollen sac sides lanate.
  2. Thyrse axes glabrous. . . . . 1. *Penstemon barrettiae*
  2. Thyrse axes glandular-pubescent at least distally, sometimes also retrorsely hairy.
    3. Cymes 2–7-flowered; staminodes 10–13 mm; stems (18–)30–80 cm; leaves 8–13 pairs, blades lanceolate, rarely linear, 23–130 × 3–20 mm . . . . 6. *Penstemon lyallii*
    3. Cymes 1-flowered; staminodes 7–16 mm; stems 3–40 cm; leaves 2–10 pairs, blades orbiculate, obovate, spatulate, oblanceolate, ovate, elliptic, broadly lanceolate, or round, 4–55(–60) × 3–17(–28) mm.

[4. Shifted to left margin.—Ed.]

4. Leaves: distals seldom distinctly smaller than proximals, all deciduous, sometimes persistent.
    5. Leaves glabrous, not glaucous; leaf blades obovate to ovate, elliptic, or lanceolate, margins ± crenulate-serrulate; thyrses cylindric to ± secund . . . . .4. *Penstemon ellipticus*
    5. Leaves glabrous or puberulent, or puberulent and glandular-pubescent, glaucous or not; leaf blades obovate to oblanceolate, ovate, lanceolate, or elliptic, margins entire, subentire, serrate, or dentate; thyrses secund. . . . . .7. *Penstemon montanus*
4. Leaves: distals usually distinctly smaller than proximals, all persistent.
    6. Leaf blades lanceolate to oblanceolate or elliptic, not glaucous. . . . . .5. *Penstemon fruticosus*
    6. Leaf blades round, elliptic, obovate, ovate, orbiculate, or spatulate, glaucous or not.
        7. Corollas blue, lavender, violet, or purple; leaves ± or not glaucous; stamens included.
            8. Thyrses 5–14 cm, ± interrupted; proximal bract margins serrate, sometimes entire; stems 10–27 cm. . . . . .2. *Penstemon cardwellii*
            8. Thyrses 1–6 cm, continuous; proximal bract margins entire; stems 4–10(–17) cm. . . . . .3. *Penstemon davidsonii*
        7. Corollas pink, red, rose red, scarlet, pinkish lavender, lavender, or purple; leaves usually glaucous; stamens: longer pair exserted (included in *P. newberryi* var. *berryi*).
            9. Verticillasters 4–12; corollas moderately to densely white-lanate internally abaxially . . . . .8. *Penstemon newberryi*
            9. Verticillasters 1–3; corollas glabrous internally or sparsely white-lanate abaxially . . . . 10. *Penstemon rupicola*

1. **Penstemon barrettiae** A. Gray in A. Gray et al., Syn. Fl. N. Amer. ed. 2, 2(1): 440. 1886 (as Pentstemon barrettae) • Barrett's beardtongue ⃞C ⃞E ⃞F

Subshrubs, sometimes cespitose. **Stems** ascending to erect, 13–40 cm, glabrous, glaucous. **Leaves** persistent, 3–5 pairs, short-petiolate or sessile, 15–80 × (5–)11–28 mm, blade elliptic to ovate, oblanceolate, or spatulate, base tapered to clasping, margins entire or irregularly serrate to serrulate, apex rounded to obtuse or acute, glabrous, glaucous. **Thyrses** continuous, secund, 7–13 cm, axis glabrous, verticillasters 6–8, cymes 1- or 2-flowered; proximal bracts ovate, 7–28 × 3–17 mm; peduncles and pedicels ascending to erect, glabrous. **Flowers:** calyx lobes ovate to lanceolate, 5–7.5 × 2.2–3.5 mm, glabrous or inconspicuously glandular proximally; corolla lilac to purple, essentially unlined internally but abaxial ridges usually lilac or white, not personate, funnelform, 33–38 mm, glabrous externally, sparsely to densely white-lanate internally abaxially, tube 6–8 mm, throat 7–10 mm diam.; stamens included, pollen sacs 1.2–1.6 mm; staminode 7–13 mm, essentially terete distally, 0.1 mm diam., tip straight, glabrous; style 20–26 mm. **Capsules** 8–10 × 5–7 mm. *2n* = 16.

Flowering Apr–Jun. Basalt cliffs, outcrops, talus slopes; of conservation concern; 70–300 m; Oreg., Wash.

Penstemon barrettiae is known from along the Columbia and Klickitat rivers in Hood River, Multnomah, and Wasco counties, Oregon, and Klickitat County, Washington.

2. **Penstemon cardwellii** Howell, Fl. N.W. Amer., 510. 1901 (as Pentstemon) • Cardwell's beardtongue ⃞E

Shrubs, sometimes cespitose. **Stems** ascending to erect, 10–27 cm, retrorsely hairy or puberulent, not glaucous. **Leaves** persistent, 6–9 pairs, distals usually distinctly smaller than proximals, short-petiolate or sessile, 15–55 × 4–16 mm, blade elliptic, base tapered, margins subentire or serrate, apex rounded to obtuse or acute, glabrous, ± glaucous. **Thyrses** ± interrupted, secund, 5–14 cm, axis retrorsely hairy proximally, retrorsely hairy and glandular-pubescent distally, verticillasters 3–7, cymes 1-flowered; proximal bracts ovate to lanceolate, 6–18 × 3–12 mm, margins serrate, sometimes entire; peduncles and pedicels ascending to erect, glandular-pubescent, sometimes also puberulent. **Flowers:** calyx lobes ovate to lanceolate, 8–14 × 2–3.1 mm, glandular-pubescent; corolla purple to violet, essentially unlined internally but abaxial ridges usually light purple or white, not personate, funnelform, (25–)30–38 mm, glabrous externally, sparsely white-lanate internally abaxially, tube 6–8 mm, throat

9–10 mm diam.; stamens included, pollen sacs 1.5–1.7 mm; staminode 11–13 mm, essentially terete distally, 0.1 mm diam., tip straight, distal 4–6 mm moderately to densely villous, hairs white or yellowish, to 1.5 mm; style 25–29 mm. **Capsules** 8–11 × 4–6 mm. *2n* = 16.

Flowering May–Aug. Rock outcrops, pumice and talus slopes, forest clearings, roadcuts; 400–1900 m; Oreg., Wash.

Penstemon cardwellii occurs primarily on the west slope of the Cascade Range from southwestern Oregon to northwestern Washington, with isolated populations east of the crest.

3. Penstemon davidsonii Greene, Pittonia 2: 241. 1892 (as Pentstemon) • Davidson's beardtongue [E]

Penstemon menziesii Hooker subsp. *davidsonii* (Greene) Piper

Subshrubs, cespitose. **Stems** ascending to erect, 4–10(–17) cm, retrorsely hairy or puberulent, not glaucous. **Leaves** persistent, 5–10 pairs, distals usually distinctly smaller than proximals, short-petiolate or sessile, (5–)10–20(–30) × 4–11 mm, blade elliptic to orbiculate, ovate, or spatulate, base tapered, margins entire or serrulate, apex rounded to obtuse or acute, glabrous or obscurely puberulent, not glaucous. **Thyrses** continuous, ± secund, 1–6 cm, axis retrorsely hairy proximally, retrorsely hairy and glandular-pubescent distally, verticillasters (1 or)2–7, cymes 1-flowered; proximal bracts round to ovate or lanceolate, 3–12 × 1–6 mm, margins entire; peduncles and pedicels ascending to erect, glandular-pubescent, sometimes also puberulent. **Flowers:** calyx lobes lanceolate to linear, 7–12(–14) × 2.5–4.3 mm, glandular-pubescent; corolla lavender to bluish lavender, violet, or purple, essentially unlined internally but abaxial ridges usually lavender or white, nearly personate, funnelform, 20–45 mm, glabrous externally, sparsely to moderately white-villous internally abaxially, tube (4–)6–12 mm, throat 7–12 mm diam.; stamens included, pollen sacs 0.9–1.6 mm; staminode 7–14 mm, slightly flattened distally, 0.1–0.2 mm diam., tip straight, distal 4–7 mm sparsely to densely villous, hairs whitish, to 1 mm; style 19–26 mm. **Capsules** 8–13 × 5–8 mm.

Varieties 3 (3 in the flora): w North America.

Penstemon davidsonii occurs mostly at high elevations in the Cascade Range and Sierra Nevada.

1. Corollas (30–)34–45 mm; leaf blades elliptic to ovate, sometimes spatulate, margins entire . . . . 3c. *Penstemon davidsonii* var. *praeteritus*
1. Corollas 20–36 mm; leaf blades spatulate to orbiculate or elliptic, margins serrulate or entire.
　2. Leaf blade margins entire, sometimes serrulate, apices rounded to obtuse . . . . 3a. *Penstemon davidsonii* var. *davidsonii*
　2. Leaf blade margins ± serrulate, apices rounded to obtuse or acute . . . . 3b. *Penstemon davidsonii* var. *menziesii*

3a. Penstemon davidsonii Greene var. **davidsonii** [E]

Leaf blades spatulate to orbiculate, margins entire, sometimes serrulate, apex rounded to obtuse. **Flowers:** corolla lavender or bluish lavender, violet, or purple, 20–36 mm, moderately white-villous internally abaxially. *2n* = 16.

Flowering May. Rock outcrops, talus slopes; 900–3700 m; B.C.; Calif., Nev., Oreg., Wash.

The ranges of var. *davidsonii* and var. *menziesii* overlap throughout much of British Columbia and Washington. Variety *davidsonii* hybridizes with var. *menziesii*, and with *Penstemon cardwellii*, *P. fruticosus*, *P. newberryi*, and *P. rupicola*.

3b. Penstemon davidsonii Greene var. **menziesii** (D. D. Keck) Cronquist in C. L. Hitchcock et al., Vasc. Pl. Pacif. N.W. 4: 379. 1959 [E]

Penstemon davidsonii subsp. *menziesii* D. D. Keck, Brittonia 8: 247. 1957; *P. menziesii* Hooker subsp. *thompsonii* Pennell & D. D. Keck

Leaf blades spatulate to elliptic, margins ± serrulate, apex rounded to obtuse or acute. **Flowers:** corolla violet, 20–35 mm, moderately white-villous internally abaxially. *2n* = 16.

Flowering Jun–Aug. Rock outcrops, ledges, talus slopes; 30–2000 m; B.C.; Oreg., Wash.

The ranges of var. *davidsonii* and var. *menziesii* overlap throughout much of British Columbia and Washington.

3c. Penstemon davidsonii Greene var. **praeteritus**
Cronquist, Leafl. W. Bot. 10: 129. 1964 E

Leaf blades elliptic to ovate, sometimes spatulate, margins entire, apex obtuse to acute. **Flowers:** corolla bluish lavender to violet, (30–)34–45 mm, sparsely white-lanate internally abaxially.

Flowering May–Jul. Rock outcrops, talus slopes; 1500–2900 m; Nev., Oreg.

Populations of var. *praeteritus* occur on isolated, relatively dry peaks in the Great Basin. It is known from the Jackson Mountains and Pine Forest and Santa Rosa ranges in Humboldt County, Nevada, and the Pueblo Mountains and Steens Mountain in Harney County, Oregon, and is disjunct to the west from the nearest populations of var. *davidsonii* by approximately 120 km.

4. Penstemon ellipticus J. M. Coulter & Fisher, Bot. Gaz. 18. 302. 1893 (as Pentstemon) • Rocky ledge beardtongue E

Penstemon davidsonii Greene var. *ellipticus* (J. M. Coulter & Fisher) B. Boivin

Subshrubs, cespitose. **Stems** ascending to erect, 5–18 cm, puberulent to retrorsely hairy, usually in lines proximally, not glaucous. **Leaves** deciduous, sometimes persistent, 2–6 pairs, distals seldom distinctly smaller than proximals, petiolate or sessile, (5–)10–42 × 4–15 mm, blade obovate to ovate, elliptic, or lanceolate, base tapered to clasping, margins ± crenulate-serrulate, apex obtuse to acute, glabrous, not glaucous. **Thyrses** continuous, cylindric to ± secund, 1–4 cm, axis glandular-pubescent, verticillasters 2–4, cymes 1-flowered; proximal bracts ovate to lanceolate, (4–)8–25 × 3–17 mm; peduncles and pedicels ascending to erect, glandular-pubescent. **Flowers:** calyx lobes lanceolate, 8–15 × (1.8–)2.5–3.7 mm, glandular-pubescent; corolla violet to purple, unlined internally but abaxial ridges usually light violet to whitish, not personate, funnelform, 27–40 mm, glabrous externally, sparsely to moderately white-lanate internally abaxially, tube 5–8 mm, throat 8–11 mm diam.; stamens included, pollen sacs 1.1–1.3 mm; staminode 9–16 mm, flattened distally, 0.3–0.4 mm diam., tip straight, glabrous or distal 1–3(–13) mm sparsely to moderately villous, hairs yellow, to 1.3 mm; style 25–28 mm. **Capsules** 8–11 × 6–8 mm. **2n** = 16.

Flowering May–Aug(–Sep). Cliffs, ledges, rock outcrops, talus slopes; 900–2700 m; Alta., B.C.; Idaho, Mont., Wash.

Penstemon ellipticus is known from the Canadian, Northern, and Middle Rocky mountains, including the Columbia Mountains.

5. Penstemon fruticosus (Pursh) Greene, Pittonia 2: 239. 1892 (as Pentstemon) • Shrubby beardtongue E

Gerardia fruticosa Pursh, Fl. Amer. Sept. 2: 423, plate 18. 1813

Shrubs or subshrubs, usually cespitose. **Stems** ascending, 13–40 cm, glabrous or retrorsely hairy, usually in lines, not glaucous. **Leaves** persistent, 2–6 pairs, distals usually distinctly smaller than proximals, petiolate or sessile, (5–)6–50(–60) × 3–12 mm, blade lanceolate to oblanceolate or elliptic, base tapered to clasping, margins entire or ± serrate, or dentate, apex obtuse to acute, glabrous, not glaucous. **Thyrses** continuous to ± interrupted, secund, 3–11 cm, axis sparsely glandular-pubescent, sometimes also retrorsely hairy proximally, verticillasters 2–7, cymes 1-flowered; proximal bracts lanceolate, 4–16 × 1–5 mm; peduncles and pedicels ascending to erect, glandular-pubescent. **Flowers:** calyx lobes lanceolate to linear, (7–)9–13(–15) × 1.8–3 mm, glandular-pubescent; corolla bluish lavender to purplish or violet, unlined internally or lined with faint bluish lavender nectar guides abaxially, not personate, funnelform, 28–48 mm, glabrous externally, sparsely to moderately white-lanate or -villous internally abaxially, tube (5–)8–10 mm, throat 8–13 mm diam.; stamens included, pollen sacs 1–1.3(–1.5) mm; staminode 7–12 mm, slightly flattened distally, 0.1–0.3 mm diam., tip straight to recurved, distal 1–4 mm sparsely to densely pubescent, hairs yellow, to 1.2 mm; style 24–27 mm. **Capsules** 8–13 × 5.5–7 mm.

Varieties 3 (3 in the flora): w North America.

Penstemon fruticosus occurs widely across the Central and Northern Rocky mountains and Cascade Ranges from southwestern Alberta and interior British Columbia to northwestern Wyoming, southern Idaho, and southern Oregon.

Drug, food, fiber, dye, and decorative uses of *Penstemon fruticosus* are reported among six tribes of Native Americans in northwestern North America (D. E. Moerman 1998).

1. Leaf margins sharply serrate to dentate, at least distally, blades oblanceolate to elliptic, 6–25 × 3–8 mm; wc Idaho, ne Oregon, se Washington . . . . .5c. *Penstemon fruticosus* var. *serratus*
1. Leaf margins entire, serrate, or dentate, rarely sharply serrate or dentate, blades lanceolate to oblanceolate or elliptic, (5–)8–50(–60) × 3–12 mm; sw Alberta, s British Columbia, Idaho, sw Montana, Oregon, Washington, nw Wyoming.

[2. Shifted to left margin.—Ed.]

2. Leaf blades lanceolate to oblanceolate or elliptic; corollas 28–40 mm; s British Columbia, Idaho, sw Montana, Oregon, Washington, nw Wyoming . . . . 5a. *Penstemon fruticosus* var. *fruticosus*
2. Leaf blades lanceolate; corollas 35–48 mm; sw Alberta, s British Columbia, n Idaho, ne Washington . . .5b. *Penstemon fruticosus* var. *scouleri*

5a. Penstemon fruticosus (Pursh) Greene var. **fruticosus** E

Leaves 15–50(–60) × 5–12 mm, blade lanceolate to oblanceolate or elliptic, margins entire, serrate, or dentate, rarely sharply serrate or dentate. **Flowers:** corolla 28–40 mm. **2n** = 16.

Flowering Apr–Jul. Rock outcrops, gravelly slopes, forest openings, roadcuts; 300–3000 m; B.C.; Idaho, Mont., Oreg., Wash., Wyo.

Variety *fruticosus* is known from the Cascade Range, Columbia Plateau, and southern part of the Northern Rocky Mountains, from southern British Columbia, through much of Washington, Idaho, and Oregon to southwestern Montana and northwestern Wyoming.

5b. Penstemon fruticosus (Pursh) Greene var. **scouleri** (Lindley) Cronquist in C. L. Hitchcock et al., Vasc. Pl. Pacif. N.W. 4: 385. 1959 E

Penstemon scouleri Lindley, Edwards's Bot. Reg. 15: plate 1277. 1829 (as Pentstemon); *P. fruticosus* subsp. *scouleri* (Lindley) Pennell & D. D. Keck

Leaves (5–)8–48 × 3–8 mm, blade lanceolate, margins entire, serrate, or dentate, rarely sharply serrate or dentate. **Flowers:** corolla 35–48 mm.

Flowering Apr–Jul. Cliffs, rock outcrops, rocky slopes, forest openings, roadcuts; 200–2200 m; Alta., B.C.; Idaho, Wash.

Variety *scouleri* is known from southwestern Alberta, southern British Columbia, northern Idaho, and northeastern Washington.

5c. Penstemon fruticosus (Pursh) Greene var. **serratus** (D. D. Keck) Cronquist in C. L. Hitchcock et al., Vasc. Pl. Pacif. N.W. 4: 385. 1959 E

Penstemon fruticosus subsp. *serratus* D. D. Keck in L. Abrams and R. S. Ferris, Ill. Fl. Pacific States 3: 765. 1951

Leaves 6–25 × 3–8 mm, blade oblanceolate to elliptic, margins sharply serrate to dentate, at least distally. **Flowers:** corolla 38–43 mm.

Flowering Apr–Jun. Rock outcrops, rocky slopes; 1400–2500 m; Idaho, Oreg., Wash.

Variety *serratus* is known from the Seven Devils Mountains of Idaho, Wallowa Mountains of Oregon, and Blue Mountains of Washington.

6. Penstemon lyallii (A. Gray) A. Gray in A. Gray et al., Syn. Fl. N. Amer. ed. 2, 2(1): 440. 1886 (as Pentstemon lyalli) • Lyall's beardtongue E

Penstemon menziesii Hooker var. *lyallii* A. Gray, Proc. Amer. Acad. Arts 6: 76. 1862 (as Pentstemon menziesii var. lyalli)

Herbs. Stems ascending to erect, (18–)30–80 cm, puberulent or retrorsely hairy proximally, puberulent or glandular-pubescent distally, not glaucous. **Leaves** deciduous, 8–13 pairs, short-petiolate or sessile, 23–130 × 3–20 mm, blade lanceolate, rarely linear, base tapered, margins entire or remotely serrate, apex acute to acuminate, rarely obtuse, glabrous or puberulent to pubescent, not glaucous. **Thyrses** interrupted, ± secund, (3–)6–15(–28) cm, axis sparsely to moderately glandular-pubescent, verticillasters 3–7, cymes 2–7-flowered; proximal bracts lanceolate to linear, (5–)15–80 × (1–)2–8(–20) mm; peduncles and pedicels spreading or ascending, glandular-pubescent. **Flowers:** calyx lobes lanceolate, 7–16 × 2–4 mm, glandular-pubescent; corolla lavender to purple, unlined internally or lined with faint lavender nectar guides abaxially, nearly personate, funnelform, 35–46 mm, glabrous externally, moderately white-lanate internally abaxially, tube 8–14 mm, throat 8–12 mm diam.; stamens included, pollen sacs 1.1–1.8 mm; staminode 10–13 mm, flattened distally, 0.2–0.3 mm diam., tip straight, glabrous; style 30–40 mm. **Capsules** 10–14 × 5–7 mm.

Flowering Jun–Aug. Rocky slopes, rock outcrops, gravel bars along streams; 600–2400 m; Alta., B.C.; Idaho, Mont., Wash.

Penstemon lyallii is known from the Central, Northern, and Canadian Rocky mountains in southwestern Alberta, southeastern British Columbia, northern Idaho, northwestern Montana, and northeastern Washington.

7. Penstemon montanus Greene, Pittonia 2: 240. 1892
(as Pentstemon) • Cordroot or mountain beardtongue
E F

Herbs or subshrubs, sometimes cespitose. **Stems** ascending to erect, 6–30 cm, pubescent, sometimes retrorsely hairy proximally and glandular-pubescent distally, not glaucous. **Leaves** deciduous, 4–7 pairs, distals seldom distinctly smaller than proximals, short-petiolate or sessile, 9–35(–56) × 7–17(–28) mm, blade obovate to oblanceolate, ovate, lanceolate, or elliptic, base tapered, cuneate, or cordate-clasping, margins entire, subentire, serrate, or dentate, apex obtuse to acute, glabrous, puberulent, or puberulent and glandular-pubescent, glaucous or not. **Thyrses** continuous, secund, 1–7 cm, axis sparsely to moderately glandular-pubescent, sometimes also pubescent proximally, verticillasters 1–5, cymes 1-flowered; proximal bracts ovate to lanceolate, 6–28 × 3–22 mm; peduncles and pedicels ascending to erect, glandular-pubescent. **Flowers:** calyx lobes lanceolate to linear, 8–15 × 1.8–3.5 mm, glandular-pubescent; corolla lavender, blue, or violet, unlined internally or lined with faint lavender or blue nectar guides abaxially, not personate, funnelform, 26–33 (–39) mm, glabrous externally, moderately to densely white-lanate internally abaxially, tube 7–9 mm, throat 8–10 mm diam.; stamens included, pollen sacs 1.1–1.6 mm; staminode 11–15 mm, slightly flattened distally, 0.1–0.3 mm diam., tip straight to recurved, glabrous or distal 3–6 mm pubescent, hairs white, to 1.4 mm; style 25–30 mm. **Capsules** 10–14 × 5–7 mm.

Varieties 2 (2 in the flora): w United States.

Penstemon montanus is known from the Central and Northern Rocky mountains.

1. Leaf blades puberulent and glandular-pubescent, not glaucous, blade margins serrate to dentate
. . . . 7a. *Penstemon montanus* var. *montanus*
1. Leaf blades glabrous or puberulent, usually glaucous, blade margins entire or subentire. . . . .
. . . . 7b. *Penstemon montanus* var. *idahoensis*

7a. Penstemon montanus Greene var. **montanus** E F

Leaf blades: margins serrate to dentate, puberulent and glandular-pubescent, not glaucous. **Flowers:** corolla densely white-lanate internally abaxially.

Flowering Jun–Aug. Rocky basalt, granite, limestone, and sandstone slopes, rock slides, cliffs; 1500–3500 m; Idaho, Mont., Utah, Wyo.

7b. Penstemon montanus Greene var. **idahoensis** (D. D. Keck) Cronquist in C. L. Hitchcock et al., Vasc. Pl. Pacif. N.W. 4: 394. 1959 E F

Penstemon montanus subsp. *idahoensis* D. D. Keck, Brittonia 8: 247. 1957

Leaf blades: margins entire or subentire, glabrous or puberulent, usually glaucous. **Flowers:** corolla moderately white-lanate internally abaxially.

Flowering Jun–Aug. Granite crevices, weathered granite; 2100–3200 m; Idaho.

Variety *idahoensis* is known from the mountains of central Idaho in Blaine, Custer, Elmore, and Valley counties.

8. Penstemon newberryi A. Gray in War Department [U.S.], Pacif. Railr. Rep. 6(3): 82, plate 14. 1858 (as Pentstemon) • Newberry's beardtongue E

Subshrubs, cespitose. **Stems** ascending to erect, (5–)10–32 cm, glabrous or retrorsely hairy, not glaucous. **Leaves** persistent, 4–9 pairs, distals distinctly to slightly smaller than proximals, short-petiolate or sessile, (5–) 9–42 × 4–14(–18) mm, blade elliptic to obovate, base tapered, cuneate, or cordate-clasping, margins ± serrulate or serrulate-denticulate to serrate, apex rounded to obtuse or acute, glabrous, glaucous. **Thyrses** continuous, ± secund, 2–20 cm, axis glandular-pubescent, verticillasters 4–12, cymes 1-flowered; proximal bracts ovate to lanceolate, 4–11(–20) × 2–8(–15) mm; peduncles and pedicels ascending, glandular-pubescent. **Flowers:** calyx lobes lanceolate, 6–10 × 2–2.5 mm, glandular-pubescent; corolla pink, red, scarlet, lavender, or purple, essentially unlined internally but abaxial ridges sometimes pink to whitish, not personate, funnelform, 22–35 mm, glabrous externally, moderately to densely white-lanate internally abaxially, tube 6–9 mm, throat 5–12 mm diam.; stamens included or longer pair exserted, pollen sacs 0.8–1.7 mm; staminode 8–14 mm, slightly flattened distally, 0.1–0.3 mm diam., tip straight, distal 2–8 mm pubescent to lanate, hairs white or yellowish, to 1 mm; style 20–24 mm. **Capsules** (8–) 10–13 × (4–)5–8 mm.

Varieties 3 (3 in the flora): w United States.

Penstemon newberryi is known from the Sierra Nevada and Cascade ranges of California, Nevada, and Oregon.

1. Corollas pink to lavender or purple, 27–35 mm; stamens included, pollen sacs 1.2–1.7 mm . . . . .8b. *Penstemon newberryi* var. *berryi*
1. Corollas red, scarlet, or purple, 22–30 mm; stamens: longer pair exserted, pollen sacs 0.8–1.3 mm.
   2. Corollas red to scarlet; distal leaves distinctly smaller than proximals . . . . .8a. *Penstemon newberryi* var. *newberryi*
   2. Corollas purple; distal leaves slightly smaller than proximals. . . . . .8c. *Penstemon newberryi* var. *sonomensis*

8a. Penstemon newberryi A. Gray var. **newberryi** E

Leaves: distals distinctly smaller than proximals. **Flowers:** corolla red to scarlet, 22–30 mm; stamens: longer pair exserted, pollen sacs 0.8–1.3 mm. 2*n* = 16.

Flowering May–Aug. Rock outcrops, talus slopes; 500–3500 m; Calif., Nev.

Variety *newberryi* is known from the Sierra Nevada from northern to south-central California and adjacent northwestern Nevada.

8b. Penstemon newberryi A. Gray var. **berryi** (Eastwood) N. H. Holmgren, Brittonia 44: 482. 1992 E

Penstemon berryi Eastwood, Bull. Torrey Bot. Club 32: 209. 1905 (as Pentstemon); *P. newberryi* subsp. *berryi* (Eastwood) D. D. Keck

Leaves: distals distinctly smaller than proximals. **Flowers:** corolla pink to lavender or purple, 27–35 mm; stamens included, pollen sacs 1.2–1.7 mm.

Flowering May–Aug. Rock outcrops, talus slopes; 900–3300 m; Calif., Oreg.

Variety *berryi* is known from the Klamath region of northwestern California (Del Norte, Glenn, Humboldt, Mendocino, Siskiyou, and Trinity counties) and southwestern Oregon (Curry, Douglas, Jackson, and Josephine counties).

8c. Penstemon newberryi A. Gray var. **sonomensis** (Greene) Jepson, Fl. W. Calif., 401. 1901 (as Pentstemon) C E

Penstemon sonomensis Greene, Pittonia 2: 218. 1891 (as Pentstemon); *P. newberryi* subsp. *sonomensis* (Greene) D. D. Keck

Leaves: distal slightly smaller than proximal. **Flowers:** corolla purple, 22–30 mm; stamens: longer pair exserted, pollen sacs 1.1–1.3 mm.

Flowering May–Aug. Rock outcrops, talus slopes; of conservation concern; 800–2300 m; Calif.

Variety *sonomensis* is known from mountain peaks in Lake, Napa, and Sonoma counties.

9. Penstemon personatus D. D. Keck, Madroño 3: 248. 1936 • Closed-throat beardtongue C E

Herbs; caudex herbaceous, rhizomelike. **Stems** ascending to erect, 30–57 cm, puberulent or retrorsely hairy, rarely glandular-pubescent distally, not glaucous. **Leaves** deciduous, 3–6(or 7) pairs, petiolate or sessile, (6–)30–65(–95) × (5–)12–50 mm, blade ovate to oblong or lanceolate, base truncate to cordate, margins entire or ± denticulate, apex obtuse to acute, puberulent to pubescent abaxially, glabrate adaxially, glaucescent. **Thyrses** interrupted, conic, 7–33 cm, axis sparsely to densely glandular-pubescent, verticillasters (2 or)3–5(–7), cymes 2–5(–7)-flowered; proximal bracts ovate to lanceolate, 7–28(–72) × 3–17(–45) mm; peduncles and pedicels ascending to erect, glandular-pubescent. **Flowers:** calyx lobes lanceolate, 3.8–6 × 1.1–2 mm, glandular-pubescent; corolla lavender to purple or white, lined internally with violet nectar guides, personate, tubular-funnelform, 20–26 mm, glabrous or sparsely glandular-pubescent externally, moderately to densely white-villous internally, tube 7–8 mm, throat 6–8 mm diam.; stamens included, pollen sacs 0.8–1.1 mm, glabrous; staminode 4–5 mm, flattened distally, 0.1–0.2 mm diam., tip straight, distal 3–4 mm densely villous, hairs yellow, to 1 mm; style 17–19 mm. **Capsules** 6–9 × 4–5 mm.

Flowering (Jun–)Jul–Aug. Sugar pine, yellow pine, California red fir, and mixed evergreen forests; of conservation concern; 1100–2100 m; Calif.

Penstemon personatus is known from the northern Sierra Nevada in Butte, Nevada, Plumas, and Sierra counties.

10. Penstemon rupicola (Piper) Howell, Fl. N.W. Amer., 510. 1901 (as Pentstemon) • Cliff beardtongue E

Penstemon newberryi A. Gray var. *rupicola* Piper, Bull. Torrey Bot. Club 27: 397. 1900 (as Pentstemon)

Subshrubs, cespitose. **Stems** ascending to erect, 3–16 cm, puberulent to pubescent, not glaucous. **Leaves** persistent, 3–8 pairs, distals usually distinctly smaller than proximals, short-petiolate or sessile, 4–22 × 3–14 mm, blade round to elliptic or spatulate, base tapered to cuneate, margins ± serrate, apex rounded to obtuse, rarely acute, glabrous or hairy, usually glaucous. **Thyrses** continuous, ± secund, 1–3 cm, axis hairy proximally, pubescent or glandular-pubescent distally, verticillasters 1–3, cymes 1-flowered; proximal bracts elliptic to ovate, 3–9 × 2–6 mm; peduncles and pedicels ascending to erect, glandular-pubescent. **Flowers:** calyx lobes lanceolate-elliptic to oblong, 6–11 × 2–2.8 mm, glandular-pubescent; corolla pink to red, pinkish lavender, or rose red, essentially unlined internally but abaxial ridges usually white, not personate, funnelform, 25–36 mm, glabrous externally, glabrous internally or sparsely white-lanate abaxially, tube 8–12 mm, throat 6–8 mm diam.; stamens: longer pair exserted, pollen sacs 1–1.5 mm; staminode 13–15 mm, slightly flattened distally, 0.1 mm diam., tip straight, glabrous or distal 1–2 mm sparsely villous, hairs yellow, to 0.8 mm; style 26–30 mm. **Capsules** 7–9 × 5–7 mm. **2n** = 16.

Flowering Jun–Jul. Cliffs, rock outcrops; 60–2400 m; Calif., Oreg., Wash.

Penstemon rupicola occurs largely in the Siskiyou Mountains and Cascade Range from northern California to northern Washington. It is partly sympatric with and forms hybrids with *P. cardwellii, P. davidsonii* var. *davidsonii,* and *P. fruticosus* var. *fruticosus* (A. D. Every 1977). Hybrids between *P. rupicola* and *P. davidsonii* var. *davidsonii* are encountered frequently in the vicinity of Crater Lake National Park and Mt. Hood (Every).

26b. PENSTEMON Schmidel subg. PENSTEMON

Flowers: pollen sacs navicular, navicular-sigmoid, saccate, or explanate, dehiscing completely or incompletely, connective splitting or not, sides glabrous or hairy, hairs not concealing surface, sutures smooth, papillate, or denticulate; staminode included to exserted, if included, length ¾+ times corolla throat. **Seeds** angled or rounded, rarely angled-elongate, reniform, or disciform, wings absent.

Species ca. 270 (229 in the flora): North America, Mexico, Central America (Guatemala).

Following the recommendation of A. D. Wolfe et al. (2006), subg. *Penstemon* is expanded here to include species of *Penstemon* not included in subg. *Dasanthera.* Molecular data indicate that some sections as usually circumscribed are paraphyletic. Absent a robust phylogeny for *Penstemon,* a morphology-based classification is followed here with minor adjustments. As circumscribed here, subg. *Penstemon* comprises 16 sections, 15 in the flora area.

Section *Leptostemon* Trautvetter includes 27 species from Mexico and Guatemala. Among its members are *Penstemon campanulatus* (Cavanilles) Willdenow, *P. gentianoides* (Kunth) Poiret, and *P. hartwegii* Bentham, three red- or purple-flowered species popular among European plant breeders in the nineteenth century and used in the development of European hybrid penstemons (R. Nold 1999; D. Lindgren and E. Wilde 2003).

Key to Sections of subg. *Penstemon*

1. Cauline leaf blade margins deeply pinnatifid to nearly pinnatisect.... 26b.8. *Penstemon* sect. *Dissecti,* p. 144
1. Cauline leaf blade margins entire or toothed, rarely laciniate-pinnatifid.
    2. Subshrubs or shrubs; staminodes glabrous; corollas pink, red, scarlet, or magenta, rarely lavender or violet.
        3. Cauline leaf blades linear and pollen sacs explanate........26b.1. *Penstemon* sect. *Ambigui,* p. 94
        3. Cauline leaf blades obovate, ovate, oblanceolate, lanceolate, elliptic, or linear; pollen sacs navicular, saccate, subexplanate, or explanate.

4. Pollen sacs subexplanate to explanate, opposite; corollas weakly ventricose; thyrses cylindric; leaves glaucous . . . . .26b.13.   *Penstemon* sect. *Petiolati*, p. 228

4. Pollen sacs saccate or navicular, parallel or divergent; corollas salverform or funnelform; thyrses secund; leaves not glaucous.

    5. Cauline leaf margins toothed or entire; corollas funnelform, glandular-pubescent internally abaxially and adaxially; cymes 1 per node; Texas . . . . 26b.2.   *Penstemon* sect. *Baccharifolii*, p. 96

    5. Cauline leaf margins entire; corollas salverform, hairy internally abaxially; cymes 2 per node; Arizona, California, Colorado, Nevada, New Mexico, Utah. . . . . .26b.3.   *Penstemon* sect. *Bridgesiani*, p. 97

[2. Shifted to left margin.—Ed.]

2. Herbs, shrubs, or subshrubs; staminodes hairy distally or glabrous; corollas usually white, lavender, lilac, blue, violet, purple, ochroleucous, yellow, orange, or crimson, if pink, red, scarlet, or magenta, herbs and/or staminodes hairy, or both.

  6. Pollen sacs saccate, dehiscing incompletely, distal $\frac{1}{5}$–$\frac{2}{3}$ indehiscent, connectives splitting, rarely not.

    7. Pollen sac sutures papillate; herbs. . . . . .26b.12.   *Penstemon* sect. *Penstemon* (in part), p. 183

    7. Pollen sac sutures denticulate; herbs or subshrubs.

      8. Herbs; pollen sac sides glabrous; staminodes glabrous, tips recurved; cauline leaf blade margins entire; thyrses interrupted . . . . .26b.5.   *Penstemon* sect. *Chamaeleon* (in part), p. 108

      8. Herbs or subshrubs; pollen sac sides glabrous or hairy; staminodes glabrous or hairy distally, tips straight; cauline leaf blade margins entire or toothed, rarely laciniate-pinnatifid; thyrses interrupted or continuous . . . . .26b.14.   *Penstemon* sect. *Saccanthera*, p. 229

  6. Pollen sacs navicular, navicular-sigmoid, sigmoid, subexplanate, or explanate, dehiscing completely, if incompletely proximal $\frac{1}{5}$–$\frac{1}{2}$ indehiscent and/or connectives not splitting.

    9. Pollen sacs dehiscing incompletely, proximal $\frac{1}{5}$–$\frac{1}{2}$ indehiscent and/or connectives not splitting.

      10. Corollas red to scarlet or orange, rarely yellow, glabrous externally. . . . . 26b.9.   *Penstemon* sect. *Elmigera*, p. 145

      10. Corollas white to pink, lavender, lilac, blue, violet, or purple, glandular-pubescent or glabrous externally, if corollas reddish pink or rose red, glandular-pubescent externally.

        11. Pollen sacs: connectives not splitting, sutures papillate, sides glabrous; corollas tubular-funnelform . . . . .26b.7.   *Penstemon* sect. *Cristati* (in part), p. 125

        11. Pollen sacs: proximal $\frac{1}{5}$–$\frac{1}{3}$ indehiscent and/or connectives not splitting, sutures denticulate, sometimes papillate or smooth, sides glabrous or hairy; corollas funnelform, ventricose, or ventricose-ampliate . . . . 26b.11.   *Penstemon* sect. *Glabri* (in part), p. 156

    9. Pollen sacs dehiscing completely (rarely connectives not splitting in *P. incertus*).

      12. Staminodes glabrous; pollen sacs parallel to divergent, sutures denticulate, teeth to 0.2–0.3 mm. . . . . .26b.5.   *Penstemon* sect. *Chamaeleon* (in part), p. 108

      12. Staminodes hairy distally, sometimes glabrous; pollen sacs opposite or divergent, rarely parallel, sutures smooth, papillate, or denticulate, teeth to 0.1–0.2 mm.

        13. Subshrubs; leaves cauline, rarely basal and cauline, blade margins entire; stems not glaucous; corollas 2-ridged abaxially . . . . 26b.4.   *Penstemon* sect. *Caespitosi*, p. 98

        13. Herbs, shrubs, or subshrubs; leaves basal and cauline, sometimes basal few or absent, blade margins entire or toothed, if shrubs or subshrubs, leaves basal and cauline or leaf blade margins toothed or stems glaucous; corollas rounded to 2-ridged abaxially.

[14. Shifted to left margin.—Ed.]

14. Stems and/or leaves and/or thyrse axes hairy; leaf blades not (or rarely) glaucous, not (or rarely) leathery.
    15. Corollas glabrous externally, ventricose or ventricose-ampliate.................... ....26b.11. *Penstemon* sect. *Glabri* (in part), p. 156
    15. Corollas glandular-pubescent externally, if glabrous then tubular, tubular-funnelform, or funnelform.
        16. Corolla throats rounded abaxially, sometimes slightly 2-ridged; staminodes: distal (40–)50–100% hairy, hairs to 4 mm, exserted, sometimes reaching orifice or included, tips straight to recurved or coiled; seeds 1.4–4.8 mm, brown, dark brown, or black ....26b.7. *Penstemon* sect. *Cristati* (in part), p. 125
        16. Corolla throats slightly to prominently 2-ridged abaxially, rarely rounded; staminodes: distal 5–60(–90)% hairy, hairs to 3 mm, sometimes glabrous, included or reaching orifice, sometimes exserted, tips straight to recurved; seeds 0.4–2(–3.5) mm, tan, brown, gray, or black ....26b.12. *Penstemon* sect. *Penstemon* (in part), p. 183
14. Stems, leaves, and thyrse axes glabrous, rarely scabrous; leaf blades glaucous, rarely not, leathery, rarely not.
    17. Shrubs or subshrubs; stamen filaments: shorter pair glandular-puberulent proximally ....26b.15. *Penstemon* sect. *Spectabiles* (in part), p. 245
    17. Herbs; stamen filaments glabrous.
        18. Leaf blade margins toothed.................26b.15. *Penstemon* sect. *Spectabiles* (in part), p. 245
        18. Leaf blade margins entire (rarely remotely and obscurely toothed in *P. parryi*).
            19. Stamens exserted; corollas red, 32–36 mm.....26b.6. *Penstemon* sect. *Coerulei* (in part), p. 110
            19. Stamens included or longer pair reaching orifice, rarely exserted; corollas lavender to blue, violet, purple, white, pink, lilac, red, scarlet, crimson, magenta, orange, yellow, or ochroleucous, 6–33 mm.
                20. Basal and proximal cauline leaf blades linear, 0.5–1.5 mm wide ....26b.7. *Penstemon* sect. *Cristati* (in part), p. 125
                20. Leaf blades ovate, obovate, lanceolate, oblanceolate, spatulate, trullate, oblong, or elliptic, rarely linear, 1–70 mm wide.
                    21. Pollen sacs 0.3–1.2 mm, opposite; corollas 7–22 mm, throats 2-ridged abaxially, sometimes rounded .....26b.12. *Penstemon* sect. *Penstemon* (in part), p. 183
                    21. Pollen sacs 0.6–3 mm, opposite, divergent, or parallel; corollas 10–48 mm, throats rounded abaxially, rarely slightly 2-ridged.
                        22. Pollens sacs navicular; corollas glabrous externally, rarely obscurely glandular (*P. carnosus*) ....26b.6. *Penstemon* sect. *Coerulei* (in part), p. 110
                        22. Pollen sacs explanate, sometimes navicular; corollas glandular-pubescent externally, if glabrous staminode papillate distally, papillae to 0.2 mm ....26b.10. *Penstemon* sect. *Gentianoides*, p. 150

26b.1. PENSTEMON Schmidel (subg. PENSTEMON) sect. AMBIGUI (Rydberg) Pennell, Contr. U.S. Natl. Herb. 20: 326, 333. 1920

Penstemon [unranked] *Ambigui* Rydberg, Fl. Colorado, 306. 1906 (as Pentstemon)

Subshrubs. Stems glabrous, glabrate, or scabrous proximally, not glaucous. **Leaves** cauline, opposite, not leathery, glabrous or scabrous, not glaucous; cauline sessile, blade linear, margins entire. **Thyrses** interrupted, cylindric to conic, axis glabrous or scabrous, cymes 2 per node; peduncles and pedicels spreading or ascending. **Flowers:** calyx lobes: margins entire or erose, broadly scarious, glabrous; corolla pink or lavender to violet, bilaterally symmetric, bilabiate, not personate, salverform or funnelform, glabrous externally, glandular-pubescent internally abaxially, throat slightly inflated, not constricted at orifice, rounded abaxially; stamens included or longer pair reaching orifice or barely exserted, filaments glabrous, pollen sacs divergent, explanate, dehiscing completely, connective splitting, sides glabrous, sutures smooth; staminode included, terete or slightly flattened distally, 0.1 mm diam., tip straight, glabrous; style glabrous. **Capsules** glabrous. **Seeds** dark brown to reddish brown, angled, 1–2.8 mm.

Species 2 (2 in the flora): w United States, n Mexico.

Section *Ambigui* includes two subshrubby species of the southwestern and south-central United States and northern Mexico.

1. Corollas salverform, (14–)16–22(–28) mm . . . . 11. *Penstemon ambiguus*
1. Corollas funnelform, 8–14 mm . . . . 12. *Penstemon thurberi*

11. Penstemon ambiguus Torrey, Ann. Lyceum Nat. Hist. New York 2: 228. 1827 (as Pentstemon ambiguum) • Gilia beardtongue F

Stems (20–)30–40(–60) cm, glabrous or scabrous proximally. **Leaves** (7–)10–25 pairs, (3–)5–30(–40) × 0.5–1(–2.5) mm, blade base tapered, margins glabrous or scabrous, apex acuminate to mucronate. **Thyrses** 6–15 cm, verticillasters 6–10, cymes 1–3-flowered; proximal bracts linear, 6–27(–33) × 0.3–1.5 mm; peduncles and pedicels glabrous or scabrous. **Flowers:** calyx lobes ovate, 1.5–3.5 × 1–1.5 mm; corolla pink (limb milky pink or milky white), lined internally abaxially with reddish purple nectar guides, salverform, (14–)16–22(–28) mm, glandular-pubescent internally in 2 lines on abaxial surface, tube 3–4 mm, throat 3–5 mm diam.; stamens included, pollen sacs 0.5–0.6 mm; staminode 7–9 mm; style 8–10 mm. **Capsules** 6–9 × 3–5 mm.

Varieties 2 (2 in the flora): w United States, n Mexico.

G. T. Nisbet and R. C. Jackson (1960) reported natural hybrids between *Penstemon ambiguus* and *P. thurberi* from Lincoln and Socorro counties, New Mexico.

Some Native American tribes used *Penstemon ambiguus* as a drug or ceremonial plant (D. E. Moerman 1998).

1. Stems scabrous proximally; leaf blade margins scabrous . . . . . 11a. *Penstemon ambiguus* var. *ambiguus*
1. Stems glabrous; leaf blade margins glabrous or scabrous . . . . 11b. *Penstemon ambiguus* var. *laevissimus*

11a. Penstemon ambiguus Torrey var. **ambiguus** E

Stems scabrous proximally. **Leaf blades:** margins scabrous. $2n = 16$.

Flowering May–Aug(–Sep). Dunes and loose sand, sandsage shrublands; 700–2100 m; Colo., Kans., Nebr., N.Mex., Okla., Tex.

11b. Penstemon ambiguus Torrey var. **laevissimus** (D. D. Keck) N. H. Holmgren, Brittonia 31: 104. 1979 F

Penstemon ambiguus subsp. *laevissimus* D. D. Keck, J. Wash. Acad. Sci. 29: 491. 1939

Stems glabrous. **Leaf blades:** margins glabrous or scabrous. $2n = 16$.

Flowering May–Aug(–Sep). Sandy soils, creosote bush, blackbrush, and sagebrush shrublands, juniper woodlands; 700–1800 m; Ariz., Nev., N.Mex., Tex., Utah; Mexico (Chihuahua).

12. Penstemon thurberi Torrey, Pacif. Railr. Rep. Parke, Bot., 15. 1856 (as Pentstemon) • Thurber's beardtongue

Penstemon thurberi var. *anestius* Reveal & Beatley

Stems 20–40 cm, glabrous or glabrate. **Leaves** 10–20 pairs, 5–32 × 0.3–2 mm, blade base tapered, margins glabrous or scabrous, apex acuminate to mucronate. **Thyrses** 2–15(–23) cm, verticillasters (2–)5–10 (–30), cymes 1- or 2-flowered; proximal bracts linear, 3–22 × 0.2–1 mm; peduncles and pedicels glabrous or scabrous. **Flowers:** calyx lobes ovate, 1–2.7 × 0.9–1.5 mm; corolla lavender to violet, lined internally abaxially with reddish purple nectar guides, funnelform, 8–14 mm, glandular-pubescent internally in 2 lines on abaxial surface, sometimes also in sinuses of adaxial lobes, tube (1.5–)2–3 mm, throat 3–5 mm diam.; stamens included or longer pair reaching orifice or barely exserted, pollen sacs 0.4–0.5 mm; staminode 5–6 mm; style 7–9 mm. **Capsules** (4–)5–8 × 3–4 mm. $2n = 16$.

Flowering Apr–Aug(–Sep). Sandy or rocky slopes, chaparral, creosote shrublands, pine-juniper woodlands; 500–1200(–2100) m; Ariz., Calif., Nev., N.Mex., Tex., Utah; Mexico (Baja California, Chihuahua).

Plants of *Penstemon thurberi* from northern Clark County, Nevada, with corollas 8–9 mm and capsules 4–5 mm have been treated as var. *anestius*.

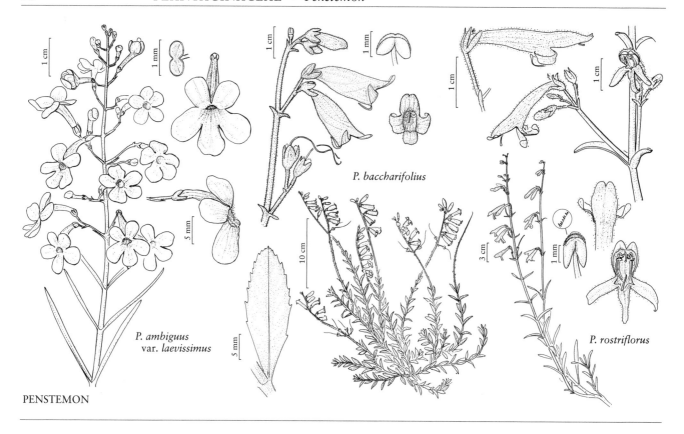

P. baccharifolius

P. ambiguus
var. laevissimus

P. rostriflorus

PENSTEMON

26b.2. PENSTEMON Schmidel (subg. PENSTEMON) sect. BACCHARIFOLII Pennell, Monogr. Acad. Nat. Sci. Philadelphia 1: 270. 1935

Subshrubs. Stems retrorsely hairy proximally, glandular-pubescent distally, not glaucous. **Leaves** cauline, opposite, leathery, glabrous or retrorsely hairy proximally, not glaucous; cauline short-petiolate or sessile, blade oblanceolate, obovate, or elliptic, margins toothed or entire. **Thyrses** interrupted, secund, axis glandular-pubescent, cymes 1 per node; peduncles and pedicels ascending to erect. **Flowers:** calyx lobes: margins entire, herbaceous, glandular-pubescent; corolla red, bilaterally symmetric, bilabiate, not personate, funnelform, glandular-pubescent externally, glandular-pubescent internally abaxially and adaxially, throat gradually inflated, not constricted at orifice, rounded abaxially; stamens included or longer pair reaching orifice, filaments glabrous, pollen sacs divergent, navicular to saccate, dehiscing completely or incompletely, sometimes distal $1/5$–$1/4$ indehiscent, connective splitting, sides glabrous, sutures papillate or denticulate, papillae or teeth to 0.1 mm; staminode included, flattened distally, 0.3–0.4 mm diam., tip straight, glabrous; style glabrous. **Capsules** glabrous. **Seeds** dark brown, angled, 1.5–2 mm.

Species 1: Texas, n Mexico.

13. Penstemon baccharifolius Hooker, Bot. Mag.
78: plate 4627. 1852 (as Pentstemon) • Baccharis-
leaf beardtongue F

Stems 12–50 cm. **Leaves:** cauline 5–15 pairs (distal ones much-reduced), 5–30(–46) × 4–15 mm, blade base tapered to acute to cuneate, margins sharply dentate, especially distally, or entire, apex obtuse to acute. **Thyrses** (2–)5–15 (–40) cm, verticillasters (2 or) 3–7(–12), cymes 1- or 2(or 3)-flowered; proximal bracts elliptic to lanceolate, 2–5 × 1–3 mm; peduncles and pedicels glandular-pubescent. **Flowers:** calyx lobes elliptic to ovate, 3.5–5 × 1.5–2.2 mm; corolla unlined internally, 22–27 mm, glandular-pubescent internally, tube 4–5 mm, throat 6–7 mm diam.; pollen sacs 1.2–1.5 mm; staminode 16–19 mm; style 18–25 mm. **Capsules** 7–10 × 5–7 mm.

Flowering Apr–Oct. Limestone cliffs, ledges, hills; 300–2000 m; Tex.; Mexico (Coahuila, San Luis Potosí).

Penstemon baccharifolius is known from the Edwards Plateau and eastern Trans-Pecos of Texas, from Bandera and Medina counties west to extreme eastern Presidio County. It also occurs in the mountains of northern Coahuila, Mexico. I. Méndez Larios and J. L. Villaseñor Ríos (2001) listed it for the state of San Luis Potosí, Mexico.

26b.3. PENSTEMON Schmidel (subg. PENSTEMON) sect. BRIDGESIANI (Rydberg) Pennell, Contr. U.S. Natl. Herb. 20: 327, 335. 1920

Penstemon [unranked] *Bridgesiani* Rydberg, Fl. Colorado, 306, 308. 1906 (as Pentstemon)

Subshrubs. Stems scabrous proximally, glabrous distally, sometimes wholly glabrous, glaucous or not. **Leaves** mostly cauline, basal usually absent or greatly reduced, opposite, leathery or not, glabrous, not glaucous; basal and proximal cauline petiolate; cauline petiolate, blade oblanceolate, lanceolate, or linear, margins entire. **Thyrses** interrupted, secund, axis glabrous proximally, glandular-pubescent distally, cymes 2 per node; peduncles and pedicels ascending. **Flowers:** calyx lobes: margins entire, herbaceous or narrowly scarious, glandular-pubescent; corolla red to scarlet, bilaterally symmetric, strongly bilabiate, not personate, salverform, glandular-pubescent externally, hairy internally abaxially, throat slightly inflated, not constricted at orifice, slightly 2-ridged abaxially; stamens: longer pair exserted (concealed by galeate adaxial lobes), filaments glabrous, pollen sacs parallel, saccate, dehiscing incompletely, distal $^2/_3$–$^3/_4$ indehiscent, connective splitting, sides glabrous, sometimes scabrellous near filament attachment, sutures denticulate, teeth to 0.1 mm; staminode included, terete, 0.1–0.2 mm diam., tip straight, glabrous; style glabrous. **Capsules** glabrous. **Seeds** light brown to brown, angled, 0.9–1.9 mm.

Species 1: w United States, nw Mexico.

Penstemon rostriflorus was included in sect. *Saccanthera* by D. D. Keck (1932). Molecular data (C. A. Wessinger et al. 2016) provided evidence for placing *P. rostriflorus* near sect. *Saccanthera*.

14. **Penstemon rostriflorus** Kellogg, Hutchings' Ill. Calif. Mag. 5: 102, fig. [p. 102]. 1860 (as Pentstemon rostriflorum) • Beak-flower beardtongue F

Penstemon bridgesii A. Gray

Stems (24–)30–100 cm. **Leaves:** basal and proximal cauline 20–52 × 3–11 mm, blade base tapered, margins entire, apex obtuse; cauline 4–12 pairs, 20–70 × 2–14 mm, blade base tapered, apex obtuse or acute. **Thyrses** (3–)6–28 cm, verticillasters (3–)6–12, cymes 2–5-flowered; proximal bracts lanceolate to linear, 6–36 × 1–6 mm; peduncles and pedicels glandular-pubescent. **Flowers:** calyx lobes ovate to lanceolate, 4–6 × 1.7–2.1 mm; corolla orangish red to scarlet, throat usually yellowish or orangish, unlined internally, 22–33 mm, sparsely white-puberulent, tube 3–5 mm, throat 4–6 mm diam.; pollen sacs 1.8–2.3(–2.5) mm; staminode 14–16 mm; style 18–26 mm, exserted. **Capsules** 7–10 × 4–5 mm. $2n = 16$.

Flowering Jul–Aug(–Sep). Rocky sagebrush shrublands, pinyon-juniper woodlands, montane forests; 1200–3200 m; Ariz., Calif., Colo., Nev., N.Mex., Utah; Mexico (Baja California).

A chromosome count of $2n = 42$ (P. G. Zhukova 1967) appears to be in error.

The Kawaiisu tribe of southern California prepared a poultice from the roots of *Penstemon rostriflorus* that was applied to swollen limbs (D. E. Moerman 1998).

26b.4. PENSTEMON Schmidel (subg. PENSTEMON) sect. CAESPITOSI (Rydberg) Pennell, Contr. U.S. Natl. Herb. 20: 327, 334. 1920

Penstemon [unranked] *Caespitosi* Rydberg, Fl. Colorado, 306. 1906 (as Pentstemon)

Subshrubs, usually cespitose. **Stems** retrorsely hairy, rarely glabrous, not glaucous. **Leaves** cauline, rarely basal and cauline (*P. retrorsus*), opposite, leathery or not, glabrous, glabrate, pubescent, scabrous, puberulent, or retrorsely hairy, not glaucous; basal and proximal cauline petiolate; cauline petiolate or sessile, blade obovate, spatulate, oblanceolate, lanceolate, elliptic, or linear, margins entire. **Thyrses** continuous or interrupted, cylindric or secund, axis retrorsely hairy, glandular-pubescent, or puberulent and glandular-pubescent, rarely glabrous, cymes 1 or 2 per node; peduncles and pedicels spreading to ascending or erect. **Flowers:** calyx tube 0.2–1 mm (1.5–2 mm in *P. retrorsus*), lobes: margins entire or erose, ± scarious or herbaceous, glandular-pubescent or retrorsely hairy, rarely glabrate; corolla lavender, blue, violet, or purple, rarely scarlet, bilaterally symmetric, bilabiate, not personate, funnelform, tubular-funnelform, salverform, ventricose-ampliate, or ampliate, glandular-pubescent externally, hairy internally abaxially, rarely glabrous, throat slightly to abruptly inflated, not constricted at orifice, 2-ridged abaxially; stamens included to exserted, filaments glabrous, rarely sparsely glandular-puberulent proximally, pollen sacs opposite, rarely divergent, navicular to subexplanate or explanate, dehiscing completely, connective splitting, sides glabrous, sutures papillate or smooth; staminode included to exserted, ± flattened distally, 0.1–0.5 mm diam., tip straight to recurved, distal (20–)50–100% hairy, hairs to 0.7–1.3 mm; style glabrous. **Capsules** glabrous. **Seeds** dark brown to black, angled to rounded, 1.1–2.1 mm.

Species 11 (11 in the flora): w United States, nw Mexico.

Most species of sect. *Caespitosi* formed a clade in the molecular study by A. D. Wolfe et al. (2006); placement of the clade within subg. *Penstemon* is not clear.

SELECTED REFERENCES Keck, D. D. 1937. Studies in *Penstemon* IV. The section *Ericopsis*. Bull. Torrey Bot. Club 64: 357–381. Nelson, A. 1937. Taxonomic studies. *Penstemon*. Publ. Univ. Wyoming 3: 101–107.

1. Corollas salverform, scarlet (throats sometimes yellow or yellow- or orange-spotted) . . . . 21. *Penstemon pinifolius*
1. Corollas tubular-funnelform, funnelform, ampliate, or ventricose-ampliate, lavender, blue, violet, or purple.

[2. Shifted to left margin.—Ed.]

2. Leaves glabrous or retrorsely hairy, hairs appressed, white, scalelike.
   3. Stems prostrate, spreading, or ascending; cauline leaves 1–5 pairs.
      4. Leaves densely retrorsely hairy, hairs appressed, white, scalelike; pollen sacs 0.8–1.2(–1.4) mm; Arizona, California, Nevada, Utah . . . . .24.   *Penstemon thompsoniae*
      4. Leaves glabrous or sparsely hairy, hairs appressed, white, scalelike abaxially, ± densely hairy, hairs appressed, white, scalelike adaxially; pollen sacs 0.6–0.8 mm; Utah . . . . 25.   *Penstemon xylus*
   3. Stems ascending to erect; cauline leaves 6–30 pairs.
      5. Leaves glabrous, rarely sparsely retrorsely hairy, hairs white, scalelike abaxially, densely puberulent, hairs white, scalelike adaxially; corollas 9–14 mm; pollen sacs 0.5–0.8 mm; Arizona . . . . 19.   *Penstemon discolor*
      5. Leaves sparsely to densely retrorsely hairy, hairs appressed, white, scalelike; corollas 14–20 mm; pollen sacs 0.9–1.4 mm; Arizona, California, New Mexico.
         6. Corollas tubular-funnelform, 14–18 mm, purple to violet; California . . . . .17.   *Penstemon californicus*
         6. Corollas ventricose-ampliate, 16–20 mm, blue to lavender; Arizona, New Mexico. . . . . .20.   *Penstemon linarioides* (in part)
2. Leaves glabrous, glabrate, or hairy, hairs pointed.
   7. Calyx tubes 1.5–2 mm; proximal bracts oblanceolate . . . . .22.   *Penstemon retrorsus*
   7. Calyx tubes 0.2–1 mm; proximal bracts oblanceolate to linear.
      8. Calyx lobes herbaceous- or narrowly scarious-margined.
         9. Stems and leaves not cinereous; leaf blades obovate to spatulate, oblanceolate, or linear; Arizona, Colorado, Utah, Wyoming . . . . .16.   *Penstemon caespitosus*
         9. Stems and leaves cinereous; leaf blades linear; Colorado . . . . 23.   *Penstemon teucrioides*
      8. Calyx lobes broadly scarious-margined.
         10. Corollas ventricose-ampliate; calyx lobe apices acute to acuminate . . . . .20.   *Penstemon linarioides* (in part)
         10. Corollas funnelform; calyx lobe apices acuminate to caudate.
            11. Leaf blades linear, rarely oblanceolate; stems ascending to erect; corollas glabrous or sparsely white-villous internally abaxially; Utah . . . . . . .15.   *Penstemon abietinus*
            11. Leaf blades elliptic to obovate, oblanceolate, or linear; stems prostrate, decumbent, ascending, or erect; corollas sparsely to densely yellow-lanate or yellow-pilose internally abaxially; Colorado, New Mexico, Utah . . . . 18.   *Penstemon crandallii*

15. **Penstemon abietinus** Pennell, Contr. U.S. Natl. Herb. 20: 376. 1920 • Fir-leaf beardtongue C E

Stems ascending to erect, 5–15 (–23) cm, retrorsely hairy or puberulent, hairs pointed. **Leaves** not leathery, puberulent proximally, hairs pointed, sometimes only along margins, glabrous distally; cauline 2–7 pairs, petiolate or sessile, 7–17 (–25) × 0.7–1.3(–1.7) mm, blade linear, rarely oblanceolate, base tapered, apex acute to mucronate. **Thyrses** continuous, secund, 2–8 cm, axis retrorsely hairy, hairs pointed, and, sometimes, sparsely glandular-pubescent, verticillasters 2–10, cymes 1–3 (or 4)-flowered, 1(or 2) per node; proximal bracts linear, 11–21 × 1–1.4 mm; peduncles and pedicels ascending to erect, retrorsely hairy, hairs pointed, and, sometimes, sparsely glandular-pubescent. **Flowers:** calyx lobes lanceolate, 4–5.5 × 1–1.6 mm, broadly scarious-margined, apex acuminate to caudate, sparsely glandular-pubescent, sometimes glabrate; corolla blue to bluish violet, lined internally abaxially with reddish violet nectar guides, funnelform, 12–15(–18) mm, glabrous or sparsely white-villous internally abaxially, tube 5–6 mm, throat gradually inflated, 6–7 mm diam., 2-ridged abaxially; stamens: longer pair reaching orifice or exserted, filaments glabrous, pollen sacs widely divergent or opposite, navicular, 0.8–1.1 mm, sutures papillate; staminode 6–8 mm, included, flattened distally, 0.3–0.4 mm diam., tip straight to recurved, distal 6–7 mm densely pilose, hairs golden yellow, to 1.1 mm; style 11–13 mm. **Capsules** 4.5–7 × 2.8–3.5 mm.

Flowering Jun–Jul. Limestone-derived soils in pinyon-juniper and oak-juniper woodlands; of conservation concern; 1800–2600 m; Utah.

Penstemon abietinus is known from Emery, Iron, Sevier, and Utah counties. It bears a strong resemblance to erect varieties of *P. crandallii* in habit, leaf shape and pubescence, and corolla shape.

16. Penstemon caespitosus Nuttall ex A. Gray, Proc. Amer. Acad. Arts 6: 66. 1862 (as Pentstemon) • Mat beardtongue E

Stems prostrate, only tip ascending, 2–8(–10) cm, not cinereous, retrorsely hairy, hairs pointed. **Leaves** not leathery, not cinereous, retrorsely hairy, hairs pointed; cauline 3–20 pairs, petiolate, 2.5–14(–21) × 1–4 mm, blade obovate to spatulate, oblanceolate, or linear, base tapered, apex rounded to acute or mucronate. **Thyrses** continuous, cylindric to secund, 0.5–3 cm, axis retrorsely hairy to retrorsely cinereous, hairs pointed, sometimes also glandular-pubescent, verticillasters 1–5, cymes 1–3-flowered, 1 per node; proximal bracts linear, 2.8–13(–17) × 0.7–3 mm; peduncles and pedicels ascending to erect, retrorsely hairy to retrorsely cinereous, hairs pointed, and glandular-pubescent. **Flowers:** calyx lobes lanceolate to linear, 4–6.5(–8.5) × 1–1.5(–1.9) mm, herbaceous- or narrowly scarious-margined, retrorsely hairy, hairs pointed, and sparsely glandular-pubescent; corolla blue to purplish lavender or lavender, lined internally abaxially with reddish violet nectar guides, funnelform, 10–17(–21) mm, yellow-villous internally abaxially, tube 5–6 mm, throat gradually inflated, 4.5–6 mm diam., 2-ridged abaxially; stamens: longer pair reaching orifice or exserted, filaments glabrous, pollen sacs opposite, navicular to explanate, 0.6–1.2 mm, sutures papillate; staminode 9–12 mm, reaching orifice, flattened distally, 0.3–0.5 mm diam., tip straight to recurved, distal 7–9 mm densely pilose, hairs golden yellow, to 1 mm; style 13–15 mm. **Capsules** 3.5–5 × 3–4 mm.

Varieties 3 (3 in the flora): w United States.

The varieties of *Penstemon caespitosus* are weakly differentiated and largely allopatric.

1. Pollen sacs 0.9–1.2 mm; leaf blades oblanceolate; Arizona, Utah . . . .
. . . . 16b. *Penstemon caespitosus* var. *desertipicti*
1. Pollen sacs 0.6–0.9 mm; leaf blades obovate, spatulate, oblanceolate, or linear; Colorado, Utah, Wyoming.
　2. Leaf blades oblanceolate to linear; pollen sacs explanate, 0.6–0.8 mm; Colorado, Utah, Wyoming . . . .
　. . . . . . . 16a. *Penstemon caespitosus* var. *caespitosus*
　2. Leaf blades obovate or spatulate; pollen sacs navicular to subexplanate, 0.8–0.9 mm; Colorado, Utah . . . .
　. . . . . 16c. *Penstemon caespitosus* var. *perbrevis*

16a. Penstemon caespitosus Nuttall ex A. Gray var. **caespitosus** E

Leaves 4–10(–21) × 1–3 mm; blade oblanceolate to linear. **Flowers:** pollen sacs explanate, 0.6–0.8 mm.

Flowering May–Jul(–Aug). Gravelly to clay soils, sagebrush shrublands, juniper woodlands, alpine meadows; 2000–2300 (–3700) m; Colo., Utah, Wyo.

Variety *caespitosus* is the northern element of the species, occurring in much of western Colorado (Chaffee, Eagle, Grand, Gunnison, Jackson, Mesa, Moffat, Park, Rio Blanco, and Routt counties), extreme northeastern Utah (Daggett, Duchesne, Rich, Uintah, and Wasatch counties), and southwestern Wyoming (Carbon, Lincoln, Sublette, Sweetwater, and Uinta counties). The variety is sympatric with var. *perbrevis* in Duchesne and Uintah counties, Utah. Indument varies considerably throughout the range of this variety, with glabrate forms not unusual (D. D. Keck 1937).

16b. Penstemon caespitosus Nuttall ex A. Gray var. **desertipicti** (A. Nelson) N. H. Holmgren, Brittonia 31: 104. 1979 E

Penstemon desertipicti A. Nelson, Univ. Wyoming Publ. Sci., Bot. 1: 130. 1926; *P. caespitosus* subsp. *desertipicti* (A. Nelson) D. D. Keck

Leaves 2.5–9 × 1–2.8 mm; blade oblanceolate. **Flowers:** pollen sacs navicular, 0.9–1.2 mm. $2n = 16$.

Flowering May–Jul. Gravelly to clay soils, sagebrush shrublands, pinyon-juniper and pine woodlands; 1500–2800 m; Ariz., Utah.

Variety *desertipicti* is known from the Aquarius, Paunsaugunt, and Sevier plateaus in south-central Utah southward to Painted Desert and Grand Canyon regions of northern Arizona. Populations have been documented in Apache, Coconino, and Mohave counties, Arizona, and Beaver, Garfield, Kane, Piute, and Wayne counties, Utah.

16c. Penstemon caespitosus Nuttall ex A. Gray var. **perbrevis** (Pennell) N. H. Holmgren, Brittonia 31: 104. 1979 [E]

Penstemon caespitosus subsp. *perbrevis* Pennell, Contr. U.S. Natl. Herb. 20: 375. 1920

Leaves 2.5–14 × 1–4 mm; blade obovate or spatulate. **Flowers:** pollen sacs navicular to subexplanate, 0.8–0.9 mm. Flowering May–Aug.

Gravelly to clay soils, sagebrush shrublands, pinyon-juniper and pine woodlands; 2100–2900 m; Colo., Utah.

Variety *perbrevis* occurs in Carbon, Duchesne, Emery, Garfield, Sanpete, Uintah, Utah, and Wasatch counties, Utah, and in Garfield County, Colorado. Some specimens from Rio Blanco County, Colorado, also approach this variety.

17. Penstemon californicus (Munz & I. M. Johnston) D. D. Keck, Bull. Torrey Bot. Club 64: 378. 1937 • California beardtongue

Penstemon linarioides A. Gray var. *californicus* Munz & I. M. Johnston, Bull. S. Calif. Acad. Sci. 23: 31. 1924

Stems ascending to erect, (5–)8–30 cm, retrorsely hairy, hairs white, scalelike. **Leaves** ± leathery or not, densely retrorsely hairy, hairs appressed, white, scalelike; cauline 10–30 pairs, petiolate, 8–16 × 1.5–2.5 mm, blade oblanceolate, base tapered, apex mucronate. **Thyrses** interrupted or continuous, secund, (1–)2–9 cm, axis retrorsely hairy with appressed, white, scalelike hairs, verticillasters 2–13, cymes 1–4-flowered, 1 per node; proximal bracts oblanceolate, 2–10 × 0.5–1.5 mm; peduncles and pedicels spreading to ascending, retrorsely hairy with appressed, white, scalelike hairs, and sparsely glandular-pubescent. **Flowers:** calyx lobes ovate, 3.5–5.5 × 1.5–2 mm, scarious-margined, retrorsely hairy with appressed, white, scalelike hairs, and sparsely glandular-pubescent; corolla purple to violet, lined internally abaxially with reddish purple nectar guides, tubular-funnelform, 14–18 mm, sparsely white-lanate internally abaxially, tube 5–6 mm, throat gradually inflated, 4.5–6 mm diam., 2-ridged abaxially; stamens included, filaments glabrous or sparsely puberulent and sometimes with appressed, white, scalelike hairs proximally, pollen sacs opposite, navicular, 0.9–1.4 mm, sutures papillate; staminode 7–9 mm, reaching orifice or exserted, flattened distally, 0.3–0.4 mm diam., tip straight to recurved, distal 4–5 mm pilose, hairs yellow, to 0.7 mm; style 9–11 mm. **Capsules** 5–7 × 4–5 mm.

Flowering May–Jul(–Aug). Sandy or gravelly soils, pine-juniper woodlands, pine forests; 1200–2300 m; Calif.; Mexico (Baja California).

In California, *Penstemon californicus* is known from the San Jacinto Mountains in Orange and Riverside counties; it also occurs in the Sierra San Pedro Mártir in Baja California. The chromosome number of $2n = 16$ reported by N. H. Holmgren (1993) is likely an error based on a count for *Scrophularia californica* reported in P. H. Raven et al. (1965) and later misattributed to *P. californicus* by D. V. Clark (1971) as $n = 48$ (the count listed in Raven et al.), and then corrected to $2n = 16$.

18. Penstemon crandallii A. Nelson, Bull. Torrey Bot. Club 26: 354. 1899 (as Pentstemon) • Crandall's beardtongue [E]

Stems prostrate, decumbent, ascending, or erect, 2–28 cm, retrorsely hairy or pubescent. **Leaves** not leathery, glabrous, glabrate, scabrous, or pubescent, hairs pointed; cauline (1 or)2–4 pairs, petiolate, (2–)4–47 × 0.5–8 mm, blade elliptic to obovate, oblanceolate, or linear, base tapered, apex acute to mucronate. **Thyrses** continuous, ± secund, 1–8(–16) cm, axis retrorsely hairy, hairs pointed and, usually, sparsely glandular-pubescent, verticillasters (1 or)2–8, cymes 1–4-flowered, 1 or 2 per node; proximal bracts oblanceolate to linear, 10–25 × 0.5–4 mm; peduncles and pedicels ascending to erect, retrorsely hairy and, usually, sparsely glandular-pubescent. **Flowers:** calyx lobes lanceolate, 4–7(–7.5) × 0.9–1.8(–2.2) mm, broadly scarious-margined, apex acuminate to caudate, retrorsely puberulent, hairs pointed, and sparsely glandular-pubescent; corolla blue to violet or lavender, lined internally abaxially with violet or reddish purple nectar guides, funnelform, 14–20 mm, sparsely to densely yellow-lanate or yellow-pilose internally abaxially, tube 6–8 mm, throat gradually inflated, 6–8 mm diam., 2-ridged abaxially; stamens: longer pair reaching orifice, filaments glabrous, pollen sacs opposite, navicular to subexplanate, 0.9–1.4 mm, sutures papillate; staminode 11–14 mm, reaching orifice or exserted, flattened distally, 0.3–0.4 mm diam., tip straight to recurved, distal 7–10 mm densely pilose, hairs golden yellow, to 1 mm; style 13–15 mm. **Capsules** 7–8 × 3.5–4.5 mm.

Varieties 5 (5 in the flora): w United States.

Penstemon crandallii comprises five partly sympatric, sometimes intergrading varieties.

1. Leaves glabrous or sparsely puberulent, especially along margins, blades obovate to elliptic; stems prostrate or decumbent; Colorado . . . . 18d. *Penstemon crandallii* var. *procumbens*
1. Leaves glabrate or scabrous to pubescent, blades linear, oblanceolate, or obovate; stems decumbent, ascending, or erect; Colorado, New Mexico, Utah.
    2. Leaf blades obovate or oblanceolate; stems decumbent to ascending.
        3. Leaves glabrate or scabrous, blades oblanceolate; Colorado, Utah. . . . . 18a. *Penstemon crandallii* var. *crandallii*
        3. Leaves ± scabrous, blades obovate or oblanceolate; Utah . . . . 18b. *Penstemon crandallii* var. *atratus*
    2. Leaf blades linear to oblanceolate; stems ascending to erect.
        4. Stems retrorsely hairy or puberulent, ascending to erect; Colorado, New Mexico. . . . . 18c. *Penstemon crandallii* var. *glabrescens*
        4. Stems puberulent to pubescent, erect; Colorado . . . . 18e. *Penstemon crandallii* var. *ramaleyi*

18a. Penstemon crandallii A. Nelson var. **crandallii** E

Penstemon suffrutescens Rydberg

Stems decumbent to ascending, 4–15 cm, retrorsely hairy. **Leaves** glabrate or scabrous, 6–26 × (0.5–)1–5 mm; blade oblanceolate.

Flowering May–Jul(–Aug). Sagebrush shrublands, pinyon-juniper and scrub oak woodlands; 1800–3400 m; Colo., Utah.

Variety *crandallii* is encountered in central and southwestern Colorado (Archuleta, Chaffee, Delta, Dolores, Eagle, Garfield, Gilpin, Grand, Gunnison, Hinsdale, Mesa, Montezuma, Montrose, Ouray, Park, Routt, and Teller counties) and San Juan County, Utah. Leaf indument is variable: typical var. *crandallii* has glabrate leaves; populations with scabrous leaves are not unusual.

18b. Penstemon crandallii A. Nelson var. **atratus** (D. D. Keck) N. H. Holmgren, Brittonia 31: 105. 1979 E

Penstemon crandallii subsp. *atratus* D. D. Keck, Bull. Torrey Bot. Club 64: 370. 1937

Stems decumbent to ascending, 2–8 cm, retrorsely hairy or puberulent. **Leaves** ± scabrous, 7–23 × 2–5 mm; blade obovate or oblanceolate. $2n = 16$.

Flowering May–Jul. Sagebrush shrublands, pinyon-juniper and scrub oak woodlands; 2100–2600 m; Utah.

Variety *atratus* apparently is limited to the La Sal Mountains of Grand County. N. H. Holmgren (1984) stated that some specimens of var. *crandallii* from the Abajo Mountains of San Juan County, Utah, approach var. *atratus* in leaf indument. Within populations of var. *atratus*, leaf indument varies from densely to sparsely puberulent.

18c. Penstemon crandallii A. Nelson var. **glabrescens** (Pennell) C. C. Freeman, J. Bot. Res. Inst. Texas 11: 287. 2017 E

Penstemon glabrescens Pennell, Contr. U.S. Natl. Herb. 20: 375. 1920; *P. coloradoensis* A. Nelson var. *glabrescens* (Pennell) A. Nelson; *P. crandallii* subsp. *glabrescens* (Pennell) D. D. Keck; *P. linarioides* A. Gray subsp. *taosensis* D. D. Keck

Stems ascending to erect, 6–28 cm, retrorsely hairy or puberulent. **Leaves** glabrate or scabrous, 6–25 × 0.5–1.5 mm; blade linear, rarely oblanceolate. $2n = 16$.

Flowering May–Jul(–Aug). Pinyon-juniper woodlands, ponderosa pine forests; 1700–3400 m; Colo., N.Mex.

Variety *glabrescens* is weakly differentiated from var. *crandallii*. Variety *glabrescens* is documented in Archuleta, Conejos, Custer, Delta, Dolores, Fremont, La Plata, Mineral, Montezuma, and Saguache counties, Colorado, and Rio Arriba and Taos counties, New Mexico.

18d. Penstemon crandallii A. Nelson var. **procumbens** (Greene) C. C. Freeman, PhytoKeys 80: 34. 2017 E

Penstemon procumbens Greene, Pl. Baker. 3: 23. 1901 (as *Pentstemon*); *P. crandallii* subsp. *procumbens* (Greene) D. D. Keck

Stems prostrate or decumbent, 2–18 cm, retrorsely hairy. **Leaves** glabrous or sparsely puberulent, especially along margins, (2–)4–24 × 2–8 mm; blade obovate to elliptic.

Flowering Jun–Jul. Sagebrush shrublands, scrub oak woodlands, spruce woodlands, talus slopes; 2000–3400 m; Colo.

Specimens referable to var. *procumbens* come from Carbon Peak, Kebler Pass, and West Elk Mountains. D. D. Keck (1937) noted the strong resemblance of the type specimen to *Penstemon caespitosus* subsp. *suffruticosus* A. Gray (= *P. xylus*). More material is needed to determine where var. *procumbens* fits in the *P. crandallii/P. caespitosus* complex.

18e. Penstemon crandallii A. Nelson var. **ramaleyi** (A. Nelson) C. C. Freeman, PhytoKeys 80: 35. 2017 E

Penstemon ramaleyi A. Nelson, Publ. Univ. Wyoming 3: 106. 1937

Stems erect, 6–25 cm, puberulent to pubescent. **Leaves** glabrate or scabrous to pubescent, 7–47 × 0.5–2 mm; blade oblanceolate to linear.

Flowering Jun–Aug. Sagebrush shrublands, pinyon-juniper and aspen woodlands; 2400–3300 m; Colo.

Variety *ramaleyi* has been largely overlooked; D. D. Keck (1937) did not account for it. The variety is similar in habit to var. *glabrescens* but has spreading hairs on the stems and, usually, on the leaves; var. *glabrescens* typically bears relatively shorter, retrorse hairs. Also, the leaves are longer and the stems more stiffly erect in var. *ramaleyi* than in var. *glabrescens*. Variety *ramaleyi* is known from Hinsdale, Mineral, and Saguache counties.

19. Penstemon discolor D. D. Keck, Bull. Torrey Bot. Club 64: 379. 1937 • Catalina beardtongue C E

Stems ascending to erect, 10–35 cm, glabrous or sparsely retrorsely hairy, hairs white, scalelike. **Leaves** leathery, glabrous, rarely sparsely retrorsely hairy, hairs white, scalelike abaxially, densely puberulent, hairs white, scalelike adaxially; cauline 10–30 pairs, petiolate, 5–18 × 1–1.5 mm, blade oblanceolate to linear, base tapered, apex mucronate. **Thyrses** continuous or interrupted, ± secund, 2–15 cm, axis glandular-pubescent, verticillasters 5–13, cymes 1–3-flowered, 1 per node; proximal bracts oblanceolate to linear, 5–12 × 0.5–1 mm; peduncles and pedicels ascending to erect, glandular-pubescent. **Flowers:** calyx lobes ovate to lanceolate, 2.5–3.5 × 1.2–1.8 mm, scarious-margined, glandular-pubescent, hairs pointed; corolla lavender to violet or purple, lined internally abaxially with reddish purple nectar guides, tubular-funnelform, 9–14 mm, densely yellow- or whitish lanate internally abaxially, tube 3–5 mm, throat gradually inflated, 3.5–5 mm diam., slightly 2-ridged abaxially; stamens included, filaments glabrous, pollen sacs opposite, navicular, 0.5–0.8 mm, sutures papillate; staminode 7–8 mm, included or reaching orifice, flattened distally, 0.2–0.4 mm diam., tip straight to recurved, distal 5–6 mm densely pilose, hairs golden yellow, to 1 mm; style 8–11 mm. **Capsules** 5–8 × 3–4.8 mm.

Flowering Jun–Aug. Pinyon-juniper and oak woodlands, granite crevices; of conservation concern; 1900–2100 m; Ariz.

Penstemon discolor is known from the Atascosa, Dragoon, Santa Catalina, and Santa Teresa mountains in Cochise, Graham, Pima, and Santa Cruz counties.

20. Penstemon linarioides A. Gray in W. H. Emory, Rep. U.S. Mex. Bound. 2(1): 112. 1859 (as Pentstemon) • Toad-flax beardtongue E

Stems ascending to erect, 8–50 cm, retrorsely hairy, hairs pointed or white, scalelike. **Leaves** not leathery, glabrous, glabrate, or sparsely to densely retrorsely hairy, hairs pointed, sometimes appressed, white, scalelike; cauline 6–12(–20) pairs, petiolate or sessile, 4–26 × 0.5–3.5 mm, blade oblanceolate to lanceolate or linear, base tapered to truncate, apex mucronate. **Thyrses** continuous or interrupted, ± secund, 3–17 cm, axis glandular-pubescent, verticillasters 3–9(–12), cymes 1 or 2(or 3)-flowered, 1(or 2) per node; proximal bracts lanceolate to linear, 6–25 × 0.7–1 mm; peduncles and pedicels ascending to erect, retrorsely hairy, hairs pointed, and sparsely glandular-pubescent. **Flowers:** calyx lobes ovate, rarely lanceolate, 4–7(–9) × 2–3 mm, broadly scarious-margined, apex acute to acuminate, glandular-pubescent, hairs pointed; corolla blue or lavender, lined internally abaxially with reddish purple nectar guides, ventricose-ampliate, 16–20 mm, moderately white- or yellowish villous internally abaxially, tube 5–6 mm, throat abruptly inflated, 5.5–8 mm diam., slightly 2-ridged abaxially; stamens: longer pair reaching orifice or slightly exserted, filaments glabrous, pollen sacs opposite, navicular, 0.9–1.3 mm, sutures papillate; staminode 6–8 mm, exserted, flattened distally, 0.3–0.5 mm diam., tip straight to recurved, distal 1–5 mm densely pilose, hairs yellow or golden yellow, to 1.2 mm, rest of distal 1–7 mm glabrous or sparsely to moderately pilose; style 10–11 mm. **Capsules** 5–9 × 3.5–5 mm.

Varieties 4 (4 in the flora): w United States.

Penstemon linarioides is widespread and highly variable. D. D. Keck (1937) recognized seven subspecies on the basis of habit, pubescence, leaf shape, and staminode bearding. Some of the variation on which those subspecies are based appears to be clinal or too variable within or among regional populations to be of taxonomic value.

1. Leaves glabrous, glabrate, retrorsely hairy, or pubescent, hairs pointed . . . . 20d. *Penstemon linarioides* var. *sileri*
1. Leaves sparsely to densely retrorsely hairy, hairs appressed, white, scalelike.
  2. Cauline leaf blades oblanceolate, distals sometimes lanceolate, 1.5–3.5 mm wide . . . . 20c. *Penstemon linarioides* var. *maguirei*
  2. Cauline leaf blades oblanceolate to linear, 0.5–2 mm wide.
    3. Staminodes: distal 3–5 mm densely pilose, hairs golden yellow or yellow, to 1 mm, rest of distal 1–4 mm sparsely to moderately pilose . . . . 20a. *Penstemon linarioides* var. *linarioides*
    3. Staminodes: distal 1 mm densely pilose, hairs yellow, to 0.8 mm, rest of distal 3–4 mm glabrous or sparsely pilose (with much shorter hairs) . . . . 20b. *Penstemon linarioides* var. *coloradoensis*

20a. Penstemon linarioides A. Gray var. **linarioides** [E]

Penstemon linarioides subsp. *compactifolius* D. D. Keck

Stems 10–50 cm. **Leaves** sparsely to densely retrorsely hairy, hairs appressed, white, scalelike, cauline 4–26 × 0.5–1.5 mm, blade linear. **Flowers:** staminode: distal 3–5 mm densely pilose, hairs golden yellow or yellow, to 1 mm, rest of distal 1–4 mm sparsely to moderately pilose. *2n* = 16.

Flowering Jun–Aug. Rocky slopes, sagebrush shrublands, pinyon-juniper woodlands; 1400–2900 m; Ariz., N.Mex.

Variety *linarioides* is known from Arizona (Apache, Cochise, Coconino, Gila, Graham, Greenlee, Navajo, and Yavapai counties) and western New Mexico (Cibola, Doña Ana, Grant, Hidalgo, Luna, McKinley, and Socorro counties). I. Méndez Larios and J. L. Villaseñor Ríos (2001) listed *Penstemon linarioides* for Baja California and Sonora, Mexico. D. D. Keck (1937) named plants with more compact, heathlike leaves, ascending stems, and decumbent rootstocks as subsp. *compactifolius*; he stated the subspecies was limited to the vicinity of Flagstaff and intergraded with both subsp. *sileri* and var. *viridis* D. D. Keck. Plants with leaves similar to subsp. *compactifolius* are encountered elsewhere and appear to be a confluent phenotypic expression of var. *linarioides*.

20b. Penstemon linarioides A. Gray var. **coloradoensis** (A. Nelson) C. C. Freeman, PhytoKeys 80: 35. 2017 [E]

Penstemon coloradoensis A. Nelson, Bull. Torrey Bot. Club 26: 355. 1899 (as Pentstemon); *P. linarioides* subsp. *coloradoensis* (A. Nelson) D. D. Keck

Stems 10–30 cm. **Leaves** moderately to densely retrorsely hairy, hairs appressed, white, scalelike, cauline 4–23 × 0.8–2 mm, blade oblanceolate to linear. **Flowers:** staminode: distal 1 mm densely pilose, hairs yellow, to 0.8 mm, rest of distal 3–4 mm glabrous or sparsely pilose (with much shorter hairs). *2n* = 16.

Flowering Jun–Jul. Sagebrush shrublands, pinyon-juniper and scrub oak woodlands; 2100–2600 m; Ariz., Colo., N.Mex.

Variety *coloradoensis* is known from northeastern Arizona (Apache and Navajo counties), southwestern Colorado (Dolores, La Plata, Montezuma, and San Miguel counties), and northwestern New Mexico (Rio Arriba and San Juan counties).

A decoction made from var. *coloradoensis* is used as a gynecological aid by the Ramah Navajo of western New Mexico (D. E. Moerman 1998).

20c. Penstemon linarioides A. Gray var. **maguirei** (D. D. Keck) A. Nelson, Publ. Univ. Wyoming 3: 105. 1937 • Maguire's beardtongue [E]

Penstemon linarioides subsp. *maguirei* D. D. Keck, Bull. Torrey Bot. Club 64: 378. 1937

Stems 15–25 cm. **Leaves** densely retrorsely hairy, hairs appressed, white, scalelike, cauline 8–23 × 1.5–3.5 mm, blade oblanceolate, distals sometimes lanceolate. **Flowers:** staminode: distal 1–2 mm densely pilose, hairs yellow, to 1 mm, rest of distal 6–7 mm sparsely pilose.

Flowering Jun–Oct. Rocky slopes and limestone cliffs in pine-juniper woodlands; 1800–2000 m; Ariz., N.Mex.

Variety *maguirei* is known from the Gila River Valley where it has been collected rarely in Greenlee County, Arizona, and Grant County, New Mexico.

20d. Penstemon linarioides A. Gray var. **sileri** A. Gray in A. Gray et al., Syn. Fl. N. Amer. 2(1): 270. 1878 (as Pentstemon) [E]

Penstemon linarioides subsp. *sileri* (A. Gray) D. D. Keck; *P. linarioides* var. *viridis* D. D. Keck

Stems 8–40 cm. **Leaves** glabrous, glabrate, retrorsely hairy, or puberulent, hairs pointed, cauline 8–26 × 0.5–2 mm, blade linear. **Flowers:** staminode: distal 3–5 mm densely pilose, hairs yellow, to 1.2 mm, rest of distal 2–3 mm sparsely to moderately pilose.

Flowering May–Sep(–Oct). Sandy to clay soils, sagebrush shrublands, oak or oak-juniper woodlands, pine forests; 1200–2400 m; Ariz., Nev., Utah.

Variety *sileri* is known from northern Arizona, southern Nevada, and southwestern Utah. Plants with glabrous or nearly glabrous leaves from northern Arizona and southern Utah have been referred to var. *viridis*. As noted by N. H. Holmgren (1984), such forms occur sporadically through much of the range of var. *sileri*, suggesting that taxonomic recognition of var. *viridis* is unwarranted. Some specimens included in var. *sileri* from northern Arizona, especially from the Kaibab Plateau, have the habit of subsp. *compactifolius* (included here in var. *linarioides*) but the leaf pubescence of var. *sileri* (or var. *viridis*). Holmgren identified three morpho-geographic races that may warrant recognition.

21. Penstemon pinifolius Greene, Bot. Gaz. 6: 218. 1881 (as Pentstemon) • Pine-leaf beardtongue [F]

Stems erect, (5–)10–50 cm, glabrous or retrorsely hairy, hairs pointed. **Leaves** leathery, sparsely retrorsely hairy, hairs pointed, especially along margins, or glabrous; cauline 10–30 pairs, sessile, 4–32 × 0.5–1.2 mm, blade linear, base tapered to clasping, apex acuminate. **Thyrses** interrupted, secund, (1–)3–15 cm, axis retrorsely hairy to puberulent, or glabrous, verticillasters 1–7, cymes 1–3-flowered, 1(or 2) per node; proximal bracts linear, 3–14 × 0.3–1 mm; peduncles and pedicels ascending to erect, glandular-pubescent, sometimes puberulent. **Flowers:** calyx lobes lanceolate, 5–7 × 1.5–2 mm, scarious-margined, glandular-pubescent, hairs pointed; corolla scarlet, throat sometimes yellow or yellow- or orange-spotted, unlined internally, salverform, 25–32 mm, yellow-lanate internally abaxially, tube 4–6 mm, throat slightly inflated, 4–5 mm diam., slightly 2-ridged abaxially; stamens exserted (often hidden by galeate adaxial lobes), filaments glabrous, pollen sacs opposite, explanate, 0.8–1 mm, sutures smooth; staminode 12–17 mm, included, slightly flattened distally, 0.1–0.2 mm diam., tip straight, distal 9–10 mm densely pilose, hairs yellow, to 1.3 mm; style 21–27 mm. **Capsules** 5–9 × 3–4 mm. $2n = 16$.

Flowering Jun–Jul(–Sep). Rocky slopes, cliffs; 1800–3100 m; Ariz., N.Mex.; Mexico (Chihuahua, Sonora).

Penstemon pinifolius has been included as an anomalous member of sect. *Leptostemon* Trautvetter; R. M. Straw (1962) expressed doubt about that association. It was allied with members of sect. *Caespitosi* in the molecular study by A. D. Wolfe et al. (2006).

22. Penstemon retrorsus Payson ex Pennell, Contr. U.S. Natl. Herb. 20: 373. 1920 • Adobe beardtongue [E]

Stems ascending to erect, 10–20 cm, retrorsely hairy, hairs pointed. **Leaves** not leathery, retrorsely hairy, hairs pointed; basal and proximal cauline 5–21 × 1–4 mm, blade spatulate to oblanceolate, base tapered, apex rounded to obtuse; cauline 2–4 pairs, petiolate or sessile, 8–30 × 2–5 mm, blade spatulate to oblanceolate, base tapered, apex rounded to obtuse, acute, or mucronate. **Thyrses** continuous, cylindric, 3–12 cm, axis retrorsely hairy, hairs pointed,

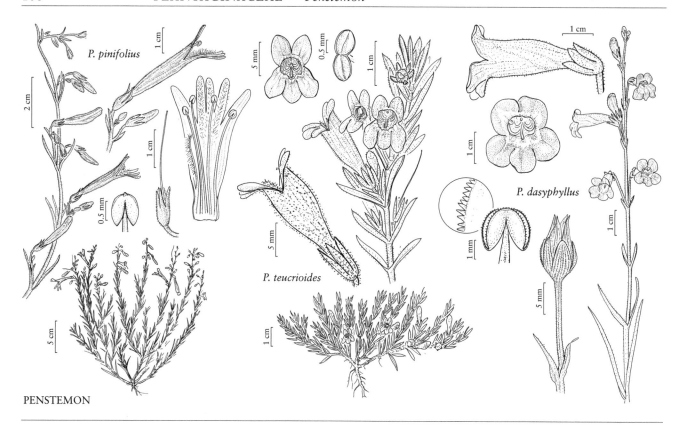

P. pinifolius

P. teucrioides

P. dasyphyllus

PENSTEMON

verticillasters 4–12, cymes 1–4-flowered, 1 or 2 per node; proximal bracts oblanceolate, 11–22 × 2–3 mm; peduncles and pedicels ascending to erect, retrorsely hairy, hairs pointed. **Flowers:** calyx tube 1.5–2 mm, lobes lanceolate to oblong, 3.5–4.5 × 0.7–1.2 mm, herbaceous- or narrowly scarious-margined, retrorsely hairy, hairs pointed; corolla bluish violet, lined internally abaxially with reddish violet nectar guides, tubular-funnelform, 16–20 mm, sparsely white-lanate internally abaxially, tube 6–7 mm, throat gradually inflated, 4.5–6 mm diam., 2-ridged abaxially; stamens included, filaments glabrous, pollen sacs opposite, subexplanate, 0.7–0.9 mm, sutures papillate; staminode 7–9 mm, included or reaching orifice, flattened distally, 0.2–0.3 mm diam., tip recurved, distal 4–6 mm pubescent, hairs yellowish orange, to 0.8 mm; style 12–15 mm. **Capsules** not seen.

Flowering May–Jun. Clay soils, sagebrush, saltbush shrublands, pinyon-juniper woodlands; 1500–2500 m; Colo.

Penstemon retrorsus is known only from Delta and Montrose counties, where populations occur on barren hills of the Mancos Shale. It is unique in having calyx tubes roughly as long as the calyx lobes.

23. **Penstemon teucrioides** Greene, Pl. Baker. 3: 23. 1901 (as Pentstemon) • Germander beardtongue [E] [F]

Stems prostrate or ascending, 2–10 cm, cinereous, retrorsely hairy, hairs pointed. **Leaves** not leathery, cinereous, densely retrorsely hairy, hairs pointed; cauline 5–9 pairs, sessile, 4–14 × 0.5–1.4 mm, blade linear, base tapered, apex mucronate, sometimes acuminate. **Thyrses** continuous, cylindric, 3–15 cm, axis retrorsely hairy to retrorsely cinereous, hairs pointed, verticillasters 1–8, cymes 1 or 2(or 3)-flowered, 1 per node; proximal bracts linear, 6–15 × 0.6–1.5 mm; peduncles and pedicels ascending to erect, retrorsely cinereous, hairs pointed, and, sometimes, sparsely glandular-pubescent. **Flowers:** calyx lobes lanceolate, (3–)4–6 × 0.7–1 mm, narrowly scarious-margined, retrorsely hairy, hairs pointed, and sparsely glandular-pubescent; corolla blue or violet, lined internally abaxially with reddish violet nectar guides, funnelform, 15–19 mm, sparsely white- or yellowish lanate internally abaxially, tube 6–7 mm, throat gradually inflated, 4–5.5 mm diam., 2-ridged abaxially; stamens: longer pair reaching orifice, filaments glabrous, pollen sacs opposite, navicular to subexplanate, 0.7–1 mm, sutures papillate; staminode 8–9 mm, included or reaching orifice, flattened distally,

0.2–0.3 mm diam., tip straight to recurved, distal 6–7 mm pilose, hairs yellowish orange, to 0.7 mm; style 13–16 mm. **Capsules** 3.3–4.5 × 2.8–3.6 mm. **2***n* = 16.

Flowering May–Jul(–Aug). Silt or gravelly slopes in sagebrush shrublands; 2200–3400 m; Colo.

Penstemon teucrioides is known from the Gunnison Basin. Populations have been documented in Chaffee, Gunnison, Hinsdale, Park, and Saguache counties. E. O. Wooton and P. C. Standley (1915) cited a specimen from Tierra Amarilla, Rio Arriba County, New Mexico, which appears to be in error. *Penstemon teucrioides* is most likely confused with *P. crandallii* but can be distinguished reliably by its densely retrorsely hairy, cinereous leaves, narrowly scarious-margined calyx lobes, and 0.7–1 mm pollen sacs. *Penstemon teucrioides* also can be mistaken for *P. caespitosus*; the two species can be distinguished by the indument of stems and leaves.

24. **Penstemon thompsoniae** (A. Gray) Rydberg, Bull. Torrey Bot. Club 36: 690. 1909 (as Pentstemon)

• Thompson's beardtongue E

Penstemon pumilus Nuttall var. *thompsoniae* A. Gray in A. Gray et al., Syn. Fl. N. Amer. 2(1): 269. 1878 (as Pentstemon); *P. thompsoniae* var. *desperatus* Neese; *P. thompsoniae* subsp. *jaegeri* D. D. Keck

Stems prostrate or ascending, 4–15(–25) cm, retrorsely hairy, hairs appressed, white, scalelike. **Leaves** not leathery, densely retrorsely hairy, hairs appressed, white, scalelike; cauline 1–5 pairs, petiolate, 6–25(–33) × 2–6.5 mm, blade obovate to spatulate, base tapered, apex mucronate, sometimes rounded or obtuse. **Thyrses** continuous, secund, (1–)2–12 cm, axis retrorsely hairy, hairs appressed, white, scalelike, sometimes also glandular-pubescent distally, verticillasters (1–)3–12, cymes 1–3(–5)-flowered, 1 per node; proximal bracts spatulate to oblanceolate, 5–19 × 1.5–4 mm; peduncles and pedicels spreading to ascending, retrorsely hairy, hairs pointed, and, sometimes, sparsely glandular-pubescent. **Flowers:** calyx lobes lanceolate, 4–6 × 1–1.7 mm, herbaceous- or narrowly scarious-margined, sparsely glandular-pubescent and retrorsely hairy, hairs appressed, white, scalelike; corolla blue to violet or purple, lined internally abaxially with reddish violet nectar guides, ampliate, 10–18 mm, yellow-lanate internally abaxially, tube 6–7 mm, throat gradually inflated, 4.5–6 mm diam., 2-ridged abaxially; stamens reaching orifice or longer pair exserted, filaments glabrous, pollen sacs divergent, navicular, 0.8–1.2(–1.4) mm, sutures papillate; staminode 8–9 mm, exserted, flattened distally, 0.2–0.3 mm diam., tip recurved, distal

6–8 mm pubescent, hairs yellow or orange, to 0.8 mm; style 10–13 mm. **Capsules** 3.5–5.5 × 3–4 mm.

Flowering May–Aug. Sandy to gravelly soils, sagebrush shrublands, pine-juniper woodlands, pine forests; 1500–3400 m; Ariz., Calif., Nev., Utah.

D. D. Keck (1937) recognized two subspecies in his treatment of *Penstemon thompsoniae*: subsp. *jaegeri*, restricted to mountains of Clark County, Nevada, which he distinguished by its few, remote stems and open inflorescences, and subsp. *thompsoniae* in Arizona, Nevada, and Utah, which he distinguished by its tufted stems and compact inflorescences. A morphologic continuum exists between the two subspecies.

Neese described var. *desperatus* from Beaver and Iron counties, Utah, and eastern Nevada, distinguishing it from var. *thompsoniae* by its longer stems and inflorescences, and smaller leaves. She later concluded that geographic variation in *Penstemon thompsoniae* was not sufficiently geographically correlated to recognize infraspecific taxa (E. C. Neese 1993).

25. **Penstemon xylus** A. Nelson, Bot. Gaz. 34: 31. 1902 (as Pentstemon) • Tushar Range beardtongue C E

Penstemon caespitosus Nuttall ex A. Gray var. *suffruticosus* A. Gray in A. Gray et al., Syn. Fl. N. Amer. 2(1): 270. 1878 (as Pentstemon)

Stems prostrate, spreading, or ascending, 6–12 cm, retrorsely hairy, hairs pointed, sometimes also appressed, white, scalelike. **Leaves** not leathery, glabrous or sparsely hairy, hairs appressed, white, scalelike abaxially, ± densely hairy, hairs appressed, white, scalelike adaxially; cauline 1–5 pairs, petiolate, 6–16 × 2–4.5 mm, blade obovate to spatulate, base tapered, apex rounded to mucronate or acute. **Thyrses** continuous, secund, 2–12 cm, axis retrorsely hairy, hairs appressed or spreading, white, scalelike, sometimes also glandular-pubescent distally, verticillasters 3–12, cymes 1–3(or 4)-flowered, 1 per node; proximal bracts spatulate to oblanceolate, 8–17 × 1.4–4 mm; peduncles and pedicels spreading to ascending, glandular-pubescent, hairs appressed, white, scalelike. **Flowers:** calyx lobes lanceolate to oblong, 3–5 × 0.7–1 mm, herbaceous-margined, glandular-pubescent and retrorsely hairy, hairs appressed, white, scalelike; corolla blue to lavender, or violet, lined internally abaxially with reddish violet nectar guides, ampliate, 12–18 mm, yellow-lanate internally abaxially, tube 5–6 mm, throat gradually inflated, 3.5–5 mm diam., 2-ridged abaxially; stamens reaching orifice or longer pair exserted, filaments glabrous, pollen sacs opposite, navicular to subexplanate, 0.6–0.8 mm, sutures papillate; staminode 7–8 mm, included, flattened distally, 0.2–0.3 mm diam., tips straight to recurved, distal 6–7 mm pubescent, hairs yellow or orange, to 0.7 mm; style 11–14 mm. **Capsules** 3–5 × 3–4 mm.

Flowering Jun–Aug. Sandy or gravelly slopes, sagebrush shrublands, aspen woodlands, pine woodlands; of conservation concern; 2100–3100 m; Utah.

Penstemon xylus is known from south-central Utah. Populations are documented in Beaver, Garfield, Iron, Piute, and Sevier counties. The placement of *P. xylus* remains unclear. D. D. Keck (1937) treated it as *P. caespitosus* subsp. *suffruticosus*. N. H. Holmgren (1979b) believed it to be more closely related to *P. thompsoniae*, which also has vestiture of appressed, white, scalelike hairs.

The correct name for *Penstemon xylus* at the rank of species, which has been a source of confusion, was discussed by C. C. Freeman (2017).

26b.5. PENSTEMON Schmidel (subg. PENSTEMON) sect. CHAMAELEON Crosswhite, Sida 2: 339. 1966

Herbs. **Stems** retrorsely hairy or puberulent, sometimes glandular-pubescent distally or glabrescent, not glaucous. **Leaves** basal and cauline, basal usually absent at anthesis, opposite, leathery or not, glabrous, puberulent, or retrorsely hairy, not glaucous; basal and proximal cauline petiolate, sometimes short-petiolate or sessile (*P. dasyphyllus*); cauline sessile, sometimes short-petiolate (*P. lanceolatus*), blade lanceolate or linear, margins entire. **Thyrses** interrupted, ± secund to cylindric, axis glandular-pubescent or glabrous, cymes 1 or 2 per node; peduncles and pedicels ascending to erect. **Flowers:** calyx lobes: margins entire or erose (*P. stenophyllus*), herbaceous or narrowly scarious, sometimes broadly scarious (*P. stenophyllus*), glandular-pubescent or glabrous; corolla lavender to blue, violet, purple, red, or scarlet, bilaterally symmetric, bilabiate, not personate, ventricose or tubular-funnelform, glandular-pubescent externally, glabrous or glandular-pubescent internally abaxially, throat abruptly or gradually inflated, not constricted at orifice, rounded to 2-ridged abaxially; stamens included to exserted, filaments glabrous, pollen sacs parallel to divergent, navicular to slightly saccate, dehiscing completely or incompletely, distal ⅕ or less sometimes indehiscent, connective splitting, sides glabrous, sutures denticulate, teeth to 0.2–0.3 mm; staminode included or exserted, flattened distally, 0.3–2.6 mm diam., tip recurved, glabrous; style glabrous. **Capsules** glabrous. **Seeds** brown to dark brown, angled, 1.9–3.2 mm.

Species 4 (3 in the flora): sw United States, Mexico.

Crosswhite circumscribed sect. *Chamaeleon* to include a group of southern species distinguished by incompletely dehiscent pollen sacs with prominently denticulate sutures. *Penstemon punctatus* Brandegee, a species not treated by Crosswhite, might belong here; the only specimens of that species seen were from Coahuila, Mexico.

SELECTED REFERENCE Crosswhite, F. S. 1966. Revision of *Penstemon* section *Chamaeleon* (Scrophulariaceae). Sida 2: 339–346.

1. Corollas scarlet to red, tubular-funnelform, glandular-pubescent internally abaxially . . . .
   . . . . 27. *Penstemon lanceolatus*
1. Corollas violet to blue, lavender, or purple, ventricose, glabrous internally.
   2. Thyrse axes glandular-pubescent . . . . 26. *Penstemon dasyphyllus*
   2. Thyrse axes glabrous . . . . 28. *Penstemon stenophyllus*

26. Penstemon dasyphyllus A. Gray in W. H. Emory,
Rep. U.S. Mex. Bound. 2(1): 112. 1859 (as Pentstemon)
• Thick-leaf beardtongue F

Stems 20–50 cm, puberulent or retrorsely hairy proximally, puberulent and glandular-pubescent distally. **Leaves** puberulent or retrorsely hairy, sometimes only along midveins and margins; basal and proximal cauline 28–68 × 3–9 mm, blade oblanceolate to lanceolate or linear, base tapered, margins entire, apex obtuse to acute; cauline 6–9 pairs, 32–95 × 4–12 mm, blade lanceolate to linear, base tapered, apex acute. **Thyrses** ± secund, 6–28 cm, axis glandular-pubescent, verticillasters 4–11, cymes 1- or 2-flowered; proximal bracts lanceolate to linear, 4–35 × 1–4 mm; peduncles and pedicels glandular-pubescent. **Flowers:** calyx lobes ovate to lanceolate, 5–8 × 2–2.8 mm, glandular-pubescent; corolla violet to lavender to purple, lined internally abaxially with purple nectar guides, ventricose, 25–35 mm, glabrous internally, tube 7–9 mm, throat abruptly inflated, 9–10 mm diam., slightly 2-ridged abaxially; stamens: longer pair reaching orifice, pollen sacs divergent, 1.8–2.2 mm, sutures denticulate, teeth to 0.3 mm; staminode 17–19 mm, included, 0.6–1 mm diam.; style 15–20 mm. **Capsules** 11–15 × 7–9 mm. $2n = 16$.

Flowering Apr–Jul(–Sep). Rocky ridges, gravelly slopes, desert grasslands; 1100–1700 m; Ariz., N.Mex., Tex.; Mexico (Chihuahua, Coahuila, Durango).

In the flora area, *Penstemon dasyphyllus* is known from southeastern Arizona (Cochise, Gila, Pima, Pinal, and Santa Cruz counties), southwestern New Mexico (Hidalgo and Luna counties), and western Texas (Brewster, Crockett, Pecos, Presidio, and Terrell counties). Populations are documented southward to northern Durango, Mexico.

27. Penstemon lanceolatus Bentham, Pl. Hartw., 22.
1839 (as Pentstemon) • Lance-leaf beardtongue

Penstemon ramosus Crosswhite

Stems 24–75(–98) cm, sparsely to densely retrorsely hairy or puberulent, usually also glandular-pubescent distally. **Leaves** retrorsely hairy; basal and proximal cauline 8–80 × 3–12 mm, blade lanceolate or spatulate, base tapered, margins entire, apex rounded to obtuse or acute; cauline 7–15 pairs, 9–85(–110) × 1–11(–22) mm, blade lanceolate to linear, base tapered to truncate, apex acute to acuminate, rarely obtuse. **Thyrses** ± secund, 5–36 cm, axis glandular-pubescent, verticillasters 3–6, cymes 1 or 2(or 3)-flowered; proximal bracts lanceolate to linear, 5–16 × 1–3 mm; peduncles and pedicels glandular-pubescent, sometimes also retrorsely hairy. **Flowers:** calyx lobes ovate to lanceolate, 2.9–7 × 1.5–2.5(–3.6) mm, glandular-pubescent; corolla scarlet to red, unlined internally or lined on abaxial surface with faint reddish purple nectar guides, tubular-funnelform, 22–35 mm, glandular-pubescent internally abaxially, tube 5–6 mm, throat gradually inflated, 6–8 mm diam., rounded abaxially; stamens exserted (hidden by galeate adaxial lobes), pollen sacs parallel to divergent, 1.5–2 mm, sutures denticulate, teeth to 0.2 mm; staminode 15–17 mm, included, 0.3–0.4 mm diam.; style 25–30 mm, usually barely exserted from galea. **Capsules** 9–14 × 6–7 mm. $2n = 16$.

Flowering May–Sep. Gravelly pinyon-juniper woodlands, pine woodlands, thorn scrub, desert grasslands; 1200–1800 m; Ariz., N.Mex., Tex.; Mexico (Aguascalientes, Chihuahua, Coahuila, Durango, Nayarit, Nuevo León, San Luis Potosí, Tamaulipas, Veracruz, Zacatecas).

Penstemon lanceolatus is known from southern Arizona (Cochise, Graham, and Greenlee counties), southern New Mexico (Doña Ana, Grant, Hidalgo, Luna, and Sierra counties), and western Texas (Brewster County).

Crosswhite published *Penstemon ramosus* as an avowed substitute for the homonym *P. pauciflorus* Greene, distinguishing *P. ramosus* from *P. lanceolatus* by the former's branched stems and relatively narrower leaves (1 mm wide versus 4–8 mm wide) with revolute margins. He mapped *P. ramosus* in southeastern Arizona and southwestern New Mexico and *P. lanceolatus* in western Texas and northern Mexico. Morphologic differences between these taxa are not consistent enough to warrant recognition of *P. ramosus* (J. L. Anderson et al. 2007).

SELECTED REFERENCE Anderson, J. L., S. Richmond-Williams, and O. Williams. 2007. *Penstemon lanceolatus* Benth. or *P. ramosus* Crosswhite in Arizona and New Mexico, a peripheral or endemic species? In: P. L. Barlow-Irick et al., eds. 2007. Southwestern Rare and Endangered Plants: Proceedings of the Fourth Conference. Fort Collins. Pp. 8–15.

28. Penstemon stenophyllus A. Gray in W. H. Emory,
Rep. U.S. Mex. Bound. 2(1): 112. 1859 (as Pentstemon)
• Sonoran beardtongue

Stems 20–90 cm, retrorsely hairy, sometimes glabrescent. **Leaves** glabrous or retrorsely hairy; basal and proximal cauline 16–80 × 3–10 mm, blade oblanceolate to lanceolate, base tapered, margins entire, apex obtuse to acute; cauline 6–15(–20) pairs, 45–150 × 1–6 mm, blade linear, base tapered, apex acute to acuminate.

Thyrses cylindric to ± secund, 10–35 cm, axis glabrous, verticillasters (2–)5–8, cymes (1 or)2–4 (or 5)-flowered; proximal bracts linear, 8–30(–80) × 0.5–2 mm; peduncles and pedicels glabrous or pedicels rarely glandular-pubescent. **Flowers:** calyx lobes elliptic to ovate, 4.2–7(–8.5) × 2–2.6(–3.5) mm, glabrous, rarely glandular-pubescent; corolla violet to blue, lavender, or purple, lined internally abaxially with reddish purple nectar guides, ventricose, 23–37 mm, glabrous internally, tube 7–10 mm, throat abruptly inflated, 9–10 mm diam., slightly 2-ridged abaxially; stamens included, pollen sacs parallel to divergent, 1.6–1.8 mm, sutures denticulate, teeth to 0.2 mm; staminode 20–23 mm, exserted, 1.8–2.6 mm diam.; style 20–24 mm. **Capsules** 9–15 × 6–9 mm.

Flowering Jul–Oct. Desert grasslands, openings in pine and pine-oak woodlands; 1200–1700 m; Ariz.; Mexico (Chihuahua, Durango, Sonora, Zacatecas).

In the flora area, *Penstemon stenophyllus* is known from the Huachuca and Patagonia mountains in Cochise and Santa Cruz counties. In Mexico, populations occur primarily along the Sierra Madre Occidental southward to central Durango at elevations to 2300 m.

26b.6. Penstemon Schmidel (subg. Penstemon) sect. Coerulei Pennell, Contr. U.S. Natl. Herb. 20: 326, 331. 1920

Herbs. Stems glabrous, sometimes scabrous, glaucous. **Leaves** basal and cauline, sometimes basal absent or reduced, opposite, leathery, glabrous, rarely scabrous, glaucous; basal and proximal cauline petiolate, sometimes short-petiolate (*P. immanifestus, P. lentus*); cauline sessile, sometimes short-petiolate, blade obovate, ovate, spatulate, trullate, oblanceolate, lanceolate, oblong, elliptic, linear, or orbiculate, margins entire. **Thyrses** continuous or interrupted, cylindric, sometimes secund, axis glabrous, rarely scabrous or glutinous, cymes 2 per node; peduncles and pedicels ascending to erect. **Flowers:** calyx lobes: margins entire or erose, herbaceous or ± scarious, glabrous, rarely scabrous, glandular, glandular-pubescent, or glutinous; corolla lavender to blue, violet, purple, or pink, rarely white or red, bilaterally symmetric, rarely nearly radially symmetric (*P. cyathophorus*), weakly, rarely strongly, bilabiate, not personate, tubular-funnelform, funnelform, or weakly ventricose, rarely tubular-salverform or weakly ampliate, glabrous externally, rarely obscurely glandular, glabrous or hairy internally abaxially, throat slightly to abruptly inflated, not constricted at orifice, rounded abaxially, rarely slightly 2-ridged; stamens included to exserted, filaments glabrous, pollen sacs opposite, rarely divergent or parallel, navicular, rarely explanate (*P. murrayanus*), dehiscing completely, connective splitting, sides glabrous, sutures papillate, sometimes smooth; staminode included to exserted, flattened distally, (0.2–)0.4–3 mm diam., tip recurved, rarely coiled or straight, distal 10–50(–70)% hairy, hairs to 2.5 mm, rarely glabrous; style glabrous. **Capsules** glabrous. **Seeds** brown or dark brown, angled, angled-elongate, or disciform, 1.8–5.4 mm.

Species 20 (20 in the flora): North America, n Mexico.

D. D. Keck (1951) proposed the section name *Anularius* for sect. *Coerulei*, believing that the name of Pennell was invalid. N. H. Holmgren (1979b) concluded that the section names of Pennell are valid and that sect. *Anularius* Keck is invalid; the new combinations by Keck in that publication lacked descriptions.

Members of sect. *Coerulei* usually are recognized by their mostly glabrous, glaucous, and fleshy herbage, their blue or pink corollas that are glabrous, rarely obscurely glandular, externally, and their relatively broad and heavily bearded staminodes. *Penstemon murrayanus* has red corollas and glabrous staminodes. Members of sect. *Coerulei* mostly occur on well-drained, sandy soils. Molecular data (C. A. Wessinger et al. 2016) support the monophyly of the section.

1. Corollas red; staminodes glabrous; pollen sacs explanate . . . . 44. *Penstemon murrayanus*
1. Corollas lavender to blue, violet, purple, or pink, rarely white; staminodes hairy; pollen sacs navicular.

[2. Shifted to left margin.—Ed.]

2. Pollen sacs parallel or divergent; stamens: 2 or 4 prominently exserted; corollas without nectar guides.
  3. Stamens: 4 exserted; pollen sacs 1.2–1.5 mm; corollas 11–14 mm . . . . 35. *Penstemon cyathophorus*
  3. Stamens: 2 exserted; pollen sacs 2–3 mm; corollas 17–24 mm . . . . 39. *Penstemon harringtonii*
2. Pollen sacs opposite or divergent; stamens included or longer pair reaching orifice to slightly exserted; corollas usually with nectar guides.
  4. Corollas 21–48 mm, ampliate, glabrous internally, throats abruptly inflated.
    5. Corollas 35–48 mm; cauline leaf blades spatulate to orbiculate . . . . 38. *Penstemon grandiflorus*
    5. Corollas 21–28 mm; cauline leaf blades lanceolate to linear . . . . 40. *Penstemon haydenii*
  4. Corollas 10–25(–28) mm, tubular-funnelform, tubular-salverform, funnelform, or ventricose, white-villous internally abaxially or glabrous, throats not abruptly inflated.
    6. Corollas tubular-salverform; thyrses interrupted. . . . . 36. *Penstemon fendleri*
    6. Corollas tubular-funnelform, funnelform, or ventricose; thyrses continuous or interrupted.
      7. Stems 3–10(–15) cm; cauline leaf pairs 1–3; thyrses 1–6 cm; verticillasters 1–5(or 6) . . . . .32. *Penstemon bracteatus*
      7. Stems (4–)7–77(–82) cm; cauline leaf pairs 1–8(or 9); thyrses (2–)3–37(–57) cm; verticillasters (2 or)3–20(–35).
        8. Thyrses secund.
          9. Cauline leaves 2–4 pairs; staminodes 8–9 mm, distal 3–4 mm sparsely to densely villous, hairs yellowish, to 1 mm; Arizona, Colorado, New Mexico, Utah . . . . 42. *Penstemon lentus* (in part)
          9. Cauline leaves 4–6 pairs; staminodes 10–13 mm, distal 4–6 mm densely villous, hairs golden yellow, to 2 mm; Colorado, New Mexico, Wyoming. . . . . 48. *Penstemon secundiflorus*
        8. Thyrses cylindric.
          10. Staminodes 2–3 mm diam.; corollas with nectar guides, sparsely to moderately white-villous internally abaxially, rarely glabrous . . . . 46. *Penstemon osterhoutii*
          10. Staminodes (0.2–)0.4–2 mm diam.; corollas with or without nectar guides, glabrous or sparsely white-villous internally abaxially.
            11. Corollas pink or white.
              12. Corollas white . . . . 42. *Penstemon lentus* (in part)
              12. Corollas pink.
                13. Leaves: basal usually reduced or absent; corollas glabrous internally; staminodes: distal 0.5–1.5 mm lanulose, hairs to 0.6 mm . . . . 37. *Penstemon flowersii*
                13. Leaves: basal not reduced; corollas sparsely white-hairy internally abaxially or glabrous; staminodes: distal 2–6 mm villous, hairs to 1 mm.
                  14. Cauline leaf apices acute to acuminate; corollas tubular-funnelform; calyx lobes ovate to lanceolate, 4–8 × 1–2.5 mm . . . . .30. *Penstemon angustifolius* (in part)
                  14. Cauline leaf apices obtuse to acute; corollas funnelform to ventricose; calyx lobes ovate, 5–8(–12) × 2–3.5 mm. . . . . 41. *Penstemon immanifestus* (in part)
            11. Corollas blue, pinkish blue, lavender-blue, lavender, violet, or purple, rarely some individuals with corollas pink.
              15. Cauline leaf blades ovate to orbiculate; corollas with nectar guides, villous internally abaxially . . . . 43. *Penstemon mucronatus*
              15. Cauline leaf blades elliptic to ovate, obovate, lanceolate, oblanceolate, or linear; corollas with or without nectar guides, glabrous or villous internally abaxially.

[16. Shifted to left margin.—Ed.]

16. Stems decumbent to ascending, (4–)7–18(–30) cm; thyrses continuous; proximal bracts oblanceolate to lanceolate; corollas (10–)12–15 mm; pollen sacs 0.7–0.8 mm...... 31. *Penstemon arenicola*
16. Stems ascending to erect, (5–)9–60(–82) cm; thyrses interrupted or continuous; proximal bracts orbiculate to ovate, lanceolate, oblong, or linear; corollas 10–22 mm; pollen sacs 0.7–1.6 mm.
    17. Proximal bracts orbiculate to ovate; corollas lavender to bluish lavender or pinkish blue, glabrous internally; capsules 12–18(–20) mm...................... 33. *Penstemon buckleyi*
    17. Proximal bracts ovate to oblong, lanceolate, or linear, rarely orbiculate; corollas blue, pinkish blue, bluish lavender, violet, or purple, hairy internally abaxially or glabrous; capsules 7–15 mm.
        18. Corollas obscurely glandular externally or glabrous; cauline leaves 1–3 pairs; thyrses 3–15 cm, continuous, sometimes interrupted ....34. *Penstemon carnosus*
        18. Corollas glabrous externally; cauline leaves 2–8 pairs; thyrses (2–)5–30(–37) cm, continuous or interrupted.
            19. Staminodes 1–2 mm diam., distal 6–8 mm villous, hairs to 2.5 mm........ ....47. *Penstemon pachyphyllus* (in part)
            19. Staminodes (0.2–)0.5–1.5 mm diam., distal 1.5–7 mm pilose to villous, hairs to 1.5 mm.
                20. Cauline leaf blades: apices obtuse to acute, sometimes rounded or mucronate; peduncles to 12–28 mm; corolla throats 5–8 mm diam.
                    21. Cymes 1–3(or 4)-flowered; cauline leaves sessile; seeds 3.5–5 mm... .... 42. *Penstemon lentus* (in part)
                    21. Cymes (1 or)2–6-flowered; cauline leaves short-petiolate or sessile; seeds 2–4 mm.
                        22. Corollas bluish lavender to lavender, 15–22 mm; peduncles to 12 mm; calyx lobes ovate, 5–8(–12) mm, apices acuminate; staminodes: distal 2–5 mm densely villous, hairs yellowish, brownish, or orangish, to 1 mm .... 41. *Penstemon immanifestus* (in part)
                        22. Corollas blue to violet, sometimes lavender, 12–20 mm; peduncles to 26 mm; calyx lobes ovate to lanceolate, (2.5–)4–7 mm, apices acute to acuminate or caudate; staminodes: distal 4–8 mm densely villous, hairs yellow to golden yellow, to 2.5 mm ....47. *Penstemon pachyphyllus* (in part)
                20. Cauline leaf blades: apices acute to acuminate, sometimes mucronate (*P. nitidus*); peduncles to 3–4 mm; corolla throats 4–6 mm diam.
                    23. Cymes 2–5-flowered, verticillasters (2–)4–10; corolla tubes 4–6 mm; stems glabrous....................................... 45. *Penstemon nitidus*
                    23. Cymes (2–)4–8(–12)-flowered, verticillasters (3–)5–15(–26); corolla tubes 5–9 mm; stems glabrous or scabrous.
                        24. Staminodes: distal 1.5–3.5 mm densely pilose, hairs to 0.6 mm; calyx lobes lanceolate, (3.2–)4.5–10 × 1.2–3.2(–3.8) mm; pollen sacs 0.7–1.3(–1.5) mm; Idaho, Nevada, Oregon, Washington... ....29. *Penstemon acuminatus*
                        24. Staminodes: distal 4–6 mm sparsely villous, hairs to 1 mm; calyx lobes ovate to lanceolate, 4–8 × 1–2.5 mm; pollen sacs (0.9–)1.1–1.5 mm; Arizona, Colorado, Kansas, Montana, Nebraska, New Mexico, North Dakota, Oklahoma, South Dakota, Utah, Wyoming.................30. *Penstemon angustifolius* (in part)

29. Penstemon acuminatus Douglas ex Lindley,
Edwards's Bot. Reg. 15: plate 1285. 1829
(as Pentstemon acuminatum) • Sand-dune
beardtongue E

Stems ascending to erect,
(9–)20–60 cm, glabrous or
± scabrous, especially distally.
Leaves basal and cauline,
glabrous, sometimes scabrous;
basal and proximal cauline
(14–)30–95 × 4–15(–24) mm,
blade oblanceolate to elliptic,
base tapered, apex obtuse to
acute or mucronate; cauline (2 or)3–6(or 7) pairs,
sessile, 18–60(–85) × (4–)8–28 mm, blade ovate to
lanceolate, base tapered to cordate-clasping, apex
acute to acuminate. **Thyrses** continuous or interrupted,
cylindric, 6–25(–30) cm, axis glabrous, rarely scabrous
or glutinous, verticillasters (5–)7–15, cymes (2–)4–8
(–12)-flowered; proximal bracts ovate to lanceolate,
14–60 × 9–30 mm, margins entire, apex obtuse to
acuminate or caudate; peduncles and pedicels glabrous
or scabrous, sometimes also glutinous, peduncles to
4 mm. **Flowers:** calyx lobes lanceolate, (3.2–)4.5–10 ×
1.2–3.2(–3.8) mm, margins entire or erose, herbaceous
or narrowly scarious, glabrous or scabrous proximally
and, sometimes, along margins, sometimes also
glutinous; corolla blue to violet or purple, with reddish
purple nectar guides, tubular-funnelform, 11–20 mm,
glabrous externally, glabrous internally, tube 5–9 mm,
throat gradually inflated, 4–6 mm diam., rounded
abaxially; stamens included or longer pair reaching
orifice to slightly exserted, pollen sacs opposite,
navicular, 0.7–1.3(–1.5) mm, sutures papillate;
staminode 6–12 mm, reaching orifice, 0.6–1 mm diam.,
tip recurved, distal 1.5–3.5 mm densely pilose, hairs
golden yellow, to 0.6 mm, hairs on proximal part of
staminode mostly along margins; style 8–15 mm.
Capsules 7–13 × 4–7 mm.

Varieties 2 (2 in the flora): w United States.

Penstemon acuminatus includes the northern var.
acuminatus, which has denser thyrses than the southern
var. *latebracteatus*. Herbage of var. *acuminatus* is
usually scabrous; herbage of var. *latebracteatus* is rarely
scabrous.

The Blackfoot tribe of the Central Rocky Mountains
use *Penstemon acuminatus* as an analgesic, antiemetic,
and gastrointestinal aid (D. E. Moerman 1998).

1. Styles 11–15 mm; corollas 14–20 mm; pollen sacs
0.8–1.3(–1.5) mm; calyx lobes 5–10 mm . . . . . .
. . . . .29a. *Penstemon acuminatus* var. *acuminatus*
1. Styles 8–11 mm; corollas 11–15 mm; pollen sacs
0.7–1 mm; calyx lobes (3.2–)4.5–7 mm . . . .
. . . . .29b. *Penstemon acuminatus* var. *latebracteatus*

29a. Penstemon acuminatus Douglas ex Lindley var.
acuminatus E

Flowers: calyx lobes 5–10 mm;
corolla 14–20 mm; pollen
sacs 0.8–1.3(–1.5) mm; style
11–15 mm.

Flowering Apr–Jun(–Jul).
Sandy soils, dunes; 60–400 m;
Oreg., Wash.

Variety *acuminatus* occurs
primarily in the Columbia
River Basin from north-central Oregon to central
Washington.

29b. Penstemon acuminatus Douglas ex Lindley var.
latebracteatus N. H. Holmgren, Brittonia 31: 232,
fig. 8. 1979 E

Flowers: calyx lobes (3.2–)
4.5–7 mm; corolla 11–15 mm;
pollen sacs 0.7–1 mm; style
8–11 mm.

Flowering May–Jun(–Jul).
Sandy soils, dunes, sagebrush
and shadscale communities;
600–1700 m; Idaho, Nev.,
Oreg.

Variety *latebracteatus* is more widely distributed
than var. *acuminatus*. Most populations occur in the
Snake River Basin of southern Idaho, the Owyhee
River Basin of southwestern Oregon, and the
Humboldt River Basin of Nevada.

30. Penstemon angustifolius Pursh, Fl. Amer. Sept.
2: 738. 1813 (as Pentstemon angustifolia) • Narrow-
leaf beardtongue E F

Stems ascending to erect,
(6–)15–45(–65) cm, glabrous
or scabrous. **Leaves** basal and
cauline, basal not reduced,
glabrous, sometimes scabrous;
basal and proximal cauline
(25–)40–90 × 2–18 mm, blade
spatulate to oblanceolate or
linear, base tapered, apex
rounded to obtuse or acute; cauline 2–8 pairs, sessile,
18–113 × 2–24(–40) mm, blade ovate to lanceolate or
linear, base tapered to cordate-clasping, apex acute to
acuminate. **Thyrses** continuous or interrupted,
cylindric, (2–)5–30(–37) cm, axis glabrous, rarely
scabrous, verticillasters (3–)5–15(–26), cymes (2–)4–8
(–10)-flowered; proximal bracts ovate to lanceolate or
linear, (3–)12–98 × (1–)2–28 mm; peduncles and
pedicels glabrous or scabrous, peduncles to 3 mm.
Flowers: calyx lobes ovate to lanceolate, 4–8 ×
1–2.5 mm, margins entire or erose, broadly scarious,

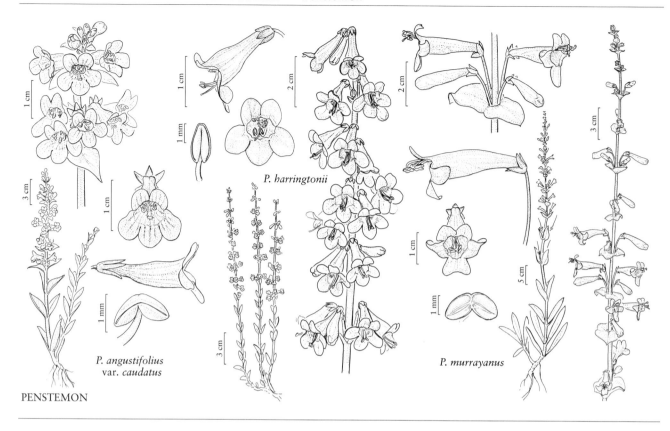

P. *harringtonii*

P. *angustifolius*
var. *caudatus*

P. *murrayanus*

PENSTEMON

glabrous or scabrous; corolla lavender, blue, or pink, with violet or reddish purple nectar guides, tubular-funnelform, 13–20(–22) mm, glabrous externally, sparsely white-pubescent internally abaxially or glabrous, tube 6–8 mm, throat gradually inflated, 4–6 mm diam., rounded abaxially; stamens included or longer pair reaching orifice, pollen sacs opposite, (0.9–)1.1–1.5 mm, sutures papillate; staminode 8–10 mm, reaching orifice, 0.9–1.2 mm diam., tip recurved, distal 4–6 mm sparsely villous, hairs golden yellow, to 1 mm, hairs on proximal part of staminode mostly along margins; style 8–12 mm. **Capsules** 9–15 × 5–7 mm.

Varieties 5 (5 in the flora): w United States.

Penstemon angustifolius comprises five varieties: two widespread in the west-central Great Plains and three with more limited, allopatric ranges.

1. Corollas pink; cauline leaf blades lanceolate to linear; Juab and Millard counties, Utah . . . .
. . . . 30c. *Penstemon angustifolius* var. *dulcis*
1. Corollas blue or lavender, rarely pink; cauline leaf blades ovate to lanceolate or linear (if corollas pink, leaf blades broader or plants not of Juab or Millard counties, Utah); Arizona, Colorado, Kansas, Montana, Nebraska, New Mexico, North Dakota, Oklahoma, South Dakota, Utah, Wyoming.

[2. Shifted to left margin.—Ed.]
2. Cauline leaf blades and proximal bracts lanceolate to linear.
   3. Stems glabrous or scabrous; corollas 14–18 mm; cauline leaves 2–5(or 6) pairs; stems (6–)10–30 cm; Colorado, Montana, Nebraska, North Dakota, South Dakota, Wyoming . . . .
   . . . . 30a. *Penstemon angustifolius* var. *angustifolius*
   3. Stems glabrous, rarely scabrous; corollas (14–)18–20(–22) mm; cauline leaves 4–7 pairs; stems (8–)15–45 cm; Colorado, Utah. . . . .
   . . . . . . 30e. *Penstemon angustifolius* var. *vernalensis*
2. Cauline leaf blades and proximal bracts ovate to lanceolate.
   4. Corollas 13–20(–22) mm, blue, rarely lavender or pink; Colorado, Kansas, Nebraska, New Mexico, Oklahoma, South Dakota, Wyoming
   . . . . 30b. *Penstemon angustifolius* var. *caudatus*
   4. Corollas 14–17 mm, blue to lavender or pink; Arizona, Colorado, New Mexico, Utah . . . . .
   . . . . 30d. *Penstemon angustifolius* var. *venosus*

30a. Penstemon angustifolius Pursh var. **angustifolius** E

Stems (6–)10–30 cm, glabrous or scabrous. **Cauline leaves** 2–5 (or 6) pairs, blade lanceolate to linear. **Thyrses:** proximal bracts lanceolate to linear. **Flowers:** corolla blue, rarely lavender or pink, 14–18 mm. $2n = 16$.

Flowering May–Jul. Sandy or gravelly soils, shortgrass prairies, sagebrush shrublands; 600–2400 m; Colo., Mont., Nebr., N.Dak., S.Dak., Wyo.

Variety *angustifolius* is known from the northern Great Plains, and it passes into var. *caudatus* in northeastern Colorado and western Nebraska.

30b. Penstemon angustifolius Pursh var. **caudatus** (A. Heller) Rydberg, Bull. Torrey Bot. Club 33: 151. 1906 (as Pentstemon) E F

Penstemon caudatus A. Heller, Minnesota Bot. Stud. 2: 34. 1898 (as Pentstemon)

Stems (10–)15–60(–65) cm, glabrous, rarely scabrous. **Cauline leaves** 3–8 pairs, blade ovate to lanceolate. **Thyrses:** proximal bracts ovate to lanceolate. **Flowers:** corolla blue, rarely lavender or pink, 13–20(–22) mm. $2n = 16$.

Flowering May–Jun. Sandy or gravelly soils, shortgrass prairies, sagebrush shrublands; 1200–2600 m; Colo., Kans., Nebr., N.Mex., Okla., S.Dak., Wyo.

Variety *caudatus* is known from the southwestern Great Plains. While mostly distinct in the southern and central parts of its range, the variety passes into var. *angustifolius* in northeastern Colorado and western Nebraska.

30c. Penstemon angustifolius Pursh var. **dulcis** Neese, Great Basin Naturalist 46: 459. 1986 • Sweet beardtongue E

Stems 27–45 cm, glabrous. **Cauline leaves** 5–8 pairs, blade lanceolate to linear. **Thyrses:** proximal bracts ovate to lanceolate. **Flowers:** corolla pink, 13–17 mm.

Flowering May–Jun. Sand dunes, sagebrush, saltbush, juniper communities; 1400–1700 m; Utah.

Variety *dulcis* is known from Juab and Millard counties.

30d. Penstemon angustifolius Pursh var. **venosus** (D. D. Keck) N. H. Holmgren, Brittonia 31: 104. 1979 E

Penstemon angustifolius subsp. *venosus* D. D. Keck, J. Wash. Acad. Sci. 29: 490. 1939

Stems 17–65 cm, glabrous. **Cauline leaves** 5–8 pairs, blade ovate to lanceolate. **Thyrses:** proximal bracts ovate to lanceolate. **Flowers:** corolla blue to lavender or pink, 14–17 mm.

Flowering Apr–Jun. Sandy soils, blackbrush, sagebrush, and pinyon-juniper woodlands; 1400–2300 m; Ariz., Colo., N.Mex., Utah.

Although variety *venosus* shares many morphologic features with var. *caudatus*, it generally exhibits more variation in corolla color than var. *caudatus*, and the varieties are largely allopatric except possibly in northcentral New Mexico.

30e. Penstemon angustifolius Pursh var. **vernalensis** N. H. Holmgren, Brittonia 31: 229, fig. 7. 1979 • Vernal narrow-leaf beardtongue E

Stems (8–)15–45 cm, glabrous, rarely scabrous. **Cauline leaves** 4–7 pairs, blade lanceolate to linear. **Thyrses:** proximal bracts lanceolate to linear. **Flowers:** corolla blue, (14–)18–20(–22) mm.

Flowering May–Jun. Sandy soils, sand dunes, sagebrush and sagebrush-juniper communities; 1500–1800 m; Colo., Utah.

Variety *vernalensis* is known from the Green River Basin in extreme northwestern Colorado (Moffat and Rio Blanco counties) and northeastern Utah (Daggett and Uintah counties). The herbage of var. *vernalensis* usually is glabrous; some populations from Browns Park in Daggett and Moffat counties have scabrous leaves, bracts, and calyces.

31. Penstemon arenicola A. Nelson, Bull. Torrey Bot. Club 25: 280. 1898 (as Pentstemon) • Red Desert beardtongue [E]

Stems decumbent to ascending, (4–)7–18(–30) cm, glabrous. **Leaves** basal and cauline, glabrous; basal and proximal cauline 21–50(–70) × 4–11 (–17) mm, blade spatulate to oblanceolate, base tapered, apex rounded to obtuse, acute, or mucronate; cauline 2–5 pairs, short-petiolate or sessile, 21–58 × 3–11 mm, blade lanceolate to oblanceolate, base tapered to clasping, apex obtuse to acute or mucronate. **Thyrses** continuous, cylindric, 3–14 cm, axis glabrous, verticillasters 4–9, cymes (1 or)2–6-flowered; proximal bracts oblanceolate to lanceolate, 13–60 × 5–16 mm; peduncles and pedicels glabrous. **Flowers:** calyx lobes lanceolate, 4–6.2(–7.5) × 0.9–1.6 mm, margins entire or erose, narrowly scarious, glabrous; corolla blue to violet, with reddish purple nectar guides, tubular-funnelform, (10–)12–15 mm, glabrous externally, sparsely white-pubescent internally abaxially or glabrous, tube 4–5 mm, throat gradually inflated, 4.5–5.5 mm diam., rounded abaxially; stamens included, pollen sacs divergent, 0.7–0.8 mm, sutures smooth or obscurely papillate; staminode 6–8 mm, included, 0.8–1 mm diam., tip recurved, distal 3–4 mm densely pubescent, hairs golden yellow, to 1 mm; style 8–10 mm. **Capsules** 6–10(–12) × 5–6 mm.

Flowering May–Jun. Sandy or shaley soils, sagebrush shrublands; 1800–2400 m; Wyo.

Penstemon arenicola occurs throughout much of the western two-thirds of Wyoming. Specimens documenting occurrence of the species in Colorado and Montana have not been seen.

32. Penstemon bracteatus D. D. Keck, Leafl. W. Bot. 1: 82. 1934 • Platy or Red Canyon beardtongue [C][E]

Plants: caudex rhizomelike. **Stems** decumbent to erect, 3–10(–15) cm, glabrous. **Leaves** basal and cauline, glabrous; basal and proximal cauline (10–)20–35 × 5–17 mm, blade obovate, base tapered, apex rounded to obtuse; cauline 1–3 pairs, sessile, 10–29 × 5–12 mm, blade ovate to elliptic, base tapered, apex rounded to mucronate. **Thyrses** continuous, cylindric, 1–6 cm, axis glabrous, verticillasters 1–5(or 6), cymes 3–5-flowered; proximal bracts ovate, 6–14 × 3–13 mm; peduncles and pedicels glabrous. **Flowers:** calyx lobes ovate to

lanceolate, (3–)4–5.5 × 1.7–2.2 mm, margins entire or erose, narrowly to broadly scarious, glabrous; corolla blue to violet or lavender, with reddish purple nectar guides, tubular-funnelform, 13–17(–19) mm, glabrous externally, sparsely white-villous internally abaxially, tube 4–5 mm, throat gradually inflated, 4–5 mm diam., rounded abaxially; stamens included, pollen sacs divergent to opposite, 0.8–1.2 mm, sutures smooth or obscurely papillate; staminode 6–8 mm, included, 0.8–1 mm diam., tip recurved, distal 4–5 mm moderately to densely pubescent, hairs golden yellow, to 1 mm; style 7–8 mm. **Capsules** 5–7 × 5–6 mm.

Flowering May–Jul. Gravelly and talus slopes, pine woodlands; of conservation concern; 2100–2500 m; Utah.

Penstemon bracteatus is known from Garfield, Iron, and Kane counties. Plants grow on limestones of the Claron Formation.

33. Penstemon buckleyi Pennell, Proc. Acad. Nat. Sci. Philadelphia 73: 486. 1922 • Buckley's beardtongue [E]

Penstemon amplexicaulis Buckley, Proc. Acad. Nat. Sci. Philadelphia 13: 461. 1862 (as amplexicaule), not Moench 1794

Stems ascending to erect, 15–55 (–82) cm, glabrous. **Leaves** basal and cauline, glabrous; basal and proximal cauline 19–116 (–150) × 3–23(–31) mm, blade spatulate to oblanceolate, base tapered, apex rounded to obtuse, acute, or mucronate; cauline 3–9 pairs, sessile, 20–95 × 10–30 mm, blade ovate to lanceolate, base clasping to cordate-clasping, apex acute. **Thyrses** interrupted, cylindric, (4–)10–35(–57) cm, axis glabrous, verticillasters 4–20(–35), cymes (2 or)3–5 (–11)-flowered; proximal bracts orbiculate to ovate, 15–70 × 5–31 mm; peduncles and pedicels glabrous. **Flowers:** calyx lobes ovate, 3.5–6 × 1.5–2.5 mm, margins entire or erose, broadly scarious, glabrous; corolla lavender to bluish lavender or pinkish blue, with reddish or reddish purple nectar guides, tubular-funnelform, (12–)14–20 mm, glabrous externally, glabrous internally, tube 5–7 mm, throat gradually inflated, 4–6 mm diam., rounded abaxially; stamens included, pollen sacs opposite, 0.8–1.2 mm, sutures papillate; staminode 7–9 mm, included or reaching orifice, 0.8–0.9 mm diam., tip recurved, distal 3–5 mm moderately to densely villous, hairs golden yellow, to 1.5 mm; style 9–11 mm. **Capsules** 12–18 (–20) × 7–11 mm. $2n = 16$.

Flowering Apr–Jun. Sandy soils, sand, sandsage, or mixed-grass prairies; 400–1300 m; Colo., Kans., Nebr., N.Mex., Okla., Tex.

Penstemon buckleyi frequently is found on aeolian sands of dune fields associated with the major streams of the southern Great Plains; it also occurs on alluvial sands and sandstone-derived soils. It has been documented in Banner and Platte counties, Nebraska, well north of its native range; those plants probably represent roadside introductions (R. B. Kaul et al. 2006).

34. Penstemon carnosus Pennell ex D. D. Keck, Ann. Carnegie Mus. 26: 329. 1937 • Fleshy beardtongue E

Stems ascending to erect, 12–35 cm, glabrous. **Leaves** basal and cauline, glabrous; basal and proximal cauline 20–110 × 14–40(–60) mm, blade spatulate to obovate, base tapered, apex rounded, mucronate, or retuse; cauline 1–3 pairs, sessile, 20–58 × 8–23 mm, blade ovate to obovate or lanceolate, base tapered to cordate-clasping, apex obtuse to mucronate or retuse. **Thyrses** continuous, sometimes interrupted, cylindric, 3–15 cm, axis glabrous, verticillasters 5–10, cymes 2–6-flowered; proximal bracts ovate to lanceolate, 11–38 × 5–13 mm; peduncles and pedicels glabrous, rarely obscurely glandular distally. **Flowers:** calyx lobes ovate to lanceolate, 4.5–8 × 1.8–2.4 mm, margins erose, broadly scarious, glabrous rarely obscurely glandular-pubescent proximally or along margins; corolla blue to violet, rarely pink, with or without reddish or purple nectar guides, tubular-funnelform, 16–20(–22) mm, obscurely glandular externally or glabrous, sparsely white-villous internally abaxially or glabrous, tube 6–7 mm, throat gradually inflated, 5–6 mm diam., rounded or slightly 2-ridged abaxially; stamens included or longer pair reaching orifice, pollen sacs opposite, 0.8–1.4 mm, sutures papillate; staminode 8–10 mm, included or reaching orifice, 0.4–1 mm diam., tip strongly recurved, distal 3–4 mm moderately to densely pubescent, hairs golden yellow, to 0.7 mm; style 8–10 mm. **Capsules** 8–12 × 6–8 mm.

Flowering May–Jul. Desert or sagebrush shrublands, pinyon-juniper woodlands; 1500–3000 m; Utah.

Penstemon carnosus is known from eastern Utah from the Uinta Basin south to the San Rafael Swell, Henry Mountains, Aquarius and Paunsaugunt plateaus, and west to the Wasatch Plateau, documented in Carbon, Emery, Garfield, Kane, Sevier, and Wayne counties. The species is confused most frequently with *P. lentus*, which occurs to the southeast in the Abajo and La Sal mountains, and *P. pachyphyllus*, which occurs to the immediate north and west. Glands on the corollas usually are obscure; they are most easily observed on flower buds.

35. Penstemon cyathophorus Rydberg, Bull. Torrey Bot. Club 31: 643. 1905 (as Pentstemon) • Sagebrush or Middle Park beardtongue E

Stems erect, (13–)19–42 cm, glabrous. **Leaves** basal and cauline, glabrous; basal and proximal cauline 13–57 × 4–19 mm, blade spatulate to oblanceolate, base tapered, apex mucronate, sometimes rounded to obtuse; cauline 3–7 pairs, sessile, 9–30 × 6–15 mm, blade oblanceolate to ovate, base clasping to cordate-clasping, apex acute to acuminate, sometimes mucronate. **Thyrses** interrupted or continuous, cylindric, 10–16 cm, axis glabrous, verticillasters (6–)8–12, cymes 2–7-flowered, proximal bracts orbiculate to ovate, 11–18 × 6–14 mm; peduncles and pedicels glabrous. **Flowers:** calyx lobes lanceolate, 4–7 × 1.1–1.9 mm, margins erose, broadly scarious, glabrous; corolla lavender to lavender-blue, without nectar guides, funnelform, nearly rotate, 11–14 mm, glabrous externally or glandular, glabrous internally, tube 6.5–8 mm, throat gradually inflated, 4–5 mm diam., rounded abaxially; stamens: 4 prominently exserted (4–5 mm) beyond limb, pollen sacs parallel, 1.2–1.5 mm, sutures smooth; staminode 6–7 mm, reaching orifice or exserted (adaxially), 0.4–0.5 mm diam., tip recurved, distal 1–2 mm densely villous, hairs golden yellow, to 0.8 mm; style 12–14 mm. **Capsules** 5–6 × 4–5 mm.

Flowering May–Jul(–Sep). Sagebrush shrublands; 2100–2900 m; Colo., Wyo.

Penstemon cyathophorus occurs mostly in Middle Park, North Park, and the southern end of the North Platte River valley in the Southern Rocky Mountains. Populations have been documented in Grand, Jackson, and Summit counties, Colorado, and Carbon County, Wyoming. Besides having all four stamens prominently exserted, *P. cyathophorus* is unique in *Penstemon* in having the distal end of the staminode situated against the adaxial surface of the throat in mature corollas; it is situated against the abaxial surface in other species of *Penstemon*.

36. Penstemon fendleri Torrey & A. Gray in War Department [U.S.], Pacif. Railr. Rep. 2(4): 168, plate 5. 1857 (as Pentstemon) • Fendler's beardtongue

Stems erect, (15–)20–55 (–60) cm, glabrous. **Leaves** basal and cauline, glabrous; basal and proximal cauline 20–100 × 4–24 mm, blade spatulate to oblanceolate, base tapered, apex rounded to obtuse or acute, sometimes mucronate; cauline 2–5 pairs, sessile, (14–)23–95 × (4–)6–31 mm, blade lanceolate or ovate to trullate, base clasping to cordate-clasping, apex obtuse to acute. **Thyrses** interrupted, cylindric, (5–)11–38 cm, axis glabrous, verticillasters (3 or)4–12, cymes 2- or 3(–5)-flowered; proximal bracts trullate to ovate, 11–70 × 7–38 mm; peduncles and pedicels glabrous. **Flowers:** calyx lobes ovate, 4.5–7 × 1.5–3.5 mm, margins entire or erose, broadly scarious, glabrous or obscurely glandular along margins distally; corolla lavender to violet or bluish, with violet or reddish purple nectar guides, tubular-salverform, 14–23(–28) mm, glabrous externally or glandular, glabrous or sparsely white-villous internally abaxially, tube 7–9 mm, throat slightly inflated, 4–6 mm diam., rounded abaxially; stamens included, pollen sacs opposite, 1–1.3 mm, sutures papillate; staminode 8–11 mm, reaching orifice, 0.8–1.6 mm diam., tip recurved, distal 1–3 mm villous, hairs golden yellow, to 1.5 mm; style 11–15 mm. **Capsules** 10–15 × 8–10 mm. $2n = 16$.

Flowering Mar–Jun(–Jul). Sandy or gravelly soils, mixed-grass, shortgrass, or sandsage prairies; 200–2300 m; Ariz., Kans., N.Mex., Okla., Tex.; Mexico (Chihuahua, Coahuila).

Penstemon fendleri occurs on mesas and plains from the southern Great Plains through western Texas, New Mexico, and southeastern Arizona into northern Mexico.

The Ramah Navajo of western New Mexico use *Penstemon fendleri* as a dermatological aid (D. E. Moerman 1998).

37. Penstemon flowersii Neese & S. L. Welsh, Great Basin Naturalist 43: 429, fig. 1. 1983 • Flowers's beardtongue C E

Stems ascending to erect, 8–25(–32) cm, glabrous. **Leaves** essentially cauline, basal usually reduced or absent, glabrous; proximal cauline (15–)20–55 × (4–)10–25 mm, blade spatulate to lanceolate, base tapered, apex rounded to obtuse or acute; cauline 4–6 pairs, short-petiolate or sessile, 30–58 × 9–26 mm, blade ovate to elliptic or lanceolate, base tapered to clasping, apex obtuse to acute. **Thyrses** ± interrupted, cylindric, 8–14 cm, axis glabrous, verticillasters 4–9, cymes 1–5-flowered; proximal bracts ovate, 16–44 × 6–24 mm; peduncles and pedicels glabrous. **Flowers:** calyx lobes lanceolate, 5–6.5 × 1.9–3 mm, margins entire or erose, broadly scarious, glabrous; corolla pink, with rose pink nectar guides, tubular-funnelform, 15–18 mm, glabrous externally, glabrous internally, tube 5–6 mm, throat gradually inflated, 4–5 mm diam., rounded abaxially; stamens included, pollen sacs opposite, 1–1.2 mm, sutures smooth; staminode 8–9 mm, reaching orifice, 0.7–0.8 mm diam., tip recurved, distal 0.5–1.5 mm sparsely lanulose, hairs yellow, to 0.6 mm; style 14–16 mm. **Capsules** 7–10 × 5–8 mm.

Flowering May–Jun. Shaley and clayey soils, shadscale shrublands; of conservation concern; 1500–1600 m; Utah.

Penstemon flowersii is known from a 200 square km area of the Duchesne River drainage between Myton and Randlett in Duchesne and Uintah counties. According to Neese and Welsh, plants grow on slopes and benches of the Uinta Formation.

38. Penstemon grandiflorus Nuttall, Cat. Pl. Upper Louisiana, no. 64. 1813 (as grandiflorum) • Shell-leaf beardtongue E

Stems erect, (40–)50–95 (–120) cm, glabrous. **Leaves** basal and cauline, glabrous; basal and proximal cauline 30–160 × 6–50 mm, blade spatulate to obovate, base tapered, apex rounded to obtuse or acute; cauline 4–8 pairs, sessile, 18–90(–110) × 15–50 mm, blade spatulate to orbiculate, base clasping, apex rounded to obtuse. **Thyrses** interrupted, cylindric, 12–30(–40) cm, axis glabrous, verticillasters 3–7(–9), cymes 2–4-flowered; proximal bracts ovate to elliptic or orbiculate, (9–)16–83 × (9–)16–54 mm; peduncles and pedicels glabrous. **Flowers:** calyx lobes ovate to lanceolate, 7–11 × 2.5–4 mm, margins entire, rarely erose, herbaceous or narrowly scarious, glabrous; corolla lavender to blue or pinkish blue, with magenta nectar guides, ampliate, 35–48 mm, glabrous externally, glabrous internally, tube 10–13 mm, throat abruptly inflated, 15–18 mm diam., rounded abaxially; stamens included, pollen sacs opposite, 2.1–2.6 mm, sutures papillate; staminode 16–21 mm, included or reaching orifice, 2–2.6 mm diam., tip recurved to coiled, distal 1–2 mm sparsely villous, hairs golden yellow, to 0.5 mm; style 19–30 mm. **Capsules** 16–20(–25) × 8–15 mm. $2n = 16$.

Flowering Apr–Jul. Sandy or calcareous soils, tallgrass, mixed-grass, and sand prairies; 200–1800 (–2400) m; Colo., Conn., Ill., Ind., Iowa, Kans., Mass., Mich., Minn., Mo., Mont., Nebr., N.Mex., N.Dak., Ohio, Okla., S.Dak., Tex., Wyo.

Widely distributed in the western Midwest and Great Plains in the United States, *Penstemon grandiflorus* is cultivated as an ornamental for its showy flowers. Reports from Connecticut (D. W. Magee and H. E. Ahles 2007), Indiana (K. Yatskievych 2000), Massachusetts (Magee and Ahles), Michigan (E. G. Voss 1972–1996), and Ohio (T. S. Cooperrider 1995) appear to be based on introductions. *Penstemon grandiflorus* has been seeded along highways in Iowa and Nebraska, where it is also native.

The validity of the name *Penstemon grandiflorus* has been debated owing to the meager diagnosis by Nuttall. The name is accepted here, making *P. bradburyi* Pursh, an illegitimate, superfluous replacement for *P. grandiflorus*.

The Dakota, Kiowa, and Sioux tribes, centered in the Great Plains, use *Penstemon grandiflorus* as an analgesic, a gastrointestinal aid, and for fevers (D. E. Moerman 1998).

39. Penstemon harringtonii Penland, Madroño 14: 153, fig. 1. 1958 • Harrington's beardtongue E F

Stems erect, (30–)40–70 cm, glabrous. **Leaves** basal and cauline, glabrous; basal and proximal cauline (15–)30–70 × (6–)13–26 mm, blade trullate to ovate or elliptic, base tapered, apex acute, obtuse, rounded, or mucronate; cauline 3–7 pairs, sessile, (10–)18–53 × (7–)12–26 mm, blade ovate to elliptic or oblanceolate, base tapered to clasping, apex obtuse to acute, rounded, or mucronate. **Thyrses** interrupted, cylindric, 13–30 cm, axis glabrous, verticillasters 6–10, cymes 1–4(–7)-flowered; proximal bracts ovate to lanceolate, 10–20 × 3–12 mm; peduncles and pedicels glabrous. **Flowers:** calyx lobes ovate to lanceolate, 3.5–7(–9) × 2–2.7 mm, margins entire or erose, broadly scarious, glabrous; corolla blue to violet, lavender, or pinkish blue, without nectar guides, funnelform, 17–24 mm, glabrous externally, glabrous internally, tube 8–11 mm, throat gradually inflated, 5–7 mm diam., rounded abaxially; stamens: 2 prominently exserted, 4–6 mm, 2 included, pollen sacs parallel or divergent, 2–3 mm, sutures smooth; staminode 7–10 mm, included or reaching orifice, 0.5–1(–1.5) mm diam., tip recurved, distal 4–6 mm densely villous, hairs golden yellow, to 1.5 mm; style 18–22 mm. **Capsules** 7–11 × 4–6 mm.

Flowering Jun–Jul. Sandy soils, sagebrush shrublands, pinyon-juniper woodlands; 2000–2900 m; Colo.

Penstemon harringtonii is known from Eagle, Garfield, Grand, Pitkin, Routt, and Summit counties in northwestern Colorado (S. S. Panjabi and D. G. Anderson 2006). It sometimes grows with *P. osterhoutii*.

SELECTED REFERENCE Panjabi, S. S. and D. G. Anderson. 2006. *Penstemon harringtonii* Penland (Harrington's Beardtongue): A Technical Conservation Assessment. Fort Collins.

40. Penstemon haydenii S. Watson, Bot. Gaz. 16: 311. 1891 (as Pentstemon haydeni) • Blowout beardtongue C E

Plants: caudex rhizomelike. **Stems** decumbent to ascending, (15–)20–48 cm, glabrous. **Leaves** essentially cauline, basal absent or reduced, glabrous; basal and proximal cauline (25–)55–130(–175) × 3–25 mm, blade linear to lanceolate, base tapered, apex acute to acuminate; cauline 2–10 pairs, sessile, 60–110(–120) × 7–30 mm, blade lanceolate to linear, base clasping, apex acuminate to long-acuminate. **Thyrses** continuous, cylindric, (2–)6–21(–34) cm, axis glabrous, verticillasters (2–)6–10(–17), cymes 1–8-flowered; proximal bracts ovate, 58–120 × 20–45 mm; peduncles and pedicels glabrous. **Flowers:** calyx lobes lanceolate to linear, 8–13 × 1–3 mm, margins entire or erose, herbaceous or scarious, glabrous; corolla lavender to bluish, usually with magenta nectar guides, ampliate, 21–28 mm, glabrous externally, glabrous internally, tube 7–9 mm, throat abruptly inflated, 9–11 mm diam., rounded abaxially; stamens included, pollen sacs opposite, 1.1–2 mm, sutures papillate; staminode 13–16 mm, included, 1.2–3 mm diam., tip recurved, distal 0.1–5 mm sparsely to densely villous, hairs golden yellow, to 1 mm, rarely glabrous; style 13–23 mm. **Capsules** 8–12 × 5–9 mm. $2n = 16$.

Flowering May–Jul(–Sep). Blowouts in sand dunes; of conservation concern; 900–2300 m; Nebr., Wyo.

Penstemon haydenii is known from the Nebraska Sandhills, where extant populations occur in Box Butte, Cherry, Garden, Hooker, Morrill, and Sheridan counties. Historic populations occurred in Thomas County (D. M. Sutherland 1988).

Penstemon haydenii was discovered in northern Carbon County, Wyoming, in the late 1990s some 300 km west of the nearest Nebraska populations; it might have been collected in Wyoming as early as 1877 (W. Fertig 2001). A morphometric analysis of Nebraska and Wyoming plants revealed differences that could justify recognition of Wyoming populations as a distinct variety (C. C. Freeman 2015). It is listed as endangered by the U.S. Department of the Interior.

SELECTED REFERENCES Fertig, W. 2001. Survey for Blowout Penstemon (*Penstemon haydenii*) in Wyoming. Laramie. Freeman, C. C. 2015. Final Report on an Assessment of the Status of Blowout Beardtongue (*Penstemon haydenii* S. Watson, Plantaginaceae) Using Molecular and Morphometric Approaches. Lawrence, Kans. Sutherland, D. M. 1988. Historical notes on collections and taxonomy of *Penstemon haydenii* S. Wats. (blowout penstemon), Nebraska's only endemic plant species. Trans. Nebraska Acad. Sci. 16: 191–194.

41. **Penstemon immanifestus** N. H. Holmgren,
    Brittonia 30: 334, fig. 1. 1978 • Steptoe Valley
    beardtongue E

Stems ascending to erect, 9–30 (–35) cm, glabrous. **Leaves** basal and cauline, basal not reduced, glabrous; basal and proximal cauline 15–60(–90) × 8–25(–39) mm, blade oblanceolate to obovate, base tapered, apex rounded to obtuse, sometimes mucronate; cauline 3–7 pairs, short-petiolate or sessile, (25–)35–80 × 10–20(–28) mm, blade lanceolate to ovate, base tapered to clasping, apex obtuse to acute. **Thyrses** continuous or interrupted, cylindric, (4–)6–20 cm, axis glabrous, verticillasters (3–)5–10(–14), cymes 2–5-flowered; proximal bracts ovate to lanceolate, (5–)13–56 × (2–)4–24 mm; peduncles and pedicels glabrous, peduncles to 12 mm. **Flowers:** calyx lobes ovate, 5–8(–12) × 2–3.5 mm, margins entire or erose, broadly scarious, apex acuminate, glabrous; corolla bluish lavender to lavender or pink, with dark reddish violet or reddish nectar guides, funnelform to ventricose, 15–22 mm, glabrous externally, sparsely white-villous internally abaxially or glabrous, tube 5–7 mm, throat gradually inflated, 5–8 mm diam., rounded abaxially; stamens included, pollen sacs opposite, (1–)1.2–1.6 mm, sutures papillate; staminode 8–10 mm, included or reaching orifice, 0.7–1 mm diam., tip strongly recurved, distal 2–5 mm densely villous, hairs yellowish, brownish, or orangish, to 1 mm; style 10–12 mm. **Capsules** 8–12 × 4–6 mm. **Seeds** 2–4 mm.

Flowering May–Jun. Sandy or sandy-loam soils, sagebrush grasslands; 1500–2200 m; Nev., Utah.

Penstemon immanifestus occurs along the western edge of the Bonneville Basin in eastern Nevada and western Utah, and in the Calcareous Mountains, central Great Basin, and Tonopah regions of eastern and central Nevada. Populations have been documented in Elko, Eureka, Lander, Nye, and White Pine counties, Nevada, and in Juab, Millard, and Tooele counties, Utah.

42. **Penstemon lentus** Pennell, Contr. U.S. Natl. Herb.
    20: 359. 1920 • Abajo beardtongue E

Stems ascending to erect, 14–40(–60) cm, glabrous. **Leaves** basal and cauline, glabrous; basal and proximal cauline 40–65(–110) × 8–20 (–30) mm, blade obovate, base tapered, apex obtuse to acute, sometimes mucronate; cauline 2–4 pairs, sessile, (9–)12–60 × 7–26 mm, blade lanceolate to ovate or elliptic, base tapered to clasping or cordate-clasping, apex obtuse to

acute, sometimes mucronate. **Thyrses** interrupted or ± continuous, secund, sometimes cylindric, 5–23 cm, axis glabrous, verticillasters 4–8, cymes 1–3(or 4)-flowered; proximal bracts ovate to lanceolate, (3–)7–38 × (1–)2–18 mm; peduncles and pedicels glabrous, peduncles to 28 mm. **Flowers:** calyx lobes ovate to lanceolate, 5–6.7(–7.2) × 2–3(–3.7) mm, margins entire or erose, broadly scarious, glabrous; corolla lavender to violet, blue, pinkish blue, or white, with or without reddish violet nectar guides, tubular-funnelform, 17–20 (–22) mm, glabrous externally, sparsely white-villous internally abaxially or glabrous, tube 5–8 mm, throat gradually inflated, 5–6 mm diam., rounded abaxially; stamens included or longer pair reaching orifice, pollen sacs opposite, 1.1–1.6 mm, sutures papillate; staminode 8–9 mm, reaching orifice or slightly exserted, 0.6–1 mm diam., tip strongly recurved, distal 3–4 mm sparsely to densely villous, hairs mostly along margins, yellowish, to 1 mm; style 10–12 mm. **Capsules** 9–12 × 5–7 mm. **Seeds** 3.5–5 mm.

Varieties 2 (2 in the flora): w United States.

Penstemon lentus is found in the Four Corners region of the Colorado Plateau. Reported differences (N. H. Holmgren 1984) between the two varieties in the sizes of the pollen sacs appear to be inconsistent and not taxonomically useful.

1. Corollas lavender to violet, blue, or pinkish blue; Arizona, Colorado, New Mexico, Utah . . . . 42a. *Penstemon lentus* var. *lentus*
1. Corollas white; San Juan County, Utah. . . . . 42b. *Penstemon lentus* var. *albiflorus*

42a. **Penstemon lentus** Pennell var. **lentus** E

Flowers: corolla lavender to violet, blue, or pinkish blue.

Flowering May–Jun. Sagebrush shrublands, pinyon-juniper and oak woodlands, ponderosa pine forests; 1300–2600 m; Ariz., Colo., N.Mex., Utah.

Variety *lentus* is known from southwestern Colorado, where it has been documented in Archuleta, Dolores, Montezuma, Montrose, and San Miguel counties; it occurs also in Apache and Navajo counties, Arizona, McKinley and San Juan counties, New Mexico, and San Juan County, Utah.

42b. Penstemon lentus Pennell var. **albiflorus**
(D. D. Keck) Reveal, Great Basin Naturalist 35: 370.
1976 [E]

Penstemon lentus subsp. *albiflorus*
D. D. Keck, Amer. Midl. Naturalist
23: 616. 1940

Flowers: corolla white.

Flowering May–Jun. Sage-
brush shrublands, pinyon-
juniper woodlands; 1900–
2600 m; Utah.

Variety *albiflorus* is known
only from the Abajo Mountains in San Juan County.

43. Penstemon mucronatus N. H. Holmgren,
Brittonia 31: 234, fig. 10. 1979 • Sheep Creek
beardtongue [E]

Penstemon pachyphyllus A. Gray
ex Rydberg var. *mucronatus*
(N. H. Holmgren) Neese

Stems ascending to erect, 10–35
(–46) cm, glabrous. **Leaves**
basal and cauline, glabrous;
basal and proximal cauline
23–65(–100) × (9–)13–40 mm,
blade oblanceolate, base
tapered, apex rounded to obtuse or acute; cauline (2–)
4–6(–9) pairs, sessile, (8–)20–30 × 8–26 mm, blade ovate
to orbiculate, base clasping, apex mucronate. **Thyrses**
interrupted, cylindric, (2–)11–18(–27)cm, axis glabrous,
verticillasters 6–9(–15), cymes 2–6-flowered; proximal
bracts orbiculate to ovate, 7–28 × 3–26 mm; peduncles
and pedicels glabrous. **Flowers:** calyx lobes ovate, (3–)
4.5–6.5 × 1.5–2.3 mm, margins entire or erose, broadly
scarious, glabrous; corolla blue to violet, with reddish
violet nectar guides, guides sometimes passing onto limb,
tubular-funnelform, (10–)13–17(–20) mm, glabrous
externally, sparsely white-villous internally abaxially,
tube (5–)7–10 mm, throat gradually inflated, 4.5–5.5
(–7) mm diam., rounded abaxially; stamens included
or longer pair reaching orifice, pollen sacs divergent
to opposite, 0.7–1.1 mm, sutures papillate; staminode
7–13 mm, reaching orifice, (0.6–)0.9–1.9 mm diam., tip
strongly recurved, distal 2–5 mm moderately to densely
pilose, hairs brownish yellow or golden yellow, to
1 mm; style (7–)9–12 mm. **Capsules** 8–10 × 6–7 mm.
2n = 16.

Flowering May–Jun. Sandy or gravelly soils, juniper
and pinyon-juniper woodlands; 1500–2500 m; Colo.,
Utah, Wyo.

Penstemon mucronatus is known from Moffat and
Rio Blanco counties, Colorado, Daggett and Uintah
counties, Utah, and Carbon and Sweetwater counties,
Wyoming. E. C. Neese (1993) considered it transitional
to *P. pachyphyllus* var. *pachyphyllus* in northeastern

Utah and treated it as a variety of that species. The
two taxa often have similar leaf morphologies;
P. mucronatus usually is easily distinguished from
P. pachyphyllus var. *pachyphyllus* by its narrower
staminode bearing shorter, less tangled hairs at the apex
and nectar guides that extend onto the corolla limbs.
Flowers of *P. mucronatus* are in many respects more
similar to those of *P. osterhoutii* than they are to those
of *P. pachyphyllus*. A chromosome count reported for
P. pachyphyllus by F. S. Crosswhite and S. Kawano
(1965) is referable to *P. mucronatus*.

44. Penstemon murrayanus Hooker, Bot. Mag. 63: plate
3472. 1836 (as Pentstemon) • Cup-leaf beardtongue
[E] [F]

Stems ascending to erect,
50–100(–150) cm, glabrous.
Leaves basal and cauline,
glabrous; basal and proximal
cauline 40–100 × (8–)14–30
mm, blade spatulate to ovate,
base tapered, apex rounded to
obtuse, rarely retuse; cauline
5–9 pairs, sessile, 12–110 ×
16–70 mm, blade oblong to ovate, base clasping to
connate-perfoliate, apex obtuse to acute. **Thyrses**
interrupted, cylindric, 15–50 cm, axis glabrous,
verticillasters 6–10(–16), cymes 1–5-flowered; proximal
bracts ovate, 9–45 × 14–40 mm; peduncles and pedicels
glabrous. **Flowers:** calyx lobes lanceolate, 4.5–7 ×
1.8–2.9 mm, margins entire, narrowly scarious,
glabrous; corolla red, without nectar guides, tubular-
funnelform, 32–36 mm, glabrous externally, glabrous
internally, tube 8–11 mm, throat abruptly inflated,
9–12 mm diam., rounded abaxially; stamens exserted,
pollen sacs opposite, explanate, 1.3–1.7 mm, sutures
smooth; staminode 18–20 mm, exserted, 0.2–0.3 mm
diam., tip straight to recurved, glabrous; style 24–31
mm. **Capsules** 9–13 × 6–10 mm.

Flowering Mar–Jun. Sandy soils, deciduous and
pine woodlands, sandhill prairies; 10–200 m; Ark., La.,
Okla., Tex.

F. S. Crosswhite and C. D. Crosswhite (1981)
hypothesized that *Penstemon murrayanus* was derived
from *P. grandiflorus*, with which it shares many
morphologic features. The two species have been
crossed artificially, yielding hybrids that once were
commercially popular (G. Viehmeyer 1958; R. Nold
1999). Molecular data support the sister relationship
of *P. murrayanus* and *P. grandiflorus* (C. A. Wessinger
et al. 2016).

45. Penstemon nitidus Douglas ex Bentham in A. P. de Candolle and A. L. P. P. de Candolle, Prodr. 10: 323. 1846 (as Pentstemon) • Wax-leaf beardtongue E

Stems ascending to erect, (5–)9–40 cm, glabrous. **Leaves** basal and cauline, glabrous; basal and proximal cauline 15–140 × 2–27 mm, blade spatulate to oblanceolate or lanceolate, base tapered, apex obtuse to acute, sometimes rounded or mucronate; cauline 2–4(–6) pairs, sessile, (11–)18–85 × (3–)5–28(–32) mm, blade ovate to lanceolate, base clasping or cordate-clasping, apex acuminate to acute, sometimes mucronate. **Thyrses** ± interrupted, cylindric, (2–)5–20 cm, axis glabrous, verticillasters (2–)4–10, cymes 2–5-flowered; proximal bracts orbiculate to ovate or lanceolate, 19–70 × 7–28 mm; peduncles and pedicels glabrous, peduncles to 3 mm. **Flowers:** calyx lobes ovate to lanceolate, 3–7.5 × (0.8–)1.2–1.8 mm, margins entire or erose, narrowly scarious, glabrous; corolla blue to violet, with reddish violet nectar guides, tubular-funnelform, (10–)13–15(–18) mm, glabrous externally, glabrous or sparsely white-villous internally abaxially, tube 4–6 mm, throat gradually inflated, 4–6 mm diam., rounded abaxially; stamens included or longer pair reaching orifice to slightly exserted, pollen sacs opposite, 0.7–1.2 mm, sutures papillate; staminode 8–11 mm, reaching orifice or slightly exserted, 0.8–1.5 mm diam., tip strongly recurved, distal 3–4 mm densely villous, hairs golden yellow, to 1.5 mm, hairs on proximal part of staminode mostly along margins; style 8–10 mm. **Capsules** 9–13 × 5–8 mm.

Varieties 2 (2 in the flora): w North America.

Some specimens of *Penstemon nitidus* from western Montana are referable to variety with difficulty.

1. Proximal bracts orbiculate to ovate, rarely lanceolate; calyx lobes ovate to lanceolate, 3–6.5 mm; Alberta, British Columbia, Manitoba, Saskatchewan, Montana, North Dakota, South Dakota, Wyoming. . .45a. *Penstemon nitidus* var. *nitidus*
1. Proximal bracts ovate to lanceolate; calyx lobes lanceolate, (3.2–)5–7.5 mm; Idaho, Montana . . . . .45b. *Penstemon nitidus* var. *polyphyllus*

45a. Penstemon nitidus Douglas ex Bentham var. **nitidus** E

Thyrses: proximal bracts orbiculate to ovate, rarely lanceolate. **Flowers:** calyx lobes ovate to lanceolate, 3–6.5 mm. $2n = 16$.

Flowering May–Jun(–Jul). Gumbo hills, shortgrass prairies, montane grasslands, roadcut banks; 200–2500 m; Alta., B.C., Man., Sask.; Mont., N.Dak., S.Dak., Wyo.

Variety *nitidus* occurs widely in the northwestern Great Plains and at low elevations in the Central, Northern, and Canadian Rocky mountains.

45b. Penstemon nitidus Douglas ex Bentham var. **polyphyllus** (Pennell) Cronquist in C. L. Hitchcock et al., Vasc. Pl. Pacif. N.W. 4: 395. 1959 E

Penstemon nitidus subsp. *polyphyllus* Pennell, Notul. Nat. Acad. Nat. Sci. Philadelphia 95: 6. 1942

Thyrses: proximal bracts ovate to lanceolate. **Flowers:** calyx lobes lanceolate, (3.2–)5–7.5 mm.

Flowering Apr–Jun. Rocky or gravelly soils, slopes and banks, montane grasslands; 400–1700 m; Idaho, Mont.

Variety *polyphyllus* is known from the upper Clark Fork River valley of east-central Idaho (Lemhi County) and west-central Montana (Beaverhead, Deer Lodge, Granite, Jefferson, Madison, Missoula, Powell, and Silver Bow counties). Compared to var. *nitidus*, it generally is taller and has relatively longer, narrower leaves, bracts, and calyx lobes.

46. Penstemon osterhoutii Pennell, Contr. U.S. Natl. Herb. 20: 358. 1920 • Osterhout's beardtongue E

Stems ascending to erect, (6–)26–77 cm, glabrous. **Leaves** basal and cauline, glabrous; basal and proximal cauline 32–150 × 9–40 mm, blade oblong to ovate, base tapered, apex obtuse to acute; cauline 4–7 pairs, sessile, (11–)18–75 × (4–)10–38 mm, blade oblong to ovate, base clasping or cordate-clasping, apex obtuse to acute. **Thyrses** interrupted, cylindric, (4–)14–37 cm, axis glabrous, verticillasters (3–)6–10(–14), cymes (1 or)2–8-flowered; proximal bracts ovate to oblong, (6–)24–43 × (3–)16–28 mm; peduncles and pedicels glabrous. **Flowers:** calyx lobes ovate, 4–6.8 × 2–3 mm, margins erose, broadly scarious, glabrous; corolla blue to violet or lavender, with reddish violet nectar guides, guides passing onto limb, tubular-funnelform to ventricose, 15–18 mm, glabrous externally, sparsely to moderately white-villous internally abaxially, rarely glabrous, tube 6–8 mm, throat gradually inflated, 7–10 mm diam., rounded abaxially; stamens included or longer pair reaching orifice, pollen sacs divergent to opposite, 1.1–1.3 mm, sutures papillate; staminode 11–13 mm, reaching orifice or slightly exserted, 2–3 mm diam., tip recurved, prominently bifurcate, distal 5–7 mm densely villous, hairs yellowish or brownish yellow, to 1.5 mm; style 10–12 mm. **Capsules** 11–13 × 5–7 mm.

Flowering May–Jun. Sandy or clayey soils, sagebrush and desert shrublands, pine-juniper woodlands; 1600–2700 m; Colo.

Populations of *Penstemon osterhoutii* are known from Eagle, Garfield, Grand, Pitkin, Rio Blanco, Routt, and Summit counties. The species shares some characteristics with *P. pachyphyllus*; it differs by larger, sky blue corollas with violet nectar guides that extend onto the corolla limbs (as in *P. mucronatus*) and corollas sparsely to moderately white-villous internally, rarely glabrous, on the abaxial surfaces.

47. **Penstemon pachyphyllus** A. Gray ex Rydberg, Fl. Rocky Mts., 770, 1066. 1917 (as Pentstemon)

• Thick-leaf beardtongue [E]

Penstemon nitidus Douglas ex Bentham var. *major* Bentham in A. P. de Candolle and A. L. P. P. de Candolle, Prodr. 10: 323. 1846 (as Pentstemon)

Stems ascending to erect, (10–)15–57 cm, glabrous. **Leaves** basal and cauline, glabrous; basal and proximal cauline 18–180 × 5–45(–53) mm, blade spatulate to lanceolate, base tapered, apex rounded to obtuse, acute, or mucronate; cauline 2–5 pairs, short-petiolate or sessile, 15–80 × 7–32 mm, blade obovate, ovate, oblanceolate, or lanceolate, base tapered, clasping, or cordate-clasping, apex acute to rounded or mucronate. **Thyrses** continuous or interrupted, cylindric, (5–)9–27 cm, axis glabrous, verticillasters (2–)4–10, cymes (1–)3–6-flowered; proximal bracts ovate to lanceolate or oblong, 9–46 × 7–15 mm; peduncles and pedicels glabrous, peduncles to 26 mm. **Flowers:** calyx lobes ovate to lanceolate, (2.5–)4–7 × 1.8–2.8 mm, margins entire or erose, broadly scarious, apex acute to acuminate or caudate, glabrous; corolla blue to violet, sometimes lavender, with or without reddish violet nectar guides, tubular-funnelform to ventricose, 12–20 mm, glabrous externally, glabrous internally or sparsely white-villous abaxially, tube 4–6 mm, throat gradually inflated, 5–6 mm diam., rounded abaxially; stamens included or longer pair reaching orifice,

pollen sacs opposite, (0.7–)1–1.5 mm, sutures smooth or papillate; staminode 8–12 mm, reaching orifice or slightly exserted, (0.2–)0.5–2 mm diam., tip recurved, prominently bifurcate, distal 4–8 mm densely villous, hairs yellow to golden yellow, to 2.5 mm; style 11–15 mm. **Capsules** 10–14 × 5–7 mm. **Seeds** 2–3 mm.

Varieties 2 (2 in the flora): w United States.

E. C. Neese (1993) considered *Penstemon mucronatus* transitional to *P. pachyphyllus* var. *pachyphyllus* in northeastern Utah. *Penstemon mucronatus* is distinguished from *P. pachyphyllus* var. *pachyphyllus* by its narrower staminodes bearing shorter, less tangled hairs, and its nectar guides extending onto the corolla limbs.

The Havasupai tribe of northwestern Arizona uses leaves of *Penstemon pachyphyllus* to make a deer call for hunting (D. E. Moerman 1998).

1. Staminodes 8–9 × 1–2 mm, distal 6–8 mm densely villous, hairs to 2.5 mm; Utah. . . . . 47a. *Penstemon pachyphyllus* var. *pachyphyllus*
1. Staminodes 10–12 × (0.2–)0.5–1 mm, distal 4–7 mm densely villous, hairs to 1.5 mm; Arizona, Nevada, Utah . . . . 47b. *Penstemon pachyphyllus* var. *congestus*

47a. **Penstemon pachyphyllus** A. Gray ex Rydberg var. **pachyphyllus** [E]

Flowers: staminode 8–9 × 1–2 mm, distal 6–8 mm densely villous, hairs to 2.5 mm.

Flowering May–Jul. Sandy to gravelly sagebrush shrublands, pine-juniper, Gambel oak, ponderosa pine, and bristlecone pine woodlands; 1300–3200 m; Utah.

Variety *pachyphyllus* is reported from Carbon, Daggett, Duchesne, Uintah, and Wasatch counties.

47b. **Penstemon pachyphyllus** A. Gray ex Rydberg var. **congestus** (M. E. Jones) N. H. Holmgren, Brittonia 31: 105. 1979 [E]

Penstemon acuminatus Douglas ex Lindley var. *congestus* M. E. Jones, Proc. Calif. Acad. Sci., ser. 2, 5: 714. 1895 (as Pentstemon); *P. pachyphyllus* subsp. *congestus* (M. E. Jones) D. D. Keck

Flowers: staminode 10–12 × (0.2–)0.5–1 mm, distal 4–7 mm densely villous, hairs to 1.5 mm. *2n* = 16.

Flowering May–Jul(–Aug). Sandy to gravelly sagebrush shrubland, pinyon-juniper, Gambel oak, ponderosa pine, and bristlecone pine woodlands; 1200–3100 m; Ariz., Nev., Utah.

Variety *congestus* is known from the Uinta Basin of northeastern Utah to the Grand Canyon Plateau of Arizona and to the Calcareous Mountains of east-central Nevada.

48. Penstemon secundiflorus Bentham in A. P. de Candolle and A. L. P. P. de Candolle, Prodr. 10: 325. 1846 (as Pentstemon) • Side-bells beardtongue E

Plants: caudex sometimes herbaceous. **Stems** ascending to erect, (15–)20–45(–50) cm, glabrous. **Leaves** basal and cauline, glabrous; basal and proximal cauline 20–80(–102) × 2–25 mm, blade spatulate to oblanceolate, base tapered, apex rounded to obtuse or acute, sometimes mucronate; cauline 4–6 pairs, sessile, (16–)20–78 × 3–24 mm, blade ovate to elliptic or lanceolate, base clasping to cordate-clasping, apex acute to acuminate. **Thyrses** interrupted to nearly continuous, secund, 6–24(–31) cm, axis glabrous, verticillasters (2 or)3–10(–12), cymes 2–7-flowered; proximal bracts ovate, 9–70 × 2–26 mm; peduncles and pedicels glabrous. **Flowers:** calyx lobes ovate, 4–7 × 2.4–4 mm, margins erose, sometimes entire, broadly scarious, glabrous; corolla blue to violet or lavender, rarely pink, with reddish or reddish purple nectar guides, tubular-funnelform to ventricose, 15–25 mm, glabrous externally, glabrous or sparsely to densely white-villous internally abaxially, tube 8–11 mm, throat gradually inflated, 4–7 mm diam., rounded abaxially; stamens included or longer pair reaching orifice, pollen sacs opposite, 1–1.4 mm, sutures papillate; staminode 10–13 mm, included or reaching orifice, 1–1.4 mm diam., tip abruptly recurved, bifurcate, distal 4–6 mm densely villous, hairs golden yellow, to 2 mm; style 9–15 mm. **Capsules** 9–12 × 7–9 mm.

Varieties 2 (2 in the flora): w United States.

1. Cauline leaf blades lanceolate to ovate; corollas light blue or lavender, rarely pink, sparsely to densely white-villous internally abaxially; Colorado, New Mexico, Wyoming . . . . .48a. *Penstemon secundiflorus* var. *secundiflorus*
1. Cauline leaf blades oblanceolate to ovate or elliptic; corollas dark blue or violet, glabrous or sparsely white-villous internally abaxially; Colorado. . . . . .48b. *Penstemon secundiflorus* var. *versicolor*

48a. Penstemon secundiflorus Bentham var. secundiflorus E

Penstemon unilateralis Rydberg

Cauline leaf blades lanceolate to ovate. **Corollas** light blue or lavender, rarely pink, sparsely to densely white-villous internally abaxially. $2n = 16$.

Flowering May–Jul. Rocky or gravelly soils, shortgrass prairies, foothills, mesas, parklands; 1500–2900 m; Colo., N.Mex., Wyo.

Variety *secundiflorus* occurs from south-central Wyoming through the Rocky Mountains to Santa Fe County, New Mexico. It is characteristic of the Rocky Mountain Front Range.

48b. Penstemon secundiflorus Bentham var. versicolor (Pennell) C. C. Freeman, PhytoKeys 80: 37. 2017 E

Penstemon versicolor Pennell, Contr. U.S. Natl. Herb. 20: 358. 1920

Cauline leaf blades oblanceolate to ovate or elliptic. **Corollas** dark blue or violet, glabrous or sparsely white-villous internally abaxially.

Flowering May–Jun. Rocky, gravelly, or sandy soils, shortgrass prairies, foothills; 1400–1700 m; Colo.

Variety *versicolor* is known from the Arkansas River drainage in south-central Colorado in Baca, Fremont, Las Animas, Otero, and Pueblo counties.

26b.7. PENSTEMON Schmidel (subg. PENSTEMON) sect. CRISTATI (Rydberg) Pennell, Contr. U.S. Natl. Herb. 20: 325, 328. 1920

Penstemon [unranked] *Cristati* Rydberg, Fl. Colorado, 306. 1906 (as Pentstemon); *Penstemon* sect. *Albidi* Pennell

Herbs, caulescent or essentially acaulescent (*P. acaulis, P. yampaënsis*). **Stems** retrorsely hairy, sometimes glandular-pubescent or glandular-villous distally or wholly, rarely glabrous, glabrate, or scabrous proximally or wholly, not glaucous. **Leaves** basal and cauline, sometimes ± cauline or ± basal, opposite, rarely whorled (*P. albidus, P. auriberbis, P. jamesii*), leathery or not, glabrous or glabrate, scabrous, retrorsely hairy, glandular-pubescent, pubescent, villous, or glandular-villous, not glaucous; basal and proximal cauline petiolate, sometimes sessile, rarely short-petiolate (*P. pumilus*); cauline sessile or short-petiolate, blade ovate, deltate-ovate, oblanceolate, lanceolate, elliptic, oblong, or linear, rarely pandurate, margins entire or toothed. **Thyrses** continuous or interrupted, cylindric, rarely ± secund to secund, axis glandular-pubescent, glandular-villous, or retrorsely hairy, rarely puberulent or glabrous, cymes 1 or 2 per node; peduncles and pedicels ascending to erect. **Flowers:** calyx lobes: margins entire, sometimes erose (*P. atwoodii, P. laricifolius, P. triflorus*), herbaceous or ± scarious, glandular-pubescent or glandular-villous, sometimes also scabrous, rarely entirely retrorsely hairy, puberulent, or glabrous; corolla white to lilac, lavender, blue, violet, purple, red, pink, or magenta, bilaterally symmetric, rarely nearly radially symmetric (*P. goodrichii*), weakly to strongly bilabiate, not personate, funnelform, tubular-funnelform, ventricose, or ampliate, glandular-pubescent externally, rarely glabrous (*P. laricifolius*), hairy internally abaxially, rarely glabrous or glandular-pubescent, throat gradually to abruptly inflated, rarely slightly inflated, constricted or not at orifice, rounded abaxially, sometimes slightly 2-ridged; stamens included, reaching orifice, or longer pair exserted, filaments glabrous, pollen sacs parallel to divergent or opposite, navicular to subexplanate or explanate, rarely navicular-sigmoid, dehiscing completely, rarely incompletely, proximal ⅕–½ indehiscent, connective splitting, rarely not, sides glabrous, sutures smooth or papillate; staminode included to exserted, flattened distally, 0.2–1.1 mm diam., tip straight to recurved or coiled, distal (40–)50–100% hairy, hairs to 4 mm; style glabrous, sometimes proximally glandular-pubescent (*P. eriantherus*). **Capsules** glabrous, rarely glandular-pubescent distally. **Seeds** brown, dark brown, or black, angled to reniform or slightly rounded, 1.5–4.8 mm.

Species 31 (31 in the flora): w North America, n Mexico.

Members of sect. *Cristati* are mostly calciphiles on soils derived from limestone, dolomite, or other calcareous substrates. Species in sect. *Cristati* generally have hairy herbage, glandular-pubescent inflorescences and flowers, and staminodes prominently bearded more than half their lengths. Two species often assigned elsewhere (*Penstemon acaulis, P. laricifolius*) are included in this section based on molecular data (A. D. Wolfe et al. 2006). D. D. Keck's (1938) sect. *Aurator*, which appears widely in the literature, is invalid, published without a Latin diagnosis.

SELECTED REFERENCE Keck, D. D. 1938. Studies in *Penstemon* VI. The section *Aurator*. Bull. Torrey Bot. Club 65: 233–255.

1. Stems to 1 cm.
    2. Basal and proximal cauline leaves 6–15(–22) × 0.6–1.3(–1.5) mm, blades linear... 49. *Penstemon acaulis*
    2. Basal and proximal cauline leaves 15–25(–35) × 1.5–2.6(–4.5) mm, blades oblanceolate to linear. . . . . .79. *Penstemon yampaënsis*
1. Stems 2–65(–100) cm.
    3. Corollas glabrous externally; calyx lobes glabrous . . . . 69. *Penstemon laricifolius*
    3. Corollas glandular-pubescent externally; calyx lobes hairy.

[4. Shifted to left margin.—Ed.]

4. Pollen sacs parallel, navicular.
    5. Staminodes prominently exserted; basal and proximal cauline leaf margins sharply serrate-dentate, rarely entire; corollas ventricose . . . . .76. *Penstemon pinorum*
    5. Staminodes included to barely exserted; basal and proximal cauline leaf margins entire or ± dentate; corollas funnelform or tubular-funnelform.
        6. Thyrses 8–30 cm, interrupted; stems (25–)30–60 cm; Arizona . . . . 58. *Penstemon distans*
        6. Thyrses 2–7(–8) cm, continuous or interrupted; stems 2–25 cm; Utah.
            7. Leaves glabrous abaxially, retrorsely hairy adaxially, usually only along midveins on distal caulines; pollen sacs 1.3–1.6 mm . . . . 62. *Penstemon franklinii*
            7. Leaves densely retrorsely hairy; pollen sacs 1.7–2 mm . . . . 74. *Penstemon nanus*
4. Pollen sacs opposite or divergent, explanate or subexplanate to navicular or navicular-sigmoid.
    8. Thyrse axes puberulent or retrorsely hairy.
        9. Leaves: basal and proximal cauline blades oblanceolate to linear, 1–4 mm wide, caulines usually arcuate; Idaho . . . . 77. *Penstemon pumilus*
        9. Leaves: basal and proximal cauline blades spatulate to oblanceolate, 2–14 mm wide, caulines not arcuate; Nevada, Utah.
            10. Cauline leaves 12–38 × 3–7 mm; stems 2–14(–20) cm, prostrate, decumbent, ascending, or erect; corollas with, rarely without, faint blue or reddish purple nectar guides; ec Nevada, wc Utah . . . . 59. *Penstemon dolius*
            10. Cauline leaves 14–23 × 1–4 mm; stems 2–6(–8) cm, ascending to erect; corollas without nectar guides; ne Utah . . . . .60. *Penstemon duchesnensis*
    8. Thyrse axes glandular-pubescent or glandular-villous, sometimes retrorsely hairy.
        11. Capsules sparsely glandular-pubescent distally; styles sparsely glandular-pubescent proximally or glabrous . . . . 61. *Penstemon eriantherus*
        11. Capsules glabrous; styles glabrous.
            12. Pollen sacs navicular or navicular-sigmoid.
                13. Corollas ventricose, 19–30 mm; staminodes prominently exserted . . . . .
                . . . . .64. *Penstemon gormanii*
                13. Corollas tubular-funnelform, 10–22(–24) mm; staminodes included to exserted.
                    14. Stems puberulent, hairs usually white, scalelike; leaves densely retrorsely hairy, hairs white, scalelike; California . . . . 73. *Penstemon monoensis*
                    14. Stems glabrous or retrorsely hairy, sometimes sparsely glandular-pubescent distally, hairs not white or scalelike; leaves glabrous, puberulent, or retrorsely hairy, hairs not white or scalelike; Colorado, New Mexico, Utah.
                        15. Pollen sacs divergent, dehiscing incompletely; corollas weakly bilabiate; Colorado, New Mexico .52. *Penstemon auriberbis*
                        15. Pollen sacs opposite, dehiscing completely; corollas bilabiate; Colorado, Utah.
                            16. Basal and proximal cauline leaf blades oblanceolate to lanceolate; corollas 10–13 mm; styles 6–8 mm; Utah . . .51. *Penstemon atwoodii*
                            16. Basal and proximal cauline leaf blades spatulate to oblanceolate or ovate; corollas 15–20(–22) mm; styles 10–12 mm; Colorado, Utah . . . . 72. *Penstemon moffatii*
            12. Pollen sacs explanate, rarely subexplanate.
                17. Corollas tubular-funnelform to funnelform.
                    18. Corollas white, sometimes tinged pink or lavender.
                        19. Leaves glabrate or puberulent to scabrous, basals and proximals (4–)7–18(–20) mm wide, blades oblanceolate or obovate to lanceolate; thyrse axes densely glandular-pubescent; Alberta, Manitoba, Ontario, Saskatchewan, Colorado, Iowa, Kansas, Minnesota, Montana, Nebraska, New Mexico, North Dakota, Oklahoma, South Dakota, Texas, Wyoming . . . . 50. *Penstemon albidus*

19. Leaves glabrous or glabrate, rarely basals retrorsely hairy along midveins, basals and proximals 1–6 mm wide, blades linear, rarely lanceolate; thyrse axes glandular-pubescent; Texas . . . . 66. *Penstemon guadalupensis*

18. Corollas blue, bluish lavender, lavender, pink, rose red, rose purple, violet, or purple.

20. Corollas pink to rose red or rose purple; California, sw Nevada . . . . .55. *Penstemon calcareus*

20. Corollas lavender, violet, purple, bluish lavender, or blue; e Nevada, Utah.

21. Staminodes exserted; corollas bilabiate, tubular-funnelform . . . . .57. *Penstemon concinnus*

21. Staminodes included; corollas weakly bilabiate, funnelform.

22. Stems glabrous or slightly retrorsely hairy proximally, glandular-pubescent distally; basal and proximal cauline leaf blades lanceolate to linear; staminodes: distal 5–7 mm densely pilose, hairs yellow, to 0.8 mm . . . . . .63. *Penstemon goodrichii*

22. Stems retrorsely hairy; basal and proximal cauline leaf blades oblanceolate to lanceolate; staminodes: distal 3–5 mm sparsely to densely pubescent, hairs yellow, to 1 mm . . . . .70. *Penstemon marcusii*

[17. Shifted to left margin.—Ed.]

17. Corollas ventricose, ventricose-ampliate, or ampliate.

23. Corollas 35–55 mm; staminodes 23–30 mm; styles 18–34 mm . . . . 56. *Penstemon cobaea*

23. Corollas 10–35 mm; staminodes 7–20 mm; styles 7–24 mm.

24. Corollas glandular-pubescent internally abaxially, sometimes also pilose.

25. Corollas 14–20(–22) mm; styles 11–13 mm; capsules 6–11 mm . . . . . .75. *Penstemon ophianthus*

25. Corollas 22–35 mm; styles 17–24 mm; capsules (7–)10–16 mm.

26. Staminodes exserted, distal 10–14 mm lanate, hairs to 3.5 mm; corolla throats abruptly inflated . . . . 67. *Penstemon jamesii*

26. Staminodes included or reaching orifice, distal 8–12 mm sparsely to moderately pilose, hairs to 1 mm; corolla throats gradually inflated . . . . 78. *Penstemon triflorus*

24. Corollas lanate or villous internally abaxially.

27. Styles 15–24 mm; corollas 18–35 mm, tubes (6–)8–13 mm; staminodes 16–20 mm.

28. Corollas 28–35 mm, ventricose-ampliate, sparsely white-lanate internally abaxially; Colorado, Utah . . . . .65. *Penstemon grahamii*

28. Corollas 18–28 mm, ventricose, densely white-lanate internally abaxially; California, Idaho, Nevada, Oregon . . . . .68. *Penstemon janishiae*

27. Styles 7–11 mm; corollas 10–22 mm, tubes 4–7 mm; staminodes 7–10 mm.

29. Staminodes reaching orifices or exserted; corollas ampliate; leaves retrorsely hairy, sometimes glabrate; Arizona, Colorado, New Mexico, Utah . . . . .54. *Penstemon breviculus*

29. Staminodes exserted or prominently exserted; corollas ventricose; leaves retrorsely hairy, glabrate, or glabrous; California, Idaho, Nevada, Oregon.

30. Corollas 10–14 mm; styles 7–9 mm; capsules 3–6 mm; California, Nevada . . . . .53. *Penstemon barnebyi*

30. Corollas 14–22 mm; styles 9–10 mm; capsules (6–)8–11 mm; Idaho, Oregon . . . . 71. *Penstemon miser*

49. Penstemon acaulis L. O. Williams, Ann. Missouri Bot. Gard. 21: 345. 1934 • Stemless beardtongue C E

Stems prostrate to ascending, to 1 cm (to 15 cm diam.), scabrous or puberulent. **Leaves** essentially basal, not leathery, scabrous; basal and proximal cauline sessile, 6–15(–22) × 0.6–1.3(–1.5) mm, blade linear, base tapered, margins entire, apex acute. **Thyrses** essentially absent, verticillasters 1, cymes 1(or 2)-flowered, 1 per node; proximal bracts linear, 5–10 × 0.6–1 mm; peduncles and pedicels glandular-pubescent and scabrous. **Flowers:** calyx lobes lanceolate, 3.5–5.5 × 0.8–1.5 mm, glandular-pubescent and scabrous; corolla lavender to blue or violet, without nectar guides, funnelform, 11–15 mm, yellowish or white-villous internally abaxially, tube 5–6 mm, throat gradually inflated, not constricted at orifice, 4.5–6 mm diam., slightly 2-ridged abaxially; stamens: longer pair reaching orifice or exserted, pollen sacs widely divergent or opposite, navicular, 0.7–0.9 mm, dehiscing completely, sutures papillate; staminode 7–9 mm, exserted, 0.4–0.5 mm diam., tip straight to slightly recurved, distal 4–7 mm densely pilose, hairs orange, to 0.8 mm; style 7–8 mm. **Capsules** 2.5–3.5 × 2.5–3 mm.

Flowering May–Jul. Semi-barren rock ledges, clayey ridges, gravelly hilltops; of conservation concern; 1500–2500 m; Utah, Wyo.

Penstemon acaulis is known from the Bridger Basin. Populations are reported from Daggett County, Utah, and Sweetwater County, Wyoming. E. C. Neese (1993) considered populations from Browns Park, Utah, to be transitional between *P. acaulis* and *P. yampaënsis*, treating the latter as *P. acaulis* var. *yampaënsis* (Penland) Neese. Some specimens from the vicinity of Clay Basin and Red Creek in northwestern Browns Park, Utah, are morphologically intermediate; other specimens from the same area can be assigned to species unambiguously. Given the morphologic distinctness of the majority of specimens, which fall into two basically discrete geographic regions, *P. acaulis* and *P. yampaënsis* are here treated as distinct, closely related species.

50. Penstemon albidus Nuttall, Gen. N. Amer. Pl. 2: 53. 1818 (as Pentstemon albidum) • White beardtongue E

Stems ascending to erect, (10–)15–50(–55) cm, retrorsely hairy proximally, glandular-pubescent distally. **Leaves** basal and cauline, not leathery, glabrate or puberulent to scabrous; basal and proximal cauline petiolate, 20–85(–110) × (4–)7–18(–20) mm, blade oblanceolate or obovate to lanceolate, base tapered, margins entire to obscurely or distinctly serrate, apex obtuse to acute; cauline 2–5(or 6) pairs, sessile or proximals short-petiolate, 25–65 × (3–)7–19(–21) mm, blade ovate to lanceolate, base tapered to clasping, margins entire or serrate to dentate, apex acute. **Thyrses** continuous or interrupted, cylindric, 4–24(–30) cm, axis densely glandular-pubescent, verticillasters (2 or)3–10, cymes 2–7-flowered, 2 per node; proximal bracts lanceolate, 17–65 × 3–17 mm; peduncles and pedicels densely glandular-pubescent. **Flowers:** calyx lobes ovate to lanceolate, 4–7 × 1.5–3 mm, glandular-pubescent; corolla white, rarely tinged pink or lavender, with red or reddish purple nectar guides, funnelform, (12–)16–20 mm, glandular-pubescent internally, tube 4–6 mm, throat gradually inflated, not constricted at orifice, (4–)6–8 mm diam., rounded abaxially; stamens included, pollen sacs opposite, explanate, 0.7–1.1 mm, dehiscing completely, sutures smooth; staminode 8–9 mm, included, 0.3–0.4 mm diam., tip straight to recurved, distal 3–6 mm sparsely to moderately villous, hairs yellowish, to 1 mm; style 9–11(–13) mm. **Capsules** 8–12 × 4–7 mm. *2n* = 16.

Flowering Apr–Jul(–Sep). Silty or gravelly soils, mixed-grass and shortgrass prairies; 300–1800 m; Alta., Man., Ont., Sask.; Colo., Iowa, Kans., Minn., Mont., Nebr., N.Mex., N.Dak., Okla., S.Dak., Tex., Wyo.

Penstemon albidus is remarkably uniform throughout its range. It probably is introduced in Ontario.

51. Penstemon atwoodii S. L. Welsh, Great Basin Naturalist 35: 378. 1976 • Atwood's beardtongue E

Stems ascending to erect, 14–35(–50) cm, glabrous or sparsely retrorsely hairy proximally, usually sparsely glandular-pubescent distally. **Leaves** basal and cauline, leathery or not, glabrous or sparsely retrorsely hairy proximally adaxially; basal and proximal cauline petiolate, 30–80(–100) × 6–12 mm, blade oblanceolate to lanceolate, base tapered, margins

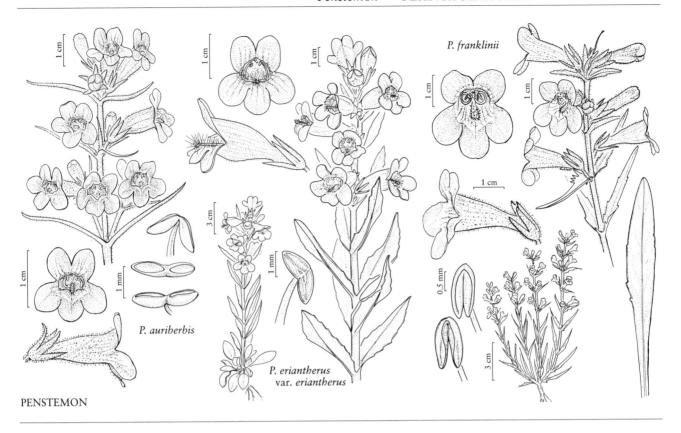

P. franklinii

P. auriberbis

P. eriantherus
var. *eriantherus*

PENSTEMON

entire or obscurely serrate, apex obtuse to acute or mucronate; cauline 3 or 4 pairs, sessile, 28–78 × 3–7 mm, blade oblanceolate to lanceolate, base attenuate to clasping, margins entire, apex obtuse to acute. **Thyrses** interrupted, cylindric, (3–)7–24 cm, axis glandular-pubescent, verticillasters 3–8, cymes 3–12-flowered, 2 per node; proximal bracts lanceolate to linear, 22–65 × 2–10 mm; peduncles and pedicels glandular-pubescent. **Flowers:** calyx lobes lanceolate, 6–8(–9) × 1.5–2 mm, glandular-pubescent; corolla blue to lavender, with dark violet nectar guides, bilabiate, tubular-funnelform, 10–13 mm, sparsely to moderately yellow-lanate internally abaxially, tube 4–5 mm, throat gradually inflated, not constricted at orifice, 3–5 mm diam., rounded abaxially; stamens reaching orifice or longer pair exserted, pollen sacs opposite, navicular, 0.9–1.5 mm, dehiscing completely, sutures papillate; staminode 6–8 mm, exserted, 0.3–0.4 mm diam., tip straight, distal 4–5 mm moderately to densely retrorsely-pubescent, hairs golden yellow, to 0.6 mm; style 6–8 mm. **Capsules** 8–12 × 4–7 mm.

Flowering May–Jun. Pinyon-juniper woodlands; 1600–2400 m; Utah.

According to Welsh, *Penstemon atwoodii* is known only from the Kaiparowits Plateau of eastern Garfield and Kane counties.

52. Penstemon auriberbis Pennell, Contr. U.S. Natl. Herb. 20: 339. 1920 • Colorado beardtongue E F

Penstemon parviflorus Pennell

Stems ascending to erect, 10–30(–35) cm, retrorsely hairy proximally, glandular-pubescent distally. **Leaves** basal and cauline, not leathery, glabrous, puberulent, or retrorsely hairy; basal and proximal cauline petiolate, (15–)30–60(–100) × (1–)2–5(–7) mm, blade lanceolate to linear, base tapered, margins entire, apex obtuse to acute; cauline 2–6 pairs, sessile, (25–)40–80 × 2–7 mm, blade lanceolate to linear, base tapered to slightly clasping, margins entire or obscurely denticulate, apex acute to acuminate. **Thyrses** continuous or interrupted, cylindric or slightly secund, (5–)7–23 cm, axis glandular-pubescent, verticillasters 3–8, cymes 2–4-flowered, 2 per node; proximal bracts lanceolate to linear, 18–100 × 2–12 mm; peduncles and pedicels glandular-pubescent. **Flowers:** calyx lobes ovate to lanceolate, (6–)7–9 × 1–2 mm, densely glandular-pubescent; corolla violet to lilac or pinkish blue, with magenta or violet nectar guides, weakly bilabiate, tubular-funnelform, (16–)18–22(–24) mm, moderately white-pilose internally abaxially, tube 4–6 mm, throat gradually inflated, not constricted at orifice, 7–9 mm diam., rounded abaxially;

stamens included, pollen sacs divergent, navicular, 1.2–1.5 mm, dehiscing incompletely, connective not splitting, sutures papillate; staminode 13–16 mm, reaching orifice or exserted, 0.6–0.9 mm diam., tip recurved, distal 10–15 mm moderately to densely villous, hairs orangish, to 2.5 mm; style 11–14 mm. **Capsules** (6–)8–10 × 3–4 mm. *2n* = 16.

Flowering May–Aug. Silty or gravelly soils, short-grass prairies, sagebrush shrublands; 1200–2500 m; Colo., N.Mex.

Penstemon auriberbis is known from Chaffee, Costilla, Custer, El Paso, Elbert, Fremont, Huerfano, Kiowa, Las Animas, Otero, and Pueblo counties, Colorado, and Colfax and Union counties, New Mexico. The species is unique in sect. *Cristati* in having incompletely dehiscing anthers.

Penstemon parviflorus is known only from the type locality in Montezuma County, Colorado (S. L. O'Kane 1988). A report for McKinley County, New Mexico, has not been confirmed. The type was collected July 1890 by Alice Eastwood, reputedly near Mancos, Colorado. F. W. Pennell (1920b) distinguished *P. parviflorus* from the more eastern *P. auriberbis* primarily by the former's shorter corollas, longer calyx lobes, and more slender stems. D. D. Keck (1938) observed that the type specimen has whorled cauline leaves, which Pennell did not mention in his description, and Keck suggested that *P. parviflorus* might be a genetically aberrant individual of *P. auriberbis* that was accidentally introduced into Montezuma County. Whorled cauline leaves are not uncommon in *P. auriberbis*, and they occur rarely in other species in sect. *Cristati*. *Penstemon parviflorus* appears to be *P. auriberbis*, and the type specimen actually may have come from south-central Colorado; Eastwood collected *P. auriberbis* in Pueblo County in 1891.

53. Penstemon barnebyi N. H. Holmgren, Brittonia 31: 226, fig. 5. 1979 • Barneby's beardtongue E

Stems ascending to erect, 5–20 (–30) cm, retrorsely hairy. **Leaves** basal and cauline, not leathery, retrorsely hairy; basal and proximal cauline petiolate, 12–55(–75) × 4–12 mm, blade elliptic to spatulate, base tapered, margins entire or obscurely and remotely serrate, apex rounded to obtuse or acute; cauline 2–5 pairs, short-petiolate or sessile, 19–65 × 2–8 mm, blade lanceolate to oblanceolate, base truncate to tapered, margins entire or remotely serrate, apex obtuse to acute. **Thyrses** continuous, cylindric, 2–10 cm, axis densely glandular-pubescent, verticillasters 3–8, cymes 3–5-flowered, 2 per node; proximal bracts lanceolate, 18–70 × 3–8 mm; peduncles and pedicels densely glandular-pubescent. **Flowers:** calyx lobes lanceolate, 4.5–7 × 1.1–1.8 mm, glandular-pubescent; corolla lavender to bluish lavender or violet, with reddish violet nectar guides, ventricose, 10–14 mm, sparsely to densely white- or yellowish lanate internally abaxially, tube 4–6 mm, throat abruptly inflated, ± constricted at orifice, 4–5 mm diam., rounded abaxially; stamens included, pollen sacs opposite, explanate, 0.6–0.9 mm, dehiscing completely, sutures smooth; staminode 9–10 mm, prominently exserted, 0.2 mm diam., tip coiled, distal 5–6 mm densely pubescent, hairs orange or orangish yellow, to 1.2 mm; style 7–9 mm. **Capsules** 3–6 × 3–5 mm.

Flowering May–Jun. Alluvial gravels, calcareous silts, sagebrush shrublands, pinyon-juniper woodlands; 1500–2500 m; Calif., Nev.

Penstemon barnebyi is known from Mono County, California, and Esmeralda, Lincoln, Mineral, Nye, and White Pine counties, Nevada.

54. Penstemon breviculus (D. D. Keck) G. T. Nisbet & R. C. Jackson, Univ. Kansas Sci. Bull. 41: 734. 1960 • Short-stem beardtongue E

Penstemon jamesii Bentham subsp. *breviculus* D. D. Keck, Bull. Torrey Bot. Club 65: 241. 1938

Stems ascending to erect, 8–20(–35) cm, retrorsely hairy. **Leaves** basal and cauline, not leathery, retrorsely hairy, sometimes glabrate; basal and proximal cauline petiolate, 35–95 × 3–18 mm, blade oblanceolate to lanceolate or linear, base tapered, margins entire or obscurely dentate, rarely prominently dentate distally, apex rounded to obtuse or acute; cauline (1 or)2–5 pairs, short-petiolate or sessile, 33–78 × 4–10 mm, blade elliptic to oblanceolate, lanceolate or linear, base tapered to clasping, margins entire or dentate, apex obtuse to acute. **Thyrses** continuous or ± interrupted, cylindric, 4–11(–18) cm, axis glandular-pubescent, verticillasters 3–5, cymes (1 or)2–6-flowered, 2 per node; proximal bracts lanceolate to linear, 16–38 × 2–8 mm; peduncles and pedicels glandular-pubescent. **Flowers:** calyx lobes lanceolate, 4.5–7.5 × 1.2–2 mm, glandular-pubescent; corolla dark blue to violet or purple, with reddish violet nectar guides, ampliate, 10–15(–18) mm, yellowish or whitish villous internally abaxially, tube 5–6 mm, throat abruptly inflated, ± constricted at orifice, 3.5–6 mm diam., rounded abaxially; stamens included, pollen sacs opposite, explanate, 0.6–1 mm, dehiscing completely, sutures smooth; staminode 7–8 mm, reaching orifice or exserted, 0.3–0.4 mm diam., tip straight to recurved, distal 5–6 mm densely pubescent, hairs orangish yellow, to 1.5 mm, and medial hairs shorter, stiffer, and retrorse; style 9–11 mm. **Capsules** 6–9 × 4.5–6 mm. *2n* = 16.

Flowering May–Jun. Sandy or clayey soils, sage-brush shrublands, pinyon-juniper woodlands, desert grasslands; 1500–2000 m; Ariz., Colo., N.Mex., Utah.

Penstemon breviculus is known from the Four Corners region. Populations have been documented in Arizona (Apache County), Colorado (Montezuma, Montrose, and San Miguel counties), New Mexico (McKinley and San Juan counties), and Utah (Grand, San Juan, and Wayne counties).

55. Penstemon calcareus Brandegee, Zoë 5: 152. 1903 (as Pentstemon) • Limestone beardtongue C E

Stems ascending to erect, 5–35 cm, puberulent. **Leaves** basal and cauline, ± leathery, ± retrorsely hairy, sometimes glabrate abaxially; basal and proximal cauline petiolate, 15–55 × 6–25 mm, blade elliptic to broadly ovate, base tapered, rarely truncate, margins entire or remotely and obscurely serrate, apex obtuse to acute; cauline 2 or 3 pairs, sessile, 25–75 × 10–21 mm, blade lanceolate to oblanceolate, rarely ovate, base truncate to subcordate-clasping, margins entire or obscurely dentate, apex obtuse to acute. **Thyrses** continuous or interrupted, cylindric, 6–18 cm, axis glandular-pubescent, verticillasters 2–8, cymes (1–)3–7-flowered, 2 per node; proximal bracts lanceolate to ovate, 13–38 × 9–20 mm; peduncles and pedicels glandular-pubescent. **Flowers:** calyx lobes lanceolate, 5–7.5 × 0.9–1.8 mm, glandular-pubescent; corolla pink to rose red or rose purple, with reddish nectar guides, tubular-funnelform, 12–17 mm, sparsely white-lanate internally abaxially, tube 4–5 mm, throat gradually inflated, not constricted at orifice, 2.5–4 mm diam., slightly 2-ridged abaxially; stamens included, pollen sacs opposite, subexplanate to explanate, 0.6–0.7 mm, dehiscing completely, sutures smooth; staminode 7–9 mm, included, 0.2–0.3 mm diam., tip straight to recurved, distal 5–6 mm densely pilose, hairs golden yellow, to 0.8 mm; style 4–6 mm. **Capsules** 6–9 × 4–5.5 mm.

Flowering Apr–May. Limestone crevices, rocky slopes, pine-juniper woodlands; of conservation concern; 1200–2000 m; Calif., Nev.

Penstemon calcareus is known from Inyo and San Bernardino counties, California, and Esmeralda County, Nevada. The later homonym, *P. calcareus* M. E. Jones, is referable to *P. petiolatus*.

56. Penstemon cobaea Nuttall, Trans. Amer. Philos. Soc., n. s. 5: 182. 1835 (as Pentstemon) • Cobaea beardtongue E

Stems retrorsely ascending to erect, (15–)25–65(–100) cm, puberulent proximally, glandular-pubescent distally. **Leaves** basal and cauline, sometimes basal absent or withering, ± leathery or not, glabrous or puberulent to pubescent; basal and proximal cauline petiolate, 35–120(–150) × 8–55(–76) mm, blade spatulate to oblanceolate or elliptic, base tapered, margins subentire or serrate to dentate, apex rounded to acute; cauline 4–8 pairs, sessile, 35–120(–150) × 10–45(–54) mm, blade ovate to lanceolate, base tapered to cordate-clasping, margins subentire to serrate or dentate, apex acute, rarely obtuse. **Thyrses** interrupted or continuous, cylindric, 10–30(–52) cm, axis densely glandular-pubescent, verticillasters 3–6(–8), cymes 2–6-flowered, 2 per node; proximal bracts ovate, 20–78 × 8–74 mm; peduncles and pedicels densely glandular-pubescent. **Flowers:** calyx lobes ovate to lanceolate, (8–)10–16 × 2.5–4.5 mm, densely glandular-pubescent; corolla white to pink or violet to purple, with reddish purple nectar guides, ventricose-ampliate, 35–55 mm, glandular-pubescent internally, tube 10–14 mm, throat abruptly inflated, slightly constricted at orifice, (15–)18–25 mm diam., rounded abaxially; stamens included, pollen sacs opposite, explanate, 1.3–1.6 mm, dehiscing completely, sutures smooth; staminode 23–30 mm, included or barely exserted, 0.6–0.9 mm diam., tip recurved, distal 20–25 mm sparsely to moderately villous, hairs white or golden yellow, to 3.5 mm; style 18–34 mm. **Capsules** 13–18 × 8–10 mm.

Varieties 2 (2 in the flora): United States.

Penstemon cobaea occurs widely in the central and southern Great Plains and into the Ozark Plateau. The species is a popular ornamental used in the breeding of the European or large-flowered hybrids of *Penstemon* (D. Lindgren and E. Wilde 2003). A chromosome count of *n* = 32 (R. Piotrowska 1934) appears doubtful.

1. Corollas white to pink . . . .
. . . . 56a. *Penstemon cobaea* var. *cobaea*
1. Corollas violet to purple . . . .
. . . . .56b. *Penstemon cobaea* var. *purpureus*

56a. Penstemon cobaea Nuttall var. **cobaea** [E]

Penstemon hansonii A. Nelson

Corollas white to pink. **2***n* = 16.

Flowering Apr–Jun. Limestone and gypsum soils, tallgrass and mixed-grass prairies; 200–1000 m; Ariz., Ark., Colo., Ill., Iowa, Kans., Mo., Nebr., N.Mex., Ohio, Okla., Tex.

Reports of var. *cobaea* from Arizona, Colorado, Illinois, New Mexico, and Ohio appear to be based on introductions, as are some reports from western Kansas and western Nebraska.

56b. Penstemon cobaea Nuttall var. **purpureus** Pennell, Proc. Acad. Nat. Sci. Philadelphia 73: 490. 1922 [E]

Penstemon cobaea subsp. *purpureus* (Pennell) Pennell

Corollas violet to purple. **2***n* = 16.

Flowering Apr–Jun. Limestone and dolomite glades, barrens, cliffs, calcareous prairies; 100–300 m; Ark., Mo., Okla.

Variety *purpureus* is concentrated in the Ozark Plateau of north-central Arkansas and south-central Missouri. Populations of *Penstemon cobaea* on the Coastal Plain region of southwestern Arkansas and southeastern Oklahoma, which are 300–400 km disjunct from the core populations of var. *purpureus*, are included here.

57. Penstemon concinnus D. D. Keck, Amer. Midl. Naturalist 23: 608. 1940 • Tunnel Springs beardtongue [E]

Stems ascending to erect, 7–18(–24) cm, retrorsely hairy. **Leaves** basal and cauline, not leathery, glabrous or sparsely retrorsely hairy on petiole and, rarely, midvein; basal and proximal cauline petiolate, 30–55(–70) × 2–5(–8) mm, blade oblanceolate to lanceolate or linear, base tapered, margins entire or obscurely and remotely dentate, apex obtuse to acute; cauline 3–5 pairs, short-petiolate or sessile, 17–60 × 1–5 mm, blade lanceolate to linear, base tapered to clasping, margins entire or remotely dentate, apex acute. **Thyrses** interrupted or continuous, cylindric, 2–11 cm, axis glandular-pubescent, verticillasters 3–6, cymes 2–6-flowered, 2 per node; proximal bracts lanceolate to linear, 12–55 × 2–9 mm; peduncles and pedicels

glandular-pubescent. **Flowers:** calyx lobes lanceolate, 6–8 × 1–1.4 mm, glandular-pubescent; corolla violet to lavender or purple, with dark violet nectar guides, bilabiate, tubular-funnelform, 8–11 mm, white-lanate internally abaxially, tube 4–5 mm, throat gradually inflated, not constricted at orifice, 3.5–4.5 mm diam., slightly 2-ridged abaxially; stamens included or longer pair reaching orifice or slightly exserted, pollen sacs opposite, explanate, 0.5–0.9 mm, dehiscing completely, sutures papillate; staminode 6–8 mm, exserted, 0.3–0.4 mm diam., tip recurved to coiled, distal 4–5 mm densely pilose, hairs white or pale yellow, to 1 mm; style 5–7 mm. **Capsules** 4–6 × 3–4 mm.

Flowering May–Jun. Gravelly soils, desert shrublands, pinyon-juniper woodlands; 1600–2200 m; Nev., Utah.

Penstemon concinnus is known from the Snake Range in Lincoln and White Pine counties, Nevada, and in the Burbank Hills, Needle Range, and Tunnel Spring and Wah Wah mountains in Beaver and Millard counties, Utah.

58. Penstemon distans N. H. Holmgren, Brittonia 32: 326, fig. 1. 1980 • Mt. Trumbull beardtongue [C] [E]

Stems ascending to erect, (25–)30–60 cm, puberulent proximally, glandular-pubescent distally. **Leaves** basal and cauline, not leathery, puberulent proximally, glandular-pubescent distally; basal and proximal cauline petiolate, 55–140 × 2–14 mm, blade oblanceolate, base tapered, margins entire or ± dentate, apex obtuse to acute; cauline 3–6 pairs, short-petiolate or sessile, 25–40(–60) × 1–3(–7) mm, blade lanceolate, base tapered to subcordate-clasping, margins entire or ± serrate, apex obtuse to acute. **Thyrses** interrupted, cylindric, 8–30 cm, axis glandular-pubescent, verticillasters 6–10, cymes 1- or 2(or 3)-flowered, 2 per node; proximal bracts lanceolate to linear, 5–35 × 1–6 mm; peduncles and pedicels glandular-pubescent. **Flowers:** calyx lobes lanceolate, 4–7 × 1–1.8 mm, glandular-pubescent; corolla blue to violet, with reddish purple nectar guides, funnelform, 16–20 mm, sparsely to moderately white- or yellow-lanate internally abaxially, tube 7–8 mm, throat gradually inflated, not constricted at orifice, 6–8 mm diam., 2-ridged abaxially; stamens included, pollen sacs parallel, navicular, 1.4–1.8 mm, dehiscing completely, sutures papillate; staminode 9–11 mm, reaching orifice or barely exserted, 0.7–0.9 mm diam., tip recurved, distal 6–8 mm densely hairy, hairs yellow-orange, to 1.4 mm; style 7–9 mm. **Capsules** 7–9 × 4–6 mm.

Flowering Apr–Jun. Limestone gravel slopes, pine-juniper woodlands; of conservation concern; 1200–1600 m; Ariz.

Penstemon distans is known from the southeastern end of the Shivwits Plateau in Mohave County.

59. Penstemon dolius M. E. Jones ex Pennell, Contr. U.S. Natl. Herb. 20: 341. 1920 • Jones's beardtongue E

Stems prostrate, decumbent, ascending, or erect, 2–14 (–20) cm, pubescent to densely retrorsely hairy. **Leaves** basal and cauline, not leathery, retrorsely hairy; basal and proximal cauline petiolate, (5–) 10–40(–55) × 4–14 mm, blade spatulate to oblanceolate, base tapered, margins entire, apex obtuse to acute, rarely rounded; cauline 1–3 pairs, sessile, 12–38 × 3–7 mm, blade oblanceolate, not arcuate, base tapered, margins entire, apex obtuse to acute. **Thyrses** continuous, cylindric, 0.5–8(–10) cm, axis retrorsely hairy, verticillasters 1–4(–7), cymes 1- or 2(–4)-flowered, 1 or 2 per node; proximal bracts oblanceolate to lanceolate, 12–34 × 2–8 mm; peduncles and pedicels retrorsely hairy, sometimes sparsely glandular-pubescent. **Flowers:** calyx lobes lanceolate, 4–6 × 1–1.9 mm, retrorsely hairy, sometimes sparsely glandular-pubescent; corolla light blue to blue or violet, with, rarely without, faint blue or reddish purple nectar guides, funnelform, 14–20 mm, glandular-pubescent externally, sparsely white-lanate internally abaxially, tube 6–7 mm, throat gradually inflated, not constricted at orifice, 4–7 mm diam., slightly 2-ridged abaxially; stamens included or longer pair reaching orifice, pollen sacs opposite, navicular, 0.9–1.2 mm, dehiscing completely, sutures papillate; staminode 7–10 mm, included or reaching orifice, 0.7–0.9 mm diam., tip recurved to coiled, distal 5–8 mm sparsely pubescent to lanate, hairs yellow or yellow-orange, to 1 mm; style 10–12 mm. **Capsules** 4–6 × 4–5 mm.

Flowering May–Jun. Gravelly soils, shadscale and sagebrush shrublands, pinyon-juniper woodlands; 1400–2100 m; Nev., Utah.

Penstemon dolius is known from east-central Nevada (Elko, Nye, and White Pine counties) and west-central Utah (Beaver, Juab, Millard, Sanpete, and Tooele counties). *Penstemon dolius* is distinguished from *P. duchesnensis* by longer stems and leaves and more sparsely bearded staminodes. Corollas of *P. dolius* usually have nectar guides; *P. duchesnensis* appears to lack nectar guides.

60. Penstemon duchesnensis (N. H. Holmgren) Neese, Great Basin Naturalist 46: 459. 1986 C E

Penstemon dolius M. E. Jones ex Pennell var. *duchesnensis* N. H. Holmgren, Brittonia 31: 219, fig. 6. 1979

Stems ascending to erect, 2–6(–8) cm, pubescent or densely retrorsely hairy. **Leaves** basal and cauline, not leathery, retrorsely hairy; basal and proximal cauline petiolate, 8–30 × 2–9 mm, blade spatulate to oblanceolate, base tapered, margins entire, apex rounded to obtuse or acute; cauline (0 or)1–3 pairs, sessile, 14–23 × 1–4 mm, blade oblanceolate to linear, not arcuate, base tapered, margins entire, apex acute. **Thyrses** continuous, cylindric, 1–4 cm, axis puberulent or retrorsely hairy, verticillasters 1–5, cymes (1 or)2-flowered, 2 per node; proximal bracts oblanceolate to lanceolate or linear, 13–24 × 1–4 mm; peduncles and pedicels retrorsely hairy, sometimes sparsely glandular-pubescent. **Flowers:** calyx lobes lanceolate, 5–8.5 × 1–2.1 mm, retrorsely hairy, sometimes sparsely glandular-pubescent; corolla light blue to blue, lavender, or violet, without nectar guides, funnelform, 14–20 mm, glabrous or sparsely white-lanate internally abaxially, tube 6–8 mm, throat gradually inflated, not constricted at orifice, 4–7 mm diam., slightly 2-ridged abaxially; stamens included or longer pair reaching orifice, pollen sacs opposite, navicular, 0.9–1.2 mm, dehiscing completely, sutures papillate; staminode 9–11 mm, included or reaching orifice, 0.5–1.1 mm diam., tip straight to recurved, distal 5–7 mm moderately pubescent to lanate, hairs yellow or yellow-orange, to 1.5 mm; style 9–12 mm. **Capsules** 4–6.5 × 4–5 mm.

Flowering May–Jun. Gravelly, sandy, and clayey hills, sagebrush shrublands, pinyon-juniper woodlands; of conservation concern; 1500–2000 m; Utah.

Penstemon duchesnensis is known only from the western Uinta Basin in the vicinity of Duchesne, Duchesne County.

61. Penstemon eriantherus Pursh, Fl. Amer. Sept. 2: 737. 1813 (as Pentstemon erianthera) • Crested-tongue beardtongue E F

Stems ascending to erect, (2–)6–60 cm, retrorsely hairy, sometimes also villous or glandular-villous, or glabrate. **Leaves** basal and cauline, not leathery, glabrous or glabrate to retrorsely hairy and, sometimes, sparsely villous abaxially, glabrous or glabrate to retrorsely hairy and, sometimes, villous or glandular-villous adaxially; basal and proximal cauline petiolate,

(14–)22–130(–180) × 3–30(–40) mm, blade spatulate to obovate, oblanceolate, elliptic, or linear, base tapered, margins entire or ± dentate to serrate, apex rounded to obtuse or acute; cauline 2–6 pairs, sessile or proximals short-petiolate, 18–90 × 3–20(–25) mm, blade oblanceolate to oblong, lanceolate, or linear, base tapered to clasping or cordate-clasping, margins entire or ± dentate to serrate, apex obtuse to acute. **Thyrses** continuous or interrupted, cylindric, (1–)2–20 (–27) cm, axis glandular-pubescent to glandular-villous, sometimes also retrorsely hairy, verticillasters 2–7(–9), cymes (1 or)2–6-flowered, 2 per node; proximal bracts oblanceolate to oblong or lanceolate, 12–75 × 2–23 mm; peduncles and pedicels glandular-villous or glandular-pubescent and, sometimes, also retrorsely hairy. **Flowers:** calyx lobes lanceolate, 5.5–13 × 1.4–3 mm, glandular-pubescent to glandular-villous; corolla lavender to violet, purple, pink, or blue, with reddish purple nectar guides, funnelform or ventricose-ampliate, 16–35(–42) mm, sparsely to densely white- or yellow-villous internally abaxially, sometimes glandular-pubescent laterally, tube 4–9 mm, throat gradually to abruptly inflated, constricted or not at orifice, 6–14 mm diam., rounded abaxially; stamens included, pollen sacs widely divergent to opposite, navicular to subexplanate or explanate, 0.8–1.9 mm, dehiscing completely, sutures smooth or papillate; staminode 12–18 mm, exserted or prominently exserted, 0.3–0.9 mm diam., tip recurved, distal 6–15 mm sparsely to densely lanate, hairs yellow, yellowish, or orangish, to 4 mm, and medial hairs shorter, stiffer, and retrorse; style 9–20 mm, glabrous or proximal 1–10 mm sparsely glandular-pubescent. **Capsules** 6–13 × 4–6 mm, sparsely glandular-pubescent distally.

Varieties 5 (5 in the flora): w, c North America.

D. D. Keck (1938) treated this complex as three species: *Penstemon cleburnei*, *P. eriantherus* (with three varieties), and *P. whitedii* (with three subspecies). Most authors followed A. Cronquist (1959), who employed a broader species concept; the approach by Cronquist is followed here.

1. Corollas constricted at orifices; styles 9–12(–14) mm; pollen sacs 0.8–1.2 mm. . . . .
. . . . .61c. *Penstemon eriantherus* var. *cleburnei*
1. Corollas not constricted at orifices; styles 9–20 mm; pollen sacs 1–1.9 mm.
    2. Stems glabrate or retrorsely hairy, rarely glandular-villous, proximally, glandular-villous distally; leaves glabrous or glabrate abaxially, glabrate or villous to glandular-villous adaxially. . . . .
    . . . . 61e. *Penstemon eriantherus* var. *whitedii*
    2. Stems retrorsely hairy and, sometimes, sparsely villous, glandular-villous, or glandular-pubescent distally; leaves retrorsely hairy and, sometimes, sparsely villous or glabrous abaxially, retrorsely hairy or puberulent and, sometimes, villous or glandular-villous adaxially.

[3. Shifted to left margin.—Ed.]
3. Corollas funnelform; staminodes sparsely lanate distally; leaves glabrous abaxially . . . .
. . . . 61b. *Penstemon eriantherus* var. *argillosus*
3. Corollas ventricose-ampliate; staminodes moderately to densely lanate distally; leaves retrorsely hairy, sparsely villous, or glabrous abaxially.
    4. Pollen sacs explanate; stems retrorsely hairy and, usually, sparsely villous or glandular-villous distally; corollas (20–)22–35(–42) mm . . . . 61a. *Penstemon eriantherus* var. *eriantherus*
    4. Pollen sacs navicular to subexplanate; stems retrorsely hairy, sometimes also glandular-pubescent distally or wholly; corollas 18–25 mm . . . . 61d. *Penstemon eriantherus* var. *redactus*

61a. Penstemon eriantherus Pursh var. **eriantherus** E F

Stems retrorsely hairy and, usually, sparsely villous or glandular-villous distally. **Leaves** retrorsely hairy and, sometimes, sparsely villous abaxially, retrorsely hairy and villous or glandular-villous adaxially. **Flowers:** corolla lavender to purple or pink, ventricose-ampliate, (20–)22–35(–42) mm, not constricted at orifice; pollen sacs explanate, 1–1.6(–1.8) mm; staminode: distal 4–5 mm densely lanate, hairs yellowish, to 4 mm; style 15–20 mm. 2*n* = 16.

Flowering Jun–Jul. Sandy, gravelly, shaley, and clayey soils, shortgrass prairies, sagebrush shrublands, stream terraces; 500–2400 m; Alta., B.C.; Colo., Idaho, Mont., Nebr., N.Dak., S.Dak., Wash., Wyo.

61b. Penstemon eriantherus Pursh var. **argillosus**
M. E. Jones, Contr. W. Bot. 12: 62. 1908 (as Pentstemon) E

Penstemon whitedii Piper subsp. *dayanus* (Howell) D. D. Keck

Stems retrorsely hairy, hairs sometimes appressed, white, scalelike, and sometimes also glandular-villous distally. **Leaves** glabrous abaxially, retrorsely hairy or puberulent adaxially, hairs sometimes appressed, white, scalelike. **Flowers:** corolla lavender to violet, funnelform, 18–28 mm, not constricted at orifice; pollen sacs navicular to subexplanate, 1.2–1.6 mm; staminode: distal 3–5 mm sparsely lanate, hairs yellow, to 1 mm; style 12–14 mm.

Flowering May–Jun. Sandy or gravelly soils, juniper woodlands; 300–1300 m; Oreg.

Variety *argillosus* is known from north-central Oregon. Populations are concentrated in the John Day and Deschutes river drainages in Crook, Gilliam, Grant, and Wheeler counties.

61c. Penstemon eriantherus Pursh var. **cleburnei** (M. E. Jones) Dorn, Vasc. Pl. Wyoming, 300. 1988 • Cleburn's beardtongue E

Penstemon cleburnei M. E. Jones, Contr. W. Bot. 12: 62. 1908 (as Pentstemon cleburni)

Stems retrorsely hairy, sometimes also villous. **Leaves** glabrate or retrorsely hairy abaxially, retrorsely hairy and, sometimes, villous adaxially. **Flowers:** corolla lavender to violet, ventricose-ampliate, 16–22(–25) mm, constricted at orifice; pollen sacs explanate, 0.8–1.2 mm; staminode: distal 2–4 mm densely lanate, hairs orangish, to 3 mm; style 9–12(–14) mm.

Flowering May–Jun(–Jul). Clayey, sandy, or gravelly soils, sagebrush shrublands; 1400–2500 m; Utah, Wyo.

Variety *cleburnei* occurs mostly in central and western Wyoming, west of the Bighorn Mountains, Casper Arch, and Laramie Mountains; some populations extend into northeastern Utah. Variety *cleburnei* and var. *eriantherus* intergrade in eastern and north-central Wyoming.

61d. Penstemon eriantherus Pursh var. **redactus** Pennell & D. D. Keck, Bull. Torrey Bot. Club 65: 252. 1938 E

Penstemon whitedii Piper subsp. *tristis* Pennell & D. D. Keck

Stems retrorsely hairy, sometimes also glandular-pubescent distally or wholly. **Leaves** glabrous or retrorsely hairy abaxially, retrorsely hairy and, sometimes, sparsely villous adaxially. **Flowers:** corolla lavender to violet, ventricose-ampliate, 18–25 mm, not constricted at orifice; pollen sacs navicular to subexplanate, 1.2–1.9 mm; staminode: distal 4–5 mm moderately lanate, hairs yellow, to 3 mm; style 9–14 mm. $2n = 16$.

Flowering May–Jul. Sandy or gravelly soils, sagebrush shrublands; 600–2700 m; Idaho, Mont., Oreg.

Variety *redactus* occurs largely in the Salmon River Basin in Idaho (Butte, Clark, Custer, Idaho, and Lemhi counties), southwestern Montana (Beaverhead, Deer Lodge, Madison, and Silverbow counties), and northeastern Oregon (Wallowa County).

61e. Penstemon eriantherus Pursh var. **whitedii** (Piper) A. Nelson, Bot. Gaz. 54: 148. 1912 (as Pentstemon erianthera) • Whited's beardtongue C E

Penstemon whitedii Piper, Bot. Gaz. 22: 490. 1896 (as Pentstemon)

Stems glabrate or retrorsely hairy, rarely glandular-villous, proximally, glandular-villous distally. **Leaves** glabrous or glabrate abaxially, glabrate or villous to glandular-villous adaxially. **Flowers:** corolla lavender to violet or blue, ventricose-ampliate, 17–22 mm, not constricted at orifice; pollen sacs navicular, 1.3–1.6 mm; staminode: distal 3–6 mm moderately lanate, hairs yellow, to 2 mm; style 9–13 mm.

Flowering May–Jun. Dry, rocky, bitterbrush and rabbitbrush shrublands; of conservation concern; 200–1200 m; Wash.

Variety *whitedii* is known from the foothills of the Cascade Range and on the plains and in valleys in the Columbia Basin. The variety has been documented in Chelan, Douglas, Franklin, Kittitas, Klickitat, and Lincoln counties.

62. Penstemon franklinii S. L. Welsh, Rhodora 95: 414, fig. 21. 1993 • Franklin's beardtongue C E F

Stems ascending to erect, 4–25 cm, retrorsely ± hairy proximally, glandular-pubescent distally. **Leaves** basal and cauline, not leathery, glabrous abaxially, retrorsely hairy adaxially, usually only along midvein on distal caulines; basal and proximal cauline petiolate, 12–73 × 2–9 mm, blade spatulate to oblanceolate, base tapered, margins entire or obscurely dentate, apex rounded to obtuse or acute; cauline 3–5 pairs, sessile, 11–48 × 2–6 mm, blade lanceolate, base clasping, margins entire or dentate proximally, apex acute to acuminate. **Thyrses** interrupted or continuous, cylindric, 2–7 cm, axis glandular-pubescent, verticillasters 3–7, cymes 3–5-flowered, 2 per node; proximal bracts lanceolate, 7–28 × 2–6 mm; peduncles and pedicels glandular-pubescent. **Flowers:** calyx lobes lanceolate, 4.8–7.5 × 0.9–1.9 mm, glandular-pubescent; corolla blue to bluish lavender or light purple, with purple nectar guides, funnelform, 14–20 mm, yellow-lanate internally abaxially, tube 4–6 mm, throat gradually inflated, not constricted at orifice, 5–7 mm diam., rounded or slightly 2-ridged abaxially; stamens: longer pair reaching orifice or exserted, pollen sacs parallel, navicular, 1.3–1.6 mm, dehiscing completely, sutures papillate; staminode 7–9 mm, included or reaching orifice, 0.7–0.9 mm diam., tip straight to recurved, distal 5–6 mm sparsely to densely pilose, hairs golden yellow, to 2 mm; style 8–10 mm. **Capsules** 6–10 × 3–5.5 mm.

Flowering May–Jun. Sagebrush shrublands, grasslands; of conservation concern; 1600–1800 m; Utah.

Penstemon franklinii is known only from Cedar Valley in Iron County. According to Welsh, morphology and geography suggest that the species is most closely related to *P. pinorum*.

63. Penstemon goodrichii N. H. Holmgren, Brittonia 30: 416, fig. 1. 1978 • Lapoint or Goodrich's beardtongue C E

Stems ascending to erect, (9–)15–35 cm, glabrous or slightly retrorsely hairy proximally, glandular-pubescent distally. **Leaves** basal and cauline, not leathery, glabrous abaxially, retrorsely hairy adaxially, sometimes only along midvein; basal and proximal cauline petiolate, 32–65(–80) × 1–6(–9) mm, blade lanceolate to linear, base tapered, margins entire or obscurely serrate, apex obtuse to acute; cauline 3–7 pairs, sessile, 30–45(–70) × 1–5(–8) mm, blade lanceolate to linear, base tapered to slightly clasping, margins entire or obscurely serrate, apex acute. **Thyrses** interrupted, cylindric, (4–)6–15 cm, axis glandular-pubescent, verticillasters 3–6(–9), cymes (1 or)2–5-flowered, 2 per node; proximal bracts lanceolate to linear, 7–20(–45) × 1–4(–8) mm; peduncles and pedicels glandular-pubescent. **Flowers:** calyx lobes elliptic-lanceolate, 4–6 × 1–1.6 mm, glandular-pubescent; corolla violet to blue or bluish lavender, with reddish purple nectar guides, weakly bilabiate, funnelform, 10–16 mm, glabrous or sparsely white-villous internally abaxially, tube 5–7 mm, throat gradually inflated, not constricted at orifice, 5–6 mm diam., rounded abaxially; stamens included, pollen sacs opposite, explanate, 0.7–0.9 mm, dehiscing completely, sutures smooth; staminode 6–8 mm, included, 0.5–0.9 mm diam., tip straight to recurved, distal 5–7 mm densely pilose, hairs yellow, to 0.8 mm; style 4–5 mm. **Capsules** 7–10 × 3–5 mm.

Flowering May–Jun. Clayey slopes, shadscale and sagebrush shrublands, juniper-mountain mahogany woodlands; of conservation concern; 1700–1900 m; Utah.

Penstemon goodrichii is known from the Lapoint-Tridell-Whiterocks vicinity in Duchesne and Uintah counties on clayey badlands associated with the Duchesne River Formation.

64. Penstemon gormanii Greene, Ottawa Naturalist 16: 39. 1902 (as Pentstemon gormani) • Gorman's beardtongue E

Stems ascending to erect, (6–)10–30(–65) cm, glabrous proximally, glandular-pubescent to glandular-villous distally. **Leaves** basal and cauline, not leathery, glabrous or distal cauline sometimes glabrous abaxially, glandular-pubescent adaxially; basal and proximal cauline petiolate, (18–)35–100 × (2–)3–10 mm, blade spatulate to oblanceolate, base tapered, margins entire or obscurely serrate, apex rounded to obtuse or acute; cauline 2–4 pairs, sessile, 15–68 × 2–16 mm, blade oblong to lanceolate, base tapered to clasping, margins entire or obscurely serrate, apex obtuse to acute. **Thyrses** continuous or ± interrupted, cylindric, (1–)3–15 cm, axis glandular-villous, verticillasters 2–6, cymes (1 or) 2–5-flowered, 2 per node; proximal bracts lanceolate to linear, (5–)10–56 × (1–)3–10 mm; peduncles and pedicels glandular-pubescent. **Flowers:** calyx lobes lanceolate, 7–10.5 × 1.4–2.2 mm, glandular-villous; corolla lavender to purple or bluish violet, with reddish purple nectar guides, ventricose, 19–30 mm, sparsely white- or yellowish villous internally abaxially, tube 5–6 mm, throat abruptly inflated, not constricted at orifice, 8–14 mm diam., rounded abaxially; stamens included, pollen sacs opposite, navicular, 1.3–1.5 mm, dehiscing completely, sutures smooth; staminode 13–17 mm, prominently exserted, 0.4–0.6 mm diam., tip straight to recurved, distal 2–3 mm sparsely lanate, hairs yellow, to 2 mm, and medial 4–8 mm sparsely to moderately pilose, hairs yellow, to 1 mm; style 11–14 mm. **Capsules** 7–10 × 4–5 mm. $2n = 16$.

Flowering Jun–Aug. Rocky slopes, sand dunes, gravelly stream terraces, clearings in forests; 400–1200 m; B.C., N.W.T., Yukon; Alaska.

Penstemon gormanii resembles *P. eriantherus* vars. *argillosus* and *whitedii*.

65. Penstemon grahamii D. D. Keck, Ann. Carnegie Mus. 26: 331. 1937 (as grahami) • Graham's or Uinta Basin beardtongue C E

Stems ascending to erect, 7–18 cm, retrorsely hairy. **Leaves** basal and cauline, or basal absent or reduced, leathery, glabrous or distal cauline sometimes glandular-pubescent along margins and midvein; basal and proximal cauline petiolate, 16–50 × 5–20 mm, blade oval to elliptic, base tapered, margins entire, apex rounded to obtuse; cauline 2 or 3 pairs, sessile or proximals short-petiolate, 25–48 × 6–15 mm, blade oblanceolate to elliptic or lanceolate, base tapered to clasping, margins entire or obscurely serrate distally, apex obtuse to acute. **Thyrses** continuous, cylindric, 4–7 cm, axis glandular-pubescent, verticillasters 4–6, cymes 2–6-flowered, 2 per node; proximal bracts lanceolate, 26–36 × 5–12 mm; peduncles and pedicels glandular-pubescent. **Flowers:** calyx lobes lanceolate, 6.2–9 × 1.7–2 mm, glandular-pubescent; corolla lavender to pink, with violet nectar guides, ventricose-ampliate, 28–35 mm, sparsely white-lanate internally abaxially, tube 9–13 mm, throat abruptly inflated, prominently constricted at orifice, 11–15 mm diam., rounded abaxially; stamens: longer pair reaching orifice or exserted, pollen sacs opposite, explanate, 0.9–1.4 mm, dehiscing completely, sutures smooth; staminode 17–20 mm, prominently exserted, 0.3–0.5 mm diam., tip recurved to coiled, distal 13–16 mm densely pilose, hairs golden yellow or orange, to 0.8 mm; style 21–24 mm. **Capsules** 8–10 × 5–6 mm.

Flowering May–Jun. Talus slopes, benches, knolls, washes in desert shrublands, pinyon-juniper woodlands; of conservation concern; 1400–2000 m; Colo., Utah.

Penstemon grahamii is known from the Uinta Basin. Populations have been documented from Raven Ridge in extreme northwestern Rio Blanco County, Colorado (S. L. O'Kane 1988), south and west in an arc across southern Uintah County, Utah, into extreme northeastern Carbon County and extreme southeastern Duchesne County. Plants grow in loose shale and shale-derived soils associated with the Green River Formation.

66. **Penstemon guadalupensis** A. Heller, Contr. Herb. Franklin Marshall Coll. 1: 92, plate 7. 1895 (as Pentstemon) • Guadalupe beardtongue [E]

Penstemon guadalupensis subsp. *ernestii* Pennell

Stems ascending to erect, 20–50 cm, retrorsely hairy. **Leaves** basal and cauline, not leathery, glabrous or glabrate, rarely basal retrorsely hairy along midvein; basal and proximal cauline petiolate, 25–95 × 1–6 mm, blade linear, rarely lanceolate, base tapered, margins entire, apex acute; cauline 6–11, sessile or proximals short-petiolate, 22–65(–95) × 1–10(–18) mm, blade lanceolate to linear, base tapered to clasping, margins entire or obscurely serrate, apex acute. **Thyrses** continuous or ± interrupted, cylindric, 3–16(–26) cm, axis glandular-pubescent, verticillasters 3–8, cymes 2–4-flowered, 2 per node; proximal bracts lanceolate, 15–48 × 3–8 mm; peduncles and pedicels glandular-pubescent. **Flowers:** calyx lobes lanceolate, 5–6.5 × 1.2–2 mm, glandular-pubescent; corolla white,

sometimes tinged pink or lavender, with reddish violet nectar guides, funnelform, 13–20 mm, glandular-pubescent and sparsely white-villous internally abaxially, tube 4–6 mm, throat gradually inflated, not constricted at orifice, 6–7 mm diam., rounded abaxially; stamens included, pollen sacs opposite, explanate, 0.8–1 mm, dehiscing completely, sutures smooth; staminode 10–14 mm, reaching orifice or exserted, 0.5–0.7 mm diam., tip straight to recurved, distal 3–7 mm sparsely pilose, hairs yellow, to 0.9 mm; style 9–12 mm. **Capsules** 7–9 × 4.5–6 mm.

Flowering Mar–Jun. Sandy or gravelly, calcareous soils, prairies; 400–800 m; Tex.

Penstemon guadalupensis is known from the southern Cross Timbers and Prairies, Edwards Plateau, and southern Rolling Plains physiographic provinces of central Texas. Pennell distinguished subsp. *ernestii* by its shorter corollas and broader leaves, but that distinction seems trivial.

67. **Penstemon jamesii** Bentham in A. P. de Candolle and A. L. P. P. de Candolle, Prodr. 10: 325. 1846 (as Pentstemon) • James's beardtongue

Stems ascending to erect, 10–45(–52) cm, glabrate or retrorsely hairy proximally, glandular-pubescent distally. **Leaves** basal and cauline, not leathery, glabrous or retrorsely hairy along midvein, especially adaxially; basal and proximal cauline petiolate, 20–80(–105) × (2–)5–10(–13) mm, blade oblanceolate to lanceolate or linear, base tapered, margins entire or serrate, apex obtuse to acute; cauline 3–6 pairs, sessile, 20–100(–110) × 5–15 mm, blade lanceolate to linear, base tapered to slightly clasping, margins entire or remotely dentate to remotely serrate, apex acute, rarely obtuse. **Thyrses** continuous or interrupted, secund, 5–20(–24) cm, axis densely glandular-pubescent, verticillasters 2–8, cymes 2–5-flowered, 2 per node; proximal bracts lanceolate, (9–)25–90 × (1–)3–15 mm; peduncles and pedicels densely glandular-pubescent. **Flowers:** calyx lobes ovate to lanceolate, 8–12 × 2–3 mm, glandular-pubescent; corolla white to lavender, pink, or violet, with magenta or violet-blue nectar guides, ventricose, 24–32 (–35) mm, moderately to densely white-pilose and glandular-pubescent internally abaxially, tube 8–10 mm, throat abruptly inflated, slightly constricted at orifice, 9–15 mm diam., rounded abaxially; stamens included, pollen sacs opposite, explanate, 1–1.2 mm, dehiscing completely, sutures smooth; staminode 14–17 mm, prominently exserted, 0.3–0.4 mm diam., tip recurved, distal 10–14 mm lanate, hairs yellow, to 3.5 mm, and medial hairs shorter, stiffer, and retrorse; style 17–18 mm. **Capsules** 10–16 × 5–7 mm. $2n = 16$.

Flowering May–Jul. Sandy, gravelly, or loamy soils, shortgrass prairies, sagebrush shrublands; 1100–2300 m; Colo., Kans., N.Mex., Tex.; Mexico (Coahuila).

Penstemon jamesii occurs on the southern High Plains and trans-Pecos regions from southeastern Colorado and southwestern Kansas through eastern New Mexico and western Texas to northern Coahuila, Mexico.

68. Penstemon janishiae N. H. Holmgren, Brittonia 31: 223, fig. 3. 1979 • Janish's beardtongue E

Stems ascending to erect, 4–20(–25) cm, retrorsely hairy. Leaves basal and cauline, not leathery, retrorsely hairy or basal and proximal cauline sometimes glabrate or glabrous abaxially; basal and proximal cauline petiolate, 15–50(–60) × 5–18(–23) mm, blade oblance-olate to elliptic, base tapered, margins entire or remotely serrate distally, apex obtuse to acute; cauline 2–4 pairs, sessile, 15–60 × 3–8 mm, blade oblanceolate to lanceolate, rarely pandurate, base tapered to clasping, margins entire or remotely serrate, apex obtuse to acute. Thyrses continuous, cylindric to ± secund, 1–13 cm, axis glandular-pubescent, verticillasters 2–6, cymes 2- or 3(or 4)-flowered, 2 per node; proximal bracts lanceolate, rarely oblanceolate, 12–42(–76) × 1–12 (–16) mm; peduncles and pedicels glandular-pubescent. Flowers: calyx lobes lanceolate, 6–10(–12) × 1.6–2.6 (–3) mm, glandular-pubescent; corolla lavender to purple or pink, with reddish purple nectar guides, ventricose, 18–28 mm, densely white-lanate internally abaxially, tube (6–)8–12 mm, throat abruptly inflated, prominently constricted at orifice, 7–12 mm diam., rounded abaxially; stamens: longer pair reaching orifice, pollen sacs opposite, explanate, 0.8–1.2 mm, dehiscing completely, sutures smooth; staminode 16–18 mm, prominently exserted, 0.2–0.3 mm diam., tip coiled, distal 8–11 mm densely pilose, hairs orange or yellowish orange, to 1.3 mm; style 15–23 mm. Capsules 7–11 × 5–7 mm.

Flowering May–Jul. Igneous clayey soils, sagebrush shrublands, juniper and pine-juniper woodlands; 1300–2300 m; Calif., Idaho, Nev., Oreg.

Penstemon janishiae occurs largely in the Great Basin physiographic region. Populations have been documented in northeastern California (Lassen and Modoc counties), southwestern Idaho (Owyhee County), northern Nevada (Elko, Eureka, Humboldt, Lander, Nye, Washoe, and White Pine counties), and southeastern Oregon (Harney County).

69. Penstemon laricifolius Hooker & Arnott, Bot. Beechey Voy., 376. 1839 (as Pentstemon) • Larch-leaf beardtongue E F

Plants cespitose. Stems erect, 10–30 cm, glabrous or puberulent. Leaves basal and cauline, not leathery, glabrous, rarely puberulent proximally, especially along margins; basal and proximal cauline sessile, 15–20(–40) × 0.5–1.5 mm, blade linear, base tapered to ± clasping, margins entire, apex acute; cauline 3–7 pairs, sessile, 11–35 × 0.3–1 mm, blade linear, base tapered to ± clasping, margins entire, apex acute. Thyrses continuous or interrupted, cylindric, 3–15 cm, axis glabrous or retrorsely hairy, verticillasters 3–7, cymes 1- or 2(or 3)-flowered, 1 or 2 per node; proximal bracts linear, 10–30 × 0.3–1 mm; peduncles and pedicels glabrous, rarely puberulent. Flowers: calyx lobes ovate, 3.5–5.8 × 1.3–2.2 mm, glabrous; corolla white or pink to violet or purple, without nectar guides, tubular-funnelform, 9–18 mm, glabrous externally, moderately yellow-villous internally abaxially, tube 4–5 mm, throat gradually inflated, not constricted, 5–6 mm diam., rounded abaxially; stamens included, pollen sacs opposite, navicular, 1–1.5 mm, dehiscing completely, sutures papillate; staminode 7–11 mm, included or reaching orifice, 0.5–0.7 mm diam., tip straight to recurved, distal 3–5 mm pilose, hairs yellow or yellow-orange, to 1 mm; style 9–12 mm. Capsules 4–5 × 2.5–4 mm.

Varieties 2 (2 in the flora): w United States.

Molecular data (A. D. Wolfe et al. 2006; C. A. Wessinger et al. 2016) place *Penstemon laricifolius* with or near sect. *Cristati*. Two varieties, differentiated by corolla color, are difficult to distinguish in the herbarium, where corollas are usually brown. Populations with a mix of white, pink, and purple corollas have been documented in Wyoming (B. L. Heidel and J. Handley 2007).

1. Corollas pink to violet or purple, 10–18 mm; Montana, Wyoming . . . .
. . . . 69a. *Penstemon laricifolius* var. *laricifolius*
1. Corollas white, 9–15 mm; Colorado, Wyoming
. . . . 69b. *Penstemon laricifolius* var. *exilifolius*

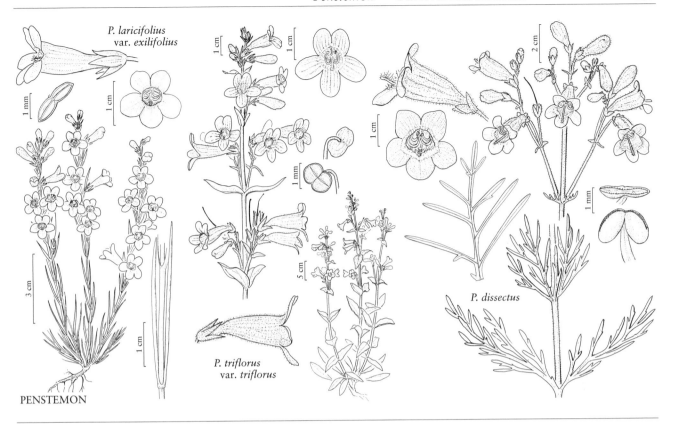

P. laricifolius var. *exilifolius*

PENSTEMON

P. triflorus var. *triflorus*

P. dissectus

69a. Penstemon laricifolius Hooker & Arnott var. **laricifolius** [E]

Flowers: corolla pink to violet or purple, 10–18 mm.

Flowering May–Aug. Sandy or gravelly sagebrush shrublands, openings in pine forests; 1500–2400 m; Mont., Wyo.

Variety *laricifolius* is known from south-central Montana and central and western Wyoming. The type material (Snake County, *Tolmie s.n.*, K) reputedly came from southern Idaho. D. D. Keck (1937) expressed doubt; no other collections are known from Idaho, so the variety is excluded from that state.

69b. Penstemon laricifolius Hooker & Arnott var. **exilifolius** (A. Nelson) Payson, Univ. Wyoming Publ. Sci., Bot. 1: 97. 1924 [E] [F]

Penstemon exilifolius A. Nelson, Bull. Torrey Bot. Club 28: 230. 1901 (as Pentstemon); *P. laricifolius* subsp. *exilifolius* (A. Nelson) D. D. Keck

Flowers: corolla white, 9–15 mm.

Flowering Jun–Aug. Sandy or gravelly grasslands, sagebrush shrublands; 1500–2900 m; Colo., Wyo.

Variety *exilifolius* is concentrated in the Laramie Basin of Wyoming and Laramie River Valley of northwestern Larimer County, Colorado. Other isolated populations are known in Wyoming (B. L. Heidel and J. Handley 2007). A specimen from southern Montana, reported by Heidel and Handley, could be this variety; the specimen has not been seen for this study.

SELECTED REFERENCE   Heidel, B. L. and J. Handley. 2007. *Penstemon laricifolius* Hook. & Arn. Ssp. *exilifolius* (A. Nels.) D. D. Keck (Larchleaf Beardtongue): A Technical Conservation Assessment. Laramie.

70. Penstemon marcusii (D. D. Keck) N. H. Holmgren, Brittonia 31: 105. 1979 • Marcus Jones's beardtongue [C] [E]

Penstemon moffatii Eastwood subsp. *marcusii* D. D. Keck, Bull. Torrey Bot. Club 65: 243. 1938, based on *P. pseudohumilis* M. E. Jones, Contr. W. Bot. 12: 65. 1908, not Rydberg 1900

Stems ascending to erect, (9–) 15–32 cm, retrorsely hairy. **Leaves** basal and cauline, not leathery, glabrous or retrorsely hairy; basal and proximal cauline petiolate, 20–85 × 4–8(–12) mm, blade oblanceolate to lanceolate, base tapered, margins entire or obscurely dentate, apex obtuse; cauline 2 or 3 pairs, sessile, 30–75 × 3–10 mm, blade lanceolate, base tapered to clasping, margins entire, apex acute. **Thyrses** interrupted, cylindric, 5–18 cm, axis glandular-pubescent, verticillasters 3–7, cymes 2–5-flowered, 2 per node; proximal bracts lanceolate, 5–38 × 2–10 mm; peduncles and pedicels glandular-pubescent. **Flowers:** calyx lobes lanceolate, (2.2–)4.5–7.5 × (0.9–)1.2–1.6 mm, glandular-pubescent; corolla violet or blue, with reddish violet nectar guides, weakly bilabiate, funnelform, 8–12 mm, sparsely yellow-lanate or glabrous internally abaxially, tube 4–5 mm, throat gradually inflated, not constricted at orifice, 2–5 mm diam., rounded abaxially; stamens included, pollen sacs opposite, explanate, 0.6–1 mm, dehiscing completely, sutures smooth; staminode 6–8 mm, included, 0.3–0.6 mm diam., tip recurved, distal 3–5 mm sparsely to densely pubescent, hairs yellow, to 1 mm; style 3–4 mm. **Capsules** 4.5–6.5 × 4–5 mm.

Flowering May–Jun. Clayey and gravelly soils, sagebrush shrublands; of conservation concern; 1600–2000 m; Utah.

Penstemon marcusii is found in the San Rafael Swell and along the eastern edge of the Wasatch Plateau in Carbon and Emery counties, Utah. The species is most similar to *P. goodrichii*, differing primarily in its puberulent stems and smaller flowers.

71. Penstemon miser A. Gray in A. Gray et al., Syn. Fl. N. Amer. ed. 2, 2(1): 441. 1886 (as Pentstemon) • Malheur beardtongue E

Stems ascending to erect, 8–28(–40) cm, retrorsely hairy. **Leaves** basal and cauline, not leathery, retrorsely hairy, hairs sometimes appressed, white, scalelike, sometimes basal and proximal cauline glabrate or glabrous abaxially; basal and proximal cauline petiolate, 15–60(–78) × 4–12(–17) mm, blade spatulate to oblanceolate, base tapered, margins entire or dentate-serrate, apex rounded to obtuse or acute; cauline 3 or 4 pairs, sessile or proximals short-petiolate, 20–50(–75) × 2–5(–10) mm, blade lanceolate to oblanceolate, base tapered to clasping, margins entire or obscurely and remotely dentate, apex obtuse to acute. **Thyrses** ± interrupted, cylindric, (3–)6–18 cm, axis glandular-pubescent and, sometimes, retrorsely hairy, verticillasters 3–6, cymes 2–5-flowered, 2 per node; proximal bracts lanceolate, 18–46 × 2–12 mm; peduncles and pedicels glandular-pubescent. **Flowers:** calyx lobes lanceolate, 5.5–8 × 1.2–2 mm, glandular-pubescent; corolla light blue to bluish purple or magenta, with reddish violet nectar guides, ventricose, 14–22 mm, sparsely white-lanate internally abaxially, tube 5–7 mm, throat abruptly inflated, not constricted at orifice, (3–)5–7 mm diam., rounded abaxially; stamens: longer pair reaching orifice, pollen sacs opposite, explanate, 0.7–1(–1.2) mm, dehiscing completely, sutures smooth; staminode 8–9 mm, exserted, 0.3–0.5 mm diam., tip straight to

recurved, distal 5–7 mm moderately to densely pilose, hairs yellow or orangish yellow, to 1.2 mm; style 9–10 mm. **Capsules** (6–)8–11 × 4.5–7 mm.

Flowering May–Jun. Clayey soils, sagebrush shrublands, pine-juniper woodlands; 800–1400 m; Idaho, Oreg.

Penstemon miser is known from Owyhee County, Idaho, and southern Baker and northern Malheur counties, Oregon.

72. Penstemon moffatii Eastwood, Zoë 4: 9. 1893 (as Pentstemon) • Moffat's beardtongue E

Penstemon moffatii subsp. *paysonii* (Pennell) D. D. Keck

Stems ascending to erect, (3–)7–30 cm, retrorsely hairy. **Leaves** basal and cauline, not leathery, retrorsely hairy; basal and proximal cauline petiolate, 15–45(–65) × 3–20(–25) mm, blade spatulate to oblanceolate or ovate, base tapered, margins entire or obscurely dentate distally, apex rounded to obtuse; cauline (1 or)2–4 pairs, sessile, 16–55 × 4–10(–15) mm, blade lanceolate to oblanceolate or linear, base tapered to clasping, margins entire or obscurely dentate distally, apex obtuse to acute. **Thyrses** continuous to ± interrupted, cylindric, 1–12 cm, axis glandular-pubescent, verticillasters (1–)3–7, cymes (1 or)2–4-flowered, 2 per node; proximal bracts ovate to lanceolate, 8–34 × 2–16 mm; peduncles and pedicels glandular-pubescent. **Flowers:** calyx lobes lanceolate, 4.8–6.2 × 1.2–1.9 mm, glandular-pubescent; corolla blue to violet, with reddish violet nectar guides, bilabiate, tubular-funnelform, 15–20(–22) mm, sparsely white-lanate or glabrous internally abaxially, tube 4–6 mm, throat gradually inflated, not constricted at orifice, 4–7 mm diam., rounded abaxially; stamens included, pollen sacs opposite, navicular, 1.1–1.4 mm, dehiscing completely, sutures papillate; staminode 7–9 mm, included or reaching orifice, 0.3–0.5 mm diam., tip straight to recurved, distal 4–5 mm sparsely pubescent, hairs yellow, to 0.8 mm; style 10–12 mm. **Capsules** 6–10 × 4–6 mm.

Flowering Apr–Jun. Shale or gravelly mesas and slopes, sagebrush shrublands, pinyon-juniper woodlands; 1300–2000 m; Colo., Utah.

Penstemon moffatii is known from Delta, Grand, Mesa, and Montrose counties, Colorado, and Garfield, Grand, and Wayne counties, Utah. Plants with oblanceolate basal leaves and sessile cauline leaves from Montrose County, Colorado, have been distinguished as subsp. *paysonii*. Plants with these characteristics are found sporadically throughout the range of *P. moffatii*, sometimes in populations with plants referable to subsp. *moffatii*.

73. Penstemon monoensis A. Heller,
Muhlenbergia 2: 246. 1906 (as Pentstemon)
• Mono beardtongue E

Stems ascending to erect, 7–35 cm, puberulent, hairs usually white, scalelike. **Leaves** basal and cauline, or basal sometimes few or absent, ± leathery, densely retrorsely hairy, hairs white, scalelike; basal and proximal cauline petiolate, (28–)50–130 × 7–40 mm, blade ovate to oblong, base tapered, margins entire or dentate (often crisped), apex obtuse to acute; cauline 2–4 pairs, short-petiolate or sessile, 35–83 × 7–26 mm, blade elliptic to deltate-ovate or lanceolate, base tapered to clasping, margins entire, apex obtuse to acute or acuminate. **Thyrses** continuous to ± interrupted, cylindric, 4–23 cm, axis glandular-pubescent, verticillasters 4–9, cymes (1 or)2–7-flowered, 2 per node; proximal bracts ovate to lanceolate, 19–61 × 15–30 mm; peduncles and pedicels glandular-pubescent. **Flowers:** calyx lobes lanceolate, 8–11 × 1.6–2 mm, glandular-pubescent; corolla pink to reddish pink or violet, with or without faint reddish nectar guides, tubular-funnelform, 14–20 mm, sparsely white-lanate internally abaxially, tube 5–7 mm, throat gradually inflated, not constricted at orifice, 4–7 mm diam., 2-ridged abaxially; stamens included, pollen sacs divergent, navicular-sigmoid, 1.1–1.4 mm, dehiscing incompletely, connective not splitting, sutures papillate; staminode 8–10 mm, included, 0.3–0.4 mm diam., tip straight to recurved, distal 5–7 mm densely pilose, hairs golden yellow, to 1 mm; style 11–13 mm. **Capsules** 6–9 × 4–5.5 mm.

Flowering Apr–May. Sandy or gravelly washes and hills, sagebrush shrublands, pinyon-juniper woodlands; 1200–2100 m; Calif.

Penstemon monoensis occurs in the Inyo and White mountains in Inyo and Mono counties. The species appears to be closely related to *P. calcareus*. They differ by their herbage indument and anther morphology.

74. Penstemon nanus D. D. Keck, Amer. Midl. Naturalist
23: 607. 1940 • Low beardtongue E

Stems ascending, 2–10(–14) cm, densely retrorsely hairy. **Leaves** basal and cauline, not leathery, densely retrorsely hairy; basal and proximal cauline petiolate, 12–35(–50) × 3–10 mm, blade ovate to lanceolate, base tapered, margins entire, rarely obscurely dentate, apex obtuse to acute; cauline 1–3 pairs, sessile, 13–34 × 2–6 mm, blade oblanceolate to lanceolate or linear, base tapered to

slightly clasping, margins entire, apex obtuse to acute. **Thyrses** continuous, cylindric, axis glandular-pubescent, 2–3(–8) cm, verticillasters 1–5(or 6), cymes 2–4-flowered, 2 per node; proximal bracts lanceolate, 11–20 × 2–6 mm; peduncles and pedicels glandular-pubescent. **Flowers:** calyx lobes ovate to lanceolate, 4–6 × 1.8–2.2 mm, glandular-pubescent; corolla violet to blue or purple, with reddish violet nectar guides, tubular-funnelform, 12–16 mm, sparsely yellow-lanate internally abaxially, tube 5–6 mm, throat gradually inflated, not constricted at orifice, 4–7 mm diam., rounded abaxially; stamens included, pollen sacs parallel, navicular, 1.7–2 mm, dehiscing completely, sutures papillate; staminode 8–9 mm, included or reaching orifice, 0.8–0.9 mm diam., tip straight, distal 5–6 mm moderately to densely pubescent, hairs orange, to 0.8 mm; style 9–11 mm. **Capsules** 4–7 × 3–4.5 mm.

Flowering May–Jun. Pinyon-juniper woodlands, desert shrublands; 1700–2300 m; Utah.

Penstemon nanus is known from the ranges and dry valleys of western Beaver and Millard counties. Populations are concentrated in and near the Burbank Hills, Confusion Range, Halfway Hills, Mountain Home Range, and Tunnel Spring and Wah Wah mountains. The parallel pollen sacs and glandular-pubescent thyrse axes of *P. nanus* distinguish it from *P. dolius*.

75. Penstemon ophianthus Pennell, Contr. U.S. Natl.
Herb. 20: 343. 1920 • Arizona beardtongue E

Penstemon jamesii Bentham subsp. *ophianthus* (Pennell) D. D. Keck

Stems ascending to erect, (10–)13–36(–40) cm, glabrous or sparsely retrorsely hairy proximally, sometimes also sparsely glandular-pubescent. **Leaves** basal and cauline, not leathery, glabrous or sparsely retrorsely hairy along midvein; basal and proximal cauline petiolate, 16–75(–120) × 6–15(–22) mm, blade oblanceolate, base tapered, margins entire or sinuate-dentate, apex obtuse to acute; cauline 2–4 pairs, short-petiolate or sessile, 25–105 × 2–12 mm, blade lanceolate to linear, base tapered, margins entire or sinuate-dentate to dentate, apex acute, rarely obtuse. **Thyrses** interrupted to compact, secund, 6–20 cm, axis densely glandular-pubescent, verticillasters 4–9, cymes (1–)3–7-flowered, 2 per node; proximal bracts lanceolate, 12–70 × 2–8 mm; peduncles and pedicels glandular-pubescent. **Flowers:** calyx lobes lanceolate, 6–9 × 1.3–2.2 mm, glandular-pubescent; corolla lavender to violet, with deep violet nectar guides, ventricose-ampliate, 14–20 (–22) mm, moderately white-pilose and glandular-pubescent internally abaxially, tube 5–8 mm, throat abruptly inflated, not constricted at orifice, 7–11 mm diam., rounded abaxially; stamens included, pollen sacs opposite, explanate, 0.8–1.2 mm, dehiscing completely, sutures smooth; staminode 11–13 mm, prominently exserted, 0.7–0.9 mm diam., tip recurved, distal 8–9 mm densely lanate, hairs yellow, to 2 mm, and medial hairs shorter, stiffer, and retrorse; style 11–13 mm. **Capsules** 6–11 × 5–6 mm.

Flowering May–Jun. Sandy, gravelly, or clayey soils, sagebrush shrublands, pinyon-juniper and oak woodlands, ponderosa pine forests; 1500–2300 m; Ariz., Colo., N.Mex., Utah.

Penstemon ophianthus is confused most often with *P. jamesii* and less often with *P. breviculus*.

76. Penstemon pinorum L. M. Shultz & J. S. Shultz, Brittonia 37: 98, fig. 1. 1985 • Pine Valley beardtongue C E

Stems ascending to erect, 13–33 cm, ± retrorsely pubescent. **Leaves** basal and cauline, not leathery, glabrous or ± retrorsely hairy along margins; basal and proximal cauline petiolate, 20–90 × (3–)6–12 mm, blade oblanceolate to lanceolate, base tapered, margins sharply serrate-dentate, rarely entire, apex obtuse to acute; cauline 2–4 pairs, sessile, (20–)30–56 × 3–6 mm, blade oblanceolate, base clasping, margins coarsely serrate-dentate or dentate, apex obtuse. **Thyrses** interrupted, cylindric, 7–15(–21) cm, axis glandular-pubescent, verticillasters 3–7, cymes (1 or)2–7-flowered, 2 per node; proximal bracts lanceolate, 18–40 × 2–5 mm; peduncles and pedicels glandular-pubescent. **Flowers:** calyx lobes lanceolate, 3.8–6 × 1.1–2 mm, glandular-pubescent; corolla blue to violet or light purple, rarely pink, with or without purple nectar guides, ventricose, 10–15 mm, yellowish lanate internally abaxially, tube 4–6 mm, throat gradually inflated, not constricted at orifice, 4–6 mm diam., slightly 2-ridged abaxially; stamens included, pollen sacs parallel, navicular, 1.2–1.5 mm, dehiscing completely, sutures papillate; staminode 7–10 mm, prominently exserted, 0.4–0.6 mm diam., tip straight to recurved, distal 5–7 mm moderately to densely pilose, hairs orange, to 0.8 mm; style 6–7 mm. **Capsules** 5–7 × 4–5 mm.

Flowering May–Jun. Pinyon-juniper woodlands, desert shrub communities; of conservation concern; 1700–2100 m; Utah.

Penstemon pinorum is known from the Pine Valley Mountains in Iron and Washington counties.

77. Penstemon pumilus Nuttall, J. Acad. Nat. Sci. Philadelphia 7: 46. 1834 (as Pentstemon pumilum) • Northern beardtongue E

Stems decumbent, ascending, or erect, 3–12 cm, retrorsely hairy. **Leaves** basal and cauline or mostly basal, not leathery, retrorsely hairy; basal and proximal cauline short-petiolate or sessile, 7–35(–50) × 1–4 mm, blade oblanceolate to linear, base tapered, margins entire, apex obtuse to acute; cauline 1–3 pairs, sessile, 12–30 × 1–4 mm, blade oblanceolate to oblong or linear, usually arcuate, base tapered, margins entire, apex acute. **Thyrses** interrupted to compact, cylindric, 1–6 cm, axis retrorsely hairy, verticillasters 1–6, cymes 1- or 2-flowered, 1 or 2 per node; proximal bracts oblanceolate to lanceolate, 7–30 × 1–3 mm; peduncles and pedicels retrorsely hairy and sparsely glandular-pubescent. **Flowers:** calyx lobes lanceolate, 4.5–7 × 1.5–2 mm, puberulent and glandular-pubescent; corolla blue to violet, with reddish purple nectar guides, tubular-funnelform, 15–20(–23) mm, glabrous or sparsely white-lanate internally abaxially, tube 4–6 mm, throat gradually inflated, not constricted at orifice, 6–8 mm diam., rounded abaxially; stamens included, pollen sacs opposite, navicular, 0.9–1.2 mm, dehiscing completely, sutures papillate; staminode 9–12 mm, included, 0.5–0.6 mm diam., tip recurved, distal 6–8 mm sparsely pilose, hairs yellow, to 0.5 mm; style 9–11 mm. **Capsules** 4–7 × 3.5–5.5 mm.

Flowering May–Jun. Gravelly soils, limestone, sagebrush shrublands; 1400–2100 m; Idaho.

Penstemon pumilus is known from east-central Idaho in Butte, Clark, Custer, and Lemhi counties. The arcuate cauline leaves with acute apices and tufted basal leaves reliably distinguish this species from the other diminutive members of sect. *Cristati*.

78. Penstemon triflorus A. Heller, Contr. Herb. Franklin Marshall Coll. 1: 92, plate 8. 1895 (as Pentstemon) • Heller's beardtongue F

Stems erect, 30–65 cm, retrorsely hairy. **Leaves** basal and cauline or, sometimes, basal absent or withering, not leathery to ± leathery, glabrous or retrorsely hairy along midvein and, sometimes, along margins; basal and proximal cauline petiolate, 22–120(–190) × 7–40(–65) mm, blade oblanceolate to elliptic, base tapered, margins entire or serrate to dentate, apex obtuse to acute; cauline 6–8(–10) pairs, sessile, 30–95(–115) × 7–30 mm, blade ovate to lanceolate to linear, base tapered to cordate clasping, margins entire or serrate, apex acute. **Thyrses** interrupted, cylindric, 5–20 cm, axis glandular-pubescent, verticillasters (3–)4–6, cymes 2–4(–6)-flowered, 2 per node; proximal bracts ovate to lanceolate, 14–50 × 6–17 mm; peduncles and pedicels glandular-pubescent. **Flowers:** calyx lobes ovate to lanceolate, 6.5–13 × 2.5–4.8 mm, glandular-pubescent; corolla pink to pinkish red or reddish violet, with red or crimson nectar guides, ventricose, 22–35 mm, glandular-pubescent internally, tube 8–11 mm, throat gradually inflated, not constricted at orifice, 10–13 mm diam., rounded abaxially; stamens: longer pair exserted, pollen sacs opposite, explanate, 0.9–1.2 mm, dehiscing completely, sutures smooth; staminode 17–20 mm, included or reaching orifice, 0.4–0.6 mm diam., tip straight to recurved, distal 8–12 mm sparsely to moderately pilose, hairs yellow, to 1 mm; style 18–24 mm. **Capsules** (7–)11–14 × 5–8.5 mm.

Varieties 2 (2 in the flora): Texas, n Mexico.

Penstemon triflorus occurs largely on the Edwards Plateau, though a few populations are known from adjacent physiographic provinces in central Texas.

1. Cauline leaf blades ovate to lanceolate, margins sharply serrate; calyx lobes 8–13 × 3–4.8 mm; corollas 30–35 mm . . . . .78a. *Penstemon triflorus* var. *triflorus*
1. Cauline leaf blades lanceolate to linear, margins entire or obscurely, rarely sharply, serrate; calyx lobes 6.5–8.5 × 2.4–3.5 mm; corollas 22–30 mm . . . . .78b. *Penstemon triflorus* var. *integrifolius*

78a. Penstemon triflorus A. Heller var. **triflorus** E F

Penstemon helleri Small

Cauline leaves: blade ovate to lanceolate, margins sharply serrate. **Flowers:** calyx lobes 8–13 × 3–4.8 mm; corolla 30–35 mm.

Flowering Mar–Jun(–Jul). Limestone prairies and bluffs; 200–600 m; Tex.

Variety *triflorus* has been documented in Bandera, Bexar, Brown, Burnet, Coleman, Gillespie, Hays, Kendall, Kerr, Real, Runnels, Schleicher, Sutton, and Uvalde counties. Plants with white or pale pink corollas, relatively short peduncles and pedicels, more continuous inflorescences, and broader leaf bases have been called *Penstemon helleri.*

78b. Penstemon triflorus A. Heller var. **integrifolius** (Pennell) Cory, Rhodora 38: 407. 1936

Penstemon triflorus subsp. *integrifolius* Pennell, Monogr. Acad. Nat. Sci. Philadelphia 1: 251. 1935

Cauline leaves: blade lanceolate to linear, margins entire or obscurely, rarely sharply, serrate. **Flowers:** calyx lobes 6.5–8.5 × 2.4–3.5 mm; corolla 22–30 mm.

Flowering Apr–May(–Jul). Limestone prairies and bluffs; 400–700 m; Tex.; Mexico (Coahuila).

Variety *integrifolius* has been documented in Concho, Crockett, Edwards, Irion, Reagan, Schleicher, Sutton, Tom Green, and Val Verde counties. Specimens from Concho and Tom Green counties with margins of cauline leaves serrate as in var. *triflorus* and smaller flowers as in var. *integrifolius* are included here.

79. Penstemon yampaensis Penland, Madroño 14: 156, fig. 2. 1958 • Yampa beardtongue C E

Penstemon acaulis L. O. Williams var. *yampaensis* (Penland) Neese

Stems prostrate to ascending, to 1 cm, scabrous or puberulent. **Leaves** essentially basal, not leathery, scabrous; basal and proximal cauline sessile, 15–25 (–35) × 1.5–2.6(–4.5) mm, blade oblanceolate to linear, base tapered, margins entire or obscurely dentate distally, apex obtuse to acute. **Thyrses** essentially absent, 0.1–0.6 cm, axis puberulent, verticillasters 1 or 2, cymes 1(–3)-flowered, 1(or 2) per node; proximal bracts oblanceolate to linear, 8–20 × 1–3 mm; peduncles and pedicels glandular-pubescent and scabrous. **Flowers:** calyx lobes lanceolate, 5–6.5 × 1–1.8 mm, glandular-pubescent and scabrous; corolla lavender to bluish lavender, without nectar guides, funnelform, 14–18 mm, yellowish or whitish villous internally abaxially, tube 5–6 mm, throat gradually inflated, not constricted at orifice, 4.5–6 mm diam., slightly 2-ridged abaxially; stamens: longer pair reaching orifice, pollen sacs widely divergent to opposite, navicular, 0.8–1.3 mm, dehiscing completely, sutures papillate; staminode 8–11 mm, prominently exserted, 0.5–0.8 mm diam., tip straight to recurved, distal 5–7 mm densely pubescent, hairs orange, to 1 mm; style 9–12 mm. **Capsules** 3.5–5 × 3–4 mm.

Flowering May–Jul. Semi-barren ledges and ridges, hilltops, pinyon-juniper woodlands, sagebrush shrublands; of conservation concern; 1800–2200 m; Colo., Utah, Wyo.

Penstemon yampaënsis is known from Moffat County, Colorado, Daggett and Uintah counties, Utah, and Sweetwater County, Wyoming. The species appears to be closely related to *P. acaulis*. Seed release appears to occur as the walls of the capsules deteriorate, rather than through an apical opening as in other *Penstemon* species. This type of dehiscence also occurs in *P. acaulis*.

26b.8. Penstemon Schmidel (subg. Penstemon) sect. Dissecti (Bentham) Pennell, Monogr. Acad. Nat. Sci. Philadelphia 1: 270. 1935　Ⓒ Ⓔ

Penstemon subsect. *Dissecti* Bentham in A. P. de Candolle and A. L. P. P. de Candolle, Prodr. 10: 322. 1846 (as Pentstemon)

Herbs. Stems retrorsely hairy, not glaucous. **Leaves** cauline, basal usually withering, opposite, not leathery, glabrous or puberulent abaxially along midrib, not glaucous; cauline short-petiolate or sessile, blade ovate, margins deeply pinnatifid to nearly pinnatisect. **Thyrses** interrupted, cylindric, axis puberulent, cymes 2 per node; peduncles and pedicels ascending. **Flowers:** calyx lobes: margins entire or erose, narrowly scarious, glandular-pubescent; corolla lavender to violet, bilaterally symmetric, bilabiate, not personate, ventricose, glandular-pubescent externally, hairy internally abaxially, throat abruptly inflated, not constricted at orifice, slightly 2-ridged abaxially; stamens included or longer pair reaching orifice, filaments glabrous, pollen sacs divergent, saccate, dehiscing incompletely, distal ⅕–¼ indehiscent, connective splitting, sides glabrous, sutures denticulate, teeth to 0.1 mm; staminode exserted, flattened distally, 0.3–0.4 mm diam., tip straight, distal 10–20% hairy, hairs to 2 mm; style sparsely glandular-pubescent proximally. **Capsules** sparsely glandular-pubescent distally. **Seeds** dark brown to black, angled, 1.5–2.4 mm.

Species 1: Georgia.

Placement of *Penstemon dissectus*, the lone member of sect. *Dissecti*, remains uncertain. It is one of two eastern North American species with saccate pollen sacs (the other being *P. multiflorus*) not allied with sect. *Saccanthera*; molecular data provide support for including it in subg. *Penstemon* (A. D. Wolfe et al. 2006).

80. **Penstemon dissectus** Elliott, Sketch Bot. S. Carolina 2: 129. 1822 (as Pentstemon dissectum) • Cut-leaf beardtongue　Ⓒ Ⓔ Ⓕ

Stems 32–75 cm. **Leaves:** cauline 7–11 pairs, 16–50 × 10–28 mm, base tapered to truncate, apex obtuse to acute. **Thyrses** 6–18 cm, verticillasters 2–4, cymes (1 or)2 or 3(or 4)-flowered; proximal bracts ovate to lanceolate, 4–10 × 3–10 mm; peduncles and pedicels retrorsely hairy. **Flowers:** calyx lobes ovate to lanceolate, 3–5 × 1.3–2.2 mm; corolla with lavender or violet nectar guides, 20–25 mm, sparsely white-lanate internally abaxially, tube 6–7 mm, throat 8–10 mm diam.; pollen sacs 1.1–1.4 mm; staminode 17–19 mm, distal 2–4 mm sparsely villous, hairs white, to 2 mm; style 14–15 mm. **Capsules** 7–9 × 4–7 mm.

Flowering Apr–May. Open, mixed oak-pine forests, rock outcrops; of conservation concern; 10–50 m; Ga.

Penstemon dissectus is known from approximately 16 counties in the Coastal Plain. The species is unique in *Penstemon* in having margins of cauline leaves one- or two-pinnatifid. Basal leaves, which usually wither by anthesis, have entire margins.

26b.9. PENSTEMON Schmidel (subg. PENSTEMON) sect. ELMIGERA (Reichenbach ex Spach) Bentham in A. P. de Candolle and A. L. P. P. de Candolle, Prodr. 10: 329. 1846 (as Pentstemon)

Elmigera Reichenbach ex Spach, Hist. Nat. Vég. 9: 299. 1840

Herbs. Stems glabrous or retrorsely hairy, glaucous or not. **Leaves** basal and cauline, opposite, leathery or not, glabrous, puberulent, or retrorsely hairy, glaucous or not; basal and proximal cauline petiolate; cauline sessile or short-petiolate, blade subcordate, ovate, spatulate, oblanceolate, lanceolate, elliptic, or linear, margins entire. **Thyrses** interrupted, ± secund to secund, rarely cylindric, axis glabrous, sometimes retrorsely hairy, cymes 2 per node, sometimes 1 per node (*P. labrosus*); peduncles and pedicels spreading to ascending or erect. **Flowers:** calyx lobes: margins entire or erose, broadly scarious, sometimes scarious (*P. labrosus*), glabrous or glandular-pubescent; corolla red to scarlet or orange, rarely yellow, bilaterally symmetric or nearly radially symmetric (*P. eatonii*), weakly to strongly bilabiate, not personate, salverform, tubular-funnelform, or tubular, glabrous externally, glabrous or hairy internally abaxially, throat slightly to gradually inflated, not constricted to constricted at orifice, rounded to slightly 2-ridged abaxially; stamens included to exserted, filaments glabrous, pollen sacs parallel, divergent, or opposite, navicular, dehiscing incompletely, proximal ¼–½ indehiscent, connective not splitting, sides glabrous or hispidulous to lanate-villous; sutures papillate or denticulate, papillae or teeth to 0.1 mm; staminode included or slightly exserted, flattened distally, 0.3–0.9 mm diam., tip straight, rarely recurved, glabrous or distal 5–20% hairy, hairs to 1.5 mm; style glabrous. **Capsules** glabrous. **Seeds** brown to dark brown, angled, 1.4–4.2 mm.

Species 8 (4 in the flora): w United States, Mexico.

Members of sect. *Elmigera* all exhibit the syndrome of characteristics usually associated with hummingbird pollination. Section *Elmigera* is not monophyletic; species in the flora area probably are related to species in sect. *Glabri* (C. A. Wessinger et al. 2016). Four species usually included in sect. *Elmigera*, but not in the flora area, are *Penstemon henricksonii* Straw, *P. imberbis* (Kunth) Trautvetter, *P. luteus* G. L. Nesom, and *P. wislizeni* (A. Gray) Straw; all occur in Mexico.

1. Corollas densely yellow-villous internally abaxially, throats constricted at orifices . . .82. *Penstemon cardinalis*
1. Corollas glabrous or sparsely white- or yellow-lanate internally abaxially, throats not constricted at orifices.
    2. Corollas nearly radially symmetric, weakly bilabiate, abaxial lobes projecting to barely spreading. . . . . 83. *Penstemon eatonii*
    2. Corollas bilaterally symmetric, strongly bilabiate, abaxial lobes spreading to reflexed.
        3. Corollas tubular-funnelform; basal and proximal cauline leaf blades (6–)12–30 (–35) mm wide, glabrous or sparsely to densely puberulent . . . . .81. *Penstemon barbatus*
        3. Corollas salverform; basal and proximal cauline leaf blades 3–11 mm wide, glabrous . . . . 84. *Penstemon labrosus*

81. Penstemon barbatus (Cavanilles) Roth, Catal. Bot.
3: 49. 1806 (as Pentstemon barbata) • Beard-lip
beardtongue F

Chelone barbata Cavanilles, Icon.
3: 22, plate 242. 1795

Stems erect, sometimes ascending, 30–100 cm, glabrous or
puberulent proximally, usually
slightly glaucous. **Leaves** glabrous or sparsely to densely
puberulent, not glaucous;
basal and proximal cauline
30–80(–140) × (6–)12–30(–35) mm, blade oblanceolate, base tapered, margins entire, apex obtuse, acute,
or acuminate; cauline 3–7(–14) pairs, sessile, 28–142 ×
1–8(–15) mm, blade lanceolate to linear, base tapered,
apex acuminate, sometimes acute. **Thyrses** ± secund,
sometimes cylindric, 11–69 cm, axis glabrous,
verticillasters 6–14(–19), cymes (1 or)2–4-flowered;
proximal bracts linear, 15–70(–126) × 1–10 mm;
peduncles and pedicels spreading to ascending, glabrous.
Flowers: calyx lobes ovate to lanceolate, (3–)4–6(–9) ×
1.9–2.8(–3.2) mm, margins entire or erose, glabrous or
obscurely glandular-pubescent; corolla red or orangish
red to crimson or scarlet, with reddish violet nectar
guides, bilaterally symmetric, strongly bilabiate, tubular-
funnelform, 26–32(–36) mm, glabrous or sparsely
white-lanate or yellow-lanate internally abaxially, tube
5–8 mm, throat 5–6 mm diam., not constricted at
orifice, slightly 2-ridged abaxially, abaxial lobes strongly
reflexed, adaxial lobes projecting; stamens exserted
(but usually concealed beneath projecting adaxial lip),
pollen sacs opposite, proximal ¼–⅓ indehiscent, 1.5–2
(–2.2) mm, sides glabrous or sparsely to moderately
lanate-villous, hairs white, to 1.5 mm, sutures papillate
or denticulate, teeth to 0.1 mm; staminode 15–17
mm, included, tip straight, glabrous; style 22–27 mm.
Capsules 10–14 × 6–8 mm.

Varieties 3 (3 in the flora): w, sc United States,
n Mexico.

Penstemon barbatus is among the more widespread
and conspicuous penstemons in the desert Southwest
and Southern Rocky mountains. Putative hybrids
reported between *P. barbatus* and *P. comarrhenus*,
P. glaber, *P. strictus*, and *P. virgatus* were summarized
by F. S. Crosswhite (1965); putative hybrids involving
P. pseudoputus and *P. putus* also are suspected among
some herbarium specimens. *Penstemon barbatus*
is widely cultivated for its showy flowers; cultivars
with yellow corollas also exist. *Penstemon barbatus*
was reported for Suffolk County, Massachusetts
(D. W. Magee and H. E. Ahles 2007); that report, or
that those plants persisted outside of cultivation, has not
been confirmed.

Various groups of Apache, Navajo, and Western
Keres of Arizona and New Mexico used *Penstemon
barbatus* for drugs, ceremonies, and decorations
(D. E. Moerman 1998).

1. Pollen sac sides sparsely to moderately lanate-
villous . . . . .81c. *Penstemon barbatus* var. *trichander*
1. Pollen sac sides glabrous.
  2. Corollas yellow-lanate internally abaxially
    . . . . .81a. *Penstemon barbatus* var. *barbatus*
  2. Corollas glabrous or sparsely white-lanate
    internally abaxially . . . .
    . . . . 81b. *Penstemon barbatus* var. *torreyi*

81a. Penstemon barbatus (Cavanilles) Roth var.
barbatus

Flowers: corolla yellow-lanate
internally abaxially; pollen sac
sides glabrous.

Flowering Jun–Sep. Pinyon-
juniper, Gambel oak, Douglas
fir, and ponderosa pine woodlands; 1200–3000(–3400) m;
Ariz., Calif., N.Mex., Tex.,
Utah; Mexico (Durango, Nuevo
León, Tamaulipas).

Populations of var. *barbatus* occur predominantly
southwest of the Mogollon Rim; outlying populations
in the flora area are known from southern California,
trans-Pecos Texas, and south-central Utah.

81b. Penstemon barbatus (Cavanilles) Roth var. **torreyi**
(Bentham) A. Gray, Proc. Amer. Acad. Arts 6: 59.
1862 (as Pentstemon) F

Penstemon torreyi Bentham in
A. P. de Candolle and
A. L. P. P. de Candolle, Prodr.
10: 324. 1846 (as Pentstemon);
P. barbatus subsp. *torreyi*
(Bentham) D. D. Keck

Flowers: corolla glabrous or
sparsely white-lanate internally
abaxially; pollen sac sides glabrous. $2n = 16$.

Flowering May–Aug(–Oct). Pinyon-juniper, Gambel
oak, ponderosa pine, and spruce-fir woodlands, montane
meadows; 1800–3200 m; Ariz., Colo., N.Mex., Tex.,
Utah; Mexico (Chihuahua).

Variety *torreyi* occurs from the Mogollon Rim
northeastward. Some specimens (for example, *D. B.
Dunn 6155*, Torrance County, New Mexico, KANU),
which otherwise match var. *torreyi*, bear anthers with a
few straight hairs 0.1–0.2 mm.

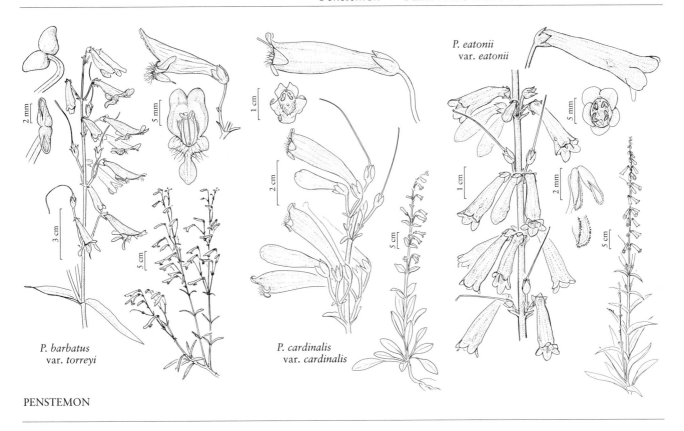

P. barbatus
var. *torreyi*

P. cardinalis
var. *cardinalis*

P. eatonii
var. *eatonii*

PENSTEMON

81c. Penstemon barbatus (Cavanilles) Roth var.
trichander A. Gray, Proc. Amer. Acad. Arts 11: 94.
1876 (as Pentstemon) E

Penstemon barbatus subsp.
trichander (A. Gray) D. D. Keck

Flowers: corolla glabrous
internally; pollen sac sides
sparsely to moderately lanate-
villous.

Flowering Jun–Aug. Pinyon-
juniper and birch-maple woodlands; 1600–2200 m; Ariz.,
Colo., N.Mex., Utah.

F. S. Crosswhite (1965) hypothesized that var.
trichander is the product of hybridization and introgression between *Penstemon barbatus* and *P. strictus*;
Crosswhite noted that var. *trichander* occurs where var.
torreyi and *P. strictus* are sympatric. Variety *trichander*
is concentrated in the Four Corners region.

82. Penstemon cardinalis Wooton & Standley,
Contr. U.S. Natl. Herb. 16: 171. 1913 (as Pentstemon)
• Cardinal or scarlet beardtongue C E F

Stems ascending to erect,
32–100 cm, glabrous, slightly
glaucous. **Leaves** glabrous, not
glaucous; basal and proximal
cauline 40–80(–150) × 14–40
(–60) mm, blade spatulate to
obovate, oblanceolate, or elliptic, base tapered, margins entire,
apex rounded to obtuse; cauline
3–7 pairs, sessile, (11–)20–135 × 9–50 mm, blade
spatulate to elliptic, ovate, or subcordate, base truncate
to broadly cordate-clasping, apex rounded to acute or
mucronate. **Thyrses** secund, (3–)6–44 cm, axis glabrous, verticillasters 3–15, cymes 2–4(–8)-flowered;
proximal bracts suborbiculate, ovate, or lanceolate,
(4–)16–60 × (2–)5–45 mm; peduncles and pedicels
erect, glabrous. **Flowers:** calyx lobes ovate, 1.6–6 ×
1.4–3 mm, margins entire or erose, glabrous; corolla red
to crimson, with reddish violet nectar guides, bilaterally
symmetric, weakly bilabiate, tubular, 22–30 mm, densely
yellow-villous internally abaxially, tube 6–8 mm, throat
6–8 mm diam., constricted at orifice, rounded abaxially,
abaxial lobes spreading to reflexed, adaxial lobes
projecting to erect; stamens included, pollen sacs
divergent, 1.5–1.9 mm, sides glabrous, sutures
denticulate, teeth to 0.1 mm; staminode 15–17 mm,

included or slightly exserted, tip straight to recurved, distal 1–3 mm sparsely lanate, hairs yellow, to 1.5 mm; style 18–28 mm, usually exserted. **Capsules** 10–15 × 5–9 mm.

Varieties 2 (2 in the flora): sc United States.

Penstemon cardinalis occurs as two weakly differentiated varieties in isolated ranges of the Southern Rocky Mountains.

1. Calyx lobes 1.6–3.5(–4) × 1.4–2 mm; cauline leaf blades elliptic to spatulate or ovate, not leathery; proximal bracts ovate to lanceolate. . . . .
. . . . 82a. *Penstemon cardinalis* var. *cardinalis*
1. Calyx lobes 4–6 × 2.3–3 mm; cauline leaf blades ovate to subcordate, sometimes elliptic, leathery; proximal bracts suborbiculate to ovate . . . .
. . . . .82b. *Penstemon cardinalis* var. *regalis*

82a. Penstemon cardinalis Wooton & Standley var. **cardinalis** [C] [E] [F]

Leaves not leathery, cauline blade elliptic to spatulate or ovate. **Thyrses:** proximal bracts ovate to lanceolate. **Flowers:** calyx lobes 1.6–3.5(–4) × 1.4–2 mm. $2n = 16$.

Flowering Jun–Jul. Canyons and rocky slopes, pinyon-juniper woodlands, pine forests; of conservation concern; 2100–2700 m; N.Mex.

Variety *cardinalis* is known from the Capitan and Sacramento mountains in Lincoln and Otero counties.

82b. Penstemon cardinalis Wooton & Standley var. **regalis** (A. Nelson) C. C. Freeman, PhytoKeys 80: 34. 2017 [C] [E]

Penstemon regalis A. Nelson, Amer. J. Bot. 21: 578. 1934; *P. cardinalis* subsp. *regalis* (A. Nelson) G. T. Nisbet & R. C. Jackson

Leaves leathery, cauline blade ovate to subcordate, sometimes elliptic. **Thyrses:** proximal bracts suborbiculate to ovate. **Flowers:** calyx lobes 4–6 × 2.3–3 mm.

Flowering May–Jun(–Aug). Rocky canyons, pine woodlands; of conservation concern; 1400–2500 m; N.Mex., Tex.

Variety *regalis* is known from the Guadalupe Mountains in New Mexico (Eddy County) and the Davis, Guadalupe, and Sierra Diablo mountains in western Texas (Culberson and Jeff Davis counties).

83. Penstemon eatonii A. Gray, Proc. Amer. Acad. Arts 8: 395. 1872 (as Pentstemon eatoni) • Eaton's or firecracker beardtongue [E] [F]

Stems ascending to erect, 40–100 cm, glabrous or retrorsely hairy, not glaucous. **Leaves** glabrous or retrorsely hairy, not glaucous; basal and proximal cauline 50–100(–200) × 15–28(–50) mm, blade obovate to elliptic, base tapered, margins entire, apex obtuse to acute; cauline 3–5 pairs, short-petiolate or sessile, 24–90(–145) × 8–28 mm, blade ovate or lanceolate, proximals sometimes oblanceolate, base tapered to cordate-clasping, apex acute, rarely obtuse. **Thyrses** secund, (9–)25–45 cm, axis glabrous or retrorsely hairy, verticillasters 4–12, cymes 1- or 2(–4)-flowered; proximal bracts lanceolate, 6–49 × 1–8 mm; peduncles and pedicels erect, sometimes ascending, glabrous or sparsely puberulent. **Flowers:** calyx lobes ovate, (2.5–)3–5(–6) × 1.8–3 mm, margins erose, rarely entire, glabrous; corolla red to scarlet, essentially without nectar guides, nearly radially symmetric, weakly bilabiate, tubular, 24–30(–33) mm, glabrous internally, tube 6–10 mm, throat 5–7(–9) mm diam., not constricted at orifice, rounded abaxially, abaxial lobes projecting to barely spreading, adaxial lobes projecting to barely spreading; stamens included, reaching orifice, or exserted, pollen sacs parallel, proximal ¼–½ indehiscent, (1.4–)1.8–2.4(–2.8) mm, sides glabrous or obscurely hispidulous with tan enations less than 0.1 mm, sutures papillate or denticulate, teeth to 0.1 mm; staminode 14–17 mm, included, tip straight, glabrous or distal 1–4 mm sparsely to densely pubescent, hairs yellow, to 1 mm; style 23–27 mm, usually exserted. **Capsules** 10–14 × 4–8 mm.

Varieties 3 (3 in the flora): w United States.

Penstemon ×*jonesii* Pennell, a putative hybrid between *P. eatonii* and *P. laevis*, is known only from Kane and Washington counties, Utah (E. C. Neese and N. D. Atwood 2003). *Penstemon* ×*crideri* A. Nelson, a putative hybrid between *P. eatonii* and *P. pseudospectabilis*, has been reported from Arizona (A. Nelson 1936). *Penstemon* ×*mirus* A. Nelson, a putative hybrid between *P. eatonii* and *P. palmeri*, has been reported from Arizona (Nelson 1938). A wild hybrid between *P. centranthifolius* and *P. eatonii* has been reported from San Bernardino County, California (Paul Wilson and M. Valenzuela 2002).

The Hopi and Kayenta Navajo of northeastern Arizona use *Penstemon eatonii* as a drug or ceremonial plant (D. E. Moerman 1998).

1. Stems glabrous, rarely obscurely retrorsely hairy at proximal nodes; leaves glabrous . . . . .83a. *Penstemon eatonii* var. *eatonii*
1. Stems retrorsely hairy; leaves retrorsely hairy.
   2. Stamens: both pairs exserted . . . . .83b. *Penstemon eatonii* var. *exsertus*
   2. Stamens included or reaching orifice, longer pair rarely exserted . . . . .83c. *Penstemon eatonii* var. *undosus*

83a. Penstemon eatonii A. Gray var. **eatonii** [E] [F]

Stems glabrous, rarely obscurely retrorsely hairy at proximal nodes. **Leaves** glabrous. **Flowers:** stamens included or reaching orifice, longer pair rarely exserted. $2n = 16$.

Flowering Mar–Jul. Sagebrush shrublands, pinyon-juniper woodlands, pine forests; 1200–2900 m; Ariz., Calif., Colo., Idaho, Nev., N.Mex., Utah, Wyo.

Some specimens of var. *eatonii* from the Schell Creek Mountains in White Pine County, Nevada, have the longer pairs of stamens prominently exserted and stems obscurely puberulent. Records from Lake County, Colorado, Oneida County, Idaho, and Carbon County, Wyoming, probably are based on plants that were seeded along highways or pipelines.

83b. Penstemon eatonii A. Gray var. **exsertus** (A. Nelson) C. C. Freeman, PhytoKeys 80: 35. 2017 [E]

Penstemon exsertus A. Nelson, Amer. J. Bot. 18: 438. 1931 (as Pentstemon); *P. eatonii* subsp. *exsertus* (A. Nelson) D. D. Keck

Stems retrorsely hairy. **Leaves** retrorsely hairy. **Flowers:** stamens: both pairs exserted. $2n = 16$.

Flowering Mar–Jun. Sagebrush shrublands, pinyon-juniper woodlands, pine forests; 600–1500 m; Ariz.

Variety *exsertus* is concentrated in central Arizona, especially from Flagstaff to Tucson. Records are known from Coconino, Gila, Greenlee, Maricopa, Mohave, Pinal, and Yavapai counties.

83c. Penstemon eatonii A. Gray var. **undosus** M. E. Jones, Proc. Calif. Acad. Sci., ser. 2, 5: 715. 1895 (as Pentstemon eatoni) [E]

Penstemon eatonii subsp. *undosus* (M. E. Jones) D. D. Keck

Stems retrorsely hairy. **Leaves** retrorsely hairy. **Flowers:** stamens included or reaching orifice, longer pair rarely exserted. $2n = 16$.

Flowering Mar–Jun. Sagebrush shrublands, pinyon-juniper woodlands, pine forests; 800–2800 m; Ariz., Calif., Colo., Nev., N.Mex., Utah.

As in var. *eatonii*, var. *undosus* rarely has the longer pairs of stamens prominently exserted and the shorter pairs reaching the orifices.

84. Penstemon labrosus (A. Gray) Hooker f., Bot. Mag. 110: plate 6738. 1884 • San Gabriel beardtongue

Penstemon barbatus (Cavanilles) Roth var. *labrosus* A. Gray in W. H. Brewer et al., Bot. California 1: 622. 1876 (as Pentstemon)

Stems erect, sometimes ascending, 30–80 cm, glabrous, slightly glaucous. **Leaves** glabrous, slightly glaucous; basal and proximal cauline 35–110 × 3–11 mm, blade oblanceolate to lanceolate, base tapered, margins entire, apex obtuse to acute; cauline 3–6 pairs, sessile or proximals short-petiolate, 27–85 × 1–6 mm, blade linear, base tapered, apex acute, rarely obtuse. **Thyrses** ± secund, 13–30 cm, axis glabrous, verticillasters 5–10, cymes 1- or 2(or 3)-flowered; proximal bracts linear, 5–30 × 1–2 mm; peduncles and pedicels ascending to erect, glabrous. **Flowers:** calyx lobes ovate to lanceolate, 3.5–6 × 1.8–2.4 mm, margins erose, glabrous; corolla red to scarlet or orange, rarely yellow, without nectar guides, bilaterally symmetric, strongly bilabiate, salverform, 30–40 mm, glabrous internally, tube 5–7 mm, throat 4–6 mm diam., not constricted at orifice, rounded abaxially, abaxial lobes spreading to reflexed, adaxial lobes projecting, galeate; stamens exserted (but concealed by projecting adaxial lip), pollen sacs opposite, proximal ¼–⅓ indehiscent, 1.6–2 mm, sides glabrous, sutures papillate; staminode 11–14 mm, included, tip straight, glabrous; style 20–28 mm, exserted. **Capsules** 7–10 × 4–6 mm. $2n = 16$.

Flowering Jun–Sep. Pinyon-juniper woodlands, pine and mixed-hardwood forests; 1200–2400 m; Calif.; Mexico (Baja California).

Penstemon labrosus is known from the Peninsular and Transverse ranges of southern California and mountain ranges of Baja California. In California, populations have been documented in Inyo, Kern, Los Angeles, Riverside, San Bernardino, San Diego, and Ventura counties.

26b.10. PENSTEMON Schmidel (subg. PENSTEMON) sect. GENTIANOIDES G. Don, Gen. Hist. 4: 640. 1837 (as Pentstemon)

Herbs. Stems glabrous, glaucous, sometimes not. **Leaves** basal and cauline, sometimes basal few (*P. subulatus*), or essentially cauline (*P. centranthifolius*), opposite, leathery, rarely not (*P. alamosensis*, *P. parryi*), glabrous, rarely scabrous, glaucous, rarely not; basal and proximal cauline petiolate; cauline sessile or short-petiolate, blade ovate, oblanceolate, lanceolate, oblong, elliptic, linear, or linear-subulate, margins entire, rarely toothed. **Thyrses** ± interrupted, cylindric or ± secund, axis glabrous, rarely scabrous, cymes 2 per node, sometimes 1 per node (*P. alamosensis*); peduncles and pedicels spreading to ascending or erect. **Flowers:** calyx lobes: entire or erose, ± scarious, glabrous or sparsely glandular or glandular-pubescent; corolla red, scarlet, crimson, magenta, pink, orange, lavender, violet, reddish violet, or purple, bilaterally symmetric or nearly radially symmetric, bilabiate or weakly bilabiate, not personate, salverform, tubular-salverform, or tubular-funnelform, sometimes ventricose, glabrous or glandular-pubescent to glandular externally, glabrous or glandular-pubescent internally abaxially or wholly, throat slightly to gradually inflated, not constricted at orifice, rounded or slightly 2-ridged abaxially; stamens included or longer pair reaching orifice, filaments glabrous, pollen sacs opposite, dehiscing completely, explanate, sometimes navicular, connective splitting, sides glabrous, sutures smooth, sometimes papillate; staminode included, flattened distally or ± terete, 0.1–1(–2.5) mm diam., tip straight or recurved, glabrous or distal 10–40(–70)% papillate or hairy, papillae or hairs to 1.5 mm; style glabrous. **Capsules** glabrous. **Seeds** brown to black, angled, 1–4 mm.

Species 11 (10 in the flora): w United States, n Mexico.

Section *Gentianoides* is not monophyletic. Molecular data (C. A. Wessinger et al. 2016) placed some of the species in a clade that includes sect. *Coerulei* and other species in a clade that includes sect. *Spectabiles*; still others may be allied with sect. *Glabri* (A. D. Wolfe et al. 2006). D. D. Keck (1937) included most of the species of sect. *Gentianoides* in subsect. *Centranthifolii* [an unpublished name] of sect. *Peltanthera* D. D. Keck, including *Penstemon cerrosensis* Kellogg, known only from Cedros Island off the west coast of Baja California, Mexico.

1. Corollas glabrous externally.
    2. Corollas salverform, weakly bilabiate.
        3. Corollas 25–33 mm, tubes 8–10 mm; cauline leaf blades ovate to lanceolate . . . .
            . . . . 86. *Penstemon centranthifolius*
        3. Corollas 18–26 mm, tubes 4–7 mm; cauline leaf blades lanceolate to linear or
            linear-subulate . . . . 91. *Penstemon subulatus*
    2. Corollas tubular-funnelform, bilabiate.
        4. Corollas glabrous or glandular-pubescent internally; peduncles and pedicels
            ascending to erect . . . . 87. *Penstemon confusus* (in part)
        4. Corollas glabrous internally; peduncles and pedicels spreading to ascending . . . .
            . . . . 90. *Penstemon patens* (in part)
1. Corollas glandular-pubescent externally.
    5. Corollas white-pilose and glandular-pubescent internally abaxially . . . . 89. *Penstemon parryi*
    5. Corollas glandular-pubescent or glabrous internally abaxially.

[6. Shifted to left margin.—Ed.]

6. Calyx lobes sparsely glandular-pubescent.
   7. Staminodes glabrous.
      8. Staminodes 7–9 mm; leaves not leathery . . . . 85. *Penstemon alamosensis*
      8. Staminodes 14–20 mm; leaves leathery . . . . 88. *Penstemon havardii*
   7. Staminodes retrorsely hairy or sparsely to densely pilose distally.
      9. Corollas nearly radially symmetric, weakly bilabiate; staminodes: distal 2–3 mm
         retrorsely hairy, hairs yellow or whitish, to 1.5 mm . . . . 92. *Penstemon superbus* (in part)
      9. Corollas bilaterally symmetric, bilabiate; staminodes: distal 2–7 mm sparsely to
         densely pilose, hairs yellow, to 1.2 mm . . . . 94. *Penstemon wrightii*
6. Calyx lobes glabrous.
   10. Staminodes: distal 2–3 mm retrorsely hairy, hairs yellow or whitish, to 1.5 mm; corollas
       nearly radially symmetric, weakly bilabiate . . . . 92. *Penstemon superbus* (in part)
   10. Staminodes glabrous or distal 1–3 mm papillate, papillae golden yellow or reddish
       yellow, to 0.2 mm; corollas bilaterally symmetric, bilabiate.
       11. Corollas red to crimson, tubular-salverform, glandular-pubescent externally and
           internally; pollen sacs explanate . . . . 93. *Penstemon utahensis*
       11. Corollas violet or reddish violet to purple or lavender, tubular-funnelform, glabrous
           or glandular-pubescent externally, glabrous or glandular-pubescent internally;
           pollen sacs navicular, sometimes explanate.
           12. Corollas glabrous or glandular-pubescent internally; pedicels ascending to
               erect . . . . 87. *Penstemon confusus* (in part)
           12. Corollas glabrous internally; pedicels spreading to ascending . . . . . 90. *Penstemon patens* (in part)

85. Penstemon alamosensis Pennell & G. T. Nisbet,
Univ. Kansas Sci. Bull. 41: 709, figs. 41, 43, 45. 1960
   • Alamo beardtongue E

Stems ascending to erect,
14–120 cm, sometimes glaucous.
Leaves not leathery, glabrous,
sometimes glaucous; basal and
proximal cauline 25–140 ×
9–45 mm, blade elliptic to
obovate or lanceolate, base
tapered, margins entire, apex
obtuse to acute; cauline 2–6
pairs, sessile or proximals short-petiolate, 10–135 ×
3–30 mm, blade oblong to oblanceolate or lanceolate,
base tapered to truncate, margins entire, apex obtuse to
acute. **Thyrses** interrupted, cylindric to ± secund, 12–34
(–90) cm, axis glabrous, verticillasters 5–15(–30), cymes
(1 or)2(–4)-flowered; proximal bracts lanceolate to
subulate, 4–15 × 1–5 mm; peduncles and pedicels
ascending to erect, peduncles glabrous, pedicels glabrous
or sparsely glandular-pubescent. **Flowers:** calyx lobes
lanceolate, 3–4.5 × 1–1.5 mm, margins entire, sparsely
glandular-pubescent; corolla red to orangish red, without nectar guides, nearly radially symmetric, weakly
bilabiate, tubular-funnelform, 19–25 mm, glandular-
pubescent externally, glandular-pubescent internally
abaxially, tube 6–8 mm, throat gradually inflated, 5–7
mm diam., rounded abaxially; stamens included, pollen
sacs explanate, 1–1.4 mm, sutures smooth; staminode
7–9 mm, flattened distally, 0.3–0.5 mm diam., tip
straight to recurved, glabrous; style 9–11 mm. **Capsules**
11–16 × 5–8 mm. **2*n* = 16.**

Flowering Apr–Jun. Limestone canyons, cliffs, and
slopes; 1300–1600 m; N.Mex., Tex.

Penstemon alamosensis is known from the
Sacramento and San Andres mountains in New Mexico
(Doña Ana, Lincoln, and Otero counties) and in the
Hueco Mountains in western Texas (El Paso County).

86. Penstemon centranthifolius (Bentham) Bentham,
Scroph. Ind., 7. 1835 (as Pentstemon centranthifolium)
   • Scarlet bugler F

Chelone centranthifolia Bentham,
Trans. Hort. Soc. London, ser. 2,
1: 481. 1835

Stems ascending to erect,
30–120 cm, glaucous. **Leaves**
glabrous, glaucous; cauline
5–11 pairs, short-petiolate or
sessile, 40–100 × 10–40 mm,
blade ovate to lanceolate, base
tapered to auriculate-clasping, margins entire, apex
rounded to acute. **Thyrses** interrupted, secund to
± cylindric, 15–60(–100) cm, axis glabrous or
obscurely scabrous, verticillasters 8–18(–22), cymes
1–5(–11)-flowered; proximal bracts ovate to lanceolate,
(3–)7–25(–87) × (1–)3–14(–30) mm; peduncles and
pedicels erect, glabrous or obscurely scabrous. **Flowers:**
calyx lobes ovate to lanceolate, 3–6 × 2–3 mm, margins
erose, glabrous; corolla scarlet, without nectar guides,
nearly radially symmetric, weakly bilabiate, salverform,
25–33 mm, glabrous externally, glabrous internally,
tube 8–10 mm, throat slightly inflated, 4.5–6 mm diam.,

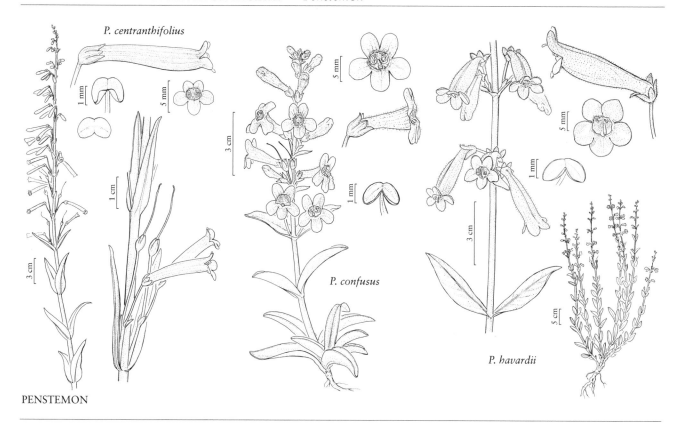

P. centranthifolius

P. confusus

P. havardii

PENSTEMON

rounded abaxially; stamens: longer pair reaching orifice, pollen sacs explanate, 0.9–1.2 mm, sutures smooth; staminode 13–14 mm, terete to slightly flattened distally, 0.1–0.3 mm diam., tip straight, glabrous; style 18–27 mm, exserted. **Capsules** 8–11 × 6–7 mm. **2*n* = 16.**

Flowering Feb–Jul. Hillsides, chaparral, oak, pinyon-juniper, and Joshua tree woodlands, coastal sage scrub, pine forests; 0–1800 m; Calif.; Mexico (Baja California).

Penstemon ×*parishii* A. Gray, a naturally occurring hybrid between *P. centranthifolius* and *P. spectabilis* (A. D. Wolfe and W. J. Elisens 1993), has been reported from Los Angeles, Riverside, San Bernardino, and San Diego counties, California (D. D. Keck 1937; Paul Wilson and M. Valenzuela 2002). A wild hybrid between *P. centranthifolius* and *P. eatonii* also was reported from San Bernardino County, California (Wilson and Valenzuela). *Penstemon* ×*dubius* Davidson was described from Mount Lowe in the San Gabriel Mountains, Los Angeles County, California, growing with *P. centranthifolius* and *P. grinnellii*. Hybridization between *P. centranthifolius* and *P. grinnellii* has been documented using allozyme and DNA data (Wolfe and Elisens 1993, 1994).

Three southern and central Californian tribes of Native Americans used *Penstemon centranthifolius* for drugs, food, and decorations (D. E. Moerman 1998).

87. **Penstemon confusus** M. E. Jones, Zoë 4: 280. 1893 (as Pentstemon) • Owens Valley beardtongue [E] [F]

Stems erect, 3–40(–44) cm, glaucous. **Leaves** glabrous or obscurely scabrous, especially along margins, glaucous; basal and proximal cauline (25–) 30–60 × 5–10(–14) mm, blade oblanceolate, base tapered, margins entire, apex obtuse to acute, sometimes mucronate; cauline 2 or 3 pairs, sessile, 15–50(–57) × 2–8(–12) mm, blade lanceolate to linear, base tapered, margins entire, apex acute. **Thyrses** ± interrupted, cylindric to ± secund, (2–)3–25(–30) cm, axis glabrous, verticillasters 3–10, cymes 1–5-flowered; proximal bracts lanceolate, 7–30 (–50) × 3–12 mm; peduncles and pedicels ascending to erect, glabrous. **Flowers:** calyx lobes ovate to obovate, rarely elliptic, (3.5–)4–6 × 1.8–3 mm, margins entire or erose, glabrous; corolla violet or reddish violet to purple, with reddish or reddish purple nectar guides, bilaterally symmetric, bilabiate, tubular-funnelform, 14–18 (–20) mm, glabrous or glandular-pubescent externally, glabrous or glandular-pubescent internally, tube 6–8 mm, throat gradually inflated, 6–7 mm diam., rounded or slightly 2-ridged abaxially; stamens included, pollen sacs navicular, sometimes explanate, 0.9–1.2 mm, sutures smooth or papillate; staminode 8–9 mm, flattened distally, 0.5–0.6 mm diam., tip recurved, distal

1–2 mm papillate, papillae golden yellow, to 0.2 mm, rarely glabrous; style 10–12 mm. **Capsules** 8–11 × 4–6 mm.

Flowering Apr–Jun. Gravelly, often calcareous, soils, sagebrush shrublands, pinyon-juniper woodlands; 1700–2200 m; Ariz., Nev., Utah.

Penstemon confusus is known from northwestern Arizona, east-central Nevada, and southeastern Utah. The species is barely sympatric with *P. utahensis* in Lincoln County, Nevada, and Washington County, Utah.

88. **Penstemon havardii** A. Gray, Proc. Amer. Acad. Arts 20: 306. 1885 (as Pentstemon havardi) • Havard's beardtongue F

Stems ascending to erect, 45–150(–200) cm, glaucous. **Leaves** glabrous, glaucous; basal and proximal cauline 75–130 × 18–50 mm, blade ovate to elliptic, base tapered, margins entire, apex rounded to obtuse; cauline 3–7 pairs, proximals usually short-petiolate, distal ones sessile, 24–145 × 12–70 mm, blade elliptic to ovate, base tapered to clasping, margins entire, apex obtuse to acute. **Thyrses** interrupted, cylindric, 14–60(–80) cm, axis glabrous, verticillasters 7–20, cymes 2–8(–18)-flowered; proximal bracts ovate to lanceolate, 15–33 × 4–20 mm; peduncles and pedicels ascending to erect, glabrous or sparsely glandular-pubescent. **Flowers:** calyx lobes elliptic to ovate, 3.5–5(–5.8) × 1.8–2.4 mm, margins entire or erose, sparsely glandular-pubescent; corolla scarlet to orange, without nectar guides, bilaterally symmetric, bilabiate, tubular-funnelform, 20–28 mm, sparsely glandular-pubescent externally, glandular-pubescent internally, tube 4–7 mm, throat gradually inflated, 5–6 mm diam., rounded abaxially; stamens included or longer pair reaching orifice, pollen sacs explanate, 1–1.2 mm, sutures smooth; staminode 14–20 mm, flattened distally, 0.3–0.6 mm diam., tip straight, glabrous; style 16–20 mm. **Capsules** 7–12 × 5–8 mm.

Flowering Apr–Oct. Dry, rocky slopes, streambeds; 1500–1800 m; Tex.; Mexico (Chihuahua, Coahuila, Nuevo León, Zacatecas).

In Texas, *Penstemon havardii* is known from Brewster, Jeff Davis, and Presidio counties. Fruiting cymes of *P. havardii* sometimes elongate to 10 cm.

89. **Penstemon parryi** (A. Gray) A. Gray in A. Gray et al., Syn. Fl. N. Amer. 2(1): 264. 1878 (as Pentstemon) • Parry's beardtongue

Penstemon puniceus A. Gray var. *parryi* A. Gray in W. H. Emory, Rep. U.S. Mex. Bound. 2(1): 113. 1859 (as Pentstemon)

Stems ascending to erect, 30–120(–160) cm, glaucous. **Leaves** glabrous, glaucous; basal and proximal cauline 36–125 × 8–25 mm, blade spatulate to oblanceolate or elliptic, base tapered, margins entire, rarely remotely and obscurely dentate, apex rounded to obtuse or acute; cauline 2–5 pairs, sessile, 23–130 × 5–25 mm, blade lanceolate to oblong, base clasping to auriculate-clasping, margins entire, rarely remotely and obscurely dentate, apex obtuse to acute. **Thyrses** interrupted, cylindric, (4–)10–60 cm, axis glabrous, verticillasters (2–)5–16, cymes 2–7-flowered; proximal bracts lanceolate, (8–)13–55 × 2–8 mm; peduncles and pedicels ascending to erect, glabrous or pedicels slightly glandular-pubescent. **Flowers:** calyx lobes ovate to lanceolate, 3–4.6 × 1.2–2 mm, apex acute to short-acuminate, glabrous or sparsely glandular-pubescent; corolla rose pink to rose magenta, with reddish purple nectar guides, bilaterally symmetric, bilabiate, ventricose, 13–22 mm, glandular-pubescent externally, white-pilose and glandular-pubescent internally abaxially, tube 3–7 mm, throat gradually inflated, 5–7 mm diam., rounded abaxially; stamens included, pollen sacs explanate, 1.1–1.4 mm, sutures smooth; staminode 10–12 mm, flattened distally, 0.7–1(–2.5) mm diam., tip straight, distal 4–5 mm retrorsely hairy, hairs yellow or whitish, to 1 mm; style 13–15 mm. **Capsules** 7–9 × 4–5 mm. $2n = 16$.

Flowering (Feb–)Mar–May. Rocky hillsides, washes, canyons, oak scrub and deserts; 200–1500 m; Ariz.; Mexico (Sonora).

Penstemon parryi is known from the scattered desert mountain ranges of southern Arizona in Cochise, Pima, Pinal, Santa Cruz, and Yavapai counties. The species resembles *P. superbus*; it is distinguished by its narrow leaves and corollas that are rose pink or rose magenta, more bilaterally symmetric, and white-pilose abaxially in the throats.

90. Penstemon patens (M. E. Jones) N. H. Holmgren, Brittonia 31: 106. 1979 • Lone pine beardtongue [E]

Penstemon confusus M. E. Jones var. *patens* M. E. Jones, Contr. W. Bot. 12: 63. 1908 (as Pentstemon); *P. confusus* subsp. *patens* (M. E. Jones) D. D. Keck

Stems ascending to erect, (15–)20–40 cm, glaucous. **Leaves** glabrous or obscurely scabrous, especially along margins, glaucous; basal and proximal cauline (30–)50–75 × 7–14 mm, blade oblanceolate, base tapered, margins entire, apex obtuse to acute, sometimes mucronate; cauline 2–4 pairs, sessile, 25–70(–90) × 4–15(–25) mm, blade lanceolate, base tapered, margins entire, apex acute. **Thyrses** ± interrupted, ± secund, 5–26 cm, axis glabrous, verticillasters 5 or 6(–10), cymes 3–7(–13)-flowered; proximal bracts lanceolate, 10–54 × 3–16 mm; peduncles and pedicels spreading to ascending, glabrous. **Flowers:** calyx lobes ovate, 3–6 (–7) × 1.9–2.6 mm, margins entire or erose, glabrous; corolla violet or reddish violet to lavender, with red or reddish purple nectar guides, bilaterally symmetric, bilabiate, tubular-funnelform, 13–17(–20) mm, glabrous externally or buds and young flowers sometimes with sessile glands distally, glabrous internally, tube 5–7 mm, throat gradually inflated, 6–7 mm diam., rounded or slightly 2-ridged abaxially; stamens included, pollen sacs navicular, 1–1.3 mm, sutures smooth or papillate; staminode 7–8 mm, flattened distally, 0.5–0.6 mm diam., tip recurved, distal 1–3 mm sparsely papillate, papillae golden yellow or reddish yellow, to 0.2 mm; style 10–12 mm. **Capsules** 8–10 × 4–6 mm. *2n* = 16.

Flowering May–Jun. Sagebrush shrublands, pinyon-juniper woodlands, pine forests; 1800–3000 m; Calif., Nev.

Penstemon patens occurs on mountain slopes surrounding the Owens Valley in Inyo and Mono counties, California, in the Huntoon Mountains in western Mineral County, Nevada, and in the East Desert and Sheep ranges in Clark County, Nevada. *Penstemon patens* usually is distinguishable from *P. confusus* and *P. utahensis* by its glabrous (versus glandular-pubescent) corollas; flower buds and young flowers of *P. patens* often bear some sessile glands distally.

91. Penstemon subulatus M. E. Jones, Contr. W. Bot. 12: 63. 1908 (as Pentstemon) • Hackberry beardtongue [E]

Stems ascending to erect, 20–60 cm, glaucous, sometimes not. **Leaves** glabrous, glaucous; basal and proximal cauline 28–100 × 2–11 mm, blade oblanceolate to elliptic, base tapered, margins entire, apex obtuse to acute; cauline 3–6 pairs, sessile, (11–)30–80(–120) × 2–9 mm, blade lanceolate to linear or linear-subulate, base clasping to auriculate-clasping, margins entire, apex rounded to acute. **Thyrses** interrupted, cylindric, 6–26 cm, axis glabrous, verticillasters 5–9, cymes 1–4-flowered; proximal bracts lanceolate to linear, 6–33 × 1–4 mm; peduncles and pedicels erect, glabrous. **Flowers:** calyx lobes ovate to lanceolate, 2–4.5 × 1.4–2.1 mm, margins entire or erose, glabrous or ± glandular along margins; corolla scarlet, without nectar guides, nearly radially symmetric, weakly bilabiate, salverform, 18–26 mm, glabrous externally, glabrous internally, tube 4–7 mm, 1.6–3 times as long as calyx lobes, throat slightly inflated, 3–5 mm diam., rounded abaxially; stamens: longer pair reaching orifice, pollen sacs explanate, 0.7–0.9 mm, sutures smooth; staminode 13–14 mm, essentially terete, 0.1 mm diam., tip straight, glabrous, rarely distal 1–2 mm sparsely papillate, papillae yellow, to 0.2 mm; style 10–19 mm. **Capsules** 6–8 × 4–5 mm. *2n* = 16.

Flowering (Feb–)Mar–May. Rocky slopes, mesas, canyons; 500–1300 m; Ariz.

Penstemon subulatus has been documented in Gila, Graham, Maricopa, Mohave, Pinal, and Yuma counties in central Arizona.

92. Penstemon superbus A. Nelson, Proc. Biol. Soc. Wash. 17: 100. 1904 (as Pentstemon) • Superb beardtongue

Penstemon puniceus A. Gray in W. H. Emory, Rep. U.S. Mex. Bound. 2(1): 113. 1859, not Lilja 1843

Stems ascending to erect, 30–140(–160) cm, glaucous. **Leaves** glabrous, glaucous; basal and proximal cauline 40–160 × 14–40 mm, blade spatulate to oblanceolate or elliptic, base tapered, margins entire, apex rounded to obtuse or acute; cauline 3–8 pairs, sessile, (17–)44–115 × (4–)23–45 mm, blade ovate to oblanceolate or lanceolate, base tapered to cordate-clasping to connate-perfoliate, margins entire, apex rounded or obtuse to acute. **Thyrses** interrupted, cylindric, (15–)30–60(–90) cm, axis glabrous, verticillasters 9–12(–20), cymes (1–)3–9-flowered; proximal bracts ovate to lanceolate, 11–55 × 4–30 mm; peduncles and pedicels ascending to erect, glabrous or pedicels sparsely glandular-pubescent. **Flowers:** calyx lobes ovate to lanceolate, 2.8–5 × 1.5–2.1 mm, margins entire or erose, sparsely glandular-pubescent proximally, sometimes glabrous; corolla orangish pink to red, without nectar guides, nearly radially symmetric, weakly bilabiate, tubular-funnelform, 17–22 mm, glandular-pubescent externally, glandular-pubescent internally abaxially, tube 5–7 mm, throat slightly inflated, 4–6 mm diam., rounded abaxially; stamens included, pollen sacs explanate, 1–1.5 mm, sutures smooth; staminode 8–11 mm, flattened distally, 0.6–1 mm diam., tip straight, distal 2–3 mm retrorsely hairy, hairs yellow or whitish, to 1.5 mm; style 10–11 mm. **Capsules** 10–13 × 5–8 mm. $2n = 16$.

Flowering (Mar–)Apr–Jun. Gravelly or rocky canyons, slopes, washes, desert grasslands, pinyon-juniper and oak woodlands; 900–1800 m; Ariz., N.Mex.; Mexico (Chihuahua, Sonora).

Penstemon superbus is known from Cochise, Gila, Graham, Greenlee, and Pima counties, Arizona, Grant and Hidalgo counties, New Mexico, and in adjacent Chihuahua and Sonora, Mexico. The species resembles *P. parryi*; it differs by broader leaves and corollas that are orangish pink to red, more nearly radially symmetric, and without white hairs abaxially in the throats. *Penstemon superbus* also resembles *P. alamosensis*, which is known from New Mexico and Texas to the east of *P. superbus*. The glabrous staminode of *P. alamosensis* distinguishes it from *P. superbus*.

93. Penstemon utahensis Eastwood, Zoë 4: 124. 1893
• Utah beardtongue E

Stems ascending to erect, 15–60 cm, glaucous. **Leaves** glabrous or obscurely scabrous, especially along margins, glaucous; basal and proximal cauline 35–80(–100) × 5–20 mm, blade oblanceolate, base tapered, margins entire, apex obtuse to acute; cauline 2–4 pairs, sessile, 15–75 × 4–26 mm, blade elliptic to lanceolate, rarely ovate, base tapered, margins entire, apex acute. **Thyrses** ± interrupted, ± secund, 11–25 cm, axis glabrous, verticillasters 5–15, cymes 1–4-flowered; proximal bracts lanceolate, 6–23 × 1–4 mm; peduncles and pedicels erect, glabrous. **Flowers:** calyx lobes ovate, 2.5–4(–5) × 1.5–4(–4.6) mm, margins erose, glabrous; corolla red to crimson, with red nectar guides, bilaterally symmetric, bilabiate, tubular-salverform, 17–22(–25) mm, glandular-pubescent externally, especially distally, glandular-pubescent internally, tube 6–9 mm, throat slightly inflated, 6–7 mm diam., rounded abaxially;

stamens included, pollen sacs explanate, 0.6–1.1 mm, sutures smooth; staminode 7–10 mm, flattened distally, 0.4–0.5 mm diam., tip recurved, glabrous, rarely distal 1 mm papillate, papillae golden yellow, to 0.1 mm; style 10–14 mm. **Capsules** 7–10 × 5–7 mm.

Flowering Apr–Jun. Sagebrush shrublands, pinyon-juniper woodlands; 400–2500 m; Ariz., Calif., Colo., Nev., Utah.

Penstemon utahensis is widespread in the Colorado River Basin from western Colorado through northwestern Arizona, southeastern California, southern Nevada, and southern Utah.

The Hopi tribe of northeastern Arizona used *Penstemon utahensis* as a ceremonial and decorative plant (D. E. Moerman 1998).

94. Penstemon wrightii Hooker, Bot. Mag.
77: plate 4601. 1851 (as Pentstemon) • Wright's beardtongue E

Stems ascending to erect, (30–)40–70(–130) cm, glaucous. **Leaves** glabrous, glaucous; basal and proximal cauline 45–125 × 12–45 mm, blade spatulate to oblanceolate or elliptic, base tapered, margins entire, apex rounded to obtuse or acute; cauline 4–6 pairs, sessile or proximals sometimes short-petiolate, (35–)50–70(–100) × 15–40(–50) mm, blade oblong to elliptic, oblanceolate, or ovate, base tapered to auriculate-clasping or cordate-clasping, margins entire, apex rounded to acute. **Thyrses** interrupted, cylindric, 10–72 cm, axis glabrous, verticillasters 6–17, cymes 3–8-flowered; proximal bracts ovate to lanceolate, 8–40 × 3–20 mm; peduncles and pedicels erect, glabrous or pedicels sparsely glandular-pubescent distally. **Flowers:** calyx lobes ovate to lanceolate, 3.5–6 × 1.5–2.2 mm, margins erose, sparsely glandular-pubescent, especially along margins; corolla red to pink, without nectar guides, bilaterally symmetric, bilabiate, tubular-funnelform to ventricose, 14–19 mm, glandular-pubescent externally, glandular-pubescent internally, tube 3–6 mm, throat abruptly inflated, 6–7 mm diam., rounded abaxially; stamens included, pollen sacs explanate, 0.9–1.2 mm, sutures smooth; staminode 8–10 mm, flattened distally, 0.6–1 mm diam., tip straight to recurved, distal 2–7 mm sparsely to densely pilose, hairs yellow, to 1.2 mm; style 8–11 mm. **Capsules** 6–12 × 4–7 mm. $2n = 16$.

Flowering Apr–Jun. Rocky canyons and hillsides; 900–1800 m; Tex.

Penstemon wrightii is known from the mountains of the Trans-Pecos of western Texas. Populations are known from Brewster, Jeff Davis, Presidio, and Reeves counties.

26b.11. Penstemon Schmidel (subg. Penstemon) sect. Glabri (Rydberg) Pennell, Contr. U.S. Natl. Herb. 20: 326, 329. 1920

Penstemon [unranked] *Glabri* Rydberg, Fl. Colorado, 306. 1906 (as Pentstemon); *Penstemon* subg. *Habroanthus* Crosswhite

Herbs. Stems glabrous, retrorsely hairy, canescent, glandular, or glandular-pubescent, rarely glutinous, glaucous or not. **Leaves** basal and cauline, sometimes basal absent or reduced, opposite, leathery or not, glabrous, glabrescent, scabrous, puberulent, retrorsely hairy, or glandular-pubescent, glaucous or not; basal and proximal cauline petiolate, sometimes short-petiolate (*P. gibbensii, P. idahoensis*); cauline sessile or petiolate, blade obovate, ovate, spatulate, oblanceolate, elliptic-lanceolate, lanceolate, elliptic, oblong, or linear, margins entire. **Thyrses** continuous or interrupted, cylindric, subsecund, or secund, axis glabrous, glandular-pubescent, puberulent, or retrorsely hairy, rarely glandular, cymes 1 or 2 per node; peduncles and pedicels ascending or erect, sometimes spreading (*P. comarrhenus, P. cyananthus, P. nudiflorus*). **Flowers:** calyx lobes: margins erose or entire, rarely lacerate (*P. hallii*), ± scarious, glabrous or glandular-pubescent, sometimes glandular, glutinous, or puberulent; corolla white to pink, lavender, blue, violet, or purple, bilaterally symmetric, weakly bilabiate to strongly bilabiate, not personate, funnelform, ventricose, or ventricose-ampliate, glabrous or glandular-pubescent externally, glabrous or hairy internally abaxially, rarely glandular-pubescent internally abaxially or adaxially, throat gradually to abruptly inflated, slightly constricted or not at orifice, rounded to 2-ridged abaxially; stamens included or longer pair reaching orifice or exserted, filaments glabrous, pollen sacs divergent to opposite, navicular, navicular-sigmoid, or sigmoid, rarely subexplanate, dehiscing incompletely proximal ⅕–⅓ indehiscent, rarely completely, connective splitting or not, sides glabrous, papillate, or hairy, sutures denticulate, teeth to 0.1–0.2 mm, sometimes papillate or smooth; staminode included to exserted, flattened distally, 0.2–1.3 mm diam., tip straight to recurved, distal 5–60(–90)% hairy or papillate, hairs or papillae to 1.5 mm, rarely glabrous; style glabrous. **Capsules** glabrous. **Seeds** tan to brown or dark brown, angled to angled-elongate, 0.9–4 mm.

Species 44 (43 in the flora): w United States, n Mexico.

Pollen sacs dehiscing from their distal ends proximally and connectives usually not splitting has traditionally been used as a synapomorphy for subg. *Habroanthus* (F. S. Crosswhite 1967). Within subg. *Habroanthus*, species with blue or violet and tubular or ventricose corollas have been included in sect. *Glabri*; species with red and tubular corollas have been placed in sect. *Elmigera*. Molecular studies (A. D. Wolfe et al. 2006; C. A. Wessinger et al. 2016) revealed that the four red-flowered species of sect. *Elmigera* in the flora area are closely related to blue-flowered species of sect. *Glabri*. *Penstemon galloensis* G. L. Nesom, known from Nuevo León, Mexico, is the only member of sect. *Glabri* occurring strictly outside the flora area.

SELECTED REFERENCES Crosswhite, F. S. 1967. Revision of *Penstemon* section *Habroanthus* (Scrophulariaceae) I: Conspectus. Amer. Midl. Naturalist 77: 1–11. Crosswhite, F. S. 1967b. Revision of *Penstemon* section *Habroanthus* (Scrophulariaceae) II: Series *Speciosi*. Amer. Midl. Naturalist 77: 12–27. Crosswhite, F. S. 1967c. Revision of *Penstemon* section *Habroanthus* (Scrophulariaceae) III: Series *Virgati*. Amer. Midl. Naturalist 77: 28–41. Moorman, M. L. 1982. Systematic Revision of *Penstemon subglaber*, *P. saxosorum*, and *P. mensarum*. M.S. thesis. University of Wyoming.

1. Pollen sac sides glabrous.
  2. Corollas glandular-pubescent externally.
    3. Stems glabrous; leaves glabrous . . . . 111. *Penstemon leiophyllus*
    3. Stems retrorsely hairy or puberulent to pubescent; leaves retrorsely hairy.
      4. Corollas 13–20 mm; pollen sacs 1–1.4 mm, navicular; styles 11–15 mm . . . .
        . . . . 120. *Penstemon parvus*
      4. Corollas 24–30 mm; pollen sacs 2–2.5 mm, sigmoid; styles 20–24 mm . . .137. *Penstemon wardii*

[2. Shifted to left margin.—Ed.]
2. Corollas glabrous externally, rarely glutinous and covered with sand.
  5. Stems glutinous and covered with sand, usually ± fistulose . . . . 96. *Penstemon ammophilus*
  5. Stems glabrous or retrorsely hairy, puberulent, pubescent, canescent, or glandular-pubescent, not glutinous and covered with sand, not fistulose.
    6. Basal and proximal cauline leaf blades linear.
      7. Stems retrorsely hairy; corollas glabrous internally . . . . 126. *Penstemon pseudoputus*
      7. Stems glabrous; corollas sparsely white-lanate internally abaxially . . . 127. *Penstemon putus*
    6. Basal and proximal cauline leaf blades ovate to elliptic, obovate, oblanceolate, lanceolate, or spatulate, rarely linear (*P. virgatus*).
      8. Stems glaucous; thyrses interrupted; peduncles and pedicels spreading to ascending; pollen sacs dehiscing completely . . . . 118. *Penstemon nudiflorus*
      8. Stems slightly or not glaucous; thyrses continuous or interrupted; peduncles and pedicels ascending to erect; pollen sacs dehiscing incompletely or completely.
        9. Thyrse axes glandular-pubescent.
          10. Corollas 14–25 mm; pollen sacs navicular to subexplanate, 1.2–1.4 mm, dehiscing completely . . . . 108. *Penstemon hallii* (in part)
          10. Corollas 25–35 mm; pollen sacs sigmoid, (1.8–)2–3 mm, dehiscing incompletely . . . . 130. *Penstemon speciosus* (in part)
        9. Thyrse axes glabrous or retrorsely hairy.
          11. Pollen sacs dehiscing incompletely, proximal ¹⁄₅–¹⁄₃ indehiscent, connectives not splitting.
            12. Calyx lobes (4–)4.5–11(–13.5) mm; pollen sacs sigmoid, (1.8–)2–3 mm.
              13. Stems glabrous; pollen sacs 2–2.5 mm, sides glabrous; staminodes: distal 5–7 mm ± pilose, hairs yellow, to 1 mm; capsules 8–10 mm. 110. *Penstemon laevis*
              13. Stems glabrous or retrorsely hairy; pollen sacs (1.8–)2–3 mm, sides glabrous or sparsely hispidulous; staminodes glabrous or distal 1–7 mm sparsely pilose, hairs yellow to yellowish orange, to 0.8 mm; capsules 10–16 mm . . . . 130. *Penstemon speciosus* (in part)
            12. Calyx lobes 2.5–5 mm; pollen sacs navicular, navicular-sigmoid, or sigmoid, 1.2–2.2 mm.
              14. Staminodes glabrous; pollen sac sides lanate, rarely glabrous. . . . . 116. *Penstemon navajoa* (in part)
              14. Staminodes pilose distally; pollen sac sides glabrous.
                15. Corollas sparsely yellow- or white-villous internally abaxially; stems glabrous; pollen sacs sigmoid, 1.6–2.2 mm. . . . . 119. *Penstemon pahutensis*
                15. Corollas glabrous internally; stems glabrous or retrorsely hairy; pollen sacs navicular-sigmoid or navicular, 1.3–1.7 mm.
                  16. Stems glabrous or retrorsely hairy; calyx lobes: margins broadly scarious, apices rounded to acute or mucronate; Idaho, Oregon . . . 125. *Penstemon perpulcher*
                  16. Stems retrorsely hairy; calyx lobes: margins narrowly scarious, apices obtuse, sometimes acute to cuspidate; Utah . . . . . 134. *Penstemon tidestromii* (in part)
          11. Pollen sacs dehiscing completely or connectives not splitting.
            17. Staminodes pilose to lanate distally.
              18. Pollen sacs 1.2–1.5 mm, sutures smooth or papillate; staminodes: distal 0.5–2 mm sparsely to moderately pilose; styles 10–14 mm. . . . . 103. *Penstemon deaveri*
              18. Pollen sacs 1.2–2.5(–2.8) mm, sutures denticulate; staminodes: distal 1–8 mm pilose or lanate; styles 11–20 mm.

19. Corollas 22–48 mm, glabrous or sparsely to moderately white-lanate internally abaxially . . . . .107. *Penstemon glaber* (in part)

19. Corollas 13–22 mm, glabrous internally.

    20. Stems glabrous . . . . 121. *Penstemon payettensis*

    20. Stems retrorsely hairy . . . . 134. *Penstemon tidestromii* (in part)

17. Staminodes glabrous.

    21. Cauline leaves 1–5 mm wide; pollen sac sides lanate, hairs to 0.9 mm, rarely glabrous; stems glabrous . . . . .116. *Penstemon navajoa* (in part)

    21. Cauline leaves 3–35(–43) mm wide; pollen sac sides glabrous or sparsely to moderately hirsute, hairs to 0.5 mm; stems glabrous or puberulent to pubescent.

        22. Pollen sac sutures denticulate; cauline leaves (6–)22–35(–43) mm wide . . . . 107. *Penstemon glaber* (in part)

        22. Pollen sac sutures smooth or papillate; cauline leaves 3–17 mm wide.

            23. Stems glabrous; corollas sparsely to densely white-villous internally abaxially; New Mexico . . . . .117. *Penstemon neomexicanus*

            23. Stems glabrous or puberulent; corollas sparsely white-villous internally abaxially, rarely glabrous; Arizona, Colorado, New Mexico, Wyoming . . . . 136. *Penstemon virgatus*

1. Pollen sac sides hairy.

    24. Corollas ± glandular-pubescent or glandular externally.

        25. Pollen sacs: proximal ⅕–¼ indehiscent.

            26. Corollas 18–33 mm, throats 6–11 mm diam.; pollen sacs (1.3–)1.5–2.4 mm; styles 12–22 mm; stems glabrous . . . . .129. *Penstemon scariosus*

            26. Corollas 14–18(–20) mm, throats 5–7 mm diam.; pollen sacs 1–1.7 mm; styles 9–16 mm; stems puberulent, pubescent, retrorsely pubescent, or glabrous.

                27. Thyrse axes glabrous; stems glabrous or ± puberulent proximally . . . . .102. *Penstemon cyanocaulis* (in part)

                27. Thyrse axes glandular-pubescent; stems pubescent to retrorsely hairy, sometimes glabrous proximally, glandular-pubescent distally. . . . . . .106. *Penstemon gibbensii*

        25. Pollen sacs: connectives not splitting.

            28. Stems 5–15(–24) cm; corollas without nectar guides . . . . 135. *Penstemon uintahensis*

            28. Stems (10–)18–90(–130) cm; corollas with violet or reddish purple nectar guides.

                29. Staminodes: distal 7–9 mm sparsely to moderately villous; styles 9–11 mm . . . . .114. *Penstemon mensarum*

                29. Staminodes: distal 2–5 mm sparsely villous; styles 14–18 mm.

                    30. Stems glabrous; corollas glabrous or ± glandular-pubescent internally abaxially; Colorado, Wyoming . . . . 128. *Penstemon saxosorum* (in part)

                    30. Stems glabrous proximally, usually sparsely glandular distally; corollas glabrous internally; Idaho, Utah, Wyoming . . . . 133. *Penstemon subglaber*

    24. Corollas glabrous externally.

        31. Stems glutinous, usually covered with sand.

            32. Corollas 20–28 mm; styles 17–20 mm; pollen sacs 2–2.5 mm, proximal ⅕ indehiscent, connectives not splitting; Montana, Wyoming . . . . 97. *Penstemon caryi* (in part)

            32. Corollas 15–21 mm; styles 12–14 mm; pollen sacs 1.5–2 mm, connectives not splitting; Idaho, Nevada, Utah . . . . 109. *Penstemon idahoensis*

        31. Stems glabrous or retrorsely hairy, puberulent, pubescent, cinereous, or glandular-pubescent.

[33. Shifted to left margin.]

33. Caudices herbaceous, rhizomelike; corollas white to lavender; proximal bracts obovate to oblanceolate; stems 4–9 cm . . . . .104. *Penstemon debilis*
33. Caudices woody; corollas blue to bluish violet, violet, or purple, rarely pinkish blue or lavender; proximal bracts ovate to elliptic, oblong, lanceolate, or linear, rarely oblanceolate; stems 5–85(–120) cm.
    34. Thyrse axes glandular or glandular-pubescent, sometimes only distally, sometimes also retrorsely hairy.
        35. Corollas funnelform, 13–16 mm; styles 8–10 mm. . . . . 123. *Penstemon penlandii*
        35. Corollas ventricose, 14–35 mm; styles 10–23 mm.
            36. Pollen sacs navicular to subexplanate, 1.2–1.4 mm, dehiscing completely; calyx lobe margins erose or lacerate; corolla tubes 2–3 mm. . . . . .108. *Penstemon hallii* (in part)
            36. Pollen sacs navicular, navicular-sigmoid, or sigmoid, 1.1–3 mm, dehiscing incompletely; calyx lobe margins erose; corolla tubes 5–13 mm.
                37. Pollen sacs sigmoid or navicular-sigmoid, sides glabrous or hispidulous to hispid or pubescent, hairs to 0.3 mm; corollas 18–35 mm, tubes 7–13 mm; leaves retrorsely hairy or glabrous; calyx lobes (4–)6–12(–13.5) mm.
                    38. Proximal bracts ovate to elliptic; pollen sacs: connectives not splitting; stems 7–27 cm. . . . . .95. *Penstemon absarokensis* (in part)
                    38. Proximal bracts lanceolate, rarely ovate or linear; pollen sacs: proximal $\frac{1}{5}$–$\frac{1}{3}$ indehiscent, connectives not splitting; stems 7–80(–90) cm.
                        39. Pollen sacs 1.8–2.2 mm, sides sparsely hispid; stems retrorsely hairy; leaves retrorsely hairy . . . . .112. *Penstemon lemhiensis*
                        39. Pollen sacs (1.8–)2–3 mm, sides glabrous or sparsely hispidulous; stems glabrous or retrorsely hairy to pubescent; leaves glabrous or retrorsely hairy . . . . .130. *Penstemon speciosus* (in part)
                37. Pollen sacs navicular, sides hispid, pubescent, or villous, hairs to 0.3–1 mm; corollas 17–33 mm, tubes 5–11 mm; leaves glabrous; calyx lobes 3.5–8 mm.
                    40. Pollen sac sides hispid or pubescent, hairs to 0.3 mm . . . . 128. *Penstemon saxosorum* (in part)
                    40. Pollen sac sides pubescent to villous, hairs to 0.7–1 mm.
                        41. Corollas 18–25 mm, tubes 5–7 mm; styles 14–17 mm; staminodes glabrous or distal 1–3 mm sparsely pilose, hairs yellow or white, to 0.5 mm. . . . . 99. *Penstemon compactus* (in part)
                        41. Corollas 27–33 mm, tubes 9–11 mm; styles 17–21 mm; staminodes: distal 1–6 mm sparsely lanate, hairs yellowish, to 0.8 mm . . . . 115. *Penstemon moriahensis* (in part)
    34. Thyrse axes glabrous, retrorsely hairy, or puberulent, not glandular or glandular-pubescent.
        42. Thyrse axes moderately to densely retrorsely hairy; corollas funnelform. . . .105. *Penstemon fremontii*
        42. Thyrse axes glabrous, puberulent, glandular-pubescent, or ± glandular; corollas ventricose, rarely ventricose-ampliate or funnelform.
            43. Pollen sac sides villous to lanate, rarely moderately pubescent, hairs to 0.7–2.5 mm.
                44. Corollas sparsely white-villous internally abaxially; staminodes glabrous . . . . .116. *Penstemon navajoa* (in part)
                44. Corollas glabrous internally; staminodes glabrous or distal 1–6 mm pilose, villous, or lanate.
                    45. Leaves: basal and proximal cauline blades oblanceolate to linear, cauline blades linear; calyx lobes (6–)7–12 mm. . . . . .97. *Penstemon caryi* (in part)
                    45. Leaves: basal and proximal cauline blades oblanceolate to obovate, sometimes lanceolate, cauline blades oblanceolate to lanceolate or linear; calyx lobes (2.2–)3–8(–9) mm.

46. Pollen sac sides villous or lanate, hairs to 2.5 mm; cauline leaf blades oblanceolate to linear; corollas blue to pinkish blue or lavender; thyrses interrupted, secund, sometimes subsecund. . . . . .98. *Penstemon comarrhenus*

46. Pollen sac sides pubescent, villous, or lanate, hairs to 1.8 mm; cauline leaf blades oblanceolate to lanceolate or linear; corollas blue, lavender, violet, or purple; thyrses continuous or interrupted, cylindric or secund.

    47. Cauline leaf blades oblanceolate to lanceolate, rarely linear, apices obtuse to acute; proximal bracts oblong to lanceolate, apices acute to acuminate; peduncles to 7 mm, pedicels 1–11 mm.

        48. Corollas 18–25 mm, tubes 5–7 mm; styles 14–17 mm; staminodes glabrous or distal 1–3 mm sparsely pilose, hairs yellow or white, to 0.5 mm . . . . . . 99. *Penstemon compactus* (in part)

        48. Corollas 27–33 mm, tubes 9–11 mm; styles 17–21 mm; staminodes: distal 1–6 mm sparsely lanate, hairs yellowish, to 0.8 mm . . . . 115. *Penstemon moriahensis* (in part)

    47. Cauline leaf blades lanceolate to linear, apices acute or acuminate; proximal bracts lanceolate to linear, apices acuminate; peduncles to 20–27 mm, pedicels 1–20 mm.

        49. Calyx lobes (4.5–)6–8(–9) × 2–3 mm, apices acuminate to caudate; pollen sac sides sparsely to densely lanate, hairs to 1 mm; staminodes: distal (2–)4–5 mm sparsely to moderately villous, hairs yellow or white, to 1 mm . . . . 131. *Penstemon strictiformis*

        49. Calyx lobes (2.5–)3–5 × 1.3–2.5 mm, apices acute, sometimes acuminate or obtuse; pollen sac sides moderately to densely lanate to villous, hairs to 1.8 mm; staminodes glabrous or distal 1–5 mm sparsely villous, hairs yellow or orange, to 0.7 mm . . . . 132. *Penstemon strictus*

[43. Shifted to left margin.—Ed.]

43. Pollen sac sides hispidulous, hispid, pubescent, hirsute, or pilose, hairs to 0.1–1 mm, rarely glabrous.

  50. Pollen sacs navicular.

    51. Corollas 24–48 mm, throats 8–13(–18) mm diam.; pollen sacs 1.4–2.5(–2.8) mm, connectives not splitting . . . . .107. *Penstemon glaber* (in part)

    51. Corollas 12–25(–30) mm, throats 4–11 mm diam.; pollen sacs 1.2–2(–2.2) mm, proximal ⅕–⅓ indehiscent, rarely only connectives not splitting.

        52. Staminodes reaching orifices or slightly exserted, distal 3–5 mm villous, hairs yellow, to 1 mm; stems erect; corolla tubes 5–8 mm, throats 6–8 mm diam. . . . . 128. *Penstemon saxosorum* (in part)

        52. Staminodes included or reaching orifices, distal 5–9 mm pilose to villous or lanate, hairs orange, yellow, or white, to 1 mm; stems decumbent, ascending, or erect; corolla tubes 5–11 mm, throats 4–11 mm diam.

            53. Cauline leaf blades ovate or oblanceolate to lanceolate, bases cordate-clasping to clasping; calyx lobe apices acute to acuminate or cuspidate; stems decumbent to erect, 5–70(–90) cm . . . . 100. *Penstemon cyananthus*

            53. Cauline leaf blades oblanceolate to lanceolate or linear, bases tapered to truncate; calyx lobe apices acuminate, sometimes caudate; stems decumbent to ascending, 8–18(–25) cm . . . . 122. *Penstemon paysoniorum*

  50. Pollen sacs sigmoid or navicular-sigmoid.

    54. Calyx lobes obovate to ovate, 2.5–4.5 mm, apices mucronate, sometimes acute to acuminate . . . . 101. *Penstemon cyaneus*

    54. Calyx lobes ovate to lanceolate, rarely obovate, 3.5–12(–13.5) mm, apices acute to acuminate, cuspidate, or caudate.

[55. Shifted to left margin.—Ed.]

55. Calyx lobes (4–)6–12(–13.5) mm, glandular or glabrous; pollen sacs (1.8–)2–3 mm.
  56. Proximal bracts ovate to elliptic; pollen sacs: connectives not splitting; stems 7–27 cm
    . . . . .95. *Penstemon absarokensis* (in part)
  56. Proximal bracts lanceolate, rarely ovate or linear; pollen sacs: proximal ⅓ indehiscent,
    connectives not splitting; stems 7–65(–90) cm. . . . . .130. *Penstemon speciosus* (in part)
55. Calyx lobes 3.5–8.5 mm, glabrous; pollen sacs 1.2–2.7(–2.8) mm.
  57. Corollas 14–18(–20) mm; pollen sacs 1.2–1.7 mm, sides sparsely to moderately pilose,
    hairs to 0.5(–0.7) mm; calyx lobes 3.5–4.5 mm. . . . . .102. *Penstemon cyanocaulis* (in part)
  57. Corollas (20–)24–35 mm; pollen sacs 1.8–2.4(–2.8) mm, sides hirsute or hispid, hairs
    to 0.3 mm; calyx lobes 4.5–8.5 mm.
    58. Corollas (20–)24–30(–32) mm; staminodes: distal 0.5–1 mm moderately pilose,
      hairs to 0.8 mm, remaining distal 3–4 mm with scattered yellow hairs; cymes
      1- or 2(or 3)-flowered . . . . 113. *Penstemon longiflorus*
    58. Corollas 29–35 mm; staminodes: distal 4–6 mm ± uniformly sparsely pilose, hairs
      to 1 mm; cymes 1–6-flowered . . . . .124. *Penstemon pennellianus*

95. **Penstemon absarokensis** Evert, Madroño 31: 140, fig. 1. 1984 • Absaroka Range beardtongue C E

Stems ascending to erect, 7–27 cm, glabrous or ± puberulent distally or at nodes, not glaucous. **Leaves** essentially cauline, basal leaves absent or reduced, ± leathery, glabrous, not glaucous; basal and proximal cauline 12–90(–105) × 3–23 mm, blade elliptic-lanceolate to oblanceolate, base tapered, margins entire, apex obtuse to acute; cauline 2–4 pairs, petiolate or sessile, 35–95(–107) × 11–30 mm, blade elliptic-lanceolate to oblanceolate, base tapered to truncate, apex obtuse to acute. **Thyrses** continuous, ± secund, 3–15 cm, axis glabrous or ± glandular, verticillasters 1–7, cymes 2–8-flowered, 2 per node; proximal bracts ovate to elliptic, 16–85 × 7–31 mm; peduncles and pedicels glabrous or ± glandular, peduncles to 15 mm, pedicels 1–3 mm. **Flowers:** calyx lobes ovate to lanceolate, 6.5–12 × 2.8–4.5 mm, margins erose, apex acuminate to caudate, ± glandular; corolla blue to violet, with violet nectar guides, ventricose, 18–33 mm, glabrous externally, glabrous internally, tube 7–9 mm, throat abruptly inflated, slightly constricted at orifice, 7–9 mm diam., rounded abaxially; stamens included or longer pair reaching orifice, pollen sacs divergent to nearly opposite, navicular-sigmoid, (1.8–)2.2–2.6 mm, dehiscing incompletely, connective not splitting, sides sparsely to moderately hispidulous or pubescent, hairs white, to 0.3 mm; sutures denticulate, teeth to 0.1 mm; staminode 13–17 mm, included, 0.5–0.9 mm diam., tip straight, glabrous or distal 1–3 mm pubescent, hairs yellow, to 0.5 mm; style 17–20 mm. **Capsules** 8–12 × 5–8 mm.

Flowering Jun–Jul. Volcanic talus slopes; of conservation concern; 1800–3000 m; Wyo.

Penstemon absarokensis is known from the Absaroka Range of northwestern Wyoming (Fremont and Park counties).

96. **Penstemon ammophilus** N. H. Holmgren & L. M. Shultz, Brittonia 34: 381, fig. 1. 1982 • Canaan Mountain beardtongue C E

Stems decumbent to ascending, 5–26 cm, usually ± fistulose, glutinous and covered with sand, not glaucous. **Leaves** basal and cauline, not leathery, glutinous and covered with sand, not glaucous; basal and proximal cauline 18–80 × 4–15 mm, blade oblanceolate, base tapered, margins entire, crisped, apex obtuse to acute; cauline 2–5 pairs, sessile, 15–75 × 2–8 mm, blade oblanceolate to lanceolate, base tapered to truncate, margins crisped, apex rounded or obtuse to acute. **Thyrses** continuous, cylindric, 4–15 cm, axis glutinous and covered with sand, verticillasters 4–10, cymes 1–6-flowered, 2 per node; proximal bracts lanceolate, 14–45 × 2–5(–8) mm; peduncles and pedicels glutinous and covered with sand. **Flowers:** calyx lobes lanceolate, 5–6.5 × 1–2 mm, glutinous and covered with sand; corolla lavender to blue-lavender, with violet nectar guides, weakly ventricose, 14–17 mm, glabrous externally, usually glutinous and covered with sand distally, especially on lobes, glabrous internally, tube 5–6 mm, throat gradually inflated, not constricted at orifice, 5–6 mm diam., slightly 2-ridged abaxially; stamens: longer pair reaching orifice, pollen sacs opposite, navicular, 0.9–1.4 mm, dehiscing completely or incompletely, connective splitting or not, sides glabrous, sutures papillate; staminode: 8–9 mm, included, 0.3–0.5 mm diam., tip straight, distal 4–5 mm sparsely papillate, papillae violet, to 0.2 mm; style 10–12 mm. **Capsules** 3.5–6.5 mm.

Flowering May–Jun. Sand dunes, ponderosa pine forests, mixed shrublands; of conservation concern; 1800–2200 m; Utah.

Penstemon ammophilus is known from Garfield, Kane, and Washington counties. Populations occur on aeolian sand, derived from the Navajo Sandstone. The fistulose stems of *P. ammophilus* are unique in *Penstemon*.

97. Penstemon caryi Pennell, Contr. U.S. Natl. Herb. 20: 354. 1920 • Cary's beardtongue E

Stems ascending to erect, 9–35 cm, glabrous, often glutinous and covered with sand, not glaucous. **Leaves** basal and cauline, ± leathery or not, glabrous, often glutinous and covered with sand, not glaucous; basal and proximal cauline 18–105 × 1–10 mm, blade oblanceolate to linear, base tapered, margins entire, apex obtuse to acute; cauline (2 or)3 or 4 pairs, petiolate or sessile, 28–110 × 1–6 mm, blade linear, base tapered to truncate, apex acute. **Thyrses** continuous, secund, 3–26 cm, axis glabrous, sometimes glutinous and covered with sand, verticillasters 3–6(–9), cymes 1–3-flowered, 2 per node; proximal bracts linear, 22–115 × 1.5–4 mm; peduncles and pedicels glabrous, sometimes glutinous and covered with sand. **Flowers:** calyx lobes lanceolate, (6–)7–12 × 2.8–4 mm, glabrous, glutinous and sometimes covered with sand; corolla blue to violet or purple, with violet nectar guides, ventricose, 20–28 mm, glabrous externally, glabrous internally, tube 8–9 mm, throat abruptly inflated, slightly constricted at orifice, (7–)10–12 mm diam., slightly 2-ridged abaxially; stamens: longer pair reaching orifice, pollen sacs divergent, navicular-sigmoid, 2–2.5 mm, dehiscing incompletely, proximal ⅕ indehiscent, connective not splitting, sides moderately villous or lanate, hairs white, to 1 mm, sutures denticulate, teeth to 0.1 mm; staminode: 11–18 mm, reaching orifice or slightly exserted, 0.6–0.8 mm diam., tip straight, distal 1–2 mm sparsely to moderately pilose, hairs yellow, to 1 mm; style 17–20 mm. **Capsules** (5–)7–11 × (4–)5–8 mm.

Flowering May–Jul(–Aug). Rock outcrops, rocky soils, sagebrush shrublands, juniper, Douglas fir, and limber pine woodlands; 1600–2900 m; Mont., Wyo.

Penstemon caryi is known from the Bighorn Mountains of Wyoming (Big Horn, Sheridan, and Washakie counties) and the Pryor Mountains of Montana (Carbon County).

98. Penstemon comarrhenus A. Gray, Proc. Amer. Acad. Arts 12: 81. 1876 (as Pentstemon) • Dusty beardtongue E

Stems ascending to erect, (18–) 40–80(–120) cm, glabrous, sometimes retrorsely hairy proximally, not glaucous. **Leaves** basal and cauline, ± leathery or not, glabrous or sparsely retrorsely hairy, rarely densely retrorsely hairy, not glaucous; basal and proximal cauline 40–80(–120) × 7–20(–30) mm, blade obovate to oblanceolate, base tapered, margins entire, apex obtuse to acute; cauline 5–7(or 8) pairs, sessile, 38–110 × 1–10 mm, blade oblanceolate to linear, base tapered, apex obtuse to acute. **Thyrses** erupted, secund, sometimes subsecund, (12–)15–32 cm, axis glabrous, verticillasters 6–12, cymes 1–2(–5)-flowered, 2 per node; proximal bracts linear, (6–)26–84 × 1–4 mm; peduncles and pedicels glabrous. **Flowers:** calyx lobes ovate, (2.2–)3.5–6(–7) × 2.1–2.7 mm, glabrous; corolla blue to pinkish blue or lavender, with violet nectar guides, ventricose, 25–35(–38) mm, glabrous externally, glabrous internally, tube 8–11(–14) mm, throat abruptly inflated, slightly constricted at orifice, 9–13 mm diam., slightly 2-ridged abaxially; stamens: longer pair exserted, pollen sacs divergent to nearly opposite, navicular-sigmoid, 2–2.5(–2.8) mm, dehiscing incompletely, proximal ⅕ indehiscent, connective not splitting, sides densely villous or lanate, hairs white, to 2.5 mm, sutures smooth, papillate, or denticulate, teeth to 0.1 mm; staminode 12–17 mm, included, 0.8–1 mm diam., tip straight, glabrous or distal 1–2 mm sparsely pilose, hairs white, to 0.5 mm; style 16–25 mm. **Capsules** (7–)10–15 × 4–5 mm.

Flowering Jun–Aug. Pinyon-juniper, Gambel oak, and ponderosa pine woodlands; (1000–)1600–2800 m; Ariz., Colo., Nev., N.Mex., Utah.

Penstemon comarrhenus usually is distinguished from *P. strictus* by its paler blue corollas, more diffuse inflorescences, and more densely lanate anthers. A putative hybrid between *P. comarrhenus* and *P. barbatus* var. *trichander* has been collected in La Plata County, Colorado (*R. E. Umber, B. E. Nelson & T. Davis 999b*, KANU, RM).

99. **Penstemon compactus** (D. D. Keck) Crosswhite, Amer. Midl. Naturalist 77: 6. 1967 • Bear River Range beardtongue [C] [E]

Penstemon cyananthus Hooker subsp. *compactus* D. D. Keck, Amer. Midl. Naturalist 23: 615. 1940

Stems decumbent to ascending, 8–26 cm, glabrous, not glaucous. **Leaves** basal and cauline, ± leathery or not, glabrous, not glaucous; basal and proximal cauline (15–)20–110 × (3–)6–14(–25) mm, blade oblanceolate, base tapered, margins entire, apex obtuse to acute; cauline (1 or)2 or 3 pairs, sessile or proximals short-petiolate, 13–55 × 5–14 mm, blade oblanceolate to lanceolate, base tapered, apex obtuse to acute. **Thyrses** continuous, cylindric to secund, 2–13 cm, axis glabrous or glandular-pubescent, verticillasters 2 or 3(–5), cymes 1–7-flowered, 2 per node; proximal bracts lanceolate to oblong, 10–45 × 4–18 mm, apex acute; peduncles and pedicels glabrous or ± glandular-pubescent, peduncles to 7 mm, pedicels 1–4 mm. **Flowers:** calyx lobes lanceolate, 5.5–8 × 1.5–2 mm, margins erose, glabrous or ± glandular-pubescent; corolla blue to violet, without nectar guides, ventricose, 18–25 mm, glabrous externally, glabrous internally, tube 5–7 mm, throat gradually inflated, not constricted at orifice, 6–10 mm diam., 2-ridged abaxially; stamens: longer pair exserted, pollen sacs divergent, navicular, 1.3–1.8 mm, dehiscing incompletely, proximal ⅓ indehiscent, connective not splitting, sides moderately pubescent to villous, hairs white, to 0.7 mm, sutures denticulate, teeth to 0.1 mm; staminode 10–12 mm, included, 0.5–0.6 mm diam., tip straight, glabrous or distal 1–3 mm sparsely pilose, hairs yellow or white, to 0.5 mm; style 14–17 mm. **Capsules** 5–7.5 × 4–5 mm. *2n* = 16.

Flowering Jun–Aug. Rocky, subalpine openings; of conservation concern; 2000–2900 m; Idaho, Utah.

Penstemon compactus is found in the northern Wasatch Range in Franklin County, Idaho, and Cache County, Utah.

100. **Penstemon cyananthus** Hooker, Bot. Mag. 75: plate 4464. 1849 (as Pentstemon) • Wasatch beardtongue [E]

Stems decumbent to erect, 5–70(–90) cm, glabrous or sparsely to densely puberulent, not glaucous. **Leaves** basal and cauline, ± leathery or not, glabrous or basal ones sparsely to densely puberulent, not glaucous; basal and proximal cauline (20–)45–160 × 3–40 (–55) mm, blade obovate to oblanceolate, base tapered,

margins entire, apex obtuse to acute; cauline 1–6 pairs, sessile or proximals short- to long-petiolate, 16–100 × 5–45(–60) mm, blade ovate or oblanceolate to lanceolate, base cordate-clasping to clasping, apex obtuse to acute. **Thyrses** continuous or interrupted, cylindric to subsecund, 2–30 cm, axis glabrous, verticillasters 2–12, cymes 3–7-flowered, 2 per node; proximal bracts ovate to lanceolate, 7–80 × (1–)3–35 mm; peduncles and pedicels glabrous, peduncles to 20 mm, pedicels 1–9 mm. **Flowers:** calyx lobes ovate to lanceolate, 3–6.5 × 1.3–2.6 mm, apex acute to acuminate or cuspidate, glabrous; corolla blue to violet, without nectar guides, ventricose, 12–25(–28) mm, glabrous externally, glabrous internally, tube 6–11 mm, throat gradually inflated, not constricted at orifice, 6–11 mm diam., 2-ridged abaxially; stamens included or longer pair reaching orifice, pollen sacs divergent, navicular, 1.2–2(–2.2) mm, dehiscing incompletely, proximal ¼–⅓ indehiscent, connective not splitting, sides hispid, hairs white, to 0.5 mm, sutures denticulate, teeth to 0.1 mm; staminode 11–14 mm, included or reaching orifice, 0.3–0.6 mm diam., tip straight to recurved, distal 6–9 mm sparsely to moderately pilose to lanate, hairs orange, yellow, or white, to 1 mm; style 9–16 mm. **Capsules** 7–15 × 5–10 mm.

Varieties 3 (3 in the flora): w United States.

Penstemon compactus and *P. longiflorus* have been treated as elements of the *P. cyananthus* complex (F. W. Pennell 1920b; D. D. Keck 1940b). All of these taxa occur from the Wasatch Range northward into the Central Rocky Mountains.

1. Cauline leaf blades lanceolate to oblanceolate; calyx lobes ovate to lanceolate, apices acute to cuspidate, rarely acuminate. . . . .
. . . . 100c. *Penstemon cyananthus* var. *subglaber*
1. Cauline leaf blades ovate to lanceolate; calyx lobes lanceolate, rarely ovate, apices acuminate, rarely acute.
  2. Verticillasters 4–12; thyrses 4–30 cm; stems ascending to erect, 20–70(–90) cm; corollas 16–25(–28) mm. . . . .
. . . . . .100a. *Penstemon cyananthus* var. *cyananthus*
  2. Verticillasters 2 or 3; thyrses 2–6 cm; stems decumbent to ascending, 5–25(–32) cm; corollas 12–16(–20) mm . . . .
. . . . 100b. *Penstemon cyananthus* var. *judyae*

100a. Penstemon cyananthus Hooker var.
   cyananthus [E]

Stems ascending to erect, 20–70 (–90) cm. **Cauline leaves:** blade ovate to lanceolate. **Thyrses** 4–30 cm, verticillasters 4–12. **Flowers:** calyx lobes lanceolate, rarely ovate, apex acuminate, rarely acute; corolla 16–25 (–28) mm.

Flowering May–Jul. Sandy or gravelly slopes, sagebrush shrublands; 1300–3200 m; Idaho, Utah, Wyo.

Variety *cyananthus* passes into var. *subglaber* in southeastern Idaho and west-central Wyoming; some specimens from that area are difficult to assign to variety with confidence.

100b. Penstemon cyananthus Hooker var. **judyae**
   N. D. Atwood in S. L. Welsh et al., Utah Fl. ed. 3,
   622. 2003 [E]

Stems decumbent to ascending, 5–25(–32) cm. **Cauline leaves:** blade ovate. **Thyrses** 2–6 cm, verticillasters 2–3. **Flowers:** calyx lobes lanceolate, apex acuminate; corolla 12–16(–20) mm.

Flowering Jun–Jul. Alpine meadows; 2900–3300 m; Utah.

Variety *judyae* is known primarily from the Wasatch Range from the vicinity of Lake Catherine (Salt Lake County) south to Mt. Timpanogos and Provo Peak (Utah County), and Mt. Nebo (Juab County). It also occurs on Duchesne Ridge at the western end of the Uinta Mountains (Wasatch County).

100c. Penstemon cyananthus Hooker var. **subglaber**
   (A. Gray) N. H. Holmgren, Brittonia 31: 105.
   1979 [E]

Penstemon fremontii Torrey & A. Gray var. *subglaber* A. Gray in A. Gray et al., Syn. Fl. N. Amer. 2(1): 262. 1878 (as Pentstemon fremonti); *P. cyananthus* subsp. *subglaber* (A. Gray) Pennell

Stems ascending to erect, 22–70 (–90) cm. **Cauline leaves:** blade lanceolate to oblanceolate. **Thyrses** 5–21 cm, verticillasters 4–11. **Flowers:** calyx lobes ovate to lanceolate, apex acute to cuspidate, rarely acuminate; corolla 16–25(–28) mm. $2n = 16$.

Flowering May–Jul. Clayey soils, sagebrush shrublands; 1400–2900 m; Idaho, Mont., Utah, Wyo.

Variety *subglaber* is easily confused with individuals of *Penstemon subglaber* with sparsely glandular-pubescent corollas; calyx lobe shape and anther morphology usually distinguish the two reliably. *Penstemon holmgrenii* S. L. Clark is an invalid name that applies here.

101. Penstemon cyaneus Pennell, Contr. U.S. Natl.
   Herb. 20: 351. 1920 • Dark blue beardtongue [E]

Stems ascending to erect, (17–)30–80 cm, glabrous, not or ± glaucous. **Leaves** basal and cauline, ± leathery or not, glabrous, ± glaucous; basal and proximal cauline 50–170 × 8–25(–35) mm, blade oblanceolate to elliptic, base tapered, margins entire, apex obtuse to acute; cauline 3–7 pairs, sessile or proximals short-petiolate, (15–)30–90(–120) × (4–)8–25 mm, blade lanceolate to ovate, base cordate-clasping or proximals tapered, apex acute. **Thyrses** continuous or interrupted, secund, (3–)7–27 cm, axis glabrous, verticillasters (3–)5–14, cymes (2 or)3–6-flowered, 2 per node; proximal bracts lanceolate, 15–65 × 2–26 mm; peduncles and pedicels glabrous, peduncles to 18 mm, pedicels 3–18 mm. **Flowers:** calyx lobes obovate to ovate, 2.5–4.5 × 2–3.4(–3.8) mm, apex mucronate, sometimes acute to acuminate, glabrous; corolla blue to violet or purple, without nectar guides, ventricose, 24–30(–34) mm, glabrous externally, glabrous internally, tube 7–10 mm, throat gradually inflated, not constricted at orifice, 8–11 mm diam., slightly 2-ridged abaxially; stamens: longer pair reaching orifice, pollen sacs divergent, sigmoid, 1.8–2.7(–3) mm, dehiscing incompletely, proximal ⅕ indehiscent, connective not splitting, sides hispid to hispidulous, hairs white or tan, to 0.2 mm, sutures denticulate, teeth to 0.1 mm; staminode 14–17 mm, included, 0.5–0.7 mm diam., tip straight, distal 4–7 mm sparsely pilose, hairs yellow, to 1 mm; style 17–23 mm. **Capsules** 9–13 × 5–7 mm.

Flowering Jun–Aug. Gravelly or sandy hillsides, sagebrush shrublands; 1400–2900 m; Idaho, Mont., Wyo.

Penstemon cyaneus occurs from the vicinity of Yellowstone National Park westward into south-central Idaho.

102. **Penstemon cyanocaulis** Payson, Bot. Gaz. 60: 380. 1915 (as Pentstemon) • Blue-stem beardtongue [E]

Stems ascending to erect, (16–)20–60 cm, glabrous or ± puberulent proximally, ± glaucous. **Leaves** basal and cauline, ± leathery or not, glabrous or proximals puberulent proximally or along midveins, ± glaucous; basal and proximal cauline (30–)50–95 × 10–15(–28) mm, blade spatulate to oblanceolate, base tapered, margins entire, usually undulate, apex rounded to obtuse; cauline 2–4 pairs, sessile, 20–73 × 4–15 mm, blade lanceolate to elliptic, base truncate to clasping, margins crisped or undulate, apex acute. **Thyrses** interrupted, secund to subsecund, (4–)8–31 cm, axis glabrous, verticillasters 5–9(–12), cymes (1 or)2–5-flowered, 2 per node; proximal bracts ovate to lanceolate, 15–50 × 3–23 mm; peduncles and pedicels glabrous, peduncles to 18 mm, pedicels 1–7 mm. **Flowers:** calyx lobes ovate, 3.5–4.5 × 1.4–2 mm, apex acute to acuminate, glabrous; corolla blue to violet, with violet nectar guides, funnelform to weakly ventricose, 14–18(–20) mm, glabrous or ± glandular externally, glabrous internally, tube 5–8 mm, throat gradually inflated, not constricted at orifice, 5–7 mm diam., rounded abaxially; stamens: longer pair reaching orifice, pollen sacs divergent, navicular-sigmoid, 1.2–1.7 mm, dehiscing incompletely, proximal ¼ indehiscent, connective not splitting, sides sparsely to moderately pilose, hairs white, to 0.5(–0.7) mm, sutures denticulate, teeth to 0.1 mm; staminode: 9–11 mm, reaching orifice, 0.2–0.4 mm diam., tip straight, distal 2–6 mm sparsely pilose, hairs golden yellow, to 0.6 mm; style 9–12 mm. **Capsules** 6–10 × 4–5 mm.

Flowering Apr–Jun. Rocky slopes, pinyon-juniper, oak, and oak-juniper woodlands; 1300–2700 m; Colo., Utah.

Penstemon cyanocaulis is known from Mesa, Montrose, and San Miguel counties, Colorado, and Carbon, Emery, Grand, and San Juan counties, Utah. It has crisped or undulate leaf margins that help distinguish it in the field.

103. **Penstemon deaveri** Crosswhite, Amer. Midl. Naturalist 77: 6, 38. 1967 • Deaver's or Mount Graham beardtongue [E]

Penstemon hallii A. Gray var. *arizonicus* A. Gray in A. Gray et al., Syn. Fl. N. Amer. 2(1): 263. 1878 (as Pentstemon), not *P. arizonicus* A. Heller 1899

Stems ascending to erect, 16–60 cm, retrorsely hairy, not glaucous. **Leaves** basal and cauline, not leathery, glabrous or proximals sometimes puberulent proximally, not glaucous; basal and proximal cauline 35–110 × 5–22 mm, blade spatulate to oblanceolate, base tapered, margins entire, apex rounded to obtuse; cauline 3–7 pairs, sessile, 20–95 × 4–16 mm, blade oblanceolate to elliptic or oblong, base tapered, apex obtuse to acute. **Thyrses** continuous, cylindric, 4–13 cm, axis puberulent, verticillasters 3–8, cymes (1 or)2–4-flowered, (1 or)2 per node; proximal bracts lanceolate, proximals 7–55 × 1–15 mm; peduncles and pedicels puberulent. **Flowers:** calyx lobes lanceolate to oblanceolate, 3–5.5 × 1.5–2.2 mm, glabrous or sparsely puberulent; corolla lavender to violet, with violet nectar guides, ventricose, 16–25 mm, glabrous externally, sparsely white-lanate internally abaxially, tube 4–5 mm, throat gradually inflated, not constricted at orifice, 6–8 mm diam., rounded abaxially; stamens: longer pair reaching orifice, pollen sacs opposite, navicular, 1.2–1.5 mm, dehiscing completely or incompletely, connective splitting or not, sides glabrous, sutures smooth or papillate; staminode 9–14 mm, exserted, 0.7–1 mm diam., tip straight to recurved, distal 0.5–2 mm sparsely to moderately pilose, hairs yellow, to 0.8 mm; style 10–14 mm. **Capsules** 9–12 × 4.5–8 mm.

Flowering Jul–Sep. Rocky slopes, pine forests, alpine meadows; 2000–3400 m; Ariz., N.Mex.

Penstemon deaveri is found in the Pinaleño and White mountains of Arizona and in the Datil, Mogollon, San Mateo, and Zuni mountains of New Mexico. The species has been documented in Apache, Graham, and Greenlee counties, Arizona, and Catron, Cibola, and Socorro counties, New Mexico. The bearded staminodes and glabrous and generally broader leaves of *P. deaveri* distinguish it from *P. virgatus*; rare individuals of *P. deaveri* have glabrous staminodes.

104. Penstemon debilis O'Kane & J. L. Anderson, Brittonia 39: 412, fig. 1. 1987 • Parachute beardtongue C E

Caudex herbaceous, rhizome-like. **Stems** ascending, 4–9 cm, puberulent to glandular-pubescent, sometimes glabrate proximally, not glaucous. **Leaves** cauline, leathery, glabrous, glaucous; cauline 1 or 2, sessile or short-petiolate, 7–28 (–32) × 3–12 mm, blade obovate to oblanceolate or elliptic, base tapered, with a white band 0.2 mm wide, apex rounded to obtuse or acute. **Thyrses** continuous, ± secund, 1–7 cm, axis puberulent to glandular-pubescent, verticillasters 2–6, cymes 1- or 2(or 3)-flowered, 2 per node; proximal bracts obovate to oblanceolate, 14–32 × 8–14 mm; peduncles and pedicels puberulent to glandular-pubescent. **Flowers:** calyx lobes lanceolate, 5.5–7 × 1.5–2.2 mm, glabrous; corolla white to lavender, without nectar guides, funnelform to weakly ventricose, (14–)17–20 mm, glabrous externally, glabrous internally, tube 6–7 mm, throat gradually inflated, not constricted at orifice, 6–8 mm diam., slightly 2-ridged abaxially; stamens included, pollen sacs divergent, navicular, 0.9–1.4 mm, dehiscing incompletely, proximal ⅕–¼ indehiscent, connective not splitting, sides sparsely hispid, hairs white, to 0.4 mm, sutures denticulate, teeth to 0.1 mm; staminode 9–10 mm, included or slightly exserted, 0.5–0.6 mm diam., tip straight, 7–9 mm sparsely pilose, hairs yellow, to 1 mm; style 9–11 mm. **Capsules** 3.5–7.5 × 2.5–5 mm.

Flowering Jun–Jul. Oil-shale talus slopes; of conservation concern; 2400–2700 m; Colo.

Penstemon debilis is known from the vicinity of Mt. Callahan, Mt. Logan, and the Roan Plateau in Garfield County. Plants grow on steep, sparsely vegetated oil-shale talus slopes of the Parachute Creek Member of the Green River Formation (S. L. O'Kane 1988; S. Spackman et al. 1997).

SELECTED REFERENCES McMullen, A. L. 1998. Factors Concerning the Conservation of a Rare Shale Endemic Plant: The Reproductive Ecology and Edaphic Characteristics of *Penstemon debilis* (Scrophulariaceae). M.S. thesis. Utah State University. Spackman, S. et al. 1997. Field Survey and Protection Recommendations for the Globally Imperiled Parachute Penstemon, *Penstemon debilis* O'Kane and Anderson. Fort Collins.

105. Penstemon fremontii Torrey & A. Gray, Proc. Amer. Acad. Arts 6: 60. 1862 (as Pentstemon fremonti) • Fremont's beardtongue E

Stems ascending to erect, 8–25 (–40) cm, densely retrorsely hairy, not glaucous. **Leaves** basal and cauline, or basal absent or reduced, not leathery, retrorsely hairy or glabrous, sometimes glabrescent, not glaucous; basal and proximal cauline 30–75(–100) × 2–18 (–27) mm, blade oblanceolate to elliptic, base tapered, margins entire, apex obtuse to acute; cauline 2–4 pairs, sessile, rarely short-petiolate, 22–77 × 3–10(–15) mm, blade lanceolate, proximals sometimes oblanceolate, base tapered to truncate, apex acute. **Thyrses** interrupted to ± continuous, cylindric, (3–)8–22(–28) cm, axis moderately to densely retrorsely hairy, verticillasters (5–)7–15, cymes (1 or)2- or 3-flowered, 2 per node; proximal bracts lanceolate, 16–50(–63) × 3–10 mm; peduncles and pedicels retrorsely hairy. **Flowers:** calyx lobes ovate, (3–)4–6.8 × 1.7–2.8 mm, glabrous, rarely ± puberulent; corolla blue to violet or purple, with violet nectar guides, funnelform, (12–)14–23(–28) mm, glabrous externally, glabrous internally, tube 6–8 mm, throat gradually inflated, not constricted at orifice, 6–9 mm diam., rounded abaxially; stamens included, pollen sacs divergent, navicular, (1–)1.2–1.5(–1.8) mm, dehiscing incompletely, proximal ¼ indehiscent, connective not splitting, sides sparsely to moderately hispid, hairs white or tan, to 0.4(–1) mm, sutures papillate or denticulate, teeth to 0.1 mm; staminode 10–12 mm, included, 0.8–1.1 mm diam., tip recurved, distal 5–7 mm sparsely pilose, hairs yellow, to 1 mm; style 12–15 mm. **Capsules** 7–10 × 5–6 mm.

Varieties 2 (2 in the flora): w United States.

1. Leaves densely retrorsely hairy, rarely glabrescent, basal and proximal cauline 2–18(–27) mm wide . . . . 105a. *Penstemon fremontii* var. *fremontii*
1. Leaves glabrous or glabrescent, except sometimes margins and midveins, basal and proximal cauline 2–12 mm wide . . . . 105b. *Penstemon fremontii* var. *glabrescens*

105a. Penstemon fremontii Torrey & A. Gray var.
      fremontii E

Leaves densely retrorsely hairy, rarely glabrescent, basal and proximal cauline 2–18(–27) mm wide.

Flowering May–Jun. Clayey or sandy soils, sagebrush shrublands; 1500–2500 m; Colo., Utah, Wyo.

Variety *fremontii* is known from northwestern Colorado, northeastern Utah, and south-central and southwestern Wyoming.

105b. Penstemon fremontii Torrey & A. Gray var.
      glabrescens Dorn & Lichvar, Madroño 37: 195,
      fig. 1. 1990 C E

Penstemon luculentus
R. L. Johnson & M. R. Stevens

Leaves glabrous or glabrescent, except sometimes margins and midveins, basal and proximal cauline 2–12 mm wide.

Flowering May–Jun(–Jul). Shaley, barren slopes, rabbitbrush shrublands; of conservation concern; 1800–2500 m; Colo.

Variety *glabrescens* is known from Garfield and Rio Blanco counties, within the range of var. *fremontii*. Records of var. *glabrescens* come from the Piceance and Roan creek drainages. R. L. Johnson et al. (2016) may be correct that var. *glabrescens* is morphologically and molecularly distinct from *Penstemon fremontii* and should be recognized as *P. luculentus*.

SELECTED REFERENCE Johnson, R. L. et al. 2016. Molecular and morphological evidence for *Penstemon luculentus* (Plantaginaceae): A replacement name for *Penstemon fremontii* var. *glabrescens*. PhytoKeys 63: 47–62.

106. Penstemon gibbensii Dorn, Brittonia 34: 334.
      1982 • Gibbens's beardtongue C E

Stems ascending to erect, 10–37 cm, pubescent to retrorsely hairy, sometimes glabrous proximally, glandular-pubescent distally, not glaucous. Leaves basal and cauline, or basal absent or reduced, not leathery, proximals glabrous or puberulent to scabrous, distals puberulent or scabrous to glandular-pubescent, not glaucous; basal and proximal cauline 15–90 × 2–7 (–8) mm, blade oblanceolate to linear, base tapered, margins entire, apex obtuse to acute; cauline 6–10 pairs, short-petiolate or sessile, 9–68 × 1–5 mm, blade

oblanceolate to linear, base tapered to truncate, apex obtuse to acute. Thyrses interrupted, secund, (2–) 5–14 cm, axis glandular-pubescent, verticillasters (2–) 5–8, cymes 1–3-flowered, 1 or 2 per node; proximal bracts lanceolate, proximals 8–43 × 1–3 mm; peduncles and pedicels glandular-pubescent. Flowers: calyx lobes ovate to lanceolate, 3.5–7(–8) × 1.8–2.5 mm, glandular-pubescent; corolla lavender to light blue or light violet, with faint reddish purple nectar guides, funnelform, (15–)16–18(–20) mm, glandular-pubescent externally, sparsely to moderately glandular-pubescent or glandular internally abaxially, tube 5.5–6.5 mm, throat gradually inflated, not constricted at orifice, 5–6 mm diam., rounded abaxially; stamens: longer pair reaching orifice, pollen sacs divergent, navicular, 1–1.4 mm, dehiscing incompletely, proximal ⅕ indehiscent, connective not splitting, sides moderately hirsute, hairs white, to 0.7 mm, sutures denticulate, teeth to 0.1 mm; staminode 8–10 mm, included, 0.6–0.7 mm diam., tip recurved, distal 1–3 mm sparsely pilose or lanate, hairs yellow or whitish, to 1 mm; style 11–16 mm. Capsules 5–8 × 3.5–5 mm.

Flowering Jun–Sep. Barren hills, pinyon-juniper woodlands, sagebrush and greasewood-saltbush shrublands; of conservation concern; 1700–2300 m; Colo., Utah, Wyo.

Penstemon gibbensii is known from fewer than ten populations in Moffat and Rio Blanco counties, Colorado, Daggett County, Utah, and Carbon and Sweetwater counties, Wyoming (S. L. O'Kane 1988; E. C. Neese and N. D. Atwood 2003). Plants occur on shales or sandstones of the Browns Park Formation and Green River Formation (B. L. Heidel 2009).

SELECTED REFERENCE Heidel, B. L. 2009. Survey and Monitoring of *Penstemon gibbensii* (Gibbens' Beardtongue) in South-central Wyoming. Laramie.

107. Penstemon glaber Pursh, Fl. Amer. Sept. 2: 738.
      1813 (as Pentstemon glabra) • Southern smooth
      beardtongue E F

Stems ascending, (10–)50–65 (–80) cm, glabrous or puberulent to pubescent, not glaucous. Leaves basal and cauline, or basal absent or reduced, not leathery, glabrous or puberulent to pubescent, not glaucous; basal and proximal cauline 20–80(–155) × 5–20(–45) mm, blade obovate to oblanceolate or lanceolate, base tapered, margins entire, usually undulate, apex obtuse to acute, sometimes mucronate; cauline (2 or)3–6(–8) pairs, sessile, 27–120(–150) × (6–)22–35(–43) mm, blade lanceolate, base truncate to cordate, apex obtuse to acute. Thyrses continuous, secund, (3–)6–26 (–30) cm, axis glabrous or puberulent, verticillasters

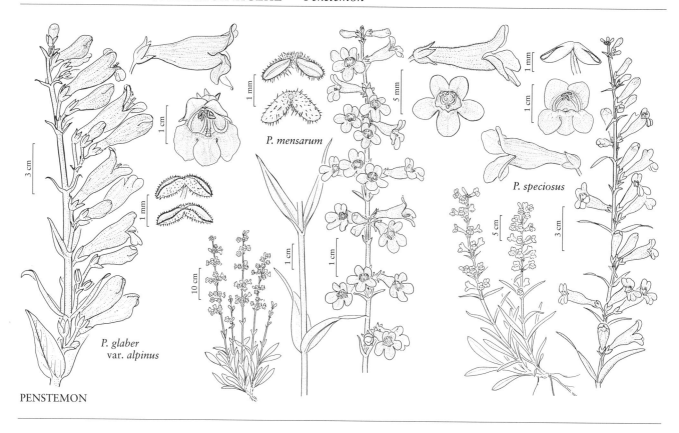

P. mensarum

P. speciosus

P. glaber
var. *alpinus*

PENSTEMON

(3–)5–12, cymes 2–4-flowered, 2 per node; proximal bracts ovate to lanceolate, 14–100 × 2–36 mm; peduncles and pedicels glabrous or puberulent, peduncles to 71 mm, pedicels 4–14 mm. **Flowers:** calyx lobes ovate to orbiculate, 2–11 × 1.3–5.5 mm, glabrous or puberulent; corolla blue to violet, with reddish purple nectar guides, ventricose, 24–48 mm, glabrous externally, glabrous internally or sparsely to moderately white-lanate internally abaxially, tube 5–12 mm, throat gradually inflated, slightly constricted at orifice, 8–13 (–18) mm diam., rounded abaxially; stamens included or longer pair reaching orifice, pollen sacs divergent, navicular, 1.4–2.5(–2.8) mm, dehiscing incompletely, connective not splitting, sides sparsely to moderately hirsute, hairs white or tan, to 0.5 mm, rarely glabrous, sutures denticulate, teeth to 0.2 mm; staminode (15–) 18–22 mm, included or barely exserted, 0.6–1.3 mm diam., tip straight to recurved, rounded to distinctly bifurcate at apex, glabrous or distal 1–2 mm sparsely lanate, hairs pale yellow, to 1.5 mm; style 18–20 mm. **Capsules** 10–17 × 5–8 mm.

Varieties 3 (3 in the flora): w United States.

The *Penstemon glaber* complex has been treated with as many as four species (F. W. Pennell 1920b) or one variable species as treated herein. M. K. W. Latady (1985) found clinal variation in most morphologic characters used to distinguish taxa in the complex and recognized only one species and two varieties: var. *glaber* (including var. *alpinus*) in the plains and mountains from southeastern Montana south to central Colorado; and var. *brandegeei*, in the mountains from central Colorado south to north-central New Mexico. A case for recognizing var. *alpinus*, as is done here, can be made on morphological, phenological, and ecological grounds.

SELECTED REFERENCE Latady, M. K. W. 1985. A Systematic Revision of *Penstemon glaber* and *P. alpinus*. M.S. thesis. University of Wyoming.

1. Calyx lobes 2–4.8 mm, apices rounded or abruptly short-acuminate; staminode apices rounded . . . .
. . . . .107a. *Penstemon glaber* var. *glaber*
1. Calyx lobes 3.8–11 mm, apices long-acuminate; staminode apices rounded or bifurcate.
  2. Staminode apices rounded or bifurcate; corollas 24–35 mm; stems glabrous or puberulent proximally . . . .
. . . . .107b. *Penstemon glaber* var. *alpinus*
  2. Staminode apices usually bifurcate; corollas 27–48 mm; stems puberulent to pubescent proximally. . .107c. *Penstemon glaber* var. *brandegeei*

107a. Penstemon glaber Pursh var. **glaber** E

Stems glabrous. **Flowers:** calyx lobes ovate to orbiculate, 2–4.8 × 1.4–3.8 mm, apex rounded or abruptly short-acuminate; corolla 24–33 mm, glabrous or sparsely to moderately white-lanate internally abaxially; staminode apex rounded. $2n = 16$.

Flowering Jul–Aug(–Sep). Sandy or gravelly short-grass prairies, pine forests; 1100–2200 m; Mont., Nebr., N.Dak., S.Dak., Wyo.

Variety *glaber* occurs from southwestern North Dakota and north-central South Dakota into the Nebraska Panhandle, southeastern Montana, and through most of eastern Wyoming, and it is replaced by var. *alpinus* and var. *brandegeei* southward along the Front Range of the Rocky Mountains.

107b. Penstemon glaber Pursh var. **alpinus** (Torrey) A. Gray, Proc. Amer. Acad. Arts. 6: 60. 1862 (as Pentstemon) E F

Penstemon alpinus Torrey, Ann. Lyceum Nat. Hist. New York 1: 35. 1823 (as Pentstemon alpina); *P. alpinus* subsp. *magnus* (Pennell) Penland

Stems glabrous or puberulent proximally. **Flowers:** calyx lobes ovate, 3.8–9.5 × 1.3–3.3 mm, apex long-acuminate; corolla 24–35 mm, sparsely to moderately white-lanate internally abaxially; staminode apex rounded or bifurcate. $2n = 16$.

Flowering Jun–Aug(–Sep). Sandy or gravelly slopes and banks in shortgrass prairies, foothills, mountains; 1500–2900 m; Colo., Nebr., N.Mex., Wyo.

Variety *alpinus* occurs in the Front Range of the Rocky Mountains on gravelly, granitic slopes, and roadcuts. The type of *Penstemon magnus* Pennell, on which *P. alpinus* subsp. *magnus* is based, was collected in Jackson County, Colorado. The specimen is in poor condition but has calyx lobes to 8 mm with apices short- to long-acuminate; it is tentatively assigned here.

107c. Penstemon glaber Pursh var. **brandegeei** (Porter) C. C. Freeman, Phytologia 60: 105. 1986 (as brandegei) E

Penstemon cyananthus Hooker var. *brandegeei* Porter in T. C. Porter and J. M. Coulter, Syn. Fl. Colorado, 91. 1874 (as Pentstemon brandegei); *P. alpinus* Torrey subsp. *brandegeei* (Porter) Penland

Stems puberulent to pubescent proximally. **Flowers:** calyx lobes ovate to orbiculate, (4–)5–11 × 1.8–5.5 mm, apex long-acuminate; corolla 27–48 mm, glabrous internally or sparsely white-lanate internally abaxially; staminode apex usually bifurcate. $2n = 16$.

Flowering Jun–Aug. Sandy or gravelly foothills, mountains; 1800–2100 m; Colo., N.Mex.

Variety *brandegeei* is known from Costilla, Custer, El Paso, Fremont, Huerfano, Las Animas, and Pueblo counties in south-central Colorado, and Colfax and Union counties in northeastern New Mexico. G. T. Nisbet and R. C. Jackson (1960) reported hybrids between it and var. *alpinus*.

108. Penstemon hallii A. Gray, Proc. Amer. Acad. Arts 6: 70. 1862 (as Pentstemon) • Hall's beardtongue E

Stems ascending to erect, (7–)9–16(–25) cm, glabrous or ± glandular-pubescent distally, not glaucous. **Leaves** basal and cauline, or basal absent or reduced, ± leathery, glabrous except for antrorsely-scabrous margins, not glaucous; basal and proximal cauline 16–85 × 3–9 mm, blade spatulate to lanceolate, base tapered, margins entire, apex rounded to obtuse or acute; cauline 2–4 pairs, sessile, 24–55 × 2–5 mm, blade lanceolate, proximals sometimes spatulate, base truncate, apex obtuse to acute. **Thyrses** continuous, cylindric to secund, (1–)2–7 cm, axis glandular-pubescent, verticillasters 1–7, cymes 1–4-flowered, 2 per node; proximal bracts lanceolate, rarely oblanceolate, 9–26 × 1–3 mm; peduncles and pedicels ascending to erect, glandular-pubescent. **Flowers:** calyx lobes obovate to elliptic, (4–)5–6.5 × 3–4 mm, margins erose or lacerate, glandular-pubescent, sometimes only proximally; corolla blue to violet or purple, with reddish purple nectar guides, ventricose, 14–25 mm, glabrous externally, glabrous or sparsely white-lanate internally abaxially, tube 2–3 mm, throat abruptly inflated, not constricted at orifice, 6–7 mm diam., rounded abaxially; stamens included, pollen sacs opposite, navicular to subexplanate,

1.2–1.4 mm, dehiscing completely, sides glabrous, rarely sparsely pilose, hairs white, to 0.8 mm near connective, sutures smooth or obscurely papillate; staminode 10–15 mm, included or reaching orifice, 0.6–1 mm diam., tip straight, distal 3–6 mm moderately to densely pubescent, hairs golden yellow, to 0.6 mm (hairs scattered proximally); style 10–13 mm. **Capsules** 6–9 × 4–7 mm.

Flowering Jul–Aug. Rocky to gravelly alpine meadows; 2800–4000 m; Colo.

Penstemon hallii is known from alpine zones on the highest peaks in Colorado.

109. Penstemon idahoensis N. D. Atwood &
S. L. Welsh, Great Basin Naturalist 48: 496, fig. 2. 1989 • Idaho beardtongue [C][E]

Stems ascending to erect, 8–20 cm, glutinous, usually covered with sand, not glaucous. **Leaves** basal and cauline, ± leathery, glutinous, usually covered with sand, not glaucous; basal and proximal cauline 35–70 × 2–5(–8) mm, blade oblanceolate, base tapered, margins entire, apex obtuse to acute; cauline 2–8 pairs, sessile, 34–55 × 2–6 mm, blade elliptic to linear, base tapered to truncate, apex obtuse to acute. **Thyrses** continuous, secund, 3–9 cm, axis glutinous, sometimes covered with sand, verticillasters 3–5, cymes 1–5-flowered, 2 per node; proximal bracts lanceolate, 15–23 × 1–3 mm; peduncles and pedicels glutinous, sometimes covered with sand. **Flowers:** calyx lobes lanceolate, 3–6 × 1.5–1.8 mm, glutinous and covered with sand; corolla violet to blue, with faint reddish purple nectar guides, ventricose-ampliate, 15–21 mm, glabrous externally, glabrous internally, tube 4–5 mm, throat gradually inflated, not constricted at orifice, 6–7 mm diam., rounded abaxially; stamens included or longer pair reaching orifice, pollen sacs divergent, navicular, 1.5–2 mm, dehiscing incompletely, connective not splitting, sides sparsely to moderately lanate, hairs white, to 0.6 mm, sutures denticulate, teeth to 0.1 mm; staminode 8–10 mm, included, 0.4–0.5 mm diam., tip straight to recurved, distal 1–3 mm sparsely lanate, hairs yellowish, to 0.6 mm; style 12–14 mm. **Capsules** 7.5–10 × 5–6 mm.

Flowering Jun–Jul. Sandy, tuffaceous outcrops, sagebrush and juniper communities; of conservation concern; 1500–1700 m; Idaho, Nev., Utah.

Penstemon idahoensis is known from the Goose Creek basin in southwestern Cassia County, Idaho, extreme northeastern Elko County, Nevada, and extreme northwestern Box Elder County, Utah. Nevada plants differ from others in having distinctly wavy leaf margins.

110. Penstemon laevis Pennell, Contr. U.S. Natl. Herb.
20: 347. 1920 • Southwestern beardtongue [E]

Stems ascending to erect, 28–90 cm, glabrous, ± glaucous. **Leaves** basal and cauline, ± leathery, glabrous, ± glaucous; basal and proximal cauline 35–120(–150) × 8–30 mm, blade obovate to oblanceolate or elliptic, base tapered, margins entire, sometimes undulate, apex obtuse; cauline 3–6 pairs, sessile or proximals sometimes short- to long-petiolate, (20–)30–120 × 6–20(–29) mm, blade oblanceolate to lanceolate or elliptic, base truncate to clasping, sometimes tapered, margins sometimes undulate, apex obtuse to acute. **Thyrses** continuous, secund, 8–48 cm, axis glabrous; verticillasters 5–8(–17), cymes 1–4-flowered, 2 per node; proximal bracts lanceolate, 10–65 × 2–15 mm; peduncles and pedicels glabrous. **Flowers:** calyx lobes ovate to lanceolate, (4–)4.5–9(–10) × 2.1–3.4(–3.8) mm, glabrous; corolla blue to bluish violet, with reddish violet nectar guides, ventricose, (20–)25–32 mm, glabrous externally, glabrous internally, tube 7–11 mm, throat abruptly inflated, slightly constricted at orifice, 8–12 mm diam., 2-ridged abaxially; stamens: longer pair reaching orifice or exserted, pollen sacs divergent, sigmoid, 2–2.5 mm, dehiscing incompletely, proximal ¼ indehiscent, connective not splitting, sides glabrous, sutures denticulate, teeth to 0.1 mm; staminode 16–19 mm, included or reaching orifice, 0.6–0.9 mm diam., tip straight to recurved, distal 5–7 mm ± pilose, hairs yellow, to 1 mm; style (17–)19–23 mm. **Capsules** 8–10 × 5–6 mm.

Flowering May–Jun. Sandy soils, sagebrush, juniper, oak-juniper, and pine-juniper communities; 1100–2000 m; Ariz., Utah.

Penstemon laevis is known from Coconino and Mohave counties, Arizona, and Garfield, Kane, and Washington counties, Utah. *Penstemon ×jonesii* Pennell, a putative hybrid between *P. eatonii* and *P. laevis*, is known from Kane and Washington counties, Utah (E. C. Neese and N. D. Atwood 2003).

111. Penstemon leiophyllus Pennell, Contr. U.S. Natl.
Herb. 20: 346. 1920 • Charleston beardtongue [E]

Stems ascending to erect, 4–55(–80) cm, glabrous, not glaucous. **Leaves** basal and cauline, not leathery, glabrous, not glaucous; basal and proximal cauline (20–)30–100 × 4–15 (–20) mm, blade oblanceolate, base tapered, margins entire, apex obtuse to acute; cauline 2–7 pairs, sessile or proximals sometimes short-petiolate, 16–80 × 2–20 mm, blade lanceolate to linear, base tapered, apex obtuse to acute. **Thyrses** interrupted or continuous, secund, (1–)3–30(–40) cm, axis glandular-pubescent, verticillasters 2–8(–12), cymes 1–3(or 4)-flowered, (1 or)2 per node; proximal bracts lanceolate, (6–)10–60 × (1–)3–24 mm; peduncles and pedicels glandular-pubescent. **Flowers:** calyx lobes lanceolate, (3.8–)4.5–8.5 × 1.6–3.8 mm, glandular-pubescent; corolla violet to blue or purple, with violet or reddish purple nectar guides, ventricose, 18–35 mm, glandular-pubescent externally, glabrous or white-lanate internally abaxially, glandular-pubescent adaxially, tube 6–10 mm, throat gradually inflated, not constricted at orifice, 5–10 mm diam., 2-ridged abaxially; stamens: longer pair reaching orifice, pollen sacs divergent to nearly opposite, navicular, 1.2–2 mm, dehiscing incompletely, proximal ⅕ indehiscent, connective not splitting, sides glabrous, sutures denticulate, teeth to 0.1 mm; staminode 10–14 mm, included, 0.4–0.5 mm diam., tip straight, distal 1–5 mm sparsely to densely pilose, hairs yellow or orangish yellow, to 1 mm; style 12–19 mm. **Capsules** 7–11 × 5–6 mm.

Varieties 3 (3 in the flora): w United States.

F. S. Crosswhite (1967) recognized three species in the *Penstemon leiophyllus* complex: *P. francisci-pennellii*, *P. keckii*, and *P. leiophyllus*. N. H. Holmgren (1984) expanded *P. leiophyllus* to include the other species as varieties. The complex is unique in the section in having patches of glandular hairs on the internal, adaxial surfaces of the corolla throats. The varieties are essentially allopatric; characters distinguishing them are minor.

1. Stems 15–55(–80) cm; cauline leaves 4–20 mm wide, blades lanceolate, usually flat and straight; corollas glabrous, rarely white-lanate, internally abaxially; Utah . . . . 111a. *Penstemon leiophyllus* var. *leiophyllus*
1. Stems 4–25(–43) cm; cauline leaves 2–5(–10) mm wide, blades lanceolate to linear, usually folded lengthwise and curved; corollas white-lanate, rarely glabrous, internally abaxially; Nevada.
2. Corollas (22–)25–35 mm; Lincoln, Nye, and White Pine counties, Nevada . . . . .111b. *Penstemon leiophyllus* var. *francisci-pennellii*
2. Corollas 18–22 mm; Clark County, Nevada . . . . 111c. *Penstemon leiophyllus* var. *keckii*

111a. Penstemon leiophyllus Pennell var. **leiophyllus** E

Stems 15–55(–80) cm. **Cauline leaves** 18–70 × 4–20 mm, blade lanceolate, usually flat and straight. **Flowers:** corolla 20–32 mm, glabrous, rarely white-lanate, internally abaxially.

Flowering Jun–Aug. Spruce forests, pine-aspen woodlands; 1700–3400 m; Utah.

Variety *leiophyllus* is known from the Markagunt and Paunsaugunt plateaus and Pine Valley Mountains (Garfield, Iron, Kane, and Washington counties). Some collections from the Pine Valley Mountains have corollas that are white-lanate internally abaxially.

111b. Penstemon leiophyllus Pennell var. **francisci-pennellii** (Crosswhite) N. H. Holmgren in A. Cronquist et al., Intermount. Fl. 4: 432. 1984 C E

Penstemon francisci-pennellii Crosswhite, Leafl. W. Bot. 10: 170. 1965

Stems 4–25(–43) cm. **Cauline leaves** 19–80 × 2–5(–10) mm, blade lanceolate to linear, usually folded lengthwise and curved. **Flowers:** corolla (22–) 25–35 mm, white-lanate, rarely glabrous, internally abaxially.

Flowering Jul–Aug. Rocky, limestone slopes in sagebrush shrublands and pine-aspen woodlands; of conservation concern; 2100–4000 m; Nev.

Variety *francisci-pennellii* is known from Lincoln, Nye, and White Pine counties. Populations have been documented in the Egan, Grant, Schell Creek, Snake, and Wilson Creek ranges.

111c. Penstemon leiophyllus Pennell var. **keckii** (Clokey) N. H. Holmgren in A. Cronquist et al., Intermount. Fl. 4: 432. 1984 C E

Penstemon keckii Clokey, Madroño 4: 128. 1937

Stems 5–24(–30) cm. **Cauline leaves** 16–56 × 2–5(–10) mm, blade lanceolate to linear, usually folded lengthwise and curved. **Flowers:** corolla 18–22 mm, white-lanate, rarely glabrous, internally abaxially.

Flowering Jun–Aug. Pine forests; of conservation concern; 900–3500 m; Nev.

Variety *keckii* is known from the Spring Mountains in Clark County.

112. Penstemon lemhiensis (D. D. Keck) D. D. Keck & Cronquist, Brittonia 8: 248. 1957 • Lemhi beardtongue E

Penstemon speciosus Douglas ex Lindley subsp. *lemhiensis* D. D. Keck, Amer. Midl. Naturalist 23: 612. 1940

Stems ascending to erect, 25–80 cm, retrorsely hairy, not glaucous. **Leaves** basal and cauline, not leathery, retrorsely hairy, not glaucous; basal and proximal cauline (35–)50–160 × (5–)8–24 mm, blade oblanceolate to lanceolate, base tapered, margins entire, apex obtuse to acute; cauline 4–8 pairs, short-petiolate or sessile, 25–85(–125) × 6–15 mm, blade elliptic to lanceolate, base tapered to truncate, apex acute, rarely obtuse. **Thyrses** interrupted or continuous, cylindric, 7–30 cm, axis retrorsely hairy proximally, mostly glandular-pubescent distally, verticillasters 3–7(–9), cymes 2–8(–10)-flowered, 2 per node; proximal bracts lanceolate, 18–95 × 5–20 mm; peduncles and pedicels retrorsely hairy. **Flowers:** calyx lobes ovate to lanceolate, (4.8–)6–11 × 2.2–3.5 mm, margins erose, glabrous or sparsely glandular, sometimes also sparsely puberulent; corolla violet to blue, without nectar guides, ventricose, 25–35 mm, glabrous externally, glabrous internally, tube 7–10 mm, throat gradually inflated, not constricted at orifice, 7–11 mm diam., 2-ridged abaxially; stamens included, pollen sacs divergent to nearly opposite, sigmoid, 1.8–2.2 mm, dehiscing incompletely, proximal ⅕–¼ indehiscent, connective not splitting, sides sparsely hispid, hairs white, to 0.3 mm, sutures denticulate, teeth to 0.1 mm; staminode 14–16 mm, included, 0.7–0.8 mm diam., tip straight, glabrous; style (11–)15–22 mm. **Capsules** 8–11 × 5–6 mm.

Flowering Jun–Jul. Montane grasslands, sagebrush shrublands, pine and pine-fir woodlands; 1200–2500 m; Idaho, Mont.

Penstemon lemhiensis is known from the Central Rocky Mountains in east-central Idaho (Lemhi County) and southwestern Montana (Beaverhead, Deer Lodge, Ravalli, and Silver Bow counties).

113. Penstemon longiflorus (Pennell) S. L. Clark, Great Basin Naturalist 35: 434. 1976 • Long-flower beardtongue E

Penstemon cyananthus Hooker subsp. *longiflorus* Pennell, Contr. U.S. Natl. Herb. 20: 353. 1920

Stems ascending to erect, (11–)20–60 cm, ± puberulent proximally, glabrous distally, not glaucous. **Leaves** basal and cauline, not leathery, proximals ± puberulent, distals glabrous, not glaucous; basal and proximal cauline (24–)40–90(–120) × 6–20(–30) mm, blade obovate, base tapered, margins entire, apex obtuse to acute; cauline 3–5 pairs, sessile, (25–)35–70 × 8–23 mm, blade ovate to lanceolate or elliptic, base tapered to clasping, apex acute, rarely obtuse. **Thyrses** interrupted, secund, 7–21 cm, axis glabrous, verticillasters 3–6, cymes 1- or 2(or 3)-flowered, 2 per node; proximal bracts ovate to lanceolate, 5–28 × 2.5–24 mm; peduncles and pedicels glabrous, peduncles to 26 mm, pedicels 1–8(–12) mm. **Flowers:** calyx lobes ovate, 4.5–7 × 2.5–3 mm, apex acute to acuminate, glabrous; corolla blue to violet, without nectar guides, ventricose, (20–)24–30(–32) mm, glabrous externally, glabrous internally, tube 8–11 mm, throat gradually inflated, not constricted at orifice, 8–10(–12) mm diam., 2-ridged abaxially; stamens: longer pair reaching orifice, pollen sacs divergent, rarely nearly opposite, sigmoid, 1.8–2.4(–2.8) mm, dehiscing incompletely, proximal ⅕ indehiscent, connective not splitting, sides sparsely hispid, hairs white, to 0.3 mm, sutures denticulate, teeth to 0.1 mm; staminode 14–16 mm, included, 0.6–1 mm diam., tip straight to recurved, distal 0.5–1 mm moderately pilose, hairs yellow, to 0.8 mm, remaining distal 3–4 mm with scattered yellow hairs; style 18–22 mm. **Capsules** 8–14 × 6–10 mm. **2***n* = 16.

Flowering Jun–Jul. Sagebrush shrublands, Gambel oak and mountain mahogany woodlands; 1800–2700 m; Utah.

Penstemon longiflorus is known from Beaver, Juab, Millard, Piute, Sevier, and Tooele counties (E. C. Neese and N. D. Atwood 2003). The species is most often confused with the partially sympatric *P. cyananthus*.

114. Penstemon mensarum Pennell, Contr. U.S. Natl. Herb. 20: 380. 1920 • Tiger or Grand Mesa beardtongue E F

Stems erect, 25–90 cm, glabrous, slightly glaucous or not. **Leaves** basal and cauline, not leathery, glabrous, not glaucous; basal and proximal cauline 28–145 × 4–30 mm, blade oblong to oblanceolate or elliptic, base tapered, margins entire, apex acute; cauline 2–4(or 5) pairs, sessile, rarely short-petiolate, 20–135 × 4–31 mm, blade lanceolate, proximals sometimes oblanceolate, base tapered to cordate-clasping, apex acute to acuminate. **Thyrses** interrupted, secund, 10–56 cm, axis ± glandular-pubescent at least distally, verticillasters 4–12, cymes (1 or)2–6-flowered, 2 per node; proximal bracts lanceolate, 10–58 × 10–85 mm; peduncles and pedicels glandular-pubescent, sometimes sparsely so. **Flowers:** calyx lobes oblong-ovate, 3.5–5.8 × 1.2–2.1 mm, glandular-pubescent; corolla blue to purple or violet, with faint reddish purple nectar guides, ventricose, 16–25 mm, ± glandular-pubescent externally, glabrous or sparsely glandular internally, tube 5–8 mm, throat gradually inflated, not constricted at orifice, 6–10 mm diam., rounded abaxially; stamens included, pollen sacs divergent to nearly opposite, navicular, 0.8–1.4 mm, dehiscing incompletely, connective not splitting, sides sparsely to moderately hispid, hairs white or yellowish, to 0.2 mm, sutures papillate or denticulate, teeth to 0.1 mm; staminode 9–11 mm, included, 0.4–0.5 mm diam., tip straight, distal 7–9 mm sparsely to moderately villous, hairs yellow, to 1 mm; style 9–11 mm. **Capsules** 8–10 × 4–6 mm.

Flowering Jun–Aug. Sagebrush shrublands, openings in aspen woodlands; 2200–3000 m; Colo.

Penstemon mensarum is known from Delta, Gunnison, Mesa, Montrose, and Pitkin counties.

115. Penstemon moriahensis N. H. Holmgren, Brittonia 30: 422, fig. 4. 1978 • Mt. Moriah beardtongue C E

Stems ascending to erect, 18–50 cm, glabrous, not glaucous. **Leaves** basal and cauline, or basal sometimes reduced, not leathery, glabrous, not glaucous; basal and proximal cauline 35–80(–120) × 4–13 mm, blade oblanceolate, base tapered, margins entire, apex rounded to obtuse, or acute; cauline 3–5 pairs, sessile or proximals sometimes short-petiolate, 20–110 × 5–18 mm, blade oblanceolate to lanceolate, rarely linear, base tapered to clasping, apex obtuse to acute. **Thyrses** interrupted or continuous, secund, 4–17 cm, axis glabrous or ± glandular distally, verticillasters 4–11, cymes 2- or 3(–5)-flowered, 2 per node; proximal bracts lanceolate, 10–50 × 4–16 mm; peduncles and pedicels glandular to glandular-pubescent, peduncles to 7 mm, pedicels 2–11 mm. **Flowers:** calyx lobes ovate to lanceolate, 5–7.5 × 2–4 mm, margins erose, glandular to glandular-pubescent; corolla blue to violet, without nectar guides, ventricose, 27–33 mm, glabrous externally, glabrous internally, tube 9–11 mm, throat gradually to abruptly inflated, not constricted at orifice, 8–12 mm diam., 2-ridged abaxially; stamens: longer pair reaching orifice, pollen sacs divergent, navicular, 2–2.8 mm, dehiscing incompletely, proximal ⅕ indehiscent, connective not splitting, sides villous, hairs white, to 1 mm, sutures denticulate, teeth to 0.1 mm; staminode 16–18 mm, included, 0.6–0.8 mm diam., tip straight, distal 1–6 mm sparsely lanate, hairs yellowish, to 0.8 mm; style 17–21 mm. **Capsules** 7–9 × 4.5–5.5 mm.

Flowering Jun–Jul. Gravelly or silty soils, subalpine sagebrush shrublands, mountain mahogany, pinyon-juniper, and pine woods; of conservation concern; 2200–3300 m; Nev.

Penstemon moriahensis is found in the northern Snake Range near Mount Moriah, White Pine County.

116. Penstemon navajoa N. H. Holmgren, Brittonia 30: 419, fig. 3. 1978 • Navajo beardtongue C E

Stems ascending to erect, 15–65 cm, glabrous, not glaucous. **Leaves** basal and cauline, not leathery, glabrous or retrorsely hairy, not glaucous; basal and proximal cauline 12–80 × 4–10(–15) mm, blade spatulate to oblanceolate or lanceolate, base tapered, margins entire, apex rounded to obtuse, or acute; cauline 3 or 4 pairs, sessile, 12–50(–75) × 1–5 mm, blade oblanceolate to linear, base tapered, apex acute to acuminate. **Thyrses** interrupted, secund, 2–25(–47) cm, axis glabrous, verticillasters 4–14, cymes 1- or 2(or 3)-flowered, 1 or 2 per node; proximal bracts linear, 6–15(–68) × 1–2 mm; peduncles and pedicels glabrous. **Flowers:** calyx lobes ovate, 2.6–4 × 1.5–2.2 mm, glabrous; corolla blue to lavender, with faint reddish purple nectar guides, ventricose, 18–23(–25) mm, glabrous externally, sparsely white-villous internally abaxially, tube 7–9 mm, throat gradually to abruptly inflated, not constricted at orifice, 7–10 mm diam., slightly 2-ridged abaxially; stamens reaching orifice or longer pair slightly exserted, pollen sacs opposite, navicular, 1.2–1.8 mm, dehiscing incompletely, proximal ⅕ sometimes indehiscent, connective not splitting, sides lanate, rarely glabrous, hairs white, to 0.9 mm, sutures denticulate, teeth to 0.1 mm; staminode 9–11 mm, reaching orifice, 0.5–0.7 mm diam., tip straight, glabrous; style 12–17 mm. **Capsules** 6.9–13 × 4.5–6.5 mm.

Flowering Jul–Aug. Rocky ponderosa pine, aspen, Douglas fir, and subalpine fir forests; of conservation concern; 2500–3200 m; Utah.

Penstemon navajoa is known from the Abajo Mountains, upper Dark Canyon, and Navajo Mountain in San Juan County.

117. Penstemon neomexicanus Wooton & Standley, Contr. U.S. Natl. Herb. 16: 172. 1913 (as Pentstemon) • New Mexico beardtongue

Stems erect, sometimes ascending, 30–63 cm, glabrous, not glaucous. **Leaves** basal and cauline, or basal absent, not leathery, glabrous or retrorsely hairy, not glaucous; basal and proximal cauline 30–90(–116) × 5–26 mm, blade oblanceolate, base tapered, margins entire, apex obtuse to acute; cauline 6–12 pairs, sessile, 25–113 × 3–13 mm, blade lanceolate to linear, base tapered to truncate, apex obtuse to acute. **Thyrses** continuous, sometimes interrupted, secund, 16–30 cm, axis glabrous, verticillasters 7–11, cymes 1–4-flowered,

(1 or)2 per node; proximal bracts lanceolate to linear, 10–72 × 1–12 mm; peduncles and pedicels glabrous. **Flowers:** calyx lobes lanceolate to elliptic, 3–5 × 2–2.6 mm, glabrous; corolla lavender to violet, with reddish purple nectar guides, ventricose, 26–34 mm, glabrous externally, sparsely to densely white-villous internally abaxially, tube 7–8 mm, throat gradually to abruptly inflated, not constricted at orifice, 7–10 mm diam., rounded abaxially; stamens: longer pair exserted, pollen sacs opposite, navicular, 1.4–1.9 mm, dehiscing completely, sides glabrous, sutures smooth or papillate; staminode 14–16 mm, reaching orifice or exserted, 1–1.2 mm diam., tip straight, glabrous; style 14–18 mm. **Capsules** 9–14 × 5–6 mm. $2n = 16$.

Flowering Jun–Aug. Slopes and clearings in pine, pine-spruce, and spruce-fir forests; 1800–2700 m; N.Mex.; Mexico (Chihuahua).

Penstemon neomexicanus is known from the Capitan, Sacramento, and White mountains in Lincoln and Otero counties. The species is also known from a single historic collection from near Colonia Garcia, Chihuahua. G. T. Nisbet and R. C. Jackson (1960) suggested that some plants from northern Lincoln and southern Torrance counties might be hybrids between *P. neomexicanus* and *P. virgatus*. Specimens from the vicinity of the Gallinas Mountains are referred here to *P. neomexicanus*.

118. Penstemon nudiflorus A. Gray, Proc. Amer. Acad. Arts 20: 306. 1885 (as Pentstemon) • Flagstaff beardtongue E

Stems erect, (35–)55–100 cm, glabrous, glaucous. **Leaves** basal and cauline, or basal absent or reduced, leathery, glabrous, glaucous; basal and proximal cauline 25–110 × (4–)10–27 mm, blade ovate to elliptic or lanceolate, base tapered, margins entire, apex obtuse to acute; cauline 3–6 pairs, sessile, (5–)15–105 × (1–)2–30 mm, blade lanceolate, base tapered to clasping, apex acute to acuminate. **Thyrses** interrupted, cylindric, 8–55 cm, axis glabrous, verticillasters 5–13, cymes 1- or 2(or 3)-flowered, (1 or)2 per node; proximal bracts lanceolate to subulate, 4–8 × 1–3 mm; peduncles and pedicels spreading to ascending, glabrous. **Flowers:** calyx lobes ovate to lanceolate, 4–5.8 × 2.5–3.2 mm, glabrous; corolla violet to lavender, with reddish purple nectar guides, ventricose-ampliate, 20–33 mm, glabrous externally, sparsely white-lanate internally abaxially, tube 7–9 mm, throat gradually to abruptly inflated, not constricted at orifice, 9–11 mm diam., rounded abaxially; stamens: longer pair exserted, pollen sacs opposite, navicular, 1.7–2.3 mm, dehiscing completely, sides glabrous, sutures denticulate, teeth to 0.1 mm; staminode 16–20 mm, reaching orifice or exserted, 0.5–0.8 mm diam., tip straight, distal 2–4 mm sparsely lanate, hairs yellow, to 1.5 mm; style 17–22 mm. **Capsules** 8–10 × 3–4 mm.

Flowering May–Jun. Rocky, basaltic soils in pinyon-juniper woodlands; 1500–2300 m; Ariz.

Penstemon nudiflorus occurs mostly along the Mogollon Rim in central Arizona in Coconino, Gila, Mohave, Navajo, and Yavapai counties.

119. Penstemon pahutensis N. H. Holmgren, Aliso 7: 351, fig. 1. 1971 • Pahute Mesa beardtongue E

Stems ascending to erect, 15–25(–35) cm, glabrous, not glaucous. **Leaves** basal and cauline, leathery or not, glabrous, sometimes glaucous; basal and proximal cauline 48–110 × 10–18 mm, blade oblanceolate to elliptic, base tapered, margins entire, apex obtuse to acute; cauline 5–8 pairs, petiolate, 30–80 (–120) × 4–18 mm, blade lanceolate to linear, base tapered, apex acute. **Thyrses** interrupted or continuous, ± secund, 13–48 cm, axis glabrous, verticillasters 6–14, cymes 2–4(–8)-flowered, 2 per node; proximal bracts lanceolate, 8–50(–100) × 2–10(–20) mm; peduncles and pedicels glabrous. **Flowers:** calyx lobes ovate to obovate, 3–5 × 1.8–3 mm, glabrous; corolla bluish lavender, with reddish purple nectar guides, ventricose, (17–)21–26 (–30) mm, glabrous externally, sparsely yellow- or white-villous internally abaxially, tube 5–8 mm, throat gradually inflated, not constricted at orifice, 6–10 mm diam., rounded abaxially; stamens: longer pair reaching orifice, pollen sacs divergent, sigmoid, 1.6–2.2 mm, dehiscing incompletely, proximal ⅓ indehiscent, connective not splitting, sides glabrous, sutures denticulate, teeth to 0.1 mm; staminode 12–15 mm, reaching orifice, 0.3–0.5 mm diam., tip straight, distal 4–6 mm densely pilose, hairs golden yellow, to 1.5 mm; style 17–20 mm. **Capsules** 8–10 × 5–7 mm. $2n = 16$.

Flowering Jun–Jul. Loose soils, rocky crevices, sagebrush shrublands, pinyon-juniper woodlands; 1600–2500 m; Calif., Nev.

Penstemon pahutensis is known from Inyo County, California, and Esmeralda and Nye counties, Nevada.

120. **Penstemon parvus** Pennell, Contr. U.S. Natl. Herb. 20: 345. 1920 • Aquarius Plateau beardtongue CE

Stems decumbent to ascending, 6–20(–25) cm, retrorsely hairy or puberulent, not glaucous. **Leaves** basal and cauline, not leathery, retrorsely hairy, not glaucous; basal and proximal cauline 8–40(–60) × 2–7 mm, blade spatulate to oblanceolate, base tapered, margins entire, apex rounded to obtuse or acute; cauline 2–4 pairs, sessile, 10–38 × 1–6 mm, blade lanceolate, base tapered, apex acute. **Thyrses** continuous to ± interrupted, secund, (2–)4–9 cm, axis retrorsely hairy proximally, mostly glandular-pubescent distally, verticillasters (3 or)4–8, cymes 1- or 2-flowered, (1 or)2 per node; proximal bracts linear, 3–26 × 0.5–2.5 mm; peduncles and pedicels glandular-pubescent, peduncles also usually retrorsely hairy. **Flowers:** calyx lobes ovate, 3–4(–5) × 2–2.9 mm, glandular-pubescent; corolla blue to violet or purple, with reddish purple nectar guides, ventricose, 13–20 mm, glandular-pubescent externally, glabrous internally, tube 4.5–6 mm, throat gradually inflated, not constricted at orifice, 4–6(–8) mm diam., slightly 2-ridged abaxially; stamens included or longer pair reaching orifice, pollen sacs opposite, navicular, 1–1.4 mm, dehiscing incompletely, connective not splitting, sides glabrous, sutures papillate; staminode 8–10 mm, reaching orifice, 0.5–0.6 mm diam., tip straight, glabrous; style 11–15 mm. **Capsules** 7–8 × 6–7 mm.

Flowering Jun–Aug. Sagebrush shrublands; of conservation concern; 2600–3500 m; Utah.

Penstemon parvus is known from the Aquarius Plateau in Garfield, Piute, and Sevier counties.

121. **Penstemon payettensis** A. Nelson & J. F. Macbride, Bot. Gaz. 62: 147. 1916 (as Pentstemon payetensis) • Payette beardtongue E

Stems ascending to erect, (15–)30–70 cm, glabrous, not glaucous. **Leaves** basal and cauline, leathery or not, glabrous, not glaucous; basal and proximal cauline 80–140 (–180) × (10–)15–35(–40) mm, blade ovate to elliptic, base tapered, margins entire, apex obtuse to acute; cauline 5–7 pairs, sessile, (28–)40–80 (–110) × 6–30(–42) mm, blade ovate to lanceolate, base tapered to cordate-clasping, apex acute. **Thyrses** continuous, sometimes interrupted proximally, cylindric, (3–)8–32 cm, axis glabrous, verticillasters 4–10(–16), cymes 2–4(–8)-flowered, 2 per node; proximal bracts ovate to lanceolate, 12–80 × 2–35 mm; peduncles and

pedicels glabrous. **Flowers:** calyx lobes lanceolate, 5–8.5(–10) × 1–2 mm, glabrous; corolla blue to violet, without nectar guides, ventricose, 18–22 mm, glabrous externally, glabrous internally, tube 6–9 mm, throat gradually inflated, not constricted at orifice, 7–10 mm diam., slightly 2-ridged abaxially; stamens: longer pair reaching orifice or slightly exserted, pollen sacs opposite, navicular, 1.2–1.9 mm, dehiscing incompletely, connective not splitting, sides glabrous, sutures denticulate, teeth to 0.1 mm; staminode: 13–15 mm, included, 0.5–0.9 mm diam., tip straight, distal 2–3 mm sparsely pilose, sometimes glabrous, hairs yellow, to 0.5 mm; style 13–17 mm. **Capsules** 8–15 × 5–8 mm. *2n* = 16.

Flowering May–Aug. Gravelly sagebrush shrublands, open coniferous forests; 1200–2600 m; Idaho, Mont., Oreg.

Penstemon payettensis is known from the Northern Rocky Mountains and eastern Columbia Plateau.

122. **Penstemon paysoniorum** D. D. Keck, Leafl. W. Bot. 5: 57. 1947 • Payson's beardtongue E

Stems decumbent to ascending, 8–18(–25) cm, glabrous, not glaucous. **Leaves** basal and cauline, leathery or not, glabrous, not glaucous; basal and proximal cauline (15–) 30–65(–95) × 2–12(–20) mm, blade oblanceolate to elliptic, rarely spatulate, base tapered, margins entire, apex obtuse to acute, rarely rounded; cauline 1–3 pairs, sessile, 20–70(–85) × 5–12 mm, blade oblanceolate to lanceolate or linear, base tapered to truncate, margins undulate, apex acute. **Thyrses** continuous, secund, (3–)5–19 cm, axis glabrous, verticillasters 2–6, cymes 2–4-flowered, 2 per node; proximal bracts lanceolate, rarely elliptic or oblanceolate, 11–55 × 5–11 mm; peduncles and pedicels glabrous or ± glandular, peduncles to 8 mm, pedicels 1–4 mm. **Flowers:** calyx lobes ovate to obovate, (3.5–) 4–7 × 1.8–2(–3) mm, apex acuminate, sometimes caudate, glabrous or sparsely glandular; corolla blue to violet or lavender, with or without faint purple nectar guides, ventricose, 14–22 mm, glabrous externally, glabrous internally, tube 5–8 mm, throat gradually inflated, not constricted at orifice, 4–6.5 mm diam., rounded abaxially; stamens: longer pair reaching orifice, pollen sacs divergent, navicular, 1.2–1.7 mm, dehiscing incompletely, proximal ⅕ indehiscent, connective not splitting, sides pilose, hairs white, to 0.3 mm, sutures papillate or denticulate, teeth to 0.1 mm; staminode 8–12 mm, included, 0.4–0.6 mm diam., tip straight, distal 5–6 mm sparsely pilose to villous, hairs yellow, to 0.5 mm; style 9–12 mm. **Capsules** 5.5–8 × 3.5–4.5 mm.

Flowering May–Jul. Shale, sandstone, and limestone outcrops, gravels, sagebrush shrublands; 1900–2600 m; Wyo.

Penstemon paysoniorum is known from southwestern Wyoming in Fremont, Lincoln, Natrona, Sublette, Sweetwater, and Uinta counties. The species can be confused with *P. scariosus* var. *garrettii*, with which it is sympatric in extreme southwestern Wyoming. A form of *P. paysoniorum* with oblanceolate to broadly elliptic basal and proximal cauline leaves is represented among herbarium specimens from south-central Sublette County and west-central and northwestern Sweetwater County, roughly in the middle of the range of the species.

123. Penstemon penlandii W. A. Weber, Phytologia 60: 459. 1986 • Penland's beardtongue C E

Stems ascending to erect, 7–25 cm, retrorsely hairy or puberulent, not glaucous. Leaves basal and cauline, not leathery, puberulent or retrorsely hairy, not glaucous; basal and proximal cauline 15–80 × 0.8–2 mm, blade linear, base tapered, margins entire, apex obtuse to acute; cauline 1–3 pairs, sessile, 10–75 × 0.5–1.5 mm, blade linear, base tapered, apex acute. Thyrses continuous, secund, 3–10 cm, axis glandular-pubescent, verticillasters 2–6, cymes 1- or 2(–4)-flowered, 2 per node; proximal bracts linear, 10–56 × 0.5–1.5 mm; peduncles and pedicels glandular-pubescent. Flowers: calyx lobes lanceolate, 4.5–5.5 × 1.8–2.5 mm, glabrous to sparsely glandular; corolla blue to bluish violet, with reddish purple nectar guides, funnelform, 13–16 mm, glabrous externally, glabrous internally, tube 5–6 mm, throat gradually inflated, not constricted at orifice, 4.5–5.5 mm diam., rounded abaxially; stamens included, pollen sacs divergent to opposite, navicular, 1.1–1.5 mm, dehiscing incompletely, connective not splitting, sides moderately hirsute, hairs white, to 0.5 mm, sutures denticulate, teeth to 0.1 mm; staminode 8–9 mm, included, 0.9–1.1 mm diam., tip straight, distal 6–7 mm densely pubescent, hairs yellowish orange, to 0.9 mm; style 8–10 mm. Capsules 6–9(–14) × 4–6 mm.

Flowering Jun–Jul. Clayey soils, sagebrush shrublands; of conservation concern; 2300–2400 m; Colo.

Penstemon penlandii is known from Middle Park in Grand County. Plants grow on strongly seleniferous shales of the Troublesome Formation (S. L. O'Kane 1988). The species is listed as endangered by the U.S. Department of the Interior.

124. Penstemon pennellianus D. D. Keck, Amer. Midl. Naturalist 23: 614. 1940 • Pennell's beardtongue E

Stems ascending to erect, 18–85 cm, glabrous, not glaucous. Leaves basal and cauline, ± leathery or not, glabrous, not glaucous; basal and proximal cauline 55–250 × 7–40 mm, blade lanceolate to elliptic, base tapered, margins entire, apex obtuse to acute; cauline 3–5 pairs, short-petiolate or sessile, 40–83 × 10–36 mm, blade ovate to lanceolate, base tapered to clasping, apex obtuse to acute. Thyrses continuous, cylindric, 4–35 cm, axis glabrous, verticillasters 4–11, cymes 1–6-flowered, 2 per node; proximal bracts ovate to lanceolate, 22–85 × 11–35 mm; peduncles and pedicels glabrous, peduncles to 20 mm, pedicels 1–6 mm. Flowers: calyx lobes ovate to lanceolate, (5.3–)6.5–8.5 × 1.6–3 mm, apex acuminate to caudate, glabrous; corolla blue to violet or purple, with reddish purple nectar guides, ventricose, 29–35 mm, glabrous externally, glabrous internally, tube 8–11 mm, throat gradually inflated, not constricted at orifice, 9–10 mm diam., slightly 2-ridged abaxially; stamens: longer pair reaching orifice, pollen sacs divergent to opposite, sigmoid, 2–2.7 mm, dehiscing incompletely, proximal ⅕–⅓ indehiscent, connective not splitting, sides sparsely hirsute, hairs white or tan, to 0.1 mm, especially near connective, sometimes merely scurfy near connective, sutures denticulate, teeth to 0.1 mm; staminode 16–20 mm, reaching orifice, 0.5–1 mm diam., tip straight, distal 4–6 mm ± uniformly sparsely pilose, hairs yellow, to 1 mm; style 20–24 mm. Capsules 10–13 × 8–10 mm.

Flowering May–Jun. Rocky ridges, sandy to rocky slopes, coniferous forests; 1300–1800 m; Oreg., Wash.

Penstemon pennellianus occurs primarily in the Blue Mountains of northeastern Oregon (Grant, Umatilla, Union, and Wallowa counties) and southeastern Washington (Asotin, Columbia, and Garfield counties).

125. Penstemon perpulcher A. Nelson, Bot. Gaz. 52: 273. 1911 (as Pentstemon) • Minidoka beardtongue E

Stems ascending to erect, (14–)30–60(–90) cm, glabrous or retrorsely hairy, not glaucous. Leaves basal and cauline, not leathery, glabrous or retrorsely hairy, not glaucous; basal and proximal cauline 50–130(–150) × 5–15(–25) mm, blade oblanceolate to elliptic, base tapered, margins entire, usually undulate, apex obtuse to acute;

cauline (2 or)3–5(–8) pairs, petiolate or sessile, 49–90 × 5–15 mm, blade lanceolate, base tapered, truncate, or clasping, margins undulate, apex acute to acuminate. **Thyrses** interrupted or continuous, cylindric to ± secund, (2–)9–20(–40) cm, axis glabrous, verticillasters (3–)5–9(–25), cymes 2–8(–10)-flowered, 2 per node; proximal bracts lanceolate, 18–50(–80) × 2–20 mm; peduncles and pedicels glabrous. **Flowers:** calyx lobes ovate, 3–5 × 1.6–2.8 mm, margins broadly scarious, apex rounded to acute or mucronate, glabrous; corolla violet to blue or purple, without nectar guides, funnelform to weakly ventricose, 18–22 mm, glabrous externally, glabrous internally, tube 6–8 mm, throat gradually inflated, not constricted at orifice, 5–8 mm diam., slightly 2-ridged abaxially; stamens included or longer pair reaching orifice, pollen sacs divergent to nearly opposite, navicular-sigmoid, 1.4–1.7 mm, dehiscing incompletely, proximal ⅕ indehiscent, connective not splitting, sides glabrous, sutures denticulate, teeth to 0.1 mm; staminode 9–12 mm, included, 0.6–1 mm diam., tip straight to recurved, distal 4–8 mm sparsely to moderately pilose, hairs yellow, to 1 mm; style 13–16 mm. **Capsules** 9–14 × 6–8 mm.

Flowering May–Jul. Sagebrush shrublands; 600–2000 m; Idaho, Oreg.

Penstemon perpulcher is common in the Snake River Plains of southern Idaho (Ada, Bannock, Boise, Canyon, Cassia, Custer, Elmore, Gooding, Minidoka, Oneida, Owyhee, Payette, and Twin Falls counties) and western Oregon (Malheur and Wallowa counties).

126. Penstemon pseudoputus (Crosswhite) N. H. Holmgren, Brittonia 31: 106. 1979 • Kaibab beardtongue E

Penstemon virgatus A. Gray subsp. *pseudoputus* Crosswhite, Amer. Midl. Naturalist 77: 35. 1967

Stems ascending to erect, (8–)15–50 cm, retrorsely hairy, not glaucous. **Leaves** basal and cauline, basal sometimes absent, not leathery, retrorsely hairy, not glaucous; basal and proximal cauline 16–60(–80) × 1–3(–4) mm, blade linear, base tapered, margins entire, apex obtuse to acute; cauline 6–11 pairs, sessile, (2–)10–75 × (0.2–)0.5–3 mm, blade linear, base tapered, apex acuminate. **Thyrses** interrupted, secund, (2–)5–25 cm, axis glabrous or puberulent, verticillasters 3–10, cymes 1- or 2-flowered, 1 or 2 per node; proximal bracts linear, 4–30 × 0.5–2 mm; peduncles and pedicels glabrous or puberulent. **Flowers:** calyx lobes ovate, sometimes obovate or truncate, 3–4.5(–5) × 1.5–2.4 mm, glabrous or sparsely puberulent; corolla bluish violet to violet, lavender, or purple, with reddish purple nectar guides, ventricose,

17–23 mm, glabrous externally, glabrous internally, tube 5–7 mm, throat gradually inflated, slightly constricted at orifice, (5–)7–9 mm diam., rounded abaxially; stamens: longer pair exserted, pollen sacs opposite, navicular, 1.2–1.5 mm, dehiscing incompletely, connective not splitting, sides glabrous, sutures papillate; staminode 12–14 mm, exserted, 0.5–0.8 mm diam., tip straight to recurved, glabrous; style 13–17 mm. **Capsules** 7–14 × 4–5 mm.

Flowering Jul–Aug. Subalpine meadows, pine forests; 2000–2900 m; Ariz., Utah.

Penstemon pseudoputus is known from the Kaibab and Walhalla plateaus of northern Coconino County, Arizona, with apparently isolated populations in Markagunt Plateau in eastern Garfield and eastern Kane counties in southwestern Utah.

127. Penstemon putus A. Nelson, Univ. Wyoming Publ. Sci., Bot. 1: 131. 1926 • Black River beardtongue E

Penstemon virgatus A. Gray subsp. *putus* (A. Nelson) Crosswhite

Stems ascending to erect, 20–45(–60) cm, glabrous, not glaucous. **Leaves** basal and cauline, leathery or not, glabrous, not glaucous; basal and proximal cauline 15–95 × 1–4(–5) mm, blade linear, base tapered, margins entire, apex acute to acuminate; cauline 5–7 pairs, sessile, 8–95 × 0.5–3(–4) mm, blade linear, base truncate, apex acute to acuminate. **Thyrses** interrupted, secund, 6–30 cm, axis glabrous, verticillasters 5–11, cymes 1–3-flowered, 1 or 2 per node; proximal bracts linear to subulate, (2–)3–30 (–50) × 0.5–2 mm; peduncles and pedicels glabrous. **Flowers:** calyx lobes ovate, 2.2–4 × 1.3–3.5 mm, glabrous; corolla white to pink, lavender, or violet, with reddish purple nectar guides, ventricose, 18–22 mm, glabrous externally, sparsely white-lanate internally abaxially, tube 4–6 mm, throat gradually inflated, not constricted at orifice, 7–10 mm diam., rounded abaxially; stamens: longer pair exserted, pollen sacs opposite, navicular, 1.2–1.6 mm, dehiscing completely or incompletely, connective splitting or not, sides glabrous, sutures papillate or denticulate, teeth to 0.1 mm; staminode 11–14 mm, exserted, 0.4–0.7 mm diam., tip straight to recurved, glabrous; style 12–17 mm. **Capsules** 6–11 × 4–6 mm. $2n = 16$.

Flowering Jun–Sep. Rocky to sandy pine forests, pinyon-juniper woodlands; 1500–2600 m; Ariz.

Penstemon putus occurs primarily along the Mogollon Rim in central and east-central Arizona. Populations have been documented in Apache, Coconino, Gila, Navajo, and Yavapai counties.

128. Penstemon saxosorum Pennell, Contr. U.S. Natl. Herb. 20: 349. 1920 • Upland beardtongue [E]

Stems erect, (10–)30–80 cm, glabrous, not glaucous. **Leaves** basal and cauline, not leathery, glabrous, not glaucous; basal and proximal cauline 26–140 × 3–19 mm, blade oblanceolate, base tapered, margins entire, apex obtuse to acute; cauline 2–5 pairs, sessile, 50–90 × (3–)6–9(–18) mm, blade oblanceolate to lanceolate, base clasping, apex acute. **Thyrses** continuous to ± interrupted, secund or cylindric, 5–40 cm, axis glabrous or ± glandular-pubescent, verticillasters (4 or) 5–9, cymes 1–5-flowered, 2 per node; proximal bracts lanceolate, 15–54 × 2–15 mm; peduncles and pedicels glabrous or glandular-pubescent, peduncles to 11 mm, pedicels 1–10 mm. **Flowers:** calyx lobes ovate, 3.5–8 × 1.8–3 mm, margins erose, glabrous or glandular-pubescent; corolla blue to bluish violet, with or without reddish purple nectar guides, ventricose, 17–25(–30) mm, glabrous or glandular-pubescent externally, glabrous or ± glandular-pubescent internally abaxially, tube 5–8 mm, throat gradually inflated, slightly constricted at orifice, 6–8 mm diam., rounded abaxially; stamens: longer pair reaching orifice or slightly exserted, pollen sacs opposite, navicular, 1.2–1.8 mm, dehiscing incompletely, connective not splitting, sides sparsely to moderately hispid or pubescent, hairs whitish, to 0.3 mm, sutures denticulate, teeth to 0.1 mm; staminode 12–15 mm, reaching orifice or slightly exserted, 0.4–0.6 mm diam., tip straight to recurved, distal 3–5 mm sparsely villous, hairs yellow, to 1 mm; style 15–18 mm. **Capsules** 8–12 × 4–6 mm. *2n* = 16.

Flowering Jun–Aug. Sagebrush shrublands, openings in pine forests; 2400–3600 m; Colo., Wyo.

Penstemon saxosorum is known in north-central and northwestern Colorado and south-central Wyoming. Populations in the Bridger Basin in Moffat County, Colorado, and Daggett and Uintah counties, Utah, lie between the main ranges of *P. saxosorum* and *P. subglaber*. Plants in those populations have pollen sacs 0.7–0.9 mm; corollas sparsely glandular and 18–20 mm; and corolla throats 5–7 mm in diameter. M. L. Moorman (1982) believed that they represented an undescribed species. Pending further study, Colorado plants are referred to *P. saxosorum* and Utah plants to *P. subglaber*.

129. Penstemon scariosus Pennell, Contr. U.S. Natl. Herb. 20: 353. 1920 • Garrett's beardtongue [E]

Stems ascending or erect, rarely decumbent, (8–)16–50(–60) cm, glabrous, not glaucous. **Leaves** basal and cauline, or basal not persisting, ± leathery or not, glabrous, not glaucous; basal and proximal cauline (18–)35–180 × (3–)6–15(–24) mm, blade oblanceolate to oblong or linear, base tapered, margins entire, apex obtuse to acute; cauline 3–5(or 6) pairs, sessile or proximals sometimes short-petiolate, 34–115 × 2–14 mm, blade oblanceolate to oblong or linear, base tapered, apex obtuse to acute. **Thyrses** interrupted or continuous, secund, (1–)6–18(–25) cm, axis glabrous or sparsely glandular-pubescent, verticillasters (2 or)3–7(–9), cymes (1 or)2–5-flowered, 2 per node; proximal bracts lanceolate to linear, 25–70(–95) × (1–)3–12(–20) mm; peduncles and pedicels glabrous or glandular-pubescent. **Flowers:** calyx lobes ovate to lanceolate, (4–)4.8–11 × 1.3–3.2(–3.7) mm, glabrous or glandular-pubescent; corolla blue to lavender, with reddish purple nectar guides, ventricose, 18–33 mm, sparsely to moderately glandular-pubescent externally, glabrous, rarely sparsely white-villous, internally abaxially, tube 6–11 mm, throat gradually inflated, slightly constricted at orifice, 6–11 mm diam., rounded abaxially; stamens: longer pair reaching orifice, pollen sacs divergent to opposite, navicular to navicular-sigmoid, (1.3–)1.5–2.4 mm, dehiscing incompletely, proximal ⅕ indehiscent, connective not splitting, sides moderately pubescent, hairs white, to 0.8 mm, sutures denticulate, teeth to 0.1 mm; staminode 11–19 mm, reaching orifice, 0.3–0.8 mm diam., tip straight to recurved, distal 3–7 mm sparsely to moderately villous, hairs yellow, to 0.8 mm; style 12–22 mm. **Capsules** 6–10 × 5–6.5 mm.

Varieties 4 (4 in the flora): w United States.

Penstemon scariosus comprises four varieties centered in the northwestern Colorado Plateau, Uinta Basin, and Wasatch Range.

1. Corollas 25–33 mm; calyx lobes 7–11 mm; pollen sacs 1.8–2.4 mm . . . . 129a. *Penstemon scariosus* var. *scariosus*
1. Corollas 18–28 mm; calyx lobes (4–)4.8–8 mm; pollen sacs (1.3–)1.5–2.2 mm.
   2. Corollas sparsely to moderately glandular-pubescent externally, limbs blue to bluish purple; styles 12–14 mm . . . . 129c. *Penstemon scariosus* var. *cyanomontanus*
   2. Corollas sparsely glandular-pubescent externally, limbs lavender to pale blue or blue; styles 13–22 mm.

[3. Shifted to left margin.—Ed.]

3. Corolla limbs lavender to pale blue; styles 18–22 mm; basal leaves seldom persisting. . . . . 129b. *Penstemon scariosus* var. *albifluvis*
3. Corolla limbs blue; styles 13–17 mm; basal leaves usually persisting. . . . . 129d. *Penstemon scariosus* var. *garrettii*

129a. Penstemon scariosus Pennell var. **scariosus** E

Leaves: basal usually persisting. **Flowers:** calyx lobes 7–11 × 2.6–3.2(–3.7) mm; corolla 25–33 mm, sparsely glandular-pubescent externally, limb blue to bluish purple; pollen sacs 1.8–2.4 mm; style 17–22 mm.

Flowering May–Jul. Sagebrush and mountain mahogany shrublands, pinyon-juniper, pine, aspen, and spruce-fir woodlands; 1600–3200 m; Colo., Utah.

Variety *scariosus* occurs mostly in the Aquarius, Fish Lake, and Wasatch plateaus. Populations are documented in Carbon, Duchesne, Emery, Grand, Juab, Piute, Sanpete, Sevier, Uintah, and Wayne counties, Utah. A collection from the East Tavaputs Plateau in western Garfield County, Colorado (*R. E. Umber, M. L. Moorman & W. Robertson 1336*, RM), appears to be this taxon, though it is somewhat disjunct from other populations. It has obscurely glandular-pubescent calyces and corollas, and pollen sacs ca. 2.4 mm.

129b. Penstemon scariosus Pennell var. **albifluvis** (England) N. H. Holmgren in A. Cronquist et al., Intermount. Fl. 4: 442. 1984 • White River beardtongue C E

Penstemon albifluvis England, Great Basin Naturalist 42: 367. 1982

Leaves: basal seldom persisting. **Flowers:** calyx lobes 5–7.3 × 2.4–3 mm; corolla 22–28 mm, sparsely glandular-pubescent externally, limb lavender to pale blue; pollen sacs (1.6–)1.8–2.2 mm; style 18–22 mm.

Flowering May–Jun. Desert shrublands, pinyon-juniper woodlands; of conservation concern; 1500–2200 m; Colo., Utah.

Variety *albifluvis* occurs in the Uinta Basin. The variety is known in Colorado only from small populations on barren rock outcrops of Green River Shale on Raven Ridge, Rio Blanco County (S. L. O'Kane 1988); Utah populations are restricted to Uintah County.

129c. Penstemon scariosus Pennell var. **cyanomontanus** Neese, Great Basin Naturalist 46: 460. 1986 C E

Leaves: basal usually persisting. **Flowers:** calyx lobes 4.8–8 × 1.3–2.2 mm; corolla 18–22 mm, sparsely to moderately glandular-pubescent externally, limb blue to bluish purple; pollen sacs 1.5–1.7 mm; style 12–14 mm.

Flowering May–Jul. Pinyon-juniper woodlands, sagebrush grasslands; of conservation concern; 1600–2700 m; Colo., Utah.

Variety *cyanomontanus* is known from the Uinta Basin in Moffat County, Colorado, and Uintah County, Utah. Populations are concentrated from the Yampa Plateau (Blue Mountain area) eastward into Dinosaur National Monument along the Yampa River.

129d. Penstemon scariosus Pennell var. **garrettii** (Pennell) N. H. Holmgren in A. Cronquist et al., Intermount. Fl. 4: 442. 1984 E

Penstemon garrettii Pennell, Contr. U.S. Natl. Herb. 20: 353. 1920

Leaves: basal usually persisting. **Flowers:** calyx lobes (4–)6–8 × 1.8–2.6 mm; corolla 18–27 mm, sparsely glandular-pubescent externally, limb blue; pollen sacs (1.3–)1.6–2.2 mm; style 13–17 mm.

Flowering May–Jul. Sagebrush shrublands, oak-maple, aspen, juniper, fir, and pine woodlands; 1700–3000 m; Colo., Utah, Wyo.

Variety *garrettii* occurs largely in the Uinta Mountains and southern Wasatch Mountains. The variety is known in Colorado from Moffat County, in Utah from Carbon, Daggett, Duchesne, Grand, Uintah, Utah, and Wasatch counties, and in Wyoming from Sweetwater and Uinta counties.

130. Penstemon speciosus Douglas ex Lindley, Edwards's Bot. Reg. 15: plate 1270. 1829 (as Pentstemon speciosum) • Royal or showy beardtongue E F

Penstemon speciosus subsp. *kennedyi* (A. Nelson) D. D. Keck

Stems decumbent to ascending or erect, 7–65(–90) cm, glabrous or retrorsely hairy to pubescent, not glaucous. **Leaves** basal and cauline, basal sometimes absent, ± leathery or not, glabrous or retrorsely hairy, not glaucous; basal and proximal cauline

(25–)30–90(–140) × (3–)6–20(–35) mm, blade oblance-olate to elliptic or lanceolate, base tapered, margins entire, apex obtuse to acute; cauline 3–8 pairs, short-petiolate or sessile, 35–70(–150) × 3–20(–30) mm, blade oblanceolate to elliptic, lanceolate, or linear, base tapered to clasping, apex obtuse to acute. **Thyrses** continuous, secund or cylindric, (1–)3–35 cm, axis glabrous or puberulent, rarely glandular or glandular-pubescent, verticillasters (2–)4–14, cymes 1–4-flowered, 2 per node; proximal bracts lanceolate, rarely ovate or linear, (15–)20–80 × (1–)2–23 mm; peduncles and pedicels ascending to erect, glabrous or retrorsely hairy, rarely glandular-pubescent, peduncles to 20 mm, pedicels 1–8 mm. **Flowers:** calyx lobes ovate to lanceolate, rarely obovate, (4–)6–11(–13.5) × 2–4 mm, margins erose, apex acute to acuminate or cuspidate, glabrous or ± glandular proximally; corolla blue to violet, with reddish purple nectar guides, ventricose, 25–35 mm, glabrous externally, glabrous, rarely sparsely white-pilose, internally abaxially, tube 8–13 mm, throat gradually inflated, slightly constricted at orifice, 7–10 (–13) mm diam., 2-ridged abaxially; stamens: longer pair reaching orifice, pollen sacs divergent, sigmoid, (1.8–)2–3 mm, dehiscing incompletely, proximal ⅓ indehiscent, connective not splitting, sides glabrous or sparsely hispidulous, hairs tan, to 0.3 mm, sutures papillate or denticulate, teeth to 0.1 mm; staminode 14–18 mm, included, 0.6–1 mm diam., tip straight to recurved, glabrous or distal 1–7 mm sparsely pilose, hairs yellow to yellowish orange, to 0.8 mm; style 18–23 mm. **Capsules** 10–16 × 6–8 mm. **2n** = 16.

Flowering May–Aug(–Sep). Sagebrush shrublands, pinyon-juniper woodlands, clearings in pine forests; 300–3300 m; Calif., Idaho, Nev., Oreg., Utah, Wash.

Penstemon speciosus is widespread and highly polymorphic. F. S. Crosswhite (1967b) examined morphologic variation and concluded it was clinal.

131. **Penstemon strictiformis** Rydberg, Bull. Torrey Bot. Club 31: 642. 1905 (as Pentstemon) • Stiff beardtongue [E]

Penstemon strictus Bentham subsp. *strictiformis* (Rydberg) D. D. Keck

Stems ascending to erect, (11–)16–55(–70) cm, glabrous, not glaucous. **Leaves** basal and cauline, ± leathery or not, gla-brous, not glaucous; basal and proximal cauline 60–90(–150) × 5–12 mm, blade oblanceolate to lanceolate, base tapered, margins entire, apex acute; cauline 3–5 pairs, sessile, 18–80 × 3–18 mm, blade lanceolate, base tapered to truncate or clasping, apex acute. **Thyrses** continuous or interrupted, ± secund,

5–40 cm, axis glabrous, verticillasters 4–8(–12), cymes 2- or 3(–5)-flowered, 2 per node; proximal bracts lanceolate, 12–46(–115) × 2–15 mm, apex acuminate; peduncles and pedicels glabrous or ± glandular, peduncles to 20 mm, pedicels 2–13(–18) mm. **Flowers:** calyx lobes ovate to lanceolate, (4.5–)6–8(–9) × 2–3 mm, apex acuminate to caudate, glabrous or ± glandular proximally; corolla lavender to blue, without nectar guides, ventricose, (19–)25–32 mm, glabrous externally, glabrous internally, tube 5–7 mm, throat gradually to abruptly inflated, not constricted at orifice, 7–10 mm diam., slightly 2-ridged abaxially; stamens: longer pair slightly exserted, pollen sacs divergent, navicular-sigmoid, 1.3–2.3 mm, dehiscing incompletely, proximal ⅕–¼ indehiscent, connective not splitting, sides sparsely to densely lanate, hairs white, to 1 mm, sutures denticulate, teeth to 0.1 mm; staminode (8–)11–15 mm, included, 0.5–0.1(–1.2) mm diam., tip straight to recurved, distal (2–)4–5 mm sparsely to moderately villous, hairs yellow or white, to 1 mm; style (11–)13–21 mm. **Capsules** 7–11 × 4–6 mm.

Flowering May–Jun. Juniper and pinyon-juniper woodlands; 1700–2400 m; Ariz., Colo., N.Mex., Utah.

Plants of *Penstemon strictiformis* from southwestern Colorado (Montezuma County) and northwestern New Mexico (San Juan County) closely match the type, with large corollas, broad calyx lobes, and large pollen sacs. Plants from southeastern Utah have relatively smaller corollas, narrower calyx lobes, and smaller pollen sacs.

Penstemon strictiformis often is confused with *P. strictus*; many specimens from southwestern Colorado are difficult to place with certainty due to morphologic intermediacy.

132. **Penstemon strictus** Bentham in A. P. de Candolle and A. L. P. P. de Candolle, Prodr. 10: 324. 1846 (as Pentstemon) • Rocky Mountain beardtongue [E]

Penstemon strictus subsp. *angustus* Pennell

Stems ascending to erect, (20–)35–70(–90) cm, glabrous or ± puberulent proximally, rarely distinctly puberulent, not glaucous. **Leaves** basal and cauline, ± leathery or not, glabrous except for antrorsely-scabrous margins or ± puberulent proximally, rarely distinctly puberulent, not glaucous; basal and proximal cauline (30–)50–150 × 5–16(–20) mm, blade oblanceolate, base tapered, margins entire, apex rounded to obtuse; cauline 4–8 pairs, sessile, 40–100 × 2–7(–10) mm, blade lanceolate to linear, base tapered, apex acuminate. **Thyrses** interrupted to ± continuous, secund, (3–)9–40 cm, axis glabrous, verticillasters 4–10(–13), cymes 1- or 2(–4)-flowered, 2 per node; proximal bracts lanceolate to linear, 10–65 × 1–8 mm, apex acuminate; peduncles and pedicels glabrous, peduncles to 27 mm, pedicels 1–20 mm. **Flowers:** calyx lobes ovate, (2.5–)3–5 × 1.3–2.5 mm, apex acute, sometimes acuminate or obtuse, glabrous; corolla purple to violet or blue, with violet nectar guides, ventricose-ampliate, (18–)24–32 mm, glabrous externally, glabrous internally, tube 4–7 mm, throat gradually inflated, slightly constricted at orifice, 6–10 (–12) mm diam., slightly 2-ridged abaxially; stamens: longer pair slightly exserted, pollen sacs divergent to opposite, navicular-sigmoid, 1.7–2.4 mm, dehiscing incompletely, proximal ¼ sometimes indehiscent, connective not splitting, sides moderately to densely lanate to villous, hairs white, to 1.8 mm, sutures denticulate, teeth to 0.1 mm; staminode 11–14 mm, included, 0.5–0.8 mm diam., tip straight, glabrous or distal 1–5 mm sparsely villous, hairs yellow or orange, to 0.7 mm; style 17–21 mm. **Capsules** 7–13 × 4–7 mm. $2n = 16$.

Flowering Jun–Aug. Gravelly sagebrush shrublands, pinyon-juniper woodlands, oak woodlands, spruce-aspen forests; 1700–3500 m; Ariz., Calif., Colo., N.Mex., Utah, Wyo.

A disjunct population of *Penstemon strictus* in Mono County, California, appears to be an introduction, and collections from some counties in east-central Colorado (Arapahoe) and northeastern New Mexico (Harding and Union) also probably are due to recent introductions. *Penstemon strictus* is a popular ornamental; it has been seeded along some highways in Arizona, Colorado, and Wyoming.

133. Penstemon subglaber Rydberg, Bull. Torrey Bot. Club 36: 688. 1909 (as Pentstemon) • Northern smooth beardtongue E

Penstemon glaber Pursh var. *utahensis* S. Watson, Botany (Fortieth Parallel), 217. 1871 (as Pentstemon), not *P. utahensis* Eastwood 1893

Stems erect, (10–)18–90(–130) cm, glabrous proximally, usually sparsely glandular distally, not glaucous. **Leaves** basal and cauline, not leathery, glabrous, not glaucous; basal and proximal cauline 35–140(–170) × 8–24 mm, blade oblanceolate, sometimes lanceolate, base tapered, margins entire, apex obtuse to acute; cauline 3–6(–11) pairs, short-petiolate or sessile, 28–95(–130) × 4–15 mm, blade oblanceolate to lanceolate, base tapered to clasping, apex acute. **Thyrses** interrupted, secund, (4–)12–38 cm, axis glabrous or glandular-pubescent, verticillasters 4–12, cymes 1–3(–6)-flowered, 2 per node; proximal bracts lanceolate, (3–)15–75 × (1–)2–22 mm;

peduncles and pedicels glandular-pubescent, sometimes sparsely so. **Flowers:** calyx lobes ovate to lanceolate, 4.5–7.5 × (1.5–)2.1–2.8 mm, glandular-pubescent; corolla blue to violet or purple, with violet nectar guides, ventricose, 19–28 mm, ± glandular-pubescent externally, glabrous internally, tube 6–10(–11) mm, throat gradually inflated, not constricted at orifice, 7–10 mm diam., rounded abaxially; stamens included or longer pair reaching orifice, pollen sacs opposite, navicular, 1.6–2 mm, dehiscing incompletely, connective not splitting, sides sparsely to moderately hispid or pubescent, hairs white, to 0.2 mm, sutures denticulate, teeth to 0.1 mm; staminode 13–16 mm, reaching orifice or slightly exserted, 0.6–0.8 mm diam., tip straight, distal 2–3 mm sparsely villous, hairs yellow, to 0.7 mm; style 14–18 mm. **Capsules** 8–12 × 5–6 mm. $2n = 16$.

Flowering Jun–Aug. Sagebrush shrublands, clearings in fir forests; 1600–3400 m; Idaho, Utah, Wyo.

Studies of morphology and flavonoids confirm the distinctness of *Penstemon subglaber* from *P. mensarum* and *P. saxosorum* (M. L. Moorman 1982). *Penstemon subglaber* occurs primarily from the Teton and Wind River mountains of western Wyoming through the Uinta and Wasatch mountains to the Tushar Mountains in south-central Utah.

134. Penstemon tidestromii Pennell, Contr. U.S. Natl. Herb. 20: 379. 1920 • Tidestrom's beardtongue E

Penstemon leptanthus Pennell

Stems ascending to erect, 25–50 cm, retrorsely hairy, not glaucous. **Leaves** basal and cauline, ± leathery or not, retrorsely hairy, not glaucous; basal and proximal cauline 40–120 × 4–33 mm, blade obovate to oblanceolate, base tapered, margins entire, apex obtuse to acute; cauline 3–5 pairs, short-petiolate or sessile, 35–70(–90) × 8–22 mm, blade lanceolate to oblanceolate, base tapered to truncate or clasping, apex obtuse to acute. **Thyrses** interrupted, secund, 9–22 cm, axis glabrous or sparsely retrorsely hairy, verticillasters 7–15, cymes 5- or 6-flowered, 2 per node; proximal bracts lanceolate, 12–50 × 3–16 mm; peduncles and pedicels glabrous or sparsely retrorsely hairy. **Flowers:** calyx lobes ovate, 2.5–4 × 1.5–1.7(–2) mm, margins narrowly scarious, apex obtuse, sometimes acute to cuspidate, glabrous; corolla blue to violet, without nectar guides, ventricose, 13–21 mm, glabrous externally, glabrous internally, tube 5–6 mm, throat gradually inflated, not constricted at orifice, 5–7 mm diam., rounded abaxially; stamens included or longer pair reaching orifice, pollen sacs divergent, navicular, 1.3–1.7 mm, dehiscing incompletely, proximal ⅓–½ sometimes indehiscent, connective not splitting, sides glabrous, sutures denticulate, teeth to 0.1 mm; staminode 8–11 mm, included, 0.6–0.8 mm diam., tip straight, distal 5–8 mm sparsely to moderately pilose, hairs yellow, to 1 mm; style 11–13 mm. **Capsules** 6–9 × 4.5–6.5 mm.

Flowering May–Jul. Sagebrush shrublands, pinyon-juniper woodlands; 1600–2500 m; Utah.

Penstemon tidestromii is known from Juab and Sanpete counties. Specimens from the San Pitch Mountains with retrorsely hairy stems, corollas obscurely glandular, and pollen sacs 1.5–1.7 mm and sparsely hispid might be hybrids or introgressants between *P. subglaber* and *P. tidestromii*. *Penstemon leptanthus*, known only from the type collection (*Ward 280*, US), seems to be morphologically confluent with *P. tidestromii*.

135. Penstemon uintahensis Pennell, Contr. U.S.
Natl. Herb. 20: 350. 1920 • Uinta Mountains beardtongue [E]

Stems ascending, 5–15(–24) cm, glabrous, not glaucous. **Leaves** basal and cauline, ± leathery or not, glabrous, not glaucous; basal and proximal cauline (12–)20–60(–100) × (2–)4–12 mm, blade oblanceolate, base tapered, margins entire, apex obtuse to acute; cauline 2–4 pairs, sessile, (10–)15–45 × 1–4(–6) mm, blade lanceolate to linear, base tapered, apex acute. **Thyrses** continuous, secund, (1–)3–15 cm, axis ± glandular-pubescent, verticillasters 3–7, cymes 1(or 2)-flowered, 2 per node; proximal bracts oblanceolate to lanceolate or linear, 9–50 × 1.5–4 mm; peduncles and pedicels glandular-pubescent, sometimes sparsely so. **Flowers:** calyx lobes ovate, (3–)4–6.5(–7) × 2.5–3.5 mm, glandular-pubescent; corolla blue to violet, without nectar guides, ventricose, 17–24 mm, glandular-pubescent externally, glabrous internally, tube 5–8 mm, throat gradually inflated, not constricted at orifice, 5–8 mm diam., rounded abaxially; stamens included or longer pair reaching orifice, pollen sacs divergent, navicular, (0.8–)1–1.4 mm, dehiscing incompletely, connective not splitting, sides hispid, hairs white, to 0.3 mm, sutures denticulate, teeth to 0.1 mm; staminode 10–12 mm, included, 0.2–0.4 mm diam., tip straight, distal 6–8 mm moderately to densely pilose, hairs yellow or yellow-orange, to 1 mm; style 10–12 mm. **Capsules** 8–10.5 × 4.5–7 mm.

Flowering Jun–Aug. Rocky subalpine and alpine meadows; 3000–3800 m; Utah.

Penstemon uintahensis is known from the eastern part of the Uinta Mountains in Daggett, Duchesne, Summit, and Uintah counties from the vicinity of Mt. Beulah east to Dyer Mountain.

136. Penstemon virgatus A. Gray in W. H. Emory, Rep.
U.S. Mex. Bound. 2(1): 113. 1859 (as Pentstemon)
• Upright blue beardtongue [E]

Stems erect, sometimes ascending, (12–)20–65(–90) cm, glabrous or puberulent, not glaucous. **Leaves** basal and cauline or, often, basal absent, not leathery, glabrous or puberulent, not or slightly glaucous; basal and proximal cauline (20–)60–114 × (3–)12–21 mm, blade oblanceolate or lanceolate, rarely linear, base tapered, margins entire, apex obtuse to acute; cauline 5–11 pairs, sessile, 40–118 × 3–17 mm, blade lanceolate, base tapered to truncate, apex obtuse to acute. **Thyrses** continuous, sometimes interrupted, secund, (3–)6–40 cm, axis glabrous or retrorsely hairy, verticillasters (3–)6–14, cymes 2–5-flowered, 2 per node; proximal bracts lanceolate to linear, 19–74(–98) × 2–10 mm; peduncles and pedicels glabrous or retrorsely hairy. **Flowers:** calyx lobes ovate to elliptic, 2.2–5 × 1.3–2.2 mm, glabrous or puberulent; corolla violet to lavender, pink-lavender, or purple, with reddish purple nectar guides, ventricose, 17–27 mm, glabrous externally, sparsely white-villous, rarely glabrous, internally abaxially, tube 5–8 mm, throat gradually to abruptly inflated, slightly constricted at orifice, 6–8 mm diam., rounded abaxially; stamens: longer pair slightly exserted, pollen sacs divergent to opposite, navicular, 1.5–1.8 mm, dehiscing completely or incompletely, connective splitting or not, sides glabrous, sutures smooth or papillate; staminode 12–14 mm, reaching orifice or exserted, 0.7–1.2 mm diam., tip straight, glabrous; style 13–15 mm. **Capsules** 9–14 × 4–8 mm.

Varieties 2 (2 in the flora): w United States.

Penstemon virgatus is treated here to include two essentially allopatric varieties. F. S. Crosswhite (1967c) circumscribed the species more broadly, including as subspecies two additional taxa treated here as species: *P. pseudoputus* and *P. putus*.

Whole plants or roots of *Penstemon virgatus* are used by the Ramah Navajo of western New Mexico as a medicine (D. E. Moerman 1998).

1. Stems puberulent; leaves ± puberulent . . . .
 . . . . 136a. *Penstemon virgatus* var. *virgatus*
1. Stems glabrous; leaves glabrous . . . .
 . . . . 136b. *Penstemon virgatus* var. *asa-grayi*

136a. Penstemon virgatus A. Gray var. **virgatus** E

Stems puberulent. **Leaves** ± puberulent. *2n* = 16.

Flowering Jun–Oct. Sandy to gravelly hillsides, oak woodlands, pine forests; 2000–2900 m; Ariz., Colo., N.Mex.

Variety *virgatus* occurs discontinuously in the mountains of north-central Arizona and of north-central and western New Mexico. Plants from Arizona, where the variety is sympatric with *Penstemon putus* (and sometimes collected with it), often are more slender and bear smaller flowers than do plants from New Mexico.

136b. Penstemon virgatus A. Gray var. **asa-grayi** (Crosswhite) Dorn, Vasc. Pl. Wyoming, 300. 1988 E

Penstemon virgatus subsp. *asa-grayi* Crosswhite, SouthW. Naturalist 10: 318. 1965 (as asa-grayii)

Stems glabrous. **Leaves** glabrous. *2n* = 16.

Flowering Jun–Sep. Sandy or gravelly roadsides and hillsides, oak woodlands, pine woodlands; 1400–3000 m; Colo., N.Mex., Wyo.

Variety *asa-grayi* is known from the Front Range of the Southern Rocky Mountains. The variety is encountered from Albany and Laramie counties in southeastern Wyoming, south to Mora County, New Mexico. A specimen from Sublette County, Wyoming, likely is from an introduction. Variety *asa-grayi* sometimes grows in mixed populations with *Penstemon strictus*, although the two species can be distinguished immediately by their pollen sacs, which are lanate and divergent in *P. strictus* and glabrous and opposite in *P. virgatus*.

The name *Penstemon unilateralis* Rydberg, a synonym of *P. secundiflorus*, was widely misapplied to this taxon in much of the 1900s; F. S. Crosswhite (1965f, 1967c) clarified and corrected the situation.

137. Penstemon wardii A. Gray, Proc. Amer. Acad. Arts 12: 82. 1876 (as Pentstemon wardi) • Ward's beardtongue E

Stems ascending to erect, (6–)15–42 cm, ± retrorsely hairy or pubescent, not glaucous. **Leaves** basal and cauline, ± leathery or not, densely retrorsely hairy, not glaucous; basal and proximal cauline 20–70 × 3–16(–22) mm, blade oblanceolate, base tapered, margins entire, apex obtuse to acute; cauline 2–5 pairs, short-petiolate or sessile, 19–72 × 3–9 mm, blade oblanceolate to lanceolate, base tapered to truncate, apex obtuse to acute. **Thyrses** continuous, secund, (4–)10–22 cm, axis retrorsely hairy to retrorsely pubescent or pubescent, verticillasters (3–)5–8, cymes 1–5-flowered, 2 per node; proximal bracts oblanceolate to lanceolate, 25–70 × 3–9 mm; peduncles and pedicels retrorsely hairy and glandular-pubescent. **Flowers:** calyx lobes ovate, 6–8.5 × 2.8–3.8 mm, glandular-pubescent; corolla blue to lavender, with or without faint lavender nectar guides, ventricose, 24–30 mm, glandular-pubescent externally, glabrous internally, tube 8–12 mm, throat gradually inflated, slightly constricted at orifice, 6–9(–12) mm diam., slightly 2-ridged abaxially; stamens: longer pair reaching orifice or slightly exserted, pollen sacs divergent, sigmoid, 2–2.5 mm, dehiscing incompletely, proximal ⅕–¼ indehiscent, connective not splitting, sides glabrous, sutures denticulate, teeth to 0.1 mm; staminode 16–18 mm, included, 0.5–0.8 mm diam., tip straight, glabrous; style 20–24 mm. **Capsules** 12–15 × 8–10 mm.

Flowering May–Jul. Clayey or loamy soils, foothills, desert shrub communities, pinyon-juniper woodlands; 1500–2000 m; Utah.

Penstemon wardii is known from the foothills surrounding Sevier Valley in Millard, Piute, Sanpete, and Sevier counties (E. C. Neese and N. D. Atwood 2003).

26b.12. Penstemon Schmidel (subg. **Penstemon**) sect. **Penstemon** E

Herbs or subshrubs. Stems glabrous, glabrate, retrorsely hairy, puberulent, or pubescent, sometimes glandular-pubescent, glandular-lanate, or glandular-villous, rarely glutinous, glaucous or not. **Leaves** basal and cauline, sometimes basals absent or poorly developed, opposite, rarely whorled (*P. deustus*), subopposite (*P. deustus, P. sudans*), or alternate (*P. gairdneri, P. seorsus*), leathery or not, glabrous, glandular, puberulent, glandular-pubescent, glandular-lanate, retrorsely hairy, pubescent, or glabrate, not glaucous or rarely so (*P. euglaucus, P. glaucinus*), sometimes

± glaucescent (*P. albomarginatus, P. griffinii, P. inflatus, P. oliganthus, P. tracyi*); basal and proximal cauline petiolate; cauline sessile or petiolate, blade obovate, ovate, triangular-ovate, spatulate, oblanceolate, lanceolate, oblong, elliptic, or linear, rarely lyrate, margins entire or toothed. **Thyrses** continuous or interrupted, cylindric to conic, rarely secund, axis glabrous or glandular-pubescent, puberulent, pubescent, or retrorsely hairy, rarely glandular-lanate, cymes 1 or 2 per node; peduncles and pedicels spreading to ascending or erect. **Flowers:** calyx lobes: margins entire, erose, or lacerate, ± scarious, sometimes herbaceous, glandular-pubescent or glabrous, rarely glabrescent, puberulent, retrorsely hairy, glutinous, or glandular-ciliolate; corolla white, ochroleucous, pink, lavender, blue, violet, purple, or yellow, bilaterally symmetric, bilabiate or strongly bilabiate, rarely weakly bilabiate (*P. tubaeflorus*), not personate, sometimes personate (*P. hirsutus, P. oklahomensis, P. tenuiflorus*) or nearly personate (*P. australis, P. degeneri, P. griffinii, P. laxiflorus, P. oliganthus*), funnelform, tubular-funnelform, tubular, ampliate, or ventricose, glabrous, glabrate, glabrescent, or glandular-pubescent externally, glabrous or hairy internally abaxially, rarely glandular-pubescent abaxially and adaxially, throat slightly to abruptly inflated to slightly ventricose, not constricted at orifice, slightly to prominently 2-ridged adaxially, rarely rounded; stamens included or longer pair reaching orifice or exserted, filaments glabrous, pollen sacs opposite or divergent, rarely parallel, navicular to subexplanate or explanate, rarely saccate, dehiscing completely, rarely incompletely, distal ⅕–⅔ sometimes indehiscent, connective splitting, rarely not, sides glabrous, rarely hairy, sutures papillate, sometimes smooth; staminode included to exserted, flattened distally, 0.2–0.8(–1.1) mm diam., tip straight to recurved, distal 5–60(–90)% hairy, hairs to 3 mm, sometimes glabrous; style glabrous. **Capsules** glabrous, rarely glandular-puberulent distally. **Seeds** tan, brown, gray, or black, angled to slightly rounded, 0.4–2(–3.5) mm.

Species 62 (62 in the flora): North America.

Elaborate subsectional classifications were presented by D. D. Keck (1945) and F. S. Crosswhite (1965b). Those are not followed; a more well-resolved phylogeny of the genus is needed.

SELECTED REFERENCES   Clements, R. K., J. M. Baskin, and C. C. Baskin. 1998. The comparative biology of the two closely-related species *Penstemon tenuiflorus* Pennell and *P. hirsutus* (L.) Willd. (Scrophulariaceae, section *Graciles*): I. Taxonomy and geographical distribution. Castanea 63: 138–153. Clements, R. K., J. M. Baskins, and C. C. Baskins. 1998b. The comparative biology of the two closely-related species *Penstemon tenuiflorus* Pennell and *P. hirsutus* (L.) Willd. (Scrophulariaceae, section *Graciles*): II. Reproductive biology. Castanea 63: 299–309. Crosswhite, F. S. 1965b. Revision of *Penstemon* section *Penstemon* Scrophulariaceae II. A western alliance in series *Graciles*. Amer. Midl. Naturalist 74: 429–442. Crosswhite, F. S. 1965c. Subdivisions of *Penstemon* section *Penstemon*. Sida 2: 160–162. Crosswhite, F. S. 1965e. Revision of *Penstemon* (Sect. *Penstemon*) Series *Graciles* (Scrophulariaceae) with a Synopsis of the Genus. M.S. thesis. University of Wisconsin. Keck, D. D. 1940b. Studies in *Penstemon* VII. The subsections *Gairdneriani, Deusti,* and *Arenarii* of the *Graciles,* and miscellaneous new species. Amer. Midl. Naturalist 23: 594–616. Keck, D. D. 1945. Studies in *Penstemon* VIII. A cyto-taxomonic account of the section *Spermunculus*. Amer. Midl. Naturalist 33: 128–206. Koelling, A. C. 1964. Taxonomic Studies in *Penstemon deamii* and Its Allies. Ph.D. dissertation. University of Illinois.

Key to *Penstemon* sect. *Penstemon* East of the Rocky Mountains

1. Pollen sacs parallel, saccate, dehiscing incompletely. . . . . 174. *Penstemon multiflorus*
1. Pollen sacs opposite or divergent, navicular, subexplanate, or explanate, dehiscing completely.
   2. Corollas glabrous externally . . . . 181. *Penstemon procerus*
   2. Corollas glandular-pubescent externally.
      3. Corollas glandular-pubescent internally, without nectar guides; pollen sacs explanate . . . . 194. *Penstemon tubaeflorus*
      3. Corollas hairy internally, not glandular-pubescent, with, sometimes without, nectar guides; pollen sacs navicular or subexplanate.
         4. Corollas personate, with or without nectar guides.

5. Corollas lavender or purplish; leaves glabrous or sparsely glandular-lanate along midveins, rarely moderately glandular-lanate . . . . 166. *Penstemon hirsutus*

5. Corollas white to ochroleucous; leaves puberulent or retrorsely hairy, or moderately to densely glandular-lanate, especially along major veins abaxially.

    6. Leaves puberulent or retrorsely hairy; cauline leaf blades lanceolate to linear; Oklahoma, Texas . . . . 175. *Penstemon oklahomensis*

    6. Leaves moderately to densely glandular-lanate, especially along major veins abaxially; cauline leaf blades ovate to lanceolate; Alabama, Kentucky, Mississippi, Tennessee. . . . . 191. *Penstemon tenuiflorus*

[4. Shifted to left margin.—Ed.]

4. Corollas not personate, sometimes nearly personate, with nectar guides.

  7. Stems glabrous or glabrate, puberulent, or retrorsely hairy.

    8. Corollas tubular to tubular-funnelform.

      9. Cauline leaf pairs (6–)8–12; corollas 15–19 mm; capsules 4–7 mm . . . . 143. *Penstemon arkansanus* (in part)

      9. Cauline leaf pairs (3–)5–7; corollas 20–28(–30) mm; capsules 8–10 mm . . . . 171. *Penstemon laxiflorus* (in part)

    8. Corollas ventricose, sometimes funnelform.

     10. Corollas (17–)20–35 mm; calyx lobes 4–9 mm.

       11. Pollen sac sides glabrous, rarely sparsely pubescent; calyx lobes 5–9 × 0.8–1.9 mm, apices acuminate; corollas pale lavender to violet . . . 148. *Penstemon calycosus*

       11. Pollen sac sides sparsely pubescent, rarely glabrous; calyx lobes 4–8 × 2–3 mm, apices acute to acuminate; corollas white, sometimes tinged lavender . . . . 155. *Penstemon digitalis*

     10. Corollas 10–22 mm; calyx lobes 1.5–6 mm.

       12. Corollas 10–14 mm, funnelform; cauline leaf margins sharply serrate, sometimes ± serrate . . . . 192. *Penstemon tenuis*

       12. Corollas 15–22 mm, ventricose; cauline leaf margins entire or denticulate, serrulate, or ± serrate.

        13. Calyx lobes 1.5–3.5(–4) mm; corollas white . . . . 152. *Penstemon deamii*

        13. Calyx lobes 3–6 mm; corollas pale lavender to violet . . . . 170. *Penstemon laevigatus*

  7. Stems retrorsely hairy and glandular-pubescent, glandular-lanate, or glandular-villous, sometimes sparsely and only distally.

    14. Corollas 28–35 mm; staminodes: distal 13–15 mm ± pilose; proximal bracts 56–130 × 23–55 mm, margins sharply serrate . . . . 187. *Penstemon smallii*

    14. Corollas 10–30 mm; staminodes: distal 4–12 mm hairy; proximal bracts 3–95 × 1–45 mm, margins entire, ± dentate, ± crenate, serrulate, or serrate.

     15. Corollas ventricose. . . . . 149. *Penstemon canescens*

     15. Corollas tubular to tubular-funnelform.

      16. Corollas 20–28(–30) mm; styles 10–20 mm; capsules 6–10 mm.

       17. Stems retrorsely hairy and glandular-lanate, glandular hairs sometimes sparse; cymes each usually with only 1 branch elongating, other branches distinctly shorter. . . . . 145. *Penstemon australis*

       17. Stems puberulent to retrorsely hairy, rarely also glandular-pubescent distally; cymes each usually with all branches elongating and of equal length . . . . 171. *Penstemon laxiflorus* (in part)

      16. Corollas 10–24 mm; styles 9–17 mm; capsules 4–8 mm.

       18. Corollas light lavender to lavender or purple; calyx lobes (1.5–)2–6 mm.

        19. Calyx lobes (1.5–)2–3.5 mm; cauline leaves 11–32 mm wide, blades ovate to lanceolate or oblanceolate . . . . 147. *Penstemon brevisepalus*

        19. Calyx lobes 4–6 mm; cauline leaves (2–)4–10(–15) mm wide, blades lanceolate to linear. . . . . 162. *Penstemon gracilis*

       18. Corollas white, sometimes tinged lavender or blue; calyx lobes (2–) 2.8–5.5 mm.

[20. Shifted to left margin.—Ed.]

20. Corollas 10–16 mm; cauline leaves 120–150 × 30–40 mm . . . . 169. *Penstemon kralii*
20. Corollas 15–22 mm; cauline leaves 22–100 × 4–24 mm.
  21. Stems retrorsely hairy, sometimes also ± glandular-pubescent distally; leaves glabrous or sparsely retrorsely hairy, especially along midveins and margins abaxially, glabrous or sparsely puberulent adaxially; cauline leaves (6–)8–12 pairs, blades lanceolate. . . . . 143. *Penstemon arkansanus* (in part)
  21. Stems retrorsely hairy and glandular-villous proximally, retrorsely hairy and glandular-pubescent distally, glandular hairs sometimes sparse; leaves sparsely to densely pubescent, sometimes also with scattered glandular hairs abaxially, sparsely pubescent adaxially; cauline leaves 4–8 pairs, blades ovate to lanceolate. . . . . 178. *Penstemon pallidus*

Key to *Penstemon* sect. *Penstemon* from the Rocky Mountains Westward

1. Subshrubs; leaves alternate, subopposite, whorled, or opposite.
  2. Leaves: margins dentate, dentate-serrate, serrate, or serrulate.
    3. Thyrse axes, peduncles, pedicels, and calyx lobes ± glandular-pubescent . . . . 154. *Penstemon deustus* (in part)
    3. Thyrse axes, peduncles, pedicels, and calyx lobes densely glandular-pubescent. . . . . 190. *Penstemon sudans*
  2. Leaves: margins entire.
    4. Corollas white to ochroleucous, sometimes tinged lavender (abaxial lobes sometimes brown) . . . . 154. *Penstemon deustus* (in part)
    4. Corollas blue to violet or purple.
      5. Corollas glandular-pubescent internally . . . . 159. *Penstemon gairdneri*
      5. Corollas glabrous internally . . . . 186. *Penstemon seorsus*
1. Herbs; leaves opposite.
  6. Corollas glabrous externally; calyx lobes glabrous, rarely puberulent, retrorsely hairy, glandular-pubescent, or glandular-ciliolate.
    7. Corollas yellow or light yellow.
      8. Corollas 6–11 mm; pollen sacs 0.4–0.5 mm; staminodes 0.3–0.4 mm diam., distal 0.5–1 mm pilose . . . . 151. *Penstemon confertus* (in part)
      8. Corollas 12–16 mm; pollen sacs 0.7–0.9 mm; staminodes 0.4–0.7 mm diam., distal 2–3 mm moderately pilose . . . . 158. *Penstemon flavescens* (in part)
    7. Corollas white, ochroleucous, pink, pinkish lavender, lavender, blue, violet, or purple.
      9. Pollen sacs dehiscing incompletely, distal ⅕–¼ indehiscent. . . . . 161. *Penstemon globosus*
      9. Pollen sacs dehiscing completely.
        10. Staminodes glabrous.
          11. Proximal bracts oval to oblong; corollas pinkish lavender, 13–18 mm . . . . 139. *Penstemon albomarginatus*
          11. Proximal bracts lanceolate to linear; corollas violet, blue, purple, lavender, or white, 7–15 mm.
            12. Corollas white or lavender . . . . 180. *Penstemon pratensis* (in part)
            12. Corollas violet to blue or purple . . . . 181. *Penstemon procerus* (in part)
        10. Staminodes hairy.
          13. Stems glabrous, glutinous, usually covered with sand; caudices rhizomelike . . . . 141. *Penstemon arenarius*
          13. Stems glabrous, retrorsely hairy, or puberulent, not glutinous and covered with sand; caudices not rhizomelike.
            14. Stems and leaves glaucous. . . . . 157. *Penstemon euglaucus*
            14. Stems and leaves not glaucous, rarely glaucescent.
              15. Proximal bracts elliptic to ovate; thyrses 2–4 cm; stems 8–12 cm. . . . . 193. *Penstemon tracyi*
              15. Proximal bracts lanceolate to linear, rarely ovate; thyrses 0.5–23(–50) cm; stems 3–70(–120) cm.

16. Calyx lobes 1–1.8 mm, apices truncate to cuspidate; corollas 7–9(–11) mm ....150. *Penstemon cinicola*
16. Calyx lobes 1.2–8 mm, apices truncate to obtuse, acute, short- to long-caudate, short- to long-cuspidate, acuminate, or mucronate; corollas 6–20 mm.
    17. Leaves essentially cauline, basal absent or poorly developed.
        18. Corollas densely yellow-villous internally abaxially; pollen sacs explanate; stems ± retrorsely hairy...................... 172. *Penstemon laxus*
        18. Corollas sparsely white-villous internally abaxially; pollen sacs navicular; stems glabrous or sparsely puberulent.............. 197. *Penstemon watsonii*
    17. Leaves basal and cauline, basal sometimes poorly developed.
        19. Corollas yellow, white, ochroleucous, pink, or lavender; calyx lobes glabrous.
            20. Pollen sacs explanate, 0.4–0.5 mm; corollas 6–11 mm ....151. *Penstemon confertus* (in part)
            20. Pollen sacs navicular, 0.6–0.8 mm; corollas 10–15 mm ....180. *Penstemon pratensis* (in part)
        19. Corollas violet to blue or purple; calyx lobes glabrous, retrorsely hairy, or puberulent.
            21. Corollas 7–11 mm, throats 2–3 mm diam.; pollen sacs explanate, 0.3–0.6 mm....... .... 181. *Penstemon procerus* (in part)
            21. Corollas 10–20 mm, throats 3–4.5 mm diam.; pollen sacs navicular, 0.5–1.1 mm ....185. *Penstemon rydbergii*

[6. Shifted to left margin.—Ed.]

6. Corollas glandular-pubescent externally, rarely glabrescent; calyx lobes glandular-pubescent, rarely glabrous or glabrescent.
22. Pollen sac sides sparsely puberulent................................. 146. *Penstemon bleaklyi*
22. Pollen sac sides glabrous.
    23. Capsules glandular-pubescent distally.........................198. *Penstemon whippleanus*
    23. Capsules glabrous.
        24. Thyrses ± conic, interrupted; peduncles and pedicels spreading to ascending or erect.
            25. Cauline leaf blades oblanceolate to lanceolate or linear, rarely ovate, margins entire; staminodes: distal 6–10 mm pilose.
                26. Pollen sacs 0.9–1.1 mm; corollas 14–19 mm; Colorado ...... 153. *Penstemon degeneri*
                26. Pollen sacs 1.3–1.6 mm; corollas 17–27 mm; New Mexico..... 168. *Penstemon inflatus*
            25. Cauline leaf blades oblong to ovate, triangular-ovate, or lanceolate, margins serrulate, serrate, or dentate, sometimes entire; staminodes: distal 1–5 mm pilose to villous.
            27. Thyrse axes sparsely to moderately glandular-pubescent; calyx lobes ovate, margins broadly scarious ....199. *Penstemon wilcoxii* (in part)
            27. Thyrse axes moderately to densely glandular-pubescent; calyx lobes ovate to lanceolate, margins narrowly to broadly scarious.
                28. Corollas 15–22 mm; staminodes: distal 1–2 mm densely pilose, hairs yellow; cauline leaf blades oblong to ovate or triangular-ovate.. 177. *Penstemon ovatus*
                28. Corollas 12–18 mm; staminodes: distal 2–3 mm sparsely to moderately pilose, hairs yellow or yellow-orange; cauline leaf blades lanceolate to oblong..............189. *Penstemon subserratus* (in part)

[24. Shifted to left margin.—Ed.]

24. Thyrses cylindric or secund, continuous or interrupted; peduncles and pedicels ascending to erect.

    29. Stems and leaves glaucous; corollas without nectar guides . . . . 160. *Penstemon glaucinus* (in part)

    29. Stems and leaves not glaucous, leaves rarely ± glaucescent (*P. griffinii, P. oliganthus*); corollas with, sometimes without, nectar guides.

        30. Caudices rhizomelike; leaves cauline; alpine talus, gravel slopes, boulder fields, 3200–4200 m . . . . 164. *Penstemon harbourii*

        30. Caudices not rhizomelike; leaves basal and cauline, basal sometimes poorly developed, rarely absent; prairies, meadows, woodlands, forests, 30–3900 m.

            31. Staminodes exserted; corollas ventricose, ventricose-ampliate, or ampliate.

                32. Corollas 13–18 mm; styles 10–13 mm; staminodes: distal 2–6 mm sparsely pilose, hairs yellow, to 0.6 mm, rarely glabrous; pollen sacs 0.8–1.2 mm. . . . . .140. *Penstemon anguineus*

                32. Corollas 20–30 mm; styles 16–19 mm; staminodes: distal 6–10 mm sparsely to moderately lanate, hairs yellow or white, to 3 mm; pollen sacs 1.2–1.5 mm . . . . .184. *Penstemon rattanii*

            31. Staminodes included or reaching orifices, rarely slightly exserted; corollas tubular, tubular-funnelform, or funnelform.

              33. Proximal cauline leaves 1–5(–7) mm wide, blades linear to narrowly elliptic or narrowly oblanceolate, margins entire.

                  34. Calyx lobes ovate to lanceolate, 2.2–4.6 × 1–2.4 mm; stems 6–20(–25) cm; corollas 11–17 mm, blue to violet. .142. *Penstemon aridus*

                  34. Calyx lobes obovate to ovate or lanceolate, 1.5–3 × 0.7–1.5 mm; stems 20–50(–65) cm; corollas 8–10 mm, violet to light blue, pink, or light purple . . . . .179. *Penstemon peckii*

              33. Proximal cauline leaves 2–55 mm wide, blades triangular to ovate, lanceolate, oblanceolate, elliptic, obovate, or spatulate, margins serrate, dentate, serrulate, denticulate, or entire.

                35. Leaves essentially cauline, basal absent or poorly developed; cauline leaf blade margins entire . . . . 183. *Penstemon radicosus*

                35. Leaves basal and cauline; cauline leaf blade margins serrate, dentate, serrulate, denticulate, or entire.

                  36. Caudices herbaceous; corollas tubular . . . . 162. *Penstemon gracilis*

                  36. Caudices not herbaceous; corollas tubular, tubular-funnelform or funnelform.

                    37. Thyrses secund; proximal bracts lanceolate to linear.

                      38. Cauline leaf blades linear; corollas densely golden-villous internally abaxially, hairs passing onto limbs (and deep into throats), 17–25 mm; Colorado, New Mexico . . . . .163. *Penstemon griffinii*

                      38. Cauline leaf blades oblanceolate to lanceolate; corollas sparsely to moderately white-villous to white-lanate internally abaxially, 11–20 mm; Arizona, New Mexico . . . . 176. *Penstemon oliganthus*

                  37. Thyrses cylindric to conic, rarely ± secund; proximal bracts ovate to lanceolate.

                    39. Corollas light yellow; leaves glabrous . . . . 158. *Penstemon flavescens* (in part)

                    39. Corollas blue, lavender, violet, bluish purple, or purple, rarely light yellow or whitish; leaves glabrous or puberulent to retrorsely hairy, pubescent, or glandular-pubescent.

[40. Shifted to left margin.—Ed.]

40. Corollas 9–12 mm, throats 2–3 mm diam.; pollen sacs 0.5–0.6 mm; thyrses 1–6 cm; Washington . . . . .196. *Penstemon washingtonensis*
40. Corollas (7–)8–23 mm, throats 3–9 mm diam.; pollen sacs 0.5–1.2 mm; thyrses 1–39 cm; widespread, including Washington.
  41. Stems and leaves glaucous . . . . 160. *Penstemon glaucinus* (in part)
  41. Stems and leaves not glaucous.
    42. Basal and proximal cauline leaf blades triangular to ovate or lanceolate; peduncles and pedicels spreading to ascending; corolla throats 4–9 mm diam. . . . . .199. *Penstemon wilcoxii* (in part)
    42. Basal and proximal cauline leaf blades ovate to obovate, lanceolate, oblanceolate, elliptic, or spatulate; peduncles and pedicels ascending to erect; corolla throats 3–6 mm diam.
      43. Staminodes: distal 4–6 mm villous or pilose.
        44. Pollen sacs 0.9–1.2 mm; cauline leaves 12–27 mm wide; peduncles and pedicels erect; corollas 16–20 mm, tube 5–7 mm; calyx lobes lanceolate . . . . .173. *Penstemon metcalfei*
        44. Pollen sacs 0.5–0.8 mm; cauline leaves 1–14 mm wide; peduncles and pedicels ascending to erect; corollas (7–)8–19 mm, tube 2–5 mm; calyx lobes ovate to lanceolate.
          45. Leaves glabrous, retrorsely hairy, or puberulent, blade margins entire, sometimes serrulate or serrate; California, Colorado, Idaho, Montana, Nevada, Oregon, Utah, Washington, Wyoming . . . . . . 167. *Penstemon humilis* (in part)
          45. Leaves glabrous, blade margins serrulate or denticulate, sometimes entire; Colorado, Wyoming . . . . .195. *Penstemon virens*
      43. Staminodes: distal 0.5–4 mm villous or pilose.
        46. Corollas essentially without nectar guides; leaves glabrous; blade margins entire . . . . 165. *Penstemon heterodoxus*
        46. Corollas with nectar guides; leaves retrorsely hairy, puberulent, or glabrous; blade margins ± serrulate, serrate, dentate, denticulate, or entire.
        47. Staminodes 9–12 mm; corollas 14–23 mm, throats 5–6 mm diam. . . . . .156. *Penstemon elegantulus* (in part)
        47. Staminodes (3–)6–10 mm; corollas 7–20 mm, throats 3–6 mm diam.
          48. Cauline leaf blade margins serrulate to serrate or dentate, sometimes entire; proximal bract margins serrate to serrulate, sometimes entire; cymes 3–11-flowered.
            49. Pollen sacs 0.6–0.8 mm; staminodes: distal 0.5–1 mm densely pilose, hairs yellow, to 0.8 mm; basal and cauline leaves retrorsely hairy or pubescent, sometimes also glandular-pubescent; cauline leaf blades triangular-ovate to lanceolate . . . . .182. *Penstemon pruinosus*
            49. Pollen sacs 0.7–1.1 mm; staminodes: distal 2–3 mm sparsely to moderately pilose, hairs yellow or yellow-orange, to 1.5 mm; basal and cauline leaves glabrous or puberulent; cauline leaf blades lanceolate to oblong . . . 189. *Penstemon subserratus* (in part)
          48. Cauline leaf blade margins entire, sometimes ± serrulate, or dentate, rarely ± serrate; proximal bract margins entire or ± serrate to dentate; cymes (1–)2–8(–11)-flowered.
            50. Pollen sacs saccate, dehiscing incompletely, distal ⅕ or less sometimes indehiscent . . . . .144. *Penstemon attenuatus* (in part)
            50. Pollen sacs navicular, subexplanate, or explanate, dehiscing completely.
              51. Staminodes 9–12 mm; pollen sacs 0.9–1.2 mm; corollas 14–23 mm . . . . 156. *Penstemon elegantulus* (in part)
              51. Staminodes (3–)6–9 mm; pollen sacs 0.5–1.2 mm; corollas 7–20 mm.

[52. Shifted to left margin.—Ed.]

52. Basal and proximal cauline leaf margins ± serrate or dentate; pollen sacs subexplanate to explanate . . . . .138.   *Penstemon albertinus*
52. Basal and proximal cauline leaf margins entire, sometimes denticulate, serrate, or serrulate; pollen sacs navicular to subexplanate.
 53. Cauline leaves 1–3 pairs, sessile, bases truncate; proximal bracts 2–7 mm wide; staminodes: distal 0.5–1.5 mm densely pilose, hairs to 0.7 mm . . . . .188.   *Penstemon spatulatus*
 53. Cauline leaves (1–)2–5(–7) pairs, sessile or proximals short- to long-petiolate, bases tapered to clasping; proximal bracts 1–15(–28) mm wide; staminodes: distal 1–4 mm sparsely to densely pilose, hairs to 1.2 mm.
  54. Stems glabrous or retrorsely hairy; cauline leaves 2–25(–38) mm wide; peduncles and pedicels erect; calyx lobe margins erose to lacerate, apices acuminate to short-caudate or long-acuminate; pollen sacs 0.6–1.2 mm; staminodes 0.3–0.4 mm diam. . . . . .144.   *Penstemon attenuatus* (in part)
  54. Stems retrorsely hairy; cauline leaves 1–11 mm wide; peduncles and pedicels ascending to erect; calyx lobe margins entire or erose, apices acute; pollen sacs 0.5–0.8 mm; staminodes 0.1–0.3 mm diam. . . . . 167.   *Penstemon humilis* (in part)

138. Penstemon albertinus Greene, Leafl. Bot. Observ. Crit. 1: 167. 1906 (as Pentstemon) • Alberta beardtongue E

Penstemon caelestinus Pennell

Herbs. Stems ascending to erect, (10–)15–38 cm, glabrous or sparsely retrorsely hairy, sometimes also sparsely glandular-pubescent. **Leaves** basal and cauline, ± leathery or not, glabrous, sometimes sparsely retrorsely hairy along midvein abaxially, not glaucous; basal and proximal cauline (13–)40–90 × (3–)6–18 mm, blade ovate to oblanceolate or lanceolate, base tapered, margins ± serrate or dentate, apex obtuse to acute; cauline 2–6 pairs, short-petiolate or sessile, 15–55 × 3–10 mm, blade lanceolate, proximals sometimes oblanceolate to oblong, base tapered or truncate, margins entire or ± serrate to ± dentate, apex obtuse to acute. **Thyrses** continuous or ± interrupted, cylindric, 4–19 cm, axis glabrous or ± glandular-pubescent, verticillasters 4–8, cymes (1 or)2–6-flowered, 2 per node; proximal bracts lanceolate, 9–40 × 3–15 mm, margins entire or ± serrate to dentate; peduncles and pedicels erect, sometimes ascending, glandular-pubescent. **Flowers:** calyx lobes ovate, 2.8–4.8 × 0.9–1.8 mm, glandular-pubescent; corolla blue to light blue or violet, with reddish purple nectar guides, funnelform, 13–20 mm, glandular-pubescent externally, sparsely white-villous internally abaxially, tube 4–5 mm, throat gradually inflated, 4–5 mm diam., 2-ridged abaxially; stamens included, pollen sacs opposite, subexplanate to explanate, 0.5–0.8 mm, dehiscing completely, connective splitting, sides glabrous, sutures smooth or papillate; staminode 8–9 mm, included, 0.2–0.3 mm diam., tip straight to recurved, distal 0.5–2 mm ± pilose, hairs golden yellow, to 0.8 mm; style 9–11 mm. **Capsules** 4–6 × 3–4 mm, glabrous. $2n = 16$.

Flowering May–Aug. Gravelly slopes, cliffs, rocky clearings, roadcuts in coniferous forests; 800–2800 m; Alta., B.C.; Idaho, Mont.

Penstemon caelestinus has been treated as a synonym of *P. albertinus* (D. D. Keck 1945; A. Cronquist 1959). D. V. Clark (1971) grew these and related species in transplant gardens and considered *P. caelestinus* to be distinct based on flower color, leaf shape and color, phenology, and habitat. A broad zone of hybridization between *P. albertinus* and *P. wilcoxii* exists in east-central Idaho and western Montana, especially in habitats disturbed by humans (Clark).

139. Penstemon albomarginatus M. E. Jones, Contr. W. Bot. 12: 61. 1908 (as Pentstemon) • White-margin beardtongue [C] [E] [F]

Herbs. Stems ascending, 10–35 cm, glabrous, glaucescent, proximal stems usually buried, rhizomelike. **Leaves** cauline, leathery, glabrous; cauline 2–5 pairs, short-petiolate or sessile, 18–60 × 5–22 mm, blade spatulate to obovate, base tapered, margins entire, apex rounded to obtuse or truncate. **Thyrses** continuous or interrupted, cylindric, 5–12 cm, axis glabrous, verticillasters 3–7, cymes (1–)3–8-flowered, 2 per node; proximal bracts oval to oblong, 22–44 × 8–18 mm, margins entire or erose; peduncles and pedicels ascending to erect, glabrous. **Flowers:** calyx lobes ovate-oblong to lanceolate, 3–5 × 1.4–2.1 mm, glandular-pubescent; corolla pinkish lavender, with reddish or violet nectar guides, funnelform, 13–18 mm, glabrous externally, moderately yellowish pilose internally abaxially, tube 4–6 mm, throat gradually inflated, 4–6 mm diam., 2-ridged abaxially; stamens included, pollen sacs opposite, explanate, 0.8–1.1 mm, dehiscing completely, connective splitting, sides glabrous, sutures smooth; staminode 5–7 mm, reaching orifice, 0.2–0.3 mm diam., tip straight to slightly recurved, glabrous; style 7–10 mm. **Capsules** 8–11 × 5–7 mm, glabrous. *2n* = 16.

Flowering Apr–Jun. Aeolian and colluvial sand, desert sand dunes and scrub; of conservation concern; 500–1800 m; Ariz., Calif., Nev.

Penstemon albomarginatus is known from northwestern Arizona (Mohave County), southeastern California (San Bernardino County), and southern Nevada (Clark and Nye counties). The species is morphologically anomalous in sect. *Penstemon* and probably belongs elsewhere in subg. *Penstemon*.

140. Penstemon anguineus Eastwood, Bull. Torrey Bot. Club 32: 208. 1905 (as Pentstemon) • Siskiyou beardtongue [E] [F]

Herbs. Stems ascending to erect, (10–)30–80 cm, glabrous, not glaucous. **Leaves** basal and cauline, not leathery, glabrous; basal and proximal cauline 30–150 × 6–40 mm, blade ovate to elliptic or oval, base tapered, margins serrate to dentate, sometimes denticulate or entire, apex obtuse to acute; cauline 2–4 pairs, sessile or proximals short-petiolate, 11–90 × 5–50 mm, blade oblanceolate to oblong or triangular-ovate, base tapered to cordate-clasping, margins serrate to denticulate, rarely entire, apex obtuse to acute. **Thyrses** interrupted, sometimes continuous, cylindric, (4–)8–40 cm, axis glandular-pubescent, verticillasters 3–10, cymes (2 or) 3–15-flowered, 2 per node; proximal bracts ovate to triangular, 10–85 × 7–40 mm, margins serrate or entire; peduncles and pedicels ascending, glandular-pubescent. **Flowers:** calyx lobes lanceolate, 4–6.5 × 0.9–2 mm, glandular-pubescent; corolla light blue to lavender, purple, or violet, with violet nectar guides, ampliate, 13–18 mm, glandular-pubescent externally, glabrous or sparsely white-lanate internally abaxially, tube 5–7 mm, throat gradually inflated, 4–6 mm diam., 2-ridged abaxially; stamens included, pollen sacs opposite, navicular, 0.8–1.2 mm, dehiscing completely, connective splitting, sides glabrous, sutures papillate; staminode 10–13 mm, exserted, 0.2–0.4 mm diam., tip straight to slightly recurved, distal 2–6 mm glabrous or sparsely pilose, hairs yellow, to 0.6 mm; style 10–13 mm. **Capsules** 5–7 × 3–4 mm, glabrous. *2n* = 16.

Flowering Jun–Aug. Rocky, clayey, or humus soils, openings in coniferous forests; 600–2000 m; Calif., Oreg.

Penstemon anguineus occurs primarily in the Cascade, Klamath, and North Coast ranges from southwestern Oregon (Curry, Douglas, Jackson, Josephine, and Klamath counties) south into northwestern California (Del Norte, El Dorado, Glenn, Humboldt, Mendocino, Siskiyou, Tehama, and Trinity counties).

141. Penstemon arenarius Greene, Pittonia 1: 282. 1889 (as Pentstemon) • Nevada sand-dune beardtongue [E]

Herbs. Caudex rhizomelike. **Stems** ascending to erect, 10–27 cm, glabrous, glutinous, usually covered with sand, not glaucous. **Leaves** cauline, leathery, glabrous, sometimes glutinous; cauline 3–5 pairs, short-petiolate or sessile, (12–)20–50(–55) × (3–)10–24 mm, blade obovate or spatulate proximally, obovate to lanceolate distally, base tapered, margins dentate, apex rounded to obtuse or acute. **Thyrses** continuous, cylindric, (1–)2–9 cm, axis glabrous or sparsely puberulent, glutinous, verticillasters 2–8, cymes 2–6-flowered, 2 per node; proximal bracts lanceolate, 16–58 × 3–12 mm, margins dentate; peduncles and pedicels erect, glabrous or puberulent, glutinous. **Flowers:** calyx lobes ovate to lanceolate, 4.5–8.4 × 1.6–3 mm, glutinous; corolla white to lavender, with reddish purple nectar guides, tubular-funnelform, 11–15 (–18) mm, glabrous externally, sparsely white-pilose internally abaxially, tube 6–7 mm, throat gradually inflated, 3–5 mm diam., rounded abaxially; stamens included, pollen sacs opposite, navicular to subexplanate, 0.8–0.9 mm, dehiscing completely, connective splitting,

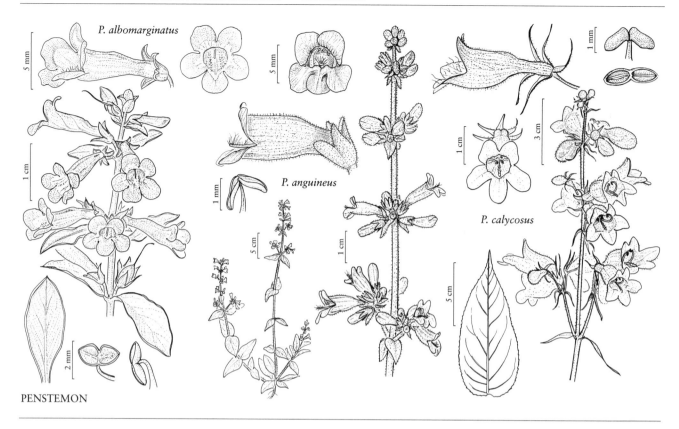

P. albomarginatus

P. anguineus

P. calycosus

PENSTEMON

sides glabrous, sutures smooth; staminode 7–10 mm, included, 0.6–0.8 mm diam., tip slightly recurved, distal 0.5–2.5 mm sparsely pilose, hairs white, to 0.5 mm; style 7–8 mm. **Capsules** 5–8 × 4–5 mm, glabrous. **2***n* = 16.

Flowering May–Jul. Loose sands, dunes, valleys, desert shrublands; 1200–1800 m; Nev.

Penstemon arenarius has been documented in Churchill, Mineral, and Nye counties. Plants usually are covered with sand because of their glutinous herbage.

142. Penstemon aridus Rydberg, Mem. New York Bot. Gard. 1: 348. 1900 (as Pentstemon) • Stiff-leaf beardtongue [E]

Herbs. Stems ascending to erect, 6–20(–25) cm, retrorsely hairy proximally, glabrous or retrorsely hairy distally, not glaucous. **Leaves** basal and cauline, not leathery, glabrous or ± puberulent; basal and proximal cauline 15–65 × 1–5 (–7) mm, blade narrowly oblanceolate to linear, base tapered, margins entire, apex obtuse to acute; cauline 2–5 pairs, short-petiolate or sessile, 7–33 × 1–3 mm, blade lanceolate to linear, base sessile to clasping, margins entire, apex obtuse to acute. **Thyrses** interrupted, cylindric, 2–11 cm, axis glandular-pubescent, sometimes also retrorsely hairy, verticillasters 2–6, cymes 1–5-flowered, 2 per node; proximal bracts lanceolate to linear, 9–32 × 1–6 mm, margins entire; peduncles and pedicels erect, ± glandular-pubescent. **Flowers:** calyx lobes ovate to lanceolate, 2.2–4.6 × 1–2.4 mm, glandular-pubescent; corolla blue to violet, with reddish purple guides, funnelform, 11–17 mm, glandular-pubescent externally, moderately whitish lanate internally abaxially, tube 3–4 mm, throat gradually inflated, 4–6 mm diam., 2-ridged abaxially; stamens included, pollen sacs divergent to opposite, subexplanate to explanate, 0.6–0.8 mm, dehiscing completely, connective splitting, sides glabrous, sutures papillate; staminode 7–8 mm, reaching orifice, 0.2–0.4 mm diam., tip recurved, distal 0.5–1 mm densely pilose, hairs golden yellow, to 0.8 mm; style 9–10 mm. **Capsules** 5–7 × 4–5 mm, glabrous.

Flowering Jun–Jul. Rocky hillsides, sagebrush shrublands, montane grasslands; 1500–2900 m; Idaho, Mont., Wyo.

Penstemon aridus is known from the Rocky Mountains in eastern Idaho and southwestern and south-central Montana and the Big Horn and northern Absaroka Mountains of Wyoming. The sometimes grasslike basal leaves distinguish *P. aridus* from *P. humilis* and *P. virens*. At lower elevations or in shaded, forest habitats, plants sometimes have broader basal leaves and tend to be much taller than typical plants; the broadly scarious-margined calyx lobes of *P. aridus* distinguish it from similar species.

143. Penstemon arkansanus Pennell, Proc. Acad. Nat. Sci. Philadelphia 73: 493. 1922 • Arkansas beardtongue E

Penstemon multicaulis Pennell; *P. pallidus* Small subsp. *arkansanus* (Pennell) R. W. Bennett; *P. wherryi* Pennell

Herbs. Stems ascending to erect, (18–)25–45(–61) cm, retrorsely hairy, sometimes also ± glandular-pubescent distally, not glaucous. **Leaves** basal and cauline, basal often withering by anthesis, not leathery, glabrous or sparsely retrorsely hairy, especially along midvein and margins, abaxially, glabrous or sparsely puberulent adaxially; basal and proximal cauline 20–95 × 5–28 mm, blade spatulate or obovate to ovate, base tapered, margins subentire to ± serrate or dentate, apex rounded to obtuse or acute; cauline (6–) 8–12 pairs, short-petiolate or sessile, 22–80 × 4–20 mm, blade lanceolate, base tapered to clasping, margins ± serrate or dentate, apex acute to acuminate. **Thyrses** interrupted, conic, 6–24 cm, axis retrorsely hairy and ± glandular-pubescent, verticillasters 4–8, cymes 3–13-flowered, 2 per node; proximal bracts lanceolate, 6–40 × 2–12 mm, margins entire or ± dentate; peduncles and pedicels ascending, retrorsely hairy and ± glandular-pubescent. **Flowers:** calyx lobes ovate, 2.8–4.5 × 1.1–2 mm, sparsely glandular-pubescent; corolla white, with reddish purple nectar guides, tubular to tubular-funnelform, 15–19 mm, glandular-pubescent externally, moderately whitish lanate internally abaxially, tube 4–6 mm, throat slightly inflated, 4–7 mm diam., 2-ridged abaxially; stamens included, pollen sacs opposite, navicular, 0.8–1 mm, dehiscing completely, connective splitting, sides glabrous, sutures papillate; staminode 10–12 mm, exserted, 0.2–0.4 mm diam., tip straight to slightly recurved, distal 5–7 mm ± villous, hairs yellow or golden yellow, to 1.2 mm; style 9–11 mm. **Capsules** 4–7 × 4–5 mm, glabrous. *2n* = 16.

Flowering Apr–Jun. Rocky oak-hickory woodlands, glades, rocky stream bottoms; 60–300 m; Ark., Mo., Okla.

Penstemon arkansanus is centered in the Interior Highlands. The species is most similar to *P. pallidus*, from which it differs by its narrower, less hairy, and more numerous cauline leaves. Some specimens from the narrow zone of sympatry in north-central Arkansas and south-central Missouri exhibit varying degrees of morphologic intermediacy. *Penstemon arkansanus* has been reported from Illinois (R. H. Mohlenbrock 1986) and Texas (S. D. Jones et al. 1997); those reports seem unlikely.

144. Penstemon attenuatus Douglas ex Lindley, Edwards's Bot. Reg. 15: plate 1295. 1830 (as Pentstemon attenuatum) • Taper-leaf beardtongue E

Herbs. Stems ascending to erect, (2–)5–70(–90) cm, glabrous or retrorsely hairy, not glaucous. **Leaves** basal and cauline, not leathery, glabrous or puberulent; basal and proximal cauline (30–)55–90(–150) × 5–20(–40) mm, blade spatulate to obovate or elliptic, base tapered, margins entire, rarely denticulate, apex rounded to obtuse or acute; cauline 2–5(–7) pairs, sessile or proximals short- to long-petiolate, 15–90(–130) × 2–25(–38) mm, blade usually lanceolate, proximals sometimes oblanceolate, base tapered to clasping, margins entire, rarely serrate or dentate, apex acute, sometimes obtuse. **Thyrses** interrupted, cylindric, 3–27 cm, axis moderately glandular-pubescent, verticillasters 2–7, cymes (1 or) 2–8-flowered, 2 per node; proximal bracts lanceolate, 15–55(–80) × 2–15(–28) mm, margins entire or ± serrate to dentate; peduncles and pedicels erect, glandular-pubescent. **Flowers:** calyx lobes ovate to obovate, ovate-oblong, or lanceolate, 3.2–7.5 × 0.9–2.6 mm, margins erose or lacerate, apex acuminate to short-caudate or long-acuminate, glandular-pubescent; corolla light blue to violet, bluish purple, or pinkish blue, rarely light yellow or whitish, with reddish purple nectar guides, tubular-funnelform, 7–20 mm, glandular-pubescent externally, moderately white-lanate internally abaxially, tube 4–5 mm, throat slightly inflated, 3–5(–6) mm diam., 2-ridged abaxially; stamens included, pollen sacs opposite, navicular to slightly saccate, 0.6–1.2 mm, dehiscing completely or incompletely, distal ⅕ or less sometimes indehiscent, connective splitting or not, sides glabrous, sutures smooth or papillate; staminode 7–9 mm, reaching orifice or slightly exserted, 0.3–0.4 mm diam., tip straight to slightly recurved, distal 1–3 mm ± pilose, hairs yellow, to 1.2 mm; style 8–12 mm. **Capsules** 4–8 × 2–5 mm, glabrous. *2n* = 48.

Varieties 4 (4 in the flora): nw United States.

1. Pollen sacs dehiscing incompletely, distal ⅕ or less usually indehiscent, connectives not splitting, sutures papillate . . . . 144b. *Penstemon attenuatus* var. *militaris*
1. Pollen sacs dehiscing completely, connectives splitting, sutures smooth, sometimes ± papillate.
　2. Corollas 7–12 mm; calyx lobes lanceolate . . . . 144c. *Penstemon attenuatus* var. *palustris*
　2. Corollas 12–20 mm; calyx lobes ovate, obovate, ovate-oblong, or lanceolate.

[3. Shifted to left margin.—Ed.]

3. Stems (22–)30–70(–90) cm; corollas light blue to violet, rarely light yellow or whitish, 14–20 mm; calyx lobe margins erose . . . .
. . . . 144a. *Penstemon attenuatus* var. *attenuatus*
3. Stems (2–)5–40 cm; corollas bluish purple, 12–16 mm; calyx lobe margins erose or lacerate . . . . . .
. . . . . . 144d. *Penstemon attenuatus* var. *pseudoprocerus*

144a. Penstemon attenuatus Douglas ex Lindley var. **attenuatus** E

Penstemon nelsoniae D. D. Keck & J. W. Thompson

Stems (22–)30–70(–90) cm. **Flowers:** calyx lobes ovate to obovate or lanceolate, margins erose; corolla light blue to violet, rarely light yellow or whitish, 14–20 mm; pollen sacs 0.8–1.2 mm, dehiscing completely, connective splitting, sutures smooth, sometimes ± papillate. *2n* = 48.

Flowering May–Aug. Gravelly or rocky clearings in pine forests, clearcut slopes, montane meadows; 500–2300 m; Idaho, Mont., Oreg., Wash.

Some specimens of var. *attenuatus* from Idaho County, Idaho, have relatively large cauline leaves with serrate to dentate margins as in *Penstemon wilcoxii*. They generally have shorter, more erect peduncles than are typical of *P. wilcoxii*. Also, the glandular hairs of the thyrse axes and corollas tend to be longer and stouter than in *P. wilcoxii*. *Penstemon nelsoniae*, which has relatively large, yellow corollas, is known only from two collections in Clallam County, Washington. It was evidently a garden adventive or originated from seeds that were sown intentionally. Forms of var. *attenuatus* with whitish or yellowish corollas occur in Idaho, Oregon, and Washington, sometimes in pure stands and sometimes in mixed stands with blue-flowered forms (D. D. Keck 1945). These plants (which have glandular-pubescent inflorescences and pollen sacs 0.9–1.1 mm) from northeastern Oregon could be the source of reports of *P. flavescens* from that state.

144b. Penstemon attenuatus Douglas ex Lindley var. **militaris** (Greene) Cronquist in C. L. Hitchcock et al., Vasc. Pl. Pacif. N.W. 4: 373. 1959 E

Penstemon militaris Greene, Leafl. Bot. Observ. Crit. 1: 166. 1906 (as Pentstemon); *P. attenuatus* subsp. *militaris* (Greene) D. D. Keck

Stems (17–)30–50(–70) cm. **Flowers:** calyx lobes lanceolate, rarely ovate, margins erose; corolla bluish purple, 10–17 mm; pollen sacs 0.6–0.8 mm, dehiscing incompletely, distal 1/5 or less usually indehiscent, connective not splitting, sutures papillate.

Flowering Jun–Aug. Rocky Douglas fir and lodgepole pine forests; 1500–2800 m; Idaho.

Variety *militaris* bears incompletely dehiscent pollen sacs, which also occur in *Penstemon globosus*. On specimens where the extent of dehiscence of the sacs is difficult to determine, the presence of prominently papillate-toothed sutures is a fairly reliable character distinguishing var. *militaris* from other varieties of *P. attenuatus*, though the teeth can be absent. On some specimens of var. *militaris*, the teeth reach nearly 0.1 mm.

144c. Penstemon attenuatus Douglas ex Lindley var. **palustris** (Pennell) Cronquist in C. L. Hitchcock et al., Vasc. Pl. Pacif. N.W. 4: 373. 1959 E

Penstemon palustris Pennell, Notul. Nat. Acad. Nat. Sci. Philadelphia 71: 8. 1941; *P. attenuatus* subsp. *palustris* (Pennell) D. D. Keck

Stems 15–40 cm. **Flowers:** calyx lobes lanceolate, margins erose; corolla bluish purple to pinkish blue, 7–12 mm; pollen sacs 0.6–0.8(–0.9) mm, dehiscing completely, connective splitting, sutures smooth, sometimes ± papillate.

Flowering Jun–Aug. Marshes, stream banks, moist to wet meadows; 400–2400 m; Oreg., Wash.

Variety *palustris* is known from the Blue Mountains of northeastern Oregon and southeastern Washington.

144d. Penstemon attenuatus Douglas ex Lindley var. **pseudoprocerus** (Rydberg) Cronquist in C. L. Hitchcock et al., Vasc. Pl. Pacif. N.W. 4: 373. 1959 E

Penstemon pseudoprocerus Rydberg, Mem. New York Bot. Gard. 1: 346. 1900 (as Pentstemon); *P. attenuatus* subsp. *pseudoprocerus* (Rydberg) D. D. Keck

Stems (2–)5–40 cm. **Flowers:** calyx lobes ovate, obovate, ovate-oblong, or lanceolate, margins erose or lacerate; corolla bluish purple, 12–16 mm; pollen sacs 0.7–0.8 mm, dehiscing completely, connective splitting, sutures smooth, sometimes ± papillate. $2n = 48$.

Flowering Jul–Aug. Moist hillsides, rocky slopes, subalpine and alpine meadows; 1600–3200 m; Idaho, Mont., Wyo.

Variety *pseudoprocerus* is concentrated in the Central Rocky Mountains in eastern Idaho, southwestern Montana, and northwestern Wyoming.

145. Penstemon australis Small, Fl. S.E. U.S., 1060, 1337. 1903 (as Pentstemon) • Southern beardtongue E

Herbs. Stems erect, 30–86 cm, retrorsely hairy and glandular-lanate, not glaucous. **Leaves** basal and cauline, not leathery, retrorsely hairy, sometimes also glandular-pubescent, rarely glabrate; basal and proximal cauline 32–130 × 7–40 mm, blade spatulate to oblanceolate, base tapered, margins entire or ± serrate, apex rounded to obtuse; cauline 5–8 pairs, sessile or proximals short-petiolate, 16–122 × 3–23 mm, blade lanceolate, base clasping or tapered, margins ± serrate, apex acute to acuminate. **Thyrses** interrupted, narrowly conic, rarely conic, 7–26 cm, axis ± glandular-pubescent, verticillasters 3–6, cymes 2–6-flowered, usually only 1 branch of each cyme elongating, others nearly sessile or distinctly shorter, 2 per node; proximal bracts lanceolate, 3–20 ×1–4 mm, margins entire; peduncles and pedicels erect, sometimes ascending, ± glandular-pubescent. **Flowers:** calyx lobes ovate to lanceolate, 4–5.2 × 1.6–2.3 mm, sparsely glandular-pubescent; corolla white to pinkish or light lavender, with dark purple nectar guides, tubular, 20–25 mm, glandular-pubescent externally, moderately white-lanate internally abaxially, tube 4–5 mm, throat abruptly inflated, 4.5–6 mm diam., 2-ridged abaxially; stamens included, pollen sacs opposite, navicular, 1.2–1.5 mm, dehiscing completely, connective splitting, sides glabrous, sutures papillate; staminode 11–14 mm, exserted, 0.4–0.7 mm diam., tip straight to slightly recurved, distal 8–10 mm densely pilose, hairs yellow, to 2 mm; style 10–12 mm. **Capsules** 6–9 × 4.5–6 mm, glabrous.

Flowering Mar–Jun. Sandy pine and oak woodlands, pine savannas, hammocks, granite hills, sandy open areas; 10–200 m; Ala., Fla., Ga., Miss., N.C., S.C., Tenn., Va.

Penstemon australis is known from the southern Atlantic Coastal Plain and eastern Gulf Coastal Plain. The species is confused most often with *P. laxiflorus*, which occurs farther west. *Penstemon australis* is distinguished from *P. laxiflorus* most readily by the stem vesture; *P. australis* usually has stems with a mix of short, eglandular hairs and much longer, glandular hairs, while *P. laxiflorus* has stems with only short, retrorse hairs or, if glandular hairs also are present, they are sparse and occur just below the inflorescences. *Penstemon australis* also tends to have narrower inflorescences due to the unequal elongation of cyme branches. Ranges of the two species overlap in southwestern Alabama, and many specimens from Baldwin, Escambia, and Mobile counties are morphologically intermediate.

Many specimens of *Penstemon* from the Appalachian Piedmont, especially in northeastern Alabama, northern Georgia, central North Carolina, and western South Carolina, combine morphological features of *P. australis* and *P. canescens*. These plants usually have equally developed cyme branches (as in *P. canescens*) but ascending or erect peduncles and pedicels (as in *P. australis*). Leaf shape generally is intermediate between *P. australis* and *P. canescens*. F. W. Pennell in the 1930s annotated most such specimens as *P. australis*. F. S. Crosswhite (1965e) also included these specimens in his concept of *P. australis*, assigning them subspecies status, though he never published the subspecies names that he used in his thesis and on herbarium sheets. The relationship of these plants remains unclear; they will key to *P. canescens* or *P. laxiflorus*.

A. E. Radford et al. (1968) gave a chromosome number of $n = 8$ for *Penstemon australis*; this does not appear to be supported by any other published chromosome count.

146. Penstemon bleaklyi O'Kane & K. D. Heil, Phytoneuron 2014-61: 1, figs. 1–3. 2014 • Bleakly's beardtongue [C] [E]

Herbs. Stems decumbent to ascending, 2–8 cm, retrorsely hairy, not glaucous. **Leaves** basal and cauline, not leathery, glabrous; basal and proximal cauline 15–35 × 4–15 mm, blade lanceolate to elliptic or oblanceolate, base tapered, margins entire or serrulate, apex obtuse to acute; cauline 3–7 pairs, petiolate, 10–20 × 3–5 mm, blade lanceolate to linear-lanceolate, base tapered to cuneate, margins entire or serrulate, apex obtuse to acute. **Thyrses** continuous, ± secund, 1–3 cm, axis glandular-pubescent, verticillasters 1 or 2, cymes (1 or)2- or 3-flowered, 1 or 2 per node; proximal bracts lanceolate, 10–20 × 4–8 mm, margins entire; peduncles and pedicels ascending to erect, glandular-pubescent. **Flowers:** calyx lobes lanceolate, 5–8 × 0.8–1.2 mm, glandular-pubescent; corolla bluish lavender to lavender-purple, with violet nectar guides, tubular-funnelform, 12–25 mm, glandular-pubescent externally, densely white-lanate internally abaxially, tube 4–6 mm, throat gradually inflated, 6–8 mm diam., scarcely 2-ridged abaxially; stamens included, pollen sacs opposite, subexplanate, 1.1–1.3 mm, dehiscing completely, connective splitting, sides sparsely puberulent, sutures smooth; staminode 14–15 mm, exserted, 0.3–0.4 mm diam., tip slightly recurved, distal 0.8–1 mm sparsely lanate, hairs yellow, to 0.4 mm; styles 10–15 mm. **Capsules** 4–7 × 3–5 mm, glabrous.

Flowering Jul. Alpine scree slopes; of conservation concern; 3800–3900 m; N.Mex.

Penstemon bleaklyi is known only from the Culebra Range of the Sangre de Cristo Mountains in Taos County.

147. Penstemon brevisepalus Pennell in J. K. Small, Man. S.E. Fl., 1204, 1508. 1933 • Short-sepal beardtongue [E]

Herbs. Stems ascending to erect, 22–60 cm, retrorsely hairy and glandular-pubescent or glandular-villous, not glaucous. **Leaves** basal and cauline, basal sometimes withering by anthesis, not leathery, sparsely to densely pubescent, usually also sparsely glandular-hairy, abaxially, glabrous or sparsely pubescent adaxially; basal and proximal cauline 20–95 × 11–28 mm, blade spatulate to obovate or ovate, base tapered, margins subentire to ± serrate, apex rounded to obtuse; cauline 4 or 5, sessile or proximals short-petiolate, 35–95 × 11–32 mm, blade ovate to lanceolate or oblanceolate, base tapered to truncate or clasping, margins ± serrate or dentate, apex obtuse to acute. **Thyrses** interrupted, conic, (4–)8–20 cm, axis sparsely to densely glandular-pubescent, verticillasters 3–8, cymes 5–16-flowered, 2 per node; proximal bracts lanceolate, 7–50 × 2–20 mm, margins entire or serrate; peduncles and pedicles spreading to ascending, sparsely to densely glandular-pubescent. **Flowers:** calyx lobes ovate, rarely lanceolate, (1.5–)2–3.5 × 0.9–2.2 mm, sparsely glandular-pubescent; corolla lavender to purple, with reddish purple nectar guides, tubular to tubular-funnelform, 15–24 mm, glandular-pubescent externally, ± whitish lanate internally abaxially, tube 4–6 mm, throat slightly inflated, 4–7 mm diam., 2-ridged abaxially; stamens included, pollen sacs opposite, navicular, 0.9–1.1 mm, dehiscing completely, connective splitting, sides glabrous, sutures papillate; staminode 12–18 mm, exserted, 0.3–0.5 mm diam., tip straight to slightly recurved, distal 8–9 mm ± villous, hairs yellow or golden yellow, to 1.5 mm; style 12–17 mm. **Capsules** 4–7 × 3–5 mm, glabrous.

Flowering May–Jun. Rocky oak-hickory forests, bluffs, roadcuts; 200–700 m; Ky., Tenn., Va., W.Va.

Penstemon brevisepalus has not been widely recognized in the literature or herbaria, usually being synonymized with *P. canescens* or *P. pallidus*. *Penstemon brevisepalus* combines features of *P. canescens* and *P. pallidus*, and its range lies between the ranges of those two species, with little overlap. In addition to calyx lobe length and corolla color, *P. brevisepalus* usually has more open inflorescences and cauline leaf margins that are more finely and regularly toothed compared to *P. pallidus*. Plants with pink- or lavender-tinged corollas occasionally occur in *P. pallidus*, and individuals with short calyx lobes (2–3 mm) also are encountered infrequently; they may be the sources of reports of *P. brevisepalus* elsewhere (for example, Georgia, Illinois, and Ohio).

148. Penstemon calycosus Small, Bull. Torrey Bot. Club 25: 470. 1898 (as Pentstemon) • Long-sepal beardtongue E F

Herbs. **Stems** erect, (40–)60–120 cm, puberulent, at least proximally, slightly glaucous or not. **Leaves** basal and cauline, basal often withering by anthesis, not leathery, glabrous; basal and proximal cauline 47–110 × 10–24 mm, blade oblanceolate, base tapered, margins entire or ± serrate, apex rounded to obtuse; cauline 3–7 pairs, sessile or proximals short-petiolate, (14–)44–155 × (2–)13–46 mm, blade lanceolate to ovate, proximals sometimes oblanceolate, base clasping or tapered, margins serrate, apex acute to acuminate. **Thyrses** interrupted, conic, (5–)14–20 cm, axis sparsely puberulent and glandular-pubescent, rarely glabrous, verticillasters 2–5, cymes (1–)5–15-flowered, 2 per node; proximal bracts ovate to lanceolate, sometimes linear, 13–80 × (2–)5–32 mm, margins serrate, sometimes entire; peduncles and pedicels spreading to ascending, sparsely glandular-pubescent. **Flowers:** calyx lobes lanceolate, 5–9 × 0.8–1.9 mm, apex acuminate, sparsely glandular-pubescent; corolla pale lavender to violet, with faint violet nectar guides, ventricose, 20–35 mm, glandular-pubescent externally, ± white-pubescent internally abaxially, tube 5–7 mm, throat abruptly inflated, 8–11 mm diam., slightly 2-ridged abaxially; stamens included or longer pair reaching orifice, pollen sacs opposite, navicular, 1.2–1.4 mm, dehiscing completely, connective splitting, sides glabrous, rarely sparsely pubescent, hairs white, to 0.2 mm, sutures papillate; staminode 11–13 mm, reaching orifice, 0.2–0.3 mm diam., tip straight to slightly recurved, distal 2–5 mm sparsely pubescent, hairs yellowish, to 1 mm; style 14–16 mm. **Capsules** 7–8 × 4–5 mm, glabrous. *2n* = 96.

Flowering May–Jun. Woodlands, meadows, rocky slopes, stream banks; 10–300 m; Ont.; Ala., Conn., D.C., Ga., Ill., Ind., Ky., Maine, Md., Mass., Mich., Minn., Mo., N.J., N.Y., N.C., Ohio, Pa., R.I., S.C., Tenn., Vt., Va.

Penstemon calycosus is concentrated in the Central Lowlands and Interior Low Plateaus; occurrences farther south and east, especially in New England, may be from introductions (F. W. Pennell 1935). The range lies almost entirely within that of *P. digitalis*, but historic ranges of both *P. calycosus* and *P. digitalis* are difficult to delimit due to both past and continuing introduction and spread of both species. Calyx lobe lengths to 12 mm are reported in the literature; those appear to be from fruiting calyces. *Penstemon calycosus* usually has glabrous anthers, but many specimens, especially from Illinois and Ohio, bear hairs on the adaxial surfaces of the anthers. Some populations from this region exhibit variation in vegetative (leaf size and bract shape and margins) and floral characters (calyx lobe shape and size, corolla color, and anther pubescence), suggesting possible hybridization with *P. digitalis*, an observation that would be consistent with the findings of A. C. Koelling (1964).

149. Penstemon canescens (Britton) Britton, Mem. Torrey Bot. Club 5: 291. 1894 (as Pentstemon) • Appalachian beardtongue E

Penstemon laevigatus Aiton var. *canescens* Britton, Mem. Torrey Bot. Club 2: 30. 1890 (as Pentstemon); *P. brittoniorum* Pennell; *P. canescens* subsp. *brittoniorum* (Pennell) Pennell

Herbs. **Stems** ascending to erect, (20–)35–80 cm, retrorsely hairy and glandular-pubescent, sometimes only glandular-pubescent distally, not glaucous. **Leaves** basal and cauline, not leathery, sparsely to densely pubescent, rarely glabrate; basal and proximal cauline 28–170 × 11–42 mm, blade orbiculate to obovate or elliptic, base tapered, margins irregularly crenate to irregularly serrate, apex rounded to obtuse; cauline 3–7 pairs, sessile or proximals short-petiolate, 28–170 × 9–40 mm, blade oblanceolate to lanceolate, proximals usually ± lyrate, base clasping or tapered, margins crenate to serrate, apex obtuse to acute. **Thyrses** interrupted, conic, (5–)8–23 cm, axis ± pubescent and glandular-pubescent, verticillasters 3–6, cymes 5–13-flowered, 2 per node; proximal bracts lanceolate, 8–60 × 2–20 mm, margins entire or ± crenate to serrate; peduncles and pedicels spreading to ascending, ± pubescent and glandular-pubescent. **Flowers:** calyx lobes lanceolate, 3.2–6(–7) × 1.4–1.8 (–2) mm, sparsely to moderately glandular-pubescent; corolla lavender to purple, with lavender or purple nectar guides, ventricose, 20–30 mm, glandular-pubescent externally, ± white-pubescent internally abaxially, tube (4–)5–8(–10) mm, throat abruptly inflated, 5–8 mm diam., slightly 2-ridged abaxially; stamens included or longer pair reaching orifice, pollen sacs opposite, navicular, 1–1.5 mm, dehiscing completely, connective splitting, sides glabrous, rarely sparsely pubescent, hairs white, to 0.2 mm, sutures papillate; staminode (11–)14–22 mm, exserted, 0.5–0.7 mm diam., tip straight to slightly recurved, distal (7–)10–12 mm ± pubescent, hairs yellowish, to 2 mm; style 13–15 mm. **Capsules** 5–7 × 4–5 mm, glabrous. *2n* = 16.

Flowering (Apr–)May–Jul(–Sep). Woodlands, thickets, cliffs, barrens; 200–1600 m; Ala., Ga., Ill., Ind., Ky., Md., N.C., Ohio, Pa., S.C., Tenn., Va., W.Va.

F. W. Pennell (1935) acknowledged that *Penstemon canescens* varies greatly in leaf size, flower size, and vestiture. He distinguished subsp. *brittoniorum* by its glabrate leaves with sharply serrate margins and dark purple corollas. Plants with those features are encountered mostly in the southern part of the range of the species, although they are essentially sympatric with subsp. *canescens* and sometimes in mixed populations with plants with the features of subsp. *canescens*.

150. Penstemon cinicola D. D. Keck, Publ. Carnegie Inst. Wash. 520: 294. 1940 • Ash beardtongue [E]

Herbs. Stems ascending, (8–)10–40 cm, glabrous or retrorsely hairy, not glaucous. **Leaves** essentially cauline, basal usually poorly developed, not leathery, glabrous; basal and proximal cauline 10–40(–65) × 1–6 mm, blade oblanceolate, base tapered, margins entire, apex obtuse to acute; cauline 3–5 pairs, sessile or proximals short-petiolate, 12–60 × 2–5(–7) mm, blade oblanceolate to lanceolate or linear, base tapered, margins entire, apex acute. **Thyrses** interrupted, cylindric, 1–14 cm, axis glabrous or ± puberulent at axils, verticillasters (1 or)2–6, cymes (1–)3–7-flowered, 2 per node; proximal bracts lanceolate to linear, 8–30 (–45) × 1–7 mm, margins entire; peduncles and pedicels erect, glabrous. **Flowers:** calyx lobes obovate to ovate, 1–1.8 × 0.7–1.1 mm, apex truncate to cuspidate, glabrous; corolla violet to blue or purple, without nectar guides, funnelform, 7–9(–11) mm, glabrous externally, moderately yellowish or white-pilose internally abaxially, tube 3–4 mm, throat slightly inflated, 2–3 mm diam., 2-ridged abaxially; stamens included or longer pair reaching orifice, pollen sacs opposite, explanate, 0.3–0.4(–0.5) mm, dehiscing completely, connective splitting, sides glabrous, sutures smooth; staminode 4–6 mm, included, 0.3–0.4 mm diam., tip straight, distal 0.5–1 mm sparsely to moderately pilose, hairs yellow, to 0.4 mm; style 5–6 mm. **Capsules** 2.5–4 × 1.8–2.5 mm, glabrous. *2n* = 16, 32.

Flowering Jun–Aug. Dry, volcanic soils in sagebrush openings in pine forests; 1000–2600 m; Calif., Oreg.

Penstemon cinicola occurs along the eastern flank of the Cascade Range in central Oregon (Crook, Deschutes, Douglas, Klamath, and Lake counties) south to northern California (Lassen, Modoc, Shasta, Siskiyou, and Tehama counties).

151. Penstemon confertus Douglas, Edwards's Bot. Reg. 15: plate 1260. 1829 (as Pentstemon confertum) • Yellow beardtongue [E]

Herbs. Stems ascending to erect, 4–40(–70) cm, glabrous or retrorsely hairy, not glaucous. **Leaves** basal and cauline, not leathery, glabrous; basal and proximal cauline 45–125 × 5–26 mm, blade elliptic to oblanceolate, base tapered, margins entire, apex obtuse to acute; cauline 4–7 pairs, sessile or proximals short-petiolate, (7–)12–86 × (2–)3–20(–25) mm, blade oblanceolate to lanceolate or elliptic, rarely ovate, base tapered to truncate, margins entire, apex obtuse to acute. **Thyrses** interrupted, cylindric, 2–25(–50) cm, axis glabrous or retrorsely hairy, verticillasters (1 or) 2–10, cymes 2–9-flowered, 2 per node; proximal bracts lanceolate, rarely ovate, 4–50 × 2–12 mm, margins entire; peduncles and pedicels erect, retrorsely hairy. **Flowers:** calyx lobes oblong to lanceolate, 3–4.8 × 0.7–1.8 mm, apex truncate to cuspidate or short- to long-caudate, glabrous; corolla yellow, rarely white, ochroleucous, pink, or lavender, without nectar guides, tubular to tubular-funnelform, 6–11 mm, glabrous externally, moderately yellowish or white-lanate internally abaxially, tube 3–4 mm, throat slightly inflated, 2–3 mm diam., 2-ridged abaxially; stamens included or longer pair reaching orifice, pollen sacs opposite, explanate, 0.4–0.5 mm, dehiscing completely, connective splitting, sides glabrous, sutures smooth; staminode 4–6 mm, included, 0.3–0.4 mm diam., tip straight, distal 0.5–1 mm pilose, hairs yellowish brown, to 0.5 mm; style 7–9 mm. **Capsules** 3–5 × 2–3.5 mm, glabrous. *2n* = 32.

Flowering May–Aug. Dry meadows, grassy slopes, openings in pine forests; 300–2400 m; Alta., B.C., Sask.; Alaska, Idaho, Mont., Oreg., Wash.

Penstemon confertus is widespread in the Central, Northern, and Canadian Rocky mountains in southern Alberta, southern British Columbia, southwestern Saskatchewan, northern Idaho, western Montana, northeastern Oregon, and eastern Washington. Isolated populations are known in northern British Columbia (Lackman Lake) and southern Alaska (Haines). E. Heitz (1927) reported a chromosome number of *2n* = ca. 16 for *P. confertus*; the voucher for that count has not been verified. Counts from reports by J. Clausen et al. (1940), D. D. Keck (1945), and R. Spellenberg (1971) all are *2n* = 32. Putative hybrids between *P. confertus* and *P. globosus* have been documented in Idaho County, Idaho (D. V. Clark 1971).

The Thompson and Okanagan-Colville tribes of southern British Columbia and northeastern Washington use *Penstemon confertus* for drugs, food, and dye (D. E. Moerman 1998).

152. **Penstemon deamii** Pennell, Monogr. Acad. Nat. Sci. Philadelphia 1: 212. 1935 • Deam's beardtongue C E

Herbs. Stems ascending to erect, 40–90 cm, retrorsely hairy proximally, usually retrorsely hairy distally, not glaucous. **Leaves** basal and cauline, basal sometimes withering by anthesis, not leathery, glabrate to puberulent; basal and proximal cauline 40–150 × 10–26 mm, blade oval to elliptic, base tapered, margins entire or ± denticulate, apex rounded to obtuse; cauline 7 or 8 pairs, sessile, 17–110 × 3–20 mm, blade oblanceolate to lanceolate, base tapered to clasping, margins entire or denticulate to serrulate, apex obtuse to acute. **Thyrses** interrupted, conic, 11–22 cm, axis glabrous or sparsely glandular-pubescent, verticillasters 4–6, cymes (2–)5–8-flowered, 2 per node; proximal bracts lanceolate, 10–15 × 3–5 mm, margins entire; peduncles and pedicels ascending to erect, glabrous or sparsely glandular-pubescent. **Flowers:** calyx lobes ovate, 1.5–3.5(–4) × 0.9–1.5 mm, sparsely glandular-pubescent; corolla white, sometimes tinged lavender or purple, with reddish purple nectar guides, ventricose, 15–20(–22) mm, glandular-pubescent externally, sparsely white-pubescent internally abaxially, tube 5–6 mm, throat abruptly inflated, 5–7 mm diam., slightly 2-ridged abaxially; stamens included, pollen sacs opposite, navicular, 0.8–1 mm, dehiscing completely, connective splitting, sides glabrous, sometimes sparsely pubescent, hairs white, to 0.2 mm, sutures papillate; staminode 8–10 mm, reaching orifice, 0.3–0.4 mm diam., tip straight to slightly recurved, distal 3–4 mm pilose, hairs yellowish, to 1 mm; style 10–12 mm. **Capsules** 5–8 × 4–5 mm, glabrous.

Flowering May–Jun. Rocky oak woods, fields; of conservation concern; 100–200 m; Ind.

A. C. Koelling (1964) assessed *Penstemon deamii* and its allies using morphological, cytological, and breeding studies. He concluded that *P. deamii* is a distinct member of the eastern North American polyploids that include *P. calycosus*, *P. digitalis*, and *P. laevigatus*. Geographically sympatric with *P. calycosus* and *P. digitalis*, *P. deamii* is distinguished from those two species by its shorter calyx lobes [1.5–3.5(–4) mm versus 4–9 mm], corollas [15–20(–22) mm versus (17–)20–35 mm], and pollen sacs (0.8–1 mm versus 1.2–1.7 mm). The species appears to be limited to Clark, Crawford, Floyd, Harrison, and Marion counties (K. Yatskievych 2000). Reports of *P. deamii* from southern Illinois appear to be based on specimens of questionable provenance or misidentifications (Koelling).

153. **Penstemon degeneri** Crosswhite, Amer. Midl. Naturalist 74: 434, fig. 4. 1965 • Degener's beardtongue C E

Herbs. Stems ascending to erect, 25–40 cm, retrorsely hairy proximally, glandular-pubescent distally, not glaucous. **Leaves** basal and cauline, basal sometimes withering by anthesis, not leathery, glabrous or retrorsely hairy proximally; basal and proximal cauline 25–70(–85) × 5–23 mm, blade ovate to lanceolate, base tapered, margins entire, apex acute; cauline 3–6 pairs, petiolate or sessile, 20–90(–110) × 2–18 mm, blade oblanceolate to lanceolate, base tapered, margins entire, apex acute. **Thyrses** interrupted, narrowly conic, 3–10 cm, axis sparsely glandular-pubescent, verticillasters 3–6, cymes 2–10-flowered, 2 per node; proximal bracts lanceolate to linear, 3–40 × 0.5–4 mm, margins entire; peduncles and pedicels ascending to erect, sparsely glandular-pubescent. **Flowers:** calyx lobes ovate to lanceolate, 3.5–7 × 1–2 mm, glandular-pubescent; corolla blue to light violet, with reddish purple nectar guides, funnelform, 14–19 mm, glandular-pubescent or glabrescent externally, glabrous or sparsely white- or yellow-villous internally abaxially, tube 5–6 mm, throat gradually inflated, 4–9 mm diam., slightly 2-ridged abaxially; stamens included, pollen sacs opposite, navicular, 0.9–1.1 mm, dehiscing completely, connective splitting, sides glabrous, sutures papillate; staminode 15–16 mm, included, 0.3–0.4 mm diam., tip straight to slightly recurved, distal 6–10 mm pilose, hairs golden, to 1.5 mm; style 13–16 mm. **Capsules** 7–10 × 2.5–4 mm, glabrous.

Flowering Jun–Aug. Pine-juniper woodlands, ponderosa pine parklands, montane grasslands; of conservation concern; 1800–2900 m; Colo.

Penstemon degeneri is known from the Arkansas River Canyon from near Salida to Cañon City, and to near Wet Mountain Valley. Most populations are in Fremont County (B. L. Beatty et al. 2004); the species is also known from Chaffee and Custer counties. Typical, large-leaved forms are unlikely to be confused with any other species in the area, but narrow-leaved forms can be mistaken for *P. griffinii*. F. S. Crosswhite (1965b) described the corollas of *P. degeneri* as bearded at the orifices with light yellow hairs. In most populations, the abaxial limbs and throats are sparsely yellow-lanate; however, pubescence can vary greatly within populations, from sparsely white-lanate only on the limbs to moderately yellow-lanate on the limbs and throats. When the throats are bearded in *P. degeneri*, the hairs extend 1–3 mm into the throats; in *P. griffinii* they extend deeper into the throats, usually 5–8 mm.

SELECTED REFERENCE Beatty, B. L., W. F. Jennings, and R. C. Rawlinson. 2004. *Penstemon degeneri* Crosswhite (Degener's Beardtongue): A Technical Conservation Assessment. Denver.

154. Penstemon deustus Douglas ex Lindley, Edwards's Bot. Reg. 16: plate 1318. 1830 (as Pentstemon deustum) • Hot-rock beardtongue E F

Subshrubs. Stems ascending to erect, (6–)15–50 cm, retrorsely hairy, sometime glabrate, not glaucous. **Leaves** cauline, opposite, subopposite, or whorled, not leathery, glabrous or retrorsely hairy; cauline 5–9 pairs, petiolate or sessile, (7–)10–50(–52) × (0.5–)1–20 mm, blade oblanceolate to ovate or spatulate, lanceolate to linear distally, base of proximals tapered, distals clasping, margins serrate to serrulate, rarely entire, apex rounded to obtuse, acute, or acuminate. **Thyrses** interrupted, cylindric, 6–30 cm, axis ± glandular-pubescent, verticillasters 5–11, cymes 1–6-flowered, 2 per node; proximal bracts lanceolate to linear, 6–40 × 1–4 mm, margins entire or serrate to serrulate; peduncles and pedicels ascending to erect, glandular-pubescent. **Flowers:** calyx lobes ovate to lanceolate, 2.3–7 × 0.6–1.9 mm, glandular-pubescent; corolla white to ochroleucous, sometimes tinged lavender, with brownish or purple nectar guides, funnelform, 9–18 mm, glabrate to glandular-pubescent externally, glandular-pubescent internally, sometimes glabrous or glabrate adaxially, tube 4–5 mm, throat slightly inflated, 2.5–5 mm diam., slightly 2-ridged abaxially; stamens: longer pair slightly exserted, pollen sacs opposite, explanate, 0.4–0.8 mm, dehiscing completely, connective splitting, sides glabrous, sutures papillate; staminode 8–9 mm, exserted, 0.2–0.3 mm diam., tip straight, glabrous or distal 0.5–2.5 mm sparsely pilose or villous, hairs yellow or white, to 0.5 mm; style 7–10 mm. **Capsules** 3–5 × 2.5–4 mm, glabrous.

Varieties 4 (4 in the flora): w United States.

Penstemon deustus comprises four marginally discrete varieties. Extreme forms of each variety are distinctive, but extensive zones of contact exist among the varieties, and populations frequently exhibit degrees of intermediacy, especially in Oregon where all four varieties occur.

The Paiute and Shoshoni tribes, centered in the Great Basin, used *Penstemon deustus* to treat dermatological, gastrointestinal, immunological, and other disorders (D. E. Moerman 1998).

1. Leaves whorled, sometimes opposite or subopposite, blades oblanceolate or lanceolate to linear, margins serrate to serrulate distally or entire. . . . . 154d. *Penstemon deustus* var. *variabilis*
1. Leaves opposite, blades ovate, spatulate, oblanceolate, or lanceolate, margins sharply to obscurely serrate, sometimes entire.
    2. Calyx lobes lanceolate, (3.5–)4–7 mm, apices acuminate; corollas 9–12 mm; cauline leaf blade margins sharply serrate. . . . . 154c. *Penstemon deustus* var. *suffrutescens*
    2. Calyx lobes ovate to lanceolate, 2.3–5(–6.3) mm, apices acute to acuminate; corollas 10–18 mm; cauline leaf blade margins obscurely to sharply serrate, sometimes entire.
        3. Corollas 12–18 mm, glandular-pubescent externally, glandular-pubescent internally, sometimes obscurely adaxially, abaxial lobes white to ochroleucous, sometimes tinged lavender. . . . . 154a. *Penstemon deustus* var. *deustus*
        3. Corollas 10–12(–15) mm, glabrate or glandular-pubescent externally, glabrous internally, or glabrate or glandular-pubescent abaxially and glabrous adaxially, abaxial lobes brown, sometimes white to ochroleucous . . . . 154b. *Penstemon deustus* var. *pedicillatus*

154a. Penstemon deustus Douglas ex Lindley var. deustus E F

Leaves opposite, cauline (7–)10–50 × (5–)8–20 mm, blade ovate, sometimes distals oblanceolate, spatulate, or lanceolate, margins sharply serrate. **Flowers:** calyx lobes ovate to lanceolate, 3–5(–6.3) × 1–1.9 mm, apex acute to acuminate; corolla 12–18 mm, glandular-pubescent externally, glandular-pubescent internally, sometimes obscurely adaxially, abaxial lobes white to ochroleucous, sometimes tinged lavender; staminode glabrous, sometimes distal 1–2 mm sparsely pilose, hairs yellow, to 0.2 mm. $2n = 16$.

Flowering May–Jul. Rocky hillsides, talus slopes, rock outcrops, gravelly banks, sagebrush shrublands, juniper woodlands, open pine forests; 400–2600 m; Idaho, Mont., Nev., Oreg., Wash., Wyo.

Variety *deustus* is widespread in the Columbia Basin and Central Rocky Mountains from northwestern and north-central Wyoming and south-central Montana to northern Idaho and western Washington and to western Oregon and northern Nevada.

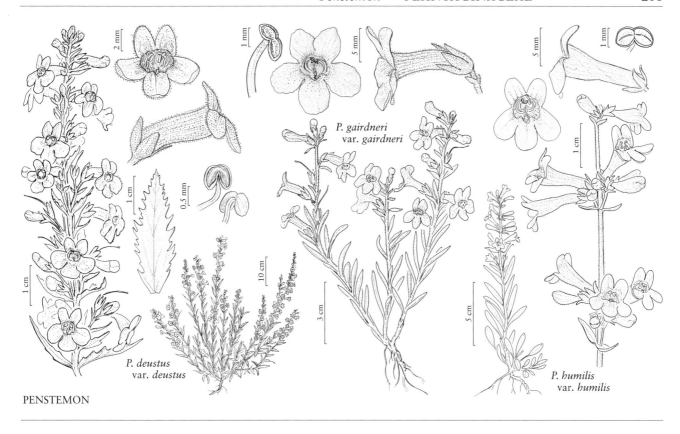

P. gairdneri
var. gairdneri

P. deustus
var. deustus

PENSTEMON

P. humilis
var. humilis

154b. Penstemon deustus Douglas ex Lindley var. **pedicillatus** M. E. Jones, Zoë 4: 281. 1893 (as Pentstemon) E

Penstemon deustus subsp. *heterander* (Torrey & A. Gray) Pennell & D. D. Keck; *P. deustus* var. *heterander* (Torrey & A. Gray) Pennell & D. D. Keck

Leaves opposite, cauline (7–) 10–35(–52) × 2.5–8(–14) mm, blade lanceolate, sometimes oblanceolate, spatulate, or ovate, margins obscurely to sharply serrate, sometimes entire. **Flowers:** calyx lobes ovate to lanceolate, 2.3–4.6 × 0.7–1.5 mm, apex acute to acuminate; corolla 10–12(–15) mm, glabrate or glandular-pubescent externally, glabrous internally, or glabrate or glandular-pubescent abaxially and glabrous adaxially, abaxial lobes brown, sometimes white to ochroleucous; staminode glabrous, sometimes distal 1–2 mm sometimes sparsely pilose, hairs yellow, to 0.5 mm.

Flowering May–Jul. Rocky and talus slopes, sagebrush shrublands, pinyon-juniper woodlands, coniferous forests; 900–3000 m; Calif., Nev., Oreg., Utah.

Fresh abaxial corolla lobes usually are brown in var. *pedicillatus*, and the abaxial lobes sometimes dry distinctly darker than the abaxial lobes.

154c. Penstemon deustus Douglas ex Lindley var. **suffrutescens** L. F. Henderson, Rhodora 33: 206. 1931 (as Pentstemon) E

Leaves opposite, cauline (8–) 13–40 × 6–14 mm, blade ovate to spatulate, oblanceolate, or lanceolate, margins sharply serrate. **Flowers:** calyx lobes lanceolate, (3.5–)4–7 × 0.6–1.5 mm, apex acuminate; corolla 9–12 mm, glandular-pubescent externally, glandular-pubescent internally, sometimes glabrous or ± glandular-pubescent adaxially, abaxial lobes white to ochroleucous, sometimes tinged lavender; staminode glabrous or distal 1.5–2.5 mm sparsely pilose, hairs yellow, to 0.4 mm.

Flowering May–Jul. Rocky and talus slopes, rock outcrops, coniferous forests; 600–2700 m; Calif., Oreg.

154d. Penstemon deustus Douglas ex Lindley var. **variabilis** (Suksdorf) Cronquist in C. L. Hitchcock et al., Vasc. Pl. Pacif. N.W. 4: 380. 1959 [C] [E]

Penstemon variabilis Suksdorf, Deutsche Bot. Monatsschr. 18: 153. 1900 (as Pentstemon); *P. deustus* subsp. *variabilis* (Suksdorf) Pennell & D. D. Keck

Leaves whorled, sometimes opposite or subopposite, cauline 11–48 × (0.5–)1–5 mm, blade oblanceolate or lanceolate to linear, margins serrate to serrulate distally or entire. **Flowers:** calyx lobes ovate to lanceolate, 2.8–4.5 × 0.8–1.6 mm, apex acute to acuminate; corolla 10–12 (–15) mm, glandular-pubescent externally, glandular-pubescent internally, abaxial lobes white to ochroleucous, sometimes tinged lavender; staminode: distal 0.5 mm sparsely villous, hairs white, to 0.5 mm. 2*n* = 32.

Flowering Jun–Jul. Gravelly slopes, streambeds; of conservation concern; 600–2600 m; Oreg., Wash.

Variety *variabilis* occurs from southern Washington to central and eastern Oregon. Populations are most concentrated in the Blue Mountains.

155. Penstemon digitalis Nuttall ex Sims, Bot. Mag. 52: plate 2587. 1825 (as Pentstemon) • Tall white beardtongue, penstémon digitale [E]

Penstemon alluviorum Pennell

Herbs. Stems erect, 25–90 cm, glabrous or sparsely retrorsely hairy, slightly glaucous or not. **Leaves** basal and cauline, basal sometimes withering by anthesis, not leathery, glabrous; basal and proximal cauline 30–180(–250) × 4–38(–70) mm, blade spatulate to obovate or lanceolate, base tapered, margins entire or ± serrate or denticulate, apex rounded to obtuse or acute; cauline 5–8 pairs, petiolate or sessile, 26–195 × 4–55 mm, blade ovate to lanceolate, base tapered to clasping, margins entire or denticulate, apex acuminate to acute. **Thyrses** interrupted, conic, 7–26(–34) cm, axis glabrous proximally, sparsely to densely glandular-pubescent distally, verticillasters (2 or)3–6, cymes (3–)5–12(–18)-flowered, 2 per node; proximal bracts lanceolate to linear, 9–105 × 1–40 mm, margins entire, sometimes serrulate; peduncles and pedicels spreading to ascending, sparsely glandular-pubescent. **Flowers:** calyx lobes ovate,

sometimes lanceolate, 4–8 × 2–3 mm, apex acute to acuminate, glandular-pubescent; corolla white, sometimes tinged lavender, with reddish purple nectar guides, ventricose, (17–)20–30 mm, glandular-pubescent externally, sparsely to moderately white-pubescent internally abaxially, tube 6–9 mm, throat abruptly inflated, 8–12 mm diam., slightly 2-ridged abaxially; stamens included or longer pair reaching orifice, pollen sacs opposite, navicular, 1.4–1.7 mm, dehiscing completely, connective splitting, sides sparsely pubescent, hairs white or purplish, to 0.6 mm, rarely glabrous, sutures papillate; staminode 13–17 mm, reaching orifice, 0.3–0.4 mm diam., tip straight to slightly recurved, distal 6–8 mm sparsely to moderately villous, hairs yellowish, to 1.5(–2) mm; style 13–18 mm. **Capsules** 8–14 × 4–6 mm, glabrous. 2*n* = 96.

Flowering Apr–Jul. Prairies, meadows, roadsides, clearings in woods; 10–500 m; N.B., N.S., Ont., Que.; Ala., Ark., Conn., Del., D.C., Ill., Ind., Iowa, Kans., Ky., La., Maine, Md., Mass., Mich., Minn., Miss., Mo., Nebr., N.H., N.J., N.Y., N.C., Ohio, Okla., Pa., R.I., S.C., S.Dak., Tenn., Tex., Vt., Va., W.Va., Wis.

Penstemon digitalis appears to be native in the central Mississippi River Basin, with human activities expanding its range, particularly eastward (F. W. Pennell 1935). It is listed in VASCAN as introduced in the four Canadian provinces where it occurs (http://data.canadensys.net/vascan/taxon/7273).

Plants resembling *Penstemon digitalis* but with smaller corollas (17–23 mm versus 23–30 mm), less glandular-pubescent inflorescences, less scarious-margined calyx lobes, and stems sometimes puberulent (versus glabrous) have been named *P. alluviorum*. Plants with those features come mostly from the southeastern Central Lowlands and north-central Coastal Plain provinces in the eastern United States in the south-central part of the range of *P. digitalis* (Alabama, Arkansas, Illinois, Indiana, Kentucky, Mississippi, Missouri, Ohio, and Tennessee). A. C. Koelling (1964) observed overlap in characters used to distinguish *P. alluviorum* from *P. digitalis* and concluded it was a small-flowered variant of *P. digitalis*.

Specimens of *Penstemon digitalis* with lanceolate and acuminate-tipped calyx lobes can be mistaken for *P. calycosus*, but calyx lobe length and, usually, the presence of hairs on the pollen sacs allow them to be accurately identified.

Penstemon digitalis is widely cultivated and spreads readily. R. R. Clinebell and P. Bernhardt (1998) found evidence that plants are self-compatible.

156. Penstemon elegantulus Pennell, Notul. Nat. Acad. Nat. Sci. Philadelphia 71: 14. 1941 • Rockvine beardtongue E

Herbs. Stems erect, 10–31 cm, retrorsely hairy, not glaucous. **Leaves** basal and cauline, not leathery, glabrous or retrorsely hairy primarily along midvein and margins; basal and proximal cauline 14–75 × 4–16 mm, blade ovate to lanceolate or elliptic, base tapered, margins entire or ± serrate distally, apex obtuse to acute; cauline 3–5 pairs, sessile or proximals short-petiolate, 15–45 × 2–8 mm, blade oblanceolate to lanceolate, base tapered, margins entire or ± serrate or ± dentate, primarily distally, apex obtuse to acute. **Thyrses** interrupted, cylindric, 4–18 cm, axis retrorsely hairy proximally, sparsely glandular-pubescent distally, verticillasters 2–5, cymes 2–6-flowered, (1 or)2 per node; proximal bracts lanceolate, 9–38 × 2–12 mm, margins entire or ± serrate distally; peduncles and pedicels ascending, sparsely glandular-pubescent. **Flowers:** calyx lobes ovate, 3–4 × 1–2 mm, sparsely glandular-pubescent; corolla violet to blue or purple, with purple nectar guides, funnelform, 14–23 mm, glandular-pubescent externally, ± white-pubescent internally abaxially, tube 4–6 mm, throat gradually inflated, 5–6 mm diam., slightly 2-ridged abaxially; stamens included, pollen sacs opposite, navicular, 0.9–1.2 mm, dehiscing completely, connective splitting, sides glabrous, sutures smooth; staminode 9–12 mm, reaching orifice, 0.4–0.6 mm diam., tip straight, distal 0.5–2 mm pilose, hairs yellow or golden yellow, to 0.8 mm; style 9–13 mm. **Capsules** 5–8 × 3.5–5 mm, glabrous.

Flowering May–Jun. Rocky, granitic meadows and hillsides; 900–1800 m; Idaho, Oreg.

Penstemon elegantulus occurs largely in the Hell's Canyon region of the Snake River in Idaho and Nez Perce counties, Idaho, and Wallowa County, Oregon. A specimen from near Silver City in Owyhee County, Idaho (*Hitchcock & Muhlick 22585*, WTU), also appears to be this species. *Penstemon elegantulus* combines morphologic features of *P. albertinus* and *P. humilis*; it generally has obscurely serrate leaves as in the former, and retrorsely hairy leaves and stems as in the latter.

157. Penstemon euglaucus English, Proc. Biol. Soc. Wash. 41: 197. 1928 • Glaucous beardtongue E

Herbs. Stems ascending to erect, 15–55 cm, glabrous, glaucous. **Leaves** basal and cauline, ± leathery or not, glabrous, glaucous; basal and proximal cauline (15–)30–115 × 4–25 mm, blade elliptic to oblanceolate or spatulate, base tapered, margins entire, apex obtuse to acute; cauline 3–6 pairs, sessile or proximals short-petiolate, (17–)40–72 × 9–28 mm, blade ovate to lanceolate, base tapered to clasping, margins entire, apex obtuse to acute. **Thyrses** continuous or interrupted, cylindric, 2–21 cm, axis glabrous, verticillasters 1–6, cymes (2 or)3–8-flowered, 2 per node; proximal bracts lanceolate, 9–45 × 4–13 mm, margins entire; peduncles and pedicels erect, glabrous. **Flowers:** calyx lobes ovate to lanceolate, 2.5–4.5 × 1–2.4 mm, glabrous; corolla violet to blue or purple, with light violet nectar guides, funnelform, 11–15 mm, glabrous externally, sparsely whitish or yellowish pubescent internally abaxially, tube 3–4 mm, throat gradually inflated, 3.5–5 mm diam., 2-ridged abaxially; stamens: longer pair exserted, pollen sacs opposite, navicular, 0.5–0.7 mm, dehiscing completely, connective splitting, sides glabrous, sutures smooth; staminode 7–8 mm, reaching orifice, 0.2–0.3 mm diam., tip straight to slightly recurved, distal 0.5–2 mm sparsely to densely pilose, hairs golden yellow, to 0.8 mm; style 8–10 mm. **Capsules** 5–6 × 3.5–4.5 mm, glabrous. **2***n* = 48.

Flowering Jun–Sep. Sandy, open pine forests, subalpine meadows; 1200–2400 m; Oreg., Wash.

Penstemon euglaucus is a hexaploid of the Cascade Range from Three Sisters, Oregon, north to Mount Adams, Washington. A chromosome count of $2n = 16$ (P. G. Zhukova 1967) is inconsistent with earlier reports of $2n = 48$ (J. Clausen et al. 1940; D. D. Keck 1945) and likely is in error.

158. Penstemon flavescens Pennell, Notul. Nat. Acad. Nat. Sci. Philadelphia 95: 4. 1942 • Tall yellow beardtongue E

Herbs. Stems erect, (6–)10–40 cm, glabrous or puberulent, not glaucous. **Leaves** basal and cauline, ± leathery or not, glabrous; basal and proximal cauline (12–)25–115 × (4–)7–22 mm, blade oblanceolate to lanceolate or elliptic, base tapered, margins entire, apex obtuse to acute; cauline 2–4 pairs, sessile, 18–65 × 3–16 mm, blade oblanceolate to lanceolate, sometimes ovate, base tapered to clasping, margins entire, apex obtuse to acute. **Thyrses** continuous or interrupted, cylindric,

1–12 cm, axis ± puberulent, rarely glandular-pubescent, verticillasters 1–4, cymes 2–10-flowered, 2 per node; proximal bracts lanceolate, sometimes ovate, 12–44 × 2–13 mm, margins entire; peduncles and pedicels erect, glabrous or ± puberulent, rarely glandular-pubescent. **Flowers:** calyx lobes ovate to lanceolate, 5–9 × 1.8–2.7 mm, glabrous, rarely glandular-pubescent; corolla light yellow, without nectar guides, funnelform, 12–16 mm, glabrous externally, rarely glandular-pubescent, sparsely to moderately yellow-pubescent internally abaxially, tube 5–7 mm, throat gradually inflated, 4–6 mm diam., 2-ridged abaxially; stamens included or longer pair reaching orifice, pollen sacs opposite, navicular to subexplanate, 0.7–0.9 mm, dehiscing completely, connective splitting, sides glabrous, sutures smooth; staminode 8–9 mm, included, 0.4–0.7 mm diam., tip straight to slightly recurved, distal 2–3 mm moderately pilose, hairs yellow or golden yellow, to 0.8 mm; style 8–11 mm. **Capsules** 5–8 × 4–5 mm, glabrous. *2n* = 48.

Flowering Jun–Aug. Open or wooded, rocky slopes; 1800–2500 m; Idaho, Mont.

Penstemon flavescens is known from the Bitterroot Mountains in east-central Idaho (Clearwater, Idaho, and Shoshone counties) and southwestern Montana (Ravalli County). A yellow-flowered, glandular-pubescent hexaploid from near Hoodoo Pass in Shoshone County, Idaho, might be a hybrid between *P. attenuatus* and *P. flavescens* (D. V. Clark 1971). Reports of *P. flavescens* from Oregon might be based on whitish or yellowish flowered forms of *P. attenuatus* var. *attenuatus* (D. D. Keck 1945).

159. Penstemon gairdneri Hooker, Fl. Bor.-Amer. 2: 99. 1838 (as Pentstemon) • Gairdner's beardtongue
E F

Subshrubs. **Stems** ascending to erect, 6–40 cm, retrorsely hairy, not glaucous. **Leaves** basal and cauline, alternate or mostly opposite, not leathery, retrorsely hairy; basal and proximal cauline 10–50(–70) × 1–4 mm, blade spatulate to oblanceolate or linear, base tapered, margins entire, revolute, apex rounded to obtuse or acute; cauline 8–12 pairs (or 13–25 leaves when alternate), sessile, 7–35 × 1–3 mm, blade oblanceolate to linear, base tapered, margins entire, apex acute. **Thyrses** interrupted, cylindric to secund, 2–12 cm, axis glandular-pubescent and retrorsely hairy, verticillasters 3–12, cymes 1(or 2)-flowered, 1 or 2 per node; proximal bracts lanceolate to linear, 5–18 × 0.5–3 mm, margins entire; peduncles and pedicels erect, glandular-pubescent and/or retrorsely hairy. **Flowers:** calyx lobes lanceolate, 5–8.5 × 1.7–2.7 mm, puberulent and glandular-pubescent; corolla violet to blue or purple,

with or without reddish nectar guides, funnelform, 15–22 mm, glandular-pubescent externally, glandular-pubescent internally, tube 3–4 mm, throat gradually inflated, 5–6 mm diam., slightly 2-ridged abaxially; stamens included or longer pair reaching orifice, pollen sacs opposite, navicular to subexplanate, 0.8–1 mm, dehiscing completely, connective splitting, sides glabrous, sutures smooth; staminode 8–11 mm, included, 0.4–0.6 mm diam., tip straight, distal 3–5 mm sparsely to densely pubescent, hairs yellow, to 1 mm; style 10–12 mm. **Capsules** 5–9 × 3–5.3 mm, glabrous. *2n* = 16.

Varieties 2 (2 in the flora): nw United States.

Following A. Cronquist (1959), two varieties are recognized here.

1. Leaves alternate; stems 6–20(–35) cm . . . .
. . . . 159a. *Penstemon gairdneri* var. *gairdneri*
1. Leaves opposite, sometimes alternate; stems 20–40 cm . . . . 159b. *Penstemon gairdneri* var. *oreganus*

159a. Penstemon gairdneri Hooker var. **gairdneri**
E F

Penstemon gairdneri subsp. *hians* (Piper) D. D. Keck

Stems 6–20(–35) cm. **Leaves** alternate.

Flowering (Apr–)May–Jun. Rock outcrops, sagebrush shrublands, prairies; 100–1800 m; Oreg., Wash.

D. D. Keck (1940b) recognized three subspecies of *Penstemon gairdneri*: subsp. *gairdneri*, subsp. *hians*, and subsp. *oreganus*. He distinguished the two alternate-leaved subspecies (subsp. *gairdneri* and subsp. *hians*) on the basis of corolla color (rose purple versus lavender-purple with blue limb, respectively) and limb size (15–20 mm diam. versus 12–14 mm diam., respectively), acknowledging considerable variation in limb size in the former.

Variety *gairdneri* occurs in eastern Oregon and central Washington. Its range overlaps with that of var. *oreganus* in southeastern Oregon.

159b. Penstemon gairdneri Hooker var. **oreganus**
A. Gray in A. Gray et al., Syn. Fl. N. Amer. ed. 2, 2(1): 441. 1886 (as Pentstemon) E

Penstemon gairdneri subsp. *oreganus* (A. Gray) D. D. Keck

Stems 20–40 cm. **Leaves** opposite, sometimes alternate.

Flowering May–Jun. Rock outcrops, sagebrush shrublands, prairies; 1000–1600 m; Idaho, Oreg.

Variety *oreganus* occurs in eastern Oregon and central Idaho.

160. **Penstemon glaucinus** Pennell, Notul. Nat. Acad. Nat. Sci. Philadelphia 71: 10. 1941 • Blue-leaf beardtongue E

Herbs. Stems ascending to erect, 26–45 cm, glabrous, glaucous. **Leaves** basal and cauline, ± leathery or not, glabrous, glaucous; basal and proximal cauline 15–60 × 7–13 mm, blade spatulate to elliptic, base tapered, margins entire, apex rounded to obtuse or acute; cauline 2–4 pairs, sessile or proximals short-petiolate, 26–55 × 7–14 mm, blade oblanceolate to oblong or lanceolate, base truncate to clasping, margins entire, apex obtuse to acute. **Thyrses** interrupted, cylindric, 8–22 cm, axis glandular-pubescent, verticillasters 2–8, cymes 3–6-flowered, 2 per node; proximal bracts lanceolate, 16–40 × 2–14 mm, margins entire; peduncles and pedicels ascending to erect, pedicels glandular-pubescent, sometimes peduncles glabrous. **Flowers:** calyx lobes ovate to lanceolate, 2.4–5.5 × 1–2 mm, glandular-pubescent; corolla violet, without nectar guides, tubular-funnelform, 12–15 mm, glandular-pubescent externally, moderately yellow-pilose internally abaxially, tube 3–4 mm, throat slightly inflated, 3.5–5 mm diam., 2-ridged abaxially; stamens included, pollen sacs opposite, navicular, 0.7–1 mm, dehiscing completely, connective splitting, sides glabrous, sutures papillate; staminode 6–8 mm, included, 0.3–0.5 mm diam., tip straight to slightly recurved, distal 1–2 mm densely pilose, hairs golden yellow, to 0.5 mm, or glabrous; style 8–10 mm. **Capsules** 5–7 × 3–4 mm, glabrous.

Flowering Jun–Aug. Clearings in pine and pine-fir forests; 1900–2100 m; Oreg.

Penstemon glaucinus is known only from Klamath and Lake counties.

161. **Penstemon globosus** (Piper) Pennell & D. D. Keck, Publ. Carnegie Inst. Wash. 520: 294. 1940 • Globe beardtongue E

Penstemon confertus Douglas var. *globosus* Piper, Bull. Torrey Bot. Club 27: 397. 1900 (as Pentstemon)

Herbs. Stems ascending to erect, (10–)20–65 cm, glabrous or sparsely retrorsely hairy distally, not glaucous. **Leaves** basal and cauline, not leathery, glabrous; basal and proximal cauline 28–160(–235) × 8–25 (–35) mm, blade oblanceolate to elliptic, base tapered, margins entire, apex obtuse to acute; cauline 2–5 pairs, sessile or proximals short-petiolate, 10–130 × 3–35 mm, blade oblong to ovate or lanceolate, base truncate to clasping, margins entire, apex obtuse to acute. **Thyrses** continuous or interrupted, cylindric, (1–)2–11(–27) cm, axis glabrous or sparsely retrorsely hairy, verticillasters 1–3(or 4), cymes (2–)4–13-flowered, 2 per node; proximal bracts lanceolate, (5–)16–66 × (1–)4–19 mm, margins entire; peduncles and pedicels erect, glabrous. **Flowers:** calyx lobes oblanceolate to lanceolate, (4–)5–9 × 1.4–3.5 mm, glabrous; corolla blue to purple or violet, without nectar guides, tubular-funnelform, 14–21 mm, glabrous externally, moderately white- or yellowish pilose internally abaxially, tube 5–6 mm, throat slightly inflated to slightly ventricose, 4–7 mm diam., 2-ridged abaxially; stamens included, pollen sacs divergent, saccate, 0.7–1.2 mm, dehiscing incompletely, distal ⅕–¼ indehiscent, connective not splitting, sides glabrous, sutures papillate; staminode 7–9 mm, included, 0.4–0.7 mm diam., tip straight to slightly recurved, distal 1–2 mm densely pilose to lanate, hairs golden yellow, to 1.4 mm, medial 2–4 mm sparsely lanate; style 8–10 mm. **Capsules** 5–7 × 3–4 mm, glabrous. $2n = 16, 32$.

Flowering May–Aug. Dry to wet meadows, moist mountain slopes; 800–3100 m; Idaho, Mont., Oreg.

The saccate anthers and relatively broad leaves of *Penstemon globosus* are diagnostic. D. V. Clark (1971) reported plants in northeastern Oregon (Baker, Union, and Wallowa counties) exhibiting degrees of morphologic intermediacy between *P. globosus* and *P. rydbergii*. These could be hybrids or introgressants involving those two species, though Clark did not observe any populations where hybridization was evident. Putative hybrids between *P. globosus* and *P. confertus* have been documented in Idaho County, Idaho (Clark).

162. **Penstemon gracilis** Nuttall, Gen. N. Amer. Pl. 2: 52. 1818 (as Pentstemon gracile) • Lilac or slender beardtongue E

Penstemon gracilis subsp. *wisconsinensis* Pennell; *P. gracilis* var. *wisconsinensis* (Pennell) Fassett

Herbs. Caudex herbaceous. **Stems** ascending to erect, (15–)20–50 cm, retrorsely hairy and glandular-pubescent, not glaucous. **Leaves** basal and cauline, not leathery, glabrous or sparsely, rarely densely, puberulent; basal and proximal cauline 25–75 × 4–15 mm, blade ovate to oblanceolate or lanceolate, base tapered, margins subentire or ± serrate, apex obtuse to acute; cauline 4–7 pairs, sessile, 25–80(–90) × (2–)4–10(–15) mm, blade lanceolate to linear, base truncate to clasping, margins entire or serrulate to serrate, apex acute to acuminate. **Thyrses** interrupted, sometimes continuous, cylindric, (3–)5–17(–21) cm, axis glandular-pubescent, verticillasters (2 or)3–5(–7), cymes 2–6-flowered, 2 per node; proximal bracts lanceolate, 5–95 × 1–12 mm, margins entire or serrulate, rarely serrate; peduncles and pedicels ascending to erect, glandular-pubescent. **Flowers:** calyx lobes ovate to lanceolate, 4–6 × 1.5–2 mm, glandular-pubescent; corolla light lavender to lavender, with violet nectar guides, tubular, 14–22 mm, glandular-pubescent externally, moderately white-pilose internally abaxially, tube 4–6 mm, throat slightly inflated, 4–6 mm diam., prominently 2-ridged abaxially; stamens included, pollen sacs divergent, navicular, 1–1.3 mm, dehiscing completely, connective splitting, sides glabrous, sutures papillate; staminode 11–12 mm, reaching orifice, 0.4–0.5 mm diam., tip slightly recurved, distal 7–9 mm densely villous, hairs golden yellow, to 1.5 mm; style 9–12 mm. **Capsules** 6–8 × 3–4 mm, glabrous. $2n = 16$.

Flowering May–Aug. Tallgrass, mixed grass, and shortgrass prairies, foothills; 300–2100 m; Alta., B.C., Man., Ont., Sask.; Colo., Ill., Ind., Iowa, Mich., Minn., Mont., Nebr., N.Mex., N.Dak., S.Dak., Wis., Wyo.

Glabrous-leaved plants are characteristic on the prairies of most of the central and northern Great Plains, and in the foothills of the Central, Northern, and Canadian Rocky mountains. Puberulent-leaved plants from the Driftless Area of Wisconsin have been named var. *wisconsinensis*. Puberulent-leaved plants also occur in Alberta and in North Dakota (specimens from Barnes, Benton, Eddy, Pierce, Ramsey, Wells, and Williams counties have been seen), sometimes with glabrous-leaved plants. *Penstemon gracilis* is introduced in Indiana (K. Yatskievych 2000).

The roots of *Penstemon gracilis* are used by the Lakota of the northern Great Plains for protection from snakebites (D. E. Moerman 1998).

163. **Penstemon griffinii** A. Nelson, Bot. Gaz. 56: 70. 1913 (as Pentstemon) • Griffin's beardtongue E

Herbs. Stems erect, 12–50 cm, retrorsely hairy proximally, becoming glandular-pubescent distally, not glaucous. **Leaves** basal and cauline, not leathery, glabrous or retrorsely hairy abaxially along midvein and glabrous adaxially, ± glaucescent; basal and proximal cauline 13–55 × 4–8(–12) mm, blade spatulate to oblanceolate or elliptic, base tapered, margins entire, apex rounded to acute; cauline 2–5(or 6) pairs, sessile, 5–40(–52) × 0.3–5 mm, blade linear, base tapered, margins entire, apex acuminate. **Thyrses** interrupted, secund, 1–14 cm, axis glandular-pubescent, verticillasters (1–)3 or 4(or 5), cymes 1–3-flowered, 1(or 2) per node; proximal bracts linear, 3–15 × 0.3–1.4 mm; peduncles and pedicels erect, sometimes ascending, glandular-pubescent. **Flowers:** calyx lobes ovate to lanceolate, 3.8–7.5 × 1.5–2.6 mm,

glabrescent or glandular-pubescent; corolla violet to blue or purple, with violet nectar guides, tubular-funnelform, 17–25 mm, glandular-pubescent externally, densely golden-villous internally abaxially, tube 4–7 mm, throat slightly inflated, 5–7 mm diam., prominently 2-ridged abaxially; stamens included, pollen sacs divergent to opposite, navicular, 1–1.3 mm, dehiscing completely, connective splitting, sides glabrous, sutures papillate; staminode 10–13 mm, included, 0.2–0.3 mm diam., tip straight to recurved, distal 7–11 mm densely pilose, hairs golden, to 1.5 mm; style 12–17 mm. **Capsules** 6–9 × 4–5 mm, glabrous.

Flowering Jun–Aug. Rocky hillsides, open coniferous forests, montane grasslands; 2500–3000 m; Colo., N.Mex.

Populations of *Penstemon griffinii* are documented in Chaffee, Conejos, Fremont, Mineral, Park, Rio Grande, and Saguache counties, Colorado, and Rio Arriba and Taos counties, New Mexico. *Penstemon griffinii* can be confused with narrow-leaved plants of *P. degeneri*, but the pubescence on the internal abaxial surfaces of the corollas of *P. griffinii* is consistently golden-lanate and extends from the abaxial limbs 5–8 mm into the throats of the corollas. In *P. degeneri*, the internal abaxial surfaces of the corollas are white- or yellow-lanate, sometimes sparsely so, and the indument barely extends into the throats of the corollas.

164. **Penstemon harbourii** A. Gray, Proc. Amer. Acad. Arts 6: 71. 1862 (as Pentstemon) • Harbour's beardtongue E

Herbs. Caudex rhizomelike. **Stems** decumbent, ascending, or weakly erect, 4–18 cm, retrorsely hairy, not glaucous. **Leaves** cauline, not leathery, glabrate, retrorsely hairy, or puberulent; cauline 2–4 pairs, petiolate or sessile, 7–28 × 3–12 mm, blade spatulate to oblanceolate, base tapered to clasping, margins entire, apex rounded or obtuse to acute. **Thyrses** interrupted or continuous, ± secund, 1–3 cm, axis ± glandular-pubescent, verticillasters 1 or 2(or 3), cymes 1- or 2-flowered, 1 or 2 per node; proximal bracts lanceolate, 17–23 × 4–10 mm, margins entire; peduncles and pedicels ascending to erect, retrorsely hairy and glandular-pubescent. **Flowers:** calyx lobes lanceolate, 6–10 × 1.3–2 mm, glandular-pubescent; corolla lavender or bluish lavender to lavender-purple, without nectar guides, bilabiate, not personate, tubular-funnelform, 15–20 mm, glandular-pubescent externally, densely white-lanate internally abaxially, tube 4–6 mm, throat gradually inflated, 5–6 mm diam., prominently 2-ridged abaxially; stamens included, pollen sacs opposite, subexplanate, 0.7–0.9 mm, dehiscing completely, connective splitting, sides glabrous, sutures smooth; staminode 8–9 mm, included, 0.3–0.4 mm diam., tip recurved, distal 4–5 mm moderately to densely lanate, hairs yellow, to 1 mm; style 11–12 mm. **Capsules** 6–8 × 4–5 mm, glabrous. $2n = 16$.

Flowering Jun–Aug. Alpine talus, gravel slopes, boulder fields; 3200–4200 m; Colo.

Penstemon harbourii is found on the high peaks of the Central Rocky Mountains. Populations have been documented in at least 17 counties. Habit, pubescence, and root characters of *P. harbourii* are somewhat anomalous in sect. *Penstemon*, prompting D. D. Keck (1945) to suggest possible ties with sect. *Caespitosi* or sect. *Cristati*. Molecular data appear to provide support for ties with sect. *Caespitosi* (A. D. Wolfe et al. 2006; C. A. Wessinger et al. 2016).

165. Penstemon heterodoxus A. Gray in A. Gray et al., Syn. Fl. N. Amer. 2(1): 269. 1878 (as Pentstemon)

• Sierra beardtongue   E

Herbs. **Stems** decumbent, ascending, or erect, 5–65 cm, glabrous or retrorsely hairy, sometimes also glandular-pubescent distally, not glaucous. **Leaves** basal and cauline, not leathery, glabrous; basal and proximal cauline 5–140 × 3–20 mm, blade spatulate to oblanceolate or elliptic, base tapered, margins entire, apex rounded to obtuse or acute; cauline 2–5 pairs, short-petiolate or sessile, 10–60(–70) × 1–14 mm, blade oblong to oblanceolate or lanceolate, base tapered to clasping, margins entire, apex obtuse to acute. **Thyrses** interrupted, cylindric, 1–12 cm, axis ± glandular-pubescent, verticillasters 1–6, cymes (1 or)2–6-flowered, 2 per node; proximal bracts lanceolate, 4–35 × 1–6 mm, margins entire; peduncles and pedicels erect, ± glandular-pubescent. **Flowers:** calyx lobes oblong to lanceolate, 4–6 × 0.7–1.2 mm, glandular-pubescent; corolla lavender to blue, violet, or purple, essentially without nectar guides, tubular-funnelform, 10–16 mm, glandular-pubescent externally, moderately yellow-pilose to yellow-lanate internally abaxially, tube 5–7 mm, throat slightly inflated, 3–5 mm diam., prominently 2-ridged abaxially; stamens: longer pair reaching orifice or slightly exserted, pollen sacs opposite, navicular, 0.6–0.8(–1) mm, dehiscing completely, connective splitting, sides glabrous, sutures smooth; staminode 6–9 mm, included, 0.3–0.6 mm diam., tip straight to recurved, glabrous or distal 1–3 mm sparsely to moderately pilose, hairs yellow, to 0.5 mm; style 9–12 mm. **Capsules** 4–6 × 2.5–3.5 mm, glabrous.

Varieties 3 (3 in the flora): w United States.

Penstemon heterodoxus comprises three partially sympatric varieties centered in the Sierra Nevada of eastern California.

1. Thyrse axes usually sparsely glandular-pubescent; n California. . . . . 165c. *Penstemon heterodoxus* var. *shastensis*
1. Thyrse axes moderately to densely glandular-pubescent; e California, wc Nevada.
  2. Stems 5–25 cm; basal and proximal cauline leaves 5–50 × 3–10 mm; verticillasters (1–)3; ec California, wc Nevada. 165a. *Penstemon heterodoxus* var. *heterodoxus*
  2. Stems 15–35 cm; basal and proximal cauline leaves (17–)20–75 × 6–18 mm; verticillasters 2–6; se California. . . . . 165b. *Penstemon heterodoxus* var. *cephalophorus*

165a. Penstemon heterodoxus A. Gray var. **heterodoxus**   E

Stems 5–25 cm. **Leaves:** basal and proximal cauline 5–50 × 3–10 mm. **Thyrses:** axis moderately to densely glandular-pubescent, verticillasters 1(–3). $2n = 16$.

Flowering Jul–Aug. Montane and alpine slopes, montane meadows, talus slopes; 2000–3900 m; Calif., Nev.

Variety *heterodoxus* is known from the Klamath Range, Sierra Nevada, and Inyo and White mountains in east-central California and west-central Nevada. Populations are partly geographically sympatric with those of var. *shastensis*, which occurs to the north, and with those of var. *cephalophorus*, which occurs to the south.

165b. Penstemon heterodoxus A. Gray var. **cephalophorus** (Greene) N. H. Holmgren, Brittonia 44: 482. 1992   E

Penstemon cephalophorus Greene, Leafl. Bot. Observ. Crit. 1: 79. 1904 (as Pentstemon); *P. heterodoxus* subsp. *cephalophorus* (Greene) D. D. Keck

Stems 15–35 cm. **Leaves:** basal and proximal cauline (17–)20–75 × 6–18 mm. **Thyrses:** axis moderately to densely glandular-pubescent, verticillasters 2–6. $2n = 16$.

Flowering Jul–Aug(–Sep). Subalpine and alpine meadows; 2100–3300 m; Calif.

Variety *cephalophorus* occurs at the southern end of the range of *Penstemon heterodoxus*; it has been documented in Alpine, Fresno, Inyo, Kern, Madera, Mono, and Tulare counties.

165c. Penstemon heterodoxus A. Gray var. **shastensis**
(D. D. Keck) N. H. Holmgren, Brittonia 44: 482.
1992 E

Penstemon shastensis D. D. Keck,
Amer. Midl. Naturalist 33: 165.
1945

Stems (10–)15–65 cm. **Leaves:**
basal and proximal cauline
30–140 × 6–20 mm. **Thyrses:**
axis usually sparsely glandular-
pubescent, verticillasters 2–6.
$2n = 32$.

Flowering Jul–Aug.   Montane meadows; 1100–
2400 m; Calif.

Variety *shastensis* occurs at the northern end of
the range of *Penstemon heterodoxus* in Butte, Lassen,
Modoc, Plumas, Shasta, and Siskiyou counties.

166. Penstemon hirsutus (Linnaeus) Willdenow, Sp.
Pl. 3: 227. 1800 (as Pentstemon) • Northeastern
beardtongue, penstémon hirsute E

Chelone hirsuta Linnaeus, Sp. Pl.
2: 611. 1753, name conserved;
C. pentstemon Linnaeus

Herbs.   Stems ascending to
erect, 30–80 cm, retrorsely
hairy and sparsely to densely
glandular-villous, not glaucous.
Leaves basal and cauline, not
leathery, glabrous or sparsely
glandular-lanate along midveins, rarely moderately
glandular-lanate; basal and proximal cauline 20–126 ×
4–16 mm, blade spatulate to oblanceolate, lanceolate,
spatulate, or elliptic, base tapered, margins entire or
serrate to dentate, apex rounded to obtuse or acute;
cauline 5–8(–10) pairs, sessile or short-petiolate,
20–130 × 2–30 mm, blade lanceolate to oblanceolate,
base clasping or tapered, margins finely to coarsely
serrate or dentate, apex acute to acuminate. **Thyrses**
interrupted, conic, 6–37 cm, axis glandular-pubescent
to glandular-lanate, verticillasters (3 or)4–8, cymes
2–11(–15)-flowered, 2 per node; proximal bracts
lanceolate to linear, 8–60 × 1–6 mm, margins entire
or ± serrate to dentate; peduncles and pedicels
ascending, glandular-pubescent to glandular-lanate.
Flowers: calyx lobes ovate to lanceolate, 4–7 ×
1.5–2.3 mm, glandular-pubescent; corolla lavender
or purplish, with or without faint violet nectar
guides, bilabiate, personate, tubular, 20–26(–28) mm,
glandular-pubescent externally, moderately to densely
white-villous internally abaxially, tube 6–7 mm, throat
slightly inflated, 6–8 mm diam., 2-ridged abaxially;
stamens included, pollen sacs opposite, navicular
to subexplanate, 0.8–1 mm, dehiscing completely,
connective splitting, sides glabrous, sutures papillate;
staminode 13–16 mm, exserted, 0.6–0.7 mm diam., tip

straight, distal 11–14 mm moderately to densely villous,
hairs yellowish, to 1.2 mm; style 16–18 mm. **Capsules**
6–9 × 5–6 mm, glabrous. $2n = 16$.

Flowering May–Jul.   Sandy or rocky woods, rocky
fields, bluffs, cliffs; 10–200 m; Ont., Que.; Conn., Del.,
D.C., Ill., Ind., Ky., Maine, Md., Mass., Mich., N.J.,
N.Y., Ohio, Pa., R.I., Tenn., Vt., Va., W.Va., Wis.

Penstemon hirsutus occurs widely in the northeastern
United States, and in southern Ontario and Quebec. Its
long, tubular, personate corollas and herbage with some
stellate hairs may indicate a close relationship with
P. oklahomensis and *P. tenuiflorus*.

167. Penstemon humilis Nuttall ex A. Gray, Proc. Amer.
Acad. Arts 6: 69. 1862 (as Pentstemon) • Low
beardtongue E F

Herbs.   Stems decumbent to
erect, (2–)5–30(–65) cm,
retrorsely hairy, not glaucous.
Leaves basal and cauline, not
leathery, glabrous or retrorsely
hairy; basal and proximal
cauline (8–)15–105 × 2–22
(–32) mm, blade obovate to
ovate or oblanceolate, base
tapered, margins entire, sometimes serrulate or serrate,
apex obtuse to acute; cauline (1 or)2–4 pairs, sessile or
proximals short-petiolate, 10–35(–45) × 1–11 mm,
blade obovate to oblanceolate or lanceolate, base
tapered to clasping, margins entire, rarely ± serrate,
apex obtuse to acute. **Thyrses** interrupted, cylindric to
± secund, (1–)3–20(–39) cm, axis moderately glandular-
pubescent, verticillasters (1–)3–9(–11), cymes (1 or)2–6
(–11)-flowered, 2 per node; proximal bracts lanceolate,
rarely ovate, 6–34(–48) × 1–11 mm, margins entire;
peduncles and pedicels ascending to erect, glandular-
pubescent and, usually, retrorsely hairy. **Flowers:** calyx
lobes ovate to lanceolate, 2.8–5(–6) ×1–1.9 mm,
margins entire or erose, apex acute, glandular-pubescent;
corolla blue to purple or violet, with reddish violet
nectar guides, bilabiate, not personate, tubular-
funnelform, (7–)8–19 mm, glandular-pubescent
externally, sparsely to moderately white- or yellow-
lanate internally abaxially, tube 2–5 mm, throat slightly
inflated, (3–)4.5–6 mm diam., 2-ridged abaxially;
stamens included, pollen sacs opposite, navicular,
0.5–0.8 mm, dehiscing completely, connective splitting,
sides glabrous, sutures smooth or papillate; staminode
(3–)7–9 mm, included, 0.1–0.3 mm diam., tip slightly
recurved, distal 1–4 mm sparsely to densely pilose,
hairs yellow or golden yellow, to 0.7 mm (medial hairs
sparser and shorter); style 6–12 mm. **Capsules** 4–6 ×
2.5–3.5 mm, glabrous.

Varieties 3 (3 in the flora): w United States.

Penstemon humilis is widespread and extremely
variable throughout its range in the western United
States.   Except for two fairly distinct elements in

Utah, most variation in the species is contained in the widespread var. *humilis*. D. D. Keck (1945) and N. H. Holmgren (1984) summarized the ranges and morphologic tendencies of some of the more distinctive phases.

1. Cauline leaf blade margins ± serrate; basal leaves (4–)7–22(–32) mm wide; sw Utah . . . . .167c. *Penstemon humilis* var. *obtusifolius*
1. Cauline leaf blade margins entire, rarely ± serrate distally; basal leaves 2–20 mm wide; California, nw Colorado, s Idaho, sw Montana, Nevada, e Oregon, Utah, Washington, w Wyoming.
  2. Basal leaves puberulent or retrorsely hairy; corollas (7–)12–19 mm; California, nw Colorado, s Idaho, sw Montana, Nevada, e Oregon, Utah, Washington, w Wyoming . . . . .167a. *Penstemon humilis* var. *humilis*
  2. Basal leaves glabrous, sometimes ± puberulent, especially along midveins and margins; corollas 10–13 mm; nc Utah . . . . 167b. *Penstemon humilis* var. *brevifolius*

167a. Penstemon humilis Nuttall ex A. Gray var. **humilis** E F

Penstemon cinereus Piper; *P. cinereus* subsp. *foliatus* D. D. Keck; *P. decurvus* Pennell ex Crosswhite; *P. humilis* var. *desereticus* S. L. Welsh

Leaves: basal leaves puberulent or retrorsely hairy, (8–)15–105 × 2–13(–18) mm, cauline blade margins entire, rarely ± serrate distally. **Flowers:** corolla (7–)12–19 mm. $2n = 16$.

Flowering May–Jul. Open, rocky slopes, hillsides, sagebrush shrublands, pine-juniper woodlands, coniferous forests, alpine meadows; 1000–3200 m; Calif., Colo., Idaho, Mont., Nev., Oreg., Utah, Wash., Wyo.

Welsh described var. *desereticus* based on material with relatively larger corollas (15–19 mm) and smaller basal leaves (8–25 × 2–6 mm) from mountain ranges in the western Bonneville Basin of Utah; E. C. Neese and N. D. Atwood (2003) considered the variety to be limited to Utah. Variety *desereticus* appears to be confluent with a form of the species in the Calcareous Mountains of eastern Nevada, which N. H. Holmgren (1984) discussed and later recognized as *Penstemon decurvus* (Holmgren 2017). *Penstemon decurvus*, described from eastern Lincoln County, Nevada, also is referable here. Plants from northeastern California, extreme southwestern Idaho, extreme northwestern Nevada, and eastern and central Oregon often have cinereous leaves and relatively longer stems, shorter calyx lobes, and, sometimes, more open inflorescences; these plants have been called *P. cinereus*. This element grades into other phases to the east and appears to be another form of highly variable

var. *humilis*. Some specimens from eastern Idaho (Clark and Fremont counties) have glabrous leaves, possibly from genetic exchange with *P. aridus*.

167b. Penstemon humilis Nuttall ex A. Gray var. **brevifolius** A. Gray in A. Gray et al., Syn. Fl. N. Amer. 2(1): 267. 1878 (as Pentstemon) E

Penstemon humilis subsp. *brevifolius* (A. Gray) D. D. Keck

Leaves: basal leaves glabrous, sometimes ± puberulent, especially along midvein and margins, (10–)15–67 × (2–)4–20 mm, cauline blade margins entire. **Flowers:** corolla 8–13 mm.

Flowering May–Jul. Sagebrush shrublands, pinyon-juniper woodlands, coniferous forests, alpine meadows; 1700–3400 m; Utah.

Variety *brevifolius* is known from the northern Wasatch Range. The variety has been documented in Box Elder, Cache, Juab, Morgan, Rich, Salt Lake, Sanpete, Sevier, Tooele, Utah, Wasatch, and Weber counties.

167c. Penstemon humilis Nuttall ex A. Gray var. **obtusifolius** (Pennell) Reveal, Great Basin Naturalist 35: 369. 1976 E

Penstemon obtusifolius Pennell, Contr. U.S. Natl. Herb. 20: 370. 1920; *P. humilis* subsp. *obtusifolius* (Pennell) D. D. Keck

Leaves: basal leaves glabrous, rarely ± puberulent along midvein and margins, (11–)25–55 (–75) × (4–)7–22(–32) mm, cauline blade margins ± serrate. **Flowers:** corolla 10–13 mm.

Flowering May–Jun. Pinyon-juniper and pine woodlands; 1500–3400 m; Utah.

Variety *obtusifolius* is known only from Washington County.

168. Penstemon inflatus Crosswhite, Amer. Midl. Naturalist 74: 436, fig. 5. 1965 • Inflated beardtongue E

Herbs. Stems ascending to erect, (10–)17–70 cm, ± retrorsely hairy proximally, also sparsely glandular-pubescent distally, not glaucous. **Leaves** basal and cauline, basal sometimes withering by anthesis, not leathery, glabrous (except for puberulent petioles and mid-veins); basal and proximal cauline (12–)25–87(–120) × (3–)5–18(–28) mm, blade lanceolate, base tapered, margins entire, apex obtuse to acute; cauline 4–6 pairs, short-petiolate or sessile, 17–98 × 2–16 mm, blade lanceolate to linear, base tapered to clasping, margins entire, apex acute. **Thyrses** interrupted, narrowly conic, (3–)11–30 cm, axis sparsely glandular-pubescent, verticillasters 3–8, cymes 2–4-flowered, 2 per node; proximal bracts lanceolate to linear, 10–45 × 1–5 mm, margins entire; peduncles and pedicels ascending, sparsely glandular-pubescent. **Flowers:** calyx lobes lanceolate, 3–6 × 1.4–2 mm, glandular-pubescent; corolla lavender to light blue, with reddish purple nectar guides, funnelform, 17–27 mm, glandular-pubescent externally, sparsely white- or yellowish lanate internally abaxially, tube 5–7 mm, throat gradually inflated, 6–7 mm diam., slightly 2-ridged abaxially; stamens included, pollen sacs opposite, navicular, 1.3–1.6 mm, dehiscing completely, connective splitting, sides glabrous, sutures papillate; staminode 12–14 mm, included, 0.3–0.4 mm diam., tip recurved, distal 7–8 mm moderately to densely pilose, hairs yellow-orange, to 1 mm; style 11–14 mm. **Capsules** 8–10 × 3.2–4 mm, glabrous. *2n* = 16.

Flowering Jun–Aug(–Oct). Hillsides, meadows, pinyon-juniper woodlands, pine and pine-Douglas fir forests; 2000–3400 m; N.Mex.

Penstemon inflatus is concentrated in the Sangre de Cristo Range of north-central New Mexico, extending southward into the Sandia-Manzano Mountains. The species has been documented in Bernalillo, Colfax, San Miguel, Santa Fe, Taos, and Torrance counties.

169. Penstemon kralii D. Estes, J. Bot. Res. Inst. Texas 6: 1, figs. 1, 2. 2012 • Kral's beardtongue E

Herbs. Stems ascending to erect, 40–70 cm, spreading-puberulent or retrorsely hairy proximally, glandular-pubescent distally, not glaucous. **Leaves** basal and cauline, not leathery, glabrate or puberulent; basal and proximal cauline 20–70 × (5–)9–20 mm, blade oblanceolate, base tapered, margins entire or ± serrate or crenate, apex obtuse to acute; cauline 6–9 pairs, sessile, 120–150 × 30–40 mm, blade oblong to lanceolate or ovate-lanceolate, base tapered to clasping, margins serrate to dentate, apex acute to acuminate. **Thyrses** interrupted, conic, 5–20 cm, axis glandular-pubescent, verticillasters 4–7, cymes 3–7-flowered, 2 per node; proximal bracts ovate-lanceolate to lanceolate, 16–60 (–80) × 3–45 mm, margins entire or serrate; peduncles and pedicels ascending, glandular-pubescent. **Flowers:** calyx lobes lanceolate, 4–5.5 × 0.8–1.2 mm, glandular-pubescent; corolla pale lavender to nearly white, with violet nectar guides, tubular-funnelform, 10–16 mm, glandular-pubescent externally, sparsely white-lanate internally abaxially, tube 6–7 mm, throat gradually inflated, 5–7 mm diam., slightly 2-ridged abaxially;

stamens included, pollen sacs opposite, navicular, 0.8–1 mm, dehiscing completely, connective splitting, sides glabrous, sutures papillate; staminode 7–8 mm, reaching orifice or slightly exserted, 0.3–0.5 mm diam., tip straight to slightly recurved, distal 4–5 mm moderately to densely pilose, hairs golden, to 1.2 mm; style 9–12 mm. **Capsules** 5–6 × 3.5–4 mm, glabrous.

Flowering May. Calcareous soils, juniper-oak-hickory woodlands; 200–300 m; Ala., Tenn.

Penstemon kralii is known from the southern Cumberland Plateau. It has been collected in Blount, Jackson, Madison, and Morgan counties, Alabama, and in Franklin County, Tennessee.

170. Penstemon laevigatus Aiton, Hort. Kew. 2: 361. 1789 (as Pentstemon laevigata) • Eastern beardtongue E

Herbs. Stems ascending to erect, 40–115 cm, glabrous or retrorsely hairy, not glaucous. **Leaves** basal and cauline, basal sometimes withering by anthesis, not leathery, glabrous or puberulent; basal and proximal cauline 40–150 × 7–48 mm, blade oblanceolate to lanceolate, base tapered, margins entire or ± serrate, apex obtuse to acute; cauline 3–7 pairs, sessile or proximals sometimes petiolate, 18–128 × 8–28 mm, blade lanceolate, proximals sometimes oblanceolate, base clasping, sometimes tapered, margins entire or ± serrate, apex acute. **Thyrses** interrupted, narrowly conic, (3–)8–20 cm, axis glabrous or retrorsely hairy and sparsely glandular-pubescent, verticillasters 2–6, cymes (2–)4–11-flowered, 2 per node; proximal bracts lanceolate to linear, 8–38 × 1–3 mm, margins entire; peduncles and pedicels spreading or ascending, glabrous or retrorsely hairy and sparsely glandular-pubescent. **Flowers:** calyx lobes ovate, 3–6 × 1.5–2.1 mm, glabrous or glandular-pubescent; corolla pale lavender to violet, with reddish purple nectar guides, ventricose, 15–22 mm, glandular-pubescent externally, sparsely to moderately white-lanate internally abaxially, tube 5–6 mm, throat abruptly inflated, 5–8 mm diam., slightly 2-ridged abaxially; stamens included or longer pair reaching orifice, pollen sacs opposite, navicular, 1–1.2 mm, dehiscing completely, connective splitting, sides glabrous, sutures papillate; staminode 10–12 mm, exserted, 0.4–0.5 mm diam., tip straight to slightly recurved, distal 5–7 mm moderately to densely pubescent, hairs yellow, to 2.3 mm; style 12–15 mm. **Capsules** 5–8 × 3.5–5 mm, glabrous.

Flowering May–Jun. Meadows, floodplain forests, fields, rock outcrops, calcareous bluffs; 10–400 m; Ala., Conn., Del., D.C., Fla., Ga., Ky., La., Md., Mass., Miss., N.J., N.C., Ohio, Pa., S.C., Tenn., Va., W.Va.

Penstemon laevigatus primarily is a species of the Piedmont and eastern Appalachians; scattered populations occur outside those regions. L. La Cour (1931) reported a chromosome count of *n* = 48 for *P. laevigatus*. Without a voucher, it is impossible to know if this count was for the species as treated here, or *P. calycosus* or *P. digitalis*, which have been treated as infraspecific taxa of *P. laevigatus*.

An infusion made from *Penstemon laevigatus* is used by the Cherokee tribe of the southeastern United States as a gastrointestinal aid (D. E. Moerman 1998).

171. Penstemon laxiflorus Pennell, Monogr. Acad. Nat. Sci. Philadelphia 1: 229. 1935 • Nodding beardtongue [E]

Herbs. Stems ascending to erect, 25–65(–70) cm, puberulent or retrorsely hairy, rarely also glandular-pubescent distally, not glaucous. **Leaves** basal and cauline, basal often withering by anthesis, not leathery, glabrous or retrorsely hairy to puberulent; basal and proximal cauline 25–90 × 8–25 mm, blade spatulate, oblanceolate, or ovate, base tapered, rarely truncate, margins subentire or serrate to dentate, apex rounded to obtuse or acute; cauline (3–)5–7 pairs, short-petiolate or sessile, (20–)30–90(–110) × (2–)5–22 mm, blade lanceolate to oblanceolate, base tapered to cordate-clasping, margins subentire or serrate to dentate, apex acute. **Thyrses** interrupted, conic, 5–26(–32) cm, axis retrorsely hairy and ± glandular-pubescent, verticillasters 3–7, cymes 2–6-flowered, branches of each cyme usually elongating, of equal length, 2 per node; proximal bracts lanceolate, 4–28 × 1–6 mm, margins entire or ± serrulate; peduncles and pedicels spreading or ascending, puberulent or retrorsely hairy and, usually, glandular-pubescent. **Flowers:** calyx lobes ovate to lanceolate, 2.5–5.5 × 2–3 mm, sparsely puberulent and glandular-pubescent; corolla white to light lavender, sometimes tinged pink, with reddish purple nectar guides, tubular, 20–28(–30) mm, glandular-pubescent externally, moderately white- or yellow-lanate internally abaxially, tube 5–7 mm, throat abruptly inflated, 4–8 mm diam., 2-ridged abaxially; stamens included, pollen sacs opposite, navicular, 1–1.3 mm, dehiscing completely, connective splitting, sides glabrous, sutures papillate; staminode 15–20 mm, exserted, 0.6–0.8 mm diam., tip straight to slightly recurved, distal 8–10 mm densely pilose, hairs yellow, to 1.5 mm; style 14–20 mm. **Capsules** 8–10 × 6–7 mm, glabrous. *2n* = 16.

Flowering Mar–Jun. Sandy or rocky open woods, tallgrass prairies, sand barrens; 10–500 m; Ala., Ark., La., Miss., Okla., Tex.

Penstemon laxiflorus is a species of the central and western Gulf Coastal Plain and southern Interior Lowland. Pennell cited one specimen each from Florida and Georgia; specimens from those states have not been confirmed. The species shares many features with *P. australis*, which occurs farther east along the southern Atlantic Coastal Plain and eastern Gulf Coastal Plain. *Penstemon laxiflorus* usually can be distinguished from *P. australis* by stem vestiture. *Penstemon laxiflorus* has stems with only short, retrorse hairs or, if glandular hairs also are present, they are sparse and occur just below the inflorescences. By contrast, *P. australis* has stems with a distinct mix of short, eglandular hairs and much longer, glandular hairs.

172. Penstemon laxus A. Nelson, Bot. Gaz. 54: 147. 1912 (as Pentstemon) • Lax beardtongue [E]

Penstemon watsonii A. Gray subsp. *laxus* (A. Nelson) D. D. Keck

Herbs. Stems ascending to erect, 30–70 cm, ± retrorsely hairy, not glaucous. **Leaves** essentially cauline, basal leaves absent or poorly developed, not leathery, glabrous, sometimes ± puberulent; cauline 4–8(–10) pairs, sessile, 30–90 × 4–17 mm, blade lanceolate, base tapered, margins entire, apex acute. **Thyrses** continuous or interrupted, cylindric, (1–)2–15 cm, axis ± retrorsely hairy, verticillasters 1–5, cymes (2–)6–13-flowered, 2 per node; proximal bracts lanceolate to linear, 8–38 × 1–8 mm, margins entire; peduncles and pedicels ascending to erect, glabrous or retrorsely hairy. **Flowers:** calyx lobes broadly ovate to obovate or elliptic, 1.7–3 × 0.9–2 mm, apex obtuse to acute or slightly cuspidate, glabrous; corolla violet to blue or purple, with faint reddish purple nectar guides, funnelform, 12–15 mm, glabrous externally, densely yellow-villous internally abaxially, tube 3–4 mm, throat gradually inflated, 3–4 mm diam., 2-ridged abaxially; stamens included or longer pair slightly exserted, pollen sacs opposite, explanate, 0.4–0.8 mm, dehiscing completely, connective splitting, sides glabrous, sutures smooth; staminode 6–8 mm, included, 0.2–0.3 mm diam., tip straight to slightly recurved, distal 2–3 mm sparsely to densely villous, hairs yellow, to 0.7 mm; style 7–9 mm. **Capsules** 4.5–6 × 2.5–3.5 mm, glabrous.

Flowering Jun–Jul. Dry meadows, sagebrush shrublands; 1700–2400 m; Idaho.

Penstemon laxus is known from south-central Idaho; populations are known in Blaine, Boise, and Elmore counties.

173. Penstemon metcalfei Wooton & Standley,
Torreya 9: 145. 1909 (as Pentstemon) • Metcalf's
beardtongue E

Penstemon puberulus Wooton &
Standley, Bull. Torrey Bot. Club
36: 112. 1909 (as Pentstemon),
not M. E. Jones 1908

Herbs. Stems ascending to erect,
17–40 cm, retrorsely hairy,
sometimes also ± glandular-
pubescent distally, not glaucous. **Leaves** basal and cauline,
not leathery, retrorsely hairy, sometimes only along
midribs and margins; basal and proximal cauline
12–130 × 10–35 mm, blade ovate to elliptic, base
tapered, margins entire or ± denticulate, apex obtuse
to acute; cauline 3 or 4 pairs, petiolate or sessile,
33–100 × 12–27 mm, blade ovate to lanceolate, base
tapered to truncate, sometimes clasping on distal ones,
margins entire or ± denticulate, apex acute. **Thyrses**
interrupted, cylindric, 5–15 cm, axis glandular-
pubescent, verticillasters 3–5, cymes 2–6-flowered, 2
per node; proximal bracts ovate to lanceolate, 18–50
× 4–22 mm, margins entire; peduncles and pedicels
erect, glandular-pubescent. **Flowers:** calyx lobes
lanceolate, 3–6.5 × 0.9–1.5 mm, glandular-pubescent;
corolla lavender to purple, with reddish purple nectar
guides, bilabiate, not personate, funnelform, 16–20 mm,
glandular-pubescent externally, densely white-lanate
internally abaxially, tube 5–7 mm, throat gradually
inflated, 5–6 mm diam., 2-ridged abaxially; stamens
included, pollen sacs opposite, navicular, 0.9–1.2 mm,
dehiscing completely, connective splitting, sides glabrous,
sutures papillate; staminode 10–12 mm, included or
reaching orifice, 0.4–0.5 mm diam., tip recurved, distal
5–6 mm ± villous, hairs golden yellow, to 1.5 mm; style
10–13 mm. **Capsules** 5–8 × 3–4 mm, glabrous.

Flowering Jul–Aug. Cliffs, slopes, pine and spruce-fir
forests; 2000–2900 m; N.Mex.

Penstemon metcalfei is known from the Black
Range in Grant and Sierra counties. It was synonymized with *P. whippleanus* by D. D. Keck (1945);
T. K. Todsen (1998) enumerated clear differences
between the two species.

SELECTED REFERENCE Todsen, T. K. 1998. *Penstemon metcalfei*
(Scrophulariaceae), a valid species. Sida 18: 621–622.

174. Penstemon multiflorus (Bentham) Chapman ex
Small, Fl. S.E. U.S., 1061. 1903 (as Pentstemon)
• Many-flower beardtongue E

Penstemon pubescens Aiton var.
multiflorus Bentham in
A. P. de Candolle and
A. L. P. P. de Candolle, Prodr.
10: 327. 1846 (as Pentstemon)

Herbs. Stems ascending to
erect, (37–)80–150 cm, glabrous, not glaucous. **Leaves**
basal and cauline, not leathery,
glabrous; basal and proximal cauline 65–160 × 15–40
mm, blade oblanceolate, base tapered, margins entire
or slightly crenulate, apex acute; cauline 3–8 pairs,
short-petiolate or sessile, (16–)40–170 × (5–)12–35
mm, blade lanceolate, base tapered to clasping,
margins entire, apex acute. **Thyrses** interrupted,
conic, 22–35(–60) cm, axis glabrous or sparsely
glandular-pubescent distally, verticillasters 3–8, cymes
5–19-flowered, 2 per node; proximal bracts lanceolate,
8–47(–60) × 3–16(–25) mm, margins entire; peduncles
and pedicels ascending, glabrous or sparsely glandular-
pubescent. **Flowers:** calyx lobes ovate, 3–5 × 1–2 mm,
glandular-pubescent; corolla white to light lavender,
without nectar guides, funnelform, 20–22 mm,
glandular-pubescent externally, glabrous internally,
tube 5–6 mm, throat abruptly inflated, 5–6 mm diam.,
2-ridged abaxially; stamens included, pollen sacs parallel,
saccate, 0.8–1 mm, dehiscing incompletely, distal
⅔ indehiscent, connective splitting, sides glabrous,
sutures papillate; staminode 10–13 mm, exserted,
0.3–0.5 mm diam., tip straight to recurved, distal 3–4
mm sparsely pubescent, hairs yellowish, to 0.3 mm;
style 7–10 mm. **Capsules** 5–8 × 4–5 mm, glabrous.

Flowering Mar–Jul(–Aug). Sandy pinelands, sandy
scrublands; 0–100 m; Ala., Fla., Ga.

Penstemon multiflorus occurs in the Coastal Plain
throughout Florida and in southern Alabama and
southern Georgia.

175. Penstemon oklahomensis Pennell, Monogr. Acad.
Nat. Sci. Philadelphia 1: 237. 1935 • Oklahoma
beardtongue E F

Herbs. Stems ascending to erect,
(15–)35–55 cm, spreading-
puberulent or retrorsely hairy
proximally, glandular-pubescent
medially and distally, not glaucous. **Leaves** basal and cauline,
basal often withering by
anthesis, not leathery, puberulent or retrorsely hairy; basal
and proximal cauline 25–80(–110) × 5–24(–32) mm,
blade spatulate to obovate, oblanceolate, or lanceolate,

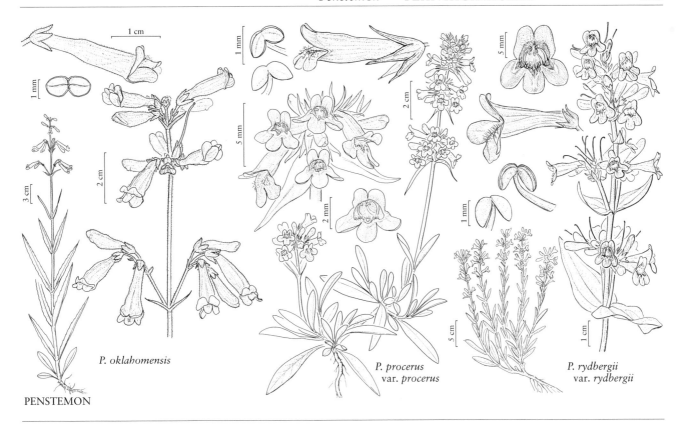

P. oklahomensis

P. procerus
var. procerus

P. rydbergii
var. rydbergii

PENSTEMON

base tapered, margins subentire or ± denticulate or serrate, apex rounded to obtuse or acute; cauline 2–9 pairs, short-petiolate or sessile, (25–)60–120 × 4–20 mm, blade lanceolate to linear, base tapered to clasping, margins entire or ± denticulate to serrate, apex acute. **Thyrses** interrupted, cylindric, (5–)8–18(–25) cm, axis glandular-pubescent, verticillasters (2 or)3–6, cymes 2–4(–6)-flowered, 2 per node; proximal bracts lanceolate to linear, 60 × 6 mm, margins entire; peduncles and pedicels ascending to erect, glandular-pubescent. **Flowers:** calyx lobes ovate to lanceolate, 5–7 × 3–4 mm, glandular-pubescent; corolla white to ochroleucous, without nectar guides, personate, tubular, 24–32 mm, glandular-pubescent externally, densely yellowish lanate internally abaxially, tube slightly differentiated from throat, 6–10 mm, throat slightly inflated, 6–8 mm diam., 2-ridged abaxially; stamens included, pollen sacs divergent, navicular to subexplanate, 1–1.5 mm, dehiscing completely, connective splitting, sides glabrous, sutures papillate; staminode 18–20 mm, included or slightly exserted, 0.4–0.6 mm diam., tip straight, distal 14–16 mm ± villous, hairs yellowish orange, to 1.5 mm; style 20–22 mm. **Capsules** 8–13 × 3–5 mm, glabrous. $2n = 16$.

Flowering Apr–Jun. Tallgrass prairies, open woods; 200–300 m; Okla., Tex.

Penstemon oklahomensis is known from 26 counties primarily in east-central Oklahoma and in Lamar County, Texas (J. N. Mink et al. 2010).

176. Penstemon oliganthus Wooton & Standley, Contr. U.S. Natl. Herb. 16: 172. 1913 (as Pentstemon)

• Apache beardtongue [E]

Penstemon pseudoparvus Crosswhite

Herbs. **Stems** erect, (6–)10–45 cm, retrorsely hairy, also glandular-pubescent distally, not glaucous. **Leaves** basal and cauline, not leathery, proximals usually retrorsely hairy proximally, especially along midveins and margins, mostly glabrate distally, distals retrorsely hairy and, sometimes, sparsely glandular-pubescent, sometimes ± glaucescent; basal and proximal cauline 15–75 × (2–)5–19 mm, blade spatulate to oblanceolate or elliptic, base tapered, margins entire, apex rounded to obtuse or acute, sometimes mucronate; cauline 2–5 pairs, sessile, 11–70 × (1–)4–15 mm, blade of proximal leaves oblanceolate to lanceolate, distal leaves lanceolate, base tapered to clasping, margins entire, rarely ± denticulate distally, apex acute to acuminate. **Thyrses** interrupted, secund, (1–)3–24(–30) cm, axis glandular-pubescent, verticillasters (2 or)3–7, cymes 1–4-flowered, (1 or)2 per node; proximal bracts lanceolate to linear, 7–38(–60) × 1–7 mm, margins entire; peduncles and pedicels erect, glandular-pubescent. **Flowers:** calyx lobes lanceolate, 2.8–6 × 0.9–2 mm, glandular-pubescent; corolla blue to violet-blue, with reddish purple nectar guides, tubular to tubular-funnelform, 11–20 mm, glandular-pubescent externally, sparsely to moderately white-villous to white-lanate internally abaxially, tube 5–7 mm, throat slightly inflated, 3–6 mm diam., 2-ridged abaxially; stamens included, pollen sacs opposite, navicular, 0.8–1.2 mm, dehiscing completely, connective splitting, sides glabrous, sutures papillate; staminode 10–12 mm, included or slightly exserted, 0.3–0.4 mm diam., tip straight to slightly recurved, distal 3–6 mm densely pilose, hairs golden, to 2 mm; style 9–13 mm. Capsules 5–7 × 3–4.5 mm, glabrous. $2n = 16$.

Flowering Jun–Aug(–Oct). Montane meadows, ciénagas, clearings in pine and spruce-fir forests; 2400–3500 m; Ariz., N.Mex.

Penstemon oliganthus is known from the Mogollon Rim (Coconino County) and White Mountains of east-central Arizona (Apache and Greenlee counties) and the Jemez and San Mateo mountains of northwestern New Mexico (McKinley, Sandoval, and Valencia counties). Crosswhite described *P. pseudoparvus* from five specimens from the Magdalena and San Mateo mountains in Socorro County, New Mexico, and more than 24 specimens of *P. oliganthus* from Arizona and New Mexico. He separated the two species based on stem indument (obscurely puberulent in *P. oliganthus* versus obviously puberulent in *P. pseudoparvus*), and flower orientation and shape (horizontal or more usually drooping and little inflated in *P. oliganthus* versus ascending and not inflated in *P. pseudoparvus*). More than 60 collections referable to these species were examined for this treatment, including the types, specimens annotated by Crosswhite, and 20 specimens from the San Mateo and Magdalena mountains in Socorro County; the two taxa appear to be indistinguishable.

177. Penstemon ovatus Douglas, Bot. Mag. 56: plate 2903. 1829 (as Pentstemon) • Ovate-leaf beardtongue E

Herbs. Stems ascending to erect, (20–)30–100 cm, puberulent to pubescent, sometimes glabrous, not glaucous. Leaves basal and cauline, not leathery, puberulent to pubescent, sometimes only along midvein and margins; basal and proximal cauline 40–150(–230) × 14–50(–70) mm, blade ovate to deltate-ovate, base truncate to tapered, margins serrate-dentate, apex obtuse to acute; cauline (3–)5–7 pairs, sessile, 25–80(–120) × 15–45(–70) mm, blade oblong to ovate or triangular-ovate, base clasping, margins serrate, apex acute. Thyrses interrupted, narrowly conic, 6–30 cm, axis densely glandular-pubescent, verticillasters 4–10, cymes 5–13-flowered, 2 per node; proximal bracts ovate to

lanceolate, 11–35(–70) × 4–30(–45) mm, margins serrate, sometimes entire; peduncles and pedicels ascending to erect, densely glandular-pubescent. Flowers: calyx lobes ovate to lanceolate, (2–)3–5 × 0.9–1.6(–1.9) mm, margins narrowly scarious, sparsely to densely glandular-pubescent; corolla violet to blue or purple, with reddish purple nectar guides, funnelform, 15–22 mm, glandular-pubescent externally, densely white-villous internally abaxially, rarely glabrous, tube 5–6 mm, throat slightly inflated, 5–7 mm diam., 2-ridged abaxially; stamens included, pollen sacs opposite, navicular, 0.8–1.1 mm, dehiscing completely, connective splitting, sides glabrous, sutures smooth or papillate; staminode 9–12 mm, reaching orifice or slightly exserted, 0.3–0.4 mm diam., tip recurved, distal 1–2 mm densely pilose, hairs yellow, to 1.5 mm; style 9–12 mm. Capsules 4–6 × 2.5–4.5 mm, glabrous. $2n = 16$.

Flowering May–Aug. Open rocky slopes and woods; 10–900 m; B.C.; Oreg., Wash.

Penstemon ovatus is known in the Cascade Mountains from Multnomah County, Oregon, to southwestern British Columbia. The species is easily confused with the more eastern *P. wilcoxii*, which has less glandular inflorescence. Herbarium specimens of *P. ovatus* also sometimes are misidentified as *P. serrulatus*, which has an eglandular inflorescence and saccate pollen sacs.

178. Penstemon pallidus Small, Fl. S.E. U.S., 1060, 1337. 1903 (as Pentstemon) • Pale beardtongue E

Herbs. Stems ascending to erect, 25–55(–65) cm, retrorsely hairy and glandular-villous proximally, retrorsely hairy and glandular-pubescent distally, glandular hairs sometimes sparse, not glaucous. Leaves basal and cauline, basal often withering by anthesis, not leathery, sparsely to densely pubescent, sometimes also with scattered glandular hairs, abaxially, sparsely pubescent adaxially; basal and proximal cauline 20–120(–180) × 5–35(–40) mm, blade spatulate to obovate or ovate, base tapered, margins subentire or ± serrate or dentate, apex rounded to obtuse or acute; cauline 4–8 pairs, sessile, 22–100 × 4–24 mm, blade ovate to lanceolate, base truncate to clasping, margins subentire or ± serrate or dentate, apex acute to acuminate. Thyrses interrupted, conic, 5–26(–30) cm, axis sparsely to densely glandular-pubescent, verticillasters 3–8, cymes 2–8(–16)-flowered, 2 per node; proximal bracts lanceolate, 7–72 × 2–22 mm, margins entire or ± dentate; peduncles and pedicels ascending, sparsely to densely glandular-pubescent. Flowers: calyx lobes ovate to lanceolate, (2–)3–5 × 1–3 mm, sparsely glandular-pubescent; corolla white, sometimes tinged lavender or blue, with reddish purple nectar guides, tubular to tubular-funnelform, 16–22 mm, glandular-pubescent externally, moderately whitish or yellowish lanate internally abaxially, tube 4–6 mm, throat slightly inflated, 4–7 mm diam., 2-ridged abaxially; stamens included, pollen sacs opposite, navicular, (0.8–)1–1.2 mm, dehiscing completely, connective splitting, sides glabrous, sutures papillate; staminode 10–12 mm, exserted, 0.2–0.3 mm diam., tip straight to slightly recurved, distal 8–9 mm moderately to densely villous, hairs yellow or golden yellow, to 1.5 mm; style 12–14 mm. **Capsules** 5–7 × 3–5 mm, glabrous. $2n = 16$.

Flowering Apr–Jun(–Jul). Tallgrass prairie, limestone and sandstone glades, barrens, rocky oak-hickory woodlands; 10–300 m; Ont.; Ala., Ark., Conn., D.C., Ga., Ill., Ind., Iowa, Kans., Ky., Maine, Md., Mass., Mich., Minn., Mo., N.H., N.J., N.Y., N.C., Ohio, Pa., R.I., Tenn., Vt., Va., W.Va., Wis.

Penstemon pallidus is concentrated in the Central Lowlands and Ozark Plateau of the central United States. F. W. Pennell (1935) believed it to be native there, having spread eastward as a result of human activities, as in Ontario.

Penstemon brevisepalus, which has been synonymized with *P. pallidus* by many authors, is treated here as a distinct species. Characters distinguishing the species are discussed under 147. *P. brevisepalus*.

Some specimens from north-central Arkansas and north-central Missouri are morphologically intermediate between *Penstemon pallidus* and *P. arkansanus*.

179. Penstemon peckii Pennell, Notul. Nat. Acad. Nat. Sci. Philadelphia 71: 12. 1941 • Peck's beardtongue E

Herbs. Stems ascending to erect, 20–50(–65) cm, glabrous or retrorsely hairy, not glaucous. **Leaves** basal and cauline, basal often poorly developed, not leathery, glabrous; basal and proximal cauline 20–75 × 2–5 (–7) mm, blade narrowly elliptic to linear, base tapered, margins entire, apex obtuse to acute; cauline 5–8 pairs, sessile or proximals long-petiolate, 15–75 × 2–8 mm, blade lanceolate to linear, base tapered, margins entire, apex acute. **Thyrses** interrupted or continuous, cylindric, 3–25 cm, axis glandular-pubescent, verticillasters 2–6 (–10), cymes 1–7-flowered, 2 per node; proximal bracts lanceolate, 6–55 × 1–8 mm, margins entire; peduncles and pedicels erect, glandular-pubescent. **Flowers:** calyx lobes obovate to ovate or lanceolate, 1.5–3 × 0.7–1.5 mm, glandular-pubescent; corolla violet to light blue, pink, or light purple, with or without pink or purple nectar guides, tubular to tubular-funnelform, 8–10 mm, glandular-pubescent externally, moderately yellow-pilose internally abaxially, tube 3–4 mm, throat slightly

inflated, 2–3 mm diam., slightly 2-ridged abaxially; stamens included, pollen sacs opposite, navicular to subexplanate, 0.4–0.5 mm, dehiscing completely, connective splitting, sides glabrous, sutures smooth; staminode 4–6 mm, included, 0.3–0.4 mm diam., tip straight, distal 0.3–0.5 mm sparsely pilose, hairs yellow, to 0.4 mm; style 7–8 mm. **Capsules** 3.5–5 × 2–3 mm, glabrous. $2n = 32$.

Flowering Jun–Aug. Dry, sandy, volcanic soils, pine forests; 900–1500 m; Oreg.

Penstemon peckii is known from the eastern slope of the Cascade Range in Crook, Deschutes, Hood River, and Jefferson counties.

180. Penstemon pratensis Greene, Leafl. Bot. Observ. Crit. 1: 165. 1906 (as Pentstemon) • White-flower beardtongue E

Herbs. Stems ascending to erect, 15–40(–65) cm, glabrous, not glaucous. **Leaves** basal and cauline, not leathery, glabrous; basal and proximal cauline (23–)40–90(–120) × 6–14(–24) mm, blade spatulate to elliptic, base tapered, margins entire, apex rounded to obtuse or acute; cauline 3–6 pairs, sessile or proximals short- to long-petiolate, 24–115 × 4–15 mm, blade oblanceolate to lanceolate, base tapered to truncate or clasping, margins entire, apex obtuse to acute. **Thyrses** interrupted, cylindric, 6–15 cm, axis glabrous, verticillasters 2–6, cymes 3–11-flowered, 2 per node; proximal bracts lanceolate, 8–50 × 1–8 mm, margins entire; peduncles and pedicels erect, glabrous. **Flowers:** calyx lobes ovate to lanceolate, 3–5.5 × 1.2–2.1 mm, apex acute to short-caudate, glabrous; corolla white or lavender, without nectar guides, funnelform, 10–15 mm, glabrous externally, moderately yellow- or white-lanate internally abaxially, tube 4–5 mm, throat slightly inflated, 3–4 mm diam., 2-ridged abaxially; stamens: longer pair reaching orifice or slightly exserted, pollen sacs opposite, navicular, 0.6–0.8 mm, dehiscing completely, connective splitting, sides glabrous, sutures smooth; staminode 6–7 mm, included, 0.3–0.4 mm diam., tip straight, distal 0.5–3 mm moderately pilose, hairs yellow, to 0.8 mm, rarely glabrous; style 7–8 mm. **Capsules** 4–7 × 3–4 mm, glabrous. $2n = 32$.

Flowering Jun–Jul. Moist meadows, stream banks, aspen woodlands; 1300–2800 m; Idaho, Nev., Oreg.

Penstemon pratensis is known from the northern Central Great Basin and Owyhee Desert regions of the northern Great Basin. The species is documented in Owyhee County, Idaho, Elko and Eureka counties, Nevada, and Harney and Malheur counties, Oregon. In Elko County, Nevada, where it occurs with *P. rydbergii* var. *oreocharis*, *P. pratensis* appears to occur mostly in the East Humboldt Range and Ruby Mountains, south and east of the Humboldt River. Most populations of *P. rydbergii* var. *oreocharis* in that county occur north and west of the Humboldt River in the Independence and Jarbidge mountains.

181. Penstemon procerus Douglas ex Graham, Edinburgh New Philos. J. 7: 348. 1829 (as Pentstemon procerum) • Small-flower beardtongue E F

Penstemon confertus Douglas var. *procerus* (Douglas ex Graham) Coville

Herbs. Stems ascending to erect, 3–55(–70) cm, glabrous or ± retrorsely hairy, not glaucous. **Leaves** basal and cauline, basal sometimes poorly developed, ± leathery or not, glabrous, puberulent, or retrorsely hairy; basal and proximal cauline 9–90(–115) × 2–19 mm, blade spatulate to oblanceolate, lanceolate, or elliptic, base tapered, margins entire, apex rounded to obtuse or acute; cauline 2–6 pairs, sessile or proximals sometimes short-petiolate, 6–85 × 1–17 mm, blade oblanceolate to spatulate, obovate, oblong, elliptic, lanceolate, or linear, base tapered to clasping, margins entire, apex rounded to obtuse or acute. **Thyrses** interrupted, cylindric, 0.5–23 cm, axis glabrous or retrorsely hairy, verticillasters 1–8(–11), cymes 3–11-flowered, 2 per node; proximal bracts lanceolate to linear, 10–50(–60) × 1–8(–16) mm, margins entire; peduncles and pedicels erect, glabrous, puberulent, or retrorsely hairy. **Flowers:** calyx lobes ovate to obovate or lanceolate, 1.2–5(–5.5) × 0.5–2 mm, apex truncate, obtuse, acute, acuminate, cuspidate, caudate, or mucronate, glabrous or puberulent to retrorsely hairy; corolla violet to blue or purple, with, sometimes without, reddish purple nectar guides, funnelform, 7–11 mm, glabrous externally, moderately white- or yellow-lanate internally abaxially, tube 2–3 mm, throat slightly inflated, 2–3 mm diam., 2-ridged abaxially; stamens included, pollen sacs opposite, explanate, 0.3–0.6 mm, dehiscing completely, connective splitting, sides glabrous, sutures smooth or papillate; staminode 4–6 mm, included, 0.2–0.3 mm diam., tip straight to slightly recurved, distal 0.5–1 mm pilose, hairs yellow or golden yellow, to 0.6 mm, rarely glabrous; style 4–7 mm. **Capsules** 3–5 × 2–4 mm, glabrous.

Varieties 6 (6 in the flora): w North America.

Penstemon procerus is a widespread, frequent, and highly variable species distributed throughout much of western North America. This variability has resulted in varied interpretations of its taxonomic units.

1. Calyx lobes (2–)3–5(–5.5) mm, apices acuminate, cuspidate, or short- to long-caudate.
  2. Verticillasters (1–)3–8(–11); stems (6–)15–55(–70) cm; thyrses (1–)5–23 cm. . . . . 181a. *Penstemon procerus* var. *procerus*
  2. Verticillasters 1 or 2(–5); stems 3–20(–28) cm; thyrses 1–5(–13) cm. . . . . 181f. *Penstemon procerus* var. *tolmiei*
1. Calyx lobes 1.2–3.2 mm, apices truncate to obtuse, acute, acuminate, mucronate, or short-caudate.
  3. Verticillasters 1 or 2(–5); stems 3–15(–20) cm; cauline leaves 1–4(–8) mm wide. . . . . 181d. *Penstemon procerus* var. *formosus*
  3. Verticillasters (1 or)2–6; stems (4–)8–35(–42) cm; cauline leaves 2–13 mm wide.
    4. Calyx lobes 1.2–2.4 mm, apices truncate to obtuse or mucronate, rarely short-caudate or acute; Nevada . . . . 181e. *Penstemon procerus* var. *modestus*
    4. Calyx lobes (1.4–)1.8–3.2 mm, apices acute to acuminate or short-caudate; California, Oregon, Utah.
      5. Pollen sacs 0.3–0.4 mm; corollas 7–10 mm; stems 10–35(–42) cm; Utah . . . . . 181b. *Penstemon procerus* var. *aberrans*
      5. Pollen sacs 0.4–0.6 mm; corollas 9–12 mm; stems 9–20(–30) cm; California, Oregon . . . . 181c. *Penstemon procerus* var. *brachyanthus*

181a. Penstemon procerus Douglas ex Graham var. **procerus** • Pincushion beardtongue E F

Stems (6–)15–55(–70) cm. **Leaves:** basal and proximal cauline 23–90(–115) × 4–19 mm; cauline 13–85 × 4–17 mm, blade lanceolate to oblanceolate, sometimes elliptic or oblong. **Thyrses** (1–)5–23 cm, verticillasters (1–)3–8(–11). **Flowers:** calyx lobes (2.5–)3–5 × 0.6–2 mm, apex acuminate or short- to long-caudate; corolla (7–)8–11 mm; pollen sacs (0.3–)0.4–0.6 mm; staminode: distal 0.5–1 mm densely pilose, hairs yellow or golden yellow, to 0.5 mm. *2n* = 16, 32.

Flowering Jun–Aug(–Sep). Sagebrush shrublands, grassy slopes, aspen woodlands, open forest slopes, subalpine and alpine meadows; (300–)600–3700 m; Alta., B.C., Man., Sask., Yukon; Alaska, Colo., Idaho, Mont., N.Dak., Oreg., Utah, Wash., Wyo.

Variety *procerus* is widespread in the Rocky Mountains from southern Colorado to the Yukon. The caudate apices of the calyx lobes sometimes are as long as the bodies of the lobes, especially in more northern populations.

181b. Penstemon procerus Douglas ex Graham var. **aberrans** (M. E. Jones) A. Nelson, Bot. Gaz. 54: 146. 1912 (as Pentstemon) [E]

Penstemon confertus Douglas var. *aberrans* M. E. Jones, Proc. Calif. Acad. Sci., ser. 2, 5: 715. 1895 (as Pentstemon); *P. procerus* subsp. *aberrans* (M. E. Jones) D. D. Keck

Stems 10–35(–42) cm. **Leaves:** basal and proximal cauline (13–)30–60 × (2–)4–13 mm; cauline 14–48 × 2–11 mm, blade lanceolate, proximals sometimes obovate. **Thyrses** 2–14 cm, verticillasters 2–6. **Flowers:** calyx lobes (1.4–)1.8–3 × 0.9–1.2 mm, apex acute to acuminate or short-caudate; corolla 7–10 mm; pollen sacs 0.3–0.4 mm; staminode: distal 0.5–0.8 mm densely pilose, hairs yellow or golden yellow, to 0.5 mm. **2*n*** = 16.

Flowering Jul–Aug. Sagebrush shrublands, meadows, clearings in spruce-fir, pine, and aspen woodlands; 2200–3400 m; Utah.

Variety *aberrans* is known from the Utah Plateau from Soldier Summit of the Wasatch Plateau to the Aquarius and Tushar plateaus.

181c. Penstemon procerus Douglas ex Graham var. **brachyanthus** (Pennell) Cronquist in C. L. Hitchcock et al., Vasc. Pl. Pacif. N.W. 4: 398. 1959 [E]

Penstemon brachyanthus Pennell, Notul. Nat. Acad. Nat. Sci. Philadelphia 71: 3. 1941; *P. procerus* subsp. *brachyanthus* (Pennell) D. D. Keck; *P. tolmiei* Hooker subsp. *brachyanthus* (Pennell) D. D. Keck

Stems 9–20(–30) cm. **Leaves:** basal and proximal cauline 10–62 × 4–12 mm; cauline 13–45 × 3–13 mm, blade obovate to oblanceolate or elliptic, distals sometimes lanceolate. **Thyrses** 2–11 cm, verticillasters (1 or)2–4. **Flowers:** calyx lobes 2–3.2 × 0.6–1.5 mm, apex acute to acuminate; corolla 9–12 mm; pollen sacs 0.4–0.6 mm; staminode: distal 0.5–0.8 mm sparsely to densely pilose, hairs golden yellow, to 0.6 mm. **2*n*** = 16.

Flowering Jun–Aug. Rocky or grassy montane meadows, open pine woodlands; 1200–2400 m; Calif., Oreg.

Variety *brachyanthus* is known from the Cascade Range from northern Oregon to northwestern California.

181d. Penstemon procerus Douglas ex Graham var. **formosus** (D. D. Keck) Cronquist in C. L. Hitchcock et al., Vasc. Pl. Pacif. N.W. 4: 398. 1959 [E]

Penstemon tolmiei Hooker subsp. *formosus* D. D. Keck, Amer. Midl. Naturalist 33: 147. 1945, based on *P. formosus* A. Nelson, Proc. Biol. Soc. Wash. 17: 100. 1904 (as Pentstemon), based on *P. pulchellus* Greene, Pittonia 3: 310. 1898 (as Pentstemon), not Lindley 1828; *P. procerus* subsp. *formosus* (D. D. Keck) D. D. Keck

Stems 3–15(–20) cm. **Leaves:** basal and proximal cauline 9–30(–48) × 2–7(–12) mm; cauline 6–20 (–45) × 1–4(–8) mm, blade lanceolate to linear, proximals sometimes spatulate. **Thyrses** 0.5–3(–12) cm, verticillasters 1 or 2(–5). **Flowers:** calyx lobes 1.4–3 × 0.5–1.5 mm, apex acute or short-caudate, rarely truncate, obtuse, or mucronate; corolla 7–10 mm; pollen sacs 0.4–0.6 mm; staminode glabrous or distal 0.5 mm sparsely pilose, hairs yellow, to 0.4 mm.

Flowering Jun–Aug. Alpine talus slopes, alpine meadows; 2200–3600 m; Calif., Nev., Oreg.

D. D. Keck (1945) mapped populations of var. *formosus* in three widely separated areas: eastern Oregon (Blue Mountains, Steens Mountain, and Wallowa Mountains), northern California in the Klamath Range, and in the southern High Sierra Nevada and ranges in the western Great Basin in east-central California and western Nevada. Populations in the northern part of the range are typical var. *formosus*, which have spatulate, oblanceolate, or lanceolate basal and proximal cauline leaves. Populations in the southern part of the range usually have linear-oblanceolate basal and proximal cauline leaves. Tall plants can be mistaken for *Penstemon cinicola*, but var. *formosus* has calyx lobes that are consistently ovate to lanceolate with apices acute to short-caudate. Calyx lobes of *P. cinicola* are obovate to ovate with apices truncate or cuspidate.

Specimens from the Toquima Range in Nye County, Nevada, have been referred to var. *formosus*. They have the short stems and capitate thyrses of that variety but bear short (1–1.5 mm), truncate or mucronate calyx lobes that are typical of var. *modestus*.

Variety *formosus* has been reported from Custer County, Idaho (S. Caicco et al. 1983). Specimens cited in that paper have not been seen, but a collection from the same area by one of the coauthors and identified as var. *formosus* (*D. Henderson 6439*, RM) appears to be representative of the cited populations. That sheet bears three short (6–8 cm) flowering stems, each with two or three pairs of elliptic leaves and a more or less capitate inflorescence, calyx lobes 2–2.7 mm with acute apices, and corollas 10–12 mm, features consistent with var. *formosus*. However, the thyrses, calyces, and corollas are glandular-pubescent, and the corolla throats are 3–4 mm in diameter, suggesting possible contact with *Penstemon humilis*; consequently, Idaho is excluded from the distribution of var. *formosus*.

181e. Penstemon procerus Douglas ex Graham var. **modestus** (Greene) N. H. Holmgren, Brittonia 31: 106. 1979 • Ruby Mountain beardtongue E

Penstemon modestus Greene, Leafl. Bot. Observ. Crit. 1: 165. 1906 (as Pentstemon); *P. procerus* subsp. *modestus* (Greene) D. D. Keck; *P. tolmiei* Hooker subsp. *modestus* (Greene) D. D. Keck

Stems (4–)8–28 cm. **Leaves:** basal and proximal cauline (12–)16–52 × 5–14 mm; cauline 11–52 × 3–12 mm, blade lanceolate to oblanceolate or elliptic. **Thyrses** 2–13 cm, verticillasters (2 or)3–5. **Flowers:** calyx lobes rarely lanceolate, 1.2–2.4 × 0.6–1.3 mm, apex truncate to obtuse or mucronate, rarely short-caudate or acute; corolla 7–10 mm; pollen sacs 0.4–0.6 mm; staminode glabrous or distal 0.5 mm sparsely pilose, hairs yellow, to 0.4 mm.

Flowering Jun–Jul. Gravelly alpine slopes; 2700–3200 m; Nev.

Variety *modestus* is known from the East Humboldt and Ruby mountains of Elko County. Specimens from the Toquima Range in Nye County, which are here referred to var. *formosus*, bear short (1–1.5 mm), truncate to mucronate calyx lobes that are typical of var. *modestus*, but they are more similar to var. *formosus* in habit and inflorescence characters.

181f. Penstemon procerus Douglas ex Graham var. **tolmiei** (Hooker) Cronquist in C. L. Hitchcock et al., Vasc. Pl. Pacif. N.W. 4: 398. 1959 E

Penstemon tolmiei Hooker, Fl. Bor.-Amer. 2: 98. 1838 (as Pentstemon); *P. procerus* subsp. *tolmiei* (Hooker) D. D. Keck

Stems 3–20(–28) cm. **Leaves:** basal and proximal cauline 15–55(–70) × 3–15 mm; cauline 11–45 × 3–13 mm, blade lanceolate to elliptic or oblong. **Thyrses** 1–5(–13) cm, verticillasters 1 or 2(–5). **Flowers:** calyx lobes (2–)3–4.8(–5.5) × 1–2 mm, apex cuspidate or short-caudate; corolla 8–11 mm; pollen sacs 0.4–0.5 mm; staminode: distal 0.5–1 mm densely pilose, hairs yellow or golden yellow, to 0.5 mm. $2n = 16$.

Flowering Jun–Sep. Rocky, montane to alpine slopes, ridges, meadows; 1200–2300 m; B.C.; Wash.

Variety *tolmiei* is known from the northern Cascade Range from east-central British Columbia through eastern Washington nearly to the Columbia River and the Olympic Mountains in Washington.

182. Penstemon pruinosus Douglas ex Lindley, Edwards's Bot. Reg. 15: plate 1280. 1829 (as Pentstemon pruinosum) • Chelan beardtongue E

Herbs. Stems ascending or erect, 8–40(–60) cm, glabrate, retrorsely hairy, or glandular-pubescent, not glaucous. **Leaves** basal and cauline, not leathery, retrorsely hairy or pubescent, sometimes also glandular-pubescent; basal and proximal cauline 25–100(–150) × (4–)9–20(–34) mm, blade elliptic to ovate, base tapered, margins ± serrate to dentate, apex obtuse to acute; cauline 2–5 pairs, sessile, 11–60 × 5–25(–30) mm, blade triangular-ovate to lanceolate, base truncate to cordate-clasping, margins serrate to dentate, apex acute, rarely obtuse. **Thyrses** interrupted, cylindric, 4–28 cm, axis densely glandular-pubescent, rarely sparsely glandular-pubescent, verticillasters 3–8, cymes 3–11-flowered, 2 per node; proximal bracts ovate to lanceolate, 8–58 × 4–25 mm, margins serrate, sometimes entire; peduncles and pedicels ascending to erect, densely glandular-pubescent. **Flowers:** calyx lobes ovate to lanceolate, 2.5–4.5(–6) × 0.9–1.5 mm, glandular-pubescent; corolla blue to lavender, purple, or violet, with purple nectar guides, funnelform, 11–16 mm, glandular-pubescent externally, sparsely white-villous internally abaxially, tube 4–5 mm, throat gradually inflated, 3–6 mm diam., rounded abaxially; stamens included or longer pair reaching orifice, pollen sacs opposite, navicular to subexplanate, 0.6–0.8 mm, dehiscing completely, connective splitting, sides glabrous, sutures smooth; staminode 7–9 mm, included or slightly exserted, 0.2–0.3 mm diam., tip recurved, distal 0.5–1 mm densely pilose, hairs yellow, to 0.8 mm; style 8–10 mm. **Capsules** 4–7 × 3–4 mm, glabrous. $2n = 16$.

Flowering Apr–Jul. Open grassy, sandy, gravelly, and rocky slopes, pine and pine-fir woodlands, sagebrush shrublands; 300–2000 m; B.C.; Wash.

Penstemon pruinosus occurs along the eastern slope of the Cascade Range in Chelan, Douglas, Grant, and Kittitas counties, Washington, east into the scablands of Adams and Franklin counties, Washington, and north into extreme southern British Columbia.

A blue dye is made from the flowers of *Penstemon pruinosus* by the Okanagan-Coville tribe of northeastern Washington (D. E. Moerman 1998).

183. Penstemon radicosus A. Nelson, Bull. Torrey Bot. Club 25: 280. 1898 (as Pentstemon) • Mat-root beardtongue E

Herbs. Stems ascending to erect, (15–)20–42 cm, retrorsely hairy or puberulent, not glaucous. **Leaves** essentially cauline, basal absent or poorly developed, not leathery, ± puberulent; cauline (4 or)5–8 pairs, sessile, (20–)30–65 × (2–)4–10(–20) mm, blade ovate to lanceolate, base tapered, margins entire, apex obtuse to acute. **Thyrses** interrupted or continuous, cylindric, 5–14 cm, axis glandular-pubescent, verticillasters (2 or)3–6, cymes 2–4(–8)-flowered, 2 per node; proximal bracts lanceolate to linear, 7–46 × 1–10(–20) mm, margins entire; peduncles and pedicels ascending to erect, glandular-pubescent. **Flowers:** calyx lobes lanceolate, 5–8 × 1.2–2.5 mm, glandular-pubescent; corolla blue to dark blue, purple, or violet, with reddish purple nectar guides, bilabiate, not personate, funnelform, 16–23 mm, glandular-pubescent externally, moderately yellowish lanate internally abaxially, tube 4.5–6 mm, throat slightly inflated, 4–6 mm diam., 2-ridged abaxially; stamens included, pollen sacs opposite, navicular to subexplanate, 0.8–1.3 mm, dehiscing completely, connective splitting, sides glabrous, sutures papillate; staminode 10–12 mm, included, 0.3–0.4 mm diam., tip straight, distal 1 mm densely pilose or lanate, hairs yellow, to 1.4 mm, remaining distal 4–5 mm sparsely pilose, hairs yellow, to 0.5 mm; style 11–14 mm. **Capsules** 5–8 × 3.5–5 mm, glabrous. *2n* = 16.

Flowering May–Jul. Rocky flats, slopes, ravines, sagebrush shrublands, pine woodlands; 1500–2400 m; Colo., Idaho, Mont., Nev., Utah, Wyo.

Penstemon radicosus is known from the Central Rocky Mountains and mountain ranges in the northern Great Basin. The species is most likely to be confused with *P. laxus* or *P. watsonii*.

184. Penstemon rattanii A. Gray, Proc. Amer. Acad. Arts 15: 50. 1879 (as Pentstemon rattani) • Rattan's beardtongue E

Herbs. Stems ascending to erect, (15–)30–120 cm, glabrous, sometimes glabrous proximally, glandular-pubescent distally, not glaucous. **Leaves** basal and cauline, not leathery, glabrous; basal and proximal cauline 30–250 × 7–50 mm, blade elliptic to lanceolate, base tapered, margins serrate to dentate, apex obtuse to acute; cauline 2–5 pairs, short-petiolate or sessile, (10–)20–75(–110) × 5–30(–38) mm, blade oblong to triangular-ovate, base truncate to cordate-clasping distally, margins serrate to dentate, apex obtuse to acute. **Thyrses** interrupted, cylindric, 2–24 cm, axis glandular-pubescent, verticillasters 2–7, cymes 3–11-flowered, 2 per node; proximal bracts triangular to lanceolate, 6–50 × 1–22 mm, margins entire or ± serrate; peduncles and pedicels ascending to erect, glandular-pubescent. **Flowers:** calyx lobes ovate to ovate-oblong or lanceolate, 4–10 × (1–)1.5–2.5 mm, glandular-pubescent; corolla lavender to purple or violet, with reddish purple nectar guides, ventricose to ventricose-ampliate, 20–30 mm, glandular-pubescent externally, sparsely to densely white-lanate internally, tube 6–7 mm, throat abruptly inflated, 8–10 mm diam., 2-ridged abaxially; stamens included, pollen sacs opposite, subexplanate, 1.2–1.5 mm, dehiscing completely, connective splitting, sides glabrous, sutures smooth or papillate; staminode 21–26 mm, exserted, 0.3–0.4 mm diam., tip straight to recurved, distal 6–10 mm sparsely to moderately lanate, hairs white or yellow, to 3 mm; style 16–19 mm. **Capsules** 5–9 × 3.5–5 mm, glabrous. *2n* = 16.

Varieties 2 (2 in the flora): w United States.

Penstemon rattanii comprises two morphologically distinct and geographically allopatric varieties confined to the Coast Range of California and Oregon.

1. Calyx lobes usually longer than capsules, lanceolate, sometimes ovate, (5.5–)6.5–10 mm, apices acute to acuminate, sometimes obtuse; nw California, sw Oregon. . . . . 184a. *Penstemon rattanii* var. *rattanii*
1. Calyx lobes usually shorter than capsules, ovate-oblong, 4–7 mm, apices obtuse to acute; wc California . . . . . . 184b. *Penstemon rattanii* var. *kleei*

184a. Penstemon rattanii A. Gray var. **rattanii** E

Flowers: calyx lobes lanceolate, sometimes ovate, (5.5–)6.5–10 × (1–)2–2.5 mm, usually longer than capsules, apex acute to acuminate, sometimes obtuse. *2n* = 16.

Flowering Jun–Jul. Grassy slopes in fir-spruce, redwood, and mixed-evergreen forests; 10–1200 m; Calif., Oreg.

Variety *rattanii* is known from the Coast Range of northwestern California (Del Norte, Humboldt, and Mendocino counties) and southwestern Oregon (Benton, Coos, Curry, and Douglas counties).

184b. Penstemon rattanii A. Gray var. **kleei** (Greene) A. Gray in A. Gray et al., Syn. Fl. N. Amer. ed. 2, 2(1): 441. 1886 (as Pentstemon rattani) • Santa Cruz Mountains beardtongue [E]

Penstemon kleei Greene, Bull. Torrey Bot. Club 10: 127. 1883 (as Pentstemon); *P. rattanii* subsp. *kleei* (Greene) D. D. Keck

Flowers: calyx lobes ovate-oblong, 4–7 × 1.5–1.8 mm, usually shorter than capsules, apex obtuse to acute. *2n* = 16.

Flowering Jun–Jul. Open, coastal forests; 400–1000 m; Calif.

Variety *kleei* is known only from the peaks of the Santa Cruz Mountains in Santa Clara and Santa Cruz counties.

185. Penstemon rydbergii A. Nelson, Bull. Torrey Bot. Club 25: 281. 1898 (as Pentstemon) • Rydberg's beardtongue [E] [F]

Herbs. Stems decumbent, ascending, or erect, 20–70 (–120) cm, glabrous or retrorsely hairy, sometimes only in lines, not glaucous. **Leaves** basal and cauline, not leathery, glabrous; basal and proximal cauline (25–)35–70(–150) × (5–)10–15 (–22) mm, blade oblanceolate to elliptic, base tapered, margins entire, apex obtuse to acute, rarely mucronate; cauline 3–6 pairs, sessile or proximals short-petiolate, 25–70(–110) × 3–24 mm, blade lanceolate to elliptic or oblong, base tapered to clasping, margins entire, apex obtuse to acute. **Thyrses** interrupted, cylindric, 3–10(–18) cm, axis retrorsely hairy or puberulent, sometimes glabrous, verticillasters (1 or)2–7, cymes (3–)5–11-flowered, 2 per node; proximal bracts lanceolate, (4–)11–52(–70) × (1–)2–18 mm, margins entire or erose to lacerate; peduncles and pedicels erect, glabrous, retrorsely hairy, or puberulent. **Flowers:** calyx lobes oblong to ovate or obovate, 3–8 × 1.2–3.5 mm, apex acute to acuminate or short- to long-caudate, glabrous or puberulent; corolla violet to blue or purple, without nectar guides, funnelform, 10–20 mm, glabrous externally, sparsely white- or yellowish villous internally abaxially, tube 3–4 mm, throat gradually inflated, 3–4.5 mm diam., 2-ridged abaxially; stamens: longer pair reaching orifice or exserted, pollen sacs opposite, navicular, 0.5–1.1 mm, dehiscing completely, connective splitting, sides glabrous, sutures smooth or ± papillate; staminode 6–7 mm, reaching orifice, 0.2–0.4 mm diam., tip straight, distal 1.5–2 mm moderately to densely pilose, hairs golden yellow, to 1 mm; style 8–10 mm. **Capsules** 4–5 × 3–4 mm, glabrous.

Varieties 3 (3 in the flora): w United States.

As treated here, *Penstemon rydbergii* includes three partially sympatric varieties. D. D. Keck (1945) treated the same taxa as four species and two subspecies (*P. hesperius*, *P. oreocharis*, *P. rydbergii* subsp. *aggregatus*, *P. rydbergii* subsp. *rydbergii*, and *P. vaseyanus*).

1. Calyx lobe margins lacerate, rarely erose, broadly scarious, apices acuminate or short- to long-caudate; proximal bract margins erose to lacerate proximally, sometimes entire; thyrse axes retrorsely hairy or puberulent . . . . 185a. *Penstemon rydbergii* var. *rydbergii*
1. Calyx lobe margins erose, sometimes lacerate, narrowly to broadly scarious, apices acute, acuminate, or short-caudate; proximal bract margins entire; thyrse axes glabrous or retrorsely hairy.
  2. Stems ± retrorsely hairy; thyrse axes glabrous or retrorsely hairy; corollas (10–)13–20 mm; pollen sacs (0.6–)0.8–1.1 mm . . . . 185b. *Penstemon rydbergii* var. *aggregatus*
  2. Stems glabrous, rarely ± retrorsely hairy; thyrse axes glabrous, rarely ± retrorsely hairy; corollas 10–14 mm; pollen sacs 0.5–0.7(–0.8) mm . . . . . 185c. *Penstemon rydbergii* var. *oreocharis*

185a. Penstemon rydbergii A. Nelson var. **rydbergii** [E] [F]

Penstemon rydbergii var. *varians* (A. Nelson) Cronquist

Stems retrorsely hairy, sometimes only in lines. **Thyrses:** axis retrorsely hairy or puberulent; proximal bract margins usually erose to lacerate proximally, sometimes entire. **Flowers:** calyx lobes (3.5–) 4–6.5 × 1.6–3.5 mm, margins lacerate, rarely erose, broadly scarious, apex acuminate or short- to long-caudate; corolla 10–15 mm; pollen sacs 0.5–0.8 mm. *2n* = 16.

Flowering Jun–Aug. Moist, grassy meadows, moist ravines, aspen thickets, rocky hillsides, subalpine parks; 1900–3300 m; Colo., N.Mex., Utah, Wyo.

Variety *rydbergii* is concentrated in the Southern Rocky Mountains from south-central Wyoming to northern New Mexico. D. D. Keck (1945) referred material from the vicinity of the Grand Canyon in Arizona and from the Uinta Mountains of northeastern Utah to var. *rydbergii*. Some Utah plants fit well with the concept of var. *rydbergii* used here; others appear to pass into *Penstemon procerus*. The Arizona plants are intermediate between var. *aggregatus* and var. *rydbergii*.

185b. Penstemon rydbergii A. Nelson var. **aggregatus** (Pennell) N. H. Holmgren, Brittonia 31: 106. 1979 [E]

Penstemon aggregatus Pennell, Contr. U.S. Natl. Herb. 20: 367. 1920; *P. rydbergii* subsp. *aggregatus* (Pennell) D. D. Keck

Stems ± retrorsely hairy. **Thyrses:** axis glabrous or retrorsely hairy; proximal bract margins entire. **Flowers:** calyx lobes (4.5–)5–8 × 1.5–2.5 mm, margins erose, rarely lacerate, narrowly to broadly scarious, apex acute to acuminate; corolla (10–)13–20 mm; pollen sacs (0.6–)0.8–1.1 mm. $2n = 32$.

Flowering Jun–Aug. Sagebrush shrublands, aspen woodlands; 1800–3200 m; Ariz., Colo., Idaho, Mont., Utah, Wyo.

Variety *aggregatus* is known from the Rocky Mountains from central Idaho and western Montana to north-central and western Wyoming, north-central Colorado, and central Utah. The variety is sympatric with var. *rydbergii* in north-central Colorado and south-central Wyoming; some plants from that area are difficult to assign to variety with confidence.

185c. Penstemon rydbergii A. Nelson var. **oreocharis** (Greene) N. H. Holmgren in A. Cronquist et al., Intermount. Fl. 4: 390. 1984 [E]

Penstemon oreocharis Greene, Leafl. Bot. Observ. Crit. 1: 163. 1906 (as Pentstemon); *P. hesperius* M. Peck; *P. vaseyanus* Greene

Stems glabrous, rarely ± retrorsely hairy. **Thyrses:** axis glabrous, rarely ± retrorsely hairy; proximal bract margins entire. **Flowers:** calyx lobes 3–5.5 × 1.2–2.2 mm, margins erose, sometimes lacerate, narrowly to broadly scarious, apex acute to short-caudate; corolla 10–14 mm; pollen sacs 0.5–0.7 (–0.8) mm. $2n = 16$.

Flowering Jun–Aug. Moist grassy meadows, stream banks, montane and subalpine forests; 600–3100 m; Calif., Idaho, Nev., Oreg., Wash.

Variety *oreocharis* is widespread in the mountains of the western United States, mostly from the crest of the Cascades and Sierra Nevada eastward, from north-central Washington through eastern Oregon to east-central California and west-central Nevada. D. D. Keck (1945) recognized *Penstemon vaseyanus*, which he distinguished from *P. oreocharis* (= var. *oreocharis*) by the former's stouter stems, leaf margins occasionally denticulate, and corollas more funnelform and white-lanate abaxially. Many plants from the

eastern flank of the Cascade Range in central Washington exhibit some or all of these features, which also are expressed among other populations of var. *oreocharis*.

Peck described *Penstemon hesperius* from plants collected in a boggy meadow in Washington County, Oregon, commenting on their tall stature and unusual habitat. D. D. Keck (1945) recognized *P. hesperius*, distinguishing it from *P. oreocharis* by its generally taller stems [50–80(–120) cm] and longer (more than 5 mm), hirtellous calyx lobes with attenuate apices. He also noted that the few specimens he had seen (from Clackamas and Washington counties, Oregon, and one from Clark or Skamania counties, Washington) came from boggy meadows. A. Cronquist (1959) synonymized *P. hesperius* with *P. rydbergii*, stating that the two species appeared to be morphologically confluent.

Plants consistent with the type of *Penstemon hesperius* were discovered along the Tualatin River in southeastern Washington County, Oregon, in 2008 and 2009. Preliminary assessments suggest that *P. hesperius* may be morphologically and genetically distinct from *P. rydbergii*; work to characterize *P. hesperius* is ongoing (G. Maffitt 2012).

SELECTED REFERENCE Maffitt, G. 2012. The Tualatin Basin *Penstemon* revisited. Bull. Amer. Penstemon Soc. 71: 48–53.

186. Penstemon seorsus (A. Nelson) D. D. Keck, Amer. Midl. Naturalist 23: 595. 1940 • Short-lobe beardtongue [E]

Penstemon linarioides A. Gray var. *seorsus* A. Nelson, Bot. Gaz. 54: 147. 1912 (as Pentstemon)

Subshrubs. Stems ascending to erect, 10–40 cm, moderately to densely retrorsely hairy, not glaucous. **Leaves** basal and cauline, opposite, rarely alternate, not leathery, ± retrorsely hairy; basal and proximal cauline 10–25(–65) × 1–2(–6) mm, blade linear, base tapered, margins entire, apex acute, rarely obtuse; cauline 7–10 pairs, sessile or proximals sometimes petiolate, 4–45 × 1–4 (–6) mm, blade linear, rarely lanceolate distally, base truncate or clasping, margins entire, usually revolute, apex acute. **Thyrses** interrupted, cylindric, 3–18 cm, axis retrorsely hairy and glandular-pubescent, sometimes only retrorsely hairy, verticillasters 3–7, cymes 1- or 2(or 3)-flowered, 2 per node; proximal bracts lanceolate to linear, 5–24 × 1–2 mm, margins entire; peduncles and pedicels ascending to erect, retrorsely hairy and glandular-pubescent. **Flowers:** calyx lobes broadly lanceolate, 4–6.5 × 1.5–3 mm, glandular-pubescent; corolla blue to violet or purple, with violet nectar guides, funnelform, (13–)15–22 mm, glandular-pubescent externally, glabrous internally, tube 5–6 mm, throat gradually inflated, 4–6(–7) mm diam., slightly 2-ridged abaxially; stamens included, pollen sacs opposite, navicular, 1.1–1.3 mm, dehiscing completely, connective splitting, sides glabrous, sutures papillate; staminode 13–17 mm, exserted, 0.5–0.9 mm diam., tip straight, distal 7–9 mm sparsely yellow-pubescent, hairs to 1 mm; style 11–13 mm. **Capsules** 6–8 × 3–5 mm, glabrous.

Flowering Jun–Jul. Sagebrush shrublands, juniper woodlands; 1000–2100 m; Idaho, Oreg.

Penstemon seorsus is known from the southern Columbia Plateau and northwestern Great Basin. The species has been documented in Adams, Owyhee, and Washington counties, Idaho, and in Baker, Grant, Harney, Jefferson, Malheur, and Wasco counties, Oregon.

187. Penstemon smallii A. Heller, Bull. Torrey Bot. Club 21: 25. 1894 (as Pentstemon) • Blue Ridge beardtongue E

Herbs. Stems ascending to erect, 35–80 cm, puberulent or retrorsely hairy proximally, retrorsely hairy and sparsely glandular-pubescent distally, not glaucous. **Leaves** basal and cauline, basal sometimes withering by anthesis, not leathery, glabrous or puberulent along midveins and, sometimes, on proximal parts of blade; basal and proximal cauline 55–170 × 15–60 mm, blade triangular-ovate to cordate or lanceolate, base tapered, margins crenate to sharply serrate, apex obtuse to acute; cauline 4–6 pairs, sessile, 66–105 × 12–48 mm, blade ovate to lanceolate, proximals sometimes lyrate, base truncate to broadly clasping, margins crenate to sharply serrate, apex acute. **Thyrses** interrupted, narrowly conic, (6–)10–28 cm, axis puberulent and glandular-pubescent, verticillasters 4–7, cymes (3–)5–12-flowered, 2 per node; proximal bracts lanceolate, 56–130 × 23–55 mm, margins sharply serrate; peduncles and pedicels ascending to erect, puberulent and glandular-pubescent. **Flowers:** calyx lobes ovate to lanceolate, 3.5–5 × 1.2–2.1 mm, glandular-pubescent; corolla lavender to violet or purple, with violet nectar guides, funnelform to ventricose-ampliate, 28–35 mm, glandular-pubescent externally, moderately white-lanate internally abaxially, tube 7–8 mm, throat abruptly inflated, 9–16 mm diam., 2-ridged abaxially; stamens included, pollen sacs opposite, navicular, 1.3–1.5 mm, dehiscing completely, connective splitting, sides glabrous, sutures papillate; staminode 15–18 mm, included, 0.4–0.6 mm diam., tip straight, distal 13–15 mm ± pilose, hairs yellow, to 2 mm; style 17–20 mm. **Capsules** 7–10 × 4–5 mm, glabrous. *2n* = 16.

Flowering Apr–Jul. Rocky slopes, bluffs, cliffs; 200–1200 m; Ala., Ga., N.C., S.C., Tenn.

Penstemon smallii is known from the southern Appalachians. Foliose inflorescence bracts, truncate or cordate cauline leaf bases, and lavender to purple corollas distinguish it from other eastern penstemons.

188. Penstemon spatulatus Pennell, Notul. Nat. Acad. Nat. Sci. Philadelphia 71: 10. 1941 • Wallowa beardtongue E

Herbs. Stems ascending to erect, 6–26 cm, retrorsely hairy or glabrate, not glaucous. **Leaves** basal and cauline, ± leathery or not, glabrous or puberulent proximally on petiole and along midvein abaxially; basal and proximal cauline 10–60 × 3–18 mm, blade spatulate to elliptic, base tapered, margins entire, sometimes ± serrulate distally, apex rounded to obtuse or acute; cauline 1–3 pairs, sessile, 14–35 × 2–8 mm, blade lanceolate, sometimes oblanceolate, base truncate, margins entire, rarely ± serrulate distally, apex obtuse to acute. **Thyrses** interrupted, cylindric, 1–13 cm, axis moderately glandular-pubescent, verticillasters 1–5, cymes 2–5-flowered, 2 per node; proximal bracts lanceolate, 8–30 × 2–7 mm, margins entire; peduncles and pedicels ascending to erect, glandular-pubescent. **Flowers:** calyx lobes ovate to lanceolate, 2.6–4.8 × 1–1.8 mm, glandular-pubescent; corolla light blue to light purple, with violet nectar guides, funnelform, 10–14 mm, glandular-pubescent externally, moderately white- or yellowish lanate internally abaxially, tube 4–5 mm, throat slightly inflated, 3.5–4.5 mm diam., 2-ridged abaxially; stamens included, pollen sacs opposite, navicular to subexplanate, 0.6–0.8 mm, dehiscing completely, connective splitting, sides glabrous, sutures papillate; staminode 6–7 mm, included, 0.3–0.4 mm diam., tip recurved, distal 0.5–1.5 mm densely pilose, hairs yellow, to 0.7 mm; style 8–10 mm. **Capsules** 4–6 × 2.5–4 mm, glabrous.

Flowering Jul–Aug. Open rocky slopes, meadows; 2300–2500 m; Oreg.

Penstemon spatulatus is known only from the Strawberry and Wallowa mountains of northeastern Oregon.

189. Penstemon subserratus Pennell, Notul. Nat. Acad. Nat. Sci. Philadelphia 71: 13. 1941 • Subserrate beardtongue E

Herbs. Stems ascending to erect, 20–80 cm, glabrous or retrorsely hairy, rarely sparsely glandular-pubescent, not glaucous. Leaves basal and cauline, not leathery, glabrous or puberulent (especially on petiole and along major veins and margins); basal and proximal cauline 30–120(–160) × 4–32 mm, blade ovate to elliptic, rarely spatulate, base tapered, margins serrate to serrulate or entire, apex obtuse to acute, rarely rounded; cauline 2–6 pairs, sessile, 20–75 × 5–27 mm, blade lanceolate to oblong, base truncate to clasping, margins serrate to serrulate, sometimes entire, apex obtuse to acute or acuminate. Thyrses interrupted, cylindric to narrowly conic, 18–37 cm, axis moderately to densely glandular-pubescent, verticillasters 5–11, cymes 3–11-flowered, 2 per node; proximal bracts ovate to lanceolate, 25–60 × 9–30 mm, margins serrate to serrulate, sometimes entire; peduncles and pedicels ascending to erect, glandular-pubescent. Flowers: calyx lobes ovate to lanceolate, 3–5 × 1.4–2.5 mm, margins narrowly to broadly scarious, glandular-pubescent to densely glandular-pubescent; corolla blue to violet or purple, with reddish purple nectar guides, funnelform, 12–18 mm, glandular-pubescent externally, moderately white-lanate internally abaxially, tube 4–5 mm, throat slightly inflated, 3–6 mm diam., 2-ridged abaxially; stamens included, pollen sacs opposite, navicular to subexplanate, 0.7–1.1 mm, dehiscing completely, connective splitting, sides glabrous, sutures papillate; staminode 8–10 mm, included, 0.3–0.5 mm diam., tip recurved, distal 2–3 mm sparsely to moderately pilose, hairs yellow or yellow-orange, to 1.5 mm; style 8–11 mm. Capsules 4–5 × 3–4 mm, glabrous. 2*n* = 32.

Flowering May–Jul. Open slopes in pine forests and pine-oak woodlands; 30–1800 m; Oreg., Wash.

Penstemon subserratus occurs along the eastern flank of the Cascade Range.

190. Penstemon sudans M. E. Jones, Contr. W. Bot. 8: 37. 1898 (as Pentstemon) • Susanville beardtongue E

Penstemon deustus Douglas ex Lindley subsp. *sudans* (M. E. Jones) Pennell & D. D. Keck

Subshrubs. Stems ascending to erect, 18–70 cm, glandular-pubescent, sometimes densely so distally, not glaucous. Leaves cauline, opposite or subopposite, not leathery, glandular-pubescent; cauline (4–)7–10 pairs, sessile or proximals

short-petiolate, 14–60 × 8–20(–30) mm, blade obovate to ovate or lanceolate, base tapered to clasping, margins dentate, dentate-serrate, or serrate, apex acute to acuminate. Thyrses interrupted or continuous, cylindric, 13–35 cm, axis densely glandular-pubescent, verticillasters 9–18, cymes 1–5-flowered, 2 per node; proximal bracts ovate to lanceolate, (7–)15–40 × (2–)8–18 mm, margins dentate to dentate-serrate; peduncles and pedicels ascending to erect, densely glandular-pubescent. Flowers: calyx lobes ovate to lanceolate, 3–6 × 0.8–1.6 mm, densely glandular-pubescent; corolla white to light lavender or light pink, with reddish violet nectar guides, funnelform, 9–11 mm, glandular-pubescent externally, glandular-pubescent internally, tube 3–4 mm, throat slightly inflated, 3–4 mm diam., rounded abaxially; stamens included, pollen sacs opposite, navicular to subexplanate, 0.5–0.7 mm, dehiscing completely, connective splitting, sides glabrous, sutures smooth; staminode 9–10 mm, included or exserted, 0.2–0.3 mm diam., tip straight, glabrous; style 7–8 mm. Capsules 4–5 × 2.5–3 mm, glabrous.

Flowering Jun–Jul. Igneous soils and crevices, basalt in sagebrush shrublands and pine woodlands; 1200–1800 m; Calif., Nev.

Penstemon sudans is known from northeastern California (Lassen, Modoc, and Placer counties); it has been reported for Washoe County, Nevada. D. D. Keck (1940b) observed that some plants of *P. deustus* from southeastern Oregon (Curry, Jackson, and Josephine counties) approach *P. sudans* in glandularity, though he considered these to be one extreme of the morphologic spectrum exhibited by *P. deustus*.

191. Penstemon tenuiflorus Pennell, Addisonia 4: 79, plate 160. 1919 • Eastern white-flower beardtongue E

Herbs. Stems ascending to erect, 20–80 cm, retrorsely hairy and ± glandular-lanate, retrorse hairs sometimes few or distal only, not glaucous. Leaves basal and cauline, not leathery, moderately to densely glandular-lanate, especially along major veins abaxially; basal and proximal cauline 40–138 × 7–37 mm, blade spatulate to oblanceolate or elliptic, base tapered, margins entire or ± serrate to dentate, apex rounded to obtuse or acute; cauline 4–7 pairs, sessile or proximals sometimes petiolate, (11–)20–110 × 3–30(–40) mm, blade ovate to lanceolate, base truncate to clasping, sometimes tapered, margins entire or crenate-serrate to dentate, apex acute to acuminate. Thyrses interrupted, conic, (3–)8–37 cm, axis moderately to densely glandular-pubescent to glandular-lanate, verticillasters 3–6(or 7), cymes 3–10(–13)-flowered, 2 per node; proximal bracts lanceolate, 6–40 × 1–10 mm, margins entire or ± serrate; peduncles and pedicels ascending, moderately to densely glandular-pubescent to glandular lanate. **Flowers:** calyx lobes ovate to broadly lanceolate, 2–5 × 1.3–2 mm, glandular-pubescent; corolla white, without nectar guides, personate, tubular, 21–28 mm, sparsely glandular-pubescent externally, moderately to densely white-villous internally abaxially, tube 6–8 mm, throat slightly inflated, 5–8 mm diam., 2-ridged abaxially; stamens included, pollen sacs opposite, navicular to subexplanate, 1–1.2 mm, dehiscing completely, connective splitting, sides glabrous, sutures papillate; staminode 13–16 mm, exserted, 0.4–0.6 mm diam., tip straight, distal 10–12 mm moderately to densely pubescent, hairs yellow, to 1.6 mm; style 16–19 mm. **Capsules** 5–7 × 3–5 mm, glabrous.

Flowering Apr–Jun. Dry woods, cedar glades, prairies, barrens, cliffs; 30–80 m; Ala., Ky., Miss., Tenn.

Penstemon tenuiflorus is concentrated in the Interior Low Plateaus region of western Kentucky, central Tennessee, and northwestern Alabama, and in the Black Belt subsection of the Coastal Plain of central Alabama and northeastern Mississippi.

R. K. Clements et al. (1998) observed that the leaves of *Penstemon tenuiflorus* bear both simple and stellate hairs, reporting only simple hairs on the leaves of closely related *P. hirsutus*. However, stellate hairs occur in the leaf axils of plants in some populations of *P. hirsutus*, and they are common on the stems. Stellate hairs also occur on other diploid, tubular-flowered eastern North American species, including *P. arkansanus*, *P. australis*, *P. brevisepalus*, *P. canescens*, *P. laxiflorus*, *P. oklahomensis*, and *P. pallidus*. Stellate hairs have not been observed on any of the polyploid, ventricose-flowered eastern North American species.

192. Penstemon tenuis Small, Fl. S.E. U.S., 1061, 1337. 1903 (as Pentstemon) • Sharp-sepal beardtongue [E]

Herbs. Stems ascending to erect, 25–85 cm, glabrate or puberulent, not glaucous. **Leaves** basal and cauline, basal often withering by anthesis, not leathery, glabrous or ± puberulent along major veins; basal and proximal cauline 65–160 × 8–48 mm, blade oblanceolate to elliptic, base tapered, margins serrate, sometimes subentire, apex obtuse to acute; cauline (3–)7 or 8 pairs, sessile, 45–140 × 6–37 mm, blade lanceolate, proximals sometimes oblanceolate, base truncate to clasping, margins sharply serrate, sometimes ± serrate, apex acute to acuminate. **Thyrses** interrupted, conic, 6–20 cm, axis glabrous or retrorsely hairy, rarely ± glandular-pubescent distally, verticillasters 3–8, cymes (3–)5–15-flowered, 2 per node; proximal bracts lanceolate, 22–100 × 6–30 mm, margins serrate; peduncles and pedicels ascending, retrorsely hairy or ± glandular-pubescent distally. **Flowers:** calyx lobes ovate to lanceolate, 3.1–6 × 1–2.2 mm, glabrous or sparsely glandular-pubescent; corolla lavender to violet, with sometimes faint reddish purple nectar guides, funnelform, 10–14 mm, sparsely glandular-pubescent externally, moderately to densely white-villous internally abaxially, tube 4–6 mm, throat moderately inflated, 4–6 mm diam., rounded abaxially; stamens included, pollen sacs opposite, navicular, 0.8–1 mm, dehiscing completely, connective splitting, sides glabrous, sutures smooth or ± papillate; staminode 8–9 mm, exserted, 0.3–0.4 mm diam., tip straight, distal 4–5 mm moderately to densely pubescent, hairs orangish, to 1 mm; style 8–10 mm. **Capsules** 4–5 × 3–4 mm, glabrous.

Flowering Apr–Jun. Alluvial soils, open floodplain forests, open hardwood flatwoods, marsh margins, low prairies, wet depressions; 0–200 m; Ark., La., Miss., Mo., Tex.

Penstemon tenuis is widespread in the western Gulf Coastal Plain and lower Mississippi River Basin.

193. Penstemon tracyi D. D. Keck, Amer. Midl. Naturalist 23: 603, fig. 2. 1940 • Tracy's beardtongue [C][E]

Herbs. Stems erect, 8–12 cm, glabrous, glaucescent. **Leaves** basal and cauline, leathery, glabrous; basal and proximal cauline 11–40 × 3–14 mm, blade spatulate to oval or orbiculate, base tapered, margins entire or ± denticulate or serrulate distally, apex rounded to obtuse; cauline 2–4, sessile, 9–20 × 4–10 mm, blade: proximals spatulate, distals obovate or elliptic, base truncate, margins entire or ± denticulate distally, apex rounded or mucronate. **Thyrses** continuous, cylindric, 2–4 cm, axis glabrous, verticillasters 2 or 3, cymes 2–5-flowered, 2 per node; proximal bracts elliptic to ovate, 6–18 × 2–10 mm, margins entire; peduncles and pedicels ascending to erect, glabrous. **Flowers:** calyx lobes ovate, 1.5–2 × 0.8–1.2 mm, glabrous or ± glandular-ciliolate; corolla white to pink, without nectar guides, funnelform, 11–15 mm, glabrous externally, densely white-villous internally abaxially, tube slightly differentiated, 3–5 mm, throat slightly inflated, 2–3 mm diam., rounded abaxially; stamens included, pollen sacs opposite, explanate, 0.4–0.5 mm, dehiscing completely, connective splitting, sides glabrous, sutures smooth; staminode 8–10 mm, reaching orifice, 0.2–0.3 mm diam., tip straight, distal 2–3 mm sparsely pilose, hairs white or yellowish, to 0.6 mm; style 9–13 mm. **Capsules** 5–8 × 3–4.5 mm, glabrous.

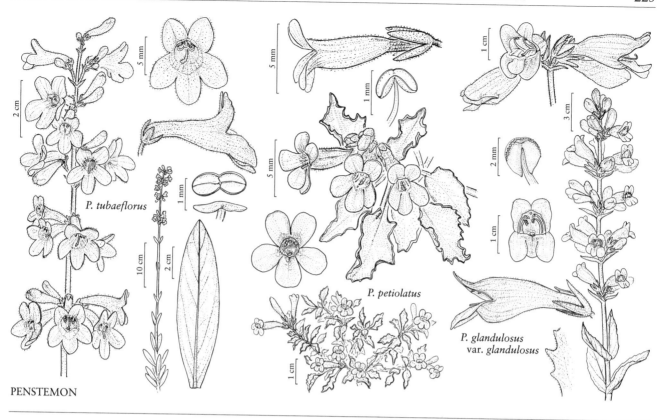

P. tubaeflorus

PENSTEMON

P. petiolatus

P. glandulosus var. glandulosus

Flowering Jul–Aug. Rock crevices; of conservation concern; 2000–2100 m; Calif.

Penstemon tracyi has been documented only in Siskiyou and Trinity counties.

194. Penstemon tubaeflorus Nuttall, Trans. Amer. Philos. Soc., n. s. 5: 181. 1835 (as Pentstemon tubaeflorum) • Tube beardtongue E F

Penstemon tubaeflorus var. *achoreus* Fernald

Herbs. Stems erect, 25–85 (–100) cm, glabrous, not glaucous. **Leaves** basal and cauline, basal often withering by anthesis, not leathery, glabrous; basal and proximal cauline 25–110 × 6–38 mm, blade spatulate to ovate, base tapered, margins entire or ± serrate, apex rounded to obtuse; cauline 5–8 (–10) pairs, sessile, 15–100(–135) × 4–20(–38) mm, blade ovate to lanceolate, base tapered to clasping, margins entire or ± serrate, apex acute to acuminate. **Thyrses** interrupted, cylindric, 8–30(–40) cm, axis glandular-pubescent, verticillasters 4–8(–12), cymes 3–9-flowered, 2 per node; proximal bracts ovate to lanceolate, 9–50 × 2–10 mm, margins entire or ± serrulate; peduncles and pedicels ascending to erect, glabrous or glandular-pubescent. **Flowers:** calyx lobes ovate, 2.5–5 × 1.5–2.5 mm, glandular-pubescent; corolla white, without nectar guides, funnelform, 15–22 mm, glandular-pubescent externally, glandular-pubescent internally, tube slightly differentiated, 5–7 mm, throat slightly inflated, 4–6 mm diam., rounded abaxially; stamens included or longer pair reaching orifice, pollen sacs divergent, explanate, 0.8–1 mm, dehiscing completely, connective splitting, sides glabrous, sutures smooth; staminode 9–11 mm, included or reaching orifice, 0.2–0.3 mm diam., tip recurved, distal 3–4 mm sparsely villous, hairs yellowish or brownish yellow, to 0.8 mm; style 10–12 mm. **Capsules** 7–10 ×3–5 mm, glabrous. *2n* = 32.

Flowering May–Jul. Tallgrass prairies, woodlands, roadsides; 20–600 m; Ont.; Ark., Conn., Ill., Ind., Iowa, Kans., Ky., La., Maine, Mass., Miss., Mo., Nebr., N.H., N.Y., Ohio, Pa., R.I., Tenn., Tex., Vt., Wis.

Penstemon tubaeflorus is concentrated in the southeastern Great Plains and Ozark Highlands, where it probably is native. *Penstemon tubaeflorus* appears to have naturalized eastward and northward in North America due to habitat disturbance and planting. Sometimes it grows in mixed populations with *P. digitalis*. Fruiting specimens of *P. tubaeflorus* often are misidentified as *P. digitalis*; the former can be distinguished by its more cylindric inflorescences with erect peduncles and smaller capsules.

195. Penstemon virens Pennell ex Rydberg, Fl. Rocky Mts., 773, 1066. 1917 (as Pentstemon) • Green beardtongue E

Herbs. Stems ascending to erect, 10–40 cm, puberulent or retrorsely hairy proximally, rarely glabrous, glandular-pubescent distally, not glaucous. **Leaves** basal and cauline, not leathery, glabrous; basal and proximal cauline 20–102 × 4–15 mm, blade lanceolate to oblanceolate or spatulate, base tapered, margins serrulate or denticulate, sometimes entire, apex obtuse to acute; cauline (2 or)3–5 pairs, sessile, 18–50(–70) ×3–14 mm, blade ovate to lanceolate, base tapered to clasping, margins serrulate or denticulate, sometimes entire, apex acute to acuminate. **Thyrses** interrupted or continuous, cylindric, 5–18 cm, axis glandular-pubescent, verticillasters 3–6(–8), cymes 2–5-flowered, 2 per node; proximal bracts ovate to lanceolate, 6–40 × 2–10 mm, margins entire or serrulate to denticulate; peduncles and pedicels ascending to erect, puberulent or glandular-pubescent. **Flowers:** calyx lobes ovate to lanceolate, 2–4.5 × 1.5–2.5 mm, glandular-pubescent; corolla blue to light purple or violet, with reddish purple to bluish purple nectar guides, funnelform, 10–16(–18) mm, glandular-pubescent externally, moderately white-pilose internally abaxially, tube 4–5 mm, throat 3–5 mm diam., 2-ridged abaxially; stamens included, pollen sacs divergent to opposite, navicular, 0.6–0.8 mm, dehiscing completely, connective splitting, sides glabrous, sutures smooth or ± papillate; staminode 8–10 mm, included or reaching orifice, 0.2–0.3 mm diam., tip straight to slightly recurved, distal 4–5 mm moderately to densely villous, hairs golden yellow, to 1.3 mm; style 8–11 mm. **Capsules** 5–7 × 2–3 mm, glabrous. $2n = 16$.

Flowering May–Aug. Rocky or gravelly soils, shortgrass prairies, foothills, and mountain meadows; 1600–2600 m; Colo., Wyo.

Penstemon virens is known in the Front Range of the Rocky Mountains from Raton Mesa in south-central Colorado to the eastern Sierra Madre and Medicine Bow and Laramie mountains of southeastern Wyoming.

196. Penstemon washingtonensis D. D. Keck, Amer. Midl. Naturalist 33: 150. 1945 • Washington beardtongue E

Herbs. Stems erect, 3–25 cm, glabrous or ± retrorsely hairy, not glaucous. **Leaves** basal and cauline, not leathery, glabrous; basal and proximal cauline (10–)17–60 × 5–18 mm, blade spatulate to oblanceolate, lanceolate, or elliptic, base tapered, margins entire, apex rounded to obtuse or acute; cauline 2–6 pairs, short-petiolate or sessile, 10–55 × 3–15 mm, blade ovate to lanceolate, base tapered to clasping, margins entire, apex obtuse to acute. **Thyrses** interrupted, cylindric, 1–6 cm, axis sparsely glandular-pubescent, sometimes also retrorsely hairy, verticillasters 1–3(or 4), cymes 2–6-flowered, 2 per node; proximal bracts lanceolate, 2–15 × 1–7 mm, margins entire; peduncles and pedicels erect, sparsely to moderately glandular-pubescent. **Flowers:** calyx lobes lanceolate, 3–4 × 0.8–1.2 mm, glandular-pubescent; corolla dark blue, sometimes whitish, without nectar guides, funnelform, 9–12 mm, glandular-pubescent externally, moderately whitish lanate internally abaxially, tube 3–4 mm, throat slightly inflated, 2–3 mm diam., 2-ridged abaxially; stamens included, pollen sacs opposite, explanate to subexplanate, 0.5–0.6 mm, dehiscing completely, connective splitting, sides glabrous, sutures smooth; staminode 5–6 mm, included, 0.2–0.3 mm diam., tip straight to slightly recurved, distal 0.5–1 mm moderately pilose, hairs golden yellow, to 0.5 mm; style 4–6 mm. **Capsules** 4–7 × 2.5–4 mm, glabrous.

Flowering Jul–Aug(–Sep). Alpine meadows, rocky slopes, grassy openings in pine-fir forests; 1800–2300 m; Wash.

Penstemon washingtonensis is known from Chelan and Okanogan counties. Except for its glandular-pubescent inflorescence, *P. washingtonensis* is morphologically similar to *P. procerus* var. *tolmiei*.

197. Penstemon watsonii A. Gray in A. Gray et al., Syn. Fl. N. Amer. 2(1): 267. 1878 (as Pentstemon watsoni) • Watson's beardtongue E

Herbs. Stems ascending to erect, 25–60 cm, glabrous or sparsely puberulent, not glaucous. **Leaves** essentially cauline, basal absent or poorly developed, not leathery, glabrous or puberulent; cauline 4–8 pairs, short-petiolate or sessile, 30–70(–80) × 8–18(–35) mm, blade oblanceolate to lanceolate, base tapered or clasping, margins entire, apex obtuse to acute. **Thyrses** continuous or interrupted, cylindric, (1.5–)5–16 cm, axis ± retrorsely hairy, verticillasters 2–6(–10), cymes (2–)6–14-flowered, 2 per node; proximal bracts lanceolate to linear, rarely ovate, 3–60 × 1–23 mm, margins entire; peduncles and pedicels ascending to erect, ± retrorsely hairy. **Flowers:** calyx lobes broadly ovate, 1.8–3(–3.5) × 1.4–2 mm, apex obtuse to acute or short-cuspidate, glabrous; corolla blue to violet or purple, with faint reddish purple nectar guides, funnelform, 12–16(–18) mm, glabrous externally, sparsely white-villous internally abaxially, tube 4–5 mm, throat gradually inflated, 3–6 mm diam., 2-ridged abaxially; stamens included or longer pair slightly exserted, pollen sacs opposite, navicular, 0.9–1 mm, dehiscing completely, connective splitting, sides glabrous, sutures papillate; staminode 8–9 mm, reaching orifice, 0.2–0.3 mm diam., tip slightly recurved, distal 3–4 mm moderately to densely villous, hairs golden yellow, to 1 mm; style 9–13 mm. **Capsules** 4–7 × 3.5–5 mm, glabrous. $2n = 16$.

Flowering May–Aug. Dry, rocky slopes, sagebrush shrublands, pine-oak and pine woodlands; 1700–3200 m; Ariz., Colo., Idaho, Nev., Utah, Wyo.

Penstemon watsonii is known from mountain ranges throughout the Intermountain Region. *Penstemon laxus*, which has been included as a variety of this species, has corollas 2–3 mm in diameter and densely white-villous internally abaxially, and is known from south-central Idaho north of the Snake River. *Penstemon watsonii* does enter extreme south-central Idaho but is not known to occur north of the Snake River.

198. Penstemon whippleanus A. Gray, Proc. Amer. Acad. Arts 6: 73. 1862 (as Pentstemon) • Whipple's beardtongue E

Herbs. Stems ascending to erect, (8–)20–65(–100) cm, glabrous or ± puberulent proximally, ± puberulent or glandular-pubescent distally, not glaucous. **Leaves** basal and cauline, not leathery, glabrous; basal and proximal cauline 40–90(–130) × (5–)10–30(–75) mm, blade ovate to lanceolate or elliptic, base tapered to cuneate, margins entire or ± repand, denticulate, or dentate, apex obtuse to acute; cauline 2–5 pairs, short-petiolate or sessile, 25–60(–85) × 3–15(–25) mm, blade lanceolate to oblanceolate, base tapered to clasping or cordate-clasping, margins entire, sometimes ± repand to denticulate, apex obtuse to acute. **Thyrses** interrupted or continuous, secund, (2–)5–35 cm, axis sparsely to densely glandular-pubescent, verticillasters 2–5(–7), cymes 2–4-flowered, 2 per node; proximal bracts lanceolate, 11–85 × 1–18 mm, margins entire or ± repand proximally; peduncles and pedicels ascending to erect, glandular-pubescent. **Flowers:** calyx lobes lanceolate, 7–10 × 1–2.2 mm, glandular-pubescent; corolla brownish purple to bluish violet, sometimes ochroleucous, lavender, or white, with alternating white or lavender and purple nectar guides in dark-colored forms, with lavender nectar guides in light-colored forms, ventricose to ventricose-ampliate, 20–27(–30) mm, glandular-pubescent externally, sparsely white-villous internally abaxially, tube 5–8 mm, throat abruptly inflated, 8–10 mm diam., 2-ridged abaxially; stamens included or longer pair reaching orifice, pollen sacs opposite, explanate, 1–1.4 mm, dehiscing completely, connective splitting, sides glabrous, sutures smooth; staminode 12–15 mm, reaching orifice or exserted,

0.6–1.1 mm diam., tip straight to slightly recurved, distal 1–3 mm sparsely to densely villous, hairs yellow, to 1 mm; style 12–16 mm. **Capsules** 6–9 × 4–5 mm, glandular-puberulent distally. $2n = 16$.

Flowering Jun–Aug(–Sep). Rocky slopes in subalpine forests, alpine meadows; 2100–3700 m; Ariz., Colo., Idaho, Mont., N.Mex., Utah, Wyo.

Penstemon whippleanus is known from forests and meadows of the Southern and Central Rocky mountains and westward into the Wasatch Mountains and Utah Plateaus. Plants with light-colored corollas often are found growing next to plants with dark-colored corollas.

199. Penstemon wilcoxii Rydberg, Bull. Torrey Bot. Club 28: 28. 1901 (as Pentstemon) • Wilcox's beardtongue E

Penstemon ovatus Douglas var. *pinetorum* Piper

Herbs. Stems ascending to erect, (20–)35–100 cm, glabrous or sparsely to densely puberulent, sometimes also sparsely glandular-pubescent, especially distally, not glaucous. **Leaves** basal and cauline, ± leathery or not, glabrous or puberulent to pubescent, especially along midveins; basal and proximal cauline (25–)35–190 × (5–)12–55 mm, blade triangular to ovate or lanceolate, base tapered, rarely truncate, margins ± serrate or dentate, apex acute, rarely obtuse; cauline 2–5 pairs, short-petiolate or sessile, 35–90 × (8–)15–50 mm, blade ovate to oblong or lanceolate, base tapered to clasping or cordate-clasping, margins serrate to dentate, apex obtuse to acute. **Thyrses** interrupted, cylindric to conic, (5–)8–30 cm, axis sparsely to moderately glandular-pubescent, verticillasters 3–7, cymes 3–11-flowered, 2 per node; proximal bracts ovate to lanceolate, (8–)18–75 × (5–)8–35 mm, margins ± serrate to dentate; peduncles and pedicels spreading to ascending, ± glandular-pubescent. **Flowers:** calyx lobes ovate, 2.4–5.5 × 0.9–1.8 mm, margins broadly scarious, glandular-pubescent; corolla blue to violet or light blue, with reddish purple nectar guides, funnelform, 13–23 mm, glandular-pubescent externally, sparsely white- or yellow-villous internally abaxially, tube 4–8 mm, throat gradually inflated, 4–9 mm diam., 2-ridged abaxially; stamens included, pollen sacs opposite, navicular to subexplanate, 0.8–1 mm, dehiscing completely, connective splitting, sides glabrous, sutures smooth or ± papillate; staminode 8–11 mm, reaching orifice, 0.2–0.3 mm diam., tip straight to recurved, distal 1–5 mm pilose to villous, hairs yellow, to 1.2 mm; style 9–13 mm. **Capsules** 4–6.5 × 3–4 mm, glabrous. $2n = 16, 32$.

Flowering May–Aug. Rocky or gravelly slopes in forests and woodlands, roadcuts; 300–2300 m; Idaho, Mont., Oreg., Wash.

Penstemon wilcoxii hybridizes with *P. albertinus* in areas of sympatry in east-central Idaho and west-central Montana (D. D. Keck 1945; A. Cronquist 1959; D. V. Clark 1971). Putative hybrids and/or introgressants exhibit varying degrees of morphologic intermediacy between the parent species. Many herbarium specimens are difficult to determine with confidence.

26b.13. PENSTEMON Schmidel (subg. PENSTEMON) sect. PETIOLATI (Rydberg) Pennell, Contr. U.S. Natl. Herb. 20: 326, 328. 1920 ⒺE

Penstemon [unranked] *Petiolati* Rydberg, Fl. Rocky Mts., 764. 1917 (as Pentstemon)

Shrubs or subshrubs. Stems glabrous or scabrous to puberulent, glaucous. **Leaves** cauline, opposite, ± leathery or not, glabrate or scabrous, glaucous; cauline petiolate or sessile, blade ovate, margins toothed. **Thyrses** continuous, cylindric, axis puberulent, cymes 2 per node; peduncles and pedicels spreading to ascending. **Flowers:** calyx lobes: margins entire, narrowly scarious, glandular-pubescent, rarely glabrous; corolla magenta to pink, bilaterally symmetric, weakly bilabiate, not personate, weakly ventricose, glandular-pubescent externally, hairy internally abaxially, glandular-pubescent adaxially, throat moderately inflated, not constricted at orifice, rounded abaxially; stamens included or longer pair slightly exserted, filaments glabrous, pollen sacs opposite, subexplanate to explanate, dehiscing completely, connective splitting, sides glabrous, sutures smooth; staminode slightly exserted, flattened distally, 0.4–0.5 mm diam., tip straight, glabrous; style glabrous. **Capsules** glabrous. **Seeds** dark brown to black, angled, 0.9–1.2 mm.

Species 1: w United States.

Penstemon petiolatus is the only member of sect. *Petiolati*. D. D. Keck (1937) noted morphologic similarities between *P. petiolatus* and *P. clevelandii* of sect. *Spectabiles*. *Penstemon petiolatus* has not been examined molecularly.

200. Penstemon petiolatus Brandegee, Bot. Gaz. 27: 455. 1899 (as Pentstemon) • Petiolate beardtongue ⒺE ⒻF

Stems 5–25(–40) cm. **Leaves:** cauline 2–4 pairs, 14–30 (–35) × 6–17 mm, base tapered to cuneate, margins dentate to serrate-dentate, apex obtuse to acute. **Thyrses** 3–6 cm, verticillasters 2–6, cymes 1–3(–5)-flowered; proximal bracts ovate to lanceolate, 6–25 × 2–14 mm; peduncles and pedicels puberulent and sparsely glandular-pubescent, especially distally. **Flowers:** calyx lobes lanceolate, 4.5–6 × 1.2–1.9(–2.1) mm; corolla with violet nectar guides, 13–17 mm, sparsely glandular-pubescent externally, sparsely whitish or yellowish villous internally abaxially, tube 5–6 mm, throat 4–5 mm diam.; stamens: pollen sacs 0.5–0.9 mm; staminode 10–11 mm; style 9–13 mm. **Capsules** 5–7 × 4–5 mm.

Flowering May–Jun. Crevices in limestone outcrops, desert shrub communities, juniper woodlands; 1000–1700 m; Ariz., Nev., Utah.

Penstemon petiolatus occurs in the Beaver Dam and Virgin mountains in northwestern Mohave County, Arizona, the Charleston Mountains and Sheep Range in Clark County, Nevada, and the Beaver Dam Mountains in southwestern Washington County, Utah. *Penstemon calcareus* M. E. Jones is an illegitimate, later homonym that applies here.

26b.14. PENSTEMON Schmidel (subg. PENSTEMON) sect. SACCANTHERA Bentham in A. P. de Candolle and A. L. P. P. de Candolle, Prodr. 10: 329. 1846 (as Pentstemon) [E]

Penstemon subg. *Saccanthera* (Bentham) A. Gray

Herbs or subshrubs. **Stems** glabrous, puberulent, or retrorsely hairy, rarely glandular-pubescent distally or wholly or retrorsely hairy, glaucous or not. **Leaves** basal and cauline, basals sometimes few, or cauline, opposite, sometimes alternate, rarely whorled or subalternate, leathery, rarely not (*P. papillatus*, *P. platyphyllus*), glabrous, glabrate, puberulent, retrorsely hairy, canescent, pilose, hirsute, or glandular-pubescent, glaucous or not; basal and proximal cauline petiolate; cauline sessile, petiolate, or short-petiolate, blade obovate, ovate, spatulate, oblanceolate, lanceolate, elliptic, oblong, or linear, margins entire or toothed, rarely laciniate-pinnatifid. **Thyrses** interrupted or continuous, cylindric, sometimes secund, axis glabrous, glandular-pubescent, puberulent, or retrorsely hairy, cymes 1 or 2 per node; peduncles and pedicels spreading to ascending or erect, rarely appressed. **Flowers:** calyx lobes: margins entire or erose, rarely lacerate (*P. serrulatus*), narrowly scarious or scarious, sometimes herbaceous, rarely broadly scarious (*P. serrulatus*), glabrous or glandular-pubescent, rarely scabrous, puberulent, pubescent, or ciliolate; corolla lavender to blue, violet, purple, reddish purple, or pinkish, bilaterally symmetric, ± bilabiate, not personate, funnelform, ventricose, ampliate, or ventricose-ampliate, glabrous or glandular-pubescent externally, glabrous or hairy internally abaxially, throat slightly to gradually inflated, not constricted at orifice, rounded to 2-ridged abaxially; stamens included or longer pair reaching orifice or exserted, filaments glabrous, rarely hairy distally, pollen sacs parallel, rarely divergent, saccate, dehiscing incompletely, distal $1/5$–$2/3$ indehiscent, connective splitting, sides glabrous or hairy, sutures denticulate, teeth to 0.3 mm; staminode included to exserted, flattened distally, 0.2–1 mm diam., tip straight, glabrous or distal 5–50(–60)% hairy, hairs to 2.5 mm; style glabrous. **Capsules** glabrous. **Seeds** tan, gray, or brown, angled, rarely patelliform, 0.5–3.4 mm.

Species 27 (27 in the flora): w North America.

D. D. Keck (1932) included most species of *Penstemon* with saccate anthers in subg. *Saccanthera*, sect. *Saccanthera*. Most members of sect. *Saccanthera* formed a clade nested in subg. *Penstemon* in the molecular study by A. D. Wolfe et al. (2006).

SELECTED REFERENCE Keck, D. D. 1932. Studies in *Penstemon*. A systematic treatment of section *Saccanthera*. Univ. Calif. Publ. Bot. 16: 367–426.

1. Leaf blade margins serrate, dentate, laciniate-dentate, or laciniate-pinnatifid.
   2. Stems viscid glandular-pubescent. . . . . .207. *Penstemon glandulosus* (in part)
   2. Stems glabrous or retrorsely hairy.
      3. Calyx lobes glabrous, glabrate, or sparsely puberulent, margins sometimes ciliate.
         4. Styles 10–16 mm; filaments glabrous . . . . .224. *Penstemon serrulatus*
         4. Styles 18–25 mm; filaments white-lanate distally, rarely glabrous. . . . .
            . . . . 227. *Penstemon venustus* (in part)
      3. Calyx lobes glandular-pubescent.
         5. Corollas ventricose-ampliate, 15–32 mm; pollen sacs 1.3–2 mm; styles 11–23 mm . . . . 220. *Penstemon richardsonii*
         5. Corollas funnelform, 13–19 mm; pollen sacs 0.9–1.3 mm; styles 8–12 mm.
            6. Leaves opposite, sometimes alternate; staminodes: distal 1–4 mm sparsely to moderately lanate, hairs white or yellowish, to 1.5 mm. . . . . . . 204. *Penstemon diphyllus*
            6. Leaves whorled, sometimes alternate; staminodes: distal 6–7 mm moderately to densely lanate, hairs yellow, to 2.5 mm . . . 226. *Penstemon triphyllus* (in part)

1. Leaf blade margins entire or subentire.
    7. Stems viscid glandular-pubescent...........................207. *Penstemon glandulosus* (in part)
    7. Stems glabrous, retrorsely hairy, puberulent, or pilose.
        8. Staminodes hairy.
            9. Stems retrorsely hairy, hairs appressed, ± scalelike........... 225. *Penstemon tiehmii* (in part)
            9. Stems glabrous, retrorsely hairy, or puberulent, sometimes glabrate or hirsute, hairs pointed.
                10. Thyrse axes glabrous or retrorsely hairy; corollas glabrous externally or lobe margins sometimes densely white-ciliate.
                    11. Corollas 14–20 mm, glabrous externally; styles 11–14 mm........ ....215. *Penstemon parvulus* (in part)
                    11. Corollas 24–38 mm, glabrous externally except lobe margins densely white-ciliate; styles 18–25 mm..................227. *Penstemon venustus* (in part)
                10. Thyrse axes glandular-pubescent; corollas glandular-pubescent externally.
                    12. Leaves hirsute..................................... 222. *Penstemon scapoides*
                    12. Leaves glabrous, retrorsely hairy, or puberulent, sometimes glabrate.
                        13. Leaves whorled, sometimes alternate, cauline blades linear, sometimes lanceolate..................... 226. *Penstemon triphyllus* (in part)
                        13. Leaves opposite, cauline blades spatulate, oblanceolate, oblong, elliptic, or lanceolate.
                            14. Corollas 15–20(–24) mm; styles 12–15 mm....... 208. *Penstemon gracilentus*
                            14. Corollas 24–35 mm; styles 22–30 mm.
                                15. Stems retrorsely hairy or puberulent; pollen sacs: distal ⅓ indehiscent..........................214. *Penstemon papillatus*
                                15. Stems glabrous proximally, sparsely glandular-pubescent distally; pollen sacs: distal ½–⅔ indehiscent.......217. *Penstemon pudicus*
        8. Staminodes glabrous.
            16. Corollas glabrous externally.
                17. Stems puberulent or retrorsely hairy, at least proximally.
                    18. Styles 11–18 mm; corollas 14–26 mm.
                        19. Pedicels puberulent or retrorsely hairy.................203. *Penstemon cusickii*
                        19. Pedicels glabrous or papillate distally.................212. *Penstemon leonardii*
                    18. Styles 18–27 mm; corollas (20–)22–38 mm.
                        20. Pollen sacs 2.2–3 mm; calyx lobes obovate to oblanceolate or ovate to lanceolate, glabrous or puberulent; staminodes included; California............................ 209. *Penstemon heterophyllus* (in part)
                        20. Pollen sacs 1.5–1.9(–2.1) mm; calyx lobes lanceolate, glabrous; staminodes reaching orifice or exserted; Utah........ 216. *Penstemon platyphyllus*
                17. Stems glabrous.
                    21. Pollen sacs: distal ½–⅔ indehiscent, sides glabrous; Nevada, Utah.
                        22. Cauline leaf pairs 2–6; pollen sacs 2.4–2.8 mm; calyx lobes (4–)5–6.3 mm; Nevada...........................206. *Penstemon floribundus*
                        22. Cauline leaf pairs 6–14; pollen sacs 1.5–2 mm; calyx lobes 1.8–3.2 mm; Utah................................. 223. *Penstemon sepalulus*
                    21. Pollen sacs: distal ⅓–½ indehiscent, sides hispidulous, pubescent, pilose, or lanate near filament attachments; California, Oregon.
                        23. Corollas 14–20 mm, ventricose; pollen sacs 1.4–1.8 mm; styles 11–14 mm.............................215. *Penstemon parvulus* (in part)
                        23. Corollas 18–38 mm, ventricose, ventricose-ampliate, or funnelform; pollen sacs 1.8–3.5 mm; styles 16–27 mm.
                            24. Leaves glabrous, glaucous; thyrses secund, axes glabrous; proximal bracts ovate to lanceolate................201. *Penstemon azureus*
                            24. Leaves glabrous or retrorsely hairy, rarely glaucous; thyrses cylindric, axes glabrous or retrorsely hairy; proximal bracts lanceolate to linear................ 209. *Penstemon heterophyllus* (in part)

[16. Shifted to left margin.—Ed.]
16. Corollas glandular-pubescent externally.
  25. Cauline leaves 0.5–1(–2) mm wide, blades linear; proximal bracts linear ......205. *Penstemon filiformis*
  25. Cauline leaves (1–)2–24 mm wide, blades ovate to obovate, spatulate, oblanceolate, lanceolate, or linear; proximal bracts ovate to oblanceolate or lanceolate, rarely linear.
    26. Leaves canescent to pilose...................................218. *Penstemon purpusii*
    26. Leaves glabrous or retrorsely hairy, sometimes glabrate or puberulent.
      27. Stems retrorsely hairy, hairs appressed, scalelike.
        28. Cauline leaves 3–7 mm wide, blades oblanceolate to lanceolate, proximals sometimes spatulate; pollen sacs 0.9–1.2 mm ....219. *Penstemon rhizomatosus*
        28. Cauline leaves 7–24 mm wide, blades ovate to spatulate or lanceolate; pollen sacs 1.1–1.6 mm ....225. *Penstemon tiehmii* (in part)
      27. Stems glabrous, retrorsely hairy, or ± puberulent, hairs pointed.
        29. Pollen sacs 1–1.5(–1.8) mm, sides glabrous.
          30. Leaves glabrous, usually glaucous; California ....202. *Penstemon caesius*
          30. Leaves retrorsely hairy, not glaucous; Nevada, Oregon.........210. *Penstemon kingii*
        29. Pollen sacs 1.5–3.2 mm, sides hispidulous, pubescent, or pilose to lanate.
          31. Stems glabrous, sometimes ± puberulent; leaves glabrous, sometimes proximals puberulent, glaucous; pollen sacs 2.4–3.2 mm ....213. *Penstemon neotericus*
          31. Stems retrorsely hairy; leaves puberulent or retrorsely hairy, sometimes glabrate, not glaucous; pollen sacs 1.5–2.8 mm.
            32. Corollas 21–35 mm; pollen sacs 1.8–2.8 mm ....211. *Penstemon laetus*
            32. Corollas 15–25 mm; pollen sacs 1.5–2 mm...............221. *Penstemon roezlii*

201. Penstemon azureus Bentham, Pl. Hartw., 327. 1849 (as Pentstemon) • Azure beardtongue [E]

Subshrubs. Stems ascending to erect, 13–50(–70) cm, glabrous, glaucous. **Leaves** basal and cauline, basal sometimes few, opposite, glabrous, glaucous; basal and proximal cauline 15–60 × 2–10(–20) mm, blade obovate to oblanceolate or linear, base tapered, margins entire, apex obtuse to acute; cauline 3–20 pairs, sessile or proximals short-petiolate, 10–90 × 2–20 mm, blade elliptic to ovate, lanceolate, or linear, base tapered to clasping or cordate-clasping, margins entire, apex obtuse to acute. **Thyrses** interrupted or continuous, secund, (3–)6–31 cm, axis glabrous, verticillasters 3–10, cymes 1–3(or 4)-flowered, 2 per node; proximal bracts ovate to lanceolate, 4–50 × 1–32 mm; peduncles and pedicels ascending to erect, glabrous. **Flowers:** calyx lobes ovate to oblong or obovate, 2.5–6.5 × 1.4–3.2 mm, glabrous; corolla violet to blue or purple, with or without purple nectar guides, funnelform to ventricose, 18–35 mm, glabrous externally, glabrous internally, tube 7–9 mm, throat gradually inflated, 8–10 mm diam., slightly 2-ridged abaxially; stamens: longer pair reaching orifice or exserted, filaments glabrous, pollen sacs parallel, 1.8–3.5 mm, distal ⅓–½ indehiscent, sides hispidulous to pilose, hairs white, to 0.8 mm near filament attachment, sutures denticulate, teeth to 0.3 mm; staminode 12–19 mm, included, 0.4–0.6 mm diam., glabrous; style 16–22 mm. **Capsules** 7–11 × 14–6 mm.

Varieties 2 (2 in the flora): w United States.

Penstemon azureus occurs from southwestern Oregon to central California.

1. Calyx lobes obovate to oblong, 2.5–4.8 × 2–3.2 mm, apices cuspidate to mucronate; cauline leaves 10–65 × 4–20 mm, blades elliptic to ovate or lanceolate .... 201a. *Penstemon azureus* var. *azureus*
1. Calyx lobes ovate, 5–6.5 × 1.4–2.3 mm, apices acuminate to caudate; cauline leaves 40–90 × 2–6 mm, blades lanceolate to linear ....201b. *Penstemon azureus* var. *angustissimus*

201a. Penstemon azureus Bentham var. **azureus** [E]

Cauline leaves 10–65 × 4–20 mm, blade elliptic to ovate or lanceolate, base tapered to clasping or cordate-clasping. **Flowers:** calyx lobes obovate to oblong, 2.5–4.8 × 2–3.2 mm, apex cuspidate to mucronate. $2n = 48$.

Flowering May–Aug. River benches, mesas, and slopes in oak, pine, or juniper woodlands; 500–2500 m; Calif., Oreg.

Variety *azureus* is known from the Cascade, Klamath, and North Coast ranges, and in the Sierra Nevada from southwestern Oregon (Curry, Jackson, and Josephine counties) south to Fresno County, California.

201b. Penstemon azureus Bentham var. **angustissimus**
A. Gray in A. Gray et al., Syn. Fl. N. Amer. 2(1): 272.
1878 (as Pentstemon) E

Penstemon azureus subsp.
angustissimus (A. Gray)
D. D. Keck

Cauline leaves 40–90 × 2–6 mm,
blade lanceolate to linear, base
tapered to clasping. **Flowers:**
calyx lobes ovate, 5–6.5 ×
1.4–2.3 mm, apex acuminate
to caudate. $2n = 48$.

Flowering May–Aug. Moist woodlands, forests;
300–700 m; Calif.

Variety *angustissimus* occurs in northern and
central California.

202. Penstemon caesius A. Gray, Proc. Amer. Acad. Arts
19: 92. 1883 (as Pentstemon) • San Bernardino
beardtongue E

Herbs. Stems erect, 14–30
(–45) cm, glabrous, sometimes
glaucous. **Leaves** basal and
cauline, opposite, glabrous,
usually glaucous; basal and
proximal cauline (7–)15–50 ×
(4–)7–20 mm, blade ovate to
deltate, base tapered to
truncate,
margins entire, apex rounded to
obtuse; cauline 1 or 2(or 3) pairs, short-petiolate or
sessile, 12–57 × 3–20 mm, blade oblanceolate, base
tapered to truncate, margins entire, apex obtuse to
acute. **Thyrses** interrupted, cylindric, 6–17 cm, axis
glandular-pubescent, verticillasters 3–5, cymes 1–3-
flowered, 2 per node; proximal bracts oblanceolate to
lanceolate, 6–28 × 1–6 mm; peduncles and pedicels
ascending, glandular-pubescent. **Flowers:** calyx lobes
oblong to lanceolate, 3.5–5 × 1.5–2.1 mm, glandular-
pubescent; corolla light purple to lavender, violet, or
blue, without nectar guides, ventricose, 17–23 mm,
glandular-pubescent externally, glabrous internally, tube
4.5–6 mm, throat slightly inflated, 4.5–7 mm diam.,
2-ridged abaxially; stamens included, filaments glabrous,
pollen sacs parallel, 1.1–1.3 mm, distal ½ indehiscent,
sides glabrous, sutures denticulate, teeth to 0.1 mm;
staminode 10–13 mm, included, 0.2–0.3 mm diam.,
glabrous; style 11–13 mm. **Capsules** (6–)8–10 ×
4–5.5 mm.

Flowering Jun–Aug. Rock slopes, openings in pine
forests; 2000–3100 m; Calif.

Penstemon caesius is known from the High Sierra
Nevada and San Bernardino and San Gabriel mountains
in southern California. The species has been documented
in Fresno, Inyo, Kern, Los Angeles, San Bernardino, and
Tulare counties.

203. Penstemon cusickii A. Gray, Proc. Amer. Acad.
Arts 16: 106. 1880 (as Pentstemon) • Cusick's
beardtongue E

Herbs or subshrubs. Stems
ascending to erect, 15–45 cm,
retrorsely hairy, hairs pointed,
not glaucous. **Leaves** cauline,
opposite or subalternate,
retrorsely hairy, hairs pointed,
not glaucous; cauline 4–8 pairs,
short-petiolate or sessile, 16–70
× 1–8(–14) mm, blade lance-
olate to linear, base tapered, margins entire, apex obtuse
to acute. **Thyrses** interrupted, cylindric, 4–15 cm, axis
puberulent or retrorsely hairy, verticillasters 3–10,
cymes 1–5(–7)-flowered, 2 per node; proximal bracts
lanceolate to linear, 7–50 × 1–6 mm; peduncles and
pedicels ascending to erect, puberulent or retrorsely
hairy. **Flowers:** calyx lobes ovate, 3–5.5 × 1.4–2.3 mm,
glabrous or sparsely puberulent; corolla blue to violet
or purple, without nectar guides, ampliate, 16–21 mm,
glabrous externally, glabrous internally, tube 4–5 mm,
throat gradually inflated, 5–6 mm diam., slightly
2-ridged abaxially; stamens included, filaments glabrous,
pollen sacs parallel, 1–1.5(–1.8) mm, distal ½–⅔
indehiscent, sides hispidulous, hairs white, to 0.1 mm
near filament attachment, sutures denticulate, teeth to
0.1 mm; staminode 10–12 mm, reaching orifice or barely
exserted, 0.5–0.7 mm diam., glabrous; style 11–13 mm.
Capsules 6–7 × 4.5–6 mm. $2n = 16$.

Flowering May–Jul. Rocky, often basaltic,
sagebrush shrublands; 600–1500 m; Idaho, Oreg.

Penstemon cusickii is known from west-central and
southwestern Idaho (Ada, Blaine, Canyon, Elmore,
Gem, Owyhee, and Washington counties) and eastern
Oregon (Baker, Grant, Harney, Malheur, Union, and
Wheeler counties).

204. Penstemon diphyllus Rydberg, Mem. New York
Bot. Gard. 1: 349. 1900 (as Pentstemon) • Two-
leaf beardtongue E

Subshrubs. Stems ascending to
erect, (10–)20–80 cm, retrorsely
hairy, not glaucous. **Leaves**
cauline, opposite, sometimes
alternate, glabrous, not glaucous; cauline (4–)7–10 pairs,
sessile or proximals short-
petiolate, 20–80 × 5–17 mm,
blade elliptic to lanceolate or
oblanceolate, base tapered, margins irregularly dentate
or serrate, apex acute. **Thyrses** interrupted, cylindric,
3–25(–50) cm, axis glandular-pubescent, verticillasters
4–10, cymes 1–11(–18)-flowered, 1 or 2 per node;
proximal bracts lanceolate, rarely ovate, 9–45 × 2–16
mm; peduncles and pedicels spreading to ascending,

glandular-pubescent. **Flowers:** calyx lobes lanceolate, (2–)3–5.2 × 0.9–1.3 mm, glandular-pubescent; corolla bluish lavender to violet or purple, with reddish purple nectar guides, funnelform, 13–19 mm, glandular-pubescent externally, glabrous internally or sparsely white-lanate abaxially, tube 5–6 mm, throat gradually inflated, 5–6 mm diam., rounded abaxially; stamens included, filaments glabrous or sparsely white-pubescent distally, pollen sacs parallel, 0.9–1.3 mm, distal ½ indehiscent, sides glabrous, sutures denticulate, teeth to 0.1 mm; staminode 9–11 mm, reaching orifice or barely exserted, 0.4–0.5 mm diam., distal 1–4 mm sparsely to moderately lanate, hairs white or yellowish, to 1.5 mm; style 8–12 mm. **Capsules** 4–6 × 2.8–4.5 mm.

Flowering Jun–Aug(–Sep). Talus slopes, cliffs, rocky banks; 1000–2300 m; Idaho, Mont., Wash.

Penstemon diphyllus occurs in northeastern Idaho, western Montana, and southeastern Washington (Whitman County).

205. Penstemon filiformis (D. D. Keck) D. D. Keck in L. Abrams and R. S. Ferris, Ill. Fl. Pacific States 3: 761. 1951 • Thread-leaf beardtongue E

Penstemon laetus A. Gray subsp. *filiformis* D. D. Keck, Univ. Calif. Publ. Bot. 16: 394. 1932

Herbs or subshrubs. Stems ascending to erect, 16–50 cm, puberulent to retrorsely hairy, hairs pointed, not glaucous. **Leaves** cauline, opposite, puberulent, hairs pointed, or glabrous, not glaucous; cauline 8–12 pairs, sessile, 12–70 × 0.5–1(–2) mm, blade linear, base not tapered, margins entire, apex acute. **Thyrses** interrupted, cylindric, 9–22 cm, axis glandular-pubescent, verticillasters 1–5, cymes 1–3-flowered, (1 or)2 per node; proximal bracts linear, 9–30 × 0.5–1 mm; peduncles and pedicels spreading to ascending, glandular-pubescent. **Flowers:** calyx lobes ovate to lanceolate, 2–4.5 × 0.7–1.7 mm, glandular-pubescent; corolla light purple to violet or purple, without nectar guides, funnelform, 13–18 mm, glandular-pubescent externally, glabrous internally, tube 3–4 mm, throat gradually inflated, 5–7 mm diam., 2-ridged abaxially; stamens included or longer pair reaching orifice, filaments glabrous, pollen sacs parallel, 1.2–1.6 mm, distal ½ indehiscent, sides pubescent, hairs white, to 0.5 mm near filament attachment, sutures denticulate, teeth to 0.2 mm; staminode 10–12 mm, included or reaching orifice, 0.3–0.5 mm diam., glabrous; style 11–15 mm. **Capsules** 5–7.5 × 2.5–4 mm. $2n = 16$.

Flowering Jun–Jul. Rocky shrublands, pine woodlands; 400–600 m; Calif.

Penstemon filiformis is known from Lake, Shasta, Siskiyou, and Trinity counties.

206. Penstemon floribundus Danley, Brittonia 37: 321, figs. 1, 2. 1985 • Cordelia beardtongue C E

Herbs. Stems ascending to erect, (5–)8–23 cm, glabrous, not glaucous. **Leaves** cauline, opposite, glabrous abaxially, glabrate or retrorsely hairy adaxially, hairs appressed, white, scalelike, not glaucous; cauline 2–6 pairs, sessile or proximals short-petiolate, 13–70 × (3–)7–17 mm, blade elliptic to lanceolate, base tapered, margins entire, apex obtuse to acute. **Thyrses** continuous, cylindric, (1–)2–12 cm, axis glabrous, verticillasters (2 or)3–7, cymes 3–5-flowered, 2 per node; proximal bracts lanceolate, 14–45 × 2–12 mm; peduncles and pedicels ascending to erect, glabrous or pedicels ± puberulent, hairs appressed, white, scalelike. **Flowers:** calyx lobes ovate, (4–)5–6.3 × 1.5–2.6 mm, glabrous or ± scabrous proximally; corolla violet to light blue, without nectar guides, abaxial ridges usually white, ventricose, 24–32 mm, glabrous externally, glabrous internally, tube 7–10 mm, throat gradually inflated, 8–13 mm diam., 2-ridged abaxially; stamens: longer pair reaching orifice or exserted, filaments glabrous, pollen sacs parallel, 2.4–2.8 mm, distal ½–⅔ indehiscent, sides glabrous, sutures denticulate, teeth to 0.1 mm; staminode 15–18 mm, included, 0.4–0.7 mm diam., glabrous; style 18–27 mm. **Capsules** 7–9 × 5–6 mm.

Flowering May–Jun. Rocky alluvium, talus slopes, desert scrub, juniper woodlands; of conservation concern; 1300–2300 m; Nev.

Penstemon floribundus is known only from the Jackson Mountains in Humboldt County.

207. Penstemon glandulosus Douglas ex Lindley, Edwards's Bot. Reg. 15: plate 1262. 1829 (as *Pentstemon glandulosum*) • Sticky-stem beardtongue E F

Herbs. Stems ascending to erect, (30–)40–100 cm, viscid glandular-pubescent, not glaucous. **Leaves** basal and cauline, opposite, viscid glandular-pubescent, not glaucous; basal and proximal cauline 25–120 (–260) × 7–20(–65) mm, blade elliptic to lanceolate, base tapered, margins entire, subentire, or irregularly serrate to dentate, apex acute to acuminate; cauline 4–10 pairs, sessile, 32–80(–120) × 12–45 mm, blade ovate to obovate, elliptic, or lanceolate, base tapered to cordate-clasping, margins entire or subentire to irregularly serrate to dentate, apex obtuse to acute or acuminate. **Thyrses** interrupted, cylindric, (5–)8–31 cm, axis viscid glandular-pubescent, verticillasters 3–7, cymes (1 or) 2–8-flowered, 2 per node; proximal bracts ovate, 18–60 × 11–44 mm; peduncles and pedicels ascending to erect, viscid glandular-pubescent. **Flowers:** calyx lobes lanceolate, 9–15 × 2.3–3.8 mm, viscid glandular-pubescent; corolla lavender to bluish lavender or purple, with dark violet nectar guides, ampliate, (28–)30–40 mm, glandular-pubescent externally, sparsely to moderately white-pubescent internally abaxially, tube 9–11 mm, throat gradually inflated, 12–15 mm diam., rounded abaxially; stamens included, filaments glabrous, pollen sacs parallel, 1.6–2.2 mm, distal ½ indehiscent, sides glabrous, sutures denticulate, teeth to 0.2 mm; staminode 22–26 mm, exserted, 0.8–1 mm diam., glabrous, rarely distal 2–4 mm sparsely pubescent, hairs yellow, to 0.4 mm; style 20–26 mm. **Capsules** 9–13 × 5–6.5 mm.

Varieties 2 (2 in the flora): nw United States.

Two allopatric varieties of *Penstemon glandulosus* have been distinguished based on basal leaf morphology. Leaf margins occasionally vary from entire to serrate or dentate among individuals in some populations, but the general pattern of leaf variation does appear to be geographically correlated.

1. Basal leaves: blade margins irregularly serrate to dentate . . .207a. *Penstemon glandulosus* var. *glandulosus*
1. Basal leaves: blade margins entire or subentire . . . . 207b. *Penstemon glandulosus* var. *chelanensis*

207a. Penstemon glandulosus Douglas ex Lindley var. **glandulosus** E F

Basal leaves: blade margins irregularly serrate to dentate.

Flowering May–Jul. Rocky sagebrush slopes and hillsides, cliffs; 200–2000 m; Idaho, Oreg., Wash.

Variety *glandulosus* is known from the Snake River Plateau of west-central Idaho, northeast Oregon, and southeast Washington.

207b. Penstemon glandulosus Douglas ex Lindley var. **chelanensis** (D. D. Keck) Cronquist in C. L. Hitchcock et al., Vasc. Pl. Pacif. N.W. 4: 388. 1959 E

Penstemon glandulosus subsp. *chelanensis* D. D. Keck in L. Abrams and R. S. Ferris, Ill. Fl. Pacific States 3: 764. 1951

Basal leaves: blade margins entire or subentire.

Flowering Apr–Jul. Rocky sagebrush slopes, hillsides, and banks; 200–2000 m; Oreg., Wash.

Variety *chelanensis* is known from the eastern slope of the Cascade Range in Benton, Chelan, Kittitas, Klickitat, and Yakima counties, Washington, and Hood River and Wasco counties, Oregon.

208. Penstemon gracilentus A. Gray in War Department [U.S.], Pacif. Railr. Rep. 6(3): 82. 1858 (as Pentstemon) • Slender beardtongue E

Subshrubs. Stems ascending to erect, 20–65 cm, glabrous or retrorsely hairy, hairs ± pointed, sometimes glaucous. **Leaves** essentially cauline, basal absent or essentially so, opposite, glabrous, sometimes glaucous distally or abaxially; cauline (3–)6–10 pairs, petiolate or sessile, (20–)40–80(–105) × (2–)4–15 mm, blade spatulate to oblanceolate or lanceolate, base tapered, margins entire, apex rounded to obtuse or acute. **Thyrses** interrupted, cylindric, (3–)7–28 cm, axis glandular-pubescent, usually sparsely so proximally, verticillasters 3–6(–9), cymes 2–4(–7)-flowered, 2 per node; proximal bracts lanceolate to linear, 14–75 × 1–6 mm; peduncles and pedicels ascending to erect, glandular-pubescent. **Flowers:** calyx lobes oblong to lanceolate, 3.5–5.5 × 0.8–1.4 mm, glandular-pubescent; corolla blue to violet or purple, with or without dark blue nectar guides, funnelform, 15–20(–24) mm, glandular-pubescent externally, glabrous internally or sparsely to densely white- or yellow-pilose abaxially, tube 6–8 mm, throat gradually inflated, 4–5 mm diam., 2-ridged abaxially; stamens included or longer pair reaching orifice, filaments glabrous, pollen sacs parallel, 1–1.3 mm, distal ½ indehiscent, sides glabrous, sutures denticulate, teeth to 0.1 mm; staminode 8–10 mm, included, 0.3–0.4 mm diam., distal 3–5 mm sparsely to moderately pubescent, hairs yellow, to 0.5 mm; style 12–15 mm. **Capsules** 5–7 × 3.5–4 mm. **2n** = 16.

Flowering Jun–Aug. Lava and granitic sand or gravel, sagebrush shrublands, juniper woodlands, pine-fir forests; 1000–3000 m; Calif., Nev., Oreg.

Penstemon gracilentus occurs from south-central Oregon (Lake County) through northeastern California and northwestern Nevada to near Lake Tahoe. D. D. Keck (1932) hypothesized that *P. gracilentus* was closely related to species of sect. *Penstemon*; his hypothesis is supported by molecular data (A. D. Wolfe et al. 2006).

209. Penstemon heterophyllus Lindley, Edwards's Bot. Reg. 22: plate 1899. 1836 (as Pentstemon heterophyllum) • Foothill beardtongue [E]

Subshrubs. Stems ascending to erect, 30–150 cm, glabrous or retrorsely hairy, hairs pointed, rarely glaucous. **Leaves** essentially cauline, basal absent or essentially so, opposite, glabrous or retrorsely hairy to puberulent, hairs pointed, rarely glaucous; cauline 5–11 pairs, petiolate or sessile, 10–100 × 0.5–12 mm, blade oblanceolate to lanceolate or linear, base tapered to clasping, margins entire, apex acute. **Thyrses** interrupted or continuous, cylindric, (6–)10–32 cm, axis glabrous or retrorsely hairy, verticillasters (3–)5–14, cymes 1(–3)-flowered, (1 or)2 per node; proximal bracts lanceolate to linear, 5–65 × 0.5–5 mm; peduncles and pedicels ascending to erect, glabrous or retrorsely hairy. **Flowers:** calyx lobes obovate to oblanceolate or ovate to lanceolate, 2.6–6(–7) × 1.3–3.3 mm, glabrous or puberulent; corolla blue to violet or purple, without nectar guides, ventricose to ventricose-ampliate, 24–38 mm, glabrous externally, glabrous internally, tube 7–10 mm, throat gradually inflated, 8–11 mm diam., 2-ridged abaxially; stamens included or longer pair reaching orifice or slightly exserted, filaments glabrous, pollen sacs parallel, 2.2–3 mm, distal ⅓–½ indehiscent, sides pubescent to lanate, hairs white, to 0.8 mm near filament attachment, sutures denticulate, teeth to 0.2 mm; staminode 16–19 mm, included, 0.4–0.6 mm diam., glabrous; style 18–27 mm. **Capsules** 6–9 × 4–5 mm. *2n* = 16, 32.

Varieties 3 (3 in the flora): California.

Penstemon heterophyllus apparently occurs only in California (see discussion under 209c. var. *purdyi*). The species includes three varieties concentrated in the Coast Ranges from Humboldt County south to San Bernardino and San Diego counties.

1. Thyrse axes glabrous; stems glabrous, rarely retrorsely hairy; calyx lobes obovate to oblanceolate, rarely lanceolate . . . .
. . . . . 209a. *Penstemon heterophyllus* var. *heterophyllus*
1. Thyrse axes retrorsely hairy; stems retrorsely hairy; calyx lobes ovate to lanceolate.
  2. Cauline leaves 15–45 × 0.5–4 mm, axillary fascicles present, sometimes absent . . . .
. . . . . . 209b. *Penstemon heterophyllus* var. *australis*
  2. Cauline leaves (25–)30–100 × 2–12 mm, axillary fascicles absent, rarely present. . . . . .
. . . . .209c. *Penstemon heterophyllus* var. *purdyi*

209a. Penstemon heterophyllus Lindley var. **heterophyllus** [E]

Stems glabrous, rarely retrorsely hairy. **Leaves:** axillary fascicles present, sometimes absent; cauline 10–55 × 0.5–6 mm. **Thyrses:** axis glabrous. **Flowers:** calyx lobes obovate to oblanceolate, rarely lanceolate.

Flowering Apr–Jul. Grasslands, chaparral, sagebrush shrublands, pine woodlands; 50–1600 m; Calif.

Variety *heterophyllus* occurs through much of western California from Humboldt County to San Diego County.

209b. Penstemon heterophyllus Lindley var. **australis** Munz & I. M. Johnston, Bull. S. Calif. Acad. Sci. 23: 40. 1924 [E]

Penstemon heterophyllus subsp. *australis* (Munz & I. M. Johnston) D. D. Keck

Stems retrorsely hairy. **Leaves:** axillary fascicles present, sometimes absent; cauline 15–45 × 0.5–4 mm. **Thyrses:** axis retrorsely hairy. **Flowers:** calyx lobes ovate to lanceolate.

Flowering Apr–Jul. Grasslands, chaparral, forest openings; 50–1700 m; Calif.

Variety *australis* occurs in west-central and southwestern California from Colusa County to San Diego County.

209c. Penstemon heterophyllus Lindley var. **purdyi**
(D. D. Keck) McMinn, Man. Calif. Shrubs, 518.
1939 E

Penstemon heterophyllus subsp.
purdyi D. D. Keck, Univ. Calif.
Publ. Bot. 16: 409, fig. 15. 1932;
P. heterophyllus subsp. *spinulosus*
(Wooton & Standley) D. D. Keck

Stems retrorsely hairy. **Leaves:**
axillary fascicles absent, rarely
present; cauline (25–)30–100 ×
2–12 mm. **Thyrses:** axis
retrorsely hairy. **Flowers:** calyx lobes ovate to
lanceolate.

Flowering Apr–Jul. Grasslands, chaparral, pine
forest openings; 50–1600 m; Calif.

Variety *purdyi* occurs primarily in northern
California from Humboldt, Lassen, and Trinity
counties to San Benito County.

The status of subsp. *spinulosus* has been problematic.
Known only from the type collection (*G. R. Vasey*,
June 1881, US) reputedly from the Santa Magdalena
Mountains of New Mexico, the subspecies is disjunct
600 km east of the nearest California populations of
Penstemon heterophyllus. F. W. Pennell examined
the type in 1941 and provided evidence that the
specimen was actually collected by Vasey in 1880 in
Marin County, California, and D. D. Keck (1932)
concluded that the only character distinguishing subsp.
spinulosus from subsp. *purdyi* (= var. *purdyi*) is the
length of the teeth along the anther sutures. This
treatment concurs with the conclusion by Pennell that
subsp. *spinulosus* is based on a mislabeled specimen and
synonymous with var. *purdyi*.

210. Penstemon kingii S. Watson, Botany (Fortieth
Parallel), 222. 1871 (as Pentstemon) • King's
beardtongue E

Herbs. Stems decumbent to
ascending, 7–25(–30) cm,
retrorsely hairy, hairs pointed,
often densely so, not glaucous.
Leaves cauline, opposite,
retrorsely hairy or puberulent,
hairs pointed, not glaucous;
cauline (2 or)3–6 pairs, short-
petiolate or sessile, 17–50(–80)
× 2–7(–10) mm, blade oblanceolate to linear, base
tapered, margins entire, apex obtuse to acute. **Thyrses**
continuous, cylindric, 3–15 cm, axis retrorsely hairy or
puberulent, verticillasters 4–10, cymes (1 or)2–4-
flowered, 2 per node; proximal bracts oblanceolate to
lanceolate, 11–40 × 2–8 mm; peduncles and pedicels
ascending to erect, peduncles puberulent, pedicels
puberulent and glandular-pubescent, at least distally.

Flowers: calyx lobes lanceolate, 4–6 × 1.4–1.9 mm,
glandular-pubescent; corolla violet to purple, reddish
purple, or blue, without nectar guides, ventricose,
15–22 mm, glandular-pubescent externally, glabrous
internally, tube 5–7 mm, throat gradually inflated,
4.5–7 mm diam., slightly 2-ridged abaxially; stamens
included or longer pair reaching orifice, filaments
glabrous, pollen sacs parallel, 1–1.5(–1.8) mm, distal
½–⅔ indehiscent, sides glabrous, sutures denticulate,
teeth to 0.1 mm; staminode 11–13 mm, reaching orifice
or barely exserted, 0.4–0.5 mm diam., glabrous; style
12–16 mm. **Capsules** 6–10 × 4–6 mm.

Flowering May–Jul. Sagebrush-juniper woodlands;
1500–2100 m; Nev., Oreg.

Penstemon kingii is known from mountain
ranges throughout central and northwestern Nevada
(Churchill, Elko, Esmeralda, Eureka, Humboldt,
Lander, Nye, and Pershing counties) and southeastern
Oregon (Malheur County). N. H. Holmgren (1984)
reported a chromosome number of $2n = 32$ for *P. kingii*;
the basis for that report has not been determined.

211. Penstemon laetus A. Gray, Proc. Boston Soc. Nat.
Hist. 7: 147. 1859 (as Pentstemon) • Western gray
beardtongue E

Herbs or subshrubs. Stems
decumbent to ascending or
erect, 15–90 cm, retrorsely
hairy, hairs pointed, not glaucous. **Leaves** cauline, opposite,
puberulent or retrorsely hairy,
hairs pointed, sometimes
glabrate, not glaucous; cauline
4–8 pairs, sessile or proximals
short-petiolate, 13–85(–110) × 2–22 mm, blade
oblanceolate to lanceolate or linear, base tapered to
truncate, margins entire, apex acute. **Thyrses** interrupted, cylindric, 10–30(–43) cm, axis glandular-
pubescent, verticillasters 6–12, cymes 1–5-flowered,
2 per node; proximal bracts lanceolate, rarely linear,
9–40 × 2–10 mm; peduncles and pedicels spreading to
ascending, glandular-pubescent. **Flowers:** calyx lobes
ovate to lanceolate, 4–11 × 1–2.8 mm, glandular-
pubescent; corolla violet to blue or purple, with or
without violet nectar guides, funnelform to ampliate,
21–35 mm, glandular-pubescent externally, glabrous
internally, tube 6–9 mm, throat gradually inflated,
8–10 mm diam., 2-ridged abaxially; stamens included
or longer pair reaching orifice, filaments glabrous,
pollen sacs parallel, 1.8–2.8 mm, distal ⅕–⅓
indehiscent, sides pubescent to lanate, hairs white, to
1 mm near filament attachment, sutures denticulate,
teeth to 0.3 mm; staminode 17–19 mm, included or
reaching orifice, 0.5–0.6 mm diam., glabrous; style
16–22 mm. **Capsules** 7–9 × 4–6 mm.

Varieties 3 (3 in the flora): w United States.

Penstemon laetus occurs from the Cascade Range of south-central Oregon south through the Sierra Nevada of California to the Western Transverse Ranges of southwestern California. The species comprises three partially sympatric varieties.

The Karuk tribe of northwestern California use an infusion of *Penstemon laetus* as a psychological aid (D. E. Moerman 1998).

1. Calyx lobes lanceolate, 7–11 × 1–2 mm . . . . . .
    . . . . 211b. *Penstemon laetus* var. *leptosepalus*
1. Calyx lobes ovate to lanceolate, 4–8.5 × 1–2.8 mm.
    2. Anthers ovate to elliptic in outline, pollen sacs 1.8–2.3 mm . . . . . 211a. *Penstemon laetus* var. *laetus*
    2. Anthers sagittate in outline, pollen sacs 2.1–2.8 mm . . . . . .211c. *Penstemon laetus* var. *sagittatus*

211a. Penstemon laetus A. Gray var. **laetus** [E]

Flowers: calyx lobes ovate to lanceolate, 4–8.5 × (1–)2–2.8 mm; anthers ovate to elliptic in outline, pollen sacs 1.8–2.3 mm. $2n = 16$.
Flowering (Mar–)Apr–Jul. Oak and pine-oak woodlands, pine forests; 400–2500 m; Calif.

Variety *laetus* occurs from the southern Cascade Range of northern California through the Sierra Nevada to the Western Transverse Ranges of southwestern California. As discussed by D. D. Keck (1932), corolla lengths and leaf sizes usually are larger among plants in the southern part of the range as compared to plants in the north.

211b. Penstemon laetus A. Gray var. **leptosepalus**
Greene ex A. Gray in A. Gray et al., Syn. Fl. N. Amer. ed. 2, 2(1): 442. 1886 (as Pentstemon) [E]

Penstemon laetus subsp. *leptosepalus* (Greene ex A. Gray) D. D. Keck

Flowers: calyx lobes lanceolate, 7–11 × 1–2 mm; anthers ovate to elliptic in outline, pollen sacs 2–2.7 mm. $2n = 16$.
Flowering May–Jul. Pine forests, pine-oak woodlands; 400–1700 m; Calif.

Variety *leptosepalus* occurs in the northern Sierra Nevada and the southern Cascade Range.

211c. Penstemon laetus A. Gray var. **sagittatus**
(D. D. Keck) McMinn, Man. Calif. Shrubs, 518. 1939 [E]

Penstemon laetus subsp. *sagittatus* D. D. Keck, Univ. Calif. Publ. Bot. 16: 395, fig. 12. 1932

Flowers: calyx lobes ovate to lanceolate, 4–6 × 1–2.3 mm; anthers sagittate in outline, pollen sacs 2.1–2.8 mm.
Flowering May–Aug(–Sep). Sagebrush shrublands, coniferous forests; 500–2300 m; Calif., Oreg.

Variety *sagittatus* occurs in the Cascade Range of northern California and south-central Oregon. In addition to its sagittate anthers, it usually has glabrate leaves, though this character varies among populations.

212. Penstemon leonardii Rydberg, Bull. Torrey Bot. Club 40: 483. 1913 (as Pentstemon leonardi)

• Leonard's beardtongue [E]

Herbs or subshrubs. Stems decumbent to ascending or erect, (5–)10–30(–47) cm, puberulent or retrorsely hairy, hairs pointed, not glaucous. **Leaves** cauline, opposite, glabrous, sometimes proximals puberulent, hairs pointed, not glaucous; cauline 2–7 pairs, short-petiolate or sessile, 11–48(–60) × 2–12 mm, blade oblanceolate to spatulate or obovate, rarely linear, base tapered, margins entire, apex rounded to obtuse or acute. **Thyrses** continuous or interrupted, cylindric to ± secund, (1–)2–15(–25) cm, axis glabrous or retrorsely hairy, verticillasters 2–8, cymes 1–4-flowered, 2 per node; proximal bracts lanceolate, 10–35 × 1–4 mm; peduncles and pedicels ascending to erect, peduncles glabrous or sparsely retrorsely hairy, pedicels glabrous or papillate distally. **Flowers:** calyx lobes ovate to lanceolate, 3–5.5 × 0.9–2.2 mm, margins entire or erose, scarious, apex acuminate to caudate, glabrous; corolla lavender to blue or violet, with or without faint lavender nectar guides, funnelform to ventricose, 14–26 mm, glabrous externally, glabrous internally, tube 5–8 mm, throat gradually inflated, 6–8(–10) mm diam., slightly 2-ridged abaxially; stamens included or longer pair reaching orifice, filaments glabrous, pollen sacs parallel, 0.9–1.9 mm, distal ½–⅔ indehiscent, sides glabrous or puberulent, hairs white, to 0.1 mm, sutures denticulate, teeth to 0.2 mm; staminode 10–15 mm, included or reaching orifice, 0.6–0.9 mm diam., glabrous; style 12–18 mm. **Capsules** 6–8 × 2.5–4.5 mm.

Varieties 3 (3 in the flora): w United States.

Penstemon leonardii is common in the Wasatch Mountains, Utah Plateaus, and mountain ranges in the eastern Basin and Range Province. Elements allied with it have been interpreted variously as one species comprising three varieties, two species (one with two varieties), or three species; the broadest concept is followed here.

1. Corollas 20–26 mm; pollen sacs (1.2–)1.4–1.9 mm; e Nevada, w Utah . . . .
    . . . . 212c. *Penstemon leonardii* var. *patricus*
1. Corollas 14–22 mm; pollen sacs 0.9–1.6 mm; nw Arizona, se Idaho, se Nevada, Utah.
    2. Corolla limbs blue to lavender; thyrses continuous; se Idaho, n Utah . . . .
        . . . . 212a. *Penstemon leonardii* var. *leonardii*
    2. Corolla limbs lavender to violet; thyrses continuous or interrupted; nw Arizona, se Nevada, sw Utah . . . .
        . . . . .212b. *Penstemon leonardii* var. *higginsii*

212a. Penstemon leonardii Rydberg var. **leonardii** [E]

Thyrses continuous. **Flowers:** corolla 14–22 mm, limb blue to lavender; pollen sacs 0.9–1.5 mm. $2n = 16$.

Flowering Jun–Jul. Rocky sites in pine-juniper and aspen-pine woodlands, shrublands; 1800–3100 m; Idaho, Utah.

Variety *leonardii* has been documented in Bear Lake County, Idaho, and Cache, Davis, Juab, Millard, Morgan, Rich, Salt Lake, Sanpete, Summit, Utah, Wasatch, and Weber counties, Utah.

212b. Penstemon leonardii Rydberg var. **higginsii** Neese, Great Basin Naturalist 46: 459. 1986 • Higgins's beardtongue [E]

Penstemon higginsii (Neese) N. H. Holmgren & N. D. Atwood

Thyrses continuous or interrupted. **Flowers:** corolla 14–18 mm, limb lavender to violet; pollen sacs 1.3–1.6 mm.

Flowering (May–)Jun–Jul. Pine-juniper, oak-maple, conifer, and aspen-conifer woodlands; 1700–3100 m; Ariz., Nev., Utah.

Variety *higginsii* is known from the Black Rock Mountains (Mohave County) of Arizona, the Clover Mountains (Lincoln County) and Snake Range (White Pine County) of Nevada, and the Bull Valley and Pine Valley mountains and Kolob Plateau (Iron and Washington counties) of Utah.

212c. Penstemon leonardii Rydberg var. **patricus** (N. H. Holmgren) Neese, Great Basin Naturalist 46: 460. 1986 • Dad's beardtongue [E]

Penstemon patricus N. H. Holmgren, Brittonia 31: 238, fig. 11. 1979

Thyrses continuous or interrupted. **Flowers:** corolla 20–26 mm, limb blue to violet or purple; pollen sacs (1.2–) 1.4–1.9 mm.

Flowering Jun–Jul. Rocky sagebrush-grasslands, sagebrush and mixed shrubland communities, pine and aspen-pine woodlands; 2100–2700 m; Nev., Utah.

Variety *patricus* is known from the Kern Mountains in White Pine County, Nevada, and the Deep Creek and House ranges in Juab, Millard, and Tooele counties, Utah. Flower features tend to be slightly larger in var. *patricus* than in var. *leonardii* and var. *higginsii*.

213. Penstemon neotericus D. D. Keck, Univ. Calif. Publ. Bot. 16: 398, fig. 13. 1932 • Plumas County beardtongue [E]

Subshrubs. **Stems** ascending to erect, 20–80 cm, glabrous, sometimes ± puberulent, hairs pointed, glaucous. **Leaves** basal and cauline, basal sometimes few, opposite, glabrous, sometimes proximals puberulent, hairs pointed, glaucous; basal and proximal cauline 15–80 × 2–10 mm, blade spatulate to oblanceolate, base tapered, margins entire, apex rounded to obtuse or acute; cauline 4–6 pairs, sessile, 20–80 × 3–9 mm, blade ovate to lanceolate, base tapered, margins entire, apex acute. **Thyrses** interrupted, cylindric, 5–25 cm, axis glandular-pubescent, verticillasters 4–6, cymes 1–3-flowered, 2 per node; proximal bracts ovate to lanceolate, 4–50 × 1–8 mm; peduncles and pedicels ascending, glandular-pubescent. **Flowers:** calyx lobes ovate to lanceolate, 3.8–6 × 1.5–2.2 mm, glandular-pubescent; corolla violet to blue or purple, without nectar guides, ventricose to ventricose-ampliate, 25–38 mm, glandular-pubescent externally, glabrous internally, tube 7–9 mm, throat gradually inflated, 8–10 mm diam., 2-ridged abaxially; stamens included, filaments glabrous, pollen sacs parallel, 2.4–3.2 mm, distal 1/5–1/3 indehiscent, sides pilose to lanate, hairs white, to 0.6 mm near filament attachment, sutures denticulate, teeth to 0.2 mm; staminode 11–15 mm, included, 0.4–0.6 mm diam., glabrous; style 15–20 mm. **Capsules** 7–9 × 4–5 mm. $2n = 64$.

Flowering May–Aug(–Oct). Volcanic soils, scrub, open pine and fir forests; 1000–2200 m; Calif.

Penstemon neotericus is known from the Cascade Range and northern Sierra Nevada of northern California (Butte, Lassen, Placer, Plumas, Shasta, Sierra, Siskiyou, Tehama, and Yuba counties). D. D. Keck (1932) used morphologic and geographic data to hypothesize that *P. neotericus*, an octoploid, arose through hybridization between *P. azureus* and *P. laetus*. J. Clausen (1933) provided cytological evidence supporting that hypothesis. A diploid chromosome number of $2n = 32$ listed by N. H. Holmgren (1993) is probably a transcription error.

SELECTED REFERENCE Clausen, J. 1933. Cytological evidence for the hybrid origin of *Penstemon neotericus* Keck. Hereditas (Lund) 18: 65–76.

214. **Penstemon papillatus** J. T. Howell, Leafl. W. Bot. 2: 119. 1938 • Inyo beardtongue E

Herbs. Stems ascending to erect, 11–48 cm, retrorsely hairy or puberulent, hairs pointed, not glaucous. **Leaves** basal and cauline, opposite, puberulent, hairs pointed, sometimes glabrate abaxially, not glaucous; basal and proximal cauline (12–)25–100 × 6–15(–25) mm, blade spatulate to oblanceolate, base tapered, margins entire, apex rounded to obtuse or acute; cauline 2–4 pairs, sessile or proximals short-petiolate, 24–70 × 7–20 mm, blade elliptic or oblanceolate to lanceolate or oblong, base tapered to cordate-clasping, margins entire, apex obtuse to acute. **Thyrses** interrupted, cylindric, 8–20 cm, axis glandular-pubescent, verticillasters 3–7, cymes 1–4-flowered, 2 per node; proximal bracts lanceolate, 13–45 × 4–12 mm; peduncles and pedicels erect, glandular-pubescent. **Flowers:** calyx lobes lanceolate, 8–10 × 1.4–2.6 mm, glandular-pubescent; corolla blue to lavender or violet, without nectar guides, ampliate, 24–30 mm, glandular-pubescent externally, glabrous internally, tube 8–10 mm, throat gradually inflated, 6–8 mm diam., 2-ridged abaxially; stamens included, filaments glabrous, pollen sacs parallel, 1.6–1.9 mm, distal 1/3 indehiscent, sides glabrous, sutures denticulate, teeth to 0.1 mm; staminode 13–15 mm, included, 0.4–0.5 mm diam., distal 5–8 mm densely pilose, hairs yellow, to 1.3 mm; style 22–25 mm. **Capsules** 7–11 × 5–6 mm.

Flowering Jun–Aug. Rocky openings, pine-juniper woodlands, pine forests; 2000–2900 m; Calif.

Penstemon papillatus is limited to the High Sierra Nevada and Inyo and White mountains in Inyo and Mono counties.

215. **Penstemon parvulus** (A. Gray) Krautter, Contr. Bot. Lab. Morris Arbor. Univ. Pennsylvania 3: 193. 1908 (as Pentstemon) E

Penstemon azureus Bentham var. *parvulus* A. Gray in A. Gray et al., Syn. Fl. N. Amer. 2(1): 272. 1878 (as Pentstemon); *P. azureus* subsp. *parvulus* (A. Gray) D. D. Keck

Herbs or subshrubs. Stems ascending to erect, (10–)15–40 cm, glabrous, glaucous. **Leaves** cauline, opposite, glabrous, sometimes proximals retrorsely hairy, hairs pointed, usually glaucous; cauline 3–8 pairs, short-petiolate or sessile, (6–)10–50 × (2–)4–9 mm, blade oblanceolate to oblong or lanceolate, base tapered to truncate or cordate, margins entire, apex obtuse to acute. **Thyrses** continuous, cylindric, 4–11 cm, axis glabrous, verticillasters 4 or 5, cymes 2–4-flowered, 2 per node; proximal bracts ovate to oblong or lanceolate, 4–34 × 2–8 mm; peduncles and pedicels ascending to erect, glabrous. **Flowers:** calyx lobes obovate to ovate or lanceolate, 3–5 × 1.5–2.5 mm, glabrous; corolla blue to violet or purple, without nectar guides, ventricose, 14–20 mm, glabrous externally, glabrous internally, tube 4–5 mm, throat gradually inflated, 5–7 mm diam., 2-ridged abaxially; stamens included, filaments glabrous, pollen sacs parallel, 1.4–1.8 mm, distal 1/3–1/2 indehiscent, sides hispidulous, hairs white, to 0.3 mm near filament attachment, sometimes sparsely so, sutures denticulate, teeth to 0.2 mm; staminode 10–12 mm, included, 0.4–0.6 mm diam., glabrous or distal 0.5 mm sparsely pubescent, hairs whitish, to 0.2 mm; style 11–14 mm. **Capsules** 5–8 × 3.5–5 mm. $2n = 32$.

Flowering Jun–Aug. Rocky, open foothills and montane forests; 500–2500(–3000) m; Calif., Oreg.

Penstemon parvulus occurs in the Klamath Ranges of northern California and southwestern Oregon and in the southern High Sierra Nevada of south-central California.

216. **Penstemon platyphyllus** Rydberg, Bull. Torrey Bot. Club 36: 690. 1909 (as Pentstemon) • Broad-leaf beardtongue E F

Penstemon heterophyllus Lindley var. *latifolius* S. Watson, Botany (Fortieth Parallel), 222. 1871 (as Pentstemon), not *P. latifolius* Hoffmannsegg 1824

Herbs or subshrubs. Stems ascending to erect, (15–)35–60 cm, retrorsely hairy, hairs pointed, at least proximally, not glaucous. **Leaves** cauline, opposite, glabrous, glaucous; cauline 4–8 pairs, short-petiolate or sessile,

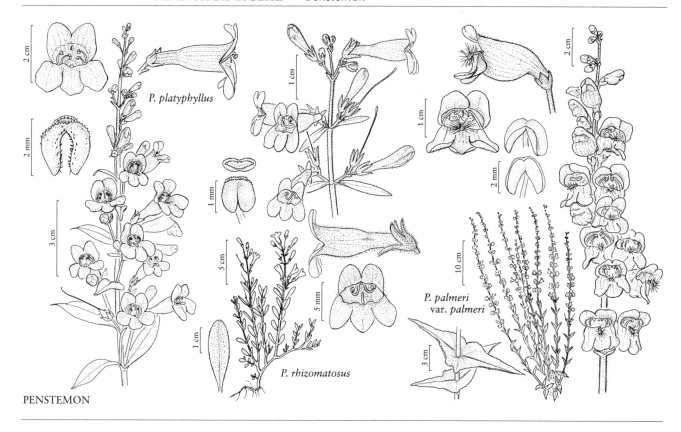

P. platyphyllus

P. rhizomatosus

P. palmeri
var. *palmeri*

PENSTEMON

(15–)20–60 × (2–)6–13(–20) mm, blade elliptic or proximals oblanceolate to lanceolate, base tapered, margins entire, apex obtuse to acute. **Thyrses** continuous, cylindric, 5–22 cm, axis glabrous or retrorsely hairy, verticillasters 4–8(–11), cymes 1- or 2(–4)-flowered, 2 per node; proximal bracts lanceolate to elliptic, 8–66 × 1–13 mm; peduncles and pedicels ascending to erect, peduncles glabrous or sparsely retrorsely hairy, pedicels papillate or glandular distally. **Flowers:** calyx lobes lanceolate, 4–6(–7) × 1.4–2.6 mm, glabrous; corolla lavender to light violet or violet, with or without faint lavender nectar guides, ventricose-ampliate, (20–)22–25(–30) mm, glabrous externally, glabrous internally, tube 6–8 mm, throat gradually inflated, 7–9 mm diam., 2-ridged abaxially; stamens: longer pair exserted, filaments glabrous, pollen sacs parallel, 1.5–1.9(–2.1) mm, distal ½–⅔ indehiscent, sides glabrous or hispidulous to lanate, hairs white, to 0.5 mm, sutures denticulate, teeth to 0.2 mm; staminode 12–14 mm, reaching orifice or exserted, 0.6–0.8 mm diam., glabrous; style 18–20 mm. **Capsules** 7–11 × 3.5–4.5 mm. $2n = 32$.

Flowering Jun–Aug. Talus slopes, rocky canyons, foothills; 1400–2400 m; Utah.

Penstemon platyphyllus is known from the Wasatch Mountains in Davis, Salt Lake, Utah, and Weber counties, with one somewhat isolated population in Indian Canyon in extreme southwestern Duchesne County. Reports of *P. platyphyllus* from White Pine County, Nevada, are based on a specimen cited by D. D. Keck (1932) that is referable to *P. leonardii* var. *patricus*, as are most collections cited by Keck from western Juab and southwestern Tooele counties, Utah.

217. **Penstemon pudicus** Reveal & Beatley, Bull. Torrey Bot. Club 98: 332. 1972 • Kawich Range beardtongue [C][E]

Herbs. Stems ascending to erect, 24–50 cm, glabrous proximally, sparsely glandular-pubescent distally, glaucous. **Leaves** basal and cauline, basal sometimes few, opposite, glabrous, glaucous; basal and proximal cauline 20–75(–90) × 4–16(–20) mm, blade oblanceolate to elliptic, base tapered, margins entire, apex rounded to obtuse; cauline 3 or 4 pairs, sessile or proximals short-petiolate, 17–80 × 3–18 mm, blade oblanceolate to oblong or lanceolate, base tapered to cordate-clasping, margins entire, apex obtuse to acute. **Thyrses** interrupted, cylindric, 5–22 cm, axis glandular-pubescent, verticillasters 3–6, cymes 1- or 2 (or 3)-flowered, 2 per node; proximal bracts lanceolate, 20–44 × 4–13 mm; peduncles and pedicels ascending to erect, glandular-pubescent. **Flowers:** calyx lobes lanceolate, 7–10 × 1.8–2.6 mm, glandular-pubescent; corolla violet to blue or lavender, without nectar guides, ventricose, (25–)30–35 mm, glandular-pubescent externally, glabrous internally, tube 10–11 mm, throat gradually inflated, 7–9 mm diam., 2-ridged abaxially; stamens included, filaments glabrous, pollen sacs parallel, 1.5–2 mm, distal ½–⅔ indehiscent, sides glabrous, sutures denticulate, teeth to 0.1 mm; staminode 19–24 mm, included, 0.4–0.6 mm diam., distal 10–14 mm moderately to densely pilose, hairs golden yellow, to 1.2 mm; style 26–30 mm. **Capsules** 7–9 × 4–5 mm.

Flowering Jun–Jul. Rocky soils, pine-juniper woodlands, mountain mahogany woodlands, subalpine sagebrush shrublands; of conservation concern; 2300–3000 m; Nev.

Penstemon pudicus is known from the Hot Creek and Kawich ranges in central Nye County.

218. Penstemon purpusii Brandegee, Bot. Gaz. 27: 455. 1899 (as Pentstemon purpusi) • Snow Mountain beardtongue E

Herbs. Stems decumbent to ascending, 8–20 cm, retrorsely hairy to pilose, hairs pointed, not glaucous. **Leaves** basal and cauline, opposite, canescent to pilose, hairs pointed, not glaucous; basal and proximal cauline 8–30 × 3–12 mm, blade round to spatulate or ovate, base tapered, margins entire or shallowly dentate or serrate, apex rounded to obtuse; cauline 2–6 pairs, short-petiolate or sessile, 15–30 × 4–10 mm, blade obovate to spatulate or oblanceolate, base tapered, margins entire, apex obtuse to acute. **Thyrses** interrupted or continuous, cylindric, 2–8(–15) cm, axis glandular-pubescent, verticillasters 2–6, cymes 1–3-flowered, 1 or 2 per node; proximal bracts lanceolate, (4–)9–20 × (0.5–)1.5–4 mm; peduncles and pedicels appressed, glandular-pubescent. **Flowers:** calyx lobes ovate to lanceolate, 5–9.5 × 1–2.2 mm, glandular-pubescent; corolla purple to violet, blue, or purple, with purple or violet nectar guides, funnelform, 25–31 mm, glandular-pubescent externally, glabrous internally, tube 8–10 mm, throat gradually inflated, 7–9 mm diam., 2-ridged abaxially; stamens: longer pair reaching orifice, filaments glabrous, pollen sacs parallel, 2.4–2.6 mm, distal ¼–⅓ indehiscent, sides hirsute, hairs white, to 0.7 mm, sutures denticulate, hairs to 0.1 mm; staminode 15–18 mm, included, 0.4–0.5 mm diam., glabrous; style 19–23 mm. **Capsules** 8–13 × 5–8 mm. **2n = 16.**

Flowering Jun–Aug. Rocky ridges, peaks, open slopes in pine and fir forests; 1500–2400 m; Calif.

Penstemon purpusii is found in the North Coast and Klamath ranges. Populations have been documented in Colusa, Glenn, Humboldt, Lake, Mendocino, Shasta, Tehama, and Trinity counties.

219. Penstemon rhizomatosus N. H. Holmgren, Brittonia 50: 162, fig. 1A–E. 1998 • Rhizome beardtongue C E F

Herbs. Stems ascending, 7–20 (–28) cm, retrorsely hairy, hairs appressed, white, scalelike, not glaucous. **Leaves** cauline, opposite, glabrous abaxially, glabrate or retrorsely hairy, hairs appressed, white, scalelike, adaxially, not glaucous; cauline 4–7 pairs, short-petiolate or sessile, 17–35 × 3–7 mm, blade oblanceolate to lanceolate, proximals sometimes spatulate, base tapered, margins entire, apex rounded to obtuse or acute. **Thyrses** interrupted, secund, 1–11 cm, axis glandular-pubescent distally, sometimes also retrorsely hairy proximally, hairs appressed, white, scalelike, verticillasters 2–5(–7), cymes 1- or 2(or 3)-flowered, 2 per node; proximal bracts lanceolate, 4–26 × 1–7 mm; peduncles and pedicels ascending to erect, glandular-pubescent. **Flowers:** calyx lobes ovate to lanceolate, 3.5–5.5(–6.5) × 1.7–2.1 mm, glandular-pubescent; corolla violet to reddish purple, without nectar guides, funnelform, 14–18 mm, glandular-pubescent externally, glabrous internally, tube 5–6 mm, throat gradually inflated, 4.5–6 mm diam., 2-ridged abaxially; stamens included or longer pair reaching orifice, filaments glabrous, pollen sacs parallel, 0.9–1.2 mm, distal ½–⅔ indehiscent, sides glabrous, sutures denticulate, teeth to 0.1 mm; staminode 7–8 mm, exserted, 0.4–0.6 mm diam., glabrous; style 10–13 mm. **Capsules** 6–9 × 4–5 mm. **2n = 16.**

Flowering Jun–Aug(–Sep). Talus slopes and crevices, limestone, pine and subalpine conifer woodlands; of conservation concern; 3000–3500 m; Nev.

Penstemon rhizomatosus is known only from the Schell Creek Range of White Pine County. The species occurs near the summits of Cave Mountain, Cleve Creek Baldy, and Taft Peak (N. H. Holmgren 1998).

220. Penstemon richardsonii Douglas ex Lindley, Bot. Reg. 13: plate 1121. 1828 (as Pentstemon) • Richardson's beardtongue E

Subshrubs. Stems ascending to erect, 20–95 cm, retrorsely hairy, hairs pointed, not glaucous. **Leaves** cauline, opposite or subopposite, glabrate or retrorsely hairy, not glaucous; cauline 5–13 pairs, short-petiolate or sessile, 13–70 × 5–30 mm, blade ovate to lanceolate, base tapered, margins laciniate-dentate to irregularly laciniate-pinnatifid, apex obtuse to acute. **Thyrses** interrupted, cylindric, 2–25(–48) cm, axis densely glandular-pubescent, verticillasters 3–7(–15), cymes 2–7-flowered, 1 or 2 per node; proximal bracts lanceolate, 7–43 × 2–16 mm; peduncles and pedicels spreading, sparsely to densely glandular-pubescent. **Flowers:** calyx lobes ovate to lanceolate, 4–7 × 1.3–2.3 mm, glandular-pubescent; corolla bluish to lavender, or pinkish, with reddish purple nectar guides, ventricose-ampliate, 15–32 mm, glandular-pubescent externally, glabrous internally or sparsely to moderately white-lanate abaxially, tube 7–12 mm, throat gradually inflated, 5–14 mm diam., rounded abaxially; stamens: longer pair reaching orifice or exserted, filaments glabrous, pollen sacs parallel, 1.3–2 mm, distal ½ indehiscent, sides glabrous, sutures denticulate, teeth to 0.2 mm; staminode 16–23 mm, exserted, 0.3–0.8 mm diam., glabrous or distal 2–8 mm sparsely lanate, hairs yellow, to 2 mm; style 11–23 mm. **Capsules** 5–9 × 4–5 mm. $2n$ = 16.

Varieties 3 (3 in the flora): w North America.

Penstemon richardsonii consists of three weakly differentiated varieties that are partially sympatric on the eastern slope of the Cascade Range.

The Okanagan-Coville tribe of northeastern Washington and the Paiute tribe of the Great Basin region used *Penstemon richardsonii* as a drug plant to treat typhoid fever and as a dermatological aid (D. E. Moerman 1998).

1. Corollas 15–20 mm; styles 11–15 mm; staminodes glabrous or sparsely hairy distally . . . . 220b. *Penstemon richardsonii* var. *curtiflorus*
1. Corollas 22–32 mm; styles 16–23 mm; staminodes with distal 2–8 mm sparsely lanate, rarely glabrous.
  2. Leaf blades ovate to lanceolate (in outline), margins laciniate-dentate to irregularly laciniate-pinnatifid . . . . 220a. *Penstemon richardsonii* var. *richardsonii*
  2. Leaf blades ovate, sometimes lanceolate, margins dentate, rarely laciniate-dentate . . . . 220c. *Penstemon richardsonii* var. *dentatus*

220a. Penstemon richardsonii Douglas ex Lindley var. **richardsonii** E

Leaves: blade ovate to lanceolate, margins laciniate-dentate to irregularly laciniate-pinnatifid. **Flowers:** corolla 22–32 mm; staminode: distal 2–8 mm sparsely lanate, rarely glabrous; style 16–23 mm.
Flowering May–Aug(–Sep). Talus slopes, cliffs, crevices, often in basalt; 30–1000 m; B.C.; Oreg., Wash.

Variety *richardsonii* occurs mostly from the Columbia River to the Okanagan Valley in British Columbia. A specimen (*A. Kellogg & W. G. Harford 672*, MO), collected in 1868 or 1869 and attributed to California, is doubtful as to locality.

220b. Penstemon richardsonii Douglas ex Lindley var. **curtiflorus** (D. D. Keck) Cronquist in C. L. Hitchcock et al., Vasc. Fl. Pacif. N.W. 4: 401. 1959 E

Penstemon richardsonii subsp. *curtiflorus* D. D. Keck, Brittonia 8: 249. 1957

Leaves: blade ovate to lanceolate, margins laciniate-dentate, sometimes irregularly laciniate-pinnatifid. **Flowers:** corolla 15–20 mm; staminode glabrous or sparsely hairy distally; style 11–15 mm.
Flowering May–Aug. Talus slopes, cliffs, crevices, basalt; 800–900 m; Oreg.

Variety *curtiflorus* is known from Gilliam, Wasco, and Wheeler counties in north-central Oregon. D. D. Keck and A. Cronquist (1957) cited a specimen from Wheeler County (*L. F. Henderson 5503*, CAS) as probably this taxon, noting the flowers were immature. A duplicate specimen at MO clearly is var. *curtiflorus*, with corollas 15–18 mm and staminodes with one or two hairs near their tips. An anomalous specimen at MO (*E. A. Purer 7811*) collected in Yakima County, Washington, appears close to var. *curtiflorus*, with corollas 18–20 mm, staminodes sparsely lanate, and leaf margins dentate. This collection was made about 120 km north of the northernmost collections of the variety.

220c. Penstemon richardsonii Douglas ex Lindley var. **dentatus** (D. D. Keck) Cronquist in C. L. Hitchcock et al., Vasc. Fl. Pacif. N.W. 4: 401. 1959 E

Penstemon richardsonii subsp. *dentatus* D. D. Keck, Brittonia 8: 250. 1957

Leaves: blade ovate, sometimes lanceolate, margins dentate, rarely laciniate-dentate. **Flowers:** corolla 22–30 mm; staminode: distal 2–8 mm sparsely lanate; style 16–22 mm.
Flowering May–Aug(–Sep). Talus slopes, cliffs, crevices, often in basalt; 300–1500 m; Oreg.

Variety *dentatus* is known from north-central and northeastern Oregon (Baker, Crook, Deschutes, Grant, Jefferson, Morrow, Umatilla, Union, Wasco, and Wheeler counties).

221. Penstemon roezlii Regel, Gartenflora 21: 239. 1872 (as Pentstemon roezli) • Roezl's beardtongue E

Penstemon laetus A. Gray subsp. *roezlii* (Regel) D. D. Keck; *P. laetus* var. *roezlii* (Regel) Jepson

Herbs or subshrubs. Stems ascending to erect, 13–50 cm, retrorsely hairy, hairs pointed, not glaucous. **Leaves** essentially cauline, opposite, retrorsely hairy, hairs pointed, sometimes glabrate, not glaucous; cauline 5–8 pairs, petiolate or sessile, 22–80 × (1–)2–9 mm, blade oblanceolate to lanceolate or linear, base tapered, margins entire, apex obtuse to acute. **Thyrses** interrupted, cylindric, 4–14 cm, axis glandular-pubescent, verticillasters (3 or)4–8 (–12), cymes 1–5-flowered, 2 per node; proximal bracts lanceolate, 5–40 × 1–4 mm; peduncles and pedicels ascending to erect, glandular-pubescent. **Flowers:** calyx lobes ovate to lanceolate, 3.8–7 × 1.4–2.6 mm, glandular-pubescent; corolla blue to violet or purple, without nectar guides, funnelform, 15–25 mm, glandular-pubescent externally, glabrous internally, tube 6–8 mm, throat gradually inflated, 6–9 mm diam., slightly 2-ridged abaxially; stamens included, filaments glabrous, pollen sacs parallel, 1.5–2 mm, distal ½ indehiscent, sides hispidulous, hairs white, to 0.5 mm near filament attachment, sutures denticulate, teeth to 0.2 mm; staminode 12–15 mm, reaching orifice, 0.5–1 mm diam., glabrous; style 13–17 mm. **Capsules** (5.5–)7–9 × (3–)4–6 mm.

Flowering May–Aug. Sagebrush shrublands, juniper woodlands, pine forests; 600–2500 m; Calif., Nev., Oreg.

Penstemon roezlii occurs from north-central Oregon (Wheeler County) to central California (Inyo and Tulare counties) and northwestern Nevada (Storey and Washoe counties). Populations are concentrated in the Cascade and Klamath ranges, the High Sierra Nevada, and the Modoc Plateau and East Cascades.

222. Penstemon scapoides D. D. Keck, Univ. Calif. Publ. Bot. 16: 379, fig. 4. 1932 • Pinyon beardtongue E

Herbs. Stems erect, 15–45 (–60) cm, glabrous, glaucous. **Leaves** basal and cauline, opposite, hirsute, hairs pointed, not glaucous; basal and proximal cauline 15–30(–60) × 4–10(–14) mm, blade ovate to elliptic, base tapered, margins entire, apex rounded to acute; cauline 2 or 3 pairs, petiolate or sessile, 8–25(–45) × 1–3 mm, blade spatulate to oblanceolate or linear, base tapered, margins entire, apex rounded to acute. **Thyrses** interrupted, cylindric, 4–24 cm, axis glandular-pubescent, verticillasters 2–4, cymes 1–4-flowered, 1 or 2 per node; proximal bracts subulate to lanceolate or linear, 4–35 × 1–3 mm; peduncles and pedicels ascending to erect, glandular-pubescent. **Flowers:** calyx lobes oblong to ovate, 3–4.5(–5) × 0.9–2 mm, glandular-pubescent; corolla lavender to violet or purple, with or without faint reddish purple nectar guides, ventricose, 25–34 mm, glandular-pubescent externally, yellow-lanate internally abaxially, tube 6–9 mm, throat gradually inflated, 6–8 mm diam., 2-ridged abaxially; stamens included, filaments glabrous, pollen sacs nearly parallel, 1.4–1.5(–1.7) mm, distal ½ indehiscent, sides glabrous, sutures denticulate, teeth to 0.1 mm; staminode 15–19 mm, included, 0.2–0.3 mm diam., distal 4–8 mm sparsely pilose, hairs yellow, to 0.7 mm; style 20–29 mm. **Capsules** 6–10 × 4–6 mm. $2n = 16$.

Flowering Jun–Aug. Sagebrush shrublands, pine-juniper and pine woodlands; 2000–3100 m; Calif.

Penstemon scapoides is known from the Inyo, Last Chance, and White mountains in Inyo County.

223. Penstemon sepalulus A. Nelson in J. M. Coulter and A. Nelson, New Man. Bot. Rocky Mt., 449. 1909 (as Pentstemon) • Little-cup beardtongue E

Penstemon azureus Bentham var. *ambiguus* A. Gray in A. Gray et al., Syn. Fl. N. Amer. 2(1): 272. 1878 (as Pentstemon), not *P. ambiguus* Torrey 1827

Herbs or subshrubs. Stems ascending to erect, 40–90 cm, glabrous, glaucous. **Leaves** cauline, opposite, glabrous, glaucous; cauline 6–14 pairs, sessile, 30–85(–102) × 3–13 mm, blade elliptic, base tapered, margins entire, apex acute. **Thyrses** interrupted, cylindric, 6–30 cm, axis glabrous, verticillasters (5–)8–14, cymes 1(or 2)-flowered, 2 per node; proximal bracts narrowly elliptic to linear, 8–45 × 1–8 mm; peduncles and pedicels ascending to erect, glabrous. **Flowers:** calyx lobes obovate to ovate or elliptic, 1.8–3.2 × 1.5–2.2 mm, glabrous or minutely ciliolate distally; corolla light lavender to violet or purple, with violet nectar guides, weakly ventricose, 22–26(–28) mm, glabrous externally, glabrous internally, tube 7–9 mm, throat gradually inflated, 7–9 mm diam., 2-ridged abaxially; stamens included or longer pair exserted, filaments glabrous, pollen sacs parallel to slightly divergent, 1.5–2 mm, distal ⅔ indehiscent, sides glabrous, sutures denticulate, teeth to 0.1 mm; staminode 15–16 mm, included, 0.5–0.6 mm diam., glabrous; style 16–20 mm. **Capsules** 7–10 × 4.5–5.5 mm.

Flowering Jun–Aug. Rocky to gravelly and talus slopes, Gambel oak, maple, and aspen woodlands; 1200–2300 m; Utah.

Penstemon sepalulus is found in the Wasatch Mountains in Juab, Sevier, Utah, and Washington counties.

224. Penstemon serrulatus Menzies ex Smith in A. Rees, Cycl. 26: Pentstemon no. 5. 1814 (as Pentstemon serrulata) • Cascade beardtongue [E]

Herbs or subshrubs. Stems ascending to erect, 13–80 cm, glabrous or retrorsely hairy, hairs pointed, not glaucous. **Leaves** cauline, opposite, glabrous, sometimes sparsely puberulent, hairs pointed, adaxially along midvein, not glaucous; cauline 3–6 pairs, petiolate proximally or sessile distally, (12–)30–80 × (7–)15–50 mm, blade ovate-oblong proximally, ovate or lanceolate distally, base truncate to tapered proximally, cordate-clasping distally, margins serrate, apex obtuse to acute. **Thyrses** continuous or interrupted, cylindric, 2–18(–28) cm, axis retrorsely hairy, verticillasters 1–3(–5), cymes 3–10-flowered, 2 per node; proximal bracts ovate, 13–35(–80) × 6–16(–36) mm; peduncles and pedicels ascending to erect, retrorsely hairy to puberulent. **Flowers:** calyx lobes obovate to lanceolate, 3.5–11 × 1–3 mm, sparsely pubescent and ciliolate along margins, sometimes glabrate; corolla blue to violet or purple, without nectar guides, weakly ventricose, 17–25(–28) mm, glabrous externally, glabrous or sparsely to densely white-lanate internally abaxially, tube 4–6 mm, throat gradually inflated, 5–8 mm diam., 2-ridged abaxially; stamens included, filaments glabrous, pollen sacs parallel, 1.1–1.6 mm, distal ½–⅔ indehiscent, sides glabrous, sutures denticulate, teeth to 0.1 mm; staminode 12–14 mm, exserted, 0.3–0.6 mm diam., distal 2–7 mm sparsely lanate, hairs yellow, to 1.5 mm; style 10–16 mm. **Capsules** 5–8 × 3–5.5 mm. *2n* = 16.

Flowering Jun–Aug(–Sep). Sandy to rocky stream banks, gullies, slopes, roadcuts, coniferous forests; 0–1800 m; B.C.; Alaska, Oreg., Wash.

Penstemon serrulatus occurs from southeastern Alaska through western British Columbia and western Washington to southwestern Oregon, largely in the Cascade and Coast ranges. Herbarium specimens sometimes are misidentified as *P. ovatus*, which has glandular-pubescent inflorescences and navicular pollen sacs.

225. Penstemon tiehmii N. H. Holmgren, Brittonia 50: 159, fig. 1F–J. 1998 • Shoshone or Tiehm's beardtongue [C][E]

Herbs. Stems ascending to erect, 6–20 cm, retrorsely hairy, hairs appressed, white, ± scalelike, not glaucous. **Leaves** cauline, opposite, retrorsely hairy, hairs white, scalelike, not glaucous; cauline 1–4 pairs, petiolate, 15–45 × 7–24 mm, blade ovate to spatulate or lanceolate, base tapered, margins entire, apex rounded to acute. **Thyrses** interrupted, cylindric, 2–7 cm, axis glandular-pubescent, sometimes also retrorsely hairy or appressed-hairy proximally, hairs white, scalelike, verticillasters 3 or 4(–7), cymes (1 or)2–5-flowered, 1 or 2 per node; proximal bracts lanceolate, 9–25 × 1–5 mm; peduncles and pedicels ascending, glandular-pubescent. **Flowers:** calyx lobes lanceolate, 4–4.8 × 1.4–2 mm, glandular-pubescent; corolla violet to lavender, without nectar guides, abaxial ridges usually white, weakly ventricose, 15–20 mm, glandular-pubescent externally, glabrous internally, tube 4–6 mm, throat gradually inflated, 4–5 mm diam., 2-ridged abaxially; stamens included or longer pair reaching orifice, filaments glabrous, pollen sacs nearly parallel to divergent, 1.1–1.6 mm, distal ⅔ indehiscent, sides glabrous, sutures denticulate, teeth to 0.1 mm; staminode 8–10 mm, exserted, 0.3–0.6 mm diam., glabrous or sparsely pilose, hairs yellow, to 0.2 mm; style 10–13 mm. **Capsules** 6–8 × 4–5 mm. *2n* = 16.

Flowering Jun–Aug. Sandy-loam soils, steep volcanic talus slopes; of conservation concern; 2300–2900 m; Nev.

Penstemon tiehmii is known only from Mount Lewis in the Shoshone Range in Lander County.

226. Penstemon triphyllus Douglas, Edwards's Bot. Reg. 15: plate 1245. 1829 (as Pentstemon triphyllum) • Whorled beardtongue [E]

Subshrubs. Stems ascending to erect, 25–80 cm, retrorsely hairy, hairs pointed, not glaucous. **Leaves** cauline, whorled, sometimes alternate, glabrate or retrorsely hairy, hairs pointed, not glaucous; cauline 15–20 pairs, sessile, (5–)10–52 × 1–6(–12) mm, blade linear, sometimes lanceolate, base tapered, margins coarsely serrate-dentate, distals sometimes entire, apex acute. **Thyrses** interrupted, cylindric, 4–35 cm, axis glandular-pubescent, verticillasters 5–10, cymes 1–4-flowered, 1 or 2(or 3) per node; proximal bracts linear, rarely lanceolate, 5–34 × 1–6 mm; peduncles and pedicels spreading to ascending, peduncles retrorsely hairy and, sometimes, glandular-pubescent distally, pedicels glandular-pubescent. **Flowers:** calyx lobes lanceolate, 3.5–5.8 × 1–2 mm, glandular-pubescent; corolla bluish lavender to lavender, pinkish, pale purple, or violet, with reddish purple nectar guides, funnelform, 13–19 mm, glandular-pubescent externally, glabrous internally or sparsely white-lanate abaxially, tube 5–7 mm, throat gradually inflated, 5–6 mm diam., rounded abaxially; stamens included, filaments glabrous or white-pubescent distally, pollen sacs parallel, 0.9–1.3 mm, distal $2/3$ indehiscent, sides glabrous, sutures denticulate, teeth to 0.1 mm; staminode 9–11 mm, exserted, 0.2–0.3 mm diam., distal 6–7 mm moderately to densely lanate, hairs yellow, to 2.5 mm; style 8–10 mm. **Capsules** 4.5–7 × 3–4.5 mm. $2n = 16$.

Flowering May–Jul. Cliffs, rocky slopes, basalt; 300–1800 m; Idaho, Oreg., Wash.

Penstemon triphyllus occurs near the Snake River in northeastern Oregon (Baker and Wallowa counties), southeastern Washington (Asotin, Franklin, Garfield, Klickitat, Okanogan, and Whitman counties), and adjacent Idaho (Adams, Idaho, Lemhi, Lewis, Nez Perce, and Valley counties), extending up the Salmon River.

227. **Penstemon venustus** Douglas ex Lindley, Edwards's Bot. Reg. 16: plate 1309. 1830 (as Pentstemon venustum) • Lovely beardtongue [E]

Subshrubs. Stems ascending to erect, 18–80 cm, glabrous or retrorsely hairy, hairs pointed, not glaucous. **Leaves** cauline, opposite, glabrous, sometimes glaucous; cauline 7–15 pairs, short-petiolate or sessile, (20–)40–100(–120) × (7–)10–30 (–38) mm, blade ovate to lanceolate, base tapered to truncate, margins subentire or serrate, apex acute. **Thyrses** interrupted, ± secund, (2–)7–40 cm, axis glabrous or retrorsely hairy, verticillasters (3–)5–14, cymes 2–7-flowered, 2 per node; proximal bracts lanceolate, (6–)23–70(–115) × (1–)4–28(–35) mm; peduncles and pedicels ascending to erect, glabrous or retrorsely hairy. **Flowers:** calyx lobes ovate to obovate or lanceolate, 2.5–6 × 1.3–1.8 mm, glabrous except for margins sometimes ciliate; corolla lavender to purple or violet, with reddish purple nectar guides, weakly ventricose, 24–38 mm, glabrous externally except lobe margins densely white-ciliate, glabrous internally, tube 7–9 mm, throat gradually inflated, 9–10 mm diam., rounded abaxially; stamens: longer pair reaching orifice or exserted, filaments white-lanate distally, rarely glabrous, pollen sacs parallel, 1.6–2.1 mm, distal $1/2$–$2/3$ indehiscent, sides hispidulous, hairs white, to 0.2 mm near point of attachment of connective, sutures denticulate, teeth to 0.2 mm; staminode 16–19 mm, reaching orifice or barely exserted, 0.4–0.5 mm diam., distal 2–5 mm sparsely to moderately lanate, hairs white, to 2 mm; style 18–25 mm. **Capsules** 6–9 × 4–5 mm. $2n = 64$.

Flowering (May–)Jun–Sep. Rock outcrops, rocky and talus slopes, especially on basalt, shrublands, forest clearings; 300–2700 m; Calif., Idaho, Oreg., Utah, Wash.

Penstemon venustus is known from the Snake River Plateau in west-central Idaho, northeastern Oregon, and southeastern Washington and is naturalized in Lassen County, California. E. C. Neese and N. D. Atwood (2003) reported *P. venustus*, apparently introduced, from Duchesne County, Utah. E. Heitz (1927) reported a chromosome number of $2n = 14$–16 for *P. venustus*; that count conflicts with all other reports for *P. venustus*.

26b.15. PENSTEMON Schmidel (subg. PENSTEMON) sect. SPECTABILES Pennell, Contr. U.S. Natl. Herb. 20: 326, 329. 1920

Herbs, shrubs, or subshrubs. Stems glabrous, glaucous, sometimes not. **Leaves** basal and cauline, basals sometimes few (*P. clevelandii*, *P. stephensii*) or absent (*P. clevelandii*), or leaves essentially cauline (*P. fruticiformis*, *P. incertus*), opposite, leathery, sometimes not (*P. pseudospectabilis*, *P. spectabilis*, *P. stephensii*), glabrous, glaucous, sometimes not (*P. clevelandii*, *P. spectabilis*, *P. stephensii*); basal and proximal cauline petiolate; cauline sessile or short-petiolate, blade cordate, ovate, triangular-ovate, triangular, oblanceolate, triangular-lanceolate, lanceolate, ovate-oblong, or linear, margins entire or toothed. **Thyrses** interrupted, rarely continuous, cylindric or secund, axis glabrous or glandular-pubescent, cymes 2 per node; peduncles and pedicels spreading to ascending or erect. **Flowers:** calyx lobes: margins entire or erose, sometimes denticulate (*P. spectabilis*), ± scarious, sometimes herbaceous (*P. pseudospectabilis*), glabrous or glandular-pubescent; corolla white to pink, rose, red, magenta, yellow, lavender, blue, violet, or purple, bilaterally symmetric, bilabiate to strongly bilabiate, not personate, ventricose or ventricose-ampliate, sometimes tubular-funnelform or funnelform, glandular-pubescent externally, sometimes glabrous, glabrous or hairy internally abaxially, glandular-pubescent on both surfaces, or glandular-pubescent on both surfaces and hairy abaxially, throat gradually to abruptly inflated, constricted or not at orifice, rounded abaxially, sometimes 2-ridged; stamens included or longer pair reaching orifice or exserted, filaments glabrous or shorter pair glandular-puberulent proximally, pollen sacs opposite, rarely divergent (*P. incertus*), navicular to subexplanate or explanate, dehiscing completely, rarely incompletely (*P. incertus*), connective splitting, rarely not (*P. incertus*), sides glabrous, sutures smooth, rarely papillate or denticulate, teeth to 0.1 mm; staminode included or exserted, flattened distally, 0.2–1.1 mm diam., tip straight to recurved, rarely coiled, glabrous or distal 10–50% hairy, sometimes proximal 10–40% hairy, hairs glandular; style glabrous, rarely sparsely glandular-pubescent proximally. **Capsules** glabrous or sparsely glandular-pubescent distally. **Seeds** brown to dark brown or black, angled, rarely ± rounded, 1–4 mm.

Species 16 (12 in the flora): w United States, n Mexico.

D. D. Keck (1937b) circumscribed sect. *Peltanthera*, a name that is illegitimate and superfluous, to include species from southwestern United States and northern Mexico with showy, white to pink, red, or purple, tubular to ventricose corollas and completely dehiscent, glabrous pollen sacs. His section encompasses species included here mostly in sect. *Spectabiles* and species treated here in sects. *Gentianoides* and *Petiolati*. Core species of sect. *Spectabiles*, those with ventricose corollas and treated by Keck in so-called subsect. *Spectabiles* (not validly published), were monophyletic in a molecular study (C. A. Wessinger et al. 2016) that included nine of the 12 species treated here in sect. *Spectabiles*. That study also provided support for inclusion of *Penstemon alamosensis* and *P. parryi*, species with tubular corollas and treated here in sect. *Gentianoides*, in sect. *Spectabiles*. However, broader taxon sampling is needed before definitive circumscription of sects. *Gentianoides* and *Spectabiles* is possible. Four species and one variety in sect. *Spectabiles* occur in Mexico: *P. angelicus* (I. M. Johnston) Moran, *P. eximius* D. D. Keck, *P. rotundifolius* A. Gray, *P. spectabilis* var. *subinteger* (D. D. Keck) C. C. Freeman, and *P. vizcainensis* Moran.

SELECTED REFERENCES Clokey, I. W. and D. D. Keck. 1939. Reconsideration of certain members of *Penstemon* subsection *Spectabiles*. Bull. S. Calif. Acad. Sci. 38: 8–13. Keck, D. D. 1937b. Studies in *Penstemon* V. The section *Peltanthera*. Amer. Midl. Naturalist 18: 790–829.

1. Cauline leaf blade margins entire.
  2. Corollas glabrous externally . . . . 232. *Penstemon fruticiformis* (in part)
  2. Corollas glandular-pubescent externally.
    3. Corollas tubular-funnelform to slightly ventricose, red to reddish purple . . . .
      . . . . .229. *Penstemon clevelandii* (in part)
    3. Corollas ventricose-ampliate, purple, violet, blue, bluish lavender, lavender, or light pink.
      4. Peduncles and pedicels glabrous; corolla tubes 6–8 mm, lengths 1–1.5 times calyx lobes . . . . 232. *Penstemon fruticiformis* (in part)
      4. Peduncles and pedicels glandular-pubescent; corolla tubes 7–11 mm, lengths 1.8–2 times calyx lobes . . . . .234. *Penstemon incertus* (in part)
1. Cauline leaf blade margins serrate to dentate, rarely ± serrulate (at least proximals, distals sometimes entire).

[5. Shifted to left margin.—Ed.]

5. Corollas tubular-funnelform or funnelform to slightly ventricose.
   6. Calyx lobes 3.2–6 mm; corolla throats 5–8 mm diam.; staminodes glabrous or distal 1–6 mm pilose, hairs yellow, to 0.8 mm, proximal 1–2 mm sometimes glandular-puberulent. . . . . .229. *Penstemon clevelandii* (in part)
   6. Calyx lobes 3–4.2(–5) mm; corolla throats 4–6 mm diam.; staminodes glabrous . . . .
     . . . . .239. *Penstemon stephensii*
5. Corollas ventricose to ventricose-ampliate.
   7. Corolla tube lengths 0.9–1.5 times calyx lobes.
     8. Cauline leaves 2–8 mm wide, blades lanceolate to linear, sometimes oblanceolate
     . . . . .232. *Penstemon fruticiformis* (in part)
     8. Cauline leaves 4–45 mm wide, blades lanceolate to ovate-oblong, ovate, or triangular.
       9. Cauline leaf blades lanceolate to ovate-oblong, bases tapered; staminodes 20–22 mm; styles 15–18 mm . . . . .233. *Penstemon grinnellii*
       9. Cauline leaf blades ovate to triangular, bases of distals auriculate-clasping or connate-perfoliate; staminodes 25–30 mm; styles 19–23 mm. . . . . .235. *Penstemon palmeri*
   7. Corolla tube lengths 1.6–2.2 times calyx lobes.
     10. Cauline leaves (1–)2.5–4(–6) mm wide, blades lanceolate to linear. . . . .
     . . . . .234. *Penstemon incertus* (in part)
     10. Cauline leaves 5–76 mm wide, blades ovate to triangular or lanceolate.
       11. Peduncles and pedicels spreading to ascending; pollen sacs navicular, sutures papillate or denticulate; corollas lavender, blue, pink, or purple. . . . . .238. *Penstemon spectabilis*
       11. Peduncles and pedicels ascending to erect; pollen sacs explanate or navicular, sutures smooth; corollas pink to rose pink, lavender, purple, or light yellow.
         12. Cauline leaf blades ovate to lanceolate; filaments glabrous.
           13. Corollas 21–32 mm; staminodes included, glabrous . . . . .231. *Penstemon floridus*
           13. Corollas (27–)30–35 mm; staminodes exserted, distal 6–7 mm pilose
           . . . . .237. *Penstemon rubicundus*
         12. Cauline leaf blades ovate to triangular or lanceolate; filaments of shorter stamen pairs glandular-puberulent proximally.
           14. Staminodes: distal 4–5 mm lanate, hairs to 2.8 mm; Arizona, California, Nevada . . . . 228. *Penstemon bicolor*
           14. Staminodes: glabrous or distal 1–4 mm sparsely to moderately lanate, hairs to 1 mm; Arizona, California, New Mexico, Utah.
             15. Corollas glandular-pubescent internally, also usually moderately white-lanate internally abaxially, throats 9–12 mm diam.; Arizona. . . . . 230. *Penstemon clutei*
             15. Corollas glandular-pubescent internally, throats 7–10 mm diam.; Arizona, California, New Mexico, Utah. . . . . 236. *Penstemon pseudospectabilis*

228. Penstemon bicolor (Brandegee) Clokey & D. D. Keck, Bull. S. Calif. Acad. Sci. 38: 12. 1939

  • Two-color beardtongue [E]

Penstemon palmeri A. Gray var. *bicolor* Brandegee, Univ. Calif. Publ. Bot. 6: 360. 1916 (as Pentstemon); *P. bicolor* subsp. *roseus* Clokey & D. D. Keck; *P. pseudospectabilis* M. E. Jones subsp. *bicolor* (Brandegee) D. D. Keck

Herbs. Stems ascending to erect, 60–150 cm, glaucous. **Leaves:** basal and proximal cauline 37–110 × 10–50 mm, blade obovate, base tapered, margins coarsely serrate to coarsely dentate, apex obtuse to acute; cauline 4–8 pairs, sessile, 40–110 × 15–55 mm, blade ovate, base connate-perfoliate, margins coarsely serrate to coarsely dentate, apex acute to acuminate. **Thyrses** interrupted, secund, 16–90 cm, axis glandular-pubescent, verticillasters 9–23, cymes 1–4-flowered; proximal bracts ovate to lanceolate, 8–25 × 3–28 mm; peduncles and pedicels ascending to erect, glandular-pubescent. **Flowers:** calyx lobes ovate, 4–6 × 2.2–3 mm, glandular-pubescent; corolla pink to rose pink, purple, or light yellow, with or without reddish or reddish purple nectar guides, strongly bilabiate, ventricose, 18–27 mm, glandular-pubescent externally, glandular-pubescent internally, and sparsely white- or yellowish lanate internally abaxially, tube 7–10 mm, length 1.7–2 times calyx lobes, throat abruptly inflated, constricted at orifice, 6–9 mm diam., rounded abaxially; stamens: longer pair reaching orifice, filaments of shorter pair glandular-puberulent prox-imally, pollen sacs explanate, 1.4–1.8 mm, sutures smooth; staminode 14–16 mm, included or exserted, 0.3–0.4 mm diam., tip recurved, distal 4–5 mm lanate, hairs yellow, to 2.8 mm, proximal 3–4 mm glandular-puberulent; style 14–16 mm, glabrous. **Capsules** 10–13 × 5–7 mm, glabrous or sparsely glandular-pubescent distally.

Flowering Mar–Jun. Gravelly soils, arroyos, road-sides, talus slopes, desert scrub, juniper woodlands; 500–1700 m; Ariz., Calif., Nev.

Penstemon bicolor is known from the Mojave Desert in the Black Mountains of northwestern Mohave County, Arizona, the Castle, Clark, and New York mountains of eastern San Bernardino County, California, and the desert mountain ranges of southern and western Clark County, Nevada. Two subspecies differing in corolla color have been recognized, with yellow-corolla forms restricted to Clark County, Nevada. Subspecies are not recognized here, following J. D. Morefield (2006).

Habitat destruction and genetic swamping caused by hybridization between *Penstemon bicolor* and *P. palmeri* threaten many populations of *P. bicolor* in the vicinity of Las Vegas, Nevada (G. Glenne 2003).

SELECTED REFERENCES Glenne, G. 2003. Reproductive Biology, Hybridization Isolating Mechanisms, and Conservation Implications of Two Rare Subspecies of *Penstemon bicolor* (Brandeg.) Clokey and Keck: Ssp. *bicolor* and ssp. *roseus* Clokey and Keck (Scrophulariaceae) in Clark County, Nevada. M.S. thesis. Utah State University. Morefield, J. D. 2006. Current Knowledge and Conservation Status of *Penstemon bicolor* (Brandegee) Clokey & Keck (Plantaginaceae), the Two-tone Beardtongue. Carson City.

229. Penstemon clevelandii A. Gray, Proc. Amer. Acad. Arts 11: 94. 1876 (as Pentstemon clevelandi)

• Cleveland's beardtongue

Herbs. Stems ascending to erect, 30–70 cm, glaucous. **Leaves:** basal and proximal cauline 15–90 × 8–35 mm, blade ovate, base truncate to tapered, mar-gins entire or coarsely dentate, apex acute; cauline 4–7 pairs, short-petiolate or sessile, 10–70 × 8–32 mm, blade cordate to triangular-lanceolate, base tapered or truncate on proximal leaves to connate-perfoliate on distal leaves, margins coarsely dentate, sometimes entire, apex acute. **Thyrses** interrupted, secund, 10–40(–65) cm, axis glabrous or glandular-pubescent, verticillasters 6–12 (–22), cymes 2–8-flowered; proximal bracts depressed-ovate to ovate or triangular, 6–45 × 11–26 mm; pedun-cles and pedicels spreading to ascending or erect, glabrous or glandular-pubescent. **Flowers:** calyx lobes ovate, 3.2–6 × 1.6–4 mm, glabrous or glandular-pubescent; corolla red to reddish purple, with or without reddish or reddish purple nectar guides, bilabiate, tubular-funnelform to slightly ventricose, 17–24 mm,

glandular-pubescent externally, glabrous internally or glandular-pubescent, sometimes also pilose abaxially, tube 6–9 mm, length 1.8–2 times calyx lobes, throat gradually inflated, slightly constricted or not at orifice, 5–8 mm diam., rounded abaxially; stamens included or longer pair reaching orifice, filaments glabrous, pollen sacs navicular or explanate, 1.2–1.7 mm, sutures smooth or denticulate, teeth to 0.1 mm; staminode 6–11 mm, included, 0.3–0.5 mm diam., tip straight to recurved, glabrous or distal 1–6 mm pilose, hairs yellow, to 0.8 mm, proximal 1–2 mm sometimes glandular-puberulent; style 11–14 mm, glabrous. **Capsules** 6–9 × 5–7 mm, glabrous. $2n = 16$.

Varieties 3 (3 in the flora): California, nw Mexico.

Penstemon clevelandii is an ancient, stable, diploid hybrid between *P. centranthifolius* and *P. spectabilis* (A. D. Wolfe et al. 1998). *Penstemon* ×*parishii* A. Gray is a naturally occurring hybrid, also with *P. centranthifolius* and *P. spectabilis* as parents; it has been reported from Los Angeles, Riverside, San Bernardino, and San Diego counties, California (D. D. Keck 1937b; Paul Wilson and M. Valenzuela 2002). Keck reported that *P. clevelandii* var. *clevelandii* could be distinguished from *P.* ×*parishii* by the former's shorter and narrower corollas, more glandular-pubescent inflorescences, and explanate pollen sacs.

1. Distal cauline leaves sessile, bases connate-perfoliate; pollen sacs navicular, sutures denticulate; peduncles, pedicels, and calyx lobes glabrous . . . . 229b. *Penstemon clevelandii* var. *connatus*
1. Distal cauline leaves short-petiolate or sessile, bases tapered or truncate; pollen sacs explanate, sutures smooth; peduncles, pedicels, and calyx lobes glandular-pubescent, rarely glabrous.
   2. Corollas glandular-pubescent internally, not pilose or sparsely white- or yellowish pilose internally abaxially; staminodes 9–11 mm . . . . 229a. *Penstemon clevelandii* var. *clevelandii*
   2. Corollas not glandular-pubescent internally, densely white- or yellowish pilose internally abaxially; staminodes 6–8 mm . . . . 229c. *Penstemon clevelandii* var. *mohavensis*

229a. Penstemon clevelandii A. Gray var. clevelandii

Distal cauline leaves short-petiolate or sessile, base tapered or truncate. **Thyrses:** peduncles and pedicels glandular-pubescent, rarely glabrous. **Flowers:** calyx lobes glandular-pubescent, rarely glabrous; corolla glandular-pubescent internally abaxially; pollen sacs explanate, sutures smooth; staminode 9–11 mm.

Flowering Feb–Jun. Rocky to sandy slopes, pinyon-juniper woodlands, scrub, chaparral; 200–1500 m; Calif.; Mexico (Baja California).

In the flora area, var. *clevelandii* is known from southern California.

229b. Penstemon clevelandii A. Gray var. **connatus** Munz & I. M. Johnston, Bull. Torrey Bot. Club 49: 357. 1923 (as clevelandi) E

Penstemon clevelandii subsp. *connatus* (Munz & I. M. Johnston) D. D. Keck

Distal cauline leaves sessile, base connate-perfoliate. **Thyrses:** peduncles and pedicels glabrous. **Flowers:** calyx lobes glabrous; corolla sparsely white- or yellowish pilose internally abaxially; pollen sacs navicular, sutures denticulate, teeth to 0.1 mm; staminode 7–11 mm.

Flowering Feb–May. Rocky to sandy slopes, pinyon-juniper woodlands, scrub, chaparral; 200–1700 m; Calif.

Variety *connatus* is known only from southern California, especially in the Cuyamaca-Lacuna, San Jacinto, and Santa Rosa mountains.

229c. Penstemon clevelandii A. Gray var. **mohavensis** (D. D. Keck) McMinn, Man. Calif. Shrubs, 511. 1939 E

Penstemon clevelandii subsp. *mohavensis* D. D. Keck, Amer. Midl. Naturalist 18: 810. 1937

Distal cauline leaves short-petiolate or sessile, base tapered or truncate. **Thyrses:** peduncles and pedicels glandular-pubescent. **Flowers:** calyx lobes glandular-pubescent; corolla densely white- or yellowish pilose internally abaxially; pollen sacs explanate, sutures smooth; staminode 6–8 mm.

Flowering Feb–May. Rocky to sandy slopes, pinyon-juniper woodlands, scrub, chaparral; 400–1500 m; Calif.

Variety *mohavensis* is known from southern California mostly in the Granite, Little San Bernardino, and Sheephole mountains.

230. Penstemon clutei A. Nelson, Amer. Bot. (Binghamton) 33: 109. 1927 • Sunset Crater beardtongue C E

Penstemon pseudospectabilis M. E. Jones subsp. *clutei* (A. Nelson) D. D. Keck

Herbs. Stems ascending to erect, 30–60 cm, glaucous. **Leaves:** basal and proximal cauline 26–56 × 7–15 mm, blade lanceolate to elliptic, base tapered, margins coarsely dentate or serrate, apex obtuse to acute; cauline 4–7 pairs, short-petiolate or sessile, 30–55 × 5–32 mm, blade ovate to lanceolate, base tapered or auriculate-clasping to connate-perfoliate on distal leaves, margins coarsely dentate, apex acute to acuminate. **Thyrses** interrupted, secund, 8–30 cm, axis glabrous or sparsely glandular-pubescent distally, verticillasters 3–10, cymes 1–3-flowered; proximal bracts ovate to triangular, 12–35 × 8–32 mm; peduncles and pedicels ascending to erect, glandular-pubescent. **Flowers:** calyx lobes ovate to elliptic, 3.5–5.5 × 2.5–3.8 mm, glandular-pubescent; corolla pink to rose pink, with reddish purple nectar guides, strongly bilabiate, ventricose, 23–27 mm, glandular-pubescent externally, glandular-pubescent internally, also usually moderately white-lanate abaxially, tube 6–10 mm, length 1.6–2 times calyx lobes, throat abruptly inflated, ± constricted at orifice, 9–12 mm diam., rounded abaxially; stamens: longer pair reaching orifice, filaments of shorter pair glandular-puberulent proximally, pollen sacs explanate, 1.3–1.6 mm, sutures smooth; staminode 16–18 mm, exserted, 0.4–0.6(–1) mm diam., tip straight, glabrous or distal 2–4 mm moderately lanate, hairs yellow, to 1 mm; style 17–22 mm, glabrous. **Capsules** 9–12 × 6–7 mm, glabrous. $2n = 16$.

Flowering Apr–Aug. Volcanic cinder fields, pine forests; of conservation concern; 1900–2600 m; Ariz.

Penstemon clutei occurs only near Indian Flat and Sunset Crater in south-central Coconino County.

231. Penstemon floridus Brandegee, Bot. Gaz. 27: 454. 1899 • Panamint beardtongue E

Herbs. Stems ascending or erect, 50–120 cm, glaucous. **Leaves:** basal and proximal cauline 45–80(–100) × 15–30 (–40) mm, blade obovate to ovate, oblanceolate, or lance-olate, base tapered, margins coarsely dentate, apex obtuse to acute; cauline 6–9 pairs, sessile or proximals short-petiolate, 50–100 × (8–)20–45 mm, blade ovate to lanceolate, base cordate-clasping or tapered, margins coarsely dentate, apex acute, sometimes obtuse. **Thyrses** interrupted, secund, 30–80 cm, axis glabrous or sparsely glandular-pubescent distally, verticillasters 9–19, cymes 1–5-flowered; proximal bracts lanceolate, 10–55 × 2–30 mm; peduncles and pedicels ascending to erect, glandular-pubescent. **Flowers:** calyx lobes ovate, 4.2–6 × 2–3 mm, glandular-pubescent; corolla lavender to light pink, rose pink, or light yellow, with reddish nectar guides, strongly bilabiate, ventricose, 21–32 mm, glandular-pubescent externally, glandular-pubescent internally, tube 6–12 mm, length 2–2.2 times calyx lobes, throat gradually to abruptly inflated, constricted or slightly so at orifice, 6–16 mm diam., rounded abaxially; stamens included, filaments glabrous, pollen sacs explanate, 1.4–1.8 mm, sutures smooth; staminode 16–19 mm, included, 0.7–0.9 mm diam., tip straight, glabrous; style 15–20 mm, glabrous. **Capsules** 8–13 × 4–7 mm, glabrous. $2n = 16$.

Varieties 2 (2 in the flora): w United States.

The two varieties of *Penstemon floridus* are sympatric in the southern Silver Peak Range and southern White Mountains, where they apparently hybridize and introgress (N. H. Holmgren 1984).

1. Corollas 24–32 mm, tubes 7–12 mm, throats abruptly inflated, constricted at orifices, (10–)11– 16 mm diam.. . . . 231a. *Penstemon floridus* var. *floridus*
1. Corollas 21–27 mm, tubes 6–9 mm, throats gradually inflated, slightly constricted at orifices, 6–10(–11) mm diam.. . . . . .231b. *Penstemon floridus* var. *austinii*

231a. Penstemon floridus Brandegee var. **floridus** E

Flowers: corolla 24–32 mm, tube 7–12 mm, throat abruptly inflated, constricted at orifice, (10–)11–16 mm diam.

Flowering Jun–Jul. Gravelly washes, canyon floors, sagebrush shrublands, pinyon-juniper woodlands; 1600–2700 m; Calif., Nev.

Variety *floridus* is known from the central Sierra Nevada, the north end of Owens Valley, and the White Mountains of Inyo and Mono counties, California, and the Silver Peak Range of Esmeralda County, Nevada.

231b. Penstemon floridus Brandegee var. **austinii** (Eastwood) N. H. Holmgren, Brittonia 31: 105. 1979 E

Penstemon austinii Eastwood, Bull. Torrey Bot. Club 32: 206. 1905 (as Pentstemon austini); *P. floridus* subsp. *austinii* (Eastwood) D. D. Keck

Flowers: corolla 21–27 mm, tube 6–9 mm, throat gradually inflated, slightly constricted at orifice, 6–10(–11) mm diam.

Flowering Jun–Jul. Gravelly washes, canyon floors, sagebrush shrublands, pinyon-juniper woodlands; 1000–2400 m; Calif., Nev.

Variety *austinii* is known from the Grapevine, Inyo, Last Chance, Panamint, and White mountains of Inyo County, California, and in Nevada from the Silver Peak Range of Esmeralda County and Shoshone Mountain and Tolicha Peak in Nye County.

232. Penstemon fruticiformis Coville, Contr. U.S. Natl. Herb. 4: 170. 1893 (as Pentstemon) • Death Valley beardtongue E

Shrubs or subshrubs. Stems ascending to erect, 30–60 cm, glaucous. **Leaves:** cauline 6–12 pairs, sessile or short-petiolate, (12–)25–65 × 2–8 mm, blade lanceolate to linear, sometimes oblanceolate, base truncate or tapered, margins entire, rarely ± serrulate distally, apex obtuse to acute. **Thyrses** interrupted, cylindric, 8–30 cm, axis glabrous, verticillasters 3–8, cymes 1–3-flowered; proximal bracts lanceolate to linear, 9–28 × 2–5 mm; peduncles and pedicels spreading to ascending, glabrous. **Flowers:** calyx lobes depressed-ovate to round or ovate, 3.5–6.5 × 2.5–5 mm, glabrous; corolla bluish lavender to lavender or light pink, with reddish purple nectar guides, strongly bilabiate, ventricose-ampliate, 22–28 mm, glabrous or glandular-pubescent externally, moderately to densely white-lanate internally abaxially, sometimes also sparsely to densely glandular-pubescent internally, tube 6–8 mm, length 1–1.5 times calyx lobes, throat abruptly inflated, constricted at orifice, 8–14 mm diam., 2-ridged abaxially; stamens included, filaments of shorter pair glandular-puberulent proximally, pollen sacs explanate to subexplanate, 1.6–2.2 mm, sutures smooth; staminode 15–19 mm, exserted, 0.6–0.8 mm diam., tip recurved, distal 8–9 mm moderately pilose to lanate, hairs yellow, to 2.5 mm; proximal 5–6 mm glandular-pubescent; style 19–24 mm, glabrous. **Capsules** 9–13 × 5–7 mm, glabrous. $2n = 16$.

Varieties 2 (2 in the flora): w United States.

Penstemon fruticiformis is known from the Mojave Desert. Specimens from the Kingston Range and the eastern flank of the southern Sierra Nevada sometimes combine features of *P. fruticiformis* and *P. incertus*.

1. Corollas glabrous externally; calyx lobes depressed-ovate to round or ovate, 3.5–5 × 3.2–5 mm . . . . .232a. *Penstemon fruticiformis* var. *fruticiformis*
1. Corollas ± glandular-pubescent externally; calyx lobes ovate, 4.5–6.5 × 2.5–4 mm . . . . 232b. *Penstemon fruticiformis* var. *amargosae*

232a. Penstemon fruticiformis Coville var.
 fruticiformis [E]

Flowers: calyx lobes depressed-ovate to round or ovate, 3.5–5 × 3.2–5 mm; corolla 24–28 mm, glabrous externally.

Flowering Apr–Jun. Gravelly washes, canyon floors, creosote shrublands, pinyon-juniper woodlands; 1000–2100 m; Calif.

Variety *fruticiformis* is known only from Inyo County. Some specimens from the Last Chance and Panamint ranges have broadly ovate to round calyx lobes and moderately glandular-pubescent corollas internally as in var. *fruticiformis*, but they also have corollas glandular-pubescent externally as in var. *amargosae*.

232b. Penstemon fruticiformis Coville var. **amargosae**
 (D. D. Keck) N. H. Holmgren, Brittonia 44: 481.
 1992 [E]

Penstemon fruticiformis subsp. *amargosae* D. D. Keck, Amer. Midl. Naturalist 18: 801. 1937

Flowers: calyx lobes ovate, 4.5–6.5 × 2.5–4 mm; corolla 22–24 mm, ± glandular-pubescent externally.

Flowering Apr–Jun. Creosote shrublands; 900–1900 m; Calif., Nev.

Variety *amargosae* is known from Inyo and San Bernardino counties, California, and Nye County, Nevada. The glandular hairs of the corollas can be sparse; they usually are easy to see on flower buds.

233. Penstemon grinnellii Eastwood, Bull. Torrey Bot.
 Club 32: 207. 1905 (as Pentstemon) • Grinnell's
 beardtongue [E]

Penstemon grinnellii subsp. *scrophularioides* (M. E. Jones) Munz; *P. grinnellii* var. *scrophularioides* (M. E. Jones) N. H. Holmgren; *P. scrophularioides* M. E. Jones

Herbs. Stems ascending to erect, 10–90 cm, glaucous or not. Leaves: basal and proximal cauline 31–90 × 11–20 mm, blade oblanceolate to elliptic, base tapered, margins finely to coarsely dentate, apex rounded to obtuse; cauline 3–6 pairs, short-petiolate or sessile, (15–)30–90 × 4–30 mm, blade lanceolate to ovate-oblong, base tapered, margins coarsely dentate, distals sometimes entire, apex obtuse or acute. Thyrses interrupted, cylindric, (5–)7–34 cm, axis sparsely

to densely glandular-pubescent, verticillasters 4–10, cymes 2–4-flowered; proximal bracts ovate to lanceolate, 5–50 × 2–18 mm; peduncles and pedicels spreading to ascending, glandular-pubescent. Flowers: calyx lobes ovate, 4.5–8.5 × (2–)2.9–3.2 mm, glandular-pubescent; corolla white to pink or light lavender to light blue or light violet, with reddish purple nectar guides, strongly bilabiate, ventricose-ampliate, 22–35 mm, glandular-pubescent externally, glandular-pubescent internally, also white-villous abaxially, tube 5–8 mm, length 0.9–1.2 times calyx lobes, throat abruptly inflated, constricted at orifice, 10–18 mm diam., rounded abaxially; stamens included, filaments of shorter pair glandular-puberulent proximally, pollen sacs navicular to subexplanate, 1.6–2.2 mm, sutures smooth; staminode 20–22 mm, exserted, 0.4–0.8 mm diam., tip recurved, distal 8–10 mm densely lanate, hairs yellowish or whitish, to 4 mm; proximal 6–8 mm densely glandular-pubescent; style 15–18 mm, glabrous. Capsules 10–14 × 5–8 mm, glabrous. 2*n* = 16.

Flowering Apr–Jul. Chaparral, foothills, pine-juniper woodlands, pine forests; 500–2700 m; Calif.

Two varieties of *Penstemon grinnellii* have been recognized. Variety *grinnellii*, with non-glaucous stems 10–60 cm and corollas white to pink and 22–30 mm, has been reported from the Western Transverse and Peninsular ranges, and in the San Bernardino, San Gabriel, and San Jacinto mountains. Variety *scrophularioides*, with glaucous stems 45–90 cm and corollas light lavender, light blue, or light violet and 26–35 mm, has been reported in the San Francisco Bay Area, southern Sierra Nevada and South Coast and Western Transverse ranges. Characteristics that distinguish the varieties overlap in some populations.

Penstemon ×*peirsonii* Munz & I. M. Johnston, a putative hybrid between *P. grinnellii* and *P. speciosus*, and *P.* ×*dubius* Davidson, a putative hybrid between *P. centranthifolius* and *P. grinnellii*, both have been reported from California (D. D. Keck 1937b).

234. Penstemon incertus Brandegee, Bot. Gaz. 27: 454.
 1899 (as Pentstemon) • Mojave beardtongue [E]

Shrubs or subshrubs. Stems ascending to erect, 22–100 cm, glaucous. Leaves: cauline 6–18 pairs, sessile, (8–)25–70 × (1–)2.5–4(–6) mm, blade lanceolate to linear, base tapered, margins entire, apex acute to acuminate. Thyrses interrupted, cylindric, 3–27 cm, axis sparsely glandular-pubescent, rarely glabrous, verticillasters (2–)4–9, cymes 1–3(–5)-flowered; proximal bracts lanceolate to linear, 3–35 × 1–5 mm; peduncles and pedicels spreading or ascending, glandular-pubescent.

Flowers: calyx lobes ovate, 3.8–6 × 2.6–4 mm, glabrous or sparsely glandular-pubescent; corolla blue to violet or purple, without nectar guides, strongly bilabiate, ventricose-ampliate, 23–32 mm, glabrous or glandular-pubescent externally, sparsely white-lanate internally abaxially, glandular-pubescent adaxially, tube 7–11 mm, length 1.8–2 times calyx lobes, throat gradually inflated, not constricted at orifice, 8–13 mm diam., slightly 2-ridged abaxially; stamens included, filaments of shorter pair glandular-puberulent proximally, pollen sacs navicular, 1.8–2.4 mm, sutures denticulate, teeth to 0.1 mm; staminode 10–13 mm, included, 0.6–0.7 mm diam., tip straight, distal 3–7 mm sparsely to densely pilose, hairs yellow, to 1 mm; proximal 1–3 mm glandular-pubescent; style 19–23 mm, glabrous. **Capsules** 10–17 × 5–7 mm, glabrous.

Flowering Apr–Jul. Sandy soils, sagebrush shrublands, Joshua tree and pinyon-juniper woodlands; 800–1800 m; Calif.

Populations of *Penstemon incertus* are concentrated in the southern Sierra Nevada and adjacent Tehachapi Mountains, with other populations scattered in mountain ranges in and near the Mojave Desert. The species is documented in Inyo, Kern, Los Angeles, Riverside, San Bernardino, San Diego, and Tulare counties. Occasional specimens from the Kingston Range and the eastern flank of the southern Sierra Nevada combine features of *P. incertus* and *P. fruticiformis*.

235. Penstemon palmeri A. Gray, Proc. Amer. Acad. Arts 7: 379. 1868 (as Pentstemon) • Palmer's or scented beardtongue E F

Herbs. Stems ascending to erect, 43–140 cm, glaucous. **Leaves:** basal and proximal cauline (28–)52–120(–145) × (5–)25–50 (–60) mm, blade ovate, base tapered, margins coarsely dentate, apex obtuse to acute; cauline 5–8 pairs, sessile or connate-perfoliate, 25–130 × 7–45 mm, blade ovate to triangular, base of distals auriculate-clasping or connate-perfoliate, margins coarsely dentate, distals sometimes entire, apex acute. **Thyrses** interrupted, rarely continuous, secund, 7–63 cm, axis glabrous or glandular-pubescent, verticillasters (4–)7–17, cymes (1 or)2–5-flowered; proximal bracts ovate to lanceolate, 8–40 × 4–38 mm; peduncles and pedicels ascending to erect, glabrous or glandular-pubescent. **Flowers:** calyx lobes ovate, (3.4–)4.2–6(–7.5) × 1.9–3.6 mm, glabrous or glandular-pubescent; corolla pale pink to white or pinkish lavender, with reddish purple nectar guides, strongly bilabiate, ventricose-ampliate, 25–38 mm, glandular-pubescent externally, glandular-pubescent internally,

sometimes also sparsely white-lanate abaxially, tube 4–8 mm, length 1–1.4 times calyx lobes, throat abruptly inflated, constricted at orifice, 12–21 mm diam., rounded abaxially; stamens included or longer pair reaching orifice, filaments of shorter pair glandular-puberulent proximally, pollen sacs navicular, 1.6–2.4 mm, sutures smooth; staminode 25–30 mm, exserted, 0.9–1.1 mm diam., tip strongly recurved to coiled, distal 8–10 mm, and especially tip, moderately to densely lanate, hairs yellow, to 3.5 mm, proximal 6–8 mm glandular-puberulent; style 19–23 mm, glabrous or sparsely glandular-pubescent proximally. **Capsules** 11–16 × 6–8 mm, sparsely glandular-pubescent distally.

Varieties 3 (3 in the flora): w United States.

Penstemon ×*bryantiae* D. D. Keck, a hybrid between *P. palmeri* and *P. spectabilis*, was described based on plants that volunteered in the Rancho Santa Ana Botanic Garden in 1935 (D. D. Keck 1937b). *Penstemon* ×*mirus* A. Nelson, a putative hybrid between *P. eatonii* and *P. palmeri*, has been reported from Arizona (A. Nelson 1938).

A poultice prepared from *Penstemon palmeri* is applied to snakebites by the Kayenta Navajo of northeastern Arizona (D. E. Moerman 1998).

1. Corolla tubes 7–8 mm; distal cauline leaves sessile, sometimes connate-perfoliate. 235c. *Penstemon palmeri* var. *macranthus*
1. Corolla tubes 4–6 mm; distal cauline leaves connate-perfoliate.
   2. Peduncles, pedicels, and calyx lobes glandular-pubescent; pollen sacs 1.8–2.4 mm . . . . 235a. *Penstemon palmeri* var. *palmeri*
   2. Peduncles, pedicels, and calyx lobes glabrous; pollen sacs 1.6–2(–2.2) mm . . . . 235b. *Penstemon palmeri* var. *eglandulosus*

235a. Penstemon palmeri A. Gray var. palmeri E F

Distal cauline leaves connate-perfoliate. **Thyrses:** peduncles and pedicels glandular-pubescent. **Flowers:** calyx lobes glandular-pubescent; corolla tube 4–6 mm; pollen sacs 1.8–2.4 mm. 2*n* = 16.

Flowering May–Jul. Washes, roadsides, canyon floors, creosote shrublands, pinyon-juniper woodlands; 800–2800 m; Ariz., Calif., Colo., Idaho, Nev., N.Mex., Utah, Wash., Wyo.

Because of its stature and showy flowers, var. *palmeri* is seeded widely in the western United States along roadsides after construction. Populations in Colorado, Idaho, Washington, and Wyoming are introduced; some in New Mexico also appear to have resulted from seeding.

235b. Penstemon palmeri A. Gray var. **eglandulosus**
(D. D. Keck) N. H. Holmgren, Brittonia 31: 105.
1979 [E]

Penstemon palmeri subsp.
eglandulosus D. D. Keck, Amer.
Midl. Naturalist 18: 797. 1937

Distal cauline leaves connate-
perfoliate. **Thyrses:** peduncles
and pedicels glabrous. **Flowers:**
calyx lobes glabrous; corolla
tube 4–6 mm; pollen sacs
1.6–2(–2.2) mm.

Flowering Jun–Jul.   Washes, roadsides, desert
shrublands, pinyon-juniper woodlands, pine woodlands; 1500–2600 m; Ariz., N.Mex., Utah.

Variety *eglandulosus* occurs from the southern end of
the Utah Plateau (Emery, Garfield, Grand, Iron, Kane,
San Juan, Washington, and Wayne counties, Utah) to
the Kaibab Plateau in northern Coconino and Mohave
counties, Arizona.   A population in Socorro County,
New Mexico, appears to have been seeded along a
roadside.

235c. Penstemon palmeri A. Gray var. **macranthus**
(Eastwood) N. H. Holmgren, Brittonia 31: 105.
1979 • Lahontan beardtongue [E]

Penstemon macranthus Eastwood,
Bull. Torrey Bot. Club 32: 207.
1905 (as Pentstemon)

Distal cauline leaves sessile,
sometimes connate-perfoliate.
Thyrses: peduncles and pedicels
glandular-pubescent. **Flowers:**
calyx lobes glandular-pubescent;
corolla tube 7–8 mm; pollen
sacs 1.9–2.1 mm.

Flowering May–Jul(–Aug).   Washes, roadsides,
canyons, pinyon-juniper and sagebrush-juniper woodlands; 1000–1400 m; Nev.

Variety *macranthus* has been documented in
Churchill, Nye, and Pershing counties.

236. Penstemon pseudospectabilis M. E. Jones, Contr.
W. Bot. 12: 66. 1908 (as Pentstemon) • Mojave
beardtongue

Herbs.   Stems ascending to
erect, (30–)60–100 cm, slightly
glaucous or not. **Leaves:** basal
and proximal cauline 20–165 ×
12–50 mm, blade ovate to
lanceolate, base tapered, margins finely to coarsely dentate,
apex obtuse to acute; cauline
4–7 pairs, sessile or proximals
short-petiolate, 45–90 × 22–76 mm, blade ovate to
triangular, base tapered or sessile on proximals,

auriculate-clasping to connate-perfoliate on distal ones,
margins finely to coarsely dentate, apex acute. **Thyrses**
interrupted or continuous, secund, (6–)12–48 cm,
axis glabrous, verticillasters 4–13, cymes 1–3(–6)-
flowered; proximal bracts ovate (connate-perfoliate) or
lanceolate (auriculate-clasping), (9–)16–55 × 10–60 mm;
peduncles and pedicels ascending to erect, glabrous or
glandular-pubescent.   **Flowers:** calyx lobes ovate,
3–5.5 × 2–3.2 mm, glabrous or glandular-pubescent;
corolla rose pink, with reddish purple nectar guides,
strongly bilabiate, ventricose, (18–)25–34 mm,
glandular-pubescent externally, glandular-pubescent
internally, tube 6–9 mm, length 1.6–2 times calyx
lobes, throat gradually inflated, ± constricted at orifice,
7–10 mm diam., rounded abaxially; stamens: longer
pair reaching orifice, filaments of shorter pair glandular-
puberulent proximally, pollen sacs explanate,
1.5–2 mm, sutures smooth; staminode 11–17 mm,
included, 0.2–0.4 mm diam., tip straight to recurved,
glabrous or distal 1–2 mm sparsely to moderately lanate,
hairs yellow, to 1 mm; style 20–25 mm, glabrous.
Capsules 7–12 × 4–6 mm, glabrous.

Varieties 2 (2 in the flora): sw United States,
n Mexico.

Two varieties, distinguished on the basis of
inflorescence indument, are largely allopatric.
Penstemon ×*crideri* A. Nelson, a putative hybrid
between *P. eatonii* and *P. pseudospectabilis*, has been
reported from Arizona (A. Nelson 1936).

1. Peduncles, pedicels, and calyx lobes glandular-
   pubescent . . . . .236a. *Penstemon pseudospectabilis*
   var. *pseudospectabilis*
1. Peduncles, pedicels, and calyx lobes glabrous
   . . . . 236b. *Penstemon pseudospectabilis*
   var. *connatifolius*

236a. Penstemon pseudospectabilis M. E. Jones var.
pseudospectabilis

Thyrses: peduncles and pedicels
glandular-pubescent. **Flowers:**
calyx lobes glandular-
pubescent.

Flowering Mar–Apr. Rocky
or gravelly washes, canyon
floors, creosote shrublands,
juniper-pine woodlands; 100–
1400 m; Ariz., Calif., Utah;
Mexico (Sonora).

Variety *pseudospectabilis* is known from southern
California (Riverside and San Bernardino counties),
western and south-central Arizona (Coconino, Mohave,
Pima, and Pinal counties), and northwestern Sonora,
Mexico.   A collection from Kane County, Utah, may
have come from an introduction (E. C. Neese and
N. D. Atwood 2003).

236b. Penstemon pseudospectabilis M. E. Jones var. **connatifolius** (A. Nelson) C. C. Freeman, PhytoKeys 80: 37. 2017　E

Penstemon connatifolius A. Nelson, Amer. J. Bot. 18: 437. 1931 (as Pentstemon); *P. pseudospectabilis* subsp. *connatifolius* (A. Nelson) D. D. Keck

Thyrses: peduncles and pedicels glabrous. **Flowers:** calyx lobes glabrous. $2n = 16$.

Flowering Feb–Jun, Oct–Nov. Rocky or gravelly slopes, washes, pine-oak woodlands; 600–2300 m; Ariz., N.Mex.

Variety *connatifolius* is known from northwestern Arizona (Cochise, Coconino, Gila, Graham, Greenlee, Maricopa, Mohave, Navajo, and Pima counties) and southwestern New Mexico (Catron, Doña Ana, Grant, Hidalgo, and Luna counties).

237. Penstemon rubicundus D. D. Keck, Amer. Midl. Naturalist 18: 802. 1937 • Wassuk beardtongue　E

Herbs. Stems erect, 50–120 cm, glaucous. **Leaves:** basal and proximal cauline 25–100 × (5–)10–35 mm, blade oblanceolate, base tapered, margins coarsely dentate, apex obtuse to acute; cauline 3–5 pairs, sessile, 25–65 × 12–25 mm, blade lanceolate, base sessile, clasping or distal ones connate-perfoliate, margins coarsely dentate, apex acute. **Thyrses** interrupted, secund, 20–60 cm, axis glabrous, sometimes sparsely glandular-pubescent distally, verticillasters 10–16, cymes 2- or 3-flowered; proximal bracts lanceolate, 15–60 × 4–22 mm; peduncles and pedicels ascending to erect, glandular-pubescent. **Flowers:** calyx lobes ovate, 4–5.5 × 2–2.7 mm, glandular-pubescent; corolla pink to rose pink, with reddish nectar guides, strongly bilabiate, ventricose to ventricose-ampliate, (27–)30–35 mm, glandular-pubescent externally, sparsely white-lanate internally abaxially, not glandular-pubescent, tube 7–10 mm, length 1.8–2 times calyx lobes, throat gradually inflated, not constricted at orifice, 10–16 mm diam., rounded abaxially; stamens included or longer pair exserted, filaments glabrous, pollen sacs navicular, 1.2–1.5 mm, sutures smooth; staminode 22–24 mm, exserted, 0.4–0.6 mm diam., tip recurved, distal 6–7 mm pilose, hairs yellow or golden yellow, to 2 mm; style 18–24 mm, glabrous. **Capsules** 7–11 × 4–6 mm, glabrous.

Flowering May–Sep. Rocky or gravelly slopes, drainages, roadsides, sagebrush or shadscale shrublands, pine-juniper woodlands; 1300–2100 m; Nev.

Penstemon rubicundus is known from the Pine Nut Mountains (Douglas County) and the Wassuk Range (Mineral County) of western Nevada.

238. Penstemon spectabilis Thurber ex A. Gray in War Department [U.S.], Pacif. Railr. Rep. 4(5): 119. 1857 (as Pentstemon) • Showy beardtongue

Herbs. Stems erect, (50–)80–120 cm, glaucous or not. **Leaves:** basal and proximal cauline 40–100 × 8–50 mm, blade oblanceolate to elliptic, base tapered, margins coarsely dentate, apex obtuse to acute; cauline 5–8 pairs, sessile, 32–120 × 10–45 mm, blade ovate to lanceolate, base sessile, proximals usually auriculate-clasping, distal ones usually cordate-clasping, margins dentate to serrate, apex acute to acuminate. **Thyrses** interrupted, cylindric, 20–80 cm, axis glabrous or glandular-pubescent distally, verticillasters 9–16, cymes (1–)3–8(–13)-flowered; proximal bracts ovate to lanceolate, 15–35 × 4–25 mm; peduncles and pedicels spreading to ascending, glabrous or glandular-pubescent. **Flowers:** calyx lobes round to ovate or lanceolate, 3.5–5.5 × 2.5–3.2 mm, glabrous or glandular-pubescent; corolla lavender, blue, pink, or purple, with reddish purple nectar guides, strongly bilabiate, ventricose-ampliate, 24–34 mm, glandular-pubescent externally, glabrous or sparsely white-lanate internally abaxially, glandular-pubescent adaxially, tube 7–9 mm, length 1.6–2 times calyx lobes, throat abruptly inflated, not constricted at orifice, 8–14 mm diam., rounded abaxially; stamens included or longer pair reaching orifice, filaments of shorter pair glandular-puberulent proximally, pollen sacs navicular, 1.8–2.4 mm, sutures papillate or denticulate, papillae or teeth to 0.1 mm; staminode 15–16 mm, included, 0.5–0.9 mm diam., tip straight, glabrous distally or distal 1–2 mm sparsely pilose, hairs yellow, to 0.5 mm [5–6 mm moderately pilose, hairs yellow, to 1.5 mm], proximal 1–2 mm glandular-pubescent; style 15–19 mm, glabrous. **Capsules** 9–14 × 6–8 mm, glabrous. $2n = 16$.

Varieties 3 (2 in the flora): w United States, nw Mexico.

Penstemon spectabilis consists of three partially sympatric varieties: var. *spectabilis* in southern California and northern Baja California, Mexico; var. *subinteger* (D. D. Keck) C. C. Freeman known only from the area of El Rosario in Baja California; and var. *subviscosus* in southern California. *Penstemon* ×*bryantiae* D. D. Keck, a hybrid between *P. palmeri* and *P. spectabilis*, was described based on plants that volunteered in the Rancho Santa Ana Botanic Garden in 1935 (D. D. Keck 1937b). *Penstemon* ×*parishii* A. Gray is a hybrid between *P. centranthifolius* and *P. spectabilis* reported from Los Angeles, Riverside, San Bernardino, and San Diego counties, California (Keck; Paul Wilson and M. Valenzuela 2002).

1. Pedicels and calyx lobes glabrous . . . .
 . . . . 238a. *Penstemon spectabilis* var. *spectabilis*
1. Pedicels and calyx lobes glandular-pubescent . . .
 . . . . 238b. *Penstemon spectabilis* var. *subviscosus*

238a. Penstemon spectabilis Thurber ex A. Gray var. spectabilis

Thyrses: axis glabrous; pedicels glabrous. **Flowers:** calyx lobes glabrous.

Flowering Apr–Jul. Sandy or gravelly slopes, washes, sagebrush shrublands, chaparral, oak woodlands; 100–1800 m; Calif., Nev.; Mexico (Baja California).

In California, var. *spectabilis* is known from the Peninsular and Transverse ranges in Los Angeles, Riverside, San Bernardino, and San Diego counties; its range extends into northern Baja California, Mexico.

238b. Penstemon spectabilis Thurber ex A. Gray var. subviscosus (D. D. Keck) McMinn, Man. Calif. Shrubs, 513. 1939 E

Penstemon spectabilis subsp. *subviscosus* D. D. Keck, Amer. Midl. Naturalist 18: 818. 1937

Thyrses: axis glabrous or glandular-pubescent distally; pedicels glandular-pubescent. **Flowers:** calyx lobes glandular-pubescent.

Flowering Apr–Jul. Sandy or gravelly slopes, washes, sagebrush shrublands, chaparral, oak woodlands; 500–2400 m; Calif.

Variety *subviscosus* is known from the Transverse Ranges. Populations are documented in Los Angeles, San Bernardino, San Diego, San Luis Obispo, and Santa Barbara counties.

239. Penstemon stephensii Brandegee, Zoë 5: 151. 1903 (as Pentstemon stephensi) • Stephens's beardtongue C E

Herbs. **Stems** ascending to erect, 30–150 cm, glaucous, sometimes not. **Leaves:** basal and proximal cauline 35–80 × 18–30 mm, blade spatulate to ovate or elliptic, base tapered, margins coarsely serrate to coarsely dentate, apex rounded to acute; cauline 4–7 pairs, sessile, 18–80 × 10–40 mm, blade ovate-oblong to triangular-ovate, base connate-perfoliate, margins finely serrate, apex obtuse to acute or acuminate. **Thyrses** interrupted, secund, 16–45 cm, axis glabrous or sparsely glandular-pubescent distally, verticillasters 8–14, cymes 2–6-flowered; proximal bracts depressed-ovate to ovate or triangular, 12–50 × 15–55 mm; peduncles and pedicels erect, sparsely glandular-pubescent. **Flowers:** calyx lobes ovate, 3–4.2(–5) × 1.5–2.5 mm, glandular-pubescent; corolla rose to magenta or pinkish lavender, without nectar guides, bilabiate, funnelform to slightly ventricose, 16–20 mm, glandular-pubescent externally, glandular-pubescent internally, tube 5–6 mm, length 1.7–2 times calyx lobes, throat gradually inflated, not constricted at orifice, 4–6 mm diam., rounded abaxially; stamens included, filaments glabrous, pollen sacs explanate, 1.1–1.5 mm, sutures smooth; staminode 9–13 mm, included, 0.2–0.3 mm diam., tip straight, glabrous; style 10–12 mm, glabrous. **Capsules** 7–10 × 4–6 mm, glabrous.

Flowering Apr–Jun. Rocky slopes, washes, rock crevices, creosote shrublands, pinyon-juniper woodlands; of conservation concern; 1200–1700 m; Calif.

Penstemon stephensii is known from the Granite, Kingston, Nopah, and Providence ranges in Inyo and San Bernardino counties.

27. TONELLA Nuttall ex A. Gray, Proc. Amer. Acad. Arts 7: 378. 1868 • [Derivation unknown; perhaps a misspelling of Latin *tenella*, delicate, alluding to filiform branches] E

Craig C. Freeman

Herbs, annual. **Stems** ascending to erect, glabrous proximally, glandular-hairy distally. **Leaves** cauline, opposite; petiole present or absent; blade not fleshy, not leathery, margins entire, crenate, dentate, or serrate. **Inflorescences** terminal, racemes; bracts present. **Pedicels** present, glandular-hairy, sometimes glabrous or glabrate; bracteoles absent. **Flowers** bisexual; sepals 5, proximally connate, calyx ± bilaterally symmetric, subrotate to campanulate, lobes triangular to lanceolate; corolla white on tube and lobes proximally, blue to violet or lavender on lobes distally, with dark violet spots internally near bases of adaxial and lateral lobes, bilaterally symmetric, weakly bilabiate, subrotate, tube base gibbous adaxially, sometimes obscurely so, not spurred abaxially, throat not densely pilose internally, lobes 5, abaxial 3, middle lobe of abaxial lip not folded lengthwise, not enclosing stamens and style, adaxial 2; stamens 4, medially adnate to corolla, equal or ± didynamous, exserted, filaments glandular; staminode 1, minute; ovary 2-locular, placentation axile; stigma linear. **Fruits** capsules, dehiscence septicidal and loculicidal. **Seeds** 2 or 4, dark brown to black, reniform to ovoid, wings absent.

Species 2 (2 in the flora): w North America.

Studies of vascular anatomy, organography, and gametophyte development (G. F. Schrock and B. F. Palser 1967) and molecular phylogenetic analyses (B. G. Baldwin et al. 2011) support the monophyly of *Tonella* and its sister-relationship to *Collinsia.*

SELECTED REFERENCE Baldwin, B. G., S. Kalisz, and W. S. Armbruster. 2011. Phylogenetic perspectives on diversification, biogeography, and floral evolution of *Collinsia* and *Tonella* (Plantaginaceae). Amer. J. Bot. 98: 731–753.

1. Corollas 5–7 × 6–12 mm; racemes: flowers 2–10 per node . . . . 1. *Tonella floribunda*
1. Corollas 2–2.5 × 2–4 mm; racemes: flowers 1–3 per node . . . . 2. *Tonella tenella*

1. **Tonella floribunda** A. Gray in W. H. Brewer et al., Bot. California 1: 556. 1876 • Many-flower tonella
E F

Annuals 7–40 cm. **Stems** erect, branched, sometimes simple, glandular-pilose. **Leaves:** petiole 0–15 mm; proximal cauline leaves simple or tripartite, margins entire or crenate to dentate or serrate, medial cauline leaves deeply tripartite, segments lanceolate-elliptic, margins entire, sometimes crenate to serrate; surfaces sparsely pilose adaxially, abaxially along midvein, and along margins, sometimes glabrous. **Racemes:** flowers 2–10 per node; bracts deeply tripartite, sometimes simple, segments linear to narrowly lanceolate or elliptic, margins entire or serrate. **Pedicels** 10–25 mm, sparsely to densely glandular-pubescent. **Flowers:** calyx subrotate, lobes narrowly triangular to lanceolate, 0.5–2 × 0.4–1 mm; lateral lobes of abaxial lip appearing closely associated with medial lobe; corolla 5–7 × 6–12 mm; style 5–7 mm. **Capsules** 2–3 × 2–3 mm. **Seeds** 4, 1.2–3 × 0.7–1 mm.

Flowering Apr–Jul. Open rocky canyons and slopes, open pine forests; 30–1600 m; Idaho, Oreg., Wash.

Populations of *Tonella floribunda* are concentrated mainly in the Snake River Canyon and close tributaries in west-central Idaho, northeastern Oregon, and southeastern Washington, with scattered, disjunct populations in northern (Shoshone County) and southwestern (Owyhee County) Idaho, and along the Columbia River below its confluence with the Snake River.

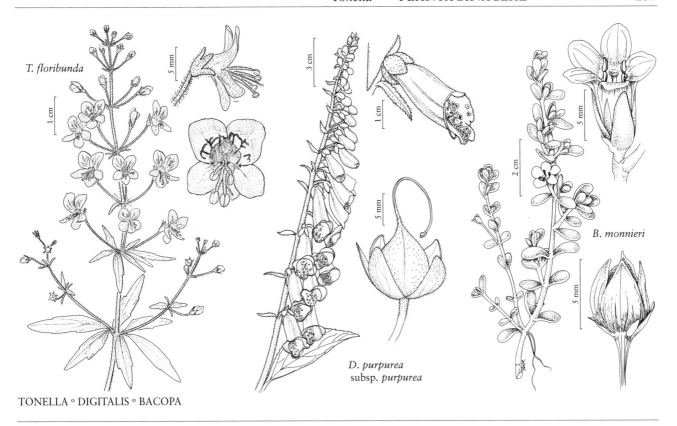

T. floribunda

D. purpurea subsp. *purpurea*

B. monnieri

TONELLA ∘ DIGITALIS ∘ BACOPA

2. **Tonella tenella** (Bentham ex A. de Candolle) A. Heller, Muhlenbergia 1: 5. 1900 • Lesser baby-innocence, small-flower tonella E

Collinsia tenella Bentham ex A. de Candolle in A. P. de Candolle and A. L. P. P. de Candolle, Prodr. 10: 593. 1846

Annuals 5–37 cm. **Stems** ascending to erect, branched, sometimes simple, glabrous, sometimes minutely pubescent distally. **Leaves:** petiole 0–15 mm; proximal cauline leaves simple, margins usually crenate, medial cauline leaves tripartite, segments narrowly elliptic to lanceolate or linear, margins entire or serrate; surfaces sparsely pilose adaxially, abaxially along midvein, and along margins, sometimes glabrous. **Racemes:** flowers 1–3 per node; bracts elliptic to lanceolate or deeply tripartite, segments narrowly elliptic to lanceolate, margins entire or crenate to serrate. **Pedicels** 8–16 mm, sparsely glandular-hairy distally, sometimes glabrous or glabrate. **Flowers:** calyx subrotate to campanulate, lobes triangular to oblong, 0.5–1.5 × 0.4–1 mm; lateral lobes of abaxial lip not closely associated with medial lobe, widely spreading and closely associated with lobes of adaxial lip; corolla 2–2.5 × 2–4 mm; style 1.5–2.5 mm. **Capsules** 1.5–3 × 1.5–2 mm. **Seeds** 2, 1.6–2 × 0.8–1 mm.

Flowering Mar–Jun. Moist, shaded canyons, moist open coniferous and deciduous forests, talus slopes; 20–1600 m; B.C.; Calif., Oreg., Wash.

Tonella tenella occurs from San Benito County, California, and the High Sierra Nevada north to Washington mostly in the Cascade and Coast ranges but extending eastward through the Columbia River Gorge. Disjunct populations occur in central Oregon, and on Saltspring Island and Mount Tzouhalmen on Vancouver Island in southwestern British Columbia.

28. DIGITALIS Linnaeus, Sp. Pl. 2: 621. 1753; Gen Pl. ed. 5, 272. 1754 • Foxglove

[Latin *digitalis*, finger of a glove, alluding to resemblance of tubular flowers to glove fingers] [I]

Kerry A. Barringer

Neil A. Harriman

Herbs [shrubs], biennial or perennial. **Stems** erect, simple or branching from base, glabrous, glabrate, pilose, or villous. **Leaves** basal and cauline, alternate, smaller distally; petiole absent [present]; blade not fleshy, not leathery, margins entire or serrate to coarsely doubly serrate. **Inflorescences** terminal, racemes, often secund; bracts present. **Pedicels** present; bracteoles usually absent. **Flowers** bisexual; sepals 5, distinct, narrowly triangular to lanceolate or ovate-lanceolate, calyx ± bilaterally symmetric, campanulate; corolla brown, yellow, pink to purple, or white, bilaterally symmetric, ± bilabiate, funnelform, tubular-funnelform, or globular to ovoid, tube base not spurred or gibbous, lobes 5, abaxial 3, adaxial 2; stamens 4, adnate to corolla, didynamous, filaments glabrous or hairy; staminode 0; ovary 2-locular, placentation axile; stigma 2-lobed or punctiform. **Fruits** capsules, dehiscence septicidal, sometimes secondarily loculicidal. **Seeds** 20–60, brown to black, prismatic or cylindric to ovoid, wings absent. $x = 28$.

Species 22 (4 in the flora): introduced; Europe, w Asia; introduced also nearly worldwide.

All species of *Digitalis* are poisonous, containing cardiac glycosides including digitoxin. In addition to the following species, *D. ferruginea* Linnaeus is sometimes found in cultivation in North America. It has yellow to yellow-brown corollas, like *D. lanata*, but the corolla tubes are elongate, not globular.

SELECTED REFERENCES Aronson, J. K. 1986. An Account of the Foxglove and Its Medicinal Uses, 1785–1985. Oxford. Werner, K. 1965. Taxonomie und Phylogenie der Gattungen *Isoplexis* Lindl. und *Digitalis* L. Repert. Spec. Nov. Regni Veg. 70: 109–135.

1. Corolla tubes globular to ovoid; leaf blade margins entire. . . . . .2. *Digitalis lanata*
1. Corolla tubes funnelform or tubular-funnelform; leaf blade margins serrate or serrate at least distally.
  2. Corolla tubes 13–15 mm; throats 5–7 mm diam. . . . . .3. *Digitalis lutea*
  2. Corolla tubes 25–60 mm; throats 14–25 mm diam.
    3. Corolla tubes pale yellow; leaf blade margins finely and evenly serrate distally. . . .
      . . . . .1. *Digitalis grandiflora*
    3. Corolla tubes purple-pink to white; leaf blade margins coarsely serrate . . . . 4. *Digitalis purpurea*

1. Digitalis grandiflora Miller, Gard. Dict. ed. 8,
Digitalis no. 4 [as magno flore], corr. 1768 • Yellow
foxglove [I]

Digitalis ambigua Murray;
D. orientalis Miller

Stems 50–100 cm, pilose to villous, hairs glandular and eglandular. **Leaves:** blade lanceolate to oblanceolate, 10–21 × 2–4 cm, margins finely and evenly serrate distally, with a glandular tip. **Inflorescences** secund, glandular-pilose, bracts 20–30 mm. **Pedicels** pendent, 5–12 mm, glandular-pilose. **Flowers:** sepals narrowly lanceolate, 8–12 × 1.5–2 mm, villous; corolla tube pale yellow, funnelform, 25–40 mm, throat 14–20 mm diam., abaxial lip pendent, pale yellow often marked by brown veins, broadly rounded to broadly triangular, 3–4 mm. **Capsules** ovoid, 8–11 mm, glandular-pilose. **Seeds** brown to black, prismatic, 1 mm, finely reticulate-alveolate. $2n = 56$ (Europe).

Flowering Mar–Jul. Disturbed sites, roadsides, old fields; 0–1000 m; introduced; Ont.; Conn., Maine, Mass., Mich., Minn., Mont., N.H., N.Y., Ohio, Vt., Wis.; e Europe; w Asia; introduced also in South America, elsewhere in Europe, elsewhere in Asia, Africa, Australia.

Plants of *Digitalis grandiflora* are occasionally found in cultivation and sometimes escape in the northeastern part of the flora area.

Although the names *Digitalis grandiflora* and *D. orientalis* were published at the same time, *D. grandiflora* has long been the preferred name for this species and has been in general use since it was published. The name *D. orientalis* Miller was long confused with the later homonym *D. orientalis* Lamarck, a synonym of *D. lanata*.

2. Digitalis lanata Ehrhart, Beitr. Naturk. 7: 152. 1792
• Grecian or woolly foxglove, digitale laineuse I

Stems 30–100 cm, glabrous or glabrate. **Leaves:** blade lanceolate to oblanceolate, 5–15 × 1–2 cm, margins entire. **Inflorescences** not secund, villous; bracts 15–30 mm. **Pedicels** spreading, 1–4 mm, villous. **Flowers:** sepals narrowly lanceolate to narrowly triangular, 8–10 × 1.5–2 mm, villous; corolla tube yellow to yellow-brown with red to brown veins, globular to ovoid, 10–15 mm, throat 10–15 mm diam., abaxial lip strongly curved, white, lingulate, 7–15 mm. **Capsules** ovoid-conical, 10–15 mm, villous. **Seeds** brown to black, prismatic, 1 mm, finely reticulate-alveolate. $2n = 56$ (Asia).

Flowering Jun–Jul. Disturbed sites, roadsides, abandoned lots; 0–1000 m; introduced; Ont., Que.; Conn., Ind., Kans., Maine, Md., Mass., Mich., Minn., Nebr., N.H., N.J., N.Y., Ohio, Pa., Vt., W.Va., Wis.; Eurasia; introduced also in South America, elsewhere in Europe, elsewhere in Asia, Africa.

Plants of *Digitalis lanata* are the principal source of the drug digitalin. *Digitalis lanata* can be confused with *D. leucophaea* Sibthorp & Smith, which is rarely cultivated and has linear bracts and smaller flowers.

3. Digitalis lutea Linnaeus, Sp. Pl. 2: 622. 1753
• Straw foxglove, digitale jaune I

Stems 50–80 cm, glabrous. **Leaves:** blade oblanceolate to narrowly lanceolate, 7–18 × 2–4 cm, margins serrate. **Inflorescences** secund, glabrous; bracts 5–15 mm. **Pedicels** spreading to slightly pendent, 3–4 mm, glabrous or sparsely pilose. **Flowers:** sepals lanceolate, 6–8 × 2–3 mm, glabrous or sparsely glandular; corolla tube pale yellow, tubular-funnelform, 13–15 mm, throat 5–7 mm diam., abaxial lip pendent or spreading, pale yellow, lingulate, 4–5 mm. **Capsules** ovoid-conical, 8–12 mm, glabrous. **Seeds** brown, prismatic, 1 mm, reticulate-alveolate. $2n = 56$ (Europe).

Flowering Jun–Jul. Disturbed sites, roadsides; 0–1000 m; introduced; B.C., Que.; Conn., Md., Mass., Mich., N.H., N.Y., Ohio, Pa., Vt.; Europe; introduced also in South America, Asia.

Digitalis lutea is easy to grow and is distinguished by its relatively small, tubular-funnelform flowers.

4. Digitalis purpurea Linnaeus, Sp. Pl. 2: 621. 1753
• Purple foxglove, digitale pourpre F I W

Subspecies 5 (1 in the flora): introduced; Europe; introduced also in Mexico, Central America, South America, Asia, Africa, Pacific Islands (New Zealand), Australia.

Digitalis purpurea was once used as a commercial source of digitalin, is widely cultivated, and has many cultivars. Some plants have been identified as European subspecies; all variability in the flora area appears to be from cultivars of subsp. *purpurea*. *Digitalis* ×*mertonensis* B. H. Buxton & C. D. Darlington (strawberry or giant foxglove) is a hybrid of *D. purpurea* with *D. grandiflora* that is sometimes cultivated.

4a. Digitalis purpurea Linnaeus subsp. **purpurea**
F I W

Stems 50–270 cm, hairy, hairs glandular and eglandular. **Leaves:** blade lanceolate to oblanceolate or obovate, 10–42 × 2–12 cm, margins coarsely serrate. **Inflorescences** secund, villous, hairs glandular and eglandular; bracts 14–50 mm. **Pedicels** pendent, 5–15 mm, villous, hairs glandular and eglandular. **Flowers:** sepals ovate-lanceolate, 9–25 × 2–13 mm, glabrous or villous, hairs glandular and eglandular; corolla tube pink-purple to white, funnelform, 25–60 mm, throat 20–25 mm diam., abaxial lip pendent, pink-purple to white, rounded, 8–10 mm. **Capsules** ovoid-conical, slightly 2-lobed, 10–17 mm, pilose. **Seeds** brown, cylindric to ovoid, 1 mm, reticulate-alveolate. $2n = 56$ (Europe).

Flowering May–Aug. Disturbed sites, roadsides, clearcuts, old fields, pastures; 0–1000 m; introduced; St. Pierre and Miquelon; B.C., N.B., Nfld. and Labr. (Nfld.), N.S., Ont., Que.; Ark., Calif., Colo., Conn., Idaho, Maine, Md., Mass., Mich., Mont., N.H., N.J., N.Y., N.C., Ohio, Oreg., Pa., Utah, Vt., Wash., W.Va., Wis., Wyo.; Europe; introduced also in Mexico, Central America, South America, Asia, Africa, Pacific Islands (New Zealand), Australia.

29. BACOPA Aublet, Hist. Pl. Guiane 1: 128, plate 49. 1775, name conserved

- Water-hyssop [Aboriginal name in French Guiana]

Adjoa Richardson Ahedor

Herbs, annual or perennial, emergent or submerged, sometimes rooting from proximal nodes. **Stems** prostrate or erect, glabrous or slightly hairy. **Leaves** cauline, opposite [whorled]; petiole obscure or absent; blade fleshy (*B. monnieri*) or not, not leathery, margins entire, crenate, or serrate. **Inflorescences** axillary, flowers solitary or in pairs; bracts absent. **Pedicels** present or absent; bracteoles present [absent]. **Flowers** bisexual; sepals 4 or 5, distinct, outer wider than inner, ovate, lanceolate, or oblong, calyx radially or bilaterally symmetric, campanulate; corolla white (sometimes with yellow throat), pink, or violet-blue, radially or bilaterally (*B. repens*) symmetric, regular or weakly bilabiate, rotate or campanulate, rarely tubular, tube base not spurred or gibbous, throat not densely pilose internally, lobes 4 or 5, abaxial 2 or 3, adaxial 2; stamens 2–4, adnate to corolla, didynamous or equal, filaments glabrous; staminode 0; ovary 2-locular, placentation axile; stigmas peltately flattened and semicapitate. **Fruits** capsules, dehiscence septicidal-loculicidal or loculicidal. **Seeds** 80–120, brownish black, ellipsoid to cylindric, reticulate, wings absent. *x* = 14.

Species ca. 70 (7 in the flora): North America, Mexico, West Indies, Central America, South America, s Europe, Asia, Africa, Indian Ocean Islands (Madagascar), Australia.

Bacopa laxiflora (Bentham) Wettstein ex Edwall, native to Central America and South America, has been introduced to Mexico and may be expected in the flora area. Plants of *B. laxiflora* resemble *B. monnieri*; they differ by erect stems and acute leaf blade apices.

SELECTED REFERENCES Barrett, S. C. H. and J. L. Strother. 1978. Taxonomy and natural history of *Bacopa* (Scrophulariaceae) in California. Syst. Bot. 3: 408–419. Schuyler, A. E. 1989. Intertidal variants of *Bacopa rotundifolia* and *B. innominata* in the Chesapeake Bay drainage. Bartonia 55: 18–22.

1. Leaf blades 1-nerved . . . . 1. *Bacopa monnieri*
1. Leaf blades (3–)5–9+-nerved.
  2. Plants aromatic; bracteoles present; corollas violet-blue with violet-blue throats. . . .2. *Bacopa caroliniana*
  2. Plants not aromatic; bracteoles absent; corollas white with yellow or white throats or
    pink with pink throats.
    3. Leaf blade bases narrowly cuneate.
      4. Leaf blade margins crenate. . . . . .4. *Bacopa egensis*
      4. Leaf blade margins entire . . . . 7. *Bacopa repens*
    3. Leaf blade bases broadly cuneate to truncate.
      5. Pedicels 6–10 mm; corollas 2–5 mm, white with white throats. . . . . 6. *Bacopa innominata*
      5. Pedicels 9–51 mm; corollas 5–14 mm, white with yellow throats.
        6. Pedicels 9–15 mm; corollas 5–10 mm . . . . 3. *Bacopa rotundifolia*
        6. Pedicels 17–51 mm; corollas 10–14 mm . . . . 5. *Bacopa eisenii*

1. **Bacopa monnieri** (Linnaeus) Wettstein in H. G. A. Engler and K. Prantl, Nat. Pflanzenfam. 67[IV,3b]: 77. 1891 (as monniera) • Brahmi, Indian pennywort, herb of grace F W

Lysimachia monnieri Linnaeus, Cent. Pl. II, 9. 1756

Perennials, sometimes annuals. Stems prostrate, 15–30 cm, glabrous. **Leaves** glabrous; blade fleshy, base narrowly cuneate, margins entire or serrate, apex obtuse, 1-nerved. **Pedicels** 5–30 mm; bracteoles present. **Flowers:** sepals 5, ovate to lanceolate, calyx radially symmetric; corolla white with yellow throat, 5–10 mm, lobes 5; stamens 4, didynamous. $2n = 64$.

Flowering Apr–Sep. Wetlands, wet sands, mud flats, riparian areas; 0–1500 m; Ala., Ariz., Calif., Fla., Ga., La., Md., Miss., N.C., Okla., S.C., Tex., Va.; Mexico; West Indies; Central America; South America; s Europe; Asia; Africa; Australia; introduced in sw Europe (Portugal, Spain), Asia (China, Taiwan).

Bacopa monnieri is thought to be native throughout much of its range, though it is weedy and cultivated. It readily colonizes irrigated fields, especially rice fields, and seeds easily get mixed with rice and are planted in new locations. *Bacopa monnieri* is introduced in parts of Europe (Portugal, Spain) and Asia (China, Taiwan). It can be propagated vegetatively by cuttings.

Bacopa monnieri is used medicinally in Asia in traditional Indian medicine (Ayurveda). It is edible and contains steroidal saponins, including bacosides, that have beneficial effects on the nervous system. Leaf, stem, and root extracts are used as cardiac and nerve tonics, sedatives, and vasoconstrictors. Leaves and stems are diuretic and used in treating constipation and indigestion. An alcohol extract of the whole plant has been used to treat Walker carcinoma and as a cardiovascular and muscle relaxer. In the United States, recent studies suggest it has potential for enhancing cognitive performance in the elderly and in the treatment of Alzheimer's disease (C. Calabres et al. 2008; S. Aguiar and T. Borowski 2013). Extracts are used in the treatment of nerve and brain disorders; they also are believed to enhance intellect and decrease fertility.

2. **Bacopa caroliniana** (Walter) B. L. Robinson, Rhodora 10: 66. 1908 • Blue water-hyssop, lemon bacopa E W

Obolaria caroliniana Walter, Fl. Carol., 166. 1788; *Hydrotrida caroliniana* (Walter) Small

Perennials, aromatic. **Stems** prostrate, 15–30 cm, hairy. **Leaves** hairy; blade base broadly cuneate, margins entire, apex obtuse. **Pedicels** 5–15 mm; bracteoles present. **Flowers:** sepals 5, ovate, calyx bilaterally symmetric; corolla violet-blue with violet-blue throat, 10–13 mm, lobes 4; stamens 2–4, didynamous.

Flowering Jun–Nov. Marshes, swamps, margins of streams, pastures; 0–300 m; Ala., Fla., Ga., La., Md., Miss., N.C., S.C., Tex., Va.

Bacopa caroliniana is used in aquascaping in freshwater aquariums. The species can be propagated vegetatively through cuttings. The leaves of *B. caroliniana* are lemon scented when crushed.

3. **Bacopa rotundifolia** (Michaux) Wettstein in H. G. A. Engler and K. Prantl, Nat. Pflanzenfam. 67[IV,3b]: 76. 1891 • Round leaf water-hyssop E F W

Monniera rotundifolia Michaux, Fl. Bor.-Amer. 2: 22. 1803; *Bacopa nobsiana* H. Mason; *B. simulans* Fernald; *Bramia rotundifolia* (Michaux) Britton; *Macuillamia rotundifolia* (Michaux) Rafinesque

Annuals [perennials]. Stems prostrate, 15–60 cm, hairy, sometimes sparsely so, glabrescent. **Leaves** hairy; blade base broadly cuneate, margins entire, apex ± rounded. **Pedicels** 9–15 mm. **Flowers:** sepals 5, ovate to oblong, calyx bilaterally symmetric; corolla white with yellow throat, 5–10 mm; stamens 2–4, equal. $2n = 56$.

Flowering May–Nov. In water or on mud in lakes, ponds, pools, in and around marshes and ditches, ephemeral pools on rock outcrops; 0–2300 m; Alta., Sask.; Ala., Ariz., Ark., Calif., Colo., Idaho, Ill., Ind., Iowa, Kans., Ky., La., Md., Minn., Miss., Mo., Mont., Nebr., N.Mex., N.C., N.Dak., Okla., S.Dak., Tenn., Tex., Va., Wis., Wyo.; introduced in Europe (Greece, Spain).

Bacopa rotundifolia is the most widespread member of the genus in the United States; it is invasive in rice fields.

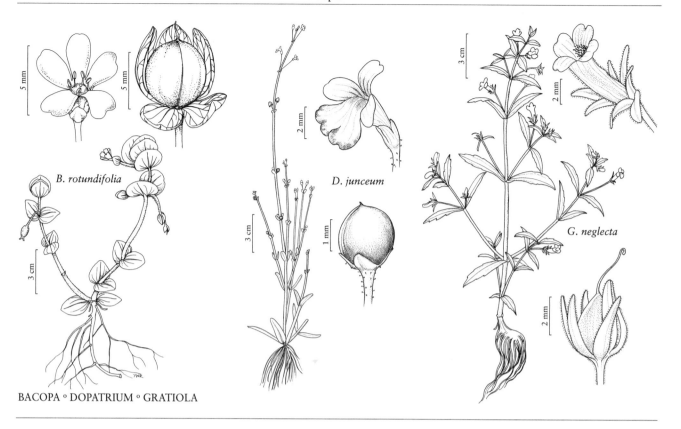

B. rotundifolia

D. junceum

G. neglecta

BACOPA ∘ DOPATRIUM ∘ GRATIOLA

4. Bacopa egensis (Poeppig) Pennell, Proc. Acad. Nat. Sci. Philadelphia 98: 96. 1946 • Brazilian water-hyssop [I]

Hydranthelium egense Poeppig in E. G. Poeppig and S. L. Endlicher, Nov. Gen. Sp. Pl. 3: 75, plate 287, figs. 1–10. 1845

Perennials. Stems prostrate, 15–30 cm, hairy. **Leaves** glabrous; blade base narrowly cuneate, margins crenate, apex rounded with a narrow petiole-like base. **Pedicels** 10–12 mm. **Flowers:** sepals 4(or 5), ovate to oblong, calyx radially symmetric; corolla white with yellow throat, 3–5 mm; stamens 3 or 4, didynamous.

Flowering Aug–Oct. Shallow, quiescent water, lakes; 0–20 m; introduced; Ark., La.; Central America (Costa Rica, Nicaragua); South America (Brazil); introduced also in w, c Africa (Republic of Cameroon, Democratic Republic of Congo, Ghana, Nigeria, Senegal).

Bacopa egensis was first collected in the United States by Josiah Hale in the early 1800s (C. E. Depoe 1969). There are subsequent collections from Arkansas and Louisiana.

SELECTED REFERENCE   Depoe, C. E. 1969. *Bacopa egensis* (Poeppig) Pennell (Scrophulariaceae) in the United States. Sida 3: 313–318.

5. Bacopa eisenii (Kellogg) Pennell, Proc. Acad. Nat. Sci. Philadelphia 98: 96. 1946 • Western hydranthele, Gila River water-hyssop [E] [W]

Ranapalus eisenii Kellogg, Proc. Calif. Acad. Sci. 7: 113. 1877

Perennials. Stems prostrate, 15–60 cm, hairy. **Leaves** hairy; blade base broadly cuneate to truncate, margins entire, apex rounded. **Pedicels** 17–51 mm. **Flowers:** sepals 5, ovate to oblong, calyx bilaterally symmetric; corolla white with yellow throat, 10–14 mm; stamens 3, equal. $2n = 56$.

Flowering May–Oct. Marshes, ponds, pools, wet soil, wet ditches, rice fields; 30–1100 m; Ariz., Calif., Nev.

Bacopa eisenii was reported by T. H. Kearney and R. H. Peebles (1960) based on a specimen labeled *B. rotundifolia* at ARIZ.

6. Bacopa innominata (M. Gómez) Alain, Revista Soc. Cub. Bot. 13: 61. 1957 • Tropical water-hyssop E W

Conobea innominata M. Gómez, Anales Soc. Esp. Hist. Nat. 23: 278. 1895; *Bacopa cyclophylla* Fernald; *B. stragula* Fernald; *Herpestis rotundifolia* C. F. Gaertner 1807, not *B. rotundifolia* (Michaux) Wettstein

Perennials. Stems prostrate, 8–14 cm, hairy. **Leaves** hairy; blade base ± truncate, margins entire, apex rounded. **Pedicels** 6–10 mm. **Flowers:** sepals 5, ovate to oblong, calyx bilaterally symmetric; corolla white with white throat, 2–5 mm; stamens 2–4, equal.

Flowering Jun–Sep. Marshy areas, alluvial deposits along streams, wet ditches, muddy shores and banks; 0–200 m; Fla., Ga., Md., N.C., S.C., Va.

Bacopa innominata differs from *B. rotundifolia* in having corollas entirely white; *B. rotundifolia* has corollas with yellow throats.

7. Bacopa repens (Swartz) Wettstein in H. G. A. Engler and K. Prantl, Nat. Pflanzenfam. 67[IV,3b]: 76. 1891 • Creeping water-hyssop I W

Gratiola repens Swartz, Prodr., 14. 1788; *Macuillamia repens* (Swartz) Pennell

Annuals or perennials. Stems erect, 15–40 cm, hairy, glabrescent. **Leaves** hairy, glabrescent; blade base narrowly cuneate, margins entire, apex rounded. **Pedicels** 1.5–15 mm. **Flowers:** sepals 4(or 5), oblong to lanceolate, calyx bilaterally symmetric; corolla pink with pink throat or white with white throat, 2–5 mm; stamens 4, didynamous. $2n = 28$.

Flowering Jul–Sep. Muddy sites, pools, rice fields; 0–100 m; introduced; Calif., Fla., La., S.C., Tex.; West Indies (Greater Antilles, Lesser Antilles, Puerto Rico); Central America; introduced also in Mexico, South America (Colombia), Asia (China).

SELECTED REFERENCE Thieret, J. W. 1970. *Bacopa repens* (Scrophulariaceae) in the conterminous United States. Castanea 35: 132–136.

30. DOPATRIUM Buchanan-Hamilton ex Bentham, Edwards's Bot. Reg. 21: sub plate 1770, genus no. 46. 1835 • [Hindi *do*, two, and *patra*, leaves, alluding to opposite leaf arrangement] I

Robert E. Preston

Herbs, annual. **Stems** erect, glabrous [sparsely glandular]. **Leaves** basal and cauline, opposite; petiole absent; blade fleshy, not leathery, margins entire. **Inflorescences** axillary, flowers solitary; bracts present. **Pedicels** present; bracteoles absent. **Flowers** bisexual; sepals 5, proximally connate, calyx bilaterally symmetric, campanulate, lobes oblong; corolla pale blue to lavender [yellow], bilaterally symmetric, bilabiate, tubular, tube base not spurred or gibbous, lobes 5, abaxial 3, adaxial 2; stamens 2, adnate to corolla, filaments glabrous; staminode 0[2]; ovary 1[2]-locular, placentation parietal; stigma slightly 2-lobed, ± peltate. **Fruits** capsules, dehiscence loculicidal. **Seeds** 50–200, dark reddish brown, oblong-ovoid, wings absent. $x = 7$.

Species ca. 14 (1 in the flora): introduced; Asia, Africa, Australia.

SELECTED REFERENCE Fischer, E. 1997. A revision of the genus *Dopatrium* (Scrophulariaceae-Gratioloideae). Nordic J. Bot. 17: 527–555.

1. Dopatrium junceum (Roxburgh) Buchanan-Hamilton ex Bentham, Scroph. Ind., 31. 1835 • Horsefly's eye F I W

Gratiola juncea Roxburgh, Pl. Coromandel 2: 16. 1799

Annuals 10–30 cm. **Leaves:** basal and proximal 4–8, blade oblong, 10–25 × 2–10 mm, apex obtuse; distal blade linear-lanceolate, 2–7 × 0.5–2 mm. **Inflorescences:** proximal flowers sessile, cleistogamous, distal pedicellate, chasmogamous; proximal bracts leaflike, distal scalelike. **Pedicels** spreading in fruit, 4–11 mm, sparsely glandular. **Flowers:** calyx 1–2 mm, lobes equal, apex obtuse; corolla 4–7 mm, tube slightly longer than calyx, abaxial lip with middle lobe longer than lateral lobes, adaxial lip erect. **Capsules** 2 mm. **Seeds** 0.4–0.5 × 0.2 mm, testa netted. $2n = 14$.

Flowering Jun–Sep. Drying wetlands, rice fields, ditches; 0–200 m; introduced; Calif., La.; Asia; Africa; Australia.

First collected as a rice field weed in Butte County, California, in 1944 (C. V. Morton and J. T. Howell 1945), *Dopatrium junceum* has spread across similar habitats in the Sacramento Valley (S. C. H. Barrett and D. E. Seaman 1980) and in Louisiana (J. W. Thieret and D. H. Dike 1969). *Dopatrium junceum* germinates in inundated wetlands in May and blooms during summer, which suggests that, at least in California, it is unlikely to spread to vernal pools and other seasonal wetlands that are inundated only during the winter rainy season.

31. GRATIOLA Linnaeus, Sp. Pl. 1: 17. 1753; Gen. Pl. ed. 5, 11. 1754 • Hedge-hyssop

[Latin *gratia*, graceful, and *-ola*, diminutive, alluding to medicinal qualities of some species]

Craig C. Freeman

Amphianthus Torrey

Herbs, annual or perennial. **Stems** decumbent to ascending or erect, glabrous or glandular-hairy. **Leaves** cauline (also basal and clustered in *G. amphiantha*), opposite, monomorphic (dimorphic in *G. amphiantha*); petiole absent; blade not fleshy, not leathery, margins entire or toothed, surfaces often glandular-punctate. **Inflorescences** axillary, flowers 1 or 2 per node; bracts present. **Pedicels** present, sometimes short; bracteoles present or absent, sepal-like. **Flowers** bisexual; sepals 5, distinct or proximally connate, calyx ± bilaterally symmetric, campanulate, lobes (sepals when distinct) linear to lanceolate, obovate, or oblong; corolla white, cream, or yellow, sometimes pinkish, tube yellowish, brownish, or purplish, often with dark veins, bilaterally symmetric, bilabiate, tubular, tube base not spurred or gibbous, lobes 5, abaxial 3, adaxial 2, nearly completely connate and appearing as 1 emarginate lobe, throat glabrous or pubescent; stamens 2, medially adnate to corolla, filaments glabrous, pollen sacs perpendicular to filaments, connective dilated; staminodes 0 or 2, minute; ovary 2-locular, placentation axile; stigma capitate or lamelliform. **Fruits** capsules, dehiscence septicidal, sometimes also loculicidal. **Seeds** ca. 200, yellow to brown or black, oblong to obpyramidal or short-cylindric, wings absent. $x = 9$.

Species ca. 30 (13 in the flora): North America, Mexico, Central America (Guatemala), South America, Europe, e Asia (Japan), Africa (Morocco), Pacific Islands (New Zealand), Australia.

Members of *Gratiola* sometimes are confused with *Lindernia* and *Mecardonia*. *Sophronanthe* often is included in *Gratiola*; molecular and morphological data support separating the two as sister genera (D. Estes 2008).

This treatment was prepared in part from information contained in a partial manuscript submitted by D. Estes.

SELECTED REFERENCE Estes, D. 2008. Systematics of *Gratiola* (Plantaginaceae). Ph.D. dissertation. University of Tennessee.

1. Capsules bilaterally symmetric; leaves dimorphic, submersed blades usually narrower than floating blades . . . . 1. *Gratiola amphiantha*
1. Capsules ± radially symmetric; leaves monomorphic, submersed blades absent or, if present, similar to emersed blades.
    2. Perennials.
        3. Bracteoles 0 or 1; leaf blades linear-lanceolate to lanceolate-ovate, 1–3 mm wide, margins entire or with 1 or 2(or 3) pairs of blunt teeth . . . . 10. *Gratiola ramosa*
        3. Bracteoles (1 or)2; if 1, leaf blades lanceolate to lanceolate-ovate, ovate, or oblong, rarely linear-lanceolate, (1–)3–8(–13) mm wide, margins with (1 or)2–4 pairs of sharp teeth.
            4. Corolla limbs yellow, rarely white or cream, veins yellow to orangish yellow, rarely white; pedicels sparsely glandular-puberulent . . . . 7. *Gratiola lutea*
            4. Corolla limbs white, veins yellow, brownish yellow, blue, or lavender; pedicels sparsely to densely glandular-puberulent to glandular-pubescent.
                5. Leaf blade margins with (1 or)2–4 pairs of teeth, blades lanceolate to lanceolate-ovate or oblong; corolla veins yellow or brownish yellow; sepals linear-lanceolate . . . . 2. *Gratiola brevifolia*
                5. Leaf blade margins with (3–)5–10(–12) pairs of teeth, blades lanceolate-ovate to ovate or oblong, rarely linear-lanceolate in submersed forms; corolla veins blue or lavender; sepals lanceolate to elliptic or ovate-lanceolate . . . . 13. *Gratiola viscidula*
    2. Annuals.
        6. Bracteoles 0; w North America.
            7. Calyces slightly bilaterally symmetric, sepals distinct, (4–)7–11 mm, lanceolate; leaf blade apices acuminate to attenuate . . . . 3. *Gratiola ebracteata*
            7. Calyces distinctly bilaterally symmetric, sepals connate proximally, lobes 3.5–6 mm, elliptic-oblanceolate; leaf blade apices obtuse to rounded . . . . 6. *Gratiola heterosepala*
        6. Bracteoles 1 or 2; e, c, or w North America.
            8. Corolla limbs yellow; capsules ovoid . . . . 11. *Gratiola torreyi*
            8. Corolla limbs white, sometimes tinged pink, lavender, or purple; capsules ovoid to short-cylindric or subglobular.
                9. Pedicels stout, 1–12 mm, lengths 0.1–0.3 times bracts . . . . 12. *Gratiola virginiana*
                9. Pedicels slender, (5–)7–45(–55) mm, lengths (0.3–)0.4–2(–2.3) times bracts.
                    10. Corollas 14–25 mm; pedicels 20–45(–55) mm . . . . 4. *Gratiola floridana*
                    10. Corollas 7–12 mm; pedicels (5–)7–35 mm.
                        11. Leaf blades 6–13(–18) mm; pedicel lengths 0.9–2(–2.3) times bracts; granite outcrops . . . . 5. *Gratiola graniticola*
                        11. Leaf blades 10–65 mm; pedicel lengths (0.3–)0.4–1(–1.6) times bracts; wet meadows, stream banks, shores of ponds, mudflats, salt marshes, crop fields, seeps, pools, and streams in limestone and dolomite glades, calcareous grasslands.
                            12. Leaf blades linear to narrowly elliptic, oblanceolate, or elliptic-obovate, rarely falcate, 3–11(–18) mm wide, margins with (1 or)2–5(–7) pairs of blunt to sharp teeth, rarely entire; wet meadows, stream banks, shores of ponds, mudflats, salt marshes, crop fields . . . . 8. *Gratiola neglecta*
                            12. Leaf blades linear to linear-lanceolate or lanceolate-elliptic, often falcate, (1–)2.5–5 mm wide, margins entire or with 1 or 2(or 3) pairs of blunt teeth distally; seeps, pools, and streams in limestone and dolomite glades, calcareous grasslands. . . . . 9. *Gratiola quartermaniae*

1. Gratiola amphiantha D. Estes & R. L. Small, Syst.
Bot. 33: 181. 2008 • Pool sprite, snorkelwort [C] [E]

Amphianthus pusillus Torrey, Ann. Lyceum Nat. Hist. New York 4: 82. 1837, not *Gratiola pusilla* Willdenow 1797 [= *Lindernia pusilla*]

Annuals. Stems erect, simple or few-branched, (7–)9–21 (–29) cm, glabrous or glabrate proximally, glabrate or glandular-puberulent distally. **Leaves:** dimorphic, submersed basal and clustered, floating paired at ends of branches to 70 mm, blade of submersed lanceolate to oblanceolate or oblong, 1–7 × 0.5–3 mm, margins entire, apex obtuse to acute, surfaces glabrous, blade of floating elliptic to ovate or nearly round, 3–10 × 2–9 mm, margins entire, apex obtuse to rounded or retuse, surfaces glabrous. **Pedicels** stout, 0.1–3 mm, length less than 0.1 times bract, glabrous; bracteoles 0. **Flowers:** sepals connate proximally, calyx lobes obovate to oblong, 0.7–1 mm; corolla 3–4 mm, tube and limb white tinged pink or purple, veins lavender or purple; style 0.5–1 mm. **Capsules** bilaterally symmetric, obcordiform, 2–3 × 3–4 mm. **Seeds** (0.3–)0.4–0.5 mm. 2*n* = 18.

Flowering Mar–Apr. Shallow, ephemeral pools on exposed granite outcrops; of conservation concern; 100–300 m; Ala., Ga., S.C.

Phylogenetic studies confirmed that *Gratiola amphiantha*, long treated in monospecific *Amphianthus*, is embedded in *Gratiola* (D. Estes and R. L. Small 2008). *Gratiola amphiantha* is listed as a federally-threatened species by the United States Fish and Wildlife Service.

2. Gratiola brevifolia Rafinesque, Atlantic J. 1: 176.
1833 • Sticky or short-leaf hedge-hyssop [E]

Perennials. Stems decumbent to ascending, simple or few- to much-branched, 20–60 cm, sparsely to moderately glandular-puberulent. **Leaves:** blade lanceolate to lanceolate-ovate or oblong, 8–25 × (1–)3–8(–13) mm, margins with (1 or)2–4 pairs of sharp teeth distally, apex acute, surfaces sparsely to densely glandular-puberulent. **Pedicels** slender, (5–)10–25(–30) mm, length 0.8–1.8 times bract, densely glandular-puberulent to glandular-pubescent; bracteoles (1 or)2, 4–4.5 mm. **Flowers:** sepals distinct, linear-lanceolate, 5–7 mm; corolla 10–13 mm, tube yellowish brown, veins yellow or brownish yellow, limb white; style 3–5 mm. **Capsules** subglobular to ovoid, 1.5–2.5 × 1–1.3 mm. **Seeds** 0.3–0.4 mm. 2*n* = 28.

Flowering Apr–Jun. Sandy pinelands, oak barrens, sandy stream banks; 0–200 m; Ala., Ark., Del., Fla., Ga., La., Okla., Tenn., Tex.

Gratiola brevifolia is morphologically variable. In parts of its range, narrow-leaved plants approach *G. ramosa* and some broad-leaved plants approach *G. viscidula*. A population of *G. brevifolia* in Delaware is disjunct from the nearest conspecific population by more than 800 km (W. M. Knapp and D. Estes 2006).

3. Gratiola ebracteata Bentham ex A. de Candolle in
A. P. de Candolle and A. L. P. P. de Candolle, Prodr. 10: 595. 1846 • Bractless hedge-hyssop [E]

Annuals. Stems decumbent to ascending or erect, simple or few-branched, (5–)15–22 cm, glabrous or glandular-puberulent distally. **Leaves:** blade linear-lanceolate to lanceolate-ovate, 7–20(–32) × 1.5–5(–7) mm, margins entire, rarely with 1 or 2 pairs of teeth distally, apex acuminate to attenuate, surfaces glabrate or glandular-pubescent. **Pedicels** stout, (3–)7–25 mm, length 0.5–2.5 times bract, glabrous or obscurely glandular-pubescent distally; bracteoles 0. **Flowers:** sepals distinct, lanceolate, (4–)7–11 mm; calyx slightly bilaterally symmetric; corolla 5–8(–10) mm, tube yellowish green or yellow, veins purple, limb white to pinkish white; style 2–3 mm. **Capsules** subglobular, 3–6 × 3.5–5 mm. **Seeds** 0.4–0.9 mm.

Flowering Apr–Sep(–Nov). Muddy to sandy stream banks, pond and lake shorelines, shallow water, wet meadows, vernal pools; 0–2100 m; B.C.; Calif., Idaho, Mont., Oreg., Wash.

4. Gratiola floridana Nuttall, J. Acad. Nat. Sci.
Philadelphia 7: 103. 1834 • Florida hedge-hyssop [E]

Annuals. Stems decumbent to ascending or erect, simple or few- to much-branched, 10–40 cm, glabrous proximally, glandular-pubescent distally. **Leaves:** blade oblanceolate to narrowly obovate or oval, 20–45 × 4–18 mm, margins with 1–6 pairs of teeth distally, apex obtuse to acute, surfaces glabrous. **Pedicels** slender, 20–45(–55) mm, length 0.8–2 times bract, glabrate or finely glandular-pubescent; bracteoles 2, 3–6 mm. **Flowers:** sepals distinct, linear to narrowly lanceolate or narrowly oblong, 3–6 mm; corolla 14–25 mm, tube yellowish green, veins reddish purple to brownish yellow, limb white; style 5–7 mm. **Capsules** ovoid, 3–6 × 2.8–3.2 mm. **Seeds** 0.6–0.8 mm.

Flowering Mar–May. Low wet woods, wooded stream banks, open meadows; 0–300 m; Ala., Fla., Ga., La., Miss., Tenn.

5. Gratiola graniticola D. Estes, J. Bot. Res. Inst. Texas 1: 166, figs. 3A,D, 9. 2007 • Granite hedge-hyssop E

Annuals. Stems erect, simple, sometimes few-branched, (7–)9–21(–29) cm, glabrous or glabrate proximally, glandular-puberulent distally. **Leaves:** blade lanceolate-ovate to oblong, not falcate, 6–13(–18) × 1–3(–5) mm, margins entire or with 1 or 2(or 3) pairs of blunt teeth distally, apex obtuse, surfaces glabrate or glandular-puberulent. **Pedicels** slender, (5–)8–17(–22) mm, length 0.9–2(–2.3) times bract, sparsely to densely glandular-puberulent; bracteoles 2, 2–4.5 mm. **Flowers:** sepals distinct, lanceolate, 2–4.5 mm; corolla 7–9 mm, tube and limb white tinged with pink or purple, veins lavender or purple; style 1.9–2.2 mm. **Capsules** subglobular to ovoid, 2.5–3.6 × 2–3.7 mm. **Seeds** 0.3–0.5 mm.

Flowering Apr–May. Seasonal pools on granite outcrops; 100–300 m; Ga., S.C.

Gratiola graniticola is known from about a dozen counties in the southeastern piedmont of northern Georgia and north-central South Carolina, where it occurs only on granite flatrocks (D. Estes and R. L. Small 2007; D. F. Brunton 2009).

6. Gratiola heterosepala H. Mason & Bacigalupi, Madroño 12: 150, figs. 1–8. 1954 • Boggs Lake hedge-hyssop C E

Annuals. Stems ascending to erect, simple or few-branched, 2–12 cm, glabrous proximally, glabrous or glandular-puberulent distally. **Leaves:** blade linear-lanceolate to oblanceolate, elliptic, or oblong, 4–8(–20) × 1.5–3.5 mm, margins entire, apex obtuse to rounded, surfaces glabrous. **Pedicels** slender, 2–14(–28) mm, length 0.9–2.3 times bract, sparsely to densely glandular-puberulent; bracteoles 0. **Flowers:** sepals connate proximally, calyx distinctly bilaterally symmetric, lobes elliptic-oblanceolate, 3.5–6 mm; corolla 4–7.5(–9) mm, tube yellow or greenish yellow, veins purple, limb white abaxially, greenish yellow adaxially; style 1.5–2 mm. **Capsules** subglobular to ovoid, 3.5–4.7 × 3–4.5 mm. **Seeds** 0.5–0.7 mm.

Flowering Apr–Sep. Shallow water and exposed mud of vernal pools, wet meadows, lake margins; of conservation concern; 0–1900 m; Calif., Oreg.

7. Gratiola lutea Rafinesque, Med. Repos., hexade 3, 2: 333. 1811 • Golden hedge-hyssop, gratiole dorée E

Gratiola aurea Muhlenberg

Perennials. Stems decumbent to ascending or erect, simple or few-branched, 5–47 cm, glabrous proximally, sparsely glandular-puberulent distally. **Leaves:** blade lanceolate-ovate to oblong or ovate-elliptic, sometimes linear-lanceolate on submerged plants, 5–26 × (1.5–)3–7(–9) mm, margins entire or with 1 or 2 pairs of teeth, apex obtuse, rarely acute, surfaces glabrous. **Pedicels** slender, (3–)5–20 mm, length 0.4–2.1 times bract, sparsely glandular-puberulent; bracteoles 2, 2–4 mm. **Flowers:** sepals distinct, linear-lanceolate to lanceolate, 3.5–7 mm; corolla 8–15 mm, tube and limb yellow, rarely white or cream, veins yellow to orangish yellow, rarely white; style 3.5–5 mm. **Capsules** ovoid, 2.2–4.5 × 2–3.5 mm. **Seeds** 0.4–0.6 mm. $2n = 28$.

Flowering May–Oct. Acidic freshwater pondshores, blackwater stream banks, cypress savannas, acidic wetlands, swamps; 0–300 m; N.B., Nfld. and Labr. (Nfld.), N.S., Ont., Que.; Conn., Del., D.C., Fla., Ga., Ill., Maine, Md., Mass., Mich., N.H., N.J., N.Y., N.C., N.Dak., Pa., R.I., S.C., Vt., Va., Wis.; introduced in e Asia (Japan).

The name *Gratiola aurea* has been used widely for *G. lutea* due to questions about the applicability of the name by Rafinesque, which predates the name by Muhlenberg by two years. F. W. Pennell (1935) reviewed the nomenclatural history, noting that the type on which the name by Rafinesque is based, a collection by A. Michaux deposited at P, had been identified by A. Gray, M. L. Fernald, and S. F. Blake as *G. neglecta*. However, Pennell determined that the collection by Michaux was a mixed gathering and that the name by Rafinesque applies to the yellow-flowered element thereof.

Gratiola lutea (as *G. aurea*) has been reported from Crittenden County, Arkansas (W. H. Wilcox 1973); it is excluded from Arkansas in recent state checklists (E. B. Smith 1991; J. L. Gentry et al. 2013). Reports from Alabama may be based on misidentified specimens. A specimen identified as *G. lutea* has been collected in Japan (D. Estes 2008), where it was presumably introduced.

8. Gratiola neglecta Torrey, Cat. Pl. New York, 89. 1819
• Clammy hedge-hyssop, gratiole négligée E F

Gratiola neglecta var. *glaberrima* (Fernald) Fernald

Annuals. Stems ascending or erect, few- to much-branched, rarely simple, (4–)10–40 cm, glabrous or glandular-puberulent proximally, glandular-puberulent distally. **Leaves:** blade linear to narrowly elliptic, oblanceolate, or elliptic-obovate, rarely falcate, 10–65 × 3–11(–18) mm, margins with (1 or)2–5(–7) pairs of blunt to sharp teeth, apex acute, surfaces glabrate or glandular-puberulent. **Pedicels** slender, 8–35 mm, length (0.3–)0.4–0.9 (–1.3) times bract, sparsely to densely glandular-puberulent; bracteoles 2, 2.5–7 mm. **Flowers:** sepals distinct, lanceolate to narrowly lanceolate-elliptic, 2.5–6 mm; corolla 7–12 mm, tube yellow, yellowish white, or yellowish green, veins brownish violet, limb white, sometimes tinged lavender; style 3–4 mm. **Capsules** ovoid, 2.6–6 × 3–5 mm. **Seeds** 0.4–0.7 mm. $2n = 14, 16$.

Flowering Mar–Oct. Wet meadows, stream banks, shorelines of ponds, mudflats, salt marshes, crop fields; 0–2400 m; Alta., B.C., Man., N.B., N.S., Ont., Que., Sask.; Ala., Ariz., Ark., Calif., Colo., Conn., Del., D.C., Ga., Idaho, Ill., Ind., Iowa, Kans., Ky., La., Maine, Md., Mass., Mich., Minn., Miss., Mo., Mont., Nebr., Nev., N.H., N.J., N.Mex., N.Y., N.C., N.Dak., Ohio, Okla., Oreg., Pa., R.I., S.C., S.Dak., Tenn., Tex., Utah, Vt., Va., Wash., W.Va., Wis., Wyo.; introduced in Europe (Finland, France).

Gratiola neglecta exhibits the widest range and broadest ecological amplitude of the North American *Gratiola* species. Besides occurring in a variety of freshwater wetland communities, it sometimes is found in salt marshes and damp or wet agricultural fields.

9. Gratiola quartermaniae D. Estes, J. Bot. Res. Inst. Texas 1: 163, figs. 3C,F, 8. 2007 • Limestone or Quarterman's hedge-hyssop E

Annuals. Stems erect, simple or few-branched, (7–)10–21 (–30) cm, glabrous or glabrate proximally, sparsely glandular-puberulent distally. **Leaves:** blade linear to linear-lanceolate or lanceolate-elliptic, often falcate, (16–)18–30(–43) × (1–) 2.5–5 mm, margins entire or with 1 or 2(or 3) pairs of blunt teeth distally, apex obtuse to acute, surfaces glabrous or glabrate. **Pedicels** slender, 7–14(–22) mm, length 0.5–1(–1.6) times bract, sparsely glandular-puberulent; bracteoles 2, 2–4.5 mm. **Flowers:** sepals distinct, lanceolate, 2.7–5 mm; corolla

7–9 mm, tube and limb white tinged with pink or purple, veins white to greenish white or lavender; style 3–4.6 mm. **Capsules** ovoid, 3.4–5.1 × 3–4.5 mm. **Seeds** 0.4–0.7 mm.

Flowering Apr–Jun. Seeps, pools, and streams in limestone and dolomite glades, alvars, calcareous grasslands; 50–300 m; Ont.; Ala., Ill., Ky., Tenn., Tex.

Populations of *Gratiola quartermaniae* are concentrated in northern Alabama and central Tennessee. Disjunct populations are known from the Edwards Plateau of Texas, Will County, Illinois, and southeastern Ontario. D. Estes and R. L. Small (2007) discussed the distribution and ecology of *G. quartermaniae*.

10. Gratiola ramosa Walter, Fl. Carol., 61. 1788
• Branched hedge-hyssop E

Perennials. Stems ascending to erect, simple or few-branched, 10–35 cm, glabrous or glabrate proximally, glandular-puberulent distally. **Leaves:** blade linear-lanceolate to lanceolate-ovate, 7–20 × 1–3mm, margins entire or with 1 or 2(or 3) pairs of blunt teeth distally, apex acute, surfaces glabrate or sparsely glandular-puberulent. **Pedicels** slender to stout, 6–17 mm, length 0.8–2 times bract, sparsely to densely glandular-puberulent; bracteoles 0 or 1, 0.8–1 mm. **Flowers:** sepals distinct, linear to narrowly lanceolate or oblong, 3–6 mm; corolla 10–14 mm, tube greenish yellow, veins yellow, brownish yellow, or purple, limb white; style 2–4 mm. **Capsules** subglobular to ovoid, 1.2–2 × 1.2–2 mm. **Seeds** 0.2–0.3 mm. $2n = 14$.

Flowering Apr–Oct. Wet pine savannas, Carolina bays, shorelines of ponds, marshes; 0–100 m; Ala., Fla., Ga., La., Md., Miss., N.C., S.C., Va.

11. Gratiola torreyi Small, Fl. S.E. U.S., 1066, 1338. 1903 • Yellow hedge-hyssop E

Gratiola flava Leavenworth ex Pennell

Annuals. Stems ascending to erect, usually much-branched from base, 5–10 cm, glabrous. **Leaves:** blade linear-lanceolate, 5–15 × 1–5 mm, margins entire or with 1 or 2 pairs of teeth distally, apex obtuse, surfaces glabrous. **Pedicels** slender, 4–10 mm, length 1–2.3 times bract, glabrous; bracteoles 1, 1.5–4 mm. **Flowers:** sepals distinct, linear to linear-lanceolate, 3–6 mm; corolla 9–12 mm, tube yellow to orangish yellow, veins yellow to greenish yellow or brownish yellow, limb yellow; style 3–5 mm. **Capsules** ovoid, 2–2.5 × 1.5–2 mm. **Seeds** 0.3–0.4 mm.

Flowering Mar–Apr. Wet, exposed soils in meadows, clearings in post oak woodlands, saline prairies; 0–100 m; La., Tex.

The name *Gratiola flava* has been widely used for *G. torreyi*. F. W. Pennell (1935) synonymized *G. torreyi* with *G. neglecta*, stating that the type of *G. torreyi* (*Wright s.n.*, NY) has white corolla lobes and yellow corolla tubes instead of entirely golden yellow corollas as reported by Small in the protologue of *G. torreyi*. The statement by Pennell appears to be in error.

M. H. MacRoberts et al. (2007) discussed historic and recent collections of *Gratiola torreyi* (as *G. flava*) from Louisiana. The provenance of collections attributed to Arkansas is equivocal (MacRoberts et al.).

Gratiola pusilla Torrey ex Bentham 1846 (not Willdenow 1797) is an illegitimate name and pertains here.

12. Gratiola virginiana Linnaeus, Sp. Pl. 1: 17; 2: 1200. 1753 • Round-fruit or Virginia hedge-hyssop E

Gratiola virginiana var. *aestuariorum* Pennell

Annuals. Stems ascending to erect, simple or few-branched, 4–50 cm, glabrous or glabrate proximally, glabrous or glandular-puberulent distally. **Leaves:** blade lanceolate to elliptic or oblong-obovate, 15–70 × 5–25 mm, margins entire or with 1–4 pairs of blunt or sharp teeth distally, apex obtuse to acute, surfaces glabrous. **Pedicels** stout, 1–12 mm, length 0.1–0.3 times bract, glabrous or sparsely glandular-puberulent; bracteoles 2, 2–6 mm. **Flowers:** sepals distinct, linear-lanceolate to lanceolate or oblong, 4–7 mm; corolla 8–15 mm, tube greenish white to greenish yellow or yellow, veins purple or brownish purple, limb white, sometimes tinged lavender; style 2–4 mm. **Capsules** subglobular, (3–)4–9 × 4–8 mm. **Seeds** 0.7–0.8 mm. $2n = 16$.

Flowering Mar–Oct. Stream banks, swamps, floodplain pools and ponds, swamps; 0–500 m; Ala., Ark., Del., D.C., Fla., Ga., Ill., Ind., Iowa, Kans., Ky., La., Md., Mich., Miss., Mo., N.J., N.C., Ohio, Okla., R.I., S.C., Tenn., Tex., Va., W.Va.; introduced in Mexico (Veracruz).

Some plants from tidal wetlands in Maryland, New Jersey, and Virginia are purportedly relatively shorter in stature and bear shorter pedicels and smaller capsules than most plants of *Gratiola virginiana*. They have been treated as var. *aestuariorum*; the distinctness of var. *aestuariorum* has not been assessed in the context of morphological variation across the range of *G. virginiana*.

13. Gratiola viscidula Pennell, Torreya 19: 145. 1919 • Short's hedge-hyssop E F

Gratiola viscosa Schweinitz ex Leconte, Ann. Lyceum Nat. Hist. New York 1: 106. 1824, not Hornemann 1807; *G. viscidula* subsp. *shortii* Pennell

Perennials. Stems decumbent, ascending, or erect, simple, sometimes few-branched, (8–)15–60 cm, sparsely to densely glandular-puberulent or glandular-pubescent proximally, glandular-puberulent distally. **Leaves:** blade lanceolate-ovate to ovate or oblong, rarely linear-lanceolate in submersed forms, 5–28 × 4–15 mm, margins with (3–)5–10(–12) pairs of sharp teeth, apex acute, rarely obtuse, surfaces sparsely to densely glandular-puberulent. **Pedicels** slender, 10–23 mm, length 0.5–2 times bract, sparsely to densely glandular-puberulent; bracteoles 2, 3.5–7 mm. **Flowers:** sepals distinct, lanceolate to elliptic or ovate-lanceolate, 3–7 mm; corolla 9–13 mm, tube yellowish brown, veins blue or lavender, limb white; style 3–4 mm. **Capsules** globular, 1–2.5 × 1–2.5 mm. **Seeds** 0.3–0.4 mm. $2n = 16$.

Flowering May–Oct. Bogs, wet meadows, floodplain wetlands, beaver ponds, stream banks, seeps, shores of ponds and lakes; 0–300 m; Del., D.C., Fla., Ga., Ky., Md., N.C., Ohio, S.C., Tenn., Va., W.Va.

D. M. Spooner (1984) examined intraspecific variation in *Gratiola viscidula* and concluded that recognition of infraspecific taxa is not warranted.

SELECTED REFERENCE Spooner, D. M. 1984. Intraspecific variation in *Gratiola viscidula* Pennell (Scrophulariaceae). Rhodora 86: 79–87.

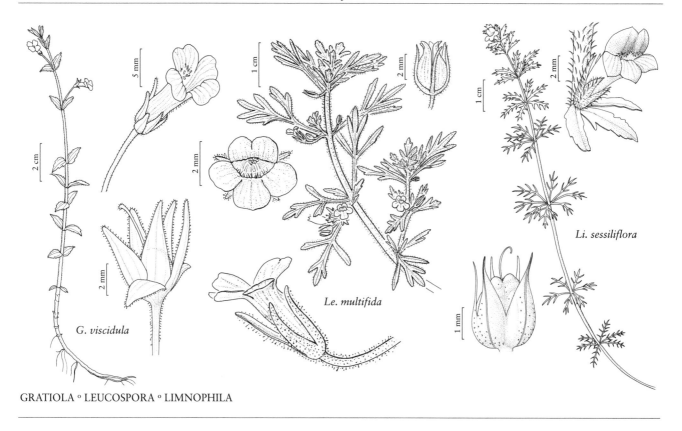

G. *viscidula*

Le. *multifida*

Li. *sessiliflora*

GRATIOLA ∘ LEUCOSPORA ∘ LIMNOPHILA

32. LEUCOSPORA   Nuttall, J. Acad. Nat. Sci. Philadelphia 7: 87. 1834   •   Paleseed [Greek *leucos*, white or clear, and *spora*, seed, alluding to transparency of matured seeds]

Craig C. Freeman

Herbs, annual. **Stems** erect or ascending, glandular-villosulous. **Leaves** cauline, opposite or whorled; petiole present; blade not fleshy, not leathery, margins pinnatifid to bipinnatifid. **Inflorescences** axillary, flowers solitary; bracts absent. **Pedicels** present; bracteoles absent. **Flowers** bisexual; sepals 5, distinct, lanceolate to linear-lanceolate, calyx obscurely bilaterally symmetric, tubular; corolla pale blue to pale lavender, bilaterally symmetric, bilabiate, tubular, tube base not spurred or gibbous, throat not densely pilose internally, lobes 5, abaxial 3, adaxial 2; stamens 4, medially adnate to corolla, didynamous, filaments glabrous; staminode 0; ovary 2-locular, placentation axile; stigma cuneate. **Fruits** capsules, dehiscence septicidal. **Seeds** 50–100, white, parallel-ridged, ovoid, wings absent.

Species 2 (1 in the flora): c, e North America, n Mexico.

The second species of *Leucospora*, as circumscribed here, is *L. coahuilensis* Henrickson, which is known from Chihuahua, Coahuila, Durango, and Zacatecas, Mexico. It differs from *L. multifida* in its longer corollas, shorter pedicels, longer and more slender capsules, and smaller seeds (J. Henrickson 1989).

Based on a molecular phylogenetic study, D. Estes and R. L. Small (2008) suggested that *Leucospora* belongs to a clade that includes *Bacopa*, *Gratiola*, *Mecardonia*, *Scoparia*, and *Stemodia*, genera traditionally placed in Gratioleae. Relationships of some species, and the circumscriptions of some genera and the tribe itself, remain uncertain due to limited sampling. *Leucospora* was found to be sister to *Stemodia verticillata* (Miller) Hassler; *Stemodia* was paraphyletic, with the type species of that genus, *S. maritima* Linnaeus, included in a second, separate clade. B. L. Turner and C. C. Cowan (1993) included *Leucospora* in *Stemodia* based on morphological evidence.

Schistophragma, another genus possibly allied with *Leucospora*, has not been examined molecularly; it shares some morphological similarities with *Leucospora*. On the basis of morphology, J. Henrickson (1989) argued that *Schistophragma* should be submerged in *Leucospora*. Pending more complete taxon sampling for Gratioleae, *Leucospora* is retained here in its traditional sense.

1. **Leucospora multifida** (Michaux) Nuttall, J. Acad. Nat. Sci. Philadelphia 7: 87. 1834 • Narrow-leaf paleseed E F

Capraria multifida Michaux, Fl. Bor.-Amer. 2: 22, plate 35. 1803; *Conobea multifida* (Michaux) Bentham; *Stemodia multifida* (Michaux) Sprengel

Stems (4–)10–30(–41) cm, usually much-branched from base; branches spreading to ascending. **Leaves** 2 or 3(or 4) per node; blade ovate to triangular-ovate in outline, rarely lanceolate in outline, 8–30 × 4–23 mm, ultimate segments lanceolate to linear. **Pedicels** spreading to ascending, (1–)2–5 mm in flower, to 7 mm in fruit, glandular-villosulous. **Flowers:** calyx lobes slightly falcate, abaxial slightly shorter than adaxial, 2.6–4 × 0.4–0.7 mm; corolla glabrous, tube 2–3 mm, throat open, yellow or greenish yellow abaxially within, lobes 1 × 1 mm; stamens included; style 0.5–1 mm. **Capsules** ovoid, 4.5 × 1.8–2.3 mm. **Seeds** 0.3–0.4 mm, 8-ribbed, ribs longitudinal, straight.

Flowering Jun–Nov. Sandy or silty shores of streams, lakes, and ponds, muddy or sandy bars, rocky intermittent stream beds, seeps, solution holes in rocks, wet places in urban areas; 0–600 m; Ont.; Ala., Ark., Fla., Ga., Ill., Ind., Iowa, Kans., Ky., La., Mich., Miss., Mo., Nebr., N.J., N.Y., N.C., Ohio, Okla., Pa., Tenn., Tex., Va.

As noted by F. W. Pennell (1935), populations of *Leucospora multifida* often are associated with calcareous substrates; populations also occur on other substrates. They occasionally are reported along railroads and in damp sites in parking lots in urban areas.

33. **LIMNOPHILA** R. Brown, Prodr., 442. 1810, name conserved • Ambulia [Greek *limne*, pool, and *philos*, loving, alluding to habitat] I

Kerry A. Barringer

Herbs, perennial [annual], paludal or aquatic, emergent. **Stems** erect [prostrate, creeping], hairy or glabrous. **Leaves** cauline, opposite or whorled, dimorphic; petiole present or absent; blade not fleshy, not leathery, margins of submerged leaves pinnatifid, of aerial leaves entire, serrate, or pinnatifid. **Inflorescences** terminal or axillary, racemes or flowers solitary, rarely paired [or spikes]; bracts present or absent. **Pedicels** present or absent; bracteoles present or absent. **Flowers** bisexual, chasmogamous or cleistogamous; sepals 5, basally connate, calyx bilaterally symmetric, tubular, lobes triangular or lanceolate [ovate]; corolla blue, blue-purple, or white, bilaterally symmetric, bilabiate, tubular or funnelform, tube base not gibbous or spurred, lobes 4 or 5, abaxial 3, adaxial 1 or 2; stamens 4, basally adnate to corolla, didynamous, filaments glabrous or hairy; staminode 0; ovary 2-locular, placentation axile; stigma capitate, 2-lobed. **Fruits** capsules, dehiscence septicidal. **Seeds** 20–200, brown to dark brown, conic to cylindric, wings absent. *x* = 17.

Species 40 (2 in the flora): introduced; Asia, Africa, Pacific Islands, Australia; introduced also in South America.

The two species recognized here could be difficult to distinguish throughout their ranges; intermediate plants, which are believed to be hybrids, have been found.

Limnophila is closely related to *Hydrotriche* Zuccarini, a small genus of aquatic plants native to Madagascar. Together they form a group sister to *Gratiola* in Gratioleae (D. Estes and R. L. Small 2008).

SELECTED REFERENCE   Philcox, D. 1970. A taxonomic revision of the genus *Limnophila* R. Br. (Scrophulariaceae). Kew Bull. 24: 101–170.

1. Pedicels 3–8 mm in flower, 5–12 mm in fruit; bracteoles 1–3.5 mm; corolla tubes 4–5(–8) mm; cleistogamous flowers absent . . . . 1. *Limnophila indica*
1. Pedicels 0–2 mm in flower, 0–4 mm in fruit; bracteoles 0–1.5 mm; corolla tubes 8–10 mm; cleistogamous flowers present, submerged . . . . 2. *Limnophila sessiliflora*

1. Limnophila indica (Linnaeus) Druce, Rep. Bot. Exch. Club Soc. Brit. Isles 3: 420. 1914 [I]

Hottonia indica Linnaeus, Syst. Nat. ed. 10, 2: 919. 1759

Perennials aquatic or paludal. **Stems:** submerged glabrous, aerial to 15 cm, glabrous or glandular-pubescent with stalked or sessile glands. **Leaves** verticillate or opposite; blade of submerged leaves broadly ovate in outline, 10–35 × 10–15 mm, segments capillary, blade of aerial leaves linear to lanceolate, 10–40 × 1–10 mm. **Inflorescences** terminal, racemes or flowers solitary, axillary. **Pedicels** 3–8 mm in flower, 5–12 mm in fruit; bracteoles 1–3.5 mm. **Flowers:** cleistogamous absent; chasmogamous aerial; calyx 3–3.5 mm, lobes triangulate, 1–2 mm; corolla tube white, 4–5 (–8) mm, lobes lavender to purple, 1–2 × 1–1.5 mm; stamens 4–5 mm; style 3–4 mm. **Capsules** compressed, 3–5 mm. **Seeds** brown, conic. $2n = 34$ (Asia).

Flowering Jun–Aug. Ponds, rice fields; 0–100 m; introduced; Fla.; Asia.

Limnophila indica is uncommon in the flora area and may not persist. The species is variable: leaf size, shape, number, and dissection can vary with light intensity, day length, and water level (D. Philcox 1970). Corolla and calyx colors vary, and the shape and size of the calyx lobes change during the development of the flower and fruit. *Limnophila indica* tends to have whorled leaves proximally and flowers in racemes, characteristics that can help distinguish it from *L. sessiliflora*. Plants sold as *L. indica* in the aquarium trade are often *L. sessiliflora*. A hybrid with *L. sessiliflora* (*L.* ×*ludoviciana* Thieret) is discussed below.

2. Limnophila sessiliflora Blume, Bijdr. Fl. Ned. Ind. 14: 749. 1826 (as Lymnophila sessiflora) [F] [I] [W]

Perennials aquatic. **Stems:** submerged glabrous, aerial to 20 cm, glabrous or pubescent with eglandular and glandular hairs. **Leaves** opposite; blade of submerged leaves ovate to broadly ovate in outline, 10–35 × 10–15 mm, segments flattened or capillary, blade of aerial leaves elliptic to lanceolate, 10–30(–80) × 2–5 (–10) mm. **Inflorescences** axillary, flowers solitary, rarely paired. **Pedicels** 0–2 mm in flower, 0–4 mm in fruit; bracteoles 0–1.5 mm. **Flowers:** cleistogamous submerged, budlike; chasmogamous aerial; calyx 3–5 mm, lobes lanceolate, 2–4 mm; corolla tube white, 8–10 mm, lobes blue-purple, 0.9–1.2 × 1–1.5 mm; stamens 4–6 mm; style 1–3 mm. **Capsules** often sulcate, slightly compressed, 2–2.5 mm. **Seeds** brown to dark brown, conic to cylindric.

Flowering Jul–Sep. Ponds, swamps, rice fields, streams; 0–500 m; introduced; Calif., Fla., Ga., La., Tex.; Asia.

Bentham (in N. Wallich 1828[–1849]) seems to have been the first to correct the spelling by Blume from *sessiflora* to *sessiliflora*, and later works have generally accepted the corrected epithet. Some works treat *Hottonia sessiliflora* Vahl as the basionym, but no homotypy has been established between the two names.

Limnophila sessiliflora is the more common species of the genus found in North America. It is a problematic weed in rice fields and shallow ponds worldwide; it has been placed on the Federal Noxious Weed List (http://www.aphis.usda.gov/plant_health/plant_pest_info/weeds/downloads/weedlist.pdf). It is also sometimes sold in the aquarium trade under the name *L. indica*.

An uncommon hybrid with *Limnophila indica* (*L.* ×*ludoviciana* Thieret) is distinguished by its ebracteolate pedicels 5–10 mm and calyx lobes 1–2 mm with eglandular trichomes (D. Philcox 1970). Some of the plants identified as the hybrid are actually *L. sessiliflora* with pedicellate flowers, a condition found throughout the range of the species.

34. MECARDONIA   Ruiz & Pavon, Fl. Peruv. Prodr., 95. 1794   •   [For Antoni de Meca-Caçador-Cardona i de Beatrin, 1726–1788, benefactor of Royal College of Surgery of Barcelona]

Adjoa Richardson Ahedor

Herpestis C. F. Gaertner

Herbs, perennial; caudex herbaceous or woody, herbage tending to dry black. **Stems** prostrate, ascending, erect, or spreading, glabrous. **Leaves** cauline, opposite; petiole present; blade not fleshy, not leathery, margins weakly or sharply serrate. **Inflorescences** axillary, flowers solitary; bracts present. **Pedicels** present; bracteoles smaller than calyx lobes, not surrounding calyx of flowers they subtend. **Flowers** bisexual; sepals 5, proximally connate, calyx bilaterally symmetric, campanulate, lobes ovate, outer lobes ± as wide as inner; corolla white or lemon yellow with purple or reddish veins in throat, bilaterally symmetric, weakly bilabiate, campanulate [salverform], tube base not spurred or gibbous, throat not densely pilose internally, lobes 5, abaxial 3, adaxial 2, abaxial lip projecting to spreading; stamens 4, medially or distally adnate to corolla, didynamous, filaments glabrous; staminode 0; ovary 2-locular, placentation axile; stigma 2-lobed. **Fruits** capsules, dehiscence septicidal, slightly loculicidal at apex. **Seeds** 40–80, black or brown, angled, wings absent. $x = 11$.

Species ca. 10 (2 in the flora): c, e United States, Mexico, Central America, South America; introduced in Asia (India, Sri Lanka), Africa (Cameroon, Sierra Leone), Australia.

SELECTED REFERENCES   Ahedor, A. R. 2007. Systematics of the *Mecardonia acuminata* (tribe Gratioleae, Plantaginaceae) Complex of Southern USA. Ph.D. dissertation. University of Oklahoma.   Ahedor, A. R. and W. J. Elisens. 2015. Morphological analyses of the *Mecardonia acuminata* (Plantaginaceae) species complex in the southeastern U.S.A. SouthE. Naturalist 14: 173–196.   Rossow, R. A. 1987. Revisión del género *Mecardonia*. Candollea 42: 431–474.

1. Stems erect; leaf blade margins weakly serrate; corollas white with purple veins in throats, adaxial lobes connate ½–⅔ lengths; seeds: radial walls of reticulum smooth . . . . 1. *Mecardonia acuminata*
1. Stems spreading, prostrate, or ascending; leaf blade margins sharply serrate; corollas lemon yellow with reddish veins in throats, adaxial lobes connate nearly to apices; seeds: radial walls of reticulum mamillate . . . . 2. *Mecardonia procumbens*

1. **Mecardonia acuminata** (Walter) Small, Fl. S.E. U.S., 1065, 1337. 1903 • Axilflower E F

Gratiola acuminata Walter, Fl. Carol., 61. 1788; *Bacopa acuminata* (Walter) B. L. Robinson

Stems erect, 10–60 cm, glandular. **Leaves** glandular; blade 9–30 × 6–10 mm, 1–2 times bracts, margins weakly serrate. **Pedicels** divaricate or ascending, 6–22 mm. **Flowers:** calyx lobes unequal; corolla white with purple veins in throat, adaxial lobes connate ½–⅔ lengths, abaxial lip not prominent; stamens distally adnate to corolla. **Capsules** 6–7 mm. **Seeds** 40–60, black, radial walls of reticulum smooth.

Varieties 3 (3 in the flora): c, e United States.

1. Stems branched from bases and diffuse; fruiting pedicels ascending, 1–2 times bracts . . . . 1c. *Mecardonia acuminata* var. *peninsularis*
1. Stems branched from middles, rarely diffuse; fruiting pedicels divaricate, ½–2 times bracts.
  2. Fruiting pedicels 17–35 mm, 1–2 times bracts . . . . 1a. *Mecardonia acuminata* var. *acuminata*
  2. Fruiting pedicels 10–18 mm, ½–1 times bracts. . . 1b. *Mecardonia acuminata* var. *microphylla*

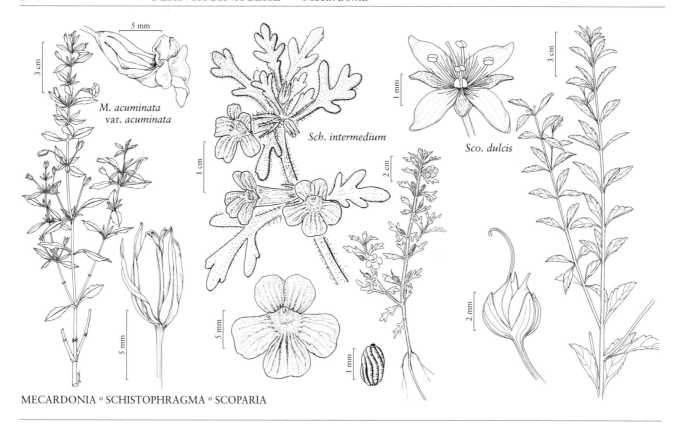

M. *acuminata*
var. *acuminata*

Sch. intermedium

Sco. dulcis

MECARDONIA ° SCHISTOPHRAGMA ° SCOPARIA

1a. Mecardonia acuminata (Walter) Small var. **acuminata** E F

Stems branched from middle, rarely diffuse. **Leaf blades** 20–30 mm. **Fruiting pedicels** divaricate, 17–35 mm, 1–2 times bracts.

Flowering Jul–Oct. Moist sand and loam, near streams, ditches, roadsides, partial shade; 0–900 m; Ala., Ark., Fla., Ga., Ill., Ind., Kans., La., Miss., Mo., N.C., Okla., S.C., Tenn., Tex., Va., W.Va.

1b. Mecardonia acuminata (Walter) Small var. **microphylla** (Rafinesque) Pennell, Torreya 22: 79. 1922 E

Ambulia rigida Rafinesque var. *microphylla* Rafinesque, Autik. Bot., 43. 1840; *Mecardonia acuminata* subsp. *microphylla* (Rafinesque) Rossow

Stems branched from middle, rarely diffuse. **Leaf blades** 20–30 mm. **Fruiting pedicels** divaricate, 10–18 mm, ½–1 times bracts.

Flowering Jul–Oct. Pinelands, swamps, low wet woods, sloughs, wet meadows, bottomland fields and wet open areas; 0–400 m; Ala., Ark., Fla., Ga., La., N.C., S.C., Tex., Va., W.Va.

1c. Mecardonia acuminata (Walter) Small var. **peninsularis** Pennell, Proc. Acad. Nat. Sci. Philadelphia 71: 237. 1920 E

Mecardonia acuminata subsp. *peninsularis* (Pennell) Rossow

Stems branched from base and diffuse. **Leaf blades** 9–14 mm. **Fruiting pedicels** ascending, 16–28 mm, 1–2 times bracts.

Flowering Jun–Oct. Moist sand and loam, near streams, ditches, roadsides, partial shade; 0–300 m; Ala., Fla., Ga., La., Miss.

2. **Mecardonia procumbens** (Miller) Small, Fl. S.E. U.S., 1065, 1338. 1903 • Baby jump-up

Erinus procumbens Miller, Gard. Dict. ed. 8, Erinus no. 6. 1768; *Bacopa procumbens* (Miller) Greenman; *Mecardonia tenuis* Small

Stems spreading, prostrate, or ascending, 10–60 cm, glandular. **Leaves** glandular; blade 10–25 × 6–12 mm, 4–8 times bracts, margins sharply serrate. **Pedicels** ascending, 7–25 mm.

Flowers: calyx lobes equal; corolla lemon yellow with reddish veins in throat, abaxial lip prominent, adaxial lobes connate nearly to apices; stamens medially adnate to corolla. **Capsules** 4–5 mm. **Seeds** 60–80, brown, radial walls of reticulum mamillate. $2n = 22, 44$.

Flowering May–Oct. Wetlands, edges of streams, springs and seeps, deserts in upland washes and canyons with moisture; 90–1600 m; Ala., Ariz., Fla., Ga., La., Miss., N.Mex., Tex.; Mexico; Central America; introduced in Asia (India, Sri Lanka), Africa (Cameroon, Sierra Leone), Australia.

35. SCHISTOPHRAGMA Bentham in S. L. Endlicher, Gen. Pl. 9: 679. 1839 • [Greek *schist*, cleft, and *phragma*, fence, alluding to incomplete septum of ovary and fruit]

Kerry A. Barringer

Herbs, annual. **Stems** erect or ascending, glandular-hairy. **Leaves** cauline, opposite; petiole present; blade not fleshy, not leathery, margins pinnatifid [entire]. **Inflorescences** axillary, flowers solitary; bracts absent. **Pedicels** present; bracteoles absent. **Flowers** bisexual; sepals 5, basally connate, calyx bilaterally symmetric, tubular, lobes narrowly triangular; corolla pink or purple, bilaterally symmetric, bilabiate, tubular, tube base not spurred or gibbous, throat not densely pilose internally, lobes 5, abaxial 3, adaxial 2; stamens 4, proximally adnate to corolla, didynamous, filaments glabrous; staminode 0; ovary incompletely 2-locular, placentation axile; stigma capitate, slightly 2-lobed. **Fruits** capsules, dehiscence septicidal. **Seeds** 30–100, yellow or brown, spirally ridged, ovoid or fusiform, wings absent. $x = 20$.

Species 3 (1 in the flora): sw United States, Mexico, Central America, South America (Colombia).

Schistophragma is related to *Leucospora*, *Limnophila*, and *Stemodia*, and shares with them distinctive, stipitate anthers and a curved, capitate and two-lobed stigma. They are all in Gratioleae. Morphological characters have not been sufficient to clarify the relationships of the genera in this tribe, and molecular data are not available for many of the species, including *S. intermedium*.

1. **Schistophragma intermedium** (A. Gray) Pennell, Notul. Nat. Acad. Nat. Sci. Philadelphia 43: 2. 1940 (as intermedia) • Harlequin spiralseed [F]

Conobea intermedia A. Gray in W. H. Emory, Rep. U.S. Mex. Bound. 2(1): 117. 1859

Stems simple or branching, 3–10 (–15) cm. **Leaves** 2 per node; blade ovate to lanceolate, 10–20 × 3–10 mm, ultimate segments ovate to obovate. **Pedicels** ascending, 1–2(–4) mm in flower, 2–5(–7) mm in fruit, glabrous or glandular-villosulous.
Flowers: calyx not falcate, abaxial lobe shorter than adaxial lobes, 2–4 × 0.3–0.5 mm, glabrous or glandular-villosulous; corolla glabrous, tube funnelform, 5–6 mm,

throat open, pale yellow, lobes spreading, violet to purple, usually with darker lines, 1.5–2 × 1–2 mm; stamens included; style 2–2.5 mm. **Capsules** green or purple, ovoid, 4–6(–8) × 1.5–2.5 mm, slightly upcurved. **Seeds** shallowly reticulate, longitudinally spiraled. $2n = 40$.

Flowering Jul–Sep. Dry, sunny slopes and washes; 1000–2500 m; Ariz., N.Mex.; Mexico (Chihuahua, Sonora).

Schistophragma intermedium is closely related to *Leucospora multifida*, which has two or three (or four) leaves per node, corolla tubes 2–3 mm, and smaller fruits with obtuse or rounded apices. *Schistophragma intermedium* might be confused with species of *Verbena* due to the pinnatifid leaves and purple tubular corollas; the flowers of *Schistophragma* are solitary in leaf axils, and the corollas are bilabiate.

36. SCOPARIA Linnaeus, Sp. Pl. 1: 116. 1753; Gen. Pl. ed. 5, 52. 1754   •   Goat-weed, sweet-broom [Latin *scopa*, broom, and *-aria*, resemblance, alluding to appearance and use]

Craig C. Freeman

Herbs [subshrubs], annual or perennial. **Stems** decumbent, spreading, ascending, or erect, glabrous or puberulent to glandular-puberulent. **Leaves** cauline, opposite, distal sometimes whorled; petiole absent or nearly so; blade not fleshy, not leathery, margins entire, crenate, dentate, or pinnately lobed, surfaces distinctly punctate. **Inflorescences** axillary, flowers 1–4 per node; bracts present. **Pedicels** present, spreading to ascending; bracteoles absent. **Flowers** bisexual; sepals 4 or 5, proximally connate, calyx radially symmetric, short-campanulate, lobes ovate to elliptic-ovate or lanceolate; corolla white, sometimes tinged pink or lavender, or yellow or orangish yellow, radially symmetric, rotate or subrotate, tube base not spurred or gibbous, throat densely pilose internally, lobes 4; stamens 4, proximally adnate to corolla, subequal, exserted, filaments glabrous; staminode 0; ovary 2-locular, placentation axile; stigma capitate. **Fruits** capsules, dehiscence septicidal and secondarily loculicidal. **Seeds** 50–200, brown to dark brown, oblong or angled, wings absent. *x* = 10.

Species ca. 10 (2 in the flora): United States, Mexico, West Indies, Central America, South America; introduced in Asia, Africa, Indian Ocean Islands (Madagascar), Australia.

SELECTED REFERENCE   Chodat, R. 1908. Étude critique des genres *Scoparia* L. et *Hasslerella* Chod. Bull. Herb. Boissier, sér. 2, 8: 1–16.

1. Corollas white, sometimes tinged pink or lavender; calyx lobes 4 . . . . 1. *Scoparia dulcis*
1. Corollas yellow or orangish yellow; calyx lobes 5 . . . . 2. *Scoparia montevidensis*

1. Scoparia dulcis Linnaeus, Sp. Pl. 1: 116. 1753
   • Licorice-weed F W

Annuals or perennials. Stems erect to ascending, usually much-branched distally, (17–) 30–100(–150) cm, glabrous or puberulent. **Leaves:** blade oblanceolate to narrowly oblanceolate or rhombic, 8–53 × 3–25 mm, base tapered to cuneate, margins crenate to dentate in distal ½. **Inflorescences:** flowers 1 or 2(or 3) per axil; bracts narrowly oblanceolate to narrowly elliptic, 4–35 mm. **Pedicels** 2–10 mm, glabrous. **Flowers:** calyx lobes 4, ovate to elliptic-ovate, 1.2–1.5 × 0.6–1 mm, margins ciliolate; corolla white, sometimes tinged pink or lavender, 2–2.5 × 3–4 mm. **Capsules** ovoid to subglobose, (1.6–)2–2.5(–4) × 1.4–2 mm. **Seeds:** 0.1–0.3 mm. *2n* = 40 (India).

Flowering May–Nov(–Jan). Marshes, wet hammocks, flatwoods, sandy woods, disturbed sites; 0–300 m; Ala., Fla., Ga., La., Miss., S.C., Tex.; Mexico; West Indies; Central America; South America; introduced in Asia, Africa, Indian Ocean Islands (Madagascar), Australia.

Scoparia dulcis is a pantropical weed. Noting that it was a widespread weed in lowland tropical America, F. W. Pennell (1935) believed that it was adventive in the United States. Most United States floras consider it to be native in the flora area, and it is treated that way here.

2. Scoparia montevidensis (Sprengel) R. E. Fries, Ark. Bot. 6(9): 22. 1907   • Broomwort I

Microcarpaea montevidensis Sprengel, Syst. Veg. 1: 42. 1824; *Scoparia flava* Chamisso & Schlechtendal; *S. montevidensis* var. *flava* (Chamisso & Schlechtendal) Chodat; *S. montevidensis* var. *glandulifera* (Fritsch) R. E. Fries

Perennials. Stems decumbent, spreading, ascending, or erect, sometimes rooting at nodes, usually much-branched proximally, 10–30 cm, glabrous or distally sparsely glandular-puberulent. **Leaves:** blade oblanceolate to narrowly oblanceolate, sometimes linear, 5–25 × 0.4–8 mm, base tapered, margins dentate to pinnately lobed, sometimes entire.

Inflorescences: flowers 1 or 2 per axil; bracts narrowly oblanceolate to linear, 2–20 mm. **Pedicels** 6–17 mm, glabrous or sparsely glandular-puberulent. **Flowers:** calyx lobes 5, ovate to lanceolate, 1.6–2.2 × 0.5–0.7 mm; corolla yellow or orangish yellow, 2.5–2.9 × 3–4.5 mm. **Capsules** ovoid, 2.2–4 × 1.3–2.5 mm. **Seeds:** faces 0.2–0.4 mm.

Flowering Apr–Jul. Sandy, disturbed areas, including vacant lots, roadsides, ballast; 0–100 m; introduced; Fla., N.C.; Mexico (Veracruz); Central America; South America.

R. Chodat (1908) noted the polymorphic nature of *Scoparia montevidensis* and emphasized habit, leaf margin, and pedicel characters in distinguishing seven varieties. These characters appear to vary freely among South American plants, making the application of infraspecific names of questionable value. The name var. *glandulifera* has been applied to some specimens in the flora area with glandular-puberulent pedicels. A specimen collected in 1874 on ballast at Kaighn Point in Camden, New Jersey (*C. F. Parker s.n.*, MO) was annotated by F. W. Pennell in 1934 as *S. flava*. He later cited that specimen, the only collection from New Jersey, as *S. montevidensis* (Pennell 1935), apparently adopting a concept of that species that included *S. flava*, an approach that is followed here.

37. SOPHRONANTHE Bentham in J. Lindley, Nat. Syst. Bot., 445. 1836 • Hedge-hyssop [Greek *sphoron*, modest, and *anthos*, flower, alluding to small flowers] E

Craig C. Freeman

Tragiola Small & Pennell

Herbs, perennial; caudex thick, hard. **Stems** ascending to erect, villous (hairs jointed), rarely glabrous. **Leaves** cauline, sometimes also basal, opposite; petiole absent, sometimes present on basal leaves; blade not fleshy, leathery, margins entire or serrate. **Inflorescences** axillary, flowers 1 or 2 per node; bracts present. **Pedicels** present or absent; bracteoles present, 2. **Flowers** bisexual; sepals 5, distinct, linear to linear-lanceolate, calyx ± bilaterally symmetric, campanulate; corolla white, sometimes tinged purple or pink, bilaterally symmetric, bilabiate to bilabiate and personate, tubular to salverform, tube base not spurred or gibbous, lobes 5, abaxial 3, adaxial 2, throat hairy; stamens 2, medially adnate to corolla, filaments glabrous, pollen sacs parallel to filaments, connective not dilated; staminodes 0 or 2, minute; ovary 2-locular, placentation axile; stigma crateriform. **Fruits** capsules, dehiscence septicidal, sometimes also loculicidal. **Seeds** ca. 200, yellow to brown or black, ovoid to short-cylindric, wings absent.

Species 2 (2 in the flora): se United States.

The two species of *Sophronanthe* often have been included in *Gratiola* or in the monospecific genera *Sophronanthe* (*S. hispida*) and *Tragiola* (*S. pilosa*). Molecular and morphological data support the monophyly of *Sophronanthe* (D. Estes 2008), which is sister to *Gratiola* in the narrow sense.

1. Leaf blades 0.7–3 mm wide, linear to lanceolate-ovate, margins prominently revolute; corollas salverform, 7–16 mm . . . . .1. *Sophronanthe hispida*
1. Leaf blades 4–11 mm wide, lanceolate to ovate, margins not or slightly revolute; corollas tubular, 5–9 mm . . . . .2. *Sophronanthe pilosa*

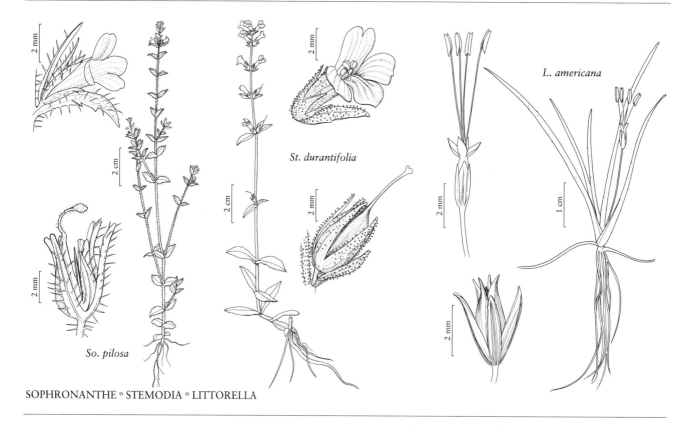

SOPHRONANTHE ° STEMODIA ° LITTORELLA

So. pilosa

St. durantifolia

L. americana

1. Sophronanthe hispida Bentham in J. Lindley, Nat. Syst. Bot., 445. 1836 • Rough hedge-hyssop E

Gratiola hispida (Bentham) Pollard

Stems ascending to erect, 10–75 cm, villous. **Leaves:** blade linear to lanceolate-ovate, 8–20 × 0.7–3 mm, base not clasping, truncate, margins entire, prominently revolute, apex acute to acuminate, surfaces villous, papillate, not distinctly punctate abaxially. **Pedicels** 0–1 mm; bracteoles 6–9 mm. **Flowers:** calyx lobes 3–7 × 0.5–1 mm; corolla salverform, 7–16 mm, tube white, veins inconspicuous or pale purple, limb white. **Capsules** conic, 4–5 × 1.8–2.5 mm. **Seeds** 0.2–0.4 mm.

Flowering May–Oct. Sandy scrub, dry to wet flatwoods, sandhills, seasonally wet clearings; 0–100 m; Ala., Fla., Ga., Miss.

Gratiola subulata Baldwin ex Bentham is an illegitimate name that pertains to *Sophronanthe hispida*.

2. Sophronanthe pilosa (Michaux) Small, Fl. S.E. U.S., 1067, 1338. 1903 • Shaggy hedge-hyssop E F

Gratiola pilosa Michaux, Fl. Bor.-Amer. 1: 7. 1803; *G. pilosa* var. *epilis* Pennell; *Tragiola pilosa* (Michaux) Small & Pennell

Stems erect, 20–65 cm, villous, rarely glabrous. **Leaves:** blade lanceolate to ovate, 8–20 × 4–13 mm, base clasping, rounded to subcordate, margins entire or serrate, not or slightly revolute, apex rounded to obtuse or acute, surfaces villous, rarely glabrous, papillate, distinctly punctate abaxially. **Pedicels** 0–3(–5) mm; bracteoles 6–9(–11) mm. **Flowers:** calyx lobes 5–7 × 0.5–1.5 mm; corolla tubular, 5–9 mm, tube yellow, veins bluish purple, limb white. **Capsules** conic, 4–5 × 2–2.8 mm. **Seeds** 0.2–0.4 mm. $2n = 22$.

Flowering May–Oct. Sandy pinelands, bogs, swamps, cypress-gum depressions, pine savannas, wet flatwoods; 0–700 m; Ala., Ark., Del., Fla., Ga., Ky., La., Md., Miss., N.J., N.C., Okla., S.C., Tenn., Tex., Va.

Plants with glabrous stems and leaves have been named *Gratiola pilosa* var. *epilis*. The combination for that taxon in *Sophronanthe* has not been made.

38. STEMODIA Linnaeus, Syst. Nat. ed. 10, 2: 1091, 1118, 1374. 1759, name conserved

• Twintips [Greek *stemon*, stamen, and *dis*, double, alluding to each stamen bearing two anthers]

Kerry A. Barringer

Herbs [**suffruticose herbs**], annual or perennial. **Stems** erect, ascending, decumbent, or prostrate, hairy [glabrous]. **Leaves** cauline, opposite or whorled; petiole absent [present]; blade not fleshy, not leathery, margins serrate or denticulate. **Inflorescences** terminal or axillary, spikes, racemes, or flowers solitary; bracts present or absent. **Pedicels** present or absent; bracteoles smaller than calyx lobes, not surrounding calyx of flower they subtend. **Flowers** bisexual; sepals 5, basally connate, calyx bilaterally symmetric [radially symmetric], tubular, lobes narrowly lanceolate to narrowly triangular, outer lobes ± as wide as inner; corolla blue-purple, lavender, or white, bilaterally symmetric, bilabiate, tubular, tube base not spurred or gibbous, lobes 5, abaxial 3, adaxial 2; stamens 4, proximally adnate to corolla, didynamous, filaments glabrous [hairy]; staminode 0 or 1, short-cylindric [filiform]; ovary 2-locular, placentation axile; stigma capitate, 2-lobed. **Fruits** capsules, dehiscence loculicidal, 4-valved. **Seeds** 10–150, light brown [black], ovoid or ellipsoid, wings absent.

Species 52 (3 in the flora): s United States, Mexico, West Indies, Central America, South America, Asia, Africa, Australia.

Stemodia is closely allied to *Leucospora*, *Limnophila*, and *Schistophragma* in Gratioleae. D. Estes and R. L. Small (2008) found *Stemodia* to be paraphyletic as now delimited. Of the species in the flora area, only *S. schottii* was sampled; it appears to be in the cluster containing *S. maritima* Linnaeus, the type of the genus.

Species of *Stemodia* can resemble other small-flowered Plantaginaceae but always can be distinguished by their distinct anther cells, their ovoid, loculicidal, four-valved capsules, and their parallel-ridged seeds. *Schistophragma* has pinnately lobed leaf blade margins, narrowly cylindric, septicidal fruits, and spirally ridged seeds. *Leucospora* is distinguished by its pinnatifid to bipinnatifid leaf blade margins, pedicels without bracteoles, and septicidal capsules. *Limnophila* grows in mud or shallow water and has pinnatifid blade margins of submersed leaves.

SELECTED REFERENCE Turner, B. L. and C. C. Cowan. 1993. Taxonomic overview of *Stemodia* (Scrophulariaceae) for North America and the West Indies. Phytologia 74: 61–103.

1. Leaf blade surfaces densely white-tomentose, bases not auriculate or clasping. . . . . 2. *Stemodia lanata*
1. Leaf blade surfaces sparsely hairy or sparsely glandular-hairy, bases auriculate or clasping.
    2. Inflorescences terminal, spikes, racemes, or flowers 1–4 per axil; corollas 5–8 mm . . .
         . . . . .1. *Stemodia durantifolia*
    2. Inflorescences axillary, flowers 1 or 2 per axil; corollas 10–13 mm . . . . 3. *Stemodia schottii*

1. Stemodia durantifolia (Linnaeus) Swartz, Observ. Bot., 240. 1791 [F]

Capraria durantifolia Linnaeus, Syst. Nat. ed. 10, 2: 1116. 1759; *Stemodia arizonica* Pennell

Stems erect or ascending, 10–100 cm, glandular- and eglandular-hairy. **Leaves** opposite or whorled proximally, opposite distally; blade lanceolate to oblong, 15–50(–70) × 3–15(–20) mm, base clasping, margins serrate distally, flat, surfaces sparsely hairy. **Inflorescences** terminal, spikes, racemes, or flowers 1–4 per axil; bracts 4–5 mm. **Pedicels** 0–8(–15) mm; bracteoles [1 or]2, 2–3.5 mm. **Flowers:** calyx lobes narrowly lanceolate, 3–5 mm; corolla blue-purple, 5–8(–10) mm, puberulent, lobes 1–2 × 1 mm; style 2–4 mm. **Capsules** 4–5 mm; style often persistent. **Seeds** ellipsoid, 2–3 mm.

Flowering year-round. Wet sand and rock in open, sometimes disturbed, sites; 0–400 m; Ariz., Calif., Fla.; Mexico; West Indies; Central America; South America.

Stemodia durantifolia might be confused with *S. maritima* Linnaeus, which is found in the Caribbean and along the coast in Mexico and elsewhere in the New World tropics. *Stemodia maritima* has corollas less than 5 mm and stalked pollen sacs. Florida plants of *S. durantifolia* are introduced.

2. Stemodia lanata Ruiz & Pavon ex Bentham in A. P. de Candolle and A. L. P. P. de Candolle, Prodr. 10: 383. 1846 • Woolly twintip

Stems prostrate or decumbent, 5–10 cm, densely white-tomentose. **Leaves** opposite; blade elliptic to obovate, 10–30 × 5–10 mm, base not auriculate or clasping, margins slightly serrate distally, slightly revolute, surfaces densely white-tomentose. **Inflorescences** axillary, flowers 1 or 2 per axil; bracts 0 mm. **Pedicels** 0 mm; bracteoles 1 or 2, 1–2 mm. **Flowers:** calyx lobes narrowly triangular, 2–3(–5) mm; corolla blue-purple to lavender, 6–7 mm, glabrous or puberulent, lobes 1.5–2 × 1.5–2 mm; style 1–2 mm. **Capsules** 2–3 mm; style not persistent. **Seeds** ellipsoid, 0.2–0.5 mm.

Flowering Apr–Jul. Sandy soils, woodland openings, disturbed sites; 0–200 m; Tex.; Mexico (Tamaulipas, Veracruz).

Stemodia lanata is sometimes called *S. tomentosa* (Miller) Greenman & C. H. Thompson, a later homonym of *S. tomentosa* G. Don 1838.

3. Stemodia schottii Holzinger, Contr. U.S. Natl. Herb. 1: 286, plate 20. 1893

Stems ascending to decumbent, 5–15(–25) cm, glandular-pubescent. **Leaves** opposite; blade obovate to oblanceolate, 10–30 × 5–15 mm, base auriculate or clasping, margins denticulate, flat, surfaces sparsely glandular-hairy. **Inflorescences** axillary, flowers 1 or 2 per axil; bracts 0 mm. **Pedicels** 2–6 mm; bracteoles 1 or 2, 3–5 mm. **Flowers:** calyx lobes narrowly triangular, 4.5–6.5 mm; corolla blue-purple or white, 10–13 mm, slightly hairy, lobes 3.4–4 × 2–4 mm; style 5–6 mm. **Capsules** 5–6 mm; style not persistent. **Seeds** ovoid, 0.2–0.5 mm.

Flowering year-round. Sandbars, moist, sandy disturbed sites; 0–50 m; Tex.; Mexico (Coahuila, Durango, Nuevo León, Tamaulipas, Veracruz).

Stemodia schottii is closely related to *S. durantifolia* and is distinguished by its smaller size and axillary flowers. In Texas, *S. schottii* is known from counties along the lower Rio Grande and from the Big Bend region.

39. LITTORELLA P. J. Bergius, Kongl. Vetensk. Acad. Handl. 29: 341. 1768 • [Latin *littora*, shores, and *-ella*, small, alluding to small lakeshore habitat]

Alexey Shipunov

Herbs, perennial; stoloniferous. **Stems** erect or creeping, glabrous. **Leaves** basal, alternate; petiole absent; blade fleshy, not leathery, margins entire. **Inflorescences** axillary, spikes; bracts present. **Pedicels** present; bracteoles absent. **Flowers** unisexual; sepals 4, nearly distinct, calyx radially symmetric, cuplike, lobes oblong; corolla semitransparent, radially symmetric, rotate, tube not spurred or gibbous, lobes 4; stamens 3 or 4, free, equal, filaments glabrous; staminode 0; ovary 1-locular, placentation basal; stigma elongate. **Fruits** nutlets. **Seeds** 1, black or brown, oblong, wings absent. $x = 6$.

Species 3 (1 in the flora): North America, South America, Europe.

According to a recent treatment (R. K. Hoggard et al. 2003), *Littorella* is a sister group to *Plantago*; this corresponds with morphological differences between these taxa and justifies recognizing separate genera.

SELECTED REFERENCE Hoggard, R. K. et al. 2003. Molecular systematics and biogeography of the amphibious genus *Littorella* (Plantaginaceae). Amer. J. Bot. 90: 429–435.

1. Littorella americana Fernald, Rhodora 20: 62. 1918

• American shoreweed, littorelle d'Amérique E F

Perennials fibrous-rooted. **Leaves** dark green or green when dry, usually arcuate, linear, or subulate, 10–40 × 1–2 mm, glabrous. **Spikes:** staminate flowers 1, peduncles 5–30 mm, glabrous; pistillate flowers 2–5, at base of plant; bracts rounded. **Flowers:** sepals 2.5–3 mm; corolla lobes erect; stamen connective to 1 mm. **Nutlets** 2.5–3 mm. **Seeds** 2–3 mm. *2n* = 12.

Flowering summer. Shorelines of lakes, ponds, and slow moving streams; 0–200 m; St. Pierre and Miquelon; N.B., Nfld. and Labr. (Nfld.), N.S., Ont., Que.; Maine, Mich., Minn., N.H., N.Y., Vt., Wis.

The relationship between *Littorella americana* and Eurasian *L. uniflora* (Linnaeus) Ascherson has been debated. Molecular data provide support for recognition of North American plants as a distinct species (R. K. Hoggard et al. 2003).

Plantago americana (Fernald) Rahn is an invalid name and pertains here.

40. PLANTAGO Linnaeus, Sp. Pl. 1: 112. 1753; Gen. Pl. ed. 5, 52. 1754 • Plantain [Latin *planta*, sole or flat, and *-ago*, resemblance, alluding to leaf shape of *P. major*]

Alexey Shipunov

Psyllium Miller

Herbs, annual or perennial, sometimes biennial [rarely suffrutescent or arborescent]; caudex usually present when perennial. **Stems** present or absent, if present, erect, glabrous or hairy. **Leaves** usually basal, usually alternate, (cauline and opposite in *P. afra*, *P. indica*, *P. sempervirens*); petiole absent or present; blade fleshy, leathery or not, margins entire or toothed. **Scapes** erect or ascending, rarely decumbent (*P. coronopus*), surpassing leaves, sometimes slightly so (*P. tweedyi*) or not (*P. major*). **Inflorescences** axillary, spikes or spiciform, dull, sometimes shiny (*P. canescens*, *P. lanceolata*, *P. media*); bracts present. **Pedicels** absent or present; bracteoles absent. **Flowers** bisexual; sepals 3 or 4, nearly distinct (abaxials connate in *P. lanceolata*), oblong, calyx radially, rarely bilaterally, symmetric, cuplike; corolla semitransparent, radially or weakly bilaterally symmetric, lateral lobes smaller, ± tubular to ± funnelform, tube base not spurred or gibbous, tube glabrous, rarely hairy (*P. coronopus*, *P. maritima*), lobes 4; stamens 2 or 4, free, equal, filaments glabrous; staminode 0; ovary 2-locular, placentation free-central, sometimes axile; stigma elongate. **Fruits** pyxides, lanceoloid, rarely ovoid (*P. macrocarpa*), dehiscence circumscissile (indehiscent or dehiscence not circumscissile in *P. macrocarpa*). **Seeds** (1 or)2–35, black or brown, sometimes dark red (*P. rhodosperma*) or yellowish brown (*P. virginica*), oblong, wings absent. *x* = 4, 5, 6.

Species ca. 210 (32 in the flora): North America, Mexico, Central America, South America, Europe, Asia, Africa, Pacific Islands (New Zealand), Australia.

Plantago lanceolata and *P. major* have become established on all continents except Antarctica. A specimen of *P. asiatica* Linnaeus (New York City, *US 295731*) is ambiguous as to locality, and there is no evidence that it is established outside of cultivation in the flora area. Among North American *Plantago*, several native species have been introduced to states or provinces outside their native range.

For species with bilaterally symmetric calyces, sepal lengths in the descriptions are for the adaxial sepals.

SELECTED REFERENCES Bassett, I. J. 1966. Taxonomy of North American *Plantago* L. section *Micropsyllium* Decne. Canad. J. Bot. 44: 467–479. Bassett, I. J. 1967. Taxonomy of *Plantago* L. in North America: Sections *Holopsyllium* Pilger, *Palaeopsyllium* Pilger, and *Lamprosantha* Decne. Canad. J. Bot. 45: 565–577. Bassett, I. J. 1973. The Plantains of Canada. Ottawa. Kuiper, P. J. C. and M. Bos, eds. 1992. *Plantago*: A Multidisciplinary Study. Berlin. Rahn, K. 1974. *Plantago* section *Virginica*: A taxonomic revision of a group of American plantains using experimental, taximetric and classical methods. Dansk Bot. Ark. 30: 1–180. Rahn, K. 1978. *Plantago* ser. *Gnaphaloides* Rahn: A taxonomic revision. Bot. Tidsskr. 73: 137–154. Rahn, K. 1979. *Plantago* ser. *Ovatae*. A taxonomic revision. Bot. Tidsskr. 74: 13–20. Ronsted, N. et al. 2002. Phylogenetic relationships within *Plantago* (Plantaginaceae): Evidence from nuclear ribosomal ITS and plastid *trn*L-F sequence data. Bot. J. Linn. Soc. 139: 323–338.

1. Leaves cauline.
   2. Perennials, sometimes woody . . . . 27. *Plantago sempervirens*
   2. Annuals.
      3. Spikes glandular-hairy; all bracts similar. . . . . 1. *Plantago afra*
      3. Spikes eglandular; proximal bracts strongly differing from distal bracts . . . . 16. *Plantago indica*
1. Leaves basal.
   4. Leaf margins usually 1- or 2-pinnatifid; scapes decumbent, sometimes erect; corolla tubes hairy . . . . 7. *Plantago coronopus*
   4. Leaf margins entire, toothed, or lobed; scapes erect or ascending; corolla tubes glabrous, rarely hairy (*P. maritima*).
      5. Annuals; leaf blades linear, narrowly lanceolate, narrowly elliptic, or almost filiform; roots taproots.
         6. Seeds (3 or)4–25(–30); corollas radially symmetric, lobes 0.5–1 mm; leaf blade surfaces glabrous or hairy.
            7. Seeds 10–25(–30), 0.5–0.8 mm . . . . 14. *Plantago heterophylla*
            7. Seeds (3 or)4–9(–12), 0.8–2.5 mm.
               8. Corolla lobes spreading or reflexed, not forming a beak; seeds (3 or)4–9(–12), 1.5–2.5 mm . . . . 8. *Plantago elongata*
               8. Corolla lobes erect, forming a beak; seeds 4, 0.8–1.3 mm . . . . 24. *Plantago pusilla*
         6. Seeds 2; corollas bilaterally or radially symmetric, lobes 1.3–3.6 mm; leaf blade surfaces lanate, sericeous, or villous, rarely glabrate or glabrous.
            9. Spikes: flowers in spirals; scapes without antrorse hairs; bracts ovate or elliptic . . . . 22. *Plantago ovata*
            9. Spikes: flowers in whorls or pairs; scapes with some antrorse hairs; bracts ovate, triangular, or almost linear.
               10. Corollas radially symmetric, lobe bases obtuse or slightly cordate.
                  11. Scapes with antrorse, long and short hairs; bract lengths 0.3–0.8 times sepals; corolla lobes 2–2.7 mm; California, Oregon . . . . 9. *Plantago erecta*
                  11. Scapes with patent, long and antrorse, short hairs; bract lengths 0.6–2.2 times sepals; corolla lobes 3–3.6 mm; New Mexico, Texas . . . . 13. *Plantago helleri*
               10. Corollas bilaterally symmetric, lobe bases slightly to deeply cordate.
                  12. Leaf blades: adaxial surfaces glabrous or sparsely villous, margins entire, rarely toothed; stems 10–40 mm.
                     13. Bract lengths 2–12 times sepals; corolla lobes: adaxials 1.4–2.3 mm, laterals symmetric; flowering spring–fall . . . . . 3. *Plantago aristata*
                     13. Bract lengths 0.4–0.8 times sepals; corolla lobes: adaxials 2.4–3 mm, laterals asymmetric; flowering summer . . . . . 32. *Plantago wrightiana*
                  12. Leaf blades: adaxial surfaces sericeous or villous, rarely lanate, margins entire or toothed; stems 0–20 mm.
                     14. Bracts ovate, lengths 0.4–0.7 times sepals; leaves 1.5–4 mm wide; anther connectives slightly elongated, apices obtuse . . . . 2. *Plantago argyrea*
                     14. Bracts triangular or ovate, lengths 0.6–2 times sepals; leaves 1–4 or 4–10 mm wide; anther connectives elongated to significantly elongated, apices acute.

15. Bract lengths 0.6–1.4 times sepals; corolla lobes 2.2–2.5 mm; leaves 4–10 mm wide, blade margins toothed, rarely entire; flowering spring . . . . 15. *Plantago hookeriana*
15. Bract lengths 1–2 times sepals; corolla lobes 1.6–2.1 mm; leaves 1–4 mm wide, blade margins entire, rarely toothed; flowering early summer . . . . .23. *Plantago patagonica*

[5. Shifted to left margin.—Ed.]

5. Perennials or annuals; leaf blades ovate, elliptic, or lanceolate, sometimes cordate-ovate, lanceolate-spatulate, linear, oblanceolate, obovate, or oval; roots taproots or fibrous.
    16. Corolla lobes usually forming a beak, erect or patent; annuals or perennials (usually without caudex).
        17. Annuals; roots taproots.
            18. Seeds: adaxial faces flat; bracts triangular, 2–3.1 mm; sepals 1.8–2.8 mm . . . . 11. *Plantago firma*
            18. Seeds: adaxial faces concave; bracts ovate or triangular, 1.6–3.2 mm; sepals 1.5–3.6 mm.
                19. Sepals 2.7–3.6 mm, apices acuminate; bracts 2.5–3.2 mm, narrowly triangular or triangular; seeds dark red .25. *Plantago rhodosperma*
                19. Sepals 1.5–2.4 mm, apices obtuse; bracts 1.6–2.4 mm, narrowly ovate or ovate; seeds brown or yellowish brown . . . . 31. *Plantago virginica*
        17. Perennials; roots taproots or fibrous.
            20. Adaxial surfaces of leaves: hairs floccose, slender, 4–6 × 0.01–0.03 mm. . . . 12. *Plantago floccosa*
            20. Adaxial surfaces of leaves: hairs not floccose, less than 2 mm long, 0.03+ mm wide.
                21. Roots fibrous; sepals 2–2.5 mm. . . . . .4. *Plantago australis*
                21. Roots taproots; sepals 2.6–3.1 mm . . . . .29. *Plantago subnuda*
    16. Corolla lobes not forming a beak, spreading or reflexed; perennials (sometimes with caudex), rarely annuals.
        22. Fruits ovoid, indehiscent or dehiscence not circumscissile. . . . . 18. *Plantago macrocarpa*
        22. Fruits lanceoloid, dehiscence circumscissile.
            23. Leaf blades linear to lanceolate, veins not conspicuous; corolla tubes hairy . . . . .20. *Plantago maritima*
            23. Leaf blades lanceolate, linear, oblanceolate, oval, cordate-ovate, lanceolate-spatulate, ovate, or elliptic, veins conspicuous; corolla tubes glabrous.
                24. Spikes grayish, whitish, or yellowish, shiny, corolla lobes of neighboring flowers often overlapping.
                    25. Sepals: adaxial 2 connate; scapes groove-angled . . . . 17. *Plantago lanceolata*
                    25. Sepals: adaxial 2 nearly distinct; scapes not groove-angled.
                        26. Leaves ascending, 6–20 mm wide, blades linear to lanceolate or oblanceolate, surfaces hairy (hairs 1 mm) or glabrate; seeds 3–7, 1–1.8 mm. . . . . .5. *Plantago canescens*
                        26. Leaves prostrate, sometimes ascending, 30–70 mm wide, blades elliptic to ovate, surfaces hairy (hairs 0.5 mm); seeds 2–4, 2 mm . . . . 21. *Plantago media*
                24. Spikes brownish or greenish, dull, corolla lobes of neighboring flowers not overlapping.
                    27. Caudices absent.
                        28. Fruits (2–)4–5 mm, dehiscing at middle; seeds 5–35, 0.5–1 mm; bracts 0.5–1 mm. . . . . .19. *Plantago major*
                        28. Fruits 4–6(–8) mm, dehiscing proximal to middle; seeds 4 or 5(–8), 1.5–2 mm; bracts 2 mm . . . . 26. *Plantago rugelii*
                    27. Caudices well developed, conspicuous.
                        29. Spikes densely flowered, rachises not clearly visible between flowers; scapes slightly surpassing leaves. . . . . 30. *Plantago tweedyi*
                        29. Spikes loosely flowered, rachises visible between flowers; scapes surpassing leaves.

[30. Shifted to left margin.—Ed.]

1. Plantago afra Linnaeus, Sp. Pl. ed. 2, 1: 168. 1762
   • Glandular plantain I

Annuals; roots taproots, slender. Stems 100–350 mm, freely branched. Leaves cauline, opposite, 30–60 × 1–4 mm; blade linear to linear-lanceolate, margins entire or slightly toothed, veins conspicuous or not, surfaces hairy. Scapes 30–50 mm, hairy. Spikes greenish or brownish, 40–65 mm, densely flowered, glandular-hairy; bracts all similar, ovate, 3–5 mm, lengths 1–1.5 times sepals. Flowers: sepals 3–3.5 mm; corolla radially symmetric, lobes reflexed, 2–3 mm, base obtuse; stamens 4. Capsules lanceoloid. Seeds 2, 2–3 mm. $2n = 12$.

Flowering summer. Disturbed habitats; 0–200 m; introduced; Mass.; s Europe.

Plantago afra is known in Massachusetts from a single collection made in 1927 in Worcester County.

Plantago psyllium Linnaeus (1762, not 1753), a rejected name, and *P. indica* Linnaeus are misapplied names that pertain here. *Plantago squalida* Salisbury is an illegitimate name that pertains here.

2. Plantago argyrea E. Morris, Bull. Torrey Bot. Club 27: 111. 1900 • Saltmeadow plantain

Annuals; roots taproots, slender. Stems 0–20 mm. Leaves 45–140 × 1.5–4 mm; blade linear, margins entire, veins conspicuous or not, surfaces sericeous, rarely lanate. Scapes 300–1300 mm, hairy, hairs antrorse, long and short. Spikes greenish or brownish, 70–200 mm, densely or loosely flowered, flowers in whorls or pairs; bracts ovate, 1.5–2 mm, length 0.4–0.7 times sepals, apex acute. Flowers: sepals 2.6–3.3 mm; corolla bilaterally symmetric, lobes reflexed, 1.7–2.6 mm, base cordate; stamens 4, connective slightly elongated, apex obtuse. Seeds 2, 2.3–2.8 mm.

Flowering summer–fall. Clearings in forests; 700–2900 m; Ariz., N.Mex.; Mexico (Baja California, Chihuahua).

3. Plantago aristata Michaux, Fl. Bor.-Amer. 1: 95. 1803
   • Largebract plantain E

Plantago patagonica Jacquin var. *aristata* (Michaux) A. Gray; *P. purshii* Roemer & Schultes var. *aristata* (Michaux) M. E. Jones

Annuals; roots taproots, slender. Stems 20–40 mm. Leaves 30–200 × 3–7 mm; blade linear or narrowly lanceolate, margins entire, rarely toothed, veins conspicuous or not, abaxial surface villous, adaxial glabrous or sparsely villous. Scapes 100–500 mm, hairy, hairs antrorse, long and short. Spikes greenish or brownish, 80–150 mm, densely flowered, flowers in whorls or pairs; bracts almost linear, 15–30 mm, length 2–12 times sepals. Flowers: sepals 2.7–3.7 mm; corolla bilaterally symmetric, lobes reflexed, adaxials 1.4–2.3 mm, laterals symmetric, base deeply cordate; stamens 4, connective elongated, apex acute. Seeds 2, 2.5–2.9 mm. $2n = 20$.

Flowering spring–fall. Roadsides, pastures, disturbed ground; 0–700 m; N.S., Ont.; Ala., Ark., Calif., Conn., Del., D.C., Fla., Ga., Ill., Ind., Iowa, Kans., Ky., La., Maine, Md., Mass., Mich., Minn., Miss., Mo., Nebr., N.H., N.J., N.Mex., N.Y., N.C., Ohio, Okla., Oreg., Pa., R.I., S.C., S.Dak., Tenn., Tex., Vt., Va., Wash., W.Va., Wis.; introduced in Central America, Europe, Asia.

Plantago aristata is similar to *P. patagonica*; the latter is distinguished by its dense, villous indument.

4. Plantago australis Lamarck in J. Lamarck and J. L. M. Poiret, Tabl. Encycl. 1: 339. 1792 • Mexican plantain

Plantago australis subsp. *hirtella* (Kunth) Rahn; *P. hirtella* Kunth; *P. hirtella* var. *galeottiana* (Decaisne) Pilger; *P. hirtella* var. *mollior* Pilger

Perennials; caudex glabrous; roots fibrous, stout. Stems 0–10 mm. Leaves 40–350 × 6–77 mm; blade elliptic to narrowly elliptic, margins entire, veins conspicuous, surfaces pilose, rarely glabrate, adaxial surface hairs not floccose, less than 2 mm long, more than 0.03 mm wide. Scapes 30–560 mm, hairy, hairs antrorse, short. Spikes greenish or brownish, 100–1000 mm, densely flowered;

bracts narrowly triangular, 1.6–4.2 mm, length 0.8–1.5 times sepals. **Flowers:** sepals 2–2.5 mm; corolla radially symmetric, lobes erect, forming a beak, 2–2.8 mm, base obtuse; stamens 4. **Seeds** 3, 1.2–2.2 mm. $2n = 24$.

Flowering summer. Open places; 0–1000 m; Ariz.; Mexico; Central America; South America.

Plantago australis occurs in Cochise, Coconino, and Pima counties. *Plantago australis* is most diverse in South America, where as many as 16 subspecies (K. Rahn 1974) may be recognized. Plants from California identified as *P. hirtella* are most likely *P. subnuda*. However, since the most important distinguishing character of *P. australis* is the absence of the developed taproot (which is fragile and often broken in herbarium specimens), all these samples require careful examination. Further research is needed also to clarify the circumscriptions of *P. australis* and *P. subnuda*.

5. **Plantago canescens** Adams, Nouv. Mém. Soc. Imp. Naturalistes Moscou 9: 233, plate 13, fig. 1. 1834
  • Gray-pubescent plantain

Plantago septata E. Morris

Perennials; caudex usually woolly; roots taproots, thick. **Stems** 0–20 mm. **Leaves** ascending, 180–250 × 6–20 mm; blade linear to lanceolate or oblanceolate, margins entire, rarely toothed, veins conspicuous, surfaces glabrate or hairy, hairs 1 mm. **Scapes** 50–230 mm, not groove-angled, hairy or glabrous. **Spikes** grayish or whitish, 80–350 mm, usually densely flowered, shiny; corolla lobes of neighboring flowers often overlapping; bracts broadly ovate, 1.8–2 mm, length 0.9–1 times sepals. **Flowers:** sepals 2 mm, adaxial 2 nearly distinct; corolla radially symmetric, lobes reflexed, 2 mm, base obtuse; stamens 4. **Seeds** 3–7, 1–1.8 mm. $2n = 12$.

Flowering summer. Grassy, gravelly, and rocky slopes, cliffs; 0–2000 m; Alta., B.C., N.W.T., Nunavut, Yukon; Alaska, Mont.; Asia.

N. N. Tzvelev (1983) recognized six subspecies (including two in North America) within *Plantago canescens*; North American material is not segregated as such here.

6. **Plantago cordata** Lamarck in J. Lamarck and J. L. M. Poiret, Tabl. Encycl. 1: 338. 1792 • Heartleaf plantain E

Perennials; caudex well developed, conspicuous, glabrous; roots fibrous, thick. **Stems** 0–20 mm. **Leaves** 100–300 × 80–200 mm; petiole to 300 mm; blade broadly oval to cordate-ovate, margins entire, veins conspicuous, laterals branching from midvein distal to base, surfaces glabrous. **Scapes** 200–300 mm, glabrous. **Spikes** brownish or greenish, 100–500 mm, loosely flowered, rachis visible between flowers; bracts round-ovate, 2 mm, length 0.8–1 times sepals. **Flowers:** sepals 2–2.5 mm; corolla radially symmetric, lobes spreading, 2–2.5 mm, base obtuse; stamens 4. **Seeds** 2–4, 2.5–3.5 mm. $2n = 24$.

Flowering late spring–early summer. Rocky or gravelly beds of shallow, slow-moving streams, sloughs, swamps; 0–200 m; Ont.; Ala., Ark., D.C., Fla., Ga., Ill., Ind., Iowa, Ky., Md., Mich., Miss., Mo., N.Y., N.C., Ohio, Tenn., Va., Wis.

Plantago cordata is listed as federally endangered in Canada and is in the Center for Plant Conservation's National Collection of Endangered Plants.

7. **Plantago coronopus** Linnaeus, Sp. Pl. 1: 115. 1753
  • Buckhorn plantain I

Plantago coronopus subsp. *commutata* (Gussone) Pilger

Annuals, sometimes biennials; roots taproots, stout. **Stems** 0–10 mm. **Leaves** 20–80(–115) × 5–15 mm; blade lanceolate, margins usually 1- or 2-pinnatifid, veins conspicuous or not, surfaces villous, hairs septate, sometimes glabrate. **Scapes** decumbent, sometimes erect, 15–150(–210) mm, villous. **Spikes** decumbent, sometimes erect, greenish, purplish, or brownish, (15–)30–300 mm, densely flowered; bracts ovate to lanceolate, 1.5–2 mm, length 0.5–0.6 times sepals. **Flowers:** sepals 2–3 mm; corolla radially symmetric, tube hairy, lobes reflexed, 1 mm, base obtuse; stamens 4. **Seeds** (2–)4 (plus 1 smaller, distal one of different shape), 1–1.5 mm. $2n = 10, 20, 30$ (all Eurasia).

Flowering summer. Moist, gravelly or sandy soils; 0–200 m; introduced; Greenland; B.C., Man.; Calif., Mass., N.J., N.Y., Oreg., Pa., Tex., Wash.; Eurasia; Africa; introduced also in s South America.

8. Plantago elongata Pursh, Fl. Amer. Sept. 2: 729. 1813
   • Prairie plantain

Plantago bigelovii A. Gray; *P. bigelovii* subsp. *californica* (Greene) Bassett; *P. elongata* subsp. *pentasperma* Bassett

Annuals; roots taproots, slender. **Stems** 0–5(–7) mm. **Leaves** 10–70 × (0.8–)1–2 mm; blade linear to almost filiform, margins entire, veins conspicuous or not, surfaces glabrous or hairy. **Scapes** 10–80 mm, glabrous or hairy. **Spikes** greenish, brownish, or gray, (30–)50–150 mm, densely or loosely flowered; bracts ovate, 2–2.5 mm, length 0.8–1.2 times sepals. **Flowers:** sepals 2–2.5 mm; corolla radially symmetric, lobes spreading or reflexed, not forming a beak, 0.5–1 mm, base obtuse; stamens 2. **Seeds** (3 or) 4–9(–12), 1.5–2.5 mm. $2n = 12, 20, 36$.

Flowering spring–early summer. Mostly moist soils; 0–2100 m; Alta., B.C., Man., Sask.; Ariz., Calif., Colo., Idaho, Kans., Minn., Mont., Nebr., N.Mex., N.Dak., Okla., Oreg., S.Dak., Tex., Utah, Wash., Wyo.; Mexico (Baja California).

Purported differences between *Plantago bigelovii* and *P. elongata* (I. J. Bassett 1966) do not appear to be taxonomically meaningful.

9. Plantago erecta E. Morris, Bull. Torrey Bot. Club 27: 118. 1900 • Dotseed plantain

Plantago patagonica Jacquín var. *californica* Greene, Man. Bot. San Francisco, 236. 1894; *P. erecta* subsp. *rigidior* Pilger; *P. hookeriana* Fischer & C. A. Meyer var. *californica* (Greene) Poe

Annuals; roots taproots, slender. **Stems** 0–10 mm. **Leaves** 50–120 × 1–4 mm; blade linear, margins entire or toothed, veins conspicuous or not, surfaces villous or lanate, rarely glabrate. **Scapes** 200–1500 mm, hairy, hairs antrorse, long and short. **Spikes** greenish or brownish, 70–150 mm, densely flowered, flowers in whorls or pairs; bracts ovate, 1–2 mm, length 0.3–0.8 times sepals. **Flowers:** sepals 2.5–3.4 mm; corolla radially symmetric, lobes reflexed, 2–2.7 mm, base obtuse or slightly cordate; stamens 4, connective elongated, apex acute. **Seeds** 2, 2.1–2.9 mm. $2n = 20$.

Flowering spring. Dunes, grassy hills and flats, clearings in woods; -50–1400 m; Calif., Oreg.; Mexico (Baja California).

10. Plantago eriopoda Torrey, Ann. Lyceum Nat. Hist. New York 2: 237. 1827 • Redwool plantain, plantain à base velue

Plantago shastensis Greene

Perennials; caudex well developed, conspicuous, brown-woolly; roots taproots, thick. **Stems** 0–20 mm. **Leaves** (30–) 50–250 × (5–)15–70 mm; blade lanceolate to elliptic, margins entire, veins conspicuous, laterals branching from base, surfaces glabrous or hairy. **Scapes** (40–)50–300 mm, glabrous or hairy. **Spikes** brownish or greenish, (25–) 80–500 mm, loosely flowered, rachis visible between flowers; bracts broadly ovate, 2–2.5 mm, length 0.8–1.2 times sepals. **Flowers:** sepals 2–2.5 mm; corolla radially symmetric, lobes reflexed, 1–1.5 mm, base obtuse; stamens 4. **Seeds** 2–4, 2–2.5 mm. $2n = 24$.

Flowering late spring–early summer. Moist meadows and prairies, wetlands, marshes, fens, ditches, stream banks, saline or alkaline soils; 0–2900 m; Alta., B.C., Man., N.W.T., Que., Sask., Yukon; Alaska, Ariz., Calif., Colo., Idaho, Iowa, Minn., Mont., Nebr., Nev., N.Mex., N.Y., N.Dak., Oreg., S.Dak., Utah, Wyo.; Mexico (Durango).

11. Plantago firma Kunze ex Walpers, Nov. Actorum Acad. Caes. Leop. Carol. Nat. Cur. 19, suppl. 1: 402. 1843 • Chilean plantain [I]

Plantago truncata Chamisso & Schlechtendal subsp. *firma* (Kunze ex Walpers) Pilger

Annuals; roots taproots, slender. **Stems** 0–10 mm. **Leaves** 20–50 × 2–15 mm; blade narrowly obovate or narrowly elliptic, margins toothed, veins conspicuous, surfaces densely pilose or glabrate. **Scapes** 10–50 mm, hairy, hairs appressed or nearly patent. **Spikes** greenish or brownish, 40–60 mm, densely flowered; bracts triangular, 2–3.1 mm, length 0.7–1.1 times sepals. **Flowers:** sepals 1.8–2.8 mm; corolla radially symmetric, lobes patent or erect, forming a beak, 1.7–3.1 mm, base obtuse; stamens 4. **Seeds** 2, 1.5–2.2 mm, adaxial face flat. $2n = 24$.

Flowering fall. Sandy soils; 0–1000 m; introduced; Calif.; South America (Chile).

Plantago firma has been known in California, especially in Marin County, since at least 1896, but the most recent collection known was made in 1957.

12. **Plantago floccosa** Decaisne in A. P. de Candolle and A. L. P. P. de Candolle, Prodr. 13(1): 723. 1852 • Floccose plantain [I]

Perennials; caudex well developed, conspicuous, glabrous or hairy; roots several taproots, fragile. **Stems** 0–30 mm. **Leaves** 50–220 × 9–60 mm; blade elliptic to narrowly elliptic, margins with inconspicuous teeth, veins conspicuous, surfaces hairy, adaxial surface hairs floccose, slender, 4–6 × 0.01–0.03 mm. **Scapes** 55–220 mm, lanate, hairs variously directed, long. **Spikes** greenish or brownish, 200–500 mm, densely flowered, flowers less crowded proximally; bracts narrowly triangular or triangular, 2–2.8 mm, lengths 0.9–1 times sepals. **Flowers:** sepals 1.9–2.7 mm; corolla radially symmetric, lobes erect, forming a beak, 2–2.9 mm, base obtuse; stamens 4. **Seeds** 3, 1.8–2.4 mm.

Flowering spring–fall. Roadsides; 0–300 m; introduced; Fla.; Mexico (Hidalgo, México, Querétaro, San Luis Potosi, Tamaulipas, Veracruz).

According to J. Burkhalter (pers. comm.), *Plantago floccosa* is well established in northwestern Florida near the border with Alabama.

13. **Plantago helleri** Small, Bull. New York Bot. Gard. 1: 288. 1899 • Heller's plantain [E]

Annuals; roots taproots, slender. **Stems** 0–12 mm. **Leaves** 50–130 × 3–7 mm; blade linear, margins entire, veins conspicuous or not, surfaces villous, adaxial rarely glabrous. **Scapes** 80–250 mm, hairy, hairs patent, long and antrorse, short. **Spikes** greenish, whitish, or brownish, 50–120 mm, densely flowered, flowers in whorls or pairs; bracts triangular, 2.5–8 mm, length 0.6–2.2 times sepals. **Flowers:** sepals 3.5–4 mm; corolla radially symmetric, lobes reflexed, 3–3.6 mm, base slightly cordate; stamens 4, connective elongated, apex acute. **Seeds** 2, 3.2–3.8 mm. $2n = 20$.

Flowering spring. Dry slopes and flats on limestone; 100–1700 m; N.Mex., Tex.

Populations of *Plantago helleri* are concentrated in central Texas, especially in the Edwards Plateau and Trans-Pecos. The species also occurs in Eddy and Lincoln counties, New Mexico.

14. **Plantago heterophylla** Nuttall, Trans. Amer. Philos. Soc., n.s. 5: 177. 1835 • Slender plantain [E]

Annuals; roots taproots, slender. **Stems** 0–10 mm. **Leaves** 30–80 × 1–4 mm; blade linear, margins lobed, rarely entire, veins conspicuous or not, surfaces hairy, sometimes glabrous. **Scapes** 5–60 mm, hairy, sometimes glabrous. **Spikes** greenish or brownish, 50–150 mm, loosely or densely flowered; bracts ovate, 2 mm, length 0.9–1.1 times sepals. **Flowers:** sepals 2 mm; corolla radially symmetric, lobes spreading, 0.5–1 mm, base obtuse; stamens 2. **Seeds** 10–25(–30), 0.5–0.8 mm. $2n = 12$.

Flowering spring–early summer. Moist sandy soils; 0–200 m; Ala., Ark., Del., Fla., Ga., Ill., Kans., Ky., La., Md., Miss., Mo., N.J., N.Y., N.C., Okla., Pa., S.C., Tenn., Tex., Va.; introduced in South America (Argentina).

15. **Plantago hookeriana** Fischer & C. A. Meyer, Index Seminum (St. Petersburg) 1838: 39. 1839 • Hooker's plantain

Plantago hookeriana var. *nuda* (A. Gray) Poe

Annuals; roots taproots, slender. **Stems** 0–10 mm. **Leaves** 60–120 × 4–10 mm; blade linear, margins toothed (teeth to 4 mm), rarely entire, veins conspicuous or not, surfaces villous or sericeous. **Scapes** 280–580 mm, hairy, hairs antrorse, long and short. **Spikes** greenish or brownish, 50–180 mm, densely flowered, flowers in whorls or pairs; bracts ovate or triangular, 1.5–6 mm, length 0.6–1.4 times sepals, apex acute or acuminate. **Flowers:** sepals 2.5–4 mm; corolla bilaterally symmetric, abaxial and lateral lobes reflexed, adaxial erect, 2.2–2.5 mm, base slightly cordate; stamens 4, connective significantly elongated, apex acute. **Seeds** 2, 2.4–3.4 mm. $2n = 20$.

Flowering spring. Sandy soils, disturbed areas; 0–1300 m; La., Miss., Tex.; Mexico (Chihuahua, Coahuila, Nuevo León, Tamaulipas).

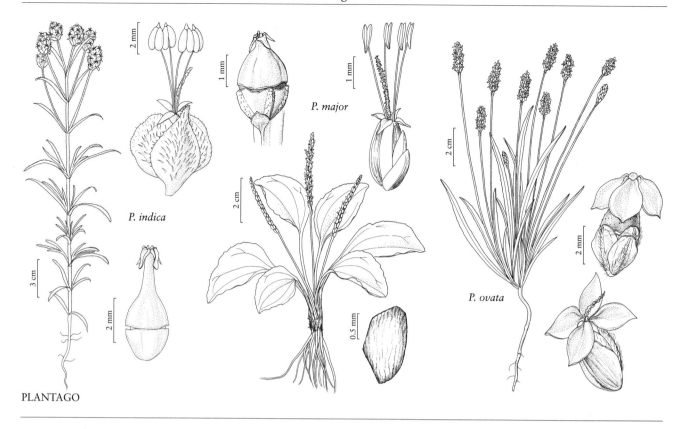

PLANTAGO

16. Plantago indica Linnaeus, Syst. Nat., ed. 10, 2: 896.
1759, legitimacy of name in question • Sand plantain, plantain des sables F I

Plantago arenaria Waldstein & Kitaibel; *P. psyllium* Linnaeus, name rejected

Annuals; roots taproots, slender. **Stems** 100–350 mm, freely branched. **Leaves** cauline, opposite, 60–80 × 1–3 mm; blade linear to linear-lanceolate, margins entire or toothed, veins conspicuous or not, surfaces hairy. **Scapes** 100–650 mm, hairy. **Spikes** greenish or brownish, (50–)150–250 mm, densely flowered, eglandular; bracts proximal strongly differing from distal, ovate, 2–5(–7) mm, length 1–1.5 times sepals, proximal bracts: apex acute. **Flowers:** sepals 2–3 mm; corolla radially symmetric, lobes reflexed, 2–4 mm, base obtuse; stamens 4. **Seeds** 2, 2–2.5 mm. $2n = 12$.

Flowering late summer–fall. Roadsides, railroads, sandy shorelines; 0–200 m; introduced; B.C., Man., Ont., Que.; Calif., Conn., Del., Ill., Ind., Iowa, Maine, Mass., Mich., Minn., Mo., N.H., N.J., N.Y., N.C., Ohio, Oreg., Pa., Vt., Va., Wash., Wis.; Eurasia.

While considering the proposal made by W. L. Applequist (2006) to reject *Plantago psyllium*, the Nomenclature Committee for Vascular Plants also decided that *P. indica* is a legitimate name (R. K. Brummitt 2009). A. B. Doweld and A. Shipunov (2017) published a proposal to reject *P. indica* in favor of *P. arenaria*. That proposal awaits a decision by that committee.

17. Plantago lanceolata Linnaeus, Sp. Pl. 1: 113. 1753
• Narrow-leaf or English plantain I

Plantago lanceolata var. *angustifolia* de Candolle; *P. lanceolata* var. *sphaerostachya* Mertens & W. D. J. Koch

Perennials; caudex hairy or glabrous; roots fibrous, slender. **Stems** 0–20 mm. **Leaves** 30–300 × 5–25(–45) mm; blade lanceolate to oblanceolate, margins entire or toothed, veins conspicuous, surfaces glabrous or sericeous. **Scapes** 300–400 mm, groove-angled, hairy. **Spikes** grayish, whitish, or yellowish, (5–)100–450(–1000) mm, usually shorter than scape, densely flowered, shiny; corolla lobes of neighboring flowers often overlapping; bracts broadly ovate, 2 mm, length 0.8–1 times sepals. **Flowers:** sepals 2–2.5 mm, adaxial 2 connate; corolla radially symmetric, lobes reflexed, 2–2.5 mm, base obtuse; stamens 4. **Seeds** (1 or)2, 2–3(–4) mm. $2n = 12$.

Flowering summer. Roadsides, trails, lawns, urban areas, other disturbed sites; 0–3200 m; introduced; Greenland; St. Pierre and Miquelon; B.C., N.B., Nfld.

and Labr. (Nfld.), N.S., Ont., P.E.I., Que.; Ala., Alaska, Ariz., Ark., Calif., Colo., Conn., Del., D.C., Fla., Ga., Idaho, Ill., Ind., Iowa, Kans., Ky., La., Maine, Md., Mass., Mich., Minn., Miss., Mo., Mont., Nebr., Nev., N.H., N.J., N.Mex., N.Y., N.C., N.Dak., Ohio, Okla., Oreg., Pa., R.I., S.C., S.Dak., Tenn., Tex., Utah, Vt., Va., Wash., W.Va., Wis., Wyo.; Europe; introduced also in Central America, South America, Asia, Africa, Pacific Islands, Australia.

Plantago lanceolata is known from historic collections in Manitoba and Saskatchewan.

The name *Plantago altissima* Linnaeus sometimes has been misapplied to North American plants of *P. lanceolata*.

18. **Plantago macrocarpa** Chamisso & Schlechtendal, Linnaea 1: 166. 1826 • Seashore plantain

Perennials; caudex well developed, conspicuous, glabrous; roots fibrous, thick. **Stems** 0–20 mm. **Leaves** (80–)100–400(–550) × (5–)10–35(–40) mm; blade oblanceolate or almost linear, margins entire, veins conspicuous, 4 or 5, surfaces glabrous. **Scapes** 300–400 mm, glabrous or sparsely hairy, becoming densely so distally. **Spikes** greenish or brownish, 350–450 mm, loosely flowered; bracts ovate to deltate, 3–4 mm, length 1.5–2.5 times sepals. **Flowers:** sepals 1.5–2 mm; corolla radially symmetric, lobes spreading, 1.5–2 mm, base obtuse; stamens 4. **Fruits** ovoid, indehiscent or dehiscence not circumscissile. **Seeds** 1 or 2, 4–5 mm. $2n = 24$.

Flowering late spring–early summer. Wet places, tidal marshes, saline areas; 0–700 m; B.C.; Alaska, Oreg., Wash.; Asia.

Plantago macrocarpa has been documented along the Pacific coast south to the mouth of the Yachats River in Oregon.

19. **Plantago major** Linnaeus, Sp. Pl. 1: 112. 1753 • Common plantain, plantain majeur [F][I]

Plantago major var. *pachyphylla* Pilger; *P. major* var. *pilgeri* Domin; *P. major* var. *scopulorum* Fries & Broberg

Perennials, sometimes annuals; caudex absent; roots fibrous, thick. **Stems** 0–20 mm. **Leaves** 20–150(–400) × 10–120(–170) mm; petiole to 200 mm; blade ovate to cordate-ovate, margins entire or toothed, veins conspicuous, surfaces glabrous or hirsute. **Scapes** 50–250(–500) mm, surpassing leaves or not, glabrous or hirsute. **Spikes** brownish or greenish, (20–)

50–300(–400) mm, densely flowered; bracts lanceolate, 0.5–1 mm, length 0.3–0.7 times sepals. **Flowers:** sepals 1.5–2 mm; corolla radially symmetric, lobes reflexed, 0.5–1 mm, base obtuse; stamens 4. **Fruits** (2–)4–5 mm, dehiscing at middle. **Seeds** 5–35, 0.5–1 mm. $2n = 12$.

Flowering summer. Roadsides, trails, stream banks, urban areas, lawns, other disturbed areas; 0–3000 m; introduced; Greenland; St. Pierre and Miquelon; Alta., B.C., Man., N.B., Nfld. and Labr., N.W.T., N.S., Ont., P.E.I., Que., Sask., Yukon; Ala., Alaska, Ariz., Ark., Calif., Colo., Conn., Del., D.C., Fla., Ga., Idaho, Ill., Ind., Iowa, Kans., Ky., La., Maine, Md., Mass., Mich., Minn., Miss., Mo., Mont., Nebr., Nev., N.H., N.J., N.Mex., N.Y., N.C., N.Dak., Ohio, Okla., Oreg., Pa., R.I., S.C., S.Dak., Tenn., Tex., Utah, Vt., Va., Wash., W.Va., Wis., Wyo.; Eurasia; introduced also in Mexico, Central America, South America, Africa, Pacific Islands, Australia.

Subspecies *intermedia* (Gilibert) Lange is often accepted by European botanists as a separate species, *Plantago uliginosa* F. W. Schmidt. Observations suggest that plants referable to this taxon may occur in the United States; it has been reported from the New England states (A. Haines 2011). Subspecies *intermedia* is distinguished by more abundant (11–35) and smaller (0.8–1 mm) seeds, ascending spikes, ovoid fruits, and elliptic or lanceolate leaf blades. Without detailed morphologic and genetic investigations of North American plants similar to the study of European plants by M. Morgan-Richards and K. Wolff (1999), it is not possible to draw any conclusions about the status and distribution of this or any other possible infraspecific taxa of *P. major* in North America.

20. **Plantago maritima** Linnaeus, Sp. Pl. 1: 114. 1753 • Goose tongue, plantain maritime

Plantago borealis Lange; *P. decipiens* Barnéoud; *P. juncoides* Lamarck; *P. juncoides* var. *californica* Fernald; *P. juncoides* var. *decipiens* (Barnéoud) Fernald; *P. juncoides* var. *glauca* (Hornemann) Fernald; *P. juncoides* var. *laurentiana* Fernald; *P. maritima* subsp. *borealis* (Lange) A. Blytt; *P. maritima* var. *californica* (Fernald) Pilger; *P. maritima* var. *decipiens* (Barnéoud) Fernald; *P. maritima* var. *glauca* Hornemann; *P. maritima* subsp. *juncoides* (Lamarck) Hultén; *P. maritima* var. *juncoides* A. Gray; *P. oliganthos* Roemer & Schultes; *P. oliganthos* var. *fallax* Fernald

Perennials, rarely annuals; caudex absent or well developed, conspicuous, glabrous or hairy; roots taproots, thick. **Stems** 0–40 mm, usually branched. **Leaves** 10–220 × (1–)10–15 mm; blade linear to lanceolate, margins entire or toothed, veins not conspicuous, surfaces glabrous, sometimes hairy. **Scapes** 40–120 mm, glabrous or hirsute. **Spikes** greenish or brownish, (15–)50–200(–290) mm, densely or loosely flowered; bracts broadly ovate, 1.5–4(–6) mm, length 0.8–1.2 times sepals. **Flowers:** sepals 1.5–3.5 mm; corolla radially symmetric, tube hairy, lobes reflexed, 1–1.5 mm, base obtuse; stamens 4. **Seeds** 1–3, 1.5–3 mm. $2n = 24$.

Flowering summer. Marine shorelines, crevices of large rocks in sea spray, coastal and inland salt marshes, alkaline and saline flats, roadsides; 0–800 m; Greenland; St. Pierre and Miquelon; Alta., B.C., Man., N.B., Nfld. and Labr., N.W.T., N.S., Nunavut, Ont., P.E.I., Que., Sask., Yukon; Alaska, Calif., Conn., Maine, Mass., N.H., N.J., N.Y., Oreg., R.I., Va., Wash.; Mexico; Central America; South America; Eurasia; Africa.

Plantago maritima has been reported from Utah; no specimen supporting that report has been found.

Since the 1930s, when *Plantago maritima* was shown to have high levels of phenotypic plasticity (J. W. Gregor and J. M. S. Lang 1950), it usually has been accepted in a broad sense. That approach is followed here, with all dwarf and loose-flowered forms (such as *P. borealis* and *P. decipiens*, respectively) included under this name.

SELECTED REFERENCE Gregor, J. W. and J. M. S. Lang. 1950. Intracolonial variation in plant size and habit in *Plantago maritima*. New Phytol. 49: 135–141.

21. **Plantago media** Linnaeus, Sp. Pl. 1: 113. 1753
   • Hoary plantain, plantain moyen [I]

Perennials; caudex glabrous or hairy; roots taproots, thick. **Stems** 0–20 mm. **Leaves** prostrate, sometimes ascending, 40–200 × 30–70 mm; blade elliptic to ovate, margins entire or toothed, veins conspicuous, surfaces hairy, hairs 0.5 mm. **Scapes** 80–300 mm, not groove-angled, hirsute. **Spikes** grayish or whitish, 100–400 mm, densely flowered, shiny; corolla lobes of neighboring flowers overlapping; bracts ovate, 1.8–2 mm, length 0.9–1 times sepals. **Flowers:** sepals 2 mm, adaxial 2 nearly distinct; corolla radially symmetric, lobes reflexed, 1.5 mm, base obtuse; stamens 4. **Seeds** 2–4, 2 mm. $2n = 12, 24$.

Flowering summer. Disturbed areas, neutral and basic soils; 0–200 m; introduced; Man., N.B., Ont., Que.; Conn., Ill., Maine, Mass., Mich., N.H., N.J., N.Y., Pa., R.I., Wis.; Eurasia.

Three varieties of *Plantago media* have been recognized in western Europe. It is not clear how names of those varieties apply to *P. media* in eastern Europe and North America.

22. **Plantago ovata** Forsskål, Fl. Aegypt.-Arab., 31. 1775
   • Desert Indian-wheat [F]

Plantago insularis Eastwood; *P. insularis* var. *fastigiata* (E. Morris) Jepson

Annuals; roots taproots, slender. **Stems** 0–30 mm, often branched. **Leaves** 10–230 × 0.5–12 mm; blade linear or narrowly elliptic, margins toothed, veins conspicuous or not, surfaces villous or lanate to sericeous. **Scapes** 10–400 mm, hairy, hairs woolly, long. **Spikes** grayish or brownish, 20–400 mm, densely flowered, flowers in spirals; bracts ovate or elliptic, 1.7–4 mm, length 0.8–1.2 times sepals, apex not reached by green nerve. **Flowers:** sepals 1.9–3.5 mm; corolla radially symmetric, lobes reflexed, 1.3–2.8 mm, base cuneate; stamens 4. **Seeds** 2, 2–2.6 mm. $2n = 8$.

Flowering spring. Sandy deserts and steppes; 0–1800 m; Ariz., Calif., Nev., Tex., Utah; Mexico (Baja California, Sonora); Eurasia; Africa.

From molecular evidence, S. C. Meyers and A. Liston (2008) suggested that *Plantago ovata* was introduced to North America during the Pleistocene. They recognized four varieties; North American specimens can be treated as two varieties based on bract and corolla color: the inland var. *fastigiata* (E. Morris) S. C. Meyers & Liston (midribs of mature flower bracts green, corolla lobes without reddish brown midribs) and the coastal var. *insularis* (Eastwood) S. C. Meyers & Liston (midribs of mature flower bracts brown, corolla lobe midribs prominent, reddish brown). Unfortunately, these features are not easily seen on many herbarium specimens, and these taxa are not recognized here.

SELECTED REFERENCE Meyers, S. C. and A. Liston. 2008. The biogeography of *Plantago ovata* Forssk. (Plantaginaceae). Int. J. Pl. Sci. 169: 954–962.

23. **Plantago patagonica** Jacquin, Icon. Pl. Rar. 2: 9, plate 306. 1795 • Woolly plantain [F]

Plantago patagonica var. *breviscapa* (Shinners) Shinners; *P. patagonica* var. *gnaphalioides* (Nuttall) A. Gray; *P. patagonica* var. *spinulosa* (Decaisne) A. Gray; *P. purshii* Roemer & Schultes var. *oblonga* (E. Morris) Shinners; *P. spinulosa* Decaisne

Annuals; roots taproots, slender. **Stems** 0–15 mm. **Leaves** (25–)50–120 × 1–4 mm; blade linear, margins entire, rarely toothed, veins conspicuous or not, surfaces villous. **Scapes** (10–)40–240(–260) mm, hairy, hairs antrorse, long and short and patent, long. **Spikes** grayish or brownish,

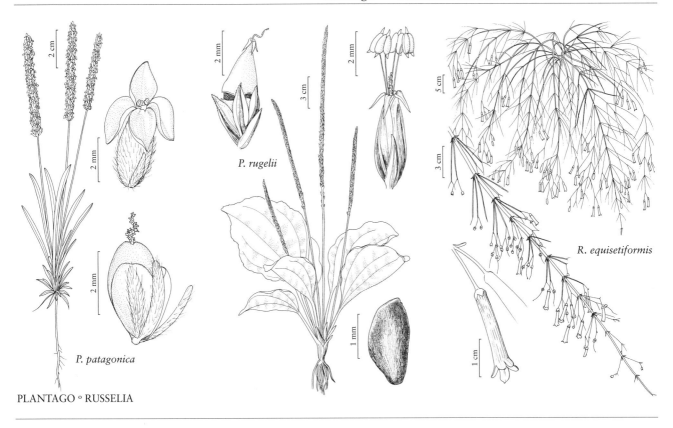

P. rugelii

R. equisetiformis

P. patagonica

PLANTAGO ∘ RUSSELIA

40–120 mm, densely flowered, flowers in whorls or pairs; bracts triangular or narrowly ovate, 2–10 mm, length 1–2 times sepals, apex acute or acuminate. **Flowers:** sepals 2.2–4.8 mm; corolla bilaterally symmetric, lobes reflexed, 1.6–2.1 mm, base cordate; stamens 4, connective usually elongated, apex acute. Seeds 2, 2.2–2.6 mm. $2n = 20$.

Flowering early summer. Dry sandy soils, grasslands, openings, disturbed areas; 0–3600 m; Alta., B.C., Man., Sask.; Ariz., Ark., Calif., Colo., Ga., Idaho, Ill., Ind., Iowa, Kans., La., Maine, Mass., Mich., Minn., Mo., Mont., Nebr., Nev., N.H., N.J., N.Mex., N.Y., N.C., N.Dak., Ohio, Okla., Oreg., S.C., S.Dak., Tenn., Tex., Utah, Vt., Va., Wash., W.Va., Wis., Wyo.; Mexico (Baja California); South America.

The circumscription of *Plantago patagonica* is treated here in accordance with K. Rahn (1978). The phenotypic plasticity of it does not allow varieties or subspecies to be recognized. Plants recognized in other floras as *P. purshii* (for example, *Flora of Indiana* and *Flora of New Mexico*) are referable to either *P. aristata* or *P. patagonica* (Rahn).

24. **Plantago pusilla** Nuttall, Gen. N. Amer. Pl. 1: 100. 1818 • Dwarf plantain [E]

Plantago hybrida W. P. C. Barton; *P. pusilla* var. *major* Engelmann

Annuals; roots taproots, slender. **Stems** 0–5 mm. **Leaves** 20–70 × 1–2 mm; blade linear, margins entire, sometimes toothed, veins conspicuous or not, surfaces hairy or glabrous. **Scapes** 15–60 mm, hairy, sometimes glabrous. **Spikes** greenish or brownish, 20–100 mm, loosely or densely flowered; bracts triangular-ovate, 1.5–2 mm, length 0.9–1.1 times sepals. **Flowers:** sepals 1.5–2 mm; corolla radially symmetric, lobes erect, forming a beak, 0.5 mm, base obtuse; stamens 2. Seeds 4, 0.8–1.3 mm. $2n = 12$.

Flowering spring–early summer. Dry to moist, sandy, alluvial soils; 0–200 m; Ala., Ark., Calif., Conn., Del., D.C., Fla., Ga., Ill., Ind., Kans., Ky., La., Md., Mass., Miss., Mo., N.J., N.Y., N.C., Okla., Oreg., Pa., R.I., S.C., Tenn., Tex., Va., Wash.

Plantago pusilla is considered to be introduced in Oregon and Washington, and possibly in California.

25. Plantago rhodosperma Decaisne in A. P. de Candolle and A. L. P. P. de Candolle, Prodr. 13(1): 722. 1852

• Redseed plantain

Annuals; roots taproots, slender. **Stems** 0–10 mm. **Leaves** 60–290 × 7–30 mm; blade obovate or elliptic, margins toothed, veins conspicuous, surfaces pilose. **Scapes** 15–150 mm, hairy, hairs patent, long. **Spikes** greenish, yellowish, or brownish, 35–300 mm, densely or loosely flowered; bracts narrowly triangular or triangular, 2.5–3.2 mm, length 0.7–1.1 times sepals. **Flowers:** sepals asymmetric, 2.7–3.6 mm, apex acuminate; corolla radially symmetric, lobes erect, forming a beak, 2–3.1 mm, base obtuse; stamens 4. **Seeds** 2, dark red, 2.3–2.8 mm, adaxial face concave. $2n = 24$.

Flowering spring. Rocky or sandy soils, grasslands, disturbed areas; 0–2600 m; Ala., Ariz., Ark., Calif., Ga., Ill., Kans., Ky., La., Miss., Mo., Nebr., N.Mex., Okla., Tenn., Tex.; n Mexico.

26. Plantago rugelii Decaisne in A. P. de Candolle and A. L. P. P. de Candolle, Prodr. 13(1): 700. 1852

• Blackseed plantain, plantain de Rugel E F

Perennials, sometimes annuals; caudex absent; roots fibrous, thick. **Stems** 0–20 mm. **Leaves** 20–150 × 10–120 mm; petiole to 200 mm; blade ovate to cordate-ovate, margins entire or toothed, veins conspicuous, surfaces glabrous or hirsute. **Scapes** 50–250 mm, glabrous or hirsute. **Spikes** brownish or greenish, 50–300 mm, densely or loosely flowered; bracts narrowly lanceolate, 2 mm, length 1–1.2 times sepals. **Flowers:** sepals 1.5–2 mm; corolla radially symmetric, lobes reflexed, 0.5–1 mm, base obtuse; stamens 4. **Fruits** 4–6(–8) mm, dehiscing proximal to middle. **Seeds** 4 or 5(–8), 1.5–2 mm. $2n = 24$.

Flowering summer. Open woods, meadows, pastures, waste places; 0–2000 m; N.B., N.S., Ont., Que.; Ala., Ark., Conn., Del., D.C., Fla., Ga., Ill., Ind., Iowa, Kans., Ky., La., Maine, Md., Mass., Mich., Minn., Miss., Mo., Mont., Nebr., N.H., N.J., N.Y., N.C., N.Dak., Ohio, Okla., Pa., R.I., S.C., S.Dak., Tenn., Tex., Vt., Va., W.Va., Wis.

Plants of *Plantago major* and *P. rugelii* are morphologically indistinguishable when young; they may be differentiated with certainty only at fruiting stage. DNA sequence data (A. Shipunov et al. 2014) confirm that the two species are distinct: *P. rugelii* is different from *P. major* by 11 substitutions in the ITS2 sequence.

27. Plantago sempervirens Crantz, Inst. Rei Herb. 2: 331. 1766   • Evergreen plantain   I

Perennials, sometimes woody; roots taproots, slender. **Stems** 100–400 mm, freely branched. **Leaves** cauline, opposite, 30–60 × 0.75–1 mm; blade linear to linear-lanceolate, margins entire, veins inconspicuous, surfaces hairy. **Scapes** 30–80 mm, hairy. **Spikes** greenish or brownish, 40–85 mm, densely flowered; bracts broadly ovate, 5–6 mm, lengths equal to sepals. **Flowers:** sepals 5–6 mm; corolla radially symmetric, lobes reflexed, 3–3.5 mm, base obtuse; stamens 4. **Seeds** 1 or 2, 2 mm. $2n = 12$.

Flowering summer–fall. Disturbed habitats; 0–200 m; introduced; Calif.; s Europe.

Plantago sempervirens is known from San Diego County; one specimen (JEPS) was collected in 2008 in Torrey Pines State Reserve.

Plantago cynops Linnaeus (1762, not 1753), a rejected name, has been misapplied to *P. sempervirens*.

28. Plantago sparsiflora Michaux, Fl. Bor.-Amer. 1: 94. 1803   • Pineland plantain   E

Perennials; caudex well developed, conspicuous, glabrous; roots taproots, thick. **Stems** 0–10 mm. **Leaves** 50–300 × 10–30 mm; blade lanceolate, margins entire, sometimes toothed, veins conspicuous, laterals branching from base, surfaces sparsely hairy. **Scapes** 75–150 mm, sparsely hairy. **Spikes** brownish or greenish, 100–450 mm, loosely flowered, rachis visible between flowers; bracts ovate, 1 mm, length 0.5–0.6 times sepals. **Flowers:** sepals 2 mm; corolla radially symmetric, lobes spreading, 1 mm, base obtuse; stamens 4. **Seeds** 2, 2 mm. $2n = 24$.

Flowering summer. Moist, sandy soils, open, undisturbed pine woods; 0–200 m; Fla., Ga., N.C., S.C.

Populations of *Plantago sparsiflora* occur mainly along the Atlantic coast from Columbia County, North Carolina, south to Volusia County, Florida.

29. Plantago subnuda Pilger, Notizbl. Königl. Bot. Gart. Berlin 5: 260. 1912 • Tall coastal plantain E

Perennials; caudex well developed, conspicuous, glabrous or hairy; roots taproots, fragile. **Stems** 0–10 mm. **Leaves** 60–360 × 15–65 mm; blade elliptic to narrowly elliptic, margins toothed, veins conspicuous, surfaces pilose, rarely glabrate, adaxial surface hairs not floccose, less than 2 mm long, more than 0.03 mm wide. **Scapes** 55–360 mm, hairy, hairs antrorse, long. **Spikes** greenish or brownish, 110–720 mm, densely or loosely flowered; bracts ovate, rarely triangular, 2.5–4 mm, length 0.8–1.3 times sepals. **Flowers:** sepals 2.6–3.1 mm; corolla radially symmetric, lobes erect, forming a beak, 2.4–2.7 mm, base obtuse; stamens 4. **Seeds** 3, 1.8–2.5 mm. $2n = 48$.

Flowering late spring–fall. Moist ground; 0–300 m; Calif., Oreg., Wash.

Plantago subnuda occurs primarily in counties along the Pacific coast from southwestern Washington to southern California.

30. Plantago tweedyi A. Gray in A. Gray et al., Syn. Fl. N. Amer. ed. 2, 2(1): 390. 1886 • Tweedy's plantain E

Perennials; caudex well developed, conspicuous, glabrous; roots taproots, thick. **Stems** 0–20 mm. **Leaves** 40–200 × 10–30 mm; blade lanceolate-spatulate to narrowly ovate, margins entire, sometimes toothed, veins conspicuous, surfaces usually glabrous. **Scapes** 25–200 mm, slightly surpassing leaves, glabrous. **Spikes** brownish or greenish, 45–250 mm, densely flowered, rachis not clearly visible between flowers; bracts broadly ovate, 2 mm, length 0.8–1 times sepals. **Flowers:** sepals 2–2.5 mm; corolla radially symmetric, lobes spreading, 1 mm, base obtuse; stamens 4. **Seeds** 3 or 4, 2–2.3 mm. $2n = 24$.

Flowering summer. Grasslands, sagebrush steppes, montane and subalpine meadows; 1600–4000 m; Ariz., Colo., Idaho, Mont., N.Mex., Utah, Wyo.

31. Plantago virginica Linnaeus, Sp. Pl. 1: 113. 1753 • Virginia plantain

Plantago virginica var. *viridescens* Fernald

Annuals; roots taproots, slender. **Stems** 0–20 mm. **Leaves** 20–120 × 4–25 mm; blade obovate or narrowly obovate, margins toothed, veins conspicuous, surfaces pilose or glabrate. **Scapes** 30–240 mm, hairy, hairs patent, long. **Spikes** greenish or yellowish, 38–260 mm, densely or loosely flowered; bracts narrowly ovate or ovate, 1.6–2.5 mm, length 0.7–1 times sepals. **Flowers:** sepals 1.5–2.4 mm, apex obtuse; corolla radially symmetric, lobes erect, forming a beak, 1.1–2.9 mm, base obtuse; stamens 4. **Seeds** 2, brown or yellowish brown, 1.6–2 mm, adaxial face deeply concave. $2n = 24$.

Flowering spring–early summer. Sandy soils, disturbed areas; 0–2300 m; Ont.; Ala., Ariz., Ark., Calif., Conn., Del., D.C., Fla., Ga., Ill., Ind., Iowa, Kans., Ky., La., Maine, Md., Mass., Mich., Minn., Miss., Mo., Nebr., N.J., N.Mex., N.Y., N.C., Ohio, Okla., Oreg., Pa., R.I., S.C., S.Dak., Tenn., Tex., Va., W.Va., Wis.; n Mexico.

32. Plantago wrightiana Decaisne in A. P. de Candolle and A. L. P. P. de Candolle, Prodr. 13(1): 712. 1852 • Wright's plantain

Annuals; roots taproots, slender. **Stems** 10–40 mm. **Leaves** 60–160 × 3–5 mm; blade linear, margins entire, veins conspicuous or not, abaxial surface villous, adaxial glabrous, rarely sparsely villous. **Scapes** 800–1600 mm, hairy, hairs antrorse, long and short. **Spikes** greenish or brownish, (20–)150–250 mm, densely flowered, flowers in whorls or pairs; bracts ovate or triangular, 1–3.5 mm, length 0.4–0.8 times sepals. **Flowers:** sepals 3–4 mm; corolla bilaterally symmetric, lobes reflexed, adaxials 2.4–3 mm, laterals asymmetric, base cordate; stamens 4, connective elongated, apex acute. **Seeds** 2, 2.8–3.2 mm. $2n = 20$.

Flowering summer. Sandy and gravelly soils, roadsides; 0–1000(–1300) m; Ala., Ariz., Ark., Fla., Ga., Kans., Ky., La., Md., Miss., N.Mex., N.C., Okla., S.C., Tenn., Tex., Va., W.Va.; Mexico (Chihuahua, Durango).

41. RUSSELIA Jacquin, Enum. Syst. Pl., 6, 25. 1760 • [For Alexander Russell, c. 1715–1768, Scottish physician and naturalist] [I]

Kerry A. Barringer

Shrubs [herbs, perennial]. Stems arching [erect or ascending], glabrous [hairy]. **Leaves** caducous, cauline, whorled or opposite; petiole absent or present; blade not fleshy, not leathery, margins dentate or entire, distal leaf blade needlelike or scalelike. **Inflorescences** axillary, cymes; bracts present. **Pedicels** present; bracteoles usually present. **Flowers** bisexual; sepals 5, basally connate, calyx bilaterally symmetric, tubular or campanulate, lobes broadly ovate [lanceolate]; corolla red, bilaterally symmetric, bilabiate, tubular [funnelform], tube base not spurred or gibbous, lobes 5, abaxial 3, adaxial 2; stamens 4, basally adnate to corolla, didynamous, filaments glabrous; staminode [0 or]1, conical [filiform]; ovary 2-locular, placentation axile; stigma subcapitate. **Fruits** capsules, dehiscence septicidal, densely packed with white, membranous hairs. **Seeds** 50–200, dark brown [black], ovoid, wings absent. *x* = 10.

Species 52 (1 in the flora): introduced, Florida; Mexico, Central America, n South America.

Russelia is unique in having capsules filled with densely packed hairs. Morphological and molecular characters suggest *Russelia* is related to *Tetranema* Bentham; both seem to be basal to the clade containing *Chelone* and *Penstemon* (D. C. Albach et al. 2005).

SELECTED REFERENCE Carlson, M. C. 1957. Monograph of the genus *Russelia*. Fieldiana, Bot. 29: 231–292.

1. **Russelia equisetiformis** Schlechtendal & Chamisso, Linnaea 6: 377. 1831 • Fountain or firecracker or coral plant [F] [I]

Russelia juncea Zuccarini

Stems striate, sometimes rooting at tips, 4–12-angled, 7–15 dm, gland-dotted. **Leaves** dimorphic; proximal: petiole 3–4 mm, in whorls of 3–6, blade ovate, 8–15 × 6–10 mm, margins dentate, surfaces glandular-punctate; distal: petiole 0–1 mm, opposite or whorled, blade narrowly triangular, needlelike or scalelike, 1–2 × 1 mm, margins entire.

Inflorescences 2-flowered; peduncles 3–4 cm. **Pedicels** 10–15 mm. **Flowers:** calyx lobes 2–3 mm, apex acute or mucronate; corolla 15–25 mm, glabrous, adaxial lip deeply notched; stamens 15–20 mm; staminode 0.5–0.7 mm, sometimes with an abortive anther. **Capsules** globular, 3–6 mm; style persistent. **Seeds** 0.3–0.4 mm, pitted. *2n* = 20.

Flowering year-round. Disturbed pinelands, palmetto scrub, old gardens; 0–100 m; introduced; Fla.; Mexico.

Russelia equisetiformis is distinguished by its tubular, red corollas and arching, ridged, striate stems with relatively small distal leaves. It has been in cultivation since the eighteenth century and is sometimes grown as an annual in temperate areas.

42. LAGOTIS Gaertner, Novi Comment. Acad. Sci. Imp. Petrop. 14(1): 533, plate 18, fig. 2. 1770 • Weasel snout [Greek *lagos*, hare, and *otos*, ear, alluding to calyx of some species]

Reidar Elven

David F. Murray

Gymnandra Pallas

Herbs, perennial; rhizomatous. **Stems** erect or ascending, glabrous. **Leaves** basal and cauline, opposite; petiole present or absent; blade not fleshy, leathery, margins entire, crenulate, dentate, or serrate. **Inflorescences** terminal, compact spikes; bracts present. **Pedicels** absent; bracteoles absent. **Flowers** bisexual; sepals 2, basally connate, calyx bilaterally symmetric, tubular, abaxial lobe enclosed by spathelike adaxial lobe; corolla blue, bilaterally symmetric, bilabiate, funnelform, tube base not spurred or gibbous, lobes 3, abaxial 2, adaxial 1; stamens 2, adnate to throat of corolla, filaments glabrous; staminode 0; ovary 2-locular, placentation axile; stigma capitate. **Fruits** drupelike. **Seeds** 2, tan, globular, wings absent. *x* = 11.

Species ca. 30 (1 in the flora): nw North America, Eurasia.

1. **Lagotis glauca** Gaertner, Novi Comment. Acad. Sci. Imp. Petrop. 14(1): 534, plate 18, fig. 2. 1770 F

Perennials 6–35 cm; rhizomes short, stout. **Stems** 1 or 2 (or 3), unbranched, fleshy. **Leaves:** basal long-petiolate; cauline short-petiolate or sessile. **Spikes** capitate to cylindric, 15–77 mm, elongating in fruit to 4–14 cm; bracts enlarging post anthesis. **Flowers:** calyx lobe margins entire or ciliate; style exserted.

Subspecies 3 (2 in the flora): nw North America, Asia (Russian Far East, Siberia).

Subspecies *minor* (Willdenow) Hultén occurs in Eurasia.

1. Basal leaf blades broadly obovate-oblanceolate, apices rounded or obtuse, sometimes subacute, margins crenate-dentate; filaments to 1.5 mm .1a. *Lagotis glauca* subsp. *glauca*
1. Basal leaf blades narrowly oblanceolate, apices obtuse or subacute, sometimes acute, margins entire or dentate to serrate; filaments 2+ mm . . . . .1b. *Lagotis glauca* subsp. *lanceolata*

1a. **Lagotis glauca** Gaertner subsp. **glauca** F

Basal leaves 6–16 cm; blade broadly obovate-oblanceolate, 35–80 × 20–65 mm, base truncate, margins crenate-dentate, apex rounded or obtuse, sometimes subacute. **Cauline leaves:** blade ovate to triangular, margins crenate-dentate. **Flowers:** filaments to 1.5 mm, often shorter than to as long as anthers. *2n* = 22.

Flowering summer. Coastal tundra meadows; 0–1100 m; Alaska; e Asia (Russian Far East).

Subspecies *glauca* occurs along the coast of the Alaska Peninsula, Aleutian Islands, Kodiak Island, the Pribilov Islands, and, intermittently, to Teller on the Seward Peninsula. Transitional basal leaf forms occur where subsp. *glauca* is sympatric with subsp. *lanceolata*; subsp. *glauca* is then distinguished in most cases by having shorter stamen filaments. Some specimens are difficult to place in either subspecies in this zone of contact.

1b. **Lagotis glauca** Gaertner subsp. **lanceolata** (Hultén) D. F. Murray & Elven, J. Bot. Res. Inst. Texas 4: 220. 2010 F

Lagotis glauca var. *lanceolata* Hultén, Fl. Kamtchatka 4: 105. 1930

Basal leaves 2–17 cm; blade narrowly oblanceolate, 12–70 × 12–30 mm, base cuneate, margins entire or dentate to serrate, apex obtuse or subacute, sometimes acute. **Cauline leaves:** blade lanceolate, margins entire or dentate to serrate. **Flowers:** filaments 2+ mm, longer than anthers. *2n* = 22.

Flowering summer. Tundra meadows, tussock tundra, seeps, moist screes, margins of shallow tundra pools; 0–1800 m; N.W.T., Yukon; Alaska; Asia (Russian Far East, Siberia).

Where subsp. *glauca* and subsp. *lanceolata* are sympatric, there is local separation by habitat preferences. Subspecies *lanceolata* occurs farther north and eastward along the Arctic coastal plain, Arctic foothills, and Brooks Range southward to the mountain ranges of interior Alaska, arctic and interior Yukon, and westernmost Northwest Territories.

W. J. Cody (1966), E. Hultén (1968), and V. V. Petrovsky (1980) all misapplied the name subsp. *minor* (Willdenow) Hultén to plants of subsp. *lanceolata*. Subspecies *minor* is typified on tetraploid (*2n* = 44) plants, ones probably from the lower Lena River, which are otherwise found in northeastern Europe, European Russia, and Siberia, but do not reach the Russian Far East or North America.

Gjaerevoll proposed, but did not validly publish, *Lagotis glauca* subsp. *stelleri* in 1967, based on *Gymnandra stelleri* Chamisso & Schlechtendal, a name itself published twice and which cannot be clearly linked to this taxon (D. F. Murray et al. 2010). The legitimate var. *lanceolata* by Hultén has a diagnosis that refers unambiguously to a northern amphi-Beringian plant.

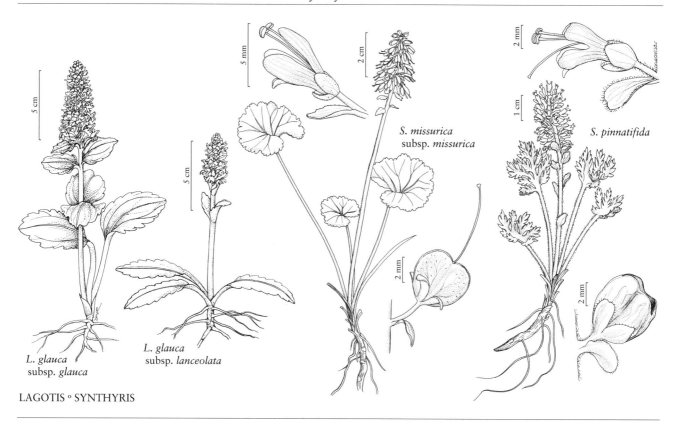

S. missurica
subsp. *missurica*

S. pinnatifida

L. glauca
subsp. *glauca*

L. glauca
subsp. *lanceolata*

LAGOTIS ∘ SYNTHYRIS

43. SYNTHYRIS Bentham in A. P. de Candolle and A. L. P. P. de Candolle, Prodr. 10: 454.
1846 • Kittentail [Greek *syn*, together, and *thyris*, valve, alluding to capsule valves
adhering below to the placentiferous axis]

Larry D. Hufford

Besseya Rydberg; *Veronica* Linnaeus subg. *Synthyris* (Bentham) M. M. Martínez Ort., Albach &
M. A. Fischer

Herbs, perennial; rhizomatous. **Stems** erect when aerial, horizontal to erect when subterranean,
glabrous. **Leaves** basal, helical; petiole present; blade not fleshy, leathery or not, margins with
simple or compound teeth and some lobes deeply incised to pinnatifid. **Inflorescences** axillary,
racemes, erect or decumbent; bracts present. **Pedicels** present; bracteoles absent. **Flowers**
bisexual; sepals 2–4(or 5), basally connate, calyx bilaterally symmetric, campanulate, lobes
ovate to lanceolate; corolla white, yellow, green, pink, reddish, lavender, or blue to bluish purple,
bilaterally symmetric, bilabiate or ± regular (unilabiate in *S. rubra*), campanulate, tubular,
ellipsoid, or rudimentary, tube base not spurred or gibbous, lobes 0 or 3 or 4(or 5), abaxial
2 or 3(or 4), adaxial 1; stamens 2, epipetalous or inserted on receptacle, filaments glabrous;
staminode 0; ovary 1-locular, placentation parietal centrally, axile distally and proximally;
stigma capitate. **Fruits** capsules, flattened, dehiscence loculicidal and basipetal over apex. **Seeds**
2–40, brown, disc- or boat-shaped (*S. cordata*, *S. reniformis*), wings absent. $x = 12$.

Species 19 (19 in the flora): North America, n Mexico.

Plants of *Synthyris* flower in the early spring and as early as snowmelt in alpine and tundra
areas, and fruits mature by mid spring. Flowers are protogynous. Species of *Synthyris* are not
consistent with Darwin's pollination syndrome that associates acropetal floral maturation with

protandry and upward movement of pollinators on the inflorescence; instead, insect visitors (primarily pollen foraging bees) show a weak tendency to move vertically among the protogynous flowers with both self- and cross-pollination (M. J. McCone et al. 1995).

Synthyris occurs in arctic and alpine tundra, grasslands, savannas, and mesophytic forests. Some species have restricted distributions, and two are of conservation concern.

Synthyris is part of a monophyletic Veroniceae that includes *Veronica* and allied genera. The closest relatives of *Synthyris* are uncertain and may be either clades of *Veronica* or Asian Veroniceae (Albach et al. 2004; D. C. Albach and H. M. Meudt 2010).

Species treated as *Besseya* by P. A. Rydberg (1903), F. W. Pennell (1933), and C. G. Schaack (1983) are monophyletic in *Synthyris*. The bilabiate flowers and more coriaceous leaves characteristic of the besseyas are derived in *Synthyris*. Dissected leaves are convergent in *Synthyris*, evolving independently in *S. pinnatifida* and the clade consisting of *S. canbyi*, *S. dissecta*, and *S. lanuginosa*, in association with shifts into alpine and treeline environments (L. Hufford and M. McMahon 2004). *Synthyris cordata* and *S. reniformis* are low elevation species of moist forests that uniquely share decumbent infructescences, an apparent specialization for seed dispersal by ants.

SELECTED REFERENCES Hedglin, F. L. 1959. A Survey of the Genus *Synthyris*. M.S. thesis. University of Washington. Hufford, L. 1992. Floral structure of *Besseya* and *Synthyris* (Scrophulariaceae). Int. J. Pl. Sci. 153: 217–229. Hufford, L. 1992b. Leaf structure of *Besseya* and *Synthyris* (Scrophulariaceae). Canad. J. Bot. 70: 921–932. Hufford, L. and M. McMahon. 2004. Morphological evolution and systematics of *Synthyris* and *Besseya* (Veroniceae): A phylogenetic analysis. Syst. Bot. 29: 716–736. Kruckeberg, A. R. and F. L. Hedglin. 1963. Natural and artificial hybrids of *Besseya* and *Synthyris* (Scrophulariaceae). Madroño 17: 109–115. Marlowe, K. and L. Hufford. 2008. Evolution of *Synthyris* sect. *Dissecta* (Plantaginaceae) on sky islands in the northern Rocky Mountains. Amer. J. Bot. 95: 381–392. Pennell, F. W. 1933. A revision of *Synthyris* and *Besseya*. Proc. Acad. Nat. Sci. Philadelphia 85: 77–106. Schaack, C. G. 1983. A Monographic Revision of the Genera *Synthyris* and *Besseya* (Scrophulariaceae). Ph.D. dissertation. University of Montana.

1. Leaves strictly annual, disintegrating in 1st year; leaf blade margins crenate or incised-crenate; corollas bilabiate, unilabiate (*S. rubra*), or absent (*S. wyomingensis*), longer than or as long as calyces, or corollas rudimentary or absent.
   2. Petals 0 or 1–4, rudimentary if present, corollas much shorter than calyces if present.
      3. Sepals 4, basal connation between abaxial and adaxial lobes on each side of flower; petals 1–4, rudimentary . . . . .17. *Synthyris rubra*
      3. Sepals 2–4, all lobes connate, if 2+ lobes then connate for at least ½ of length on abaxial side; petals 0. . . . . 19. *Synthyris wyomingensis*
   2. Petals 3 or 4(or 5), corollas nearly as long as or longer than calyces.
      4. Corollas blue, bluish purple, lavender, or reddish . . . . 1. *Synthyris alpina*
      4. Corollas white, yellow, or pink.
         5. Corollas yellow.
            6. Basal veins of leaves extending into distal ½ of blade, lateral veins 3–6 on each side of midvein; corolla tubes conspicuous; stamens epipetalous; ovaries puberulent to villous . . . . .3. *Synthyris bullii*
            6. Basal veins of leaves extending through basal ½ of blade, lateral veins 5–12 on each side of midvein; corolla tubes absent; stamens inserted on receptacles, abaxial and adaxial petal lips basally adnate to stamens; ovaries glabrous or sparsely hairy at apex . . . . .16. *Synthyris ritteriana*
         5. Corollas pink to white.
            7. Corollas 2–3 mm longer than calyces; ovaries puberulent to villous at apex; Sierra Blanca Range, New Mexico . . . . 10. *Synthyris oblongifolia*
            7. Corollas 0–2 mm longer than calyces; ovaries glabrous; Arizona, Colorado, New Mexico, Wyoming. . . . . .12. *Synthyris plantaginea*
1. Leaves persistent, some withering in 2d year as new leaves expand; leaf blade margins dentate, crenate, incised-crenate, laciniate, pinnately lobed, or ± palmately lobed; corollas ± regular, much longer than calyces.

[8. Shifted to left margin.—Ed.]

8. Racemes decumbent, sterile bracts usually 0.
   9. Leaf blades ovate to ovate-cordate; corollas glabrous . . . . .5. *Synthyris cordata*
   9. Leaf blades cordate to reniform; corollas puberulent-villous in throat. . . . . .15. *Synthyris reniformis*
8. Racemes erect, sterile bracts 2 or 3+.
   10. Petal apices laciniate.
      11. Racemes ovate-spatulate, largest sterile bracts less than 1 cm; Idaho . . . . . . .13. *Synthyris platycarpa*
      11. Racemes fan-shaped, largest sterile bracts 2+ cm; nw Oregon, w Washington . . .
      . . . . .18. *Synthyris schizantha*
   10. Petal apices entire or erose.
      12. Leaf blade margins laciniate, ± incised-crenate, dentate, or ± palmately lobed.
         13. Capsules hairy; Northwest Territories, Yukon, Alaska. . . . . .2. *Synthyris borealis*
         13. Capsules glabrous, glabrescent, or sparsely hairy; nw United States.
            14. Leaf blades less than 25 mm wide; corolla tubes inconspicuous. . . .14. *Synthyris ranunculina*
            14. Leaf blades 25+ mm wide; corolla tubes conspicuous.
               15. Leaf blades ovate or cordate, margins ± incised-crenate or laciniate to
                  ± pinnatifid, sometimes ± palmately lobed; Montana . . . . . .4. *Synthyris canbyi* (in part)
               15. Leaf blades orbiculate, reniform, or cordate, margins ± laciniate or
                  ± incised-crenate to dentate, sometimes palmately lobed; California,
                  Idaho, Montana, Oregon, Utah, Washington.
                    16. Leaf blade margins ± laciniate, sometimes palmately lobed or
                      incised-crenate; Utah. . . . . 7. *Synthyris laciniata*
                    16. Leaf blade margins ± incised-crenate to dentate; California, Idaho,
                      Montana, Oregon, Washington. . . . . .9. *Synthyris missurica*
      12. Leaf blade margins pinnately lobed or 1–3-pinnatifid.
         17. Leaf blade surfaces canescent or tomentose.
            18. Leaf blade surfaces tomentose; adaxial petals flat; ec Idaho, sw Montana
            . . . . .6. *Synthyris dissecta* (in part)
            18. Leaf blade surfaces canescent; adaxial petals not flat; nw Washington. . .
            . . . . 8. *Synthyris lanuginosa*
         17. Leaf blade surfaces glabrous, sparsely hairy, puberulous, or villous.
            19. Capsules hairy.
               20. Ovaries sparsely hairy at apices; flowers 10–50; Montana. . .4. *Synthyris canbyi* (in part)
               20. Ovaries pilose to tomentose; flowers 10–30; ec Idaho, Montana. . . .
               . . . . .6. *Synthyris dissecta* (in part)
            19. Capsules glabrous.
               21. Leaf blade margins ± pinnatifid, teeth apices obtuse to rounded,
                 surfaces sparsely hairy; Montana. .4. *Synthyris canbyi* (in part)
               21. Leaf blade margins 1- or 2-pinnatifd, teeth apices obtuse to acute,
                 surfaces glabrous or villous; Idaho, Utah, Wyoming . . . . 11. *Synthyris pinnatifida*

1. **Synthyris alpina** A. Gray, Amer. J. Sci. Arts, ser. 2,
34: 251. 1862 • Alpine kittentail or besseya [E]

Besseya alpina (A. Gray) Rydberg; *Veronica besseya* M. M. Martínez Ort. & Albach

Leaves strictly annual, disintegrating in 1st year; blade ovate to slightly cordate, sometimes oblong, 25+ mm wide, leathery, base cuneate, truncate, cordate, or lobate, margins crenate, teeth apices acute, surfaces glabrous or sparsely hairy; basal veins extending into distal ½ of blade, lateral veins 3–6 on each side of midvein. **Racemes** erect, to 10 cm in fruit; sterile bracts 5–14, ovate-spatulate, largest 1+ cm; flowers 100+, densely aggregated (separating in fruit). **Sepals** 4. **Petals** (3 or) 4(or 5), apex entire or erose; corolla blue, bluish purple, lavender, or reddish, bilabiate, ± rudimentary, tubular, longer than calyx, glabrous, lateral and abaxial petals of abaxial lip connate ½+ their lengths, tube absent. **Stamens** inserted on receptacle. **Ovaries:** ovules 17–40. **Capsules** glabrous.

Flowering May–Aug; fruiting Jun–Sep. Stony alpine meadows or fellfields; 2800–4300 m; Colo., N.Mex., Utah, Wyo.

Synthyris alpina is a host plant for checkerspot butterfly (*Euphydryas anicia*), which consumes leaves and sequesters iridoid glycosides (K. M. L'Empereur and F. R. Stermitz 1990). Flowering in plants of *S. alpina* begins at the margins of melting snow banks.

2. **Synthyris borealis** Pennell, Proc. Acad. Nat. Sci. Philadelphia 85: 88, fig. 1. 1933 • Northern kittentail E

Veronica alaskensis M. M. Martínez Ort. & Albach

Leaves persistent, some withering in 2d year as new leaves expand; blade ovate to cordate, 25+ mm wide, not leathery, base lobate, margins laciniate, teeth apices rounded, surfaces villous, hairs prominent on margins; basal veins extending into distal ½ of blade, lateral veins 2–4 on each side of midvein. **Racemes** erect, to 15 cm in fruit; sterile bracts 3+, ovate-spatulate, largest 1+ cm; flowers 15–50, loosely aggregated. **Sepals** 4. **Petals** (3 or)4(or 5), apex entire or erose; corolla blue, ± regular, campanulate, much longer than calyx, glabrous, tube conspicuous. **Stamens** epipetalous. **Ovaries:** ovules 10–16. **Capsules** hairy. *2n* = 24.

Flowering May–Jul; fruiting May–Sep. Tundra heaths, fellfields, talus slopes; 200–2500 m; N.W.T., Yukon; Alaska.

Synthyris borealis is distributed primarily in unglaciated portions of the Northwest Territories, Yukon, and Alaska.

3. **Synthyris bullii** (Eaton) A. Heller, Muhlenbergia 1: 4. 1900 • Bull's kittentail or coraldrops E

Gymnandra bullii Eaton in A. Eaton and J. Wright, Man. Bot. ed. 8, 259. 1840; *Besseya bullii* (Eaton) Rydberg; *Veronica bullii* (Eaton) M. M. Martínez Ort. & Albach; *Wulfenia bullii* (Eaton) Barnhart

Leaves strictly annual, disintegrating in 1st year; blade elliptic or ovate to broadly ovate, 25+ mm wide, leathery, base rounded to lobate or cordate, margins crenate, teeth apices obtuse to rounded, surfaces hairy; basal veins extending into distal ½ of blade, lateral veins 3–6 on each side of midvein. **Racemes** erect, to 40 cm in fruit; sterile bracts 10–30, ovate-spatulate, largest 1+ cm; flowers 100+, densely aggregated (separating in fruit). **Sepals** 4. **Petals** (3 or)4, apex entire or erose; corolla yellow, bilabiate, tubular to ellipsoid, longer than calyx, hairy, lateral and abaxial petals connate ca. ½ their lengths, tube conspicuous. **Stamens** epipetalous. **Ovaries** puberulent to villous; ovules 17–40. **Capsules** hairy.

Flowering Apr–Jun; fruiting May–Jul. Sandy prairies, sandy secondary deciduous forests; 100–400 m; Ill., Ind., Iowa, Mich., Minn., Ohio, Wis.

Synthyris bullii is protected throughout its range and may be extirpated in Ohio (NatureServe, www. natureserve.org).

4. **Synthyris canbyi** Pennell, Proc. Acad. Nat. Sci. Philadelphia 85: 93. 1933 • Canby's kittentail E

Veronica canbyi (Pennell) M. M. Martínez Ort. & Albach

Leaves persistent, some withering in 2d year as new leaves expand; blade ovate or cordate, 25+ mm wide, not leathery, base cordate to lobate, margins ± incised-crenate or laciniate to ± pinnatifid, sometimes ± palmately lobed, teeth apices obtuse to rounded, surfaces sparsely hairy; basal veins extending into distal ½ of blade, lateral veins 2–4 on each side of midvein. **Racemes** erect, to 16 cm in fruit; sterile bracts 2+, ovate-spatulate, sometimes 1+ cm; flowers 10–50, loosely aggregated. **Sepals** 4. **Petals** (3 or)4(or 5), apex entire or erose; corolla blue, ± regular, campanulate, much longer than calyx, glabrous, tube conspicuous. **Stamens** epipetalous. **Ovaries** glabrous or sparsely hairy at apex; ovules 10–16. **Capsules** glabrous or sparsely hairy.

Flowering Jun–Jul; fruiting Jun–Aug. Alpine ridges, scree slopes; 2100–3200 m; Mont.

Synthyris canbyi is known from the Mission, Rattlesnake, and Swan mountain ranges of northwest Montana. Flowering in *S. canbyi* begins at the margins of melting snow banks. Specimens from the southern end of the Mission Range appear intermediate between *S. canbyi* and *S. dissecta*.

5. **Synthyris cordata** (A. Gray) A. Heller, Muhlenbergia 1: 4. 1900 • Snow queen E

Synthyris reniformis (Douglas ex Bentham) Bentham var. *cordata* A. Gray in W. H. Brewer et al., Bot. California 1: 571. 1876; *Veronica californica* M. M. Martínez Ort. & Albach

Leaves persistent, some withering in 2d year as new leaves expand; blade ovate to ovate-cordate, 25+ mm wide, leathery, base lobate to cordate, margins incised-crenate, teeth apices acute to obtuse, surfaces glabrous or hairy; basal veins extending into distal ½ of blade, lateral veins 2–4 on each side of midvein. **Racemes** decumbent, to 14 cm in fruit; sterile bracts usually 0; flowers 5–10, loosely aggregated. **Sepals** 4. **Petals** (3 or)4(or 5), apex entire or erose; corolla blue, ± regular, campanulate, much longer than calyx, glabrous, tube conspicuous. **Stamens**

epipetalous. **Ovaries:** ovules 4. **Capsules** sparsely hairy on margins.

Flowering Feb–Apr; fruiting Mar–Jun. Forests; 100–1000 m; Calif., Oreg.

6. Synthyris dissecta Rydberg, Bull. Torrey Bot. Club 36: 691. 1909 • Cut-leaf kittentail E

Synthyris pinnatifida S. Watson var. *canescens* (Pennell) Cronquist; *Veronica dissecta* (Rydberg) M. M. Martínez Ort. & Albach

Leaves persistent, some withering in 2d year as new leaves expand; blade oblong-ovate to ovate, 25+ mm wide, not leathery, base lobate, margins 1- or 2-pinnatifid, ultimate lobes oblanceolate to lanceolate or linear, teeth apices obtuse, surfaces glabrous, sparsely hairy, puberulous, sparsely villous, or tomentose; basal veins extending through proximal ½ of blade, lateral veins 2–4 on each side of midvein. **Racemes** erect, to 15 cm in fruit; sterile bracts 3+, ovate-spatulate, largest 1+ cm; flowers 10–30, loosely aggregated. **Sepals** 4. **Petals** (3 or)4(or 5), apex entire or erose, adaxials flat; corolla blue, ± regular, campanulate, much longer than calyx, glabrous, tube conspicuous. **Stamens** epipetalous. **Ovaries** pilose to tomentose; ovules 10–16. **Capsules** hairy.

Flowering May–Jul; fruiting Jun–Sep. Treeline to alpine tundra; 1600–3000 m; Idaho, Mont.

Synthyris dissecta is found in east-central Idaho and southwestern Montana.

7. Synthyris laciniata (A. Gray) Rydberg, Mem. New York Bot. Gard. 1: 353. 1900 • Cut-leaf kittentail E

Synthyris pinnatifida S. Watson var. *laciniata* A. Gray in A. Gray et al., Syn. Fl. N. Amer. 2(1): 286. 1878; *Veronica utahensis* M. M. Martínez Ort. & Albach

Leaves persistent, some withering in 2d year as new leaves expand; blade ± orbiculate, reniform, or cordate, 25+ mm wide, not leathery, base cordate or lobate, margins ± laciniate, sometimes palmately lobed or incised-crenate, teeth apices obtuse to rounded, surfaces glabrous or sparsely hairy; basal veins usually extending into distal ½ of blade, lateral veins 2–4 on each side of midvein. **Racemes** erect, to 20 cm (usually less than 25 cm in fruit); sterile bracts 3+, ovate-spatulate, largest 1+ cm; flowers 15–40, loosely aggregated. **Sepals** 4. **Petals** (3 or)4(or 5), apex entire or erose; corolla blue, ± regular, campanulate, much longer than calyx, glabrous, tube conspicuous. **Stamens** epipetalous. **Ovaries:** ovules 10–16. **Capsules** glabrous.

Flowering late May–Aug; fruiting Jun–Oct. Subalpine meadows, alpine tundra, fellfields; 2900–4000 m; Utah.

Synthyris laciniata is found only in high mountain areas of central to west-central and southern Utah. Flowering in *S. laciniata* begins at the margins of melting snow banks.

8. Synthyris lanuginosa (Piper) Pennell & J. W. Thompson, Proc. Acad. Nat. Sci. Philadelphia 85: 93. 1933 • Olympic cut-leaf or woolly kittentail E

Synthyris pinnatifida S. Watson subsp. *lanuginosa* Piper, Contr. U.S. Natl. Herb. 11: 504. 1906; *S. pinnatifida* var. *lanuginosa* (Piper) Cronquist; *Veronica dissecta* (Rydberg) M. M. Martínez Ort. & Albach subsp. *lanuginosa* (Piper) M. M. Martínez Ort. & Albach

Leaves persistent, some withering in 2d year as new leaves expand; blade oblong-ovate to ovate, 25+ mm wide, not leathery, base lobate, margins pinnately lobed or 2- or 3-pinnatifid, teeth apices obtuse, surfaces canescent; basal veins extending through proximal ½ of blade, lateral veins 2–4 on each side of midvein. **Racemes** erect, to 15 cm in fruit; sterile bracts 3+, ovate-spatulate, largest 1+ cm; flowers 10–30, loosely aggregated. **Sepals** 4. **Petals** (3 or)4(or 5), apex entire or erose, adaxials not flat; corolla blue, ± regular, campanulate, much longer than calyx, glabrous, tube conspicuous. **Stamens** epipetalous. **Ovaries:** ovules 10–16. **Capsules** hairy. $2n = 24$.

Flowering May–Aug; fruiting Jun–Sep. Alpine fellfields, talus slopes; 1600–2200 m; Wash.

Synthyris lanuginosa is restricted to the Olympic Mountains of northwestern Washington.

9. Synthyris missurica (Rafinesque) Pennell, Proc. Acad. Nat. Sci. Philadelphia 85: 89. 1933 • Mountain kittentail E F

Veronica missurica Rafinesque, Amer. Monthly Mag. & Crit. Rev. 2: 175. 1818, based on *V. reniformis* Pursh, Fl. Amer. Sept. 1: 10. 1813, not Rafinesque 1808

Leaves persistent, some withering in 2d year as new leaves expand; blade orbiculate to reniform, 25+ mm wide, not leathery, base cordate to lobate, margins ± incised-crenate to dentate, teeth apices rounded to acute or acuminate, surfaces glabrous, rarely sparsely hairy; basal veins extending into distal ½ of blade, lateral veins 2–4 on each side of midvein. **Racemes** erect, to 33 cm in fruit; sterile bracts 3+, ovate-spatulate, largest 1+ cm; flowers 12–100, loosely aggregated.

Sepals 4. **Petals** (3 or)4(or 5), apex entire or erose; corolla blue, ± regular, campanulate, much longer than calyx, glabrous or puberulent, tube conspicuous. **Stamens** epipetalous. **Ovaries:** ovules 10–16. **Capsules** glabrous or glabrescent.

Subspecies 3 (3 in the flora): w United States.

1. Racemes usually 20–100-flowered, to 33 cm at end of flowering; ovaries glabrous; capsules glabrous . . . . 9b. *Synthyris missurica* subsp. *major*
1. Racemes usually 12–45-flowered, to 25 cm at end of flowering; ovaries sparsely hairy, especially along margins; capsules glabrescent.
  2. Leaf blade margins regularly toothed; racemes to 20 cm at end of flowering . . . .
         . . . . 9a. *Synthyris missurica* subsp. *missurica*
  2. Leaf blade margins regularly to irregularly toothed; racemes to 25 cm at end of flowering . . . . . 9c. *Synthyris missurica* subsp. *stellata*

9a. Synthyris missurica (Rafinesque) Pennell subsp. **missurica** [E] [F]

Leaves 35–80 mm during flowering, margins regularly toothed. **Racemes** to 20 cm at end of flowering, longer in fruit, usually 12–35-flowered. **Ovaries** sparsely hairy, especially along margins. **Capsules** glabrescent.

Flowering late Mar–Jul; fruiting Apr–Aug. Moist forests; 300–2900 m; Calif., Idaho, Mont., Oreg., Wash.

9b. Synthyris missurica (Rafinesque) Pennell subsp. **major** (Hooker) Pennell, Proc. Acad. Nat. Sci. Philadelphia 85: 91. 1933 [E]

Synthyris reniformis (Douglas ex Bentham) Bentham var. *major* Hooker, Hooker's J. Bot. Kew Gard. Misc. 5: 257. 1853; *Veronica missurica* Rafinesque subsp. *major* (Hooker) M. M. Martínez Ort. & Albach

Leaves 65–125 mm during flowering, margins regularly toothed. **Racemes** to 33 cm at end of flowering, longer in fruit, usually 20–100-flowered. **Ovaries** glabrous. **Capsules** glabrous.

Flowering Mar–May; fruiting Apr–Jul. Moist forest slopes; 300–1100 m; Idaho, Wash.

Subspecies *major* occurs most commonly in the Clearwater River drainage but also on Craig Mountain near Lewiston, Idaho, and Kamiak Butte near Pullman, Washington.

9c. Synthyris missurica (Rafinesque) Pennell subsp. **stellata** (Pennell) Kartesz & Gandhi, Phytologia 67: 464. 1989 [E]

Synthyris stellata Pennell, Proc. Acad. Nat. Sci. Philadelphia 85: 89. 1933; *Veronica missurica* Rafinesque subsp. *stellata* (Pennell) M. M. Martínez Ort. & Albach

Leaves 45–105 mm during flowering, margins regularly to irregularly toothed. **Racemes** to 25 cm at end of flowering, longer in fruit, usually 24–45-flowered. **Ovaries** sparsely hairy, especially along margins. **Capsules** glabrescent.

Flowering Mar–May; fruiting Mar–Jul. Moist forests; 30–1000 m; Oreg., Wash.

Subspecies *stellata* is known from the Columbia River Gorge and at higher elevations in Washington in environments that tend to be drier than those at the base of the Columbia Gorge.

10. Synthyris oblongifolia (Pennell) L. Hufford & M. McMahon, Syst. Bot. 29: 735. 2004 • Eggleaf kittentail or coraldrops [C] [E]

Besseya oblongifolia Pennell, Proc. Acad. Nat. Sci. Philadelphia 85: 101. 1933; *Veronica oblongifolia* (Pennell) M. M. Martínez Ort. & Albach

Leaves strictly annual, disintegrating in 1st year; blade narrowly ovate to ovate or oblong-ovate, 25+ mm wide, slightly leathery, base obtuse, truncate, rounded, or cordate, margins crenate, teeth apices acute to obtuse, surfaces glabrous or sparsely hairy; basal veins extending through proximal ½ of blade, lateral veins 5–12 on each side of midvein. **Racemes** erect, to 30 cm in fruit; sterile bracts 11–20, ovate-spatulate, largest 1+ cm; flowers 100+, densely aggregated (separating in fruit). **Sepals** 3(or 4). **Petals** 3 (abaxial petals sometimes with numerous subsidiary lobes), apex entire or erose; corolla pink, bilabiate, tubular to ellipsoid, 2–3 mm longer than calyx, glabrous, lateral and abaxial petals of abaxial lip connate ½+ their lengths, abaxial and adaxial lips sometimes basally adnate slightly to stamen filaments, tube absent. **Stamens** inserted on receptacle. **Ovaries** puberulent to villous at apex; ovules 17–40. **Capsules** densely hairy.

Flowering May–Sep; fruiting Jun–Oct. Alpine and subalpine meadows; of conservation concern; 2900–3700 m; N.Mex.

Synthyris oblongifolia is endemic to the Sierra Blanca Range in Lincoln and Otero counties.

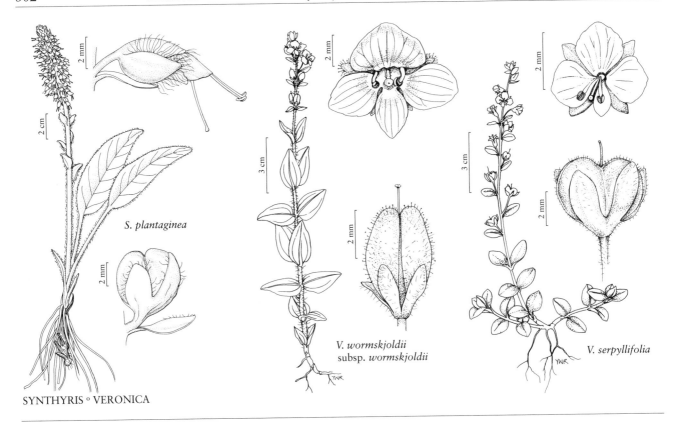

S. plantaginea

V. wormskjoldii
subsp. wormskjoldii

V. serpyllifolia

SYNTHYRIS ∘ VERONICA

11. Synthyris pinnatifida S. Watson, Botany (Fortieth Parallel), 227, plate 22, figs. 1, 2 [as pinnata]. 1871 • Feather-leaf kittentail E F

Veronica paysonii (Pennell & L. O. Williams) M. M. Martínez Ort. & Albach

Leaves persistent, some withering in 2d year as new leaves expand; blade oblong-ovate to ovate, 25+ mm wide, not leathery, base lobate, margins 1- or 2-pinnatifid, teeth apices obtuse to acute, surfaces glabrous or villous; basal veins extending through proximal ½ of blade, lateral veins 2–4 on each side of midvein. **Racemes** erect, to 30 cm in fruit; sterile bracts 3+, ovate-spatulate, largest 1+ cm; flowers 10–40, loosely aggregated. **Sepals** 4. **Petals** (3 or)4(or 5), apex entire or erose; corolla blue, ± regular, campanulate, much longer than calyx, glabrous, tube conspicuous. **Stamens** epipetalous. **Ovaries:** ovules 10–16. **Capsules** glabrous.

Flowering Apr–Aug; fruiting May–Oct. Forest openings, subalpine meadows, alpine tundra; 2100–3800 m; Idaho, Utah, Wyo.

Flowering in plants of *Synthyris pinnatifida* begins at the margins of melting snow banks.

12. Synthyris plantaginea (E. James) Bentham in A. P. de Candolle and A. L. P. P. de Candolle, Prodr. 10: 455. 1846 • Foothills kittentail F

Veronica plantaginea E. James, Trans. Amer. Philos. Soc., n. s. 2: 173. 1825; *Besseya plantaginea* (E. James) Rydberg

Leaves strictly annual, disintegrating in 1st year; blade narrowly to broadly ovate, 25+ mm wide, leathery, base obtuse to rounded or lobate, margins crenate, teeth apices acute to obtuse, surfaces sparsely hairy to villous; basal veins extending through proximal ½ of blade, lateral veins 5–12 on each side of midvein. **Racemes** erect, to 40 cm in fruit; sterile bracts 9–31, ovate-spatulate, largest 1+ cm; flowers 100+, densely aggregated (separating in fruit). **Sepals** 4. **Petals** (3 or)4(or 5), apex entire or erose; corolla pink to white, bilabiate, ellipsoid, 0–2 mm longer than calyx, glabrous or sparsely hairy, lateral and abaxial petals of abaxial lip connate ½+ their lengths, tube absent. **Stamens** inserted on receptacle. **Ovaries** glabrous; ovules 17–40. **Capsules** glabrous.

Flowering May–Jun; fruiting May–Aug. Montane to subalpine meadows, open, montane, conifer forests; 1800–3400 m; Ariz., Colo., N.Mex., Wyo.; Mexico (Chihuahua).

Both diploid (2*n* = 24) and tetraploid (2*n* = 48; described as *Besseya gooddingii* Pennell) populations have been discovered among southern populations of *Synthyris plantaginea* in Arizona (C. G. Schaack 1983). *Synthyris plantaginea* is a host for the checkerspot butterfly (*Euphydryas anicia*), which sequesters iridoid glycosides as it consumes the leaves (K. M. L'Empereur and F. R. Stermitz 1990).

13. Synthyris platycarpa Gail & Pennell, Amer. J. Bot. 24: 40. 1937 • Evergreen kittentail E

Veronica idahoensis M. M. Martínez Ort. & Albach

Leaves persistent, some withering in 2d year as new leaves expand; blade broadly cordate or reniform to nearly orbiculate, 25+ mm wide, chartaceous, base cordate to lobate, margins crenate or incised-crenate, teeth apices rounded, surfaces glabrous or hairy; basal veins extending into distal ½ of blade, lateral veins 2–4 on each side of midvein. **Racemes** erect, to 35 cm in fruit; sterile bracts 2, ovate-spatulate, largest less than 1 cm; flowers 15–35, loosely aggregated. **Sepals** 4. **Petals** (3 or)4(or 5), apex laciniate; corolla blue or lavender, suffused with white, ± regular, campanulate, much longer than calyx, glabrous, tube conspicuous. **Stamens** epipetalous. **Ovaries:** ovules 4. **Capsules** glabrous.

Flowering May–Jun; fruiting Jun–Jul. Forest openings of upper *Thuja-Pseudotsuga* zone; 500–800(–2000) m; Idaho.

Synthyris platycarpa is known from the mountains above the Selway River and North Fork of the Clearwater River in Nez Perce-Clearwater National Forests in Clearwater and Idaho counties.

14. Synthyris ranunculina Pennell, Proc. Acad. Nat. Sci. Philadelphia 85: 92. 1933 • Charleston Mountains kittentail C E

Veronica ranunculina (Pennell) M. M. Martínez Ort. & Albach

Leaves persistent, some withering in 2d year as new leaves expand; blade reniform to orbiculate, less than 25 mm wide, chartaceous, base cordate to lobate, margins incised-crenate, teeth apices rounded, surfaces glabrous; basal veins extending into distal ½ of blade, lateral veins 2–4 on each side of midvein. **Racemes** erect, to 10 cm in fruit; sterile bracts 3+, ovate-spatulate, largest less than 1 cm; flowers 5–15, loosely aggregated. **Sepals** 4. **Petals** (3 or)4 (or 5), apex entire or erose; corolla blue, ± regular,

campanulate, much longer than calyx, glabrous, tube inconspicuous. **Stamens** epipetalous. **Ovaries:** ovules 10–16. **Capsules** glabrous.

Flowering May–Jul; fruiting May–Aug. Rocky slopes; of conservation concern; 3000–3700 m; Nev.

Synthyris ranunculina is known from the Spring (Charleston) Mountains, Clark County.

15. Synthyris reniformis (Douglas ex Bentham) Bentham in A. P. de Candolle and A. L. P. P. de Candolle, Prodr. 10: 454. 1846 • Round-leaf kittentail, snow queen E

Wulfenia reniformis Douglas ex Bentham, Scroph. Ind., 46. 1835; *Veronica regina-nivalis* M. M. Martínez Ort. & Albach

Leaves persistent, some withering in 2d year as new leaves expand; blade cordate to reniform, 25+ mm wide, leathery, base cordate to lobate, margins incised-crenate, teeth apices obtuse to rounded, surfaces glabrous or hairy; basal veins extending into distal ½ of blade, lateral veins 2–4 on each side of midvein. **Racemes** decumbent, to 18 cm in fruit; sterile bracts usually 0; flowers 5–10, loosely aggregated. **Sepals** 4. **Petals** (3 or)4(or 5), apex entire or erose; corolla blue, ± regular, campanulate, much longer than calyx, puberulent-villous in throat, tube conspicuous. **Stamens** epipetalous. **Ovaries:** ovules 4. **Capsules** sparsely hairy.

Flowering Mar–Jun; fruiting Apr–Aug. Forests; 0–1500 m; Calif., Oreg., Wash.

16. Synthyris ritteriana Eastwood, Proc. Calif. Acad. Sci., ser. 3, 1: 123, plate 9, fig. 2. 1898 • Ritters' kittentail E

Besseya ritteriana (Eastwood) Rydberg; *Veronica ritteriana* (Eastwood) M. M. Martínez Ort. & Albach

Leaves strictly annual, disintegrating in 1st year; blade broadly elliptic to lanceolate, 25+ mm wide, leathery, base truncate to tapered, margins crenate, teeth apices acute to rounded, surfaces hairy; basal veins extending through proximal ½ of blade, lateral veins 5–12 on each side of midvein. **Racemes** erect, to 40 cm in fruit; sterile bracts 10–40, ovate-spatulate, largest 1+ cm; flowers 100+, densely aggregated (separating in fruit). **Sepals** 4. **Petals** 3(or 4), apex entire or erose; corolla yellow, bilabiate, ellipsoid, longer than calyx, puberulent on margins, lateral and abaxial petals of abaxial lip connate ½+ their lengths, abaxial and adaxial petal lips basally adnate to stamens, tube absent. **Stamens** inserted on receptacle, but abaxial and adaxial petal lips basally adnate to stamens. **Ovaries** glabrous or sparsely hairy at apex; ovules 17–40. **Capsules** glabrous.

Flowering May–Aug; fruiting Jun–Oct. Moist subalpine or alpine meadows; 2100–3800 m; Colo.

Synthyris ritteriana is known from the San Juan and La Plata mountains, and adjacent areas of Gunnison and Montrose counties on San Juan and Hansen tuffs.

17. Synthyris rubra (Douglas ex Hooker) Bentham in A. P. de Candolle and A. L. P. P. de Candolle, Prodr. 10: 455. 1846 • Red kittentail or coraldrops E

Gymnandra rubra Douglas ex Hooker, Fl. Bor.-Amer. 2: 103, plate 172. 1838; *Besseya rubra* (Douglas ex Hooker) Rydberg; *Veronica rubra* (Douglas ex Hooker) M. M. Martínez Ort. & Albach

Leaves strictly annual, dis-integrating in 1st year; blade oblong-ovate to ovate, 25+ mm wide, leathery, base cuneate, truncate, or rounded to lobate, sometimes cordate, margins crenate, teeth apices acute to rounded, surfaces sparsely hairy; basal veins extending into distal ½ of blade, lateral veins 3–6 on each side of midvein. **Racemes** erect, to 45 cm in fruit; sterile bracts 10–30, ovate-spatulate, largest 1+ cm; flowers 100+, densely aggregated (separating in fruit). **Sepals** 4, basal connation between abaxial and adaxial lobes on each side of flower. **Petals** 1–4, apex entire or erose; corolla reddish, bluish purple, green, or white, bilabiate or unilabiate, rudimentary, much shorter than calyx, glabrous, lateral and abaxial petals of abaxial lip connate ½+ their lengths, abaxial and adaxial petal lips basally adnate to stamens, tube absent. **Stamens** inserted on receptacle. **Ovaries:** ovules 17–40. **Capsules** densely hairy.

Flowering Mar–Jun; fruiting Apr–Jul. Grasslands, open coniferous forests; 200–1700 m; Idaho, Mont., Oreg., Wash.

Natural hybrids of *Synthyris missurica* and *S. rubra* occur near Kamiah, Idaho (A. R. Kruckeberg and F. L. Hedglin 1963).

18. Synthyris schizantha Piper, Bull. Torrey Bot. Club 29: 223. 1902 • Fringed kittentail E

Veronica schizantha (Piper) M. M. Martínez Ort. & Albach

Leaves persistent, some withering in 2d year as new leaves expand; blade ovate to reniform or orbiculate, 25+ mm wide, chartaceous, base cordate to lobate, margins incised-crenate, teeth apices obtuse to rounded, surfaces ± hairy; basal veins extending into

distal ½ of blade, lateral veins 2–4 on each side of midvein. **Racemes** erect, to 35 cm in fruit; sterile bracts 2, fan-shaped, largest 2+ cm; flowers 15–80, loosely aggregated. **Sepals** 4. **Petals** (3 or)4(or 5), apex laciniate; corolla blue to lavender with veins deeper colored (except tube yellowish white), ± regular, campanulate, much longer than calyx, glabrous, tube conspicuous. **Stamens** epipetalous. **Ovaries:** ovules 2–7. **Capsules** glabrous or sparsely hairy along margins.

Flowering May–Jun; fruiting May–Aug. Moist slopes, forest edges; 900–1400 m; Oreg., Wash.

Synthyris schizantha is known from the southern Olympic Mountains in Grays Harbor County and the Cascade Mountains in Lewis County, Washington, and in the vicinity of Saddle Mountain in Clatsop County, Oregon.

19. Synthyris wyomingensis (A. Nelson) A. Heller, Muhlenbergia 1: 5. 1900 • Wyoming kittentail E

Wulfenia wyomingensis A. Nelson, Bull. Torrey Bot. Club 25: 281. 1898; *Besseya wyomingensis* (A. Nelson) Rydberg; *Veronica wyomingensis* (A. Nelson) M. M. Martínez Ort. & Albach

Leaves strictly annual, dis-integrating in 1st year; blade broadly to narrowly lanceolate to ovate, 25+ mm wide, leathery, base cuneate, rounded, lobate, or cordate, margins crenate or incised-crenate, teeth apices acute to rounded, surfaces hairy; basal veins extending into distal ½ of blade, lateral veins 3–8 on each side of midvein. **Racemes** erect, to 47 cm in fruit; sterile bracts 10–25, ovate to spatulate, largest 1+ cm; flowers 100+, densely aggregated (separating in fruit). **Sepals** 2–4, all lobes connate, if 2+ lobes, then connate for at least ½ of length on abaxial side. **Petals** 0. **Stamens** inserted on receptacle. **Ovaries:** ovules 17–40. **Capsules** densely hairy.

Flowering Apr–Jul; fruiting May–Aug. Montane, subalpine, and alpine grasslands, open, coniferous forests; 1000–3700 m; Alta., B.C., Sask.; Colo., Idaho, Mont., Nebr., S.Dak., Utah, Wyo.

Synthyris wyomingensis is highly variable over its wide elevational and geographic ranges. Populations in the northern and western part of the range have bluish purple stamens; those in the southeastern part of the range have white to whitish yellow stamens. Tetraploid ($2n = 48$) populations are known only from the Bridger Mountains, Montana; populations from elsewhere are diploid $2n = 24$ (C. G. Schaack 1983).

44. **VERONICA** Linnaeus, Sp. Pl. 1: 9. 1753; Gen. Pl. ed. 5, 10. 1754 • Speedwell [Late Latin form of Greek *Berenike/Pherenike*, *phero*, bearer, and *nike*, victory, probably alluding to Saint Veronica]

Dirk C. Albach

Herbs or subshrubs (*V. fruticans*) [**shrubs**], perennial, biennial (*V. triphyllos*), or annual; rhizomes creeping or absent. **Stems** creeping to erect, glabrous or hairy, glandular or eglandular. **Leaves** cauline, rarely in basal rosettes, opposite, sometimes alternate distally, rarely proximally (*V. fruticans*); petiole absent or present; blade not fleshy, not leathery (except *V. fruticans*), margins entire, dentate, serrate, 3–5-pinnately lobed, 3–7 pinnatifid, or ± palmatifid. **Inflorescences** terminal and/or axillary, racemes; bracts present, sometimes leaflike. **Pedicels** absent or present; bracteoles absent. **Flowers** bisexual; sepals 4(or 5), basally connate, calyx bilaterally symmetric, rotate to campanulate, lobes linear to suborbiculate or triangular-cordate; corolla light to deep blue with lighter or reddish center, violet, lavender, or purple to pink with darker center, or white, bilaterally symmetric, weakly bilabiate, rotate or short-tubular, sometimes campanulate, tube base not spurred or gibbous, lobes 4, abaxial 3, adaxial 1; stamens 2, basally adnate to corolla, filaments glabrous; staminode 0; ovary 2-locular, placentation axile; stigma pointed to capitate. **Fruits** capsules, dehiscence loculicidal. **Seeds** (1–)4–140, yellow to brown, rarely blackish (*V. triphyllos*), planoconvex to urn-shaped, wings absent. $x = 9$.

Species ca. 450 (34 in the flora): North America, Mexico, Central America, South America, Eurasia, Africa, Atlantic Islands, Pacific Islands, Australia.

Veronica includes many horticultural and weedy plants. Introduction and subsequent naturalization of species is often a possibility. The classification by D. C. Albach et al. (2004b) recognized 13 subgenera within *Veronica*; this was reduced to 12 by P. J. Garnock-Jones et al. (2007). Of these 12, ten are represented in the flora area. Species in this treatment are ordered by their placement in these subgenera as follows: subg. *Pseudolysimachium* (W. D. J. Koch) Buchenau (species 1 and 2, type *V. spicata*), subg. *Veronica* Linnaeus (species 3–10, type *V. officinalis*), subg. *Beccabunga* (Hill) M. M. Martínez Ortega, Albach & M. A. Fischer (species 11–17, type *V. beccabunga*), subg. *Pellidosperma* (E. B. J. Lehman) M. M. Martínez Ortega, Albach & M. A. Fischer (species 18, type *V. praecox* Allioni), subg. *Cochlidiosperma* (Reichenbach) M. M. Martínez Ortega, Albach & M. A. Fischer (species 19–22, type *V. hederifolia*), subg. *Chamaedrys* (W. D. J. Koch) M. M. Martínez Ortega, Albach & M. A. Fischer (species 23–26, type *V. chamaedrys*), subg. *Stenocarpon* (Borissova) M. M. Martínez Ortega, Albach & M. A. Fischer (species 27, type *V. ciliata* Fischer), subg. *Pocilla* (Dumortier) M. M. Martínez Ortega, Albach & M. A. Fischer (species 28–33, type *V. agrestis*), and subg. *Pentasepalae* (Bentham) M. M. Martínez Ortega, Albach & M. A. Fischer (species 34, type *V. teucrium*).

One of the subgenera, subg. *Synthyris* (Bentham) M. M. Martínez Ortega, Albach & M. A. Fischer, includes species treated here as the genus *Synthyris*. Other taxa included in *Veronica* but sometimes recognized at the generic rank are *Pseudolysimachion* (W. D. J. Koch) Opiz (D. C. Albach 2008) and *Hebe* Commerson ex Jussieu (P. J. Garnock-Jones et al. 2007).

Subgenus *Pseudoveronica* J. B. Armstrong includes *Veronica franciscana* Eastwood [= *Hebe franciscana* (Eastwood) Souster] and *V. speciosa* R. Cunningham ex A. Cunningham [= *H. speciosa* (R. Cunningham ex A. Cunningham) Andersen], which have been introduced to California as ornamentals that might escape.

Characters known to vary in *Veronica* are the indument (to densely white tomentose in *V. incana* Linnaeus), leaf shape, and flower color. Among many of the creeping species of the genus (for example, *V. agrestis*, *V. filiformis*, *V. hederifolia*, *V. persica*), most leaflike organs on the stems are bracts. The genus also includes a number of ornamentally grown plants, most of them more or less closely related to *V. teucrium*. They differ mainly in habit and leaf shape and should key out close to *V. chamaedrys*, *V. officinalis*, or *V. teucrium*, but none of them have been reported to persist outside their native range.

SELECTED REFERENCES Albach, D. C. and H. M. Meudt. 2010. Phylogeny of *Veronica* in the southern and northern hemispheres based on plastid, nuclear ribosomal and nuclear low-copy DNA. Molec. Phylogen. Evol. 54: 457–471.   Albach, D. C., P. Schönswetter, and A. Tribsch. 2006. Comparative phylogeography of the *Veronica alpina* complex in Europe and North America. Molec. Ecol. 15: 3269–3286.

1. Racemes 1–25, axillary.
    2. Stems hairy, rarely glabrate or glabrous; leaf blades 1–3 times as long as wide, rarely to 4 times as long (*V. teucrium*); meadows, forests, and other dry habitats.
        3. Leaf blades elliptic-ovate, ovate, ovate-oblong, or obovate, bases cuneate to attenuate or obtuse.
            4. Pedicels 8–10 mm, longer than subtending bracts; arctic . . . . .8. *Veronica grandiflora*
            4. Pedicels (0.5–)1–2(–3) mm, shorter than subtending bracts; not arctic . . . . .9. *Veronica officinalis*
        3. Leaf blades ± ovate, ovate-orbiculate, or oblong-ovate, rarely lanceolate, bases truncate to ± cordate.
            5. Stems ascending, rarely erect, (7–)10–30(–50) cm, hairs along stem in 2 prominent lines; calyces 4-lobed; styles (2.5–)4–5 mm. . . . . .26. *Veronica chamaedrys*
            5. Stems erect, rarely ascending, (15–)30–70(–100) cm, hairs evenly distributed around stem; calyces 5-lobed; styles 5–6 mm . . . . 34. *Veronica teucrium*
    2. Stems glabrous or, if hairy, then only distally; leaf blades 1.5–10 times as long as wide; wet meadows, marshes, or aquatic habitats.
        6. Petioles 2–6(–10) mm.
            7. Leaf blades widest near or distal to middle; styles (1.3–)1.5–3(–3.5) mm . . . .
            . . . . 12. *Veronica beccabunga*
            7. Leaf blades widest proximal to middle; styles (1.7–)2.5–3.5(–4) mm . . . . 13. *Veronica americana*
        6. Petioles 0(–5) mm or proximals to 8 mm and distals 0 mm.
            8. Capsules compressed in cross section, apices emarginate by ⅓ length; leaf blades narrowly lanceolate or linear, rarely elliptic. . . . . .10. *Veronica scutellata*
            8. Capsules ± compressed or ± round in cross section, apices rounded or ± emarginate; leaf blades ovate-oblong, ovate, oblong, linear-lanceolate, lanceolate, elliptic, oblong-ovate, or oblong-lanceolate.
                9. Corollas white to pale pink; calyx lobe apices obtuse; stamens 5 mm; pedicels equal to ± shorter than subtending bract; leaf margins entire or subentire . . . . .15. *Veronica catenata*
                9. Corollas pale lilac to pale blue or lavender, rarely white; calyx lobe apices acute; stamens 2–3.5 mm; pedicels equal to longer than subtending bract; leaf margins crenate, serrulate, or denticulate to subentire, sometimes serrate or ± undulate (*V. undulata*).
                    10. Pedicels suberect or arcuate-erect to subpatent; racemes 5–10 mm diam., axes glabrous, rarely glandular-hairy . . . . 14. *Veronica anagallis-aquatica*
                    10. Pedicels patent; racemes 10–15 mm diam., axes sparsely glandular-hairy, rarely glabrous. . . . . 16. *Veronica undulata*
1. Racemes 1(–20), terminal, sometimes also with 1–4 axillary racemes, flowers sometimes appearing solitary.
    11. Perennials; leaf blades (5–)7–40(–150) mm (usually at least 3 mm longer than bracts); bracts 1–9(–15) mm.
        12. Corollas campanulate; racemes (50–)70–200(–300) mm, (50–)100–300-flowered.
            13. Stems usually with 10–20 nodes, (50–)70–120(–150) cm; leaf blade margins serrate to biserrate, apices acute. . . . . .1. *Veronica longifolia*
            13. Stems usually with 4–10 nodes, 5–45(–60) cm; leaf blade margins shallowly crenate to shallowly serrate to subentire, apices obtuse to rounded . . . . 2. *Veronica spicata*
        12. Corollas rotate; racemes (± loose), 5–100(–130) mm, (1–)2–30(–60)-flowered.

14. Capsules wider than long; stems with scattered eglandular hairs only, often also with glandular hairs, sometimes glabrate . . . . 11. *Veronica serpyllifolia*

14. Capsules ca. as long as or longer than wide; stems eglandular- or glandular-hairy.

 15. Subshrubs; stems densely branched from woody base . . . . 27. *Veronica fruticans*

 15. Herbs; stems unbranched or sparsely branched.

  16. Stamens 1–2.3 mm; corollas 2.5–11 mm diam.; styles 0.5–4(–6) mm.

   17. Raceme axes eglandular-hairy or glabrate; pedicels 1.5–2(–4) mm; styles 0.5–1.5(–2) mm. . . . . 3. *Veronica alpina*

   17. Raceme axes glandular-hairy; pedicels 2–10(–15) mm; styles 0.8–4(–6) mm . . . . 4. *Veronica wormskjoldii*

  16. Stamens 4–8 mm; corollas (8–)10–13 mm diam.; styles 3–10 mm.

   18. Leaf blade surfaces glabrous or glabrate . . . . 5. *Veronica cusickii*

   18. Leaf blade surfaces hairy.

    19. Leaf blade margins entire; stems glandular-hairy . . . . 6. *Veronica copelandii*

    19. Leaf blade margins dentate or serrate; stems eglandular-hairy . . . . 7. *Veronica stelleri*

[11. Shifted to left margin.—Ed.]

11. Mostly annuals (*V. triphyllos* sometimes biennial, *V. filiformis* perennial); leaf blades (2–)5–28(–35) mm; bracts 3–25 mm (if bracts less than 9 mm, leaf blades usually less than 10 mm).

20. Pedicels 0–4(–5) mm.

 21. Corollas white or pale pink; leaf blades 3–10 times as long as wide, margins entire or dentate distally . . . . 17. *Veronica peregrina*

 21. Corollas ± blue; leaf blades 1–2.5 times as long as wide, margins crenate-serrate, pinnatifid, or subpalmatifid.

  22. Leaf margins crenate-serrate. . . . . 23. *Veronica arvensis*

  22. Leaf margins, at least distal leaves, pinnatifid or ± palmatifid.

   23. Corollas 1.5–3 mm diam.; styles 0.2–0.6 mm, stigmas white . . . . 24. *Veronica verna*

   23. Corollas 4–6 mm diam.; styles 0.8–1.5 mm, stigmas violet . . . . 25. *Veronica dillenii*

20. Pedicels (2–)4–30 mm.

 24. Leaf blade margins 3–5(–7)-palmatifid . . . . 18. *Veronica triphyllos*

 24. Leaf blade margins serrate, dentate, crenate-serrate, or 3–9-lobed.

  25. Styles (1.5–)2–4 mm; pedicels (12–)15–30(–38) mm; corollas ± blue, 8–14 mm diam.

   26. Capsules reticulate with prominent veins, sinus angle (80–)90–120(–150)°; pedicels (12–)15–27(–38) mm, length 1–2(–3) times subtending bracts . . . . 29. *Veronica persica*

   26. Capsules absent or almost smooth, sinus angle 50–90°; pedicels (15–)20–30 mm, length 2–5 times subtending bracts . . . . 31. *Veronica filiformis*

  25. Styles 0.2–2 mm; pedicels (2–)3–15(–30) mm; corollas ± blue, white, pale lilac, pinkish, or pale violet, 2–12 mm diam.

   27. Stems erect.

    28. Bract margins: proximals serrate, distals sometimes entire; leaf blades (6–)12–28(–35) mm; styles (0.7–)1–1.3(–1.7) mm. . . . . 32. *Veronica argute-serrata*

    28. Bract margins entire; leaf blades (3–)4–12(–20) mm; styles 0.4–0.8(–2) mm. . . . . 33. *Veronica biloba*

   27. Stems creeping, decumbent, or ascending, rarely erect in young plants (*V. sublobata*).

    29. Leaf blade margins dentate to serrate; capsules ± compressed or ± round in cross section, apices emarginate.

     30. Calyx lobes broadly ovate, 2.6–3.8 mm wide; capsules densely eglandular-hairy (hairs less than 0.1 mm), sometimes also glandular-puberulent (hairs 0.2–0.3 mm), rarely glabrous; corollas intense dark to bright blue, rarely pale lilac or white . . . . 28. *Veronica polita*

     30. Calyx lobes linear-lanceolate, 2.2–2.6 mm wide; capsules sparsely glandular-hairy; corollas white or pale pinkish or pale blue . . . 30. *Veronica agrestis*

[29. Shifted to left margin.—Ed.]

29. Leaf blade margins 3–5(–9)-lobed; capsules ± round in cross section, apices ± emarginate.
    31. Corollas white; calyx lobes not ciliate . . . . .22. *Veronica cymbalaria*
    31. Corollas blue or pale violet to whitish; calyx lobes ciliate.
        32. Corollas deep blue with white center; calyx lobe abaxial surfaces puberulent, rarely glabrous; pedicel length 1–2 times calyx . . . . 21. *Veronica triloba*
        32. Corollas bright blue with bright white center or pale violet to whitish; calyx lobe abaxial surfaces sparsely hairy or glabrous; pedicel length 2–4(–6) times calyx.
            33. Corollas bright blue with bright white center, 5–7(–9) mm diam.; styles (0.6–)0.7–0.9(–1.1) mm; pedicel length 2–3 times calyx, eglandular-hairy in single line adaxially . . . . 19. *Veronica hederifolia*
            33. Corollas pale violet to whitish, 4–5(–6) mm diam.; styles 0.2–0.7 mm; pedicel length 3–4(–6) times calyx, densely eglandular-hairy adaxially . . . . .20. *Veronica sublobata*

1. Veronica longifolia Linnaeus, Sp. Pl. 1: 10.
1753 • Long-leaved speedwell, véronique à longues feuilles [I]

Veronica maritima Linnaeus

Perennials. Stems erect, usually with 10–20 nodes, (50–)70–120(–150) cm, proximal ½ glabrous, distal ½ eglandular-hairy with some glandular hairs intermixed. **Leaves:** blade narrowly ovate-lanceolate, almost linear, or narrowly ovate, (60–)80–120(–150) × 10–30(–35) mm, base cuneate, truncate, or cordate, margins serrate to biserrate, apex acute, surfaces sparsely or densely eglandular-hairy, rarely glabrous. **Racemes** 1–7(–9), terminal and occasionally axillary, 100–200 (–300) mm, 150–300-flowered, axis eglandular- and glandular-hairy; bracts linear to subulate, 2–6 mm. **Pedicels** erect, (1–)1.5–3(–5) mm, shorter than subtending bract, densely eglandular-hairy, sometimes also glandular-hairy. **Flowers:** calyx lobes 2–2.5 (–4) mm, apex acute, eglandular-hairy, sometimes glandular-hairy; corolla blue, campanulate, longer than wide, 5–7 mm diam.; stamens 5–6 mm; style (5–)6–8 (–10) mm. **Capsules** ± compressed in cross section, ovoid, 2.5–3 × 2.5–3 mm, apex emarginate, glabrous. **Seeds** 2–40, light brown, ellipsoid, flat, 0.7–1.2 × 0.5–0.8 mm, 0.2–0.3 mm thick, smooth. *2n* = 34, 68 (Eurasia).

Flowering Jun–Aug(–Sep). Moist (swampy) habitats, roadsides; 0–800 m; introduced; Alta., B.C., Man., N.B., Nfld. and Labr. (Nfld.), N.W.T., N.S., Ont., P.E.I., Que., Sask., Yukon; Alaska, Conn., Fla., Ill., Ind., Iowa, Ky., Maine, Md., Mass., Mich., Minn., Mo., Mont., N.H., N.J., N.Y., N.Dak., Ohio, Pa., R.I., Vt., Wash., W.Va., Wis.; Eurasia; introduced also in South America, Pacific Islands (New Zealand).

Reports of *Veronica bachofenii* Heuffel, *V. grandis* Fischer ex Sprengel (= *V. daurica* Steven), or *V. spuria* Linnaeus, seem always to refer to garden varieties of *V. longifolia*. Garden varieties often differ from typical plants in having glandular hairs and broader leaves, which may indicate past hybridization with other species.

Subgenus *Pseudolysimachium* includes *Veronica longifolia* and *V. spicata*, which are naturalized in the flora area, but also other species and hybrids of these, which are commonly grown ornamentally and may escape from cultivation. All of these plants should easily key out with *V. longifolia* and *V. spicata* but differ in specifics of the species descriptions.

2. Veronica spicata Linnaeus, Sp. Pl. 1: 10. 1753
• Spiked speedwell, véronique en épi [I]

Perennials. Stems ascending, usually with 4–10 nodes, 5–45 (–60) cm, proximally eglandular- or glandular-hairy, sometimes glabrous, distally eglandular- and, usually, glandular-hairy. **Leaves:** blade ovate-oblong to narrowly oblong-lanceolate, (30–)40–70(–80) × 5–20 mm, base long-cuneate, margins shallowly crenate to shallowly serrate to subentire, apex obtuse to rounded, surfaces densely glandular- or eglandular-hairy, rarely glabrate. **Racemes** 1(–7), terminal, sometimes with lateral ones, (50–)70–130(–170) mm, (50–)100–300-flowered, axis eglandular- and glandular-hairy; bracts oblong to lanceolate, 3–5 mm. **Pedicels** suberect to patent, 0.5–1(–2) mm, shorter than subtending bract, eglandular- and glandular-hairy. **Flowers:** calyx lobes 2–3 mm, ciliate, apex obtuse, glandular-hairy, rarely eglandular-hairy; corolla blue, campanulate, longer than wide, 5–6 mm diam., lobes 3–4.5 × 1.5–2 mm; stamens 5 mm; style 8 mm. **Capsules** ± compressed in cross section, broadly ovoid, 2–3 × 2–3 mm, apex obtuse, densely glandular-puberulent. **Seeds** 3–40, light brown, ellipsoid, flat, 0.6–1.2 × 0.5–0.8 mm, 0.2–0.4 mm thick, smooth. *2n* = 34, 68 (Europe).

Flowering Jun–Sep. Dry grasslands; 0–700 m; introduced; Ont., Que.; Conn., N.H., N.Y.; Eurasia.

Veronica spicata is widely distributed in horticulture and a multitude of cultivars is available. Some specimens may also be derived from a related species, *V. barrelieri* Schott ex Roemer & Schultes, differentiated from *V. spicata* by glabrous calyx lobes, and in its typical variety, by eglandular pubescence. Another closely related species, *V. incana* Linnaeus, has a dense white woolly indumentum and is native to northern Asia and eastern Europe; it has not escaped from cultivation in the flora area.

3. **Veronica alpina** Linnaeus, Sp. Pl. 1: 11. 1753
   • Alpine speedwell, véronique alpine

Veronica alpina subsp. *australis* (Wahlenberg) Á. Löve & D. Löve; *V. alpina* var. *australis* Wahlenberg; *V. pumila* Allioni

Perennials. Stems ascending, often dark bluish distally, sparsely branched at base, 5–15(–25) cm, eglandular-hairy. **Leaves:** blade ovate or oblong-elliptic, 7–30 × 4–20 mm, base cuneate, margins indistinctly dentate or entire, apex short-acuminate, abaxial surface glabrate, adaxial ± hairy. **Racemes** 1, terminal, 5–20 mm, to 30 mm in fruit, (1–)3–20-flowered, axis eglandular-hairy; bracts oblanceolate, 4–5 mm. **Pedicels** ascending to erect, 1.5–2(–4) mm, shorter than subtending bract, eglandular-hairy. **Flowers:** calyx lobes 3–4 mm, ciliate, apex obtuse or acuminate, eglandular-hairy; corolla sky blue or blue-violet, sometimes white, rotate, 2.5–5.5 mm diam.; stamens 2–2.3 mm; style 0.5–1.5(–2) mm. **Capsules** compressed in cross section, obovoid or oblong-obovoid, 4.5–7.5 × 3.5–5.5 mm, ca. as long as wide, apex shallowly emarginate, eglandular-hairy or glabrate. **Seeds** 9–53, brown to yellow, ellipsoid-oblong, flat, 0.7–1.2 × 0.4–1 mm, 0.1 mm thick, smooth. $2n = 18$.

Flowering (Jun–)Jul–Aug. Slopes, moist rocks, hillocks, moist alpine and subalpine meadows; 0–1000 m; Greenland; Nunavut, Que.; Eurasia.

4. **Veronica wormskjoldii** Roemer & Schultes in J. J. Roemer et al., Syst. Veg. 1: 101. 1817 (as wormskjoldi) • American alpine speedwell [E] [F]

Veronica alpina Linnaeus subsp. *wormskjoldii* (Roemer & Schultes) Elenevsky; *V. alpina* var. *wormskjoldii* (Roemer & Schultes) Hooker

Perennials. Stems erect or ascending, light grayish green distally, unbranched, (3–)8–50 cm, sparsely to densely villous-hirsute, sometimes also glandular-hairy. **Leaves:** blade elliptic to lanceolate or oblong-ovate, 8–40 × 5–20 mm, base cuneate, margins entire, dentate, or serrate, apex short-acuminate, surfaces sparsely to densely villous-hirsute or glabrous. **Racemes** 1, terminal, 5–40 mm, to 60(–150) mm in fruit, (2–)5–25-flowered, axis densely villous-hirsute and ± glandular-hairy; bracts linear to lanceolate, 1–8 mm. **Pedicels** erect, 2–10(–15) mm, ca. equal to subtending bract, densely villous-hirsute or glandular-hairy. **Flowers:** calyx lobes (2.5–)3–5.5(–7) mm, apex obtuse or acuminate, glabrous or densely glandular-hairy; corolla deep blue-violet or violet-purple to deep blue, rotate, 3–11 mm diam.; stamens 1–2 mm; style 0.8–4(–6) mm. **Capsules** compressed in cross section, oblong-obovoid, 4–6(–8) × (2.8–)4–5.5 mm, ca. as long as wide, apex emarginate, usually densely glandular-hairy. **Seeds** 10–50, straw colored, ellipsoid, flat, 0.7–1 × 0.4–1 mm, 0.05–1 mm thick, very minutely striate.

Subspecies 2 (2 in the flora): North America.

Veronica wormskjoldii occurs in two cytotypes, a diploid ($2n = 18$) in western North America and a tetraploid ($2n = 36$) in eastern North America and Greenland. As demonstrated by D. C. Albach et al. (2006), the tetraploid is likely to be a hybrid of the diploid *V. alpina* and *V. wormskjoldii*. In that publication, species rank was used for simplicity, not as a taxonomic conclusion. However, subsequent morphological analyses in conjunction with preparing this treatment revealed that the large variation in the diploid *V. wormskjoldii*, also supported by the large number of varieties established within that taxon, make it nearly impossible to differentiate the two North American taxa morphologically. In cases where two taxa are morphologically undifferentiable, but karyologically and genetically distinct and geographically clearly separate, the rank of subspecies seems more appropriate.

1. Pedicels 2–5 mm; e North America . . . .
    . . . . 4a. *Veronica wormskjoldii* subsp. *wormskjoldii*
1. Pedicels (2–)5–10(–15) mm; w North America
    . . . . 4b. *Veronica wormskjoldii* subsp. *nutans*

4a. **Veronica wormskjoldii** Roemer & Schultes subsp. **wormskjoldii** • Eastern American alpine speedwell, véronique de Wormskjöld [E] [F]

Veronica alpina Linnaeus var. *terrae-novae* Fernald

Pedicels 2–5 mm. $2n = 36$.

Flowering Jul–Aug. Moist meadows, stream banks, bogs, moist open slopes; 0–1500 m; Greenland; Nfld. and Labr., Nunavut, Ont., Que.; Maine, N.H.

4b. Veronica wormskjoldii Roemer & Schultes subsp. **nutans** (Bongard) Pennell, Rhodora 23: 15. 1921 • Western American alpine speedwell [E]

Veronica nutans Bongard, Mem. Acad. Imp. Sci. St.-Petersbourg, Ser. 6, Sci. Math. 2: 157. 1832; *V. alpina* Linnaeus var. *alterniflora* Fernald; *V. alpina* var. *cascadensis* Fernald; *V. alpina* var. *geminiflora* Fernald; *V. alpina* var. *nutans* (Bongard) Ledebour; *V. alpina* var. *unalaschcensis* Chamisso & Schlechtendal; *V. wormskjoldii* subsp. *alterniflora* (Fernald) Pennell

Pedicels (2–)5–10(–15) mm. *2n* = 18.

Flowering Jun–Aug. Moist alpine meadows, heathlands, and scree, bogs, fens, stream banks, pond and lake margins; 0–3500 m; Alta., B.C., N.W.T., Nunavut, Yukon; Alaska, Ariz., Calif., Colo., Idaho, Mont., Nev., N.Mex., Oreg., Utah, Wash., Wyo.

Hybrids of subsp. *nutans* with *Veronica stelleri* show morphological intermediacy. Especially plants on the Alaskan coast may have serrate leaves as in *V. stelleri* and larger corollas as in *V. wormskjoldii*. The degree and direction of introgression are unknown.

5. Veronica cusickii A. Gray in A. Gray et al., Syn. Fl. N. Amer. 2(1): 288. 1878 • Cusick's speedwell [E]

Veronica allenii Greenman

Perennials. Stems erect, unbranched or sparsely branched at base, (5–)10–15(–30) cm, finely glandular-hairy. **Leaves:** blade elliptic-ovate, (5–)10–25 × 5–10 mm, base cuneate, margins entire or subentire, apex short-acuminate, surfaces glabrous or glabrate. **Racemes** 1, terminal, (15–)40–80(–130) mm, (4–)8–30(–40)-flowered, axis finely glandular-hairy; bracts lanceolate, 3–7(–11) mm. **Pedicels** erect, 3–9 mm, equal to or ± longer than subtending bract, finely glandular-hairy. **Flowers:** calyx lobes 2–3 mm, apex acute to obtuse, finely glandular-hairy; corolla deep blue-violet, rarely white, rotate, (8–)10–13 mm diam.; stamens 4–8 mm; style (5–)6–9(–10) mm, conspicuously exerted. **Capsules** compressed in cross section, oblong-obovoid, 4–6 × 3.5–5.5 mm, ca. as long as wide, apex emarginate, finely glandular-hairy. **Seeds** 28–32, straw colored or dark brown, ovoid, flat, 0.6 × 0.4 mm, 0.1 mm thick, rough. *2n* = 18, 72.

Flowering Jun–Aug. Gravelly soils, openings in coniferous forests and in subalpine and alpine meadows; 1500–3200 m; B.C.; Calif., Idaho, Mont., Oreg., Wash.

Veronica cusickii is sister to *V. copelandii* and closely related to *V. wormskjoldii* subsp. *nutans* (D. C. Albach et al. 2006).

6. Veronica copelandii Eastwood, Bot. Gaz. 41: 288, fig. 2. 1906 (as copelandi) • Copeland's speedwell [E]

Perennials. Stems ascending, unbranched, 5–15 cm, densely glandular-hairy. **Leaves:** blade oblong-elliptic, (5–)10–15(–35) × 4–8 mm, base cuneate, margins entire, apex short-acuminate, surfaces hairy. **Racemes** 1, terminal, distalmost leaves often with 1 or 2 axillary flowers, 10–80 mm, (3–)5–15-flowered, axis glandular-hairy; bracts lanceolate, 3–5 mm. **Pedicels** erect, 5–8 mm, equal to ± longer than subtending bract, densely glandular-hairy. **Flowers:** calyx 4(or 5)-lobed, lobes (1–)2–3 mm, apex obtuse, glandular-hairy; corolla pale blue to purple, rotate, 8–10 mm diam.; stamens 4–5 mm; style 7 mm. **Capsules** compressed in cross section, broadly oblong, 5–6 × 3.5 mm, longer than wide, apex emarginate, glandular-hairy. **Seeds** number unknown, brown, ovoid, flat, 1–1.2 × 0.7–1.1 mm, thickness and texture unknown. *2n* = 18.

Flowering Jul–Aug. Subalpine meadows, alpine slopes; 1900–2500 m; Calif.

Veronica copelandii is sister to *V. cusickii* in the phylogenetic analysis by D. C. Albach et al. (2006). It occurs in the Klamath Ranges of northwestern California.

7. Veronica stelleri Link, Jahrb. Gewächsk. 1(3): 40. 1820 • Steller's speedwell

Veronica stelleri var. *glabrescens* Hultén; *V. wormskjoldii* Roemer & Schultes var. *stelleri* (Link) S. L. Welsh

Perennials. Stems erect to ascending, unbranched, (5–)10–20(–25) cm, eglandular-hairy. **Leaves:** blade ovate, 15–30 × 10–20 mm, base rounded, margins dentate or serrate, apex subacute to subobtuse, surfaces eglandular-hairy. **Racemes** 1, terminal, 10–60 mm, to 100 mm in fruit, 5–15-flowered, axis hairy; bracts lanceolate, 2–5 mm. **Pedicels** erect, 2–11 mm, equal to longer than subtending bract, crisp-hairy. **Flowers:** calyx lobes 4–5 mm, eglandular-ciliate, apex acuminate, glabrous or sparsely hairy; corolla pale blue or violet, rotate, 8–10 mm diam.; stamens 7 mm; style 3–6 mm. **Capsules** compressed in cross section, oblong, 5–6(–8) × 3–4.5 mm, longer than wide, apex ± emarginate, glabrous proximally, hairy distally. **Seeds** number unknown, brown, ovoid, flat, 1–1.5 × 0.7–1.2 mm, thickness and texture unknown. *2n* = 18 (Japan, Russia).

Flowering Jul–Sep. Dry to moderately moist slopes, meadows in alpine regions, moraines; 0–1800 m; Alaska; Asia (Japan, Korea, Russia).

Veronica stelleri may be difficult to differentiate from 4b. *V. wormskjoldii* subsp. *nutans* where sympatric.

8. **Veronica grandiflora** Gaertner, Novi Comment. Acad. Sci. Imp. Petrop. 14(1): 531, plate 18, fig. 1. 1770
   • Large-flowered speedwell

Veronica aphylla Linnaeus var. *grandiflora* (Gaertner) Vahl; *V. grandiflora* var. *minor* Hultén

Perennials. Stems prostrate to erect, unbranched, (3–)5–15 (–20) cm, viscid-hairy. **Leaves:** blade elliptic-ovate, 20–45 × 15–30 mm, 1–3 times as long as wide, base cuneate, margins ± crenate-serrate, apex acute to rounded, abaxial surface with articulate hairs or glabrous, adaxial hirsute, or surfaces sparsely ciliate only. **Racemes** 1–4, axillary, 50–150 mm, 3–8-flowered, axis viscid-hairy; bracts oblong, 3–5 mm. **Pedicels** suberect, 8–10 mm, longer than subtending bract, viscid-hairy. **Flowers:** calyx lobes 4 mm, apex obtuse, hairy; corolla bright blue to violet-blue, 8–9 mm diam.; stamens 7 mm; style 7–9 mm. **Capsules** compressed in cross section, obovoid, 9–11 × 7–8 mm, apex emarginate, sparsely hairy. **Seeds** unknown. $2n$ = ca. 48–50 (Russia).

Flowering Jul–Aug. Meadows, between stones, slopes, alpine slopes; 0–600 m; Alaska; Asia (Japan, Russia).

9. **Veronica officinalis** Linnaeus, Sp. Pl. 1: 11. 1753
   • Common speedwell, véronique officinale ⊡

Perennials. Stems creeping, decumbent to ascending distally, 10–40(–50) cm, densely eglandular-hairy, rarely glabrate or glabrous. **Leaves:** blade ovate, ovate-oblong, or obovate, (8–)20–35(–50) × (3–)10–20 (–30) mm, 1.5–2 times as long as wide, base attenuate to cuneate or obtuse, margins dentate, denticulate, or serrulate, apex obtuse to rounded, surfaces sparsely to densely eglandular-hairy, rarely glabrate. **Racemes** 1–4(–8), axillary, 40–60 mm, 10–35(–40)-flowered, axis densely to sparsely eglandular-hairy, sometimes also glandular-hairy, rarely glabrate; bracts linear-lanceolate or narrowly ovate, (1.5–)3–4(–5) mm. **Pedicels** erect, (0.5–)1–2(–3) mm, shorter than subtending bract, densely eglandular-hairy, rarely glandular-hairy or glabrate. **Flowers:** calyx 4(or 5)-lobed, lobes 2–3.5 (–4) mm, apex obtuse, glandular- and eglandular-hairy;

corolla pale blue-lilac to pale pink, rarely white with pink veins, 6–8 mm diam.; stamens 2.8–3.4 mm; style 2.5–3(–4) mm. **Capsules** compressed in cross section, triangulate-obdeltoid to obcordiform, 3–5 × 4–5 mm, apex rounded or truncate, rarely emarginate, glandular-hairy, rarely glabrous. **Seeds** 9–24, brown to yellow, ellipsoid, ovoid, or obovoid, flat, 0.8–1.6 × (0.3–) 0.6–1.2 mm, 0.2–0.3 mm thick, smooth. $2n$ = 18, 36 (Europe).

Flowering (Apr–)Jun–Jul(–Aug). Forests, meadows, pastures, scrub, ruderal places; 0–1600 m; introduced; Greenland; St. Pierre and Miquelon; B.C., N.B., Nfld. and Labr. (Nfld.), N.S., Ont., P.E.I., Que.; Calif., Conn., Del., D.C., Ga., Idaho, Ill., Ind., Iowa, Kans., Ky., Maine, Md., Mass., Mich., Minn., Mont., N.H., N.J., N.Y., N.C., N.Dak., Ohio, Oreg., Pa., R.I., S.C., S.Dak., Tenn., Vt., Va., Wash., W.Va., Wis., Wyo.; Eurasia; introduced in s South America (Chile).

Veronica ×*tournefortii* Villars [*V. officinalis* var. *tournefortii* (Villars) Dumortier] refers to the hybrid between *V. officinalis* and *V. allionii* Villars in Europe and is not applicable to any North American plant.

10. **Veronica scutellata** Linnaeus, Sp. Pl. 1: 12. 1753
    • Marsh speedwell, véronique en écusson

Veronica connata Rafinesque; *V. scutellata* var. *villosa* Schumacher

Perennials. Stems ascending, (5–)15–45(–80) cm, glabrous, sometimes sparsely eglandular-hairy distally. **Leaves:** petiole 0 mm; blade narrowly lanceolate or linear, rarely elliptic, (10–)30–50(–80) × (2–)3–7(–14) mm, 6–10 times as long as wide, base cuneate, margins remote- and fine-dentate, apex acute, abaxial surface glabrous, adaxial glabrous or sparsely eglandular-hairy. **Racemes** 1–5 (–10), axillary, 40–90(–150) mm, (2–)5–20-flowered, axis glabrous, sometimes sparsely eglandular-hairy; bracts lanceolate, 1.5–3 mm. **Pedicels** patent or ± curled, (6–)10–15(–17) mm, longer than subtending bracts, glabrous. **Flowers:** calyx lobes (1–)2–3 mm, apex obtuse to acute, glabrate; corolla pale lilac, pale sky blue, or whitish, with pink or dark blue narrow stripes, 2.5–7 mm diam.; stamens 4–5 mm; style (1–)2–4 mm. **Capsules** compressed in cross section, ovoid or reniform, 2.5–5 × (3–)4–6 mm, apex emarginate by ⅓ length, glabrous. **Seeds** 5–20, pale brownish, globular or ovoid, flat, 1.2–1.9 × 1–1.5 mm, 0.2 mm thick, smooth. $2n$ = 18.

Flowering May–Sep. Moist meadows, marshes, shallows, forests, steppes, fens, stream banks, lakeshores; 0–1800(–2400) m; St. Pierre and Miquelon; Alta., B.C., Man., N.B., Nfld. and Labr., N.W.T., N.S.,

Nunavut, Ont., P.E.I., Que., Sask., Yukon; Alaska, Ariz., Calif., Colo., Conn., D.C., Idaho, Ill., Ind., Iowa, La., Maine, Md., Mass., Mich., Minn., Mont., Nev., N.H., N.J., N.Y., N.C., N.Dak., Ohio, Oreg., Pa., R.I., Tenn., Vt., Va., Wash., W.Va., Wis., Wyo.; Eurasia; Africa (Algeria); introduced in s South America (Argentina).

Alpine dwarf forms of *Veronica scutellata* tend to have relatively broad leaves and may therefore be easily confused with other species.

11. Veronica serpyllifolia Linnaeus, Sp. Pl. 1: 12. 1753

• Thyme-leaved speedwell F

Veronica humifusa Dickson; *V. serpyllifolia* var. *decipiens* B. Boivin; *V. serpyllifolia* subsp. *humifusa* (Dickson) Syme; *V. serpyllifolia* var. *humifusa* (Dickson) Vahl; *V. tenella* Allioni

Perennials. Stems creeping to ascending, 5–40 cm, scattered eglandular hairs only, often also with glandular hairs, sometimes glabrate. **Leaves:** blade oblong-lanceolate to ovate, 8–25 × 5–13 mm, 1.5–2.5 times as long as wide, base cuneate, margins subentire or serrulate-crenate, apex rounded to short-acuminate, surfaces glabrate. **Racemes** 1, terminal, 50–100 mm, 10–30(–60)-flowered, axis eglandular- and glandular-hairy; bracts oblong, 4–7 mm. **Pedicels** erect, 2–5 mm, 4–6 mm in fruit, shorter than subtending bract in flower, eglandular- and, sometimes, glandular-hairy. **Flowers:** calyx lobes 2–3 mm, apex acute, ciliate; corolla white, blue, or pink, with purple or dark blue veins (except on abaxial lobe), rotate, 5–8 mm diam.; stamens 2.5–3 mm; style 2–4 mm. **Capsules** compressed in cross section, obcordiform, 2.5–3.5 × 4–5.5 mm, wider than long, apex acutely emarginate, glandular-ciliate, otherwise glabrate. **Seeds** 50–72, brown, ellipsoid, flat, 0.6–1.1 × 0.4–1 mm, 0.2 mm thick, smooth. $2n = 14$.

Flowering Apr–Aug. River banks, wet places, moist meadows, shady forests; 0–3300 m; Alta., B.C., N.B., Nfld. and Labr., N.S., Ont., P.E.I., Que., Sask.; Ala., Alaska, Ariz., Ark., Calif., Colo., Conn., Del., D.C., Ga., Idaho, Ill., Ind., Iowa, Kans., Ky., Maine, Md., Mass., Mich., Minn., Mo., Mont., Nev., N.H., N.J., N.Mex., N.Y., N.C., N.Dak., Ohio, Okla., Oreg., Pa., R.I., S.C., Tenn., Utah, Vt., Va., Wash., W.Va., Wis., Wyo.; Mexico (Baja California); South America; Eurasia; Australia.

Alpine plants of *Veronica serpyllifolia* with bright blue, larger corollas, and smaller, hairier raceme axes have been treated at various ranks under the epithet *humifusa*. The characters are labile and show intergradations with the typical plants. *Veronica serpyllifolia* is widespread; it is not clear where it is native.

12. Veronica beccabunga Linnaeus, Sp. Pl. 1: 12. 1753

• Brooklime, véronique beccabunga I

Subspecies 3 (1 in the flora): introduced; Eurasia, Africa; introduced also in s South America.

Subspecies *abscondita* M. A. Fischer is found from eastern Turkey to western Iran, while the native range of subsp. *muscosa* (Korshinsky) Elenevsky extends from eastern Iran to southern China.

12a. Veronica beccabunga Linnaeus subsp. beccabunga I

Veronica baxteri House

Perennials. Stems decumbent to ascending proximally, creeping distally, 10–60 cm, glabrous. **Leaves:** petiole 2–4(–8) mm; blade broadly oblong, ovate, or subelliptic, widest near or distal to middle, (10–)15–35 (–65) × (6–)8–20(–33) mm, 1.5–2 times as long as wide, base rounded-truncate to cuneate, margins denticulate, subserrulate, or subentire, rarely crenulate, apex obtuse-rounded, shining, surfaces glabrous. **Racemes** 2–6(–12), axillary, 30–50 mm, (5–)10–15(–22)-flowered, axis glabrous; bracts lanceolate, 3–6 mm. **Pedicels** subpatent, arcuate-erect, or suberect, (2.5–)3–7(–10) mm, equal to subtending bract, glabrous. **Flowers:** calyx lobes 2–2.5 mm, apex acute, glabrous; corolla deep or bright, rarely pale, blue with white center, 5–6 mm diam.; stamens 1.4–2.8 mm; style (1.3–)1.5–3(–3.5) mm. **Capsules** slightly compressed in cross section, broadly ellipsoid or globular, 2.5–4(–5.5) × 3–4(–4.5) mm, apex not or ± emarginate, glabrous. **Seeds** 38–70(–110), brown, broadly ellipsoid, flat, 0.4–0.6 × 0.3–0.5 mm, 0.2–0.3 mm thick, smooth. $2n = 18, 36$.

Flowering May–Sep(–Oct). Streams, marshes, ditches, wet meadows, running water, disturbed sites; 0–2300 m; introduced; B.C., Ont., Que.; Calif., Conn., Ill., Maine, Md., Mass., Mich., Mo., Nev., N.J., N.Y., Ohio, Pa., Va., W.Va., Wis.; Eurasia; Africa; introduced in s South America (Argentina, Chile).

For the flora area, *Veronica beccabunga* was first recorded in 1876 (New York, D. H. Les and R. L. Stuckey 1985); it seems to have been introduced multiple times independently.

SELECTED REFERENCE Les, D. H. and R. L. Stuckey. 1985. The introduction and spread of *Veronica beccabunga* (Scrophulariaceae) in eastern North America. Rhodora 87: 503–515.

13. Veronica americana Schweinitz ex Bentham in
A. P. de Candolle and A. L. P. P. de Candolle, Prodr.
10: 468. 1846 • American brooklime, véronique
d'Amérique

Veronica beccabunga Linnaeus var.
americana Rafinesque

Perennials. Stems decumbent or
ascending, 5–50 cm, glabrous.
Leaves: petiole 2–6(–10) mm;
blade ovate to lanceolate,
widest proximal to middle,
(5–)30–50(–100) × (3–)7–20
(–30) mm, 2–4 times as long
as wide, base truncate, rounded or almost cordate,
abruptly turning into petioles, margins entire or serrate,
apex obtuse-orbicular or acute, surfaces glabrous.
Racemes 1–8, axillary, 30–100 mm, (3–)10–30-
flowered, axis glabrous; bracts lanceolate, 3–6 mm.
Pedicels patent, 3–10(–12) mm, equal to ± longer than
subtending bract, glabrous. **Flowers:** calyx lobes 2–5
(–6) mm, apex acute, glabrous; corolla pink or sky
blue, 4–10 mm diam.; stamens 2–4 mm; style (1.7–)
2.5–3.5(–4) mm. **Capsules** slightly compressed in
cross section, globular, 2–4 × 3–5 mm, apex not or
± emarginate, glabrous. **Seeds** 20–30, brownish,
ellipsoid to ovoid, indistinct-compressed or planoconvex, 0.5–0.7 × 0.3–0.6 mm, thickness varies due
to compression in capsule, indistinctly wrinkled or
± rugose. $2n = 36$.

Flowering May–Aug. Slowly flowing waters, banks,
sand bars, gravel flood plains, moist soils, springs,
ditches, swamps, marshes; 0–3600 m; Alta., B.C.,
Man., N.B., Nfld. and Labr. (Nfld.), N.W.T., N.S., Ont.,
P.E.I., Que., Sask., Yukon; Alaska, Ariz., Ark., Calif.,
Colo., Conn., Del., Idaho, Ill., Ind., Iowa, Kans., Ky.,
Maine, Md., Mass., Mich., Minn., Mo., Mont., Nebr.,
Nev., N.H., N.J., N.Mex., N.Y., N.C., N.Dak., Ohio,
Okla., Oreg., Pa., R.I., S.C., S.Dak., Tenn., Tex., Utah,
Vt., Va., Wash., W.Va., Wis., Wyo.; Mexico; Asia
(Japan, Russia).

Although difficult to separate, *Veronica beccabunga*
and *V. americana* are maintained as species here.
Evidence suggesting separate species status is
different ploidy level (the tetraploid level is rare in
V. beccabunga) and the occurrence of *V. americana* in
more natural habitats.

14. Veronica anagallis-aquatica Linnaeus, Sp. Pl. 1: 12.
1753 (as anagall. ▽) • Water speedwell, véronique
mouron-d'eau ⊡

Veronica anagallis-aquatica var.
terrea Farwell; *V. brittonii* Porter;
V. comosa Richter; *V. glandifera*
Pennell; *V. micromera* Wooton &
Standley

Annuals or perennials. Stems
erect or prostrate basally,
(20–)30–100(–170) cm, usually
thick-fleshy, glabrous. **Leaves:**
petiole 0–2(–8) mm (basal lateral branches usually
distinctly petiolate to 8 mm) proximally, 0 mm distally;
blade (of proximal leaves) ovate, elliptic, or oblong,
(15–)30–80(–145) × (7–)10–30(–45) mm, 1.5–3 times
as long as wide, base obtuse proximally to cordate-
amplexicaul in middle and ± cuneate distally, margins
± serrulate or denticulate, apex acute, surfaces
glabrous, rarely glandular-hairy. **Racemes** 6–25,
axillary, 50–100(–150) mm, 5–10 mm diam., (20–)
30–40(–60)-flowered, axis glabrous, rarely glandular-
hairy; bracts linear to lanceolate, 2–5 mm, apex acute.
Pedicels suberect or arcuate-erect to subpatent, curved
upwards in fruit, 3–7(–10) mm, longer than subtending
bracts, glabrous or hairy. **Flowers:** calyx lobes 2–3 mm,
apex acute, glabrous, rarely hairy; corolla lavender to pale
blue, rarely pale lilac, (4–)6–8 mm diam.; stamens 3–3.5
mm; style 1.5–2.5 mm. **Capsules** slightly compressed
in cross section, globular, (2.5–)3–3.5(–4) × 2.5–3.2
(–4) mm, apex rounded or ± emarginate, glabrous or
sparsely short glandular-hairy. **Seeds** 40–77, yellow-
brown, ellipsoid to subglobular, planoconvex, (0.3–)
0.5–0.7 × 0.3–0.5 mm, 0.2–0.3 mm thick, smooth.
$2n = 36$.

Flowering Mar–Sep(–Nov). Stream margins, ditches,
banks, springs, swamps, wet meadows; 0–4000 m;
introduced; Alta., B.C., Ont.; Ariz., Calif., Colo., Conn.,
Del., D.C., Fla., Ga., Ill., Ind., Iowa, Kans., Ky., Maine,
Mich., Mo., Mont., Nebr., Nev., N.J., N.Mex., N.Y.,
N.C., N.Dak., Ohio, Okla., Oreg., Pa., R.I., S.Dak.,
Tenn., Tex., Utah, Va., Wash., W.Va., Wis., Wyo.;
Eurasia; Africa; Atlantic Islands; likely introduced also
in Mexico, Central America, South America.

Veronica anagallis-aquatica is widespread. It is not
clear whether it is native to the flora area; it is certainly
introduced in some states and commonly dispersed by
human activity. The species varies with water availability. It is frequently confused with *V. catenata*.
Ecological differences (see discussion under 15.
V. catenata) may suggest that *V. anagallis-aquatica*
does not occur in most parts of Canada and Alaska
from which the species is reported but is present in some
states excluded from the distribution area due to the lack
of herbarium specimens seen (for example, Alabama,

Arkansas, Idaho, Louisiana, Maryland, Massachusetts, Mississippi, South Carolina, and Vermont). Hybrids are frequent in Europe (*V.* ×*lackschewitzii* J. Keller) and have been reported for California and Nebraska (L. R. Heckard and P. Rubtzoff 1977). The sterile hybrids never form capsules and bear relatively long inflorescences. Another closely related species, *V. anagalloides* Gussone (= *V. salina* Schur), is not present in North America.

15. Veronica catenata Pennell, Rhodora 23: 37. 1921
    • Chain speedwell

Veronica catenata var. *glandulosa* (Farwell) Pennell; *V. connata* Rafinesque subsp. *glaberrima* Pennell

Annuals or perennials. Stems erect or ascending, 15–60 (–80) cm, glabrous or glandular-hairy distally. **Leaves:** petiole 0 mm; blade oblong-ovate to oblong-lanceolate, (5–)25–50(–100) × 4–15(–30) mm, 2.5–5 times as long as wide, base amplexical-truncate or amplexical-subcordate, margins entire or subentire, apex acute, surfaces glabrous. **Racemes** 10–25, axillary, 100–160 mm, 15–25-flowered, axis glabrous or glandular-hispid; pedicels less than 3 per cm, glabrous or scarcely to densely short glandular-hairy, rarely completely glabrous; bracts oblong, 3–5 mm, apex obtuse. **Pedicels** patent, (3–)5–10 mm, equal to or ± shorter than subtending bract, glabrous or glandular-hairy. **Flowers:** calyx lobes 2.5–3 mm, apex obtuse, glabrous or glandular-hairy; corolla white to pale pink with darker veins not reaching margins, 4–5 mm diam.; stamens 5 mm; style (1.3–)1.5–2.5 mm. **Capsules** ± compressed in cross section, subglobular, 2.5–3(–3.5) × 3–4 mm, apex emarginate, sinus 0.1–0.3 mm, glabrous or glandular-hairy. **Seeds** 26–123, yellow-brown, ellipsoid to subglobular, planoconvex, 0.4–0.7 × 0.3–0.5 mm, 0.2–0.3 mm thick, smooth. $2n = 36$.

Flowering Jun–Oct. Wet places, rarely running water, lakeshores, ditches, muddy places, stream channels; 0–2500 m; Alta., B.C., Man., Ont., Que., Sask.; Calif., Idaho, Ill., Ind., Iowa, Kans., Maine, Mass., Mich., Minn., Mo., Mont., Nebr., Nev., N.Mex., N.Y., N.Dak., Ohio, Okla., Oreg., Pa., S.Dak., Tenn., Tex., Vt., Va., Wash., W.Va., Wis., Wyo.; Europe.

Veronica catenata seems to be a relative of *V. anagallis-aquatica* that is more cold-adapted, as seen in Europe (R. Götte 2007), and native to the flora area. Although no specimens were seen it may be found in Alaska, Connecticut, and Kentucky.

16. Veronica undulata Wallich in W. Roxburgh, Fl. Ind. 1: 147. 1820 • Wavy-leaved water speedwell ⊡

Annuals or perennials. Stems erect or prostrate basally, 10–100 cm, glabrous at least proximally, glandular-hairy distally. **Leaves:** petiole 0–5 mm; blade elliptic to ovate, sometimes ovate-oblong or linear-lanceolate, rarely lanceolate, 20–50(–100) × 5–20(–25) mm, 2.5–4 times as long as wide, base attenuate, upwards amplexicaul, margins subentire or crenate to serrate or ± undulate, apex acute to obtuse, surfaces glabrous. **Racemes** 6–25, axillary, 50–220 mm, 10–15 mm diam., 10–100-flowered, axis sparsely glandular-hairy, rarely glabrous; pedicels 2–6 per cm; bracts linear to lanceolate, 3–6 mm, apex acute. **Pedicels** patent, 3–5 mm, equal to subtending bract, sparsely glandular-hairy. **Flowers:** calyx lobes 1.5–2(–3) mm, apex acute, sparsely glandular-hairy, rarely glabrous; corolla pale blue or pale lilac, rarely white, 2.5–5 mm diam.; stamens 2 mm; style 0.9–2 mm. **Capsules** slightly compressed in cross section, obcordiform to globular, 2–3 × 2–3 mm, apex rounded or ± emarginate, sparsely glandular-hairy. **Seeds** 30–40, ochre, broadly ellipsoid, ± flat, convex on both sides, 0.5–0.6 × 0.4–0.5 mm, thickness and texture unknown. $2n = 54$ (Asia).

Flowering (Mar–)Apr–Sep. Disturbed, wet places, ditches, or swamps; 0–200 m; introduced; Ala., Oreg., Wash.; Asia.

In the flora area, *Veronica undulata* was introduced near ports (Mobile, Alabama, and Portland, Oregon) before 1900 via ship ballast from trade with Asia, did not spread much, and may not have persisted.

17. Veronica peregrina Linnaeus, Sp. Pl. 1: 14. 1753
    • Purslane speedwell, véronique voyageuse

Veronica peregrina var. *laurentiana* Victorin & J. Rousseau; *V. peregrina* subsp. *xalapensis* (Kunth) Pennell; *V. peregrina* var. *xalapensis* (Kunth) Pennell; *V. sherwoodii* M. Peck; *V. xalapensis* Kunth

Annuals. Stems erect or ascending, (2.5–)4–25(–35) cm, glabrous or densely glandular-hairy. **Leaves:** blade oblanceolate proximally, narrowly oblong distally, 5–28 (–35) × 2–6(–10) mm, 3–10 times as long as wide, base cuneate, margins entire or dentate distally, apex acute, surfaces glabrous or densely glandular-hairy. **Racemes** 1, terminal, sometimes 1 or 2 axillary, 20–200 mm, (2–)5–40-flowered, axis glabrous or densely glandular-hairy; bracts spatulate to linear-lanceolate, 3–22 mm. **Pedicels** erect, 0.2–1(–2) mm, much shorter than subtending bract, glabrous or densely glandular-hairy.

Flowers: calyx lobes (2–)3–6 mm, 0.9–2 mm wide, apex obtuse to acute, glabrous or densely glandular-hairy; corolla white or pale pink, 2–5 mm diam.; stamens 1 mm; style 0.1–0.5 mm. **Capsules** strongly compressed in cross section, obcordiform, 2.5–5 × 2.5–6 mm, apex ± emarginate, glabrous. **Seeds** 12–140, yellow or pale brown, oblong, flat, 0.6–1.6 × 0.4–0.9 mm, 0.1–0.2 mm thick, smooth. $2n = 52$.

Flowering (Feb–)Mar–Jun(–Nov). Moist waste lands, gardens, roadsides, stream banks, pond shorelines, vernal pools, other cultivated land; 0–700(–3000) m; Alta., B.C., Man., N.B., Nfld. and Labr. (Nfld.), N.W.T., N.S., Ont., P.E.I., Que., Sask., Yukon; Ala., Alaska, Ariz., Ark., Calif., Colo., Conn., Del., D.C., Fla., Ga., Idaho, Ill., Ind., Iowa, Kans., Ky., La., Maine, Md., Mass., Mich., Minn., Miss., Mo., Mont., Nebr., Nev., N.H., N.J., N.Mex., N.Y., N.C., N.Dak., Ohio, Okla., Oreg., Pa., R.I., S.C., S.Dak., Tenn., Tex., Utah, Vt., Va., Wash., W.Va., Wis., Wyo.; Mexico (Baja California, Baja California Sur, Chihuahua, México, Sonora); Central America; South America; Eurasia; Australia.

Stem indument of *Veronica peregrina* has been used to distinguish var. *xalapensis* with a glandular-hairy stem (F. W. Pennell 1935), a variety that seems to be confined to drier places than the type variety. Variety *laurentiana* is also differentiated from var. *peregrina* in having a glandular-hairy stem, fleshier leaves, shorter stamens, smaller corollas, and slightly differently shaped capsules (Frère Marie-Victorin and J. Rousseau 1940).

18. **Veronica triphyllos** Linnaeus, Sp. Pl. 1: 14. 1753
    • Fingered speedwell I

Annuals, sometimes biennials. **Stems** erect or ascending, often blue tinged, (3–)5–15(–20) cm, densely hairy. **Leaves:** blade: proximal ovate to triangular, 5–7 × 3–6 mm, distal 8–18 × 8–18 mm, base rounded, margins (proximal) coarsely crenate or (distal) 3–5(–7)-palmatifid, apex obtuse, surfaces sparsely hairy, often glandular. **Racemes** 1, terminal, sometimes also 1 or 2 axillary, 30–120(–150) mm, 6–15(–25)-flowered, axis densely eglandular- and glandular-hairy; proximal bracts 5-palmatifid, distal 3-fid, 4–18 mm. **Pedicels** patent or arcuate-ascending, 4–11(–20) mm, ± equal to subtending bract, length 1–2 times calyx, densely hairy. **Flowers:** calyx lobes 4–6(–11) mm, apex obtuse, glandular-hairy; corolla deep blue, 5–10 mm diam.; stamens 0.9–1.7 mm; style (0.5–)0.7–1.5(–2) mm. **Capsules** ± inflated basally, compressed distally in cross section, obcordiform, 4–5(–10) × 4.5–6.5(–8) mm, apex emarginate, glandular-hairy. **Seeds** 14–22(–35), dark brown or blackish, subglobular, cymbiform, 1–2.2 × 0.7–2 mm, 0.4–0.9 mm thick, ± rugulose to cristate. $2n = 14$ (Eurasia).

Flowering (Feb–)Mar–May(–Jul). Pine forests, stony pastures, rocky banks, sandy fields, gardens, roadsides, rarely calcareous soils; 100–1000 m; introduced; Calif., Idaho, Kans., Mo., Okla., Oreg., Wash.; Eurasia.

19. **Veronica hederifolia** Linnaeus, Sp. Pl. 1: 13. 1753
    (as hederaefolia) • Ivy-leaved speedwell F I

Annuals. **Stems** decumbent to ascending, 5–40(–50) cm, eglandular-hairy. **Leaves:** blade suborbiculate, (3–)5-lobed, central lobe usually overtopping lateral ones, (5–)7–15(–20) × 8–16(–25) mm, base truncate, margins (3–)5-lobed, apex acute, surfaces sparsely eglandular-hairy. **Racemes** 1–10, terminal, 50–400(–500) mm, 5–20-flowered, axis eglandular-hairy; bracts suborbiculate, (3–)5-lobed, central lobe usually overtopping lateral ones, (5–)7–15(–20) mm. **Pedicels** patent or deflexed, (5–)9–15(–20) mm, equal to or ± shorter than subtending bract, length 2–3 times calyx, eglandular-hairy (in single line adaxially). **Flowers:** calyx lobes 5–6(–7) mm, apex acute, abaxial surface usually glabrous, ciliate with 25–35 hairs per side; corolla bright blue with bright white center, 5–7(–9) mm diam.; stamens sky blue, 0.7–1.2 mm; style (0.6–)0.7–0.9(–1.1)mm. **Capsules** ± round in cross section, ovoid, 3–4 × 4.5–6 mm, apex ± emarginate, glabrate. **Seeds** 1–4, bright yellow, ellipsoid to subglobular, urn-shaped, 2.3–3.3 × 2–3 mm, 1–2.2 mm thick, weakly cristate or rugose, ± smooth. $2n = 54$ (Eurasia).

Flowering (Feb–)Mar–Jun. Fields, lawns, gardens, ruderal places, vineyards, open forests, shady rocky places, dunes; 0–2000 m; introduced; B.C., Ont.; Ala., Ark., Calif., Conn., Del., D.C., Fla., Ga., Ill., Ind., Kans., Ky., La., Md., Mich., Mo., Nebr., N.J., N.Y., N.C., Ohio, Okla., Oreg., Pa., S.C., S.Dak., Tenn., Utah, Va., Wash., W.Va.; Eurasia.

20. **Veronica sublobata** M. A. Fischer, Oesterr. Bot. Z. 114: 201, 227, figs. 3c, 4c. 1967 • False ivy-leaved speedwell F I

Veronica hederifolia Linnaeus subsp. *lucorum* (Klett & Richter) Hartl; *V. hederifolia* var. *lucorum* Klett & Richter

Annuals. **Stems** decumbent to ascending, sometimes erect in young plants, 5–40(–50) cm, eglandular-hairy. **Leaves:** blade ovate to broadly ovate, 5(–7)-lobed, central lobe longer than wide and broadest, (7–)9–15(–25) × (7–)9–17(–27) mm, base truncate, margins 5(–7)-lobed, apex obtuse to rounded, surfaces sparsely

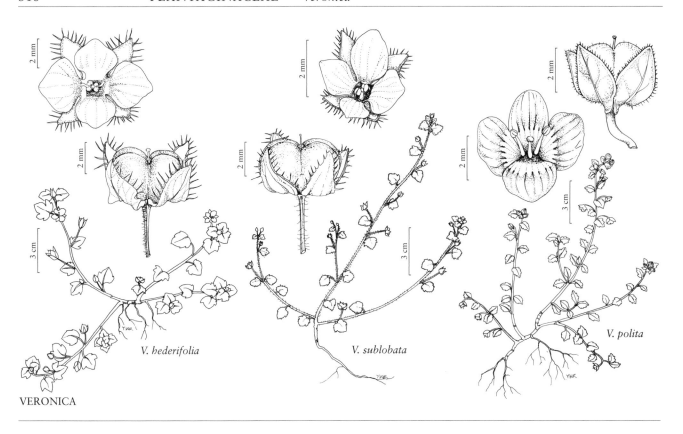

V. *hederifolia*

V. *sublobata*

V. *polita*

VERONICA

eglandular-hairy.   **Racemes** 1–10, terminal, 50–400 (–500) mm, 5–20-flowered, axis eglandular-hairy; bracts broadly ovate, 5(–7)-lobed, central lobe longer than wide and broadest, (7–)9–15(–25) mm. **Pedicels** patent or deflexed, (7–)10–18(–24) mm, ± longer than subtending bract, length 3–4(–6) times calyx, eglandular-hairy (densely adaxially).   **Flowers:** calyx lobes 3–4 mm, apex acute, sparsely hairy on abaxial surface or glabrous, ciliate with 25–35 hairs per side; corolla pale violet to whitish, 4–5(–6) mm diam.; stamens pale violet, 0.4–0.8 mm; style 0.2–0.7 mm. **Capsules** ± round in cross section, ovoid, 2–3 × 3.5–5 mm, apex ± emarginate, glabrate. **Seeds** 1–4, bright reddish brown, broadly ovoid to subglobular, urn-shaped, 2.2–2.7 × 2–2.4 mm, 1–1.8 mm thick, transverse ribs inconspicuous, thinner towards margin, margins strongly involute. $2n = 36$ (Europe).

Flowering Mar–May. Moist forests, damp, shady ruderal places, thickets, gardens, montane fields; 0–1000 m; introduced; Ala., Ky., Md., Mo., Ohio, Utah, Va.; Europe.

Veronica sublobata is closely related to V. *hederifolia* and has been frequently synonymized with or treated as a subspecies of the latter. It has rarely been reported in the flora area and is surely more widespread than indicated above. It seems to be more rare than V. *hederifolia* with only about a dozen verified records from seven states. However, its frequent occurrence in cultivated habitats in Europe makes it likely that it has been introduced to the flora area on multiple occasions.

21. **Veronica triloba** (Opiz) Opiz, Naturalientausch 11: 467. 1826 • Trilobed speedwell [I]

Veronica hederifolia Linnaeus var. *triloba* Opiz, Hesperus 1815: 327. 1815 (as hederaefolia)

Annuals. Stems decumbent, (5–)10–30 cm, ± eglandular-hairy. **Leaves:** blade broadly ovate to suborbiculate, 3(–5)-lobed, central lobe ¼–½ times length of whole leaf, (3–)5–10 (–15) × (4–)6–12(–18) mm, ± fleshy, base truncate to subcordate, margins 3(–5)-lobed, apex acute, surfaces sparsely eglandular-hairy. **Racemes** 1–10, terminal, (50–)100–300 mm, 5–20-flowered, axis sparsely eglandular-hairy; bracts broadly ovate to suborbiculate, 3(–5)-lobed, (3–)5–10(–15) mm. **Pedicels** patent, deflexed in fruit, (2–)4–8(–10) mm, shorter than subtending bract, length 1–2 times calyx, glabrous abaxially, hairy adaxially. **Flowers:** calyx lobes 4–5 (–6) mm, apex acute, puberulent on abaxial surface, rarely glabrous, densely ciliate with 35–60 per side; corolla deep blue with white center, 3–5 mm diam.; stamens dark blue, length unknown; style 0.5–0.9 (–1.1) mm. **Capsules** ± round in cross section, ovoid, 2.5–3(–3.5) × 4–5 mm, apex ± emarginate, glabrous. **Seeds** (1–)4, dull yellowish brown, ellipsoid, often elongate-cuboidal, deeply urn-shaped (elaiosome inside), 2.2–2.8 × 1.5–2.2 mm, 1–1.4 mm thick, cristate to strongly rugose at margins. $2n = 18$ (Eurasia).

Flowering (Feb–)Mar–May(–Jun). Disturbed sites; 100–200 m; introduced; Calif., Ohio; Europe (Balkans); Asia (Turkey).

Veronica triloba is known in the flora area from three recent collections in California and Ohio; it is possibly more widespread than reported here. It is closely related to *V. hederifolia*.

22. Veronica cymbalaria Bodard, Mém. Véronique Cymb., 3. 1798 • Pale speedwell I

Annuals. **Stems** decumbent to ascending, (5–)10–30 cm, sparsely subvillous to glabrate. **Leaves:** blade suborbiculate, 6–20 × 9–20 mm, base subcordate, truncate, or cuneate, margins 5–9-lobed ⅙–⅓ of width of blade, lateral lobes ± narrower, apex acute, surfaces eglandular-hairy, rarely also glandular-hairy or glabrate. **Racemes** 1–10, terminal, (50–)100–300 mm, 5–20-flowered, axis eglandular-hairy, rarely also glandular-hairy or glabrate; bracts suborbiculate, 6–20 mm, 5- to 9-lobed for ⅙–⅓ of width of leaf blade, lateral lobes ± narrower. **Pedicels** patent or recurved, 5–15(–30) mm, equal in flower to ± longer than subtending bract in fruit, eglandular- or glandular-pilose or glabrous. **Flowers:** calyx lobes erect-patent, (3–)4–5 mm, apex obtuse, abaxial surfaces eglandular- or glandular-hairy, rarely glabrous, not ciliate; corolla white, 6–12 mm diam.; stamens 1–2 mm; style 1–2 mm. **Capsules** ± round in cross section, ovoid to ellipsoid, 2.5–3.5 × 3.5–4.5 mm, apex ± emarginate, eglandular- or glandular-pilose. **Seeds** 1–4, brownish orange, subglobular, urn-shaped (elaiosome inside), (2–)2.5–3.1 × (1.7–)2–2.5(–2.8) mm, 1.1–1.7 mm thick, cristate. **2n = 36, 54** (Eurasia).

Flowering Mar–May(–Jun). Lawns; 20 m; introduced; La.; Eurasia (Mediterranean).

The occurrence of *Veronica cymbalaria* in other states is possible. It has been reported from Maryland; a herbarium specimen could not be located. The report seems to be based on incorrect synonymy (C. Frye, pers. comm.).

23. Veronica arvensis Linnaeus, Sp. Pl. 1: 13. 1753 • Wall or corn speedwell, véronique des champs I

Annuals. **Stems** erect to ascending, 1–30(–40) cm, glandular- or eglandular-hairy. **Leaves** in 3–6 pairs per stem; blade oblong to broadly ovate, (2–)5–14(–35) × (2–)3–10(–18) mm, 2.5 times as long as wide, base truncate, margins crenate-serrate, apex obtuse to rounded, surfaces sparsely eglandular-hairy, rarely also glandular-

hairy. **Racemes** 1(–3), terminal, 20–150 mm, (5–)15–40(–60)-flowered, axis densely eglandular- and glandular-hairy; bracts linear-oblong, (3.5–)4–6 (–10) mm. **Pedicels** erect, 0–4 mm, shorter than subtending bract, densely eglandular- and glandular-hairy. **Flowers:** calyx lobes (2–)3.5–5(–6) mm, 0.8–2 mm wide, apex acute, glandular-hairy; corolla sky blue to intense blue, 2–4 mm diam.; stamens 0.3–0.5 mm; style (0.2–)0.4–0.6(–1) mm. **Capsules** compressed in cross section, obcordiform, 2–4 × 2.5–5 mm, apex markedly emarginate, glandular-ciliate, otherwise glabrous. **Seeds** 10–30, yellowish, ovoid, flat, 0.7–1.7 × 0.4–1 mm, 0.2–0.4 mm thick, rugose. **2n = 16.**

Flowering (Feb–)Mar–Jun(–Oct). Disturbed sites, lawns, fields, open forests, scrub, grasslands, rocky sites, coasts; 0–2900 m; introduced; Greenland; B.C., N.B., Nfld. and Labr. (Nfld.), N.S., Ont., P.E.I., Que., Yukon; Ala., Alaska, Ariz., Ark., Calif., Colo., Conn., Del., D.C., Fla., Ga., Idaho, Ill., Ind., Iowa, Kans., Ky., La., Maine, Md., Mass., Mich., Minn., Miss., Mo., Mont., Nebr., Nev., N.H., N.J., N.Mex., N.Y., N.C., Ohio, Okla., Oreg., Pa., R.I., S.C., S.Dak., Tenn., Tex., Utah, Vt., Va., Wash., W.Va., Wis., Wyo.; Eurasia; Africa; Atlantic Islands; Pacific Islands; Australia; introduced also in Mexico, Central America.

24. Veronica verna Linnaeus, Sp. Pl. 1: 14. 1753 • Spring speedwell I

Annuals. **Stems** erect or ascending, (1–)3–15(–20) cm, at least distally glandular- and eglandular-hairy. **Leaves:** blade lanceolate to ovate, (4–)6–13 × (2.5–)4–10 mm, 1.3–1.6 times as long as wide, base cuneate, proximal margins coarsely crenate-serrate, distal pinnatifid to subpalmatifid, lobes 3–7, central largest, lateral linear to lanceolate, apex of central lobe obtuse, lateral ± acute, surfaces sparsely glandular-hairy. **Racemes** 1–3, terminal and axillary, 20–80 mm, (5–)15–40(–60)-flowered, axis eglandular- and glandular-hairy; bracts proximalmost often 3-fid, others linear-lanceolate, 3–5(–8) mm. **Pedicels** erect, (0.4–)1–3 mm, shorter than subtending bract, length ⅓–½ times calyx, eglandular- and glandular-hairy. **Flowers:** calyx lobes (2–)3–5 (–7) mm, apex acute, eglandular- and glandular-hairy; corolla sky to pale blue, 1.5–3 mm diam.; stamens 0.2–0.6 mm; style 0.2–0.6 mm, stigma white. **Capsules** compressed in cross section, obcordiform, 2.5–3.5 × 3.5–5 mm, apex emarginate, eglandular-hairy. **Seeds** 8–20(–26), yellowish, ellipsoid, flat, 0.9–1.6 × 0.6–1.3 mm, 0.2 mm thick, smooth to ± rugulose. **2n = 16** (Eurasia).

Flowering (Mar–)Apr–Jun(–Aug). Open pine and oak forests, rocky and sandy steppes, pastures, meadows; 300–2600 m; introduced; B.C., N.B., N.S., Ont., P.E.I.; Idaho, Ind., Mass., Mich., Minn., Mont., N.Y., Oreg., Wash., Wis., Wyo.; Eurasia.

Specimens of *Veronica verna* from Alberta have not been verified.

25. Veronica dillenii Crantz, Stirp. Austr. Fasc. ed. 2, 2: 352. 1769 • Dillenius speedwell ☐

Annuals. Stems erect to ascending, (8–)10–20(–40) cm, glandular- and eglandular-hairy. **Leaves:** blade lanceolate to ovate, 7–19(–21) × (3–)5–12 mm, 1–2 times as long as wide, base cuneate, margins (proximal) crenate-dentate or (distal) ± palmatifid, lobes 3–7+, apex of central lobe obtuse, lateral ± acute, surfaces glandular- and eglandular-hairy. **Racemes** 1, terminal, 60–180 mm, 15–50(–120)-flowered, axis eglandular- and glandular-hairy; proximal bracts similar to leaves, distal ones linear-lanceolate, 3–5 mm. **Pedicels** erect, 2–4(–5) mm, shorter than subtending bract, length ½–1 times calyx, eglandular- and glandular-hairy. **Flowers:** calyx lobes 3.5–6 mm, apex acute, eglandular- and glandular-hairy; corolla deep blue, 4–6 mm diam.; stamens 0.8–1.5 mm; style 0.8–1.5 mm, stigma violet. **Capsules** compressed in cross section, narrowly obcordiform, 3.5–4.5 × 4–6 mm, apex emarginate, angle of sinus ca. 90°, ciliate with glandular and eglandular hairs. **Seeds** 10–28, yellowish, ellipsoid, flat, 0.9–1.6 × 0.7–1.3 mm, 0.2–0.3 mm thick, smooth. $2n$ = 16 (Eurasia).

Flowering (Mar–)Apr–Jul. Open pine and oak forests, rocky, dry, sandy slopes; 500–2200 m; introduced; Ill., Ind., Mich., N.Y., Va., Wis.; Eurasia.

Veronica dillenii is closely related to *V. verna* but with larger flowers; it may have been overlooked and may be distributed more widely. Most herbarium specimens of *V. dillenii* blacken when dry due to the presence of aucubin, which distinguishes them from *V. verna*.

26. Veronica chamaedrys Linnaeus, Sp. Pl. 1: 13. 1753 • Germander speedwell, véronique petit-chêne

Perennials. Stems ascending, rarely erect, (7–)10–30(–50) cm, densely eglandular-hairy, hairs along stem in 2 prominent lines. **Leaves:** blade narrowly ovate to ovate-orbiculate, (10–)12–30 (–42) × (6–)10–22(–30) mm, 1–2 times as long as wide, base truncate to ± cordate, margins crenate to deeply incised, apex obtuse, surfaces variably hairy. **Racemes** 1–4, axillary, 40–100(–200) mm,

15–40(–60)-flowered, axis eglandular-hairy, sometimes also glandular-hairy; bracts linear-elliptic, 3–7 mm. **Pedicels** suberect, (3–)5–8(–10) mm, equal to or shorter than subtending bract, eglandular- and glandular-hairy. **Flowers:** calyx 4-lobed, lobes 2–8 mm, apex acute, eglandular- and, sometimes, glandular-hairy; corolla blue obscure darker nerves and sometimes whitish margin, (6–)10–14(–17) mm diam.; stamens 4.5–6.5 mm; style (2.5–)4–5 mm. **Capsules** strongly compressed in cross section, obcordiform to obdeltoid, (2–)3.5–4(–5) × (3.5–)4–5(–5.5) mm, apex ± emarginate, eglandular-hairy. **Seeds** (2–)12–20(–28), yellow, ellipsoid, flat, 1.1–1.7 × 0.6–1.5 mm, 0.2–0.4 mm thick, smooth to subrugose. $2n$ = 16, 32 (Eurasia).

Flowering Apr–Jun(–Oct). Rich soils, deciduous forests, forest edges, roadsides, chaparral, scrub, meadows, lawns; 0–2200 m; Alta., B.C., N.B., Nfld. and Labr. (Nfld.), N.S., Ont., P.E.I., Que.; Alaska, Conn., D.C., Idaho, Ill., Ind., Maine, Md., Mass., Mich., Mo., Mont., N.H., N.J., N.Y., N.C., Ohio, Oreg., Pa., R.I., Vt., Va., Wash., W.Va., Wis.; Eurasia; introduced in South America (Argentina).

The description provided here for *Veronica chamaedrys* is solely for the tetraploid cytotype, most probably the exclusive cytotype in the flora area and in central and western Europe. The diploid cytotype is so far only known from eastern and southeastern Europe (K. E. Bardy et al. 2010). A significant change in morphology can occur in shaded habitats, in which especially the petiole can be elongated beyond the range given.

It is unclear whether *Veronica chamaedrys* is introduced throughout the flora area; it may be native in northeastern areas of North America.

27. Veronica fruticans Jacquin, Enum. Stirp. Vindob. 2: 200. 1762 • Rock speedwell ☐E

Perennials. Stems ascending to erect, densely branched from woody base, (5–)10–15(–30) cm, eglandular-hairy, sometimes glandular-hairy. **Leaves:** blade obovate to ovate or spatulate, sometimes suborbiculate proximally, (7–)8–20(–25) × (2–)3–6(–7) mm, mostly shorter to equal to internodes, base cuneate, margins entire or ± crenate or serrate, apex acute, surfaces glabrous or glabrate. **Racemes** 1–10(–20), terminal, rarely with 1–4 axillary, 20–40 mm, (1–)4–10(–18)-flowered, axis glabrate to sparsely eglandular-hairy; bracts linear or linear-lanceolate or long-ovate, (1–)3.5–9(–15) mm. **Pedicels** erect, (2–)5–7(–15) mm, equal to ± longer than subtending bracts, eglandular-hairy. **Flowers:** calyx 4(or 5)-lobed, lobes (2.5–)4.5–6(–8) mm, apex obtuse, eglandular-hairy, sometimes also glandular-hairy; corolla intense blue, sometimes with reddish or white center, rarely white, rotate, (6–)10–14(–15) mm diam.; stamens 5–7 mm; style (3–)5–7(–8.5) mm. **Capsules** ± compressed in cross section, ovoid, (5–)6–8(–9) × (3–)4–4.5(–5.5) mm, longer than wide, apex attenuate to apiculate and acute, rarely rounded, not emarginate, eglandular-hairy, sometimes mixed with larger glandular hairs. **Seeds** 35–62, orange, ellipsoid, flat, 0.9–1.5 × (0.4–)0.9–1.2 mm, 0.2–0.3 mm thick, smooth. $2n = 16$.

Flowering (May–)Jun–Aug. Fissures, rocky places, scree; 0–1000 m; Greenland.

28. **Veronica polita** Fries, Novit. Fl. Svec., 63. 1819 • Gray speedwell F I

Annuals. Stems creeping to decumbent, 5–20(–40) cm, eglandular-hairy. **Leaves:** blade suborbiculate to ovate, rarely oblong-ovate, (3–)6–11(–14) × (3–)4–9(–10) mm, base truncate, margins serrate to dentate, teeth 2–4 per side, apices usually rounded, apex acute to obtuse, abaxial surface densely hairy, adaxial sparsely hairy to glabrate. **Racemes** 1–4, terminal, 50–200(–400) mm, 5–20-flowered, axis sparsely to densely hairy or glabrate; bracts suborbiculate to ovate, (3–)6–11(–14) mm. **Pedicels** usually semicircularly recurved, (3–)6–13(–15) mm, 0.5–1.5 times subtending bract, densely eglandular-hairy. **Flowers:** calyx lobes broadly ovate, (3–)4–6(–7) mm, 2.6–3.8 mm wide, apex acute, eglandular-hairy, rarely glandular-hairy; corolla intense dark to bright blue, rarely pale lilac or white, with darker nerves, abaxial ½ often brighter to whitish, (3–)4–7(–8) mm diam.; stamens 0.5 mm; style (0.5–)1–1.6 (–1.8) mm. **Capsules** ± round in cross section, obcordiform, (2.5–)3–4(–4.5) × (3.5–)4–6(–6.3) mm, apex emarginate, sinus angle 20–60(–80)°, densely eglandular-hairy, sometimes also glandular-puberulent, rarely glabrous. **Seeds** (7–)16–24(–30), pale yellow, ellipsoid to ovoid, deeply cymbiform, 0.9–1.6 × (0.5–)0.8–1.3 mm, 0.4–0.8 mm thick, rugose-cristate. $2n = 14$ (Eurasia).

Flowering (Feb–)Mar–Jun(–Jul). Fields, ruderal places, calcareous soils, lawns; 0–600 m; introduced; Man., N.B., Ont.; Ala., Ariz., Ark., Colo., Conn., D.C., Fla., Ill., Ind., Iowa, Kans., Ky., La., Maine, Md., Mass., Mich., Mo., Nebr., N.Y., N.C., Ohio, Okla., Pa., Tenn., Tex., Va., W.Va., Wis.; Eurasia; nw Africa (Algeria, Morocco); introduced also in Mexico, s South America.

Plants of *Veronica polita* are similar to those of the more frequent *V. persica* and are probably frequently overlooked.

29. **Veronica persica** Poiret in J. Lamarck et al., Encycl. 8: 542. 1808 • Large field or bird's-eye speedwell, véronique de Perse F I

Veronica persica var. *aschersoniana* (E. B. J. Lehmann) Drabble & J. E. Little; *V. persica* var. *corrensiana* (E. B. J. Lehmann) Stroh; *V. tournefortii* C. C. Gmelin subsp. *aschersoniana* E. B. J. Lehmann; *V. tournefortii* subsp. *corrensiana* E. B. J. Lehmann

Annuals. Stems creeping to decumbent, 10–50(–60) cm, eglandular-hairy. **Leaves:** blade suborbiculate, broadly ovate, or broadly lanceolate, (6–)9–18(–30) × (5–)8–15(–20) mm, base truncate, margins serrate, apex acute, surfaces sparsely eglandular-hairy. **Racemes** 1–6, terminal, 100–500 (–600) mm, 5–30-flowered, axis eglandular-hairy; bracts suborbiculate or broadly ovate or broadly lanceolate, (6–)9–18(–25) mm. **Pedicels** spreading, deflexed in fruit, (12–)15–27(–38) mm, length 1–2 (–3) times subtending bract, densely eglandular-hairy. **Flowers:** calyx lobes (4.5–)5.5–8(–9.5) mm, (1.7–)2.4–3.6(–4.2) mm wide, apex acuminate, eglandular-hairy; corolla intense bright blue, 8–14 mm diam.; stamens 1.2 mm; style (1.5–)2–2.8(–3.2) mm. **Capsules** compressed in cross section, broadly obcordiform, 4–6 × (5–)6–8.5(–9.5) mm, apex acute, sinus angle (80–)90–120(–150)°, reticulate with prominent veins, ± sparsely to densely eglandular- and/or glandular-hairy or glabrate. **Seeds** (10–)12–18(–20), pale brownish yellow, ellipsoid to globular, cymbiform, (1.3–)1.4–2.3 (–2.5) × (0.8–)0.9–1.6(–1.9) mm, 0.5–1 mm thick, cristate-rugose. $2n = 28$.

Flowering (Jan–)Apr–Jul(–Dec). Roadsides, lawns, fields, waste places; 0–500(–2000) m; introduced; Alta., B.C., Man., N.B., Nfld. and Labr. (Nfld.), N.W.T., N.S., Ont., P.E.I., Que., Sask.; Ala., Alaska, Ariz., Ark., Calif., Colo., Conn., Del., D.C., Fla., Ga., Idaho, Ill., Ind., Iowa, Kans., Ky., La., Maine, Md., Mass., Mich., Minn., Miss., Mo., Mont., Nebr., Nev., N.H., N.J., N.Mex., N.Y., N.C., Ohio, Okla., Oreg., Pa., R.I., S.C., S.Dak., Tenn., Tex., Utah, Vt., Va., Wash., W.Va., Wis., Wyo.; sw Asia; introduced also in Mexico (Michoacán, Veracruz), Central America, South America, Eurasia, e Asia, Africa, Atlantic Islands, Pacific Islands, Australia.

The names *Veronica buxbaumii* Tenore and *V. tournefortii* C. C. Gmelin (not *V. ×tournefortii* Villars) have been used for *V. persica*.

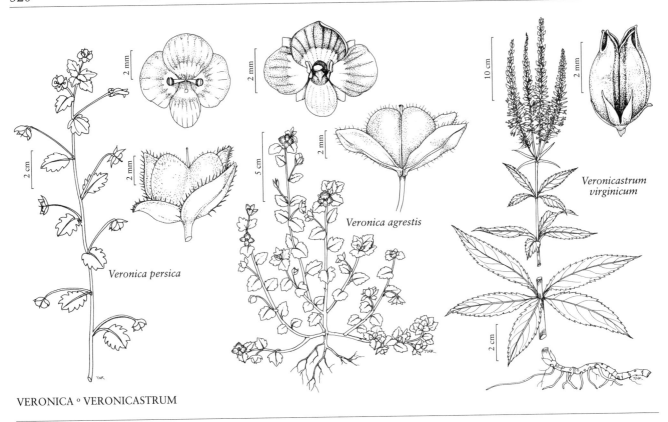

Veronica persica

Veronica agrestis

Veronicastrum virginicum

VERONICA ∘ VERONICASTRUM

30. Veronica agrestis Linnaeus, Sp. Pl. 1: 13. 1753

• Field speedwell, véronique agreste [F][I]

Annuals. Stems creeping to decumbent, 5–25 cm, hairy. **Leaves:** blade elliptic-ovate to oblong, (6–)8–16(–19) × (3–)4–10(–13) mm, base truncate, margins serrate, teeth 4–6(–8) per side, apex obtuse to acute, abaxial surface ± more densely hairy than adaxial. **Racemes** 1–5, terminal, 50–250 mm, 5–15-flowered, axis eglandular-hairy, sometimes glandular-hairy; bracts elliptic-ovate to oblong, (6–)9–16(–19) mm. **Pedicels** deflexed, (5–)6–10(–15) mm, ± shorter than subtending bracts, densely eglandular- and, sometimes, glandular-hairy distally. **Flowers:** calyx lobes linear-lanceolate, (5–)6–7 mm, 2.2–2.6 mm wide, apex rounded, sparsely eglandular- or glandular-hairy; corolla white or pale pinkish or pale blue, 4–5(–6) mm diam.; stamens 0.5–1 mm; style (0.6–)0.9–1.1(–1.2) mm. **Capsules** ± compressed in cross section, ± subglobular, 3.5–4.5(–4.7) × 4.5–6(–6.3) mm, apex emarginate, sparsely glandular-hairy. **Seeds** 6–14, yellow to ochre, globular, cymbiform, 1.3–2.1 × 1–1.6 mm, 0.6–1.1 mm thick, strongly rugose (transversely ribbed). $2n = 28$ (Europe).

Flowering May–Oct. Light, noncalcareous, moist soils, gravelly soils; (0–)300–800 m; introduced; Nfld. and Labr. (Nfld.), N.S., Que.; D.C., Fla., Ill., La., Mass., N.Y., Pa., Tex., Vt.; Europe.

Veronica agrestis is similar to the more frequent *V. persica* and probably frequently overlooked and to be expected elsewhere. However, it should be noted that it is less common than *V. persica* or *V. polita* and always in more humid habitats.

31. Veronica filiformis Smith, Trans. Linn. Soc. London 1: 195. 1791 • Slender speedwell, véronique filiforme [I]

Perennials. Stems creeping, 5–50 cm, eglandular- and glandular-hairy. **Leaves:** blade suborbiculate, 4–13 × 4–14 mm, base cordate, margins crenate-serrate, teeth (2 or)3–5(or 6) per side, apex rounded, surfaces sparsely eglandular- and/or glandular-hairy. **Racemes** 1–6, terminal, 50–500 mm, 10–20-flowered, axis sparsely eglandular- and/or glandular-hairy; bracts suborbiculate, 4–13 mm. **Pedicels** patent or recurved, (15–)20–30 mm, length 2–5 times subtending bract, eglandular- and glandular-hairy. **Flowers:** calyx lobes 4–7 mm, 1.5–2.5 mm wide, apex rounded, sparsely glandular-hairy to glabrate; corolla blue-violet to pale blue and white, 8–14 mm diam.; stamens 2.1–2.5 mm; style 3–4 mm. **Capsules** compressed in cross section, subglobular, 3.5–5 × 5.5–6.5 mm, apex emarginate, sinus angle 50–90°, absent or almost smooth, glabrous or glandular-ciliate. **Seeds** (2–)6–14(–18), ochre, ovoid, flat to ± excavate, 1.3–1.7 × 1–1.3 mm, 0.6–0.8 mm thick, rugose to rugulose. $2n = 14$ (Europe).

Flowering (Mar–)Apr–Jun(–Aug). Moist lawns and meadows, waste fields, stream banks; 0–1000 m; introduced; B.C., Ont., Que.; Calif., Conn., Del., D.C., Idaho, Md., Mich., N.Y., Ohio, Oreg., Pa., Vt., Wash., W.Va.; Eurasia.

Veronica filiformis is self-incompatible. In the flora area, plants seldom produce capsules; it is dispersed by stem cuttings. The species is native to subalpine meadows in the Caucasus and northern Turkey.

32. Veronica argute-serrata Regel & Schmalhausen, Trudy Imp. S.-Peterburgsk. Bot. Sada 5: 626. 1878 (as argute serrata) • Sawtooth speedwell ☐

Veronica bornmuelleri Hausknecht

Annuals. Stems erect, (7–)10–20 (–30) cm, eglandular- and/or glandular-hairy. **Leaves:** blade lanceolate or elliptic, (6–)12–28 (–35) × 3–12(–20) mm, base cuneate, margins serrate, teeth 4–7(–12) per side, apex acute, surfaces sparsely glandular-hairy or glabrate. **Racemes** 1–8, terminal, usually reduced and replaced by 2 pseudodichotomous racemes with additional axillary racemes, 80–150 mm, (5–)10–25(–35)-flowered, axis eglandular- and/or glandular-hairy; bracts lanceolate, 9–12 mm, proximals with margins serrate, distals sometimes entire. **Pedicels** erect-patent to spreading proximally, deflexed distally, (3–)6–9(–12) mm, shorter than subtending bract, densely glandular- and eglandular-hairy. **Flowers:** calyx lobes (5–)6–8.5(–10) mm, apex acuminate-acute, pairs connate 0.8–1.1 mm, glandular- and eglandular-hairy, rarely sparsely hairy; corolla mostly intense blue with white center, 4–6 mm diam.; stamens 2.5 mm; style (0.7–)1–1.3(–1.7) mm. **Capsules** slightly compressed in cross section, 3.5–5 × (4–)5–8 mm, apex emarginate, sinus angle 30–45(–60)°, densely eglandular- and glandular-hairy, rarely eglandular-hairy. **Seeds** 4–10, yellowish brown, oblong to ovoid, cymbiform, 1.5–2.5 × 0.8–1.5 mm, 0.8 mm thick, cristate, dorsally reticulate-verrucate. $2n =$ 18, 42 (Asia).

Flowering Apr–Jun. Bare soils, oak and juniper forests, steppes, alpine meadows, fields, stony slopes, streams, gardens, ruins, ruderal places, calcareous and siliceous soils; (600–)900–3000 m; introduced; B.C.; Colo., Idaho, Ill., Kans., Mont., Nebr., Nev., N.Mex., N.Y., Oreg., Utah, Wash., Wyo.; Eurasia.

Veronica argute-serrata has been recorded mostly under the name *V. biloba* and occasionally under *V. campylopoda* Boissier. Most of the records for *V. biloba* likely belong under this name. *Veronica biloba* is also found in the flora area, sometimes mixed with *V. argute-serrata*; *V. campylopoda* has not been recorded from the flora area. A chromosome count of $2n = 18$ (C. R. Bell 1965) for American material is suspicious because $2n = 42$ has been reported multiple times in Turkish plants, and a base chromosome number of $x = 9$ is not known in related species.

33. Veronica biloba Linnaeus, Mant. Pl. 2: 172. 1771 • Bilobed speedwell ☐

Annuals. Stems erect, (1–)2–10 (–18) cm, eglandular-hairy. **Leaves:** blade ovate to lanceolate-ovate, (3–)4–12(–20) × (1–)2–6(–12) mm, base cuneate, margins remotely serrate, rarely dentate, apex acute, surfaces glabrate. **Racemes** 1–8, terminal, 20–80 mm, 5–15(–30)-flowered, axis eglandular- and glandular-hairy; bracts lanceolate, 6–9 mm, margins entire, apex acute. **Pedicels** straight or scarcely deflexed, erect-spreading to horizontally spreading in fruit, (2–)3–8(–11) mm, shorter than subtending bract, eglandular-hairy, sometimes glandular-hairy, rarely glabrous. **Flowers:** calyx lobes (3–)4–7(–9) mm, margins entire, apex acuminate-acute, pairs connate 0.5–1.5 mm, puberulent or glabrate; corolla pale to clear blue, 2–4 mm diam.; stamens 1–2 mm; style 0.4–0.8(–2) mm. **Capsules** slightly compressed in cross section, obcordiform, (2.4–)2.8–4(–4.5) × (3–)3.5–5(–6) mm, apex emarginate, eglandular- and/or glandular-hairy, rarely glabrous. **Seeds** (2–)4–6(–10), bright yellow, oblong, cymbiform, (1.4–)1.7–2(–2.2) × (0.7–)1–1.3 mm, 0.7 mm thick, smooth. $2n = 28$ (Asia).

Flowering Apr–Jul. Bare soils, open forests, scree, mountain and alpine meadows, humid places near streams, waste fields, ruins, calcareous and non-calcareous soils; (400–)1300–3000 m; introduced; B.C.; Mont., Utah; Asia.

Most reports of *Veronica biloba* pertain to *V. argute-serrata*. There are gatherings from potentially mixed populations (for example, Mt. Baldy, British Columbia, *Lomer 97-118*, UBC). Studies on phenotypic plasticity in these two species are necessary to ascertain the diagnostic characters.

Veronica intercedens Bornmüller, native to western Asia, was recently reported from Idaho. It is most similar to *V. biloba*, distinguished by a large calyx with broadly ovate lobes (3–5 mm versus 2–3.5 mm), the pairs joined by more than one half.

34. Veronica teucrium Linnaeus, Sp. Pl. ed. 2, 1: 16. 1762 • Large speedwell [1]

Veronica austriaca Linnaeus subsp. *teucrium* (Linnaeus) D. A. Webb; *V. austriaca* var. *teucrium* (Linnaeus) O. Bolòs & Vigo

Perennials. Stems erect, rarely ascending, (15–)30–70(–100) cm, eglandular-hairy, hairs evenly distributed around stem. **Leaves:** blade ovate or oblong-ovate, rarely lanceolate, (20–)30–55(–70) × (6–)15–25 (–45) mm, 2–3(–4) times as long as wide, base ± cordate, margins dentate to serrate to crenate-incised, rarely subentire, apex acute, abaxial surface hairy, adaxial glabrous. **Racemes** 2–4(–8), axillary, 60–150 mm, 50–150-flowered, axis hairy; bracts linear-lanceolate to linear, 4–8(–14) mm. **Pedicels** erect, 3–7(–14) mm, equal to subtending bract, eglandular-hairy. **Flowers:** calyx 5-lobed, abaxial lobes 3–4 mm, adaxial 1–1.5 mm, apex acute, glabrous or hairy; corolla bright blue, rarely pinkish or white, (9–)12–18 mm diam.; stamens 4.5–6 mm; style 5–6 mm. **Capsules** ± compressed to ± round in cross section, broadly obovoid or obcordiform, (3–)4–5.5 × (2.5–)3.5–5 mm, apex not deeply, narrowly emarginate, glabrous or sparsely hairy. **Seeds** (2–)8–18, brown, widely ellipsoid to widely obovoid, flat, 1.2–2.1 × 1–1.5 mm, 0.2–0.4 mm thick, smooth. $2n$ = 48, 64 (Europe).

Flowering May–Jul. Dry meadows, forest edges, scrub, calcareous soils; 0–1000 m; introduced; Nfld. and Labr. (Nfld.), Ont., Que.; Conn., D.C., Ill., Ind., Maine, Md., Mass., Mich., Mo., N.H., N.J., N.Y., Ohio, Pa., S.Dak., Vt., Wis.; Europe.

Veronica teucrium has often been considered an infraspecific taxon of *V. austriaca*; many records of *V. teucrium* have been reported under the name *V. austriaca*. The occurrence of *V. austriaca* in the narrow sense in the flora area could be neither supported nor excluded; it differs from *V. teucrium* mainly in distal leaves of the vegetative shoot linear and entire.

Related species have been introduced for horticultural reasons but apparently have not escaped. *Veronica satureiifolia* Poiteau & Turpin was reported from Missouri by G. Yatskievych [1999–2013; as *V. scheereri* (J.-P. Brandt) Holub] outside cultivation based on a single specimen and was similarly found in MIN but likely not naturalized. The name *V. latifolia* Linnaeus, which has at times been used for *V. teucrium*, is now considered ambiguous and should not be used (R. K. Brummitt 2007). Reports of *V. teucrium* in Saskatchewan have not been verified.

45. VERONICASTRUM Heister ex Fabricius, Enum., 111. 1759 • [Genus *Veronica* and Latin *-astrum*, resembling]

Craig C. Freeman

Herbs, perennial; rhizomatous. **Stems** erect, sparsely to densely villous proximally, sometimes glabrous distally. **Leaves** cauline, whorled, rarely opposite; petiole present; blade not fleshy, not leathery, margins serrate to doubly serrate. **Inflorescences** terminal, spikelike racemes; peduncle absent; bracts present. **Pedicels** present; bracteoles present. **Flowers** bisexual; sepals (4 or)5, proximally connate, calyx bilaterally symmetric, cylindric, lobes lanceolate; corolla white or pinkish, weakly bilaterally symmetric, weakly bilabiate, tubular-salverform, tube base not spurred or gibbous, lobes 4, abaxial 3, adaxial 1; stamens 2, proximally adnate to corolla, filaments pubescent proximally; staminode 0; ovary 2-locular, placentation axile; stigma capitate. **Fruits** capsules, dehiscence loculicidal, sometimes also septicidal, becoming 2- or apparently 4-valved distally. **Seeds** 10–30, brown or reddish brown, ellipsoid, wings absent. x = 17.

Species ca. 20 (1 in the flora): c, e North America, e Asia.

Veronicastrum appears to be part of a grade of genera in Veroniceae that is ancestral to *Veronica* (D. C. Albach et al. 2004).

1. Veronicastrum virginicum (Linnaeus) Farwell, Druggist's Circ. 61: 231. 1917 • Culver's-root or -physic E F

Veronica virginica Linnaeus, Sp. Pl. 1: 9. 1753

Stems unbranched or branched distally, 80–200 cm. **Leaves:** proximal leaves: withering, petiole 2–4 mm, blade lanceolate to broadly lanceolate or elliptic, (40–)70–140 × 10–36 mm; distal leaves: petiole 0.1–3 mm, blade lanceolate to narrowly lanceolate or elliptic, 20–40 × 3–10 mm. **Racemes** 1–8(–12), continuous, cylindric, 6–35 cm; bracts leaflike, smaller distally, (4–)12–75 × 1–12 mm; cymes 1- or 2-flowered. **Pedicels** ascending, 0.3–1.2 mm, glabrous; bracteoles linear-lanceolate to linear. **Flowers:** calyx glabrous, lobes 1.2–3 × 0.5–1 mm, abaxial 2 shorter than abaxial (2 or)3; corolla 4–5.5(–6.5) mm, glabrous externally, obscurely pubescent internally, tube not differentiated from throat, 1–1.3 mm diam., lobes spreading, broadly ovate to triangular, 1.2–2.2 mm, abaxial 3 narrower than adaxial 1; stamens long-exserted, filaments 7–9 mm; nectariferous ring at base of ovary; style 7–9 mm. **Capsules** ovoid to ellipsoid, 2.5–4.5(–5.2) × 1.8–2.3 mm, glabrous. **Seeds** 0.3–0.7 × 0.2–0.4 mm. $2n = 34$.

Flowering Jun–Aug. Dry to mesic forests, tallgrass prairies, thickets, oak savannas; 0–300 m; Man., N.S., Ont.; Ala., Ark., Conn., Del., D.C., Fla., Ga., Ill., Ind., Iowa, Kans., Ky., La., Maine, Mass., Mich., Minn., Miss., Mo., Nebr., N.J., N.Y., N.C., N.Dak., Ohio, Okla., Pa., S.C., S.Dak., Tenn., Tex., Vt., Va., W.Va., Wis.

The roots and rhizomes of *Veronicastrum virginicum* were used widely by Native Americans as an emetic and cathartic (D. E. Moerman 1998). The pharmacologic properties of *V. virginicum* have been studied and promoted since the early 1800s (K. Kindscher 1992). It is grown widely as an ornamental and often escapes from cultivation. Populations in Nova Scotia are introduced.

SCROPHULARIACEAE Jussieu

• Figwort Family

Richard K. Rabeler

Craig C. Freeman

Wayne J. Elisens

Shrubs, subshrubs, trees, or herbs, annual, biennial, or perennial, not fleshy [fleshy], autotrophic. **Stems** prostrate, ascending, pendent, or erect. **Leaves** deciduous, semipersistent, or persistent, basal and cauline or cauline, opposite or alternate, simple; stipules absent or present (most *Buddleja, Emorya, Limosella*); petiole present or absent; blade fleshy or not, leathery or not, margins entire to subentire, undulate, toothed, lobed, divided, or incised. **Inflorescences** terminal, subterminal, or axillary, racemes, cymes, panicles, or thyrses (or combinations thereof), spikes, fascicles, or flowers 1(or 2). **Flowers** bisexual or unisexual (some *Buddleja*), perianth and androecium hypogynous; sepals 4 or 5, ± distinct (*Capraria*), connate proximally, or to past middle (*Limosella*), calyx radially or bilaterally symmetric; petals 4 or 5, proximally connate, corolla radially or bilaterally symmetric, regular or bilabiate, rotate to salverform, tubular, funnelform, or campanulate; stamens mostly 4 or 5(–8 in *Myoporum*), adnate to corolla, didynamous or equal, staminode 0 or 1; pistil 1, 2-carpellate, ovary superior, 2- or 4-locular (partition incomplete and ovary 1-locular distally in *Limosella*), placentation axile (free-central in *Limosella*, apical in *Myoporum*); ovules anatropous or hemitropous (*Buddleja*), unitegmic, tenuinucellate; style 1; stigma 1, sometimes 2-lobed. **Fruits** capsules, dry and dehiscence septicidal or loculicidal, or fleshy and drupelike (*Bontia, Myoporum*) or berries (some *Buddleja*), [schizocarps]. **Seeds** 1–300, white, yellow, orangish, brown, or black, ovoid, oblong-ovoid, conic, ellipsoid, L-shaped, angled, cylindric, threadlike, or fusiform; embryo straight or slightly curved, endosperm abundant or not.

Genera ca. 60, species ca. 1700 (9 genera, 45 species in the flora): nearly worldwide except boreal and arctic North America and Asia, tropical Africa, Antarctica.

As noted in the introduction to this volume, the authors follow a narrow circumscription of Scrophulariaceae. Molecular studies (especially R. G. Olmstead et al. 2001 and B. Oxelman et al. 2005) have shown that some of the taxa commonly included in earlier treatments of Scrophulariaceae are better placed elsewhere, particularly in Orobanchaceae (hemiparasitic members, for example, Rhinantheae Lamarck & de Candolle) and Plantaginaceae

324

(for example, *Penstemon*). Oxelman et al. and D. C. Tank et al. (2006) both recognized eight tribes in their analyses of the Scrophulariaceae. Five of the eight, Buddlejeae Bartling (*Buddleja*, *Emorya*), Leucophylleae Miers (*Capraria*, *Leucophyllum*), Limoselleae Dumortier (*Limosella*), Myoporeae Reichenbach (*Bontia*, *Myoporum*), and Scrophularieae Dumortier (*Scrophularia*, *Verbascum*), are represented in the flora area.

Inclusion of Myoporaceae (three or four genera and 125 species, in the sense of A. Cronquist 1981) is warranted based on its close similarity to Leucophylleae. Morphological similarities have led workers either to propose transferring Leucophylleae from Scrophulariaceae to Myoporaceae (C. J. Niezgoda and A. S. Tomb 1975) or to question the validity of recognizing Myoporaceae as distinct from Scrophulariaceae (J. Henrickson and L. D. Flyr 1985). R. G. Olmstead et al. (2001) confirmed this closeness in their molecular survey, with *Leucophyllum* and *Myoporum* clustering together and forming a clade sister to Buddlejaceae; Leucophylleae and Myoporeae also clustered together in the studies by B. Oxelman et al. (2005) and E. Gándara and V. Sosa (2013).

Members of Buddlejaceae (10 genera, ca. 150 species, in the sense of A. Cronquist 1981) often have been included in Loganiaceae (G. K. Rogers 1986) or treated as a separate family. Cronquist noted that the four-lobed corolla may be the primitive condition for the Scrophulariales and that Buddlejaceae is clearly not primitive. B. Oxelman et al. (1999) showed that Buddlejaceae is monophyletic; R. G. Olmstead et al. (2001) and Oxelman et al. (2005) both found that *Buddleja* clustered within Scrophulariaceae in the strict sense; that interpretation is followed here. J. H. Chau et al. (2017) found that Buddlejeae is monophyletic.

SELECTED REFERENCES   Chau, J. H. et al. 2017. Phylogenetic relationships in tribe Buddlejeae (Scrophulariaceae) based on multiple nuclear and plastid markers. Bot. J. Linn. Soc. 184: 137–166. Chinnock, R. J. 2007. *Eremophila* and Allied Genera: A Monograph of the Myoporaceae. Dural, New South Wales. Fischer, E. 2004. Scrophulariaceae. In: K. Kubitzki et al., eds. 1990+. The Families and Genera of Vascular Plants. 14+ vols. Berlin, etc. Vol. 7, pp. 333–432. Flyr, L. D. 1970b. A Systematic Study of the Tribe Leucophylleae (Scrophulariaceae). Ph.D. dissertation. University of Texas. Henrickson, J. and L. D. Flyr. 1985. Systematics of *Leucophyllum* and *Eremogeton* (Scrophulariaceae). Sida 11: 107–172. Karrfalt, E. E. and A. S. Tomb. 1983. Air spaces, secretory cavities, and the relationship between Leucophylleae (Scrophulariaceae) and Myoporaceae. Syst. Bot. 8: 29–32. Niezgoda, C. J. and A. S. Tomb. 1975. Systematic palynology of tribe Leucophylleae (Scrophulariaceae) and selected Myoporaceae. Pollen & Spores 17: 495–516. Olmstead, R. G. et al. 2001. Disintegration of the Scrophulariaceae. Amer. J. Bot. 88: 348–361. Oxelman, B., M. Backlund, and B. Bremer. 1999. Relationships of the Buddlejaceae s.l. investigated using parsimony jackknife and branch support analysis of chloroplast *ndh*F and *rbc*L sequence data. Syst. Bot. 24: 164–182. Oxelman, B. et al. 2005. Further disintegration of Scrophulariaceae. Taxon 54: 411–425. Theisen, I. and E. Fischer. 2004. Myoporaceae. In: K. Kubitzki et al., eds. 1990+. The Families and Genera of Vascular Plants. 14+ vols. Berlin, etc. Vol. 7, pp. 289–292. Zona, S. 1998. The Myoporaceae in the southeastern United States. Harvard Pap. Bot. 3: 171–179.

1. Herbs or subshrubs.
    2. Plants aquatic or paludal; stems prostrate or ascending; leaves basal; stolons present
        . . . . 7. *Limosella*, p. 338
    2. Plants terrestrial; stems erect; leaves basal and cauline or cauline; stolons absent.
        3. Leaves opposite. . . . . 8. *Scrophularia*, p. 339
        3. Leaves alternate.
            4. Leaf surfaces with punctate glands and internal secretory oil cavities; inflorescences axillary, racemes. . . . . 4. *Capraria*, p. 334
            4. Leaf surfaces without punctate glands and internal secretory oil cavities; inflorescences terminal, spikes, racemes, or panicles. . . . . 9. *Verbascum*, p. 343
1. Shrubs or trees.
    5. Sepals 4; stipules usually present or as stipular lines.
        6. Calyx lobes linear-subulate; corollas long-tubular. . . . . 1. *Emorya*, p. 326
        6. Calyx lobes ovate to lanceolate; corollas campanulate-rotate, salverform, funnelform, or tubular . . . . 2. *Buddleja*, p. 327
    5. Sepals 5; stipules absent.

[7. Shifted to left margin.—Ed.]

1. EMORYA Torrey in W. H. Emory, Rep. U.S. Mex. Bound. 2(1): 121, plate 36. 1859

• Emory bush [For William Hemsley Emory, 1811–1877, commander of Texas-Mexico boundary survey]

Eliane Meyer Norman

Shrubs; stolons absent. **Stems** erect, appressed-tomentulose. **Leaves** persistent, cauline, opposite, decussate; stipules present; petiole present; blade not fleshy, subleathery or thinner, margins sinuate-dentate. **Inflorescences** terminal or subterminal, thyrselike; bracts present. **Pedicels** present; bracteoles present. **Flowers** bisexual; sepals 4, calyx radially symmetric, tubular, lobes linear-subulate; petals 4, corolla yellow or greenish yellow, radially symmetric, long-tubular, 5–6 times tube diam., abaxial lobes 2, adaxial 2; stamens 4, adnate to sinus or medially in tube, equal, filaments glabrous, staminode 0; ovary 2-locular, placentation axile; stigma subglobose. **Fruits** capsules, narrowly ovoid, dehiscence septicidal for ½ their lengths, loculicidal distally. **Seeds** 100–150, yellowish brown, lingulate, wings present. *x* = 19.

Species 2 (1 in the flora): Texas, n Mexico.

Emorya rinconensis Mayfield [*Buddleja rinconensis* (Mayfield) J. H. Chau] is endemic to Coahuila, Mexico.

SELECTED REFERENCE Norman, E. M. and R. J. Moore. 1968. Notes on *Emorya* (Loganiaceae). SouthW. Naturalist 13: 137–142.

1. Emorya suaveolens Torrey in W. H. Emory, Rep. U.S. Mex. Bound. 2(1): 121, plate 36. 1859 [F]

Buddleja normaniae J. H. Chau

Plants to 2 m, older branches terete, younger ones sub-quadrangular. **Stems** much-branched; bark grayish or light brown, splitting. **Leaves:** petiole 5–10 mm; blade ovate-deltate, 2–7 × 1.3–4 cm, base hastate, margins irregularly cleft, slightly revolute, apex acute or obtuse, abaxial surface appressed stellate-tomentose, adaxial glabrate. **Inflorescences** 5–12 cm, cyme-bearing nodes 3–5, cymes 2 per node, subtended by small leaves proximally. **Pedicels** 2–5 mm; bracteoles 1–3, linear. **Flowers** slightly fragrant; calyx tomentulose externally, tube 5–7 mm, lobes marcescent, 4–6 mm; corolla puberulent externally, glabrous internally, tube 30–35 mm, 2–3 mm diam. at base, 4–5 mm in distal ½, lobes ± erect, 2.5–5 × 3–4 mm, apex rounded or obtuse; filaments flat, 15–20 mm, anthers slightly exserted or included, 2–2.5 × 1 mm; ovary conical, 5–7 mm, tomentulose; style filiform, 25–30 mm, surpassing stamens, stigma 0.5–0.6 mm. **Capsules** 10–12 × 3–4 mm, hairy, hairs stellate and glandular. **Seeds** 2–2.5 × 0.4–0.7 mm. *2n* = 38.

Flowering Jun–Oct; fruiting Jul–Nov. Limestone cliffs or canyons in chaparral with *Juniperus, Quercus intricata, Fraxinus greggii, Senegalia berlandieri,* and *Nolina;* 600–700 m; Tex.; Mexico (Coahuila, Nuevo León).

The exact location where Parry collected the type material for *Emorya suaveolens* is not known. The author believes that all of Parry's collections of *Emorya* are from one locality. Based on available collections from the area through which he traveled, the most likely locations are the canyons of Boquillas, Maravillas, or San Vincente.

In Mexico, *Emorya suaveolens* occurs at elevations to 2400 m.

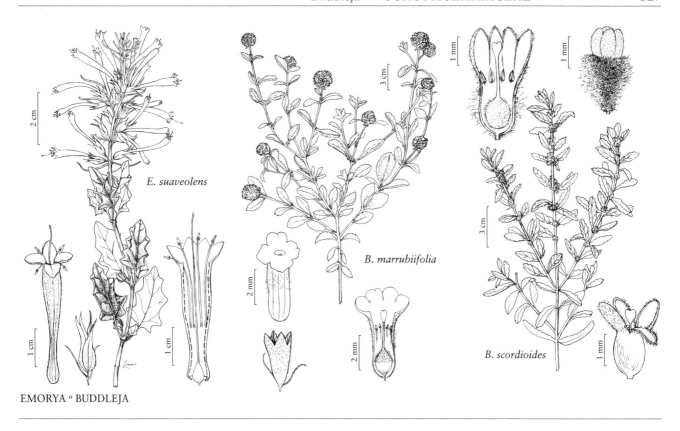

E. suaveolens

B. marrubiifolia

B. scordioides

EMORYA ∘ BUDDLEJA

2. BUDDLEJA Linnaeus, Sp. Pl. 1: 112. 1753; Gen. Pl. ed. 5, 51. 1754 • Butterfly bush
[For Adam Buddle, 1660–1715, English botanist, vicar of Farmbridge]

Eliane Meyer Norman

Chilianthus Burchell; *Nicodemia* Tenore

Shrubs or trees; stolons absent. **Stems** erect or pendent, hairy when young, hairs usually stellate
and glandular. **Leaves** persistent, semipersistent, or deciduous, cauline, opposite, decussate;
stipules present (*B. davidii*) or as stipular lines or absent; petiole present or absent; blade
not fleshy, subleathery or not, margins crenate, dentate, undulate, serrate, sinuate-dentate,
subentire, or entire, sometimes involute. **Inflorescences** terminal, rarely axillary, cymes in heads,
pseudoverticillate spikes, or panicles; bracts present. **Pedicels** often absent; bracteoles present
or absent. **Flowers** bisexual, sometimes functionally dioecious, rarely trioecious; sepals 4, calyx
radially symmetric, campanulate, lobes ovate to lanceolate; petals 4, corolla yellow, yellowish
orange, greenish yellow, purple, pink, or white, sometimes with yellow-orange eye or white
throat, radially symmetric, campanulate-rotate, salverform, funnelform, or tubular; stamens 4,
adnate to corolla tube, included, rarely exserted, equal, filaments glabrous, staminode 0; ovary
2- or 4-locular, placentation axile; stigma capitate, clavate, or clavate and slightly 2-lobed. **Fruits**
berries or capsules, ovoid or globular to cylindric, oblong, ellipsoid, or subglobular, dehiscence
septicidal and/or loculicidal, rarely indehiscent. **Seeds** 20–100+, yellow to brown, tetrahedral,
fusiform, or threadlike (*B. davidii*), usually obcompressed, wings present or absent. $x = 19$.

Species ca. 100 (10 in the flora): North America, Mexico, Central America, South America,
Asia, e Africa, Indian Ocean Islands (Comoro, Madagascar, Mascarene Islands); introduced in
Europe, elsewhere in Africa, Pacific Islands (Hawaii, New Zealand), Australia.

Buddleja species, especially Asiatic ones, are cultivated widely for their attractive flowers and leaves and their pleasant floral scent. Some plants readily persist and spread after cultivation; one of these, *B. davidii*, is considered an invasive of concern in some parts of the world (N. G. Tallent-Halsell and M. S. Watt 2009). Some species are used as remedies in folk medicine (E. M. Norman 2000). Almost all buddlejas thrive in sunny disturbed habitats.

The *Buddleja* species of Madagascar and nearby islands have berries rather than capsules and have been segregated by some authors into *Nicodemia*; African taxa that have flowers with open corollas with well-exserted stamens rather than tubular flowers with included, sessile or subsessile stamens have been referred to *Chilianthus*. Here both are treated in *Buddleja*.

J. F. Chau et al. (2017) conducted a phylogenetic study of Buddlejeae with the five genera of the tribe all treated within seven sections of *Buddleja*. This treatment is consistent with that study except for recognition of *Emorya*.

SELECTED REFERENCE   Norman, E. M. 2000. Buddlejaceae. In: Organization for Flora Neotropica. 1968+. Flora Neotropica. 121+ nos. New York. No. 81, pp. 1–225.

1. Inflorescences broadly paniculate; corollas white; stamens exserted from corolla tube. . . . . 7. *Buddleja saligna*
1. Inflorescences narrowly paniculate, spicate, racemose, or capitate; corollas yellow, yellowish orange, greenish yellow, orange, purple, violet, or pink, rarely white, sometimes with yellow-orange eye or white throat; stamens included in corolla tube.
   2. Corollas tubular or salverform, tubes 7–14 mm.
      3. Leaf blades orbiculate to elliptic; corollas greenish yellow . . . . 2. *Buddleja indica*
      3. Leaf blades ovate, ovate-lanceolate, or elliptic; corollas yellowish orange, purple, violet, pink, or white, sometimes with yellow-orange eye or white throat.
         4. Corollas yellowish orange; stamens inserted near throat . . . . 4. *Buddleja madagascariensis*
         4. Corollas purple, violet, pink, or white, sometimes with yellow-orange eye or white throat; stamens inserted medially or 3–5 from base.
            5. Cymes almost encircling inflorescence axis; corollas straight . . . . 1. *Buddleja davidii*
            5. Cymes secund; corollas curved. . . . . 3. *Buddleja lindleyana*
   2. Corollas funnelform, salverform, or campanulate-rotate, tubes 1.5–5 mm.
      6. Leaf blade adaxial surfaces glabrous or glabrescent.
         7. Shrubs, 3–15 dm; cymes 6–12-flowered; corolla lobes orbiculate . . . . 6. *Buddleja racemosa*
         7. Shrubs or trees, 15–40 dm; cymes 10–35-flowered; corolla lobes ovate . . . 9. *Buddleja sessiliflora*
      6. Leaf blade surfaces glandular-tomentose (grayish).
         8. Inflorescences capitate; corollas deep yellow (turning orange) . . . . 5. *Buddleja marrubiifolia*
         8. Inflorescences pseudoverticillate spikes; corollas light yellow, lemon yellow, or greenish yellow.
            9. Corolla tubes 1.5–2 mm, lobes oblong. . . . . 8. *Buddleja scordioides*
            9. Corolla tubes 4–5 mm, lobes orbiculate. . . . . 10. *Buddleja utahensis*

1. Buddleja davidii Franchet, Nouv. Arch. Mus. Hist. Nat., sér. 2, 10: 65. 1887–1888 (as Budleia davidi)

• Summer-lilac ⓘ Ⓦ

Shrubs, 5–30 dm. **Stems** branched, tomentose. **Leaves** often with auriculate stipules; petiole 5 mm; blade ovate to ovate-lanceolate, 10–15 × 2–4 cm, base attenuate or cuneate, margins serrate or subentire, apex acute or acuminate, abaxial surface tomentose, adaxial glabrous or glabrescent. **Inflorescences** terminal, tapered-spicate or narrowly paniculate thyrses, 10–20 × 2–4 cm, cymes almost encircling inflorescence axis, 10–25 pairs, 3–30-flowered. **Pedicels** essentially absent; bracteoles present. **Flowers** fragrant; calyx sparsely hairy externally, tube 1.5–2 mm, lobes 0.5–1.5 mm; corolla straight, violet or pink, rarely white, often with yellow-orange eye, salverform, tube 8–10 mm, lobes suborbiculate, 2–3 × 2–3 mm; stamens inserted medially in corolla tube, included in tube; ovary ovoid, 2 mm, glabrous or puberulent; stigma clavate, 1 mm. **Fruits** capsules, brown, narrowly ellipsoid, 5–9 × 1.5–2 mm, glabrous or puberulent, dehiscence primarily septicidal. **Seeds** threadlike, 3–4 × 0.5 mm, wings long. *2n* = 76.

Flowering May–Oct; fruiting Jul–Nov. Roadsides, railroad embankments, quarries, streambeds, landslide scars, sandy lakeshores, disturbed sites; 0–1300 m;

introduced; B.C., Ont.; Ala., Calif., Conn., Del., D.C., Ga., Ill., Ky., Md., Mass., Mich., Mo., N.J., N.Y., N.C., Ohio, Oreg., Pa., R.I., S.C., Tenn., Va., Wash., W.Va.; Asia (China); introduced also in Central America, South America, Europe, Africa, Pacific Islands (Hawaii, New Caledonia, New Zealand), Australia.

Buddleja davidii has been designated as a noxious weed in Oregon and Washington; it is a serious invader also in England, New Zealand, and Australia. It can often form dense thickets and produce abundant seeds (N. G. Tallent-Halsell and M. S. Watt 2009).

SELECTED REFERENCE Tallent-Halsell, N. G. and M. S. Watt. 2009. The invasive *Buddleja davidii* (butterfly bush). Bot. Rev. (Lancaster) 75: 292–325.

2. **Buddleja indica** Lamarck in J. Lamarck et al., Encycl. 1: 513. 1785 (as Budleia) • Oak-leaved butterfly bush, parlor-oak [I]

Shrubs, 5–20 dm. Stems branched, brown-tomentose when young, glabrescent. Leaves: stipules absent; petiole 3–10 mm; blade orbiculate to elliptic, 3–6 × 2–5 cm, base cuneate, margins crenate to undulate, apex rounded, surfaces glabrous except veins abaxially. Inflorescences axillary, rarely terminal, paniculate, 2–3.5 × 1–2 cm, cymes 1–3, 5–7-flowered. Pedicels essentially absent; bracteoles present. Flowers not fragrant; calyx tomentose externally, tube 1.5–2.5 mm, lobes 1–2 mm; corolla greenish yellow, salverform, tube 7–8 mm, lobes broadly ovate, 2–2.5 × 1.2–1.5 mm; stamens inserted near orifice of corolla tube, included in tube; ovary ovoid or ellipsoid, 1–1.5 mm, tomentose; stigma capitate, 0.5 mm. Fruits berries, white, ovoid, 12–15 × 5–7 mm, glandular-pubescent, indehiscent. Seeds ovoid, 2–2.5 × 1.5–2 mm, wings absent. *2n* = 76.

Flowering and fruiting year-round. Scrub, roadsides; 0–10 m; introduced; Fla.; Indian Ocean Islands (Comoro Islands, Madagascar, Mascarene Islands).

Buddleja indica was first collected in Florida in 1998; it is now found in Broward, Martin, and Palm Beach counties. It is often cultivated, especially in greenhouses in temperate regions.

3. **Buddleja lindleyana** Fortune, Edwards's Bot. Reg. 30(misc.): 25. 1844 (as Buddlea) • Lindley's butterfly bush [I]

Shrubs, 10–30 dm. Stems branched, tomentulose. Leaves: stipules absent; petiole 1–5 mm; blade elliptic to ovate, 2–10 × 1–4 cm, base acute, margins entire or sinuate-dentate, apex acuminate, abaxial surface appressed-hairy, adaxial glabrate. Inflorescences terminal, pendent, tapering-spicate thyrses, 4–20 × 2–4 cm, cymes secund, 10–20 pairs, 1–5-flowered. Pedicels essentially absent; bracteoles absent or present. Flowers fragrant; calyx densely hairy externally, tube 1.5–3 mm, lobes 0.5–1 mm; corolla curved, purple with white throat, tubular, tube 11–14 mm, lobes suborbiculate, 2–3 × 2–3 mm; stamens inserted 3–5 mm distal to base of corolla tube, included in tube; ovary ovoid, 1.5–2.5 mm, glabrate; stigma clavate, 1–1.5 mm. Fruits capsules, brown, ellipsoid, 4–6 × 2–2.5 mm, glabrous or sparsely glandular, dehiscence primarily septicidal. Seeds tetrahedral, 0.7–1 × 0.5–0.7 mm, wings narrow at edges. *2n* = 38.

Flowering Jun–Oct; fruiting Aug–Nov. Roadsides, railroad embankments; 10–200 m; introduced; Ala., Ariz., Fla., Ga., La., Miss., N.C., S.C., Tex.; Asia (China, Japan); introduced also in West Indies (Guadeloupe, Jamaica, Martinique).

4. **Buddleja madagascariensis** Lamarck in J. Lamarck et al., Encycl. 1: 513. 1785 (as Budleia) • Smoke bush [I]

Nicodemia madagascariensis (Lamarck) R. Parker

Shrubs, climbing, sarmentose, 20–50 dm. Stems branched, tomentose. Leaves often with globular stipular lines; petiole 5–15 mm; blade ovate to elliptic, 8–15 × 2–5 cm, base rounded, margins entire, apex acuminate, abaxial surface tomentose, adaxial glabrous or glabrate. Inflorescences terminal, tapering-spicate thyrses, 5–25 × 3–5 cm, cymes 10–25 pairs, 5–15-flowered. Pedicels essentially absent; bracteoles present. Flowers slightly fragrant; calyx tomentose externally, tube 2–3 mm, lobes 0.5–1 mm; corolla yellowish orange, salverform, tube 7–9 mm, lobes suborbiculate, 2.5–3.5 × 2–3 mm; stamens inserted proximal to throat of corolla tube, included in tube; ovary subglobular, 1–1.5 mm, tomentose at tip; stigma clavate, 1–1.5 mm. Fruits berries, bluish purple, globular, 2.5–5 × 2.5–5 mm, papillose, indehiscent. Seeds ovoid, 0.6–0.9 × 0.5–0.6 mm, wings absent. *2n* = 38.

Flowering Mar–May; fruiting not known in continental United States. Roadsides; 10–20 m; introduced; Fla.; Indian Ocean Islands (Madagascar); introduced also in Mexico, West Indies (Cuba, Puerto Rico), South America (Argentina, Uruguay), Europe (Greece), Africa (Republic of South Africa), Indian Ocean Islands (Mauritius, Réunion), Pacific Islands (Fiji, Hawaii, New Caledonia, New Zealand), Australia.

Buddleja madagascariensis is considered a serious invader in Hawaii at 900–1200 m, where it sets abundant fruits (Hawaii Invasive Species Council, https://dlnr.hawaii.gov/hisc/info/species/smoke-bush/); this contrasts with the report by A. J. M. Leeuwenberg (1979) that he had observed specimens with fruits only from Crete, Madagascar, and Mauritius.

5. **Buddleja marrubiifolia** Bentham in A. P. de Candolle and A. L. P. P. de Candolle, Prodr. 10: 441. 1846 • Woolly butterfly bush [F]

Shrubs, 5–20 dm. Stems much-branched, tomentose. Leaves: stipular lines faint; petiole 3–5 mm; blade grayish, ovate to rhombic, 1–6 × 1–2.5 cm, base cuneate, acute, or attenuate, margins crenate, apex obtuse or rounded, surfaces densely tomentose. Inflorescences terminal, capitate, 1–1.5 × 0.8–1.2 cm, cymes 1, 5–25-flowered. Pedicels essentially absent; bracteoles absent. Flowers fragrant; calyx tomentulose externally, tube 2–3.5 mm, lobes 1–1.5 mm; corolla deep yellow (turning orange), salverform, tube 3–4.5 mm, lobes orbiculate, 1–1.5 ×1.5–2 mm; stamens inserted in distal ⅓ of corolla tube, included in tube; ovary ovoid, 1–1.5 mm, distal ½ glandular-tomentose; stigma clavate, slightly 2-lobed at apex, 0.6–1 mm. Fruits capsules, brown, oblong-ovoid, 3–4 × 1.5–2 mm, glandular-tomentulose at apex, dehiscence septicidal and loculicidal. Seeds ellipsoid, 1–1.3 × 0.2–0.3 mm, wings short at chalazal end. $2n = 38$.

Flowering Feb–Aug; fruiting Jul–Oct. Limestone cliffs and canyons; 600–1300 m; Tex.; Mexico (Chihuahua, Durango, Hidalgo, Nuevo León, San Luis Potosí, Tamaulipas, Zacatecas).

Buddleja marrubiifolia is found with *Agave lechuguilla*, *Dasylirion leiophyllum*, and *Fouquieria splendens*. In Mexico, it is found growing at elevations to 2300 m. Although cultivated in Arizona and Hawaii, it does not appear to have escaped in either state.

6. **Buddleja racemosa** Torrey in W. H. Emory, Rep. U.S. Mex. Bound. 2(1): 121. 1859 (as Buddleia) • Wand butterfly bush [E]

Shrubs, often pendent, lax, 3–15 dm. Stems much-branched, glandular-tomentose when young. Leaves: stipular lines faint; petiole 5–20 mm; blade ovate-oblong or lanceolate, 3–10 × 1.5–4 cm, base hastate, cuneate, or truncate, margins irregularly dentate, crenate, or subentire, apex acute, abaxial surface appressed-tomentose or with stellate hairs, always with glandular hairs, adaxial glabrous or glabrate. Inflorescences terminal, racemose, 3–10 × 1.5–4 cm, cymes capitate, 4–28 pairs, 6–12-flowered. Pedicels absent; bracteoles absent. Flowers slightly fragrant; calyx tomentose and glandular externally, tube 1–1.7 mm, lobes 0.5–0.8 mm; corolla pale yellow, funnelform, salverform, or campanulate-rotate, tube 1.5–2.4 mm, lobes orbiculate, 0.7–1 × 1–1.5 mm; stamens inserted in distal ⅓ of corolla tube, included in tube; ovary ovoid, 0.7–1.2 mm, glandular-tomentose; stigma clavate, slightly 2-lobed at apex, 0.5–0.7 mm. Fruits capsules, brown, ovoid, 2–2.5 × 1.7–2 mm, glandular-tomentose, dehiscence septicidal and loculicidal. Seeds ellipsoid, 0.5–0.6 × 0.2–0.3 mm, wings absent.

Varieties 2 (2 in the flora): Texas.

1. Leaves sparsely stellate-pubescent abaxially . . . .
   . . . . 6a. *Buddleja racemosa* var. *racemosa*
1. Leaves appressed-stellate-tomentose abaxially . .
   . . . . 6b. *Buddleja racemosa* var. *incana*

6a. **Buddleja racemosa** Torrey var. **racemosa** [E]

Leaves sparsely stellate-pubescent abaxially. $2n = 38$.

Flowering May–Aug; fruiting Jul–Oct. Limestone cliffs, along streams; 200–800 m; Tex.

Variety *racemosa* is found on the eastern end of Edwards Plateau.

6b. **Buddleja racemosa** Torrey var. **incana** Torrey in W. H. Emory, Rep. U.S. Mex. Bound. 2(1): 121. 1859 (as Buddleia) [E]

Leaves appressed-stellate-tomentose abaxially.

Flowering May–Aug; fruiting Jul–Oct. Limestone cliffs, along streams; 200–800 m; Tex.

Variety *incana* is found on the southwestern part of Edwards Plateau.

7. Buddleja saligna Willdenow, Enum. Pl. 1: 159. 1809
  • Squarestem butterfly bush [I]

Shrubs or trees, 5–50 dm. Stems much-branched, lepidote when young. **Leaves:** stipular lines faint; petiole 2–10 mm; blade narrowly elliptic to linear, 6–15 × 0.4–3 cm, base cuneate or decurrent, margins entire, apex acute to acuminate, abaxial surface appressed-tomentose, adaxial glabrate. **Inflorescences** axillary or terminal, broadly paniculate, 8–14 × 10–16 cm, cymes 5–7 pairs, 15–60-flowered, cymules 3–5-flowered. **Pedicels** present; bracteoles present or absent. **Flowers** fragrant; calyx lepidote externally, tube 0.5–1 mm, lobes 0.2–0.3 mm; corolla white, campanulate-rotate, tube 1–1.2 mm, lobes suborbiculate, 1–1.5 × 0.7–1 mm; stamens inserted medially in corolla tube, exserted from tube; ovary ovoid, 0.5–0.8 mm, sparsely lepidote; stigma capitate, 0.2 mm. **Fruits** capsules, brown, oblong, 1.5–2.5 × 0.8–1 mm, sparsely lepidote, dehiscence primarily septicidal. **Seeds** ovoid, 1–1.4 × 0.4–0.6 mm, wings short at both ends.

Flowering May–Sep; fruiting Jul–Oct. Roadsides, disturbed chaparral, banks of canyons; 400–1200 m; introduced; Calif.; Africa (Republic of South Africa); introduced also in Pacific Islands (Hawaii), Australia.

Buddleja saligna is established in Los Angeles County, where it was first collected in 1965 (*Fuller 13379*, CDA). It was originally reported for California under the illegitimate name *Chilianthus oleaceous* Burchell.

8. Buddleja scordioides Kunth in A. von Humboldt et al., Nov. Gen. Sp. 2(fol.): 278; 2(qto.): 345; plate 183. 1818 • Escobilla butterfly bush [F]

Shrubs, 3–12 dm. Stems much-branched, tomentose and glandular. **Leaves:** stipular lines faint; petiole 0 mm; blade grayish, oblong to linear, 1–3 × 0.3–0.8 cm, base cuneate or attenuate, margins coarsely crenate, apex obtuse, surfaces glandular-tomentose. **Inflorescences** terminal, pseudoverticillate spikes, 2–10 × 0.4–0.7 cm, cymes 3–15 pairs, 6–15-flowered. **Pedicels** absent; bracteoles absent. **Flowers** fragrant; calyx thickly tomentose externally, tube 1.5–1.8 mm, lobes 0.2–0.5 mm; corolla lemon yellow or greenish yellow, funnelform, tube 1.5–2 mm, lobes oblong, 1–1.5 × 0.5–0.9 mm; stamens inserted at orifice of corolla tube, included in tube; ovary globular, 0.5–1 mm, glandular-tomentose; stigma clavate, slightly 2-lobed at apex, 0.3–0.5 mm. **Fruits** capsules, grayish green, subglobular,

1.5–2.5 × 1.5 mm, densely glandular-tomentulose at apex, indehiscent. **Seeds** ovoid, 0.7–0.8 × 0.5–0.6 mm, wings absent. $2n = 38$.

Flowering May–Aug; fruiting Sep–Nov. Roadsides, rangeland, thorn scrub; 600–1400 m; Ariz., N.Mex., Tex.; Mexico.

Buddleja scordioides is weedy and grows on limestone or gypsum soils in the trans-Pecos region of Texas with *Flourensia cernua*, *Larrea tridentata*, and *Senegalia greggii*, and in semixeric areas in central Arizona and southeastern New Mexico. It occurs also in Mexico, throughout the Chihuahuan Desert and as far south as México, at elevations to 2500 m.

9. Buddleja sessiliflora Kunth in A. von Humboldt et al., Nov. Gen. Sp. 2(fol.): 278; 2(qto.): 345; plate 182. 1818
  • Rio Grande butterfly bush

Buddleja pringlei A. Gray

Shrubs or trees, 15–40 dm. Stems much-branched, tomentose when young. **Leaves:** basal: stipular lines evident; petiole 10–40 mm; blade ovate, 9–23 × 5–14 cm, base attenuate to obtuse, decurrent, margins serrate or double-serrate, apex acute or acuminate, abaxial surface tomentulose, adaxial glabrate; cauline: stipular lines evident; petiole 0–10 mm; blade lanceolate or narrowly elliptic, 5–15 × 1.5–3 cm, base attenuate, decurrent, margins entire or irregularly serrate, apex acute, abaxial surface tomentulose, adaxial glabrate. **Inflorescences** terminal, pseudoverticillate spikes, 6–25 × 1–3 cm, cymes 5–20 pairs, 10–35-flowered. **Pedicels** absent; bracteoles absent. **Flowers** malodorous; calyx tomentose externally, tube 1.7–3 mm, lobes 1–1.8 mm; corolla yellow, funnelform, tube 3.2–4.2 mm, lobes ovate, 1–1.5 × 1.2–1.7 mm; stamens inserted 1 mm proximal to orifice of corolla tube, included in tube; ovary ovoid, 1–1.5 mm, distal ½ tomentose; stigma clavate, 1.2–1.4 mm. **Fruits** capsules, brown, cylindric, 4–4.5 × 1.7–2 mm, tomentose at apex, dehiscence septicidal and loculicidal. **Seeds** fusiform, 0.8–1.2 × 0.3–0.4 mm, wings short. $2n = 76$.

Flowering Jan–Apr; fruiting Mar–Jun. Tamaulipan thorn scrub, riparian woodlands, roadsides; 10–1400 m; Ariz., Tex.; Mexico.

Buddleja sessiliflora is found in association with *Dodonaea viscosa*, *Fouquieria splendens*, and *Parkinsonia aculeata* in Arizona and with *Celtis laevigata*, *Fraxinus berlandieriana*, and *Prosopis glandulosa* in southernmost Texas. It is widespread in Mexico from Sonora and Tamaulipas to the Isthmus de Tehuantepec, except in the Chihuahuan Desert, at elevations to 2800 m.

10. Buddleja utahensis Coville, Proc. Biol. Soc. Wash. 7: 69. 1892 (as Buddleia) • Panamint or Utah butterfly bush E

Buddleja marrubiifolia Bentham var. *utahensis* (Coville) M. E. Jones

Shrubs, 3–10 dm. **Stems** much-branched, gray-tomentose. **Leaves:** stipular lines faint; petiole 0 mm; blade grayish, linear to oblong, 1.5–3.5 × 0.3–0.5 cm, base attenuate, margins involute, crenate to undulate, apex rounded, surfaces densely glandular-tomentose. **Inflorescences** terminal, pseudoverticillate spikes, 4–12 × 0.8–1.3 cm, cymes 3–7 pairs, 5–15-flowered. **Pedicels** absent; bracteoles present or absent. **Flowers** slightly fragrant; calyx glandular-tomentose externally, tube 2.5–3.5 mm, lobes 1–2 mm; corolla light yellow, salverform, tube 4–5 mm, lobes orbiculate, 1–1.2 × 1.5–1.7 mm; stamens inserted 1.5–1.7 mm proximal to orifice of corolla tube, included in tube; ovary ovoid, 1–1.5 mm, distal ½ tomentulose; stigma clavate, slightly 2-lobed at apex, 0.5–1 mm. **Fruits** capsules, brown, ovoid, 2–2.5 × 1.3–1.5 mm, tomentulose near apex, dehiscence septicidal and loculicidal. **Seeds** ellipsoid, 0.4–0.5 × 0.2–0.3 mm, wings absent.

Flowering Apr–Aug; fruiting Jun–Sep. Limestone outcrops in pinyon-juniper and Joshua tree woodlands; 700–2000 m; Ariz., Calif., Nev., Utah.

3. LEUCOPHYLLUM Humboldt & Bonpland, Pl. Aequinoct. 2: 95, plate 109. 1812

• Barometer-bush, silver-leaf, cenizo [Greek *leukos*, white, and *phyllon*, leaf]

James Henrickson

Guy L. Nesom

Shrubs; stolons absent. **Stems** erect, well branched, densely silvery gray-tomentose, hairs conic to cylindric, dendritic, sometimes appressed-stellate, glabrescent, [glabrous and sessile-glandular]. **Leaves** persistent, cauline, alternate to opposite or subopposite; stipules absent; petiole usually present; blade plane or ± conduplicate, not fleshy, not leathery, margins entire. **Inflorescences** axillary, flowers solitary; bracts absent. **Pedicels** present; bracteoles absent. **Flowers** bisexual; sepals 5, calyx radially symmetric, campanulate, lobes linear to oblong-lanceolate or oblong; petals 5, corolla purplish lavender to violet or blue, rarely white, with white in proximal throat, usually marked with yellow-brown, orange, or purple-violet spots, slightly bilaterally symmetric, funnelform-campanulate, abaxial lobes 3, adaxial 2, adaxial lip with lobes slightly smaller than abaxial; stamens 4, adnate to proximal ¼–⅓ of corolla tube, didynamous, filaments glabrous or pilose near base, staminode 0; ovary 2-locular, placentation axile; stigma 2-lobed. **Fruits** capsules, oblong-ovoid, dehiscence initially septicidal to near base, eventually loculicidal halfway to base. **Seeds** 15–25, yellowish brown, irregularly ovoid, minutely reticulate, wings absent. *x* = 17.

Species 15 (3 in the flora): sc United States, Mexico.

Leucophyllum is closely related to *Eremogeton* Standley & L. O. Williams of southern Mexico and Central America and *Capraria* of tropical southern Florida, Mexico, the West Indies, and Central and South America. Studies by E. Gándara and V. Sosa (2013) show both *Capraria* and *Eremogeton* imbedded within *Leucophyllum* in chloroplast and chloroplast-dominated phylogenies, but not in their nuclear (ITS) phylogeny (pers. comm.), leaving the relationship of these taxa unresolved.

The relationship of *Leucophyllum* with Myoporaceae often has been suggested (L. D. Flyr 1970b; C. J. Niezgoda and A. S. Tomb 1975; E. E. Karrfalt and Tomb 1983). Molecular studies recognize Myoporeae and Leucophylleae as sister tribes within the redrawn Scrophulariaceae (R. G. Olmstead et al. 2001; B. Oxelman et al. 2005).

Leucophyllum zygophyllum I. M. Johnston is mapped in Hidalgo County, Texas (B. L. Turner et al. 2003); the voucher, deposited at TEX, was most likely collected in Mexico.

1. Leaves: abaxial surfaces silvery gray, adaxial more greenish; plants 5–20(–30) dm; sc to trans-Pecos Texas. . . . . 1. *Leucophyllum frutescens*
1. Leaf surfaces equally silvery gray; plants 2–10(–15) dm; trans-Pecos Texas.
   2. Young stems and leaves densely and evenly silvery, hairs stellate, closely and tightly appressed, densely overlapping; plants often appearing thorny . . . . 2. *Leucophyllum minus*
   2. Young stems and leaves irregularly canescent-tomentose, hairs conic to cylindric, dendritic, uneven in height; plants not appearing thorny. . . . . 3. *Leucophyllum candidum*

1. Leucophyllum frutescens (Berlandier) I. M. Johnston, Contr. Gray Herb. 70: 89. 1924 • Texas barometer-bush or silver-leaf, purple sage F W

Terania frutescens Berlandier, Mem. Comis. Limites, 4. 1832

Shrubs erect, not intricately branched, rounded, 5–20(–30) dm, not appearing thorny. **Young stems** densely canescent-tomentose, hairs conic to cylindric, dendritic, uneven in height. **Leaves** alternate, rarely opposite; petiole 1–2 mm; blade obovate to oblong-obovate or obovate-orbiculate, 10–25(–35) mm, base cuneate, midvein and major lateral veins raised abaxially, abaxial surface silvery gray, adaxial more greenish, hairs conic to cylindric, dendritic, uneven in height. **Flowers:** calyx lobes oblong-lanceolate, 3–5 mm; corolla rose lavender to light violet, pink, and rose pink, rarely white, campanulate, 18–26 mm, tube not notably narrowed. $2n = 34$.

Flowering (Mar–)May–Sep(–Oct). Rocky and gravelly hillsides, talus, arroyos, ridges, flats, roadcuts, clay dunes, scrub, chaparral, thorn scrub, riparian communities; 10–1200 m; Tex.; Mexico (Coahuila, Nuevo León, Tamaulipas).

Leucophyllum frutescens is widely cultivated; horticultural varieties differ in habit, vestiture, and corolla color. The plants are cold hardy and can withstand moderate frosts. As in most or all *Leucophyllum* species, plants usually flower in response to rain.

In Texas, *Leucophyllum frutescens* is known from much of the southwestern half of the state.

2. Leucophyllum minus A. Gray in W. H. Emory, Rep. U.S. Mex. Bound. 2(1): 115. 1859 • Big Bend barometer-bush or silver-leaf

Shrubs erect, intricately branched, compact, 2–8(–15) dm, often appearing thorny. **Young stems** densely and evenly silvery stellate, hairs closely and tightly appressed, overlapping, appearing stellate, actually compressed-dendritic with radii extending from multiple levels, typically with one series of radii at tip. **Leaves** alternate, crowded in axillary fascicles or on compressed lateral

shoots; petiole (0.5–)1–3(–4) mm; blade oblanceolate or spatulate to obovate-orbiculate, (2–)3–10(–16) mm, base cuneate-attenuate, surfaces equally silvery gray, hairs stellate, closely and tightly appressed, overlapping. **Flowers:** calyx lobes linear to oblong-lanceolate, (2–)3–4.5 mm; corolla lavender to purple or blue, rarely white, campanulate, (12–)18–25 mm, abruptly ampliate distal to narrow tube. $2n = 34$.

Flowering (May–)Jun–Aug(–Nov). Limestone ridges, slopes, and ledges, gravel, clayey, sandy hills, gravelly washes; 900–1700 m; N.Mex., Tex.; Mexico (Chihuahua, Coahuila).

In Texas, *Leucophyllum minus* is known from the Big Bend region westward through the Trans-Pecos.

3. Leucophyllum candidum I. M. Johnston, J. Arnold Arbor. 22: 120. 1941 • Brewster County barometer-bush

Leucophyllum violaceum Pennell

Shrubs erect, not intricately branched, rounded, 3–10(–15) dm, not appearing thorny. **Young stems** irregularly canescent-tomentose, hairs conic to cylindric, dendritic, uneven in height. **Leaves** alternate or subopposite, sometimes opposite near stem apices; petiole 1–3(–6) mm; blade broadly obovate to obovate-orbiculate, reniform, or ovate, 6–10(–16) mm, base rounded to cuneate, veins not prominently raised abaxially, surfaces densely equally silvery gray-tomentose or irregularly canescent-tomentose, hairs conic to cylindric, dendritic, uneven in height. **Flowers:** calyx lobes oblong to oblong-lanceolate, 2.5–5(–6) mm; corolla dark to light violet-purple, campanulate, (10–)12–22(–25) mm, tube not notably narrowed. $2n = 68$.

Flowering (Apr–)May–Sep. Limestone hills, canyons, steep slopes, gravelly slopes, flats, roadsides; 600–1200 m; Tex.; Mexico (Chihuahua, Coahuila, Durango, Zacatecas).

Leucophyllum candidum is known in the flora area only from Brewster County.

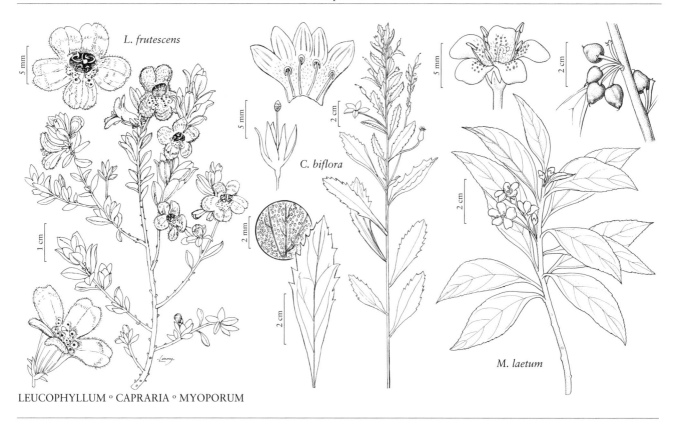

LEUCOPHYLLUM ∘ CAPRARIA ∘ MYOPORUM

4. CAPRARIA Linnaeus, Sp. Pl. 2: 628. 1753; Gen Pl. ed. 5, 276. 1754 • Goat's head [Latin *capri*, goat, and *-arius*, pertaining to, alluding to consumption by goats] ☐

Justin K. Williams

Pogostoma Schrader; *Xuarezia* Ruiz & Pavon

Subshrubs; stolons absent, taproot woody. **Stems** erect, glabrous or glabrate to hirsute [densely glandular-pubescent]. **Leaves** deciduous, mostly cauline, alternate; stipules absent; petiole absent; blade not fleshy, not leathery, margins serrate distally, surfaces with punctate glands and internal secretory oil cavities. **Inflorescences** axillary, racemes; bracts absent. **Pedicels** present; bracteoles absent. **Flowers** bisexual; sepals 5, ± distinct, lanceolate, calyx radially symmetric, rotate; petals [4]5, corolla white, sometimes with scattered purple spots inside, radially or bilaterally symmetric, rotate or tubular-funnelform, abaxial lobes 3, adaxial 2; stamens 4 or 5, adnate to base of corolla tube, didynamous or equal, filaments glabrous, staminode 0; ovary 2-locular, placentation axile; stigma ellipsoid [reniform]. **Fruits** capsules, elliptic, dehiscence loculicidal. **Seeds** 90–120, brown, ovoid, wings absent. x = 14 or 15.

Species 4 (2 in the flora): introduced; Mexico, West Indies, Central America, South America, Pacific Islands; introduced also in Africa (Cape Verde Islands, Ghana), Indian Ocean Islands (Mauritius).

Capraria is unique in Scrophulariaceae in having alternate leaves with surfaces with punctate glands and with internal oil-secreting cavities.

A molecular study by E. Gándara and V. Sosa (2013) showed that, depending on how *Leucophyllum* is defined, *Capraria* may either be considered nested within a broadly defined *Leucophyllum* or retained as a separate genus, as treated here.

SELECTED REFERENCES Sprague, T. A. 1921. A revision of the genus *Capraria*. Bull. Misc. Inform. Kew 1921: 205–212. Williams, J. K. 2004. A revision of *Capraria* (Scrophulariaceae). Lundellia 7: 53–78.

1. Stems and leaf blades hirsute to glabrate or glabrous; leaf blades spatulate; flowers bilaterally symmetric, 10–13 mm; corollas white with scattered purple spots inside, tubular-funnelform, villous inside; stamens 4(or 5), didynamous; styles included. . . . . 1. *Capraria biflora*
1. Stems and leaf blades glabrous; leaf blades lanceolate; flowers radially symmetric, 8–10 mm; corollas white, rotate, glabrous; stamens 5, equal; styles exserted . . . . 2. *Capraria mexicana*

1. Capraria biflora Linnaeus, Sp. Pl. 2: 628. 1753 F I

Stems branched, 4–20 dm, hirsute to glabrate. **Leaves:** blade spatulate, 30–80 × 5–35 mm, glabrous or moderately hirsute. **Pedicels** 5–22 mm, glabrous or glandular-pubescent. **Flowers** bilaterally symmetric, 10–13 mm; sepals 4–7 mm, glabrous; corolla white, with purple spots inside, tubular-funnelform, villous inside; stamens 4(or 5), didynamous; ovary glabrous; style included, 3–5 mm, glabrous or sparsely pilose. **Seeds** 0.4–0.5 × 0.3–0.4 mm. $2n$ = 28, 60 (Africa).

Flowering fall–spring. Beaches, dunes, empty lots, roadways, streams; 0–10 m; introduced; Fla.; Mexico; West Indies; Central America; South America (e of the Andes); Pacific Islands (Galapagos Islands).

In the United States, *Capraria biflora* grows only in the southern tip of Florida, where it is widely distributed on the southern quarter of the peninsula as well as throughout the Florida Keys. It is commonly cultivated throughout the world for its purported healing properties.

2. Capraria mexicana Moricand ex Bentham in A. P. de Candolle and A. L. P. P. de Candolle, Prodr. 10: 429. 1846 I

Stems branched, 5–20 dm, glabrous. **Leaves:** blade lanceolate, 40–100 × 6–22 mm, glabrous. **Pedicels** 5–12 mm, glabrous. **Flowers** radially symmetric, 8–10 mm; sepals 3–5 mm, glabrous; corolla white, rotate, glabrous; stamens 5, equal; ovary glabrous; style exserted, 3–5 mm, glabrous. **Seeds** to 0.4 × 0.2 mm.

Flowering fall–spring. Beaches, roadsides, streams, disturbed areas; 0–700 m; introduced; Tex.; Mexico (Guanajuato, Querétaro, San Luis Potosí, Tamaulipas, Veracruz); Central America (Belize).

Known in the flora area only since 1993 (J. Ideker 1996b), *Capraria mexicana* grows in a handful of populations in two counties of southern Texas (Cameron and Starr). Reports suggest that a flood extirpated the only known population in Starr County (A. Richardson and K. King 2006). Although rare in Texas, this species is widespread throughout the northeastern half of Mexico.

5. MYOPORUM Solander ex G. Forster, Fl. Ins. Austr., 44. 1786 • [Greek *myo*, to shut, and *poros*, hole, alluding to transparent spots on leaves closed with pellucid substance] I

Robert E. Preston

Shrubs or trees; stolons absent. **Stems** ascending to prostrate, glabrous. **Leaves** persistent, cauline, alternate, rarely opposite, with embedded, translucent glands; stipules absent; petiole present [absent]; blade fleshy or not, leathery or not, margins entire or serrate. **Inflorescences** axillary, in clusters [flowers solitary]; bracts absent. **Pedicels** present; bracteoles absent. **Flowers** bisexual; sepals 5, calyx radially or bilaterally symmetric, campanulate, lobes oblong; petals 5, corolla white [pale purple], spotted with purple [orange-brown to yellow or unspotted], radially or bilaterally symmetric, campanulate, abaxial lobes 3, adaxial 2, lobes glabrous [pubescent] abaxially, variously pubescent adaxially; stamens 4–8, adnate to corolla, equal, filaments hairy [glabrous], staminode 0; ovary [2–]4[–6]-locular, placentation apical; stigma capitate or 2–5-lobed. **Fruits** drupelike capsules, ovoid to globular, fleshy [dry]. **Seeds** 1–4, white to pale brown, ovoid to oblong or ellipsoid, wings absent. x = 18.

Species ca. 30 (3 in the flora): introduced, California; Indian Ocean Islands, Pacific Islands, Australia; introduced also in s South America.

Myoporum species have been introduced for ground cover or hedges in coastal or low rainfall regions of many countries, especially *M. insulare* R. Brown, *M. laetum*, and *M. montanum* R. Brown. The gall-inducing thrip *Klambothrips myopori*, native to Australia, has recently been introduced to California and is causing substantial damage to *Myoporum* species used in landscaping (L. A. Mound and D. C. Morris 2007). The leaves and fruits of *M. laetum* and other *Myoporum* species are toxic to livestock; they contain ngaione, a furanoid sesquiterpene that causes photosensitization and liver damage (G. S. Richmond and E. L. Ghisalberti 1995; K. Parton and A. N. Bruere 2002).

1. Shrubs, prostrate; leaf blades narrowly oblanceolate, 2–4 cm; capsules white to pale brown . . . . 3. *Myoporum parvifolium*
1. Shrubs or trees, broadly spreading; leaf blades narrowly elliptic to lanceolate, 5–14 cm; capsules pale to dark reddish purple.
    2. Leaf blades lanceolate, margins finely serrate distal to middle, glands conspicuous; flowers 2–4 per axil; anthers well exserted from tube; ovaries smooth . . . . 1. *Myoporum laetum*
    2. Leaf blades narrowly elliptic, tapering proximally and distally, margins entire, glands inconspicuous; flowers 6–8 per axil; anthers slightly exserted from tube; ovaries rugose . . . . .2. *Myoporum acuminatum*

1. Myoporum laetum G. Forster, Fl. Ins. Austr., 44. 1786 • Ngaio tree F I W

Shrubs or trees, broadly spreading, 30–100 dm. **Stems** ascending to prostrate, much branched; twig tips and young leaves bronze green, sticky. **Leaves:** blade bright green, lanceolate, 5–12.5 × 1.5–3 cm, margins finely serrate distal to middle, embedded glands conspicuous. **Flowers** 2–4 per axil; corolla white with purple spots on lobes and distal tube, tube 3.5–4.5 mm, lobes equal, 4–5.5 mm, densely long-hairy adaxially; anthers well exserted from tube; ovary smooth. **Capsules** pale to dark reddish purple, ovoid, 5–10 mm. **Seeds** oblong, 3–3.5 mm. $2n$ = 108 (New Zealand).

Flowering (Jan–)Mar–Aug. Open areas in grasslands, scrub, riparian habitats, generally coastal; 0–500 m; introduced; Calif.; Pacific Islands (New Zealand); introduced also in s South America (Argentina, Uruguay).

Myoporum laetum is commonly cultivated in coastal areas of California. Although first collected outside of cultivation in 1949, it was not recognized as an introduced element of local and regional floras until the 1970s. It has naturalized mostly in southern California to San Luis Obispo County with some populations north along the coast to the San Francisco Bay area.

Myoporum insulare R. Brown, also cultivated in California, is similar to *M. laetum*, and some reports of *M. laetum* are possibly *M. insulare*. *Myoporum insulare* has leaves that are lighter green when young,

and the translucent glands of the mature leaves are less conspicuous. The flowers are slightly smaller with anthers that are only slightly exserted from the tubes, and the fruits are smaller and globular.

2. Myoporum acuminatum R. Brown, Prodr., 515. 1810 • Waterbush I W

Shrubs or trees, broadly spreading, 20–100 dm. **Stems** ascending to prostrate, much branched; twig tips and young leaves green to blackish, sticky. **Leaves:** blade dark green, narrowly elliptic, tapering proximally and distally, 5–14 × 1–3 cm, margins entire [obscurely serrate distal to middle], embedded glands inconspicuous. **Flowers** 6–8 per axil; corolla white with purple spots on lobes and distal tube, tube 3–4 mm, lobes equal, 3–4.5 mm, long-hairy adaxially; anthers slightly exserted from tube; ovary rugose. **Capsules** dark purple, ovoid, 4–7 mm. **Seeds** ovoid-oblong, 2.2–2.5 mm.

Flowering Feb–May. Coastal sage scrub and chaparral; 0–100 m; introduced; Calif.; Australia.

Cultivated in coastal southern California, *Myoporum acuminatum* is established in nearby wildlands.

Myoporum montanum R. Brown, also cultivated in California, is similar to *M. acuminatum*, and some reports of *M. acuminatum* are possibly *M. montanum*. *Myoporum montanum* has narrower leaves (5–20 mm wide), anthers that are included in the tubes, smooth ovaries, and pink to light purple fruits.

3. Myoporum parvifolium R. Brown, Prodr., 516. 1810 • Creeping myoporum [I] [W]

Shrubs, prostrate, to 5 dm. **Stems** prostrate, much branched, 15 dm, often rooting at nodes; twig tips and young leaves green, not sticky. **Leaves:** blade green, narrowly oblanceolate, 2–4 × 0.5 cm, margins entire or sparsely serrate distal to middle, embedded glands inconspicuous. **Flowers** 1–3 per axil; corolla white, purple-spotted at bases of lobes, tube 2.5–3 mm, lobes equal, 3–4 mm, sparsely hairy adaxially; anthers well exserted from tube; ovary smooth. **Capsules** white to pale brown, globular, 5–7 mm, fleshy. **Seeds** ovoid to ellipsoid, 2 mm.

Flowering Apr–Oct. Vacant lots, open, mesic areas in chaparral; 100–300 m; introduced; Calif.; Australia.

Myoporum parvifolium is widely cultivated as a ground cover in the southwestern United States; it appears to be established in canyons of urban southern California.

---

6. BONTIA Linnaeus, Sp. Pl. 2: 638 [as 938]. 1753; Gen. Pl. ed. 5, 285. 1754 • [For Jacobus Bontius, 1592–1631, Dutch physician and botanist in Java]

Scott Zona

Shrubs or trees, all parts bearing spherical resin cavities; stolons absent. **Stems** ascending, glabrous. **Leaves** persistent, cauline, alternate; stipules absent; petiole present; blade not fleshy, leathery, margins entire (except in seedling leaves, which are distally serrate). **Inflorescences** axillary, 2-flowered or flowers solitary; bracts absent. **Pedicels** present; bracteoles absent. **Flowers** bisexual; sepals 5, calyx radially symmetric, funnelform, lobes broadly ovate; petals 5, corolla buff or coppery brown to greenish tan, bilaterally symmetric, tubular, abaxial lobes 3, strongly recurved and bearing a conspicuous longitudinal crest of violet hairs, adaxial 2; stamens 4, adnate to corolla, didynamous, filaments sparsely hairy, staminode 0; ovary 2-locular, placentation axile; stigma minutely 2-fid. **Fruits** drupelike capsules, ovoid, fleshy, indehiscent. **Seeds** 1–4, brown, L-shaped and terete, wings absent. $x = 18$.

Species 1: Florida, West Indies, South America.

Bontia is monospecific and the only genus of Myoporeae (the former Myoporaceae) native to the Americas. Phylogenetic analysis suggests that it was derived from within the Australian genus *Eremophila* R. Brown and achieved its current distribution via long-distance dispersal (S. A. Kelchner et al. 2001).

1. Bontia daphnoides Linnaeus, Sp. Pl. 2: 638 [as 938]. 1753 • White alling [F]

Plants 1–5 m. **Bark** tan, smooth in young shoots, furrowed in older shoots. **Leaves** spirally arranged, concolorous, 65–130 × 13–19 mm; petiole 2–7 mm; blade narrowly elliptic, midvein conspicuous, apex acute. **Flowers:** calyx lobes 3 × 3 mm, margins ciliate, apex acuminate; corolla 15–20 mm; stamens exserted, assurgent against adaxial lip of corolla, anther dehiscence introrse; style elongate. **Fruits** greenish yellow, 12–15 × 10–12 mm, pedicel and style persistent, exocarp glossy, glabrous, mesocarp corky, endocarp woody. $2n = 36$.

Flowering Nov–Jun. Coastal thickets, littoral or mangrove forests; 0–10 m; Fla.; West Indies; South America.

Bontia daphnoides was first discovered in Florida in the early 1990s (S. Zona 1998). Field studies are needed to determine whether flowers are pollinated by bees (as postulated by Zona), hummingbirds (R. J. Chinnock 2007), both, or neither. The fruits float and are adapted to dispersal by water. *Bontia daphnoides* has potential as an ornamental shrub, especially because it tolerates alkaline soil, salt, and drought. It is used medicinally throughout much of the Caribbean (E. S. Ayensu 1981b).

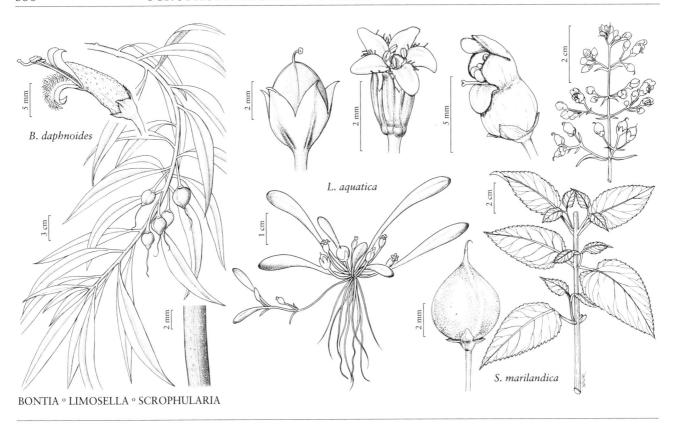

B. daphnoides

L. aquatica

S. marilandica

BONTIA ° LIMOSELLA ° SCROPHULARIA

7. LIMOSELLA Linnaeus, Sp. Pl. 2: 631. 1753; Gen. Pl. ed. 5, 280. 1754 • Mudwort

[Latin *limosus*, full of mud, and *-ella*, diminutive, alluding to habitat]

Kerry A. Barringer

Herbs, annual [perennial], aquatic or paludal; stolons present. **Stems** stoloniferous, prostrate or ascending, glabrous. **Leaves** basal, opposite, sometimes rosulate; stipules absent; petiole present or absent; blade fleshy or not, leathery or not, margins entire. **Inflorescences** axillary, flowers solitary; bracts absent. **Pedicels** present; bracteoles present or absent. **Flowers** bisexual; chasmogamous flowers emergent; sepals [4]5, calyx radially or slightly bilaterally symmetric, campanulate, lobes lanceolate to triangular; petals [4]5, corolla white, pink, pale purple, or lavender, slightly bilaterally symmetric, campanulate, not saccate, abaxial lobes 3, adaxial 2; stamens 4, adnate near mouth of corolla, didynamous or equal, filaments glabrous, staminode 0; ovary 2-locular proximally, 1-locular distally, placentation free-central; stigma capitate or slightly 2-lobed. **Fruits** capsules, ellipsoid to globular, dehiscence septicidal. **Seeds** 3–200, yellow to dark brown, fusiform or cylindric, wings absent.

Species 15 (2 in the flora): nearly worldwide.

Interpretations of morphological and molecular data place *Limosella* in Scrophulariaceae near the African genera formerly classified as Manuleae Bentham & Hooker f. (P. Kornhall and B. Bremer 2004). *Limosella* is in need of revision; the characters often used for identification are variable or difficult to observe.

SELECTED REFERENCE Lourteig, A. 1964. Étude sur *Limosella* L. Publ. Comité Natl. Franç. Rech. Antarct. Biol. 1: 165–173.

1. Corolla lobes acute; stamens attached at different levels; styles 0.2–0.4 mm; leaf blades spatulate to linear; young leaves with auriculate sheaths at bases . . . . 1. *Limosella aquatica*
1. Corolla lobes rounded; stamens attached at same level; styles 1 mm; leaf blades linear to filiform; young leaves with tapering sheaths at bases . . . . 2. *Limosella australis*

1. Limosella aquatica Linnaeus, Sp. Pl. 2: 631. 1753

• Limosella aquatique F

Limosella pubiflora Pennell

Annuals solitary or mat-forming. **Leaves** clustered, usually rosulate; blade spatulate to linear, (1.5–)3–10 (–20) × 0.5–3 cm, young leaves auriculate at base. **Pedicels** spreading in fruit, (0.04–) 2–6 cm. **Flowers:** calyx often purple spotted between lobes, lobes 1 mm, tube 1–2 mm; corolla white to pink, rarly pale purple abaxially, 2–3 mm, lobes acute; stamens attached at different levels; style 0.2–0.4 mm, entire or 2-lobed. **Capsules** ellipsoid to globular, (2–)2.5–4 mm. *2n* = 40.

Flowering Apr–Oct. Periodically flooded mud of streams and ponds, estuaries, vernal pools; 0–3200 m; Greenland; Alta., B.C., Man., Nfld. and Labr., N.W.T., Nunavut, Ont., P.E.I., Que., Sask., Yukon; Alaska, Ariz., Ark., Calif., Colo., Idaho, Minn., Mo., Mont., Nebr., Nev., N.Mex., N.Dak., Oreg., S.Dak., Utah, Wash., Wyo.; Mexico; Europe; Asia.

Limosella aquatica grows primarily in fresh water; it has also been found in brackish and alkaline water. Leaves are usually spatulate; narrow-leaved plants occur throughout the range. Narrow-leaved plants in the southwest have been called *L. pubiflora*; there are no consistent characteristics to distinguish *L. pubiflora* from *L. aquatica*. Narrow-leaved plants in the western United States have been called *L. acaulis* Sessé & Mociño; these specimens do not match true *L. acaulis*. Clasping leaf bases are best seen on young leaves.

2. Limosella australis R. Brown, Prodr., 443. 1810

• Limosella à feuilles subulées

Limosella subulata E. Ives; *L. tenuifolia* J. P. Wolff

Annuals mat-forming. **Leaves** clustered or scattered along stolons; blade linear to filiform, 1–3.5 × 0.1–0.2 cm, young leaves with tapering sheaths at base. **Pedicels** recurved in fruit, 1–2 cm. **Flowers:** calyx not purple spotted, lobes 0.5–1 mm, tube 1–2 mm; corolla white to pale lavender, 2–4 mm, lobes rounded; stamens attached at same level; style 1 mm, capitate. **Capsules** globular, 2–3 mm. *2n* = 20.

Flowering May–Aug(–Oct). Muddy or sandy pond or stream margins, often in brackish water; 0–10 m; St. Pierre and Miquelon; N.B., Nfld. and Labr., N.S., P.E.I., Que.; Conn., Del., Maine, Md., Mass., N.H., N.J., N.Y., N.C., Pa., R.I., Va.; South America; Africa; Pacific Islands (New Zealand); Australia.

In the flora area, *Limosella australis* is native to eastern North America and is rare in much of its range. Reports of *L. australis* from California are likely based on misidentified material.

8. SCROPHULARIA Linnaeus, Sp. Pl. 2: 619. 1753; Gen. Pl. ed. 5, 271. 1754 • Figwort [Association with the disease scrofula by the doctrine of signatures]

Kim R. Kersh

Subshrubs or herbs, annual or perennial; caudex herbaceous or woody; stolons absent. **Stems** erect, glabrous, glabrate, glandular-pubescent, puberulent, or villous. **Leaves** persistent, cauline, opposite, decussate; stipules absent; petiole present; blade not fleshy (fleshy in *S. desertorum*), not leathery, margins serrulate, serrate, dentate, or incised. **Inflorescences** terminal, panicles, sometimes axillary, cymes (*S. peregrina*); bracts present. **Pedicels** present; bracteoles present. **Flowers** bisexual; sepals 5, calyx radially symmetric, campanulate, lobes triangular-ovate to lanceolate; petals 5, corolla ± dark, often bicolored, red, black-red, brown-red, brick red, purple-red, cream, or green, often paler abaxially, bilaterally symmetric, proximally inflated, distally constricted, abaxial lobes 1, lateral 2, adaxial 2; stamens 4, adnate to corolla near base, subdidynamous, filaments glandular-puberulent, staminode (0 or)1, clavate to flabellate, sometimes rudimentary, scalelike or awnlike; ovary 2-locular, placentation axile; stigma capitate or 2-lobed. **Fruits** capsules, ± pear-shaped, dehiscence septicidal. **Seeds** 100–130+, black to brown, oblong-ovoid, rugose, wings absent.

Species 150+ (11 in the flora): North America, n Mexico, West Indies, Europe, Asia (sw China), n Africa.

Chromosome numbers of $2n$ = 18, 24, 26, 36, 40, 50, 52, 56, 58, 78, 80, 84, 86, 90, 92, and 96 have been reported for *Scrophularia*. Base chromosome numbers 7, 9, 10, 12, and 13 have been proposed for *Scrophularia* (see C. Carlbom 1969); a hypothesis for one distinct base number has yet to be presented.

Scrophularia nodosa Linnaeus, native to Europe, is discussed under 7. *S. marilandica*. European species *S. aquatica* Linnaeus (later referred to as *S. umbrosa* Dumortier), *S. auriculata* Linnaeus, and *S. canina* Linnaeus were reported on ballast in New Jersey, New York, and Pennsylvania in the 1870s and 1880s and are not known to have persisted.

North American scrophularias can become especially abundant in areas with human disturbance. Native American uses of *Scrophularia* were reported by D. E. Moerman (1998).

Scrophularias are visited, and presumably pollinated, by bees, wasps, and other insects. *Scrophularia macrantha*, with relatively showy, large, red corollas, is often cited as hummingbird-visited, and other species are hummingbird-visited as well; *S. montana* is a major nectar resource for migrant *Selasphorus* hummingbirds (D. Heinemann 1992).

Morphological plasticity in *Scrophularia* produces variation so wide as to make characters that have been used to define and separate some taxa of limited use. The most distinct species are of relatively narrow distribution.

SELECTED REFERENCES Carlbom, C. 1969. Evolutionary relationships in the genus *Scrophularia* L. Hereditas (Lund) 61: 287–301. Shaw, R. J. 1962. The biosystematics of *Scrophularia* in western North America. Aliso 5: 147–178. Stiefelhagen, H. 1910. Systematische und pflanzengeographische Studien zur Kenntnis der Gattung *Scrophularia*. Bot. Jahrb. Syst. 44: 406–496.

1. Inflorescences axillary, cymes...1. *Scrophularia peregrina*
1. Inflorescences terminal, panicles.
    2. Pedicels densely villous ....2. *Scrophularia villosa*
    2. Pedicels glabrate or stipitate-glandular.
        3. Corollas bright red, (10–)13–21 mm.............................3. *Scrophularia macrantha*
        3. Corollas black-red, brown-red, brick red, red, purple-red, green, or cream, 5–12 (–14) mm.
            4. Corollas black-red, throats constricted...........................4. *Scrophularia atrata*
            4. Corollas brown-red, brick red, red, purple-red, green, or cream, throats open or narrow.
                5. Herbage light gray-green (fresh leaves fleshy); corollas bicolored, sometimes unicolored, paler abaxially (abaxial lobe often cream to white) .... 5. *Scrophularia desertorum*
                5. Herbage light green to dark green (fresh leaves not fleshy); corollas unicolored or ± bicolored, paler abaxially.
                    6. Leaf blade margins regularly serrulate or serrate.
                        7. Pedicels relatively stout; petiole lengths $\frac{1}{12}$–$\frac{1}{6}$ blades; blades 8.5–14 cm, lengths 3–4.5 times widths, bases truncate or cuneate, margins regularly serrulate .... 6. *Scrophularia montana*
                        7. Pedicels slender; petiole lengths $\frac{1}{3}$–$\frac{1}{2}$ blades; blades 10–19(–25) cm, lengths 2–3 times widths, bases rounded to cordate, margins regularly serrate .... 7. *Scrophularia marilandica*
                    6. Leaf blade margins serrate, dentate, and/or incised.
                      8. Herbage puberulent or glabrate; leaf blades 4–7(–8) cm.
                        9. Herbage puberulent, more densely distally; petiole lengths $\frac{1}{6}$–$\frac{1}{3}$ leaf blades ....8. *Scrophularia parviflora*
                        9. Herbage glabrate; petiole lengths $\frac{1}{3}$–$\frac{1}{2}$ leaf blades ....9. *Scrophularia laevis*
                      8. Herbage glandular-pubescent, rarely glabrate; leaf blades 8–17(–20) cm.
                        10. Staminodes yellow-green, sometimes green to brown-red, flabellate, lengths less than widths ....10. *Scrophularia lanceolata*
                        10. Staminodes red, brown-red, or green, clavate to obovate, lengths greater than widths...................11. *Scrophularia californica*

1. **Scrophularia peregrina** Linnaeus, Sp. Pl. 2: 621. 1753 [I]

Herbs, annual, 1.5–10 dm; herbage bright yellow-green, glabrate or glandular-pubescent. **Leaves:** petiole length ⅓ blade; blade deltate to ovate, 5–10 cm, length 1.2–2 times width, base cordate to truncate, margins dentate. **Inflorescences** axillary, cymes. **Pedicels** slender, glabrate or stipitate-glandular. **Flowers:** corolla dark brown-red or purple-red, usually unicolored, 6–9 mm, throat narrow; staminode obovate to orbiculate, length greater than width. $2n = 36$.

Flowering Feb–May. Gardens, disturbed areas immediately adjacent; 400 m; introduced; Calif.; Europe.

Scrophularia peregrina was introduced in Rancho Santa Ana Botanic Garden for biosystematic studies in the 1950s. It is abundant and weedy in the Garden and immediately adjacent in uncultivated and non-irrigated areas; it is considered likely to spread, with potential negative impacts on the native flora, especially in southern California (T. S. Ross and S. Boyd 1996).

2. **Scrophularia villosa** Pennell, Publ. Field Mus. Nat. Hist., Bot. Ser. 5: 223. 1923 [C]

Subshrubs, 12–18 dm; herbage light green to dark green, glabrate or glandular-pubescent. **Leaves:** petiole length ⅙–⅓ blade; blade deltate to ovate or lanceolate, 8–15 cm, length 1–2 times width, base cordate to truncate, margins serrate to dentate. **Pedicels** slender to moderately stout, densely villous. **Flowers:** corolla dark brown-red, unicolored, sometimes ± bicolored, paler abaxially, 8–11 mm, throat narrow; staminode awnlike, rudimentary, or lacking. $2n = $ ca. 94–96.

Flowering Feb–May. Canyon bottoms, coastal scrub, chaparral; of conservation concern; 0–400 m; Calif.; Mexico (Baja California).

Scrophularia villosa occurs on San Clemente and Santa Catalina islands, California, and Guadalupe Island, Mexico.

3. **Scrophularia macrantha** Greene ex Stiefelhagen, Bot. Jahrb. Syst. 44: 461, 483. 1910 [C][E]

Scrophularia neomexicana R. J. Shaw

Herbs, perennial, 4–11 dm; herbage light green to dark green, glabrate. **Leaves:** petiole length ⅓–½ blade; blade ovate to lanceolate, 5–8(–11) cm, length 2–2.5 times width, base truncate, margins serrate to dentate. **Pedicels** relatively stout, glabrate or stipitate-glandular. **Flowers:** corolla bright red, unicolored, (10–)13–21 mm, throat narrow; staminode obovate, length greater than width. $2n = 92$.

Flowering Jul–Oct. Rocky slopes, canyon bottoms, pinyon-juniper woodlands, lower montane coniferous forests; of conservation concern; 2000–2500 m; N.Mex.

Scrophularia macrantha has been called *S. coccinea* A. Gray, an illegitimate later homonym of *S. coccinea* Linnaeus. *Scrophularia macrantha* is known from Grant and Luna counties.

4. **Scrophularia atrata** Pennell, Proc. Acad. Nat. Sci. Philadelphia 99: 172. 1947 [C][E]

Herbs, perennial, 10–15 dm; herbage light green to dark green, glandular-pubescent. **Leaves:** petiole length ⅓–⁷⁄₁₂ blade; blade deltate or ovate to lanceolate, 5–9(–11) cm, length 1.1–2 times width, base cordate to truncate, margins serrate to dentate. **Pedicels** slender, glabrate or stipitate-glandular. **Flowers:** corolla black-red, unicolored, sometimes ± bicolored, paler abaxially, 9–11 mm, throat constricted; staminode obovate, length greater than width, apex acute. $2n = $ ca. 92–96.

Flowering Apr–Jul. Calcareous, diatomaceous soils; of conservation concern; 0–400 m; Calif.

Scrophularia atrata is known from San Luis Obispo and Santa Barbara counties.

5. **Scrophularia desertorum** (Munz) R. J. Shaw, Aliso 5: 174. 1962 [E]

Scrophularia californica Chamisso & Schlechtendal var. *desertorum* Munz, Aliso 4: 99. 1958

Herbs, perennial, 7–12 dm; herbage light gray-green, glabrate or glandular-pubescent. **Leaves:** petiole length ⅙–½ blade; blade lanceolate, 4–8 (–12) cm, length 2.2–3.5 times width, fleshy fresh, base truncate to cuneate, margins serrate to dentate. **Pedicels** slender, glabrate or stipitate-glandular. **Flowers:** corolla brick red to cream, bicolored, sometimes unicolored, paler abaxially, abaxial lobe often cream to white, 7–9 mm, throat relatively open; staminode clavate, length greater than width. $2n = 96$.

Flowering Apr–Aug. Rocky slopes, canyons, gravelly washes; 800–3300 m; Calif., Nev.

6. **Scrophularia montana** Wooton, Bull. Torrey Bot. Club 25: 308. 1898 [E]

Herbs, perennial, 10–15 dm; herbage dark green, finely glandular-pubescent. **Leaves:** petiole length $\frac{1}{12}$–$\frac{1}{6}$ blade; blade lanceolate, 8.5–14 cm, length 3–4.5 times width, base truncate or cuneate, margins regularly serrulate. **Pedicels** relatively stout, stipitate-glandular. **Flowers:** corolla brown-red or purple-red to green, unicolored, sometimes ± bicolored, paler abaxially, 6–10 mm, throat ± open; staminode obovate, length usually greater than width. $2n = $ ca. 70–76.

Flowering Jul–Sep. Mountainous areas, open woodlands, forest edges; 2200–3200 m; N.Mex.

7. **Scrophularia marilandica** Linnaeus, Sp. Pl. 2: 619. 1753 [E][F][W]

Herbs, perennial, 10–30 dm; herbage light green to dark green, sparsely glandular-pubescent. **Leaves:** petiole length $\frac{1}{3}$–$\frac{1}{2}$ blade; blade ovate to lanceolate, 10–19(–25) cm, length 2–3 times width, base rounded to cordate, margins regularly serrate. **Pedicels** slender, glabrate or stipitate-glandular. **Flowers:** corolla brown-red, unicolored, sometimes ± bicolored, paler abaxially, 5–8 mm, throat narrow; staminode clavate, length greater than width. $2n = $ ca. 86.

Flowering Jul–Sep. Open woods; 0–1200 m; N.B., Nfld. and Labr. (Nfld.), N.S., Ont., P.E.I., Que.; Ala., Ark., Conn., Del., D.C., Fla., Ga., Ill., Ind., Iowa, Kans., Ky., La., Maine, Md., Mass., Mich., Minn., Miss., Mo., Nebr., N.H., N.J., N.Y., N.C., Ohio, Okla., Pa., R.I., S.C., S.Dak., Tenn., Tex., Vt., Va., W.Va., Wis.

Scrophularia marilandica has been confused with the European *S. nodosa*. Some North American floristic works have attempted to separate *S. nodosa* from *S. marilandica* using various characteristics including relatively shorter and narrowly wing-margined petioles,

shorter leaves, stems acutely angled, corollas with green bases and tips of lobes, and nodular rhizomes. Nevertheless, the two cannot be reliably distinguished. *Scrophularia nodosa* has been reported from New Brunswick, Newfoundland, Nova Scotia, Prince Edward Island, and Quebec, and from Connecticut, Iowa, Massachusetts, Minnesota, New Jersey, New York, Pennsylvania, Rhode Island, and Washington (cultivated at the University of Washington Medicinal Herb Garden and present on and near the campus). Reports are generally based on collections from ballast in the 1870s and 1880s, other introductions of European plants associated with gardens and medicinal uses, and attempts to apply characters such as those above to North American plants. Plants in Newfoundland have generally been referred to as *S. nodosa*. While M. L. Fernald (1933) considered *S. nodosa* native in Newfoundland, later workers have treated *S. nodosa* as introduced in Newfoundland (for example, H. J. Scoggan 1978–1979).

Plants from maritime Canada that have been treated as *Scrophularia nodosa*, alternately as native to North America and as introduced from Europe, are treated here provisionally as of North American provenance and included in *S. marilandica*.

Although European plants of *Scrophularia nodosa* may have occasionally been introduced to North America, the extent to which they may have persisted locally is difficult to assess due to morphological variability and resemblance to North American plants. Linnaeus, while recognizing both *S. marilandica* and *S. nodosa*, stated that the distinction is scarcely merited. A review of literature also suggests that the two species may well be one.

8. **Scrophularia parviflora** Wooton & Standley, Contr. U.S. Natl. Herb. 16: 173. 1913 [E]

Herbs, perennial, 5–11 dm; herbage light green to dark green, puberulent, more densely distally. **Leaves:** petiole length $\frac{1}{6}$–$\frac{1}{3}$ blade; blade lanceolate to ovate, 4–7(–8) cm, length 2–3 times width, base truncate, margins dentate. **Pedicels** slender, glabrate or stipitate-glandular. **Flowers:** corolla red to green, unicolored or ± bicolored, paler abaxially, 6–8 mm, throat ± open; staminode obovate, length usually greater than width. $2n = 92$.

Flowering Jul–Oct. Coniferous forests, riparian areas; 1500–3100 m; Ariz., N.Mex.

9. **Scrophularia laevis** Wooton & Standley, Contr. U.S. Natl. Herb. 16: 173. 1913 [C][E]

Herbs, perennial, 4–10 dm; herbage light green, glabrate. **Leaves:** petiole length ⅓–½ blade; blade lanceolate to ovate, 5–7 cm, length 2–2.5 times width, base truncate, margins dentate. **Pedicels** slender, glabrate or stipitate-glandular. **Flowers:** corolla red to green, unicolored or ± bicolored, paler abaxially, 7–11 mm, throat open; staminode orbiculate, length equal to width.

Flowering Jul–Sep. Moist canyons; of conservation concern; 2100–2600 m; N.Mex.

Scrophularia laevis is known from the Organ Mountains in Doña Ana County.

10. **Scrophularia lanceolata** Pursh, Fl. Amer. Sept. 2: 419. 1813 [E][W]

Herbs, perennial, 8–15 dm; herbage light green to dark green, glabrate or glandular-pubescent. **Leaves:** petiole length ⅙–⅓ blade; blade ovate to lanceolate, 9–15(–20) cm, length 1.5–3 times width, base truncate to slightly rounded, margins serrate to dentate. **Pedicels** moderately stout, glabrate or stipitate-glandular. **Flowers:** corolla brown-red to green, unicolored or ± bicolored, paler abaxially, 8–11(–14) mm, throat relatively open; staminode yellow-green, sometimes green to brown-red, flabellate, length less than width. $2n = 96$.

Flowering May–Jul. Damp places in open woods, riparian areas, meadows, canyons; 0–2800 m; Alta., B.C., N.B., N.S., Ont., Que., Sask.; Calif., Colo., Conn., Del., Idaho, Ill., Ind., Iowa, Kans., Maine, Md., Mass., Mich., Minn., Mo., Mont., Nebr., Nev., N.H., N.J., N.Mex., N.Y., N.Dak., Ohio, Okla., Oreg., Pa., R.I., S.Dak., Utah, Vt., Va., Wash., W.Va., Wis., Wyo.

11. **Scrophularia californica** Chamisso & Schlechtendal, Linnaea 2: 585. 1827 (as Scrofularia) [F]

Scrophularia californica subsp. *floribunda* (Greene) R. J. Shaw; *S. californica* var. *floribunda* Greene; *S. californica* var. *oregana* (Pennell) B. Boivin; *S. multiflora* Pennell; *S. oregana* Pennell

Herbs, perennial, 8–12 dm; herbage light green to dark green, glandular-pubescent. **Leaves:** petiole length ⅓–½ blade; blade deltate or ovate to lanceolate, 8–17 cm, length 1.5–3 times width, base cordate to truncate, sometimes cuneate, margins serrate or dentate, sometimes incised. **Pedicels** slender to moderately stout, glabrate or stipitate-glandular. **Flowers:** corolla red, brown-red, or green, unicolored or ± bicolored, paler abaxially, 8–12 mm, throat narrow to open; staminode red, brown-red, or green, clavate to obovate, length greater than width, apex sometimes acute. $2n =$ ca. 90–96.

Flowering Mar–Jul. Relatively damp places on slopes, chaparral, woodlands, roadsides; 0–2500 m; B.C.; Ariz., Calif., Oreg., Wash.; Mexico (Baja California).

Subspecies (or var.) *floribunda*, characteristically from inland and drier areas, may be part of a complex of infraspecific diversity within *Scrophularia californica* needing investigation.

9. VERBASCUM Linnaeus, Sp. Pl. 1: 177. 1753; Gen. Pl. ed. 5, 83. 1754 • Mullein, molène [Ancient Latin name used by Pliny, probably corruption of *barbascum*, bearded, alluding to dense tomentum, or *barbarum*, medicinal plaster, alluding to use of some species] [I]

Guy L. Nesom

Herbs, annual, biennial, or perennial; stolons absent. **Stems** erect, glabrous, glabrate, puberulent, hirsute, tomentose, or floccose, sometimes glabrescent, stipitate-glandular or eglandular. **Leaves** basal and cauline, alternate; stipules absent; petiole present or absent; blade not fleshy, leathery or not, margins entire, serrate, crenate, dentate, sinuate, or lobed. **Inflorescences** terminal, spikes, racemes, or panicles, flowers solitary or clustered in fascicles; bracts present. **Pedicels** present; bracteoles present or absent. **Flowers** bisexual; sepals 5, calyx radially or bilaterally symmetric, campanulate, lobes linear-oblong to elliptic or triangular; petals 5, corolla yellow to

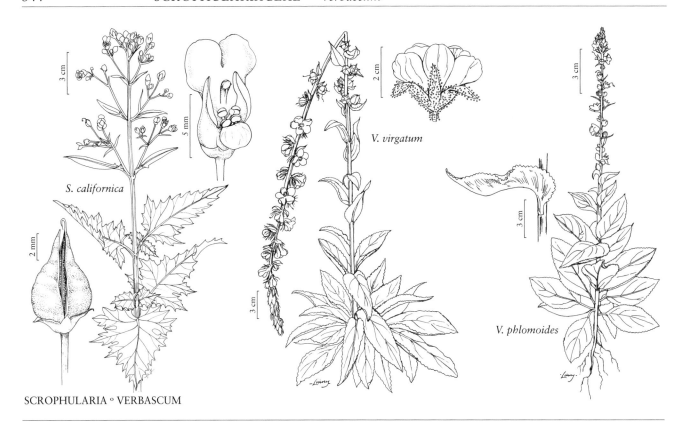

S. californica

V. virgatum

V. phlomoides

SCROPHULARIA ⁰ VERBASCUM

orange, white, purple, cream, or pink, sometimes with purple center or red-tinged tips, radially or bilaterally symmetric, rotate with petals spreading to deflexed or shallowly cupulate, abaxial lobes 3, adaxial 2; stamens [4]5, adnate to base of corolla, equal to subequal or anterior pair longer, filaments densely villous, sometimes anterior pair glabrous, staminode 0[1]; ovary 2-locular, placentation axile; stigma capitate or spatulate. **Fruits** capsules, ovoid to ellipsoid-ovoid, broadly ellipsoid, ovoid-globular, or subglobular, dehiscence septicidal. **Seeds** 50–300, tan or brown to orangish, conic to cylindric, usually pitted and rugose, appearing transversely and longitudinally ribbed, wings absent. $x = 6$.

Species 300–360 (12 in the flora): introduced; Europe, Asia, ne Africa; introduced also in Mexico, South America, elsewhere in Africa (Tunisia), Pacific Islands, Australia.

Chromosome counts of $2n = 18, 26, 30, 32, 34, 36, 40, 44, 45, 48, 50, 64$, and 66 have been reported for species of *Verbascum*. A base number is not clear, but $x = 6$ appears to be a reasonable fit ($2n = 18, 30, 36, 48$, and 66 would thus be polyploid levels). In any case, the range of numbers reflects both polyploidy and dysploidy.

Stamen number, placenta structure (entire and sessile versus two-fid and stalked), capsule shape, and number of flowers per node/bract have been used as diagnostic characters to distinguish *Celsia* Linnaeus and *Staurophragma* Fischer & C. A. Meyer as segregates from *Verbascum* (S. Murbeck 1925; A. Huber-Morath 1973; F. A. Karaveliogullari and Z. Aytac 2008). Huber-Morath (1978) included *Celsia* and *Staurophragma* within *Verbascum*. All species treated here are *Verbascum* in the strict sense.

Taxonomic treatments of *Verbascum* have been inconsistent in use of rank for infraspecific taxa; both varieties and subspecies appear in this treatment as well. Since all of our taxa are introduced to North America, further study of the species in their native environs is warranted before any uniformity can be established. In any case, use of one rank or the other does not imply a biological or evolutionary difference.

Hybrids are commonly produced among many combinations of parents. *Verbascum thapsus*, which is the most common and widespread of the species in the flora area, is known to have formed four hybrids in North America; see species discussions for nothospecies and purported parental species. *Verbascum ×ramigerum* Link ex Schrader [*V. densiflorum* × *V. lychnitis*] has been reported in Europe; it and other combinations may be expected in the flora area.

Plants of *Verbascum* collected as weeds in a lawn in Duluth, Minnesota, in 2001 and 2002 (*Schimpf DJS318, DJS327*, MIN), have been identified as *V. chaixii* Villars (http://plants. usda.gov); they perhaps are hybrids with *V. nigrum* as one of the parents (Dirk Albach, pers. comm.). The leaves are glabrate, the basal ones ovate-elliptic, subentire to shallowly crenate, and subpetiolate, and the cauline ones mostly non-clasping; the inflorescences are stipitate-glandular, unbranched, with flowers in loosely overlapping clusters of three to six; corollas are white to yellowish with pellucid glands; and staminal filaments all are villous with violet hairs. Without an unequivocal identification and evidence that they have become naturalized, they are not formally treated here.

SELECTED REFERENCES Murbeck, S. 1933. Monographie der Gattung *Verbascum*. Acta Univ. Lund, n. s. 29: 1–630. Murbeck, S. 1936. Nachträge zur Monographie der Gattung *Verbascum*. Acta Univ. Lund, n. s. 32: 1–46. Murbeck, S. 1939. Weitere Studien über die Gattungen *Verbascum* und *Celsia*. Acta Univ. Lund, n. s. 35: 1–71.

1. Flowers solitary in axils at least distally.
    2. Corollas purple to violet; bracteoles 0; cauline leaves abruptly smaller distally . . . . . .
    . . . . 1. *Verbascum phoeniceum*
    2. Corollas yellow, sometimes white or pink (*V. blattaria*); bracteoles 0 or 2; cauline leaves gradually smaller distally.
        3. Pedicels 5–11(–15) mm; bracteoles 0; stems and leaves glabrous or glabrate . . . .
        . . . . 2. *Verbascum blattaria*
        3. Pedicels (0–)1–3 mm; bracteoles 2; stems and leaves densely stipitate-glandular, sometimes also sparsely hirsute-villous . . . . .3. *Verbascum virgatum*
1. Flowers mostly in clusters of 2–10.
    4. Inflorescences unbranched, sometimes branched from proximal nodes.
        5. Basal and proximal cauline leaf blades: bases shallowly cordate to nearly truncate, surfaces sparsely tomentose to glabrate, abaxial soon glabrescent, sometimes both glabrate; filaments villous, hairs purple to violet . . . . 9. *Verbascum nigrum* (in part)
        5. Basal and proximal cauline leaf blades: bases attenuate, surfaces densely and persistently tomentose; filaments villous or glabrous, hairs yellowish to whitish.
            6. Flower clusters loosely overlapping . . . . 7. *Verbascum bombyciferum*
            6. Flower clusters densely overlapping or remote proximally.
                7. Cauline leaves: bases not decurrent, rarely slightly so . . . . .6. *Verbascum phlomoides*
                7. Cauline leaves: bases decurrent.
                    8. Corollas 14–20(–30) mm diam., pellucid glands relatively numerous; anthers yellow; stigmas capitate; pedicels mostly or completely adnate to rachis. . . . . .4. *Verbascum thapsus*
                    8. Corollas 30–55 mm diam., pellucid glands absent or relatively few; anthers orange; stigmas spatulate; pedicels free or adnate to rachis at base. . . . . 5. *Verbascum densiflorum* (in part)
    4. Inflorescences freely branched, sometimes unbranched (*V. densiflorum, V. nigrum*) or branched from proximal nodes and forming panicles (*V. nigrum*).
        9. Leaf surfaces persistently tomentose.
            10. Filaments: proximals glabrous, distal pairs villous; cauline leaves: bases long-decurrent . . . . 5. *Verbascum densiflorum* (in part)
            10. Filaments villous; cauline leaves: bases not decurrent or short-decurrent.
                11. Leaf blade margins sinuate to incised or incised-lobed; flowers remote, clustered, rarely solitary; pedicels 2–5 mm; filament hairs violet to purple
                . . . . 11. *Verbascum sinuatum*
                11. Leaf blade margins entire or minutely serrate-dentate; flowers loosely overlapping, clustered; pedicels (3–)5–12 mm; filament hairs whitish to yellowish . . . . .12. *Verbascum speciosum*

[9. Shifted to left margin.—Ed.]

9. Leaf surfaces glabrate, sometimes abaxial surfaces sparsely persistently hairy and soon glabrescent.
    12. Basal and proximal cauline leaves shallowly cordate or nearly truncate to a narrow petiole 5–15(–20) mm; filament hairs purple to violet; inflorescences unbranched, sometimes branched from proximal nodes, narrowly conic panicles . . . . . . 9. *Verbascum nigrum* (in part)
    12. Basal leaves sessile or basally attenuate and sessile, less commonly with petiolar regions 10–50 mm, cauline leaves sessile; filament hairs yellow to whitish; inflorescences freely branched, loosely conic to broadly cylindric panicles, sometimes unbranched.
        13. Cauline leaves: bases not clasping or slightly so, surfaces thinly tawny- to gray-tomentose, abaxial soon glabrescent, adaxial persistent, closely adherent; bracts linear to narrowly lanceolate, 8–15 mm; pedicels 6–11 mm; corollas white, sometimes yellowish . . . . 8. *Verbascum lychnitis*
        13. Cauline leaves: bases distinctly clasping or subclasping, surfaces densely and loosely white-floccose, glabrescent, especially abaxially, sometimes thin-persistent adaxially; bracts linear, 3–5 mm; pedicels (1–)2–5(–7) mm; corollas yellow . . . . .
         . . . . 10. *Verbascum pulverulentum*

1. Verbascum phoeniceum Linnaeus, Sp. Pl. 1: 178. 1753 • Purple mullein [I]

Subspecies 2 (1 in the flora): introduced; Europe, Asia.

1a. Verbascum phoeniceum Linnaeus subsp. **phoeniceum** [I]

Biennials or perennials. Stems 30–100 cm, stipitate-glandular, sometimes also sparsely puberulent. **Leaves:** surfaces stipitate-glandular, sometimes also sparsely puberulent; basal and proximal cauline with petiole 5–40 mm; blade ovate to elliptic-ovate, ovate-lanceolate, or oblong, 5–15 × 3–7.5(–10) cm, base subrounded to broadly cuneate; cauline not clasping, abruptly smaller distally, base not decurrent, margins crenate, sinuate, or subentire, apex of distal cauline and floral bracts long-acuminate. **Inflorescences** unbranched, rarely branched from proximal nodes, narrowly cylindric, flowers remote, solitary in axils at least distally; rachis stipitate-glandular, without other vestiture or sparsely puberulent; bracts ± lanceolate, 4–7 mm, base not decurrent, apex acute to acute-acuminate, stipitate-glandular, without other vestiture or sparsely puberulent. **Pedicels** free, 6–20(–25) mm; bracteoles 0. **Flowers:** calyx 4–7 mm, stipitate-glandular, without other vestiture or sparsely puberulent, lobes elliptic; corolla purple to violet, 20–30(–35) mm diam., pellucid glands absent; filaments villous, hairs violet; stigma capitate. **Capsules** ovoid, 6–8 mm, glabrous or sparsely glandular. *2n* = 32, 36.

Flowering May–Aug. Fields, roadsides, disturbed sites; 100–300 m; introduced; Ky., N.Y., Ohio; Europe; Asia.

Plants of *Verbascum phoeniceum* with yellow corollas, mostly from Greece, have been named subsp. *flavidum* (Boissier) Bornmueller.

2. Verbascum blattaria Linnaeus, Sp. Pl. 1: 178. 1753 • Moth mullein, molène blattaire [I][W]

Annuals or biennials. Stems 60–150 cm, glabrous or glabrate. **Leaves:** surfaces glabrous or glabrate; basal and proximal cauline with petiole 1–2 mm; blade obovate to oblanceolate, oblong, or oblong-lanceolate, 4–12(–20) × 1.5–5 cm, base subrounded to broadly cuneate; cauline subclasping, gradually smaller distally, base not decurrent, margins coarsely and regularly crenate-dentate to dentate or pinnately dentate-lobed, apex of distal cauline and floral bracts acute. **Inflorescences** unbranched, rarely branched from proximal nodes, narrowly cylindric, flowers remote, solitary in axils at least distally; rachis stipitate-glandular, without other vestiture; bracts ovate to lanceolate-ovate, 7–10(–15) mm, base decurrent, apex acute to short-acuminate, stipitate-glandular. **Pedicels** free or adnate to rachis at base, 5–11(–15) mm; bracteoles 0. **Flowers:** calyx 5–7 mm, stipitate-glandular, lobes linear-lanceolate to narrowly oblong; corolla purple in bud, becoming yellow, yellow-orange, yellow with purple center, white, cream with red-tinged tips, or pink, 25–35 mm diam., pellucid glands absent or relatively few; proximal filaments hairy, hairs purple, distal pair villous, hairs white and purple or violet; stigma spatulate, base decurrent. **Capsules** subglobular, 5–8 mm, sparsely stipitate-glandular apically. *2n* = 18, 30, 32.

Flowering May–Jul(–Oct). Stream banks, lake edges, ditches, dry hills, railroad rights-of-way, orchards, prairies, open oak woods, rocky meadows, roadsides, fields, disturbed sites; 0–1300 m; introduced; B.C., N.B., Ont., Que.; Ala., Ariz., Ark., Calif., Colo., Conn., Del., D.C., Fla., Ga., Idaho, Ill., Ind., Iowa, Kans., Ky., La., Maine, Md., Mass., Mich., Miss., Mo., Mont., Nebr., Nev., N.H., N.J., N.Mex., N.Y., N.C., N.Dak., Ohio, Okla., Oreg., Pa., R.I., S.C., S.Dak., Tenn., Tex., Utah, Vt., Va., Wash., W.Va., Wis.; Europe; Asia; introduced also in Pacific Islands (Hawaii, New Zealand), Australia.

Verbascum ×pterocaulon Franchet is a hybrid between *V. blattaria* and *V. thapsus.*

3. Verbascum virgatum Stokes in W. Withering, Bot. Arr. Brit. Pl. ed. 2, 1: 227. 1787 • Wand mullein, molène en baguette F I W

Biennials. Stems 50–100 cm, densely stipitate-glandular, sometimes also sparsely hirsute-villous. **Leaves:** surfaces densely stipitate-glandular, sometimes also sparsely hirsute-villous; basal and proximal cauline with petiole 1–2 mm; blade elliptic to elliptic-obovate, 8–20(–30) × 2.5–8(–15) cm, base subrounded to broadly cuneate; cauline not clasping, gradually smaller distally, base not decurrent, margins coarsely crenate to crenulate, apex of distal cauline and floral bracts acute to obtuse. **Inflorescences** unbranched, narrowly cylindric, flowers remote, solitary in axils at least distally, sometimes 1(–5) at proximal nodes; rachis densely stipitate-glandular, sometimes also sparsely hirsute-villous with simple hairs; bracts linear-lanceolate, 8–20 mm, base not decurrent, apex long-acuminate, densely stipitate-glandular, sometimes also sparsely hirsute-villous with simple hairs. **Pedicels** free, (0–)1–3 mm; bracteoles 2. **Flowers:** calyx 4–9 mm, densely stipitate-glandular, sometimes also sparsely hirsute-villous with simple hairs, lobes ovate-lanceolate to triangular or narrowly lanceolate; corolla yellow, (25–)30–40 mm diam., pellucid glands absent or relatively few; proximal filaments glabrous at least distally, distal pair villous, hairs purplish to violet or whitish; stigma capitate. **Capsules** ovoid-globular to subglobular, 6–10 mm, stipitate-glandular.

Flowering Apr–Jun(–Oct). Fields, roadsides, disturbed sites; 10–2000 m; introduced; B.C., N.S., Ont., Que.; Ala., Ariz., Calif., Fla., Ga., Idaho, Ill., Ind., La., Nev., N.Mex., N.Y., N.C., Ohio, Pa., S.C., Tex., Utah; Europe; Asia; introduced also in Mexico (Coahuila), South America (Argentina, Chile), s Asia (India), Pacific Islands (Hawaii, New Zealand), Australia.

The occurrence of *Verbascum virgatum* in Nova Scotia may be historic; Ruth collected specimens from 1940 through 1960 from East Chester, Sydney, and Wolfville; it apparently has not been seen there subsequently. It also may be historic in British Columbia.

4. Verbascum thapsus Linnaeus, Sp. Pl. 1: 177. 1753 • Common mullein, grande molène I W

Subspecies 3 (1 in the flora): introduced; Europe; introduced also in South America, Asia, Pacific Islands, Australia.

4a. Verbascum thapsus Linnaeus subsp. thapsus I W

Verbascum schraderi G. Meyer; *V. simplex* Hoffmannsegg & Link

Annuals or biennials. Stems 30–200 cm, densely and persistently tomentose, eglandular. **Leaves:** surfaces densely and persistently tomentose, eglandular; basal and proximal cauline with petiole 5–25 mm; blade broadly elliptic to elliptic-ovate, elliptic-obovate, or oblong, 7–30(–45) × 3.5–10(–14) cm, base gradually attenuate; cauline not clasping, gradually smaller distally, base decurrent, margins subentire or crenulate, apex of distal cauline and floral bracts acute to obtuse. **Inflorescences** unbranched, sometimes branched from proximal nodes, narrowly cylindric, flowers densely overlapping, in clusters of 2–7; rachis densely and persistently tomentose, eglandular; bracts ovate to lanceolate, 12–18 mm, base decurrent, apex acuminate, densely and persistently tomentose, eglandular. **Pedicels** mostly or completely adnate to rachis, 1–5 mm; bracteoles 2. **Flowers:** calyx (5–)8–12 mm, densely and persistently tomentose, eglandular, lobes lanceolate to triangular; corolla yellow, 14–20(–30) mm diam., pellucid glands relatively numerous; proximal filaments glabrous, distal pair villous, hairs white or yellow, anthers yellow; stigma capitate. **Capsules** elliptic-ovoid to ovoid, 7–10 mm, tomentose. $2n = 32, 36.$

Flowering May–Sep. Lake edges, stream banks, bottomlands, prairies, pastures, fields, woods edges, rocky slopes, railroad embankments, roadsides; 0–2300(–2700) m; introduced; Greenland; Alta., B.C., Man., N.B., Nfld. and Labr. (Nfld.), N.S., Ont., P.E.I., Que., Sask.; Ala., Alaska, Ariz., Ark., Calif., Colo., Conn., Del., D.C., Fla., Ga., Idaho, Ill., Ind., Iowa, Kans., Ky., La., Maine, Md., Mass., Mich., Minn., Miss., Mo., Mont., Nebr., Nev., N.H., N.J., N.Mex., N.Y., N.C., N.Dak., Ohio, Okla., Oreg., Pa., R.I., S.C., S.Dak., Tenn., Tex., Utah, Vt., Va., Wash., W.Va., Wis., Wyo.; Europe; introduced also in South America (Argentina, Chile), Asia, Pacific Islands (Hawaii, New Zealand), Australia.

Verbascum thapsus is known to have formed four hybrids in North America: *V. ×kerneri* Fritsch [*V. thapsus × V. phlomoides*]; *V. ×pterocaulon* Franchet [*V. thapsus × V. blattaria*]; *V. ×spurium* W. D. J. Koch [*V. thapsus × V. lychnitis*]; and *V. ×humnickii* Franchet [*V. thapsus × V. densiflorum*].

Subspecies *crassifolium* (Lamarck) Murbeck and subsp. *giganteum* (Willkomm) Nyman have densely villous abaxial stamens and occupy restricted ranges in Europe.

5. **Verbascum densiflorum** Bertoloni, Rar. Lig. [Ital.] Pl. 3: 52. 1810 • Denseflower mullein ☐

Verbascum thapsiforme Schrader

Biennials. Stems 30–120 cm, densely and persistently tomentose, eglandular. **Leaves:** surfaces densely and persistently tomentose, eglandular; basal and proximal cauline with petiole 10–30 mm; blade obovate to oblong-ovate, 5–25(–30) × 4–8(–12) cm, base attenuate; cauline not clasping, gradually smaller distally, base long-decurrent, margins crenate, apex of distal cauline and floral bracts long-acuminate. **Inflorescences** unbranched, narrowly cylindric, sometimes branched from proximal nodes, forming a panicle, flowers densely overlapping, in clusters of 2–8; rachis densely and persistently tomentose, eglandular; bracts ovate, 15–40 mm, base decurrent, apex long-acuminate, densely and persistently tomentose, eglandular. **Pedicels** free or adnate to rachis at base, 3–15 mm; bracteoles 2. **Flowers:** calyx 5–12 mm, densely and persistently tomentose, eglandular, lobes ovate-lanceolate to lanceolate; corolla yellow, 30–55 mm diam., pellucid glands absent or relatively few; proximal filaments glabrous, distal pair villous, hairs white or yellow, anthers orange; stigma spatulate, base decurrent. **Capsules** elliptic-ovoid, 5–8 mm, tomentose.

Flowering May–Aug. Fields, roadsides, disturbed sites; 50–300 m; introduced; Iowa, Mass., Mich., Mo., Wis.; Europe; Asia; introduced also in South America (Chile), Australia.

Verbascum ×*humnickii* Franchet is a hybrid between *V. densiflorum* and *V. thapsus.*

6. **Verbascum phlomoides** Linnaeus, Sp. Pl. 2: 1194. 1753 • Orange mullein, molène faux-phlomis ☐ ☐ ☐

Biennials. Stems (30–)50–200 cm, densely and persistently tomentose, eglandular. **Leaves:** surfaces densely and persistently tomentose, eglandular; basal and proximal cauline with petiole 40–80 mm; blade ovate-lanceolate to ovate-elliptic or oblong, (10–)15–25(–35) × 4–10(–15) cm, base attenuate; cauline subauriculate-clasping, gradually smaller distally, base not decurrent, rarely slightly so, margins entire or shallowly crenate, apex of distal cauline and floral bracts caudate-acuminate to short-acuminate. **Inflorescences** unbranched, narrowly cylindric, flowers densely overlapping or remote proximally, in clusters of 2–9; rachis densely and persistently tomentose, eglandular; bracts ovate-lanceolate, 9–15 mm, base short-decurrent or not at all, apex acute to short-acuminate, densely and persistently tomentose, eglandular. **Pedicels** adnate to rachis at base, 2–8(–15) mm; bracteoles 2. **Flowers:** calyx 5–12 mm, densely and persistently tomentose, eglandular, lobes lanceolate to triangular; corolla yellow, 30–55 mm diam., pellucid glands absent or relatively few; proximal filaments glabrous at least distally, distal pair villous, hairs white or yellow; stigma spatulate, base decurrent. **Capsules** elliptic-ovoid, 5–8 mm, tomentose. $2n = 32$.

Flowering Jun–Aug. Fields, roadsides, disturbed sites; 0–600 m; introduced; Alta., B.C., Man., Ont., P.E.I., Que., Sask.; Ark., Colo., Conn., Del., D.C., Ga., Ill., Ind., Iowa, Ky., Maine, Md., Mass., Mich., Minn., Mo., N.J., N.Y., N.C., Ohio, Oreg., Pa., R.I., S.C., Tenn., Vt., Va., Wash., W.Va., Wis.; Europe; Asia; introduced also in South America (Ecuador), Pacific Islands (New Zealand).

In the flora area, *Verbascum phlomoides* is known from a single location each in Manitoba (near Roseisle) and Saskatchewan (near Moose Jaw). The record for Washington possibly is only a waif (King County, Seattle, in waste ground, introduced from Europe, 12 September 1936, *W. J. Eyerdam s.n.*, SMU), because it apparently has not been recorded there since.

Verbascum ×*kerneri* Fritsch is a hybrid between *V. phlomoides* and *V. thapsus.*

7. **Verbascum bombyciferum** Boissier, Diagn. Pl. Orient. 1(4): 52. 1844 • Turkish or silver mullein ☐

Biennials. Stems 60–200 cm, densely and persistently tomentose, eglandular. **Leaves:** surfaces densely and persistently tomentose, eglandular; basal and proximal cauline with petiole 15–40 mm; blade broadly elliptic to lanceolate-oblong, 25–35 × 15–25 cm, base attenuate; cauline slightly auriculate-clasping, gradually smaller distally, base not decurrent, margins obscurely crenate or entire, apex of distal cauline and floral bracts acute. **Inflorescences** unbranched or branched from proximal nodes, narrowly cylindric, flowers loosely overlapping, in clusters of 2–8; rachis densely and persistently tomentose, eglandular; bracts ovate to lanceolate-triangular, 7–12 mm, base not decurrent, apex acuminate, densely and persistently tomentose, eglandular. **Pedicels** free, 2–5 mm; bracteoles 2. **Flowers:** calyx 6–10 mm, densely and persistently tomentose, eglandular, lobes lanceolate to narrowly lanceolate; corolla yellow, (20–)30–40 mm diam., pellucid glands absent; filaments villous, hairs yellowish to yellowish white, or 2 proximal glabrous distally or completely; stigma spatulate, base decurrent. **Capsules** ovoid to subglobular, 5–8 mm, tomentose.

Flowering Jun–Sep. Grassy, rocky benches, streambeds; 300–500 m; introduced; Calif.; Asia (Turkey); introduced also in Europe (England, Germany).

Verbascum bombyciferum is naturalized in Sonoma County, escaped from ornamental plantings in 1976 at a residence on the Pepperwood Preserve (F. Hrusa et al. 2002). Photos of the population (http://www. calflora.org) show plants (intermixed with typical *V. thapsus*) with a dense, persistent, bright white vestiture, spikes unbranched or proximally few-branched and 1–2 m, the floral clusters thick and somewhat remotely arranged, yellow corollas with yellowish to yellowish white filament hairs, and broadly elliptic, basally attenuate leaves densely and persistently tomentose on both surfaces. Internet photos confirm the identification as *V. bombyciferum* and indicate that the Calflora photos show plants just beginning to flower, as the plants potentially elongate proximally and the spikes may develop lateral branches, although the central one usually remains dominant.

Verbascum bombyciferum of Sonoma County has been identified previously (F. Hrusa et al. 2002) as *V. olympicum* Boissier, and that name has correspondingly been registered in other literature. *Verbascum bombyciferum* (as well as *V. olympicum*) is endemic in native range to Mount Olympus (now known as Uludağ) in northwestern Turkey.

8. **Verbascum lychnitis** Linnaeus, Sp. Pl. 1: 177. 1753
   • White mullein [I] [W]

Subspecies 2 (1 in the flora): introduced; Europe, Asia.

Verbascum lychnitis is recognized by its bicolored leaves, sessile and non-clasping cauline leaves, freely branched inflorescences with loosely overlapping flower clusters, relatively long pedicels, and relatively small, white corollas. Subspecies *moenchii* (C. F. Schultz) Holub & F. Mlady was recognized in 1978, referring to a white-flowered variant in central Europe.

8a. **Verbascum lychnitis** Linnaeus subsp. **lychnitis**
   [I] [W]

Biennials. Stems 50–150 cm, thinly tawny- to gray-tomentose, glabrescent, eglandular. **Leaves:** surfaces thinly tawny- to gray-tomentose, abaxial soon glabrescent, adaxial persistent, closely adherent, eglandular; basal leaves sessile or basally attenuate and sessile, less commonly with petiolar region 10–50 mm, cauline sessile; blade obovate to elliptic-obovate, (8–)10–15(–30) × 3–7(–11) cm, base attenuate; cauline not clasping or slightly so, gradually smaller distally, base not decurrent, margins coarsely to shallowly crenate-serrate or subentire, apex of distal cauline and floral bracts long-acuminate. **Inflorescences** freely branched, loosely conic to broadly cylindric, elliptic, or ovate panicle, flowers loosely overlapping, sometimes barely remote, in clusters of 2–5; rachis thinly tawny- to gray-tomentose, glabrescent, persistent and closely adherent on abaxial leaf surfaces, not completely obscuring epidermis, eglandular; bracts linear to narrowly lanceolate, 8–15 mm, base not decurrent, apex acute, thinly tawny- to gray-tomentose, glabrescent, persistent and closely adherent on abaxial leaf surfaces, not completely obscuring epidermis, eglandular. **Pedicels** free, 6–11 mm; bracteoles 2. **Flowers:** calyx 2.5–4 mm, thinly tawny- to gray-tomentose, glabrescent, persistent and closely adherent on abaxial leaf surfaces, not completely obscuring epidermis, eglandular, lobes lanceolate; corolla white, sometimes yellowish, 12–20 mm diam., pellucid glands absent; filaments villous, hairs yellow to whitish; stigma capitate. **Capsules** ovoid-ellipsoid, 4–5 mm, tomentose. 2*n* = 32, 34.

Flowering Jun–Aug. Sandy fields, vacant lots, roadsides, disturbed sites; 50–300 m; introduced; Ont.; Colo., Conn., Del., Iowa, Md., Mass., Mich., Mo., N.H., N.J., N.Y., Pa., R.I., Vt., Va., W.Va.; Europe; Asia.

J. K. Small (1933) listed *Verbascum lychnitis* (without further taxonomic restriction) as occurring in North Carolina, but no specimen has yet been located.

Verbascum ×*spurium* W. D. J. Koch is a hybrid between *V. lychnitis* and *V. thapsus*.

9. Verbascum nigrum Linnaeus, Sp. Pl. 1: 178. 1753
  • Black mullein [I]

Perennials. Stems 50–120 cm, sparsely tomentose to glabrate, glabrescent, eglandular. **Leaves:** surfaces sparsely tomentose to glabrate, abaxial soon glabrescent, sometimes both glabrate, eglandular; basal and proximal cauline with petiole 5–15(–20) mm; blade lanceolate to ovate or oblong, 12–30 × 5–12(–15) cm, base shallowly cordate to nearly truncate; cauline not clasping or distal ones clasping, gradually smaller distally, base not decurrent, margins crenate, apex of distal cauline and floral bracts acute. **Inflorescences** unbranched, sometimes branched from proximal nodes, narrowly conic panicle, flowers loosely overlapping, in clusters of 5–10; rachis sparsely tomentose to glabrate, glabrescent, thinly tomentose on abaxial leaf surfaces, not completely obscuring epidermis, sometimes glabrate on both surfaces, eglandular; bracts linear, 4–7(–15) mm, base not decurrent, apex acute, sparsely tomentose to glabrate, glabrescent, thinly tomentose on abaxial leaf surfaces, not completely obscuring epidermis, sometimes glabrate on both surfaces, eglandular. **Pedicels** free, 5–12(–15) mm; bracteoles 2. **Flowers:** calyx 3–4.5 mm, sparsely tomentose to glabrate, glabrescent, thinly tomentose on abaxial leaf surfaces, not completely obscuring epidermis, sometimes glabrate on both surfaces, eglandular, lobes linear-oblong to linear-lanceolate; corolla yellow to cream, 18–25 mm diam., pellucid glands relatively numerous; filaments villous, hairs purple to violet; stigma capitate. **Capsules** ovoid-ellipsoid to ellipsoid-obovoid, 4–5 mm, tomentose.

Flowering Jun–Aug. Fields, roadsides, disturbed sites; 100–300(–700) m; introduced; Alta., Ont., Sask.; Ill., Mass., Minn., N.H., N.J., Pa., Wis.; Europe; Asia.

Verbascum nigrum is recognized by its petiolate, basally cordate to truncate basal leaves, bicolored or dark on both surfaces, dark-colored stems, usually unbranched inflorescences, relatively small flowers, and narrow calyx lobes. Plants in the flora area appear to be subsp. *nigrum*, with a mostly unbranched inflorescence; subsp. *abietinum* (Borbás) I. K. Ferguson, found in Germany, has freely branched inflorescences.

10. Verbascum pulverulentum Villars, Prosp. Hist. Pl. Dauphiné, 22. 1779 • Broad-leaf mullein [I]

Verbascum floccosum Waldstein & Kitaibel

Biennials. Stems 50–150(–200) cm, densely and loosely white-floccose, glabrescent, eglandular. **Leaves:** surfaces densely and loosely white-floccose, glabrescent, especially abaxially, sometimes thin-persistent adaxially, eglandular; basal and proximal cauline sessile; blade obovate to oblong-ovate, 12–30 (–40) × 5–10(–15) cm, base attenuate; cauline clasping or subclasping, gradually smaller distally, base not decurrent, margins crenate or subentire, apex of distal cauline and floral bracts acuminate. **Inflorescences** freely branched, broadly elliptic to ovate panicle, flowers becoming remote in fruit, in clusters of 2–5; rachis densely and loosely white-floccose on both surfaces, easily separating, glabrescent, eglandular; bracts linear, 3–5 mm, base not decurrent, apex acute, densely and loosely white-floccose on both surfaces, easily separating, glabrescent, eglandular. **Pedicels** free, (1–)2–5(–7) mm; bracteoles 2. **Flowers:** calyx 2–3.5 mm, densely and loosely white-floccose on both surfaces, easily separating, glabrescent, eglandular, lobes linear-lanceolate; corolla yellow, 18–20 mm diam., pellucid glands numerous; filaments villous, hairs white; stigma capitate. **Capsules** ellipsoid-globular, 3–5(–8) mm, glabrescent.

Flowering Jul–Sep. Fields, roadsides, disturbed sites; 20–30 m; introduced; Wash.; Europe; introduced also in Pacific Islands (New Zealand).

Verbascum pulverulentum is recognized by its densely white-floccose vestiture (separating easily in clumps from the stems and leaf surfaces), non-decurrent cauline leaves, freely branched inflorescences with remote fruiting clusters, and relatively small flowers.

Verbascum pulverulentum was discovered in the flora area in 1999 as a weed in the Washington Park Arboretum, Seattle (A. L. Jacobsen et al. 2001). These plants were clearly seen as growing outside of cultivation at the time, and plants were collected again in 2005, suggesting that *V. pulverulentum* persists at the arboretum site.

11. Verbascum sinuatum Linnaeus, Sp. Pl. 1: 178. 1753 • Wavy-leaf mullein [I]

Varieties 2 (1 in the flora): introduced; Europe, Asia; introduced also in Africa, Australia.

11a. Verbascum sinuatum Linnaeus var. **sinuatum** [I]

Biennials. Stems 50–100 cm, persistently and loosely tawny-tomentose, usually eglandular. Leaves: surfaces persistently and loosely tawny-tomentose, usually eglandular; basal and proximal cauline with petiole 5–25 mm; blade oblong to obovate, oblong-obovate, or broadly lanceolate, (10–)15–25(–35) × 4–8(–12) cm, base attenuate; cauline subauriculate-clasping, gradually smaller distally, base short-decurrent, margins sinuate to incised or incised-lobed, apex of distal cauline and floral bracts acute. Inflorescences freely branched, broadly elliptic to ovate panicle, flowers remote, in clusters of 2–7, rarely solitary; rachis persistently and loosely tawny-tomentose, abaxial leaf epidermis evident, usually eglandular; bracts cordate-deltate, 3–8 mm, base not decurrent, apex cuspidate, persistently and loosely tawny-tomentose, abaxial leaf epidermis evident, usually eglandular. Pedicels free, 2–5 mm; bracteoles 2. Flowers: calyx 3–5 mm, persistently and loosely tawny-tomentose, abaxial leaf epidermis evident, usually eglandular, lobes ovate-lanceolate to lanceolate; corolla yellow, 15–30 mm diam., pellucid glands relatively numerous; filaments villous, hairs violet to purple; stigma capitate. Capsules broadly ellipsoid to subglobular, 3–5 mm, tomentose. $2n = 30$.

Flowering Jun–Sep. Fields, roadsides, disturbed sites; 100–300 m; introduced; Md., N.J., N.Y., Pa.; Europe; Asia; introduced also in Africa (Tunisia), Australia.

Variety *sinuatum* is recognized by its persistent tomentum, sinuate to lobed leaf margins, subauriculate and short-decurrent cauline leaves, and freely branched inflorescences with distantly remote flower clusters. Plants recognized as var. *adenosepalum* Murbeck occur from Turkey to Iran; they differ from var. *sinuatum* in having non-decurrent cauline leaves and glandular calyx lobes. *Verbascum gaillardotii* Boissier of the eastern Mediterranean region (Lebanon, Palestine, Syria, Turkey) has sometimes been recognized as *V. sinuatum* subsp. *gaillardotii* (Boissier) Bornmueller; compared to typical *V. sinuatum*, it has narrower bracts and bracteoles, slightly smaller corollas, and four (or five) stamens [versus (four or) five in *V. sinuatum*].

12. Verbascum speciosum Schrader, Index Seminum (Göttingen) 2: 22, plate 16. 1811 • Showy mullein [I]

Subspecies 2 (1 in the flora): introduced; Europe.

It is not clear that subsp. *speciosum* is the best identification for plants in the flora area, if the infraspecific distinction is to be made. Subspecies *speciosum* is characterized by leaves with grayish, rather harsh tomentum and capsules 3–6 mm; subsp. *megaphlomos* (Boissier & Heldreich) Nyman is characterized by leaves with a white or yellowish, soft, thick tomentum and capsules 5–7 mm. Distinctly elongate basal and proximal cauline leaves are often found in *Verbascum speciosum*.

12a. Verbascum speciosum Schrader subsp. **speciosum** [I]

Biennials. Stems 50–200 cm, closely and persistently tomentose, often yellowish, eglandular. Leaves: surfaces closely and persistently tomentose, often yellowish, abaxial epidermis obscured, eglandular; basal and proximal cauline with petiole 15–50 mm; blade broadly oblanceolate to elliptic-obovate or oblong-lanceolate, (12–)25–42 × 4–10 cm, base attenuate; cauline auriculate-subclasping, gradually smaller distally, base not decurrent, margins entire or minutely serrate-dentate, apex of distal cauline and floral bracts caudate to long-acuminate. Inflorescences freely branched, broadly elliptic or ovate panicle, flowers loosely overlapping, in clusters of 5–8; rachis closely and persistently tomentose, often yellowish, densely obscuring abaxial leaf epidermis, eglandular; bracts broadly ovate-lanceolate to lanceolate, (5–)8–15(–20) mm, base not decurrent, apex acuminate, closely and persistently tomentose, often yellowish, densely obscuring abaxial leaf epidermis, eglandular. Pedicels free, (3–)5–12 mm; bracteoles 2. Flowers: calyx 3–5 mm, closely and persistently tomentose, often yellowish, densely obscuring abaxial leaf epidermis, eglandular, lobes narrowly lanceolate; corolla yellow, 18–30 mm diam., pellucid glands absent; filaments villous, hairs whitish to yellowish; stigma spatulate, base decurrent. Capsules ovoid-oblong to oblong-ovate, 4–7 mm, tomentose. $2n = 36$.

Flowering May–Aug. Fields, roadsides, disturbed sites; 100–600 m; introduced; Calif., Ill.; Europe.

There is a historic record of subsp. *speciosum* collected at Portland, Oregon, in 1909 (*Suksdorf s.n.*, WS).

LINDERNIACEAE Borsch, Kai Müller & Eb. Fischer
• False-pimpernel Family

Deborah Q. Lewis

Richard K. Rabeler

Craig C. Freeman

Wayne J. Elisens

Herbs [**subshrubs**], annual or perennial, not fleshy [fleshy], autotrophic. **Stems** prostrate or repent to ascending or erect, often rooting at nodes, 4-angled or weakly so (*Micranthemum*). **Leaves** cauline, or basal and cauline [basal], opposite, rarely whorled, simple; stipules absent; petiole present or absent; blade not fleshy [fleshy], leathery or not, margins entire, undulate, or toothed. **Inflorescences** terminal, racemelike, or axillary, flowers solitary (*Micranthemum*). **Pedicels** present; bracteoles absent. **Flowers** bisexual, perianth and androecium hypogynous; sepals 4 or 5, connate or partially so proximally, calyx radially or bilaterally symmetric; petals 4 or 5, connate, corolla bilaterally symmetric, bilabiate or unilabiate, cylindric, campanulate, or rotate; stamens 2 or 4, adnate to corolla throat, didynamous or equal, abaxial filaments geniculate; staminodes 0 or 2; pistil 1, 2-carpellate, ovary superior, 2-locular (sometimes incompletely in *Micranthemum*), placentation axile; ovules orthotropous, unitegmic, tenuinucellate; style 1; stigma 1, 2-lobed, capitate or clavate or neither (*Micranthemum*). **Fruits** capsules, dehiscence septicidal, poricidal (later septicidal and loculicidal), or irregular. **Seeds** 15–600, white, yellow, or gold, cylindric, ellipsoid, oblong to narrowly obconic, or irregularly angled, wings absent; embryo straight, endosperm rudimentary.

Genera 20, species ca. 250 (3 genera, 10 species in the flora): North America, Mexico, West Indies, Central America, South America, Europe, Asia, Africa, Pacific Islands, Australia; pantropical to temperate, mostly Old World.

The genera here included in Linderniaceae were placed traditionally in Scrophulariaceae tribe Gratioleae D. Don (for example, F. W. Pennell 1935 and most subsequent authors). Molecular analyses by D. C. Albach et al. (2005), B. Oxelman et al. (2005), and R. Rahmanzadeh et al. (2005) have contributed to the redefinition of Scrophulariaceae and to the recognition of Linderniaceae, including *Micranthemum*.

E. Fischer et al. (2013) presented a phylogeny for Linderniaceae based on broad taxon sampling and concluded that, while the family is monophyletic, realignments within the family are necessary. As an example, some species placed traditionally in *Lindernia* (including three of those occurring in the flora area) were found to be included within segregated genera.

In addition to the molecular data supporting the monophyly of this clade, Linderniaceae are recognized morphologically as a distinct lineage in the Lamiales by possessing abaxial geniculate filaments, which are expressed in the species of the flora area either in stamens or staminodes. They also lack iridoid compounds and a type of protein body present in the genera previously allied with those now in Linderniaceae (D. C. Albach et al. 2005; R. Rahmanzadeh et al. 2005; D. C. Tank et al. 2006).

1. Corolla tubes 21–36(–40) mm; calyces winged . . . . .2. *Torenia*, p. 357
1. Corolla tubes 0.3–10 mm; calyces not winged.
    2. Sepals 5; petals 5; stamens 4 and staminodes 0 or stamens 2 and staminodes 2 . . . . . 1. *Lindernia*, p. 353
    2. Sepals 4; petals 4; stamens 2, staminodes 0 . . . . .3. *Micranthemum*, p. 358

1. LINDERNIA Allioni, Mélanges Philos. Math. Soc. Roy. Turin 3: 178, plate 5, fig. 1. 1766 • False-pimpernel, blue moneywort [For Franz Balthasar von Lindern, 1682–1755, French botanist and physician]

Deborah Q. Lewis

Bonnaya Link & Otto; *Ilysanthes* Rafinesque; *Vandellia* P. Browne ex Linnaeus

Annuals or perennials. Stems erect to prostrate, glabrous or sparsely stipitate-glandular. **Leaves** cauline, or basal and cauline, opposite; petiole present or absent; blade leathery or not, margins entire, remotely toothed, undulate, serrate, or serrulate. **Inflorescences** terminal, racemelike; bracts present. **Flowers:** sepals 5, ± connate, calyx radially to bilaterally symmetric, campanulate, not winged, lobes narrowly lanceolate or elliptic to triangular-ovate; petals 5, corolla white or purplish or bluish with darker purple, violet, or blue markings, bilaterally symmetric, bilabiate, cylindric, tube 0.3–10 mm, abaxial lobes 3, adaxial 1, emarginate to entire [2-lobed], forming a projecting adaxial lip; stamens 2 or 4, didynamous when 4, abaxial filaments 0 if stamens 2, papillate proximally, glabrous distally, appendage present, adaxial filaments glabrous, appendage absent; staminodes 0 or 2, geniculate, papillate proximally, glabrous distally, appendage present or absent. **Capsules:** dehiscence septicidal. **Seeds** 150–600, yellowish gold to reddish gold, cylindric, ellipsoid, oblong, ovoid, or irregularly angled. x = 7, 8, 9.

Species 30 or ca. 100 (6 in the flora): North America, Mexico, West Indies, Central America, South America, Europe, Asia, Africa, Pacific Islands, Australia; tropical and temperate regions worldwide.

F. W. Pennell (1935) merged *Bonnaya*, *Ilysanthes*, and *Vandellia* with *Lindernia*, and most subsequent authors have followed this broad generic concept, including about 100 species, with sections being used to account for the breadth of variation within the genus; for example, D. Q. Lewis (2000) recognized three sections for the North American species. Molecular studies led E. Fischer et al. (2013) to conclude that *Lindernia* as it has been treated is paraphyletic, and they proposed a new phylogeny of Linderniaceae. In their concept, *Lindernia* would include 30 species. In particular, this impacts the North American species as follows: *L. antipoda* = *B. antipoda*, *L. ciliata* = *B. ciliata*, and *L. crustacea* = *Torenia crustacea*.

Lindernia procumbens (Krocker) Borbás has been reported from North America [F. W. Pennell (1935) as *L. pyxidaria* Linnaeus; http://plants.usda.gov; and elsewhere] based on specimens that resemble this European species. These reports should be referred to *L. dubia,* since the specimens examined have the two fertile stamens of *L. dubia* instead of the four of *L. procumbens. Lindernia diffusa* (Linnaeus) Wettstein has been reported from South Carolina (http://plants.usda.gov) and Texas (P. J. Mahan, http://www.oldthingsforgotten.com/bearcreekpark/bearcreekpark.htm). The South Carolina report is based on a specimen of *L. dubia* that had been mislabeled in annotation, and the one from Texas on a misidentification of *L. crustacea.*

SELECTED REFERENCES   Berger, B. A. 2005. Character Polymorphism and Taxonomy of the *Lindernia dubia* Complex (Scrophulariaceae). M.S. thesis. University of Oklahoma.   Lewis, D. Q. 2000. A revision of the New World species of *Lindernia* (Scrophulariaceae). Castanea 65: 93–122.

1. Stamens 4; sepals connate ½–¾ lengths . . . . 6. *Lindernia crustacea*
1. Stamens 2; sepals connate to ¼ lengths.
   2. Capsules narrowly lanceoloid to narrowly ellipsoid or cylindric; staminodes without appendages and distal segments.
      3. Leaf blade margins shallowly serrate to subentire . . . . .4. *Lindernia antipoda*
      3. Leaf blade margins sharply aristate-serrate . . . . .5. *Lindernia ciliata*
   2. Capsules ellipsoid, ovoid, or subglobular; staminodes with appendages and distal segments.
      4. Stems repent or prostrate, matted; distal leaves well developed or slightly reduced; leaf blades orbiculate, widely elliptic, or widely ovate . . . . 1. *Lindernia grandiflora*
      4. Stems erect, ascending, or prostrate, not matted; distal leaves well developed or much reduced; leaf blades spatulate, oblanceolate, obovate, elliptic, ovate, lanceolate, or linear-subulate.
         5. Perennials; leaves basal and cauline or, rarely, basal leaves absent, thick, leathery . . . . .2. *Lindernia monticola*
         5. Annuals; leaves cauline, thin, not leathery . . . . 3. *Lindernia dubia*

1. **Lindernia grandiflora** Nuttall, Gen. N. Amer. Pl. 2: 43. 1818 • Savanna false-pimpernel E W

Ilysanthes grandiflora (Nuttall) Bentham

Perennials. Stems repent or prostrate, matted, rooting at proximal or most nodes, 2–30 (–43) cm. **Leaves** cauline, thick; petiole absent; blade orbiculate, widely elliptic, or widely ovate, 2–16 × 1.5–16 mm, palmately 3–7-veined, leathery, margins entire, remotely toothed, or undulate; distal well developed or slightly reduced. **Pedicels** 6–38 mm, 2–5 times subtending leaves. **Flowers:** sepals 2–5.8 mm, connate to ⅛ lengths; corolla tube and adaxial lip lavender or blue to white, abaxial lobes white with violet to blue spots or streaks, tube 5–9 mm, adaxial lip ¼ abaxial; stamens 2; staminodes each with appendage and distal segment. **Capsules** ellipsoid, sometimes obliquely or narrowly ovoid, 2.3–6.9 × 1.2–2.5 mm. **Seeds** 6-angled, strongly ribbed.

Flowering year-round. Swamps, low woods and grasslands, wet depressions, ditches and along edges of streams and ponds, usually in sandy soils; 0–50 m; Fla., Ga.

Lindernia grandiflora has been introduced into the horticultural trade in the southern United States as a container plant or groundcover for wet areas or in bogs or water gardens. It is sometimes sold under the name "Ilysanthes floribunda," which has no standing in botanical nomenclature, or as blue moneywort or angel's tears.

2. **Lindernia monticola** Nuttall, Gen. N. Amer. Pl. 2: Add. 1818 • Piedmont false-pimpernel E

Ilysanthes monticola (Nuttall) Rafinesque; *I. saxicola* (M. A. Curtis) Chapman; *Lindernia saxicola* M. A. Curtis

Perennials. Stems erect to prostrate, not matted, sometimes rooting at proximal nodes, 6–41 cm. **Leaves** basal and cauline or, rarely, basal leaves absent, most larger cauline leaves basally disposed, thick; petiole absent; blade (basal) oblanceolate, obovate, or elliptic, 5–28(–35) × 2–15 mm; blade (cauline) ovate, elliptic, oblanceolate, or linear-subulate, 1–30 × 0.2–8(–11) mm, palmately 3–5-veined, pinnately veined, or 1-nerved, leathery, margins entire or remotely toothed; distal much reduced. **Pedicels** 2–40(–52) mm, 1–10 times subtending leaves. **Flowers:** sepals 1.1–4.3 mm, connate to ⅛ lengths; corolla tube and adaxial lip lavender or blue to white, abaxial lobes white with purple to blue markings, tube 5–10 mm, adaxial lip ½ abaxial; stamens 2; staminodes each with appendage and distal segment. **Capsules** narrowly ellipsoid to ovoid, (1.8–)3.4–7.1 × 1–2.1 mm. **Seeds** usually 6-angled, not strongly ribbed.

Flowering Mar–Oct. Shallow, moist soil pockets on granite outcrops, granite or calcareous rock in streams, sand in pine-flatwoods; 50–300 m; Ala., Fla., Ga., N.C., S.C.

Plants of *Lindernia monticola* are almost exclusively found in intermittent pools in depressions or cracks on granite outcrops of the southeastern United States, where they typically show the basal rosette (heterophyllous) form. When it occurs in other habitats, the basal rosette is less distinct or absent, the plants having more or less homophyllous leaves.

3. **Lindernia dubia** (Linnaeus) Pennell, Monogr. Acad. Nat. Sci. Philadelphia 1: 141. 1935 • Yellowseed false-pimpernel, lindernie douteuse [F] [W]

Gratiola dubia Linnaeus, Sp. Pl. 1: 17. 1753; *Ilysanthes dubia* (Linnaeus) Barnhart; *I. inaequalis* (Walter) Pennell; *I. riparia* Rafinesque; *Lindernia anagallidea* (Michaux) Pennell; *L. dubia* var. *anagallidea* (Michaux) Cooperrider; *L. dubia* var. *inundata* (Pennell) Pennell; *L. dubia* var. *riparia* (Rafinesque) Fernald

Annuals. **Stems** erect, ascending, or prostrate, not matted, usually rooting at proximal nodes, 1.5–27 (–38) cm. **Leaves** cauline, thin; petiole absent; blade spatulate, lanceolate, oblanceolate, elliptic, ovate, or obovate, (1–)5–37 × (0.5–)3–18 mm, palmately 3–5-veined or 1-nerved, not leathery, margins entire or remotely, sometimes coarsely, toothed; distal well developed or much reduced. **Pedicels** 0.5–31 mm, ½–5 times subtending leaves. **Flowers** chasmogamous or cleistogamous; chasmogamous: sepals 0.7–6.1 mm, connate to ⅛ lengths; corolla tube and adaxial lip lavender or blue to white, abaxial lobes white with purple to blue markings, tube 2.5–8 mm, adaxial lip ½ abaxial; stamens 2; staminodes each with appendage and distal segment. **Capsules** ellipsoid, often obliquely, sometimes ovoid or subglobular, 1.4–6.3(–7.5) × 1.2–3.3 mm. **Seeds** usually 6-angled, usually ribbed. $2n = 18$.

Flowering year-round. Wet ditches, meadows, borders of ponds, lakes, streams, moist to wet disturbed habitats; 0–2600 m; B.C., N.B., N.S., Ont., P.E.I., Que.;

Ala., Ariz., Ark., Calif., Colo., Conn., Del., D.C., Fla., Ga., Idaho, Ill., Ind., Iowa, Kans., Ky., La., Maine, Md., Mass., Mich., Minn., Miss., Mo., Nebr., Nev., N.H., N.J., N.Mex., N.Y., N.C., N.Dak., Ohio, Okla., Oreg., Pa., R.I., S.C., S.Dak., Tenn., Tex., Vt., Va., Wash., W.Va., Wis.; Mexico; West Indies; Central America; South America; introduced in Europe, Asia, Africa, Australia.

In Asia, *Lindernia dubia* is considered a noxious weed in rice paddies, where it has become resistant to some commonly used herbicides.

Lindernia dubia shows extreme morphological plasticity, especially in vegetative characters. This has led to the naming of species and varieties that have been accepted or not in recent treatments (for example, D. Q. Lewis 2000). B. A. Berger (2005) examined variation within the *L. dubia* complex and concluded that the recognition of these taxa is unwarranted.

Variety *inundata*, an estuarine form from the intertidal zone along the Atlantic Coast, continues to be recognized in several databases. However, W. R. Ferren Jr. and A. E. Schuyler (1980) described the clinal variation in these intertidal populations, ranging from typical *Lindernia dubia* to this form, with such variation sometimes evident on submerged and emergent parts of the same plant.

4. **Lindernia antipoda** (Linnaeus) Alston in H. Trimen et al., Handb. Fl. Ceylon 6: 214. 1931 • Sparrow false-pimpernel [I]

Ruellia antipoda Linnaeus, Sp. Pl. 2: 635. 1753; *Bonnaya antipoda* (Linnaeus) Druce

Annuals. **Stems** erect to decumbent, not matted, usually rooting at proximal nodes, 3–17 cm. **Leaves** cauline, thin; petiole absent; blade obovate, oblanceolate, or elliptic, distal sometimes linear, (2–)10–25 × (0.4–)3.5–9 mm, pinnately veined to 1-nerved, not leathery, margins shallowly serrate to subentire; distal well developed or much reduced. **Pedicels** 3–15 mm, ½–3 times subtending leaves. **Flowers:** sepals 5–6 mm, connate to ¼ lengths; corolla white, lavender, or pink, abaxial lobes sometimes with purple markings, tube 5.5–8 mm, adaxial lip equal to abaxial; stamens 2; staminodes without appendage and distal segment. **Capsules** narrowly lanceoloid, 8–14 × 1.2–3.8 mm. **Seeds** irregularly oblong, ovoid, or rhombic, not ribbed. $2n = 36$ (India).

Flowering Jun–Sep. Moist to wet roadsides, disturbed areas, pine flatwoods; 0–50 m; introduced; La.; Asia; Australia; introduced also in Mexico, South America (Venezuela).

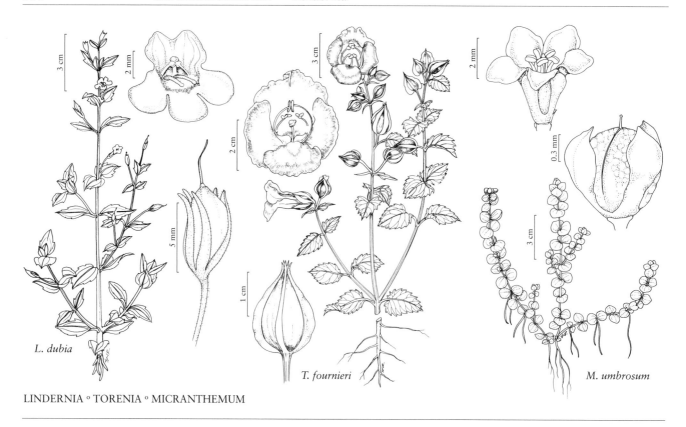

L. dubia

T. fournieri

M. umbrosum

LINDERNIA ° TORENIA ° MICRANTHEMUM

Lindernia antipoda has only recently been introduced in the Americas; it was first collected in 1979 (*Allen, Vincent & Erbe 9259*, LAF) from Livingston Parish, Louisiana (K. A. Vincent 1981). It has subsequently been documented from Mexico and Venezuela.

5. Lindernia ciliata (Colsmann) Pennell, Brittonia 2: 182. 1936 • Fringed false-pimpernel [I]

Gratiola ciliata Colsmann, Prodr. Descr. Gratiol., 14. 1793; *Bonnaya ciliata* (Colsmann) Sprengel

Annuals. **Stems** erect to decumbent, not matted, sometimes rooting at proximal nodes, 3–20 cm. **Leaves** cauline, thin; petiole absent; blade oblong to lanceolate, distal sometimes linear, 4–45 × 0.7–12 mm, not leathery, margins sharply aristate-serrate; distal much reduced. **Pedicels** 2–11 mm, ½–1[–3] times subtending leaves. **Flowers:** sepals 4–6 mm, connate to ¼ lengths; corolla white, pink, or lavender, abaxial lobes sometimes with darker pink or purple markings, tube 5–7 mm, adaxial lip ¾ abaxial; stamens 2; staminodes without appendage and distal segment. **Capsules** narrowly lanceoloid to narrowly ellipsoid or cylindric, 12–14 × 1–3.2 mm. **Seeds** ovoid to ellipsoid or oblong, irregularly angled, not ribbed.

Flowering Jun–Nov. Moist, sandy disturbed areas; 0–50 m; introduced; Fla.; Asia; Australia.

A specimen collected in Hillsborough County in 1992 (*Holland & Donovan s.n.*, USF) represents the first record of *Lindernia ciliata* in the flora area.

6. Lindernia crustacea (Linnaeus) F. Mueller, Syst. Census Austral. Pl., 97. 1882/1883 • Malaysian false-pimpernel [I]

Capraria crustacea Linnaeus, Mant. Pl. 1: 87. 1767; *Torenia crustacea* (Linnaeus) Chamisso & Schlechtendal

Annuals. **Stems** erect to ascending or prostrate, not matted, usually rooting at proximal nodes, 2–26 cm. **Leaves** cauline, thin; petiole present or absent; blade ovate, elliptic, or suborbicular, distal sometimes lanceolate to nearly linear, 2–28 × 0.5–17 mm, pinnately veined or distal sometimes 1-nerved, not leathery, margins serrate, serrulate, crenate, or coarsely toothed, rarely nearly entire; distal well developed or much reduced. **Pedicels** 2–36 mm, ½–4 times subtending leaves. **Flowers:** sepals 1.8–5.9 mm, connate ½–¾ lengths; corolla tube and adaxial lip lavender, pink, or blue to white, abaxial lobes white with purple to blue markings, tube 3–9 mm, adaxial lip ¾–1 times abaxial; stamens 4; staminodes absent. **Capsules** broadly ellipsoid to globular, 1.4–5.5 × 1.7–3.5 mm. **Seeds** cylindric, ellipsoid, or irregularly oblong, not ribbed. $2n$ = 42 (India).

Flowering year-round. Wet ditches and depressions, moist to wet disturbed habitats; 0–100 m; introduced; Ala., Ark., Fla., Ga., La., Miss., N.C., S.C., Tex.; Asia (China, India, Japan, Malaysia); Africa; Pacific Islands (Philippines); Australia; introduced also in Mexico, West Indies, Central America, South America, Asia (Indonesia), elsewhere in Pacific Islands.

Lindernia crustacea is common and weedy, as well as used medicinally, in much of the Eastern Hemisphere. It is difficult to assess the native range of the species; here the author follows E. Fischer et al. (2013). The first collection in the flora area was in 1916 in Lee County, Florida (*Standley 390*, MO). The species is apparently continuing to spread in the southeastern United States.

2. TORENIA Linnaeus, Sp. Pl. 2: 619. 1753; Gen. Pl. ed. 5, 270. 1754 • Wishbone flower, bluewings [For Olaf Toren, 1718–1753, Swedish clergyman and naturalist with Swedish East India Company] ⓘ

Deborah Q. Lewis

Annuals [perennials]. Stems erect, sometimes prostrate, glabrous or hairy. **Leaves** cauline, opposite; petiole present; blade not leathery, margins serrate [crenate]. **Inflorescences** terminal, racemelike; bracts present. **Flowers:** sepals 5, connate ½+ length, calyx bilaterally to radially symmetric, ellipsoid, winged, lobes triangular-ovate; petals 5, corolla usually white, violet, purple, or blue [red, pink, or yellow], bilaterally symmetric, bilabiate, campanulate, tube 21–36(–40) mm, abaxial lobes 3, adaxial 2, connate, forming erect to spreading adaxial lip; stamens 4, didynamous, abaxial filaments glabrous, appendage absent [present], adaxial filaments glabrous, appendage absent; staminode 0. **Capsules:** dehiscence initially poricidal, later loculicidal and septicidal. **Seeds** 250–400, white to gold, cylindric. x = [8]9.

Species ca. 50 (1 in the flora): introduced; Asia, Africa; introduced also in Mexico, West Indies, Central America, South America, Europe, Australia.

In addition to the two widely cultivated taxa, *Torenia fournieri* and *T. fournieri* × *T. concolor* Lindley, *T. thouarsii* (Chamisso & Schlechtendal) Kuntze is found as a weedy species introduced in Central America and South America.

1. **Torenia fournieri** Linden ex E. Fournier, Ill. Hort. 23: 129, plate 249. 1876 Ⓕ ⓘ

Stems 2–38 cm. **Leaves:** petiole 3–21 mm; blade lanceolate to ovate or distal sometimes linear, 3–46 × 2–24 mm; distal well developed or greatly reduced. **Pedicels** 5–24 mm, 0.7–1.5 times subtending leaves. **Flowers:** sepals 11–18 mm; corolla adaxial lip equal to abaxial. **Capsules** narrowly ellipsoid, 8–12 × 1.5–3 mm. **Seeds** 0.5–0.6 × 0.3–0.4 mm. $2n$ = 18 (India).

Flowering May–Oct. Lawns, compost piles, roadsides, disturbed places, persisting after cultivation; 20–1200 m; introduced; Ala., Fla., Iowa, La., N.C.; Asia; introduced also in Mexico, Central America, South America, Europe, Australia.

Torenia fournieri and its hybrid with *T. concolor*, often called "Torenia hybrida" in the horticultural trade, are popular bedding plants throughout much of North America. The hybrid is sterile, reportedly producing neither seeds nor viable pollen (http://www.health.gov.au/internet/ogtr/publishing.nsf/Content/torenia-3/$FILE/biologytorenia08.pdf). J. D. Pittillo and A. E. Brown (1988) published the first report of *T. fournieri* as a waif in the flora area, from Jackson County, North Carolina.

3. MICRANTHEMUM Michaux, Fl. Bor.-Amer. 1: 10, plate 2. 1803, name conserved • Mudflower [Greek *micros*, small, and *anthemom*, flower]

Brian R. Keener

Hemianthus Nuttall

Annuals [perennials], forming mats. **Stems** prostrate to weakly ascending, glabrous. **Leaves** cauline, opposite or whorled; petiole absent or present; blade not leathery, margins entire. **Inflorescences** axillary, solitary flowers; bracts absent. **Flowers:** sepals 4, connate proximally, calyx nearly radially symmetric, campanulate, not winged, lobes lanceolate or oblanceolate, equal or unequal; petals 4, corolla white, weakly to strongly bilaterally symmetric, weakly bilabiate and ± rotate or unilabiate, tube 0.3–0.8 mm, abaxial lobes 1 or 3, lateral 0 or 2, adaxial 0 or 1, adaxial lip sometimes prominent; stamens 2, filaments gibbous or bearing a process near base, glabrous, appendage present; staminode 0. **Capsules:** dehiscence irregular. **Seeds** 15–20, yellowish, ellipsoid or oblong to narrowly obconic.

Species 14 (3 in the flora): s, e United States, West Indies, Central America, South America.

Morphological data (corolla and stamen structures) can be interpreted as providing support for recognizing the genus *Hemianthus* (which would include *Micranthemum glomeratum* and *M. micranthemoides*), as was done by J. K. Small (1933) and F. W. Pennell (1935).

1. Leaf blades orbiculate, ovate, or obovate; corollas shorter than or equal to calyces. . . . . .1. *Micranthemum umbrosum*
1. Leaf blades elliptic to oblanceolate; corollas longer than calyces.
    2. Seeds ellipsoid; flowers chasmogamous. . . . . .2. *Micranthemum glomeratum*
    2. Seeds oblong to narrowly obconic; flowers cleistogamous. . . . . .3. *Micranthemum micranthemoides*

1. Micranthemum umbrosum (J. F. Gmelin) S. F. Blake, Rhodora 17: 131. 1915 • Shade mudflower, baby's tears F W

Globifera umbrosa J. F. Gmelin, Syst. Nat. 1: 32. 1791 (as umbros)

Leaves opposite; petiole 0–0.4 mm; blade orbiculate, ovate, or obovate, 2–9 × 1.5–7 mm. **Pedicels** 0.2–0.9 mm. **Flowers** chasmogamous; calyx symmetric, 1.2–1.5 mm, tube 0.2–0.4 mm, lobes 0.9–1.1 mm; corolla slightly bilaterally symmetric, weakly bilabiate, ± rotate, 1.3–1.5 mm, shorter than or equal to calyx, tube 0.7–1 mm, abaxial lobe 1, not forming a prominent lip, 0.8–1 mm, lateral 2, 0.6–0.7 mm, adaxial 0.4–0.5 mm. **Capsules** 1.1–1.2 × 0.9–1 mm. **Seeds** oblong to narrowly obconic, 0.3–0.4 × 0.1–0.2 mm.

Flowering late summer–fall. Swamps, margins of ponds, lakes, and sluggish streams, wet ditches; 0–200 m; Ala., Ark., Fla., Ga., La., Miss., N.C., S.C., Tenn., Tex., Va.; West Indies; Central America; South America.

2. Micranthemum glomeratum (Chapman) Shinners, Sida 1: 252. 1964 • Manatee mudflower E W

Micranthemum nuttallii A. Gray var. *glomeratum* Chapman, Fl. South. U.S. ed. 2 repr. 2, 690. 1892; *Hemianthus glomeratus* (Chapman) Pennell

Leaves opposite or whorled; petiole 0–0.2 mm; blade elliptic to oblanceolate, 2.5–8.5 (–15) × 0.5–3.1 mm. **Pedicels** 0.5–4 mm. **Flowers** chasmogamous; calyx irregularly symmetric, tube deeply lobed abaxially, 1.1–1.4 mm, tube 0.5–0.8 mm, lobes 0.5–0.7 mm; corolla strongly bilaterally symmetric, unilabiate or nearly so, 1.7–1.8 mm, longer than calyx, tube 0.5–0.8 mm, abaxial lobes 3, forming a prominent lip, 0.9–1 mm, lateral 0, adaxial (0–)0.7–0.8 mm. **Capsules** 1.1–1.3 × 0.9–1.2 mm. **Seeds** ellipsoid, 0.2–0.3 × 0.05–0.1 mm.

Flowering spring–late summer. Margins of lakes and streams, sandy soils; 0–30 m; Fla.

Micranthemum glomeratum primarily occurs in the central part of the peninsula; it has not been found in the extreme southern counties. In the northern panhandle region, there is an outlier record from Gadsden County.

3. **Micranthemum micranthemoides** (Nuttall) Wettstein in H. G. A. Engler and K. Prantl, Nat. Pflanzenfam. 67[IV,3b]: 77. 1891 • Nuttall's mudflower E

Hemianthus micranthemoides Nuttall, J. Acad. Nat. Sci. Philadelphia 1: 119, plate 6, fig. 2. 1817

Leaves opposite or whorled; petiole 0–0.5 mm; blade elliptic to oblanceolate, 3–6 × 1.2–3 mm. **Pedicels** 0.6–2 mm. **Flowers** cleistogamous; calyx irregularly symmetric, tube deeply lobed abaxially, 1.2–1.5 mm, tube 0.7–1 mm, lobes 0.5–0.6 mm; corolla strongly bilaterally symmetric, unilabiate or nearly so, 1.8–2 mm, longer than calyx, tube 0.3–0.5 mm, abaxial lobes 3, forming a prominent lip, 1.4–1.5 mm, lateral 0, adaxial (0–)0.4–0.6 mm. **Capsules** 1.1–1.3 × 0.9–1 mm. **Seeds** oblong to narrowly obconic, 0.3–0.4 × 0.2–0.3 mm.

Flowering summer–fall. Fresh water tidal areas in sandy, gravelly substrates; 0–10 m; Del., D.C., Md., N.J., N.Y., Pa., Va.

The range given for *Micranthemum micranthemoides* is historic; it is presumed extinct in the wild but may still be available in the aquarium plant trade. The most recently collected specimen examined by the author was gathered in 1941 by M. L. Fernald (*Fernald & Long 13754*, GH) along the Chickahominy River near Cypress Bank Landing, Virginia.

Micranthemum nuttallii A. Gray is an older, but illegitimate name that pertains here; it was an avowed substitute for *Hemianthus micranthemoides*.

PEDALIACEAE R. Brown

• Sesame Family

Richard K. Rabeler

Craig C. Freeman

Wayne J. Elisens

Herbs [shrubs or trees], annual [perennial], not fleshy [± fleshy], autotrophic, stipitate-glandular, hairs mucilaginous. **Stems** erect or arching [prostrate]. **Leaves** cauline, opposite, rarely alternate distally, simple, 3-lobed, or 3-foliolate [digitate]; stipules absent; petiole present; blade not fleshy [± fleshy], not leathery, margins entire, toothed, or lobed. **Inflorescences** terminal, racemes or axillary, flowers solitary; bracts present. **Flowers** bisexual, perianth and androecium hypogynous; sepals 5, proximally connate, calyx ± bilaterally symmetric; petals 5, connate, corolla bilaterally symmetric, bilabiate [lobes subequal], funnelform; stamens 4, adnate to corolla base, didynamous, glabrous, filaments glabrous; staminode 0 or 1; pistil 1, 2-carpellate, ovary superior, 2- or 4-locular, often with false septa proximally, especially in fruit, placentation axile; ovules anatropous, unitegmic, tenuinucellate; style 1; stigma 1, 2-lobed, lobes spreading, filiform. **Fruits** capsules, dehiscence loculicidal, sometimes incomplete, with or without 2, subapical horns. **Seeds** [10–]30–60, white, brown, or black, obovoid to pyriform or ovoid; embryo straight, endosperm sparse.

Genera ca. 13, species ca. 79 (2 genera, 2 species in the flora): introduced; s Asia, Africa, Australia; introduced also in South America, Europe, elsewhere in Asia, n Africa, Pacific Islands.

Often combined with the chiefly New World Martyniaceae, Pedaliaceae are an Old World family distinguished by mucilaginous hairs, nectaries on the pedicels, axile placentation, colpate pollen, and fruits without an enlarged, woody endocarp (H.-D. Ihlenfeldt 1967, 2004). Pedaliaceae appear to be most closely related to members of Acanthaceae and Bignoniaceae. Molecular study supports results of morphological studies (R. G. Olmstead et al. 2001) and indicates that Martyniaceae and Pedaliaceae are nested in different clades.

The center of diversity of Pedaliaceae is sub-Saharan Africa, where all but one genus are native. Both genera introduced to the flora area belong to Sesameae (Endlicher) Meisner, one of three tribes recognized (H.-D. Ihlenfeldt 2004).

SELECTED REFERENCES Ihlenfeldt, H.-D. 1967. Über die Abgrenzung und die natürliche Gleiderung der Pedaliaceae R. Br. Mitt. Staatsinst. Allg. Bot. Hamburg 12: 43–128. Ihlenfeldt, H.-D. 2004. Pedaliaceae. In: K. Kubitzki et al., eds. 1990+. The Families and Genera of Vascular Plants. 14+ vols. Berlin, etc. Vol. 7. Pp. 307–322. Manning, S. D. 1991. The genera of Pedaliaceae in the southeastern United States. J. Arnold Arbor. Suppl. Ser. 1: 313–347.

1. Capsules with 2, subapical horns; plants foul smelling; staminodes 0(or 1) . . . . 1. *Ceratotheca*, p. 361
1. Capsules without horns; plants not foul smelling; staminodes 1 . . . . 2. *Sesamum*, p. 361

1. CERATOTHECA Endlicher, Linnaea 7: 5. 1832 • [Greek *keratos*, horned, and *theke*, case, alluding to barbed fruit] [I]

Kerry A. Barringer

Annuals foul smelling. **Stems** erect or arching, puberulent, hairs glandular and eglandular. **Leaves** opposite proximally, opposite to subopposite distally. **Inflorescences** terminal, racemes [axillary, flowers solitary]. **Pedicels** often without paired nectaries at base; bracteoles present or absent. **Flowers:** calyx cupulate, lobes narrowly triangular; corolla purple to white; staminode 0(or 1), cylindric; style base not persistent. **Capsules** flattened, style base not enlarged, with 2, subapical horns. **Seeds** [20–]30–50, black [to brown], obovoid to pyriform, wings absent. $x = 16$.

Species 5 (1 in the flora): introduced, Florida; c, se Africa.

1. **Ceratotheca triloba** (Bernhardi) E. Meyer ex Hooker f., Bot. Mag. 114: plate 6974. 1888 • Wild or white or South African foxglove [F] [I]

Sporledera triloba Bernhardi, Linnaea 16: 42. 1842; *Ceratotheca lamiifolia* (Engler) Engler; *Sesamum lamiifolium* Engler

Stems branched proximally, slightly 4-angled, 50–150 cm. **Leaves:** petiole 2–10(–15) cm, hairy; blade broadly ovate to cordate or triangular, 3–10 (–20) × 3–10 cm, base rounded to cordate, apex obtuse to acute. **Inflorescences** secund. **Pedicels** 2–4 mm; bracteoles 0–2, linear. **Flowers** drooping; sepals often purple distally, 7–10 mm; corolla 25–40 mm, glandular-pubescent externally, strongly curved at base, mouth 15 mm diam. **Capsules** 20–30 × 4–5 mm. **Seeds** 2–4 mm, smooth. $2n = 32$ (Africa).

Flowering Jul–Nov. Disturbed sites; 0–100 m; introduced; Fla.; s South Africa (Natal).

Plants of *Ceratotheca triloba* resemble those of *Digitalis* with their tall, herbaceous habit, secund racemes curved toward the tip, and drooping, tubular flowers. Capsules of *C. triloba* are four-locular and have two subapical, hornlike appendages, basal leaves sometimes three-foliolate, and corolla lobes relatively large and rounded. Plants escaped from cultivation are known from Highlands and Lake counties.

Another species, *Ceratotheca sesamoides* Endlicher, distinguished by smaller flowers, dense indument, and all leaves simple, is sometimes cultivated in the flora area.

2. SESAMUM Linnaeus, Sp. Pl. 2: 634. 1753; Gen. Pl. ed. 5, 282. 1754 • [Greek *sésamon* (Arabic *simsim*), literally meaning plant oil] [I]

Kerry A. Barringer

Annuals not foul smelling. **Stems** erect [prostrate], hairy [glabrous], hairs glandular and eglandular. **Leaves** opposite proximally, opposite or alternate distally. **Inflorescences** axillary, flowers solitary. **Pedicels** often with paired nectaries at base; bracteoles present or absent. **Flowers:** calyx tubular or cupulate, lobes linear to narrowly triangular; corolla white to pale pink or pale purple, often with darker veins proximally; staminode 1, filiform; style base persistent.

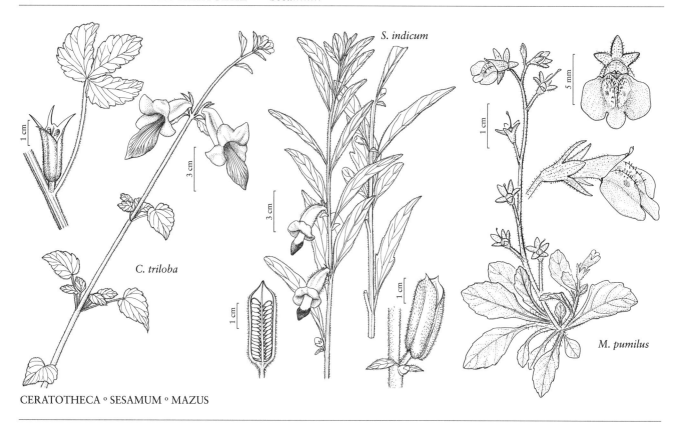

S. indicum

C. triloba

M. pumilus

CERATOTHECA ∘ SESAMUM ∘ MAZUS

Capsules not flattened, style base enlarged, forming a persistent beak, lateral horns absent. **Seeds** [10]40–60[–80], white, brown, or black, ovoid [obovoid], wings absent or present. *x* = 16.

Species 21 (1 in the flora): introduced; Asia, Africa; tropical areas; introduced also in South America, Europe, elsewhere in Asia, n Africa, Pacific Islands, Australia.

SELECTED REFERENCES Bedigian, D., ed. 2011. Sesame: The Genus *Sesamum*. Boca Raton.   Ihlenfeldt, H.-D. and U. Grabow-Seidensticker. 1979. The genus *Sesamum* and the origin of cultivated sesame. In: K. Kunkel, ed. 1979. Taxonomic Aspects of African Economic Botany. Las Palmas de Gran Canaria. Pp. 53–60.

1. **Sesamum indicum** Linnaeus, Sp. Pl. 2: 634. 1753, name conserved • Sesame, benne F I

Stems simple or branched proximally, 4-angled, grooved, 30–100(–150) cm. **Leaves:** petiole 5–12 cm; blade narrowly lanceolate to ovate, 5–20 × 1–5 cm, base cuneate, apex acute, glabrate adaxially. **Pedicels** 3–5 mm; bracteoles 0–2, linear. **Flowers** drooping; sepals 4–8 mm; corolla 2–4 cm, glandular-pubescent externally, tube slightly curved at base, mouth 10 mm diam. **Capsules** 15–30 × 5–7 mm. **Seeds** 2–3 mm, smooth. *2n* = (18), 32, 52 (Asia).

Flowering Jun–Aug.   Disturbed sites; 0–1000 m; introduced; Ala., Calif., Fla., Ga., La., Mass., Mo., N.J., N.Y., Ohio, Pa., S.C., Tex., Wis.; origin unknown, probably s Asia (India); introduced also in South America, Europe, elsewhere in Asia, n Africa, Pacific Islands, Australia.

Sesamum indicum is the source of sesame seeds and sesame oil. It has been in cultivation for perhaps more than 5000 years; it is difficult to determine its native range. Sesame seeds are a component of many bird seed mixes, so plants often occur near feeders. The name *S. orientale* Linnaeus, which pertains here, has been rejected.

MAZACEAE Reveal

• Mazus Family

Richard K. Rabeler

Craig C. Freeman

Herbs, annual or perennial, not fleshy, autotrophic. **Stems** horizontal to ascending or erect, 4-angled. **Leaves** usually cauline and basal [basal only], opposite, simple; stipules absent; petiole present or absent; blade not fleshy, not leathery, margins entire, toothed, or divided. **Inflorescences** scapose or terminal racemes. **Flowers** bisexual, perianth and androecium hypogynous; sepals 5, proximally connate, calyx radially symmetric; petals 5, proximally connate, corolla bilaterally symmetric, bilabiate, tubular; stamens 4, adnate to corolla throat, didynamous, staminode 0; pistil 1, 2-carpellate, ovary superior, 2-locular, placentation axile; ovules anatropous, unitegmic, tenuinucellate; style 1; stigma 1, capitate and slightly 2-lobed. **Fruits** capsules, dehiscence loculicidal [fleshy, berrylike (*Lancea*)]. **Seeds** 20–40, yellowish brown, ellipsoid; embryo straight, endosperm sparse.

Genera 3, species ca. 36 (1 genus, 2 species in the flora): introduced; e Asia, Australia; introduced also in Central America, South America, Europe, Africa.

Mazaceae, which includes *Dodartia* Linnaeus, *Lancea* Hooker f. & Thomson, and *Mazus*, was described in 2011 to accommodate these genera traditionally allied with *Mimulus* in the Scrophulariaceae. P. M. Beardsley and R. G. Olmstead (2002) suggested that the three genera belonged in a more broadly circumscribed Phrymaceae. Two subfamilies were proposed, but not published, to accommodate the alignment of *Lancea* and *Mazus* in a clade sister to the rest of the Phrymaceae. D. C. Albach et al. (2009) suggested that, on the basis of monophyly, *Lancea* and *Mazus* could not be included in the Phrymaceae or any other then-circumscribed family.

1. MAZUS   Loureiro, Fl. Cochinch. 2: 385. 1790   •   [Greek *mazos*, breast, alluding to two ridges on abaxial lip of corolla or to nipplelike tubercles at inner throat of corolla in *M. pumilus*]   ①

Brian R. Keener

Hornemannia Willdenow

Annuals or perennials, stolons absent or present. **Stems** glabrous or hairy. **Racemes:** bracts absent. **Pedicels** present; bracteoles present. **Flowers:** calyx campanulate to funnelform, lobes lanceolate to slightly ovate; corolla purplish to violet and white or white with throats usually spotted yellow, abaxial lobes 3, forming a prominent, projecting lip, adaxial 2; abaxial throat lined with clavate trichomes; stamens: filaments glabrous, anthers often loosely connate in pairs. **Capsules** obovoid. **Seeds:** wings absent. *x* = 20.

Species ca. 33 (2 in the flora): introduced; e Asia; introduced also in Central America, South America, Europe, Africa, Australia.

1. Stolons absent; corollas 7–9 mm . . . . .1. *Mazus pumilus*
1. Stolons present; corollas 13–22 mm . . . . .2. *Mazus miquelii*

1. Mazus pumilus (Burman f.) Steenis, Nova Guinea, n. s. 9: 31. 1958   •   Japanese mazus   F I W

Lobelia pumila Burman f., Fl. Indica, 186, plate 60, fig. 3. 1768; *Lindernia japonica* Thunberg; *Mazus japonicus* (Thunberg) Kuntze; *M. rugosus* Loureiro

Annuals, 2–20 cm; stolons absent. **Stems** erect, ascending, or prostrate, sometimes branched near base, hairy. **Leaves:** basal leaves: petiole 2–13 mm; blade ovate to oblanceolate, 5–30 × 2–15 mm, faces glabrous; cauline leaves: petiole 0–8 mm, ciliate or glabrous; blade obovate to spatulate, 8–30 × 3–20 mm, base long-cuneate, usually ciliate near base, faces glabrous or abaxially hairy along veins, adaxially hairy proximally. **Racemes** scapose or terminating erect stems, 0.5–12 cm. **Pedicels** 2–10 mm, hairy or glabrous; bracteoles narrowly triangular, 1–1.5 mm, glabrous. **Flowers:** calyx 5–7 mm, sparsely hairy, tube 2–3 mm, lobes 3–4 mm; corolla 7–9 mm, tube 4–5 mm, abaxial lobes 3–4 mm, adaxial 2–3 mm. **Capsules** 2–3 × 1.8–2.2 mm. **Seeds** 0.4–0.6 × 0.3–0.4 mm. *2n* = 40 (Asia).

Flowering spring–fall. Lawns, roadsides, stream banks, lakeshores, shady, moist areas; 0–700 m; introduced; Ala., Ark., Del., D.C., Fla., Ga., Ill., La., Mass., Miss., Mo., N.Y., N.C., Ohio, Okla., Oreg., Pa., S.C., Tenn., Tex., Va., Wash., W.Va., Wis.; e Asia; introduced also in Central America, South America, Europe, Africa, Australia.

The description and illustration of the type by Burman do not seem to match this species. Steenis examined specimens used by Burman and determined that they are indeed referable to *Mazus pumilus*; his combination has priority over *M. japonicus*.

2. Mazus miquelii Makino, Bot. Mag. (Tokyo) 16: 162. 1902   •   Creeping or Miquel's mazus   I

Vandellia japonica Miquel, Ann. Mus. Bot. Lugduno-Batavi 2: 118. 1865, not *Mazus japonicus* (Thunberg) Kuntze 1891; *M. reptans* N. E. Brown

Perennials, 5–35 cm; stolons present. **Stems** prostrate, rooting adventitiously, branched near base, glabrous. **Leaves:** petiole 0 mm; basal leaves: blade oblanceolate, 10–40 × 3–15 mm, faces glabrous; cauline leaves: blade obovate to spatulate, 8–30 × 4–12 mm, base long-cuneate, faces glabrous. **Racemes** scapose or terminating stolons, 5–15 cm. **Pedicels** 2–10 mm, glabrous; bracteoles narrowly triangular, 1.5–2 mm, glabrous. **Flowers:** calyx 4.5–7 mm, glabrous, tube 2.5–4 mm, lobes 2.5–4 mm; corolla 13–22 mm, tube 6–7.5 mm, abaxial lobes 7–13 mm, adaxial 6–8 mm. **Capsules** 3.5–5 × 3–3.5 mm. **Seeds** 0.4–0.6 × 0.3–0.4 mm. *2n* = 20 (Asia).

Flowering spring–late summer. Lawns, shady, moist areas; 10–500 m; introduced; Del., La., Md., Mich., N.J., N.Y., N.C., Pa., S.C., Tenn., Va., W.Va.; e Asia; introduced also in Europe, Australia.

Mazus miquelii is often sold as *M. reptans* in the horticulture trade. Corollas of cultivated forms are sometimes solid white except for yellow spots in the throat; these may escape in the flora area.

PHRYMACEAE Schauer
• Monkeyflower Family

Richard K. Rabeler

Craig C. Freeman

Wayne J. Elisens

Herbs, subshrubs, or shrubs, annual or perennial, aquatic or terrestrial, sometimes fleshy, autotrophic. **Stems** erect or ascending to prostrate, 4-angled, sometimes winged. **Leaves** deciduous or persistent, basal and cauline or all cauline, rarely subrosulate or rosulate (*Erythranthe*), opposite, or alternate distally, simple; stipules absent; petiole present or absent; blade fleshy, semi-fleshy, or not, not leathery, margins entire or toothed. **Inflorescences** terminal and axillary racemes or flowers solitary (*Glossostigma*, some *Erythranthe*, some annual plants); flowers erect to nodding or strongly reflexed and appressed to inflorescence axis (*Phryma*). **Flowers** bisexual, perianth and androecium hypogynous; sepals (3 or)4 or 5, proximally connate, calyx radially or bilaterally symmetric; petals 3–5, proximally connate, corolla bilaterally symmetric, rarely nearly radially in reduced forms, strongly to weakly bilabiate, rarely nearly regular, salverform to tubular-funnelform, funnelform, campanulate, or compressed; stamens (2–)4, adnate to corolla, didynamous [both pairs of equal length in autogamous forms], staminode 0; pistil 1, 2-carpellate, ovary superior, (1 or)2-locular, placentation axile, basal (*Phryma*), or parietal (*Diplacus*, *Mimetanthe*); ovules anatropous or orthotropous (*Phryma*), unitegmic, tenuinucellate; style 1; stigma 1, 2-lobed. **Fruits** capsules, dehiscence loculicidal [septicidal or irregular], or achenes [berry]. **Seeds** 1–2000, yellowish brown or brown, narrowly ellipsoid, slightly flattened bilaterally; embryo straight, endosperm sparse.

Genera 13, species ca. 200 (6 genera, 139 species in the flora): North America, Mexico, Central America, w South America (primarily Andean), s Asia (India), se Asia, e Africa, Indian Ocean Islands (Madagascar), Pacific Islands (New Zealand), Australia; introduced in Europe, s Africa.

Over one half of the species in Phrymaceae are members of *Diplacus* and *Erythranthe* and together include over 160 species; all other genera each have seven or fewer species.

Until recently, Phrymaceae consisted only of *Phryma leptostachya*, a taxonomically isolated species of eastern North America and eastern Asia. Molecular studies have established a relationship not with the Verbenaceae, as was earlier postulated (see H. L. Whipple 1972;

R. Venkata Ramana et al. 2000), but rather with *Mimulus* and other genera, suggesting that Phrymaceae should be enlarged. The sequence of genera in Phrymaceae here follows the phylogeny proposed by P. M. Beardsley and R. G. Olmstead (2002).

One of the major lineages of Phrymaceae is primarily a Southern Hemisphere group ranging from Australia and New Zealand to southeastern and south Asia (India), Madagascar, and South Africa. *Mimulus* in the narrow sense, including the two endemic North American species, is part of this group, which includes 24 species in seven genera. The largest major lineage includes 158 species in five genera from North America, South America, and southeast Asia. This lineage includes two genera from Mexico and Central America: *Hemichaena* Bentham, which is sister to the North American *Diplacus* and *Mimetanthe*, and *Leucocarpus* D. Don, which is sister to the American and Asian *Erythranthe*.

SELECTED REFERENCES   Barker, W. R. et al. 2012. A taxonomic conspectus of Phrymaceae: A narrowed circumscription for *Mimulus*, new and resurrected genera, and new names and combinations. Phytoneuron 2012-39: 1–60.   Beardsley, P. M. and R. G. Olmstead. 2002. Redefining Phrymaceae: The placement of *Mimulus*, tribe Mimuleae, and *Phryma*. Amer. J. Bot. 89: 1093–1102.   Cantino, P. D. 2004. Phrymaceae. In: K. Kubitzki et al., eds. 1990+. The Families and Genera of Vascular Plants. 14+ vols. Berlin etc. Vol. 7, pp. 323–326.   Fischer, E. 2004. Scrophulariaceae. In: K. Kubitzki et al., eds. 1990+. The Families and Genera of Vascular Plants. 14+ vols. Berlin etc. Vol. 7, pp. 333–432.

1. Plants aquatic or semi-aquatic, mat-forming.
    2. Stems functionally stolons, leaves on stolons; sepals 3; leaf blades fleshy . . . . 2. *Glossostigma*, p. 369
    2. Stems prostrate to decumbent or erect; sepals 5, sometimes 3 in reduced forms; leaf blades sometimes thickened or semi-fleshy . . . . 4. *Erythranthe* (in part), p. 372
1. Plants terrestrial or, if semi-aquatic, not mat-forming.
    3. Flowers strongly reflexed and appressed to inflorescence axes in fruit; fruits achenes; bracteoles present . . . . 3. *Phryma*, p. 370
    3. Flowers lateral or erect to nodding, not strongly reflexed and appressed in fruit; fruits capsules; bracteoles absent.
        4. Fruit apices rounded to truncate; placentation axile.
            5. Leaf venation brochidodromous; stamens adnate to middle of corolla . . . . 1. *Mimulus*, p. 366
            5. Leaf venation acrodromous (veins usually basal only, sometimes basal and suprabasal); stamens adnate proximal to middle of corolla . . . . . 4. *Erythranthe* (in part), p. 372
        4. Fruit apices attenuate; placentation parietal.
            6. Pedicels ± equal to or slightly longer than calyces; calyx lobe midveins low-rounded, not wing-angled; fruit walls densely pustulate-glandular . . . . 5. *Mimetanthe*, p. 425
            6. Pedicels nearly absent or shorter than calyces, rarely ± equal to or slightly longer than calyces; calyx lobe midveins angled or wing-angled; fruit walls smooth, eglandular . . . . .6. *Diplacus*, p. 426

1. MIMULUS Linnaeus, Sp. Pl. 2: 634. 1753; Gen. Pl. ed. 5, 283. 1754 • Monkeyflower [Latin *mimulus*, diminutive of *mimus*, comic or mimic actor, alluding to monkey-faced corolla of some species]

Guy L. Nesom

Herbs, perennial, terrestrial or semi-aquatic. **Stems** ascending to erect, 4-angled, winged, sometimes narrowly, glabrous. **Leaves** basal and cauline, opposite; petiole present or absent; blade sometimes semi-fleshy, margins dentate to subentire, venation brochidodromous. **Inflorescences** axillary, flowers solitary at medial to distal nodes; bracts absent. **Pedicels** present, shorter or longer than calyces; bracteoles absent. **Flowers** mostly erect, not strongly reflexed and appressed in fruit; sepals 5, calyx bilaterally symmetric, tubular, tube midveins wing-angled, lobes deltate to triangular; petals deciduous, 5, corolla blue to violet, purplish, or light pink, rarely white [yellow in *M. bracteosus*], bilaterally symmetric, strongly bilabiate, dorsoventrally compressed, abaxial lobes 3, adaxial 2; stamens 4, adnate to middle of corolla, didynamous, filaments glabrous; ovary 2-locular, placentation axile; stigma bilamellate. **Fruits** capsules, apex rounded to truncate, wall smooth, eglandular, dehiscence loculicidal. **Seeds** 500–1000, brown, narrowly ellipsoid, flattened, wings absent. $x = 8, 11$.

Species 8 (2 in the flora): North America, s, se Asia (se China, India, Indochina), s Africa, Indian Ocean Islands (Madagascar), Australia; introduced in w Europe.

A strictly defined *Mimulus* is sister to the radiation of Phrymaceae that has occurred primarily in Australia (P. M. Beardsley and R. G. Olmstead 2002; Beardsley et al. 2004; Beardsley and W. R. Barker 2005; Barker et al. 2012). In addition to those in North America, six *Mimulus* species are natives of the Eastern and Southern hemispheres; these include *M. aquatilis* A. R. Bean (northeastern Australia), *M. bracteosus* P. C. Tsoong (southeastern China), *M. gracilis* R. Brown (Australia), *M. madagascariensis* Bentham (Madagascar), *M. orbicularis* Wallich ex Bentham (mainland Southeast Asia), and *M. strictus* Bentham (Africa, India). Only *M. gracilis* and *M. ringens* have been subject to molecular analysis, but morphological similarities among these species indicate that they are monophyletic (Barker et al.).

The clade including strictly defined *Mimulus* is sister to the essentially western American monkeyflower species, which are treated here within separate genera. These can be divided into two groups: those with axile placentation and with long pedicels (*Erythranthe*) versus those with parietal placentation and short pedicels (*Diplacus* and *Mimetanthe*). Molecular data confirm that *Erythranthe* and *Diplacus/Mimetanthe* are in separate lineages and are not the closest relative of each other. Closest to *Diplacus* and *Mimetanthe* is the Mexican-Central American *Hemichaena* Bentham, also with parietal placentation. Closest to *Erythranthe* is the monospecific *Leucocarpus* D. Don of Central America to northern South America, which has axile placentation.

Mimulus alatus and *M. ringens* maintain strongly distinct morphologies over a broad region of sympatry, but the two have been reported to hybridize naturally along the Patapsco River in Maryland (D. R. Windler et al. 1976). In a mixed population, about 40 percent of the individuals apparently were F_1 hybrids; consistently high pollen stainability of the parental plants suggests that backcrossing was not occurring. The two species differ in chromosome number (*M. alatus*, $2n = 22$; *M. ringens*, $2n = 16, 24$), but chromosome numbers were not determined in this study.

SELECTED REFERENCES Beardsley, P. M. et al. 2004. Patterns of evolution in western North American *Mimulus* (Phrymaceae). Amer. J. Bot. 91: 474–489. Grant, A. L. 1924. A monograph of the genus *Mimulus*. Ann. Missouri Bot. Gard. 11: 99–388.

1. Leaves sessile; blade bases auriculate-clasping or subclasping; stems not or narrowly winged; fruiting pedicels 10–17 mm or 20–45(–60) mm, longer than calyces . . . . .1. *Mimulus ringens*
1. Leaves petiolate; blade bases rounded to cuneate; stems narrowly winged; fruiting pedicels 5–14(–30) mm, shorter than calyces . . . . .2. *Mimulus alatus*

1. Mimulus ringens Linnaeus, Sp. Pl. 2: 634. 1753
[E] [F]

Stems ascending to erect, not or narrowly winged on angles, (20–)40–130 cm. **Leaves** sessile; blade oblanceolate to narrowly oblong or elliptic-lanceolate, 25–80(–150) × 6–20(–35) mm, base auriculate-clasping or subclasping, margins bluntly serrate to shallowly crenulate, apex acute. **Fruiting pedicels** 10–17 or 20–45(–60) mm, longer than calyces. **Calyces** cylindric, 8–16(–20) mm, lobes triangular, 1.5–6 mm, apex acute to subulate or aristate, ciliolate. **Corollas** purplish blue to light blue or pinkish or nearly white, tube 20–30(–35) mm, throats nearly closed, palate puberulent. **Capsules** ovoid, 10–12 mm. $2n = 16, 24$.

Varieties 2 (2 in the flora): North America; introduced in w Europe.

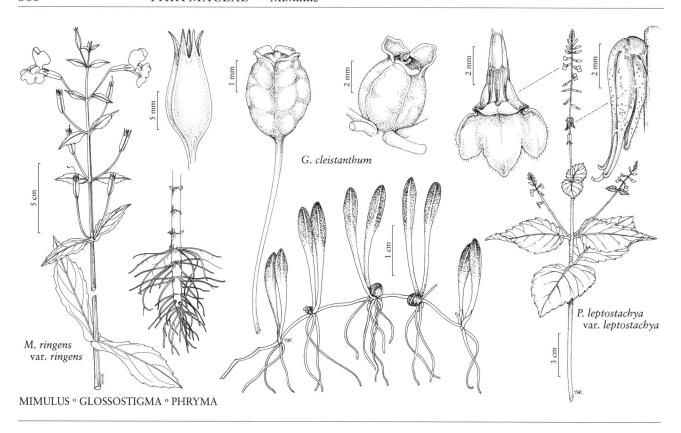

M. *ringens*
var. *ringens*

G. *cleistanthum*

P. *leptostachya*
var. *leptostachya*

MIMULUS ∘ GLOSSOSTIGMA ∘ PHRYMA

1. Leaf blades 50–150 mm, internodes 30–70 mm; fruiting pedicels 20–45(–60) mm; calyces 10–16 (–20) mm, lobes 2–6 mm; moist to wet habitats in w, e North America....1a. *Mimulus ringens* var. *ringens*
1. Leaf blades 25–50 mm, internodes 15–25 mm; fruiting pedicels 10–17 mm; calyces 8–10 mm, lobes 1.5–2 mm; freshwater tidal wetlands of Quebec, Maine, Vermont . . . . 1b. *Mimulus ringens* var. *colpophilus*

1a. Mimulus ringens Linnaeus var. **ringens**

• Allegheny or square-stemmed monkeyflower, marsh skullcap, mad-dog weed, mimule à fleurs entrouvertes E F

Mimulus ringens var. *congesta* Farwell; *M. ringens* var. *minthodes* (Greene) A. L. Grant

Leaves: blade 50–150 mm, internodes 30–70 mm. **Fruiting pedicels** 20–45(–60) mm. **Calyces** 10–16(–20) mm, lobes 2–6 mm.

Flowering Jun–Sep. Fens, seepage slopes, wet meadows, marshy shores, swamp borders, wet woods; 0–500(–1000) m; Alta., Man., N.B., N.S., Ont., P.E.I., Que., Sask.; Ala., Ark., Calif., Colo., Conn., Del., D.C., Ga., Idaho, Ill., Ind., Iowa, Kans., Ky., La., Maine, Md., Mass., Mich., Minn., Miss., Mo., Mont., Nebr., N.H., N.J., N.Y., N.C., N.Dak., Ohio, Okla., Oreg., Pa., R.I., S.C., S.Dak., Tenn., Vt., Va., Wash., W.Va., Wis.; introduced in w Europe (England).

Variety *ringens* in Alberta, California, Colorado, Idaho, Montana, Oregon, and Washington occurs in very scattered, often apparently natural habitats; with more or less certainty, these plants probably are introduced. The basis for all accounts from Colorado apparently is a single specimen collected near Denver in 1895 (*Bethel s.n.*, COLO); there is no indication in recent literature that *Mimulus ringens* has persisted there. It has been included in the Texas flora based on a listing by S. L. Hatch et al. (1990); the basis for that is obscure, and *M. ringens* has not been otherwise reported or documented from Texas.

1b. Mimulus ringens Linnaeus var. **colpophilus** Fernald, Rhodora 34: 119. 1932 • Estuary monkeyflower E

Leaves: blade 25–50 mm, internodes 15–25 mm. **Fruiting pedicels** 10–17 mm. **Calyces** 8–10 mm, lobes 1.5–2 mm.

Flowering Jun–Sep. Freshwater tidal shores, mud flats, meadows, wet places; 0–10 m; Que.; Maine, Vt.

Variety *colpophilus* is restricted to St. Lawrence River estuaries in Quebec and nearby drainages in Maine and Vermont.

F. W. Pennell (1935) speculated that the flowers might be cleistogamous; this apparently has not been verified. Tentative evidence supporting the distinction of var. *colpophilus* was presented by J. L. Stone and B. A. Drummond (2006); most other floristic summaries have treated it as an informal variant of *Mimulus ringens*.

2. Mimulus alatus Aiton, Hort. Kew. 2: 361. 1789

• Sharp-wing monkeyflower [E]

Stems erect, narrowly winged on angles, 30–70 cm. **Leaves** petiolate; blade oblong-lanceolate to broadly lanceolate or ovate, 50–80(–150) × 25–40 mm, base rounded to cuneate, margins coarsely serrate, apex acute-acuminate. **Fruiting pedicels** 5–14(–30) mm, shorter than calyces. **Calyces** broadly cylindric, 12–18 mm, lobes deltate, 0.8–2.5 mm, apex obtuse-aristate, -apiculate, or -subulate, ciliate or glabrous. **Corollas** blue to light violet or pinkish, rarely white, tube 20–28 mm, throats closed, palate villous. **Capsules** ovate-oblong to ovoid, 8–11 mm. $2n = 22$.

Flowering Jun–Sep(–Nov). Floodplain forests, swamps, stream banks, marshy shores, ditches; 0–500 m; Ont.; Ala., Ark., Conn., Del., D.C., Fla., Ga., Ill., Ind., Iowa, Kans., Ky., La., Md., Mass., Mich., Miss., Mo., Nebr., N.J., N.Y., N.C., Ohio, Okla., Pa., S.C., Tenn., Tex., Va., W.Va.

2. GLOSSOSTIGMA Wight & Arnott, Nova Acta Phys.-Med. Acad. Caes. Leop.-Carol. Nat. Cur. 18: 355. 1836, name conserved • Mud mat [Greek *glossa*, tongue, and *stigma*, spot, alluding to ligulate stigma] [I]

Donald H. Les

Robert S. Capers

Herbs, annual or perennial, aquatic, mat-forming; stolons horizontal. **Stems** functionally stolons. **Leaves** on stolons, opposite; petiole absent [present]; blade fleshy, margins entire, obscurely uninerved (midvein). **Inflorescences** axillary, flowers solitary; bracts absent. **Pedicels** absent or present, longer or shorter than calyces; bracteoles absent. **Flowers** [monomorphic] dimorphic (perianth), erect or sessile; sepals 3[4], calyx radially or bilaterally symmetric, campanulate or urceolate, lobes oblong, adaxial lobe longer; petals 0 or [3–]5, corolla fugacious, white [suffused with blue, pink, or purple], bilaterally symmetric, weakly bilabiate, short-funnelform, abaxial lobes 3, adaxial 2, smaller; stamens 2[4], equal [didynamous], filaments glabrous; staminode 0; ovary 2-locular, placentation axile; stigma broad, flaplike, or relatively small. **Fruits** capsules, dehiscence loculicidal. **Seeds** 12–73[100], brown, flattened, wings absent. $x = 5$.

Species 7 or 8 (1 in the flora): introduced; Asia, Africa, Pacific Islands (New Zealand), Australia.

Chromosome counts from New Zealand (E. J. Beuzenberg and J. B. Hair 1983; P. J. de Lange et al. 2004) indicate a base number of $x = 5$ for *Glossostigma*. An anomalous count ($2n = 32$) has been reported for *G. diandrum* (Linnaeus) Kuntze from Bangladesh (see M. O. Rahman 2006).

SELECTED REFERENCE Les, D. H., R. S. Capers, and N. P. Tippery. 2006. Introduction of *Glossostigma* (Phrymaceae) to North America: A taxonomic and ecological overview. Amer. J. Bot. 93: 927–939.

1. Glossostigma cleistanthum W. R. Barker, J. Adelaide Bot. Gard. 15: 72. 1992 • Cleistogamous mud mat F I

Roots (3–)5(or 6) per node. **Stolons** creeping along substrate surface, internodes 1–18 mm. **Leaves:** blade erect, green or yellow-green, narrowly spatulate, 4–11 mm emersed, (7–)9–25(–57) mm submersed, base with a pair of longitudinal lacunae, glossy. **Pedicels** 4–15 mm (chasmogamous), 0–2 mm (cleistogamous). **Flowers** dimorphic, chasmogamous and/or cleistoga-mous emersed, cleistogamous submersed; calyx green, red, or brown, campanulate, 1 mm (chasmogamous), urceolate, 1.5–3 mm (cleistogamous), lobes spreading, equal to or shorter than tube, obtuse; corolla 5-lobed (chasmogamous), tube 0.6 mm, abaxial lobe 0.5–0.7 mm, adaxial lobes 0.2 mm, lateral lobes 0.3–0.5 mm, lobes distinct, obtuse, roundish, spreading or corolla rudimentary (cleistogamous); stamens exserted 0.1–0.2 mm; stigma 1 mm, exserted, inrolled over anthers distally, straightening when touched, adaxial surface papillate, receptive (chasmogamous), or smaller, elliptic, non-tactile (cleistogamous). **Capsules** oblong (chasmogamous) or globular (cleistogamous), 1.4–1.8 mm. **Seeds** oblong, 0.4–0.5 mm. *2n* = 50 (New Zealand).

Flowering spring–fall; fruiting summer–winter. Fresh, oligotrophic waters to 26 dm, margins of coastal plain lakes, ponds, and watercourses, on sand or sandy substrates mixed with clay, gravel, or silt; 10–200 m; introduced; Conn., Del., Md., Mass., N.J., Pa., R.I.; Pacific Islands (New Zealand); Australia.

North American plants initially were misidentified as *Glossostigma diandrum* due to the presence of chasmogamous flowers, which had not been reported in *G. cleistanthum*. Field and genetic studies confirmed that all known cleistogamous and chasmogamous North American plants belonged to *G. cleistanthum* (D. H. Les et al. 2006).

When submersed, *Glossostigma cleistanthum* is perennial (perhaps also annual) and can remain green throughout winter, even under ice. Submersed plants produce only cleistogamous flowers but longer leaves and stolon internodes. Emersed plants have shorter leaves and stolons and are annual in the flora area. They produce cleistogamous and/or chasmogamous flowers; however, the latter are observed infrequently in North America.

In chasmogamous flowers, the apex of the flaplike stigma curls inward to cover the anthers but quickly straightens, displacing the receptive surface against the adaxial corolla when contacted; it then slowly recurls to cover the anthers. Despite this presumed adaptation for outcrossing, autogamy occurs frequently by means of the cleistogamous flowers. Seed set and germination typically are high. Seed production is prodigious, reaching densities of 23,000 seeds m².

Glossostigma is regulated as a noxious aquatic weed even in the states of Oklahoma and Washington, which are quite remote from any of the known populations. Though small, these plants are seriously invasive because they can become dense (to 25,000 plants m²) in some sites, displacing imperiled native species. Sterile or cleistogamous plants easily avoid detection because they often resemble masses of seedlings. Dispersal occurs by transport of seeds (in mud or feathers of waterfowl), or locally by dislodged fruit-bearing plants, which float. Chasmogamous flowers exhibit splash-cup seed dispersal. The Delaware and Maryland attributions are based on photographs sent to us by Wayne Longbottom (W. M. Knapp et al. 2011). The presence of *G. cleistanthum* in New York is probable but not documented.

The use of *Glossostigma* as an aquarium plant (C. Kasselmann 2003) likely is associated with its introduction to North America. The first documented North American specimens were collected from a tidal freshwater wetland in New London County, Connecticut, in 1992.

3. PHRYMA Linnaeus, Sp. Pl. 2: 601. 1753; Gen. Pl. ed. 5, 262. 1754 • Lopseed [Derivation unknown]

Jay B. Walker

Herbs, perennial, terrestrial; rhizomes fleshy, vertical. **Stems** erect, terete proximally, 4-angled distally, glabrous or puberulent. **Leaves** cauline, opposite, decussate; petiole present; blade not fleshy, margins crenate-serrate, venation pinnate. **Inflorescences** terminal, sometimes axillary at distal nodes, spikelike racemes, long-pedunculate, elongate; bracts present. **Pedicels** present, shorter than calyx; bracteoles present. **Flowers** opposite, proximalmost sometimes subopposite, erect in bud, becoming horizontal to ascending at anthesis and strongly reflexed and appressed in fruit; sepals 5, calyx bilaterally symmetric, narrowly campanulate, lobes triangular or subulate-uncinate; petals 5, corolla lavender to purple to white, bilaterally symmetric, bilabiate, narrowly campanulate, abaxial lobes 3, adaxial 2; stamens 4, included, filaments glabrous, staminode 0; ovary 1-locular, placentation basal; stigma 2-lobed. **Fruits** achenes, enclosed in accrescent calyx. **Seeds** 1, yellowish brown or brown, narrowly ellipsoid, wings absent. $x = 14$.

Species 1: c, e North America, e Asia.

The name lopseed alludes to the fruit appearing to lop or hang as it matures, reflexes, and becomes appressed to the infructescence axis.

Phryma has often been included in Verbenaceae based on presumed affinities with members of Lantaneae Endlicher, particularly *Stachytarpheta* (H. L. Whipple 1972). Others have placed it in Phrymaceae based on its unusual gynoecium with only one ovule becoming a seed. Molecular evidence (P. M. Beardsley and R. G. Olmstead 2002; B. Oxelman et al. 2005) suggests that *Phryma* is allied with genera traditionally placed in the Scrophulariaceae (*Mazus*, *Mimulus*, and others). This evidence has led to an expanded circumscription of Phrymaceae, followed here.

SELECTED REFERENCES Nie, Z. L. et al. 2006. Evolution of biogeographic disjunction between eastern Asia and eastern North America in *Phryma*. Amer. J. Bot. 93: 1343–1356. Venkata Ramana, R. et al. 2000. Embryology of *Phryma leptostachya* L. (Verbenaceae) with considerations of its systematic status and affinities. Feddes Repert. 111: 231–248. Whipple, H. L. 1972. Structure and systematics of *Phryma leptostachya*. J. Elisha Mitchell Sci. Soc. 88: 1–17.

1. **Phryma leptostachya** Linnaeus, Sp. Pl. 2: 601. 1753
   • American lopseed, phryma à épis grèles [F]

Varieties 2 (1 in the flora): c, e North America, Asia (Himalayas to e China, Japan, Russian Far East, Taiwan).

Phryma leptostachya is a classic example of the eastern Asian/eastern North American disjunction pattern, which has long fascinated botanists. Two varieties are traditionally recognized: var. *leptostachya* in North America and var. *asiatica* H. Hara in eastern Asia. They are distinguished on minor differences in leaf size, shape of adaxial lip of the corolla, and length of the adaxial, spinulose calyx lobes (Nie Z. L. et al. 2006). Despite the morphological similarities, recent analyses demonstrate significant molecular divergence between the two varieties (N. S. Lee et al. 1996; Xiang Q. Y. et al. 2000; Nie et al. 2006, 2009b) and a North American origin of the genus with subsequent dispersal to eastern Asia approximately three to six mya has been hypothesized (Xiang et al.; Nie et al. 2006).

Ground roots of *Phryma leptostachya* and extracts of the entire plant have been used as insecticides in Asia (G. M. Hocking 1997), where it has been used also to treat fevers, ulcers, ringworms, scabies, boils, carbuncles, and cancers (D. F. Austin 2004). In the New World, the Ojibwa reportedly used it in the treatment of sore throat and rheumatism (Austin), as did the Chippewa (F. Densmore 1928).

1a. **Phryma leptostachya** Linnaeus var. **leptostachya**
[E] [F]

Phryma leptostachya var. *confertifolia* Fernald

Stems reddish, sometimes green, simple or branched, 4–10 dm, swollen distal to each node, some hairs glandular. **Leaves:** petiole 2–5 cm; blade ovate to ovate-lanceolate, 5–16 × 3–10 cm, base cuneate to truncate, apex acute to acuminate, surfaces glabrous or puberulent. **Inflorescences:** flowers opposite or subopposite; proximal bracts cordate, leaflike, distal usually present, subulate, 1/3–1/2 distance between base of inflorescence and proximal flowers. **Pedicels** to 1 mm; bracteoles 3, persistent, subulate. **Flowers:** calyx green, 5–8 mm, glabrous, abaxial lobes 2, triangular, 0.1–0.5 mm, adaxial 3, subulate-uncinate and indurate in fruit, 2–3 mm; corolla glabrous externally, puberulent on abaxial lip at mouth, tube 8–12 mm, abaxial lobes white to lavender, 2–4 mm, apex rounded, adaxial lavender to purple, 2–4 mm, apex acute, usually emarginate; ovule 1; style slightly exserted, glabrous; stigma lobes unequal. **Achenes** strongly reflexed and appressed to infructescence axis, 3–4 mm. $2n = 28$.

Flowering May–Aug; fruiting Jul–Oct. Moist woods, edges of woods; 10–1500 m; Man., N.B., Ont., Que.; Ark., Conn., Del., D.C., Fla., Ga., Ill., Ind., Iowa, Kans., Ky., La., Maine, Md., Mass., Mich., Minn., Miss., Mo., Nebr., N.H., N.J., N.Y., N.C., N.Dak., Ohio, Okla., Pa., R.I., S.C., S.Dak., Tenn., Tex., Vt., Va., W.Va., Wis., Wyo.

Variety *leptostachya* ranges from eastern North America to easternmost Wyoming and eastern Texas and Manitoba. *Phryma* had previously been reported as occurring as a garden waif in central California (C. Best et al. 1996); it has since been rejected as being naturalized in that state (http://ucjeps.berkeley.edu/cgi-bin/get_cpn.pl?37935). Fruit dispersal is presumably via animals, with attachment facilitated by the uncinate lobes of the persistent calyx (P. D. Cantino 2004).

4. ERYTHRANTHE Spach, Hist. Nat. Vég. 9: 312. 1840 • Stalked monkeyflower

[Greek *erythros*, red, and *anthe*, bloom, alluding to corolla color of type species, *E. cardinalis*]

Guy L. Nesom

Naomi S. Fraga

Mimulus Linnaeus sect. *Erythranthe* (Spach) Greene; *Mimulus* subg. *Synplacus* A. L. Grant

Subshrubs or herbs, annual (fibrous-rooted or taprooted) or perennial (rhizomatous), terrestrial or semi-aquatic. **Stems** prostrate to decumbent or erect, terete or 4-angled, glabrous or hairy to glandular-hairy. **Leaves** basal and cauline, basal, cauline, or basal not persistent, rarely subrosulate (*E. linearifolia*) or rosulate (*E. primuloides*); petiole present or absent; blade sometimes thickened or semi-fleshy, not leathery, margins toothed, subentire, or entire, venation acrodromous (veins basal only, sometimes basal and suprabasal). **Inflorescences** of axillary flowers at medial to distal nodes or at all nodes or rarely solitary; bracts absent or present when distal leaves much reduced. **Pedicels** present, usually distinctly longer in fruit than calyces; bracteoles absent. **Flowers** erect, spreading, or nodding; sepals 5 or 3 (in reduced forms), calyx bilaterally or radially symmetric, tubular, sometimes inflated and sagittally compressed with abaxial lobes characteristically upcurving and closing throat or not, lobes 5, rarely 3, all usually ca. equal size or adaxial slightly longer, mostly triangular to deltate, rarely reduced and barely evident; petals 5, corolla deciduous, rarely marcescent, pink, red, crimson, magenta, lavender, purple, or yellow, rarely white, bilaterally symmetric, sometimes ± radially symmetric, ± bilabiate, sometimes regular, sometimes sagittally compressed, abaxial lobes 3, adaxial 2; stamens 4, adnate proximal to middle of corolla, didynamous, filaments glabrous or hairy, staminode 0; ovary 2-locular, placentation axile; stigma bilamellate. **Fruits** capsules, apex rounded to truncate, dehiscence loculicidal partly to base along both sutures. **Seeds** 100–2000, brown, narrowly ellipsoid, flattened bilaterally, not winged. $x = 7, 8, 9$.

Species 114 (86 in the flora): North America, Mexico, Central America, South America, e, se Asia; introduced in Europe, e Asia (Japan), s Africa, Pacific Islands (New Zealand).

Erythranthe includes the mostly western American and Asian species with axile placentation and pedicels distinctly longer than the calyces; these were formerly treated within *Mimulus* in the broad sense. The species of *Erythranthe* have been divided into 12 sections (W. R. Barker et al. 2012), closely corresponding to the cladistic patterns in the molecular-based phylogenetic analysis by P. M. Beardsley et al. (2004): sect. *Achlyopitheca* G. L. Nesom & N. S. Fraga (species 1–3), sect. *Paradantha* (A. L. Grant) G. L. Nesom & N. S. Fraga (species 4–19), sect. *Monantha* G. L. Nesom & N. S. Fraga (species 20, 21), sect. *Monimanthe* (Pennell) G. L. Nesom & N. S. Fraga (species 22–24), sect. *Erythranthe* (species 25–31), sect. *Alsinimimulus* G. L. Nesom & N. S. Fraga (species 32), sect. *Simigemma* G. L. Nesom & N. S. Fraga (species 33), sect.

Mimulosma G. L. Nesom & N. S. Fraga (species 34–51), sect. *Mimulasia* G. L. Nesom & N. S. Fraga (species 52), sect. *Exigua* G. L. Nesom & N. S. Fraga (species 53), and sect. *Simiolus* (Greene) G. L. Nesom & N. S. Fraga (species 54–86).

Recent taxonomic studies have clarified species definitions, and to some extent relationships, in sect. *Achlyopitheca* (G. L. Nesom 2012g), sect. *Erythranthe* (P. M. Beardsley et al. 2003), sect. *Mimulosma* (M. L. Carlson 2002; J. B. Whittall et al. 2006; Nesom 2012h), sect. *Simiolus* (G. R. Campbell 1950; Nesom 2012i), and sect. *Paradantha* (N. S. Fraga 2012). A secondary radiation of species has occurred in Andean South America; C. von Bohlen (1995) studied the Chilean species. A taxonomic summary of the primarily Central American genera *Hemichaena* Bentham and *Leucocarpus* D. Don, which are closely related to *Diplacus* and *Erythranthe*, respectively, was provided by Nesom (2011d).

A revision of *Mimulus* in the broad sense was provided by A. L. Grant (1924), followed by a major update and overview by F. W. Pennell (1951) that covered the majority of the western North American species of *Erythranthe*. A floristic treatment of *Mimulus* in the broad sense for California by D. M. Thompson (1993) recognized fewer species of *Erythranthe* than did Pennell, especially in sects. *Achlyopitheca*, *Mimulosma*, and *Simiolus*. The present treatment is more similar to that by Pennell.

A number of the species of *Erythranthe* currently serve as experimental organisms for various laboratories studying the genetic and genomic basis of adaptation and speciation (M. R. Dudash et al. 2005; C. A. Wu et al. 2008), and sect. *Simiolus* was the object of intensive investigation from about 1950 until the 1990s by R. K. Vickery and his students, who made hundreds of chromosome counts and interspecific and infraspecific crosses and from those data drew corresponding inferences about isolating mechanisms. Most studies have centered on *E. guttata* and close relatives (sect. *Simiolus*), and *E. cardinalis* and close relatives (sect. *Erythranthe*), focusing on mating system evolution, reproductive isolation, hybridization and hybrid incompatibility, genetic architecture of floral traits, corolla morphology and pollination, inbreeding phenomena, and epistasis.

The relative position of stigma and anthers is indicated in the descriptions by the terms herkogamous (stigma distinctly above the level of the anthers) and plesiogamous (stigma and anthers in essentially the same level). Herkogamous corollas are generally assumed to be outcrossing except in the relatively smaller corollas where, even if herkogamous, crowding of the stigma and anthers may facilitate occasional autogamy.

SELECTED REFERENCES Argue, C. L. 1986. Some taxonomic implications of pollen and seed morphology in *Mimulus hymenophyllus* and *M. jungermannioides* and comparisons with other putative members of the *M. moschatus* alliance (Scrophulariaceae). Canad. J. Bot. 64: 1331–1337. Beardsley, P. M., A. Yen, and R. G. Olmstead. 2003. AFLP phylogeny of *Mimulus* section *Erythranthe* and the evolution of hummingbird pollination. Evolution 57: 1397–1410. Campbell, G. R. 1950. *Mimulus guttatus* and related species. Aliso 2: 319–335. Carlson, M. L. 2002. Evolution of Mating System and Inbreeding Depression in the *Mimulus moschatus* (Scrophulariaceae) Alliance. Ph.D. dissertation. University of Alaska. Fraga, N. S. 2012. A revision of *Erythranthe montioides* and *Erythranthe palmeri* (Phrymaceae), with descriptions of five new species from California and Nevada, USA. Aliso 30: 49–68. Nesom, G. L. 2011d. Recognition and synopsis of *Mimulus* sect. *Tropanthus* and sect. *Leucocarpus* (Phrymaceae). Phytoneuron 2011-28: 1–8. Nesom, G. L. 2012g. Taxonomic summary of *Erythranthe* sect. *Achlyopitheca* (Phrymaceae). Phytoneuron 2012-42 1–4. Nesom, G. L. 2012h. Taxonomy of *Erythranthe* sect. *Mimulosma* (Phrymaceae). Phytoneuron 2012-41: 1–36. Nesom, G. L. 2012i. Taxonomy of *Erythranthe* sect. *Simiola* (Phrymaceae) in the USA and Mexico. Phytoneuron 2012-40: 1–123. Nesom, G. L. 2014b. Taxonomy of *Erythranthe* sect. *Erythranthe* (Phrymaceae). Phytoneuron 2014-31: 1–41. Nesom, G. L. 2017. Taxonomic review of the *Erythranthe moschata* complex (Phrymaceae). Phytoneuron 2017-17: 1–29. Vickery, R. K. 1978. Case studies in the evolution of species complexes in *Mimulus*. Evol. Biol. 11: 404–506. Whittall, J. B. et al. 2006. The *Mimulus moschatus* alliance (Phrymaceae): Molecular and morphological phylogenetics and their conservation implications. Syst. Bot. 31: 380–397.

1. Fruiting calyces inflated and sagittally compressed with abaxial lobes characteristically upcurving and closing throat or not; corollas yellow, often red-spotted; leaf blades palmately, sometimes subpinnately, veined.
    2. Plants rhizomatous.
        3. Corolla lobes fimbriate; stems procumbent; plants mat-forming.
            4. Leaf blade surfaces villous-hirsute, hairs whitish, thickened, flattened, stiff, gland-tipped, fruiting calyces villous-hirsute, fruiting pedicels and distal stems stipitate-glandular; corollas: abaxial limbs spreading . . . . .84. *Erythranthe parvula*
            4. Leaf blade surfaces: abaxials glabrous, adaxials sometimes moderately villosulous, fruiting calyces glabrous or sparsely villosulous-glandular, hairs vitreous, flattened, eglandular or minutely gland-tipped, fruiting pedicels and stems glabrous; corollas: abaxial limbs strongly reflexed . . . . .85. *Erythranthe chinatiensis*
        3. Corolla lobes not fimbriate (entire or apically notched); stems decumbent, ascending, or erect to suberect, sometimes procumbent; plants not mat-forming.
            5. Rhizomes prolifically produced, filiform.
                6. Leaf blade surfaces hirtellous to softly hirsute . . . . .61. *Erythranthe corallina*
                6. Leaf blade surfaces glabrous, glabrate, pilose, villous, glandular-villous, or stipitate-glandular.
                    7. Flowers 6–16, from proximal to distal nodes; corolla tube-throats 10–15 mm; fruiting pedicels (25–)40–75 mm; stems usually erect, 20–50 cm . . . . 62. *Erythranthe utahensis*
                    7. Flowers 1–3(–5), from distal nodes; corolla tube-throats 9–11 or 15–28 mm; fruiting pedicels 10–35(–40) mm; stems erect to erect-ascending, procumbent, or decumbent to decumbent-ascending, 2–35 cm.
                        8. Stems procumbent or decumbent to decumbent-ascending, 3–10 cm; plants forming matted colonies; leaf blades 3–12 mm, margins entire, mucronulate, or barely denticulate; corolla tube-throats 15–18 mm . . . . .59. *Erythranthe caespitosa*
                        8. Stems erect to erect-ascending, 2–35 cm; plants solitary to colonial; leaf blades 5–35(–55) mm, margins shallowly dentate to denticulate; corolla tube-throats 9–11 mm or 15–28 mm.
                            9. Corolla tube-throats 15–28 mm, exserted 5–10 mm beyond calyx margin. . . . . .58. *Erythranthe tilingii*
                            9. Corolla tube-throats 9–11 mm, exserted 0–1(–2) mm beyond calyx margin. . . . . 60. *Erythranthe minor*
            5. Rhizomes usually 1–few, usually broader than filiform.
                10. Calyx throats not closing at maturity.
                    11. Corollas weakly bilabiate, tube-throats 6–8 mm, exserted 1–3 mm beyond calyx margin, limbs expanded 5–8 mm; flowers plesiogamous. . . . . .54. *Erythranthe geyeri*
                    11. Corollas strongly bilabiate, tube-throats 10–14 mm, exserted 5–8 mm beyond calyx margin, limbs expanded 10–15 mm; flowers herkogamous. . . . . .57. *Erythranthe michiganensis*
                10. Calyx throats closing at maturity.
                    12. Leaf blades oblong-elliptic to oblong-lanceolate, usually 3–4 times longer than wide, bases attenuate, margins evenly, shallowly dentate or crenate to mucronate or mucronulate; plants glabrous . . . . . .69. *Erythranthe scouleri*
                    12. Leaf blades ovate to ovate-elliptic, broadly elliptic, suborbicular to broadly ovate-triangular or ovate-lanceolate, usually 1–2 times longer than wide, bases rounded, cuneate, subcordate, or truncate, margins dentate (*E. decora*) or crenate to dentate or toothed, rarely lobed or dissected; plants glabrate or hairy, rarely glabrous.

13. Leaf blades ovate-triangular to ovate-lanceolate, bases rounded to truncate to shallowly cuneate, palmately (3–)5–7-veined; corolla tube-throats 18–26 mm; stems minutely hirtellous, sometimes glabrate. . . . . 68. *Erythranthe decora*

13. Leaf blades ovate-elliptic to ovate or suborbicular, bases rounded, cuneate, truncate, or subcordate, usually subpinnately 5–7-veined; corolla tube-throats (10–)12–20 mm or (14–)16–24 mm; stems villous-glandular to hirtellous, hirsutulous, or pilose-hirsutulous.

    14. Stems (6–)15–65(–80) cm; fruiting pedicels, fruiting calyces, and distal stems vestiture variable, not puberulent-glandular; corolla tube-throats (10–)12–20 mm, exserted 3–5 mm beyond calyx margin; fruiting calyces 11–17(–20) mm. . . . . 63. *Erythranthe guttata*

    14. Stems (25–)50–120(–160) cm; fruiting pedicels, fruiting calyces, and distal stems densely puberulent, hairs a mix of crinkly and minutely stipitate-glandular; corolla tube-throats (14–)16–24 mm, exserted (8–)10–15 mm beyond calyx margin; fruiting calyces 15–22(–25) mm . . . . 64. *Erythranthe grandis*

[2. Shifted to left margin.—Ed.]

2. Plants fibrous-rooted or taprooted, without rhizomes or stolons (sometimes rooting at nodes).

15. Flowers herkogamous or plesiogamous, chasmogamous; corolla tube-throats (4–)7–23 mm, exserted (0–)3–8 mm (sometimes 1 mm in smallest corollas of *E. microphylla*) beyond fruiting calyx margin.

  16. Stems moderately to densely villous-glandular, at least proximally.

    17. Stems, leaves, and fruiting pedicels moderately to densely villous-glandular, without eglandular hairs, fruiting calyces densely hirtellous, sometimes sparsely stipitate-glandular, densely villous at sinuses; fruiting pedicels 15–45 mm; flowers herkogamous . . . . 71. *Erythranthe marmorata*

    17. Stems, leaves, fruiting pedicels, and fruiting calyces moderately villous-glandular with gland-tipped hairs, mixed hirtellous and stipitate-glandular, or moderately to densely stipitate-glandular, sometimes hirtellous-eglandular; fruiting pedicels 7–17 mm; flowers plesiogamous or herkogamous.

      18. Corollas tube-throats 11–20 mm, exserted 4–8 mm beyond calyx margin; flowers herkogamous; coastal and near-coastal localities in Monterey, San Luis Obispo, and Santa Cruz counties, California. . . . . 65. *Erythranthe arenicola*

      18. Corollas tube-throats 8–12 mm, exserted 1–2 mm beyond calyx margin; flowers plesiogamous; Yellowstone National Park, Wyoming . . . . 66. *Erythranthe thermalis*

  16. Stems delicately, sparsely stipitate-glandular, short villous-glandular, glabrous, or glabrate, sometimes hirtellous or distals puberulent-glandular.

    19. Corollas pale yellow, palate dark yellow, drying blue-green; s Arizona, New Mexico . . . . 67. *Erythranthe unimaculata*

    19. Corollas, including palate, usually yellow, sometimes with a median splotch or red-spotted; British Columbia, California, Idaho, Nevada, Oregon, Washington.

      20. Leaf blade margins bipinnately dissected, ca. 5–12(–15) primary divisions on each side. . . . . 75. *Erythranthe filicifolia*

      20. Leaf blade margins entire or toothed to sometimes lobed at base.

        21. Cauline leaf blades 5–15(–30) × 1–5 mm.

          22. Cauline leaf blades 5–15(–30) mm; plants glabrous or stems, leaves, and fruiting calyces minutely stipitate-glandular, hairs 0.05–0.1 mm, at least just above nodes; fruiting calyces 6–13 mm; corollas without a large red splotch; Colusa, Glenn, Lake, Mendocino, Napa, and Sonoma counties, California . . . . . . 73. *Erythranthe nudata*

          22. Cauline leaf blades 7–10 mm; plants glabrous; fruiting calyces (4–)5–6 mm; corollas with a large red splotch at base of proximal middle lip; Plumas County, California. . . . . 74. *Erythranthe percaulis*

21. Cauline leaf blades (3–)10–50 × 3–25 mm.

    23. Distal cauline leaf blades connate-perfoliate, disclike; stems and leaf surfaces glaucous . . . . .72. *Erythranthe glaucescens*

    23. Distal cauline leaves subclasping or narrowly perfoliate, not connate-perfoliate, not disclike; stems and leaf surfaces not glaucous.

        24. Flowers usually from distal nodes; corolla tube-throats (6–)8–16(–20) mm, exserted (1–)2–6(–8) mm beyond calyx margin, limbs expanded 8–25 mm; stems and fruiting pedicels: distals hirtellous or mixed hirtellous and stipitate-glandular, sometimes only short villous-glandular, stems sometimes glabrous below inflorescence; British Columbia, California, Idaho, Nevada, Oregon, Washington . . . . 70. *Erythranthe microphylla*

        24. Flowers usually evenly distributed from proximal to distal nodes; corolla tube-throats 7–10(–12) mm, exserted 1–3 mm beyond calyx margin, limbs expanded 8–12 mm; stems and fruiting pedicels short, delicately stipitate-glandular, distals minutely puberulent-glandular; Amador, Calaveras, El Dorado, Placer, Tehama, and Tuolumne counties, California . . . . 78. *Erythranthe pardalis*

[15. Shifted to left margin.—Ed.]

15. Flowers plesiogamous, chasmogamous or cleistogamous; corolla tube-throats 4–14 mm, exserted 0–5 mm beyond fruiting calyx margin.

  25. Flowers chasmogamous; corolla limbs expanded 2–14 mm, tube-throats 4–14 mm.

    26. Distal leaves closely paired and auriculate-clasping; stems usually glabrous.

        27. Flowers from remote distal nodes; stems glabrous, sometimes minutely hirtellous in inflorescence, hairs deflexed, eglandular; fruiting calyces (7–)9–14 mm; British Columbia, California, Idaho, Montana, Nevada, Oregon, Utah, Washington, Wyoming . . . . .79. *Erythranthe arvensis* (in part)

        27. Flowers from distal nodes or all nodes; stems glabrous or sparsely, minutely stipitate-glandular; fruiting calyces 7–18(–20) mm; Arizona, California, Colorado, Nevada, New Mexico, Texas, Utah.

            28. Fruiting calyx throats not closing; flowers often produced from all nodes; stems, fruiting pedicels, and calyces glabrous or calyces minutely scabrous-hirtellous; fruiting calyces 7–11 mm; Texas. . . . . 56. *Erythranthe inamoena*

            28. Fruiting calyx throats closing; flowers at distal nodes; stems, fruiting pedicels, and fruiting calyces sparsely stipitate-glandular; fruiting calyces (8–)14–18(–20) mm; Arizona, California, Colorado, Nevada, New Mexico, Texas, Utah . . . . .81. *Erythranthe cordata* (in part)

    26. Distal leaves not paired and not auriculate-clasping; stems glabrous or sparsely hirtellous and/or finely villosulous-glandular.

        29. Calyx lobes usually 3 or 3 and 5 on same plant, if 5 then with 2 interpolated lobes much smaller than abaxial pair . . . . 86. *Erythranthe calciphila*

        29. Calyx lobes 5, ca. equal size or adaxial slightly longer.

            30. Leaves as long as wide or wider than long, blades elliptic-ovate to broadly ovate, suborbicular, or depressed-ovate, margins irregularly dentate to dentate-serrate or nearly lacerate-dentate, commonly doubly toothed; plants commonly producing tiny cleistogamous flowers on branches separate from those with larger flowers; corolla tube-throats (5–)8–12 mm . . . . .76. *Erythranthe nasuta* (in part)

            30. Leaves longer than wide, blades elliptic to elliptic-obovate, oblanceolate, or oblong, margins narrowly pinnately lobed or dissected, sometimes merely shallowly toothed; plants producing flowers of only one size; corolla tube-throats 4–6 mm . . . . 77. *Erythranthe laciniata*

25. Flowers cleistogamous; corolla limbs not or barely expanded or only 1–3 mm, tube-throats 4–12 mm.

  31. Flowers and fruits subsessile; fruiting pedicels shorter than or equal to subtending leaves . . . . 80. *Erythranthe brachystylis*

  31. Flowers and fruits distinctly pedicellate; fruiting pedicels longer than subtending leaves.

    32. Distal leaves: petioles 0 mm, blade surfaces glabrous, sometimes hirtellous; stems villous-glandular just above nodes, sometimes hirtellous distally; fruiting calyces with adaxial lobe usually distinctly longer than abaxial, slightly falcate . . . . .76. *Erythranthe nasuta* (in part)

    32. Distal leaves: petioles 0 mm or 1–4 mm, blade surfaces glabrous or villous on one or both surfaces; stems usually glabrous; fruiting calyces with adaxial lobe not distinctly longer than abaxial (except in *E. regni*), not falcate.

      33. Fruiting calyces glabrous, (5–)7–10 mm.

        34. Stems 15–45 cm, terete, sometimes becoming slightly fistulose; leaf blades 15–20(–50) × 15–25(–50) mm; fruiting pedicels 15–30 mm; corolla tube-throats 9–12 mm, exserted 3–5 mm beyond calyx margin; Arizona . . . . .55. *Erythranthe regni*

        34. Stems 2–8 cm, 4-angled; leaf blades 5–11 × 3–9 mm; fruiting pedicels 6–14 mm; corolla tube-throats 4–6 mm, exserted 0.5–1 mm beyond calyx margin; Colorado. . . . . .83. *Erythranthe hallii*

      33. Fruiting calyces stipitate-glandular, hirtellous, or hirsutulous, sometimes sparsely glandular or mixed glandular-hirsutulous, (7–)9–18(–20) mm.

        35. Fruiting calyces stipitate-glandular, (8–)14–18(–20) mm . . . .
. . . . .81. *Erythranthe cordata* (in part)

        35. Fruiting calyces usually minutely hirtellous, (7–)9–14 mm.

          36. Abaxial surfaces of distal and bracteal leaves densely villous, hairs long, sometimes vitreous, flattened, eglandular, multicellular; middle and distal cauline leaf blades depressed-ovate or broadly orbicular to nearly reniform; petioles 0 mm; stems, leaves, and fruiting calyces usually green; fruiting calyx throats closing or not, remaining open; stems erect to ascending; plants sometimes rooting at proximal cauline nodes if decumbent . . .79. *Erythranthe arvensis* (in part)

          36. Leaves glabrous, proximals sometimes sparsely villous; middle and distal cauline leaf blades ovate to ovate-lanceolate; petioles: distals 1–4 mm; stems, leaves, and fruiting calyces usually dark purplish; fruiting calyx throats closing; stems erect; plants not rooting at proximal nodes . . . . 82. *Erythranthe charlestonensis*

1. Fruiting calyces not inflated and sagittally compressed, abaxial lobes not upcurving; corollas yellow, pink, red, crimson, magenta, lavender, or purple, rarely white; leaf blades palmately or pinnately veined.

  37. Plants rhizomatous and/or stoloniferous, perennial.

    38. Corollas red to purple or orange, rarely white; petioles 0 mm; blades: bases clasping to subclasping.

      39. Stems scandent to pendent, stoloniferous; fruiting pedicels 10–30(–40) mm . . . . .31. *Erythranthe eastwoodiae*

      39. Stems erect to ascending or decumbent, rhizomatous, without stolons; fruiting pedicels (25–)30–120(–150) mm.

        40. Corollas purple or light pink, rarely crimson, pale violet, white, pinkish white, yellowish white, or lavender; anthers included; leaf blade margins denticulate, subentire, or entire.

          41. Corollas purple, rarely crimson, pale violet, white, pinkish white, yellowish white, or lavender; calyx tubes 12–15(–17) × 9–12 mm; widespread in nw North America . . . . .26. *Erythranthe lewisii*

          41. Corollas light pink; calyx tubes 14–19 × 6–8 mm; Sierra Nevada, California and Nevada . . . . 27. *Erythranthe erubescens*

40. Corollas orange-red to scarlet, deep, dull, or red-orange, or crimson, rarely yellow or yellow-tinged; anthers exserted; leaf blade margins dentate to serrate.
  42. Corolla tube-throats tubular, exserted 13–25 mm beyond calyx margin. . . . . 30. *Erythranthe verbenacea*
  42. Corolla tube-throats funnelform or tubular, exserted 2–12 mm beyond calyx margin.
    43. Leaf blade surfaces adaxially ± glandular-villous to glabrate on veins and lamina; fruiting calyces 17–28(–30) mm, lobes 4–7 mm, ovate to ovate-deltate, lobe apices attenuate-acute; corolla tube-throats (15–)20–30 mm; California, Nevada, Oregon, (5–)50–2300(–2800) m. . . . . 28. *Erythranthe cardinalis*
    43. Leaf blade surfaces adaxially glabrous or minutely sessile- or stipitate-glandular along veins; fruiting calyces (27–)29–34 mm, lobes 7–10 mm, ovate, lobe apices abruptly attenuate to linear-caudate; corolla tube-throats 29–36 mm; Arizona, 2100–3300 m . . . . 29. *Erythranthe cinnabarina*
38. Corollas yellow to orange-yellow, often red-spotted or -striped or brown-spotted; petioles 0 mm or 0.5–5(–20) mm; blades: bases subclasping to clasping or not clasping.
  44. Stems with shortened internodes; leaves all basal or near basal (sometimes proximal cauline).
    45. Leaf blades oblanceolate to elliptic-obovate; corollas densely hirsute on abaxial side of opening . . . . 20. *Erythranthe primuloides*
    45. Leaf blades linear to narrowly oblanceolate; corollas loosely hirsute on abaxial side of opening . . . . 21. *Erythranthe linearifolia*
  44. Stems with evident internodes; leaves cauline, basal usually not persistent.
    46. Stems erect to erect-ascending, eglandular; corollas ± bilabiate . . . . . 52. *Erythranthe dentata*
    46. Stems erect, ascending, or decumbent to procumbent or prostrate, glandular; corollas weakly bilabiate to nearly radially symmetric (strongly bilabiate in *E. jungermannioides*).
      47. Calyx lobes 1–2 mm; anthers glabrous; styles scabrous; stolons forming overwintering turions; plants usually on cliff faces . . . 34. *Erythranthe jungermannioides*
      47. Calyx lobes 2–9 mm; anthers glabrous or hairy; styles glabrous; stolons without turions; plants usually of habitats other than cliff faces.
        48. Cauline leaf blades 30–70 mm; fruiting pedicels (15–)22–50 mm; calyx lobes 5–9 mm, linear-lanceolate to narrowly triangular . . . . .47. *Erythranthe ptilota*
        48. Cauline leaf blades 10–40(–50) mm; fruiting pedicels 4–25 mm; calyx lobes 2–4 mm, triangular to linear-lanceolate or narrowly triangular-acuminate.
          49. Stems (2–)5–20 cm, usually erect, nodes 2–4(or 5); widely distributed . . . . 45. *Erythranthe moschata*
          49. Stems 7–45 cm, usually sprawling-decumbent, nodes (2–)4–15+; North Feather River, California . . . . .46. *Erythranthe willisii*
[37. Shifted to left margin.—Ed.]
37. Plants taprooted or fibrous-rooted, annual.
  50. Leaves: petioles 0 mm, blades: bases subclasping to clasping.
    51. Corollas bicolored (3 lobes of abaxial limbs yellow or medial lobe yellow and 2 lateral lobes maroon, 2 lobes of adaxial limbs maroon). . . . . 19. *Erythranthe shevockii*
    51. Corollas not bicolored (lobes yellow, pinkish, or purple).
      52. Corollas yellow . . . . 7. *Erythranthe carsonensis* (in part)
      52. Corollas white to light lavender, pinkish, rosy, or purplish.

53. Corollas white to lavender, pinkish, or rosy, palate and abaxial limb with or without small, reddish spots, yellow ridges present; corolla limbs not expanded . . . . .25. *Erythranthe parishii*
53. Corollas pink to purple, palate and abaxial limb sometimes with yellow markings, yellow ridges absent; corolla limbs expanded 3–7 mm.
    54. Corollas strongly bilabiate, adaxial lips darker than abaxials; corolla limbs expanded 7–10 mm . . . . .16. *Erythranthe purpurea*
    54. Corollas weakly bilabiate, adaxial lips not darker than abaxials; corolla limbs expanded 3–7 mm .17. *Erythranthe androsacea*
[50. Shifted to left margin.—Ed.]
50. Leaves: petioles 0 mm or to 30(–35) mm, blades: bases not clasping or subclasping.
  55. Corollas light lavender to purple, abaxial lobes and palates with yellow patches, tube-throats 1.5–2.5 mm; stigmas persistent in fruit; capsules distinctly exserted . . . . .53. *Erythranthe exigua*
  55. Corollas yellow to pink, lavender, purple, magenta, red, or white, tube-throats 3–17 mm; stigmas not persistent; capsules included to slightly exserted.
    56. Petioles 2–3 mm, laterally compressed, deeply saccate at base, usually containing a lenticular propagule . . . . .33. *Erythranthe gemmipara*
    56. Petioles 0–30 mm, not compressed, not saccate at base, not containing a lenticular propagule.
      57. Fruiting calyx margins subtruncate, lobes reduced.
        58. Fruiting calyx lobes: 1 usually slightly longer; corollas yellow, abaxial limbs with a large maroon splotch; stems glandular-puberulent; petioles 1–20(–30) mm . . . . .32. *Erythranthe alsinoides*
        58. Fruiting calyx lobes subequal; corollas pale pink to rose pink, rose red, rose purple, or purple to magenta, throat sometimes yellow, abaxial limbs without a large maroon splotch; stems glabrous; petioles 0–5 mm.
          59. Corollas pale pink to rose pink or purple to magenta, throats sometimes yellow, tube-throats 5–9 mm, limbs expanded 5–6 mm; flowers plesiogamous; fruiting pedicels 5–15 mm . . . . .1. *Erythranthe inconspicua*
          59. Corollas rose red or pale pink to rose purple, throats sometimes yellow, tube-throats 8–12 mm, limbs expanded 7–12 mm; flowers herkogamous; fruiting pedicels 6–7 mm or 10–23 mm.
            60. Fruiting pedicels 6–7 mm, shorter than subtending leaves . . . . . .2. *Erythranthe grayi*
            60. Fruiting pedicels 10–23 mm, longer than subtending leaves . . . . .3. *Erythranthe acutidens*
      57. Fruiting calyx margins distinctly lobed, lobes pronounced.
        61. Fruiting calyces strongly angled, ribs corky, lobes spreading.
          62. Corollas yellow, adaxial lip white, sometimes yellow . . . . .23. *Erythranthe bicolor*
          62. Corollas pink to red, rose red, red-purple, purple, or lavender.
            63. Corolla tube-throats 3–7 mm, throats usually light purple to lavender (similar in color to rest of corolla); fruiting pedicels 4–12 mm . . . . .22. *Erythranthe breweri*
            63. Corolla tube-throats 8–12(–17) mm, throats dark red-purple (darker than rest of corolla); fruiting pedicels 9–30 mm . . .24. *Erythranthe filicaulis*
        61. Fruiting calyces weakly or strongly angled, not corky, lobes erect.
          64. Corollas pink, purple, yellow, rose lavender, or (in *E. barbata* and *E. shevockii*) bicolored dark maroon and yellow.
            65. Calyx lobe margins ciliate.
              66. Corollas: 2 adaxial lobes much reduced, smaller than 3 abaxials . . . . .8. *Erythranthe gracilipes*
              66. Corollas: abaxial and adaxial lobes ± same size.
                67. Fruiting calyces becoming red-angled or red; corollas weakly bilabiate, limbs expanded 3–5 mm, lobes entire or weakly notched . . . . .5. *Erythranthe rubella*
                67. Fruiting calyces sometimes red-spotted on ribs, becoming straw colored; corollas strongly bilabiate, limbs expanded 5–17 mm, lobes deeply notched.

68. Fruiting calyces minutely glandular, sometimes red-spotted . . . . .9. *Erythranthe sierrae*
68. Fruiting calyces glabrous, sometimes red-spotted on ribs, becoming straw colored. . . . . .14. *Erythranthe palmeri*

65. Calyx lobe margins glabrous.
69. Fruiting pedicels erect to ascending.
70. Corollas deep pink to purple, abaxial limbs with 2 yellow ridges, limbs expanded 7–15 mm. . . . . . . .11. *Erythranthe discolor* (in part)
70. Corollas pale pink to pink, palate with a broad yellow patch covering ridges and lateral areas, limbs expanded 16–25 mm . . . . .13. *Erythranthe rhodopetra*
69. Fruiting pedicels ascending to often spreading horizontally.
71. Corollas bicolored (2 lobes of adaxial lips maroon, 3 lobes of abaxial limbs yellow or medial lobe yellow and 2 lateral lobes maroon) . . . . .12. *Erythranthe barbata* (in part)
71. Corollas not bicolored (lobes pinkish to purple).
72. Styles distally pubescent; stigmas distinctly shorter than corolla tubes . . . . 15. *Erythranthe diffusa*
72. Styles glabrous; stigmas equal in length to corolla tubes or exserted . . . . 18. *Erythranthe hardhamiae*

[64. Shifted to left margin.—Ed.]

64. Corollas yellow or (in color morph of *E. calcicola*) white, sometimes pinkish or flesh colored.
73. Stems erect; leaves: blades linear to lanceolate, sometimes ovate or spatulate, 0.5–8(–10) mm wide, margins entire, rarely toothed, petioles 0(–2) mm.
74. Abaxial corolla limb with one large central red spot, if red spot absent then throat mottled red.
75. Corolla tube-throats 4–6 mm, indistinct from throat, palates glabrous or sparsely bearded. . . . . 6. *Erythranthe suksdorfii*
75. Corolla tube-throats (5–)7–11 mm, distinct from abruptly expanding throat, palates densely bearded . . . . 7. *Erythranthe carsonensis* (in part)
74. Abaxial corolla limb red-spotted, but not with a single large red spot.
76. Corollas yellow, tube-throats funnelform, adaxial surfaces red-tinged . . . . . .
. . . . .11. *Erythranthe discolor* (in part)
76. Corollas yellow, white, or bicolored (abaxial limbs yellow with red spots, adaxials maroon-purple), tube-throats funnelform to cylindric, adaxial surfaces not red-tinged.
77. Corollas yellow or white, tube-throats 6–13 mm; calyx ribs strongly angled . . . . 4. *Erythranthe calcicola*
77. Corollas yellow or bicolored (abaxial limbs yellow with red spots, adaxials maroon-purple), tube-throats (5–)6–15 mm; calyx ribs weak.
78. Corolla tube-throats cylindric to funnelform, lateral lobes entire or shallowly notched, palates glabrous or sparsely bearded; calyx margins ciliate. . . . . 10. *Erythranthe montioides*
78. Corolla tube-throats cylindric, lateral lobes 2-fid, palates bearded; calyx margins glabrous . . . . .12. *Erythranthe barbata* (in part)
73. Stems prostrate to procumbent-trailing, decumbent, ascending, or erect; leaves: blades lanceolate to oblanceolate, elliptic, oblong, ovate, obovate, or deltate, (1–)2–30 mm wide, margins toothed, sometimes entire, petioles (0–)1–30 mm.
79. Cauline leaves: petioles 1–30(–35) mm, blades ovate, sometimes deltate, ovate-lanceolate, broadly lanceolate, elliptic-ovate, or triangular, palmately or pinnately veined, bases rounded to cuneate, truncate, or cordate.
80. Stems prostrate to ascending-erect, sharply bent at basal nodes; fruiting pedicels usually closely paired . . . . .36. *Erythranthe hymenophylla*
80. Stems erect to prostrate, decumbent, or ascending, straight (if erect) or geniculate at nodes; fruiting pedicels not paired.

81. Stems and pedicels villous or villous-glandular, pedicels sometimes stipitate-glandular, hairs 0.5–1.2(–2) mm; leaf blades pinnately to subpinnately veined, sometimes ± palmately veined.
 82. Stems erect to ascending; fruiting calyces greenish; styles hispid-hirtellous; Oregon . . . . 35. *Erythranthe washingtonensis* (in part)
 82. Stems erect to prostrate, decumbent, or ascending; fruiting calyces usually red-spotted; styles glabrous; widespread.
  83. Corolla tube-throats (4–)5–10 mm, limbs expanded 3–4 mm diam.; flowers plesiogamous . . . . 49. *Erythranthe floribunda* (in part)
  83. Corolla tube-throats 9–12 mm, limbs expanded 10–18 mm diam.; flowers herkogamous . . . . 50. *Erythranthe geniculata*
81. Stems and fruiting pedicels stipitate-glandular, sometimes puberulent-glandular to villous-glandular, hairs 0.05–0.5 mm; leaf blades palmately veined.
 84. Corolla tube-throats 8–12(–14) mm.
  85. Stems moderately puberulent-glandular to villous-glandular, hairs 0.1–0.8 mm, flattened, vitreous, leaves moderately puberulent-glandular to villous-glandular, fruiting calyces densely stipitate-glandular; stems terete; styles hispid-hirtellous . . . .
  . . . . 35. *Erythranthe washingtonensis* (in part)
  85. Stems and leaves sparsely sessile- to subsessile-glandular, hairs 0.05–0.2 mm, fruiting calyces sparsely stipitate-glandular or glabrous; stems 4-angled; styles glabrous . . . . 38. *Erythranthe ampliata*
 84. Corolla tube-throats 5–8 mm.
  86. Petioles (5–)8–25 mm; blades 4–12(–17) mm; fruiting pedicels 10–25(–38) mm; fruiting calyces 5–6(–7) mm, sparsely stipitate-glandular to sparsely hirtellous; corolla tube-throats 7–8 mm . . . . 37. *Erythranthe patula*
  86. Petioles 3–5(–8) mm; blades 4–20 mm; fruiting pedicels 6–13 mm; fruiting calyces 4–5 mm, densely invested with tiny, waxy-white, eglandular, papillose hairs between angles; corolla tube-throats 5–7 mm . . . . 39. *Erythranthe taylorii*

[79. Shifted to left margin.—Ed.]

79. Cauline leaves: petioles 0 mm or 1–10(–15) mm, blades elliptic to oblong, lanceolate, oblanceolate, ovate, or obovate, palmately 3–5-veined, sometimes 1-veined, subpalmately veined, or with 1–3 distal vein pairs diverging pinnately, bases attenuate, sometimes cuneate, obtuse, rounded, subauriculate, truncate, or cordate.
 87. Stems villous-glandular, hairs 0.2–1(–1.5) mm.
  88. Midcauline leaves: petioles 1–12 mm; corolla tube-throats (4–)5–10 mm . . . .
  . . . . 49. *Erythranthe floribunda* (in part)
  88. Midcauline leaves: petioles 0 mm or 1–10(–15) mm; corolla tube-throats 9–16 mm.
   89. Midcauline leaves: petioles 0 mm or proximals 1–3(–5) mm; corollas red-dotted but without discrete red splotches or white patches, tube-throats 9–12(–14) mm; fruiting calyces: ribs angled, lobes erect . . . . 48. *Erythranthe arenaria*
   89. Midcauline leaves: petioles 5–10(–15) mm; corollas: base of each lobe with a prominent maroon blotch, abaxial limbs with white patch at 2 sinus bases, tube-throats 12–16 mm; fruiting calyces: ribs rounded-thickened, lobes often incurved . . . . 51. *Erythranthe norrisii*
 87. Stems sessile-glandular, minutely stipitate-glandular, or puberulent, sometimes minutely hirtellous, hairs 0.1–0.3 mm.
  90. Petioles 2–9 mm; fruiting pedicels divergent-arcuate.
   91. Leaves basal and cauline; petioles 2–9 mm, distinctly 3-veined, winged; fruiting pedicels 12–38 mm; tube-throats, lobes, and palate ridges yellow . . . . 40. *Erythranthe pulsiferae*
   91. Leaves usually cauline; petioles 4–8 mm, 1-veined, not winged; fruiting pedicels 9–17 mm; tube-throats yellow, lobes (limbs) pink or white with pink distal borders, palate ridges yellow . . . . 41. *Erythranthe trinitiensis*

[90. Shifted to left margin.—Ed.]

90. Petioles 0 mm or 1–3 mm; fruiting pedicels straight.
  92. Fruiting calyces 8–12 mm, short stipitate-glandular or sessile-glandular; fruiting pedicels 11–28 mm; leaves basal and cauline. . . . . .44. *Erythranthe latidens*
  92. Fruiting calyces 5–11 mm, sparsely, minutely hirtellous, eglandular, sometimes sparsely sessile-glandular; fruiting pedicels 5–11 mm or 7–18 mm; leaves usually cauline, basal usually deciduous by flowering.
    93. Fruiting calyces 5–6 mm; fruiting pedicels 5–11 mm; corolla tube-throats 3.5–5 mm, not exserted beyond calyx margins; distal leaves: petioles 1–3 mm . . .
. . . . .42. *Erythranthe breviflora*
    93. Fruiting calyces 7–11 mm; fruiting pedicels 7–18 mm; corolla tube-throats 5–8 mm, exserted 1–3 mm beyond calyx margins; distals leaves: petioles 0 mm
. . . . .43. *Erythranthe inflatula*

1. Erythranthe inconspicua (A. Gray) G. L. Nesom, Phytoneuron 2012-39: 34. 2012 E

Mimulus inconspicuus A. Gray in War Department [U.S.], Pacif. Railr. Rep. 4(5): 120. 1857

Annuals, fibrous-rooted. **Stems** erect to ascending, simple or branched from base, 4-angled, 3–16 cm, glabrous. **Leaves** basal and cauline; petiole: proximals 1–5 mm, mid cauline and distals 0 mm; blade palmately 3-veined, broadly elliptic to ovate or broadly ovate, 6–20 × 6–12 mm, base rounded to cordate, margins subentire to denticulate, apex obtuse to acute or acuminate, surfaces sparsely villous. **Flowers** plesiogamous, 1–12, from proximal to distal nodes. **Fruiting pedicels** 5–15 mm, glabrous. **Fruiting calyces** campanulate, 6–9 mm, margins subtruncate, glabrous, lobes reduced, subequal. **Corollas** pale pink to rose pink or purple to magenta, throat sometimes yellow, lobes sometimes yellowish with pale rose spots, bilaterally symmetric, weakly bilabiate; tube-throat cylindric, 5–9 mm, exserted beyond calyx margin; limb expanded 5–6 mm. **Styles** glabrous. **Anthers** included, minutely villous-hirsute. **Capsules** included, 4–9 mm.

Flowering Apr–Jun(–Jul). Steep, north- or northwest-facing slopes, canyon walls, moist talus, granitic sand on outcrops, moist gravelly open spots, sandy lakeshores, hillside streams or seeps, riparian woodlands, grassy slopes, gray pine, yellow pine, yellow pine-Kellogg oak, chaparral, *Pseudotsuga-Pinus-Cornus*, canyon live oak woodlands; 200–2100 m; Calif.

Erythranthe inconspicua occurs in Sierran counties from Kern north to El Dorado and then is apparently disjunct further northward to Butte County. A record from Los Angeles County (*Bigelow s.n.*, 14 May 1854, the type of the species) is probably mislabeled, as other collections by Bigelow on the same day are from Calaveras County.

2. Erythranthe grayi (A. L. Grant) G. L. Nesom, Phytoneuron 2012-39: 34. 2012 C E

Mimulus grayi A. L. Grant, Ann. Missouri Bot. Gard. 11: 203, plate 6. 1925

Annuals, fibrous-rooted. **Stems** erect, simple or branched from base, weakly 4-angled, 8–20 cm, glabrous. **Leaves** basal and cauline; petiole 0 mm; blade palmately 3–5-veined, broadly ovate, 7–18 × 5–12 mm, base rounded, margins denticulate, apex acute, surfaces glabrous, rarely sparsely puberulent. **Flowers** herkogamous, 2–20, from proximal to distal nodes. **Fruiting pedicels** 6–7 mm, shorter than subtending leaves, glabrous. **Fruiting calyces** campanulate, 9–11 × 5–6 mm, margins subtruncate, glabrous, sometimes densely papillate at flowering with tiny, 1-celled, eglandular hairs, these apparently deciduous by fruiting, lobes reduced, subequal. **Corollas** rose red, throat pink lined with rose red and a yellow patch, abaxial ridges yellow, bilaterally symmetric, weakly bilabiate; tube-throat cylindric-funnelform, 8–11 mm, exserted beyond calyx margin; limb expanded 7–10 mm. **Styles** glabrous. **Anthers** included, minutely villous-hirsute. **Capsules** included, 5–9 mm.

Flowering May–Jul(–Oct). Drying pond beds, creek banks, yellow pine, yellow pine-*Libocedrus* woodlands; of conservation concern; 1000–1900 m; Calif.

In addition to the features noted in the key and descriptions, the fruiting calyces of *Erythranthe grayi* are distinctly more inflated than those of *E. acutidens* and *E. inconspicua*. G. L. Nesom (2012g) maintained *E. grayi* as distinct from *E. acutidens*, relying primarily on fruiting pedicel length (see key above), but the two have nearly identical ranges (Tuolumne County south to Kern County), and study of additional collections suggests that only a single species may be represented.

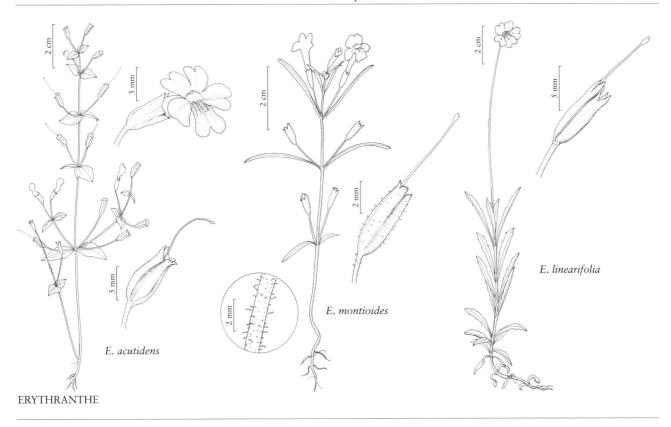

ERYTHRANTHE

E. acutidens

E. montioides

E. linearifolia

3. **Erythranthe acutidens** (Greene) G. L. Nesom,
Phytoneuron 2012-39: 34. 2012 [C] [E] [F]

Mimulus acutidens Greene, Bull. Calif. Acad. Sci. 1: 117. 1885; *M. inconspicuus* A. Gray var. *acutidens* (Greene) A. Gray

Annuals, fibrous-rooted. **Stems** erect to ascending-erect, simple or diffusely branched, 4-angled, 7–20 cm, glabrous. **Leaves** basal and cauline; petiole 0 mm; blade palmately 3–5 veined, ovate to broadly ovate, 10–20 × 7–11 mm, base rounded to truncate, margins serrate-denticulate, apex acute, surfaces glabrous. **Flowers** herkogamous, 2–20, from proximal to distal nodes. **Fruiting pedicels** divergent-arcuate, sometimes becoming deflexed, 10–23 mm, longer than subtending leaves, glabrous. **Fruiting calyces** campanulate, 7–9 mm, margins subtruncate, glabrous, lobes reduced, subequal. **Corollas** pale pink to rose purple, throat yellow or deep pink externally with 2 yellow spots below abaxial limb, bilaterally symmetric, weakly bilabiate; tube-throat cylindric-funnelform, 9–12 mm, exserted beyond calyx margin; limb expanded 9–12 mm. **Styles** glabrous. **Anthers** included, minutely villous-hirsute. **Capsules** included, (4–)5–10 mm.

Flowering Apr–Jul. Grassy slopes, sandy terraces, marshy places, lakeshores, creek sides, seep edges, shaded canyon slopes, roadcuts and roadsides, woodlands of *Pseudotsuga menziesii-Quercus chrysolepis*, oak-gray pine, or interior live oak; of conservation concern; 200–2000 m; Calif.

Mimulus acutidens Greene is heterotypic with the South American *M. acutidens* Reiche, which is a later homonym.

4. **Erythranthe calcicola** N. S. Fraga & D. A. York,
Aliso 30: 54, figs. 12–16. 2012 • Limestone
monkeyflower [C] [E]

Annuals, taprooted. **Stems** erect, simple or branched from basal nodes, 2–10(–15) cm, sparsely glandular-pubescent, internodes elongate, distinct. **Leaves** cauline, basal not persistent; petiole 0–1 mm; blade palmately 3-veined (in broader ones), lanceolate to ovate, 3–25 × 2–8(–10) mm, base attenuate, margins entire, sometimes toothed, apex acute to obtuse, surfaces sparsely glandular-pubescent. **Flowers** herkogamous, sometimes plesiogamous, 1–16, from distal or medial to distal nodes. **Fruiting pedicels** (3–)5–20 mm. **Fruiting calyces** campanulate to widely cylindric, 5–7 mm, margins distinctly toothed or lobed, sparsely glandular-pubescent, ribs strongly angled, lobes pronounced, erect. **Corollas** yellow or white and throat yellow, throat and limb red-spotted, bilaterally symmetric, weakly bilabiate; tube-throat cylindric, 6–13 mm, exserted beyond calyx margin; limb expanded 3–7(–9) mm, lobes notched, abaxial limb sparsely bearded or glabrous. **Styles** glabrous. **Anthers** included, glabrous. **Capsules** included, 4–6 mm.

Flowering Apr–Jun. Creosote bush and Joshua tree woodlands, usually on carbonate substrate, primarily on talus slopes; of conservation concern; 900–2200 m; Calif., Nev.

Erythranthe calcicola was previously included in the broader concept of *E. montioides*; the former can be distinguished by its glandular herbage, broader lanceolate to ovate leaves, strongly angled calyces, and smaller flowers. The species is restricted to the northern Mojave Desert and southwestern Great Basin and occurs primarily on talus slopes on substrates derived from carbonate rock.

5. **Erythranthe rubella** (A. Gray) N. S. Fraga, Phytoneuron 2012-39: 35. 2012 • Redstem or little redstem monkeyflower

Mimulus rubellus A. Gray in W. H. Emory, Rep. U.S. Mex. Bound. 2(1): 116. 1859; *M. gratioloides* Rydberg

Annuals, taprooted. **Stems** erect, simple, sometimes branched from basal nodes, 3–32 cm, minutely puberulent. **Leaves** cauline, basal not persistent; petiole 0 mm; blade palmately 3-veined (in broader ones), linear to elliptic, 5–22(–30) × 1–5 mm, base narrowed, margins entire, sometimes toothed, apex acute to obtuse, surfaces minutely puberulent. **Flowers** herkogamous, sometimes plesiogamous, 1–106, from distal or medial to distal nodes. **Fruiting pedicels** 2–18 mm. **Fruiting calyces** becoming red-angled or red, campanulate to nearly cylindric, 4–9 mm, margins distinctly toothed or lobed, glabrous or minutely puberulent, ribs thickened, lobes pronounced, erect, margins ciliate. **Corollas** yellow and abaxial limb and throat red dotted or pink to purple and throat yellow, bilaterally symmetric, weakly bilabiate; tube-throat cylindric, 4–10 mm, exserted beyond calyx margin; limb expanded 3–5 mm, lobes entire or weakly notched, abaxial limb glabrous. **Styles** glabrous. **Anthers** included, glabrous. **Capsules** included, 3–8 mm.

Flowering Apr–Jun. Open slopes and washes; 300–3000 m; Ariz., Calif., Colo., Nev., N.Mex., Tex., Utah, Wyo.; Mexico (Baja California, Sonora).

6. **Erythranthe suksdorfii** (A. Gray) N. S. Fraga, Phytoneuron 2012-39: 35. 2012 • Suksdorf's monkeyflower E

Mimulus suksdorfii A. Gray in A. Gray et al., Syn. Fl. N. Amer. ed. 2, 2(1): 450. 1886

Annuals, taprooted, densely compact. **Stems** erect, simple or many-branched from basal nodes, 0.5–10(–13) cm, minutely puberulent, internodes shortened, not evident. **Leaves** cauline, basal not persistent; petiole 0 mm; blade 1-veined or palmately 3-veined (in broader ones), linear to lanceolate or ovate, 2–20(–25) × 0.5–4 mm, base attenuate, margins entire, apex acute to obtuse, surfaces minutely puberulent. **Flowers** herkogamous or plesiogamous, 1–72, from distal or medial to distal nodes. **Fruiting pedicels** 2–10 mm. **Fruiting calyces** cylindric to urn-shaped, 3–6 mm, margins distinctly toothed or lobed, glabrous or minutely puberulent, ribs thickened, lobes pronounced, erect. **Corollas** yellow, throat sometimes with red markings, abaxial limb sometimes with a red dot, bilaterally symmetric, weakly bilabiate; tube-throat cylindric, indistinct from throat, 4–6 mm, exserted beyond calyx margin; limb expanded 2–3 mm, lobes entire or weakly notched, palate glabrous or sparsely bearded. **Styles** glabrous. **Anthers** included, glabrous. **Capsules** included, 3–6 mm.

Flowering Apr–Aug. Moist, generally loamy to clay soils in open areas; 700–4200 m; B.C.; Ariz., Calif., Idaho, Mont., Nev., N.Mex., Oreg., Utah, Wash., Wyo.

7. **Erythranthe carsonensis** N. S. Fraga, Aliso 30: 59, figs. 17–21. 2012 • Carson Valley monkeyflower C E

Mimulus rubellus A. Gray var. *latiflorus* S. Watson

Annuals, taprooted, densely compact. **Stems** erect, simple or branched from basal nodes, 2–7(–8) cm, minutely glandular, internodes shortened, not evident. **Leaves** cauline, basal not persistent; petiole 0 mm; blade palmately 3-veined (in broader ones), linear to spatulate, (3–)5–23 × 1–5 mm, base truncate to truncate-cordate, clasping, margins entire, apex acute to obtuse, surfaces minutely glandular. **Flowers** herkogamous, 1–35, from distal or medial to distal nodes. **Fruiting pedicels** (3–)5–14 mm. **Fruiting calyces** campanulate to widely urceolate, 4–7 mm, margins distinctly toothed or lobed, minutely glandular, ribs thickened, lobes pronounced, erect. **Corollas** yellow, palate red-dotted and 1 large central spot, bilaterally symmetric, strongly bilabiate; tube-throat cylindric, distinct from abruptly expanding throat, (5–)7–11 mm, exserted beyond calyx margin; limb expanded 7–12(–15) mm, each lobe 2-fid, palate densely bearded. **Styles** glabrous. **Anthers** included, glabrous. **Capsules** included, 3–6 mm.

Flowering Mar–Jun. Openings in sage brush/bitterbrush scrub in sand of decomposed granite; of conservation concern; 1400–1800 m; Calif., Nev.

Erythranthe carsonensis is restricted to the Carson Valley, Eagle Valley, and Washoe Valley region of Nevada and adjacent California, with one known disjunct occurrence about 58 km to the north in Nevada. The species was previously included in the broader concept of *E. montioides* but can be distinguished by its much branched and compact habit, linear to spatulate leaves with clasping bases, calyx with glabrous margins, and larger corolla with one large red spot in the center. *Erythranthe carsonensis* has been impacted by agriculture, urbanization, and other anthropogenic changes.

8. **Erythranthe gracilipes** (B. L. Robinson) N. S. Fraga, Phytoneuron 2012-39: 35. 2012 • Slender-stalked monkeyflower C E

Mimulus gracilipes B. L. Robinson, Proc. Amer. Acad. Arts 26: 176. 1891

Annuals, taprooted. **Stems** erect, simple or branched, 5–15 cm, sparsely glandular-puberulent. **Leaves** cauline, basal not persistent; petiole 0 mm; blade 1-veined or palmately 3-veined (in broader ones), linear to lanceolate, (3–)7–13 × 1–3 mm, base attenuate, margins entire, sometimes toothed, apex acute to obtuse, surfaces sparsely glandular-puberulent. **Flowers** herkogamous, 1–15, from distal or medial to distal nodes. **Fruiting pedicels** 8–30 mm. **Fruiting calyces** becoming red to straw colored, campanulate, 4–7 mm, margins distinctly toothed or lobed, sparsely glandular-puberulent, ribs thickened, lobes pronounced, erect, margins ciliate. **Corollas** pink to rose lavender, throat deep pink to purple with 2 yellow longitudinal ridges, bilaterally symmetric, strongly bilabiate, 2 adaxial lobes much reduced, smaller than 3 abaxials; tube-throat cylindric, 7–10 mm, exserted beyond calyx margin; limb expanded 5–9 mm, abaxial lobes rounded and shallowly 2-fid, abaxial limb glabrous or sparsely bearded. **Styles** glabrous. **Anthers** included, glabrous. **Capsules** included, 4–5 mm.

Flowering Apr–May. Open areas in thin decomposed granite soils, often on edges of large granite boulders; of conservation concern; 500–1300 m; Calif.

Erythranthe gracilipes is endemic to the foothills of the central Sierra Nevada in Fresno, Madera, and Mariposa counties and apparently is most abundant after fire.

9. **Erythranthe sierrae** N. S. Fraga, Aliso 30: 67, figs. 29–31. 2012 • Sierra Nevada monkeyflower C E

Annuals, taprooted. **Stems** erect, simple or branched from basal nodes, (3–)4–20 cm, minutely glandular. **Leaves** cauline, basal not persistent; petiole 0 mm; blade 1-veined or palmately 3-veined (in broader ones), linear to oblanceolate, 3–23 × 1–10 mm, base attenuate, margins entire, sometimes toothed, apex acute, surfaces minutely glandular. **Flowers** herkogamous, 1–38, from distal or medial to distal nodes. **Fruiting pedicels** 10–33(–40) mm. **Fruiting calyces** sometimes red-spotted, campanulate to cylindric, 4–8 mm, margins distinctly toothed or lobed, minutely glandular, ribs weak, lobes pronounced, erect, margins ciliate. **Corollas** pale pink, abaxial limb with 2 yellow ridges, bilaterally symmetric, strongly bilabiate; tube-throat funnelform, 5–15 mm, exserted beyond calyx margin; limb expanded 5–17 mm, lobes deeply notched, abaxial limb glabrous or sparsely puberulent. **Styles** glabrous. **Anthers** included, glabrous. **Capsules** included, 4–9 mm.

Flowering Apr–Jul. Foothill oak woodlands, mixed coniferous forests; of conservation concern; 200–2300 m; Calif.

Erythranthe sierrae is endemic to the southern Sierra Nevada in Fresno, Kern, and Tulare counties. The species was previously treated as *E. palmeri* but can be distinguished by having leaf margins that are often toothed, pale pink corollas, and white stamens. *Erythranthe palmeri* has entire leaf margins that are never toothed, deep pink to purple corollas, and yellow stamens.

10. **Erythranthe montioides** (A. Gray) N. S. Fraga, Phytoneuron 2012-39: 35. 2012 • Montia-like monkeyflower C E F

Mimulus montioides A. Gray, Proc. Amer. Acad. Arts. 7: 380. 1868

Annuals, taprooted. **Stems** erect, simple or branched from basal nodes, 2–15 cm, glabrous or minutely puberulent, internodes elongate, distinct. **Leaves** cauline, basal not persistent; petiole 0 mm; blade 1-veined or palmately 3-veined (in broader ones), linear to lanceolate, (3–)4–25 × 0.5–2 mm, base attenuate, margins entire, apex acute, surfaces glabrous or minutely puberulent. **Flowers** herkogamous, 1–6, from medial to distal nodes. **Fruiting pedicels** (4–)5–20 mm. **Fruiting calyces** becoming straw colored, campanulate, 5–7 mm, margins distinctly toothed or lobed, ciliate, glabrous or minutely puberulent, ribs weak, lobes pronounced, erect. **Corollas** yellow, abaxial limb red-spotted, bilaterally symmetric, strongly bilabiate; tube-throat cylindric to funnelform, 6–10 mm, exserted beyond calyx margin; limb expanded 5–12 mm, lateral lobes entire or shallowly notched, palate glabrous or sparsely bearded. **Styles** glabrous. **Anthers** included, glabrous. **Capsules** included, 5–6 mm.

Flowering Jun–Aug. Sandy opening in mixed coniferous and lodgepole forests; of conservation concern; 1900–2900 m; Calif.

Erythranthe montioides has been previously treated as a highly polymorphic species with a relatively broad geographic range but is now recognized as narrowly endemic to Fresno and north-central Tulare counties. The following species were segregated from *E. montioides*: *E. barbata*, *E. calcicola*, *E. carsonensis*, and *E. discolor*.

11. **Erythranthe discolor** (A. L. Grant) N. S. Fraga, Phytoneuron 2012-39: 35. 2012 • Party-colored monkeyflower C E

Mimulus discolor A. L. Grant, Ann. Missouri Bot. Gard. 11: 257. 1925

Annuals, taprooted. **Stems** erect, simple or branched from basal nodes, 5–12 cm, sparsely glandular-pubescent, internodes elongate, distinct. **Leaves** cauline, basal not persistent; petiole 0–2 mm; blade 1-veined or palmately 3-veined (in broader ones), linear to lanceolate or ovate, 5–15 × 1–4 mm, base attenuate, margins entire, apex acute, surfaces sparsely glandular-pubescent. **Flowers** herkogamous, 1–8, from distal or medial to distal nodes. **Fruiting pedicels** erect to ascending, 6–11(–32) mm. **Fruiting calyces** sometimes red-dotted on ribs, campanulate, 4–8 mm, margins distinctly toothed or lobed, sparsely glandular-pubescent, ribs weak, lobes pronounced, erect, margins glabrous. **Corollas** yellow and tube-throat adaxial surface red-tinged, abaxial limb red-spotted or deep pink to purple and abaxial limb with 2 yellow ridges, bilaterally symmetric, strongly bilabiate; tube-throat funnelform, 9–13 mm, exserted beyond calyx margin; limb expanded 7–15 mm, lobes notched, adaxial limb glabrous or sparsely bearded. **Styles** glabrous. **Anthers** included, glabrous. **Capsules** included, 5–8 mm.

Flowering May–Jun. Moist open areas on gentle slopes in desert chaparral and pine transition areas; of conservation concern; 1400–2700 m; Calif.

Erythranthe discolor has two distinctive floral morphs; one is yellow with red spots on the palate, the other is pink with two yellow ridges on the palate. Populations can be monomorphic (usually yellow) or mixed, with the yellow morph most often in higher frequency.

Erythranthe discolor was placed previously in synonymy with *E. montioides*, and the pink form of *E. discolor* has commonly been confused with *E. palmeri*. *Erythranthe discolor* is known to hybridize with *E. barbata* and is restricted to Kern and Tulare counties in the southern Sierra Nevada.

12. **Erythranthe barbata** (Greene) N. S. Fraga, Phytoneuron 2012-39: 35. 2012 • Bearded monkeyflower E

Mimulus barbatus Greene, Bull. Calif. Acad. Sci. 1: 9. 1884; *M. deflexus* S. Watson

Annuals, taprooted. **Stems** erect, simple or branched from basal nodes, 2–14.5 cm, sparsely glandular-pubescent, internodes elongate, distinct. **Leaves** cauline, basal not persistent; petiole 0 mm; blade 1-veined or palmately 3-veined (in broader ones), linear to lanceolate, 5–15 × 0.5–2 mm, base attenuate, margins entire, apex acute to obtuse, surfaces sparsely glandular-pubescent. **Flowers** herkogamous, 1–30, from distal or medial to distal nodes. **Fruiting pedicels** ascending to often spreading horizontally, (5–)9–25 mm. **Fruiting calyces** sometimes red-dotted on ribs, campanulate, 3–6 mm, margins distinctly toothed or lobed, glabrous, sparsely glandular-pubescent, ribs weak, lobes pronounced, erect, margins glabrous. **Corollas** bicolored (abaxial limb yellow with red spots, adaxial maroon-purple) or yellow, bilaterally symmetric, strongly bilabiate; tube-throat cylindric, (5–)8–15 mm, exserted beyond calyx margin; limb expanded 6–15 mm, lateral lobes 2-fid, palate bearded. **Styles** glabrous. **Anthers** included, glabrous. **Capsules** included or equal to calyx, 3–5 mm.

Flowering May–Aug. Open areas in pine forests, edges of meadows and ephemeral streams; 1800–3400 m; Calif.

Erythranthe barbata previously has been placed in synonymy with *E. montioides* but differs from it in having each corolla lobe deeply notched and a consistently bearded palate. In *E. montioides*, each corolla lobe is entire or shallowly notched, and the palate is glabrous or sparsely bearded. *Erythranthe barbata* is most abundant in Tulare County but also occurs in immediately adjacent Inyo and Kern counties.

13. **Erythranthe rhodopetra** N. S. Fraga, Aliso 30: 66, figs. 26–28. 2012 • Red Rock Canyon monkeyflower [C] [E]

Annuals, taprooted. Stems erect, simple or branched from basal nodes, 6–7(–21) cm, minutely puberulent. Leaves cauline, basal not persistent; petiole 0 mm; blade palmately 3-veined, linear to oblanceolate, 5–22 × 1–7(–10) mm, base attenuate, margins entire, apex acute to obtuse, surfaces minutely puberulent. Flowers herkogamous, 1–46, from distal or medial to distal nodes. Fruiting pedicels erect to ascending, 10–33(–40) mm. Fruiting calyces becoming reddish, campanulate to cylindric, 5–10 mm, margins distinctly toothed or lobed, minutely puberulent, ribs thickened, lobes pronounced, erect, margins glabrous. Corollas pale pink to pink, throat broad, yellow, palate with a broad yellow patch covering ridges and lateral areas, bilaterally symmetric, strongly bilabiate; tube-throat cylindric to funnelform, 9–15 mm, exserted beyond calyx margin; limb expanded 16–25 mm, lobes notched, abaxial limb glabrous or sparsely bearded. Styles glabrous. Anthers included, glabrous. Capsules included, 4–9 mm.

Flowering Mar–May. Washes in Mojave Desert scrub; of conservation concern; 600–900 m; Calif.

Erythranthe rhodopetra is endemic to the El Paso Mountains in Kern County. The species was treated previously as part of *E. palmeri* but is distinguished by having a wider limb (16–25 mm) than *E. palmeri* (8–15 mm) and pale pink flowers with a broad yellow palate and orifice. In contrast, *E. palmeri* has deep pink corollas with two yellow ridges on the palate.

14. **Erythranthe palmeri** (A. Gray) N. S. Fraga, Phytoneuron 2012-39: 35. 2012 • Palmer's monkeyflower [E]

Mimulus palmeri A. Gray, Proc. Amer. Acad. Arts 12: 82. 1876

Annuals, taprooted. Stems erect, simple or branched from basal nodes, 4–17 cm, minutely puberulent. Leaves cauline, basal not persistent; petiole 0 mm; blade pinnately veined, palmately 3-veined (in broader ones), linear to oblanceolate, (3–)4–17 × 1–4 mm, base attenuate, margins entire, apex acute to obtuse, surfaces minutely puberulent. Flowers herkogamous, 1–36, from distal or medial to distal nodes. Fruiting pedicels spreading horizontally, 5–33 mm. Fruiting calyces sometimes red-spotted on ribs, becoming straw colored, cylindric, 4–8 mm, margins distinctly toothed or lobed, glabrous, ribs weak, lobes pronounced, erect, margins ciliate. Corollas pink to purple, abaxial limb with 2 yellow ridges, bilaterally symmetric, strongly bilabiate; tube-throat funnelform, 6–15 mm, exserted beyond calyx margin; limb expanded 8–15 mm, lobes deeply notched, abaxial limb sparsely bearded. Styles glabrous. Anthers included, glabrous. Capsules included, 4–8 mm.

Flowering Apr–Jun. Moist areas in openings in pine forest and desert chaparral transitions; 900–2200 m; Calif.

Erythranthe palmeri has been confused with other closely related species, including *E. diffusa*, *E. discolor* (pink form), *E. rhodopetra*, and *E. sierrae*. It was previously thought to be a widely distributed species because of this taxonomic confusion but now is regarded as endemic to the Transverse Range in the San Bernardino and San Gabriel mountains.

15. **Erythranthe diffusa** (A. L. Grant) N. S. Fraga, Phytoneuron 2012-39: 35. 2012 • Palomar monkeyflower [C]

Mimulus diffusus A. L. Grant, Ann. Missouri Bot. Gard. 11: 254. 1925; *M. grantianus* Eastwood

Annuals, taprooted. Stems erect, simple or branched from basal nodes, 4–20 cm, minutely puberulent. Leaves cauline, basal not persistent; petiole 0 mm; blade 1-veined or palmately 3-veined (in broader ones), lanceolate to ovate, 3–20 × 1–6 mm, base attenuate, margins entire, sometimes toothed, apex acute, surfaces minutely puberulent. Flowers herkogamous, 1–36, from distal or medial to distal nodes. Fruiting pedicels ascending to spreading horizontally, (2–)12–60(–68) mm. Fruiting calyces sometimes red-dotted, campanulate, 3–6 mm, margins distinctly toothed or lobed, minutely puberulent, ribs weak, lobes pronounced, erect, margins glabrous. Corollas pink to purple, abaxial limb with 2 yellow palate ridges, bilaterally symmetric, weakly bilabiate; tube-throat funnelform, (3–)6–10 mm, exserted beyond calyx margin; limb expanded 3–14 mm, lobes notched, adaxial limb glabrous, sometimes sparsely bearded. Styles distally pubescent. Stigmas distinctly shorter than corolla tube. Anthers included, glabrous. Capsules included, 5–8 mm.

Flowering Apr–Jun. Moist areas in openings of chaparral, dry meadows in pine and oak woodlands; of conservation concern; 300–1800 m; Calif.; Mexico (Baja California).

Erythranthe diffusa has been included previously in *E. palmeri* but can be distinguished morphologically by glabrous margins on the calyx and pubescence on the distal end of the style. It occurs in Orange, Riverside, and San Diego counties as well as close to the California border in Baja California.

16. Erythranthe purpurea (A. L. Grant)
N. S. Fraga, Phytoneuron 2012-39: 35. 2012
• Purple monkeyflower C

Mimulus purpureus A. L. Grant, Ann. Missouri Bot. Gard. 11: 255, plate 5, fig. 2. 1925; *M. purpureus* var. *pauxillus* A. L. Grant

Annuals, taprooted. **Stems** erect, simple or branched from basal nodes, 3–10 cm, minutely puberulent. **Leaves** cauline, basal not persistent; petiole 0 mm; blade palmately 3–5-veined, elliptic to lanceolate, 4–15 × 1–5 mm, base truncate to truncate-cordate, clasping, margins entire, sometimes toothed, apex acute to obtuse, surfaces minutely puberulent. **Flowers** herkogamous, 1–22, from distal or medial to distal nodes. **Fruiting pedicels** ascending to often spreading horizontally, 13–57(–70) mm. **Fruiting calyces** becoming reddish, campanulate, 3–8 mm, margins distinctly toothed or lobed, minutely puberulent, ribs thickened, lobes pronounced, erect, margins glabrous. **Corollas** pink to purple, adaxial limb darker than abaxial, abaxial limb with yellow markings, bilaterally symmetric, strongly bilabiate; tube-throat cylindric to funnelform, 7–13 mm, exserted beyond calyx margin; limb expanded 7–10 mm, bilabiate, 3 abaxial lobes notched, 2 adaxial nearly entire, abaxial limb sparsely bearded. **Styles** glabrous. **Anthers** included, glabrous. **Capsules** included, 3–8 mm.

Flowering May–Jun. Moist openings along streams, swales, and depressions, pine duff in yellow pine forests, margins of dry meadows; of conservation concern; 1900–2800 m; Calif.; Mexico (Baja California).

Erythranthe purpurea is restricted to the San Bernardino Mountains in San Bernardino County and is disjunct in the Sierra San Pedro Mártir in Baja California, Mexico.

17. Erythranthe androsacea (Curran ex Greene)
N. S. Fraga, Phytoneuron 2012-39: 35. 2012
• Rock-jasmine monkeyflower E

Mimulus androsaceus Curran ex Greene, Bull. Calif. Acad. Sci. 1: 121. 1885; *M. palmeri* A. Gray var. *androsaceus* (Curran ex Greene) A. Gray

Annuals, taprooted. **Stems** erect, simple or branched from basal nodes, 0.5–10 cm, glabrous or minutely puberulent. **Leaves** cauline, basal not persistent; petiole 0 mm; blade 1-veined or palmately 3-veined (in broader ones), linear to lanceolate or ovate, 2–8 × 1–5 mm, base attenuate, sometimes clasping, margins entire, sometimes toothed, apex acute, surfaces glabrous or minutely puberulent. **Flowers** herkogamous, sometimes plesiogamous, 1–22, from distal or medial to distal nodes. **Fruiting pedicels** ascending to often spreading horizontally, 5–30 mm. **Fruiting calyces** sometimes red-dotted, campanulate, 3–7 mm, margins distinctly toothed or lobed, glabrous or minutely puberulent, ribs thickened, lobes pronounced, erect, margins glabrous. **Corollas** pink to purple, palate sometimes with yellow or pink markings, without yellow palate ridges, bilaterally symmetric, weakly bilabiate; tube-throat cylindric, 3–8 mm, exserted 2–3 mm beyond calyx margin; limb expanded 3–7 mm, lobes entire, sometimes notched, adaxial limb glabrous, sometimes sparsely bearded. **Styles** glabrous. **Anthers** included, glabrous. **Capsules** included, 3–5 mm.

Flowering Feb–May. Moist open areas on gentle slopes; 100–3600 m; Calif.

Erythranthe androsacea occurs primarily in the southern half of California but is scattered as far north as Lake and Modoc counties.

18. Erythranthe hardhamiae N. S. Fraga, Aliso 30:
64, figs. 23–25. 2012 • Hardham's monkeyflower C E

Annuals, taprooted. **Stems** erect, simple or branched from basal nodes, (2–)3–13 cm, glabrous or minutely puberulent. **Leaves** cauline, basal not persistent; petiole 0 mm; blade palmately 3-veined (in broader ones), linear to oblanceolate, 2–12 × 1–5 mm, base attenuate, margins entire, sometimes toothed, apex acute, surfaces glabrous or minutely puberulent. **Flowers** herkogamous, 1–12, from distal or medial to distal nodes. **Fruiting pedicels** ascending to often spreading horizontally, 10–60 mm. **Fruiting calyces** becoming reddish, sometimes red-spotted, campanulate to cylindric, 4–6 mm, margins distinctly toothed or lobed, glabrous or minutely puberulent, ribs weak, lobes pronounced, erect, margins glabrous. **Corollas** pink to purple, abaxial limb with 2 yellow palate ridges, bilaterally symmetric, strongly bilabiate; tube-throat funnelform, 5–10 mm, exserted 2–5 mm beyond calyx margin; limb expanded 7–13 mm, lobes notched, abaxial limb sparsely bearded. **Styles** glabrous. **Stigmas** equal in length to corolla tube or exserted. **Anthers** included or slightly exserted, glabrous. **Capsules** included, 4–5 mm.

Flowering Mar–Apr. Sandy soils near sandstone outcrops and chaparral; of conservation concern; 300–800 m; Calif.

Erythranthe hardhamiae is endemic to the central coast region in Monterey and San Luis Obispo counties. The species was previously included in *E. palmeri* but can be distinguished by having a wider limb

(16–25 mm) than *E. palmeri* (8–15 mm) and pale pink flowers with a broad yellow palate and orifice. In contrast, *E. palmeri* has deep pink flowers with two yellow ridges on the palate.

19. Erythranthe shevockii (Heckard & Bacigalupi) N. S. Fraga, Phytoneuron 2012-39: 35. 2012

• Kelso Creek or Shevock's monkeyflower [C] [E]

Mimulus shevockii Heckard & Bacigalupi, Madroño 33: 271, figs. 1, 2. 1986

Annuals, taprooted. **Stems** erect, simple or branched from basal nodes, 2–12 cm, minutely puberulent or glabrous. **Leaves** cauline, basal not persistent; petiole 0 mm; blade palmately 3-veined (in broader ones), lanceolate to ovate, 3–10 × 1–5 mm, base truncate to truncate-cordate, clasping, margins entire, apex acute to obtuse, surfaces minutely puberulent or glabrous. **Flowers** herkogamous, 1–16, from distal or medial to distal nodes. **Fruiting pedicels** ascending to often spreading horizontally, 10–22 mm. **Fruiting calyces** red-spotted or red, campanulate, 4–7 mm, margins distinctly toothed or lobed, minutely puberulent or glabrous, ribs weak, lobes pronounced, erect, margins glabrous. **Corollas** maroon and yellow, 2 lateral lobes maroon, 1 much larger central lobe yellow (red-spotted), 2 adaxial lobes maroon, bilaterally symmetric, strongly bilabiate; tube-throat cylindric, 8–12 mm, exserted 2–3 mm beyond calyx margin; limb expanded 8–15 mm, central lobe 2-fid, abaxial limb sparsely villous-bearded. **Styles** glabrous. **Anthers** included, glabrous. **Capsules** included, 5–6 mm. *2n* = 32.

Flowering Mar–May. Level openings in juniper and Joshua tree woodlands; of conservation concern; 900–1400 m; Calif.

Erythranthe shevockii is known only from the southernmost Sierra Nevada in Kern County. It (as *Mimulus shevockii*) is in the Center for Plant Conservation's National Collection of Endangered Plants.

20. Erythranthe primuloides (Bentham) G. L. Nesom & N. S. Fraga, Phytoneuron 2012-39: 35. 2012

• Primrose monkeyflower [E]

Mimulus primuloides Bentham, Scroph. Ind., 29. 1835; *M. nevadensis* Gandoger; *M. pilosellus* Greene; *M. primuloides* var. *minimus* M. Peck; *M. primuloides* var. *pilosellus* (Greene) Smiley

Perennials, rhizomatous or stoloniferous, mat-forming, rhizomes or stolons flagelliform. **Stems** erect to ascending, usually simple, 2–10(–20) cm, villous, internodes shortened. **Leaves** all basal or near basal, often rosulate; petiole 0 mm; blade palmately 3-veined, oblanceolate to elliptic-obovate, 7–40 × 4–12 mm, base cuneate to attenuate, margins entire, distally denticulate to dentate, or sharply serrate-dentate, apex acute to obtuse, abaxial surface glabrous, adaxial glabrous or glabrate to sparsely to densely long-villous, eglandular. **Flowers** herkogamous, 1. **Fruiting pedicels** 30–110 (–130) mm, glabrous or sparsely stipitate-glandular near base. **Fruiting calyces** tubular-campanulate, weakly or not inflated, 6–8 mm, glabrous. **Corollas** yellow to orange-yellow, usually brown-spotted abaxially, base of each abaxial lobe usually with a larger reddish brown spot, bilaterally or nearly radially symmetric, weakly bilabiate or nearly regular, densely hirsute on abaxial side of opening; tube-throat narrowly campanulate, 15–20 mm, exserted beyond calyx margin; lobes broadly obovate-oblong, apex rounded- or truncate-notched, throat open, palate densely villous, abaxial ridges prominent. **Styles** glabrous. **Anthers** slightly exserted, margins ciliate, glabrous. **Capsules** included, 6–7 mm. *2n* = 34.

Flowering Jun–Aug. Wet meadows, seeps, streamsides; 600–3400 m; Ariz., Calif., Idaho, Mont., Nev., N.Mex., Oreg., Utah, Wash.

Flowers in *Erythranthe primuloides* and *E. linearifolia* characteristically appear to be scapose, but the scapes are pedicels arising from axils of greatly foreshortened stems. Occasionally in both species the internodes may lengthen somewhat, and the leaves are not so densely clustered at the base of the stems.

In northern Klamath, western Deschutes, and eastern Douglas counties, Oregon, an area within the range of typical populations, *Erythranthe primuloides* has distinctively large corollas (limbs 10–15 mm wide). Apparent clones of large-flowered and smaller-flowered plants sometimes grow in close proximity or even intermixed, appearing as two different entities.

21. **Erythranthe linearifolia** (A. L. Grant) G. L. Nesom & N. S. Fraga, Phytoneuron 2012-39: 35. 2012 • Threadleaf primrose monkeyflower E F

Mimulus primuloides Bentham var. *linearifolius* A. L. Grant, Ann. Missouri Bot. Gard. 11: 246. 1925; *M. linearifolius* (A. L. Grant) Pennell; *M. primuloides* subsp. *linearifolius* (A. L. Grant) Munz

Perennials, rhizomatous, densely cespitose, forming large patches and turfs 0.3–1 m diam. **Stems** erect to ascending, simple, 2–10 cm, sparsely hirsute and stipitate-glandular, internodes shortened. **Leaves** basal or near basal, sometimes proximal cauline, subrosulate; petiole 0 mm; blade 1-veined or palmately 3-veined, linear to narrowly oblanceolate, 15–50 × 1.5–5 mm, base long-cuneate, often subclasping, margins entire, dentate-serrate, or distally dentate, apex acute, surfaces glabrous or adaxial sparsely short-pilose, eglandular. **Flowers** herkogamous, 1. **Fruiting pedicels** (40–)65–85 (–120) mm, glabrous or sparsely stipitate-glandular near base. **Fruiting calyces** winged- or plicate-angled, tubular-campanulate, weakly or not inflated, 9–10(–12) mm, glabrous. **Corollas** yellow, red-spotted or -striped, bilaterally symmetric, weakly bilabiate, loosely hirsute on abaxial side of opening; tube-throat narrowly campanulate, 18–22 mm, exserted beyond calyx margin; lobes broadly obovate-oblong, apex rounded- or truncate-notched, throat open. **Styles** glabrous. **Anthers** included or slightly exserted, margins ciliate, glabrous. **Capsules** included, 6–7 mm.

Flowering Jul–Sep. Wet banks, *Darlingtonia* seeps and bogs, seepages in serpentine talus; 600–2800 m; Calif.

Erythranthe linearifolia is endemic to serpentine substrates in Shasta, Siskiyou, and Trinity counties; typical *E. primuloides* occurs in the same area but not on serpentine. *Erythranthe linearifolia* is distinct from *E. primuloides* especially in its narrow leaves and cespitose habit. A collection from Tulare County appears to be *E. linearifolia* (*Shevock 10597*, CAS), but this appears to be far out of range and the voucher should be reexamined; it probably is better identified as an unusual collection of *E. primuloides*.

22. **Erythranthe breweri** (Greene) G. L. Nesom & N. S. Fraga, Phytoneuron 2012-39: 36. 2012 • Brewer's monkeyflower E F

Eunanus breweri Greene, Bull. Calif. Acad. Sci. 1: 101. 1885; *Mimulus breweri* (Greene) Coville; *M. rubellus* A. Gray var. *breweri* (Greene) Jepson

Annuals, taprooted. **Stems** erect, simple or branched from basal nodes, 2–20 cm, densely glandular-puberulent, hairs 0.05–0.1 mm, gland-tipped. **Leaves** cauline, basal not persistent; petiole 1–4 mm; blade 1-veined or palmately 3-veined (in broader ones), linear-oblanceolate to linear-lanceolate or linear, (5–)10–15(–20) × 1–2(–5) mm, base attenuate, margins entire or remotely mucronulate distally, apex acute, surfaces densely glandular-puberulent, hairs 0.05–0.1 mm, gland-tipped. **Flowers** herkogamous, 1–8, from medial to distal nodes. **Fruiting pedicels** 4–12 mm, densely glandular-puberulent, hairs 0.05–0.1 mm, gland-tipped. **Fruiting calyces** red-dotted, strongly angled, cylindric-campanulate, 4–7 mm, margins distinctly toothed or lobed, densely glandular-puberulent, hairs 0.05–0.1 mm, gland-tipped, ribs corky, lobes pronounced, spreading. **Corollas** pink to red, rose red, red-purple, or lavender, throat usually light purple to lavender (similar in color to rest of corolla), palate ridges yellow, abaxial limb with deep pink markings, bilaterally symmetric, strongly bilabiate; tube-throat cylindric-funnelform, 3–7 mm, exserted 0–1 mm beyond calyx margin; limb expanded 2–3 mm, lobes notched or entire. **Styles** glabrous. **Anthers** included, glabrous. **Capsules** included, 3–7 mm.

Flowering Jun–Aug(–Sep). Seeps and springs, damp rocks, vernal creek beds, cliffs, granite outcrops, rocky ridges, gravelly areas, meadow edges, stream edges; (700–)1300–3500 m; Alta., B.C.; Ariz., Calif., Idaho, Mont., Nev., Oreg., Utah, Wash., Wyo.

23. **Erythranthe bicolor** (Hartweg ex Bentham) G. L. Nesom & N. S. Fraga, Phytoneuron 2012-39: 36. 2012 • Yellow and white monkeyflower E

Mimulus bicolor Hartweg ex Bentham, Pl. Hartw., 328. 1849

Annuals, taprooted. **Stems** erect, simple or branched from basal nodes, 4–30 cm, densely glandular-puberulent, hairs 0.05–0.1 mm, gland-tipped. **Leaves** cauline, basal not persistent; petiole 0 mm or short-attenuate; blade 1-veined or palmately 3-veined (in broader ones), narrowly elliptic to linear-lanceolate or linear-oblanceolate, 10–30 × 2–6 mm at mid stem, base

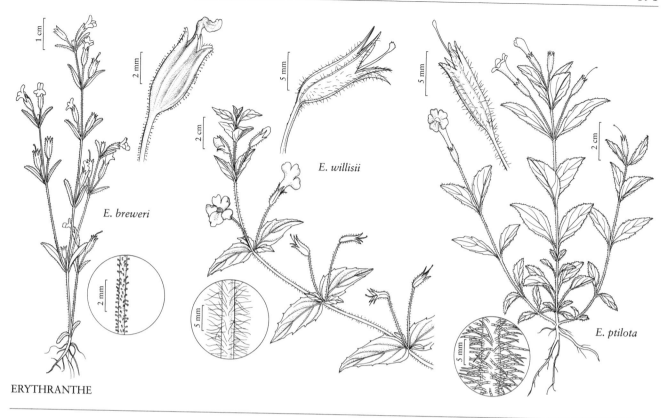

E. breweri

E. willisii

E. ptilota

ERYTHRANTHE

attenuate, margins entire or remotely shallowly dentate to mucronulate, teeth 1–4 per side, apex acute, surfaces densely glandular-puberulent, hairs 0.05–0.1 mm, gland-tipped. **Flowers** herkogamous, 2–8, from medial to distal nodes. **Fruiting pedicels** (5–)7–20(–30) mm, densely glandular-puberulent, hairs 0.05–0.1 mm, gland-tipped. **Fruiting calyces** red-dotted, strongly angled, cylindric-campanulate, (5–)7–12 mm, margins distinctly toothed or lobed, densely glandular-puberulent, hairs 0.05–0.1 mm, gland-tipped, ribs corky, lobes pronounced, spreading. **Corollas** yellow, adaxial lip white, sometimes yellow, throat and abaxial limb red-spotted, bilaterally symmetric, weakly bilabiate; tube-throat funnelform, 8–12 mm, exserted beyond calyx margin; limb broadly expanded, lobes 2-fid, palate villous. **Styles** minutely glandular. **Anthers** included, margins ciliate, glabrous. **Capsules** included, 4–6 mm.

Flowering Apr–Jun. Moist banks, serpentine and granite outcrops, seepage areas, volcanic mudflows, open red clay exposures; (100–)400–1700(–2500) m; Calif.

Erythranthe bicolor ranges from Shasta and Trinity counties south to Tulare County; the identity of apparently disjunct populations in San Bernardino County needs to be reexamined.

24. Erythranthe filicaulis (S. Watson) G. L. Nesom & N. S. Fraga, Phytoneuron 2012-39: 36. 2012 · Slender-stemmed monkeyflower C E

Mimulus filicaulis S. Watson, Proc. Amer. Acad. Arts 26: 125. 1891; *M. biolettii* Eastwood

Annuals, taprooted. **Stems** erect, simple or branched at basal nodes, 4–30 cm, densely glandular-puberulent, hairs 0.05–0.1 mm, gland-tipped. **Leaves** cauline, basal usually not persistent; petiole 0 mm; blade 1-veined or palmately 3-veined, linear to oblanceolate, 6–15 × (1–)2–3 mm, base attenuate, margins entire to remotely mucronulate, teeth 1–3 per side, apex acute, surfaces densely glandular-puberulent, hairs 0.05–0.1 mm, gland-tipped. **Flowers** herkogamous, 2–8, from medial to distal nodes. **Fruiting pedicels** 9–30 mm, densely glandular-puberulent, hairs 0.05–0.1 mm, gland-tipped. **Fruiting calyces** reddish brown-spotted, strongly angled, cylindric-campanulate, 7–9 mm, margins distinctly toothed or lobed, densely glandular-puberulent, hairs 0.05–0.1 mm, gland-tipped, ribs corky, lobes pronounced, spreading. **Corollas** pink to purple, throat dark red-purple, palate ridges yellow, bilaterally or nearly radially symmetric, weakly bilabiate or nearly regular; tube-throat cylindric-funnelform, 8–12(–17) mm, exserted beyond calyx margin; limb expanded 10–14 mm, lobes notched, palate villous. **Styles** glabrous. **Anthers** included, hirtellous. **Capsules** included, 5–7 mm. $2n = 16$.

Flowering May–Jul. Moist open areas on gentle slopes, meadows, roadsides, gravelly soils; of conservation concern; 800–1800 m; Calif.

Erythranthe filicaulis is known only for certain from Mariposa and Tuolumne counties. *Kappler 1691* (UCLA) from Madera County also may be this species.

25. Erythranthe parishii (Greene) G. L. Nesom & N. S. Fraga, Phytoneuron 2012-39: 37. 2012

• Parish's monkeyflower

Mimulus parishii Greene, Bull. Calif. Acad. Sci. 1: 108. 1885

Annuals, taprooted or fibrous-rooted. **Stems** erect, usually simple, (3–)10–85 cm, villous-glandular to stipitate-glandular. **Leaves** cauline; petiole 0 mm; blade palmately 3-veined, oblanceolate to narrowly ovate or oblong, (8–)15–75 × 3–17 mm, base slightly narrowed, subclasping to clasping, margins distally denticulate to irregularly dentate, apex acute to obtuse, surfaces villous-glandular to stipitate-glandular. **Flowers** herkogamous, 4–12, from medial to distal nodes. **Fruiting pedicels** 15–20 mm. **Fruiting calyces** slightly ridge-angled, cylindric-campanulate, not inflated, 8–13 mm, villous-glandular to stipitate-glandular. **Corollas** white to light lavender, pinkish, or rosy, palate ridges yellow-striped, palate and abaxial limb with or without small, reddish spots, bilaterally symmetric, weakly bilabiate; tube-throat cylindric, 8–10 mm, exserted beyond calyx margin; lobe apex truncate to rounded, slightly erose, throat open. **Styles** glabrous. **Anthers** included, glabrous. **Capsules** included, 6–10 mm.

Flowering May–Aug. Wet, sandy streamsides, rocky riverbeds, canyon drainages; 400–2300 m; Calif., Nev.; Mexico (Baja California).

Erythranthe parishii occurs in the southern Sierra Nevada (Kern and Tulare counties), mountain ranges of southern California (five counties), desert mountains (Inyo County), and in Clark and southern Nye counties, Nevada. It is the only annual species of sect. *Erythranthe.*

26. Erythranthe lewisii (Pursh) G. L. Nesom & N. S. Fraga, Phytoneuron 2012-39: 36. 2012

• Great purple or Lewis's monkeyflower E

Mimulus lewisii Pursh, Fl. Amer. Sept. 2: 427, plate 20. 1813; *M. lewisii* var. *tetonensis* A. Nelson

Perennials, rhizomatous. **Stems** erect, usually simple, (15–)25–60(–75) cm, stipitate-glandular to glandular-villous. **Leaves** cauline; petiole 0 mm; blade palmately veined, elliptic to ovate, ovate-lanceolate, or broadly oblanceolate, (10–)25–75(–90) × 5–35 mm, base rounded to cuneate, subclasping, margins denticulate, subentire, or entire, apex acute, surfaces stipitate-glandular to glandular-villous. **Flowers** herkogamous, 2–6(–10), axillary at leafy medial to distal nodes. **Fruiting pedicels** (25–)35–70 mm, stipitate-glandular to glandular-villous. **Fruiting calyces** broadly cylindric-campanulate, not inflated, 15–22 mm, stipitate-glandular to glandular-villous, tube 12–15(–17) × 9–12 mm. **Corollas** purple, rarely crimson, pale violet, white, pinkish white, yellowish white, or lavender, sometimes lined with red dots, bilaterally symmetric, strongly bilabiate; tube-throat funnelform, 22–28 mm, exserted beyond calyx margins; lobe apex usually truncate to shallowly convex, shallowly retuse, throat open. **Styles** glabrous. **Anthers** included, white-villous, thecae spreading. **Capsules** included, 6–11 mm. $2n = 16$.

Flowering Jun–Sep. Stream banks, springs, wet meadows, subalpine slopes, talus, crevices, ditches; 600–2900(–3200) m; Alta., B.C.; Alaska, Calif., Colo., Idaho, Mont., Nev., Oreg., Utah, Wash., Wyo.

Erythranthe lewisii in California occurs in Modoc, Siskiyou, and Trinity counties. California records from Shasta County and south previously identified as *E. lewisii* are identified here as *E. erubescens.* The record for *E. lewisii* in Alaska is documented by this collection: Hyder [noted on handwritten label as "New to Alaska, Standley"], damp land, 27 June 1924, *K. Whited 1291* (MO).

Apparent exceptions to the characteristic flower color are these: white to lavender in Nevada (Clark County, Charleston Mountains, *Train 2068*, MO); pinkish white in Washington (Skamania and Yakima counties, Mt. Paddo, *Suksdorf 5779*, MO); white or tinged with yellow, in Wyoming (Teton County, as described by Nelson in the protologue of *Mimulus lewisii* var. *tetonensis*).

27. **Erythranthe erubescens** G. L. Nesom, Phytoneuron 2014-31: 12, figs. 11–13. 2014 • California blushing monkeyflower E

Perennials, rhizomatous. **Stems** erect, usually simple, 25–90 cm, stipitate-glandular to glandular-villous. **Leaves** cauline; petiole 0 mm; blade palmately veined, elliptic to ovate, ovate-lanceolate, or lanceolate, (20–)30–90 × 5–25(–35) mm, base rounded to cuneate, subclasping, margins denticulate, subentire, or entire, apex acute, surfaces stipitate-glandular to glandular-villous. **Flowers** herkogamous, 2–8, axillary at leafy medial to distal nodes. **Fruiting pedicels** 45–90 mm. **Fruiting calyces** cylindric-campanulate, not inflated, 15–22 mm, stipitate-glandular to glandular-villous, tube 14–19 × 6–8 mm, lobes subequal to distinctly unequal, ovate, apex linear-caudate. **Corollas** light pink, darker pink stripes down middle of each lobe, abaxial 3 lobes with a white basal patch, palate ridges yellow, bilaterally symmetric, strongly bilabiate; tube-throat funnelform, 20–30 mm, exserted beyond calyx margins; lobe apex usually truncate, shallowly retuse, throat open. **Styles** glabrous. **Anthers** included, white-villous, thecae spreading. **Capsules** included, 7–13 mm. $2n = 16$.

Flowering Jul–Aug. Springs and seeps, meadows, cliffs, steep rocky slopes, ridges; (1400–)1800–3000(–3500) m; Calif., Nev.

Erythranthe erubescens was long identified as *E. lewisii* but is distinct in its light pink corollas (versus mostly magenta-rose to purplish in *E. lewisii*), more broadly cylindric calyx tube [14–19 × 6–8 mm versus 12–15(–17) × 9–12 mm], and its geographic range in the Sierra Nevada of California (versus widespread from southern Alaska south to northwestern California, northern Utah, eastern Nevada, and northern Colorado in *E. lewisii*). The two are genetically isolated and phylogenetically distinct (see summary of evidence in G. L. Nesom 2014b).

In California, *Erythranthe erubescens* ranges from Modoc, Plumas, and Tehama counties south to Fresno County; in Nevada, it is known only from Washoe County and Carson City.

28. **Erythranthe cardinalis** (Douglas ex Bentham) Spach, Hist. Nat. Vég. 9: 313. 1840 • Scarlet monkeyflower

Mimulus cardinalis Douglas ex Bentham, Scroph. Ind., 28. 1835; *M. cardinalis* var. *exsul* Greene; *M. cardinalis* var. *griseus* Greene; *M. cardinalis* var. *rigens* Greene

Perennials, rhizomatous. **Stems** usually erect to ascending, freely branched, 25–75 cm, ± glandular-villous or glabrate. **Leaves** usually cauline; petiole 0 mm; blade palmately veined, elliptic-ovate to ovate or obovate, 20–90(–110) × 10–38(–60) mm, thick, base acuminate, subclasping, margins irregularly serrate to dentate, apex acute, surfaces ± glandular-villous to glabrate. **Fruiting calyces** cylindric to cylindric-campanulate, not inflated, 17–28(–30) mm, hispid-hirsute to hirsute, lobes 4–7 mm, ovate to ovate-deltate, apex attenuate-acute. **Flowers** herkogamous, 2–12, axillary at leafy medial to distal nodes. **Fruiting pedicels** 30–90(–120) mm. **Corollas** scarlet to orange-red, rarely yellow, throat yellowish, red-striped, palate red, yellow-villous, not spotted or striped, bilaterally symmetric, strongly bilabiate; tube-throat funnelform, (15–)20–30 mm, exserted beyond calyx margin; throat open. **Styles** glabrous. **Anthers** exserted, white-villous, thecae spreading. **Capsules** included, 10–16 mm. $2n = 16$.

Flowering May–Sep. Moist to wet places along streams, lakes, creek beds, canyon, arroyo, and ravine bottoms, around springs and seepage areas, flood plains, moist clearings and woods edges; (5–)50–2300(–2800) m; Calif., Nev., Oreg., Wash.; Mexico (Baja California).

Yellow-flowered populations of *Erythranthe cardinalis* occur on Cedros Island, Baja California, and in the Siskiyou Mountains of Oregon (R. K. Vickery 1992). They have been documented also on Santa Cruz Island, California.

Populations of *Erythranthe cardinalis* in King and Yakima counties, Washington, are introduced and naturalized.

29. **Erythranthe cinnabarina** G. L. Nesom, Phytoneuron 2014-31: 16, figs. 16, 17. 2014 • Arizona big red monkeyflower E

Perennials, rhizomatous. **Stems** usually erect to ascending, freely branched, 25–60 cm, glabrous. **Leaves** usually cauline; petiole 0 mm; blade palmately veined, elliptic to oblong-elliptic, elliptic-lanceolate, or broadly lanceolate, 60–125 × 25–46 mm, base narrowly auriculate, clasping to subclasping, margins shallowly dentate, teeth sharp-pointed, apex acute, adaxial surface glabrous or minutely sessile- or stipitate-glandular along veins, lamina glabrous. **Flowers** herkogamous, 2–4(–8), axillary at leafy distal nodes. **Fruiting pedicels** 50–95 mm. **Fruiting calyces** cylindric-campanulate, not inflated, (27–)29–34 mm, minutely stipitate- or sessile-glandular, lobes 7–10 mm, ovate, apex abruptly attenuate to linear-caudate. **Corollas** deep orange, dull orange, red-orange, or deep scarlet, throat yellow-orange, dark red stripes leading onto basal part of lobes, not spotted, palate ridges red, bilaterally symmetric, strongly bilabiate; tube-throat tubular, 29–36 mm, exserted 7–12 mm beyond calyx margin; throat open, palate ridges densely short-villous, hairs yellowish. **Styles** glabrous. **Anthers** exserted, white-villous, thecae spreading. **Capsules** included, 14–18 mm. **2*n*** = 16.

Flowering Jun–Aug(–Sep). Canyons, ravines, streambeds and margins, riparian vegetation, mixed conifer forest; 2100–3300 m; Ariz.

Erythranthe cinnabarina is similar to typical *E. cardinalis* in its spreading anther thecae, relatively short-exserted corolla tube, and its reflexing corolla lobes but distinct in its generally larger leaves with reduced vestiture, fewer flowers, larger calyx and corolla, apically caudate calyx lobes, and its separate geographical range.

Erythranthe cinnabarina occurs in Cochise County (Chiricahua Mountains), Graham County (Pinaleño Mountains), and Pima County (Santa Catalina Mountains). *Erythranthe verbenacea*, with which it sometimes has been confused, occurs at lower elevations (350–2600 m) and ranges over most of the state (Apache, Cochise, Coconino, Gila, Graham, La Paz, Maricopa, Mohave, Pima, Pinal, Santa Cruz, and Yavapai counties). *Erythranthe cinnabarina* apparently occurs alone (without *E. verbenacea*) in the Pinaleño Mountains and in the Chiricahua Mountains, but both species have been abundantly documented in the Santa Catalina Mountains, where they sometimes closely co-occur in areas of elevational overlap (for example, at Marshall Gulch, about 2500 m; at Bear Wallow Campground, about 2600 m).

30. **Erythranthe verbenacea** (Greene) G. L. Nesom & N. S. Fraga, Phytoneuron 2012-39: 37. 2012

• Crimson monkeyflower

Mimulus verbenaceus Greene, Leafl. Bot. Observ. Crit. 2: 2. 1909; *M. cardinalis* Douglas ex Bentham var. *verbenaceus* (Greene) Kearney & Peebles; *M. lugens* Greene

Perennials, rhizomatous. **Stems** erect to decumbent, usually simple, weakly 4-angled, 20–60 cm, ± glandular-villous. **Leaves** cauline; petiole 0 mm; blade palmately 3–5-veined, elliptic to obovate, rhombic-ovate, or broadly spatulate, 50–75 × 15–26(–30) mm, base subcordate, subclasping, margins coarsely serrate, sometimes only distally, apex acute to obtuse, surfaces ± glandular-villous. **Flowers** herkogamous, 2–12, axillary at leafy medial to distal nodes. **Fruiting pedicels** 45–90(–150) mm. **Fruiting calyces** campanulate, weakly inflated, 20–28 mm, sparsely glandular-villosulous to stipitate-glandular, lobes triangular to ovate-triangular, apex linear-triangular. **Corollas** crimson, often yellow-tinged, palate ridges dark red, bilaterally symmetric, strongly bilabiate; tube-throat tubular, 25–35 mm, exserted 13–25 mm beyond calyx margin; abaxial limb spreading, adaxial erect, lobe apex truncate, often emarginate, throat open, palate ridges densely short-villous. **Styles** glabrous. **Anthers** exserted, white-villous, thecae reflexed 45°. **Capsules** included, 15–22 mm. **2*n*** = 16.

Flowering Jun–Sep. Stream edges and beds, flood plains, around seeps and springs, canyon bottoms, moist cliff crevices and ledges; 300–2600 m; Ariz., Utah; Mexico (Baja California, Chihuahua, Durango, Sinaloa, Sonora).

R. K. Vickery (1992) noted that yellow-flowered morphs of *Erythranthe verbenacea* occur in a population at Vasey's Paradise in the Grand Canyon (Coconino County), 32 miles downstream from Lees Ferry.

Populations of *Erythranthe verbenacea* in the Oak Creek Canyon area in southern Coconino County, Arizona, have leaves with a narrow, lateral, undulating, purple stripe across the mid lamina. The coloration is retained even in dried specimens.

In Utah, *Erythranthe verbenacea* is known only from the Zion Canyon area.

Molecular (P. M. Beardsley et al. 2003) and morphological (G. L. Nesom 2014b) data indicate that *Erythranthe verbenacea* is sister to *E. eastwoodiae*.

31. **Erythranthe eastwoodiae** (Rydberg) G. L. Nesom & N. S. Fraga, Phytoneuron 2012-39: 36. 2012

• Eastwood's monkeyflower E

Mimulus eastwoodiae Rydberg, Bull. Torrey Bot. Club 40: 483. 1913

Perennials, stoloniferous, sometimes also rhizomatous. **Stems** scandent to pendent, usually simple, 5–30(–40) cm, villous-glandular to minutely stipitate-glandular, hairs often a mixture of longer and much shorter ones, gland-tipped. **Leaves** cauline; petiole 0 mm; blade palmately 3-veined, flabellate distally to obovate to oblanceolate or elliptic, (5–)13–40(–55) × 8–20(–25) mm, largest near mid stem or distally, thick, base cuneate to rounded, subclasping, margins coarsely serrate on distal ½, apex acute, surfaces villous-glandular to minutely stipitate-glandular, hairs often a mixture of longer and much shorter ones, gland-tipped. **Flowers** herkogamous, 2–8, axillary at leafy medial to distal nodes. **Fruiting pedicels** 10–30(–40) mm. **Fruiting calyces** cuneate-cylindric to cylindric, weakly or not inflated, 15–23(–27) mm, glabrous or minutely stipitate-glandular to sparsely glandular-villosulous, lobes triangular-acuminate. **Corollas** scarlet to orange-red or orange, palate red, not spotted or striped, bilaterally symmetric, strongly bilabiate; tube-throat narrowly funnelform, 20–30 mm, exserted 5–15 mm beyond calyx margin; throat open, palate puberulent. **Styles** glabrous. **Anthers** exserted, villous, thecae reflexed 45°. **Capsules** included, 6–10 mm. $2n = 16$.

Flowering May–Sep(–Nov). Seepages in sandstone overhangs, cave roofs, walls, crevices, and cliff bases, pinyon-juniper woodlands; 900–2000 m; Ariz., Colo., N.Mex., Utah.

The range of *Erythranthe eastwoodiae* appears to be essentially contiguous with that of *E. verbenacea* in the Grand Canyon region, but there is no evidence of hybridization.

Erythranthe eastwoodiae (as *Mimulus eastwoodiae*) is in the Center for Plant Conservation's National Collection of Endangered Plants.

32. Erythranthe alsinoides (Douglas ex Bentham) G. L. Nesom & N. S. Fraga, Phytoneuron 2012-39: 37. 2012 • Wing-stem or chickweed monkeyflower E

Mimulus alsinoides Douglas ex Bentham, Scroph. Ind., 29. 1835

Annuals, fibrous-rooted. **Stems** erect, usually simple, (0.5–)2–6(–15) cm, glandular-puberulent, hairs 0.1–0.2 mm, gland-tipped, nodes 2(or 3), usually red-tinged. **Leaves** basal and cauline; petiole 1–20(–30) mm, distinctly 3-veined (winged); blade palmately 3-veined, lanceolate-ovate to ovate, elliptic, or suborbicular, 3–18(–32) × 3–12(–25) mm, base cuneate to truncate, margins dentate to denticulate or subentire, apex acute to obtuse or rounded, surfaces glandular-puberulent, hairs 0.1–0.2 mm, gland-tipped. **Flowers** herkogamous, 1–4(–8), from distal or medial to distal nodes. **Fruiting pedicels** 15–32 mm, glandular-puberulent, hairs 0.1–0.2 mm, gland-tipped. **Fruiting calyces** purplish, slightly ridge-angled, campanulate-cylindric, weakly or not inflated, 5–8 mm, margins subtruncate, sparsely minutely stipitate-glandular, lobes 4, (0–)0.5–1 mm, sometimes barely evident, 1 lobe usually slightly longer, margins appearing subtruncate, shallowly convex to rounded-mucronulate. **Corollas** yellow, abaxial limb with a large maroon splotch, also

red-spotted, bilaterally symmetric, ± bilabiate; tube-throat funnelform-cylindric, 6–9 mm, exserted beyond calyx margin; throat open, palate villous, abaxial ridges low. **Styles** glabrous. **Anthers** included, glabrous. **Capsules** included, (3–)5–7 mm.

Flowering Mar–Jun. Open, rocky slopes, cliff faces, bluffs, mossy rock crevices, ledges, moist rocks, roadsides, along wet paths and trails; 10–1900 m; B.C.; Calif., Idaho, Oreg., Wash.

Erythranthe alsinoides is distinct in its short, erect stems with few nodes, small, mostly ovate to elliptic-ovate, petiolate leaves, minutely stipitate-glandular vestiture, small corollas with a prominent maroon splotch on the abaxial limb, small, non-inflated mature calyces and, most especially, by its nearly truncate calyx margin. *Erythranthe pulsiferae* is superficially similar to *E. alsinoides* but has larger calyces borne on divergent-arcuate pedicels, smaller leaf blades with attenuate to cuneate bases, and the corolla limbs are smaller and lack a prominent maroon splotch.

33. Erythranthe gemmipara (W. A. Weber) G. L. Nesom & N. S. Fraga, Phytoneuron 2012-39: 37. 2012 • Rocky Mountain or petiole-purse monkeyflower C E

Mimulus gemmiparus W. A. Weber, Madroño 21: 423, fig. 1. 1972 (as *Minulus*)

Annuals, taprooted. **Stems** erect, straight at nodes, simple, 1–10 cm, glabrous. **Leaves** cauline; petiole 2–3 mm, laterally compressed, base deeply saccate, usually containing a lenticular propagule; blade emerging from bulbils, palmately veined, elliptic-ovate to ovate, 2–8 (–10) × 2–5(–7) mm, base truncate to shallowly cordate, margins entire or remotely denticulate, apex obtuse to rounded, surfaces glabrous. **Flowers** herkogamous, (1 or)2–12, from medial or medial to distal nodes. **Fruiting pedicels** 4–6 mm, slightly longer than calyx, glabrous. **Fruiting calyces** strongly angled, subcampanulate, weakly inflated, 3–4 mm, margins distinctly toothed or lobed, glabrous, lobes pronounced, erect, incurved-triangular. **Corollas** yellow, palate yellow, not spotted or striped, bilaterally symmetric, weakly bilabiate; tube-throat broadly cylindric-funnelform, 3–4 mm, exserted beyond calyx margin; lobes subequal, oblong-obovate, throat open, palate puberulent, abaxial ridges low. **Styles** glabrous. **Anthers** included, glabrous. **Capsules** unknown. $2n = 16$.

Flowering Jul–Aug(–Sep). Granitic seeps, thin soils over bedrock cliff bases, crevices, ledges, talus, among rocks and boulders, Douglas fir, spruce-fir, and aspen forests; of conservation concern; 2600–3700 m; Colo.

Erythranthe gemmipara is known only from Grand, Jefferson, Larimer, and Park counties in north-central Colorado. Flowers in this species are uncommon, and seed set has not been observed in natural populations; reproduction in nature appears to be solely asexual via overwintering propagules (bulbils) formed in leaf axils. Two meristems are initiated in each axil. The proximal meristem produces a pair of starch-thickened storage leaves, a rudimentary axis, and a distal pair of preformed leaf primordia that enclose the shoot apical meristem. Root primordia are present within the first node of the bulbil. The petiole of the subtending leaf expands laterally and folds adaxially to enclose the developing bulbil, and entangled trichomes along the petiole margins secure it following leaf abscission and dispersal. The leaf blades commonly are deciduous, leaving the bulbil still attached (M. R. Beardsley 1997).

SELECTED REFERENCES Beardsley, M. R. 1997. Colorado's Rare Endemic Plant, *Mimulus gemmiparus*, and Its Unique Mode of Reproduction. M.S. thesis. Colorado State University. Moody, A., P. K. Diggle, and D. A. Steingraeber. 1999. Developmental analysis of the evolutionary origin of vegetative propagules in *Mimulus gemmiparus*. Amer. J. Bot. 86: 1512–1522.

34. Erythranthe jungermannioides (Suksdorf) G. L. Nesom, Phytoneuron 2012-39: 38. 2012 • Liverwort monkeyflower E

Mimulus jungermannioides Suksdorf, Deutsche Bot. Monatsschr. 18: 154. 1900

Perennials, stoloniferous, stolons thin, forming overwintering turions. **Stems** decumbent to procumbent, simple or branching near base, 5–38(–60) cm, densely glandular-villous, hairs 0.5–1.2(–1.5) mm, gland-tipped, internodes evident. **Leaves** cauline, basal not persistent; petiole 2–5(–20) mm; blade subpalmately to pinnately veined, broadly ovate to broadly lanceolate, 7–35(–40) × 8–25 mm, base rounded, margins sharply, irregularly dentate to denticulate, apex acute to obtuse, surfaces glandular-villous. **Flowers** herkogamous, 2 or 3, from medial to distal nodes. **Fruiting pedicels** 15–35 mm, densely glandular-villous, hairs 0.5–1.2(–1.5) mm, gland-tipped. **Fruiting calyces** plicate-angled, cylindric-urceolate, weakly inflated, 6–12 mm, densely glandular-villous, hairs 0.5–1.2(–1.5) mm, gland-tipped, lobes 1–2 mm, apex rounded to mucronate. **Corollas** yellow, with scattered red spots, palate ridges with 2 white patches at tips, bilaterally symmetric, strongly bilabiate; tube-throat funnelform, (12–)16–20(–24) mm, exserted beyond calyx margin; limb expanded 8–10 mm, lobes obovate-oblong, apex rounded to truncate. **Styles** scabrous. **Anthers** included, glabrous. **Capsules** included, 5–9 mm. $2n = 32$.

Flowering May–Jul(–Aug). Basalt crevices in seepage zones in vertical cliff faces and canyon walls; 100–400(–1200) m; Oreg., Wash.

The occurrence of *Erythranthe jungermannioides* in the Columbia River Gorge of Klickitat County, Washington (the only record from the state of Washington), is based on an imprecise, unconfirmed observation from the early 1990s (Washington National Heritage Program 2005).

35. Erythranthe washingtonensis (Gandoger) G. L. Nesom, Phytoneuron 2012-39: 39. 2012 (as washingtoniensis) • John Day or Washington monkeyflower E

Mimulus washingtonensis Gandoger, Bull. Soc. Bot. France 66: 218. 1919

Annuals, fibrous-rooted or filiform-taprooted. **Stems** erect to ascending, straight or geniculate at nodes, usually many-branched, terete, 5–25 cm, moderately puberulent-glandular to villous-glandular, hairs 0.1–0.8 mm, flattened, sometimes vitreous, distinctly multicellular, gland-tipped. **Leaves** cauline, basal not persistent; petiole 2–14 mm; blade palmately veined, deltate or ovate to ovate-lanceolate, 4–16(–23) × 2–11(–16) mm, base rounded to cuneate or truncate, margins denticulate or entire, apex acute, surfaces moderately puberulent-glandular to villous-glandular, hairs 0.1–0.8 mm, flattened, sometimes vitreous, distinctly multicellular, gland-tipped. **Flowers** herkogamous, 1–6, from proximal to distal nodes. **Fruiting pedicels** divergent at nearly right angles, 20–50 mm, densely, minutely stipitate-glandular. **Fruiting calyces** greenish, ridge-angled, tubular, weakly inflated, 6–8 mm, margins distinctly toothed or lobed, densely, minutely stipitate-glandular, lobes pronounced, erect. **Corollas** yellow with small reddish brown dots, abaxial limb with 2 white patches (abaxial ridges), bilaterally symmetric, strongly bilabiate; tube-throat funnelform, 8–10 mm, exserted beyond calyx margin; limb expanded 7–10 mm, lobes obovate-oblong, apex rounded to rounded-cuneate. **Styles** hispid-hirtellous. **Anthers** included, glabrous. **Capsules** included, 5–8.5 mm. $2n = 32$.

Flowering May–Sep. Shallow basalt gravel in narrow channels and intermittent streams, sandy stream banks, open slopes, rocky shelves near seeps; 700–1300 m; Oreg.

Erythranthe washingtonensis is considered to be extirpated in Washington by the Washington Natural Heritage Program.

36. **Erythranthe hymenophylla** (Meinke) G. L. Nesom, Phytoneuron 2012-39: 38. 2012 • Thin-sepal monkeyflower C E

Mimulus hymenophyllus Meinke, Madroño 30: 147, fig. 1. 1983

Annuals, filiform-taprooted. **Stems** prostrate to ascending-erect, sharply bent at basal nodes, simple or few-branched, 5–25 cm, glandular-puberulent to glandular-villous, hairs 0.1–0.8 mm, vitreous, flattened, multicellular, gland-tipped. **Leaves** basal and cauline, largest at mid stem; petiole 6–30 mm; blade pinnately veined, broadly lanceolate to ovate, 10–35 × 10–30 mm, distinctly membranous, base cuneate to shallowly cordate, margins coarsely dentate to shallowly denticulate or entire, apex acute to obtuse, surfaces glandular-puberulent to glandular-villous, hairs 0.1–0.8 mm, vitreous, flattened, multicellular, gland-tipped, glandular. **Flowers** herkogamous, 1–6, from proximal to distal nodes. **Fruiting pedicels** divergent at right angles from stem, usually closely paired, 10–45 mm, negatively phototropic, causing capsules to be pressed against a cliff face or crevice at time of dehiscence, glandular-puberulent to glandular-villous, hairs 0.1–0.8 mm, vitreous, flattened, multicellular, gland-tipped. **Fruiting calyces** angled, tubular-campanulate, slightly inflated, 5–7 mm, margins distinctly toothed or lobed, sparsely stipitate-glandular, lobes pronounced, erect. **Corollas** light yellow, throat and abaxial lobes red- or purple-spotted, sometimes with small white patches, bilaterally symmetric, weakly bilabiate; tube-throat funnelform, 10–14 mm, exserted beyond calyx margin; lobes obovate-oblong, apex rounded to truncate or notched. **Styles** glabrous. **Anthers** included, glabrous. **Capsules** included, 3–6 mm. *2n* = 32.

Flowering Apr–Aug(–Sep). Steep, seasonally moist, basalt cliffs with west or southwest exposure, mesic coniferous forests; of conservation concern; 800–1300 m; Idaho, Mont., Oreg.

In the protologue, R. J. Meinke observed that plants of *Erythranthe hymenophylla* have reflexed fruiting pedicels that increase seed dispersal back onto the vertical cliff wall, the characteristic habitat of the species. The hanging habit of *E. hymenophylla* is reflected in a sharp (90° to 180°) bend in the basal nodes and the long pedicels that are closely paired and divergent in parallel at about right angles from the stem. The species also is characterized by it very short calyx to corolla length, relatively short capsules, and large seeds.

37. **Erythranthe patula** (Pennell) G. L. Nesom, Phytoneuron 2012-39: 39. 2012 • Stalk-leaf monkeyflower E

Mimulus patulus Pennell, Proc. Acad. Nat. Sci. Philadelphia 99: 162. 1947

Annuals, fibrous-rooted or filiform-taprooted. **Stems** erect to ascending, straight or geniculate at nodes, usually simple, (3–)5–15(–24) cm, stipitate-glandular, hairs 0.2–0.5 mm, gland-tipped. **Leaves** cauline, basal not persistent; petiole (5–)8–25 mm; blade palmately 3-veined, deltate or ovate to ovate-lanceolate, 4–12(–17) × 3–10(–14) mm, base rounded to cuneate-truncate, margins usually denticulate, apex acute to obtuse, surfaces stipitate-glandular, hairs 0.2–0.5 mm, gland-tipped. **Flowers** herkogamous, 1–10, from proximal to distal nodes. **Fruiting pedicels** 10–25 (–38) mm, stipitate-glandular, hairs 0.2–0.5 mm, gland-tipped. **Fruiting calyces** tubular, weakly or not inflated, 5–6(–7) mm, margins distinctly toothed or lobed, sparsely stipitate-glandular to sparsely hirtellous, lobes pronounced, erect. **Corollas** yellow, abaxial limb usually with a few red or brownish dots, radially or bilaterally symmetric, regular or weakly bilabiate; tube-throat funnelform, 7–8 mm, exserted beyond calyx margin; lobes oblong, apex rounded to truncate. **Styles** glabrous. **Anthers** included, glabrous. **Capsules** included, 4–6 mm. *2n* = 32.

Flowering Apr–May(–Aug). Ephemeral seeps, springs, rocky stream banks, moist basalt, fine gravel on bedrock, muddy hillside seeps, crevices; 200–1900 (–2900) m; Alta., B.C.; Idaho, Mont., Oreg., Wash., Wyo.

Erythranthe patula is distinctive with its long-petiolate leaves with ovate blades and its small, weakly bilabiate to nearly radially symmetric corollas. Vesture may include only minute, stipitate-glandular hairs or it may be an intergrading mix of stipitate-glandular hairs and minute (0.1–0.2 mm), sharp-pointed, eglandular hairs. Plants may have stipitate-glandular pedicels and calyces but hirtellous, eglandular stems, or they may have stipitate-glandular stems and pedicels but hirtellous calyces.

38. Erythranthe ampliata (A. L. Grant) G. L. Nesom, Phytoneuron 2012-39: 38. 2012 • Nez Perce monkeyflower E

Mimulus ampliatus A. L. Grant, Ann. Missouri Bot. Gard. 11: 214. 1925

Annuals, fibrous-rooted or filiform-taprooted. **Stems** erect to ascending, straight or geniculate at nodes, usually many-branched, 4-angled, 5–17 cm, sparsely sessile- to subsessile-glandular, hairs 0.05–0.2 mm, gland-tipped. **Leaves** cauline, basal not persistent; petiole 8–20 mm; blade palmately veined, broadly ovate to lanceolate, 8–25 × 5–19 mm, base cuneate, margins dentate to coarsely denticulate, apex acute to obtuse, surfaces sparsely sessile- to subsessile-glandular, hairs 0.05–0.2 mm, gland-tipped. **Flowers** herkogamous, 1–10, from proximal to distal nodes. **Fruiting pedicels** 10–22 mm, sparsely sessile- to subsessile-glandular, hairs 0.05–0.2 mm, gland-tipped. **Fruiting calyces** tubular-campanulate, not or weakly inflated, 6–8 mm, margins distinctly toothed or lobed, minutely, sparsely stipitate-glandular or glabrous, lobes pronounced, erect. **Corollas** deep yellow, sometimes with small white patches, abaxial limb with a few brownish dots, bilaterally symmetric, strongly bilabiate; tube-throat broadly funnelform, 8–12(–14) mm, exserted beyond calyx margin; lobes obovate-oblong, apex rounded to truncate. **Styles** glabrous. **Anthers** included, glabrous. **Capsules** included, 5–6 mm.

Flowering Jun–Jul. Basalt outcrops, seepy roadcuts, grassland seeps; 900–1700 m; Idaho, Mont.

Erythranthe ampliata is known from Clearwater, Lewis, and Nez Perce counties in Idaho and six western counties in Montana.

39. Erythranthe taylorii G. L. Nesom, Phytoneuron 2013-43: 6, figs. 5–7. 2013 (as taylori) • Taylor's or Shasta limestone monkeyflower C E

Annuals, filiform-taprooted. **Stems** erect, straight at nodes, simple or few-branched from base, 5–10 cm, sparsely eglandular-villous proximally, becoming sparsely short stipitate-glandular distally. **Leaves** usually cauline, basal not persistent, largest at mid stem or basal and mid stem to nearly even-sized; petiole 3–5(–8) mm; blade often purple adaxially, palmately 3–5-veined, broadly ovate to elliptic-ovate or ovate-lanceolate, 4–20 × 4–12 mm, base rounded to truncate, margins serrate-dentate, teeth 2–4 per side, shallow, apex rounded to obtuse or acute, surfaces: distals moderately short-stipitate-glandular. **Flowers** herkogamous, sometimes plesiogamous, 2–6(–8), from proximal to distal nodes. **Fruiting pedicels** divergent to arcuate-divergent, 6–13 mm. **Fruiting calyces** wing-angled, tubular-campanulate, 4–5 mm, margins distinctly toothed or lobed, densely invested with tiny, waxy-white, eglandular, papillose hairs between angles, lobes pronounced, erect. **Corollas** yellow, throat ceiling sometimes red-spotted or -lined, abaxial limb yellow or with 1 or 2 red splotches, bilaterally symmetric, bilabiate; tube-throat funnelform, 5–7 mm, exserted beyond calyx margin. **Styles** glabrous. **Anthers** included, glabrous. **Capsules** included, 3–4 mm.

Flowering Feb–May. Crevices in limestone cliff faces and outcrops; of conservation concern; 900–1100 m; Calif.

Erythranthe taylorii is known only from the Shasta Lake region of northwestern Shasta County. Its broad, distinctly bilabiate corollas and ovate leaf blades with palmate venation are similar to those of species of the northern group of sect. *Mimulosma*, the "Columbia River clade" (J. B. Whittall et al. 2006) of Idaho, Montana, Oregon, Washington, and Wyoming, particularly to the Idaho endemic *E. ampliata. Erythranthe taylorii* is distinct from *E. ampliata* in its larger, papillose calyces, shorter fruiting pedicels, corollas with shorter tube-throats, and shorter capsules. Considerable corolla color variation exists in *E. taylorii* in the occurrence and density of red dots and lines on the throat ceiling and larger red splotches on the abaxial limb.

40. Erythranthe pulsiferae (A. Gray) G. L. Nesom, Phytoneuron 2012-39: 39. 2012 • Pulsifer's or candelabrum monkeyflower E

Mimulus pulsiferae A. Gray, Proc. Amer. Acad. Arts 11: 98. 1876

Annuals, shallowly fibrous-rooted. **Stems** erect, straight at nodes, simple or few-branched at base, 5–12 (–18) cm, minutely stipitate-glandular, hairs 0.1–0.3 mm, gland-tipped. **Leaves** basal and cauline; petiole 2–9 mm, distinctly 3-veined, 2-winged; blade palmately 3-veined, elliptic-oblong to ovate or oblanceolate, 3–14(–23) × 2–9(–15) mm, base cuneate to attenuate, margins denticulate to entire, apex acute to obtuse, surfaces minutely stipitate-glandular, hairs 0.1–0.3 mm, gland-tipped. **Flowers** herkogamous, 1–9, from medial to distal nodes. **Fruiting pedicels** divergent-arcuate, 12–38 mm, minutely stipitate-glandular, hairs 0.1–0.3 mm, gland-tipped. **Fruiting calyces** cylindric, ± inflated, 7–10 mm, margins distinctly toothed or lobed, minutely stipitate-glandular, hairs 0.1–0.3 mm, gland-tipped, lobes pronounced, erect. **Corollas** yellowish, tube-throat, palate ridges, and limb yellow to pale yellow, abaxial limb red-dotted or not, bilaterally symmetric, weakly bilabiate; tube-throat funnelform, 6–9 mm, exserted beyond calyx margin; lobes broadly obovate-suborbicular, apex rounded. **Styles** glabrous. **Anthers** included, glabrous. **Capsules** included, 5–8 mm. $2n = 32$.

Flowering Apr–Jul. Damp depressions, moist gravel, rocky flats, granite outcrops, wet meadows, lava beds, vernal pools, forest openings, commonly in or near coniferous forests, chaparral-live oak woodlands; 50–1300(–2500) m; Calif., Oreg., Wash.

Erythranthe pulsiferae is characterized by minutely stipitate-glandular vestiture (lacking villous hairs), elongate internodes, persistent basal leaves, small, palmately veined, cauline leaves with short, three-veined petioles and elliptic-oblong to ovate or oblanceolate blades, divergent-arcuate pedicels, and small, all yellow, weakly bilabiate corollas.

41. **Erythranthe trinitiensis** G. L. Nesom, Phytoneuron 2013-43: 1, figs. 1–3. 2013 • Trinity Mountains or pink-margined monkeyflower C E

Annuals, shallowly fibrous-rooted. **Stems** erect, straight at nodes, simple or branched at base, 5–15 cm, puberulent, hairs gland-tipped, glands often dark. **Leaves** mostly cauline, largest at mid stem; petiole 4–8 mm, 1-veined, not winged; blade palmately 3-veined, ovate to elliptic-ovate, 6–17 × 4–9 mm, base attenuate, margins entire or dentate-serrate, teeth 1–2 per side, shallow, apex acute, surfaces puberulent, hairs gland-tipped, glands often dark, adaxial sometimes sparsely villous-glandular. **Flowers** herkogamous, 1–12, from proximal to distal nodes. **Fruiting pedicels** divergent-arcuate, 9–17 mm. **Fruiting calyces** oblong-ovoid, distinctly inflated, 8–10 mm, margins distinctly toothed or lobed, puberulent, hairs gland-tipped, glands often dark, lobes pronounced, erect. **Corollas** yellow and light pink to white, tube-throat yellow (inner and outer surfaces), lobes (limb) pink or white with pink distal borders, palate ridges yellow, throat floor and ridges weakly red-spotted, bilaterally symmetric, weakly bilabiate; tube-throat funnelform, 7–10 mm, exserted beyond calyx margin. **Styles** glabrous. **Anthers** included, glabrous. **Capsules** included, 6–8 mm.

Flowering Jun–Jul(–Aug). Seeps over serpentine, wet meadows, roadsides; of conservation concern; 1300–2000 m; Calif.

Erythranthe trinitiensis is similar to *E. pulsiferae* in its narrow leaves, glandular-puberulent vestiture, and weakly bilabiate corollas; it differs in its early-shed basal leaves, cauline leaves with one-veined petioles, and yellow corolla tubes and throats with pink lobes or lobe margins. The species is known only in Humboldt, Siskiyou, and Trinity counties.

42. **Erythranthe breviflora** (Piper) G. L. Nesom, Phytoneuron 2012-39: 38. 2012 • Short-flower monkeyflower E

Mimulus breviflorus Piper, Bull. Torrey Bot. Club 28: 45. 1901

Annuals, shallowly fibrous-rooted. **Stems** ascending, geniculate at nodes, branched at proximal and medial nodes, 4–15 cm, minutely stipitate-glandular, hairs 0.1–0.3 mm, gland-tipped, sometimes minutely hirtellous, hairs sharp-pointed, eglandular. **Leaves** usually cauline, basal usually deciduous by flowering; petiole 1–3 mm; blade palmately 3-veined, narrowly ovate or narrowly lanceolate to elliptic or elliptic-lanceolate, largest 5–15 × 2–6 mm, relatively even-sized, or slightly reduced distally, base attenuate, margins entire, mucronulate, or denticulate, apex acute to obtuse, surfaces minutely stipitate-glandular, hairs 0.1–0.3 mm, gland-tipped, sometimes minutely hirtellous, hairs sharp-pointed, eglandular. **Flowers** plesiogamous, 10–20, from medial to distal nodes. **Fruiting pedicels** straight, 5–11 mm, minutely stipitate-glandular, hairs 0.1–0.3 mm, gland-tipped, sometimes minutely hirtellous, hairs sharp-pointed, eglandular. **Fruiting calyces** winged, plicate-angled, campanulate becoming ovoid-ellipsoid to campanulate, distinctly inflated, 5–6 mm, margins distinctly toothed or lobed, sparsely, minutely hirtellous, eglandular, sometimes sparsely sessile-glandular, lobes pronounced, erect. **Corollas** yellow, red-spotted or striped, bilaterally symmetric, weakly bilabiate; tube-throat cylindric to narrowly funnelform, 3.5–5 mm, not exserted beyond calyx margin; limb barely widened, lobes broadly obovate, apex rounded. **Styles** glabrous. **Anthers** included, glabrous. **Capsules** included, 4–6 mm.

Flowering May–Jul. Stream and lake sides, gravel bars, springs, moist slopes, damp swales between dunes, along trails; 700–2300 m; B.C.; Calif., Idaho, Mont., Nev., Oreg., Wash.

43. Erythranthe inflatula (Suksdorf) G. L. Nesom, Phytoneuron 2012-39: 38. 2012 • Disappearing monkeyflower E

Mimulus inflatulus Suksdorf, Werdenda 1: 38. 1927; *M. evanescens* Meinke

Annuals, fibrous-rooted or filiform-taprooted. **Stems** erect to ascending, straight or geniculate at nodes, simple or branched at proximal and medial nodes, 6–20(–25) cm, minutely stipitate-glandular, hairs 0.1–0.3 mm, glandtipped. **Leaves** usually cauline, basal usually deciduous by flowering; petiole: proximals 1–3 mm, distals 0 mm; blade palmately 3–5-veined, narrowly ovate or narrowly lanceolate to elliptic or elliptic-lanceolate, largest 8–18(–30) × (1–)3–7 mm, relatively even-sized, or slightly reduced distally, base attenuate to obtuse or rounded, margins entire, mucronulate, or denticulate, apex acute to obtuse, surfaces minutely stipitateglandular, hairs 0.1–0.3 mm, gland-tipped. **Flowers** plesiogamous, 10–20, from medial to distal nodes. **Fruiting pedicels** straight, 7–18 mm, minutely stipitateglandular, hairs 0.1–0.3 mm, gland-tipped. **Fruiting calyces** winged, plicate-angled, maturing ovoid-ellipsoid to campanulate or broadly urceolate, distinctly inflated, 7–11 mm, margins distinctly toothed or lobed, sparsely, minutely hirtellous, eglandular, lobes pronounced, erect. **Corollas** yellow to pale yellow, sparsely red-spotted or not, bilaterally symmetric, weakly bilabiate; tube-throat cylindric, 5–8 mm, exserted 1–3 mm beyond calyx margin; limb barely widened, lobes broadly obovate, apex rounded or mucronate. **Styles** glabrous. **Anthers** included, glabrous. **Capsules** included, 5–9 mm.

Flowering Jun–Jul. Drying edges, banks, and beds of summer-dry watercourses, near drying edges of small lakes or impoundments, often among rocks and shoreline detritus, occasionally in moist protected areas beneath low shrubs; 1200–1700 m; Calif., Idaho, Nev., Oreg.

No natural occurrences of *Erythranthe inflatula* are known from Washington; the type collection from Klickitat County is from a cultivated plant.

Morphological and molecular data (R. J. Meinke 1995; P. M. Beardsley et al. 2004) indicate that *Erythranthe inflatula* originated as a hybrid between *E. breviflora* and *E. latidens*. Its geography and biology suggest that it is reproductively stable. The putative parents are geographically and ecologically separated for most of their ranges, and the range of *E. inflatula* is considerably broader than the relatively small region where the parents are sympatric. In the region of sympatry, however, *E. inflatula* may be difficult to distinguish from one or both of its putative parents.

G. L. Nesom (2012g) was not able to find morphology that would distinguish the recently described *Mimulus evanescens* from *E. inflatula*.

44. Erythranthe latidens (A. Gray) G. L. Nesom, Phytoneuron 2012-39: 38. 2012 • Broad-tooth monkeyflower

Mimulus inconspicuus A. Gray var. *latidens* A. Gray in A. Gray et al., Syn. Fl. N. Amer. ed. 2, 2(1): 450. 1886; *M. latidens* (A. Gray) Greene

Annuals, fibrous-rooted or filiform-taprooted. **Stems** ascending to ascending-erect, geniculate at nodes, usually many-branched from base, 3–10(–25) cm, short stipitate-glandular or sessileglandular, hairs 0.1–0.3 mm, gland-tipped. **Leaves** basal and cauline, largest at base or near mid stem, sometimes unreduced in size up to distalmost nodes; petiole 0 mm; cauline blade palmately 3(–5)-veined, ovate to ovatelanceolate, 8–26(–35) mm, base abruptly cuneate to rounded, sometimes subauriculate, margins entire or barely mucronulate to shallowly dentate-mucronulate, teeth or mucronulae 1–3 per side, apex acute to rounded, surfaces short stipitate-glandular or sessileglandular, hairs 0.1–0.3 mm, gland-tipped. **Flowers** plesiogamous, (1–)3–12, from medial to distal nodes. **Fruiting pedicels** straight, 11–28 mm, short stipitateglandular or sessile-glandular, hairs 0.1–0.3 mm, gland-tipped. **Fruiting calyces** purplish, prominently 5-angled, tubular-campanulate or ovoid-ellipsoid, strongly inflated, 8–12 mm, margins distinctly toothed or lobed, short stipitate-glandular or sessile-glandular, hairs 0.1–0.3 mm, gland-tipped, lobes pronounced, erect. **Corollas** white to pinkish or flesh colored, rarely yellowish, throat and abaxial lobes red-spotted, nearly radially symmetric; tube-throat cylindric, 5–6(–8) mm, exserted beyond calyx margin; limb barely widened, lobes broadly obovate, apex rounded. **Styles** glabrous. **Anthers** included, glabrous. **Capsules** included, 6–7 mm.

Flowering Apr–Jun. Drained flats or slopes subject to vernal inundation, depressions in open fields, bare clay soils, vacant lots, roadsides; 10–800 m; Calif., Oreg.; Mexico (Baja California).

The distinction between *Erythranthe latidens* and *E. inflatula* sometimes seems arbitrary, perhaps because of gene exchange where they are sympatric in northwestern California. *Erythranthe latidens* in Oregon is known only from southern Harney County.

45. Erythranthe moschata (Douglas ex Lindley) G. L. Nesom, Phytoneuron 2012-39: 38. 2012 • Muskflower, musk plant, mimule musqué E

Mimulus moschatus Douglas ex Lindley, Bot. Reg. 13: plate 1118. 1828; *Erythranthe inodora* (Greene) G. L. Nesom; *E. moniliformis* (Greene) G. L. Nesom; *M. crinitus* A. L. Grant; *M. guttatus* de Candolle var. *moschatus* (Douglas ex Lindley) Provancher; *M. inodorus* Greene; *M. leibergii* A. L. Grant; *M. macranthus* Pennell; *M. moniliformis* Greene; *M. moschatus* var. *longiflorus* A. Gray; *M. moschatus* var. *moniliformis* (Greene) Munz; *M. moschatus* var. *pallidiflorus* Suksdorf

Perennials, rhizomatous, rooting at proximal nodes. **Stems** erect, sometimes ascending to decumbent, simple or branched, (2–)5–20 cm, nodes 2–4(or 5), glabrate to glandular-villous, hairs 0.5–2 mm, gland-tipped, internodes evident. **Leaves** usually cauline, basal not persistent, distinctly separated; petiole 0 mm or (0.5–)1–5(–10) mm; blade pinnately veined, oblong-ovate to ovate, (10–)15–40(–50) × 5–25 mm, base obtuse-cuneate to truncate, rounded or subcordate, subclasping to sessile, margins coarsely serrate-dentate to denticulate or subentire, apex acute to obtuse, surfaces glabrate to glandular-villous. **Flowers** herkogamous, 1–8, from medial to distal nodes. **Fruiting pedicels** (7–)10–25 mm, glabrate to glandular-villous. **Fruiting calyces** ridge- to wing-angled, campanulate to cylindric-campanulate, weakly or not inflated, 6–13 mm, villous to glandular-villous, lobes erect to spreading-recurving, strongly unequal to subequal, triangular to linear-lanceolate or narrowly triangular-acuminate, 2–4 mm, apex acute to obtuse. **Corollas** yellow, throat with fine red to blackish or brown lines extending onto lobes, red to brown dots in throat and lobes present or absent, bilaterally or nearly radially symmetric, bilabiate or nearly regular; tube-throat narrowly funnelform, 11–18 mm, exserted beyond calyx margin; lobes oblong-obovate, apex rounded to notched. **Styles** glabrous. **Anthers** included, glabrous or slightly hirtellous to scabrous. **Capsules** included, 6–8 mm. *2n* = 32.

Flowering May–Aug. Springs and seeps, creek edges, moist meadows, ditches, along trails, roadsides, rocky ridges, granite outcrops, shaded and wet places in sagebrush, aspen, fir, spruce-fir, lodgepole pine forests, meadows; (300–)400–3100 m; St. Pierre and Miquelon; B.C., N.B., Nfld. and Labr. (Nfld.), N.S., Ont., P.E.I., Que.; Calif., Colo., Conn., Idaho, Maine, Mass., Mich., Mont., Nev., N.H., N.J., N.Y., Oreg., Pa , R.I., Utah, Vt., Va., Wash., W.Va., Wis., Wyo.; introduced in South America (Chile), Europe, e Asia (Japan), Pacific Islands (New Zealand), Australia.

Earlier segregation of *Erythranthe moniliformis* as distinct from *E. moschata* (for example, G. L. Nesom 2012g) emphasized a primarily erect habit and tendency toward sessile to subsessile and more densely arranged cauline leaves in *E. moniliformis* versus a decumbent to procumbent habit and consistently petiolate leaves on longer internodes in *E. moschata*. Discontinuities in morphology, geography, and ecology were not confirmed in later study by Nesom (2017). Rhizomes with small, tuberlike swellings can be observed over the whole *moschata/moniliformis* range, and there apparently are no consistent distinctions in vesture and corolla size.

Mimulus acutidens Reiche (1911), a later homonym of *M. acutidens* Greene, pertains here.

46. Erythranthe willisii G. L. Nesom, Phytoneuron 2017-17: 7, figs. 14–22. 2017 • Willis's monkeyflower C E F

Perennials, rhizomatous, rarely rooting at proximal nodes, usually forming large colonies, rhizomes white, usually highly branching. **Stems** usually sprawling-decumbent, branched, sometimes simple, 7–45 cm, nodes (2–)4–15+, densely glandular-villous, hairs 1–2 mm, glandular, internodes evident. **Leaves** usually cauline, basal not persistent, distinctly separated; petiole 0 mm, sometimes 1–2 mm at proximal nodes; blade bicolored, purplish abaxially, pinnately veined, ovate to elliptic-ovate, midcauline 10–35 × 6–18 mm, base rounded to subcordate, margins coarsely serrate-dentate to denticulate or subentire, apex short-attenuate to acute, obtuse, or rounded, surfaces densely glandular-villous, hairs 1–2 mm, gland-tipped. **Flowers** herkogamous, (4–)8–30+, from medial to distal nodes, sometimes from all nodes. **Fruiting pedicels** 4–20(–25) mm, densely glandular-villous, hairs 1–2 mm, gland-tipped. **Fruiting calyces** ridge- to wing-angled, campanulate to cylindric-campanulate, weakly inflated, 7–10 mm, densely glandular-villous, lobes erect to slightly spreading, unequal, triangular to linear-lanceolate, 2–4 mm, apex acuminate-apiculate. **Corollas** yellow, throat, tube, and proximal portion of abaxial 3 lobes with fine, red to brownish lines, weakly bilaterally or nearly radially symmetric, weakly bilabiate or nearly regular; tube-throat narrowly funnelform, 12–15 mm, exserted beyond calyx margin; limb 9–12 mm wide (pressed), lobes oblong-obovate, apex rounded to notched. **Styles** glabrous. **Anthers** included, glabrous or finely hirtellous to scabrous. **Capsules** included, 4–5 mm.

Flowering May–Sep. Seepage, drainage margins, moist soils, talus, cracks and crevices, soils deprived from serpentine; of conservation concern; (500–)700–900 m; Calif.

Erythranthe willisii is narrowly endemic over serpentine along the North Fork Feather River (including the North Branch) in Plumas County. In the original description, its range was said to include serpentine localities in closely adjacent areas of east-central Butte, Plumas, and northwestern Yuba counties, but subsequent field work has shown that these peripheral populations are *E. moschata*, and that *E. willisii* occurs only in the bottom of the Serpentine Canyon area. The most consistent and recognizable features of *E. willisii* are the long, sprawling stems often spread over a large area, sometimes reaching at least 45 cm and often with many crowded nodes, sessile or subsessile leaves with rounded to subcordate bases, and short pedicels, characteristically no longer than the subtending leaves (except sometimes the distal ones where subtending leaves are distinctly reduced in size). It is possible that stem growth in *E. willisii* is indeterminate versus determinate in *E. moschata*. Sessile to subsessile leaves occur in *E. moschata*, especially in the California Sierra Nevada, but petiole length and leaf base shape are variable within populations; lack of petioles and a rounded/subcordate base are fixed characters in *E. willisii* (as they are also in *E. ptilota*). Although large colonies of *E. moschata* are sometimes encountered, the individual plants tend to be erect (in California) and with few distal flowers. In the field, the dense vesture of *E. willisii* is a prominent feature, but this is harder to distinguish in pressed specimens, and there is a strong tendency for purple abaxial leaf coloration in *E. willisii*. Phenology and flower morphology of *E. willisii* and *E. moschata* appear to be similar, but *E. moschata* in north-central California does not occur at as low elevations as *E. willisii*.

47. Erythranthe ptilota G. L. Nesom, Phytoneuron 2017-17: 4. 2017 • Wing-leaf monkeyflower E F

Mimulus moschatus Douglas ex Lindley var. *sessilifolius* A. Gray in A. Gray et al., Syn. Fl. N. Amer. ed. 2, 2(1): 447. 1886

Perennials, rhizomatous, sometimes rooting at proximal nodes. **Stems** prostrate, sometimes decumbent to ascending, few-branched, 20–80 cm, villous, hairs 1–2 mm, eglandular, sometimes mixed with much shorter stipitate-glandular ones, internodes evident. **Leaves** cauline, basal not persistent, often congested; petiole 0 mm, rarely 1–2(–3) mm; blade pinnately veined, oblong-lanceolate, 30–70 × 10–22 mm, base rounded, margins denticulate to dentate, apex acute, surfaces villous, hairs 1–2 mm, eglandular, sometimes mixed with much shorter stipitate-glandular ones. **Flowers** herkogamous, 4–10, from medial to distal nodes. **Fruiting pedicels** (15–)22–50 mm, villous, hairs 1–2 mm, eglandular, sometimes mixed with much shorter stipitate-glandular ones. **Fruiting calyces** wing- or plicate-angled, cylindric-campanulate, weakly inflated, 10–12 mm, villous-glandular, hairs gland-tipped, lobes distinctly spreading, strongly unequal, linear-lanceolate to narrowly triangular, 5–9 mm, apex long acuminate-apiculate. **Corollas** yellow, throat with fine blackish or brownish lines on all sides, weakly bilaterally or nearly radially symmetric, weakly bilabiate or nearly regular; tube-throat narrowly campanulate, 15–18 mm, exserted beyond calyx margin; lobe apex rounded. **Styles** glabrous. **Anthers** included, finely hirtellous to hispidulous. **Capsules** included, 6–8 mm.

Flowering (May–)Jun–Aug. Creek banks, gravel bars, flood plains, shallow ditches and natural drainages, swales, damp banks, wet sand, moist soils in coniferous woods, marshes, bogs; 0–1000(–1900) m; B.C.; Calif., Oreg., Wash.

Erythranthe ptilota is recognized by its prostrate to decumbent or decumbent-ascending habit, large, consistently sessile leaves, densely villous vesture, long pedicels, large calyces and corollas, hispid-hirtellous anthers, and particularly by its long, strongly unequal, linear-triangular calyx lobes usually distally falcate. Leaf bases typically are truncate to rounded or subcordate. Rarely the leaves are short-petiolate, but in such cases, the distinctive leaf bases, vesture, calyx morphology, and pubescent anthers are diagnostic. *Erythranthe ptilota* is widely sympatric with *E. moschata* but usually occurs at lower elevations and characteristically in wetter habitats. The epithet *ptilota* (Greek *ptilotos*, winged) alludes to a fancied winglike aspect of the pairs of sessile leaves.

A population system of *Erythranthe ptilota*-like plants occurs in southern California, about 480 km disjunct from the main range of the species. These plants have the prostrate habit, large leaves, long pedicels, and large corollas of *E. ptilota*, but the calyx lobes are variable in length and usually do not show the characteristic attenuate-apiculate apices. The southern California plants are identified here as *E. moschata*.

Erythranthe ptilota is a new name at specific rank for *Mimulus moschatus* var. *sessilifolius* [not *E. sessilifolia* (Maximowicz) G. L. Nesom].

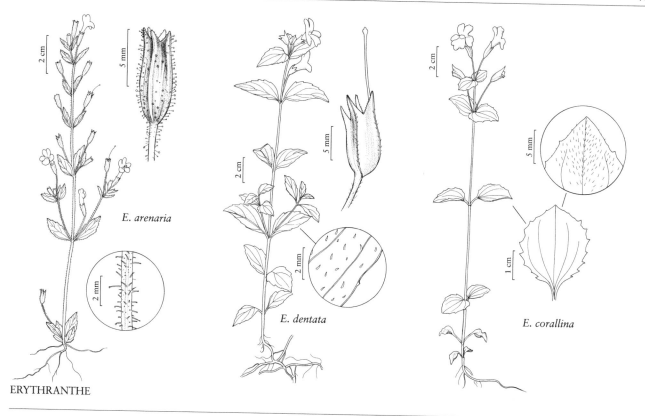

E. arenaria

E. dentata

E. corallina

ERYTHRANTHE

48. Erythranthe arenaria (A. L. Grant) G. L. Nesom, Phytoneuron 2012-39: 38. 2012 • Sand-loving monkeyflower E F

Mimulus arenarius A. L. Grant, Ann. Missouri Bot. Gard. 11: 215. 1925; *M. floribundus* Douglas ex Lindley var. *subulatus* A. L. Grant; *M. multiflorus* Pennell; *M. subulatus* (A. L. Grant) Pennell; *M. trisulcatus* Pennell

Annuals, fibrous-rooted or filiform-taprooted. **Stems** erect to ascending, straight or geniculate at nodes, simple or branched, 5–20 cm, villous-glandular, hairs 0.2–0.8 mm, gland-tipped. **Leaves** basal and cauline; petiole 0 mm or proximals 1–3(–5) mm; blade 1-veined or palmately 3-veined, elliptic to narrowly elliptic, ovate-elliptic, or ovate-lanceolate, 5–12(–17) × 3–7 mm, base rounded to cuneate-attenuate, margins entire or sparsely dentate to serrate, apex acuminate to acute or obtuse, surfaces villous-glandular, hairs 0.2–0.8 mm, gland-tipped. **Flowers** herkogamous, 1–22, from proximal to distal nodes. **Fruiting pedicels** divergent-arcuate, 10–23 mm, villous-glandular, hairs 0.2–0.8 mm, gland-tipped. **Fruiting calyces** usually red-dotted, narrowly campanulate, not or weakly inflated, 5–7(–9) mm, margins distinctly toothed or lobed, villous-glandular, ribs angled, lobes pronounced, erect. **Corollas** yellow, abaxial limb red-dotted, bilaterally symmetric, weakly bilabiate; tube-throat funnelform, 9–12(–14) mm, exserted beyond calyx margin; lobes broadly obovate, apex rounded. **Styles** glabrous. **Anthers** included, glabrous. **Capsules** included, 4–7 mm. $2n = 32$.

Flowering May–Sep. Sandy flats, bars, gullies, washes, trails, roadcuts, seasonal creek beds and drainages, rocky slopes, seepy loam, ditches, lake edges, meadows, openings in pine-fir and pine-oak woodlands; (100–)500–2600(–2800) m; Calif.

Erythranthe arenaria is known from a cluster of six counties of the central Sierra Nevada: Fresno, Madera, Mariposa, Merced, Tulare, and Tuolumne.

Most plants of *Erythranthe arenaria* have relatively even-sized cauline leaves, all sessile to proximally subsessile. Plants in the Yosemite area with persistent basal leaves that are short-petiolate, ovate with a cuneate base, and relatively larger than the more distal cauline ones, and possibly related to *E. arenaria*, have been named *Mimulus floribundus* var. *subulatus*. These might be construed as showing the influence of *E. geniculata*, but that species occurs only at the lower range of elevation of *E. arenaria*, while plants referable to *M. floribundus* var. *subulatus* occur at least to 2300 m and also have the erect habit characteristic of *E. arenaria*. These variants should be investigated, especially in the Yosemite area where they appear to be relatively common, with the possibility that they indeed represent a distinct entity.

49. **Erythranthe floribunda** (Lindley) G. L. Nesom, Phytoneuron 2012-39: 38. 2012 • Many-flowered monkeyflower

Mimulus floribundus Lindley, Bot. Reg. 13: plate 1125. 1828; *M. deltoideus* Gandoger; *M. floribundus* var. *membranaceus* (A. Nelson) A. L. Grant; *M. membranaceus* A. Nelson; *M. peduncularis* Douglas ex Bentham; *M. serotinus* Suksdorf

Annuals, fibrous-rooted or filiform-taprooted. **Stems** erect to decumbent, sometimes procumbent-trailing, straight or geniculate at nodes, simple or many-branched, 3–22(–40) cm, villous-glandular, hairs greatly variable in length and density, gland-tipped, sometimes 0.2–0.5 mm, sparsely stipitate-glandular. **Leaves** cauline, basal mostly deciduous by flowering; petiole 1–12 mm; blade pinnately to subpalmately veined, ovate, (3–)8–25(–35) × (1–)5–18 (–26) mm, base cuneate to truncate or cordate, margins serrate to sparsely dentate, apex acute, surfaces villous-glandular, hairs greatly variable in length and density, gland-tipped, sometimes 0.2–0.5 mm, sparsely stipitate-glandular. **Flowers** plesiogamous, 1–20, from proximal to distal nodes. **Fruiting pedicels** 5–20(–26) mm, villous-glandular, hairs greatly variable in length and density, gland-tipped, sometimes 0.2–0.5 mm, sparsely stipitate-glandular. **Fruiting calyces** greenish or purplish to red-dotted, cylindric, ± inflated, 4–7 mm, margins distinctly toothed or lobed, villous-glandular, lobes pronounced, erect. **Corollas** yellow, abaxial limb red-dotted, bilaterally symmetric, weakly bilabiate; tube-throat funnelform-cylindric, (4–)5–10 mm, exserted slightly beyond calyx margin or not; limb expanded 3–4 mm diam., lobes usually oblong, apex notched. **Styles** glabrous. **Anthers** included, glabrous. **Capsules** included, 4–7 mm. 2*n* = 32.

Flowering (May–)Jun–Aug(–Sep). Under overhangs, moist roofs of caves, wet rock crevices, cliff faces, wet cliff bases, below waterfalls, seeps, springs, humus and moist soils over rocks and slabs, moist slopes, ditches and pond edges, wet edges of creeks and rivers, drying mud on margins of wetland depressions, creek beds, wet or swampy meadows, along trails, in lodgepole pine, ponderosa pine, ponderosa pine-Douglas fir, and spruce-fir woodlands; (100–)1800–2600(–3100) m; Alta., B.C.; Ariz., Ark., Calif., Colo., Idaho, Mont., Nev., N.Mex., Oreg., S.Dak., Utah, Wash., Wyo.; Mexico (Baja California, Baja California Sur, Chihuahua, Sinaloa, Sonora).

Some plants identified here as *Erythranthe floribunda* in Arizona and southwestern New Mexico are distinctive in their prominently inflated calyces, sessile to subsessile leaves with attenuate bases and palmately three- to five-veined venation, and much-elongated pedicels (20–43 mm); numerous intermediates in Arizona make it difficult to conclude that the variants represent an entity discontinuous from plants of typical morphology. The variant morphology has not been observed among Mexican populations. Further discussion of this situation was given by G. L. Nesom (2012h).

Erythranthe floribunda has been documented from 12 counties in northern Arkansas (Carroll, Cleburne, Crawford, Franklin, Izard, Johnson, Logan, Newton, Pope, Searcy, Stone, and Washington), where it occurs at 300–500 m. The unpublished name *Mimulus floribundus* subsp. *moorei* Iltis appears in various checklists in reference to the Arkansas plants, but there appears to be no basis for treating them as distinct from the rest of the species. Elsewhere in the main range (western states), scattered variants extremely reduced in size, leaves, flowers, and overall stature appear to be at the lower limits of the species rather than taxonomically distinct.

50. **Erythranthe geniculata** (Greene) G. L. Nesom, Phytoneuron 2012-39: 38. 2012 • Bent-stem or Dudley's monkeyflower E

Mimulus geniculatus Greene, Bull. Calif. Acad. Sci. 1: 280. 1885; *M. dudleyi* A. L. Grant; *M. floribundus* Douglas ex Lindley var. *geniculatus* (Greene) A. L. Grant

Annuals, fibrous-rooted or filiform-taprooted. **Stems** ascending to decumbent or prostrate, geniculate at nodes, simple or diffusely branched, 5–60 cm, moderately villous, hairs 0.8–2 mm, multicellular, eglandular and also 0.1–0.3 mm, stipitate-glandular. **Leaves** basal and cauline, basal usually deciduous by flowering; petiole 2–10(–35) mm; blade pinnately to subpinnately veined, broadly ovate or elliptic-ovate to triangular, 8–35 × 5–30 mm, base cuneate to rounded or subcordate, margins serrate or dentate, teeth 3–10 per side, apex acute to obtuse or rounded, surfaces moderately villous, hairs 0.8–2 mm, multicellular, eglandular and also 0.1–0.3 mm, stipitate-glandular. **Flowers** herkogamous, (1–)6–20, from all or medial to distal nodes. **Fruiting pedicels** 12–26 (–55) mm, moderately villous, hairs 0.8–2 mm, multicellular, eglandular and also 0.1–0.3 mm, stipitate-glandular. **Fruiting calyces** red-spotted, campanulate-cylindric, weakly inflated, (5–)6–8 mm, margins distinctly toothed or lobed, sparsely to moderately villous-glandular, ribs shallowly wing-angled, lobes pronounced, erect to spreading or spreading-recurving. **Corollas** yellow, without white patches, throat red-spotted, spots concentrated or becoming coalescent into a somewhat discrete splotch at base of each of 3 abaxial lobes and sometimes 2 adaxial, bilaterally symmetric, ± bilabiate; tube-throat cylindric, 9–12 mm, exserted beyond calyx margin; limb expanded 10–18 mm diam. **Styles** glabrous. **Anthers** included, glabrous. **Capsules** included, 4–6(–7) mm. **2n** = 32.

Flowering (Mar–)Apr–Jul. Granite crevices, canyon slopes, talus, crevices in volcanic outcrops, edges of boulders, roadsides, damp sandy soils, sandy water edges, gravelly soils and creek bottoms; 200–900 (–1200) m; Calif.

Erythranthe geniculata is known from an apparently disjunct cluster of populations in Butte, Sutter, and Yuba counties and then from Tuolumne and Stanislaus counties south to Kern County.

Erythranthe geniculata, compared to *E. floribunda*, has larger, chasmogamous, and allogamous flowers. The anther pairs of *E. geniculata* are at different levels, and the stigma is slightly above the adaxial anther pair; in *E. floribunda* both anther pairs and the stigma are at the same level.

Erythranthe arenaria, *E. geniculata*, and *E. norrisii* constitute a group of apparently closely related species within sect. *Mimulosma* endemic along the Sierra Nevada. All have ovate-petiolate leaves (only the basal ones are sometimes ovate in *E. arenaria*) with pinnate to subpinnate venation. The more widespread *E. floribunda*, which is part of the above group, also is similar, but all three endemics have larger corollas with the tube exserted at greater length beyond the calyx margin.

51. Erythranthe norrisii (Heckard & Shevock)
G. L. Nesom, Phytoneuron 2012-39: 39. 2012

• Norris's monkeyflower C E

Mimulus norrisii Heckard & Shevock, Madroño 32: 179, figs. 1, 2. 1985

Annuals, fibrous-rooted or filiform-taprooted. **Stems** ascending to erect-ascending, geniculate at nodes, usually branched from proximal nodes, 2–15(–25) cm, villous-glandular. **Leaves** basal and cauline; petiole 5–10(–15) mm; blade palmately 3–5-veined, sometimes with 1–3 distal vein pairs diverging pinnately, elliptic to elliptic-obovate, 20–35 × 10–20 mm, base usually attenuate, margins subentire to distally denticulate, apex acute to obtuse, surfaces villous-glandular. **Flowers** herkogamous, 1–5, from medial to distal nodes. **Fruiting pedicels** 20–35(–50) mm, villous-glandular. **Fruiting calyces** red-dotted, campanulate, weakly inflated, 4–6 mm, margins distinctly toothed or lobed, villous-glandular, ribs rounded-thickened, lobes pronounced, erect, often incurved, linear-oblong to oblong-lanceolate,

apex rounded to blunt. **Corollas** yellow, base of each lobe with a prominent maroon splotch, abaxial limb with white patch at 2 sinus bases, weakly bilaterally or radially symmetric, weakly bilabiate or regular; tube-throat cylindric-funnelform, 12–16 mm, exserted beyond calyx margin; limb expanded 15–30 mm, lobes oblong-obovate to orbicular-obovate, apex rounded-truncate. **Styles** glabrous. **Anthers** included, glabrous. **Capsules** usually slightly exserted, 4–6 mm. **2n** = 32.

Flowering Mar–May. Steep marble outcrops in soil pockets, moss covered marble and quartzite ledges, cracks, fractures, weathered faces, chamise chaparral or blue oak woodlands; of conservation concern; 300–1300 m; Calif.

Erythranthe norrisii is known only from the Kaweah River drainage; most populations are in Sequoia National Park in Tulare County. The species is characterized by its short-petiolate leaves with attenuate bases, very large corollas with red splotches at the base of each lobe and two white patches on the abaxial limb, and very short, purple-dotted calyces with rounded-thickened ribs and linear-oblong lobes incurved in fruit. The capsules often are slightly exserted.

52. Erythranthe dentata (Nuttall ex Bentham)
G. L. Nesom, Phytoneuron 2012-39: 41. 2012

• Coastal or tooth-leaf monkeyflower E F

Mimulus dentatus Nuttall ex Bentham in A. P. de Candolle and A. L. P. P. de Candolle, Prodr. 10: 372. 1846

Perennials, rhizomatous. **Stems** erect to erect-ascending, simple or few-branched, 15–40 cm, coarsely pilose to hirsute-pilose, glabrescent, internodes evident. **Leaves** cauline; petiole (0–)2–12 mm, not winged, distally sometimes sessile, subclasping; blade pinnately veined, lanceolate to ovate-lanceolate or elliptic-ovate, 25–75 mm, thick, base rounded to cuneate, margins coarsely dentate to serrate, apex acute to obtuse, surfaces coarsely pilose to hirsute-pilose, glabrescent. **Flowers** herkogamous, 1–5, from distal nodes. **Fruiting pedicels** 12–25(–50) mm. **Fruiting calyces** narrowly campanulate, not or weakly inflated, 9–14 mm, villous-hirsute on ribs. **Corollas** yellow, palate and throat brown to reddish brown-spotted, bilaterally symmetric, ± bilabiate; tube-throat funnelform, 15–26 mm, exserted beyond calyx margin; throat open, palate villous, abaxial ridges low. **Styles** glabrous. **Anthers** included, short villous-hirsute. **Capsules** included, 8–9 mm.

Flowering May–Aug. Stream banks; 20–400 m; B.C.; Calif., Oreg., Wash.

Erythranthe dentata is distinctive and rarely misidentified.

53. Erythranthe exigua (A. Gray) G. L. Nesom &
N. S. Fraga, Phytoneuron 2012-39: 42. 2012
· San Bernardino Mountains monkeyflower C

Mimulus exiguus A. Gray, Proc.
Amer. Acad. Arts 20: 307. 1885

Annuals, taprooted. **Stems**
erect, simple, sometimes
branched near base, 2–10 cm,
minutely stipitate-glandular.
Leaves cauline; petiole 0 mm;
blade 1-veined, obovate-oblong
to narrowly elliptic, ovate, or
narrowly ovate, 3–6 mm, base rounded to truncate
or cuneate, margins entire or shallowly dentate, apex
rounded, surfaces minutely stipitate-glandular. **Flowers**
plesiogamous, (1 or)2–6, from distal or medial to distal
nodes. **Fruiting pedicels** divergent-spreading, 15–20
mm, minutely stipitate-glandular. **Fruiting calyces**
campanulate, 2–2.5 mm, minutely stipitate-glandular.
Corollas light lavender to purple, abaxial lobe and
palate ridges with yellow patches, bilaterally symmetric,
± bilabiate; tube-throat narrowly funnelform-cylindric,
1.5–2.5 mm, exserted 0.5 mm beyond calyx margin;
lobes spreading. **Styles** glabrous. **Stigmas** persistent in
fruit. **Anthers** included, glabrous. **Capsules** distinctly
exserted, 3–4 mm.

Flowering Jun–Jul. Gentle slopes, along small
streams, vernal creeks, pebble plains, openings in Jeffrey
pine-juniper forests, runoff areas, vernal depressions,
roadsides; of conservation concern; 1800–2400
(–2600) m; Calif.; Mexico (Baja California).

Plants of *Erythranthe exigua* are diminutive annuals
with few nodes and greatly reduced leaves, corollas,
and calyces, wide spreading pedicels, and lavender
flowers with small but bilabiate limbs. The species is
known only from the San Bernardino Mountains of San
Bernardino County and in adjacent Mexico.

54. Erythranthe geyeri (Torrey) G. L. Nesom,
Phytoneuron 2012-39: 43. 2012 · Geyer's
monkeyflower, mimule de James

Mimulus geyeri Torrey in
J. N. Nicollet, Rep. Ill. Map
Hydrogr. Basin Upper Mississippi,
157. 1843; *M. glabratus* Kunth
var. *fremontii* (Bentham)
A. L. Grant; *M. glabratus* var.
jamesii (Torrey & A. Gray ex
Bentham) A. Gray; *M. glabratus*
var. *oklahomensis* Fassett;
M. jamesii Torrey & A. Gray ex Bentham; *M. jamesii* var.
fremontii Bentham

Perennials, rhizomatous, rooting at nodes. **Stems**
decumbent-ascending to ascending or erect-ascending,
branched, (3–)10–40 cm, glabrous. **Leaves** cauline,
basal not persistent; petiole 3–10(–20) mm or 0 mm

distally; blade palmately 3–5-veined, suborbicular to
depressed-ovate or broadly elliptic-ovate to reniform,
6–25 mm, relatively even-sized or largest often at
mid stem, bracteal reduced, base cuneate to truncate
or subcordate, margins shallowly dentate to crenate-
dentate, teeth 3–7(–10) per side, apex rounded, adaxial
surface of distals sparsely short villous-glandular
or glabrous. **Flowers** plesiogamous, 2–8(–12), from
distal nodes, sometimes from most nodes, very loosely
racemose. **Fruiting pedicels** 18–30 mm, sparsely
short villous-glandular or glabrous. **Fruiting calyces**
obtriangular to broadly obtriangular or deeply cupulate,
inflated, sagittally compressed, (7–)8–12 mm, sparsely
short villous-glandular or glabrous, throat not closing,
lateral lobes shallowly convex-mucronulate, adaxial
ovate with apex rounded. **Corollas** yellow, sparsely red-
dotted or not, bilaterally symmetric, weakly bilabiate;
tube-throat cylindric-funnelform, 6–8 mm, exserted
1–3 mm beyond calyx margin; limb expanded 5–8 mm.
Styles glabrous. **Anthers** included, glabrous. **Capsules**
included, (4.5–)5–8 mm. **2***n* = 30.

Flowering May–Aug(–Oct). Edges of flowing streams,
marsh edges, drainage ditches, seepage areas, springs,
muddy or moist banks; 200–2500 m; Alta., Man., Ont.,
Que., Sask.; Ariz., Colo., Ill., Iowa, Kans., Mich., Minn.,
Mo., Nebr., N.Mex., Okla., Pa., S.Dak., Tex., Wis.,
Wyo.; Mexico (Chihuahua, Coahuila, Distrito Federal,
Durango, Hidalgo, Nuevo León, México, Querétaro,
San Luis Potosí, Sonora, Veracruz, Zacatecas).

Erythranthe geyeri has commonly been regarded as
conspecific with *E. glabrata* (Kunth) G. L. Nesom (as
Mimulus glabratus var. *jamesii*), but typical *E. glabrata*
has a different chromosome number and distinct
morphology and its range does not reach the United
States. In Mexico, the two species are broadly sympatric
without intermediates. An allozyme study of the
M. glabratus complex (R. K. Vickery 1990) indicated
that the Great Plains populations of *E. geyeri* are distinct
from those in New Mexico and Mexico, corresponding
to a difference in pedicel vesture.

55. Erythranthe regni G. L. Nesom, Phytoneuron
2012-40: 24. 2012 · King of Arizona monkeyflower
C E

Annuals, fibrous-rooted, sometimes rooting at proximal nodes.
Stems erect to ascending-erect,
branched, sometimes becoming
slightly fistulose, 15–45 cm,
glabrous. **Leaves** basal and
cauline; petiole: proximals
5–25(–30) mm, mid cauline and
distals not connate, 0 mm; blade
palmately 5–7-veined, proximal sometimes subpinnate,
proximals ovate to depressed-orbicular, 15–20(–50) ×
15–25(–50) mm, medials and distals broadly depressed-
ovate to obtriangular or flabellate, 15–35 mm, largest

basal or at mid stem with distal slightly reduced, base attenuate-cuneate, margins shallowly serrate-dentate, sometimes irregularly, to mucronulate or apiculate, teeth (3–)5–7 per side, rarely subentire, apex rounded, surfaces glabrous. **Flowers** plesiogamous, 6–16, from all nodes or medial to distal, cleistogamous. **Fruiting pedicels** 15–30 mm, longer than subtending leaves, glabrous. **Fruiting calyces** sparsely purple-dotted, broadly campanulate-cylindric, inflated, sagittally compressed, 7–9 mm, glabrous, throat not closing, adaxial lobe longest. **Corollas** yellow, not red-dotted, bilaterally or radially symmetric, bilabiate or regular; tube-throat cylindric-funnelform, 9–12 mm, exserted 3–5 mm beyond calyx margin; limb expanded 1–1.5 mm. **Styles** glabrous. **Anthers** included, glabrous. **Capsules** included, 4–5 mm.

Flowering Mar–May. Moist to wet, sandy loam soils; of conservation concern; 800–1000 m; Ariz.

Erythranthe regni is endemic to the Kofa Mountains of Yuma County; all collections have been made from the Kofa Game Refuge (Kofa National Wildlife Refuge). Because its calyces remain open at maturity, this species is hypothesized to be most closely related to *E. geyeri*, from which it differs by its erect habit, apparently annual duration, larger leaves, purple-dotted calyces, and corollas with longer tube-throat and barely bilabiate limb. Geography and other morphology, however, suggest that its evolutionary origins are closer to *E. guttata*.

56. **Erythranthe inamoena** (Greene) G. L. Nesom, Phytoneuron 2012-39: 44. 2012 • Texas monkeyflower

Mimulus inamoenus Greene, Pittonia 5: 137. 1903; *M. jamesii* (Torrey & A. Gray ex Bentham) A. Gray var. *texensis* A. Gray

Annuals, fibrous-rooted, rooting at proximal nodes, sometimes forming matlike colonies. **Stems** decumbent to decumbent-ascending basally, becoming fully erect at least in inflorescence, simple, sometimes few-branched from proximal nodes, usually distinctly fistulose, 10–30 cm, glabrous. **Leaves** basal and cauline, basal sometimes deciduous by flowering; petiole: basal and proximals to mid cauline 10–70 mm, distals 0 mm; blade palmately veined, ovate to broadly ovate or elliptic-ovate, becoming subreniform distally, (5–)15–35(–60) mm, distal closely paired, auriculate-subclasping, base truncate to subcordate, margins dentate-serrate to shallowly dentate, teeth 5–11 per side, apex obtuse to rounded, surfaces glabrous. **Flowers** plesiogamous, (6–)8–18(–24), often produced from all nodes, loosely to densely racemose, chasmogamous.

Fruiting pedicels 9–20 mm, glabrous, rarely sparsely stipitate-glandular. **Fruiting calyces** purple-spotted, sometimes greenish, broadly cylindric-campanulate, inflated, sagittally compressed, 7–11 mm, glabrous, rarely minutely scabrous-hirtellous or sparsely stipitate-glandular, throat not closing, abaxial lobe slightly upcurving 10–45°, spreading 45°, or sometimes deflexed 40°. **Corollas** yellow, red-spotted, bilaterally symmetric, bilabiate; tube-throat cylindric, 7–11 mm, exserted (1–)2–3(–4) mm beyond calyx margin; limb slightly expanded. **Styles** glabrous. **Anthers** included, glabrous. **Capsules** included, 4.5–6 mm. **2***n* = 60.

Flowering Jan–Apr(–Sep). Edge of seeps and creeks, mud or gravel, shallow running water, wet crevices, canyon drainages; 100–2400 m; Tex.; Mexico (Chihuahua, Coahuila).

Erythranthe inamoena is distinctive in its completely glabrous herbage, small corollas, flowers in racemes mostly at distal nodes and with reduced bracts, short and open-throated fruiting calyces, erect and fistulose stems, and apparent annual duration (fibrous-rooted but usually rooting at proximal cauline nodes).

Presumably because of its autogamous reproduction, *Erythranthe inamoena* has been confused with *E. cordata*, especially in the trans-Pecos region of Texas, where the two are sympatric. In contrast to *E. cordata*, *E. inamoena* usually has glabrous pedicels and calyces, flowers often produced from all nodes, smaller calyces [(7–)8–11 mm] that do not close at maturity, and corollas with a shorter tube-throat (7–11 mm).

Some populations in Brewster, Presidio, and Val Verde counties are identified here as *Erythranthe inamoena* (based on proximal-to-distal distribution of flowers and the short mature calyces with open throat) but have sparsely stipitate-glandular pedicels and calyces.

57. **Erythranthe michiganensis** (Pennell) G. L. Nesom, Phytoneuron 2012-39: 44. 2012 • Michigan monkeyflower C E

Mimulus glabratus Kunth subsp. *michiganensis* Pennell, Monogr. Acad. Nat. Sci. Philadelphia 1: 119. 1935; *M. glabratus* var. *michiganensis* (Pennell) Fassett; *M. michiganensis* (Pennell) Posto & Prather

Perennials, rhizomatous, usually producing numerous leafy stolons from basal nodes, rooting at distal nodes, sometimes forming mats. **Stems** ascending-erect or basally decumbent, becoming erect in inflorescence, branched, 12–50(–70) cm, glabrous or minutely hirtellous and stipitate-glandular. **Leaves** cauline, basal not persistent; petiole 1–5(–15) mm, mid stem and distals 0 mm; blade palmately 3–5-veined, broadly ovate to broadly ovate-elliptic or suborbicular, 8–30 mm, relatively even-sized or diminishing in size distally, bracteal reduced, slightly falcate, base truncate to cuneate, margins evenly or unevenly dentate-serrate to dentate, teeth 3–8 per side, apex usually rounded, surfaces glabrous or minutely hirtellous and stipitate-glandular. **Flowers** herkogamous, 2–14, mostly from distal nodes or medial to distal nodes. **Fruiting pedicels** 10–25 mm, villous-glandular to minutely villosulous-glandular. **Fruiting calyces** cylindric-campanulate, inflated, sagittally compressed, 7–10 mm, puberulous to softly hirtellous, mixed with longer, stipitate-glandular hairs, throat not closing, adaxial lobe 2 times longer than others, slightly upcurving. **Corollas** yellow, sometimes faintly red-spotted, bilaterally symmetric, strongly bilabiate; tube-throat cylindric-campanulate, 10–14 mm, exserted 5–8 mm beyond calyx margin; limb expanded 10–15 mm. **Styles** glabrous. **Anthers** included, glabrous. **Capsules** unknown. $2n$ = (28)30.

Flowering Jun–Aug(–Oct). Cold calcareous springs, seeps, depressions, streams, alkaline shorelines at mouths of small drainages, steep moraine slopes, bluff bases, commonly in northern white cedar swamps; of conservation concern; 500–600 m; Mich.

Based on data from allozyme and RAPD studies, morphology, and crossing studies, A. L. Posto and L. A. Prather (2003) raised *Erythranthe* [*Mimulus*] *michiganensis* to specific rank. The species is endemic to a small area in the Mackinac Straits and Grand Traverse regions of Michigan; it is known to be extant at 15 sites and apparently is extinct at three previously known sites. Plants of all but one of the populations are essentially pollen-sterile and reproduce through rhizomes. However, the didynamous stamens and stigma positioned above the adaxial anther pair, along with the relatively large corollas with broadly expanded limb, are reflective of an allogamous breeding system, which seems unusual in view of its apparent sterility. The narrow geographic distribution of *E. michiganensis* lies within the wider range of *E. geyeri*, and the two are known to co-occur at two sites, apparently without hybridization or morphologically intermediate individuals.

Erythranthe michiganensis (as *Mimulus michiganensis*) is in the Center for Plant Conservation's National Collection of Endangered Plants.

SELECTED REFERENCE Posto, A. L. and L. A. Prather. 2003. The evolutionary and taxonomic implications of RAPD data on the genetic relationships of *Mimulus michiganensis* (comb. et stat. nov.: Scrophulariaceae). Syst. Bo t. 28: 172–178.

58. Erythranthe tilingii (Regel) G. L. Nesom, Phytoneuron 2012-39: 44. 2012 • Tiling's monkeyflower E

Mimulus tilingii Regel, Gartenflora 18: 321, plate 631. 1869 (as tilingi); *M. caespitosus* (Greene) Greene var. *implexus* (Greene) M. Peck; *M. implexus* Greene; *M. implicatus* Greene; *M. langsdorffii* Donn ex Greene var. *tilingii* (Regel) Greene; *M. lucens* Greene; *M. veronicifolius* Greene

Perennials, rhizomatous, solitary to weakly colonial, rhizomes forming a mass, yellowish, branching, filiform. **Stems** erect-ascending, usually freely branched, 2–35 cm, glabrous or sparsely stipitate-glandular to short glandular-villous. **Leaves** cauline; petiole 0–25 mm, distals 0 mm; blade palmately 3–5-veined, ovate to lanceolate-triangular or narrowly lanceolate (broadly ovate in large-leaved forms), 5–35(–55) mm, base cuneate to attenuate, margins irregularly denticulate, apex acute to obtuse or rounded, surfaces glabrous, sparsely stipitate-glandular to short glandular-villous, glabrate, or sparsely to moderately villous, hairs thick-vitreous, eglandular. **Flowers** herkogamous, 1–3(–5), from distal nodes. **Fruiting pedicels** 15–35(–40) mm, sparsely stipitate-glandular to short glandular-villous. **Fruiting calyces** usually purple-tinged and purple-dotted, broadly campanulate, inflated, sagittally compressed, 11–15 mm, glabrous or sparsely stipitate-glandular to short glandular-villous, villous at sinuses, throat closing, lobes broadly ovate, abaxial usually longer than lateral, adaxial at least 2 times as long as others. **Corollas** yellow, red-dotted, bilaterally symmetric, bilabiate; tube-throat broadly funnelform, 15–28 mm, exserted 5–10 mm beyond calyx margin; limb expanded 14–30 mm. **Styles** hirtellous. **Anthers** included, glabrous. **Capsules** included, 5–7 mm. $2n$ = 28, 56.

Flowering Jul–Sep. Seeps, springs, stream banks, shallow rivulets, cliff bases, ledges and crevices, steep gravelly slopes, wet meadows; 1400–3400 m; Alta.; Ariz., Calif., Idaho, Mont., Nev., Oreg., Utah.

Plants of *Erythranthe tilingii* are characterized by their relatively low stature and stems arising from a system of thin rhizomes and producing mostly one to three large flowers each; they usually occur at relatively high elevations. *Erythranthe tilingii* sometimes has been considered to include one or several infraspecific entities; from within this taxonomic amalgam, four distinct species are recognized here: *E. caespitosa*, *E. corallina*, *E. minor*, and *E. tilingii*. *Euphrasia corallina* and *E. minor* probably are more closely related to *E. guttata*. The populations identified here as *E. tilingii* from northeastern Oregon northeast to Alberta and southeast to Utah may prove to be a separate (undescribed) species.

Erythranthe tilingii in the strict sense is relatively widespread over the western United States and is sympatric with *E. caespitosa* and *E. corallina*. Leaves in *E. tilingii* are variable in size, and particularly in Idaho, they may approach the small size of those of *E. caespitosa*, but the leaf margins of *E. tilingii* are distinctly toothed, and the stems are taller and more erect. Across the range of the species, plants sometimes produce very large leaves, but these often occur on plants with characteristically smaller leaves. This wide variability in size apparently does not occur in *E. caespitosa*.

59. **Erythranthe caespitosa** (Greene) G. L. Nesom, Phytoneuron 2012-39: 43. 2012 • Olympic monkeyflower E

Mimulus scouleri Hooker var. *caespitosus* Greene, Pittonia 2: 22. 1889; *M. caespitosus* (Greene) Greene; *M. tilingii* Regel var. *caespitosus* (Greene) A. L. Grant

Perennials, rhizomatous, rooting at proximal nodes, sometimes producing creeping, small-leaved runners, forming matted colonies, rhizomes filiform. **Stems** procumbent or decumbent to decumbent-ascending, delicate, usually in masses, terete or flattish, branched, 3–10 cm, glabrous, minutely hirtellous, or stipitate-glandular. **Leaves** basal and cauline; petiole: proximals 2–5 mm, distals 0 mm; blade often purple beneath, palmately 3-veined, orbicular to narrowly elliptic or ovate, proximals usually sublyrate, 3–12 mm, becoming larger distally, base cuneate to a short petiole, margins entire, mucronulate, or barely denticulate, apex obtuse, surfaces sparsely to moderately puberulent, hairs minute, stipitate-glandular. **Flowers** herkogamous, 1–3, from distal nodes, commonly solitary. **Fruiting pedicels** 10–30(–40) mm, sparsely to moderately villous, hairs short, gland-tipped, sometimes hirtellous. **Fruiting calyces** broadly campanulate, inflated, sagittally compressed, 7–15 mm, glabrous, minutely hirtellous, or stipitate-glandular, throat closing, proximalmost lobe pair upcurving, distalmost 3–5 mm, prominently protruding. **Corollas** yellow, dark red-spotted, bilaterally symmetric, bilabiate; tube-throat broadly funnelform to cylindric-funnelform, 15–18 mm, exserted beyond calyx margin; abaxial limb with deflexed-spreading lobes, adaxial with ascending lobes, palate partially closed. **Styles** minutely hirtellous. **Anthers** included, glabrous. **Capsules** included, 4–5 mm.

Flowering Jul–Sep. Alpine meadows and slopes, stream banks, wet rocks in streams, wet crevices, talus; 1100–2000(–2300) m; B.C.; Wash.

Erythranthe caespitosa is endemic to northwestern and central Washington (Cascade and Olympic mountains) and adjacent British Columbia (Cascades, Selkirk Mountains and Chilliwack Valley, Coast Mountains). The plants have consistently small leaves with subentire margins, and the stems are consistently procumbent to decumbent-ascending, usually forming matted colonies. *Erythranthe caespitosa* and *E. tilingii* appear to be sympatric in counties of northwestern Washington, but this needs to be verified in the field.

60. **Erythranthe minor** (A. Nelson) G. L. Nesom, Phytoneuron 2012-39: 44. 2012 • Colorado monkeyflower E

Mimulus minor A. Nelson, Proc. Biol. Soc. Wash. 17: 178. 1904; *M. alpinus* (Piper) Piper; *M. langsdorffii* Donn ex Greene var. *alpinus* Piper; *M. langsdorffii* var. *minor* (A. Nelson) Cockerell

Perennials, rhizomatous, colonial, rhizomes forming a mass, branching, filiform. **Stems** erect to erect-ascending, branched, 5–20 cm, densely minutely hirtellous and eglandular or with a mixture of hirtellous and gland-tipped hairs. **Leaves** basal and cauline; petiole 0 mm or proximals 1–3 mm; blade palmately 3-veined, broadly ovate to elliptic-ovate or lanceolate, 8–25 × 5–15 mm, base cuneate to truncate, margins shallowly dentate to denticulate, apex acute to obtuse, surfaces glabrous. **Flowers** herkogamous, 1–3, from distal nodes. **Fruiting pedicels** 10–20 mm, densely minutely hirtellous and eglandular or with a mixture of hirtellous and gland-tipped hairs. **Fruiting calyces** nodding 80–100°, not purple-dotted, cylindric-campanulate, inflated, sagittally compressed, 10–13 mm, densely minutely hirtellous and eglandular or with a mixture of hirtellous and gland-tipped hairs, throat closing. **Corollas** yellow, not red-dotted, bilaterally symmetric, bilabiate; tube-throat tubular-funnelform, 9–11 mm, exserted 0–1(–2) mm beyond calyx margin. **Styles** sparsely hirtellous. **Anthers** included, glabrous. **Capsules** included, 5–8 mm.

Flowering Jul–Aug(–Sep). Stream and lake edges, intermittent subalpine water courses, roadside ditches, subalpine to alpine; 3000–3700 m; Colo., N.Mex.

The corollas of *Erythranthe minor* are shorter than those of typical *E. tilingii*, and the two species are allopatric. Corollas of *E. tilingii* rarely may be equally as short as those of *E. minor* but are produced in scattered localities on plants that are depauperate in other ways. The range of *E. minor* is primarily in Colorado apparently extending southward into the Wheeler Peak area of Taos County, New Mexico. Attribution of its range into the La Sal Mountains of east-central Utah has been based on misidentifications of *E. guttata*; the distinction between *E. guttata* and *E. minor* in Colorado also needs clarification.

Mimulus luteus Linnaeus var. *alpinus* A. Gray (1863, the type from Colorado) is an illegitimate name for *Erythranthe minor*, preceded by *M. luteus* var. *alpinus* Lindley (1827).

61. **Erythranthe corallina** (Greene) G. L. Nesom, Phytoneuron 2012-39: 44. 2012 • Coralline monkeyflower E F

Mimulus corallinus Greene, Erythea 4: 21. 1896; *M. minusculus* Greene; *M. tilingii* Regel var. *corallinus* (Greene) A. L. Grant

Perennials, rhizomatous, rhizomes often forming a mass, usually branching, filiform. **Stems** usually erect to ascending-erect, few-branched, 6–25(–38) cm, moderately hirsute to hirtellous, hairs deflexed. **Leaves** basal and cauline, becoming larger distally or even-sized; petiole 0 mm or proximals 1–15 mm; blade palmately 5-veined, ovate to broadly ovate, 15–45 mm, base mostly truncate to shallowly cordate, margins sharply dentate-serrate, apex obtuse, surfaces hirtellous to softly hirsute, hairs ascending, straight, dull gray, sharp-pointed, thick-walled, eglandular. **Flowers** herkogamous, 1–3(–6), commonly solitary or from distal nodes. **Fruiting pedicels** (10–)25–75 mm, glabrous or puberulent proximally, hairs stipitate-glandular. **Fruiting calyces** sometimes purple-spotted, broadly cylindric-campanulate, inflated, sagittally compressed, 11–15 mm, glabrous, throat not closing, proximal lobe pair slightly upcurving. **Corollas** yellow, red-spotted, bilaterally symmetric, bilabiate; tube-throat narrowly funnelform to broadly cylindric, 13–20 mm, exserted beyond calyx margin; limb expanded 12–22 mm. **Styles** sparsely hirtellous. **Anthers** included, glabrous. **Capsules** included, stipitate, 7–10 mm. *2n* = 48, 56.

Flowering (May–)Jun–Aug. Creek banks, moraine water courses, bogs, marshes, wet meadows, roadside ditches; (1400–)1700–2700(–3000) m; Calif., Nev.

Erythranthe corallina is a morphologically consistent entity that occurs only in the Sierra Nevada of California and adjacent Nevada (Washoe County and Carson City). Its chromosome number is reported as *2n* = 48 and 56, compared to *2n* = 28 and 56 in *E. tilingii*; identities of the *E. corallina* vouchers should be rechecked and additional counts made, since the occurrence of such wide dysploidy seems unlikely. Compared to the leaf blades of *E. tilingii* in the strict sense, those of *E. corallina* are relatively broader (broadly ovate to orbicular-ovate), the plants generally taller, and long-

pedicellate flowers occasionally are produced from mid stem or even proximal nodes. The hirsutulous to hirsute vestiture of eglandular hairs on both leaf surfaces is a reliably diagnostic feature and usually easily observed with a 10× lens.

Some plants of *Erythranthe corallina* from San Bernardino County, California, produce decumbent-ascending stems (4–10 cm) and ovate-triangular leaves (blade 5–10 × 3–6 mm), but the dense system of filiform rhizomes, flowers one to three, and hirtellous foliar vestiture serve to identify them.

62. **Erythranthe utahensis** (Pennell) G. L. Nesom, Phytoneuron 2012-39: 44. 2012 • Utah monkeyflower E

Mimulus glabratus Kunth subsp. *utahensis* Pennell, Monogr. Acad. Nat. Sci. Philadelphia 1: 123. 1935

Perennials, rhizomatous, rooting at nodes, rhizomes filiform. **Stems** erect, some-times decumbent-ascending proximally, simple or few-branched, 20–50 cm, glabrous or sparsely stipitate-glandular in inflorescence. **Leaves** basal and cauline, even-sized or largest near mid stem; petiole 0 mm or proximalmost 2–10 mm; blade palmately 3–5-veined, orbicular or suborbicular to broadly elliptic, broadly ovate, or depressed-ovate, 20–40(–75) × 12–35 (–40) mm, base usually truncate to broadly cuneate, margins entire or subentire to mucronulate, shallowly dentate, or denticulate, apex rounded, surfaces glabrous or glabrate to sparsely stipitate-glandular and sparsely pilose, hairs thin-walled, abaxial often glaucous. **Flowers** herkogamous, 6–16, from proximal to distal nodes, in a loose raceme, distal bracts becoming much reduced. **Fruiting pedicels** (25–)40–75 mm, stipitate-glandular to short villous, hairs gland-tipped. **Fruiting calyces** broadly ovate-cylindric, inflated, sagittally compressed, (10–)11–17(–20) mm, stipitate-glandular or minutely hirtellous or a mixture, hairs sometimes also longer, thin-walled, eglandular or glandular, throat not closing, adaxial lobe slightly longer, triangular-blunt. **Corollas** yellow, abaxial limb prominently darker yellow, sparsely purple-spotted, bilaterally symmetric, weakly bilabiate; tube-throat narrowly funnelform to broadly cylindric, 10–15 mm, exserted 5–8 mm beyond calyx margin; limb expanded 12–20 mm. **Styles** hirtellous. **Anthers** included, glabrous. **Capsules** included, 5–7 mm. *2n* = 28, 30.

Flowering (Feb–)May–Aug(–Oct). Drainage ditches, springs, seeps, wet meadows, margins of ponds and small streams, marshy areas; 1400–2500 m; Calif., Colo., Nev., Utah.

Erythranthe utahensis is characterized by its erect stems, prolifically produced filiform rhizomes, basal leaves short-petiolate to subsessile and cauline sessile, blades suborbicular to broadly ovate or depressed ovate with thin-walled villous-glandular hairs on both surfaces, margins subentire, proximal pedicels elongating to 75 mm, and calyces open at maturity.

63. **Erythranthe guttata** (de Candolle) G. L. Nesom, Phytoneuron 2012-39: 43. 2012 • Common or seep monkeyflower

Mimulus guttatus de Candolle, Cat. Pl. Hort. Monsp., 127. 1813; *M. clementinus* Greene; *M. equinus* Greene; *M. glabratus* Kunth var. *adscendens* A. Gray; *M. grandiflorus* Howell; *M. guttatus* subsp. *haidensis* Calder & Roy L. Taylor; *M. guttatus* var. *puberulus* (Greene ex Rydberg) A. L. Grant; *M. hirsutus* Howell; *M. langsdorffii* Donn ex Greene var. *argutus* Greene; *M. langsdorffii* var. *californicus* Jepson; *M. langsdorffii* var. *guttatus* (de Candolle) Jepson; *M. langsdorffii* var. *minimus* J. K. Henry; *M. langsdorffii* var. *platyphyllus* Greene; *M. lyratus* Bentham; *M. paniculatus* Greene; *M. petiolaris* Greene; *M. prionophyllus* Greene; *M. puberulus* Greene ex Rydberg; *M. rivularis* Nuttall

Perennials, rhizomatous, sometimes rooting at proximal nodes. **Stems** erect to ascending-erect, branched distally, sometimes fistulose, to 10 mm wide, pressed, (6–)15–65(–80) cm, villous-glandular or moderately to densely hirtellous, hairs eglandular or glandular and eglandular. **Leaves** basal and cauline or basal not persistent; petiole 0 mm or proximals 1–95 mm; blade subpinnately, sometimes palmately, 5–7-veined, ovate-elliptic to ovate or suborbicular, 4–125 mm, 1–2 times longer than wide, gradually or abruptly reduced in size distally, base rounded to cuneate to truncate, margins crenate to coarsely dentate, proximally shallowly toothed to irregularly small-lobed or lyrate-dissected, apex rounded to obtuse, surfaces glabrous. **Flowers** herkogamous, (1–)3–20(–28), from distal nodes, sometimes in relatively compact racemes with reduced bracts. **Fruiting pedicels** 15–40(–60) mm, villous-glandular or moderately to densely hirtellous, hairs eglandular or glandular and eglandular. **Fruiting calyces** nodding, usually without red markings, ovate-campanulate, inflated, sagittally compressed, 11–17(–20) mm, villous-glandular or moderately to densely hirtellous, hairs eglandular or glandular and eglandular, throat closing. **Corollas** yellow, red-dotted, bilaterally symmetric, bilabiate; tube-throat funnelform, (10–)12–20 mm, exserted 3–5 mm beyond calyx margin; limb expanded 12–24 mm. **Styles** minutely hirsutulous to villosulous. **Anthers** included, glabrous. **Capsules** included, 7–11(–12) mm. $2n = 28, 56$.

Flowering Apr–Sep. Springs and seeps, marshes, beaver dams, along rivers, streams, and irrigation canals, loamy soils in conifer forests, wet and damp meadows, wet roadsides; 20–3200(–3700) m; Alta., B.C., N.B., N.W.T., Sask., Yukon; Alaska, Ariz., Calif., Colo., Conn., Idaho, Mich., Mont., Nebr., Nev., N.Mex., N.Y., Oreg., Pa., S.Dak., Tex., Utah, Wash., Wyo.; Mexico (Baja California, Baja California Sur, Chihuahua, Coahuila, Nayarit, Sonora); introduced in Europe.

Erythranthe guttata is markedly variable in stature, leaf shape, vesture, flower size, and the separation distance between anthers and stigma; it ranges from subalpine and near-alpine habitats into desert situations where water is available.

In all of Colorado, the Four Corners area, and north-central New Mexico, the vesture of stems and calyces is consistently densely hirsute-hirtellous, without glandular hairs. Plants with similar vesture also occur in British Columbia, Oregon, and Washington, and in scattered localities elsewhere. In northwestern Arizona, California, Nevada, and southern Oregon, vesture is consistently villous-glandular, without eglandular hairs. Elsewhere in the geographic range the vesture is a mix of hirsute-hirtellous (eglandular) and villous-glandular hairs. Other morphological variants and patterns, as well as variation in ploidy level, within *Erythranthe guttata* were discussed by G. L. Nesom (2012i).

Plants of *Erythranthe guttata* with extremely large corollas have been frequently collected on the Aleutian Islands, Kodiak Island, and in other Alaskan localities (for example, Admiralty Island, Amakuk, Juneau, and Yakutat Bay). Corolla tube-throats are 19–26 mm, and the limbs are expanded to 18–25 mm. The type collection of *E. guttata* is one of these plants, and the name *E. guttata* may prove to apply most appropriately only to Alaskan populations. Diploids and tetraploids appear to be sympatric in Alaska.

Mimulus guttatus subsp. *haidensis* was described as an endemic subalpine race that occurs in and along the flanks of the Queen Charlotte Mountains on Graham Island and Moresby Island. The subspecies was distinguished on the basis of its hirtellous vesture, but plants of similar hirtellous vesture occur over the whole range of the species. A tetraploid chromosome number ($2n = 56$) was reported for subsp. *haidensis* from a total of five localities on Graham Island and Moresby Island (J. A. Calder and R. L. Taylor 1968, vol. 2), and diploids ($2n = 28$) were documented from one locality on each of the two islands. At least one of the diploids has densely hirtellous stems, pedicels, and calyces, matching the morphology of subsp. *haidensis*.

Erythranthe guttata is naturalized in Europe and has been introduced to the northeastern United States (Connecticut, Michigan, New York, and Pennsylvania) and eastern Canada (New Brunswick).

64. **Erythranthe grandis** (Greene) G. L. Nesom, Phytoneuron 2012-39: 43. 2012 • Magnificent monkeyflower E

Mimulus guttatus de Candolle var. *grandis* Greene, Man. Bot. San Francisco, 277. 1894; *M. grandis* (Greene) A. Heller; *M. guttatus* subsp. *litoralis* Pennell; *M. langsdorffii* Donn ex Greene var. *grandis* (Greene) Greene; *M. procerus* Greene

Perennials, rhizomatous, sometimes rooting at proximal nodes. **Stems** erect, sometimes decumbent basally, branched, often fistulose, (25–)50–120(–160) cm, densely hirsutulous to softly hirtellous-puberulent to pilose-hirsutulous, hairs usually crinkly, and eglandular or with a mixture of hirtellous-puberulent and stipitate-glandular hairs, sometimes ± stipitate-glandular or glandular-villous without hirtellous-puberulent hairs. **Leaves** basal and cauline, basal usually not persistent, bracteate in inflorescence; petiole 10–80 mm, gradually reduced distally; blade subpinnately, sometimes palmately, 5–7-veined, ovate to broadly elliptic, 25–60 × 20–40(–60) mm, usually 1–2 times longer than wide, base truncate or truncate-cuneate to subcordate, margins crenulate to dentate, proximally sometimes sublyrate, apex rounded to obtuse, surfaces of distals densely hirsutulous to softly hirtellous-puberulent to pilose-hirsutulous, hairs usually crinkly, and eglandular or with a mixture of hirtellous-puberulent and stipitate-glandular hairs, sometimes ± stipitate-glandular or glandular-villous without hirtellous-puberulent hairs. **Flowers** herkogamous, 8–26, mostly from distal nodes, usually in bracteate racemes. **Fruiting pedicels** 10–35 mm, densely hirsutulous to softly hirtellous-puberulent to pilose-hirsutulous, hairs usually crinkly, and eglandular or with a mixture of hirtellous-puberulent and stipitate-glandular hairs, sometimes ± stipitate-glandular or glandular-villous without hirtellous-puberulent hairs. **Fruiting calyces** straight-erect or nodding 45–100°, ovate-campanulate, inflated, sagittally compressed, 15–22(–25) mm, densely hirsutulous to softly hirtellous-puberulent to pilose-hirsutulous, hairs usually crinkly, and eglandular or with a mixture of hirtellous-puberulent and stipitate-glandular hairs, sometimes ± stipitate-glandular or glandular-villous without hirtellous-puberulent hairs, throat closing. **Corollas** yellow, red-dotted within, bilaterally symmetric, bilabiate; tube-throat broadly funnelform, (14–)16–24 mm, exserted (8–)10–15 mm beyond calyx margin; limb broadly expanded. **Styles** hirtellous. **Anthers** included, glabrous. **Capsules** included, 8–12 mm. *2n* = 28.

Flowering (Apr–)May–Jul(–Sep). Beaches, dunes, coastal bluffs, wet cliff faces, mud flats and seeps, marshes, drainage ditches, creeks, rarely in coastal sage scrub; 0–200(–800) m; Calif., Oreg.

The densely, evenly puberulent vestiture of pedicels, calyces, and distal stems usually is diagnostic, especially in combination with the large flowers (corollas and mature calyces) and tall stature. Plants from scattered collections are much shorter than normal but have large corollas and characteristic vestiture.

Erythranthe grandis characteristically occurs in coastal localities from southern California to northern Oregon but also is found in inland localities and habitats near the coast but well away from salt spray.

65. **Erythranthe arenicola** (Pennell) G. L. Nesom, Phytoneuron 2012-39: 43. 2012 • Beach monkeyflower C E

Mimulus guttatus de Candolle subsp. *arenicola* Pennell, Proc. Acad. Nat. Sci. Philadelphia 99: 166. 1947

Annuals, fibrous-rooted or slender-taprooted, rarely rooting at nodes. **Stems** erect, rarely prostrate to prostrate-ascending, few-branched, 3–17 cm, moderately villous-glandular, hairs gland-tipped, or mixed hirtellous and stipitate-glandular. **Leaves** basal and cauline; petiole: basal 2–8 mm or mid and distals absent; blade palmately 3–5-veined, suborbicular to broadly ovate or depressed-ovate, 5–17 × 6–15 mm, base truncate or truncate-cuneate to subcordate, margins subentire or crenulate, apex rounded to obtuse, surfaces moderately villous-glandular, hairs gland-tipped, or mixed hirtellous and stipitate-glandular. **Flowers** herkogamous, 1–6, at distal nodes, chasmogamous. **Fruiting pedicels** 9–17 mm, moderately villous-glandular, hairs gland-tipped, or mixed hirtellous and stipitate-glandular. **Fruiting calyces** nodding, ovoid-campanulate, inflated, sagittally compressed, 9–16 mm, moderately villous-glandular, hairs gland-tipped, or mixed hirtellous and stipitate-glandular, throat closing. **Corollas** yellow, red-dotted, bilaterally symmetric, bilabiate; tube-throat funnelform, 11–20 mm, exserted 4–8 mm beyond calyx margin; limb expanded 10–18 mm. **Styles** hirtellous. **Anthers** included, glabrous. **Capsules** included, 5–12 mm.

Flowering Apr–Aug. Sandy beaches, especially in moist hollows among dunes, sea cliff bases, chaparral near beaches, mudstone outcrops; of conservation concern; 0–100 m; Calif.

F. W. Pennell (1947, 1951) considered *Erythranthe arenicola* an endemic of Monterey County, but plants from adjacent San Luis Obispo and Santa Cruz counties also belong here. Most of the localities are at seaside, but some are more than a mile inland. *Erythranthe arenicola* is hypothesized here to be a derivative of *E. guttata* or *E. grandis*, retaining the herkogamous breeding system of its putative ancestor but reduced in size and duration.

66. **Erythranthe thermalis** (A. Nelson) G. L. Nesom, Phytoneuron 2012-39: 44. 2012 • Yellowstone monkeyflower [C] [E]

Mimulus thermalis A. Nelson, Bull. Torrey Bot. Club 27: 269. 1900

Annuals, taprooted, rarely with a basal, runnerlike stem. **Stems** erect, simple or branched from basal nodes, 1.5–10(–15) cm, villous-glandular proximally, moderately to densely stipitate-glandular, sometimes hirtellous and eglandular distally. **Leaves** basal and cauline, cauline 2–5 pairs; petiole: basal and proximal cauline 3–20 mm, distals 0 mm; blade palmately 3–5-veined, suborbicular to ovate, depressed-ovate, ovate-deltate, or reniform, 4–15(–20) × 4–20 mm, base cuneate to truncate or subcordate, margins evenly crenate-dentate to subentire, apex acute to obtuse or rounded, surfaces moderately to densely stipitate-glandular, sometimes hirtellous and eglandular. **Flowers** plesiogamous, 1–5(–9), usually at distal nodes, chasmogamous. **Fruiting pedicels** 7–12 mm, moderately to densely stipitate-glandular, sometimes hirtellous and eglandular. **Fruiting calyces** ovate-campanulate, inflated, sagittally compressed, 8–11 mm, moderately to densely stipitate-glandular, sometimes hirtellous and eglandular, throat closing, adaxial lobe longer than others. **Corollas** yellow, red-dotted or not, bilaterally symmetric, strongly bilabiate; tube-throat funnelform, 8–12 mm, exserted 1–2 mm beyond calyx margin; limb expanded 12–15 mm, throat open, palate villous. **Styles** hirtellous. **Anthers** included, glabrous. **Capsules** included, 5–6 mm. $2n = 28$.

Flowering Mar–Aug. Hot, shallow, quick-drying soils around thermal pools and vents; of conservation concern; 2200–2600 m; Wyo.

Erythranthe thermalis is endemic to Yellowstone National Park in northwestern Wyoming. The species is recognized by its annual duration (without rhizomes), reduced stature and leaf size, and short, but broad-limbed, corollas with autogamous fertilization. Typical *E. guttata* (rhizomatous, herkogamous) also grows in the immediately surrounding areas but apparently not in hot soils. Each species maintains distinctions in growth form, phenology and mating system in common garden experiments (Y. Lekberg et al. 2012).

SELECTED REFERENCE Lekberg, Y. et al. 2012. Phenotypic and genetic differentiation among yellow monkeyflower populations from thermal and non-thermal soils in Yellowstone National Park. Oecologia 170: 111–122.

67. **Erythranthe unimaculata** (Pennell) G. L. Nesom, Phytoneuron 2012-39: 44. 2012 • Green-palate monkeyflower

Mimulus unimaculatus Pennell, Notul. Nat. Acad. Nat. Sci. Philadelphia 43: 5. 1940

Annuals, shallowly fibrous-rooted or slender-taprooted, sometimes rooting at proximal nodes. **Stems** erect or basally ascending-erect, simple or few-branched, becoming fistulose in larger plants, (2–)10–30(–100) cm, delicately short glandular-villous to stipitate-glandular, often glabrous below inflorescence. **Leaves** usually cauline, basal sometimes persistent, distal connate-perfoliate, often bractlike; petiole: proximals 4–10(–15) mm, distals 0 mm; blade: proximals sometimes subpinnately veined, usually with (1 or)2 pairs arising from midvein above base, becoming palmately veined distally, ovate-lanceolate to ovate or broadly ovate-elliptic, mid cauline 12–40(–50) × 10–25(–45) mm, base rounded to truncate or cuneate, margins shallowly dentate-serrate to serrate, teeth 7–12 per side, apex rounded to obtuse, surfaces glabrous, sometimes with sharp-pointed hirtellous, vitreous-flattened, or gland-tipped hairs. **Flowers** herkogamous, (1–)3–14, usually from mid stem and distally, chasmogamous. **Fruiting pedicels** 10–40 mm, delicately short glandular-villous to stipitate-glandular, often glabrous below inflorescence. **Fruiting calyces** nodding 30–90°, broadly campanulate, inflated, sagittally compressed, 9–13(–15) mm, sparsely glandular-villous to stipitate-glandular, throat closing, adaxial lobe distinctly longer. **Corollas** pale yellow, palate and abaxial throat dark yellow, drying blue-green, red-spotted, bilaterally symmetric, bilabiate; tube-throat funnelform to subfunnelform, (7–)9–14 mm, exserted 3–4 mm beyond calyx margin; limb broadly expanded (8–17 mm pressed), palate densely bearded. **Styles** hirtellous. **Anthers** included, glabrous. **Capsules** included, 5–8 mm. $2n = 28$.

Flowering Jan–Jun. Stream and canal sides, pool edges, canyon bottoms, sand, gravel, and mud, riparian habitats, pine-oak forests; 200–2000 m; Ariz., N.Mex.; Mexico (Chihuahua, Sonora).

Erythranthe unimaculata is recognized by its annual duration (fibrous-rooted, without stolons or rhizomes), delicate-glandular vestiture, mostly sessile to subsessile, often widely spaced leaves, closed fruiting calyces, and relatively large, pale yellow to nearly white corollas with a dark yellow palate that commonly dries blue-green.

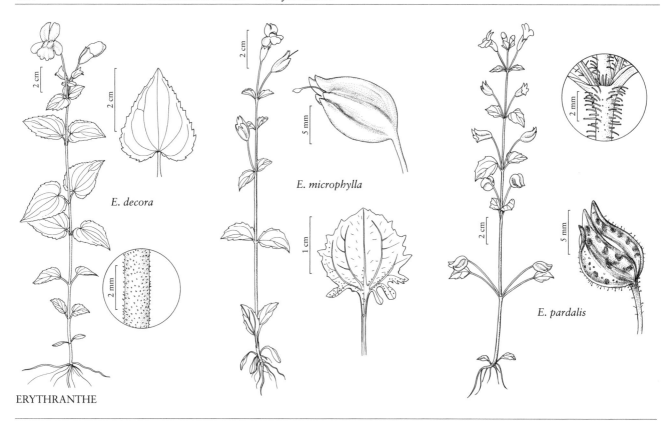

E. decora

E. microphylla

E. pardalis

ERYTHRANTHE

The breeding system is allogamous. Plants commonly are single-stemmed and usually erect but sometimes produce decumbent-ascending stems branching from the base, these sometimes rooting at proximal nodes.

Erythranthe unimaculata is known from Cochise, Gila, Pima, and Santa Cruz counties, Arizona, and from Doña Ana County, New Mexico.

68. Erythranthe decora (A. L. Grant) G. L. Nesom, Phytoneuron 2012-39: 43. 2012 • Mannered monkeyflower [E] [F]

Mimulus guttatus de Candolle var. *decorus* A. L. Grant, Ann. Missouri Bot. Gard. 11: 173, plate 5, fig. 1. 1925; *M. decorus* (A. L. Grant) Suksdorf

Perennials, rhizomatous, sometimes producing numerous, long, remotely leafy runners from basal cauline nodes. **Stems** erect, simple, 20–100 cm, distals densely, minutely hirtellous. **Leaves** cauline, basal usually not persistent; petiole: proximals 8–25 mm, midcauline 3–5 mm, distalmost 0 mm; blade palmately (3–)5–7-veined, broadly ovate-triangular to ovate-lanceolate, 20–50(–60) × 10–30(–40) mm, usually 1–2 times longer than wide, base rounded to truncate to shallowly cuneate, margins dentate, teeth sharp-pointed,

apex acute, surfaces sometimes glabrate, distals densely, minutely hirtellous. **Flowers** herkogamous, (1 or)2–7 (–14), from distal nodes. **Fruiting pedicels** 18–35 (–40) mm, distals densely, minutely hirtellous, hairs often slightly deflexed. **Fruiting calyces** green, sometimes red-spotted, ovoid, inflated, sagittally compressed, 15–19 mm, distals densely, minutely hirtellous, throat closing, lobe margins (sinuses) villous. **Corollas** yellow to chrome yellow, throat floor and tube usually red-spotted, bilaterally symmetric, bilabiate; tube-throat broadly funnelform, 18–26 mm, exserted (8–)10–15 mm beyond calyx margin; limb expanded 22–30 mm. **Styles** minutely, prominently hirsutulous to villosulous. **Anthers** included, glabrous. **Capsules** included, 8–10 mm.

Flowering May–Aug. River banks, streamsides, dripping banks, moist meadows; 1000–1600 m; B.C.; Idaho, Oreg., Wash.

Erythranthe decora is distinct in its uniformly ovate to ovate-lanceolate leaf blades with truncate bases and regularly toothed margins, relatively long internodes, rhizomatous habit, mostly unbranched stems, often with leafy runners from basal nodes, large corollas, hairy styles, and minutely hirtellous stems, pedicels, calyces, and leaf surfaces. The densely produced rhizomes suggest a relationship with the *E. tilingii* group. A population in Clearwater County, Idaho, appears to be long-disjunct from the main range of the species in northwestern Oregon and the western half of Washington.

69. **Erythranthe scouleri** (Hooker) G. L. Nesom, Phytoneuron 2012-39: 44. 2012 • Scouler's monkeyflower [C] [E]

Mimulus scouleri Hooker, Fl. Bor.-Amer. 2: 100. 1838; *M. guttatus* de Candolle subsp. *scouleri* (Hooker) Pennell

Perennials, rhizomatous, producing long, sparsely leafy runners from basal nodes. **Stems** erect, simple or few-branched, 15–80 cm, glabrous. **Leaves** cauline, basal usually not persistent; petiole: proximals and medial 10–25 mm (gradually merging into blade), distals 0 mm; blade palmately (3–)5–7-veined to subpinnate, (proximal to medial) oblong-elliptic to oblong-lanceolate, 25–60 × 8–18 mm, usually 3–4 times longer than wide, base attenuate, margins evenly, shallowly dentate or crenate to mucronate or mucronulate, teeth 10–20 per side, sometimes more deeply toothed at base, apex obtuse to acute, surfaces glabrous. **Flowers** herkogamous, (1 or)2–8, from distal nodes. **Fruiting pedicels** 20–25 mm, glabrous. **Fruiting calyces** ovoid, inflated, sagittally compressed, 13–14 mm, glabrous, throat closing. **Corollas** yellow, without red markings, bilaterally symmetric, bilabiate; tube-throat funnelform, 20–24 mm, exserted 10–15 mm beyond calyx margin; limb expanded 22–30 mm. **Styles** minutely, prominently hirsutulous to villosulous. **Anthers** included, glabrous. **Capsules** unknown.

Flowering May–Aug. Salt marshes, river banks; of conservation concern; 0–20 m; Oreg., Wash.

Erythranthe scouleri is distinctive in its oblong-elliptic leaves with long-tapering bases and closely toothed margins, completely glabrous vestiture, long, leafy runners from basal cauline nodes, large corollas with broad limbs, and prominently hairy styles. Several features suggest a close relationship to *E. decora*, particularly its very large corollas, hairy styles, closely toothed leaf margins, tall, simple, and erect stems, numerous runners, and its geographic range. All collections apparently have been made near the mouth of the Columbia River in Clatsop and Columbia counties, Oregon (G. L. Nesom 2013d). Recent observations (Alexander John Wright, pers. comm.; photos http://www.inaturalist.org/observations/3537008) indicate that it also occurs in Wahkiakum County, Washington, in the Julia Butler Hansen Refuge.

70. **Erythranthe microphylla** (Bentham) G. L. Nesom, Phytoneuron 2012-39: 44. 2012 • Bentham's monkeyflower [E] [F]

Mimulus microphyllus Bentham in A. P. de Candolle and A. L. P. P. de Candolle, Prodr. 10: 371. 1846; *M. glareosus* Greene; *M. guttatus* de Candolle var. *depauperatus* (A. Gray) A. L. Grant; *M. guttatus* var. *insignis* Greene; *M. guttatus* var. *microphyllus* (Bentham) Pennell; *M. guttatus* var. *platycalyx* (Pennell) D. W. Taylor; *M. langsdorffii* Donn ex Greene var. *insignis* (Greene) Greene; *M. langsdorffii* var. *microphyllus* (Bentham) A. Nelson & J. F. Macbride; *M. luteus* Linnaeus var. *depauperatus* A. Gray; *M. nasutus* Greene var. *insignis* (Greene) A. L. Grant; *M. platycalyx* Pennell

Annuals, fibrous-rooted. **Stems** erect, simple, sometimes many-branched from basal cauline nodes, terete, sometimes distinctly 4-angled, (3–)5–30(–45) cm, glabrous below inflorescence, sometimes distals hirtellous, hairs commonly deflexed, or mixed hirtellous and stipitate-glandular, sometimes only short villous-glandular, hairs gland-tipped. **Leaves** basal and cauline, basal sometimes absent at flowering; petiole: basal or proximals to medials 3–25(–35) mm, distals 0 mm (then blade subclasping to narrowly perfoliate); blade often purplish, palmately 3–5-veined, ovate or ovate-lanceolate to elliptic-ovate, suborbicular, or depressed-ovate, (3–)10–35 × 3–25 mm, base rounded to truncate or subcordate, margins shallowly crenate to sharply crenate-serrate, teeth 5–10 per side, basal and proximal often irregularly incised near petiole and sublyrate, apex acute to obtuse-rounded, surfaces glabrous or sparsely to moderately hirtellous, eglandular. **Flowers** herkogamous, 1–8(–14), usually from distal nodes, chasmogamous. **Fruiting pedicels** 8–30(–50) mm, distals hirtellous, hairs commonly deflexed, or mixed hirtellous and stipitate-glandular, sometimes only short villous-glandular, hairs gland-tipped. **Fruiting calyces** nodding 30–90°, sometimes red-tinged or -dotted, ovoid-campanulate to broadly cylindric-campanulate, inflated, sagittally compressed, (7–)9–16(–20) mm, minutely hirtellous, hairs sometimes reduced to basal cells, or glabrous, throat strongly to weakly closing. **Corollas** yellow to golden yellow or orangish yellow, usually red-spotted, abaxial limb sometimes with a large red splotch, bilaterally symmetric, bilabiate; tube-throat broadly funnelform, (6–)8–16(–20) mm, exserted (1–)2–6(–8) mm beyond calyx margin; limb expanded 8–25 mm, palate villous. **Styles** sparsely hirtellous. **Anthers** included, glabrous. **Capsules** included, 6–9(–11) mm. *2n* = 28, 56.

Flowering Mar–Jul. Rock depressions, rocky ridges, cliff faces, roadcuts, wet meadows, seeps, stream banks, drying ponds, ephemeral stream channels, vernal springs over serpentine, roadsides and roadside ditches, dry banks, lava soils, loam, clay, gravel, yellow pine, oak-pine, mixed oak woodlands, oak-chaparral; 20–1700 (–2600) m; B.C.; Calif., Idaho, Nev., Oreg., Wash.

Erythranthe microphylla is characterized by its annual duration (fibrous-rooted), usually simple stems, relatively widely spaced leaves, glabrous or hirtellous vestiture, open corollas, and calyces closing at the throat. Even in the smallest corollas, the stigma is positioned above the adaxial anther pair, indicating that all are primarily allogamous. Some plants have basal and proximal cauline leaves with exaggeratedly and irregularly toothed-incised margins (especially in Lake and Napa counties, California, as in the types of *Mimulus glareosus* and *M. guttatus* var. *insignis*, respectively), but a similar tendency can be seen over most of the geographic range. Plants of *E. microphylla* vary greatly in height, leaf size, and flower size (the larger flowers approaching the size of those in *E. grandis* and *E. decora*), yet all seem to be within the expression of a single species.

An inversion sequence of chromosome 8 (the DIV1 region) is perfectly correlated with the life history features of at least four species of sect. *Simiolus* (D. B. Lowry and J. H. Willis 2010). One sequence occurs in *Erythranthe guttata* and *E. grandis*, which are perennial and rhizomatous, occur in habitats with year-round moisture, and flower relatively late in the season, while the opposite sequence occurs in *E. nasuta* and *E. microphylla*, which are annual and slender-taprooted or fibrous-rooted, occur in quickly drying habitats, and flower in early season. The inversion, with its tightly linked, locally adaptive alleles, contributes to isolating mechanisms between species with contrasting sequences and preserves the constellation of features that makes each a recognizable entity.

Hybridization and apparent introgression occur among *Erythranthe guttata*, *E. microphylla*, and *E. nasuta*, yet each remains distinct in duration, perennating morphology, and habit. *Erythranthe arvensis* (annual) also is a member of this gene-sharing group. A whole-genome analysis by A. D. Twyford and J. Friedman (2015) showed that populations of *E. guttata* and *E. microphylla* cluster together within each of several geographically delimited regions; they interpreted their trees as showing phylogenetic relationships and concluded that *E. guttata* and *E. microphylla* are conspecific. Other evidence (G. L. Nesom 2014c, 2014d), particularly the DIV1 inversion sequence, indicates that *E. guttata* and *E. grandis* are most closely related to each other and were derived from an ancestor of annual duration. Thus, *E. microphylla* and *E. nasuta* are morphologically and phylogenetically distinct from *E. guttata*, despite extensive gene flow where they are sympatric with it.

71. **Erythranthe marmorata** (Greene) G. L. Nesom, Phytoneuron 2012-39: 44. 2012 • Stanislaus or Whipple's monkeyflower [C] [E]

Mimulus marmoratus Greene, Erythea 3: 73. 1895; *M. whipplei* A. L. Grant

Annuals, taprooted. **Stems** erect, simple or branched from base, 7–28 cm, usually moderately to densely villous-glandular, without eglandular hairs. **Leaves** usually cauline or basal persistent; petiole: proximals to medials 7–15 mm, distals 0 mm; blade palmately (3–)5-veined, ovate or broadly ovate to elliptic-ovate or depressed-ovate, (10–)15–30 × 6–15 mm, base truncate to shallowly cuneate, margins shallowly to coarsely dentate, apex acute, surfaces usually moderately to densely villous-glandular, without eglandular hairs. **Flowers** herkogamous, (1–)2–6, axillary from middle to distal nodes, chasmogamous. **Fruiting pedicels** 15–45 mm, usually moderately to densely villous-glandular, without eglandular hairs. **Fruiting calyces** sharply nodding, usually densely purple-spotted, broadly campanulate, inflated, sagittally compressed, 9–12 mm, densely hirtellous, sometimes sparsely stipitate-glandular, densely villous at sinuses, throat closing, adaxial lobe ca. 2 times length of others. **Corollas** yellow, throat red-spotted, abaxial limb base with a large red splotch, bilaterally symmetric, bilabiate; tube-throat narrowly cylindric-funnelform, 10–12 mm, exserted 4–5 mm beyond calyx margin; limb abruptly expanded 14–20 mm. **Styles** glabrous or sparsely hirtellous. **Anthers** included, glabrous. **Capsules** included, 6–9 mm.

Flowering Mar–May. Habitat unknown, not over serpentine; of conservation concern; 100–900 m; Calif.

Erythranthe marmorata is recognized by its erect, taprooted habit and annual duration, villous-glandular vestiture, ovate-petiolate leaves, flowers from middle to distal nodes, long, narrow corolla tube-throat abruptly flaring into a broad limb, abaxial middle corolla lobe with a large red splotch, and fruiting calyces dark-spotted and sharply nodding. The species is known from foothills in Calaveras, Stanislaus, and Tuolumne counties in the Stanislaus River drainage and from Amador County in the Mokelumne River drainage. A collection from Fresno County appears to be somewhat disjunct from the main range, and the plants are more densely villous than characteristic elsewhere, but their identification as *E. marmorata* otherwise seems secure.

Erythranthe marmorata (previously identified as *Mimulus whipplei*) had been considered extremely rare or even perhaps extinct. See G. L. Nesom (2013d) for citations of recent collections.

72. Erythranthe glaucescens (Greene) G. L. Nesom, Phytoneuron 2012-39: 43. 2012 • Shieldbract monkeyflower E

Mimulus glaucescens Greene, Bull. Calif. Acad. Sci. 1: 113. 1885; *M. guttatus* de Candolle var. *glaucescens* (Greene) Jepson

Annuals, slender-taprooted or fibrous-rooted, rarely with runners from basal nodes. **Stems** erect, simple or branched, terete, sometimes 4-angled distally, (5–)30–60(–80) cm, glabrous, glaucous. **Leaves** basal and cauline; petiole: basal and proximal cauline as long as or much longer than blade, slender, sometimes pubescent or villous, distals absent; blade palmately 3–5-veined, (proximal) ovate to ovate-elliptic or orbicular-ovate, sometimes subcordate, 10–50 mm, midcauline to distal orbicular, 5–45 mm wide, distinctly connate-perfoliate, disclike distally, base rounded to subcordate, margins: proximals denticulate to dentate or coarsely, irregularly toothed, sometimes lobed at base, distals nearly entire or toothed, teeth scattered, small, apex rounded, surfaces glabrous, glaucous. **Flowers** herkogamous, 1–16, from distal nodes, sometimes from nearly all, chasmogamous. **Fruiting pedicels** 10–50 mm, glabrous, glaucous. **Fruiting calyces** broadly campanulate, inflated, sagittally compressed, 7–16 mm, glabrous, glaucous, throat closing. **Corollas** yellow, sometimes with a median splotch, abaxial limb densely dark yellow, others much lighter, throat floor and tube red-dotted, bilaterally symmetric, bilabiate; tube-throat funnelform, 12–23 mm, exserted 4–8 mm beyond calyx margin; limb expanded 14–36 mm. **Styles** minutely hirtellous-puberulent. **Anthers** included, glabrous. **Capsules** included, 5–11 mm. *2n* = 28.

Flowering Mar–May(–Jun). Seepage areas, wet rocks, moist cliffs, pool edges, gravelly stream banks, serpentine outcrops, roadsides and roadcuts, low pastures, riparian woodlands, blue oak woodlands, chaparral, grasslands; 80–900(–1100) m; Calif.

Plants from one locality in Butte County are unusual in producing filiform, small-leaved runners from basal cauline nodes. *Erythranthe glaucescens* is known only from Butte and Tehama counties.

73. Erythranthe nudata (Curran ex Greene) G. L. Nesom, Phytoneuron 2012-39: 44. 2012 • Bare monkeyflower E

Mimulus nudatus Curran ex Greene, Bull. Calif. Acad. Sci. 1: 114. 1885

Annuals, taprooted or fibrous-rooted. **Stems** erect or ascending, simple or few-branched from basal nodes, branches mostly reddish purple, (5–)9–30 cm, glabrous or minutely stipitate-glandular, hairs 0.05–0.1 mm, at least just above nodes. **Leaves** cauline on wide internodes; petiole 5–30 mm, distals 0 mm; blade 1-veined or palmately 3–5-veined, proximals lanceolate or oblong-lanceolate to ovate, narrowly spatulate, or oblanceolate, distals usually linear, not perfoliate, 5–15(–30) × 1–5 mm, base attenuate, margins denticulate to proximally dentate-lobed, apex acute, surfaces glabrous or minutely stipitate-glandular, hairs 0.05–0.1 mm, at least just above nodes. **Flowers** herkogamous, (2–)4–8, usually in proximal or medial to distal axils, chasmogamous. **Fruiting pedicels** erect in flower, spreading to divaricate, rarely recurved, in fruit, 10–35 mm, glabrous or minutely stipitate-glandular, hairs 0.05–0.1 mm, at least just above nodes. **Fruiting calyces** ovate-campanulate, inflated, sagittally compressed, 6–13 mm, glabrous or minutely stipitate-glandular, hairs 0.05–0.1 mm, at least just above nodes, throat closing, abaxial lobe upcurving over lateral ones, nearly closing orifice. **Corollas** yellow, without a large red splotch, throat floor and tube red-spotted, bilaterally symmetric, bilabiate; tube-throat cylindric-funnelform, 8–12 mm, exserted 2–4 mm beyond calyx margin; limb expanded 8–12 mm. **Styles** glabrous. **Anthers** included, glabrous. **Capsules** included, 6–7 mm.

Flowering Apr–Jun. Open gravelly seeps on serpentine outcrops, serpentine crevices, springs, streamsides, gravelly creek beds, roadside drainages and swales; 200–700 m; Calif.

Erythranthe nudata is distinct in its annual duration, few, inconspicuous, and narrow leaves, long and spreading-divaricate pedicels, and large corollas. The plants apparently are restricted to serpentine substrate and known only from Colusa, Glenn, Lake, Mendocino, Napa, and Sonoma counties.

74. Erythranthe percaulis G. L. Nesom, Phytoneuron 2013-70: 1, figs. 1–5. 2013 • Serpentine Canyon monkeyflower C E

Annuals, fibrous-rooted. **Stems** erect or slightly ascending from base, often purplish, simple or few-branched from basal nodes, 7–28 cm, glabrous. **Leaves** basal and cauline; petiole: basal and proximalmost cauline 5–10 mm or cauline 0 mm; blade often spreading at right angles to stem, purple, palmately 3–5-veined, narrowly ovate, rhombic-elliptic, ovate to lanceolate, or elliptic-lanceolate to ovate or oblong-ovate, 7–10 mm, cauline even-sized or slightly smaller distally, 4–10 mm, base truncate to attenuate, margins entire or proximals shallowly sinuate, serrations 2–4, shallow, apex rounded, surfaces glabrous. **Flowers** herkogamous, 8–12, usually on distal ⅔ of stem, not clustered, chasmogamous. **Fruiting pedicels** 15–35 mm, glabrous. **Fruiting calyces** sharply wing-angled, urceolate to urceolate-campanulate, inflated, sagittally compressed, (4–)5–6 mm, glabrous, throat closing. **Corollas** yellow, throat floor with a few red dots, proximal middle lip base with a larger red splotch, bilaterally symmetric, bilabiate; tube-throat cylindric-funnelform, 4–6 mm, exserted 2–3 mm beyond calyx margin; limb expanded 4–5 mm, palate ridges yellow, densely hairy. **Styles** glabrous. **Anthers** included, glabrous. **Capsules** included, 3 mm.

Flowering May–Jun. Soil pockets, crevices, and boulders on serpentine cliffs, slopes, and roadcuts; of conservation concern; 2800 m; Calif.

Erythranthe percaulis was described from only the type collection from Serpentine Canyon of the Feather River in Plumas County, but the type locality has recently been relocated and the population determined to comprise many thousands of individuals (S. Schoenig 2016). Plants are characterized by their completely glabrous vestiture, terete and mostly simple stems, small leaves on relatively widely spaced nodes, small calyces, and small, yellow corollas with herkogamous arrangement of stigma and anthers.

SELECTED REFERENCE Schoenig, S. 2016. Rediscovery of *Erythranthe percaulis* (Phrymaceae) in the Feather River Canyon. Phytoneuron 2016-69: 1–14.

75. Erythranthe filicifolia (Sexton, K. G. Ferris & Schoenig) G. L. Nesom, Phytoneuron 2013-80: 1. 2013 • Fernleaf monkeyflower C E

Mimulus filicifolius Sexton, K. G. Ferris & Schoenig, Madroño 60: 237, figs. 2–4. 2013

Annuals, fibrous-rooted. **Stems** erect or slightly ascending from base, sometimes purplish, simple or few-branched from proximal nodes, 3–38 cm, glabrous. **Leaves** basal and cauline; petiole (proximal and proximal to mid cauline), distal bracteate or absent, ovate, margins entire; blade: divisions 1-veined, oblong-lanceolate to ovate in outline, 3–68 mm, slightly thickened, base (at proximalmost division) truncate, margins (of divisions) entire, bipinnately dissected, ca. 5–12(–15) primary divisions on each side, ultimate division nearly linear, apex obtuse to acute, surfaces glabrous. **Flowers** plesiogamous, 4–10(–12), on foreshortened distal nodes, usually clustered, chasmogamous, sometimes cleistogamous. **Fruiting pedicels** 6–14 mm, glabrous. **Fruiting calyces** urceolate to urceolate-campanulate, inflated, sagittally compressed, 8–11 mm, glabrous, throat closing. **Corollas** yellow, proximal lip sometimes red-splotched, palate ridges yellow, bilaterally symmetric, bilabiate; tube-throat cylindric-funnelform, 4–8 mm, exserted 0–2 mm beyond calyx margin or not; palate ridges densely villous. **Styles** glabrous. **Anthers** included, glabrous. **Capsules** included, 3–8 mm.

Flowering Apr–Jun(–Sep). Slow-draining, ephemeral seeps on exfoliating granite slabs, over basalt; of conservation concern; 300–1600 m; Calif.

Erythranthe filicifolia is known only from localities in Butte and Plumas counties.

76. Erythranthe nasuta (Greene) G. L. Nesom, Phytoneuron 2012-39: 44. 2012 • Calyx-nose monkeyflower

Mimulus nasutus Greene, Bull. Calif. Acad. Sci. 1: 112. 1885; *M. bakeri* Gandoger; *M. cuspidatus* Greene; *M. erosus* Greene; *M. guttatus* de Candolle var. *gracilis* (A. Gray) G. R. Campbell; *M. guttatus* var. *nasutus* (Greene) Jepson; *M. langsdorffii* Donn ex Greene var. *nasutus* (Greene) Jepson; *M. luteus* Linnaeus var. *gracilis* A. Gray; *M. puncticalyx* Gandoger; *M. sookensis* B. G. Benedict, Modliszewski, Sweigart, N. H. Martin, Ganders & John H. Willis; *M. subreniformis* Greene

Annuals, fibrous-rooted or slender-taprooted. **Stems** erect to ascending-erect or decumbent, simple or branched from proximal nodes, 4-angled, sometimes shallowly 4-winged, thin-wiry, or fistulose, 2–35 (–100) cm, glabrous except for a consistently small, villous-glandular area just above nodes, sometimes hirtellous distally. **Leaves** basal and cauline or basal not persistent; petiole: proximals to medials 3–35 mm, base narrowly flanged, distals 0 mm; blade ± red tinged abaxially or purple-spotted, palmately 3–5-veined, elliptic-ovate to broadly ovate, suborbicular, or depressed-ovate, (5–)10–49(–80) × (3–)10–25(–60) mm, as long as wide or wider than long, proximals largest and persistent, base cuneate to truncate or subcordate, margins irregularly dentate to dentate-serrate or nearly lacerate-dentate, commonly doubly toothed, main teeth 4–9 per side, sometimes sublacerate to sublyrate basally, apex acute to obtuse, surfaces glabrous, sometimes hirtellous, hairs dull, terete, sharp-pointed, eglandular. **Flowers** plesiogamous, (1 or)2–12(–20), from distal nodes, sometimes from medial to distal, chasmogamous or cleistogamous. **Fruiting pedicels** (3–)7–20(–40) mm, longer than subtending leaves, glabrate, sometimes glandular-villous adaxially at axils. **Fruiting calyces** nodding 30–180°, usually purple-tinged or -spotted, ovoid-campanulate, inflated, sagittally compressed, (5–)10–15(–19) mm, glabrous or minutely hirtellous to appressed-hirtellous, minutely short-ciliate at sinuses, throat closing, adaxial lobe usually longer than abaxial, slightlty falcate. **Corollas** yellow, throat usually red-spotted, abaxial limb base usually with a red splotch, bilaterally symmetric, weakly bilabiate; tube-throat broadly cylindric, (5–)8–12 mm, exserted (0–)1–2 mm beyond calyx margin; limb expanded 6–12 mm. **Styles** minutely scabrous or glabrous. **Anthers** included, glabrous. **Capsules** included, (4–)5–9(–10) mm. $2n$ = 26, 28, 56.

Flowering (Mar–)Apr–Jun(–Jul). Cliff faces, ledges, crevices, and bases, wet rocks in rivers, streamsides, sand bars, mossy seeps, wet clay banks, moist fields, sandy soils, depressions over granite, roadsides; (0–) 600–2300(–3200) m; B.C.; Ariz., Calif., Idaho, Nev., N.Mex., Oreg., Utah, Wash.; Mexico (Baja California, Sonora).

Erythranthe nasuta is characterized by its annual duration (fibrous-rooted), four-angled stems, broadly ovate leaves commonly with irregularly toothed margins, calyces with longish, protruding adaxial lobes, and short corollas (all autogamous, chasmogamous or cleistogamous). Flowers may vary significantly in size, even on a single plant. Plants commonly produce tiny cleistogamous flowers on branches (usually at the base of the plant) separate from those with larger flowers. At least the distal and bracteal leaves consistently have hirtellous to hirsutulous adaxial surfaces, even in the smallest of plants. Glandular vestiture is produced only in the axils. Some plants, apparently at the upper limits of populational variability, produce thick-fistulose stems to 100 cm, large leaves (to 80 × 60 mm), and large fruiting calyces (16–19 mm).

Plants described as *Mimulus sookensis* (B. G. Benedict 1993; Benedict et al. 2012) are tetraploid and have been hypothesized to have arisen as alloploids between *Erythranthe nasuta* and *E. microphylla* in two or more independent events. In morphology, however, they are indistinguishable from *E. nasuta* and are treated here within it. In contrast, experimental hybrids and naturally occurring *nasuta-microphylla* hybrids apparently of contemporary origin are intermediate in morphology (see review by G. L. Nesom 2013e). The tetraploid occurs from northern California to southwestern British Columbia (A. L. Sweigart et al. 2008; Benedict et al. 2012).

77. **Erythranthe laciniata** (A. Gray) G. L. Nesom, Phytoneuron 2012-39: 44. 2012 • Cut-leaf monkeyflower E

Mimulus laciniatus A. Gray, Proc. Amer. Acad. Arts 11: 98. 1876; *M. eisenii* Kellogg

Annuals, slender-taprooted or fibrous-rooted. **Stems** erect, simple or branched from base, 3–38 cm, glabrous or sparsely hirtellous, finely villosulous-glandular above nodes. **Leaves** cauline, basal deciduous by flowering; petiole 1–35 mm, distals 0 mm; blade 1-veined or palmately 3-veined, elliptic to elliptic-obovate, oblanceolate, or oblong, 3–55 mm, longer than wide, base attenuate, margins narrowly pinnately lobed or dissected, sometimes merely shallowly toothed, apex acute to obtuse, surfaces glabrate. **Flowers** plesiogamous, 2–8, from medial to distal nodes, chasmogamous, sometimes cleistogamous. **Fruiting pedicels** nodding 30–140° at calyx base, 5–25 mm. **Fruiting calyces** red-spotted, cylindric-campanulate, inflated, sagittally compressed, 8–10 mm, glabrate, throat closing, lobes ca. equal size or adaxial slightly longer. **Corollas** yellow, throat red-spotted, abaxial limb of larger usually with 1 large red splotch, bilaterally symmetric, ± bilabiate; tube-throat funnelform, 4–6 mm, exserted 1–2 mm beyond calyx margin; limb expanded 5–6 mm. **Styles** glabrous. **Anthers** included, glabrous. **Capsules** included, stipitate, 5–7 mm. $2n$ = 28.

Flowering Apr–Jul(–Aug). Cracks, depressions, and seeps in granite outcrops, ledges, talus and scree, rocky streamsides, rocky slopes, roadsides, intermittent drainages; 900–2300(–3300) m; Calif.

Erythranthe laciniata is known from Amador County south to Kern County.

As in *Erythranthe nasuta*, the adaxial calyx lobe in *E. laciniata* tends to be narrowly lanceolate to triangular (noselike) and perceptibly falcate, curving slightly upward both in flower and in fruit. The adaxial lobe is not so prominently protruding as it often is in *E. nasuta*.

Corolla size is variable in *Erythranthe laciniata*, but the size of those with an open throat (versus much reduced in size and apparently cleistogamous) is not strongly correlated with size of the individual plant, and all on one plant are about the same size (compare with *E. nasuta*). Corollas on some plants, however, are all or nearly all greatly reduced and apparently cleistogamous. Fertilization in even the larger corollas apparently is autogamous; the anther pairs are slightly separated or equal in level, and the stigma is in the middle of the anthers or at the level of the adaxial pair.

78. **Erythranthe pardalis** (Pennell) G. L. Nesom, Phytoneuron 2012-39: 44. 2012 • Pennell's panther

E F

Mimulus pardalis Pennell, Proc. Acad. Nat. Sci. Philadelphia 99: 164. 1947; *M. cupriphilus* Macnair; *M. guttatus* de Candolle var. *cupriphilus* (Macnair) D. W. Taylor; *M. guttatus* var. *pardalis* (Pennell) D. W. Taylor

Annuals, fibrous-rooted or taprooted. **Stems** decumbent-ascending, erect distally, simple, sometimes branched from proximal to medial nodes, 5–30 cm, short, delicately stipitate-glandular, distals minutely puberulent-glandular, hairs 0.1–0.4 mm (to 1 mm on proximal portions of stems), gland-tipped. **Leaves** usually cauline, basal usually not persistent; petiole: proximals and medials 8–20 mm, distalmost 1–2 mm; blade palmately 3-veined, usually ovate or broadly ovate to depressed-ovate, proximals and medials 7–22 × 6–18 mm, sometimes largest at mid stem, base rounded or cuneate to gradually attenuate, margins shallowly dentate-serrate, teeth 2 or 3(–5) per side mostly distally, apex obtuse to obtuse-acuminate, surfaces sparsely villous to puberulent-glandular, hairs vitreous, gland-tipped, sometimes glabrous. **Flowers** plesiogamous, 2–12, usually evenly distributed from proximal to distal nodes, chasmogamous, anther pairs in larger corollas slightly separated, stigma at level of distal pair, or both anther pairs and stigma at same level; in smaller corollas without expanded limb and barely exserted beyond calyx margin, both anther pairs and stigma at same level. **Fruiting pedicels** 10–35 mm, short, delicately stipitate-glandular, distals minutely puberulent-glandular, hairs 0.1–0.4 mm, gland-tipped. **Fruiting calyces** nodding 45–180°, consistently dark purple-spotted, cylindric-campanulate, inflated, sagittally compressed,

8–11 mm, glabrous or sparsely puberulent-glandular, sometimes minutely hirtellous, throat closing. **Corollas** yellow, throat floor sometimes red-spotted, bilaterally symmetric, bilabiate; tube-throat narrowly funnelform to cylindric, 7–10(–12) mm, exserted 1–3 mm beyond calyx margin; limb expanded 8–12 mm, palate villous. **Styles** glabrous. **Anthers** included, glabrous. **Capsules** included, stipitate, 4–6 mm. *2n* = 28.

Flowering (Mar–)Apr–May. Crevices of serpentine rock, stony red soils, red clay, among boulders, along streams, ditches, tailings at copper mines; 100–700 m; Calif.

The relative constancy of *Erythranthe pardalis* in morphology suggests that genetic influence from other species is slight. It is recognized by its annual duration and relatively delicate habit, ovate to depressed-ovate leaves toothed mostly on the distal margins, small flowers produced from all nodes (proximal to distal), dark-spotted calyces, and stipitate-glandular cauline and foliar vestiture. While the corolla limbs are distinctly expanded, the tubes are only slightly exserted from the calyx, and the flowers apparently are plesiogamous. The epithet *pardalis* alludes to the dark-spotted calyx.

Plants of *Erythranthe pardalis* occur primarily on serpentine rocks and soils but also grow on copper tailings at mine sites. The species is known from Amador, Calaveras, El Dorado, Placer, Tehama, and Tuolumne counties. The plants in Tehama County, geographically and ecologically disjunct from the main range, were recorded as growing in basalt crevices.

79. **Erythranthe arvensis** (Greene) G. L. Nesom, Phytoneuron 2012-39: 43. 2012 • Villous-bracted monkeyflower F

Mimulus arvensis Greene, Pittonia 1: 37. 1887; *M. guttatus* de Candolle subsp. *arvensis* (Greene) Munz; *M. guttatus* var. *arvensis* (Greene) A. L. Grant; *M. guttatus* subsp. *micranthus* (A. Heller) Munz; *M. langsdorffii* Donn ex Greene var. *arvensis* (Greene) Jepson; *M. longulus* Greene; *M. micranthus* A. Heller

Annuals, taprooted or fibrous-rooted, sometimes rooting at proximal cauline nodes if decumbent. **Stems** erect to decumbent-ascending, simple or branched from proximal to medial nodes, usually 4-angled, fistulose to very narrow, 5–70 cm, glabrous, sometimes minutely hirtellous in inflorescence, hairs deflexed, eglandular. **Leaves** basal and cauline or basal not persistent, often largest at mid stem or above, reduced in size distally; petiole 3–20(–90) mm, distals 0 mm; blade palmately 3–5-veined, ovate to orbicular, orbicular-ovate, oblong-ovate, or (middle and distal cauline) broadly orbicular

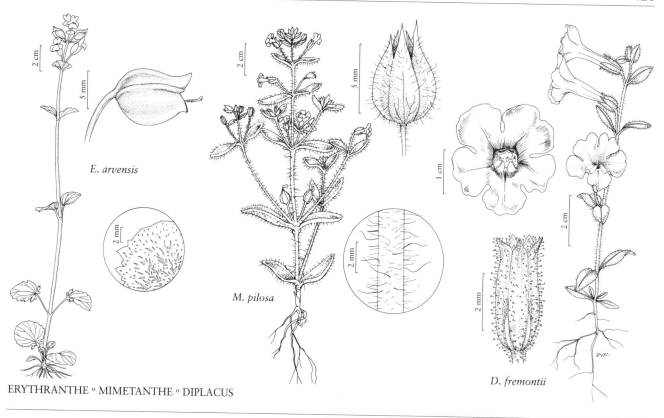

E. arvensis

M. pilosa

D. fremontii

ERYTHRANTHE ° MIMETANTHE ° DIPLACUS

to depressed-ovate or nearly reniform, (5–)10–35(–45) × 6–26(–50) mm, distal closely paired, auriculate-subclasping, base rounded to truncate, subcordate, or shallowly cordate, margins denticulate or subentire to distinctly dentate, on larger plants proximal characteristically lacerate-lobed to pinnatifid at margin base, apex rounded, surfaces glabrous except for bracts densely villous abaxially, sometimes also adaxially, hairs long, sometimes vitreous, flattened, eglandular, multicellular. **Flowers** plesiogamous, 3–8(–16), from remote distal nodes, chasmogamous or cleistogamous. **Fruiting pedicels** 5–40(–90) mm, longer than subtending leaves, glabrous. **Fruiting calyces** red-dotted or not, ovate-campanulate, inflated, sagittally compressed, (7–)9–14 mm, minutely hirtellous, throat closing or not, remaining open, lobes upcurving weakly, adaxial lobe not distinctly longer than abaxial, not falcate. **Corollas** yellow, usually red-spotted, weakly bilaterally or nearly radially symmetric, weakly bilabiate or nearly regular; tube-throat cylindric-funnelform, (7–)8–12 mm, exserted (0–)1–2(–3) mm beyond calyx margin; limb expanded 5–10 mm. **Styles** glabrous. **Anthers** included, glabrous. **Capsules** included, stipitate, (5–)6–7 mm. $2n = 28$.

Flowering Apr–Jun(–Jul). Hills, ridges, clay banks, stream banks, moist woods; 30–1900(–2300) m; B.C.; Calif., Idaho, Mont., Nev., Oreg., Utah, Wash., Wyo.; Mexico (Baja California).

Erythranthe arvensis usually is easily recognized, characterized by its annual duration (but commonly rooting at proximal cauline nodes, suggestive of a rhizomatous habit), glabrous stems with nodes relatively few and remotely spaced, depressed-ovate leaves with margins often sublyrate (lacerate-lobed to subpinnatifid) at the base, distal leaves and bracts densely villous with vitreous eglandular hairs, other leaves glabrous, and corollas varying in size from relatively small but perhaps chasmogamous (the type of *Mimulus arvensis*) to even smaller (cleistogamous; the type of *M. micranthus*). The breeding system is consistently autogamous. The relatively short and even-sized calyx lobes that do not turn upward to close the orifice have been considered diagnostic of *E. arvensis*. This feature is evident in some plants, but others (perhaps reflecting gene flow from other species) have a longer adaxial calyx lobe and abaxial lobes that turn upward variably.

80. **Erythranthe brachystylis** (Edwin) G. L. Nesom, Phytoneuron 2012-39: 43. 2012 • Short-pedicel monkeyflower C E

Mimulus brachystylis Edwin, Leafl. W. Bot. 7: 137. 1954

Annuals, fibrous-rooted, sometimes taprooted, apparently sometimes producing thin runners from basal nodes. **Stems** erect, simple or branched from proximal to medial nodes, 4-angled, filiform to slightly thickened, not distinctly fistulose, 6–22 cm, glabrous. **Leaves** basal and cauline; petiole: proximals 1–8 mm, distals 0 mm; blade palmately 3–5-veined, ovate to depressed-ovate or suborbicular, 10–40 × 6–25 mm, base truncate to subcordate, margins undulate, subentire, or weakly, irregularly dentate, apex rounded, surfaces: proximals and medials glabrous, distals villous, hairs thin-walled, flattened, vitreous and sharp-pointed, eglandular. **Flowers** plesiogamous, 4–10, from medial to distal nodes, cleistogamous. **Fruiting pedicels** 5–10 mm in proximal axils, shorter than or equal to subtending leaves, 1–5 mm distally and flowers and fruits appearing sessile or subsessile, glabrous. **Fruiting calyces** red-tinged to sparsely purple-dotted or not, broadly elliptic-ovoid, inflated, sagittally compressed, 10–13 mm, minutely hirtellous, throat not or slightly closing. **Corollas** yellow, without red markings, weakly bilaterally or nearly radially symmetric, weakly bilabiate or nearly regular; tube-throat narrowly cylindric, 7–9 mm, exserted 0–1 mm beyond calyx margin; limb expanded 3 mm. **Styles** glabrous. **Anthers** included, glabrous. **Capsules** included, stipitate, 4–5 mm.

Flowering Jun–Aug. Around springs, steep slopes; of conservation concern; 2100 m; Nev.

Erythranthe brachystylis is closely similar to *E. arvensis*. Plants of both are annual in duration and produce depressed-ovate leaves, the distal with vitreous-villous surfaces, and tiny corollas barely exserted from the calyx and probably cleistogamous. Vesture of the distal leaves includes an admixture of eglandular sharp-pointed hairs, sometimes encountered in *E. arvensis*, though not typical, perhaps reflecting introgression from *E. nasuta*.

The distinction of *Erythranthe brachystylis* from *E. arvensis* is primarily in its foreshortened pedicels and more inflated fruiting calyces. The fruiting calyces appear to be subsessile or on pedicels shorter or only equaling the subtending leaves. The difference is essentially qualitative but produces a distinctive aspect. *Erythranthe brachystylis* is known only from the type collection in Nye County, a region where *E. arvensis* has not been documented.

81. **Erythranthe cordata** (Greene) G. L. Nesom, Phytoneuron 2012-39: 43. 2012 • Tinytooter monkeyflower E

Mimulus cordatus Greene, Leafl. Bot. Observ. Crit. 2: 5. 1909; *M. maguirei* Pennell

Annuals, fibrous-rooted, sometimes producing leafy runners from basal nodes, stems often rooting at proximal nodes and appearing rhizomelike. **Stems** usually erect, usually simple, usually fistulose, 12–40(–100) cm, sparsely stipitate-glandular, hairs fine, gland-tipped. **Leaves** basal and cauline, basal persistent; petiole: basal and proximals 6–20(–40) mm, midcauline to distals 0 mm; blade not connate, palmately 3–5(–7)-veined, orbicular to broadly elliptic-ovate or oblong-elliptic, cauline becoming broadly ovate to narrowly reniform, basal and mid cauline 15–30(–50) mm, gradually reduced in size distally to 6 mm, basal largest, distal closely paired, auriculate-subclasping, base cuneate to truncate or shallowly cordate, margins shallowly, evenly to unevenly dentate, apex obtuse to rounded, surfaces glabrous. **Flowers** plesiogamous, (5–)10–16, at distal nodes, in bracteate racemes, chasmogamous or cleistogamous. **Fruiting pedicels** 10–30(–45) mm, longer than subtending leaves, minutely stipitate-glandular. **Fruiting calyces** nodding 45–90°, not red-dotted, broadly elliptic-ovoid, inflated, sagittally compressed, (8–)14–18(–20) mm, glabrous or sparsely stipitate-glandular to hirsutulous, sometimes mixed glandular-hirsutulous, throat closing, adaxial lobe not distinctly longer than abaxial, not falcate. **Corollas** yellow, red-spotted, abaxial limb deeper yellow, weakly bilaterally or radially symmetric, weakly bilabiate or regular; tube-throat sometimes tubular and not opening (cleistogamous), 8–14 mm, exserted 1–3 mm beyond calyx margin; limb not expanded or expanded 9–14 mm. **Styles** glabrous. **Anthers** included, glabrous. **Capsules** included, stipitate, 5–7 mm. $2n = 60$.

Flowering (Jan–)Mar–Jun(–Nov). Springs, seeps, stream edges, muddy banks, flood plains, marshes and swamps, wash bottoms, wet depressions, wet places among boulders; (600–)800–2400(–3000) m; Ariz., Calif., Colo., Nev., N.Mex., Tex., Utah.

Erythranthe cordata is characterized by its fibrous-rooted habit (annual in duration, without rhizomes but commonly rooting at the proximal nodes), short corollas and autogamous reproduction (anthers and stigma at the same level), closed calyces, sparsely villous-glandular vesture (lacking hirtellous, eglandular hairs), and stems commonly fistulose in larger plants. The short corollas and other features of autogamous reproduction of *E. cordata* are diagnostic and prominent. Plants of *E. cordata* are highly variable in size, from tiny fibrous-rooted plants with nearly filiform stems to much larger individuals with fistulose stems rooting at proximal nodes.

Erythranthe cordata and *E. nasuta* are sympatric in Arizona, southeastern New Mexico, and southern Utah, and small plants of each species may be similar in aspect, both with cleistogamous flowers and reduced vestiture. *Erythranthe nasuta* can be recognized by its distal and bracteal leaves with hirtellous to hirsutulous adaxial surfaces; a 10× lens usually is required to see this feature, and it sometimes is most obvious around the leaf margins.

The common name of *Erythranthe cordata* alludes to a fancied resemblance of the corollas to the horn of a diminutive trumpet.

82. Erythranthe charlestonensis G. L. Nesom,
Phytoneuron 2012-40: 80. 2012 • Charleston Mountains monkeyflower [E]

Annuals, fibrous-rooted, usually dark purplish. Stems erect, simple, weakly 4-angled, 4–16 (–24) cm, slender, glabrous or sparsely villous-glandular near nodes. Leaves basal and cauline, basal usually persistent, largest at mid stem or above, cauline relatively few on long internodes; petiole: proximals to distals 1–4 mm; blade palmately 3-veined, ovate to ovate-lanceolate, 5–16 (–20) × 3–11 mm, base truncate to subcordate, margins shallowly, evenly crenulate to serrate-dentate or denticulate, apex acute to obtuse, surfaces: proximals glabrous or sparsely villous, distals and bracteals sparsely hirtellous or glabrous abaxially or along distal margin, sparsely villous adaxially, hairs vitreous, flattened, eglandular. Flowers plesiogamous, (1–)4–7, usually from all nodes, usually beginning about mid stem, cleistogamous. Fruiting pedicels deflexed 45–90° at calyx, 6–19 mm, longer than subtending leaves, glabrous or sparsely villous-glandular. Fruiting calyces sometimes purple-dotted, broadly elliptic-ovoid, inflated, sagittally compressed, 10–13 mm, minutely hirtellous, sometimes also sparsely glandular, throat closing, adaxial lobe not distinctly longer than abaxial, not falcate. Corollas yellow, sparsely red-dotted, bilateral or nearly radially symmetric, bilabiate o r nearly regular; tube-throat narrowly cylindric, 4–6 mm, exserted 0.5–1 mm beyond calyx margin; limb barely expanded. Styles glabrous. Anthers included, glabrous. Capsules included, stipitate, 6–8 mm.

Flowering Apr–Jul(–Aug). Grassy slopes, damp soils, moist rocks; (900–)1700–2400(–2800) m; Ariz., Nev.

Erythranthe charlestonensis is characterized by its annual duration, autogamous reproduction, small stature, commonly purplish color, regularly ovate, short-petiolate leaves with shallowly crenate margins

and (distal leaves) sparsely villous or glabrous adaxial surfaces, hirtellous calyces without a prominently longer adaxial lobe, and lack of glandular hairs. It is endemic to the Charleston (Spring) Mountains of Clark County, Nevada, and one locality (Union Pass) in Mohave County, Arizona.

83. Erythranthe hallii (Greene) G. L. Nesom,
Phytoneuron 2012-39: 43. 2012 • Hall's monkeyflower [E]

Mimulus hallii Greene, Bull. Calif. Acad. Sci. 1: 113. 1885; *M. guttatus* de Candolle var. *hallii* (Greene) A. L. Grant

Annuals, fibrous-rooted, sometimes apparently rooting at proximal nodes if stems proximally decumbent. Stems erect, simple, 4-angled, 2–8 cm, slender, glabrous. Leaves basal and cauline or basal deciduous, largest at mid stem or above, cauline relatively few on long internodes; petiole: basal and proximals to midcauline 1–4 mm, distals 0 mm; blade palmately 3-veined, ovate to ovate-lanceolate, 5–11 × 3–9 mm, base truncate to cuneate, margins very shallowly dentate or denticulate, apex acute to obtuse, surfaces glabrous or distals and bracteals sparsely villous, hairs vitreous, flattened, eglandular, multicellular. Flowers plesiogamous, (1–)4–10, sometimes from all nodes, usually beginning about mid stem, cleistogamous. Fruiting pedicels usually deflexed 90° at calyx, 6–14 mm, longer than subtending leaves. Fruiting calyces sometimes red-dotted, broadly elliptic-ovoid, inflated, sagittally compressed, (5–)7–10 mm, glabrous, throat closing, adaxial lobe not distinctly longer than abaxial, not falcate. Corollas yellow, usually red-dotted, bilaterally or nearly radially symmetric, bilabiate or nearly regular; tube-throat narrowly cylindric, 4–6 mm, exserted 0.5–1 mm beyond calyx margin; limb barely expanded. Styles glabrous. Anthers included, glabrous. Capsules included, 4–6 mm. $2n = 32$.

Flowering May–Aug. Ledges, seeps, along streams, wet meadows; 1900–3200 m; Colo.

Erythranthe hallii is known from Boulder, Clear Creek, Fremont, Grand, Jefferson, Larimer, Routt, and Saguache counties. The Colorado population system is morphologically and geographically coherent.

Erythranthe hallii is similar to *E. arvensis*; both have four-angled stems, autogamous reproduction, a tendency to root at basal nodes and distally, and both have bracteal leaves villous with vitreous, flattened, eglandular, multicellular hairs, although this vestiture is barely developed and often absent in *E. hallii*. The only reported chromosome number from the Colorado plants ($2n = 32$) also appears to be distinct among possible relatives of *E. hallii*.

84. Erythranthe parvula (Wooton & Standley) G. L. Nesom, Phytoneuron 2012-39: 44. 2012 • Southwestern mat monkeyflower

Mimulus parvulus Wooton & Standley, Contr. U.S. Natl. Herb. 16: 171. 1913

Perennials, rhizomatous, sometimes rooting at nodes, mat-forming. **Stems** procumbent, branched, 5–15 cm, stipitate-glandular distally. **Leaves** cauline; petiole 1–4 mm; blade palmately 3-veined, ovate to orbicular-ovate or depressed-ovate, 3–11 × 3–9 mm, base truncate to cuneate, margins shallowly denticulate to dentate, teeth 3–5 per side, apex acute to obtuse, surfaces villous-hirsute, hairs whitish, thickened, flattened, stiff, gland-tipped. **Flowers** plesiogamous, 2–8, axillary at distal nodes. **Fruiting pedicels** 7–15 mm, minutely stipitate-glandular. **Fruiting calyces** nodding 45–90°, 5-lobed, cylindric-ovoid, inflated, sagittally compressed, 4–5 mm, villous-hirsute, throat closing. **Corollas** yellow, red-dotted, bilaterally symmetric, bilabiate; tube-throat funnelform, 6–8 mm, exserted beyond calyx margin; limb expanded 4–6 mm, abaxial limb spreading, lobes fimbriate. **Styles** glabrous. **Anthers** included, glabrous. **Capsules** included, 3–4 mm. $2n = 32$.

Flowering Apr–Sep. Wet vertical rock faces, ledges, and rocky slopes, seepy wash banks; 500–2400 (–3400) m; Ariz., N.Mex.; Mexico (Sonora).

All monkeyflowers with laciniate-lobed corollas have generally been identified as *Mimulus dentilobus* B. L. Robinson & Fernald, but these occur as three, morphologically distinct, widely allopatric population systems, each of which is regarded as a separate species (G. L. Nesom 2012g). *Erythranthe dentiloba* (B. L. Robinson & Fernald) G. L. Nesom, which is endemic to Mexico, is the only one of the three with an allogamous breeding system. *Erythranthe parvula* is restricted to Arizona and New Mexico except for one locality in northern Sonora, Mexico. *Erythranthe chinatiensis* is the third species of the group.

85. Erythranthe chinatiensis G. L. Nesom, Phytoneuron 2012-40: 86, figs. 12–14. 2012 • Chinati Mountains monkeyflower [C][E]

Perennials, rhizomatous, sometimes rooting at nodes, mat-forming. **Stems** procumbent, branched, 5–20 cm, glabrous. **Leaves** cauline; petiole 2–10 (–20) mm; blade palmately 3–5(–7)-veined, ovate to broadly ovate or orbicular-ovate, 4–15 (–22) × 4–15(–18) mm, base truncate to cuneate, margins shallowly denticulate or merely mucronate to mucronulate, teeth 3–6 per side,

apex acute to obtuse, surfaces glabrous, adaxial sometimes moderately villosulous, hairs vitreous, flattened, eglandular or minutely gland-tipped. **Flowers** plesiogamous, 2–8, axillary at distal nodes. **Fruiting pedicels** 10–20 mm, glabrous. **Fruiting calyces** nodding 45–90°, 5-lobed, ellipsoid, inflated, sagittally compressed, 5–6 mm, glabrous or sparsely villosulous-glandular, throat closing. **Corollas** yellow, red-dotted, bilaterally symmetric, bilabiate; tube-throat funnelform, 7–8 mm, exserted beyond calyx margin; limb expanded 6–7 mm, abaxial limb strongly reflexed, lobes fimbriate. **Styles** glabrous. **Anthers** included, glabrous. **Capsules** included, 4–5 mm.

Flowering Feb–Sep. Seeps in vertical cliff faces, wet bluffs; of conservation concern; 600–1900(–2300) m; Tex.

Erythranthe chinatiensis is similar to *E. parvula* in its prostrate habit, five-lobed calyces, and fimbriate corolla lobes. It differs from the latter in its nearly glabrous leaves and strongly reflexed abaxial corolla lip. *Erythranthe chinatiensis* is known only from Presidio County but should be expected to occur also in adjacent Chihuahua, Mexico.

86. Erythranthe calciphila (Gentry) G. L. Nesom, Phytoneuron 2012-39: 43. 2012 • Mexican moss monkeyflower

Mimulus calciphilus Gentry, Madroño 9: 21. 1947; *M. minutiflorus* R. K. Vickery

Annuals, usually fibrous-rooted, rarely rooting at proximal nodes. **Stems** erect, sometimes decumbent-ascending, branched, 4–30 cm, minutely stipitate-glandular, also delicately villosulous-glandular along whole length. **Leaves** basal and cauline; petiole 1–3 mm, cauline blade slightly or hardly reduced in size from basal, becoming subsessile to sessile (1–3 pairs of cauline leaves); blade palmately 3(–5)-veined, orbicular-ovate to oblong-ovate, 7–28 × 5–22 mm, base truncate to shallowly cuneate, margins shallowly dentate to denticulate, teeth 3–6 per side, apex rounded to obtuse, surfaces villous, hairs thin-walled, vitreous, eglandular or minutely gland-tipped, usually also minutely stipitate-glandular. **Flowers** plesiogamous, 1–6(–10), axillary at all nodes, chasmogamous. **Fruiting pedicels** 15–30 (–55) mm, minutely stipitate-glandular, sometimes minutely hirtellous and minutely stipitate-glandular, sometimes short glandular-villous. **Fruiting calyces** nodding 90° at maturity, usually 3(–5)-veined, ovoid, inflated, sagittally compressed, 6–10 mm, minutely stipitate-glandular, throat closing, lobes usually 3 or 3 and 5 on same plant, if 5 then with 2 interpolated lobes much smaller than abaxial pair. **Corollas** light yellow, red-spotted, weakly bilaterally symmetric, ± bilabiate; tube-throat narrowly funnelform, 5–7 mm, exserted 1–2 mm beyond calyx margin; limb expanded 2–4 mm. **Styles** glabrous. **Anthers** included, glabrous. **Capsules** included, 2–4 mm. $2n$ = 30, 32.

Flowering Mar–Sep(–Nov). Rocky knobs, moist boulders, wet rock faces, roadcuts, seepages, springs, with moss, usually in pine or pine-oak woods; 1800–2500 m; Ariz.; Mexico (Chihuahua, Durango, Sinaloa, Sonora).

Erythranthe calciphila is recognized by its annual duration (fibrous-rooted), short, erect stems with few, even-sized leaves (the basal often persistent), delicate stipitate-glandular vesture, three-lobed calyces that are relatively large in fruit, small corollas, and autogamous reproduction. Plants rarely root at proximal nodes. In Arizona, at the northernmost extremity of its range, *E. calciphila* occurs in the Chiricahua, Huachuca, and Mule mountains of Cochise County.

5. MIMETANTHE Greene, Bull. Calif. Acad. Sci. 1: 181. 1885 • Mimic monkeyflower [Greek *mimos*, imitator, and *anthe*, flower, alluding to *Mimulus*-like corolla]

Guy L. Nesom

Herpestis C. F. Gaertner sect. *Mimuloides* Bentham in A. P. de Candolle and A. L. P. P. de Candolle, Prodr. 10: 394. 1846; *Mimulus* Linnaeus sect. *Mimuloides* (Bentham) Bentham & Hooker f.

Herbs, annual, terrestrial. **Stems** erect, terete, prominently glandular-villous. **Leaves** mostly cauline, opposite; petiole present or absent; blade not fleshy, not leathery, margins entire, venation acrodromous. **Inflorescences** axillary, flowers paired at nodes, becoming racemelike when nodes condensed; bracts absent. **Pedicels** present, ± equal to or slightly longer than calyces; bracteoles absent. **Flowers** mostly erect, not strongly reflexed and appressed in fruit; sepals 5, calyx bilaterally symmetric, tubular, lobes triangular-lanceolate to deltate-lanceolate, midvein low-rounded, not wing-angled; petals marcescent to fugacious, 5, corolla yellow with 2 purple spots on abaxial lip, bilaterally symmetric, weakly to strongly bilabiate, narrowly tubular-funnelform, abaxial lobes 3, adaxial 2; stamens (2 or)4, didynamous, filaments glabrous; ovary 2-locular, placentation parietal; stigma bilamellate. **Fruits** capsules, apex attenuate, walls densely pustulate-glandular, dehiscence loculicidal along distal ⅓–½ of both sutures. **Seeds** 75–100, brown, narrowly ellipsoid, flattened, wings absent.

Species 1: w United States, nw Mexico.

Mimetanthe has been recognized by some authors (A. L. Grant 1924; R. F. Hoover 1970; N. H. Holmgren 1984) but not others (F. W. Pennell 1951; P. A. Munz 1959; D. M. Thompson 1993, 2005). Bentham originally described the species in *Herpestis*; *Herpestis* is now regarded as a synonym of *Bacopa* (Plantaginaceae).

The ovary and fruit morphology of *Mimetanthe pilosa* (apically attenuate fruits without prismatic or angled walls, parietal placentation) are synapomorphic with *Diplacus*, and *M. pilosa* could justifiably be included as a basal element, sister to the rest of *Diplacus*. Fusion of parietal placentae, glandular seeds, and glandular fruit walls in *M. pilosa* are specializations (C. L. Argue 1980, 1986), and pollen and floral morphology have been noted as distinctions by previous botanists. In treating *M. pilosa* as the monotypic *Mimulus* sect. *Mimuloides*, A. Gray et al. (1886, vol. 2) emphasized the parietal placentation and the deeply divided, essentially unangled calyx.

1. Mimetanthe pilosa (Bentham) Greene, Bull. Calif. Acad. Sci. 1: 181. 1885 • False, downy, or snouted monkeyflower, hairy mimetanthe F

Herpestis pilosa Bentham, Compan. Bot. Mag. 2: 57. 1836; *Mimulus exilis* Durand & Hilgard; *M. pilosus* (Bentham) S. Watson

Annuals fibrous-rooted or taprooted, often becoming woody; stems, leaves, and pedicels villous with mixture of minute (0.1–0.2 mm) stipitate- or sessile-glandular hairs and longer (0.5–1.5 mm), multicellular, flattened, vitreous hairs. **Stems** simple or much-branched, 2–35 cm. **Leaves** 12–50(–90) × 3–15 mm; petiole 0–1 mm; blade narrowly elliptic to oblanceolate or oblong-oblanceolate. **Pedicels** 9–25 (–35) mm. **Flowers:** fruiting calyx erect, swollen-ovoid, 5–8 mm, lobes strongly unequal, 3–4 mm, adaxial longest; corolla 6–10 mm, tube slightly exserted from calyx, expanding abruptly into limb, palate sparsely pubescent; stamens included, proximal pair sometimes reduced or essentially absent; style sessile-glandular. **Capsules** included or slightly exserted, ovoid-fusiform, 4–7 mm, minutely and densely pustulate-glandular.

Flowering Apr–Aug(–Sep). Sandy and gravelly stream banks, sandy streambeds and washes, sandstone seeps; (500–)700–2000(–3000) m; Ariz., Calif., Idaho, Nev., Oreg., Utah, Wash.; Mexico (Baja California, Baja California Sur).

6. DIPLACUS Nuttall, Ann. Nat. Hist. 1: 137. 1838 • Sessile monkeyflower [Greek *dis*-, two, and *plakos*, placenta, alluding to splitting of capsule into valves bearing parietal placentae]

Guy L. Nesom

Melissa C. Tulig

Eunanus Bentham; *Mimulus* Linnaeus sect. *Diplacus* (Nuttall) A. Gray

Herbs, subshrubs, or shrubs, annual or perennial, terrestrial. **Stems** erect, terete, rarely 4-sided (*D. pictus*), glabrous, puberulent, hirtellous, hirsutulous, glandular-puberulent, or glandular-pubescent to glandular-villous with gland-tipped hairs. **Leaves** basal and cauline, or basal deciduous by flowering; petiole absent or present; blade not fleshy, leathery or not, margins toothed to subentire or entire, plane or revolute, venation palmate to subpinnate. **Inflorescences** axillary, flowers usually at medial to distal nodes, 0 or 1(or 2) in each axil, thus 1 or 2(–4) flowers at each node; bracts absent or present. **Pedicels** nearly absent or shorter than calyces, rarely ± equal to or slightly longer than calyces; bracteoles absent. **Flowers** erect to lateral or nodding, not strongly reflexed and appressed in fruit; sepals 5, calyx symmetric, tubular, lobes triangular to deltate, rarely reduced or barely evident, midvein angled or wing-angled; petals 5, corolla marcescent (caducous in *D. mohavensis*, *D. pictus*), blue, pink, red, magenta, lavender, purple, purplish brown, orange, or yellow, rarely white, sometimes multicolor, bilaterally symmetric, rarely radially symmetric, weakly to strongly bilabiate or not, rarely ± rotate, funnelform, or salverform, abaxial lobes 3, adaxial 2; stamens 4, didynamous, filaments glabrous or hairy; staminode 0; ovary 2-locular, placentation parietal; stigma 2-lobed. **Fruits** capsules, ovoid or asymmetrically ovoid to lanceoloid or nearly cylindric, sometimes longitudinally compressed, apex attenuate, walls smooth, eglandular, dehiscence loculicidal along both sutures at least ½ to base, indehiscent (species 29–38), or tardily dehiscent (*D. pictus*). **Seeds** 100–2000, yellow to olive green or dark brown, ovoid to oblong, flattened, wings absent. $x = 8, 9, 10$.

Species 49 (48 in the flora): w United States, nw Mexico.

Diplacus includes western North American species with parietal placentation and pedicels usually distinctly shorter than the calyces, which were formerly treated within *Mimulus* in the broad sense. The closest relatives of *Diplacus* apparently are *Mimetanthe* and the Mexican-Central American *Hemichaena* Bentham, which also are characterized by parietal placentation. *Diplacus* has been divided into six sections (W. R. Barker et al. 2012; G. L. Nesom 2013b), corresponding to morphological patterns and cladistic patterns in the molecular-based phylogenetic analysis by P. M. Beardsley et al. (2004). The species are ordered here following that system as follows: sect. *Eunanus* (Bentham) G. L. Nesom & N. S. Fraga (species 1–27, type *D. nanus*), sect. *Pseudoenoe* (A. L. Grant) G. L. Nesom & N. S. Fraga (species 28, type *D. pictus*), sect. *Cleisanthus* (J. T. Howell) G. L. Nesom & N. S. Fraga (species 29–33, type *D. douglasii*), sect. *Oenoe* (A. Gray) G. L. Nesom & N. S. Fraga (species 34–38, type *D. tricolor*), and sect. *Diplacus* (species 39–48, type *D. aurantiacus*).

Evolutionary radiation in *Diplacus* has been restricted to the United States and Mexico, in contrast to *Erythranthe*, the other large segregate from *Mimulus*. The only species of *Diplacus* that does not occur in the United States is an endemic of Baja California on Cedros Island, *D. stellatus* Kellogg, a member of sect. *Diplacus* (M. C. Tulig and G. L. Nesom 2012). D. M. Thompson (2005) treated *D. stellatus* as a synonym of his broadly conceived *D. aurantiacus*.

A revision of *Mimulus* in the broad sense was provided by A. L. Grant (1924), followed by a major update and overview by F. W. Pennell (1951) that covered the majority of the western North American species of *Diplacus*. The present treatment closely follows that of D. M. Thompson (2005); a major area of taxonomic difference is in sect. *Diplacus*, where he treated most of the variation as within two species (*D. aurantiacus* and *D. clevelandii*), recognizing a number of varieties within *D. aurantiacus* that are here treated at specific rank. The detailed revision of Thompson forms a substantial basis of the treatment here.

Plants in sect. *Diplacus* are subshrubs or shrubs (except for the herbaceous *Diplacus clevelandii*) and are the only perennials (except for *D. rupicola*) in *Diplacus*. They also are distinctive in their glutinous, revolute-margined leaves, often with fascicles of smaller leaves in the axils of main shoots, and minutely glandular styles. Molecular data indicate that sect. *Diplacus* is derived from annual, herbaceous ancestors.

SELECTED REFERENCES Ezell, W. L. 1971. Biosystematics of the *Mimulus nanus* Complex in Oregon. Ph.D. dissertation. Oregon State University. McMinn, H. 1951. Studies in the genus *Diplacus*, Scrophulariaceae. Madroño 11: 33–128. Michener, D. C. 1983. Systematic and ecological wood anatomy of Californian Scrophulariaceae. I. *Antirrhinum, Castilleja, Galvezia*, and *Mimulus* sect. *Diplacus*. Aliso 10: 471–487. Tulig, M. C. 2000. Morphological Variation in *Mimulus* Section *Diplacus* (Scrophulariaceae). M.S. thesis. California State Polytechnic University. Tulig, M. C. and G. L. Nesom. 2012. Taxonomic overview of *Diplacus* sect. *Diplacus* (Phrymaceae). Phytoneuron 2012-45: 1–17.

1. Subshrubs, shrubs, or herbs, perennial; leaf blade margins plane or revolute; calyces 17–40 mm.
    2. Herbs, rhizomatous; corollas bright golden yellow . . . . 39. *Diplacus clevelandii*
    2. Subshrubs or shrubs, not stoloniferous or rhizomatous; corollas red, maroon, scarlet, orange-red, orange, pale orange, dull orange, pale yellow-orange, cream, yellow, pale yellow, or nearly white, sometimes golden yellow (*D. aridus*).
        3. Corollas red, scarlet, orange-red, orange, or maroon.
            4. Calyces glandular-puberulent and short glandular-villous to hirsute-villous . . . . 45. *Diplacus rutilus*
            4. Calyces glabrous.
                5. Leaf blades linear-oblong or narrowly elliptic to narrowly lanceolate or oblanceolate, apices acute; corolla limbs 15–23 mm diam. . . . . 46. *Diplacus puniceus*
                5. Leaf blades elliptic to broadly elliptic-oblanceolate or elliptic-lanceolate, apices acute to obtuse or rounded; corolla limbs 12–16 mm diam. . . . 47. *Diplacus parviflorus*

3. Corollas pale yellow, yellow, golden yellow, light orange, pale yellow-orange, orange, orange-yellow, dull orange, cream, or nearly white.
  6. Abaxial leaf surfaces hairy.
    7. Calyces glabrous or hirtellous and/or minutely stipitate-glandular; leaf blade surfaces glabrous or abaxial usually sparsely to densely hairy, hairs branched, adaxial usually without unbranched hairs . . . . 42. *Diplacus aurantiacus*
    7. Calyces densely glandular-puberulent or glandular-pubescent to short glandular-villous or hirsute-villous; leaf blade abaxial surfaces moderately villous, hairs unbranched, vitreous or densely hairy, hairs branched, adaxial glabrous or glabrescent.
      8. Corollas usually pale yellow or cream to yellow, limbs 20–30 mm diam.; calyces 28–40 mm; leaf blade abaxial surfaces moderately villous, hairs unbranched, vitreous . . . . 43. *Diplacus calycinus*
      8. Corollas light orange to pale yellow-orange, limbs (25–)28–40 mm diam.; calyces 22–32 mm; leaf blade abaxial surfaces densely hairy, hairs branched. . . . . 44. *Diplacus longiflorus*
  6. Abaxial leaf surfaces glabrous.
    9. Corolla lobes entire; calyces 35–40 mm, tubes distinctly dilated distally; stems 80–330 mm . . . . 48. *Diplacus aridus*
    9. Corolla lobes apically incised ¼–½ length, appearing 2-lobed; calyces 20–30 mm, tubes slightly dilated distally; stems 300–800(–1200) mm.
      10. Leaf blades elliptic or oblong-elliptic to elliptic-lanceolate or elliptic-oblanceolate, 4–17 mm wide; corolla limbs 30–45 mm diam.; c Sierra Nevada. . . . . 40. *Diplacus grandiflorus*
      10. Leaf blades narrowly elliptic to narrowly lanceolate or narrowly elliptic-oblong, 3–9 mm wide; corolla limbs 20–25 mm diam.; coastal counties. . . . . .41. *Diplacus linearis*
1. Herbs, annual, rarely perennial (*D. rupicola*); leaf blade margins plane; calyces usually 2–18 mm, (6–)9–25(–31) mm in *D. bolanderi*, *D. brevipes*, *D. traskiae*, and *D. tricolor*.
  11. Herbs, perennial; calyx lobes unequal; corolla limbs pinkish white to nearly white . . . . .38. *Diplacus rupicola*
  11. Herbs, annual; calyx lobes subequal or unequal; corolla limbs not pinkish white to nearly white.
    12. Corollas salverform-rotate, throats dark purplish brown without internal or external markings, palate ridges absent.
      13. Corolla lobes purplish brown basally with red veins; leaf blades narrowly elliptic to elliptic-lanceolate, margins entire; stems terete; stigma lobes subequal. . . . . 27. *Diplacus mohavensis*
      13. Corolla lobes white with purplish brown veins; leaf blades elliptic to elliptic-ovate or obovate, margins crenate; stems 4-sided; stigma lobes unequal. . . . . . .28. *Diplacus pictus*
    12. Corollas bilabiate, sometimes not, throats whitish, yellow, pink, magenta, red, or purple, palate ridges present, sometimes absent (*D. congdonii*).
      14. Corollas yellow or predominantly yellow, sometimes with red, reddish, reddish brown, maroon, or purple spots or lines.
        15. Calyces (7–)10–25(–31) mm; corolla tube-throats (10–)15–30(–34) mm; leaves 7–90(–125) mm. . . . . .7. *Diplacus brevipes*
        15. Calyces (3–)5–14 mm; corolla tube-throats 5–23 mm; leaves 2–30(–55) mm.
          16. Stems 5–15(–20) mm; leaf blade margins usually ciliate along proximal ½; corolla tube-throats 5–10 mm. . . . . 34. *Diplacus pygmaeus*
          16. Stems 10–200(–240) mm; leaf blade margins not ciliate; corolla tube-throats (8–)10–23 mm.
            17. Flowers 1 per node . . . . 2. *Diplacus vandenbergensis*
            17. Flowers 2 per node, or 1 or 2 per node on 1 plant.

18. Stigmas exserted or at opening of corolla tube-throat......
...................................23. *Diplacus mephiticus* (in part)
18. Stigmas included.
  19. Calyx lobes subequal; corolla throats with dark maroon stripes; s Sierra Nevada, (1200–)1500–3300 m.......
.................................12. *Diplacus bicolor* (in part)
  19. Calyx lobes unequal; corolla throats usually with 6–8 narrow reddish spots or lines in arc on abaxial lip around mouth and throat floor with reddish spots, colored palate ridges; Mohave Desert, (600–)800–1700(–1800) m....
.................................25. *Diplacus parryi* (in part)

[14. Shifted to left margin.—Ed.]

14. Corollas shades of red to purple, sometimes white or magenta with yellow abaxial lobes or palate.
  20. Corolla lips: abaxial essentially absent or smaller than adaxial; calyces distinctly asymmetrically attached to pedicels.
    21. Corolla lips: abaxial absent or nearly absent.........................29. *Diplacus douglasii*
    21. Corolla lips: abaxial present, usually smaller than adaxial.
      22. Calyces 18–21 mm; Santa Catalina Island.........................33. *Diplacus traskiae*
      22. Calyces 5–16(–17) mm; Santa Catalina and Santa Cruz islands (*D. brandegeei*) or mainland.
        23. Palate ridges absent or purple; corolla throats not golden yellow at base
................................30. *Diplacus congdonii*
        23. Palate ridges golden yellow; corolla throats golden yellow at base.
          24. Corolla limbs 10–18 mm diam.; pedicels usually twisting, inverting calyx in fruit; leaf margins not ciliate; California (mainland) and Oregon.........................31. *Diplacus kelloggii*
          24. Corolla limbs 5–11 mm diam.; pedicels not twisting, inverting calyx in fruit; leaf margins usually ciliate in proximal ½; California (Santa Catalina and Santa Cruz islands).........................32. *Diplacus brandegeei*
  20. Corolla lips: abaxial and adaxial of ± equal size; calyces symmetrically, sometimes slightly asymmetrically, attached to pedicels.
    25. Plants acaulescent or short-caulescent; stems 30–70 mm; basal leaves densely clustered; corolla tube-throats 15–60 mm.
      26. Corollas not bicolored or tricolored (throat and limb usually magenta to purple with a dark maroon-purple spot or elongate blotch at base of each abaxial lobe, palate ridges and bases of abaxial lips yellow)................35. *Diplacus angustatus*
      26. Corollas bicolored or tricolored.
        27. Corollas bicolored....................................36. *Diplacus pulchellus*
        27. Corollas tricolored.................................37. *Diplacus tricolor* (in part)
    25. Plants caulescent; stems 10–900(–1200) mm; basal leaves seldom densely clustered; corolla tube-throats 6–30 mm.
      28. Leaf blade apices (at least distal leaves) abruptly acuminate or cuspidate to long-tapering or long-acuminate.
        29. Corollas not bilabiate; calyces inflated in fruit, lobes strongly unequal
................................9. *Diplacus bigelovii* (in part)
        29. Corollas bilabiate; calyces not inflated in fruit, lobes subequal or slightly unequal.
          30. Leaf surfaces glabrous or sparsely glandular-puberulent, or adaxial minutely stipitate-glandular.
            31. Corolla tube-throats 13–16(–19) mm, limbs 16–24 mm diam.; leaves (10–)15–25(–35) mm; Malheur County, Oregon, and Ada and Owyhee counties, Idaho........................ 17. *Diplacus cusickii*
            31. Corolla tube-throats 8–12 mm, limbs 10–16 mm diam.; leaves 10–15(–25) mm; Crook, Deschutes, Jefferson, Klamath, Lake, and Wheeler counties, Oregon......................... 19. *Diplacus deschutesensis*
          30. Leaf surfaces densely glandular-villous.

32. Stems (10–)30–240(–350) mm, usually simple or few-branched; calyces 7–12 mm; corolla tube-throats 13–16(–19) mm, limbs 14–26 mm diam.; capsules 10–17 mm; California, Idaho, Oregon, Washington. . . . . .18. *Diplacus cusickioides*

32. Stems 20–140 mm, usually many-branched; calyces 7–9(–10) mm; corolla tube-throats 9–11 mm, limbs 12–15 mm diam.; capsules 6–8 mm; Nevada . . . . .20. *Diplacus ovatus*

[28. Shifted to left margin.—Ed.]

28. Leaf blade apices rounded, rounded-acute, obtuse, or acute.

33. Capsules indehiscent; corolla lobes each with discrete, dark spot at base; flowers 1 per node . . . . 37. *Diplacus tricolor* (in part)

33. Capsules dehiscent; corolla lobes not each with discrete, dark spot at base; flowers 1, 2, or 1 or 2 per node on 1 plant.

34. Flowers 1 per node.

35. Corolla tube-throats 7–10 mm, limbs 4–7 mm diam. . . . . 3. *Diplacus rattanii*

35. Corolla tube-throats 9–23 mm, limbs 8–26 mm diam.

36. Corollas: palate ridges yellow, throat floor glabrous or puberulent at mouth . . . . .1. *Diplacus fremontii*

36. Corollas: palate ridges white or yellow fading distally to white, throat villous at mouth.

37. Corolla lobes with dark red-purple midveins extending from throat, lobes not each with dark spot at base, throat ceilings glabrous; styles glandular-puberulent; abaxial stigma lobes 1.5 times adaxials. . . . . . 4. *Diplacus viscidus*

37. Corolla lobes without dark midveins, lobes dark at base, throat ceilings villous-pilose; styles glabrous or sparsely eglandular-puberulent; abaxial stigma lobes 3–4 times adaxials .5. *Diplacus compactus*

34. Flowers 2 per node, or 1 or 2 per node on 1 plant.

38. Stigmas exserted or at opening of corolla tube-throats.

39. Corolla limbs not bilabiate; anthers glabrous.

40. Corollas: palate ridges and throat floors yellow, throats usually with large dark spot on each side of mouth on lateral walls; stigma lobes equal; calyces not inflated in fruit, ribs inconspicuous . . . . 6. *Diplacus johnstonii*

40. Corollas: palate ridges or throat floors white, throats without dark spots; stigma lobes unequal, abaxials 2–3 times adaxials; calyces inflated in fruit, ribs prominent . . . . .8. *Diplacus bolanderi*

39. Corolla limbs bilabiate, rarely not; anthers hairy.

41. Leaf blade margins crenate to serrulate, at least on biggest leaves; pedicels 2–7(–10) mm in fruit . . . . .16. *Diplacus clivicola*

41. Leaf blade margins entire; pedicels 1–3 mm in fruit.

42. Corolla limbs 5–7 mm diam., tube-throats 7–10 mm, tubes 0.8–1.2 mm diam. at filament insertion; capsules 4–6 mm . . . . 24. *Diplacus jepsonii*

42. Corolla limbs (5–)7–14(–15) mm diam., tube-throats 8–14(–15) mm, tubes 1.1–1.9 mm diam. at filament insertion; capsules 5–12 mm.

43. Leaf surfaces glandular-pubescent and (at least along veins) viscid-villous. . . . . .23. *Diplacus mephiticus* (in part)

43. Leaf surfaces minutely glandular-puberulent.

44. Distal leaves: blades oblanceolate to elliptic-lanceolate, 2–4 mm wide; corolla tube-throats 11–15 mm, throat floors villous with hairs extending onto adaxial lip; capsules 8–12 mm . . . . .21. *Diplacus nanus*

44. Distal leaves: blades obovate to broadly oblanceolate, 5–8(–12) mm wide; corolla tube-throats 8–10 mm, throat floors glabrous; capsules 5–8(–9) mm . . . . . 22. *Diplacus cascadensis*

[38. Shifted to left margin.—Ed.]

38. Stigmas included.
    45. Stems glandular-puberulent; calyx lobes unequal, adaxial longer . . . . .25. *Diplacus parryi* (in part)
    45. Stems glandular-puberulent to glandular-pubescent or glandular-villous, sometimes glabrous or glabrate proximally; calyx lobes subequal.
        46. Corolla tube-throats 6–8 mm, limbs 3–5 mm diam. . . . . 15. *Diplacus leptaleus*
        46. Corolla tube-throats (8–)9–22(–25) mm, limbs 6–24 mm diam.
            47. Calyx lobes shallowly triangular to broadly ovate; herbage not drying dark
            . . . . 26. *Diplacus torreyi*
            47. Calyx lobes triangular or narrowly triangular; herbage usually drying dark.
                48. Calyx tubes not strongly plicate, ribs thin, not strongly raised . . . .
                . . . . .12. *Diplacus bicolor* (in part)
                48. Calyx tubes strongly plicate, ribs thickened, strongly raised.
                    49. Nodes 4–15(–20); calyx ribs narrow, intercostal areas green to purple, not membranous . . . . .14. *Diplacus graniticola*
                    49. Nodes 3–6; calyx ribs broad, intercostal areas whitish, membranous.
                        50. Calyces not inflated in fruit; abaxial stigma lobes 1.5 times adaxial . . . . 13. *Diplacus layneae*
                        50. Calyces inflated in fruit; stigma lobes equal to subequal.
                            51. Leaf apices abruptly acuminate, acute-acuminate, or cuspidate to long-tapering or long-acuminate . . . . .9. *Diplacus bigelovii* (in part)
                            51. Leaf apices acute or rounded.
                                52. Stems and leaves glandular-puberulent; corolla tube-throats 10–14 mm, limbs 6–10 mm diam.; palate ridges yellow . . . . 10. *Diplacus thompsonii*
                              52. Stems glandular-villous; corolla tube-throats (10–)13–22(–25) mm, limbs 14–23 mm diam.; palate ridges white . . . . .11. *Diplacus constrictus*

1. **Diplacus fremontii** (Bentham) G. L. Nesom, Phytoneuron 2012-39: 28. 2012 • Fremont's monkeyflower [F]

Eunanus fremontii Bentham in A. P. de Candolle and A. L. P. P. de Candolle, Prodr. 10: 374. 1846 (as fremonti); *Mimulus fremontii* (Bentham) A. Gray; *M. subsecundus* A. Gray

Herbs, annual. **Stems** erect, 10–200(–240) mm, glandular-puberulent or glandular-pubescent. **Leaves** basal and cauline, basal in rosette, cauline reduced distally; petiole absent; blade narrowly elliptic, sometimes obovate to oblanceolate, 2–30(–55) × 1–10(–16) mm, margins entire, sometimes crenate to serrate, plane, apex rounded to acute, surfaces: proximals glabrous, distals glandular-puberulent or glandular-pubescent. **Pedicels** 1–4(–7 on proximal) mm

in fruit. **Flowers** 1 per node, chasmogamous. **Calyces** symmetrically attached to pedicels, inflated in fruit, 5–14 mm, glandular-puberulent to glandular-pubescent or ribs almost tomentose and viscid, lobes subequal, apex rounded and apiculate or acute, intercostal areas white. **Corollas** magenta to dark reddish purple, throat often darker near mouth, palate ridges yellow at mouth, throat floor glabrous or minutely puberulent, tube-throat 9–23 mm, limb 8–26 mm diam., not bilabiate. **Anthers** included, glabrous, rarely minutely puberulent. **Styles** minutely puberulent. **Stigmas** included, lobes subequal. **Capsules** 6.5–13(–14) mm. $2n = 16$.

Flowering Mar–Jun. Soft, sandy soils along washes, flood plains, areas of water runoff, sandy hilltops and flats; 100–2100 m; Calif.; Mexico (Baja California).

Diplacus fremontii occurs from Monterey and San Benito counties south to San Diego County, east to Kern County and adjacent Inyo County, and in Baja California.

2. Diplacus vandenbergensis (D. M. Thompson)
G. L. Nesom, Phytoneuron 2012-47: 2. 2012
• Vandenberg monkeyflower E

Mimulus fremontii (Bentham) A. Gray var. *vandenbergensis* D. M. Thompson, Syst. Bot. Monogr. 75: 134, fig. 53G, plate 6b. 2005

Herbs, annual. **Stems** erect, 10–200(–240) mm, glandular-puberulent or glandular-pubescent. **Leaves** basal and cauline, basal in rosette, cauline reduced distally; petiole absent; blade narrowly elliptic, sometimes obovate to oblanceolate, 2–30(–55) × 1–10(–16) mm, margins entire, sometimes crenate to serrate, plane, not ciliate, apex rounded to acute, surfaces: proximals glabrous, distals glandular-puberulent or glandular-pubescent. **Pedicels** 1–4(–7 on proximal) mm in fruit. **Flowers** 1 per node, chasmogamous. **Calyces** inflated in fruit, 5–14 mm, glandular-puberulent to glandular-pubescent or ribs almost tomentose and viscid, lobes subequal, apex rounded and apiculate or acute, intercostal areas white. **Corollas** yellow with reddish brown spots near mouth, palate ridges yellow, throat floor puberulent, tube-throat (9–)10–23 mm, limb 8–26 mm diam., not bilabiate. **Anthers** included, glabrous, rarely minutely puberulent. **Styles** minutely puberulent. **Stigmas** included, lobes equal. **Capsules** 6.5–13(–14) mm. *2n* = 16.

Flowering Apr–Jun. Sandy open or disturbed areas, among shrubs; 70–150 m; Calif.

Diplacus vandenbergensis is endemic to Santa Barbara County, mostly on the north side of the city of Lompoc (La Purisima Mission State Historic Park; also on and near Vandenberg Air Force Base) plus one other locality about 10 kilometers farther west (Santa Ynez Valley, 8 miles west of Buellton). Thompson described *Mimulus fremontii* var. *vandenbergensis* as a geographically distinct variant of *M. fremontii*. The discontinuous morphological difference, allopatric/parapatric geographical distribution, and lack of intergrading populations support recognition of the plants with yellow corollas at specific rank. Plants with yellow corollas are found in *D. bicolor*, *D. mephiticus*, and *D. parryi*, but these morphs are found through most of the geographic ranges of these taxa and sometimes are intermixed in populations.

3. Diplacus rattanii (A. Gray) G. L. Nesom, Phytoneuron 2012-39: 29. 2012 • Rattan's monkeyflower E

Mimulus rattanii A. Gray, Proc. Amer. Acad. Arts 20: 307. 1885 (as rattani); *M. rattanii* subsp. *decurtatus* (A. L. Grant) Munz

Herbs, annual. **Stems** erect, 10–180(–230) mm, densely glandular-pubescent and viscid. **Leaves** basal and cauline, basal in rosette, cauline gradually reduced distally; petiole absent; blade obovate to narrowly elliptic, 3–46(–70) × 1–20(–25) mm, margins entire or crenate, plane, apex rounded or obtuse, surfaces: proximals glabrate, distals glandular-pubescent and viscid. **Pedicels** 1–3(–6) mm in fruit. **Flowers** 1 per node, chasmogamous. **Calyces** symmetrically attached to pedicels, inflated in fruit, 5–10 mm, glandular-pubescent and viscid, lobes subequal, apex obtuse, often apiculate, ribs dark green to purplish, intercostal areas whitish. **Corollas** pink to magenta, throat floor with 3 dark purple lines meeting abaxial lip lobes, palate ridges yellow, tube-throat 7–10 mm, limb 4–7 mm diam., not bilabiate. **Anthers** nearly exserted, glabrous. **Styles** eglandular-puberulent. **Stigmas** nearly exserted, lobes unequal, abaxial 5–7 times adaxial. **Capsules** 7–11(–12) mm. *2n* = 16.

Flowering Apr–Jul. Recently burned or cleared areas, sandhills, sandstone outcrops, sandy gravel and loam, decomposed granite, serpentine-derived soils, open chaparral, chaparral margins, open yellow pine-manzanita woodlands; 300–1300 m; Calif.

Diplacus rattanii occurs mostly in near-coastal localities from Glenn and Lake counties south to Ventura County.

4. Diplacus viscidus (Congdon) G. L. Nesom, Phytoneuron 2012-39: 29. 2012 • Sticky monkeyflower E

Mimulus viscidus Congdon, Erythea 7: 187. 1900; *M. fremontii* (Bentham) A. Gray var. *viscidus* (Congdon) Jepson; *M. subsecundus* A. Gray var. *viscidus* (Congdon) A. L. Grant

Herbs, annual. **Stems** erect, (30–)60–370 mm, densely glandular-pubescent with viscid hairs. **Leaves** usually cauline, relatively even-sized or largest proximally and gradually reduced distally; petiole absent; blade obovate to narrowly elliptic, (4–)8–54 (–70) × (2–)3–23 mm, margins entire or serrate, plane, apex obtuse to rounded, surfaces: proximals glabrous abaxially, distals glandular-pubescent. **Pedicels** 1–4 (–5) mm in fruit. **Flowers** 1 per node, chasmogamous. **Calyces** symmetrically attached to pedicels, inflated in fruit, (7–)8–15 mm, villous, hairs eglandular, lobes subequal, apex acute to attenuate, ribs and intercostal areas often reddish. **Corollas** lavender to magenta with diffuse dark markings on sides of darker tube-throat and with dark red-purple midveins on lobes extending from throat, lobes not dark at base, floor white or yellow, fading to white at mouth, palate ridges white or yellow fading to white distally, throat ceiling glabrous, tube-throat 10–20 mm, limb 8–20 mm diam., not bilabiate. **Anthers** included, ciliate. **Styles** glandular-puberulent. **Stigmas** included, lobes unequal, abaxial 1.5 times adaxial. **Capsules** 7–11 mm. *2n* = 16.

Flowering Apr–Jul. Chaparral clearings and openings; 90–1300 m; Calif.

D. M. Thompson (2005) noted that *Diplacus viscidus* and *D. compactus* (as *Mimulus viscidus* var. *compactus*) are parapatric and may intergrade in central Mariposa County. The two taxa are distinguished by the presence or absence of dark stripes on the corolla lobe midveins, which are evident even on herbarium specimens. Thompson found that the two remained distinct when grown together in the greenhouse.

Diplacus viscidus is known from Amador, Calaveras, Eldorado, Mariposa, Merced, and Tuolumne counties; *D. compactus* continues south through Fresno, northern Kern, Madera, Mariposa, and Tulare counties.

5. **Diplacus compactus** (D. M. Thompson) G. L. Nesom, Phytoneuron 2012-47: 1. 2012 • Compact monkeyflower [E]

Mimulus viscidus Congdon var. *compactus* D. M. Thompson, Syst. Bot. Monogr. 75: 129. 2005

Herbs, annual. **Stems** erect, 20–280 mm, densely glandular-pubescent and viscid. **Leaves** usually cauline, gradually reduced distally; petiole absent; blade obovate to narrowly elliptic, 4–40 × 0.7–20 mm, margins entire or serrate, plane, apex obtuse to rounded, surfaces: proximals glabrous abaxially, distals glandular-pubescent. **Pedicels** 1–4(–5) mm in fruit. **Flowers** 1 per node, chasmogamous. **Calyces** symmetrically attached to pedicels, inflated in fruit, 6–10(–12) mm, villous, hairs eglandular, lobes unequal, apex acute to attenuate, ribs and intercostal areas often reddish. **Corollas** lavender to magenta or red with diffuse dark markings on sides of darker tube-throat, lobes dark at base, without radiating dark lines, floor and palate ridges white or yellow fading distally to white, palate ridges and throat ceiling villous-pilose, tube-throat 10–20 mm, limb 8–20 mm diam., not bilabiate. **Anthers** included, ciliate. **Styles** glabrous or sparsely eglandular-puberulent. **Stigmas** included, lobes unequal, abaxial 3–4 times adaxial. **Capsules** 6–9 mm. *2n* = 16.

Flowering May–Jul. Hillsides, washes, recently burned areas, soil and scree banks, granitic sand, chaparral openings, gray pine-blue oak woodlands; 300–1000 m; Calif.

6. **Diplacus johnstonii** (A. L. Grant) G. L. Nesom, Phytoneuron 2012-39: 29. 2012 • Johnston's monkeyflower [E]

Mimulus johnstonii A. L. Grant, Ann. Missouri Bot. Gard. 11: 280. 1925

Herbs, annual. **Stems** erect, (10–)30–200(–300) mm, densely glandular-puberulent. **Leaves** basal and cauline, gradually reduced distally; petiole absent, bases of larger leaves often with petiole-like extensions; blade obovate or oblanceolate, sometimes elliptic, (4.5–)7–25(–32) × 2–12(–15) mm, margins entire, plane, apex rounded to acute, surfaces densely glandular-puberulent. **Pedicels** 1–4(–5) mm in fruit. **Flowers** 2 per node, or 1 or 2 per node on 1 plant, chasmogamous. **Calyces** symmetrically attached to pedicels, not inflated in fruit, (6–)7–11 mm, glandular-puberulent, lobes unequal, apex acute to acuminate, ribs inconspicuous, intercostal areas reddish. **Corollas** magenta, darker and more reddish in throat and, often, along narrow radiating lines extending from throat onto midveins of lobes, throat usually with a large dark spot on each side of mouth on lateral walls, palate ridges and throat floor yellow with reddish spots, tube-throat 9–15 mm, limb 10–15 mm diam., not bilabiate. **Anthers** included, glabrous. **Styles** densely glandular-puberulent distally. **Stigmas** exserted or at opening of corolla tube-throat, lobes equal. **Capsules** 7–12 mm. *2n* = 16.

Flowering May–Aug. Steep, unstable scree slides, talus slopes, gravel slides, cracks in granite cliffs, ridges, washes, steep sand and gravel slopes, canyon bottoms, gravelly road banks, recent burns, desert scrub, chaparral, juniper, pinyon-juniper, lodgepole pine, yellow pine, Jeffrey pine, and Jeffrey pine-western white pine-fir woodlands; (1000–)1300–2900 m; Calif.

Populations of *Diplacus johnstonii* occur in Los Angeles, Riverside, San Bernardino, Santa Barbara, and Ventura counties. The populations in northwestern Los Angeles, Santa Barbara, and Ventura counties were noted by D. M. Thompson (2005) to be intermediate between *D. constrictus* and *D. johnstonii*.

7. **Diplacus brevipes** (Bentham) G. L. Nesom, Phytoneuron 2012-39: 28. 2012 • Wide-throat yellow monkeyflower

Mimulus brevipes Bentham, Scroph. Ind., 28. 1835

Herbs, annual. **Stems** erect, (25–)50–800(–1000) mm, usually glandular-puberulent or glandular-pubescent, viscid. **Leaves** basal and cauline, basal usually in rosette, cauline gradually reduced distally; petiole often present proximally, usually absent distally; blade linear-lanceolate, elliptic, narrowly oblanceolate, or lanceolate, sometimes ovate or obovate, 7–90 (–125) × 1–40(–48) mm, margins serrate or entire, plane, apex rounded to acute, surfaces: proximals glabrate, distals glandular-puberulent or glandular-pubescent. **Pedicels** 2–10(–17 at proximalmost node) mm in fruit. **Flowers** 2 per node, or 1 or 2 per node on 1 plant, chasmogamous. **Calyces** inflated in fruit, (7–)10–25(–31) mm, glandular-pubescent and viscid, lobes unequal, apex acute to acuminate, ribs green, sometimes purplish, intercostal areas white. **Corollas** yellow, usually with reddish brown spots, palate ridges yellow, tube-throat (10–)15–30(–34) mm, limb 11–30 mm diam., not bilabiate. **Anthers** included, glabrous. **Styles** glandular-puberulent. **Stigmas** included, lobe unequal, abaxial 1.5–2 times adaxial. **Capsules** (7–)8–14(–17) mm. **2*n*** = 16.

Flowering Apr–Jul. Openings in chaparral or coastal sage scrub, recently burned or mechanically disturbed areas; 30–1800(–2200) m; Calif.; Mexico (Baja California).

Diplacus brevipes occurs in the southern quarter of California. It is distinctive in its relatively large, yellow corollas, linear-lanceolate leaves, and relatively long internodes.

8. **Diplacus bolanderi** (A. Gray) G. L. Nesom, Phytoneuron 2012-39: 28. 2012 • Bolander's monkeyflower E

Mimulus bolanderi A. Gray, Proc. Amer. Acad. Arts 7: 381. 1868; *M. platylaemus* Pennell

Herbs, annual. **Stems** erect, 20–900(–1200) mm, usually glandular-pubescent and viscid. **Leaves** basal and cauline, basal in rosette, cauline gradually reduced distally; petiole present proximally, usually absent distally; blade ovate, obovate, elliptic, or oblanceolate, 5–60(–85) × 1.5–25(–32) mm, margins entire or largest irregularly serrate, plane, apex rounded to obtuse, surfaces: proximals nearly glabrous,

distals glandular-pubescent. **Pedicels** 2–5(–8) mm in fruit. **Flowers** 2 per node, or 1 or 2 per node on 1 plant (1 per node on depauperate plants), chasmogamous. **Calyces** symmetrically attached to pedicels, inflated in fruit, (7–)9–25(–27) mm, glandular-pubescent and viscid, lobes often spreading in fruit, unequal if calyx large, apex acute to attenuate, ribs prominent, purplish, intercostal areas white. **Corollas** magenta, throat often darkening near floor, without dark spots, palate ridges or floor white, usually magenta-speckled, tube-throat 10–30 mm, limb 6–20 mm diam., not bilabiate. **Anthers** included, glabrous. **Styles** glandular-puberulent. **Stigmas** exserted, lobes unequal, abaxial 2–3 times adaxial. **Capsules** 8–20 mm. **2*n*** = 16.

Flowering Apr–Jul. Granite outcrops, slopes with decomposed granite soils; (100–)300–1700(–2000) m; Calif., Oreg.

Diplacus bolanderi is scattered and uncommon through most of California (with concentrations in the southern North Coast Ranges and Central Sierra) and apparently is disjunct in Jackson County, Oregon.

Diplacus bolanderi is highly variable in height, leaf size, corolla size, and calyx size. Identity is unmistakable in larger plants; the inflated calyx with lobes strongly unequal in length usually is helpful in identification of smaller plants.

9. **Diplacus bigelovii** (A. Gray) G. L. Nesom, Phytoneuron 2012-39: 28. 2012 • Bigelow's monkeyflower E

Eunanus bigelovii A. Gray in War Department [U.S.], Pacif., Railr. Rep. 4(5): 121. 1857; *Mimulus bigelovii* (A. Gray) A. Gray

Herbs, annual, herbage usually drying dark. **Stems** erect, (10–) 20–250(–320) mm, nodes 3–6, internodes 1–6 mm, glandular-pubescent to glandular-villous. **Leaves** usually cauline, relatively even-sized or reduced distally; petiole absent, bases of largest leaves often long-tapered to petiole-like extensions; blade obovate, elliptic, or oblanceolate, (5–)7–35(–50) × (2–)3–18 (–26) mm, margins entire, rarely toothed, plane, apex abruptly acuminate, acute-acuminate, or cuspidate to long-tapering or long-acuminate, surfaces glandular-pubescent. **Pedicels** 1–4(–8) mm in fruit. **Flowers** 2 per node, or 1 or 2 per node on 1 plant, chasmogamous. **Calyces** symmetrically attached to pedicels, inflated in fruit, 6–13(–15) mm, glandular-pubescent, tube strongly plicate, lobes slightly recurved, narrowly triangular, subequal, often slightly indurate, apex acuminate to attenuate, ribs broad, darkened, blackish, thickened, strongly raised, intercostal areas whitish, membranous. **Corollas** magenta with dark reddish spot on each side of mouth on interior lateral walls of throat, usually with reddish lines extending from throat onto midveins of lobes, throat floor yellow with reddish speckling and variable reddish markings, palate ridges yellow, tube-throat (9–)12–22 mm, limb 12–24 mm diam., not bilabiate. **Anthers** included, glabrous, sometimes ciliate. **Styles** glandular-puberulent. **Stigmas** included, lobes equal. **Capsules** (6–)7–13(–15) mm.

Varieties 2 (2 in the flora): w United States.

Diplacus bigelovii is distributed in southeastern California from southern Mono County south to Imperial and San Diego counties through southern Nevada into Washington County, Utah, and La Paz and Mohave counties, Arizona. The relatively sharp line dividing the two varieties roughly follows the Inyo-San Bernardino county line, then cuts across Clark County, Nevada, and Mojave County, Arizona.

Diplacus bigelovii can generally be recognized by its relatively large, nearly radially symmetric corollas, included stigmas, and inflated mature calyces with lobes of unequal length and apices acuminate-attenuate. The two varieties have distinctive leaf shapes; D. M. Thompson (2005) reported them as very closely parapatric and exhibiting limited intergradation near their contiguous occurrence.

1. Leaf blades: distals gradually narrower than proximals, apices long-tapering or long-acuminate; internodes: proximals usually longer than distals . . . . 9a. *Diplacus bigelovii* var. *bigelovii*
1. Leaf blades: distals usually relatively broader than proximals, apices abruptly acute-acuminate, sometimes cuspidate; internodes usually subequal . . . . 9b. *Diplacus bigelovii* var. *cuspidatus*

9a. Diplacus bigelovii (A. Gray) G. L. Nesom var. **bigelovii** ⌐E¬

Stem internodes: proximals usually longer than distals. **Leaf blades:** proximals usually obovate, distals gradually narrower than proximals, apices of proximals usually rounded, distals long-tapering or long-acuminate.

Flowering Feb–Jun(–Sep). Washes and adjacent hillsides; 100–1700(–2000) m; Ariz., Calif., Nev.

9b. Diplacus bigelovii (A. Gray) G. L. Nesom var. **cuspidatus** (A. L. Grant) G. L. Nesom, Phytoneuron 2012-39: 28. 2012 ⌐E¬

Mimulus bigelovii (A. Gray) A. Gray var. *cuspidatus* A. L. Grant, Ann. Missouri Bot. Gard. 11: 279. 1925; *M. bigelovii* var. *panamintensis* Munz; *M. spissus* A. L. Grant

Stem internodes usually subequal. **Leaf blades:** proximals obovate to oblanceolate, distals usually relatively broader, apices of proximals abruptly acute-acuminate, distals abruptly acuminate, sometimes cuspidate.

Flowering Feb–Jun(–Sep). Washes, dry hillsides, limestone outcrops, scree slopes, periphery of trees and shrubs; (10–)100–2300(–3300) m; Ariz., Calif., Nev., Utah.

Leaves of var. *cuspidatus* sometimes are more congested near stem tips under conditions of severe drought stress. Seeds from plants identified as *Mimulus spissus*, with leaves and flowers densely crowded at least distally, were greenhouse grown by D. M. Thompson (2005), who found that they produced plants identical to var. *cuspidatus*.

10. Diplacus thompsonii G. L. Nesom, Phytoneuron 2013-46: 1, figs. 2, 3. 2013 • Thompson's monkeyflower ⌐E¬ ⌐F¬

Herbs, annual, herbage usually drying dark. **Stems** erect, 30–150 mm, nodes 3–6, internodes 1–6 mm, glandular-puberulent. **Leaves** usually cauline, relatively even-sized; petiole absent, proximal base narrowed to petiole-like extension; blade lanceolate or narrowly lance-olate to elliptic-lanceolate, 7–30 × 2–7 mm, margins entire, plane, apex acute, surfaces glandular-puberulent. **Pedicels** 1–3 mm in fruit. **Flowers** 2 per node, or 1 or 2 per node on 1 plant, chasmogamous. **Calyces** symmetrically attached to pedicels, inflated in fruit, 7–9 mm, glandular-villous, tube strongly plicate, lobes narrowly triangular, subequal, apex acuminate to attenuate, ribs broad, darkened, blackish, thickened, strongly raised, intercostal areas whitish, membranous. **Corollas** dark magenta to red-purple, throat floor yellowish and red-dotted, with a red-purple midvein radiating onto base of each lobe of proximal lip, palate ridges yellow, tube-throat 10–14 mm, limb 6–10 mm diam., not bilabiate. **Anthers** included, ciliate. **Styles** puberulent. **Stigmas** included, lobes subequal. **Capsules** 7–9 mm. $2n = 16$.

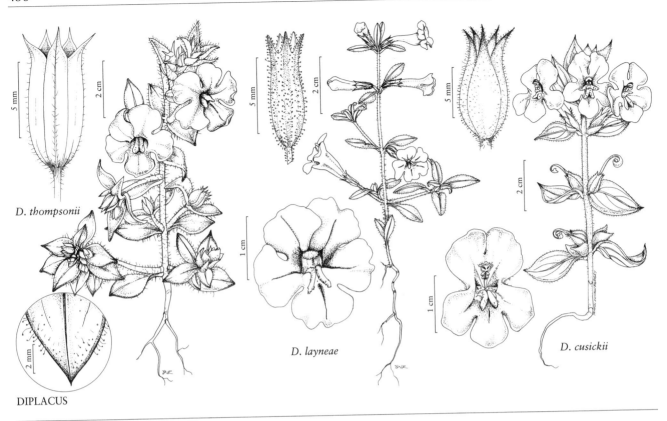

D. *thompsonii*

DIPLACUS

D. *layneae*

D. *cusickii*

Flowering May–Jul. Granitic soils and sands, sandy loam, outwash slopes, dry sagebrush flats, pinyon and Jeffrey pine woodlands; 1800–2600(–3000) m; Calif.

Diplacus thompsonii is abundant along the eastern side of the Sierra Nevada in Inyo and southern Mono counties. D. M. Thompson (2005) identified these populations as intermediate between *Mimulus nanus* var. *mephiticus* and *M. bigelovii*; diagnostic features are more similar to those of the latter, especially the inflated and strongly wing-angled calyx and the non-bilabiate corollas. The range of *D. thompsonii* is within the northern limit of *D. bigelovii* var. *cuspidatus*, which is different in leaf morphology.

11. **Diplacus constrictus** (A. L. Grant) G. L. Nesom, Phytoneuron 2012-39: 28. 2012 • Dense-fruited monkeyflower [E]

Mimulus subsecundus A. Gray subsp. *constrictus* A. L. Grant, Ann. Missouri Bot. Gard. 11: 287. 1925; *M. constrictus* (A. L. Grant) Pennell; *M. viscidus* Congdon subsp. *constrictus* (A. L. Grant) Munz

Herbs, annual, herbage usually drying dark. **Stems** erect, (10–) 20–240(–350) mm, nodes 3–6, internodes 1–6 mm, glandular-villous. **Leaves** basal and cauline, relatively even-sized; petiole indistinct; blade obovate, oblance-olate, or narrowly elliptic, (3.5–)5–32(–47) × 3–15 (–18) mm, margins entire or toothed, plane, apex acute or rounded, surfaces: proximals glabrous, distals glandular-pubescent. **Pedicels** 0.5–3(–4) mm in fruit. **Flowers** 2 per node, or 1 or 2 per node on 1 plant, chasmogamous. **Calyces** symmetrically attached to pedicels, inflated in fruit, (5–)7–12(–15) mm, glandular-pubescent to glandular-villous, tube strongly plicate, lobes triangular, subequal, apex acute, ribs broad, darkened, blackish, thickened, strongly raised, inter-costal areas whitish, membranous. **Corollas** magenta or pinkish to red-purple, throat floor whitish with dark lines or streaks, often yellowish deep inside throat, never at mouth, palate ridges white, tube-throat (10–)13–22 (–25) mm, limb 14–23 mm diam., not bilabiate. **Anthers** included, ciliate. **Styles** glandular-puberulent. **Stigmas** included, lobes equal. **Capsules** (7–)8–12(–13.5) mm. $2n = 16$.

Flowering May–Aug. Disturbed areas with concentrated runoff from rains on, or just above, verges of roadside banks; 800–2100(–2400) m; Calif.

Diplacus constrictus is endemic to Kern, Los Angeles, Tulare, and Ventura counties.

12. Diplacus bicolor (A. Gray) Hrusa, Phytoneuron 2014-17: 1. 2014 • Whitney's monkeyflower Ⓔ

Eunanus bicolor A. Gray, Proc. Amer. Acad. Arts 7: 381. 1868, not *Mimulus bicolor* Hartweg ex Bentham 1849; *M. nanus* Hooker & Arnott var. *bicolor* (A. Gray) A. Gray; *M. whitneyi* A. Gray

Herbs, annual, herbage usually drying dark. **Stems** erect or ascending, 10–140(–220) mm, glandular-puberulent to glandular-pubescent. **Leaves** usually cauline, relatively even-sized; petiole absent; blade narrowly elliptic to narrowly oblanceolate, linear-lanceolate, or linear, (4–)7–23(–34) × 1–5(–8) mm, margins entire, plane, not ciliate, apex acute, surfaces glandular-puberulent. **Pedicels** 1–3(–4) mm in fruit. **Flowers** 2 per node, or 1 or 2 per node on 1 plant, chasmogamous. **Calyces** symmetrically attached to pedicels, not inflated in fruit, (3–)4–8(–10) mm, glandular-puberulent, tube not strongly plicate, lobes triangular, subequal, apex acute-apiculate, ribs narrow, darkened, purplish, thin, not strongly raised, intercostal areas pale green. **Corollas:** (a) magenta, darkening toward mouth and within tube, often nearly obscuring dark longitudinal stripes within, throat floor or at least palate ridges yellow, or (b) yellow, usually with maroon stripes in throat extending onto lobe bases, sometimes with maroon blotches on adaxial lateral walls of throat, colored palate ridges ending in throat, tube-throat (10–)13–18(–20) mm, limb 10–19 mm diam., not bilabiate. **Anthers** included, ciliate. **Styles** glandular-puberulent. **Stigmas** included, lobes equal. **Capsules** (4.5–)6–10(–13) mm. $2n = 16$.

Flowering May–Sep. Disturbed areas, water runoff areas, granitic soils, edges of granite outcrops; (1200–)1500–3300 m; Calif.

Diplacus bicolor is endemic to Fresno, Kern, and Tulare counties in the southern Sierra Nevada. The combination *D. whitneyi* (A. Gray) G. L. Nesom is illegitimate, as *Mimulus whitneyi* A. Gray 1886, is a replacement name based on *Eunanus bicolor* A. Gray 1868.

Diplacus bicolor, D. bigelovii, D. constrictus, D. graniticola, D. layneae, and *D. thompsonii* appear to be closely related species, sometimes intergrading where sympatric. Plants of each species often produce flowers at all nodes and have dark magenta corollas with nearly regular to weakly bilabiate limbs and villous vestiture. Specimens of each species dry to a dark color.

13. Diplacus layneae (Greene) G. L. Nesom, Phytoneuron 2012-39: 29. 2012 • Layne's monkeyflower Ⓔ Ⓕ

Eunanus layneae Greene, Bull. Calif. Acad. Sci. 1: 104. 1885; *Mimulus brachiatus* Pennell; *M. layneae* (Greene) Jepson

Herbs, annual, herbage usually drying dark. **Stems** erect, 30–160(–300) mm, nodes 3–6, glandular-puberulent to glandular-pubescent, hairs 0.2–0.8 mm. **Leaves** usually cauline, relatively even-sized; petiole weakly delimited; blade elliptic to narrowly elliptic, oblanceolate, elliptic-oblanceolate, or elliptic-lanceolate, 8–27(–35) × 2–8 mm, margins entire, rarely toothed, plane, apex rounded to obtuse, surfaces: proximals often glabrate, distals glandular-puberulent or glandular-pubescent. **Pedicels** 2–4(–5) mm in fruit. **Flowers** 2 per node, or 1 or 2 per node on 1 plant, chasmogamous. **Calyces** symmetrically attached to pedicels, not inflated in fruit, (5–)6–8(–9) mm, glandular-puberulent to glandular-pubescent, tube strongly plicate, lobes triangular, subequal, apex acute, ribs broad, darkened, blackish, thickened, strongly raised, intercostal areas whitish, membranous. **Corollas** pinkish or pale to dark magenta or red-purple, each lobe usually with a faint to dark medial line extending ½ or less to tip, throat floor yellowish near base, mostly white with red-purple dots near mouth, palate ridges white, tube-throat 10–15 mm, limb (8–)10–16 mm diam., not bilabiate. **Anthers** included, ciliate. **Styles** glandular-puberulent. **Stigmas** included, lobes unequal, abaxial 1.5 times adaxial. **Capsules** 6–10(–13) mm. $2n = 16$.

Flowering May–Aug. Road banks, serpentine, granitic sand, red clay, lava beds and volcanic soils, openings in chaparral, shallow dry streambeds or stream banks, burned or otherwise disturbed open areas; (100–)400–2400 m; Calif.

D. M. Thompson (2005) observed that two forms of *Diplacus layneae* co-occur from the Yosemite National Park area southward; one of these is recognized here as *D. graniticola*.

14. **Diplacus graniticola** Schoenig, Phytoneuron 2017-24: 1, figs. 1, 3–10. 2017 • Granite-crack monkeyflower E

Herbs, annual, herbage usually drying dark. **Stems** erect, 60–120 (–150) mm, nodes 4–15(–20), internodes shorter than leaves, glandular-villous with gland-tipped hairs 1–1.6 mm. **Leaves** usually cauline, relatively even-sized; petiole weakly delimited; blade usually lanceolate to ovate-lanceolate, 20–40 × 4–12 mm, margins entire, rarely toothed, plane, apex rounded to obtuse or acute, surfaces: proximals often glabrate abaxially, distals glandular-villous. **Pedicels** 1–3 mm in fruit. **Flowers** 2 per node, or 1 or 2 per node on 1 plant, chasmogamous. **Calyces** symmetrically attached to pedicels, not inflated in fruit, 8–12 mm, glandular-villous, tube strongly plicate, lobes triangular, subequal, apex acute, ribs narrow, darkened, blackish, thickened, strongly raised, intercostal areas green to purple, not membranous. **Corollas** nearly white or pale lavender to pinkish or pale to dark magenta, each lobe with a dark medial line extending nearly to tip, throat with a dark red or purple splotch at junction of each abaxial lobe and adjacent lateral lobe, throat floor sometimes with 2 adjacent white splotches at lateral lobe bases, palate ridges yellow, tube-throat 15–20 mm, limb 10–16 mm diam., bilabiate. **Anthers** included, ciliate. **Styles** glandular-puberulent. **Stigmas** included, lobes unequal, abaxial 1.5 times adaxial. **Capsules** 6–10 mm. $2n = 16$.

Flowering Apr–Sep. Granite cracks and crevices; 300–2100 m; Calif.

Diplacus graniticola occurs in the Sierra Nevada from Tuolumne County to northern Tulare County. These plants previously were identified within *D. layneae*, with which they are partially sympatric; where these two occur together, *D. layneae* often grows in granite-derived sand and gravel immediately adjacent to the granite rock habitat of *D. graniticola*.

15. **Diplacus leptaleus** (A. Gray) G. L. Nesom, Phytoneuron 2012-39: 29. 2012 • Least-flowered monkeyflower E

Mimulus leptaleus A. Gray, Proc. Amer. Acad. Arts 11: 96. 1876

Herbs, annual. **Stems** erect to ascending, 10–140 mm, glandular-puberulent to short glandular-villous. **Leaves** usually cauline, relatively even-sized; petiole absent; blade oblanceolate to linear, (5–)6–24 (–34) × (0.5–)1–4(–7.5) mm, margins entire, plane, apex rounded to acute, surfaces glandular-puberulent. **Pedicels** 1(–3) mm in fruit. **Flowers** 2 per node, or 1 or 2 per node on 1 plant, chasmogamous. **Calyces** symmetrically attached to pedicels, not inflated in fruit, (2.5–)3–5.5(–6.5) mm, glandular-puberulent, lobes subequal, apex acute, ribs darkened, blackish, intercostal areas purplish. **Corollas** magenta, sometimes white, throat whitish, often speckled with dark spots, palate ridges weak, whitish to pinkish, tube-throat 6–8 mm, limb 3–5 mm diam., not bilabiate. **Anthers** included, glabrous. **Styles** sparsely glandular-puberulent. **Stigmas** included, lobes unequal, abaxial nearly 2 times adaxial. **Capsules** (3–)4–6(–6.5) mm. $2n = 16$.

Flowering Jun–Aug. Developing granitic soils, boulders of granite outcrops, disturbed areas, water runoff areas; (1800–)2100–3400 m; Calif., Nev.

Diplacus leptaleus occurs in the Sierra Nevada from Nevada County south to Tulare County, California, and in southern Washoe County, Nevada. *Diplacus leptaleus* is the only species of *Diplacus* that commonly produces white flowers. Populations rarely are fixed for white corollas; they are commonly fixed for magenta corollas, especially at higher elevations (D. M. Thompson 2005).

16. **Diplacus clivicola** (Greenman) G. L. Nesom, Phytoneuron 2012-39: 28. 2012 • Slope monkeyflower E

Mimulus clivicola Greenman, Erythea 7: 119. 1899

Herbs, annual. **Stems** erect, (10–)20–180 mm, glandular-puberulent to short glandular-villous. **Leaves** usually cauline, relatively even-sized; petiole absent, base sometimes petiole-like; blade narrowly elliptic, sometimes broadly elliptic to obovate or oblanceolate, (2.5–)6–20(–26) × (1–)2–10(–14) mm, margins crenate to serrulate or entire, plane, apex rounded or acute, surfaces glandular-puberulent. **Pedicels** 2–7(–10) mm in fruit. **Flowers** 2 per node, or 1 or 2 per node on 1 plant, chasmogamous. **Calyces** symmetrically attached to pedicels, not inflated in fruit, (5–)7–8 mm, glandular-puberulent, lobes subequal, apex acute, ribs green, intercostal areas whitish. **Corollas** rose pink to purplish, limb often pale, especially abaxial lip, abaxial lip often purple-dotted near base, markings often coalescing and forming broken lines radiating toward each lobe, tube yellow, palate ridges yellow with magenta speckling, confluent and extending onto abaxial lip base, tube-throat (8–)11–12 mm, limb 7–12 mm diam., bilabiate. **Anthers** included, ciliate. **Styles** glandular-puberulent distally. **Stigmas** exserted, lobes subequal. **Capsules** 8–13 mm. $2n = 16$.

Flowering May–Aug. Bluffs, disturbed slopes, well-developed loam soils, vegetation openings; 500–1200 (–2000) m; Idaho, Mont., Oreg.

Diplacus clivicola is known from northern Idaho and immediately adjacent Montana and Oregon. It is similar to typical *D. nanus* in its strongly bilabiate corollas; it differs in its slightly toothed leaf blade margins, relatively long pedicels, and calyces with cuneate bases.

17. **Diplacus cusickii** (Greene) G. L. Nesom, Phytoneuron 2012-39: 28. 2012 • Cusick's monkeyflower E F

Eunanus cusickii Greene, Pittonia 1: 36. 1887; *Mimulus cusickii* (Greene) Rattan

Herbs, annual. **Stems** erect to erect-ascending, 10–80 mm, distal internodes 2–20 mm, minutely glandular-puberulent. **Leaves** basal and cauline or usually cauline, relatively even-sized or gradually larger distally; petiole absent, proximal base short petiole-like, 1–5 mm; blade ovate to broadly elliptic-ovate, (10–)15–25(–35) × 4–17 mm, margins entire, plane, apex acuminate, surfaces glabrous or adaxial minutely glandular-puberulent. **Pedicels** 1–1.5 mm in fruit. **Flowers** 2 per node, or 1 or 2 per node on 1 plant, chasmogamous. **Calyces** symmetrically attached to pedicels, not inflated in fruit, 7–10 mm, glabrous or minutely stipitate-glandular, lobes unequal, apex linear-acuminate, sharp-pointed, ribs green distally, intercostal areas whitish. **Corollas** magenta or rose purple, tube yellow, throat yellow, throat and distal tube red-spotted on floor, palate ridges yellow, tube-throat 13–16(–19) mm, limb 16–24 mm diam., bilabiate. **Anthers** included, glabrous or sparsely hirsutulous. **Styles** pubescent, at least on distal ½. **Stigmas** exserted, lobes subequal. **Capsules** 10–15 mm.

Flowering May–Jul. Slopes, canyons, washes, ditches, sand talus, diatomaceous slopes, basalt outcrops, black volcanic gravel, volcanic ash and sand, sagebrush areas; 800–1000 m; Idaho, Oreg.

Diplacus cusickii is endemic to northern Malheur County, Oregon, and along the Snake River in Ada and Owyhee counties, Idaho. Its narrow geographic range reflects the segregation of the more widely distributed *D. cusickioides*.

D. M. Thompson (2005) noted that collections in northern Malheur County, Oregon, were intermediate between *Mimulus cusickii* and *M. nanus*; the type of *M. cusickii* is from this area and is among the narrowly endemic, supposedly putative intermediates

(G. L. Nesom 2013c). These plants have abruptly and sharply acuminate leaf apices like the more widespread form traditionally identified as *Diplacus cusickii*; they differ in having glabrous leaf surfaces. Typical *D. nanus* occurs in close sympatry, without intergradation, with the populations in northern Malheur County. Because of their distinctive morphology and coherent geography, the northern Malheur County plants are reasonably recognized as a distinct species. The more widely distributed form formerly identified as *D. cusickii* now is identified as *D. cusickioides*. Populations of *D. cusickii* in the narrow sense along the Snake River in Ada and Owyhee counties, Idaho, may have dispersed there from the Oregon center. Some plants of *D. cusickioides* in the Leslie Gulch area of east-central Malheur County have somewhat reduced vestiture, approaching that of *D. cusickii*.

18. **Diplacus cusickioides** G. L. Nesom, Phytoneuron 2013-65: 6, figs. 3, 4. 2013 E

Herbs, annual. **Stems** erect to erect-ascending, (10–)30–240 (–350) mm, distal internodes 1–5 mm, densely glandular-villous. **Leaves** basal and cauline or usually cauline, relatively even-sized or gradually larger distally; petiole absent, proximal base short petiole-like, 1–5 mm; blade ovate to broadly elliptic-ovate, (10–)15–25(–35) × 4–16 mm, margins entire, plane, apex abruptly acuminate, surfaces densely glandular-villous. **Pedicels** 1–1.5 mm in fruit. **Flowers** 2 per node, or 1 or 2 per node on 1 plant, chasmogamous. **Calyces** symmetrically attached to pedicels, not inflated in fruit, 7–12 mm, glabrous or minutely stipitate-glandular, lobes unequal, apex linear-acuminate, sharp-pointed, ribs green distally, intercostal areas whitish. **Corollas** magenta or rose purple, tube yellow, throat usually yellow, throat and distal tube red-spotted on floor, palate ridges yellow, tube-throat 13–16(–19) mm, limb 14–26 mm diam., bilabiate. **Anthers** included, glabrous or sparsely hirsutulous. **Styles** puberulent, at least on distal ½. **Stigmas** exserted, lobes subequal (herkogamous). **Capsules** 10–17 mm.

Flowering May–Aug. Lava formations, steep slopes, roadsides, volcanic gravels, scree, ash; (400–)600–1500(–2000) m; Calif., Idaho, Oreg., Wash.

Diplacus cusickioides occurs in western Idaho, eastern Oregon, Klickitat County, Washington, and apparently in a disjunct population system in Modoc County, California, on the east side of the Warner Mountains.

19. **Diplacus deschutesensis** G. L. Nesom, Phytoneuron 2013-65: 8, fig. 5. 2013

• Deschutes monkeyflower [E]

Herbs, annual. **Stems** erect to erect-ascending, 40–150 mm, distal internodes 1–2 mm, minutely glandular-puberulent. **Leaves** usually cauline, relatively even-sized or gradually larger distally; petiole absent, proximal base short petiole-like; blade broadly ovate or obovate to elliptic-ovate or elliptic-oblanceolate, 10–15(–25) × 4–13 mm, margins entire, plane, apex acuminate or cuspidate, surfaces glabrous or sparsely glandular-puberulent. **Pedicels** 1–1.5 mm in fruit. **Flowers** usually from proximalmost to distal nodes, 2 per node, or 1 or 2 per node on 1 plant, chasmogamous. **Calyces** symmetrically attached to pedicels, not inflated in fruit, 7–8 mm, glandular-puberulent, lobes subequal, apex linear-acuminate, ribs green distally, intercostal areas whitish. **Corollas** light pink to magenta or rose purple, usually with a darker narrow line extending from throat onto each lobe midvein, throat yellow, palate ridges yellow, tube-throat 8–12 mm, limb 10–16 mm diam., bilabiate. **Anthers** included, glabrous or sparsely hispidulous. **Styles** puberulent, at least on distal ½. **Stigmas** exserted, lobes subequal, abaxial slightly longer. **Capsules** 7–9 mm. $2n = 16$.

Flowering Jun–Aug. Sandy and ashy soils, pumice sand and gravel, red clay slopes, hillsides, roadsides, bare areas, sagebrush, sagebrush-juniper, juniper, yellow pine and lodgepole pine forests; 700–1500 m; Oreg.

Diplacus deschutesensis is endemic to Crook, Deschutes, Jefferson, Klamath, Lake, and Wheeler counties of central Oregon. D. M. Thompson (2005) regarded these plants as a zone of stabilized hybrids, intermediate between *Mimulus cusickii* and typical *M. nanus*, the range just outside and west of the wider range of typical *M. cusickii*. In an earlier study that included both of the latter species, W. L. Ezell (1971, and by annotation in 1987) identified the same set of plants simply as *M. cusickii*, not associating them at all with *M. nanus*. A. L. Grant (1924, and by annotation of MO collections) identified them variously as either *M. cusickii* or *M. ovatus*. Thompson did not say what features of intermediacy he observed in the putative hybrids, but he did note that they produced leaves with acuminate-cuspidate apices and that they would key to *M. cusickii*.

Leaves of *Diplacus deschutesensis* are broad with abruptly and sharply acuminate apices like those of *D. cusickii*, and the corolla coloration also is similar. The flowers (calyx length, corolla tube-throat length, limb width) and capsules of *D. deschutesensis* are considerably smaller, and the distal leaves are smaller with glabrous surfaces.

20. **Diplacus ovatus** (A. Gray) G. L. Nesom, Phytoneuron 2012-39: 29; 2012-47: 3. 2012

• Steamboat or eggleaf monkeyflower [C][E]

Mimulus bigelovii A. Gray var. *ovatus* A. Gray in A. Gray et al., Syn. Fl. N. Amer. ed. 2, 2(1): 445. 1886; *M. ovatus* (A. Gray) N. H. Holmgren

Herbs, annual. **Stems** erect to ascending, 20–140 mm, distal internodes 1–3 mm, glandular-pubescent and short glandular-villous. **Leaves** usually cauline, relatively even-sized or slightly reduced distally; petiole absent, base sometimes tapered to narrow, petiole-like extension; blade obovate to broadly oblanceolate, 13–33 × 5–12(–16) mm, margins entire, plane, apex acuminate, surfaces densely glandular-villous. **Pedicels** 2–3(–5) mm in fruit. **Flowers** 2 per node, or 1 or 2 per node on 1 plant, chasmogamous. **Calyces** symmetrically attached to pedicels, not inflated in fruit, 7–9(–10) mm, coarsely glandular-pubescent, lobes subequal, apex lanceolate to narrowly lanceolate, ribs purplish, intercostal areas white. **Corollas** magenta to red-purple with a yellow patch on palate, sometimes yellow with a red-brown patch, palate ridges orange-yellow, tube-throat 9–11 mm, limb 12–15 mm diam., bilabiate. **Anthers** exserted, sparsely hirsutulous. **Styles** glandular-puberulent. **Stigmas** exserted, lobes subequal. **Capsules** 6–8 mm.

Flowering Apr–Jun(–Aug). Dry to moist, often barren, loose, sandy to gravelly slopes, andesite or rhyolite deposits, sandy alkaline valley floors, roadsides, washes, sagebrush, pinyon-juniper, open yellow pine woodlands; of conservation concern; 1300–1900(–2400) m; Nev.

Mimulus ovatus was treated as a distinct species by N. H. Holmgren (1984); the plants were considered by D. M. Thompson (2005) to be hybrids between *M. nanus* var. *mephiticus* and *M. cusickii*, and he placed the name as a synonym of *M. cusickii*. *Diplacus ovatus* is known only from Carson City, Douglas, and southern Washoe counties.

21. **Diplacus nanus** (Hooker & Arnott) G. L. Nesom, Phytoneuron 2012-39: 29. 2012 • Dwarf or purple monkeyflower [E]

Mimulus nanus Hooker & Arnott, Bot. Beechey Voy., 378. 1839; *M. tolmiei* (Bentham) Rydberg

Herbs, annual. **Stems** erect, 30–120 mm, minutely glandular-puberulent. **Leaves** basal and cauline, relatively even-sized; petiole absent; blade narrowly elliptic to oblanceolate, ovate, obovate, or elliptic-lanceolate, (1–)3–30(–50) × (0.4–)0.7–8(–20) mm, margins entire, plane, apex rounded or obtuse, surfaces minutely glandular-puberulent. **Pedicels** 1–3 mm in fruit. **Flowers** 2 per node, or 1 or 2 per node on 1 plant, chasmogamous. **Calyces** symmetrically attached to pedicels, not inflated in fruit, 6–9 mm, minutely glandular-puberulent, lobes subequal, apex acute-apiculate, acuminate, or attenuate, ribs dark green or reddish, intercostal areas whitish. **Corollas** magenta to purplish, dark line often extending onto each abaxial lip lobe from throat, palate ridges yellow with red-purple speckling and border, throat floor villous with hairs extending onto abaxial lip, tube 1.1–1.9 mm diam. at filament insertion, tube-throat 11–15 mm, limb 8–14 mm diam., usually, rarely not, bilabiate. **Anthers** included or exserted, ciliate. **Styles** glandular-puberulent or glandular-pubescent. **Stigmas** exserted, lobes equal. **Capsules** 8–12 mm. $2n = 16$.

Flowering (Apr–)May–Jul. Openings in sagebrush, disturbed slopes, granite outcrops; (300–)1100–2300 (–2900) m; Calif., Idaho, Mont., Nev., Oreg., Wash., Wyo.

Diplacus nanus is broadly distributed through northern California, southern Idaho, and eastern Oregon, with stations in Ravalli County, Montana, and Park County, Wyoming, and scattered localities in Washington.

Diplacus nanus is generally recognized by its strongly bilabiate corollas with purplish (not yellow) tubes and two dark purple patches along the sides of the throats. The glandular-puberulent vestiture of *D. nanus* contrasts with the glandular-pubescent and viscid-villous vestiture (with hairs much longer) of *D. mephiticus*.

W. L. Ezell (1971) noted that in the Siskiyou Mountains of Josephine County, Oregon, and adjacent Siskiyou and Trinity counties, California, corollas of *Diplacus nanus* do not have clearly differentiated abaxial and adaxial lips.

22. **Diplacus cascadensis** G. L. Nesom, Phytoneuron 2013-65: 13, figs. 8, 9. 2013 • Cascades monkeyflower E

Herbs, annual. **Stems** erect to ascending-erect, 20–100 mm, distal internodes 1–4 mm, short glandular-villous to glandular-puberulent. **Leaves** usually cauline, relatively even-sized; petiole present proximally, absent distally; blade elliptic-spatulate to obovate or broadly oblanceolate, 10–22 × 2–10 mm, margins entire, plane, apex obtuse to rounded-acute, surfaces minutely glandular-puberulent. **Pedicels** 1–3 mm in fruit. **Flowers** 2 per node, or 1 or 2 per node on 1 plant, chasmogamous. **Calyces** symmetrically attached to pedicels, not inflated in fruit, (4–)5–7 mm, minutely glandular-puberulent, lobes subequal, apex acute to acuminate, ribs dark green

or reddish, intercostal areas whitish. **Corollas** magenta to purplish, usually with a darker narrow line extending from throat onto each lobe midvein, palate ridges yellow with red spots, throat floor glabrous, tube 1.1–1.9 mm diam. at filament insertion, tube-throat 8–10 mm, limb 7–11 mm diam., bilabiate. **Anthers** (distal pair) exserted, minutely viscid-villosulous. **Styles** apparently glabrous. **Stigmas** exserted, lobes usually subequal. **Capsules** 5–8(–9) mm. $2n = 16$.

Flowering Jun–Aug. Open pumice flats, scree slopes, sandy soils, juniper-sagebrush, juniper, pine-juniper, yellow pine, lodgepole pine forests; 1400–2400 (–2600) m; Oreg.

Diplacus cascadensis is known from Deschutes, Klamath, and Lake counties. Plants of this species have been identified as *D. nanus* (similar in its purplish leaves congested on crowded distal nodes, minutely glandular-puberulent vestiture, and purplish and strongly bilabiate corollas), but they differ from *D. nanus* in their broader distal leaves, shorter calyces, shorter corollas with glabrous throats and magenta tubes, and shorter capsules.

23. **Diplacus mephiticus** (Greene) G. L. Nesom, Phytoneuron 2012-39: 29. 2012 • Skunky monkeyflower E

Mimulus mephiticus Greene, Bull. Calif. Acad. Sci. 1: 9. 1884; *M. coccineus* Congdon; *M. coccineus* var. *wolfii* (Eastwood) D. W. Taylor; *M. densus* A. L. Grant; *M. nanus* Hooker & Arnott var. *mephiticus* (Greene) D. M. Thompson; *M. reifschneiderae* Edwin; *M. stamineus* A. L. Grant; *M. washoensis* Edwin; *M. wolfii* Eastwood

Herbs, annual. **Stems** erect, (20–)30–150(–180) mm, glandular-pubescent and viscid-villous. **Leaves** usually cauline, relatively even-sized; petiole absent; blade ovate to oblong or narrowly elliptic to narrowly lanceolate, narrowly oblong, or linear, 10–25 × 1–5 mm, margins entire, plane, not ciliate, apex rounded to obtuse or acute, surfaces usually glandular-pubescent and (at least along veins) viscid-villous. **Pedicels** 1–3 mm in fruit. **Flowers** 2 per node, or 1 or 2 per node on 1 plant, chasmogamous. **Calyces** symmetrically attached to pedicels, slightly inflated in fruit, (3–)4–7(–9) mm, glandular-pubescent and viscid-villous (at least along veins), lobes subequal, apex acute to acuminate, ribs dark green to purplish, intercostal areas whitish. **Corollas** of 2 color forms: (a) dark magenta, purplish, or reddish with palate ridges or whole throat floor yellow, red- or purple-dotted, lateral lobes yellowish inside and (b) yellow with red or purple spots on floor, tube-throat 8–12(–15) mm, tube 1.3–1.9 mm diam. at filament insertion, limb 5–12 (–15) mm diam., bilabiate. **Anthers** exserted, short-hirsute. **Styles** glabrous or sparsely glandular-puberulent. **Stigmas** exserted or at opening of corolla tube-throat, lobes subequal to unequal, abaxial to 2 times adaxial. **Capsules** 5–8 mm.

Flowering (May–)Jun–Aug(–Sep). Openings in sage-brush, disturbed slopes, granite outcrops, serpentine substrates, gravelly and sandy soils, sandy moraines, pumice flats, gravelly washes, meadows, shadscale and sagebrush communities, pinyon-juniper and ponderosa pine woodlands; 1300–3700 m; Calif., Nev.

Diplacus mephiticus occurs in eastern California and west-central Nevada.

Various synonyms treated here are in agreement with D. M. Thompson (2005). *Mimulus coccineus* (mostly from Eldorado to Tulare counties, California, and, apparently, including *Eunanus angustifolius* Greene from Mt. Rose, Nevada) includes relatively small, tufted plants at high elevations with relatively small calyces and relatively small, dark red-purple, strongly bilabiate corollas with prominently exserted stamens. *Mimulus densus* (mostly in Nevada and in Lassen, Nevada, and Plumas counties, California) includes taller plants at lower elevations with a strong tendency to produce populations with all individuals with larger, yellow, nearly regular corollas with more nearly included stamens. Typical *Diplacus mephiticus* has moderate-sized plants at medium elevations with magenta, bilabiate corollas.

The specific epithet *mephiticus* alludes to the musky odor of the plants; this has also been noted in plants of *Mimulus coccineus* and *M. densus*. *Diplacus nanus*, in which D. M. Thompson (2005) included *D. mephiticus* as a variety, apparently does not produce a mephitic odor. *Diplacus cusickii* also produces a mephitic odor (W. L. Ezell 1971).

The later homonym *Mimulus angustifolius* (Greene) A. L. Grant 1925, not Hochstetter ex Richard 1850, based on *Eunanus angustifolius*, pertains here.

24. **Diplacus jepsonii** (A. L. Grant) G. L. Nesom, Phytoneuron 2012-39: 29. 2012 • Jepson's monkeyflower E

Mimulus jepsonii A. L. Grant, Ann. Missouri Bot. Gard. 11: 306. 1925; *M. microcarpus* Pennell; *M. nanus* Hooker & Arnott var. *jepsonii* (A. L. Grant) D. M. Thompson

Herbs, annual. **Stems** erect, 10–150 mm, minutely glandular-puberulent. **Leaves** usually cauline, relatively even-sized; petiole absent; blade narrowly elliptic-oblanceolate to narrowly oblanceolate,

6–22 × 1–5(–7) mm, margins entire, plane, apex acute to obtuse, surfaces glandular-puberulent to glandular-pubescent. **Pedicels** 1 mm in fruit. **Flowers** 2 per node, or 1 or 2 per node on 1 plant, chasmogamous. **Calyces** symmetrically attached to pedicels, not inflated in fruit, 3–5 mm, minutely puberulent, lobes subequal, apex acute to acuminate, ribs dark green to purplish, intercostal areas whitish. **Corollas** lavender-purple to rose purple, throat darker, palate ridges yellow, red-spotted, tube 0.8–1.2 mm diam. at filament insertion, tube-throat 7–10 mm, limb 5–7 mm diam., bilabiate. **Anthers** exserted, sparsely hispidulous. **Styles** glandular-puberulent or glandular-pubescent. **Stigmas** exserted, lobes equal. **Capsules** 4–6 mm. $2n = 16$.

Flowering Jun–Aug. Gentle slopes, sandy and volcanic soils, meadows in spruce-fir forests, openings among pines and in chaparral, shallow drainage areas, disturbed open areas; (900–)1100–2500(–2800) m; Calif., Nev., Oreg.

D. M. Thompson (2005) treated *Diplacus jepsonii* as conspecific with *D. nanus* [as W. L. Ezell (1971) had done earlier] because of putative intergrades. In its typical form, *D. jepsonii* is immediately distinct in its narrower leaves, stems with well-separated proximal nodes, and smaller corollas with nearly filiform tubes. R. J. Meinke (1992), in a field and greenhouse study not cited by Thompson, found that *D. jepsonii* and *D. nanus* are distinct in morphology as well as in habitat and that the two do not grow intermixed in nature.

Diplacus jepsonii is known only from north-central California, southern Washoe County, Nevada, and southern Oregon. Putative outliers in the central Sierra Nevada are dubiously identified.

25. **Diplacus parryi** (A. Gray) G. L. Nesom & N. S. Fraga, Phytoneuron 2012-39: 27. 2012 • Parry's monkeyflower E F

Mimulus parryi A. Gray, Proc. Amer. Acad. Arts 11: 97. 1876; *M. spissus* A. L. Grant var. *lincolnensis* Edwin

Herbs, annual. **Stems** erect, 10–120(–170) mm, finely and minutely glandular-puberulent. **Leaves** usually cauline, relatively even-sized; petiole absent; blade narrowly elliptic to sublinear or oblanceolate, sometimes obovate, (5–)8–25(–31) × (1–)2–9(–12) mm, margins entire, plane, not ciliate, apex: proximals usually rounded, distals usually acute, surfaces glandular-puberulent. **Pedicels** (1.5–)2–4(–9) mm in fruit. **Flowers** 2 per node, or 1 or 2 per node on 1 plant, chasmogamous. **Calyces** symmetrically attached to pedicels, not inflated in fruit, (5–)7–12(–13) mm, glandular-puberulent, lobes unequal, adaxial longer, apex broadly rounded to

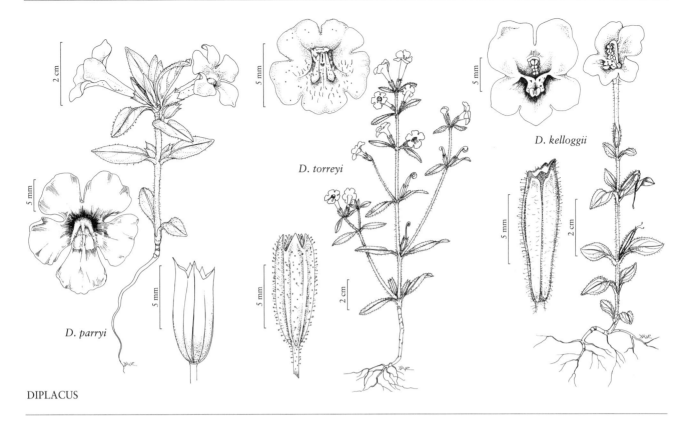

DIPLACUS

acute, often apiculate, ribs often dark purple, intercostal areas purplish or white. **Corollas** of 2 color forms: (a) magenta, ± deepening at mouth, usually with 6–8 darker spots in arc on abaxial lip around mouth, throat floor yellow to whitish with reddish speckling and (b) yellow with 6–8 narrow reddish spots or lines in arc on abaxial lip around mouth and reddish speckling on throat floor, palate ridges yellow extending onto lip, tube-throat (10–)12–18(–20) mm, limb 11–17.5 (–20) mm diam., not bilabiate. **Anthers** included, glabrous. **Styles** glandular-puberulent. **Stigmas** included, lobes equal. **Capsules** (5.5–)6.5–10.5 mm. $2n = 16$.

Flowering Apr–Jun(–Jul). Banks, gravel bars, washes, sandy ravines, rocky hillsides, ledges and bases of limestone ledges and boulders, clay loam-basalt, bare areas, often with *Coleogyne* and *Larrea*, sagebrush and pinyon-juniper; (600–)800–1700(–2200) m; Ariz., Calif., Nev., Utah.

Diplacus parryi has a limited range, primarily in the Mohave Desert in four states: Arizona (Mohave County), California (Inyo County, where apparently disjunct, in pinyon-juniper woodlands, and at higher than typical elevation), Nevada (Clark and Lincoln counties), and Utah (Washington County). The glandular-puberulent vestiture and unequal calyx lobes are diagnostic.

26. **Diplacus torreyi** (A. Gray) G. L. Nesom, Phytoneuron 2012-39: 32. 2012 • Torrey's monkeyflower E F

Mimulus torreyi A. Gray, Proc. Amer. Acad. Arts 11: 97. 1876

Herbs, annual, herbage not drying dark. **Stems** erect, (20–) 40–380(–530) mm, glandular-puberulent to glandular-pubescent. **Leaves** usually cauline, gradually reduced distally; petiole poorly delimited; blade elliptic or oblanceolate, (3–)6–40(–48) × (0.5–)1–14 (–19) mm, margins entire, plane, apex rounded, surfaces: proximals glabrate, distals glandular-puberulent or glandular-pubescent. **Pedicels** 1–3(–6) mm in fruit. **Flowers** 2 per node, or 1 or 2 per node on 1 plant, chasmogamous. **Calyces** symmetrically attached to pedicels, not inflated in fruit, (2–)5–10(–12) mm, glandular-puberulent to glandular-pubescent, lobes shallowly triangular to broadly ovate, subequal, apex obtuse, rarely apiculate, ribs purplish. **Corollas:** tube magenta, limb magenta to pale rose, rarely nearly white, abaxial lip usually paler, palate ridges yellow, tube-throat (8–)9–18(–20) mm, limb 8–20 mm diam., weakly bilabiate. **Anthers** included, sometimes slightly ciliate. **Styles** glandular-puberulent. **Stigmas** included, lobes unequal, abaxial 2 times adaxial. **Capsules** (5–)6–11(–13) mm. $2n = 20$.

Flowering May–Aug. Moist sand in channel beds and flood plains, exposed slopes along streams, roadside ditches, serpentine slopes, volcanic outcrop margins, gravelly openings, rocky serpentine soils, ridges in yellow pine and Jeffrey pine forests, openings in yellow pine, lodgepole pine, red fir, and mixed conifer woodlands, chaparral edges, meadows, recent burns, roadsides, road banks; (100–)700–2000(–2400) m; Calif.

Diplacus torreyi occurs from Lassen and Shasta counties to Fresno County in the Sierra Nevada and Cascade Range.

Diplacus torreyi is recognized by its erect, usually single-stemmed habit (usually with the proximal one or two internodes elongate), obovate to oblanceolate leaf blades, calyces with membranous walls, thin ribs, and shallowly triangular to broadly ovate lobes, and weakly bilabiate corollas with relatively narrower limbs, usually drying a light color with yellow palate ridges.

27. **Diplacus mohavensis** (Lemmon) G. L. Nesom, Phytoneuron 2012-39: 29. 2012 • Mohave monkeyflower C E

Mimulus mohavensis Lemmon, Bot. Gaz. 9: 142. 1884; *Eunanus mohavensis* (Lemmon) Greene

Herbs, annual. **Stems** erect, (10–)20–100(–140) mm, terete. **Leaves** basal and cauline, relatively even-sized; petiole absent; blade narrowly elliptic to elliptic-lanceolate, (6–)7–27 × 1.2–8(–10) mm, margins entire, plane, apex obtuse to acute or acuminate, surfaces green, often red-purple tinted, usually glabrous, veins and margins glandular-puberulent or ciliate. **Pedicels** 2–5(–6) mm in fruit. **Flowers** 2 per node, or 1 or 2 per node on 1 plant, chasmogamous. **Calyces** asymmetrically attached to pedicel, inflated in fruit, 7–15(–16) mm, glabrous or with glandular-puberulent veins, lobes unequal, apex acuminate, ribs and intercostal areas purplish brown. **Corollas** salverform-rotate, throat dark purplish brown without internal or external markings, floor purplish brown-pilose, lobes purplish brown basally with red veins, palate ridges absent, tube-throat 9–15(–18) mm, limb 8–11(–14) mm diam., not bilabiate. **Anthers** included, glabrous or with a few hairs at base of flower pair. **Styles** sparsely glandular-puberulent. **Stigmas** included, lobes subequal. **Capsules** (7–)8–13 mm. $2n = 16$.

Flowering Apr–May. Gravelly hillsides and slopes, limestone, granite, fine gravel in wash bottoms and edges, commonly with *Larrea*; of conservation concern; 600–900 m; Calif.

Diplacus mohavensis is known from San Bernardino County.

Diplacus mohavensis is similar to *D. pictus* in features of corolla morphology and color patterning, and the pair sometimes has been segregated as *Mimulus* sect. *Mimulastrum* A. Gray (for example, by D. M. Thompson 2005). Molecular data (P. M. Beardsley et al. 2004) indicate that *D. mohavensis* arose from within sect. *Eunanus*. It is distinct from other species of sect. *Eunanus* (and similar to *D. pictus*) in its salverform-rotate corollas with an abrupt tube-throat transition and vein-patterned limb. In *D. mohavensis*, the limb is purplish brown basally with red, irregularly patterned veins fading into a wide, whitish distal border; in *D. pictus*, the limb is all white, and the purplish brown vein patterning is more regular and not fading distally.

28. **Diplacus pictus** (Curran ex Greene) G. L. Nesom, Phytoneuron 2012-39: 30. 2012 • Calico monkeyflower C E

Eunanus pictus Curran ex Greene, Bull. Calif. Acad. Sci. 1: 106. 1885; *Mimulus pictus* (Curran ex Greene) A. Gray

Herbs, annual. **Stems** erect to ascending, 20–380 mm, 4-sided, glandular-pubescent. **Leaves** basal and cauline, relatively even-sized; petiole absent; blade elliptic to elliptic-ovate or obovate, 7–45(–57) × 3–20 (–32) mm, margins crenate, plane, apex obtuse to rounded, surfaces glandular-pubescent. **Pedicels** 1.5–6 (–7) mm in fruit. **Flowers** 2 per node, or 1 or 2 per node on 1 plant, chasmogamous, sometimes cleistogamous. **Calyces** asymmetrically attached to pedicel, not inflated in fruit, 6–18 mm, densely glandular-pubescent, lobes unequal, adaxial longest, apex obtuse, ribs green, intercostal areas whitish. **Corollas** salverform-rotate, throat dark purplish brown without internal or external markings, lobes white with bold, purplish brown veins, palate ridges absent, tube-throat 6.5–18 mm, limb 4–16 mm diam., not bilabiate. **Anthers** included, glabrous. **Styles** glandular-puberulent. **Stigmas** included, lobes unequal, abaxial 6–8 times adaxial. **Capsules** (5.5–)7–17 mm. $2n = 16$.

Flowering Mar–May. Rocky granitic slopes, granite rocks and outcrops, sandy granitic soils, blue oak, blue oak-grey pine, Douglas oak, Douglas oak-canyon live oak, and gray pine-Douglas oak woodlands, steep canyon slopes with box elder; of conservation concern; 100–1300 m; Calif.

Diplacus pictus is known from Kern and Tulare counties in the Sierra Nevada foothills and Tehachapi Mountains.

Diplacus pictus and *D. mohavensis* have been treated together as *Mimulus* sect. *Mimulastrum*; molecular data (P. M. Beardsley et al. 2004) indicate that the two species are independently derived and that the similarities in corolla morphology are convergent. *Diplacus pictus* produces cleistogamous flowers in drought conditions.

29. **Diplacus douglasii** (Bentham) G. L. Nesom, Phytoneuron 2012-39: 32. 2012 • Brownies, purple mouse-ears [E]

Eunanus douglasii Bentham in A. P. de Candolle and A. L. P. P. de Candolle, Prodr. 10: 374. 1846; *Mimulus cleistogamus* J. T. Howell; *M. douglasii* (Bentham) A. Gray

Herbs, annual. **Stems** erect, 3–40(–80) mm, glandular-puberulent and/or glandular-pubescent. **Leaves** usually basal, reduced distally; petiole absent, larger with petiole-like extension; blade obovate to elliptic, 5–28(–35) × (2–)3–10(–12) mm, margins entire or crenate, plane, not ciliate, apex obtuse, surfaces: proximals glabrate, distals glandular-pilose adaxially. **Pedicels** (1–)2–4 mm in fruit, usually twisting to invert calyx. **Flowers** 2 per node, or 1 or 2 per node on 1 plant, chasmogamous or cleistogamous. **Calyces** distinctly asymmetrically attached to pedicel, not inflated in fruit, (6–)8–14(–15) mm, sparsely glandular-pilose, lobes subequal, apex appearing acute due to their continuity with ribs, ribs green to purplish, intercostal areas whitish. **Corollas:** throat magenta to dark purple, inside with longitudinal gold markings basally, adaxial lip magenta, palate ridges golden yellow, tube-throat (15–)20–41(–45) mm, limb 10–14 mm diam., abaxial lip essentially absent. **Anthers** (distal pair) nearly exserted, glabrous or glabrate. **Styles** pubescent distally. **Stigmas** exserted, lobes unequal, abaxial 10–20 times adaxial. **Capsules** (2.5–)3–6.5 mm. **2*n*** = 18.

Flowering Feb–Apr(–May). Gentle to moderately steep slopes, upper banks of small creeks and ditches; 50–1200 m; Calif., Oreg.

Diplacus douglasii occurs in southwestern Oregon and the northern two-thirds of California.

Diplacus douglasii is distinct in the complete or nearly complete lack of abaxial corolla lobes. *Diplacus brandegeei*, *D. congdonii*, *D. kelloggii*, and *D. traskiae*, the other species of sect. *Cleisanthus*, also show a distinct tendency toward reduction of the abaxial corolla lobes, and all species of the group except *D. brandegeei* produce markedly elongate corolla tubes.

30. **Diplacus congdonii** (B. L. Robinson) G. L. Nesom, Phytoneuron 2012-39: 32. 2012 • Congdon's monkeyflower [E]

Mimulus congdonii B. L. Robinson, Proc. Amer. Acad. Arts 26: 175. 1891

Herbs, annual. **Stems** erect to ascending, 0–100(–120) mm, glandular-puberulent and/or glandular-pubescent. **Leaves** usually cauline, relatively even-sized; petiole absent, larger with petiole-like extension; blade oblanceolate to elliptic, (4–)8–32(–37) × (1.5–)2.5–14(–18) mm, margins entire or crenate, plane, apex obtuse, surfaces glabrous or puberulent and/or pilose adaxially. **Pedicels** (1–)2–5 mm in fruit, usually twisting to invert calyx. **Flowers** 2 per node, or 1 or 2 per node on 1 plant, chasmogamous or cleistogamous. **Calyces** distinctly asymmetrically attached to pedicel, inflated in fruit, 5–14 mm, sparsely pilose, lobes subequal, apex obtuse, ribs green to purplish, intercostal areas whitish. **Corollas:** throat whitish to dark magenta, not golden yellow at base, lobes magenta, without markings or with dark magenta dots and, sometimes, some yellow at bases of abaxial lobes, palate ridges absent or purple, tube-throat 8–30 mm, limb 3–9 mm diam., bilabiate, abaxial lobe usually smaller than adaxial. **Anthers** (distal pair) exserted, glabrous. **Styles** glandular-pubescent. **Stigmas** exserted, lobes unequal, abaxial 3–4 times adaxial. **Capsules** 4–8.5 mm, indehiscent. **2*n*** = 18.

Flowering Feb–May. Serpentine soils, periphery of granite outcrops, disturbed hillsides, soil from decomposed granite, near water runoff areas, away from seeps or other areas with prolonged surface moisture; 100–1100(–1700) m; Calif.

Diplacus congdonii occurs in the Sierra Nevada Foothills and mountain ranges along the coast.

31. **Diplacus kelloggii** (Curran ex Greene) G. L. Nesom, Phytoneuron 2012-39: 32. 2012 • Kellogg's monkeyflower [E] [F]

Eunanus kelloggii Curran ex Greene, Bull. Calif. Acad. Sci. 1: 100. 1885; *Mimulus kelloggii* (Curran ex Greene) Curran ex A. Gray

Herbs, annual. **Stems** erect to ascending, 10–310(–370) mm, glandular-puberulent and/or glandular-pubescent. **Leaves** usually basal, sometimes basal and cauline, relatively even-sized; petiole absent, larger with petiole-like extension; blade obovate to elliptic, (4–)6–40(–52) × (2–)3–17(–26) mm, margins entire or crenate, plane, apex obtuse, surfaces often pubescent. **Pedicels** 2–6 (–10) mm in fruit, usually twisting to invert calyx. **Flowers** 2 per node, or 1 or 2 per node on 1 plant, chasmogamous. **Calyces** distinctly asymmetrically attached to pedicel, not inflated in fruit, (7–)8–16(–17) mm, densely glandular-puberulent or glandular-pubescent, lobes subequal, apex obtuse, ribs purplish, intercostal areas whitish. **Corollas:** throat dark purple, golden yellow at base with reddish speckling, limb magenta to reddish purple, lateral adaxial lobes each with a dark purple basal spot, palate ridges golden yellow, tube-throat (13–)20–45(–50) mm, limb 10–18 mm diam., bilabiate, abaxial lip smaller than adaxial. **Anthers** (distal pair) exserted, glabrous. **Styles** densely glandular-puberulent distally. **Stigmas** exserted, lobes unequal, abaxial 4–5 times adaxial. **Capsules** 6–12(–13) mm, indehiscent. $2n = 18$.

Flowering Mar–Jun. Near water runoff areas, away from seeps or other areas with prolonged moisture; 50–1500 m; Calif., Oreg.

Diplacus kelloggii occurs in southwestern Oregon and broadly in northern California.

32. Diplacus brandegeei (Pennell) G. L. Nesom, Phytoneuron 2012-54: 1. 2012 • Santa Cruz Island or broadleaf monkeyflower C

Mimulus brandegeei Pennell, Proc. Acad. Nat. Sci. Philadelphia 99: 170. 1947 (as brandegei); *M. latifolius* A. Gray 1876, not *Diplacus latifolius* Nuttall 1838

Herbs, annual. **Stems** erect, 10–100 mm, glandular-puberulent to glandular-pubescent. **Leaves** mostly cauline, relatively even-sized; petiole absent, larger with petiole-like extension; blade narrowly to broadly ovate to broadly elliptic, (4–)8–39 × (2–)4–23 mm, margins entire or crenate, plane, proximal ½ usually ciliate, apex obtuse, surfaces glandular-pubescent. **Pedicels** 1.5–4 mm in fruit, not twisting to invert calyx. **Flowers** 2 per node, or 1 or 2 per node on 1 plant, chasmogamous. **Calyces** distinctly asymmetrically attached to pedicel, inflated in fruit, 8–13 mm, glandular-pubescent, lobes subequal, apex obtuse, ribs green, intercostal areas pale green to whitish. **Corollas:** throat magenta, golden yellow at base, limb magenta, palate ridges golden yellow, purple-speckled, tube-throat 12–19 mm, limb 5–11 mm diam., bilabiate, abaxial lip smaller than adaxial. **Anthers** (distal pair) nearly exserted, glabrous. **Styles** pubescent distally. **Stigmas** exserted, lobes unequal, abaxial 3–4 times adaxial. **Capsules** (5–)6–11 mm, indehiscent.

Flowering Mar–Apr(–May). Rocky, brushy slopes; of conservation concern; 100–300 m; Calif.; Mexico (Baja California).

Diplacus brandegeei is known only from Santa Catalina (Los Angeles County), Santa Cruz (Santa Barbara County), and Guadalupe (Baja California) islands. The earliest name for the species, *Mimulus latifolius* A. Gray, was first transferred to *Diplacus* as *D. latifolius* (A. Gray) G. L. Nesom, but that name is a later homonym of *D. latifolius* Nuttall.

33. Diplacus traskiae (A. L. Grant) G. L. Nesom, Phytoneuron 2012-39: 32. 2012 • Santa Catalina Island monkeyflower E

Mimulus traskiae A. L. Grant, Publ. Field Mus. Nat. Hist., Bot. Ser. 5: 226. 1923

Herbs, annual. **Stems** erect, 80–120 mm, glandular-pubescent. **Leaves** cauline, nearly even-sized or enlarging distally; petiole absent or with petiole-like extension; blade ovate to obovate, 12–41 × 5–21 mm, margins entire or crenate, plane, usually ciliate proximally, apex obtuse, surfaces glandular-pubescent. **Pedicels** 3–5 mm in fruit. **Flowers** 2 per node, or 1 or 2 per node on 1 plant, chasmogamous. **Calyces** distinctly asymmetrically attached to pedicel, not inflated in fruit, 18–21 mm, glandular-pubescent, lobes subequal, apex acute, ribs usually green, intercostal areas whitish. **Corollas:** throat magenta, abaxial lip magenta, adaxial lip whitish, palate ridges not seen, tube-throat 20–23 mm, limb 4–5 mm diam., bilabiate, lobes of abaxial lip smaller than adaxial. **Anthers** (distal pair) nearly exserted, glabrous. **Styles** puberulent distally. **Stigmas** exserted, lobes unequal, abaxial 6–8 times adaxial. **Capsules** not seen.

Flowering Mar–Apr. Rocky, brushy slopes; 0–100 m; Calif.

Diplacus traskiae is known only from Santa Catalina Island. According to the California Native Plant Society Online Inventory of Rare and Endangered Plants, ed. 8 (http://www.cnps.org/cnps/rareplants/inventory/), it (as *Mimulus traskiae*) is possibly extirpated.

34. Diplacus pygmaeus (A. L. Grant) G. L. Nesom, Phytoneuron 2012-39: 30. 2012 • Egg Lake monkeyflower E

Mimulus pygmaeus A. L. Grant, Ann. Missouri Bot. Gard. 11: 312. 1925

Herbs, annual, usually acaulescent. **Stems** erect, 5–12(–20) mm, glandular-puberulent. **Leaves** basal or basal and cauline, relatively even-sized; petiole absent; blade narrowly elliptic to oblanceolate, 2–15 × 0.5–3.5 mm, margins entire or toothed, plane, proximal ½ usually ciliate, apex rounded, surfaces glandular-puberulent. **Pedicels** 0–1 mm in fruit. **Flowers** 1 per node, usually 1–5 per plant, chasmogamous. **Calyces** slightly asymmetrically attached to pedicel, inflated in fruit, 3.5–8 mm, glandular-puberulent, lobes subequal, apex rounded, ribs reddish purple, intercostal areas white. **Corollas:** throat, limb, and palate ridges yellow, central abaxial lobe and throat floor sparsely red-spotted, tube-throat 5–10 mm, limb 2–5 mm diam., bilabiate. **Anthers** included, glabrous. **Styles** sparsely glandular-puberulent distally. **Stigmas** usually exserted, lobes subequal. **Capsules** 2–4 mm, indehiscent. $2n = 20$.

Flowering May–Jun. Vernally flooded swales, mud flats flanking streams, stream banks, low spots in meadows; 1100–1800 m; Calif., Oreg.

Diplacus pygmaeus occurs in northeastern California and south-central Oregon.

35. **Diplacus angustatus** (A. Gray) G. L. Nesom, Phytoneuron 2012-39: 30. 2012 • Narrow-leaved or purple-lip pansy monkeyflower [E]

Eunanus coulteri Harvey & A. Gray ex Bentham var. *angustatus* A. Gray, Proc. Amer. Acad. Arts 7: 381. 1868; *E. angustatus* (A. Gray) Greene; *Mimulus angustatus* (A. Gray) A. Gray; *M. tricolor* Hartweg ex Lindley var. *angustatus* (A. Gray) A. Gray

Herbs, annual, acaulescent or caulescent. **Stems** erect, 0 or 20–70 mm, eglandular-puberulent. **Leaves:** basal densely clustered; petiole absent; blade narrowly lanceolate or elliptic to linear, (3.5–)5–36 × 0.5–3.8 (–5) mm, margins entire, plane, proximal ½+ ciliate, apex obtuse, surfaces glabrous. **Pedicels** 0–1 mm in fruit. **Flowers** 1 per node, chasmogamous. **Calyces** slightly asymmetrically attached to pedicel, inflated in fruit, (4–)7–14(–17) mm, pilose, lobes subequal, apex obtuse, ribs becoming reddish, intercostal areas pale green to whitish. **Corollas:** throat and limb magenta to purple with a dark maroon-purple spot or elongate blotch at base of each abaxial lobe, palate ridges and base of abaxial lip yellow, tube-throat 20–60 mm, limb 10–20 mm diam., bilabiate, lobes equal. **Anthers** (distal pair) sometimes nearly exserted, hirsute with 2 apical tufts. **Styles** glandular-puberulent. **Stigmas** usually exserted, lobes subequal. **Capsules** 2–4.5 mm, indehiscent. $2n = 18$.

Flowering Mar–May(–Jun). Vernally flooded depressions or swales, summits of mesas or foothills; 300–1200 m; Calif.

Diplacus angustatus occurs in northern California in the Sierra Nevada and across the Sacramento Valley in the North Coast Range (Lake, Mendocino, and Napa counties); it also is known from one area in northwestern Shasta County and has a disjunct series of populations in Fresno and Tulare counties.

36. **Diplacus pulchellus** (Drew ex Greene) G. L. Nesom, Phytoneuron 2012-39: 30. 2012 • Pansy or yellow-lip pansy monkeyflower [E]

Eunanus pulchellus Drew ex Greene, Pittonia 2: 104. 1890; *Mimulus pulchellus* (Drew ex Greene) A. L. Grant

Herbs, annual, acaulescent or caulescent. **Stems** erect, 0 or 30–50(–60) mm, glandular-puberulent. **Leaves:** basal densely clustered; petiole absent; blade linear, (6–)8–35 × 1–4.5(–9) mm, margins entire, plane, proximal ½+ ciliate, apex obtuse, surfaces glabrous. **Pedicels** 0–2 mm in fruit. **Flowers** 1 per node, chasmogamous. **Calyces** slightly asymmetrically attached to pedicel, inflated in fruit, (5–)7–15 mm, pilose, lobes subequal, apex obtuse, ribs reddish purple, intercostal areas pale green to whitish. **Corollas** bicolored, adaxial lip lavender to magenta-purple, adaxial throat very dark purple, abaxial lip or at least central lobe golden yellow, abaxial lip base and palate ridges sparsely red-dotted, 2 adaxial lobes and 2 lateral lobes of abaxial lip with dark maroon purple bands at base and extending into and merging within throat, palate ridges yellow, tube-throat (15–)20–40 mm, limb 10–18 mm diam., bilabiate, lips equal. **Anthers** (distal pair) sometimes nearly exserted, with apical tufts of short, eglandular hairs. **Styles** glandular-puberulent. **Stigmas** usually exserted, lobes subequal. **Capsules** (2–) 3–5.5 mm, indehiscent. $2n = 18$.

Flowering Apr–Jul. Vernally flooded depressions or swales, ruts, seepage areas on gentle slopes; 600–2000 m; Calif.

Diplacus pulchellus is known from Calaveras, Mariposa, and Tuolumne counties.

37. **Diplacus tricolor** (Hartweg ex Lindley) G. L. Nesom, Phytoneuron 2012-39: 30. 2012 • Tricolor monkeyflower [E]

Mimulus tricolor Hartweg ex Lindley, J. Hort. Soc. London 4: 222. 1849

Herbs, annual, acaulescent or caulescent. **Stems** erect or ascending, 10–140(–170) mm, densely glandular-puberulent. **Leaves** basal densely clustered; petiole absent; blade narrowly elliptic to oblanceolate, (5–)8–45(–60) × (1–)3–12 (–20) mm, margins entire, sometimes toothed, plane, not ciliate, apex obtuse, surfaces glandular-puberulent. **Pedicels** 1–3(–5) mm in fruit. **Flowers** 1 per node, chasmogamous. **Calyces** slightly asymmetrically attached to pedicel, not inflated in fruit, (6–)11–23 mm, densely glandular-puberulent, lobes subequal, apex obtuse, ribs often purplish proximally, intercostal areas whitish. **Corollas** tricolored, limb and throat magenta to purple, each lobe with a discrete, dark maroon-purple blotch at base, all 3 blotches of abaxial lip round and not usually extending into throat, palate ridges yellow, flanked with white, sometimes purple-spotted, tube-throat (13–)15–50 mm, limb 7–21 mm diam., bilabiate, lobes equal. **Anthers** included, with apical tufts of short, eglandular hairs. **Styles** usually glandular-puberulent. **Stigmas** nearly exserted, lobes subequal. **Capsules** (2–)3–8(–10) mm, indehiscent. $2n = 18$.

Flowering Mar–Jul(–Aug). Vernally flooded depressions in grasslands, low spots and ditches in and around agricultural fields; 50–1500 m; Calif., Oreg.

Diplacus tricolor occurs in northwestern and south-central Oregon and from there across a disjunction to central California as far as Kern County.

38. **Diplacus rupicola** (Coville & A. L. Grant) G. L. Nesom & N. S. Fraga, Phytoneuron 2012-39: 27. 2012 • Death Valley monkeyflower, rock midget E

Mimulus rupicola Coville & A. L. Grant, J. Wash. Acad. Sci. 26: 99, fig. s.n. [p. 100]. 1936

Herbs, perennial, with woody caudex. **Stems** erect to ascending, sometimes pendent, 10–170 mm, densely and finely glandular-puberulent. **Leaves** usually basal rosettes and proximal cauline, relatively even-sized; petiole absent, base gradually narrowed to broad, petiole-like extension; blade oblanceolate, (10–)18–60(–80) × (1.5–)3–15(–26) mm, margins entire, plane, not ciliate, apex acute, surfaces glandular-puberulent. **Pedicels** 1–3 mm in fruit. **Flowers** 1 or 2 per node, chasmogamous. **Calyces** asymmetrically attached to pedicel, not inflated in fruit, 8–18 mm, densely glandular-puberulent, lobes unequal, apex acuminate, ribs green, intercostal areas pale green. **Corollas:** limb pinkish white to nearly white with a large magenta-purple round or 2-lobed blotch at base of each lobe, throat and palate ridges golden yellow with magenta speckling, palate ridges short-pilose, throat glabrous, tube-throat 17–35 mm, limb 8–21 mm diam., not bilabiate. **Anthers** included, glabrous or slightly puberulent at base. **Styles** glandular-puberulent. **Stigmas** included, lobes equal. **Capsules** 3–8 mm, indehiscent until senescence of pedicel, then opening along both sutures only after wetting. $2n = 16$.

Flowering Feb–Jun. Crevices in limestone cliffs and walls, limestone ridge tops and slopes, wash edges, gravelly slopes, canyon sides; 300–1800 m; Calif.

Diplacus rupicola is known from Inyo County.

39. **Diplacus clevelandii** (Brandegee) Greene, Erythea 4: 22. 1896 (as clevelandi) • Cleveland's bush monkeyflower

Mimulus clevelandii Brandegee, Gard. & Forest 8: 134, fig. 20. 1895 (as clevelandi)

Herbs, perennial, sometimes rhizomatous. **Stems** erect, (200–)300–950(–1250) mm, glandular-villous. **Leaves** basal and cauline, relatively even-sized; petiole absent; blade lanceolate to oblong-lanceolate or elliptic-lanceolate, 20–110(–130) × 5–33(–40) mm, margins dentate to serrate, plane or revolute, apex acute, surfaces finely pubescent-glandular. **Pedicels** 2–4(–7) mm in fruit. **Flowers** 2 per node, chasmogamous. **Calyces** not inflated in fruit, 20–35(–37) mm, densely glandular-pubescent, tube slightly dilated distally, lobes subequal, apex acute. **Corollas** bright golden yellow, often with reddish brown speckling on throat floor, palate ridges yellow, tube-throat 21–40 mm, limb 25–30 mm diam., nearly regular, lobes obovate-oblong. **Anthers** included, glabrous. **Styles** minutely glandular. **Stigmas** included, lobes unequal, abaxial 2 times longer than adaxial. **Capsules** 8–15(–17) mm. $2n = 20$.

Flowering Apr–Jun(–Jul). Dry, rocky openings in chaparral, roadcuts; 900–1500 m; Calif.; Mexico (Baja California).

Diplacus clevelandii is restricted to Orange, Riverside, and San Diego counties and in adjacent Mexico.

40. **Diplacus grandiflorus** Groenland, Rev. Hort. 6: 402, fig. 136. 1857 • Sierra bush monkeyflower E

Diplacus glutinosus (J. C. Wendland) Nuttall var. *grandiflorus* Lindley & Paxton; *D. longiflorus* Nuttall var. *grandiflorus* (Groenland) Jepson; *Mimulus aurantiacus* Curtis var. *grandiflorus* (Lindley & Paxton) D. M. Thompson; *M. bifidus* Pennell

Subshrubs. Stems erect, 400–700 mm, minutely hirtel-lous to hirsutulous with slightly deflexed, eglandular hairs. **Leaves** cauline, relatively even-sized; petiole absent; blade elliptic or oblong-elliptic to elliptic-lanceolate or elliptic-oblanceolate, (10–)20–55 × 4–17 mm, margins entire, serrulate, or mucronulate, plane or revolute, apex obtuse or rounded to acute, surfaces glabrous. **Pedicels** 5–12 mm in fruit. **Flowers** 2 per node, chasmogamous. **Calyces** not inflated in fruit, 20–30 mm, glabrous except for lobe apices, tube slightly dilated distally, lobes subequal, apex acute. **Corollas** pale yellow or nearly white to pale orange or light yellow-orange, not spotted or striped, palate ridges yellow to golden yellow, tube-throat 35–45 mm, limb 30–45 mm diam., bilabiate, lobes each apically incised ¼–½ length, appearing 2-lobed. **Anthers** included, glabrous. **Styles** sparsely minutely glandular. **Stigmas** included, lobes equal. **Capsules** 15–25 mm. $2n$ = 20.

Flowering Apr–Jun. Rock walls, dry rocky soils; 100–1600 m; Calif.

Diplacus grandiflorus (central Sierran) and *D. linearis* (coastal) are remarkably similar; it seems likely that they are vicariants. Leaves of *D. linearis* are narrower, and the nodes tend to be considerably more crowded. Corollas of *D. linearis* have narrower limbs, and both pairs of anthers and the stigma are at the same level and relatively deeply included. In *D. grandiflorus*, the anther pairs are separated, and the stigma is above the distal anther pair at or near the throat opening.

41. **Diplacus linearis** (Bentham) Greene, Pittonia 2: 156. 1890 • Monterey monkeyflower E

Mimulus linearis Bentham, Scroph. Ind., 27. 1835; *Diplacus fasciculatus* (Pennell) McMinn; *D. longiflorus* Nuttall var. *linearis* (Bentham) McMinn; *M. bifidus* Pennell subsp. *fasciculatus* Pennell; *M. glutinosus* J. C. Wendland var. *linearis* (Bentham) A. Gray; *M. longiflorus* (Nuttall) A. L. Grant var. *linearis* (Bentham) A. L. Grant

Subshrubs. Stems erect, 300–800(–1200) mm, minutely hirtellous to hirsutulous with slightly deflexed, eglandular hairs. **Leaves** cauline, relatively even-sized; petiole absent; blade narrowly elliptic to narrowly lanceolate or narrowly elliptic-oblong, 12–37 × 3–9 mm, margins entire, serrulate, or mucronulate, plane or revolute, apex usually obtuse to rounded, surfaces glabrous. **Pedicels** 3–10 mm in fruit. **Flowers** (1 or)2 per node, chasmogamous. **Calyces** not inflated in fruit, 20–28 mm, glabrous, tube slightly dilated distally, lobes unequal to subequal, apex acute. **Corollas** yellow-orange to dull orange, not spotted or striped, palate ridges yellow, tube-throat 35–45 mm, limb 20–25 mm diam., bilabiate, lobes oblong, each apically incised ¼–½ length, appearing 2-lobed. **Anthers** included, glabrous. **Styles** minutely glandular. **Stigmas** included, lobes equal. **Capsules** 18–30 mm. $2n$ = 20.

Flowering Apr–Jun. Dry hillsides, rock outcrops; 100–300 m; Calif.

The coastal *Diplacus linearis* (as *Mimulus bifidus* subsp. *fasciculatus*) was allied by F. W. Pennell (1947) with the Sierran *D. grandiflorus* (as *M. bifidus* subsp. *typicus*) as a narrower-leaved and smaller-flowered subspecies; see discussion concerning its distribution under 40. *D. grandiflorus*. The two were considered synonymous by D. M. Thompson (2005). They are distinct in geography, ecology, and morphology.

42. **Diplacus aurantiacus** (Curtis) Jepson, Man. Fl. Pl. Calif., 919. 1925 • Orange-bush monkeyflower E

Mimulus aurantiacus Curtis, Bot. Mag. 10: plate 354. 1796; *Diplacus glutinosus* (J. C. Wendland) Nuttall; *D. glutinosus* var. *aurantiacus* (Curtis) Lindley; *M. glutinosus* J. C. Wendland

Subshrubs or shrubs. Stems erect to ascending, 500–1200 (–1500) mm, minutely hirtellous-hirsutulous and minutely stipitate-glandular. **Leaves** usually cauline, relatively even-sized; petiole absent or indistinct; blade narrowly oblong to narrowly elliptic-lanceolate proximally to lanceolate to elliptic-lanceolate or elliptic distally, 15–60(–75) × 2–20 mm, margins entire or shallowly serrate, plane or revolute, apex obtuse to rounded, surfaces glabrous or abaxial sparsely to densely hairy, hairs branched, adaxial usually without unbranched hairs. **Pedicels** 4–13 mm in fruit. **Flowers** 2(–4) per node, chasmogamous. **Calyces** not inflated in fruit, 18–30 mm, glabrous or minutely hirtellous and/or minutely stipitate-glandular, tube slightly dilated distally, lobes subequal to unequal, apex acute, ribs green, intercostal areas light green. **Corollas** yellow-orange to orange, not spotted or striped, palate ridges yellow to golden yellow or orange, tube-throat 25–30 mm, limb 20–30 mm diam., bilabiate to nearly rotate, lobes oblong, apex of adaxial 2 each shallowly, asymmetrically incised. **Anthers** exserted (at throat), glabrous. **Styles** sparsely glandular. **Stigmas** exserted, lobes equal. **Capsules** 18–31 mm. $2n$ = 20.

Flowering Apr–Aug. Sand dunes and bluffs, dry hillsides, grassy slopes, road banks, stream banks, basaltic knolls, rocky slopes and outcrops, open pine forests, coastal scrub; 0–700(–1000) m; Calif., Oreg.

Diplacus aurantiacus occurs from southwestern Oregon (Curry County) southward to Santa Barbara County, California.

43. **Diplacus calycinus** Eastwood, Bot. Gaz. 41: 287. 1906 • Rock bush monkeyflower E

Diplacus longiflorus Nuttall var. *calycinus* (Eastwood) Jepson; *Mimulus longiflorus* (Nuttall) A. L. Grant subsp. *calycinus* (Eastwood) Munz; *M. longiflorus* var. *calycinus* (Eastwood) A. L. Grant

Subshrubs. Stems erect, 150–1500 mm, glandular-pubescent to viscid-villous. **Leaves** cauline, relatively even-sized; petiole absent; blade elliptic-lanceolate or lanceolate to narrowly lanceolate or linear-lanceolate, sometimes narrowly oblong, 20–75(–100) × 4–20 (–28) mm, margins entire or shallowly crenate, plane or revolute, apex acute to obtuse, abaxial surfaces moderately villous, hairs unbranched, vitreous, adaxial glabrous. **Pedicels** 3–5 mm in fruit. **Flowers** 2 per node, chasmogamous. **Calyces** not inflated in fruit, 28–40 mm, densely glandular-pubescent to short glandular-villous, tube slightly dilated distally, lobes unequal, apex acute, ribs green, intercostal areas light green. **Corollas** usually pale yellow or cream to yellow, not spotted or striped, palate ridges yellow to golden yellow, tube-throat 35–42 mm, limb 20–30 mm diam., bilabiate, lobes oblong, apex of adaxial 2 each shallowly, asymmetrically incised. **Anthers** exserted, glabrous. **Styles** minutely glandular. **Stigmas** exserted, lobes equal. **Capsules** 25–35 mm. $2n = 20$.

Flowering Apr–Jul. Granite outcrops, boulders, rocky gullies; (300–)700–2200 m; Calif.

Although first described as a separate species, *Diplacus calycinus* has more recently been treated at subspecific or varietal rank (A. L. Grant 1924; F. W. Pennell 1951; P. A. Munz and D. D. Keck 1973). D. M. Thompson (2005) included both *D. calycinus* and *D. longiflorus* within his concept of *Mimulus aurantiacus* var. *pubescens* (Torrey) D. M. Thompson. He did not reference the study of sect. *Diplacus* by M. C. Tulig (2000), but results from the Tulig morphometric analyses indicated that *D. calycinus* is distinct from *D. longiflorus*, especially in corolla length, corolla tube length, and style length.

The type of *Diplacus calycinus* is from Tulare County, and the concept of the species is perhaps best restricted to the Sierran population system in Fresno, Kern, and Tulare counties, disjunct from *D. longiflorus*, which occurs primarily in coastal counties. The Sierran system is characterized by distinct abaxial leaf vestiture; the hairs are unbranched, broad, and vitreous, compared to the branched, thinner, and dull hairs of *D. longiflorus*. Plants of *D. calycinus* parapatric with *D. longiflorus* also show a tendency toward the characteristic vestiture and also have lighter-colored (but more variable in color) corollas with narrower but slightly shorter tubes. Intergradation between *D. calycinus* and *D. longiflorus* occurs in the region connecting the San Bernardino and San Gabriel mountains in San Bernardino County.

44. Diplacus longiflorus Nuttall, Ann. Nat. Hist. 1: 139. 1838 (as longiflora) • Southern bush monkeyflower F

Diplacus arachnoideus Greene; *D. glutinosus* (J. C. Wendland) Nuttall var. *pubescens* Torrey; *Mimulus aurantiacus* Curtis var. *pubescens* (Torrey) D. M. Thompson; *M. longiflorus* (Nuttall) A. L. Grant

Subshrubs. Stems erect, 300–1000(–2000) mm, glandular-puberulent and short-villous. **Leaves** usually cauline, relatively even-sized; petiole absent; blade elliptic to lanceolate, elliptic-lanceolate, or elliptic-oblanceolate, 25–65(–80) × 4–15(–25) mm, margins entire or serrate, revolute, apex acute to obtuse, abaxial surfaces densely hairy, hairs branched, adaxial glabrescent. **Pedicels** 5–16 mm in fruit. **Flowers** 2 per node, chasmogamous. **Calyces** not inflated in fruit, 22–32 mm, glandular-puberulent and short glandular-villous to hirsute-villous, tube slightly dilated distally, lobes unequal, apex acute, ribs green, intercostal areas light green. **Corollas** light orange to pale yellow-orange, palate ridges orangish, tube-throat 34–45 mm, limb (25–)28–40 mm diam., bilabiate, lobes oblong, apex of adaxial 2 each shallowly incised. **Anthers** included, glabrous. **Styles** minutely glandular. **Stigmas** included, lobes equal. **Capsules** 18–28 mm. $2n = 20$.

Flowering (Mar–)Apr–Jul. Rocky hillsides and slopes, talus, chaparral, live oak woodlands; (50–)100–1300(–1800) m; Calif.; Mexico (Baja California).

Diplacus longiflorus occurs in southwestern California and northeastern Baja California.

Plants and populations intermediate between *Diplacus longiflorus* and *D. puniceus* are found where their ranges meet in Los Angeles, Orange, Riverside, and San Diego counties. The intermediate morphology and geography indicate that these are hybrids (as has been hypothesized by, for example, M. A. Streisfeld and J. R. Kohn 2005; D. M. Thompson 2005; M. C. Tulig and G. L. Nesom 2012), which have been identified as *D.* ×*australis* (McMinn ex Munz) Tulig. Streisfeld and Kohn found that in San Diego County, *D. longiflorus* and *D. puniceus* are discrete in morphology and separated in geography, with a narrow zone of hybrids and putative introgressants between.

Plants identified as *Diplacus* ×*lompocensis* McMinn (as species) occur where the geographic ranges of *D. aurantiacus* and *D. longiflorus* meet in Santa Barbara County and southern San Luis Obispo County; these plants have floral features intermediate between these two species. Stable populations of the putative hybrid are found throughout this region, although at either end of its distribution, the populations may more closely resemble the nearer parent. Considering that both

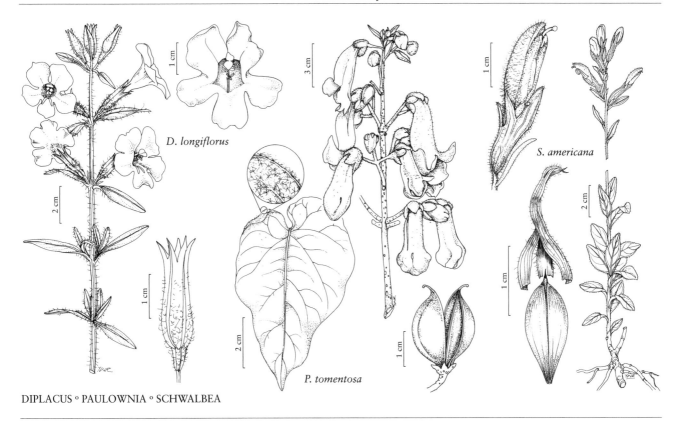

D. longiflorus

S. americana

P. tomentosa

DIPLACUS ∘ PAULOWNIA ∘ SCHWALBEA

D. *aurantiacus* and D. *longiflorus* are morphologically consistent across broad regions, D. ×*lompocensis* is perhaps best interpreted as a zone of introgression.

Diplacus ×*australis* and D. ×*lompocensis* are similar to D. *longiflorus* as well as to each other in most features; they are easily separated only by geographic range. Diplacus *longiflorus* is distinct from both in its larger corolla features and, frequently, calyx indument.

45. Diplacus rutilus (A. L. Grant) McMinn, Madroño 11: 83. 1951 • Santa Susana bush monkeyflower E

Mimulus longiflorus (Nuttall) A. L. Grant var. *rutilus* A. L. Grant, Ann. Missouri Bot. Gard. 11: 333. 1925; *Diplacus longiflorus* Nuttall var. *rutilus* (A. L. Grant) McMinn

Subshrubs. Stems erect, 300–1000(–2000) mm, glandular-puberulent and short-villous. **Leaves** usually cauline, relatively even-sized; petiole absent; blade elliptic to lanceolate, elliptic-lanceolate, or elliptic-oblanceolate, 25–65(–80) × 4–15(–25) mm, margins entire or serrate, revolute, apex acute to obtuse, abaxial surfaces densely hairy, hairs branched, adaxial glabrescent. **Pedicels** 5–16 mm in fruit. **Flowers** 2 per node, chasmogamous. **Calyces** not inflated in fruit, 22–32 mm, glandular-puberulent and short glandular-villous to hirsute-villous, tube slightly dilated distally,

lobes unequal, apex acute, ribs green, intercostal areas light green. **Corollas** red to orange-red, throat whitish at least on floor, palate ridges orange or white with orange crest, tube-throat 34–45 mm, limb (25–)28–40 mm diam., bilabiate, lobes oblong, apex of adaxial 2 each shallowly, asymmetrically incised. **Anthers** included, glabrous. **Styles** minutely glandular. **Stigmas** included, lobes equal. **Capsules** 18–28 mm.

Flowering Mar–Jun. Chaparral; 400–600 m; Calif.

H. E. McMinn (1951), M. C. Tulig and G. L Nesom (2012), and Nesom (2013c) treated *Diplacus rutilus* at specific rank, but P. A. Munz and D. D. Keck (1973) followed the original assessment of A. L. Grant (1924) in treating it as a variety of the light orange- to pale yellow-orange-flowered D. *longiflorus*, while R. M. Beeks (1962) and D. M. Thompson (2005) regarded this taxon as only a variant of D. *longiflorus*, without formal rank. It is maintained here as a distinct, red-flowered species localized in Los Angeles County, California. *Diplacus rutilus* occurs in pockets in a strip from Whittier and Pomona through North Pasadena westward to near the Ventura County line (the Santa Susana area being the type locality), a distance of almost 60 miles.

Very few of these red-flowered collections can be separated in any feature except corolla color from typical *Diplacus longiflorus*. Red corollas have not been observed in D. *longiflorus* outside of Los Angeles County, and D. *rutilus* might be interpreted as reflecting local introgression in this area from D. *puniceus*, but the distinctive and tightly coherent geographical distribution of these red-flowered plants and their apparent absence elsewhere in the area where *D.* ×*australis* occurs suggest that the origin of *D. rutilus* is different from that of the highly variable *D.* ×*australis*.

46. **Diplacus puniceus** Nuttall, Ann. Nat. Hist. 1: 137. 1838 (as punicea) • Red bush monkeyflower

Diplacus glutinosus (J. C. Wendland) Nuttall var. *puniceus* (Nuttall) Bentham; *Mimulus aurantiacus* Curtis var. *puniceus* (Nuttall) D. M. Thompson; *M. glutinosus* J. C. Wendland var. *puniceus* (Nuttall) A. Gray; *M. puniceus* (Nuttall) Steudel

Subshrubs or shrubs. Stems erect to ascending-erect or sprawling, 200–1500(–2000) mm, glabrous. **Leaves** cauline, relatively even-sized; petiole absent; blade linear-oblong or narrowly elliptic to narrowly lanceolate or oblanceolate, 10–60 × 1–11(–15) mm, margins entire or serrate, plane or revolute, apex acute, surfaces glabrous. **Pedicels** 5–25 mm in fruit. **Flowers** 2(–4) per node, chasmogamous. **Calyces** not inflated in fruit, 17–25 mm, glabrous, lobes unequal, apex acute, ribs green, intercostal areas light green. **Corollas** deep red, orange-red, or orange to maroon, throat sometimes orangish, palate ridges red, rarely yellow, tube-throat 27–35 mm, limb 15–23 mm diam., bilabiate, lobes oblong, apex truncate, sometimes notched. **Anthers** (distal pair) exserted, glabrous. **Styles** minutely glandular. **Stigmas** exserted, lobes equal. **Capsules** 14–22 mm. **2*n*** = 20.

Flowering Jan–Jul(–Oct). Rocky hillsides, boulders, moist hillsides, canyons, wash bottoms, roadsides, chaparral; (20–)100–700 m; Calif.; Mexico (Baja California).

Diplacus puniceus occurs in southwestern California and northeastern Baja California. Hybrids are common with *D. longiflorus* and have been called *D.* ×*australis*.

47. **Diplacus parviflorus** Greene, Pittonia 1: 36. 1887 • Island bush monkeyflower E

Mimulus aurantiacus Curtis var. *parviflorus* (Greene) D. M. Thompson; *M. flemingii* Munz

Subshrubs or shrubs. Stems erect to ascending-erect, 300–1200 mm, glabrous. **Leaves** cauline, relatively even-sized; petiole absent; blade elliptic to broadly elliptic-oblanceolate or elliptic-lanceolate, 15–60 × 4–21 mm, margins entire or serrate, plane or revolute, apex acute to obtuse or rounded, surfaces glabrous. **Pedicels** 6–15 mm in fruit. **Flowers** 2 per node, chasmogamous. **Calyces** not inflated in fruit, 18–25 mm, glabrous, tube slightly dilated distally, lobes unequal, apex acute, ribs green, intercostal areas light green. **Corollas** deep red to scarlet, throat sometimes orange, not spotted or striped, palate ridges red to orange-red, tube-throat 27–33 mm, limb 12–16 mm diam., bilabiate, lobes oblong, each truncate-entire to slightly emarginate. **Anthers** exserted, glabrous. **Styles** minutely glandular. **Stigmas** exserted, lobes equal. **Capsules** 14–21 mm. **2*n*** = 20.

Flowering Mar–Aug. Hillsides, canyons, rocky slopes and walls, bluffs, sea cliffs; 10–400 m; Calif.

Diplacus parviflorus is known from four of the Channel Islands (Anacapa, San Clemente, Santa Cruz, and Santa Rosa).

Hybrids with *Diplacus longiflorus* occur on Santa Cruz Island; A. L. Grant (according to label data of collections) found these to be fairly common on open hillsides near Friar's Harbor and Valdez, where the two species grew near each other though apparently separated in habitat, with typical *D. parviflorus* mostly in the canyons and *D. longiflorus* on open hillsides. She noted that the apparent hybrids were variable in all possible combinations of features of the leaves, calyces, and corollas, including color.

Mimulus parviflorus (Greene) A. L. Grant 1925, not Lindley 1825, pertains here.

48. **Diplacus aridus** Abrams, Bull. Torrey Bot. Club 32: 540. 1905 • Low bush monkeyflower E

Mimulus aridus (Abrams) A. L. Grant; *M. aurantiacus* Curtis var. *aridus* (Abrams) D. M. Thompson

Subshrubs. Stems erect, 80–330 mm, glabrous. **Leaves** cauline, relatively even-sized; petiole absent or short; blade narrowly ovate to elliptic-lanceolate, elliptic, or oblanceolate, 20–50 × 3–15 mm, margins dentate to serrate, plane or revolute, apex obtuse to acute, surfaces glabrous. **Pedicels** 3–10 mm in fruit. **Flowers** 2 per node, chasmogamous. **Calyces** inflated in fruit, 35–40 mm, glabrous, tube distinctly dilated distally, lobes subequal, apex acute, ribs green, intercostal areas light green. **Corollas** pale yellow to cream or nearly white, pale orange, golden yellow, or orange-yellow, not spotted or striped, palate ridges golden yellow, throat internally glabrous, tube-throat 37–52 mm, limb 15–22 mm diam., not bilabiate, lobes entire. **Anthers** exserted (at throat), glabrous. **Styles** minutely glandular. **Stigmas** exserted, lobes equal. **Capsules** 20–36 mm. **2*n*** = 20.

Flowering Apr–Jun. Rock walls, dry rocky soils; 200–1500 m; Calif.

Diplacus aridus is known from Imperial and San Diego counties.

PAULOWNIACEAE Nakai

• Paulownia Family

Craig C. Freeman

Richard K. Rabeler

Wayne J. Elisens

Trees, not fleshy, autotrophic. **Leaves** deciduous [persistent], cauline, opposite, rarely in whorls of 3, simple; stipules absent; petiole present; blade not fleshy, not leathery, margins entire or shallowly 3(–5)-lobed and usually serrate to dentate on leaves of young plants. **Inflorescences** terminal thyrses of (1–)3–8(–11)-flowered cymes. **Flowers** bisexual, perianth and androecium hypogynous; sepals 5, connate, calyx bilaterally symmetric; petals 5, connate, corolla bilaterally symmetric, bilabiate, funnelform; stamens 4, proximally adnate to corolla, didynamous, staminode 0; pistil 1, 2-carpellate, ovary superior, 2-locular, placentation axile; ovules anatropous, unitegmic, tenuinucellate; style 1; stigma 1. **Fruits** capsules, dehiscence loculicidal. **Seeds** ca. 2000, brown, ellipsoid; embryo straight, endosperm present.

Genus 1, species 7 (1 in the flora): introduced; e Asia; introduced also in Europe.

Paulownia has been allied with Bignoniaceae or Scrophulariaceae based on morphological, anatomical, and embryological evidence (D. H. Campbell 1930; J. E. Armstrong 1985). Molecular phylogenetic evidence showed that it does not belong with Bignoniaceae or Scrophulariaceae; instead it is an isolated lineage near Lamiaceae, Orobanchaceae, and Phrymaceae (R. G. Olmstead et al. 2001; B. Oxelman et al. 2005).

SELECTED REFERENCES Campbell, D. H. 1930. The relationships of *Paulownia*. Bull. Torrey Bot. Club 57: 47–50. Hu, S. Y. 1959. A monograph of the genus *Paulownia*. Quart. J. Taiwan Mus. 12: 1–54. Hu, S. Y. 1961. The economic botany of the paulownias. Econ. Bot. 15: 11–27. Liang, Z. Y. and Chen Z. Y. 1997. Studies of the cytological taxonomy of the genus *Paulownia*. J. Huazhong Agric. Univ. 16: 609–613. Millsaps, V. 1936. The structure and development of the seed of *Paulownia tomentosa* Steud. J. Elisha Mitchell Sci. Soc. 52: 56–75.

1. PAULOWNIA Siebold & Zuccarini, Fl. Jap. 1: 25. 1835 • [For Anna Paulowna Romanov, 1795–1865, Grand Duchess of Russia and daughter of Czar Paul I, Hereditary Princess of the Netherlands] □

Craig C. Freeman

Thyrses: bracts absent. **Pedicels** present; bracteoles absent. **Flowers:** calyx campanulate, lobes ovate to broadly ovate or oblong; corolla lavender, pinkish purple, or purple externally, whitish or yellowish internally on palate and lined with reddish purple nectar guides, abaxial lobes 3, adaxial 2; stamens: filaments glabrous; stigma capitate. **Seeds:** margins winged, wings clear or tan. $x = 20$.

Species 7 (1 in the flora): introduced; Asia (China); introduced also in Europe, e Asia.

Paulownias long have held mythical, spiritual, cultural, and economic significance in China and Japan (Hu S. Y. 1959, 1961). The wood of some species is highly prized in Asia; *Paulownia tomentosa* is grown in the United States in plantations for wood that is exported to Japan.

The fossil record provides evidence of *Paulownia* in North America and Europe during the Tertiary (C. J. Smiley 1961).

1. **Paulownia tomentosa** (Thunberg) Steudel, Nomencl. Bot. ed. 2, 2: 278. 1841 • Empress-tree, princess-tree, royal paulownia F I W

Bignonia tomentosa Thunberg in J. A. Murray, Syst. Veg. ed. 14, 563. 1784

Stems to 20 m; crown spreading; bark grayish brown, thin, slightly fissured; twigs usually viscid when young, pith hollow or chambered, lenticels conspicuous, white. **Leaves** usually largest on sprouts, smallest near inflorescences; petiole (1.5–)5–21(–36) cm, floccose or sparsely to densely lanate, rarely glabrate; blade ovate to broadly ovate, (6–)14–40(–50) × (4–)8–30(–56) cm, base cordate, sometimes truncate, apex acute, acuminate, or cuspidate, floccose or sparsely to densely lanate abaxially, sometimes only along veins, glabrate or sparsely lanate adaxially. **Thyrses** 18–40 cm; verticillasters 6–10, interrupted, conical, axis glabrous or sparsely lanate; peduncle ascending, 0–5 cm, glabrous or sparsely to densely lanate. **Pedicels** 0.8–1.5 cm, brownish lanate, jointed at apex with flower. **Flowers** pendent, fragrant; calyx tube 6–9 mm, brownish lanate, lobes 7–9 × 5–8 mm, leathery, brownish lanate, margins entire, apex obtuse to acute; corolla 4–6 cm, glandular-pubescent externally, glabrous internally, tube 1–1.5 cm, slightly downcurved, throat gradually inflated, 1.5–2 cm diam., palate 2-ridged, abaxial lobes projecting-spreading, adaxial reflexed; stamens included; pollen sacs opposite, white, tan, or brown, flattened, 1.8–2.3 mm, dehiscing across connective, glabrous; styles 24–30 mm, sparsely hairy proximally. **Capsules** persistent, ovoid, 2.5–5 × 1.5–2.6 cm, cartilaginous, viscid, floccose distally. **Seeds** fusiform, 3–4 mm, surrounded by prominent wings, margins irregular, wings reticulate. $2n = 40$.

Flowering Apr–Jun. Roadsides, riparian areas, forest margins, old homesteads, strip mines, soil banks, clearcuts; 0–1200 m; introduced; Ala., Ark., Conn., Del., D.C., Fla., Ga., Ill., Ind., Ky., La., Md., Mass., Miss., Mo., N.J., N.Y., N.C., Ohio, Okla., Oreg., Pa., R.I., S.C., Tenn., Tex., Va., Wash., W.Va.; Asia (China); introduced also in Europe, elsewhere in e Asia (Japan, Korea).

Paulownia tomentosa was brought to the United States around 1844 from plants introduced into cultivation in Europe (Hu S. Y. 1961); it has been planted as an ornamental and shade tree in southern Canada, eastern and northwestern United States, Europe, Japan, and Korea. Trees are fast-growing and produce sturdy, light wood. Princess-tree is a prolific seed producer, resprouts from suckers when cut, and can become an aggressive invader in disturbed habitats. In North America, it shows a tendency to invade natural habitats more frequently than do plants in central Europe (E. Franz 2007); it is increasingly invading natural forests in the southern Appalachian Mountains. *Paulownia tomentosa* was designated in 2004 as an exotic with a severe threat to escape in Tennessee (Tennessee Exotic Pest Plant Council 2004); it is prohibited in the nursery trade in Connecticut (Connecticut Invasive Plants Council, http://cipwg.uconn.edu/invasive_plant_list/). The broad wings of the seeds aid in wind dissemination and apparently facilitate imbibition (R. Vujičić et al. 1993). A count of 2000 seeds per capsule, which appears widely in the literature, apparently came from R. S. Walker (1919).

The type and abundance of hairs of *Paulownia tomentosa* varies greatly with phenology and location on plants. Dendritic hairs occur on leaves and in inflorescences; they are dense on pedicels and calyces. Their abundance on leaves varies, and there they may occur mixed with eglandular or glandular, erect, unbranched, multicellular hairs. Leaf vestiture has been used to distinguish two varieties: var. *tomentosa*, with leaf blades densely lanate abaxially, and var. *tsinlingensis* (Pai) Gong Tong, with leaf blades glabrous or sparsely lanate abaxially when young (Hong D. Y. et al. 1998). The validity of these varieties cannot be evaluated from North American material alone.

OROBANCHACEAE Ventenat

• Broomrape Family

Craig C. Freeman

Richard K. Rabeler

Wayne J. Elisens

Herbs, rarely subshrubs or shrubs, annual, biennial, or perennial, sometimes fleshy, hemiparasitic or holoparasitic (without chlorophyll) [autotrophic]. **Stems** subterranean or aerial; aerial stems prostrate to decumbent, ascending, or erect [viny]. **Leaves** deciduous, cauline or basal and cauline, rarely basal only or absent, sometimes scales, opposite, alternate, whorled, or spiral, simple; stipules absent; petiole present or absent; blade usually not fleshy or leathery, rarely fleshy, leathery, or chartaceous, margins entire, toothed, or lobed. **Inflorescences** terminal and/or axillary, racemes, panicles, spikes, corymbs, or flowers 1 or 2. **Flowers** bisexual, perianth and androecium hypogynous; sepals (0 or)2–5(–8), connate, calyx radially or bilaterally symmetric; petals [4 or]5, connate, corolla bilaterally symmetric, bilabiate or strongly bilabiate, tubular, funnelform, campanulate, salverform, or club-shaped, sometimes cylindric, subrotate, or curved; stamens (2 or)4, adnate to corolla tube, didynamous, subequal, or equal, staminodes 0 or 2; pistil 1, 2[or 3]-carpellate, ovary superior, 1- or 2-locular, placentation axile, sometimes parietal; ovules anatropous or campylotropous-like (*Rhinanthus*), unitegmic, tenuinucellate; style 1; stigma 1. **Fruits** capsules, dehiscence loculicidal and/or septicidal or indehiscent (*Conopholis*). **Seeds** 1–2500(–5000), brown or black, sometimes tan, white, yellow, amber, or gray, ovoid to ellipsoid, reniform, globular, oblong, or angled; embryo straight, endosperm present.

Genera ca. 100, species ca. 2000 (27 genera, 292 species in the flora): nearly worldwide; especially in warm temperate regions.

Orobanchaceae are now defined to include both the holoparasitic members traditionally included in the family (A. Cronquist 1981) and the hemiparasitic genera formerly included in Scrophulariaceae. Although multiple research groups focus on members of the Orobanchaceae, a widely accepted infrafamilial classification of the family in the sense of Angiosperm Phylogeny Group (2016) has not yet appeared.

The classification by J. R. McNeal et al. (2013), who found that Orobanchaceae comprise six clades, is followed herein (their named clades are roughly equivalent to tribes). The autotrophic *Lindenbergia* Lehmann (12 species in the Old World) corresponds to the basal clade sister to the rest of the clades. Species in our region are distributed among the remaining five clades: Cymbarieae D. Don (genus 1), Orobancheae Lamarck & de Candolle (genera 2–6), Rhinantheae Lamarck & de Candolle (genera 7–12), Buchnereae Bentham (genera 13 and 14), and Pedicularideae Duby (genera 15–27). Within the family, genera are arranged alphabetically within tribes, or within Pedicularideae, in subgroups within the tribe.

Parasitic plants attach to their hosts via haustoria (L. J. Irving and D. D. Cameron 2009). Haustoria are produced by both hemiparasitic and holoparasitic Orobanchaceae (E. Fischer 2004). In hemiparasitic taxa, haustoria usually tap their host's xylem, mostly taking up water, mineral nutrients, and nitrogen from their host, and sometimes also carbon. Holoparasitic taxa derive all of their growth requirements predominantly from the host's phloem (Irving and Cameron).

Parasitism has evolved once in the family (N. D. Young et al. 1999; J. R. McNeal et al. 2013); holoparasitism has arisen independently three times from the hemiparasitic condition (J. R. Bennett and S. Mathews 2006; McNeal et al.).

Some Orobanchaceae are serious pests, primarily on legume and grain crops in warmer and drier areas, especially in sub-Saharan Africa. *Striga* is a particularly serious pest that parasitizes mostly monocots; *S. gesnerioides* attacks eudicots (K. I. Mohamed et al. 2006). *Orobanche* parasitizes eudicot crops primarily in temperate parts of the world (E. S. Teryokhin 1997). All *Striga* species and non-native species of *Orobanche* in the flora area are listed on the Federal Noxious Weed List (http://www.aphis.usda.gov/plant_health/plant_pest_info/weeds/downloads/weedlist.pdf) in the United States.

SELECTED REFERENCES Bennett, J. R. and S. Mathews. 2006. Phylogeny of the parasitic plant family Orobanchaceae inferred from Phytochrome A. Amer. J. Bot. 93: 1039–1051. McNeal, J. R. et al. 2013. Phylogeny and origins of holoparasitism in Orobanchaceae. Amer. J. Bot. 100: 971–983. Park, J. M. et al. 2008. A plastid phylogeny of the non-photosynthetic parasitic *Orobanche* (Orobanchaceae) and related genera. J. Pl. Res. 121: 365–376. Teryokhin, E. S. 1997. Weed Broomrapes: Systematics, Ontogenesis, Biology, Evolution. Landshut. Thieret, J. W. 1971. The genera of Orobanchaceae in the southeastern United States. J. Arnold Arbor. 52: 404–434. Wolfe, A. D. et al. 2005. Phylogeny and biogeography of Orobanchaceae. Folia Geobot. 40: 115–134. Young, N. D., K. E. Steiner, and C. W. dePamphilis. 1999. The evolution of parasitism in Scrophulariaceae/Orobanchaceae: Plastid gene sequences refute an evolutionary transition series. Ann. Missouri Bot. Gard. 86: 876–893.

1. Plants holoparasitic, achlorophyllous.
    2. Corollas salverform; annuals . . . . 14. *Striga* (in part), p. 508
    2. Corollas short-tubular, tubular, or funnelform; perennials or annuals.
        3. Flowers cleistogamous and chasmogamous; petals 5 (appearing as 4); stems absent . . . . 4. *Epifagus*, p. 463
        3. Flowers chasmogamous; petals 5; stems present.
            4. Capsules indehiscent; calyces divided abaxially, not divided adaxially; stamens exserted . . . . 3. *Conopholis*, p. 461
            4. Capsules dehiscent; calyces divided roughly uniformly abaxially and adaxially; stamens included.
                5. Corollas tinged pink to purple, yellow, or blue, pallid proximally; palatal folds present (longitudinal folds in abaxial side of tube); calyces narrowly campanulate to campanulate; roots short, sometimes coralloid . . . . . . 6. *Orobanche*, p. 467
                5. Corollas dark red or purple, sometimes yellow; palatal folds absent; calyces cup-shaped; roots absent.
                    6. Inflorescences dense spikes; pedicels absent, bracteoles absent; corollas short-tubular . . . . 2. *Boschniakia*, p. 460
                    6. Inflorescences compact or open racemes; pedicels present, bracteoles present, rarely absent; corollas funnelform . . . . 5. *Kopsiopsis*, p. 464
1. Plants hemiparasitic, chlorophyllous.
    7. Corollas bilabiate, adaxial lips not galeate, cucullate, or beaked.
        8. Corollas salverform.
            9. Corollas purple, blue-purple, blue, violet, rosy, or white; filaments pilose . . . . 13. *Buchnera*, p. 506
            9. Corollas red, brownish red, or purple, rarely white or yellow; filaments glabrous . . . . 14. *Striga* (in part), p. 508

8. Corollas tubular, campanulate, or subrotate.
  10. Leaves whorled . . . . .18. *Brachystigma*, p. 559
  10. Leaves alternate, opposite, or subopposite.
    11. Leaves alternate . . . . 16. *Agalinis* (in part), p. 534
    11. Leaves opposite or subopposite.
      12. Corollas pale pink to rose purple or purple, rarely white; leaf blade
        margins entire, rarely proximally cleft, pinnatifid, or 2-pinnatifid . . .
        . . . . 16. *Agalinis* (in part), p. 534
      12. Corollas yellow or bright orange; leaf blade margins toothed or
        irregularly lobed, pinnatifid, or 2-pinnatifid, sometimes entire.
        13. Corollas bright orange, tubular; stamens equal . . . . .20. *Macranthera*, p. 561
        13. Corollas yellow, campanulate; stamens didynamous, subequal, or
          equal.
          14. Anthers villous . . . . 17. *Aureolaria*, p. 555
          14. Anthers glabrous.
            15. Calyx lobes ovate to oblong-ovate; stamens didynamous
            . . . . 19. *Dasistoma*, p. 560
            15. Calyx lobes linear to lanceolate; stamens equal to
            subequal . . . . .21. *Seymeria*, p. 562

[7. Shifted to left margin.—Ed.]
7. Corollas strongly bilabiate or bilabiate, adaxial lips galeate, cucullate, or beaked.
  16. Perennials, caudices woody or fleshy.
    17. Bracteoles present; sepals 5 . . . . 1. *Schwalbea*, p. 459
    17. Bracteoles absent; sepals 2, 4, or 5.
      18. Cauline leaves decussate . . . . 7. *Bartsia*, p. 488
      18. Cauline leaves alternate, rarely whorled.
        19. Pollen sacs equal; corollas: adaxial lips sometimes with an upcurved or
          coiled beak . . . . 15. *Pedicularis*, p. 510
        19. Pollen sacs unequal; corollas: adaxial lips straight, rarely hooked . . . . . .
        . . . . 22. *Castilleja* (in part), p. 565
  16. Annuals, rarely biennials, caudices absent.
    20. Cauline leaves opposite, sometimes subopposite or alternate.
      21. Leaf blade margins entire, sometimes margins of distal leaves proximally
        toothed; calyx lobes subulate; seeds 1–4 . . . . 10. *Melampyrum*, p. 502
      21. Leaf blade margins toothed; calyx lobes deltate, triangular, or lanceolate; seeds
        (2–)10–450.
        22. Calyces ovate to suborbiculate, flattened laterally, accrescent in fruit. . . .
        . . . . .12. *Rhinanthus*, p. 504
        22. Calyces tubular to campanulate, not flattened laterally, not accrescent in
          fruit.
          23. Anther mucros unequal; capsule dehiscence septicidal. . . . . .9. *Euphrasia*, p. 492
          23. Anther mucros equal or absent; capsule dehiscence loculicidal.
            24. Filaments glabrous; inflorescences spikelike racemes. . . . . .8. *Bellardia*, p. 490
            24. Filaments papillose; inflorescences unilateral racemes . . . . .11. *Odontites*, p. 503
    20. Cauline leaves alternate, proximals rarely subopposite to opposite.
      25. Stamens 2.
        26. Leaf blades: margins of proximals 3-lobed, margins of distals entire . . . .
        . . . . 24. *Cordylanthus* (in part), p. 669
        26. Leaf blades: margins entire or pinnately 5–11-lobed.
          27. Leaf blade margins entire or pinnately 5- or 7-lobed; pollens sacs
            approximate, connectives not elongate . . . . .23. *Chloropyron* (in part), p. 666
          27. Leaf blade margins pinnately 8–11-lobed; pollens sacs separate,
            connectives elongate . . . . .25. *Dicranostegia*, p. 679
      25. Stamens 4.

[28. Shifted to left margin.—Ed.]
28. Corollas: adaxial lips ± straight, openings directed forward, rarely beaked, bent, or hooked at tip and openings directed downward; stigmas capitate or 2-lobed.
   29. Cauline leaves alternate; pollen sacs 2 . . . . 22. *Castilleja* (in part), p. 565
   29. Cauline leaves: proximals usually subopposite to opposite, distals alternate; pollen sacs 1 . . . . .27. *Triphysaria*, p. 684
28. Corollas: adaxial lips rounded at apex, sometimes obscurely so, openings directed downward; stigmas not or slightly expanded.
   30. Sepals 4, calyces tubular . . . . .26. *Orthocarpus*, p. 680
   30. Sepals 2, calyces spathelike.
      31. Leaf blade margins entire; corollas: middle lobes of abaxial lips not revolute; saline marshes, alkaline flats . . . . .23. *Chloropyron* (in part), p. 666
      31. Leaf blade margins entire or 3–7-lobed; corollas: middle lobes of abaxial lips tightly revolute; sagebrush scrub, chaparral, woodlands, forests . . . . 24. *Cordylanthus* (in part), p. 669

1. SCHWALBEA Linnaeus, Sp. Pl. 2: 606. 1753; Gen. Pl. ed. 5, 265. 1754 • Chaffseed
[For Christian Georg Schwalbe, eighteenth-century medical botanist] C E

Bruce A. Sorrie

Herbs, perennial; hemiparasitic, caudex knotty, semiwoody. **Stems** erect, not fleshy, densely pubescent. **Leaves** cauline, alternate; petiole absent; blade not fleshy, not leathery, margins entire. **Inflorescences** terminal, racemes; bracts absent. **Pedicels** present; bracteoles present. **Flowers:** sepals 5, calyx bilaterally symmetric, tubular, abaxial lobes long-triangular, larger and with 2 lateral lobes and larger central lobe 3-notched at summit, adaxial short-triangular; petals 5, corolla pale yellow, strongly suffused with purple distally, strongly bilabiate, tubular, abaxial lobes 3, adaxial 2, adaxial lip ± galeate; stamens 4, equal, filaments pilose; staminode 0; ovary 2-locular, placentation axile; stigma capitate. **Capsules:** dehiscence septicidal and loculicidal. **Seeds** ca. 300, brown or amber, linear, flattened, ± twisted, wings present. *x* = 18.

Species 1: e United States.

SELECTED REFERENCE Kirkman, L. K., M. B. Drew, and D. Edwards. 1998. Effects of experimental fire regimes on the population dynamics of *Schwalbea americana* L. Pl. Ecol. 137: 115–137.

1. **Schwalbea americana** Linnaeus, Sp. Pl. 2: 606. 1753
C E F

Schwalbea americana var. *australis* (Pennell) Reveal & C. R. Broome; *S. australis* Pennell

Perennials blackening upon drying. **Stems** simple or branched from near base, 3–8 dm. **Leaves:** blade lanceolate to narrowly elliptic, 20–50 × 6–10 mm, gradually smaller distally, 3-nerved, surfaces densely pubescent. **Pedicels** ascending, 1–5 mm; bracteoles 2, linear, 6–12 mm, arising just proximal to calyx. **Flowers:** sepals 15–22 mm; calyx prominently 10–12-ribbed, oblique, glandular-pubescent, lobes unequal; petals suffused with purple in distal ½ and on veins, 22–35 mm; stamens included in adaxial lip; style included, 2–3 mm, glabrous. **Capsules** brown, ellipsoid to ovoid, 10–12 mm, glabrous. **Seeds** 2–2.5 mm, wings membranous. *2n* = 36.

Flowering Apr–Aug. Savannas, flatwoods, ecotones of depression ponds and streamheads, moist glacial outwash plains, pine-oak woodlands, longleaf pine, other pines northward and inland; of conservation concern; 0–400 m; Ala., Conn., Del., Fla., Ga., Ky., La., Md., Mass., Miss., N.J., N.Y., N.C., S.C., Tenn., Va.

Schwalbea americana is among the more fire-dependent species in the flora, responding favorably to annual burns. Documented host plants include *Aletris farinosa, Carphephorus odoratissimus, Gaylussacia dumosa, Ilex glabra,* and *Pityopsis graminifolia.* It has declined greatly since the advent of widespread fire suppression and is now federally listed as endangered; extant populations occur in Florida, Georgia, Louisiana, New Jersey, North Carolina, and South Carolina. Chaffseed was reported from eastern Texas by D. S. Correll and M. C. Johnston (1970); no voucher has been seen.

Schwalbea americana is in the Center for Plant Conservation's National Collection of Endangered Plants.

2. BOSCHNIAKIA   C. A. Meyer ex Bongard, Mém. Acad. Imp. Sci. Saint-Pétersbourg, Sér. 6, Sci. Math. 2: 159.   1832   •   Ground-cone [For Alexander Karlovich Boschniak, 1786–1831, Russian botanist]

L. Turner Collins

Alison E. L. Colwell

George Yatskievych

Herbs, monocarpic perennial; achlorophyllous, holoparasitic, with rhizomelike vegetative structure, surfaces tessellate or with irregular scaly plates, roots absent. **Stems** erect, fleshy, glabrous. **Leaves** cauline, spiral, proximally imbricate, less so distally; petiole absent; blade stiffly chartaceous, margins entire. **Inflorescences** terminal, dense spikes; bracts present. **Pedicels** absent; bracteoles absent. **Flowers:** sepals 4 or 5, calyx bilaterally symmetric, cup-shaped, lobes triangular-acuminate; petals 5, corolla dark red or dark purple, strongly bilabiate, cucullate, short-tubular, palatal folds absent, abaxial lobes 0 or 3, adaxial 2, adaxial lip entire or with shallow notch; stamens 4, didynamous, included, filaments glabrous; staminode 0; ovary 1-locular, placentation parietal; stigma 2–4-lobed, broadly clavate-crateriform or nearly capitate. **Capsules:** dehiscence loculicidal. **Seeds** 2000–2500, light tan or brown, irregularly columnar or oblong-ellipsoid, wings absent.

Species 2 (1 in the flora): nw North America, Asia (Bhutan, China, Nepal, Russian Far East).

Boschniakia has not been well studied using modern systematic techniques. The overall relationship of the genera of holoparasites in Orobanchaceae is currently under review.

Boschniakia himalaica Hooker f. & Thomson is distributed in southern China and the Himalayan countries of Asia. It is parasitic on species of *Rhododendron*.

SELECTED REFERENCE   Zhang, Zhi Y. 1987. A taxonomic study of the genus *Boschniakia*. Acta Bot. Yunnan. 9: 289–296.

1. **Boschniakia rossica** (Chamisso & Schlechtendal) B. Fedtschenko in B. A. Fedtschenko and A. F. Flerow, Fl. Evropeiskoi Ross., 896. 1910 • Northern ground-cone, poque [F]

Orobanche rossica Chamisso & Schlechtendal, Linnaea 3: 132. 1828

Stems 1–3, dark wine red or purple, unbranched, 9–38 cm, with rhizomelike cylindric to globular base 10–25(–40) mm diam. **Leaves** yellow or wine red; blade triangular, lanceolate, or ovate, 3–10 mm. **Spikes** dark purple or red, 10–25 mm diam.; bracts yellow or yellow-tipped, sometimes translucent, triangular, lanceolate, or ovate, margins erose or ciliolate, apex acute or acuminate, rarely rounded. **Flowers:** calyx purple or red-brown, cleft adaxially, 3–6 mm; corolla 8–13 mm, glabrous except margins, tube base ± inflated, abaxial lobes 0–1 mm, margins white-ciliate; stamens included or exserted; ovary ovoid; style base persistent on fruit. **Capsules** subglobular, 5–6 × 5–6 mm, apex beaked. **Seeds** 0.5–0.7 × 0.2–0.3 mm, reticulate, foveate.

Flowering Jun–Aug. Forests, muskegs, bogs, coastal and interior along streams and lakeshores; 0–1500 m; Alta., B.C., N.W.T., Yukon; Alaska; Asia (n China, Russian Far East).

The confirmed host of *Boschniakia rossica* is *Alnus*; unconfirmed hosts include *Betula* and *Vaccinium*; dubious hosts include *Chamaedaphne*, *Picea*, and *Salix*.

Boschniakia rossica has a rather uniform morphology in North America; however, var. *flavida* Y. Zhang & J. Y. Ma has been described in a region of China.

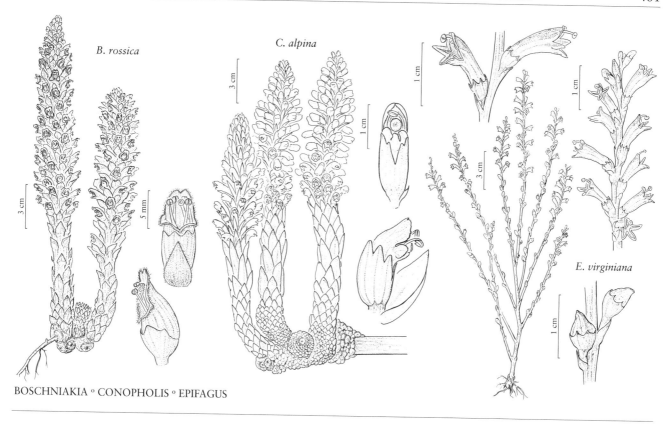

B. rossica

C. alpina

E. virginiana

BOSCHNIAKIA ∘ CONOPHOLIS ∘ EPIFAGUS

3. CONOPHOLIS Wallroth, Orobanches Gen. Diask., 78. 1825 • Cancer- or squaw-root [Greek *conos*, cone, and *pholis*, scale, alluding to conelike inflorescences]

L. Turner Collins

Alison E. L. Colwell

George Yatskievych

Herbs, perennial; achlorophyllous, holoparasitic, with perennial, tuberlike underground vegetative structure attached to host root, dark brown or black, with a sclerenchymatous, knobby surface, roots absent. **Stems** erect, yellow, brown, or black, fleshy, glabrous. **Leaves** cauline, spiral, tightly imbricate; petiole absent; blade stiffly chartaceous, margins entire or minutely erose or ciliate. **Inflorescences** terminal, compact racemes; bracts present. **Pedicels** present; bracteoles present, sometimes absent. **Flowers:** sepals 2, 4, or 5, calyx bilaterally symmetric, tubular, divided abaxially, not divided adaxially, lobes lanceolate; petals 5, corolla yellow, strongly bilabiate, tubular, arching forward, abaxial lobes [1–]3, adaxial 2, adaxial lip notched; stamens 4, didynamous, exserted, filaments glabrous or hairy at base; staminode 0; ovary 1-locular, placentation parietal; stigma disciform-crateriform or 2–4-lobed. **Capsules:** indehiscent. **Seeds** 300–600, glossy tan, brown, or black, irregularly ellipsoid or angled, wings absent.

Species 3 (2 in the flora): sw, e North America, Mexico, Central America.

Five species have been described in the genus; most authors accept *Conopholis* as including only from one to three species (R. R. Haynes 1971; A. G. Rodrigues et al. 2011, 2013). All taxa in the genus are strikingly similar morphologically. Differences in scale leaves, floral bracts, flowers, vesture, and plant size have been used to separate species and varieties; the variability of these characters within taxa occasionally exceeds between-taxon differences. There is considerable overlap in any character set used to separate taxa. J. W. Thieret (1971) suggested that all taxa could be considered conspecific. Molecular and morphological studies by Rodrigues et al. (2011, 2013) provided strong evidence that three species should be recognized. The third taxon, *C. panamensis* Woodson, is known only from Central America. Hosts are all members of *Quercus* (Fagaceae).

Conopholis fruits develop a woody texture at maturity. Although members of the family are generally stated to produce capsular fruits, those of *Conopholis* generally remain indehiscent until degraded by either physical abrasion or invasion by predatory insects, such as ants. Seed dispersal tends to be of two types: short-range dispersal by the action of gravity, wind, and/or water following the breach of the wall; and longer-range dispersal following ingestion of the plants by various mammals.

SELECTED REFERENCES Haynes, R. R. 1971. A monograph of *Conopholis*. Sida 4: 246–264. Rodrigues, A. G. et al. 2013. Morphometric analysis and taxonomic revision of the North American holoparasitic genus *Conopholis* (Orobanchaceae). Syst. Bot. 38: 795–804. Rodrigues, A. G., A. E. L. Colwell, and S. Stefanović. 2011. Molecular systematics of the parasitic genus *Conopholis* (Orobanchaceae) inferred from plastid and nuclear sequences. Amer. J. Bot. 98: 896–908.

1. Bracts glabrous, lanceolate, mostly concealing calyces, margins entire or minutely erose, not ciliate; leaves glabrous or minutely hairy along margins, triangular to broadly lanceolate; calyces 3–8 mm, lobe apices acute; corollas 8–14 mm; anthers glabrous . . . . 1. *Conopholis americana*
1. Bracts glandular-pubescent, narrowly lanceolate, mostly not concealing calyces, margins entire, glandular-ciliate; leaves mostly glandular-pubescent, triangular to narrowly lanceolate; calyces 6–9 mm, lobe apices obtuse; corollas 14–20 mm; anthers sparsely pilose or glabrous . . . . 2. *Conopholis alpina*

1. **Conopholis americana** (Linnaeus) Wallroth, Orobanches Gen. Diask., 78. 1825 • American cancer-root, conopholis d'Amérique

Orobanche americana Linnaeus, Mant. Pl. 1: 88. 1767

Stems 6–20 cm. Leaves triangular to broadly lanceolate, 5.5–18 × 5–12 mm, glabrous or minutely hairy along margins. Bracts mostly concealing calyces, lanceolate, 5–18 × 2–8 mm, margins entire or minutely erose, not ciliate, glabrous, veins visible or obscure. Pedicels 0–4(–6) mm; bracteoles 2. Flowers: calyx 3–8 mm, lobe margins entire or erose, not ciliate, apex acute; corolla 8–14 mm; anthers glabrous. Capsules 5–15 × 5–12 mm; styles persistent.

Flowering Feb–Jun. Under oaks in moist deciduous or mixed deciduous-coniferous woods; 30–600 m; Man., N.S., Ont., Que.; Ala., Conn., Del., D.C., Fla., Ga., Ill., Ind., Ky., Maine, Md., Mass., Mich., N.H., N.J., N.Y., N.C., Ohio, Pa., R.I., S.C., Tenn., Vt., Va., W.Va., Wis.; s Mexico.

A. G. Rodrigues et al. (2013) noted the existence of a single historical specimen from Jeff Davis County, Texas, and suggested that this represented a historical disjunction. However, it is also possible that the specimen in question was mislabeled.

V. Baird and J. L. Riopel (1986) documented the dispersal of *Conopholis americana* in the eastern United States by deer and rodents.

SELECTED REFERENCE Baird, V. and J. L. Riopel. 1986. Life history studies of *Conopholis americana* (Orobanchaceae). Amer. Midl. Naturalist 116: 140–151.

2. **Conopholis alpina** Liebmann, Förh. Skand. Naturf. Mötet 4: 184. 1847 • Bear corn F

Conopholis alpina var. *mexicana* (A. Gray ex S. Watson) R. R. Haynes; *C. mexicana* A. Gray ex S. Watson

Stems 11–33 cm. Leaves triangular to narrowly lanceolate, 12–22 × 3–9 mm, mostly glandular-pubescent. Bracts mostly not concealing calyces, narrowly lanceolate, 12–22 × 2–7 mm, margins entire, glandular-ciliate, glandular-pubescent, veins not evident. Pedicels 0–3(–4) mm; bracteoles 0–2. Flowers: calyx 6–9 mm, lobe apex obtuse; corolla 14–20 mm; anthers sparsely pilose or glabrous. Capsules 8–15 × 6–12 mm; styles mostly deciduous.

Flowering Feb–Jul. Oak woodlands, mixed montane forests; 1500–2800 m; Ariz., N.Mex., Tex.; Mexico.

R. R. Haynes (1971) recognized two varieties within *Conopholis alpina* based on purported differences in foliar texture, relative bract length, and glandularity, as well as semi-distinct ranges. However, the molecular and morphological analyses by A. G. Rodrigues et al. (2013) resulted in a reevaluation of taxon circumscriptions and redeterminations of selected specimens. The present treatment follows that of Rodrigues et al. in not recognizing infraspecific taxa within *C. alpina*.

B. R. McKinney and J. D. Villalobos (2014) documented the seeds of *Conopholis alpina* in the diet of Mexican black bears. The foraging behavior of bears in the western United States has given rise to one of the vernacular names, bear corn.

4. EPIFAGUS Nuttall, Gen. N. Amer. Pl. 2: 60. 1818, name conserved • Beechdrops, cancer-root, épifage [Greek *epi-*, upon, and Latin *fagus*, beech, alluding to host plant]

L. Turner Collins

Alison E. L. Colwell

George Yatskievych

Herbs, annual; achlorophyllous, holoparasitic, roots coralloid, from among scales of cormlike structure. **Stems** absent. **Leaves** absent. **Inflorescences** terminal, open panicles of racemes; bracts present. **Pedicels** present; bracteoles present. **Flowers:** sepals 5, calyx ± bilaterally symmetric, obliquely campanulate, lobes toothlike. **Chasmogamous flowers:** petals 5 (appearing as 4), corolla white or pale yellow with purple to reddish purple or reddish brown stripes, bilabiate, tubular, abaxial lobes 3, adaxial 1 (by connation); stamens 4, didynamous, filaments densely pilose at base, sparsely pilose distally; staminode 0; ovary 1-locular, placentation parietal; stigma capitate or ca. 2-lobed. **Cleistogamous flowers:** corolla not persistent, purplish red and white, rarely pale yellow, bilaterally symmetric, calyptriform; stamens included, anthers connate, adnate to stigma; corolla, stamens, and style shed as unit. **Capsules:** dehiscence loculicidal. **Seeds** 500–600, pale amber, narrowly ovoid to oblong-ellipsoid, not or only slightly flattened, wings absent.

Species 1: c, e North America, ne Mexico.

Epifagus is the only genus among the holoparasites to produce cleistogamous flowers. The relative occurrence of cleistogamous and chasmogamous flowers varies from plant to plant.

The cormlike structure of *Epifagus* is entirely underground and differs from the stems of *Boschniakia*, *Conopholis*, and *Kopsiopsis* by having both scales and coralloid roots. The racemes emerge directly from the base; there is no aerial stem. Floral buds occur in the axils of all bracts, often even those on the cormlike base. Other genera of holoparasites have a well-differentiated vegetative stem that supports the inflorescence.

SELECTED REFERENCE Thieret, J. W. 1969d. Notes on *Epifagus*. Castanea 34: 397–402.

1. Epifagus virginiana (Linnaeus) W. P. C. Barton, Comp. Fl. Philadelph. 2: 50. 1818 (as virginianus) F

Orobanche virginiana Linnaeus, Sp. Pl. 2: 633. 1753

Racemes usually with 1–12 erect or ascending branches, yellow, blackish purple, purplish brown, yellowish brown, or brown, (3.5–)15–45(–54) cm, arising from underground, ovoid, vegetative, cormlike structure covered with imbricate scale leaves and short, coralloid, adventitious, yellow to brownish orange roots; bracts triangular to ovate or ± awl-shaped, 2–5 mm. **Pedicels** 1–3 mm; bracteoles 2, adnate to base of calyx, triangular-ovate, 1–2 mm. **Flowers:** calyx 2–3 mm, lobes acuminate, 1 mm. **Chasmogamous flowers:** appearing bisexual, rarely producing fruit; corolla ± laterally compressed, 8–12 mm, abaxial lip with lobes erect to ± spreading, apex acute, adaxial lobe erect, slightly incurved apically, margins entire or notched; stamens included or barely exserted; stigma usually exserted. **Cleistogamous flowers:** corolla 2–3 mm. **Capsules** obliquely ovoid, laterally compressed, ± reniform in silhouette, thickest proximally, 3–4 × 3–3.5 mm, 2(or 3)-valved. **Seeds** 50 × 20 μm, testa striate-reticulate.

Flowering Jul–Oct. Mesic deciduous forests, mixed broadleaf-conifer forests; 0–1000 m; N.B., N.S., Ont., P.E.I., Que.; Ala., Ark., Conn., Del., Fla., Ga., Ill., Ind., Ky., La., Maine, Md., Mass., Mich., Miss., Mo., N.H., N.J., N.Y., N.C., Ohio, Okla., Pa., R.I., S.C., Tenn., Tex., Vt., Va., W.Va., Wis.; Mexico (Hidalgo, Tamaulipas).

Epifagus virginiana is an obligate parasite of the American beech, *Fagus grandifolia* (Fagaceae). Its range is coincident with that of its host, including disjunct occurrences in the highlands of Hidalgo and Tamaulipas, Mexico (J. W. Thieret 1969d, 1971). Different degrees of anthocyanin production among populations have led to the description of two trivial color forms. Local common names such as cancer-root, clap-wort, and flux-plant allude to the historical belief that the plant was efficacious medicinally for a number of ailments (Thieret 1971).

Epifagus americana Nuttall is an illegitimate name for *E. virginiana*.

5. KOPSIOPSIS (Beck) Beck in H. G. A. Engler, Pflanzenr. 96[IV,261]: 304. 1930

• Ground-cone, poque [Genus *Kopsia* and Greek *-opsis*, resemblance]

L. Turner Collins

Alison E. L. Colwell

George Yatskievych

Orobanche Linnaeus sect. *Kopsiopsis* Beck, Biblioth. Bot. 19: 74, 85. 1890

Herbs, perennial; achlorophyllous, holoparasitic, with a tuberlike underground vegetative structure attached to host root, surface divided into polygonal plates, roots absent. **Stems** erect, fleshy, glabrous. **Leaves** cauline, alternate; petiole absent; blade stiffly chartaceous, margins entire or slightly erose. **Inflorescences** terminal, compact or open racemes; bracts present. **Pedicels** present; bracteoles present, rarely absent. **Flowers:** perianth persistent; sepals (0 or)2–5, calyx bilaterally symmetric, cup-shaped, lobes attenuate, linear-subulate, or filiform; petals 5, corolla dark red, purple, or yellow, strongly bilabiate, funnelform, palatal folds absent, abaxial lobes 3, adaxial 2, adaxial lip ± galeate; stamens 4, didynamous, included, filaments with tuft of hair at base, villous or glabrous distally; staminode 0; ovary 1-locular, placentation parietal; stigma obscurely 2–4-lobed, crateriform. **Capsules:** dehiscence loculicidal. **Seeds** 100–500, light tan or brown, irregularly globular or ovoid to oblong-ellipsoid, prismatic, not or slightly flattened, wings absent.

Species 2 (2 in the flora): w North America, nw Mexico.

Kopsiopsis was first recognized by Beck but was treated as part of *Boschniakia* by many subsequent authors. Zhang Zhi Y. (1987) and Yu W. B. (2013) presented further morphological evidence for separating the two genera. *Kopsiopsis* differs from *Boschniakia* in the following traits: inflorescences compact or open racemes versus dense spikes; pedicels present versus absent; bracteoles present (rarely absent) versus absent; corollas funnelform versus short-tubular, constricted above ovary versus corolla base inflated; leaves spatulate versus triangular or lanceolate; and, seeds 1.5–3 mm versus 0.5–0.7 mm. Molecular phylogenetic analysis of the family (J. R. McNeal et al. 2013) supports *Boschniakia* as sister to a lineage that includes the genera *Conopholis*, *Epifagus*, and *Kopsiopsis*, and as-yet unpublished molecular data suggest that *Kopsiopsis* may be sister to the southern Mexican genus *Eremitilla* Yatskievych & J. L. Contreras (S. Mathews, pers. comm.).

SELECTED REFERENCE   Gilkey, H. M. 1945. *Boschniakia* in the western United States. Oregon State Monogr., Stud. Bot. 9: 7–15.

1. Bracts spatulate or broadly oblanceolate, apices obtuse or truncate, rolled adaxially; stems stout, 10–30 mm diam.; anther bases mucronulate; corolla lobe margins glabrous, sometimes ciliate; inflorescences compact or open racemes; leaf blades lanceolate, ovate, or broadly triangular, apices obtuse, margins entire, sometimes ± erose; parasitic on *Arbutus* and *Arctostaphylos* . . . . 1. *Kopsiopsis strobilacea*
1. Bracts narrowly spatulate or lanceolate, apices ± acute, rarely obtuse, slightly rolled adaxially; stems slender, 5–17 mm diam.; anther bases rounded; corolla lobe margins minutely ciliate; inflorescences compact racemes; leaf blades triangular-obovate or rhombic, apices obtuse or ± acute, margins finely erose; parasitic on *Gaultheria shallon* . . . . .2. *Kopsiopsis hookeri*

1. Kopsiopsis strobilacea (A. Gray) Beck in H. G. A. Engler, Pflanzenr. 96[IV,261]: 306. 1930

• California groundcone   F

Boschniakia strobilacea A. Gray in War Department [U.S.], Pacif. Railr. Rep. 4(5): 118. 1857

Tuberlike bases 25–55(–90) mm diam., surface coarsely tessellate. **Stems** dark red-brown or purple, sometimes yellow, (100–)140–260(–300) mm, stout, 10–30 mm diam. **Leaves** imbricate; blade lanceolate, ovate, or broadly triangular, 7–12 × 7–15 mm, margins entire, sometimes ± erose, apex obtuse. **Inflorescences** compact or open racemes, 2–5 cm diam.; bracts spreading, dark purple or yellow, spatulate or broadly oblanceolate, 8–17(–20) × 8–15(–19) mm, margins entire or ± erose, frequently white-translucent or lavender, apex obtuse or truncate, rolled adaxially. **Pedicels** 0–1.5(–3.5) mm; bracteoles 0–2. **Flowers** dark purple, sometimes yellow, 15–20 mm; calyx cup 1–3 mm deep, lobes sometimes deciduous, 0–5, attenuate to filiform, equal to or shorter than cup; corolla lobe margins white, lavender, or pinkish, glabrous, sometimes ciliate, abaxial lip as long as or slightly shorter than adaxial lip, lobes 3–5 mm, apex acute, obtuse, or rounded, adaxial lip margins inrolled; filaments with tuft of hair at base, sparsely villous distally, anthers villous, ± exserted, base mucronulate, sterile appendage (connective) minute or absent; style equal to or shorter than stamens; stigma capitate or 2–4-lobed; carpels 4; placentae (3 or)4. **Capsules** (3 or)4-valved. **Seeds** 2–3 mm.

Flowering Apr–Jul. Coastal mountains, open woodlands, chaparral, dry forests with *Arbutus* and *Arctostaphylos*; 300–3000 m; Calif., Oreg.; Mexico (Baja California).

Kopsiopsis strobilacea is parasitic on *Arbutus menziesii* and inland species of *Arctostaphylos*. Some populations in Marin, Siskiyou, and Trinity counties, California, and Jackson and Josephine counties, Oregon, where the ranges of the two species would allow contact, present characters intermediate with *K. hookeri*. The intermediate conditions observed were corolla lobes more or less ciliate versus glabrous and floral bracts narrower (almost lanceolate versus spatulate); intermediate specimens examined are at CAS, GH, NY, ORE, OSC, and RSA. Although the two species are almost completely allopatric, morphological evidence points to the possibility that some genetic exchange has occurred between the two species. The intermediate conditions were observed only in individuals within populations of *K. strobilacea* in the northern portion of the geographical range. No intermediate characters were observed in *K. hookeri*.

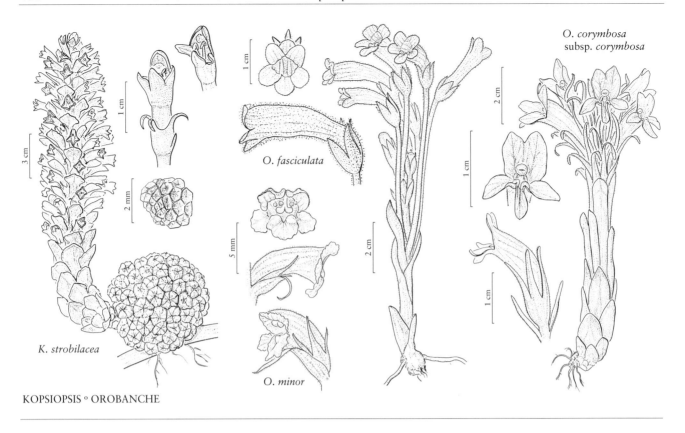

O. corymbosa
subsp. *corymbosa*

O. fasciculata

K. strobilacea

O. minor

KOPSIOPSIS ° OROBANCHE

2. **Kopsiopsis hookeri** (Walpers) Govaerts, World Checkl. Seed Pl. 2: 14. 1996 • Small or Hooker's groundcone E

Boschniakia hookeri Walpers, Repert. Bot. Syst. 3: 479. 1844, based on *Orobanche tuberosa* Hooker, Fl. Bor.-Amer. 2: 92, plate 168. 1838, not Vellozo 1829

Tuberlike bases 20–50 mm diam., surface finely tessellate. **Stems** dark red-brown, purple, or yellow, 55–160 mm, slender, 5–17 mm diam. **Leaves** tightly to loosely imbricate; blade triangular-obovate or rhombic, 6–12 × 5–12 mm, margins finely erose, apex obtuse or ± acute. **Inflorescences** compact racemes, 6–10 cm diam.; bracts erect, sometimes spreading, purple or yellow, narrowly spatulate or lanceolate, 7–11(–14) × 6–10 mm, margins erose, frequently white-translucent, apex obtuse or acute, slightly rolled adaxially. **Pedicels** 0–1.5 mm; bracteoles 2. **Flowers** dark purple, pink, or frequently yellow, 10–15 mm; calyx cup 1–3 mm deep, lobes persistent, 2–4, attenuate or linear-subulate, equal to cup; corolla lobe margins white, lavender, wine red, pink, or yellow, minutely ciliate, abaxial lip shorter than adaxial, lobes 3–5 mm, apex apiculate, adaxial lip margins inrolled; filaments with tuft of hair at base, glabrous distally, anthers glabrous or sparsely villous after anthesis, included, base rounded, sterile appendage

(connective) short; style equal to or longer than stamens; stigma obscurely 2- or 3-lobed; carpels 3; placentae 3. **Capsules** 3-valved. **Seeds** 1.5–2 mm.

Flowering Apr–Jul. Sandy coastal areas in thickets; 0–700 m; B.C.; Calif., Oreg., Wash.

Kopsiopsis hookeri is a coastal species parasitic on *Gaultheria shallon* (Ericaceae) and confined to the range of the host species. It has been reported rarely on *Alnus*, *Arbutus*, and *Arctostaphylos*. Reports on *Pinus* are clearly spurious. It is often confused with *K. strobilacea* because of their close morphological similarity, but *K. hookeri* is a much smaller plant.

The range of *Kopsiopsis hookeri* is from southern Oregon and northwestern California to the Queen Charlotte Islands (Haida Gwaii) in British Columbia, with apparently disjunct populations in isolated locations in California. A few historic collections at CAS and UC from southern California are morphologically indistinguishable from typical *K. hookeri*, indicating that this taxon may have once occurred south of its present distributional boundary. Populations represented by specimen records from Marin (1925, 1944, 1958), Mendocino (1966), and Monterey (1957) counties may still be extant and indicate the need for additional studies to determine the current range of the species. A single specimen from Los Angeles County may be mislabeled, as this area is outside the range of the host taxon.

No morphological evidence was observed in the disjunct populations to suggest genetic influence from *Kopsiopsis strobilacea*; refer to 1. *K. strobilacea* for further discussion.

6. OROBANCHE Linnaeus, Sp. Pl. 2: 632. 1753; Gen. Pl. ed. 5, 281. 1754

• Broomrape, cancer-root [Greek *orobos*, a kind of vetch, and *anchein*, to strangle, alluding to host plant and parasitic habit]

L. Turner Collins

Alison E. L. Colwell

George Yatskievych

Aphyllon Mitchell; *Myzorrhiza* Philippi

Herbs, annual, rarely perennial; achlorophyllous, holoparasitic, lacking a rhizomelike or cormlike underground vegetative structure; roots short, sometimes coralloid. **Stems** erect, white or yellow, rarely purple, fleshy, glabrous or puberulent, at least distally. **Leaves** cauline, spiral, imbricate at least proximally; petiole absent; blade fleshy or not, not leathery, margins entire, erose, or erosulate. **Inflorescences** terminal, spikes, spikelike racemes, racemes, panicles, or corymbs, sometimes solitary flowers or fascicles (*O. fasciculata, O. uniflora*); bracts present. **Pedicels** present or absent distally; bracteoles present, sometimes absent. **Flowers:** sepals 4 or 5 (5d sometimes vestigial), calyx ± bilaterally or ± radially symmetric, narrowly campanulate to campanulate, lobes linear-subulate to lanceolate-attenuate or narrowly triangular-acuminate; petals 5, corolla tinged pink to purple, yellow, or blue, pallid proximally, bilabiate, tubular, usually constricted above ovary, curved or bent forward, palatal folds present (longitudinal folds in abaxial side of tube), lobes loosely ascending to recurved (not cucullate), abaxial lobes 3, adaxial 2; stamens 4, didynamous, included, filaments glabrous or pubescent proximally; staminode 0; ovary 1-locular (sometimes irregularly 2- or 4-locular by intrusion of placentae in sect. *Gymnocaulis*), placentation parietal; stigma 2–4-lobed, sometimes very shallowly so, bilamellate, broadly clavate to crateriform-peltate, or nearly capitate. **Capsules:** dehiscence loculicidal. **Seeds** 500–2000(–5000), tan to dark brown, rarely black, irregularly globular or ovoid to oblong-ellipsoid, prismatic, wings absent. $x = 19, 24$.

Species ca. 150 (17 in the flora): North America, Mexico, Central America, South America, Europe, Asia, n Africa; introduced widely.

As noted by J. W. Thieret (1971), systematists have generally accepted the classification by G. Beck (1930) of a broadly circumscribed *Orobanche* comprising four sections, all of which are represented in the flora area. However, J. Holub (1990) and some other authors have questioned whether this classification accurately reflects the phylogeny and taxonomic complexity of the group and have proposed recognizing the sections of Beck as separate genera. More recently, cytological and molecular phylogenetic studies have added new data to the discussion but have not resulted in a well-supported, revised classification. G. M. Schneeweiss et al. (2004b) reviewed a large series of new and earlier chromosome counts and concluded that sect. *Orobanche* has a base number of $x = 19$, whereas members of the other sections are characterized by $x = 24$. As summarized by J. M. Park et al. (2008), molecular data have resulted in discordant phylogenies, depending on taxon sampling and whether plastid or nuclear markers were sequenced, but have suggested that five lineages should be recognized within the traditional *Orobanche*. The fifth lineage has been segregated by some authors as the monospecific *Boulardia latisquama* F. W. Schultz [*O. latisquama* (F. W. Schultz) Battandier], native to the Iberian Peninsula and Morocco. The molecular data support the group as monophyletic whether treated as one genus or several. Although the authors retain *Orobanche* in the broad sense, it seems inevitable that *Orobanche* will be split into three or more genera once the number of well-supported segregates is resolved. Most recently, A. C. Schneider (2016) has made a case for splitting all of the species native to the New World [sect. *Gymnocaulis* Nuttall and sect. *Nothaphyllon* (A. Gray) Heckard] into two sections of a single generic segregate, *Aphyllon* Mitchell, and has published new combinations for the constituent taxa.

As outlined in the key to species below, the two Old World sections present as adventives in our region are notable for their usually four-lobed or -toothed calyces and spicate inflorescences. Members of the two New World sections have five-lobed calyces and a variety of inflorescence types. Section *Orobanche* includes 120 to 130 species and is widespread in Europe, Asia, and Africa. The flowers lack bracteoles, the calyces are deeply two-parted with one or both halves toothed or lobed, and the corollas are usually yellow, red, or brown. The 12 to 16 species of sect. *Trionychon* Wallroth (*Phelipanche* Pomel) are most diverse in the Mediterranean region; their flowers have a small pair of bracteoles adnate to the calyx bases, the calyces are four-toothed or -lobed, and the corollas are usually purple or blue.

In the New World, the small sect. *Gymnocaulis* [sect. *Euanoplon* (Endlicher ex Walpers) Thieret; *Aphyllon*] comprises two or more species widespread in the flora area and is notable for its condensed inflorescence axes shorter than the elongate pedicels, flowers lacking bracteoles, and calyces shallowly to moderately lobed. About 15 members of sect. *Nothaphyllon* [sect. *Myzorrhiza* (R. A. Philippi) Beck-Mannagetta; *Myzorrhiza* R. A. Philippi] are widespread from temperate North America south to Guatemala, with a few disjunct species in western South America. They are characterized by usually elongate inflorescence axes usually much longer than the pedicels, flowers with a pair of bracteoles, and deeply five-lobed calyces.

Several of the Old World species are introduced widely, including the species treated below. They parasitize various crop plants, principally tomatoes (*Solanum lycopersicum*), tobacco (*Nicotiana tabacum*), clovers (*Trifolium* spp.), and hemp (*Cannabis sativa*) in the United States and are considered agricultural pests. In response, the Federal Government has placed all of the non-native taxa of *Orobanche* on the United States Department of Agriculture's noxious weed list (http://www.aphis.usda.gov/plant_health/plant_pest_info/weeds/downloads/weedlist.pdf), which regulates their movement into, out of, and within the United States. Several states also regulate various non-native broomrapes as noxious weeds.

In 2014, an infestation of *Orobanche aegyptiaca* Persoon [*Phelipanche aegyptiaca* (Persoon) Pomel] was discovered in a tomato field in Solano County, California, the first report of this taxon from the flora area. Egyptian broomrape is native to the Middle East and adjacent portions of Europe and Asia but is a widely distributed weed in Europe, Asia, and Africa, and has also been reported from Cuba (R. Oviedo Prieto et al. 2012). Its principal negative economic impact is to crop species in the Solanaceae (tomatoes, potatoes, tobacco), but, like *O. ramosa*, it has a broad host range, including a variety of other crops. Although the appearance of the species in North America is not unexpected, a full treatment is not provided here, because there is no certainty that it will become established. Vigorous efforts have been undertaken to eradicate plants at the sole known location. *Orobanche aegyptiaca* is similar to *O. ramosa* in its branched, glandular-pubescent stems and general floral morphology, but it differs from that species in its usually more robust habit (stems to 40 cm), with larger corollas (20–35 mm) that are often darker purple, and by having densely villous versus glabrous anthers.

Plants of *Orobanche hederae* Duby (ivy broomrape), another Eurasian species in sect. *Orobanche*, have been documented since 2000 parasitizing planted *Hedera* (Araliaceae) in Alameda County, California, on the campus of the University of California-Berkeley. It can be difficult to distinguish from the morphologically variable *O. minor*, differing in its often slightly more lax inflorescences and corollas with a constriction on the abaxial side just behind the throat. For recognition, most botanists focus on the difference in hosts, as *O. hederae* parasitizes species of *Hedera*. The taxon is excluded from the flora for the present, because its host specificity has limited its ability to spread beyond the immediate vicinity of the initial infection site.

The native North American species have had a negligible impact on agriculture. *Orobanche cooperi* and *O. riparia* (as *O. ludoviciana*) have been reported on cultivated tomato crops in southern California and on tobacco crops in the Ohio River Valley, respectively, with *O. cooperi* having required control measures (G. H. Starr 1943; S. Wilhelm et al. 1958).

Some of the western North American taxa were recorded as a minor, but apparently widely utilized, food source for western American Indians and were also used medicinally, mainly for treating colds and other pulmonary ailments (D. E. Moerman 1998).

The North American *Orobanche* species are more host-specific than has been reported previously (A. C. Schneider et al. 2016b), and understanding of host relationships continues to be refined. Many collectors merely note the nearest living plant as the host species or have neglected to note host species entirely, thus perpetuating imprecise knowledge of host relationships in this group. In this treatment, a host species is noted as a primary host when it has been confirmed multiple times by excavation and is frequently reported as the host. Less frequent, but reliable, reports of other hosts are noted as occasional hosts. Localized shifts in host preference in *Orobanche* species may reflect an indication of lineage divergence.

SELECTED REFERENCES Achey, D. M. 1933. A revision of the section *Gymnocaulis* of the genus *Orobanche*. Bull. Torrey Bot. Club 60: 441–451. Munz, P. A. 1930. The North American species of *Orobanche*, sect. *Myzorrhiza*. Bull. Torrey Bot. Club 57: 611–624.

1. Calyces bilaterally symmetric, sometimes strongly so, 2- or 4-lobed (rarely with an additional vestigial abaxial lobe); inflorescences spikes or spikelike racemes; pedicels 0–1 mm (rarely to 30 mm in proximalmost flowers).
   2. Calyces deeply divided into 2 lateral lobes (rarely with an additional vestigial abaxial lobe), lobes entire or asymmetrically divided into 2 teeth or short lobes, these much shorter than tube; bracteoles 0 [sect. *Orobanche*]. . . . . .1. *Orobanche minor*
   2. Calyces divided into 4 subequal lobes, lobes entire, slightly shorter to slightly longer than tube; bracteoles 2 [sect. *Trionychon*]. . . . . .2. *Orobanche ramosa*
1. Calyces ± radially symmetric or weakly bilaterally symmetric, 5-lobed; inflorescences fascicles, corymbs, panicles, or racemes, sometimes spikelike, or of solitary flowers; pedicels 1+ mm (except in distalmost flowers of a few species).
   3. Pedicels (8–)10–110(–170) mm, as long as or longer than plant axis, distal sometimes shorter (*O. fasciculata*); inflorescences of solitary flowers or fascicles, irregular corymbs, or short racemes of (1–)6–15(–20) flowers; bracteoles 0 [sect. *Gymnocaulis*].
      4. Flowers 1 or 2(–4); pedicels much longer than plant axis. . . . . 3. *Orobanche uniflora*
      4. Flowers (1–)6–15(–20); proximal pedicels as long as or ± longer than plant axis, distals sometimes shorter . . . . 4. *Orobanche fasciculata*
   3. Pedicels 0–30 mm (to 35 mm proximally), shorter than plant axis; inflorescences racemes, sometimes spikelike, corymbs, or panicles (flowers numerous, rarely 10 or fewer in depauperate plants); bracteoles 2 [sect. *Nothaphyllon*].
      5. Inflorescences corymbs or panicles, sometimes racemes.
         6. Inflorescences panicles, open or dense and pyramidal; anthers glabrous or sparsely pubescent; corollas less than 20 mm; palatal folds not prominent, pale or light yellow.
            7. Inflorescences dense, pyramidal, dark purple-brown, imbricately branched; calyces divided into 5 unequal lobes, cleft to base only on adaxial side, otherwise deeply lobed; filaments glabrous . . . . 5. *Orobanche bulbosa*
            7. Inflorescences open, cylindric (axis visible between flowers), ochraceous, red-brown, purple or purple streaked, yellow, or cream-white, loosely branched, rarely simple; calyces divided into 5 subequal lobes (lobes slightly shorter than to ca. as long as tube); filaments with ring of hairs at base. . . . . .6. *Orobanche pinorum*

6. Inflorescences corymbs, sometimes racemes; anthers woolly, tomentose, pubescent, or glabrous; corollas (8–)12–50(–55) mm; palatal folds prominent, yellow, sometimes cream to lemon or white.

    8. Pedicels 2–10 mm (to 35 mm proximally); corollas (8–)15–25(–30) mm, tubes white to cream, sometimes pale purplish tinged distally, sometimes with purple veins, adaxial lips 4–6(–9) mm; inflorescences compact, corymbs to subracemose . . . . 7. *Orobanche robbinsii*

    8. Pedicels (0–)3–20(–40) mm; corollas (15–)18–50(–55) mm, tubes white to grayish white, cream, ± pink, ± purple, or pinkish to purplish tinged, rarely brick red, sometimes with darker veins, adaxial lips 5–15(–18) mm; inflorescences corymbs, subcorymbose, racemes, or subcapitate.

        9. Corollas (15–)18–34 mm, lips 5–9 mm, erect, sometimes spreading or reflexed; palatal folds glabrous (with blisterlike swellings); Great Basin . . . . .8. *Orobanche corymbosa*

        9. Corollas 22–50(–55) mm, lips 8–15 mm, ± spreading; palatal folds glabrous (without blisterlike swellings), sometimes pubescent; w of Cascade and Sierra mountains. . . . . .9. *Orobanche californica*

[5. Shifted to left margin.—Ed.]

5. Inflorescences racemes or spikelike racemes.

  10. Abaxial corolla lobe apices rounded or obtuse.

    11. Corolla lips internally maroon or reddish purple, sometimes with maroon or reddish purple stripes, veins, or blotches; palatal folds glabrous; s California, Nevada, Oregon . . . . 12. *Orobanche parishii*

    11. Corolla lips internally pink or purple, sometimes white with purple veins, rarely light yellow; palatal folds pubescent; absent in California, widespread to the n and e.

      12. Corollas 14–20 mm, lips 3–6 mm; anthers glabrous or with a few woolly hairs along sutures . . . . 10. *Orobanche ludoviciana*

      12. Corollas 22–36 mm, lips 5–12 mm; anthers woolly . . . . .11. *Orobanche multiflora*

  10. Abaxial corolla lobe apices acute, pointed, or with an apiculate tooth.

    13. Corollas 15–32(–33) mm, lips 4–9(–10) mm.

      14. Corolla tubes purple or lavender, rarely white, tinged with purple, lips dark purple to lavender, with darker purple veins, lobes often with apiculate teeth; anthers usually with inconspicuous stalked glands . . . . 13. *Orobanche cooperi*

      14. Corolla tubes white or pale yellow to pale pink, lips white to pale pink, often with darker pink veins, lobes without apiculate teeth; anthers without stalked glands. . . . . .17. *Orobanche vallicola*

    13. Corollas 12–22 mm, lips 3–6 mm.

      15. Corollas 12–16(–18) mm, tubes dark purple, sometimes yellow to white abaxially, adaxial lips 3–5 mm. . . . . .14. *Orobanche valida*

      15. Corollas (13–)15–20(–22) mm, tubes white, distally sometimes tinged with purple or pink, or with dark purple veins, adaxial lips 4–6 mm.

        16. Corollas 15–20(–22) mm, tubes white, adaxial lips dark purple, sometimes lavender; Colorado Plateau and Great Basin Desert; host *Gutierrezia* . . . . 15. *Orobanche arizonica*

        16. Corollas (13–)15–22 mm, tubes white, distally often tinged with purple or pink, or with dark purple veins, adaxial lips lavender or dark purple; riparian habitats; host *Ambrosia*, *Dicoria*, or *Xanthium* . . . . 16. *Orobanche riparia*

1. **Orobanche minor** Smith, Engl. Bot. 6: plate 422. 1797 • Common or lesser broomrape, hellroot F I W

Orobanche columbiana H. St. John & English

Plants simple, (8–)12–55 (–70) cm, slender, base sometimes abruptly enlarged. **Roots** usually conspicuous (often forming a globular mass), very slender, usually branched. **Leaves** several to numerous, loosely ascending to spreading, imbricate only near stem base; blade lanceolate to oblong-ovate or triangular-ovate, 6–20 mm, margins entire, apex acute to acuminate, surfaces moderately to densely glandular-pubescent. **Inflorescences** spikes, reddish brown to purple or yellow, simple, sparsely to densely glandular-pubescent; flowers numerous, axis visible between flowers; bracts slightly reflexed, narrowly lanceolate, 6–17 mm, apex attenuate, glandular-pubescent. **Pedicels** 0–0.8 mm (rarely to 30 mm in proximalmost flowers); bracteoles 0. **Flowers:** calyx yellow or brownish red to brownish purple, strongly bilaterally symmetric, (6–)8–12 mm, deeply divided into 2 lateral lobes (rarely with an additional vestigial abaxial lobe), lobes entire or asymmetrically divided into 2 teeth or short lobes, these much shorter than tube, lanceolate to subulate-attenuate, ± glandular-villous; corolla 10–19 mm, tube white to pale yellow, not or only slightly constricted above ovary, ± curved, glandular-puberulent; palatal folds prominent, yellow to nearly white, usually glabrous; lips similar in color to tube, more commonly purplish tinged and/or veined, sometimes more strongly so externally, abaxial lip spreading abruptly from base, 3–4 mm, lobes broadly ovate to ± semiorbiculate (this sometimes difficult to observe because of the crinkled, erose-crenulate margins and overlapping sinuses), apex rounded or shallowly emarginate, adaxial lip erect or curved outward at tip, 3–5 mm, lobes shallow, ± semiorbiculate, apex broadly rounded; filaments sparsely pubescent, distal hairs gland-tipped, anthers included, glabrous or tomentulose. **Capsules** ovoid to oblong-ovoid, 5–9 mm. **Seeds** 0.2–0.4 mm. $2n = 38$.

Flowering Apr–Jul. Old fields, forest margins, woodland openings, railroad embankments, roadsides, pastures, crop fields, orchards, gardens, lawns, disturbed areas, greenhouses; 0–300 m; introduced; Del., D.C., Fla., Ga., Idaho, Md., N.J., N.Y., N.C., Oreg., Pa., S.C., Tex., Vt., Va., Wash., W.Va.; Eurasia; n Africa; introduced widely.

Orobanche minor has been documented most frequently parasitizing introduced clovers (mainly *Trifolium arvense* and *T. repens*), and collected rarely on *Crotalaria* (J. W. Thieret 1971) and *Vicia*. It also has been recorded, at least historically, on a variety of cultivated hosts in the region, including hemp, carrots (*Daucus carota*), tobacco, geraniums (*Pelargonium* spp.), and *Petunia* spp. Allegedly, the species is toxic to livestock (Thieret). The sole specimen from Idaho (*J. A. Allen s.n.*, 1875, NY) lacks locality data; if the provenance is correct, the elevational range would be extended upward.

European authors have recognized a number of infrataxa and segregates; for example, F. J. Rumsey and S. L. Jury (1991) provisionally accepted four varieties of *Orobanche minor* as occurring in the British Isles. However, they noted that little is known about cytological and morphological variation within the complex. Thus, it seems inappropriate to apply an infraspecific classification to the North American plants.

A single historical specimen (*J. C. Nelson 3337*, 25 August 1920, GH) collected from ship's ballast in the Linnton area of Portland, Oregon, is an unusually stout plant with apparently pale corollas and filaments relatively densely pubescent toward their bases. This plant may represent a record of *Orobanche loricata* Reichenbach, a European species that parasitizes mainly *Picris* and other Asteraceae, and does not affect any crop plants. However, specimen condition precludes definitive determination, and the label does not list a host species. Other materials from Oregon have the typical morphology of *O. minor*.

2. **Orobanche ramosa** Linnaeus, Sp. Pl. 2: 633. 1753 • Hemp or branching broomrape I W

Plants few- to many-branched, (5–)10–45 cm, slender, base usually only slightly enlarged. **Roots** often inconspicuous (sometimes forming a globose mass), very slender, usually branched. **Leaves** relatively few, not imbricate, appressed to loosely ascending or spreading; blade lanceolate to narrowly ovate, 4–7 mm, margins entire, apex acute, surfaces glandular-pubescent or glabrous. **Inflorescences** spikes or spikelike racemes, pale tan to yellow or brownish purple, sometimes with a brownish cast, simple, densely glandular-pubescent; flowers numerous; bracts ± ascending, lanceolate, 4–6 mm, apex acute, glandular-pubescent. **Pedicels** 0–1 mm; bracteoles 2. **Flowers:** calyx purple or yellow, sometimes brownish, bilaterally symmetric, 4.5–6.5 mm, divided into 4 subequal lobes, lobes triangular-acuminate, entire, glandular-pubescent; corolla 10–15(–17) mm, tube white or light purple to bluish purple, slightly constricted above ovary, slightly to moderately curved forward, glandular-pubescent; palatal folds prominent, white, pubescent; lips similar in color to tube or slightly darker purple to bluish purple, sometimes with ± darker veins (these often externally darker), abaxial lip slightly reflexed, 3–5 mm, lobes oblong-ovate, apex rounded to bluntly pointed, adaxial lip usually spreading to ± recurved near tip, 3–5 mm, lobes broadly ovate to semiorbiculate, apex rounded to bluntly pointed; filaments glabrous or sparsely glandular-pubescent, anthers included, glabrous or nearly so. **Capsules** broadly ovoid to broadly oblong-ellipsoid, 5–9 mm. **Seeds** 0.2–0.5 mm. $2n = 24$.

Flowering Feb–Apr, Jul–Sep. Roadsides, crop fields, lawns, disturbed areas, greenhouses; 0–300 m; introduced; Calif., Ill., Ky., N.J., Tex., Va.; Eurasia; Africa; introduced widely.

The identity of the sole historical collection of *Orobanche ramosa* from North Carolina (Roan Mtn., *J. Ball s.n.*, 17 September 1884, E) remains tentative (L. J. Musselman 1984). Many other historic infestations of this weedy species in the region have been eradicated. However, in recent decades the species appears to be spreading in central and eastern Texas using disturbed roadsides as avenues for dispersal.

This destructive agricultural weed has a large host range. L. J. Musselman and K. C. Nixon (1981) documented plants within a single population in Texas attached to the roots of ten different hosts in eight angiosperm families: Apiaceae, Asteraceae, Caryophyllaceae, Fabaceae, Geraniaceae, Malvaceae, Onagraceae, and Verbenaceae. Other voucher specimens have recorded an even more diverse assemblage of hosts, mainly eudicots and occasionally even monocots. Historically, the principal crop species affected in the region have been hemp, tobacco, and tomato, but an infestation in a commercial greenhouse in New York impacted coleus (*Plectranthus*) production in the 1920s. Currently, the main negative economic impact of *Orobanche ramosa* is in tomato fields in California.

As many as 20 microspecies have been described as segregates from *Orobanche ramosa* (F. J. Rumsey and S. L. Jury 1991). However, A. O. Chater and D. A. Webb (1972), who chose to circumscribe *O. ramosa* broadly as a single species with three subspecies, noted that the characters used to discriminate among these taxa exhibit nearly continuous variation and are often quite variable within populations. In their treatment, subsp. *ramosa* is the widespread weedy taxon that has emigrated to the New World, and the other two infrataxa [subsp. *mutelii* (F. J. Schultz) Coutinho and subsp. *nana* (Reuter) Coutinho] are uncommon, non-weedy taxa endemic to southern Europe that differ from subsp. *ramosa* in perianth details. Further studies are needed before an infraspecific classification can be supported.

As noted by H. J. Scoggan (1978–1979), *Orobanche purpurea* Jacquin (yarrow broomrape), was collected in Huron County, Ontario, in 1895 by Jasi A. Morton but has not been rediscovered in the flora area since the initial collection. The voucher specimens lack habitat data, but in its native range the species parasitizes mainly open grassland and woodland taxa of *Achillea* and *Artemisia* (Asteraceae). The species resembles *O. ramosa*, differing in its usually simple stems, longer corollas (18–25 mm), and calyces with slightly longer, more narrowly triangular lobes. Although its native distribution in Europe, Asia, and northern Africa is broad, C. A. J. Kreutz (1995) noted that it is rare in much of the range, and F. J. Rumsey and S. L. Jury (1991) were alarmed at its historical decline in Great Britain.

3. **Orobanche uniflora** Linnaeus, Sp. Pl. 2: 633. 1753
   • One-flowered or small cancer-root, one-flowered or naked broomrape, orobanche uniflore

Anoplanthus uniflorus (Linnaeus) Endlicher; *Aphyllon uniflorum* (Linnaeus) Torrey & A. Gray; *Thalesia uniflora* (Linnaeus) Britton

Plants simple or few-branched basally, sometimes forked medially, 3.5–18(–25) cm (including pedicels), stem portion 1–5(–7) cm, slender, base not enlarged. **Roots** inconspicuous, slender or stout, unbranched or few-branched. **Leaves** relatively few, loosely imbricate or more remote, loosely ascending to spreading; blade oblong-lanceolate to awl-shaped, 2–10 mm, margins entire, often inrolled, apex acuminate, surfaces glabrous. **Inflorescences** of solitary flowers or fascicles of 2(–4) at stem or branch tips, white to cream, sometimes purple tinged, simple, glabrous; bracts loosely ascending and erect, oblanceolate to broadly ovate, obovate, rhombic, or awl-shaped, 5–12 mm, apex acute to acuminate, glabrous, rarely glandular-pubescent distally. **Pedicels** (8–)20–110(–170) mm, much longer than plant axis; bracteoles 0. **Flowers:** calyx white to straw colored or light yellow, or pale to dark, dull purple, sometimes brownish, ± radially symmetric, 4–12(–15) mm, divided into 5 subequal lobes, lobes slightly shorter than to 2 times as long as tube, ± triangular or subulate-triangular, sparsely to moderately glandular-pubescent; corolla (11–)15–30(–35) mm, tube white to purple, blue, or yellow, sometimes lighter proximally or with light or darker purple or blue veins, slightly to moderately constricted above ovary, ± bent forward, glandular-pubescent; palatal folds ± prominent, bright yellow, glandular- and/or eglandular-pubescent, sometimes glabrescent; lips white, yellow, purple, or blue, sometimes with light or darker purple or blue veins, abaxial lip slightly to moderately spreading, sometimes ± recurved distally, (1–)2–6(–9) mm, lobes oblong-obovate to nearly round, apex rounded to bluntly pointed or shallowly emarginate (sometimes with 2 notches), adaxial lip slightly to moderately spreading, sometimes recurved, 2–6(–9) mm, lobes broadly oblong-ovate to oblong-semiorbiculate, apex rounded, rarely bluntly pointed or shallowly emarginate; filaments glabrous, anthers included, glabrous or villous-tomentose. **Capsules** ovoid to oblong-ovoid, 4–8(–11) mm. **Seeds** 0.1–0.4 mm.

Subspecies 2 (2 in the flora): North America, Mexico.

Orobanche uniflora forms a polymorphic complex that requires more detailed study. The detection of broad-scale patterns of morphological variation is confounded by the differentiation among local races. D. M. Achey (1933) recognized five varieties, and K. C. Watson (1975), in her unpublished thesis, revised the classification to three subspecies. The present treatment, which accepts only the two major infraspecific variants as subspecies, should be considered highly tentative.

1. Calyx lobes slightly shorter than to only slightly longer than tubes, narrowly to broadly lanceolate-triangular; corollas white to pale yellow, sometimes pale purplish tinged and/or with light purple veins. . . . . 3a. *Orobanche uniflora* subsp. *uniflora*
1. Calyx lobes ca. 2 times as long as tubes, subulate-triangular; corollas purple to blue (often with a white throat), yellow, or white, often with darker purple or blue veins. . . . . .3b. *Orobanche uniflora* subsp. *occidentalis*

3a. Orobanche uniflora Linnaeus subsp. **uniflora**

Orobanche terrae-novae Fernald; *O. uniflora* var. *terrae-novae* (Fernald) Achey

Plants simple, rarely few-branched basally, often forked medially, (including pedicels) (4–)7–18(–25) cm, stems and inflorescence axes only 1–5(–7) cm. **Leaves and bracts** usually 3–7 per stem or branch. **Pedicels** (8–)30–110(–170) mm. **Calyx lobes** narrowly to broadly lanceolate-triangular, 4–12(–15) mm, slightly shorter than to only slightly longer than tube, narrowed relatively evenly from base to tip. **Corollas** white to pale yellow, sometimes pale purplish tinged and/or with light purple veins, (11–)15–25(–30) mm, throat 2–4 mm diam., palatal folds slender, elongate; limb 3–6(–9) mm. $2n = 36, 72$.

Flowering Apr–Jul. Old fields, upland hardwood forests, woodlands, thickets, hardwood hammocks, glades, bluffs, bogs, stream banks, riverbanks, gardens; 0–400(–2000) m; St. Pierre and Miquelon; B.C., N.B., Nfld. and Labr. (Nfld.), N.S., Ont., P.E.I., Que.; Alaska, Ark., Conn., Del., D.C., Fla., Ill., Ind., Iowa, Kans., Ky., Maine, Md., Mass., Mich., Minn., Miss., Mo., Mont., Nebr., N.H., N.J., N.Y., N.C., N.Dak., Ohio, Okla., Pa., R.I., S.C., Tenn., Tex., Vt., Va., W.Va., Wis.; Mexico (Oaxaca).

Subspecies *uniflora*, the widespread eastern subspecies, occurs as far west as the eastern edge of the Great Plains southward to eastern Texas, with disjunct populations in Alaska, British Columbia, and Montana. A single specimen parasitic on *Baccharis* (*Camp 2451*, NY) documents a major disjunct occurrence in Oaxaca, Mexico. The name *Orobanche terrae-novae* corresponds to a local race in Newfoundland with exceptionally deep corolla lobes bearing slightly shorter cilia.

Many specimens lack information on plants being parasitized or evidence of haustorial connections. The primary hosts of subsp. *uniflora* in temperate North America are perennial members of the Asteraceae, especially selected species of *Eurybia*, *Rudbeckia*, *Solidago*, *Symphyotrichum*, and rarely *Hypochaeris*. However, occasional and likely spurious host records are *Betula* and *Ostrya* (Betulaceae), *Rhododendron* (Ericaceae), *Quercus* (Fagaceae), and *Potentilla* (Rosaceae).

3b. Orobanche uniflora Linnaeus subsp. **occidentalis** (Greene) Abrams ex Ferris, Contr. Dudley Herb. 5: 99. 1958 E

Aphyllon uniflorum (Linnaeus) Torrey & A. Gray var. *occidentale* Greene, Man. Bot. San Francisco, 285. 1894; *A. inundatum* Suksdorf; *A. minutum* Suksdorf; *A. purpureum* (A. Heller) Holub; *A. sedi* Suksdorf; *Orobanche porphyrantha* Beck; *O. sedi* (Suksdorf) Fernald; *O. uniflora* var. *minuta* (Suksdorf) Beck; *O. uniflora* var. *occidentalis* (Greene) Roy L. Taylor & Macbryde; *O. uniflora* var. *purpurea* (A. Heller) Achey; *O. uniflora* var. *sedi* (Suksdorf) Achey; *Thalesia minuta* (Suksdorf) Rydberg; *T. purpurea* A. Heller; *T. sedi* (Suksdorf) Rydberg

Plants simple or few-branched basally, sometimes also forked medially, (including pedicels) 3.5–18(–20) cm, stems and inflorescence axes only 1–5(–7) cm. **Leaves and bracts** usually 1–5 per stem or branch. **Pedicels** (8–)20–110(–140) mm. **Calyx lobes** subulate-triangular, 5–15 mm, ca. 2 times as long as tube, tapered abruptly above a broad base to tip, sparsely to moderately glandular-pubescent adaxially. **Corollas** purple to blue (often with a white throat), yellow, or white, often with darker purple or blue veins, (13–)15–30(–35) mm, throat (1–)2–4(–5) mm diam., palatal folds slender to relatively broad, elongate or short; limb (1–)2–5(–8) mm.

Flowering Apr–Aug. Sagebrush, hardwood and coniferous woodlands and forests, moist rock ledges, stream banks, meadows; 10–3200 m; Alta., B.C., Man., Sask.; Ariz., Calif., Colo., Idaho, Mont., Nev., N.Mex., Oreg., Utah, Wash., Wyo.

Subspecies *occidentalis* has been divided into four infraspecific taxa in older literature. Two main groups of plants are the predominant hosts: Crassulaceae (*Sedum*) and Saxifragaceae (*Lithophragma, Saxifraga, Tellima*). However, several specimens have recorded or suggested parasitism on a variety of plants, even including mosses (*Polytrichum*), lycophytes (*Selaginella*), and ferns (*Cryptogramma*). Among the miscellaneous angiosperm reports are: Apiaceae (*Lomatium, Perideridia*), Asteraceae (*Antennaria, Artemisia, Balsamorhiza, Coreopsis, Erigeron, Eriophyllum, Packera, Pseudognaphalium, Solidago*), Caryophyllaceae (*Eremogone*), Liliaceae in the sense of this flora (*Maianthemum*), Poaceae (*Bromus, Poa*), Polemoniaceae (*Leptodactylon*), Ranunculaceae (*Ranunculus*), and Rubiaceae (*Galium*).

Based on two unusual specimens from Mohave County, Arizona, in which the proximal flowers were partially buried in loose substrate, B. D. Parfitt and M. L. Butterwick (1981) suggested that subsp. *occidentalis* might be capable of producing cleistogamous flowers. This condition has not been documented elsewhere among the holoparasitic taxa of Orobanchaceae in the region, except in *Epifagus*.

4. Orobanche fasciculata Nuttall, Gen. N. Amer. Pl. 2: 59. 1818 • Clustered broomrape F

Anoplanthus fasciculatus (Nuttall) Walpers; *Anoplon fasciculatum* (Nuttall) G. Don; *Aphyllon fasciculatum* (Nuttall) Torrey & A. Gray; *Orobanhe fasciculata* var. *franciscana* Achey; *O. fasciculata* var. *lutea* (A. Gray) Achey; *O. fasciculata* var. *subulata* Goodman; *Phelypaea fasciculata* (Nuttall) Sprengel; *Thalesia fasciculata* (Nuttall) Britton

Plants branched proximally and/or distally, rarely simple, 6–25(–35) cm (including pedicels), stem portion 1.5–15(–22) cm, slender to moderately stout, base slightly enlarged. **Roots** inconspicuous, slender or stout, unbranched or few-branched. **Leaves** few to several, erect or reflexed; blade oblong-ovate to ovate-triangular or awl-shaped, (4–)6–12(–15) mm, margins entire, apex acute or acuminate, surfaces glandular-pubescent distally. **Inflorescences** fascicles, irregular corymbs, or short racemes of (1–)6–15(–20) flowers at stem or branch tips, light yellow to yellow or tinged pinkish to reddish purple, simple, densely glandular-pubescent, sometimes glabrescent proximally; bracts erect or ± spreading, oblanceolate to oblong, lanceolate, or awl-shaped, sometimes ovate, 7–12 mm, apex acute, sometimes acuminate, moderately to densely glandular-pubescent. **Pedicels** 10–70(–150) mm, proximal as long as or ± longer than plant axis, distal sometimes shorter; bracteoles 0. **Flowers:** calyx light yellow to

orangish yellow, tan, or grayish tan, often purplish tinged distally, sometimes entirely pinkish purple to reddish purple or dark purple, ± radially symmetric, (4–)6–12(–18) mm, divided into 5 subequal lobes, lobes shorter than to slightly longer than tube, triangular to subulate-triangular, moderately to densely glandular-pubescent; corolla (11–)14–30(–38) mm, tube white to cream or yellow, purplish tinged, or pinkish purple to reddish purple, sometimes with darker pink, purple, or brown veins, slightly to moderately constricted above ovary, ± bent forward, glabrate or glandular-pubescent; palatal folds ± prominent, usually yellow, moderately to densely glandular-pubescent; lips yellow or pinkish purple to reddish purple, rarely white, sometimes with darker purple veins, abaxial lip ± spreading, 3–6 (–9) mm, lobes oblong-obovate to nearly round, sometimes oblong-elliptic, apex rounded or ± pointed, sometimes shallowly emarginate, adaxial lip slightly to moderately spreading or recurved, (2–)3–6(–9) mm, lobes oblong-ovate to nearly round, sometimes oblong-elliptic, apex rounded or ± pointed; filaments glabrous, anthers included, glabrous or villous-tomentose. **Capsules** ovoid to oblong-ovoid, 6–12 mm. **Seeds** 0.2–0.5 mm. $2n = 48$.

Flowering Apr–Aug. Sagebrush, chaparral, upland prairies, dunes, desert scrub, rocky slopes, hardwood and coniferous woodlands and forests, thickets, alpine meadows, roadsides, gardens; 150–3300 m; Alta., B.C., Man., Ont., Sask., Yukon; Alaska, Ariz., Calif., Colo., Idaho, Ill., Ind., Iowa, Kans., Mich., Minn., Mont., Nebr., Nev., N.Mex., N.Dak., Okla., Oreg., S.Dak., Tex., Utah, Wash., Wis., Wyo.; Mexico (Baja California, Chihuahua).

Similar to *Orobanche uniflora*, *O. fasciculata* forms a polymorphic complex that may involve cryptic species. However, unlike *O. uniflora*, infraspecific taxa in *O. fasciculata* lack strong correlations with morphology, geography, and host ranges; they are not recognized here. Previously, D. M. Achey (1933) separated the species into three varieties based mainly on plant color and flower size. In her unpublished thesis, K. C. Watson (1975) expanded this to four subspecies but circumscribed her taxa differently. Both authors noted morphological overlap among taxa.

Of particular interest is a series of populations from California and adjacent Oregon [always parasitic on *Galium* (Rubiaceae)] to which Watson applied the manuscript name "subsp. *uniflorioides*" and that, in many ways, are morphologically intermediate between *Orobanche fasciculata* and *O. uniflora*. Recently, A. E. L. Colwell et al. (2017) segregated these under the name *Aphyllon epigalium* Colwell & A. C. Schneider. They are distinctive in having typically two to four flowers per stem, usually cream to yellow corollas (sometimes tinged with pink or purple), including the palatal folds, and glandular (versus ciliolate) corolla margins. Colwell et al. went further in subdividing their new species into two subspecies, segregating plants with somewhat smaller, cream-colored corollas having at most slightly recurved lips as subsp. *notocalifornicum* A. C. Schneider & Colwell and retaining plants with somewhat larger, yellow corollas with spreading lobes as subsp. *epigalium*. This treatment is tentative, pending further research into the population genetics within the entire complex.

Most references suggest that *Orobanche fasciculata* uses a broad range of hosts. However, there are four main genera of host plants: *Artemisia* (Asteraceae), *Phacelia* (Hydrophyllaceae), *Eriodictyon* (Namaceae), and *Eriogonum* (Polygonaceae). As noted above, a morphologically distinctive set of populations parasitizes *Galium*. Other less commonly reported hosts include *Ericameria* and *Eriophyllum* (Asteraceae), *Atriplex* and *Grayia* (Chenopodiaceae), *Convolvulus* (Convolvulaceae), *Arctostaphylos* (Ericaceae), *Mirabilis* (Nyctaginaceae), *Pinus* (Pinaceae), grasses (Poaceae), *Delphinium* (Ranunculaceae), *Adenostoma*, *Prunus*, and *Purshia* (Rosaceae), and *Vitis* (Vitaceae). Some minor hosts are listed based only on specimen label data and require confirmation.

5. Orobanche bulbosa Beck, Biblioth. Bot. 19: 83. 1890 • Chaparral broomrape

Phelypaea tuberosa A. Gray, Proc. Amer. Acad. Arts 7: 371. 1868 (as Phelipaea), not *Orobanche tuberosa* Vellozo 1829; *Aphyllon tuberosum* (A. Gray) A. Gray; *Myzorrhiza tuberosa* (A. Gray) Rydberg

Plants simple, sometimes branched, 8–30 cm, stout, base enlarged. **Roots** relatively conspicuous (forming a globular mass), slender or stout, branched or unbranched. **Leaves** numerous, large and imbricate proximally, appressed or slightly spreading; blade lanceolate, 5–10 mm, margins entire, apex acute, surfaces glabrous. **Inflorescences** dense, pyramidal, thyrsoid panicles, dark purple-brown, imbricately branched, cinereous- to ferruginous-puberulent, hairs eglandular; flowers numerous; bracts strongly reflexed, lanceolate-subulate, 3–5 mm, apex acuminate, puberulent. **Pedicels** 0–5 mm, much shorter than plant axis, sometimes adnate to stem for a portion of their length or flattened; bracteoles 2. **Flowers:** calyx dark purple, weakly bilaterally symmetric, 6–11 mm, divided into 5 unequal lobes, cleft to base on adaxial side, otherwise deeply lobed, lobes attenuate, puberulent; corolla 10–18 mm, tube dark purple, wine colored, or dark gray, sometimes dark pink, slightly constricted above ovary, ± straight, puberulent; palatal folds not prominent, pale yellow, glabrous; lips purple

to dark purple or wine colored, sometimes dark pink or with darker pink to purple veins, sometimes internally variegated with white, dark gray externally, abaxial lip spreading, 2–4 mm, lobes narrowly oblong or oblong-lanceolate, apex acute, often bluntly pointed, adaxial lip erect, spreading at tip, 2–4 mm, lobes triangular to oblong-triangular, apex acute, sometimes with a minute tooth; filaments glabrous, anthers included, minutely apiculate at base, glabrous or sparsely pubescent. **Capsules** narrowly ovoid, 5–6 mm. **Seeds** 0.3–0.5 mm. $2n = 48$.

Flowering mid Apr–Jul. Chaparral; 150–2000 m; Calif.; Mexico (Baja California).

Orobanche bulbosa is endemic to chaparral in California and northern Baja California, Mexico, distributed throughout the range of its host, *Adenostoma fasciculatum* (Rosaceae).

6. Orobanche pinorum Geyer ex Hooker, Hooker's J. Bot. Kew Gard. Misc. 3: 297. 1851 • Conifer broomrape [E]

Aphyllon pinorum (Geyer ex Hooker) A. Gray; *Myzorrhiza pinorum* (Geyer ex Hooker) Rydberg

Plants simple or few-branched from near base, 10–40 cm, slender, base enlarged. **Roots** conspicuous (usually forming an irregularly globular mass), stout, unbranched. **Leaves** numerous, appressed proximally, spreading distally; blade lanceolate to oblong-ovate or triangular-ovate, 6–20 mm, margins entire, apex acute to acuminate, surfaces usually glabrous. **Inflorescences** open, cylindric panicles, ochraceous, red-brown, purple or purple streaked, yellow, or cream-white, loosely branched, rarely simple, cinereous glandular-puberulent; flowers numerous, widely spaced proximally, clustered distally; bracts reflexed, narrowly lanceolate, 3–6 mm, apex acuminate, moderately glandular-puberulent. **Pedicels** 0–2(–6) mm, much shorter than plant axis; bracteoles 2. **Flowers:** calyx yellow, brown, or purple, ± radially symmetric, 5–8 mm, divided into 5 subequal lobes, lobes slightly shorter than to ca. as long as tube, subulate, puberulent; corolla 13–19 mm, tube white, cream, or light yellow, sometimes reddish brown to purple tinged or with reddish brown or purple veins, strongly constricted above ovary, bent forward, glandular-pubescent; palatal folds not prominent, pale or light yellow, glabrous; lips externally white, cream, or light yellow, sometimes reddish or purplish tinged distally or with reddish brown to purple veins, internally reddish brown to purple, sometimes pale with reddish brown or purple veins, abaxial lip spreading, 3–4 mm, lobes narrowly lanceolate, apex rounded, adaxial lip spreading, 3–4 mm, lobes lanceolate, apex rounded, acute, or emarginate; filaments with ring of hairs at base, anthers included or 1 pair exserted, glabrous or sparsely pubescent. **Capsules** ovoid, 6–7 mm. **Seeds** 0.3–0.5 mm. $2n = 48$.

Flowering Jul–Sep. Dry coniferous forests, rocky slopes; 200–2500 m; B.C.; Calif., Idaho, Nev., N.Mex., Oreg., Wash.

Orobanche pinorum is unusual in appearance as the stem and inflorescence axis are often markedly darker in color than the flowers.

Orobanche pinorum is chiefly host-specific on *Holodiscus discolor* (Rosaceae) and is largely sympatric with that species. There are a few occurrences outside the range of *H. discolor* where it is reported on other species of *Holodiscus*. Unverified reports on herbarium sheets of parasitism on various conifers exist. Disjunct locations in Nevada and New Mexico are noteworthy, because they suggest that this species may be found in appropriate habitat in the intervening Great Basin territory.

7. **Orobanche robbinsii** Heckard ex Colwell & Yatskievych, Phytoneuron 2016-58: 2, fig. 1A–D. 2016 [C] [E]

Aphyllon robbinsii (Heckard ex Colwell & Yatskievych) A. C. Schneider

Plants simple, sometimes few-branched from base, 4–15 (–26) cm, stout, base enlarged in robust specimens. **Roots** usually relatively conspicuous, slender, unbranched or with short bifurcations. **Leaves** numerous, appressed; blade broadly rounded or deltate, 3–8 mm, margins entire or erosulate, apex rounded to subacute, surfaces glabrous. **Inflorescences** compact corymbs (sometimes subracemose in robust specimens), pallid, usually infused with purple, often branched, densely glandular-puberulent; flowers numerous; bracts erect to slightly reflexed, ± lanceolate to oblanceolate, 5–10 mm, apex retuse or erosulate, sometimes obtuse to acute, glandular-pubescent. **Pedicels** 2–10 mm (to 35 mm proximally), shorter than plant axis; bracteoles 2. **Flowers:** calyx pallid to dark purple, weakly bilaterally symmetric, 6–14 mm, deeply divided into 5 subequal lobes, lobes subulate to narrowly spatulate, densely glandular-pubescent; corolla (8–)15–25(–30) mm, tube white to cream, sometimes pale purplish tinged distally, sometimes with purple veins, slightly constricted above ovary, slightly dilated distally, straight or ± curved forward, glandular-puberulent; palatal folds prominent, cream to lemon, glabrous; lips white to cream, sometimes purplish tinged distally (especially internally and on adaxial lip), sometimes with purple veins, abaxial lip erect to slightly spreading, 4–6(–9) mm, lobes oblong to oblong-ovate, apex rounded to truncate, adaxial lip erect to barely spreading, 4–6(–9) mm, lobes oblong to lanceolate, apex rounded to truncate, sometimes shallowly emarginate; filaments glabrous at base, anthers included or barely exserted, glabrous or with few long hairs. **Capsules** ovoid to cylindric-ovoid, 8–10 mm. **Seeds** 0.3–0.5 mm. $2n = 48$.

Flowering Apr–Jul. Rocky seaside bluffs, ancient shell mounds and sand dunes, eroding cliff slide areas; of conservation concern; 0–100 m; Calif.

Orobanche robbinsii is distributed from Marin to San Luis Obispo counties. It is parasitic on *Eriophyllum staechadifolium*, with single reports on *Artemisia pycnocephala* (Asteraceae) and *Phacelia californica* (Hydrophyllaceae).

8. **Orobanche corymbosa** (Rydberg) Ferris, Contr. Dudley Herb. 5: 99. 1958 • Flat-top broomrape [E] [F]

Myzorrhiza corymbosa Rydberg, Bull. Torrey Bot. Club 36: 696. 1909; *Aphyllon corymbosum* (Rydberg) A. C. Schneider; *Orobanche californica* Chamisso & Schlechtendal var. *corymbosa* (Rydberg) Munz

Plants simple, rarely branched, 5–12(–18) cm, usually slender, base sometimes enlarged. **Roots** usually inconspicuous (sometimes forming an irregular mass), slender, unbranched or with short bifurcations. **Leaves** few to several, appressed; blade broadly lanceolate or ovate, 5–10 mm, margins entire or slightly erose, apex acute or obtuse, surfaces often glandular-pubescent. **Inflorescences** racemelike to elongate corymbs or short racemes (sometimes corymbose), rose, pink, purple, or white, simple, densely glandular-pubescent; flowers numerous (rarely 10 or fewer in depauperate plants); bracts slightly reflexed, narrowly lanceolate or almost linear, 7–11 mm, apex acute, glandular-pubescent. **Pedicels** 3–20 mm proximally, 0 mm distally, shorter than plant axis; bracteoles 2. **Flowers:** calyx lavender, pink, white, or yellow, sometimes burgundy, ± radially symmetric, 12–24 mm, deeply divided into 5 subequal (reflexed or contorted) lobes, lobes linear-subulate, densely glandular-pubescent; corolla (15–)18–34 mm, tube white to grayish white, pale pink to pink, or pale purple to purple, rarely brick red, sometimes with darker pink to purple veins, slightly constricted above ovary, bent forward, glandular-pubescent or glabrate; palatal folds prominent, yellow, glabrous (with blisterlike swellings); lips white to ± pink, pale purple to purple, brick red, or pinkish red, sometimes with darker pink to purple veins or internally darker, abaxial lip spreading, 5–9 mm, lobes oblong to oblong-lanceolate, apex acute or rounded, adaxial lip erect or reflexed, 5–9 mm, lobes oblong, apex rounded, truncate, or emarginate, rarely acute; filaments glabrous, anthers included, tomentose, sometimes glabrous (subsp. *mutabilis*). **Capsules** ovoid to oblong-ovoid, 5–13 mm. **Seeds** 0.3–0.5 mm.

Subspecies 2 (2 in the flora): w North America.

The range of *Orobanche corymbosa* includes the Great Basin Desert, Intermountain Region, Columbia Plateau, and contiguous areas. The two subspecies are sympatric through most of the geographic range of the species. Areas of sympatry exist with several other species of *Orobanche*: in the west with *O. californica*; in the north and east with *O. ludoviciana*; and in the south with *O. arizonica* and *O. multiflora*. Subspecies *mutabilis* tends to produce more racemose inflorescences, while subsp. *corymbosa* typically produces more compact corymbose inflorescences. This suggests a genetic affinity between *O. corymbosa* and members of the *O. californica* and *O. ludoviciana* groups. The southern members of subsp. *corymbosa* in California and Nevada appear to intergrade with *O. californica* across the Sierra Nevada. However, there are exceptional individuals in every population, so seasonal weather may cause variations in plant morphology.

Inflorescence differences could also be associated with the ploidy differences reported in *Orobanche corymbosa* (L. R. Heckard and T. I. Chuang 1975). Ploidy and morphological instability may indicate significant introgression or a hybrid origin of *O. corymbosa*.

Both subspecies share a somewhat ampliate corolla tube, which is a good field character for this species when compared with species with which it has an overlapping range. *Orobanche corymbosa* is parasitic on species of *Artemisia* (Asteraceae), principally *A. tridentata*, but has occasionally been reported on *Iva* (Asteraceae) and *Atriplex* and *Sarcobatus* (Chenopodiaceae).

1. Inflorescences simple corymbs or short corymbose racemes; calyces (13–)15–24 mm, equal to or shorter than corollas; corollas glandular-pubescent; anthers tomentose . . . .
. . . . 8a. *Orobanche corymbosa* subsp. *corymbosa*
1. Inflorescences short racemes or racemelike to elongate corymbs; calyces 12–18 mm, sometimes equal to or longer than corollas; corollas glabrate or slightly glandular-pubescent; anthers tomentose or glabrous . . . .
. . . . 8b. *Orobanche corymbosa* subsp. *mutabilis*

8a. Orobanche corymbosa (Rydberg) Ferris subsp. **corymbosa** E F

Plants simple, sometimes branched. **Inflorescences** simple corymbs or short corymbose racemes. **Calyces** (13–)15–24 mm, equal to or shorter than corolla. **Corollas** rose, pink, or purple with darker veins, glandular-pubescent; anthers tomentose.

Flowering Jun–Sep. Sandy soils and rocky slopes in open sagebrush communities and subalpine meadows; 100–3200 m; Ariz., Calif., Idaho, Mont., Nev., N.Mex., Oreg., Utah, Wash., Wyo.

Subspecies *corymbosa* is most often mistaken for *Orobanche californica* with which it was formerly associated taxonomically. The two species are very similar morphologically; *Orobanche corymbosa*, with only two subspecies, has much less morphological variability. Except for a small area of sympatry in eastern California, the two species are otherwise allopatric.

8b. Orobanche corymbosa (Rydberg) Ferris subsp. **mutabilis** Heckard, Canad. J. Bot. 56: 187. 1978 E

Aphyllon corymbosum (Rydberg) A. C. Schneider subsp. *mutabile* (Heckard) A. C. Schneider

Plants simple, rarely branched. **Inflorescences** short racemes or racemelike to elongate corymbs. **Calyces** 12–18 mm, sometimes equal to or longer than corolla. **Corollas** rose, pink, pale purple, or white with darker veins, glabrate or slightly glandular-pubescent; anthers tomentose or glabrous.

Flowering Jun–Sep. Sandy soils and rocky slopes in open sagebrush communities; 300–3200 m; Alta., B.C., Man.; Calif., Nev., N.Mex., Wyo.

Populations of subsp. *mutabilis* are more abundant in the northern portion of the range; populations elsewhere appear to be somewhat isolated. The more elongated inflorescence seems to be the distinguishing character, although this is often difficult to determine. The corollas often display less pigmentation. Anthers vary from quite woolly in some populations to completely glabrous in other populations. This variation could be the result of geographic isolation or hybridization.

9. **Orobanche californica** Chamisso & Schlechtendal, Linnaea 3: 134. 1828 • California broomrape [F]

Aphyllon californicum (Chamisso & Schlechtendal) A. Gray; *Myzorrhiza californica* (Chamisso & Schlechtendal) Rydberg; *Phelypaea californica* (Chamisso & Schlechtendal) G. Don

Plants simple or branched proximally, 4–35(–40) cm, slender or stout, base sometimes enlarged. **Roots** inconspicuous (rarely forming a bulbous mass), slender or stout, unbranched, sometimes branched (subsp. *grandis*). **Leaves** numerous, appressed; blade broadly ovate, triangular, deltate, lanceolate, or oblong, 4–12 mm, margins entire or erosulate, apex obtuse or rounded, surfaces glabrous. **Inflorescences** corymbs (sometimes subcapitate), sometimes racemes or subcorymbose racemes, dark purple, reddish purple, pinkish, or pallid cream to nearly white, simple, sometimes inconspicuously branched, densely glandular-puberulent; flowers numerous (rarely 10 or fewer in depauperate plants); bracts appressed to spreading, ± lanceolate to oblanceolate, 5–15 mm, apex acute, glandular-pubescent. **Pedicels** 0–20(–25) mm, shorter than plant axis; bracteoles 2. **Flowers:** calyx pallid to dark purple, pink, yellow, or white, ± weakly bilaterally symmetric, 8–20(–27) mm, deeply divided into 5 (sometimes reflexed or contorted) lobes, lobes subulate to linear-subulate, gradually attenuate, glandular-pubescent; corolla 22–50(–55) mm, tube white or cream to pinkish or purplish tinged or pink to purple, sometimes with darker veins, constricted above ovary, curved forward, sparsely to moderately glandular-puberulent; palatal folds prominent, yellow, glabrous (lacking blisterlike swellings), sometimes pubescent; lips white or cream to pinkish or purplish tinged (then sometimes appearing reddish brown in herbarium specimens) or pink to purple, sometimes with darker veins or dark purple distally, abaxial lip widely spreading, 8–15 mm, lobes ± lanceolate to ± oblong, lanceolate-subulate, narrowly triangular, or lanceolate-ovate, apex ± acute to rounded or obtuse, sometimes retuse or emarginate, adaxial lip ± spreading, 10–15(–18) mm, lobes broadly deltate to ovate or oblong, apex bluntly pointed to rounded, obtuse, acute, shallowly retuse, erosulate, emarginate, shallowly notched, or erose; filaments glabrous, anthers included, densely villous on sutures, rarely glabrous. **Capsules** ovoid to cylindric-ovoid, 10–12 mm. **Seeds** 0.4–0.6 mm.

Subspecies 6 (6 in the flora): w North America, nw Mexico.

Plants of *Orobanche californica* occur almost entirely in and west of the Cascade-Sierra Nevada-Peninsular ranges from British Columbia south to the Sierra San Pedro Mártir in Baja California. Host plants are various perennial members of Asteraceae.

L. R. Heckard (1973) discussed the difficulty of presenting a classification for *Orobanche californica* based on the few morphological features that must be used for taxonomic delineation and the sometimes baffling disjunctions in the distribution of the variants. He recognized six subspecies based on geographic variations that demonstrate the considerable variability within the species. The following key to subspecies is adapted from the key by Heckard.

1. Corollas dark purple or red-purple distally; calyx lobes and bracts purple tinged, drying blackish purple.... 9a. *Orobanche californica* subsp. *californica*
1. Corollas pallid, pinkish, pale lavender, whitish, or yellowish distally, often with darker veins; calyx lobes pallid, pinkish tinged, whitish, yellow, or buff, sometimes purplish or tinged pinkish, purple, or brownish, bracts pallid to pinkish tinged, drying brown.
 2. Corolla tubes gradually widening toward throat, corollas pallid or yellow distally, sometimes pinkish tinged................
  .... 9f. *Orobanche californica* subsp. *feudgei*
 2. Corolla tubes abruptly widening toward throat, corollas pallid, pale pink, or pale lavender, sometimes yellowish tinged distally.
  3. Corollas 35–50 mm, abaxial lobes lanceolate to lanceolate-ovate, 5–7 mm wide ... 9e. *Orobanche californica* subsp. *grandis*
  3. Corollas 25–35(–40) mm, abaxial lobes lanceolate to lanceolate-subulate, narrowly lanceolate, or lanceolate-oblong to narrowly triangular, 2–4(–5) mm wide.
   4. Plants (8–)10–35 cm, portions proximal to inflorescences 6–15 cm; inflorescences racemes, sometimes subcorymbose, 5–20 cm ....
    ....9c. *Orobanche californica* subsp. *jepsonii*
   4. Plants 4–10(–15) cm, portions proximal to inflorescences 1–8(–12) cm; inflorescences corymbs, sometimes subcorymbose or subcapitate racemes, 2–6 cm.
    5. Plants: portions proximal to inflorescences 1–3(–4) cm; abaxial corolla lobes lanceolate to lanceolate-subulate, 2–3 mm wide, apices acute ... 9b. *Orobanche californica* subsp. *grayana*
    5. Plants: portions proximal to inflorescences 3–8(–12) cm; abaxial corolla lobes lanceolate-oblong to narrowly triangular, 3–4 mm wide, apices bluntly acute or obtuse, sometimes retuse or emarginate ...
     .... 9d. *Orobanche californica* subsp. *condensa*

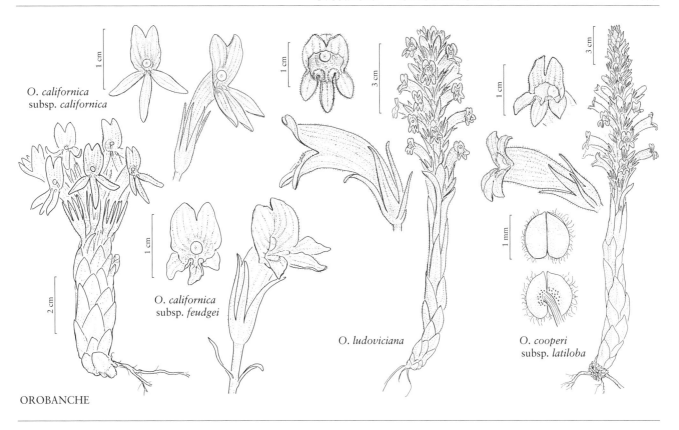

O. californica
subsp. *californica*

O. californica
subsp. *feudgei*

O. ludoviciana

O. cooperi
subsp. *latiloba*

OROBANCHE

9a. Orobanche californica Chamisso & Schlechtendal
subsp. **californica** E F

Aphyllon violaceum Eastwood;
Myzorrhiza violacea (Eastwood)
Rydberg; *Orobanche comosa*
Hooker var. *violacea* (Eastwood)
Jepson; *O. grayana* Beck var.
nelsonii Munz; *O. grayana* var.
violacea (Eastwood) Munz

Plants simple or branched from
below ground, 5–20(–27) cm,
portion proximal to inflorescence 1–14 cm, slender.
Inflorescences corymbs, sometimes subcorymbose or
subcapitate racemes, 4–8 cm; bracts purple tinged,
drying blackish purple. **Pedicels** 0–10 mm. **Calyces**
8–17(–24) mm, lobes purple tinged, drying blackish
purple, subulate to linear-subulate, 6–20 mm. **Corollas**
dark purple or red-purple distally, (22–)25–40(–55) mm,
tube slender, abruptly widening toward throat; throat
8–10 mm wide at base of lobes; lips 10–14(–18) mm,
abaxial lobes lanceolate to oblong, 3–4(–5) mm wide,
apex acute to obtuse, adaxial lobes oblong-ovate, apex
broadly obtuse to acute, shallowly retuse, or erosulate.
2*n* = 48.

Flowering (Mar–)May–Sep(–Oct). Rocky banks,
sand flats, tidal marshes, coastal prairies and slopes;
0–100 m; B.C.; Calif., Oreg., Wash.

Subspecies *californica* is uncommon on rocky banks,
sand flats, and slopes near the sea from southwestern

British Columbia, San Juan Islands, and adjacent
mainland Washington, and to coastal prairie and bluff
tops in California from Humboldt County south to San
Luis Obispo County. Available habitat in the southern
portion of the range is adversely affected by infestations
of *Carpobrotus edulis* (Aizoaceae).

The primary host for subsp. *californica* is *Grindelia*
(Asteraceae). Reports on *Ambrosia chamissonis*,
Erigeron glaucus, and *Heterotheca villosa* need
confirmation.

Subspecies *californica* is fairly discrete and easily
recognizable throughout most of its range. The deep
lavender to reddish purple exterior of the corolla, long
corolla tube, and long lips are unique to this taxon.
The maritime habitat is shared with several other taxa
in California: *Orobanche californica* subsp. *grandis*,
O. parishii subsp. *brachyloba*, and *O. robbinsii*;
however, the microhabitats and primary hosts of each
differ markedly, with subsp. *californica* the only one of
these primarily on coastal prairie, especially blufftop
(marine terrace) habitat, and primarily parasitic on
Grindelia.

Morphological variation within subsp. *californica* is
partially correlated with geography. Along the central
California coast, plants usually have corollas 35–55 mm
with broad lobes, apically rounded, and deep purple.
The plants of coastal Oregon, Washington, and British
Columbia have shorter and narrower corollas (25–35
mm), often with pointed lobes and slightly paler color.

9b. Orobanche californica Chamisso & Schlechtendal subsp. **grayana** (Beck) Heckard, Madroño 22: 54. 1973 E

Orobanche grayana Beck, Biblioth. Bot. 19: 79. 1890, based on *O. comosa* Hooker, Fl. Bor.-Amer. 2: 92. 1837, not Wallroth 1822; *Aphyllon californicum* (Chamisso & Schlechtendal) A. Gray subsp. *grayanum* (Beck) A. C. Schneider; *Myzorrhiza grayana* (Beck) Rydberg

Plants usually branched proximally, 4–10 cm, portion proximal to inflorescence 1–3(–4) cm, slender. **Inflorescences** corymbs, sometimes subcorymbose racemes, 3–6 cm; bracts pallid to pinkish tinged, drying brown. **Pedicels** 5–20 mm. **Calyces** 11–15 mm, lobes pallid, sometimes purplish, linear-subulate, (7–)9–13 (–16) mm. **Corollas** pallid to pinkish or pale lavender, often with lavender veins, (25–)28–33 mm; tube slender, abruptly widening toward throat; throat 8–10 mm wide at base of lobes; lips 10–12 mm, abaxial lobes lanceolate to lanceolate-subulate, 2–3 mm wide, apex acute, adaxial lobes oblong-ovate, apex narrowly rounded, shallowly retuse, or erosulate. $2n = 48$.

Flowering (Jun–)Aug–Sep. Moist meadows and stream margins; (50–)300–2100 m; Calif., Oreg., Wash.

Subspecies *grayana* is distributed in the Cascade-Sierra Nevada ranges, Coast Ranges (central California and southern Oregon), and mountains of the Columbia Plateau in Oregon, from Klickitat County, Washington, south to Tuolumne County, California. It is rare throughout the range, possibly locally extirpated in portions of the range in Oregon.

The hosts are primarily species of *Aster* and *Erigeron* and occasionally *Grindelia*. Other reported non-Asteraceae hosts are unlikely.

9c. Orobanche californica Chamisso & Schlechtendal subsp. **jepsonii** (Munz) Heckard, Madroño 22: 57. 1973 E

Orobanche grayana Beck var. *jepsonii* Munz, Bull. Torrey Bot. Club 57: 617, plate 38, fig. 10. 1931; *Aphyllon californicum* (Chamisso & Schlechtendal) A. Gray subsp. *jepsonii* (Munz) A. C. Schneider

Plants simple or branched proximally, (8–)10–35 cm, portion proximal to inflorescence 6–15 cm, slender. **Inflorescences** racemes, sometimes subcorymbose, 5–20 cm; bracts pallid to pinkish tinged, drying brown. **Pedicels** 3–20 mm. **Calyces** (10–)15–24(–27) mm, lobes pallid or pinkish tinged, subulate to linear-subulate,

(8–)10–20(–25) mm. **Corollas** whitish to pinkish distally, often with veins of lips deep rose or purplish, (25–)30–35(–40) mm; tube slender, abruptly widening toward throat; throat 8–10 mm wide at base of lobes; lips 10–12(–15) mm, abaxial lobes usually narrowly lanceolate, 3–5 mm wide, apex acute, adaxial lobes deltate, apex acute, often erosulate or emarginate. $2n = 48$.

Flowering May–Oct. Bottomlands, dry hillsides; 100–2000 m; Calif., Oreg.

Subspecies *jepsonii* is distributed primarily in the Central Valley and surrounding mountains from the Cascade Range of Plumas and Shasta counties in the Sierra Nevada to Kern County and the central California Coast Ranges from Santa Cruz to Tehama counties.

The most frequently reported hosts for subsp. *jepsonii* are *Baccharis*, *Grindelia*, and *Solidago* (Asteraceae). Reports of *Eriophyllum* are unconfirmed, and reports of non-Asteraceae hosts such as *Eriogonum* and *Rubus* are highly unlikely.

Subspecies *jepsonii* is circumscribed similarly to what Munz described as *Orobanche grayana* var. *jepsonii*. It includes foothill and montane plants that resemble subsp. *grayana* in size and corymbose inflorescence, albeit with a deeper colored corolla and a preference for dry, rocky habitats.

L. R. Heckard (1973) indicated that subsp. *jepsonii* and subsp. *grayana* are very similar both geographically and morphologically. The two subspecies are separated by minor differences in plant size, corolla color and shape, and habitat.

9d. Orobanche californica Chamisso & Schlechtendal subsp. **condensa** Heckard, Madroño 22: 59, fig. 1I–L. 1973 E

Aphyllon californicum (Chamisso & Schlechtendal) A. Gray subsp. *condensum* (Heckard) A. C. Schneider

Plants simple or sparsely branched below ground, 5–10 (–15) cm, portion proximal to inflorescence 3–8(–12) cm, slender or stout. **Inflorescences** corymbs, sometimes subcorymbose or subcapitate racemes, 2–6 cm; bracts pallid to pinkish tinged, drying brown. **Pedicels** 5–12 mm. **Calyces** 10–16(–20) mm, lobes whitish or yellow, tinged with purple, linear-subulate, 7–13 mm. **Corollas** sordid white to yellowish distally, 25–35 mm; tube relatively broad, usually abruptly widening slightly toward throat; throat 6–8 mm wide; lips 8–10(–12) mm, abaxial lobes lanceolate-oblong to narrowly triangular, 3–4 mm wide, apex bluntly acute or obtuse, sometimes retuse or emarginate, adaxial lobes broadly oblong, apex obtuse, rounded, or emarginate, rarely acute. $2n = 48$.

Flowering May–Jun. Sandy or gravelly soils of dry bottomlands; 100–500 m; Calif.

Subspecies *condensa* is found in the interior South Coast Ranges of southern California from San Benito to Santa Barbara counties. It appears to be host specific on *Heterotheca villosa* (Asteraceae).

9e. Orobanche californica Chamisso & Schlechtendal subsp. **grandis** Heckard, Madroño 22: 60, figs. 1P–R, 3A, 4E. 1973　E

Aphyllon californicum (Chamisso & Schlechtendal) A. Gray subsp. *grande* (Heckard) A. C. Schneider

Plants simple, rarely branched from base, 8–35 cm, proximal portion to inflorescence (5–)10–20 cm, stout. **Inflorescences** corymbs, sometimes subcapitate to subcorymbose racemes, 3–15 cm; bracts pallid to pinkish tinged, drying brown. **Pedicels** 5–15(–25) mm. **Calyces** 12–20(–25) mm, lobes buff or yellow, tinged pink or brownish, linear-subulate, 15–20 mm. **Corollas** pallid to pinkish tinged, drying reddish brown, 35–50 mm; tube broad, abruptly widening toward throat; throat 9–10 mm wide; lips 12–14 mm, abaxial lobes lanceolate to lanceolate-ovate, 5–7 mm wide, apex acute, adaxial lobes ovate, apex obtuse, shallowly notched, erose.

Flowering Apr–Oct(–Nov). Sandy soils or lee of foredunes near coast; 0–150 m; Calif.

Subspecies *grandis* is commonly found near the ocean shore from Los Angeles to San Mateo counties and on Santa Rosa Island. It is uncommon, and its habitat is impacted by *Carpobrotus* (Aizoaceae) invasions. The primary host is *Corethrogyne filaginifolia* (Asteraceae), but it can also utilize *Grindelia* (Asteraceae). Other hosts reported are *Adenostoma fasciculatum* (Rosaceae), *Artemisia*, *Isocoma veneta*, and *Heterotheca* (Asteraceae).

Plants included in subsp. *grandis* have been identified variously in the past. The northern coastal specimens were aligned with subsp. *californica*, whereas the southern collections were identified as subspp. *feudgei* or *jepsonii*. Subspecies *grandis* is roughly sympatric with other coastal subspecies, but its unique morphology and its preferences for sand dune habitat and *Corethrogyne filaginifolia* as a host make it distinctive.

9f. Orobanche californica Chamisso & Schlechtendal subsp. **feudgei** (Munz) Heckard, Madroño 22: 62. 1973　F

Orobanche grayana Beck var. *feudgei* Munz, Bull. Torrey Bot. Club 57: 616, plate 38, fig. 8. 1931; *Aphyllon californicum* (Chamisso & Schlechtendal) A. Gray subsp. *feudgei* (Munz) A. C. Schneider

Plants usually simple proximally, 10–30(–40) cm, portion proximal to inflorescence (6–)10–18 cm, stout. **Inflorescences** corymbs, sometimes subcorymbose or subcapitate racemes, sometimes racemose, 4–12 cm; bracts pallid to pinkish tinged, drying brown. **Pedicels** 5–15(–20) mm. **Calyces** 16–22 mm, lobes pallid, sometimes tinged with reddish purple or reddish brown, linear-subulate, (12–)15–20 mm. **Corollas** pallid or yellow distally, sometimes pinkish tinged, often with darker veins, 25–35 mm; tube broad, gradually widening toward throat; throat 8–10 mm wide; lips (8–)10–12 mm, abaxial lobes narrowly oblong, 5–7 mm wide, apex subacute or obtuse, sometimes shallowly retuse, adaxial lobes broadly oblong and rounded, apex sometimes emarginate or shallowly retuse. $2n = 48$.

Flowering May–Jul. Washes, open sagebrush flats and slopes; 800–2600 m; Calif.; Mexico (Baja California).

Subspecies *feudgei* occurs from the Piute Mountains of Kern County and the vicinity of Mount Pinos in Ventura County, California, south to the Sierra San Pedro Mártir in Baja California.

Subspecies *feudgei* is apparently host-specific on *Artemisia tridentata* (Asteraceae).

Plants with the morphological features defining subsp. *feudgei* correlate fairly well with a distinct ecology and geography. The geographical area occupied by subsp. *feudgei* coincides with the southeastern portion of the range of *Orobanche californica*.

L. R. Heckard (1973) noted the morphological similarities of subsp. *feudgei* to *Orobanche corymbosa*.

10. Orobanche ludoviciana Nuttall, Gen. N. Amer. Pl. 2: 58. 1818 • Louisiana broomrape　E F W

Aphyllon ludovicianum (Nuttall) A. Gray; *Myzorrhiza ludoviciana* (Nuttall) Rydberg; *Orobanche ludoviciana* var. *arenosa* (Suksdorf) Cronquist; *O. multiflora* Nuttall var. *arenosa* (Suksdorf) Munz

Plants simple or few-branched, 7–40(–54) cm, usually stout, base enlarged in robust specimens. **Roots** inconspicuous to conspicuous (often forming an amorphous mass), slender, usually branched. **Leaves**

several to numerous, appressed; blade lanceolate to lanceolate-ovate, 5–10 mm, margins entire, apex acute, surfaces sometimes glandular-pubescent. **Inflorescences** spikelike racemes, purple, lavender, or pallid distally, rarely yellow, sometimes branched, glandular-pubescent; flowers numerous; bracts ± reflexed, lanceolate, 8–12 mm, apex acute or attenuate, densely glandular-pubescent. **Pedicels** 0–15 mm, much shorter than plant axis; bracteoles 2. **Flowers:** calyx purple, lavender, or pallid, weakly bilaterally symmetric, 8–14 mm, deeply divided into 5 lobes, lobes lanceolate-subulate, glandular-pubescent; corolla 14–20 mm, tube white to pallid or cream, sometimes pinkish or light purplish tinged distally, sometimes with purple veins, constricted above ovary, slightly curved forward, glandular-pubescent; palatal folds prominent, yellow, pubescent; lips externally white to pallid or cream, sometimes pinkish or light purplish tinged, internally pink or purple, sometimes white with purple veins, rarely light yellow, abaxial lip spreading, 3–5 mm, lobes oblong-lanceolate, apex obtuse or rounded, adaxial lip erect or slightly reflexed, 4–6 mm, lobes ovate, sometimes deltate, apex rounded or obtuse to bluntly pointed; filaments glabrous or pilose at base, anthers included or slightly exserted, glabrous or with few woolly hairs along sutures. **Capsules** ovoid, 6–13 mm. **Seeds** 0.3–0.5 mm. $2n = 48$.

Flowering Apr–Aug. Prairies, sand hills, sand dunes, eroded ground, glades, roadsides; 0–2500 m; Alta., B.C., Man., Sask.; Colo., Idaho, Ill., Kans., Minn., Mo., Mont., Nebr., N.Mex., N.Dak., Okla., S.Dak., Tex., Wash., Wis., Wyo.

Orobanche ludoviciana is one of the most widely distributed species of *Orobanche* in North America. It commonly occurs in wind and water eroded habitats, principally in the Great Plains of North America and contiguous areas.

Throughout most of its range, *Orobanche ludoviciana* parasitizes *Grindelia squarrosa* and several species of *Artemisia*. However, at the southern limits of the range in Texas, it has been reported on *Baccharis*, *Haploësthes*, *Heterotheca*, and *Thelesperma* (Asteraceae); in Canada, *Heterotheca villosa* is an important host. The reports on cultivated crops (tomato and tobacco) are the result of misidentifications by P. A. Munz (1930) and should be attributed to *O. cooperi* and *O. riparia*.

The binomial *Orobanche ludoviciana* has often been broadly applied to several taxa in western states, including several taxa treated herein as species.

P. A. Munz (1930) inadvertently used a specimen of *Orobanche riparia* to describe and illustrate the corollas of *O. ludoviciana* as having pointed corolla lobes. This led him to include several western taxa that have pointed corolla lobes within *O. ludoviciana*, including taxa treated here as species: *O. cooperi*, *O. riparia*, and *O. valida*. L. T. Collins et al. (2009) clarified this issue, pointing out that the corolla lobes are in fact rounded.

11. Orobanche multiflora Nuttall, Proc. Acad. Nat. Sci. Philadelphia 4: 22. 1848 • Many-flower broomrape [E]

Aphyllon multiflorum (Nuttall) A. Gray; *Myzorrhiza multiflora* (Nuttall) Rydberg; *Orobanche ludoviciana* Nuttall subsp. *multiflora* (Nuttall) L. T. Collins ex H. L. White & W. C. Holmes; *O. ludoviciana* var. *multiflora* (Nuttall) Beck

Plants simple or few-branched, 7–27 cm, relatively slender, base usually not enlarged. **Roots** usually inconspicuous, slender, unbranched or branched. **Leaves** numerous, appressed; blade lanceolate or broadly triangular, 3–10 mm, margins entire, sometimes ciliate, apex acute, surfaces glabrous. **Inflorescences** racemes, sometimes thyrsoid, pale yellow, white, or tan proximally, purple or pale lavender dis-tally, sometimes branched, densely glandular-pubescent, appearing whitish or canescent, sometimes with axillary branches; flowers numerous; bracts erect to reflexed, ± lanceolate, 11–20 mm, apex acute or acuminate, densely glandular-pubescent. **Pedicels** 0–5 mm, much shorter than plant axis; bracteoles 2. **Flowers:** calyx pallid externally, purple internally, weakly bilaterally symmetric, 15–21 mm, deeply divided into 5 lobes, lobes lanceolate-attenuate, densely glandular-pubescent; corolla 22–36 mm, tube white to pallid or cream, sometimes pinkish or light purplish tinged, rarely light yellow distally, sometimes with purple veins, constricted above ovary, only slightly bent forward, densely pubescent; palatal folds prominent, yellow, densely pubescent; lips externally white to pallid or cream, sometimes pinkish or light purplish tinged, internally pink or purple, sometimes white with purple veins, rarely light yellow, abaxial lip spreading, 5–9 mm, lobes broadly lanceolate, apex obtuse or rounded, adaxial lip erect or reflexed, 6–12 mm, lobes oblong, apex rounded; filaments pilose with ring of hairs at insertion, anthers included, woolly. **Capsules** ovoid, 8–12 mm. **Seeds** 0.4–0.5 mm. $2n = 48$.

Flowering Aug–Sep. Arid grasslands, semideserts, open woodlands; 0–2500 m; Colo., N.Mex., Tex.

Orobanche multiflora is parasitic mainly on *Gutierrezia* and occasionally *Heterotheca* (Asteraceae). A population in southern Texas is parasitic on *Varilla texana* (Asteraceae).

There is considerable confusion about what constitutes *Orobanche multiflora*. It has been interpreted as both a variety and subspecies of *O. ludoviciana*. The much larger flowers (22–36 mm versus 14–20 mm) set it apart from *O. ludoviciana*. P. A. Munz (1930) described four varieties of *O. multiflora*: vars. *arenosa*, *multiflora* (as *typica*), *pringlei*, and *xanthocroa*. Variety *xanthocroa* was based on a specimen of *Conopholis*. Variety *arenosa* is treated in synonymy under *O. ludoviciana*. Variety *pringlei* was based on a few specimens from northeastern Mexico that appear to represent an undescribed species.

The distribution range is not well defined in the literature. The authors consider *Orobanche multiflora* to have a much more restricted range than previously indicated, occurring mostly in Colorado, New Mexico, and adjacent northwestern Texas, with disjunct populations in southern Texas that are apparently this species. Specimen data indicate that the species distribution is most likely east of the Continental Divide; plants from west of the continental divide with this binomial are probably misidentified.

Plants from five counties in extreme southern Texas, near the Mexican border, appear to be a disjunct variant of *Orobanche multiflora* that flowers in spring. The flowers are about the same length, but the corolla tube is much narrower. Margins of the petals are often undulate and are an unusual shade of blue-purple. The pubescence at the base of the filaments is also reduced. Host and habitat differences also raise questions about the classification of these plants.

12. Orobanche parishii (Jepson) Heckard, Madroño 22: 66. 1973 • Parish's broomrape

Orobanche californica Chamisso & Schlechtendal var. *parishii* Jepson, Man. Fl. Pl. Calif., 952. 1925; *Aphyllon parishii* (Jepson) A. C. Schneider

Plants simple, rarely branched, 5–22(–35) cm, stout, base not enlarged. **Stems** thickened, fleshy. **Roots** inconspicuous (often very short and knobby), slender, usually unbranched. **Leaves** numerous, appressed; blade broadly ovate, 7–12 mm, margins entire, apex obtuse, surfaces glandular-pubescent. **Inflorescences** spikelike racemes, pallid creamy or yellow, purplish tinged, simple, sometimes branched, densely glandular-pubescent; flowers numerous; bracts erect to spreading, narrowly lanceolate-ovate, 10–12(–19) mm, apex acute, glandular-pubescent. **Pedicels** 0–10(–12) mm, much shorter than plant axis; bracteoles 2. **Flowers:** calyx white or pinkish, or purple tinged, often pallid, radially or weakly bilaterally symmetric, 7–18 mm, deeply divided into 5 lobes, lobes subulate to attenuate, densely glandular-pubescent; corolla 15–25 mm, tube white to pallid, yellow, or buff, slightly constricted above ovary, straight to curved forward, glandular-pubescent; palatal folds prominent, yellow, glabrous; lips externally white to pallid, yellow, or buff, sometimes slightly reddish tinged distally, internally maroon or reddish purple, sometimes with maroon or reddish purple stripes, veins, or blotches, abaxial lip erect to spreading or recurved, 4–8 mm, lobes narrowly ovate to oblong, apex rounded, blunt, retuse, or erosulate, adaxial lip erect to ± spreading, 4–8 mm, lobes oblong or oblong-ovate, apex rounded, truncate, retuse, or erosulate; filaments sparsely pilose at base, sometimes glandular hairs present near connective, anthers included, moderately woolly or glabrous. **Capsules** oblong-ovoid, 7–12 mm. **Seeds** 0.3–0.5 mm.

Subspecies 2 (2 in the flora): w United States, nw Mexico.

Orobanche parishii is distributed mainly in the southern High Sierra Nevada, Tehachapi Mountains, Central Coast, southwestern California, Desert, Inyo, and White mountains, and Channel Islands.

The hosts of *Orobanche parishii* include several herbs and shrubs of Asteraceae.

1. Corolla lobes 18–25 mm, lips 6–8 mm; calyces (7–)10–15(–18) mm, lobes subulate-attenuate; anthers moderately woolly on sutures; stigma lobes spreading, not recurved, thin; mainland range of species . . . . 12a. *Orobanche parishii* subsp. *parishii*
1. Corolla lobes 15–20(–24) mm, lips 4–6(–7) mm; calyces 9–11 mm, lobes subulate; anthers glabrous, rarely sparsely woolly; stigma lobes strongly recurved, thick; seaside plants of Channel Islands and mainland San Diego County, California . . . . 12b. *Orobanche parishii* subsp. *brachyloba*

12a. Orobanche parishii (Jepson) Heckard subsp. parishii

Plants (10–)15–22(–35) cm. **Inflorescences** 5–8 cm. **Calyces** radially symmetric, (7–)10–15 (–18) mm, lobes subulate-attenuate, (6–)9–13 mm, apex often recurved or twisted. **Corolla lobes** internally usually with maroon or reddish purple stripes or veins distally, 18–25 mm, lips 6–8 mm; anthers moderately woolly on sutures; stigma unequally 2-lobed or not, lobes spreading, not recurved, thin. $2n = 48$.

Flowering Mar–Sep. Open chaparral, scrub, pinyon-juniper woodlands, rocky soils; 200–2700 m; Calif., Nev., Oreg.; Mexico (Baja California).

Confirmed frequently reported hosts are *Artemisia dracunculus*, *Baccharis sergilioides*, *Corethrogyne filaginifolia*, and *Hazardia squarrosus* (Asteraceae). Infrequently reported hosts are *A. nova*, *A. tridentata*, *Ericameria nauseosa*, *Eriophyllum confertiflorum*, *E. lanatum*, and *Iva axillaris* (Asteraceae).

The diversity of reported hosts and range of elevation, habitat, and morphology reported for subsp. *parishii* suggests that it encompasses multiple lineages or that it comprises one or more hybrids with unstable host preferences. Some specimens cited by L. R. Heckard (1973) were previously misidentified as *Orobanche ludoviciana*.

12b. Orobanche parishii (Jepson) Heckard subsp. **brachyloba** Heckard, Madroño 22: 68, figs. 2J, 3N. 1973

Aphyllon parishii (Jepson) A. C. Schneider subsp. *brachylobum* (Heckard) A. C. Schneider

Plants (5–)7–18(–25) cm. **Inflorescences** 3–10(–15) cm. **Calyces** slightly bilaterally symmetric, 9–11 mm, lobes subulate, 7–10 mm, apex ascending. **Corolla lobes** internally usually with maroon or reddish purple blotches or relatively uniformly colored, lacking maroon or reddish purple stripes or veins distally, 15–20 (–24) mm, lips 4–6(–7) mm; anthers glabrous, rarely sparsely woolly; stigma 2-lobed, lobes strongly recurved, thick. *2n* = 96.

Flowering May–Sep. Beaches, sandy soils near coasts; 0–200 m; Calif.; Mexico (Baja California).

Subspecies *brachyloba* is found primarily on the Channel, San Nicholas, and Santa Catalina islands. It is rarely found on coastal sites of the mainland in San Diego County. The most frequently reported host is *Isocoma menziesii* (Asteraceae). Reports on *Atriplex californica* (Chenopodiaceae) and *Eriogonum latifolium* (Polygonaceae) are unconfirmed.

13. Orobanche cooperi (A. Gray) A. Heller, Cat. N. Amer. Pl., 7. 1898 • Cooper's or desert broomrape [F] [W]

Aphyllon cooperi A. Gray, Proc. Amer. Acad. Arts 20: 307. 1885; *Myzorrhiza cooperi* (A. Gray) Rydberg; *Orobanche ludoviciana* Nuttall var. *cooperi* (A. Gray) Beck

Plants simple, branched, or multiple stems from host attachment, 5–45 cm, stout, sometimes slender, base usually enlarged. **Roots** usually relatively conspicuous (often forming an irregular mass), slender, branched. **Leaves** numerous, appressed; blade ± lanceolate to broadly ovate, 6–12 mm, margins entire, apex acute or obtuse, surfaces sometimes glandular-pubescent. **Inflorescences** spikelike racemes, purple, usually branched, densely glandular-pubescent, sometimes appearing canescent; flowers numerous; bracts erect to reflexed, lanceolate to linear, 5–12 mm, apex acuminate, obtuse, or acute,

glandular-pubescent. **Pedicels** 0–30 mm, much shorter than plant axis; bracteoles 2. **Flowers:** calyx dark purple or lavender, weakly bilaterally symmetric, 8–12 mm, deeply divided into 5 lobes, lobes lanceolate-acute to -attenuate, densely glandular-pubescent; corolla 15–32 mm, tube purple or lavender, rarely white, tinged with purple, constricted above ovary, bent forward, ± glandular-pubescent; palatal folds prominent, yellow, densely pubescent; lips dark purple to lavender, with darker purple veins, abaxial lip spreading or slightly reflexed, 3–9 mm, lobes oblong-lanceolate to narrowly ovate, apex acute, often with apiculate tooth, adaxial lip erect, reflexed, or revolute, 6–10 mm, lobes ± triangular, rarely truncate, apex acute, often with apiculate teeth; filaments glabrous, anthers included, glabrous, sparsely villous, pubescent, or tomentulose along sutures, usually also with inconspicuous stalked glands (these minute, appearing peglike under magnification). **Capsules** ovoid, 6–12 mm. **Seeds** 0.2–0.5 mm.

Subspecies 3 (3 in the flora): sw, sc United States, n Mexico.

The hosts for *Orobanche cooperi* are shrubs of *Ambrosia* and *Viguiera* (Asteraceae).

L. R. Heckard and T. I. Chuang (1975) mentioned an undescribed polyploid variant (*2n* = 96) with smaller, shorter-lobed corollas and peltate, bowl-shaped stigmas occurs on the same hosts in southern California.

1. Corollas 15–18(–22) mm, lips 3–6 mm, adaxial lips reflexed or revolute; anthers glabrous or tomentulose from pollen sacs, stalked glands often few on dorsal surfaces or absent; Chihuahuan Desert . . . . 13c. *Orobanche cooperi* subsp. *palmeri*
1. Corollas (15–)18–32 mm, lips 5–10 mm, adaxial lips erect or reflexed; anthers glabrous, sparsely villous, or pubescent, stalked glands present on dorsal surfaces, sometimes obscure, rarely absent; Sonoran Desert and adjacent portions of California and Nevada.
  2. Corollas (15–)18–22 mm, adaxial lips erect, lobes with or without apiculate teeth . . . . 13a. *Orobanche cooperi* subsp. *cooperi*
  2. Corollas 22–32 mm, adaxial lips erect or reflexed, lobes with apiculate teeth . . . . 13b. *Orobanche cooperi* subsp. *latiloba*

13a. Orobanche cooperi (A. Gray) A. Heller subsp. **cooperi**

Plants simple or branched, base sometimes enlarged. **Leaves** sometimes imbricate proximally; blade broadly lanceolate, 7–11 mm, apex obtuse. **Inflorescences:** bracts 8–11(–13) mm, apex not or only slightly reflexed. **Corollas** (15–)18–22 mm; palatal folds pubescent; abaxial lip 4–7 mm, adaxial lip erect, 5–9 mm, lobes with or without apiculate tooth. **Anthers** glabrous or sparsely villous, stalked glands present on dorsal surface, sometimes obscure, rarely absent. **Stigmas** peltate, crateriform, or bilaminar, rarely 2-lobed. $2n$ = 48, 72, 96.

Flowering Jan–Apr. Sandy desert, dry washes; -50–1000 m; Ariz., Calif., Nev.; Mexico (Baja California, Sonora).

Subspecies *cooperi* is most abundant in the Sonoran Desert. A single specimen collected recently in Santa Cruz County, Arizona (*Carnahan 1365*, ARIZ, MO) is anomalous in habitat (oak woodland), elevation (1525 m), flowering time (August), and apparent host (*Artemisia*). The specimen has pale flowers but otherwise keys closest to subsp. *cooperi*.

13b. **Orobanche cooperi** (A. Gray) A. Heller subsp. **latiloba** (Munz) L. T. Collins, Phytoneuron 2015-48: 15. 2015 [F]

Orobanche ludoviciana Nuttall var. *latiloba* Munz, Bull. Torrey Bot. Club 57: 621, plate 39, fig. 18. 1931; *Aphyllon cooperi* A. Gray subsp. *latilobum* (Munz) A. C. Schneider; *O. multicaulis* Brandegee

Plants usually branched, base usually enlarged. **Leaves** rarely imbricate proximally; blade broadly lanceolate, 8–12 mm, apex obtuse. **Inflorescences:** bracts 7–11 mm, apex acuminate or obtuse, sometimes slightly reflexed. **Corollas** 22–32 mm; palatal folds densely villous; abaxial lip 5–9 mm, adaxial lip erect or reflexed, 6–10 mm, lobes with apiculate tooth. **Anthers** pubescent, stalked glands present on dorsal surface, rarely absent. **Stigmas** bilaminar-rhomboid, sometimes peltate or crateriform. $2n$ = 48, 96.

Flowering Jan–Apr. Warm sandy deserts and dry washes of Sonoran Desert; -50–1000 m; Ariz., Calif., Nev.; Mexico (Baja California, Baja California Sur, Sonora).

13c. **Orobanche cooperi** (A. Gray) A. Heller subsp. **palmeri** (Munz) L. T. Collins, Phytoneuron 2015-48: 16. 2015

Orobanche multicaulis Brandegee var. *palmeri* Munz, Bull. Torrey Bot. Club 57: 613, plate 38, fig. 2. 1931; *Aphyllon cooperi* A. Gray subsp. *palmeri* (Munz) A. C. Schneider

Plants sometimes branched, base sometimes enlarged. **Leaves** imbricate proximally; blade narrowly lanceolate to broadly ovate, 6–10 mm,

apex acute. **Inflorescences:** bracts 7–12 mm, apex acute to acuminate, strongly reflexed. **Corollas** 15–18(–22) mm; palatal folds densely villous distally; abaxial lip 3–5 mm, adaxial lip reflexed or revolute, 4–6 mm, lobes with or without apiculate tooth, sometimes truncate. **Anthers** glabrous or tomentulose along sutures, stalked glands often few on dorsal surface or absent. **Stigmas** bilaminar-rhomboid, peltate, or crateriform, rarely 2-lobed. $2n$ = 48.

Flowering Aug–May. Volcanic mountains, sand dunes, dry washes, hillsides; 100–1800 m; N.Mex., Tex.; Mexico (Chihuahua, Coahuila, Durango, Hidalgo, San Luis Potosí).

Subspecies *palmeri* is found principally in the Chihuahuan Desert from southern New Mexico and western Texas southward to central Mexico. The known hosts are a few species of *Viguiera* (Asteraceae).

Subspecies *palmeri* is sometimes confused with the Mexican species, *Orobanche dugesii* (S. Watson) Munz. An issue regarding the separation of subsp. *palmeri* from *O. dugesii*, raised by G. C. de Rzedowski (1998), was addressed by L. T. Collins and G. Yatskievych (2015), who maintained the two taxa as separate entities.

14. **Orobanche valida** Jepson, Madroño 1: 255. 1929
   • Rock Creek broomrape [E]

Aphyllon validum (Jepson) A. C. Schneider; *Orobanche ludoviciana* Nuttall var. *valida* (Jepson) Munz

Plants simple or with a few short branches, (6–)10–30 cm, relatively slender, base sometimes slightly enlarged. **Roots** inconspicuous, slender, unbranched or branched. **Leaves** few, imbricate proximally, erect; blade narrowly to broadly ovate, 5–7 mm, margins entire, apex acute to obtuse, surfaces glabrous. **Inflorescences** spikelike racemes, creamy white proximally, dark purple distally, simple, densely glandular-pubescent, sometimes also sparsely pilose proximally (subsp. *howellii*); flowers numerous; bracts ascending to recurved, narrowly lanceolate-acuminate to lanceolate-subulate, 5–10 mm, apex acuminate, densely glandular-pubescent. **Pedicels** 0–5 mm (to 10 mm proximally), much shorter than plant axis; bracteoles 2. **Flowers:** calyx dark purple externally, weakly bilaterally symmetric, (4–)6–10 (–11) mm, deeply divided into 5 lobes, lobes linear-subulate, densely glandular-pubescent; corolla 12–16 (–18) mm, tube dark purple, sometimes yellow to white abaxially, constricted above ovary, bent forward, glandular-pubescent; palatal folds prominent, pale yellow to white, glabrous or puberulent; lips dark purple, abaxial lip usually white to pale lavender with purple veins, abaxial lip spreading abruptly from base, 4–5 mm, lobes narrowly oblong to oblong-triangular, apex acute or bluntly pointed, adaxial lip erect or reflexed distally, 3–5 mm, lobes triangular, apex acute; filaments glabrous or with a few hairs at base, anthers included, glabrous or pilose. **Capsules** narrowly ovoid to oblong-ovoid, 6–9 mm. **Seeds** 0.2–0.4 mm.

Subspecies 2 (2 in the flora): California.

Orobanche valida is the rarest species in the genus in North America, occurring in two disjunct sets of populations on a few rocky mountain slopes of granite and volcanic rock in the San Gabriel and central North Coast Range mountains. It is also the most narrowly endemic species of *Orobanche*. The populations of the two subspecies are separated by several hundred kilometers. The host species are also unique when compared with other *Orobanche* species; *O. valida* is parasitic on *Garrya* (Garryaceae) and has also been reported on *Eriodictyon* (Hydrophyllaceae) and *Quercus* (Fagaceae). P. A. Munz (1930) treated *O. valida* as a variety of *O. ludoviciana*, but L. R. Heckard and L. T. Collins (1982) demonstrated its status as a species and described an additional subspecies. Because of its restricted distribution and relative rarity, *O. valida* should be considered for addition to California's list of plants of conservation concern.

SELECTED REFERENCE Heckard, L. R. and L. T. Collins. 1982. Taxonomy and distribution of *Orobanche valida* (Orobanchaceae). Madroño 29: 95–100.

1. Corollas 12–14 mm, lips and distal tubes puberulent, glabrous or sparsely puberulent at constriction and proximally; palatal folds glabrous; anthers glabrous or with a few glandular hairs near connective; filaments glabrous . . . . 14a. *Orobanche valida* subsp. *valida*
1. Corollas 14–16(–18) mm, lips and tubes densely glandular-pilose throughout; palatal folds puberulent; anthers pilose; filaments sparsely pilose at bases . . . 14b. *Orobanche valida* subsp. *howellii*

14a. Orobanche valida Jepson subsp. **valida** C E

Plants 15–30 cm, base usually not swollen. **Bracts, bracteoles, and calyces** glandular-puberulent, hairs 0.1 mm. **Corollas** 12–14 mm; palatal folds glabrous; lips and distal tube glandular-puberulent, glabrous or sparsely puberulent at constriction and proximally; filaments glabrous, anthers glabrous or with a few glandular hairs near connective. $2n = 48$.

Flowering May–Aug. Rocky slopes of loose decomposed granite; of conservation concern; 1200–2000 m; Calif.

Subspecies *valida* is restricted to a few mountain slopes in the San Gabriel and Topatopa mountains of southern California. It is parasitic on *Garrya* (Garryaceae) and is associated with chaparral shrubs including *Ceanothus* (Rhamnaceae), *Cercocarpus* (Rosaceae), *Eriogonum* (Polygonaceae), and *Quercus* (Fagaceae).

14b. Orobanche valida Jepson subsp. **howellii** Heckard & L. T. Collins, Madroño 29: 98, fig. 1A–E. 1982 • Howell's broomrape E

Aphyllon validum (Jepson) A. C. Schneider subsp. *howellii* (Heckard & L. T. Collins) A. C. Schneider

Plants (6–)10–20 cm, base often slightly enlarged. **Bracts, bracteoles, and calyces** densely glandular-pilose, hairs 0.2–0.4 mm. **Corollas** 14–16(–18) mm; palatal folds puberulent; lips and tube densely glandular-pilose throughout, hairs 0.4–0.7 mm; filaments sparsely pilose at base, anthers pilose. $2n = 48$.

Flowering Jun–Sep. Rocky volcanic, serpentine slopes, open chaparral; 1200–1700 m; Calif.

Subspecies *howellii* is restricted to Glenn, Lake, Mendocino, Napa, and Tehama counties in the northern and central Coast Ranges, geographically disjunct from subsp. *valida*. The confirmed hosts are species of *Garrya*. Reports of *Eriodictyon* and *Quercus* as hosts need confirmation. Further research on this rare subspecies is warranted to establish its range and conservation status.

15. Orobanche arizonica L. T. Collins, Phytoneuron 2015-48: 16, figs. 2, 5, 6A, 7. 2015 E

Aphyllon arizonicum (L. T. Collins) A. C. Schneider

Plants simple, rarely branched, sometimes with several stems from host attachment, 5–35 cm, relatively slender, base not enlarged, glabrous proximally, glandular-pubescent distally. **Roots** inconspicuous, slender, usually unbranched. **Leaves** numerous, sometimes imbricate distally, appressed; blade lanceolate to broadly ovate, 4–10 mm, margins entire, apex acute, surfaces glabrous. **Inflorescences** spikelike racemes, dark purple, sometimes pale lavender, sometimes branched, densely glandular-pubescent, often viscid; flowers numerous; bracts erect to recurved, narrowly oblong-lanceolate, 5–11 mm, apex acute, glandular-pubescent. **Pedicels** 0–5 mm (15 mm proximally), much shorter than plant axis; bracteoles 2. **Flowers:** calyx lavender to purple externally, weakly bilaterally symmetric, 8–12 (–14) mm, deeply divided into 5 lobes, lobes lanceolate-linear to linear-subulate, densely glandular-pubescent; corolla 15–20(–22) mm, tube white, constricted above ovary, slightly bent forward, glandular-puberulent or pubescent; palatal folds prominent, yellow, villous distally, hairs eglandular; lips dark purple, abaxial lip sometimes lavender, abaxial lip erect to slightly spreading, 3–4 mm, lobes linear, apex acute, adaxial lip erect to ± spreading, 4–6 mm, lobes triangular-acute to narrowly oblong-triangular, apex obtuse-rounded; filaments glabrous or with a few scattered hairs, anthers included, glabrous or sparsely woolly along sutures. **Capsules** ovoid, 6–10 mm. **Seeds** 0.3–0.5 mm. $2n$ = 48.

Flowering Jun–Jul. Red sands, high deserts, pinyon-juniper woodlands; 1000–3000 m; Ariz., Nev., N.Mex., Utah.

Orobanche arizonica is rather common in the high desert of northern Arizona and southern Utah. It is parasitic on *Gutierrezia sarothrae* (Asteraceae) and possibly other species of the genus.

Orobanche arizonica occurs in the Colorado Plateau and southern Great Basin Desert and contiguous areas. It is often confused with *O. ludoviciana* or is considered a small-flowered *O. cooperi*. The smaller flowers, summer flowering period, host, and ecological setting separate it from *O. cooperi* of the warm Sonoran Desert (L. T. Collins and G. Yatskievych 2015).

16. Orobanche riparia L. T. Collins, J. Bot. Res. Inst. Texas 3: 7, fig. 1A,B. 2009 • River broomrape E

Aphyllon riparium (L. T. Collins) A. C. Schneider; *Myzorrhiza riparia* (L. T. Collins) Weakley

Plants simple or few- to many-branched, 5–35 cm, stout, sometimes slender, base enlarged in robust specimens. **Roots** inconspicuous or conspicuous (often forming an amorphous mass), slender, branched or unbranched. **Leaves** numerous, appressed; blade broadly ovate to ovate-triangular, narrower distally, 6–9 mm, margins entire, apex acute or obtuse, surfaces glabrous. **Inflorescences** spikelike racemes, purple, lavender, or pallid, sometimes branched, glandular-pubescent, often ± viscid; flowers numerous; bracts usually reflexed, lanceolate, 8–15 mm, apex acute, glandular-pubescent. **Pedicels** 0–10(–12) mm, much shorter than plant axis; bracteoles 2. **Flowers:** calyx purple, rarely pale lavender externally, weakly bilaterally symmetric, 7–11(–13) mm, deeply divided into 5 lobes, lobes lanceolate-linear to linear-subulate, densely glandular-pubescent; corolla (13–)15–22 mm, tube white, distally often tinged with purple or pink, or with dark purple veins, constricted above ovary, slightly to moderately bent forward, glandular-puberulent to -pubescent; palatal folds prominent, yellow, pubescent; lips internally ± purple

or lavender, often with darker veins, abaxial lip erect to slightly spreading, 3–4 mm, lobes narrowly oblong-triangular, apex acute, adaxial lip erect to ± spreading, 4–6 mm, lobes triangular, apex acute; filaments glabrous or pubescent at base, anthers included, glabrous or sparsely pubescent. **Capsules** ovoid, 7–10 mm. **Seeds** 0.3–0.5 mm. $2n$ = 48.

Flowering Aug–Sep. Stream banks, sand bars, flood plains; 100–1500 m; Colo., D.C., Ill., Ind., Kans., Ky., Mo., Nebr., N.Mex., Ohio, Okla., Tenn., Tex., Va., W.Va.

Orobanche riparia was formerly included in *O. ludoviciana* (P. A. Munz 1930). The two species can be distinguished on the basis of morphology, habitat, host associations, and phenology. *Orobanche riparia* appears to be more closely allied to *O. cooperi* than *O. ludoviciana*. The eastern and western populations show slight differences in color intensity and degree of pubescence.

Orobanche riparia is the only species of *Orobanche* that exclusively parasitizes annual hosts, including *Ambrosia trifida*, *Dicoria canescens*, and *Xanthium strumarium* (Asteraceae), and rarely *Nicotiana tabacum* (Solanaceae) (L. T. Collins et al. 2009). Its riparian distribution, entirely on sandbars, sandy banks, and silt deposits mainly of the Mississippi, Ohio, Platte, and Rio Grande rivers and tributaries, is likewise unique. Extant populations are most abundant along the Ohio and Platte rivers but are imperiled by habitat destruction along stream banks.

17. Orobanche vallicola (Jepson) Heckard, Madroño 22: 64. 1973 • Hillside broomrape E

Orobanche comosa Hooker var. *vallicola* Jepson, Man. Fl. Pl. Calif., 952. 1925; *Aphyllon vallicola* (Jepson) A. C. Schneider; *O. californica* Chamisso & Schlechtendal var. *claremontensis* Munz

Plants simple or few-branched, 7–40 cm, stout, base usually enlarged. **Roots** usually inconspicuous, slender, branched or unbranched. **Leaves** numerous, erect; blade broadly triangular proximally, grading to lanceolate distally, 7–10 mm, margins entire, apex obtuse, acute on distal leaves, surfaces glabrous. **Inflorescences** racemes, sometimes spikelike, pink, lavender, or yellow, sometimes branched, glandular-puberulent; flowers numerous; bracts usually reflexed toward tips, narrowly triangular to lanceolate-linear or subulate, 5–10 mm, apex attenuate, glandular-puberulent. **Pedicels** 5–25 mm, much shorter than plant axis; bracteoles 2. **Flowers:** calyx pale or pinkish tinged, ± radially symmetric, (6–)9–15(–20) mm, deeply divided into 5 lobes, lobes narrowly subulate, glandular-puberulent; corolla 17–28(–33) mm, tube white to pale yellow or pale pink, slightly constricted above ovary, straight or slightly bent forward, sparsely glandular-pubescent or glabrate; palatal folds prominent, yellow, glabrous; lips white to pale pink, often with darker pink veins, abaxial lip usually widely spreading, 5–9(–10) mm, lobes narrowly oblong-triangular to lanceolate, apex acute, adaxial lip usually widely spreading, 5–9 (–10) mm, lobes triangular or triangular-ovate, apex acute; filaments glabrous, anthers included, glabrous or ± villous along sutures. **Capsules** ovoid to cylindric-ovoid, 10–13 mm. **Seeds** 0.4–0.5 mm.

Flowering (Mar–)May–Nov. Woodlands, thickets, openings, lowland valleys and foothills; 0–400 m; Calif.

Orobanche vallicola is rare but occurs occasionally in widely scattered localities in cismontane California from Trinity County south to Los Angeles County.

The most frequently reported host for *Orobanche vallicola* is *Sambucus* (Adoxaceae), but *O. vallicola* has also been reported on *Baccharis douglasii* and *Ericameria nauseosa* (Asteraceae), *Pyrus* (Rosaceae), *Quercus agrifolia* (Fagaceae), and *Symphoricarpos albus* (Caprifoliaceae) (L. R. Heckard 1973).

7. BARTSIA Linnaeus, Sp. Pl. 2: 602. 1753; Gen. Pl. ed. 5, 262. 1754, name conserved • [For Johann Bartsch, 1709–1738, German physician]

Christopher P. Randle

Simon Uribe-Convers

Herbs, perennial; hemiparasitic, caudex woody. **Stems** erect, not fleshy, pilose and eglandular at base, hirsute and glandular at apex. **Leaves** basal (scalelike) and cauline (expanded), decussate; petiole absent; blade not fleshy, subleathery or not, margins crenate to serrate. **Inflorescences** terminal, racemes; bracts present. **Pedicels** present; bracteoles absent. **Flowers:** sepals 4, calyx bilaterally symmetric, tubular, lobes triangular; petals 5, corolla violet to yellow, strongly bilabiate, funnelform, abaxial lobes 3, adaxial 2, adaxial lip galeate; stamens 4, didynamous, filaments minutely pubescent; staminode 0; ovary 2-locular, placentation axile; stigma subcapitate. **Capsules:** dehiscence loculicidal. **Seeds** ca. 50, white, fusiform-cylindric, wings present. $x = 12$.

Species 1: ne North America, Europe.

Until recently, *Bartsia* included 49 species distributed in North America, South America, Eurasia, and Africa (U. Molau 1990). However, phylogenetic analyses of nuclear and chloroplast genes, obtained using both traditional and high-throughput sequencing, indicate that *Bartsia* so circumscribed is polyphyletic (S. Uribe-Convers and D. C. Tank 2015, 2016; Uribe-Convers et al. 2016). The type species, *B. alpina*, is recovered on a branch with no other species of *Bartsia*; therefore, *Bartsia* is herein treated as monospecific following Uribe-Convers and Tank (2016). The remaining species of *Bartsia* have been placed in *Bellardia*, *Hedbergia* Molau, and a new South American genus, *Neobartsia* Uribe-Convers & Tank. Of these, only *Bellardia* occurs in the flora area.

SELECTED REFERENCES Molau, U. 1990. The genus *Bartsia* (Scrophulariaceae-Rhinanthoideae). Opera Bot. 102: 5–99. Uribe-Convers, S. and D. C. Tank. 2016. Phylogenetic revision of the genus *Bartsia* (Orobanchaceae): Disjunct distributions correlate to independent lineages. Syst. Bot. 41: 672--684.

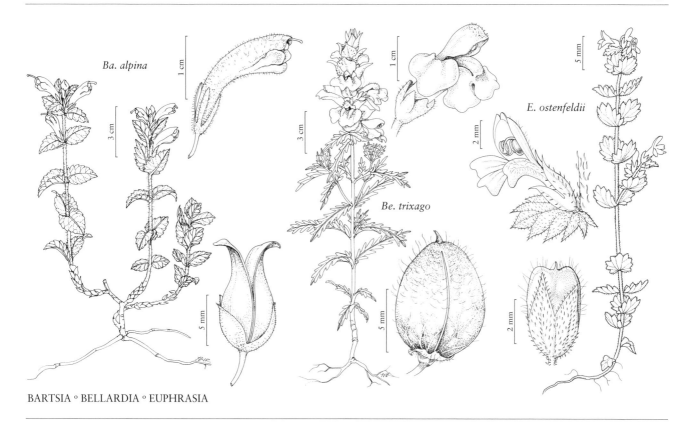

Ba. alpina

E. ostenfeldii

Be. trixago

BARTSIA ∘ BELLARDIA ∘ EUPHRASIA

1. Bartsia alpina Linnaeus, Sp. Pl. 2: 602. 1753, name conserved • Velvet bells, alpine bartsia, tornarsup-narsue, djaevelens blomster, bartsie alpine, alpenhelm ☐F

Bartsia alpina var. *jensenii* Lange; *B. alpina* var. *pallida* Wormskjold ex Lange

Perennials simple or branched, 10–30 cm. **Leaves** 4–10 pairs, divaricate; blade ovate, (5–)10–25 × (6–)9–17 mm, rugose, abaxial surface glabrescent to hirsute, adaxial glabrescent. **Inflorescences** with 2–8 pairs of flowers; bracts resembling foliage leaves, distal ones violet. **Pedicels** 2–4 mm. **Flowers** ± divaricate, 15–20 mm; calyx green, often with violet markings, 5–9 mm, divided less than ½ length, hirsute; corolla scarcely curved, pilose, galea 3–5 mm, divided from abaxial lip less than ¼ length of corolla, abaxial lip 2–3 mm; stamens included, ⅓–½ length of corolla from base, extending to abaxial lip; anthers white, equal, apex mucronate, villous; style 14–20 mm, stigma ± exserted. **Capsules** ovoid, terete, 6–10 mm. **Seeds** 1.4–2 mm. $2n = 24$.

Flowering Jun–Aug. Subarctic and arctic tundra, lakeshores, fens; 0–1000 m; Greenland; Man., Nfld. and Labr., N.W.T., Nunavut, Ont., Que.; Europe.

Bartsia alpina has been reported to parasitize a wide range of hosts, including members of Ericaceae, Fabaceae, and Poaceae (U. Molau 1990). Outside the flora area, it occurs throughout northern Europe and in disjunct populations in the Alps and Pyrenees.

Bartsia alpina has been divided into varieties based primarily on variation in the color of the corolla. The most common form has violet corollas; other populations throughout the range occasionally include individuals with pale violet to yellow corollas. U. Molau (1990) reported that these forms are likely an expression of phenotypic variation induced by infection by either a gall fly or imperfect fungus.

8. BELLARDIA   Allioni, Fl. Pedem. 1: 61.   1785   •   Lineseed [For Carlo Antonio Lodovico Bellardi, 1741–1826, professor of botany at University of Turin]

Elizabeth H. Zacharias

Bartsia Linnaeus sect. *Bellardia* (Allioni) Molau; *Parentucellia* Viviani

Herbs, annual [perennial]; hemiparasitic. **Stems** erect, not fleshy, retrorsely short-strigose or glandular-hairy. **Leaves** cauline, opposite, sometimes some subopposite or alternate; petiole absent; blade not fleshy, not leathery, margins coarsely crenate-dentate. **Inflorescences** terminal, spikelike racemes; bracts present. **Pedicels** present; bracteoles absent. **Flowers:** sepals 4, calyx radially or bilaterally symmetric, not flattened laterally, tubular or campanulate, not accrescent in fruit, lobes triangular or lanceolate; petals 5, corolla white with purple galea, yellow, or red-purple, strongly bilabiate, tubular-funnelform, abaxial lobes 3, adaxial 2, adaxial lip galeate; stamens 4, didynamous, filaments glabrous, anther mucros equal or absent; staminode 0; ovary 2-locular, placentation axile; stigma clavate, capitate, or ± 2-lobed. **Capsules:** dehiscence loculicidal. **Seeds** 150–450, white, reddish brown with age, ellipsoid or ellipsoid-oblong, wings absent. $x = 12$.

Species 48 (3 in the flora): introduced; Europe; introduced also in s South America, Asia, Africa, Atlantic Islands, Pacific Islands, s Australia.

Bellardia trixago has been included in a monospecific section of *Bartsia* in the broad sense (U. Molau 1990). Morphology and molecular phylogenetic analyses indicate a close relationship among *Bellardia*, *Parentucellia*, and New World species of *Bartsia* (A. D. Wolfe et al. 2005; J. R. Bennett and S. Mathews 2006; S. Uribe-Convers and D. C. Tank, http://2010. botanyconference.org/engine/search/index.php?func=detail&aid=685; A. Scheunert et al. 2012). Molecular phylogenetic analyses suggest that South American *Bartsia* are nested within a highly supported clade including *Bellardia* and *Parentucellia* (Scheunert et al.). An expanded *Bellardia* includes *P. latifolia* (Linnaeus) Caruel, *P. viscosa* (Linnaeus) Caruel, and the South American species of *Bartsia*. After this treatment had been completed, the author became aware of the alternative view presented by Uribe-Convers and Tank (2016), where all of the South American *Bartsia* taxa are transferred to the new genus *Neobartsia* Uribe-Convers & Tank, *Parentucellia* is recognized, and *Bellardia* includes only *B. trixago* and *B. viscosa*.

1. Corollas white with purple galea; calyx lobes unequal. . . . . 2. *Bellardia trixago*
1. Corollas red-purple or yellow; calyx lobes ± equal.
   2. Corollas red-purple; leaf blades 4–10(–12) mm. . . . . 1. *Bellardia latifolia*
   2. Corollas yellow; leaf blades (10–)20–46 mm. . . . . .3. *Bellardia viscosa*

1. Bellardia latifolia (Linnaeus) Cuatrecasas, Trab. Mus. Ci. Nat. Barcelona 12: 428. 1929 • Broadleaf glandweed or parentucellia ☐

Euphrasia latifolia Linnaeus, Sp. Pl. 2: 604. 1753; *Parentucellia latifolia* (Linnaeus) Caruel

Stems simple or with few ascending branches proximally, 6–30 cm, glandular-hairy. **Leaves** 4 or 5 pairs, ascending or divaricate, glandular-hairy; blade ovate, 4–10(–12) × 2–7 mm, margins purple, apex acute. **Spikelike racemes** 2–18 cm; flowers 1–12 pairs, interrupted proximally, dense distally, glandular-hairy; peduncle absent; bracts foliaceous, 7–9 × 6–7 mm, margins of distal bracts with 1 pair of teeth. **Pedicels** 0.5–2 mm, glabrous. **Flowers:** calyx tubular, 6–12 mm, tube 4–9.5 mm, hairy or glandular-hairy, lobes ± equal, lanceolate, 2–4 × 1.5–2 mm, herbaceous, margins entire, apex acute, hairy or glandular-hairy; corolla red-purple, 8–18 mm, glandular-hairy externally, throat with 2 yellow, ridgelike appendages, abaxial lobes spreading, adaxial projecting; stamens included, pollen sacs yellow, 1–2 mm, mucronate distally, glabrous, dehiscing longitudinally in distal ¾–⅘; style 4–6 mm, glabrous; stigma 2-lobed. **Capsules** 6–12 × 2–3 mm, glabrous. **Seeds** 0.3–0.5 mm, smooth. $2n = 48$ (Europe).

Flowering Apr–May. Pastures, roadsides, open areas; 0–100 m; introduced; Calif.; Europe; n Africa; Atlantic Islands (Canary Islands); introduced also in South America, w Asia, s Australia.

Bellardia latifolia may be found also in the Canary Islands and as an introduction to South America.

2. Bellardia trixago (Linnaeus) Allioni, Fl. Pedem. 1: 61. 1785 • Mediterranean lineseed F I

Bartsia trixago Linnaeus, Sp. Pl. 2: 602. 1753

Stems simple or with few ascending branches, (5–)10–70 cm, retrorsely short-strigose. **Leaves** 12–22 pairs, ascending or divaricate, antrorsely short-strigose, glandular-hairy; blade oblong-lanceolate or linear, (10–)14–90[–95] × 2–20 mm, margins green, apex acute. **Spikelike racemes** 1–14 cm; flowers 2–16(–25) pairs, dense, not interrupted proximally, glandular-hairy; peduncle absent; bracts foliaceous, 8–23 × 3–11 mm, margins of proximal bracts ± coarsely dentate, margins of distal bracts ± entire. **Pedicels** 1–2 mm, hairy. **Flowers:** calyx tubular, 7–9 mm, tube 4–6 mm, hairy or glandular-hairy, lobes unequal, triangular, 0.5–7 × 2–4 mm, herbaceous, margins entire, apex acute, glandular-hairy; corolla white with purple galea, 18–25(–30) mm, sparsely glandular-hairy externally, throat with 2 inflated lines between lateral and central lobes, abaxial lobes spreading, adaxial projecting; stamens included, pollen sacs yellow, 2–2.5 mm, mucronate distally, brownish villous proximally, scarcely hairy distally, dehiscing longitudinally in distal ³⁄₄–⁴⁄₅; style 14–20 mm, puberulent; stigma clavate, capitate, or ± 2-lobed. **Capsules** 7–10 × 4.5–8 mm, setose or villous, some hairs glandular. **Seeds** 0.5–1 mm, longitudinally ridged or smooth. *2n* = 24 (Spain).

Flowering Mar–Jun. Disturbed grasslands, roadsides, fields, serpentine grasslands; 0–900 m; introduced; Calif., La., Tex.; Europe; n Africa; introduced also in s South America, w Asia, s Africa, Atlantic Islands (Canary Islands), s Australia.

Although known in California since at least 1889 (*Greene s.n.*, 1889, UC), *Bellardia trixago* has appeared in the southeastern United States much more recently. First collected in Texas in 1970, *B. trixago* was documented in Louisiana in 2007, its easternmost locality in the flora area, and it seems to be moving eastward and southeastward more quickly than northward (J. R. Singhurst et al. 2012). *Bellardia trixago* rarely forms dense populations in the flora area.

3. Bellardia viscosa (Linnaeus) Fischer & C. A. Meyer, Index Seminum (St. Petersburg) 2: 4. 1836 • Yellow glandweed or parentucellia or bartsia, sticky parentucellia I

Bartsia viscosa Linnaeus, Sp. Pl. 2: 602. 1753; *Parentucellia viscosa* (Linnaeus) Caruel

Stems simple or with few ascending branches, (8–)12–50 (–100) cm, densely glandular-hairy. **Leaves** 6–30 pairs, ascending or divaricate, glandular-hairy; blade oblong-lanceolate, (10–)20–46 × 3–18 mm, margins green, apex acute. **Spikelike racemes** 3–30 cm; flowers 4–35 pairs, interrupted proximally, dense distally, glandular-hairy; peduncle absent; bracts foliaceous, 15–25 × 3–6 mm, margins crenate-dentate. **Pedicels** 1 mm, glandular-hairy. **Flowers:** calyx campanulate, 10–16 mm, tube 8–10 mm, glandular-hairy, lobes ± equal, narrowly lanceolate, 4–10 × 1.5–2 mm, herbaceous, margins entire, apex acute, glandular-hairy; corolla yellow, 20–25 mm, glandular-hairy externally, throat with 2 inflated lines between lateral and central lobes, abaxial lobes spreading, adaxial projecting; stamens included, pollen sacs yellow, 1–2 mm, glabrous proximally, villous distally around pore, dehiscing longitudinally in distal ³⁄₄–⁴⁄₅; style 10–20 mm, short-strigose; stigma capitate. **Capsules** 6–10 × 2–4 mm, long-strigose. **Seeds** 0.2–0.3 mm, smooth. *2n* = 48.

Flowering Apr–Sep. Wetland prairies, fields, pastures, beach foreshores, roadsides; 0–1500 m; introduced; B.C.; Ariz., Ark., Calif., La., Miss., Okla., Oreg., Tex., Wash.; Europe; introduced also in s South America, e Asia, n Africa, Atlantic Islands, Pacific Islands (Hawaii, New Zealand), s Australia.

9. EUPHRASIA Linnaeus, Sp. Pl. 2: 604. 1753; Gen. Pl. ed. 5, 263. 1754 • Eyebright [Greek *euphraino*, to delight, alluding to supposed improvements in vision from the application of *E. officinalis*]

Galina Gussarova

Herbs, annual [perennial]; hemiparasitic. **Stems** erect, not fleshy, retrorsely hairy. **Leaves** cauline, opposite; petiole absent or nearly so; blade sometimes fleshy, not leathery, margins crenate, serrate, dentate, or incised. **Inflorescences** terminal, subcapitate to diffuse spikes, flowers solitary in each axil; bracts present, subopposite or irregularly alternate. **Pedicels** present or absent; bracteoles absent. **Flowers:** sepals 4, calyx bilaterally symmetric, not flattened laterally, tubular, not accrescent in fruit, lobes triangular; petals 5, corolla white or cream to purple, lilac, violet, brownish purple, or yellow with violet veins (often without violet veins in *E. subarctica*) and yellow spot in throat and on abaxial lip, adaxial lip sometimes lilac, purple, or yellow, contrasting with rest of corolla, bilabiate, funnelform, abaxial lobes 3, emarginate, adaxial 2, adaxial lip cucullate; stamens 4, didynamous, filaments glabrous, anther mucros unequal; staminode 0; ovary 2-locular, placentation axile; stigma capitate. **Capsules:** dehiscence septicidal, opening in distal ½, margins ciliate, sometimes glabrous or short-ciliate (*E. salisburgensis*). **Seeds** 10–18, grayish, fusiform, wings absent. $x = 11$.

Species ca. 350 (18 in the flora): North America, South America, Eurasia, Africa (Morocco), Atlantic Islands (Azores, Iceland), Pacific Islands (Oceania), Australia.

Euphrasia is distributed throughout temperate regions of the northern and southern hemispheres, with one transtropical connection across the high mountains of Oceania. All *Euphrasia* species in the flora area belong to sect. *Euphrasia*, the largest of 15 sections in the genus (G. Gussarova et al. 2008). The section is noted for its taxonomic complexity. Major factors explaining the complex patterns of variation in sect. *Euphrasia* include interspecific hybridization, recent radiations, parasitism, polyploidization, and breeding system transitions (T. Karlsson 1974, 1986; P. F. Yeo 1978; G. C. French et al. 2008; R. A. Ennos et al. 2012). All North American species are tetraploids except, probably, *E. disjuncta*, *E. farlowii*, *E. oakesii*, *E. randii*, and *E. subarctica*. Of 18 species in the flora, nine are considered endemic. To facilitate identification of the species, the aggregate species concept is sometimes used to show relationships among the so-called more conventional, discrete taxa (Gussarova 2005). Following that concept, four aggregate species are represented in the flora area and their associated species treated here are: *E. borealis* (F. Townsend) Wettstein agg. (species 3 and 4), *E. nemorosa* agg. (species 6–10), *E. oakesii* agg. (species 13–15), and *E. minima* Jacquin ex de Candolle agg. (species 16 and 17).

The key makes no provision for hybrids, which often have intermediate features or a mixture of character states typical of their putative parental species. For a reliable identification, multiple representative plants should be collected and examined from a population to account for individual variation in key characters. The bracts referred to in the key are proximal bracts, which are most representative of the variation along the stem. The teeth of the bracts become progressively more acute up the stem as well as on any one leaf from apex to base. To characterize teeth shape, the key uses the most basal ones. Sessile glands are often present on the abaxial surface of bracts; they have no taxonomic value. Glandular (on short versus long stalks) and eglandular hairs are used in characterizing indumentum of different species.

Two infraspecific levels are used in this treatment. This is consistent with the usage in other *Euphrasia* treatments (for example, P. F. Yeo 1978). Subspecific rank is used for geographic or ecological integrity; varietal rank is used to indicate transitions between the extreme forms without any geographic or ecological distinction.

SELECTED REFERENCES Fernald, M. L. and K. M. Wiegand. 1915. The genus *Euphrasia* in North America. Rhodora 17: 181–201. Gussarova, G. et al. 2008. Molecular phylogeny and biogeography of the bipolar *Euphrasia* (Orobanchaceae): Recent radiations in an old genus. Molec. Phylogen. Evol. 48: 444–460. Sell, P. D. and P. F. Yeo. 1970. A revision of the North American species of *Euphrasia* L. (Scrophulariaceae). Bot. J. Linn. Soc. 63: 189–234.

1. Bracts oblong to lanceolate, lengths 2.5+ times widths; capsule margins eciliate or short-ciliate . . . . .18. *Euphrasia salisburgensis*
1. Bracts ovate, obovate, deltate, suborbiculate, elliptic, oval, oblong-ovate, or oblong, lengths not more than 2 times widths; capsule margins conspicuously long-ciliate.
    2. Bract surfaces hairy, hairs glandular (with stalks sometimes flexuous, 3–6-celled, 0.2–0.6 mm).
        3. Stems simple, sometimes with 1 or 2 pairs of branches; corollas 3–4 mm, white to yellow; Alberta, British Columbia, Manitoba, Northwest Territories, Saskatchewan, Yukon, Alaska, Montana . . . . 1. *Euphrasia subarctica* (in part)
        3. Stems simple or with 1–6 pairs of branches; corollas 4–5.5 mm, white; Newfoundland and Labrador, Nova Scotia, Nunavut, Quebec, Maine . . . . 2. *Euphrasia disjuncta* (in part)
    2. Bract surfaces glabrous or hairy, hairs eglandular or glandular (with stalks 1- or 2-celled, 0.1–0.2 mm).
        4. Bract surfaces hairy, hairs glandular.
            5. Inflorescences beginning at nodes (5–)7–9(–11); corollas 8–10 mm; capsules narrowly oblong, apices truncate to retuse; usually pastures, scrub, marshy places . . . . 5. *Euphrasia stricta* (in part)
            5. Inflorescences beginning at nodes 2–4; corollas 5–8(–10) mm; capsules oblong to elliptic or obovate, apices truncate, retuse, or emarginate; usually coastal.
                6. Bract teeth much longer than wide, apices acute to acuminate; corollas 6–8(–13) mm . . . . 3. *Euphrasia arctica* (in part)
                6. Bract teeth as long as or slightly longer than wide, apices subacute to acute, rarely aristate; corollas 5–8 mm . . . . 4. *Euphrasia frigida*
        4. Bract surfaces glabrous or hairy, hairs eglandular.
            7. Corollas yellow.
                8. Corollas 3–4 mm; leaf blades oblanceolate to broadly ovate; bract surfaces sparsely hirsute . . . . 1. *Euphrasia subarctica* (in part)
                8. Corollas 4–5.5 mm; leaf blades ovate to orbiculate; bract surfaces densely hirsute . . . . 12. *Euphrasia mollis*
            7. Corollas white, cream, lilac, violet, purple, or brownish purple.
                9. Bracts elliptic to ovate, bases strongly cuneate, tooth apices acute, sometimes aristate . . . . 11. *Euphrasia hudsoniana*
                9. Bracts oval, ovate to obovate, suborbiculate, deltate, or oblong-ovate, bases round, truncate, or cuneate, tooth apices obtuse, acute, or subacute, rarely aristate.
                    10. Corollas 6–11(–13) mm.
                        11. Inflorescences beginning at nodes 3–5 . . . . 3. *Euphrasia arctica* (in part)
                        11. Inflorescences beginning at nodes (5–)7–9(–11) . . . . . . 5. *Euphrasia stricta* (in part)
                    10. Corollas 2.5–7.5(–8.5) mm.
                        12. Sinuses between teeth rounded, bract tooth apices obtuse or acute; corollas purple or lilac with darker lines . . . . . . 13. *Euphrasia oakesii* (in part)
                        12. Sinuses between teeth acute, bract tooth apices obtuse, subacute, or acute, rarely aristate; corollas white, cream, lilac, violet, or brownish purple, rarely purple.
                            13. Inflorescences capitate, beginning at nodes 4 or 5; calyx lobes falcate, apices obtuse . . . . 13. *Euphrasia oakesii* (in part)
                            13. Inflorescences sparsely or densely spicate, beginning at nodes 2–12; calyx lobes straight, apices acute.
                                14. Inflorescences densely spicate, sometimes 4-angled . . . . .8. *Euphrasia tetraquetra*
                                14. Inflorescences sparsely spicate, not 4-angled.

[15. Shifted to left margin.—Ed.]

15. Inflorescences beginning at nodes 2–5; stems with 0–2 pairs of branches; arctic regions, sometimes non-arctic regions.
     16. Capsules 6–8 mm, elliptic to obovate, equal to or slightly longer than calyces, apices emarginate . . . . 16. *Euphrasia wettsteinii*
     16. Capsules 4–6 mm, oblong to narrowly elliptic, shorter than or equal to calyces, apices retuse . . . . 17. *Euphrasia vinacea*
15. Inflorescences beginning at nodes 3–12; stems with 0–7 pairs of branches; non-arctic regions or, if arctic, inflorescences usually beginning at nodes 5+.
     17. Bracts suborbiculate, broadly ovate, or oblong-ovate, surfaces ± densely hirsute, teeth shorter than or as long as wide, apices obtuse or subacute to acute.
         18. Stems to 12(–15) cm; bract teeth as long as wide . . . . 9. *Euphrasia ostenfeldii*
         18. Stems to 27(–37) cm; bract teeth shorter than wide . . . . 10. *Euphrasia suborbicularis*
     17. Bracts ovate to obovate, oblong, or oval, surfaces glabrous or hirsute, teeth longer than or as long as wide, apices obtuse to acute, sometimes aristate.
         19. Corollas 4.5–7.5(–8.5) mm.
             20. Corollas 5–7.5(–8.5) mm; bracts not glossy, surfaces glabrous or hirsute, tooth apices acute, sometimes aristate . . . . .6. *Euphrasia nemorosa*
             20. Corollas 4.5–6.5 mm; bracts glossy, surfaces glabrous, tooth apices acute . . . . 7. *Euphrasia micrantha*
         19. Corollas 3–5.5 mm.
             21. Stems branched from middle and/or distal cauline nodes, branches erect; cauline internode lengths (1–)2–5 times subtending leaves; inflorescences: proximal internode lengths 0.8–3(–4) times bracts.
                22. Stems simple, sometimes branched, branches 1 or 2 pairs; corollas 3–4 mm, white to yellow; Alberta, British Columbia, Manitoba, Northwest Territories, Saskatchewan, Yukon, Alaska, Montana . . . . . . 1. *Euphrasia subarctica* (in part)
                22. Stems simple or branched, branches 1–6 pairs; corollas 4–5.5 mm, white; Newfoundland and Labrador, Nova Scotia, Nunavut, Quebec, Maine . . . . 2. *Euphrasia disjuncta* (in part)
             21. Stems branched from basal, middle, and/or distal cauline nodes, branches ascending; cauline internode lengths 1–3(–5) times subtending leaves; inflorescences: proximal internode lengths 0.8–1.5 times bracts.
                23. Cauline internode lengths 1–3(–5) times subtending leaves; bracts 2–7 mm, abaxial surfaces setulose on veins, adaxial puberulent . . . . . . .14. *Euphrasia randii*
                23. Cauline internode lengths 1–2 times subtending leaves; bracts 2–4 mm, surfaces coarsely, densely hirsute . . . . 15. *Euphrasia farlowii*

1. **Euphrasia subarctica** Raup, Rhodora 36: 87, plate 278. 1934 • Arctic eyebright E

Euphrasia arctica Lange ex Rostrup var. *dolosa* (B. Boivin) B. Boivin; *E. disjuncta* Fernald & Wiegand var. *dolosa* B. Boivin; *E. pennellii* Callen var. *incana* Callen

Stems simple, sometimes branched, to 30 cm; branches 1 or 2 pairs, erect, from distal cauline nodes; cauline internode lengths (1–)2–5 times subtending leaves. **Leaves:** blade oblanceolate to broadly ovate, 2–6(–11) mm, margins crenate, teeth 1–3(–5) pairs, apices obtuse to subacute. **Inflorescences** sparsely spicate, not 4-angled, beginning at node 3 or 4(–6); proximal internode lengths 0.8–3(–4) times bracts; bracts green, often purplish adaxially, broader than leaves, ovate to obovate, length not more than 2 times width, 3–10 mm, base cuneate, surfaces sparsely hirsute and hairs eglandular or ± densely, sometimes sparsely, glandular-pilose and hairs glandular, stalks sometimes flexuous, 3–6-celled, 0.2–0.6 mm, teeth 2–6 pairs, longer than wide, apices subacute to acute, rarely aristate, sinuses between teeth acute. **Flowers:** calyx lobes straight, apex acute; corolla white to yellow, seldom with violet veins, 3–4 mm, lips ± equal. **Capsules** elliptic to oblong, (2.5–)5–6.5 mm, apex truncate to retuse.

Flowering summer. Sandy, gravelly, or damp grassy places, stream banks, shores, thickets, heathlands, tundra; 0–2500 m; Alta., B.C., Man., N.W.T., Sask., Yukon; Alaska, Mont.

The holotype (*Raup & Abbe 4633*, GH) fixes the name *Euphrasia subarctica* to individuals with long, glandular hairs. Eglandular individuals are sometimes found growing mixed together with typical, glandular *E. subarctica* in Alaska and northwestern Canada. Similarly, eglandular and long glandular-pubescent individuals are intermixed in populations of otherwise morphologically similar, but not identical, *E. disjuncta*, including the specimens of its paratype (*Fernald & Wiegand 6166*, CAN). Due to the clear tendency for the glandular forms to be more frequent in the west, while eglandular ones are more typical in the east, as well as other morphological differences, *E. subarctica* and *E. disjuncta* are treated as separate species here. However, it is likely that they represent a single lineage with a disjunct distribution, which could be a result of postglacial colonization from two separate refugia in western and eastern North America. Co-occurrence of glandular and eglandular states is documented in other diploids [for example, *E. picta* Wimmer in the Austrian Alps (P. F. Yeo 1978)]. Occasional individuals of *E. subarctica* with yellow corollas may be the result of introgression from *E. mollis*.

2. Euphrasia disjuncta Fernald & Wiegand, Rhodora 17: 190. 1915 • Polar eyebright, euphraise à aires disjointes E

Euphrasia arctica Lange ex Rostrup var. *disjuncta* (Fernald & Wiegand) Cronquist

Stems simple or branched, to 25 cm; branches 1–6 pairs, erect, from middle and distal cauline nodes; cauline internode lengths (2.5–)3–5 times subtending leaves. **Leaves:** blade broadly ovate, suborbiculate, or oval, 3–9(–11.5) mm, margins crenate to crenate-serrate, teeth 1–4 (or 5), apices subacute to acute. **Inflorescences** sparsely spicate, not 4-angled, beginning at node 3–8; proximal internode lengths 1–3 times bracts; bracts purplish adaxially, broader than leaves, ovate to obovate, length not more than 2 times width, 5–8 mm, base cuneate, surfaces hirsute and hairs eglandular or pubescent and hairs glandular, stalks sometimes flexuous, 3–6-celled, 0.2–0.6 mm, teeth (2 or)3 or 4(or 5) pairs, longer than wide, apices acute, sinuses between teeth acute. **Flowers:** calyx lobes straight, apex acute; corolla white, 4–5.5 mm, lips ± equal. **Capsules** oblong to elliptic, 4–5 mm, apex truncate to retuse.

Flowering mid summer–fall. Damp woods, gravel and ledges near streams, bogs; 0–600 m; Nfld. and Labr., N.S., Nunavut, Que.; Maine.

The circumscription of *Euphrasia disjuncta* follows that of E. Hultén (1941–1950, vol. 7, 1968) and P. D. Sell and P. F. Yeo (1970). According to the protologue, *E. disjuncta* was understood as having a disjunct distribution in western and eastern North America. Western populations (Alberta, Northwest Territories, and Alaska) are treated here as the chiefly glandular-pubescent *E. subarctica*; eastern populations are treated as typically eglandular *E. disjuncta*. The holotype (*Fernald, Wiegand & Darlington 6169*, GH) comprises only eglandular individuals and thus unambiguously assigns the name. Both glandular and eglandular forms are found growing mixed together, similar to the situation in *E. subarctica*; see additional comments under 1. *E. subarctica*.

3. Euphrasia arctica Lange ex Rostrup, Bot. Tidsskr. 4: 47. 1870 • Arctic eyebright

Stems simple or branched, to 25(–35) cm; branches 1–5(or 6) pairs, from proximal, middle, or distal cauline nodes; cauline internode lengths 2–4 times subtending leaves. **Leaves:** blade usually broadly ovate, 3–16 mm, margins serrate, teeth 1–6 pairs, apices acute. **Inflorescences** beginning at node 3–5; bracts green or purplish adaxially, broader than leaves, broadly ovate or deltate, length not more than 2 times width, (4–)5–9(–11) mm, surfaces glabrous or setose and hairs eglandular, sometimes pubescent and hairs glandular, stalks 1- or 2-celled, 0.1–0.2 mm, teeth 3–6 pairs, as long as or much longer than wide, apices acute to acuminate. **Flowers:** corolla white, sometimes lilac, adaxial lip lilac, 6–11(–13) mm, abaxial lip exceeding adaxial. **Capsules** oblong to elliptic or obovate, 4.5–8 mm, apex truncate to retuse.

Subspecies 5 (2 in the flora): n North America, Europe, Atlantic Islands (Iceland).

The other three subspecies of *Euphrasia arctica* occur in Europe (P. F. Yeo 1978).

1. Stem bases flexuous or decumbent; branches 1 or 2(–5) pairs, usually flexuous; bract surfaces glabrous or hairy, hairs eglandular bristles, teeth not much longer than wide; corollas 7–11(–13) mm . . . . 3a. *Euphrasia arctica* subsp. *arctica*
1. Stem bases erect; branches 1–5(or 6) pairs, usually straight; bract surfaces hairy, hairs glandular and eglandular bristles, teeth much longer than wide; corollas 6–8(–10) mm . . . . 3b. *Euphrasia arctica* subsp. *borealis*

3a. Euphrasia arctica Lange ex Rostrup subsp. **arctica**

Stems: base flexuous or decumbent; branches 1 or 2(–5) pairs, usually flexuous. **Inflorescences:** bract surfaces glabrous or hairy, hairs eglandular bristles, teeth not much longer than wide. **Flowers:** corolla 7–11(–13) mm.

Flowering early summer–fall. Grassy meadows, pastures, roadsides, coastal regions; 0–300 m; Nfld. and Labr. (Nfld.), Ont.; Europe (Faroe, Orkney, and Shetland islands).

Subspecies *arctica* is represented in continental North America by as yet only a few collections (*M. J. Oldham 29317*, TRTE; *M. J. Oldham 32017*, MICH; *M. O. Malte N943/29*, CAN).

3b. Euphrasia arctica Lange ex Rostrup subsp. **borealis** (F. Townsend) Yeo, Bot. J. Linn. Soc. 64: 358. 1971 • Boreal eyebright

Euphrasia rostkoviana Hayne forma *borealis* F. Townsend, Bot. Exch. Club Brit. Isles Rep. 1: 307. 1891; *E. ×aequalis* Callen; *E. borealis* (F. Townsend) Wettstein; *E. ×vestita* Callen

Stems: base erect; branches 1–5(or 6) pairs, usually straight. **Inflorescences:** bract surfaces hairy, hairs glandular and eglandular bristles, teeth much longer than wide. **Flowers:** corolla 6–8(–10) mm. $2n = 44$ (Europe).

Flowering early summer–fall. Meadows, pastures, roadsides, dunes, wood margins, scrub, grassy places, rocky shorelines, open, moist calcareous barrens; 0–300 m; Greenland; B.C., N.B., Nfld. and Labr., N.S., Ont., P.E.I., Que.; Maine; Europe (British Isles, Norway); Atlantic Islands (Iceland).

4. Euphrasia frigida Pugsley, J. Linn. Soc., Bot. 48: 490. 1930 • Cold-weather eyebright, euphraise arctique [E]

Stems simple or branched, to 14 cm; branches 1 or 2 pairs, from middle cauline nodes; cauline internode lengths 2–3(–6) times leaves or, in very compact forms, lengths 0.5–1 times leaves. **Leaves:** blade oblanceolate to ovate, 2–9 (–12) mm, margins crenate-dentate to serrate-dentate, teeth 1 or 2(–4) pairs, apices obtuse to acute. **Inflorescences** beginning at node 2–4; bracts green, sometimes purplish, broader than leaves, ovate to obovate, length not more than 2 times width,

(2.5–)4–11(–14) mm, surfaces sparsely to moderately hirsute and glandular-pubescent and hairs glandular, stalks 1- or 2-celled, 0.1–0.2 mm, teeth 2–5 pairs, as long as or slightly longer than wide, apices subacute to acute, rarely aristate. **Flowers:** corolla white, sometimes tinged lilac, 5–8 mm, abaxial lip exceeding adaxial. **Capsules** oblong to elliptic or obovate, 6–8 mm, apex retuse or emarginate. $2n = 44$.

Flowering summer. Wet or dry, rocky, grassy, or sandy places, usually coastal; 0–1000 m; Greenland; Nfld. and Labr., Que.

H. W. Pugsley (1933b) chose a glandular-pubescent specimen from Greenland (Ujarasguit, Godthaabs Fjord, W. Greenland, *S. Hansen s.n.*, 13 VIII 1885, C) as the lectotype of *Euphrasia frigida*, fixing application of the name *E. frigida* to the glandular-pubescent form with medium-sized corollas that occurs in Greenland, Newfoundland, and Quebec (P. D. Sell and P. F. Yeo 1970). See 16. *E. wettsteinii* for the widely distributed and circumpolar, small-flowered, eglandular form that has traditionally been treated as *E. frigida* (G. Gussarova 2005).

Euphrasia latifolia Pursh and *E. latifolia* Pursh ex Wettstein, names sometimes considered to be later homonyms of *E. latifolia* Linnaeus and that apply here, were never formally published.

5. Euphrasia stricta J. P. Wolff ex J. F. Lehmann, Prim. Lin. Fl. Herbipol., 43. 1809 • Drug eyebright, euphraise dressée [I]

Stems branched, rarely simple, to 30 cm; branches 1–5 pairs, from middle and distal cauline nodes; cauline internode length 0.5–2 times subtending leaves. **Leaves:** blade narrowly oblong to ovate-oblong, 2–9(–12) mm, margins serrate, teeth 1–3 pairs, apices subacute to aristate. **Inflorescences** beginning at node (5–)7–9(–11); bracts green, sometimes purple adaxially, broader than leaves, oblong to ovate, length not more than 2 times width, 5–10 mm, surfaces glabrous or sparsely to densely hirsute and hairs eglandular, sometimes glandular-pubescent and hairs glandular, stalks 1- or 2-celled, 0.1–0.2 mm, teeth 4 or 5 pairs, longer than wide, apices acute, sometimes aristate. **Flowers:** corolla white, usually suffused with lilac, 8–10 mm, abaxial lip exceeding adaxial. **Capsules** narrowly oblong, 4.5–6 mm, apex truncate to retuse.

Varieties 7 (2 in the flora): introduced; Europe, Asia (w Siberia).

Occurrence of *Euphrasia stricta* in Alberta is based on the report by G. C. D. Griffiths (2002); no specimen has been located. Griffiths listed *E. arctica* subsp. *borealis* as new for Alberta, but both morphological and habitat descriptions fit *E. stricta* more closely. If true, then this introduced species has crossed the continent; all earlier records are from eastern Canada and New England (S. R. Downie and J. McNeill 1988).

SELECTED REFERENCE Downie, S. R. and J. McNeill. 1988. Description and distribution of *Euphrasia stricta* in North America. Rhodora 90: 223–231.

1. Bract surfaces glabrous or sparsely to densely hirsute, hairs eglandular . . . . 5a. *Euphrasia stricta* var. *stricta*
1. Bract surfaces hairy, hairs glandular . . . . 5b. *Euphrasia stricta* var. *brevipila*

5a. Euphrasia stricta J. P. Wolff ex J. F. Lehmann var. stricta

Euphrasia condensata Jordan; *E. rigidula* Jordan

Bracts: surfaces glabrous or sparsely to densely hirsute, hairs eglandular. $2n$ = 44 (Germany).

Flowering early summer–fall. Pastures, scrub, marshy places; 0–200 m; introduced; N.B., N.S., Ont., P.E.I., Que.; Ill., Maine, Mass., Mich., Minn., N.H., N.Y., Pa., R.I., Vt., Wis.; Europe; Asia (w Siberia).

5b. Euphrasia stricta J. P. Wolff ex J. F. Lehmann var. brevipila (Burnat & Gremli) Hartl in G. Hegi et al., Ill. Fl. Mitt.-Eur. ed. 2, 6(1:5): 356. 1972

Euphrasia brevipila Burnat & Gremli in A. Gremli, Excursionsfl. Schweiz ed. 5, 329. 1885

Bracts: surfaces hairy, hairs glandular.

Flowering early summer–fall. Pastures, scrub, marshy places; 0–200 m; introduced; N.B., N.S., Ont., P.E.I., Que.; Ill., Maine, Mass., Mich., N.H., N.Y., Pa., R.I., Vt., Wis.; Europe; Asia (w Siberia).

6. Euphrasia nemorosa (Persoon) Wallroth, Annus Bot., 81. 1815 • Common eyebright, euphraise des bois

Euphrasia officinalis Linnaeus var. *nemorosa* Persoon, Syn. Pl. 2: 149. 1806; *E. canadensis* F. Townsend

Stems simple or branched, to 25(–42) cm; branches 1–7 pairs, from middle or distal cauline nodes; cauline internode lengths 2–3 times subtending leaves. **Leaves:** blade oval to oblong or ovate, 2–10 mm, margins serrate, teeth 1–5(or 6) pairs, apices subacute, sometimes aristate. **Inflorescences**

sparsely spicate, not 4-angled, beginning at node (7–)10–12; bracts green, sometimes purplish adaxially, broader than leaves, oblong to ovate, length not more than 2 times width, 5–8 mm, not glossy, margins ciliate, surfaces glabrous or hirsute and hairs eglandular, teeth 4–6 pairs, longer than wide, apices acute to long-aristate. **Flowers:** corolla white to lilac, 5–7.5(–8.5) mm, lips ± equal. **Capsules** narrowly oblong, 4–5.5 mm, apex truncate to retuse. $2n$ = 44 (England).

Flowering early summer–fall. Dry or moist, grassy, gravelly, or sandy places, pastures, roadsides, cliff ledges, wood margins, coastal or inland; 0–1300 m; introduced; St. Pierre and Miquelon; Alta., B.C., N.B., N.S., Ont., P.E.I., Que.; Alaska, Maine, Mich., Minn., N.H., Oreg., Vt., Wash.; n, c Europe.

7. Euphrasia micrantha Reichenbach, Fl. Germ. Excurs., 358. 1830 • Northern eyebright

Euphrasia gracilis (Fries) Fries

Stems simple or branched, to 18 cm; branches 1–4 pairs, from middle or distal cauline nodes; cauline internode lengths 2–3 times leaves. **Leaves:** blade ovate or elliptic, 3–10 mm, margins serrate, teeth 1–5 pairs, apices subacute to acute. **Inflorescences** sparsely spicate, not 4-angled, beginning at node 7–11; bracts green, purple, or blackish purple, broader than leaves, oblong to ovate, length not more than 2 times width, 4–8 mm, glossy, margins ciliate, surfaces glabrous and hairs eglandular, teeth 3–5 pairs, longer than wide, apices acute. **Flowers:** corolla white to violet, 4.5–6.5 mm, lips ± equal. **Capsules** oblong or elliptic, 4–5.5 mm, apex truncate to retuse. $2n$ = 44 (Iceland).

Flowering summer–fall. Grassy places, roadsides, abandoned railways and gravel pits, disturbed shorelines, wood margins; 0–1200 m; introduced; N.S., Ont.; Mass.; Europe.

8. Euphrasia tetraquetra (Brébisson) Arrondeau, Bull. Soc. Polymath. Morbihan 1862: 96. 1863 • Maritime eyebright

Euphrasia officinalis Linnaeus subvar. *tetraquetra* Brébisson, Fl. Normandie ed. 2, 183. 1849

Stems branched, sometimes simple, to 12(–18) cm; branches 1–3(–5), from middle and distal, sometimes basal, nodes; cauline internode lengths 0.8–1.5 times subtending leaves. **Leaves:** blade ± ovate, 2–8(–11) mm, margins serrate-dentate, teeth 2 or 3(–6) pairs, apices subacute to acute. **Inflorescences**

densely spicate, sometimes 4-angled, beginning at node 3–5(–7); bracts light to dark green, broader than leaves, ovate to obovate, length not more than 2 times width, (4–)5–7(–9) mm, glossy, surfaces glabrous or hairy and hairs eglandular bristles on veins, teeth 4 or 5 pairs, as long as or longer than wide, apices acute, sometimes aristate, sinuses between teeth acute. **Flowers:** calyx lobes straight, apex acute; corolla white, adaxial lip sometimes tinged lilac, 4–5 mm, abaxial lip exceeding adaxial. **Capsules** oblong, 3–6 mm, apex truncate. $2n = 44$ (England).

Flowering summer. Grassy and gravelly places, coastal; 0–200 m; introduced; N.S., Ont., Que.; Maine; n Europe.

9. **Euphrasia ostenfeldii** (Pugsley) Yeo, Bot. J. Linn. Soc. 64: 359. 1971 • Ostenfeld's eyebright [F]

Euphrasia curta (Fries) Wettstein var. *ostenfeldii* Pugsley, J. Bot. 71: 308. 1933

Stems simple or branched, to 12(–15) cm; branches 1–4(–6) pairs, from middle and distal cauline nodes; cauline internode lengths 0.5–3(–5) times subtending leaves. **Leaves:** blade suborbiculate, ovate, or oblong-ovate, 2–10(–14) mm, base round to cuneate, margins crenate to crenate-serrate, teeth 1–4(or 5) pairs, apices obtuse to subacute. **Inflorescences** sparsely spicate, not 4-angled, beginning at node 4–7(–9); proximal internode lengths 0.8–1.5 (–2.5) times bracts; bracts ± strongly flushed purple adaxially or on both surfaces, broader or as broad as leaves, suborbiculate, broadly ovate, or oblong-ovate, length not more than 2 times width, 4–8(–10) mm, base round, truncate, or broadly cuneate, surfaces ± densely hirsute and hairs eglandular, teeth 3–5 pairs, as long as wide, apices subacute to acute. **Flowers:** corolla white, adaxial lip sometimes lilac, (3.5–)4.5–6 mm, lips ± equal. **Capsules** elliptic-oblong to oblong, 4–5.5(–6) mm, apex retuse to emarginate. $2n = 44$ (Europe).

Flowering summer. Grasslands, dry or barren areas on limestone, stony and sandy places, rock ledges, near sea; 0–400 m; Nfld. and Labr., Ont., Que.; Mich., Minn.; n Europe.

10. **Euphrasia suborbicularis** P. D. Sell & Yeo, Feddes Repert. Spec. Nov. Regni Veg. 64: 203. 1962 • Roundleaf eyebright, euphraise suborbiculaire [E]

Stems branched, sometimes simple, to 27(–37) cm; branches 1–5 pairs, from distal cauline nodes; cauline internode lengths 1–3 times subtending leaves. **Leaves:** blade suborbiculate, rhombic, or obovate, 3–12 mm, margins crenate-serrate to serrate, teeth 1–4 pairs, apices obtuse to acute. **Inflorescences** sparsely spicate, not 4-angled, beginning at node 5–9; proximal internode lengths 0.8–2.5 times bracts; bracts green, broader than leaves, suborbiculate, length not more than 2 times width, 5–15 mm, surfaces hirsute and hairs eglandular, teeth 4 or 5(or 6) pairs, shorter than wide, apices obtuse or acute. **Flowers:** corolla white, adaxial lip sometimes tinged lilac, 5–7 mm, lips ± equal. **Capsules** oval, (4–)5–6.5 mm, apex retuse to emarginate.

Flowering summer. Turf and peat on calcareous barrens, rock and gravel near shorelines, sparsely wooded beach ridges; 0–200 m; Nfld. and Labr., N.S., Ont., Que.

11. **Euphrasia hudsoniana** Fernald & Wiegand, Rhodora 17: 194. 1915 • Hudson Bay eyebright [E] [F]

Stems simple or branched, to 35 cm; branches 1–5 pairs, from distal cauline nodes; cauline internode lengths (1–)2–5 times subtending leaves. **Leaves:** blade lanceolate to ovate, 5–12(–16) mm, base strongly cuneate, margins serrate to incised-serrate, teeth 1–5 pairs, apices subacute to acute. **Inflorescences** beginning at node 2–9; proximal internode lengths 1–2.5 times bracts, distal shorter than bracts; bracts green, sometimes purplish adaxially, broader than leaves, elliptic to ovate, length not more than 2 times width, 7–15(–20) mm, base strongly cuneate, surfaces ± hirsute and hairs eglandular, teeth 2–6(or 7) pairs, as long as or longer than wide, apices acute, sometimes aristate. **Flowers:** corolla white, adaxial lip tinged lilac, 5–6.5 mm, lips ± equal. **Capsules** narrowly oblong, 4–7.5 mm, apex truncate to emarginate.

Flowering summer. Gravelly, sandy, grassy, and rocky places, often calcareous, coastal or shores of streams and lakes; 0–200 m; Alta., Man., Nfld. and Labr. (Labr.), Ont., Que.; Mich., Minn.

Euphrasia hudsoniana has been reported from Alberta; the voucher specimen supporting that occurrence has not been verified by the author.

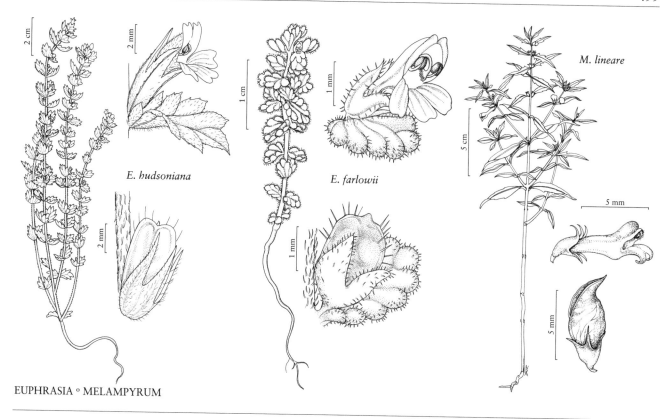

E. hudsoniana

E. farlowii

M. lineare

EUPHRASIA ∘ MELAMPYRUM

The status of var. *contracta* P. D. Sell & Yeo and var. *ramosior* P. D. Sell & Yeo is unclear. Some specimens identified as var. *contracta* represent precocious states of var. *hudsoniana*. Variety *ramosior* specimens look very different from typical var. *hudsoniana*, resembling the hybrid *Euphrasia nemorosa* × *E. hudsoniana*. But variation in their characters is consistent, particularly their minute corollas and type of pubescence, which is not typical of *Euphrasia* hybrids.

12. Euphrasia mollis (Ledebour) Wettstein, Monogr. Euphrasia, 141. 1896 • Subalpine eyebright

Euphrasia officinalis Linnaeus var. *mollis* Ledebour, Fl. Ross. 3: 263. 1849; *E. arctica* Lange ex Rostrup var. *mollis* (Ledebour) S. L. Welsh

Stems simple, rarely branched, to 16 cm; branches 1 or 2 pairs, from basal cauline nodes; cauline internode lengths (1–) 2–6 times subtending leaves. **Leaves:** blade ovate to orbiculate, 2–6 mm, margins incised-crenate, teeth 1–4(or 5) pairs, apices obtuse to subacute. **Inflorescences** beginning at node 2–5, proximal internode lengths 0.5–1.5(–2.5) times bracts; bracts green, sometimes purplish adaxially, broader than leaves, ovate to obovate, length not more than 2 times width, 4–6(–8) mm, surfaces densely hirsute and hairs eglandular, teeth 4 or 5(or 6) pairs, as long as

wide, apices acute. **Flowers:** corolla yellow, adaxial lip sometimes lilac, 4–5.5 mm, abaxial lip slightly exceeding adaxial. **Capsules** oblong to elliptic, 4–5(–6) mm, apex slightly retuse.

Flowering summer. Meadows, low swampy tundra, dry mountain ridges; 0–500 m; Alaska; e Asia (Japan, Russian Far East [Commander, Kamchatka, and Kuril islands]).

13. Euphrasia oakesii Wettstein, Monogr. Euphrasia, 142, plate 4, figs. 211–215; plate 12, fig. 6. 1896 • Oakes's eyebright, euphraise d'Oakes' E

Stems simple, sometimes branched, to 8(–12) cm; branches 1 or 2 pairs, from basal cauline nodes; cauline internode lengths 2–3 times subtending leaves. **Leaves:** blade obovate or orbiculate, 2–6(–8) mm, base cuneate, margins incised-crenate to incised-dentate, teeth 1–3(or 4) pairs, apices obtuse or acute. **Inflorescences** capitate, beginning at node 4 or 5; proximal internode lengths 0.5–1(–1.5) times bracts; bracts green or with slight bronze tinge or suffused with purple adaxially especially near margins, broader than leaves, ovate to obovate, length not more than 2 times width, 3–7 mm, base cuneate, surfaces hirsute and hairs eglandular, teeth 2–4 pairs, as long as wide, apices obtuse or acute, sinuses between teeth rounded or acute. **Flowers:** calyx lobes falcate, apex obtuse; corolla brownish purple, purple, or lilac with darker lines, abaxial side of tube and base of abaxial lip sometimes yellow, 3–4 mm, lips equal. **Capsules** obovate or broadly elliptic, 4–4.5 mm, apex emarginate.

Varieties 2 (2 in the flora): e North America.

1. Corollas purple or lilac with darker lines; bracts adaxially suffused with purple especially near margins; leaf teeth apices obtuse, sinuses between teeth rounded . . . . . . 13a. *Euphrasia oakesii* var. *oakesii*
1. Corollas brownish purple; bracts green or adaxially slightly bronze tinged; leaf teeth apices acute, sinuses between teeth acute. . . . . 13b. *Euphrasia oakesii* var. *williamsii*

13a. Euphrasia oakesii Wettstein var. **oakesii** [E]

Leaves: teeth apices obtuse, sinuses between teeth rounded. **Bracts** adaxially suffused with purple especially near margins. **Corollas** purple or lilac with darker lines.

Flowering summer. Open gravelly or rocky places; 0–1500 m; Nfld. and Labr., Que.; Maine, N.H.

13b. Euphrasia oakesii Wettstein var. **williamsii** (B. L. Robinson) Gussarova, J. Bot. Res. Inst. Texas 11: 289. 2017 [E]

Euphrasia williamsii B. L. Robinson, Rhodora 3: 272. 1901; *E. williamsii* var. *vestita* Fernald & Wiegand

Leaves: teeth apices acute, sinuses between teeth acute. **Bracts** green or adaxially slightly bronze tinged. **Corollas** brownish purple.

Flowering summer. Open gravelly or rocky places; 0–1500 m; Nfld. and Labr., Que.; Maine, N.H.

14. Euphrasia randii B. L. Robinson, Rhodora 3: 273. 1901 • Small eyebright [E]

Euphrasia purpurea Reeks ex Fernald & Wiegand [not Desfontaines] var. *randii* (B. L. Robinson) Fernald & Wiegand; *E. randii* var. *reeksii* Fernald

Stems simple or branched, to 23(–40) cm; branches 1–4(or 5) pairs, ascending, from middle and distal cauline nodes; cauline internode lengths 1–3(–5) times subtending leaves. **Leaves:** petiole 0–10(–13) mm; blade long-orbiculate to triangular-ovate or oval, 2–10(–13) mm, margins crenate to incised-crenate, teeth 1–4(–6) pairs, apices obtuse to subacute. **Inflorescences** sparsely spicate, not 4-angled, beginning at node 3–7(–10); proximal internode lengths 0.8–1.5 times bracts; bracts green or suffused with purple, as broad as leaves, ovate or oval, length not more than 2 times width, 2–7 mm, base round, abaxial surface setulose on veins, adaxial puberulent and hairs eglandular, teeth 3–5(or 6) pairs, longer than wide, apices obtuse to acute, sinuses between teeth acute. **Flowers:** calyx lobes straight, apex acute; corolla white or cream, abaxial lip lilac or purple, adaxial lilac, 2.5–4.5 mm, lips equal. **Capsules** oval to oblong or obovate, 2.5–5 mm, apex retuse to emarginate.

Flowering summer. Damp habitats near sea, sand, salt marshes, *Sphagnum* marshes, hummocks, ledges, grassy slopes, non-calcareous soils; 0–600 m; N.B., Nfld. and Labr., N.S., P.E.I., Que.; Maine.

Euphrasia purpurea Reeks ex Fernald & Wiegand is an illegitimate name that pertains here.

15. Euphrasia farlowii (B. L. Robinson) Gussarova, J. Bot. Res. Inst. Texas 11: 290. 2017 • Farlow's eyebright [E] [F]

Euphrasia randii B. L. Robinson var. *farlowii* B. L. Robinson, Rhodora 3: 274. 1901; *E. purpurea* Reeks ex Fernald & Wiegand [not Desfontaines] var. *farlowii* (B. L. Robinson) Fernald & Wiegand

Stems simple or branched, to 9(–12) cm; branches 1–3 pairs, ascending, from basal cauline nodes; cauline internode lengths 1–2 times subtending leaves. **Leaves:** blade orbiculate to triangular-ovate or oval, 2–4(–6) mm, margins crenate to incised-crenate, teeth 1–4 pairs, apices obtuse to subacute. **Inflorescences** sparsely spicate, not 4-angled, beginning at node 4–6, proximal internode lengths 1–1.5 times bracts; bracts green or suffused with purple, as broad as leaves, ovate or oval, length not more than 2 times width, 2–4 mm, base round, surfaces coarsely, densely hirsute with short-eglandular hairs, teeth 3–5 pairs, as long as wide, apices obtuse to acute, sinuses between teeth acute. **Flowers:** calyx lobes straight, apex acute; corolla white or cream, rarely purple, adaxial lip lilac or purple, 2.5–4.5 mm, lips ± equal. **Capsules** oval to oblong or obovate, 2.5–4 mm, apex retuse to emarginate.

Flowering mid summer–fall. Dry, grassy habitats on sandstone or limestone barrens, rocks, ledges, sandy beaches; 0–300 m; N.B., Nfld. and Labr. (Nfld.), N.S., Ont., P.E.I.; Maine.

Euphrasia farlowii differs from *E. randii* by its compact growth with condensed (versus elongated) cauline internodes and growth on calcareous soils versus non-calcareous soils (G. Gussarova 2017).

16. Euphrasia wettsteinii Gussarova, Bot. Zhurn. (Moscow & Leningrad) 90: 1103. 2005 • Wettstein's eyebright

Stems simple or branched, to 20 cm; branches 1 or 2 pairs, from basal cauline nodes; cauline internode lengths 1.5–4 times subtending leaves. **Leaves:** blade ovate to cuneate-obovate, 2–10 mm, margins crenate-dentate to serrate-dentate, teeth 1 or 2(–4) pairs, apices obtuse to acute. **Inflorescences** sparsely spicate, not 4-angled, beginning at node 2–5; proximal internode lengths 2–4 times subtending bracts; bracts green, sometimes purplish adaxially, broader than leaves, broadly ovate, length not more than 2 times width, (2–)4–13 mm, surfaces glabrous, moderately hirsute, or with bristles on veins and hairs eglandular, teeth 3–6 pairs, as long as or slightly longer than wide, apices subacute to acute. **Flowers:** calyx lobes straight, apex acute; corolla white, sometimes tinged lilac, 5–7 mm, lips ± equal. **Capsules** elliptic to obovate, 6–8 mm, equal to or slightly longer than calyx, apex emarginate.

Flowering summer. Tundra, grassy meadows, rocky and gravelly shores, beaches, stream flood plains; 0–200 m; Greenland; Nfld. and Labr., Nunavut, Ont., Que.; Eurasia.

Euphrasia wettsteinii applies to eglandular populations formerly included by some authors in *E. frigida*; see 4. *E. frigida* for details of the lectotypification that fixed the application of *E. frigida*.

17. Euphrasia vinacea P. D. Sell & Yeo, Feddes Repert. Spec. Nov. Regni Veg. 64: 203. 1962 • Glacier eyebright E

Stems branched, sometimes simple, to 8 cm; branches 1 pair, from basal cauline nodes; cauline internode lengths 1.5–5 times subtending leaves. **Leaves:** blade obovate to ovate or oblong, 2–6.5 mm, margins crenate to crenate-serrate, teeth 1–3 pairs, apices obtuse. **Inflorescences** sparsely spicate, not 4-angled, beginning at node 2 or 3(–5); proximal internode lengths 2–3 times

bracts, distal shorter than bracts; bracts green, often purple on veins abaxially and on surface adaxially, broader than leaves, obovate to ovate, length not more than 2 times width, 4–8 mm, surfaces sparsely glabrate or hirsute and hairs eglandular, teeth 3–5 pairs, as long as wide, apices acute, sometimes slightly aristate. **Flowers:** calyx lobes straight, apex acute; corolla white, adaxial lip sometimes lilac, 4–6 mm, lips ± equal. **Capsules** oblong to narrowly elliptic, 4–6 mm, shorter than or equal to calyx, apex retuse.

Flowering summer. Peat, gravel, clay; 0–200 m; Man., Nfld. and Labr., N.S., Ont., Que.

18. Euphrasia salisburgensis Funck ex Hoppe, Bot. Taschenb. 1794, 190. 1794

Varieties 3 (1 in the flora): Newfoundland, Europe (Ireland, nw Turkey).

18a. Euphrasia salisburgensis Funck ex Hoppe var. **hibernica** Pugsley, J. Linn. Soc., Bot. 48: 533, plate 37, figs. f, g. 1930

Stems simple or branched, to 10 cm; branches 1–6 pairs, from middle cauline nodes; cauline internode lengths 1 times subtending leaves. **Leaves:** blade narrowly oblong or narrowly oblanceolate, 2–8(–10) mm, margins sparsely dentate, teeth 1 or 2(or 3) pairs, apices acute. **Inflorescences** beginning at node 5–10; proximal internode lengths 0.8–1.5 times subtending bracts; bracts green, sometimes purplish, slightly broader than leaves, oblong to lanceolate, length 2.5+ times width, 4–8 mm, surfaces glabrous, teeth 1–3 pairs, longer or much longer than wide, apices narrowly acute, sometimes aristate. **Flowers:** corolla white, adaxial lip lilac, 5–7 mm, abaxial lip slightly exceeding or equal to adaxial. **Capsules** oblong to elliptic-oblong, 4–5 mm, margins eciliate or short-ciliate, apex truncate, retuse, or emarginate. $2n = 44$ (Ireland).

Flowering mid summer–fall. Short turf on shallow soils over limestone; 0–200 m; Nfld. and Labr. (Nfld.); w Europe (Ireland).

10. MELAMPYRUM   Linnaeus, Sp. Pl. 2: 605. 1753; Gen. Pl. ed. 5, 264. 1754, name conserved   •   Cow-wheat [Greek *melam-* (combining form of *melas* before *b* and *p*), black, and *pyros*, wheat, alluding to color of seeds]

Christopher P. Randle

Herbs, annual; hemiparasitic. **Stems** erect, not fleshy, puberulent. **Leaves** cauline, ± decussate; petiole absent or minute; blade not fleshy, not leathery, margins entire, sometimes margins of distal leaves proximally toothed. **Inflorescences** axillary, flowers 2 per axil; bracts present. **Pedicels** absent or minute; bracteoles absent. **Flowers:** sepals 4, calyx bilaterally symmetric, campanulate, lobes subulate; petals 5, corolla white, sometimes tinged with pink, strongly bilabiate, scarcely curved, abaxial lobes 3, adaxial 2, adaxial lip galeate; stamens 4, didynamous, filaments glabrous; staminode 0; ovary 2-locular, placentation axile; stigma ± capitate. **Capsules:** dehiscence loculicidal. **Seeds** 1–4, tan drying black, ellipsoid, wings absent. $x = 9$.

Species 20–35 (1 in the flora): North America, Eurasia.

Melampyrum occupies a broad range of habitats and infects a taxonomically and ecologically diverse range of hosts. No recent comprehensive taxonomic treatment of *Melampyrum* exists. Species number estimates range from 20 (K. J. Kim and S. M. Yun 2012) to 35 (E. Fischer 2004). A molecular phylogenetic study of the Rhinantheae (in the sense of Fisher) supports the monophyly of *Melampyrum*, placing it sister to the remaining Rhinantheae (J. Těšitel et al. 2010). *Melampyrum* is unusual in Orobanchaceae in that the seeds are relatively large and few per capsule. The seeds bear an elaiosome that has been linked to dispersal by ants (W. Gibson 1993).

1. **Melampyrum lineare** Desrousseaux in J. Lamarck et al., Encycl. 4: 22. 1797 • Narrowleaf cow-wheat, mélampyre linéaire E F W

Melampyrum latifolium Muhlenberg ex Britton; *M. lineare* var. *americanum* (Michaux) Beauverd; *M. lineare* var. *latifolium* (Muhlenberg ex Britton) Beauverd; *M. lineare* var. *pectinatum* (Pennell) Fernald

Annuals 5–40 cm; branches 1–3(–5) pairs, opposite, rarely with secondary branches. **Leaves:** blade linear to ovate, 13–55 × 2–22 mm, puberulent. **Inflorescences:** flower pairs 4–10; bracts resembling foliage leaves or with 1 or 2 pairs of proximal teeth. **Pedicels** 0–3(–5) mm. **Flowers:** calyx 3–5 × 2–3 mm, lobes equal to tube, often reflexed, adaxial slightly longer than abaxial; corolla 10–14 × 3–4 mm, abaxial lip with palate yellow and divided by a longitudinal groove, sinuses 1–2 mm, lanose within; anthers included in adaxial lip, hairy. **Capsules** 6–9 × 3–5 mm, ± falcate, compressed. **Seeds** 2–4 × 0.5–1 mm, testa smooth. $2n = 18$.

Flowering May–Sep; fruiting Jul–Oct. Coniferous and deciduous forests, sandy glades, gravelly terraces, heaths, rocky barrens, coastal headlands, dry meadows, peatlands, fens, roadsides; 0–2000(–3000) m; St. Pierre and Miquelon; Alta., B.C., Man., N.B., Nfld. and Labr. (Nfld.), N.S., Ont., P.E.I., Que., Sask.; Conn., Del., Ga., Idaho, Ill., Ind., Ky., Maine, Md., Mass., Mich., Minn., Mont., N.H., N.J., N.Y., N.C., Ohio, Pa., R.I., S.C., Tenn., Vt., Va., Wash., W.Va., Wis.

Melampyrum lineare has been divided into four varieties on the basis of vegetative traits: leaf shape and margin, internode length, and degree of branching (or bushiness). Without exception, circumscriptions are overlapping for these traits, none of which is strictly diagnostic. Authors are also not in agreement about which varieties to recognize. Geographic ranges of the varieties are largely confluent; much of the morphological diversity of *M. lineare* is present throughout the range.

Melampyrum lineare is an obligate parasite insofar as it does not flower or fruit without a host (J. E. Cantlon et al. 1963). It parasitizes primarily woody plants, including *Acer saccharum, Pinus banksiana, P. resinosa, P. strobus, P. sylvestris, Populus grandidentata, P. tremuloides, Quercus rubra,* and *Vaccinium angustifolium.*

11. ODONTITES Ludwig, Inst. Regn. Veg. ed. 2, 120. 1757 • [Greek *odontos*, tooth, and *-ites*, connection or association, alluding to traditional use to treat toothaches] □

Christopher P. Randle

Herbs, annual; hemiparasitic. **Stems** erect, not fleshy, villous. **Leaves** cauline, opposite; petiole absent; blade not fleshy, not leathery, margins coarsely serrate. **Inflorescences** terminal, loose, unilateral racemes; bracts present. **Pedicels** present; bracteoles absent. **Flowers:** sepals 4, calyx ± symmetric, not flattened laterally, campanulate, not accrescent in fruit, lobes deltate, subequal; petals 5, corolla purple to pink, dorsally obcompressed, strongly bilabiate, tubular to funnelform, abaxial lobes 3, adaxial 2, adaxial lip galeate; stamens 4, slightly didynamous, filaments papillose proximally, with spiral hairs at apex and club-shaped hairs at connective, anther mucros equal; staminode 0; ovary 2-locular, placentation axile; stigma capitate. **Capsules:** dehiscence loculicidal. **Seeds** 20–32, white, fusiform, wings absent. $x = 9, 10$.

Species 26 (1 in the flora): introduced; w Eurasia, n Africa.

Molecular phylogenetic analysis indicates a close relationship between *Odontites* and six other genera, of which *Bartsia*, *Bellardia*, and *Euphrasia* occur in the flora area (J. R. Bennett and S. Mathews 2006; J. Těšitel et al. 2010; S. Uribe-Convers and D. C. Tank 2016). *Odontites* has been subdivided into five genera, four of which consist of one or two geographically limited species (M. Bollinger 1996); *Odontites* in the broad sense, as treated here, is much more diverse and geographically widespread.

SELECTED REFERENCE Bollinger, M. 1996. Monographie der Gattung *Odontites* (Scrophulariaceae) sowie der verwandten Gattungen *Macrosyringion*, *Odontitella*, *Bornmuellerantha* und *Bartsiella*. Willdenowia 26: 37–168.

1. Odontites vulgaris Moench, Methodus, 439. 1794
• Red bartsia F I W

Euphrasia odontites Linnaeus, Sp. Pl. 2: 604. 1753

Stems 15–50 cm; unbranched or branched, branches 3–12 pairs, arising near base of stem, ascending. **Leaves:** blade lanceolate, 20–50 × 5–10 mm, marginal teeth 3–5, apex blunt, surfaces hispid. **Racemes:** 4–10 pairs of flowers; bracts foliaceous, 5–15 × 2–4 mm. **Pedicels** 2–3 mm. **Flowers** 8–10 mm; calyx often with purple markings, 4–6 mm, lobes less than ½ length of whole, puberulent; corolla tube light purple, glabrescent, lobes pink, villous, abaxial lobes 3–4 × 2–3 mm, adaxial lip 4–5 mm; anthers slightly exserted, yellow, equal in size and length; style exserted, 8–10 mm. **Capsules** compressed-ellipsoid, 7–8 × 3–4 mm. **Seeds** 1–2 mm, reticulate. $2n = 18, 20$.

Flowering Jun–Aug. Roadsides, fields, disturbed areas; 0–700 m; introduced; St. Pierre and Miquelon; Alta., Man., N.B., Nfld. and Labr. (Nfld.), N.S., Ont., P.E.I., Que., Sask.; Maine, Mass., Mich., N.H., N.Y., Vt., W.Va., Wis.; Eurasia.

Accurate identification of species of *Odontites* has been made difficult by a combination of complex taxonomic history, extreme variability in morphological characters across geographic ranges, and phenotypic plasticity (M. Bollinger 1996). The native range of *O. vulgaris* spans the Iberian Peninsula to Scandinavia and into Siberia. The expansion of its range in the New World is likely to continue into maritime grasslands (N. M. Hill and C. S. Blaney 2009) and elsewhere (Randolph County, West Virginia; *Grafton s.n.*, WVA).

Largely due to taxonomic confusion among species that co-occur in its native range, specimens in the New World have been identified as either *Odontites ruber* Gilibert, *O. serotinus* Dumortier, or *O. vernus* Dumortier. The subspecific names of *O. serotinus* have been synonymized with names of other species, and *O. serotinus*, an illegitimate name, was treated as a synonym of *O. vulgaris* by M. Bollinger (1996). *Odontites vulgaris* differs from *O. vernus* in ploidy level, the latter being tetraploid ($4n = 40$), and by degree of branching (Bollinger), the latter with fewer branches occurring at acute angles. Branching characters are difficult to discern in preserved material; specimens from North America appear to be *O. vulgaris*. Reported chromosome counts indicate New World *Odontites* specimens are diploid (E. H. Moss and J. G. Packer 1983); undocumented tetraploids may also exist.

Red bartsia is classified as a prohibited noxious weed in the Canadian Seeds Act and Regulations.

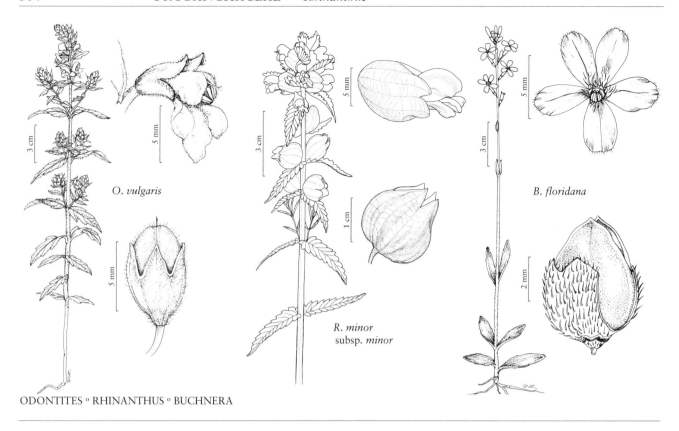

O. *vulgaris*

R. *minor*
subsp. *minor*

B. *floridana*

ODONTITES ○ RHINANTHUS ○ BUCHNERA

12. RHINANTHUS Linnaeus, Sp. Pl. 2: 603. 1753; Gen. Pl. ed. 5, 263. 1754 • Yellow rattle [Greek *rhis*, nose or snout, and *anthos*, flower, alluding to protruding adaxial lip of corolla]

Gordon C. Tucker

Bradley M. Daugherty

Herbs, annual; hemiparasitic. **Stems** erect, not fleshy, glabrous, hairy on 2 opposite sides. **Leaves** cauline, opposite; petiole present or absent; blade not fleshy to ± fleshy, not leathery, margins coarsely toothed distally. **Inflorescences** terminal, racemes, ± secund; bracts present. **Pedicels** present; bracteoles absent. **Flowers:** sepals 4, calyx bilaterally symmetric, flattened laterally, ovate to suborbiculate, accrescent in fruit, lobes deltate; petals 5, corolla yellow or yellowish [bronze to bluish], compressed, strongly bilabiate, narrowly campanulate, abaxial lobes 3, adaxial 2, adaxial lip galeate, ovate, obtuse, entire, with subapical tooth on each side; stamens 4, didynamous, filaments lanate; staminode 0; ovary 2-locular, placentation axile; stigma capitate. **Capsules:** dehiscence loculicidal longitudinally. **Seeds** 2–10[+], brown, kidney-shaped, flattened, wings present (absent in some *R. major*). $x = 11$.

Species ca. 50 (2 in the flora): North America, Europe, n Asia.

Although the taxonomy of *Rhinanthus* in the Old World is complex, the taxa found in North America are reasonably distinct. The subapical teeth on the adaxial corolla lip have been described as galea or nipples; the term teeth is used in this account. Leaf measurements are for mid-stem leaves; calyx features are for post-anthesis calyces.

The European species *Rhinanthus serotinus* (Schönheit ex Halácsy & H. Braun) Oborný has been reported from Maine (E. Hultén and M. Fries 1986); the authors have found no specimens to substantiate this report. If it were to be encountered in North America, it would key to *R. major*, from which it can be distinguished by having narrowly triangular bracts and glabrous calyces as in *R. minor*.

1. Corollas 15–20 mm; teeth of galea of corolla 1.5–2.5 mm; calyces villous, glabrescent . . .
  . . . . 1. *Rhinanthus major*
1. Corollas 13–15 mm; teeth of galea of corolla (0.4–)0.6–1 mm; calyces glabrous, scabrid, or puberulent . . . . 2. *Rhinanthus minor*

1. **Rhinanthus major** Linnaeus, Amoen. Acad. 3: 53. 1756 • European yellow rattle [I]

Rhinanthus alectorolophus (Scopoli) Pollich

Stems without black streaks, simple or branched, 15–40(–80) cm. **Leaves:** blade lanceolate to ovate, (3.5–)5–10 mm wide, margins crenate-serrate. **Bracts** rhombic-triangular, puberulent, teeth not bristle-tipped, basal teeth scarcely longer than others. **Calyces** round-ovate in fruit, 15–20 mm, initially villous, hairs white, partly glabrescent in fruit. **Corollas** 15–20 mm; tube slightly curved; throat closed; teeth of galea 1.5–2.5 mm. **Capsules** 8–12 mm. **Seeds** 2–6, 4–5 mm.

Flowering summer. Grasslands, clearings; 0–100 m; introduced; Mass., N.Y.; Europe; Asia.

The name *Rhinanthus alectorolophus* has been applied to *R. major* by some authors (for example, A. Haines 2011) on the grounds that *R. major* is an ambiguous name (R. Soó and D. A. Webb 1972). Variants with wingless seeds are found throughout the range of *R. major*.

2. **Rhinanthus minor** Linnaeus, Amoen. Acad. 3: 54. 1756 • Little yellow rattle [F][W]

Alectorolophus minor (Linnaeus) Dumortier

Stems sometimes with black streaks, simple or slightly branched, 5–20(–55) cm. **Leaves:** blade ovate-oblong to linear-lanceolate or ± oblong, (2–)5–15 mm wide, margins crenate-serrate or coarsely dentate. **Bracts** triangular, glabrous or puberulent, sometimes scabrid, teeth bristle-tipped or not, basal teeth slightly longer than others. **Calyces** 13–16 mm, glabrous, scabrid, or puberulent. **Corollas** 13–15 mm; tube straight; throat ± open; teeth of galea (0.4–)0.6–1 mm. **Capsules** 10–12 mm. **Seeds** 8–10, 4 mm.

Subspecies 4 (2 in the flora): North America, Europe.

1. Leaves: blades ovate-oblong to linear-lanceolate, margins crenate-serrate (distal leaves sometimes dentate); teeth of galea of corolla bluish or bluish gray; stems glabrous . . . .
  . . . . 2a. *Rhinanthus minor* subsp. *minor*
1. Leaves: blades ± oblong, margins coarsely dentate; teeth of galea of corolla whitish; stems hairy on 2 opposite sides . . . .
  . . . . 2b. *Rhinanthus minor* subsp. *groenlandicus*

2a. **Rhinanthus minor** Linnaeus subsp. **minor** • Petit rhinanthe [F][I]

Rhinanthus borealis (Sterneck) Chabert subsp. *kyrollae* (Chabert) Pennell; *R. crista-galli* Linnaeus var. *fallax* W. D. J. Koch ex E. F. Linton; *R. stenophyllus* (Schur) Schinz & Thellung

Stems 5–20(–55) cm, glabrous. **Leaves:** blade dark green, ovate-oblong to linear-lanceolate, margins crenate-serrate, distal leaves sometimes dentate. **Bracts** longer than to slightly shorter than calyx, glabrous, sometimes somewhat scabrid, teeth bristle-tipped. **Corollas** 13–15 mm; throat ± open; teeth of galea bluish or bluish gray, apex rounded, acute, or truncate-acute. $2n = 22$.

Flowering summer. Clearings, meadows, rocky slopes, open, grassy slopes at edges of mixed woods, roadsides, often on calcareous soils or rocks; 0–600 (–2700) m; introduced; Greenland; St. Pierre and Miquelon; Alta., B.C., Man., N.B., Nfld. and Labr., N.W.T., N.S., Ont., P.E.I., Que., Sask.; Alaska, Ariz., Colo., Conn., Idaho, Maine, Mass., Mich., Mont., N.Mex., N.Y., Oreg., Wash., Wis.; Europe.

The name *Rhinanthus crista-galli* has been misapplied to *R. minor* by numerous European and American authors. All reports of *R. crista-galli* from eastern North America appear to be based on subsp. *minor*. Reports of subsp. *minor* from North Dakota and Rhode Island are probably erroneous; confirming specimens could not be found. Populations of subsp. *minor* usually are associated with some type of anthropogenic disturbance.

2b. Rhinanthus minor Linnaeus subsp. **groenlandicus**
(Chabert) Neuman, Bot. Not. 1905: 257. 1905
　　· Arctic yellow rattle, rhinanthe du Groenland

Rhinanthus groenlandicus Chabert,
Bull. Herb. Boissier 7: 515. 1899;
Alectorolophus groenlandicus
(Chabert) Ostenfeld; *R. arcticus*
(Sterneck) Pennell; *R. borealis*
(Sterneck) Chabert; *R. minor*
subsp. *borealis* (Sterneck) Á. Löve

Stems 13–30 cm, hairy on 2
opposite sides. **Leaves:** blade
bright yellowish green, ± oblong, margins coarsely
dentate. **Bracts** longer than calyx, glabrous or
puberulent, teeth not bristle-tipped. **Corollas** 15 mm;
throat open; teeth of galea whitish, apex rounded to
blunt. $2n = 22$.

Flowering summer. Meadows, clearings, forests,
shores, slopes, sandy or rocky soils, alpine and subalpine
habitats, dry hillsides, wet banks, floodplain woods,
swamps, edges of lakes, damp roadsides, stream banks,
margins of muskeg, calcareous substrates; 0–2100 m;
Greenland; Alta., B.C., Man., N.B., Nfld. and Labr.,
N.W.T., N.S., Nunavut, Ont., Que., Yukon; Alaska,
N.H., N.Y., Oreg., Wash.; n Europe.

Reports of subsp. *groenlandicus* from Maine,
Montana, and Vermont appear to be erroneous; the
authors have seen no specimens from those states.

13. BUCHNERA Linnaeus, Sp. Pl. 2: 630. 1753; Gen. Pl. ed. 5, 278. 1754 · Bluehearts [For Andreas Elias von Büchner, 1701–1769, physician]

Bruce A. Sorrie

Herbs, biennial or perennial [annual]; hemiparasitic, caudex ca. as wide as stem, semiwoody.
Stems erect, not fleshy, glabrate or hispid-hirsute. **Leaves** basal and cauline or cauline, opposite
or subopposite; petiole absent or nearly so; blade not fleshy, not leathery, margins entire, dentate,
or crenate. **Inflorescences** terminal, spikes or spikelike racemes; bracts present. **Pedicels** absent
or present; bracteoles present. **Flowers:** sepals 5, calyx nearly radially symmetric, tubular, lobes
narrowly triangular; petals 5, corolla purple, blue-purple, blue, violet, rosy, or white, bilabiate,
salverform, pilose within, abaxial lobes 3, adaxial 2; stamens 4, subequal, filaments pilose;
staminode 0; ovary 2-locular, placentation axile; stigma short-cylindric. **Capsules:** dehiscence
loculicidal. **Seeds** ca. 250, dark brown to blackish, cylindric-hexahedral, slightly broader at one
end, wings absent. $x = 20$.

Species ca. 100 (3 in the flora): North America, Mexico, West Indies, Central America, South
America, Asia (Malesia), Africa, Australia.

In some publications, the honoree of the name *Buchnera* is given as Johann Gottfried Büchner
(1695–1749), German botanist. C. Linnaeus (1738) explicitly stated that the honoree is A. E.
von Büchner. Linnaeus omitted this information in 1753 and 1754, thus perhaps opening the
door to erroneous etymology.

SELECTED REFERENCE　Philcox, D. 1966. Revision of the New World species of *Buchnera* L. (Scrophulariaceae). Kew Bull.
18: 275–315.

1. Leaves: blades linear to linear-lanceolate, not smaller distally; calyx tubes prominently
　 10-nerved . . . . .1. *Buchnera obliqua*
1. Leaves: blades narrowly ovate, lanceolate, or narrowly oblanceolate, much smaller distally;
　 calyx tubes obscurely 10-nerved.
　　2. Leaves: larger blades narrowly ovate to lanceolate, major veins 3, minor veins (0–)2,
　　　 apex acute, margins irregularly dentate, teeth 2–3 mm; corolla lobes 5–8 mm; calyces
　　　 6–8 mm. . . . . .2. *Buchnera americana*
　　2. Leaves: larger blades narrowly oblanceolate, lanceolate, or broadly linear, major veins
　　　 1, minor veins (0–)2, apex usually obtuse or rounded, margins entire or crenate, teeth
　　　 0.5–1.5 mm; corolla lobes 2–5 mm; calyces 4.5–5.5 mm. .3. *Buchnera floridana*

1. Buchnera obliqua Bentham in A. P. de Candolle and A. L. P. P. de Candolle, Prodr. 10: 498. 1846

Buchnera arizonica (A. Gray) Pennell; *B. pilosa* Bentham var. *arizonica* A. Gray

Perennials; not blackening upon drying. **Stems** simple or branched distal to middle, 2–6 dm, spreading- or ascending-hispid, hairs pustular-based. **Leaves** not smaller distally; blade obscurely 3-nerved, linear to linear-lanceolate, 20–50 × 2–6 mm, margins entire or irregularly dentate, teeth 0.5–2 mm, apex acute, surfaces short-hispid. **Spikes:** bracts ovate-lanceolate, 5–9 mm. **Pedicels** 0–1 mm; bracteoles 5–9 mm. **Flowers:** calyx 7–9 mm, tube prominently 10-nerved, ascending-hispid, hairs pustular-based; corolla 10–15 mm, hairy externally, lobes 3–5 mm; style included, 2–3 mm. **Capsules** blackish, ovoid, 4.5–6 mm, glabrous. **Seeds** 0.6–0.8 mm.

Flowering Aug–Sep. Streamsides in oak woodlands; 1100–2000 m; Ariz.; Mexico; Central America (Guatemala, Honduras); South America (Ecuador).

Buchnera obliqua was first collected in the flora area in the Huachuca Mountains in August 1882; it was next collected in 1993, also from the Huachuca Mountains.

2. Buchnera americana Linnaeus, Sp. Pl. 2: 630. 1753 • American bluehearts E

Biennials; blackening upon drying. **Stems** simple or branched distally, 3–9 dm, spreading-hirsute proximally, appressed-hirsute or glabrous distally. **Leaves** much smaller distally; larger blade: major veins 3, minor veins (0–)2, narrowly ovate to lanceolate, 25–65 × 5–18 mm, margins irregularly dentate, teeth 2–3 mm, apex acute, surfaces short-hispid. **Spikes:** bracts ovate-lanceolate, 4–6 mm. **Pedicels** 1–1.5 mm; bracteoles 2–3 mm. **Flowers:** calyx 6–8 mm, tube obscurely 10-nerved, ascending- to appressed-hispid, hairs often pustular-based; corolla 15–21 mm, glabrate externally, lobes 5–8 mm; style included, 1–2 mm. **Capsules** blackish, ovoid, 6–8 mm, glabrate. **Seeds** 0.6–0.8 mm. $2n = 40$.

Flowering May–Oct. Moist to dry prairies, prairie openings, barrens, glades, pine savannas, interdune pannes; 20–400 m; Ont.; Ala., Ark., Del., D.C., Fla., Ga., Ill., Ind., Kans., Ky., La., Md., Mich., Miss., Mo., N.J., N.Y., N.C., Ohio, Okla., Pa., S.C., Tenn., Tex., Va.

Buchnera americana has declined significantly in the past century and now is of conservation concern in most states east of the Mississippi River and in Ontario; its current stronghold is in Kansas, Missouri, and Oklahoma. In greenhouse studies, *B. americana* and *B. floridana* parasitized a variety of grass and tree species (*Celtis, Fraxinus, Liquidambar, Liriodendron, Nyssa, Paspalum, Pinus, Quercus*); natural hosts remain largely undocumented (L. J. Musselman and W. F. Mann 1977, 1978). It is nearly restricted to older geological regions away from the coastal plain, primarily in circumneutral to high pH soils; there are records from eastern Texas-central Louisiana, southeastern Louisiana-southern Mississippi, and a few records from northwestern Florida, all apparently in acidic soils. There appears to be no morphological intergradation with *B. floridana* in those areas, and the occurrence of *B. americana* there is puzzling.

3. Buchnera floridana Gandoger, Bull. Soc. Bot. France 66: 217. 1919 • Florida bluehearts F

Buchnera breviflora Pennell

Biennials; blackening upon drying. **Stems** simple or branched distally, 2–8 dm, appressed-hirsute or spreading-hirsute proximally, appressed-hirsute or glabrous distally. **Leaves** much smaller distally; larger blade: major vein 1, minor veins (0–)2, narrowly oblanceolate, lanceolate, or broadly linear, 25–60 × 3–12 mm, margins entire or crenate, teeth 0.5–1.5 mm, apex obtuse or rounded, sometimes acute, surfaces glabrate or appressed-pilose. **Spikes:** bracts ovate-lanceolate, 1.5–4 mm. **Pedicels** 0.5–1 mm; bracteoles 1–2 mm. **Flowers:** calyx 4.5–5.5 mm, tube obscurely 10-nerved, ascending- to appressed-hispid, hairs often pustular-based; corolla 10–16 mm, glabrate externally, lobes 2–5 mm; style included, 1–2 mm. **Capsules** blackish, ovoid, 4.5–6.5 mm, glabrate. **Seeds** 0.5–0.7 mm.

Flowering May–Oct. Moist pine savannas, flatwoods, streamhead ecotones, seepage slopes, pitcher-plant bogs; 0–150 m; Ala., Ark., Fla., Ga., La., Miss., N.C., S.C., Tex.; Mexico (Yucatán); West Indies (Bahamas, Cuba, Dominican Republic, Jamaica); Central America (Belize).

Buchnera floridana is mostly restricted to the Coastal Plain, occurring primarily in fire-maintained, pine-graminoid ecosystems in strongly acidic soils. Collections also have been made from higher pH soils of chalk prairies in Alabama and Mississippi and from limestone substrates in Florida and Texas; in the Caribbean, it regularly occurs in limestone soils. The situation in Texas is unusual in that *B. floridana* occurs far inland as well as on the Coastal Plain; most Texas specimens of *Buchnera* belong to *B. floridana*; *B. americana* is known from counties along the Red River and in the pinelands of eastern Texas.

Some specimens of *Buchnera floridana* from southern Florida and the Keys have slender leaves and reduced indument on the calyx, which are typical characteristics of *B. longifolia* Kunth (= *B. elongata* Swartz; both names have been applied to Florida plants) of the Caribbean, Central America, and South America; they lack the glabrous corollas, long calyces, and large inflorescence bracts of *B. longifolia*. The presence of one or more *B. longifolia* characteristics in southern Florida plants may be the result of past contact between the two species; there is no evidence that the two come into contact in Florida at present.

14. **STRIGA** Loureiro, Fl. Cochinch. 1: 22. 1790 • Witchweed [Latin *strigosus*, slender, alluding to habit] ☐

Kamal I. Mohamed

Lytton J. Musselman

Herbs, annual; chlorophyllous or achlorophyllous, hemiparasitic or holoparasitic, haustoria either single and relatively large, or multiple, smaller, and formed on secondary roots. **Stems** erect, sometimes fleshy, hispid, puberulent, or glabrous. **Leaves** cauline, opposite, subopposite, or alternate; petiole absent; blade not fleshy, not leathery, margins entire. **Inflorescences** terminal, racemes or spikes; bracts present. **Pedicels** present or absent; bracteoles present. **Flowers:** sepals 5(–8), calyx radially or bilaterally symmetric, tubular, lobes lanceolate or subulate; petals 5, corolla red, brownish red, or purple, rarely white or yellow, bilabiate, salverform, abaxial lobes 3, adaxial 2; stamens 4, didynamous, filaments glabrous; staminode 0; ovary 2-locular, placentation axile; stigma capitate. **Capsules:** dehiscence loculicidal. **Seeds** 400–600, brown or black, ovoid, wings absent.

Species ca. 40 (2 in the flora): introduced; s Asia, Africa, Australia.

Striga produces leaves of different sizes; typical proximal leaves are scalelike, and mid-stem leaves are larger. *Striga* is distinguished from its close relative *Buchnera* by its bilabiate corolla with an abruptly bent tube, one pollen sac, and glabrous filaments. *Buchnera* has a bilabiate corolla with a straight or slightly curved tube, two pollen sacs, and pilose filaments. *Striga* has been divided into three sections based on the number of ribs on the calyx tube (R. Wettstein 1891–1893): sect. *Pentapleurae* Wettstein with five, sect. *Polypleurae* Wettstein with ten, and sect. *Tetrasepalum* Engler with 15.

Thirty-four species and subspecies of witchweeds occur in Africa; 22 are endemic (K. I. Mohamed et al. 2001). All *Striga* species parasitize hosts in the Poaceae except *S. gesnerioides*, which grows on hosts in Acanthaceae, Convolvulaceae, Euphorbiaceae, Fabaceae, and Solanaceae. *Striga asiatica*, *S. aspera* Bentham, *S. forbesii* Bentham, *S. gesnerioides*, and *S. hermonthica* (Delile) Bentham are of economic importance. Crops most affected by *Striga* include *Digitaria exilis* (fonio), *Oryza* subspp. (upland rice), *Pennisetum glaucum* (bulrush millet), *Sorghum vulgare* (sorghum), and *Zea mays* (maize). *Striga gesnerioides* is a serious pest on *Vigna unguiculata* (cowpea, Fabaceae) and a minor pest on other dicot crops. All species of witchweed are listed as noxious weeds by the United States Department of Agriculture and 11 state governments. New infestations of quarantine pests in the United States, such as witchweeds, should be reported to the State Plant Health Director in the appropriate state (http://www.aphis.usda.gov/services/report_pest_disease/report_pest_disease.shtml).

1. Calyx ribs 10; mid-stem leaf blades ascending or spreading, linear or narrowly elliptic, 20–50 mm; bracts linear, longer than calyces; corollas red, rarely yellow, with yellow throats; parasitic on Poaceae . . . . .1. *Striga asiatica*
1. Calyx ribs 5; mid-stem leaf blades appressed, lanceolate, 3–7 mm; bracts lanceolate, shorter than calyces; corollas brownish red or purple, rarely white; parasitic on dicots . . . . 2. *Striga gesnerioides*

1. **Striga asiatica** (Linnaeus) Kuntze, Revis. Gen. Pl. 2: 466. 1891 F I W

Buchnera asiatica Linnaeus, Sp. Pl. 2: 630. 1753

Annuals 15–35 cm; hemiparasitic. **Taproots** slender, not fleshy; secondary roots present; haustoria multiple, small, globular. **Flowering stems** drying green, simple or branched medially, obtusely square, not fleshy, antrorsely scabrous-hispid. **Leaves:** basal blade opposite, lanceolate scales, mid-stem blade alternate, ascending or spreading, linear or narrowly elliptic, 20–50 × 2–4 mm, surfaces scabrous-hispid. **Inflorescences** racemes, lax; flowers alternate; bracts linear, 20–35 × 1–2 mm, longer than calyx, surfaces hispid. **Flowers:** sepals 5(–8), 7–8 mm; calyx: tube 5 mm, ribs 10, hispid, teeth 5 and equal or 6–8 and subequal, lanceolate or sublanceolate, 2–3 mm; corolla red, rarely yellow, throat yellow, glandular-pubescent, tube bent, expanded distally, 15 mm, abaxial lobes 12 mm wide, adaxial 5 mm wide; style 7–10 mm, rolled in when dry, sparsely hairy or glabrous. **Capsules** oblong, 5–7 × 2–3 mm. $2n$ = 24, 40 (Nigeria).

Flowering Jul–Aug. Parasitic on Poaceae; chiefly in maize fields; 20–90 m; introduced; N.C., S.C.; Africa.

Striga asiatica was discovered in southern North Carolina and adjacent South Carolina in 1956 (H. R. Garriss and J. C. Wells 1956). At one time, 38 counties in the Carolinas were infested (R. E. Eplee 1981). Through a comprehensive control program and quarantine measures, the original infested area has been reduced by 99%, with only about 2135 acres remaining in 2009 in five counties (Bladen, Cumberland, Pender, Robeson, and Sampson) in North Carolina and no acres infested in South Carolina (R. Iverson et al. 2011). Additional information on *S. asiatica*, maps of infected areas, and the eradication program in the Carolinas can be found in Iverson et al. The authors treat *S. asiatica* in the narrow sense; reports of *S. asiatica* from southern Asia require further study. They may refer to *S. lutea* Loureiro or *S. hirsuta* Bentham, species included by some (http://www.cabi.org/isc/datasheet/51786) in *S. asiatica*.

Use of the name *Striga lutea* by C. J. Saldanha (1963) and C. E. Smith (1966) for plants in the flora area is a misapplication. Saldanha rejected *S. asiatica*, believing that its Linnaean basionym could not be lectotypified; see F. N. Hepper (1974) for a counterargument and lectotypification.

Studies on climatic requirements and potential for spread indicate that *Striga asiatica* could invade new areas as a result of global warming (D. T. Patterson et al. 1982; K. I. Mohamed et al. 2006, 2007).

2. **Striga gesnerioides** (Willdenow) Vatke, Oesterr. Bot. Z. 25: 11. 1875 • Cowpea or tobacco witchweed F I W

Buchnera gesnerioides Willdenow, Sp. Pl. 3: 338. 1800

Annuals 15–30 cm; holoparasitic. **Taproots** stout, fleshy; secondary roots absent; haustoria single, large, globular. **Flowering stems** drying black or brown, simple or branched from base, obtusely square or terete, fleshy, puberulent, pilose, or glabrous. **Leaves:** blade opposite, appressed, lanceolate, scalelike, 3–7 × 2 mm, surfaces puberulent. **Inflorescences** spikes, lax, sometimes congested; flowers opposite, rarely alternate; bracts lanceolate, 5–7 × 1–2 mm, shorter than calyx, surfaces glabrous or puberulent. **Flowers:** sepals 5, 4–7 mm; calyx: tube 3–5 mm, ribs 5, scarious between ribs, teeth 5, unequal, subulate, 1–2 mm, adaxial shorter; corolla brownish red or purple, rarely white, sparsely pubescent or glabrous, tube bent, dilated distally above calyx, 8–12 mm, abaxial lobes 6 mm wide, adaxial 3–4 mm wide; style 5 mm, curved out, glabrous. **Capsules** oblong or ovoid, 4–5 × 2–3 mm. $2n$ = 40 (Nigeria).

Flowering Aug–Oct. Parasitic on dicots; heavily disturbed phosphate mines, mine reclamation sites; 0–50 m; introduced; Fla.; Asia (Arabian Peninsula); Africa.

Worldwide, host-specific strains of *Striga gesnerioides* occur on *Lepidagathis* Willdenow (Acanthaceae), *Ipomoea* and *Merremia* (Convolvulaceae), *Euphorbia* (Euphorbiaceae), *Indigofera*, *Tephrosia*, and *Vigna* (Fabaceae), and *Nicotiana* (Solanaceae) (K. I. Mohamed et al. 2001); some of these may be potential hosts for this species in the flora area.

Striga gesnerioides was discovered in Florida in 1979 as a parasite on *Indigofera hirsuta* (hairy indigo), an African species planted for phosphate mine reclamation. *Alysicarpus ovalifolius* (alyce clover) is also attacked in the field but with much less frequency (L. Herbaugh et al. 1980; L. J. Musselman et al. 1980). *Striga gesnerioides* is known from Citrus, Hillsborough, Lake, Orange, Polk, Seminole, and Volusia counties (http://www.plantatlas.usf.edu/). Greenhouse experiments conducted

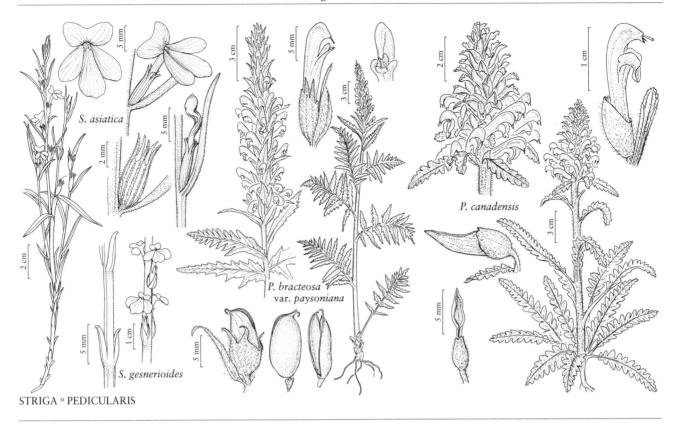

S. asiatica

S. gesnerioides

P. bracteosa
var. paysoniana

P. canadensis

STRIGA ° PEDICULARIS

in Florida on over 125 known potential hosts of *S. gesnerioides* (Herbaugh et al.) showed that it poses little threat to American agriculture (Musselman et al.). In addition to hairy indigo and alyce clover, the only other hosts reported in the Florida study were *Jacquemontia tamnifolia*, *Helianthus annuus* (sunflower), and *Ipomoea batatas* (sweet potato) (Herbaugh et al.). In an experiment conducted by Musselman and C. Parker (1981) on more than 30 potential hosts, the American strain of *S. gesnerioides* failed to grow on any of them except hairy indigo, showing the strict specificity of this strain.

15. PEDICULARIS Linnaeus, Sp. Pl. 2: 607. 1753; Gen. Pl. ed. 5, 266. 1754

• Lousewort, fernflower [Latin *pediculus*, louse, alluding to belief that livestock feeding on *P. palustris* developed lice]

Bruce W. Robart

Elephantella Rydberg; *Pediculariopsis* Á. Löve & D. Löve

Herbs, perennial [annual]; hemiparasitic, rhizomatous, caudex woody or fleshy, scaly. **Stems** erect, fleshy, glabrous, hispid, villous, or woolly. **Leaves:** basal rosette [absent], petiole present, blade not fleshy, not leathery, margins entire, serrate, or crenate; cauline alternate, rarely whorled [nearly opposite], sometimes absent, petiole present or absent, blade chartaceous, margins entire, serrate, or crenate. **Inflorescences** terminal or axillary, racemes; bracts present. **Pedicels** present; bracteoles absent. **Flowers:** sepals 2 or [4]5, calyx bilaterally symmetric, campanulate or tubular, lobes triangular, spatulate, or filiform; petals 5, corolla pink, purple, red, yellow, or white, strongly bilabiate, cylindric to funnelform, abaxial lobes 3, adaxial 2, adaxial lip galeate, enclosing anthers and style, beaked or beakless, margins entire medially and distally, with 1 set of teeth medially and entire distally, with 1 set of teeth medially and distally, or entire medially and with 1 set of teeth distally; stamens 4, didynamous, filaments glabrous or hairy, pollen sacs equal; staminode 0; ovary 2-locular, placentation axile; stigma capitate. **Capsules:** dehiscence loculicidal. **Seeds** 5–100, dark gray, brown, or tan, ovoid, wings absent. $x = 8$.

Species 400–600 (37 in the flora): North America, Mexico, South America, Europe, s, c Asia, Atlantic Islands (Iceland).

E. Hultén (1968) listed three *Pedicularis* taxa in his *Flora of Alaska and Neighboring Territories* that do not occur in the flora area; *P. amoena* Adams ex Steven, *P. kanei* Durand subsp. *adamsii* (Hultén) Hultén, and *P. villosa* Ledebour ex Sprengel are found on the Chukchi Peninsula of Siberia.

The most common chromosome number for *Pedicularis* is $2n = 16$. Two species are reported to have a diploid number of $2n = 12$, and at least five species have a number of $2n = 14$. See reviews in Jie C. et al. (2004) and M. I. S. Saggoo and D. K. Srivastava (2009).

Infrageneric classification of *Pedicularis* is both difficult and incomplete. The author has not attempted to present one for the North American taxa. Inconsistencies result from using either floral morphology or phyllotaxy to construct major subgeneric groups (C. Steven 1823; G. Bentham 1835; C. J. Maximowicz 1888; D. Prain 1890; G. Bonati 1918; Tsoong P. C. 1955); the latter author proposed 130 series to accommodate the worldwide variation in *Pedicularis*. Recognizing that floral form probably evolved in parallel among groups, Li H. L. (1948) employed phyllotaxy to erect three informal subgeneric taxa (greges) for the *Pedicularis* of China, including grex Allophyllum, species with alternate leaves; grex Cyclophyllum, species with opposite or verticillate arrangement of leaves; and grex Poecilophyllum, species with alternate to subopposite leaves on the same plant. However, molecular phylogenies (Yang F. S. et al. 2003; R. H. Ree 2005; B. W. Robart et al. 2015; Yu W. B. et al. 2015) indicate inconsistencies even in the treatment by Li, although the study by Yu et al., which included 257 species, did show that grex Cyclophyllum is monophyletic.

Pediculariopsis was described to account for species (for example, *Pedicularis verticillata*) that exhibit a different base chromosome number ($x = 6$ versus $x = 8$). That concept has not been accepted by others, because there is little morphological support.

The use of infraspecific ranks in *Pedicularis* follows T. F. Stuessy (1990), who argued that the subspecific rank is appropriate when several obvious morphological differences are associated with an allopatric or peripatric geographical pattern, whereas the varietal rank is appropriate when one to few differences are associated with more or less geographic overlap. Unfortunately, *Pedicularis* specialists have not used these levels consistently. The subspecies rank is used here, except in *P. bracteosa*, *P. centranthera*, and *P. contorta*.

Pedicularis is a common element of arctic and alpine habitats in the Northern Hemisphere, with the greatest concentration of species occurring in Asia.

SELECTED REFERENCES Bonati, G. 1918. Le Genre *Pedicularis* L. Morphologie, Classification, Distribution Géographique, Èvolution et Hybridation. Nancy. Ree, R. H. 2005. Phylogeny and the evolution of floral diversity in *Pedicularis* (Orobanchaceae). Int. J. Pl. Sci. 166: 595–613. Robart, B. W. et al. 2015. Phylogeny and biogeography of North American and Asian *Pedicularis* (Orobanchaceae). Syst. Bot. 40: 229–258. Steven, C. 1823. Monographia *Pedicularis*. Mém. Soc. Imp. Naturalistes Moscou 6: 1–60. Tsoong, P. C. 1955. A new system for the genus *Pedicularis*. Acta Phytotax. Sin. 4: 71–147. Tsoong, P. C. 1956. A new system for the genus *Pedicularis* (continued). Acta Phytotax. Sin. 5: 41–73. Yu, W. B. et al. 2015. Towards a comprehensive phylogeny of the large temperate genus *Pedicularis* (Orobanchaceae), with an emphasis on species from the Himalaya-Hengduan Mountains. B. M. C. Pl. Biol. 15: 176.

1. Racemes verticillate; cauline leaves whorled.
   2. Galeas beaked; basal leaf blades 15–40 mm . . . . 8. *Pedicularis chamissonis*
   2. Galeas beakless; basal leaf blades 10–20 mm . . . . 37. *Pedicularis verticillata*
1. Racemes simple or paniculate; cauline leaves alternate.

[3. Shifted to left margin.—Ed.]
3. Calyx lobes 2(–4).
    4. Galeas beaked.
        5. Beaks sickle-shaped. . . . . 32. *Pedicularis racemosa*
        5. Beaks straight.
            6. Calyx lobe apices distally serrate . . . . .21. *Pedicularis lanceolata*
            6. Calyx lobe apices distally entire . . . . 23. *Pedicularis lapponica*
    4. Galeas beakless.
        7. Galea margins 1-toothed medially, entire distally . . . . 28. *Pedicularis parviflora*
        7. Galea margins entire or 1-toothed medially, 1-toothed distally.
            8. Galea margins 1-toothed medially.
                9. Galeas 6.5–9 mm. . . . . 26. *Pedicularis palustris*
                9. Galeas 3–6.5 mm. . . . . 29. *Pedicularis pennellii*
            8. Galea margins entire medially.
                10. Racemes paniculate, or buds present in cauline leaf axils.
                    11. Basal leaves 0; cauline leaf blades undivided . . . . 1. *Pedicularis angustifolia*
                    11. Basal leaves 2 or 3; cauline leaf blades undivided or 1- or 2-pinnatifid
                         . . . . .19. *Pedicularis labradorica*
                10. Racemes simple.
                    12. Cauline leaf blades 1-pinnatifid, lobe margins 1- or 2-serrate . . . 5. *Pedicularis canadensis*
                    12. Cauline leaf blades undivided, lobe margins 2-crenate. . . . . 10. *Pedicularis crenulata*
3. Calyx lobes 5.
    13. Galeas beaked.
        14. Galea beaks coiled.
            15. Galea beak apices surrounded by abaxial lips . . . . 9. *Pedicularis contorta*
            15. Galea beak apices not surrounded by abaxial lips.
                16. Galea beaks abruptly upcurved; beaks 3–6 mm. . . . . .2. *Pedicularis attollens*
                16. Galea beaks gradually upcurved; beaks 5–18 mm . . . . .16. *Pedicularis groenlandica*
        14. Galea beaks straight.
            17. Beaks 0.8–2.5 mm; bracts undivided.
                18. Corolla tubes and galeas yellow. . . . . 4. *Pedicularis bracteosa* (in part)
                18. Corolla tubes and galeas white . . . . 18. *Pedicularis howellii*
            17. Beaks 2–8 mm; bracts undivided or ± lobed.
                19. Galeas 4–6.5 mm. . . . . .25. *Pedicularis ornithorhynchos*
                19. Galeas 7–10 mm . . . . .27. *Pedicularis parryi*
    13. Galeas beakless.
        20. Galea margins entire medially, 1-toothed distally.
            21. Calyces glabrous.
                22. Bracts 2-pinnatifid to midrib . . . . .36. *Pedicularis sylvatica*
                22. Bracts undivided proximally, 1-pinnatifid ½ to midrib distally, or undivided
                    with or without long auricles, or 1-pinnatifid.
                    23. Pedicels 2.5–5 mm; corolla tubes 11–13 mm. . . . . . .22. *Pedicularis langsdorffii* (in part)
                    23. Pedicels 1–2.5 mm; corolla tubes 9–11 mm. . . . . 35. *Pedicularis sudetica* (in part)
            21. Calyces ± tomentose, hispid-glandular, hispid to hirsute, white- or yellowish
              white-lanate, sparsely pilose, or ± woolly.
                24. Corolla tubes yellow to light or greenish yellow.
                    25. Corollas 14–19 mm. . . . . .15. *Pedicularis furbishiae*
                    25. Corollas 22–30 mm. . . . . .30. *Pedicularis procera* (in part)
                24. Corolla tubes red, violet-red, lavender, purple, magenta, or pink, sometimes
                white.
                  26. Basal leaf blades 150–250 mm . . . . .30. *Pedicularis procera* (in part)
                  26. Basal leaf blades 5–110 mm.

27. Basal leaves: margins of adjacent lobes nonoverlapping to extensively overlapping.
    28. Bracts undivided or 1- or 2-auricled, sometimes 1-pinnatifid . . . . 11. *Pedicularis cystopteridifolia*
    28. Bracts 2-pinnatifid, not auricled .31. *Pedicularis pulchella*
27. Basal leaves: margins of adjacent lobes nonoverlapping or slightly overlapping distally.
    29. Corollas 11–19 mm, galea apices nearly straight to arching slightly over abaxial lips . . . . 17. *Pedicularis hirsuta*
    29. Corollas 16–25 mm, galea apices strongly arching over abaxial lips.
        30. Pedicels 2.5–5 mm . . . . .22. *Pedicularis langsdorffii* (in part)
        30. Pedicels 1–2.5 mm . . . . 35. *Pedicularis sudetica* (in part)
[20. Shifted to left margin.—Ed.]
20. Galea margins entire medially and distally.
    31. Racemes not exceeding basal leaves.
        32. Galeas 13–15 mm . . . . .7. *Pedicularis centranthera*
        32. Galeas 5–12 mm . . . . 34. *Pedicularis semibarbata*
    31. Racemes exceeding basal leaves.
        33. Galea apices straight.
            34. Corolla abaxial lips 3–7 mm . . . . 3. *Pedicularis aurantiaca*
            34. Corolla abaxial lips 8–15 mm . . . . 12. *Pedicularis densiflora*
        33. Galea apices arching over or beyond abaxial lips.
            35. Calyces glabrous or ciliate.
                36. Galeas yellow . . . . 4. *Pedicularis bracteosa* (in part)
                36. Galeas yellow proximally, dark red to purple distally . . . . 14. *Pedicularis flammea*
            35. Calyces hispid to tomentose, hirsute, or densely woolly.
                37. Racemes capitate, each 2–8-flowered, galea apices arching beyond abaxial lips . . . . 6. *Pedicularis capitata*
                37. Racemes simple, each 6–100-flowered, galea apices arching over or beyond abaxial lips.
                    38. Galeas bicolored . . . . .24. *Pedicularis oederi*
                    38. Galeas concolored.
                      39. Corolla tubes light to dark yellow or dark blood red.
                          40. Pedicels 0.5–1 mm . . . . 4. *Pedicularis bracteosa* (in part)
                          40. Pedicels 1–3.5 mm . . . . .33. *Pedicularis rainierensis*
                      39. Corolla tubes pink or reddish purple, rarely white.
                          41. Bracts undivided . . . . .13. *Pedicularis dudleyi*
                          41. Bracts 1-pinnatifid distally . . . . .20. *Pedicularis lanata*

1. Pedicularis angustifolia Bentham, Pl. Hartw., 22. 1839 • Mogollon Mountain lousewort [C]

Pedicularis angustissima Greene

Plants 35–55 cm. **Leaves:** basal 0; cauline 10–20, blade linear to narrowly lanceolate, 15–70 × 1–6 mm, undivided, margins serrate, surfaces glabrous. **Racemes** paniculate or simple or buds present in cauline leaf axils, 3–15, each 2–12-flowered; bracts linear to narrowly lanceolate, 2–70 × 1–6 mm, undivided, proximal margins entire, distal serrate, surfaces glabrous. **Pedicels** 1.5–4.5 mm. **Flowers:** calyx 5.5–8.5 mm, glabrous, lobes 2, triangular, 0.5–1 mm, apex entire, glabrous; corolla 12–20 mm, tube yellow, 4–10 mm; galea yellow, 8–11 mm, beakless, margins entire medially, 1-toothed distally, apex arching over abaxial lip; abaxial lip yellow, 6–8 mm.

Flowering Jul–Aug. Moist forested ridges and slopes; of conservation concern; 2000–3000 m; N.Mex.; Mexico (Chihuahua, Durango, Michoacán).

Plants of *Pedicularis angustifolia* have two-lobed calyces, undivided linear leaves, and branched or unbranched inflorescences. The beakless galea has a single tooth on each abaxial margin at the distal tip of the galea apex.

2. Pedicularis attollens A. Gray, Proc. Amer. Acad. Arts 7: 384. 1868 • Little elephant's head, elephant snouts, Attoll lousewort, woolly mammoth E

Elephantella attollens (A. Gray) A. Heller

Plants 15–78 cm. **Leaves:** basal 5–25, blade elliptic, 60–150 (or 200–250) × 3–23 mm, 1- or 2-pinnatifid, margins of adjacent lobes nonoverlapping, serrate, surfaces glabrous or scattered glands; cauline 2–20, blade elliptic, 5–50(–100) × 1–5 mm, undivided or 1(or 2)-pinnatifid, margins of adjacent lobes nonoverlapping, serrate, surfaces glabrous. **Racemes** simple, 1–3, exceeding basal leaves, each 10–50-flowered; bracts lanceolate to triangular, 5–10 × 3–10 mm, pinnatifid, margins entire, surfaces glabrous or tomentose. **Pedicels** 1.2–1.6 mm. **Flowers:** calyx 4–5 mm, glabrous or tomentose, lobes 5, triangular, 2–2.5 mm, apex entire, glabrous; corolla 6–8 mm, tube pink, rarely white, 3–6 mm; galea white or pink with 2 purple spots or stripes, 1–2 mm, beaked, beak coiled, 3–6 mm, base curving, margins entire medially and distally, apex not surrounded by abaxial lip, axis of coil nearly vertical; abaxial lip pendulous, white or pink with purple stripe, 4–5.5 mm. $2n = 16$.

Subspecies 2 (2 in the flora): w United States.

The flowers of *Pedicularis attollens*, like those of *P. groenlandica*, resemble an elephant's head, and A. Heller placed them both in *Elephantella*. The short, upturned beak, in contrast to the long, more horizontal downturned beak of *P. groenlandica*, is a distinguishing feature of *P. attollens*. Whereas *P. groenlandica* occurs across much of western and arctic North America, *P. attollens* is found primarily in the Cascade Range of central and southern Oregon and the Sierra Nevada of California and Nevada. It is also reported from the Klamath Range to the west and the White and Sweetwater mountains and the Warner Range to the east of the Sierra Nevada.

1. Basal leaves 1-pinnatifid . . . .
. . . . 2a. *Pedicularis attollens* subsp. *attollens*
1. Basal leaves 2-pinnatifid . . . .
. . . . 2b. *Pedicularis attollens* subsp. *protogyna*

2a. Pedicularis attollens A. Gray subsp. **attollens** E

Leaves: basal 1-pinnatifid.

Flowering Jul–Sep. Alpine bogs, wet meadows; 1200–4000 m; Calif., Nev., Oreg.

2b. Pedicularis attollens A. Gray subsp. **protogyna** Pennell, Proc. Acad. Nat. Sci. Philadelphia 99: 175. 1947 E

Leaves: basal 2-pinnatifid.

Flowering Jun–Aug. Alpine bogs, wet meadows; 1200–4000 m; Calif., Oreg.

Subspecies *protogyna* is found in the southern Cascade Range of Oregon and Lassen County, California.

3. Pedicularis aurantiaca (E. F. Sprague) Monfils & Prather, Madroño 54: 311. 2008 • Indian warrior E

Pedicularis densiflora Bentham subsp. *aurantiaca* E. F. Sprague, Aliso 4: 130. 1958

Plants 10–50 cm. **Leaves:** basal 1–10, blade lanceolate, 30–200 × 20–70 mm, 2-pinnatifid, margins of adjacent lobes nonoverlapping or extensively overlapping distally, 1-serrate, surfaces glabrous, hispid, or downy, abaxial veins downy; cauline 4–20, blade lanceolate, 15–250 × 5–100 mm, 2-pinnatifid, margins of adjacent lobes nonoverlapping or extensively overlapping distally, serrate, surfaces glabrous or hispid to downy. **Racemes** simple, 1–5, exceeding basal leaves, each 10–50-flowered; bracts lanceolate, 10–35 × 3–5 mm, undivided to pinnatifid distally, proximal margins entire, surfaces glabrous. **Pedicels** 2–4 mm. **Flowers:** calyx 12–24 mm, downy, lobes 5, narrowly triangular, 5.5–7.5 mm, apex entire, ciliate; corolla 23–43 mm, tube dark red, purple, or orange-yellow, 9–17 mm; galea dark red, purple, or orange-yellow, 14–26 mm, beakless, margins entire medially and distally, apex straight; abaxial lip dark red, purple, or orange-yellow, 3–7 mm.

Flowering Apr–Jun. Mixed coniferous forests; 600–2100 m; Calif., Oreg.

Pedicularis aurantiaca is distinguished from *P. densiflora* by smaller abaxial lip lobes and longer calyces.

4. Pedicularis bracteosa Bentham in W. J. Hooker, Fl. Bor.-Amer. 2: 110. 1838 • Bracted or towering lousewort E F

Plants 20–80 cm. **Leaves:** basal 0–10, blade lanceolate, 20–120 × 10–60 mm, 1- or 2-pinnatifid, margins of adjacent lobes nonoverlapping or slightly overlapping distally, 1- or 2-serrate, surfaces glabrous; cauline 4–10, blade lanceolate, 10–270 × 8–150 mm, undivided or 1- or 2-pinnatifid, margins of adjacent lobes nonoverlapping, serrate to 2-serrate, surfaces glabrous or scattered glandular. **Racemes** simple, 1–4, exceeding basal leaves, each 15–75-flowered; bracts lanceolate or subulate to trullate, 10–20 × 2–10 mm, undivided, proximal margins entire, distal entire or serrate, surfaces glabrous, hispid, or tomentose. **Pedicels** 0.5–1 mm. **Flowers:** calyx 7–15 mm, glabrous or tomentose, lobes 5, triangular or filiform, 1–10 mm, apex entire, glabrous or ciliate; corolla 14–27 mm, tube yellow or dark blood red, 6–12 mm; galea yellow to yellow tinged with red, purple tinged with yellow, or dark blood red, 6–15 mm, beakless or beaked, beak straight, 0.8–2.5 mm, margins entire medially and distally, apex arching over abaxial lip; abaxial lip expanded, yellow, yellow tinged with purple, or dark blood red, 4.5–6.5 mm.

Varieties 8 (8 in the flora): w North America.

Pedicularis bracteosa is found in subalpine habitats across much of western North America, occurring throughout the Rocky Mountains from central British Columbia and Alberta to northern New Mexico, as well as the Coast Range south to northern California.

1. Galeas beaked.
  2. Calyces tomentose . . . .
    . . . . 4c. *Pedicularis bracteosa* var. *canbyi*
  2. Calyces glabrous . . . .
    . . . . 4h. *Pedicularis bracteosa* var. *siifolia*
1. Galeas beakless.
  3. Calyx lobes filiform.
    4. Corollas: tubes yellow, galeas yellow to yellow tinged with red, or purple tinged with yellow, abaxial lips yellow to yellow tinged with purple . . . .
        . . . . .4a. *Pedicularis bracteosa* var. *bracteosa*
    4. Corollas: tubes, galeas, and abaxial lips dark blood red . . . .
        . . . . . 4b. *Pedicularis bracteosa* var. *atrosanguinea*
  3. Calyx lobes triangular.
    5. Galeas 10–15 mm. . . . .
       . . . . 4g. *Pedicularis bracteosa* var. *paysoniana*
    5. Galeas 7–11 mm.

[6. Shifted to left margin.—Ed.]
6. Galea apices acute; British Colombia, Idaho, Washington . . . . . . 4e. *Pedicularis bracteosa* var. *latifolia*
6. Galea apices obtuse; Blue and Wallowa mountains, Cascade Range, Oregon and Washington.
  7. Calyces slightly hispid; Cascade Range, Oregon and Washington . . . .
    . . . . .4d. *Pedicularis bracteosa* var. *flavida*
  7. Calyces tomentose; Blue and Wallowa mountains, Oregon, and Washington. . . . . . .
    . . . . .4f. *Pedicularis bracteosa* var. *pachyrhiza*

4a. Pedicularis bracteosa Bentham var. **bracteosa** E

Flowers: calyx hispid, lobes filiform; corolla: tube yellow; galea yellow to yellow tinged with red, or purple tinged with yellow, 8–13 mm, beakless, apex obtuse; abaxial lip yellow to yellow tinged with purple.

Flowering Jun–Aug. Edges and openings in coniferous forests; 600–2200 m; Alta., B.C.; Idaho, Mont., Wash.

4b. Pedicularis bracteosa Bentham var. **atrosanguinea** (Pennell & J. W. Thompson) Cronquist in C. L. Hitchcock et al., Vasc. Pl. Pacif. N.W. 4: 356. 1959 • Dark-flowered lousewort E

Pedicularis atrosanguinea Pennell & J. W. Thompson, Bull. Torrey Bot. Club 61: 443. 1934

Flowers: calyx hispid, lobes filiform; corolla: tube dark blood red; galea dark blood red, 8–13 mm, beakless, apex obtuse; abaxial lip dark blood red. $2n = 16$.

Flowering Jul–Aug. Subalpine edges and openings in coniferous forests; 1200–1600 m; B.C.; Wash.

Variety *atrosanguinea* is only known from the Olympic Mountains and mountains of Vancouver Island. It has solid, blood red corollas as opposed to the mostly yellow to partly purple corollas of var. *bracteosa*. Populations with yellow corollas, identified as var. *bracteosa*, have been collected from the southern Olympic Mountains in the vicinity of Mt. Steel. Both varieties are easily distinguishable by their filiform calyx lobes compared to the triangular lobes of other varieties.

4c. Pedicularis bracteosa Bentham var. **canbyi**
(A. Gray) Cronquist in C. L. Hitchcock et al., Vasc. Pl.
Pacif. N.W. 4: 356. 1959 • Canby's lousewort [E]

Pedicularis canbyi A. Gray in
A. Gray et al., Syn. Fl. N. Amer.
ed. 2, 2(1): 454. 1886

Flowers: calyx tomentose, lobes
triangular; corolla: tube yellow;
galea yellow, 7–10 mm, beaked,
apex acute; abaxial lip yellow.
Flowering late Jun–Aug.
Moist openings in coniferous
forests; 2200–2400 m; Idaho, Mont.

Variety *canbyi* occurs in the northern Bitterroot
Range and through contiguous ranges almost to
Wyoming. Variety *canbyi* and var. *siifolia* both have
beaked galeas and are distinguishable to some degree by
the presence or absence of calyx hairs, respectively. In
some regions, such as in the northern Bitterroot Range,
the calyx of var. *canbyi* is tomentose; in the Sawtooth
Mountains, var. *siifolia* has glabrous calyces. In the
Pioneer Mountains and the middle of the Bitterroot
Range, where the ranges of the two varieties overlap,
calyx hairiness varies among populations.

4d. Pedicularis bracteosa Bentham var. **flavida**
(Pennell) Cronquist in C. L. Hitchcock et al., Vasc. Pl.
Pacif. N.W. 4: 356. 1959 [E]

Pedicularis flavida Pennell, Bull.
Torrey Bot. Club 61: 445. 1934

Flowers: calyx slightly hispid,
lobes triangular; corolla: tube
yellow; galea yellow, 7–11 mm,
beakless, apex obtuse; abaxial
lip yellow.
Flowering late Jun–Sep.
Moist, grassy, alpine meadows,
edges in coniferous forests; 1400–2400 m; Calif.,
Oreg., Wash.

The tip of the galea is more rounded in var. *flavida*
than in plants of var. *latifolia*. Corollas of var. *flavida*
are yellow and never bicolored like those of var. *latifolia*.
Variety *flavida* occurs in the Cascade Range from
southern Washington to northern California.

4e. Pedicularis bracteosa Bentham var. **latifolia**
(Pennell) Cronquist in C. L. Hitchcock et al.,
Vasc. Pl. Pacif. N.W. 4: 358. 1959 • Wide-leaved
lousewort [E]

Pedicularis latifolia Pennell, Bull.
Torrey Bot. Club 61: 448. 1934;
P. paddoensis Pennell;
P. thompsonii Pennell

Flowers: calyx slightly hispid,
lobes triangular; corolla: tube
yellow; galea yellow to yellow
tinged with red, 7–11 mm,
beakless, apex acute; abaxial lip
yellow.
Flowering Jun–Aug. Open subalpine coniferous
forests; 1100–1900 m; B.C.; Idaho, Wash.

Variety *latifolia* occurs throughout the Cascade
Range to Mt. Adams, Washington, where it intergrades
with var. *flavida*, and extends to eastern Washington
and northern Idaho, overlapping somewhat with the
distribution of var. *bracteosa*.

Pedicularis rainierensis sometimes grows adjacent to
this variety, and Mt. Rainier Park personnel (unpubl.)
have noted putative interspecific hybrids between the
two species.

4f. Pedicularis bracteosa Bentham var. **pachyrhiza**
(Pennell) Cronquist in C. L. Hitchcock et al., Vasc.
Pl. Pacif. N.W. 4: 356. 1959 • Blue Mountains
lousewort [E]

Pedicularis pachyrhiza Pennell,
Bull. Torrey Bot. Club 61: 445.
1934

Flowers: calyx tomentose, lobes
triangular; corolla: tube yellow;
galea yellow, 8–9 mm, beakless,
apex obtuse; abaxial lip yellow.
Flowering Jul–Aug. Damp
openings and edges in coniferous
forests; 1000–2200 m; Oreg., Wash.

Variety *pachyrhiza* occurs in the Blue and Wallowa
mountains of eastern Washington and Oregon; except
for its wider leaf lobes, it is indistinguishable from
var. *flavida* or var. *latifolia*.

4g. Pedicularis bracteosa Bentham var. **paysoniana** (Pennell) Cronquist in C. L. Hitchcock et al., Vasc. Pl. Pacif. N.W. 4: 358. 1959 • Payson's lousewort E F

Pedicularis paysoniana Pennell, Bull. Torrey Bot. Club 61: 446. 1934; *P. bracteosa* subsp. *paysoniana* (Pennell) W. A. Weber

Flowers: calyx tomentose, lobes triangular; corolla: tube yellow; galea yellow, 10–15 mm, beakless, apex obtuse, sometimes acute; abaxial lip yellow.

Flowering Jun–Aug. Moist alpine slopes, grassy meadows, coniferous forests; 2200–3600 m; Colo., Idaho, Mont., N.Mex., Utah, Wyo.

The long, open throat of the galea of var. *paysoniana* raises the summit of the galea well above the abaxial lip as compared to the other varieties.

4h. Pedicularis bracteosa Bentham var. **siifolia** (Rydberg) Cronquist in C. L. Hitchcock et al., Vasc. Pl. Pacif. N.W. 4: 356. 1959 • Smoothflower lousewort E

Pedicularis siifolia Rydberg, Bull. Torrey Bot. Club 34: 35. 1907

Flowers: calyx glabrous, margins sometimes ciliate, lobes triangular; corolla: tube yellow; galea yellow, 6–11 mm, beaked, apex acute; abaxial lip yellow.

Flowering late Jun–Aug. Moist openings in coniferous forests; 2100–2700 m; Idaho, Mont., Wash.

Variety *siifolia* is recorded from southeastern Washington (F. W. Pennell 1934); the author has not seen specimens from that area.

5. Pedicularis canadensis Linnaeus, Mant. Pl. 1: 86. 1767 • Canadian lousewort, wood betony, pédiculaire du Canada F

Pedicularis canadensis var. *dobbsii* Fernald; *P. canadensis* subsp. *fluviatilis* (A. Heller) W. A. Weber; *P. canadensis* var. *fluviatilis* (A. Heller) J. F. Macbride

Plants 4–50 cm. **Leaves:** basal 2–20, blade lanceolate, 15–100 × 3–40 mm, 1- or 2-pinnatifid, margins of adjacent lobes nonoverlapping or slightly overlapping distally, 1- or 2-serrate, surfaces glabrous or hispid; cauline 1–10, blade lanceolate, 10–70 × 5–20 mm, 1-pinnatifid, margins of adjacent lobes nonoverlapping or slightly overlapping distally, 1- or 2-serrate, surfaces glabrous or hispid. **Racemes** simple, 1–5, exceeding basal leaves, each 10–40-flowered; bracts lanceolate, ovate, spatulate, or trullate, 5–40 × 1–10 mm, undivided proximally, undivided to 1-pinnatifid distally, proximal margins entire, distal 1- or 2-serrate, sometimes crenate, surfaces glabrous or hispid to tomentose. **Pedicels** 1.5–2 mm. **Flowers:** calyx 7–12 mm, glabrous, hispid, or tomentose, lobes 2, triangular, 0.5–2 mm, apex entire, glabrous or ciliate; corolla 18–25 mm, tube yellow, 8–15 mm; galea yellow, yellow with red veins, or red, sometimes purple, 10–14 mm, beakless, margins entire medially, 1-toothed distally, apex arching over abaxial lip; abaxial lip expanded, yellow or white, 6–7 mm. $2n = 16$.

Flowering Apr–Jun. Deciduous forests, forest edges, prairies, alpine wet meadows; 70–2500 m; Man., N.B., Ont., Que.; Ala., Ark., Colo., Conn., Del., D.C., Fla., Ga., Ill., Ind., Iowa, Kans., Ky., La., Maine, Md., Mass., Mich., Minn., Miss., Mo., Nebr., N.H., N.J., N.Mex., N.Y., N.C., N.Dak., Ohio, Okla., Pa., R.I., S.C., S.Dak., Tenn., Tex., Vt., Va., W.Va., Wis.; Mexico.

Pedicularis canadensis is chiefly of the Midwestern prairies; in eastern states and provinces, it grows in forest openings and along forest edges. It also inhabits alpine wet meadows at higher elevations along the Front Range of the Rocky Mountains. Corollas may be yellow, yellow with red veins, red (forma *praeclara* A. H. Moore), or bicolored with a yellow corolla tube, yellow abaxial lip, and red galea (forma *bicolor* Farwell).

6. Pedicularis capitata Adams, Nouv. Mém. Soc. Imp. Naturalistes Moscou 5: 100. 1817 • Capitate lousewort, pédiculaire capitée F

Plants 2–13.5 cm. **Leaves:** basal 2 or 3, blade ovate or elliptic, 10–40 × 5–20 mm, 2(or 3)-pinnatifid, margins of adjacent lobes nonoverlapping or extensively overlapping distally, serrate, surfaces glabrous or hirsute; cauline 0–2, blade elliptic, 5–25 × 5–10 mm, 2-pinnatifid, margins of adjacent lobes nonoverlapping or extensively overlapping distally, serrate, surfaces scattered hispid. **Racemes** simple, capitate, 1 or 2, exceeding basal leaves, each 2–8-flowered; bracts subulate to narrowly lanceolate, 10–20(–50) × 6–8 mm, undivided proximally, undivided or 1- or 2-pinnatifid distally, proximal margins entire, distal serrate, surfaces sparsely tomentose. **Pedicels** 2–3 mm. **Flowers:** calyx 10–15 mm, hirsute, lobes 5, triangular to spatulate, 4–6 mm, apex serrate, glabrous; corolla 19–40 mm, tube light yellow, sometimes cream to pink, 5–20 mm; galea light yellow, sometimes cream to pink, apically sometimes diffuse purple, 12–20 mm, beakless, margins

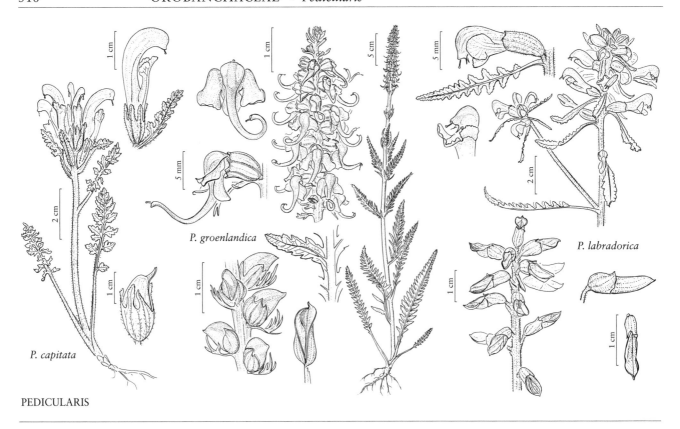

P. groenlandica

P. labradorica

P. capitata

PEDICULARIS

entire medially and distally, apex arching beyond abaxial lip; abaxial lip light yellow, sometimes cream to pink, 10–15 mm. $2n$ = 16 (Asia).

Flowering Jun–Aug. Moist arctic and alpine tundras, heathlands, alpine slopes; 10–2100 m; Greenland; Alta., B.C., N.W.T., Nunavut, Que., Yukon; Alaska; n Asia.

Inflorescences of *Pedicularis capitata* display no more than five clustered flowers; the long, vertical corolla tube and curved galea that arches over the appressed, broad lobes of the abaxial lip are distinctive. The tip of the galea can be the same color as the corolla tube and abaxial lip but sometimes is a diffuse pink to purple and not distinctly bicolored as in the galeas of *P. flammea* or *P. oederi*, with which it is often confused.

7. **Pedicularis centranthera** A. Gray in W. H. Emory, Rep. U.S. Mex. Bound. 2(1): 120. 1859 • Dwarf lousewort E

Plants 4–12 cm. **Leaves:** basal 6–8, blade elliptic or spatulate, 35–120 × 10–30 mm, undivided or 1- or 2-pinnatifid, margins of adjacent lobes nonoverlapping or extensively overlapping distally, entire or 1- or 2-serrate, surfaces glabrous or scattered abaxial glands; cauline 0–4, blade elliptic, sometimes lanceolate, 20–110 × 5–30 mm, 1- or 2-pinnatifid, margins of adjacent lobes overlapping distally, 1- or 2-serrate, surfaces glabrous.

Racemes simple, 1–4, not exceeding basal leaves, each 8–14-flowered; bracts spatulate, 40–60 × 3–6 mm, undivided proximally, undivided or 1-pinnatifid distally, proximal margins entire, distal 1- or 2-serrate, surfaces glabrous, sometimes arachnoid along main veins. **Pedicels** 1–5 mm. **Flowers:** calyx 17–22 mm, glabrous, lobes 5, narrowly triangular, 4–7 mm, apex entire or serrate, glabrous or ciliate; corolla 28–40 mm, tube white or pale purple, 15–30 mm; galea white or pale purple, apically sometimes dark violet to purple, 13–15 mm, beakless, margins entire medially and distally, apex arching over abaxial lip; abaxial lip purple, 1–4 mm.

Varieties 2 (2 in the flora): w United States.

The leaves of *Pedicularis centranthera* exceed the inflorescence, giving the impression that the cauline leaves are basal. Proximalmost basal leaves are distinct: brown, membranous, and spatulate with undivided and entire margins. *Pedicularis semibarbata* has similar basal leaves.

Pedicularis centranthera occurs in evergreen forests, often under pinyon pine, juniper, ponderosa pine, and yellow pine.

1. Bracts arachnoid and/or margins ciliate . . . .
. . . . .7a. *Pedicularis centranthera* var. *centranthera*
1. Bracts glabrous . . . .
. . . . 7b. *Pedicularis centranthera* var. *exulans*

7a. Pedicularis centranthera A. Gray
var. **centranthera** E

Inflorescences: bracts arachnoid and/or margins ciliate.

Flowering Apr–Jun. Under pinyon pines, juniper woodlands; 1300–2500 m; Ariz., Calif., Colo., Nev., N.Mex., Oreg., Utah.

S. L. Welsh et al. (2008) reported var. *centranthera* from dry conifer and scrub communities including pinyon-juniper, *Shepherdia*, oak, serviceberry, mountain mahogany, Douglas fir, and ponderosa pine-manzanita communities.

7b. Pedicularis centranthera A. Gray var. **exulans**
M. Peck, Torreya 28: 56. 1928 E

Inflorescences: bracts glabrous.

Flowering Apr–Jun. Open woodlands; 1300–2500 m; Oreg.

Peck based var. *exulans* on plants with glabrous or glabrate bracts.

8. Pedicularis chamissonis Steven, Mém. Soc. Imp.
Naturalistes Moscou 6: 20, plate 4, fig. 1. 1822
• Chamisso's lousewort

Plants 10–25 cm. **Leaves:** basal 5–15, blade lanceolate, 15–40 × 10–20 mm, 2-pinnatifid, margins of adjacent lobes non-overlapping or slightly overlapping distally, serrate, surfaces glabrous; cauline whorled, whorls 1–4, leaves per whorl 3–5, blade lanceolate, 10–60 × 10–30 mm, 1- or 2-pinnatifid, margins of adjacent lobes nonoverlapping or slightly overlapping distally, entire or serrate, surfaces glabrous. **Racemes** verticillate, 1–3, exceeding basal leaves, each 8–20-flowered; bracts lanceolate to subulate, 10–30 × 5–10 mm, 1-pinnatifid or undivided, margins serrate to 2-serrate, surfaces glabrous. **Pedicels** 1.5–2.5 mm. **Flowers:** calyx 6–8 mm, glabrous, lobes 5, triangular, 0.5–1.4 mm, apex entire, ciliate; corolla 18–20 mm, tube white to pink, 12–15 mm; galea purple, 6–8 mm, beaked, beak straight, 1–3 mm, margins entire medially and distally, apex extending beyond abaxial lip; abaxial lip pink or purple, 4–5 mm.

Flowering Jul–Sep. Arctic meadows, tundras; 10–300 m; Alaska; e Asia.

Pedicularis chamissonis differs from *P. verticillata* by having leaves and flowers about 1.5 times larger with a galea that is extended into a conical beak, leaves lanceolate versus elliptic, and lateral lobes of the abaxial lip more than twice the size and more pendulous.

Pedicularis chamissonis is chiefly an Asian species, found on the Kamchatka Peninsula, Kuril Islands, Sakhalin Island, and Japan; the only North American populations occur in the Aleutian Islands of Alaska.

9. Pedicularis contorta Bentham in W. J. Hooker, Fl.
Bor.-Amer. 2: 108. 1838 • Coil-beaked or coiled lousewort E

Plants 10–50 cm. **Leaves:** basal 3–10, blade lanceolate, 30–100 × 5–30 mm, 1- or 2-pinnatifid, margins of adjacent lobes non-overlapping, entire or serrate, surfaces glabrous; cauline 2–10, blade elliptic or triangular to lanceolate, 5–70 × 10–25 mm, 1-pinnatifid, margins of adjacent lobes nonoverlapping, entire or serrate, surfaces glabrous. **Racemes** simple, 1–6, exceeding basal leaves, each 12–40-flowered; bracts trullate to obtrullate, 5–18 × 2–15 mm, pinnatifid, margins entire, surfaces glabrous. **Pedicels** 1.5–5.5 mm. **Flowers:** calyx 5–9 mm, glabrous, lobes 5, triangular, 1–3 mm, apex entire, glabrous; corolla 7–13 mm, tube white or cream to yellowish or pink to pinkish purple, 4–7 mm; galea white or cream to yellowish with purple spots at base, or pink to pinkish purple, 2–5.5 mm, beaked, beak coiled, 5–9 mm, base curving, margins entire medially and distally, apex surrounded by abaxial lip, axis of coil nearly vertical; abaxial lip surrounding beak, white or cream to yellowish, or pink to pinkish purple, 5–8 mm.

Varieties 3 (3 in the flora): w North America.

The coiled beak that extends directly downward from the top of the galea is distinctive in *Pedicularis contorta*. Seen from the side, the large, upward-expanded, lateral lobes of the abaxial lip surround the beak, concealing it from view. The flowers of *P. contorta* are very similar to those of *P. racemosa*; however, the leaves of *P. racemosa* are undivided, the beak curves down and to one side, and the inflorescence often branches, forming a panicle of short racemes with long internodes between flowers.

Flower and calyx color and width of the subtending floral bracts delimit three varieties in *Pedicularis contorta*; corolla color is difficult to discern in herbarium specimens. The following key is adapted from R. N. Reese (1984).

1. Corolla tubes, galeas, and abaxial lips white or cream to yellowish, galeas with purple spots at base; calyces pale green . . . . 9a. *Pedicularis contorta* var. *contorta*
1. Corolla tubes, galeas, and abaxial lips pink to pinkish purple; calyces green or reddish with purple spots.
  2. Calyces green; bracts 5–8 mm wide, proximal margins ciliate . . . . 9b. *Pedicularis contorta* var. *ctenophora*
  2. Calyces reddish with purple spots; bracts 2–4 mm wide, proximal margins glabrous . . . . 9c. *Pedicularis contorta* var. *rubicunda*

9a. Pedicularis contorta Bentham var. contorta E

Inflorescences: bracts 3–15 mm wide, proximal margins glabrous. **Flowers:** calyx pale green; corolla: tube white or cream to yellowish; galea white or cream to yellowish with purple spots at base; abaxial lip white or cream to yellowish. $2n = 16$.

Flowering Jun–Aug. Alpine meadows, glades, open subalpine coniferous forest; 1400–3000 m; Alta., B.C.; Calif., Idaho, Mont., Oreg., Utah, Wash., Wyo.

9b. Pedicularis contorta Bentham var. ctenophora (Rydberg) A. Nelson & J. F. Macbride, Bot. Gaz. 61: 44. 1916 E

Pedicularis ctenophora Rydberg, Bull. Torrey Bot. Club 24: 293. 1897

Inflorescences: bracts 5–8 mm wide, proximal margins ciliate. **Flowers:** calyx green; corolla: tube pink to pinkish purple; galea pink to pinkish purple; abaxial lip pink to pinkish purple. $2n = 16$.

Flowering Jul–Aug. Dry slopes; 2300–3100 m; Idaho, Mont., Wyo.

9c. Pedicularis contorta Bentham var. rubicunda R. N. Reese, Brittonia 36: 63, figs. 1–3. 1984 E

Inflorescences: bracts 2–4 mm wide, proximal margins glabrous. **Flowers:** calyx reddish with purple spots; corolla: tube pink to pinkish purple; galea pink to pinkish purple; abaxial lip pink to pinkish purple. $2n = 16$.

Flowering Jun–Aug. Open subalpine forests, meadows, dry rocky exposures, deep loam soils, rocky well-drained soils; 2000–2900 m; Idaho, Mont.

Reese noted that the center of distribution of var. *rubicunda* was Clearwater County, Idaho, and adjacent Ravalli County, Montana. Intermediates are common between var. *contorta* and var. *rubicunda* in areas of sympatry where the same species of *Bombus* pollinate both varieties during foraging (Reese).

10. Pedicularis crenulata Bentham in A. P. de Candolle and A. L. P. P. de Candolle, Prodr. 10: 568. 1846
  • Meadow lousewort E

Plants 10–40 cm. **Leaves:** basal 8–10, blade narrowly elliptic to linear, 15–40 × 3–6 mm, undivided, margins of adjacent lobes nonoverlapping, 2-crenate, surfaces glabrous; cauline 10–40, blade linear to narrowly oblanceolate, 10–60 × 2–6 mm, undivided, margins of adjacent lobes nonoverlapping, 2-crenate, surfaces glabrous. **Racemes** simple, 1–10, exceeding basal leaves, each 10–50-flowered; bracts triangular or lanceolate to oblanceolate, 10–15 × 3–4 mm, undivided, proximal margins entire, distal crenate to 2-crenate, surfaces glabrous. **Pedicels** 3–3.5 mm. **Flowers:** calyx 8.5–11 mm, hirsute along veins or glabrous, lobes 2, triangular, 0.5–1 mm, apex entire, glabrous or ciliate; corolla 20–26 mm, tube light pink, rarely white, 12–15 mm; galea reddish violet, sometimes white, 8–11 mm, beakless, margins entire medially, 1-toothed distally, apex arching over abaxial lip; abaxial lip reddish violet, sometimes white, 4–8 mm.

Flowering May–Aug. Moist grassy, sagebrush basins, alpine meadows; 1500–3200 m; Calif., Colo., Mont., Nebr., Nev., N.Mex., Wyo.

Pedicularis crenulata is abundant in moist swales of alpine to subalpine sagebrush meadows of the mid to southern Rocky Mountains, as well as California and Nevada, where large populations with their reddish violet flowers create a distinctive swath across the otherwise green landscape. The undivided, nearly linear leaves with two-crenate margins are unique among North American species of *Pedicularis*. The galea bears a single apical tooth on each abaxial margin at the distal tip. Other species with the combination of two calyx lobes and undivided, linear to nearly linear leaves are *P. angustifolia* and *P. racemosa*.

11. Pedicularis cystopteridifolia Rydberg, Mem. New York Bot. Gard. 1: 365. 1900 • Fern-leaved lousewort E

Plants 10–50 cm. **Leaves:** basal 2–10, blade elliptic to lanceolate, 20–90 × 5–15 mm, 2(or 3)-pinnatifid, margins of adjacent lobes nonoverlapping to extensively overlapping distally, serrate, surfaces glabrous; cauline 2–8, blade triangular to lanceolate, 20–120 × 5–20 mm, 1- or 2-pinnatifid, margins of adjacent lobes nonoverlapping or extensively overlapping distally, serrate, surfaces glabrous or unevenly hispid to tomentose. **Racemes** simple, 1–2, exceeding basal leaves, each 10–40-flowered; bracts trullate to obtrullate or subulate to trullate, 10–25 × 2–5 mm, undivided or 1- or 2-auricled, sometimes 1-pinnatifid, proximal margins entire, distal entire or serrate, surfaces tomentose. **Pedicels** 1–3 mm. **Flowers:** calyx 8–12 mm, tomentose, lobes 5, narrowly triangular, 3–4 mm, apex entire, ciliate; corolla 20–26 mm, tube red or pink, 13–15 mm; galea red or pink, 7–11 mm, beakless, margins entire medially, 1-toothed distally, apex arching over abaxial lip; abaxial lip red or pink, 6–7.5 mm. 2*n* = 32.

Flowering May–Aug. Rocky alpine tundras, meadows; 2100–3100 m; Mont., Wyo.

Pedicularis cystopteridifolia occurs only in the Rocky Mountains of Montana and Wyoming. As the specific epithet implies, the leaves strongly resemble those of the fern *Cystopteris fragilis*. Although not sympatric, this species could easily be misidentified as *P. sudetica* subsp. *scopulorum*, which has flowers of a similar shape and color, and leaves that are also two-pinnatifid. The secondary leaf lobes of *P. cystopteridifolia*, however, are much larger, longer, more deeply incised, and more heavily toothed, making them appear more finely dissected than the linear to deltate secondary and smaller toothed lobes of *P. sudetica*. Many of the adjacent leaf lobes of *P. cystopteridifolia* also overlap, whereas the lobes of *P. sudetica* are more widely spaced and therefore not overlapping. The galea of *P. cystopteridifolia* is also more highly domed and broader, and the leaves are a paler shade of green in contrast to the dark green leaves of *P. sudetica* subsp. *scopulorum*.

12. Pedicularis densiflora Bentham in W. J. Hooker, Fl. Bor.-Amer. 2: 110. 1838 • Indian warrior

Plants 10–50 cm. **Leaves:** basal 1–10, blade lanceolate, 30–200 × 20–70 mm, 2(or 3)-pinnatifid, margins of adjacent lobes nonoverlapping or extensively overlapping distally, 1-serrate, surfaces glabrous, hispid, or downy; cauline 4–20, blade lanceolate, 15–250 × 5–100 mm, 2-pinnatifid, margins of adjacent lobes nonoverlapping or extensively overlapping distally, serrate, surfaces glabrous, hispid, or downy. **Racemes** simple, 1–5, exceeding basal leaves, each 10–50-flowered; bracts lanceolate to trullate, 10–35 × 3–5 mm, undivided or 1-pinnatifid, proximal margins entire, distal 1- or 2-serrate, surfaces glabrous. **Pedicels** 2–4 mm. **Flowers:** calyx 9–18 mm, downy to tomentose, lobes 5, triangular, 3–4 mm, apex entire, ciliate; corolla 23–43 mm, tube dark red, purple, or orange-yellow, rarely white, 8–18 mm; galea dark red, purple, or orange-yellow, rarely white, 15–25 mm, beakless, margins entire medially and distally, apex straight; abaxial lip dark red, purple, or orange-yellow, rarely white, 8–15 mm. 2*n* = 16.

Flowering Feb–May. Mixed coniferous forests; 30–3100 m; Calif., Oreg.; Mexico (Baja California).

Scarlet corollas with an undomed, toothless galea and two- or three-pinnatifid leaves are diagnostic of *Pedicularis densiflora*. This species occurs in forested subalpine regions of southern Oregon, western slopes of the Sierra Nevada, and the Coast Ranges of California south to Baja California. Herbarium records indicate northern populations of *P. densiflora* occur at higher elevations than do more southern populations.

13. Pedicularis dudleyi Elmer, Bot. Gaz. 41: 316. 1906 • Dudley's lousewort C E

Plants 10–30 cm. **Leaves:** basal 2–12, blade elliptic to lanceolate, 30–260 × 40–60 mm, 2-pinnatifid, margins of adjacent lobes nonoverlapping or slightly overlapping distally, serrate, surfaces glabrous; cauline 1–5, blade elliptic to lanceolate, 70–120 × 10–40 mm, 1- or 2-pinnatifid, margins of adjacent lobes nonoverlapping or slightly overlapping distally, serrate, surfaces glabrous. **Racemes** simple, 1–3, exceeding basal leaves, each 6–20-flowered; bracts lanceolate, 5–15 × 3–5 mm, undivided, proximal margins entire, distal serrate, surfaces glabrous. **Pedicels** 2–3 mm. **Flowers:** calyx 10–14 mm, tomentose, lobes 5, triangular,

5–7 mm, apex entire, glabrous; corolla 18–22 mm, tube pinkish, rarely white, 8–13 mm; galea pink to lavender, rarely white, 10–11 mm, beakless, margins entire medially and distally, apex arching over abaxial lip; abaxial lip lavender, 4–7 mm. $2n = 16$.

Flowering Apr–Jun. Coastal chaparral or forests, riparian sites in coastal redwood forests; of conservation concern; 10–300 m; Calif.

Shapes of flowers and leaves of *Pedicularis dudleyi* are similar to those of *P. semibarbata*; *P. dudleyi* has larger vegetative features. Floral features that set *P. dudleyi* apart include a 10–14 mm calyx and pink to purple corolla versus a 7–9 mm calyx and pale yellow corolla in *P. semibarbata*.

Pedicularis dudleyi occurs in the coastal mountains of central California in Monterey, San Luis Obispo, San Mateo, and Santa Cruz counties.

14. Pedicularis flammea Linnaeus, Sp. Pl. 2: 609. 1753
• Red-tipped lousewort, redrattle, pédiculaire flammée

Plants 1.5–20 cm. **Leaves:** basal 2–5, blade lanceolate, 5–25 × 2–6 mm, 1-pinnatifid, margins of adjacent lobes nonoverlapping or slightly overlapping distally, 2-serrate, surfaces glabrous; cauline 3–6, blade lanceolate, 5–30 × 3–10 mm, 1-pinnatifid, margins of adjacent lobes nonoverlapping or slightly overlapping distally, 2-serrate, surfaces glabrous. **Racemes** simple, 1 or 2, exceeding basal leaves, each 5–10-flowered; bracts lanceolate, 10–15 × 1–2 mm, pinnatifid, proximal margins entire, distal 1- or 2-serrate, surfaces glabrous. **Pedicels** 3–10 mm. **Flowers:** calyx 4.5–7 mm, glabrous, lobes 5, triangular, 1.5–2.5 mm, apex entire, glabrous; corolla 12–15 mm, tube yellow, 8–10 mm; galea bicolored, yellow proximally, dark red to purple distally, 4–5 mm, beakless, margins entire medially and distally, apex arching slightly over abaxial lip; abaxial lip yellow, 1.5–3 mm. $2n = 16$.

Flowering Jul. Wet meadows, along streams, tundras, flood plains; 0–2500 m; Greenland; Alta., Man., Nfld. and Labr., N.W.T., Nunavut, Ont., Que.; Europe (Norway, Sweden); Asia; Atlantic Islands (Iceland).

Pedicularis flammea is often confused with *P. oederi*, which also has red- or purple-tipped galeas. Compared to other *Pedicularis* species, including *P. oederi*, flowers of *P. flammea* are much smaller.

Pedicularis flammea is a boreal-arctic species found only in northern parts of many Canadian provinces.

15. Pedicularis furbishiae S. Watson, Proc. Amer. Acad. Arts 17: 375. 1882 • Furbish lousewort C E

Plants 40–90 cm. **Leaves:** basal 4, blade lanceolate to elliptic, 70–130 × 35–50 mm, 2-pinnatifid, margins of adjacent lobes nonoverlapping or slightly overlapping distally, serrate, surfaces hispid; cauline 7, blade lanceolate to elliptic, 20–90 × 8–35 mm, 1- or 2-pinnatifid, margins of adjacent lobes nonoverlapping or slightly overlapping distally, serrate, surfaces hispid. **Racemes** simple or paniculate, 1–4, exceeding basal leaves, each 3–30-flowered; bracts trullate, 8–13 × 7–10 mm, undivided or pinnatifid, margins serrate to 2-serrate, surfaces hispid. **Pedicels** 1–3 mm. **Flowers:** calyx 5–12 mm, hispid-glandular, lobes 5, narrowly triangular, 3–4.5 mm, apex entire or dentate, glabrous; corolla 14–19 mm, tube yellow, 8–10 mm; galea yellow, apically sometimes tinged red, 6–8.5 mm, beakless, margins entire medially, 1-toothed distally, apex arching slightly over abaxial lip; abaxial lip yellow with apex sometimes tinged red, 7–7.5 mm.

Flowering Jul–Sep. Riverbanks; of conservation concern; 100–300 m; N.B.; Maine.

Discovered in 1880, and at one time believed extinct, *Pedicularis furbishiae* was rediscovered in 1974 during an environmental impact survey for a proposed dam on the St. John's River and thereafter was placed on the Federal Register under the Endangered Species Act (L. W. Macior 1981). Metapopulation dynamics suggest that an ecologically intact watershed is required for long-term persistence (E. S. Menges 1990). A recovery strategy has been adopted for this species in New Brunswick (Furbish's Lousewort Recovery Team 2006; Environment Canada 2010). *Pedicularis furbishiae* is in the Center for Plant Conservation's National Collection of Endangered Plants.

SELECTED REFERENCES Environment Canada. 2010. Recovery Strategy for the Furbish's Lousewort (*Pedicularis furbishiae*) in Canada. Ottawa. Furbish's Lousewort Recovery Team. 2006. Recovery Strategy for Furbish's Lousewort (*Pedicularis furbishiae*) in New Brunswick. Fredericton. Macior, L. W. 1981. The Furbish lousewort: Weed, weapon, or wonder? Amer. Biol. Teacher 43: 323–326. Menges, E. S. 1990. Population viability analysis of an endangered plant. Conservation Biol. 4: 52–62.

16. Pedicularis groenlandica Retzius, Fl. Scand. Prodr. ed. 2, 145. 1795 • Elephant's head, pédiculaire du Groenland [E] [F]

Elephantella groenlandica (Retzius) Rydberg; *Pedicularis groenlandica* subsp. *surrecta* (Bentham) Pennell

Plants 10–60 cm. **Leaves:** basal 5–20, blade lanceolate, 20–150 × 5–25 mm, 1-pinnatifid or slightly 2-pinnatifid, margins of adjacent lobes nonoverlapping, 1- or 2-serrate, surfaces glabrous; cauline 3–31, blade lanceolate, 10–150 × 1–25 mm, 1-pinnatifid, margins of adjacent lobes nonoverlapping, serrate, surfaces glabrous. **Racemes** simple, 1 or 2, exceeding basal leaves, each 20–75-flowered; bracts linear to trullate, 5–10 × 2–10 mm, undivided to pinnatifid, margins entire, serrate, or 2-serrate, surfaces glabrous. **Pedicels** 0.5–1 mm. **Flowers:** calyx 3–5 mm, glabrous or hispid, lobes 5, deltate, 0.5–1.5 mm, apex entire, glabrous; corolla 5–8 mm, tube purple, rarely white, 3–5 mm; galea pink to purple, rarely white, 1.5–3 mm, beaked, beak coiled, 5–18 mm, base curving, margins entire medially and distally, apex not surrounded by abaxial lip, axis of coil nearly horizontal; abaxial lip pendulous, purple, rarely white, 2–5 mm. $2n = 16$.

Flowering Jun–Sep. Montane and alpine to arctic bogs, fens, marshes, and forested swamps, seepage areas, stream banks, fens, clay gravel flood plains of rivers; 600–3500 m; Greenland; Alta., B.C., Man., Nfld. and Labr. (Labr.), Nunavut, Ont., Que., Sask., Yukon; Alaska, Ariz., Calif., Colo., Idaho, Mont., Nev., N.Mex., Oreg., Utah, Wash., Wyo.

Pedicularis groenlandica has a domed galea, a long, curved beak, and relatively large lateral lobes of the abaxial lip that remarkably resemble the head, trunk, and ears of an elephant. *Pedicularis groenlandica* is the most widely distributed *Pedicularis* species in North America; it occurs from the southern Sierra Nevada and southern Rocky Mountains well into the Arctic and Greenland.

Pedicularis attollens and *P. groenlandica* are sympatric in the Sierra Nevada and the Cascades of central and southern Oregon and can be difficult to distinguish. *Pedicularis groenlandica* is distinguished by the size and orientation of the beak and color of the corolla; the beak is over twice the length of the beak of *P. attollens* and is oriented more horizontally compared to the upturned beak of *P. attollens*. Beak orientation is very difficult to ascertain in herbarium material due to flattening; beak length is a better character on herbarium specimens. The corollas of *P. groenlandica* are purple with lighter purple to pinkish abaxial lips, whereas the corollas of *P. attollens* vary from pink to light pink often with conspicuous dark purple spots on the galeas. White-flowered forms of *P. groenlandica* (forma *pallida* Lepage) are occasionally seen.

17. Pedicularis hirsuta Linnaeus, Sp. Pl. 2: 609. 1753 • Hairy lousewort, pédiculaire hirsute

Plants 3–15 cm. **Leaves:** basal 3–10, blade elliptic to lanceolate, 5–20 × 2–5 mm, 1(or 2)-pinnatifid, margins of adjacent lobes nonoverlapping or slightly overlapping distally, serrate, surfaces glabrous; cauline 3–10, blade lanceolate to oblanceolate, 5–30 × 1–7 mm, 1-pinnatifid, margins of adjacent lobes nonoverlapping or slightly overlapping distally, serrate, surfaces glabrous. **Racemes** simple, 1–10, exceeding basal leaves, each 5–25-flowered; bracts lanceolate, 5–15 × 1–5 mm, 1-pinnatifid ½–¾ to midrib, proximal margins entire, distal serrate, surfaces densely wooly. **Pedicels** 2–5 mm. **Flowers:** calyx 7–10 mm, densely woolly, lobes 5, triangular, 1–4 mm, apex entire or serrate, glabrous or ciliate; corolla 11–19 mm, tube pale pink, sometimes white, 6–14 mm; galea pale pink, sometimes white, 4–7 mm, beakless, margins entire medially, 1-toothed distally, apex nearly straight to arching slightly over abaxial lip; abaxial lip pale pink, sometimes white, 4–5 mm. $2n = 16$.

Flowering Jul–Aug. Tundra bogs, stream and lake banks; 0–500 m; Greenland; Nfld. and Labr. (Labr.), N.W.T., Nunavut, Que.; Alaska; Eurasia.

Pedicularis hirsuta can be difficult to distinguish from *P. lanata*. Unlike *P. lanata*, the galea subapex of *P. hirsuta* is toothed. Floral surfaces of *P. lanata* are sparsely to densely hirsute; those of *P. hirsuta* are glabrous. On fresh specimens, the roots of *P. hirsuta* are pale yellow; those of *P. lanata* are bright yellow.

18. Pedicularis howellii A. Gray, Proc. Amer. Acad. Arts 20: 307. 1885 • Howell's lousewort [E]

Plants 15–40 cm. **Leaves:** basal 1–4, blade elliptic to lanceolate, 20–60 × 15–40 mm, undivided or 1-pinnatifid, margins of adjacent lobes nonoverlapping or slightly overlapping distally, serrate, surfaces glabrous; cauline 10–12, blade ovate to lanceolate, 25–60 × 15–40 mm, undivided or 1-pinnatifid, sometimes auricled, margins of adjacent lobes nonoverlapping or slightly overlapping distally, entire or serrate to crenate, surfaces glabrous or scattered woolly along main vein. **Racemes** simple, 1–8, exceeding basal leaves, each 15–40-flowered; bracts trullate to cordate, 6–8 × 4–8 mm, undivided, proximal margins entire, distal entire, surfaces glabrous or tomentose. **Pedicels** 2.5–3 mm. **Flowers:** calyx 6–6.5 mm, tomentose, lobes 5, triangular, 1.5–2 mm, apex entire, ciliate; corolla 10–13 mm, tube white, 6–8 mm; galea white, apically sometimes tinged with red to violet, 3.5–5 mm, beaked, beak straight, 1–2 mm, margins entire medially and distally, apex extending beyond abaxial lip; abaxial lip white, 1.5–2 mm.

Flowering Jun–Aug. Alpine forest clearings and edges; 1100–2000 m; Calif., Oreg.

Pedicularis howellii is found in the Siskiyou Mountains along the California/Oregon border; it has undivided distal leaves. The division of the proximal leaves into irregular and asymmetric lobes, sometimes appearing auricled, is a unique feature of this species.

19. **Pedicularis labradorica** Wirsing, Eclog. Bot., [2], plate 10. 1778 • Labrador lousewort, pédiculaire du Labrador F

Pedicularis labradorica var. *sulphurea* Hultén

Plants 2–25 cm. **Leaves:** basal 2 or 3, blade lanceolate, 10–20 × 2–3 mm, 1- or 2-pinnatifid, margins of adjacent lobes nonoverlapping, serrate, surfaces glabrous; cauline 1–4, blade linear to lanceolate, 10–50 × 2–10 mm, undivided or 1- or 2-pinnatifid, margins of adjacent lobes nonoverlapping, serrate to 2-serrate, surfaces glabrous or sparsely downy to hispid. **Racemes** paniculate or buds present in cauline leaf axils, 1–8, exceeding basal leaves, each 5–20-flowered; bracts linear to narrowly lanceolate, 7–15 × 1–2 mm, undivided or 1-pinnatifid, proximal margins entire, distal 1- or 2-serrate, sometimes crenulate, surfaces glabrous or hispid. **Pedicels** 0.5–2 mm. **Flowers:** calyx 5–8 mm, glabrous, lobes 2, triangular, 0.5–1.5 mm, apex entire, sometimes slightly bifurcate, glabrous; corolla 12–18 mm, tube deep yellow, 7–10 mm; galea dark yellow or yellow tinged with purple or spotted, 5–9 mm, beakless, margins entire medially, 1-toothed distally, apex arching over abaxial lip; abaxial lip dark yellow, 5–7 mm. $2n = 16$.

Flowering Jun–Aug. Open forests, tundras, heathlands, rocky slopes, muskegs; 300–1100 m; Greenland; Alta., B.C., Man., Nfld. and Labr. (Labr.), N.W.T., Nunavut, Ont., Que., Sask., Yukon; Alaska; Asia (China, Russia).

The flowers of *Pedicularis labradorica* are usually yellow or dark yellow, and the galea is tinged distally with red or purple; sometimes, the yellow color of the tube abruptly transitions into red or purple. Hultén based var. *sulphurea* on the solid yellow color variant from the Yukon.

20. **Pedicularis lanata** Willdenow ex Chamisso & Schlechtendal, Linnaea 2: 583. 1827 • Woolly lousewort, pédiculaire laineuse

Subspecies 4 (1 in the flora): n North America, Eurasia.

E. Hultén (1967) listed three subspecies of *Pedicularis lanata*, subspp. *adamsii* (Hultén) Hultén, *dasyantha* (Hadač) Hultén, and *pallasii* (Vvedensky) Hultén, as occurring in Eurasia; only subsp. *lanata* is found in North America.

20a. **Pedicularis lanata** Willdenow ex Chamisso & Schlechtendal subsp. **lanata**

Pedicularis kanei Durand

Plants 6–20 cm. **Leaves:** basal 2–20, blade lanceolate to oblanceolate, 10–30 × 2–5 mm, 1- or 2-pinnatifid, margins of adjacent lobes nonoverlapping, serrate, surfaces glabrous; cauline 2–20, blade lanceolate to oblanceolate, 10–40 × 2–5 mm, 1- or 2-pinnatifid, margins of adjacent lobes nonoverlapping, serrate, surfaces glabrous or scattered woolly. **Racemes** simple, 1–4, exceeding basal leaves, each 12–100-flowered; bracts subulate, 10–40 × 2–5 mm, undivided or 1-pinnatifid, proximal margins entire, distal serrate, proximal surfaces densely woolly, distal glabrous or slightly to densely woolly. **Pedicels** 1–2 mm. **Flowers:** calyx 5–6 mm, densely woolly, lobes 5, triangular, 1–2 mm, apex entire, long-ciliate; corolla 14–22 mm, tube pink or reddish purple, 10–15 mm; galea concolored, pink to reddish purple, 4–7 mm, beakless, margins entire medially and distally, apex arching over abaxial lip; abaxial lip pink or reddish purple, 4–7 mm. $2n = 16$.

Flowering May–Aug. Grassy alpine and arctic tundras, fellfields, rocky slopes; 20–2100 m; Greenland; Alta., B.C., N.W.T., Nunavut, Que., Yukon; Alaska, Mont.; n Asia.

Although *Pedicularis lanata* has priority, *P. kanei* (1856) is a synonym that is used commonly for this species. In fact, E. Hultén (1968) in his list of synonyms of *P. kanei* annotated *P. lanata* as of American authors. In the same treatment, Hultén listed *P. kanei* subsp. *adamsii*, but his distribution map did not show it occurring in North America. H. J. Scoggan (1978–1979) listed *P. lanata* forma *alba* Cody based on its white flowers.

The densely woolly inflorescence and bright yellow taproot of *Pedicularis lanata* are diagnostic. *Pedicularis hirsuta*, also with a densely woolly inflorescence and similar corolla color, can be confused with *P. lanata*; however, the galea margins of *P. hirsuta* are minutely toothed, and the taproots are pale.

21. Pedicularis lanceolata Michaux, Fl. Bor.-Amer. 2: 18. 1803 • Swamp lousewort E

Plants 20–100 cm. **Leaves:** basal 0; cauline 10–30, blade lanceolate, 20–100 × 10–30 mm, 1-pinnatifid, margins of adjacent lobes nonoverlapping, 2-serrate, surfaces hispid. **Racemes** paniculate, 1, each 10–20-flowered; bracts lanceolate to trullate, 5–10 × 3–5 mm, undivided or 1-pinnatifid and 1- or 2-auricled, proximal margins entire, distal 1- or 2-serrate, surfaces glabrous or hispid. **Pedicels** 1–1.5 mm. **Flowers:** calyx 7–12 mm, glabrous or hispid, lobes 2, trullate, ovate, elliptic, or triangular, 2.5–3.5 mm, apex serrate, glabrous, sometimes ciliate; corolla 16–22 mm, tube white, cream, or light yellow, 8–12 mm; galea white, cream, or light yellow, 8–12 mm, beaked, beak straight, 0.5–2.5 mm, margins entire medially and distally, apex extending over abaxial lip; abaxial lip white, cream, or light yellow, 7–10 mm. 2*n* = 16.

Flowering Aug–Oct. Wet meadows, fens, springs, moist prairies, swamps; 10–1100 m; Man., Ont.; Ark., Conn., Del., Ga., Ill., Ind., Iowa, Ky., Md., Mass., Mich., Minn., Mo., Nebr., N.J., N.Y., N.C., N.Dak., Ohio, Pa., S.Dak., Tenn., Va., W.Va., Wis.

The long corolla tubes of *Pedicularis lanceolata* are uncharacteristically nectarless, and only late season pollen-foraging worker bumblebees pollinate this species (L. W. Macior 1969). The uniquely hinged abaxial lip covering the opening of the galea is an adaptation to allow only worker bumblebees access to the anthers, as they must learn to push it aside during foraging.

SELECTED REFERENCE Macior, L. W. 1969. Pollination adaptation in *Pedicularis lanceolata*. Amer. J. Bot. 56: 853–859.

22. Pedicularis langsdorffii Fischer ex Steven, Mém. Soc. Imp. Naturalistes Moscou 6: 49, plate 9, fig. 2. 1822 (as langsdorfii) • Langsdorff's lousewort

Plants 4–30 cm. **Leaves:** basal 0–10, blade elliptic, 10–30 × 3–15 mm, 2-pinnatifid, margins of adjacent lobes nonoverlapping or slightly overlapping distally, serrate, surfaces glabrous; cauline 1–4, blade elliptic, 10–40 × 2–10 mm, 1- or 2-pinnatifid, margins of adjacent lobes nonoverlapping or slightly overlapping distally, serrate, sometimes crenate, surfaces glabrous, sometimes sparsely tomentose. **Racemes** simple, 1–6, exceeding basal leaves, each 10–50-flowered; bracts subulate or linear, 5–25 × 1–10 mm, undivided or 1-pinnatifid, proximal margins entire, distal serrate, surfaces glabrous or sparsely tomentose to tomentose. **Pedicels** 2.5–5 mm.

Flowers: calyx 6–11 mm, glabrous or ± tomentose, lobes 5, triangular, 2–5 mm, apex entire or serrate to dentate, glabrous; corolla 17–25 mm, tube pink or lavender, 11–13 mm; galea pink or lavender, 6–12 mm, beakless, margins entire medially, 1-toothed distally, apex strongly arching over abaxial lip; abaxial lip pink or lavender, 5–8 mm.

Subspecies 2 (2 in the flora): n North America, Asia.

Pedicularis langsdorffii may be mistaken for *P. hirsuta*, *P. lanata*, and *P. sudetica*, which have similar growth forms and habitat requirements. *Pedicularis langsdorffii* generally has larger, pink to lavender corollas with toothed galeas that strongly arch over the abaxial lips in contrast to the smaller, toothless, pink corollas and slightly arching galeas of *P. lanata*. *Pedicularis sudetica* has up to five cauline leaves or lacks them. The straight, smaller galeas and pale pink or white corollas of *P. hirsuta* differentiate it from *P. langsdorffii*.

1. Bracts and calyces glabrous. . . . .
. . . . .22a. *Pedicularis langsdorffii* subsp. *langsdorffii*
1. Bracts and calyces ± tomentose . . . .
. . . . 22b. *Pedicularis langsdorffii* subsp. *arctica*

22a. Pedicularis langsdorffii Fischer ex Steven subsp. langsdorffii

Racemes: bracts glabrous. **Flowers:** calyx glabrous. 2*n* = 16.

Flowering Jun–Aug. Arctic, grassy slopes, moist areas, upland tundras, rocky alpine tundras, stream banks; 0–1000 m; Alaska; Asia.

In addition to its glabrous inflorescence, E. Hultén (1968) contrasted the shorter subtending floral bracts in the upper part of the inflorescence of subsp. *langsdorffii* to the longer, more prominent bracts of subsp. *arctica*. While some specimens can be clearly distinguished using bract size, many others cannot.

22b. Pedicularis langsdorffii Fischer ex Steven subsp. arctica (R. Brown) Pennell ex Hultén, Ark. Bot., n. s. 7: 122. 1968 (as langsdorfii)

Pedicularis arctica R. Brown, Chlor. Melvill., 22. 1823; *P. langsdorffii* var. *arctica* (R. Brown) Polunin

Racemes: bracts ± tomentose. **Flowers:** calyx ± tomentose. 2*n* = 16.

Flowering Jun–Jul. Moist heath, grassy or rocky meadows, dry arctic and alpine tundras, moist stream banks; 0–1800 m; Greenland; Alta., B.C., N.W.T., Nunavut, Yukon; Alaska; Asia.

Citing N. Polunin (1940), S. G. Aiken et al. (2007) reiterated that subsp. *arctica* grades into *Pedicularis hirsuta*. L. W. Macior (1975) showed partial fertility between *P. langsdorffii* and *P. lanata*, another species with densely lanate inflorescences. The apparent absence of strong reproductive barriers suggests the status of these taxa as species might be questioned and may explain the apparent character overlap between subsp. *arctica* and subsp. *langsdorffii*.

23. Pedicularis lapponica Linnaeus, Sp. Pl. 2: 609. 1753
· Lapland lousewort, pédiculaire de Laponie

Plants 5–15 cm. **Leaves:** basal 0–4, blade lanceolate, 6–25 × 3–13 mm, 1-pinnatifid, margins of adjacent lobes nonoverlapping, serrate, surfaces glabrous; cauline 3–7, blade lanceolate, 10–35 × 2–6 mm, 1(or 2)-pinnatifid, margins of adjacent lobes nonoverlapping, serrate, surfaces glabrous. **Racemes** simple, capitate, 1–3, exceeding basal leaves, each 6–12-flowered; bracts linear to triangular, 6–9 × 1–3 mm, undivided or 1-pinnatifid, proximal margins entire, distal serrate, surfaces glabrous. **Pedicels** 1–2 mm. **Flowers:** calyx 4–5.5 mm, glabrous, lobes 2, deltate, 0.2–1 mm, apex entire, glabrous; corolla 11–17 mm, tube yellow, 6–8 mm; galea yellow, 5–9 mm, beaked, beak straight, 0.5–2 mm, margins entire medially and distally, apex extending over abaxial lip; abaxial lip yellow, 4–7 mm. 2*n* = 16.

Flowering Jun–Aug. Arctic-alpine tundras, heathlands, subarctic, moist hummocky tundras, hummocks, open white spruce and tamarack forests; 50–1200 m; Greenland; Man., Nfld. and Labr. (Labr.), N.W.T., Nunavut, Ont., Que., Yukon; Alaska; Europe; Asia.

W. J. Cody (2000) described the beak of *Pedicularis lapponica* as toothed, but this is a misinterpretation of its irregular fimbriate apex that sometimes appears to be toothed. Basal leaves are usually not present on herbarium material, but if present, they are often larger than the cauline leaves but otherwise similar in form.

24. Pedicularis oederi Vahl ex Hornemann, Fors. Oecon.
Plantel. ed. 2, 580. 1806 · Oeder's lousewort

Pedicularis oederi var. *albertae* (Hultén) B. Boivin

Plants 2–15 cm. **Leaves:** basal 2–10, blade lanceolate, 10–70 × 3–15 mm, 1-pinnatifid, margins of adjacent lobes nonoverlapping or extensively overlapping distally, 2-serrate, surfaces glabrous or slightly tomentose; cauline 1–5, blade lanceolate, 15–50 × 3–20 mm, 1-pinnatifid, margins of adjacent lobes nonoverlapping or extensively overlapping distally, 1- or 2-serrate, surfaces glabrous or slightly tomentose. **Racemes** simple, 1 or 2, exceeding basal leaves, each 10–50-flowered; bracts linear to lanceolate, 5–20 × 1–2 mm, undivided or 1-pinnatifid, proximal margins entire, distal serrate, surfaces glabrous or tomentose. **Pedicels** 2–5 mm. **Flowers:** calyx 8–11 mm, tomentose, lobes 5, triangular, 1–3 mm, apex entire or serrate, glabrous, sometimes ciliate; corolla 15–24 mm, tube yellow, 9–15 mm; galea bicolored, yellow proximally, brown or red distally, 6–9 mm, beakless, margins entire medially and distally, apex arching slightly over abaxial lip; abaxial lip yellow, 4–5 mm. 2*n* = 16 (Asia).

Flowering Jun–Aug. Arctic and alpine tundras; 500–3700 m; Alta., B.C., N.W.T., Yukon; Alaska, Mont., Wyo.; Eurasia.

Pedicularis oederi is known from the mountains of Europe, Asia, and western North America.

Two arctic species can be easily confused with *Pedicularis oederi*. *Pedicularis flammea* also has bicolored flowers that are yellow with galeas that are red- or purple-tipped, but the flowers of *P. oederi* are twice the size of those of *P. flammea*. The flowers of *P. capitata* are sometimes also yellow but may or may not be bicolored. If bicolored, the color is more diffuse and lighter than that of either *P. flammea* or *P. oederi*. The flowers of *P. capitata* are also larger than those of *P. oederi*. In addition, the inflorescences of *P. capitata* usually have no more than five flowers clustered at the tips, while those of *P. oederi* have at least 10 to 50 flowers along at least one third their lengths. H. J. Scoggan (1978–1979) listed *P. oederi* var. *albertae* based upon its densely woolly inflorescence.

25. Pedicularis ornithorhynchos Bentham in
W. J. Hooker, Fl. Bor.-Amer. 2: 108. 1838
· Bird's-beak or ducksbill lousewort E F

Plants 10–25 cm. **Leaves:** basal 2–10, blade lanceolate to oblanceolate, 15–80 × 3–10 mm, 1- or 2-pinnatifid, margins of adjacent lobes nonoverlapping or slightly overlapping distally, entire or serrate, surfaces glabrous; cauline 0–4, blade lanceolate, 5–40 × 3–15 mm, 1- or 2-pinnatifid, margins of adjacent lobes nonoverlapping or slightly overlapping distally, serrate, surfaces glabrous. **Racemes** simple, 1–5, exceeding basal leaves, each 4–15-flowered; bracts trullate, sometimes lanceolate, 5–13 × 1–3 mm, ± lobed, margins entire or serrate, surfaces glabrous or tomentose. **Pedicels** 3–6 mm. **Flowers:** calyx 6.5–9 mm, tomentose, lobes 5, triangular, 2.5–4 mm, apex entire or serrulate, glabrous or ciliate; corolla 12–15 mm, tube lavender, 8–9 mm; galea lavender, 4–6.5 mm, beaked, beak

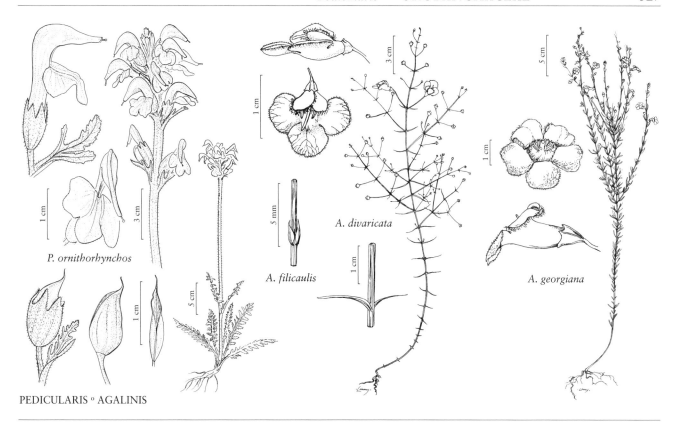

P. ornithorhynchos

A. divaricata

A. filicaulis

A. georgiana

PEDICULARIS ° AGALINIS

straight, 2–6.5 mm, margins entire medially and distally, apex extending beyond abaxial lip; abaxial lip pink, 6–8 mm. $2n = 16$.

Flowering Jul–Sep. Moist alpine meadows, heathlands, tundras; 200–2400 m; B.C.; Alaska, Wash.

Pedicularis ornithorhynchos is commonly misspelled in the literature as *P. ornithorhyncha*. As this is the orthography used in floras since C. L. Hitchcock et al. (1955–1969, vol. 4), it also appears on recent herbarium specimens. How this error arose is unclear because the same misspelling also appears on specimens older than 1959. Another occasional misspelling is *ornithorhynchus*.

The galeas of *Pedicularis ornithorhynchos* taper into long conical, uncoiled beaks that are very conspicuous above the horizontally expanded lateral lobes of the abaxial lips. No more than 15 flowers are borne on the capitate heads, with the internodes greatly expanding as the fruits develop. Compared to the basal leaves, the cauline leaves of the inflorescence are very few and much shorter, but otherwise similar in form. This alpine and tundra species occurs in the Alaskan panhandle and Coast Mountains as well as the coast ranges of mainland British Columbia south into the northern Cascade Range as far south as Mount Rainier; it is also recorded from alpine areas on Vancouver Island and the Haida Gwaii (Queen Charlotte Islands), British Columbia.

Pedicularis ornithorhynchos is pollinated by bumblebees that hang inverted from the galea and cause release of pollen by wing-muscle vibrations (L. W. Macior 1973).

26. **Pedicularis palustris** Linnaeus, Sp. Pl. 2: 607. 1753 • European purple or marsh lousewort, marsh redrattle, pédiculaire des marais

Subspecies 4 (1 in the flora): e Canada, Eurasia.

Four subspecies of *Pedicularis palustris* have been designated in Europe; North American material appears to be subsp. *palustris*.

The overall form and habit of *Pedicularis palustris* are very similar to those of *P. pennellii*. Inflorescences of both species consist of paniculate racemes, and galea margins have two sets of paired teeth, one set medial and the other distal.

26a. **Pedicularis palustris** Linnaeus subsp. **palustris**

Plants 8–45 cm. **Leaves:** basal 1–4, blade linear to lanceolate, 6–20 × 2–4 mm, 1-pinnatifid, margins of adjacent lobes nonoverlapping, serrate, surfaces glabrous; cauline 0–10, blade lanceolate to triangular, 6–65 × 1–20 mm, 1- or 2-pinnatifid, margins of adjacent lobes nonoverlapping, serrate, surfaces glabrous or hispid to tomentose. **Racemes** simple or paniculate, 1–6, exceeding basal leaves, each 10–20-flowered; bracts triangular to deltate, 7–20 × 3–8 mm, 1- or 2-pinnatifid, margins serrate, surfaces glabrous or hispid. **Pedicels** 1–4 mm. **Flowers:** calyx 5–10 mm, glabrous or hispid

to tomentose, lobes 2(–4), round, oval, or trullate to deltate, 1–5 mm, apex dentate, glabrous; corolla 16–20 mm, tube white or light purple, 7–11 mm; galea purple, 6.5–9 mm, beakless, margins 1-toothed medially and distally, apex arching over abaxial lip and upturned; abaxial lip pink to purple, 5–9 mm. $2n = 16$ (Europe).

Flowering Jun–Aug. Marshes, bogs, swamps, ditches; 10–2600 m; Nfld. and Labr. (Nfld.), N.S., Que.; Eurasia.

27. Pedicularis parryi A. Gray, Amer. J. Sci. Arts, ser. 2, 34: 250. 1862 • Parry's lousewort E

Plants 7–65 cm. **Leaves:** basal 4–20, blade elliptic or lanceolate to oblanceolate, 10–70 × 3–15 mm, 1-pinnatifid, margins of adjacent lobes nonoverlapping or slightly overlapping distally, 1- or 2-serrate, surfaces glabrous; cauline 0–20, blade lanceolate, 10–50 × 2–10 mm, 1-pinnatifid, margins of adjacent lobes nonoverlapping or slightly overlapping distally, serrate, surfaces glabrous or tomentose. **Racemes** simple, 1–10, exceeding basal leaves, each 5–50-flowered; bracts subulate to trullate, 10–25 × 8–10 mm, undivided or ± lobed, proximal margins entire, distal serrate or crenate, surfaces glabrous or tomentose. **Pedicels** 2–3 mm. **Flowers:** calyx 6–10 mm, glabrous or tomentose, lobes 5, triangular to deltate, 1–3 mm, apex entire, glabrous; corolla 14–22 mm, tube white, yellowish, or light purple to purple, 7–15 mm; galea white, yellowish, or light purple to purple, 7–10 mm, beaked, beak straight, 5–8 mm, margins entire medially and distally, apex extending beyond abaxial lip; abaxial lip white, yellowish, or light purple to purple, 4–9 mm. $2n = 16$.

Subspecies 3 (3 in the flora): w United States.

Pedicularis parryi is characterized by a domed galea with a conical beak and light green calyces and floral bracts with dark purple stripes. The beaked galea extends well beyond the expanded lobes of the abaxial lip.

Pedicularis parryi is an alpine complex. Corolla color, number of cauline leaves, and bract margin variation are important features differentiating infraspecific taxa within *P. parryi* (G. D. Carr 1971, from which the following key is adapted).

SELECTED REFERENCE Carr, G. D. 1971. Taxonomy of *Pedicularis parryi* (Scrophulariaceae). Brittonia 23: 280–291.

1. Corolla tubes ± purple; calyces tomentose. 27c. *Pedicularis parryi* subsp. *purpurea*
1. Corolla tubes white or yellowish; calyces glabrous.
  2. Cauline leaves (0 or)1–6; bract margins entire or ± crenate . . . . 27a. *Pedicularis parryi* subsp. *parryi*
  2. Cauline leaves 5–20; bract margins ± serrate . . . . 27b. *Pedicularis parryi* subsp. *mogollonica*

27a. Pedicularis parryi A. Gray subsp. **parryi** E

Leaves: cauline (0 or)1–6. **Bracts:** margins entire or ± crenate. **Flowers:** calyx glabrous; corolla tube white or yellowish; galea white or yellowish; abaxial lip white or yellowish.

Flowering Jun–Aug. Barren well-drained rocky slopes; 2400–4000 m; Ariz., Colo., N.Mex., Utah, Wyo.

27b. Pedicularis parryi A. Gray subsp. **mogollonica** (Greene) G. D. Carr, Brittonia 23: 288. 1971 E

Pedicularis mogollonica Greene, Leafl. Bot. Observ. Crit. 1: 151. 1905; *P. parryi* var. *mogollonica* (Greene) Dorn

Leaves: cauline 5–20. **Bracts:** margins ± serrate. **Flowers:** calyx glabrous; corolla tube white or yellowish; galea white or yellowish; abaxial lip white or yellowish.

Flowering Jun–Sep. Boggy open meadows, along streams and lakeshores; 2300–3500 m; Ariz., Colo., N.Mex., Utah.

27c. Pedicularis parryi A. Gray subsp. **purpurea** (Parry) G. D. Carr, Brittonia 23: 289. 1971 E

Pedicularis parryi var. *purpurea* Parry, Amer. Naturalist 8: 214. 1874

Leaves: cauline (0 or)1–8. **Bracts:** margins entire. **Flowers:** calyx tomentose; corolla tube light purple to purple; galea ± purple; abaxial lip ± purple.

Flowering Jun–Aug. Somewhat barren well-drained rocky slopes, gravel meadows along streams and lakeshores; 1800–3200 m; Idaho, Mont., Utah, Wyo.

28. Pedicularis parviflora Smith in A. Rees, Cycl. 26: Pedicularis no. 4. 1813 • Small-flowered lousewort, pédiculaire parviflore E

Pedicularis macrodontis Richardson

Plants 7–65 cm. **Leaves:** basal 0–2, blade elliptic, 3–7 × 2–5 mm, 1- or 2-pinnatifid, margins of adjacent lobes nonoverlapping or slightly overlapping distally, dentate, surfaces glabrous; cauline 0–12, blade lanceolate or elliptic to deltate, 5–50 × 3–20 mm,

1- or 2-pinnatifid, margins of adjacent lobes nonoverlapping or slightly overlapping distally, serrate, surfaces glabrous. **Racemes** simple or paniculate, 1–4, exceeding basal leaves, each 3–12-flowered; bracts deltate to trullate, 5–30 × 3–15 mm, 1- or 2-pinnatifid, margins serrate, surfaces glabrous or slightly arachnoid. **Pedicels** 1–2.5 mm. **Flowers:** calyx 5.5–8 mm, glabrous, lobes 2(–4), deltate, 2–3.5 mm, apex pinnatifid, sometimes 2-fid into triangular lobes, entire, glabrous; corolla 8–16 mm, tube light pink to purple, 5–11 mm; galea purple, 3–6 mm, beakless, margins 1-toothed medially, entire distally, apex arching slightly over abaxial lip; abaxial lip pink to purple, sometimes purple-spotted, 2–5 mm.

Flowering Jun–Aug. Muskegs, boggy flood plains, gravel stream bars, moist meadows, sedge meadows, fens, bogs, black spruce-tamarack wetlands; 0–900 m; Alta., B.C., Man., N.W.T., Nunavut, Ont., Sask., Yukon; Alaska.

Pedicularis parviflora belongs to a complex of taxa [including *P. parviflora* var. *macrodontis* (Richardson) S. L. Welsh and *P. pennellii*] that have traditionally been treated as species, subspecies, or varieties. Two features unite this group: a highly branched paniculate raceme and a tooth on each medial margin of the galea covered with pyriform glands on the inner surface. Lack of apical teeth sets *P. parviflora* apart from *P. palustris* and *P. pennellii*, which have both sets of teeth.

The distinction of *Pedicularis macrodontis* is not clear. With a galea that lacks apical teeth, it is clearly associated with *P. parviflora*, but there are no unique characters to set it apart as a distinct species; all foliar and floral features are very similar. A number of intermediate specimens were seen, suggesting a lack of reproductive barriers; treatment as a variety of *P. parviflora*, following S. L. Welsh (1974), may be warranted, but further research is required before recognizing it as a taxon.

29. **Pedicularis pennellii** Hultén, Fl. Aleut. Isl., 300, plate 14. 1937 • Pennell's lousewort

Pedicularis parviflora Smith subsp. *pennellii* (Hultén) Hultén; *P. pennellii* subsp. *insularis* Calder & Roy L. Taylor; *P. pennellii* var. *insularis* (Calder & Roy L. Taylor) B. Boivin

Plants 4–30 cm. **Leaves:** basal 0–2, blade elliptic, 1–10 × 1–5 mm, 1-pinnatifid, margins of adjacent lobes nonoverlapping or slightly overlapping distally, dentate, surfaces glabrous; cauline 0–5, blade lanceolate or elliptic to deltate, 7–30 × 1–25 mm, 1- or 2-pinnatifid, margins of adjacent lobes nonoverlapping or slightly overlapping distally, serrate, surfaces glabrous. **Racemes** simple or paniculate, 1–3, exceeding basal leaves, each 8–20-flowered; bracts deltate,

5–20 × 10–15 mm, 1- or 2-pinnatifid nearly to midrib, lobes sometimes laciniate, margins serrate, surfaces glabrous. **Pedicels** 1–1.5 mm. **Flowers:** calyx 3.5–8.5 mm, glabrous, lobes 2(–4), triangular, 2–6 mm, apex dentate to 2-dentate, glabrous; corolla 8–16 mm, tube light pink to purple, 5–9 mm; galea bicolored, yellow with purple spots proximally, purple distally, 3–6.5 mm, beakless, margins 1-toothed medially and distally, apex straight to arching slightly over abaxial lip; abaxial lip yellow to pink with purple spots, 3–8 mm. **2n** = 16 (Asia).

Flowering Jun–Aug. Moist tundras, stream banks, wet terraces, willow thickets, wet meadows, fens, bogs; 0–500 m; B.C.; Alaska; Asia.

Pedicularis palustris and *P. pennellii* are extensively branched, with branching in *P. pennellii* more compact and subequal, making it appear shrubby. The branches of *P. palustris*, in contrast, become progressively shorter distally on the stem, giving the plant a pyramidal appearance.

Calder and Taylor recognized subsp. *insularis* by its smaller or sometimes absent apical teeth on the galea margins. Boivin subsequently reduced it to a variety. Given this minor difference, this taxon (found only on Haida Gwaii [the Queen Charlotte Islands] of British Columbia) is not recognized here.

30. **Pedicularis procera** A. Gray, Amer. J. Sci. Arts, ser. 2, 34: 251. 1862 • Giant or Gray's lousewort E

Plants 75–150 cm. **Leaves:** basal 2–4, blade lanceolate, 150–250 × 80–120 mm, 2-pinnatifid, margins of adjacent lobes nonoverlapping, entire or serrate, surfaces glabrous or hirsute; cauline 4–10, blade triangular to lanceolate, 60–300 × 5–90 mm, undivided or 1- or 2-pinnatifid, margins of adjacent lobes nonoverlapping, 2-serrate, surfaces glabrous. **Racemes** simple, sometimes paniculate, 1–3, exceeding basal leaves, each 10–50-flowered; bracts linear or narrowly lanceolate to subulate, 15–80 × 3–8 mm, undivided, proximal margins entire, distal entire or serrate, surfaces hispid to tomentose. **Pedicels** 0–1 mm. **Flowers:** calyx 10–15 mm, hispid to hirsute, lobes 5, triangular, 4–7 mm, apex entire, ciliate; corolla 22–30 mm, tube light yellow, greenish yellow, or light pink, 10–15 mm; galea light yellow, greenish yellow, or light pink, with purple to red veins, 9–15 mm, beakless, margins entire medially, 1-toothed distally, apex arching over abaxial lip; abaxial lip light yellow or light pink, with purple veins, 9–15 mm. **2n** = 32.

Flowering Jul–Aug. Moist alpine meadows, aspen groves; 2400–4000 m; Ariz., Colo., N.Mex., S.Dak., Utah, Wyo.

Although *Pedicularis grayi* A. Nelson appears in older floras, the name is superfluous and illegitimate. *Pedicularis procera* Adams ex Steven 1822 is invalid.

Pedicularis procera is the tallest species of *Pedicularis* in North America. Because the leaves closely resemble those of *P. bracteosa*, smaller plants can be easily mistaken for this species.

31. Pedicularis pulchella Pennell, Notul. Nat. Acad. Nat. Sci. Philadelphia 95: 7, fig. [p. 9 (right)]. 1942

• Mountain lousewort E

Plants 6–12 cm. Leaves: basal 15–20, blade lanceolate to elliptic, 10–20 × 5–10 mm, 2- or 3-pinnatifid, margins of adjacent lobes extensively overlapping, serrate, surfaces glabrous; cauline 4–10, blade elliptic, 10–25 × 3–7 mm, 2- or 3-pinnatifid, margins of adjacent lobes extensively overlapping throughout, serrate, apex cuspidate, surfaces glabrous or tomentose. Racemes simple, 1–4, exceeding basal leaves, each 8–50-flowered; bracts lanceolate or elliptic to trullate, 4–25 × 3–6 mm, 2-pinnatifid, adjacent margins extensively overlapping, proximal margins entire, distal cuspidate, surfaces glabrous or tomentose. Pedicels 4–10 mm. Flowers: calyx 8–12 mm, tomentose, lobes 5, triangular, 1.5–3 mm, apex entire or pinnatifid to serrate, glabrous; corolla 17–27 mm, tube violet-red, 9–10 mm; galea violet-red, 7–15 mm, beakless, margins entire medially, 1-toothed distally, apex arching over abaxial lip; abaxial lip violet-red, 6–9 mm.

Flowering Jun–Jul. Gravel fields and slopes at or above tree lines; 2700–3000 m; Mont., Wash., Wyo.

Pedicularis pulchella occurs in the Absaroka and Beartooth ranges of southwestern Montana and northwestern Wyoming, the Anaconda and Madison ranges of Montana, and one site in the Cascade Range of Washington. The overlapping adjacent lobes of its two-, or sometimes three-, pinnatifid leaves are a characteristic feature of *P. pulchella*.

32. Pedicularis racemosa Douglas ex Bentham in W. J. Hooker, Fl. Bor.-Amer. 2: 108. 1838

• Sickletop lousewort, parrot's-beak E

Plants 0.5–15 cm. Leaves: basal 0; cauline 8–25, blade linear or narrowly lanceolate, 10–80 × 3–15 mm, undivided, adjacent margins nonoverlapping, 1- or 2-serrate, surfaces glabrous. Racemes simple, 1–4, each 3–25-flowered; bracts lanceolate, 5–40 × 3–10 mm, undivided, proximal margins entire, distal 1- or 2-serrate,

surfaces glabrous. Pedicels 1–3.5 mm. Flowers: calyx 4.5–7 mm, glabrous, lobes 2, triangular to deltate, 0.5–1 mm, apex entire, glabrous; corolla 10–15 mm, tube white or light pink, 6–9 mm; galea white or light pink, 4–6 mm, beaked, beak sickle-shaped, 5–8 mm, margins entire medially and distally, apex not extending beyond abaxial lip; abaxial lip white or light pink, 4–5 mm. 2n = 16.

Subspecies 2 (2 in the flora): w North America.

Pedicularis racemosa occurs in the Rocky Mountains from southern Canada to New Mexico, in the Cascade Range from British Columbia to California, and in mountainous areas of Arizona.

Herbarium sheets of *Pedicularis racemosa* are sometimes misidentified as *P. contorta* or *P. groenlandica*, but the calyx has two lobes, not five as in the latter species. The uncoiled beak also bends downward from an undomed galea, which the large lateral lobes of the abaxial lip often conceal, whereas the coiled beaks of *P. contorta* and *P. groenlandica* are very noticeable.

1. Corolla tubes light pink; cauline leaf blades narrowly elliptic-lanceolate, widest at middles ....32a. *Pedicularis racemosa* subsp. *racemosa*
1. Corolla tubes white; cauline leaf blades linear to linear-lanceolate, widest near bases or at middles .... 32b. *Pedicularis racemosa* subsp. *alba*

32a. Pedicularis racemosa Douglas ex Bentham subsp. racemosa E

Leaves: cauline leaf blade narrowly elliptic-lanceolate, widest at middles. Flowers: corolla tube, galea, and abaxial lip light pink.

Flowering Jun–Sep. Subalpine coniferous forests and glades, forest edges; 1000–2500 m; B.C.; Calif., Oreg., Wash.

The range of subsp. *racemosa* is much more limited than that of subsp. *alba*. C. L. Hitchcock et al. (1955–1969, Vol. 4) recognized these taxa as varieties and noted that var. *racemosa* occurs in the Cascades and northern Sierra Nevada and westward, whereas var. *alba* was mostly east of this region. However, L. W. Macior (1973) noted that *Pedicularis racemosa* had white flowers on Mt. Rainier, a mountain peak of the Cascade Range. In the Cascade Range sympatry is considerable as well as in the Wallowa Mountains of northeastern Oregon.

32b. Pedicularis racemosa Douglas ex Bentham subsp. **alba** Pennell, Proc. Acad. Nat. Sci. Philadelphia 99: 176. 1947 [E]

Pedicularis racemosa var. *alba* (Pennell) Cronquist

Leaves: cauline leaf blade linear to linear-lanceolate, widest near bases or at middles. **Flowers:** corolla tube, galea, and abaxial lip white.

Flowering Jun–Aug. Subalpine coniferous forests and glades, forest edges; 700–3100 m; Alta.; Ariz., Calif., Colo., Idaho, Mont., N.Mex., Oreg., Utah, Wash., Wyo.

33. Pedicularis rainierensis Pennell & F. A. Warren, Bull. Torrey Bot. Club 55: 317. 1928 • Mt. Rainier lousewort [C][E]

Plants 10–40 cm. **Leaves:** basal 2–5, blade lanceolate, 40–80 × 3–20 mm, 2-pinnatifid, margins of adjacent lobes nonoverlapping or slightly overlapping distally, serrate to 2-serrate, surfaces glabrous; cauline 2–6, blade lanceolate, 15–70 × 5–20 mm, 1- or 2-pinnatifid, margins of adjacent lobes nonoverlapping or slightly overlapping distally, serrate, surfaces glabrous. **Racemes** simple, 1–4, exceeding basal leaves, each 10–50-flowered; bracts lanceolate or subulate to trullate, 10–15 × 1–2 mm, undivided or pinnatifid, proximal margins entire, distal serrate, surfaces glabrous or tomentose. **Pedicels** 1–3.5 mm. **Flowers:** calyx 7.5–11 mm, hispid to tomentose, lobes 5, linear to narrowly triangular, 4–5 mm, apex entire, glabrous or ciliate; corolla 16–19 mm, tube light or dark yellow, 8–10 mm; galea light or dark yellow, 8–9 mm, beakless, margins entire medially and distally, apex arching beyond abaxial lip; abaxial lip light or dark yellow, 4–5 mm.

Flowering Jun–Aug. Moist alpine grassy meadows, gravelly slopes; of conservation concern; 1200–2000 m; Wash.

Pedicularis rainierensis is known from Mt. Rainier and the Crystal Mountain area. The species is easily confused with *P. bracteosa* var. *latifolia*, which often occurs in the same meadows. While the sizes and shapes of their flowers are nearly indistinguishable, *P. rainierensis* is a much smaller plant with leaves only about three fourths the size, proximal leaf lobes less than one fifth the size, inflorescences about one half the length, and the number of flowers greatly reduced in comparison to those of *P. bracteosa* var. *latifolia*.

34. Pedicularis semibarbata A. Gray, Proc. Amer. Acad. Arts 7: 385. 1868 • Pinewoods lousewort [E]

Plants 1–6 cm. **Leaves:** basal 3–5, blade lanceolate or spatulate, 20–90 × 5–30 mm, undivided or 1- or 2-pinnatifid, margins of adjacent lobes nonoverlapping or slightly overlapping distally, entire or dentate, surfaces glabrous or tomentose; cauline 1 or 2, blade lanceolate, 25–80 × 5–10 mm, 1- or 2-pinnatifid, margins of adjacent lobes nonoverlapping or slightly overlapping distally, serrate to dentate, surfaces glabrous or tomentose. **Racemes** simple, 1–5, not exceeding basal leaves, each 4–20-flowered; bracts lanceolate to oblanceolate, 30–90 × 5–40 mm, 1- or 2-pinnatifid nearly to midrib, margins serrate to dentate, surfaces glabrous. **Pedicels** 2–4 mm. **Flowers:** calyx 7–9 mm, glabrous or tomentose along veins, lobes 5, narrowly triangular, 1.5–5 mm, apex entire, glabrous or ciliate; corolla 12–25 mm, tube light green or pale yellow, sometimes cream, 7–13 mm; galea concolored, light green or pale yellow, sometimes cream, 5–12 mm, beakless, margins entire medially and distally, apex nearly straight to arching slightly over abaxial lip; abaxial lip yellow, sometimes cream, 4–7 mm. $2n = 16$.

Subspecies 2 (2 in the flora): w United States.

The basal and cauline leaves of *Pedicularis semibarbata* are distinctly one- or two-pinnatifid into deep pinnae and narrow subpinnae with serrate margins. The leaves and bracts far exceed the length of the inflorescence, often concealing it. Obvious spatulate, tan-colored, undivided, and membranous leaves are proximal to the divided basal leaves, but they are not as conspicuous as those of *P. centranthera*, a species with a similar growth form. Proximal floral bracts of *P. semibarbata* are similar to basal and cauline leaves, whereas in some specimens the distal bracts are spatulate and either once-divided or merely serrate at the apex. Surfaces of the corolla tube and sometimes the galea as well are hispid.

Pedicularis semibarbata grows under ponderosa pine, incense cedar, sugar pine, and white fir, primarily in the southern Cascade Range and Sierra Nevada, in the San Bernardino and San Gabriel mountains of California, and the Mount Charleston region of Nevada.

The flowers of *Pedicularis semibarbata* in Yosemite National Park were pollinated only by *Osmia tristella* (L. W. Macior 1977).

1. Corollas 12–20(–22) mm; leaf lobes 6–20 × 3–14 mm; outside of Mount Charleston region, Nevada . . . . 34a. *Pedicularis semibarbata* subsp. *semibarbata*
1. Corollas 20–25 mm; leaf lobes 3–12 × 1–8 mm; within Mount Charleston region, Nevada . . . . 34b. *Pedicularis semibarbata* subsp. *charlestonensis*

34a. Pedicularis semibarbata A. Gray subsp. **semibarbata** E

Leaves: lobes 6–20 × 3–14 mm. **Flowers:** corolla 12–20(–22) mm.

Flowering May–Jul. Dry, open pine forests; 1500–3500 m; Calif., Nev., Oreg.

34b. Pedicularis semibarbata A. Gray subsp. **charlestonensis** (Pennell & Clokey) Clokey, Madroño 8: 60. 1945 E

Pedicularis semibarbata var. *charlestonensis* Pennell & Clokey, Bull. S. Calif. Acad. Sci. 38: 6. 1939

Leaves: lobes 3–12 × 1–8 mm. **Flowers:** corolla 20–25 mm.

Flowering May–Jul. Dry open pine forest; 1500–3500 m; Nev.

Subspecies *charlestonensis* is known from the Mount Charleston region. Pennell and Clokey described this taxon based on its longer corolla and less pinnatifid leaves. Clokey later cited its restricted range in treating it as a subspecies. Examination of specimens of both subspecies confirmed that corolla length is a good discriminator, but leaf dissection is a less reliable character. A better character is the size of the primary leaf lobes; the lobes of subsp. *charlestonensis* are consistently smaller than those of subsp. *semibarbata* with much less overlap in size between the two.

35. Pedicularis sudetica Willdenow, Sp. Pl. 3: 209. 1800
  • Sudeten lousewort

Plants 2–45 cm. **Leaves:** basal 1–20, blade elliptic to lanceolate, 10–110 × 3–26 mm, 1- or 2-pinnatifid, margins of adjacent lobes nonoverlapping or slightly overlapping distally, serrate, surfaces glabrous; cauline 0–5, blade lanceolate to elliptic, 20–90 × 2–20 mm, 1- or 2-pinnatifid, margins of adjacent lobes nonoverlapping or slightly overlapping distally, 1- or 2-serrate, surfaces glabrous, some hairs along veins on abaxial surface. **Racemes** simple, 1–4, exceeding basal leaves, each 10–50-flowered; bracts linear to subulate or trullate, 2–15 × 1–4 mm, undivided with or without long auricles, or 1-pinnatifid, margins entire, serrate, or serrulate, surfaces glabrous, white- or yellowish white-lanate, or sparsely pilose. **Pedicels** 1–2.5 mm. **Flowers:** calyx

7–13 mm, glabrous, white-lanate, yellowish white-lanate, or sparsely pilose, lobes 5, subulate or triangular, 1.5–5 mm, apex entire, crenulate, or serrulate, glabrous, sometimes ciliate; corolla 16–21 mm, tube pink, purple, or magenta, 9–11 mm; galea purple, magenta, or bicolored, 7–12 mm, beakless, margins entire medially, 1-toothed or entire distally, apex arching over abaxial lip; abaxial lip white or pink with purple spots, purple, or magenta, 4–8 mm.

Subspecies 6 (5 in the flora): North America, Eurasia.

Pedicularis sudetica is a difficult complex. Hultén employed the degree of lobing on subtending floral bracts, length of corolla tubes, and inflorescence vesture in recognizing eight infraspecific taxa. U. Molau and D. F. Murray (1996) emphasized presence or absence of spots on the abaxial lip, inflorescence vesture, length of petioles and calyx, and ecological features to define four species, subsuming several of Hultén's other subspecific taxa into these species. According to Molau and Murray, *P. sudetica* in the narrow sense is a morphologically distinct, disjunct taxon endemic to the Sudeten Mountains of central Europe, but part of the broader circumscription by Hultén (1961, 1964), which is treated here as the sixth subspecies.

U. Molau and D. F. Murray (1996) did not include subsp. *scopulorum* in their analysis, noting only that teeth are virtually absent on the galea of *Pedicularis scopulorum*. Presence or absence of apical teeth on the margins of the galea often distinguishes species in other *Pedicularis* taxa. Because teeth are sometimes present in this taxon, it is treated here as one of the five North American subspecies in the broad sense of *P. sudetica*. A recent molecular study confirms its close relationship to other members of the complex (B. W. Robart et al. 2015).

When comparing the two alternative taxonomies, the treatment by U. Molau and D. F. Murray (1996) is easier to apply. Close inspection of specimens identified as these species often reveals combinations of traits attributable to more than one taxon; Molau and Murray reported finding hybridization common between *Pedicularis albolabiata* and *P. arctoeuropaea*, and *P. albolabiata* and *P. pacifica* where habitats overlap, indicating that reproductive isolation is not complete. Recognition of these taxa as varieties may be more appropriate considering the broad geographic overlap where the few distinguishing features tend to intergrade. They are treated here, however, as subspecies. The following key is modified from Molau and Murray.

SELECTED REFERENCE Molau, U. and D. F. Murray. 1996. Taxonomic revision of the *Pedicularis sudetica* complex (Scrophulariaceae): The Arctic species. Symb. Bot. Upsal. 31: 33–46.

1. Calyx lobe margins entire distally.
    2. Calyx lobes subulate; Alberta, British Columbia, Manitoba, Northwest Territories, Nunavut, Ontario, Yukon, Alaska. . . . .
        . . . 35c. *Pedicularis sudetica* subsp. *interior* (in part)
    2. Calyx lobes triangular; Colorado, New Mexico, Wyoming. . . . .
        . . . . . . .35e. *Pedicularis sudetica* subsp. *scopulorum*
1. Calyx lobe margins crenulate or serrulate distally.
    3. Corollas: abaxial lips purple-spotted.
        4. Corollas: galeas bicolored; abaxial lips white, rarely pink; calyces and bracts glabrous or white-lanate. . . . .
            . . . . . 35a. *Pedicularis sudetica* subsp. *albolabiata*
        4. Corollas: galeas not bicolored; abaxial lips magenta; calyces and bracts yellowish white-lanate, rarely glabrous . . . .
            . . .35b. *Pedicularis sudetica* subsp. *arctoeuropaea*
    3. Corollas: abaxial lips not purple-spotted.
        5. Calyces and bracts white-lanate, calyx lobe margins entire or serrulate distally . . . . . .
            . . . . .35c. *Pedicularis sudetica* subsp. *interior* (in part)
        5. Calyces and bracts glabrous or sparsely pilose, calyx lobe margins serrulate distally . . . .35d. *Pedicularis sudetica* subsp. *pacifica*

35a. Pedicularis sudetica Willdenow subsp. **albolabiata** Hultén, Svensk Bot. Tidskr. 55: 203. 1961

Pedicularis albolabiata (Hultén) Kozhevnikov

Racemes: bracts glabrous or white-lanate. **Flowers:** calyx glabrous or white-lanate, lobes triangular, margins crenulate distally; corolla: galea bicolored, base pink, margins entire medially, 1-toothed distally; abaxial lip white, rarely pink, purple-spotted.

Flowering Jun–Aug. Wet tundras, moist thickets, alpine-montane taigas; 0–1200 m; Greenland; N.W.T., Nunavut, Que., Yukon; Alaska; e Asia (Russian Far East, Siberia).

35b. Pedicularis sudetica Willdenow subsp. **arctoeuropaea** Hultén, Svensk Bot. Tidskr. 55: 202. 1961

Pedicularis arctoeuropaea (Hultén) Molau & D. F. Murray

Racemes: bracts yellowish white-lanate, rarely glabrous. **Flowers:** calyx yellowish white-lanate, rarely glabrous, lobes triangular, margins crenulate distally; corolla: galea not bicolored, margins entire medially, 1-toothed distally; abaxial lip magenta, purple-spotted.

Flowering Jun–Aug. Tundra flood plains and terraces; 0–1100 m; N.W.T., Nunavut, Yukon; Alaska; Eurasia.

35c. Pedicularis sudetica Willdenow subsp. **interior** (Hultén) Hultén, Svensk Bot. Tidskr. 55: 203. 1961

Pedicularis sudetica var. *interior* Hultén, Fl. Alaska Yukon 9: 1417. 1949; *P. interior* (Hultén) Molau & D. F. Murray; *P. sudetica* var. *gymnocephala* Trautvetter; *P. sudetica* subsp. *interioroides* Hultén

Racemes: bracts white-lanate. **Flowers:** calyx white-lanate, lobes subulate, margins entire or serrulate distally; corolla: galea purple to magenta, margins entire medially, 1-toothed distally; abaxial lip pale purple to magenta, purple-spotted or not. $2n = 16$ (Asia).

Flowering Jun–Aug. Taiga bogs and fens, boreal thickets, alpine and arctic tundras; 0–2100 m; Alta., B.C., Man., N.W.T., Nunavut, Ont., Que., Yukon; Alaska; Asia.

35d. Pedicularis sudetica Willdenow subsp. **pacifica** Hultén, Svensk Bot. Tidskr. 55: 203. 1961

Pedicularis pacifica (Hultén) Kozhevnikov; *P. sudetica* var. *pacifica* (Hultén) S. L. Welsh

Racemes: bracts glabrous or sparsely pilose. **Flowers:** calyx glabrous or sparsely pilose, lobes subulate, margins serrulate distally; corolla: galea purple to magenta, margins entire medially, 1-toothed distally; abaxial lip purple to magenta, not purple-spotted.

Flowering Jun–Aug. Arctic coasts, alpine meadows, stream banks, taigas; 0–900 m; Alaska; Asia.

35e. Pedicularis sudetica Willdenow subsp. **scopulorum** (A. Gray) Hultén, Svensk Bot. Tidskr. 58: 437. 1964 ⒺE

Pedicularis scopulorum A. Gray in A. Gray et al., Syn. Fl. N. Amer. 2(1): 308. 1878

Racemes: bracts white-lanate. **Flowers:** calyx white-lanate, lobes triangular, margins entire distally; corolla: galea purple to magenta, margins entire medially, 1-toothed or entire distally; abaxial lip pale purple to magenta, not purple-spotted.

Flowering Jul–Aug. Alpine tundra meadows; 3100–3700 m; Colo., N.Mex., Wyo.

Subspecies *scopulorum* can be mistaken for *Pedicularis cystopteridifolia*, which has similar flowers and growth form. The leaves of subsp. *scopulorum*, however, are not finely dissected like those of *P. cystopteridifolia*.

36. Pedicularis sylvatica Linnaeus, Sp. Pl. 2: 607. 1753 • Redrattle I

Plants 3–10 cm. **Leaves:** basal 2–7, blade elliptic, 9–20 × 4–7 mm, 1- or 2-pinnatifid, margins of adjacent lobes nonoverlapping or slightly overlapping distally, dentate, surfaces glabrous; cauline 2–6, blade elliptic to triangular, 5–16 × 1–5 mm, 1- or 2-pinnatifid, margins of adjacent lobes nonoverlapping or slightly overlapping, dentate, surfaces glabrous. **Racemes** simple, 1–5, exceeding basal leaves, each 2–10-flowered; bracts trullate to triangular, 3–15 × 6–11 mm, 2-pinnatifid, margins serrate, surfaces glabrous. **Pedicels** 0.7–3 mm. **Flowers:** calyx 9–14 mm, glabrous, lobes 5, triangular, 1.5–3.5 mm, apex pinnatifid, entire, glabrous; corolla 24–30 mm, tube pink, 15–25 mm; galea pink, 7–11 mm, beakless, margins entire medially, 1-toothed distally, apex arching slightly over abaxial lip; abaxial lip pink, 5.5–9 mm. *2n* = 16 (Europe).

Flowering Jul–Aug. Wet meadows; 30–80 m; introduced; Nfld. and Labr. (Nfld.); Europe.

Pedicularis sylvatica is considered by L. Brouillet et al. (http://data.canadensys.net/vascan/) to be introduced in the flora area.

37. Pedicularis verticillata Linnaeus, Sp. Pl. 2: 608. 1753 • Whorled lousewort

Pediculariopsis verticillata (Linnaeus) Á. Löve & D. Löve

Plants 5–25 cm. **Leaves:** basal 5–25, blade elliptic, 10–20 × 3–8 mm, 1- or 2-pinnatifid, margins of adjacent lobes nonoverlapping or slightly overlapping distally, serrate, surfaces glabrous or tomentose; cauline whorled, whorls 1 or 2, leaves per whorl 3–5, blade lanceolate, 5–30 × 3–10 mm, 1- or 2-pinnatifid, margins of adjacent lobes nonoverlapping or slightly overlapping, serrate, surfaces glabrous. **Racemes** verticillate, 1–10, exceeding basal leaves, each 10–40-flowered; bracts lanceolate to trullate, 4–20 × 2–5 mm, undivided or 1- or 2-pinnatifid, margins serrate, surfaces glabrous or sparsely tomentose. **Pedicels** 0.7–2.5 mm. **Flowers:** calyx 3.5–10 mm, glabrous or pilose, lobes 5, deltate, 0.3–1 mm, apex entire, glabrous or ciliate; corolla 10–18 mm, tube light or dark pink, 5–8 mm; galea light or dark pink or purple, 5–7 mm, beakless, margins entire medially and distally, apex arching slightly over abaxial lip; abaxial lip light pink with purple veins and/or purple spots or purple with dark veins, 5–7 mm. *2n* = 12.

Flowering Jun–Sep. Arctic meadows and thickets, muskegs, arctic and alpine tundras; 0–1800 m; B.C., N.W.T., Yukon; Alaska; Europe; Asia.

In the flora area, *Pedicularis verticillata* and *P. chamissonis* are the only species of *Pedicularis* with whorled leaves.

16. AGALINIS Rafinesque, New Fl. 2: 61. 1837, name conserved • False foxglove [Greek *aga-*, very or much, and genus *Linum*, alluding to resemblance of stems and leaves]

Judith M. Canne-Hilliker†

John F. Hays

Tomanthera Rafinesque, name rejected

Herbs, annual, rarely perennial (*A. linifolia*); hemiparasitic, rarely rhizomatous (*A. linifolia*). **Stems** erect, rarely leaning, rarely fleshy, glabrous, hispid, scabrous, scabridulous, glabrate, sericeous, or papillate. **Leaves** cauline, opposite or subopposite, rarely alternate (*A. filifolia*); petiole absent; blade not leathery, rigid, rarely fleshy, margins entire, rarely proximally cleft, pinnatifid, or 2-pinnatifid. **Inflorescences** terminal, racemes or panicles, rarely racemiform or spikelike; bracts present. **Pedicels** present; bracteoles absent. **Flowers:** sepals 5, calyx radially, rarely bilaterally, symmetric, campanulate to hemispheric, turbinate, or funnelform, lobes obtusely to acutely subulate or deltate, acute- to acuminate-triangular, or lanceolate; petals 5, corolla pale pink to rose-purple or purple, rarely white, throats usually with 2 abaxial yellow lines and red spots within, bilabiate, funnelform to campanulate, abaxial throat gibbous, rounded, or straight, abaxial lobes 3, adaxial 2; stamens 4, didynamous; filaments glabrate to lanate; staminode 0; ovary 2-locular, placentation axile; stigma linear, rarely 2-lobed (*A. neoscotica*). **Capsules:** dehiscence loculicidal. **Seeds** 60–600, yellow, tan, brown, or black, angled, wings absent. *x* = 13.

Species ca. 60 (34 in the flora): c, e North America, Mexico, West Indies, Central America, South America.

F. W. Pennell (1929, 1935) placed *Agalinis auriculata* and *A. densiflora* in *Tomanthera* based on their retrorsely hispid stems, calyx lobes longer than the tube, leaf blades lanceolate to ovate and lobed, anthers of the proximal pair of stamens shorter than anthers of the distal pair, and absence of yellow lines in the corolla. With the exception of hispid stems, these characteristics occur in other species of *Agalinis* and/or do not occur with regularity in both *A. auriculata* and *A. densiflora*.

Agalinis is an attractive component of the fall flora; corollas last less than a day, usually falling in early afternoon. Some species of *Agalinis* darken when dried, some turning nearly black. The stem leaves of some species produce axillary fascicles. Some have incrustations of silica, which appear as smooth, gray or whitish, marbled patches or striations on leaves and stems. Siliceous hairs are also common.

For descriptions presented here, corolla length is the distance from the base of the sinus between the abaxial calyx lobes to the apex of the extended (not reflexed) mid abaxial corolla lobe. Corolla throat is the flared portion of the corolla between the cylindrical tube, which is held within the calyx, and the bases of the corolla lobes. Because the delicate corollas are easily distorted by pressing, it is particularly useful to note in the field whether the adaxial corolla lobes are reflexed or project flatly forward or arch forward, whether the two adaxial lobes are obviously shorter than the three abaxial lobes, and whether the adaxial side of the corolla throat is shorter than, or about equal to, the abaxial side of the throat. The interior of the distal corolla throat at the bases of the adaxial corolla lobes should be examined for a band of relatively long hairs.

Molecular phylogenetic studies support monophyly for 24 of the 29 taxa investigated (J. B. Pettengill and M. C. Neel 2008). *Agalinis acuta* was not differentiated from *A. decemloba* by molecular or morphological data (Pettengill and Neel 2008, 2011); *A. decemloba* includes *A. acuta* as treated here. *Agalinis paupercula* clustered with *A. purpurea* in molecular phylogenetic analyses and is treated here as a variety of *A. purpurea*. *Agalinis tenella* clustered with *A. decemloba*; *A. tenella* is treated here as a species based on morphological data.

SELECTED REFERENCES Canne, J. M. 1979. A light and scanning electron microscope study of seed morphology in *Agalinis* (Scrophulariaceae) and its taxonomic significance. Syst. Bot. 4: 281–296. Canne, J. M. and C. M. Kampny. 1991. Taxonomic significance of leaf and stem anatomy of *Agalinis* (Scrophulariaceae) from the U.S.A. and Canada. Canad. J. Bot. 69: 1935–1950. Pennell, F. W. 1929. *Agalinis* and allies in North America: II. Proc. Acad. Nat. Sci. Philadelphia 81: 111–249. Pettengill, J. B. and M. C. Neel. 2008. Phylogenetic patterns and conservation among North American members of the genus *Agalinis* (Orobanchaceae). B. M. C. Evol. Biol. 8: 264.

1. Bracts longer than pedicels.
    2. Leaf blade margins entire or mid to distal with 1 or 2 proximal lobes, or pinnatifid distally or 2-pinnatifid; bract margins pinnatifid or with proximal lobes; branches retrorsely short-sericeous and hispid; calyx lobes (5–)7–13 mm.
        3. Calyx tubes hirsute; leaf blades lanceolate, margins entire or mid to distal with 1 or 2 proximal lobes . . . . 3. *Agalinis auriculata*
        3. Calyx tubes densely, finely scabridulous and hispid; leaf blades triangular-ovate, margins pinnatifid distally or 2-pinnatifid . . . . 7. *Agalinis densiflora*
    2. Leaf blade margins entire, rarely proximalmost 3-cleft; bract margins entire; branches glabrous, glabrate, or scabridulous to scabrous; calyx lobes 0.2–8 mm.

[4. Shifted to left margin.—Ed.]

4. Leaf blades fleshy, adaxial surfaces with sessile, dome-shaped hairs.
  5. Calyx lobes narrowly lanceolate, 2.7–7 mm; abaxial corolla lobes glabrous externally; capsules obovoid to oblong, 10–12 mm . . . . 5. *Agalinis calycina*
  5. Calyx lobes deltate, 0.5–2 mm; abaxial corolla lobes pilose externally; capsules subglobular to globular, 5–7 mm.
    6. Stems 70–160 cm, (perennials); corollas 25–35 mm, lobes 6–14 mm; styles 15–28 mm; pedicels 5–25 mm . . . . .20. *Agalinis linifolia*
    6. Stems 5–75 cm, (annuals); corollas 7.5–21 mm, lobes 2.5–6 mm; styles 6–13 mm; pedicels 2.5–11 mm . . . . .21. *Agalinis maritima*
4. Leaf blades not fleshy, adaxial surfaces scabridulous to scabrous.
  7. Corollas: abaxial lobes glabrous externally; capsules ovoid-oblong; calyx lobes triangular-lanceolate or lanceolate.
    8. Leaves: blades linear to filiform, 0.4–1.5(–2) mm wide, surfaces scabrous; axillary fascicles well developed; pedicels 4–20 mm. . . . . .2. *Agalinis aspera*
    8. Leaves: blades narrowly lanceolate to elliptic-lanceolate, sometimes proximalmost 3-cleft, 2–6(–7) mm wide, surfaces scabridulous to slightly scabrous; axillary fascicles absent, rarely present; pedicels 1–5(–6) mm . . . . 17. *Agalinis heterophylla*
  7. Corollas: abaxial lobes pilose externally; capsules globular; calyx lobes deltate-subulate, triangular, or triangular-subulate, rarely lanceolate or triangular-lanceolate.
    9. Pedicels 6–25 mm; corollas: adaxial lobes projected over distal anthers, throats glabrous within across bases and sinus of adaxial lobes . . . . 33. *Agalinis tenuifolia* (in part)
    9. Pedicels 0.5–8 mm; corollas: adaxial lobes spreading, erect, or reflexed-spreading, rarely projected distal to corolla mouth, throats villous within across bases and sinus of adaxial lobes.
      10. Calyx lobes deltate-subulate, 0.2–1 mm; leaf blades filiform, 0.2–0.8 mm wide; inflorescences racemiform, with pseudoterminal flowers . . . . 26. *Agalinis plukenetii*
      10. Calyx lobes triangular, triangular-subulate, triangular-lanceolate, or lanceolate, 0.6–4(–5) mm; leaf blades ± linear to linear-filiform, linear-lanceolate, or linear-elliptic to elliptic, 0.4–5 mm wide; inflorescences racemes.
        11. Leaf blades linear-elliptic to elliptic; calyx lobes triangular-lanceolate; styles included; stems 5–30(–47) cm . . . . .23. *Agalinis neoscotica*
        11. Leaf blades ± linear to linear-filiform or linear-lanceolate; calyx lobes triangular to triangular-subulate or lanceolate; styles exserted, rarely included; stems (7–)20–200 cm.
          12. Leaves: axillary fascicles well developed, shorter than or equal to subtending leaves; branches moderately to often copiously scabrous or scabridulous on faces and angles.
            13. Corollas 22–36 mm, lobes 5–12 mm; leaf blades 15–40(–50) mm; styles 14–22 mm, strongly exserted; stems (10–)50–200 cm . . . . .10. *Agalinis fasciculata*
            13. Corollas 12–14 mm, lobes 3–5 mm; leaf blades 10–15(–20) mm; styles 4–7.7 mm, included or slightly exserted; stems 20–40(–50) cm . . . . 15. *Agalinis georgiana*
          12. Leaves: axillary fascicles absent or shorter than subtending leaves; branches glabrous, glabrate, or scabridulous to sparsely scabrous.
            14. Branches ascending, slightly quadrangular-ridged distally; leaf blades linear to linear-filiform, 0.5–1.4 mm wide; calyx lobes triangular-subulate, keeled; seeds black. . . . . 16. *Agalinis harperi*
            14. Branches ascending, spreading, or arching, quadrangular-ridged or winged distally; leaf blades narrowly linear, linear, or linear-lanceolate, 0.5–4(–5) mm wide; calyx lobes triangular, lanceolate, or triangular-subulate, not keeled; seeds dark brown . . . . .28. *Agalinis purpurea*

1. Bracts shorter than, sometimes equal to, pedicels.
    15. Leaves slightly divergent or appressed to strongly ascending, blades subulate to narrowly triangular, 0.5–3 mm.
        16. Inflorescences racemes, sometimes with 1–3 mm, floriferous axillary branches; pedicels 1–3(–4) mm; branches stiffly ascending, quadrangular; corolla throats pilose externally . . . . 1. *Agalinis aphylla*
        16. Inflorescences paniculate with solitary flowers; pedicels 3–10(–12) mm; branches widely and laxly spreading, nearly terete; corolla throats glabrous externally. . . 11. *Agalinis filicaulis*
    15. Leaves erect, erect-ascending, ascending, spreading-ascending, recurved, spreading, arching, or reflexed, rarely appressed or ascending-appressed, blades subulate, filiform, linear, linear-filiform, linear-spatulate, linear-elliptic, linear-lanceolate, elliptic, spatulate, or lanceolate, (1–)4–70 mm.
        17. Abaxial corolla lobes pilose externally.
            18. Leaves (alternate, blades fleshy), axillary fascicles equal to or longer than subtending leaves . . . . 12. *Agalinis filifolia*
            18. Leaves (opposite, blades not fleshy), axillary fascicles absent or shorter than subtending leaves.
                19. Corollas: adaxial lobes projected over distal anthers, throats glabrous within across bases and sinus of adaxial lobes; leaf blades narrowly linear to linear-lanceolate; capsules globular; seeds tan to brown . . . 33. *Agalinis tenuifolia* (in part)
                19. Corollas: adaxial lobes recurved, reflexed, or reflexed-spreading, throats villous within across bases and sinus of adaxial lobes; leaf blades filiform to narrowly linear; capsules elliptic-ovate or globular-ovoid; seeds black or yellowish tan.
                    20. Inflorescences racemes, sometimes interrupted by short floriferous branches, flowers 2 per node; capsules elliptic-ovate . . . . 29. *Agalinis setacea*
                    20. Inflorescences racemiform or racemose-paniculate, flowers 1 per node; capsules globular-ovoid.
                        21. Calyx lobes triangular-lanceolate; branches spreading-ascending; seeds yellowish tan; pedicels spreading-ascending . . . . 14. *Agalinis gattingeri*
                        21. Calyx lobes deltate-subulate; branches widely and laxly ascending; seeds black; pedicels widely spreading . . . . 19. *Agalinis laxa*
        17. Abaxial corolla lobes glabrous externally.
            22. Pedicels scabrous; leaves: axillary fascicles usually ½+ length of subtending leaves, sometimes absent.
                23. Corollas: lobes unequal, 2–6 mm, adaxial arched over anthers; calyx lobes triangular-subulate, recurved, 0.6–1.5 mm. 18. *Agalinis homalantha*
                23. Corollas: lobes equal, 6–12 mm, adaxial spreading; calyx lobes subulate, erect, 0.1–0.6 mm . . . . .27. *Agalinis pulchella*
            22. Pedicels glabrous or ± scabridulous; leaves: axillary fascicles absent or shorter than subtending leaves.
                24. Corolla throats glabrous externally, sometimes pilose externally proximal to sinuses.
                    25. Calyces: tubes 1.5–3 mm, lobes 0.1–0.5 mm, deltate; capsules 3–5 mm; styles 6–9(–10) mm; branches widely spreading . . . . 8. *Agalinis divaricata*
                    25. Calyces: tubes 3–5.5 mm, lobes 0.4–2.5 mm, triangular-subulate to subulate or triangular-lanceolate; capsules 5–8 mm; styles 12–17 mm; branches ascending-spreading.
                        26. Leaf blades filiform to narrowly linear; corollas: lobes unequal, adaxial arched over anthers; calyx lobes triangular-subulate to subulate, 0.4–0.8(–1.3) mm . . . . 9. *Agalinis edwardsiana*
                        26. Leaf blades linear; corollas: lobes equal, adaxial erect to spreading; calyx lobes triangular-lanceolate, 0.8–2.5 mm . . . . .31. *Agalinis strictifolia*
                24. Corolla throats pilose externally.

[27. Shifted to left margin.—Ed.]

27. Leaves: proximal and mid reflexed or recurved, distal spreading; anthers 2–3.8 mm; styles 11–16.5 mm; corolla throats glabrous within across bases of adaxial lobes, sparsely villous at sinus.
  28. Calyces hemispheric-campanulate; inflorescences racemes; corollas 17–30 mm. . . . . 4. *Agalinis caddoensis*
  28. Calyces funnelform-obconic; inflorescences racemiform-paniculate; corollas 15–24 mm. . . . . 22. *Agalinis navasotensis*
27. Leaves erect, erect-ascending, ascending-appressed, spreading, or spreading-ascending; anthers 0.6–2.6 mm; styles 3–12 mm; corolla throats villous within across bases and sinus of adaxial lobes.
  29. Branches stiffly arching-ascending, ascending, erect-ascending, or spreading-ascending, quadrangular; capsules globular to oblong; leaves erect to erect-ascending, or ascending, or proximal slightly spreading, distal sometimes ascending-appressed.
    30. Leaf blades subulate, elliptic, or filiform; anthers 1.8–2.6 mm; corollas 15–25 mm, pink to dark pink, with 2 yellow lines and red spots present in abaxial throat, lobes 4–7(–9) mm. . . . . 25. *Agalinis oligophylla*
    30. Leaf blades linear-elliptic, narrowly spatulate, linear-spatulate, or linear; anthers 0.6–2.1 mm; corollas 8–16(–17) mm, whitish to pink, with 2 yellow lines and red or pink spots pale or absent in abaxial throat, lobes 3–5 mm.
      31. Calyx lobes deltate, 0.2–0.5 mm; leaf blades linear-elliptic to narrowly spatulate; anthers 1.5–2.1 mm; stems 30–80(–100) cm . . . . 24. *Agalinis obtusifolia*
      31. Calyx lobes triangular-subulate, 0.3–1.2 mm; leaf blades linear; anthers 0.6–1.2 mm; stems 10–50(–60) cm. . . . . 30. *Agalinis skinneriana*
  29. Branches: proximal arching upward, others spreading-ascending to laxly and widely spreading, subterete to quadrangular-ridged; capsules ovoid, obovoid-oblong, ovoid-globular, or obovoid; leaves spreading to spreading-ascending.
    32. Calyx lobes subulate to triangular or lanceolate, (0.2–)0.5–2.5 mm.
      33. Branches ascending to spreading, quadrangular-ridged; capsules ovoid; calyx lobes subulate to triangular, (0.2–)0.5–1.5(–2) mm; styles (5–)7–12 mm . . . . 6. *Agalinis decemloba*
      33. Branches laxly and widely spreading, proximal arching upward, strongly quadrangular; capsules obovoid; calyx lobes lanceolate, 1.3–2.5 mm; styles 3–5 mm . . . . 34. *Agalinis viridis*
    32. Calyx lobes deltate-subulate or subulate, 0.2–0.5 mm.
      34. Corolla lobes 3–4.5 mm; anthers 1.2–1.8 mm; styles 5–7 mm; leaf blades linear to linear-spatulate; capsules obovoid-oblong; seeds golden brown . . . . 13. *Agalinis flexicaulis*
      34. Corolla lobes 4–7.3 mm; anthers 1.7–2.5 mm; styles 7–9 mm; leaf blades filiform to linear-filiform; capsules ovoid-globular; seeds yellow . . . . 32. *Agalinis tenella*

1. **Agalinis aphylla** (Nuttall) Rafinesque, New Fl. 2: 65. 1837 • Scaleleaf false foxglove E F

Gerardia aphylla Nuttall, Gen. N. Amer. Pl. 2: 47. 1818; *Agalinis microphylla* Rafinesque; *A. oligophylla* Pennell var. *pseudaphylla* Pennell; *A. pseudaphylla* (Pennell) Shinners; *G. pseudaphylla* (Pennell) Pennell

Stems simple or branched, 40–140 cm; branches stiffly ascending, quadrangular, siliceous-ridged, glabrate or scabridulous. **Leaves** appressed or slightly divergent; blade subulate to narrowly triangular, 0.5–3 × 0.2–1 mm, margins entire, siliceous, adaxial surface scabridulous; axillary fascicles absent. **Inflorescences** racemes, sometimes with 1–3 mm, floriferous branches, flowers 1 or 2 per node; bracts shorter than or both shorter and longer than, sometimes equal to, pedicels. **Pedicels** ascending, 1–3(–4) mm, glabrous. **Flowers:** calyx hemispheric, tube (1.5–)2–3.5 mm, glabrous, lobes subulate, 0.2–0.6 mm; corolla pale to dark pink, with 2 yellow lines and dark pink spots in abaxial throat, 12–20 mm, throat pilose externally and villous within across bases and sinus of adaxial lobes, lobes: abaxial spreading, adaxial reflexed-spreading, 4–6 mm, glabrous externally; proximal anthers parallel to filaments, distal perpendicular to filaments, pollen sacs (1.6–)2–3 mm; style exserted, 7–9 mm. **Capsules** globular, 3–5.5 mm. **Seeds** yellowish tan, 0.7–1.4 mm. $2n = 26$.

Flowering early Sep–mid Oct. Bogs, wet flatwoods, hydric savannas, seepage slopes; 0–100 m; Ala., Fla., Ga., La., Miss., N.C., S.C.

Plants of *Agalinis aphylla* grow in more mesic portions of wetlands and rarely grow in the deep, saturated, mucky soil where *A. harperi* and *A. linifolia* are found. The ecotone between these communities is marked by the distribution of *A. aphylla*, which often forms a distinctive border between inundated and saturated to mesic soil. The pedicels, buds, and capsules of *A. aphylla* dry reddish brown to black and contrast markedly with the much paler greenish tan stems and branches. Corollas dry brownish pink, and styles are often pilose.

2. **Agalinis aspera** (Douglas ex Bentham) Britton in N. L. Britton and A. Brown, Ill. Fl. N. U.S. ed. 2, 3: 209. 1913 • Rough false foxglove E

Gerardia aspera Douglas ex Bentham in A. P. de Candolle and A. L. P. P. de Candolle, Prodr. 10: 517. 1846; *Agalinis greenei* Lunell

Stems simple or branched, 7–100 cm; branches ascending, terete, ridged, scabridulous proximally, angular, ridged, scabrous distally, sometimes sparingly. **Leaves** ascending; blade linear to filiform, 15–50(–60) × 0.4–1.5(–2) mm, margins entire, siliceous, adaxial surface scabrous; axillary fascicles well developed, shorter to longer than subtending leaves. **Inflorescences** racemes, flowers 1 or 2 per node; bracts longer than pedicels. **Pedicels** ascending-spreading, 4–20 mm, glabrous or proximally scabridulous. **Flowers:** calyx campanulate, tube (3–)3.5–5 mm, glabrous, lobes triangular-lanceolate, (1.5–)2–4 mm; corolla pink, with 2 yellow lines and dark red spots in abaxial throat, 15–28 mm, throat pilose externally and glabrous within across bases and sinus of adaxial lobes, lobes: abaxial projecting to spreading, adaxial spreading to erect, 3–7 mm, glabrous externally; proximal anthers parallel to filaments, distal perpendicular to filaments, pollen sacs 1.8–3 mm; style included or exserted, 8–15 mm. **Capsules** ovoid-oblong, 6.5–11 mm. **Seeds** dark brown, 0.9–2 mm. *2n* = 26.

Flowering Jul–Sep. Tallgrass or loess hills or upland prairies, dry rocky or sandy soils over limestone, limestone bluffs, gravelly moraines; 200–400 m; Man.; Ill., Iowa, Kans., Minn., Miss., Mo., Nebr., N.Dak., Okla., S.Dak., Wis.

Plants of *Agalinis aspera* have strongly siliceous hairs that are rough to the touch and appear marbled. The style is usually pilose.

3. **Agalinis auriculata** (Michaux) S. F. Blake, Rhodora 20: 71. 1918 • Earleaf false foxglove C E

Gerardia auriculata Michaux, Fl. Bor.-Amer. 2: 20. 1803; *Otophylla auriculata* (Michaux) Small; *Tomanthera auriculata* (Michaux) Rafinesque

Stems simple or branched, (18–)30–100 cm; branches spreading-ascending, obtusely angular, retrorsely short-sericeous and hispid. **Leaves** spreading; blade lanceolate, 12–60 × (2–)5–20(–25) mm, margins and midveins entire or mid to distal ones with 1 or 2 proximal lobes, hispid, adaxial surface scabrous; axillary fascicles absent. **Inflorescences** spikelike racemes, flowers 1 or 2 per node; bracts longer than pedicels, margins with 1 or 2 proximal lobes. **Pedicels** ascending, 0.5–3 mm, hispid. **Flowers:** calyx campanulate, tube (2–)3–9 mm, hirsute, lobes ovate-lanceolate, (5–)7–13 mm, unequal; corolla pink, usually without 2 yellow lines and with dark pink spots in abaxial throat, 16–30 mm, throat pilose externally, villous within across bases and sinus of adaxial lobes, lobes: abaxial projecting to spreading, adaxial erect to recurved, 4–8 mm, abaxial sparsely pilose externally, adaxial glabrous externally; proximal anthers parallel to filaments, distal perpendicular to filaments, pollen sacs 1.2–3 mm; style exserted, 15–16 mm. **Capsules** ovoid, 7–20 mm. **Seeds** brown, 1–2 mm. *2n* = 26.

Flowering Aug–Sep. Seasonally wet meadows, mesic prairies, glades, roadsides, fallow fields; of conservation concern; 30–500 m; Ala., Ark., D.C., Ill., Ind., Iowa, Kans., Ky., Md., Mich., Minn., Miss., Mo., N.J., Ohio, Okla., Pa., S.C., Tenn., Tex., Va., W.Va., Wis.

Agalinis auriculata is rare throughout its relatively broad range and has been the focus of recent field studies. It is probably extirpated in the District of Columbia, Michigan, New Jersey, Texas, and West Virginia; it was rediscovered in Pickens County, Alabama, in 2007 and Lewis County, Kentucky, in 1998. The species is considered critically imperiled in at least 11 states and imperiled in another five and is most abundant in Illinois, eastern Iowa, and northern Missouri.

Agalinis auriculata is in the Center for Plant Conservation's National Collection of Endangered Plants.

4. Agalinis caddoensis Pennell, Proc. Acad. Nat. Sci. Philadelphia 73: 519. 1922 • Caddo false foxglove C E

Gerardia caddoensis (Pennell) Pennell

Stems widely branched, 20–60 cm; branches ascending, angular distally, scabridulous. **Leaves** spreading or reflexed or recurved; blade filiform, 15–35 × 0.3–0.6 mm, margins entire, siliceous, adaxial surface scabrellous; axillary fascicles absent. **Inflorescences** racemes, flowers 1 or 2 per node; bracts equal to or shorter than pedicels. **Pedicels** spreading-ascending, 7–22 mm, proximally scabridulous. **Flowers:** calyx hemispheric-campanulate, tube 3.5–5.5 mm, glabrous, lobes triangular-subulate, 0.8–1.7 mm; corollas rose purple, with 2 yellow lines and red spots in abaxial throat, 17–30 mm, throat pilose externally and glabrous within across bases of adaxial lobes, lobes spreading, 6–9(–11) mm, glabrous externally; proximal anthers parallel to filaments, distal perpendicular to filaments, pollen sacs 2.5–3.8 mm; style exserted, 12–16.5 mm. **Capsules** unknown. **Seeds** unknown.

Flowering Oct. Oak woods, dry loamy soils; of conservation concern; 50–100 m; La.

Agalinis caddoensis is known only from two collections made by F. W. Pennell in 1913 and is most similar morphologically to *A. navasotensis*; see 22. *A. navasotensis* for a comparison. *Agalinis caddoensis* should be expected in northwestern Louisiana and eastern Texas and likely flowers in September as does *A. navasotensis*.

5. Agalinis calycina Pennell, Proc. Acad. Nat. Sci. Philadelphia 81: 141. 1929 • Leoncita false foxglove C

Gerardia calycina (Pennell) Pennell

Stems branched, 40–62 cm; branches ascending, terete to ± quadrangular, glabrous or sparsely scabridulous. **Leaves** spreading; blade narrowly linear, 12–28(–35) × 0.8–1.5 mm, fleshy, margins entire, often siliceous, adaxial surface with siliceous, sessile, dome-shaped hairs; axillary fascicles shorter than subtending leaves. **Inflorescences** racemes or panicles, flowers 1 or 2 per node; bracts longer than pedicels. **Pedicels** ascending, 5–9 mm, glabrous. **Flowers:** calyx elongate-campanulate, tube 4.5–6 mm, glabrous, lobes narrowly lanceolate, 2.7–7 mm; corolla pink, with 2 yellow lines and dark red spots in abaxial throat, 17–23 mm, throat pilose

externally and glabrous within across bases and sinus of adaxial lobes, lobes spreading, 4–6 mm, glabrous externally; proximal anthers parallel to filaments, distal parallel or oblique to filaments, pollen sacs 3–3.2 mm; style included or exserted, 13–15 mm. **Capsules** obovoid to oblong, 10–12 mm. **Seeds** brown, 1–2 mm.

Flowering Aug–Sep. Wet alkaline, saline meadows and marshes associated with springs and seeps; of conservation concern; 800–1100 m; N.Mex., Tex.; Mexico (Coahuila).

Within the flora area, *Agalinis calycina* is known only from desert, alkaline-saline, wet meadows and marshes. It is most closely related to *A. heterophylla*, from which it differs by its narrowly linear leaves, longer pedicels, shorter corolla lobes, included to slightly exserted style, and larger seeds and capsules. In Texas, *A. calycina* is known only from Pecos County in the trans-Pecos region.

6. Agalinis decemloba (Greene) Pennell, Bull. Torrey Bot. Club 40: 434. 1913 • Ten-lobed false foxglove C E

Gerardia decemloba Greene, Pittonia 4: 51, plate 9. 1899; *Agalinis acuta* Pennell; *G. acuta* (Pennell) Pennell

Stems simple or branched, 6–60 cm; branches ascending to spreading, quadrangular-ridged, wingless distally, glabrous or sparsely scabridulous. **Leaves** ascending to spreading; blade linear-filiform to linear, (6–)8–25 × 0.4–1.5 mm, margins entire, often siliceous, adaxial surface finely scabrous; axillary fascicles absent. **Inflorescences** racemes, short to elongate, interrupted, flowers 1 or 2 per node; bracts usually shorter than, sometimes both shorter and longer than, pedicels. **Pedicels** ascending to ± spreading, (3–)6–26 mm, glabrous. **Flowers:** calyx obconic to hemispheric, tube (1.3–)2–4.5(–5.5) mm, glabrous, lobes subulate to triangular, (0.2–)0.5–1.5(–2) mm; corollas pale pink to pink, usually with 2 yellow lines and red spots in abaxial throat, (5–)6–14.5 mm, throat pilose externally and villous within across bases and sinus of adaxial lobes, lobes: abaxial spreading, adaxial erect to recurved, (2–)3–6(–7) mm, glabrous externally; proximal anthers parallel to filaments, distal perpendicular to filaments, pollen sacs 0.8–2.2 mm; style exserted, (5–)7–12 mm. **Capsules** ovoid, 3.5–5 mm. **Seeds** yellowish tan, 0.3–1.2 mm. $2n = 26$.

Flowering Aug–mid Oct. Sand plain grasslands, dry prairie remnants, dry roadsides, cemeteries, margins and openings in mesic to dry mixed woodlands, serpentine grasslands; of conservation concern; 0–700 m; Ala., Conn., D.C., Ky., Md., Mass., N.Y., N.C., Pa., R.I., S.C., Tenn., Va.

J. B. Pettengill and M. C. Neel (2011) provided morphological and molecular evidence that showed *Agalinis decemloba* and *A. acuta* are conspecific. Their data also indicated that *A. tenella* is most closely related to *A. decemloba* (including *A. acuta*) and may merit infraspecific status within *A. decemloba*. The authors maintain *A. tenella* based on morphological characters that include: larger corollas with proportionally larger recurved lobes; larger anthers; longer styles and stigma; and larger plants with many laxly spreading branches than *A. decemloba*. *Agalinis decemloba* is rare, and populations show extreme variation in numbers of plants produced per year. The authors agree with Pettengill and Neel that *A. decemloba*, as circumscribed here to include the federally listed endangered *A. acuta*, is threatened and deserves protection. Study of a population of *A. decemloba* (as *A. acuta*) in Massachusetts showed these plants were self-compatible (Neel 2002). *Agalinis decemloba* is distinguished from *A. obtusifolia* by the following characteristics: stems and branches flexible and drying green versus stiff and drying stramineous, leaves linear to linear-filiform versus linear-elliptic to spatulate, bracts shorter than to slightly longer than pedicels versus much shorter than pedicels, calyx lobes subulate to triangular versus deltate, and corollas usually with two yellow lines and red spots in abaxial throat versus two yellow lines and pink spots absent or pale in abaxial throat.

Agalinis decemloba is in the Center for Plant Conservation's National Collection of Endangered Plants.

7. **Agalinis densiflora** (Bentham) S. F. Blake, Rhodora 20: 71. 1918 • Osage false foxglove E

Gerardia densiflora Bentham, Compan. Bot. Mag. 1: 206. 1836; *Otophylla densiflora* (Bentham) Small; *Tomanthera densiflora* (Bentham) Pennell

Stems simple or branched, 15–82 cm; branches ascending, subterete, retrorsely short-sericeous and hispid. **Leaves** spreading-ascending; blade triangular-ovate, 16–35 × 10–45 mm, margins 2-pinnatifid with 1–3 pairs of lobes, pinnatifid distally, margins and midvein hispid, siliceous, adaxial surface scabrous or glabrous; axillary fascicles absent. **Inflorescences** spikelike racemes, dense, flowers 1 or 2 per node; bracts longer than pedicels, margins pinnatifid. **Pedicels** erect, 0.5–2 mm, scabridulous. **Flowers:** calyx campanulate, tube 4–7 mm, densely, finely scabridulous and hispid, lobes lanceolate, 7–11 mm, unequal; corolla pink to pale pink, usually without 2 yellow lines and with red spots in abaxial throat, 18–33 mm, throat pilose externally and villous within at sinus and/or across bases of adaxial lobes, lobes: abaxial spreading, adaxial erect, 5.5–11 mm, abaxial sparsely pilose externally, adaxial shorter than abaxial, glabrous externally; proximal anthers parallel to filaments, distal oblique or perpendicular to filaments, pollen sacs 1.5–3.8 mm; style exserted, 16–26 mm. **Capsules** ovoid-obovoid, 7–10 mm. **Seeds** dark brown to black, 1.8–3 mm. **2*n*** = 26.

Flowering Aug–Sep. Prairies, grassy roadsides, pastures, well-drained calcareous soils; 100–900 m; Kans., Okla., Tex.

Agalinis densiflora is distinguished from other species of the genus by the pinnatifid leaves, yellow color of the proximal, tubular portion of the corolla, and the asymmetric calyx. The adaxial wall of the calyx tube is flat, the abaxial wall is convex, and the filaments of the shorter pair of stamens are narrower than those of the longer filaments (as the filaments are in *A. auriculata*).

Agalinis densiflora is known from north-central Texas northward to northeastern Kansas.

8. **Agalinis divaricata** (Chapman) Pennell, Bull. Torrey Bot. Club 40: 437. 1913 • Pineland false foxglove E F

Gerardia divaricata Chapman, Fl. South. U.S., 299. 1860

Stems simple or branched, 20–50(–60) cm; branches widely spreading, quadrangular-ridged, scabridulous or glabrate. **Leaves** spreading to ± reflexed at mid stem, distalmost sometimes ascending; blade linear-filiform, 10–30 × 0.2–0.6 mm, margins entire, adaxial surface glabrous or slightly scabrous; axillary fascicles absent. **Inflorescences** racemes, flowers 1 or 2 per node; bracts shorter than pedicels. **Pedicels** spreading to slightly spreading-ascending, 12–28 mm, glabrous. **Flowers:** calyx campanulate to funnelform, tube 1.5–3 mm, glabrous, lobes spreading, deltate, 0.1–0.5 mm; corolla pale to dark pink, rarely white, without 2 yellow lines and red spots in abaxial throat, 14–20 mm, throat glabrous externally and within across bases and sinus of adaxial lobes, lobes: abaxial spreading, adaxial projected flatly over anthers, 2.5–6 mm, unequal, abaxial 4–6 mm, adaxial 2.5–3 mm, glabrous externally; proximal anthers parallel to filaments, distal perpendicular to filaments, pollen sacs 1.4–2.6 mm; style exserted, 6–9(–10) mm. **Capsules** globular, 3–5 mm. **Seeds** yellowish brown to tan, 0.5–0.7 mm. **2*n*** = 28.

Flowering early Sep–mid Oct. Dry open pine savannas, dry oak openings, dry sandy ground, mesic flatwoods, margins of bogs and seepage slopes, dry roadsides; 0–70 m; Ala., Fla., Ga.

Agalinis divaricata is largely a Floridian species, just entering southeasternmost Alabama and southwestern Georgia. In Florida, it is found from the Panhandle to the northern peninsula.

9. Agalinis edwardsiana Pennell, Proc. Acad. Nat. Sci. Philadelphia 73: 522. 1922 • Plateau false foxglove E

Agalinis edwardsiana var. *glabra* Pennell; *Gerardia edwardsiana* (Pennell) Pennell; *G. edwardsiana* subsp. *glabra* (Pennell) Pennell

Stems branched, 37–80 cm; branches ascending-spreading, terete proximally, obtusely quadrangular-ridged distally, glabrate or sparsely scabridulous. **Leaves** mostly spreading, proximal often ± reflexed; blade filiform to narrowly linear, 10–36 × 0.3–1 mm, margins entire, adaxial surface scabridulous; axillary fascicles absent or shorter than subtending leaves. **Inflorescences** racemes, flowers 1 or 2 per node; bracts shorter than pedicels. **Pedicels** ascending, 9–30 mm, scabridulous. **Flowers:** calyx hemispheric, tube 3–5 mm, glabrous, lobes triangular-subulate to subulate, 0.4–0.8(–1.3) mm; corolla pink to rose pink, with 2 yellow lines and red spots in abaxial throat, 14.5–24 mm, throat glabrous externally, sometimes pilose proximal to sinuses, glabrous within across bases of adaxial lobes, sparsely villous at sinus, lobes: abaxial spreading to recurved, adaxial arched over anthers, 4.5–9 mm, unequal, abaxial 5.5–9 mm, adaxial 4.5–6 mm, glabrous externally; proximal and distal anthers perpendicular to filaments, pollen sacs 3–4 mm; style exserted, 13–17 mm. **Capsules** globular to obovoid-globular, 5–7 mm. **Seeds** dark brown, 0.8–1.8 mm. $2n = 26$.

Flowering Aug–Oct. Grasslands, open rocky slopes, calcareous, clay or sandy soils; 200–600 m; Tex.

Agalinis edwardsiana is found mostly on the Edwards Plateau. The species is distinguished from *A. strictifolia* by the nearly straight to rounded abaxial corolla throat of *A. edwardsiana* and the often strongly upcurved abaxial corolla throat of *A. strictifolia*.

10. Agalinis fasciculata (Elliott) Rafinesque, New Fl. 2: 63. 1837 • Fascicled false foxglove

Gerardia fasciculata Elliott, Sketch Bot. S. Carolina 2: 115. 1822; *G. purpurea* Linnaeus var. *fasciculata* (Elliott) Chapman

Stems simple or branched, (10–)50–200 cm; branches ascending, distal portions often arching, subterete proximally to ± quadrangular-ridged distally, moderately to copiously scabrous. **Leaves** spreading; blade narrowly to broadly linear, 15–40(–50) × 1–3 (–4) mm, margins entire, adaxial surface scabrous; axillary fascicles shorter than or equal to subtending leaves. **Inflorescences** racemes, elongate, flowers 2 per node; bracts longer than pedicels. **Pedicels** ascending, 2–5(–6) mm, scabrous. **Flowers:** calyx long-campanulate, tube 3–5 mm, glabrous or scabrous, lobes triangular to triangular-subulate, 0.6–2 mm; corolla pale to dark pink, with 2 yellow lines and dark red spots in abaxial throat, 22–36 mm, throat pilose externally and villous within across bases and sinus of adaxial lobes, lobes: abaxial spreading, adaxial reflexed-spreading, 5–12 mm, pilose externally; proximal anthers parallel to filaments, distal perpendicular to filaments, pollen sacs 2.2–4.6 mm; style strongly exserted, 14–22 mm. **Capsules** globular, 4.5–6 mm. **Seeds** blackish, 0.7–1 mm. $2n = 28$.

Flowering year-round. Dry or wet sandy ground, waste places, roadsides, ditches, pastures, fallow fields, alluvial ground of streams, brackish soils, thickets, lawns; 0–300 m; Ala., Ark., Del., Fla., Ga., Ill., Ind., Kans., Ky., La., Md., Miss., Mo., N.J., N.Y., N.C., Okla., Pa., S.C., Tenn., Tex.; West Indies (Cuba, Puerto Rico).

Although uncommon in the northern and northwestern portions of its range, *Agalinis fasciculata* is the most common and weedy species of the genus in the southeast, and also the largest in North America (especially in peninsular Florida around the strands), with a massive root system and stems with basal diameters to 25 mm. Plants in southern Florida flower intermittently year-round. It is one of the more variable species of the genus in the flora area in size of stems and leaves, and overall indument. See discussion under 28. *A. purpurea* for characters differentiating it and *A. fasciculata*.

11. **Agalinis filicaulis** (Bentham) Pennell, Bull. Torrey Bot. Club 40: 438. 1913 • Wire-stemmed false foxglove [E] [F]

Gerardia aphylla Nuttall var. *filicaulis* Bentham, Compan. Bot. Mag. 1: 174. 1836; *G. filicaulis* (Bentham) Chapman

Stems simple or branched, 20–40(–50) cm; branches widely and laxly spreading, nearly terete, papillate proximally to glabrate distally, sometimes glaucous. **Leaves** appressed to strongly ascending; blade subulate to narrowly triangular, 0.8–2.1 × 0.4–0.7 mm, margins entire, siliceous, abaxial midvein scabrous, adaxial surface glabrate; axillary fascicles absent. **Inflorescences** paniculate, flowers 1 per node; bracts shorter than pedicels. **Pedicels** ascending, 3–10 (–12) mm, glabrous. **Flowers:** calyx campanulate, tube 1.5–2.1 mm, glabrous, lobes triangular, siliceous at tips, 0.1–0.3 mm; corolla pink, without 2 yellow lines and red spots in abaxial throat, 8–11 mm, throat glabrous externally and glabrous within across bases and sinus of adaxial lobes, lobes: abaxial spreading, adaxial arching forward and overlapped resembling a keel, 1.6–4.2 mm, unequal, abaxial 2.5–4.2 mm, adaxial 1.6–2.5 mm, glabrous externally; proximal anthers parallel to filaments, distal perpendicular to filaments, pollen sacs 0.9–1.6 mm; style exserted, 6.5–7.2 mm. **Capsules** globular, 3–4 mm. **Seeds** tan, 0.5–1 mm. $2n = 28$.

Flowering mid Sep–early Nov. Mesic to moist longleaf pine forests, moist prairies, shallow bogs, moist roadsides, remnant wiregrass communities, borrow pits; 0–20 m; Ala., Fla., Ga., La., Miss.

Plants of *Agalinis filicaulis* are among the more diminutive in the genus. The corolla with short, arching adaxial lobes that appear keeled is unique in *Agalinis*. This along with the weak, glaucous stems and small, solitary flowers makes it immediately recognizable.

12. **Agalinis filifolia** (Nuttall) Rafinesque, New Fl. 2: 65. 1837 • Florida false foxglove [E]

Gerardia filifolia Nuttall, Gen. N. Amer. Pl. 1: 48. 1818

Stems often leaning, branched, 30–70 cm; branches ascending to widely spreading, subterete proximally to obtusely quadrangular distally, glabrous or scabridulous. **Leaves** alternate, spreading; blade filiform, widened slightly distally, 10–20(–30) × 0.2–0.6(–1) mm, fleshy, margins entire, adaxial surface sparsely scabrous; axillary fascicles often equal to or longer than subtending leaves. **Inflorescences** racemes, elongate, flowers 1 or 2 per node; bracts shorter than pedicels. **Pedicels** spreading-ascending, 9–33 mm, glabrous. **Flowers:** calyx campanulate, tube 3–5 mm, glabrous, lobes subulate, 0.3–1.5 mm; corolla dark pink to nearly rose, with 2 yellow lines and purple spots in abaxial throat, 15–30 mm, throat pilose externally and villous within across bases and sinus of adaxial lobes, lobes: abaxial spreading, adaxial reflexed-spreading, 5–10 mm, pilose externally; proximal anthers parallel to filaments, distal perpendicular to filaments, pollen sacs 2.5–3.7 mm; style exserted, 10–20 mm. **Capsules** globular, 3.9–5.3 mm. **Seeds** nearly black, 0.4–0.5 mm. $2n = 28$.

Flowering Jun–Nov. Xeric, sandy, open pine forests, open coastal scrub habitats, dunes, open areas of pine flatwoods, hydric soils of pine flatwoods; 0–30 m; Ala., Fla., Ga.

Agalinis filifolia is largely a species of Florida, where it is found in xeric, open, sandy, upland sites throughout the state. The species will, however, tolerate hydric conditions in open, pine flatwoods of southeastern Florida and is more common in south-central Florida than in any other part of its range. In the southernmost portion of its range, it flowers from early June through November, with different populations showing distinct phenologies. *Agalinis filifolia* is uncommon to rare in the Florida Panhandle and in southeastern Georgia, reaching its westernmost distribution along the remaining coastal scrub habitats of southern Baldwin County, Alabama.

13. **Agalinis flexicaulis** Hays, J. Bot. Res. Inst. Texas 4: 1, fig. 1. 2010 • Hampton false foxglove [C] [E]

Stems branched, 30–90 cm; branches spreading-ascending, larger branches widely spreading and decumbent-ascending, quadrangular, glabrous or scabridulous on angles. **Leaves** spreading to spreading-ascending; blade linear to linear-spatulate, 7–13 × 0.8–1.5 mm, margins entire, midvein scabrous; axillary fascicles absent. **Inflorescences** paniculate-racemiform, flowers 1 or 2 per node; bracts shorter than pedicels. **Pedicels** spreading, 4–20 mm, glabrous. **Flowers:** calyx elongate-campanulate, tube 2–3.2 mm, glabrous, lobes deltate-subulate, 0.2–0.5 mm; corolla pink to rose, without 2 yellow lines and sometimes with dark pink spots in abaxial throat, (8–)10–12 mm, throat pilose externally and villous within across bases and sinus of adaxial lobes, lobes: abaxial spreading, adaxial reflexed-spreading, 3–4.5 mm, glabrous externally; proximal anthers parallel to filaments, distal perpendicular to filaments, pollen sacs 1.2–1.8 mm; style exserted, 5–7 mm. **Capsules** obovoid-oblong, 3.8–4.5 mm. **Seeds** golden brown, 0.5–0.6 mm.

Flowering Sep–late Oct. Mesic to moist soils, open, wiregrass-dominated longleaf pine systems, savannas, prairies, seepage slopes, depressed wetlands, disturbed ground; of conservation concern; 0–60 m; Fla.

Agalinis flexicaulis is of conservation concern because of its limited distribution. The species is currently known from Bradford County.

Plants of *Agalinis flexicaulis* are distinguished by their weakly ascending main branches, which become more drooping and lax as they mature. It is most easily confused with the more widespread *A. obtusifolia* and *A. tenella*. *Agalinis flexicaulis* is differentiated from these species by characters in the key.

14. **Agalinis gattingeri** (Small) Small in N. L. Britton and A. Brown, Ill. Fl. N. U.S. ed. 2, 3: 213. 1913 • Midwest false foxglove [E]

Gerardia gattingeri Small, Fl. S.E. U.S., 1078, 1338. 1903

Stems simple or branched, 10–60 cm; branches spreading-ascending, quadrangular-ridged, glabrate, scabridulous, or scabrous. **Leaves** spreading or arching; blade narrowly linear to filiform, 13–30(–40) × 0.4–1.4 mm, not fleshy, margins entire, adaxial surface finely scabrous; axillary fascicles absent. **Inflorescences** racemiform, flowers 1 per node, sometimes with pseudoterminal flowers on lateral branches; bracts shorter than pedicels. **Pedicels** spreading-ascending, (4–)8–35 mm, glabrous. **Flowers:** calyx turbinate to hemispheric, tube 2–4 mm, glabrous, sometimes hairy, lobes triangular-lanceolate, 0.5–1.4(–2.6) mm; corolla pink to pale purple, with 2 yellow lines and dark pink spots in abaxial throat, 7–17 mm, throat pilose externally and villous within across bases and sinus of adaxial lobes, lobes: abaxial spreading, adaxial recurved, 3–5 mm, abaxial pilose externally, adaxial sparsely pilose or glabrous externally; proximal anthers parallel to filaments, distal perpendicular to filaments, pollen sacs 1.3–2.5 mm; style exserted, 7–13 mm. **Capsules** globular-ovoid, 3.5–5 mm. **Seeds** yellowish tan, 0.7–1.2 mm. *2n* = 26.

Flowering mid Aug–Oct. Dry roadsides, open woodlands, forest margins, mesic prairies, glades, bluffs, exposed ridges, alvars, often in cherty limestone, or sandy, rocky soils; 0–500 m; Man., Ont.; Ala., Ark., Ill., Ind., Iowa, Kans., Ky., La., Mich., Minn., Miss., Mo., Nebr., Ohio, Okla., Pa., Tenn., Tex., Wis.

Plants of *Agalinis gattingeri* are most often confused with those of *A. skinneriana* and *A. tenuifolia*. They can be distinguished by features discussed under 33. *A. tenuifolia*. *Agalinis gattingeri* is also confused with *A. skinneriana* from which it can be separated by the mostly solitary flowers on lateral branches of *A. gattingeri* versus the well-formed central raceme of *A. skinneriana*; pink-purple corollas of *A. gattingeri* versus the pale pink to nearly white corollas of *A. skinneriana*; the flexible, well-branched stems of *A. gattingeri* versus the strict, brittle, mostly simple to few-branched stems of *A. skinneriana*; and the pilose abaxial corolla lobes in *A. gattingeri* versus the glabrous external corolla lobes of *A. skinneriana*. Isolated populations of *A. gattingeri* on the islands of Georgian Bay, Ontario, have calyx lobes to 2.6 mm and hairs on the calyx tube, characteristics not seen elsewhere in the species.

15. **Agalinis georgiana** (C. L. Boynton) Pennell, Bull. Torrey Bot. Club 40: 427. 1913 • Boynton's false foxglove [C][E][F]

Gerardia georgiana C. L. Boynton, Biltmore Bot. Stud. 1: 148. 1902

Stems simple or branched, 20–40(–50) cm; branches ascending, quadrangular-ridged, scabridulous on faces and angles. **Leaves** spreading-ascending; blade linear, widened distally, 10–15(–20) × 0.4–1.5 mm, not fleshy, margins entire, midvein sometimes abaxially scabrous, adaxial surface scabrous; axillary fascicles shorter than or equal to subtending leaves. **Inflorescences** racemes, elongate, flowers 1 or 2 per node; bracts longer than pedicels. **Pedicels** ascending, 0.5–3.4 mm, glabrous. **Flowers:** calyx hemispheric, tube 2.5–3 mm, sometimes scabrous proximally, lobes triangular, 1–1.4 mm; corolla pale pink, without 2 yellow lines and red spots, 12–14 mm, throat pilose externally and villous within across bases and sinus of adaxial lobes, lobes: abaxial spreading, adaxial reflexed-spreading, 3–5 mm, abaxial pilose externally, adaxial glabrous externally; proximal anthers parallel to filaments, distal perpendicular to filaments, pollen sacs 1.2–1.7 mm; style included or slightly exserted, 4–7.7 mm. **Capsules** globular, 3.5–5 mm. **Seeds** dark brown, 0.4–0.6 mm.

Flowering late Aug–late Sep(–Oct). Dry or moist longleaf pine savannas, oak openings, wiregrass-dominated seepage slopes, margins of bogs; of conservation concern; 0–80 m; Ala., Fla., Ga.

Although *Agalinis georgiana* is listed as a synonym of *A. fasciculata* in some databases and floras, differences in anatomy and characters used in the key suggest that *A. georgiana* is distinct.

Agalinis georgiana is rare and known from southeastern Alabama, the northwestern panhandle of Florida, and south-central Georgia.

16. Agalinis harperi Pennell in J. K. Small, Fl. Miami, 167, 200. 1913 • Harper's false foxglove

Agalinis delicatula Pennell; *A. pinetorum* Pennell; *A. pinetorum* var. *delicatula* (Pennell) Pennell; *Gerardia harperi* (Pennell) Pennell; *G. pulchella* Pennell; *G. pulchella* var. *delicatula* (Pennell) Pennell

Stems simple or branched, 50–80 cm; branches ascending, subterete, ridged proximally, quadrangular-ridged distally, glabrous or scabridulous. **Leaves** spreading-ascending, widely spaced; blade linear to linear-filiform, 10–30 × 0.5–1.4 mm, not fleshy, margins entire, adaxial surface scabrous; axillary fascicles absent or shorter than subtending leaves. **Inflorescences** racemes, elongate, flowers 2 per node, nodes widely spaced; bracts much longer than pedicels. **Pedicels** erect-ascending, 0.5–4(–5) mm, glabrous. **Flowers:** calyx funnelform-obconic, tube 4–6 mm, glabrous, lobes triangular-subulate, keeled, 1.3–2.8 mm; corolla pink to rose pink, with 2 yellow lines and dark purple spots in abaxial throat, 15–23 mm, throat pilose externally and villous within across bases and sinus of adaxial lobes, lobes: abaxial spreading, adaxial reflexed-spreading, 6–7(–8) mm, abaxial pilose externally, adaxial glabrous externally; proximal anthers parallel to filaments, distal perpendicular to filaments, pollen sacs 2–3 mm; style exserted, 10–16 mm. **Capsules** globular, 4–5(–6) mm. **Seeds** black, 0.6–1 mm. $2n = 28$.

Flowering late Sep–Oct. Wet to mesic savannas, bogs, wet prairies, borders of fresh to brackish marshes, roadsides and ditches, drier sites near these habitats; 0–100 m; Ala., Fla., Ga., La., Miss., N.C., S.C., Tex.; West Indies (Bahamas).

The authors have not seen specimens from southeastern coastal Virginia; *Agalinis harperi* is expected there. The species is known from Liberty and Newton counties, Texas.

17. Agalinis heterophylla (Nuttall) Small in N. L. Britton and A. Brown, Ill. Fl. N. U.S. ed. 2, 3: 209. 1913 • Prairie false foxglove [E]

Gerardia heterophylla Nuttall, Trans. Amer. Philos. Soc., n. s. 5: 180. 1835

Stems branched, 40–100 cm; branches spreading-ascending, obtusely quadrangular-ridged, glabrous or scabridulous, sometimes slightly glaucous. **Leaves** spreading-ascending (primary branches) to erect or ascending (secondary branches); blade narrowly lanceolate to elliptic-lanceolate, 10–40(–50) × 2–6(–7) mm, not fleshy, margins

of proximalmost sometimes 3-cleft, distal entire, midvein sometimes abaxially scabrous, adaxial surface scabridulous to slightly scabrous; axillary fascicles absent, rarely developed. **Inflorescences** racemes, elongate, flowers 2 per node; bracts longer than pedicels. **Pedicels** ascending, 1–5(–6) mm, glabrous. **Flowers:** calyx campanulate, tube 3–5(–7) mm, glabrous, lobes lanceolate, 3–8 mm, unequal; corolla pale to dark pink, with 2 yellow lines and dark purple spots in abaxial throat, 20–32 mm, throat pilose externally and glabrous within across bases of adaxial lobes, sparsely villous at sinus, lobes: abaxial spreading, adaxial spreading-reflexed, 6–9 mm, glabrous externally; proximal anthers parallel to filaments, distal perpendicular to filaments, pollen sacs 2.5–4 mm; style exserted, 11–24 mm. **Capsules** ovoid-oblong, 5–9 mm. **Seeds** dark brown, 0.7–1.1 mm. $2n = 28$.

Flowering late Aug–early Oct. Moist open sites, moist prairies, margins of mesic to wet forests, fallow fields, roadsides, ditches, margins of marshes and ponds, disturbed sandy soils; 0–300 m; Ala., Ark., Kans., La., Miss., Mo., N.C., Okla., Tenn., Tex.

Agalinis heterophylla is the most common species of *Agalinis* in Oklahoma and is common in southern Arkansas, eastern Texas, and throughout Louisiana, except the extreme southeastern portion of the state (J. E. Williams 1973; K. A. Vincent 1982). The species occurs sporadically in eastern Mississippi and is rare in Alabama, Missouri, and Tennessee. It differs from the closely related, and rare, *A. calycina* by wider leaves, shorter pedicels, larger corollas, and smaller capsules. It differs from *A. auriculata*, which has auriculate leaves, retrorse hairs on stems and branches, spikelike racemes, and ovate-lanceolate calyx lobes.

18. Agalinis homalantha Pennell, Proc. Acad. Nat. Sci. Philadelphia 73: 525. 1922 • San Antonio false foxglove [E]

Agalinis nuttallii Shinners; *Gerardia homalantha* (Pennell) Pennell; *G. longifolia* Nuttall 1835, not *A. longifolia* Rafinesque 1837

Stems simple or branched, 30–100 cm; branches ascending, obtusely quadrangular proximally, quadrangular-ridged distally, scabrous. **Leaves** spreading-ascending; blade narrowly linear to linear, 14–35(–40) × 0.6–1.5 mm, not fleshy, margins entire, abaxial midvein scabrous, adaxial surface scabrous; axillary fascicles usually to ½+ as long as subtending leaves, sometimes absent. **Inflorescences** racemes, flowers 1 or 2 per node; bracts shorter than or both shorter and longer than pedicels. **Pedicels** ascending, often arching upwards distally, 6–30 mm, scabrous, sometimes only proximally. **Flowers:** calyx funnelform to obconic, tube 3–5 mm, glabrous, lobes recurved, triangular-subulate, 0.6–1.5 mm; corolla dark pink, with 2 yellow lines and dark purple spots in abaxial throat, 15–26 mm, throat pilose externally and glabrous within across bases of adaxial lobes, sparsely villous at sinus, lobes: abaxial projecting-spreading, adaxial arched over anthers, 2–6 mm, unequal, abaxial 4–6 mm, adaxial 2–4 mm, glabrous externally; proximal anthers perpendicular to filaments, distal oblique or perpendicular to filaments, pollen sacs 3–4 mm; style exserted, 12–18(–20) mm. **Capsules** globular, 5–7 mm. **Seeds** dark brown to nearly black, 0.8–1.6 mm. $2n = 26$.

Flowering late Aug–early Oct. Dry, open woodlands, dry to xeric sandy terrace communities above streams, dry roadsides, open sandy habitats; 0–300 m; Ark., La., Miss., Okla., Tex.

Agalinis homalantha is a common component of dry to xeric, open communities in Oklahoma and Texas and to a lesser extent in Louisiana. In Arkansas, populations are found from Fort Smith along the Arkansas River to the Mississippi River and southward to Bolivar County, Mississippi, where they are often associated with sandy terraces just downstream from dams and levees. Flooding of the sandbanks often carries seeds of *A. homalantha* well away from the streams, where it thrives in recently disturbed, weedy, sandy areas (J. F. Hays 1998).

Agalinis homalantha is distinguished from *A. tenuifolia* by its scabrous branches and pedicels; short, arched adaxial corolla lobes; prominent, wider yellow lines in the corolla; larger anthers with longer awns; and larger seeds.

19. **Agalinis laxa** Pennell, Bull. Torrey Bot. Club 40: 431. 1913 • Long-pedicelled false foxglove [E] [F]

Gerardia laxa (Pennell) Pennell

Stems simple or branched, 20–100 cm; branches widely and laxly ascending, terete proximally, terete or bluntly angular-ridged distally, glabrous or scabridulous. **Leaves** spreading; blade filiform, (7–)10–30 × 0.3–1 mm, not fleshy, margins entire, finely scabrous or nearly glabrous, adaxial surface finely scabrous or glabrate; axillary fascicles absent. **Inflorescences** racemose-paniculate, flowers 1 per node, some flowers pseudoterminal; bracts much shorter than pedicels. **Pedicels** widely spreading, 15–40(–47) mm, glabrous. **Flowers:** calyx narrowly campanulate, tube 2.5–4 mm, glabrous, lobes deltate-subulate, 0.3–1 mm; corolla pale pink to pink, with 2 yellow lines and red spots in abaxial throat, 8–16.5 mm,

throat pilose externally and villous within across bases and sinus of adaxial lobes, lobes: abaxial spreading, adaxial reflexed-spreading, 2.7–6.4(–7) mm, abaxial pilose externally, adaxial glabrous externally; proximal anthers parallel to filaments, distal perpendicular to filaments, pollen sacs 1.3–2.8 mm; style exserted, 5–12 mm. **Capsules** globular-ovoid, 3.7–5 mm. **Seeds** black, 0.4–0.7 mm. $2n = 28$.

Flowering Sep–Oct. Dry to mesic pinelands, pine and oak savannas, sand hills, openings in saw palmetto flatwoods, sandy soils; 0–60 m; Fla., Ga., S.C.

Agalinis laxa occurs in northeastern South Carolina and is expected to occur in southeastern North Carolina; it is most common near the coast in Georgia and northeastern Florida but is found sporadically on the coastal plain as far south as Hernando County, Florida. In the field, *A. laxa* is most easily confused with *A. tenella*; both have laxly spreading branches, widely spaced leaves, long pedicels subtended by much shorter bracts, and inflorescences with some pseudoterminal flowers. *Agalinis laxa* is most easily differentiated from *A. tenella* by pedicels to 40 mm in fruit, shorter and narrower corollas, straight corolla throats, pilose abaxial corolla lobes, and black seeds. *Agalinis tenella* has larger gibbous corollas, glabrous abaxial corolla lobes, and golden yellow seeds.

20. **Agalinis linifolia** (Nuttall) Britton in J. K. Small, Fl. Miami, 167. 1913 • Coastal false foxglove [E]

Gerardia linifolia Nuttall, Gen. N. Amer. Pl. 2: 47. 1818

Stems simple or branched, 70–160 cm; branches stiffly erect or arching-ascending, terete, faintly ridged, glabrous. **Leaves** strongly-ascending; blade linear-lanceolate, 10–50 × 1–3 mm, fleshy, margins entire, adaxial surface with sessile, dome-shaped hairs; axillary fascicles absent. **Inflorescences** racemes, elongate, flowers 1 or 2 per node, some flowers pseudoterminal; bracts both shorter and longer than, or longer than, pedicels. **Pedicels** strongly ascending, 5–25 mm, glabrous. **Flowers:** calyx narrowly campanulate, tube 4–6.5 mm, glabrous, lobes deltate, 0.5–0.9 mm; corolla pink, without 2 yellow lines and with dark pink spots in abaxial throat, 25–35 mm, throat pilose externally and villous within across bases and sinus of adaxial lobes, lobes: abaxial spreading, adaxial reflexed-spreading, 6–14 mm, pilose or sparsely so externally; proximal and distal anthers parallel to filaments, pollen sacs 2.8–5 mm; style exserted, 15–28 mm. **Capsules** subglobular to globular, 5–7 mm. **Seeds** dark brown to blackish, 1.1–1.5 mm. $2n = 28$.

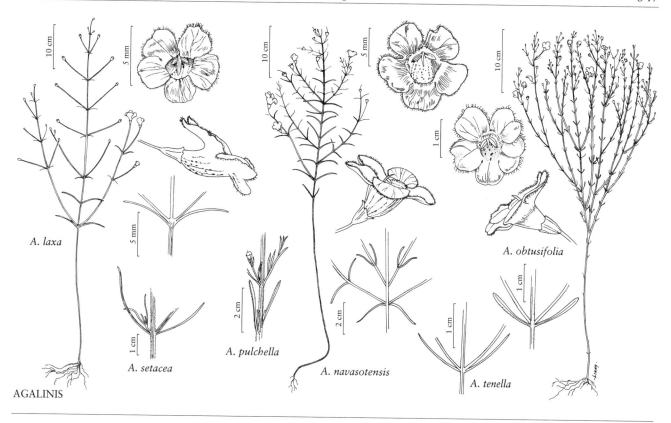

A. laxa

A. setacea

A. pulchella

A. navasotensis

A. tenella

A. obtusifolia

AGALINIS

Flowering late Aug–early Oct. Wet savannas and prairies, depressed areas of mesic savannas or open flatwoods, bogs, margins and centers of cypress domes, margins of ponds, lakes, brackish and freshwater marshes, open stream banks, roadsides, ditches; 0–100 m; Ala., Del., Fla., Ga., La., Md., Miss., N.C., S.C., Va.

Agalinis linifolia is the only perennial member of the genus in North America. It and *A. maritima* are the only obligate wetland species of *Agalinis* in eastern North America; both grow in standing water.

21. Agalinis maritima (Rafinesque) Rafinesque, New Fl. 2: 62. 1837 • Saltmarsh false foxglove

Gerardia maritima Rafinesque, Med. Repos., hexade 2, 5: 361. 1808

Stems erect, elongate main stem overtopping lateral branches, simple or branched, 5–75 cm; branches spreading-ascending to upward arching, terete, ridged proximally, obtusely quadrangular, ridged distally, ± fleshy, glabrous. **Leaves** spreading; blade linear to linear-elliptic, 10–40 × 0.5–3.5 mm, fleshy, wrinkled when dry, margins entire, adaxial surface with sessile, dome-shaped hairs; axillary fascicles absent or shorter than subtending leaves. **Inflorescences** racemes, elongate, flowers 2 per node, distal nodes usually widely spaced; bracts both longer and shorter than, or longer than, pedicels. **Pedicels** spreading-ascending, 2.5–11 mm, glabrous, ± fleshy. **Flowers:** calyx obconic to hemispheric, tube 2–5 mm, glabrous, lobes deltate, 0.7–2 mm, fleshy; corolla pink to rosy pink, with 2 yellow lines and red spots in abaxial throat, 7.5–21 mm, throat pilose externally and villous, glabrate, or glabrous within across bases and sinus of adaxial lobes, lobes: abaxial spreading, adaxial reflexed, 2.5–6 mm, abaxial pilose externally, adaxial glabrous externally; proximal anthers parallel to filaments, distal parallel or perpendicular to filaments, pollen sacs 1.2–2.3 mm; style included or exserted, 6–13 mm. **Capsules** globular, 5–7 mm. **Seeds** dark brown, 0.6–1.6 mm. $2n = 28$.

Varieties 2 (2 in the flora): e, s North America, Mexico, West Indies.

Some mixing of characters between varieties occurs along the coasts of Virginia and northern North Carolina.

1. Calyx lobes erect, apices truncate or acute; filaments glabrous or glabrate; corolla throats glabrous or glabrate within across bases and sinus of adaxial lobes, or sparsely villous within at lateral-lobe sinuses . . . . 21a. *Agalinis maritima* var. *maritima*
1. Calyx lobes divergent, apices obtuse to acute; filaments lanate; corolla throats villous within across bases and sinus of adaxial lobes .21b. *Agalinis maritima* var. *grandiflora*

21a. Agalinis maritima (Rafinesque) Rafinesque var. **maritima** E

Stems simple or branched, 5–40 cm. **Flowers:** calyx tube 2–3 mm, lobes erect, apex truncate or acute; corolla 7.5–12 mm, throat glabrous or glabrate within across bases and sinus of adaxial lobes, or sparsely villous at lateral-lobe sinuses; anthers and filaments glabrous or glabrate; style included, 6–9 mm.

Flowering Jul–Sep. Salt marshes, salt flats, tidal flats, edges of saline or brackish pools, ditches; 0–10 m; N.S.; Conn., Del., Maine, Md., Mass., N.H., N.Y., N.C., R.I., Va.

Variety *maritima* is known historically from New Brunswick; it is considered extirpated in that province.

21b. Agalinis maritima (Rafinesque) Rafinesque var. **grandiflora** (Bentham) Pennell, Proc. Acad. Nat. Sci. Philadelphia 81: 154. 1929

Gerardia maritima Rafinesque var. *grandiflora* Bentham, Compan. Bot. Mag. 1: 208. 1836

Stems branched, 35–75 cm. **Flowers:** calyx tube 3–5 mm, lobes divergent, apex obtuse to acute; corolla 12–19(–21) mm, throat villous within across bases and sinus of adaxial lobes; anthers and filaments lanate; style exserted, 10–14 mm.

Flowering year-round. Saline coastal areas, salt marshes, edges of saline or brackish pools, open salt terns, edges of ditches, salt flats, and tidal flats; 0–10 m; Ala., Fla., Ga., La., Miss., N.C., S.C., Tex., Va.; e Mexico; West Indies.

22. Agalinis navasotensis Dubrule & Canne-Hilliker, Sida 15: 426, figs. 1–7. 1993 • Navasota false foxglove C E F

Stems branched, 25–80 cm; branches spreading-ascending, nearly terete proximally, obtusely quadrangular-ridged distally, glabrous or scabridulous distally. **Leaves** proximal to mid reflexed or recurved, distal spreading; blade filiform, (11–)17–30(–40) × 0.5–1.2 mm, not fleshy, margins entire, siliceous, abaxial midvein scabridulous, adaxial surface scabridulous; axillary fascicles absent. **Inflorescences** racemiform-paniculate, flowers 1 or 2 per node; bracts both longer and shorter than, or shorter than, pedicels. **Pedicels** ascending-

spreading, (2–)6–25 mm, scabridulous proximally or glabrous. **Flowers:** calyx funnelform-obconic, tube 2.2–4.6 mm, glabrous, lobes triangular-subulate to subulate, 0.5–1.5 mm; corolla pink to rose, with 2 yellow lines and red spots in abaxial throat, 15–24 mm, throat pilose externally and glabrous within across bases of adaxial lobes, sparsely villous at sinus, lobes spreading, 5–7 mm, equal, glabrous externally; proximal anthers parallel to filaments, distal perpendicular to filaments, pollen sacs 2–3.2 mm; style exserted, 11–15 mm. **Capsules** ovoid to obovoid-oblong, (4–)6–7 mm. **Seeds** dark brown, 0.8–2.3 mm. $2n = 26$.

Flowering Sep–Oct. Rocky prairie remnants on sandstone outcrops, sandy clay soils of longleaf pine savannas; of conservation concern; 90–100 m; Tex.

Agalinis navasotensis is known from Grimes and Tyler counties in eastern Texas; it should be looked for elsewhere in eastern Texas and northwestern Louisiana. *Agalinis navasotensis* differs from *A. caddoensis* by subtleties of its calyx shape, leaf length, corolla length, inflorescence form, and offset pollen sacs. Additional collections may show that *A. navasotensis* and the morphologically similar, but poorly known, *A. caddoensis* of western Louisiana are conspecific.

Agalinis navasotensis is in the Center for Plant Conservation's National Collection of Endangered Plants.

23. Agalinis neoscotica (Greene) Fernald, Rhodora 23: 139. 1921 • Nova Scotia false foxglove E

Gerardia neoscotica Greene, Leafl. Bot. Observ. Crit. 2: 106. 1910; *Agalinis paupercula* (A. Gray) Britton var. *neoscotica* (Greene) Pennell & H. St. John; *A. purpurea* (Linnaeus) Pennell var. *neoscotica* (Greene) B. Boivin

Stems simple or branched, 5–30(–47) cm; branches spreading-ascending, obtusely quadrangular proximally to quadrangular with wings on angles distally, glabrous or sparsely scabridulous. **Leaves** spreading; blade linear-elliptic to elliptic, 6–40 × 0.8–5 mm, not fleshy, margins entire, abaxial midvein sparsely scabrous, adaxial surface scabrous; axillary fascicles absent or shorter than subtending leaves. **Inflorescences** racemes, flowers 2 per node; bracts longer than pedicels. **Pedicels** spreading-ascending, (0.8–)1.5–5.3(–6.5) mm, glabrous. **Flowers:** calyx hemispheric-campanulate, tube 1.5–3 mm, glabrous, lobes triangular-lanceolate, 1.1–4(–5) mm, unequal, mid adaxial shortest; corolla pale to dark pink, with 2 yellow lines and red spots in abaxial throat, or lines faint or absent, (8–)10–15 mm, throat pilose externally and villous within across bases and sinus of adaxial lobes, lobes: abaxial slightly spreading to projected, adaxial slightly spreading or projected distal to corolla mouth, 1.6–6 mm, unequal, pilose or densely so externally; proximal and distal anthers parallel to filaments, pollen sacs 1–2 mm; style included, (3–)4–7.5 mm. **Capsules** globular, 5–6 mm. **Seeds** brown, 0.9–1.5 mm. $2n = 28$.

Flowering late Jul–early Sep. Sandy or peaty soils, margins of pools, lakes, bogs, estuaries, and marshes, sand flats, dune hollows, ditches; 0–10 m; N.B., N.S.; Maine.

Agalinis neoscotica is rare in Maine, occurring only in Washington County on shores and tidal pools, and in adjacent eastern New Brunswick; it is most abundant in southwestern Nova Scotia and is reported from neighboring Grand Manan Island and Sable Island on the extreme northeastern edge of the range of *Agalinis*.

Features that characterize *Agalinis neoscotica* are narrow tubular corolla throats with red spots in two rows on the two yellow lines (the lines sometimes absent); corolla lobes that either project forward or are only slightly spreading; relatively small anthers, all held parallel to the filaments; two-lobed stigmas, obvious in live plants but rarely on dried specimens; both anthers and stigma included within the corolla throat; calyx in which the middle adaxial lobe is shortest and curved toward the corolla; bracts longer than both the pedicels and flowers they subtend and, sometimes, longer than the main stem leaves; and leaves with narrowed bases that terminate at a creaselike abscission zone. *Agalinis neoscotica* has a mixed mating system (H. M. Stewart et al. 1996). Self-pollination occurs in bud but delayed pollen germination allows for out-crossing. Stewart and J. M. Canne (1998) presented data that showed *A. neoscotica* flower development and morphology differ from those of *A. purpurea*.

24. Agalinis obtusifolia Rafinesque, New Fl. 2: 64. 1837 • Savanna false foxglove E F

Agalinis parvifolia (Chapman) Small; *Gerardia parvifolia* Chapman; *G. setacea* J. F. Gmelin var. *parvifolia* Bentham

Stems simple or branched, 30–80(–100) cm; branches erect-ascending to arching-ascending, quadrangular, with siliceous ridges on angles and, often, faces proximal to leaves, glabrous, sometimes scabridulous on ridges and at nodes. **Leaves** erect to erect-ascending; blade linear-elliptic to narrowly spatulate, most widened distally, 10–20 × 0.4–1.5 mm, margins entire, adaxial surface scabrous; axillary fascicles absent. **Inflorescences** racemiform-paniculate, flowers 1 or 2 per node, interrupted by short multinoded branches bearing pseudoterminal flowers subtended by

tiny bractlets; bracts shorter than pedicels. **Pedicels** spreading-ascending, 4–25 mm, glabrous, rarely scabridulous. **Flowers:** calyx hemispheric, tube 2–3 mm, glabrous, lobes deltate, 0.2–0.5 mm; corolla pink, with 2 yellow lines and pink spots pale or absent in abaxial throat, 12–15(–17) mm, throat pilose externally and villous within across bases and sinus of adaxial lobes, lobes: abaxial spreading, adaxial reflexed-spreading, 3–5 mm, glabrous; proximal anthers parallel to filaments, distal perpendicular to filaments, pollen sacs 1.5–2.1 mm; style exserted, 5–7 mm. **Capsules** globular to oblong, 4–5 mm. **Seeds** pale yellowish brown, 0.6–0.8(–1) mm. $2n = 26$.

Flowering Sep–late Oct. Mesic to dry savannas, dry roadsides with native vegetation, open rocky ground, open pine flatwoods, cutover and edges of pine plantations, margins of bogs and seepage slopes; 0–100 m; Ala., Del., Fla., Ga., Ky., La., Md., Miss., N.C., Pa., S.C., Tenn., Va.

Agalinis obtusifolia is usually found in dry habitats and less frequently in more hydric conditions. The following suite of characters is useful for differentiating *A. obtusifolia* from *A. decemloba*, *A. flexicaulis*, and *A. skinneriana*: stems and branches stiffly erect, brittle, prominently siliceous-ridged, stramineous when dried; leaves obtusely and narrowly elliptic-spatulate, margins scabrous and revolute, stramineous when dried; calyx lobes minute and deltoid; corolla lacking internal markings or markings faint; and inflorescences of interrupted racemes with secondary and tertiary branches bearing pseudoterminal flowers. Characters useful for differentiating *A. obtusifolia* and *A. tenella* are discussed under 32. *A. tenella*.

F. W. Pennell (1929) used the name *Agalinis erecta* (J. F. Gmelin) Pennell for *A. obtusifolia*. Pennell based his combination on *Anonymos erecta* Walter, an invalid name.

25. Agalinis oligophylla Pennell, Bull. Torrey Bot. Club 40: 432. 1913 • Ridgestem false foxglove E F

Gerardia aphylla Nuttall var. *grandiflora* Bentham, Compan. Bot. Mag. 1: 174. 1836; *G. microphylla* Small

Stems simple or branched, 30–90(–110) cm; branches spreading-ascending, quadrangular, with siliceous ridges on angles, scabridulous. **Leaves** ascending; blade subulate, elliptic, or filiform, (1–)4–13 × 0.3–1.1 mm, margins entire, heavily siliceous, scabridulous, adaxial surface siliceous, scabridulous; axillary fascicles absent. **Inflorescences** racemiform-paniculate, with short multinoded floriferous branches, some with pseudoterminal flowers, flowers 1 or 2

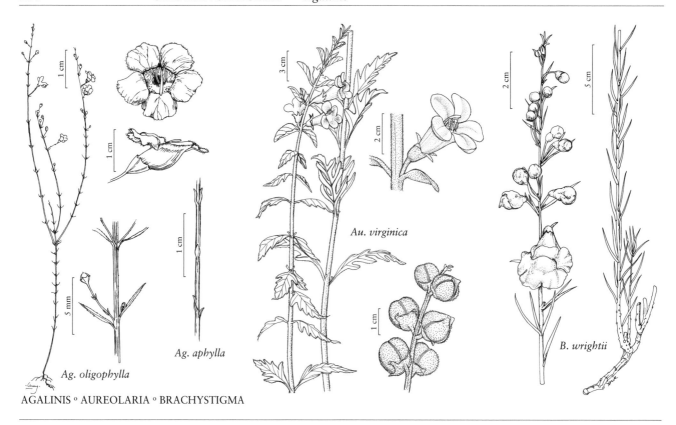

Au. virginica

Ag. aphylla

Ag. oligophylla

B. wrightii

AGALINIS ° AUREOLARIA ° BRACHYSTIGMA

per node; bracts shorter than or sometimes equal to pedicels. **Pedicels** ascending, 4–16 mm, ± scabridulous. **Flowers:** calyx campanulate to funnelform, tube 2.5–3.2 mm, glabrous, lobes subulate to triangular-subulate, 0.1–1 mm; corolla pink to dark pink, with 2 yellow lines and red spots in abaxial throat, 15–25 mm, throat pilose externally and villous within across bases and sinus of adaxial lobes, lobes: abaxial spreading, adaxial reflexed-spreading, 4–7(–9) mm, glabrous externally; proximal anthers parallel to filaments, distal oblique or perpendicular to filaments, pollen sacs 1.8–2.6 mm; style exserted, 6–11 mm. **Capsules** globular, 4–6 mm. **Seeds** tan to yellow, 0.5–0.9 mm. $2n = 26$.

Flowering late Sep–mid Nov. Moist to dry longleaf pine savannas, edges of pine plantations, dry roadsides, chalky outcrops, seepage slopes of clay roadsides; 0–300 m; Ala., Ark., La., Miss., Tenn., Tex.

Plants of *Agalinis oligophylla* are rare in the easternmost portions of the range; they are abundant farther west into Texas. *Agalinis oligophylla* is often the last species of *Agalinis* to begin flowering, usually in late September, sometimes not until mid October and often remains in flower well into November. Unbranched plants of *A. oligophylla* can be confused with *A. aphylla*, with which it often shares the habitat. *Agalinis oligophylla* can be distinguished from *A. aphylla* by the longer leaves, more diffusely branched inflorescence, glabrous style, and the much later flowering period of *A. oligophylla*. Also, on the older buds of *A. oligophylla* the corolla is globular, but

in *A. aphylla* the corolla on older buds is obovoid. Corollas of *A. oligophylla* are asymmetric; the adaxial side of the corolla throat is conspicuously shorter than the upcurved abaxial side of the corolla. The buds, pedicels, and capsules of *A. oligophylla* turn dark brown or blackish, and corollas turn dark purplish or pinkish brown when dried. These dark structures contrast strikingly with the pale green to gray (or reddish) branches and leaves.

26. **Agalinis plukenetii** (Elliott) Rafinesque, New Fl. 2: 63. 1837 (as plukeneti) • Plukenet's false foxglove E

Gerardia plukenetii Elliott, Sketch Bot. S. Carolina 2: 114. 1822

Stems often leaning, simple or branched, 30–100 cm, bushy; branches spreading-ascending, subterete proximally to quadrangular-ridged distally, glabrous or sparsely scabridulous distally. **Leaves** widely spreading to slightly ascending; blade filiform, 18–45 × 0.2–0.8 mm, not fleshy, margins entire, adaxial surface scabridulous; axillary fascicles absent. **Inflorescences** racemiform, flowers 1 per node, some flowers pseudoterminal; bracts longer than pedicels. **Pedicels** spreading-ascending, 3–8 mm, glabrous. **Flowers:** calyx hemispheric, tube 3–5 mm, glabrous, lobes deltate-subulate, 0.2–1 mm; corolla dark pink to rose pink, with 2 yellow lines and dark pink spots in abaxial throat, 18–30 mm, throat pilose externally and villous within across bases and sinus of adaxial lobes, lobes: abaxial spreading, adaxial reflexed-spreading, 6–10 mm, abaxial pilose externally, adaxial glabrous externally; proximal anthers parallel to filaments, distal perpendicular to filaments, pollen sacs 2.2–4 mm; style exserted, 11–17 mm. **Capsules** globular, 4–5 mm. **Seeds** dark brown to black, 0.5–0.8 mm. $2n = 28$.

Flowering late Sep–early Nov. Dry to xeric, sandy, gravelly or clay roadsides, pine-oak forests, margins of savannas, disturbed ground; 0–300 m; Ala., Fla., Ga., La., Miss., S.C., Tenn.

Agalinis plukenetii can be bushy, relatively large, and showy. *Agalinis plukenetii* is a common component of dry to xeric roadsides in the southern portions of its range. It readily colonizes open, dry ground with very little vegetation. It was reported in South Carolina (F. W. Pennell 1935) based on a specimen that has little label data; its occurrence there is doubtful.

27. Agalinis pulchella Pennell, Bull. Torrey Bot. Club 40: 428. 1913 • Beautiful false foxglove [E] [F]

Gerardia pulcherrima Pennell

Stems branched, 50–120 cm; branches spreading-ascending, quadrangular-ridged, scabrous. **Leaves** spreading-ascending; blade filiform, 16–40 × 0.4–1 mm, margins entire, midvein harshly scabrous, adaxial surface scabrous; axillary fascicles: length ½–1 times subtending leaves. **Inflorescences** racemiform-paniculiform, flowers 1 per node, some flowers pseudoterminal; bracts shorter than pedicels. **Pedicels** spreading-ascending, 12–50 mm, scabrous. **Flowers:** calyx hemispheric, tube 3–4 (–5) mm, glaucous, lobes erect, subulate, 0.1–0.6 mm; corolla dark pink to rose, with 2 yellow lines and purple spots in abaxial throat, 22–33 mm, throat sparsely pilose externally and glabrous within across bases of adaxial lobes, sparsely villous at sinus, lobes: abaxial reflexed-spreading, adaxial spreading, 6–12 mm, equal, glabrous externally; proximal anthers parallel to filaments, distal perpendicular to filaments, pollen sacs 2.5–3.8 mm; style strongly exserted, 9–18 mm. **Capsules** globular, 4–6 mm. **Seeds** black, 0.5–0.8 mm. $2n = 26$.

Flowering Sep–early Oct. Dry, open pine savannas, open pine-oak sandhills, dry upslope areas of wiregrass-dominated mesic prairies, chalky glades or roadsides, dry sandy or clay roadsides beside existing or remnant savannas; 0–100 m; Ala., Fla., Ga., La., Miss., Tex.

Agalinis pulchella is an uncommon species in the easternmost area of its range and is common only westward in southeastern Texas.

28. Agalinis purpurea (Linnaeus) Pennell, Bull. Torrey Bot. Club 40: 126. 1913 • Purple false foxglove, gérardie pourpre [E]

Gerardia purpurea Linnaeus, Sp. Pl. 2: 610. 1753

Stems simple or branched, 7–120 cm; branches ascending, spreading, or arching, quadrangular-ridged to winged distally, glabrate or scabridulous to sparsely scabrous, especially on angles or also on faces near nodes. **Leaves** spreading to ascending; blade narrowly linear to linear or linear-lanceolate, 7.5–50 × 0.5–4(–5) mm, not fleshy, margins entire, abaxial midvein sometimes scabrous, adaxial surface scabrous; axillary fascicles absent or shorter than subtending leaves. **Inflorescences** racemes, short to elongate, flowers 1 or 2 per node; bracts longer than pedicels. **Pedicels** spreading-ascending, 1–5(–6) mm, glabrous, rarely scabrous. **Flowers:** calyx tubular-campanulate to hemispheric, tube (2–)3–5 mm, glabrous, rarely scabrous on major veins, lobes triangular, lanceolate, or triangular-subulate, not keeled, 0.8–3(–5) mm; corolla pink to rosy pink, with 2 yellow lines and red spots in abaxial throat or these pale or absent, 9–36 mm, throat pilose externally and villous within across bases and sinus of adaxial lobes, lobes: abaxial spreading to reflexed or projected to somewhat spreading (small-flowered plants), adaxial erect to reflexed, 2.5–11 mm, pilose externally; proximal anthers parallel to filaments, distal oblique or perpendicular to filaments (parallel in small-flowered plants), pollen sacs 0.9–4.5 mm; style included or exserted, 6.5–21 mm. **Capsules** globular, 4–6(–7) mm. **Seeds** dark brown, 0.7–1.4 mm. $2n = 28$.

Varieties 2 (2 in the flora): c, e North America.

Reports of *Agalinis purpurea* from the West Indies have not been confirmed.

1. Corollas 18–36 mm, lobes (4–)5–11 mm; styles strongly exserted; distal anthers perpendicular to filaments; stems 40–120 cm, branches ascending, spreading, or arching . . . . 28a. *Agalinis purpurea* var. *purpurea*
1. Corollas 9–20 mm, lobes 2.5–5(–6) mm; styles included or slightly exserted; distal anthers oblique or parallel to filaments; stems 7–60 cm, simple or branches ascending . . . . .28b. *Agalinis purpurea* var. *parviflora*

28a. Agalinis purpurea (Linnaeus) Pennell var. **purpurea** E

Agalinis keyensis Pennell; *A. purpurea* var. *carteri* Pennell; *A. virgata* Rafinesque; *Gerardia purpurea* Linnaeus var. *carteri* (Pennell) Pennell; *G. racemulosa* Pennell

Stems 4–120 cm; branches ascending, spreading, or arching. **Flowers:** calyx tube 3–5 mm; corolla 18–36 mm, lobes spreading to reflexed, (4–)5–11 mm; distal anthers exserted, perpendicular to filaments, pollen sacs 2–4.5 mm; style strongly exserted.

Flowering Aug–Oct. Moist sandy soils at margins of ponds, lakes, streams, bogs, freshwater and brackish marshes, ditches, pine barrens, meadows, prairies, coastal swales, dry soils, margins of pine plantations, inundated marl prairies; 0–1100 m; Ont.; Ala., Ark., Conn., Del., D.C., Fla., Ga., Ill., Ind., Iowa, Kans., Ky., La., Md., Mass., Mich., Minn., Miss., Mo., Nebr., N.J., N.Y., N.C., Ohio, Pa., R.I., S.C., Tenn., Tex., Va., W.Va., Wis.

Reports of var. *purpurea* from Maine have not been verified by the authors.

Variety *purpurea* is most easily confused with *Agalinis harperi* but differs from *A. harperi* by characters used in the key and by having leaves usually as long as, or longer than, the internodes they subtend. Plants of *A. harperi* have leaves shorter than the elongated internodes. The internodes immediately proximal to the racemes are especially elongated in *A. harperi*, imparting a much more open appearance than the often bushy and leafy var. *purpurea*. The narrowly linear-leaved plants of var. *purpurea* found in brackish marshes and pine barrens appear more leafy than *A. harperi* because of many axillary fascicles and/or axillary branchlets. The calyx lobes of var. *purpurea* are variable in shape and length but are not keeled like those of *A. harperi*.

Plants of var. *purpurea* usually are easily distinguished from *Agalinis fasciculata* by their glabrate stems and few to no axillary fascicles versus the often harshly, and more uniformly, scabrous stems and abundant axillary fascicles of *A. fasciculata*. Variety *purpurea* has spreading to broadly arching branches; branches of *A. fasciculata* are ascending from a clearly defined elongated main stem and the terminal racemes arch downward as fruits mature. Some plants of var. *purpurea* in Arkansas, Louisiana, and Texas are more pubescent than usual and can be confused with *A. fasciculata*. The larger, recurved trichomes are concentrated along the angles of the main stem and largest branches on *A. purpurea*. On *A. fasciculata* the larger trichomes occur on both the angles and faces of the main stem and largest branches. Both taxa may have scabrous branches, especially distally. It is best to evaluate indument on the mid region of the main stem and largest branches when comparing *A. fasciculata* and *A. purpurea*.

28b. Agalinis purpurea (Linnaeus) Pennell var. **parviflora** (Bentham) B. Boivin, Phytologia 22: 337. 1972 E

Gerardia purpurea Linnaeus var. *parviflora* Bentham, Compan. Bot. Mag. 1: 208. 1836; *Agalinis paupercula* (A. Gray) Britton; *A. paupercula* var. *borealis* Pennell; *G. paupercula* (A. Gray) Britton; *G. paupercula* subsp. *borealis* (Pennell) Pennell; *G. purpurea* var. *paupercula* A. Gray

Stems 7–60 cm; branches ascending. **Flowers:** calyx tube 2–3.5 mm; corolla 9–20 mm, lobes projecting to spreading, 2.5–5(–6) mm; distal anthers included or exserted, oblique or perpendicular to filaments, pollen sacs 0.9–2 mm; style included or slightly exserted.

Flowering Jul–Sep. Moist sandy, clay, or peaty soils at margins of ponds, lakes, streams, bogs, freshwater or brackish marshes, ditches, roadsides; 0–300 m; Man., N.B., N.S., Ont., Que.; Conn., Del., D.C., Ill., Ind., Iowa, Maine, Mass., Mich., Minn., N.H., N.J., N.Y., Ohio, Pa., R.I., Vt., Wis.

Plants of var. *purpurea* are shorter and less branched in the northernmost part of their range, and the flowers are not notably smaller. In the Great Lakes region, New England states, and neighboring areas of Canada, var. *parviflora* has smaller, narrower corollas with included stamens and style, and ascending branches. Length and form of the calyx lobes are of no value for differentiation of varieties within *Agalinis purpurea*.

29. Agalinis setacea (J. F. Gmelin) Rafinesque, New Fl. 2: 64. 1837 • Threadleaf false foxglove E F

Gerardia setacea J. F. Gmelin, Syst. Nat. 2: 928. 1792; *Agalinis holmiana* (Greene) Pennell; *G. holmiana* Greene

Stems branched, 15–72 cm, bushy; branches spreading-ascending, terete to quadrangular proximally, quadrangular-ridged distally, glabrous or sparsely scabridulous. **Leaves** spreading or arching; blade filiform to sometimes narrowly linear, 10–45 × 0.2–1(–1.5) mm, not fleshy, margins entire, abaxial midvein occasionally scabridulous, adaxial surface finely scabrous; axillary fascicles absent or shorter than subtending leaves. **Inflorescences** racemes, sometimes interrupted by floriferous branches, flowers 2 per node; bracts shorter than or nearly equal to pedicels.

Pedicels ascending-spreading, (5–)10–35 mm, glabrous. **Flowers:** calyx turbinate-obconic, tube 2.2–4.4 mm, glabrous, lobes triangular-subulate, 0.3–0.9 mm; corolla pink to rosy pink, with 2 yellow lines and red spots in abaxial throat, (9–)13–22 mm, throat pilose externally and villous within across bases and sinus of adaxial lobes, lobes: abaxial spreading to reflexed, adaxial reflexed, 3.5–7 mm, pilose externally throughout or only proximally; proximal anthers parallel to filaments, distal perpendicular to filaments, pollen sacs 2–3.2 mm; style exserted, 8–15 mm. **Capsules** elliptic-ovate, 3–4 mm. **Seeds** black, 0.4–0.7 mm. *2n* = 28.

Flowering Aug–Oct. Dry sandy soils, pine savannas, margins of pine-oak woodlands and scrub, sand hills, roadside embankments; 0–500 m; Ala., Del., D.C., Fla., Ga., Md., N.J., N.Y., N.C., S.C., Tenn., Va.

Agalinis setacea is most common in North Carolina, South Carolina, and Virginia, and is much more rare or localized in other states.

Agalinis setacea is largely replaced toward the south and west on the coastal plain by *A. plukenetii*. Plants similar to *A. setacea* but with more slender leaves and stems, longer internodes, and racemes similar to those in *A. plukenetii* occur from the mid-western peninsula of Florida near Tampa east to Lake Wales area south to Sebring. F. W. Pennell (1920, 1929) referred these plants to *A. stenophylla* Pennell. That species may warrant recognition based on the examination of specimens from Tampa and along the Lake Wales Ridge.

30. **Agalinis skinneriana** (Alph. Wood) Britton in N. L. Britton and A. Brown, Ill. Fl. N. U.S. ed. 2, 3: 212. 1913 • Skinner's false foxglove E

Gerardia skinneriana Alph. Wood, Class-book Bot. ed. 2, 408. 1847

Stems simple or branched, 10–50(–60) cm; branches ascending, sharply quadrangular, siliceous ridges and/or wings on angles distally, scabrous or scabridulous mostly on angles, especially proximally. **Leaves:** proximal slightly spreading, distal ascending-appressed; blade linear, 5–20(–25) × 0.5–1.5 mm, not fleshy, margins entire, abaxial midvein scabrous, adaxial surface scabrous; axillary fascicles absent. **Inflorescences** racemes, flowers 2 per node on central terminal raceme; bracts both longer and shorter than, or shorter than, pedicels, secondary branches with shorter pedicels, bracts shorter than pedicels. **Pedicels** erect-ascending, (3–)5–17(–30) mm, glabrous or sparsely scabridulous. **Flowers:** calyx hemispheric, tube 2–4.5 mm, glabrous, lobes triangular-subulate, 0.3–1.2 mm; corolla whitish to pale pink or pink, with 2 pale yellow lines and red spots in abaxial throat or lines absent, 8–16 mm, throat pilose externally

and villous within across bases and sinus of adaxial lobes, lobes: abaxial spreading, adaxial spreading to reflexed, 3–4 mm, glabrous externally; proximal anthers parallel to filaments, distal perpendicular to filaments, pollen sacs 0.6–1.2 mm; style exserted, 4–8 mm. **Capsules** globular, 3.5–5 mm. **Seeds** yellow, 0.6–1 mm. *2n* = 26.

Flowering Aug–early Oct. Dry to mesic areas in sand and loess hill prairies, bluff prairies, dolomite glades, dunes, open woods, areas of low and/or sparse herbaceous cover; 10–300 m; Ont.; Ark., Ill., Ind., Iowa, Kans., La., Md., Mich., Miss., Mo., Ohio, Wis.

Agalinis skinneriana is rare throughout its relatively wide range. Specimens of *A. skinneriana* are often misidentified as *A. gattingeri* or *A. tenuifolia*, which also have elongated pedicels. Characters that differ between *A. gattingeri* and *A. skinneriana* are given in the discussion of 14. *A. gattingeri*. Although *A. skinneriana* and *A. tenuifolia* both have low wings on the stem angles, the flowers of the two species differ. The corolla throat is villous within at the bases of the reflexed adaxial lobes in *A. skinneriana*, while the corolla throat is glabrous within at the bases of forward-projecting adaxial lobes of *A. tenuifolia*. The abaxial corolla lobes are glabrous externally in *A. skinneriana*, but the abaxial lobes of *A. tenuifolia* are pilose externally. The main stem on plants of *A. skinneriana* is often simple or has few branches, and the leaves are strongly ascending to appressed on secondary branches. Stems of *A. tenuifolia* are usually much branched and the leaves are spreading, arching, or reflexed, sometimes ascending, but not appressed. Plants of *A. tenuifolia* may have axillary fascicles; fascicles are absent in *A. skinneriana*.

Agalinis skinneriana is in the Center for Plant Conservation's National Collection of Endangered Plants.

31. **Agalinis strictifolia** (Bentham) Pennell, Proc. Acad. Nat. Sci. Philadelphia 73: 520. 1922 • Stiffleaf false foxglove

Gerardia strictifolia Bentham, Compan. Bot. Mag. 1: 209. 1836

Stems branched, 30–100 cm; branches ascending-spreading, subterete to obtusely angular, finely ridged, glabrous or sparsely scabridulous. **Leaves** spreading, proximal sometimes reflexed, distal appressed; blade linear, 8–35 × 0.5–2 mm, margins entire, siliceous, adaxial surface finely scabrous; axillary fascicles absent. **Inflorescences** racemes, flowers 1 or 2 per node; bracts shorter than pedicels. **Pedicels** spreading, often upcurved distally, (6–)10–30 mm, glabrous. **Flowers:** calyx turbinate-campanulate, tube 3–5.5 mm, glabrous, lobes triangular-lanceolate, 0.8–2.5 mm; corolla pink, with 2 yellow lines and red spots in abaxial throat, 15–35 mm, throat strongly upcurved, glabrous externally and within across bases and sinus of adaxial lobes, lobes: abaxial spreading, adaxial erect to spreading, 4.5–8(–10) mm, equal, glabrous externally; proximal anthers parallel to filaments, distal perpendicular to filaments, pollen sacs 3–4 mm; style exserted, 12–17 mm. **Capsules** globular-ovate to oblong, 5–8 mm. **Seeds** dark brown to black, 0.5–1 mm. $2n = 26$.

Flowering mid Jul–Oct. Grasslands, old fields, pastures, open mesquite plains, dunes along Gulf Coast; 0–200 m; Tex.; Mexico.

Within the flora area, *Agalinis strictifolia* and *A. edwardsiana* are restricted to Texas. They are distinguished by characters given in the key and those mentioned in the discussion for 9. *A. edwardsiana*.

32. **Agalinis tenella** Pennell, Bull. Torrey Bot. Club 40: 434. 1913 • Delicate false foxglove [E] [F]

Gerardia tenella (Pennell) Pennell

Stems simple or branched, 25–90 cm; branches laxly and widely spreading, subterete proximally, quadrangular-ridged distally, glabrous or scabridulous. **Leaves** spreading; blade filiform to linear-filiform, 6–25 × 0.2–1(–1.5) mm, margins entire, often siliceous, abaxial midvein sometimes scabridulous, adaxial surface finely scabrous; axillary fascicles absent. **Inflorescences** racemes, flowers 2 per node, sometimes with branches bearing pseudoterminal flowers; bracts shorter than pedicels. **Pedicels** spreading, (7–)10–25(–30) mm, glabrous. **Flowers:** calyx obconic to hemispheric, tube 2–3 mm, glabrous, lobes subulate, 0.2–0.4 mm; corolla pink to rosy pink, with 2 yellow lines and red spots in abaxial throat, 10–17 mm, throat pilose externally and villous within across bases and sinus of adaxial lobes, lobes: abaxial spreading, adaxial erect to strongly recurved, 4–7.3 mm, glabrous externally; proximal anthers parallel to filaments, distal perpendicular to filaments, pollen sacs 1.7–2.5 mm; style exserted, 7–9 mm. **Capsules** ovoid-globular, 3–3.7 mm. **Seeds** yellow, 0.7–1 mm. $2n = 26$.

Flowering Sep–Oct. Dry sandy to mesic pine savannas, margins of pine plantations, mixed open woodlands and oak-pine scrub, open pine-palmetto palm woodlands, ditch banks, sandy embankments, roadsides; 0–70 m; Ala., Fla., Ga., S.C.

Agalinis tenella is common in the panhandle of Florida, adjacent Alabama, and east-central Georgia, and occasional in southern South Carolina. J. B. Pettengill and M. C. Neel (2011) showed *A. tenella* to be most closely related to the rare *A. decemloba*. *Agalinis tenella*

is distinguished from *A. decemloba* (including *A. acuta*) by the laxly and widely spreading branching habit, generally much shorter calyx lobes, larger corollas, and truncate capsules of *A. tenella*. *Agalinis tenella* can also be confused with *A. obtusifolia*; *A. tenella* has well-formed racemes usually with two flowers per node, while the inflorescence of *A. obtusifolia* is paniculate and has short, slender branches each appearing to terminate in a pedicelled flower subtended by tiny bractlets.

33. **Agalinis tenuifolia** (Vahl) Rafinesque, New Fl. 2: 64. 1837 • Slenderleaf false foxglove, gérardie à feuilles ténues [E]

Gerardia tenuifolia Vahl, Symb. Bot. 3: 79. 1794; *Agalinis besseyana* (Britton) Britton; *A. tenuifolia* var. *leucanthera* (Rafinesque) Pennell; *A. tenuifolia* var. *macrophylla* (Bentham) S. F. Blake; *A. tenuifolia* var. *parviflora* (Nuttall) Pennell; *A. tenuifolia* var. *polyphylla* (Small) Pennell; *G. besseyana* Britton; *G. tenuifolia* subsp. *leucanthera* (Rafinesque) Pennell; *G. tenuifolia* subsp. *macrophylla* (Bentham) Pennell; *G. tenuifolia* subsp. *parviflora* (Nuttall) Pennell; *G. tenuifolia* subsp. *polyphylla* (Small) Pennell

Stems simple or branched, 10–100 cm; branches ascending to spreading, quadrangular, sharply ridged to winged distally, glabrate, sometimes scabrous. **Leaves** spreading, sometimes arching, ascending, or reflexed; blade narrowly linear to linear-lanceolate, 10–70 × 0.3–6 mm, not fleshy, margins entire, adaxial surface scabrous; axillary fascicles absent or shorter than subtending leaves. **Inflorescences** racemes, elongate, flowers 2 per node; bracts shorter than, or longer than, or both shorter and longer than, pedicels. **Pedicels** ascending-spreading, some upcurved distally, 6–25 mm, glabrous. **Flowers:** calyx obconic to hemispheric, tube 2.3–5.5 mm, glabrous, lobes subulate to triangular-subulate, 0.3–2 mm; corolla pink to rose purple, with 2 yellow lines and red spots in abaxial throat, 7–23 mm, throat pilose externally and glabrous within across bases and sinus of adaxial lobes, lobes: abaxial projected or spreading, adaxial projected over distal anthers, 2–8 mm, abaxial pilose externally, adaxial glabrous externally or pilose proximally; proximal anthers perpendicular or oblique to filaments, distal perpendicular and vertical to filaments, pollen sacs 1–4 mm; style exserted, 6.7–18 mm. **Capsules** globular, 4–7 mm. **Seeds** tan to brown, 0.5–1.5 mm. $2n = 28$.

Flowering (late Jul–)Aug–Nov. Wet to dry roadsides, ditches, margins of streams and ponds, borders of woodlands, dry to moist prairies, fallow fields, railroad embankments, rocky cliff faces and bluffs; 0–1600 m;

Man., N.B., N.S., Ont., P.E.I., Que.; Ala., Ark., Colo., Conn., Del., D.C., Fla., Ga., Ill., Ind., Iowa, Kans., La., Maine, Md., Mass., Mich., Miss., Mo., Nebr., N.H., N.J., N.Mex., N.Y., N.C., N.Dak., Ohio, Okla., Pa., R.I., S.C., S.Dak., Tenn., Tex., Vt., Va., W.Va., Wis., Wyo.

Populations of *Agalinis tenuifolia* in New Brunswick, Nova Scotia, and possibly Prince Edward Island in Canada are presumed introduced.

Agalinis tenuifolia is the most widespread and morphologically variable species of the genus in the flora area. Infraspecific taxa have been recognized based on differences in sizes of corollas, calyx lobes, anthers, capsules, and leaves; presence or absence of axillary fascicles; density of indument on stamens; branches ascending versus spreading; and even the stoutness of reticulations on seed coats. These characters intergrade within and among populations and occur in many other combinations in addition to those described, making these infraspecific taxa arbitrary and inconsistent with plants in the field. Pressed specimens of *A. tenuifolia* are often confused with *A. gattingeri* from which they differ by lacking a villous band of trichomes within the corolla at the bases of the adaxial corolla lobes present in *A. gattingeri*; projecting adaxial corolla lobes versus erect to recurved lobes in *A. gattingeri*; elongate racemes with two flowers per node versus one flower per node, often appearing to terminate branches in *A. gattingeri*; and low wings of tissue on the branch angles that are absent or less pronounced in *A. gattingeri*.

34. Agalinis viridis (Small) Pennell, Proc. Acad. Nat. Sci. Philadelphia 73: 521. 1922 • Green false foxglove [E]

Gerardia viridis Small, Bull. Torrey Bot. Club 25: 619. 1898

Stems branched, 15–70 cm; branches laxly and widely spreading, proximal arching upward, subterete proximally, quadrangular, with siliceous ridges and wings on angles distally, glabrous, sometimes scabridulous. **Leaves** spreading to spreading-ascending; blade linear, 8–30 × 0.5–2.3(–3) mm, not fleshy, margins entire, siliceous, adaxial surface finely scabrous; axillary fascicles absent. **Inflorescences** racemiform, with lateral branches bearing solitary flowers, flowers 1 per node; bracts both shorter and longer than, or shorter than, pedicels. **Pedicels** spreading to spreading-ascending or arching, 3–14 mm, glabrous. **Flowers:** calyx campanulate, tube 3–5.5 mm, glabrous, lobes lanceolate, 1.3–2.5 mm; corolla pale pink (sometimes nearly translucent), with 2 yellow or pale lines and pale purple spots in abaxial throat, 8–12 mm, throat pilose externally and villous within across bases and sinus of adaxial lobes, lobes: abaxial spreading, adaxial reflexed-spreading, 3–5 mm, glabrous externally; proximal anthers parallel to filaments, distal perpendicular or oblique to filaments, pollen sacs 0.8–1.5 mm; style exserted, 3–5 mm. **Capsules** obovoid, 4–6(–7) mm. **Seeds** tan to pale brown, 0.8–1 mm.

Flowering late Aug–Oct. Mesic to wet areas of prairies, moist to wet savannas, moist roadsides, mesic to wet edges of recently clear-cut forests, dry or wet roadsides; 0–100 m; Ala., Ark., La., Miss., Mo., Okla., Tex.

Agalinis viridis is tolerant of mowing and grazing; plants severed at mid stem will flower vigorously from the proximal nodes

17. AUREOLARIA Rafinesque, New Fl. 2: 58. 1837 • Yellow false foxglove [Latin *aureolus*, golden, and *-arius*, possession, alluding to corolla]

Jeffery J. Morawetz

Herbs, annual or perennial; hemiparasitic, caudex knotted. **Stems** erect, not fleshy, glabrous, sparsely to densely puberulent, densely villous, or glandular-pubescent. **Leaves** basal and cauline, opposite; petiole present; blade leathery, not fleshy, margins entire, pinnatifid, or 2-pinnatifid. **Inflorescences** terminal, loose racemes; bracts present. **Pedicels** present; bracteoles absent. **Flowers:** sepals 5, calyx radially symmetric, campanulate or turbinate, lobes linear to deltate, sometimes lanceolate; petals 5, corolla yellow, bilabiate, campanulate, abaxial lobes 3, adaxial 2; stamens 4, didynamous, filaments glabrous or ciliate, anthers villous; staminode 0; ovary 2-locular, placentation axile; stigma truncate. **Capsules:** dehiscence loculicidal. **Seeds** 300–500, brown to blackish, ovoid to deltoid, wings present (absent in *A. pedicularia*). *x* = 14.

Species 8 (7 in the flora): c, e North America, n Mexico.

Molecular phylogenetic analysis supports a close relationship of *Aureolaria* with *Agalinis*, *Esterhazya* J. C. Mikan, and *Seymeria* (J. R. Bennett and S. Mathews 2006). Further evidence for these relationships is similarities in floral morphology among these genera.

Aureolaria greggii (S. Watson) Pennell occurs broadly throughout northern Mexico.

SELECTED REFERENCES Musselman, L. J. 1969. Observations on the life history of *Aureolaria grandiflora* and *Aureolaria pedicularia* (Scrophulariaceae). Amer. Midl. Naturalist 82: 307–311. Pennell, F. W. 1928. *Agalinis* and allies in North America: I. Proc. Acad. Nat. Sci. Philadelphia 80: 339–449.

1. Calyx lobe margins crenate or pinnatifid; plants annual.
　　2. Leaf blade margins 2-pinnatifid, sometimes undivided or basal leaves less deeply incised; calyx tubes campanulate; corolla tubes floccose .1. *Aureolaria pectinata*
　　2. Leaf blade margins pinnatifid or 2-pinnatifid (first sinus deeper than second when both present); calyx tubes turbinate; corolla tubes glabrous or sparsely pubescent . . . . .2. *Aureolaria pedicularia*
1. Calyx lobe margins entire; plants perennial.
　　3. Stems and leaf blade surfaces sparsely to densely brown-pubescent; capsules densely pubescent . . . . .3. *Aureolaria virginica*
　　3. Stems and leaf blade surfaces glabrous, puberulent, or white-puberulent; capsules glabrous.
　　　　4. Leaf blade margins proximally serrate to pinnatifid, entire on distal ¼–½ . . . . . .4. *Aureolaria patula*
　　　　4. Leaf blade margins entire, serrate, or shallowly to deeply pinnatifid.
　　　　　　5. Stems and leaf blade surfaces puberulent or white-puberulent . . . . .5. *Aureolaria grandiflora*
　　　　　　5. Stems and leaf blade surfaces glabrous.
　　　　　　　　6. Leaf blade margins entire or serrate, rarely proximal leaves pinnatifid . . . . .6. *Aureolaria levigata*
　　　　　　　　6. Leaf blade margins pinnatifid or basal leaves 2-pinnatifid, rarely entire . . . . .7. *Aureolaria flava*

1. **Aureolaria pectinata** (Nuttall) Pennell, Bull. Torrey Bot. Club 40: 414. 1913 • Combleaf yellow false foxglove E

Gerardia pedicularia Linnaeus var. *pectinata* Nuttall, Gen. N. Amer. Pl. 2: 48. 1818; *G. pectinata* (Nuttall) Bentham

Annuals. Stems branched, 3–12 dm, densely white-villous, glandular-pubescent. **Leaves:** petiole 1–8 mm; blade lanceolate, (5–)10–50 × 6–26 mm, margins 2-pinnatifid, sometimes undivided or basal less deeply incised, glandular-pubescent. **Bracts** leaflike, 10–16 × 4–10 mm, margins pinnatifid to 2-pinnatifid. **Pedicels** 5–20 mm, villous, glandular-pubescent. **Flowers:** calyx villous and glandular-pubescent, tube campanulate, 3–8 mm, lobes linear, 7–15 × 1–3 mm, margins pinnatifid; corolla tube 25–40 mm, gibbous, floccose, lobes 7–11 × 6–15 mm; filaments 12–18 mm, ciliate; style 30–50 mm. **Capsules** pyriform, 11–15 × 5–7 mm, hispid.

Flowering Jun–Sep. Sandy soils, oak-pine scrubs; 0–700 m; Ala., Ark., Fla., Ga., Ky., La., Miss., Mo., N.C., Okla., S.C., Tenn., Tex., Va.

Aureolaria pectinata was treated at species rank by F. W. Pennell; some authors have included it within *A. pedicularia* (for example, R. P. Wunderlin 1998). Although *A. pectinata* and *A. pedicularia* bear a close resemblance, they can be separated reliably and recognized as species. While Pennell (1935) recognized four infraspecific taxa, these were distinguished on subtle characteristics that are difficult to discern on herbarium specimens (for example, distal leaves and bracts not excessively smaller versus distal leaves and bracts excessively smaller). No infraspecific taxa are recognized here.

2. Aureolaria pedicularia (Linnaeus) Rafinesque ex Pennell, Bull. Torrey Bot. Club 40: 412. 1913 • Fernleaf yellow false foxglove [E]

Gerardia pedicularia Linnaeus, Sp. Pl. 2: 611. 1753; *Aureolaria pedicularia* var. *ambigens* (Fernald) Farwell; *A. pedicularia* var. *austromontana* Pennell; *A. pedicularia* var. *intercedens* Pennell; *G. pedicularia* var. *ambigens* Fernald; *G. pedicularia* var. *austromontana* (Pennell) Fernald; *G. pedicularia* var. *intercedens* (Pennell) Fernald

Annuals. Stems simple or branched, 4–10 dm, glabrous, white-pubescent, or sparsely glandular-pubescent. **Leaves:** petiole 3–5 mm; blade lanceolate, 12–70 × 8–30 mm, margins pinnatifid (less than ½ to midvein) to 2-pinnatifid, first sinus deeper than second when both present, surfaces glabrous, rarely white-pubescent. **Bracts** leaflike, 9–15 × 4–6 mm, margins pinnatifid to (rarely) 2-pinnatifid. **Pedicels** (2–)8–13 mm, pubescent to glandular-pubescent. **Flowers:** calyx pubescent and/or glandular-pubescent, tube turbinate, 4–8 mm, lobes linear, sometimes narrowly lanceolate, 4–12 × 1–3 mm, margins crenate to pinnatifid; corolla tube 17–32 mm, glabrous or sparsely pubescent, lobes 8–16 × 5–13 mm; filaments 15–22 mm, ciliate; style 25–27 mm. **Capsules** ellipsoid, 10–13 × 5–6 mm, hispid. **2***n* = 28.

Flowering Jun–Sep. Roadsides, oak woods, pine barrens, clearings, dry, sandy soils; 10–900 m; Ont.; Conn., Del., D.C., Ga., Ill., Ind., Ky., Maine, Md., Mass., Mich., Minn., N.H., N.J., N.Y., N.C., Ohio, Pa., R.I., S.C., Tenn., Vt., Va., W.Va., Wis.

Species of *Aureolaria* typically parasitize oaks (*Quercus* spp.); two studies have documented *A. pedicularia* parasitizing non-oak species: *Pinus taeda* and *Vaccinium arboreum* (L. J. Musselman and H. E. Grelen 1979), and unidentified Ericaceae (C. R. Werth and J. L. Riopel 1979). In a study of seed coat morphology of *Aureolaria* and some related genera, *A. pedicularia* was documented to be the only species in the genus with unwinged seeds (J. M. Canne 1980). F. W. Pennell (1935) recognized five infraspecific taxa within *A. pedicularia* based on subtle characters (for example, stem distally closely pubescent, not or only slightly glandular, and leaves puberulent, scarcely or not glandular, versus stem distally glandular-pubescent to hirsute, and leaves glandular puberulent to pubescent). He recognized these infraspecific taxa at both the varietal and subspecific ranks (at different times). No infraspecific taxa are recognized here.

3. Aureolaria virginica (Linnaeus) Pennell, Bull. Torrey Bot. Club 40: 409. 1913 • Downy yellow false foxglove [E] [F]

Rhinanthus virginicus Linnaeus, Sp. Pl. 2: 603. 1753 (as virginica); *Aureolaria dispersa* (Small) Pennell; *A. microcarpa* Pennell; *Gerardia virginica* (Linnaeus) Britton, Sterns & Poggenburg

Perennials. Stems simple or branched, 5–15 dm, sparsely to densely brown-pubescent. **Leaves:** petiole 2–5 mm; blade lanceolate to oblong, 24–140 × 8–31 mm, margins serrate to pinnatifid, surfaces brown-pubescent. **Bracts** leaflike, 11–32 × 3–9 mm, margins entire or crenate. **Pedicels** 3–5 mm, pubescent. **Flowers:** calyx densely pubescent, tube campanulate, 5–6 mm, lobes deltate, 4–9 × 2–3 mm, margins entire; corolla tube 25–35 mm, glabrous, lobes 9–14 × 10–15 mm; filaments 9–14 mm, ciliate; style 21–33 mm. **Capsules** pyriform, 8–15 × 6–8 mm, densely pubescent. **2***n* = 26.

Flowering Jun–Sep. Oak and pine woods and slopes; 0–700 m; Ont.; Ala., Conn., Del., D.C., Fla., Ga., Ind., Ky., La., Md., Mass., Mich., Miss., N.H., N.J., N.Y., N.C., Ohio, Pa., R.I., S.C., Tenn., Tex., Vt., Va., W.Va.

This treatment follows G. L. Nesom and L. E. Brown (1998) and R. P. Wunderlin (1998) in including *Aureolaria dispersa* and *A. microcarpa* as synonyms of *A. virginica*; it is impossible to distinguish these taxa from *A. virginica*.

4. Aureolaria patula (Chapman) Pennell, Proc. Acad. Nat. Sci. Philadelphia 71: 271. 1920 • Spreading yellow false foxglove [E]

Dasystoma patulum Chapman, Bot. Gaz. 3: 10. 1878 (as patula); *Gerardia patula* (Chapman) Chapman ex A. Gray

Perennials. Stems branched, 5–9 dm, glabrous. **Leaves:** petiole 3–30 mm; blade lanceolate, 30–130 × 5–87 mm, margins proximally serrate to pinnatifid, distal ¼–½ entire, surfaces glabrous. **Bracts** lanceolate, 15–36 × 6–7 mm, margins entire. **Pedicels** 11–23 mm, sparsely villous to puberulent. **Flowers:** calyx glabrous or sparsely pubescent, tube campanulate, 3–5 mm, lobes lanceolate, 7–9 × 1–2 mm, margins entire; corolla tube 18–27 mm, glabrous, lobes 11–12 × 10–13 mm; filaments 15–28 mm, glabrous or ciliate; style 25–30 mm. **Capsules** ovoid to pyriform, 10–11 × 5–10 mm, glabrous.

Flowering Aug–Oct. Limestone substrates, edges of lakes or streams; 200–300 m; Ala., Ga., Ky., Tenn.

Aureolaria patula is poorly known and has the most restricted distribution within *Aureolaria*. Little information is available, and few specimens exist. It is state-listed as threatened in Tennessee and of special concern in Kentucky.

5. **Aureolaria grandiflora** (Bentham) Pennell, Rhodora 20: 135. 1918 • Large-flower yellow false foxglove E

Gerardia grandiflora Bentham, Compan. Bot. Mag. 1: 206. 1836; *Aureolaria grandiflora* var. *cinerea* Pennell; *A. grandiflora* var. *pulchra* Pennell; *A. grandiflora* var. *serrata* (Bentham) Pennell; *G. grandiflora* var. *cinerea* (Pennell) Cory; *G. grandiflora* var. *pulchra* (Pennell) Fernald

Perennials. Stems simple or branched, 5–15 dm, white-puberulent. **Leaves:** petiole 8–18 mm; blade lanceolate, 50–105 × 30–40 mm, margins pinnatifid and serrate, rarely serrate and not pinnatifid, surfaces puberulent. **Bracts** leaflike, 15–22 × 3–7 mm, margins serrate to weakly pinnatifid at base. **Pedicels** 4–11 mm, puberulent. **Flowers:** calyx puberulent, tube campanulate, 6–10 mm, lobes linear to lanceolate, 3–12 × 1–4 mm, margins entire; corolla tube 30–40 mm, glabrous, lobes 7–11 × 10–16 mm; filaments 13–27 mm, ciliate; style 29–35 mm. **Capsules** pyriform, 10–16 × 8–12 mm, glabrous.

Flowering Jul–Oct. Maple, pine, and oak woodlands, roadsides; 0–400 m; Ont.; Ark., Ill., Ind., Iowa, Kans., La., Mo., Okla., Tex., Wis.

J. C. McFeeley and E. P. Roberts (1974) documented *Aureolaria grandiflora*, typically a parasite of oaks, parasitizing *Juniperus virginiana* in addition to oak species. F. W. Pennell (1935) recognized four infraspecific taxa within *A. grandiflora*, differentiated on subtle characteristics (for example, bract margins serrate versus more deeply cut, leaf blade margins more versus less pinnatifid). Additionally, he recognized these infraspecific taxa at both the varietal and subspecific ranks (at different times). No infraspecific taxa are recognized here.

There is a historic record of *Aureolaria grandiflora* from Houston County, Minnesota, collected in 1899.

6. **Aureolaria levigata** (Rafinesque) Rafinesque, New Fl. 2: 59. 1837 • Entire-leaf yellow false foxglove E

Gerardia levigata Rafinesque, Ann. Nat. 1: 13. 1820

Perennials. Stems simple or branched, 4–25 dm, glabrous, not glaucous. **Leaves:** petiole 2–3 mm; blade lanceolate to oblong, 15–127 × 4–34 mm, margins entire or serrate, rarely (proximal) pinnatifid, surfaces glabrous. **Bracts** leaflike, 8–15 × 2–5 mm, margins entire. **Pedicels** 1–3 mm, glabrous. **Flowers:** calyx glabrous, tube campanulate, 5–7 mm, lobes lanceolate to narrowly deltate, 5–6 × 2–3 mm, margins entire; corolla tube 19–30 mm, glabrous, lobes 9–11 × 11–16 mm; filaments 12–20 mm, ciliate; style 21–29 mm. **Capsules** ovoid to pyriform, 8–12 × 6–7 mm, glabrous. $2n = 26$.

Flowering Jul–Oct. Rocky wooded slopes, mixed hardwoods; 10–1800 m; Ala., Ga., Ky., Md., Miss., N.C., Ohio, Pa., S.C., Tenn., Va., W.Va.

Aureolaria levigata is disjunct in southwestern Mississippi. The affinity between the floras of the Appalachian and Ozark mountains and the loess hills of southwestern Mississippi is well documented (H. R. Delcourt and P. A. Delcourt 1975).

7. **Aureolaria flava** (Linnaeus) Farwell, Rep. (Annual) Michigan Acad. Sci. 20: 188. 1919 • Smooth yellow false foxglove E

Gerardia flava Linnaeus, Sp. Pl. 2: 610. 1753; *Aureolaria calycosa* (Mackenzie & Bush) Pennell; *A. flava* var. *macrantha* Pennell; *A. flava* var. *reticulata* (Rafinesque) Pennell; *G. calycosa* (Mackenzie & Bush) Fernald; *G. flava* var. *macrantha* (Pennell) Fernald; *G. flava* var. *reticulata* (Rafinesque) Cory

Perennials. Stems branched, 4–22 dm, glabrous, glaucescent. **Leaves:** petiole 7–25 mm; blade lanceolate, 60–160 × 12–60 mm, margins pinnatifid or basal leaves 2-pinnatifid, rarely entire, surfaces glabrous. **Bracts** leaflike, 10–48 × 2–5 mm, margins entire or pinnatifid. **Pedicels** 7–12 mm, glabrous. **Flowers:** calyx glabrous, tube campanulate, 4–8 mm, lobes linear to narrowly deltate, 4–7 × 1–2 mm, margins entire; corolla tube 25–38 mm, glabrous, lobes 5–9 × 6–15 mm; filaments 12–26 mm, glabrous; style 27–31 mm. **Capsules** ovoid to pyriform, 11–13 × 9–11 mm, glabrous. $2n = 24$.

Flowering Jul–Oct. Oak or oak-pine woods, roadsides; 0–1500 m; Ont.; Ala., Ark., Conn., Del., Fla., Ga., Ill., Ind., Ky., La., Maine, Md., Mass., Mich., Miss., Mo., N.H., N.J., N.Y., N.C., Ohio, Pa., R.I., S.C., Tenn., Tex., Vt., Va., W.Va.

Despite the nuanced variation within this broadly distributed species recognized at both the varietal and subspecific ranks by F. W. Pennell (1935), R. P. Wunderlin (1998) recognized a broad *Aureolaria flava* with no varietal designations; that concept is followed here.

18. BRACHYSTIGMA Pennell, Proc. Acad. Nat. Sci. Philadelphia 80: 432. 1928

• Desert foxglove [Greek *brachys*, short, and *stigma*, stigma]

Christopher P. Randle

Herbs, perennial; hemiparasitic, caudex woody. **Stems** erect, not fleshy, hirsutulous. **Leaves** cauline, in whorls of 3; petiole absent; blade not fleshy, not leathery, margins entire. **Inflorescences** terminal, racemes; bracts present. **Pedicels** present; bracteoles absent. **Flowers:** sepals 5, calyx nearly radially symmetric, broadly campanulate, lobes deltate; petals 5, corolla yellow, bilabiate, subrotate, abaxial lobes 3, adaxial 2; stamens 4, didynamous, filaments glabrescent proximally, villous distally; staminode 0; ovary 2-locular, placentation axile; stigma clavate. **Capsules:** dehiscence loculicidal. **Seeds** 100+, dark brown, ellipsoid, wings absent or present.

Species 1: sw United States, n Mexico.

Brachystigma is monospecific and narrowly restricted to dry mountain slopes of southeastern Arizona, southwestern New Mexico, and northern Mexico. F. W. Pennell (1928) differentiated it from the similar genus *Agalinis* by its yellow corollas, glabrous anthers, capitate stigmas, more acute capsules, and winged seeds. Phylogenetic analysis of three chloroplast genes supports this distinction (M. C. Neel and M. P. Cummings 2004). Further, *Brachystigma* may be differentiated from closely related genera *Aureolaria*, *Dasistoma*, and *Seymeria* by its leaves arranged in whorls of three, a characteristic unique in Orobanchaceae.

SELECTED REFERENCE Pennell, F. W. 1928. *Agalinis* and allies in North America: I. Proc. Acad. Nat. Sci. Philadelphia 80: 339–449.

1. Brachystigma wrightii (A. Gray) Pennell, Proc. Acad. Nat. Sci. Philadelphia 80: 433. 1928 • Arizona desert foxglove [F]

Gerardia wrightii A. Gray in W. H. Emory, Rep. U.S. Mex. Bound. 2(1): 118. 1859; *Agalinis wrightii* (A. Gray) Tidestrom; *Dasistoma wrightii* (A. Gray) Wooton & Standley

Perennials unbranched or branching at caudex. **Stems** slender, 30–55 cm. **Leaves:** blade filiform-linear, 30–60 × 2–3 mm, proximals longer than distals. **Inflorescences** racemes of paired flowers, occupying to ⅔ height of stem, bracts resembling foliage leaves. **Pedicels** 11–30 mm. **Flowers** 10–26 × 7–16 mm; calyx 3–5 × 4–7 mm, lobes broadly deltate, 1–1.5 mm; corolla abruptly inflating just beyond calyx, 24–30 mm diam., externally pubescent, internally glabrous, lobes orbiculate; filaments 8–11 mm, anthers included, base bluntly sagittate, glabrous, locules equal, 5 mm; ovary ovoid, 4–5 × 2–4 mm; style incurving at maturity, 5–6 mm. **Capsules** with persistent calyces, globular-ovoid, 8–10 × 5–8 mm. **Seeds** reticulate, 2 mm.

Flowering Aug–Oct; fruiting Oct–Dec. Dry mountain slopes, often in oak chaparral; 1200–2000 m; Ariz., N.Mex.; Mexico (Chihuahua, Sonora).

19. DASISTOMA Rafinesque, J. Phys. Chim. Hist. Nat. Arts 89: 99. 1819 • Mullein-foxglove [Greek *dasys*, hairy, and *stoma*, mouth, alluding to lanate throat of corolla] E

Craig C. Freeman

Herbs, annual or biennial, possibly sometimes perennial; hemiparasitic. **Stems** erect, not fleshy, puberulent, retrorsely puberulent, or pubescent. **Leaves** loosely basal and cauline, opposite; petiole present; blade not fleshy, not leathery, margins pinnatifid to 2-pinnatifid proximally, sometimes entire distally. **Inflorescences** terminal or terminal and axillary, spikelike racemes; bracts present. **Pedicels** present; bracteoles absent. **Flowers:** sepals 5, calyx weakly bilaterally symmetric, campanulate, lobes ovate to oblong-ovate; petals 5, corolla yellow, bilabiate, campanulate, abaxial lobes 3, adaxial 2; stamens 4, didynamous, filaments lanate, anthers glabrous; staminode 0; ovary 2-locular, placentation axile; stigma capitate or 2-lobed. **Capsules:** dehiscence loculicidal. **Seeds** 50–100, tan or brown, angled, wings absent.

Species 1: e, nc United States.

SELECTED REFERENCE Pennell, F. W. 1928. *Agalinis* and allies in North America: I. Proc. Acad. Nat. Sci. Philadelphia 80: 339–449.

1. **Dasistoma macrophyllum** (Nuttall) Rafinesque, New Fl. 2: 67. 1837 (as Dasistema macrophylla) E F

Seymeria macrophylla Nuttall, Gen. N. Amer. Pl. 2: 49. 1818

Stems branched distally, rarely unbranched, square to round, 80–200 cm. **Leaves:** basal and cauline: petiole (5–)15–100 mm; blade ovate to ovate-lanceolate, 15–35 × 6–22 cm, segment margins irregularly serrate to irregularly dentate, surfaces scabrid-puberulent to pubescent; distal: petiole 2–20(–35) mm; blade lanceolate to narrowly lanceolate, 7–21 × 2–14 cm, segment margins entire or crenate, irregularly serrate, or irregularly dentate, surfaces scabrid-puberulent to pubescent. **Racemes** 1–8, congested to interrupted, cylindric, 7–35 cm; bracts leafy, gradually smaller distally, 9–65 × 7–35 mm, margins entire or irregularly serrate to irregularly dentate, apex acute; cymes 1-flowered.

Pedicels erect, 1–4 mm, puberulent to retrorsely puberulent. **Flowers:** calyx puberulent to pubescent, lobes (2–)4–5.5 × 2.5–4.5 mm, margins entire, apex acute to obtuse, membranous; corolla 15–16 mm, glabrous externally, densely lanate internally; tube 7–9 mm; throat inflated, 5–7 mm diam.; lobes spreading, ovate-reniform, 6–7 mm; stamens distinct, reaching orifice or barely exserted, pollen sacs 2–2.8 mm. **Capsules** ovoid to globular-ovoid, 8–12 × 7–9 mm, glabrous. **Seeds** reticulate, 1.8–2.5 × 1–2 mm.

Flowering Jun–Sep. Dry to mesic forests and woodlands, brushy tallgrass prairies, stream banks; 60–400 m; Ala., Ark., Ga., Ill., Ind., Iowa, Kans., Ky., La., Mich., Miss., Mo., Nebr., Ohio, Okla., Pa., S.C., Tenn., Tex., Va., W.Va., Wis.

In the greenhouse, *Dasistoma macrophyllum* formed haustorial connections with 18 species of gymnosperm and angiosperm trees, but it also could remain autotrophic (L. J. Musselman and W. F. Mann 1979); there is no evidence of autotrophy in the wild.

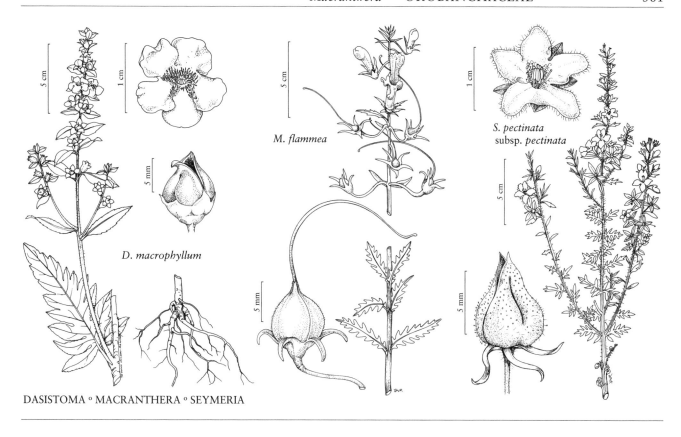

D. *macrophyllum*

M. *flammea*

S. *pectinata*
subsp. *pectinata*

DASISTOMA ∘ MACRANTHERA ∘ SEYMERIA

20. MACRANTHERA Nuttall ex Bentham, Edwards's Bot. Reg. 21: sub plate 1770, [6]. 1835 • Flame or hummingbird flower [Greek *macros*, long, and *antheros*, anther, alluding to long-exserted stamens] E

Bruce A. Sorrie

Herbs, perennial or biennial; hemiparasitic. **Stems** erect, not fleshy, retrorsely puberulent. **Leaves** basal and cauline, opposite; petiole present; blade not fleshy, not leathery, margins pinnatifid on proximal leaves, toothed on distal leaves. **Inflorescences** terminal, racemes; bracts present. **Pedicels** present; bracteoles absent. **Flowers:** sepals 5, calyx radially symmetric, campanulate to short-tubular, lobes linear; petals 5, corolla bright orange, bilabiate, tubular, abaxial lobes 3, adaxial 2; stamens 4, equal, filaments pilosulous; staminode 0; ovary 2-locular, placentation axile; stigma linear-clavate. **Capsules:** dehiscence loculicidal. **Seeds** ca. 110, brown to blackish, irregularly triangular, wings present (with tuberculate crest). *x* = 13.

Species 1: se United States.

SELECTED REFERENCES Alford, J. D. and L. C. Anderson. 2002. Taxonomy and morphology of *Macranthera flammea* (Orobanchaceae). Sida 20: 189–204. Musselman, L. J. 1972. Root parasitism of *Macranthera flammea* and *Tomanthera auriculata*. J. Elisha Mitchell Sci. Soc. 88: 58–60.

1. Macranthera flammea (W. Bartram) Pennell, Bull. Torrey Bot. Club 40: 124. 1913 [E] [F]

Gerardia flammea W. Bartram, Travels Carolina, 412. 1791

Perennials or biennials monocarpic, blackening upon drying. **Stems** virgately branched distally, 4-angled, 8–35 dm. **Leaves:** petiole winged, to 2 cm; blade lanceolate to narrowly ovate, 8–15 × 2–6 cm, smaller distally, surfaces glabrate. **Racemes** 8–36 cm. **Pedicels** deflexed-spreading proximally, strongly upcurved distally. **Flowers:** calyx tube 2–4 mm, retrorsely puberulent, lobes 7.5–15 mm, longer than tube; corolla 20–25 mm, exterior densely glandular-mealy, lobes 3–4.5 mm, shorter than tube, abaxial lobes reflexed-spreading, adaxial erect; stamens long-exserted, filaments orange, 15–46 mm; style long-exserted, 28–36 mm, glabrous. **Capsules** brown, ovoid, 9.5–13 mm, densely puberulent. **Seeds** 2.5–3 mm, wings 2–5. $2n = 26$.

Flowering Aug–Oct. Periodically burned streamheads and ecotones, baygall ecotones, seepage slopes, margins of shrub-tree bogs, cypress-gum depressions; 0–100 m; Ala., Fla., Ga., La., Miss.

Macranthera flammea is one of the more striking plants in the flora area due to its remarkable height (for an herb) and numerous, brilliant orange flowers; it is best able to compete with associated shrubs and trees by flowering prolifically following fire and attracting hummingbirds. There is precise coincidence in the flowering of this species and the arrival of ruby-throated hummingbirds (*Archilochus colubris*) prior to their trans-Gulf migration to Central America (A. L. Pickens 1927).

---

21. SEYMERIA Pursh, Fl. Amer. Sept. 2: 736. 1813, name conserved • Blacksenna [For Henry Seymer, 1714–1785, British collector]

Christopher P. Randle

Afzelia J. F. Gmelin 1792, name rejected

Herbs, annual; hemiparasitic. **Stems** erect, not fleshy, glabrous or villous to scabrid. **Leaves** cauline, opposite; petiole absent; blade not fleshy, not leathery, margins entire, irregularly lobed, pinnatifid, or 2-pinnatifid. **Inflorescences** axillary, flowers 2 per axil; bracts present. **Pedicels** present; bracteoles absent. **Flowers:** sepals 5, calyx ± radially symmetric, campanulate, lobes linear to lanceolate; petals 5, corolla yellow often with red, maroon, or purple markings in throat, bilabiate, campanulate with spreading lobes, abaxial lobes 3, adaxial 2; stamens 4, equal to subequal, filaments hairy proximally, glabrous or tomentose to lanate distally, anthers glabrous; staminode 0; ovary 2-locular, placentation axile; stigma simple. **Capsules:** dehiscence loculicidal. **Seeds** 30–40, brown, globular-ovoid or irregular, wings present or absent. $x = 13$.

Species 17 (5 in the flora): s United States, Mexico, West Indies.

The phylogenetic position of *Seymeria* remains unclear. N. D. Young et al. (1999) placed *Seymeria* in an unresolved clade with *Agalinis* and *Macranthera*, among other genera. M. C. Neel and M. P. Cummings (2004) recovered a clade including *Aureolaria*, *Brachystigma*, *Dasistoma*, and *Seymeria*. A. D. Wolfe et al. (2005) placed *Seymeria* sister to *Agalinis*. J. R. Bennett and S. Mathews (2006) found that *Seymeria* is in a clade including *Agalinis* and *Aureolaria*, and also *Esterhazya* J. G. Mikan.

Seymeria is one of the few North American root-parasitic genera of economic importance in that *S. cassioides* is a parasite of pine species used for lumber in the southeastern United States (L. J. Musselman 1996).

The common name blacksenna is derived from the superficial resemblance to *Senna* (Fabaceae).

SELECTED REFERENCES Pennell, F. W. 1925. The genus *Afzelia*: A taxonomic study in evolution. Proc. Acad. Nat. Sci. Philadelphia 77: 335–373. Turner, B. L. 1982. Revisional treatment of the Mexican species of *Seymeria*. Phytologia 51: 403–423.

1. Corollas externally pubescent or tomentose.
    2. Filaments tomentose to lanate distally, anthers dehiscing to ¼ lengths; capsules densely tomentose or pubescent to glabrescent; seeds: wings present. . . . . 4. *Seymeria pectinata*
    2. Filaments glabrous distally, anthers dehiscing ½+ lengths; capsules pubescent; seeds: wings absent . . . . 5. *Seymeria bipinnatisecta*
1. Corollas externally glabrous.
    3. Leaves: blade margins 2-pinnatifid, pinnules filiform, surfaces not scabrid; capsules pyriform . . . . 1. *Seymeria cassioides*
    3. Leaves: blade margins entire, 3-lobed, irregularly pinnatifid, or slightly 2-pinnatifid, pinnules linear to lanceolate, surfaces scabrid; capsules ± falcate.
        4. Leaves: blade margins entire, 3-lobed, or irregularly pinnatifid, surfaces minutely scabrid; pedicels 6–10 mm; capsules glabrous. . . . . 2. *Seymeria falcata*
        4. Leaves: blade margins pinnatifid or slightly 2-pinnatifid, surfaces strongly scabrid; pedicels 1.5–4 mm; capsules glandular-pubescent . . . . 3. *Seymeria scabra*

1. **Seymeria cassioides** (J. F. Gmelin) S. F. Blake, Rhodora 17: 134. 1915 • Yaupon blacksenna, senna seymeria

Afzelia cassioides J. F. Gmelin, Syst. Nat. 2: 927. 1792

Stems pubescent to villous, eglandular. **Leaves:** blade margins 2-pinnatifid, pinnules filiform, surfaces not scabrid. **Pedicels** 3–6 mm. **Flowers:** calyx lobes linear, margins entire; corolla yellow, with purple markings in throat, externally glabrous, internally pubescent proximal to adaxial sinus, between lobes, and in a ring at stamen insertion; filaments glabrous distally, anthers dehiscing to ¼ length. **Capsules** symmetric, pyriform, glabrous. **Seeds** ovoid, reticulate, wings absent. $2n = 26$.

Flowering and fruiting Sep–Oct. Dry to moist pine forests, pine savannas, sandhills and scrub of coastal plains; 0–300 m; Ala., Ark., Fla., Ga., La., Miss., N.C., S.C., Tenn., Tex., Va.; West Indies (Bahamas).

Seymeria cassioides is an obligate parasite of 11 pine species in the southeastern United States; it is also capable of growing on western pines and other conifers (W. F. Mann and L. J. Musselman 1980). Infestations of pine plantations are not common; in sufficient numbers, *S. cassioides* may produce chlorosis and mortality in cultivated *Pinus elliottii* and *P. taeda* (H. E. Grelen and Mann 1973).

2. **Seymeria falcata** B. L. Turner, Phytologia 51: 412. 1982

Varieties 2 (1 in the flora): Texas, n Mexico.

Seymeria falcata has been divided into two somewhat intergrading varieties, var. *falcata* and var. *uncinata* B. L. Turner, the latter known from Chihuahua and Coahuila, Mexico.

2a. **Seymeria falcata** B. L. Turner var. **falcata**

Stems glabrous or puberulent, glandular. **Leaves:** blade margins entire, 3-lobed, or irregularly pinnatifid, pinnules linear, surfaces minutely scabrid. **Pedicels** 6–10 mm. **Flowers:** calyx lobes linear to lanceolate, margins entire; corolla yellow, with or without purple markings in throat, externally glabrous, internally pubescent in a ring at stamen insertion; filaments glabrous distally, anthers dehiscing to ¼ length. **Capsules** symmetric, falcate, glabrous. **Seeds** irregularly shaped, wings absent.

Flowering and fruiting Aug–Sep. Rocky slopes and ridges; 1200–1700 m; Tex.; Mexico (Coahuila).

In the flora area, var. *falcata* is known from the Dead Horse Mountains of Brewster County (J. Fenstermacher 2006).

3. **Seymeria scabra** A. Gray in W. H. Emory, Rep. U.S. Mex. Bound. 2(1): 118. 1859 • Limpia blacksenna

Stems strongly scabrid, glandular. **Leaves:** blade margins pinnatifid or slightly 2-pinnatifid, pinnules lanceolate to linear, surfaces strongly scabrid. **Pedicels** 1.5–4 mm. **Flowers:** calyx lobes lanceolate, margins denticulate; corolla yellow, rarely with reddish streaks, externally glabrous, internally pubescent proximal to adaxial sinus, between lobes, and in a ring at stamen insertion; filaments glabrous distally, anthers dehiscing ½+ length. **Capsules** asymmetric, ± falcate, glandular-pubescent. **Seeds** globular to ovoid, wings absent.

Flowering and fruiting Aug–Oct. Desert, semi-desert, exposed calcareous and gypseous soils; 1400–2500 m; Tex.; Mexico (Chihuahua, Coahuila, Nuevo León).

In the flora area, *Seymeria scabra* is known from Brewster, Culberson, Jeff Davis, Pecos, and Presidio counties.

4. **Seymeria pectinata** Pursh, Fl. Amer. Sept. 2: 737. 1813 • Comb seymeria, combleaf senna E F

Afzelia pectinata (Pursh) Kuntze

Stems pubescent to villous, eglandular. **Leaves:** blade margins pinnatifid to 2-pinnatifid, pinnules lanceolate, surfaces not scabrid. **Pedicels** 3–5 mm. **Flowers:** calyx lobes lanceolate; corolla yellow, sometimes with maroon spots on adaxial lobes, externally pubescent, internally pubescent between lobes and in a ring at stamen insertion; filaments tomentose to lanate distally, anthers dehiscing to ¼ length. **Capsules** symmetric, pyriform, glabrescent to densely tomentose. **Seeds** irregularly shaped, wings present. $2n = 26$.

Subspecies 2 (2 in the flora): se United States.

1. Stems pubescent to villous; capsules densely tomentose . . . . . . 4a. *Seymeria pectinata* subsp. *pectinata*
1. Stems pubescent to puberulous; capsules glabrescent to pubescent . . . . 4b. *Seymeria pectinata* subsp. *peninsularis*

4a. **Seymeria pectinata** Pursh subsp. **pectinata** E F

Stems pubescent to villous. **Capsules** densely tomentose.

Flowering and fruiting Jul–Oct. Pine-oak sandhills and scrub; 0–300 m; Ala., Fla., Ga., La., Miss., N.C., S.C.

Plants of subsp. *pectinata* have been reported to parasitize *Aristida stricta* (Poaceae) and *Quercus laevis* (Fagaceae) (L. J. Musselman and W. F. Mann 1977). Greenhouse experiments demonstrate that plants of *Seymeria pectinata* are capable of parasitizing economically important southeastern trees (notably excluding *Pinus*) that are not known natural hosts, indicating pathogen potential and/or an incomplete host record for wild-collected plants (Musselman and Mann 1978).

4b. **Seymeria pectinata** Pursh subsp. **peninsularis** (Pennell) Pennell, Monogr. Acad. Nat. Sci. Philadelphia 1: 411. 1935 E

Afzelia pectinata (Pursh) Kuntze var. *peninsularis* Pennell, Proc. Acad. Nat. Sci. Philadelphia 71: 265. 1920

Stems pubescent to puberulous. **Capsules** glabrescent to pubescent.

Flowering and fruiting Jun–Sep. Hardwood hammocks, pine flatwoods; 0–20 m; Fla.

5. **Seymeria bipinnatisecta** Seemann, Bot. Voy. Herald, 323, plate 59. 1856 • Eagle Pass blacksenna

Seymeria bipinnatisecta var. *texana* A. Gray; *S. havardii* (Pennell) Pennell; *S. texana* (A. Gray) Pennell

Stems puberulous to villous, glandular. **Leaves:** blade margins 2-pinnatifid, pinnules lanceolate, surfaces not scabrid. **Pedicels** 4–7 mm. **Flowers:** calyx lobes lanceolate, margins slightly dentate; corolla yellow, externally tomentose, internally glabrous except pubescent in a ring at stamen insertion; filaments glabrous distally, anthers dehiscing ½+ length. **Capsules** symmetric, pyriform, pubescent. **Seeds** ovoid, wings absent.

Flowering and fruiting Jul–Oct. Rocky calcareous soils, shrubby grasslands, oak-juniper woodlands, mesic canyon bottoms, rock outcrops; 200–600 m; Ariz., Tex.; Mexico (Coahuila, Durango, Sinaloa, Sonora).

B. L. Turner (1982) circumscribed *Seymeria bipinnatisecta* as a variable species ranging widely from the Arizona and Texas borders through much of northwestern Mexico. In the flora area, the species extends into the Sierra Madre of southeastern Arizona and the Edwards Plateau of central and southwestern Texas. Morphological variation in this range is continuous with that of the Mexican specimens. Gray admitted that he had not seen Mexican specimens of *S. bipinnatisecta* but created var. *texana* based on perceived differences from the illustration of leaf and calyx incision, and of pedicel length by Seemann. Examination of specimens for this treatment indicates that the full range of diversity attributable to Mexican specimens for these characters can be found also in specimens of the Edwards Plateau. Most of the specimens collected in Texas are from the central Edwards Plateau; one clearly belonging to this species from Val Verde County on the Rio Grande (*W. R. Carr 12406*, TEX) is known; the range may not be discontinuous between Mexican and Edwards Plateau populations.

22. CASTILLEJA Mutis ex Linnaeus f., Suppl. Pl., 47, 293. 1782 • Indian paintbrush, painted-cup, owl's-clover [For Domingo Castillejo, 1744–1793, Spanish botanist]

J. Mark Egger   Peter F. Zika

Barbara L. Wilson   Richard E. Brainerd

Nick Otting

Euchroma Nuttall; *Oncorhynchus* Lehmann

Herbs or subshrubs, rarely shrubs, annual or perennial, sometimes biennial; hemiparasitic, caudex woody or fleshy, taprooted to fibrous-rooted or rhizomatous. **Stems** 1–200, strongly decumbent to erect, sometimes prostrate or sprawling, frequently with leafy axillary shoots, not fleshy, glabrous or pubescent, hairs eglandular or stipitate-glandular, unbranched, rarely ± to much-branched or stellate. **Leaves** mostly deciduous, cauline, mostly alternate, reduced proximally or forming basal rosettes in a few species; petiole absent or present; blade fleshy or not, leathery or not, margins entire or pinnately to ± palmately divided. **Inflorescences** terminal, compact to elongate spikes or secund, sometimes non-secund, racemes; bracts present, often brightly colored distally or throughout. **Pedicels** absent or present; bracteoles absent. **Flowers:** sepals 4, calyx entirely green or more often conspicuously colored distally, usually as in bracts, contrasting with bracts in a few species, radially or bilaterally symmetric, tubular, lobed distally in usually diagnostic patterns: either cleft into 4 subequal lobes (called segments in some floras), or abaxial and adaxial clefts deeper than 2 lateral clefts, or lateral clefts absent and calyx 2-lobed, or lateral clefts slightly deeper than abaxial and adaxial (*C. plagiotoma*); petals 5, corolla white to pale greenish proximally, usually becoming green, white, yellow, orange, red, pink, or purple distally, tubular proximally, strongly bilabiate distally and divided into adaxial beak of 2 lobes joined to tip, often with contrasting colors on margins, abaxial lip with 3 highly variable lobes, consisting of either greatly reduced, ± incurved, usually greenish teeth, or subpetaloid, contrastingly colored teeth, or shallow to deep, often more brightly colored pouches with small, apical, often contrastingly colored teeth, adaxial lip ± straight, beaked, opening directed forward; stamens 4, didynamous, filaments glabrous or spreading-hairy (*C. exserta*), pollen sacs 2, unequal; staminode 0; ovary 2-locular, placentation axile; stigma capitate, entire, or 2-lobed. **Capsules:** dehiscence loculicidal. **Seeds** 20–100, mostly stramineous, ovoid, pyramidal, irregularly oblong-ovoid, or trapezoidal, shiny, wings absent. $x = 12$.

Species ca. 200 (119 in the flora): North America, Mexico, Central America, South America, n Eurasia; introduced in West Indies, Pacific Islands.

Castilleja can be distinguished most easily from *Orthocarpus* in the strict sense by the structure of the flowers. The adaxial lip apices of *Castilleja* are more or less straight to slightly curved (hooked in *C. chlorotica*, *C. exserta*, *C. mexicana*, and *C. sessiliflora*) and never hooded, with the opening directed forward; in *Orthocarpus* the adaxial lip is rounded (straight in *O. cuspidatus*) and hooded, with the opening directed downward. In addition, *Castilleja* has a capitate to two-lobed stigma, while that of *Orthocarpus* is simple and unexpanded. An annual habit can be found in some *Castilleja* and all *Orthocarpus*. *Castilleja* has a base chromosome number of $x = 12$; *Orthocarpus* is $x = 14$. Presumed aneuploid reductions to $x = 10$ are known from three annual *Castilleja* species (two of these from Mexico), and a reduction to $x = 11$ is known from one species (T. I. Chuang and L. R. Heckard 1993).

Polyploidy is widespread in *Castilleja*, with about half of all species exhibiting at least one level of infra-specific polyploidy, including $n = 24, 36, 48, 60, 72$ (L. R. Heckard 1968; Heckard and T. I. Chuang 1977; Chuang and Heckard 1982).

Many species of *Castilleja* are fairly well defined morphologically and straightforward to identify using the keys; others have complex patterns of variation within local populations and across edaphic and geographic discontinuities. Some species groups, such as those surrounding *C. affinis*, *C. miniata*, and *C. pallida*, are especially challenging taxonomically, probably due at least in part to allopolyploidy (L. R. Heckard 1968; Heckard and T. I. Chuang 1977). Natural interspecific hybridization, though usually limited to individual plants or local populations, is well documented in the genus (N. H. Holmgren 1971; Heckard and Chuang). Hybrid swarms are observed in some regions, complicating identification (M. Ownbey 1959; J. M. Egger 1994), particularly among coastal populations from southeastern Alaska to central California. Homoploid hybrid speciation has also been proposed in at least one species in this genus (D. L. Clay et al. 2012).

Many of the characters used in the keys and descriptions are gathered from fresh plants. Floral characters were assessed at anthesis; one should avoid immature corollas as well as those persisting after flowering. Good collections require notes on variation in the fresh colors of bracts, calyces, and corollas. Pressing several examples of dissected flowers and bracts, separate from the plant, allows much easier measurement of corolla parts and calyx clefts, which are diagnostic for many taxa. Because these characters are difficult to assess once dried, herbarium material can be difficult to identify. Bract and flower coloration can be distinctive, but color variants are common in many taxa. Inflorescence color is usually a result of bright, non-green coloration in at least the distal portions of the bracts and calyces, although occasionally the corolla color contributes significantly to the coloration of the inflorescence or is more prominent than the colors of the other structures. The amount of bright distal coloration, especially on the bracts, is highly variable between species, within species, and even in individual plants. When only the color of the distal bracts or the distal portion of the bracts is reported here, the proximal bracts and/or the proximal part of the distal bracts are colored like the leaves. Also, in cases where bracts can be brightly colored either throughout or only distally in the same species, the colored throughout statement is presented first, followed by a statement such as "...or proximally greenish, distally as stated above." When the abaxial lip is described as reduced, it is a small green knob consisting of three incurved lobes and is not inflated. In relation to the leaf margins, two independently varying sub-traits are described. The authors contrast wavy (crisped) margins with plane (non-crisped) margins, and flat is used to contrast with involute or revolute margins.

In this treatment, the length of the corolla lobes and tube is measured from the point of contact in the notch between the abaxial lip and adaxial lip (beak). The junction of the lips is easier to find when the flower is fresh. Similarly, presence or absence of wavy leaf margins is easy to assess in the field and should be noted on herbarium labels, but this character is often obscured in herbarium material. While critical to species identification in only a few cases, woodiness or the subshrub habit and the presence of a woody caudex or rhizome should be noted in the field, as these characters are rarely evident in dried specimens.

Species of *Castilleja* occur in a wide range of habitats, from sea level to high alpine and from salt marshes and freshwater wetlands to tundra and deserts. Diversity is highest in grasslands, subalpine and alpine parklands, and other meadows, and is lower in densely forested regions. Some species tolerate a wide range of substrates, while others are specialists on soils derived from limestone, basalt, sandstone, granite, or serpentine.

All species of *Castilleja* are presumed to be root-hemiparasites, with most species studied experimentally growing more vigorously in the presence of a host (L. R. Heckard 1962). While some species of the Intermountain Region, such as *C. chromosa*, *C. flava*, and *C. linariifolia*, are closely associated with species of *Artemisia* (N. H. Holmgren 1984), host specificity is not well documented for many species. Some species occur primarily with Asteraceae hosts (especially *Artemisia*), Fabaceae (especially *Lupinus*), Poaceae (especially bunchgrasses), and Polygonaceae (especially *Eriogonum*) (M. Ownbey 1959; Holmgren; T. I. Chuang and Heckard 1993b; J. M. Egger, unpubl.). Many species of *Castilleja* will flower in greenhouse garden studies when grown on *Helianthus* and on mat-forming *Raoulia* Hooker f. ex Raoul species, both Asteraceae (Heckard; Egger, unpubl.).

Species with red to red-orange inflorescences often are visited and pollinated by nectar-seeking hummingbirds. These and other flower colors also attract nectar-gathering insects, mostly hymenopterans (especially *Bombus*) (J. M. Egger, unpubl.). Field observations confirm at least the occasional occurrence of selective, trap-line foraging on species of *Castilleja* by Hymenoptera. A few species, such as *C. cryptantha*, are self-pollinating (W. J. Duffield 1972). Some species are important food plants for various Lepidoptera larvae, especially *Euphydryas* (for example, P. R. Ehrlich et al. 1975) and *Platyptilla* (W. H. Lange 1950), and as nectar sources for hawkmoths (*Hyles*) (F. S. Crosswhite and C. D. Crosswhite 1970). Some species are known to serve as alternate or intermediate hosts for certain rust fungi (*Cronartium*, *Endocronartium*) (for example, D. B. O. Savile 1968c) and for hummingbird flower mites (Mesostigmata: Ascidae) (A. J. Heyneman et al. 1991). The leaves, stems, and inflorescences of many species are adult food sources for a variety of insects, including Coleoptera, Heteroptera, and Homoptera. Coleopteran species are especially common on a wide variety of species of *Castilleja*, especially in the inflorescences, and field studies show preliminary evidence of species-specific relationships between plant and beetle species. Species of *Castilleja* are grazed by Canada geese and by mammals, including voles, packrats, porcupines, deer, and introduced cattle, goats, and sheep.

Several species of *Castilleja* are used as ornamentals (J. Borland 1994; T. Luna 2005), as natural dyes, and by Native Americans for medicinal, practical, and ceremonial purposes (D. E. Moerman 1998).

A number of *Castilleja* species contain alkaloids, including some assimilated from parasitized hosts via haustorial bridges (for example, F. R. Stermitz and G. H. Harris 1987). Various Lepidopterans utilizing species of *Castilleja* as larval food plants either sequester or excrete the alkaloid chemicals (Stermitz et al. 1986, 1986b). The presence of alkaloids in the bracts and calyces of at least some species increases lifetime seed production through decreased herbivory and increased visitation by pollinators (L. S. Adler 2000). Some species of *Castilleja* are reported to absorb and concentrate selenium, producing potentially toxic effects in grazing animals (I. Rosenfeld and O. A. Beath 2013).

Many species of *Castilleja* decline when grazed by domestic livestock, especially on islands. Grazing by feral animals resulted in the extinction of one species (*C. guadalupensis* Brandegee) that occurred outside the flora area and the near extinction of several others, including the federally listed *C. grisea* and *C. mollis*. Plants are damaged both by cropping of inflorescences and by trampling of the often brittle stems. Removal of grazing pressures usually results in recovery of the affected populations. Numerous species in the flora area are of conservation concern, and many others are restricted in range and habitat and should be monitored. Most species in the flora area are endemic. A few species are peripheral in North America or extend into the mountains of northern and central Mexico or into the boreal and arctic regions of northeastern Asia.

Aside from limited taxonomic treatments in regional floras and works addressing particular species groups, the first comprehensive modern revision of the taxonomy and phylogenetic classification of *Castilleja* was the generic realignment and synopsis of the subtribe Castillejinae proposed by T. I. Chuang and L. R. Heckard (1991). In that work, *Castilleja* was considerably expanded, with the fragmentation of the traditional genus *Orthocarpus* and the movement of *Orthocarpus* sect. *Castillejoides* A. Gray into *Castilleja*, based on morphological congruities, base chromosome numbers, and other factors. Another monospecific genus, *Gentrya* Breedlove & Heckard, was also moved to *Castilleja*. Chuang and Heckard provided a detailed summary of earlier attempts at infrageneric classification in *Castilleja* and proposed the beginnings of a revised infrageneric classification, but their treatment in this regard was provisional and incomplete, primarily addressing the placement of the newly included groups and the species most closely related to them morphologically. More recently, additional comprehensive phylogenetic analyses of subtribe Castillejinae by D. C. Tank and R. G. Olmstead (2008) and Tank et al. (2009) largely confirmed the generic realignment by Chuang and Heckard but determined that two additional monospecific genera from Mexico, *Ophiocephalus* Wiggins and *Clevelandia* Greene, were well nested within a monophyletic *Castilleja*. However, no infrageneric classification for *Castilleja* was proposed at that time, due to discordance between previous taxonomic hypotheses and the results of genetic studies to date. Since the phylogenetic classification of *Castilleja* is the subject of ongoing research, no taxonomic structure for the North American species is proposed here.

SELECTED REFERENCES Gray, A. 1880b. Revision of the genus *Castilleja*. Amer. J. Sci. Arts, ser. 2, 34: 335–339. Hersch-Green, E. I. 2012. Polyploidy in Indian paintbrush (*Castilleja*; Orobanchaceae) species shapes but does not prevent gene flow across species boundaries. Amer. J. Bot. 99: 1680–1690. Holmgren, N. H. 1971. A taxonomic revision of the *Castilleja viscidula* group. Mem. New York Bot. Gard. 21: 1–63. Pennell, F. W. 1934b. *Castilleja* in Alaska and northwestern Canada. Proc. Acad. Nat. Sci. Philadelphia 86: 517–540.

1. Plants annual, sometimes biennial, with thin taproot or fibrous root system . . . . Group 1, p. 569
1. Plants perennial, sometimes biennial, with a ± woody caudex, sometimes from rhizomes.
   2. Woody shrubs or subshrubs; stem surfaces often nearly obscured by dense white
     (yellowish) hairs . . . . Group 2, p. 570
   2. Herbs or subshrubs, or shrubs collected without diagnostic proximal parts; stem
     surfaces not obscured by dense white (yellowish) hairs (except *C. arachnoidea*,
     C. nivea, *C. schizotricha*, which are herbaceous).
      3. Calyces cleft into 4 subequal lobes . . . . Group 3, p. 571
      3. Calyces cleft unequally in various ways, never divided into 4 subequal lobes.
        4. Stems with branched or stellate hairs, at least in part . . . . Group 4, p. 571
        4. Stems glabrous or hairy but lacking branched or stellate hairs.
          5. Leaf margins conspicuously wavy (perpendicular to plane of blade;
           sometimes obscure in pressed material); stems with dense stipitate-
           glandular hairs, often mixed with eglandular hairs . . . . Group 5, p. 572
          5. Leaf margins plane, sometimes ± wavy; stems stipitate-glandular or not.
            6. Corolla beaks relatively short, abaxial lip lengths 50–100% beak
             lengths. . . . . Group 6, p. 573
            6. Corolla beaks longer, abaxial lip lengths less than 50% beak lengths.
              7. Majority of leaves with 3–9 lobes, proximal leaves sometimes
               entire . . . . Group 7, p. 575
              7. Majority of leaves entire, sometimes distal leaves with 3–5(–7)
               lobes.
                8. Inflorescences yellow, yellow-green, yellow-orange, or white,
                 sometimes tinged with pale red or purple . . . . Group 8, p. 577
                8. Inflorescences usually reddish orange, bright red, pink-purple,
                 or purplish red . . . . Group 9, p. 579

Group 1. Plants annual, sometimes biennial, with thin taproot or fibrous root system.

1. Abaxial and adaxial calyx clefts conspicuously deeper than lateral calyx clefts, lateral clefts absent or $^1\!/_{20}$–$^1\!/_4$ depth of abaxial and adaxial clefts; bracts distally red to red-orange, yellow, green, or white, rarely pink, peach, or magenta.
   2. Bracts 0-lobed, apices acuminate or spatulate; mostly west of Continental Divide.
      3. Bracts plane-margined, narrowly lanceolate-acuminate, sometimes narrowly oblong or spatulate distally, apices red to orange, sometimes bright yellow; widespread in w North America . . . . 68. *Castilleja minor*
      3. Bracts wavy-margined, spatulate distally, apices white, sometimes very pale yellow, often aging dull pink to dull red-purple; Animas Valley, sw New Mexico . . . . 79. *Castilleja ornata*
   2. Bracts lobed or 0-lobed, apices not acuminate or spatulate; east of Continental Divide.
      4. Bracts greenish throughout; corollas conspicuously decurved distally; sw Texas . . . . .66. *Castilleja mexicana* (in part)
      4. Bracts greenish proximally, contrastingly colored distally; corollas ± straight, sometimes slightly curved; widespread east of Continental Divide, north to se Canada.
         5. Leaves and bracts deeply divided into (3–)5–9 lobes; widespread east of Continental Divide, north to sc Canada . . . . 20. *Castilleja coccinea*
         5. Leaves and bracts 0(–5)-lobed; se United States.
            6. Bracts distally pink to reddish, rarely white, peach, magenta, or yellow; basal rosettes of leaves usually absent; meadows, forest openings, various substrates including limestone; Arkansas, Louisiana, Oklahoma, and Texas, introduced sporadically along highways in Alabama and Florida . . . . .45. *Castilleja indivisa*
            6. Bracts distally bright yellow, rarely light orange; basal rosettes of leaves usually present; forest openings on dolomite limestone substrates; Bibb County, Alabama . . . . 49. *Castilleja kraliana*
1. Abaxial and adaxial calyx clefts subequal to or only slightly deeper than lateral calyx clefts, lateral clefts well developed and at least $^1\!/_3$ depth of abaxial and adaxial clefts; bracts green throughout or distally reddish purple, pink-purple, pink, dull reddish brown or brownish, cream, or white, rarely red, red-orange, or yellow.
   7. Distal bracts essentially unicolored, usually greenish to dull reddish brown, distal portions never strongly contrasting with proximal portions (except *C. lasiorhyncha* with distal margin sometimes white on immature bracts).
      8. Bracts and leaves 0-lobed . . . . .10. *Castilleja campestris*
      8. Bracts lobed; leaves lobed, sometimes 0-lobed.
         9. Corolla beaks obscured by dense, spreading hairs; distal margins of immature bracts white to rarely cream, with a tuft of white hairs; Cuyamaca, San Bernardino, San Jacinto, and Santa Rosa mountains, sw California . . . 52. *Castilleja lasiorhyncha*
         9. Corolla beaks puberulent; distal margins of immature bracts green or brownish, rarely white (*C. lacera*), lacking tufted hairs; widespread w North America.
            10. Stigmas generally exserted, equal to or longer than corolla beak; abaxial corolla lips strongly inflated, 4–10 mm wide, 3–6 mm deep.
               11. Corollas 10–22 mm, abaxial lips 4–8 mm wide . . . . 50. *Castilleja lacera*
               11. Corollas (15–)20–28 mm, abaxial lips 8–10 mm wide . . . . . . 99. *Castilleja rubicundula*
            10. Stigmas included within corolla beaks; abaxial corolla lips moderately inflated, 2–4(–5) mm wide, 1–3 mm deep.

12. Leaves linear to lanceolate, lobes linear; bract lobes linear or lanceolate, apices acute to acuminate; corollas white or yellow, usually with obscure, dull red to deep purple spots near bases of abaxial lips; widespread in w North America but not coastal and absent from Puget Trough region and Vancouver Island. . . . . 110. *Castilleja tenuis*

12. Leaves ovate to lanceolate, lobes linear to lanceolate; bract lobes lanceolate, apices broadly acute or obtuse; corollas distinctly bicolored, beaks white, sometimes faint purple, abaxial lips pale lemon yellow, unspotted; s Vancouver Island and islands bordering Haro Strait region of Puget Trough, British Columbia and Washington. . . . . 115. *Castilleja victoriae*

[7. Shifted to left margin.—Ed.]

7. Distal bracts bicolored, usually greenish proximally, distal portions strongly contrasting with proximal portions.

13. Corolla beaks prominently hooked and densely villous-hairy; filaments with long, soft, spreading hairs . . . . 32. *Castilleja exserta*

13. Corolla beaks ± straight and inconspicuously puberulent; filaments glabrous.

14. Inflorescences 1–2 cm wide; corolla tubes not expanded distally; abaxial lips inconspicuous, pouches 2 mm wide.

15. Bracts 3(–5)-lobed, distally white to sometimes pale yellow or pale pink-purple; widespread in w coastal North America, including s California . . . . .7. *Castilleja attenuata*

15. Bracts (3–)5-lobed, distally pink to purplish red; s Coast Ranges, s San Joaquin Valley and s Sierra Nevada foothills, California . . . . 9. *Castilleja brevistyla*

14. Inflorescences 1–4 cm wide; corolla tubes expanded distally; abaxial lips conspicuous, pouches 3–7 mm wide.

16. Abaxial lips of corollas yellow, bract lobes lanceolate to oblong; near the coast; British Columbia to California. . . . . 2. *Castilleja ambigua* (in part)

16. Abaxial lips of corollas not yellow or yellow distally (except *C. ambigua*), bract lobes linear or narrowly oblong; coastal and inland; California.

17. Bracts green to purple proximally, purple to red-purple or white distally . . . . 26. *Castilleja densiflora* (in part)

17. Bracts green to reddish brown proximally, sometimes with purple midrib, white to pale yellow or pale purple distally.

18. Beaks (4–)5–7 mm; stems lacking stipitate-glandular hairs; coastal grasslands, San Luis Obispo County, California . . . . . . 26. *Castilleja densiflora* (in part)

18. Beaks 1–5(–5.5) mm; stems with minute stipitate-glandular hairs; Sierra Nevada foothills and Napa County, California.

19. Calyces 8–14 mm; beaks 1–4 mm; plants 0.6–2.2 dm; Napa County, California . . . . 2. *Castilleja ambigua* (in part)

19. Calyces 15–25 mm; beaks (3–)4–5(–5.5) mm; plants 1.5–4.5 dm; Sierra Nevada foothills, California .60. *Castilleja lineariloba*

Group 2. Woody shrubs or subshrubs; stem surfaces often nearly obscured by dense white (yellowish) hairs.

1. Stems unbranched or sparsely branched on distal ½, hairs unbranched to strongly branched; s Channel Islands and California mainland to w Texas.

2. Proximal stems or branches usually with dense leafy axillary shoots; stem hairs strongly branched; corollas 16–27 mm, beaks (7–)8.5–14 mm; California . . . . 34. *Castilleja foliolosa* (in part)

2. Proximal stems or branches sometimes with dense leafy axillary shoots; stem hairs unbranched to moderately branched; corollas 23–35(–42) mm, beaks 11–22 mm; Arizona, New Mexico, sw Texas. . . . . 51. *Castilleja lanata* (in part)

1. Stems much-branched, including distal ½, hairs stellate or weakly branched; Channel Islands, California.

3. Corolla beaks 7–9 mm; bracts and calyces distally pale yellow to green; bracts (3–)5–7-lobed; stem hairs stellate; San Clemente Island. . . . . .40. *Castilleja grisea* (in part)

3. Corolla beaks 11–14 mm; bracts and calyces distally yellow to red; bracts (0–)3-lobed; stem hairs weakly branched; n Channel Islands, absent on San Clemente Island. . . . .

. . . . .43. *Castilleja hololeuca* (in part)

Group 3. Plants perennial; calyx lobes subequal.

1. Stem hairs usually conspicuously matted and reflexed-spreading to ± appressed.
    2. Plants 1–4 dm; stems erect; montane to subalpine meadows; ne Arizona, s Colorado, nw New Mexico . . . . 61. *Castilleja lineata* (in part)
    2. Plants 0.5–2 dm, sometimes to 3 dm (*C. arachnoidea*); stems ascending to erect, sometimes proximally decumbent; alpine to subalpine meadows; n California, Montana, Nevada, s Oregon, nw Wyoming.
        3. Corolla beaks 6–8 mm; bracts greenish to pale yellow-green or very pale dull purplish; c, sw Montana, nw Wyoming . . . . 75. *Castilleja nivea*
        3. Corolla beaks (2–)3–5 mm; bracts pale yellow or soft greenish to dull red or soft purple, including intermediate shades; California, Nevada, Oregon.
            4. Hairs unbranched; plants 0.6–3 dm; leaves (1–)2–4(–6) cm; bracts dull yellow, dull or brick red, green, dull orange, or pink . . . . 6. *Castilleja arachnoidea*
            4. Hairs branched; plants 0.8–1.5 dm; leaves 0.5–2 cm; bracts various shades of soft purple . . . . 104. *Castilleja schizotricha*
1. Stem hairs not conspicuously matted and spreading to reflexed (sometimes appearing more appressed if handled or pressed).
    5. Stems ashy gray; distal portions of bracts appearing dusty with dense, short stipitate-glandular hairs, many with a nodulose to pillarlike, crystallized, usually pigmented exudate, papillose at 40×; San Bernardino Mountains, s California . . . . 18. *Castilleja cinerea*
    5. Stems not ashy gray; distal portions of bracts not appearing dusty and, when stipitate-glandular hairs are present, lacking any crystallized exudate; n California, Idaho, Montana, Nevada, Oregon, Utah, Wyoming.
        6. Herbs (0.7–)1.2–3.5(–4.4) dm; stems ascending to erect, sometimes short-decumbent; bracts pale yellowish green, sometimes with pale purplish to brownish cast, especially with age; leaves 1–5.5(–8) cm; stems with eglandular hairs . . . . . 87. *Castilleja pilosa*
        6. Herbs 0.4–1.5(–1.8) dm; stems decumbent at least proximally; bracts brownish green or whitish green to dull red or purple; leaves 0.5–3.2 cm; stems usually with glandular hairs.
            7. Corolla abaxial lips conspicuously exserted from calyces, teeth pink-purple to red-purple, erect and appressed to corolla beaks, beaks green; stigmas green; bracts mostly pink-purple to red-purple; alpine Wallowa Mountains, Oregon . . . . 100. *Castilleja rubida*
            7. Corolla abaxial lips slightly exserted or included within calyces, teeth white, cream, yellow, or pink, erect to slightly spreading; stigmas blackish; bracts brownish green or whitish green to dull greenish purple, red-purple, or pink-purple; e California, Nevada, Utah.
                8. Corollas 10–16(–19) mm, tubes 8–13 mm, beaks 3–5.5 mm, scarcely exceeding abaxial lips; leaves and bracts not fleshy; subalpine and alpine, not usually associated with hot springs; California, Nevada, Utah . . . . . . 72. *Castilleja nana*
                8. Corollas 18–22(–24) mm, tubes 13–18 mm, beaks 4.5–6.5 mm, noticeably exceeding abaxial lips; leaves and bracts often fleshy; mounded, alkaline hot spring outlets; ec Nevada . . . . 102. *Castilleja salsuginosa*

Group 4. Plants perennial; calyces unequally cleft; stems with at least some branched hairs.

1. Calyx abaxial clefts clearly shallower than laterals; bracts uniformly colored, green or yellowish green except for white hairs; sw California . . . . 88. *Castilleja plagiotoma* (in part)
1. Calyx abaxial clefts clearly deeper than laterals; bracts distally colored differently from and more conspicuously than proximally, never uniformly green or yellowish green; widespread, including sw California.

[2. Shifted to left margin.—Ed.]

2. Stems not white-woolly to grayish, hairs sparse to ± dense.
    3. Leaves 0–5-lobed; short, leafy axillary shoots often conspicuous; calyx lobe apices acute
       to obtuse or rounded; coastal Santa Barbara and San Luis Obispo counties, California
       . . . . 1. *Castilleja affinis* (in part)
    3. Leaves 0-lobed, sometimes with 3–5 lobes distally; short, leafy axillary shoots usually
       lacking or inconspicuous; calyx lobe apices acute to acuminate; c, nw California to
       sw Oregon. . . . . 90. *Castilleja pruinosa* (in part)
2. Stems white-woolly to grayish, hairs dense.
    4. Leaves elliptic-oblong to obovate, 0-lobed; bracts 0–3-lobed, distally pale to bright
       yellow, sometimes brownish orange; Santa Rosa Island. . . . . 70. *Castilleja mollis*
    4. Leaves linear to lanceolate, sometimes narrowly oblong or narrowly oblanceolate,
       distal leaves sometimes 0–5-lobed; bracts 0–7-lobed, distally red, orange, or yellow;
       Channel Islands and adjacent sw North America.
       5. Stems white-felted, with long, intertwined, sparsely branched hairs; leaves 0-lobed;
          n Channel Islands, California . . . . .43. *Castilleja hololeuca* (in part)
       5. Stems ashy gray or white-woolly with shorter, much-branched or stellate hairs;
          leaves 0–7-lobed; Oregon to Texas, including s Channel Islands of California.
          6. Distal portions of bracts pale yellow to green, hairs stellate; bracts
             (3–)5–7-lobed; San Clemente Island, California . . . . .40. *Castilleja grisea* (in part)
          6. Distal portions of bracts bright red to bright yellow (yellow-green), hairs much-
             branched; bracts 0–5-lobed; Santa Catalina Island and mainland of California,
             Arizona, New Mexico, w Texas.
             7. Lateral calyx clefts absent or inconspicuous.
                8. Proximal stems or branches usually with dense leafy axillary shoots,
                   these overwintering; corollas 16–27 mm, beaks (7–)8.5–14 mm;
                   California . . . . 34. *Castilleja foliolosa* (in part)
                8. Proximal stems or branches lacking dense leafy axillary shoots; corollas
                   23–35(–42) mm, beaks 11–22 mm; Arizona to Texas . . . . . 51. *Castilleja lanata* (in part)
             7. Lateral calyx clefts 20–25% of calyx length.
                9. Calyx lobes ovate to narrowly triangular, apices obtuse to acute;
                   abaxial corolla lips deep green to ± black; bracts (0–)3(–5)-lobed; stem
                   hairs branched or often mixed with shorter, unbranched, stipitate-
                   glandular ones; San Gabriel Mountains, sw California . . . . .38. *Castilleja gleasoni*
                9. Calyx lobes lanceolate to triangular, apices acute to acuminate;
                   abaxial corolla lips deep green; bracts 0–5-lobed; stem hairs branched;
                   c, nw California to sw Oregon . . . . 90. *Castilleja pruinosa* (in part)

Group 5. Plants perennial; calyces unequally cleft; stems glabrous or pubescent with unbranched
    hairs; margins of leaves wavy; stems stipitate-glandular.

1. Corollas conspicuously decurved distally; bracts usually entirely green.
    2. Corolla beaks exserted, adaxial lips ± exserted, tubes included within calyx;
       c Oregon. . . . . 13. *Castilleja chlorotica*
    2. Corollas (including majority of tubes) conspicuously exserted from calyx; sw Texas
       . . . . .66. *Castilleja mexicana* (in part)
1. Corollas straight to slightly curved distally; bracts usually not entirely green, at least
   distally.
    3. Distal margins of central bract lobes, sometimes also side lobes, with multiple shallow
       teeth; plants coastal or near coastal. . . . . .117. *Castilleja wightii* (in part)
    3. Distal margins of bract lobes entire, sometimes with a few shallow teeth; plants usually
       not coastal or near coastal (sometimes near coastal in *C. martini* var. *martini*).

[4. Shifted to left margin.—Ed.]
4. Bracts distally greenish to yellow-green or white, rarely red or pale pink.
  5. Bracts distally white to cream, sometimes shaded with pale, dull pink or pale yellow (sharply differentiated from proximal coloration), central lobe apices broadly rounded to truncate; 400–800 m. . . . . .119. *Castilleja xanthotricha* (in part)
  5. Bracts distally greenish, yellow-green, cream, or yellow, rarely red to pale pink (sometimes gradually differentiated from proximal coloration), central lobe apices acute to acuminate or obtuse, rarely narrowly rounded; 1400–3200 m.
    6. Corollas (20–)22–30 mm, beaks 8–11(–12) mm; calyces 17–21(–23) mm; Blue and Strawberry mountains, Oregon. . . . . .37. *Castilleja glandulifera*
    6. Corollas 16–22(–25) mm, beaks 5–8(–9) mm; calyces (10–)14–18 mm; mountains of e Oregon, Nevada, s Idaho. . . . . 116. *Castilleja viscidula*
4. Bracts distally red to red-orange (proximal bracts often entirely greenish in *C. disticha*), rarely yellowish to greenish or other color variants.
  7. Corollas 15–24(–26) mm, beaks 6–10 mm; plants 0.8–2.5(–5) dm.
    8. Lateral calyx clefts 3–5(–7) mm; calyx lobe apices acute, rarely obtuse; leaves 1.2–4.8 cm; usually granitic substrates; Sierra Nevada of California and Nevada
    . . . . 4. *Castilleja applegatei* (in part)
    8. Lateral calyx clefts 1.5–4 mm; calyx lobe apices obtuse to rounded; leaves 1–2(–2.5) cm; serpentine substrates; Siskiyou Mountains of nw California and sw Oregon
    . . . . 8. *Castilleja brevilobata* (in part)
  7. Corollas (20–)24–35(–45) mm, beaks 9–20 mm; plants 1–8 dm.
    9. Lateral calyx clefts (2–)3–8 mm, abaxials (4–)8–16(–19) mm; adaxial surfaces of corolla beaks usually greenish.
      10. Abaxial calyx clefts subequal to or slightly shallower than adaxials; sagebrush valleys to subalpine slopes; w United States . . . . 4. *Castilleja applegatei* (in part)
      10. Abaxial calyx clefts deeper than adaxials; upper montane to lower subalpine; mountains of c, ec Nevada. . . . . .27. *Castilleja dissitiflora* (in part)
    9. Lateral calyx clefts 0.5–3(–5) mm, abaxials 3.4–8 mm; adaxial surfaces of corolla beaks usually reddish to yellowish green (or greenish in *C. martini* var. *clokeyi*).
      11. Bracts 0(–3)-lobed, all but distalmost entirely greenish. . . . . .28. *Castilleja disticha*
      11. Bracts (0–)3–5(–7)-lobed, most contrastingly colored other than green distally.
        12. Corolla beaks equal to or longer than tubes; leaves green to sometimes purple; bract lobe apices obtuse to rounded. . . . . 64. *Castilleja martini*
        12. Corolla beaks usually shorter than tubes; leaves gray-green, sometimes green; bract lobe apices acute, sometimes obtuse. . . . . .71. *Castilleja montigena*

Group 6. Plants perennial; calyces unequally cleft; stems glabrous or pubescent with unbranched hairs; margins of leaves plane, sometimes ± wavy; stems sparsely to not at all stipitate-glandular; corolla beaks to 2 times length of abaxial lips.

1. Corollas conspicuously curved distally, at least distal portions of tubes conspicuously exserted from calyces.
  2. Lateral calyx clefts 1.5–6 mm; bracts greenish throughout; annuals or short-lived perennials; sw Texas . . . . .66. *Castilleja mexicana* (in part)
  2. Lateral calyx clefts 5–15 mm; bracts green to purplish throughout, sometimes reddish brown, pink, or lavender throughout, or distally white or pale yellow, sometimes distally dull pink, pink, salmon, orangish, pale pink-orange, buff, or cream; perennials; widespread in wc United States and sc Canada . . . . .106. *Castilleja sessiliflora* (in part)
1. Corollas straight or slightly curved, tubes ± included within calyces.
  3. Leaves 0-lobed (except sometimes distally, just below inflorescences); boreal and arctic.
    4. Bracts yellow, yellow-green, or pale whitish throughout, sometimes with dull reddish-purplish wash proximally, especially with age . . . . .81. *Castilleja pallida*
    4. Bracts purple to lavender, pink-purple, or reddish purple throughout, sometimes proximally purple to pink-purple or reddish purple, distally whitish or pale pink.

5. Plants 0.5–3 dm; stems unbranched . . . . 30. *Castilleja elegans*
5. Plants (2.5–)3–5(–6) dm; stems unbranched to often branched distally . . . . . 95. *Castilleja raupii*
[3. Shifted to left margin.—Ed.]
3. Leaves lobed, at least distally; w North America south of boreal-arctic region (except
   C. *hyperborea*).
   6. Lateral lobes of cauline leaves ± divaricate and often abruptly up-curved from plane of
      main leaf blade; nw Canada, Alaska . . . . 44. *Castilleja hyperborea*
   6. Lateral lobes of cauline leaves ascending to erect or ± spreading and gradually or not at
      all up-curved from plane of main leaf blade; extreme sw Canada, w United States.
      7. Calyx lobe apices acute (to rarely obtuse in C. *thompsonii*); xeric sites, usually with
         sagebrush.
         8. Corollas: abaxial lips scarcely expanded, glabrous or obscurely puberulent,
            50–70% as long as corolla beak . . . . 111. *Castilleja thompsonii*
         8. Corollas: abaxial lips slightly but noticeably pouched, puberulent, 70–100%
            as long as corolla beak.
            9. Lateral calyx clefts 5–10 mm, lobes linear . . . . 77. *Castilleja oresbia*
            9. Lateral calyx clefts 0.5–4.3(–6) mm, lobes lanceolate to triangular . . . 80. *Castilleja pallescens*
      7. Calyx lobe apices obtuse to rounded or acute, rarely truncate; ± mesic sites, not
         usually associated with sagebrush.
         10. Corollas 24–45(–50) mm, beaks and abaxial lips conspicuously exserted from
            calyces . . . . 48. *Castilleja kerryana*
         10. Corollas 12–25(–28) mm, beaks and abaxial lips included to scarcely and
            inconspicuously exserted from calyces, usually only stigmas and distal portions
            of beaks emergent.
            11. Central lobes of distal bracts rounded, sometimes truncate.
               12. Calyces 20–30 mm; plants (1–)1.5–5(–6) dm . . . . 25. *Castilleja cusickii*
               12. Calyces (12–)13–23(–25) mm; plants 0.5–1(–2) dm . . . . 92. *Castilleja pulchella*
            11. Central lobes of distal bracts acute to acuminate or obtuse, rarely narrowly
               rounded.
               13. Corollas 14–16 mm; vicinity of Mt. Rainier, Washington . . . . . 24. *Castilleja cryptantha*
               13. Corollas 16–25 mm; California, Nevada, Oregon.
                  14. Bracts pale cream to pale yellow, pale yellow-green, or greenish,
                     often weakly infused with dull purple, sometimes other shades of
                     above colors; Oregon.
                     15. Bracts pale greenish or pale yellow-green throughout, or
                        proximally greenish or pale yellow-green, distally yellow
                        to whitish, sometimes pink-purple, or pale, dull purplish,
                        sometimes aging pink or yellow, often infused with light purple,
                        rarely pink; Blue and Wallowa mountains, ne Oregon . . . . .
                        . . . . 17. *Castilleja chrysantha*
                     15. Bracts pale cream to pale greenish yellow throughout, often
                        partly to entirely suffused with dull reddish purple to maroon,
                        especially proximally, along veins, and with age, sometimes
                        distal apices pale dullish red; Cascade Range, Klamath
                        County, s Oregon . . . . 21. *Castilleja collegiorum*
                  14. Bracts green to brownish or purplish proximally, pink to purple
                     or magenta distally, rarely white; California, closely adjacent
                     Nevada.
                     16. Corolla beaks adaxially white or white with pale salmon
                        margins; abaxial lips pale green to yellow-green, with white
                        distal teeth; stigmas pale green; Mt. Lassen region, Cascade
                        Range, California . . . . 53. *Castilleja lassenensis*
                     16. Corolla beaks adaxially green with red margins; abaxial lips
                        greenish with pink-purple distal teeth; stigmas greenish to
                        deep bluish purple; Sierra Nevada, California and adjacent
                        Nevada. . . . . 55. *Castilleja lemmonii*

Group 7. Plants perennial; calyces unequally cleft; stems glabrous or pubescent with unbranched hairs; margins of leaves plane, sometimes ± wavy; stems sparsely to not at all stipitate-glandular; corolla beaks 2+ times length of abaxial lips; most leaves divided, sometimes proximal leaves 0-lobed.

1. Corolla beaks clearly ½+ as long as corollas.
  2. Corolla beaks 22–39 mm; abaxial calyx clefts much (11+ mm) deeper than adaxials; se Arizona, sw New Mexico . . . . 84. *Castilleja patriotica*
  2. Corolla beaks 8–24 mm; abaxial calyx clefts subequal to or slightly (less than 2 mm) shallower than adaxials; Pacific Northwest and California to Idaho and Utah.
    3. Shrubs or subshrubs; stems white-woolly; n Channel Islands, California . . . .
       . . . . .43. *Castilleja hololeuca* (in part)
    3. Herbs; stems not white-woolly; mainland of w North America.
      4. Bracts lanceolate to oblong; corolla beaks 8–15 mm; se Idaho, ne Nevada, w Utah . . . . .3. *Castilleja angustifolia* (in part)
      4. Bracts obovate to ovate or orbicular; corolla beaks 14–24 mm; Cascade Range and west, s British Columbia to Oregon.
        5. Stems mostly glabrous; bracts mostly obovate to orbicular, lobes mostly arising at or above mid length, short and medium length, lanceolate to triangular, apices acute; Coast Range, Clatsop County, Oregon, and Pacific County, Washington . . . . 12. *Castilleja chambersii*
        5. Stems with short, soft, spreading hairs; bracts ovate to orbicular, lobes arising below mid length, long, linear to linear-lanceolate, apices rounded to acute; Cascade Range and Columbia River Gorge, s British Columbia to Oregon . . . . 101. *Castilleja rupicola*
1. Corolla beaks ca. ½ or less than ½ as long as corollas.
  6. Abaxial corolla lips usually not pouched, teeth prominent, petaloid, spreading to erect, not completely greenish nor incurved.
    7. Bracts and calyces distally yellow, sometimes orange-yellow. . . . . 19. *Castilleja citrina*
    7. Bracts and calyces distally various shades of red, red-orange, orange, purple, or pink, rarely yellowish or white.
      8. Bracts and calyces distally red to reddish orange or pale orange. . . . . . . 59. *Castilleja lindheimeri*
      8. Bracts and calyces distally purple to pink-purple or pink, light orange, or light yellow, rarely white . . . . .94. *Castilleja purpurea*
  6. Abaxial corolla lips either ± pouched and with small, erect teeth or apiculations, or teeth reduced, not petaloid, not spreading, entirely greenish and incurved.
    9. Bracts green or yellow-green throughout, or proximally green to yellow-green, distally yellow to sometimes pale yellow-orange, hairs white; corolla tubes 5–14 mm.
      10. Calyces divided more deeply abaxially than laterally; ne Arizona, s Colorado, nw New Mexico. . . . . .61. *Castilleja lineata* (in part)
      10. Calyces only slightly divided abaxially, more deeply divided laterally; sw California . . . . 88. *Castilleja plagiotoma* (in part)
    9. Bracts at least distally red, rose, purple, yellow, or orange, rarely green, white, or cream; corolla tubes 8–24 mm.
      11. Stems glabrate, sometimes puberulent proximally.
        12. Bracts distally red to red-orange, very rarely yellowish.
          13. Stems few to several, clustered, from a woody caudex; bracts lacking a yellowish medial band; Rocky Mountains, Montana and Wyoming
             . . . . .23. *Castilleja crista-galli*
          13. Stems usually solitary from slender, creeping rhizomes; bracts usually with a narrow, distinct yellow medial band; Cascade Range, Oregon and s Washington . . . . 108. *Castilleja suksdorfii*
        12. Bracts distally pink-purple to red-purple, lilac purple, rose red, pink, or white, rare individuals red to orange-red in populations of purple plants.

14. Stems glabrate or puberulent, 0.7–2 dm; corollas 20–25 mm; alpine; s Rocky Mountains, sc Colorado, n New Mexico, se Utah . . . . . . 41. *Castilleja haydenii*

14. Stems glabrate proximally, hairy distally, (0.6–)1–4(–5) dm; corollas 12–30 mm; subalpine to lower alpine; widespread in mountains of nw North America . . . . 82. *Castilleja parviflora*

[11. Shifted to left margin.—Ed.]

11. Stems hairy.
    15. Adaxial calyx clefts deeper than abaxials.
        16. Corolla beaks 50–70% as long as tubes; not associated with sagebrush; Rocky Mountains, c Idaho and adjacent Montana . . . . 22. *Castilleja covilleana*
        16. Corolla beaks 85–100% as long as tubes; usually associated with sagebrush; widespread in w United States.
            17. Bracts distally pink, magenta, pink-purple, reddish pink, pale yellow, pale yellow-orange, pale orange, or white, rarely reddish or orange-red; calyces 13–25(–28) mm; corolla beaks 8–15 mm; s Idaho, sw Montana, e Nevada, se Oregon, sw South Dakota, w Utah, n Wyoming. . . . . 3. *Castilleja angustifolia* (in part)
            17. Bracts distally bright red to scarlet or orange-red, rarely yellowish to dull orange or pink; calyces (17–)20–27 mm; corolla beaks (9–)10–18 mm; Arizona, e California, w Colorado, e Oregon, s Idaho, Nevada, nw New Mexico, Utah, s, w Wyoming. . . . . 15. *Castilleja chromosa*
    15. Abaxial and adaxial calyx clefts subequal, adaxials only slightly deeper, or abaxials deeper than adaxials.
        18. Corolla beaks 7–22 mm.
            19. Corollas 18–24 mm; bracts (0–)3–5-lobed; stems with mixture of long, soft, spreading, eglandular hairs and shorter stipitate-glandular hairs; leaf margins ± wavy, sometimes plane; substrates serpentine; Siskiyou Mountains, nw California, sw Oregon . . . . 8. *Castilleja brevilobata* (in part)
            19. Corollas 17–40(–45) mm; bracts 0–7(–11)-lobed; stems with short to long, stiff to soft, spreading hairs, sometimes mixed with stipitate-glandular hairs; leaf margins usually plane, often mixed with wavy-margined leaves (*C. affinis*, *C. hispida*); substrates serpentine or not; widespread in w North America.
                20. Stems with dense, whitish, short hairs; plants decumbent, often sprawling, distally ascending, 0.7–1.5(–2.2) dm; inflorescences 2.5–10 cm; leaves gray-green, sometimes green or reddish purple; Nevada to New Mexico
                . . . . 103. *Castilleja scabrida*
                20. Stems with long, soft to stiff hairs and short-glandular or eglandular hairs; plants erect or ascending, 1.3–6 dm; inflorescences 3–30 cm; leaves green, purplish, or red-brown; Alberta and British Columbia to Arizona.
                    21. Stems unbranched or often branched at proximal nodes; short, leafy axillary shoots common, usually conspicuous; lateral lobes of leaves almost as wide as mid blade; w California. . . . . 1. *Castilleja affinis* (in part)
                    21. Stems unbranched; short, leafy axillary shoots absent or, when present, inconspicuous; lateral lobes of leaves much narrower than mid blade; nw United States, sw Canada . . . . 42. *Castilleja hispida* (in part)
        18. Corolla beaks 4–8(–12) mm.
            22. Central bract lobes acute to acuminate, sometimes narrowly obtuse (*C. flava*).
                23. Inflorescences a variable and often contrasting mixture of bright yellow and bright red; Sierra Nevada, California and adjacent Nevada . . . .
                . . . . 86. *Castilleja peirsonii* (in part)
                23. Inflorescences soft yellow to pale yellow-green, sometimes tinged with dull reddish or orange; Great Basin and Rocky Mountains.
                    24. Herbs 1.5–5.5(–7.5) dm; sagebrush or montane; Oregon to Montana
                    . . . . 33. *Castilleja flava* (in part)
                    24. Herbs 0.8–1.5 dm; alpine to subalpine; c Rocky Mountains, Colorado, disjunct in se Montana . . . . 91. *Castilleja puberula*

[22. Shifted to left margin.—Ed.]

22. Central bract lobes rounded to obtuse, sometimes truncate.
    25. Bracts usually distally bright red to red-orange.
        26. Leaf margins plane; inflorescences 2.5–7.5(–15) cm; bract lobes arising above mid length except on proximal bracts; 1700–3400 m; Sierra Nevada, California and adjacent Nevada . . . . 86. *Castilleja peirsonii* (in part)
        26. Leaf margins plane or wavy; inflorescences 3–25(–30) cm; bract lobes arising above or below mid length; 0–1200(–1700) m; w of Sierra Nevada, California and adjacent sw Oregon.
            27. Stem hairs usually eglandular; corolla beaks subequal to or longer than tube; corollas 17–40 mm; abaxial and adaxial calyx clefts 6–22 mm; substrates serpentine or not; w California . . . . 1. *Castilleja affinis* (in part)
            27. Stem hairs stipitate-glandular; corolla beaks subequal to or shorter than tube; corollas 15–24(–26) mm; abaxial and adaxial calyx clefts 5.5–8.5 mm; substrates serpentine; nw California, sw Oregon . . . . 8. *Castilleja brevilobata* (in part)
    25. Bracts distally yellow, whitish, purplish, dull red, pinkish, or dull orange.
        28. Abaxial corolla lip teeth reduced to minute apiculations; s Sierra Nevada, California . . . . 89. *Castilleja praeterita*
        28. Abaxial corolla lip teeth not reduced to minute apiculations; s British Columbia, California, Oregon, Washington.
            29. Inflorescences bright yellow or light yellow, usually becoming deep golden yellow; bracts (0–)5–9(–13)-lobed, lobes arising above mid length; Puget Trough, British Columbia and Washington, and Willamette Valley, Oregon . . . . 57. *Castilleja levisecta*
            29. Inflorescences red, orange, yellow, or white, never becoming deep golden yellow; bracts (0–)3–7-lobed, proximal lobes arising below mid length (except in *C. brevilobata*); California, Oregon, absent in Puget Trough and Willamette Valley.
                30. Corolla beaks puberulent, eglandular; stem hairs usually eglandular; leaves 2–13 cm; w California . . . . 1. *Castilleja affinis* (in part)
                30. Corolla beaks puberulent, stipitate-glandular; stem hairs mostly stipitate-glandular; leaves 0.8–6 cm; California, Oregon.
                    31. Bracts distally red, red-orange, or scarlet, sometimes orange or yellow; corollas 15–24(–26) mm, beaks ca. 50% as long as corollas; plants 1–5 dm; substrates of serpentine origin; nw California, sw Oregon . . . . 8. *Castilleja brevilobata* (in part)
                    31. Bracts distally white to cream, rarely pale yellow or dull, pale pink; corollas 17–23 mm, beaks ca. 33% as long as corollas; plants 1–2(–3.8) dm; substrates of basaltic origin; nc Oregon . . . . 119. *Castilleja xanthotricha* (in part)

Group 8. Plants perennial; calyces unequally cleft; stems glabrous or pubescent with unbranched hairs; margins of leaves plane; stems sparsely or not stipitate-glandular; corolla beaks 2+ times length of abaxial lips; most leaves entire; inflorescences yellow, yellow-green, pale yellow-orange, pale reddish orange, or white, rarely pale reddish.

1. Corolla tubes 24–45 mm, usually partly exserted; corollas strongly curved distally . . . . 106. *Castilleja sessiliflora* (in part)
1. Corolla tubes usually less than 25 mm, mostly included within calyces; corollas straight or slightly curved.
    2. Stems solitary, sometimes few, proximally creeping and becoming rhizomatous, attached to a remote caudex; wet meadows or bogs, sometimes near hot springs.
        3. Calyces 9–17 mm; corolla tubes 7–12 mm; serpentine bogs and wetlands; nw California, sw Oregon . . . . 29. *Castilleja elata* (in part)
        3. Calyces 15–22 mm; corolla tubes 11–19 mm; non-serpentine wet meadows, sometimes near hot springs; Idaho, Montana, Oregon, Wyoming . . . . 39. *Castilleja gracillima*

[2. Shifted to left margin.—Ed.]

2. Stems few to many, rarely solitary, rhizomatous or not; wet or dry meadows.

     4. Stems loosely lanate . . . . .36. *Castilleja genevieveana*

     4. Stems glabrous or variously hairy, never lanate.

         5. Leaves densely crowded; stems usually much-branched, usually with short, leafy axillary shoots; 0–700 m; coastal; c, n California . . . . .117. *Castilleja wightii* (in part)

         5. Leaves not densely crowded; stems branched or unbranched, lacking short, leafy axillary shoots; 0–4300 m; inland (except for *C. miniata* var. *dixonii, C. unalaschcensis*); w North America.

             6. Stems usually unbranched; plants 1–3(–4) dm.

                 7. Stem hairs including stipitate-glandular ones both in inflorescence and proximal to it.

                     8. Bracts 3–5-lobed, lateral lobes medium to long; quartzite substrates; near summit of Mt. Harrison, Albion Mountains, Idaho. . . . . .14. *Castilleja christii*

                     8. Bracts 0(–5)-lobed, rarely with 1 or 2 pairs of short, usually distal lobes; serpentine substrates, sometimes not; Cascade Mountains, s British Columbia and Washington . . . . 31. *Castilleja elmeri* (in part)

                 7. Stem hairs eglandular below inflorescence, only stipitate-glandular in inflorescence.

                     9. Stems several to many, usually short-decumbent near base, becoming ascending-erect distally; bracts often conspicuously dull reddish brown or reddish purple proximally, especially with age; upper subalpine to alpine; Rocky Mountains . . . . .76. *Castilleja occidentalis*

                     9. Stems solitary to few, erect to ascending; bracts not conspicuously dull reddish brown or reddish purple proximally; usually montane to subalpine; nw Arizona, sc Utah.

                           10. Corollas 17–25 mm; stems ascending, sometimes erect; inflorescences pale yellow, rarely yellow-orange; Aquarius Plateau, sc Utah . . . . 5. *Castilleja aquariensis*

                           10. Corollas 21–30(–35) mm; stems erect, sometimes ascending; inflorescences whitish to pale yellow, pale yellow-orange, pale reddish orange, or salmon; Kaibab Plateau, nw Arizona. . . 47. *Castilleja kaibabensis*

             6. Stems branched or unbranched; plants (1.4–)2.5–6(–8) dm.

                 11. Calyx lobe apices acute to rounded or obtuse; lateral lobe apices of bracts, when present, acute to obtuse or narrowly rounded.

                     12. Bracts usually divided, with 1–3 pairs of medium-length lateral lobes.

                         13. Stems with short, scabrid hairs below inflorescence; calyces 15–25 mm; corollas 21–27 mm; nw United States, adjacent sw Canada . . . . 63. *Castilleja lutescens*

                         13. Stems with long, soft, spreading hairs; calyces 13–20 mm; corollas (15–)17–27 mm; White Mountains, ec Arizona . . . . 69. *Castilleja mogollonica*

                     12. Bracts entire or shallowly divided, with 1 or 2 pairs of short lateral lobes.

                         14. Lateral calyx clefts 1–4 mm; bracts distally usually whitish to pale yellow (pale pink), 0–3(–5)-lobed; montane and subalpine meadows and slopes; Rocky Mountains and eastward . . . . .105. *Castilleja septentrionalis*

                         14. Lateral calyx clefts 4–10 mm; bracts distally usually yellow to yellow-orange (whitish), (0–)3–5-lobed; open coastal habitats, lower montane to lower alpine meadows in mountainous regions; British Columbia, Yukon, Alaska . . . . .114. *Castilleja unalaschcensis*

                 11. Calyx lobe apices usually acute to acuminate; lateral lobe apices of bracts, when present, usually acute to acuminate.

[15. Shifted to left margin.—Ed.]

15. Abaxial calyx clefts deeper than adaxials.
    16. Herbs (2.3–)3–8 dm; leaves usually greenish, unaffected by pubescence; grasslands, balds, and open pine forests; s British Columbia, n Idaho, nc, ne Washington...... 11. *Castilleja cervina*
    16. Herbs 1.5–4(–5.5) dm; leaves often grayish from pubescence; primarily sagebrush slopes and flats; Colorado, s, c Idaho, w Montana, Nevada, Utah, Wyoming .... 33. *Castilleja flava* (in part)
15. Abaxial calyx clefts subequal to adaxials.
    17. Bracts pale yellow to cream, sometimes greenish proximally; Sonoma County, California . . . . 113. *Castilleja uliginosa*
    17. Bracts at least distally wide range of colors, rarely yellow or yellow-orange (except for *C. miniata* var. *fulva*); widespread in w North America, not in Sonoma County, California.
        18. Bracts at least distally pink, magenta, or purple, sometimes buff, dull yellow, cream, or light yellow-orange; serpentine bogs and wetlands; Siskiyou Mountains, nw California and sw Oregon. . . . . 29. *Castilleja elata* (in part)
        18. Bracts at least distally yellow to tawny yellow, pale yellow-orange, or pale reddish; non-serpentine substrates; n British Columbia and adjacent Alberta and Yukon . . . . 67. *Castilleja miniata* (in part)

Group 9. Plants perennial; calyces unequally cleft; stems glabrous or pubescent with unbranched hairs; margins of leaves plane to sometimes slightly wavy; stems stipitate-glandular or not; corolla beaks 2+ times length of abaxial lips; most leaves entire; inflorescences mostly bright reddish, orange-red, pink-purplish, or reddish purple.

1. Corolla tubes 24–45 mm, distal portion often exserted from calyces; corollas strongly curved distally . . . . 106. *Castilleja sessiliflora* (in part)
1. Corolla tubes usually 30 mm or less, mostly included within calyces; corollas straight or slightly curved.
    2. Stems woolly to lanate, hairs usually obscuring surfaces.
        3. Bracts usually entire, apices rounded distally, sometimes with 1 or 2 pairs of short, lateral lobes usually arising at or above mid length. . . . . 46. *Castilleja integra*
        3. Bracts usually deeply divided, apices acute to rounded distally, a pair of much longer lateral lobes often originating from well below mid length.
            4. Calyx lobes truncate, rounded, emarginate, or divided 0–5 mm into obtuse or rounded lateral lobes; stems densely lanate to woolly, hairs branched or unbranched; s Arizona, s New Mexico, sw Texas. . . . . 51. *Castilleja lanata* (in part)
            4. Calyx lobes divided 5–7 mm into linear to lanceolate lateral lobes, apices ± acute; stems moderately lanate, hairs unbranched; Animas Valley and south, sw New Mexico. . . . . .112. *Castilleja tomentosa*
    2. Stems glabrous or hairy, not woolly or lanate, hairs, if any, not obscuring surfaces.
        5. Lateral calyx clefts 0–0.5 mm.
            6. Abaxial calyx clefts conspicuously deeper than adaxials; bract apices acute . . . . .109. *Castilleja tenuiflora* (in part)
            6. Calyx clefts usually subequal; bract apices rounded to obtuse or truncate.
                7. Calyces distally deep green, deep purple, or blackish; stems ± straight; corolla beaks usually included in calyces; se Arizona . . . . 74. *Castilleja nervata*
                7. Calyces distally bright red; stems subtly wavy, curving back and forth slightly throughout their length; corolla beaks exserted from calyces; sw Texas . . . . 98. *Castilleja rigida*
        5. Lateral calyx clefts usually 1–12 mm (sometimes less than 1 mm in *C. subinclusa* and *C. leschkeana* of California).
            8. Bracts at least distally pink, pink-purple, red-purple, or deep purple, rarely whitish.
                9. Plants 0.8–1.5(–2) dm; stems usually unbranched; sc or sw Utah.

10. Calyces 12–18 mm, abaxial and adaxial clefts 5.5–8.5(–10) mm; stems several to many; alpine; Tushar Mountains, sc Utah . . . . 83. *Castilleja parvula*

10. Calyces 16.5–27 mm, abaxial and adaxial clefts (8–)10–15(–18) mm; stems solitary, sometimes few; pine forests; Aquarius, Markagunt, and Paunsaugunt plateaus, Utah . . . . 96. *Castilleja revealii*

9. Plants (1–)2–7(–8) dm; stems branched or unbranched; w United States, sw Canada, not sc or sw Utah (except *C. rhexiifolia*).

11. Corolla beaks 12–20 mm . . . . 1. *Castilleja affinis* (in part)

11. Corolla beaks 6–16 mm.

12. Stems branched or unbranched; calyx lobe apices acute to acuminate; leaf apices acute to acuminate; se Alberta and se British Columbia or Siskiyou Mountains.

13. Bracts at least distally pink, magenta, or purple, sometimes distally buff, dull yellow, cream, or light yellow-orange; serpentine bogs and wetlands, sometimes gravelly river bars; Siskiyou Mountains, nw California and sw Oregon . . . .
. . . . 29. *Castilleja elata* (in part)

13. Bracts at least distally deep reddish or crimson, rarely magenta or dull orange; gravel river flats, moist thickets; sw Alberta, se British Columbia . . . . .93. *Castilleja purpurascens*

12. Stems usually unbranched; calyx lobe apices obtuse to rounded, sometimes ± acute (*C. rhexiifolia*); leaf apices acute; Rocky Mountains, s British Columbia, Oregon, Washington.

14. Stems pubescent, hairs moderately dense and of medium length, mixed with shorter eglandular and stipitate-glandular ones; bracts 0(–5)-lobed; 600–2600 m; Wenatchee Mountains, Washington, northward in Cascade Mountains to s British Columbia . . . . 31. *Castilleja elmeri* (in part)

14. Stems proximally glabrous to glabrate, distally sparsely pubescent, hairs ± long, soft, eglandular, sometimes stipitate-glandular; bracts (0–)3–5(–7)-lobed; 1800–4000 m; widespread in Rocky Mountains, disjunct in ne Oregon, ne Washington . . . . .97. *Castilleja rhexiifolia*

[8. Shifted to left margin.—Ed.]

8. Bracts at least distally red to red-orange, sometimes some populations with a few plants with yellow, white, or pink inflorescences.

15. Leaves cupulate, sometimes obscurely so on distal portion of stem, sometimes ± fleshy, apices truncate or obtuse to broadly rounded; coastal; c, n California to Curry County, Oregon.

16. Corollas 19–30 mm, beaks 8.5–15 mm; c California .54. *Castilleja latifolia*

16. Corollas 28–45 mm, beaks 15–25 mm; nw California to Curry County, Oregon
. . . . .65. *Castilleja mendocinensis*

15. Leaves not cupulate, sometimes thickened but not fleshy, apices acute to acuminate or obtuse, rarely rounded (except in *C. litoralis*, *C. wightii*); coastal to inland; w North America.

17. Plants 0.8–2.5 dm; calyces brightly and conspicuously colored for at least distally ²⁄₃ and often throughout, providing much of inflorescence color, clefts subequal; alpine to subalpine; Wallowa Mountains, ne Oregon . . . . .35. *Castilleja fraterna*

17. Plants 1–10(–20) dm, usually 2+ dm; calyces often less colorful, usually providing a less significant portion of inflorescence color, except when abaxial clefts deeper than adaxials, clefts subequal or abaxials deeper or adaxials ± deeper; sea level to subalpine; w North America, including Wallowa Mountains, Oregon.

18. Abaxial calyx clefts deeper than adaxials.

19. Stem hairs at least partially stipitate-glandular; California . . . . . . .107. *Castilleja subinclusa*

19. Stem hairs not stipitate-glandular; Arizona to Texas.

20. Lateral calyx clefts 0(–2.5) mm; bracts 0-lobed, rarely with 1 pair of lateral lobes arising near apex; stems densely short-pubescent. . . . . .109. *Castilleja tenuiflora* (in part)

20. Lateral calyx clefts 1.5–7 mm; bracts usually with 1–3 pairs of lateral lobes arising above or below mid length; stems glabrous to sometimes ± pubescent.

    21. Abaxial calyx clefts 3–5 times deeper than adaxials; bracts usually 3-lobed; widespread in w North America (parapatric with *C. wootonii* in New Mexico) . . . . .58. *Castilleja linariifolia*

    21. Abaxial calyx clefts 1.2–2 times deeper than adaxials; bracts usually 3–5-lobed; New Mexico, w Texas. . . . . .118. *Castilleja wootonii*

[18. Shifted to left margin.—Ed.]

18. Abaxial calyx clefts subequal to or shallower than adaxials.

  22. Lateral calyx clefts 7–11 mm; se Alaska . . . . .16. *Castilleja chrymactis*

  22. Lateral calyx clefts 1–8 mm; widespread including se Alaska.

    23. Abaxial calyx clefts shallower than adaxials; mountains of c, e Nevada. . . . . .27. *Castilleja dissitiflora* (in part)

    23. Calyx clefts subequal; w North America.

      24. Calyx lobe apices rounded to obtuse, rarely acute.

        25. Pedicels 0 mm or nearly so; range not limited to immediate coast.

          26. Stem hairs usually not stipitate-glandular; abaxial and adaxial calyx clefts 6–22 mm; California w of Sierra Nevada. . . . . 1. *Castilleja affinis* (in part)

          26. Stem hairs usually stipitate-glandular; abaxial and adaxial calyx clefts 5–14 mm; near coast of c, n California or Cascade Range from s British Columbia to c Washington.

            27. Bracts 0(–5)-lobed; corolla tubes 13–18 mm, beaks 8–15 mm; montane to subalpine; Wenatchee Mountains and Cascade Range, c, n Washington and s British Columbia. . . . . 31. *Castilleja elmeri* (in part)

            27. Bracts (0–)3–5(–7+)-lobed; corolla tubes 10–12 mm, beaks 11.5–18 mm; coastal and near-coastal bluffs, slopes, and canyons; n, c California. . . . . .117. *Castilleja wightii* (in part)

        25. Pedicels 0–6 mm; range at immediate coast.

          28. Stems usually much-branched, usually with prominent leafy axillary shoots, pubescent, with hairs including stipitate-glandular ones; corollas 20–30 mm; c, n California. . . . . .117. *Castilleja wightii* (in part)

          28. Stems often branched, usually lacking prominent leafy axillary shoots, usually glabrous or glabrate, often puberulent distally, hairs, when present, usually not stipitate-glandular; corollas 25–46 mm; nw California to sw British Columbia, including coastal Puget Sound.

            29. Corollas 23–38(–40) mm, beaks 10–16 mm, adaxial surfaces inconspicuously puberulent; lateral calyx clefts 1–3(–5) mm, 5–12% of calyx length; far nw California to Pacific County, Washington. . . . . 62. *Castilleja litoralis* (in part)

            29. Corollas 25–46 mm, beaks 11–25 mm, adaxial surfaces ± conspicuously puberulent; lateral calyx clefts 2–7 mm, 15–30% of calyx length; Clatsop County, Oregon, northward to s British Columbia . . . . 67. *Castilleja miniata* (in part)

      24. Calyx lobe apices acute to acuminate, rarely obtuse.

        30. Bract veins usually yellow, conspicuously contrasting with base color; calyces mostly yellow, with scarlet to red or red-orange apices, these matching distal coloration of bracts; stems without stipitate-glandular hairs; sky island mountain ranges, c, e Arizona, w New Mexico . . . . . .73. *Castilleja nelsonii*

        30. Bract veins not yellow, not conspicuously contrasting with base color; calyces not mostly yellow, not with scarlet to red or red-orange apices; stems with or without stipitate-glandular hairs; widespread, including Arizona, New Mexico.

[31. Shifted to left margin.—Ed.]

31. Plants 1.3–5 dm; stems usually unbranched; leaves usually 0-lobed proximally, 3-lobed distally . . . . 42. *Castilleja hispida* (in part)

31. Plants 1.2–10 dm; stems usually branched; leaves usually 0-lobed, rarely 3–5-lobed distally.

    32. Bracts cuneate to obovate-truncate; vicinity of Pt. Reyes Peninsula, California . . . 56. *Castilleja leschkeana*

    32. Bracts lanceolate to ovate, obovate, oblong-ovate, or oblong; widespread in w North America.

        33. Inflorescences 2–4.5 cm, bracts 0(–3)-lobed; Organ Mountains of s New Mexico . . . . 78. *Castilleja organorum*

        33. Inflorescences 2.5–21 cm, bracts 0–5(–7)-lobed; widespread in w North America.

            34. Corolla beaks usually 14–25 mm; inflorescences usually bearing a thin coat of whitish, powdery exudate . . . . 67. *Castilleja miniata* (in part)

            34. Corolla beaks 8–16 mm; inflorescences lacking a whitish exudate.

                35. Bracts usually 3–5-lobed; coastal, 0–200 m; far nw California to far sw Washington . . . . 62. *Castilleja litoralis* (in part)

                35. Bracts usually 3-lobed; not coastal, 1400–2600 m; c, e Oregon and adjacent Idaho and Nevada . . . . 85. *Castilleja peckiana*

1. Castilleja affinis Hooker & Arnott, Bot. Beechey Voy., 154. 1833 • California or coast paintbrush [F]

Herbs, perennial, 1.4–6 dm; caudex woody; with a taproot. **Stems** few to many, erect to ascending, unbranched or branched at proximal nodes, sometimes with short, leafy axillary branches, hairs sparse to dense, spreading, short and long, soft to stiff, unbranched, sometimes branched, eglandular, sometimes stipitate-glandular. **Leaves** green or purplish, sometimes red-brown, linear or linear-oblong to broadly lanceolate, 2–13 cm, not or ± fleshy, margins wavy or plane, flat or involute, 0–5-lobed, sometimes with small secondary lobes, short, leafy axillary shoots common, usually conspicuous, apex acuminate or acute to rounded; lobes spreading, linear to lanceolate, lateral lobes almost as wide as mid blade, apex acute to rounded. **Inflorescences** 3–25(–30) × 1.5–5 cm; bracts proximally green or deep purple, distally red, crimson, scarlet, pink, pinkish purple, pinkish red, or yellow, sometimes rose magenta, red-orange, or orange, oblanceolate or obovate to oblong or lanceolate, (0–)3–5(–7)-lobed; lobes spreading to ascending, linear to obovate, long, proximal lobes at or arising below mid length, apices acute to obtuse, center lobe sometimes rounded. **Pedicels** 0 mm or nearly so. **Calyces** proximally pale or green, distally as in distal portion of bracts, 14–35 mm; abaxial and adaxial clefts 6–22 mm, 33–50% of calyx length, deeper than laterals, lateral 2–7 mm, 15–25% of calyx length; lobes lanceolate or oblong, apex acute to obtuse or rounded, sometimes curved upward. **Corollas** straight or ± curved, 17–40 mm; tube 10–15 mm; beak long-exserted to subequal to calyx, adaxially green, 7–20 mm, puberulent, eglandular;

abaxial lip deep green to reddish, brown, or deep purple, reduced, inconspicuous, included in calyx, 1.5–3 mm, 15–25% as long as beak; teeth ascending, green, 0.5–2 mm. $2n = 48, 72, 96$.

Varieties 3 (3 in the flora): California, nw Mexico.

Castilleja affinis is highly variable and one of the more common paintbrushes at lower elevations in California, west of the Sierra Nevada, from the northern coast south to northern Baja California. Some recent authors (for example, M. Wetherwax et al. 2012) include *C. litoralis* of the Pacific Northwest coast as a subspecies of *C. affinis*, but due to the high polyploid nature of *C. litoralis* and its significantly closer morphological resemblance to *C. miniata* var. *dixonii*, that treatment is not followed here. Also see the comments under 62. *C. litoralis*.

1. Stem hairs branched; distal portion of bracts usually pink to pinkish purple or pinkish red; coastal San Luis Obispo and Santa Barbara counties. . . . . 1b. *Castilleja affinis* var. *contentiosa*

1. Stem hairs unbranched, rarely branched; distal portion of bracts usually bright red, red-orange, crimson, scarlet, or yellow; more widespread in w California.

    2. Distal portions of bracts bright red to red-orange, crimson, or scarlet, rarely yellow, orange, rose, magenta, or pinkish red; inflorescences (2.5–)3–5 cm wide; chaparral slopes, openings, open woods, coastal scrub, stabilized dunes; much of w California . . . . 1a. *Castilleja affinis* var. *affinis*

    2. Distal portions of bracts yellow, rarely pink or pale red-orange, often becoming reddish tinted after anthesis; inflorescences 1.5–2.5 cm wide; serpentine substrates; San Francisco Bay region . . . . 1c. *Castilleja affinis* var. *neglecta*

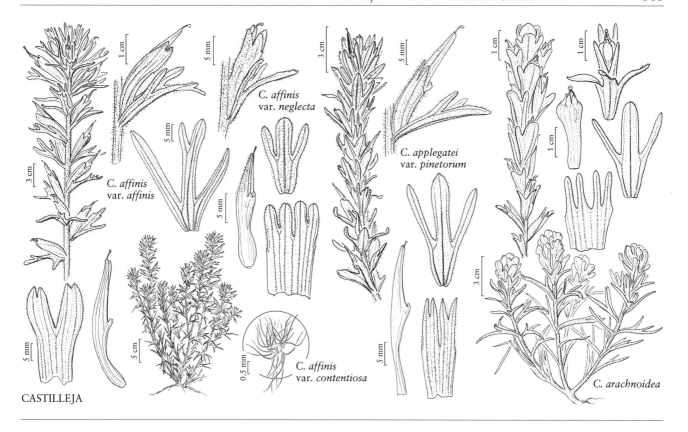

C. affinis
var. *neglecta*

C. affinis
var. *affinis*

C. applegatei
var. *pinetorum*

C. affinis
var. *contentiosa*

CASTILLEJA

C. arachnoidea

1a. Castilleja affinis Hooker & Arnott var. **affinis** [F]

Castilleja affinis subsp. *insularis* (Eastwood) Munz; *C. californica* Abrams; *C. douglasii* Bentham; *C. douglasii* subsp. *insularis* (Eastwood) Pennell; *C. inflata* Pennell; *C. wightii* Elmer subsp. *anacapensis* (Dunkle) Pennell; *C. wightii* subsp. *inflata* (Pennell) Munz

Stems: hairs sparse or dense, unbranched, rarely branched. **Leaves** not or ± fleshy. **Inflorescences** (2.5–) 3–5 cm wide; bracts proximally green or deep purple, distally bright red to red-orange, crimson, or scarlet, rarely yellow, orange, rose, magenta, or pinkish red, 5(–7)-lobed. **Calyces** colored as bracts, or yellow with red (or light orange or yellow) lobes, sometimes white with colored lobes, 20–35 mm. **Corollas** (21–) 25–40 mm; beak usually long-exserted, sometimes slightly so, 12–20 mm.

Flowering Feb–Aug(–Oct). Chaparral slopes, openings, open woods, coastal scrub, stabilized dunes; 0–1900 m; Calif.; Mexico (Baja California).

Plants of var. *affinis* of the immediate coast have more or less fleshy leaves. Populations with somewhat inflated and distinctively colored calyces, with yellow tubes and red lobes, are found from Point Reyes south to San Mateo County and have been called *Castilleja inflata*. A similar situation occurs on the northern Channel Islands, where plants with shorter corollas and paler inflorescences were named subsp. *insularis*. Their indument includes some branched hairs, which might cause them to be identified as the strictly mainland var. *contentiosa*. Hybrids with *C. wightii* are known from Marin County, California.

1b. Castilleja affinis Hooker & Arnott var. **contentiosa** (J. F. Macbride) Bacigalupi, Leafl. W. Bot. 10: 286. 1966 [E] [F]

Castilleja douglasii Bentham var. *contentiosa* J. F. Macbride, Contr. Gray Herb. 65: 44. 1922

Stems: hairs sparse proximally, becoming fairly dense distally, branched. **Leaves** often ± fleshy. **Inflorescences** 2–4 cm wide; bracts proximally greenish to dull purplish, distally pink to pinkish purple or pinkish red, sometimes yellow, pinkish white, red, or pale orange, (0–)3–5-lobed. **Calyces** colored as bracts or paler near base, 15–25 mm. **Corollas** 25–30 mm; beak subequal with or exserted from calyx, 14–20 mm.

Flowering Feb–Aug(–Sep). Coastal scrub, stabilized dunes, canyons; 0–500 m; Calif.

Variety *contentiosa* is restricted to the immediate coast in southern San Luis Obispo and northern Santa Barbara counties, where it replaces var. *affinis*. The distinctive branched hairs impart a grayish cast to the plants, and this variety is further distinguished by the commonly pink to pink-purple or purplish red inflorescences. Variety *affinis* usually has unbranched hairs and reddish to red-orange inflorescences. Transitional plants exist a few miles inland from the coastal populations of var. *contentiosa*. This form has also been confused with *Castilleja miniata*, from which it differs in numerous characters, including pubescence and leaf and calyx morphologies, as well as with the insular *C. mollis*, which is distinguished by its largely decumbent stems and broad, rounded, usually unlobed leaves and bracts.

1c. Castilleja affinis Hooker & Arnott var. **neglecta** (Zeile) J. M. Egger, Phytologia 90: 66. 2008

• Tiburon paintbrush [C] [E] [F]

Castilleja neglecta Zeile in W. L. Jepson, Man. Fl. Pl. Calif., 936. 1925 (as Castilleia); *C. affinis* subsp. *neglecta* (Zeile) T. I. Chuang & Heckard

Stems: hairs dense, sometimes sparse, unbranched, rarely branched. **Leaves** not fleshy. **Inflorescences** 1.5–2.5 cm wide; bracts proximally pale yellow or yellow-green, distally yellow, rarely pink or pale red-orange, often becoming reddish tinted after anthesis, 3-lobed. **Calyces** green proximally, yellow in outer ½, after anthesis often whitish or reddish distally, 14–20 mm. **Corollas** 17–25 mm; beak subequal to or slightly exserted from calyx, 7–11 mm.

Flowering Mar–Jul. Serpentine meadows and slopes; of conservation concern; 0–300 m; Calif.

Variety *neglecta* is endemic to Marin, Napa, and Santa Clara counties. Its habitat is threatened by urbanization and recreational uses. Within its limited range, it is the only representative of *Castilleja affinis* on serpentine substrates and is recognized by its relatively narrow, yellow inflorescences.

2. Castilleja ambigua Hooker & Arnott, Bot. Beechey Voy., 154. 1833 • Johnny-nip, paintbrush owl's-clover [E]

Herbs, annual, sometimes biennial, 0.6–3.5 dm; with fibrous roots. **Stems** few to many, sometimes solitary (var. *meadii*), ascending or erect, often decumbent proximally, unbranched or branched above base, hairs sparse to moderately dense, spreading, long, soft, eglandular, mixed with short stipitate-glandular hairs. **Leaves** green or brownish in upland forms, linear-lanceolate to widely lanceolate, linear, elliptic, obovate, or oblong, rarely ovate or cup-shaped, (0.6–)0.8–5 cm, fleshy or not, margins plane, flat, 3–5(–7)-lobed, apex rounded to obtuse or acuminate, abaxial surface often stipitate-glandular, adaxial sometimes shiny, glabrous; lobes ascending to erect, linear to lanceolate, apex acute or obtuse. **Inflorescences** 1.5–9(–13) × 1–4 cm; bracts proximally green, rarely brownish purple, distally white, cream, pink, or purple on apices, lanceolate, oblong, or ovate, (0–)3–5(–7)-lobed; lobes ascending or divaricate-ascending, oblong to linear, short to long, arising below or above mid length, central lobe apex usually rounded to truncate, others acute to obtuse. **Calyces** green to pale yellowish green or tipped with white or purple, 8–23 mm; abaxial and adaxial clefts 5–9 mm, 33–50 (–67)% of calyx length, lateral 2–5.5 mm, 20–40% of calyx length; lobes linear or narrowly lanceolate to oblong or ± triangular, apex acuminate or narrowly acute to obtuse or rounded. **Corollas** straight, 14–24 mm; tube 11–21 mm, expanded distally; abaxial lip and beak usually exserted, beak straight or slightly curved, adaxially white, yellow, or pink, sometimes green or purplish, (1–)4–7 mm, sparsely to densely short-hairy; abaxial lip pale to bright yellow, sometimes becoming pink, orange, or red after anthesis, with red-brown or purple spots at base of each tooth and sometimes at base of each pouch, conspicuous, pouches 3, ± prominent, divergent, saccate, 2–7 × 3–7 mm, (33–)60–75(–90)% as long as beak; teeth erect, sometimes spreading, white, pink, purple, or green, often with whitish bases, 1–3 mm. **Filaments** glabrous. $2n = 24$.

Varieties 4 (4 in the flora): w North America.

Castilleja ambigua is a complex species, treated here with four varieties, though many localized variations exist among populations, and these are still incompletely understood.

1. Bract lobes linear; stems erect; moist inland meadows and vernal pools on volcanic substrates; c Napa County, California . . . . .2d. *Castilleja ambigua* var. *meadii*
1. Bract lobes linear to oblong; stems ± decumbent, at least proximally; coastal salt marshes, margins of brackish estuaries, coastal sandy bluffs, mesic to ± xeric inland grasslands; British Columbia to California.
   2. Bracts pink to purple distally; stems unbranched or few-branched from mid stem; salt marshes; Humboldt Bay region, nw California. . . . . .2b. *Castilleja ambigua* var. *humboldtiensis*
   2. Bracts white distally, rarely pink or cream; stems often branched from base; salt marshes, sandy coastal bluffs, inland grasslands; British Columbia to California.

[3. Shifted to left margin.—Ed.]

3. Corolla beaks usually white or yellow, abaxial lips yellow; salt marshes, sandy coastal bluffs, inland grasslands; s British Columbia to c California. . . . . 2a. *Castilleja ambigua* var. *ambigua*

3. Corolla beaks usually pink to purplish, abaxial lips yellow, becoming white and then red or soft pink-purple; grassy coastal bluffs and adjacent sand dunes; vicinity of Monterey Bay, California . . . . 2c. *Castilleja ambigua* var. *insalutata*

2a. Castilleja ambigua Hooker & Arnott var. ambigua E

Orthocarpus castillejoides Bentham

Stems ± short-decumbent proximally, becoming ascending to erect, often branched. **Leaves** widely lanceolate, sometimes linear or narrowly oblong, rarely ovate, not fleshy, apex rounded. **Bracts** proximally greenish, rarely dull brownish, distally white, rarely fading pinkish, on lobe apices; lobes ascending, linear to oblong, 12–15 mm, usually arising near or above mid length. **Calyces** with all 4 clefts subequal; lateral clefts 2–3 mm, 20% of calyx length. **Corollas** 14–21 mm; beak pale to bright yellow or white, rarely pink, 4–6 mm; abaxial lip pale to bright yellow; teeth white, rarely pink or reddish purple.

Flowering Apr–Aug(–Oct). Sandy coastal bluffs, inland grasslands, upper margins of salt marshes; 0–500 m; B.C.; Calif., Oreg., Wash.

Variety *ambigua* is highly variable among populations. Much of the variation tends towards two ecotypes. One ecotype has thin leaves with narrow, often acute lobes and more or less spreading-ascending stems. Plants of this ecotype inhabit sandy coastal bluffs and short-grass, herbaceous meadows, often well inland, in Marin, Napa, and Sonoma counties in California. Plants of a second ecotype have fleshy leaves and bracts with more or less obtuse lobes and more upright stems. These are found in *Salicornia*-dominated coastal salt marshes and range from southern Vancouver Island to San Francisco Bay. Plants found around Tomales Bay, Marin County, may have pink beaks. Variety *ambigua* occasionally hybridizes with *Castilleja exserta* and reportedly with *C. rubicundula*.

2b. Castilleja ambigua Hooker & Arnott var. humboldtiensis (D. D. Keck) J. M. Egger, Phytologia 90: 67. 2008 • Humboldt Bay owl's-clover C E

Orthocarpus castillejoides Bentham var. *humboldtiensis* D. D. Keck, Proc. Calif. Acad. Sci., ser. 4, 16: 536. 1927; *Castilleja ambigua* subsp. *humboldtiensis* (D. D. Keck) T. I. Chuang & Heckard

Stems ± short-decumbent proximally, becoming erect to ascending, unbranched or few-branched from mid stem. **Leaves** narrowly oblong to linear, oblong, or lanceolate, cup-shaped, ± fleshy, apex rounded. **Bracts** proximally green or brownish purple, distally pink to purple on lobe apices; lobes ascending, oblong, 20–23 mm, usually arising above mid length. **Calyces** with all 4 clefts subequal; lateral clefts 3 mm, ca. 15% of calyx length. **Corollas** 21–24 mm; beak pink or purplish, 5–7 mm; abaxial lip yellow, becoming orange or red-purple after anthesis; teeth deep pinkish purple, often with whitish bases.

Flowering Apr–Jul(–Aug). Upper margins of *Salicornia* salt marshes, saline flats; of conservation concern; 0–10 m; Calif.

Variety *humboldtiensis* has purplish flowers and bract apices and is limited to the upper edge of *Salicornia*-dominated salt marshes in Humboldt and Mendocino counties, where it replaces var. *ambigua*. Most populations are around Humboldt Bay. Reports of var. *humboldtiensis* from Tomales Bay in Marin County are based on plants of the salt marsh form of var. *ambigua* and have yellowish flowers. Variety *humboldtiensis* is threatened by coastal development.

2c. Castilleja ambigua Hooker & Arnott var. insalutata (Jepson) J. M. Egger, Phytologia 90: 67. 2008 C E

Orthocarpus castillejoides Bentham var. *insalutatus* Jepson, Man. Fl. Pl. Calif., 944. 1925 (as castilleioides); *Castilleja ambigua* subsp. *insalutata* (Jepson) T. I. Chuang & Heckard

Stems ± short-decumbent proximally, becoming erect to ascending, much-branched from base. **Leaves** lanceolate to elliptic to ovate or obovate, distal ones cup-shaped, not fleshy, apex rounded to obtuse. **Bracts** proximally green or reddish brownish, distally whitish, rarely pinkish or cream, on lobe apices; lobes ascending, oblong, 12–14 mm, arising above mid length. **Calyces** with all 4 clefts subequal; lateral clefts 5–5.5 mm, 40% of calyx length. **Corollas** 15–22 mm; beak pink to purple, 4–6 mm; abaxial lip yellow, becoming pink to red after anthesis; teeth deep to bright pinkish purple, often with whitish bases.

Flowering Apr–Jul(–Aug). Grassy coastal bluffs, adjacent dunes; of conservation concern; 0–100 m; Calif.

Variety *insalutata* is now endemic to coastal habitats in northern Monterey County. This variety is also threatened by development, and past populations in Alameda County appear to be extirpated. Reports of var. *insalutata* from San Luis Obispo County are referable to other species.

2d. Castilleja ambigua Hooker & Arnott var. **meadii** J. M. Egger & Ruygt, Phytoneuron 2012-68: 2, figs. 1, 3–7, 9[left]. 2012 • Mead's owl's-clover C E

Stems erect, unbranched, sometimes with a few divaricate-ascending branches from proximal ½ of stem above base. **Leaves** linear or linear-lanceolate to lanceolate, 1 mm wide at base, not fleshy, apex acuminate. **Bracts** proximally pale greenish, distally white on lobe apices, often becoming entirely greenish with age; lobes divaricate-ascending, linear, 8–14 mm, usually arising below mid length. **Calyces** with all 4 clefts subequal or lateral clefts shallower; lateral clefts 2–3 mm, 25% of calyx length. **Corollas** 14–21 mm; beak pale, off-white, pale yellow, green with margins off-white, or yellow, sometimes orange, 1–4 mm; abaxial lip pale yellow; teeth white to green.

Flowering Apr–Jun(–Jul). Seasonally wet meadows with volcanic substrates in oak-pine woodlands or chaparral, shallow vernal pools, ephemeral stream margins; of conservation concern; 400–500 m; Calif.

Variety *meadii* is limited to vernally wet habitats, growing over rocks of the Sonoma Volcanic Formation in central Napa County near Atlas Peak. All known populations are under private ownership, and the variety is of conservation concern due to its very limited range. For the present, all populations except one are protected by conservation easements. Variety *meadii* often grows alongside, but does not hybridize with, *Castilleja attenuata* and *C. densiflora*. Ongoing study of annual species of *Castilleja* suggests this variety is genetically distinctive and may deserve full species status (S. J. Jacobs et al. 2018).

3. Castilleja angustifolia (Nuttall) G. Don, Gen. Hist. 4: 616. 1837/1838 • Northwestern or narrow-leaved paintbrush E

Euchroma angustifolia Nuttall, J. Acad. Nat. Sci. Philadelphia 7: 46. 1834

Herbs, perennial, 0.9–3.8(–4) dm; from a woody caudex; with a taproot. **Stems** few to many, ascending to erect, branched, especially near base, sometimes unbranched, hairs sparse to dense, spreading to retrorse, long, sometimes short, soft to stiff, usually mixed with short-glandular ones, sometimes viscid. **Leaves** brown or purplish, sometimes green, linear to lanceolate or broadly lanceolate, 1.2–7 (–7.5) cm, not fleshy, margins plane, sometimes ± wavy, involute or flat, (0–)3–5-lobed, rarely with secondary lobes, apex acuminate to rounded; lobes spreading, oblong or lanceolate to linear-lanceolate, apex acute to rounded. **Inflorescences** 2.5–20 × 1.5–5 cm; bracts proximally greenish or dull purplish, distally pink, magenta, pink-purple, reddish pink, pale yellow, pale yellow-orange, pale orange, or white, rarely reddish or orange-red, lanceolate to oblong, 3–5(–9)-lobed, sometimes with secondary lobes; lobes spreading or ascending, oblanceolate or linear, proximal lobes often much longer than distal, proximal lobes arising below or a little above mid length, apex acute to rounded. **Calyces** proximally green, yellow, brown, or purple, lobes colored as bract lobes, sometimes with a yellow band between proximal and distal parts, 13–25(–28) mm; abaxial clefts 3–8 mm, adaxial 5–9(–12) mm, clefts 30–50% of calyx length, deeper than laterals, lateral (1–)1.5–4(–5) mm, 10–25% of calyx length; lobes lanceolate to oblong, abaxials wider than adaxials, apex acute to rounded. **Corollas** straight, 18–27(–32) mm; tube 8–17 mm; beak usually long-exserted, adaxially green or pink, 8–15 mm; abaxial lip deep green, reduced, inconspicuous, 1–2.5 mm, 5–20% as long as beak; teeth incurved to ascending, deep green, 0.5–1.5 mm.

Varieties 3 (3 in the flora): w United States.

Much confusion exists concerning *Castilleja angustifolia* and the closely related *C. chromosa*. Sometimes *C. chromosa* is treated as a variety of *C. angustifolia*, using the name *C. angustifolia* var. *dubia*. The latter name is used here to represent a different assemblage of plants, not including *C. chromosa*. At other times, *C. chromosa* is synonymized completely under *C. angustifolia*. However, the two species are in most cases easily separable, and where they are sympatric there is little evidence of intergradation. Both *C. angustifolia* var. *dubia* and *C. chromosa* are accepted here. See additional comments under 3b. *C. angustifolia* var. *dubia* and 15. *C. chromosa*.

1. Bracts distally usually pink to pink-purple; s Idaho, sw Montana, se Oregon, nw Wyoming . . . . 3a. *Castilleja angustifolia* var. *angustifolia*
1. Bracts distally yellow, yellow-orange, pale orange, white, pink, reddish pink, or magenta; se Idaho, e Nevada, sw South Dakota, w Utah, ec Wyoming.
   2. Bracts distally usually yellow to pale orange or white; ec Wyoming, adjacent sw South Dakota. . . . . 3b. *Castilleja angustifolia* var. *dubia*
   2. Bracts distally usually yellow, yellow-orange, white, pink, or reddish pink; se Idaho, e Nevada, w Utah . . . . 3c. *Castilleja angustifolia* var. *flavescens*

3a. Castilleja angustifolia (Nuttall) G. Don var. **angustifolia** E

Stems: hairs sparse to moderately dense, spreading to retrorse, long, soft to stiff, usually mixed with short-glandular ones. **Bracts** distally pink or pink-purple, sometimes pale orange, yellowish, or white, 3–5-lobed. **Calyces** 13–20 mm; abaxial clefts 3–6 mm, adaxial 5–8 mm. **Corollas** 18–25 mm.

Flowering (Apr–)May–Jul. Dry sagebrush slopes and flats; 1300–2200 m; Idaho, Mont., Oreg., Wyo.

Variety *angustifolia* has pink to purple inflorescences over most of its range from southeastern Oregon and southern Idaho to southwestern Montana and northwestern Wyoming. At the eastern edge of its distribution, the colors vary more, possibly representing hybridization or contact with the more eastern var. *dubia*, which usually has yellow inflorescences. Plants of the western slope of the Big Horn Mountains, Wyoming, are morphologically intermediate between these two varieties. In the same area, var. *angustifolia* occasionally hybridizes with *Castilleja flava* var. *flava*.

3b. Castilleja angustifolia (Nuttall) G. Don var. **dubia** A. Nelson, Bull. Torrey Bot. Club 29: 404. 1902 E

Castilleja dubia (A. Nelson) A. Nelson

Stems: hairs fairly dense, spreading, short and long, soft to stiff, eglandular. **Bracts** distally yellow to pale orange or white, sometimes red, 3–9-lobed, sometimes with secondary lobes. **Calyces** 18–22 mm; abaxial clefts 5–7 mm, adaxial 7–8 mm. **Corollas** 18–27 mm.

Flowering Jun. Dry sagebrush slopes and flats in mountains, high plains; 1500–2400 m; S.Dak., Wyo.

Variety *dubia* is often confused with *Castilleja chromosa* in herbaria, floras, and databases, but they are not synonymous. Where *C. angustifolia* and *C. chromosa* are sympatric in Wyoming, the specimens show little evidence of hybridization. Variety *dubia*, found primarily in southwestern South Dakota and east-central Wyoming, can be recognized by its usually narrower, yellow to pale orange inflorescences, as well as by its shorter corollas, while *C. chromosa* has wider, red inflorescences and longer corollas. Occasional hybrids between var. *dubia* and *C. sessiliflora* are known from northeastern Wyoming.

3c. Castilleja angustifolia (Nuttall) G. Don var. **flavescens** (Pennell ex Edwin) N. H. Holmgren in A. Cronquist et al., Intermount. Fl. 4: 488. 1984 E

Castilleja flavescens Pennell ex Edwin, Leafl. W. Bot. 9: 45. 1959

Stems: hairs fairly dense, spreading, long, soft, with scattered, shorter, sometimes stipitate-glandular, ones. **Bracts** distally yellow, yellow-orange, white, pink, reddish pink, or magenta, usually variable within a population, 3–5-lobed. **Calyces** 21–28 mm; abaxial clefts 7–8 mm, adaxial 7–12 mm. **Corollas** 27–32 mm.

Flowering (Mar–)Apr–Jul. Dry sagebrush slopes and flats, often rocky; 2100–3100 m; Idaho, Nev., Utah.

Most plants of var. *flavescens* have yellow to pale orange inflorescences, but in some populations, such as near Wells, Nevada, the inflorescence can vary to pink, reddish pink, white, or magenta. Variety *flavescens* is associated with *Artemisia arbuscula*, its likely host plant, and is often found at higher elevations than *Castilleja chromosa* (N. H. Holmgren 1984). The variety is found in southeastern Idaho, eastern Nevada, and western Utah. Recent collections from extreme eastern Mono and Modoc counties suggest that var. *flavescens* may also occur in California, but the identity of these populations has yet to be fully verified.

4. Castilleja applegatei Fernald, Erythea 6: 49. 1898 (as Castilleia) • Applegate's paintbrush E F

Herbs, perennial, (0.8–)1–5 (–6) dm; from a branched, woody caudex; with a taproot. **Stems** few to many, erect, ascending, or decumbent, unbranched or branched, hairs sparse to dense, spreading, long, soft to ± stiff, eglandular, mixed with shorter stipitate-glandular ones. **Leaves** green to purplish or brown, linear to broadly lanceolate, sometimes ovate, 1–6 cm, not fleshy, margins wavy, involute, (0–)3(–5)-lobed, apex rounded or narrowly acute to acuminate; lobes spreading or ascending, lanceolate to broadly lanceolate, often short, apex acute to rounded. **Inflorescences** 2–12(–21 in fruit) × 1–5 cm; bracts proximally green to dull purplish brown, distally red, red-orange, or scarlet, sometimes orange, white, or yellow, rarely with a narrow yellowish band medially, lanceolate, broadly lanceolate, oblong, or lanceolate-ovate, 3–5(–7)-lobed; lobes spreading to ascending, linear, sometimes expanded near tip, long, arising from ca. mid length, central lobe apex obtuse to rounded, sometimes expanded, others acute, rounded, obtuse, or acuminate. **Calyces** proximally green or whitish, sometimes yellow, lobes colored as bract lobes, sometimes with a yellow band between proximal and distal portions, 13–25 mm; abaxial and adaxial clefts 4–14 mm, 33–50% of calyx length, deeper than laterals, lateral 2.5–8 mm, 12–50% of calyx length; lobes lanceolate-acuminate, narrowly oblong, or narrowly triangular, apex acuminate, acute, or obtuse. **Corollas** straight or curved in proximal ⅓, 16–35(–41) mm; tube 9–22 mm; beak usually long-exserted, adaxially green, yellow-green, or yellow, 6–20 mm; abaxial lip deep green to whitish or yellow, reduced, inconspicuous, protuberant, thickened, included or exserted, 1–3 mm, 5–20% as long as beak; teeth ascending or incurved, deep green to yellow, 0.5–1 mm. *2n* = 24, 48.

Varieties 4 (4 in the flora): w United States.

Castilleja applegatei is a widespread and often common species, with complex patterns of variation and several common but inconstant color forms, especially in var. *pinetorum*. *Castilleja disticha* and *C. martini*, although sometimes included as subspecies within *C. applegatei*, are treated as species here. Both are more morphologically divergent and more easily distinguished from typical *C. applegatei* than are the four varieties accepted here. Variety *pinetorum* is the most widespread form and also occurs over a wider range of elevations than the other three varieties, which are primarily montane to subalpine and occasionally alpine.

1. Corollas 25–41 mm, beaks 9–20 mm, tubes 15–22 mm; leaves usually 0–3-lobed.
  2. Corollas 34–41 mm; herbs 1–2.5(–3.5) dm; 2100–2800 m; Crater Lake to Newberry Crater, Cascade Range, Oregon . . . . .4a. *Castilleja applegatei* var. *applegatei*
  2. Corollas 25–35 mm; herbs 1.2–6 dm; 300–3600 m; California to Oregon and Idaho . . . . . . .4c. *Castilleja applegatei* var. *pinetorum*
1. Corollas 16–31 mm, beaks 6–14 mm, tubes 9–14 mm; leaves usually 3–5-lobed.
  3. Corolla beaks 6–10 mm; abaxial and adaxial calyx clefts (4–)5–8(–10) mm . . . . .4b. *Castilleja applegatei* var. *breweri*
  3. Corolla beaks 9–14 mm; abaxial and adaxial calyx clefts 7–14 mm . . . . 4d. *Castilleja applegatei* var. *viscida*

4a. **Castilleja applegatei** Fernald var. **applegatei** ☐E☐

Herbs 1–2.5(–3.5) dm. **Stems** usually branched. **Leaves** linear to lanceolate, sometimes ovate, 1.3–4.1 cm, 0–3(–5)-lobed, apex narrowly acute to acuminate. **Inflorescences** 3–8 × 2–5 cm; bracts distally usually red to red-orange or scarlet, 3(–5)-lobed, apex acute to acuminate. **Calyces** green, sometimes yellow, with red lobes, sometimes with a yellow band below lobes, 18–25 mm; abaxial and adaxial clefts 13 mm, 50% of calyx length, lateral 5–8 mm, 30% of calyx length; lobe apex acute to obtuse. **Corollas** 34–41 mm; tube 17–22 mm; beak 14–15 mm. *2n* = 48.

Flowering Jun–Aug. Dry rocky volcanic slopes, open conifer forests, subalpine; 2100–2800 m; Oreg.

Variety *applegatei* is restricted to volcanic substrates in the higher elevations of the Cascade Range in south-central Oregon between Crater Lake and Newberry Crater.

4b. **Castilleja applegatei** Fernald var. **breweri** (Fernald) N. H. Holmgren in A. Cronquist et al., Intermount. Fl. 4: 486. 1984 • Brewer's paintbrush ☐E☐

Castilleja breweri Fernald, Erythea 6: 49. 1898 (as Castilleia); *C. applegatei* subsp. *pallida* (Eastwood) T. I. Chuang & Heckard; *C. applegatei* var. *pallida* (Eastwood) N. H. Holmgren; *C. breweri* var. *pallida* Eastwood; *C. glandulifera* Pennell subsp. *pallida* (Eastwood) Pennell

Herbs (0.8–)1–3(–4.8) dm. **Stems** branched from base. **Leaves** linear to broadly lanceolate, 1.2–4.8 cm, (0–)3(–5)-lobed, apex acute, rarely obtuse. **Inflorescences** 2–12 × 2–3.5 cm; bracts distally red, scarlet, or pale orange, rarely yellow, 3-lobed, apex acute to rounded, central lobe sometimes expanded. **Calyces** green to yellow, lobe apices red to yellow, 13–15 mm; abaxial and adaxial clefts (4–)5–8(–10) mm, 33% of calyx length, lateral 3–5(–7) mm, 10–25% of calyx length; lobe apex acute, rarely obtuse. **Corollas** 16–22 mm; tube 11–13 mm; beak 6–10 mm. *2n* = 48.

Flowering Jun–Aug(–Sep). Dry rocky slopes, flats, open montane forests to alpine, granitic substrates; 1800–3600 m; Calif., Nev.

Variety *breweri* is found in California in the high Sierra Nevada and in the mountains of adjacent Nevada. Hybrids with *Castilleja chromosa* and *C. miniata* var. *miniata* have both been found in Inyo County, California.

4c. Castilleja applegatei Fernald var. **pinetorum**
(Fernald) N. H. Holmgren in A. Cronquist et al.,
Intermount. Fl. 4: 486. 1984 • Wavy-leaved
paintbrush E F

Castilleja pinetorum Fernald,
Erythea 6: 50. 1898 (as Castilleia);
C. applegatei var. *fragilis* (Zeile)
N. H. Holmgren; *C. applegatei*
subsp. *pinetorum* (Fernald)
T. I. Chuang & Heckard;
C. wherryana Pennell

Herbs 1.2–6 dm. **Stems** unbranched, sometimes branched.
Leaves linear to broadly lanceolate, 1–6 cm,
0–3(–5)-lobed, apex narrowly acute to acuminate.
Inflorescences 4–12(–21 in fruit) × 1–4 cm; bracts
distally red, red-orange, or scarlet, sometimes orange,
pale orange, or yellow, rarely with a narrow yellowish
band medially, 3(–7)-lobed, central lobe apex obtuse,
others obtuse to acute. **Calyces** proximally green,
yellow, or whitish with green veins, lobe apices colored
as bracts, sometimes with a yellow band between green
and other color, 15–25 mm; abaxial and adaxial clefts
4–9 mm, 33–50% of calyx length, lateral 2.5–5 mm,
12–20% of calyx length; lobe apex acute to acuminate.
Corollas 25–35 mm; tube 15–20 mm; beak 9–20 mm.
$2n$ = 24, 48.

Flowering (Apr–)May–Aug. Open forests, rocky
slopes, canyons, sagebrush hillsides; 300–3600 m;
Calif., Idaho, Nev., Oreg.

Variety *pinetorum* is the most widespread and
common representative of the *Castilleja applegatei*
complex. It replaces var. *breweri* in the Sierra Nevada
at low to moderate elevations. On the western flank
of Steens Mountain, Harney County, Oregon, var.
pinetorum occurs almost exclusively in a yellow-bracted
form and has been confused with *C. glandulifera*.
Variety *pinetorum* hybridizes with *C. peckiana* in
Harney County, Oregon.

4d. Castilleja applegatei Fernald var. **viscida** (Rydberg)
Ownbey in C. L. Hitchcock et al., Vasc. Pl. Pacif. N.W.
4: 299. 1959 E

Castilleja viscida Rydberg, Bull.
Torrey Bot. Club 34: 38. 1907

Herbs 1–3.1 dm. **Stems** unbranched, sometimes branched.
Leaves broadly lanceolate,
1–4.5 cm, 3–5-lobed, apex
acute. **Inflorescences** 2–12 ×
1–4 cm; bracts distally red, red-
orange, or orange, 3–5-lobed,
apex acute. **Calyces** proximally green to yellowish, lobe
apices colored as bracts, 14–24 mm; abaxial and adaxial
clefts 7–14 mm, 50% of calyx length, lateral 7–8 mm,

33–50% of calyx length; lobe apex acuminate. **Corollas**
26–31 mm; tube 9–14 mm; beak 9–14 mm. $2n$ = 24.

Flowering May–Aug. Dry rocky slopes, alpine and
open subalpine forests; 2100–3400 m; Idaho, Utah,
Wyo.

Variety *viscida* occurs in the mountains of southeastern
Idaho, Utah, and extreme southwestern Wyoming.

5. Castilleja aquariensis N. H. Holmgren, Bull. Torrey
Bot. Club 100: 87, fig. 3. 1973 • Aquarius Plateau
paintbrush C E

Herbs, perennial, 1.1–3.9 dm;
from a woody caudex; with
numerous, thickened roots.
Stems few to several, ascending
to erect, unbranched, hairs moderately dense, often retrorse,
short, ± stiff, stipitate-glandular
distally only in inflorescence.
Leaves appressed-ascending,
green to purplish, linear to narrowly lanceolate, distal
sometimes broadly lanceolate, (1–)2.5–4(–5.5) cm, not
fleshy, margins plane, involute, 0(–5)-lobed, apex
rounded to acuminate; lobes ascending, linear to
narrowly lanceolate, apex acute. **Inflorescences** 2–7 ×
1.5–2 cm; bracts proximally pale green to pale yellow-
green, distally pale to bright yellow or cream, rarely pale
orange, elliptic, narrowly ovate, elliptic-oblong, broadly
lanceolate, or ovate, (0–)3(–5)-lobed; lobes ascending or
spreading, narrowly lanceolate, short, arising near tip
on distal bracts, central lobe apex rounded to truncate,
lateral ones acute to rounded. **Calyces** proximally green
to yellowish, distal ½ yellow, (16–)18–25 mm; abaxial
and adaxial clefts (8–)9.5–12.5 mm, 50% of calyx
length, deeper than laterals, lateral 0.5–5(–6) mm,
5–20% of calyx length; lobes lanceolate to ovate, apex
acute to obtuse. **Corollas** straight, 17–25 mm; tube
9–13 mm; beak exserted, adaxially green, 6–8(–12) mm,
margins red or reddish brown, short-hairy; abaxial lip
green, reduced, sometimes exserted, 1.5–2 mm, 10–15%
as long as beak, glabrous; teeth erect, green,
0.5–1.5 mm.

Flowering Jun–Aug. Meadows with sagebrush,
openings in spruce-fir forests; of conservation concern;
2900–3400 m; Utah.

Castilleja aquariensis is endemic to the Aquarius
Plateau in the mountains of south-central Utah. Its
meadow habitats were severely degraded by livestock
grazing, and at one time the species was a candidate for
listing under the Endangered Species Act of the United
States. It is still a species of management concern.

Castilleja aquariensis is in the Center for Plant
Conservation's National Collection of Endangered
Plants.

6. **Castilleja arachnoidea** Greenman, Bot. Gaz. 53: 510.
1912 • Cobwebby or cotton paintbrush E F

Castilleja arachnoidea subsp.
shastensis Pennell; *C. eastwoodiana*
Pennell; *C. filifolia* Eastwood; *C.
payneae* Eastwood; *C. pumicicola*
Pennell

Herbs, perennial, 0.6–2(–3)
dm; from branched, sometimes
elongate, creeping basal stems;
with a taproot. **Stems** solitary
or few to many, decumbent-ascending, unbranched,
sometimes branched, hairs dense, reflexed-spreading,
matted, long, soft, unbranched, eglandular, white-
woolly. **Leaves** green to purple or steel gray, lanceolate-
linear or narrowly lanceolate, (1–)2–4(–6) cm, not
fleshy, margins plane, involute, 3(–5)-lobed, apex
acute; lobes spreading, linear to lanceolate, apex acute
to rounded. **Inflorescences** erect to ascending, 3–12 ×
1–2.5 cm; bracts proximally greenish to dull reddish
brown, rarely dull red or dull orange, distally dull red,
deep rusty red, or yellow to pale yellow, sometimes deep
pink, dull rose, pale salmon, or dull light orange, rarely
green, dull reddish, or dull orange throughout, aging
browner and/or more orange, lanceolate or elliptic-
lanceolate to obovate, (0–)3–5-lobed; lobes usually
ascending, linear, lanceolate, oblanceolate, or spatulate,
long, arising near or below mid length, central lobe apex
rounded to truncate, others obtuse to rounded. **Calyces**
colored as bracts, proximal part sometimes paler, (10–)
12–19 mm; abaxial and adaxial clefts 4–8(–10) mm,
33–50% of calyx length, all 4 clefts subequal; lobes
linear to lanceolate, apex acute. **Corollas** straight or
± curved, 12–20 mm; tube 9–15 mm; beak subequal to
calyx or slightly exserted, adaxially green or yellow-
green, (2–)3–5 mm, ± densely puberulent, hairs often
crisped; abaxial lip yellow, greenish, pink, or red-violet,
moderately conspicuous, exserted to barely included in
calyx; pouches shallow, prominent, 2–4 mm, ca. 100%
as long as beak; teeth erect, red, pink, or pale yellow,
1 mm. $2n = 24$.

Flowering (May–)Jun–Aug(–Sep). Pumice flats, sandy,
gravelly, or rocky slopes, ridges and open summits;
1300–3300 m; Calif., Nev., Oreg.

Castilleja arachnoidea is a remarkably variable species
in form and color, but inflorescence colors are often
uniform within a population. It is sometimes confused
with *C. schizotricha*, but the two differ in stature, lobing
of the leaves and bracts, bract coloration, and inflation
of the abaxial lip. They are easily separated by the
simple hairs and pale yellow to brick red inflorescences
of *C. arachnoidea*, compared to the pink-purple to
deep purple inflorescences and branched hairs of
C. schizotricha.

The names *Castilleja eastwoodiana* and *C. filifolia*
represent yellow color forms of *C. arachnoidea*. A race
on the pumice plains of Mt. Shasta in the Cascade Range
in California is distinctive.

7. **Castilleja attenuata** (A. Gray) T. I. Chuang &
Heckard, Syst. Bot. 16: 656. 1991 • Valley tassels,
narrow-leaved owl's-clover F

Orthocarpus attenuatus A. Gray in
War Department [U.S.], Pacif.
Railr. Rep. 4(5): 121. 1857

Herbs, annual, 1–4.5 dm; with
fibrous roots. **Stems** solitary,
erect to ascending, unbranched,
sometimes branched from base,
hairs spreading, short and
medium, ± stiff, eglandular.
Leaves green to purple-tinged, linear to linear-lanceolate,
(1–)2–8 cm, not fleshy, margins plane, flat, 0(–5)-lobed,
apex acuminate; lobes spreading to ascending, filiform
to lanceolate, apex sometimes acuminate. **Inflorescences**
(1.5–)2–10(–19, –30 in fruit) × 1–2 cm; bracts proximally
green to pale brown, rarely dull reddish brown, distally
white on apices, sometimes pale yellow or pale pink-
purplish on apices, rarely greenish or dull reddish brown
throughout, lanceolate or lanceolate-elliptic, 3(–5)-lobed;
lobes ascending, linear to lanceolate, long, arising below
mid length, apex acuminate, acute, or obtuse. **Calyces**
colored as bracts, (8–)15–23 mm; abaxial and adaxial
clefts (4–)6–8 mm, abaxial ca. 50% of calyx length,
adaxial ca. 75% of calyx length, lateral 3–3.5 mm, 33%
of calyx length; lobes linear to narrowly triangular,
apex acute to acuminate. **Corollas** straight, 10–25 mm;
tube 9–20 mm, not expanded distally; beak exserted,
straight, adaxially white, light yellow, or greenish,
3–5 mm, inconspicuously puberulent; abaxial lip white or
yellow with deep brown to purple spots, often becoming
pink, slightly inflated, exserted or not, pouches 3,
2 mm wide, 1–1.5 mm deep, 3–4 mm, 75–80% as long
as beak; teeth erect, white, pale yellow, or pink, 0.5–1.2
mm. **Filaments** glabrous. $2n = 24$.

Flowering (Feb–)Mar–Jun. Grasslands, pastures,
moist margins of springs and streams, damp rocky
slopes; 0–2100 m; B.C.; Ariz., Calif., Oreg., Wash.;
Mexico (Baja California); South America (Chile).

Castilleja attenuata is a common and widespread
species, ranging from southwestern Canada to northern
Baja California, with several disjunct populations in
central Chile. It is sensitive to competition from weeds.
Disjunct populations in the Rincon Mountains in
Pima County, Arizona, often have pink bracts but are
otherwise typical.

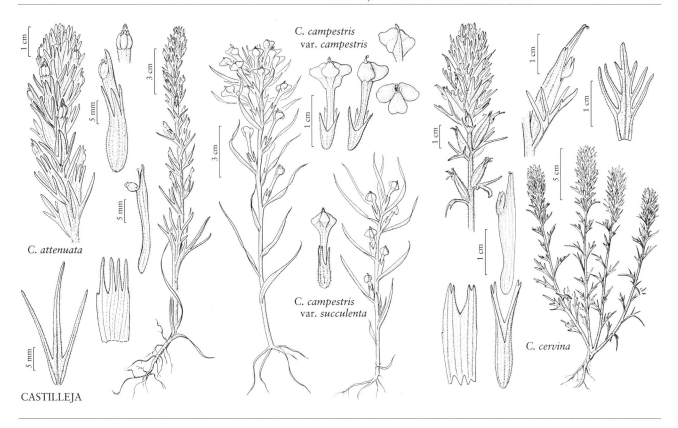

C. attenuata

C. campestris
var. campestris

C. campestris
var. succulenta

C. cervina

CASTILLEJA

8. Castilleja brevilobata Piper, Proc. Biol. Soc. Wash. 33: 104. 1920 • Short-lobed paintbrush E

Castilleja hispida Bentham subsp. *brevilobata* (Piper) T. I. Chuang & Heckard

Herbs, perennial, 1–5 dm; from a branched, woody caudex; with a taproot. **Stems** solitary or few, erect or ascending, unbranched, sometimes branched, hairs spreading, short, medium, and long, soft, short and medium ones short stipitate-glandular. **Leaves** green or ± yellow, lanceolate, elliptic, or oblong to narrowly ovate, 1–2(–2.5) cm, not fleshy, margins plane or wavy, involute, (0–)3–5(–7)-lobed, apex rounded to acute; lobes ascending to erect, linear to lanceolate, apex rounded to acute. **Inflorescences** 3–20 × 2–3.5 cm; bracts proximally greenish to dull brown, distally red, orange-red, or scarlet, sometimes orange or yellow, broadly lanceolate or oblong, (0–)3–5-lobed; lobes ascending, broadly to narrowly lanceolate, short, arising above mid length, apex acute, obtuse, or rounded. **Calyces** green or whitish with green veins, lobes colored as bract lobes or paler, 14–30 mm; abaxial and adaxial clefts 5.5–8.5 mm, 30–40% of calyx length, deeper than laterals, lateral 1.5–4 mm, 20–25% of calyx length; lobes oblong to narrowly triangular, apex obtuse to rounded. **Corollas** straight, 15–24 (–26) mm; tube 12–16 mm; beak exserted, abaxial lip equal to calyx; beak adaxially green or ± yellow-green,

7–10 mm, puberulent, stipitate-glandular; abaxial lip deep green, reduced, rounded, 1–2 mm, 10–25% as long as beak; teeth incurved to erect, light green, 0.5–1 mm. $2n = 24$.

Flowering Mar–Aug. Dry savannas, rocky slopes and open conifer forests, on serpentine; 200–1700 m; Calif., Oreg.

Castilleja brevilobata is endemic to dry serpentine openings in the Siskiyou Mountains of southwestern Oregon and adjacent California. Although sometimes treated as part of *C. applegatei* or *C. hispida*, its morphology does not suggest a close connection with either. This species occasionally hybridizes with *C. pruinosa* in Del Norte County, California.

9. Castilleja brevistyla (Hoover) T. I. Chuang & Heckard, Syst. Bot. 16: 656. 1991 • Short-styled owl's-clover E

Orthocarpus brevistylus Hoover, Four Seasons 2(4): 14. 1968

Herbs, annual, (0.6–)1–4.3 dm; with fibrous roots. **Stems** solitary or few, erect, unbranched, sometimes branched, hairs spreading, short and long, soft and stiff, eglandular. **Leaves** green to purplish, linear to linear-lanceolate, (0.8–)2–6(–8.7) cm, not fleshy, margins plane, flat, 3–5-lobed, apex acuminate; lobes ascending, linear to narrowly lanceolate, long, apex acuminate to rounded. **Inflorescences** 5–25 (longer in fruit) × 1–2 cm; bracts proximally greenish to dull reddish brown, distally pink, lavender, magenta, purple-red, or white on apices, narrowly lanceolate, (3–)5-lobed; lobes ascending, linear to narrowly lanceolate with slightly widened apices, medium length to long, arising near or below mid length, apex acute. **Calyces** colored as bracts, sometimes proximally yellow, 15–20 mm; abaxial and adaxial clefts 7–9.5 mm, abaxial ca. 33% of calyx length, adaxial ca. 66% of calyx length, deeper than laterals, lateral 3–5.5 mm, 33% of calyx length; lobes ± linear, slender, all 4 similar, apex acute. **Corollas** straight, 15–30 mm; tube 14–23 mm, not expanded distally, majority of it exserted from calyx; beak straight, adaxially white or pink (drying purple), 4–6 mm, pubescent; abaxial lip ± inconspicuous, exserted, pouches 3, 2 mm wide, 1–1.5 mm deep, 3–5 mm, 50–70% as long as beak, white, yellow, or pink with large deep purple, red, or brown spot on each pouch at or extending below middle; teeth erect, white, yellow, or pink, 1–1.5 mm. **Filaments** glabrous. $2n = 48$.

Flowering Mar–Jun. Arid grasslands in hilly country, sagebrush or alkaline flats; 50–1200 m; Calif.

Castilleja brevistyla is endemic to the foothills of the southern Sierra Nevada and the southern reaches of the Inner South Coast Ranges, in Kern and adjacent counties. It often grows with other annuals, including *C. attenuata*, *C. densiflora*, and *C. exserta*. The similar *C. attenuata* has three-lobed bracts and leaves, while *C. brevistyla* has mostly five-lobed parts, and also differs in corolla structure and spotting. Hybrids between *C. brevistyla* and *C. exserta* are known from Kern County.

10. **Castilleja campestris** (Bentham) T. I. Chuang & Heckard, Syst. Bot. 16: 656. 1991 • Field owl's-clover E F

Orthocarpus campestris Bentham, Pl. Hartw., 329. 1849

Herbs, annual, (0.3–)0.9–3 dm; with fibrous roots. **Stems** solitary or few, erect, unbranched, sometimes branched, glabrous or hairs sparse proximally, spreading, long, soft distally. **Leaves** green, sometimes purple-tinged, linear to narrowly lanceolate, (0.4–)1–4 cm, not fleshy, thin and flexible or thick and ± brittle, margins plane, involute, 0-lobed, apex acuminate. **Inflorescences** 2–15 × 2–3 cm; bracts green throughout, sometimes purplish tinged, linear to narrowly lanceolate, 0-lobed,

apex acute. **Calyces** proximally green, pale, or purplish, distally purple or green, 5.5–11 mm; abaxial and adaxial clefts 2–5 mm, 30–50% of calyx length, deeper than laterals, lateral 1.5–2.5 mm, 25–30% of calyx length; lobes linear-lanceolate to narrowly triangular, apex acute to acuminate. **Corollas** straight, 11–26 mm; tube 10–22 mm; abaxial lip, beak, and part of tube exserted from calyx; beak adaxially white or yellow, 3–6 mm; abaxial lip yellow, sometimes orange, paler near teeth, prominent, deeply saccate with 3 lobes at ca. 90° from one another, pouches 6–10 mm wide, 3–4 mm deep, abruptly widening from tube, sacs ± round or ± rounded-triangular from above, villous within at base of teeth, 4–5 mm (2–4 mm from sinus), 70–90% as long as beak, puberulent; teeth erect, white, sometimes with deeply colored bases, 0.5–2 mm. $2n = 24$.

Varieties 2 (2 in the flora): w United States.

Castilleja campestris is sometimes confused with *Triphysaria eriantha* subsp. *eriantha*, but *C. campestris* has two-celled anthers and entire leaves, while *T. eriantha* has one pollen sac per stamen and strongly divided leaves.

1. Leaves and bracts linear; leaves thin, flexible; bracts shorter than or ± equal to flowers; corollas light to bright yellow; proximal pollen sacs to ⅓ length of distal . . . . 10a. *Castilleja campestris* var. *campestris*
1. Leaves and bracts lanceolate; leaves thick, ± brittle; bracts longer than flowers; corollas usually yellow to orange; proximal pollen sacs ca. ½ length of distal . . . . 10b. *Castilleja campestris* var. *succulenta*

10a. **Castilleja campestris** (Bentham) T. I. Chuang & Heckard var. **campestris** E F

Herbs (0.3–)1–3 dm. **Leaves** linear, thin, flexible, 0.8–4 cm. **Inflorescences**: bracts linear, shorter than or ± equal to flowers. **Calyces** 7–11 mm; abaxial and adaxial clefts 3–6 mm, 45–75% of calyx length. **Corollas** light to bright yellow, 14–26 mm; tube 13–22 mm. **Anthers**: proximal pollen sac to ⅓ length of distal.

Flowering (Mar–)Apr–Jul(–Aug). Vernal pools, moist ground; 0–2100 m; Calif., Oreg.

Variety *campestris* is widespread from central California to east-central Oregon, but much of its vernal pool and wet meadow habitat has been lost to agricultural development.

10b. Castilleja campestris (Bentham) T. I. Chuang & Heckard var. **succulenta** (Hoover) J. M. Egger, Phytologia 90: 68. 2008 [C][E][F]

Orthocarpus campestris Bentham var. *succulentus* Hoover, Leafl. W. Bot. 1: 228. 1936; *Castilleja campestris* subsp. *succulenta* (Hoover) T. I. Chuang & Heckard; *C. succulenta* (Hoover) D. W. Taylor; *O. succulentus* (Hoover) Hoover

Herbs 0.9–1.4 dm. **Leaves** lanceolate, thick, ± brittle, (0.4–)1–2.5 cm. **Inflorescences:** bracts lanceolate, longer than flowers. **Calyces** 5.5–9(–11) mm; abaxial and adaxial clefts 2–3 mm, 30–35% of calyx length. **Corollas** yellow, sometimes orange, 11–18 mm; tube 10–13 mm. **Anthers:** proximal pollen sac ca. ½ length of distal.

Flowering Apr–May. Vernal pools, moist ground; of conservation concern; 0–800 m; Calif.

Variety *succulenta* is rare and limited to the northeastern San Joaquin Valley, where it is endangered by habitat loss to urbanization and agriculture. It is listed as a threatened species under the Endangered Species Act of the United States.

11. Castilleja cervina Greenman, Bot. Gaz. 25: 269. 1898 (as Castilleia) • Deer paintbrush [E][F]

Herbs, perennial, (2.3–)3–8 dm; from a woody caudex; with a taproot. **Stems** solitary or few to many, erect, branched distally, sometimes unbranched or with short, leafy branches in axils of leaves, glabrous, sometimes hairy, hairs moderately dense, spreading, short, soft, eglandular. **Leaves** green, sometimes purplish, linear or narrowly lanceolate, 1–9 cm, not fleshy, margins plane, involute, 3–5(–9)-lobed, apex acuminate to narrowly acute; lobes spreading, linear, arising near or below mid length, apex acute, rarely obtuse. **Inflorescences** 3–15(–20) × 1.5–3 cm; bracts proximally pale greenish to pale yellowish green, distally white, cream, or pale yellow, sometimes pale greenish, pale yellow-green, or pale yellow throughout, lanceolate to narrowly elliptic, 3–5(–7)-lobed; lobes ascending or spreading, linear to very narrowly linear, long, arising above or below mid length, apex acute to obtuse. **Calyces** green, rarely pale purple, lobes white, cream, or pale yellow, 18–24(–27) mm; abaxial clefts (6–)8–16 mm, adaxial 4–10(–14) mm, abaxial 45–60% of calyx length, adaxial 15–40% of calyx length, deeper than laterals, lateral 1–4 mm, 5–15% of calyx length; lobes narrowly triangular to linear, apex acute. **Corollas** straight or ± curved, 16–25 mm; tube 15–17 mm; beak exserted from calyx, adaxially green,

5–7 mm; abaxial lip deep green, reduced, exserted through abaxial cleft, 1–4 mm, 40–60% as long as beak; teeth erect, white, 0.5–1 mm. $2n = 24$.

Flowering (May–)Jun–Jul. Open pine forests and grasslands, rocky balds and dry subalpine meadows; 500–2000 m; B.C.; Idaho, Wash.

Castilleja cervina is found across northern Washington, from the eastern slope of the Cascade Range, east to northern Idaho and north to southern British Columbia. Plants at higher elevations are dwarfed. Reports from Montana need verification.

12. Castilleja chambersii J. M. Egger & Meinke, Brittonia 51: 445, fig. 1. 1999 • Chambers's paintbrush [C][E]

Herbs, perennial, 1.5–3.7 dm; from a thick, woody caudex; with a taproot. **Stems** solitary or few to several, ascending, sometimes short-decumbent and rooting, branched from near base, sometimes distally, glabrous or glabrate with hairs very sparse, ± appressed, very short, soft, eglandular. **Leaves** green, often brown- or purple-tinged, oblong to lanceolate to narrowly ovate or ovate, (1–)2–5.8 cm, not fleshy, margins plane, ± involute, deeply 3–7(–11)-lobed, sometimes with secondary lobes, apex narrowly acute; lobes erect to ascending, lanceolate to linear-lanceolate to oblanceolate, apex acute. **Inflorescences** 3–15 × 2–4.5 cm; bracts proximally greenish, distally bright red, scarlet, or pale reddish orange, rarely orange-yellow, often fading to pale yellowish orange with age, obovate to orbicular, fan-shaped, (3–)5–9(–13)-lobed, sometimes with secondary lobes; lobes erect or ascending, lanceolate to triangular, short and medium length, usually arising at or above mid length, rarely below, apex acute. **Calyces** proximally green, sometimes purple to brown, distally colored as bracts, 20–30 mm; abaxial and adaxial clefts 7–14 mm, ca. 33% of calyx length, deeper than laterals, lateral 2–4 mm, 10–15% of calyx length; lobes triangular, barely longer than wide, apex acute or acuminate to obtuse. **Corollas** straight, 30–45 mm; tube 14–19 mm; beak long-exserted, adaxially green or yellow-green to brownish, 18–24 mm; abaxial lip deep green, reduced, 1–3 mm, 10% as long as beak; teeth incurved, greenish to dull purplish, 0.5–1.5 mm. $2n = 24$.

Flowering Jun–Aug. Turf or crevices on rocky slopes and benches over basalt, se to sw aspect, sun or partial shade; of conservation concern; 600–1000 m; Oreg., Wash.

Castilleja chambersii is limited to the summits of three volcanic peaks in the northern Coast Range of Clatsop County, Oregon, and at one similar area in nearby Pacific County, Washington. It is similar to *C. rupicola*,

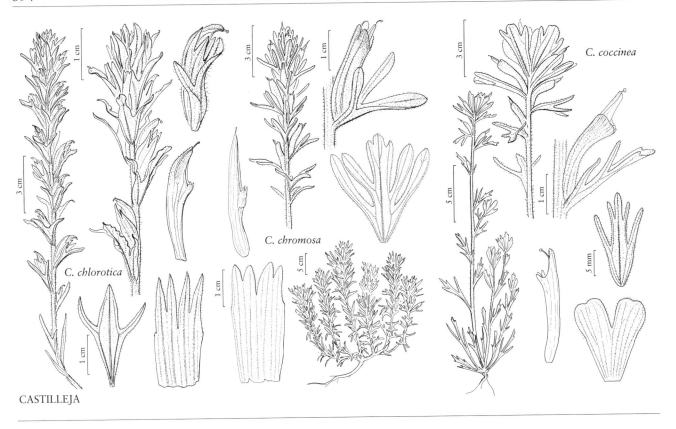

C. coccinea

C. chromosa

C. chlorotica

CASTILLEJA

and the two likely share a common ancestor. Disturbance and erosion from logging and road construction represent significant threats to *C. chambersii*. Populations of *C. chambersii* often grow near and even alongside *C. hispida*, but hybrids are rare.

13. Castilleja chlorotica Piper, Proc. Biol. Soc. Wash. 33: 104. 1920 • Green-tinged or Gearheart Mountain paintbrush E F

Herbs, perennial, 0.8–3.1 dm; from a woody caudex; with a taproot. Stems several to many, erect to ascending, unbranched, sometimes branched near base, hairs spreading, long, soft, eglandular, mixed with dense, short stipitate-glandular ones. Leaves green, narrowly to broadly lanceolate or oblong-lanceolate, 9–35 cm, not fleshy, margins wavy, involute, 0(–3)-lobed, distal sometimes 3-lobed, apex narrowly acuminate; lobes ascending or spreading, linear to lanceolate, apex acute to obtuse. Inflorescences 3–9(–18) × 2–3 cm; bracts green to pale green to rarely dull purplish brown throughout, distally rarely with pale yellow apices, narrowly lanceolate to narrowly elliptic to sometimes broadly lanceolate, 3-lobed, wavy-margined; lobes spreading or ascending, narrowly lanceolate, medium length, arising at or above mid length, sometimes wavy-margined, apex obtuse to acute. Calyces green or yellowish green, 15–19 mm; abaxial and adaxial clefts 9–11 mm, 60% of calyx length, deeper than laterals, lateral 3.5–7 mm, 15–25% of calyx length; lobes broadly or narrowly triangular, taller than wide, apex acute to obtuse. Corollas conspicuously decurved distally, 18–22 mm; tube 11–14 mm; beak exserted, adaxially green to yellow-green or yellow, 6–8 mm; abaxial lip green, reduced, fleshy, included, 1.5–2 mm, 25% as long as beak, sparsely hairy, hairs spreading; teeth ascending, green, 0.5–1 mm.

Flowering Jun–Aug. Dry open pine forests, often with sagebrush understory, rocky ridges and summits, montane to subalpine; 2000–2500 m; Oreg.

Castilleja chlorotica is an uncommon to rare endemic on dry slopes in Deschutes, Klamath, Lake, and possibly Crook counties in central Oregon. The hooked corolla beak and greenish bracts, often aging purplish distally, help distinguish it from similar species such as *C. glandulifera* and yellow forms of *C. applegatei* var. *pinetorum*.

14. Castilleja christii N. H. Holmgren, Bull. Torrey Bot. Club 100: 91, fig. 5. 1973 • John Christ's paintbrush [C] [E]

Herbs, perennial, 1.4–3 dm; from a woody caudex; with a taproot. **Stems** several, erect to ascending, unbranched, sometimes branched, glabrous or hairs spreading, short and long, ± stiff, sometimes stipitate-glandular especially distally. **Leaves** green, narrowly to broadly lanceolate, (1–)2–5(–6) cm, not fleshy, margins plane, flat or involute, (0–)3(–5)-lobed, apex acute to rounded; lobes erect or ascending, linear, lanceolate, or triangular, sometimes very small, apex acute to rounded. **Inflorescences** 3–6 × 2–4 cm; bracts proximally greenish, distally pale orange or pale yellow, sometimes red-orange, lanceolate or narrowly elliptic to ovate, sometimes obovate, 3–5-lobed; lobes ascending, linear, medium length to long, arising at or below mid length, apex acute to obtuse. **Calyces** colored as bracts, 17–22(–24) mm; abaxial clefts 9–12(–13) mm, adaxial 7–11 mm, clefts 50% of calyx length, deeper than laterals, lateral 2–6.5(–8.5) mm, 20–50% of calyx length; lobes lanceolate to narrowly lanceolate, apex acute to obtuse. **Corollas** straight, 20–30 mm; tube 12–19 mm; beak exserted, adaxially green, 7–12 mm; abaxial lip green, reduced, included or visible through cleft, 1.5–2 mm, 10–20% as long as beak; teeth incurved, deep green, 1.5 mm.

Flowering Jun–Jul. Gentle slopes, mostly northern aspect, in herbaceous or grassy subalpine to alpine meadows, sagebrush openings and swales, snowbank communities, over quartzite; of conservation concern; 2100–2900 m; Idaho.

Castilleja christii is endemic to subalpine meadows near the summit of Mt. Harrison in the Albion Mountains, Cassia County. Morphologically, it most closely resembles the widespread *C. hispida* var. *acuta*, but a recent molecular study (D. L. Clay et al. 2012) presents clear evidence for a homoploid hybrid origin for the species, incorporating portions of the genomes of *C. linariifolia* and *C. miniata*. This is the first documented case of homoploid origin in *Castilleja*.

Castilleja christii is in the Center for Plant Conservation's National Collection of Endangered Plants.

15. Castilleja chromosa A. Nelson, Bull. Torrey Bot. Club 26: 245. 1899 (as Castilleia) • Desert paintbrush [E] [F]

Castilleja ewanii Eastwood; *C. martini* Abrams subsp. *ewanii* (Eastwood) Munz; *C. martini* var. *ewanii* (Eastwood) N. H. Holmgren

Herbs, sometimes subshrubs, perennial, 1.5–3.5(–4.5) dm; from a woody caudex; with a taproot. **Stems** several to many, ascending to erect, often grayish, unbranched, rarely branched, sometimes with short, leafy axillary branches, hairs spreading-erect, long, stiff, eglandular, sometimes also with shorter, stipitate-glandular ones. **Leaves** gray-green, linear, lanceolate, or oblanceolate, sometimes broadly lanceolate, (1.5–)2.5–6(–7) cm, not fleshy, margins plane, involute, (0–)3–5(–7)-lobed, sometimes with secondary lobes, apex acuminate to obtuse; lobes spreading, linear, apex acuminate. **Inflorescences** 2.5–15 (much longer in fruit) × 1.5–5.5 cm; bracts proximally greenish to dull purplish, distally bright red to scarlet or orange-red, rarely yellowish to dull orange or pink, narrowly to broadly linear or lanceolate, narrowly ovate, or oblong-lanceolate, (0–)3–7-lobed, rarely with secondary lobes; lobes spreading, linear to oblong, sometimes oblanceolate, often expanded near tip, long, proximal lobes arising below mid length, apex rounded or obtuse to sometimes acute. **Calyces** colored as bracts, sometimes with broad yellow band below colored lobe apices, (17–)20–27 mm; abaxial clefts 4–10 mm, adaxial 6–12 mm, abaxial ca. 30% of calyx length, adaxial ca. 40% of calyx length, deeper than laterals, lateral 1–4 mm, ca. 15% of calyx length; lobes oblong or ovate to narrowly triangular or lanceolate, apex obtuse to rounded. **Corollas** straight or ± curved, 18–35(–40) mm; tube 8–15 mm; beak short- or long-exserted, adaxially green to yellow-green, (9–)10–18 mm; abaxial lip deep green, reduced, thickened, included to exserted, 2–3 mm, ca. 20% as long as beak; teeth incurved, deep green, 0.5–1 mm. $2n$ = 24, 48.

Flowering Jun–Aug. Dry sagebrush slopes and flats, pinyon-juniper stands, blackbrush, open yellow pine forests; 500–3200 m; Ariz., Calif., Colo., Idaho, Nev., N.Mex., Oreg., Utah, Wyo.

Castilleja chromosa is sometimes confused with 3b. *C. angustifolia* var. *dubia* (see discussion there). *Castilleja chromosa* retains its distinctive morphology across its wide range and is a characteristic species of much of the southwestern United States. Where it overlaps with *C. angustifolia*, the two are distinguished by inflorescence color and width and by the lengths of the calyx, corolla, and corolla beak. In the broad region of their sympatry, there is little evidence of intergradation, except in a few sites in Elko County, Nevada, and in southern Wyoming. Throughout southern Idaho and northeastern Nevada the range of the two overlap with little or no intergradation. At high elevations in Montrose County, Colorado, *C. chromosa* has narrower leaves and a longer and silkier pubescence, especially in the inflorescence. Apparent hybrids between *C. chromosa* and *C. flava* var. *rustica* are known from Custer County, Idaho, and hybrids with *C. linariifolia* are known from Montrose County, Colorado.

16. **Castilleja chrymactis** Pennell, Proc. Acad. Nat. Sci. Philadelphia 86: 539. 1934 [E]

Herbs, perennial, 2.8–6.6 dm; from a woody caudex; with a taproot. **Stems** several to many, erect to ascending, unbranched or branched, sometimes with short, leafy axillary shoots, glabrous or hairs sparse proximally, distally spreading-erect, long, soft, eglandular, mixed with shorter ones. **Leaves** green or purple-tinged, sometimes deep purple, linear-lanceolate to lanceolate or ovate, (1.5–)5–11.7 cm, not fleshy, margins plane, sometimes ± wavy, flat or involute, 0–3(–5)-lobed, apex acute to obtuse, sometimes rounded; lobes erect or ascending, linear, apex acute or acuminate. **Inflorescences** 3.5–18 × 3–7.5 cm; bracts proximally pale green to dull purplish, distally bright red or scarlet to yellow, orange, yellow-orange, and salmon, sometimes tricolored with central band of deeper purple or deeper shade of the far-distal color, broadly lanceolate, oblong, elliptic, or oblong, 0–3(–9)-lobed; lobes ascending, lanceolate, triangular, or oblong, short, arising at or above mid length, central lobe apex rounded, lateral ones acute to obtuse. **Calyces** colored as bracts, sometimes more deeply pigmented and contrasting with bract coloration, 24–31 mm; abaxial and adaxial clefts 12–19 mm, 45–65% of calyx length, deeper than laterals, lateral 7–11 mm, 25–32% of calyx length; lobes lanceolate, narrowly triangular, or almost oblong, apex acute, sometimes obtuse or acuminate. **Corollas** straight to slightly curved, 26–43 mm; tube 16–24 mm; beak exserted, adaxially green to yellow, 9–19 mm; abaxial lip whitish or dark green, reduced, not exserted, sometimes visible through front calyx cleft, 1.5–2 mm, 20% as long as beak, glabrous or sparsely hairy, hairs spreading; teeth reduced, incurved, deep green, 0.5–1 mm.

Flowering (Jan–)Feb–Sep(–Oct). Glacial outwash plains, river flats, moist or wet openings, beach meadows, roadsides; 0–20 m; Alaska.

Castilleja chrymactis is restricted to the coast of southeastern Alaska, from Glacier Bay to Yakutat. In some respects, it resembles a hybrid swarm between *C. miniata* and *C. unalaschcensis.* If *C. chrymactis* originated as a hybrid, at present it has formed a number of populations and has a stable morphology that differs from both putative parents, especially in calyx structure. Reports of its introduction to Point Reyes, California (T. I. Chuang and L. R. Heckard 1993b), are based on *C. leschkeana,* a distinctive species known only from two records near the type locality and presumed to be extinct.

17. **Castilleja chrysantha** Greenman, Bot. Gaz. 48: 146. 1909 • Yellowish or common wallowa paintbrush [E]

Castilleja ownbeyana Pennell

Herbs, perennial, (0.5–)1–2(–5) dm; from a woody caudex; with a taproot. **Stems** few to many, erect or ascending, often decumbent at base, unbranched, sometimes branched, hairs spreading, long, soft, mixed with dense, shorter stipitate-glandular ones. **Leaves** green or purple-tinged, sometimes deep purple, broadly lanceolate, sometimes linear-lanceolate or narrowly oblong, 1.5–4.8 cm, not fleshy, margins plane, flat or involute, 0–3-lobed, apex acute; lobes ascending or spreading, linear to narrowly linear or narrowly lanceolate, short, apex acuminate. **Inflorescences** 3–17 × 1–3 cm; bracts greenish or pale yellow-green throughout, or proximally greenish or pale yellow-green, distally pale yellow to whitish, sometimes pink-purple, or pale, dull purplish, sometimes aging pink or yellow, often infused with light purple, rarely pink, ovate to broadly acute, (0–)3-lobed; lobes ascending, linear-lanceolate, oblong, or triangular, short, arising below mid length, apex acute to obtuse. **Calyces** proximally green, pale with green veins, purple-tinged green, or purple, distally pale yellow, white, or purplish, 12–20 mm; abaxial and adaxial clefts 8–12 mm, 50% of calyx length, deeper than laterals, lateral 0.5–3 mm, 5–15% of calyx length; lobes short-triangular, apex obtuse, rounded, or truncate. **Corollas** straight, (16–)20–25 mm; tube 15–18 mm; beak slightly exserted, adaxially green to yellow, 5.5–8.5 mm; abaxial lip white, green and white, pink, or purple, ± prominent, appressed (proximally scarcely or not pouched), 3–5 mm, 67% as long as beak; teeth erect, white or pink with some purple or red, 1.5–2.5 mm.

Flowering May–Oct. Flat, mesic meadows, dry talus and ridges, montane to alpine; 1100–2800 m; Oreg.

Castilleja chrysantha is endemic to the mountains of northeastern Oregon, and its patterns of variation need further study. Most plants in the Blue Mountains are taller, more erect, and tolerate lower elevation habitats than populations around the type locality in the Wallowa Mountains. Plants with purplish inflorescences and longer hairs were described as *C. ownbeyana* and tend to favor drier talus and ridges than plants like the type, found in flat, mesic, montane to subalpine meadows. Hybrids between *C. chrysantha* and *C. fraterna* occur in Wallowa County. In the Wallowa Mountains, a recurrent and variable hybrid form between *C. chrysantha* and *C. rhexiifolia* was described as *C. wallowensis* Pennell.

18. **Castilleja cinerea** A. Gray, Proc. Amer. Acad. Arts 19: 93. 1883 (as Castilleia) • Ash gray paintbrush C E

Orthocarpus cinereus (A. Gray) Jepson

Herbs, perennial, 0.5–3 dm; from a woody caudex; with a taproot. **Stems** several to many, erect to ascending, or decumbent, inflorescence erect in high-elevation form, unbranched, sometimes branched, hairs dense, spreading, ashy gray, short and long, soft, mixed with short stipitate-glandular ones. **Leaves** green, brown, purple, or deep gray, linear or narrowly to broadly lanceolate to sometimes ovate, 0.7–3 cm, not fleshy, margins plane, slightly involute, 0(–3)-lobed, apex acuminate; lobes ascending to spreading, linear to lanceolate, apex acuminate. **Inflorescences** 1–8.5 × 2–5 cm; bracts proximally greenish or deep reddish purple, distally burnt orange, sometimes yellow or deep red to deep burgundy, proximal sometimes lanceolate with narrow lobes, distal or all bracts broadly lanceolate to oblong or slightly oblanceolate, (0–)3–5-lobed, appearing dusty with dense, short stipitate-glandular hairs, many with a nodulose to pillarlike, crystallized, usually pigmented exudate, papillose at 40×; lobes ascending-spreading, oblong or oblanceolate, short, arising above mid length, central lobe apex rounded, often expanded, rounded, or truncate, lateral ones acute to rounded. **Calyces** colored as bracts, sometimes whitish proximally, 1.5–20 mm (shorter in upper elevation form); abaxial and adaxial clefts 3.5–8 mm, 30–50% of calyx length, all 4 clefts subequal; lobes linear to narrowly oblong or oblanceolate, apex obtuse to rounded, densely stipitate-glandular. **Corollas** straight, 12–18 mm; tube 9–14 mm; beak included or tip just barely exserted, adaxially green or pale yellow to deep burgundy, 3–5 mm; abaxial lip green, burgundy, or reddish purple (in high-elevation form), little inflated, small, included, 2 mm, to 20% as long as beak; teeth incurved, green, 0.2–0.5 mm.

Flowering May–Aug(–Oct). Dry rocky slopes, ridges, and flats, pebble plains, sagebrush openings, open conifer forests; of conservation concern; 1800–3100 m; Calif.

Castilleja cinerea is endemic to the higher elevations of the San Bernardino Mountains, San Bernardino County.

Most plants are upright to ascending and have yellow to yellow-orange inflorescences, with occasional plants ranging to dull red, especially with age. On Sugarloaf Mountain, mostly above 2700 m, is a distinctive form with consistently reddish purple to burgundy inflorescences and a strongly decumbent growth form.

Castilleja cinerea is most often associated with and likely parasitic on *Artemisia nova* and *Eriogonum* species. *Castilleja cinerea* is known from few populations and is threatened by livestock grazing, development, and vehicle use. It is listed as a threatened species under the Endangered Species Act of the United States.

The crystalline exudate associated with the stipitate-glandular pubescence of the distal portion of the bracts is unique in the genus.

19. **Castilleja citrina** Pennell, Proc. Acad. Nat. Sci. Philadelphia 73: 532. 1922 • Lemon paintbrush E

Castilleja purpurea (Nuttall) G. Don var. *citrina* (Pennell) Shinners

Herbs, perennial, 1.5–3.5 dm; from a slender, woody caudex; with a taproot or stout, branched roots. **Stems** few to many, erect to ascending, unbranched or branched, hairs appressed to ± ascending, matted, long, soft, mixed with short-glandular ones, denser distally, sometimes obscuring surface. **Leaves** green, sometimes brown, linear to linear-lanceolate, 3–7 cm, not fleshy, margins plane, involute, 3–7-lobed, apex acute to rounded; lateral lobes spreading, narrowly linear, apex acuminate. **Inflorescences** (3.5–)8–20 × 2–5.5 cm; bracts proximally greenish, distally bright yellow, sometimes pale yellow to pale orange, sometimes aging white to pink, narrowly to broadly lanceolate or oblong, 3–5(–7)-lobed; lobes spreading to ascending, linear, long, arising at or below mid length, apex obtuse to rounded. **Calyces** green, pale green, or whitish, lobes colored as bracts, 12–18 mm; abaxial and adaxial clefts 8–15 mm, 60–75% of calyx length, deeper than laterals, lateral 2.5–3 mm, 20% of calyx length; lobes linear to very narrowly triangular (equilaterally triangular if very short), apex obtuse to rounded. **Corollas** straight but curved at tip, 30–41 mm; tube 21–26 mm; beak slightly to long-exserted, adaxially green to yellowish, sometimes aging pinkish, 10–15 mm; abaxial lip white or yellow, sometimes partly green, darkening with age, only slightly inflated, exserted from abaxial cleft, 3–7 mm, ca. 50% as long as beak; teeth prominent, petaloid, spreading, yellow, 2.5–6 mm. $2n = 24$.

Flowering (Mar–)Apr–May. Calcareous prairies, sandy fields, gravelly limestone hillsides, limestone outcrops, mesquite, juniper, oak-juniper, and post oak woodlands, roadsides; 300–800 m; Kans., Okla., Tex.

Although the range of *Castilleja citrina* overlaps the range of its close relatives, *C. lindheimeri* and *C. purpurea*, in central Texas, *C. citrina* extends considerably farther to the west and northwest of the others (G. L. Nesom and J. M. Egger 2014). The inflorescences of *C. citrina* are mostly pale to bright lemon yellow or occasionally brassy yellow. The color and usually more elongate abaxial corolla lip separate it from *C. lindheimeri* and *C. purpurea*. *Castilleja citrina* is similar to some color forms of *C. sessiliflora*, but the more conspicuously curved corolla of *C. sessiliflora* is usually exserted far above the calyx and is often white to pale pink, rather than yellow.

20. **Castilleja coccinea** (Linnaeus) Sprengel, Syst. Veg. 2: 775. 1825 (as Castilleia) • Painted-cup, red or scarlet paintbrush E F

Bartsia coccinea Linnaeus, Sp. Pl. 2: 602. 1753; *Castilleja ludoviciana* Pennell

Herbs, annual or biennial, 1.2–5(–7) dm; with fibrous roots. **Stems** solitary or few, erect to ascending, unbranched, rarely branched, hairs spreading, long, soft, eglandular, others shorter and glandular. **Leaves** green to sometimes brownish, rosette leaves persisting or withered at anthesis, oblanceolate to oblong-lanceolate, 1–3 cm, usually 0-lobed; cauline leaves linear-lanceolate to oblong-lanceolate, 2–7(–8) cm, not fleshy, margins plane, sometimes wavy, involute, 3–7(–9)-lobed, apex acute to acuminate; cauline leaves: lobes spreading to ascending, linear to oblong, (2–)5–25 mm, apex acuminate. **Inflorescences** 2–35 (longest in fruit) × 1.5–6 cm; bracts proximally greenish to brownish green, distally red, sometimes orange, yellow, or white, lanceolate to obovate, 3(–5)-lobed; lobes spreading-ascending, lanceolate and longer on proximal bracts, becoming oblanceolate, shorter, and wider on distal bracts, arising from distal ¾ of blade, apex acute to rounded. **Calyces** colored as bracts, 17–23(–28) mm; abaxial clefts 6–10 mm, adaxial 5–8 mm, clefts 33–50% of calyx length, deeper than laterals, lateral 0–1 mm, 0–5% of calyx length; lobes oblong to triangular, apex rounded to truncate, sometimes emarginate. **Corollas** straight, 18–31 mm; tube 15–18 mm; abaxial lip ± exserted, beak exserted, often whitish, yellowish, or faint dull reddish proximally, adaxially green distally, 7–10 mm; abaxial lip green to yellowish, apex pink to yellow, reduced, ± protuberant, 1.5–4 mm, 20–33% as long as beak; teeth erect, green or yellow, apices white, yellow, or pink, 2–4(–5) mm. *2n* = 24, 46, 48.

Flowering Jan–Sep. Damp or wet meadows, roadsides, prairies, swamps, peatlands, ditches, thickets, dunes, jack pine flats, rocky forests, ledges, sandstone, limestone, or granite; 0–1200 m; Man., Ont., Sask.; Ala., Ark., Conn., Del., Fla., Ga., Ill., Ind., Iowa, Kans., Ky., La., Maine, Md., Mass., Mich., Minn., Miss., Mo., N.H., N.J., N.Y., N.C., Ohio, Okla., Pa., R.I., S.C., Tenn., Tex., Va., W.Va., Wis.

Castilleja coccinea usually has red bracts. Forms with white or yellow bracts have been named but are scattered across the range of the species. However, yellow-bracted forms become markedly more common in populations in the northern portion of the range, especially in the upper midwestern region. Populations of *C. coccinea* seem ephemeral, disappearing from one site after a few years and appearing in another. It is rare in much of the eastern portion of its range, and apparently is extirpated in Louisiana, Maine, Massachusetts, and New Hampshire.

21. **Castilleja collegiorum** J. M. Egger, Phytoneuron 2015-33: 1, figs. 1–3, 9[left]. 2015 • Collegial paintbrush C E

Herbs, perennial, 1.1–2.8 dm; from a woody caudex; with a taproot. **Stems** few to many, erect or ascending, short-decumbent at base, unbranched, hairs dense, spreading to erect, ± short, soft, usually stipitate-glandular, longer ones sometimes eglandular. **Leaves** pale green to dull reddish maroon, linear to linear-lanceolate, 0.8–3.5 cm, not fleshy, margins plane, slightly involute, 0–3(–5)-lobed, apex acuminate; lateral lobes ascending to spreading, linear-lanceolate, usually arising from distal ½ of blade, usually narrower than central lobe, apex acuminate to acute. **Inflorescences** 10–40(–80 with age) × 0.5–2.5 cm; bracts pale cream to pale greenish yellow throughout, often partly to entirely suffused with dull reddish purple to maroon, especially proximally, along veins, and with age, sometimes distal apices pale, dullish red, lanceolate to ovate, usually 3-lobed, central lobe sometimes with short teeth; lobes spreading-ascending, linear-lanceolate, short to medium length, arising at or above mid length, apex acute. **Calyces** pale cream to pale greenish yellow, sometimes reddish violet to maroon on distal segments and/or with a thin vertical strip of pale reddish violet to maroon along veins, 11–20 mm; abaxial and adaxial clefts 7–12 mm, 60% of calyx length, deeper than laterals, lateral 0.5–1 mm, 5–10% of calyx length; lobes triangular, apex acute to obtuse. **Corollas** straight, 16.5–25 mm; tube 12–18 mm; beak scarcely exserted, adaxially pale green to yellowish, 3–7 mm; abaxial lip green, not inflated, grooved, 2.5 mm, 33–50% as long as beak; teeth slightly incurved, white, 1 mm.

Flowering late Jun–Jul. Hummocks and margins of moist to wet meadows; of conservation concern; 1700–1800 m; Oreg.

Castilleja collegiorum is endemic to a large meadow system in the southern Cascade Range of Klamath County. It is similar to *C. cryptantha* in Washington and *C. lemmonii* in California but differs from both in structural details of the inflorescence, calyx, bracts, and leaves.

22. **Castilleja covilleana** L. F. Henderson, Bull. Torrey Bot. Club 27: 353. 1900 (as Castilleia) • Coville's or cushion or Rocky Mountain paintbrush E

Castilleja multisecta A. Nelson

Herbs, perennial, 1–3(–4) dm; from a stout, woody caudex; with a taproot. **Stems** several, erect or ascending, unbranched, sometimes branched, hairs sparse, spreading or retrorse, moderately long, soft, crisped, eglandular, mixed with shorter ones. **Leaves** green, brown, or purple, linear to lanceolate, 1.5–6 cm, not fleshy, margins plane, rarely ± wavy, involute, 3–5(–7)-lobed, apex acuminate to acute; lobes widely spreading, deep, linear to narrowly lanceolate, lateral lobes not much narrower than central one, apex acuminate. **Inflorescences** 3–22 (longer in fruit) × 1–5 cm; bracts proximally greenish to deep purplish, distally bright red or scarlet, sometimes orange to pale yellow, linear, (3–)5–7-lobed, sometimes with secondary lobes; lobes spreading, linear to oblanceolate, long, arising below mid length, apex acute to obtuse. **Calyces** colored as bracts, sometimes paler proximally, 15–26(–33) mm; abaxial clefts 4–7.4 mm, adaxial 7–10 mm, abaxial 21–37% of calyx length, adaxial 28–51% of calyx length, deeper than laterals, lateral 1.5–3.5(–5) mm, 8–18% of calyx length; lobes lanceolate to triangular, apex acute. **Corollas** straight, 20–35 mm; tube 13–23 mm; beak and usually part of abaxial lip exserted; beak adaxially green, 6.9–13 mm; abaxial lip deep green, reduced, thickened, included or visible through front calyx cleft, 1.7–3.8 mm, 10–25% as long as beak; teeth incurved, yellow or deep green, 0.7–1.7 mm. $2n = 48$.

Flowering Jun–Aug. Rocky slopes, ledges, talus, ridges, open conifer forests, moist to dry substrates, montane to alpine; 1200–3100 m; Idaho, Mont.

Castilleja covilleana is endemic to the mountains of central Idaho and adjacent southwestern Montana. A collection of unusually tall plants was described as *C. multisecta*. *Castilleja covilleana* is closely related to *C. rupicola*.

23. **Castilleja crista-galli** Rydberg, Mem. New York Bot. Gard. 1: 355. 1900 • Cock's-comb paintbrush E

Herbs, perennial, 1–5 dm; from a woody caudex; with a taproot. **Stems** few to several, erect or ascending, unbranched or branched, sometimes with short, leafy axillary shoots, hairy, sometimes glabrate proximally, hairs spreading to retrorse, medium length to long, soft, eglandular, often mixed distally with shorter stipitate-glandular ones. **Leaves** green, linear to narrowly lanceolate, 2–8 cm, not fleshy, margins plane, involute, 0–5-lobed, apex acute; lateral lobes spreading, linear, apex acuminate. **Inflorescences** 3–6(–11) × 1.5–6.5 cm; bracts proximally greenish, distally red, red-orange, or orange, sometimes yellow or dull salmon, narrowly to broadly lanceolate or oblong, 3–5-lobed; lobes ascending-spreading, linear-lanceolate, long, arising below mid length, central lobe apex rounded to obtuse, lateral ones acute. **Calyces** colored as bracts, (20–)25–35 mm; abaxial and adaxial clefts (6–)10–17 mm, 50% of calyx length, deeper than laterals, lateral (1–)3–6(–10) mm, 35% of calyx length; lobes slender, triangular, apex acute. **Corollas** straight, (25–)30–40 (–45) mm; tube 15–20 mm; abaxial lip visible through front cleft, beak long-exserted from calyx; beak adaxially green or yellow-green, 16–21 mm; abaxial lip proximally white or yellow-green, distally green, reduced, usually visible in front cleft, 3 mm, 20% as long as beak; teeth incurved to ascending, green, 1 mm. $2n = 96$.

Flowering Jun–Aug. Rocky slopes, talus, ridges, dry to moist, open, conifer forests, montane meadows; 1500–2900 m; Idaho, Mont., Wyo.

Castilleja crista-galli is found in the Rocky Mountains of southwestern Montana and northwestern Wyoming. The extent of its distribution into adjacent Idaho is unresolved, in part because it is frequently confused with either *C. linariifolia* or *C. miniata*. *Castilleja crista-galli* appears to be morphologically intermediate between them, leading to speculation that it might be an allopolyploid derivative. A DNA study (S. Matthews and M. Lavin 1998) showed little support for a hybrid origin. *Castilleja crista-galli* may be separated with some difficulty from the other two species by the presence of at least some short hairs on the stems and the frequently three- to five-parted leaves. *Castilleja linariifolia* and *C. miniata* both usually have subglabrous stems and entire leaves, sometimes three-parted distally, near the inflorescence.

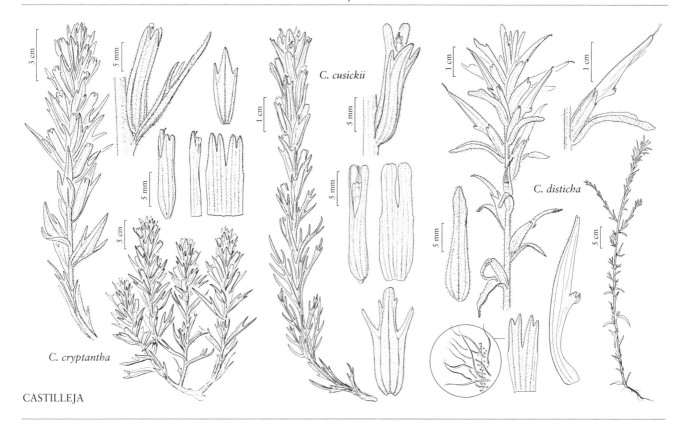

C. cusickii

C. disticha

C. cryptantha

CASTILLEJA

24. Castilleja cryptantha Pennell & G. N. Jones, Proc. Biol. Soc. Wash. 50: 208. 1937 • Mt. Rainier or obscure paintbrush C E F

Herbs, perennial, (0.8–)1–1.9 dm; from a woody caudex; with a taproot. **Stems** few to several, erect or ascending, unbranched, hairs spreading, long, soft, eglandular, mixed with short stipitate-glandular ones. **Leaves** green, often with brown or purple veins, narrowly to broadly lanceolate, 1.5–4 cm, not fleshy, margins plane, ± involute, 0–3-lobed, apex acute to acuminate; lobes spreading-ascending, narrowly lanceolate, apex acute to ± obtuse. **Inflorescences** (2.5–)3–6 × 1–2 cm; green to dull brown or dull reddish purple throughout, or proximally green to dull brown or dull reddish purple, distally yellow on apices, broadly lanceolate to ovate, (0–)3-lobed; lobes ascending, narrowly lanceolate, long or short, arising near mid length, apex acute or acuminate. **Calyces** proximally green or pale with green veins, lobes yellow, sometimes becoming deep red with age, 12–15 mm; abaxial and adaxial clefts 3–7 mm, 25–50% of calyx length, deeper than laterals, lateral 1–3(–4) mm, 8–20% of calyx length; lobes triangular, adaxial segments longer than abaxials, apex acute or obtuse. **Corollas** straight, 14–16 mm; tube 11–14 mm; whole corolla included within calyx; beak adaxially pale yellow, 1–2 mm; abaxial lip deep green, slightly inflated, 4–5 mm, 67% as long as beak; teeth ascending, pale, 1.5–2 mm. $2n = 24$.

Flowering Jul–Aug. Mesic to moist flat subalpine meadows and turf, to tree line; of conservation concern; 1500–2000 m; Wash.

Castilleja cryptantha is endemic to the vicinity of Mt. Rainier in the Cascade Range, with most populations found within Mt. Rainier National Park. Unlike most species of *Castilleja*, it is apparently self-pollinating (W. J. Duffield 1972); the small flowers are entirely enclosed within the yellowish calyces, which tend to grow deep reddish as they age. The purplish brown bracts are also unusual in the genus.

25. Castilleja cusickii Greenman, Bot. Gaz. 25: 267. 1898 (as Castilleia) E F

Castilleja lutea A. Heller

Herbs, perennial, (1–)1.5–5 (–6) dm; from a small, woody caudex; with a taproot or stout, branched roots. **Stems** solitary or few to several, erect, often decumbent at base, unbranched, sometimes branched, hairs spreading, long, soft, eglandular, mixed with shorter, sometimes stiff, stipitate-glandular ones. **Leaves** green, sometimes with prominent red-purple veins, lanceolate-linear to broadly lanceolate, 2.5–4.5(–7) cm, not fleshy, margins plane, flat, (0–)3–5(–9)-lobed, apex acuminate to acute; lobes spreading-ascending, narrowly lanceolate, arising at or above mid length, apex acute. **Inflorescences** 3.5–26 × 1.5–3.5 cm; bracts pale green, pale greenish yellow, or pale yellow throughout, or proximally pale green, pale greenish yellow, pale yellow, or reddish purple, distally white, yellow, pink, dull purple, or dull reddish purple, sometimes with a purple band below that, sometimes with veins darker than background color, lanceolate to oblong, 0–5(–9)-lobed; lobes ascending or spreading, linear to narrowly lanceolate, often short, arising above mid length, central lobe apex rounded, lateral ones acute to rounded. **Calyces** green, pale green, or pale yellow, lobes yellow to pale yellow, 20–30 mm; abaxial and adaxial clefts 6.6–14 mm, 33–50% of calyx length, deeper than laterals, lateral 0–4(–4.5) mm, 0–15% of calyx length; lobes broadly rounded to triangular, apex obtuse to rounded. **Corollas** straight, 19–25(–28) mm; tube (13–)15–20 mm; abaxial lip included, beak not or slightly exserted; beak adaxially green to yellow-green, 4–7 mm; abaxial lip yellow, prominent, inflated, 2–4.5 mm, ca. 75% as long as beak; teeth incurved to erect, white, 1.3–2.1 mm. *2n* = 24.

Flowering Apr–Aug. Moist meadows, swales, grasslands in sagebrush steppes, occasionally to subalpine; 500–2500 m; Alta., B.C.; Idaho, Mont., Nev., Oreg., Wash., Wyo.

Castilleja cusickii includes two morphologically distinct variants. One variant has a wider inflorescence, broader bracts, and wider bract lobes. These bracts often have highly contrasting purple veins, and some populations also have a purplish wash on the bracts. These plants have a range on the western edge of typical *C. cusickii*, though overlapping with more typical forms in western Idaho and adjacent northeastern Oregon and southeastern Washington. The name *C. lutea* is available for these variants, and both consistent morphological differences and preliminary results from ongoing genetic research on the plants (D. C. Tank, pers. comm.) indicate its resurrection may be justified. A second variant is distinguished by a pale pink-purple wash on the bracts and a narrow distribution in the meadows of the Bear River Valley, bordering northwestern Utah and adjacent southeastern Idaho. Only bract coloration appears to distinguish this form from typical *C. cusickii*, and the distinctive bract coloration is only more or less consistent in these populations.

Typical *Castilleja cusickii* appears to hybridize with *C. gracillima* in the Logan Valley, Grant County, Oregon. It is also reported by M. Ownbey (1959) to hybridize with *C. miniata* and *C. rhexiifolia*.

26. **Castilleja densiflora** (Bentham) T. I. Chuang & Heckard, Syst. Bot. 16: 656. 1991 • Dense-flowered owl's-clover

Orthocarpus densiflorus Bentham, Scroph. Ind., 13. 1835

Herbs, annual, 0.7–4.7 dm; with fibrous roots. **Stems** solitary, erect, branched, sometimes unbranched, glabrous or glabrate proximally, pubescent distally, hairs moderately dense, spreading, short to long, soft, eglandular, often mixed with short stipitate-glandular ones (except var. *obispoënsis*). **Leaves** pale green, linear to broadly lanceolate, linear-lanceolate, or ovate, 1.4–9 cm, not fleshy, margins plane, flat or slightly involute, (0–)3–5-lobed, apex acuminate; lobes ascending, linear to narrowly to sometimes broadly lanceolate, apex acuminate to acute. **Inflorescences** 1–16(–20) × 2.5–4 cm; bracts proximally greenish to deep purple, distally white, or pink to pink-purple or reddish purple on apices, if white sometimes aging pink, lanceolate, 3–5-lobed; lobes ascending, linear to oblanceolate, long, arising below mid length, apex acute to acuminate. **Calyces** colored as bracts, 5–20 mm; abaxial and adaxial clefts 4.7–15 mm, 33–90% of calyx length, slightly deeper than laterals or all 4 clefts subequal, lateral 3–8 mm, 33–60% of calyx length; lobes linear to narrowly oblanceolate, apex acute. **Corollas** straight, 14–29 mm; tube expanded distally; abaxial lip and beak exserted from or equal to calyx; beak straight, adaxially pink, purple, or white, if white often aging light pink, (4–)5–7 mm, densely puberulent, hairs often stipitate-glandular; abaxial lip proximally white or pink to deep purple, expanded part white throughout, or proximally white or purple, or green becoming light pink with age, distally white or yellow (sometimes becoming orange with age), purple or maroon spots or blotches on each lobe, inflated, lobes 3, pouches gradually (to abruptly) widened, 4–6 mm wide, 2–3 mm deep, deeper than tall, 3–7 mm, 80–100% as long as beak; teeth erect, pink, white (often turning pink with age), chartreuse, or purple, sometimes with deep purple spot at base, 1–2.5 mm. **Filaments** glabrous. *2n* = 24.

Varieties 3 (3 in the flora): California, Mexico (Baja California).

Castilleja densiflora is often confused with *C. exserta*, and both species are broadly overlapping in both range and habitat, often occurring in close proximity. However, intermediates are remarkably rare. The two are most easily separated by the structure and pubescence of the corollas. In addition to the characters mentioned in the key, *C. densiflora* usually has a bilobed stigma that is exserted from the apex of the corolla with a more or less vertical orientation, while that of *C. exserta* emerges horizontally and is capitate. These differences are remarkably consistent.

1. Bracts distally white, rarely pale yellow; coastal grasslands; San Luis Obispo County, sc California . . . 26c. *Castilleja densiflora* var. *obispoënsis*
1. Bracts distally pink to pink-purple to red-purple, rarely white; near-coastal and interior grasslands; widespread in w California.
   2. Abaxial lips of corollas appearing slightly inflated, pouches widening gradually, longer than deep, 3–7 mm; calyces 8–20 mm . . . . 26a. *Castilleja densiflora* var. *densiflora*
   2. Abaxial lips of corollas appearing moderately inflated, pouches widening abruptly, deeper than long, 4–5 mm; calyces 5–11 mm . . . . 26b. *Castilleja densiflora* var. *gracilis*

26a. Castilleja densiflora (Bentham) T. I. Chuang & Heckard var. **densiflora** E

Herbs 0.9–4.7 dm. **Stems:** hairs spreading, long, soft or proximal ones stiff. **Leaves** linear to broadly lanceolate or ovate, 1.5–8 cm, 0–5-lobed. **Inflorescences** 2.5–16(–20) cm; bracts distally pink to pink-purple or reddish purple on apices, sometimes white distally, if white sometimes aging pink. **Calyces** 8–20 mm; abaxial and adaxial clefts 6.5–15 mm, 75% of calyx length, often deeper than laterals, lateral 6.5–8 mm, 40–60% of calyx length. **Corollas** 18–29 mm; beak adaxially pink or purple; abaxial lip proximally white, yellow, or purple, often with purple spot in each crease, pouches widened gradually, appearing slightly inflated, longer than deep, 3–7 mm.

Flowering Mar–May. Grasslands, forest openings, meadows, roadsides, sometimes on serpentine; 0–2100 m; Calif.

Variety *densiflora* usually has a pink to purple inflorescence, although rare individuals in a population may be white; in the San Francisco Bay region, purely white-bracted populations are interspersed among typical populations with pink-purple bracts. This variety differs from the other two varieties in the structure of the abaxial lip of the corolla. In addition, it is distinguished from var. *gracilis* by a less exserted corolla within a slightly longer calyx and subtending bracts. Within the range of the species, var. *densiflora* extends a little farther north, and var. *gracilis* is more common to the south. Confusing intermediates are occasionally found, suggesting possible hybridization with either *Castilleja ambigua* var. *ambigua* or *C. attenuata*. A peculiar form from Inverness, Marin County, was named *Orthocarpus noctuinus* Eastwood and treated later by some as a hybrid with *C. rubicundula* var. *lithospermoides*.

26b. Castilleja densiflora (Bentham) T. I. Chuang & Heckard var. **gracilis** (Bentham) J. M. Egger, Phytologia 90: 68. 2008

Orthocarpus gracilis Bentham, Scroph. Ind., 12. 1835; *Castilleja densiflora* subsp. *gracilis* (Bentham) T. I. Chuang & Heckard; *O. densiflorus* Bentham var. *gracilis* (Bentham) D. D. Keck

Herbs 1–3.2 dm. **Stems:** hairs spreading, short and medium length, soft. **Leaves** linear to narrowly lanceolate, 1.5–9 cm, (0–)3–5-lobed. **Inflorescences** 1–13 cm; bracts distally pink to pink-purple or reddish purple on apices, rarely white, if white, sometimes aging pink to deep red. **Calyces** 5–11 mm; abaxial and adaxial clefts 4.7–10 mm, 90% of calyx length, all 4 clefts subequal, lateral 3–6 mm, 50% of calyx length. **Corollas** 17–20 mm; beak adaxially pink or purple, sometimes white; abaxial lip base white or pink to deep purple, expanded lip white with purple spots or purple proximally, white distally, sometimes all white, sometimes with distal yellow band and/or deep purple spots, deeply 3-lobed, appearing moderately inflated, lobes ± abrupt, deeper than long, 4–5 mm.

Flowering Mar–May(–Jun). Grasslands, chaparral, savannas, stream or pool margins; 0–1600 m; Calif.; Mexico (Baja California).

Variety *gracilis* usually has a pink to purple inflorescence, although rare individuals in a population may be white. It has a shorter calyx than var. *densiflora* and therefore a more exserted corolla, and also differs in the more divergent angle of the abaxial lip, so that the pouches of the abaxial lip are as deep as they are long. Although var. *gracilis* occurs sporadically within the range of var. *densiflora*, it is much more common in the southern portion of the range of the species.

26c. Castilleja densiflora (Bentham) T. I. Chuang & Heckard var. **obispoënsis** (D. D. Keck) J. M. Egger, Phytologia 90: 69. 2008 • San Luis Obispo owl's-clover C E

Orthocarpus densiflorus Bentham var. *obispoënsis* D. D. Keck, Proc. Calif. Acad. Sci., ser. 4, 16: 539. 1927; *Castilleja densiflora* subsp. *obispoënsis* (D. D. Keck) T. I. Chuang & Heckard

Herbs 0.7–1.2 dm. **Stems:** hairs spreading, medium length, ± stiff. **Leaves** linear-lanceolate, 1.4–6 cm, 0(–5)-lobed. **Inflorescences** 1.5–7 cm; bracts distally white on apices, rarely pale yellow. **Calyces** (8–)15–20 mm; abaxial clefts 7–10.5 mm, adaxial 12 mm, abaxial 33–50% of calyx length, adaxial ca. 50% of calyx length, all 4 clefts subequal, lateral 5–7 mm, 33% of calyx length. **Corollas** 14–22 mm; beak adaxially white, aging light pink; abaxial lip proximally creamy white or yellow-green (becoming light pink with age), distally yellow (becoming orange with age), with small purple spot, pouches appearing moderately inflated, widened gradually, longer than deep, deeply pouched (less so than var. *gracilis*), 4.5–7 mm.

Flowering Apr–May. Coastal grasslands, springs, roadsides, pastures; of conservation concern; 0–500 m; Calif.

Variety *obispoënsis* is white-bracted and is mostly limited to the coastal plain in San Luis Obispo County, where it is threatened by development. It can resemble *Castilleja attenuata*, but the spikes of var. *obispoënsis* are considerably wider, and the abaxial lip of the corolla is more inflated and conspicuous.

27. **Castilleja dissitiflora** N. H. Holmgren, Mem. New York Bot. Gard. 21(4): 46, figs. 6–8. 1971 [E]

Herbs, perennial, 1.8–4(–5) dm; from a woody caudex; with a taproot. **Stems** several to many, erect to ascending, unbranched, sometimes branched, hairs spreading, long, soft, mixed with shorter stipitate-glandular ones. **Leaves** green, linear to narrowly or broadly lanceolate, (1–)3–5(–6) cm, not fleshy, margins wavy (obscure on many pressed specimens), involute, usually 0–3(–5)-lobed, apex broadly acute to rounded; lobes widely spreading, linear to narrowly lanceolate, apex acute to acuminate. **Inflorescences** 2.5–10 × 2–5.5 cm; bracts proximally greenish, distally red to red-orange, narrowly lanceolate to broadly lanceolate, 3–5-lobed; lobes spreading, distal, if present, ascending, linear or narrowly oblanceolate to triangular, proximals long, arising below mid blade, distals short, sometimes mere teeth, near apex of central lobe, sometimes wavy-margined, apex obtuse to rounded, sometimes acute. **Calyces** whitish with green veins or green, sometimes purple, distally same color as bracts, sometimes with yellowish band below colored apices, 20–26(–29) mm; abaxial clefts (8–)13–16(–19) mm, adaxial 7–12(–14) mm, clefts 35–50% of calyx length, deeper than laterals, lateral 2–6(–8) mm, 10–30% of calyx length; lobes linear-lanceolate to lanceolate, apex obtuse to acute. **Corollas** straight to slightly curved, 24–38 mm; tube 14–21 mm; beak exserted from calyx, adaxially green, 11–16.5(–18) mm; abaxial lip green, reduced, visible or not through deep front cleft in calyx, 2 mm, 13% as long as beak; teeth incurved, green, 1 mm. **2n** = 48.

Flowering Jun–Aug. Sagebrush slopes often rocky, montane to subalpine; 1900–3300 m; Nev.

Castilleja dissitiflora is endemic to several mountain ranges in central and eastern Nevada, in the upper montane and lower subalpine zones. It has the deep abaxial calyx cleft of *C. linariifolia* and the stipitate-glandular, wavy-margined leaves of *C. applegatei* var. *pinetorum*. Based on morphological data, Holmgren suggested that it is an allopolyploid derived from hybridization of *C. applegatei* var. *pinetorum* and *C. linariifolia*. His proposal is plausible and should be further tested. *Castilleja dissitiflora* is a tetraploid, while both putative parental species have at least some diploid populations.

28. **Castilleja disticha** Eastwood, Proc. Calif. Acad. Sci., ser. 3, 2: 289. 1902 (as Castilleia) [E] [F]

Castilleja applegatei Fernald subsp. *disticha* (Eastwood) T. I. Chuang & Heckard

Herbs, perennial, 1.4–8 dm; from a woody caudex; with a taproot. **Stems** many, erect, unbranched or branched, sometimes with small, leafy axillary branches, hairs spreading, long, soft, mixed with shorter stipitate-glandular ones. **Leaves** green or purple-tinged, lanceolate to linear-oblong, 0.8–6.1 cm, not fleshy, margins wavy, involute, 0(–3)-lobed, apex rounded to acute; lobes spreading-ascending, narrowly lanceolate, apex acute to acuminate. **Inflorescences** 3–34 × 1.5–3.5 cm; flowers usually distichous, remote, except distalmost; bracts proximally greenish, distally greenish, reddish, orange-red, or dull orange, rarely yellow, proximal bracts frequently greenish throughout, lanceolate to narrowly ovate or narrowly oblong-elliptic, 0(–3)-lobed, proximal wavy-margined; lobes ascending, triangular, short, arising near apex, apex acute to obtuse. **Calyces** proximally whitish, green, or purple, distally colored as bracts, 9–18 mm; abaxial clefts 6–6.5 mm, adaxial 7–9 mm, clefts 33+% of calyx length, deeper than laterals, lateral 1.5–4 mm, ca. 10% of calyx length; lobes linear-lanceolate, apex acute. **Corollas** ± curved, 25–38 mm; tube 14–18 mm; beak and often abaxial lip exserted; beak adaxially dull orange or dull red, rarely yellow, 14–19 mm; abaxial lip slightly curved, green, red, whitish, or yellow, not inflated, 2–2.5 mm, 15% as long as beak; teeth incurved, deep green to reddish, 1–1.5 mm. **2n** = 24.

Flowering (May–)Jun–Aug(–Sep). Open conifer forests, rocky or sandy slopes, montane to subalpine; 1600–3000 m; Calif.

Castilleja disticha is limited to the central and southern Sierra Nevada. Although similar to *C. applegatei* and *C. martini*, *C. disticha* is distinctive and unique in its long, colorful, and highly exserted corollas with the beak exceeding the tube in length, calyces more brightly colored than the often unlobed bracts, and tall, strongly distichous growth form.

29. Castilleja elata Piper, Smithsonian Misc. Collect. 50: 201. 1907 • Siskiyou paintbrush [E]

Castilleja miniata Douglas ex Hooker subsp. *elata* (Piper) Munz

Herbs, perennial, 1.9–6.2 dm; from a remote woody caudex; with a taproot. **Stems** solitary, sometimes few, proximally creeping, becoming rhizomatous, distally ascending to erect, unbranched or branched, glabrous proximally, glabrate distally, hairs ascending, medium length, soft, eglandular. **Leaves** widely spaced on stem, green, linear to lanceolate or narrowly oblong, 1.3–7.2 cm, not fleshy, margins plane, ± revolute, 0-lobed, apex broadly acute to acuminate, distalmost sometimes obtuse. **Inflorescences** 2–13.5 × 1.5–4.5 cm; bracts proximally greenish, rarely pink, distally pink, magenta, or purple, sometimes distally buff, dull yellow, cream, or light yellow-orange, ovate to elliptic or lanceolate, 0–5-lobed, sometimes also with 3 small teeth on center lobe; lobes ascending or ± spreading, narrowly lanceolate to oblong, medium length, arising near or above mid length, apex acuminate to acute, sometimes obtuse. **Calyces** whitish, green, or pink, pale colored ones tending to age pink, lobes as in bracts, 9–17 mm; abaxial and adaxial clefts 4.5–8 mm, 50% of calyx length, deeper than laterals, lateral 2–3 mm, 17–22% of calyx length; lobes linear to oblanceolate, apex acute. **Corollas** straight, 15–25 mm; tube 7–12 mm; beak and abaxial lip exserted above calyx lobes or ± pendently exserted from abaxial calyx cleft; beak adaxially green to yellowish, 6–11 mm; abaxial lip white to green, reduced, inconspicuous, pouches 3, small, 0.5–1.5 mm, 10–15% as long as beak; teeth ascending, white to deep green, 0.5–1 mm. $2n = 24$.

Flowering May–Aug. Serpentine bogs and wetlands; 50–1900 m; Calif., Oreg.

Castilleja elata is endemic to the Siskiyou Mountains of northwestern California and adjacent Oregon. Although often treated as a subspecies of *C. miniata*, it differs from that species in its shorter corolla beaks and distinctive bimodal coloration, with some populations exclusively pale yellow to pale orange and others pink-purple to magenta, as well as its specialized habitat in serpentine wetlands, where it often grows alongside *Darlingtonia californica*. *Castilleja miniata* grows on more mesic to moderately xeric substrates in the general vicinity of *C. elata* but with no sign of intergradation between the two species. The origin and significance of the two discrete color forms of *C. elata* deserve further study.

30. Castilleja elegans Malte, Rhodora 36: 187. 1934 • Elegant paintbrush

Castilleja pallida (Linnaeus) Sprengel subsp. *elegans* (Malte) Pennell

Herbs, perennial, 0.5–3 dm; from a woody caudex; with a slender taproot. **Stems** few to several, ascending to erect, or slightly decumbent at base, unbranched, hairy, sometimes glabrate near base, hairs erect-ascending, whitish or yellowish, long, soft, eglandular, mixed with shorter stipitate-glandular ones, especially distally. **Leaves** green to sometimes purple-tinged, lanceolate to narrowly oblong, (1.2–)2–6(–9) cm, not fleshy, margins plane, sometimes ± wavy, flat to ± involute, 0-lobed, sometimes 3-lobed distally, immediately below inflorescence, apex acuminate to caudate; lobes ascending to erect, linear to lanceolate, lateral lobes narrower than central, apex acuminate. **Inflorescences** 1.5–9 × 1.5–3 cm; bracts purple to pink-purple or reddish purple throughout, sometimes proximally purple to pink-purple or reddish purple, distally whitish or pale pink, oblong to lanceolate or narrowly ovate, (0–)3–7-lobed; lobes spreading-ascending, linear-lanceolate, narrow, short or long, arising near or above mid length, center lobe apex rounded to sometimes acute, lateral ones acute. **Calyces** yellow to yellow-green, rarely purplish to dull red throughout, 19–25 mm; abaxial and adaxial clefts 10–17 mm, 60–70% of calyx length, deeper than laterals, lateral (2–)5–8 mm, 20–25% of calyx length; lobes narrowly oblong to lanceolate, apex rounded or obtuse, sometimes acute. **Corollas** straight, 18–28 mm; tube 17–20 mm; abaxial lip generally visible between calyx lobes, sometimes exserted above them, beak exserted; beak adaxially green, yellowish green, or purplish green, 4–8 mm; abaxial lip purple or magenta, medium sized, often visible through or above abaxial cleft, pouches 3, slightly inflated, 3–5.5 mm, 50+% as long as beak; teeth ascending, purple or magenta, 1–1.5 mm. $2n = 24$.

Flowering Jun–Aug. Meadows, tundra, fell fields, moraines, talus, rocky slopes, shrub thickets, gravel bars, lakeshores; 0–2000 m; Man., N.W.T., Nunavut, Que., Yukon; Alaska; e Asia (Russian Far East).

Castilleja elegans is a characteristic species of the arctic and arctic-alpine regions. Its eastern and western limits are poorly understood, due to confusion with the similar species *C. raupii* to the east and south, and with *C. rubra* (Drobow) Rebristaya in eastern Asia. All three are part of a morphologically variable complex of recently evolved and poorly differentiated entities informally known as the *C. pallida* complex, after the first-described member of the group. Some reports of *C. elegans* in eastern Canada are here referred to

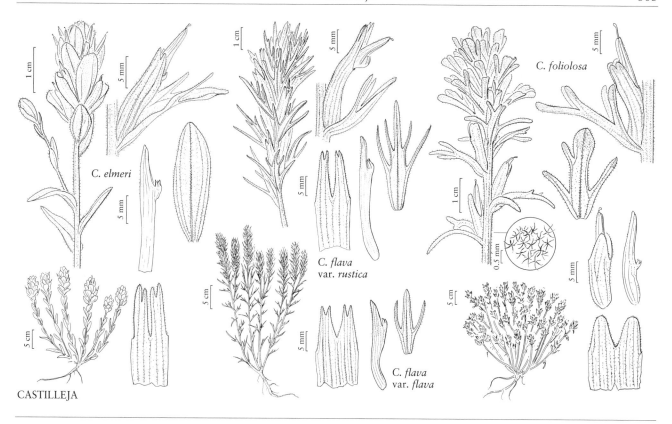

C. elmeri

C. flava
var. *rustica*

C. flava
var. *flava*

C. foliolosa

CASTILLEJA

C. septentrionalis. A recent collection of *C. elegans* from the northern Canadian Rocky Mountains is apparently disjunct, though it may also suggest its wider presence in Alberta.

31. **Castilleja elmeri** Fernald, Erythea 6: 51. 1898 (as Castilleia) • Elmer's paintbrush [E] [F]

Herbs, perennial, 1.1–4.2 dm; from a many-headed, woody caudex; with a stout taproot. **Stems** solitary or few to many, erect to ascending, sometimes slightly curved at base, unbranched, rarely branched, hairs moderately dense, spreading, medium length, soft, mixed with shorter eglandular and stipitate-glandular ones, at least on distal ½ of stem. **Leaves** green, rarely purple-tinged, linear-lanceolate, sometimes linear or lanceolate, 1.3–6.5 cm, not fleshy, margins plane, flat or involute, entire, apex acute. **Inflorescences** 2.5–9 × 1.5–3 cm; bracts red, crimson, scarlet, pink, magenta, red-orange, burnt orange, orange, pale yellow, or whitish throughout, or proximally greenish, distally as stated above, oblong, narrowly obovate, elliptic-oblong, or narrowly ovate, 0(–5)-lobed, rarely with 1 or 2 pairs of short, usually distal lobes; lobes ascending, lanceolate, very short, arising from distal edge, apex rounded to obtuse. **Pedicels** 0 mm. **Calyces** proximally green to

pale green, distally colored as bracts, 15–25 mm; abaxial and adaxial clefts 5–14 mm, 33–50% of calyx length, deeper than laterals, lateral 1–4 mm, 6–20% of calyx length; lobes lanceolate to narrowly triangular, apex rounded to obtuse. **Corollas** straight, 20–33 mm; tubes 13–18 mm; beak, and sometimes abaxial lip, partially to fully exserted; beak adaxially green to yellowish, 8–15 mm; abaxial lip incurved, green, thickened, 2–3 mm, 20–33% as long as beak; teeth ascending or incurved, green, 0.5–1 mm. **2n = 48.**

Flowering May–Aug. Moist to dry rocky slopes, meadows, swales, sagebrush steppes, open conifer forests, montane to subalpine, often on serpentine; 600–2600 m; B.C.; Wash.

The bracts and inflorescences of *Castilleja elmeri* can be red, crimson, scarlet, pink, magenta, red-orange, burnt orange, orange, pale yellow, or whitish. While the coloration is highly variable between local populations, it is usually uniform and consistent within them. It is found primarily in the Wenatchee Mountains of Washington, where it is often on serpentine, but it ranges northward in the Cascades into extreme southern British Columbia, apparently on non-serpentine substrates. Hybrids with *C. miniata* var. *miniata*, *C. parviflora* var. *albida*, and *C. thompsonii* have been found in the eastern Cascades of Washington, and a similar hybrid swarm with *C. thompsonii* is reported from southern British Columbia.

32. Castilleja exserta (A. Heller) T. I. Chuang & Heckard, Syst. Bot. 16: 657. 1991 • Purple owl's-clover, escobita

Orthocarpus exsertus A. Heller, Muhlenbergia 1: 109. 1904

Herbs, annual, 0.1–4.5 dm; with fibrous roots. **Stems** solitary, erect to ascending, unbranched or diffusely branched from near base, hairs spreading, medium length and long, stiff, mixed with short stipitate-glandular ones. **Leaves** green, sometimes purple to brownish, linear or ovate to orbicular in outline, (0.8–)1–5(–7.7) cm, not fleshy, margins plane, involute, (0–)3–9(–11)-lobed, sometimes with secondary lobing, apex acuminate to rounded or acute; lobes spreading, filiform or linear to narrowly spatulate, apex acute to acuminate or rounded. **Inflorescences** 1.5–20 × 2–4 cm; bracts proximally greenish, dark purple, brownish purple, or white, distally pink, lavender, magenta, light purple, or white on lobe apices, lanceolate to elliptic or narrowly ovate, (3–)5(–9)-lobed, often with 2–4 secondary lobes; lobes ascending to spreading, linear to filiform or narrowly spatulate, medium length to long, arising below or above mid length, apex rounded to acute. **Calyces** colored as bracts, 10–26 mm; abaxial clefts 4–12 mm, adaxial 9–18 mm, abaxial ca. 50% of calyx length, adaxial ca. 67% of calyx length, at least adaxial deeper than others, lateral 2.5–9 mm, 15–45% of calyx length; lobes linear (to narrowly oblanceolate), apex rounded to acute. **Corollas** straight, 12–33 mm; tube 7–20 mm; beak slightly exserted, hooked near apex, adaxially pink-purple to magenta, rarely white, 5–13 mm, margins colored as bracts, densely villous-hairy; abaxial lip proximally pink, purple, or magenta, rarely yellow or white, with maroon or deep purple distal to that and white to yellow or pink distally, often with purple spots, distal pale color often aging deep pink or deep red, strongly inflated, pouches 3, 3–8 mm wide, 3–4 mm deep, 3–9 mm, 67–80% as long as beak; teeth erect, white, yellow, or purple, often with purple or maroon spots, 0.5–2 mm. **Filaments** with spreading, long, soft hairs. $2n = 24$.

Varieties 3 (3 in the flora): w United States, nw Mexico.

In addition to the characters in the key, *Castilleja exserta* is distinguished from the similar *C. densiflora* by its conspicuously hairy and apically hooked beak. As a result, the capitate stigma is exserted more or less horizontally from the corolla beak. In contrast, *C. densiflora* has an unhooked, inconspicuously puberulent beak, from which the stigma emerges more vertically. *Castilleja exserta* hybridizes with *C. attenuata*

in southern California and with *C. lineariloba* in central California, and it reportedly crosses occasionally with *C. densiflora* in southern California.

1. Abaxial lips of corollas deep red-purple proximally, with distal ¼ bright yellow to yellow-orange, becoming orange or red after anthesis; w Mojave Desert, California . . . . 32c. *Castilleja exserta* var. *venusta*
1. Abaxial lips of corollas variably colored, proximally pink to deep red-purple or white, with distal ¼–⅓ white, yellow, or yellow-orange; widespread in Arizona and California, rare in New Mexico or limited to the coastal regions of California.
   2. Bracts usually less than 5 mm wide, lobes filiform to linear; inland and near-coastal grasslands and meadows; Arizona, w California, sw New Mexico . . . . .32a. *Castilleja exserta* var. *exserta*
   2. Bracts 5–7 mm wide, lobes linear to narrowly spatulate; coastal dunes and bluffs; California . . . . .32b. *Castilleja exserta* var. *latifolia*

32a. Castilleja exserta (A. Heller) T. I. Chuang & Heckard var. exserta

Orthocarpus purpurascens Bentham 1835 [not *Castilleja purpurascens* Greenman 1906]; *O. purpurascens* var. *pallidus* D. D. Keck

Herbs 0.1–4.5 dm. **Inflorescences** 1.5–20 cm; bracts: distal apices lavender, magenta, pink, or white, usually less than 5 mm wide, (3–)5(–9)-lobed; lobes filiform to linear, 2 × to 1 mm. **Calyces** 10–26 mm; abaxial clefts 6.5–12 mm, adaxial 12–18 mm, abaxial ca. 50% as long as calyx, adaxial ca. 67% of calyx length, lateral 5.5–9 mm, 33% of calyx length. **Corollas** 12–33 mm; tube 22–25 mm; beak 8–13 mm; abaxial lip variably colored, proximally pink to deep red-purple or white, distal ¼ white, yellow, or yellow-orange, 4.5–9 mm.

Flowering Jan–Jun(–Aug). Grasslands, chaparral, dunes, sandy flats, headlands, open oak forests, coastal scrub, rocky slopes, dry sand, desert washes; 0–1600 m; Ariz., Calif., N.Mex.; Mexico (Baja California).

Variety *exserta* is the most common and widespread *Castilleja* in California. In large masses, it is a showy spectacle in springtime. Pink to purple forms are most common, but color morphs with white bracts are often found intermingled in small numbers. In addition, there are often small differences among populations, especially in corolla color and markings. Early records from Oregon and Washington are either garden plants or waifs, not native populations, and none has persisted.

32b. Castilleja exserta (A. Heller) T. I. Chuang & Heckard var. **latifolia** (S. Watson) J. M. Egger, Phytologia 90: 70. 2008 E

Orthocarpus purpurascens Bentham var. *latifolius* S. Watson, Botany (Fortieth Parallel), 458. 1871; *Castilleja exserta* subsp. *latifolia* (S. Watson) T. I. Chuang & Heckard

Herbs 0.6–3.3 dm. **Inflorescences** 1.5–12 cm; bracts: distal apices pink to pale purple, sometimes white, 5–7 mm wide, 3–5(–7)-lobed; lobes linear to narrowly spatulate, less than 2 × 1–2 mm. **Calyces** 17.5–23 mm; abaxial clefts 4–6 mm, adaxial 9–16 mm, abaxial 40–50% of calyx length, adaxial 50–70% of calyx length, lateral 2.5–3 mm, 15% of calyx length. **Corollas** 20–27 mm; tube 15–20 mm; beak 5–7 mm; abaxial lip pink proximally, distal ⅓ white to yellow, becoming deep pink after anthesis, 3–5 mm.

Flowering (Mar–)Apr–Aug. Coastal grasslands, dunes, sandy flats, headlands, sandy to rocky slopes; 0–500 m; Calif.

Variety *latifolia* is a close associate of *Artemisia pycnocephala* along the coast of central and northern California. Sand dune populations in the vicinity of Humboldt Bay and occasionally elsewhere are unique in having almost uniformly greenish bracts, usually with white apices, and a less-inflated abaxial lip on the corollas. The inflorescences of this form also become anthcyanic very rapidly, with much of the coloration fading quickly to deep, dull reddish brown, even while flowering continues. These distinctive plants are the subject of ongoing research. The northernmost population of this form of var. *latifolia* occurs north of Crescent City, in northern Del Norte County, and this variety should be looked for in adjacent southwestern Oregon.

32c. Castilleja exserta (A. Heller) T. I. Chuang & Heckard var. **venusta** (A. Heller) J. M. Egger, Phytologia 90: 70. 2008 E

Orthocarpus venustus A. Heller, Muhlenbergia 2: 141. 1906; *Castilleja exserta* subsp. *venusta* (A. Heller) T. I. Chuang & Heckard; *O. purpurascens* Bentham var. *ornatus* Jepson

Herbs 0.4–2.3 dm. **Inflorescences** 2–12.5 cm; bracts: distal apices pink, light purple, or yellow- or orange-spotted, 4–5.5 mm, (3–)5(–7)-lobed; lobes linear to narrowly spatulate, 2 × 0.8–1.2 mm. **Calyces** 16–19 mm; abaxial clefts 6.5–9.5 mm, adaxial 9–13 mm, clefts 40–70% of calyx length, lateral 5.5–8 mm, 34–45% of calyx length. **Corollas** 20–27 mm; tube 13–15 mm; beak 7–10 mm; abaxial lip deep red-purple proximally, distal ¼ bright yellow to yellow-orange, becoming orange or red after anthesis, 4–7 mm.

Flowering (Mar–)Apr–May(–Jul). Dry sand, desert washes, creosote scrub, saltbush flats; 300–1500 m; Calif.

Variety *venusta* is endemic to the western Mojave Desert region of southern California. Typical var. *exserta* is found east and west of it, in less arid grasslands. In var. *venusta*, the corolla is consistently different in color pattern and intensity from var. *exserta*.

33. Castilleja flava S. Watson, Botany (Fortieth Parallel), 230. 1871 (as Castilleia) • Yellow paintbrush E F

Herbs, perennial, 1.5–5.5 (–7.5) dm; from a woody caudex; with a taproot. **Stems** several to many, erect or ascending, often grayish, unbranched or often branched distally, glabrate proximally or hairy, especially distally, rarely glabrous, hairs sparse, spreading to appressed, usually curly, ± short, soft, sometimes mixed with sparse, short stipitate-glandular ones. **Leaves** often grayish or purplish, linear to narrowly lanceolate or narrowly oblong, (1–)2.5–5(–6.7) cm, not fleshy, margins plane or wavy, involute, deeply (0–)3–5(–7)-lobed, apex acute to acuminate; lobes divaricate to spreading or ascending, linear, arising at or below mid length, apex acute to acuminate. **Inflorescences** 3.5–20(–29) × 1–4 cm; bracts pale green to pale yellow throughout, or proximally pale green to pale yellow, distally pale yellow to bright yellow, sometimes light orange, light red, or bright red, lanceolate to narrowly oblong or narrowly elliptic, slightly broader than leaves, usually 3–5-lobed; lobes ± spreading, linear or narrowly lanceolate, short or long, proximal or all bracts arising near or below mid length, apex acute, sometimes narrowly obtuse. **Calyces** proximally green, sometimes purple or pinkish, distally yellow to light orange, dull green, or colored as bracts, 11–28 mm; abaxial and adaxial clefts 4–16(–17) mm, 30–60 (–95)% of calyx length, deeper than laterals, lateral 0.5–7 mm, 4–30% of calyx length; lobes linear to lanceolate or narrowly to broadly triangular, apex acute to acuminate. **Corollas** straight or ± curved, 13–30 mm; tube 12–16 mm; beak, sometimes abaxial lip, slightly to strongly exserted, corolla often curved and exserted through abaxial cleft; beak adaxially green, (5–)6–12 mm; abaxial lip green or yellow, reduced, exserted or more commonly visible in cleft in calyx, 1–3.5 mm, 20–33(–50)% as long as beak; teeth ascending, green to yellow, 0.5–2 mm. $2n = 48$.

Varieties 2 (2 in the flora): w United States.

Both varieties of *Castilleja flava* are characteristic plants of sagebrush-dominated communities throughout its wide range, from valleys to moderate elevations in the mountains, sometimes reaching the lower subalpine. The species almost always occurs in close association with shrubby species of *Artemisia*.

1. Corolla beaks 5–12 mm, abaxial lips 1–2 mm; calyx abaxial clefts 8–16 mm, adaxial 4–8(–15.5) mm; n Great Basin, c Rocky Mountains . . . . 33a. *Castilleja flava* var. *flava*
1. Corolla beaks 5–8(–9.5) mm, abaxial lips (1.5–)2–3.5 mm; calyx abaxial and adaxial clefts subequal, (6–)7–13(–17) mm; mountains of ne Oregon to c Idaho . . . . .33b. *Castilleja flava* var. *rustica*

33a. Castilleja flava S. Watson var. **flava** E F

Leaves 0–5-lobed. **Bracts** distally pale to bright yellow, sometimes to light orange; lobes long. **Calyces** proximally green, sometimes purple, distally yellow to light orange, 13–28 mm; abaxial clefts 8–16 mm, adaxial 4–8(–15.5) mm, abaxial ca. 60% of calyx length, adaxial 30–50% of calyx length, slightly to much more deeply abaxially in front than adaxially, lateral 1.5–4 mm, 5–15% of calyx length; lobes lanceolate to narrowly triangular. **Corollas** 13–30 mm; beak 5–12 mm; abaxial lip 1–2 mm.

Flowering Jun–Aug(–Sep). Montane to subalpine; 900–3000 m; Colo., Idaho, Mont., Nev., Utah, Wyo.

Variety *flava* often differs from var. *rustica* in the color of the bracts, as well as in the structure of the calyx and corolla. They also have different ranges, with var. *rustica* replacing var. *flava* in southwestern Idaho and northeastern Oregon. The latter is found in the northern Great Basin and central Rocky Mountains. Variety *flava* hybridizes with *Castilleja viscidula* in Elko County, Nevada. Hybrids between *C. angustifolia* var. *angustifolia* and var. *flava* are known from the western slopes of the Big Horn Mountains, Wyoming. M. Ownbey (1959) reported from the same region a hybrid swarm between var. *flava* and *C. linariifolia*.

33b. Castilleja flava S. Watson var. **rustica** (Piper) N. H. Holmgren in A. Cronquist et al., Intermount. Fl. 4: 484. 1984 • Rustic or rural paintbrush E F

Castilleja rustica Piper, Bull. Torrey Bot. Club 27: 398. 1900 (as Castilleia)

Leaves 3–5(–7)-lobed. **Bracts** distally pale to bright yellow or light orange, light red, or bright red; lobes short. **Calyces** proximally pinkish, purplish, or green, distally dull green or colored as bract lobes, 11–25 mm; abaxial and adaxial clefts (6–)7–13(–17) mm, 50–60(–95)% of calyx length, subequal, lateral 0.5–7 mm, 4–30% of calyx length; lobes linear or narrowly to broadly triangular. **Corollas** 15–27 mm; beak 5–8(–9.5) mm; abaxial lip (1.5–)2–3.5 mm.

Flowering Jun–Aug. Mostly montane; 1100–2300 m; Idaho, Mont., Oreg.

Variety *rustica* is almost always closely associated with *Artemisia tridentata*, which is very likely its primary host species. It is found in central Idaho, Montana, and northeastern Oregon. Reports of var. *rustica* from British Columbia are referred to *Castilleja cervina*. Apparent hybrids between this variety and *C. chromosa* are known from Custer County, Idaho.

34. Castilleja foliolosa Hooker & Arnott, Bot. Beechey Voy., 154. 1833 • Woolly or felt paintbrush F

Herbs or subshrubs, sometimes shrubs, perennial, 1.2–6 dm; from a woody caudex; with a taproot. **Stems** many, ascending, white or grayish due to hairs branched, much-branched proximally, unbranched to sometimes branched on distal ½, also usually with short axillary branches, base often shrubby and with marcescent leaves of previous year, hairs dense, appressed to spreading, medium length, soft, much-branched, eglandular, white-woolly, obscuring surface. **Leaves** whitish or grayish, linear to narrowly lanceolate or narrowly oblong, 0.9–5 cm, not fleshy, margins plane, involute, 0–3-lobed, apex obtuse to rounded; lobes divergent, spreading, linear, apex acuminate to acute. **Inflorescences** 2.5–20 × 1.5–3 cm; bracts proximally green to dull pale purplish, distally red, scarlet, rose, yellow, orange, or cream, sometimes pink or white, lanceolate to oblanceolate, (0–)3–5-lobed; lobes spreading, or distal ones erect, often oblanceolate, if 5-lobed, distal lobes usually very shallow, long or distal short, proximal bracts arising below mid length, distal ones arising near apex, apex rounded to truncate.

Calyces proximally green, apices colored as bracts, 14–20(–22) mm; abaxial and adaxial clefts 3.5–7 (–9) mm, 33–40% of calyx length, deeper than laterals, lateral 0(–1) mm, 0(–7)% of calyx length; lobes broadly triangular, apex of lobes and segments broadly rounded to truncate, segments, if present, obtuse. **Corollas** straight or slightly curved, 16–27 mm; tube (8–)10–12 (–14) mm; beak exserted or partially so, adaxially green, (7–)8.5–14 mm; abaxial lip deep green, reduced, 1–3 mm, 20–33% as long as beak; teeth incurved, reduced, green, 0.5 mm. **2n = 24.**

Flowering Jan–Jul. Chaparral, dry rocky slopes, coastal scrub, open forests; 0–1900 m; Calif.; Mexico (Baja California).

Castilleja foliolosa inhabits chaparral from northern California southward to northern Baja California, west of the Sierra Nevada crest. The inflorescence is usually red or red-orange but can vary to white, yellow, pink, or purple. It is present on Santa Catalina Island in the southern Channel Islands, Los Angeles County, but is replaced in the northern Channel Islands by *C. hololeuca*, a related shrubby species. *Castilleja foliolosa* hybridizes with *C. martini* var. *martini* in San Diego County.

35. **Castilleja fraterna** Greenman, Bot. Gaz. 48: 147. 1909 • Fraternal paintbrush C E

Herbs, perennial, 0.8–2.5 dm; from a woody caudex; with a taproot. **Stems** few to many, ± curved at base, ascending or erect distally, unbranched, hairs moderately dense, spreading, medium length and long, soft, mixed with shorter stipitate-glandular ones. **Leaves** green, sometimes purple-tinged, lanceolate, broadly lanceolate, or narrowly ovate, 1–4.3 cm, not fleshy, margins plane, sometimes ± wavy, ± involute, 0–3-lobed, apex acute to acuminate; lobes ascending, often narrowly oblanceolate, apex acute to acuminate. **Inflorescences** 3–8.5 × 1–4 cm; bracts greenish or bright to sometimes dull red, sometimes scarlet, orange, or pale pink throughout, or proximally greenish, distally as above, ovate to elliptic, (0–)3-lobed; lobes ascending, lanceolate to narrowly triangular, short, arising above mid length, apex acute to obtuse or acuminate. **Calyces** brightly, conspicuously colored for at least distal 2/3, often throughout, providing much of inflorescence coloration, proximally pale whitish to pale pink, distally colored as in distal portion of bracts, sometimes colored throughout as in distal portion of bracts, 15–30 mm; abaxial and adaxial clefts 7–14 mm, 33–50% of calyx length, deeper than laterals, lateral 1–5 mm, 5–15% of calyx length; lobes narrowly to broadly triangular, apex rounded to acute. **Corollas** straight or slightly curved, 20–40 mm; tube 27 mm; abaxial lip often visible in abaxial cleft, sometimes

exserted, beak usually exserted from calyx; beak adaxially green to yellow, 8–14 mm; abaxial lip red, black, green, white, or green and white, ± prominent, slightly rounded, 1.5–5 mm, ca. 33% as long as beak; teeth ascending, green, bright red, pink, white, or yellow, 0.5–3 mm.

Flowering Jul–Aug(–Sep). Moist or dry rocky slopes and flats, ridges, talus, dwarf willow mats, subalpine to alpine, rarely along stream channels at lower elevations, over sedimentary rocks, often limy; of conservation concern; 2000–2900 m; Oreg.

Castilleja fraterna is endemic to one ridge system in the Wallowa Mountains of northeastern Oregon. A parallel ridge system in the range has a second Wallowa Mountains endemic, *C. rubida*. *Castilleja fraterna* is colored similarly to *C. miniata*, but the petaloid teeth of the abaxial corolla lip are more like *C. chrysantha*. It is possible that *C. fraterna* was derived through hybridization, though its chromosome number is unknown. It resembles the newly described *C. kerryana* in Montana.

36. **Castilleja genevieveana** G. L. Nesom, Phytologia 72: 210. 1992 (as genevievana) • Genevieve's paintbrush

Herbs, perennial, 1.5–4.5 dm; from a woody caudex; with a taproot. **Stems** several, erect or ascending, unbranched or often branched proximally, loosely lanate, hairs spreading to retrorse, flattened, often matted, long, soft, eglandular. **Leaves** green to gray-green, linear-lanceolate to narrowly oblong, 1.5–7.5 cm, not fleshy, margins plane, involute, 0(–3)-lobed, apex rounded; lobes spreading, linear or filiform, originating from distal 1/3 of blade, apex acute. **Inflorescences** 2.5–10 (–20 in fruit) × 1.5–2.5 cm; bracts proximally pale greenish to green, distally yellow, yellow-orange, or orange, ovate to broadly lanceolate or broadly oblanceolate-obovate, 0–3-lobed; lobes ascending to erect, narrowly lanceolate, medium length, arising near or above mid length, apex rounded to acute. **Calyces** colored as bracts, 23–27 mm; abaxial and adaxial clefts 10–15 mm, 33–50% of calyx length, deeper than laterals, lateral 4–9 mm, 15–33% of calyx length; lobes oblong to lanceolate or narrowly triangular, apex acute to rounded. **Corollas** straight or slightly curved, 24–37 mm; tube 15–22 mm; slightly shorter than calyx or usually with beak, sometimes abaxial lip, exserted beyond calyx apices; beak adaxially green to yellowish, 7–15 mm; abaxial lip green or yellowish, reduced, ± inconspicuous, slightly pouched, 1.6–6 mm, less than 50% as long as beak; teeth spreading, yellow or yellow-green, 1–4 mm.

Flowering Mar–Jun.   Rocky slopes, open stands of juniper, over limestone; 500–900 m; Tex.; Mexico (Coahuila).

Castilleja genevieveana is known from eight counties in southwestern Texas and from a small area of adjacent Coahuila, Mexico.   The species is likely related to the widespread *C. integra* and geographically replaces it.

37. **Castilleja glandulifera** Pennell, Notul. Nat. Acad. Nat. Sci. Philadelphia 74: 8. 1941 • Glandular paintbrush [E]

Herbs, perennial, 1–3 dm; from a woody caudex; with a taproot. **Stems** few to many, erect or ascending, sometimes decumbent, unbranched or often branched proximally, hairs spreading, medium length and long, soft, mixed with more abundant stipitate-glandular ones.   **Leaves** green, linear-lanceolate to sometimes narrowly oblong or narrowly oblanceolate, 0.7–3.7 cm, not fleshy, margins wavy, involute, 0(–5)-lobed, apex acute; lateral lobes ascending to erect, narrowly lanceolate to narrowly oblong, usually narrower than center lobe, apex acute.   **Inflorescences** 2.5–10 × 2–5 cm; bracts proximally pale green to pale yellow, distally yellow, whitish, pink, dull red, or purple on apices (sometimes gradually differentiated from proximal coloration), lanceolate, broadly lanceolate, or oblong, 3–5(–7)-lobed, sometimes with secondary lobes; lobes ascending to spreading, linear, sometimes rounded, medium length or distal short, arising near mid length, apex acute to rarely obtuse.   **Calyces** proximally green or pale, distally colored as bracts, 17–21(–23) mm; abaxial and adaxial clefts 4–8 mm, 33–50% of calyx length, deeper than laterals, lateral 2–6 mm, 15–33% of calyx length; lobes linear, narrowly lanceolate, or narrowly triangular to oblong, apex acute.   **Corollas** straight or slightly curved, (20–)22–30 mm; tube 15 mm; abaxial lip usually hidden or just visible in abaxial calyx notch, not exserted/longer than calyx, beak exserted; beak straight or slightly curved, adaxially green, 8–11(–12) mm; abaxial lip deep green to yellow, reduced, slightly pouched, 1–2.5 mm, to 20% as long as beak; teeth incurved, green to yellow, 0.5–1 mm.

Flowering Jun–Aug. Dry sagebrush steppes, gravelly or rocky slopes, talus, open conifer forests, subalpine; 1400–2500 m; Oreg.

Castilleja glandulifera is endemic to the upper elevations of the Blue and Strawberry mountains of northeastern Oregon, as well as a few adjacent minor ranges.   It is related to *C. applegatei* and *C. viscidula*, which are the source of reports of *C. glandulifera* in the Wallowa Mountains and on Steens Mountain. Inflorescences of *C. glandulifera* are usually white to pale yellow, but in the area around Marble Creek Pass in Baker County, they are multicolored, with a variety of reddish shades mixed in among the yellowish plants. *Castilleja glandulifera* and *C. viscidula* share a glandular pubescence, divided leaves, and usually yellowish inflorescences.   *Castilleja glandulifera* is distinguished from *C. viscidula* by its taller stature, longer corolla beak, and more deeply divided leaves and bracts with linear to linear-lanceolate lobes.   *Castilleja glandulifera* differs from *C. applegatei* by its unusual leaves and bracts as well as by its habitat and narrower and somewhat shorter corolla beak.

38. **Castilleja gleasoni** Elmer, Bot. Gaz. 39: 51. 1905 (as Castilleia) • Mt. Gleason paintbrush [C] [E]

Castilleja pruinosa Fernald subsp. *gleasoni* (Elmer) Munz

Herbs or subshrubs, perennial, 3–8 dm; from a woody caudex; with a taproot. **Stems** several to many, erect to ascending, branched, sometimes unbranched, hairs moderately dense, spreading, ± matted, ash gray, branched, sometimes unbranched, medium length, soft, often mixed with shorter, unbranched, stipitate-glandular ones, not obscuring surface.   **Leaves** ash gray, linear-lanceolate to narrowly oblong or narrowly oblanceolate, 2–6 cm, not fleshy, margins wavy, involute, 0(–3)-lobed, apex acute to obtuse; lobes spreading, narrowly lanceolate, apex acute to obtuse. **Inflorescences** 10–15(–30) cm; bracts proximally greenish to dull brownish purple, distally red to deep red or red-orange, lanceolate to oblong or narrowly ovate, (0–)3(–5)-lobed; lobes ascending to spreading, narrowly oblong to narrowly lanceolate, medium length, arising below mid length, center lobe apex obtuse or toothed, lateral ones acute.   **Calyces** proximally pale to green, sometimes dull purple, often paler or greener than bracts, distal ¼ or less colored as bract lobes, 12–17 mm; abaxial and adaxial clefts 3–8 mm, 40% of calyx length, deeper than laterals, lateral 0.5–3 mm, 20–25% of calyx length; lobes ovate to narrowly triangular, apex obtuse to acute.   **Corollas** ± straight, 18–30 mm; tube 9–15 mm; beak exserted, adaxially yellow, 9–15(–20) mm; abaxial lip spreading, deep green to ± black, reduced, 0.5–2 mm, 20% as long as beak; teeth incurved, green, 0.2–1 mm. **2***n* = 72.

Flowering May–Jun.   Ledges, rocky slopes, open yellow pine forests, montane chaparral or sagebrush; of conservation concern; 900–2200 m; Calif.

Castilleja gleasoni is an uncommon plant endemic to the upper elevations of the San Gabriel Mountains of Los Angeles County.   It was treated as a polyploid derivative of *C. affinis* var. *affinis* (as subsp. *affinis*) × *C. foliolosa* (T. I. Chuang and L. R. Heckard 1993b), a hypothetical ancestry supported by a chromosome number of $2n = 72$. Others have placed it as a subspecies of *C. pruinosa*, a similar species. However, the morphology of *C. gleasoni* also suggests it could have originated as a cross between the diploid *C. foliolosa* and a tetraploid form of *C. martini*; careful morphological and molecular analyses are needed to determine its true ancestry.

Castilleja gleasoni is in the Center for Plant Conservation's National Collection of Endangered Plants.

39. Castilleja gracillima Rydberg, Bull. Torrey Bot. Club 34: 39. 1907 • Slender paintbrush E

Herbs, perennial, 2–5.5 dm; from a remote woody caudex; with a taproot. Stems solitary, sometimes few, proximally creeping, becoming rhizomatous, erect to ascending distally, unbranched, sometimes branched, often glabrate proximally, hairy distally, hairs spreading, medium length and long, soft, mixed with much shorter stipitate-glandular ones near inflorescence. Leaves green to purplish, linear-lanceolate to broadly lanceolate or narrowly oblong, 1.2–7.1 cm, not fleshy, margins plane, sometimes ± wavy, slightly involute, 0-lobed, apex acute to acuminate. Inflorescences 4.5–18 × 1.5–6 cm; bracts white, cream, pale yellow, pink, salmon, orange, or dull red throughout, or proximally greenish, distally as above, broadly lanceolate to oblong, 0–3(–5)-lobed; lobes ascending to erect, lanceolate to triangular, often short, arising above mid length, apex obtuse or rounded, sometimes acute or acuminate. Calyces colored as bracts, pigmentation often confined to lobes, 15–22 mm; abaxial and adaxial clefts 7–14 mm, 40–50% of calyx length, deeper than laterals, lateral 2–6.5 mm, 10–20% of calyx length; lobes narrowly triangular to narrowly lanceolate, apex acute (to sometimes acuminate in Logan Valley). Corollas ± straight, 19–30 mm; tube 11–19 mm; beak exserted from calyx, adaxially green, 7.5–11 mm; abaxial lip deep green, reduced, 1–2 mm, 20% as long as beak; teeth incurved to erect, green, 0.5–1 mm. $2n = 48$.

Flowering May–Aug. Mesic or wet, usually flat meadows, sometimes near hot springs, along stream banks, montane; 1600–2500 m; Idaho, Mont., Oreg., Wyo.

Castilleja gracillima populations are centered around the Greater Yellowstone region, but its range extends sporadically west to central Oregon. It is sometimes confused with *C. miniata*, but differs from that species in its floral dimensions, mostly single-stemmed growth form, primarily white, yellow, or pinkish orange bract coloration, puberulent stems, and weakly rhizomatous habit. Where the two grow in the same general region,

there is no clear evidence of hybridization; however, *C. cusickii* and *C. gracillima* form an extensive, sporadically intergrading population in the Logan Valley, Grant County, Oregon.

Plants attributed to this species from the Rocky Mountain trench near the head of the Columbia River in southeastern British Columbia or adjacent Alberta are a combination of several other species, especially *Castilleja lutescens* and *C. miniata*.

40. Castilleja grisea Dunkle, Bull. S. Calif. Acad. Sci. 42: 31. 1943 • San Clemente Island paintbrush C E

Castilleja hololeuca Greene subsp. *grisea* (Dunkle) Munz

Subshrubs or shrubs, perennial, 4–6(–10) dm, to 20 dm wide; from a woody caudex; with thick, woody roots. Stems many, spreading and ascending, much-branched, with short, leafy axillary shoots, proximal stems 1+ cm wide, hairs dense, spreading, matted, white, medium length, soft, stellate, sometimes mixed with short stipitate-glandular ones especially near inflorescence, partially obscuring surface. Leaves ash gray, linear to narrowly lanceolate, 1–6.5 cm, not fleshy, margins plane or slightly wavy, involute, 0–7-lobed, apex rounded to obtuse, surfaces almost obscured by hairs, at least when young; lobes spreading to ascending, linear to narrowly oblong, apex acute to acuminate. Inflorescences (3–)7.5–11.5 × 1.5–3.5 cm; bracts proximally greenish, distally ± pale yellow, proximals sometimes greenish throughout, linear to narrowly lanceolate or narrowly oblong, (3–)5–7-lobed; lobes spreading, linear to narrowly oblong, long, proximal lobes arising below mid length, apex obtuse to acute. Calyces colored as bracts, 10–20 mm; abaxial and adaxial clefts 5.5–7 mm, 50% of calyx length, deeper than laterals, lateral 0–1 mm, 0–8% of calyx length; lobes oblong, apex truncate to rounded or obtuse, rarely emarginate, inner surface glandular. Corollas curved in proximal ⅓, 15–25 mm; tube 11 mm; abaxial lip not exserted, sometimes visible in abaxial cleft, beak exserted; beak adaxially green or dull yellow, 7–9 mm; abaxial lip deep green, reduced, 1–2 mm, less than 25% as long as beak; teeth incurved, green, 0.7–1 mm. $2n = 24$.

Flowering Feb–Jun. Coastal terraces and slopes, cliffs and canyon walls, coastal sage scrub; of conservation concern; 0–500 m; Calif.

Castilleja grisea is endemic to a small portion of San Clemente Island in Los Angeles County, which is managed by the U.S. Navy. It was one of the first plants listed as endangered under the Endangered Species Act of the United States, and its near extinction was due to grazing by feral goats. Plants responded well when the goats were removed, spreading from their last refuge on inaccessible cliffs onto gentle terrain at the southern end of the island, near the type locality, and are now locally fairly common. Ungrazed old shrubs can reach 1 m in height and 2 m in breadth and often have a thick woody trunk. The stellate pubescence is unusual in *Castilleja*.

Castilleja grisea is in the Center for Plant Conservation's National Collection of Endangered Plants.

41. Castilleja haydenii (A. Gray) Cockerell, Bull. Torrey Bot. Club 17: 37. 1890 (as Castilleia haydeni) • Hayden's paintbrush E

Castilleja pallida (Linnaeus) Sprengel var. *haydenii* A. Gray in A. Gray et al., Syn. Fl. N. Amer. 2(1): 297. 1878 (as Castilleia haydeni)

Herbs, perennial, 0.7–2 dm; from a woody caudex; with a taproot. **Stems** few to many, spreading to ascending, unbranched except for short, leafy shoots in axils of leaves, glabrate to distally puberulent, hairs sparse to dense, spreading, ± short, soft, sometimes stipitate-glandular. **Leaves** green to purple, linear or narrowly lanceolate to narrowly oblong, (1.1–)2–8 cm, not fleshy, margins plane, involute, (0–)3–7(–9)-lobed, apex acute to obtuse; lobes spreading, linear to narrowly lanceolate, apex acuminate. **Inflorescences** 2.5–5.5 × 1.5–2.5 cm; bracts rose purple, magenta, lilac, or crimson throughout, or proximally greenish to dull purplish, distally as above, lanceolate to ovate or broadly elliptic, 3–7(–13)-lobed; lobes spreading to erect, linear to narrowly lanceolate, long or short, proximal arising below mid length, central lobe apex rounded to acute, lateral ones acute. **Calyces** colored as bracts, 12–26 mm; abaxial and adaxial clefts 5–10 mm, 50% of calyx length, deeper than laterals, lateral 0.2–6 mm, 25% of calyx length; lobes triangular, apex acute. **Corollas** straight or slightly curved, 20–25 mm; tube 13–15 mm; beak exserted, sometimes part of abaxial lip equal to or exceeding calyx; beak adaxially green, 6–8 mm; abaxial lip greenish at base, becoming white to rose pink on apices, reduced, slightly pouched, 2–3 mm, 33–50% as long as beak; teeth ascending, white, pink, or green, 1–2.2 mm. *2n* = 24.

Flowering Jul–Sep. Rocky slopes, meadows, fellfields, alpine; 3200–4300 m; Colo., N.Mex., Utah.

Castilleja haydenii is endemic to high-elevation slopes in the Rocky Mountains of southern Colorado and northern New Mexico. It is known in Utah from a single collection from high elevations in the La Sal Mountains. Reports of this species elsewhere are usually attributable to *C. rhexiifolia*. Its affinities are uncertain. Some features are shared with *C. rhexiifolia*, but in other ways it resembles species such as *C. lemmonii* of

the Sierra Nevada. Plants in the northwestern portions of its range tend to have less divided leaves. *Castilleja haydenii* occasionally hybridizes with *C. occidentalis* where the two commingle in the lower alpine zone.

42. Castilleja hispida Bentham in W. J. Hooker, Fl. Bor.-Amer. 2: 105. 1838 • Hispid or harsh or bristly paintbrush E F

Herbs, perennial, 1.3–5(–6) dm; from a woody caudex; with a taproot. **Stems** few to many, erect or ascending, unbranched, sometimes with inconspicuous, short, leafy axillary shoots, hairs spreading to erect, long, soft to stiff, mixed with shorter stipitate-glandular ones. **Leaves** green, sometimes purple-tinged, margins sometimes red-brown, linear or narrowly to broadly lanceolate to narrowly oblong, oblanceolate, or ovate, 1–8.5 cm, not fleshy, margins plane or wavy, involute or flat, (0–)3–5(–7)-lobed, apex acute to rounded or acuminate; lobes ascending, linear to narrowly lanceolate, much narrower than mid blade, apex acute to rounded. **Inflorescences** 3–16(–30 in fruit) × 2–5 cm; bracts proximally greenish to dull reddish purple, distally red to orange or yellow, sometimes crimson, scarlet, orange-red, red-orange, or burnt orange, often becoming paler and/or duller with age, lanceolate to oblong, ovate, or obovate, 3–5(–11)-lobed; lobes spreading to ascending, linear to oblong or narrowly lanceolate, medium length or long, arising at or above mid length, central lobe apex obtuse to rounded or truncate, sometimes emarginate, truncate, or acute, lateral ones acute to obtuse. **Calyces** colored as bracts, sometimes with a yellow band proximal to red to orange apices, or ca. ½ yellowish and ½ reddish, 12–35 mm; abaxial and adaxial clefts 7–12 mm, 33–65% of calyx length, deeper than laterals, lateral 2–7 mm, 15–30% of calyx length; lobes triangular, linear, or oblong to lanceolate, apex acute or obtuse to rounded. **Corollas** slightly curved, 17–38 mm; tube 12–18 mm; beak exserted, adaxially green, sometimes yellowish, rarely red-brown, 9–20 mm; abaxial lip ascending, deep green, reduced, curved, 0.5–3 mm, to 10–33% length of beak; teeth incurved, reduced, green, 0.5–1.2 mm. *2n* = 24, 48, 96.

Varieties 2 (2 in the flora): w North America.

Castilleja hispida is likely related to *C. chromosa*, which replaces it geographically to the southeast.

1. Leaves (0–)3–5(–7)-lobed; calyx lobe apices obtuse to rounded; lateral calyx clefts 2–7 mm; Cascade Range and west to coast . . . . 42a. *Castilleja hispida* var. *hispida*
1. Leaves 0–3(–5)-lobed; calyx lobe apices acute; lateral calyx clefts 4.5–7 mm; e slope of Cascade Range to Alberta . . . . . . 42b. *Castilleja hispida* var. *acuta*

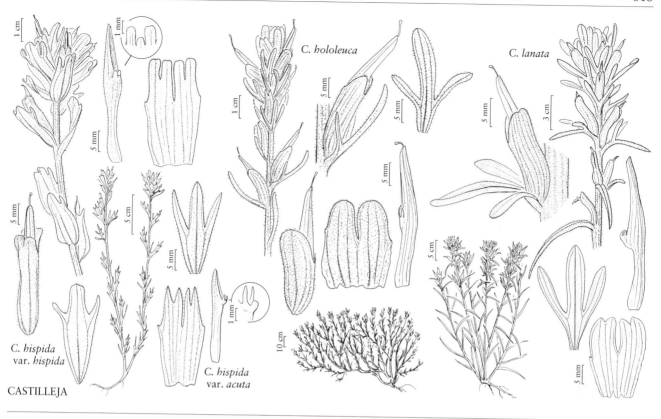

C. hololeuca

C. lanata

C. hispida
var. *hispida*

C. hispida
var. *acuta*

CASTILLEJA

42a. Castilleja hispida Bentham var. **hispida** [E] [F]

Castilleja hispida subsp. *abbreviata* (Fernald) Pennell

Leaves linear or lanceolate to narrowly oblong or ovate, (0–)3–5(–7)-lobed, apex rounded to acute. **Bracts** distally red of various shades or yellow, sometimes orange or burnt orange, older bracts sometimes aging orange to yellow on plants with red bracts; lobes linear to oblong, medium length, apex rounded, sometimes truncate or acute. **Calyces** 12–35 mm; lateral clefts 2–7 mm; lobes oblong to lanceolate, apex obtuse to rounded. **Corollas** 17–38 mm; abaxial lip 0.5–1.5 mm, to 10–20% length of beak. $2n = 24, 48$.

Flowering (Mar–)Apr–Aug. Grassy or rocky slopes, coastal bluffs, dry prairies or open forests, valleys, montane, occasionally to subalpine; 0–2100 m; Alta., B.C.; Idaho, Mont., Oreg., Wash.

Variety *hispida* is fairly common in the lowlands and islands of western Washington, northwestern Oregon, and southwestern British Columbia, and it extends into the montane and subalpine zones of the Cascade Range and Olympic Mountains. It also ranges eastward sporadically on both sides of the United States-Canada border to northwestern Montana and adjacent southern Alberta. Variety *hispida* forms occasional hybrids with *Castilleja levisecta* in Thurston County, Washington, and with *C. chambersii* in Clatsop County, Oregon.

42b. Castilleja hispida Bentham var. **acuta** (Pennell) Ownbey in C. L. Hitchcock et al., Vasc. Pl. Pacif. N.W. 4: 309. 1959 [E] [F]

Castilleja hispida subsp. *acuta* Pennell, Notul. Nat. Acad. Nat. Sci. Philadelphia 74: 11. 1941; *C. taedifera* Pennell

Leaves narrowly to broadly lanceolate, sometimes oblanceolate, 0–3(–5)-lobed, apex acute to acuminate. **Bracts** distally red to orange, sometimes crimson, scarlet, orange-red, or red-orange, lobes or proximal ones at least becoming paler and/or duller with age, orange ones becoming yellow, red ones sometimes becoming pale purple; lobes broadly linear to narrowly lanceolate, long, central lobe apex obtuse to rounded or truncate, sometimes emarginate, lateral ones acute to obtuse. **Calyces** 19–27 mm; lateral clefts 4.5–7 mm; lobes oblong, triangular, or linear, apex acute. **Corollas** 25–33 mm; abaxial lip 1–3 mm, 33% length of beak. $2n = 96$.

Flowering (Mar–)Apr–Sep. Sagebrush steppes, grassy slopes, dry open forests, ledges, talus; 500–2200 m; Idaho, Mont., Oreg., Wash.

Variety *acuta* sporadically replaces var. *hispida* on the eastern slope of the Cascade Range and extends eastward from there to southwestern Montana and northeastern Oregon. Variety *acuta* hybridizes with *Castilleja lutescens* in Garfield County, Washington.

43. Castilleja hololeuca Greene, W. Amer. Sci. 3: 3. 1886 (as Castilleia) • Channel Islands paintbrush E F

Castilleja lanata A. Gray subsp. *hololeuca* (Greene) T. I. Chuang & Heckard

Shrubs or subshrubs, 3–10 dm; from a woody caudex; with thick, woody roots. **Stems** many, erect to spreading, much-branched with many short axillary shoots, proximal stems reaching 1+ cm wide, white-felted, hairs dense, spreading to ± appressed, intertwined, long, soft, slightly branched, obscuring surface. **Leaves** ash gray, leaves of previous year persisting on proximal stem, linear, older leaves sometimes broadly linear, 1–5 cm, not fleshy, margins plane, involute, 0-lobed, apex obtuse; lobes ascending, linear-lanceolate, apex obtuse to rounded, sometimes acute. **Inflorescences** 3–16 × 2–4 cm; bracts pale gray-green throughout, or proximally pale gray-green, distally red or yellow, proximal linear, distal broader, often oblanceolate to obovate, (0–)3-lobed; lobes spreading, linear to oblong, short to long, arising below mid length, apex rounded to truncate or acute. **Calyces** colored as bracts, with conspicuous whitish veins, 14–18(–20) mm; abaxial and adaxial clefts 5–12 mm, 30–67% of calyx length, deeper than laterals, lateral 0 mm or nearly so, ca. 0% of calyx length; lobes linear-lanceolate, apex rounded, truncate, or emarginate. **Corollas** slightly curved in proximal ⅓, 14–26(–31) mm; tube 9–17 mm; abaxial lip not exserted, beak exserted; beak adaxially green to yellowish, 11–14 mm; abaxial lip ascending, deep green, reduced, 2–3 mm, less than 33% as long as beak; teeth reduced to apiculations, deep green, 1–1.5 mm.

Flowering Mar–Oct. Coastal sage scrub, chaparral slopes and flats, ledges, forest edges; 0–500 m; Calif.

Castilleja hololeuca is endemic to the four major islands of the northern Channel Islands. It is similar to *C. lanata*, a mostly Mexican species that reaches Arizona, New Mexico, and southwest Texas. However, *C. hololeuca* appears to be more closely related to *C. foliolosa*, a species of the California mainland. These three species, and *C. galehintoniae* G. L. Nesom of Nuevo León, all share a distinctive, thick indument of pale, matted, and often branched hairs on the stems and leaves. Large individuals of *C. hololeuca* form thick, woody trunks. On Anacapa Island, closest to the mainland, red-flowered forms of *C. hololeuca* predominate. On Santa Cruz Island, to the west, both red- and yellow-bracted forms are fairly common, usually in separate populations. Further offshore, on San Miguel and Santa Rosa islands, most plants are yellow to, occasionally, peach in coloration. Reports of this species from Santa Barbara and Santa Catalina Islands are referable to other species, mostly *C. foliolosa*. Populations of *C. hololeuca* historically declined from grazing by introduced game and livestock species. The plants are recovering well, following removal of the grazing animals, and are now common in many areas, especially on San Miguel Island.

44. Castilleja hyperborea Pennell, Proc. Acad. Nat. Sci. Philadelphia 86: 532. 1934 • Northern paintbrush

Castilleja kuschei Eastwood; *C. villosissima* Pennell

Herbs, perennial, 0.7–2.6 dm; from a many-headed, short, woody caudex; with a taproot. **Stems** few, ascending, unbranched, sometimes branched, glabrate to sparsely hairy proximally, becoming densely hairy distally, proximal hairs appressed, white, short, soft, eglandular, distal hairs widely spreading to erect, white or yellowish, long to very long, soft, eglandular. **Leaves** green to sometimes red-brown or dull purplish, linear to linear-lanceolate, 1.5–4 cm, not fleshy, margins plane, sometimes ± wavy, involute, (0–)3(–5)-lobed, apex acute to acuminate; lobes linear or linear-lanceolate, laterals ± divaricate, often abruptly up-curved from plane of main leaf blade, proximals usually 33–50% as long as leaf, apex acuminate to acute. **Inflorescences** 2–11 × 1–2.5 cm; bracts greenish or pale yellow, rarely purplish or deep red, often with dull purplish veins on proximal bracts, or proximally greenish or pale yellow, distally whitish, yellow, or orange-yellow, proximal sometimes linear-lanceolate, distal broadly lanceolate to oblong or obovate, (0–)3–5(–7)-lobed; lobes ascending to erect, linear to lanceolate, short to medium length, arising above mid length, proximal bract apex acute, distal obtuse, rounded, or truncate. **Calyces** colored as bracts, proximally often paler, usually yellow throughout, 10–20 mm; abaxial and adaxial clefts 7–14 mm, 40–50% of calyx length, deeper than laterals, lateral 0.5–1.5 mm, 5–15% of calyx length; lobes triangular, apex acute to rounded. **Corollas** slightly curved, 10–22 mm; tube 8–14 mm; teeth of abaxial lip sometimes slightly exserted, beak exserted or subequal; beak adaxially green to pale yellow, 5–8 mm; abaxial lip deep yellow or green, reduced, forming a small pouch visible through front cleft, 3–5 mm, 50–75% as long as beak; teeth ascending to erect, white or yellow, 1 mm. $2n = 24$.

Flowering May–Aug(–Sep). Dry to mesic rocky tundra, slopes, ridges, barrens and meadows, openings in boreal forests, arctic to alpine; 0–2200 m; B.C., N.W.T., Yukon; Alaska; e Asia (Russian Far East).

Castilleja hyperborea is widespread across boreal, alpine, and arctic habitats in western Canada and Alaska, extending into the Russian Far East. A rare form with light purple bracts was named as *C. kuschei*. Another form with particularly long hairs found in the Kluane Lake region of the southern Yukon was described as *C. villosissima* and is usually found on calcareous substrates.

45. **Castilleja indivisa** Engelmann, Boston J. Nat. Hist. 5: 255. 1845 • Texas paintbrush

Herbs, annual, (0.5–)1–4.5 dm; with a taproot or branched root system. **Stems** solitary or few, erect, unbranched or 1–4 branches from proximal ½, hairs spreading, long, soft, often mixed with shorter stipitate-glandular ones. **Leaves** green, not forming a distinct basal rosette, sometimes relatively dense proximally with short internodes, similar in size and shape to more distal cauline leaves, narrowly lanceolate to linear-lanceolate, (1.5–)2–8(–9) cm, not fleshy, margins wavy, sometimes plane, involute, 0(–5)-lobed, apex acute; lobes erect, linear or filiform to narrowly lanceolate, apex acute to acuminate. **Inflorescences** 2–16(–20) × 3–5.5 cm; bracts proximally greenish, distally scarlet or bright red, sometimes deep to pale pink, peach, yellow, white, or magenta, sometimes with a white to rarely yellow medial band between green and brightly colored distal portion, proximal narrowly lanceolate, distal shorter and oblong-obovate, broadly obovate, or obtrullate, 0(–5)-lobed; lobes erect or ascending, triangular, short, arising above mid length, proximal bract apex acute, distal obtuse, rounded, or truncate. **Calyces** proximally light green, distally red, pale pink, or white, rarely pale yellow, usually paler than bracts, often with a white to yellow medial band between green and brightly colored distal portion, 16–31 mm; abaxial and adaxial clefts 6–10 mm, 25–33% of calyx length, deeper than laterals, lateral 0(–0.2) mm, 0(–4)% of calyx length; lobes expanded distally, apices much wider than narrow calyx tube, apex rounded, truncate, or emarginate. **Corollas** curved in proximal ⅓, 15–29 mm; tube 2–3.5 mm; whole corolla included or beak partly exserted, abaxial lip included; beak adaxially green, yellow, or pink, 4–10 mm; abaxial lip green, white, or yellow, reduced, pouches 3, 2 mm, 25–30% as long as beak; teeth erect, green, white, or yellow, 0.5 mm. *2n* = 24.

Flowering Jan–Jun(–Dec). Grasslands, pastures, dunes, oak savannas, limestone glades, open woodlands, roadsides, often in sand or clay; 0–400 m; Ala., Ark., Fla., La., Okla., Tex.; Mexico (Aguascalientes, Chihuahua).

Castilleja indivisa is native in Texas and adjacent states. In Mexico it is rare, with collections only from two states; these are likely waifs. This species is possibly extirpated from Arkansas. Records from Alabama (starting in 1995) and Florida (starting in 1961) are adventive populations, often on roadsides, and in some cases spreading from ornamental highway plantings. *Castilleja indivisa* usually has bright red bract apices and red, white, or pale pink calyx apices, but many color variants are found in nature and in cultivation, including individuals with the distal portion of the bracts colored white, pink, pale yellow, peach, or, very rarely, magenta. Uniformly white-bracted populations occur on the margins of tidal salt marshes in a small area of Nueces County, Texas, between Aransas Pass and Port Aransas. These populations likely deserve nomenclatural recognition, due to their combination of consistent coloration and unique habitat. While the main bloom period is in the spring, summer rains often allow continuing or renewed flowering during virtually any month of the year. Occasionally, plants show variation in leaf lobing; this likely reflects introgression from the *C. purpurea* complex, at least in some cases, such as in Coleman and McCullough counties, Texas.

46. **Castilleja integra** A. Gray in W. H. Emory, Rep. U.S. Mex. Bound. 2(1): 119. 1859 • Entire-leaved or southwestern paintbrush

Castilleja elongata Pennell; *C. gloriosa* Britton; *C. integra* var. *gloriosa* (Britton) Cockerell

Herbs, sometimes subshrubs, perennial, 0.9–5(–10) dm; from a woody caudex; with a taproot or stout, branched roots. **Stems** solitary or few to several, erect to ascending, less commonly bent at base, unbranched, sometimes branched distally, hairs spreading to appressed, fairly short, soft and moderately dense, matted, unbranched, not quite obscuring surface. **Leaves** green to purplish, linear to lanceolate or narrowly oblong, (1–)2–7(–9) cm, not fleshy, margins plane, sometimes ± wavy, involute, 0-lobed distally, sometimes 3-lobed, apex acute to acuminate, sometimes rounded. **Inflorescences** 2–12(–15) × 1.5–4 cm; bracts red to red-orange or orange throughout, sometimes crimson, cerise, pale salmon, or pale yellow throughout, or proximally pale green to straw colored, distally colored as above, proximal sometimes narrowly lanceolate, others elliptic to narrowly elliptic, oblong, obovate, or oblanceolate, 0–3(–5)-lobed; lobes ascending, lanceolate, short, arising at or above mid length, central lobe apex obtuse to rounded, lateral ones acute. **Calyces** colored as bracts, (18–)21–35(–38) mm; abaxial and adaxial clefts (6–)9–16(–18) mm, 25–33% of calyx length, deeper than laterals, lateral (2–)4–14(–16) mm, 10–15% of calyx length; lobes lanceolate or triangular, apex acute to rounded. **Corollas** straight or slightly curved, (21–)25–45(–50) mm; tube 17–30(–33) mm; beak subequal to calyx or strongly exserted, adaxially green, (8–)10–17(–18) mm; abaxial lip deep green, reduced, usually visible in front cleft, 1–2.8 mm, 20% as long as beak; teeth incurved, green or yellow, 0.5–1.5 mm. $2n = 24, 48$.

Flowering (Jan–)Mar–Oct. Dry rocky slopes and flats, grasslands, open forests, ledges, road banks, valleys, subalpine; (600–)1000–3300 m; Ariz., Colo., N.Mex., Tex.; Mexico (Chihuahua, Coahuila, Durango, Nuevo León, Sonora).

Castilleja integra is widespread and common in the southwestern United States. It is important to Native Americans for dyes, ceremonies, as a food preservative, and as a medicine. It is sometimes confused with *C. lanata* or *C. miniata* but has a distinctive combination of entire, narrow, strongly involute leaves, soft-tomentose pubescence of unbranched hairs, and usually entire bracts, sometimes with one pair of short lobes from the middle. The leaf margins of *C. integra* are usually plane, but some populations in the Chisos Mountains of Texas are wavy margined. These plants are on the higher slopes of the Chisos Mountains in montane thorn-oak vegetation, and they are also often taller and have longer, more frequently lobed leaves. These variant populations have been called *C. elongata*, and they deserve further study. *Castilleja integra* is typically a species of dry grasslands and open forests at moderate elevations. Occasional hybrids with *C. linariifolia* are known from Montrose County, Colorado.

The *Castilleja elongata* form of *C. integra* is in the Center for Plant Conservation's National Collection of Endangered Plants.

47. **Castilleja kaibabensis** N. H. Holmgren, Bull. Torrey Bot. Club 100: 89, fig. 4. 1973 • Kaibab paintbrush [C] [E]

Herbs, perennial, 1.5–3(–4) dm; from a woody caudex; with a taproot. **Stems** solitary or few, sometimes several, erect or ascending, unbranched, rarely branched from base or from proximal inflorescence nodes, hairs spreading or reflexed, medium length, ± stiff, distally mixed with scattered, very short stipitate-glandular ones in inflorescences. **Leaves** green, often with a dull red-purplish cast, linear to narrowly lanceolate or narrowly oblong, (0.8–)1.5–4 cm, not fleshy, margins plane, involute, 0(–3)-lobed, apex acuminate to acute; lobes spreading-ascending, narrowly lanceolate, apex acute. **Inflorescences** 2–6.5(–9) × 1.5–3.5 cm; bracts yellow,

pale orange, salmon, pink, dull brick red, or reddish orange throughout, rarely red, or proximally pale greenish, distally colored as above, narrowly lanceolate to lanceolate, 0–3(–7)-lobed; lobes ascending to erect, linear to lanceolate, short to medium length, arising at or above mid length, central lobe apex obtuse, lateral ones obtuse to acute. **Calyces** colored as bracts, 20–27(–30) mm; abaxial and adaxial clefts 6.5–13(–18) mm, 40–60% of calyx length, deeper than laterals, lateral (2–)4–7.5(–10) mm, 25% of calyx length; lobes linear to narrowly lanceolate, apex rounded to acute. **Corollas** straight or slightly curved, 21–30(–35) mm; tube 13–19 mm; beak subequal to calyx or exserted, adaxially green, 8–14 mm; abaxial lip whitish to pink with deep green teeth, reduced, not exserted, sometimes visible through front calyx cleft, 0.5–2 mm, 5–15% as long as beak; teeth incurved, dark green, 0.2–1 mm. $2n = 24$.

Flowering Jul–Sep. Dry or exposed sites in subalpine meadows, low ridges and crests, openings in spruce-fir-aspen forests, fine silts and clay over limestone; of conservation concern; 2400–2800 m; Ariz.

Castilleja kaibabensis is endemic to a few large meadow systems on the Kaibab Plateau, north of the North Rim of the Grand Canyon in Coconino County. Recreational activity, grazing, and road construction remain concerns for the management of this species.

Castilleja kaibabensis is in the Center for Plant Conservation's National Collection of Endangered Plants.

48. **Castilleja kerryana** J. M. Egger, Phytoneuron 2013-21: 2, figs. 1–6, 10. 2013 • Kerry's paintbrush [C] [E]

Herbs, perennial, 0.5–19 dm; from a branching, woody caudex; with a taproot. **Stems** few to many, short-decumbent, becoming upright-ascending, unbranched, hairs moderately dense, spreading to erect, medium length and long, soft, mixed with shorter stipitate-glandular ones. **Leaves** greenish to dull purplish brown, linear to linear-lanceolate or narrowly lanceolate or narrowly oblong, 1–7 cm, not fleshy, margins plane, involute, (0–)3-lobed, apex acute, distalmost obtuse to rounded; lobes erect, linear-lanceolate to narrowly long-triangular, usually shorter and narrower than central lobe, usually arising at or above mid length, apex acute to rounded. **Inflorescences** 2.5–8 × 2–5(–8) cm at midflowering; bracts red to red-orange, crimson, or magenta throughout, sometimes green, pale orange, rose red, salmon, or yellow throughout, or proximally greenish or dull reddish, distally colored as above, lanceolate to elliptic or oblong, 3(–5)-lobed; lobes ascending to erect, linear to lanceolate or narrowly oblong, short or medium length, arising near or above mid length, central lobe apex rounded to acute, lateral ones acute to obtuse. **Calyces** colored as bracts, 24–44 mm; abaxial and adaxial clefts 9–18 mm, 30–45% of calyx length, deeper than laterals, lateral 1–4 mm, 5–10% of calyx length; lobes lanceolate to ovate, apex rounded to acute. **Corollas** straight or slightly curved, 24–45(–50) mm; tube 22–38 mm; beak and abaxial lips conspicuously exserted; beak adaxially green to yellow, 8–11 mm; abaxial lip ascending, green, sometimes yellowish or black, slightly inflated, visible through front cleft and subequal to calyx, 4–6.5 mm, 33–50% as long as beak; teeth ascending, pink, white, or rose red, 1–4 mm.

Flowering Jul–Aug(–Sep). Moist to dry, rocky or gravelly slopes and ridges, scree, talus, krummholz, turf or fellfields, snowfield or meltwater margins, subalpine to alpine, Cambrian limestone; of conservation concern; 2100–2900 m; Mont.

Castilleja kerryana is endemic to alpine peaks and ridges around the Scapegoat Plateau of Lewis and Clark County, in the eastern Rocky Mountains of central-western Montana. It is closely associated with and probably parasitic on *Dryas hookeriana*.

49. **Castilleja kraliana** J. R. Allison, Castanea 66: 159, fig. 3. 2001 • Kral's paintbrush [C][E]

Herbs, annual or biennial, (1.6–)2–4(–4.9) dm; with fibrous roots. **Stems** solitary or few, erect, from a basal rosette that usually persists into anthesis, unbranched or few-branched from mid stem, glabrate proximally, hairy distally, hairs moderately dense, spreading, long, soft, mixed with short stipitate-glandular ones. **Leaves:** basal forming a persistent rosette, often withered at anthesis, usually suffused to varying degrees with red, sometimes green, oblanceolate to oblong-oblanceolate, 2–6 cm, not fleshy, margins plane, flat, 0-lobed, apex obtuse to broadly rounded to sometimes acute or subtruncate; cauline green, sometimes proximal suffused with red or purple, linear to linear-lanceolate or linear-oblanceolate, 1.1–6.5 cm, not fleshy, margins plane, flat, 0–3(–5)-lobed, apex usually acute, rarely obtuse; lobes erect to ascending, narrowly oblong to linear or narrowly lanceolate, to 1.8 cm, apex acute. **Inflorescences** (2–)4–24(–33) × 2–4 cm; bracts proximally greenish, distally bright yellow, rarely light orange, proximal narrowly lanceolate, leaflike, distal oblong to broadly rounded-oblanceolate, continuing to grow after anthesis and becoming greener and more leaflike, lateral lobes, if present, growing even more than central lobe, 0–5-lobed; lobes erect, triangular or lanceolate, sometimes linear, short or medium length, arising above mid length, proximal bract apex acute,

distal rounded, rarely central lobe shallowly toothed. **Calyces** proximally whitish or pale green, often aging purple, distally yellow, (13–)15–18 mm; abaxial and adaxial clefts 4–7 mm, 40–50% of calyx length, deeper than laterals, lateral 0–0.2 mm, ca. 0% of calyx length; lobes curved, oblong, apex rounded to truncate, sometimes emarginate. **Corollas** straight or slightly curved, (15–)17–23 mm; tube 8–11 mm; abaxial lip rarely exserted, beak exserted; beak adaxially green or yellow-green, 5–7 mm, margins yellow or yellow-green; abaxial lip ascending, deep green, slightly inflated, 2–3 mm, 20–30% as long as beak, apex rounded; teeth ascending, green, whitish, or yellow, 0.5–2 mm.

Flowering (Mar–)Apr–Jun(–Jul). Full sun or partial shade, outcrops and glades, forest edges, river bluffs, Ketona dolomite; of conservation concern; 90–150 m; Ala.

Castilleja kraliana is restricted to dolomite glades in Bibb County, which is a local center of endemism near the Cahaba River. *Castilleja kraliana* has mostly entire, bright yellow bracts and calyces 13–18 mm, and differs from the related *C. coccinea*, which has deeply lobed and red, rarely yellow, bracts and calyces 17–28 mm.

50. **Castilleja lacera** (Bentham) T. I. Chuang & Heckard, Syst. Bot. 16: 657. 1991 • Cut-leaved paintbrush [E]

Orthocarpus lacerus Bentham, Pl. Hartw., 329. 1849

Herbs, annual, 0.5–4 dm; with fibrous roots. **Stems** solitary, erect, unbranched or branched, hairs spreading, long, soft, scattered among more numerous, medium length, stipitate-glandular ones. **Leaves** green or purplish, linear to narrowly lanceolate, 1–5 cm, not fleshy, margins plane, flat, 0–5(–7)-lobed, apex acuminate; lobes spreading to ascending, linear, apex acuminate to acute. **Inflorescences** (1.5–)3–14 × 2–3 cm; bracts green throughout, sometimes proximally green, distally white on apices, lanceolate to ovate, 3–7-lobed; lobes spreading to ascending, linear to narrowly lanceolate, long, arising below mid length, apex obtuse to acute. **Calyces** light green, lobes green, 7–13 mm; abaxial and adaxial clefts 3.5–8 mm, 50–67% of calyx length, lateral 2.5–5 mm, ca. 40% of calyx length; lobes narrowly to broadly lanceolate, apex acute to acuminate. **Corollas** straight, 10–22 mm; tube 8–15 mm; abaxial lip and beak exserted; beak adaxially yellow to greenish, 3–6 mm, densely puberulent; abaxial lip yellow with purple dots at base, inflated, pouches 3, central pouch slightly 2-lobed, pouches 4–8 mm wide, 3–6 mm deep, side pouches curving up a little at tip, 2–5 mm, 75–95% as long as beak; teeth erect, white or yellow, 0.5–2 mm. **Stigmas** equal to or slightly exserted from beak. $2n = 22, 24$.

Flowering (Mar–)Apr–Jul(–Aug). Grasslands, meadows, moist flats, vernal pool margins, moist forest openings, serpentine slopes and ledges, roadsides; 0–2700 m; Calif., Oreg.

Castilleja lacera is found in a wide range of elevations in the central and northern Sierra Nevada region and in the Siskiyou Mountains region of northwestern California and southwestern Oregon. Reports from the Coast Ranges north of the San Francisco Bay region and south of the Siskiyou region in western California are referable to other yellow-flowered annuals, including *C. ambigua*, *C. rubicundula* var. *lithospermoides*, *Triphysaria eriantha* subsp. *eriantha*, and *T. versicolor* subsp. *faucibarbata*. Although most similar to *C. rubicundula*, *C. lacera* is somewhat smaller in stature and flower size. It is also easily confused with yellow-flowered populations of *C. tenuis*, which has smaller flowers and an included stigma. Two chromosome numbers are known for this species, the more northern populations being diploid, and those to the south having an apparently aneuploid count of $2n = 22$, which is unique in the genus.

51. **Castilleja lanata** A. Gray in W. H. Emory, Rep. U.S. Mex. Bound. 2(1): 118. 1859 • Woolly paintbrush [F]

Subshrubs or herbs, perennial, 1.8–9(–10) dm; from a woody caudex; with a thick, woody taproot or system to thick, branched roots. **Stems** few to many, erect to ascending, unbranched, sometimes branched distally, sometimes with leafy axillary shoots, lanate to woolly, hairs dense, spreading to weakly appressed, white to yellowish, medium length, soft, moderately branched or unbranched in different populations, sometimes mixed, eglandular, rarely short stipitate-glandular, mostly obscuring surface. **Leaves** green to purple, often hidden by matted, white hairs, linear to narrowly oblong or narrowly lanceolate, (1–)2–7(–10) cm, not fleshy, margins plane, flat, sometimes involute, 0–3-lobed, apex rounded, sometimes acute; lobes spreading, linear or narrowly oblong, apex rounded to acute. **Inflorescences** 2–19 × 1.5–5 cm; bracts proximally pale greenish to greenish gray, distally bright red to orange-red, sometimes pinkish, magenta, or salmon, rarely yellow, oblong to narrowly lanceolate or narrowly oblanceolate, deeply 0–3(–5)-lobed; lobes spreading, oblanceolate or linear, long, arising below mid length, distal bract apex usually rounded. **Calyces** proximally whitish to green, central band whitish, distal ⅓ colored as bract lobes, sometimes lighter or deeper, 15–29 mm; abaxial and adaxial clefts 6–14 mm, ca. 50% of calyx length, deeper than laterals, lateral 0–5 mm, 0–5(–15)% of calyx length; lobes oblong to broadly triangular, apex rounded, truncate, or emarginate. **Corollas** curved in proximal ⅓, 23–35 (–42) mm; tube 12–17 mm; beak exserted, adaxially greenish to yellowish, 11–22 mm; abaxial lip dark green, reduced, 1–4 mm, ca. 20% as long as beak; teeth incurved, reduced, green, 0.5–2.5 mm. $2n = 24$.

Flowering year-round. Dry rocky slopes, ridges, canyons, flats, valleys, montane; 300–2300 m; Ariz., N.Mex., Tex.; Mexico (Baja California, Chihuahua, Coahuila, Durango, Nuevo León, San Luis Potosí, Sonora, Tamaulipas, Zacatecas).

Most records of *Castilleja lanata* are from the southern third of Arizona, New Mexico, and southwestern Texas. Plants sometimes referred to *C. lanata* from the Channel Islands of California are assigned here to *C. hololeuca*, an insular endemic. In northern Mexico, *C. lanata* extends from San Luis Potosí and Nuevo León westward to Sonora, with a disjunct distribution in central Baja California, where it overlaps the southern limits of a related species, *C. foliolosa*. *Castilleja lanata* inhabits a wide spectrum of soils and elevations and does not co-occur with the closely related *C. galehintoniae* in Nuevo León, where the latter is limited to limestone and gypsum substrates in a narrow elevational range. Hybrids of *C. lanata* with *C. sessiliflora* are known from Pecos County, Texas.

52. **Castilleja lasiorhyncha** (A. Gray) T. I. Chuang & Heckard, Syst. Bot. 16: 657. 1991 • San Bernardino Mountains paintbrush [C] [E]

Orthocarpus lasiorhynchus A. Gray, Proc. Amer. Acad. Arts 12: 82. 1876

Herbs, annual, (0.6–)1–3(–4) dm; with fibrous roots. **Stems** solitary, erect, unbranched or branched, hairs spreading, medium length and long, soft to ± stiff, mixed with shorter stipitate-glandular ones. **Leaves** green to purple, linear to narrowly lanceolate, 0.5–4.2 cm, not fleshy, margins plane, flat, 0(–5)-lobed, apex acuminate to acute; lobes ascending to erect, linear to narrowly lanceolate, apex acuminate. **Inflorescences** 1.5–15 × 1.5–3 cm; bracts green throughout, or proximally green, distally white to rarely cream on apices, with a tuft of erect, white, soft hairs, especially when immature, narrowly lanceolate to narrowly ovate, 3–5-lobed; lobes ascending, linear to lanceolate, long, proximal lobes arising near base, apex obtuse to acute. **Calyces** light green, lobes deep green, sometimes purple, 5.5–12 mm; abaxial and adaxial clefts 2.5–6.5 mm, 30–50% of calyx length, lateral 2–4.5 mm, 30–40% of calyx length; lobes linear to narrowly lanceolate or triangular, apex acuminate or acute. **Corollas** straight or slightly curved distally, 12–25 mm; tube 18 mm; abaxial lip and beak exserted; beak adaxially white or pale yellow, 3.5–9 mm, hairs dense, spreading, medium length, obscuring surface; abaxial lip yellow, inflated, abruptly expanded, obpyramidal, pouches 3, central pouch slightly 2-lobed, 4–5 mm deep, 3–8 mm, 75–90% as long as beak; teeth erect, whitish to pale yellow, 1–2 mm. $2n = 24$.

Flowering Jun–Jul. Springs, moist or wet meadows, flats, open forests; of conservation concern; 1000–2500 m; Calif.

Most populations of *Castilleja lasiorhyncha* are in the San Bernardino Mountains, with a few records in the adjacent Peninsular Ranges immediately to the south. The distal tufts of soft, pale hairs on the immature bracts are apparently unique in the genus.

53. Castilleja lassenensis Eastwood, Leafl. W. Bot. 2: 244. 1940 • Mt. Lassen paintbrush E

Herbs, perennial, 0.8–2.1 dm; from a woody caudex; with slender, branching roots. **Stems** few to many, decumbent-based to erect, unbranched except for short, leafy axillary shoots, hairs sparse, spreading, long, soft, mixed with short stipitate-glandular ones. **Leaves** green to purple (sometimes different on stems of same plant), linear-lanceolate, 1–3.5 cm, not fleshy, margins plane, flat, 0(–3)-lobed, apex acute; lobes spreading-ascending, linear-lanceolate to lanceolate, apex acute. **Inflorescences** 2–4.5 × 1.5–2 cm; bracts proximally greenish to dull brownish purple, distally pink to magenta or reddish purple, ovate, broadly lanceolate, or oblong, (0–)3–5-lobed; lobes ascending to erect, lanceolate, medium length, arising near or above mid length, apex acute to rounded. **Calyces** proximally brown or dull magenta, sometimes green, distally colored as bracts, 11–19 mm; abaxial and adaxial clefts 9–10 mm, 55–60% of calyx length, deeper than laterals, lateral 2.5–5 mm, 15–25% of calyx length; lobes oblong, apex rounded. **Corollas** straight, 18–20 mm; tube 14–15 mm; abaxial lip sometimes partly exserted, beak usually exserted; beak adaxially white, 7–10 mm, margins white or pale salmon; abaxial lip pale green to yellow-green, inflated, pouches 3, shallow, central pouch shallowly grooved, visible through front cleft, 6–7 mm, 60–70% as long as beak; teeth erect, white, 1–2 mm. **Stigmas** pale green. $2n = 24$.

Flowering Jul–Sep. Moist to wet meadows, shorelines, subalpine, volcanic soils; 1500–2500 m; Calif.

Castilleja lassenensis is restricted to the Mt. Lassen area in Shasta and Tehama counties. See the discussion of 55. *C. lemmonii* for information on the relationship between these two species.

54. Castilleja latifolia Hooker & Arnott, Bot. Beechey Voy., 154. 1833 • Monterey paintbrush E

Herbs or subshrubs, perennial, 2–6 dm; from a woody caudex and sometimes from a woody proximal stem; with a taproot. **Stems** many, spreading to erect, much-branched, with numerous short, leafy axillary shoots, hairs moderately dense, spreading, medium length to long, stiff to soft, shorter stipitate-glandular ones. **Leaves** gray-green becoming ± purple to sometimes green as hairs are lost, oblong to lanceolate-oblong or broadly lanceolate, 0.5–2 cm, ± fleshy, cupulate, sometimes obscurely so on distal portion of stem, margins plane, sometimes ± wavy, involute, 0–3-lobed, apex truncate or broadly rounded to obtuse; lobes erect to ascending, oblong, apex rounded. **Inflorescences** 2.5–20 × 1.5–5 cm; bracts proximally green to dull, deep brownish purple, distally bright red, red-orange, or orange, sometimes yellow to yellow-orange, oblong or broadly lanceolate to widely obovate or ovate, often cup-shaped, center lobe often expanded distally, 0–3(–5)-lobed, often wavy-margined; lobes ascending, oblong, short or long, arising near or above mid length, central lobe apex mostly rounded to truncate, sometimes with 5 or so very shallow teeth. **Calyces** proximally light green to sometimes purple, distally colored as bracts, 15–25 mm; abaxial and adaxial clefts 6–9.5 mm, 33–50% of calyx length, deeper than laterals, lateral 1–3 mm, ca. 12% of calyx length; lobes broadly triangular to oblong, apex rounded to obtuse, rarely acute. **Corollas** slightly curved, 19–30 mm; tube 8.5–15 mm; beak exserted, adaxially green, 8.5–15 mm; abaxial lip ascending, deep green, reduced, 0.5–1 mm, 5–10% as long as beak; teeth incurved, reduced, green or white, 0.2–0.5 mm. $2n = 24$.

Flowering Feb–Oct. Coastal dunes and scrub, chaparral, grasslands, sandy bluffs; 0–500 m; Calif.

Castilleja latifolia is endemic to the central California coast, especially around Monterey Bay. Around Half Moon Bay in San Mateo County, it apparently forms hybrids with *C. affinis* var. *affinis*. Records of this species from north of San Francisco and south of Monterey County are referable to other species.

55. Castilleja lemmonii A. Gray in A. Gray et al., Syn. Fl. N. Amer. 2(1): 297. 1878 (as Castilleia lemmoni) • Lemmon's paintbrush E

Castilleja culbertsonii Greene

Herbs, perennial, 0.8–2.5 dm; from a woody caudex; with slender, branching roots. **Stems** few to many, decumbent-based to erect, unbranched except for short, leafy axillary shoots, hairs sparse, spreading, medium length to long, soft and dense, short to medium length, stipitate-glandular. **Leaves** green or gray-green to purple (sometimes different on stems of same plant), linear-lanceolate, distal sometimes broadly lanceolate, 0.5–4 cm, not fleshy, margins plane, involute, 0(–3)-lobed, apex acute to acuminate; lobes ascending, linear to narrowly lanceolate, apex acute to acuminate, rarely obtuse. **Inflorescences** 2–12 × 1–3 cm; bracts greenish to dull purplish or brownish throughout, or proximally greenish to dull purplish, distally pink to purple or magenta, rarely white, ovate, broadly lanceolate, or oblong, (0–)3–5-lobed; lobes ascending to erect, lanceolate, medium length, arising above mid length, apex acute to rounded. **Calyces** proximally brown or dull magenta, sometimes green, distally colored as bracts, 12.5–18 mm; abaxial and adaxial clefts 5.5–10.5 mm, 40–65% of calyx length, deeper than laterals, lateral 0.5–2 mm, 5–15% of calyx length; lobes oblong, apex rounded. **Corollas** slightly curved, 16–21 mm; tube 10–16 mm; abaxial lip sometimes partly exserted, beak usually exserted; beak adaxially green, 6–7 mm, margins red; abaxial lip greenish, inflated, pouches 3, shallow, central pouch shallowly grooved, visible through front cleft, 3–4 mm, 60% as long as beak; teeth erect, violet-purple or pink, 1–2.5 mm. **Stigmas** greenish to deep bluish purple. $2n = 24$.

Flowering Jun–Aug. Moist to wet meadows and flats, shorelines, open conifer forests, subalpine and alpine, often over granite; 1500–3700 m; Calif., Nev.

As delimited here, *Castilleja lemmonii* is endemic to the highlands of the Sierra Nevada in California and in adjacent Washoe County, Nevada. It differs from *C. lassenensis*, a plant of volcanic highlands around Mt. Lassen, which has consistently white corollas. Corollas are usually pink to purplish in *C. lemmonii*. *Castilleja lemmonii* also tends to have somewhat shorter lateral calyx clefts, though the two species overlap slightly in this character.

56. Castilleja leschkeana J. T. Howell, Leafl. W. Bot. 5: 91. 1948 • Leschke's paintbrush C E

Herbs, perennial, 10 dm; from a woody caudex; with a taproot. **Stems** several, erect, decumbent at base, unbranched proximally, branched distally, glabrous proximally, hairy distally, hairs sparse, short, stiff. **Leaves** green, proximal linear, reduced, middle and distal lanceolate to narrowly elliptic or ovate, proximal 0.5–1.5 cm, middle 6–7 cm (0.8–1.5 cm wide), distal 4–6 cm (2.5–2.3 cm wide), not fleshy, margins plane to sometimes ± wavy, ± involute, 0–3-lobed, apex acute; lobes ascending, lanceolate, apex acute or acuminate. **Inflorescences** 10–14+ × 6 cm; bracts proximally dull greenish to dull brownish, distally pale orange to reddish, cuneate to obovate-truncate, 3–5(–9)-lobed, with white, stiff hairs mostly along veins; lobes ascending to erect, broadly lanceolate or oblong, medium length, arising at or above mid length, center lobe apex rounded, sometimes toothed, others acute to sometimes rounded. **Calyces** distally reddish, 20–28 mm; abaxial and adaxial clefts 12–15 mm, 50–60% of calyx length, deeper than laterals, lateral (0–)2–6 mm, 0–25% of calyx length; lobes lanceolate, apex acute. **Corollas** straight, 25–30 mm; tube 12–15 mm; beak subequal or ± exserted, adaxially green, 12–15 mm; abaxial lip deep green, reduced, strongly saccate-corrugated, 1.5 mm, 10% as long as beak; teeth incurved, reduced, green, 1 mm.

Flowering Jun. Dune swales, swampy ground, near coast; of conservation concern; 0–100 m; Calif.

Castilleja leschkeana was based on a 1947 specimen from a swale near the radio station on Point Reyes Peninsula in Marin County. It was found again in 1960, but the species has not been seen or collected since then. The type specimen of *C. leschkeana* was identified as the Alaskan species *C. chrymactis* by T. I. Chuang and L. R. Heckard (1993), who considered the California population to be an accidental introduction. However, the flowers of *C. leschkeana* are shorter than those of *C. chrymactis*, with a shorter beak that is scarcely exserted. The bract blade and bract lobes of *C. leschkeana* are shorter, and the pubescence of its bracts is very different, short-hispid and more prominent along the veins than on the blades, contrasting with the longer and soft-villous pubescence of *C. chrymactis*. M. Wetherwax et al. (2012) agreed that the Point Reyes specimen was misidentified as *C. chrymactis*, but they were hesitant to place it in any accepted California taxon. *Castilleja leschkeana* is here recognized as another narrow endemic, worthy of conservation concern if it is relocated.

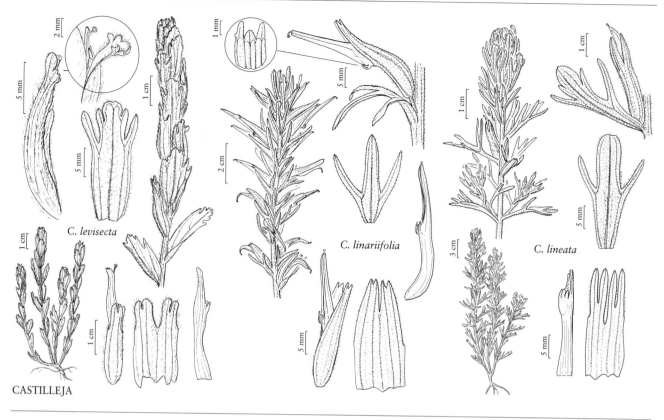

C. *levisecta*

CASTILLEJA

C. *linariifolia*

C. *lineata*

57. Castilleja levisecta Greenman, Bot. Gaz. 25: 268. 1898 (as Castilleia) • Golden paintbrush C E F

Herbs, perennial, 1–5 dm; from a woody caudex; with a taproot. **Stems** few to many, erect, ± decumbent or creeping at base, unbranched, sometimes branched, hairs spreading, medium length and long, soft, mixed with short stipitate-glandular ones. **Leaves** green to purple or brown-tinged, linear-lanceolate proximally, oblong-ovate or -obovate distally, 0.8–5.2 cm, not fleshy, margins plane, distalmost sometimes ± wavy, involute, 3–7(–11)-lobed, apex obtuse; lobes erect to ascending, linear to linear-spatulate, very short, toothlike, usually arising from distal ⅓ of blade, apex rounded. **Inflorescences** 2.5–25 × 1–4 cm; bracts bright yellow throughout, or proximally greenish, distally bright yellow, sometimes deep yellow-orange, especially with age, oblong, elliptic, or obtuse to ovate, (0–)5–9(–13)-lobed, sometimes wavy-margined; lobes erect to ascending, oblong, short to medium length, arising above mid length, central lobe apex rounded, lateral ones rounded to acute. **Calyces** distally yellow, 13–22 mm; abaxial and adaxial clefts 4–9.5 mm, 30–40% of calyx length, deeper than laterals, lateral 2.5–4.5 mm, ca. 25% of calyx length; lobes linear to narrowly oblong or narrowly lanceolate, apex obtuse, sometimes rounded to acute. **Corollas** straight or slightly curved, 17–28 mm; tube 12–15 mm; beak exserted, adaxially green or greenish yellow, 6–8 mm; abaxial lip yellow or greenish, reduced, not inflated, 2–3 mm, 25–33(–50)% as long as beak; teeth ascending to erect, yellow, 0.5–1.5 mm. $2n = 24$.

Flowering Apr–Jul(–Sep). Dry meadows, upland prairie remnants, sunny headlands and coastal bluffs, rocky islands, often over glacial outwash or deposits; of conservation concern; 0–400 m; B.C.; Oreg., Wash.

Castilleja levisecta is listed as threatened in the United States and endangered in Canada, where it is extremely rare. Most of its grassland habitat has been altered by development in the Puget Trough, and there are historical stations in the metro areas of what are now Victoria, Portland, and Seattle. For several decades, *C. levisecta* was considered extirpated from Oregon. However, recent reintroduction programs in Oregon and Washington have been very successful at reestablishing this species at several sites in the Willamette Valley. The bright yellow inflorescences often gradually age to a golden yellow color, unique in the genus.

Castilleja levisecta is in the Center for Plant Conservation's National Collection of Endangered Plants.

58. Castilleja linariifolia Bentham in A. P. de Candolle and A. L. P. P. de Candolle, Prodr. 10: 532. 1846 (as lineariaefolia) • Wyoming or narrow-leaved or long-leaf paintbrush E F

Castilleja trainii Edwin

Herbs, perennial, 1.8–10(–20) dm; from a woody caudex; with a taproot or branching roots. Stems few to many, ascending to erect, much-branched, glabrous proximally, hairy distally, sometimes glabrous or hairy throughout, hairs sparse, sometimes dense, spreading to slightly retrorse, short to long, ± stiff, eglandular. Leaves ± yellow to gray-green, sometimes becoming ± purple, linear to narrowly lanceolate, (1–)2–10 cm, not fleshy, rarely ± thickened and fleshy, margins plane, involute, 0–3(–5)-lobed, apex acute or acuminate; lobes spreading, linear, apex acute to acuminate. Inflorescences 4–20 × 2–7.5 cm; bracts red to red-orange throughout, sometimes pale green, yellow, magenta, pink-purple, or white throughout, or proximally pale greenish to straw colored, distally colored as above, lanceolate to narrowly lanceolate or narrowly oblong, 3(–5)-lobed; lobes spreading to ascending, narrowly lanceolate to linear, shorter than central lobe, arising in proximal ⅓, apex obtuse to acuminate. Calyces proximally greenish, whitish, or yellowish, distally colored as bracts, 18–30(–35) mm; abaxial clefts 10–20(–22) mm, adaxial 2–6(–12) mm, abaxial ca. 70% of calyx length, adaxial ca. 20–25% of calyx length, deeper than laterals, lateral 1.5–5 (–6) mm, 12–17% of calyx length; lobes curved slightly toward adaxial side, narrowly oblong to narrowly lanceolate, apex acute. Corollas often slightly curved, 25–45 mm; tube 11–22(–25) mm; beak exserted, longer than calyx lobes, usually projecting through abaxial cleft; beak adaxially yellow-green or green, 9–21 (–24) mm; abaxial lip deep green, reduced, often visible in exserted sideways through abaxial calyx cleft, 0.5–3 mm, 10–15% as long as beak; teeth incurved, green or whitish, 5–2(–3) mm. *2n* = 24, 48.

Flowering Apr–Oct. Sagebrush steppes, grasslands, dry rocky slopes and flats, open forests, talus, lowlands to montane, occasionally subalpine; 600–3400 m; Ariz., Calif., Colo., Idaho, Mont., Nev., N.Mex., Oreg., Utah, Wyo.

Castilleja linariifolia is widespread in the western United States and is important to Native Americans as a source of dyes and for medicinal and ceremonial purposes (D. E. Moerman 1998). It is the state flower of Wyoming. *Castilleja linariifolia* is closely associated with, and undoubtedly parasitic on, the roots of sagebrush, *Artemisia tridentata*. The bracts are usually red to red-orange, with uncommon yellow-bracted variants. On the eastern slope of the Sierra Nevada in California, most populations have pink-purple to magenta bracts, and at least one population has almost pure white bracts. Plants associated with hot springs in the Ash Meadows area of Nye County, Nevada, are unusually tall and have somewhat fleshy leaves. *Castilleja linariifolia* stems are usually glabrous, but pubescent stems are sporadic across its range. However, in the Mt. Charleston area of the Spring Mountains, in southern Nevada, most plants are pubescent, ranging from an inconspicuous layer to a fairly dense and obvious indument. These plants have been called forma *omnipubescens* Pennell. Elsewhere, especially in central and northern Arizona, individuals with short-pubescent stems are intermingled with more typical plants. In addition, late-blooming forms associated with hot spring sites in the eastern Mojave Desert are particularly thick-stemmed and vigorous. Hybrids between *C. linariifolia* and *C. scabrida* are known from Garfield County, Utah. *Castilleja linariifolia* is parapatric with the similar *C. wootonii* in central New Mexico. *Castilleja linariifolia* is reported to hybridize with *C. chromosa*, *C. flava*, and *C. miniata*. A possible hybrid with *C. septentrionalis* was named *C. ×cognata* Greene.

59. Castilleja lindheimeri A. Gray in A. Gray et al., Syn. Fl. N. Amer. 2(1): 298. 1878 (as Castilleia) • Lindheimer's paintbrush E

Castilleja purpurea (Nuttall) G. Don var. *lindheimeri* (A. Gray) Shinners

Herbs, perennial, 1.5–3 dm; from a woody caudex; with a taproot. Stems several, erect or ascending, branched or unbranched, sometimes with axillary tufts of leaves, hairs spreading to ± appressed, ± matted, short, soft, eglandular. Leaves green to purple, linear to narrowly lanceolate, 1.3–7 cm, not fleshy, margins plane, sometimes ± wavy, involute, 3–5-lobed, apex acute to obtuse; lobes ascending, linear to narrowly lanceolate, apex acute. Inflorescences 4.5–18 × 1.5–3.5 cm; bracts proximally pale greenish to dull greenish brown, distally orange, reddish orange, or pale orange, sometimes reddish, lanceolate to oblong, 3(–5)-lobed; lobes ascending to spreading, linear to broadly lanceolate or oblanceolate, proximal ones usually arising below mid length, rarely at or above mid length, apex obtuse to acute. Calyces proximally greenish or pale, distally colored as bracts, 23–33 mm; abaxial and adaxial clefts 10–15 mm, 40–50% of calyx length, deeper than laterals, lateral 7–11 mm, 20–35% of calyx length; lobes oblong to broadly linear, central lobe apex obtuse to rounded, lateral ones acute to rounded. Corollas ± curved, 30–40 mm; tube 20–27 mm; beak, sometimes teeth of abaxial lip, exserted; beak adaxially green to yellowish, 8–15 mm; abaxial lip green, reduced, 2–3.5 mm, 20–25% as long as beak; teeth prominent, petaloid, spreading to erect, colored as in distal portion of bracts, 0.5–3 mm.

Flowering Jan–May. Rocky slopes, ridges, grasslands, pastures, open forests, roadsides, sometimes over limestone or granite; 200–800 m; Tex.

Castilleja lindheimeri is endemic to the Edwards Plateau region. Unlike its close relatives, *C. citrina* and *C. purpurea*, most plants of *C. lindheimeri* have orange to reddish orange inflorescences, with smaller numbers varying to red. The leaves are also often less divided than in either *C. citrina* or *C. purpurea*.

60. **Castilleja lineariloba** (Bentham) T. I. Chuang & Heckard, Syst. Bot. 16: 657. 1991 • Thin-lobed owl's clover E

Orthocarpus linearilobus Bentham, Pl. Hartw., 330. 1849; *O. mariposanus* Congdon

Herbs, annual, 1.5–4.5 dm; with fibrous roots. **Stems** solitary or few, erect, unbranched or branched, hairs spreading, short to long, ± stiff, mixed with short stipitate-glandular ones. **Leaves** green, linear to narrowly lanceolate or narrowly oblong, 2–5.7 cm, not fleshy, margins plane, flat, 3–7(–9)-lobed, apex acuminate to acute; lobes spreading to ascending, linear to narrowly oblong, apex acuminate. **Inflorescences** 2–14 × 1–4 cm; bracts greenish throughout, or proximally greenish, distally white, cream, pale pink, or pale purple on apices, linear-lanceolate, 5–7(–9)-lobed, sometimes with secondary lobes; lobes ascending to spreading, linear to narrowly oblanceolate, long, arising all along leaf axis, apex acute to obtuse. **Calyces** colored as bracts, 15–25 mm; all 4 clefts subequal, 7–11 mm, 50–67% of calyx length; lobes linear, apices often slightly expanded, apex obtuse to acute. **Corollas** straight or slightly curved, 12–25 mm; tube 9–14 mm, expanded distally; abaxial lip sometimes slightly exserted, never hidden by slender calyx lobes, beak exserted; beak straight, adaxially white or lilac pink, 3–5.5 mm, inconspicuously puberulent; abaxial lip proximally white, distally yellow, with purple or red-brown spots, conspicuous, pouches 3, inflated, 4–6 mm, 4–5 mm wide, 2 mm deep, longer than deep, 1.5–4 mm, 90% as long as beak; teeth erect, white, usually with purple spot at base, 0.5–1 mm. **Filaments** glabrous. $2n = 20$.

Flowering Apr–Jun(–Jul). Grasslands, moist meadows, swales, shores, forest openings; 0–1600 m; Calif.

Castilleja lineariloba is endemic to the western slope of the Sierra Nevada. Its chromosome number is $2n = 20$, an apparent aneuploid reduction and documented by numerous counts. This diploid number is shared only with two very distantly related annual species endemic to central Mexico, *C. gracilis* Bentham and *C. tenuifolia* M. Martens & Galeotti.

61. **Castilleja lineata** Greene, Pittonia 4: 151. 1900 (as Castilleia) • Linear-lobed or lineated paintbrush E F

Herbs, perennial, 1–4 dm; from a woody caudex; with a taproot. **Stems** few to many, erect or strongly ascending, unbranched except for small, leafy axillary shoots, hairs spreading-reflexed to ± appressed, matted, long, soft, with much shorter stipitate-glandular ones, white-woolly. **Leaves** green, linear to narrowly oblong or narrowly lanceolate, 1.3–5 cm, not fleshy, margins plane, sometimes wavy, involute, 3–7-lobed, apex acute; lobes divergent, spreading-ascending, linear, apex acute to acuminate. **Inflorescences** 5–22 × 1–4.5 cm; bracts green to yellow-green throughout, or proximally green to yellow-green, distally yellow to sometimes pale yellow-orange, narrowly lanceolate to narrowly oblong or broadly lanceolate, 3(–7)-lobed; lobes ascending to spreading, linear to oblong, medium length to long, arising near or below mid length, central lobe apex rounded to obtuse, lateral ones acute. **Calyces** colored as bracts, 15–20 mm; abaxial and adaxial clefts 5.5–8 mm, 30–50% of calyx length, ± deeper than laterals, sometimes appearing subequal in pressed specimens, lateral 5–6 mm, ca. 33% of calyx length; lobes linear to narrowly lanceolate, sometimes expanded towards apices, apex acute. **Corollas** straight or slightly curved, 14–22 mm; tube 7–14 mm; beak tip barely exserted from calyx; beak adaxially greenish, 4–7 mm; abaxial lip green to yellow, reduced, 1–4 mm, usually less than 67% as long as beak; teeth erect, white to yellow, 1–2.5 mm. $2n = 24$.

Flowering Jun–Aug. Dry to moist slopes and meadows, shores, open conifer forests, montane to alpine; 2100–3800 m; Ariz., Colo., N.Mex.

Castilleja lineata is restricted to the mountains of northeastern Arizona, southern Colorado, and northwestern New Mexico. It is uncommon throughout its range and is without apparent close relatives. The Navajo used *C. lineata* as a medicinal plant and for its sweet nectar (D. E. Moerman 1998).

62. Castilleja litoralis Pennell, Proc. Acad. Nat. Sci.
Philadelphia 99: 183. 1947 [E]

Castilleja affinis Hooker & Arnott
subsp. *litoralis* (Pennell)
T. I. Chuang & Heckard;
C. wightii Elmer subsp. *litoralis*
(Pennell) Munz

Herbs, perennial, 1–9 dm;
from a woody caudex; with a
taproot. **Stems** few to many,
usually decumbent proximally,
becoming ascending-erect, sometimes ascending,
branched, sometimes with small, leafy axillary shoots,
glabrate or ± pubescent distally, hairs sparse to moderately
dense, spreading to ± appressed, short, soft,
sometimes mixed with short-glandular ones below
inflorescence. **Leaves** green, lanceolate to oblong or
narrowly ovate, (0.5–)3–8 cm, sometimes thickened,
not fleshy, margins plane, sometimes ± wavy, flat to
involute, 0(–3)-lobed, apex acute to rounded; lobes
ascending or spreading, linear, narrowly lanceolate
to oblong or triangular, short, apex acute to obtuse.
Inflorescences 2.5–21 × 3–5 cm; bracts proximally green,
distally bright red to crimson or orange-red, sometimes
orange or pale yellow-orange, oblong to narrowly ovate
or narrowly obovate, (0–)3–5-lobed, sometimes with a
pair of small teeth; lobes ascending, linear to oblong,
medium length, arising in middle ⅓, central lobe apex
obtuse to rounded or truncate, lateral ones ± acute.
Pedicels 0–6 mm. **Calyces** colored as bracts, 17–25
(–30) mm; abaxial and adaxial clefts (5–)7–15
(–18) mm, 33–55% of calyx length, deeper than
laterals, lateral 1–3(–5) mm, 5–10% of calyx length;
lobes broadly triangular to oblong, apex obtuse to
acute or rounded. **Corollas** straight or slightly curved,
23–38(–40) mm; tube 10–20 mm; abaxial lip often
visible through front cleft, very rarely almost exserted,
beak exserted; beak adaxially green or yellowish,
10–16 mm, surface inconspicuously puberulent;
abaxial lip ascending, green, reduced, 1–2.5 mm,
10–20% as long as beak; teeth erect or incurved, green
or white, 1–2 mm. **2n = 120, 144.**

Flowering (Apr–)May–Aug(–Sep). Steep rocky
slopes, headlands, ledges, sea cliffs, coastal scrub, dune
swales, roadcuts; 0–200 m; Calif., Oreg., Wash.

Castilleja litoralis never ranges more than one to
two kilometers from the sea, from Humboldt County,
California, north to Pacific County, Washington, near
the mouth of the Columbia River. It is a high polyploid
complex, possibly incorporating the genomes of several
species, including *C. affinis*, *C. miniata*, and possibly
C. hispida. The coastal *C. miniata* var. *dixonii* is very
similar ecologically and morphologically but replaces
C. litoralis from southwestern Washington to southern
British Columbia. Compared to *C. litoralis*, *C. miniata*

var. *dixonii* usually has somewhat longer corollas and
corolla beaks, the latter with a more conspicuously
puberulent surface and deeper lateral calyx clefts.
Castilleja litoralis has been included as a subspecies of
C. affinis by some (for example, M. Wetherwax et al.
2012), but the morphological resemblance to that species
is far more tenuous than it is to *C. miniata* var. *dixonii*.
Considering their very similar morphologies, along
with the fact that both *C. litoralis* (2n = 120, 144) and
C. miniata var. *dixonii* (2n = 96, 144) apparently combine
multiple genomes, strongly suggest that they would best
be treated as a single entity. Should they be combined at
the species level following additional research, the name
C. dixonii has priority.

Castilleja litoralis is often associated with salal,
Gaultheria shallon, on which it is likely parasitic.

63. Castilleja lutescens (Greenman) Rydberg, Mem.
New York Bot. Gard. 1: 359. 1900 • Yellowish
paintbrush [E]

Castilleja pallida (Linnaeus)
Sprengel var. *lutescens* Greenman,
Bot. Gaz. 25: 265. 1898 (as
Castilleia)

Herbs, perennial, 1.4–6 dm;
from a woody caudex; with a
taproot. **Stems** few to several,
erect or ascending, sometimes
decumbent at base, unbranched
or branched, hairs spreading to retrorse, short, scabrid
below inflorescence, sometimes becoming medium
length to long, soft to stiff, mixed with short stipitate-
glandular ones in inflorescence. **Leaves** green, sometimes
purplish, linear to lanceolate or narrowly oblong,
1–8.5 cm, not fleshy, margins plane, flat or involute
lengthwise, 0(–5)-lobed, apex acute; lobes ascending,
linear, central one sometimes shallowly toothed, apex
acute to obtuse. **Inflorescences** 3–14 × 1.5–3 cm; bracts
greenish throughout, or proximally greenish, distally
pale to bright yellow or whitish, rarely pale orangish,
lanceolate to oblong, (0–)3–7-lobed; lobes ascending
to erect, linear to lanceolate or narrowly oblong,
medium length, arising at or above mid length, central
lobe apex obtuse to rounded, lateral ones acute.
Calyces colored as bracts, 15–25 mm; abaxial and
adaxial clefts 6–13 mm, 50% of calyx length, deeper
than laterals, lateral 1–7 mm, 15% of calyx length;
lobes narrowly triangular to linear, apex acute. **Corollas**
straight or slightly curved, 21–27 mm; tube 14–16 mm;
beak partly exserted, adaxially green, 7–12 mm; abaxial
lip ascending, green, reduced, 2–4 mm, 25–50% as long
as beak; teeth erect or incurved, sometimes spreading,
green or white, 0.7–2.5 mm. **2n = 48, 96.**

Flowering May–Aug. Grasslands, open conifer
forests, moist meadows, rocky slopes, valleys, montane;
600–1900 m; Alta., B.C.; Idaho, Mont., Oreg., Wash.

Castilleja lutescens is found east of the Cascade Range in Oregon and Washington and ranges east to western Montana and the adjacent interior of western Canada. Its inflorescences vary in color from entirely greenish to white or yellowish, and the stature of the plants is also variable, trending from moderate and compact in grasslands to taller in more forested situations. In the Blue Mountains of Garfield County, Washington, occasional hybrids form between *C. lutescens* and *C. hispida* var. *acuta*, which often both occur in the same vicinity.

L. R. Heckard (1968) reported a chromosome count of ca. $2n = 120$ from a population in Montana.

64. Castilleja martini Abrams, Bull. S. Calif. Acad. Sci. 1: 69. 1902 • Camp Martin paintbrush

Castilleja applegatei Fernald subsp. *martini* (Abrams) T. I. Chuang & Heckard

Herbs, perennial, 1.5–5(–8) dm; from a woody caudex; with a taproot. **Stems** several to many, erect or ascending, sometimes curved at base, unbranched or branched, sometimes with short, leafy axillary shoots, hairs spreading, short or long and distally short, stiff or soft, stipitate-glandular. **Leaves** green to sometimes purple, narrowly oblong or lanceolate to sometimes linear or narrowly ovate, (0.8–)1–5(–6.5) cm, not fleshy, margins wavy, flat or involute, 0–5-lobed, apex rounded or obtuse, rarely acute; lobes ascending-spreading or divaricate, linear to narrowly oblong, apex acute. **Inflorescences** 2.5–25 × 1.5–4 cm; bracts proximally greenish, distally red, orange-red, or pale orange, rarely yellow, lanceolate or broadly lanceolate to oblong, (0–)3–5(–7)-lobed, plus 1–3 shallow teeth on apex of central lobe (var. *clokeyi*); lobes ascending or spreading, linear, lanceolate, or oblanceolate, medium length or long, arising at or near mid length, central lobe apex obtuse to rounded, lateral ones acute. **Calyces** proximally pale green, green, or whitish, sometimes with a yellow central band, distally colored as bracts, 14–26(–28) mm; abaxial and adaxial clefts 3–10 mm, 20–50% of calyx length, deeper than laterals, lateral 1–5 mm, 15–20% of calyx length; lobes lanceolate to broadly lanceolate, oblong, or ovate, apex acute, obtuse, or rounded. **Corollas** straight to slightly curved, (19–)22–45 mm; tube 11–22 mm; abaxial lip rarely exserted, beak exserted; beak adaxially green to yellowish, 9–21(–23) mm; abaxial lip green or deep green with white teeth or red with green teeth, reduced, not inflated, often visible through front cleft, 1–2 mm, ca. 10% as long as beak; teeth incurved, green to white, 0.5–1 mm. $2n = 24, 48, 72$.

Castilleja martini is sometimes treated as a subspecies of *C. applegatei*, but there are relatively consistent morphological discontinuities between these two variable complexes. See notes under 4. *C. applegatei*.

Varieties 2 (2 in the flora): w United States, nw Mexico.

1. Calyx lobes broadly lanceolate to ovate or oblong, apices acute, obtuse, or rounded, abaxial and adaxial clefts 20–44% of calyx lengths; w California . . . . 64a. *Castilleja martini* var. *martini*
1. Calyx lobes lanceolate to broadly lanceolate, apices acute to sometimes obtuse, abaxial and adaxial clefts 40–50% of calyx lengths; sw Nevada, adjacent e California . . . . 64b. *Castilleja martini* var. *clokeyi*

64a. Castilleja martini Abrams var. **martini**

Castilleja gyroloba Pennell; *C. roseana* Eastwood

Herbs 2–8 dm. **Stems:** hairs long, sometimes short, soft. **Leaves** 0(–5)-lobed. **Bracts** distally red, orange-red, or pale orange, rarely yellow, broadly lanceolate to oblong, 0–5-lobed; lobes linear or often narrowly oblanceolate, long. **Calyces** proximally pale green or green, sometimes with a yellow central band, 14–26 mm; abaxial and adaxial clefts 3–8 mm, 20–44% of calyx length; lobes broadly lanceolate to ovate or oblong, apex acute, obtuse, or rounded. **Corolla beaks** 9–18(–23) mm; abaxial lip deep green with white teeth or red with green teeth. $2n = 24, 48, 72$.

Flowering Mar–Sep. Chaparral, dry rocky or gravelly slopes, open conifer forests, sagebrush steppes; 100–2900 m; Calif.; Mexico (Baja California).

Variety *martini* has a wide distribution across California west of the Sierra Nevada crest and extends south to the Sierra San Pedro Mártir in northern Baja California. It occurs from low elevations to the montane zone.

64b. Castilleja martini Abrams var. **clokeyi** (Pennell) N. H. Holmgren, Mem. New York Bot. Gard. 21(4): 55. 1971 (as martinii) • Clokey's paintbrush E

Castilleja clokeyi Pennell, Proc. Acad. Nat. Sci. Philadelphia 89: 420. 1938

Herbs 1.5–4(–6) dm. **Stems:** hairs short or long and distally short, stiff, sometimes with short, leafy axillary shoots. **Leaves** 3–5-lobed. **Bracts** distally red or orange-red, lanceolate, 3–7-lobed, often with 1–3 shallow teeth on apex of central lobe; lobes lanceolate to oblanceolate, medium length. **Calyces** proximally whitish, 14–21(–28) mm; abaxial and adaxial clefts 7–10 mm, 40–50% of calyx length; lobes lanceolate to broadly lanceolate, apex acute to sometimes obtuse. **Corolla beaks** 11–21 mm; abaxial lip green. $2n = 48$.

Flowering Jun–Aug. Dry rocky or gravelly slopes, open conifer forests, montane to subalpine; 1500–3200 m; Calif., Nev.

Variety *clokeyi* has a distinct distribution in the arid mountain ranges surrounding the Death Valley region of California and adjacent southwestern Nevada, where it usually occurs near timberline.

65. Castilleja mendocinensis (Eastwood) Pennell, Proc. Acad. Nat. Sci. Philadelphia 99: 184. 1947 • Mendocino Coast paintbrush C E

Castilleja latifolia Hooker & Arnott subsp. *mendocinensis* Eastwood, Leafl. W. Bot. 1: 238. 1936

Herbs, perennial, 1.7–6.5 dm; from a woody caudex; with a taproot. **Stems** few to many, decumbent to ascending, much-branched, with leafy axillary shoots, villous, hairs spreading, long, stiff to soft, eglandular, mixed with short-glandular ones. **Leaves** gray-green becoming ± purple, or green, ± cup-shaped, oblong to narrowly elliptic or suborbicular, 0.5–2(–5) cm, ± fleshy, cupulate throughout, sometimes obscurely so on distal portion of stem, margins plane, flat to involute, 0–3-lobed, apex rounded; lobes ascending, oblong to rounded, apex truncate or rounded, sometimes ± acute. **Inflorescences** 5–23 × 1.5–3.5 cm; bracts proximally greenish, distally bright red or red-orange, sometimes orange, oblong, ovate, or widely cuneate to widely obovate to suborbiculate, sometimes cup-shaped, 0–3-lobed, sometimes with 3 additional shallow teeth at tip of central lobe; lobes erect, oblong to broadly triangular, short, arising above mid length, apex truncate, rounded, or obtuse, lateral ones sometimes acute. **Calyces** colored as bracts, often with a yellow central band, 20–31 mm; abaxial and adaxial clefts 8–12 mm, ca. 50% of calyx length, deeper than laterals, lateral 2–6 mm, 10–15% of calyx length; lobes oblong to broadly or narrowly triangular, apex obtuse or rounded, sometimes acute. **Corollas** straight or slightly curved, 28–45 mm; tube 18–20 mm; beak exserted, adaxially green or yellow-green, 15–25 mm; abaxial lip deep green, reduced, visible in front cleft, 1–1.5 mm, 10% as long as beak; teeth incurved, green, 0.5–1 mm. $2n = 72$.

Flowering Apr–Aug. Coastal scrub, headlands, sea bluffs, over sandstone or serpentine; of conservation concern; 0–100 m; Calif., Oreg.

Castilleja mendocinensis is a coastal plant from Mendocino County, California, northward to Curry County, Oregon. Its close relative, *C. latifolia*, occurs in similar habitats south of San Francisco Bay. There is one known population in Oregon, and a number of California localities are threatened by coastal development.

66. Castilleja mexicana (Hemsley) A. Gray, Proc. Amer. Acad. Arts 21: 404. 1886 (as Castilleia) • Mexican paintbrush F

Orthocarpus mexicanus Hemsley, Biol. Cent.-Amer., Bot. 2: 463, plate 63A, figs. 1–6. 1882; *Castilleja tortifolia* Pennell

Herbs, annual or short-lived perennial, 0.6–3 dm; from a woody caudex; with a slender taproot. **Stems** solitary or few, erect to erect-ascending, sometimes slightly curved at base, branched at base or unbranched, hairs spreading, long, stiff to soft, mixed with shorter stipitate-glandular ones. **Leaves** brown or purplish, sometimes green, linear to narrowly oblong, 1–5 cm, not fleshy, margins wavy, mostly involute, to flat, 3–5-lobed, apex acute; lobes spreading, linear, apex acute to obtuse. **Inflorescences** 2–17 × 1.5–6.5 cm; bracts greenish throughout, narrowly lanceolate (ovate in outline), 3–5(–7)-lobed; lobes spreading, linear to narrowly oblong, long, arising from ⅓–⅔ blade length, wavy-margined, apex rounded or obtuse to acute. **Calyces** proximally brownish green, purplish, or green, lobes tipped with same color as bracts, 18–28 mm; abaxial and adaxial clefts 6–14 mm, 33–50% of calyx length, deeper than laterals, lateral 1.5–6 mm, 8–20% of calyx length; lobes lanceolate to narrowly triangular, abaxials wider than adaxials, apex acute. **Corollas** straight proximally, conspicuously decurved distally, 35–60 mm; tube 26–46 mm; abaxial lip, beak, and majority of tube exserted; beak yellow to yellowish green, sometimes purplish tipped or drying pinkish, 9–15 mm; abaxial lip light yellow to whitish, prominent, not inflated, 4–8 mm, 50–75% as long as beak; teeth spreading-ascending, yellowish, 3–6 mm. $2n = 24$.

Flowering Feb–Oct. Dry rocky slopes, grasslands, pinyon-juniper stands; 1200–2100 m; Tex.; Mexico (Aguascalientes, Chihuahua, Coahuila, Durango, Nuevo León, San Luis Potosí, Tamaulipas, Zacatecas).

Castilleja mexicana occurs in the northern third of Mexico and reaches the flora area only in southwestern Texas. Its conspicuous corollas can be either yellow or, less commonly, white. In both cases, the flowers turn soft pink-purple with age. Texas populations of this species are yellow flowered, and the white-flowered morph appears to occur only in northeastern Mexico.

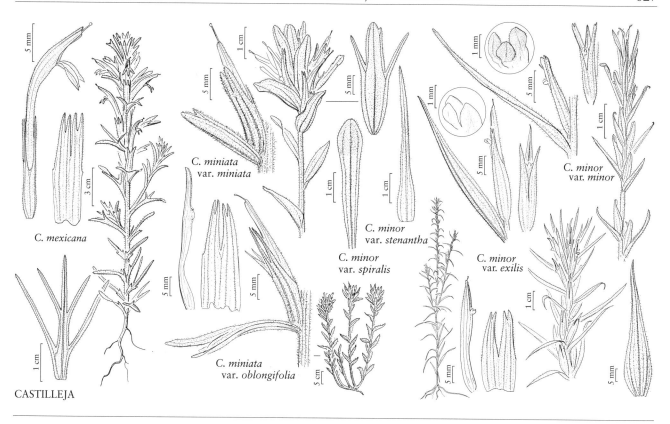

C. mexicana

C. miniata
var. miniata

C. miniata
var. oblongifolia

C. minor
var. spiralis

C. minor
var. stenantha

C. minor
var. exilis

C. minor
var. minor

CASTILLEJA

There is some indication of additional morphological differences between these color morphs that may justify varietal segregation. *Castilleja mexicana* is sometimes confused with the closely related *C. sessiliflora*, due to their conspicuous, distally curved, usually strongly exserted corollas, but the two species remain distinct.

67. **Castilleja miniata** Douglas ex Hooker, Fl. Bor.-Amer. 2: 106. 1838 • Scarlet or common or giant red paintbrush F

Herbs, perennial, 1.2–8(–10) dm; from a woody caudex (or slender rooting rhizomes in var. *dixonii*, var. *miniata*); with a taproot or with slender, branched roots from a rhizome. **Stems** few to many, erect to ascending, rarely proximally decumbent or creeping and rooting at nodes, usually branched, glabrous, glabrate, or hairy, hairs spreading to ± retrorse, short to long, soft to stiff, rarely stipitate-glandular. **Leaves** green to purple, linear to lanceolate, narrowly elliptic, narrowly oblong, or ovate, (1.5–)3–8(–9.5) cm, thin and not fleshy or slightly to moderately thickened and slightly fleshy, margins plane, rarely wavy, flat to involute, whole leaf sometimes recurved downward, 0(–5)-lobed, apex acute to obtuse, sometimes rounded; lobes ascending-spreading, narrowly lanceolate, apex acute. **Inflorescences** 3–15(–22) × 1.5–5.5 cm, often bearing a thin, white, powdery exudate, especially on bract surfaces; bracts greenish, scarlet, red, red-orange, or pale orange throughout, sometimes pink, magenta, pink-purple, yellow, greenish yellow, white, or salmon throughout, or proximally greenish, distally colored as above, lanceolate to oblong-ovate, 0–5(–7)-lobed, central lobes sometimes distally apiculate; lobes erect, linear to lanceolate, oblong, or oblanceolate, short or medium length, arising near or above mid length, central lobe apex obtuse, rounded, or truncate, lateral ones rounded to acute or acuminate. **Pedicels** 0–5 mm. **Calyces** colored as bracts, 15–38 mm; abaxial and adaxial clefts 4–24 mm, 35–70% of calyx length, deeper than laterals, lateral (1–)3–8(–12) mm, 5–30% of calyx length; lobes linear or narrowly lanceolate to narrowly triangular, apex acute to acuminate or obtuse. **Corollas** slightly curved, 20–48 mm; tube 12–26 mm; abaxial lip usually not exserted, though often visible in front calyx cleft, beak partially to fully exserted; beak adaxially green to yellow-green or whitish, (9–)14–25 mm; abaxial lip incurved or ascending, deep green or green, sometimes deep purple or yellowish, reduced, not inflated, visible in front cleft, 0.5–3.5 mm, 5–20% as long as beak (to ca. 33% as long as beak in some populations of var. *miniata*); teeth incurved or erect, green or white, 0.7–1.5 mm. $2n = 24, 48, 72, 96, 120, 144$.

Varieties 4 (4 in the flora): w North America, nw Mexico.

Castilleja miniata is widely recognized as the common scarlet paintbrush. It is highly variable and has five levels of polyploidy. Nonetheless, it remains fairly well defined morphologically across its wide range. Native Americans use it medicinally. A probable hybrid with *C. septentrionalis* from southern Nevada was named *C. ×porterae* Cockerell.

1. Bracts distally pale yellowish, pale orange, or salmon, sometimes pale red; n, c British Columbia, adjacent wc Alberta, s Yukon . . . . .67c. *Castilleja miniata* var. *fulva*
1. Bracts distally scarlet, red, or orange-red, sometimes orange, pink, pink-purple, magenta, pale orange, yellow, greenish yellow, or white; widespread in much of w North America south of the arctic.
  2. Stems glabrous or glabrate proximally, hairy medially and distally; leaves appressed-ascending, ascending, or spreading, often thickened and slightly fleshy; calyx lobe apices acuminate; Cuyamaca, Laguna, San Bernardino and e San Gabriel mountains of sw California . . . . 67d. *Castilleja miniata* var. *oblongifolia*
  2. Stems glabrous or sparsely hairy proximally and medially, subglabrous or sparsely hairy distally; leaves spreading to ascending, thin and not fleshy (except sometimes in var. *dixonii*); calyx lobe apices acute to acuminate or obtuse; widespread in w North America.
    3. Leaves linear to lanceolate, narrowly elliptic, or narrowly oblong, apices acute to acuminate, sometimes obtuse; calyx lobe apices acute, sometimes acuminate or obtuse; widespread in w North America, except near coast . . . . 67a. *Castilleja miniata* var. *miniata*
    3. Leaves narrowly lanceolate to ovate, narrowly elliptic, or narrowly oblong, apices rounded to obtuse, sometimes acute to acuminate; calyx lobe apices acute to obtuse; coastal, from extreme nw Oregon northward to British Columbia. . . . . 67b. *Castilleja miniata* var. *dixonii*

67a. Castilleja miniata Douglas ex Hooker var. **miniata** E F

Castilleja hyetophila Pennell; *C. inconstans* Standley

Herbs 1.6–6.5 dm; with a taproot or with slender, branched roots from a rhizome. **Stems** glabrous or sparsely hairy proximally and medially, subglabrous or sparsely hairy distally, hairs spreading, short to long, soft, eglandular, rarely stipitate-glandular. **Leaves** spreading to ascending, linear to lanceolate, narrowly elliptic, or narrowly oblong, thin and not fleshy,

margins plane, rarely wavy, flat to slightly involute, leaf sometimes recurved downward, 0(–3)-lobed, apex acute to acuminate, sometimes obtuse. **Bracts** distally scarlet, red, orange-red, or pale orange, sometimes pink, pink-purple, magenta, pale orange, yellow, greenish yellow, or white, 0–3(–5)-lobed, central lobe with 3–5 small teeth at apex; central lobe apex obtuse, rounded to truncate, lateral ones acute to rounded. **Calyces** 15–36 mm; abaxial and adaxial clefts 8–20 mm, ca. 40% of calyx length, subequal or abaxial longer, lateral 1–9 mm, 15–20% of calyx length; lobes narrowly lanceolate to narrowly triangular or linear, apex acute, sometimes acuminate or obtuse. **Corollas** 20–48 mm; beak 9–25 mm; abaxial lip deep green or deep purple, 0.5–3.5 mm, 10–15% as long as beak, to ca. 33% as long as beak in some populations. $2n = 24$, 48, 72, 96, 120.

Flowering Mar–Sep(–Oct). Moist or wet meadows, stream banks, shores, open forests, rocky slopes, roadsides, forest edges; 0–3700 m; Alta., B.C., Man., Ont., Sask.; Alaska, Ariz., Calif., Colo., Idaho, Mont., Nev., N.Mex., Oreg., Utah, Wash., Wyo.

Variety *miniata* is much more widespread than the other varieties of *Castilleja miniata*. Occasional hybrids are encountered between this variety and a wide range of species. Apparent hybrid swarms with *C. rhexiifolia* and/or *C. septentrionalis* are occasional in the Rocky Mountains. In the Pacific Northwest and across the prairie provinces of Canada, populations with relatively short inflorescences and pink bracts deserve further study. Likewise, populations of pink to pink-purple plants with shorter corollas and corolla beaks are found on boggy ground in the Peace River region of northwestern Alberta and adjacent British Columbia. A slightly rhizomatous form from the coast of British Columbia to southeastern Alaska was named *C. hyetophila*. The precise relationship of the rhizomatous form with var. *dixonii* and with *C. miniata* as a whole is still unclear and merits comprehensive study.

67b. Castilleja miniata Douglas ex Hooker var. **dixonii** (Fernald) A. Nelson & J. F. Macbride, Bot. Gaz. 65: 70. 1918 • Dixon's paintbrush E

Castilleja dixonii Fernald, Erythea 7: 122. 1899 (as Castilleia); *C. miniata* subsp. *dixonii* (Fernald) Kartesz

Herbs 1.2–4.7 dm; with a taproot or with slender, branched roots from decumbent proximal stems. **Stems** glabrous or glabrate proximally, sometimes sparsely hairy distally, hairs spreading to ± retrorse or appressed, short, soft, rarely mixed with short stipitate-glandular ones near inflorescence. **Leaves** spreading to ascending, narrowly lanceolate to oblong, sometimes narrowly elliptic to ovate, thin and not fleshy or often slightly to moderately thickened and slightly fleshy, margins plane, flat to slightly involute, 0(–5)-lobed, apex acute to obtuse or rounded, sometimes acuminate. **Bracts** distally scarlet, red, orange-red, or light red, sometimes with medial region ± darkened, 0–5-lobed; lobe apex obtuse to rounded or acute. **Calyces** 17–28 mm; abaxial and adaxial clefts 7–15 mm, 35–55% of calyx length, abaxial slightly deeper than adaxial or subequal, lateral 2–7 mm, 15–30% of calyx length; lobes linear or narrowly lanceolate to narrowly triangular, apex acute, sometimes obtuse. **Corollas** 25–46 mm; beak 11–25 mm, adaxial surface ± conspicuously puberulent; abaxial lip green, 1–2 mm, 5–10% as long as beak. *2n* = 96, 144.

Flowering May–Sep. Rocky islands, moist sand, vegetated dunes, sea bluffs, headlands, coastal scrub, upper margin of tidal flats; 0–200 m; B.C.; Oreg., Wash.

Variety *dixonii* is found on the outer coast, from Clatsop County, Oregon, to Vancouver Island and the mainland shore of southwestern British Columbia, as well as along the shores and some of the islands of Puget Sound. Variety *dixonii* can be difficult to separate morphologically from var. *miniata*, but var. *dixonii* occurs strictly in low-elevation, sandy, coastal habitats and differs somewhat inconsistently in a number of minor traits, especially in the leaves. Variety *miniata* usually occurs at considerably higher elevations in this region. Variety *dixonii* is also very similar to, if not conspecific with, *Castilleja litoralis*, which has a more or less shorter corollas and corolla beaks and other minor differences. The corolla beaks of var. *dixonii* have more or less longer hairs and are more obviously puberulent. It is tempting to merge *C. litoralis* into var. *dixonii*, especially since both entities are known to be high polyploids, but this is premature. Despite the attention of many botanists, the species of *Castilleja* along the Pacific coast are still in need of comprehensive study and delimitation. Also see the comments under 62. *C. litoralis*.

67c. Castilleja miniata Douglas ex Hooker var. **fulva** (Pennell) J. M. Egger, Phytologia 90: 72. 2008 [E]

Castilleja fulva Pennell, Proc. Acad. Nat. Sci. Philadelphia 86: 540. 1934

Herbs 4–5 dm; with a taproot. **Stems** glabrous or glabrate proximally, hairy distally, hairs sparse, spreading, long, soft to ± stiff, short stipitate-glandular. **Leaves** narrowly lanceolate to narrowly ovate, thin and not fleshy, margins plane, flat, usually 0-lobed, apex acuminate to acute. **Bracts**

distally pale yellowish, pale orange, or salmon, sometimes pale red, 0–3(–7)-lobed, central lobe sometimes distally apiculate; central lobe apex rounded, lateral ones acute to acuminate. **Calyces** 15–21 mm; abaxial and adaxial clefts 10 mm, 50–70% of calyx length, abaxial slightly deeper than adaxial, lateral 1–6 mm, 5–30% of calyx length; lobes narrowly lanceolate to narrowly triangular, apex acute to acuminate. **Corollas** 20–27 mm; beak 9–11 mm; abaxial lip green, 1–1.5 mm, 10–20% as long as beak. *2n* = 96.

Flowering Jun–Aug. Damp low thickets, grassy verges, roadsides, forest edges; 600–1600 m; Alta., B.C., Yukon.

Variety *fulva* is found widely in central and northern British Columbia, as well as in the Peace River valley in adjacent west-central Alberta and southern Yukon, substantially but not entirely replacing var. *miniata* in these regions. The inflorescences are consistently yellow to pale tawny orange within populations and have a distinctive morphology, though overlapped by the wide range of variation in var. *miniata*. It is not known if the distinctive coloration of var. *fulva* represents introgression from *Castilleja unalaschcensis*, or if these populations might have diverged in isolation on a nunatak during the last extensive glaciation, as postulated by F. W. Pennell (1934b).

67d. Castilleja miniata Douglas ex Hooker var. **oblongifolia** (A. Gray) Munz, Bull. S. Calif. Acad. Sci. 31: 69. 1932 [F]

Castilleja oblongifolia A. Gray in A. Gray et al., Syn. Fl. N. Amer. 2(1): 296. 1878 (as Castilleia)

Herbs 2.4–8 dm; with a taproot. **Stems** glabrous or glabrate proximally, hairy medially and distally, hairs moderately dense, appressed to weakly spreading, short to long, soft, eglandular to sometimes stipitate-glandular. **Leaves** appressed-ascending, ascending, or spreading, narrowly lanceolate to narrowly oblong or ovate, slightly to moderately thickened and slightly fleshy, sometimes thin and not fleshy, margins plane, rarely wavy, flat, 0(–3)-lobed, apex acute. **Bracts** distally red, red-orange, pale orange, pink, salmon, or reddish magenta, 0–5-lobed; lobe apex acute to acuminate. **Calyces** 19–38 mm; abaxial and adaxial clefts 4–24 mm, 40% of calyx length, subequal or abaxial longer, lateral 2–12 mm, 10–25% of calyx length; lobes linear to narrowly lanceolate, apex acuminate. **Corollas** 28–48 mm; beak 15–25 mm; abaxial lip dark green or yellowish to whitish, 1–2.5 mm, 5–15% as long as beak.

Flowering Jul–Sep(–Oct). Stream banks, seeps and damp ground in open conifer forests, roadside banks; 600–2800 m; Calif.; Mexico (Baja California).

Variety *oblongifolia* largely replaces var. *miniata* from the upper elevations of the San Bernardino and eastern San Gabriel mountains in San Bernardino County, as well as the Cuyamaca and Laguna mountains in San Diego County, and it extends southward into the mountains of northwestern Baja California, particularly in the Sierra Juarez. Variety *oblongifolia* rarely has oblong leaves, but it differs from var. *miniata* in its densely puberulent stems and leaves, unusually long, narrow, acuminate calyx lobes, usually ascending-appressed leaves that are often brittle and slightly fleshy, and usually pale orange-red inflorescences. While they are occasionally found singly in populations of the nominate variety, the combination of these otherwise unusual characters is unique in var. *oblongifolia*.

68. **Castilleja minor** (A. Gray) A. Gray in W. H. Brewer et al., Bot. California 1: 573. 1876 (as Castilleia)
• Seep or thread-torch paintbrush F

Castilleja affinis Hooker & Arnott var. *minor* A. Gray in W. H. Emory, Rep. U.S. Mex. Bound. 2(1): 119. 1859

Herbs, annual, 2–10(–15) dm; with a short taproot or small, fibrous root system. **Stems** solitary or few, erect, unbranched, rarely branched distally, hairs sparse to dense, spreading, sometimes shaggy (var. *minor*), short to long, soft to stiff, eglandular and/or sparsely to densely stipitate-glandular. **Leaves** green or purple to ± gray, linear to lanceolate, 2–10 cm, not fleshy, margins plane, sometimes wavy, ± involute, 0-lobed, apex acuminate to acute, sometimes obtuse. **Inflorescences** 5–40 × 1–4 cm; bracts proximally greenish, distally red, red-orange, or pale orange, rarely yellow, on apices, narrowly lanceolate, sometimes narrowly oblong to spatulate distally, 0-lobed, plane-margined, apex acuminate (oblong to narrowly spatulate in var. *spiralis*). **Calyces** green or yellowish green, 13–27(–28) mm; abaxial and adaxial clefts 6–15 mm, 33–75% of calyx length, deeper than laterals, lateral 0.5–4 mm, 5–20% of calyx length; lobes linear to narrowly triangular or narrowly lanceolate, apex acute or acuminate. **Corollas** straight, 13–39 mm; tube 11–16(–20) mm; beak partially to completely exserted, sometimes included, adaxially yellow, pale orange to red-orange, reddish brown, green, or white, 5–15(–20) mm; abaxial lip yellow, white, red, deep red, red-violet, or green, colored as or strongly contrasting with rest of corolla, small but jutting out at 90° from axis of corolla, often readily visible through abaxial cleft, 1–3 mm, 5–25% as long as beak; teeth spreading to strongly incurved, green, white, yellow, red, or red-purple, 0.2–1 mm. $2n = 24$.

Varieties 4 (4 in the flora): w North America, nw Mexico.

Castilleja minor is a widespread specialist of seeps, saline shores, and wet ground at moderate elevations. California has many populations on serpentine substrates. Most of the varieties have distinct ranges with little overlap, with the exception of the edaphic obligate var. *spiralis*, which is restricted to serpentine substrates.

1. Corolla beaks 5–8(–10) mm, usually ⅓ or less of length of corollas, included to more often partially exserted from calyces.
  2. Abaxial lips of corollas red to reddish purple; leaves linear or linear-lanceolate, soft; mountains of c, e Arizona and adjacent New Mexico southward . . .68a. *Castilleja minor* var. *minor*
  2. Abaxial lips of corollas whitish, pale green, or pale yellowish; leaves linear-lanceolate to lanceolate, coarse; widespread in Great Basin region . . . . .68b. *Castilleja minor* var. *exilis*
1. Corolla beaks 8–15(–20) mm, usually ⅓+ of length of corollas, exserted from calyces.
  3. Proximal bracts narrowly lanceolate or oblanceolate, distals oblong to narrowly spatulate, apices rounded or obtuse, rarely acute; abaxial lips of corollas red. . . . . 68c. *Castilleja minor* var. *spiralis*
  3. Bracts linear to narrowly lanceolate, apices acuminate; abaxial lips of corollas whitish, rarely greenish or pale yellow (sometimes red in c Arizona) . . . 68d. *Castilleja minor* var. *stenantha*

68a. **Castilleja minor** (A. Gray) A. Gray var. **minor** F

Herbs 7.2 dm. **Stems:** hairs fairly short, soft, moderately to densely stipitate-glandular. **Leaves** linear or linear-lanceolate, soft, margins plane. **Bracts** distally red, linear to linear-lanceolate, apex acuminate. **Calyces** 13–21 mm; abaxial clefts 9–11 mm, adaxial 6–8 mm, clefts 40–75% of calyx length, lateral 3–4 mm, 15–25% of calyx length; lobes linear, apex acuminate. **Corollas** 14–21(–30) mm; beak included to more often partially exserted from calyces, adaxially yellow to greenish, 5–8 mm; abaxial lip red to reddish purple.

Flowering Apr–Sep. Wet, sometimes alkaline sites, including marshes, hot springs, seeps, stream banks; 700–2500 m; Ariz., N.Mex.; Mexico (Chihuahua, Sonora).

Variety *minor* is found in central and southeastern Arizona and adjacent New Mexico southward into northern Chihuahua and Sonora, Mexico. Compared to similar var. *exilis*, plants of var. *minor* tend to be more slender, with narrower, softer-textured, linear-acuminate leaves, red-purple abaxial corolla lips, and with less dense and less conspicuous stem pubescence.

68b. Castilleja minor (A. Gray) A. Gray var. **exilis** (A. Nelson) J. M. Egger, Phytologia 90: 72. 2008 • Alkaline or annual paintbrush, Great Basin thread-torch E F

Castilleja exilis A. Nelson, Proc. Biol. Soc. Wash. 17: 100. 1904, based on *C. stricta* Rydberg, Mem. New York Bot. Gard. 1: 354. 1900, not Bentham 1846

Herbs 2–8(–11) dm. **Stems:** hairs (sparse to) dense, moderately long, stiff to ± soft, eglandular or sparsely stipitate-glandular. **Leaves** linear-lanceolate to lanceolate, coarse, margins plane. **Bracts** distally red to pale orange, sometimes yellow, linear-lanceolate to lanceolate, apex narrowly acute to acuminate. **Calyces** 14–20(–25) mm; abaxial and adaxial clefts 6–12(–15) mm, 33–67% of calyx length, lateral 1–3 mm, 20% of calyx length; lobes linear or narrowly triangular, apex acute or acuminate. **Corollas** 13–25(–30) mm; beak included to more often partially exserted from calyces, adaxially green, yellowish, pale orange, or pale reddish, rarely whitish, 5–7(–10) mm; abaxial lip whitish, pale green, or pale yellowish.

Flowering Apr–Oct(–Nov). Alkaline marshes, hot springs, seeps, stream banks, shores, dune swales; 400–2500 m; B.C.; Ariz., Calif., Colo., Idaho, Mont., Nev., N.Mex., Oreg., Utah, Wash., Wyo.

Variety *exilis* is the most widespread variety of *Castilleja minor*, and it essentially replaces var. *minor* from the northern portions of New Mexico and Arizona northward, throughout much of the Intermountain West. Compared to var. *minor*, the plants of var. *exilis* tend to be wider and more robust, with broader, coarser, often lanceolate and acuminate-tipped leaves, greenish abaxial corolla lips, and coarser, spreading stem pubescence.

68c. Castilleja minor (A. Gray) A. Gray var. **spiralis** (Jepson) J. M. Egger, Phytologia 90: 74. 2008 E F

Castilleja spiralis Jepson, Fl. W. Calif., 412. 1901 (as Castilleia); *C. minor* subsp. *spiralis* (Jepson) T. I. Chuang & Heckard; *C. stenantha* A. Gray subsp. *spiralis* (Jepson) Munz

Herbs 5–7 dm. **Stems:** hairs dense, short, soft, mixed with shorter stipitate-glandular ones. **Leaves** linear or narrowly lanceolate, soft, margins plane or wavy. **Bracts** distally red, proximal narrowly lanceolate or oblanceolate, distal oblong to narrowly spatulate, apex rounded or obtuse, rarely acute. **Calyces** 15–27 mm; abaxial and adaxial clefts 7–9 mm, 35–50% of calyx length, lateral 0.5–2.5 mm, 5–25% of calyx length; lobes linear to narrowly lanceolate, apex acute to acuminate. **Corollas** 20–35 mm; beak and often abaxial lip exserted; beak adaxially yellowish or reddish brown, 8–15 mm; abaxial lip red.

Flowering Jun–Aug. Stream banks, damp flats and seeps, over serpentine; 0–3000 m; Calif.

Variety *spiralis* is limited to a few highly serpentine sites in central California. Its range slightly overlaps var. *stenantha*, but within populations there is little or no intergradation in the distinguishing characters. Variety *spiralis* is separable from the other varieties of *Castilleja minor* by a combination of obtuse to rounded bract apices, bright red, somewhat up-curved abaxial corolla lips, and relatively long corolla beaks.

68d. Castilleja minor (A. Gray) A. Gray var. **stenantha** (A. Gray) J. M. Egger, Phytologia 90: 73. 2008 • Thread-torch, California thread-torch F

Castilleja stenantha A. Gray in A. Gray et al., Syn. Fl. N. Amer. 2(1): 295. 1878 (as Castilleia)

Herbs 2.8–10(–15) dm. **Stems:** hairs sparse to dense, short to long, soft, sparsely stipitate-glandular. **Leaves** linear-lanceolate to lanceolate, soft, margins plane. **Bracts** distally red to pale orange, linear to narrowly lanceolate, apex acuminate. **Calyces** 14–22(–28) mm; abaxial and adaxial clefts 7–15 mm, 50–60% of calyx length, lateral 1–1.5(–3) mm, 10–20% of calyx length; lobes linear to narrowly triangular, apex acute to acuminate. **Corollas** 23–39 mm; beak exserted, adaxially light reddish brown, yellowish, pale orange, or reddish orange, 8–15(–20) mm; abaxial lip whitish, rarely greenish or pale yellow (sometimes red in in c Arizona).

Flowering Feb–Sep. Riparian zones, chaparral, wet cliffs, often over serpentine; 0–2500 m; Ariz., Calif.; Mexico (Baja California).

Variety *stenantha* is most common in central and southern California west of the Sierra Nevada crest, southward to at least the Sierra San Pedro Mártir in Baja California. It is associated with serpentine, granite, and other substrates.

69. Castilleja mogollonica Pennell, Notul. Nat. Acad. Nat. Sci. Philadelphia 237: 1. 1951 • Mogollon or White Mountains paintbrush [C] [E]

Herbs, perennial, 2.5–5 dm; from a woody caudex; with a taproot. **Stems** few to several, ascending to erect, unbranched or branched distally, sometimes with a few small, leafy axillary shoots, hairs spreading, long, soft, eglandular. **Leaves** green to purple-tinged, sometimes purple, narrowly to broadly lanceolate, 2.5–5 cm, not fleshy, margins plane, sometimes ± wavy, flat, 0(–5)-lobed, apex acuminate; lobes ascending-spreading, linear to narrowly lanceolate, apex acute. **Inflorescences** 3–15 × 2–4 cm; bracts proximally greenish, distally yellow-green, green, pale yellow, cream, or pale orange, often tinged with bright orange along margins, aging dull pink, broadly lanceolate to oblong or obovate, (0–)3–5(–7)-lobed; lobes ascending to erect, linear to narrowly lanceolate, medium length, arising above or below mid length, central lobe apex rounded to obtuse, lateral ones usually acute. **Calyces** colored as bracts, 13–20 mm; abaxial and adaxial clefts 5–9 mm, 33–50% of calyx length, deeper than laterals, lateral 1–2 mm, 7–25% of calyx length; lobes oblong, apex acute to rounded. **Corollas** straight, (15–)17–27 mm; tube 11–12 mm; beak exserted, adaxially green, 9–10 mm; abaxial lip green, reduced, visible in front cleft, 0.5–2 mm, 10–20% as long as beak; teeth incurved to erect, green, 0.5–1 mm. *2n* = 24, 48.

Flowering Jun–Sep. Subalpine wet meadows and springs, mixed conifer forests, volcanic soils; of conservation concern; 2600–2900 m; Ariz.

Castilleja mogollonica is endemic to the Mogollon Rim in the White Mountains of Apache County. It is frequently confused with the widespread *C. septentrionalis* but is amply distinct therefrom. This species faces threats from grazing, road building, and recreational activities. It occasionally hybridizes with *C. nelsonii.*

70. Castilleja mollis Pennell, Proc. Acad. Nat. Sci. Philadelphia 99: 185. 1947 • Soft-leaved paintbrush [C] [E]

Herbs or subshrubs, perennial, 3–4 dm; from a woody caudex; with woody roots. **Stems** few to several, ± prostrate, sometimes ascending, much-branched, with dense, matlike growth form, often with short, leafy axillary shoots, hairs dense, tangled, short to long, fairly stiff, branched, sometimes glandular, white-woolly. **Leaves** green to deep purple, narrowly elliptic to oblong, ovate, or obovate, 1–3 cm, ± fleshy, margins plane or ± wavy, flat, 0-lobed, apex rounded, rarely acute. **Inflorescences** usually erect, 2.5–8 × 1.5–4 cm; bracts proximally greenish, distally pale to bright yellow, sometimes brownish orange, sometimes with brownish orange medial band, oblong, elliptic, or obovate, ± cup-shaped, ± fleshy, 0–3-lobed; lobes erect, oblong, short, arising near tip, central lobe apex rounded to truncate, sometimes crenate or with obtuse teeth, lateral ones obtuse. **Calyces** colored as bracts, lacking orange central band, 16–23 mm; abaxial clefts 9.5–14 mm, adaxial 8 mm, abaxial 50–67% of calyx length, adaxial 35–45% of calyx length, deeper than laterals, lateral 4.5–5 mm, 20–25% of calyx length; lobes oblong to triangular, abaxials sometimes wider than adaxials, apex acute to rounded, inner surface glabrous. **Corollas** straight, 17–26 mm; tube 12–13 mm; beak often slightly exserted, adaxially green to yellow-green, 11–13 mm; abaxial lip green, reduced, 1.5–3.5 mm, 10–20% as long as beak; teeth incurved, reduced, green, 0.5–1 mm. *2n* = 24.

Flowering Apr–Aug. Sandy openings in coastal scrub, thin sandy soils over limestone terraces, north- or northwest-facing sandy bluffs, dunes; of conservation concern; 0–100 m; Calif.

Castilleja mollis is federally listed as a threatened species under the Endangered Species Act of the United States. It is endemic to the coastal terraces of the northern portion of Santa Rosa Island, Santa Barbara County, in the northern Channel Islands of southern California. It is recorded historically from San Miguel Island. Much of the available low-elevation habitat on Santa Rosa Island was degraded by trampling and grazing of introduced ungulates, which also resulted in the apparent loss of a natural population on the western end of San Miguel Island, last seen in the 1930s and now believed extirpated. Reports of *C. mollis* from the Oso Flaco Lake area of the mainland, in San Luis Obispo County, are based on populations of *C. affinis* var. *contentiosa. Castilleja mollis* is most closely related to *C. latifolia* of the central California coast and a sister species to the north, *C. mendocinensis.*

71. Castilleja montigena Heckard, Syst. Bot. 5: 83, fig. 17 [center]. 1980 • Heckard's paintbrush [E]

Herbs or subshrubs, perennial, 1.5–4.5 dm; from a woody caudex; with a taproot. **Stems** few to several, decumbent to erect, sometimes leaning, unbranched or often much-branched distally, with a few short, leafy axillary shoots, hairs spreading, short, soft, stipitate-glandular, mixed with long-spreading, eglandular ones. **Leaves** gray-green, sometimes green, lanceolate-linear to narrowly lanceolate, 1–6.5 cm, not fleshy, margins plane, sometimes wavy, flat to involute, 0(–3)-lobed, apex acuminate; lobes spreading-ascending, linear to narrowly lanceolate, apex acute. **Inflorescences** 3–30 × 3–4 cm; bracts proximally green to dark purplish, distally red to crimson, sometimes pale salmon, linear-lanceolate to broadly lanceolate, 3–5-lobed; lobes spreading, linear, long, arising below mid length, apex acute, sometimes obtuse. **Calyces** colored as bracts, 15–20 mm; abaxial clefts 3.4–6.2 mm, adaxial 4.5–9 mm, clefts 25–33% of calyx length, deeper than laterals, lateral 0.5–2 mm, 5–10% of calyx length; lobes narrowly triangular, often slightly unequal, apex acute. **Corollas** straight or slightly curved, 20–40 mm; tube 15–23 mm; abaxial lip exserted to included, beak much exserted; beak adaxially yellow-green to reddish, 9–18 mm; abaxial lip green, reduced, 0.5–1.5 mm, 5–20% as long as beak; teeth incurved, green, (0–)0.5–1.5 mm. $2n = 48, 72$.

Flowering May–Aug. Dry rocky slopes, ledges, open conifer forests, thickets, washes; 1900–2900 m; Calif.

Castilleja montigena is endemic to the northeastern portion of the San Bernardino Mountains of southern California. In the field, this species is consistently and relatively easily distinguished from nearby populations of *C. martini* var. *martini*, which it essentially replaces in the northeastern portion of the San Bernardino Mountains. It is apparently of allopolyploid hybrid origin between *C. martini* var. *martini* and *C. chromosa*, which approaches its range from the adjacent Mojave Desert.

72. **Castilleja nana** Eastwood, Proc. Calif. Acad. Sci., ser. 3, 2: 289. 1902 (as Castilleia) • Alpine or dwarf paintbrush E

Castilleja lapidicola A. Heller

Herbs, perennial, 0.4–1.7 dm; from a woody caudex; with a taproot. **Stems** few to several, ascending to decumbent-based, unbranched, hairs spreading, long, soft to stiff, mixed with shorter stipitate-glandular ones. **Leaves** green to deep purple, linear to narrowly lanceolate, (0.5–)1–2.5(–3.1) cm, not fleshy, margins plane, flat to slightly involute, 3(–7)-lobed, apex acuminate; lobes ascending-erect, linear to filiform, apex acute to obtuse. **Inflorescences** 15–20 × 1.5–3 cm; bracts greenish or deep purplish throughout, or proximally greenish or deep purplish, distally white, pink, magenta, reddish purple, purple, or pale yellow, distal coloration often limited to apices and margins, broadly lanceolate or narrowly to broadly elliptic, 0–3(–7)-lobed; lobes spreading, linear to narrowly oblanceolate, long, arising near mid length, apex acute or obtuse. **Calyces** green to purple, margins green, white, or pink, 10–19 mm; abaxial, adaxial, and lateral clefts 3.5–7(–10) mm, 33–55% of calyx length,

all 4 clefts subequal; lobes linear, lanceolate-elliptic, or narrowly triangular, apex acute to rounded. **Corollas** straight, 10–16(–19) mm; tube 8–13 mm, with patches of blackish coloration on either side of distal portion; beak subequal to calyx or exserted, adaxially green, yellow, whitish, or pink, 3–5.5 mm, scarcely exceeding abaxial lip, margins brown or burgundy, sometimes pink; abaxial lip pale yellow, white, green, or purple, inflated, pouched, 2–5 mm, 65–95% as long as beak; teeth erect to slightly spreading, white, yellow, or pink, 0.5–2.1 mm. **Stigmas** black. $2n = 24$.

Flowering Jun–Aug. Rocky or gravelly slopes, talus, ridges, fellfields, subalpine and alpine, often over granite; 1900–4300 m; Calif., Nev., Utah.

Castilleja nana is limited to high elevations in the Sierra Nevada of California and the Great Basin ranges in central Nevada and western Utah. Plants with pink to purple inflorescences were described as *C. lapidicola*. Localized hybrid swarms between *C. nana* and *C. viscidula* are known from several mountain ranges in central and northern Nevada. *Castilleja nana* is sometimes confused with higher elevation forms of *C. pilosa* but can usually be distinguished from that species by the blackish patches on the sides of the corolla tube. Divergent populations in the central Sierra Nevada and adjacent White Mountains deserve further study.

73. **Castilleja nelsonii** Eastwood, Proc. Amer. Acad. Arts 44: 579. 1909 • Arizona or southern mountains paintbrush

Castilleja austromontana Standley & Blumer

Herbs, perennial, 2.5–8(–10) dm; from a woody caudex; with a taproot or branched root system. **Stems** few to many, ascending to erect, unbranched or often strongly and diffusely branched distally, hairs sparse to dense, spreading to matted, long proximally on stem, becoming puberulent distally, ± stiff, eglandular, often mixed with retrorse shorter ones. **Leaves** green, linear-lanceolate or narrowly to broadly lanceolate, 2–6.5(–8) cm, not fleshy, margins plane, flat to involute, 0(–3)-lobed, apex acute; lobes ascending, lanceolate, apex acute to obtuse. **Inflorescences** (2.5–)5–15 × 2–4.5 cm; bracts proximally greenish, distally scarlet to red or orange-red, rarely yellow or crimson, veins usually yellow or yellow-green, contrasting conspicuously with base color, lanceolate or elliptic to oblanceolate or obovate, 0–3(–5)-lobed; lobes ascending, lanceolate to triangular, medium length, arising above mid length, apex rounded to obtuse. **Calyces** mostly yellowish throughout, with a thin reddish apex, 15–27 mm; abaxial clefts (5–)9–11 mm, adaxial 4.5–9.5 mm, clefts 25–50% of calyx length, deeper than laterals, lateral 2–4 mm, 10–20% of calyx length;

lobes linear-lanceolate to triangular, apex acute to acuminate, rarely ± obtuse. **Corollas** slightly curved, 15–35 mm, subequal to calyx or beak partially to strongly exserted; tube 15–17 mm; beak adaxially yellowish green, 10–16 mm; abaxial lip green, reduced, ± pouched, 0.5–1.5 mm, 4–10% as long as beak; teeth incurved, deep green, 0.7–1 mm. $2n = 24$.

Flowering Jun–Oct. Rocky slopes, meadows, riparian zones, moist ground in open forests, montane to subalpine; 1900–3100 m; Ariz., N.Mex.; Mexico (Chihuahua, Durango, Nayarit, Sonora).

Castilleja nelsonii is fairly common in the upper elevations of the so-called sky island ranges from central and eastern Arizona to adjacent New Mexico, southward into the Sierra Madre Occidental, at least as far south as southern Chihuahua, where the type collection was obtained on Cerro Mohinora. Although it was long known in the United States as *C. austromontana*, the name *C. nelsonii* has priority. Some specimens from southern Coconino County, Arizona, approach *C. miniata*, but most material is easily separable. *Castilleja nelsonii* occasionally hybridizes with *C. mogollonica* in Apache County, Arizona, near the border of the range of the former.

74. **Castilleja nervata** Eastwood, Proc. Amer. Acad. Arts 44: 574. 1909 [C]

Castilleja cruenta Standley

Herbs or subshrubs, perennial, 3–6(–10) dm; from a woody caudex; with a taproot. **Stems** solitary or few, ascending to erect, straight, unbranched or branched, hairs dense, spreading, long, stiff, eglandular, mixed with deflexed, short stipitate-glandular ones. **Leaves** green to purple-tinged, linear-lanceolate to broadly lanceolate or oblanceolate, 1.5–8 cm, not fleshy, margins plane, involute, 0-lobed, apex rounded to acute. **Inflorescences** 3–16 × 2–4.5 cm; bracts proximally green to deep purple, distally red to red-orange or orange, sometimes with a pale medial band, lanceolate to oblanceolate to ovate or obovate, 0(–3)-lobed; lobes when present upright, ± triangular, short, arising from distal portion, apex obtuse to rounded or truncate. **Calyces** proximally light green, distally deep green, deep purple, or blackish, 15–24 mm; abaxial and adaxial clefts 5–9 mm, 25–45% of calyx length, deeper than laterals, lateral 0–0.5 mm, 0–3% of calyx length; lobes oblong, sometimes emarginate, apex obtuse, rounded, or obliquely truncate. **Corollas** straight, 15–24 mm; tube 11–13 mm; subequal to calyx or tip slightly exserted; beak adaxially whitish to pinkish, 5–11 mm; abaxial lip pale to deep green, reduced, 1–2 mm, 10–20% as long as beak; teeth incurved, green, 1–2.5 mm. $2n = 24, 48$.

Flowering Aug–Sep. Dry south-facing rocky slopes with scattered bunchgrasses and oaks, cliff bases, mesa tops, open pine-oak woodlands, rocky savannas; of conservation concern; 1800–2400 m; Ariz.; Mexico (Aguascalientes, Chihuahua, Colima, Durango, Guerrero, Jalisco, México, Michoacán, Morelos, Nayarit, Oaxaca, Puebla, Sonora, Veracruz).

Castilleja nervata is a common and widespread Mexican species, from northern Chihuahua and Sonora south to central Oaxaca. It has medicinal value to indigenous peoples of the Sierra Madre. In the flora area, *C. nervata* is known from a single, presumably extant population in the Chiricahua Mountains in Cochise County, though there are historical records from the Rincon and Santa Rita mountains. The sole recently verified population is on private property and is endangered.

75. **Castilleja nivea** Pennell & Ownbey, Notul. Nat. Acad. Nat. Sci. Philadelphia 227: 2. 1950 • Snow or snowy paintbrush [E]

Herbs, perennial, 0.5–1.6 dm; from a woody caudex; with a taproot. **Stems** few to several, erect to ascending, decumbent at base, unbranched except for small, leafy axillary shoots, hairs weakly spreading to appressed, ± matted, especially distally on stem, fairly short and sparse proximally, longer and denser distally, soft, eglandular, becoming woolly, often obscuring surface. **Leaves** gray with hairs, surface green to purple, linear to narrowly lanceolate, 1–3.8 cm, not fleshy, margins plane, sometimes ± wavy, involute, 0–3-lobed, apex acute; lobes ascending-spreading, linear to narrowly lanceolate, apex acute to obtuse. **Inflorescences** 2.5–6 × 1–2.5 cm; bracts greenish to pale yellow-green or very pale, dull purplish throughout, lanceolate to oblong, (0–)3(–5)-lobed; lobes usually ascending, linear, medium length to long, arising near mid length, apex acute. **Calyces** yellow, color mostly obscured by whitish hairs, 15–22 mm; abaxial, adaxial, and lateral clefts (5.5–)7–12 mm, 35–55% of calyx length, often appearing shorter because matted hairs stitch proximal part of clefts shut, all 4 clefts subequal; lobes broadly linear, apex acute. **Corollas** straight, 18–25 mm; tube 3.5–5.5 mm; subequal to calyx, or beak and sometimes abaxial lip exserted; beak adaxially yellow, 6–8 mm, hairs moderately long, matted on midline, very short-glandular on sides; abaxial lip green, inconspicuous, slightly pouched, 3.5–5.5 mm, 60–90% as long as beak; teeth erect, white or yellow, 0.5–3 mm. $2n = 24$.

Flowering Jun–Aug. Gravelly slopes and flats, turf and fellfields, mostly alpine; 1700–3600 m; Mont., Wyo.

Castilleja nivea is endemic to alpine habitats in the mountains of northwestern Wyoming and adjacent Montana. It forms occasional hybrids with *C. pulchella*, which often shares its habitat, as on the Beartooth Plateau in northwestern Wyoming.

76. **Castilleja occidentalis** Torrey, Ann. Lyceum Nat. Hist. New York 2: 230. 1827 • Western or western yellow paintbrush E

Herbs, perennial, 0.7–2(–3) dm; from a small, woody caudex; with a taproot. **Stems** several to many, erect or ascending, usually short-decumbent at base, unbranched, sometimes glabrous proximally, hairs spreading, long, soft, mixed with medium length to short stipitate-glandular ones only in inflorescence. **Leaves** green to deep purple, linear-lanceolate to broadly lanceolate (to linear on sterile shoots), 1.5–4(–5.5) cm, not fleshy, margins plane, flat, prominently veined, 0–3(–5)-lobed, apex acute to rounded; lobes ascending, lanceolate, apex acute. **Inflorescences** 2–7(–10) × 1–3.5 cm; bracts greenish to pale greenish yellow throughout, often aging dull reddish brown or reddish purple proximally, rarely dull reddish brown throughout, or proximally green, dull reddish brown, or reddish purple, distally greenish white, yellow, or cream, broadly lanceolate to widely oblong to ovate, 0–3(–7)-lobed; lobes ascending, triangular to lanceolate, medium length, usually arising at or above mid length, rarely just below, central lobe apex obtuse to rounded, others acute. **Calyces** proximally green to purple, distally colored as bracts, 12–20 mm; abaxial and adaxial clefts 5–9(–10) mm, 40–50% of calyx length, deeper than laterals, lateral 1–3(–4.5) mm, 5–20% of calyx length; lobes lanceolate to triangular, apex acute, obtuse, or rounded. **Corollas** straight, 16–25 mm; tube 9–15 mm; teeth and part of abaxial lip sometimes exserted, beak exserted; beak adaxially green, (2.5–)5–9 mm; abaxial lip green, reduced, often visible through abaxial cleft, slightly pouched, 1.5–3 mm, 25–50% as long as beak; teeth incurved to ascending, white, sometimes green, 0.7–2 mm. $2n$ = 24, 48.

Flowering Jun–Sep. Meadows, gravel slopes, talus, ridges, mostly upper subalpine to alpine; 1500–4300 m; Alta., B.C.; Colo., Mont., N.Mex., Utah.

Castilleja occidentalis has a wide distribution in the southern Rocky Mountains, a gap in its distribution in Wyoming, and reappears in Montana and the Canadian Rockies. Bract lobing and color vary considerably but without correlation to geography.

Although *C. occidentalis* resembles an alpine form of *C. septentrionalis*, *C. occidentalis* is missing in several regions in the distribution of the latter, even when extensive areas of suitable habitat are available, and has a discrete range.

77. **Castilleja oresbia** Greenman, Bot. Gaz. 48: 147. 1909 • Pale wallowa paintbrush E

Herbs, perennial, 0.9–3 dm; from a woody caudex; with a stout taproot. **Stems** few to several, erect or ascending, sometimes decumbent at base, unbranched or branched, hairs usually retrorse, medium length, ± soft, eglandular, mixed with very short-glandular ones, sometimes with spreading, long, soft ones. **Leaves** green to purple, linear to lanceolate, 2–7 cm, not fleshy, margins plane, involute, 3–5(–7)-lobed, apex acuminate to acute; lobes spreading, linear to sometimes narrowly lanceolate, apex acute. **Inflorescences** 2.5–18 × 1–3.5 cm; bracts pale green to yellow-green or pale, dull reddish brown throughout, or proximally so colored but changing gradually to cream or yellowish on distal margins, narrowly to broadly lanceolate, (3–)5–7(–9)-lobed; lobes ascending, linear, long, proximal lobes arising below mid length, central lobe apex obtuse, others acute. **Calyces** colored as bracts, 10–25 mm; abaxial and adaxial clefts 6–7 mm, 30–60% of calyx length, deeper than laterals, lateral 5–10 mm, 40–50% of calyx length; lobes linear, apex acute. **Corollas** straight, 21–36 mm; tube 16–20 mm; teeth of abaxial lip often exserted, beak exserted; beak adaxially green, 4.2–5.5 mm; abaxial lip green to purple, distally white, conspicuous, slightly but noticeably pouched, often visible through front cleft, 3–5 mm, 67–100% as long as beak, puberulent; teeth erect, white, 1.8–2.1 mm.

Flowering May–Aug. Dry slopes and plains, sagebrush meadows, grasslands, openings in conifer forests; 900–2200 m; Idaho, Oreg.

Castilleja oresbia is endemic to eastern Oregon and adjacent Idaho. It is easily confused with both varieties of *C. pallescens*, which also occur in sagebrush habitats. *Castilleja oresbia* has longer calyx lobes and softer pubescence than *C. pallescens* var. *pallescens*, although some transitional specimens are found. *Castilleja oresbia* has a combination of longer calyx lobes, longer pubescence, and obscurely nerved bracts, which usually serve to separate it from *C. pallescens* var. *inverta*. All three have different, though somewhat overlapping, ranges. *Castilleja oresbia* occasionally hybridizes with *C. peckiana* in Grant County, Oregon.

78. Castilleja organorum Standley, Muhlenbergia 5: 86. 1909 • Organ Mountains paintbrush [C] [E]

Herbs or subshrubs, perennial, 2.7–8 dm; from a small, woody caudex; with thick, woody roots. **Stems** several to many, erect to sprawling, usually profusely branched, including many small, leafy axillary shoots, hairs dense, retrorse, medium length, stiff, eglandular, distally spreading, long, soft, sometimes matted, very short-glandular. **Leaves** green, linear-lanceolate, distally sometimes broadly lanceolate, 2–4.5 cm on main branch, 0.5–1 cm on proximal part of side branches, not fleshy, margins plane, involute, sometimes flat, 0-lobed, apex acute to acuminate. **Inflorescences** 2–4.5(–14 in fruit) × 1.5–4 cm; bracts proximally greenish, distally red to reddish orange, broadly lanceolate to oblong, 0(–3)-lobed; lobes ascending, lanceolate, short, arising above mid length, apex acute to obtuse. **Calyces** proximally pale green to pale yellow-green, distally pale red to red-orange above middle, 12.5–20.5 mm; abaxial and adaxial clefts 6–9 mm, 33–50% of calyx length, deeper than laterals, lateral (1.5–)3–4 mm, 20–35% of calyx length; lobes lanceolate or broadly triangular, apex acute. **Corollas** slightly curved, 15–24 mm; tube 10–13 mm; beak exserted, adaxially green, 6–10 mm; abaxial lip green, reduced, slightly pouched, sometimes visible in front cleft, 0.5–1.5 mm, 15–20% as long as beak; teeth incurved, white or green, 0.4–0.7 mm.

Flowering Jun–Oct. Rocky slopes, shaded canyons, riparian zones, open conifer forests, sun or partial shade; of conservation concern; 1500–2500 m; N.Mex.

Castilleja organorum is endemic to the Organ Mountains of Doña Ana County. Reports from the mountains of the Mogollon Rim of east-central Arizona and adjacent New Mexico are based on specimens of *C. nelsonii*. *Castilleja organorum* is grouped with *C. linariifolia* by some authors (for example, G. L. Nesom 1992c), but it has subequal abaxial and adaxial calyx clefts and is more likely closely related to *C. integra*. *Castilleja organorum* differs from the latter species in its loose, often profusely branched habit, more compact inflorescences, and usually smaller corollas. In Fillmore Canyon, *C. integra*, *C. lanata*, and *C. organorum* are all found, but each is in a different habitat, and there is no sign of hybridization.

79. Castilleja ornata Eastwood, Proc. Amer. Acad. Arts 44: 571. 1909 • Ornate paintbrush [C] [F]

Herbs, annual, 1.7–3.5(–5) dm; with a thin taproot or fibrous root system. **Stems** solitary or few to several, erect or ascending, often branched low on stem, unbranched distally, hairs appressed or retrorse, medium length, soft, eglandular, mixed with shorter stipitate-glandular ones. **Leaves** green or purple-tinged, proximal forming a rosette, linear-lanceolate to oblong or oblanceolate, 2–4 cm, not fleshy, clasping, margins wavy, sometimes plane, involute, 0-lobed, apex acuminate, acute, or obtuse. **Inflorescences** 3–24 × 1.5–3 cm; bracts proximally green, distally white, sometimes very pale yellow, often aging dull pink or dull red-purple, spatulate, 0-lobed, sometimes seeming lobed due to wavy margins, apex obtuse to rounded. **Calyces** green throughout or distal margin white aging pink, 15–17 mm; abaxial and adaxial clefts 6–14 mm, 35–45% of calyx length, deeper than laterals, lateral 0(–0.7) mm, 0(–5)% of calyx length; lobes short-triangular, abaxial segments longer than adaxials, apex acute to obtuse or rounded. **Corollas** slightly curved, 22–24 mm; tube 10–13 mm; beak exserted, adaxially green, 5–10 mm; abaxial lip pale greenish, reduced, pouches 3, 0.5–1.5 mm, 5–10% as long as beak; teeth slightly incurved, reduced, pale greenish to white, 0.3–0.7 mm. $2n = 24$.

Flowering Jul–Sep. Seasonally damp ground, dry or sandy grasslands; of conservation concern; 1500–2100 m; N.Mex.; Mexico (Chihuahua, Durango).

Castilleja ornata is known from Chihuahua and northern Durango, Mexico, but much of its seasonally moist grassland habitat is now altered by grazing or agriculture, and there are no recent sightings of the species south of the United States border. There is a recently discovered population in southwestern New Mexico, in the southern Animas Valley of Hidalgo County. While very rare, *C. ornata* lacks federal protection. The small Animas Valley population is the last known extant occurrence, and this population was reduced to two individuals in a census conducted in 2017 (D. Roth, pers. comm.). The species appears to be critically endangered globally and in need of conservation management.

The inflorescences of *Castilleja ornata* have pale greenish bracts with white apices when young, but the apices often become pale pink to dull reddish with age. Its pubescence, wavy-margined leaves, and unusual bract color also distinguish *C. ornata*. *Castilleja exserta* and *C. minor* are the only other annual paintbrushes in New Mexico and differ from *C. ornata* by the color of their floral bract apices, which are usually pink to red-purple in *C. exserta* and bright red in *C. minor*.

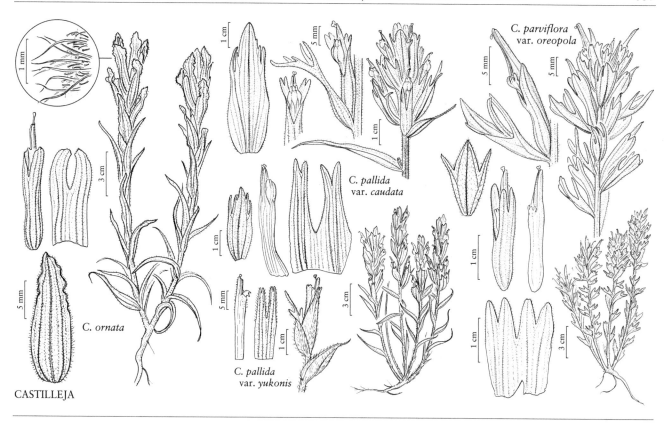

C. ornata

C. pallida var. yukonis

C. pallida var. caudata

C. parviflora var. oreopola

CASTILLEJA

80. Castilleja pallescens (A. Gray) Greenman, Bot. Gaz. 25: 266. 1898 (as Castilleia) • Pale paintbrush [E]

Orthocarpus pallescens A. Gray, Amer. J. Sci. Arts, ser. 2, 34: 339. 1862

Herbs, perennial, (0.4–)1–3 dm; from a woody caudex; with a taproot. **Stems** few to many, erect to ascending, decumbent at base, unbranched, sometimes branched, hairs moderately to very dense, retrorsely curved to appressed, short, ± stiff, eglandular. **Leaves** purple-tinged or deep purple, sometimes green, linear to narrowly lanceolate, 1–4(–5) cm, not fleshy, margins plane, sometimes ± wavy, ± involute, (0–)3–5(–7)-lobed, apex acute; lobes spreading or ascending-spreading, linear, apex acute. **Inflorescences** (1.5–)4–8(–12) × 1.5–5.5 cm; bracts pale green to yellow-green or reddish purple throughout, or proximally pale green to yellow-green, distally white to cream or pale yellowish, sometimes pink to reddish purple, lanceolate to linear-lanceolate, elliptic, or ovate, 3–5(–9)-lobed; lobes spreading to ascending, linear, long, arising along distal ⅔, apex acute to obtuse. **Calyces** colored as bracts, sometimes distally purple with age, 11–25(–27) mm; abaxial and adaxial clefts 7–13.6 mm,

40–50% of calyx length, deeper than laterals, lateral 0.5–4.3(–6) mm, (0–)5–25% of calyx length; lobes lanceolate to triangular, apex usually triangular or acute, rarely ± obtuse. **Corollas** straight, 13–23(–27) mm; tube 10–20 mm; subequal to calyx or beak slightly exserted; beak adaxially whitish or buff, rarely pink to pink-purple, 3.5–8 mm; abaxial lip proximally green, white, purple, or purplish brown, distally white, yellow, green, pink, or reddish, prominent, pouched, pouches pleated, longer than deep, gradually expanded, 2.5–8 mm, 70–100% as long as beak, puberulent; teeth erect to spreading, pink, cream, or white, sometimes with a yellow spot proximally, 1.5–3.5 mm. $2n = 24, 48.$

Varieties 2 (2 in the flora): w United States.

Castilleja pallescens occurs from valleys to alpine ridges and summits throughout its range, usually in sagebrush communities, but at higher elevations it is also found on dry sites associated with other plant species. The alpine plants are greatly reduced in stature.

1. Bracts not rigid, veins inconspicuous, usually same color as surfaces; herbs (0.5–)1–3 dm; ne Idaho, sw Montana, nw Wyoming . . . . 80a. *Castilleja pallescens* var. *pallescens*
1. Bracts rigid, veins prominent, pale and contrasting with color of surfaces; herbs 0.4–1.2(–1.7) dm; s, se Idaho, ne Nevada, Oregon. . . . . 80b. *Castilleja pallescens* var. *inverta*

80a. Castilleja pallescens (A. Gray) Greenman var. **pallescens** E

Herbs (0.5–)1–3 dm. **Leaves:** margins plane, sometimes ± wavy, (0–)3–5(–7)-lobed; lobes spreading. **Bracts** distally pale greenish to yellow-green, whitish, or pale pink, rarely purplish, lanceolate to elliptic or ovate, not rigid, veins inconspicuous, usually same color as surface. **Calyces** colored as bracts, sometimes distally purple with age, 11–23 mm; abaxial clefts 8–13.6 mm, adaxial 7–8.2 mm, lateral (0.9–)2.4–4.3 mm; lobes triangular to lanceolate. **Corollas** subequal to calyx or beak slightly exserted; beak adaxially whitish; abaxial lip green or white, sometimes distally pink or reddish; teeth white or pink. *2n* = 24, 48.

Flowering Jun–Aug. Grasslands, dry rocky slopes, sagebrush steppes, open forests, meadows, fellfields, valleys, alpine; 1500–2900 m; Idaho, Mont., Wyo.

Variety *pallescens* is found in northeastern Idaho, southwestern Montana, and northwestern Wyoming.

80b. Castilleja pallescens (A. Gray) Greenman var. **inverta** (A. Nelson & J. F. Macbride) Edwin, Leafl. W. Bot. 9: 72. 1959 • Dwarf pale or foothills paintbrush E

Castilleja fasciculata A. Nelson var. *inverta* A. Nelson & J. F. Macbride, Bot. Gaz. 55: 381. 1913; *C. inverta* (A. Nelson & J. F. Macbride) Pennell & Ownbey

Herbs 0.4–1.2(–1.7) dm. **Leaves:** margins plane, 0–3 (–7)-lobed; lobes ascending-spreading. **Bracts** distally pale green or cream, sometimes pale yellowish or reddish purple to deep purple, lanceolate to linear-lanceolate, rigid, veins prominent, pale and contrasting with color of surfaces. **Calyces** purple to green with white or pink margins and veins, 15–25(–27) mm; abaxial and adaxial clefts 7–12 mm, lateral 0.5–3(–6) mm; lobes linear-lanceolate or narrowly triangular. **Corollas** subequal to calyx or beak sometimes exserted, abaxial lip sometimes partially exserted; beak adaxially whitish, buff, or pink; abaxial lip proximally green, purple, or purplish brown, distally yellow, white, or pink; teeth white, white with yellow spot proximally, cream, or pink. *2n* = 24.

Flowering May–Aug. Dry slopes, ridges, and plains, often with sagebrush, talus, ledges, seasonally damp meadows; 1500–3200 m; Idaho, Nev., Oreg.

The range of var. *inverta* somewhat overlaps that of the similar *Castilleja oresbia*, but the two are separable by the characters of the key. The inflorescences of both species are usually pale straw yellow to pale green.

In Lemhi County, Idaho, primarily on alpine ridges along the summit of the Lemhi Range, plants apparently referable to var. *inverta* occur in large numbers. Almost all of them have deep reddish purple inflorescences, with a few rare variants of pale pink or pale yellow. This population system occurs on the border of the ranges of var. *inverta* and var. *pallescens*, and the entire complex deserves further study.

81. Castilleja pallida (Linnaeus) Sprengel, Syst. Veg. 2: 774. 1825 (as Castilleia) • Pale paintbrush F

Bartsia pallida Linnaeus, Sp. Pl. 2: 602. 1753

Herbs, perennial, sometimes biennial, 1–4.1(–5.5) dm; from a woody caudex; with a taproot. **Stems** few to many, erect or ascending, unbranched or branched, glabrate proximally or hairy, hairs usually spreading to weakly appressed, whitish or yellowish, short to long, stiff to ± soft, eglandular to rarely stipitate-glandular. **Leaves** green to red-brown or deep purple, linear-lanceolate to lanceolate or linear, (1.5–)5–10.5 cm, not fleshy, margins plane, sometimes ± wavy, flat to involute, 0-lobed, sometimes 3–5-lobed distally immediately below inflorescence, apex acuminate, caudate, or acute, sometimes obtuse; lobes ascending-spreading, linear to narrowly lanceolate, sometimes with earlike appendages, short, apex acute to obtuse. **Inflorescences** 2–16 × 1–2.5 cm; bracts yellow, yellow-green, or pale whitish throughout, sometimes with dull reddish-purplish wash proximally, especially with age, proximal few lanceolate, most broadly lanceolate to ovate or lanceolate to oblong, 0–5(–7)-lobed, sometimes central lobe with a few small teeth; lobes ascending to erect, linear, sometimes expanded distally, short, arising near or above mid length, central lobe apex obtuse to rounded or truncate, lateral ones acute to obtuse. **Calyces** proximally pale green, red-brown, or purple, distally colored as bracts, 12–18.5 mm; abaxial clefts 6–9 mm, adaxial 6–12.1 mm, clefts 45–60% of calyx length, deeper than laterals, lateral 0.3–4(–7) mm, 2–40% of calyx length; lobes linear or triangular to ovoid, apex acute to obtuse or rounded. **Corollas** straight or slightly curved, 12–26(–28) mm; tube 10–20 mm; abaxial lip sometimes exserted, beak usually exserted; beak adaxially green to yellowish, 3.5–8 mm; abaxial lip proximally green or purple, distally white, purple, yellow to greenish yellow, cream, orange-brown, or reddish, ± inconspicuous, slightly to moderately pouched, 3–6 mm, 50–75% as long as beak; teeth erect to curved, white, cream, or yellow, (0.8–)1.2–2 mm. *2n* = 48, 72.

Varieties 6 (2 in the flora): n North America, n Asia.

Castilleja pallida as interpreted here is one of the most wide-ranging species in the genus, extending from the Kola Peninsula of the western Palearctic eastward across arctic and boreal Asia to similar latitudes in the western Nearctic region. Accounts of this species in the eastern Nearctic and farther south in North America are here attributed to *C. septentrionalis*, which the authors regard as a separate species. Throughout its enormous range, *C. pallida* is extremely complex, with several levels of polyploidy documented and numerous localized races of highly variable and inconstant forms. On the other hand, some plants from as far west as the arctic Ural Mountains are virtually identical in morphology to plants of var. *caudata* in central Alaska. *Castilleja pallida* is treated differently in regional floras, as one highly variable species or as a complex of numerous species, with or without infraspecific taxa. Despite some recent collections, *C. pallida* remains an incompletely known entity, and a fairly conservative approach to its delimitation is preferred until its variations and relationships to closely related species such as *C. elegans* and *C. raupii*, as well as to the numerous named forms in the Palearctic, are understood. Access to critical type material from Russian herbaria remains problematic as well. A full examination of types and a comprehensive genetic and morphological survey of specimens across the range are needed for a satisfactory treatment of the *C. pallida* complex.

Only vars. *caudata* and *yukonis* in North America are accepted tentatively here, yet it is possible that var. *pallida* or another east-Asian form occurs in the general area of Nome, Alaska. Some collections from northwestern Alaska appear to be distinct from var. *caudata*, but their precise identity is not established. Even the morphological boundaries between vars. *caudata* and *yukonis* are problematic and variable, especially in the Kluane Lake region of the southern Yukon and adjacent Alaska, as well as in other regions where the two come in contact. The type of *Castilleja annua* consists of poor material, and the description is based on plastic traits, so here it is treated as a synonym of var. *caudata*. However, additional research may yield better characters and a rationale to distinguish it.

1. Lateral calyx clefts (1.5–)3–4(–7) mm, lobes linear, sometimes triangular; leaves lanceolate to linear-lanceolate, rarely linear; stems glabrate or hairy . . . . 81a. *Castilleja pallida* var. *caudata*
1. Lateral calyx clefts 0.3–2(–4) mm, lobes triangular to ovoid; leaves linear to linear-lanceolate; stems hairy . . . . 81b. *Castilleja pallida* var. *yukonis*

81a. **Castilleja pallida** (Linnaeus) Sprengel var. **caudata** (Pennell) B. Boivin, Naturaliste Canad. 79: 320. 1952 • Pale or yellow paintbrush F

Castilleja pallida subsp. *caudata* Pennell, Proc. Acad. Nat. Sci. Philadelphia 86: 524. 1934; *C. annua* Pennell; *C. caudata* (Pennell) Rebristaya; *C. pallida* subsp. *auricoma* Pennell

Herbs biennial or short-lived perennial. **Stems** glabrate or hairs appressed proximally, spreading distally, medium length proximally, longer distally, eglandular. **Leaves** lanceolate to linear-lanceolate, rarely linear, (1.5–)5–7(–9) cm, apex acuminate, caudate, or acute, sometimes obtuse; lobe apex acute to obtuse. **Inflorescences** 2–16 cm; bracts yellow, cream, or greenish yellow throughout, sometimes proximally green or dull purple, proximal few lanceolate, most broadly lanceolate to ovate. **Calyces** proximally pale green to red-brown or purple, distally colored as bracts, 12–13 mm; abaxial clefts 7–7.8 mm, adaxial 8.4–12.1 mm, clefts 50–60% of calyx length, lateral (1.5–)3–4(–7) mm, 20–40% of calyx length; lobes linear, sometimes triangular. **Corollas** slightly curved, 22–26 mm; tube 14–16 mm; beak margins white and often also deep purple or red-brown, or yellowish; abaxial lip with 1–3 of the colors green, purple, white, cream, or rose red, as a solid color or in 2 or 3 horizontal layers, 3–4 mm. $2n = 72$.

Flowering Jun–Aug(–Sep). Moist tundra, peatlands, stream banks, forests, bluffs, terraces, roadsides; 0–1700 m; N.W.T., Nunavut, Yukon; Alaska; ne Asia.

Variety *caudata* is both widespread and variable throughout much of Alaska and enters western and northern Yukon. The precise eastern extent of its range is incompletely known, especially since it is often similar to and easily confused with both var. *yukonis* and northern variants of *Castilleja elegans* and *C. raupii*. It apparently occurs in far northeastern Asia, but its status and distribution there is incompletely understood. See additional notes for this form under 81. *C. pallida*.

81b. **Castilleja pallida** (Linnaeus) Sprengel var. **yukonis** (Pennell) J. M. Egger, Phytologia 90: 76. 2008 • Yukon paintbrush E F

Castilleja yukonis Pennell, Proc. Acad. Nat. Sci. Philadelphia 86: 531. 1934; *C. muelleri* Pennell

Herbs perennial. **Stems:** hairs spreading, ± matted, long, eglandular, mixed with retrorse, short ones, rarely short stipitate-glandular ones. **Leaves** linear to linear-lanceolate, 1.5–10.5 cm, apex acuminate; lobe apex acute. **Inflorescences** 3–7 cm; bracts yellowish throughout, sometimes orangish or reddish, sometimes green with margins yellow or cream, lanceolate to oblong. **Calyces** proximally pale green, distally light yellow, (13–)16–18.5 mm; abaxial and adaxial clefts 6–9 mm, 45–55% of calyx length, lateral 0.3–2(–4) mm, 2–20% of calyx length; lobes triangular to ovoid. **Corollas** straight, 12–22(–28) mm; tube 10–20 mm; beak margins light yellow; abaxial lip greenish yellow, orange-brown, or red, 4–6 mm. $2n = 48$.

Flowering Jun–Aug. Dry grasslands, bluffs, eskers, pingos, alluvial plains, shores, open slopes, gravelly roadsides; 600–1100 m; N.W.T., Yukon; Alaska.

Variety *yukonis* is primarily limited to central and southern Yukon, though it also likely occurs at least sporadically in the adjacent portions of eastern Alaska. Reports of this variety from Asia are doubtful. Its status in the Mackenzie River delta region in the Northwest Territories is still unclear, and plants from that region may be referable to variably colored populations of *Castilleja raupii*. See additional notes for this form under 81. *C. pallida*.

82. Castilleja parviflora Bongard, Mém. Acad. Imp. Sci. St.-Pétersbourg, sér. 6, Sci. Math. 2(2): 158. 1832

• Mountain or rosy or small-flowered paintbrush

E F

Herbs, perennial, (0.6–)1–4(–5) dm; from a woody caudex; with a taproot or stout, branched roots. **Stems** several or many, erect or ascending, unbranched except for short, leafy axillary shoots, glabrate proximally, hairy distally, hairs sparse, spreading, ± matted, long, soft, minute-glandular. **Leaves** green or gray-green to purple-tinged or deep purple, often blackening on drying, narrowly to broadly lanceolate or elliptic, rarely linear, 1.5–5 cm, not fleshy, margins plane, sometimes ± wavy, flat, (0–)3–9-lobed, apex acute to acuminate or obtuse; lobes spreading or ascending, linear, sometimes lanceolate, much narrower than terminal lobe, evenly spaced, short, apex acute. **Inflorescences** 2–16 × 1–3.5 cm; bracts proximally greenish, dull, deep purple, or reddish purple, distally pink, pink-purple, magenta, deep rose, crimson, cream, or white, sometimes red, pale orange, or red-orange, lanceolate to broadly elliptic or ovate, 3–7-lobed; lobes spreading to ascending, linear, lanceolate, or lanceolate-acuminate, short to medium length, arising at or near mid length, apex obtuse to acute, central lobes sometimes rounded. **Calyces** colored as bracts, 12–28 mm; abaxial and adaxial clefts 6–15 mm, 40–70% of calyx length, deeper than laterals, lateral 1–8 mm, 10–35% of calyx length; lobes narrowly to broadly triangular, sometimes distally expanded and flaring, petaloid, apex obtuse or acute, sometimes rounded. **Corollas** straight or slightly curved,

12–30 mm; tube 8–19 mm; beak exserted or subequal to calyx, adaxially green-yellowish or red, 5.5–11 mm; abaxial lip green, brown, or yellow, sometimes purple, reduced, slightly or not inflated and pouched, 1–3 mm, 20–45% as long as beak; teeth erect, green, white, yellow, pink, or red, 0.5–2 mm. $2n = 24, 48$.

Varieties 4 (4 in the flora): w North America.

Castilleja parviflora is a complex, geographically widespread, and often misunderstood species ranging from southeastern Alaska through much of British Columbia, southwestern Yukon, and the Rocky Mountains of extreme western Alberta and southward in the Cascade Range to central Oregon.

1. Corollas (18–)20–30 mm; calyces 20–28 mm; leaves (0–)3(–5)-lobed; Oregon, Washington. . . . . 82d. *Castilleja parviflora* var. *oreopola*
1. Corollas 12–20(–25) mm; calyces 12–20(–28) mm; leaves (0–)3–9-lobed; Washington to Alaska, Alberta, and Yukon.
  2. Bracts distally white to cream, sometimes suffused with pink to purple; herbs 0.6–2.7 dm; n Cascade Range, Washington and s British Columbia . . . . 82b. *Castilleja parviflora* var. *albida*
  2. Bracts distally pink-purple, magenta, deep rose, or crimson, rarely white; herbs 1.2–5 dm; Olympic Mountains, Washington, w Canada, se Alaska.
    3. Leaves (3–)5–9-lobed; corolla beaks 5.5–7 mm; Alberta, British Columbia, Yukon, Alaska . . . . 82a. *Castilleja parviflora* var. *parviflora*
    3. Leaves (0–)3(–5)-lobed; corolla beaks (5.5–)7–9(–11) mm; s Vancouver Island, British Columbia, Olympic Mountains, Washington . . . . 82c. *Castilleja parviflora* var. *olympica*

82a. Castilleja parviflora Bongard var. **parviflora** E

Castilleja henryae Pennell

Herbs 1.3–3(–5) dm. **Leaves** narrowly to broadly lanceolate or elliptic, margins plane, (3–)5–9-lobed, apex acute; lobes spreading, linear. **Bracts** distally magenta to pink-purple, sometimes pale orange, pink, or whitish, (3–)5(–7)-lobed; lobes lanceolate-acuminate, arising near or above mid length. **Calyces** colored as bracts, 12–17(–20) mm; abaxial and adaxial clefts 6–11 mm, lateral 2–6 mm, 20–35% of calyx length; lobes narrowly triangular, apex acute. **Corollas** 12–20(–25) mm; tube 10–12 mm; beak exserted, 5.5–7 mm; abaxial lip green; teeth green.

Flowering Jun–Sep. Mesic to wet meadows, stream banks, rocky slopes, ridges, ledges, scree, heathlands, subalpine to lower alpine; 150–2300 m; Alta., B.C., Yukon; Alaska.

Variety *parviflora* has relatively long hairs, many-lobed leaves, and a short corolla. Compared to the other varieties, it has a more boreal range, extending north to extreme southeastern Alaska and the Kluane Lake region of extreme southwestern Yukon. Specimens from northern and central British Columbia belong to this variety, but its precise southern limits and potential overlap with var. *albida* in southern British Columbia are still incompletely known. Plants presumably assignable to var. *parviflora* are also found in the Canadian Rocky Mountains in the vicinity of Banff and Jasper National parks, but they were until recently rarely collected and deserve further study.

82b. Castilleja parviflora Bongard var. **albida** (Pennell) Ownbey in C. L. Hitchcock et al., Vasc. Pl. Pacif. N.W. 4: 317. 1959 • Pale small-flowered paintbrush E

Castilleja oreopola Greenman subsp. *albida* Pennell, Proc. Acad. Nat. Sci. Philadelphia 99: 188. 1947

Herbs 0.6–2.7 dm. **Leaves** broadly, sometimes narrowly, lanceolate to elliptic, margins plane to ± wavy, (0–)3–5(–7)-lobed, apex acute to obtuse; lobes ascending, lanceolate. **Bracts** distally white to cream, sometimes suffused with pink to purple, 3–5(–7)-lobed; lobes lanceolate, arising at or above mid length. **Calyces** green to purple or red, distally white to pink, 12–17(–21) mm; abaxial clefts (6–)7–8 mm, adaxial 8–10 mm, abaxial ca. 40% of calyx length, adaxial ca. 33% of calyx length, lateral 2–6 mm, 15–25% of calyx length; lobes triangular, apex acute, sometimes obtuse. **Corollas** 12–20(–25) mm; tube 8–11 mm; subequal to calyx or beak exserted, 5.5–8 mm; abaxial lip green, brown, or yellow; teeth white to yellow, sometimes pinkish.

Flowering (Jun–)Jul–Sep. Moist to wet meadows, snowmelt streams, receding shorelines, subalpine to lower alpine; 1200–2500 m; B.C.; Wash.

Variety *albida* is found in the Cascade Range of southern British Columbia southward to the Wenatchee Mountains of Washington. In central British Columbia, the point of transition into var. *parviflora* still needs definition. Variety *albida* is characterized by whitish to cream bracts, although some plants in Okanogan County, Washington, have pink or purple bracts in mixed populations with white-bracted plants.

82c. Castilleja parviflora Bongard var. **olympica** (G. N. Jones) Ownbey in C. L. Hitchcock et al., Vasc. Pl. Pacif. N.W. 4: 317. 1959 • Olympic Mountains paintbrush C E

Castilleja olympica G. N. Jones, Bot. Surv. Olympic Penins., 231. 1936; *C. oreopola* Greenman subsp. *olympica* Pennell

Herbs 1.2–3.2 dm. **Leaves** broadly lanceolate, rarely linear, margins plane, (0–)3(–5)-lobed, apex narrowly acute to acuminate; lobes spreading, linear to lanceolate, lateral lobes nearly as long as terminal. **Bracts** distally pink-purple, magenta, deep rose, or crimson, rarely white, 3–7-lobed; lobes lanceolate, arising near to above mid length. **Calyces** deep purple with magenta or light pink lobes, 13–20(–28) mm; abaxial clefts 6.5–12 mm, adaxial 7–15 mm, lateral 2–8 mm, 10–35% of calyx length; lobes narrowly triangular, apex acute to obtuse. **Corollas** 12–20(–25) mm; tube 8.5–19 mm; beak exserted, (5.5–)7–9(–11) mm; abaxial lip green; teeth green.

Flowering Jul–Sep. Dry to moist meadows, forest openings, ridges, subalpine to lower alpine; of conservation concern; 1000–1600 m; B.C.; Wash.

Variety *olympica* is essentially endemic to the upper elevations of the Olympic Mountains, in northwestern Washington, though a handful of collections from high elevations on Vancouver Island, British Columbia, are also referable to this form.

82d. Castilleja parviflora Bongard var. **oreopola** (Greenman) Ownbey in C. L. Hitchcock et al., Vasc. Pl. Pacific N.W. 4: 317. 1959 • Magenta paintbrush E F

Castilleja oreopola Greenman, Bot. Gaz. 25: 264. 1898 (as Castilleia)

Herbs 1–4 dm. **Leaves** broadly lanceolate to rarely linear, margins plane to ± wavy, (0–)3(–5)-lobed, apex acuminate to acute; lobes spreading or ascending, linear, very narrow. **Bracts** distally magenta to pink or pink-purple, sometimes red, red-orange, deep rose, or crimson, rarely whitish, 3–7-lobed; lobes linear, arising below or above mid length. **Calyces** colored as bracts, 20–28 mm; abaxial clefts 8–13 mm, adaxial 10–14 mm, lateral 1–3 mm, 12–20% of calyx length; lobes narrowly to broadly triangular, distally expanded and flaring, petaloid, apex obtuse to acute or rounded. **Corollas** (18–)20–30 mm; tube 12–19 mm; beak exserted, 8–11 mm; abaxial lip green, sometimes purple; teeth green, red, or white.

Flowering Jun–Sep. Dry to moist meadows, ridges, pumice, subalpine to lower alpine; 1500–2200 m; Oreg., Wash.

Variety *oreopola* is restricted to the Cascade Range from central Oregon, near the Three Sisters peaks, to just north of Mt. Rainier, in Washington. It is replaced to the north in the Wenatchee Mountains and the North Cascade Range by var. *albida*. Variety *oreopola* usually has pink to purple inflorescences, with occasional variants of pinkish red, pinkish orange, or white. Its colorful displays are a conspicuous element of the subalpine meadows and slopes of this region. Reports of var. *oreopola* in the Cascade Mountains of southern British Columbia seem plausible but have yet to be fully verified.

83. Castilleja parvula Rydberg, Bull. Torrey Bot. Club 34: 40. 1907 • Tushar Mountains paintbrush C E

Herbs, perennial, 1–2 dm; from a woody caudex; with a taproot or stout, branched roots. Stems several to many, decumbent to ascending, unbranched except for small, leafy axillary shoots, hairs retrorse, short proximally, spreading, longer distally, soft, stipitate-glandular. Leaves green to blackish, proximalmost small and scalelike, linear to narrowly or broadly lanceolate, 1–3(–4) cm, not fleshy, margins plane, involute, 0(–3)-lobed, apex obtuse to rounded; lobes ascending, linear or short-lanceolate, very small, apex acute to obtuse. Inflorescences 4–6.5 × 1–3 cm; bracts proximally greenish to deep purple near base, distally magenta, deep pink, or red, broadly lanceolate to elliptic, oblong, or ovate, 0–5-lobed; lobes ascending to erect, triangular to oblong, short, arising near apex, central lobe apex rounded, lateral ones acute to rounded. Calyces colored as bracts, 12–18 mm; abaxial and adaxial clefts 5.5–8.5(–10) mm, ca. 50% of calyx length, deeper than laterals, lateral 1–3(–5) mm, ca. 25% of calyx length; lobes broadly linear or narrowly triangular, apex acute to obtuse, sometimes rounded. Corollas straight, 16–24 mm; tube 10–13(–15) mm; beak exserted, adaxially green, 4.5–8(–9) mm; abaxial lip green to deep purple, reduced, 1–3 mm, 40–45% as long as beak; teeth incurved, greenish, (0.5–)1–2.5 mm. $2n = 24$.

Flowering Jun–Aug. Gravelly meadows, rocky slopes, talus, ridges, krummholz zone or alpine; of conservation concern; 2700–3700 m; Utah.

Castilleja parvula is limited to the upper elevations of the Tushar Mountains. Morphologically, it appears to be a species derived from the widespread Rocky Mountains species *C. rhexiifolia*.

84. Castilleja patriotica Fernald, Proc. Amer. Acad. Arts 40: 56. 1904 • Patriotic paintbrush F

Castilleja blumeri Standley; *C. patriotica* var. *blumeri* (Standley) Kearney & Peebles

Herbs, perennial, 1.7–6 dm; from a slender, woody caudex; with a taproot. Stems few to many, erect or ascending, much-branched with many short, leafy axillary shoots, hairs recurved to retrorse, short, ± stiff, eglandular, often with some spreading, long ones. Leaves green, linear-lanceolate to lanceolate, 1–3.5 cm, not fleshy, margins plane, sometimes ± wavy, flat to involute, (0–)3–7-lobed, apex acute to obtuse; lateral lobes spreading, linear, much shorter and narrower than terminal lobe, apex acute. Inflorescences 3–17 × 2–7.5 cm; bracts green throughout, sometimes distalmost proximally green, distally red or red-orange on apices, lanceolate or broadly lanceolate, (0–)3–5-lobed; lobes spreading, linear or narrowly lanceolate, short, arising near or below mid length, central lobe apex obtuse to rounded, lateral ones acute. Calyces proximal ⅓ green, distal ⅔ red to pale red-orange, rarely yellow, (17–)25–35(–40) mm; abaxial clefts 14–29 mm, 67–75% of calyx length, adaxial 2.5–7.3 mm, 5–15% of calyx length, deeper than laterals, lateral 0–0.3(–2) mm, 0–5% of calyx length; lobes broadly triangular, entire or shallowly cleft, segments, if present, often differing in length, apex acute, obtuse, or rounded. Corollas slightly curved, (24–)27–43 mm; tube 9–20 mm; beak longer than calyx, exserted through abaxial cleft, adaxially green to yellowish, 22–39 mm; abaxial lip deep green, inflated, tight, 1–2.5 mm, 5–10% as long as beak; teeth often spreading, green or white, 0.5–1.6 mm. $2n = 24$.

Flowering Jul–Oct. Dry slopes and flats, open pine forests, ledges, mostly montane; 2000–3000 m; Ariz., N.Mex.; Mexico (Chihuahua, Durango, Sinaloa, Sonora).

Castilleja patriotica is broadly distributed in the northern Sierra Madre Occidental in Mexico, south to northern Durango and Sinaloa. It is replaced in central and southern Mexico by a related species, *C. pectinata* M. Martens & Galeotti. *Castilleja patriotica* reaches the Chiricahua and Huachuca mountains of southeastern Arizona and the adjacent Animas Mountains of southwestern New Mexico. A report from the Atascosa Mountains of southern Arizona needs verification. The numerous, narrow axillary leaves and the narrow leaf lobes combine to give this plant a curious frilly appearance. The narrow, widely divaricate lateral leaf lobes are unique in the flora area. Short, compact plants with smaller flowers were named *C. blumeri*. *Castilleja patriotica* has medicinal value to native peoples of the northern Sierra Madre.

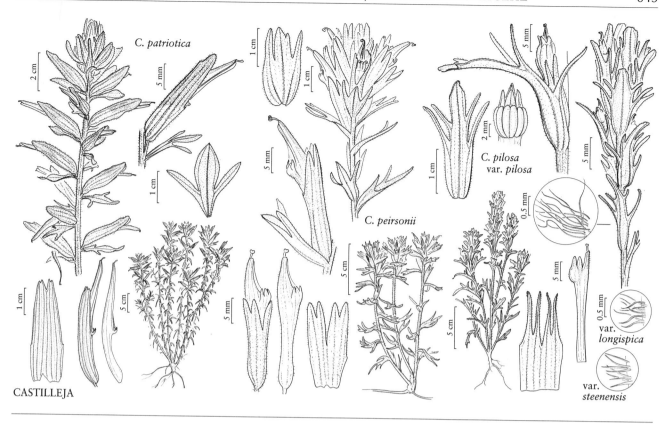

C. patriotica

C. peirsonii

C. pilosa var. pilosa

var. *longispica*

var. *steenensis*

CASTILLEJA

85. Castilleja peckiana Pennell, Notul. Nat. Acad. Nat. Sci. Philadelphia 74: 9. 1941 • Peck's paintbrush [E]

Herbs, perennial, (1.8–)2.4–6 dm; from a woody caudex; with a stout taproot. **Stems** few to many, erect or ascending, often branched distally, sometimes unbranched, sometimes with short axillary shoots, proximal hairs retrorse to appressed, short to moderately long, distal hairs spreading, longer, soft, often mixed with short stipitate-glandular ones. **Leaves** green, linear-lanceolate, rarely broadly lanceolate, (1.2–)2.5–8(–9) cm, not fleshy, margins plane, flat or involute, 0(–3)-lobed, apex acute to rounded; lobes ascending-spreading, narrowly lanceolate to linear, apex acute or obtuse. **Inflorescences** (2–)4–17 × 1.5–3 cm; bracts proximally greenish, distally red, orange-red, or orange, proximal sometimes lanceolate, distal broadly lanceolate to ovate, (0–)3(–7)-lobed; lobes spreading to ascending, linear to lanceolate, long, arising near or below mid length, central lobe apex rounded, lateral ones acute. **Calyces** proximally pale yellow or greenish, distally colored as bracts, (15–)18–28 mm; abaxial and adaxial clefts 6–12 mm, 40–45% of calyx length, deeper than laterals, lateral 2.5–8 mm, 20–30% of calyx length; lobes linear to narrowly lanceolate, apex acute to acuminate. **Corollas** straight, 23–30(–35) mm; tube 12–20 mm; beak exserted, adaxially green, 8–12(–14) mm; abaxial lip pale to deep green, reduced, rounded, 0.5–1.5 mm, 10–20% as long as beak; teeth erect to incurved, green, (0.4–)0.7–1.2(–2) mm. $2n = 72, 96,$ ca. 120.

Flowering (Apr–)May–Aug. Open conifer forests, sagebrush slopes, riparian meadows, shores; 1400–2600 m; Idaho, Nev., Oreg.

When describing *Castilleja peckiana*, Pennell noted that its variation approached *C. hispida* on one hand and *C. miniata* on the other, and it is likely of allopolyploid derivation. L. R. Heckard (1968) found chromosome numbers of *n* = 36, 48, and ca. 60. He hypothesized that *C. chromosa*, *C. hispida* var. *acuta*, and *C. miniata* were likely involved in its ancestry, and possibly *C. pruinosa* as well. Heckard suggested subsequent introgression among the derived forms introduced further complexity. Though complex, these forms are self-perpetuating and appear morphologically stable within their range.

86. Castilleja peirsonii Eastwood, Leafl. W. Bot. 1: 175. 1935 (as peirsoni) • Peirson's paintbrush [E] [F]

Castilleja carterae Beane

Herbs, perennial, 0.6–3(–4) dm; from a woody caudex; with a taproot. **Stems** few to several, erect or ascending, unbranched, sometimes branched, hairs dense distally, less so proximally, spreading, long, soft, eglandular, also mixed with short stipitate-glandular ones distally. **Leaves** green to purple, narrowly to broadly lanceolate to oblong, (0.7–)1.5–4.2(–5) cm, not fleshy, margins plane, flat or involute, (0–)3(–5)-lobed, apex acuminate; lobes ascending-spreading, narrowly lanceolate to linear, apex acute. **Inflorescences** 2.5–7.5(–15) × 1.5–3 cm; bracts proximally greenish to dull reddish, distally bright red, orange, or pale orange, sometimes yellowish or dull red, broadly lanceolate to oblong, 3–5-lobed; lobes ascending, linear to lanceolate, ± long, arising above mid length, sometimes below mid length on proximal bracts, central lobe apex acute, rarely narrowly obtuse, lateral ones acute. **Calyces** proximally light green or yellow, distally yellow or colored as bracts, 12–20 mm; abaxial and adaxial clefts 5–9 mm, 40–50% of calyx length, deeper than laterals, lateral (1.5–)3–5.8 mm, 10–30% of calyx length; lobes oblong, apex acute to obtuse. **Corollas** slightly curved, 15–28 mm; tube 13–15 mm; beak exserted, adaxially yellow or yellow-green, 7–8 mm; abaxial lip yellow or deep green, reduced, slightly inflated, pouched, protruding out abaxial cleft, 1–2.5 mm, 13–33% as long as beak; teeth erect or curved, yellow or green, 0.7–1 mm. $2n = 24$.

Flowering Jun–Sep. Moist to wet meadows, stream banks, lakeshores, montane to alpine; 1700–3400 m; Calif., Nev.

Castilleja peirsonii is endemic to the higher elevations in the Sierra Nevada of California and in Tahoe Meadows in adjacent Washoe County, Nevada. Plants with yellow bracts in the southern portion of the range were named *C. carterae*. *Castilleja peirsonii* has shorter, wider corolla beaks than the related *C. parviflora*. In addition to the corolla shape differences, *C. peirsonii* also has red to yellow bracts and fairly bright yellow corollas, especially on the beaks, while *C. parviflora* has purple, pink, or white bracts and greenish corolla tubes and dorsal beak surfaces, with the beak margins pink, purple, or white. Reports from outside the Sierra Nevada in California and immediately adjacent Nevada are misidentifications. *Castilleja peirsonii* sometimes hybridizes with *C. lemmonii* in meadows where both species often occur in large numbers.

87. Castilleja pilosa (S. Watson) Rydberg, Mem. New York Bot. Gard. 1: 361. 1900 • Hairy paintbrush [E] [F]

Orthocarpus pilosus S. Watson, Botany (Fortieth Parallel), 231, 459. 1871

Herbs, perennial, (0.7–)1.2–3.5 (–4.4) dm; from a woody caudex; with a stout taproot. **Stems** several to many, ascending to erect, sometimes short-decumbent, branched or unbranched, sometimes with short, leafy axillary shoots, hairs moderately dense, retrorse or curved to spreading, straight, curly, or ± wavy, medium length to long, soft to stiff, eglandular. **Leaves** green to purple, linear to lanceolate, 1–5.5(–8) cm, not fleshy, margins plane to ± wavy, involute, 0–5(–7)-lobed, apex acuminate to obtuse; lobes widely spreading to ascending-spreading, linear to filiform, apex acute or obtuse. **Inflorescences** (2–)3.5–16 × 1–3.5 cm; bracts light green, green, yellow-green, light purple, purple, light dusky pink, salmon, or reddish brown throughout, or these colors proximally, distally or on distal margins white, pale yellow, yellow, pale salmon, or buff, sometimes becoming reddish purple with age, lanceolate to broadly lanceolate or narrowly ovate, 3–5(–9)-lobed, often wavy-margined; lobes spreading to erect or ascending, linear to oblanceolate, short to long, arising near or above mid length, sometimes wavy-margined, central lobe apex obtuse to rounded or truncate, sometimes acute, lateral ones acute to obtuse. **Calyces** proximally whitish to pale green, distally green, whitish, pink, or yellowish, 9–28 mm; all 4 clefts subequal, 2–12 mm, 45–55% of calyx length; lobes linear to narrowly lanceolate or narrowly triangular, rarely deltoid, apex acute, rarely rounded. **Corollas** straight, 14–23 mm; tube 10–15; abaxial lip sometimes partially exserted, beak exserted; beak adaxially green or yellow-green, 3–7 mm; abaxial lip proximally green, pale yellow, pale or bright pink, or deep purple, distally white to pink or purplish, inflated, pouches 3, deeply furrowed, 2.5–8 mm, 50–100% as long as beak; teeth erect, green, white, buff, pink, or pale yellow, 1–2 mm. $2n = 24, 48, 96$.

Varieties 3 (3 in the flora): w United States.

Castilleja pilosa is a widespread and variable complex similar in growth form and coloration to *C. pallescens* and related species. However, *C. pilosa* is distinguished with relative ease by its subequally divided calyces. Plants of the *C. pallescens* complex have very shallow lateral calyx lobes.

1. Calyces 9–15(–20) mm; c, se Idaho, nw Wyoming, adjacent Montana. . . 87b. *Castilleja pilosa* var. *longispica*
1. Calyces (11–)14–28 mm; ne California, nw Nevada, e Oregon.
   2. Stem hairs spreading, ± wavy or curly, soft, (0.4–)0.8–1.3(–1.5) mm; sagebrush steppes, rocky slopes, ridges, seasonally moist meadows or pools, conifer forests, montane to subalpine; 500–2000(–3500) m; ne California, nw Nevada, e Oregon . . . . 87a. *Castilleja pilosa* var. *pilosa*
   2. Stem hairs retrorse or spreading, ± curved, stiff, 0.5–0.8(–1) mm; rocky slopes, ledges, dry meadows, sagebrush steppes, subalpine to alpine, volcanic soils; (1900–)2000–2900 m; Steens Mountain, se Oregon . . . . 87c. *Castilleja pilosa* var. *steenensis*

87a. Castilleja pilosa (S. Watson) Rydberg var. **pilosa**
E F

Castilleja pilosa subsp. *jusselii* (Eastwood) Munz; *C. psittacina* (Eastwood) Pennell

Stems: hairs spreading, ± wavy or curly, soft, (0.4–) 0.8–1.3(–1.5) mm. **Leaf lobes** widely spreading to ascending-spreading. **Inflorescences** (2–) 3.5–15 cm; bracts proximally green, sometimes pinkish, reddish brown, or light purple with margins or distal portions pale yellow or whitish, (3–)5(–9)-lobed; lobes spreading to erect, medium length to long, arising near to above mid length. **Calyces** (11–)14–28 mm; abaxial and adaxial clefts 5–7 mm, 45–50% of calyx length, lateral 5–7 mm, 45–55% of calyx length; lobes narrowly lanceolate. **Corolla beaks** 3–5 mm; abaxial lip green, sometimes pink or pale yellow, 3.5–8 mm, 70–100% as long as beak; teeth green, white, pink, buff, or pale yellow. $2n = 24$.

Flowering May–Aug. Sagebrush steppes, rocky slopes, ridges, seasonally moist meadows or pools, conifer forests, montane to subalpine; 500–2000 (–3500) m; Calif., Nev., Oreg.

Variety *pilosa* is found in sagebrush steppes and communities in the surrounding mountains from northeastern California through much of eastern Oregon, and northwestern Nevada. The inflorescences are almost always pale yellowish white when young and often turn dull reddish purple with age, especially proximally. This variety is sometimes confused with the high-elevation species *Castilleja nana*; it can usually be distinguished by the absence of blackish patches on the sides of the corolla tube, which are present in *C. nana*.

87b. Castilleja pilosa (S. Watson) Rydberg var. **longispica** (A. Nelson) N. H. Holmgren in A. Cronquist et al., Intermount. Fl. 4: 480. 1984
• White paintbrush E F

Castilleja longispica A. Nelson, Bull. Torrey Bot. Club 26: 480. 1899

Stems: hairs spreading or retrorse and curved, soft. **Leaf lobes** ascending-spreading. **Inflorescences** 2–16 cm; bracts proximally light green, or yellow-green, sometimes becoming pink or dull purple, with margins or distal portions pale yellow, sometimes soft pink or buff, 3–5(–9)-lobed; lobes erect, sometimes spreading, short to medium length, arising near or above mid length. **Calyces** 9–15(–20) mm; abaxial and adaxial clefts 2–7 mm, 40–45% of calyx length, lateral 2–5.5 mm, 40–50% of calyx length; lobes linear or narrowly triangular. **Corolla beaks** 4–7 mm; abaxial lip proximally green or pale yellow, sometimes dull purple, distally white to sometimes pink, 2.5–5 mm, 50–85% as long as beak; teeth white to pale yellow, rarely pink. $2n = 24, 48, 96$.

Flowering Jun–Aug. Moist to dry meadows, sagebrush steppes, woodlands, rocky slopes, ridges, montane; 1700–2000 m; Idaho, Mont., Wyo.

Variety *longispica* is distributed across central and southeastern Idaho, as well as adjacent Montana and Wyoming. This variety is often somewhat variable in color within populations, ranging from pale greenish yellow to various shades of pale pink to pale purple.

87c. Castilleja pilosa (S. Watson) Rydberg var. **steenensis** (Pennell) N. H. Holmgren in A. Cronquist et al., Intermount. Fl. 4: 480. 1984 • Steen's Mountain paintbrush E F

Castilleja steenensis Pennell, Notul. Nat. Acad. Nat. Sci. Philadelphia 74: 4. 1941

Stems: hairs retrorse or spreading, ± curved, stiff, 0.5–0.8 (–1) mm. **Leaf lobes** widely spreading. **Inflorescences** 2.5–10 cm; bracts proximally greenish, light dusky pink, or salmon, with margins or distal portions the same color or yellow or white, 3–5-lobed; lobes spreading to ascending, short to long, arising below mid length. **Calyces** (12–)14–20(–28) mm; abaxial and adaxial clefts 5–12 mm, 45–50% of calyx length, lateral 5–12 mm, 45–50% of calyx length; lobes linear to lanceolate. **Corolla beaks** 4.5–6.5 mm; abaxial lip purple or pink, sometimes pale yellow, distally sometimes white, 3.5–6 mm, 75–100% as long as beak; teeth white or pale yellow.

Flowering May–Aug. Rocky slopes, ledges, dry meadows, sagebrush steppes, subalpine to alpine, volcanic soils; (1900–)2000–2900 m; Oreg.

Variety *steenensis* is endemic to the upper elevations of Steens Mountain in Harney County. It appears distinct when seen in its dwarfed form on the highest slopes; some transitional plants exist at lower elevations where it encounters var. *pilosa*. Genetic studies of this complex are in progress.

88. **Castilleja plagiotoma** A. Gray, Proc. Amer. Acad.
    Arts 19: 93. 1883 (as Castilleia) • Mojave Desert
    paintbrush E F

Herbs, perennial, 2.3–5.3(–6) dm; from a woody caudex; with a taproot. Stems several, erect to weakly ascending, often leaning on nearby shrubs, much-branched, often with short, leafy axillary shoots, hairs sparse, spreading, short or long, soft, branched or unbranched, eglandular. Leaves gray-green or green to purple to ± deep red, linear or linear-lanceolate, 2–5 cm, not fleshy, margins plane to ± wavy, flat or involute, 3–5(–7)-lobed, apex acuminate to acute; lobes spreading-ascending, sometimes widely so, narrowly linear, apex acute or obtuse. Inflorescences 3–20 × 0.5–1.5 cm; bracts green throughout, sometimes tinged dull purple, proximal sometimes linear-lanceolate, distal or all bracts oblong, 3–5(–7)-lobed; lobes spreading to ascending, linear to oblanceolate, short to long, proximal lobes arising near mid length, sometimes wavy-margined, central lobe apex rounded to truncate, lateral ones obtuse to rounded. Calyces light green, cream, or light yellow, often appearing white from dense white-woolly pubescence, 10–18 mm; abaxial clefts 5–6.5 mm, closed by intertwined hairs and appearing 2 mm deep, adaxial 2.2–4.2 mm, abaxial ca. 25% of calyx length, adaxial ca. 20% of calyx length, shallower (or appearing much shallower), than laterals, lateral 4.6–7.5 mm, ca. 50% of calyx length; lobes: abaxial segments broad, paddle-shaped, 7 mm, inner surface of abaxial segments densely white-woolly, apex rounded to broadly obtuse, adaxial narrowly triangular, 5 mm, apex acute. Corollas straight, 13–20 mm; tube 5–7 mm; beak short-exserted, sometimes shorter than abaxial calyx segments, adaxially yellow, rarely greenish, 7–10 mm; abaxial lip pale green to yellowish, reduced, 0.5–1.5 mm, 7–17% as long as beak; teeth reduced to apiculations, pale green to yellowish, 0.5–0.8 mm.

Flowering Mar–Jul. Dry flats, rocky, sandy, or clayey slopes, ridges, sagebrush steppes, chaparral, desert scrub, pinyon woodlands; 200–2500 m; Calif.

Castilleja plagiotoma is unique in the genus in the structure of its calyces, with the lateral clefts deeper than the median clefts. The relative lengths of the calyx lobes are also unique, with the abaxial lobes exceeding the adaxial lobes by 2–3 mm. The abaxial lobes also bear a dense indument of whitish, branched hairs. The uniformly greenish bracts are uncommon among perennial species of *Castilleja*. *Castilleja plagiotoma* is scattered in the western Mojave Desert and the hills adjacent to the southwestern San Joaquin Valley, California. While not of immediate conservation concern, this species is uncommon and increasingly threatened by recreational vehicles, livestock grazing, residential development, and resource extraction.

Castilleja plagiotoma is most often associated with and is likely parasitic on *Eriogonum fasciculatum* var. *polifolium*, but it is also often observed with *Artemisia tridentata* or other species. It is a known larval host plant for the butterfly, *Euphydryas editha* subsp. *erlichii*.

89. **Castilleja praeterita** Heckard & Bacigalupi,
    Madroño 20: 209, fig. 1. 1970 • Salmon Creek
    paintbrush C E

Herbs, perennial, (1–)1.6–4.5 dm; from a woody caudex; with a taproot. Stems several to many, ascending to erect, ± decumbent at base, branched or unbranched, sometimes with short, leafy axillary shoots, hairs spreading to ascending, long, soft to ± stiff, mixed with short-eglandular ones. Leaves purple or green, linear-lanceolate to lanceolate, 3–5 cm, not fleshy, margins plane, partly involute, 3–5-lobed, apex acuminate, acute, or rounded; lobes sometimes divergent, spreading-ascending, linear, apex obtuse to rounded. Inflorescences (2.5–)5–15 × 1.5–2 cm; bracts proximally green to dull purplish, distally pale reddish purple, dull red, pale salmon, pale orange, or bright to pale yellow, oblong, 3(–5)-lobed; lobes spreading to ascending, linear to oblanceolate, distal pair, if present, short and toothlike, short to long, arising at or above mid length, center lobe apex rounded to truncate, lateral lobes obtuse to rounded. Calyces proximally usually whitish, distally colored as bracts, 13–18 mm; abaxial and adaxial clefts 5–6(–9) mm, ca. 50% of calyx length, deeper than laterals, lateral 0.5–1.2 mm, ca. 18% of calyx length; lobes ± hemispheric, segments often curved outwards, exposing corollas, apex rounded to obtuse, rarely acute. Corollas straight, 11–16 mm; tube 10–13 mm; beak included or tip exserted; beak adaxially green, 4–5 mm; abaxial lip deep green, reduced, with narrow pouches, 2.5–3 mm, 30–45% as long as beak; teeth reduced to minute apiculations, dark green, 0.5–1 mm. $2n = 24$.

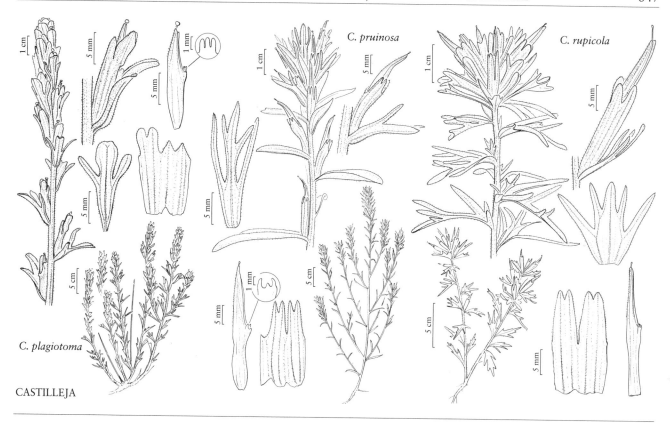

C. pruinosa

C. rupicola

C. plagiotoma

CASTILLEJA

Flowering Jun–Aug. Dry gravelly meadows and flats, with *Artemisia rothrockii*, often over granite; of conservation concern; 2200–3400 m; Calif.

Castilleja praeterita is endemic to subalpine, sagebrush-dominated meadows in the southeastern Sierra Nevada of Inyo and Tulare counties. It is closely associated with and likely parasitic on *Artemisia rothrockii*. Inflorescence coloration varies by population, with some having only yellow-bracted plants, while others are pale orange to pale red with occasional pale yellow variants. Yet other populations have only pale salmon-colored bracts.

90. Castilleja pruinosa Fernald, Erythea 6: 50. 1898 (as Castilleia) • Frosted or pruinose paintbrush E F

Herbs to subshrubs, perennial, (1.2–)2–8 dm; from a woody caudex; with a taproot. **Stems** several to many, erect or ascending, branched or unbranched, sometimes with inconspicuous, short, leafy axillary shoots, hairs ± dense, spreading to appressed, whitish or ashy gray, long, soft, much-branched, rarely unbranched, eglandular or obscurely glandular. **Leaves** green, linear-lanceolate to narrowly or broadly lanceolate, 2–8 cm, not fleshy, margins plane to sometimes wavy, flat or involute, 0-lobed, sometimes 3–5-lobed distally, apex obtuse to rounded, sometimes acute; lobes sometimes divergent, ascending-spreading, linear, apex acute. **Inflorescences** 3–20 × 1.5–4 cm; bracts proximally greenish, distally abruptly red, sometimes yellow, salmon, or orange-red, often with a narrow yellow medial band, lanceolate, broadly lanceolate, or oblong, 0–5-lobed; lobes spreading to ascending, linear, long, arising near mid length, apex obtuse to acute. **Calyces** proximally green to yellow, distally and abruptly colored as in distal bracts, often with a central yellow band, 13–28 mm; abaxial clefts 5–13 mm, adaxial 8–14 mm, clefts 33–70% of calyx length, deeper than laterals, lateral 3–9 mm, 20–25% of calyx length; lobes lanceolate to triangular, apex acute to acuminate. **Corollas** slightly curved, 20–51 mm; tube 12–21 mm; beak long-exserted, adaxially green, yellow, or red, 12–33 mm; abaxial lip deep green, reduced, 1–2.5 mm, ca. 10% as long as beak; teeth incurved to erect, white, yellow, or green, 0.5–2 mm. $2n = 48$.

Flowering Mar–Aug. Dry flats, rocky slopes, talus, thickets, open forests and forest edges, usually over serpentine; 50–2600 m; Calif., Oreg.

Castilleja pruinosa is a common and variable paintbrush at moderate elevations in the southern half of the Cascade Range, occurring westward in Oregon and in adjacent northwestern California. It is also found in the northern Sierra Nevada. The often dense pubescence of branched hairs distinguishes it from other similar species within its range. *Castilleja gleasoni* was sometimes included within *C. pruinosa*, but *C. gleasoni* is likely of hybrid origin and occurs only in the San

Gabriel Mountains of southern California. There are significant morphological differences between *C. gleasoni* and *C. pruinosa*. *Castilleja pruinosa* forms occasional hybrids with *C. applegatei* var. *pinetorum* and with *C. brevilobata*. In some populations in Douglas and eastern Lane counties, Oregon, *C. pruinosa* forms introgressive swarms with *C. hispida* var. *hispida*.

91. Castilleja puberula Rydberg, Bull. Torrey Bot. Club 31: 644. 1905 • Alpine or short-flowered paintbrush C E

Herbs, perennial, 0.8–1.5 dm; from a woody caudex; with a taproot. **Stems** few to several, erect or ascending, unbranched, sometimes branched, hairs retrorse, short, soft, eglandular. **Leaves** green to purple, linear to linear-lanceolate, 2–3.3 cm, not fleshy, margins plane, involute, 0–3(–5)-lobed, apex acuminate; lobes ascending-spreading to widely spreading, narrowly linear, apex acute. **Inflorescences** 4–5 × 1.5–2.5 cm; bracts greenish throughout, or proximally greenish, distally bright yellow, yellow-green, or yellow-orange on apices, narrowly lanceolate to linear-lanceolate, 0–5(–7)-lobed; lobes spreading, filiform to linear, long, proximal lobes arising in proximal ½, apex acute to acuminate. **Calyces** light green to yellowish, margins sometimes yellow, 10–17 mm; abaxial clefts 8–9 mm, adaxial 2–3 mm, abaxial ca. 50% of calyx length, adaxial ca. 20% of calyx length, deeper than laterals, lateral 1.9–2.5 mm, 12–20% of calyx length; lobes narrowly triangular, apex acute to acuminate. **Corollas** straight or slightly curved, 18–21 mm; tube 13–15 mm; beak exserted, adaxially yellow to yellow-green, 6–8 mm; abaxial lip green, reduced, visible in front cleft, slightly pouched, 2.5–3.5 mm, 35–60% as long as beak; teeth erect, yellow, 1 mm. *2n* = 24.

Flowering (Feb–)Jul–Aug. Moist meadows, stream banks, mesic rocky slopes, tundra, subalpine and alpine; of conservation concern; 2700–3900 m; Colo., Mont.

Castilleja puberula is a near-endemic of alpine communities in the Rocky Mountains of Colorado in Boulder, Clear Creek, Gilpin, Grand, Larimer, and Park counties. A highly disjunct population occurs near the summit of Mt. Jefferson on the Continental Divide in Beaverhead County, Montana. *Castilleja puberula* appears to be a high-elevation isolate from the widespread *C. flava*, which is common at lower elevations in the Intermountain Region. A single case of hybridization with *C. septentrionalis* is known from Clear Creek County.

92. Castilleja pulchella Rydberg, Bull. Torrey Bot. Club 34: 40. 1907 • Beautiful or showy paintbrush E

Herbs, perennial, 0.5–1(–2) dm; from a woody caudex; with a slender taproot. **Stems** few to several, erect or ascending, usually decumbent at base, unbranched, sometimes branched, hairs spreading, short and long, soft, mostly glandular. **Leaves** green to deep purple, linear to broadly lanceolate, 1–3.5(–5) cm, not fleshy, margins plane or wavy, slightly involute, (0–)3–5-lobed, apex acuminate to acute, sometimes obtuse; lateral lobes ascending-spreading, sometimes widely spreading, narrowly lanceolate to linear, mostly short, apex acute. **Inflorescences** 2–7 × 1.5–3 cm; bracts yellow-green, yellow, pinkish, pale reddish, or purple, sometimes distal margins pale white, oblong to broadly lanceolate to ovate, 0–5-lobed; lobes spreading, linear to narrowly lanceolate, medium length to long, usually arising above mid length, rarely from below mid length, center lobe apex rounded to truncate, lateral ones acute to obtuse. **Calyces** colored as bracts, sometimes strongly bicolored green or proximally whitish and distally as bract lobes, (12–)13–23(–25) mm; abaxial and adaxial clefts 5–11 mm, 45–55% of calyx length, deeper than laterals, lateral (0.5–)1–3(–5) mm, 5–20% of calyx length; lobes broadly triangular, apex obtuse to rounded, rarely acute. **Corollas** straight or slightly curved, 17–22(–25) mm; tube 11–16 mm; subequal to calyx or beak, sometimes abaxial lip, exserted; beak adaxially yellow to green, 4–6(–7) mm; abaxial lip green or yellow, reduced, often visible through front cleft, prominently pouched, thickened, 3–4(–5) mm, 50–67% as long as beak; teeth erect, white, yellow, pink, or purple, 1.5–3 mm. *2n* = 24.

Flowering (May–)Jun–Aug. Moist meadows, turf, rocky slopes and flats, talus, fellfields, subalpine to alpine; 1800–3500 m; Idaho, Mont., Utah, Wyo.

Castilleja pulchella is a mostly alpine species of the mountains of western Montana and adjacent Idaho and northwestern Wyoming, as well as in the Uinta Mountains of northeastern Utah. It is similar to and likely shares ancestry with *C. chrysantha* of the mountains of northeastern Oregon. *Castilleja pulchella* is variable in color, with inflorescences ranging from pale yellow to purplish, often within the same population. Some lower elevation populations are known. These plants are considerably taller, and they tend to have only yellowish inflorescences. Where the two occur together, *C. pulchella* occasionally forms hybrids with *C. nivea*.

93. **Castilleja purpurascens** Greenman, Bot. Gaz.
42: 146. 1906 [C] [E]

Herbs, perennial, (1.5–)2–4 dm; from a woody caudex; with branched, woody roots. **Stems** few or several, ascending to erect, branched or unbranched, shiny proximally, glabrous proximally, hairy distally, hairs spreading, long, soft, eglandular, sometimes sparsely glandular. **Leaves** deep purple to green, linear-lanceolate to narrowly lanceolate, rarely broadly lanceolate, 1.5–4.5 cm, not fleshy, margins plane, flat or slightly involute, 0(–3)-lobed, apex usually acuminate or acute; lobes upright or ascending, lanceolate, apex acute. **Inflorescences** 2–12 × 2–6 cm; bracts deep reddish or crimson, rarely magenta or dull orange, or proximally greenish near base, distally colored as above, oblong to broadly lanceolate, often spreading from base and exposing calyces, 0–3(–5)-lobed; lobes ascending to erect, linear, short to medium length, arising near or above mid length, apex acute to obtuse. **Calyces** colored as bracts, 18–28 mm; abaxial clefts 5–6 mm, adaxial 8–15 mm, clefts 28–50% of calyx length, deeper than laterals, lateral 1.5–6 mm, 5–25% of calyx length; lobes long-triangular, apex acute. **Corollas** straight to slightly curved, 20–37 mm; tube 12–21 mm; abaxial lip sometimes exserted, beak exserted; beak adaxially green, 10–16 mm; abaxial lip green, sometimes yellow, usually reduced, protruding through abaxial cleft, sometimes a little pouched, 3-lobed, 2–6 mm, 17–45% as long as beak; teeth erect, green, whitish, or yellow, 1–1.5 mm. *2n* = 48.

Flowering (May–)Jun–Jul. Gravel river flats, moist thickets; of conservation concern; 1200–2200 m; Alta., B.C.

Castilleja purpurascens is a species of gravelly flood plains and riverbanks at moderate elevations in the vicinity of the Kicking Horse River in Yoho National Park, British Columbia, and in immediately adjacent Alberta. While this tetraploid species may be derived from hybridization between *C. miniata* and *C. rhexiifolia*, its combination of traits is unique, and the species forms morphologically consistent populations limited to a habitat not particularly favored by either putative parent species. Very occasional hybrids are known with both *C. lutescens* and *C. miniata*.

94. **Castilleja purpurea** (Nuttall) G. Don, Gen. Hist.
4: 615. 1837/1838 • Prairie or purplish paintbrush [E]

Euchroma purpurea Nuttall, Trans. Amer. Philos. Soc., n.s. 5: 180. 1835; *Castilleja williamsii* Pennell

Herbs, perennial, 1.5–3(–4) dm; from a woody caudex; with a taproot. **Stems** few to many, erect to ascending, branched, sometimes unbranched, hairs fairly dense, spreading to appressed, white, fairly short, soft, ± felty, eglandular, sometimes mixed sparsely with short stipitate-glandular ones, sometimes obscuring surface. **Leaves** green to purple, linear to narrowly oblong-lanceolate, 2–7(–9) cm, not fleshy, margins plane, sometimes slightly wavy, involute, 3–7-lobed, apex narrowly acute to rounded; lobes spreading, linear, apex obtuse or acute. **Inflorescences** 2.6–16 × 2–4 cm; bracts proximally greenish to deep greenish purple, distally purple, magenta, reddish, pink, or rose, rarely white, cream, light yellow, or dull orangish, proximal linear to lanceolate, distal oblong, 3–7(–9)-lobed; lobes spreading to ascending, linear to oblanceolate, long, arising from distal ⅔, center lobe apex obtuse to rounded, lateral ones acute to rounded. **Calyces** colored as bracts, (20–)25–34 mm; abaxial and adaxial clefts (10–)13–22 mm, 50–60% of calyx length, deeper than laterals, lateral 7–16 mm, 35–45% of calyx length; lobes broadly linear to long-triangular or oblong, apex acute to obtuse. **Corollas** slightly curved, 25–40 mm; tube 16–22 mm; beak exserted, adaxially green, 9–18 mm; abaxial lip green to purple-red, reduced, not strongly pouched, ± protruding, 4–5 mm, 33–50% as long as beak; teeth prominent, petaloid, spreading to erect, colored as in distal portion of bracts, 3–4 mm.

Flowering Mar–May. Rocky slopes, ledges, prairies, woodlands, thickets, roadsides, often sandy or limy soils; 200–600 m; Kans., Mo., Okla., Tex.

Castilleja purpurea is a common species of eastern Oklahoma and central Texas, with a few records from adjacent southeastern Kansas and southwestern Missouri. It often provides beautiful, multicolored displays in the meadows within its range. *Castilleja citrina* and *C. lindheimeri* are closely related species sometimes regarded as varieties of *C. purpurea*. Hybrids and hybrid swarms between *C. indivisa* and *C. purpurea* have been observed at some localities where they are sympatric.

95. Castilleja raupii Pennell, Proc. Acad. Nat. Sci. Philadelphia 86: 528, fig. [p. 529]. 1934 • Raup's paintbrush E

Herbs, perennial, (2.5–)3–5(–6) dm; from a woody caudex; with a taproot. **Stems** few to many, erect to ascending, unbranched or often branched distally, glabrous proximally or hairy, hairs sparse, retrorse, short, ± stiff proximally, distally ± dense, spreading, longer, soft, eglandular throughout. **Leaves** purple to green, linear to lanceolate-linear, 2–8(–13) cm, not fleshy, margins plane to slightly wavy, flat or slightly involute, 0-lobed, sometimes 3-lobed distally, immediately below inflorescence, apex narrowly acuminate to acute; lobes ascending, linear, apex acute. **Inflorescences** 2.5–12 × 1.5–3.5 cm; bracts lavender, pink-purple, or reddish purple throughout, sometimes distally white, cream, or pale pink, lanceolate, oblong, broadly elliptic, or ovate, 0–5(–7)-lobed; lobes spreading to ascending, linear to narrowly lanceolate, short, arising above mid length, central lobe apex acute to obtuse, lateral ones acute. **Calyces** colored as bracts, 13–20(–25) mm; abaxial and adaxial clefts 6–11 mm, 50–60% of calyx length, deeper than laterals, lateral 0.5–3.5(–7) mm, ca. 25–30% of calyx length; lobes lanceolate-linear, apex narrowly acute or acuminate. **Corollas** straight or slightly curved, 15–20(–26) mm; tube 10–13 mm; beak and abaxial lip ± exserted; beak adaxially green, 4–6.5 mm; abaxial lip purple, magenta, or red, slightly inflated, pouches 3, 2–5 mm, 75–80% as long as beak; teeth erect, magenta, pink, or red, 1–1.5 mm. $2n = 72$.

Flowering Jun–Aug. Damp openings, thickets, stream banks, hummocks in peatlands, meadows, tundra, sandy or gravelly, calcareous ridges and roadsides; 0–2400 m; Alta., B.C., Man., N.W.T., Nunavut, Ont., Que., Sask., Yukon; Alaska.

Castilleja raupii is a member of the difficult *C. pallida* species complex. *Castilleja raupii* is widespread from southern Alaska to northern Alberta, northwards to the shore of the Arctic Sea, and east to the eastern shore of Hudson Bay. Reports from northeastern Russia are mostly referable to the similar Asian species, *C. rubra* (Drobow) Rebristaya. *Castilleja raupii* is sometimes confused with the partially sympatric *C. elegans*, but *C. raupii* is a taller plant with often somewhat branched, ascending-erect stems and is often found on more mesic substrates than *C. elegans*. The inflorescences of *C. raupii* are usually pink to pink-purple, but more variably colored populations are reported from the delta region of the Mackenzie River in the Northwest Territories of Canada.

96. Castilleja revealii N. H. Holmgren, Bull. Torrey Bot. Club 100: 87, fig. 2. 1973 • Reveal's paintbrush C E

Castilleja parvula Rydberg var. *revealii* (N. H. Holmgren) N. D. Atwood

Herbs, perennial, 0.8–1.5(–2) dm; from a woody caudex; with a taproot. **Stems** solitary, sometimes few, erect or ascending, sometimes decumbent, unbranched, glabrous proximally, hairy distally, hairs retrorse or spreading, rarely appressed, short to medium length, soft, eglandular. **Leaves** green, green tinged with purple, or dull purple, linear to narrowly lanceolate, 1.7–3.5 (–4.5) cm, not fleshy, margins plane, slightly involute, 0(–3)-lobed, apex acute; lobes ascending, linear, mostly short, apex acute. **Inflorescences** 3.5–6.5 × 2–5 cm; bracts magenta, rose red, lavender, or pink throughout, broadly lanceolate, oblong, or ovate, (0–)3(–5)-lobed; lobes ascending to erect, linear to lanceolate, short, arising above, sometimes slightly below, mid length, center lobe apex rounded, lateral ones acute. **Calyces** colored as bracts, 16.5–22(–27) mm; abaxial and adaxial clefts (8–)10–15(–18) mm, 33–40% of calyx length, deeper than laterals, lateral (1.5–)3–6(–8) mm, 20–30% of calyx length; lobes narrowly lanceolate to elliptic, apex acute, sometimes obtuse. **Corollas** straight or slightly curved, 19–25(–27.5) mm; tube 9–16 mm; beak exserted, adaxially green, 7–10(–11.5) mm; abaxial lip deep green or deep purple, slightly inflated, rounded, 1.5–2.5 mm, 20–30% as long as beak; teeth usually ± incurved, magenta or deep green, 0.6–1.4(–2) mm. $2n = 24$.

Flowering May–Jul. Dry, gravelly openings and barrens in pine forests, limestone; of conservation concern; 2400–3100 m; Utah.

Castilleja revealii is a narrow endemic of limestone gravel barrens on the Markagunt and Paunsaugunt plateaus in and adjacent to the rim of Bryce Canyon National Park and Cedar Breaks National Monument in south-central Utah. It is also known from a single collection from the southwestern edge of the Aquarius Plateau. These xeric barrens have an extremely high amount of reflective sunlight and support few other plants.

97. Castilleja rhexiifolia Rydberg, Mem. New York Bot. Gard. 1: 356. 1900 (as rhexifolia) • Rhexia-leaved or rosy paintbrush E

Castilleja oregonensis Gandoger

Herbs, perennial, (1–)2.5–6(–8) dm; from a woody caudex; with a taproot. **Stems** few to several, erect or ascending, unbranched, sometimes branched, proximally glabrous to glabrate, distally sparsely hairy, hairs spreading, ± long, soft, eglandular, sometimes stipitate-glandular. **Leaves** green or purple-tinged, linear, narrowly to broadly lanceolate, oblong, or ovate, 3–6(–7) cm, not fleshy, margins plane, flat to slightly involute, prominently veined, 0(–3)-lobed, apex acute to rarely obtuse; lobes ascending or spreading, linear to filiform, much smaller than mid blade, short, apex acute. **Inflorescences** 2.5–15 × 1.5–4.5 cm; bracts pink-purple, red-purple, purple, or crimson throughout, rarely reddish, yellowish, or white throughout, or proximally greenish, dull brownish purple, or deep purple, distally colored as above, broadly lanceolate to ovate or obovate, (0–)3–5(–7)-lobed; lobes ascending to erect, linear or lanceolate-acuminate, short, arising near or above mid length, central lobe apex obtuse to broadly rounded, lateral ones acute to sometimes rounded. **Calyces** proximally green, purplish, or whitish, distally colored as bracts, 15–25 mm; abaxial and adaxial clefts 8–12(–15) mm, 40–50% of calyx length, deeper than laterals, lateral 2–5(–8) mm, 15–25(–30)% of calyx length; lobes oblong to triangular or ovate, apex obtuse to rounded, sometimes acute. **Corollas** straight, 15–30(–36) mm; tube (11–)12–22(–24) mm; beak exserted, adaxially green, yellow, or tinged with red, 7–12 mm; abaxial lip deep green, reduced, 1.5–3.5 mm, to 33% as long as beak; teeth incurved, green or white, 0.5–2 mm. $2n$ = 24, 48, 96.

Flowering (Mar–)May–Sep. Moist meadows, open forests, slopes and ridges, sun or shade, subalpine to alpine; 1800–4000 m; Alta., B.C.; Colo., Idaho, Mont., N.Mex., Oreg., Utah, Wash., Wyo.

Castilleja rhexiifolia is a characteristic and common paintbrush in the Rocky Mountains region, from central Alberta to northern New Mexico. Disjunct populations also occur in northeastern Oregon, in the Blue, Ochoco, and Wallowa mountains, as well as near the summit of Abercrombie Mountain in the Selkirk Mountains of northeastern Washington. The inflorescence colors are variable. Typical plants are red-purple. However, many other shades of pink-white, pink-purple, and crimson are common as well. Many populations contain a wide range of color variants. Intergradation is occasionally seen with *C. miniata*, *C. occidentalis*, and *C. septentrionalis*. However, for the most part, these four

species remain distinct, and their ranges are not entirely overlapping. In the Wallowa Mountains, a recurrent and variable hybrid form between *C. chrysantha* and *C. rhexiifolia* was described as *C. wallowensis* Pennell. In the north Cascade Range of Washington, plants of *C. elmeri* with crimson inflorescences are sometimes identified as *C. rhexiifolia*. In Glacier National Park, Montana, in the vicinity of Logan Pass, *C. rhexiifolia* frequently hybridizes with *C. occidentalis*, creating a number of hybrid morphologies. Both of the parents are common in the area. Similar hybrids are also found in the Canadian Rocky Mountains.

98. Castilleja rigida Eastwood, Proc. Amer. Acad. Arts 44: 575. 1909 • Rigid paintbrush

Castilleja latebracteata Pennell

Herbs or subshrubs, perennial, 1.4–7.5 dm; from a woody caudex; with a strongly woody taproot or branched, woody roots. **Stems** few to many, erect to basally ascending, sometimes with subtle, curvy zigzags, unbranched, sometimes branched, proximally or with short, leafy axillary shoots, hairs loosely spreading, long, soft, ± wavy, eglandular and shorter stipitate-glandular. **Leaves** green, linear-oblong to oblong-lanceolate, narrowly lanceolate, or narrowly oblanceolate, 2–5.5 cm, not fleshy, margins plane, involute, 0-lobed, apex acute, distalmost obtuse to rounded. **Inflorescences** 2–7.5 × 2–3.5 cm; bracts proximally greenish, distally abruptly rose red, bright red, or red-orange, narrowly to broadly oblong-obovate to obovate or ovate-elliptic, 0-lobed; apex rounded to obtuse. **Calyces** colored as bracts, sometimes with a yellow central band, 20–32(–36) mm; abaxial and adaxial clefts (5–)7–14 mm, ca. 40% of calyx length, deeper than laterals, lateral 0–0.5 mm, 0–7% of calyx length; lobes poorly developed, if present, rounded, apex rounded to truncate, sometimes emarginate. **Corollas** straight or slightly curved, (24–)28–36(–42) mm; tube 15–22 mm; beak exserted, adaxially green to yellow-green, (8–)10–14 mm; abaxial lip dark green, reduced, 0.5–1.5 mm, 7–12% as long as beak; teeth reduced, green to yellowish green, 0.5–1.5 mm.

Flowering Mar–Jul(–Nov). Desert scrub, rocky slopes, ledges, ridges, flats, swales, roadsides, commonly with *Agave lechuguilla*; 300–1500 m; Tex.; Mexico (Chihuahua, Coahuila, Durango, Nuevo León, Zacatecas).

Castilleja rigida is endemic to desert scrub on limestone deposits in the Chihuahuan Desert region and the adjacent western edge of the Edwards Plateau. It is almost always closely associated with, and most likely parasitic on, *Agave lechuguilla*. The majority of its range is in the Mexican states of Chihuahua, east of the Sierra Madre Occidental, as well as in Coahuila and Nuevo León, where it is a characteristic component of the regional flora, but also extends into a few counties in southwestern Texas. It is sometimes confused with the similar species, *C. nervata*, which is widespread in the sierras of western Mexico and is known from a single extant population in the western Chiricahua Mountains of southeastern Arizona. *Castilleja nervata* and *C. rigida* are largely parapatric in Mexico.

99. **Castilleja rubicundula** (Jepson) T. I. Chuang & Heckard, Syst. Bot. 16: 658. 1991 [E]

Orthocarpus rubicundulus Jepson, Man. Fl. Pl. Calif., 943. 1925

Herbs, annual, 0.6–6 dm; with a taproot or branched root system. **Stems** solitary, erect, unbranched, sometimes branched, hairs spreading, short, soft, often mixed with stipitate-glandular ones. **Leaves** green to purple-tinged or dark red-brown, linear-lanceolate or distal lanceolate, 2–8(–9) cm, not fleshy, margins plane, flat to slightly curved up, 0–7-lobed, apex acute to acuminate; lobes widely spreading or ascending-spreading, linear, apex acute. **Inflorescences** 2.5–24 × 3–4 cm; bracts green throughout, lanceolate to ovate, 5–9-lobed; lobes ascending, linear-lanceolate or lanceolate, medium length, arising near mid length, apex acute to acuminate. **Calyces** green, 8–14 mm; abaxial and adaxial clefts 3–6 mm, 30–50% of calyx length, all 4 clefts subequal; lobes linear, apex acuminate to acute. **Corollas** straight, (15–)20–28 mm; tube 8–24 mm; abaxial lip, beak, and proximal part of corolla tube exserted; beak adaxially white, rarely very pale yellow or pale pink-purple, 5–7 mm, inconspicuously puberulent; abaxial lip white, fading to pink to pink-purple, or yellow, fading to white, rarely then to pink or pink-purple, both forms often with purple or red dots at base, inflated, prominent, pouches 3, 8–10 mm wide, 4–6 mm deep, 4–6 mm, 80–100% as long as beak; teeth erect, white or yellow, 0.5 mm. **Stigmas** ± exserted, as long as or slightly longer than beak and visible from abaxial side. $2n = 24$.

Varieties 2 (2 in the flora): w United States.

Castilleja rubicundula is separated into two varieties on the basis of flower color, as well as subtle differences in the pouching of the abaxial corolla lip. The ranges of the two varieties are broadly overlapping, but they never grow in the same location. Few intermediate forms have been recorded, though a very unusual population system exists in Santa Clara County, within which virtually all plants exhibit a tricolored corolla sequence, with yellow flowers aging to white and then pink or pink-purple, with all three colors visible on mature inflorescences.

1. Abaxial lips of corollas white, in most populations quickly fading to pink or pink-purple .99a. *Castilleja rubicundula* var. *rubicundula*
1. Abaxial lips of corollas light to bright yellow, in some populations quickly fading to white, then rarely to pink or pink-purple . . . . .99b. *Castilleja rubicundula* var. *lithospermoides*

99a. **Castilleja rubicundula** (Jepson) T. I. Chuang & Heckard var. **rubicundula** • Pink cream-sacs [C][E]

Orthocarpus bicolor A. Heller; *O. lithospermoides* Bentham var. *bicolor* (A. Heller) Jepson

Herbs 0.6–6 dm. **Leaves** 0–7-lobed, lobes widely spreading. **Bracts** 5–9-lobed. **Corollas:** abaxial lip white, in most populations quickly fading to pink or pink-purple, usually purple-dotted at base, teeth white.

Flowering (Apr–)May–Jun(–Jul). Grasslands, damp meadows, springs, woodland edges, dry rocky slopes, coastal valleys and foothills, often on serpentine; of conservation concern; 0–1000 m; Calif.

Variety *rubicundula* grows in the northern and western portions of the Sacramento Valley, the North Coast Range, and San Francisco Bay area of California, often on serpentine substrates. It is far less common than var. *lithospermoides* and often occurs in somewhat more mesic conditions. Populations in the north-central Sacramento Valley contain unusually robust plants with especially large flowers, and were named *Orthocarpus bicolor*. Variety *rubicundula* is of conservation concern, threatened by habitat loss to development, agriculture, and grazing.

99b. **Castilleja rubicundula** (Jepson) T. I. Chuang & Heckard var. **lithospermoides** (Bentham) J. M. Egger, Phytologia 90: 78. 2008 • Cream-sacs [E]

Orthocarpus lithospermoides Bentham, Scroph. Ind., 13. 1835; *Castilleja rubicundula* subsp. *lithospermoides* (Bentham) T. I. Chuang & Heckard

Herbs 0.8–4.4 dm. **Leaves** 0–5(–7)-lobed, lobes ascending-spreading. **Bracts** 5–7(–9)-lobed. **Corollas:** abaxial lip light to bright yellow, in most populations quickly fading to white and in a few populations then to pink or pink-purple, small purple or reddish dots near base, teeth yellow or white.

Flowering Apr–Jun(–Jul). Grasslands, damp flats, springs, vernal pools, roadbanks, coastal valleys and foothills, sometimes on serpentine; 0–2100 m; Calif., Oreg.

Variety *lithospermoides* is found in open grassland valleys along the North Coast and west of the Sierra Nevada in central and north-central California, especially in the regions west of the Sacramento Valley. It is known in southern Oregon from a single collection by T. J. Howell in the late 1800s and is apparently extirpated from that state. Variety *lithospermoides* is similar to and sometimes confused with *Castilleja lacera* or with *Triphysaria versicolor* subsp. *faucibarbata*. A peculiar form from Inverness, Marin County, was named *Orthocarpus noctuinus* Eastwood and later treated by some as a hybrid with *C. densiflora*.

100. Castilleja rubida Piper, Bull. Torrey Bot. Club 27: 398. 1900 (as Castilleia) • Purple alpine paintbrush C E

Herbs, perennial, 0.5–1.5 dm; from a woody caudex; with a taproot. Stems several, decumbent, or ascending, unbranched, hairs moderately dense, spreading, short and long, soft, eglandular and glandular. Leaves green to purple, linear to narrowly lanceolate, 0.7–3.2 cm, not fleshy, margins plane, slightly involute, 3–5-lobed, apex narrowly acute to acuminate; lobes ascending-spreading, narrowly linear to filiform, often curling, often short, apex acute or obtuse. Inflorescences 2.5–6 × 1–2 cm; bracts purple, deep burgundy, or lavender throughout, rarely pink or yellowish white throughout, sometimes pink or dull whitish on distal margins and apices, oblong, 3–5(–7)-lobed; lobes spreading, linear, medium length, proximal lobes arising below mid length, center lobe apex rounded to obtuse, lateral ones acute to obtuse. Calyces colored as bracts, 10–12 mm; all 4 clefts subequal, 3.5–6.5 mm, 35–55% of calyx length; lobes broadly linear or linear-triangular, apex obtuse to acute. Corollas straight, 12–15 mm; tube 14–16 mm; abaxial lip and beak exserted; beak adaxially green, 5–6 mm; abaxial lip colored as distal portion of bracts, prominent, pouches 3, central one grooved, pouches not strongly inflated, 4–5 mm, 80–100% as long as beak; teeth erect, appressed to beak, colored as distal portions of bracts, 1.5–2.5 mm. Stigmas green.

Flowering Jul–Aug. Rocky slopes, ledges, dry to moist gravelly flats and ridges, alpine, limestone, rarely on river cobbles at lower elevations; of conservation concern; 2200–3000 m; Oreg.

Castilleja rubida is a rare alpine species endemic to a few limestone peaks in the Wallowa Mountains of northeastern Oregon, entirely within the Eagle Cap Wilderness Area. It is likely derived from the *C. nana* complex, found in the mountains of eastern California and Nevada, but it is amply distinct. Due to its very limited range and small population numbers, *C. rubida* is a species of conservation concern.

101. Castilleja rupicola Piper, Erythea 6: 45. 1898 (as Castilleia) • Cliff paintbrush E F

Herbs, perennial, (0.8–)1–2 (–3) dm; from a woody caudex; with a taproot. Stems many, decumbent to ascending, unbranched, sparsely pubescent, hairs spreading, wavy, fairly short, soft, eglandular, sometimes glabrous proximally. Leaves purple to green, narrowly, rarely broadly, lanceolate, 1.4–4 cm, not fleshy, margins plane, flat to involute, (0–)3–5(–7)-lobed, apex acute to acuminate; lobes divergent, spreading-ascending, linear, long, not much narrower than mid blade, often with secondary lobes, creating little frilly fans, apex acute or obtuse. Inflorescences 2–6 × 2–3.5 cm; bracts proximally greenish or deep purple near base, distally red, scarlet, or crimson to red-orange, rarely orange, salmon, pink, or yellowish white, ovate to orbicular in outline, 5(–9)-lobed; lobes spreading, linear to linear-lanceolate, long, arising below mid length, apex acute to rounded. Calyces proximally purple, green, or whitish, distally colored as bract lobes, 15–25 mm; abaxial and adaxial clefts 8 mm, ca. 40–50% of calyx length, deeper than laterals, lateral 1–5 mm, 10–20% of calyx length; lobes triangular, apex obtuse or acute. Corollas straight or slightly curved, 25–35(–45) mm; tube 9–15 mm; beak exserted, adaxially green, purplish, or yellow-green, 14–22 mm; abaxial lip deep green, reduced, 0.5–2 mm, 6–12% as long as beak; teeth incurved to erect, green, 0.5 mm.

Flowering (May–)Jun–Sep. Sunny rocky slopes, scree, talus, ledges, fellfields, subalpine to alpine; (200–)1000–2500 m; B.C.; Oreg., Wash.

Castilleja rupicola is usually found in the subalpine and lower alpine zones in the Cascade Range from extreme southern British Columbia south to northern Douglas County, Oregon. Though it can be numerous where it occurs, the species as a whole is uncommon. One atypical population occurs in a moist, shaded, mossy, north-facing ravine on the Oregon side of the Columbia River Gorge, at less than 250 m. These plants often bear secondary divisions on deeply dissected leaves and bracts.

102. Castilleja salsuginosa N. H. Holmgren, Bull. Torrey Bot. Club 100: 83, fig. 1. 1973 • Monte Neva paintbrush [C] [E]

Herbs, perennial, (0.5–)0.8–1.4 (–1.8) dm; from a woody caudex; with a taproot with yellow root hairs. **Stems** several, erect, usually decumbent at base, unbranched, sometimes branched, sometimes with short, leafy axillary shoots, hairs spreading, short, rather stiff, some glandular. **Leaves** purplish brown with a grayish cast (due to adhering soil particles and salt crystals), linear to narrowly lanceolate, 1.5–2.5(–3) cm, fleshy, margins plane, sometimes wavy, involute, 0–3(–5)-lobed, apex acute; lobes spreading, linear to narrowly lanceolate, apex obtuse. **Inflorescences** 3–10 × 1.5–5 cm; bracts proximally purplish, deep burgundy, lavender, dull reddish, or deep purple, distally greenish, white, cream, or pink on margins and apices, oblong, 3(–5)-lobed; lobes ascending, ± linear, medium length, arising above mid length, central lobe apex rounded to obtuse, expanded distally, lateral ones acute. **Calyces** proximally whitish, distally purple to sometimes pink, margins white or cream, 16–20 mm; abaxial and adaxial clefts 5–8.5 mm, 20–45% of calyx length, all 4 clefts subequal; lobes linear or narrowly lanceolate, apex obtuse to rounded. **Corollas** straight or slightly curved, 18–22(–24) mm; tube 13–18 mm; beak, sometimes abaxial lip, exserted; beak adaxially purplish brown, 4.5–6.5 mm, conspicuously exceeding abaxial lip, margins reddish or colored as bracts, apices white or cream; abaxial lip reddish purple with green in a distal band or along grooves, gradually inflated, grooved, (2–)3–4(–4.5) mm, 67% as long as beak; teeth erect to slightly spreading, white to cream, often with purple spot, 1.4–2(–2.5) mm. **Stigmas** blackish. $2n = 24$.

Flowering Jun–Jul. Damp alkaline clay, hummocks, sparsely vegetated stream banks draining hot springs; of conservation concern; 1800–2000 m; Nev.

Castilleja salsuginosa is endemic to a single site in White Pine County, where it is limited to the harsh alkaline soils of travertine hot springs. This population is threatened by habitat degradation from livestock, as well as by water developments affecting the hydrology of the hot spring system. *Castilleja salsuginosa* is closely related to *C. nana* and *C. pilosa*, but genetic studies of the trio are inconclusive so far. Two populations of very similar but slightly smaller-flowered plants occur around other hot springs in adjacent Eureka County. While they resemble *C. salsuginosa* superficially, recent morphometric studies of one of these populations indicate that they may be worthy of nomenclatural recognition, separate from *C. salsuginosa*.

Castilleja salsuginosa is in the Center for Plant Conservation's National Collection of Endangered Plants.

103. Castilleja scabrida Eastwood, Bull. Torrey Bot. Club 29: 523. 1902 (as Castilleia) [E]

Herbs, perennial, 0.7–1.5(–2.2) dm; from a woody caudex; with a taproot or stout, branched roots. **Stems** several, decumbent, often sprawling, distally ascending, unbranched unless injured, rarely with small, leafy axillary shoots, hairs spreading, whitish, short, ± stiff, eglandular. **Leaves** gray-green, sometimes green or reddish purple, reduced and scalelike on proximal 10–25% of stem, linear, lanceolate, or narrowly elliptic, 1.5–4(–5) cm, not fleshy, margins plane, flat or involute, 0–3(–5)-lobed, apex acute; lobes spreading or spreading-ascending, linear or lanceolate, apex acute. **Inflorescences** 2.5–10 × 2.5–5 cm; bracts proximally green to deep greenish purple, distally bright red, sometimes brick red or orange-red, linear to lanceolate, 3–5(–7)-lobed; lobes spreading, linear-lanceolate to lanceolate, sometimes expanded near tip, long, arising near base and more distally, apex acute to obtuse. **Calyces** colored as bracts, 18–33 mm; abaxial clefts 6–12 mm, adaxial 8–15 mm, clefts 40–60% of calyx length, deeper than laterals, lateral 2–6 mm, 15–25% of calyx length; lobes lanceolate or narrowly triangular, apex acute to acuminate. **Corollas** straight or slightly curved, 25–40(–45) mm; tube 11–24 mm; beak exserted, adaxially green to yellow, 10–17(–20) mm; abaxial lip green, reduced, visible through front cleft, 1.5–2.5 mm, 10–15% as long as beak; teeth incurved, green, 0.5–1 mm. $2n = 24$.

Varieties 2 (2 in the flora): w United States.

Castilleja scabrida consists of two varieties distinguished by range and habitat differences, as well as by the morphological characters of the key. Both varieties are often confused with the similar and broadly sympatric *C. chromosa* but can be differentiated from the latter most easily by the presence of scalelike vestigial leaves on the proximal stems of *C. scabrida*.

1. Leaf lobes lanceolate, sometimes linear-lanceolate; stems and leaves appearing grayish; substrates sandstone or clay . . . . .103a. *Castilleja scabrida* var. *scabrida*
1. Leaf lobes linear; stems and leaves greenish, rarely reddish purple to grayish green; substrates calcareous . . . . 103b. *Castilleja scabrida* var. *barnebyana*

103a. Castilleja scabrida Eastwood var. **scabrida** E

Castilleja zionis Eastwood

Stems green but appearing grayish from pubescence. **Leaves** green but appearing grayish from pubescence, 0(–5)-lobed, lobes spreading, lanceolate, sometimes linear-lanceolate. **Bracts:** distally bright red, sometimes orange-red; lobe apex acute to obtuse, sometimes expanded near tip. **Calyces:** lateral clefts 2–4 mm. **Corollas:** tube 11–24 mm; beak adaxially green.

Flowering Mar–Sep. Sandy or rocky slopes, ledges, washes, sometimes on clay or cryptogamic soils, mostly sandstone; 1200–2800 m; Ariz., Colo., N.Mex., Utah.

Variety *scabrida* inhabits sandstone-derived soils in high deserts and the surrounding mountains, in Utah and adjacent western Colorado as well as a handful of records from northeastern Arizona and northwestern New Mexico. Variety *scabrida* is related to, and often confused with, *Castilleja chromosa*, but the latter species is usually a sagebrush associate, while *C. scabrida* shows no special differences for *Artemisia*. Variety *scabrida* occasionally hybridizes with *C. linariifolia*.

103b. Castilleja scabrida Eastwood var. **barnebyana** (Eastwood) N. H. Holmgren in A. Cronquist et al., Intermount. Fl. 4: 488. 1984 • Barneby's paintbrush E

Castilleja barnebyana Eastwood, Leafl. W. Bot. 3: 88. 1941

Stems greenish. **Leaves** greenish, rarely reddish purple to grayish green, 0–3(–5)-lobed, lobes spreading-ascending, sometimes divergent, linear. **Bracts:** distally bright red, sometimes orange-red or brick red; lobe apex acute. **Calyces:** lateral clefts 3–6 mm. **Corollas:** tube (12–)14–20 mm; beak adaxially green to yellow.

Flowering May–Jul. Sun or shade, rocky slopes, ledges, ridges, sagebrush steppes, limestone; 2100–2600 m; Nev., Utah.

Variety *barnebyana* grows on limestone substrates in the mountains of eastern Nevada and western Utah. It is frequently parasitic on *Petrophyton caespitosum*, a mat-forming plant from which the stems of the *Castilleja* often emerge, though the relationship is not obligate.

104. Castilleja schizotricha Greenman, Bot. Gaz. 53: 511. 1912 • Split-haired paintbrush C E

Castilleja arachnoidea Greenman subsp. *schizotricha* (Greenman) Pennell

Herbs, perennial, 0.8–1.5 dm; from a woody caudex; with a taproot. **Stems** few to many, ascending to erect, unbranched, hairs dense, appressed-ascending, matted, long, soft, branched, eglandular, ± white-woolly, obscuring surface. **Leaves** gray-green with hairs, surface green to purple, linear-lanceolate, 0.5–2 cm, not fleshy, margins plane, ± involute, 0(–3)-lobed, apex acute to acuminate, sometimes rounded; lobes sometimes divergent, spreading-ascending, narrowly linear, apex acute. **Inflorescences** 3–8 × 1–2 cm; bracts purple, lavender, pinkish, or dusty red throughout, sometimes greenish throughout or proximally greenish, distally colored as above, obscured by hairs, lanceolate, 3(–5)-lobed; lobes ascending to erect, linear to oblanceolate, short, arising near or above mid length, apex obtuse. **Calyces** colored as bracts, 11–18 mm; abaxial and adaxial clefts 3.5–6 mm, 33–50% of calyx length, all 4 clefts subequal, lateral 4–6 mm, 33–50% of calyx length; lobes linear, apex acute. **Corollas** straight, 15–20 mm; tube 8–9 mm; beak included to slightly exserted, adaxially purple or pink, 3.9–5 mm, densely puberulent with white, woolly hairs; abaxial lip deep purple, inconspicuous, pouched, pouches shallow, 3–5 mm, 80–100% as long as beak; teeth upright-ascending, reduced, appearing white, 1.5–2 mm.

Flowering Jul–Aug. Meadows, ledges, rocky or gravelly slopes, subalpine to alpine, marble, granite, or serpentine substrates; of conservation concern; 1600–2600 m; Calif., Oreg.

Castilleja schizotricha is a rare species restricted to high elevations in the mountains in Siskiyou and Trinity counties in northwestern California and in Jackson and western Klamath counties in southwestern Oregon. See the discussion of 6. *C. arachnoidea* for morphological differences between it and *C. schizotricha*, as the two are sometimes confused. Though their ranges overlap, they do not appear to hybridize.

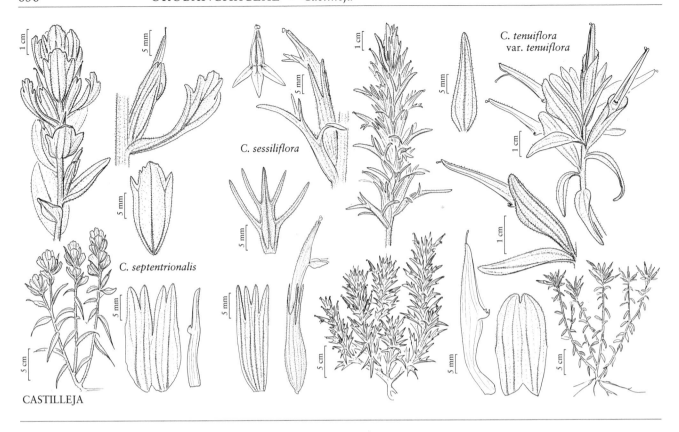

C. sessiliflora

C. tenuiflora
var. tenuiflora

C. septentrionalis

CASTILLEJA

105. Castilleja septentrionalis Lindley, Bot. Reg.
11: plate 925. 1825 • Northern or sulfur or
yellow paintbrush E F

Castilleja pallida (Linnaeus)
Sprengel subsp. *septentrionalis*
(Lindley) Scoggan; *C. rhexiifolia*
Rydberg var. *sulphurea* (Rydberg)
N. D. Atwood; *C. sulphurea*
Rydberg

Herbs, perennial, (1.5–)2.5–5.5
(–7) dm; from a woody caudex;
with a taproot or system of
slender to stout, branched roots. **Stems** few to several,
erect to ascending or ± decumbent at base, unbranched
or branched, sometimes with short, leafy axillary
shoots, glabrous or glabrate proximally, sometimes
hairy, hairs spreading to retrorsely curved, short, ± stiff
throughout, spreading, longer, soft to ± stiff distally,
often glandular and short-glandular. **Leaves** green
to purplish, linear-lanceolate or narrowly to broadly
lanceolate, 2–7(–8) cm, not fleshy, margins plane, flat
to involute, prominently veined, 0(–3)-lobed, apex
acute to acuminate; lobes ascending, linear to narrowly
lanceolate, short, apex acute or obtuse. **Inflorescences**
2.1–11 × 1.5–4 cm; bracts proximally greenish to pale
yellow-green, sometimes brownish purple, distally white
to cream or light yellow, sometimes light pink or buff,
sometimes aging pink or reddish, broadly lanceolate to
oblong or obovate, sometimes lanceolate, 0–3(–5)-lobed,
sometimes with irregular teeth at apex; lobes erect,
triangular, short, arising at or above mid length, center
lobe apex obtuse to rounded, sometimes acute, lateral
ones acute. **Calyces** colored as bracts, 13–23(–28) mm;
abaxial clefts (6–)8–13 mm, adaxial (5–)6–10(–11)
mm, clefts (25–)35–50% of calyx length, deeper than
laterals, lateral 1–4 mm, 5–25% of calyx length; lobes
triangular to lanceolate, abaxial and adaxial segments
often differing in length or width, apex obtuse to acute.
Corollas ± straight, (16–)18–30 mm; tube 10–20 mm;
teeth of abaxial lip rarely exserted, beak exserted; beak
adaxially green or yellowish, 6–12 mm; abaxial lip
green or whitish, reduced, slightly inflated, 1.5–3 mm,
25–30% as long as beak; teeth incurved to erect, green
or white, 0.5–2 mm. $2n = 24, 48, 96$.

Flowering (Apr–)May–Sep.   Moist meadows,
peatlands, open forests, gravel, rocky slopes, ridges,
slides, lowlands to subalpine; 0–3700 m; Alta., B.C.,
N.B., Nfld. and Labr., N.W.T., Nunavut, Ont., Que.;
Colo., Idaho, Maine, Mich., Minn., Mont., N.H.,
N.Mex., S.Dak., Utah, Vt., Wyo.

Castilleja septentrionalis is one of the most widespread
species in the flora area. It occurs throughout much
of the Rocky Mountain region, from New Mexico
northward into southern Canada and sporadically
eastward to the Atlantic coast, as well as in the higher
mountains and notches of New England. Rocky
Mountain plants are commonly called *C. sulphurea*,
but there is broad morphological overlap between these

plants and those in New England and eastern Canada, where *C. septentrionalis* was first described, and they are here considered synonymous. Varietal segregates may eventually prove to be appropriate, especially in the Canadian Rocky Mountains and in southeastern Canada and the adjacent United States, but this should await a comprehensive and detailed review across the full range of the species.

The relationship of *Castilleja septentrionalis* with the largely alpine species, *C. occidentalis*, has been the subject of much discussion, but the two have discrete ranges and usually remain reasonably easy to separate. The same cannot be said for plants from the Canadian Rocky Mountains, where some specimens in relatively low-elevation, riverine flood plains and lake basins seem transitional to the boreal-arctic *C. pallida*.

Castilleja septentrionalis occasionally forms localized hybrid swarms with its close relatives, *C. miniata* and *C. rhexiifolia*, where the morphological boundaries between the species seem to disintegrate. However, most populations of all three species are distinct and easily recognizable, and each has a different range, despite extensive sympatry. A possible hybrid with *C. linariifolia* bears the name *C. ×cognata* Greene, and a probable hybrid with *C. miniata* from southern Nevada is known as *C. ×porterae* Cockerell.

106. Castilleja sessiliflora Pursh, Fl. Amer. Sept. 2: 738. 1813 F

Herbs, perennial, 1–4 dm; from a branching, woody caudex; with a taproot. Stems few to many, ascending to erect, often decumbent at base, unbranched, sometimes branched, hairs spreading, sometimes matted, short to medium length, ± soft, eglandular, often with a layer of minute-glandular hairs, sometimes woolly. Leaves green to purple, or grayish with dust and hairs, linear to narrowly lanceolate, (1–)2–5(–6) cm, not fleshy, margins plane, involute, (0–)3–5-lobed, apex acuminate to acute; lobes divergent, spreading, linear, apex acute. Inflorescences 3–18 × 2.5–6.5 cm; bracts green to purplish throughout, sometimes reddish brown, pink, or lavender throughout, or distally white or pale yellow, sometimes distally dull pink, pink, salmon, orangish, pale pink-orange, buff, or cream, lanceolate, similar to distal leaves, 3(–5)-lobed; lobes spreading, linear-lanceolate, long, arising at or below mid length, apex acute to acuminate, sometimes obtuse. Calyces colored as bracts, sometimes proximally white, 20–40 mm; abaxial and adaxial clefts 12–20 mm, 40–60% of calyx length, deeper than laterals, lateral 5–15 mm, 10–25% of calyx length; lobes linear, apex acute to acuminate. Corollas strongly curved distally, 35–55 mm; tube

24–45 mm; abaxial lip, beak, and distal portion of tube exserted; beak adaxially green, yellow, pinkish, purplish, or whitish, 9–15 mm; abaxial lip green, pale green, or purple, protruding, shelflike, 4–8 mm, 50–70% as long as beak; teeth spreading, white, pale yellow, pink, or purple, 3–4 mm. 2*n* = 24.

Flowering (Feb–)Mar–Aug(–Oct). Dry mixed grass and shortgrass prairies, prairie sandhills, sandsage plains, sand prairies, rocky or sandy slopes, bluffs, open forests, or desert scrub, limestone, sandstone, gypsum, granite, other bedrock types; 0–2300 m; Alta., Man., Sask.; Ariz., Colo., Ill., Iowa, Kans., Minn., Mo., Mont., Nebr., N.Mex., N.Dak., Okla., S.Dak., Tex., Wis., Wyo.; Mexico (Coahuila, Nuevo León, Tamaulipas).

Castilleja sessiliflora ranges across the Great Plains from southern Canada to northern Mexico, where it is apparently rare. In Texas and northern Mexico, its range overlaps with the similar *C. mexicana*. Most populations of *C. sessiliflora*, especially north of Texas, have white to pale yellow inflorescences; in southwestern Texas they are more variable in color, with pink-purple plants often predominating locally. Those plants with pink-purple inflorescences were named forma *purpurina* by F. W. Pennell. In the limestone deserts of southern New Mexico and southeastern Arizona, the inflorescences are often a pale pink-orange, but these are intermingled with more typical greenish white plants. Occasional hybrids between *C. angustifolia* var. *dubia* and *C. sessiliflora* are known from northeastern Wyoming.

107. Castilleja subinclusa Greene, Pittonia 4: 2. 1899 (as Castilleia)

Herbs, perennial, 1.8–12 dm; from a woody caudex; with a taproot. Stems solitary or few to many, erect to ascending, often leaning on nearby shrubs (var. *jepsonii*), unbranched or branched, often with small, leafy axillary shoots, hairs matted to spreading, short and long, soft, mixed with short stipitate-glandular ones. Leaves green, pale gray-green, sometimes dull reddish purple, linear to narrowly, sometimes broadly, lanceolate, 2–10.5 cm, not fleshy, margins plane, sometimes ± wavy, flat or involute, 0(–3)-lobed, apex acuminate to acute; lobes spreading, linear to short-lanceolate, often arising near mid length, apex acute to obtuse. Inflorescences 3–30 × 1–7 cm; bracts green or dull purple throughout, or proximally green or dull purple, distally bright red to orange-red, sometimes salmon, orange, or yellow, usually linear to lanceolate, distal sometimes broadly lanceolate to oblong or elliptic, 0(–3)-lobed; lobes spreading to erect, linear to oblong or triangular, short or medium length, arising near or above mid length, apex acute to obtuse. Calyces proximally pale green, rarely pale purple, distally bright red or red-orange, rarely orange or yellow, or bright red or red-orange throughout, 20–42 mm; abaxial clefts 13–27 mm, adaxial 7–17 mm, abaxial 40–70% of calyx length, adaxial 12–33% of calyx length, deeper than laterals, lateral 0–7.5 mm, 0–35% of calyx length; lobes linear to narrowly triangular, strongly curved away from stem proximally and distally obviously curved toward stem, apex acute to acuminate to obtuse. **Corollas** curved proximally, straight to slightly curved distally, 25–57 mm; tube 15–29 mm; beak subequal to or exceeding calyx but abaxial lip, beak, and often part of tube usually exserted out abaxial cleft; beak adaxially green, yellow-green, or yellow, 13–21 mm, margins red, orange, or yellow; abaxial lip green, yellow, reddish, blackish, or deep red-purple, reduced, protuberant, forming a platform, 1–2 mm, 5–25% as long as beak; teeth ascending, green, purple sometimes with reddish tip, yellow, or blackish, 0.5–1 mm. *2n* = 24, 48, 72, 96.

Varieties 3 (3 in the flora): California, nw Mexico.

Castilleja subinclusa is divided into three varieties with somewhat discontinuous ranges. Identification of the varieties is often difficult when comparing only a single trait but is more easily accomplished when the characters are considered in a suite. The three varieties are also separable by range and habitat.

1. Corolla beaks adaxially bright yellow, sometimes greenish yellow near apex, margins yellow, rarely orange or red, strongly contrasting with color of calyces; stems and leaves greenish, sometimes dull reddish; lateral calyx clefts 0–4 mm; coastal scrub communities; coastal c California . . . . .107b. *Castilleja subinclusa* var. *franciscana*
1. Corolla beaks adaxially green to yellow-green, rarely yellow, margins red to orange, rarely yellow, not strongly contrasting with color of calyces; stems and leaves greenish, pale gray-green, or dull reddish; lateral calyx clefts 3–7.5 mm; dry, sandy or rocky slopes, desert scrub, foothill woodlands, chaparral, brushy openings in pine-oak woods; w foothills of Sierra Nevada in c California or sw California.
   2. Stems and leaves greenish, sometimes dull reddish; bracts (0–)3-lobed; brushy openings in pine-oak woods; mostly in w foothills of Sierra Nevada in c California . . . . 107a. *Castilleja subinclusa* var. *subinclusa*
   2. Stems usually dull reddish; leaves usually pale gray-green to ± ash-colored; bracts 0(–3)-lobed; dry, sandy or rocky slopes, desert scrub, foothill woodlands, chaparral; sw California . . . . .107c. *Castilleja subinclusa* var. *jepsonii*

107a. Castilleja subinclusa Greene var. **subinclusa** [E]

Herbs 4–4.5 dm. **Stems** greenish, sometimes dull reddish, hairs spreading, short to long. **Leaves** green, linear-lanceolate to narrowly lanceolate, margins plane; lobe apex acute. **Inflorescences** 6–21 × 2–4 cm; bracts green to purple through-out, or distally bright red to orange or salmon, (0–)3-lobed. **Calyces** proximally green to purple, distally red to orange, 20–30 mm; abaxial clefts 15–20 mm, adaxial 12–13 mm, lateral 3–5 mm; lobe apex acute to acuminate. **Corollas** 31–42 mm; beak exceeding calyx but beak, abaxial lip, and often part of tube usually exserted out abaxial cleft; beak adaxially green to yellow-green, rarely yellow, margins red to orange, rarely yellow, not strongly contrasting with color of calyces; abaxial lip green; teeth green to purple, sometimes red at tip, 0.6 mm. *2n* = ca. 72.

Flowering Apr–Jul. Brushy openings in pine-oak woods; 200–2200 m; Calif.

Variety *subinclusa* is endemic to openings in the western foothills of the Sierra Nevada from Placer County south to Calaveras County in central California. It is geographically replaced by var. *franciscana* to the west and by var. *jepsonii* to the south and southwest.

107b. Castilleja subinclusa Greene var. **franciscana** (Pennell) G. L. Nesom, Phytologia 73: 409. 1992

• Franciscan paintbrush [C] [E]

Castilleja franciscana Pennell, Proc. Acad. Nat. Sci. Philadelphia 99: 188. 1947; *C. subinclusa* subsp. *franciscana* (Pennell) T. I. Chuang & Heckard

Herbs 2.8–6.3 dm. **Stems** greenish, sometimes reddish, hairs matted to spreading, medium length or long. **Leaves** green, sometimes dull reddish, linear-lanceolate to narrowly lanceolate, margins plane, sometimes ± wavy; lobe apex acute. **Inflorescences** 5–17 × 1.5–5 cm; bracts green or purple throughout, or distally bright red to orange-red, rarely yellow, 0(–3)-lobed. **Calyces** proximally green, pale green, or pale purple, distal ²⁄₃ red to orange, 35–42 mm; abaxial clefts 20–27 mm, adaxial 11–13 mm, lateral 0–4 mm; lobe apex acute to obtuse. **Corollas** 38–57 mm; beak subequal to or exceeding calyx, beak and abaxial lip exserted through abaxial cleft; beak adaxially bright yellow, sometimes greenish yellow near apex, margins yellow, rarely orange or red, strongly contrasting with color of calyx; abaxial lip green or yellow; teeth green or yellow, 0.5–1 mm. *2n* = 24, 48, 72, 96.

Flowering Mar–Jul. Coastal scrub; of conservation concern; 0–600 m; Calif.

Variety *franciscana* is found in the coastal counties of northern and central California, from Humboldt County southward to Santa Clara and Santa Cruz counties. It is most common in the San Francisco Bay region. With its bright yellow corollas, it is the most distinctive of the three varieties.

107c. Castilleja subinclusa Greene var. **jepsonii** (Bacigalupi & Heckard) J. M. Egger, Phytologia 90: 79. 2008 • Jepson's paintbrush

Castilleja jepsonii Bacigalupi & Heckard, Leafl. W. Bot. 10: 282. 1966

Herbs 1.8–12 dm. **Stems** usually dull reddish, hairs spreading, short or long. **Leaves** pale gray-green to ± ash-colored, rarely dull reddish purple, linear to linear-lanceolate, margins plane; lobe apex acute to obtuse. **Inflorescences** 3–30 × 1–7 cm; bracts distally or entirely red to red-orange, rarely yellow, 0(–3)-lobed. **Calyces** red or red-orange throughout, or proximally green to purple, distally red to red-orange, rarely yellow, 20–38 mm; abaxial clefts 13–23 mm, adaxial 7–17 mm, lateral 3–7.5 mm, lobe apex acute to acuminate. **Corollas** 25–50 mm; beak exceeding calyx, beak, and abaxial lip, often part of tube usually exserted out abaxial cleft; beak adaxially green to yellow-green, rarely yellow, margins red to orange, rarely yellow, not strongly contrasting with color of calyces; abaxial lip green, reddish, blackish, or deep red-purple; teeth green, purplish red, or blackish, 0.5–1 mm. *2n* = 24, 48.

Flowering Mar–Jul. Dry, sandy or rocky slopes, desert scrub, foothill woodlands, chaparral; 300–1900 m; Calif.; Mexico (Baja California).

Variety *jepsonii* is a chaparral associate found at moderate elevations in California, from the southwestern Sierra Nevada and Transverse ranges southward into northern Baja California, to at least as far as the Sierra San Pedro Mártir. It differs from var. *subinclusa* in its habitat and distribution, as well as the differences listed in the key.

108. Castilleja suksdorfii A. Gray, Proc. Amer. Acad. Arts 22: 311. 1887 • Suksdorf's paintbrush E

Herbs, perennial, 3–5(–8) dm; from slender, creeping rhizomes. **Stems** solitary, sometimes few, erect from a slender, creeping base, unbranched, glabrate or hairs spreading, long, soft to ± stiff and shorter, stipitate-glandular. **Leaves** green, distal sometimes red-tipped, linear-lanceolate, sometimes distal broadly lanceolate or ovate,

1.2–8.9 cm, not fleshy, margins plane, flat or slightly involute, 0–5(–7)-lobed, apex acute to acuminate, sometimes obtuse or rounded; lateral lobes spreading-ascending or widely spreading, linear, often much narrower than mid blade, apex acute. **Inflorescences** 2.5–9(–11) × 2–5.5 cm; bracts proximally greenish, distally abruptly red to orange-red, often with a yellow, rarely purplish, medial band, narrowly lanceolate to ovate, 3–7(–11)-lobed; lobes spreading to erect, linear, narrowly lanceolate, or narrowly oblanceolate, long, arising below mid length, apex acute to obtuse. **Calyces** colored as bracts, 20–30 mm; abaxial and adaxial clefts 11–18 mm, 50–75% of calyx length, deeper than laterals, lateral 8–12 mm, 30–50% of calyx length; lobes linear, apex acute. **Corollas** ± curved, 30–50 mm; tube 11–18 mm; beak exserted, adaxially green, 18–20 mm; abaxial lip deep green, reduced, 1 mm, 10% as long as beak; teeth ascending, deep green, 1 mm. *2n* = 36.

Flowering Jun–Sep. Mesic to wet meadows, marshes, peatlands, springs, stream margins, montane to subalpine; 1000–2200 m; Oreg., Wash.

Castilleja suksdorfii is endemic to wet habitats in the Cascade Range from the Goat Rocks Wilderness Area in Yakima County, Washington, south to the vicinity of Crater Lake National Park in Klamath County, Oregon. Reports of this species farther north in Washington and southern British Columbia are referable to *C. rupicola*. *Castilleja suksdorfii* is a polyploid species and may be of hybrid origin.

109. Castilleja tenuiflora Bentham, Pl. Hartw., 22. 1839 • Santa Catalina paintbrush F

Varieties 3 (1 in the flora): sw United States, Mexico.

Castilleja tenuiflora is common and widespread across the mountains of Mexico, especially in pine-oak-madrone communities at middle elevations, as far south as Oaxaca, where it is found west and north of the Tehuantepec lowlands. There are two varieties of *C. tenuiflora* endemic to Mexico, while the typical variety crosses into the mountains of southeast Arizona and southwest New Mexico. Considerable local and regional variation exists in *C. tenuiflora*, but most of this appears to be racial in nature, and additional named varieties are likely not justified. While also commonly herbaceous, *C. tenuiflora* often forms large, multi-stemmed, subshrub plants with a woody base and ascending to strongly erect and often branched stems. It is valued in Mexican traditional medicine and is under study for potentially useful compounds (M. Jiménez et al. 1995; P. M. Sanchez et al. 2013).

109a. Castilleja tenuiflora Bentham var. **tenuiflora** F

Castilleja laxa A. Gray; *C. setosa* Pennell

Herbs or subshrubs, perennial, 1.8–6 dm; from a woody caudex; with a taproot. **Stems** few to many, erect or ascending, unbranched, sometimes branched, hairs dense, spreading to ± reflexed, short, ± stiff, eglandular. **Leaves** erect to ± reflexed, green to sometimes purple, lanceolate to linear-lanceolate, 2–3(–4) cm, not fleshy, margins plane, flat to slightly involute, 0(–3)-lobed, apex acute to acuminate; lobes spreading, linear, narrowly triangular, short, apex acute. **Inflorescences** 3–11 × (2.5–)5–7 cm; bracts red or green throughout, rarely light orange or light yellow throughout, or proximally greenish, distally colored as above, less showy than calyces, lanceolate, 0(–3)-lobed; lobes erect, linear or lanceolate-acuminate, short, arising near apex, apex acute. **Calyces** entirely red or proximally green, distally red, light orange, or light yellow, 27–35(–40) mm; abaxial clefts 12–18 mm, adaxial 8–14 mm, abaxial 55–60% of calyx length, adaxial 20–35% of calyx length, deeper than laterals, lateral 0(–2.5) mm, 0(–7)% of calyx length; lobes linear to lanceolate to broadly triangular, apex rounded to emarginate or acute. **Corollas** straight or slightly curved, 36–47 mm; tube 18–25 mm; abaxial lip often exserted, beak exserted; beak adaxially whitish, yellow to green, or reddish, 15–26(–30) mm; abaxial lip green to yellow or white, reduced, protruding through abaxial cleft, 1–2 mm, ca. 8% as long as beak; teeth incurved, white or deep green, 0.5–1 mm. *2n* = 24, 48.

Flowering Mar–Nov(–Dec). Open pine-oak woodlands, rocky slopes, ledges, chaparral, quartzite and other bedrock types; 700–2500 m; Ariz., N.Mex.; Mexico.

Variety *tenuiflora* is fairly common in the southern Sky Island ranges of southeastern Arizona and occurs in the Animas Mountains of Hidalgo County, southwestern New Mexico. It is also found throughout much of Mexico, where it is the most common species of *Castilleja*.

110. Castilleja tenuis (A. Heller) T. I. Chuang & Heckard, Syst. Bot. 16: 658. 1991 • Hairy owl's clover, slender paintbrush E

Orthocarpus tenuis A. Heller, Muhlenbergia 1: 45. 1904; *O. falcatus* Eastwood; *O. hispidus* Bentham 1835 [not *Castilleja hispida* Bentham 1838]

Herbs, annual, 0.45–5.2 dm; with a slender taproot or branched root system. **Stems** solitary, erect, unbranched or with few upright branches, hairs spreading, long, soft, mixed with shorter, eglandular and glandular ones. **Leaves** green to brown, proximal linear, distal lanceolate, 0.7–4(–8) cm, not fleshy, margins plane, flat, 0–3(–5)-lobed, apex acuminate; lobes ascending-spreading, very long linear, apex acute. **Inflorescences** 2–25 × 1–3 cm; bracts green, sometimes proximally green, distally dull brownish to deep purplish brown, lanceolate to narrowly ovate or ovate, (3–)5–7(–9)-lobed; lobes ascending to erect, linear or narrowly lanceolate, long, proximal lobes arising below mid length, apex acute to acuminate. **Calyces** green to brownish, margins sometimes deep purple or brown, 6–12 mm; abaxial and adaxial clefts 2–5 mm, 33–50% of calyx length, deeper than laterals, lateral 2–3.5 mm, 30–40% of calyx length; lobes linear to narrowly lanceolate, apex acute to acuminate. **Corollas** straight to ± curved distally, 12–20 mm; tube 9–14 mm; abaxial lip and beak exserted; beak adaxially white or pale yellowish, 3.5–5 mm, inconspicuously puberulent; abaxial lip white or yellow, 3 small red-brown dots near base, inflated, pouches 3, 2–4 mm wide, 2 mm deep, 2–4 mm, 50–70% as long as beak; teeth erect, white or yellow, 0.5–1 mm. **Stigmas** included within beak. *2n* = 24, 48.

Flowering Feb–Sep. Moist flats, vernal pools, springs, damp meadows and ditches, riparian zones, sometimes over serpentine; 200–2800 m; B.C.; Calif., Idaho, Nev., Oreg., Wash.

Castilleja tenuis is restricted to the east side of the Cascade Range in British Columbia and Washington and also occurs west of the Cascade-Sierra axis in the more arid terrain of California and Oregon. Plants in the Umpqua and Willamette valleys of western Oregon are often taller, more robust, and with slightly larger flowers than is typical in other regions. There are two color forms, with white or yellow corollas, but most individual populations are consistently unicolored. More investigation is needed to determine if corolla color is influenced by genetic and/or environmental factors. There is some evidence (T. I. Chuang and L. R. Heckard 1982) that the colors are correlated to chromosome number, with the white-flowered plants being diploid, while the yellow-flowered plants are tetraploid; however, there are exceptions. D. D. Keck (1927) cited some evidence for seasonal change, with yellow flowers occurring early, replaced with white flowers later in the season. However, no unequivocal evidence exists to support this hypothesis, and the apparently complete absence of yellow-flowered plants in Oregon makes this an unlikely explanation. Cropping by grazing animals results in occasional plants that branch. Plants with slightly curved corolla beaks were described as *Orthocarpus falcatus* but have no geographic integrity or taxonomic significance. *Castilleja tenuis* was collected as a waif in Skagway, Alaska, a century ago.

111. Castilleja thompsonii Pennell, Proc. Acad. Nat. Sci. Philadelphia 99: 178. 1947 (as thompsoni) • Thompson's paintbrush E F

Castilleja villicaulis Pennell & Ownbey

Herbs, perennial, 0.8–4 dm; from a woody caudex; with a taproot. **Stems** few to many, erect or ascending, unbranched or branched, hairs spreading, long, stiff, sometimes soft (especially in higher elevations), eglandular, mixed with shorter stipitate-glandular ones. **Leaves** green to purple or reddish brown, linear to narrowly oblong or linear-lanceolate, 1.4–7.4 cm, not fleshy, margins plane to ± wavy, involute or flat, 3(–7)-lobed, apex acuminate; lobes spreading-ascending, linear, short to long moving up leaf axis, apex acute or obtuse. **Inflorescences** 2.5–14 × 1–4 cm; bracts greenish to pale yellow or reddish brown throughout, or proximally greenish to dull reddish purple, or ruddy brown, distally greenish to yellow-green or yellow, often aging dull reddish to dull purplish, lanceolate to oblong to ovate, 3–5(–9)-lobed; lobes spreading to ascending, linear to narrowly lanceolate, long, proximal lobes arising below mid length, apex acute to obtuse. **Calyces** colored as bracts, 12–25 mm; abaxial and adaxial clefts 4–8 mm, 20–60% of calyx length, deeper than laterals, lateral (0–)1–3 mm, 7–25% of calyx length; lobes linear, lanceolate, or triangular, apex acute, sometimes obtuse. **Corollas** straight, 18–21 mm; tube 11–16 mm; subequal to calyx, sometimes beak exserted; beak adaxially green, 5–7(–8) mm; abaxial lip white, often proximally reddish, prominent, scarcely expanded, ± cylindric, 2.5–4(–5) mm, 50–70% as long as beak, glabrous or obscurely puberulent; teeth incurved to erect, white, 2.5 mm. $2n = 24, 48$.

Flowering Apr–Aug(–Sep). Dry slopes, ridges, scabland lithosol soils, meadows, sagebrush steppes, valleys, montane to alpine; 200–2100 m; B.C.; Oreg., Wash.

Castilleja thompsonii is a characteristic species of the sagebrush communities on the eastern slope of the Cascade Range in Washington, and in the high deserts of the Columbia Basin. Historically, its range approached but apparently never entered Idaho in the Spokane River valley, but much of its habitat in that area is now converted to agriculture or suburban development or overwhelmed by non-native, invasive plants. *Castilleja thompsonii* occurs in a few sites in the Okanogan Valley region of southern British Columbia and at one site on the eastern slopes of the Cascade Range in Wasco County, Oregon. A distinctive form from the subalpine and alpine zones of Mt. Adams, in the southern Cascade Range of Washington, was named *C. villicaulis*. This form may merit varietal status under *C. thompsonii*. While both names were described in the same paper, *C. thompsonii* is the name used in all regional floras since their publication, after *C. villicaulis* was reduced to synonymy by M. Ownbey (1959).

112. Castilleja tomentosa A. Gray in W. H. Emory, Rep. U.S. Mex. Bound. 2(1): 118. 1859 • Tomentose paintbrush C

Herbs or subshrubs, perennial, 1.3–5 dm; from a woody caudex; with a taproot. **Stems** few to many, ascending to erect, unbranched or branched, with short, leafy axillary shoots, moderately lanate, hairs prostrate to spreading, whitish, unbranched, short, fairly soft, eglandular. **Leaves** green, linear to narrowly lanceolate, (0.8–)3–5 cm, not fleshy, margins plane, strongly involute, 0–3(–5)-lobed, apex acute to rounded; lobes spreading, linear, short, apex acute. **Inflorescences** 5–20 × 0.5–2.5 cm; bracts proximally dull brownish to deep greenish purple, distally red, red-orange, or orange, lanceolate or oblong to obovate, deeply 3(–5)-lobed; lobes ascending, linear to lanceolate, long, arising below mid length, central lobe apex rounded to obtuse, others acute. **Calyces** colored as bracts, (10–)13–19 mm; abaxial and adaxial clefts 4–8(–11) mm, 33–50% of calyx length, deeper than laterals, lateral 5–7 mm, ca. 25% of calyx length; lobes linear to lanceolate, apex acute. **Corollas** straight or slightly curved, 12–20 mm; tube 13–15 mm; beak exserted or ± equal to calyx, adaxially pale green, 8–11.5 mm; abaxial lip green or red-violet, inconspicuous, slightly pouched, 1.5–2 mm, ca. 10–20% as long as beak; teeth incurved, pink to pale yellow or deep green, 1 mm.

Flowering Jun–Oct. Dry Chihuahuan grasslands; of conservation concern; 1300–1700 m; N.Mex.; Mexico (Sonora).

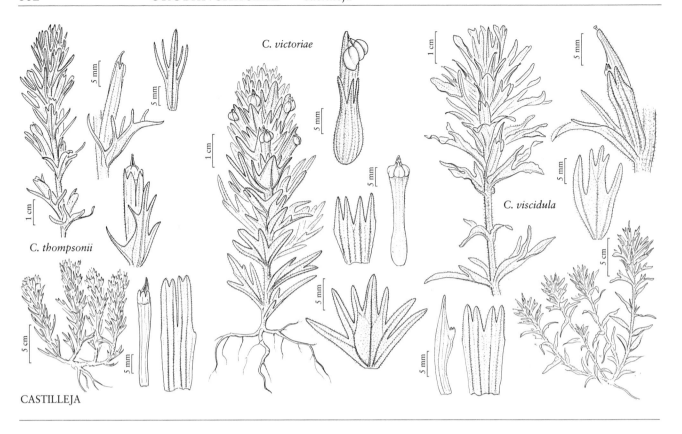

C. victoriae

C. thompsonii

C. viscidula

CASTILLEJA

In the United States, *Castilleja tomentosa* is known from a number of recently discovered populations in and near the southern Animas Valley, Hidalgo County, where it is found in *Bouteloua gracilis* and *Sporobolus airoides* grasslands. All known populations are small, and this species should be considered globally endangered. The only recorded Mexican station was the type locality from 1851 near Mabibi in adjacent northern Sonora. A. Eastwood (1909) believed *C. tomentosa* was a synonym of *C. integra*, but that species has mostly entire bracts, while the bracts of *C. tomentosa* are deeply lobed; the two also have different patterns of coloration and pubescence. T. I. Chuang annotated the holotype sheet of *C. tomentosa* as *C. lanata*, but *C. tomentosa* calyces have fairly deep lateral lobes, unlike the emarginate to very shallowly notched lobes of *C. lanata*.

113. Castilleja uliginosa Eastwood, Leafl. W. Bot. 3: 117. 1942 • Pitkin Marsh paintbrush C E

Herbs, perennial, 3–5 dm; from a woody caudex; rhizomatous. **Stems** several, decumbent proximally and sometimes becoming weakly rhizomatous, becoming ascending to erect, unbranched or often branched proximally, hairs spreading, short and long, soft, mixed eglandular and stipitate-glandular. **Leaves** green, lanceolate or broadly lanceolate, 3–5 cm, much reduced distally, not fleshy, margins plane, sometimes ± wavy, flat, 0-lobed, apex acute to rounded. **Inflorescences** 10–30 × 3–4 cm; bracts pale yellow to cream throughout, or proximally pale greenish near base, distally pale yellow to cream, broadly lanceolate, 3-lobed; lobes ascending, narrowly lanceolate, long, arising near mid length, apex narrowly acute or acuminate. **Calyces** colored as bracts, 20–25 mm; abaxial and adaxial clefts 7–14 mm, ca. 50% of calyx length, deeper than laterals, lateral 5–6 mm, ca. 33% of calyx length; lobes linear, apex acute, ciliate. **Corollas** slightly curved, 22–30 mm; tube 11–15 mm; beak partly exserted, adaxially whitish, yellowish, or greenish, 10–15 mm; abaxial lip green, small, protuberant, 3 mm, ca. 20% as long as beak; teeth erect to incurved, green, to 1 mm.

Flowering May–Jun. Margins of wet meadows, marshes, and wet thickets; of conservation concern; 40–60 m; Calif.

Castilleja uliginosa is endemic to Pitkin Marsh in Sonoma County. It differs from *C. miniata* by its uniformly pale yellow inflorescences and pubescent stems, as well as by its disjunct, low-elevation habitat. Much of its available habitat was destroyed by development, and it is apparently now extirpated from the wild. It survives in the form of tissue clones from the last wild plant, backcrossed with *C. miniata* by L. R. Heckard in the 1980s and maintained since then in the greenhouses at the University of California at Berkeley.

114. Castilleja unalaschcensis (Chamisso & Schlechtendal) Malte, Rhodora 36: 187. 1934 (as unalaschkensis) • Coastal paintbrush

Castilleja pallida (Linnaeus) Sprengel var. *unalaschcensis* Chamisso & Schlechtendal, Linnaea 2: 581. 1827

Herbs, perennial, 2–8 dm; from a short, stout, scaly, many-headed, woody caudex; with rhizomes or branching, woody roots. **Stems** few to many, erect to ascending, unbranched or branched distally, glabrate proximally, hairy distally, hairs sparse, spreading to erect, long, soft, eglandular, rarely stipitate-glandular. **Leaves** green, sometimes red-purplish, lanceolate-linear to lanceolate, elliptic-ovate, or ovate, (1.2–)3.5–10 cm, not fleshy, margins plane, sometimes ± wavy, flat to slightly involute, 0(–3)-lobed, apex acuminate to acute; lobes ascending, lanceolate, apex acute to acuminate. **Inflorescences** 3.5–18 × 2–6.5 cm; bracts proximally green, distally yellow, pale yellow, whitish, yellow-green, or pale orange, often aging reddish or pink, lanceolate, elliptic, ovate, oblanceolate, or oblong to narrowly obovate, (0–)3–5-lobed; lobes ascending, linear to oblong, short, arising above mid length, apex acute to obtuse. **Calyces** colored as bracts, 15–28 mm; abaxial cleft 12–16 mm, adaxial 8–11.5 mm, clefts 40–50% of calyx length, deeper than laterals, lateral 4–10 mm, 20–35% of calyx length; lobes oblong to lanceolate, apex acute to obtuse or rounded. **Corollas** straight or slightly curved, 18–28 mm; tube 16–19 mm; beak partially to completely exserted, abaxial lip not exserted; beak adaxially green, 6–13 mm; abaxial lip deep green, reduced, inconspicuous, 1–2 mm, ca. 20% as long as beak; teeth incurved, deep green or white, 0.5–1 mm. $2n$ = ca. 96.

Flowering May–Sep. Meadows, dunes, thickets, upper margins of tidal flats, river flats, gravel bars, tundra, open forests, roadsides, coastal to alpine; 0–2100 m; B.C., Yukon; Alaska; Asia (Russian Far East).

Castilleja unalaschcensis is a characteristic meadow species across much of the southern Alaskan coast, including the Aleutian Islands, as well as on the Queen Charlotte Islands (Haida Gwaii) and the mainland of western British Columbia and southern Yukon. It is common in the coastal littoral zone and ascends to lower alpine meadow communities. Morphologically, it appears to be a sister species to *C. septentrionalis* of the Rocky Mountains and boreal sites in eastern North America. A weakly differentiated interior form was described as subsp. *transnivalis* Pennell.

115. Castilleja victoriae Fairbarns & J. M. Egger, Madroño 54: 335, figs. 1, 3 [lower right]. 2008 • Victoria's owl's-clover [C][E][F]

Herbs, annual, 0.2–2 dm; with fibrous roots. **Stems** solitary, erect, unbranched, sometimes branched, hairs spreading, long, soft, mixed with short stipitate-glandular ones. **Leaves** usually brownish throughout, sometimes greenish proximally, brownish distally, margins deep brown, lanceolate to narrowly ovate, 0.5–2.7 cm, not fleshy, margins plane, involute, 0–3(–5)-lobed, apex acute; lobes spreading-ascending, linear to lanceolate, apex acute, obtuse, or rounded. **Inflorescences** 1–5 × 1.5–3 cm; bracts dull reddish brown throughout, or proximally dull greenish, distally dull reddish brown, narrowly ovate, 3–7-lobed; lobes ascending, lanceolate, medium length, arising near or above mid length, apex broadly acute or obtuse. **Calyces** colored as bracts, 8–12 mm; abaxial and adaxial clefts 5–6 mm, 50% of calyx length, deeper than laterals, lateral 3–4 mm, 30–40% of calyx length; lobes linear-lanceolate, apex acute. **Corollas** straight, 10–18 mm; tube 9–13 mm; abaxial lip and beak exserted; beak adaxially white, sometimes faintly diffuse purple, 3 mm, inconspicuously puberulent, hairs short stipitate-glandular; abaxial lip yellowish, cream, or soft pale yellow, lacking spots, moderately conspicuous, pouches 3, conspicuous, 1–3 mm deep, central one furrowed, conspicuous, 2–2.5 mm, 60–75% as long as beak; teeth erect, white or cream, 0.5–0.8 mm. **Stigmas** included in beak. $2n$ = 24.

Flowering (Apr–)May–Jul. Vernal pools, springs, windswept, thin-soiled rocky knolls and benches by sea, gneiss; of conservation concern; 0–10 m; B.C.; Wash.

Castilleja victoriae was first collected in 1893 but not described until 2008. It is restricted to the southern-most tip of Vancouver Island, near Oak Bay, and to several small adjacent islands within a 30 km radius in British Columbia and Washington. Its primary habitat is small depressions and vernal pools on gentle gradients within 100 m of the sea, making it particularly vulnerable to development and recreational conflicts. Never common, several historic populations near Victoria were extirpated before the species was recognized. The majority of the global population is found on Trial Island. All known extant populations would be inundated by projected sea level increases and storm surges due to climate change. It should be considered globally endangered.

116. Castilleja viscidula A. Gray in A. Gray et al., Syn. Fl. N. Amer. 2(1): 297. 1878 (as Castilleia) • Sticky paintbrush [E] [F]

Herbs, perennial, 0.5–3(–4) dm; from a woody caudex; with a taproot. **Stems** several, ascending to erect, decumbent at base, unbranched, sometimes branched, hairs spreading, long, soft, mixed with shorter stipitate-glandular ones. **Leaves** green to brown, linear, linear-lanceolate, lanceolate, or oblong (narrowly ovate nearing inflorescence), 1–4(–5) cm, not fleshy, margins wavy, flat or involute, (0–)2(–5)-lobed, apex acute; lobes ascending-spreading, oblong to narrowly lanceolate, apex acute or obtuse. **Inflorescences** 2–14 × 1–3.5 cm; bracts proximally greenish to greenish brown, distally pale yellow, cream, or yellow-green, sometimes yellow-orange or red (sometimes gradually differentiated from proximal coloration), lanceolate, broadly lanceolate, or oblong, 3(–5)-lobed, proximal wavy-margined; lobes ascending, linear to narrowly lanceolate, long, arising near or above mid length, sometimes wavy-margined, apex acute to rounded. **Calyces** colored as bracts or proximally paler, (10–)14–18 mm; abaxial and adaxial clefts (4–)5–9 mm, 30–40% of calyx length, deeper than laterals, lateral (1–)2–6 mm, ca. 25% of calyx length; lobes narrowly ovate to lanceolate, linear, or narrowly lanceolate, apex acute to obtuse. **Corollas** straight, 16–22(–25) mm; tube 10–15 mm; beak exserted, straight to sometimes curved, adaxially green to yellow, 5–8(–9) mm; abaxial lip green or yellow, sometimes deep purple, reduced, inconspicuous, often visible in abaxial cleft, 1–2 mm, 20% as long as beak; teeth erect, green to white, sometimes yellow or pink, 0.5–1 mm. $2n = 24, 72$.

Flowering Jun–Aug. Dry to mesic sagebrush steppes, rocky slopes, ledges, open woodlands, montane to subalpine; 2000–3200 m; Idaho, Nev., Oreg.

Castilleja viscidula is a member of the complex including *C. applegatei* and *C. martini*, centered in California. *Castilleja viscidula* favors isolated mountain ranges, from the Wallowa and, possibly, the Blue mountains of northeastern Oregon, eastward into southwestern Idaho, and southward into central Nevada. Most populations are greenish yellow, but in one portion of the Wallowa Mountains, reddish bracted plants are common. Many yellowish bracted populations in the same mountain range surround this reddish population. Intermediate color forms are rarely encountered. Most ranges where *C. viscidula* occurs have generated slightly differing local races, demonstrating some reproductive isolation and divergence. In addition, hybrid swarms between this species and *C. nana* are known from several mountain ranges in central and northern Nevada, and an apparent hybrid with *C. flava* var. *flava* is known from the Independence Mountains of northern Nevada.

117. Castilleja wightii Elmer, Bot. Gaz. 41: 322. 1906 (as Castilleia) • Wight's paintbrush [E]

Castilleja wightii subsp. *rubra* Pennell

Herbs or subshrubs, perennial, 3–8 dm; from a woody caudex; with woody, branching roots. **Stems** few to many, ascending to erect, much-branched, usually with prominent leafy axillary shoots, hairs spreading, long, soft, mixed with shorter, stipitate-glandular ones. **Leaves** green-tinged or ± purple, sometimes yellow-green tinged, densely crowded, linear, lanceolate, oblong, elliptic, or ovate, 1–7 cm, thickened, not usually fleshy, margins plane to wavy, involute or flat, 0–3(–5)-lobed, apex broadly rounded to obtuse or acute; lobes spreading-ascending, linear, narrowly lanceolate, triangular, or rounded, apex acute, obtuse, or rounded. **Inflorescences** 3–22 × 1.5–3.5 cm; bracts proximally green to deep purple, rarely light tan, distally red, scarlet, rose, red-orange, or yellow, sometimes orange, dull brownish orange, pale pinkish tan, yellow aging white, yellow aging pink, red with pink apices, magenta, or white, sometimes with a yellow to deep purple medial band, lanceolate, oblong, or narrowly ovate, (0–)3–5(–7+)-lobed, distal margins of central lobe and sometimes also side lobes with multiple shallow teeth, proximal often wavy-margined; lobes erect or ascending, linear to lanceolate or oblanceolate, short, arising in distal ⅔, apex obtuse to acute, central lobe often rounded to truncate. **Pedicels** 0 mm or nearly so. **Calyces** colored as bracts, 15–28 mm; abaxial and adaxial clefts 8–11 mm, 33–50% of calyx length, deeper than laterals, lateral 2–5.5 mm, 5–30% of calyx length; lobes lanceolate to oblong to broadly triangular, apex acute, obtuse, or rounded. **Corollas** slightly curved proximally, 20–30 mm; tube 10–12 mm; distal portion of beak exserted, abaxial lip included; beak adaxially green to yellow, 11.5–18 mm; abaxial lip deep green, sometimes to very deep purple, reduced, included, usually not visible through front cleft, 1–2 mm, 15–25% as long as beak; teeth erect, green to sometimes pink, 0.5–1.5 mm. $2n = 24, 48$.

Flowering Mar–Aug(–Nov). Coastal scrub, damp thickets, stream banks, sea bluffs, canyon slopes, roadsides; 0–700 m; Calif.

Castilleja wightii is found along or near the central and northern coast of California. Historical collections of *C. wightii* from Curry County, Oregon, are referable to other species, but it should be sought in the area. Reports from the south-central coast of California are referable to other species, particularly *C. affinis*. *Castilleja wightii* appears to intergrade with *C. latifolia* and perhaps *C. affinis* south of San Francisco. Despite much attention from botanists, the perennial paintbrushes along the coast between Monterey and San Francisco can be perplexing and difficult to identify. This situation is likely the result of introgression, but this complex is in need of meticulous genetic and morphological analysis. North of San Francisco, *C. wightii* is straightforward to recognize, with its abundantly stipitate-glandular stems and leaves. In addition, the leaves are often crowded on the stems, which bear axillary shoots. Most populations have either red or yellow bracts, with only occasional individual plants of the other color. Yellow populations are found primarily in Marin and southern Sonoma counties southward and are gradually replaced by red populations from northern Sonoma County northward. Mixed color populations occur in a few places, especially in San Mateo County.

118. Castilleja wootonii Standley, Muhlenbergia 5: 84. 1909 • Wooton's paintbrush [E]

Castilleja ciliata Pennell

Herbs, perennial, 1.6–6.5 dm; from a woody caudex; with a woody taproot or branching roots. Stems solitary or few to many, erect, unbranched to much-branched, glabrous or hairy proximally and/or distally, hairs sparse to dense, spreading to erect, short to fairly long, soft, eglandular. Leaves green, narrowly lanceolate to linear-lanceolate, linear, or narrowly elliptic, 2–8 cm, not fleshy, margins plane, sometimes ± wavy, flat or involute, 0(–5)-lobed, apex acuminate; lobes widely spreading, linear to narrowly lanceolate, apex acute. Inflorescences 3–16 × 2–3.5 cm; bracts proximally greenish, distally red to orange-red, sometimes with a purplish medial band, lanceolate to broadly lanceolate or ovate in outline, (0–)3–5 (–7)-lobed, sometimes also with 4 small teeth; lobes ascending, linear-lanceolate, long, arising above or below mid length, apex acuminate, acute, or obtuse. Calyces proximally green, distally red, 20–25 mm; abaxial clefts 11–14(–17) mm, adaxial 8–9 mm, abaxial 50–60% of calyx length, adaxial 35–40% of calyx length, deeper than laterals, lateral 5–7 mm, 10–15% of calyx length; lobes lanceolate to narrowly triangular, apex acute to acuminate. Corollas slightly to moderately curved distally, 25–37 mm; tube 16–20 mm; beak exserted 10–16 mm beyond calyx, adaxially green to yellowish, 11–25 mm; abaxial lip green or red, small, inconspicuous, visible through cleft of calyx, 2 mm, 15–20% as long as beak; teeth incurved, green or red, 0.7–1.5 mm.

Flowering Jun–Sep. Grasslands, rocky slopes, ledges, canyons, open forests, montane to subalpine; 2000–3700 m; N.Mex., Tex.

Castilleja wootonii is endemic to the White Mountains (Sierra Blanca) in south-central New Mexico and to the Mt. Livermore massif of western Texas. It should be sought in the intervening Guadalupe Mountains. Based on morphology, *C. wootonii* appears to be a southern derivative of *C. linariifolia.*

119. Castilleja xanthotricha Pennell, Notul. Nat. Acad. Nat. Sci. Philadelphia 74: 5. 1941 • John Day or yellow-hairy paintbrush [E]

Herbs, perennial, 1–2(–3.8) dm; from a woody caudex; with a taproot. Stems few to several, ± decumbent to erect or ascending, unbranched, sometimes with short, leafy axillary shoots, hairs erect to spreading, long, soft, eglandular, mixed with short stipitate-glandular ones. Leaves green, linear, lanceolate to broadly lanceolate, oblong, or cuneate, 0.8–5 cm, not fleshy, margins plane to wavy, involute, 0–5-lobed, apex acute, sometimes rounded; lobes spreading, linear, arising below mid length, nearly as broad as center lobe, apex acute. Inflorescences 3–14 × 1.5–4.5 cm; bracts proximally greenish, rarely dull reddish purple, distally white to cream, rarely pale yellow or dull, pale pink (sharply differentiated from proximal coloration), lanceolate or oblong to narrowly ovate, (3–)5–7-lobed; lobes ascending, linear to obovate, ± broadened distally, medium, long, proximal lobes arising below mid length, central lobe apex broadly rounded to truncate, others acute to rounded. Calyces colored as bracts, 15–26 mm; abaxial and adaxial clefts 3.5–7 mm, 25–50% of calyx length, deeper than laterals, lateral 2–5 mm, 12–25% of calyx length; lobes linear, oblong, or narrowly triangular, center lobe apex usually rounded, lobes acute to rounded. Corollas curved, 17–23 mm; tube 15–19 mm; beak exserted, adaxially green, 5–8(–9) mm, puberulent, stipitate-glandular; abaxial lip deep purple (color sometimes visible through calyx), green, pinkish, or pale yellow, ± prominent, slightly inflated, usually hidden in calyx, sometimes right at top of calyx, 2 mm, ca. 50% as long as beak; teeth ascending, whitish, yellowish, pink, or green, 1–1.5 mm. $2n = 48$.

Flowering Apr–Jul. Arid, rocky, sandy, or clay slopes of basaltic origin, sagebrush steppes; 400–800 m; Oreg.

Castilleja xanthotricha is endemic to moderate elevations in the sagebrush hills of the John Day River drainage in north-central Oregon. N. H. Holmgren (1971) hypothesized that this tetraploid species is of allopolyploid hybrid origin between *C. glandulifera* and *C. oresbia.*

23. CHLOROPYRON Behr, Proc. Calif. Acad. Sci. 1: 61. 1855 • [Greek *chloros*, green or yellow-green, and *pyros*, fire, hence red or yellow, alluding to yellowish green plants]

Kerry A. Barringer

Herbs, annual; hemiparasitic. **Stems** erect, spreading, or decumbent, not fleshy, puberulent, hispid, or villous, sometimes glandular-hairy or glabrescent. **Leaves** cauline, alternate; petiole absent; blade not fleshy, not leathery, margins entire or pinnately 5- or 7-lobed. **Inflorescences** terminal, spikes; bracts present. **Pedicels** absent; bracteoles absent. **Flowers:** sepals 2, calyx bilaterally symmetric, spathelike, lobes narrowly triangular to triangular; petals 5, corolla white, yellow, pink, or lavender, often marked or tinted with pink to purple-red lines or spots, strongly bilabiate, club-shaped, abaxial lobes 3, middle lobe erect, not revolute, adaxial 2, adaxial lip galeate, rounded at apex, opening downward; stamens 2 or 4, didynamous, filaments glabrous or sparsely pilose, pollen sacs approximate, connective not elongate; staminodes 0 or 2, peglike; ovary 2-locular, placentation axile; stigma slightly expanded at apex. **Capsules:** dehiscence loculicidal. **Seeds** 8–40, brown to dark brown, ovoid to ± reniform, wings absent. *x* = 7.

Species 4 (4 in the flora): w United States, nw Mexico.

Species of *Chloropyron* grow almost exclusively in saline and alkaline habitats.

Chloropyron is similar to *Cordylanthus* and *Dicranostegia* and, like them, has upright flowers in which the abaxial corolla lip is usually held close to the galeate adaxial lip. The flowers appear to be in bud even when they are fully open.

T. I. Chuang and L. R. Heckard (1973) placed species of *Chloropyron* in *Cordylanthus* as subg. *Hemistegia* (A. Gray) Jepson. With evidence from molecular data, D. C. Tank et al. (2009) have shown that *Chloropyron* is closely related to *Dicranostegia* and that both genera form a clade that is sister to *Castilleja* and *Triphysaria*, while *Cordylanthus* in the narrow sense is somewhat unresolved at the base of the subtribe Castillejinae clade.

SELECTED REFERENCE Chuang, T. I. and L. R. Heckard. 1973. Taxonomy of *Cordylanthus* subgenus *Hemistegia* (Scrophulariaceae). Brittonia 25: 135–158.

1. Stamens 4, staminodes 0; bract margins entire or distally 2-toothed . . . . 1. *Chloropyron maritimum*
1. Stamens 2, staminodes 2; bract margins pinnately lobed.
   2. Leaf blades 1–2 mm wide, linear to linear-lanceolate; styles puberulent . . . . . . 4. *Chloropyron tecopense*
   2. Leaf blades 2–10 mm wide, narrowly lanceolate to lanceolate; styles glabrous.
      3. Calyces 15–20 mm; seeds without abaxial crest; stems puberulent or hispid. . . .2. *Chloropyron molle*
      3. Calyces 12–15 mm; seeds with abaxial crest; stems sparsely pilose or glabrescent
       . . . . 3. *Chloropyron palmatum*

1. Chloropyron maritimum (Nuttall ex Bentham) A. Heller, Muhlenbergia 3: 133. 1907 [F]

Cordylanthus maritimus Nuttall ex Bentham in A. P. de Candolle and A. L. P. P. de Candolle, Prodr. 10: 598. 1846

Stems erect, spreading, or decumbent, 10–40 cm, puberulent or villous, sometimes glabrescent, hairs glandular and eglandular. **Leaf blades** narrowly lanceolate, 5–30 × 2–8 mm, margins entire. **Spikes** 2–9 cm; bracts often purple distally, lanceolate to oblong-lanceolate, 15–30 mm, margins entire or distally 2-toothed. **Flowers:** calyx 15–25 mm; corolla white to pale yellow or pale pink, 15–25 mm, lobes 4–5 mm, often marked with red-brown or purple-red lines; stamens 4, proximal with 2 pollen sacs, distal pair with 1 pollen sac and 1 infertile appendage; staminodes 0. **Capsules** narrowly ovoid, 6–10 mm. **Seeds** 10–40, dark brown, ovoid to reniform, 1–3 mm, without abaxial crest.

Subspecies 3 (3 in the flora): w United States, nw Mexico.

There are intermediates between the subspecies (T. I. Chuang and L. R. Heckard 1973).

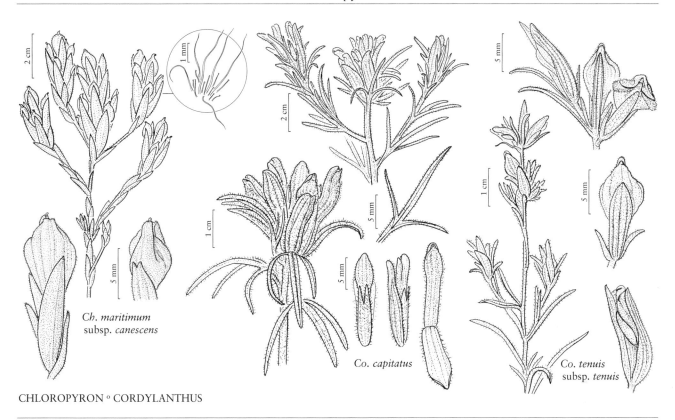

CHLOROPYRON ° CORDYLANTHUS

Ch. maritimum
subsp. *canescens*

Co. capitatus

Co. tenuis
subsp. *tenuis*

1. Seeds 25–40, 1–1.5 mm; galeas yellow to yellow-green.... 1b. *Chloropyron maritimum* subsp. *canescens*
1. Seeds 10–20, 1.5–3 mm; galeas purple-red, brown-red, or pink.
   2. Stem branches: distal usually overtopping central spike; galeas purple-red or brown-red .... 1a. *Chloropyron maritimum* subsp. *maritimum*
   2. Stem branches not overtopping central spike; galeas pink or purple-red ....
       ....1c. *Chloropyron maritimum* subsp. *palustre*

1a. Chloropyron maritimum (Nuttall ex Bentham) A. Heller subsp. **maritimum** • Saltmarsh bird's-beak C̲

Stems 15–30 cm, puberulent and often villous or short-pilose; branches decumbent to ascending, distals usually overtopping central spike. **Spikes:** bract margins entire or slightly notched. **Flowers:** corolla tube and abaxial lip pale yellow to white, galea purple-red or brown-red. **Seeds** 15–20, 1.5–2.5 mm. $2n = 30$.

Flowering May–Oct. Coastal salt marshes; of conservation concern; 0–10 m; Calif.; Mexico (Baja California).

Plants intermediate between the coastal subsp. *maritimum* and the inland subsp. *canescens* have been found in southern California.

1b. Chloropyron maritimum (Nuttall ex Bentham) A. Heller subsp. **canescens** (A. Gray) Tank & J. M. Egger, Syst. Bot. 34: 188. 2009 • Alkali bird's-beak E̲ F̲

Cordylanthus canescens A. Gray, Proc. Amer. Acad. Arts 7: 383. 1868; *C. maritimus* Nuttall ex Bentham subsp. *canescens* (A. Gray) T. I. Chuang & Heckard

Stems 15–40 cm, canescent, villous, hairs spreading; branches erect or ascending, distals overtopping central spike. **Leaves** 15–30 mm. **Spikes:** bract margins entire or distally 2-toothed. **Flowers:** corolla tube and abaxial lip white to pale pink, galea yellow to yellow-green. **Seeds** 25–40, 1–1.5 mm. $2n = 30$.

Flowering Jun–Sep. Inland alkaline flats; 600–1900 m; Calif., Nev., Oreg., Utah.

Subspecies *canescens* is the most widespread of the subspecies. Plants are always erect and have ascending branches. The stems are villous with conspicuous, straight, spreading white hairs.

1c. Chloropyron maritimum (Nuttall ex Bentham) A. Heller subsp. **palustre** (Behr) Tank & J. M. Egger, Syst. Bot. 34: 188. 2009 • Point Reyes bird's-beak
C E

Chloropyron palustre Behr, Proc. Calif. Acad. Sci. 1: 61. 1855; *Cordylanthus maritimus* Nuttall ex Bentham subsp. *palustris* (Behr) T. I. Chuang & Heckard

Stems 10–30 cm, puberulent and pilose or glabrescent; branches absent or ascending, distals not overtopping central spike. **Spikes:** bract margins entire or slightly notched. **Flowers:** corolla tube and abaxial lip white to pale yellow, galea pink or purple-red. **Seeds** 10–20, 2–3 mm. $2n = 30$.

Flowering May–Oct. Coastal salt marshes; of conservation concern; 0–10 m; Calif., Oreg.

Subspecies *palustre* grows just above the high tide line in salt marshes from Coos County, Oregon, to San Luis Obispo County, California.

2. Chloropyron molle (A. Gray) A. Heller, Muhlenbergia 3: 134. 1907 C E

Cordylanthus mollis A. Gray, Proc. Amer. Acad. Arts 7: 384. 1868

Stems erect or spreading, 10–40 cm, puberulent or hispid, hairs glandular and eglandular. **Leaf blades** narrowly lanceolate to lanceolate, 5–30 × 2–10 mm, margins entire or 7-lobed, lateral veins conspicuous. **Spikes** 2–15 cm; bracts sometimes purple distally, lanceolate to ovate, 15–25 mm, margins pinnately 3–7-lobed. **Flowers:** calyx 15–20 mm; corolla greenish white, white, or pale yellow, 15–20 mm, lobes 5–7 mm, often marked with dark red (especially toward base); stamens 2, each with 2 pollen sacs; staminodes 2. **Capsules** narrowly ovoid, 6–10 mm. **Seeds** 20–30, dark brown, ovoid to reniform, 1–3 mm, without abaxial crest.

Subspecies 2 (2 in the flora): California.

The coastal and inland subspecies intergrade.

1. Branches from distal to base; seeds 2–3 mm; corolla tubes densely tomentose . . . .
. . . . 2a. *Chloropyron molle* subsp. *molle*
1. Branches from near base; seeds 1–1.5 mm; corolla tubes sparsely tomentose. . . . .
. . . . 2b. *Chloropyron molle* subsp. *hispidum*

2a. Chloropyron molle (A. Gray) A. Heller subsp. **molle**
C E

Stems puberulent, hairs soft, sometimes stiff; branches from distal to base, ascending. **Spikes** 5–15 cm. **Flowers:** corolla tube densely tomentose. **Seeds** 2–3 mm. $2n = 28$.

Flowering Jul–Nov. Coastal salt marshes; of conservation concern; 0–10 m; Calif.

Plants of subsp. *molle* appear to lose their leaves earlier than those of subsp. *hispidum*.

2b. Chloropyron molle (A. Gray) A. Heller subsp. **hispidum** (Pennell) Tank & J. M. Egger, Syst. Bot. 34: 188. 2009 C E

Cordylanthus hispidus Pennell, Proc. Acad. Nat. Sci. Philadelphia 99: 192. 1947; *C. mollis* A. Gray subsp. *hispidus* (Pennell) T. I. Chuang & Heckard

Stems hispid, hairs stiff; branches from near base, spreading. **Spikes** 2–6 cm. **Flowers:** corolla tube sparsely tomentose. **Seeds** 1–1.5 mm. $2n = 28$.

Flowering Jun–Jul. Salt marshes, alkaline flats; of conservation concern; 0–150 m; Calif.

Subspecies *hispidum* is usually found inland from the coast in Kern, Merced, and Solano counties. Plants tend to be slightly shorter, retain their leaves longer, and have more deeply incised leaf margins than those of subsp. *molle*. Hispid indument, once thought to characterize subsp. *hispidum*, also occurs on plants of subsp. *molle*.

3. Chloropyron palmatum (Ferris) Tank & J. M. Egger, Syst. Bot. 34: 189. 2009 [C] [E]

Adenostegia palmata Ferris, Bull. Torrey Bot. Club 45: 420. 1918; *Cordylanthus carnulosus* Pennell; *C. palmatus* (Ferris) J. F. Macbride; *C. palmatus* subsp. *carnulosus* (Pennell) Munz

Stems erect or spreading, 10–30 cm, sparsely pilose or glabrescent, hairs glandular. **Leaf blades** narrowly lanceolate to lanceolate, 7–20 × 3–7 mm, margins entire or pinnately 5-lobed, lateral veins conspicuous. **Spikes** 5–15 cm; bracts often red distally, narrowly ovate to ovate, 12–20 mm, margins pinnately 3–7-lobed. **Flowers:** calyx 12–15 mm; corolla white to pale lavender, 12–20 mm, lobes 4–5 mm, often with pale lavender spots at base of abaxial lobe; stamens 2, each with 2 pollen sacs; staminodes 2. **Capsules** narrowly ovoid, 6–7 mm. **Seeds** 14–18, brown to dark brown, ± reniform, 2.5–3 mm, with abaxial crest. $2n = 42$.

Flowering Jun–Aug. Alkaline flats; of conservation concern; 10–150 m; Calif.

Chloropyron palmatum is threatened by agriculture and urbanization (T. I. Chuang and L. R. Heckard 1973) in Fresno, Madera, San Joaquin, and Yolo counties. Inflorescence bracts are not palmate but are more deeply incised than those of *C. molle*.

4. Chloropyron tecopense (Munz & J. C. Roos) Tank & J. M. Egger, Syst. Bot. 34: 189. 2009 • Tecopa bird's-beak [E]

Cordylanthus tecopensis Munz & J. C. Roos, Aliso 2: 233. 1950

Stems erect to spreading, 5–60 cm, sparsely puberulent, hairs glandular and eglandular. **Leaf blades** linear to linear-lanceolate, 5–15 × 1–2 mm, margins entire. **Spikes** 2–15 cm; bracts often purple distally, linear to narrowly lanceolate, 10–15 mm, margins pinnately 3-lobed near middle. **Flowers:** calyx 10–13 mm; corolla pale lavender to pale lilac, 10–15 mm, lobes 4–5 mm, occasionally marked with red-brown or purple-red lines; stamens 2, each with 2 pollen sacs; staminodes 2. **Capsules** narrowly ellipsoid, 5–7 mm. **Seeds** 8–10, brown, ± reniform, 2–3 mm, without abaxial crest. $2n = 28$.

Flowering Jul–Oct. Alkaline meadows and flats; 100–900 m; Calif., Nev.

Chloropyron tecopense has a delicate appearance because of their slender branches with relatively small, appressed leaves. The species is found in Inyo and eastern San Bernardino counties in California and in Esmeralda County in western Nevada.

24. CORDYLANTHUS Nuttall ex Bentham in A. P. de Candolle and A. L. P. P. de Candolle, Prodr. 10: 597. 1846, name conserved • Bird's-beak [Greek *kordyle*, club, and *anthos*, flower, alluding to somewhat clavate corolla]

Kerry A. Barringer

Adenostegia Bentham, name rejected

Herbs, annual; hemiparasitic. **Stems** erect or ascending, rarely decumbent, not fleshy, hairy or glabrous. **Leaves** cauline, alternate; petiole absent; blade not fleshy, not leathery, margins entire or 3–7-lobed. **Inflorescences** terminal, spikes or flowers solitary, often capitate; bracts present. **Pedicels** absent; bracteoles absent. **Flowers:** sepals 2, calyx bilaterally symmetric, spathelike, lobes triangular; petals 5, corolla yellow, yellow-green, purple, pink, or red, strongly bilabiate, club-shaped, abaxial lobes 3, middle lobe tightly revolute, tip distinctly folded inside-out, adaxial 2, adaxial lip galeate, rounded at apex, opening downward; stamens (2 or)4, didynamous, filaments hairy or glabrous; staminode 0; ovary 2-locular, placentation axile; stigma slightly expanded at apex. **Capsules:** dehiscence loculicidal. **Seeds** 4–25, pale brown to dark brown, ovoid to reniform, wings absent. $x = 6, 7$.

Species 13 (13 in the flora): w United States, nw Mexico.

Cordylanthus is similar to *Orthocarpus* but differs in having spathelike calyces deeply cut along one side and narrowly lanceolate, entire bracteoles subtending each flower. Also, the abaxial lip of *Orthocarpus* is spreading; it usually is appressed to the adaxial lip in *Cordylanthus*. The mature flowers often appear to be buds.

Cordylanthus is sometimes defined to include *Dicranostegia* and *Chloropyron* as subgenera (T. I. Chuang and L. R. Heckard 1986). D. C. Tank et al. (2009) found that those genera are a sister group of *Triphysaria*. The monograph by Chuang and Heckard provided the basis for most of the species delineation presented here.

SELECTED REFERENCE Chuang, T. I. and L. R. Heckard. 1986. Systematics and evolution of *Cordylanthus* (Scrophulariaceae—Pedicularieae). Syst. Bot. Monogr. 10: 1–105.

1. Stamens 2, filaments glabrous; calyx apices 2-fid, cleft 3–5 mm. . . . . 1. *Cordylanthus capitatus*
1. Stamens 4, filaments hairy; calyx apices entire or 2-fid, cleft 0–3 mm.
    2. Bract margins 5–7-lobed, flabelliform; corollas 8–9 mm . . . . .9. *Cordylanthus pringlei*
    2. Bract margins entire, 3-lobed, or (4 or)5–9-lobed, not flabelliform; corollas 10–30 mm.
        3. Fertile pollen sacs 1 per filament.
            4. Calyx apices 2-fid; bracts green or purple distally; abaxial corolla lips 7–10 mm, slightly spreading. . . . . 4. *Cordylanthus laxiflorus*
            4. Calyx apices entire; bracts white to cream distally; abaxial corolla lips 3–5 mm, appressed to adaxial . . . . 5. *Cordylanthus nevinii*
        3. Fertile pollen sacs 2 per filament.
            5. Stems decumbent, 5–10(–15) cm; corollas white with purple veins. . . .6. *Cordylanthus nidularius*
            5. Stems erect or ascending, 10–150 cm; corollas pale or bright yellow, yellow, yellow-green, pale pink, lavender-pink, pink, purple, purple-pink, or purple-red.
                6. Bract margins (4 or)5–9-lobed.
                    7. Calyx tubes 0 mm; fertile pollen sacs unequal. . . . . 10. *Cordylanthus ramosus*
                    7. Calyx tubes 1–4 mm; fertile pollen sacs equal.
                        8. Corollas 10–20 mm, throats 4–6 mm diam.; calyx tubes 1–3 mm; capsules 7–10 mm. . . . . .2. *Cordylanthus eremicus*
                        8. Corollas 15–30 mm, throats 6–8 mm diam.; calyx tubes 3–4 mm; capsules 10–15 mm. . . . . 13. *Cordylanthus wrightii*
                6. Bract margins entire or 3-lobed.
                    9. Calyx tubes 5–8 mm; corolla abaxial lips shorter than and not appressed to adaxial . . . . 7. *Cordylanthus parviflorus*
                    9. Calyx tubes 0–2.5 mm; corolla abaxial lips ca. equal to and appressed to adaxial.
                        10. Calyx apices 2-fid, clefts 2–3 mm; bracteole margins pinnately lobed. . . . . 3. *Cordylanthus kingii*
                        10. Calyx apices entire or 2-fid, clefts 0.5–1 mm; bracteole margins entire or toothed.
                            11. Calyx tubes 1–2 mm; inflorescences (2–)5–15-flowered; corollas with abaxial, U-shaped, purple markings. . . . . 11. *Cordylanthus rigidus*
                            11. Calyx tubes 0–1 mm; inflorescences 2–7-flowered, or flowers solitary; corollas streaked and spotted with maroon or marked with purple.
                                12. Stems densely puberulent and glandular-puberulent, and pilose; abaxial corolla lips 5–10 mm . . . . 8. *Cordylanthus pilosus*
                                12. Stems glabrous or sparsely glandular-puberulent, puberulent, and/or pilose; abaxial corolla lips 4–6 mm . . . . 12. *Cordylanthus tenuis*

1. Cordylanthus capitatus Nuttall ex Bentham in A. P. de Candolle and A. L. P. P. de Candolle, Prodr. 10: 597. 1846 • Yakima bird's-beak [E] [F]

Cordylanthus nevadensis Edwin

Stems erect or ascending, 10–50 cm, densely glandular-pubescent and pilose. **Leaves** glandular-pubescent and pilose; proximal 20–40(–50) mm, margins 3-lobed, lobes 1–2 mm wide; distal 10–30 × 1 mm, margins entire. **Inflorescences** capitate spikes, 2–5-flowered; bracts 1–5, 10–20 mm, margins 3-lobed, lobes green and purple, linear to linear-lanceolate. **Pedicels:** bracteoles 12–18 mm, margins entire. **Flowers:** calyx 10–15 mm, tube 2–4 mm, apex 2-fid, cleft 3–5 mm; corolla purple-red, apex yellow, 10–20 mm, throat 4–6 mm diam., abaxial lip 3–5 mm, ca. equal to and appressed to adaxial; stamens 2, filaments glabrous, fertile pollen sacs 1 per filament, vestigial pollen sacs present. **Capsules** ovoid to oblong-ovoid, 5–7 mm. **Seeds** 4–6, dark brown, narrowly ovoid to reniform, 2–2.5 mm, reticulate. $2n = 26$.

Flowering Jul–Oct. Openings in conifer woodlands and juniper scrub; 1400–2300 m; Calif., Idaho, Nev., Oreg., Wash.

Cordylanthus capitatus is often associated with *Artemisia tridentata* and may be hemiparasitic on it.

Cordylanthus capitatus is distinguished by having only two anthers, each with one fertile pollen sac; infertile vestiges of the second pollen sac usually are present. The species can be recognized by its short-capitate inflorescences, the single, spathelike, deeply cleft calyx lobe, and the purple-red corollas with yellow tips.

2. Cordylanthus eremicus (Coville & C. V. Morton) Munz, Man. S. Calif. Bot., 483, 601. 1935 • Desert bird's-beak [E]

Adenostegia eremica Coville & C. V. Morton, J. Wash. Acad. Sci. 22: 161, fig. [p. 162]. 1932; *Cordylanthus ramosus* Nuttall ex Bentham subsp. *eremicus* (Coville & C. V. Morton) Munz

Stems erect or ascending, 10–80 cm, puberulent, glabrescent. **Leaves** puberulent, sometimes scabrous; proximal 10–40 mm, margins 3–7-lobed, lobes 1 mm wide; distal 5–25 × 1 mm, margins entire. **Inflorescences** capitate spikes, 3–14-flowered; bracts 5–10, 5–20 mm, margins 5–7-lobed, lobes purple or yellow-green, linear to filiform. **Pedicels:** bracteoles 10–20 mm, margins entire. **Flowers:** calyx 10–20 mm, tube 1–3 mm, apex entire or 2-fid, cleft 0–0.5 mm; corolla pink to lavender-pink, usually spotted with purple, 10–20 mm, throat 4–6 mm diam., abaxial lip pink or yellow, 3–6 mm, shorter than and appressed to adaxial; stamens 4, filaments hairy, fertile pollen sacs 2 per filament, equal. **Capsules** oblong-lanceoloid, 7–10 mm. **Seeds** 10–15, pale brown, ovoid, 1.5–2 mm, reticulate.

Subspecies 2 (2 in the flora): California.

Cordylanthus eremicus is similar to *C. wrightii*, which also has relatively short, dense spikes and inflorescence bracts palmately three- to seven-lobed. *Cordylanthus eremicus* can be distinguished from *C. wrightii* by its gray to white hairs.

1. Bracts: lobes purple distally, apices rounded to retuse; calyx tubes 2–3 mm; filaments hairy throughout . . . 2a. *Cordylanthus eremicus* subsp. *eremicus*
1. Bracts: lobes green to yellow-green distally, apices acute; calyx tubes 1 mm; filaments hairy distally . . . . 2b. *Cordylanthus eremicus* subsp. *kernensis*

2a. Cordylanthus eremicus (Coville & C. V. Morton) Munz subsp. **eremicus** [E]

Cordylanthus bernardinus Munz

Inflorescences: bracts puberulent, lobes purple distally, apex slightly rounded to retuse. **Flowers:** calyx tube 2–3 mm; corolla lavender to pink, blotched purple, lips yellow; filaments hairy throughout. **Seeds** minutely papillate. $2n = 26$.

Flowering Aug–Sep. Dry, rocky openings in sagebrush scrub and pinyon-juniper woodlands; 1000–2800 m; Calif.

2b. Cordylanthus eremicus (Coville & C. V. Morton) Munz subsp. **kernensis** T. I. Chuang & Heckard, Syst. Bot. Monogr. 10: 91, figs 3c², 8k,l, 9d, 36i. 1986 • Kern Plateau bird's-beak [E]

Inflorescences: bracts sparsely pilose or glabrous, lobes green to yellow-green distally, apex acute. **Flowers:** calyx tube 1 mm; corolla pink to lavender, lips pink; filaments hairy distally. **Seeds** not papillate.

Flowering Jun–Sep. Dry, rocky slopes in pine-juniper woodlands; 2100–3000 m; Calif.

Subspecies *kernensis* is known from the Kern Plateau in Inyo and Tulare counties. It can resemble *Cordylanthus ramosus*, which has calyces without a tube and yellow, not pink, corollas.

3. Cordylanthus kingii S. Watson, Botany (Fortieth Parallel), 233, 460, plate 22, figs. 3–6. 1871 • King's bird's-beak E

Stems erect, 10–80 cm, puberulent, pubescent, or densely villous. **Leaves** glandular-pubescent; proximal 10–40 mm, margins 3–5-lobed, lobes linear to filiform, 1–3 mm wide; distal 10–25 × 1 mm, margins entire. **Inflorescences** capitate spikes, 2–12-flowered, or flowers solitary; bracts 1–6, 10–40 mm, margins 3-lobed, lobes purple distally, narrowly lanceolate, linear, or filiform. **Pedicels:** bracteoles 10–40 mm, margins pinnately lobed. **Flowers:** calyx 15–30(–40) mm, tube 1–2.5 mm, apex 2-fid, cleft 2–3 mm; corolla lavender-pink to purple-red or yellow with purple veins and markings, 15–25(–30) mm, throat 5–7 mm diam., abaxial lip 5–10 mm, ca. equal to and appressed to adaxial; stamens 4, filaments hairy, fertile pollen sacs 2 per filament, equal. **Capsules** oblong-lanceoloid, 6–12 mm. **Seeds** 15–20, light brown, ovoid to reniform, 2–2.5 mm, reticulate and papillate.

Subspecies 3 (3 in the flora): w United States.

1. Bracts 3–6, lobes narrowly lanceolate to linear
. . . . 3c. *Cordylanthus kingii* subsp. *helleri*
1. Bracts 1–3, lobes linear to filiform.
  2. Corollas 15–25 mm; stems puberulent to pubescent; spikes 20–40 mm, 3–5(–8)-flowered. . . . . . 3a. *Cordylanthus kingii* subsp. *kingii*
  2. Corollas 25–30 mm; stems densely villous; spikes 40–50 mm, 8–12-flowered. . . . .
. . . . 3b. *Cordylanthus kingii* subsp. *densiflorus*

3a. Cordylanthus kingii S. Watson subsp. **kingii** E

Stems puberulent to pubescent, hairs glandular and eglandular. **Proximal leaves** 20–40 mm, margins 3–5-lobed, lobes filiform. **Spikes** 3–5(–8)-flowered, 20–40 mm; bracts 1–3, lobes linear to filiform. **Corollas** 15–25 mm. $2n = 26$.

Flowering Jun–Aug. Dry, rocky openings in pinyon-juniper woodlands and sagebrush scrub; 1400–2600 m; Nev., Utah.

3b. Cordylanthus kingii S. Watson subsp. **densiflorus** T. I. Chuang & Heckard, Syst. Bot. Monogr. 10: 81, figs. 9g, 31r. 1986 E

Cordylanthus kingii var. *densiflorus* (T. I. Chuang & Heckard) N. D. Atwood

Stems densely villous, hairs glandular and eglandular. **Proximal leaves** 20–40 mm, margins 3–5-lobed, lobes linear to filiform. **Spikes** 8–12-flowered, 40–50 mm; bracts 1–3, lobes filiform. **Corollas** 25–30 mm. $2n = 26$.

Flowering Jun–Aug. Dry openings in pinyon-juniper woodlands and sagebrush scrub; 1500–2700 m; Utah.

3c. Cordylanthus kingii S. Watson subsp. **helleri** (Ferris) T. I. Chuang & Heckard, Syst. Bot. Monogr. 10: 81. 1986 • Heller's bird's-beak E

Adenostegia helleri Ferris, Bull. Torrey Bot. Club 45: 417, plate 10, fig. 17, plate 12, fig. 2. 1918; *Cordylanthus helleri* (Ferris) J. F. Macbride; *C. kingii* var. *helleri* (Ferris) N. H. Holmgren

Stems densely villous, hairs glandular. **Proximal leaves** 10–25 mm, margins 3-lobed, lobes linear. **Spikes** 2–4-flowered, or flowers solitary, 20–30 mm; bracts 3–6, lobes narrowly lanceolate to linear. **Corollas** 15–25 mm. $2n = 26$.

Flowering Jun–Sep. Open pinyon-juniper woodlands and sagebrush scrub; 1200–3200 m; Calif., Nev.

4. Cordylanthus laxiflorus A. Gray in W. H. Emory, Rep. U.S. Mex. Bound. 2(1): 120. 1859 (as Cordylanthu) • Nodding bird's-beak

Stems erect, 30–90 cm, hirsute to pilose, hairs glandular and eglandular. **Leaves** densely pilose; proximal 5–20 mm, margins 3-lobed, lobes 1 mm wide; distal 5–15 × 1 mm, margins entire. **Inflorescences** spikes, 1(–4)-flowered; bract 1, 5–7 mm, 3-lobed, lobes green or purple, narrowly lanceolate. **Pedicels:** bracteoles 10–20 mm, margins entire. **Flowers:** calyx 10–17 mm, tube 2 mm, apex 2-fid, cleft 0.5–1 mm; corolla bright yellow, 15–20 mm, throat 3–5 mm diam., abaxial lip 7–10 mm, shorter than adaxial and slightly spreading; stamens 4, filaments hairy, fertile pollen sacs 1 per filament, vestigial pollen sacs present. **Capsules** oblong-lanceoloid, 7–8 mm. **Seeds** 15–20, light brown, reniform, 1.5–2 mm, reticulate and papillate. $2n = 26$.

Flowering Jul–Oct. Rocky slopes, mesas, often limestone; 1000–1900 m; Ariz., N.Mex.; Mexico (Sonora).

Cordylanthus laxiflorus is recognized by its bright yellow corollas with adaxial lip shorter than the galea. *Cordylanthus parviflorus* also has the abaxial lip shorter than the adaxial; it has pink to purple corollas.

5. Cordylanthus nevinii A. Gray, Proc. Amer. Acad. Arts 17: 229. 1882 • Nevin's bird's-beak

Stems ascending, 20–80 cm, densely puberulent and glandular-puberulent, sometimes pilose. Leaves densely glandular-puberulent; proximal 5–30 mm, margins 3-lobed, lobes 1 mm wide; distal 5–20 × 1 mm, margins entire. Inflorescences spikes, 2(or 3)-flowered, or flowers solitary; bracts 1–3, 5–10 mm, margins 3-lobed, lobes white to cream distally, linear or slightly spatulate, apex rounded to retuse. Pedicels: bracteoles 10–15 mm, margins entire. Flowers: calyx 10–15 mm, tube 1 mm, apex entire; corolla white with purple veins, apex yellow, 10–18 mm, throat 5–8 mm diam., abaxial lip 3–5 mm, ca. equal to and appressed to adaxial; stamens 4, filaments hairy, fertile pollen sacs 1 per filament, vestigial pollen sacs present. Capsules ovoid-oblong, 5–8 mm. Seeds 6–10, dark brown, ovoid, 2.5–3 mm, striate. $2n = 28$.

Flowering Jul–Sep. Openings, hillsides, pine-oak woodlands; 1400–2600 m; Ariz., Calif.; Mexico (Baja California).

Cordylanthus nevinii has relatively short inflorescence bracts, similar to those of *C. rigidus* subsp. *brevibracteatus*, which is sympatric. *Cordylanthus nevinii* can be distinguished by having only one fertile pollen sac per stamen and flowers much longer than the inflorescence bracts.

The corolla of *Cordylanthus nevinii* is very wide at the middle just proximal to the base of the adaxial lip. The flowers appear to have a pair of lateral pouches. This appearance is enhanced because the pouches are white, in contrast to the purple-veined adaxial lip.

6. Cordylanthus nidularius J. T. Howell, Leafl. W. Bot. 3: 207. 1943 • Bird-on-nest, Mount Diablo bird's-beak C E

Stems decumbent, 5–10(–15) cm, glandular-puberulent and pilose. Leaves glandular-puberulent and pilose; proximal 10–30 mm, margins entire or 3-lobed, lobes 1–2 mm wide; distal 10–20 × 1 mm, margins entire. Inflorescences spikes, 2- or 3-flowered, or flowers solitary; bracts 2 or 3, 10–15 mm, margins 3-lobed, lobes purple distally, linear to filiform. Pedicels: bracteoles 10–15 mm, margins entire. Flowers: calyx 10–16 mm, tube 0 mm, apex entire; corolla white with purple veins, 10–15 mm, throat 5–7(–9) mm diam., abaxial lip 3–5 mm, ca. equal to and appressed to adaxial; stamens 4, filaments hairy, fertile pollen sacs 2 per filament, equal. Capsules oblong-ovoid, 5–7 mm. Seeds 7–10, dark brown, ovoid to reniform, 1.5–2 mm, irregularly striate. $2n = 28$.

Flowering Jul–Aug. Dry, open serpentine in chaparral; of conservation concern; 600–800 m; Calif.

Cordylanthus nidularius is the only member of the genus with decumbent stems. The species is similar to *C. tenuis*; it differs in its white (versus pale yellow) corollas and decumbent (versus erect to ascending) stems. It grows on the eastern slopes of Mt. Diablo.

7. Cordylanthus parviflorus (Ferris) Wiggins, Contr. Dudley Herb. 1: 174. 1933 • Purple or sticky bird's-beak E

Adenostegia parviflora Ferris, Bull. Torrey Bot. Club 45: 409, plate 10, fig. 8, plate 11, fig. 4. 1918; *Cordylanthus glandulosus* Pennell & Clokey

Stems erect to ascending, 20–60 cm, hirsute and densely glandular-pubescent. Leaves glandular-pubescent and hirsute; proximal 10–30 mm, margins 3-lobed, lobes 1 mm wide; distal 5–20 × 1 mm, margins entire. Inflorescences spikes, 2-flowered, or flowers solitary; bracts 1(or 2), 5–15 mm, margins 3-lobed, lobes sometimes purplish distally, linear-lanceolate. Pedicels: bracteoles 10–12 mm, margins entire. Flowers: calyx 10–15 mm, tube 5–8 mm, apex 2-fid, cleft 1 mm; corolla pale pink to purple-pink with darker veins, 15–20 mm, tube 10–15 mm, throat 5–7 mm diam., abaxial lip 5–6 mm, shorter than and not appressed to adaxial; stamens 4, filaments hairy, fertile pollen sacs 2 per filament, equal. Capsules oblong-lanceoloid, 7–9 mm. Seeds 15–20, dark brown, ovoid to reniform, 1.5–2 mm, reticulate and papillate. $2n = 26$.

Flowering Aug–Oct. Dry, rocky slopes, sagebrush scrub, pinyon-juniper woodlands, Joshua tree woodlands; 700–2200 m; Ariz., Calif., Idaho, Nev., Utah.

Cordylanthus parviflorus is similar to *C. laxiflorus*, which also has unequal corolla lips. It can be distinguished from *C. laxiflorus* by its pink to purple (versus yellow) corollas and stamens with two pollen sacs.

8. Cordylanthus pilosus A. Gray, Proc. Amer. Acad. Arts 7: 382. 1868 • Hairy bird's-beak E

Stems erect or ascending, 10–90(–120) cm, densely puberulent and glandular-puberulent, and pilose. Leaves puberulent and pilose; proximal 15–40 mm, margins 3-lobed, lobes 0.1–0.5 mm wide; distal 10–25 × 0.1–0.3 mm, margins entire. Inflorescences spikes, 2- or 3-flowered, or flowers solitary; bracts 1–4, 10–20 mm, margins entire or 3-lobed, lobes green or purple distally, linear-lanceolate. Pedicels: bracteoles 15–20 mm, margins entire or toothed. Flowers: calyx 15–20 mm, tube 0–1 mm, apex 2-fid, cleft 0.5–1 mm; corolla pale yellow to yellow-green, streaked and spotted with maroon, 15–20 mm, tube 5–10 mm, throat 5–8 mm diam., abaxial lip 5–10 mm, ca. equal to and appressed to adaxial; stamens 4, filaments hairy, fertile pollen sacs 2 per filament, equal. Capsules oblong-ovoid, 6–10 mm. Seeds 10–18, dark brown, ovoid, 1.5–2.5 mm, striate. $2n = 28$.

Subspecies 2 (2 in the flora): California.

1. Inflorescence bracts entire . . . .
. . . . 8a. *Cordylanthus pilosus* subsp. *pilosus*
1. Inflorescence bracts 3-lobed . . . .
. . . . 8b. *Cordylanthus pilosus* subsp. *trifidus*

8a. Cordylanthus pilosus A. Gray subsp. pilosus E

Cordylanthus diffusus Pennell; *C. pilosus* subsp. *diffusus* (Pennell) Munz

Proximal leaves 1–4 mm wide, margins entire or lobed distally. Inflorescences: bracts entire.

Flowering Jul–Sep. Open woodlands and chaparral, sometimes on serpentine; 60–1500 m; Calif.

Subspecies *pilosus* grows in the Central Valley and Coast Ranges. Plants growing on serpentine appear depauperate, generally have fewer flowers per cluster, and are usually less hairy. Plants that appear to be intermediates between subspp. *pilosus* and *trifidus* occur in the Sacramento Valley.

8b. Cordylanthus pilosus A. Gray subsp. trifidus (B. L. Robinson & Greenman) T. I. Chuang & Heckard, Syst. Bot. Monogr. 10: 68. 1986 • Hansen's bird's-beak E

Cordylanthus pilosus var. *trifidus* B. L. Robinson & Greenman, Bot. Gaz. 22: 168. 1896; *Adenostegia hansenii* Ferris; *C. hansenii* (Ferris) J. F. Macbride; *C. pilosus* subsp. *hansenii* (Ferris) T. I. Chuang & Heckard

Proximal leaves 1–2 mm wide, margins entire or 3-lobed. Inflorescences: bracts 3-lobed.

Flowering Jul–Aug. Open woodlands; 500–1000 m; Calif.

Subspecies *trifidus* grows in the foothills of the Sierra Nevada. T. I. Chuang and L. R. Heckard (1986) recognized subsp. *hansenii* as distinct based on its shallowly cleft inflorescence bracts and robust distal branches. These characteristics do not distinguish subsp. *hansenii* but occur occasionally in plants of both subsp. *hansenii* and subsp. *trifidus*. Also, the plants they separated as subsp. *hansenii* occur in two disjunct populations, separated geographically by their subsp. *trifidus*. Combining the subspecies creates a geographically, ecologically, and morphologically coherent group.

9. Cordylanthus pringlei A. Gray, Proc. Amer. Acad. Arts 19: 94. 1883 • Pringle's bird's-beak C E

Stems erect or ascending, 30–120(–150) cm, glabrous or puberulent. Leaves puberulent or glabrous; proximal 10–40 mm, margins 3-lobed, lobes 1–2 mm wide; distal 5–20 × 1 mm, margins entire. Inflorescences capitate spikes, 2–4-flowered, 15–20 mm; bracts 1–3, flabelliform, 5–8 mm, margins 3–7-lobed, lobes green, narrowly ovate. Pedicels: bracteoles 8–10 mm, margins entire. Flowers: calyx 8–10 mm, tube 0 mm, apex 2-fid, cleft 0.5–1 mm; corolla pale yellow to yellow with purple markings, 8–9 mm, throat 4 mm diam., adaxial lip 3–4 mm, ca. equal to and appressed to adaxial; stamens 4, filaments hairy, fertile pollen sacs 2 per filament, unequal. Capsules oblong-ovoid, 5–8 mm. Seeds 4–6, dark brown, ovoid to narrowly reniform, 2.5–3 mm, striate. $2n = 28$.

Flowering Jul–Sep. Dry openings in chaparral and mixed-evergreen forests; of conservation concern; 300–1900 m; Calif.

Cordylanthus pringlei grows in the Coast Range of California. The species is distinctive because of its flabelliform inflorescence bracts and relatively short corollas.

10. Cordylanthus ramosus Nuttall ex Bentham in
A. P. de Candolle and A. L. P. P. de Candolle, Prodr.
10: 597. 1846 • Bushy bird's-beak E

Adenostegia ciliosa Rydberg;
A. ramosa (Nuttall ex Bentham)
Greene; *Cordylanthus ramosus* var.
puberulus J. F. Macbride; *C.
ramosus* subsp. *setosus* Pennell

Stems erect, 10–30(–90) cm,
puberulent, sometimes pilose.
Leaves puberulent; proximal
10–40 mm, margins 3–5-lobed,
lobes 1–2 mm wide; distal 10–15 × 0.5–1 mm, margins
entire. **Inflorescences** capitate spikes, 3–7-flowered;
bracts 1–7, 10–20 mm, margins 5–7-lobed, lobes green,
sometimes purple distally, filiform. **Pedicels:** bracteoles
10–20 mm, margins entire. **Flowers:** calyx 15–20
mm, tube 0 mm, apex entire or 2-fid, cleft 0–1 mm;
corolla pale yellow, spotted and streaked with purple,
10–20 mm, throat 4–6 mm diam., abaxial lip 3–5 mm,
ca. equal to and appressed to adaxial; stamens 4,
filaments hairy distally, fertile pollen sacs 2 per filament,
unequal. **Capsules** oblong-lanceoloid, 8–10 mm.
Seeds 10–20, light brown, narrowly ovate, 1.5–2 mm,
reticulate. $2n = 24$.

Flowering late Jun–Aug. Rocky, alkaline soils,
sagebrush scrub; 400–2900 m; Calif., Colo., Idaho,
Mont., Nev., Oreg., Utah, Wyo.

11. Cordylanthus rigidus (Bentham) Jepson, Fl. W. Calif.
ed. 2, 387. 1911

Adenostegia rigida Bentham in
J. Lindley, Nat. Syst. Bot., 445.
1836

Stems ascending, 30–100
(–150) cm, pubescent, hirsute,
glabrescent, puberulent,
sparsely hispid, or downy.
Leaves puberulent and often
hirsute; proximal 10–40 mm,
margins 3-lobed, lobes 1–3 mm wide; distal 10–30 ×
0.5–1.5 mm, margins entire or 3-lobed. **Inflorescences**
capitate spikes, 2–15-flowered; bracts 5–10, 5–20 mm,
margins entire or 3-lobed, lobes purplish black, ivory,
or green distally, linear-lanceolate to linear or slightly
spatulate. **Pedicels:** bracteoles 14–20 mm, margins
entire or toothed. **Flowers:** calyx 10–20 mm, tube 1–2
mm, apex entire or 2-fid, cleft 0–1 mm; corolla pale
yellow with purple veins and an abaxial, U-shaped,
purple mark, 8–20 mm, throat 5–10 mm diam.,
abaxial lip 6–10 mm, ca. equal to and appressed to
adaxial; stamens 4, filaments hairy, fertile pollen sacs
2 per filament, subequal. **Capsules** oblong-ovoid,
8–12 mm. **Seeds** 10–25, dark brown, oblong-ovoid,
1.5–2 mm, striate. $2n = 28$.

Subspecies 4 (4 in the flora): California, nw Mexico.
Cordylanthus rigidus is variable, with many locally
distinctive populations that intergrade with each other
and with nearby species.

1. Proximal bracts 5–12 mm, shorter than flowers
. . . . . . 11b. *Cordylanthus rigidus* subsp. *brevibracteatus*
1. Proximal bracts 10–20 mm, equal to or longer
than flowers.
  2. Proximal bracts: midlobes linear, 0.5–1 mm
wide, dark purple distally . . . .
. . . . 11d. *Cordylanthus rigidus* subsp. *setigerus*
  2. Proximal bracts: midlobes linear-lanceolate
or lanceolate, 1–2.5 mm wide, green or red
distally.
    3. Stems pubescent and hirsute, hairs stiff;
proximal bracts: midlobe apices retuse . . .
. . . . 11a. *Cordylanthus rigidus* subsp. *rigidus*
    3. Stems downy to puberulent, hairs soft;
proximal bracts: midlobe apices acute to
retuse . . . .
. . . . . . 11c. *Cordylanthus rigidus* subsp. *littoralis*

11a. Cordylanthus rigidus (Bentham) Jepson subsp.
rigidus E

Cordylanthus compactus Pennell;
C. ferrisianus Pennell; *C. littoralis*
(Ferris) J. F. Macbride subsp.
platycephalus (Pennell) Munz;
C. platycephalus Pennell; *C. rigidus*
(Pennell) D. W. Taylor; *C. rigidus*
var. *ferrisianus* (Pennell) D. W.
Taylor

Stems pubescent and hirsute,
often glabrescent, hairs stiff. **Spikes** 5–15-flowered;
proximal bracts 15–20 mm, longer than flowers,
midlobes green or red distally, linear-lanceolate,
1–1.5 mm wide, apex rounded to retuse. **Flowers:**
corolla 12–20 mm.

Flowering Jul–Sep. Open, montane woodlands,
chaparral margins; 300–2700 m; Calif.

Subspecies *rigidus* is the most widespread of the
subspecies, showing great variability in height, leaf
shape, and amount of pubescence. It often can be
recognized by its inflorescence bracts, which are rounded
to retuse and longer than the flowers.

11b. Cordylanthus rigidus (Bentham) Jepson subsp. **brevibracteatus** (A. Gray) Munz, Aliso 4: 98. 1958 [E]

Cordylanthus filifolius Nuttall ex Bentham var. *brevibracteatus* A. Gray in W. H. Brewer et al., Bot. California 1: 622. 1876; *Adenostegia rigida* Bentham var. *brevibracteata* (A. Gray) Greene; *C. rigidus* var. *brevibracteatus* (A. Gray) J. F. Macbride

Stems puberulent and sparsely hispid. **Spikes** 2–12-flowered; proximal bracts 5–12 mm, shorter than flowers, midlobes green or red distally, linear-lanceolate, 1–1.5 mm wide, apex retuse to slightly rounded. **Flowers:** corolla 8–17 mm.

Flowering Jul–Sep. Rocky opening in pine woodlands, pine-juniper woodlands, sagebrush scrub; 800–2600 m; Calif.

Subspecies *brevibracteatus* occurs on the Kern Plateau and resembles subsp. *rigidus*, found to the north in the Sierra Nevada. Both have flowers in compact heads; plants of subsp. *brevibracteatus* differ in having proximal bracts 5–12 mm.

11c. Cordylanthus rigidus (Bentham) Jepson subsp. **littoralis** (Ferris) T. I. Chuang & Heckard, Syst. Bot. Monogr. 10: 42. 1986 [C][E]

Adenostegia littoralis Ferris, Bull. Torrey Bot. Club 45: 413, plate 10, fig. 13, plate 12, fig. 1. 1918; *Cordylanthus littoralis* (Ferris) J. F. Macbride; *C. rigidus* var. *littoralis* (Ferris) Jepson

Stems downy to puberulent, hairs soft. **Spikes** 5–8-flowered; proximal bracts 15–20 mm, equal to or longer than flowers, midlobes green or red distally, lanceolate, 1.5–2.5 mm wide, apex acute to retuse. **Flowers:** corolla 15–20 mm.

Flowering Jul–Sep. Dunes; of conservation concern; 0–200 m; Calif.

Subspecies *littoralis* grows only in the coastal dunes of the central coast in Monterey and Santa Barbara counties. It intergrades with subsp. *rigidus*. It is not known if this subspecies is genetically distinct or if its distinguishing features are caused by the unusual habitat where the plants grow.

11d. Cordylanthus rigidus (Bentham) Jepson subsp. **setigerus** T. I. Chuang & Heckard, Syst. Bot. Monogr. 10: 43, figs. 3k, 4n,o, 7i, 10g, 18a–g. 1986

Stems hispid. **Spikes** 5–13-flowered; proximal bracts 10–20 mm, equal to or longer than flowers, midlobes dark purple distally, linear, 0.5–1 mm wide, apex retuse. **Flowers:** corolla 13–17 mm.

Flowering Jun–Sep. Chaparral, sagebrush scrub, openings in oak woodlands and conifer woodlands; 200–2200 m; Calif.; Mexico (Baja California).

12. Cordylanthus tenuis A. Gray, Proc. Amer. Acad. Arts 7: 383. 1868 • Slender bird's-beak [E][F]

Adenostegia tenuis (A. Gray) Greene

Stems erect to ascending, 20–80(–120) cm, glabrous or sparsely glandular-puberulent, puberulent, and/or pilose. **Leaves** puberulent, often pilose, or glabrous; proximal 20–60 mm, margins entire or 3-lobed, lobes 0.5–1 mm wide; distal 10–40 × 0.3–2 mm, margins entire. **Inflorescences** spikes, 2–7-flowered, or flowers solitary; bracts 1–4, 5–20 mm, margins entire or 3-lobed, lobes green or purple distally, narrowly lanceolate to linear-lanceolate, apex rounded. **Pedicels:** bracteoles 10–20 mm, margins entire or toothed. **Flowers:** calyx 10–20 mm, tube 0 mm, apex 2-fid, cleft 1 mm; corolla pale yellow, marked with purple along veins and galea, 10–20 mm, throat 6–8 mm diam., abaxial lip 4–6 mm, ca. equal to and appressed to adaxial; stamens 4, filaments hairy, fertile pollen sacs 2 per filament, equal. **Capsules** narrowly ovoid, 5–10 mm. **Seeds** 6–16, dark brown, ovoid to rhomboid, 1.5–2.5 mm, striate.

Subspecies 5 (5 in the flora): w United States.

1. Leaf lobes filiform; stems glabrous proximally . . . . 12c. *Cordylanthus tenuis* subsp. *brunneus*
1. Leaves or leaf lobes linear to linear-lanceolate; stems puberulent, glandular-puberulent, glandular-pubescent, and/or pilose proximally.
  2. Bracts densely hirsute. . . . . .12b. *Cordylanthus tenuis* subsp. *barbatus*
  2. Bracts hirsute or pilose.
    3. Bracts entire . . . . 12a. *Cordylanthus tenuis* subsp. *tenuis*
    3. Bracts 3-lobed, sometimes entire.
      4. Leaves yellow-green; inflorescences 4–6-flowered . . . . .12d. *Cordylanthus tenuis* subsp. *pallescens*
      4. Leaves green to gray-green; inflorescences 1–3-flowered . . . . 12e. *Cordylanthus tenuis* subsp. *viscidus*

12a. Cordylanthus tenuis A. Gray subsp. **tenuis** Ⓔ Ⓕ

Cordylanthus bolanderi (A. Gray) Pennell; *C. pilosus* A. Gray subsp. *bolanderi* (A. Gray) Munz; *C. pilosus* var. *bolanderi* A. Gray

Stems puberulent to glandular-pubescent proximally. **Leaves** green, linear, entire. **Inflorescences** 1–3-flowered; bracts green, entire, hirsute with a few long hairs near margins. **Flowers:** corolla 10–20 mm.

Flowering Jul–Sep. Openings in conifer woodlands; 300–2600 m; Calif.

Subspecies *tenuis* grows in the foothills of the Sierra Nevada with a disjunct population in the Klamath Range. Some plants from the central Klamath Range are similar to subsp. *viscidus*, which has three-lobed inflorescence bracts.

12b. Cordylanthus tenuis A. Gray subsp. **barbatus** T. I. Chuang & Heckard, Syst. Bot. Monogr. 10: 58, figs. 3l³, 10k, 22h–n. 1986 • Fresno County bird's-beak Ⓔ

Stems puberulent, glandular-puberulent and often pilose. **Leaves** green, entire or 3-lobed, lobes linear to linear-lanceolate. **Inflorescences** 3–7-flowered, flowers in dense clusters; bracts green, 3-lobed, hirsute with long hairs. **Flowers:** corolla 15–18 mm. $2n = 28$.

Flowering Jul–Aug. Open, mixed deciduous forests; 1300–2400 m; Calif.

Subspecies *barbatus* is known from Fresno County. The long, dense hairs on the inflorescence bracts help to identify it.

12c. Cordylanthus tenuis A. Gray subsp. **brunneus** (Jepson) Munz, Aliso 4: 98. 1958 Ⓔ

Cordylanthus pilosus A. Gray var. *brunneus* Jepson, Man. Fl. Pl. Calif., 946. 1925; *C. brunneus* (Jepson) Pennell; *C. capillaris* Pennell; *C. tenuis* subsp. *capillaris* (Pennell) T. I. Chuang & Heckard

Stems glabrous proximally, glandular-puberulent distally. **Leaves** green, sometimes tinged purple, entire or 3-lobed, lobes filiform. **Inflorescences** 1- or 2-flowered, flowers in loose clusters; bracts green to purple, entire or 3-lobed, puberulent, often glabrous distally, without long hairs. **Flowers:** corolla 12–16 mm.

Flowering Jun–Jul. Serpentine in mixed evergreen forests and chaparral; 200–1400 m; Calif.

Subspecies *brunneus* is a serpentine endemic with glabrous or slightly puberulent stems and leaves with filiform lobes. T. I. Chuang and L. R. Heckard (1986) recognized subsp. *capillaris* as a distinct subspecies, closely related to subsp. *brunneus* but distinguished by glabrous stems and three-lobed proximal bracts. These characteristics are not reliable, varying even on a single plant. When combined, the two form a coherent subspecies distinguished by filiform leaf lobes, tendency to grow on serpentine, and distribution.

12d. Cordylanthus tenuis A. Gray subsp. **pallescens** (Pennell) T. I. Chuang & Heckard, Syst. Bot. Monogr. 10: 55. 1986 • Pale bird's-beak Ⓒ Ⓔ

Cordylanthus pallescens Pennell, Proc. Acad. Nat. Sci. Philadelphia 99: 197. 1947

Stems sparsely puberulent and/or glandular-puberulent and, often, pilose. **Leaves** yellow-green, entire or 3-lobed, lobes linear to linear-lanceolate. **Inflorescences** 4–6-flowered; bracts yellow-green, 3-lobed or entire, sparsely pilose with long hairs. **Flowers:** corolla 10–15 mm. $2n = 28$.

Flowering Jun–Sep. Open volcanic alluvium; of conservation concern; 900–1200 m; Calif.

Subspecies *pallescens* grows in Siskiyou County near Black Butte and the town of Weed (B. L. Wilson et al. 2014). It is similar to subsp. *viscidus*, differing in its yellow-green color and inflorescences with four to six pale flowers.

SELECTED REFERENCE Wilson, B. L. et al. 2014. Identification and taxonomic status of *Cordylanthus tenuis* ssp. *pallescens* (Orobanchaceae). Madroño 61: 64–76.

12e. Cordylanthus tenuis A. Gray subsp. **viscidus** (Howell) T. I. Chuang & Heckard, Syst. Bot. Monogr. 10: 56. 1986 Ⓔ

Adenostegia viscida Howell, Fl. N.W. Amer., 537. 1901; *Cordylanthus viscidus* (Howell) Pennell

Stems densely glandular-puberulent and pilose. **Leaves** green to gray-green, 3-lobed, lobes linear. **Inflorescences** loosely 1–3-flowered; bracts gray-green, 3-lobed, pilose with long hairs. **Flowers:** corolla 10–18 mm. $2n = 28$.

Flowering Jul–Sep. Serpentine in pine woodlands; 200–2000 m; Calif., Oreg.

13. Cordylanthus wrightii A. Gray in W. H. Emory, Rep. U.S. Mex. Bound. 2(1): 120. 1859

Adenostegia wrightii (A. Gray) Greene

Stems erect or ascending, 20–90 cm, puberulent and glandular-puberulent or scabrous, often glabrate. **Leaves** puberulent or glandular-puberulent, often glabrescent; proximal 20–35 mm, margins 3–7 lobed, lobes 0.5–1.5 mm wide; distal 10–20 mm, margins entire or 3-lobed. **Inflorescences** capitate spikes, 2–12-flowered, 20–40 mm; bracts 2–10, green or purple distally, 15–30 mm, margins (4 or)5–9-lobed, lobes linear. **Pedicels:** bracteoles 15–30 mm, margins toothed or lobed. **Flowers:** calyx lanceolate, 20–25 mm, tube 3–4 mm, apex 2-fid, cleft 1–2 mm; corolla bright or pale yellow or lavender-pink with purple markings, 20–30 mm, throat 6–8 mm diam., abaxial lip 5–8 mm, ca. equal to and appressed to adaxial; stamens 4, filaments hairy, fertile pollen sacs 2 per filament, equal. **Capsules** oblong, 10–15 mm. **Seeds** 15–20, brown, oblong-ovoid, 2–2.5 mm, reticulate.

Subspecies 3 (3 in the flora): w United States, n Mexico.

1. Spikes 2–4-flowered; bracts 2 or 3 . . . .
. . . . 13a. *Cordylanthus wrightii* subsp. *wrightii*
1. Spikes 5–12-flowered; bracts 4–10.
  2. Bracteole margins pinnately lobed; stems puberulent and glandular-puberulent to glabrate . . . .
      . . . . . 13b. *Cordylanthus wrightii* subsp. *kaibabensis*
  2. Bracteole margins toothed distally; stems scabrous. . . . .
      . . . . . . 13c. *Cordylanthus wrightii* subsp. *tenuifolius*

13a. Cordylanthus wrightii A. Gray subsp. wrightii

Cordylanthus wrightii var. *pauciflorus* Kearney & Peebles

Stems puberulent to sparsely puberulent and/or glandular-puberulent. **Leaves** 3–5-lobed. **Spikes** 2–4-flowered, 20–30 mm; bracts 2 or 3. **Pedicels:** bracteoles 15–30 mm, margins toothed distally. **Flowers:** corolla bright yellow or lavender-pink with purple markings, 20–25 mm. $2n = 26$.

Flowering Jul–Sep. Dry openings, pinyon-juniper woodlands, sagebrush scrub; 1600–2200 m; Ariz., Colo., N.Mex., Tex., Utah; Mexico (Chihuahua).

The inflorescence bracts of subsp. *wrightii* are three- to nine-lobed; northern populations tend to have three-lobed bracts; southerly populations have bracts with five to nine lobes.

13b. Cordylanthus wrightii A. Gray subsp. kaibabensis T. I. Chuang & Heckard, Syst. Bot. Monogr. 10: 87, fig. 35i–n. 1986 • Kaibab Plateau bird's-beak [E]

Stems puberulent and glandular-puberulent to glabrate. **Leaves** 5–7-lobed. **Spikes** 5–12-flowered, 25–40 mm; bracts 4–10. **Pedicels:** bracteoles 20–25 mm, margins pinnately lobed, lobes 5–7. **Flowers:** corolla pale yellow with purple markings, 20–30 mm.

Flowering Aug–Sep. Rocky openings, pine-juniper woodlands; 1900–2000 m; Ariz.

Subspecies *kaibabensis* is known from the Kaibab Plateau. Chuang and Heckard suggested that it may be derived from a past hybridization of subsp. *tenuifolius* and *Cordylanthus kingii* subsp. *kingii*.

13c. Cordylanthus wrightii A. Gray subsp. tenuifolius (Pennell) T. I. Chuang & Heckard, Syst. Bot. Monogr. 10: 86. 1986 [E]

Cordylanthus tenuifolius Pennell, Notul. Nat. Acad. Nat. Sci. Philadelphia 43: 9. 1940

Stems scabrous. **Leaves** 3–7-lobed. **Spikes** 5–12-flowered, 25–30 mm; bracts 4–10. **Pedicels:** bracteoles 20–25 mm, margins toothed distally. **Flowers:** corolla pale yellow with purple markings, 20–25 mm. $2n = 26$.

Flowering Jul–Oct. Open, rocky sites, pinyon-juniper woodlands; 1300–2000 m; Ariz., N.Mex.

In areas where their distributions overlap, subspp. *tenuifolius* and *wrightii* intergrade in the shape of the bracteoles and overall indument.

25. DICRANOSTEGIA (A. Gray) Pennell, Proc. Acad. Nat. Sci. Philadelphia 99: 189. 1947

• [Greek *dicranos*, two-headed, and *stegos*, sheath or cover, alluding to two-lobed calyx]

C

Kerry A. Barringer

Cordylanthus Nuttall ex Bentham [unranked] *Dicranostegia* A. Gray, Proc. Amer. Acad. Arts 19: 95. 1883

Herbs, annual; hemiparasitic. **Stems** erect or ascending, rarely prostrate, not fleshy, pubescent, hirsute, or glabrous. **Leaves** cauline, alternate; petiole absent; blade not fleshy, not leathery, margins pinnately 8–11-lobed. **Inflorescences** terminal, spikes; bracts present. **Pedicels** absent; bracteoles absent. **Flowers:** sepals 2, calyx bilaterally symmetric, spathelike, lobes triangular; petals 5, corolla white to yellow, marked with purple, strongly bilabiate, club-shaped, often gibbous at base of lobes, abaxial lobes 3, adaxial lobes 2, adaxial lip galeate; stamens 2, filaments glabrous or proximally pubescent, pollen sacs separate, connective elongate; staminodes 2, filiform; ovary 2-locular, placentation axile; stigma not expanded. **Capsules:** dehiscence loculicidal. **Seeds** 5–20, brown, ovoid, wings absent. *x* = 8.

Species 1: California, nw Mexico.

Dicranostegia is sometimes classified as *Cordylanthus* subg. *Dicranostegia* (A. Gray) T. I. Chuang & Heckard. D. C. Tank et al. (2009) have shown that *Dicranostegia* is closely related to *Chloropyron* and that both genera form a clade that is sister to *Castilleja* and *Triphysaria*.

Like *Chloropyron* and *Cordylanthus*, *Dicranostegia* has closed, club-shaped flowers with a spathelike calyx. Flowers are subtended by a single, elongate bracteole and pinnate bracts. *Dicranostegia* differs in having pinnately lobed leaves, deeply two-parted calyces that are about half as long as the corollas, and stamens with pollen sacs held apart on relatively short connectives.

SELECTED REFERENCE Chuang, T. I. and L. R. Heckard. 1975. Taxonomic status of *Cordylanthus* (subg. *Dicranostegia*) *orcuttianus* (Scrophulariaceae). Madroño 23: 88–95.

1. Dicranostegia orcuttiana (A. Gray) Pennell, Proc. Acad. Nat. Sci. Philadelphia 99: 190. 1947 • Orcutt's or Baja bird's-beak C F

Cordylanthus orcuttianus A. Gray, Proc. Amer. Acad. Arts 19: 95. 1883

Annuals 10–50 cm, puberulent and hispid to hirsute, hairs white; branches present. **Leaves:** blade ovate, 2–6(–8) × 3–5 cm, 8–11-lobed. **Spikes** 2–10 cm; bracts red to maroon distally, oblong-lanceolate, 15–25 mm, 3–7-lobed. **Pedicels:** bracteole 1, lanceolate to narrowly triangulate pinnate-lobed. **Flowers:** calyx 6–8 mm, 2-fid, lobes 4–6 mm; corolla 18–25 mm, abaxial lip not spreading, often with purple spots at base, adaxial lip yellow distally; stamens each with 2 pollen sacs widely separated on curved connectives, staminodes sometimes with sterile anthers. **Capsules** 10–12 mm. **Seeds** 1–1.5 mm, reticulate, with abaxial crest. 2*n* = 32.

Flowering Mar–Aug. Coastal scrub; of conservation concern; 10–400 m; Calif.; Mexico (Baja California).

Dicranostegia orcuttiana is found along the west coast of Baja California Peninsula, Mexico, and in the flora area in San Diego County, where it is threatened by urbanization.

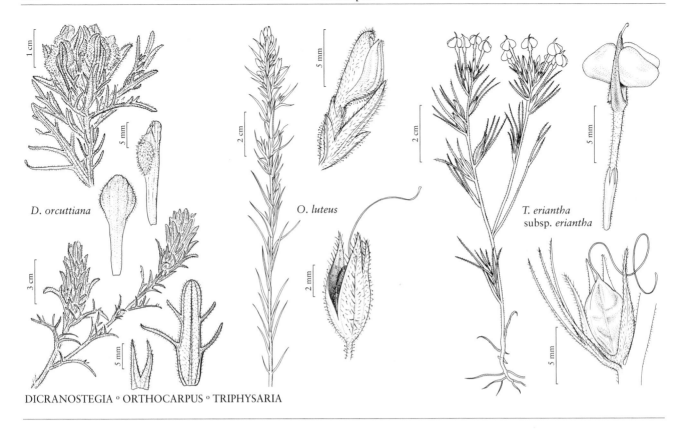

D. orcuttiana

O. luteus

T. eriantha
subsp. *eriantha*

DICRANOSTEGIA ∘ ORTHOCARPUS ∘ TRIPHYSARIA

26. **ORTHOCARPUS** Nuttall, Gen. N. Amer. Pl. 2: 56. 1818 • Owl's-clover [Greek *orthos*, straight, and *carpos*, fruit, alluding to distinctness from *Melampyrum*, which has oblique fruits] E

Kerry A. Barringer

Herbs, annual; hemiparasitic. **Stems** erect or ascending, not fleshy, glabrous or hairy. **Leaves** cauline, alternate; petiole absent; blade not fleshy, not leathery, margins entire or 3–7-lobed. **Inflorescences** terminal, spikes or racemes; bracts present. **Pedicels** absent or present; bracteoles absent. **Flowers:** sepals 4, calyx symmetric, tubular, lobes triangular; petals 5, corolla purple-pink, yellow, or white, strongly bilabiate, club-shaped, often gibbous at base of abaxial lip, abaxial lobes 0 or 3, adaxial 2, adaxial lip galeate, rounded at apex, sometimes obscurely so, opening downward; stamens 4, didynamous, filaments glabrous or hairy; staminode 0; ovary 2-locular, placentation axile; stigma linear, not expanded at apex. **Capsules:** dehiscence loculicidal. **Seeds** 5–35, light to dark brown or yellow, ovoid to reniform or oblong, wings absent. $x = 7$.

Species 9 (9 in the flora): w North America.

Orthocarpus is related to *Cordylanthus* and can be distinguished by its four-lobed, tubular calyces (D. C. Tank et al. 2009). Some species placed traditionally in *Orthocarpus* were transferred to *Castilleja* and *Triphysaria* by T. I. Chuang and L. R. Heckard (1991).

SELECTED REFERENCES Chuang, T. I. and L. R. Heckard. 1992. A taxonomic revision of *Orthocarpus* (Scrophulariaceae—tribe Pedicularieae). Syst. Bot. 17: 560–582. Keck, D. D. 1927. A revision of the genus *Orthocarpus*. Proc. Calif. Acad. Sci., ser. 4, 16: 517–571.

1. Bract margins 5–7-lobed; pedicels 2–3 mm; corollas: adaxial lips bearded at tip . . . . 1. *Orthocarpus barbatus*
1. Bract margins entire or 3(–5)-lobed; pedicels 0–2 mm; corollas: adaxial lips not bearded at tip.

[2. Shifted to left margin.—Ed.]

2. Bracts: midlobes ovate or obovate, 4–10 mm wide, apices obtuse, cuspidate.
    3. Corollas: adaxial lips not hooked at tip. . . . . .3. *Orthocarpus cuspidatus*
    3. Corollas: adaxial lips notably hooked at tip.
        4. Corollas 25–30 mm, adaxial lips glabrous, abaxial lobes rounded . . . . 6. *Orthocarpus pachystachyus*
        4. Corollas 10–20 mm, adaxial lips puberulent, abaxial lobes triangular.
            5. Distal leaf margins entire; calyces 5–7 mm; capsules 4–5 mm, apices retuse
                . . . . .4. *Orthocarpus imbricatus*
            5. Distal leaf margins 3–7 lobed; calyces 8–12 mm; capsules 6–7 mm, apices
                acute . . . . .8. *Orthocarpus tenuifolius*
2. Bracts: midlobes triangular, 0.7–4 mm wide, apices acute or acuminate, not cuspidate.
    6. Corollas: adaxial lips golden yellow, tips minutely hooked . . . . 5. *Orthocarpus luteus*
    6. Corollas: adaxial lips yellow or purple-pink to white, tips notably hooked.
        7. Distal leaf margins entire; seeds 3–6; corollas: abaxial lips with triangular lobes
            . . . . .9. *Orthocarpus tolmiei*
        7. Distal leaf margins 3-lobed; seeds 10–26; corollas: abaxial lips with lobes 0.
            8. Bracts: midlobes 3–4 mm wide, lateral lobes narrowly triangular; capsule
                apices retuse; seeds 10–15 . . . . .2. *Orthocarpus bracteosus*
            8. Bracts: midlobes 1–2 mm wide, lateral lobes linear; capsule apices cuspidate;
                seeds 18–26 . . . . .7. *Orthocarpus purpureoalbus*

1. **Orthocarpus barbatus** J. S. Cotton, Bull. Torrey Bot. Club 29: 574. 1902 • Grand Coulee owl's-clover C E

Annuals 8–27 cm. **Stems** erect or ascending, hispid to pilose and puberulent. **Leaves** 20–40 mm; blade: proximal margins entire, distal 3–5-lobed. **Inflorescences** racemes, 2–11 cm; bracts 10–20 mm, margins 5–7-lobed in distal ⅔, proximal not resembling distal leaves, midlobe yellow-green to white distally, triangular, 6–8 mm wide, apex acute, not cuspidate, lateral lobes narrowly triangular. **Pedicels** 2–3 mm. **Flowers:** calyx 8–10 mm; corolla 10–12 mm, slightly longer than bracts, abaxial lip inflated, lobes 3, triangular, adaxial lip pale yellow to yellow, 4–5 mm, 1–1.5 mm longer than abaxial, glabrous or sparsely puberulent, tip not hooked, bearded. **Capsules** 7–10 mm, apex acute. **Seeds** 18–27, light brown, narrowly ovoid to reniform, 0.8–1 mm, reticulate, ridged. *2n* = 28.

Flowering May–Jun. Sagebrush meadows and slopes; of conservation concern; 300–900 m; B.C.; Wash.

Orthocarpus barbatus is endangered in British Columbia. It is distinguished by its yellow corollas, yellow-green bracts, sometimes tipped with white, and the tuft of hairs at the tip of the adaxial corolla lip.

2. **Orthocarpus bracteosus** Bentham, Scroph. Ind., 13. 1835 • Rosy owl's-clover E

Orthocarpus bracteosus var. *albus* D. D. Keck

Annuals 6–41 cm. **Stems** erect, minutely scabrous and puberulent. **Leaves** 15–40 mm; blade: proximal margins entire, distal 3-lobed. **Inflorescences** racemes or spikes, (1.5–)3–20 cm; bracts 10–20 mm, margins entire or 3(or 5)-lobed in distal ⅔, proximal not resembling distal leaves, midlobe green or purple distally, triangular, 3–4 mm wide, apex acute, not cuspidate, lateral lobes narrowly triangular. **Pedicels** (0–)1–1.5 mm. **Flowers:** calyx 6–10 mm; corolla 12–20 mm, longer than bracts, abaxial lip inflated, lobes 0, adaxial lip purple-pink to white, 4–6 mm, equal to abaxial, glabrous or puberulent, tip notably hooked, not bearded. **Capsules** 5–7 mm, apex retuse. **Seeds** 10–15, curved, light brown, oblong, 18–25 mm, reticulate, ridged. *2n* = 30.

Flowering late May–Oct. Seasonally wet meadows; 10–2000 m; B.C.; Calif., Oreg., Wash.

Orthocarpus bracteosus is endangered in British Columbia and Washington. It resembles *O. purpureoalbus*; the midlobe of the bracts is 3–4 (versus 1–2) mm wide. Also, the leaves of *O. purpureoalbus* tend to be dark green; those of *O. bracteosus* are usually lighter green or yellow-green.

3. Orthocarpus cuspidatus Greene, Pittonia 4: 101.
1899 E

Orthocarpus pachystachyus
A. Gray var. *cuspidatus* (Greene)
Jepson

Annuals 10–40 cm. **Stems** erect
to ascending, puberulent to
scabrous. **Leaves** 10–50 mm;
blade: proximal margins entire,
distal 3-lobed. **Inflorescences**
racemes, 2–10 cm; bracts 10–20
mm, margins entire or 3-lobed from base, proximal
not resembling distal leaves, midlobe purple-pink
distally, ovate, 6–9 mm wide, apex obtuse, cuspidate,
lateral lobes lanceolate. **Pedicels** 2 mm. **Flowers:** calyx
7–10 mm; corolla 9–25 mm, equal to or longer than
bracts, abaxial lip slightly inflated, lobes 3, triangular,
adaxial lip purple-pink, 3–8 mm, equal to or 1–5 mm
longer than abaxial, densely puberulent, tip not hooked,
not bearded. **Capsules** 6–8 mm, apex retuse. **Seeds**
1–12, dark brown, ovoid to reniform, 2 mm, shallowly
reticulate, not ridged. $2n = 28$.

Subspecies 3 (3 in the flora): w United States.

1. Corollas 16–25 mm, abaxial lips 7–8 mm......
.... 3a. *Orthocarpus cuspidatus* subsp. *cuspidatus*
1. Corollas 9–18 mm, abaxial lips 3–6 mm.
  2. Corollas 14–18 mm, longer than bracts,
    abaxial lips 5–6 mm..................
    ..... 3b. *Orthocarpus cuspidatus* subsp. *copelandii*
  2. Corollas 9–14 mm, equal to bracts, abaxial
    lips 3–4 mm........................
    .... 3c. *Orthocarpus cuspidatus* subsp. *cryptanthus*

3a. Orthocarpus cuspidatus Greene subsp. **cuspidatus**
• Siskiyou Mountain owl's-clover   E

Flowers: corolla 16–25 mm,
longer than bracts, abaxial lip
7–8 mm, pouches 4–5 mm
deep.
  Flowering Jun–Aug. Open,
grassy to rocky slopes;
700–2200 m; Calif., Oreg.
  In California, subsp.
cuspidatus is known from the
northwestern part of the state and the Cascade Ranges
plus in southwestern Oregon.

3b. Orthocarpus cuspidatus Greene subsp. **copelandii**
(Eastwood) T. I. Chuang & Heckard, Syst. Bot. 17: 576.
1992 • Copeland's owl's-clover   E

Orthocarpus copelandii Eastwood,
Bot. Gaz. 41: 288, fig. 1. 1906 (as
copelandi)

Flowers: corolla 14–18 mm,
longer than bracts, abaxial lip
5–6 mm, pouches 3 mm deep.
  Flowering Jun–Aug. Open,
grassy or rocky slopes;
1200–2500 m; Calif., Oreg.
  In California, subsp. *copelandii* is known from the
Klamath Ranges, High North Coast Ranges, Cascade
Ranges, and northern High Sierra Nevada plus in
southwestern Oregon.

3c. Orthocarpus cuspidatus Greene subsp. **cryptanthus**
(Piper) T. I. Chuang & Heckard, Syst. Bot. 17: 576.
1992 E

Orthocarpus cryptanthus Piper,
Smithsonian Misc. Collect. 50:
200. 1907; *O. copelandii*
Eastwood var. *cryptanthus* (Piper)
D. D. Keck

Flowers: corolla 9–14 mm,
equal to bracts, abaxial lip
3–4 mm, pouches 1–2 mm
deep.
  Flowering Jun–Aug. Meadows, open sagebrush
communities; 1500–3200 m; Calif., Nev., Oreg.
  Subspecies *cryptanthus* has a more extensive range
in northern California than the other subspecies and
occurs in adjacent Oregon and western Nevada.

4. Orthocarpus imbricatus Torrey ex S. Watson,
Botany (Fortieth Parallel), 458. 1871 • Mountain
owl's-clover   E

Orthocarpus olympicus Elmer

Annuals (5–)10–35 cm. **Stems**
erect or ascending, puberulent
to sparsely puberulent. **Leaves**
20–50 mm; blade margins
entire. **Inflorescences** spikes,
2–8 cm; bracts 10–15 mm,
margins 3-lobed in proximal ½,
proximal not resembling distal
leaves, midlobe purple-pink distally, obovate, 6–10 mm
wide, apex obtuse, cuspidate, lateral lobes triangular.
Pedicels 0–1 mm. **Flowers:** calyx 5–7 mm; corolla
10–15 mm, slightly longer than bracts, abaxial lip
inflated, lobes 3, triangular, adaxial lip purple-pink,
3–4 mm, 0–0.5 mm longer than abaxial, puberulent, tip
hooked, not bearded. **Capsules** 4–5 mm, apex retuse.
Seeds 3–5, dark brown, reniform, 1.5–2 mm, reticulate,
not ridged. $2n = 28$.

Flowering Jun–Sep. Montane meadows, grassy or rocky slopes; 1100–2400 m; B.C.; Calif., Oreg., Wash.

In northern California, *Orthocarpus imbricatus* is known from the Klamath, North Coast, and Cascade ranges.

5. **Orthocarpus luteus** Nuttall, Gen. N. Amer. Pl. 2: 57. 1818 • Yellow owl's-clover E F

Orthocarpus strictus Bentham

Annuals 10–40(–60) cm. **Stems** erect to ascending, pubescent and pilose. **Leaves** 15–50 mm; blade: proximal margins entire, distal entire or 3-lobed. **Inflorescences** spikes, (2–)5–20 cm; bracts 10–20 mm, margins 3-lobed in distal ½, proximal resembling distal leaves, midlobe green distally, triangular, 2–3 mm wide, apex acuminate, not cuspidate, lateral lobes narrowly lanceolate. **Pedicels** 0 mm. **Flowers:** calyx 5–8 mm; corolla 10–15 mm, longer than bracts, abaxial lip inflated, lobes 3, rounded, adaxial lip golden yellow, 2–4 mm, equal to abaxial, glandular-puberulent, tip minutely hooked, not bearded. **Capsules** 5–7 mm, apex retuse. **Seeds** 20–35, yellow or dark brown, ovoid, 1.2–1.5 mm, reticulate, longitudinally ridged. $2n = 28$.

Flowering Jun–Sep. Grasslands, sagebrush scrub, mountain meadows, disturbed grounds; (200–)400–3000 m; Alta., B.C., Man., Ont., Sask.; Ariz., Calif., Colo., Idaho, Minn., Mont., Nebr., Nev., N.Mex., N.Dak., Oreg., S.Dak., Utah, Wash., Wyo.

Orthocarpus luteus can be distinguished by its golden yellow corollas, with minutely hooked tips on adaxial lips. The Ontario population is not native. The Blackfeet used the leaves of *O. luteus* to dye skins, hair, and feathers red to reddish brown (A. Johnston 1987).

6. **Orthocarpus pachystachyus** A. Gray in A. Gray et al., Syn. Fl. N. Amer. 2(1): 300. 1878 • Shasta owl's-clover C E

Annuals 10–25 cm. **Stems** erect or ascending, scabrous, sparsely soft-hairy, and puberulent. **Leaves** 30–50 mm; blade: proximal margins entire, distal 3–5-lobed. **Inflorescences** spikes, 5–10 cm; bracts broadly ovate, 20–30 mm, margins 3-lobed in distal ⅔, proximal not resembling distal leaves, midlobe purple-pink distally, ovate, 7–10 mm wide, apex obtuse, cuspidate, lateral lobes narrowly triangulate. **Pedicels** 0–1 mm. **Flowers:** calyx 15–20 mm; corolla 25–30 mm, longer than bracts, abaxial lip slightly inflated, lobes 3, rounded,

adaxial lip purple-pink, 8–10 mm, 2–3 mm longer than abaxial, glabrous, tip notably hooked, not bearded. **Capsules** 5–7 mm, apex obtuse. **Seeds** 10–15, light brown to brown, ovoid, 3 mm, reticulate, not ridged. $2n = 28$.

Flowering May–Jun. Sagebrush scrub; of conservation concern; 500–1000 m; Calif.

Orthocarpus pachystachyus is known from the eastern Klamath Range and adjacent areas of the western Cascades in Siskiyou County.

7. **Orthocarpus purpureoalbus** A. Gray ex S. Watson, Botany (Fortieth Parallel), 458. 1871 (as purpureo-albus) E

Annuals (5–)15–45 cm. **Stems** erect, scabrous and densely puberulent. **Leaves** 15–35 mm; blade: proximal margins entire, distal 3-lobed. **Inflorescences** spikes, 10–25 cm; bracts 10–20 mm, margins 3-lobed to near base, proximal resembling distal leaves, midlobe green distally, triangular, 1–2 mm wide, apex acute, not cuspidate, lateral lobes linear. **Pedicels** 1 mm. **Flowers:** calyx 5–8 mm; corolla 15–20 mm, longer than bracts, abaxial lip inflated, lobes 0, adaxial lip purple-pink, 4–5 mm, 1–1.5 mm longer than abaxial, puberulent, tip notably hooked, not bearded. **Capsules** 6–9 mm, apex cuspidate. **Seeds** 18–26, dark brown, ovoid to reniform, 1.5–2.5 mm, deeply reticulate, longitudinally ridged. $2n = 28$.

Flowering Jul–Sep. Sagebrush meadows, openings in pinyon-juniper woodlands; 1500–3000 m; Ariz., Colo., N.Mex., Utah.

Stems and leaves of *Orthocarpus purpureoalbus* are often conspicuously colored red to dark purple.

8. **Orthocarpus tenuifolius** (Pursh) Bentham, Scroph. Ind., 12. 1835 • Thin-leaved owl's-clover E

Bartsia tenuifolia Pursh, Fl. Amer. Sept. 2: 429. 1813

Annuals 8–35 cm. **Stems** erect or ascending, pilose and puberulent. **Leaves** 10–50 mm; blade: proximal margins entire, distal 3–7-lobed. **Inflorescences** racemes, 2–6 cm; bracts 10–20 mm, margins 3–5-lobed in proximal ½, proximal not resembling distal leaves, midlobe purple-pink distally, ovate, 4–6 mm wide, apex obtuse, cuspidate, lateral lobes narrowly triangular. **Pedicels** 1–2 mm. **Flowers:** calyx 8–12 mm; corolla 14–20 mm, equal to or slightly longer than bracts, abaxial lip inflated, lobes 3, triangular, adaxial lip purple-pink, 5 mm, 1 mm longer than abaxial, puberulent, tip notably hooked, not bearded. **Capsules** 6–7 mm, apex acute. **Seeds** 15–20, dark brown, ovoid to reniform, 1–1.5 mm, shallowly reticulate, not ridged. **2n** = 28.

Flowering May–Aug. Moist fields, grasslands, rocky slopes; (300–)600–2500 m; B.C.; Idaho, Mont., Oreg., Wash., Wyo.

Orthocarpus tenuifolius is one of the plants first collected by Meriwether Lewis during the Corps of Discovery Expedition.

9. Orthocarpus tolmiei Hooker & Arnott, Bot. Beechey Voy., 379. 1839 [E]

Annuals 10–35 cm. **Stems** erect, scabrous and sparsely puberulent. **Leaves** 15–50 mm; blade margins entire. **Inflorescences** spikes or racemes, 2–15 cm; bracts 10–20 mm, margins entire or 3-lobed in proximal ½, proximal resembling distal leaves, midlobe yellow-green to purple distally, narrowly triangular, 0.7–1 mm wide, apex acute, not cuspidate, lateral lobes linear to narrowly triangular. **Pedicels** 0–2 mm. **Flowers:** calyx 4–8 mm; corolla 10–17 mm, longer than bracts, abaxial lip conspicuously inflated, lobes 3, triangular, adaxial lip yellow or purple-pink, 4–5 mm, 1–2 mm longer than abaxial, glabrous or puberulent, tip notably hooked, not bearded. **Capsules** 4–6 mm, apex retuse. **Seeds** 3–6, light brown, narrowly ellipsoid, 1.8–2.5 mm, shallowly reticulate, not ridged. **2n** = 14.

Orthocarpus tolmiei is similar to both *O. luteus* and *O. purpureoalbus* but usually can be distinguished by its highly branched stems.

Subspecies 2 (2 in the flora): w United States.

1. Corollas yellow, adaxial lips sparsely puberulent; calyces glabrous or sparsely glandular-puberulent . . . . 9a. *Orthocarpus tolmiei* subsp. *tolmiei*
1. Corollas purple-pink, adaxial lips puberulent; calyces densely glandular-puberulent . . . . 9b. *Orthocarpus tolmiei* subsp. *holmgreniorum*

9a. Orthocarpus tolmiei Hooker & Arnott subsp. **tolmiei** • Tolmie's owl's-clover [E]

Flowers: calyx sparsely glandular-puberulent; corolla yellow, adaxial lip glabrous or sparsely puberulent.

Flowering Jun–Aug. Sagebrush meadows, open slopes; 2000–3200 m; Idaho, Utah, Wyo.

9b. Orthocarpus tolmiei Hooker & Arnott subsp. **holmgreniorum** T. I. Chuang & Heckard, Syst. Bot. 17: 565, fig. 1G–J. 1992 • Holmgrens's owl's-clover [E]

Flowers: calyx densely glandular-puberulent; corolla purple-pink, adaxial lip puberulent.

Flowering Jun–Aug. Sagebrush meadows, open slopes; 1700–2800 m; Idaho, Utah.

27. TRIPHYSARIA Fischer & C. A. Meyer, Index Seminum (St. Petersburg) 2: 52. 1836 • False owl's-clover [Greek *tria*, three, and *physarion*, small bladder, alluding to three pouches of abaxial corolla lip] [E]

Elizabeth H. Zacharias

Orthocarpus Nuttall subg. *Triphysaria* (Fischer & C. A. Meyer) D. D. Keck

Herbs, annual; hemiparasitic. **Stems** erect, not fleshy, glabrous or hairy. **Leaves** cauline, usually subopposite to opposite proximally, alternate distally; petiole absent; blade not fleshy, not leathery, margins entire (proximal cauline leaves) or pinnatifid, rarely bipinnatifid. **Inflorescences** terminal, spikelike racemes; bracts present. **Pedicels** present; bracteoles absent. **Flowers:** sepals 4, calyx bilaterally symmetric, tubular or campanulate, lobes triangular to narrowly lanceolate; petals 5, corolla yellow, white, purple, or white, fading to rose pink, strongly bilabiate, tubular and club-shaped, abaxial lobes 3, saccate, throat folded proximal to abaxial lips (except in

T. pusilla), adaxial 2, adaxial lip beaked, ± straight, opening directed forward; stamens 4, didynamous, filaments glabrous, pollen sac 1; staminode 0; ovary 2-locular, placentation axile; stigma capitate, 2-lobed, or filiform to subcapitate. **Capsules:** dehiscence loculicidal. **Seeds** 10–100, dark brown, ovoid to ellipsoid, wings absent. $x = 11$.

Species 5 (5 in the flora): w North America; introduced in Asia (China).

Taxa of *Triphysaria* have often been placed in *Orthocarpus*; both genera have adaxial corolla lips that are open at the tips (versus connate and folded downward forming a hood), expanded stigmas that are either capitate or two-lobed (versus unexpanded), and terminal attachments of the ovules to the placentas (versus lateral attachments). D. C. Tank and R. G. Olmstead (2008) found strong molecular support for their sister relationship. *Triphysaria* is distinguished from *Castilleja* by stamens with a single pollen sac (versus two), the throats of the corollas folded proximal to the abaxial corolla lips (except in *T. pusilla*), and the base chromosome number of 11.

Triphysaria pusilla has been introduced in China, where the name *T. chinensis* (D. Y. Hong) D. Y. Hong has been incorrectly applied to it (T. I. Chuang and L. R. Heckard 1991).

1. Stems: branches decumbent to ascending; corollas 4–7 mm, beaks hooked, abaxial lobes 0.5–1 mm . . . . 4. *Triphysaria pusilla*
1. Stems: branches ascending; corollas 6–25 mm, beaks not hooked, abaxial lobes 1.5–5 mm.
  2. Stamens exserted; corollas creamy white, rarely pale yellow . . . . 2. *Triphysaria floribunda*
  2. Stamens included; corollas yellow, rarely yellow and white, or white, fading to rose pink.
    3. Stems sparsely pubescent distally; corolla beaks ± yellow, white, or rose pink . . . . 5. *Triphysaria versicolor*
    3. Stems puberulent to glandular-puberulent or pubescent to glandular-pubescent distally; corolla beaks dark purple, rarely yellow.
      4. Corollas 10–25 mm, abaxial lobes 2–5 mm . . . . 1. *Triphysaria eriantha*
      4. Corollas 10–12 mm, abaxial lobes 1.5–2 mm . . . . 3. *Triphysaria micrantha*

1. **Triphysaria eriantha** (Bentham) T. I. Chuang & Heckard, Syst. Bot. 16: 660. 1991 • Johnny-tuck E F

Orthocarpus erianthus Bentham, Scroph. Ind., 12. 1835 (as eriantha)

Stems simple or with 1–10 ascending branches proximally, 2–37 cm, glabrous proximally, puberulent to glandular-puberulent distally. **Leaves** puberulent to glandular-puberulent; proximal cauline: blade linear, 5–25 mm; cauline: blade ± ovate or obovate, 8–50 mm, base sessile, margins pinnatifid, rarely bipinnatifid, lateral lobes 2–11. **Spikelike racemes** interrupted, dense distally, 1–24 cm; peduncle absent; bracts pinnatifid, rarely bipinnatifid, ± ovate, 2–30 mm, lateral lobes 2–8. **Pedicels** 0.3–0.5 mm, glabrous. **Flowers:** calyx 6–13 mm, puberulent to glandular-puberulent, tube 4–7 mm, lobes triangular to narrowly lanceolate, 1–4 × 1–1.5 mm; corolla yellow, rarely yellow and white, or white, fading to rose pink, 10–25 mm, densely hairy, beak dark purple, not hooked, abaxial lobes spreading, 2–5 mm, throat abruptly indented, forming a fold under abaxial corolla lip, adaxial lobes projecting; stamens included, pollen sac yellow, 1.5–2 mm, glabrous, dehiscing longitudinally; style 12–22 mm, glabrous; stigma capitate. **Capsules** 4–8 × 2.5–4.5 mm, glabrous. **Seeds** 30–100, ovoid, 0.5–1 mm.

Subspecies 2 (2 in the flora): w United States.

1. Corollas yellow, rarely yellow and white. . . . . 1a. *Triphysaria eriantha* subsp. *eriantha*
1. Corollas white, fading to rose pink . . . . 1b. *Triphysaria eriantha* subsp. *rosea*

1a. Triphysaria eriantha (Bentham) T. I. Chuang & Heckard subsp. **eriantha** E F

Orthocarpus erianthus Bentham var. *gratiosus* Jepson & J. P. Tracy

Flowers: corolla yellow, rarely yellow and white. *2n* = 22.

Flowering Mar–Jun. Grasslands, foothills, pastures, roadsides, moist flats; 0–1600 m; Calif., Oreg.

Subspecies *eriantha* was collected in Washington in 1889 near Seattle; this appears to have been an introduction that did not persist. A recent report of this taxon from British Columbia remains to be confirmed.

1b. Triphysaria eriantha (Bentham) T. I. Chuang & Heckard subsp. **rosea** (A. Gray) T. I. Chuang & Heckard, Syst. Bot. 16: 660. 1991 E

Orthocarpus erianthus Bentham var. *roseus* A. Gray in W. H. Brewer et al., Bot. California 1: 578. 1876

Flowers: corolla white, fading to rose pink. *2n* = 22.

Flowering Mar–Jun. Coastal grassy fields, bluffs, roadsides, dunes; 0–400 m; Calif.

While most populations of subsp. *rosea* occur on the coast or coastal bluffs, it is known also from sites about 15 miles inland.

2. Triphysaria floribunda (Bentham) T. I. Chuang & Heckard, Syst. Bot. 16: 660. 1991 • San Francisco owl's-clover C E

Orthocarpus floribundus Bentham, Scroph. Ind., 12. 1835

Stems simple or with 1–15 ascending branches proximally, 6–24(–30) cm, glabrous proximally, sparsely retrorsely short-strigose distally. **Leaves** glabrous or sparsely puberulent; proximal cauline: blade linear to linear-lanceolate, 6–14 mm; cauline: blade ± ovate, 10–56 mm, base sessile, margins pinnatifid, rarely bipinnatifid, lateral lobes 4–8. **Spikelike racemes** interrupted, dense distally, 1–5.5 cm; peduncle absent; bracts pinnatifid, ± ovate, 4–12 mm, lateral lobes 2–6. **Pedicels** 0–0.3 mm, glabrous. **Flowers:** calyx 4–6 mm, glabrous, tube 2–4 mm, lobes triangular to narrowly lanceolate, 1.3–2 × 0.8–1 mm; corolla creamy white, rarely pale yellow, 6–14 mm, glabrous, beak light green or white, not hooked, abaxial lobes spreading, 1.5–3 mm, throat abruptly indented, forming a fold under abaxial corolla lip, adaxial lobes projecting; stamens exserted, pollen sac yellow, 1.3–1.5 mm, glabrous, dehiscing longitudinally; style 5.5–6 mm, glabrous; stigma capitate. **Capsules** 4.5–6.5 × 3–3.5 mm, glabrous. **Seeds** 20–35, ovoid, 0.6–1.3 mm. *2n* = 22.

Flowering Apr–Jun. Coastal or valley grasslands, grassy and rocky hills, sandy soils, road edges, coastal scrub, serpentine slopes; of conservation concern; 0–200 m; Calif.

Triphysaria floribunda is known from the San Francisco Bay area. The species is threatened by grazing, non-native plants, and trampling (California Native Plant Society, http://www.rareplants.cnps.org); only 14 of the 150 specimens for *T. floribunda* in the California Consortium of Herbaria portal (http://ucjeps.berkeley.edu/consortium) were collected after 1971.

3. Triphysaria micrantha (Greene ex A. Heller) T. I. Chuang & Heckard, Syst. Bot. 16: 660. 1991 • Purple-beak owl's-clover E

Orthocarpus micranthus Greene ex A. Heller, Muhlenbergia 2: 251. 1906; *O. erianthus* Bentham var. *micranthus* (Greene ex A. Heller) Jepson

Stems simple or with 1–4 ascending branches proximally, (3.5–)5–15(–19) cm, glabrous proximally, pubescent to glandular-pubescent distally. **Leaves** pubescent to glandular-pubescent; proximal cauline: blade filiform to linear, (4–)5–25 mm; cauline: blade ± elliptic to obovate, 10–25 mm, base sessile, margins usually pinnatifid, lateral lobes 2–6(or 7). **Spikelike racemes** interrupted, open distally, 0.5–11 cm; peduncle absent; bracts pinnatifid, ± obovate, 7–13(–15) mm, lateral lobes 2–4. **Pedicels** 0.5–0.8 mm, glabrous, rarely sparsely puberulent. **Flowers:** calyx 6–7 mm, pubescent to glandular-pubescent, tube (2–)3–5 mm, lobes narrowly lanceolate, 1–3 × 0.8–1.3 mm; corolla yellow, rarely yellow and white, 10–12 mm, hairy, beak dark purple, rarely yellow, not hooked, abaxial lobes spreading, 1.5–2 mm, throat abruptly indented, forming a fold under abaxial corolla lip, adaxial lobes projecting; stamens included, pollen sac yellow, 1–1.3 mm, glabrous, dehiscing longitudinally; style 6–9 mm, glabrous; stigma 2-lobed. **Capsules** 3–5.5 × 3–4 mm, glabrous. **Seeds** (30–)40–80, ovoid to ellipsoid, 0.8–1 mm. *2n* = 22.

Flowering Mar–Apr. Grasslands, gravelly hilltops, oak woodlands; 50–800 m; Calif.

Triphysaria micrantha is known from central California from the coast to the Sierra Nevada.

4. Triphysaria pusilla (Bentham) T. I. Chuang & Heckard, Syst. Bot. 16: 661. 1991 • Dwarf owl's-clover or Johnny-tuck E

Orthocarpus pusillus Bentham, Scroph. Ind., 12. 1835; *O. chinensis* D. Y. Hong

Stems simple or with 1–6 decumbent to ascending branches proximally, (2–)4–15 (–30) cm, glabrous proximally, retrorsely short-strigose distally. **Leaves** puberulent; proximal cauline: blade linear, 5–30 mm; cauline: blade ± ovate to obovate, 5–30 mm, base sessile, margins pinnatifid, rarely bipinnatifid, lateral lobes 2–8, threadlike. **Spikelike racemes** interrupted, open distally, (1.5–) 2.5–15 cm; peduncle absent; bracts pinnatifid, rarely bipinnatifid, ± obovate, 5–19 mm, lateral lobes 2–6 (–10), threadlike. **Pedicels** 0.3–1 mm, glabrous. **Flowers:** calyx (3–)5–7 mm, puberulent, tube 2–3 mm, lobes triangular to narrowly lanceolate, (1.5–)2–3.5 × 1.5–2 mm; corolla dark purple, rarely yellow, 4–7 mm, glabrate, beak dark purple, hooked, abaxial lobes spreading, 0.5–1 mm, throat not abruptly indented, not forming a fold under abaxial corolla lip, adaxial lobes projecting; stamens included, pollen sac yellow, 0.7–0.8 mm, glabrous, dehiscing longitudinally; style 2.5–5 mm, glabrous; stigma filiform or subcapitate. **Capsules** 3.5–5 × 2.5–3.5 mm, glabrous. **Seeds** 10–30, ovoid to ellipsoid, 0.7–1.2 mm. $2n = 22$.

Flowering Apr–Jun. Grasslands, lawns, pastures, roadsides, edges of vernal pools, woodlands; 0–1400 m; B.C.; Calif., Oreg., Wash.; introduced in Asia (China).

5. Triphysaria versicolor Fischer & C. A. Meyer, Index Seminum (St. Petersburg) 2: 52. 1836 • Yellow-beak or smooth owl's-clover E

Orthocarpus erianthus Bentham var. *versicolor* (Fischer & C. A. Meyer) Jepson; *O. versicolor* (Fischer & C. A. Meyer) Greene

Stems simple or with 1–20 ascending branches, 5–60 (–80) cm, glabrous proximally, sparsely pubescent distally. **Leaves** glabrous or sparsely pubescent; proximal cauline: blade linear, 3–30(–80) mm; cauline: blade ± ovate or obovate, 15–90 mm, base sessile, margins pinnatifid, rarely bipinnatifid, lateral lobes 3–9. **Spikelike racemes** interrupted, dense distally, 2–30 cm; peduncle absent; bracts pinnatifid, ± ovate or obovate, 6–22 mm, lateral lobes 2–7. **Pedicels** 1–2 mm, glabrous. **Flowers:** calyx 5–11 mm, glabrous or sparsely puberulent, tube 4–8 mm, lobes narrowly triangular, 1.5–5 × 1–2 mm; corolla yellow, rarely yellow and white, or white, fading to rose pink, 12–25(–27) mm, densely short-hairy, beak ± yellow, white, or rose pink, not hooked, abaxial lobes spreading, 2–4 mm, throat abruptly indented, forming a fold under abaxial corolla lip, adaxial lobes projecting; stamens included, pollen sac yellow, 1.7–3 mm, glabrous, dehiscing longitudinally; style 10–20 mm, glabrous; stigma capitate. **Capsules** 6–9(–10) × 3.5–5 mm, glabrous. **Seeds** 15–60, ovoid, 0.6–1 mm.

Subspecies 2 (2 in the flora): w North America.

1. Corollas white, fading to rose pink, beaks white or rose pink....5a. *Triphysaria versicolor* subsp. *versicolor*
1. Corollas yellow, rarely yellow with white, beaks yellow or white . . . . 5b. *Triphysaria versicolor* subsp. *faucibarbata*

5a. Triphysaria versicolor Fischer & C. A. Meyer subsp. **versicolor** E

Orthocarpus faucibarbatus A. Gray subsp. *albidus* D. D. Keck

Flowers: corolla white, fading to rose pink, beak white or rose pink. $2n = 22$.

Flowering Apr–Aug. Grasslands, vernally moist seeps to dry grassy meadows, headlands, rock outcrops, coastal prairies; 0–700 m; B.C.; Calif., Oreg.

Subspecies *versicolor* is critically imperiled in British Columbia, threatened by development pressure, habitat alteration, and competition from introduced species. The Canadian populations are over 400 km north of the nearest United States populations, in Oregon. This disjunction may be a result of the discontinuity of available habitat, introduction since glaciation, or survival of populations in glacial refugia (J. L. Penny and G. W. Douglas 1999).

5b. Triphysaria versicolor Fischer & C. A. Meyer subsp. **faucibarbata** (A. Gray) T. I. Chuang & Heckard, Syst. Bot. 16: 661. 1991 E

Orthocarpus faucibarbatus A. Gray in War Department [U.S.], Pacif. Railr. Rep. 4(5): 121. 1857

Flowers: corolla yellow, rarely yellow with white, beak yellow or white. $2n = 22$.

Flowering Apr–Jun. Grasslands; 0–600 m; Calif., Oreg.

Literature Cited

Robert W. Kiger, Editor

This is a consolidated list of all works cited in volume 17, whether as selected references, in text, or in nomenclatural contexts. In citations of articles, both here and in the taxonomic treatments, and also in nomenclatural citations, the titles of serials are rendered in the forms recommended in G. D. R. Bridson and E. R. Smith (1991). When those forms are abbreviated, as most are, cross references to the corresponding full serial titles are interpolated here alphabetically by abbreviated form. In nomenclatural citations (only), book titles are rendered in the abbreviated forms recommended in F. A. Stafleu and R. S. Cowan (1976–1988) and Stafleu et al. (1992–2009). Here, those abbreviated forms are indicated parenthetically following the full citations of the corresponding works, and cross references to the full citations are interpolated in the list alphabetically by abbreviated form. Two or more works published in the same year by the same author or group of coauthors will be distinguished uniquely and consistently throughout all volumes of *Flora of North America* by lower-case letters (b, c, d, ...) suffixed to the date for the second and subsequent works in the set. The suffixes are assigned in order of editorial encounter and do not reflect chronological sequence of publication. The first work by any particular author or group from any given year carries the implicit date suffix "a"; thus, the sequence of explicit suffixes begins with "b". There may be citations in this list that have dates suffixed "b," "c," "d," etc. but that are not preceded by citations of "[a]," "b," and/or "c," etc. works for that year. In such cases, the missing "[a]," "b," and/or "c," etc. works are ones cited (and encountered first from) elsewhere in the *Flora* that are not pertinent in this volume.

Abh. Naturwiss. Naturwiss. Verein Hamburg = Abhandlungen aus dem Gebiete der Naturwissenschaften herausgegeben von dem Naturwissenschaftlichen Verein in Hamburg.

Abrams, L. and R. S. Ferris. 1923–1960. Illustrated Flora of the Pacific States: Washington, Oregon, and California. 4 vols. Stanford. (Ill. Fl. Pacific States)

Achey, D. M. 1933. A revision of the section *Gymnocaulis* of the genus *Orobanche*. Bull. Torrey Bot. Club 60: 441–451.

Acta Bot. Neerl. = Acta Botanica Neerlandica.

Acta Bot. Yunnan. = Acta Botanica Yunnanica. [Yunnan Zhiwu Yanjiu.]

Acta Phytotax. Sin. = Acta Phytotaxonomica Sinica. [Chih Wu Fen Lei Hsüeh Pao.]

Acta Soc. Bot. Poloniae = Acta Societatis Botanicorum Poloniae.

Acta Univ. Lund. = Acta Universitatis Lundensis. Nova Series. Sectio 2, Medica, Mathematica, Scientiae Rerum Naturalium. [Lunds Universitets Årsskrift N.F., Avd. 2.]

Addisonia = Addisonia; Colored Illustrations and Popular Descriptions of Plants.

Adler, L. S. 2000. Alkaloid uptake increases fitness in a hemiparasitic plant via reduced herbivory and increased pollination. Amer. Naturalist 156: 92–99.

Advances Bot. Res. = Advances in Botanical Research.

Aguiar, S. and T. Borowski. 2013. Neuropharmacological review of the nootropic herb *Bacopa monnieri*. Rejuvenation Res. 16: 313–326.

Ahedor, A. R. 2007. Systematics of the *Mecardonia acuminata* (tribe Gratioleae, Plantaginaceae) Complex of Southern USA. Ph.D. dissertation. University of Oklahoma.

Ahedor, A. R. and W. J. Elisens. 2015. Morphological analyses of the *Mecardonia acuminata* (Plantaginaceae) species complex in the southeastern U.S.A. SouthE. Naturalist 14: 173–196.

Aiken, S. G. et al. 2007. Flora of the Canadian Arctic Archipelago.... Ottawa. [CD-ROM.]

Aiton, W. 1789. Hortus Kewensis; or, a Catalogue of the Plants Cultivated in the Royal Botanic Garden at Kew. 3 vols. London. (Hort. Kew.)

Albach, D. C. 2008. Further arguments for the rejection of paraphyletic taxa: *Veronica* subgen. *Pseudolysimachium* (Plantaginaceae). Taxon 57: 1–6.

Albach, D. C. et al. 2004. Evolution of Veroniceae: A phylogenetic perspective. Ann. Missouri Bot. Gard. 91: 275–302.

Albach, D. C. et al. 2004b. A new classification of the tribe Veroniceae—Problems and a possible solution. Taxon 53: 429–452.

Albach, D. C. et al. 2009. Phylogenetic placement of *Triaenophora* (formerly Scrophulariaceae) with some implications for the phylogeny of Lamiales. Taxon 58: 749–756.

Albach, D. C. and H. M. Meudt. 2010. Phylogeny of *Veronica* in the southern and northern hemispheres based on plastid, nuclear ribosomal and nuclear low-copy DNA. Molec. Phylogen. Evol. 54: 457–471.

Albach, D. C., H. M. Meudt, and B. Oxelman. 2005. Piecing together the "new" Plantaginaceae. Amer. J. Bot. 92: 297–315.

Albach, D. C., P. Schonswetter, and A. Tribsch. 2006. Comparative phylogeography of the *Veronical alpina* complex in Europe and North America. Molec. Ecol. 15: 3269–3286.

Alex, J. F. 1962. The taxonomy, history, and distribution of *Linaria dalmatica*. Canad. J. Bot. 40: 295–307.

Alford, J. D. and L. C. Anderson. 2002. Taxonomy and morphology of *Macranthera flammea* (Orobanchaceae). Sida 20: 189–204.

Allioni, C. 1785. Flora Pedemontana sive Enumeratio Methodica Stirpium Indigenarum Pedemontii. 3 vols. Turin. (Fl. Pedem.)

Ambio = Ambio; Journal of the Human Environment, Research and Management.

Amer. Biol. Teacher = American Biology Teacher.

Amer. Bot. (Binghamton) = The American Botanist; A Monthly Journal for the Plant Lover.

Amer. Forests = American Forests [title varies].

Amer. J. Bot. = American Journal of Botany.

Amer. J. Sci. Arts = American Journal of Science, and Arts.

Amer. Midl. Naturalist = American Midland Naturalist; Devoted to Natural History, Primarily That of the Prairie States.

Amer. Monthly Mag. & Crit. Rev. = American Monthly Magazine and Critical Review.

Amer. Naturalist = American Naturalist....

Amer. Nurseryman = American Nurseryman.

Amoen. Acad.—See: C. Linnaeus 1749[–1769]

Anales Soc. Esp. Hist. Nat. = Anales de la Sociedad Española de Historia Natural.

Anderson, J. L., S. Richmond-Williams, and O. Williams. 2007. *Penstemon lanceolatus* Benth. or *P. ramosus* Crosswhite in Arizona and New Mexico, a peripheral or endemic species? In: P. L. Barlow-Irick et al., eds. 2007. Southwestern Rare and Endangered Plants: Proceedings of the Fourth Conference. Fort Collins. Pp. 8–15.

Andersson, S. 2006. On the phylogeny of the genus *Calceolaria* (Calceolariaceae) as inferred from ITS and plastid *matK* sequences. Taxon 55: 125–137.

Angiosperm Phylogeny Group. 2016. An update of the Angiosperm Phylogeny Group classification for the orders and families of flowering plants: APG IV. Bot. J. Linn. Soc. 181: 1–20.

Ann. Bot. (Oxford) = Annals of Botany. (Oxford.)

Ann. Carnegie Mus. = Annals of the Carnegie Museum.

Ann. Lyceum Nat. Hist. New York = Annals of the Lyceum of Natural History of New York.

Ann. Missouri Bot. Gard. = Annals of the Missouri Botanical Garden.

Ann. Mus. Bot. Lugduno-Batavi = Annales Musei Botanici Lugduno-Batavi.

Ann. Mus. Natl. Hist. Nat. = Annales du Muséum National d'Histoire Naturelle. ["National" dropped after vol. 5.]

Ann. Nat. = Annals of Nature; or, Annual Synopsis of New Genera and Species of Animals, Plants, &c. Discovered in North America.

Ann. Nat. Hist. = Annals of Natural History; or, Magazine of Zoology, Botany and Geology.

Ann. Roy. Bot. Gard. Calcutta = Annals of the Royal Botanic Garden, Calcutta.

Annus Bot.—See: C. F. W. Wallroth 1815

Applequist, W. L. 2006. Proposal to reject the names *Plantago psyllium* and *P. cynops* (Plantaginaceae). Taxon 55: 235–236.

Aquatic Bot. = Aquatic Botany; International Scientific Journal Dealing with Applied and Fundamental Research on Submerged, Floating and Emergent Plants in Marine and Freshwater Ecosystems.

Argue, C. L. 1980. Pollen morphology in the genus *Mimulus* (Scrophulariaceae) and its taxonomic significance. Amer. J. Bot. 67: 68–87.

Argue, C. L. 1986. Some taxonomic implications of pollen and seed morphology in *Mimulus hymenophyllus* and *M. jungermannioides* and comparisons with other putative members of the *M. moschatus* alliance (Scrophulariaceae). Canad. J. Bot. 64: 1331–1337.

Ark. Bot. = Arkiv för Botanik.

Armstrong, J. E. 1985. The delimitation of Bignoniaceae and Scrophulariaceae based on floral anatomy, and the placement of problem genera. Amer. J. Bot. 72: 755–766.

Arnold, R. M. 1981. Weeds that ride the rails. Nat. Hist. 90: 59–65.

Arnold, R. M. 1982. Floral biology of *Chaenorrhinum minus* (Scrophulariaceae) a self-compatible annual. Amer. Midl. Naturalist 108: 317–324.

Aronson, J. K. 1986. An Account of the Foxglove and Its Medicinal Uses, 1785–1985. Oxford.

Atlantic J. = Atlantic Journal, and Friend of Knowledge.

Aublet, J. B. 1775. Histoire des Plantes de la Guiane Françoise.... 4 vols. Paris. [Vols. 1 and 2: text, paged consecutively; vols. 3 and 4: plates.] (Hist. Pl. Guiane)

Auk = Auk; a Quarterly Journal of Ornithology.

Austin, D. F. 2004. Florida Ethnobotany. Boca Raton.

Austral. Syst. Bot. = Australian Systematic Botany.

Autik. Bot.—See: C. S. Rafinesque 1840

Avrorin, N. A., ed. 1967. Plantarum in Zonam Polarem Transportatio. II. Leningrad.

Ayensu, E. S. 1981b. Medicinal Plants of the West Indies. Algonac, Mich.

B. M. C. Evol. Biol. = B M C Evolutionary Biology.

Baird, V. and J. L. Riopel. 1986. Life history studies of *Conopholis americana* (Orobanchaceae). Amer. Midl. Naturalist 116: 140–151.

Bakshi, T. S. and R. T. Coupland. 1960. Vegetative propagation in *Linaria vulgaris*. Canad. J. Bot. 38: 243–249.

Baldwin, B. G. et al., eds. 2012. The Jepson Manual: Vascular Plants of California, ed. 2. Berkeley.

Baldwin, B. G., S. Kalisz, and W. S. Armbruster. 2011. Phylogenetic perspectives on diversification, biogeography, and floral evolution of *Collinsia* and *Tonella* (Plantaginaceae). Amer. J. Bot. 98: 731–753.

Bangladesh J. Pl. Taxon. = Bangladesh Journal of Plant Taxonomy.

Bardy, K. E. et al. 2010. Disentangling phylogeography, polyploid evolution and taxonomy of a woodland herb (*Veronica chamaedrys* group, Plantaginaceae) in southeastern Europe. Molec. Phylogen. Evol. 57: 771–786.

Barker, W. R. et al. 2012. A taxonomic conspectus of Phrymaceae: A narrowed circumscription for *Mimulus*, new and resurrected genera, and new names and combinations. Phytoneuron 2012-39: 1–60.

Barlow-Irick, P. L. et al., eds. 2007. Southwestern Rare and Endangered Plants: Proceedings of the Fourth Conference. Fort Collins.

Barrett, S. C. H. and D. E. Seaman. 1980. The weed flora of Californian rice fields. Aquatic Bot. 9: 351–376.

Barrett, S. C. H. and J. L. Strother. 1978. Taxonomy and natural history of *Bacopa* (Scrophulariaceae) in California. Syst. Bot. 3: 408–419.

Barton, W. P. C. 1818. Compendium Florae Philadelphicae.... 2 vols. Philadelphia. (Comp. Fl. Philadelph.)

Bartonia = Bartonia; a Botanical Annual.

Bartram, W. 1791. Travels through North and South Carolina, Georgia, East and West Florida, the Cherokee Country, the Extensive Territories of the Muscogulges, or Creek Confederacy, and the Country of the Chactaws.... Philadelphia. (Travels Carolina)

Bassett, I. J. 1966. Taxonomy of North American *Plantago* L. section *Micropsyllium* Decne. Canad. J. Bot. 44: 467–479.

Bassett, I. J. 1967. Taxonomy of *Plantago* L. in North America: Sections *Holopsyllium* Pilger, *Palaeopsyllium* Pilger, and *Lamprosantha* Decne. Canad. J. Bot. 45: 565–577.

Bassett, I. J. 1973. The Plantains of Canada. Ottawa.

Beardsley, M. R. 1997. Colorado's Rare Endemic Plant, *Mimulus gemmiparus*, and Its Unique Mode of Reproduction. M.S. thesis. Colorado State University.

Beardsley, P. M. et al. 2004. Patterns of evolution in western North American *Mimulus* (Phrymaceae). Amer. J. Bot. 91: 474–489.

Beardsley, P. M. and W. R. Barker. 2005. Patterns of evolution in Australian *Mimulus* and related genera (Phrymaceae ≈ Scrophulariaceae): A molecular phylogeny using chloroplast and nuclear sequence data. Austral. Syst. Bot. 18: 61–73.

Beardsley, P. M. and R. G. Olmstead. 2002. Redefining Phrymaceae: The placement of *Mimulus*, tribe Mimuleae, and *Phryma*. Amer. J. Bot. 89: 1093–1102.

Beardsley, P. M., A. Yen, and R. G. Olmstead. 2003. AFLP phylogeny of *Mimulus* section *Erythranthe* and the evolution of hummingbird pollination. Evolution 57: 1397–1410.

Beatty, B. L., W. F. Jennings, and R. C. Rawlinson. 2004. *Penstemon degeneri* Crosswhite (Degener's Beardtongue): A Technical Conservation Assessment. Denver.

Beck, G. 1930. Orobanchaceae. In: H. G. A. Engler, ed. 1900–1953. Das Pflanzenreich.... 107 vols. Berlin. Vol. 96[IV,261], pp. 1–348.

Bedigian, D., ed. 2011. Sesame: The Genus *Sesamum*. Boca Raton.

Beeks, R. M. 1962. Variation and hybridization in southern California populations of *Diplacus* (Scrophulariaceae). Aliso 5: 83–122.

Beitr. Naturk.—See: J. F. Ehrhart 1787–1792

Bell, C. R. 1965. In: Documented plant chromosome numbers 65: 3. Sida 2: 168–170.

Benedict, B. G. 1993. Biosystematics of the Annual Species *Mimulus guttatus* Species Complex in British Columbia. M.S. thesis. University of British Columbia.

Benedict, B. G. et al. 2012. *Mimulus sookensis* (Phrymaceae), a new allotetraploid species derived from *Mimulus guttatus* and *Mimulus nasutus*. Madroño 59: 29–43.

Bennett, J. R. and S. Mathews. 2006. Phylogeny of the parasitic plant family Orobanchaceae inferred from phytochrome A. Amer. J. Bot. 93: 1039–1051.

Bentham, G. 1835. Scrophularineae Indicae. London. (Scroph. Ind.)

Bentham, G. 1839[–1857]. Plantas Hartwegianas Imprimis Mexicanas.... London. [Issued by gatherings with consecutive signatures and pagination.] (Pl. Hartw.)

Bentham, G. 1844[–1846]. The Botany of the Voyage of H.M.S. Sulphur, under the Command of Captain Sir Edward Belcher...during the Years 1836–1842. 6 parts. London. [Parts paged consecutively.] (Bot. Voy. Sulphur)

Bentham, G. 1876. Scrophularineae. In: G. Bentham and J. D. Hooker. 1862–1883. Genera Plantarum ad Exemplaria Imprimis in Herbariis Kewensibus Servata Definita. 3 vols. London. Vol. 2, pp. 913–980.

Bentham, G. and J. D. Hooker. 1862–1883. Genera Plantarum ad Exemplaria Imprimis in Herbariis Kewensibus Servata Definita. 3 vols. London.

Berger, B. A. 2005. Character Polymorphism and Taxonomy of the *Lindernia dubia* Complex (Scrophulariaceae). M.S. thesis. University of Oklahoma.

Berlandier, J. L. [1832.] Memorias de la Comision de Limites.... [Matamoros.] (Mem. Comis. Limites)

Bertoloni, A. 1803–1810. Rariorum Liguriae [Italiae] Plantarum. 3 vols. Genoa and Pisa. (Rar. Lig. [Ital.] Pl.)

Best, C. et al. 1996. A Flora of Sonoma County: Manual of the Flowering Plants and Ferns of Sonoma County, California. Sacramento.

Beuzenberg, E. J. and J. B. Hair. 1983. Contributions to a chromosome atlas of the New Zealand flora—25 miscellaneous species. New Zealand J. Bot. 21: 13–20.

Beverley, R. 1705. The History and Present State of Virginia, in Four Parts. London.

Biblioth. Bot. = Bibliotheca Botanica; original Abhandlungen aus dem Gesammtgebiete der Botanik.

Bijdr. Fl. Ned. Ind.—See: C. L. Blume 1825–1826[–1827]

Biltmore Bot. Stud. = Biltmore Botanical Studies; a Journal of Botany Embracing Papers by the Director and Associates of the Biltmore Herbarium.

Biochem. Syst. & Ecol. = Biochemical Systematics and Ecology.

Biol. Cent.-Amer., Bot.—See: W. B. Hemsley 1879–1888

Blume, C. L. 1825–1826[–1827]. Bijdragen tot de Flora van Nederlandsch Indië. 17 parts. Batavia. [Parts paged consecutively.] (Bijdr. Fl. Ned. Ind.)

Bodard, P. H. H. [1798]. Mémoire sur la Véronique Cymbalaire. Pisa. (Mém. Véronique Cymb.)

Bohlen, C. von. 1995. El género *Mimulus* L. (Scrophulariaceae) en Chile. Gayana, Bot. 52: 7–28.

Boissier, P. E. 1842–1859. Diagnoses Plantarum Orientalium Novarum. 3 vols. in 19 parts. Geneva etc. [Parts paged independently.] (Diagn. Pl. Orient.)

Bol. Soc. Bot. México = Boletín de la Sociedad Botánica de México.

Bollinger, M. 1996. Monographie der Gattung *Odontites* (Scrophulariaceae) sowie der verwandten Gattungen *Macrosyringion, Odontitella, Bornmuellerantha* und *Bartsiella*. Willdenowia 26: 37–168.

Bonati, G. 1918. Le Genre *Pedicularis* L. Morphologie, Classification, Distribution Géographique, Évolution et Hybridation. Nancy.

Borland, J. 1994. Growing Indian paintbrush. Amer. Nurseryman 179(6): 49–53.

Boston J. Nat. Hist. = Boston Journal of Natural History.

Bot. Arr. Brit. Pl. ed. 2—See: W. Withering 1787–1792

Bot. Beechey Voy.—See: W. J. Hooker and G. A. W. Arnott [1830–]1841

Bot. California—See: W. H. Brewer et al. 1876–1880

Bot. Electronic News = Botanical Electronic News. [Electronic journal.]

Bot. Gaz. = Botanical Gazette; Paper of Botanical Notes.

Bot. J. Linn. Soc. = Botanical Journal of the Linnean Society.

Bot. Jahrb. Syst. = Botanische Jahrbücher für Systematik, Pflanzengeschichte und Pflanzengeographie.

Bot. Mag. = Botanical Magazine; or, Flower-garden Displayed.... [Edited by Wm. Curtis.] [With vol. 15, 1801, title became Curtis's Botanical Magazine; or....]

Bot. Mag. (Tokyo) = Botanical Magazine. [Shokubutsu-gaku Zasshi.] (Tokyo.)

Bot. Not. = Botaniska Notiser.

Bot. Reg. = Botanical Register....

Bot. Res. J. = Botany Research Journal.

Bot. Rev. (Lancaster) = Botanical Review, Interpreting Botanical Progress.

Bot. Surv. Olympic Penins.—See: G. N. Jones 1936

Bot. Taschenb.—See: D. H. Hoppe 1790–1849

Bot. Tidsskr. = Botanisk Tidsskrift.

Bot. Voy. Herald—See: B. Seemann 1852–1857

Bot. Voy. Sulphur—See: G. Bentham 1844[–1846]

Bot. Zhurn. (Moscow & Leningrad) = Botanicheskii Zhurnal. (Moscow and Leningrad.)

Botanical Society of America. 2001. Botany 2001 Abstracts. Columbus.

Botanical Society of America. 2014. Botany 2014 Abstracts. Boise.

Botany (Fortieth Parallel)—See: S. Watson 1871

Bragg, T. B. and J. Stubbendieck, eds. 1989. Proceedings of the Eleventh North American Prairie Conference. Prairie Pioneers: Ecology, History and Culture. Lincoln, Nebr.

Brébisson, L. A. de. 1849. Flore de la Normandie, ed. 2. Caen and Paris. (Fl. Normandie ed. 2)

Bremer, B. et al. 1994. *rbc*L sequences support exclusion of *Retzia, Desfontainia,* and *Nicodemia* (Buddlejaceae) from the Gentianales. Pl. Syst. Evol. 190: 213–230.

Bremer, B. et al. 2002. Phylogenetics of asterids based on 3 coding and 3 non-coding chloroplast DNA markers and the utility of non-coding DNA at higher taxonomic levels. Molec. Phylogen. Evol. 24: 274–301.

Brewer, W. H. et al. 1876–1880. Geological Survey of California.... Botany.... 2 vols. Cambridge, Mass. (Bot. California)

Bridson, G. D. R. 2004. BPH-2: Periodicals with Botanical Content. 2 vols. Pittsburgh.

Bridson, G. D. R. and E. R. Smith. 1991. B-P-H/S. Botanico-Periodicum-Huntianum/Supplementum. Pittsburgh.

Brit. Herb.—See: J. Hill 1756[–1757]

Britton, N. L. and A. Brown. 1913. An Illustrated Flora of the Northern United States, Canada and the British Possessions from Newfoundland to the Parallel of the Southern Boundary of Virginia, and from the Atlantic Ocean Westward to the 102d Meridian..., ed. 2. 3 vols. New York. (Ill. Fl. N. U.S. ed. 2)

Brittonia = Brittonia; a Journal of Systematic Botany....

Brown, R. 1810. Prodromus Florae Novae Hollandiae et Insulae van-Diemen.... London. (Prodr.)

Brown, R. 1823. Chloris Melvilliana. A List of Plants Collected in Melville Island...in the Year 1820.... London. [Preprint with independent pagination from W. E. Parry. 1824. A Supplement to the Appendix of Captain Parry's Voyage.... London.] (Chlor. Melvill.)

Brummitt, R. K. 2007. Report of the Nomenclature Committee for Vascular Plants: 59. Taxon 56: 1289–1296.

Brummitt, R. K. 2009. Report of the Nomenclatural Committee for Vascular Plants: 60. Taxon 58: 280–292.

Brummitt, R. K. and C. E. Powell, eds. 1992. Authors of Plant Names. A List of Authors of Scientific Names of Plants, with Recommended Standard Forms of Their Names, Including Abbreviations. Kew.

Brunton, D. F. 2009. Noteworthy collections: South Carolina. Castanea 74: 183–184.

Bull. Acad. Imp. Sci. Saint-Pétersbourg = Bulletin de l'Académie Impériale des Sciences de Saint Pétersbourg.

Bull. Amer. Penstemon Soc. = Bulletin of the American Penstemon Society.

Bull. Bot. Surv. India = Bulletin of the Botanical Survey of India.

Bull. Calif. Acad. Sci. = Bulletin of the California Academy of Sciences.

Bull. Calif. Dept. Agric. = Bulletin of the California Department of Agriculture.

Bull. Herb. Boissier = Bulletin de l'Herbier Boissier.

Bull. Misc. Inform. Kew = Bulletin of Miscellaneous Information, Royal Gardens, Kew.

Bull. Mus. Natl. Hist. Nat., B, Adansonia = Bulletin du Muséum National d'Histoire Naturelle. Section B, Adansonia: Botanique Phytochimie.

Bull. New York Bot. Gard. = Bulletin of the New York Botanical Garden.

Bull. S. Calif. Acad. Sci. = Bulletin of the Southern California Academy of Sciences.

Bull. Soc. Bot. France = Bulletin de la Société Botanique de France.

Bull. Soc. Polymath. Morbihan = Bulletin de la Société Polymathique du Morbihan.

Bull. Torrey Bot. Club = Bulletin of the Torrey Botanical Club.

Burman, N. L. 1768. Flora Indica ... Nec Non Prodromus Florae Capensis. Leiden and Amsterdam. (Fl. Indica)

Caicco, S., J. Civille, and D. M. Henderson. 1983. Noteworthy collections: Idaho. Madroño 30: 64.

Calabres, C. et al. 2008. Effects of standardized *Bacopa monnieri* extract on cognitive performance, anxiety, and depression in the elderly: A randomized, double-blind, placebo-controlled trial. J. Altern. Complement. Med. 14: 707–713.

Calder, J. A. and R. L. Taylor. 1968. Flora of the Queen Charlotte Islands. 2 vols. Ottawa.

Callihan, R. H., S. L. Carson, and R. T. Dobbins. 1995. NAWEEDS, Computer-aided Weed Identification for North America. Illustrated User's Guide plus Computer Floppy Disk. Moscow, Idaho.

Campbell, D. H. 1930. The relationships of *Paulownia*. Bull. Torrey Bot. Club 57: 47–50.

Campbell, G. R. 1950. *Mimulus guttatus* and related species. Aliso 2: 319–335.

Canad. J. Bot. = Canadian Journal of Botany.

Canad. J. Pl. Sci. = Canadian Journal of Plant Science.

Candolle, A. P. de. 1813b. Catalogus Plantarum Horti Botanici Monspeliensis.... Montpellier, Paris, and Strasbourg. (Cat. Pl. Hort. Monsp.)

Candolle, A. P. de and A. L. P. P. de Candolle, eds. 1823–1873. Prodromus Systematis Naturalis Regni Vegetabilis.... 17 vols. Paris etc. [Vols. 1–7 edited by A. P. de Candolle, vols. 8–17 by A. L. P. P. de Candolle.] (Prodr.)

Candollea = Candollea; Organe du Conservatoire et du Jardin Botaniques de la Ville de Genève.

Canne, J. M. 1979. A light and scanning electron microscope study of seed morphology in *Agalinis* (Scrophulariaceae) and its taxonomic significance. Syst. Bot. 4: 281–296.

Canne, J. M. 1980. Seed surface features in *Aureolaria, Brachystigma, Tomanthera,* and certain South American *Agalinis* (Scrophulariaceae). Syst. Bot. 5: 241–252.

Canne, J. M. and C. M. Kampny. 1991. Taxonomic significance of leaf and stem anatomy of *Agalinis* (Scrophulariaceae) from the U.S.A. and Canada. Canad. J. Bot. 69: 1935–1950.

Cantino, P. D. 2004. Phrymaceae. In: K. Kubitzki et al., eds. 1990+. The Families and Genera of Vascular Plants. 14+ vols. Berlin etc. Vol. 7, pp. 323–326.

Cantino, P. D., R. M. Harley, and S. J. Wagstaff. 1992. Genera of Labiatae: Status and classification. In: R. M. Harley et al., eds. 1992. Advances in Labiate Science. Kew. Pp. 511–522.

Cantlon, J. E., E. J. C. Curtis, and W. M. Malcolm. 1963. Studies of *Melampyrum lineare*. Ecology 44: 466–474.

Carlbom, C. 1969. Evolutionary relationships in the genus *Scrophularia* L. Hereditas (Lund) 61: 287–301.

Carlson, M. C. 1957. Monograph of the genus *Russelia*. Fieldiana, Bot. 29: 231–292.

Carlson, M. L. 2002. Evolution of Mating System and Inbreeding Depression in the *Mimulus moschatus* (Scrophulariaceae) Alliance. Ph.D. dissertation. University of Alaska.

Carnicero, P. et al. 2017. Different speciation types meet in a Mediterranean genus: The biogeographic history of *Cymbalaria* (Plantaginaceae). Taxon 66: 393–407.

Carr, G. D. 1971. Taxonomy of *Pedicularis parryi* (Scrophulariaceae). Brittonia 23: 280–291.

Caryologia = Caryologia; Giornale di Citologia, Citosistematica e Citogenetica.

Castanea = Castanea; Journal of the Southern Appalachian Botanical Club.

Castellanos, M. C. et al. 2006. Anther evolution: Pollen presentation strategies when pollinators differ. Amer. Naturalist 167: 289–296.

Castellanos, M. C., Paul Wilson, and J. D. Thompson. 2004. 'Anti-bee' and 'pro-bird' changes during the evolution of hummingbird pollination in *Penstemon* flowers. J. Evol. Biol. 17: 876–885.

Cat. N. Amer. Pl.—See: A. A. Heller 1898

Cat. Pl. Hort. Monsp.—See: A. P. de Candolle 1813b

Cat. Pl. New York—See: J. Torrey 1819

Cat. Pl. Upper Louisiana—See: T. Nuttall 1813

Catal. Bot.—See: A. W. Roth 1797–1806

Cavanilles, A. J. 1791–1801. Icones et Descriptiones Plantarum, Quae aut Sponte in Hispania Crescunt, aut in Hortis Hospitantur. 6 vols. Madrid. (Icon.)

Cent. Pl. I—See: C. Linnaeus 1755

Cent. Pl. II—See: C. Linnaeus [1756]

Chapman, A. W. 1860. Flora of the Southern United States.... New York. (Fl. South. U.S.)

Chapman, A. W. 1892. Flora of the Southern United States..., ed. 2 reprint 2. New York, Cincinnati, and Chicago. (Fl. South. U.S. ed. 2 repr. 2)

Chari, J. and Paul Wilson. 2001. Factors limiting hybridization between *Penstemon spectabilis* and *Penstemon centranthifolius*. Canad. J. Bot. 79: 1438–1448.

Chater, A. O. and D. A. Webb. 1972. *Orobanche*. In: T. G. Tutin et al., eds. 1964–1980. Flora Europaea. 5 vols. Cambridge. Vol. 3, pp. 286–293.

Chau, J. H. et al. 2017. Phylogenetic relationships in tribe Buddlejeae (Scrophulariaceae) based on multiple nuclear and plastid markers. Bot. J. Linn. Soc. 184: 137–166.

Chazelles de Prizy, L. M. 1789–1790. Supplément au Dictionnaire des Jardiniers.... 2 vols. Metz. (Suppl. Dict. Jard.)

Chinnock, R. J. 2007. *Eremophila* and Allied Genera: A Monograph of the Myoporaceae. Dural, New South Wales.

Chlor. Melvill.—See: R. Brown 1823

Chodat, R. 1908. Étude critique des genres *Scoparia* L. et *Hasslerella* Chod. Bull. Herb. Boissier, sér. 2, 8: 1–16.

Chromosome Bot. = Chromosome Botany.

Chuang, T. I. and L. R. Heckard. 1973. Taxonomy of *Cordylanthus* subgenus *Hemistegia* (Scrophulariaceae). Brittonia 25: 135–158.

Chuang, T. I. and L. R. Heckard. 1975. Taxonomic status of *Cordylanthus* (subg. *Dicranostegia*) *orcuttianus* (Scrophulariaceae). Madroño 23: 88–95.

Chuang, T. I. and L. R. Heckard. 1982. Chromosome numbers in *Orthocarpus* and related monotypic genera (Scrophulariaceae: subtribe Castillejinae). Brittonia 34: 89–101.

Chuang, T. I. and L. R. Heckard. 1986. Systematics and evolution of *Cordylanthus* (Scrophulariaceae—Pedicularieae). Syst. Bot. Monogr. 10: 1–105.

Chuang, T. I. and L. R. Heckard. 1991. Generic realignment and synopsis of subtribe Castillejinae (Scrophulariaceae—tribe Pediculareae). Syst. Bot. 16: 644–666.

Chuang, T. I. and L. R. Heckard. 1992. A taxonomic revision of *Orthocarpus* (Scrophulariaceae—tribe Pediculareae). Syst. Bot. 17: 560–582.

Chuang, T. I. and L. R. Heckard. 1993. Chromosome numbers of neotropical *Castilleja* (Scrophulariaceae: tribe Pediculareae) and their taxonomic implications. Ann. Missouri Bot. Gard. 80: 974–986.

Chuang, T. I. and L. R. Heckard. 1993b. *Castilleja.* In: J. C. Hickman, ed. 1993. The Jepson Manual. Higher Plants of California. Berkeley, Los Angeles, and London. Pp. 1016–1024.

Clark, D. V. 1971. Speciation in *Penstemon.* Ph.D. dissertation. University of Montana.

Class-book Bot. ed. 2—See: A. Wood 1847

Clausen, J. 1933. Cytological evidence for the hybrid origin of *Penstemon neotericus* Keck. Hereditas (Lund) 18: 65–76.

Clausen, J., D. D. Keck, and W. M. Hiesey. 1940. Experimental studies on the nature of species. I. Effect of varied environments on western North American plants. Publ. Carnegie Inst. Wash. 520.

Clay, D. L. et al. 2012. Homoploid hybrid speciation in a rare endemic *Castilleja* from Idaho (*Castilleja christii,* Orobanchaceae). Amer. J. Bot. 99: 1976–1990.

Clements, R. K., J. M. Baskin, and C. C. Baskin. 1998. The comparative biology of the two closely-related species *Penstemon tenuiflorus* Pennell and *P. hirsutus* (L.) Willd. (Scrophulariaceae, section *Graciles*): I. Taxonomy and geographical distribution. Castanea 63: 138–153.

Clements, R. K., J. M. Baskin, and C. C. Baskin. 1998b. The comparative biology of the two closely-related species *Penstemon tenuiflorus* Pennell and *P. hirsutus* (L.) Willd. (Scrophulariaceae, section *Graciles*): II. Reproductive biology. Castanea 63: 299–309.

Clinebell, R. R. and P. Bernhardt. 1998. The pollination ecology of five species of *Penstemon* (Scrophulariaceae) in the tallgrass prairie. Ann. Missouri Bot. Gard. 85: 126–136.

Clokey, I. W. and D. D. Keck. 1939. Reconsideration of certain members of *Penstemon* subsection *Spectabiles.* Bull. S. Calif. Acad. Sci. 38: 8–13.

Cody, W. J. 1996. Flora of the Yukon Territory. Ottawa.

Collins, L. T., A. E. L. Colwell, and G. Yatskievych. 2009. *Orobanche riparia,* a new species from the American Midwest. J. Bot. Res. Inst. Texas 3: 3–11.

Collins, L. T. and G. Yatskievych. 2015. *Orobanche arizonica* sp. nov. and nomenclatural changes in *Orobanche cooperi.* Phytoneuron 2015-48: 1–19.

Colsmann, J. [1793.] Prodromus Descriptionis Gratiolae.... Copenhagen. (Prodr. Descr. Gratiol.)

Colwell, A. E. L., K. C. Watson, and A. C. Schneider. 2017. A new species of *Aphyllon* (Orobanchaceae) parasitic on *Galium* in the western USA. Madroño 64: 99–107.

Comp. Fl. Philadelph.—See: W. P. C. Barton 1818

Compan. Bot. Mag. = Companion to the Botanical Magazine....

Conservation Biol. = Conservation Biology; Journal of the Society for Conservation Biology.

Conservation Genet. = Conservation Genetics.

Consp. Regn. Veg.—See: H. G. L. Reichenbach 1828

Contr. Bot. Lab. Morris Arbor. Univ. Pennsylvania = Contributions from the Botanical Laboratory and the Morris Arboretum of the University of Pennsylvania.

Contr. Dudley Herb. = Contributions from the Dudley Herbarium of Stanford University.

Contr. Gray Herb. = Contributions from the Gray Herbarium of Harvard University. [Some numbers reprinted from (or in?) other periodicals, e.g. Rhodora.]

Contr. Herb. Franklin Marshall Coll. = Contributions from the Herbarium of Franklin and Marshall College.

Contr. Inst. Bot. Univ. Montréal = Contributions de l'Institut Botanique de l'Université de Montréal.

Contr. U.S. Natl. Herb. = Contributions from the United States National Herbarium.

Contr. W. Bot. = Contributions to Western Botany.

Cooperrider, T. S. 1969. Notes on *Chelone* in southeastern United States. Castanea 34: 223–225.

Cooperrider, T. S. 1970. Chromosome numbers in *Chelone* (Scrophulariaceae). Brittonia 22: 175–183.

Cooperrider, T. S. 1995. The Dicotyledoneae of Ohio: Linaceae through Campanulaceae. Columbus. [The Vascular Flora of Ohio, vol. 2(2).]

Correll, D. S. and M. C. Johnston. 1970. Manual of the Vascular Plants of Texas. Renner, Tex.

Cosacov, A. et al. 2009. New insights into the phylogenetic relationships, character evolution, and phytogeographic patterns of *Calceolaria* (Calceolariaceae). Amer. J. Bot. 96: 2240–2255.

Coulter, J. M. and A. Nelson. 1909. New Manual of Botany of the Central Rocky Mountains (Vascular Plants). New York. (New Man. Bot. Rocky Mt.)

Crantz, H. J. N. von. 1766. Institutiones Rei Herbariae.... 2 vols. Vienna. (Inst. Rei Herb.)

Crantz, H. J. N. von. 1769. Stirpium Austriarum Fasciculus I [–VI]. 2 vols. in 6 fascs. Vienna. [Vols. paged consecutively.] (Stirp. Austr. Fasc. ed. 2)

Crawford, P. T. 2003. Biosystematics of North American Species of *Nuttallanthus* (Lamiales). Ph.D. dissertation. University of Oklahoma.

Crawford, P. T. and W. J. Elisens. 2006. Genetic variation and reproductive system among North American species of *Nuttallanthus* (Plantaginaceae). Amer. J. Bot. 93: 582–591.

Cronquist, A. 1959. *Penstemon.* In: C. L. Hitchcock et al. 1955–1969. Vascular Plants of the Pacific Northwest. 5 vols. Seattle. Vol. 4, pp. 365–411.

Cronquist, A. 1981. An Integrated System of Classification of Flowering Plants. New York.

Cronquist, A. et al. 1972–2017. Intermountain Flora. Vascular Plants of the Intermountain West, U.S.A. 7 vols. in 9. New York and London. (Intermount. Fl.)

Crosswhite, F. S. 1965. Hybridization of *Penstemon barbatus* (Scrophulariaceae) of section *Elmigera* with species of a section *Habroanthus*. SouthW. Naturalist 10: 234–237.

Crosswhite, F. S. 1965b. Revision of *Penstemon* section *Penstemon* Scrophulariaceae II. A western alliance in series *Graciles*. Amer. Midl. Naturalist 74: 429–442.

Crosswhite, F. S. 1965c. Subdivisions of *Penstemon* section *Penstemon*. Sida 2: 160–162.

Crosswhite, F. S. 1965d. Against proliferation of superfluous names by giving formal status to spelling variants. Taxon 16: 109–112.

Crosswhite, F. S. 1965e. Revision of *Penstemon* (Sect. *Penstemon*) Series *Graciles* (Scrophulariaceae) with a Synopsis of the Genus. M.S. thesis. University of Wisconsin.

Crosswhite, F. S. 1965f. Concerning *Penstemon unilateralis* and *Penstemon secundiflorus*. SouthW. Naturalist 10: 318.

Crosswhite, F. S. 1966. Revision of *Penstemon* section *Chamaeleon* (Scrophulariaceae). Sida 2: 339–346.

Crosswhite, F. S. 1967. Revision of *Penstemon* section *Habroanthus* (Scrophulariaceae) I: Conspectus. Amer. Midl. Naturalist 77: 1–11.

Crosswhite, F. S. 1967b. Revision of *Penstemon* section *Habroanthus* (Scrophulariaceae) II: Series *Speciosi*. Amer. Midl. Naturalist 77: 12–27.

Crosswhite, F. S. 1967c. Revision of *Penstemon* section *Habroanthus* (Scrophulariaceae) III: Series *Virgati*. Amer. Midl. Naturalist 77: 28–41.

Crosswhite, F. S. and C. D. Crosswhite. 1966. Insect pollination of *Penstemon* series *Graciles* (Scrophulariaceae) with notes on *Osmis* and other Megachilidae. Amer. Midl. Naturalist 76: 450–467.

Crosswhite, F. S. and C. D. Crosswhite. 1970. Pollination of *Castilleja sessiliflora* in southern Wisconsin. Bull. Torrey Bot. Club 97: 100–105.

Crosswhite, F. S. and C. D. Crosswhite. 1981. Hummingbirds as pollinators of flowers in the red-yellow segment of the color spectrum, with special reference to *Penstemon* and the 'open habitat'. Desert Pl. 3: 156–170.

Crosswhite, F. S. and S. Kawano. 1965. Chromosome numbers in *Penstemon* (Scrophulariaceae) I: New mitotic counts. Rhodora 67: 187–190.

Cycl.—See: A. Rees [1802–]1819–1820

Dansk Bot. Ark. = Dansk Botanisk Arkiv Udgivet af Dansk Botanisk Forening.

Datwyler, S. L. and A. D. Wolfe. 2004. Phylogenetic relationships and morphological evolution in *Penstemon* subg. *Dasanthera* (Veronicaceae). Syst. Bot. 29: 165–170.

Davidse, G., M. Sousa S., A. O. Chater, et al., eds. 1994+. Flora Mesoamericana. 5+ vols. in parts. Mexico City, St. Louis, and London.

Davis, P. H., ed. 1965–2000. Flora of Turkey and the East Aegean Islands. 11 vols. Edinburgh.

de Lange, P. J., B. G. Murray, and P. M. Datson. 2004. Contributions to a chromosome atlas of the New Zealand flora—38. Counts for 50 families. New Zealand J. Bot. 42: 873–904.

Delcourt, H. R. and P. A. Delcourt. 1975. The Blufflands: Pleistocene pathway to the Tunica Hills. Amer. Midl. Naturalist 94: 385–400.

Densmore, F. 1928. Uses of plants by the Chippewa Indians. Rep. (Annual) Bur. Amer. Ethnol. 44: 275–397.

dePamphilis, C. W., N. D. Young, and A. D. Wolfe. 1997. Evolution of plastid gene *rps*2 in a lineage of hemiparasitic and holoparasitic plants: Many losses of photosynthesis and complex patterns of rate variation. Proc. Natl. Acad. Sci. U.S.A. 94: 7367–7372.

DePoe, C. E. 1969. *Bacopa egensis* (Poeppig) Pennell (Scrophulariaceae) in the United States. Sida 3: 313–318.

Descr. Pl. Nouv.—See: É. P. Ventenat [1800–1803]

Desert Pl. = Desert Plants; a Quarterly Journal Devoted to Broadening Our Knowledge of Plants Indigenous or Adaptable to Arid and Sub-arid Regions.

Deutsche Bot. Monatsschr. = Deutsche botanische Monatsschrift; zugleich Organ der Bot. Vereine in Hamburg und Nürnberg und der Thüring. bot. Gesellschaft "Irmischie" zu Arnstadt.

Diagn. Pl. Orient.—See: P. E. Boissier 1842–1859

Dieringer, G. and L. Cabrera R. 2002. The interaction between pollinator size and the bristle staminode of *Penstemon digitalis* (Scrophulariaceae). Amer. J. Bot. 89: 991–997.

Don, G. 1831–1838. A General History of the Dichlamydeous Plants.... 4 vols. London. (Gen. Hist.)

Dorn, R. D. 1988. Vascular Plants of Wyoming. Cheyenne. (Vasc. Pl. Wyoming)

Doweld, A. B. and A. Shipunov. 2017. Proposal to reject the name *Plantago indica* (Plantaginaceae). Taxon 66: 205–206.

Downie, S. R. and J. McNeill. 1988. Description and distribution of *Euphrasia stricta* in North America. Rhodora 90: 223–231.

Druggist's Circ. = Druggist's Circular.

Dudash, M. R., C. J. Murren, and D. E. Carr. 2005. Using *Mimulus* as a model system to understand the role of inbreeding in conservation and ecological approaches. Ann. Missouri Bot. Gard. 92: 36–51.

Duffield, W. J. 1972. Pollination ecology of *Castilleja* in Mount Rainier National Park. Ohio J. Sci. 72: 110–114.

Dumortier, B. C. J. 1827. Florula Belgica, Operis Majoris Prodromus.... Tournay. (Fl. Belg.)

Eastwood, A. 1909. Synopsis of the Mexican and Central American species of *Castilleja*. Contr. Gray Herb. 36: 563–591.

Eaton, A. and J. Wright. 1840. North American Botany..., ed. 8. Troy, N.Y. (Man. Bot. ed. 8)

Eclog. Bot.—See: A. L. Wirsing 1778

Ecology = Ecology, a Quarterly Journal Devoted to All Phases of Ecological Biology.

Econ. Bot. = Economic Botany; Devoted to Applied Botany and Plant Utilization.

Edinburgh New Philos. J. = Edinburgh New Philosophical Journal.

Edwards's Bot. Reg. = Edwards's Botanical Register....

Egger, J. M. 1994. New natural hybrid combinations and comments on interpretation of hybrid populations in *Castilleja* (Scrophulariaceae). Phytologia 77: 381–389.

Ehrhart, J. F. 1787–1792. Beiträge zur Naturkunde.... 7 vols. Hannover and Osnabrück. (Beitr. Naturk.)

Ehrlich, P. R. et al. 1975. Checkerspot butterflies: A historical perspective. Science, ser. 2, 188: 221–228.

Ejeta, G. and J. Gressel. 2007. Integrating New Technologies for *Striga* Control: Towards Ending the Witch-hunt. Singapore and Hackensack.

Elisens, W. J. 1985. Monograph of the Maurandyinae (Scrophulariaceae-Antirrhineae). Syst. Bot. Monogr. 5: 1–97.

Elisens, W. J. and A. D. Nelson. 1993. Morphological and isozyme divergence in *Gambelia* (Scrophulariaceae): Species delimitation and biogeographic relationships. Syst. Bot. 18: 454–468.

Elliott, S. [1816–]1821–1824. A Sketch of the Botany of South-Carolina and Georgia. 2 vols. in 13 parts. Charleston. (Sketch Bot. S. Carolina)

Emory, W. H. 1857–1859. Report on the United States and Mexican Boundary Survey, Made under the Direction of the Secretary of the Interior. 2 vols. in parts. Washington. (Rep. U.S. Mex. Bound.)

Encycl.—See: J. Lamarck et al. 1783–1817

Endlicher, S. L. 1836–1840[–1850]. Genera Plantarum Secundum Ordines Naturales Disposita. 18 parts + 5 suppls. in 6 parts. Vienna. [Paged consecutively through suppl. 1(2); suppls. 2–5 paged independently.] (Gen. Pl.)

Engl. Bot.—See: J. E. Smith et al. 1790–1866

Engler, H. G. A., ed. 1900–1953. Das Pflanzenreich.... 107 vols. Berlin. [Sequence of vol. (Heft) numbers (order of publication) is independent of the sequence of series and family (Roman and Arabic) numbers (taxonomic order).] (Pflanzenr.)

Engler, H. G. A. and K. Prantl, eds. 1887–1915. Die natürlichen Pflanzenfamilien.... 254 fascs. Leipzig. [Sequence of fasc. (Lieferung) numbers (order of publication) is independent of the sequence of division (Teil) and subdivision (Abteilung) numbers (taxonomic order).] (Nat. Pflanzenfam.)

Ennos, R. A. et al. 2012. Process-based species action plans: An approach to conserve contemporary evolutionary processes that sustain diversity in taxonomically complex groups. Bot. J. Linn. Soc. 168: 194–203.

Enum.—See: P. C. Fabricius 1759

Enum. Pl.—See: C. L. Willdenow 1809–1813[–1814]

Enum. Stirp. Vindob.—See: N. J. Jacquin 1762

Enum. Syst. Pl.—See: N. J. Jacquin 1760

Environment Canada. 2010. Recovery Strategy for the Furbish's Lousewort (*Pedicularis furbishiae*) in Canada. Ottawa.

Eplee, R. E. 1981. *Striga*'s status as a plant parasite in the United States. Pl. Dis. Reporter 65: 951–954.

Erbar, C. and P. Leins. 2004. Callitrichaceae. In: K. Kubitzki et al., eds. 1990+. The Families and Genera of Vascular Plants. 14+ vols. Berlin etc. Vol. 7, pp. 50–56.

Erythea = Erythea; a Journal of Botany, West American and General.

Estes, D. 2008. Systematics of *Gratiola* (Plantaginaceae). Ph.D. dissertation. University of Tennessee.

Estes, D. and R. L. Small. 2007. Two new spcies of *Gratiola* (Plantaginaceae) from eastern North America and an updated circumscription for *Gratiola neglecta*. J. Bot. Res. Inst. Texas 1: 149–170.

Estes, D. and R. L. Small. 2008. Phylogenetic relationships of the monotypic genus *Amphianthus* (Plantaginaceae tribe Gratioleae) inferred from chloroplast DNA sequences. Syst. Bot. 33: 176–182.

Every, A. D. 1977. Biosystematics of *Penstemon* Subgenus *Dasanthera*—A Naturally Hybridizing Species Complex. Ph.D. dissertation. University of Washington.

Evol. Biol. = Evolutionary Biology.

Evolution = Evolution; International Journal of Organic Evolution.

Excursionsfl. Schweiz ed. 5—See: A. Gremli 1885

Ezell, W. L. 1971. Biosystematics of the *Mimulus nanus* Complex in Oregon. Ph.D. dissertation. Oregon State University.

Fabricius, P. C. 1759. Enumeratio Methodica Plantarum Horti Medici Helmstadiensis.... Helmstedt. (Enum.)

Fassett, N. C. 1951b. *Callitriche* in the New World. Rhodora 53: 137–155, 161–182, 185–194, 209–222.

Feddes Repert. = Feddes Repertorium.

Feddes Repert. Spec. Nov. Regni Veg. = Feddes Repertorium Specierum Novarum Regni Vegetabilis.

Fedtschenko, B. A. and A. F. Flerow. 1910. Flora Evropeiskoi Rossii.... St. Petersburg.

Fenstermacher, J. 2006. *Seymeria falcata* (Scrophulariaceae), a new record for Texas and the United States. Sida 22: 811–812.

Fernald, M. L. 1905b. Some recently introduced weeds. Trans. Mass. Hort. Soc. 1905: 11–22.

Fernald, M. L. 1933. Recent discoveries in the Newfoundland flora. Rhodora 35: 1–16 etc.

Fernald, M. L. 1950. Gray's Manual of Botany, ed. 8. New York.

Fernald, M. L. and K. M. Wiegand. 1915. The genus *Euphrasia* in North America. Rhodora 17: 181–201.

Fernández-Mazuecos, M., J. L. Blanco-Pastor, and P. Vargas. 2013. A phylogeny of toadflaxes (*Linaria* Mill.) based on nuclear internal transcribed spacer sequences: Systematic and evolutionary consequences. Int. J. Pl. Sci. 174: 234–249.

Ferren, W. R. Jr. and A. E. Schuyler. 1980. Intertidal vascular plants of river systems near Philadelphia. Proc. Acad. Nat. Sci. Philadelphia 132: 86–120.

Fertig, W. 2001. Survey for Blowout Penstemon (*Penstemon haydenii*) in Wyoming. Laramie.

Field & Lab. = Field & Laboratory.

Fieldiana, Bot. = Fieldiana: Botany.

Fischer, E. 1997. A revision of the genus *Dopatrium* (Scrophulariaceae-Gratioloideae). Nordic J. Bot. 17: 527–555.

Fischer, E. 2004. Scrophulariaceae. In: K. Kubitzki et al., eds. 1990+. The Families and Genera of Vascular Plants. 14+ vols. Berlin etc. Vol. 7, pp. 333–432.

Fischer, E., B. Schäferhoff, and K. Müller. 2013. The phylogeny of Linderniaceae—The new genus *Linderniella*, and new combinations within *Bonnaya*, *Craterostigma*, *Lindernia*, *Micranthemum*, *Torenia* and *Vandellia*. Willdenowia 43: 209–238.

Fl. Aegypt.-Arab.—See: P. Forsskål 1775

Fl. Alaska Yukon—See: E. Hultén 1941–1950

Fl. Aleut. Isl.—See: E. Hultén 1937

Fl. Amer. Sept.—See: F. Pursh [1813]1814

Fl. Belg.—See: B. C. J. Dumortier 1827

Fl. Bor.-Amer.—See: A. Michaux 1803; W. J. Hooker [1829–] 1833–1840

Fl. Carniol.—See: J. A. Scopoli 1760

Fl. Carol.—See: T. Walter 1788

Fl. Cochinch.—See: J. de Loureiro 1790

Fl. Colorado—See: P. A. Rydberg 1906

Fl. Franç. ed. 3—See: J. Lamarck and A. P. de Candolle 1805–1815

Fl. Germ. Excurs.—See: H. G. L. Reichenbach 1830[–1832]

Fl. Ind.—See: W. Roxburgh 1820–1824

Fl. Indica—See: N. L. Burman 1768

Fl. Ins. Austr.—See: G. Forster 1786

Fl. Jap.—See: P. F. von Siebold and J. G. Zuccarini 1835 [–1870]

Fl. Kamtchatka—See: E. Hultén 1927–1930

Fl. Miami—See: J. K. Small 1913b

Fl. N.W. Amer.—See: T. J. Howell 1897–1903

Fl. Normandie ed. 2—See: L. A. de Brébisson 1849

Fl. Pedem.—See: C. Allioni 1785

Fl. Peruv. Prodr.—See: H. Ruiz López and J. A. Pavon 1794

Fl. Rocky Mts.—See: P. A. Rydberg 1917

Fl. Ross.—See: C. F. von Ledebour [1841]1842–1853

Fl. S.E. U.S.—See: J. K. Small 1903

Fl. Scand. Prodr. ed. 2—See: A. J. Retzius 1795

Fl. South. U.S.—See: A. W. Chapman 1860

Fl. South. U.S. ed. 2 repr. 2—See: A. W. Chapman 1892

Fl. W. Calif.—See: W. L. Jepson 1901

Fl. W. Calif. ed. 2—See: W. L. Jepson 1911

Flexner, S. B. and L. C. Hauck, eds. 1987. The Random House Dictionary of the English Language, ed. 2 unabridged. New York.

Flyr, L. D. 1970b. A Systematic Study of the Tribe Leucophylleae (Scrophulariaceae). Ph.D. dissertation. University of Texas.

Folia Geobot. = Folia Geobotanica; a Journal of Plant Ecology and Systematics.

Förh. Skand. Naturf. Mötet = Förhandlingar vid (ved) de Skandinaviska Naturforskare Mötet.

Fors. Oecon. Plantel. ed. 2—See: J. W. Hornemann 1806

Forsskål, P. 1775. Flora Aegyptiaco-Arabica. Copenhagen. (Fl. Aegypt.-Arab.)

Forster, G. 1786. Florulae Insularum Australium Prodromus. Göttingen. (Fl. Ins. Austr.)

Fosberg, F. R. 1962. Miscellaneous notes on Hawaiian plants—3. Occas. Pap. Bernice Pauahi Bishop Mus. 23: 29–44.

Fraga, N. S. 2012. A revision of Erythranthe montioides and Erythranthe palmeri (Phrymaceae), with descriptions of five new species from California and Nevada, USA. Aliso 30: 49–68.

Franz, E. 2007. From ornamental to detrimental? The incipient invasion of Central Europe by Paulownia tomentosa. Preslia 79: 377–389.

Freeman, C. C. 2015. Final Report on an Assessment of the Status of Blowout Beardtongue (Penstemon haydenii S. Watson, Plantaginaceae) Using Molecular and Morphometric Approaches. Lawrence, Kans.

Freeman, C. C. 2017. Nomenclatural novelties and notes in Penstemon (Plantaginaceae). PhytoKeys 80: 33–39.

Freeman, C. E. et al. 2003. Inferred phylogeny in Keckiella (Scrophulariaceae) based on noncoding chloroplast and nuclear ribosomal DNA sequences. Syst. Bot. 28: 782–790.

Freeman, C. E. and R. Scogin. 1999. Potential utility of chloroplast trnL (UAA) gene intron sequences for inferring phylogeny in Scrophulariaceae. Aliso 18: 141–159.

French, G. C. et al. 2008. Genetics, taxonomy and the conservation of British Euphrasia. Conservation Genet. 9: 1547–1562.

Fries, E. M. 1814–1824. Novitiae Florae Svecicae.... 7 parts. Lund. [Parts paged consecutively.] (Novit. Fl. Svec.)

Fuller, T. C. 1967. Weed pest detection and identification. Bull. Calif. Dept. Agric. 56(2): 99–101.

Furbish's Lousewort Recovery Team. 2006. Recovery Strategy for Furbish's Lousewort (Pedicularis furbishiae) in New Brunswick. Fredericton.

Gaertner, P. G., B. Meyer, and J. Scherbius. 1799–1802. Oekonomisch-technische Flora der Wetterau. 3 vols. Frankfurt am Main. [Vols. 1 and 2 paged independently, vol. 3 in 2 parts paged independently.] (Oekon. Fl. Wetterau)

Gándara, E. and V. Sosa. 2013. Testing the monophyly and position of the North American shrubby desert genus Leucophyllum (Scrophulariaceae: Leucophylleae). Bot. J. Linn. Soc. 171: 508–518.

Ganders, F. R. and G. R. Krause. 1986. Systematics of Collinsia parviflora and C. grandiflora (Scrophulariaceae). Madroño 33: 63–70.

Gard. & Forest = Garden and Forest; a Journal of Horticulture, Landscape Art and Forestry.

Gard. Dict. ed. 8—See: P. Miller 1768

Gard. Dict. Abr. ed. 4—See: P. Miller 1754

Garnock-Jones, P. J., D. C. Albach, and B. G. Briggs. 2007. Botanical names in Southern Hemisphere Veronica (Plantaginaceae): Sect. Detzneria, sect. Hebe, and sect. Labiatoides. Taxon 56: 571–582.

Garriss, H. R. and J. C. Wells. 1956. Parasitic herbaceous annual associated with corn disease in North Carolina. Pl. Dis. Reporter 40: 837–839.

Gartenflora = Gartenflora; Monatsschrift für deutsche und schweizerische Garten- und Blumenkunde.

Gayana, Bot. = Gayana. Botánica.

Gen. Hist.—See: G. Don 1831–1838

Gen. N. Amer. Pl.—See: T. Nuttall 1818

Gen. Pl.—See: S. L. Endlicher 1836–1840[–1850]

Gen. Pl. ed. 5—See: C. Linnaeus 1754

Gentry, J. L. et al. 2013. Atlas of the Vascular Plants of Arkansas. Fayetteville, Ark.

Ghebrehiwet, M. 2001. Taxonomy, phylogeny, and biogeography of Kickxia and Nanorrhinum (Scrophulariaceae). Nordic J. Bot. 20: 655–690.

Ghebrehiwet, M., B. Bremer, and M. Thulin. 2000. Phylogeny of the tribe Antirrhineae (Scrophulariaceae) based on morphological and ndhF sequence data. Pl. Syst. Evol. 220: 223–239.

Gilkey, H. M. 1945. Boschniakia in the western United States. Oregon State Monogr., Stud. Bot. 9: 7–15.

Glenne, G. 2003. Reproductive Biology, Hybridization Isolating Mechanisms, and Conservation Implications of Two Rare Subspecies of Penstemon bicolor (Brandeg.) Clokey and Keck: Ssp. bicolor and ssp. roseus Clokey and Keck (Scrophulariaceae) in Clark County, Nevada. M.S. thesis. Utah State University.

Gmelin, J. F. 1791[–1792]. Caroli à Linné...Systema Naturae per Regna Tria Naturae.... Tomus II. Editio Decima Tertia, Aucta, Reformata. 2 parts. Leipzig. (Syst. Nat.)

Gormley, I. C., D. Bedigian, and R. G. Olmstead. 2015. Phylogeny of Pedaliaceae and Martyniaceae and the placement of *Trapella* in Plantaginaceae. Syst. Bot. 40: 259–268.

Götte, R. 2007. Flora im östlichen Sauerland. Arnsberg.

Govaerts, R. 1995+. World Checklist of Seed Plants. 3+ vols. in parts. Antwerp. (World Checkl. Seed Pl.)

Grant, A. L. 1924. A monograph of the genus *Mimulus*. Ann. Missouri Bot. Gard. 11: 99–388.

Gray, A. 1880b. Revision of the genus *Castilleja*. Amer. J. Sci. Arts, ser. 2, 34: 335–339.

Gray, A. et al. 1886. Synoptical Flora of North America: The Gamopetalae, Being a Second Edition of Vol. i Part ii, and Vol. ii Part i, Collected. 2 vols. New York, London, and Leipzig. [Reissued 1888 as Smithsonian Misc. Collect. 591.] (Syn. Fl. N. Amer. ed. 2)

Gray, A., S. Watson, B. L. Robinson, et al. 1878–1897. Synoptical Flora of North America. 2 vols. in parts and fascs. New York etc. [Vol. 1(1,1), 1895; vol. 1(1,2), 1897; vol. 1(2), 1884; vol. 2(1), 1878.] (Syn. Fl. N. Amer.)

Greene, E. L. 1894. Manual of the Botany of the Region of San Francisco Bay.... San Francisco. (Man. Bot. San Francisco)

Greene, E. L. [1901.] Plantae Bakerianae. 3 vols. [Washington.] (Pl. Baker.)

Gregor, J. W. and J. M. S. Lang. 1950. Intra-colonial variation in plant size and habit in *Plantago maritima*. New Phytol. 49: 135–141.

Grelen, H. E. and W. F. Mann. 1973. Distribution of senna seymeria *(Seymeria cassioides)*. Econ. Bot. 27: 339–342.

Gremli, A. 1885. Excursionsflora für die Schweiz, ed. 5. Aarau. (Excursionsfl. Schweiz ed. 5)

Griffiths, G. C. D. 2002. A second species of eyebright (*Euphrasia*, Scrophulariaceae) in Alberta. Bot. Electronic News 299: 46.

Grisebach, A. H. R. 1843–1845[–1846]. Spicilegium Florae Rumelicae et Bithynicae.... 2 vols. in 6 parts. Braunschweig. [Vols. paged independently but parts numbered consecutively.] (Spic. Fl. Rumel.)

Gussarova, G. 2005. Synopsis of the genus *Euphrasia* L. (Scrophulariaceae) of Russia and adjacent states (within the limits of the former USSR). [In Russian.] Bot. Zhurn. (Moscow & Leningrad) 90: 1087–1115.

Gussarova, G. et al. 2008. Molecular phylogeny and biogeography of the bipolar *Euphrasia* (Orobanchaceae): Recent radiations in an old genus. Molec. Phylogen. Evol. 48: 444–460.

Haines, A. 2011. New England Wildflower Society's Flora Novae Angliae.... New Haven.

Handb. Fl. Ceylon—See: H. Trimen et al. 1893–1931

Harley, R. M. et al. 2004. Labiatae. In: K. Kubitzki et al., eds. 1990+. The Families and Genera of Vascular Plants. 14+ vols. Berlin etc. Vol. 7, pp. 167–275.

Harley, R. M., T. Reynolds, and S. Atkins, eds. 1992. Advances in Labiate Science. Kew.

Harvard Pap. Bot. = Harvard Papers in Botany.

Haston, E. M. et al. 2007. A linear sequence of Angiosperm Phylogeny Group II families. Taxon 56: 7–12.

Hatch, S. L., K. N. Gandhi, and L. E. Brown. 1990. Checklist of the Vascular Plants of Texas. College Station, Tex.

Hawk, J. L. and V. J. Tepedino. 2007. The effect of staminode removal on female reproductive success in a Wyoming population of the endangered blowout penstemon, *Penstemon haydenii* (Scrophulariaceae). Madroño 54: 22–26.

Haynes, R. R. 1971. A monograph of *Conopholis*. Sida 4: 246–264.

Hays, J. F. 1998. *Agalinis* (Scrophulariaceae) in the Ozark Highlands. Sida 18: 555–577.

Hays, J. F. 2002. *Agalinis* (Scrophulariaceae) of the East Gulf Coastal Plain. M.S. thesis. University of Louisiana, Monroe.

Heckard, L. R. 1962. Root parasitism in *Castilleja*. Bot. Gaz. 124: 21–29.

Heckard, L. R. 1968. Chromosome numbers and polyploidy in *Castilleja* (Scrophulariaceae). Brittonia 20: 212–226.

Heckard, L. R. 1973. A taxonomic reinterpretation of the *Orobanche californica* complex. Madroño 22: 41–70.

Heckard, L. R. and T. I. Chuang. 1975. Chromosome numbers and polyploidy in *Orobanche* (Orobanchaceae). Brittonia 27: 179–186.

Heckard, L. R. and T. I. Chuang. 1977. Chromosome numbers, polyploidy, and hybridization in *Castilleja* (Scrophulariaceae) of the Great Basin and Rocky Mountains. Brittonia 29: 159–172.

Heckard, L. R. and L. T. Collins. 1982. Taxonomy and distribution of *Orobanche valida* (Orobanchaceae). Madroño 29: 95–100.

Heckard, L. R. and P. Rubtzoff. 1977. Additional notes on *Veronica anagallis-aquatica* × *catenata* (Scrophulariaceae). Rhodora 79: 579–582.

Hedglin, F. L. 1959. A Survey of the Genus *Synthyris*. M.S. thesis. University of Washington.

Hegelmaier, C. F. 1864. Monographie der Gattung *Callitriche*. Stuttgart.

Hegi, G. et al. 1936–1987. Illustrierte Flora von Mitteleuropa, ed. 2. 6 vols. in 14. Munich, Berlin, and Hamburg. (Ill. Fl. Mitt.-Eur. ed. 2)

Heidel, B. L. 2009. Survey and Monitoring of *Penstemon gibbensii* (Gibbens' Beardtongue) in South-central Wyoming. Laramie.

Heidel, B. L. and J. Handley. 2007. *Penstemon laricifolius* Hook. & Arn. Ssp. *exilifolius* (A. Nels.) D. D. Keck (Larchleaf Beardtongue): A Technical Conservation Assessment. Laramie.

Heinemann, D. 1992. Resource use, energetic profitability, and behavioral decisions in migrant rufous hummingbirds. Oecologia 90: 137–149.

Heinrich, B. 1975. Bee flowers: A hypothesis on flower variety and blooming times. Evolution 29: 325–334.

Heitz, E. 1927. Ueber multiple und aberrante Chromosomenzahlen. Abh. Naturwiss. Naturwiss. Verein Hamburg 21: 47–57.

Heller, A. A. 1898. Catalogue of North American Plants North of Mexico, Exclusive of the Lower Cryptogams. [Lancaster, Pa.] (Cat. N. Amer. Pl.)

Hemsley, W. B. 1879–1888. Biologia Centrali-Americana.... Botany.... 5 vols. London. (Biol. Cent.-Amer., Bot.)

Henrickson, J. 1989. A new species of *Leucospora* (Scrophulariaceae) from the Chihuahuan Desert of Mexico. Aliso 12: 435–439.

Henrickson, J. and L. D. Flyr. 1985. Systematics of *Leucophyllum* and *Eremogeton* (Scrophulariaceae). Sida 11: 107–172.

Hepper, F. N. 1974. Parasitic witchweed: *Striga asiatica* versus *S. lutea* (Scrophulariaceae). Rhodora 76: 45–47.

Herbaugh, L., N. Upton, and R. E. Eplee. 1980. *Striga gesnerioides* in the United States of America. Proc. S. Weed Sci. Soc. 33: 187–190.

Heredity = Heredity; an International Journal of Genetics.

Hersch-Green, E. I. 2012. Polyploidy in Indian paintbrush (*Castilleja*; Orobanchaceae) species shapes but does not prevent gene flow across species boundaries. Amer. J. Bot. 99: 1680–1690.

Hesperus = Hesperus; oder Belehrung und Unterhaltung für die Bewohner des Oesterreichischen Kaiserstaates....

Heyneman, A. J. et al. 1991. Host plant discrimination: Experiments with hummingbird flower mites. In: P. W. Price et al., eds. 1991. Plant-Animal Interactions: Evolutionary Ecology in Tropical and Temperate Regions. New York. Pp. 455–485.

Hickman, J. C., ed. 1993. The Jepson Manual. Higher Plants of California. Berkeley, Los Angeles, and London.

Hileman, L. C., E. M. Kramer, and D. A. Baum. 2003. Differential regulation of symmetry genes and the evolution of floral morphologies. Proc. Natl. Acad. Sci. U.S.A. 100: 12814–12819.

Hilgardia = Hilgardia; a Journal of Agricultural Science.

Hill, J. 1756[–1757]. The British Herbal: An History of Plants and Trees.... 52 fascs. London. [Fascicles paged and plates numbered consecutively.] (Brit. Herb.)

Hist. Nat. Vég.—See: É. Spach 1834–1848

Hist. Pl. Guiane—See: J. B. Aublet 1775

Hitchcock, C. L. et al. 1955–1969. Vascular Plants of the Pacific Northwest. 5 vols. Seattle. [Univ. Wash. Publ. Biol. 17.] (Vasc. Pl. Pacif. N.W.)

Hoggard, R. K. et al. 2003. Molecular systematics and biogeography of the amphibious genus *Littorella* (Plantaginaceae). Amer. J. Bot. 90: 429–435.

Holmgren, N. H. 1971. A taxonomic revision of the *Castilleja viscidula* group. Mem. New York Bot. Gard. 21: 1–63.

Holmgren, N. H. 1979b. New penstemons (Scrophulariaceae) from the Intermountain Region. Brittonia 31: 217–242.

Holmgren, N. H. 1984. Scrophulariaceae. In: A. Cronquist et al. 1972–2017. Intermountain Flora. Vascular Plants of the Intermountain West, U.S.A. 7 vols. in 9. New York and London. Vol. 4, pp. 344–506.

Holmgren, N. H. 1993. *Penstemon*. In: J. C. Hickman, ed. 1993. The Jepson Manual. Higher Plants of California. Berkeley, Los Angeles, and London. Pp. 1050–1062.

Holmgren, N. H. 2017. *Penstemon*—Update to 1984 treatment in Intermountain Flora volume 4. In: A. Cronquist et al. 1972–2017. Intermountain Flora. Vascular Plants of the Intermountain West, U.S.A. 7 vols. in 9. New York and London. Vol. 7, pp. 161–197.

Holub, J. 1990. Some taxonomic and nomenclatural changes within *Orobanche* s.l. (Orobanchaceae). Preslia 62: 193–198.

Hong, D. Y. et al. 1998. Scrophulariaceae. In: Wu Z. and P. H. Raven, eds. 1994–2013. Flora of China. 25 vols. Beijing and St. Louis. Vol. 18, pp. 1–212.

Hooker, W. J. [1829–]1833–1840. Flora Boreali-Americana; or, the Botany of the Northern Parts of British America.... 2 vols. in 12 parts. London, Paris, and Strasbourg. (Fl. Bor.-Amer.)

Hooker, W. J. and G. A. W. Arnott. [1830–]1841. The Botany of Captain Beechey's Voyage; Comprising an Account of the Plants Collected by Messrs Lay and Collie, and Other Officers of the Expedition, during the Voyage to the Pacific and Bering's Strait, Performed in His Majesty's Ship Blossom, under the Command of Captain F. W. Beechey...in the Years 1825, 26, 27, and 28. 10 parts. London. [Parts paged and plates numbered consecutively.] (Bot. Beechey Voy.)

Hooker's J. Bot. Kew Gard. Misc. = Hooker's Journal of Botany and Kew Garden Miscellany.

Hoover, R. F. 1970. The Vascular Plants of San Luis Obispo County, California. Berkeley.

Hoppe, D. H. 1790–1849. Botanisches Taschenbuch.... 23 vols. Regensburg, Nuremburg, and Altdorf. [Vols. numbered by year, 1790–1811, and 1849, the latter edited by A. E. Fürnrohr.] (Bot. Taschenb.)

Hornemann, J. W. 1806. Forsøg til en Dansk Oeconomisk Plantelaere, ed. 2. Copenhagen. (Fors. Oecon. Plantel. ed. 2)

Hort. Berol.—See: C. L. Willdenow 1803–1816

Hort. Kew.—See: W. Aiton 1789

Howell, T. J. 1897–1903. A Flora of Northwest America. 1 vol. in 8 fascs. Portland. [Fascs. 1–7 (text) paged consecutively, fasc. 8 (index) independently.] (Fl. N.W. Amer.)

Hoyt, C. A. and J. Karges. 2014. Proceedings of the Sixth Symposium on the Natural Resources of the Chihuahuan Desert Region, October 14–17, 2004. Fort Davis.

Hrusa, F. et al. 2002. Catalogue of non-native vascular plants occurring spontaneously in California beyond those addressed in The Jepson Manual—Part I. Madroño 49: 61–98.

Hu, S. Y. 1959. A monograph of the genus *Paulownia*. Quart. J. Taiwan Mus. 12: 1–54.

Hu, S. Y. 1961. The economic botany of the paulownias. Econ. Bot. 15: 11–27.

Huber-Morath, A. 1973. *Verbascum* L. s.l. (incl. *Celsia* L. et *Staurophragma* Fisch. & Mey.). Bauhinia 5: 7–16.

Huber-Morath, A. 1978. *Verbascum*. In: P. H. Davis, ed. 1965–2000. Flora of Turkey and the East Aegean Islands. 11 vols. Edinburgh. Vol. 6, pp. 461–603.

Hufford, L. 1992. Floral structure of *Besseya* and *Synthyris* (Scrophulariaceae). Int. J. Pl. Sci. 153: 217–229.

Hufford, L. 1992b. Leaf structure of *Besseya* and *Synthyris* (Scrophulariaceae). Canad. J. Bot. 70: 921–932.

Hufford, L. and M. McMahon. 2004. Morphological evolution and systematics of *Synthyris* and *Besseya* (Veroniceae): A phylogenetic analysis. Syst. Bot. 29: 716–736.

Hultén, E. 1927–1930. Flora of Kamtchatka and the Adjacent Islands.... 4 vols. Stockholm. [Vols. designated as numbers of Kongl. Svenska Vetenskapsakad. Handl., ser. 3.] (Fl. Kamtchatka)

Hultén, E. 1937. Flora of the Aleutian Islands and Westernmost Alaska Peninsula with Notes on the Flora of Commander Islands.... Stockholm. (Fl. Aleut. Isl.)

Hultén, E. 1941–1950. Flora of Alaska and Yukon. 10 vols. Lund and Leipzig. [Vols. paged consecutively and designated as simultaneous numbers of Lunds Univ. Årsskr. (= Acta Univ. Lund.) and Kungl. Fysiogr. Sällsk. Handl.] (Fl. Alaska Yukon)

Hultén, E. 1967. Comments on the flora of Alaska and Yukon. Ark. Bot., n. s. 7: 1–147.

Hultén, E. 1968. Flora of Alaska and Neighboring Territories: A Manual of the Vascular Plants. Stanford.

Hultén, E. 1973. Supplement to Flora of Alaska and Neighboring Territories. A study in the flora of Alaska and the transberingian connections. Bot. Not. 126: 459–512.

Hultén, E. and M. Fries. 1986. Atlas of North European Vascular Plants North of the Tropic of Cancer. 3 vols. Königstein.

Humboldt, A. von and A. J. Bonpland. [1805–]1808–1809 [–1817]. Plantae Aequinoctiales.... 2 vols. in 17 parts. Paris and Tübingen. [Parts numbered consecutively.] (Pl. Aequinoct.)

Humboldt, A. von, A. J. Bonpland, and C. S. Kunth. 1815[1816]–1825. Nova Genera et Species Plantarum Quas in Peregrinatione Orbis Novi Collegerunt, Descripserunt.... 7 vols. in 36 parts. Paris. (Nov. Gen. Sp.)

Hutchings' Ill. Calif. Mag. = Hutchings' Illustrated California Magazine.

Icon.—See: A. J. Cavanilles 1791–1801

Icon. Pl. ed. Keller—See: C. C. Schmidel 1762–1771

Icon. Pl. Rar.—See: N. J. Jacquin 1781–1793[–1795]

Iconogr. Bot. Pl. Crit.—See: H. G. L. Reichenbach 1823–1832

Ideker, J. 1996b. *Capraria mexicana* (Scrophulariaceae), an endangered addition to the United States flora. Sida 17: 523–526.

Ihlenfeldt, H.-D. 1967. Über die Abgrenzung und die natürliche Gliederung der Pedaliaceae R. Br. Mitt. Staatsinst. Allg. Bot. Hamburg 12: 43–128.

Ihlenfeldt, H.-D. 2004. Pedaliaceae. In: K. Kubitzki et al., eds. 1990+. The Families and Genera of Vascular Plants. 14+ vols. Berlin etc. Vol. 7, pp. 307–322.

Ihlenfeldt, H.-D. and U. Grabow-Seidensticker. 1979. The genus *Sesamum* and the origin of cultivated sesame. In: K. Kunkel, ed. 1979. Taxonomic Aspects of African Economic Botany. Las Palmas de Gran Canaria. Pp. 53–60.

Ill. Fl. Mitt.-Eur. ed. 2—See: G. Hegi et al. 1936–1987

Ill. Fl. N. U.S. ed. 2—See: N. L. Britton and A. Brown 1913

Ill. Fl. Pacific States—See: L. Abrams and R. S. Ferris 1923–1960

Ill. Hort. = L'Illustration Horticole....

Index Seminum (Göttingen) = Index Seminum Horti Academici Goettingensis Anno...Collecta.

Index Seminum (St. Petersburg) = Index Seminum, Quae Hortus Botanicus Imperialis Petropolitanus pro Mutua Commutatione Offert.

Inst. Bot.—See: V. Petagna 1785–1787

Inst. Regn. Veg. ed. 2—See: C. G. Ludwig 1757

Inst. Rei Herb.—See: H. J. N. von Crantz 1766

Int. J. Pl. Sci. = International Journal of Plant Sciences.

Intermount. Fl.—See: A. Cronquist et al. 1972–2017

Invasive Pl. Sci. Managem. = Invasive Plant Science and Management.

Irving, L. J. and D. D. Cameron. 2009. You are what you eat: Interactions between root parasite plants and their hosts. Advances Bot. Res. 50: 87–138.

Iverson, R. et al. 2011. Overview and status of the witchweed (*Striga asiatica*) eradication program in the Carolinas, 2011. In: A. Leslie and R. Westbrooks, eds. 2011. Invasive Plant Management Issues and Challenges in the United States: 2011 Overview. Washington. Pp. 51–68.

Ives, J. C. 1861. Report upon the Colorado River of the West, Explored in 1857 and 1858 by Lieutenant Joseph C. Ives.... 5 parts, appendices. Washington. (Rep. Colorado R.)

J. Acad. Nat. Sci. Philadelphia = Journal of the Academy of Natural Sciences of Philadelphia.

J. Adelaide Bot. Gard. = Journal of the Adelaide Botanic Gardens.

J. Altern. Complement. Med. = Journal of Alternative and Complementary Medicine.

J. Arnold Arbor. = Journal of the Arnold Arboretum.

J. Bot. = Journal of Botany, British and Foreign.

J. Bot. Res. Inst. Texas = Journal of the Botanical Research Institute of Texas.

J. Chem. Ecol. = Journal of Chemical Ecology.

J. Elisha Mitchell Sci. Soc. = Journal of the Elisha Mitchell Scientific Society.

J. Ethnopharmacol. = Journal of Ethnopharmacology; Interdisciplinary Journal Devoted to Bioscientific Research on Indigenous Drugs.

J. Evol. Biol. = Journal of Evolutionary Biology.

J. Hort. Soc. London = Journal of the Horticultural Society of London.

J. Huazhong Agric. Univ. = Journal of Huazhong (Central China) Agricultural University.

J. Jap. Bot. = Journal of Japanese Botany.

J. Linn. Soc., Bot. = Journal of the Linnean Society. Botany.

J. Phys. Chim. Hist. Nat. Arts = Journal de Physique, de Chimie, d'Histoire Naturelle et des Arts.

J. Pl. Res. = Journal of Plant Research. [Shokubutsu-gaku zasshi.]

J. Roy. Microscop. Soc. London = Journal of the Royal Microscopical Society of London....

J. Torrey Bot. Soc. = Journal of the Torrey Botanical Society.

J. Wash. Acad. Sci. = Journal of the Washington Academy of Sciences.

Jacobs, S. J. et al. 2018. Incongruence in molecular species delimitation schemes: What to do when adding more data is difficult. Molec. Ecol. 27: 2397–2413.

Jacobsen, A. L., F. C. Weinmann, and P. F. Zika. 2001. Noteworthy collections in Washington. Madroño 48: 213–214.

Jacquin, N. J. 1760. Enumeratio Systematica Plantarum, Quas in Insulis Caribaeis Vicinaque Americes Continente Detexit Novas.... Leiden. (Enum. Syst. Pl.)

Jacquin, N. J. 1762. Enumeratio Stirpium Plerarumque, Quae Sponte Crescunt in Agro Vindobonensi, Montibusque Confinibus. Vienna. (Enum. Stirp. Vindob.)

Jacquin, N. J. 1781–1793[–1795]. Icones Plantarum Rariorum. 3 vols. in fascs. Vienna etc. [Vols. paged independently, plates numbered consecutively.] (Icon. Pl. Rar.)

Jahrb. Gewächsk. = Jahrbücher der Gewächskunde.

Jepson, W. L. 1901. A Flora of Western Middle California.... Berkeley. (Fl. W. Calif.)

Jepson, W. L. 1911. A Flora of Western Middle California..., ed. 2. San Francisco. (Fl. W. Calif. ed. 2)

Jepson, W. L. [1923–1925.] A Manual of the Flowering Plants of California.... Berkeley. (Man. Fl. Pl. Calif.)

Jie, C. et al. 2004. Karyotypes of thirteen species of *Pedicularis* (Orobanchaceae) from the Hengduan Mountains region, NW Yunnan, China. Caryologia 57: 337–347.

Jiménez, M. et al. 1995. Iridoid glycoside constituents of *Castilleja tenuiflora*. Biochem. Syst. & Ecol. 23: 455–456.

Johnson, R. L. et al. 2016. Molecular and morphological evidence for *Penstemon luculentus* (Plantaginaceae): A replacement name for *Penstemon fremontii* var. *glabrescens*. PhytoKeys 63: 47–62.

Johnston, A. 1987. Plants and the Blackfoot. Lethbridge, Alta.

Jones, G. N. 1936. A Botanical Survey of the Olympic Peninsula. Seattle. [Univ. Wash. Publ. Biol. 5.] (Bot. Surv. Olympic Penins.)

Jones, S. D., J. K. Wipff, and P. M. Montgomery. 1997. Vascular Plants of Texas: A Comprehensive Checklist Including Synonymy, Bibliography, and Index. Austin.

Judd, W. S. and R. G. Olmstead. 2004. A survey of tricolpate (eudicot) phylogeny. Amer. J. Bot. 91: 1627–1644.

Junghans, T. and E. Fischer. 2007. Aspects of dispersal in *Cymbalaria muralis* (Scrophulariaceae). Bot. Jahrb. Syst. 127: 289–298.

Kalmiopsis = Kalmiopsis; Journal of the Native Plant Society of Oregon.

Karaveliogullari, F. A. and Z. Aytac. 2008. Revision of the genus *Verbascum* L. (Group A) in Turkey. Bot. Res. J. 1: 9–32.

Karlsson, T. 1974. Recurrent ecotypic variation in Rhinantheae and Gentianaceae in relation to hemiparasitism and mycotrophy. Bot. Not. 127: 527–539.

Karlsson, T. 1986. The evolutionary situation of *Euphrasia* in Sweden. Symb. Bot. Upsal. 27: 61–71.

Karrfalt, E. E. and A. S. Tomb. 1983. Air spaces, secretory cavities, and the relationship between Leucophylleae (Scrophulariaceae) and Myoporaceae. Syst. Bot. 8: 29–32.

Kartesz, J. T. 1987. A Flora of Nevada. Ph.D. thesis. 3 vols. University of Nevada, Reno.

Kasselmann, C. 2003. Aquarium Plants. Melbourne, Fla.

Kaul, R. B., D. Sutherland, and S. B. Rolfsmeier. 2006. The Flora of Nebraska. Lincoln.

Kearney, T. H. and R. H. Peebles. 1960. Arizona Flora, ed. 2. Berkeley.

Keck, D. D. 1927. A revision of the genus *Orthocarpus*. Proc. Calif. Acad. Sci., ser. 4, 16: 517–571.

Keck, D. D. 1932. Studies in *Penstemon*. A systematic treatment of section *Saccanthera*. Univ. Calif. Publ. Bot. 16: 367–426.

Keck, D. D. 1936. Studies in *Penstemon*. II. The section *Hesperothamnus*. Madroño 3: 200–219.

Keck, D. D. 1936b. Studies in *Penstemon* III. Madroño 3: 248–250.

Keck, D. D. 1937. Studies in *Penstemon* IV. The section *Ericopsis*. Bull. Torrey Bot. Club 64: 357–381.

Keck, D. D. 1937b. Studies in *Penstemon* V. The section *Peltanthera*. Amer. Midl. Naturalist 18: 790–829.

Keck, D. D. 1938. Studies in *Penstemon* VI. The section *Aurator*. Bull. Torrey Bot. Club 65: 233–255.

Keck, D. D. 1940b. Studies in *Penstemon* VII. The subsections *Gairdneriana*, *Deusti*, and *Arenarii* of the *Graciles*, and miscellaneous new species. Amer. Midl. Naturalist 23: 594–616.

Keck, D. D. 1945. Studies in *Penstemon* VIII. A cytotaxonomic account of the section *Spermunculus*. Amer. Midl. Naturalist 33: 128–206.

Keck, D. D. 1951. *Penstemon*. In: L. Abrams and R. S. Ferris. 1923–1960. Illustrated Flora of the Pacific States: Washington, Oregon, and California. 4 vols. Stanford. Vol. 3, pp. 733–770.

Keck, D. D. and A. Cronquist. 1957. Studies in *Penstemon* IX. Notes on northwestern American species. Brittonia 8: 247–250.

Kelchner, S. A. 2001. The Caribbean *Bontia daphnoides* and its Australian family Myoporaceae (Lamiales): Evidence of an extreme dispersal event from morphological data and *rpl*16 intron sequences. In: Botanical Society of America. 2001. Botany 2001 Abstracts. Columbus. P. 92.

Kew Bull. = Kew Bulletin.

Kiger, R. W. and D. M. Porter. 2001. Categorical Glossary for the Flora of North America Project. Pittsburgh.

Kim, K. J. and S. M. Yun. 2012. A new species of *Melampyrum* (Orobanchaceae) from southern Korea. Phytotaxa 42: 48–50.

Kindscher, K. 1992. Medicinal Wild Plants of the Prairie: An Ethnobotanical Guide. Lawrence, Kans.

Kirkman, L. K., M. B. Drew, and D. Edwards. 1998. Effects of experimental fire regimes on the population dynamics of *Schwalbea americana* L. Pl. Ecol. 137: 115–137.

Knapp, W. M. et al. 2011. Floristic discoveries in Delaware, Maryland, and Virginia. Phytoneuron 2011-64: 1–26.

Knapp, W. M. and D. Estes. 2006. *Gratiola brevifolia* (Plantaginaceae) new to the flora of Delaware, the Delmarva Peninsula, and the mid-Atlantic. Sida 22: 825–829.

Knowlton, C. H. and W. Deane. 1923. Reports on the flora of the Boston district—XL. Rhodora 25: 60–67.

Koch, W. D. J. [1835–]1837–1838. Synopsis Florae Germanicae et Helveticae.... 1 vol. in 2 sects. and index. Frankfurt am Main. [Index paged independently.] (Syn. Fl. Germ. Helv.)

Koelling, A. C. 1964. Taxonomic Studies in *Penstemon deamii* and Its Allies. Ph.D. dissertation. University of Illinois.

Kongl. Vetensk. Acad. Handl. = Kongl[iga]. Vetenskaps Academiens Handlingar.

Kornhall, P. and B. Bremer. 2004. New circumscription of the tribe Limoselleae (Scrophulariaceae) that includes taxa of the tribe Manuleae. Bot. J. Linn. Soc. 146: 453–467.

Kornhall, P., N. Heidari, and B. Bremer. 2001. Selagineae and Manuleeae, two tribes or one? Phylogenetic studies in the Scrophulariaceae. Pl. Syst. Evol. 228: 199–218.

Kral, R. 1955. Populations of *Linaria* (Scrophulariaceae) in northeastern Texas. Field & Lab. 23: 74–77.

Kreutz, C. A. J. 1995. *Orobanche*: Die Sommerwurzarten Europas.... 1. Mittel- und Nordeuropa. Maastricht.

Kruckeberg, A. R. and F. L. Hedglin. 1963. Natural and artificial hybrids of *Besseya* and *Synthyris* (Scrophulariaceae). Madroño 17: 109–115.

Kubitzki, K. et al., eds. 1990+. The Families and Genera of Vascular Plants. 14+ vols. Berlin etc.

Kuiper, P. J. C. and M. Bos, eds. 1992. *Plantago:* A Multidisciplinary Study. Berlin.

Kunkel, K., ed. 1979. Taxonomic Aspects of African Economic Botany. Las Palmas de Gran Canaria.

Kuntze, O. 1891–1898. Revisio Generum Plantarum Vascularium Omnium atque Cellularium Multarum.... 3 vols. Leipzig etc. [Vol. 3 in 3 parts paged independently; parts 1 and 3 unnumbered.] (Revis. Gen. Pl.)

L'Empereur, K. M. and F. R. Stermitz. 1990. Iridoid glycoside content of *Euphydryas anicia* (Lepidoptera: Nymphalidae) and its major host plant, *Besseya plantaginea* (Scrophulariaceae), at a high plains Colorado site. J. Chem. Ecol. 16: 187–197.

La Cour, L. 1931. Improvements in everyday technique in plant cytology. J. Roy. Microscop. Soc. London 51: 119–126.

Lamarck, J. et al. 1783–1817. Encyclopédie Méthodique. Botanique.... 13 vols. Paris and Liège. [Vols. 1–8, suppls. 1–5.] (Encycl.)

Lamarck, J. and A. P. de Candolle. 1805–1815. Flore Française, ou Descriptions Succinctes de Toutes les Plantes Qui Croissent Naturellement en France..., ed. 3. 5 tomes in 6 vols. Paris. [Tomes 1–4(2), vols. 1–5, 1805; tome 5, vol. 6, 1815.] (Fl. Franç. ed. 3)

Lamarck, J. and J. Poiret. 1791–1823. Tableau Encyclopédique et Méthodique des Trois Règnes de la Nature. Botanique.... 6 vols. Paris. [Vols. 1–2 = tome 1; vols. 3–5 = tome 2; vol. [6] = tome 3. Vols. paged consecutively within tomes.] (Tabl. Encycl.)

Lange, R. S. and P. E. Scott. 1999. Hummingbird and bee pollination of *Penstemon pseudospectabilis*. J. Torrey Bot. Soc. 126: 99–106.

Lange, W. H. 1950. Biology and systematics of plume moths of the genus *Platyptilia* in California. Hilgardia 19: 561–668.

Lansdown, R. V. 2006. Notes on the water-starworts (*Callitriche*) recorded in Europe. Watsonia 26: 105–120.

Lansdown, R. V. and M. J. M. Christenhusz. 2011. *Callitriche*. In: G. Davidse et al., eds. 1994+. Flora Mesoamericana. 5+ vols. in parts. Mexico City, St. Louis, and London. Vol. 5(1), pp. 18–22.

Lansdown, R. V. and C. E. Jarvis. 2004. Linnaean names in *Callitriche* L. (Callitrichaceae) and their typification. Taxon 53: 169–172.

Latady, M. K. W. 1985. A Systematic Revision of *Penstemon glaber* and *P. alpinus*. M.S. thesis. University of Wyoming.

Lawson, H. R., V. J. Tepedino, and T. L. Griswold. 1989. Pollen collectors and other insect visitors to *Penstemon haydenii* S. Wats. In: T. B. Bragg and J. Stubbendieck, eds. 1989. Proceedings of the Eleventh North American Prairie Conference. Prairie Pioneers: Ecology, History and Culture. Lincoln, Nebr. Pp. 233–235.

Leafl. Bot. Observ. Crit. = Leaflets of Botanical Observation and Criticism.

Leafl. W. Bot. = Leaflets of Western Botany.

Ledebour, C. F. von. [1841]1842–1853. Flora Rossica sive Enumeratio Plantarum in Totius Imperii Rossici Provinciis Europaeis, Asiaticis, et Americanis Hucusque Observatarum.... 4 vols. Stuttgart. (Fl. Ross.)

Lee, N. S. et al. 1996. Molecular divergence between disjunct taxa in eastern Asia and eastern North America. Amer. J. Bot. 83: 1373–1378.

Leeuwenberg, A. J. M. 1979. The Loganiaceae of Africa. XVIII. *Buddleja* II. Revision of the African and Asiatic species. Meded. Landbouwhoogeschool 79: 1–163.

Lehmann, J. F. 1809. Primae Lineae Florae Herbipolensis.... Herbipoli. (Prim. Lin. Fl. Herbipol.)

Lehnhoff, E. A. 2008. Invasiveness of Yellow Toadflax (*Linaria vulgaris*) Resulting from Disturbance and Environmental Conditions. Ph.D. thesis. Montana State University.

Leins, P. and C. Erbar. 2004. Hippuridaceae. In: K. Kubitzki et al., eds. 1990+. The Families and Genera of Vascular Plants. 14+ vols. Berlin etc. Vol. 7, pp. 163–166.

Lekberg, Y. et al. 2012. Phenotypic and genetic differentiation among yellow monkeyflower populations from thermal and non-thermal soils in Yellowstone National Park. Oecologia 170: 111–122.

Les, D. H., R. S. Capers, and N. P. Tippery. 2006. Introduction of *Glossostigma* (Phrymaceae) to North America: A taxonomic and ecological overview. Amer. J. Bot. 93: 927–939.

Les, D. H. and R. L. Stuckey. 1985. The introduction and spread of *Veronica beccabunga* (Scrophulariaceae) in eastern North America. Rhodora 87: 503–515.

Leslie, A. and R. Westbrooks, eds. 2011. Invasive Plant Management Issues and Challenges in the United States: 2011 Overview. Washington.

Lewis, D. Q. 2000. A revision of the New World species of *Lindernia* (Scrophulariaceae). Castanea 65: 93–122.

Li, Bo et al. 2016. A large-scale chloroplast phylogeny of the Lamiaceae sheds new light on its subfamilial classification. Sci. Rep. 6: 34343.

Li, H. L. 1948. A revision of the genus *Pedicularis* in China, part I. Proc. Acad. Nat. Sci. Philadelphia 100: 205–378.

Liang, Z. Y. and Chen Z. Y. 1997. Studies of the cytological taxonomy of the genus *Paulownia*. J. Huazhong Agric. Univ. 16: 609–613.

Lindgren, D. and E. Wilde. 2003. Growing Penstemons: Species, Cultivars, and Hybrids. Haverford.

Lindley, J. 1836. A Natural System of Botany.... London. (Nat. Syst. Bot.)

Linnaea = Linnaea; ein Journal für die Botanik in ihrem ganzen Umfange.

Linnaeus, C. 1737[1738]. Hortus Cliffortianus.... Amsterdam.

Linnaeus, C. 1749[–1769]. Amoenitates Academicae seu Dissertationes Variae Physicae, Medicae Botanicae.... 7 vols. Stockholm and Leipzig. (Amoen. Acad.)

Linnaeus, C. 1753. Species Plantarum.... 2 vols. Stockholm. (Sp. Pl.)

Linnaeus, C. 1754. Genera Plantarum..., ed. 5. Stockholm. (Gen. Pl. ed. 5)

Linnaeus, C. 1755. Centuria I. Plantarum.... Uppsala. (Cent. Pl. I)

Linnaeus, C. [1756.] Centuria II. Plantarum.... Uppsala. (Cent. Pl. II)

Linnaeus, C. 1758[–1759]. Systema Naturae per Regna Tria Naturae..., ed. 10. 2 vols. Stockholm. (Syst. Nat. ed. 10)

Linnaeus, C. 1762–1763. Species Plantarum..., ed. 2. 2 vols. Stockholm. (Sp. Pl. ed. 2)

Linnaeus, C. 1766–1768. Systema Naturae per Regna Tria Naturae..., ed. 12. 3 vols. Stockholm. (Syst. Nat. ed. 12)

Linnaeus, C. 1767[–1771]. Mantissa Plantarum. 2 parts. Stockholm. [Mantissa [1] and Mantissa [2] Altera paged consecutively.] (Mant. Pl.)

Linnaeus, C. f. 1781[1782]. Supplementum Plantarum Systematis Vegetabilium Editionis Decimae Tertiae, Generum Plantarum Editionis Sextae, et Specierum Plantarum Editionis Secundae. Braunschweig. (Suppl. Pl.)

Little, R. J. 1980. Floral Mimicry between Two Desert Annuals: *Mohavea confertiflora* (Scrophulariaceae) and *Mentzelia involucrata* (Loasaceae). Ph.D. thesis. Claremont Graduate School.

Loureiro, J. de. 1790. Flora Cochinchinensis.... 2 vols. Lisbon. [Vols. paged consecutively.] (Fl. Cochinch.)

Lourteig, A. 1964. Étude sur *Limosella* L. Publ. Comité Natl. Franç. Rech. Antarct. Biol. 1: 165–173.

Lowry, D. B. and J. H. Willis. 2010. A widespread chromosomal inversion polymorphism contributes to a major life-history transition, local adaptation, and reproductive isolation. PLOS Biol. 8(9): e1000500. doi:10.1371/journal.pbio.1000500.

Ludwig, C. G. 1757. Institutiones Historico-physicae Regni Vegetabilis..., ed. 2. Leipzig. (Inst. Regn. Veg. ed. 2)

Luna, T. 2005. Propagation protocol for Indian paintbrush, *Castilleja* species. Native Pl. J. 6: 63–68.

Lundellia = Lundellia; Journal of the Plant Resources Center of the University of Texas at Austin.

Macior, L. W. 1969. Pollination adaptation in *Pedicularis lanceolata*. Amer. J. Bot. 56: 853–859.

Macior, L. W. 1973. The pollination ecology of *Pedicularis* on Mount Rainier. Amer. J. Bot. 60: 863–871.

Macior, L. W. 1975. The pollination ecology of *Pedicularis* (Scrophulariaceae) in the Yukon Territory. Amer. J. Bot. 62: 1065–1072.

Macior, L. W. 1977. The pollination ecology of *Pedicularis* (Scrophulariaceae) in the Sierra Nevada of California. Bull. Torrey Bot. Club 104: 148–154.

Macior, L. W. 1981. The Furbish lousewort: Weed, weapon, or wonder? Amer. Biol. Teacher 43: 323–326.

MacRoberts, M. H. et al. 2007. *Minuartia drummondii* (Caryophyllaceae) and *Gratiola flava* (Plantaginaceae) rediscovered in Louisiana and *Gratiola flava* historically in Arkansas. J. Bot. Res. Inst. Texas 1: 763–767.

Madroño = Madroño; Journal of the California Botanical Society [from vol. 3: a West American Journal of Botany].

Maffitt, G. 2012. The Tualatin Basin *Penstemon* revisited. Bull. Amer. Penstemon Soc. 71: 48–53.

Magee, D. W. and H. E. Ahles. 2007. Flora of the Northeast: A Manual of the Vascular Flora of New England and Adjacent New York, ed. 2. Amherst.

Man. Bot. ed. 8—See: A. Eaton and J. Wright 1840

Man. Bot. San Francisco—See: E. L. Greene 1894

Man. Calif. Shrubs—See: H. McMinn 1939

Man. Fl. Pl. Calif.—See: W. L. Jepson [1923–1925]

Man. S. Calif. Bot.—See: P. A. Munz 1935

Man. S.E. Fl.—See: J. K. Small 1933

Mann, W. F. and L. J. Musselman. 1980. Senna seymeria parasitizes western conifers. Econ. Bot. 33: 338–339.

Manning, S. D. 1991. The genera of Pedaliaceae in the southeastern United States. J. Arnold Arbor., suppl. ser. 1: 313–347.

Mant. Pl.—See: C. Linnaeus 1767[–1771]

Marie-Victorin, Frère and J. Rousseau. 1940. Nouvelles entités de la flore phánerogamique du Canada oriental. Contr. Inst. Bot. Univ. Montréal 36: 1–74.

Marlowe, K. and L. Hufford. 2008. Evolution of *Synthyris* sect. *Dissecta* (Plantaginaceae) on sky islands in the northern Rocky Mountains. Amer. J. Bot. 95: 381–392.

Marx, H. E. et al. 2010. A molecular phylogeny and classification of Verbenaceae. Amer. J. Bot. 97: 1647–1663.

Matthews, S. and M. Lavin. 1998. A biosystematic study of *Castilleja crista-galli* (Scrophulariaceae): An allopolyploid origin reexamined. Syst. Bot. 23: 213–230.

Maximowicz, C. J. 1888. Diagnoses plantarum novarum Asiaticarum. Bull. Acad. Imp. Sci. Saint-Pétersbourg 32: 427–629.

McCone, M. J. et al. 1995. An exception to Darwin's syndrome: Floral position, protogyny, and insect visitation in *Besseya bullii*. Oecologia 101: 68–74.

McCully, M. E. and H. M. Dale. 1961. Heterophylly in *Hippuris*, a problem in identification. Canad. J. Bot. 39: 1099–1116.

McDade, L. A., T. F. Daniel, and C. A. Kiel. 2008. Toward a comprehensive understanding of phylogenetic relationships among lineages of Acanthaceae s.l. (Lamiales). Amer. J. Bot. 95: 1136–1152.

McFeeley, J. C. and E. P. Roberts. 1974. *Aureolaria grandiflora* var. *serrata*, parasite of *Juniperus virginiana*. Pl. Dis. Reporter 58: 773.

McKinney, B. R. and J. D. Villalobos. 2014. Preliminary report on Maderas del Carmen black bear study, Coahuila, México. In: C. A. Hoyt and J. Karges, eds. 2014. Proceedings of the Sixth Symposium on the Natural Resources of the Chihuahuan Desert Region, October 14–17, 2004. Fort Davis. Pp. 250–259.

McMinn, H. 1939. An Illustrated Manual of California Shrubs.... San Francisco. (Man. Calif. Shrubs)

McMinn, H. 1951. Studies in the genus *Diplacus*, Scrophulariaceae. Madroño 11: 33–128.

McMullen, A. L. 1998. Factors Concerning the Conservation of a Rare Shale Endemic Plant: The Reproductive Ecology and Edaphic Characteristics of *Penstemon debilis* (Scrophulariaceae). M.S. thesis. Utah State University.

McNeal, J. R. et al. 2013. Phylogeny and origins of holoparasitism in Orobanchaceae. Amer. J. Bot. 100: 971–983.

Med. Repos. = Medical Repository.

Meded. Landbouwhoogeschool = Mededeelingen van de Landbouwhoogeschool te Wageningen.

Meinke, R. J. 1992. Differentiating *Mimulus jepsonii* and *M. nanus* in south-central Oregon: A problem in applied systematics. Kalmiopsis 2: 10–16.

Meinke, R. J. 1995. *Mimulus evanescens* (Scrophulariaceae): A new annual species from the northern Great Basin. Great Basin Naturalist 55: 249–257.

Mélanges Philos. Math. Soc. Roy. Turin = Mélanges de Philosophie et de Mathématique de la Société Royale de Turin.

Mém. Acad. Imp. Sci. St.-Pétersbourg, Sér. 6, Sci. Math. = Mémoires de l'Académie Impériale des Sciences de St.-Pétersbourg. Sixième Série. Sciences Mathématiques, Physiques et Naturelles.

Mem. Comis. Limites—See: J. L. Berlandier [1832]

Mem. New York Bot. Gard. = Memoirs of the New York Botanical Garden.

Mém. Soc. Imp. Naturalistes Moscou = Mémoires de la Société Impériale des Naturalistes de Moscou.

Mem. Torrey Bot. Club = Memoirs of the Torrey Botanical Club.

Mém. Véronique Cymb.—See: P. H. H. Bodard [1798]

Méndez Larios, I. and J. L. Villaseñor Ríos. 2001. La familia Scrophulariaceae en México: Diversidad y distribución. Bol. Soc. Bot. México 69: 101–121.

Menges, E. S. 1990. Population viability analysis of an endangered plant. Conservation Biol. 4: 52–62.

Merriam-Webster. 1988. Webster's New Geographical Dictionary. Springfield, Mass.

Methodus—See: C. Moench 1794

Meyers, S. C. and A. Liston. 2008. The biogeography of *Plantago ovata* Forssk. (Plantaginaceae). Int. J. Pl. Sci. 169: 954–962.

Michaux, A. 1803. Flora Boreali-Americana.... 2 vols. Paris and Strasbourg. (Fl. Bor.-Amer.)

Michener, D. C. 1963. Systematic and ecological wood anatomy of Californian Scrophulariaceae. I. *Antirrhinum, Castilleja, Galvezia,* and *Mimulus* sect. *Diplacus.* Aliso 10: 471–487.

Michener, D. C. 1981. Wood and leaf anatomy of *Keckiella* (Scrophulariaceae): Ecological considerations. Aliso 10: 39–57.

Michener, D. C. 1982. Studies on the Evolution of *Keckiella* (Scrophulariaceae). Ph.D. dissertation. Claremont Graduate School.

Miller, M. 2001. Stewardship Account for Winged Water Starwort *Callitriche marginata.* Prepared for the B.C. Conservation Data Centre and the Garry Oak Ecosystem Recovery Team. Victoria.

Miller, N. G. 2001. The Callitrichaceae in the southeastern United States. Harvard Pap. Bot. 5: 277–301.

Miller, P. 1754. The Gardeners Dictionary.... Abridged..., ed. 4. 3 vols. London. (Gard. Dict. Abr. ed. 4)

Miller, P. 1768. The Gardeners Dictionary..., ed. 8. London. (Gard. Dict. ed. 8)

Millsaps, V. 1936. The structure and development of the seed of *Paulownia tomentosa* Steud. J. Elisha Mitchell Sci. Soc. 52: 56–75.

Mink, J. N., J. R. Singhurst, and W. C. Holmes. 2010. *Penstemon oklahomensis* (Scrophulariaceae) in Texas. J. Bot. Res. Inst. Texas 4: 471–472.

Minnesota Bot. Stud. = Minnesota Botanical Studies.

Mistretta, O. and R. Scogin. 1989. Foliar flavonoid aglycones of the genus *Keckiella* (Scrophulariaceae). Biochem. Syst. & Ecol. 17: 455–457.

Mitt. Staatsinst. Allg. Bot. Hamburg = Mitteilungen aus dem Staatsinstitut für Allgemeine Botanik in Hamburg.

Moench, C. 1794. Methodus Plantas Horti Botanici et Agri Marburgensis.... Marburg. (Methodus)

Moerman, D. E. 1998. Native American Ethnobotany. Portland.

Mohamed, K. I. et al. 2006. Global invasive potential of ten parasitic witchweeds and related Orobanchaceae. Ambio 35: 1–10.

Mohamed, K. I. et al. 2007. Genetic diversity of *Striga* and implications for control and modeling future distributions. In: G. Ejeta and J. Gressel, eds. 2007. Integrating New Technologies for *Striga* Control: Towards Ending the Witch-hunt. Singapore and Hackensack. Pp. 71–84.

Mohamed, K. I., L. J. Musselman, and C. R. Riches. 2001. The genus *Striga* in Africa (Scrophulariaceae). Ann. Missouri Bot. Gard. 88: 60–103.

Mohlenbrock, R. H. 1986. Guide to the Vascular Flora of Illinois, rev. ed. Carbondale.

Molau, U. 1988. Scrophulariaceae–part I. Calceolarieae. In: Organization for Flora Neotropica. 1968+. Flora Neotropica. 121+ nos. New York. No. 47, pp. 1–326.

Molau, U. 1990. The genus *Bartsia* (Scrophulariaceae-Rhinanthoideae). Opera Bot. 102: 5–99.

Molau, U. and D. F. Murray. 1996. Taxonomic revision of the *Pedicularis sudetica* complex (Scrophulariaceae): The Arctic species. Symb. Bot. Upsal. 31: 33–46.

Molec. Biol. Evol. = Molecular Biology and Evolution.

Molec. Ecol. = Molecular Ecology.

Molec. Phylogen. Evol. = Molecular Phylogenetics and Evolution.

Monogr. Acad. Nat. Sci. Philadelphia = Monographs, Academy of Natural Sciences of Philadelphia.

Monogr. Euphrasia—See: R. Wettstein 1896b

Moody, A., P. K. Diggle, and D. A. Steingraeber. 1999. Developmental analysis of the evolutionary origin of vegetative propagules in *Mimulus gemmiparus.* Amer. J. Bot. 86: 1512–1522.

Moorman, M. L. 1982. Systematic Revision of *Penstemon subglaber, P. saxorum,* and *P. mensarum.* M.S. thesis. University of Wyoming.

Morefield, J. D. 2006. Current Knowledge and Conservation Status of *Penstemon bicolor* (Brandegee) Clokey & Keck (Plantaginaceae), the Two-tone Beardtongue. Carson City.

Morgan-Richards, M. and K. Wolff. 1999. Genetic structure and differentiation of *Plantago major* reveals a pair of sympatric sister species. Molec. Ecol. 8: 1027–1036.

Morton, C. V. and J. T. Howell. 1945. A genus new to North America. Leafl. W. Bot. 4: 164–165.

Moss, E. H. and J. G. Packer. 1983. Flora of Alberta: A Manual of Flowering Plants, Conifers, Ferns, and Fern Allies Found Growing without Cultivation in the Province of Alberta, Canada, ed. 2. Toronto.

Mound, L. A. and D. C. Morris. 2007. A new thrips pest of *Myoporum* cultivars in California, in a new genus of leaf-galling Australian Phlaeothripidae (Thysanoptera). Zootaxa 1495: 35–45.

Mueller, F. J. H. 1882. Systematic Census of Australian Plants...Part i. Vasculares. Melbourne. (Syst. Census Austral. Pl.)

Muhlenbergia = Muhlenbergia; a Journal of Botany.

Munz, P. A. 1926. The Antirrhinoideae-Antirrhineae of the New World. Proc. Calif. Acad. Sci., ser 4, 15: 323–397.

Munz, P. A. 1930. The North American species of *Orobanche*, sect. *Myzorrhiza*. Bull. Torrey Bot. Club 57: 611–624.

Munz, P. A. 1935. A Manual of Southern California Botany.... San Francisco. (Man. S. Calif. Bot.)

Munz, P. A. 1959. A California Flora. Berkeley and Los Angeles.

Munz, P. A. and D. D. Keck. 1973. A California Flora and Supplement. Berkeley and Los Angeles.

Murbeck, S. 1925. Monographie der Gattung *Celsia*. Acta Univ. Lund., n. s. 22: 1–239.

Murbeck, S. 1933. Monographie der Gattung *Verbascum*. Acta Univ. Lund., n. s. 29: 1–630.

Murbeck, S. 1936. Nachträge zur Monographie der Gattung *Verbascum*. Acta Univ. Lund., n. s. 32: 1–46.

Murbeck, S. 1939. Weitere Studien über die Gattungen *Verbascum* und *Celsia*. Acta Univ. Lund., n. s. 35: 1–71.

Murray, D. F., R. Elven, and K. N. Gandhi. 2010. A new combination in *Lagotis* (Plantaginaceae). J. Bot. Res. Inst. Texas 4: 219–220.

Murray, J. A. 1784. Caroli à Linné Equitis Systema Vegetabilium.... Editio Decima Quarta.... Göttingen. (Syst. Veg. ed. 14)

Musselman, L. J. 1969. Observations on the life history of *Aureolaria grandiflora* and *Aureolaria pedicularis* (Scrophulariaceae). Amer. Midl. Naturalist 82: 307–311.

Musselman, L. J. 1972. Root parasitism of *Macranthera flammea* and *Tomanthera auriculata*. J. Elisha Mitchell Sci. Soc. 88: 58–60.

Musselman, L. J. 1984. An unusual specimen of *Orobanche* from North Carolina collected by John Ball in 1884. Castanea 49: 91–93.

Musselman, L. J. 1996. Parasitic weeds of the southern United States. Castanea 61: 271–292.

Musselman, L. J., R. E. Eplee, and P. F. Sand. 1980. Cowpea striga (*Striga gesnerioides*). Weeds Today 11: 14–15.

Musselman, L. J. and H. E. Grelen. 1979. A population of *Aureolaria pedicularia* (L.) Raf. (Scrophulariaceae) without oaks. Amer. Midl. Naturalist 102: 175–177.

Musselman, L. J. and W. F. Mann. 1977. Host plants of some Rhinanthoideae (Scrophulariaceae) of eastern North America. Pl. Syst. Evol. 127: 45–53.

Musselman, L. J. and W. F. Mann. 1978. Root Parasites of Southern Forests. New Orleans. [U.S.D.A. Forest Serv., Gen. Techn. Rep. SO-20.]

Musselman, L. J. and W. F. Mann. 1979. Haustorial frequency of some root parasites in culture. New Phytol. 83: 479–483.

Musselman, L. J. and K. C. Nixon. 1981. Branched broomrape (*Orobanche ramosa*) in Texas. Pl. Dis. 65: 752–753.

Musselman, L. J. and C. Parker. 1981. Studies on indigo witchweed, the American strain of *Striga gesnerioides* (Scrophulariaceae). Weed Sci. 29: 594–596.

N. E. Naturalist = Northeastern Naturalist.

Nakai, T. 1949. Classes, ordinae, familiae, subfamiliae, tribus, genera nova quae attinent ad plantas Koreanas. J. Jap. Bot. 24: 8–14.

Nat. Hist. = Natural History; the Magazine of the American Museum of Natural History.

Nat. Pflanzenfam.—See: H. G. A. Engler and K. Prantl 1887–1915

Nat. Syst. Bot.—See: J. Lindley 1836

Native Pl. J. = Native Plants Journal.

Naturaliste Canad. = Naturaliste Canadien. Bulletin de Recherches, Observations et Découvertes se Rapportant à l'Histoire Naturelle du Canada.

Neel, M. C. 2002. Conservation implications of the reproductive ecology of *Agalinis acuta* (Scrophulariaceae). Amer. J. Bot. 89: 972–980.

Neel, M. C. and M. P. Cummings. 2004. Section-level relationships of North American *Agalinis* (Orobanchaceae) based on DNA sequence analysis of three chloroplast gene regions. B. M. C. Evol. Biol. 4: 1–12.

Neese, E. C. 1993. *Penstemon*. In: S. L. Welsh et al., eds. 1993. A Utah Flora, ed. 2. Provo. Pp. 659–683.

Neese, E. C. 1993b. *Collinsia*. In: J. C. Hickman, ed. 1993. The Jepson Manual. Higher Plants of California. Berkeley, Los Angeles, and London. Pp. 1024–1027.

Neese, E. C. and N. D. Atwood. 2003. *Penstemon*. In: S. L. Welsh et al., eds. 2003. A Utah Flora, ed. 3. Provo. Pp. 614–635.

Nelson, A. 1936. Rocky Mountain Herbarium studies. IV. Amer. J. Bot. 23: 265–271.

Nelson, A. 1937. Taxonomic studies. *Penstemon*. Publ. Univ. Wyoming 3: 101–107.

Nelson, A. 1938. Rocky Mountain Herbarium studies. V. Amer. J. Bot. 25: 114–118.

Nelson, A. D. 1995. Polyploid Evolution in *Chelone* (Scrophulariaceae). Ph.D. dissertation. University of Oklahoma.

Nelson, A. D. and W. J. Elisens. 1999. Polyploid evolution and biogeography in *Chelone* (Scrophulariaceae): Morphological and isozyme evidence. Amer. J. Bot. 86: 1487–1501.

Nesom, G. L. 1992c. A new species of *Castilleja* (Scrophulariaceae) from southcentral Texas with comments on other Texas taxa. Phytologia 72: 209–230.

Nesom, G. L. 2011d. Recognition and synopsis of *Mimulus* sect. *Tropanthus* and sect. *Leucocarpus* (Phrymaceae). Phytoneuron 2011-28: 1–8.

Nesom, G. L. 2012g. Taxonomic summary of *Erythranthe* sect. *Achlyopitheca* (Phrymaceae). Phytoneuron 2012-42: 1–4.

Nesom, G. L. 2012h. Taxonomy of *Erythranthe* sect. *Mimulosma* (Phrymaceae). Phytoneuron 2012-41: 1–36.

Nesom, G. L. 2012i. Taxonomy of *Erythranthe* sect. *Simiola* in the USA and Mexico. Phytoneuron 2012-40: 1–123.

Nesom, G. L. 2013b. Taxonomic notes on *Diplacus* (Phrymaceae). Phytoneuron 2013-66: 1–8.

Nesom, G. L. 2013c. Three new species of *Diplacus* (Phrymaceae), primarily from Oregon. Phytoneuron 2013-65: 1–18.

Nesom, G. L. 2013d. New distribution records for *Erythranthe* (Phrymaceae). Phytoneuron 2013-67: 1–15.

Nesom, G. L. 2013e. The taxonomic status of *Mimulus sookensis* (Phrymaceae) and comments on related aspects of biology in species of *Erythranthe*. Phytoneuron 2013-69: 1–18.

Nesom, G. L. 2014b. Taxonomy of *Erythranthe* sect. *Erythranthe* (Phrymaceae). Phytoneuron 2014-31: 1–41.

Nesom, G. L. 2014c. Updated classification and hypothetical phylogeny of *Erythranthe* sect. *Simiola* (Phrymaceae). Phytoneuron 2014-81: 1–6.

Nesom, G. L. 2014d. Further observations on relationships in the *Erythranthe guttata* group (Phrymaceae). Phytoneuron 2014-93: 1–8.

Nesom, G. L. 2017. Taxonomic review of the *Erythranthe moschata* complex (Phrymaceae). Phytoneuron 2017-17: 1–29.

Nesom, G. L. and L. E. Brown. 1998. Annotated checklist of the vascular plants of Walker, Montgomery, and San Jacinto counties, east Texas. Phytologia 84: 107–153.

Nesom, G. L. and J. M. Egger. 2014. Review of the *Castilleja purpurea* complex (Orobanchaceae). Phytoneuron 2014-15: 1–16.

New Fl.—See: C. S. Rafinesque 1836[–1838]

New Man. Bot. Rocky Mt.—See: J. M. Coulter and A. Nelson 1909

New Phytol. = New Phytologist; a British Botanical Journal.

New Zealand J. Bot. = New Zealand Journal of Botany.

New Zealand Veterin. J. = New Zealand Veterinary Journal.

Newsom, V. M. 1929. A revision of the genus *Collinsia* (Scrophulariaceae). Bot. Gaz. 87: 260–301.

Nicollet, J. N. 1843. Report Intended to Illustrate a Map of the Hydrographical Basin of the Upper Mississippi River. Washington. (Rep. Ill. Map Hydrogr. Basin Upper Mississippi)

Nie, Z. L. et al. 2006. Evolution of biogeographic disjunction between eastern Asia and eastern North America in *Phryma*. Amer. J. Bot. 93: 1343–1356.

Nie, Z. L. et al. 2009b. AFLP analysis of *Phryma* (Phrymaceae) disjunct between eastern Asia and eastern North America. Acta Bot. Yunnan. 31: 289–295.

Nielson, R. L. 1998. The Reproductive Biology and Ecology of *Penstemon harringtonii* (Scrophulariaceae): Implications for Conservation. M.S. thesis. Utah State University.

Niezgoda, C. J. and A. S. Tomb. 1975. Systematic palynology of tribe Leucophylleae (Scrophulariaceae) and selected Myoporaceae. Pollen & Spores 17: 495–516.

Nisbet, G. T. and R. C. Jackson. 1960. The genus *Penstemon* in New Mexico. Univ. Kansas Sci. Bull. 41: 691–759.

Nold, R. 1999. *Penstemon*. Portland.

Nomencl. Bot. ed. 2—See: E. G. Steudel 1840–1841

Nordenskiöld, H. 1957. Hybridization experiments in the genus *Luzula*. III. The subg. *Pterodes*. Bot. Not. 110: 1–16.

Nordic J. Bot. = Nordic Journal of Botany.

Norman, E. M. 2000. Buddlejaceae. In: Organization for Flora Neotropica. 1968+. Flora Neotropica. 121+ nos. New York. No. 81, pp. 1–225.

Norman, E. M. and R. J. Moore. 1968. Notes on *Emorya* (Loganiaceae). SouthW. Naturalist 13: 137–142.

Notizbl. Königl. Bot. Gart. Berlin = Notizblatt des Königlichen botanischen Gartens und Museums zu Berlin.

Notul. Nat. Acad. Nat. Sci. Philadelphia = Notulae Naturae of the Academy of Natural Sciences of Philadelphia.

Nouv. Arch. Mus. Hist. Nat. = Nouvelles Archives du Muséum d'Histoire Naturelle.

Nouv. Mém. Soc. Imp. Naturalistes Moscou = Nouveaux Mémoires de la Société Impériale des Naturalistes de Moscou.

Nov. Actorum Acad. Caes. Leop.-Carol. Nat. Cur. = Novorum Actorum Academiae Caesareae Leopoldinae-Carolinae Naturae Curiosorum.

Nov. Gen. Sp.—See: A. von Humboldt et al. 1815[1816]–1825

Nov. Gen. Sp. Pl.—See: E. F. Poeppig and S. L. Endlicher 1835–1845

Nova Acta Phys.-Med. Acad. Caes. Leop.-Carol. Nat. Cur. = Nova Acta Physico-medica Academiae Caesareae Leopoldino-Carolinae Naturae Curiosorum Exhibentia Ephemerides, sive Observationes Historias et Experimenta....

Nova Guinea = Nova Guinea; a Journal of Botany, Zoology, Anthropology, Ethnography, Geology and Palaeontology of the Papuan Region.

Nova Hedwigia = Nova Hedwigia. Zeitschrift für Kryptogamenkunde.

Novi Comment. Acad. Sci. Imp. Petrop. = Novi Commentarii Academiae Scientiarum Imperalis Petropolitanae.

Novit. Fl. Svec.—See: E. M. Fries 1814–1824

Novon = Novon; a Journal for Botanical Nomenclature.

Nuttall, T. 1813. A Catalogue of New and Interesting Plants Collected in Upper Louisiana.... London. (Cat. Pl. Upper Louisiana)

Nuttall, T. 1818. The Genera of North American Plants, and Catalogue of the Species, to the Year 1817.... 2 vols. Philadelphia. (Gen. N. Amer. Pl.)

O'Kane, S. L. 1988. Colorado's rare flora. Great Basin Naturalist 48: 434–484.

Observ. Bot.—See: A. J. Retzius [1779–]1791; O. P. Swartz 1791

Occas. Pap. Bernice Pauahi Bishop Mus. = Occasional Papers of the Bernice Pauahi Bishop Museum of Polynesian Ethology and Natural History.

Oekon. Fl. Wetterau—See: P. G. Gaertner et al. 1799–1802

Oesterr. Bot. Z. = Oesterreichische botanische Zeitschrift. Gemeinütziges Organ für Botanik.

Ohio J. Sci. = Ohio Journal of Science.

Oikos = Oikos; Acta Oecologica Scandinavica.

Olmstead, R. G. et al. 1992. Monophyly of the Asteridae and identification of their major lineages inferred from DNA sequences of *rbc*L. Ann. Missouri Bot. Gard. 79: 249–265.

Olmstead, R. G. et al. 1993. A parsimony analysis of the Asteridae sensu lato based on *rbc*L sequences. Ann. Missouri Bot. Gard. 80: 700–722.

Olmstead, R. G. et al. 2001. Disintegration of the Scrophulariaceae. Amer. J. Bot. 88: 348–361.

Olmstead, R. G. et al. 2009. A molecular phylogeny and classification of Bignoniaceae. Amer. J. Bot. 96: 1731–1743.

Olmstead, R. G. and P. A. Reeves. 1995. Evidence for the polyphyly of the Scrophulariaceae based on chloroplast *rbc*L and *ndh*F sequences. Ann. Missouri Bot. Gard. 82: 176–193.

Opera Bot. = Opera Botanica a Societate Botanice Lundensi.

Oregon State Monogr., Stud. Bot. = Oregon State Monographs. Studies in Botany.

Organization for Flora Neotropica. 1968+. Flora Neotropica. 121+ nos. New York.

Orobanches Gen. Diask.—See: C. F. W. Wallroth 1825

Ottawa Naturalist = Ottawa Naturalist; Transactions of the Ottawa Field-Naturalists' Club.

Oviedo Prieto, R. et al. 2012. Lista nacional de especies de plantas invasoras y potencialmente invasoras en la Repúlica de Cuba–2011. Bissea 6: 22–96.

Ownbey, M. 1959. *Castilleja.* In: C. L. Hitchcock et al. 1955–1969. Vascular Plants of the Pacific Northwest. 5 vols. Seattle. Vol. 4, pp. 295–326.

Oxelman, B. et al. 2005. Further disintegration of Scrophulariaceae. Taxon 54: 411–425.

Oxelman, B., M. Backlund, and B. Bremer. 1999. Relationships of the Buddlejaceae s.l. investigated using parsimony jackknife and branch support analysis of chloroplast *ndh*F and *rbc*L sequence data. Syst. Bot. 24: 164–182.

Oyama, R. K. and D. A. Baum. 2004. Phylogenetic relationships of North American *Antirrhinum* (Veronicaceae). Amer. J. Bot. 91: 918–925.

Pacif. Railr. Rep.—See: War Department 1855–1860

Pacif. Railr. Rep. Parke, Bot.—See: J. Torrey 1856

Panjabi, S. S. and D. G. Anderson. 2006. *Penstemon harringtonii* Penland (Harrington's Beardtongue): A Technical Conservation Assessment. Fort Collins.

Parfitt, B. D. and M. L. Butterwick. 1981. Noteworthy collections: *Orobanche uniflora* L. subsp. *occidentalis.* Madroño 28: 37–38.

Park, J. M. et al. 2008. A plastid phylogeny of the non-photosynthetic parasitic *Orobanche* (Orobanchaceae) and related genera. J. Pl. Res. 121: 365–376.

Parton, K. and A. N. Bruere. 2002. Plant poisoning of livestock in New Zealand. New Zealand Veterin. J. 50(3,suppl.): 22–27.

Patterson, D. T. et al. 1982. Temperature responses and potential for spread of witchweed (*S. lutea* and *S. asiatica*) in the United States. Weed Sci. 30: 87–93.

Patterson, D. T. et al. 1989. Composite List of Weeds. Champaign.

Paxton's Fl. Gard. = Paxton's Flower Garden.

Pennell, F. W. 1919. Scrophulariaceae of the local flora. II. Torreya 19: 143–152.

Pennell, F. W. 1920. Scrophulariaceae of the southeastern United States. Proc. Acad. Nat. Sci. Philadelphia 71: 224–291.

Pennell, F. W. 1920b. Scrophulariaceae of the central Rocky Mountain states. Contr. U.S. Natl. Herb. 20: 313–381.

Pennell, F. W. 1920c. Scrophulariaceae of Colombia. Proc. Acad. Nat. Sci. Philadelphia 72: 136–188.

Pennell, F. W. 1921. Scrophulariaceae of the West Gulf states. Proc. Acad. Nat. Sci. Philadelphia 73: 459–536.

Pennell, F. W. 1925. The genus *Afzelia:* A taxonomic study in evolution. Proc. Acad. Nat. Sci. Philadelphia 77: 335–373.

Pennell, F. W. 1928. *Agalinis* and allies in North America: I. Proc. Acad. Nat. Sci. Philadelphia 80: 339–449.

Pennell, F. W. 1929. *Agalinis* and allies in North America: II. Proc. Acad. Nat. Sci. Philadelphia 81: 111–249.

Pennell, F. W. 1933. A revision of *Synthyris* and *Besseya.* Proc. Acad. Nat. Sci. Philadelphia 85: 77–106.

Pennell, F. W. 1934. Scrophulariaceae of the northwestern United States—II. *Pedicularis* of the group Bracteosae. Bull. Torrey Bot. Club 61: 441–448.

Pennell, F. W. 1934b. *Castilleja* in Alaska and northwestern Canada. Proc. Acad. Nat. Sci. Philadelphia 86: 517–540.

Pennell, F. W. 1935. The Scrophulariaceae of eastern temperate North America. Monogr. Acad. Nat. Sci. Philadelphia 1.

Pennell, F. W. 1940. Scrophulariaceae of trans-Pecos Texas. Proc. Acad. Nat. Sci. Philadelphia 92: 289–308.

Pennell, F. W. 1946. Reconsideration of the *Bacopa-Herpestis* problem of the Scrophulariaceae. Proc. Acad. Nat. Sci. Philadelphia 98: 83–98.

Pennell, F. W. 1947. Some hitherto undescribed Scrophulariaceae of the Pacific states. Proc. Acad. Nat. Sci. Philadelphia 99: 155–171.

Pennell, F. W. 1951. Scrophulariaceae. In: L. Abrams and R. S. Ferris. 1923–1960. Illustrated Flora of the Pacific States: Washington, Oregon, and California. 4 vols. Stanford. Vol. 3, pp. 686–859.

Penny, J. L. and G. W. Douglas. 1999. Status of Bearded Owl-clover in British Columbia. Victoria. [Wildlife Bull. B-89.]

Perret, M. et al. 2013. Temporal and spatial origin of Gesneriaceae in the New World inferred from plastid DNA sequences. Bot. J. Linn. Soc. 171: 61–79.

Persoon, C. H. 1805–1807. Synopsis Plantarum.... 2 vols. Paris and Tubingen. (Syn. Pl.)

Petagna, V. 1785–1787. Institutiones Botanicae.... 5 vols. Naples. (Inst. Bot.)

Petrovsky, V. V. 1980. *Lagotis.* In: A. I. Tolmatchew, ed. 1960–1987. Flora Arctica URSS. 10 vols. in 12. Moscow and Leningrad. Vol. 8(2), pp. 267–269.

Pettengill, J. B. and M. C. Neel. 2008. Phylogenetic patterns and conservation among North American members of the genus *Agalinis* (Orobanchaceae). B. M. C. Evol. Biol. 8: 264.

Pettengill, J. B. and M. C. Neel. 2011. A sequential approach using genetic and morphological analyses to test species status: The case of United States federally endangered *Agalinis acuta* (Orobanchaceae). Amer. J. Bot. 98: 859–871.

Pflanzenr.—See: H. G. A. Engler 1900–1953

Philbrick, C. T. 1989. Systematic Studies of North American *Callitriche* (Callitrichaceae). Ph.D. dissertation. University of Connecticut.

Philbrick, C. T., R. A. Aakjar, and R. L. Stuckey. 1998. Invasion and spread of *Callitriche stagnalis* (Callitrichaceae) in North America. Rhodora 100: 25–38.

Philbrick, C. T. and D. H. Les. 2000. Phylogenetic studies in *Callitriche:* Implications for interpretation of ecological, karyological and pollination system evolution. Aquatic Bot. 68: 123–141.

Philcox, D. 1966. Revision of the New World species of *Buchnera* L. (Scrophulariaceae). Kew Bull. 18: 275–315.

Philcox, D. 1970. A taxonomic revision of the genus *Limnophila* R. Br. (Scrophulariaceae). Kew Bull. 24: 101–170.

Phytologia = Phytologia; Designed to Expedite Botanical Publication.

Phytomorphology = Phytomorphology; an International Journal of Plant Morphology.

Phytopathology = Phytopathology; Official Organ of the American Phytopathological Society.

Pickens, A. L. 1927. Unique method of pollination by the ruby-throat. Auk 44: 24–27.

Piotrowska, R. 1934. Przyczynek do historji rozwoju pylników i pyklu u niektórych przedstawicieli rodaju *Penstemon* Dougl. Acta Soc. Bot. Poloniae 11: 485–500.

Pittillo, J. D. and A. E. Brown. 1988. Additions to the vascular flora of the Carolinas. III. J. Elisha Mitchell Sci. Soc. 104: 1–18.

Pl. Aequinoct.—See: A. von Humboldt and A. J. Bonpland [1805–]1808–1809[–1817]

Pl. Baker.—See: E. L. Greene 1901

Pl. Biol. (Stuttgart) = Plant Biology. [Stuttgart.]

Pl. Coromandel—See: W. Roxburgh 1795–1820

Pl. Dis. = Plant Disease; International Journal of Applied Plant Pathology.

Pl. Dis. Reporter = Plant Disease Reporter.

Pl. Ecol. = Plant Ecology.

Pl. Hartw.—See: G. Bentham 1839[–1857]

Pl. Spec. Biol. = Plant Species Biology; an International Journal.

Pl. Syst. Evol. = Plant Systematics and Evolution.

PLOS Biol. = PLOS Biology. [Electronic journal.]

Poeppig, E. F. and S. L. Endlicher. 1835–1845. Nova Genera ac Species Plantarum Quas in Regno Chilensi Peruviano et in Terra Amazonica.... 3 vols. Leipzig. (Nov. Gen. Sp. Pl.)

Porter, T. C. and J. M. Coulter. 1874. Synopsis of the Flora of Colorado.... Washington. (Syn. Fl. Colorado)

Posto, A. L. and L. A. Prather. 2003. The evolutionary and taxonomic implications of RAPD data on the genetic relationships of *Mimulus michiganensis* (comb. et stat. nov: Scrophulariaceae). Syst. Bot. 28: 172–178.

Prain, D. 1890. The species of *Pedicularis* of the Indian Empire and its frontiers. Ann. Roy. Bot. Gard. Calcutta 3: 1–62.

Preslia = Preslia. Věstník (Časopis) Československé Botanické Společnosti.

Price, P. W. et al., eds. 1991. Plant-Animal Interactions: Evolutionary Ecology in Tropical and Temperate Regions. New York.

Prim. Lin. Fl. Herbipol.—See: J. F. Lehmann 1809

Proc. Acad. Nat. Sci. Philadelphia = Proceedings of the Academy of Natural Sciences of Philadelphia.

Proc. Amer. Acad. Arts = Proceedings of the American Academy of Arts and Sciences.

Proc. Biol. Soc. Wash. = Proceedings of the Biological Society of Washington.

Proc. Boston Soc. Nat. Hist. = Proceedings of the Boston Society of Natural History.

Proc. Calif. Acad. Sci. = Proceedings of the California Academy of Sciences.

Proc. Natl. Acad. Sci. U.S.A. = Proceedings of the National Academy of Sciences of the United States of America.

Proc. S. Weed Sci. Soc. = Proceedings, Southern Weed Science Society.

Proc. U.S. Natl. Mus. = Proceedings of the United States National Museum.

Prodr.—See: R. Brown 1810; A. P. de Candolle and A. L. P. P. de Candolle 1823–1873; O. P. Swartz 1788

Prodr. Descr. Gratiol.—See: J. Colsmann [1793]

Prodr. Fl. Hispan.—See: H. M. Willkomm and J. M. C. Lange 1861–1880

Prosp. Hist. Pl. Dauphiné—See: D. Villars 1779

Publ. Carnegie Inst. Wash. = Publications of the Carnegie Institution of Washington.

Publ. Comité Natl. Franç. Rech. Antarct. Biol. = Publications de la Comité National Français des (pour les) Recherches Antarctiques. Biologie.

Publ. Field Mus. Nat. Hist., Bot. Ser. = Publications of the Field Museum of Natural History. Botanical Series.

Publ. Univ. Wyoming = Publications, University of Wyoming.

Pugsley, H. W. 1933b. The euphrasias of Iceland and the Faroes. J. Bot. 71: 303–309.

Pursh, F. [1813]1814. Flora Americae Septentrionalis; or, a Systematic Arrangement and Description of the Plants of North America. 2 vols. London. (Fl. Amer. Sept.)

Quart. J. Taiwan Mus. = Quarterly Journal of the Taiwan Museum. [Tai Wan Shǐng Li Po Wu Kuan Shi K'an.]

Radford, A. E., H. E. Ahles, and C. R. Bell. 1968. Manual of the Vascular Flora of the Carolinas. Chapel Hill.

Rafinesque, C. S. 1828[–1830]. Medical Flora; or, Manual of the Medical Botany of the United States of North America. 2 vols. Philadelphia.

Rafinesque, C. S. 1836[–1838]. New Flora and Botany of North America.... 4 parts. Philadelphia. [Parts paged independently.] (New Fl.)

Rafinesque, C. S. 1840. Autikon Botanikon. 3 parts. Philadelphia. [Parts paged consecutively.] (Autik. Bot.)

Rahman, M. O. 2006. Scrophulariaceous taxa in Bangladesh. Bangladesh J. Pl. Taxon. 13: 139–154.

Rahmanzadeh, R. et al. 2005. The Linderniaceae and Gratiolaceae are further lineages distinct from the Scrophulariaceae. Pl. Biol. (Stuttgart) 7: 67–78.

Rahn, K. 1974. *Plantago* section *Virginica*: A taxonomic revision of a group of American plantains using experimental, taximetric and classical methods. Dansk Bot. Ark. 30: 1–180.

Rahn, K. 1978. *Plantago* ser. *Gnaphaloides* Rahn: A taxonomic revision. Bot. Tidsskr. 73: 137–154.

Rar. Lig. [Ital.] Pl.—See: A. Bertoloni 1803–1810

Raven, P. H. and D. I. Axelrod. 1978. Origin and relationships of the California flora. Univ. Calif. Publ. Bot. 72: 1–134.

Raven, P. H., D. W. Kyhos, and A. J. Hill. 1965. Chromosome numbers of spermatophytes, mostly Californian. Aliso 6(1): 105–113.

Ree, R. H. 2005. Phylogeny and the evolution of floral diversity in *Pedicularis* (Orobanchaceae). Int. J. Pl. Sci. 166: 595–613.

Rees, A. [1802–]1819–1820. The Cyclopaedia; or, Universal Dictionary of Arts, Sciences, and Literature.... 39 vols. in 79 parts. London. [Pages unnumbered.] (Cycl.)

Reese, R. N. 1984. A new variety of *Pedicularis contorta* (Scrophulariaceae) endemic to Idaho and Montana. Brittonia 36: 63–66.

Refulio-Rodriguez, N. F. and R. G. Olmstead. 2014. Phylogeny of Lamiidae. Amer. J. Bot. 101: 287–299.

Reichenbach, H. G. L. 1823–1832. Iconographia Botanica seu Plantae Criticae. 10 vols. Leipzig. [Vols. 6 and 7 each published in two half-centuries paged independently.] (Iconogr. Bot. Pl. Crit.)

Reichenbach, H. G. L. 1828. Conspectus Regni Vegetabilis.... Leipzig. (Consp. Regn. Veg.)

Reichenbach, H. G. L. 1830[–1832]. Flora Germanica Excursoria.... 2 parts. Leipzig. [Parts paged consecutively.] (Fl. Germ. Excurs.)

Rejuvenation Res. = Rejuvenation Research.

Rep. (Annual) Bur. Amer. Ethnol. = Annual Report of the Bureau of American Ethnology.

Rep. (Annual) Michigan Acad. Sci. = Report (Annual) of the Michigan Academy of Science, (Arts, and Letters).

Rep. Bot. Exch. Club Soc. Brit. Isles = (Report,) Botanical Exchange Club and Society of the British Isles.

Rep. Colorado R.—See: J. C. Ives 1861

Rep. Ill. Map Hydrogr. Basin Upper Mississipp—See: J. N. Nicollet 1843

Rep. U.S. Mex. Bound.—See: W. H. Emory 1857–1859

Repert. Bot. Syst.—See: W. G. Walpers 1842–1847

Repert. Spec. Nov. Regni Veg. = Repertorium Specierum Novarum Regni Vegetabilis.

Retzius, A. J. [1779–]1791. Observationes Botanicae.... 6 vols. Leipzig. (Observ. Bot.)

Retzius, A. J. 1795. Florae Scandinaviae Prodromus..., ed. 2. Leipzig. (Fl. Scand. Prodr. ed. 2)

Rev. Hort. = Revue Horticole; Journal d'Horticulture Pratique.

Reveal, J. L. 2011. Summary of recent systems of angiosperm classification. Kew Bull. 66: 5–48.

Revis. Antirrhineae—See: D. A. Sutton 1988

Revis. Gen. Pl.—See: O. Kuntze 1891–1898

Revista Soc. Cub. Bot. = Revista de la Sociedad Cubana de Botánica; Organo Oficial del Jardin Botánico de la Universidad de la Habana.

Rhodora = Rhodora; Journal of the New England Botanical Club.

Richardson, A. and K. King. 2006. *Capraria mexicana* (Scrophulariaceae) in Cameron County, Texas: Rediscovered in the United States. Sida 22: 1237–1238.

Richmond, G. S. and E. L. Ghisalberti. 1995. Cultural, food, medicinal uses, and potential applications of *Myoporum* species (Myoporaceae). Econ. Bot. 49: 276–285.

Robart, B. W. et al. 2015. Phylogeny and biogeography of North American and Asian *Pedicularis* (Orobanchaceae). Syst. Bot. 40: 229–258.

Rodrigues, A. G. et al. 2013. Morphometric analysis and taxonomic revision of the North American holoparasitic genus *Conopholis* (Orobanchaceae). Syst. Bot. 38: 795–804.

Rodrigues, A. G., A. E. L. Colwell, and S. Stefanović. 2011. Molecular systematics of the parasitic genus *Conopholis* (Orobanchaceae) inferred from plastid and nuclear sequences. Amer. J. Bot. 98: 896–908.

Roemer, J. J., J. A. Schultes, and J. H. Schultes. 1817[–1830]. Caroli a Linné...Systema Vegetabilium...Editione XV.... 7 vols. Stuttgart. (Syst. Veg.)

Rogers, G. K. 1986. The genera of Loganiaceae in the southeastern United States. J. Arnold Arbor. 67: 143–185.

Ronsted, N. et al. 2002. Phylogenetic relationships within *Plantago* (Plantaginaceae): Evidence from nuclear ribosomal ITS and plastid *trn*L-F sequence data. Bot. J. Linn. Soc. 139: 323–338.

Rosenfeld, I. and O. A. Beath. 2013. Selenium: Geobotany, Biochemistry, Toxicity, and Nutrition. San Diego.

Ross, T. S. and S. Boyd. 1996. Noteworthy collections. California. Madroño 43: 432–436.

Rossow, R. A. 1987. Revisión del género *Mecardonia*. Candollea 42: 431–474.

Roth, A. W. 1797–1806. Catalecta Botanica.... 3 parts. Leipzig. [Parts paged independently.] (Catal. Bot.)

Rothmaler, W. 1943. Zur Gliederung der Antirrhineae. Feddes Repert. Spec. Nov. Regni Veg. 52: 16–39.

Rousseau, J., M. Gauvreau, and C. Morin. 1939. Bibliographie des travaux botaniques contenus dans les "Memoires et Comptes Rendus de la Société Royale du Canada," de 1882 à 1936 inclusivement. Contr. Inst. Bot. Univ. Montréal 33.

Roxburgh, W. 1795–1820. Plants of the Coast of Coromandel.... 3 vols. in parts. London. [Volumes paged independently, plates numbered consecutively.] (Pl. Coromandel)

Roxburgh, W. 1820–1824. Flora Indica; or Descriptions of Indian Plants.... 2 vols. Serampore. (Fl. Ind.)

Ruiz López, H. and J. A. Pavon. 1794. Flora Peruvianae, et Chilensis Prodromus.... Madrid. (Fl. Peruv. Prodr.)

Rumsey, F. J. and S. L. Jury. 1991. An account of *Orobanche* L. in Britain and Ireland. Watsonia 18: 257–295.

Rydberg, P. A. 1903. Some generic segregations. Bull. Torrey Bot. Club 30: 271–281.

Rydberg, P. A. 1906. Flora of Colorado.... Fort Collins. (Fl. Colorado)

Rydberg, P. A. 1917. Flora of the Rocky Mountains and Adjacent Plains. New York. (Fl. Rocky Mts.)

Rzedowski, G. C. de. 1998. Orobanchaceae. In: J. Rzedowski and G. C. de Rzedowski, eds. 1991+. Flora del Bajío y de Regiones Adyacentes. 99+ fascs. Pátzcuaro. Fasc. 69, pp. 1–12.

Rzedowski, J. and G. C. de Rzedowski, eds. 1991+. Flora del Bajío y de Regiones Adyacentes. 99+ fascs. Pátzcuaro.

Saggoo, M. I. S. and D. K. Srivastava. 2009. Meiotic studies in some species of *Pedicularis* L. from cold desert regions of Himachal Pradesh, India (North-West Himalaya). Chromosome Bot. 4: 83–86.

Saldanha, C. J. 1963. The genus *Striga* Lour. (Scrophulariaceae) in western India. Bull. Bot. Surv. India 5: 67–70.

Sanchez, P. M. et al. 2013. In vivo anti-inflammatory and anti-ulcerogenic activities of extracts from wild growing and in vitro plants of *Castilleja tenuiflora* Benth. (Orobanchaceae). J. Ethnopharmacol. 150: 1032–1037.

Saner, M. A. et al. 1995. The biology of Canadian weeds. 105. *Linaria vulgaris* Mill. Canad. J. Pl. Sci. 75: 525–537.

Savile, D. B. O. 1968b. The rusts of Cheloneae (Scrophulariaceae): A study in the co-evolution of hosts and parasites. Nova Hedwigia 15: 369–392.

Savile, D. B. O. 1968c. Some fungal parasites of Scrophulariaceae. Canad. J. Bot. 46: 461–471.

Savile, D. B. O. 1979. Fungi as aids in higher plant classification. Bot. Rev. (Lancaster) 45: 377–503.

Schaack, C. G. 1983. A Monographic Revision of the Genera *Synthyris* and *Besseya* (Scrophulariaceae). Ph.D. dissertation. University of Montana.

Schäferhoff, B. et al. 2010. Towards resolving Lamiales relationships: Insights from rapidly evolving chloroplast sequences. B. M. C. Evol. Biol. 10: 352.

Scheunert, A. et al. 2012. Phylogeny of tribe Rhinantheae (Orobanchaceae) with a focus on biogeography, cytology and re-examination of generic concepts. Taxon 61: 1269–1285.

Schmid, R. 1976. Fly pollination of *Penstemon davidsonii* and *P. procerus* (Scrophulariaceae). Madroño 23: 400–401.

Schmidel, C. C. 1762–1771. Icones Plantarum et Analyses Partium Aeri.... [Nuremberg.] (Icon. Pl. ed. Keller)

Schneeweiss, G. M. et al. 2004b. Chromosome numbers and karyotype evolution in holoparasitic *Orobanche* (Orobanchaceae) and related genera. Amer. J. Bot. 91: 439–448.

Schneider, A. C. 2016. Resurrection of the genus *Aphyllon* for New World broomrapes (*Orobanche* s.l., Orobanchaceae). PhytoKeys 75: 107–118.

Schneider, A. C. et al. 2016b. Extensive cryptic host-specific diversity among western hemisphere broomrapes (*Orobanche* s.l., Orobanchaceae). Ann. Bot. (Oxford) 118: 1101–1111.

Schoenig, S. 2016. Rediscovery of *Erythranthe percaulis* in the Feather River Canyon. Phytoneuron 2016-69: 1–14.

Schotsman, H. D. 1967. Les *Callitriches:* Espèces de France et Taxa Noveaux d'Europe. Paris.

Schotsman, H. D. 1985. Une nouvelle espèce de *Callitriche* (Callitrichaceae) de Papoua-Nouvelle-Guinée: *C. cycloptera* Schotsm. Bull. Mus. Natl. Hist. Nat., B, Adansonia 7: 115–121.

Schrock, G. F. and B. F. Palser. 1967. Floral development, anatomy, and embryology in *Collinsia heterophylla* with some notes on ten other species of *Collinsia* and on *Tonella tenella*. Bot. Gaz. 128: 83–104.

Schuyler, A. E. 1989. Intertidal variants of *Bacopa rotundifolia* and *B. innominata* in the Chesapeake Bay drainage. Bartonia 55: 18–22.

Schwarzbach, A. E. and L. A. McDade. 2002. Phylogenetic relationships of the mangrove family Avicenniaceae based on chloroplast and nuclear ribosomal DNA sequences. Syst. Bot. 27: 84–98.

Sci. Rep. = Scientific Reports.

Science = Science; an Illustrated Journal [later: a Weekly Journal Devoted to the Advancement of Science].

Scoggan, H. J. 1978–1979. The Flora of Canada. 4 parts. Ottawa. [Natl. Mus. Nat. Sci. Publ. Bot. 7.]

Scopoli, J. A. 1760. Flora Carniolica.... Vienna. (Fl. Carniol.)

Scroph. Ind.—See: G. Bentham 1835

Seemann, B. 1852–1857. The Botany of the Voyage of H.M.S. Herald...during the Years 1845–51. 10 parts. London. [Parts paged consecutively.] (Bot. Voy. Herald)

Sell, P. D. and P. F. Yeo. 1970. A revision of the North American species of *Euphrasia* L. (Scrophulariaceae). Bot. J. Linn. Soc. 63: 189–234.

Shaw, R. J. 1962. The biosystematics of *Scrophularia* in western North America. Aliso 5: 147–178.

Shipunov, A., K. Aquilar, and H. J. Lee. 2014. Phylogeny of plantains (*Plantago* L., Plantaginaceae). In: Botanical Society of America. 2014. Botany 2014 Abstracts. Boise. Pp. 238–239.

Sida = Sida; Contributions to Botany.

Siebold, P. F. von and J. G. Zuccarini. 1835[–1870]. Flora Japonica.... 2 vols. in 30 parts. Leiden. [Vols. paged and parted independently, plates numbered consecutively.] (Fl. Jap.)

Singhurst, J. R. et al. 2012. *Bellardia trixago* (Orobanchaceae): 40 years of range expansion in Texas and a first report from Louisiana. Phytoneuron 2012-4: 1–4.

Sketch Bot. S. Carolina—See: S. Elliott [1816–]1821–1824

Skottsberg, C. 1912. *Tetrachondra patagonica* n. sp. und die systematische Stellung der Gattung. Bot. Jahrb. Syst. 48(suppl.): 17–26.

Small, J. K. 1903. Flora of the Southeastern United States.... New York. (Fl. S.E. U.S.)

Small, J. K. 1913b. Flora of Miami.... New York. (Fl. Miami)

Small, J. K. 1933. Manual of the Southeastern Flora, Being Descriptions of the Seed Plants Growing Naturally in Florida, Alabama, Mississippi, Eastern Louisiana, Tennessee, North Carolina, South Carolina and Georgia. New York. (Man. S.E. Fl.)

Smiley, C. J. 1961. A record of *Paulownia* in the Tertiary of North America. Amer. J. Bot. 48: 175–179.

Smith, C. E. 1966. Identity of witchweed in the southeastern United States. Rhodora 68: 167.

Smith, E. B. 1991. An Atlas and Annotated List of the Vascular Plants of Arkansas, ed. 2. Fayetteville, Ark.

Smith, J. E. et al. 1790–1866. English Botany; or, Coloured Figures of British Plants.... 36 vols. + 5 suppls. London. (Engl. Bot.)

Smithsonian Misc. Collect. = Smithsonian Miscellaneous Collections.

Soltis, D. E. et al. 2011. Angiosperm phylogeny: 17 genes, 640 taxa. Amer. J. Bot. 98: 704–730.

Soó, R. and D. A. Webb. 1972. *Rhinanthus*. In: T. G. Tutin et al., eds. 1964–1980. Flora Europaea. 5 vols. Cambridge. Vol. 3, pp. 276–280.

SouthE. Naturalist = Southeastern Naturalist.

SouthW. Naturalist = Southwestern Naturalist.

Sp. Pl.—See: C. Linnaeus 1753; C. L. Willdenow et al. 1797–1830

Sp. Pl. ed. 2—See: C. Linnaeus 1762–1763

Spach, É. 1834–1848. Histoire Naturelle des Végétaux. Phanérogames.... 14 vols., atlas. Paris. (Hist. Nat. Vég.)

Spackman, S. et al. 1997. Field Survey and Protection Recommendations for the Globally Imperiled Parachute Penstemon, *Penstemon debilis* O'Kane and Anderson. Fort Collins.

Spangler, R. E. and R. G. Olmstead. 1999. Phylogenetic analysis of the Bignoniaceae based on the cpDNA gene sequences *rbc*L and *ndh*F. Ann. Missouri Bot. Gard. 86: 33–46.

Spellenberg, R. 1971. In: IOPB chromosome number reports XXXII. Taxon 20: 349–356.

Speta, F. 1980. Die Gattungen *Chaenorhinum* (DC.) Reichenb. und *Microrrhinum* (Endl.) Fourr. im östlichen Teil ihrer Areale (Balkan bis Indien). Stapfia 7: 1–72.

Spic. Fl. Rumel.—See: A. H. R. Grisebach 1843–1845[–1846]

Spooner, D. M. 1984. Intraspecific variation in *Gratiola viscidula* Pennell (Scrophulariaceae). Rhodora 86: 79–87.

Sprague, T. A. 1921. A revision of the genus *Capraria*. Bull. Misc. Inform. Kew 1921: 205–212.

Sprengel, K. [1824–]1825–1828. Caroli Linnaei...Systema Vegetabilium. Editio Decima Sexta.... 5 vols. Göttingen. [Vol. 4 in 2 parts paged independently; vol. 5 by A. Sprengel.] (Syst. Veg.)

Stafleu, F. A. et al. 1992–2009. Taxonomic Literature: A Selective Guide to Botanical Publications and Collections with Dates, Commentaries and Types. Supplement. 8 vols. Königstein.

Stafleu, F. A. and R. S. Cowan. 1976–1988. Taxonomic Literature: A Selective Guide to Botanical Publications and Collections with Dates, Commentaries and Types, ed. 2. 7 vols. Utrecht etc.

Stamp, N. E. 1984. Effect of defoliation by checkerspot caterpillars (*Euphydryas phaeton*) and sawfly larvae (*Macrophya nigra* and *Tenthredo grandis*) on their host plants (*Chelone* spp.). Oecologia 63: 275–280.

Stamp, N. E. 1987. Availability of resources for predators of *Chelone* seeds and their parasitoids. Amer. Midl. Naturalist 117: 265–279.

Stapfia = Stapfia; Publikation der Botanischen Arbeitsgemeinschaft am O. Ö. Landesmuseum, Linz.

Starr, G. H. 1943. A new parasite on tomatoes. Phytopathology 33: 257–258.

Stebbins, G. L. 1974. Flowering Plants: Evolution above the Species Level. Cambridge, Mass.

Stermitz, F. R. et al. 1986. *Euphydryas anicia* (Lepidoptera: Nymphalidae) utilization of iridoid glycosides from *Castilleja* and *Besseya* (Scrophulariaceae) host plants. J. Chem. Ecol. 12: 1459–1468.

Stermitz, F. R. et al. 1986b. Iridoids and alkaloids from *Castilleja* host plants for *Platyptilia pica*. Rhexifoline content of *P. pica*. Biochem. Syst. & Ecol. 14: 499–506.

Stermitz, F. R. and G. H. Harris. 1987. Transfer of pyrrolizidine and quinolizidine alkaloids to *Castilleja* (Scrophulariaceae) hemiparasites from composite and legume host plants. J. Chem. Ecol. 13: 1917–1925.

Steudel, E. G. 1840–1841. Nomenclator Botanicus Enumerans Ordine Alphabetico Nomina atque Synonyma tum Generica tum Specifica..., ed. 2. 2 vols. Stuttgart and Tübingen. (Nomencl. Bot. ed. 2)

Steven, C. 1823. Monographia *Pedicularis*. Mém. Soc. Imp. Naturalistes Moscou 6: 1–60.

Stewart, H. M. and J. M. Canne. 1998. Floral development of *Agalinis neoscotica, Agalinis paupercula* var. *borealis,* and *Agalinis purpurea* (Scrophulariaceae): Implications for taxonomy and mating system. Int. J. Pl. Sci. 159: 418–439.

Stewart, H. M., S. C. Stewart, and J. M. Canne. 1996. Mixed mating system in *Agalinis neoscotica* (Scrophulariaceae) with bud pollination and delayed pollen germination. Int. J. Pl. Sci. 157: 501–508.

Stiefelhagen, H. 1910. Systematische und pflanzengeographische Studien zur Kenntnis der Gattung *Scrophularia*. Bot. Jahrb. Syst. 44: 406–496.

Stirp. Austr. Fasc. ed. 2—See: H. J. N. von Crantz 1769

Stone, J. L. and B. A. Drummond. 2006. Rare estuary monkeyflower in Merrymeeting Bay is genetically distinct. N. E. Naturalist 13: 179–190.

Straw, R. M. 1955. Hybridization, homogamy, and sympatric speciation. Evolution 9: 441–444.

Straw, R. M. 1956. Adaptive morphology of the *Penstemon* flower. Phytomorphology 6: 112–119.

Straw, R. M. 1956b. Floral isolation in *Penstemon*. Amer. Naturalist 90: 47–53.

Straw, R. M. 1962. The penstemons of Mexico. II. *Penstemon hartwegii, Penstemon gentianoides*, and their allies. Bol. Soc. Bot. México 27: 1–36.

Straw, R. M. 1963. Bee-fly pollination of *Penstemon ambiguus*. Ecology 44: 818–819.

Straw, R. M. 1966. A redefinition of *Penstemon* (Scrophulariaceae). Brittonia 18: 80–95.

Streisfeld, M. A. and J. R. Kohn. 2005. Contrasting patterns of floral and molecular variation across a cline in *Mimulus aurantiacus*. Evolution 59: 2548–2559.

Stuessy, T. F. 1990. Plant Taxonomy: The Systematic Evaluation of Comparative Data. New York.

Suppl. Dict. Jard.—See: L. M. Chazelles de Prizy 1789–1790

Suppl. Pl.—See: C. Linnaeus f. 1781[1782]

Sutherland, D. M. 1988. Historical notes on collections and taxonomy of *Penstemon haydenii* S. Wats. (blowout penstemon), Nebraska's only endemic plant species. Trans. Nebraska Acad. Sci. 16: 191–194.

Sutton, D. A. 1988. A Revision of the Tribe Antirrhineae. London. (Revis. Antirrhineae)

Svensk Bot. Tidskr. = Svensk Botanisk Tidskrift Utgifven af Svenska Botaniska Föreningen.

Swartz, O. P. 1788. Nova Genera & Species Plantarum seu Prodromus.... Stockholm, Uppsala, and Åbo. (Prodr.)

Swartz, O. P. 1791. Observationes Botanicae.... Erlangen. (Observ. Bot.)

Sweigart, A. L., N. H. Martin, and J. H. Willis. 2008. Patterns of nucleotide variation and reproductive isolation between a *Mimulus* allotetraploid and its progenitor species. Molec. Ecol. 17: 2089–2100.

Symb. Bot.—See: M. Vahl 1790–1794

Symb. Bot. Upsal. = Symbolae Botanicae Upsalienses; Arbeten från Botaniska Institutionen i Uppsala.

Syn. Fl. Colorado—See: T. C. Porter and J. M. Coulter 1874

Syn. Fl. Germ. Helv.—See: W. D. J. Koch [1835–]1837–1838

Syn. Fl. N. Amer.—See: A. Gray et al. 1878–1897

Syn. Fl. N. Amer. ed. 2—See: A. Gray et al. 1886

Syn. Pl.—See: C. H. Persoon 1805–1807

Syst. Bot. = Systematic Botany; Quarterly Journal of the American Society of Plant Taxonomists.

Syst. Bot. Monogr. = Systematic Botany Monographs; Monographic Series of the American Society of Plant Taxonomists.

Syst. Census Austral. Pl.—See: F. J. H. Mueller 1882

Syst. Nat.—See: J. F. Gmelin 1791[–1792]

Syst. Nat. ed. 10—See: C. Linnaeus 1758[–1759]

Syst. Nat. ed. 12—See: C. Linnaeus 1766[–1768]

Syst. Veg.—See: J. J. Roemer et al. 1817[–1830]; K. Sprengel [1824–]1825–1828

Syst. Veg. ed. 14—See: J. A. Murray 1784

Tabl. Encycl.—See: J. Lamarck and J. Poiret 1791–1823

Takhtajan, A. L. 1980. Outline of the classification of flowering plants (Magnoliophyta). Bot. Rev. (Lancaster) 46: 225–359.

Takhtajan, A. L. 1997. Diversity and Classification of Flowering Plants. New York.

Takhtajan, A. L. 2009. Flowering Plants, ed. 2. New York.

Tallent-Halsell, N. G. and M. S. Watt. 2009. The invasive *Buddleja davidii* (butterfly bush). Bot. Rev. (Lancaster) 75: 292–325.

Tank, D. C. et al. 2006. Review of the systematics of Scrophulariaceae s.l. and their current disposition. Austral. Syst. Bot. 19: 289–307.

Tank, D. C., J. M. Egger, and R. G. Olmstead. 2009. Phylogenetic classification of subtribe Castillejinae (Orobanchaceae). Syst. Bot. 34: 182–197.

Tank, D. C. and R. G. Olmstead. 2008. From annuals to perennials: Phylogeny of subtribe Castillejinae (Orobanchaceae). Amer. J. Bot. 95: 608–625.

Taxon = Taxon; Journal of the International Association for Plant Taxonomy.

Tennessee Exotic Pest Plant Council. 2004. Invasive exotic pest plants in Tennessee—2004. Wildland Weeds 2004(fall): 13–16.

Tepedino, V. J. et al. 2007. Pollination biology of a disjunct population of the endangered sandhill endemic *Penstemon haydenii* S. Wats. (Scrophulariaceae) in Wyoming. Pl. Ecol. 193: 59–69.

Tepedino, V. J., W. R. Bowlin, and T. L. Griswold. 2006. Pollination biology of the endangered blowout penstemon (*Penstemon haydenii* S. Wats.: Scrophulariaceae) in Nebraska. J. Torrey Bot. Soc. 133: 548–559.

Tepedino, V. J., S. D. Sipes, and T. L. Griswold. 1999. The reproductive biology and effective pollinators of the endangered beardtongue *Penstemon penlandii* (Scrophulariaceae). Pl. Syst. Evol. 219: 39–54.

Teryokhin, E. S. 1997. Weed Broomrapes: Systematics, Ontogenesis, Biology, Evolution. Landshut.

Těšitel, J. et al. 2010. Phylogeny, life history evolution and biogeography of the rhinanthoid Orobanchaceae. Folia Geobot. 45: 347–367.

Theisen, I. and E. Fischer. 2004. Myoporaceae. In: K. Kubitzki et al., eds. 1990+. The Families and Genera of Vascular Plants. 14+ vols. Berlin etc. Vol. 7, pp. 289–292.

Thieret, J. W. 1969d. Notes on *Epifagus*. Castanea 34: 397–402.

Thieret, J. W. 1970. *Bacopa repens* (Scrophulariaceae) in the conterminous United States. Castanea 35: 132–136.

Thieret, J. W. 1971. The genera of Orobanchaceae in the southeastern United States. J. Arnold Arbor. 52: 404–424.

Thieret, J. W. and D. H. Dike. 1969. *Dopatrium junceum* (Scrophulariaceae) in Louisiana. Sida 3: 448.

Thompson, D. M. 1988. Systematics of *Antirrhinum* (Scrophulariaceae) in the New World. Syst. Bot. Monogr. 22: 1–142.

Thompson, D. M. 1993. *Mimulus*. In: J. C. Hickman, ed. 1993. The Jepson Manual. Higher Plants of California. Berkeley, Los Angeles, and London. Pp. 1037–1046.

Thompson, D. M. 2005. Systematics of *Mimulus* subgenus *Schizoplacus* (Scrophulariaceae). Syst. Bot. Monogr. 75: 1–213.

Thomson, J. D. et al. 2000. Pollen presentation and pollination syndromes, with special reference to *Penstemon*. Pl. Spec. Biol. 15: 11–29.

Thorne, R. F. 1983. Proposed new realignments in the angiosperms. Nordic J. Bot. 3: 85–117.

Thorne, R. F. 1992b. Classification and geography of the flowering plants. Bot. Rev. (Lancaster) 58: 225–348.

Thorne, R. F. 2000b. The classification and geography of the flowering plants: Dicotyledons of the class Angiospermae (subclasses Magnoliidae, Ranunculidae, Caryophylliidae, Dilleniidae, Rosidae, Asteridae, and Lamiidae). Bot. Rev. (Lancaster) 66: 441–647.

Thorne, R. F. and J. L. Reveal. 2007. An updated classification of the class Magnoliopsida ("Angiospermae"). Bot. Rev. (Lancaster) 73: 67–182.

Todsen, T. K. 1998. *Penstemon metcalfei* (Scrophulariaceae), a valid species. Sida 18: 621–622.

Tolmatchew, A. I., ed. 1960–1987. Flora Arctica URSS. 10 vols. in 12. Moscow and Leningrad.

Torrey, J. 1819. A Catalogue of Plants, Growing Spontaneously within Thirty Miles of the City of New York. Albany. (Cat. Pl. New York)

Torrey, J. 1856. Explorations and Surveys for a Railroad Route from the Mississippi River to the Pacific Ocean. Routes in California...Explored by Lieutenant John G. Parke.... Botanical Report. Washington. [Preprinted from War Department [U.S.]. 1855–1860. Pacif. Railr. Rep. 12 vols. in 13. Washington. Vol 7(3), pp. [3]–28. 1857.] (Pacif. Railr. Rep. Parke, Bot.)

Torreya = Torreya; a Monthly Journal of Botanical Notes and News.

Trab. Mus. Ci. Nat. Barcelona = Trabajos del Museo de Ciencias Naturales de Barcelona.

Trans. Amer. Philos. Soc. = Transactions of the American Philosophical Society Held at Philadelphia for Promoting Useful Knowledge.

Trans. Hort. Soc. London = Transactions, of the Horticultural Society of London.

Trans. Linn. Soc. London = Transactions of the Linnean Society of London.

Trans. Mass. Hort. Soc. = Transactions of the Massachusetts Horticultural Society.

Trans. Nebraska Acad. Sci. = Transactions, Nebraska Academy of Sciences.

Travels Carolina—See: W. Bartram 1791

Trimen, H., J. D. Hooker, and A. H. G. Alston. 1893–1931. A Hand-book to the Flora of Ceylon.... 6 vols. London. (Handb. Fl. Ceylon)

Trop. Woods = Tropical Woods....

Trudy Imp. S.-Peterburgsk. Bot. Sada = Trudy Imperatorskago S.-Peterburgskago Botanicheskago Sada.

Tsoong, P. C. 1955. A new system for the genus *Pedicularis*. Acta Phytotax. Sin. 4: 71–147.

Tsoong, P. C. 1956. A new system for the genus *Pedicularis* (continued). Acta Phytotax. Sin. 5: 41–73.

Tulig, M. C. 2000. Morphological Variation in *Mimulus* Section *Diplacus* (Scrophulariaceae). M.S. thesis. California State Polytechnic University.

Tulig, M. C. and G. L. Nesom. 2012. Taxonomic overview of *Diplacus* sect. *Diplacus* (Phrymaceae). Phytoneuron 2012-45: 1–17.

Tunbridge, N. D., C. Sears, and E. Elle. 2011. Variation in floral morphology and ploidy among populations of *Collinsia parviflora* and *Collinsia grandiflora*. Botany (Ottawa) 89: 19–33.

Turner, B. L. 1982. Revisional treatment of the Mexican species of *Seymeria*. Phytologia 51: 403–423.

Turner, B. L. et al. 2003. Atlas of the Vascular Plants of Texas. 2 vols. Fort Worth. [Sida Bot. Misc. 24.]

Turner, B. L. and C. C. Cowan. 1993. Taxonomic overview of *Stemodia* (Scrophulariaceae) for North America and the West Indies. Phytologia 74: 61–103.

Tutin, T. G. et al., eds. 1964–1980. Flora Europaea. 5 vols. Cambridge.

Twyford, A. D. and J. Friedman. 2015. Adaptive divergence in the monkey flower *Mimulus guttatus* is maintained by a chromosomal inversion. Evolution 69: 1476–1486.

Tzvelev, N. N. 1980. *Hippuris*. In: A. I. Tolmatchew, ed. 1960–1987. Flora Arctica URSS. 10 vols. in 12. Moscow and Leningrad. Vol. 8(2), pp. 57–61.

Tzvelev, N. N. 1983. Plantaginaceae. In: A. I. Tolmatchew, ed. 1960–1987. Flora Arctica URSS. 10 vols. in 12. Moscow and Leningrad. Vol. 8(2), pp. 16–25.

Univ. Calif. Publ. Bot. = University of California Publications in Botany.

Univ. Kansas Sci. Bull. = University of Kansas Science Bulletin.

Univ. Wyoming Publ. Sci., Bot. = University of Wyoming Publications in Science. Botany.

University of Chicago Press. 1993. The Chicago Manual of Style, ed. 14. Chicago.

Uribe-Convers, S., M. L. Settles, and D. C. Tank. 2016. A phylogenetic approach based on PCR target enrichment and high throughput sequencing: Resolving the diversity within the South American species of *Bartsia* (Orobanchaceae). PLOS ONE 11:e0148203. doi:10.1371-journal.pone.0148203.

Uribe-Convers, S. and D. C. Tank. 2016. Phylogenetic revision of the genus *Bartsia* (Orobanchaceae): Disjunct distributions correlate to independent lineages. Syst. Bot. 41: 672–684.

Utah Fl. ed. 3—See: S. L. Welsh et al. 2003

Vahl, M. 1790–1794. Symbolae Botanicae.... 3 vols. Copenhagen. (Symb. Bot.)

Vargas, P. et al. 2004. Molecular evidence for naturalness of genera in the tribe Antirrhineae (Scrophulariaceae) and three independent evolutionary lineages from the New World and the Old. Pl. Syst. Evol. 249: 151–172.

Vasc. Pl. Pacif. N.W.—See: C. L. Hitchcock et al. 1955–1969

Vasc. Pl. Wyoming—See: R. D. Dorn 1988

Venkata Ramana, R. et al. 2000. Embryology of *Phryma leptostachya* L. (Verbenaceae) with considerations of its systematic status and affinities. Feddes Repert. 111: 231–248.

Ventenat, É. P. [1800–1803.] Description des Plantes Nouvelles et Peu Connues Cultivés dans le Jardin de J. M. Cels.... 10 parts. Paris. [Plates numbered consecutively.] (Descr. Pl. Nouv.)

Vickery, R. K. 1978. Case studies in the evolution of species complexes in *Mimulus*. Evol. Biol. 11: 404–506.

Vickery, R. K. 1990. Close correspondence of allozyme groups to geographic races in the *Mimulus glabratus* complex (Scrophulariaceae). Syst. Bot. 15: 481–496.

Vickery, R. K. 1992. Pollinator preferences for yellow, orange, and red flowers of *Mimulus verbenaceus* and *M. cardinalis*. Great Basin Naturalist 52: 145–148.

Viehmeyer, G. 1958. Reversal of evolution in the genus *Penstemon*. Amer. Naturalist 92: 129–137.

Villars, D. 1779. Prospectus de l'Histoire des Plantes de Dauphiné.... Grenoble. (Prosp. Hist. Pl. Dauphiné)

Vincent, K. A. 1981. *Lindernia antipoda* (L.) Alston and *Veronica cymbalaria* Bod. (Scrophulariaceae): New to North America; *V. hederaefolia* L.: New to Louisiana. Sida 9: 185–187.

Vincent, K. A. 1982. Scrophulariaceae of Louisiana. M.S. thesis. University of Southwestern Louisiana.

Voss, E. G. 1972–1996. Michigan Flora.... 3 vols. Bloomfield Hills and Ann Arbor.

Vujičič, R., D. Grubišič, and R. Konjevič. 1993. Scanning electron microscopy of the seed coat in the genus *Paulownia* (Scrophulariaceae). Bot. J. Linn. Soc. 111: 505–511.

Vujnovic, K. and R. W. Wein. 1997. The biology of Canadian weeds. 106. *Linaria dalmatica* (L.) Mill. Canad. J. Pl. Sci. 77: 483–491.

W. Amer. Sci. = West American Scientist.

W. N. Amer. Naturalist = Western North American Naturalist.

Wagstaff, S. J. 2004. Tetrachondraceae. In: K. Kubitzki et al., eds. 1990+. The Families and Genera of Vascular Plants. 14+ vols. Berlin etc. Vol. 7, pp. 441–444.

Wagstaff, S. J. and R. G. Olmstead. 1997. Phylogeny of Labiatae and Verbenaceae inferred from *rbc*L sequences. Syst. Bot. 22: 165–179.

Walker, R. S. 1919. The *Paulownia tomentosa* tree. Amer. Forests 25: 1485–1486.

Walker-Larsen, J. and L. D. Harder. 2001. Vestigial organs as opportunities for functional innovation: The example of the *Penstemon* staminode. Evolution 55: 477–487.

Wallich, N. 1828[–1849]. A Numerical List of Dried Specimens of Plants, in the East India Company's Museum Collected under the Superintendence of Dr. Wallich of the Company's Botanic Garden at Calcutta.... London.

Wallroth, C. F. W. 1815. Annus Botanicus.... Halle. (Annus Bot.)

Wallroth, C. F. W. 1825. Orobanches Generis Diaskene.... Frankfurt am Main. (Orobanches Gen. Diask.)

Walpers, W. G. 1842–1847. Repertorium Botanices Systematicae.... 6 vols. Leipzig. (Repert. Bot. Syst.)

Walter, T. 1788. Flora Caroliniana, Secundum Systema Vegetabilium Perillustris Linnaei Digesta.... London. (Fl. Carol.)

War Department [U.S.] 1855–1860. Reports of Explorations and Surveys, to Ascertain the Most Practicable and Economical Route for a Railroad from the Mississippi River to the Pacific Ocean. Made under the Direction of the Secretary of War, in 1853[–1856].... 12 vols. in 13. Washington. (Pacif. Railr. Rep.)

Ward, S. M. et al. 2009. Hybridization between invasive populations of Dalmatian toadflax *(Linaria dalmatica)* and yellow toadflax *(Linaria vulgaris)*. Invasive Pl. Sci. Managem. 2: 369–378.

Washington Natural Heritage Program. 2005. Field Guide to Selected Rare Plants of Washington. Olympia.

Watson, K. C. 1975. Systematics of *Orobanche* Section *Gymnocaulis* (Orobanchaceae). M.S. thesis. University of California, Chico.

Watson, S. 1871. United States Geological Expolration [sic] of the Fortieth Parallel. Clarence King, Geologist-in-charge. [Vol. 5] Botany. By Sereno Watson.... Washington. [Botanical portion of larger work by C. King.] [Botany (Fortieth Parallel)]

Watsonia = Watsonia; Journal of the Botanical Society of the British Isles.

Weed Sci. = Weed Science; Journal of the Weed Science Society of America.

Welsh, S. L. 1974. Anderson's Flora of Alaska and Adjacent Parts of Canada. Provo.

Welsh, S. L. et al., eds. 1993. A Utah Flora, ed. 2. Provo.

Welsh, S. L. et al., eds. 2003. A Utah Flora, ed. 3. Provo. (Utah Fl. ed. 3)

Welsh, S. L. et al., eds. 2008. A Utah Flora, ed. 4. Provo.

Werdenda = Werdenda. Beiträge zur Pflanzenkunde.

Werner, K. 1965. Taxonomie und Phylogenie der Gattungen *Isoplexis* Lindl. und *Digitalis* L. Repert. Spec. Nov. Regni Veg. 70: 109–135.

Werth, C. R. and J. L. Riopel. 1979. A study of the host range of *Aureolaria pedicularia* (L.) Raf. (Scrophulariaceae). Amer. Midl. Naturalist 102: 300–306.

Wessinger, C. A. et al. 2016. Multiplexed shotgun genotyping resolves species relationships within the North American genus *Penstemon*. Amer. J. Bot. 103: 912–922.

Wetherwax, M., T. I. Chuang, and L. R. Heckard. 2012. *Castilleja*. In: B. G. Baldwin et al., eds. 2012. The Jepson Manual: Vascular Plants of California, ed. 2. Berkeley. Pp. 956–964.

Wetherwax, M. and D. M. Thompson. 2012. *Antirrhinum*. In: B. G. Baldwin et al., eds. 2012. The Jepson Manual: Vascular Plants of California, ed. 2. Berkeley. Pp. 1002–1004.

Wettstein, R. 1891–1893. Scrophulariaceae. In: H. G. A. Engler and K. Prantl, eds. 1887–1915. Die natürlichen Pflanzenfamilien.... 254 fascs. Leipzig. Fascs. 65, 67, 83 [IV,3b], pp. 39–107.

Wettstein, R. 1896b. Monographie der Gattung *Euphrasia*.... Leipzig. (Monogr. Euphrasia)

Whipple, H. L. 1972. Structure and systematics of *Phryma leptostachya*. J. Elisha Mitchell Sci. Soc. 88: 1–17.

Whittall, J. B. et al. 2006. The *Mimulus moschatus* alliance (Phrymaceae): Molecular and morphological phylogenetics and their conservation implications. Syst. Bot. 31: 380–397.

Widrlechner, M. P. 1983. Historical and phenological observations on the spread of *Chaenorhinum minus* across North America. Canad. J. Bot. 61: 179–187.

Wilcox, W. H. 1973. A survey of the vascular flora of Crittenden County, Arkansas. Castanea 38: 286–297.

Wildland Weeds = Wildland Weeds; a Quarterly Publication of the Florida Exotic Pest Plant Council.

Wilhelm, S., L. Benson, and J. Sagen. 1958. Studies in the control of broomrape on tomatoes. Pl. Dis. Reporter 42: 645–651.

Willdenow, C. L. 1803–1816. Hortus Berolinensis.... 2 vols. in 10 fascs. Berlin. [Fascs. and plates numbered consecutively.] (Hort. Berol.)

Willdenow, C. L. 1809–1813[–1814]. Enumeratio Plantarum Horti Regii Botanici Berolinensis.... 2 parts + suppl. Berlin. [Parts paged consecutively.] (Enum. Pl.)

Willdenow, C. L., C. F. Schwägrichen, and J. H. F. Link. 1797–1830. Caroli a Linné Species Plantarum.... Editio Quarta.... 6 vols. Berlin. [Vols. 1–5(1), 1797–1810, by Willdenow; vol. 5(2), 1830, by Schwägrichen; vol. 6, 1824–1825, by Link.] (Sp. Pl.)

Williams, J. E. 1973. *Agalinis* and Related Genera in Oklahoma. M.S. thesis. University of Oklahoma.

Williams, J. K. 2004. A revision of *Capraria* (Scrophulariaceae). Lundellia 7: 53–78.

Willkomm, H. M. and J. M. C. Lange. 1861–1880. Prodromus Florae Hispanicae.... 3 vols. in 9 parts. Stuttgart. (Prodr. Fl. Hispan.)

Wilson, B. L. et al. 2014. Identification and taxonomic status of *Cordylanthus tenuis* ssp. *pallescens* (Orobanchaceae). Madroño 61: 64–76.

Wilson, Paul et al. 2004. A multivariate search for pollination syndromes among penstemons. Oikos 104: 345–361.

Wilson, Paul et al. 2007. Constrained lability in floral evolution: Counting convergent origins of hummingbird pollination in *Penstemon* and *Keckiella*. New Phytol. 176: 883–890.

Wilson, Paul and M. Valenzuela. 2002. Three naturally occurring *Penstemon* hybrids. W. N. Amer. Naturalist 62: 25–31.

Windler, D. R., B. E. Wofford, and M. L. Bierner. 1976. Evidence of natural hybridization between *Mimulus ringens* and *Mimulus alatus* (Scrophulariaceae). Rhodora 78: 641–649.

Wirsing, A. L. 1778. Eclogae Botanicae.... Nuremberg. (Eclog. Bot.)

Withering, W. 1787–1792. A Botanical Arrangement of British Plants..., ed. 2. 3 vols. in 4. Birmingham, London, and Edinburgh. (Bot. Arr. Brit. Pl. ed. 2)

Wolfe, A. D. et al. 1997. Using restriction-site variation of PCR-amplified cpDNA genes for phylogenetic analysis of tribe Cheloneae (Scrophulariaceae). Amer. J. Bot. 84: 555–564.

Wolfe, A. D. et al. 2002. A phylogenetic and biogeographic analysis of the Cheloneae (Scrophulariaceae) based on ITS and *matK* sequence data. Syst. Bot. 27: 138–148.

Wolfe, A. D. et al. 2005. Phylogeny and biogeography of Orobanchaceae. Folia Geobot. 40: 115–134.

Wolfe, A. D. et al. 2006. Phylogeny, taxonomic affinities, and biogeography of *Penstemon* (Plantaginaceae) based on ITS and cpDNA sequence data. Amer. J. Bot. 93: 1699–1713.

Wolfe, A. D. and C. W. dePamphilis. 1998. The effect of relaxed functional constraints on the photosynthetic gene *rbc*L in photosynthetic and nonphotosynthetic parasitic plants. Molec. Biol. Evol. 15: 1243–1258.

Wolfe, A. D. and W. J. Elisens. 1993. Diploid hybrid speciation in *Penstemon* (Scrophulariaceae) revisited. Amer. J. Bot. 80: 1082–1094.

Wolfe, A. D. and W. J. Elisens. 1994. Nuclear ribosomal DNA restriction-site variation in *Penstemon* section *Peltanthera* (Scrophulariaceae): An evaluation of diploid hybrid speciation and evidence for introgression. Amer. J. Bot. 81: 1627–1635.

Wolfe, A. D., Xiang Q. Y., and S. R. Kephart. 1998. Diploid hybrid speciation in *Penstemon* (Scrophulariaceae). Proc. Natl. Acad. Sci. U.S.A. 95: 5112–5115.

Wood, A. 1847. A Class-book of Botany..., ed 2. Boston and Claremont, N.H. (Class-book Bot. ed. 2)

Wooton, E. O. and P. C. Standley. 1915. Flora of New Mexico. Contr. U.S. Natl. Herb. 19.

World Checkl. Seed Pl.—See: R. Govaerts 1995+

Wortley, A. H., D. J. Harris, and R. W. Scotland. 2007. On the taxonomy and phylogenetic position of *Thomandersia*. Syst. Bot. 32: 415–444.

Wu, C. A. et al. 2008. *Mimulus* is an emerging model system for the integration of ecological and genomic studies. Heredity 100: 220–230.

Wu, Z. and P. H. Raven, eds. 1994–2013. Flora of China. 25 vols. Beijing and St. Louis.

Wunderlin, R. P. 1998. Guide to the Vascular Plants of Florida. Gainesville.

Xia, Z., Wang Y.-Z., and J. F. Smith. 2009. Familial placement and relations of *Rehmannia* and *Triaenophora* (Scrophulariaceae s.l.) inferred from five gene regions. Amer. J. Bot. 96: 519–530.

Xiang, Q. Y. et al. 2000. Timing the eastern Asian-eastern North American floristic disjunction: Molecular clock corroborates paleontological estimates. Molec. Phylogen. Evol. 15: 462–472.

Yang, F. S., Wang X. Q., and Hong D. Y. 2003. Unexpected high divergence in nrDNA ITS and extensive parallelism in floral morphology of *Pedicularis* (Orobanchaceae). Pl. Syst. Evol. 240: 91–105.

Yatskievych, G. 1999–2013. Steyermark's Flora of Missouri. 3 vols. Jefferson City.

Yatskievych, K. 2000. Field Guide to Indiana Wildflowers. Bloomington.

Yeo, P. F. 1978. A taxonomic revision of *Euphrasia* in Europe. Bot. J. Linn. Soc. 77: 223–334.

Young, N. D., K. E. Steiner, and C. W. dePamphilis. 1999. The evolution of parasitism in Scrophulariaceae/Orobanchaceae: Plastid gene sequences refute an evolutionary transition series. Ann. Missouri Bot. Gard. 86: 876–893.

Yousefi, N., S. Zarre, and G. Heubl. 2016. Molecular phylogeny of the mainly Mediterranean genera *Chaenorhinum*, *Kickxia* and *Nanorrhinum* (Plantaginaceae, tribe Antirrhineae), with focus on taxa in the Flora Iranica region. Nordic J. Bot. 34: 455–463.

Yu, W. B. 2013. Nomenclatural clarifications for names in *Boschniakia*, *Kopsiopsis* and *Xylanche* (Orobanchaceae). Phytotaxa 77: 40–42.

Yu, W. B. et al. 2015. Towards a comprehensive phylogeny of the large temperate genus *Pedicularis* (Orobanchaceae), with an emphasis on species from the Himalaya-Hengduan Mountains. B. M. C. Pl. Biol. 15: 176.

Zhang, Zhi Y. 1987. A taxonomic study on the genus *Boschniakia* (Orobanchaceae). Acta Bot. Yunnan. 9: 289–296.

Zhukova, P. G. 1967. Karyology of some plants cultivated in the arctic-alpine botanical garden. In: N. A. Avrorin, ed. 1967. Plantarum in Zonam Polarem Transportatio. II. Leningrad. Pp. 139–149.

Zoë = Zoë; a Biological Journal.

Zona, S. 1998. The Myoporaceae in the southeastern United States. Harvard Pap. Bot. 3: 171–179.

Zootaxa = Zootaxa; an International Journal of Zootaxonomy.

Index

Names in *italics* are synonyms, casually mentioned hybrids, or plants not established in the flora. Page numbers in **boldface** indicate the primary entry for a taxon. Page numbers in *italics* indicate an illustration. Roman type is used for all other entries, including author names, vernacular names, and accepted scientific names for plants treated as established members of the flora.